Advances in Intelligent Systems and Computing

Volume 1274

AF814982

The series "Advances in Intelligent Systems and Computing" contains publications on theory, applications, and design methods of Intelligent Systems and Intelligent Computing. Virtually all disciplines such as engineering, natural sciences, computer and information science, ICT, economics, business, e-commerce, environment, healthcare, life science are covered. The list of topics spans all the areas of modern intelligent systems and computing such as: computational intelligence, soft computing including neural networks, fuzzy systems, evolutionary computing and the fusion of these paradigms, social intelligence, ambient intelligence, computational neuroscience, artificial life, virtual worlds and society, cognitive science and systems, Perception and Vision, DNA and immune based systems, self-organizing and adaptive systems, e-Learning and teaching, human-centered and human-centric computing, recommender systems, intelligent control, robotics and mechatronics including human-machine teaming, knowledge-based paradigms, learning paradigms, machine ethics, intelligent data analysis, knowledge management, intelligent agents, intelligent decision making and support, intelligent network security, trust management, interactive entertainment, Web intelligence and multimedia.

The publications within "Advances in Intelligent Systems and Computing" are primarily proceedings of important conferences, symposia and congresses. They cover significant recent developments in the field, both of a foundational and applicable character. An important characteristic feature of the series is the short publication time and world-wide distribution. This permits a rapid and broad dissemination of research results.

**** Indexing: The books of this series are submitted to ISI Proceedings, EI-Compendex, DBLP, SCOPUS, Google Scholar and Springerlink ****

More information about this series at http://www.springer.com/series/11156

Qi Liu · Xiaodong Liu ·
Tao Shen · Xuesong Qiu
Editors

The 10th International Conference on Computer Engineering and Networks

Set 1

Springer

Editors
Qi Liu
Science and Technology
Nanjing University of Information
Nanjing, Jiangsu, China

Xiaodong Liu
School of Computing
Edinburgh Napier University
Edinburgh, UK

Tao Shen
School of Electrical Engineering
University of Jinan
Jinan, Shandong, China

Xuesong Qiu
Institute of Network Technology
Beijing University of Posts
and Telecommunications
Beijing, Beijing, China

ISSN 2194-5357 ISSN 2194-5365 (electronic)
Advances in Intelligent Systems and Computing
ISBN 978-981-15-8461-9 ISBN 978-981-15-8462-6 (eBook)
https://doi.org/10.1007/978-981-15-8462-6

This Springer imprint is published by the registered company Springer Nature Singapore Pte Ltd.
The registered company address is: 152 Beach Road, #21-01/04 Gateway East, Singapore 189721, Singapore

Preface

This conference proceedings is a collection of the papers accepted by the CENet 2020—the 10th International Conference on Computer Engineering and Networks held on October 16–18 2020, in Xi'an, China.

This proceedings contains the five parts: Part I Artificial Intelligence and Applications (51 papers); Part II Communication System Detection, Analysis and Application (32 papers); Part III Information Security and Cybersecurity (25 papers), Part IV Intelligent System and Engineering (17 papers), Part V Internet of Things and Smart Systems (61 papers) and Part VI Medical Engineering And Information Systems (13 papers).

Each part can be used as an excellent reference by industry practitioners, university faculties, research fellows and undergraduates as well as graduate students who need to build a knowledge base of the most current advances and state of practice in the topics covered by this conference proceedings. This will enable them to produce, maintain and manage systems with high levels of trustworthiness and complexity.

Thanks to the authors for their hard work and dedication as well as the reviewers for ensuring the selection of only the highest-quality papers; their efforts made the proceedings possible.

Contents

Artificial Intelligence and Applications

A Unified Framework for Micro-video BackGround Music Automatic Matching

Zongzhi Chai[1], Haichao Zhang[1], Yuehua Li[1], Qianxi Yang[1], Tianyi Li[2], Fan Zhang[1], and Jinpeng Chen[1(✉)]

[1] School of Software Engineering, Beijing University of Posts and Telecommunications, Beijing 100786, China
jpchen@bupt.edu.cn
[2] University of Pittsburgh, Pittsburgh, USA
til50@pitt.edu

Abstract. The current widely spread of micro-form video is undeniable. However, neither the music-orienting nor the video-orienting way to make a video can totally express the idea of the video maker because of the fixed music types and the cost of time to find a proper music. Based on the deep learning method, this paper studies the automatic matching algorithm of background music for micro-videos which analyzes the background information and the emotions of characters in micro-videos and establishes the model to select the proper background music according to the video contents. Experiments are carried out on the database obtained from TikTok and the result shows that the current Micro-Video Background Music Automatic Matching model in this paper is effective.

Keywords: Micro-video · Deep learning · Video feature extraction · Chorus intercepting · Background music matching

1 Introduction

In recent years, with the popularity of the global mobile Internet and the development of network technology, micro videos have become a part of daily life for many people. More and more people are trying to make micro videos. But the traditional way of making micro videos with fixed music, limits this trend. In the current market, innovations in short video production and video filtering processing are imminent. Therefore, we have studied a method based on video content to add suitable background music to fill the gap in the current market demand. In this article, we propose a MVBGMAMA method to solve the above problems, and automatically select and match the appropriate background music based on the scene settings and character emotions in the short video. When analyzing the characteristics of the video, we can analyze the key frames of the video, and then use the scene recognition technology of pictures and facial expression recognition technology to label the scenes and emotional

Z. Chai and H. Zhang—These authors contributed to the work equally and should be regarded as co-first authors.

Q. Liu et al. (Eds.): CENet 2020, AISC 1274, pp. 3–10, 2021.
https://doi.org/10.1007/978-981-15-8462-6_1

tags of the video. Specifically, we first learn a mapping function (V2T) from video to text labels to fully extract the features of the video. By jointly minimizing the reconstruction error and the classification loss, these two features are merged in a supervised manner to arrive at a final label. The obtained tags are then matched with the music tags to select one or more pieces of music that match the video content. When adding background music, we automatically analyze the matching background music and intercept the chorus part of the music.

2 Related Work

2.1 Neural Network Model

In recent years, the development of CNN significantly facilitates the procedure and the implementation of image scene recognition [1, 2]. Currently, the design-related thoughts of CNN model basically develop towards the direction of deep network and more convolution calculation [3]. Residual Network Model (Resnet) has also achieved high accuracy in face recognition and scene recognition problem [3]. There is no obvious improvement in the accuracy of FER in the process of completely removing the full connection layer of the Xception model [4] and the process of depth wise separable solutions based on it. As for scene recognition, Bolei Zhou et al. have created great progress and effects on places data set [5]. AlexNet [1], GoogLeNet [6], VGG [7] and other models are used to verify, and the average accuracy of top-5 is 85.07% [5].

2.2 Music Chorus Interception

The duration of microform videos is generally limited within 15–20 s, but a song is normally about 3 min, and therefore music intercepting is required. There are two available methods to solve it: the first one relies on manual annotation; the second one uses big data of user's preference.

The advantage of the manual annotation is accurate and simple, yet it fails to label a large number of songs in a mass amount. So, there is limited music available for download at present in any music platforms which depends on this method. Another method is based on the big data of the user's playing preferences. According to statistics, a large number of users will choose to repeatedly play in the chorus part, or even directly switch to the chorus part. Through the observation of user behaviors, this method can analyze where the chorus begins, but this method requires adequate user data, and it is not possible to identify the less popular song chorus part.

3 Proposed Framework

As shown in Fig. 1, emotions and background information of the characters from the micro-videos are generated, all of which constitute the basic information of the video.

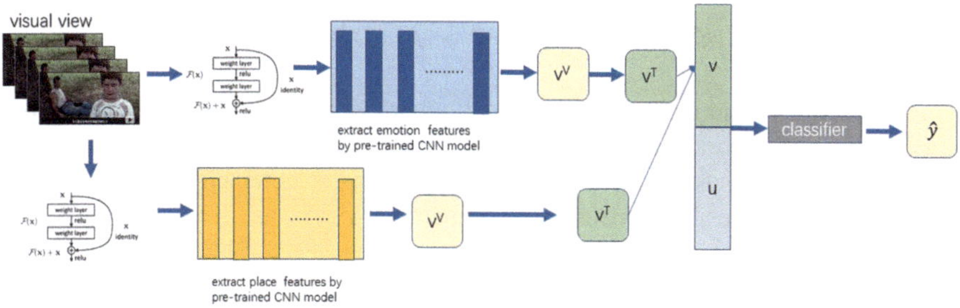

Fig. 1. An example of micro-video on TikTok.

3.1 Facial Expression Recognition

Among the seven facial expression features [8], happy, sad and angry are the three most distinguishable emotional features. In MVBGMAMA, the recognition accuracy of the above three facial expressions should be comparatively high. Our model transforms the input image layer by layer into a smaller but deeper feature map. With the increasing network depth, what this model identifies can be more complex.

According to the facial expression recognition model, the algorithm selects the keyframe expression features and processes the keyframe expression information to get the final video facial expression features.

3.2 Scene Identification

For this research topic, the scene information of the video is an important factor that affects the background music, so the training pictures that need to be selected for the public data set must contain the scene information. So we chose MIT's Places365-Standard data set [5] as the scene training data set, a total of 365 labeled diversified concept scenes. The analysis based on the above data, combined with diversified means to extract corresponding features, can achieve the purpose of training a linear support vector machine (SVM) classifier.

3.3 Intercepting Chorus of the Music

Chorus Identification. Following the experience, the chorus of a song is often repeated several times in the song. Despite different lyrics, the melody maintains basically the same, and the length of the chorus is also stable. Therefore, this characteristic is the criterion for determining the chorus of the algorithm. The reliability of the algorithm depends to a large extent on the correct determination of non-chorus.

Comparison. The FFT transform is applied to transform the original time-domain waveform to the frequency domain, after which a spectrum is obtained. The conversion method is as follows:

$$X(k) = \sum_{n}^{N-1} x(n)W_N^{nk} \cdots k = 0, 1, 2 \ldots N - 1 \quad (1)$$

The spectrum chart is three-dimensional, where X coordinate is time, Y coordinate is frequency, and Z coordinate is energy; then a series of maximum points are obtained from the spectrum chart, before the list of landmarks is obtained, and then the landmarks of the two music segments are compared one by one. The landmark is constructed as follows:

$$L = MAX(|X(k)|)^T \quad (2)$$

It is because the part of the chorus signal holds high energy, that energy works as also the most important evaluation standard. In the process, the larger the difference, the higher the score. This process is divided into X_1.

At the same time, the position, duration, and interval between repetitions of chorus are stable. Suppose the chorus appears in the $T_i - T_{i+1}$ time period with the probability of P_i, and appears in the duration of T_i with the probability of P_j, and will appear again after the duration of T_k with the probability of P_k (P_i, P_j, T_i, T_{i+1}, and T_j are counted Method), then if the start and end time of the possible chorus paragraph is between T_i and T_{i+1}, the duration is close to T_j, and the interval is close to the interval with the highest probability, then it is more likely to be the chorus of the song. Therefore, this item can also be used as a measurement scoring standard.

If the process score is x_2, then

$$x_2 = P_i * P_j * P_k\left(1 - \frac{|t_1 - T_j|}{T_j}\right)\left(1 - \frac{|t_2 - T_k|}{T_k}\right) \quad (3)$$

where t_1 is the duration of the possible chorus, and t_2 is the interval between possible choruses.

Overall score is:

$$\text{Mark} = w_1 * X_1 + w_2 * X_2 \quad (4)$$

w_1 is the score weight of X_1, w_2 is the score weight of X_2. This weight is assigned as the weight corresponding to the better test result in the experimental test.

3.4 Matching

The algorithm cuts out several frames of the video and performs facial expression recognition and scene recognition on these frames. After receiving facial expression labels, the algorithm will select the most frequent facial expression labels in all frames as the facial expression information of the video; for these scene information labels; the algorithm will select the scene labels with the highest accuracy as the information of the video according to the recognition accuracy of the scene labels.

The current research applies multiple classification logistic regression to classify facial expression tags and background tags. The classification result is one of the 11 preset video features: work, classic, driving, pub, traveling, morning, walking, afternoon tea, studying, night, work-out.

With the final classification of video, the music with the same label can be identified in the music set according to the classification label, and complete the work of matching background music according to the video features.

4 Experiments

4.1 Dataset

Facial Expression and Scene Recognition: CK+ [9] dataset includes 123 subjects and 593 image sequences. Places365 is a subset of Places2, with 1.8 million images per month from 365 scene categories.

Music Dataset. Using web crawler technology, 1133 songs with top popularity and free downloads are crawled from domestic platforms. There are 1056 complete documents and label attributes.

4.2 Evaluation Scheme

Facial Expression and Scene Recognition. In the experiment, CK+ dataset and Places dataset are randomly divided into the training set, verification set, and test set, and each kind of training set, verification set and test set is distributed in a partition as 8:1:1.

4.3 Baselines

This paper applies a series of classification algorithms to our model to predict the categories of facial expression tags and scene tags extracted from the video, to generate the final video features. Therefore, it forms several different models according to the classification algorithm. As shown below: MVBGMAMA multiple classification logistic regression, MVBGMAMA -Naive Bayesian, MVBGMAMA -SVM, verify the accuracy of our MVBGMAMA model for the classification of facial expression tags and scene tags extracted from video.

4.4 Parameter Settings

Neural Network. This paper takes Resnet as a reference and remains its full connection layer. ResNet50 is selected for its proper depth, to optimize the performance of Resnet50, we use the idea proposed in Aggregated Residual Transformations, which changes some hyper-parameters on the basis of ResNet-50 and gets better performance in image feature extracting.

4.5 Results and Analysis

Face Expression Recognition. For the CK+ [9] expression dataset, 327 sequence pictures with expression labels were selected as the expression dataset.

The accuracy of the training model has reached 66.24%. The recognition accuracy of happiness, anger, and sadness has reached 68.65% on average (the recognition accuracy of happiness has reached 87%), which can be used to obtain the facial expressions of video characters.

Scene Recognition. The results of taking Place365 data set as the training set and test set are shown in Table 1. This paper uses the class score averaged over 10-crops of each testing image to classify. Here it also fine-tunes the ResNet on Places365-standard, for 10 crop average it has 85.08% on the validation set and 85.07% on the test set for top-5 accuracy.

Table 1. Places365 dataset's performance in each model.

	Validation set of Places365		Test set of Places365	
	Top-1 acc.	Top-5 acc.	Top-1 acc.	Top-5 acc.
GoogleNet	53.63%	83.88%	53.59%	84.01%
VGG-16	55.24%	84.91%	55.19%	85.01%
ResNet50	54.74%	85.08%	54.65%	85.07%

Editing of Chorus' Climax. In this paper, 1056 pieces of music were re-identified, including 500 pieces of choruses and 556 pieces of non-choruses. The actual outcome is shown in Table 2.

The comprehensive accuracy rate is 77.65%, and the recall rate of the counter example is 79.6%. Therefore, it is reasonable to say that the method is effective.

Table 2. Identification of chorus parts of several music clips.

		Predicted value		
		Chorus	Non-chorus	Total
Actual value	Chorus	377	123	500
	Non-chorus	113	443	556
	Total	450	566	1056

Matching. The current study divides the video features into 11 categories. Each video can generate facial expression labels and scene labels through the model. It has conducted experiment to three classifiers with the above 80% data set, and detect the classification accuracy with 20% data set.

According to Table 3, our MVBGMAMA model has the highest accuracy in classifying video feature labels to generate video features, and it achieves a wonderful performance in matching background music according to video features.

Table 3. Accuracy under different models

	MVBGMAMA multiple classification logistic regression	MVBGMAMA -Naive Bayesian	MVBGMAMA -SVM
Dataset	83.74	77.32	72.43

The study also verifies the model through some videos on micro-video software such as TikTok. Since there are more than half of the background music that can match the micro-video, a video usually corresponds to one type of music. Therefore, first, the original micro-video is uploaded to MVBGMAMA, and then compares the background music automatically generated by MVBGMAMA with the original music. If the two are the same type of music, it is considered that the match is successful. Finally, 1054 micro-videos are verified, and 823 music & video are matched, with an accuracy of 78.08%.

Some videos (including background music) have been collected from micro-video software such as TikTok. Features of background music in the video collection are consistent with the video content. The research uses software to extract the features of the original music and the algorithm matching music. If the two music features are similar, the matching is successful. Finally, 1054 micro-videos are verified, and as a result, 823 music & video are matched, with an accuracy of 78.08%.

5 Conclusion

This paper proposes a new multi-view automatic background music recommendation method. In view of the tedious process and low efficiency of micro-video background music selection, this paper puts forward the idea and solution of automatically obtaining video feature information and matching appropriate music chorus. Based on the statistical conclusion, the existing algorithm has been improved to avoid a large number of unnecessary comparisons and conflicts, and to fill the gap in the field of automatic chorus recognition technology. The experimental results indicate that this method is not only suitable for the test data set, but also for the simulation of practical application scenarios.

Acknowledgment. This work is supported by Beijing Natural Science Foundation under Grant No. 4194086, National Natural Science Foundation of China under Grant No. 61702043, and Research Innovation Fund for College Students of Beijing University of Posts and Telecommunications.

References

1. Krizhevsky, A., Sutskever, I., Hinton, G.E.: Imagenet classification with deep convolutional neural networks. In: Advances in Neural Information Processing Systems (2012)
2. Russakovsky, O., Deng, J., Su, H., Krause, J., Satheesh, S., et al.: Imagenet large scale visual recognition challenge. Int. J. Comput. Vis. **115**(3), 211–252 (2015)
3. He, K., Zhang, X., Ren, S., Sun, J.: Deep residual learning for image recognition. In: Proceedings of the IEEE Conference on Computer Vision and Pattern Recognition, pp. 770–778 (2016)
4. Chollet, F.: Xception: deep learning with depthwise separable convolutions. In: IEEE Conference on Computer Vision and Pattern Recognition (CVPR). IEEE (2017)
5. Zhou, B.L., Lapedriza, A., Khosla, A., Oliva, A., Torralba, A.: Places: a 10 million image database for scene recognition. IEEE Trans. Pattern Anal. Mach. Intell. **40**(6), 1452–1464 (2017)
6. Szegedy, C., Liu, W., Jia, Y., Sermanet, P., Reed, S., et al.: Going deeper with convolutions. In: Proceedings of CVPR (2015)
7. Simonyan, K., Zisserman, A.: Very deep convolutional networks for large-scale image recognition. In: International Conference on Learning Representations (2014)
8. Bruce, V., Young, A.: Understanding face recognition. Br. J. Psychol. **77**(3), 305–327 (1986)
9. Lucey, P., Cohn, J.F., Kanade, T., Saragih, J., Ambadar, Z., et al.: The extended Cohn-Kanade dataset (CK+): a complete dataset for action unit and emotion-specified expression. In: IEEE Computer Society Conference on Computer Vision and Pattern Recognition Workshops (CVPRW), pp. 94–101. IEEE (2010)

YOLO_v3-Based Pulmonary Nodules Recognition System

Wenhao Deng[4], Zhiqiang Wang[1,2,3(✉)], Xiaorui Ren[2], Xusheng Zhang[2], Bing Wang[4], and Tao Yang[3]

[1] State Information Center, Beijing, China
[2] Beijing Electronic Science and Technology Institute, Beijing, China
wangzq@besti.edu.cn
[3] Key Lab of Information Network Security, Ministry of Public Security, Shanghai, China
[4] Guangdong Hangyu Satellite Technology Co., Ltd., Shantou, China

Abstract. The incidence of pneumonia and lung cancer is high in China. Lung CT images are widely used in the screening and adjuvant treatment of lung diseases due to their advantages of a thin layer, high definition, and low noise. The manual film-reading method has strong subjectivity, depends on the doctor's experience, and now hospitals produce a large number of lung CT images every day. The manual film-reading method is inefficient. The machine can help doctors to do lesion screening, auxiliary diagnosis, and treatment. Conventional machine learning methods have been applied to the recognition of pulmonary nodules in CT images. Still, these methods need to manually extract features that are not comprehensive or proper, resulting in misdiagnosis and missed diagnosis. These methods enable to rapidly detect lung nodules, which is terrible for patients to receive treatment in a timely manner. With the development of deep learning technology, Convolutional Neural Networks (CNNs) have been widely used in recognizing images, such as facial identification, character recognition, car plate recognition, etc. Its ability to solve computer vision problems has won the approval in some natural scene tasks. In this paper, the YOLO_v3 is used for feature extraction and classification, and residual network (ResNet), one of the classical CNN models, is used to decrease the false-positive rate, which improves the detection accuracy of pulmonary nodules in CT images.

Keywords: YOLO · ResNet · Medical imaging analysis · Pulmonary nodules

1 Introduction

As air quality worsens, lung cancer has become one of the most malignant tumors threatening human life. Early discovery and treatment can significantly improve the survival rate. Most of the early symptoms of lung cancer in the medical image manifest as isolated pulmonary nodules. In CT images, the pulmonary nodule usually shows as the approaching circle round or low-contrast spot. Traditionally, CT images need doctors to read, and this method is of subjectivity.

Moreover, everyday hospitals produce many medical images, which brings doctors intense pressure, which may increase the misdiagnosis and missed diagnosis. With

Q. Liu et al. (Eds.): CENet 2020, AISC 1274, pp. 11–19, 2021.
https://doi.org/10.1007/978-981-15-8462-6_2

computer technology development, it has become a trend to apply machine learning to analyze medical images. However, the traditional object detection method based on machine learning mainly relies on extracting features manually. The machine learning algorithm obtains these features and learns the feature values. However, the artificially predefined features are not comprehensive and have considerable limitations. Proposing an efficient pulmonary nodes detection method has become a research hotspot.

With the development of machine learning technology, deep learning is got more concerned by researchers. Deep Convolution Neural Network (DCNN) has a rich network structure. Its advantage is that its multi-layer structure has automatic learning, and it can learn multi-level features. To some certain extent, the deeper the convolution layer is, the more perceptual it is, and the learning of more abstract features can be realized. Compared with conventional machine learning, deep learning does not need to extract features manually but automatically. We only need to input the original image to the system, and the system can train and conduct predictive learning and get better results. DCNN has achieved great success in speech recognition and computer vision. DCNN has been used to detect pulmonary nodules, but the existing model based on deep learning has a high false-positive rate.

To solve this problem, this paper mainly studies the automatic recognition method about pulmonary nodules in CT images, and in this paper, we design and realize the recognition model of pulmonary nodules based on the YOLO_v3 and ResNet. The innovations and contributions of this paper are as follows:

(1) The pulmonary nodules in CT images are recognized by using the YOLO_v3 object detection algorithm. YOLO_v3 can recognize objects on three scales.
(2) Using ResNet to decrease the false positive rate can effectively improve the accuracy of pulmonary nodule recognition in CT images.

The other parts of this paper are organized as follows: In Sect. 2, we introduce the current research status; In Sect. 3, we introduce the structure of the pulmonary nodules recognition system designed in this paper; In Sect. 4, we conduct experiments on the pulmonary nodules recognition system and give the experimental results; and in Sect. 5, we summarize the whole paper and proposes the future research direction.

2 Research Status

At present, researchers have done a lot of research on medical image detection, especially on the application of conventional machine learning and deep learning technology in the analysis, processing, and recognition of medical images.

As for machine learning, Literature [1] uses machine learning to train CT images, designs the identification model. However, traditional machine learning needs to extract features manually, and the predefined features are not comprehensive or appropriate.

In the field of image recognition and computer vision, deep learning has achieved more and more significant results. Researchers have applied deep learning to the analysis, processing, and recognition of medical images. Literature [2] classifies pulmonary nodules by deep learning features and uses the principle of stacked generalization and sparse coding to establish the deep learning network. Features of the region of interest are extracted from the image to identify the internal feature that is most suitable for classification. Literature [3] analyzes the ability of the deep structural algorithm to automatically extract features in the CT image and compares its performance with that of traditional manual feature extraction computer-aided diagnosis system. Three kinds of deep structure algorithms based on multi-channel ROI are designed and implemented: convolution neural network, depth belief network, and overlaying denoising autoencoder. Experimental results show that the deep structure algorithm based on automatic feature extraction can achieve better results in the diagnosis of pulmonary nodules. A lung tumor detection method based on deep learning is proposed in Literature [4]. CNN is used to extract the features of the lung tumor image of patients, and the possible position of the tumor in the image is predicted by combining with the region proposal network. At the same time, the suggestion box is generated. The objective area is classified, and the position of the suggestion box is fine-tuned by using the learned features. This method does not need to design the target features manually.

Literature [5] uses bioinformatics and CNN in the field of machine learning algorithms to analyze medical images, the basic mathematical model is constructed by utilizing a convolution neural network algorithm, and the mathematical model is realized by Python2.7. Literature [6] applies deep convolutional neural networks to the recognition of pulmonary nodules, and this structure can learn features representation of lung nodules layer by layer and train the adaptive reverse transmission neural network. This structure fuses the decisions of three classifiers to identify nodules. In the Literature [7], VGG-16, the classic deep convolution neural network, is used to extract the features of brain MRI images by using the method of transfer learning. Finally, the extracted features are classified by the support vector machine. In Literature [8], faster R-CNN is used for pulmonary nodule recognition, which successfully solves the problem of subjective features extraction relying on doctors' experience and obtains more objective and accurate results. Literature [9] proposes a pulmonary nodule recognition system based on deep learning. According to the three-dimensional feature of the CT image, they use 3D Faster R-CNN to extract features and predict candidate nodules; and then 3D CNN is used to remove false-positive nodules position. Experiments are carried out on LUNA. Compared with traditional diagnosis and treatment methods, the correct identification rate is significantly improved, but the recognition speed is slow. Literature [10] proposes to use the YOLO_v2 to identify pulmonary nodules, but it is hard to recognize small objects for YOLO_v2, and the precision of identifying pulmonary nodules in CT images will be decreased.

The above research verifies the advantages of deep learning in medical image processing, analysis, and recognition. It also verifies the high precision of deep learning in medical image processing but fails to resolve problems of high false-positive rate and slow recognition speed at the same time. To solve this problem, this paper designs a lung nodule recognition model based on YOLO v3 and ResNet.

3 The Structure of the Recognition System

In this section, we propose an objective recognition system based on YOLO_v3 [11, 12] and ResNet [13].

The system accepts the CT image of the lung as the input and identifies the position of the pulmonary nodule preliminarily through the YOLO_v3 object recognition algorithm, and then use ResNet to reduce the false-positive effect caused by YOLO_v3. The system model is shown in Fig. 1.

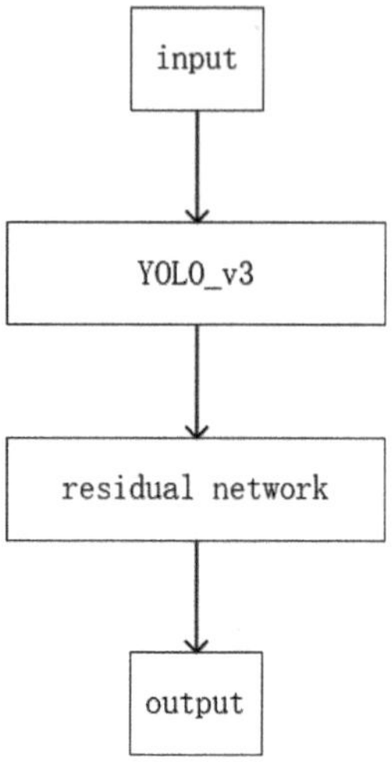

Fig. 1. The system model.

3.1 YOLO_v3

At present, in the field of object recognization, two kinds of deep learning methods are used. One is the two-stage object recognition algorithm. This algorithm needs to extract candidate boxes first and then classifies them by CNN, such as R-CNN; the other is a stage object detection algorithm, which does not need to generate candidate boxes. It directly converts the target border location into the regression problem, such as the YOLO. Compared with the two-stage target detection algorithm, one stage target detection algorithm has faster computing speed.

Since proposed, YOLO has been improved to the third version. Compared to previous versions, YOLO_v3 contains a more efficient classifier—darknet-53 without the full connection layer. Darknet-53 has higher classification accuracy and faster computation speed. And YOLO_v3 output predictions in three scales. It can recognize objects with different sizes. The structure of YOLO_v3 is shown in Fig. 2.

The main component of YOLO_v3 is the first 52 layers of darknet-53, the component is used for features extraction and use extensive skip-connections, with a precision similar to that of ResNet-152 and ResNet-101. Due to the small number of network layers, the computation speed is faster than that of ResNet-152 and ResNet-101.

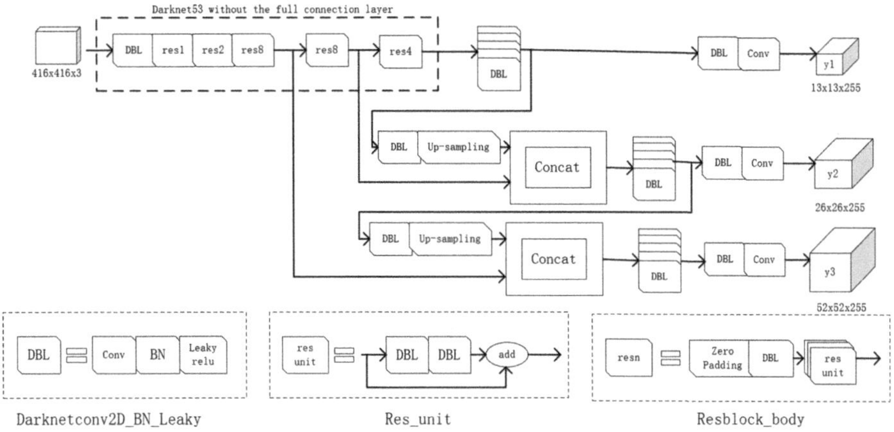

Fig. 2. The structure of YOLO_v3.

After five times of subsampling, a 13 × 13 feature map is generated, and then a 13 × 13 × 255 feature map is generated by DBLs and the full convolution layer, which is the output on the first scale y1. After four times of subsampling in darknet-53, tensor splicing was carried out with the 13 × 13 feature map extracted by darknet-53 and DBLs to produce a 26 × 26 feature map. The subsequent procedure is the same as the generation of y1, and then the output y2 on the second scale was output. The generation of y3 is similar to that of y2. YOLO_v3 produces feature maps on three different scales that can be adapted to targets of different sizes.

DBL is the basic component of YOLO_v3, including a convolution layer, a BN, and a Leaky_relu. Res unit is responsible for adding up the values extracted with DBL and output the results. Resn is a large component in YOLO_v3, which refers to the residual structure of ResNet. Darknet-19 in YOLO_v2 is improved to darknet-53 to make the result converge. The specific content of ResNet is introduced in the second part of this section.

3.2 ResNet

After the prediction results are generated by YOLO_v3, the ResNet is used for false-positive screening to obtain better accuracy.

In theory, the deeper the network level is, the higher the training accuracy is. But in fact, with the increase of network layers, the training errors will first decrease, then increase. The neural network optimized by residual structure is used to establish a direct correlation channel between input and output, so that the parametric layer can use more resources to learn the residual between input and output, and help resolve the vanishing gradient problem in the plain neural network, so as to decrease false-positive rate. The structure of ResNet is shown in Fig. 3.

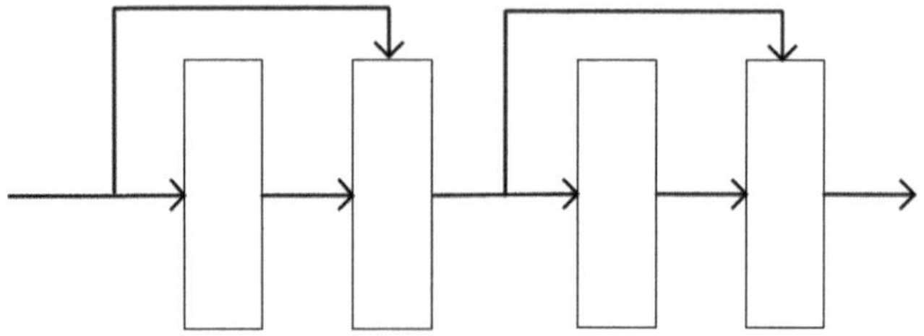

Fig. 3. The structure of ResNet.

4 Experiment and Evaluation

4.1 The Process of Objective Recognition

The pulmonary nodules detection system designed in this paper can be roughly divided into three stages: image pre-processing, preliminary recognition of pulmonary nodule positions, and false-positive reduction.

It is implemented in Python and adopts the deep learning framework Keras based on Theano. In the image pre-processing stage, we use a script to process the CT image, and the output of the script processing is a PNG image. In the stage of preliminary recognition, the positions of lung nodules are preliminarily located by the YOLO_v3, and then the system uses the ResNet to eliminate the false position of lung nodule produced in the second stage, and more accurate results are produced. And an HDF file is produced by training the model. And then, we test the system by repeating the above three stages. Finally, we get an excel file, displaying the coordinates of pulmonary nodules in each test image.

4.2 The Result and Evaluation

The model in this paper combinates YOLO_v3 with ResNet, and adopts the LUNA16 with 888 CT images, which is a subset of the largest public lung nodule data set LIDC-IDRI. The slice thickness of the CT image in LUNA16 is 3 mm, which can achieve a better recognition effect. We divide these image files into a train set and a test set. The images in the train set is used to train the model designed in this paper, and files in the test set is used to test the accuracy of the model after training. We use 800 CT images to train the model. The performance of this system is compared with that of previous methods in recognizing pulmonary nodules. The results are shown in Table 1.

Table 1. Comparison results

Model	Speed	Accuracy	Method of features extraction
The conventional machine learning methods	Slow	Slow	Manual
R-CNN-based pulmonary nodules recognition system	Higher	Higher	Automatic
YOLO_v3-based pulmonary nodules recognition system	Highest	Highest	Automatic

5 Summarizes

Facing the situation of the explosion of medical image data, doctors are under increasing pressure to read CT images. At present, there are machine reading methods to relieve doctors' burdens. In this paper, we analyze the related research status at home and abroad about the recognition of pulmonary nodules in CT images. We analyze the advantages and disadvantages of the method based on the traditional machine learning method and the method based on deep learning.

The traditional machine learning method needs to extract features manually, which has subjectivity and low accuracy. At present, the two-stage algorithm—such as fast R-CNN and R-FCN—has been used in object recognition based on deep learning, but the speed of this method to recognition is slow. This paper proposes a pulmonary nodules recognition system based on YOLO_v3. YOLO_v3 is used to locate positions of pulmonary nodules in this system, and the recognition speed is faster. By using ResNet to decrease the rate of false-positive, the accuracy of pulmonary nodule prediction is improved. By experiment, we verify that the design has higher accuracy, which is helpful in improving the efficiency of lung disease diagnosis.

Although the location of pulmonary nodules in CT images can be recognized rapidly by the system, there is no research on the classification of pulmonary lesions. In the future, we will focus on the classification of lung lesions, so as to further reduce the burdens of doctors, improve the diagnosis speed of lung diseases, and buy more time for patients.

Acknowledgment. This research was financially supported by the National Key Research and Development Plan (2018YFB1004101), Key Lab of Information Network Security, Ministry of Public Security (C19614), Special fund on education and teaching reform of Besti (jy201805), the Fundamental Research Funds for the Central Universities (328201910), China Postdoctoral Science Foundation (2019M650606), 2019 Beijing Common Construction Project-Teaching Reform and Innovation Project for Universities in Beijing, key laboratory of network assessment technology of Institute of Information Engineering, Chinese Academy of Sciences. The authors gratefully acknowledge the anonymous reviewers for their valuable comments.

Author Contributions. Wenhao Deng and Zhiqiang Wang conceived and designed the framework and the algorithm; Xiaorui Ren and Xusheng Zhang performed the experiments; Bing Wang analyzed the data; Tao Yang wrote the paper.

Conflicts of Interest. The authors declare no conflict of interest.

References

1. Liu, Y., Cheng, Y.Y., He, R.M., et al.: A machine learning-based model for the differentiation of metastatic lymph nodes in nasopharyngeal carcinoma. Chin. J. Med. Phys.
2. Jia, T., Zhang, H., Bai, Y.K.: Benign and malignant lung nodule classification based on deep learning feature. J. Med. Imaging Health Inf. **5**(8), 1936–1940 (2015)
3. Sun, W.Q., Zheng, B., Qian, W.: Automatic feature learning using multi-channel ROI based on deep structured algorithms for computerized lung cancer diagnosis. Comput. Biol. Med. **89**, 530–539 (2017)
4. Chen, Q.R., Xie, S.P.: A lung tumor detection based on deep learning. Comput. Technol. Dev. **28**(4), 201–204 (2018)
5. Li, L.S., Huang, D., Sun, H.Y., et al.: Accurate analysis of medical images of breast cancer based on convolutional neural network algorithm, no. 4, pp. 28–29 (2018). https://doi.org/10.3969/j.issn.1672-528x.2018.04.027
6. Xie, Y.T., Zhang, J.P., Xia, Y., et al.: Fusing texture, shape, and deep model-learned information at decision level for automated classification of lung nodules on chest CT. Inf. Fusion **42**, 102–110 (2018)
7. Zhang, S., Zhang, R.: A feature extraction method of medical image based on deep learning. J. Taiyuan Univ. (Nat. Sci. Ed.) **37**(3), 69–73 (2019). https://doi.org/10.14152/j.cnki.2096-191x.2019.03.014
8. Tiangong University. A detection of pulmonary nodules based on Reception and Faster R-CNN: CN201910220570.4, 28 June 2019
9. Liu, D., Wang, Y., Xu, H.: Deep learning-based medical images pulmonary nodules recognition system. Microelectron. Comput. **36**(5), 5–9 (2019)
10. Li, X.Z., Jin, W., Li, G., et al.: Asymmetric convolutional nucleus YOLO V2-based CT images pulmonary nodules recognition system. Chin. J. Biomed. Eng. **38**(4), 401–408 (2019). https://doi.org/10.3969/j.issn.0258-8021.2019.04.003
11. Wu, Y.L., Zhang, D.X.: A review of target detection algorithms based on deep learning. China Comput. Commun. (12), 46–48 (2019)
12. Lin, J.W.: A review of YOLO image detection technology. J. Fujian Comput. **35**(9), 80–83 (2019). https://doi.org/10.16707/j.cnki.fjpc.2019.09.026
13. Wang, Z., He, W.: Application of deep residual network in diagnosis of pneumoconiosis. Chin. J. Ind. Med. (1), 31–33 (2019)
14. Weng, S., Xu, X., Li, J., Wong, S.T.: Combining deep learning and coherent anti-Stokes Raman scattering imaging for automated differential diagnosis of lung cancer. J. Biomed. Opt. **22**(10), 106017 (2017)
15. Wang, C.M., Elazab, A., Wu, J.H., et al.: Lung nodule classification using deep feature fusion in chest radiography. Comput. Med. Imaging Graph.: Off. J. Comput. Med. Imaging Soc. **57**, 10–18 (2017)
16. van Tulder, G., de Bruijne, M.: Combining generative and discriminative representation learning for lung CT analysis with convolutional restricted boltzmann machines. IEEE Trans. Med. Imaging **35**(5), 1262–1272 (2016)
17. Li, C.H., Zhang, H., Shen, H.J.: Application of K-means unsupervised machine learning algorithm in Cardiac CT Image segmentation. Comput. Knowl. Technol. **15**(1), 212–213 (2019)
18. Beijing University of Technology: a method for detecting pulmonary nodules in medical images based on machine learning: CN201810352482.5 (2018)

19. Xue, C.Q., Liu, X.W., Deng, J., et al.: Advances of deep learning in medical imaging of brain tumors. Chin. J. Med. Imaging Technol. **35**(12), 1813–1816 (2019). https://doi.org/10.13929/j.1003-3289.201904061
20. Dai, J.H.: Application of deep learning in medical Image analysis. Digital Space (1), 32 (2020)
21. Xi, Z.H., Hou, C.Y., Yuan, K.P.: Residual network-based medical image super-resolution reconstruction. Comput. Eng. Appl. **55**(19), 191–197 (2019). https://doi.org/10.3778/j.issn.1002-8331.1806-0243
22. Yolo v3 in yolo series [deep analysis]. https://blog.csdn.net/leviopku/article/details/82660381

An Optimized Hybrid Clustering Algorithm for Mixed Data: Application to Customer Segmentation of Table Grapes in China

Yue Li[1], Xiaoquan Chu[1], Xin Mou[1], Dong Tian[1], Jianying Feng[1], and Weisong Mu[1,2(✉)]

[1] College of Information and Electrical Engineering, China Agricultural University, Beijing 100083, China
wsmu@cau.edu.cn
[2] Key Laboratory of Viticulture and Enology, Ministry of Agriculture, Beijing, China

Abstract. Customer segmentation based on mixed variables is an important research direction of current market segmentation methods. However, using the hybrid clustering method to divide consumer groups will overestimate the clustering contribution of categorical variables and shield numerical variables that have important marketing significance. In this study, we improve the hybrid clustering algorithm and design an indicator of assessing the customer groups' differences. Firstly, the coefficient of Features Discrepancy between Segments (FDS) and three variable weighting strategies are designed. Then, the optimal optimization scheme is determined based on three hybrid clustering algorithms and evaluation indicators. The results show that the variability weighting method can effectively improve the problem that categorical variables dominate hybrid clustering. The clustering performance and stability of PAM method are the best among the three hybrid clustering algorithms. Finally, clustering the consumer consumption values and the basic population information through the weighted PAM method to verify this method's effectiveness. This study provides a practical application value for the improvement of existing technologies in customer segmentation methods. It also offers the marketing suggestions for the table grape operators.

Keywords: Data mining · Marketing research · Customer segmentation · Balance mixed data · Table grapes

1 Introduction

Consumers' preferences and purchasing behaviors for table grapes reflect the table grape market [1–3]. To improve market competitiveness, enterprises must understand consumers' purchasing intentions [4]. Taking an effective customer segmentation method to provide different products and service solutions for different customer groups, which has important theoretical significance and application value [5, 6]. Therefore, establishing the customer-centered customer segmentation method of table grapes will become the strategic need for agricultural product customer management.

Q. Liu et al. (Eds.): CENet 2020, AISC 1274, pp. 20–32, 2021.
https://doi.org/10.1007/978-981-15-8462-6_3

In the market segmentation, clustering analysis is widely used to segment the customer groups. Due to the segmentation variables' diversity, there is no optimal clustering algorithm for all kinds of business scenarios [7, 8]. Besides the influence of clustering algorithms, data types also impact customer segmentation [9, 10]. Researchers usually use a single variable type when they select the segmentation variables. It will not be accurate enough to describe the customer groups. Nowadays, customer segmentation research based on mixed data is a practical value direction [11, 12]. The hybrid segmentation methods usually involve different segmentation criteria, which contribute to describing more certain customer groups [13, 14].

The difficulty of hybrid clustering algorithms is to balance the feature differences of mixed data. At present, there are many studies on the dissimilarity and weights of mixed variables [15–17]. For example, Modha calculated the weights based on the distortion of different variable features and proposed the weighted K-means algorithm [18]. Although only numerical variables were used in his research, weighting based on the distortion can be referenced in the hybrid clustering algorithms. Sangam proposed a new hybrid dissimilarity coefficient for the K-prototypes algorithm. He used the weighted Hamming distance to calculate categorical variables [19]. Diez encoded categorical variables into numerical variables by using the polar and spherical coordinates [20]. Foss used the KAMILA (KAy-means for MIxed LArge data) method to balance the numerical and categorical variables' contributions by using the Gaussian polynomials [21]. In these studies, partition-based clustering algorithms are the most widely-used [22–24]. However, it is still a challenge to find a suitable dissimilarity measure to deal with mixed data.

At present, the current literature on customer consumption behaviors and values of scale-based data is not mature. The research on customer segmentation of the grape industry even lacks in China. In this paper, we optimize the hybrid clustering algorithm and apply it to table grapes' customer segmentation. Firstly, the weighting method for feature variables is proposed. The aim is to make all segmentation variables present the differences in the central characteristics. Secondly, we design the cluster evaluation indicator, which plays an essential auxiliary role in measuring the optimal number of clusters. Finally, a customer group of table grapes is divided into several smaller groups based on the weighted PAM clustering algorithm. The method proposed in this paper can effectively balance the contribution of all clustering variables. It also provides the marketing references for table grape manufacturers.

2 Related Works

2.1 Dissimilarity of Mixed Data

Suppose there are n objects in the sample data set, and every object has p attributes, denoted as $x_1 = (x_{11}, x_{12}, \ldots, x_{1p}), \ldots, x_n = (x_{n1}, x_{n2}, \ldots, x_{np})$. Where x_{ij} is the value of object x_i in the jth attribute. The set of x_i $(i = 1, 2, \ldots, n)$ is called the data matrix. The matrix formed by the distance between the objects is the dissimilarity matrix [25]. In this paper, the Gower distance is used to calculate the dissimilarity of mixed data.

(1) Numerical variable: The Manhattan method (Eq. (1)) is used to calculate the dissimilarity when all numerical variables are normalized.

$$d_{ij} = \sum_{k=1}^{p} \left| x_{ik} - x_{jk} \right| \tag{1}$$

where d_{ij} represents the distance between i and j objects.

(2) Ordered categorical variable: Assuming that the variable has M states, a sort 1, 2, …, M is defined for the ordered categorical. The ordered categorical variables are transformed into the numerical attributes, and the normalized values are used as numerical variables to calculate the dissimilarity. The data is normalized to $(k - 1)/(M - 1)$, where k is the ordinal order (k = 1, 2, 3, …, M).

(3) Nominal categorical variable: The Dice method is used to calculate the dissimilarity, denoted as:

$$d_{ij} = \frac{2\left| x_i \cup x_j \right|}{\left| x_i \right| + \left| x_j \right|} \tag{2}$$

where $\left| x_i \cup x_j \right|$ represents the number of different categories in x_i and x_j, $|x_i|$ and $|x_j|$ are the lengths of x_i and x_j, respectively.

2.2 Feature Extraction of Mixed Data

To prevent information from overlapping in the variable features, Exploratory Factor Analysis (EFA) method is used to extract the features from the numerical data. It is a common method for the feature extraction of scale questions, which can objectively determine the comprehensive indicators [26, 27]. Figure 1 is the factor load map of original numerical variables. Table 1 shows the final segmentation variables.

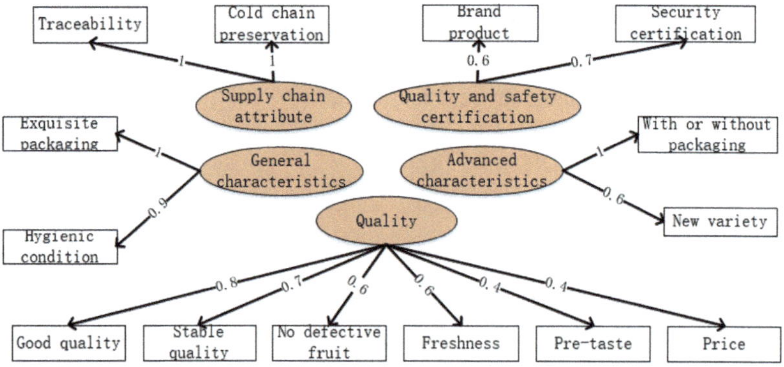

Fig. 1. The factor load map of original clustering variables.

Table 1. The final clustering variables for customer segmentation.

Number	Variables	Data type
a	Quality	Numerical
b	Supply chain attribute	Numerical
c	Advanced characteristics	Numerical
d	General characteristics	Numerical
e	Quality and safety certification	Numerical
f	City	Ordered-categorical
g	Gender	Nominal-categorical
h	Age	Ordered-categorical
i	Education	Ordered-categorical
j	Occupation	Nominal-categorical
k	Income	Ordered-categorical
l	Family size	Ordered-categorical

2.3 Hybrid Clustering Algorithms

In this paper, we choose three clustering algorithms to determine the method that is suitable for this study, including the partition-based (Partitioning Around Medoids, PAM), hierarchical-based (AGglomerative NESting, AGNES), and density-based (Adaptive Density Peak Clustering, ADP) clustering algorithms [28, 29].

PAM Algorithm Description

Algorithm 1: PAM

Input: The number of clusters k and the data set containing n objects.
Output: The data with cluster labels.
1: Select k objects arbitrarily as the initial cluster centers;
2: **Repeat** Every object is assigned to the cluster represented by its nearest center according to the dissimilarity matrix;
3: **Repeat** Select an unselected center O_i;
4: **Repeat** Select an unselected non-center O_h;
5: Calculate the total cost of replacing O_i with O_h and record it in S;
6: **Until** All non-centers are selected;
7: **Until** All centers are selected;
8: **If** S is smaller, then O_h replaces O_i;
9: **Until** The clusters are not redistributed.

AGNES Algorithm Description

Algorithm 2: AGNES

Input: The number of end condition clusters k and the data set containing n objects.
Output: The data with cluster labels.
1: Every object is treated as the initial cluster;
2: **Repeat**
3: Merge the clusters according to the closest data in two clusters;
4: **Until** The number of defined clusters is reached.

ADP Algorithm Description. ADP is a heuristic clustering algorithm that can identify the cluster allocation from the estimated local density. It is based on the assumption that the cluster center is surrounded by neighbor points with relatively low local density, and has relatively large distance from other data points with higher local density. Therefore, ADP can automatically find the cluster centers from the two-dimensional decision map. The parameters include the local density $f(x)$ (Eq. (3)), and the distance from data points with higher density (Eq. (4)).

$$\hat{f}(x_i; h_1, \ldots, h_p) = n^{-1}(\Pi_{l=1}^{p} h_l)^{-1} \cdot \sum_{j=1}^{n} K(\frac{x_{i1} - x_{j1}}{h_1}, \ldots, \frac{x_{ip} - x_{jp}}{h_p}) \tag{3}$$

$$\hat{\delta}(x_i) = \min_{j:\hat{f}(x_i) < \hat{f}(x_j)} d(x_i, x_j) \tag{4}$$

where the data set is expressed as $\{x_1, x_2, \ldots, x_n\}$, x_i is the p-dimension vector. h is the adjustment parameter. K is the number of clusters. $d(x_i, x_j)$ is the distance between x_i and x_j. The cluster centers are selected by the decision map $(\hat{f}(x_i), \hat{\delta}(x_i))$. Each of the remaining data points is assigned to the cluster to which the closest data point with higher density.

3 The Proposed Method

3.1 Design of the Evaluation Indicator

In the hybrid clustering algorithms, the distortion of categorical variables is generally greater than numerical variables. Although the data is normalized, categorical variables still have the higher potential weights, resulting in the data dissimilarity will be dominated by categorical variables. In our study, we design an indicator that can balance the feature distortion of mixed variables—the coefficient of Features Discrepancy between Segments (FDS). FDS can quantify the difference of each segmentation variable in each customer group. The construction process of FDS indicator is shown in Fig. 2, and FDS is described as follows:

Input: The labeled data and the number of clusters k.
Output: FDS value for each segmentation variable.

For each segmentation variable X_i
If X_i is the numerical variable:
Count the average ($\overline{X}_{ik}$) of X_i in each cluster k;
Normalized: $\vec{X}'_{ik} = \vec{X}_{ik} / \max \vec{X}_i$;
The variance of $\vec{X}'_i$ as the FDS value of X_i.
Else (X_i is the categorical variable):
For each category X_{ij} (j is the number of categories of variable X_i)
Calculate the frequency (count(X_{ijk})) of X_{ij} in each cluster k;
Normalized: $\overrightarrow{\text{count}}'(X_{ijk}) = \overrightarrow{\text{count}}(X_{ijk}) / \max \overrightarrow{\text{count}}(X_{ij})$;
The variance of $\overrightarrow{\text{count}}'(X_{ij})$ as the FDS value of X_{ij}.
End
The average of all variances as the FDS value of X_i.
End

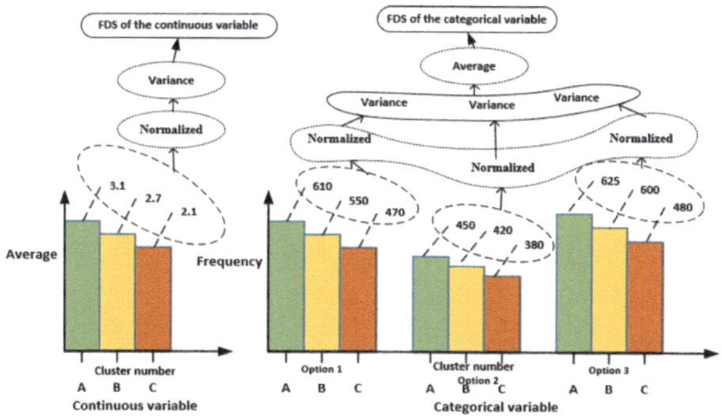

Fig. 2. The construction process of the FDS indicator.

3.2 Design of Variable Weighting Methods

Based on the Clustering Results. In order to show the difference of each segmentation variable, the spatial scale of variable with small distortion is enlarged, and the spatial scale of variable with large distortion is reduced. The most direct approach is as follows: Firstly, the clustering results are obtained by using the unweighted hybrid clustering algorithm. Then the reciprocal of FDS value is used as the variable weight. This method is an improved scheme for controlling the clustering process based on the clustering results, it called the "FDS method".

Based on the Variability. To uniformly quantify the variability of mixed variables, the Gower dissimilarity of each variable is used to measure the variance. Firstly, we obtain the dissimilarity matrix by calculating the Gower distance of each segmentation variable. Then, the variance of dissimilarity matrix is used to represent the variable variability. The reciprocal of variance is used as the variable weight. This method uses the idea of the variability in the variables, it called the "variability method".

Based on the Information Entropy. In addition to the above two methods, the weighting method based on the information entropy (it called the "entropy weight method") is widely-used for balancing the weights of mixed variables. It avoids the subjectivity of calculating weights and provides the objective basis for the comprehensive evaluation of multiple indicators [30, 31]. The entropy weight method is used to obtain the information entropy of each variable. The smaller information entropy is, the greater weight is.

4 Evaluation and Selection of Hybrid Clustering Algorithms

4.1 Evaluation of Different Weighting Methods

In order to demonstrate the effects of three weighting methods on the hybrid clustering, the segmentation variables are recombined to construct 3937 mixed data sets. PAM algorithm is used to explore the effects of three weighting schemes. Figure 3 shows the frequency distribution of FDS values. In Fig. 3, the frequency histogram is used as the basic object. The dotted line (FDS mean value) is used as the reference for the comparison of three weighting methods.

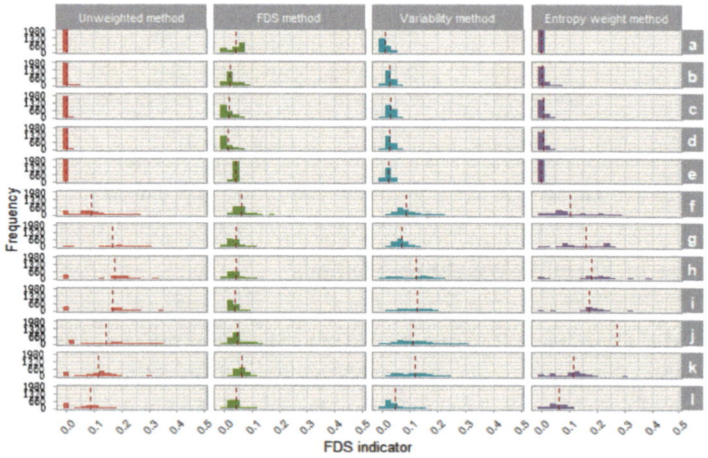

Fig. 3. The distribution of FDS values in the clustering results.

Figure 3 shows that the FDS distribution obtained through the entropy weight method is basically similar to the unweighted method. The improvement effects about the FDS and variability methods are similar in numerical variables. While in categorical variables, FDS values of the variability method are generally larger than the FDS method. The aim of this paper is to make differences in numerical variables more obvious. The variability method does not lose too much differences in categorical variables while increasing the differences in numerical variables. In summary, the variability method is the best for the hybrid clustering.

4.2 Selection of Hybrid Clustering Algorithms

To compare the performance of the three hybrid clustering algorithms, the contour coefficient and FDS indicators are used to evaluate the clustering results. In the customer segmentation, the contour coefficient can also be used to select the optimal number of clusters. The clustering results are displayed from two aspects: the relationship between the contour coefficient and the number of clusters (Fig. 4), the relationship between the FDS and the number of clusters (Fig. 5).

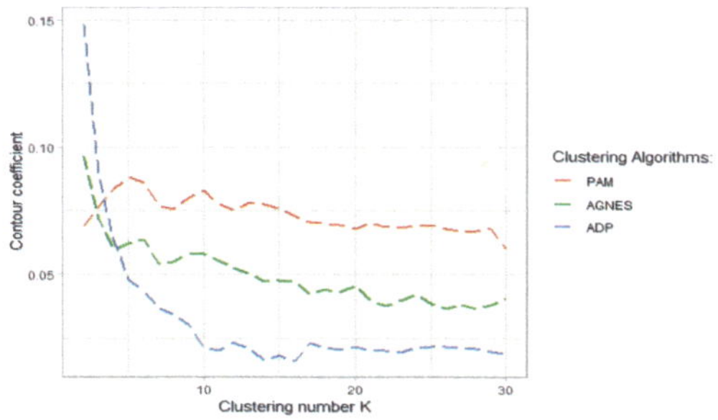

Fig. 4. Contour coefficients of three clustering methods on the number of clusters.

Figure 4 shows that when the number of clusters is 2, ADP method has the largest contour coefficient. As the number of clusters increases, the contour coefficient of ADP method decreases rapidly. Although PAM method cannot reach the maximum contour coefficient, it has the largest contour coefficient when the number of clusters is 5. PAM method can maintain a basically stable decline compared with other two algorithms. Figure 5 shows that FDS values of numerical variables increase generally with the number of clusters, and FDS values of categorical variables decrease generally. ADP method has an unstable FDS trend. FDS values of PAM and AGNES methods are relatively stable. In summary, PAM is the optimal hybrid clustering algorithm.

5 Case Study of the Optimized Hybrid Clustering Algorithm

5.1 Comparison of the Optimized Hybrid Clustering Algorithm

The experimental data used in this paper is from the questionnaire "2017 Table Grape Consumption Market Survey in China". The consumption values and consumer

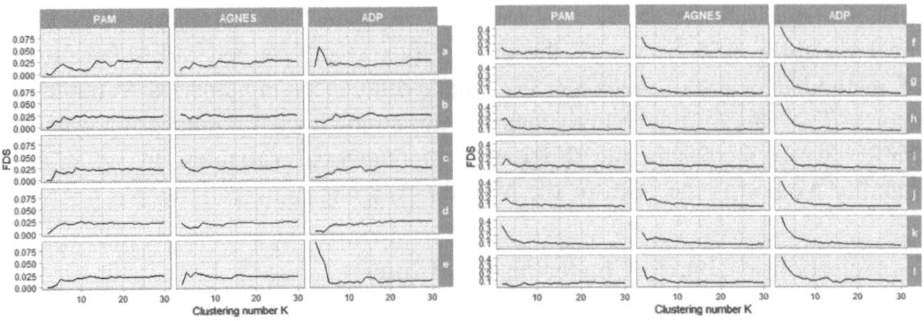

Fig. 5. FDS values of all variables on the number of clusters.

demographic characteristics are used as the mixed data for grape customer segmentation. To compare the clustering effects of PAM algorithm before and after optimization, the clustering results are evaluated by using the contour coefficient and FDS indicators. Figure 6 shows the relationship between the contour coefficient and the number of clusters. Figure 7 shows the relationship between the FDS and the number of clusters.

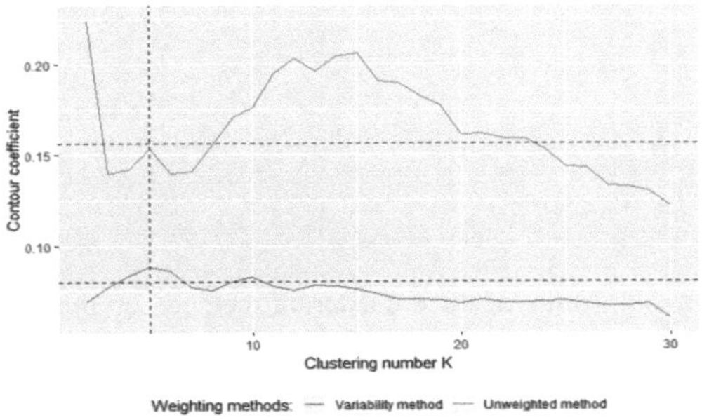

Fig. 6. The relationship between the contour coefficient and the number of clusters.

Figure 6 shows that the unweighted method has the largest contour coefficient when the number of clusters is 2. The fluctuation of the contour coefficient is unstable. However, the optimal number of clusters is 5 determined by the variability method. As the number of clusters increases, the trend of the contour coefficient decreases steadily. In Fig. 7, when the number of clusters is 2, only the variable *g* has a higher FDS by using the unweighted method. Therefore, it may be misleading to use the contour coefficient as an indicator for the unweighted hybrid clustering algorithm. By comparing Fig. 6 and Fig. 7, in the case of balancing the weights, we can obtain the consistent optimal cluster numbers by using the contour coefficient and the FDS. FDS indicator plays an important auxiliary role in selecting the optimal number of clusters.

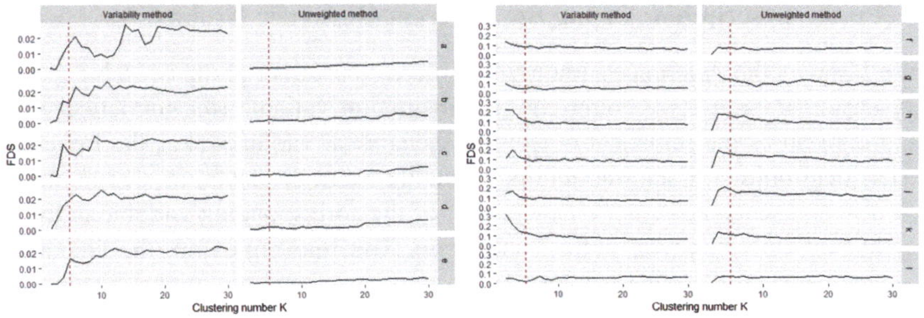

Fig. 7. The relationship between the FDS and the number of clusters.

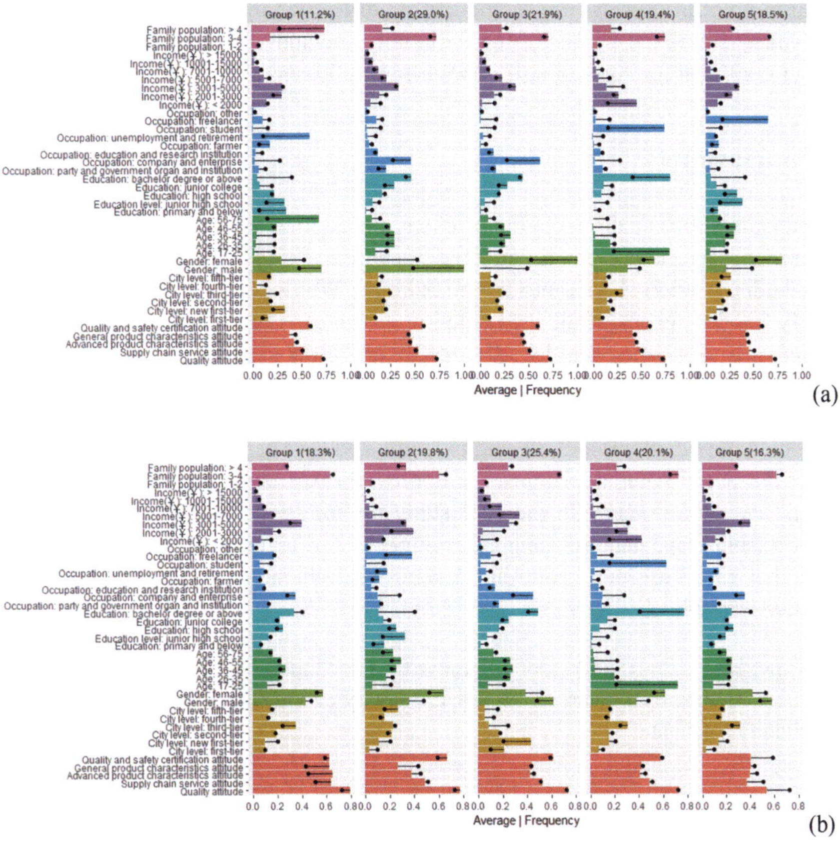

Fig. 8. Comparison of the differences in segmentation variables: (a) Before optimization; (b) After optimization.

5.2 Profile Description Among the Different Customer Groups

Figure 8 shows the differences among the different customer groups of table grapes. In every customer group, we count each segmentation variable (the numerical variable takes the mean as the statistical object, the categorical variable takes the frequency as the statistical object). Each variable is normalized and displayed in a histogram. The black dot represents the center characteristic of the corresponding variable in the grape retail market (the surveyed grape customers). The line represents the characteristic difference between the obtained customer group center and the market center.

From Fig. 8, we can understand the differences between each customer group and the grape market. We can also describe the differences among the different customer groups. Figure 8(a) indicates that each numerical center in each customer group differs very little from the grape market, so it does not satisfy the aim of the hybrid clustering algorithm optimization. However, the central difference of each numerical variable is significantly improved by using the variability method (Fig. 8(b)). Categorical variables in each customer group still remain high differences.

6 Conclusions

This paper explores solutions to the problem of customer segmentation for mixed data. The numerical variables are constructed based on the EFA method, which are integrated with the original data into the mixed data. We first design the FDS indicator to quantify the differences of all segmentation variables. Then, three weighting methods are proposed to balance the distortion of mixed variables. Finally, FDS and contour coefficient are used to evaluate the clustering performance of the three algorithms. The clustering results show that the variability weighting method can effectively improve the problem that the hybrid clustering is dominated by categorical variables. Compared with AGNES and ADP algorithms, the improved PAM algorithm has better clustering stability and performance. In the case study section, we divide the customer groups based on table grape data by using the weighted PAM algorithm. The clustering results ensure that each customer group has a significant characteristic difference in all segmentation variables.

Acknowledgment. This work was supported by the Chinese Agricultural Research System (CARS-29); and the open funds of the Key Laboratory of Viticulture and Enology, Ministry of Agriculture, PR China.

References

1. Feng, J., Wang, X., Fu, Z., Mu, W.: Assessment of consumers' perception and cognition toward table grape consumption in China. Br. Food J. **116**(4), 611–628 (2014)
2. Behroozi-Khazaei, N., Maleki, M.R.: A robust algorithm based on color features for grape cluster segmentation. Comput. Electron. Agric. **142**(Part A), 41–49 (2017)

3. Luo, L., Tang, Y., Chen, X., Zhang, P., Lu, Q., et al.: A vision methodology for harvesting robot to detect cutting points on peduncles of double overlapping grape clusters in a vineyard. Comput. Ind. **99**, 130–139 (2018)
4. Yeung, R., Morris, J., Yee, W.: The effects of risk-reducing strategies on consumer perceived risk and on purchase likelihood a modelling approach. Br. Food J. **112**(2–3), 306–322 (2010)
5. Francis, I.L., Williamson, P.O.: Application of consumer sensory science in wine research. Aust. J. Grape Wine Res. **21**, 554–567 (2015)
6. Vanneschi, L., Popovic, A., Castelli, M., Horn, D.M.: An artificial intelligence system for predicting customer default in e-commerce. Expert Syst. Appl. **104**, 1–21 (2018)
7. Morris, K., McNicholas, P.D.: Clustering, classification, discriminant analysis, and dimension reduction via generalized hyperbolic mixtures. Comput. Stat. Data Anal. **97**, 133–150 (2016)
8. Lin, T.C., Zulvia, F.E., Tsai, C.Y., Kuo, R.J.: A hybrid metaheuristic and kernel intuitionistic fuzzy c-means algorithm for cluster analysis. Appl. Soft Comput. **67**, 299–308 (2018)
9. Ji, J., Bai, T., Zhou, C., Ma, C., Wang, Z.: An improved k-prototypes clustering algorithm for mixed numeric and categorical data. Neurocomputing **120**, 590–596 (2013)
10. Chiewchanwattana, S., Wangchamhan, T., Sunat, K.: Efficient algorithms based on the k-means and chaotic league championship algorithm for numeric, categorical, and mixed-type data clustering. Expert Syst. Appl. **90**, 146–167 (2017)
11. Sağlam, B., Salman, F.S., Sayın, S., Türkay, M.: A mixed-integer programming approach to the clustering problem with an application in customer segmentation. Eur. J. Oper. Res. **173**(3), 866–879 (2006)
12. McNicholas, P.D., Browne, R.P.: Model-based clustering, classification, and discriminant analysis of data with mixed type. J. Stat. Plann. Infer. **142**(11), 2976–2984 (2012)
13. Ahmad, A., Dey, L.: A k-means type clustering algorithm for subspace clustering of mixed numeric and categorical datasets. Pattern Recogn. Lett. **32**(7), 1062–1069 (2011)
14. Cankaya, I., Kocak, O.M., Aljobouri, H.K., Algin, O., Jaber, H.A.: Clustering fMRI data with a robust unsupervised learning algorithm for neuroscience data mining. J. Neurosci. Methods **299**, 45–54 (2018)
15. Del Coso, C., Fustes, D., Dafonte, C., Novoa, F.J., Rodriguez-Pedreira, J.M., et al.: Mixing numerical and categorical data in a self-organizing map by means of frequency neurons. Appl. Soft Comput. **36**, 246–254 (2015)
16. Amiri, L., Khazaei, M., Ganjali, M.: Mixtures of general location model with factor analyzer covariance structure for clustering mixed type data. J. Appl. Stat. **46**(11), 2075–2100 (2019)
17. de Amorim, R.C., Makarenkov, V.: Applying subclustering and L-p distance in Weighted K-Means with distributed centroids. Neurocomputing **173**(Part 3), 700–707 (2016)
18. Modha, D.S., Spangler, W.S.: Feature weighting in k-means clustering. Mach. Learn. **52**(3), 217–237 (2003)
19. Sangam, R.S., Om, H.: An equi-biased k-prototypes algorithm for clustering mixed-type data. Sadhana: Acad. Proc. Eng. Sci. **43**(3), 37–48 (2018)
20. Diez, J., Barcelo-Rico, F.: Geometrical codification for clustering mixed categorical and numerical databases. J. Intell. Inf. Syst. **39**(1), 167–185 (2012)
21. Foss, A., Markatou, M., Ray, B., Heching, A.: A semiparametric method for clustering mixed data. Mach. Learn. **105**(3), 419–458 (2016)
22. Ericson, K., Pallickara, S.: On the performance of high dimensional data clustering and classification algorithms. Future Generations Comput. Syst. **29**(4), 1024–1034 (2013)
23. Zhu, X.: Variable diagnostics in model-based clustering through variation partition. J. Appl. Stat. **45**(16), 2888–2905 (2018)

24. Kuo, R.J., Nguyen, T.P.Q.: Partition-and-merge based fuzzy genetic clustering algorithm for categorical data. Appl. Soft Comput. **75**, 254–264 (2019)
25. Reybod, A., Etminan, J., Mohammadpour, A.: A Pitman measure of similarity in k-means for clustering heavy-tailed data. Commun. Stat.-Simul. Comput. **48**(6), 1595–1605 (2019)
26. Rodrigues, M.T.A., Freitas, M.H.G., Pádua, F.L.C., Gomes, R.M., Carrano, E.G.: Evaluating cluster detection algorithms and feature extraction techniques in automatic classification of fish species. Pattern Anal. Appl. **18**(4), 783–797 (2014)
27. Kamalha, E., Kiberu, J., Nibikora, I., Mwasiagi, J.I., Omollo, E.: Clustering and classification of cotton lint using principle component analysis, agglomerative hierarchical clustering, and k-means clustering. J. Nat. Fibers **15**(3/4), 425–435 (2018)
28. Jiang, J.C., Xia, Z.Y.: Cluster partition-based communication of multiagents: the model and analyses. Adv. Eng. Softw. **42**(10), 807–814 (2011)
29. Sun, H., Wang, J., Li, T., Li, K., Qi, Q., et al.: Density cluster-based approach for controller placement problem in large-scale software defined networkings. Comput. Netw. **112**, 24–35 (2017)
30. Chang, Y., Hung, W., Lee, E.S.: Weight selection in W-K-means algorithm with an application in color image segmentation. Comput. Math. Appl.: Int. J. **62**(2), 668–676 (2011)
31. Liu, X., Yang, J., Wu, J., Zhou, M., Chen, Y.: Evidential reasoning approach with multiple kinds of attributes and entropy-based weight assignment. Knowl.-Based Syst. **163**, 358–375 (2019)

Prediction of Shanghai and Shenzhen 300 Index Based on BP Neural Network

Hong Liu[1], Nan Ge[1], Bingbing Zhao[3], Yi-han Wang[2], and Fang Xia[1](✉)

[1] School of Health Management, Changchun University of Chinese Medicine, Changchun 130117, China
xiafang425@126.com
[2] Jilin University School of Public Health, Changchun 130021, China
[3] Unit 93313 of PLA, Changchun 130062, China

Abstract. The Shanghai and Shenzhen 300 Index covers the epitome of 60% of the Shanghai and Shenzhen stock markets, and is a high-level summary of the Shanghai and Shenzhen stock markets. If we can predict the closing price of the Shanghai and Shenzhen 300 Index, it will guide the investment of the Shanghai and Shenzhen stock markets, promote the sound development of the stock market, and improve the overall economic strength of the country. For the study of nonlinear systems, neural networks have unique advantages, especially BP neural networks, which can train, learn and predict complex data. In this paper, BP neural network is used to establish the BP neural network model of the Shanghai and Shenzhen 300 Index. The opening price, the highest price and the lowest price of the Shanghai and Shenzhen 300 Index are used as input variables, and the closing price is taken as the output variable to study and predict the closing of the Shanghai and Shenzhen 300 Index. Through the model results, it is found that the BP neural network model can better simulate the Shanghai and Shenzhen 300 Index and achieve the effect of predicting the Shanghai and Shenzhen 300 Index. Applying BP neural network model to nonlinear estimation of the Shanghai and Shenzhen 300 Index, solving nonlinear problems, can promote the healthy development of China's stock market.

Keywords: BP neural network · Prediction · Shanghai and Shenzhen 300 · Stock index

1 Introduction

The CSI 300 Index is a financial indicator released by the Shanghai and Shenzhen Stock Exchanges on April 8, 2008 to reflect the large and medium-sized constituent stocks of the Shanghai and Shenzhen Stock Exchanges. The Shanghai and Shenzhen 300 Index is mainly composed of the top 300 stocks of the Shanghai and Shenzhen stock exchanges through a complex algorithm, reflecting the overall summary of the Shanghai and Shenzhen large and medium-sized cities, known as the barometer of the Shanghai and Shenzhen stock markets.

In 1943, psychologist W.S. McCulloch and mathematical logician W. Pitts established a neural network and mathematical model called the MP model [1]. They proposed

Q. Liu et al. (Eds.): CENet 2020, AISC 1274, pp. 33–41, 2021.
https://doi.org/10.1007/978-981-15-8462-6_4

a formal mathematical description of neurons and a network structure method through the MP model, which proved that a single neuron can perform logical functions, thus creating an era of artificial neural network research; Hebb proposed the famous Hebb learning rule, that is, the learning process occurred on the synapse. The larger the product of the activation values between two adjacent parties, the greater the adjustment of the connection weight [2]. Rumelhart, Hinton, Williams developed the BP algorithm, and the BP neural network can be processed by practice. Complex nonlinear problems have been widely used, and until now the field of artificial intelligence neural networks is still developing [3].

In 1996, Wei Meng and others used Hopfield networks and multi-layered structured forward nerves. The network solves the problem [4]. Li Hongyi and others used BP nerve. The network predicts the remote sensing background value of the ecological environment. The data verification shows that the prediction result of BP neural network is feasible. Wang Y. proposed to use GA genetic algorithm to optimize the network structure of BP network. The results show that the error of GA-optimized network prediction model is smaller in the stock price prediction process [5]; Xu C. selected various transaction data (including the highest price, the lowest price, the opening price, and the volume of 20 kinds of data) with the code of 600000 for one year as the input variables of BP neural network, and applied these variables to the neural network. Training and learning, and obtained higher prediction accuracy [6]. In 2010, Zheng X based on BP neural network model, using PSO algorithm to BP neural network Optimized, optimized neural network prediction ability enhanced [7]. In 2017, Li, H. and others used genetics. The algorithm optimizes the BP neural network. The experimental results show that the optimized BP neural network model can obtain higher prediction accuracy and faster convergence speed [8].

Based on the above theory, this paper establishes the Shanghai and Shenzhen 300 Index BP neural network model to predict the Shanghai and Shenzhen 300 Index. While pointing out the direction for most investors, it will promote the overall level of the national economy and contribute to the development of the country.

2 BP Neural Network Theory

The neural network model is a nonlinear dynamic system suitable for dealing with problems such as multi-dimensional, inaccurate and fuzzy information. First, the neural network model does not obey the normal distribution requirements for input variables, and can guarantee more input variables. Secondly, the neural network model has the adaptability, learning ability and mapping ability. The neural network model itself has a learning function, which can simulate the human brain neural network for parallel computing in seemingly irregular data, and form a self-algorithm data by learning. Inductive. The neural network model is fault-tolerant. Compared with the traditional regression model, the neural network model can tolerate certain errors in the data [9]. The lack of a small part of the data does not affect the overall fitting effect of the model, and the adaptability is extremely strong.

The neural network model can be divided into various models according to the topology of the network connection, the characteristics of neurons, learning rules, etc., which are mainly divided into reverse transmission network, perceptron, Hopfield network, Boltzmann machine and so on [10]. The corresponding neural network models have different characteristics and deal with different problems. The most effective application is the reverse transfer network model, BP neural network model. The BP neural network model overcomes the characteristics of data linearization and ensures that multiple input variables are simultaneously incorporated into the neural network model. The BP neural network model mainly includes three major components: input layer, hidden layer and output layer. In each layer of the neural network, there are nodes connected to another layer, called neurons. The neuron passes the information of this layer to the next layer through the excitation function, and becomes the input variable of the next layer [11]. The complexity of the hidden layer determines the advantages of the BP neural network model - it can handle more complex multi-structure problems. This paper chooses BP neural network model as the theoretical basis to construct the Shanghai and Shenzhen 300 index neural network model.

The BP neural network model (see Fig. 1) belongs to the multi-layer forward network. Firstly, the system randomly selects the network weights and thresholds of [−1, 1] for training, and sets the excitation function in advance. The transfer function used by BP neural network neuron (see Fig. 2) is usually Sigmoid. Differentiable functions.

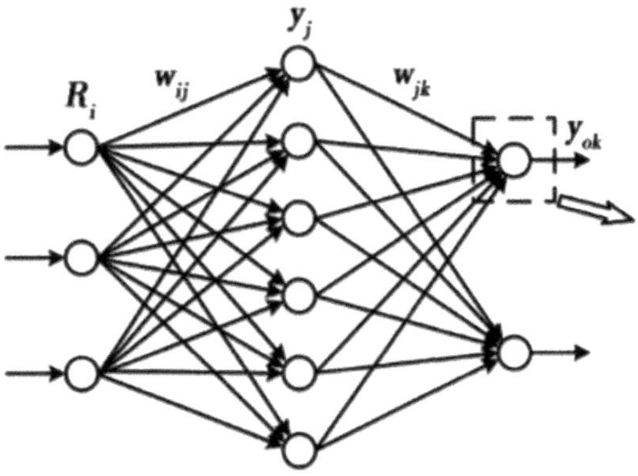

Fig. 1. BP neural network.

Its expression is:

$$S(x) = \frac{1}{1 + e^{-x}} \tag{1}$$

The main algorithms of BP neural network model are information forward propagation and error back propagation. The information forward propagation means that the input variables are added by the weights, and then the excitation function is used as the input variable of the next layer of neurons, and the output values are sequentially calculated according to the layers. Error backpropagation means that the output value

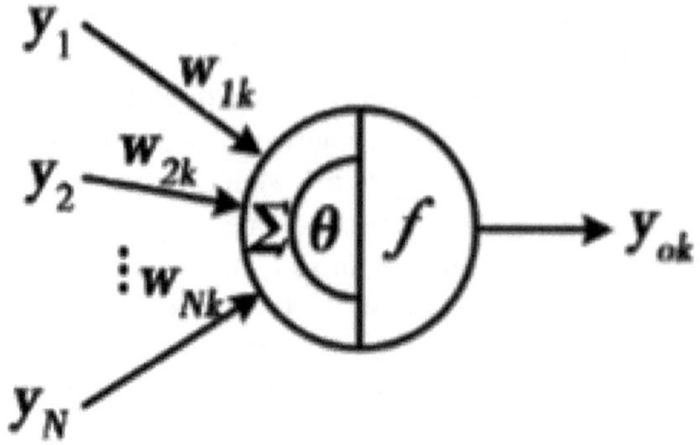

Fig. 2. Single neuron structure.

calculated by the BP neural network model is compared with the true value. If the error is smaller than the previously specified fault tolerance range, the weight and threshold of each layer of the BP neural network are from the output layer to the input. The order of the layers is updated sequentially, and the weights and thresholds are re-established. The BP neural network model will iterate through the process until the error of the output value is less than the error set in advance. The mathematical expression for this output mode is as follows:

$$\begin{cases} net_i = \sum_{i=1}^{n} xW_{ij} \\ O_j = f(net_j) \end{cases} \tag{2}$$

$$\begin{cases} net_k = \sum_{j=1}^{N} O_j W_{jk} \\ y_k = f(net_k) \end{cases} \tag{3}$$

Where netj and netk represent the net input of the hidden layer node and the output layer node in the model, xi and yi represent the input of the model and the output of the model, respectively, and M and N respectively represent the number and implied of the input layer nodes in the model. The number of layer nodes, Wij and Wjk represent the weights of the connection between the input layer, the hidden layer and the output layer in the model, Oj represents the output produced by the hidden layer nodes in the model, f represents the function, and f uses the sigmoid function.

3 Establishment of BP Neural Network Model

3.1 Selection of BP Neural Network Data

This paper uses the Choice financial terminal to export the K-line data of the Shanghai and Shenzhen 300 Index from January 4, 2005 to March 6, 2019. The data includes the opening price, the highest price, the lowest price, and the closing price. The closing price is taken as the output variable as the main research and forecasting variable. According to the rule of thumb, the opening price, the highest price and the lowest price are selected as the input of the BP neural network model. Variable. The prediction of variables has many uncertainties. Choosing too many input variables will affect the overall calculation speed of the model and the overall fit of the model. The daily closing price can be regarded as the opening price of the new day. According to the time series model, Out, the previous variable of the predictor is predicted as the main variable, and the highest price and the lowest price are updated in real time every day [12]. The prediction of high-latitude nonlinear time series is a very important variable and can contain the latest information, so The above three variables are used as input variables for the closing network BP network model of the Shanghai and Shenzhen 300 Index. Since the highest price and the lowest price are also variables that are updated in real time, it is not good to choose when to point to the highest price or the lowest price [13]. Therefore, this also requires the investor to have a certain level of professional skill, which makes the prediction result more accurate.

3.2 Data Preprocessing

In order to eliminate the influence of different dimensions and different magnitudes on the fitting results, the data was preprocessed by min-max normalization method.

$$x' = b + (a - b)\frac{x - x_{\min}}{x_{\max} - x_{\min}} \tag{4}$$

The selected data is mapped between [0, 1] by the above formula. A total of 3443 data were selected, and 2500 data were selected as the training set of BP neural network. The BP neural network was trained and trained through the training set. The remaining 943 data were used as the test set of BP neural network to test the accuracy of fitting BP neural network degree.

3.3 Establishment of Closing Price BP Neural Network Model

This paper uses R language to process the collected data and establish a closing price BP neural network model. The BP neural network model mainly consists of three parts: input layer, hidden layer and output layer. The number of nodes of the input layer and the output layer is relatively easy to determine, generally the number of input data and output data. In this paper, the opening price, the highest price and the lowest price are taken as output data. Therefore, three nodes are established in the input layer of the BP neural network, and the closing price is used as the output data, and an output node is established in the output layer [14]. There is still no clear theory for the number of layers in the hidden layer and the number of nodes in each hidden layer. In this paper, the hidden layer is positioned in two layers, and the first hidden layer contains five neurons. The second hidden layer contains 3 neurons. The BP neural network model established by this method has the smallest error and the model learning effect is the best.

The BP neural network topology of the closing price established is shown in Fig. 3:

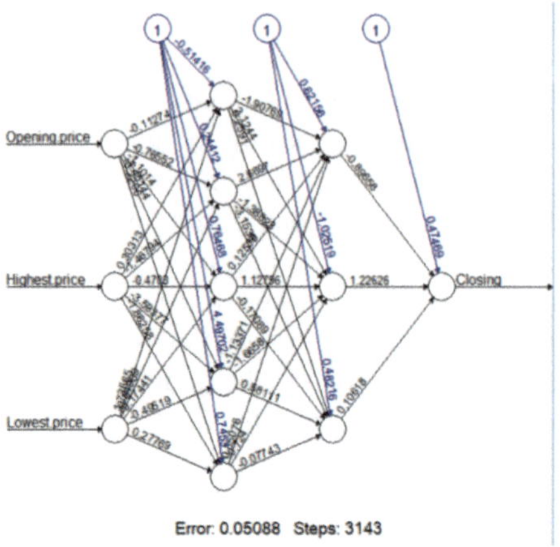

Fig. 3. Closing price BP neural network model.

In order to test the fitting degree of the closing price BP neural network model generated by R language, the test set variables are input into the BP neural network model to determine whether the model fits well or not. From the BP neural network topology map, which shows in Fig. 4, the error of the closing price BP neural network model is 0.047401. The test set data is input into the BP neural network, and the training value is obtained and compared with the real value. The error is found to be small and the fault tolerance range of the model is satisfied. The model can make a closing price forecast. Figure 5 shows the comparison of the Model's predicted and closing prices, as well as the full data line of the closing and forecasting values.

Closing	Prediction	Serial
0.03256661	0.03458837	1
0.03540304	0.03576251	2
0.03532595	0.03547854	3
0.03552954	0.03521157	4
0.03397791	0.03558643	5
0.04004214	0.03989372	6
0.04307031	0.04329424	7
0.04144553	0.04043332	8
0.04516749	0.04535295	9
0.03657517	0.03751739	10
0.03238871	0.03270687	11
0.02820028	0.02882803	12
0.02454355	0.02517909	13
0.03164946	0.02961739	14
0.03614625	0.03590049	15
0.02146004	0.02481474	16
0.02451786	0.02410946	17
0.02260647	0.02543391	18
0.01307329	0.0155831	19
0.006222365	0.01154424	20

Fig. 4. Model forecast and closing price comparison table.

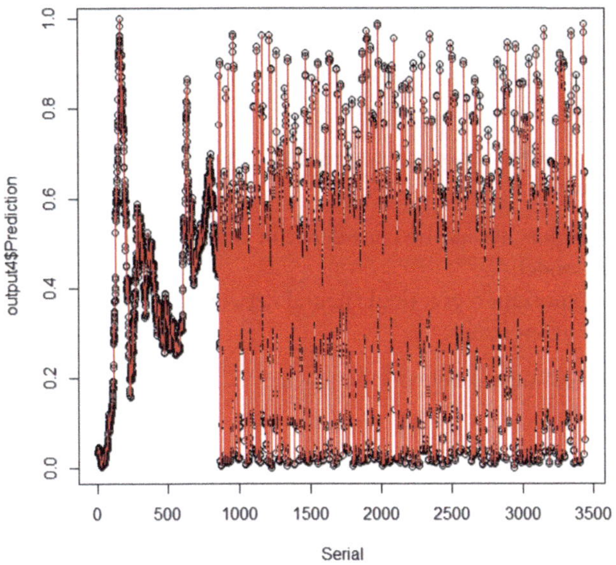

Fig. 5. Model predictive value and closing price chart.

4 Discussion

This paper mainly applies the BP neural network model to construct the Shanghai-Shenzhen 300 Index closing price forecasting model, and divides the selected data into the training layer and the prediction layer. The BP neural network established by comparing the error between the training layer and the prediction layer data is verified. The reliability of the network model finally established the Shanghai and Shenzhen 300 index prediction model and forecasted the Shanghai and Shenzhen 300 Index.

From the experimental results of this paper, the short-term prediction of the closing price of the stock index using BP neural network is feasible, and the prediction effect is also good. However, because the forecasting of its own requirements is very strict, and the stock market is a very complex comprehensive system, coupled with the relatively short history of China's stock market development, there are still many irregularities, speculation is divided into serious, irrational investment exists in large numbers. Therefore, it is basically impossible to make accurate predictions on stock indexes or stocks. This paper hopes to make some valuable explorations and attempts in stock price forecasting through hard work. At the beginning of this work, there are still many problems to be further explored. The shortcomings of this paper are mainly:

(1) Only the application model is used to study the CSI 300 Index, and the applicability of the model in different markets is not discussed. For example, the Shanghai Stock Index or the Shenzhen Stock Index is selected separately.
(2) This paper mainly uses the quantitative data in the historical data of the stock market for modeling analysis. There is basically no reference to some qualitative factors affecting the stock price fluctuation, and the stock market is a complex

dynamic nonlinear system, which integrates the world. Economic factors, national policies, economic environment, investor psychology and many other qualitative factors. Therefore, this has weakened the prediction accuracy of this paper to a certain extent.

Based on the current situation, we need to pay attention to the following issues in future research work:

(1) Qualitative factors should be considered in future studies to reduce the prediction error of the model.
(2) Parameter optimization of a single neural network model requires more effort. How to find for a given data

Finding the best neural network parameters is for further study.

5 Conclusion

The stock market is a complex nonlinear market. The traditional statistical methods are not ideal for dealing with nonlinear problems. It is difficult to realize the complicated stock market forecast. Through the above theory and experiment, this paper constructs the Shanghai-Shenzhen 300 Index BP neural network model by using the neural network to process nonlinear data. It can be seen from the experimental data that the BP neural network model of the Shanghai and Shenzhen 300 Index has the characteristics of fast convergence and high prediction accuracy in the prediction, and is particularly effective for the short-term prediction effect of the stock index. The combination of BP neural network and stock market can extract the information that has value to investors, which proves that BP neural network is an effective method in the forecast of stock market, which is feasible.

References

1. Hall, D.M., Shroff, F.R.: The study of dental ultrastructure by freeze-etching. Arch. Oral Biol. **13**(6), 703–704 (1968)
2. Steedman, W.M., Molony, V., Iggo, A.: Nociceptive neurones in the superficial dorsal horn of cat lumbar spinal cord and their primary afferent inputs. Exp. Brain Res. **58**(1), 171–182 (1985)
3. Acheroy, M.P.J., Mees, W.: Target identification by means of adaptive neural networks in thermal infrared images. In: Proceedings of SPIE - The International Society for Optical Engineering, pp. 121–132 (1991)
4. Meng, W., Huang, W., Cai, X.: Problems of adjustment in surveying by methods of neural network. Acta Geodaetica Cartogr. Sin. **4**, 298–302 (1996)
5. Wang, Y., Zhang, J., Zhao, Y., Wang, Y.: Application of elevator group control system based on genetic algorithm optimize BP fuzzy neural network. In: 2008 7th World Congress on Intelligent Control and Automation, NJ, USA, pp. 8702–8705. IEEE (2008)

6. Xu, C.: Optimization analysis of dynamic sample number and hidden layer node number based on BP neural network. In: Proceedings of the Eighth International Conference on Bio-Inspired Computing: Theories and Applications (BIC-TA), 2013, pp. 687–695. Springer, Heidelberg (2013)
7. Zheng, X., Liu, J., Wei, Z., Song, B., Feng, M.: The exploratory research of a Novel gyroscope based on superfluid Josephson effect. In: IEEE/ION Position, Location and Navigation Symposium, NJ, USA, pp. 14–19. IEEE (2010)
8. Li, H., Ding, X.: Research on artificial neural network intrusion detection photochemistry based on the improved wavelet analysis and transformation. In: AIP Conference Proceedings, NY, USA, vol. 1820, pp. 080018-1–080018-5 (2017). AIP Publishing (2017)
9. Boser, B.E., Sackinger, E., Bromley, J., Le Cun, Y., Jackel, L.D.: An analog neural network processor with programmable topology. IEEE J. Solid-State Circuits **26**(12), 2017–2025 (1991)
10. Barra, A., Bernacchia, A., Santucci, E., Contucci, P.: On the equivalence of Hopfield networks and Boltzmann machines. Neural Netw. **34**(5), 1–9 (2012)
11. Beckwith, K., Hawley, J.F., Krolik, J.H.: The influence of magnetic field geometry on the evolution of black hole accretion flows: similar disks, drastically different jets. Astrophys J. **678**(2), 1180–1199 (2008)
12. Martel, A.R., Menanteau, F., Tozzi, P., Ford, H.C., Infante, L.: The Host galaxies and environment of Chandra-selected active galactic nuclei in the deep ACS GTO cluster fields. Astrophys. J. Suppl. Ser. **168**(1), 19–57 (2007)
13. Sun, Q., Che, W.G., Wang, H.L.: Bayesian regularization BP neural network model for the stock price prediction. In: Sun, F., Li, T., Li, H. (eds.) Foundations and Applications of Intelligent Systems. Advances in Intelligent Systems and Computing, pp. 521–531. Springer, Heidelberg (2014)
14. Wang, H., Song, G.: Fault detection and fault tolerant control of a smart base isolation system with magneto-rheological damper. Smart Mater. Struct. **20**(8), 085015 (2011)

Application of Time-Frequency Features and PSO-SVM in Fault Diagnosis of Planetary Gearbox

Qing Zhang, Heng Li(✉), and Shuaihang Li

School of Mechanical Engineering, Tongji University, Shanghai 201804, China
tj_liheng@126.com

Abstract. Planetary gearboxes of cranes are often in high-speed, heavy-duty operating environments, and critical components such as gears and bearings are prone to failure. Diagnosing the fault information of key components in time can effectively avoid losses and accidents caused by mechanical equipment's shutdown. Aiming at the problems of strong time-varying and unstable vibration signals of planetary gearbox, a fault diagnosis method based on time-frequency features and PSO-SVM is proposed. First, the STFT is used to process the planetary gearbox's vibration signal to obtain the time-frequency spectrum characteristics. Then, use a support vector machine to process the time-frequency spectrum features to identify the fault type. Simultaneously, the PSO algorithm is used to optimize the support vector machine's parameters to improve the classification and recognition ability. The analysis experiment and algorithm test of the planetary gearbox's vibration signal verify that the method can extract effective characteristic parameters from non-stationary and time-varying signals and can quickly and accurately identify the fault type of the planetary gearbox.

Keywords: Time-frequency feature · PSO · SVM · Planetary gearbox · Fault diagnosis · Equipment safety protection

1 Introduction

Planetary gearboxes have been widely used in the metallurgical and mining industry, hoisting and transport industry, aerospace industry, etc. due to their compact structure, large transmission speed ratio, and strong load-carrying capacity. The planetary gearbox is the most crucial link in the mechanical transmission system. If any part of it fails, it is easy to trigger a chain reaction of failure, or even lead to safety disasters. For the question of gearbox fault diagnosis, the key is to monitor vibration signals by installing sensors and analyze these signals to obtain the health state [1]. The planetary gearbox's motion is compound, and its vibration signal is more complex, non-stationary, nonlinear, easy to occur signal modulation [2]. Therefore, it is necessary to select an

This work is supported by the National Key Research and Development Program of China [2018YFC0808902].

Q. Liu et al. (Eds.): CENet 2020, AISC 1274, pp. 42–51, 2021.
https://doi.org/10.1007/978-981-15-8462-6_5

appropriate signal processing method for the vibration signal, combined with a reliable state recognition strategy to diagnose the planetary gearbox's fault.

For the fault diagnosis of complex vibration signals, using different methods to process the signals to obtain feature vectors for diagnosis is the key of research. Zhu et al. [3] proposed an improved empirical wavelet transform method to demodulate the signal of planetary gearbox, identify the fault characteristic frequency by analyzing the sensitive component, and acquire the diagnosis result by analyzing the spectrum manually. However, the analyzing process requires a wealth of theoretical knowledge. Bi et al. [4], based on the frequency slice wavelet transform to diagnose the faults of planetary gearbox, the experimental results show that time-frequency analysis can more effectively extract features from complex working conditions. However, this method also requires a large amount of prior knowledge, which is difficult to be applied in practice. Lou et al. [5] applied PSO-SVM to gearbox fault diagnosis, combined with time and frequency domain characteristic parameters, and received better results. Nevertheless, the object of this method is limited to a few kinds of fixed axle gearbox faults. Li et al. [6] combined the multifractal spectrum and SVM to proceed with the fault diagnosis and verified that the method could better recognize the rate in the sun gear fault diagnosis. However, the method does not consider the fault diagnosis of all key components.

This paper presents a new method of fault diagnosis. First, the vibration signal is analyzed by STFT, and the valid feature parameters are extracted from the time-frequency spectrum. Secondly, the PSO-SVM classifier is constructed, which use PSO to optimize the classification performance. Finally, the classifier is automatically trained by inputting the feature parameters of the training set, and then the diagnosis results of each principal component can be obtained by inputting the test set data into the classifier.

2 Time-Frequency Spectrum Features Based on STFT

The analysis of time domain signals usually uses Fourier transform (FT) to obtain frequency domain information, and the FT performs well in the analysis of stationary signals. However, for the non-stationary signal of planetary gearbox whose frequency changes with time, it's necessary to acquire both time and frequency information. This paper uses STFT to analyze the signal of the planetary gearbox, and then extracts the characteristic values from the time-frequency spectrum for fault diagnosis.

2.1 STFT

Short-time Fourier transform (STFT) is a convenient time-frequency analysis method for processing non-stationary signals whose frequency changes with time. It uses signals in a time window to represent the signal characteristics at a certain moment [7]. That is, the original signal function is divided into many time periods by the window function, and then the FT is performed on the divided signal function of each time period. For discrete vibration signals, the calculation formula of STFT is as follows:

$$STFT\{x[n]\}(m,\omega) \equiv X(m,\omega) = \sum_{n=-\infty}^{\infty} x[n]\omega[n-m]e^{-j\omega n} \tag{1}$$

where $\omega[n-m]$ is the window function centered on m, and $x[n]$ is the time-domain signal to be transformed.

In the process of STFT, the length of the window determines the relative resolution of time and frequency domain. The longer the window, the higher the frequency resolution; on the contrary, the shorter the window, the better the time resolution. That is to say, in the STFT, the time resolution and frequency resolution cannot have both and have to choose between them according to actual needs.

In the fault diagnosis of gearboxes, the fault characteristic frequency of bearings and gears is usually taken as the reference to determine the window length and sample length. The calculation formula of the characteristic frequency of the bearing is as follows:

$$f_{BPI} = \frac{1}{2}Z\left(1 + \frac{d}{D}cos\alpha\right)f_H \tag{2}$$

$$f_{BPO} = \frac{1}{2}Z\left(1 - \frac{d}{D}cos\alpha\right)f_H \tag{3}$$

$$f_{BS} = \frac{1}{2}\frac{D}{d}\left(1 - \left(\frac{d}{D}\right)^2 \cos^2\alpha\right)f_H \tag{4}$$

where f_{BPI}, f_{BPO}, f_{BS} are respectively the frequency of the roller passing the inner ring, the frequency of the roller passing the outer ring and the rotation frequency of the roller, $f_H = n/60$ is the inner ring rotation frequency, n is the rotating speed, d and D are respectively the diameter of the roller and the pitch circle, Z is the number of rollers, and α is the contact angle.

In a planetary gearbox, the repetition frequency of the locally damaged gears is related to the speed of the gearbox, the number of planet gear and the number of teeth of each gear [8]. The calculation formula of the fault characteristic frequency of the sun gear and planet gear are as follows:

$$f_S = \frac{f_m}{Z_S}N \tag{5}$$

$$f_{p1} = \frac{f_m}{Z_p} \tag{6}$$

where f_m represents meshing frequency, Z_S is teeth number of sun gear, N represents number of planet gear and Z_p is teeth number of planet gear.

2.2 Time-Frequency Spectrum Feature Data Extraction

In time domain and frequency domain analysis, RMS, standard deviation and kurtosis of time domain or frequency domain are often used as characteristic values [5]. In this paper, the statistical characteristic values of the time and frequency domains are extended to the time-frequency domain, which are shown in Table 1.

Table 1. Statistical characteristic values.

	Equation		
RMS	$X_{rms} = \sqrt{\frac{1}{N}\sum_{i=1}^{N} x_i^2}$	$F_{rms} = \sqrt{\frac{1}{mn}\sum_{i=1}^{n}\sum_{j=1}^{m}(f(x_i, y_i))^2}$	(7)
Standard deviation	$\sigma = \sqrt{\frac{1}{N}\sum_{i=1}^{N}(x_i - \bar{x})^2}$	$\sigma = \sqrt{\frac{1}{mn}\sum_{i=1}^{n}\sum_{j=1}^{m}(f(x_i, y_i) - \bar{f})^2}$	(8)
Kurtosis	$C_q = \frac{\sum_{i=1}^{N}(\lvert x_i\rvert - \bar{x})^4}{N \cdot X_{rms}^4}$	$C_q = \frac{\sum_{i=1}^{n}\sum_{j=1}^{m}(\lvert f(x_i, y_i)\rvert - \bar{f})^4}{N \cdot F_{rms}^4}$	(9)
Skewness	$C_w = \frac{\sum_{i=1}^{N}(x_i - \bar{x}\rvert)^3}{N \cdot X_{rms}^3}$	$C_w = \frac{\sum_{i=1}^{n}\sum_{j=1}^{m}(\lvert f(x_i, y_i) - \bar{f}\rvert)^3}{N \cdot F_{rms}^3}$	(10)
Margin factor	$C_e = \frac{X_{rms}}{\bar{x}}$	$C_e = \frac{F_{rms}}{\bar{f}}$	(11)
Crest factor	$I_p = \frac{X_p}{X_{rms}}$	$I_p = \frac{F_p}{F_{rms}}$	(12)
Pulse factor	$C_f = \frac{X_p}{\bar{x}}$	$C_f = \frac{F_p}{\bar{f}}$	(13)

The time-frequency spectrum of the signal is obtained via STFT, then calculating the statistical values via equations in Table 1 to obtain a set of eigenvalues that can reflect the time-frequency characteristics of the original signal.

3 SVM Optimized by Particle Swarm

3.1 SVM and PSO

The basic method of support vector machine [9] is: Find a hyperplane ($\vec{\omega} \cdot \vec{x} + b = 0$) in the sample space of the training set $D = \{(\vec{x}_1, y_1), (\vec{x}_2, y_2), \ldots, (\vec{x}_m, y_m)\}, y_i \in \{-1, +1\}$, which divides the samples into different categories and meets the maximum distance, as shown in Fig. 1. The sample points that make up $\vec{\omega} \cdot \vec{x} + b = \pm 1$ are called support vectors, the sum of the distances between the two types of support vectors to the hyperplane is called the margin, and $\vec{\omega}$ and b satisfy $\vec{\omega} \cdot \vec{x} + b \geq 1$.

For linearly inseparable problems, vectors are mapped to a higher dimensional linearly separable space through a kernel function, and a maximum-margin hyperplane is established in this space. In this paper, the Gaussian kernel function is adopted, also known as radial basis function (RBF), and is defined as:

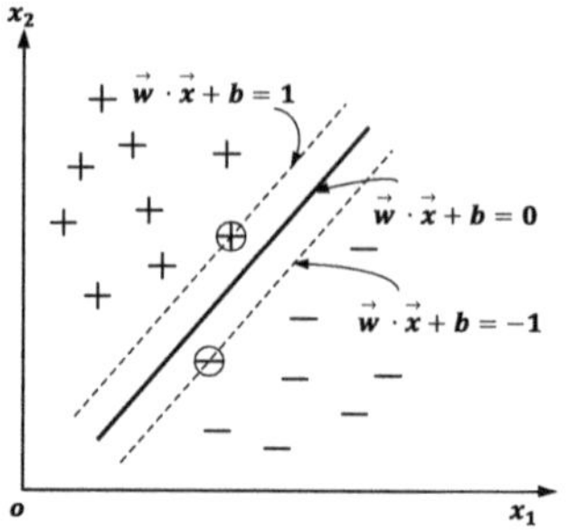

Fig. 1. The method of SVM.

$$k(x,x') = e^{-\frac{||x-x'||^2}{2\sigma^2}} \tag{14}$$

where σ is the bandwidth of the Gaussian core.

The Gaussian kernel function has only one parameter σ. In addition, the penalty parameter C also has a great impact on the classification results of SVM. As a result, this paper optimizes σ and C to improve the classification performance.

In the particle swarm optimization (PSO) algorithm, each potential solution is a particle ("bird") in the search space, and each particle has a fitness value determined by the optimization function. Each particle also has a velocity that determines its direction and distance of its flight, and follows the current optimal particle in the search space. After the PSO is initialized as a group of random particles, it iterates repeatedly until the optimal solution is found. In each iteration, the particle updates itself through two "extreme values": one is the optimal solution found by the particle itself, which is called the individual extreme value or P_{best}; and the other extreme value is currently found by the entire population the optimal solution, which is called the global extreme value or g_{best}. The equations of the PSO algorithm is as follows:

$$v_{id} = \omega v_{id} + c_1 rand_1 (p_{id} - x_{id}) + c_2 rand_2 \left(p_{gd} - x_{id}\right) \tag{15}$$

$$x_{id} = x_{id} + v_{id} \tag{16}$$

where w is inertia weight, c_1 and c_2 are acceleration constants, $rand_1$ and $rand_2$ represents two random numbers that vary in the range [0, 1].

The ith particle is represented as $X_i = (x_{i1}, x_{i2}, \ldots, x_{iD})$, and the best position it has experienced (with the best fitness value) is recorded as $P_i = (p_{i1}, p_{i2}, \ldots, p_{iD})$, which is also called as p_{best}. The best position experienced by all particles in the population is represented by the symbol P_g. The velocity of particle ith is represented by $V_i = (v_{i1}, v_{i2}, \ldots, v_{iD})$.

In this paper, the PSO algorithm is used to optimize the classification performance of the SVM, the penalty parameter C and the kernel function parameters σ of the SVM are used as the optimization parameters, and the optimization goal is the accuracy rate of the SVM classification.

3.2 Fault Diagnosis Based on PSO-SVM

The fault diagnosis model based on PSO-SVM is shown in Fig. 2. The original vibration signal is obtained by sampling the normal working condition of the planetary gearbox and several different fault types. A STFT is performed on the vibration signal to obtain a time-frequency spectrum, from which the statistical characteristic values of the time-frequency domain are extracted. The feature values and corresponding label of different faults are organized into a data set, and the data set is divided into a training set and a test set. The feature values and class labels of the training set are used to train the SVM model, and the PSO is used to optimize the SVM model to achieve the optimal SVM parameters and classification accuracy. The PSO-optimized SVM model is used in the test set samples to verify the performance of the classification model.

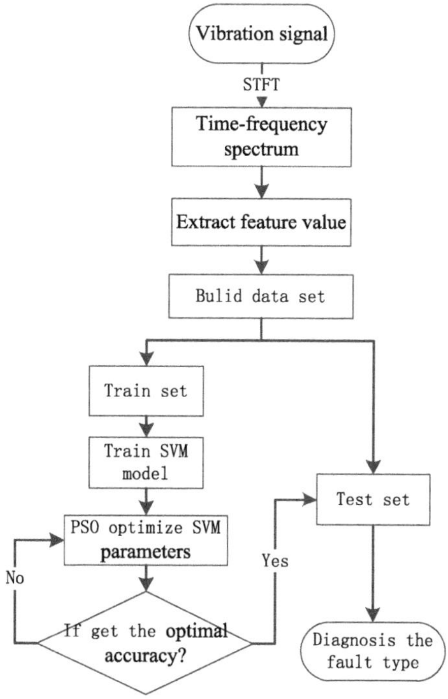

Fig. 2. Fault diagnosis model based on PSO-SVM.

4 Experiment

Experiment 1: Bearing Data Set of Case Western Reserve University.

The failures collected by this data set are the inner ring, outer ring and rolling element failures of different diameter rolling bearings (SKF6205) of drive end. In this paper, the sampling frequency is 12 kHz, the motor speed is 1792 r/min, the motor load is 0. The inner ring fault, the rolling element fault and the outer ring fault at 6 o'clock direction of 0.1778 mm, 0.3556 mm, 0.5334 mm respectively, total 9 types of faults are selected in this paper.

According to Eqs. (2)–(4), the characteristic frequency of bearing fault is calculated as shown in Table 2. According to the characteristic frequency, the sample length is determined to be 0.085 s (1024 data points), and the width of the STFT window is 0.01 s (128 data points).

Table 2. Fault characteristic frequency of rolling bearing SKF6205.

	Inner ring	Outer ring	Roller
Characteristic frequency/Hz	162.7	107.3	70.0
Characteristic period/s	0.0061	0.0093	0.0143

From the bearing data set, 100 samples of each 9 types of failures and normal state were selected as the total data set. Perform a STFT on each sample, and then calculate the 7 statistical characteristics for each sample. Take 60% samples as the training set, and the remaining 40% as the test set. Training the PSO-SVM model by the training set data. After obtaining a PSO-optimized SVM model, use it to make predictions on the test set. The prediction results obtained are shown in Table 3.

Table 3. Experiment results of data set 1.

	Normal bearing	0.1778 mm diameter bearing	0.3556 mm diameter bearing	0.5334 mm diameter bearing	Total
Recognition rate	100%	100%	94.09%	100%	98.5%

Experiment 2: Planetary gearbox experimental table for grab ship unloader.

The data set is the experimental data of the bearings and planetary gears collected by the experimental table shown in Fig. 3. The bearing type is 6205. In this paper, five types of faults are selected: the input shaft bearing inner ring failure, outer ring failure, the sun gear tooth break, the sun gear tooth root crack, and the planet gear tooth break. Among them, the bearing fault and the sun gear tooth root crack is obtained by wire cutting, and the broken teeth of the sun gear and the planet gear are completely broken teeth. When collecting data, the sampling frequency of the experimental platform is 10 kHz, the working load is 0 kg, and the corresponding speed measured by the tachometer is 376 rmp.

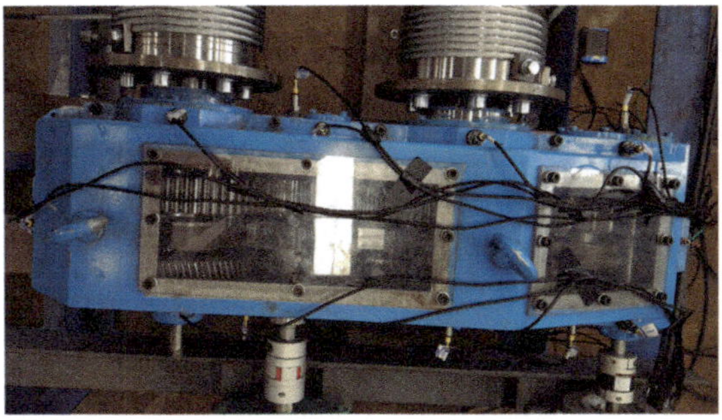

Fig. 3. Planetary gearbox experimental table.

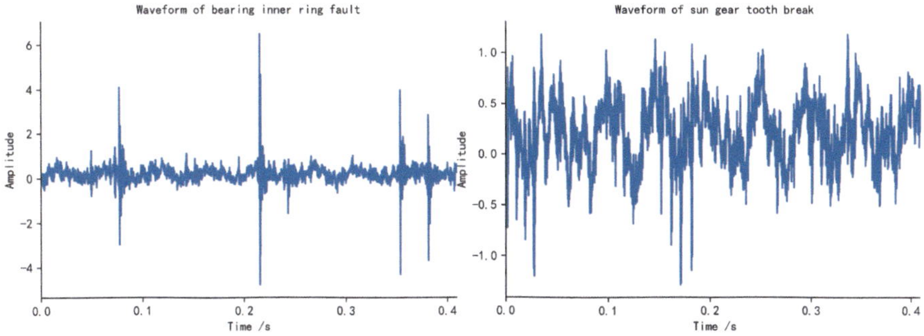

Fig. 4. Time-domain waveform diagram of different faults.

According to Eqs. (2)–(6), the characteristic frequency of the bearings and gears is calculated and shown in Table 4. According to this result, the sample length is determined to be 4096 data points, and the STFT window length is 512 data points.

Table 4. Characteristic frequency of bearing and gear of experiment table.

	Bearing inner ring	Bearing outer ring	Sun gear	planet gear
Characteristic frequency/Hz	36.1077	23.8923	0.7556	0.6612
characteristic period/s	0.0277	0.0418	1.324	1.513

A total of 192 samples of each category were obtained from the data set of the planetary gearbox experimental bench, 32 for each category. 60% of data set was taken as the training set, and 40% was the test set. The time-domain waveform diagram

(Fig. 4) of the planetary gearbox and the time-frequency spectrum diagram after the STFT are shown in Fig. 5. The comparison of the characteristic values under different faults obtained after preprocessing is shown in Fig. 6. Entering the data set into the PSO-SVM model, the final prediction result is 84.44%.

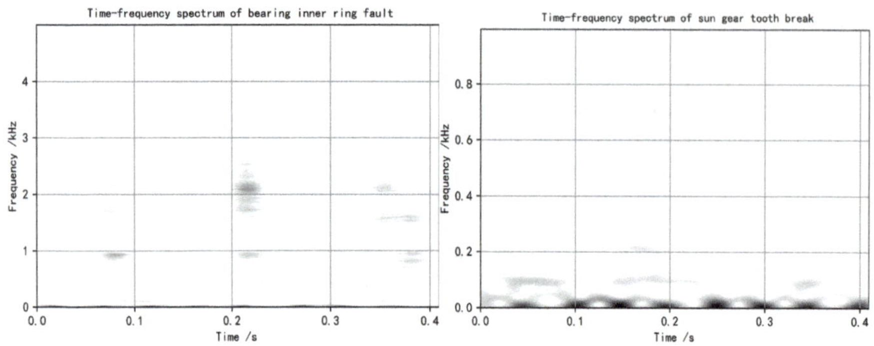

Fig. 5. Time-frequency spectrum diagram of different faults.

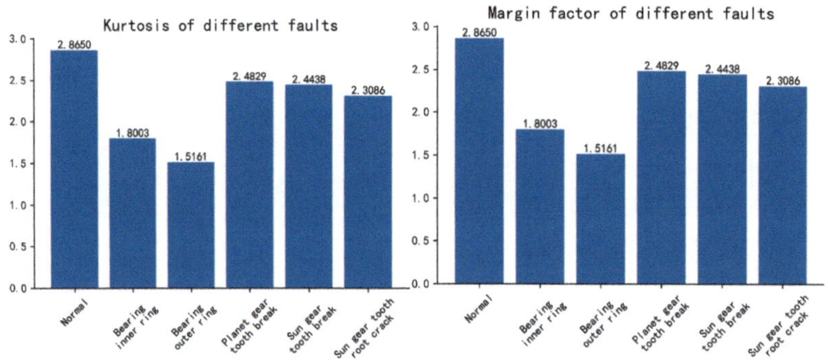

Fig. 6. Characteristic values of different faults.

5 Conclusions

A new fault diagnosis method for planetary gearbox is proposed considering the time-varying working conditions and complex vibration signals. In this paper, by calculating the fault characteristic frequency of bearings and gears, the parameters that ensure the best performance of STFT are obtained. After the STFT of the original signal, 7 kinds of characteristic values of time-frequency domain are obtained. Then, the PSO-optimized SVM automatically obtains the fault diagnosis results. Two sets of experiments verify that the method has a valid accuracy.

References

1. Goyal, D.: Condition monitoring parameters for fault diagnosis of fixed axis gearbox: a review. Arch. Comput. Methods Eng. **24**(3), 543–556 (2017)
2. Lei, Y.G.: Research advances of fault diagnosis technique for planetary gearboxes. J. Mech. Eng. **47**(19), 59–67 (2011)
3. Zhu, W.Y.: Fault diagnosis of planetary gearbox based on improved empirical wavelet transform. Chin. J. Sci. Instrument **37**(10), 2193–2201 (2016)
4. Bi, J.W.: Fault diagnosis of planetary gearbox based on wavelet transform of frequency slice. Mach. Des. Manuf. (1), 29–32 (2016)
5. Lou, H.W.: Research of gear box fault diagnosis based on PSO-SVM. Mech. Sci. Technol. Aerosp. Eng. **33**(9), 1364–1367 (2014)
6. Li, D.D.: Diagnosis and research of wind turbine planetary gearbox faults based on multifractal spectrum support vector machine (SVM). Power Syst. Prot. Control **45**(11), 43–48 (2017)
7. Li, H.: Fault diagnosis method for rolling bearings based on STFT and convolution network. J. Vib. Shock **37**(19), 124–131 (2018)
8. Feng, Z.P.: Vibration spectral characteristics of localized gear fault of planetary gearboxes. Proc. CSEE **33**(05), 119–127+19 (2013)
9. Wu, S.: Bearing fault diagnosis based on multiscale permutation entropy and support vector machine. Entropy **14**(8), 1343–1356 (2012)

Wind Power Variable-Pitch DC Servo System Based on Fuzzy Adaptive Control

Weicai Xie[1,3], Shibo Liu[2,3](✉), Yaofeng Wang[2,3], Hongzhi Liao[2,3], and Li He[2,3]

[1] Hunan Institute of Engineering, Hunan Key Laboratory of Wind Turbines and Control, Xiangtan, China

[2] Hunan Collaborative Innovation Center for Wind Power Equipment and Power Conversion, Xiangtan, China
925640210@qq.com

[3] Xiangdian Wind Energy Co, Ltd., Xiangtan, China

Abstract. The paper aims to expand a way of applying the fuzzy adaptive control strategy by analyzing the advantages and disadvantages of traditional PID control strategies. In view of the high requirements of the variable-pitch servo system for megawatt-level wind turbines, combined with the control characteristics of series-excited DC servo motors, a wind power variable-pitch DC servo system based on fuzzy adaptation was designed. The superiority and reliability of the fuzzy adaptive PI control method are verified by using simulation tools to carry out simulation experiments, and then the experiments are carried out by DC series servo motor and servo driver. In summary, the performance of the wind power variable pitch DC servo system is excellent.

Keywords: Fuzzy control · DC pitch servo system · Simulation analysis · Experiment

1 Introduction

At present, most of the control systems in the industry are mostly PID control, but many of the control objects are non-linear. The control effect of PID control is not satisfactory, and it can easily produce overshoots, shocks and other undesirable conditions. Therefore, in order to solve this situation thoroughly, experts and scholars will effectively combine the current traditional PID control with some more intelligent control methods, and then solve the shortcomings of insensitive parameter setting due to the traditional control method, so that it can achieve a better control effect [1].

In order to solve the defects of PID control, this paper proposes a fuzzy adaptive PI control method, which combines fuzzy control and PID control method. It has the advantages of small overshoot, fast rise time, strong robustness and good dynamic characteristics of fuzzy control, but also maintains the dynamic tracking characteristics of PID control and the quality of stable accuracy. The wind power variable pitch DC series motor speed control servo system adopts a double closed-loop control system composed of a current loop and a speed loop, in which the current loop is an inner loop. Because the outer speed control is a non-linear time-varying system, the

Q. Liu et al. (Eds.): CENet 2020, AISC 1274, pp. 52–59, 2021.
https://doi.org/10.1007/978-981-15-8462-6_6

traditional PID control method is adopted. In summary, the fuzzy adaptive PI control algorithm is used for this design [2].

2 Design of Fuzzy Adaptive PI Control System

At present, the algorithm cannot be changed, if the proportional coefficient, integral coefficient and differential coefficient are all determined in the classic PID control algorithm. Since the speed regulation system of the speed loop of the DC series motor is a non-linear and variable system, the PID control of the classic control system cannot achieve the optimal control effect. On the contrary, fuzzy control has the characteristics of flexible control and strong robustness. Combined with the high-precision control of PID control system, a new type of fuzzy adaptive PI control strategy can be optimized. The system performs online tuning of PI parameters according to different deviations and deviation change rates. At the same time, the system will use two-dimensional fuzzy controller. Therefore, the structure of the discussion is shown (see Fig. 1).

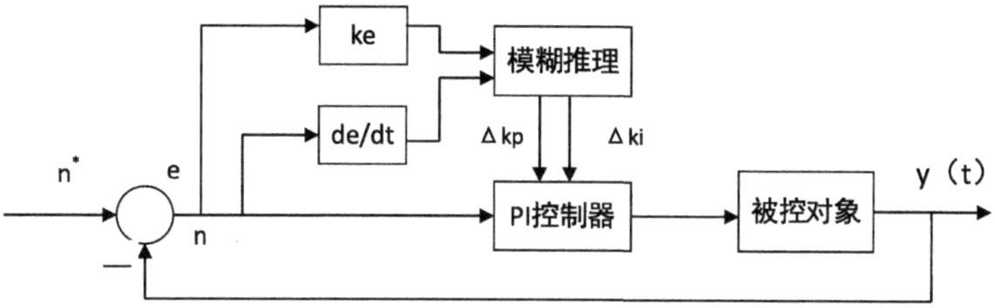

Fig. 1. Schematic diagram of fuzzy PI control.

2.1 Fuzzy Adaptive PI Parameter Tuning

Tuning the fuzzy Adaptive parameter based on PI algorithm. First, calculating the error and error rate of change, the second is to use fuzzy rules to carry out fuzzy inference, and the last is to debug parameters in real time according to the fuzzy matrix table. At the same time, the establishment of fuzzy rule table is based on summing up the experience of experts and technical knowledge in engineering design. By designing the core of the fuzzy controller, the fuzzy control rule table of Δkp and Δki under different e and ec can be obtained (see Table 1 and Table 2).

It can be seen from Table 1 and Table 2 that the adaptive correction of KP and Ki is performed after the fuzzy rule table is established. At this time, the domain of the fuzzy set is error e and error ec. During the online operation, the control system completes the online self-correction for PI parameters by processing, looking up tables and operations on the results of fuzzy logic rules.

Table 1. Δ*kp* fuzzy control rule table.

Δ*kp* (*e* \ *ec*)	NB	NM	NS	ZO	PS	PM	PB
NB	PB	PB	PM	PS	PS	ZO	ZO
NM	PB	PB	PM	PS	PS	ZO	NS
NX	PM	PB	PM	PS	ZO	NS	NS
Z0	PM	PM	PS	ZO	NS	NM	NM
PS	PS	PS	ZO	NS	NS	NM	NM
PM	NS	ZO	NS	NM	NM	NM	NB
PB	ZO	ZO	NM	NM	NM	NB	NB

Table 2. Δ*ki* fuzzy control rule table.

Δ*ki* (*e* \ *ec*)	NB	NM	NS	ZO	PS	PM	PB
NB	NB	NB	NM	NM	NS	ZO	ZO
NM	NB	NB	NM	NM	NS	ZO	ZO
NX	NB	NM	NM	NS	ZO	PS	PS
Z0	NM	NM	NS	ZO	PS	PM	PM
PS	NM	NS	ZO	PS	PS	PM	PB
PB	ZO	ZO	PM	PM	PM	PB	PB
PB	ZO	ZO	PM	PM	PM	PB	PB

3 Simulation of Variable Pitch DC Series Excitation Servo Control System

The servo speed control system of the variable pitch DC series motor is a double closed-loop control system, in which the speed loop is the outer one and the current loop is the inner one. By tracking the dynamic performance of the system, the speed outer loop can suppress the wide range of speed fluctuations, and the stability and dynamic operation of the speed control system are greatly affected. The control strategy of the speed outer loop is the fuzzy adaptive PI control principle. The fixed parameters are set by changes in the peripheral environment, and then it can be automatically changed in fuzzy control and PI control. The stability and rapidity of the speed regulation system can be well optimized by the above control strategy, and it can also effectively improve the control accuracy of the entire pitch DC series motor servo speed regulation system [5, 6]. The simulation model of the servo speed regulation system of the variable pitch DC series excitation motor is established, through the control principle of the double closed-loop speed regulation of the DC series excitation motor, which is shown in Fig. 2.

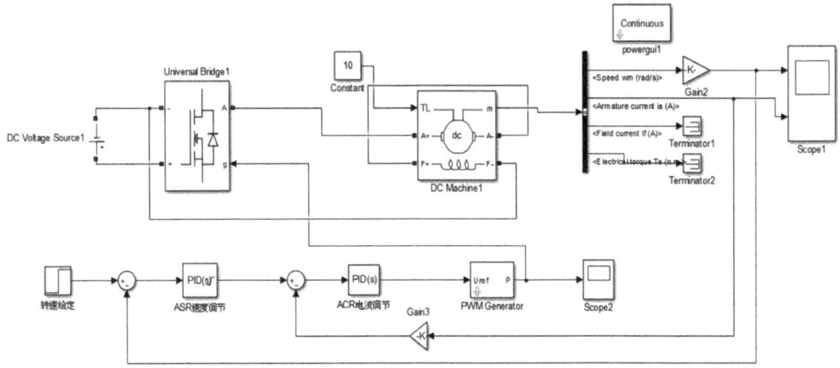

Fig. 2. Simulation of DC servo motor pitch control system.

The specific implementation process is to subtract the actual speed of the feedback from the given speed to obtain the speed deviation, and then the value is used as the input of the speed regulator, while the output I of the speed regulator is the given value of the current regulator, and then subtracts the measured current from this value to obtain the current deviation as the input to the current loop. The input of the PWM controller is the output of the current regulator, and the terminal voltage of the DC motor is changed by changing the duty cycle of the PWM, thereby completing the speed control of the DC series motor.

3.1 Simulation Results and Analysis

Before the simulation, the proportional and integral parameters of the fuzzy adaptive PI controller need to be initialized. Firstly, the fuzzy adaptive PI parameters are proportional parameter and integral parameter: kp = 0.3, ki = 0.00038. Then set the parameters in the control of the current loop: kp = 9, ki = 6.5, kd = 5. Finally, the traditional PID control strategy is compared with the fuzzy adaptive PI control strategy, and the two control strategies are simulated with the same parameters. The simulation parameter has been given: the given speed n = 2000 r/min. The simulated waveforms are shown in Fig. 3 and Fig. 4.

It can be seen from the simulated waveforms of Figs. 3 and 4. In the case of setting the speed n = 2000 r/min, the speed control system of the motor based on fuzzy PI

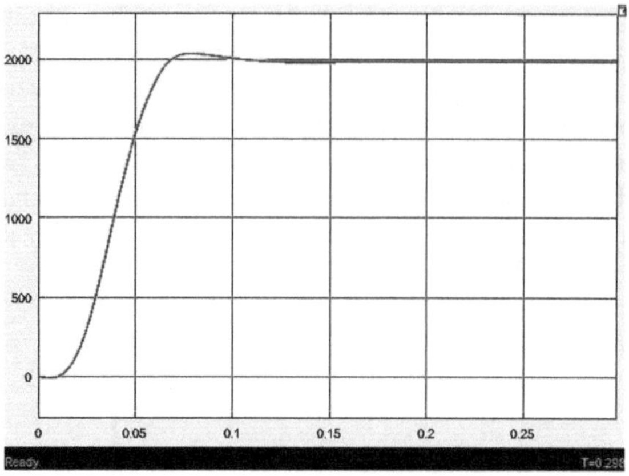

Fig. 3. Fuzzy adaptive PI speed waveform.

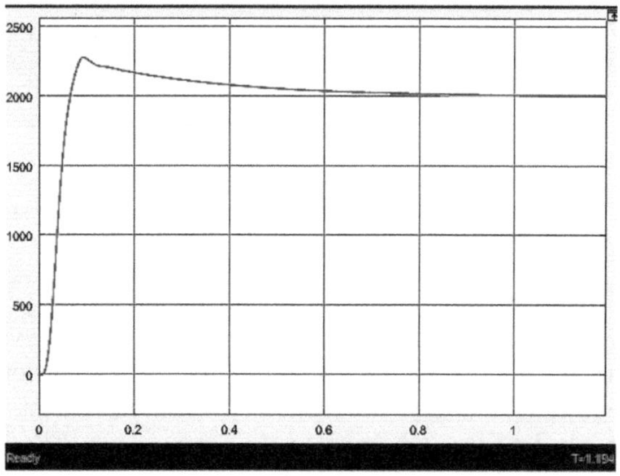

Fig. 4. Traditional PID control speed waveform.

control strategy runs smoothly with little fluctuation. In the case of setting the speed to 2000 r/min, the system responds faster than the ordinary PID control. The system enters a stable state at 0.1 s, the speed is stable, and there is not much fluctuation. The speed curve of the motor under conventional PID control fluctuates slightly, the recovery time is slow, and it does not enter a stable state until 0.8 s. The comparison of simulation results proves the superiority and reliability of the fuzzy adaptive PI control method.

4 Experimental Analysis

The current national standard for experimental methods is GB/T1311-2008. The experimental project for the wind power variable-pitch DC servo motor is basically the same as the ordinary DC motor experimental project. The motor load experiment starts with 1/10 of the rated load of the proto type and measures the input voltage, input current, and input power. By gradually adjusting the size of the load, the various electrical parameters of the prototype at different loads are recorded, and the efficiency value of the motor at different loads is calculated. The collated experimental data is shown in Fig. 5 and Fig. 6.

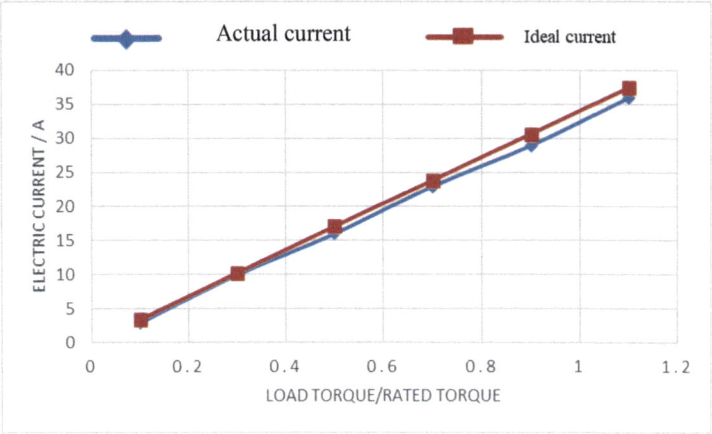

Fig. 5. Comparison of actual current and ideal current with load changes.

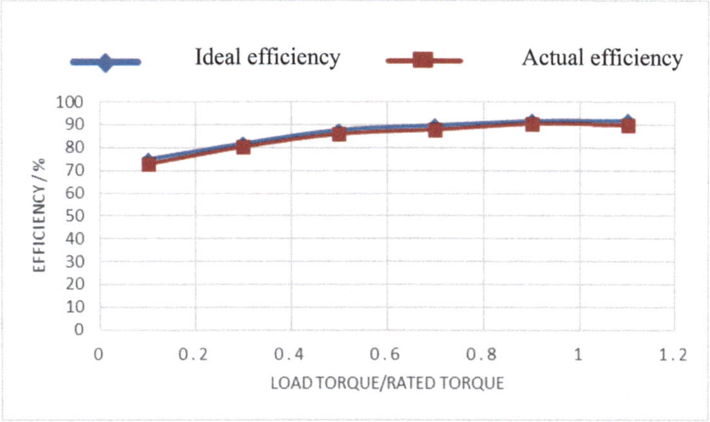

Fig. 6. Comparison of actual efficiency and ideal efficiency with load changes.

The result of simulation experiment is basically consistent with the result of design calculation, by comparing the measured date in the Simulink with the calculated data, which meets the requirements of the variable pitch servo drive of a certain type of wind turbine.

5 Conclusions

The paper proposes a control method of the wind power variable-pitch DC servo system. Aiming at the defects of traditional control methods, the paper combined fuzzy control method and PID control method, and proposes a fuzzy adaptive PI control method. It has the advantages of small overshoot, fast rise time, strong robustness and good dynamic characteristics of fuzzy control, but also maintains the characteristics of good dynamic tracking and stable accuracy of PID control. In the design process of fuzzy adaptive PI control system, the simulation model of the DC series motor servo speed control system was established by using Matlab/Simulnk, and the double closed-loop servo speed control system based on fuzzy adaptive PI control was simulated. Experiments have been carried out by running DC series servo motors and servo driver. The performance of simulation and experiment is relatively good, but there is still a lot of room for optimization, which provides a foundation for further optimization of the wind power variable pitch servo system.

Fund Project. Research project funded by the Hunan Provincial Department of Education (18A348), (18K092); Project funded by the Hunan Science and Technology Program (2016GK2018).

References

1. Qian, M., Xu, M.Q., Mi, Z.N.: Analysis of power MOSFET parallel drive characteristics. Semicond. Technol. **11**, 951–956 (2007)
2. Wei, Y.P., Guan, Q., Yang, L.: DSP minimum system design and realization of basic algorithm. Exp. Sci. Technol. **6**, 117–119 (2006)
3. Zou, Y.H.: Research on speed control system of permanent magnet brushless DC motor based on fuzzy control. Harbin Institute of Technology (2009)
4. Huang, L.S.: Research on three-phase interleaved parallel BUCK-BOOST bidirectional DC/DC converter. Hunan University of Science and Technology (2015)
5. Li, C.: Research on stepless speed regulation system of brushless DC motor based on fuzzy adaptive PI control. Nanjing Agricultural University (2011)
6. Zhao, L.X.: Design of DC motor controller based on fuzzy control. Dalian University of Technology (2014)
7. Tong, S., Li, H.X.: Fuzzy adaptive sliding-mode control for MIMO non-linear systems. IEEE Trans. Fuzzy Syst. **11**(3), 354–360 (2003)
8. Wang, R., Yu, F., Wang, J.: Adaptive non-backstepping fuzzy control for a class of uncertain nonlinear systems with unknown dead-zone input. Asian J. Control **17**(4), 1394–1402 (2015)
9. Xing, L.T., Wen, C.Y., Liu, Z.T., et al.: Event-triggered adaptive control for a class of uncertain nonlinear systems. IEEE Trans. Autom. Control **62**(4), 2071–2076 (2017)
10. Xing, L., Wen, C., Liu, Z., Su, H., Cai, J.: Event-triggered adaptive control for a class of uncertain nonlinear systems. IEEE Trans. Autom. Control **62**(4), 2071–2076 (2016)
11. Polycarpou, M.M., Mears, M.J.: Stable adaptive tracking of uncertain systems using nonlinearly parametrized on-line approximators. Int. J. Control **70**(3), 363–384 (1998)

12. Wang, M., Zhang, Z.: Globally adaptive asymptotic tracking control of non-linear systems using nonlinearly parameterized fuzzy approximator. J. Franklin Inst. **352**(7), 2783–2795 (2015)
13. Ma, L., Huo, X., Zhao, X., Ong, G.: Fuzzy adaptive tracking control for a class of uncertain switched nonlinear systems with multiple constraints: a small-gain approach. Int. J. Fuzzy Syst. **21**(8), 2609–2624 (2019)
14. Liu, Y.J., Tong, S., Chen, C.P.: Fuzzy adaptive control via observer design for uncertain nonlinear systems with unmodeled dynamics. IEEE Trans. Fuzzy Syst. **21**(2), 275–288 (2012)
15. Passino, K.M., Yurkovich, S., Reinfrank, M.: Fuzzy Control, vol. 42, pp. 15–21. Addison-wesley, Menlo Park (1998)
16. Chen, B., Liu, X., Liu, K., Lin, C.: Direct fuzzy adaptive control of nonlinear strict-feedback systems. Automatica **45**(6), 1530–1535 (2009)
17. Lin, F.J.: Fuzzy adaptive model-following position control for ultrasonic motor. IEEE Trans. Power Electron. **12**(2), 261–268 (1997)
18. Wang, W.Y., Chien, Y.H., Li, I.H.: An on-line robust and adaptive T-S fuzzy-neural controller for more general unknown systems. Int. J. Fuzzy Syst. **10**(1), 24–43 (2008)
19. Tong, S., Liu, C., Li, Y.: Fuzzy-adaptive decentralized output-feedback control for large-scale nonlinear systems with dynamical uncertainties. IEEE Trans. Fuzzy Syst. **18**(5), 845–861 (2010)
20. Cox, E.: Fuzzy adaptive systems. IEEE Spectr. **30**(2), 27–31 (1993)
21. Jwo, D.J., Yang, C.F., Chuang, C.H., et al.: Performance enhancement for ultra-tight GPS/INS integration using a fuzzy adaptive strong tracking unscented Kalman filter. Nonlinear Dyn. **73**(1–2), 377–395 (2013)
22. Niknam, T., Firouzi, B.B., Ostadi, A.: A new fuzzy adaptive particle swarm optimization for daily Volt/Var control in distribution networks considering distributed generators. Appl. Energy **87**(6), 1919–1928 (2010)
23. Elmas, C., Deperlioglu, O., Sayan, H.H.: Fuzzy adaptive logic controller for DC-DC converters. Expert Syst. Appl. **36**(2p1), 1540–1548 (2009)
24. Wang, C.X., Bai, X.M., Lv, F.D.: Design on embedded driving and control system for direct current servo motor, pp. 1307–1312. Trans Tech Publications (2013)
25. Yoshda, S.: Variable speed-variable pitch controllers for aero-servo-elastic simulations of wind turbine support structures. J. Fluid Sci. Technol. **6**(3), 300–312 (2011)

Artificial Topographic Structure of CycleGAN for Stroke Patients' Motor Imagery Recognition

Fenqi Rong[1], Tao Sun[2(✉)], Fangzhou Xu[3], Yang Zhang[4], Fei Lin[3], Xiaofang Wang[3], and Lei Wang[3]

[1] School of Electrical Engineering and Automation, Qilu University of Technology (Shandong Academy of Sciences), Jinan 250353, Shandong, China
[2] Qilu University of Technology (Shandong Academy of Sciences), Jinan 250353, Shandong, China
tutorsun@163.com
[3] School of Electronic and Information Engineering (Department of Physics), Qilu University of Technology (Shandong Academy of Sciences), Jinan 250353, Shandong, China
[4] Department of Physical Medicine and Rehabilitation, Qilu Hospital, Cheeloo College of Medicine, Shandong University, Jinan 250012, Shandong, China

Abstract. Motor imagery-based Brain-Computer Interface (MI-BCI) has already become one of the hottest research fields and has made great achievement in stroke rehabilitation. Because it is difficult for stroke patients to complete a series of motor imagery tasks, the phenomenon of insufficient data in signal processing and application is often faced. At present, the most effective way to overcome the challenge is to expand the amount of data for analysis. In this paper, we used CycleGAN for data augmentation on brain activities. First, electroencephalogram data is transformed into brain topographic map through EEG2Image. After that, we train a CycleGAN neural network. The topographic maps of healthy people and stroke patients are put into the CycleGAN model for training at the same time. In the process of training, let healthy people learn the features of the stroke patients topographic maps and generate the artificial stroke patients topographic maps. The generated topographic map can be very intuitive to see the similarity with the original topographic map. Experiments demonstrate that the method effectively generate artificial topographic maps to expand stroke dataset. And these topographic maps can be applied to data processing and analysis of MI-BCI.

Keywords: Motor imagery · Stroke · Topographic maps · CycleGAN

1 Introduction

Brain-Computer Interface (BCI) provides an approach for people to interact with the environment and control the external equipment directly without depending on peripheral nerves and muscles [1]. Various of applications, which include brain typing, personal recognition, visual image generation using brainwaves, disease identification,

Q. Liu et al. (Eds.): CENet 2020, AISC 1274, pp. 60–67, 2021.
https://doi.org/10.1007/978-981-15-8462-6_7

stroke rehabilitation and even military have been extensively explored [2]. The whole BCI system includes signal acquisition, signal processing and application (Fig. 1).

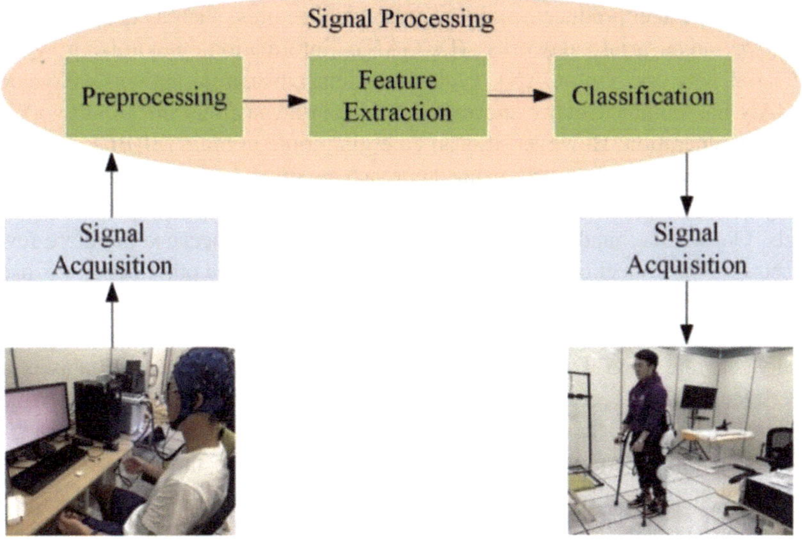

Fig. 1. The schematic of an EEG based MI-BCI system.

There are two main types of BCI signal acquisition methods: invasive and non-invasive. Electrocorticography (ECoG) based invasive BCI system and Electroencephalogram (EEG) based no-invasive BCI system are essential input signal [3].

Motor imagery-based BCI (MI-BCI) is a crucial orientation of BCI system research. It can effectively improve the ability of stroke patients to communicate with the outside world [4]. Traditional methods have been widely used in various phases of MI-BCI system, such as wavelet transform (WT), common spatial pattern (CSP), support vector machine (SVM), etc. Recently, deep learning proposed by Geoffrey Hinton has been widely used in many aspects, but only a small amount of studies has been carried out to explore the application of deep learning for MI-BCI classification. In 2014, a deep belief network (DBN) was used as a classifier to applied for two class MI-BCI classification and convolutional deep belief networks was applied to extract features from EEG signal [5, 6]. In a paper by Shiu Kumar et al., CSP-DNN framework was used as a classifier [7]. In 2017, an end-to-end deep learning way was proposed. This approach uses long short-term memory (LSTM) convolutional neural network (CNN) to classify raw EEG data without any pre-processing [8].

The key part of MI-BCI is how to extract features effectively and classify accurately. Before that, acquiring EEG signal is also a crucial step. But MI-BCI task is a tedious experimental process, which is difficult for stroke patients to complete. Therefore, it is difficult to carry out research in the field of deep learning.

In 2014, Ian J. Goodfellow et al. proposed a structure named generative adversarial networks (GAN) [9]. Gan adopts an unsupervised learning method, which can be used not only in unsupervised learning but also in semi-supervised learning. As a new deep learning network, GAN has been widely concerned once it was raised. At the same time, GAN has also produced many popular architectures, such as deep convolutional generative adversarial networks (DCGAN), information maximizing generative adversarial networks (InfoGAN), cycle-consistent adversarial networks (CycleGAN), etc. GAN has obtained many successful applications, such as image, artwork, music and video generation. However, it is an emerging topic in brain MI-BCI. Some studies have shown that Gan is a promising approach to solve problem related to BCI processing. At present, the application of Gan in EEG can be roughly divided into two aspects. On the one hand, it is used to generate EEG data because there are few EEG data sets and the collection steps are complicated [10]. On the other hand, it is used as a filter to filter out redundant information in EEG data [11].

It is feasible to generate EEG data by using GAN. In this paper, CycleGAN will be used to generate EEG Topographic maps. And these artificial EEG datasets can be applied for signal processing such as especially to solve the problem which are insufficient data and data corruption.

The remaining parts of this paper are included: Sect. 2 introduces some theories and the structure of this model. Results and discussion are shown and analyzed in Sect. 3. Finally, Sect. 4 includes conclusion of this paper and prospect or vision for the future.

2 Methods

2.1 Experimental Data

The proposed scheme has been evaluated on private dataset collected in the Department of Physical Medicine & Rehabilitation, Qilu hospital, Cheeloo College of medicine, Shandong University. The EEG dataset includes trials of 5 healthy subjects and 6 stroke patients. In the experiment, everyone has done 70 MI tasks under visual cues, including 60 imaging tasks of left and right hand and 10 gripping tasks (the subject imaged holding the limb tightly). EEG data have been recorded from 64 electrodes with a sampling frequency of 1000 Hz.

2.2 EEG2Image

EEG2Image can transform EEG activities into a train of multi-spectral images [12]. The procedure is as follows. First, mu (8–13 Hz) and beta (18–26 Hz) are the two main frequency bands of cortical oscillations related to MI. The sum of squared absolute in two frequency bands was calculated respectively and each electrode is described by two scalar values [11, 13]. Then, determining the coordinate positions of 60 electrodes in a 4 * 4 grid and plotting photos of the two frequency bands and combine them into one (Fig. 2). The transformed pictures can be used in the next experiment.

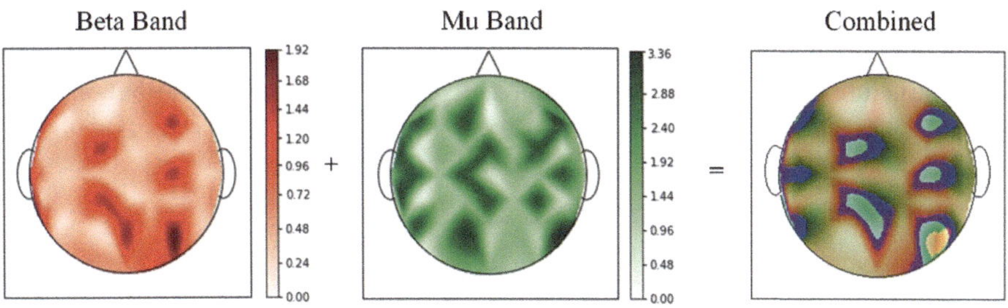

Fig. 2. Topographic map with two frequency bands.

2.3 GAN

The GAN framework consists of two opposing networks trying to outplay each other [9]. GAN has two networks: the generator (G) and the discriminator (D). Noise Z is input into G. And G tries the best to generate fake samples. The D exerts its utmost effort to distinguish between real and fake data. In general, the G generates samples that are not judged as fake by the D, and the D ties to identify the fake data generated by the G as much as possible, as shown in Fig. 3. The application of Gan in image generation is very mature [14]. And WGAN has been recently taken into consideration for the EEG data generation [10, 15].

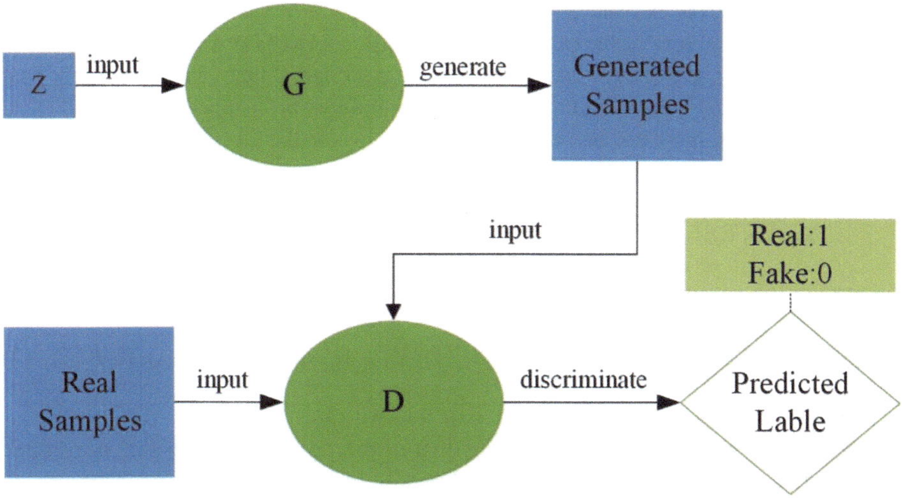

Fig. 3. Adversarial training procedure.

2.4 CycleGAN

CycleGAN, which was raised by Zhu et al. in 2017, is a popular structure for the image-to-image translation of unpaired image [16]. CycleGAN adopts an autoencoder-like structure to solve the problem of getting paired images. The structure of CycleGAN is shown in Fig. 4. G and P are two generators, D_s and D_t are two discriminators. The aim of this structure is that using G to map source distribution S to target distribution T [11].

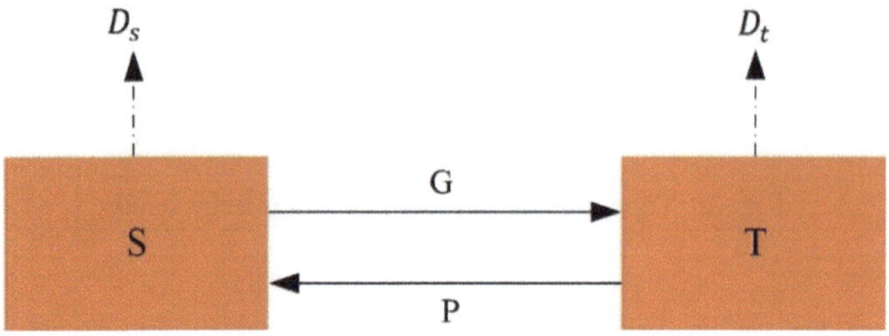

Fig. 4. The structure of CycleGAN.

CycleGAN, specifically, has two loops. The first loop is S → G(S) → P(G(S)) and the adversarial loss is defined as:

$$L_{GAN}(G, D_t, S, T) = E_{t \sim pdata(t)}[\log D_T(t)] + E_{s \sim pdata(s)}[\log(1 - D_T(G(s)))] \tag{1}$$

The goal of G is generating G(S) and let G(S) is similar to the samples in target distribution T. While, D_t aims to distinguish between G(S) generated by G and to control samples from domain S.

The adversarial loss of the other loop T → P(T) → G(P(T)) is in the similar to (1).

In this theory, CycleGAN can learn mappings G and P that can respectively generate samples identically distributed as T and S and can be mapped back.

Based on this, there are also cycle consistency loss in the CycleGAN structure. It is defined as:

$$L_{cyc}(G, P) = E_{s \sim pdata(s)}\left[\|P(G(s)) - s\|_1\right] + E_{t \sim pdata(t)}\left[\|P(G(t)) - t\|_1\right] \tag{2}$$

Finally, the total loss is as follows:

$$L(G, P, D_S, D_T) = L_{GAN}(G, D_T, S, T) + L_{GAN}(P, D_S, T, S) + \lambda L_{cyc}(G, P) \tag{3}$$

where λ controls the relative importance of the two objectives.

In this work, the two generators networks of the proposed CycleGAN have three parts: encoder, converter and decoder, as shown in Fig. 5(a). First, the encoder aims to extract features through the convolution layer. Then, the converter transforms the feature vector of a topographic map from the S to the T (or T to S) according to these features. The decoder finally converts the feature vector into a topographic map. The discriminator belongs to the convolutional network. As shown in Fig. 5(b), the discriminator consists of five convolutional layers. The first four convolutional layers extract image features, and the last convolutional layer determines whether the image is true or false.

The topographic maps of healthy people and stroke patients are input into the CycleGAN structure at the same time (the datasets of stroke patients are used as the control signals). In the game of generators and discriminators, the signals of healthy people learn the characteristics of stroke patients' signals, thereby generating more patients' signals to expand datasets of the stroke dataset.

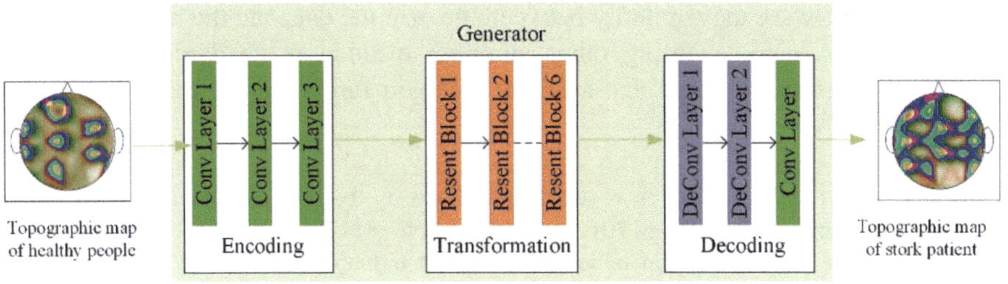

(a) The network structure of generator.

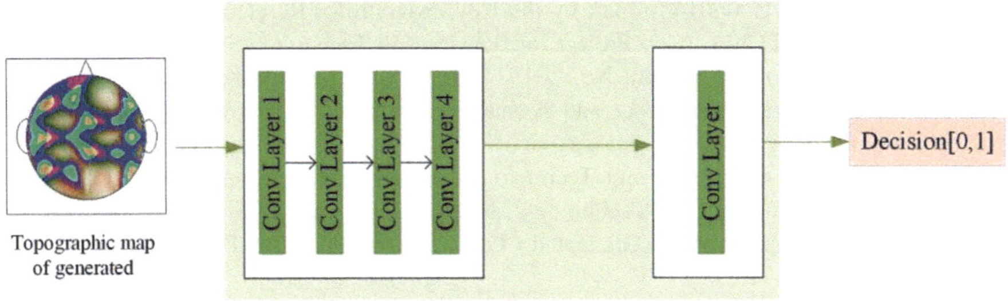

(b) The network structure of discriminator.

Fig. 5. The whole network structure of generator and discriminator.

3 Result and Discussion

This paper highlight that CycleGAN can learn the main features to output samples similar to the original data. Through this model, we have successfully transformed the EEG topographic maps of 5 healthy people into EEG topographic maps of stroke patients, thus expanding the dataset of stroke patients. In the process of training, we can judge the training result by observing the loss of generator and discriminator. After topographic maps generation, the similarity between the original data and the artificial data generation can be visualized.

The generated data can be further processed as the follow-up work of this study. According to the follow-up work, the required channels can be selected to reduce the longer calculation time and the generation of redundant data. And more network structures should be tried to achieve the simplest structure to generate the most similar data.

All in all, CycleGAN can generate higher quality brain topographic maps to prepare for the next study.

4 Conclusion

This paper mainly describes how to generate artificial dataset from CycleGAN to expand the amount of data for analysis. Firstly, the EEG data is transformed into topographic maps by using EEG2Image. Thereby, it can be seen from the generated topographic map that CycleGAN can effectively generate new data similar to the original data. Moreover, compared with other studies, the proposed method in this paper can directly see the similarity between the original data and the new data through the topographic map, avoiding other complex evaluation standards. The research should be continued to prove that artificial data generation can be applied to the BCI system.

Acknowledgement. This research is supported in part by the National Natural Science Foundation of China under Grant No. 61701270 and Grant No. 81472159, in part by the Program for Youth Innovative Research Team in University of Shandong Province in China under Grant No. 2019KJN010, in part by the Key Research and Development Plan of Shandong Province under Grant No. 2017G006014, in part by the Key Program for Research and Development of Shandong Province in China (Key Project for Science and Technology Innovation, Department and City Cooperation) under Grant No. 2019TSLH0315, in part by the Project of Shandong Province Higher Educational Science and Technology Program under Grant No. J18KA345, in part by the Jinan Program for Development of Science and Technology, in part by the Jinan Program for Leaders of Science and Technology, in part by the Young Doctor Cooperation Foundation of Qilu University of Technology (Shandong Academy of Sciences), and in part by the Shandong Province Post Graduated Education Innovation Program under Grant No. 5DYY16032.

References

1. Nicolasalonso, L.F., Gomezgil, J.: Brain computer interfaces, a review. Sensors **12**(2), 1211–1279 (2012)
2. Ang, K.K., Guan, C.: Brain-computer interface in stroke rehabilitation. J. Comput. Sci. Eng. **7**(2), 139–146 (2013)
3. Xu, F., Zhou, W., Zhen, Y., Yuan, Q., Wu, Q.: Using fractal and local binary pattern features for classification of ECoG motor imagery tasks obtained from the right brain hemisphere. Int. J. Neural Syst. **26**(6), 1650022 (2016)
4. Uehara, T., Sartori, M., Tanaka, T., Fiori, S.: Robust averaging of covariances for EEG recordings classification in motor imagery brain-computer interfaces. Neural Comput. **29**(6), 1631–1666 (2017)
5. Ren, Y., Wu, Y.: Convolutional deep belief networks for feature extraction of EEG signal. In: International Joint Conference on Neural Network 2014, Beijing, China, pp. 2850–2853 (2014)
6. Kumar, S., Sharma, A., Mamun, K.: A deep learning approach for motor imagery EEG signal classification. In: 2016 3rd Asia-Pacific World Congress on Computer Science and Engineering 2016, Nadi, Fiji, pp. 34–39 (2016)
7. An, X., Kuang, D., Guo, X., Zhao, Y., He, L.: A deep learning method for classification of EEG data based on motor imagery. In: International Conference on Intelligent Computing, 2014, Taiyuan, China, pp. 203–210 (2014)

8. Wang, P., Jiang, A., Liu, X., Shang, J., Zhang, L.: LSTM-based EEG classification in motor imagery tasks. IEEE Trans. Neural Syst. Rehabil. Eng. **26**(11), 2086–2095 (2018)
9. Goodfellow, I.J., Pouget-Abadie, J., Mirza, M., Xu, B., Warde-Farley, D., Ozair, S., Courville, A., Bengio, Y.: Generative adversarial nets. In: Neural Information Processing Systems, pp. 2672–2680 (2014)
10. Luo, Y.: EEG data augmentation for emotion recognition using a conditional wasserstein GAN. In: International Conference of the IEEE Engineering in Medicine and Biology Society 2018, Honolulu, HI, USA, pp. 2535–2538 (2018)
11. Yao, Y., Plested, J., Gedeon, T.: A feature filter for EEG using Cycle-GAN structure. In: International Conference on Neural Information Processing 2018, Siem Reap, Cambodia, pp. 567–576 (2018)
12. Siddharth, S., Jung, T.P., Sejnowski, T.J.: Utilizing deep learning towards multi-modal biosensing and vision-based affective computing. IEEE Trans. Affect. Comput. (2019)
13. Bashivan, P., Rish, I., Yeasin, M.: Learning representations from EEG with deep recurrent-convolutional neural networks. arXiv:1511.06448 (2015)
14. Mao, X., Li, Q., Xie, H., Lau, R.Y., Wang, Z., Smolley, S.P.: Least squares generative adversarial networks. In: International Conference on Computer Vision, pp. 2813–2821 (2017)
15. Hartmann, K.G., Schirrmeister, R.T., Ball, T.: EEG-GAN: generative adversarial networks for electroencephalograhic (EEG) brain signals. arXiv:1806.01875 (2018)
16. Zhu, J., Park, T., Isola, P., Efros, A.A.: Unpaired image-to-image translation using cycle-consistent adversarial networks. In: International Conference on Computer Vision, pp. 2242–2251 (2017)

Rail Defect Detection Method Based on BP Neural Network

Qinhua Xu[1], Qinjun Zhao[1], Liguo Wang[2], and Tao Shen[1(✉)]

[1] School of Electrical Engineering, University of Jinan, Jinan 250022, China
cse_st@ujn.edu.cn

[2] Jinan Working Business Section, China Railway Jinan Group Co., Ltd., Jinan 250001, China

Abstract. Ultrasonic flaw detection is the most commonly used method in rail defect detection, and the defect judgment in flaw detection data is the most important link. The B-scan data of ultrasonic flaw detection vehicle is used for data analysis and recovery. A set of rail defect identification model based on BP neural network is established according to the channel distribution characteristics, digital combination characteristics and the arrangement and combination characteristics in time sequence of the recovered data. This method takes the 1–6 channel detection data as an example, and trains the model by extracting the rail head nuclear defect as the sample and making the training set, Experiments show that the model can effectively identify the defect data by setting a suitable threshold value, and give the location of the rail defect, help the flaw detection workers to improve the efficiency of rail defect detection, greatly reduce the labor intensity, and avoid the missed judgment of the rail defect.

Keywords: Ultrasonic rail flaw detection vehicle · B-scan data · Rail defect detection · BP neural network

1 Introduction

At present, the development of China's high-speed railway is in a rapid growth stage. The intensive operation of high-speed trains and heavy loads have brought serious damage to the rails. This RCF (kind of rolling contact fatigue) caused by wheel rolling is very dangerous for the operation of railway traffic [1]. At present, in the railway flaw detection department of our country, the commonly used rail defect detection methods mostly adopt the ultrasonic flaw detection equipment to collect the data on the line, and then determine the rail defect on the terminal screen by manual discrimination. Although the application of nondestructive testing in railway rail has been very mature at present, the data defect judgment after the collection of flaw detection equipment always depends on the professional experience and interpretation ability of flaw detection workers, and will consume a lot of time and cost. Therefore, under the current tense railway maintenance and management conditions, it is particularly important to improve the efficiency of flaw detection by computer judgment and defect identification.

Q. Liu et al. (Eds.): CENet 2020, AISC 1274, pp. 68–78, 2021.
https://doi.org/10.1007/978-981-15-8462-6_8

In the research of rail defect detection system, Sperry company of the United States is at the forefront. It integrates the defect detection system into a large rail flaw detection vehicle. In the detection process, when a defect or a cluster defect is encountered, the flaw detection vehicle will automatically stop detection and conduct visual or human verification [2]. However, the detection sensitivity of general large-scale flaw detection vehicles is not high, and the detection cycle is long, which affects the normal operation of the line. The French state-owned railway company has installed high-speed track detection cameras to detect the visual rail surface defects [3]. Liu et al. from Beijing Jiaotong University have also studied the rail surface defects in terms of machine vision [4]. In theoretical research, early Wang extracted the feature quantity according to the characteristics of the defect on the B-scan image, and realized classification of rail defect based on support vector machine [5]. Zhongbei University and Fuzhou University set up the method of defect identification by extracting the characteristic parameters of ultrasonic A-scan data [6, 7]. Wu et al. established the optimization method of relevant parameters by extracting the contour features of rail defects [8]. Sun et al. of the orbital research institute directly regard the B-image as a binary sparse matrix composed of 16 channels, and use CNN's mature application in the image field to directly identify the damaged image [9]. In practical application, it is a very troublesome problem to transform a large number of flaw detection data into images.

There are several reasons for using computer to judge whether there is defect in the data of flaw detection vehicle, which has not been applied to the actual flaw detection work. The first is that when the rail is exposed to the external environment, the location and causes of the rail defect are complex and diverse, resulting in various types of rail defect. The other is that the same rail defect echo may be diversified, the exact expression of echo signal is difficult to find, and the traditional method is difficult to extract features for recognition and classification [10]. Thirdly, the normalization and standardization of operation equipment of flaw detection personnel will affect the quality of flaw detection data. Therefore, although many researchers have done a lot of theoretical research in this field, there is still no reliable and widely popularized computer detection method for rail defect in the current flaw detection system.

In this paper, after decompressing, analyzing and recovering the B-scan data, it is directly used as the input of neural network. Aiming at the rail defect type of rail head nuclear defect, a suitable BP neural network is specially designed, and a large number of data are collected to train the model, which makes the model have the ability of rail defect detection, and the pulse count output is used as the location information of rail defect. After the optimization, the requirements of the model for the inspectors are not high, and the rail defect detection meets the requirements of the manual analysis, especially in the speed of rail defect identification is far higher than the way of manual judgment. This method will be of great significance to reduce the working pressure of railway flaw detection personnel, improve the detection rate of rail defect and increase the safety guarantee coefficient of railway operation in China.

2 Preprocessing of B-Scan Data

2.1 Formation Principle of B-Scan Data

At present, in ultrasonic flaw detection equipment, B-scan image is widely used by flaw detection personnel because of its intuitive and obvious defect characteristics, while A-scan data is mostly used to assist in the judgment of rail defect. Figure 1 is a schematic diagram of nuclear defect detected by flaw detection equipment. When the ultrasonic pulse emitted by the probe is interrupted by damaged crack, slotted interface, screw hole and other media, the echo will be reflected, and the reflected signal will be received and amplified by the ultrasonic sensor, compared with the threshold value of the probe gate. If the echo voltage amplitude exceeds the current threshold value of the probe gate, the reflected signal will be recorded as the effective byte, and the flaw detection equipment will roll with the wheel, the effective signals in the echo are recorded in turn, and the B-scan image is formed after a series of drawing processing.

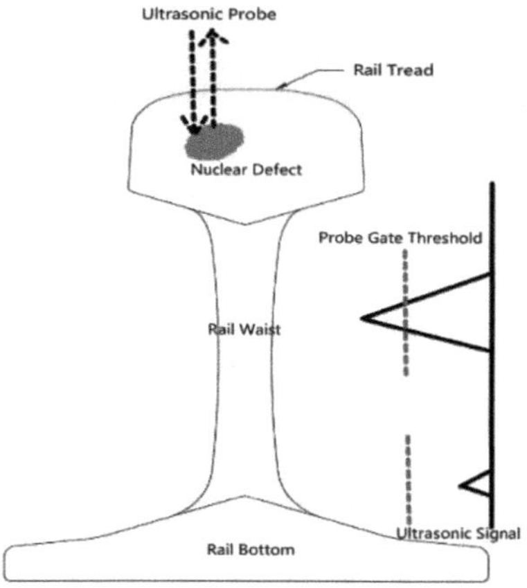

Fig. 1. A schematic diagram of nuclear defect detected by flaw detection equipment.

For the format of B-scan data, the first four bytes of B-scan data of each pulse are pulse count, the middle is channel number + mask + data, and the end is 0xff. Bit 1 in the mask indicates that the bit corresponds to a byte of B-scan data. The position of this data in the sampling point is clear. If the bit is 0 in the data, it indicates that it is an effective signal and needs to be mapped. Bit 1 indicates no signal, as shown in Fig. 2, a typical rail head nuclear defect is in the schematic diagram, B-scan image and B-scan data respectively. The unprocessed B-scan data can be regarded as a long sparse matrix along the rail direction.

The directly acquired B-scan data is encrypted and processed by the manufacturer of flaw detection equipment. Normally, it can only be viewed through B-scan image. If the B-scan data is directly processed as an image, the processing amount of the image is quite large and the processing process is quite complex. In this paper, the B-scan data is decompressed, analyzed and recovered, and it is processed into a digital matrix of 1–9

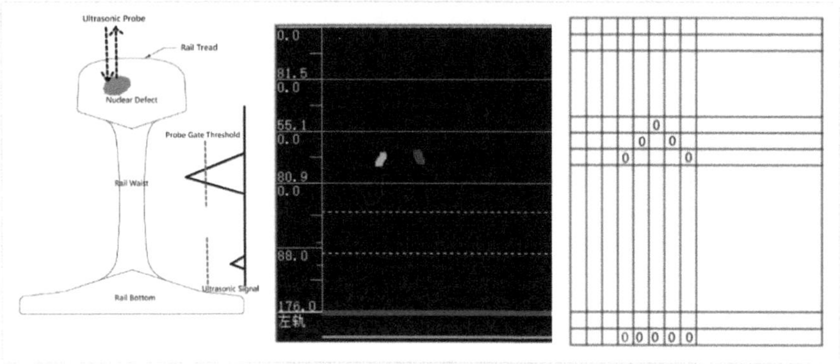

Fig. 2. Schematic diagram of B-scan data generation.

channels distributed according to the pulse time sequence. The processed B-scan data can be directly used as the input of neural network to complete the training target of rail defect detection model.

2.2 Recovery Processing of Flaw Detection Data

Because of reducing the file size and saving storage space, the original B-scan data is compressed to some extent. Before drawing, the original data of the channel needs to be recovered. For example, we need to restore channel 1–6 to 6 bytes, channel 7–8 to 15 bytes, and channel 9 to 10 bytes. This process requires shift traversal. The number of bytes sampled by 7–9 channel is greater than 8. We can define a USHORT value of 8000, and the number of bytes sampled by 1–6 channel is less than 8. We can define a USHORT value of 80 for shift control.

The following example shows how to recover the original B-scan data. The storage format of B-scan data in memory is as shown in Fig. 3. Among them, the pulse count represents the sampling time sequence, the channel number represents the different ultrasonic probes for detecting the rail damage information, and the mask represents the position of the data in the sampling point. If a bit in the data is 0, it indicates that it is an effective signal, which needs to be mapped 1 means no echo and no drawing is needed, so the B-scan image is actually a binary sparse matrix superimposed by 18 channels. It can be seen from the generation principle of B-scan image that the rail flaw detection data under each pulse has obvious digital combination characteristics.

Channel **08** is channel 9 of left track. We first define an array of 10 bytes. The initial value is 0xff. The mask **40 00** indicates the location of valid data. The PC is in small end mode. The value after reading is **00 40**. Start 0x8000 from 0 and perform bit by bit and operation with **00 40** successively. If the result is not 0, it means that there is a data in this bit, and then move it to the right to continue and operate. The above data are bitwise and until the 10th result is not 0, indicating that 0xe7 is the 10th data. The bit of bit 1 in the mask corresponds to one data. To sum up, channel **08** can be recovered: 0xff 0xff 0xff 0xff 0xff 0xff 0xff 0xff 0xff 0xe7.

[00 00 00 01]	[08 40 00 E7]	[FF]
Current pulse count Pulse count starts from **0** at the beginning of each B-scan	**08** for channel number **4000** for mask **E7** for real data	Data at each pulse ends with **FF**

Fig. 3. Example of original storage format of B-scan image.

1–6 channel defines a 6-byte array, and then carries out bitwise sum of 0x80 and mask, and then moves one bit to the right to recover the original data of 6 bytes.

3 Establish Rail Defect Detection Model

3.1 BP Neural Network Model

BP (back propagation) neural network is the most traditional neural network, which uses back propagation algorithm. Through the training of sample data, it constantly corrects the network weight and threshold value to make the error function descend along the negative gradient direction and approach the expected output, that is to say, by making the "Deviation" as small as possible, it makes the neural network model fit the data set as well as possible (Fig. 4).

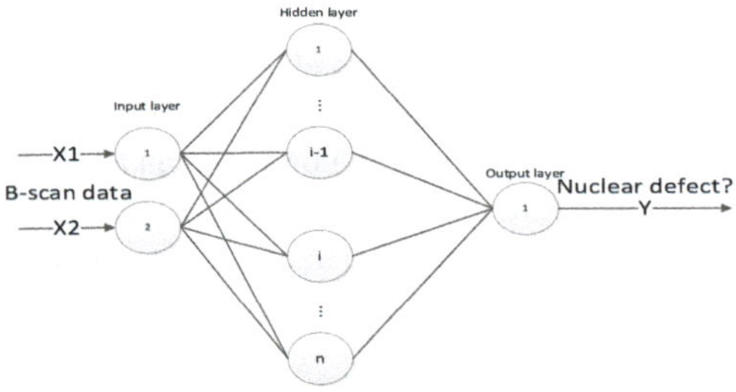

Fig. 4. BP neural network structure diagram.

When the tested B-scan data is input, in this model, each neuron receives input signals (B-scan data) from the previous neuron, and each signal is transmitted through a weighted connection. The neuron adds these signals to get a total input value, then compares the total input value with the neuron's threshold value, and then passes an "activation function" Process to get the final output. Here, the final output is the prediction result of the input B-scan data. If the output result is close to 1, it is determined to contain the data of rail head damage, and the corresponding pulse count output of the damage data is the location information of the damage. Correspondingly, if the final output result is close to 0, then it means that this is normal B-scan data without returning the location information. The acquisition of weight parameters is the process of optimizing the neural network model with training set.

The operation process of damage pattern classification based on BP neural network is as follows: first, prepare the data, decompress, analyze and recover the acquired B-scan data, obtain the original B-scan data of 18 channels of left and right rail and make the data set. Then the ultrasonic B-scan data is input into the rail damage detection system. The data can be directly input as neural network. After the trained neural network model, the classification results of the damage are determined according to the judgment rules, and finally the location information of the damage data is output. The flow chart is as follows (Fig. 5):

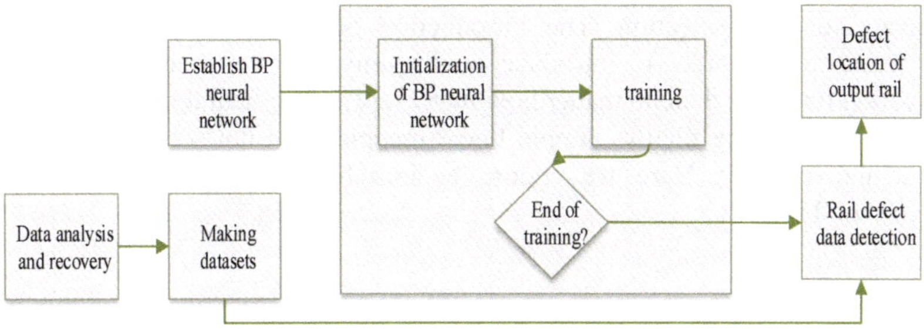

Fig. 5. Flow chart of rail defect detection model.

3.2 Structure Design of BP Neural Network

Because of the characteristics of channel distribution in different defect types of ultrasonic B-scan data, we designed different neural network structures for different defect types and trained them separately. In this paper, we design the following neural network for the example of rail head nuclear defect:

1) Number of input layer nodes. The number of nodes in the input layer depends on the dimension of the input vector. In the recovered ultrasonic B-scan data, the information related to rail head nuclear injury exists in channel 1–6, and the B-scan data under each pulse count in channel 1–6 is composed of six hexadecimal numbers. After we use the decimal number to replace it, it is composed of 12 numbers. After many experiments, we finally choose 15 sequences as a unit of defect identification, so the number of input nodes of the neural network is: $12 * 15 = 180$.
2) Number of output layer nodes. The number of neurons in the output layer also needs to be determined by the abstract model from the actual problem. Here, the output result is whether the rail defect exists or not. Therefore, the output layer

adopts one output node. 0 represents the normal B-scan data, and 1 represents the existence of such rail defect.

3) Hidden layers of neural network. BP neural network can contain one or more hidden layers. However, it has been proved theoretically that the network of a single hidden layer can realize any nonlinear mapping by appropriately increasing the number of neuron nodes. For most cases, a single hidden layer can meet the requirements. However, if there are many samples, adding a hidden layer can significantly reduce the network size. In the experiment, we choose a single hidden layer.
4) Number of hidden layer nodes. The number of hidden layer nodes has a great influence on the performance of BP neural network. Generally, more hidden layer nodes can bring better performance, but it may lead to too long training time. Here, we use the empirical formula to give the number of hidden layer nodes: $\sqrt{m+n}+a$ (m and n are the number of neurons in the output layer and the input layer respectively. a is the number between [0, 10]). According to this formula, the number of hidden layer nodes is set to 18.
5) Activate function selection. The introduction of activation function is a very important step in order to introduce nonlinearity into the model. If there is no activation function, then no matter how many layers your neural network has, it will be a linear mapping finally. Simple linear mapping can not solve the problem of linear indivisibility. Here we choose to introduce the *Relu* function: $f(u) = \max(0, u)$ (Fig. 6).

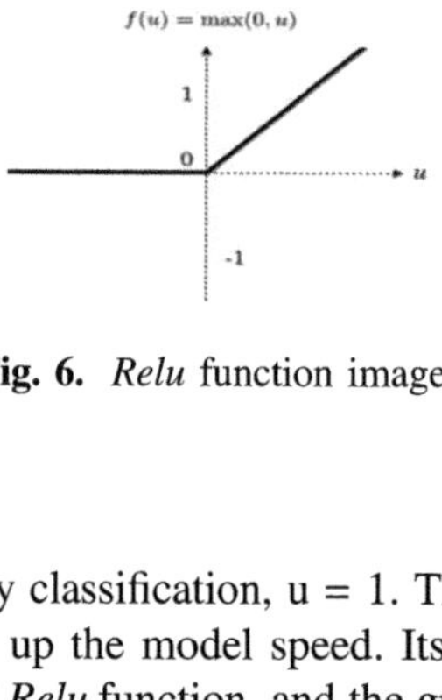

Fig. 6. *Relu* function image.

Because our output is binary classification, u = 1. The calculation of *Relu* function is simple, which can speed up the model speed. Its output result is either 1 or 0. There are only two kinds of *Relu* function, and the gradient will not be close to 0 at the edge like sigmoid. When there are many layers, the gradient may disappear, so the model cannot converge.

6) Initial weight determination. BP neural network uses iterative updating method to determine the weight, and needs an initial value, but too large or too small will affect the output results, here we choose the initial value of normal distribution for initialization.
7) Cost Function. We need to use cost function to measure how close the predicted output value is to the actual value. Suppose there are m samples in the training set, y represents the actual value in the original training sample, that is, the standard

answer, i represents the i th sample, h_θ represents the y value predicted by the parameters θ and x. Through training and learning, the whole model needs to find the weight parameter that makes the cost function J the lowest. When J is the lowest, that is, when the model fitting and prediction ability is the best, the total cost function is deduced:

$$J(\theta) = -\frac{1}{m}\left[\sum_{i=1}^{m}(y^i logh_\theta(x^i) + (1-y^i)\log(1-h_\theta(x^i))\right]$$

4 Training and Testing

The training and testing data are from RT18-D double track rail ultrasonic flaw detection vehicle, which is collected in Jinan Working Business Section, Jinan Railway Bureau. More than 5000 B-scan image samples were collected, including more than 1100 rail head nuclear defects. The schematic diagram of the rail head nuclear damage in the B-scan image is as follows. The echo of channel 1–6 in the B-scan image is in the upper half, where the rail head nuclear defect is marked with a white box, and the normal echo is not marked in this range.

All the collected B-scan data are stored in Excel file after decompressing, analyzing and recovering. After all the data are fully scrambled, 90% of them are training sets and the remaining 10% are test sets. The positive and negative samples in the training set are distinguished by the color marking method. The sequence with rail head nuclear defect is marked in green, the sequence with normal echo is marked in red, and the sequence data without echo is null. When loading data, the null value is filled with a single character (Fig. 7).

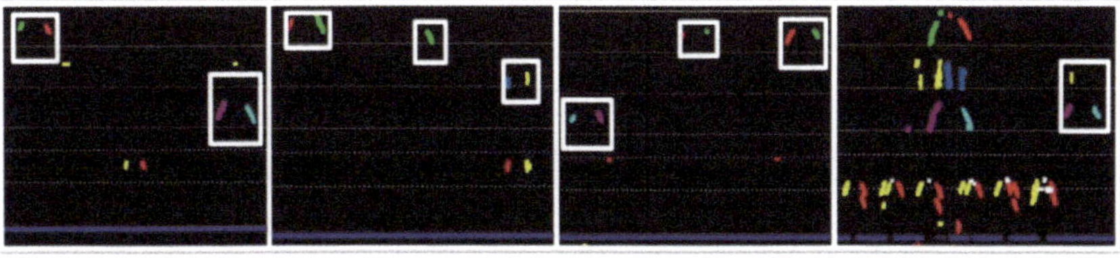

Fig. 7. B-scan of rail head nuclear defect.

It can be seen from Fig. 8 that in the training process, the cost function rapidly decreases and converges with the increase of calculation times, which means that the difference between the actual value in the training set sample and the predicted value of the model becomes smaller and smaller with the training of the model, and the prediction ability of the model approaches the actual value. After about 11000 times of training, the training is stopped and the model training is completed.

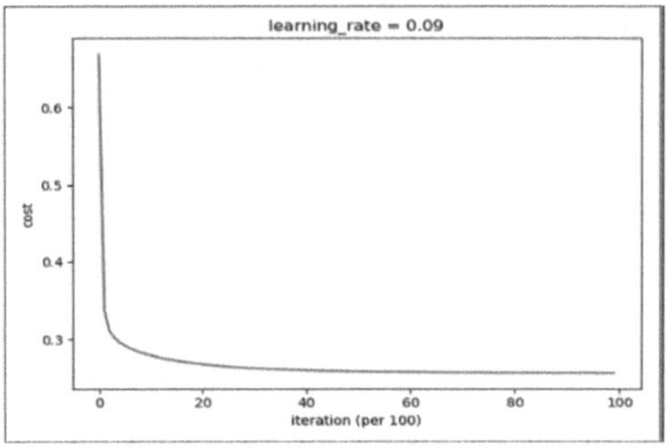

Fig. 8. Convergence graph of cost function with training times.

In the experiment, the BP neural network model is tested on 500 sample test sets. When the threshold value is 0.55, the accuracy of identifying the damage sequence reaches 95.81, and the threshold value is 0.65, The recognition accuracy of the damage sequence has reached 96.13%. When the threshold is increased, the false detection rate will also be increased, and the number of suspected defects will be increased. Therefore, it is necessary to carefully select the appropriate threshold to ensure the accuracy and reduce the number of suspected defects.

In order to further verify the recognition ability of the network model for the rail head nuclear damage, another 100 B-scan data samples of Yucheng calibration line in Jinan Working Business Section are taken, including 20 samples of rail head nuclear damage, 80 samples of normal echo, and the output value of the model is directly recorded in the test. The test results are randomly selected and 20 samples are as follows (Table 1):

Table 1. Inspection statistics of rail head nuclear defect samples.

Sample number	Whether it is rail head nuclear defect	Expected output	Actual output
1	Yes	1	1
2	Yes	1	1
3	Yes	1	1
4	Yes	1	1
5	Yes	1	1
6	Yes	1	1
7	No	0	0
8	No	0	0
9	No	0	0
10	No	0	0
11	No	0	0
12	No	0	0
13	No	0	0
14	No	0	0

(continued)

Table 1. (*continued*)

Sample number	Whether it is rail head nuclear defect	Expected output	Actual output
15	No	0	0
16	No	0	0
17	No	0	0
18	No	0	0
19	Yes	1	1
20	No	0	0

It can be seen from the test results that the predicted value of the rail defect detection model based on BP neural network is completely consistent with the actual value of the sample. After many inspections and tests, as long as the appropriate threshold value is set, the rail head nuclear defect can be detected accurately.

5 Conclusion

A method of rail defect detection based on BP neural network is provided. The model can be transplanted to other types of rail defect detection after proper modification besides detecting rail head nuclear defect. The detection accuracy reaches the standard of manual judgment of rail defects, and the detection speed is much faster than that of manual. In the later stage, more rail defect samples can be used to improve the model.

Acknowledgement. This work is supported by Key R&D projects of Shandong Province under grant 2019GNC106093, Shandong Agricultural machinery equipment R&D innovation plan project under grant 2018YF011, and Key R&D projects of Shandong Province under grant 2019JZZY021005.

References

1. Dong, Y.: Nondestructive testing method and its application in rail flaw detection. Theor. Res. Urban Constr. **013**(001), 78–79 (2018)
2. Zhao, Y., Chen, J., Sun, J.: The application and progress of nondestructive testing technique and system for the rail in-service. Nondestr. Test. **36**(3), 58–64 (2014)
3. Sun, C., Zhang, Y.: Research on automatic early warning method for rail flaw based on intelligent identification and periodic detection. J. China Railway Soc. **40**(11), 140–146 (2018)
4. Liu, Z., Wang, W., Wang, P.: Design of machine vision system for inspection of rail surface defects. J. Electron. Meas. Instrum. **24**(11), 1012–1017 (2010)
5. Wang, H., Li, C.: Real-time automatic ultrasonic rail damage detection and classification system based on SVM. J. Univ. Chin. Acad. Sci. **26**(4), 517–521 (2009)
6. Huang, S.: Research on some key techniques of intelligent defect recognition in the ultrasonic A-scan testing. Fuzhou University (2013)
7. Li, W.: Research on ultrasonic testing of rail defects and positioning. Zhongbei University (2013)

8. Wu, F., Li, Q., Li, S.: Train rail defect classification detection and its parameters learning method. Measurement **151**, 107246 (2020)
9. Sun, C., Liu, J., Qin, Y.: Intelligent detection method for rail flaw based on deep learning. China Railway Sci. **39**(5), 51–57 (2018)
10. Alahakoon, S., Sun, Y., Spiryagin, M.: Rail flaw detection technologies for safer, reliable transportation: a review. J. Dyn. Syst. Meas. Control **140**(2), 020801(2018)

Apple Grading Method Based on GA-SVM

Zheng Xu[1], Qinjun Zhao[1], Yang Zhang[1], Yuhua Zhang[2], and Tao Shen[1(✉)]

[1] School of Electrical Engineering, University of Jinan, Jinan 250000, China
cse_st@ujn.edu.cn, rongqing1986@163.com
[2] National Engineering Research Center for Agricultural Products Logistics, Jinan 250103, China

Abstract. Apple's online grading technology is an important part of Apple's commercialization. In order to further improve the efficiency of apple grading, In this paper, we study Fuji apple grading method based on machine vision and support vector machine model optimized by genetic algorithm. An apple image acquisition system was built, and the median filtering method was used to remove the noise of the image, then the target region was obtained using canny edge detection algorithm combined with morphological processing. Finally, serving as input, the roundness of apple, mean value of H component in the HSI color space, and the surface defect as the characteristic parameters was sent into the GA-SVM model. The experimental results show that the classification accuracy rate is 92.3%.

Keywords: Machine vision · Image processing · Genetic algorithm · Support vector machine

1 Introduction

The good development of agriculture is an important cornerstone of the country, and the quality of agricultural products is the key factor to measure its value. As one of the most important fruit trees cultivated in temperate areas, apple is widely loved for its rich nutritional value and sweet taste. As the largest apple production and consumption country in the world, the position of apple industry in the national economic development is self-evident. According to the data newly revised by the agricultural census of the National Bureau of statistics, China's apple production has been on the rise for seven years, including 40.93 million tons in 2016, 41.93 million tons in 2017 and 39.23 million tons in 2018 [1]. China's apple production has always accounted for more than 50% of the world's apple production, and the proportion is still increasing year by year. At the same time, China's apple export volume still maintains the first position in the world. Only in 2018, China's apple export volume was 1115000 tons, but in stark contrast to that, China's apple export volume accounted for only 1.83%–3.15% of the total output, while the world's apple export volume accounted for about 8.30% of the total output [2]. This is because China's post production processing technology and automatic grading and screening capacity are relatively weak, which to some extent restricts the export of Chinese apples.

Q. Liu et al. (Eds.): CENet 2020, AISC 1274, pp. 79–89, 2021.
https://doi.org/10.1007/978-981-15-8462-6_9

Since the 1970s, the hardware technology and image processing algorithm of computer equipment have made great progress, and machine vision technology has penetrated into various fields [3]. In recent years, the application of machine vision technology in fruit quality detection has been widely studied and applied. Moallem et al. [4] used a variety of segmentation methods to refine the defect area, and then extracted the texture and geometry of the defect area, then used SVM, MLP and K-Nearest Neighbor (KNN) to classify it, SVM classifier achieved the best results in 120 different golden crown apple images finally. Huang et al. [5] proposed an image segmentation method based on the three-layer Canny operator to obtain apple contours in order to overcome the influence of light, and then used the particle swarm optimization SVM model to analyze the fruit shape, texture and color distribution of apples. Yu et al. [6] Selected the S component in the HSI color space combined with the Otsu algorithm to extract the apple contour, and used the value of H component as the characteristic parameter of apple color grading. The support vector machine was used to classify the apple, and the accuracy rate is 89%.

In this paper, the shape, color and surface defects of apple are classified by machine vision and digital image processing. Firstly, the apple region is obtained by image preprocessing, including background segmentation, median filtering, edge detection and morphological processing. Then, the shape, color and defect characteristics of apple are extracted. Finally, genetic algorithm is used to optimize support vector machine model for classification.

2 Image Processing

The original image obtained through the apple image acquisition system is shown in Fig. 1. The background has chain, tray and other content that we are not interested in. If the original image is processed and analyzed directly, it will cause inaccurate feature extraction and classification. Therefore, we need to preprocess the original image to get the apple region that we are interested in.

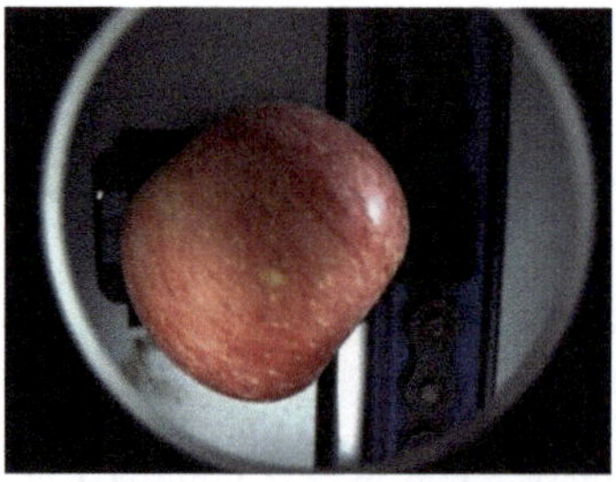

Fig. 1. Original image.

2.1 Noise Removal

In the process of image acquisition and processing, some noise may be generated in the sensor and scanner circuits, which will lead to the uncertainty decrease of image quality. If the noise is not removed before image segmentation, more unexpected objects may be detected. There are usually three traditional noise removal algorithms, namely median filter, mean filter and gaussian filter. Because salt-and-pepper noise is easy to be produced in the experiment, so we choose median filter in this paper.

2.2 Background Segmentation

The widely used background segmentation algorithms include threshold segmentation and edge detection algorithm. The segmentation process includes the selection and separation of color channels, conversion of binary images, color channel reorganization, etc. Its segmentation speed is slow because of the complexity of calculation. For the image we used in this paper, the background of the image is obviously different from that of the apple region. We find that the value of R component is greater than the G and B component in apple region by extracting the RGB channels respectively. Comparing the pixel values of RGB channels, we can remove the background by selecting the region with the large R component pixel value in the image, as shown in Fig. 2.

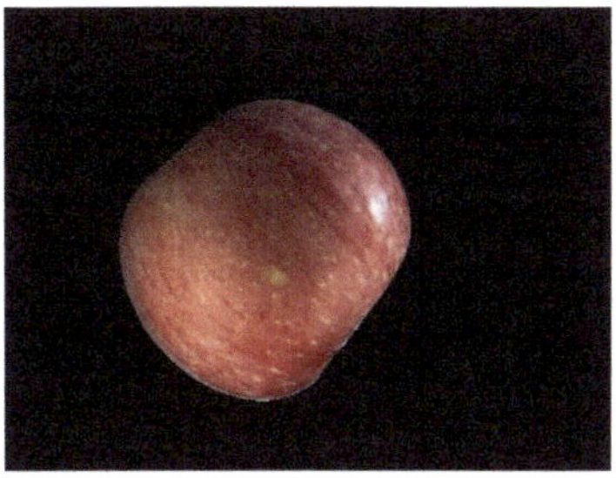

Fig. 2. Apple image after background segmentation.

3 Feature Extraction

Selecting appropriate feature parameters is very important for feature extraction and classification. In this section, firstly, I-component gray-scale image is separated for edge detection, then deleting small area objects of the binary image by using morphological operation to achieve a better denoising effect. Finally, shape, color and defect features of apples are extracted as the input of GA-SVM model.

3.1 Contour Extraction

We can get the apple region we are interested in clearly after background segmentation. In this section, edge detection algorithm combined with morphological processing is used to obtain the external contour of apple region. In this paper, Canny operator [7] with better robustness is selected for contour extraction. The steps of contour extraction based on Canny operator are as follows:

(1) Smoothing the image with Gaussian filtering. Edge detection algorithm is fundamentally based on derivative calculation of image strength. Derivative calculation is very sensitive to noise. Both noise and edge are concentrated on high-frequency signal. If not improved by filter in advance, the noise will be amplified after derivative calculation, so that more false edges are detected, which is not conducive to edge extraction. Therefore, the image is smoothed and denoised by the first derivative of Gaussian function, that is

$$g(x,y) = \frac{1}{2\pi\sigma^2}\exp\left(-\frac{x^2+y^2}{2\sigma^2}\right) * f(x,y) \quad (1)$$

Among them, σ is the variance of Gaussian function, which is used to control the smoothness; $f(x,y)$ is gray image; (x,y) is the position of pixels; $g(x,y)$ is the gray image after smoothing.

(2) Finite difference of the first-order derivative is used to calculate the magnitudes and directions of the gradient of pixels in the image. The edge of an image is a collection of pixels with large change of gray. It has two attributes: magnitude and direction. The pixels along the edge change smoothly, and the pixels perpendicular to the edge change violently. Since the image is a discrete two-dimensional function, we use the first-order finite difference of 2 * 2 to approximate the partial derivatives $g_x'(x,y)$ and $g_y'(x,y)$ of *x and y*:

$$G_x(x,y) \approx g_x'(x,y) = \frac{1}{2}[g(x,y+1) - g(x,y) + g(x+1,y+1) - g(x+1,y)] \quad (2)$$

$$G_y(x,y) \approx g_y'(x,y) = \frac{1}{2}[g(x+1,y) - g(x,y) + g(x+1,y+1) - g(x,y+1)] \quad (3)$$

According to the gradient of *x and y*, the magnitude and direction of the pixel's gradient can be calculated by the following formulas:

$$G(x,y) = \sqrt{G_x(x,y)^2 + G_y(x,y)^2} \quad (4)$$

$$\theta(x,y) = \arctan\left(G_y(x,y)/G_x(x,y)\right) \quad (5)$$

where $G_x(x,y)$ and $G_y(x,y)$ represents the gradient of the pixel in the horizontal and vertical directions respectively; $G(x,y)$ represents the magnitude of the pixel's gradient; $\theta(x,y)$ represents the direction of the pixel's gradient.

(3) Non-maximum suppression of the gradient magnitude. In the matrix of gradient's magnitude, the pixel with large value is not necessarily an edge pixel, so it needs to suppress non maximum value along the direction of gradient, that is to find the local maximum value of the pixel, and set the corresponding gray of the non-maximum value to 0. Firstly, compare the pixel's gradient to be judged with the gradient of the two 8-neighborhood pixels along the direction of gradient. If the amplitude of the pixel's gradient is not larger than that of two adjacent pixels along the gradient direction, then the gradient of this pixel is not the local maximum value and needs to be suppressed, in other words, this pixel is not an edge pixel.
(4) Detect and join edges with double-threshold algorithm. After non-maximum suppression of the gradient's magnitude, the preserved pixels can more accurately represent the edge of image, but some pixels generated by noise interference will also be preserved. A high threshold value can be used to get the edge image and remove some false edge, but the edge image generated by high threshold may not be closed. Therefore, it is also necessary to set a low threshold value. Pixels whose gradient amplitude is less than the low threshold value are regarded as non-edge pixels. If the gradient amplitude of the pixel's gradient is between the high threshold and the low threshold, it is necessary to determine whether there are pixels higher than the high threshold value in the 8-neighborhood space of the pixel. If there are pixels, they are regarded as edge pixels.

The canny edge detection results are shown in Fig. 3 (a). The contour image of Apple region can be obtained by removing the speckles and filling the holes, as shown in Fig. 3 (b).

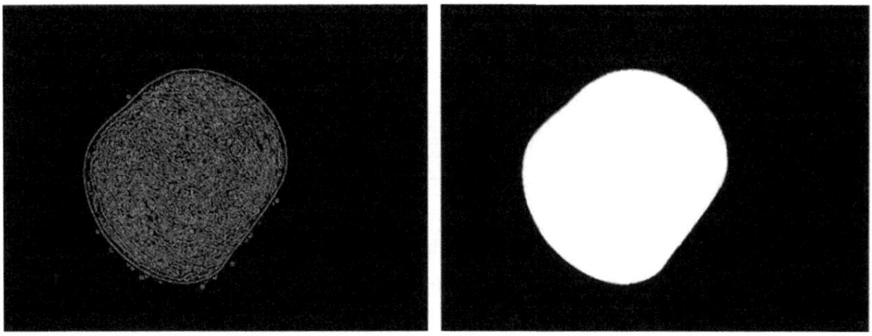

(a) (b)

Fig. 3. (a) Canny edge detection result (b) Morphological processed image.

3.2 Extraction of Shape Features

The roundness of apple is one of the important reference indexes in apple grading. The rounder the apple, the better the visual impression. In this paper, roundness is selected as the evaluation index of apple shape. The definition of roundness is:

$$e = \frac{4\pi S}{L^2} \tag{6}$$

where S is the area of apple fruit; L is the perimeter of apple fruit outline.

It should be noted that the area of apple can be obtained by calculating the number of pixels in apple fruit area; If only the number of boundary pixels of apple fruit is used to represent the perimeter of apple, it works at 0–90°, but it will be underestimated seriously in other directions. Therefore, this paper uses the algorithm of extracting Freeman chain code [8] to calculate the perimeter of apple contour, this is a series of steps that encode the boundary of an object into pixels to pixels around the object. Therefore, we simplify the binary image into a simple sequence of numbers.

3.3 Extraction of Color Features

Color is one of the important indexes of apple appearance quality grading, and apple products with uniform color have higher value. Because the I component of HSI color space has nothing to do with the color information of image, and the H and S components are closely related to the way people feel color, this paper chooses HSI color space to extract color features. It is found that the H-value of each grade of apple is distributed between 0–60°, among which the H-value of super grade apple has a peak value between 0–15°, the H-value of first grade apple has a peak value between 15–30°, the H-value of second grade apple has a peak value between 25–40°, and the H-value of extra grade apple has a peak value between 40–60°, as shown in Fig. 4 is the color histogram of different grades of apple. In order to improve the accuracy of color classification, the color of apples at all levels is divided into four intervals according to every 15°, and then the average value of each interval is used as the characteristic parameter of color classification.

3.4 Extraction of Defect Features

Apple surface damage and defects will seriously affect the quality of fruit inside and outside, which is a major difficulty in fruit classification. For the sample of apple used in this paper, in the I component of HSI color space, the gray value of the defect part is concentrated between 20–70. Therefore, the ratio between the number of pixels in this range and the number of pixels in the fruit area is selected as the defect characteristic parameter in this paper, which can be obtained by the following formula:

$$S = \frac{d}{t} \tag{7}$$

where d is the number of pixels with gray value between 20–70; t is the total number of pixels in apple fruit area.

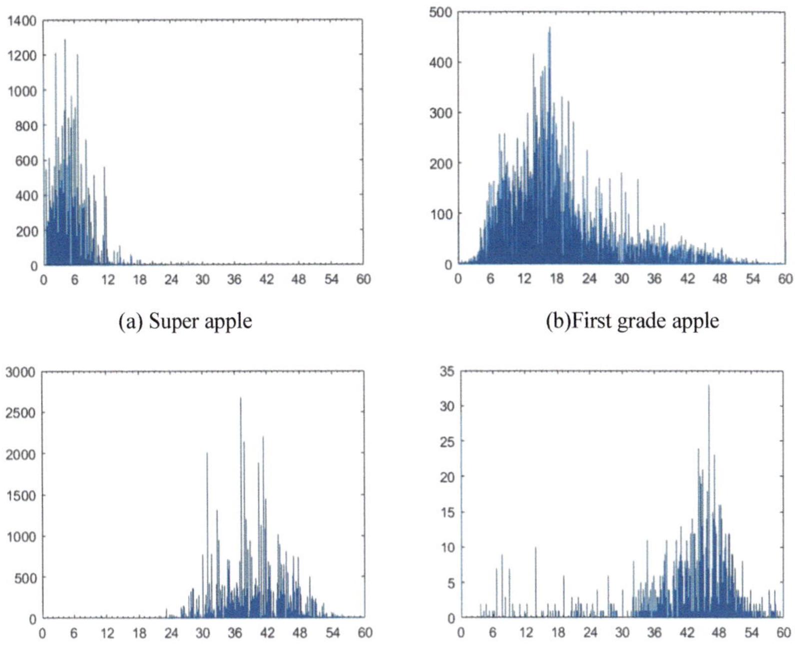

(a) Super apple (b)First grade apple

(c)Second grade apple (d)Outer apple

Fig. 4. H component histogram of apples (horizontal axis represents H value and vertical axis represents the times of H value).

4 SVM Classification Model Based on Genetic Algorithm

SVM is a machine learning method proposed by Vapnik in the mid-1990s on the basis of statistical theory. It has unique advantages in solving small sample, nonlinear and high-dimensional classification [9]. Assume that a training sample set $A = \{(x_i, y_i)i = 1, 2, \cdots, \mathrm{n}\}$, where $x_i \in R^{\mathrm{d}}$ is the input sample, n is the number of samples, $y_i \in \{1, -1\}$ is the category. The optimization problem with the largest sample classification interval can be expressed as follows:

$$\begin{cases} \min\left(\frac{1}{2}\left\|\vec{\omega}\right\|^2 + c\sum_{i=1}^{n}\xi_i\right) \\ s.\,t.\;\; y_i(\vec{\omega}\cdot\vec{\mathrm{x}}_i + b) - 1 + \xi_i \geq 0 \\ (i = 1, 2, \cdots, n), \xi_i \geq 0 \end{cases} \tag{8}$$

where c is the penalty factor, specified by users; ξ_i is the slack variable.

As shown in formula (8), this is a convex quadratic programming problem constrained by inequality. The Lagrange multiplier method can be introduced to obtain its dual problem. The kernel function of support vector machine can transform the non-linear separable samples into the linear separable feature space. Different kernel functions produce different hyperplanes. Considering that the number of samples and the feature dimension of samples are small, and compared with other polynomial kernel functions, the parameters of Gaussian radial basis function are less, so this paper chooses radial basis function as the kernel function of SVM, the formula is:

$$K(x_i, y_j) = \exp\left(-\frac{\|x - x_i\|^2}{g}\right) \tag{9}$$

In this way, the linear non separable problem is transformed into the linear separable problem in high dimensional space, that is:

$$\begin{cases} \max\limits_{\lambda} Q(\lambda) = \sum\limits_{i=1}^{n} \lambda_i - \frac{1}{2} \sum\limits_{i,j=1}^{n} \lambda_i \lambda_j y_i y_j K(x_i, y_j) \\ \text{s. } t. \quad \sum\limits_{i} \lambda_i y_i = 0\ , \qquad 0 \leq \lambda_i < c \end{cases} \tag{10}$$

where λ is Lagrange multiplier.

Therefore, to obtain a better performance SVM classifier model, we need to get the best penalty factor c and the parameter g of Gaussian radial basis function. In this paper, genetic algorithm is used to optimize the parameters of SVM classification model. The specific steps are as follows:

(1) The penalty factor c and the parameter g of Gaussian radial basis function are binary coded, and then initial the population: in this paper, the maximum evolution algebra is 100, the population number is 20, the crossover probability is 0.4, and the mutation probability is 0.01;
(2) Determine the fitness function and calculate the fitness value of each individual. In this paper, the accuracy of the cross validation set is used as the fitness;
(3) Determine whether the termination condition is met. If it is, stop the calculation and bring the optimal parameters into SVM model. Otherwise, perform crossover and mutation operations to generate a new generation of population return step (2).

5 Results and Analysis

Before the experiment, we artificially grade 215 apples, including 60 super grade apples, 60 first grade apples, 45 second grade apples and 50 outer apples, and divide them into training set and test set, as shown in Table 1, the training set contains 163 apples, and the test set contains 52 apples:

Table 1. Training and test sets.

Grade	Training set	Test set
Super	45	15
First	45	15
Second	35	10
Outer	38	12
Total	163	52

On the MATLAB platform, the feature vectors of apples in the training set and test set are extracted respectively, and the GA-SVM model is established and trained by using libsvm software package. The result of genetic algorithm optimizing the support vector is shown in Fig. 5.

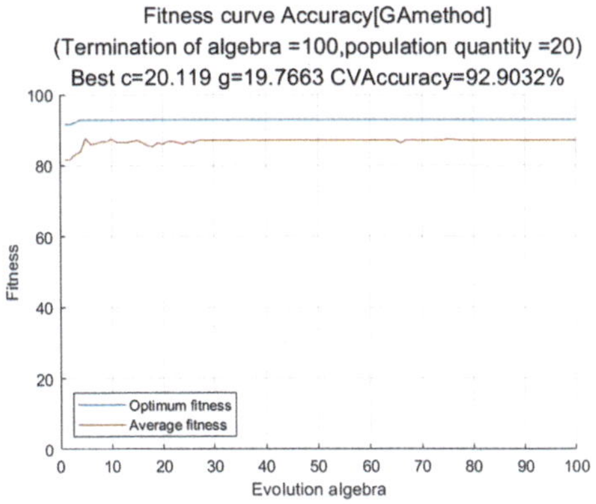

Fig. 5. Optimization of c and g by genetic algorithm.

It can be seen from Fig. 5 that after optimization by genetic algorithm, penalty parameter c = 20.119 and Gaussian radial basis function parameter g = 19.7663. The 52 feature vectors in the test set are sent to the optimized classification model for prediction. The classification results are shown in Fig. 6. The horizontal axis represents 52 test samples of four grades, the vertical axis represents classification labels, and the 0/1/2/3 represents super/first/second/substandard apples respectively.

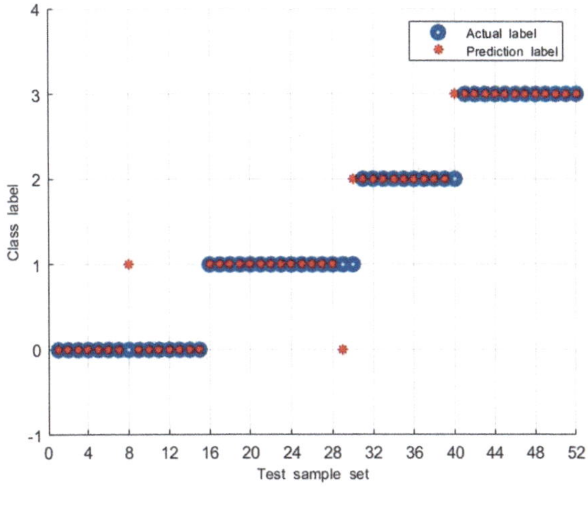

Fig. 6. Prediction results of GA-SVM model.

The classification accuracy of SVM model optimized by genetic algorithm is 92.3%, and the classification accuracy of different grades of apples is shown in Table 2.

Table 2. Classification results.

Category	Test set	Prediction results	Accuracy
Super	15	14	93.3%
First	15	13	86.7%
Second	10	9	90.0%
Outer	12	12	100.0%
Total	52	48	92.3%

As can be seen from the above analysis, this SVM model optimized by genetic algorithm can achieve the classification of apple's external quality, but there is a certain probability of misclassification, which may be caused by the following reasons: (1) the number of samples used in the experiment is relatively small; (2) we evaluate apple quality based on only one image, it is necessary to evaluate apple quality based on different images which are captured in different directions, to check the apple in all directions.

6 Conclusion

In this paper, the outline of apple is obtained by Canny edge detection algorithm with morphological processing, and the roundness, color, defects of apple are extracted as feature parameters. In order to improve the accuracy of classification, genetic algorithm is used to optimize the parameters c and g of SVM model, and the average classification accuracy rate is 92.3%. It can be seen that the GA-SVM method has certain feasibility and high engineering application value for apple's external quality classification.

Acknowledgement. This work is supported by Key R & D projects of Shandong Province under grant 2019GNC106093, Shandong Agricultural machinery equipment R & D innovation plan project under grant 2018YF011, Key R & D projects of Shandong Province under grant 2019JZZY021005.

References

1. Yang, J., Meng, X.N., Xin, L.: Analytical report on China's apple market from 2017 to 2019. Deciduous Fruits **51**(5), 5–7 (2019)
2. Zhang, B.: Analysis of China's apple industry's output, processing and trade status in the past 7 years. China Fruit Tree (004), 106–108 (2018)
3. Ren, Y.X., Shan, Z.D., Zhang, J., et al.: Research progress on computer vision technique in fruit quality inspection. J. Agric. Sci. Technol. (Beijing) **14**(1), 98–103 (2012)

4. Moallem, P., Serajoddin, A., Pourghassem, H.: Computer vision-based apple grading for golden delicious apples based on surface features. Inf. Process. Agric. **4**(1), 33–40 (2017)
5. Huang, C., Fei, J.Y.: Online apple grading based on decision fusion of image features. Trans. Chin. Soc. Agric. Eng. **33**(1), 285–291 (2017)
6. Yu, M., Li, X., Yang, H.C.: Research on apple grading algorithm based on image recognition. Autom. Instrum. (7), 39–43 (2019)
7. Li, M., Yan, J.H., Li, G., et al.: Self-adaptive Canny operator edge detection technique. J. Harbin Eng. Univ. **29**(9), 1002–1007 (2008)
8. Freeman, H.: On the encoding of arbitrary geometric configurations. IRE Trans. Electron. Comput. **10**(2), 260–268 (1961)
9. Vapnik, V.: The Nature of Statistical Learning Theory. Springer, Heidelberg (2013)

Deep Spatio-Temporal Dense Network for Regional Pollution Prediction

Qifan Wu[1], Qingshan She[1](✉), Peng Jiang[1], Xuhua Yang[2], Xiang Wu[3], and Guang Lin[4]

[1] School of Automation, Hangzhou Dianzi University, Hangzhou 310018, China
qsshe@hdu.edu.cn
[2] College of Computer Science and Technology, Zhejiang University of Technology, Hangzhou 310014, China
[3] Department of Optical Science and Engineering, Fudan University, Shanghai 200433, China
[4] Zhejiang Provincial Environmental Monitoring Center, Hangzhou 310000, China

Abstract. In most cities, the monitoring stations are sparse, and air pollution is affected by various internal and external factors. To promote the performance of regional pollution prediction, a deep spatio-temporal dense network model is proposed in this paper. First, the inverse distance weighted (IDW) space interpolation algorithm is used to construct historical emission data with insufficient station records. Based on the properties of spatial-temporal data, a deep spatial-temporal dense network model is designed to predict air pollution in each region. Finally, several experiments are conducted on the real dataset of Hangzhou. The results show that combing IDW spatial interpolation and deep spatial-temporal dense network model can effectively predict the regional air pollution and achieve superior performance compared with ARIMA, CNN, ST-ResNet, CNN-LSTM.

Keywords: Air pollution prediction · Space interpolation · Deep spatial-temporal dense network

1 Introduction

Affecting the health of billions of people, air pollution has raised a hot topic of concern in Chinese society today. Getting the reliable and accurate prediction of region air pollution is helpful to inform people of the pollution situation in advance and improve city's transportation infrastructure design for relevant government. Therefore, accurate region air pollution forecasting helps to reduce the amount of time cost, economic losses and carbon emission.

Currently, the models for predicting air pollution can roughly be categorized into three classes: deterministic models, statistical methods and deep learning models. Instead of traditional deterministic methods, the researchers also employed deep learning techniques to analyze and forecast air quality in recent years [1]. Deep learning extracts the inherent features in layers through a multilayer structure [2–4], which can

Q. Liu et al. (Eds.): CENet 2020, AISC 1274, pp. 90–99, 2021.
https://doi.org/10.1007/978-981-15-8462-6_10

automatically learn hierarchical features without relying on historical data information. Due to the excellent performance of Long Short-Term Memory (LSTM) [5] in time series prediction, LSTM can handle time-related pollutant data well. Li et al. [6] considered the spatiotemporal correlation and proposed an extended LSTM model (Long Short-Term Memory Neural Network Extended, LSTME), which uses LSTM to automatically extract features from historical air pollution data, and fuses meteorological data and time stamp data to improve prediction performance. Zhao et al. [7] have proposed a Long Short-Term Memory Fully Connected (LSTM-FC) neural network that uses historical air quality data, meteorological data, weather forecast data, and time data to predict PM2 at a single site for 48 h Concentration. It is verified on the Beijing dataset that it has better performance than ANN and LSTM. In terms of hybrid models, Qin et al. [8] integrated big data based on CNN and LSTM structures. It uses CNN as the basic layer to automatically extract input data features, and considers long-term prediction with LSTM as the output layer. In addition, many deep learning frameworks based on big data have been proposed and applied to air pollution prediction [9]. To predict the concentration of pollutants at multiple sites in urban areas, Wang et al. [10] have proposed a deep multi-task learning framework based on a gated recurrent unit (GRU) that can simultaneously predict the concentration of multiple pollutants. The prediction of multiple regions is affected by many complex factors, such as spatiotemporal correlation and external factors. Liang et al. [11] used a Recurrent Neural Network (RNN) based on Multi-Level Attention (MLA) to predict the values of multiple monitoring sites and tested it on the Beijing dataset, which was better than other basic models.

However, the challenges facing regional air pollution prediction research can be summarized as follows: (1) Data sparsity. It is difficult to predict air pollution in urban areas based on limited monitoring stations. Most existing research on spatial interpolation is based on geostatistical methods [12]. (2) Spatiotemporal dependencies. Air pollution in each area is affected by time and space. It is necessary to consider the spatiotemporal dependence in the modeling of spatiotemporal network. To address these problems, a deep spatio-temporal dense network model is proposed for regional pollution prediction. The IDW space interpolation algorithm [13] is first used to construct historical emission data with insufficient station records. Considering the properties of spatial-temporal data, a deep spatiotemporal dense network model is constructed to comprehensively predict air pollution in various regions in which the number of network layers is deepened by densely connected units and the spatiotemporal features are extracted by the LSTM network model. Several experiments are conducted on the real dataset of Hangzhou to demonstrate the effectivity of the proposed method.

2 Related Work

2.1 Spatial Interpolation

Spatial interpolation is a method to estimate the data of unplanned sites based on the observations of the site. We use the classic IDW method to generate weights for the data of surrounding monitoring sites, and then obtain the average monitoring data of the area. The main idea of inverse distance weighting is to obtain a weighted average from known data points to assign values to unknown data points. The weight is the inverse of geographical distance. The longer the geographical distance, the lower the weight. The formula of the inverse distance weighting method is:

$$z_0 = \sum_{i=1}^{n} \frac{1}{(D_i)^p} z_i \left[\sum_{i=1}^{n} \frac{1}{(D_i)^p} \right]^{-1} \tag{1}$$

$$D_i = \sqrt{(x_0 - x_i)^2 + (y_0 - y_i)^2} \tag{2}$$

where z_0 represents the estimated value, z_i is the attribute value of the point i, P is the power parameter, and D_i is the Euclidean distance between samples.

The steps for filling missing values in the area are described as follows: First, an arranged monitoring station is selected as the center, and then a weight is assigned to each monitoring station in the area according to the distance from the surrounding monitoring station to the target monitoring station. Next, a regression operation is performed to obtain average monitoring data. Finally, the IDW approach is employed to fill the data of the unarranged site area.

2.2 Deep DenseNet

DenseNet [14] is a recently proposed convolutional neural network, regarded as an extension of ResNet [15], which also uses the concept of residuals. Each layer of DenseNet receives the input of all the previous layers and feeds its output to all subsequent layers, which greatly reduces the network parameters and strengthens the feature propagation. In DenseNet model, each layer in the dense block can benefit from low-dimensional and high-dimensional features, reducing the risk of gradient explosion or disappearance.

Suppose the convolutional network consists of L layers, $H_\ell(\cdot)$ and ℓ represent the final non-linear operation of each layer, and the number of layers respectively. $H_\ell(\cdot)$ can be a compound function such as Batch Normalization (BN) [16], Rectified Linear Unit (ReLU), Pooling, Conv, etc. In DenseNet, each layer is connected to the previous layer as much as possible. Its output can be expressed as $\mathrm{x}_\ell = H_\ell([\mathrm{x}_0, \mathrm{x}_1, \ldots \mathrm{x}_{\ell-1}])$. DenseNet is mainly composed of several dense blocks and a transition layer. The transition layer refers to the convolution layer and pooling layer added between densely connected blocks. With such a structure, DenseNet can build a sufficiently flexible deep neural network. The structure of Deep DenseNet is shown in Fig. 1.

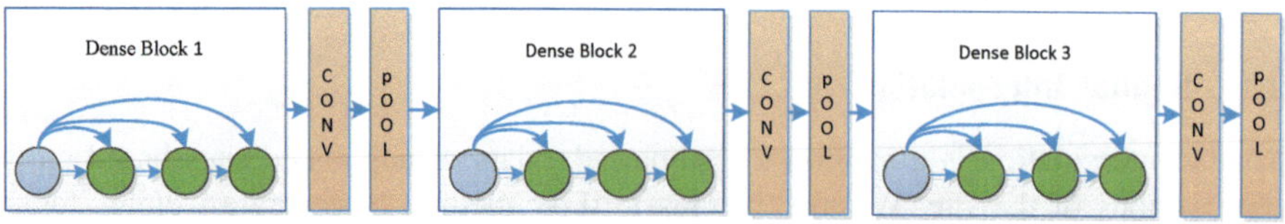

Fig. 1. DenseNet consists of 3 dense blocks.

2.3 Spatiotemporal Dependencies

The existing research for air pollution usually is time-series prediction problem, which do not take spatiotemporal dependence into account. Figure 2 (a) is the air quality index between 9:00 and 16:00 on January 1, 2018. The curve shows the time correlation in the time series. It can be seen that the air quality index of the nearest time interval is more relevant than the farthest air quality index, showing the proximity characteristics of the air quality index. Figure 2 (b) shows the change of the air quality index 7 days a week, showing the periodicity of the air quality index. Figure 2 (c) shows the change of the air quality index within 2 months. As the weather warms, the air pollution decreases, which reflects the trend characteristics.

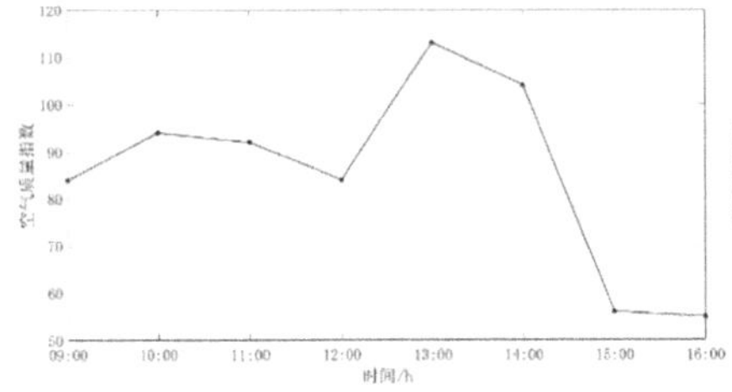

(a) Closeness

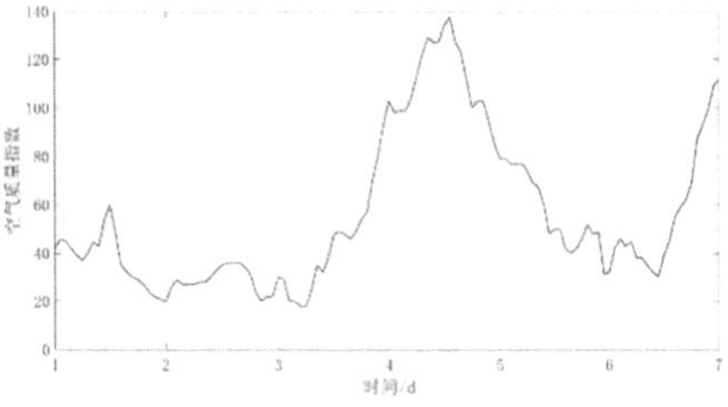

(b) Period

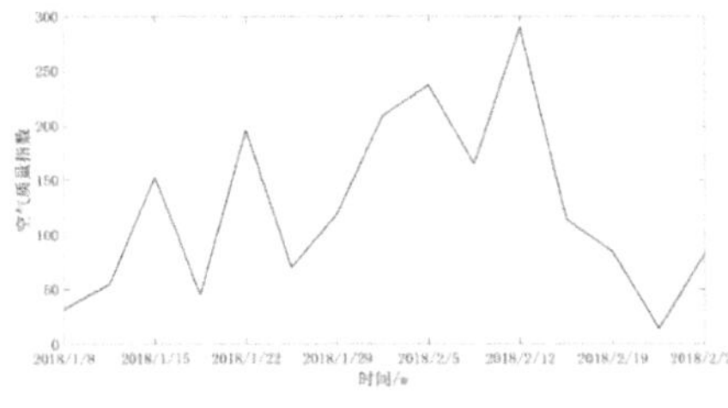

(c) Trend

Fig. 2. Temporal dependencies.

3 Our Method

The overall framework is mainly composed of three parts: data source, feature extraction and deep space-time dense network prediction model as shown in Fig. 3. After the data is obtained, E_{ext} is used to represent the historical data features and E_{ext} represents the external features of the current time t, including road network E_{road}, weather $E_{weather}$, and number of Points of interest (POI) E_{POI}. The deep spatiotemporal dense connection network mainly includes a convolutional network layer, a dense connection network layer and a long-term and short-term memory neural network layer. The system is divided into two phases, including offline training and online prediction. In offline training, a city's collection records are input to and output from a regional air pollution calculation model. These historical regional air pollution sequences are used to train deep space-time dense network models. In the online prediction, starting from the calculation of historical regional air pollution, the learned deep space-time dense model is used to predict the air pollution in the future region.

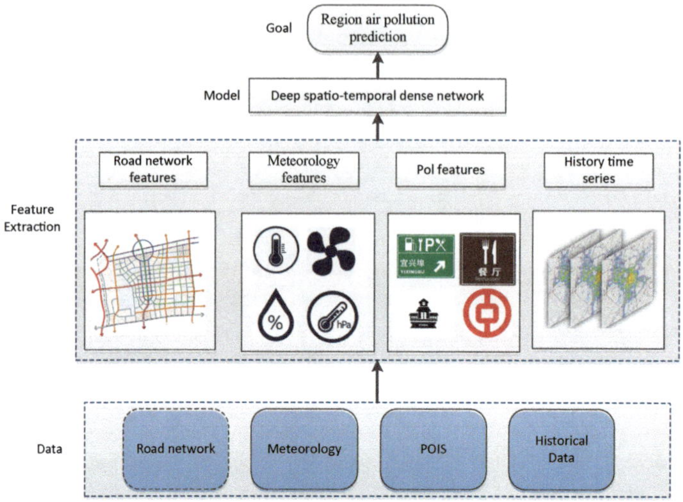

Fig. 3. System framework.

The data source mainly considers the historical data characteristics and external characteristics of air pollution affecting the target area. Three external factors are considered:

(1) Meteorological characteristics. This paper extracts 5 types of meteorological characteristics: Obtain temperature, humidity, wind level, wind direction and rainfall data every hour from the weather website.
(2) Road network characteristics. The road network consists of road sections, which are connected to each other in the form of a map. Each road segment is a directed edge with two end points. A series of intermediate points describe the road segment, the extracted features of road network include the type and length of the road.
(3) POI characteristics. POI are functional areas of the city. We divided ten categories according to their types: station, factory, service area, shop, company, bank, school, hotel, entertainment, tourist attraction. This paper counts the number of 10 types of interest points within a 200-m radius at the ends of each road section as features.

All historical observations obtained by spatial interpolation are put into the spatiotemporal component, and time series with different time delays are connected in series according to the time characteristics to construct the closeness, period and trend of time. For proximity, periodicity and trending features, they are first input to the conv layer Conv1 to extract spatial features, then sent to the densely connected network layer DenseNet, and finally LSTM is used to learn temporal features. This part will integrate closeness X_c, period X_p and trend X_s to obtain spatiotemporal characteristics X_{des}.

For external features (road network, weather, and POIs), feature vectors are obtained from the external data set at prediction time t and converted into binary vectors, and then input to the fully connected layer FC after feature extraction to obtain the external feature vector X_{ect}. Finally, the output of the two parts of the spatiotemporal feature vector X_{des} and the external feature vector X_{ect} is weighted and sent to the activation function for prediction. The model network structure is shown in Fig. 4.

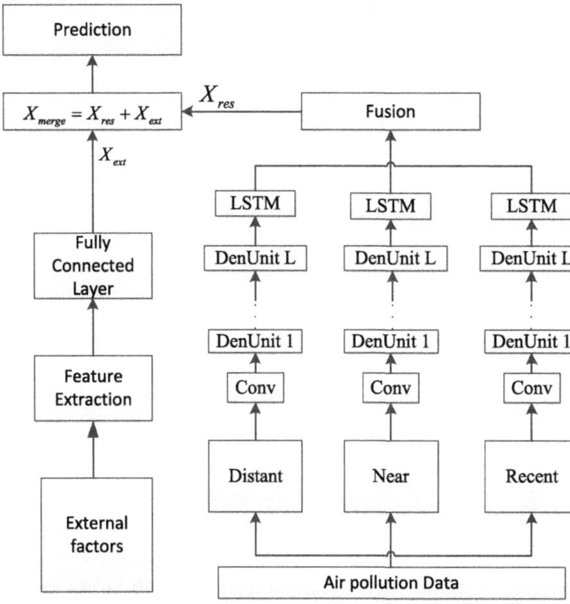

Fig. 4. The proposed network model.

In the convolutional layer, we use the convolution operations to capture spatial dependencies. The input of the convolutional layer is usually a tensor $f(W * X + b)$ where f represents the activation function, $*$ is the convolution operation, W and b are the weight and offset values. We input closeness, period, and trend into similar convolutional layers. Through convolution operations, their outputs can be written in the following forms: $X_c^{(1)} = f(\sum_{j=1}^{l_c} W_{cj}^{(1)} * x_{t-j} + b_c^{(1)})$, $X_p^{(1)} = f(\sum_{j=1}^{l_p} W_{pj}^{(1)} * x_{t-j.p} + b_p^{(1)})$, $X_s^{(1)} = f(\sum_{j=1}^{l_s} W_{sj}^{(1)} * x_{t-j.s} + b_s^{(1)})$ where $W^{(1)}$, $b^{(1)}$ are the parameters in the layer 1, respectively, c, p, s represent closeness, period and trend respectively, $X_c^{(1)}$, $X_p^{(1)}$, $X_s^{(1)}$ is the output of layer 1 in closeness, period and trend.

In order to capture closeness, period, and trend features at the same time, a convolutional layer is used after fusion. The convolutional layer is good at fusing data with similar data structures. Fusion can be expressed as:

$$X^{(1)} = f(W_c^{(2)} * X_c^{(1)} + W_p^{(2)} * X_p^{(1)} + W_s^{(2)} * X_s^{(1)} + b^{(2)}) \tag{3}$$

The output of three spatiotemporal features and the output of external features are combined by weights to predict the value of time. The predicted value can be expressed by and can be defined as:

$$\hat{X}_t = \tanh(X_{st} \cdot W_{st} + X_{ext} \cdot W_{ext}) \tag{4}$$

where W_{st} and W_{ext} are the weights, and tanh is the hyperbolic tangent function.

4 Experiments

4.1 Dataset Description

The experiments use real data sets from Hangzhou, including road network, point of interest data, meteorological data and air quality data. The road network data and points of interest data are from Baidu map. The distribution of meteorological data and air quality data comes from Hangzhou Meteorological Bureau, China's air quality monitoring platform. The time span is from January 1, 2018 to April 30, 2018, with an interval of 1 h.

4.2 Parameter Setting

The convolution layers of Conv1 and the residual units employ 64 filters with size 3×3, and Conv2 uses a convolution with 2 filters with size 3×3. The learning rate is set to 0.001. The batch size is 32. We select 90% of the training dataset for training each model, and the remaining 10% is chosen as the validation set, which is used to early-stop training algorithm for each model with the best validation score. Afterwards, we continue to train the model on the full training data for a fixed epoch number. There are 5 extra hyper parameters in our model, of which p and s are empirically fixed to one-day ($p = 1$) and one-week ($s = 7$), respectively. For lengths of the three dependent sequences, we set them as: $len_c\{0, 1, 2, 3, 4, 5, 6, 7\}$, $len_p\{0, 1, 2, 3, 4, 5, 6, 7\}$, $len_s\{0, 1, 2, 3, 4, 5, 6, 7\}$. To verify the effectiveness of the algorithm, all the following experimental results are the average values obtained after 10 times.

In order to evaluate the effectiveness of the proposed algorithm, this paper uses $\hat{X}_t$ to represent the predicted air quality level, X_t to represent the real air quality level. For a one-hour forecast, the sum of the accuracy of each type of prediction is calculated as the total accuracy A = $\sum i(\hat{X}_i == X_i) / \sum iX_i$. For predictions within 7–12, 13–24 h, the category that appears the most during this time period is taken as the real value, and its accuracy can be calculated. We also calculate the arithmetic mean F = $\sum jF1_j/u$.

For evaluation, we compare our model with the following baselines:

1) ARIMA [17]: Auto-Regressive Integrated Moving Average (ARIMA) is a well-known model for understanding and predicting future values in a time series.
2) Convolutional neural network [18]: CNN is a deep learning method used to deal with sequence dependence problems. It is often used in time series prediction, natural language processing and other scenarios.
3) CNN-LSTM [19]: Convolution LSTM is a kind of LSTM, which is good at capturing spatio-temporal features.
4) ST-ResNet [20]: Deep residual network enables convolution neural networks to have very deep layer structures, can be used for spatio-temporal prediction, which shows state-of-the-art results on the region vehicle emission prediction.

4.3 Results

Model Comparison. Table 1 shows the experimental results of the five models. From the prediction results of each model, ARIMA has the lowest accuracy, with a value of only 0.752. CNN, ST-ResNet and CNN-LSTM achieved similar accuracy of 0.915, 0.922 and 0.932 respectively. Deep learning can learn features well in the prediction of air pollution and improve the performance of model prediction. Compared with the four models, the proposed algorithm has the highest accuracy rate, improved to 0.954, which is higher than ST-ResNet, because it takes the influence of time and space characteristics of pollution into account. In terms of long-term prediction, the CNN-LSTM model combined with the LSTM and the proposed model achieved higher accuracy rates of 0.965 and 0.972, respectively. It shows the accuracy of LSTM in long-term prediction. The proposed algorithm is slightly better than CNN-LSTM, which improves 0.07. Combining the accuracy rates and, we can find that the proposed model has the best performance in predicting regional pollution.

Table 1. Model comparison.

Model	A	F
ARIMA	0.752	0.739
CNN	0.915	0.923
CNN-LSTM	0.922	0.965
ST-ResNet	0.932	0.921
Our method	**0.954**	**0.972**

Influence of External Factors. In order to assess the impact of external factors, this chapter compares the effects of individual external features and combinations on the accuracy of predictions. $F_{features} = \{F_{road}, F_{poi}, F_{meteo}\}$, $F_i \in F_{features}$ represents external characteristics. After adding road network feature F_{road}, point of interest feature

F_{poi}, and meteorological feature F_{meteo} to the model. it can be seen from Table 2 that the accuracy of the algorithm is 0.921, 0.941, and 0.936. Compared with models that do not take into account external characteristics, A has been improved to a certain extent, indicating that both the point of interest information and the meteorological information are helpful for the prediction of air pollution. Adding the features of the point of interest improves the model's performance the most, an increase of about 0.02. In the combination feature model, the accuracy of the combination of POI and weather is the highest, which A is 0.948, which is slightly better than the combination of road network features and weather features. When combining all features, A and F the model have significantly improved compared to models without considering external features, which illustrates the superiority of considering external features.

Table 2. Influence of external factors.

Factors	A	F
None	0.924	0.933
$F_{road}(F_1)$	0.921	0.938
$F_{poi}(F_2)$	0.941	0.952
$F_{meteo}(F_3)$	0.936	0.944
$F_1 + F_2$	0.947	0.945
$F_1 + F_3$	0.937	0.954
$F_2 + F_3$	0.948	0.968
$F_1 + F_2 + F_3$	0.952	0.975

5 Conclusion

This paper presents a regional air pollution prediction model based on a deep spatiotemporal dense network, which predicts the air quality of each area in the city based on historical air quality data, road network data, and meteorological data. Experiments were performed on a real dataset in Hangzhou, and the results show that the performance of the proposed model is superior to the other state of art algorithms. The proposed model has only been tested on real datasets in Hangzhou. Although it has obtained good performance, it is not convincing enough. In the future, experiments can be performed on more urban datasets to verify the performance of the model.

Acknowledgments. This work is supported by National Natural Science Foundation of China under Grant 61871427, National Key R&D Program of China under Grant 2016YFC0201400, Provincial Key R&D Program of Zhejiang Province under Grant 2017C03019 and International Science and Technology Cooperation Program of Zhejiang Province for Joint Research in High-tech Industry under Grant 2016C54007.

References

1. Qi, Z., Wang, T., Song, G., et al.: Deep air learning: interpolation, prediction, and feature analysis of fine-grained air quality. IEEE Trans. Knowl. Data Eng. **30**(12), 2285–2297 (2018)
2. Douarre, C., Schielein, R., Frindel, C., et al.: Deep learning-based root-soil segmentation from X-ray tomography. bioRxiv: 71662 (2016)
3. Chauhan, S., Vig, L.: Anomaly detection in ECG time signals via deep long short-term memory networks. In: The Proceedings of 2015 IEEE International Conference on Data Science and Advanced Analytics, Paris, France, p. 7344872 (2015)
4. Zhan, X., Li, Y., Li, R., et al.: Stock price prediction using time convolution long short-term memory network. In: Proceedings of the 11th International Conference on Knowledge Science, Engineering and Management, Changchun, China, pp. 461–468 (2018)
5. Hochreiter, S., Schmidhuber, J.: Long short-term memory. Neural Comput. **9**(8), 1735–1780 (1997)
6. Li, X., Peng, L., Yao, X., et al.: Long short-term memory neural network for air pollutant concentration predictions: method development and evaluation. Environ. Pollut. **231**, 997–1004 (2017)
7. Zhao, J., Deng, F., Cai, Y., et al.: Long short-term memory-Fully connected (LSTM-FC) neural network for PM2. 5 concentration prediction. Chemosphere **220**, 486–492 (2019)
8. Qin, D., Yu, J., Zou, G., et al.: A novel combined prediction scheme based on CNN and LSTM for urban pm 2.5 concentration. IEEE Access **7**(1), 20050–20059 (2019)
9. Kök, İ., Şimşek, M.U., Özdemir, S.: A deep learning model for air quality prediction in smart cities. In: IEEE International Conference on Big Data (Big Data), Boston, MA, USA, pp. 1983–1990 (2017)
10. Wang, B., Yan, Z., Lu, J., et al.: Deep multi-task learning for air quality prediction (2018)
11. Liang, Y., Ke, S., Zhang, J., et al.: GeoMAN: multi-level attention networks for geo-sensory time series prediction. In: Proceedings of the Twenty-Seventh International Joint Conference on Artificial Intelligence, Stockholm, Sweden, pp. 3428–3434 (2018)
12. LeCun, Y., Bengio, Y.: Convolutional networks for images, speech, and time series. In: The Handbook of Brain Theory and Neural Networks, vol. 3361, no. 10 (1995)
13. Abdel-Hamid, O., Mohamed, A., Jiang, H., et al.: Convolutional neural networks for speech recognition. IEEE/ACM Trans. Audio Speech Lang. Process. **22**(10), 1533–1545 (2014)
14. Qian, Y., Woodland, P.C.: Very deep convolutional neural networks for robust speech recognition. In: Proceedings of IEEE Spoken Language Technology Workshop (SLT), San Diego, CA, USA, pp. 481–488 (2016)
15. Spanhol, F.A., Oliveira, L.S., Petitjean, C., et al.: Breast cancer histopathological image classification using convolutional neural networks. In: International Joint Conference on Neural Networks (IJCNN), Vancouver, BC, Canada, pp. 2560–2567 (2016)
16. Ioffe, S., Szegedy, C.: Batch normalization: Accelerating deep network training by reducing internal covariate shift. arXiv preprint arXiv:1502.03167 (2015)
17. Kumar, U., Jain, V.K.: ARIMA forecasting of ambient air pollutants (O_3, NO, NO_2 and CO). Stoch. Env. Res. Risk Assess. **24**(5), 751–760 (2010)
18. Fukushima, K.: Neocognitron: a self-organizing neural network model for a mechanism of pattern recognition unaffected by shift in position. Biol. Cybern. **36**(4), 193–202 (1980)
19. Shi, X.J., Chen, Z.R., Wang, H., et al.: Convolutional LSTM network: a machine learning approach for precipitation nowcasting. In: Proceedings of Advances in Neural Information Processing Systems, Montreal, Canada, vol. 28, pp. 802–810 (2015)
20. Zhang, J., Zheng, Y., Qi, D., et al.: Predicting citywide crowd flows using deep spatio-temporal residual networks. Artif. Intell. S2058406355 (2018)

Research on the Relationship Among Electronic Word-of-Mouth, Trust and Purchase Intention-Take JingDong Shopping E-commerce Platform as an Example

Chiu-Mei Chen[1], Kun-Shan Zhang[2(✉)], and Hsuan Li

[1] University of Zhaoqing, Zhaoqing 526061, China
[2] Peking University, Beijing 100871, China
zhang_kun_shan@163.com

Abstract. With the rapid development of the Internet and the continuous improvement of online shopping facilities, online shopping has been accepted by contemporary people. Taking JD e-commerce platform as an example, this study explores the relationship among three variables: electronic word-of-mouth, trust and online consumers' purchase intention, and puts forward the following research hypotheses: (1) electronic word-of-mouth has a positive impact on the trust of online potential consumers, (2) electronic word-of-mouth has a positive impact on the purchase intention of online potential consumers, (3) trust has a positive impact on the purchase intention of online potential consumers. This study uses the quantitative questionnaire survey method and SPSS software to analyze the data, expecting that the research results can promote the operation managers to pay attention electronic word-of-mouth, develop and improve the consumer feedback system, establish the good reputation of enterprises, and promote the operation performance of e-commerce platform.

Keywords: Electronic word-of-mouth · Trust · Purchase intention · E-commerce

1 Introduction

With the rapid development of the Internet, people's daily life has been popularized and facilitated. According to the 44th statistical report on the development of Internet in China issued by China Internet Information Center in August 2019 [1], as of June 2019, the number of Internet users in China has reached 854.49 million, and the Internet penetration rate has reached 61.2%. Compared with June 2018, the number of Internet users has increased by 52.83 million. The growth rate is up to 6.59%, and the Internet penetration rate is up 3.5% year on year, as shown in Fig. 1. Therefore, the number of Internet users in China is still increasing year by year, and the network penetration rate is also increasing year by year.

Q. Liu et al. (Eds.): CENet 2020, AISC 1274, pp. 100–104, 2021.
https://doi.org/10.1007/978-981-15-8462-6_11

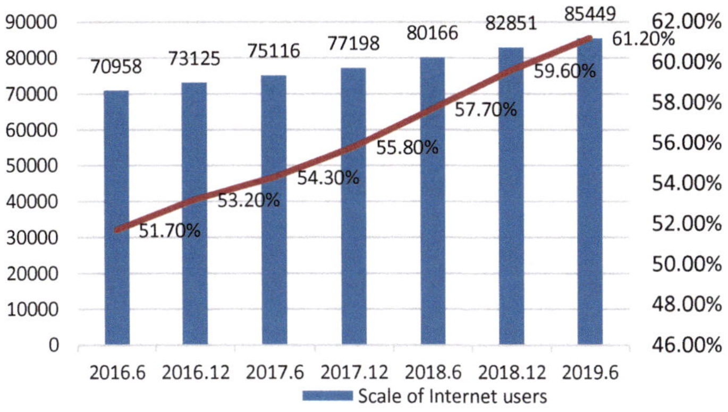

Fig. 1. Scale of Internet users and Internet penetration rate in China.

On the other hand, from June 2016 to June 2019, the survey results on the scale and utilization rate of online shopping users show that the number and utilization rate of online shopping users in China are increasing year by year, as shown in Fig. 2. Such a trend shows that China's Internet is becoming more and more livable, the convenience and popularity of the Internet have changed the shopping habits of Chinese people, making the consumption behavior of online shopping of Chinese people increasing and frequency this will also promote the expansion and improvement of the network market.

Fig. 2. Scale and utilization rate of online shopping in China.

JingDong e-commerce platform is China's online and offline retail group, which is positioned as a service enterprise based on supply chain technology. At present, the group's business involves retail, digital technology, logistics, technical services, health, insurance, logistics real estate, cloud computing, AI, overseas and other fields, among which the core business is retail, digital technology, logistics and technical services.

The online sales volume of electronic goods of this platform has a certain scale ratio in China's online shopping market. Therefore, this study chooses JD e-commerce platform as a case study to explore whether the electronic word-of-mouth of this platform will affect the trust of online consumers on the platform, and whether the electronic word-of-mouth and trust will affect the purchase intention of online consumers. It is expected that the results of this study will enable online shoppers to clearly understand the real feelings of consumers, spread and attract more potential online consumers, enable their brands to be trusted by consumers, towards stable growth and sustainable operation, and establish more perfect online management and operation strategies.

2 Literature Review

2.1 Electronic Word-of-Mouth

Gelb and Sundaram (2002) [2], Hennig-Thurau, Gwinner, Walsh, and Gremler (2004) [3] proposed that customers take the initiative and actively share their own experience, opinions and relevant knowledge on specific topics, or collect product information and topic discussion provided by other consumers, as well as the communication behavior of emotional cognition caused in the process of interaction with enterprises, which is called electronic word-of-mouth. There are many kinds of ways to spread electronic word-of-mouth, mainly through e-mail, news media, network communities (facebook, chat room, skype, line, twitter, we chat), etc. (Gelb and Sundaram, 2002 [2]; Wang, Ou, and Chen, 2019 [4]).

2.2 Trust

"Trust" has been widely used in sociology, psychology, marketing and other different fields. Many scholars have studied it and defined it differently. Lee and Turbon (2001) [5] defined trust as a belief, feeling or expectation rooted in personality and formed in the process of personal psychological development. The definition of trust is quite complex. At present, there is not a unified definition that can cover the meaning of trust in different fields and levels. This paper takes e-commerce as the research background, so this paper mainly refers to the definition of consumer trust in the field of marketing. Moorman (1993) [6] pointed out that trust depends on the willingness of trading partners, and the truster has confidence in the trustee. Hsu (2017) [7] said that trust is an important mechanism in the relationship between the buyer and the seller, because it can reduce the uncertainty of interaction and enhance the expectation of successful purchase.

2.3 Purchase Intention

Purchase intention is the probability of consumers willing to take specific purchase behavior. Dodds and Monroe (1991) [8] believed that purchase intention refers to the subjective probability or possibility of consumers to purchase a specific product, while some scholars also believed that purchase intention is the purchase plan of consumers for a specific product. Wang Meng (2018) [9] believes that the purchase intention is the subjective desire of consumers to implement the purchase behavior and a psychological attitude before taking the purchase behavior.

2.4 Electronic Word-of-Mouth, Trust and Purchase Intention

In the empirical study, You Qiuping (2019) [10] explored the impact of electronic word-of-mouth and perceived security on online group buying intention. The empirical results show that e-word-of-mouth has a positive impact on online group buying intention; e-word-of-mouth has a positive impact on perceived security; and the intermediary effect of perceived security on the relationship between electronic word-of-mouth and online group buying intention. Zhang Gaoliang et al. (2014) [11] pointed out that consumers' perception of the website's security, reputation, evaluation of consumers in the website and the usefulness of such elements as the quality of goods (services) was positively correlated with consumers' trust. Chai Kangwei et al. (2019) [12] confirmed that online word-of-mouth will affect the decision to buy. Xu Shiying (2018) [13] proposed online reviews for research and found that while online reviews affect consumer trust, trust can also improve consumer purchase intention by reducing consumer perceived risk.

3 Research Design

3.1 Research Framework

Based on the above-mentioned literature, the following research model and hypotheses are proposed in this study. The research model is shown in Fig. 3.

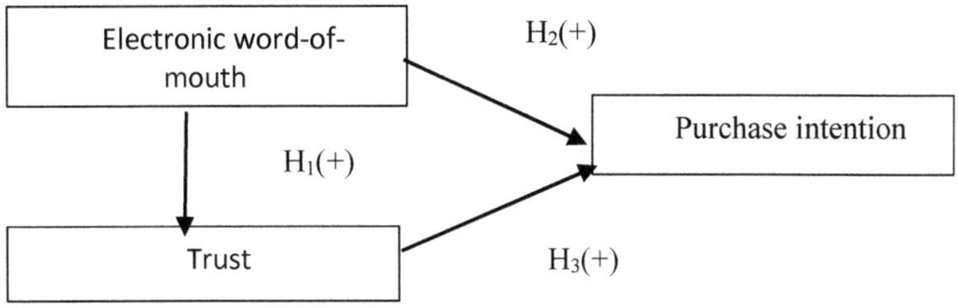

Fig. 3. Research model.

Research hypothesis:

H1: electronic word-of-mouth has a positive impact on the trust of potential consumers on the Internet.

H2: electronic word-of-mouth has a positive impact on the purchase intention of potential online consumers.

H3: trust has a positive impact on the purchase intention of potential online consumers.

3.2 Research Method

This study is expected to use a quantitative questionnaire survey method, which is distributed online to consumers who have ordered goods on the JingDong Shopping E-commerce Platform, to investigate their cognition of the relationship between electronic word-of-mouth, trust and purchase intention, and further use SPSS software for data analysis and regression model verification.

References

1. China Internet Network Information Center. The 44th China statistical report on internet development. http://www.cac.gov.cn/2019-08/30/c_1124938750.htm. 30 Aug 2019
2. Gelb, B.D., Sundaram, S.: Adapting to word of mouse. Bus. Horiz. **45**(4), 21–25 (2002)
3. Hennig-Thurau, T., Gwinner, K.P., Walsh, G., Gremler, D.D.: Electronic word-of mouth via consumer-opinion platforms: What motivates consumers to articulate themselves on the internet? J. Interact. Mark. **18**(1), 38–52 (2004)
4. Wang, W.T., Ou, W.M., Chen, W.Y.: The impact of inertia and user satisfaction on the continuance intentions to use mobile communication applications: a mobile service quality perspective. Int. J. Inf. Manage. **44**, 178–193 (2019)
5. Lee, M.K.O., Turban, E.: A trust model for consumer internet shopping. Int. J. Electr. Comm. **6**(1), 75–91 (2001)
6. Moorman, C., Deshpande, R., Zaltman, G.: Factors affecting trust in market research relationships. J. Mark. **57**(1), 81–101 (1993)
7. Hsu, L.C.: Investigating community members' purchase intention on Facebook fan page: from a dualistic perspective of trust relationships. Ind. Manag. Data Syst. **117**(5), 766–800 (2017)
8. Dodds, W.B., Monroe, K.B., Grewal, D.: Effects of price, brand, and store information on buyers' product evaluations. J. Mark. Res. **18**(3), 307–319 (1991)
9. Wang, M.: Research on the influence of online reviews on the purchase intention of hotel customers. Chongqing Jiaotong University (2018)
10. You, Q.: Master's thesis of master's in-service class of School of management. Daye University (2019)
11. Zhang, G., Zhu, W., Zhu, J.: The formation mechanism of consumer trust under the business network group buying mode. Ergonomics **20**(05), 50–55 (2014)
12. Chai, K., Ou, W., Cai, Z., Gu, Q.: Diagnosis of the relationship between shopping website brand image and online word-of-mouth: a domestic empirical study. Manag. Inf. Calculation **8**(Special 1), 24–34 (2019)
13. Xu, S.: Research on the Influence of Quality Characteristics of Online Reviews on Online First Purchase Intention of Economic Hotels. Jilin University (2018)

An Effective Face Color Comparison Method

Yuanyong Feng[1], Jinrong Zhang[2], Zhidong Xie[1], Fufang Li[1(✉)], and Weihao Lu[1,2]

[1] School of Computer Science and Cyber Engineering, Guangzhou University, Guangzhou 510006, China
liff@gzhu.edu.cn
[2] School of Mechanical and Electrical Engineering, Guangzhou University, Guangzhou 510006, China

Abstract. With the development of image processing technology and application, the recognition requirements for facial details are getting higher and higher. This paper presents a face color comparison method, which can compare the facial color difference quickly. At the same time, through experiments, this paper determined the formula of color difference used by the method and obtained the good accuracy of the method in accordance with human vision, which shows that the method has certain practicability in facial color comparison.

Keywords: Facial color · Color comparison · Image processing

1 Introduction

With the development of color difference formula, CIEDE2000 color difference formula is proposed by the International Commission on illumination (CIE), which is considered to be the best color difference formula at present, and its consistency with human vision exceeds 95% [1]. However, almost all color difference formulas are only used to evaluate the color difference of large single sample or surface [2], and more and more practical applications need to predict the color difference between photographic images [3]. Current color difference formulas do not take into account the necessary components to evaluate the stereoscopic changes in these images [4].

In the photos of human faces, the color composition is relatively complex, including dark areas such as eyebrows and pupils, and light areas such as whites of the eyes and teeth. It is difficult to compare facial colors with the color difference formula for a single color. Therefore, this paper proposes a face color comparison method, which take multi-point samples of the face first, and then calculate the color difference according to the sampling points, so as to obtain a face color comparison result closer to the visual perception of human vision. By comparison with the visual data of the experimental participants, the face color comparison method proposed in this paper is close to the actual viewing effect, so that the method has a certain practical value.

Q. Liu et al. (Eds.): CENet 2020, AISC 1274, pp. 105–112, 2021.
https://doi.org/10.1007/978-981-15-8462-6_12

2 Face Sampling Based on Dlib Library

Dlib is a general-purpose cross-platform library written in the C++ programming language, being able to deal with threads, linear algebra, statistical machine learning, image processing, numerical optimization, etc. [5].

By using the key point detection function of Dlib library, the experimenter can conveniently obtain 68 key points covering facial features and facial contour, which are evenly distributed and cover most areas with concentrated facial colors.

With the help of Dlib library, the face color comparison method takes the color from the key points of people's faces (as is shown in Fig. 1), which is used as the sample data for the face color comparison.

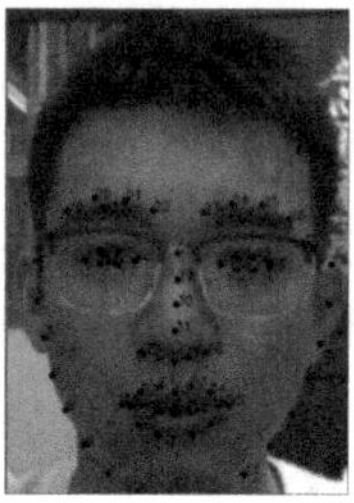

Fig. 1. Face feature points.

3 Color Space and Color Difference Formula

Color space is the organization of colors. With the help of color space and the test of physical equipment, the fixed description and digital representation of colors can be obtained. A color model is an abstract mathematical model that describes a color through a set of Numbers.

3.1 RGB Color Space

At present, the RGB color space is the most widely used color space. Other color Spaces need to be converted based on it. However, RGB color space also has the defect of inaccurate measurement of color difference. This means, the linear representation of color space is not consistent with the non-linear perception of human eyes, resulting in a big difference between the calculated value of color difference and the intuitive perception of human vision [6].

For the two colors whose coordinates are $p_1 = (R_1,G_1,B_1)$ and $p_2 = (R_2,G_2,B_2)$ in the RGB color space, there are three main measurement formulas for calculating the color difference, as following.

Simple RGB Color Difference Formula. It takes the spatial distance of two colors in the three-dimensional coordinate system established with RGB as the axis as the color difference measurement value, which needs to be established on the basis that RGB color space is uniform color space.

$$D(p_1,p_2) = \sqrt{(R_1 - R_2)^2 + (G_1 - G_2)^2 + (B_1 - B_2)^2} \quad (1)$$

According to this definition, all colors with same color difference value show the same perceptual difference. Obviously, based on the human perception difference of different colors, the calculation method of this color difference is different from human visual perception.

RGB Weighted Color Difference Formula. In view of the defect of the above formula, many literatures solve the problem of uneven color space by using the method of adding weighted coefficient. w_R is the weighted coefficient of the red component, and the other coefficients are the w_G and w_B.

$$D(p_1,p_2) = \sqrt{w_R(R_1 - R_2)^2 + w_G(G_1 - G_2)^2 + w_B(B_1 - B_2)^2} \quad (2)$$

According to the different sensitivity of human vision to the three basic colors, the weighting coefficient is defined to compensate for the non-uniformity of human vision. In most literatures, the (w_R,w_G,w_B) is (3,4,2). Many experiments have proved that the weighted color difference programmed algorithm can correct the error of the simple RGB color difference algorithm to a certain extent, but it is not always better than the non-weighted algorithm.

RGB Angular Distance Color Difference Formula. The formula was proposed by [7]. The main idea of this formula is to compare the angular difference between colors. The effect calculated with this formula is better than those results using the above two methods to a certain extent.

$$D(p_1,p_2) = 1 - \left[1 - \frac{2}{\pi}\cos^{-1}\left(\frac{p_1 \cdot p_2}{|p_1||p_2|}\right)\right]\left[1 - \frac{|p_1 - p_2|}{\sqrt{3 \cdot 255^2}}\right] \quad (3)$$

3.2 HSV Color Space

HSV color space use H(Hue), S(Saturation) and V(Value) to describe color. The hue is described by the Angle, and the wavelength of the light wave will lead to the color difference. The intensity of colors is expressed by saturation, which is determined by the purity of the wavelength of visible light. The brightness of objects is expressed by value, which is closely related to the broadcast energy [6].

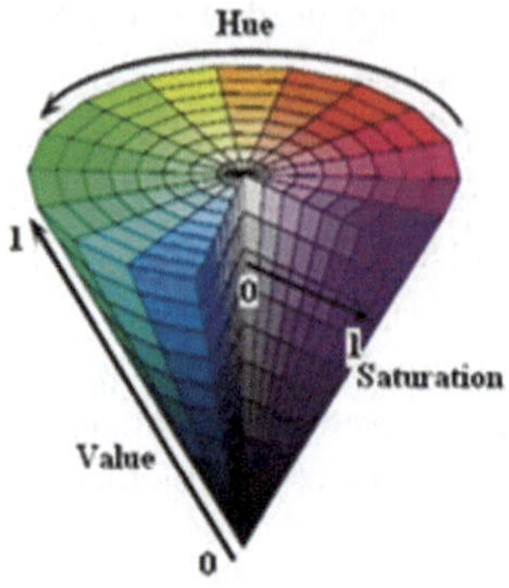

Fig. 2. HSV visual model.

A visual model of the HSV is the cone (as is shown in Fig. 2). In this representation, hue is represented as the angle around the central axis of the cone, the saturation is expressed as the distance from the center of the cone's cross section to this point, and value is represented as the distance from the center of the cone's cross section to the vertex.

The color difference measurement in HSV color space can be calculated by using Euclidean distance method [8]. The color difference of HSV calculated by Euclidean distance formula with HSV conical space coordinates, and the equations is as follows.

Color gamut conversion formula:

$$x = r \times \mathrm{V} \times \mathrm{S} \times \cos\mathrm{H}, y = r \times \mathrm{V} \times \mathrm{S} \times \sin\mathrm{H}, \mathrm{z} = \mathrm{h} \times (1 - \mathrm{V}) \tag{4}$$

Color difference calculation method:

$$D(p_1, p_2) = \sqrt{(x_1 - x_2)^2 + (x_2 - y_2)^2 + (z_1 - z_2)^2} \tag{5}$$

3.3 Lab Color Space

The International Commission on illumination (CIE) published the L × A × B color space and the color difference formula within the color space in 1976, which is also known as the color difference formula CIE L × A × B. As the color space of CIE LAB is designed to fit human visual perception, its color difference formula can better reflect the degree of color change of human visual perception [9].

CIE1976 color difference formula is defined as:

$$\Delta E^*_{ab} = \sqrt{(L^*_2 - L^*_1)^2 + (a^*_2 - a^*_1)^2 + (b^*_2 - b^*_1)^2} \tag{6}$$

4 Facial Color Comparison Method

4.1 Facial Color Comparison Method

This method (as is shown in Fig. 3) locates N facial sampling points by the Dlib library. The sampling effect is limited by the recognition. When the face in picture is blocked or only shows little part of its side, this method lost its use.

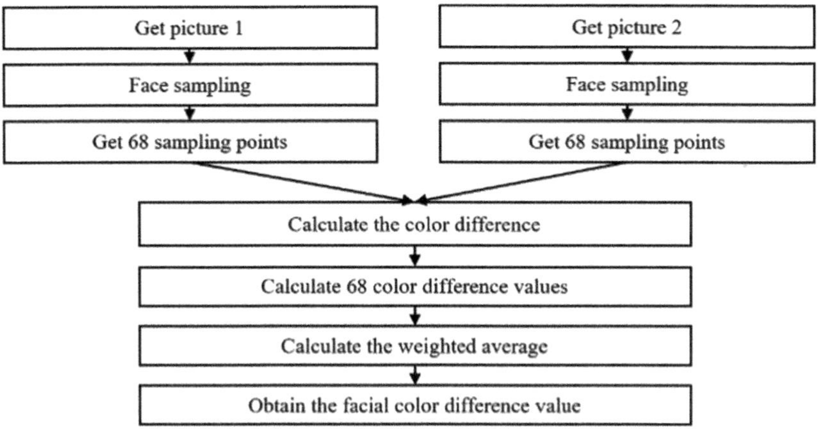

Fig. 3. The facial color comparison algorithm flow.

First, two sets C_1, C_2 of points representing contours of facial features are sampled correspondingly from each photo to be compared. Each point's color is expressed as,

$$C_j(i),\ 0 < i \leq N, \tag{7}$$

where j is index of the target photos, *i* is the index of point, and N is the contour set size.

Then calculate the color difference for each group of colors in the same position. The color difference calculation function is denoted as:

$$CCF(color_1, color_2), \tag{8}$$

where each parametric *color* is defined as in previous Sect. 3.

Next the color difference of each pair points in the same position is,

$$CD(i) = CCF[C_1(i), C_2(i)], 0 < i \leq N \tag{9}$$

Finally, the color difference values are weighted and averaged to obtain the color difference value of facial colors in the two sets of pictures under this method,

$$FCD_{face1,face2} = \sum\nolimits_{i=0}^{N} CD(i) \Big/ N \qquad (10)$$

4.2 Color Difference Formula Selection

Due to the difference between different color difference formulas in approaching human visual experience, this paper uses several commonly used color space and its widely used color difference formulas to carry out experiments, and compares the effect of multiple color difference formulas in face color comparison.

5 Experiments

5.1 Experimental Setting

The experiment used 1000 images (part of which are as shown in Fig. 4), most are of human face. There were 20 participants involved. Each of them watched 100 random groups of experimental samples, and selected the image whose face color was most similar with original image. At the same time, the program used different color difference formulas to mark the image, which had the lowest color difference between the faces and the original image, by means of the facial color comparison method proposed above. Finally, the data are summarized to calculate the probability that the method has the same result as the marking result of human vision.

Fig. 4. Example images.

Since CIE94 and CIE2000 color difference formulas require parameters such as reflection coefficient, there is no data for human face in existing studies, so they are not used in this experiment. RGB weighted color difference formula, HSV color difference formula and CIELAB color difference formula were used in this experiment. The color values shall be converted to the corresponding gamut for each color space before calculating.

In order to minimize the impact of irrelevant factors on the color difference ratio experiment, the experiment used the same display. The brightness, color temperature and other settings in the display were kept consistent, and the light in the laboratory remained constant, too. The comparison program (as Fig. 5) was executed 5 times, and the largest number of pictures was taken as the final result.

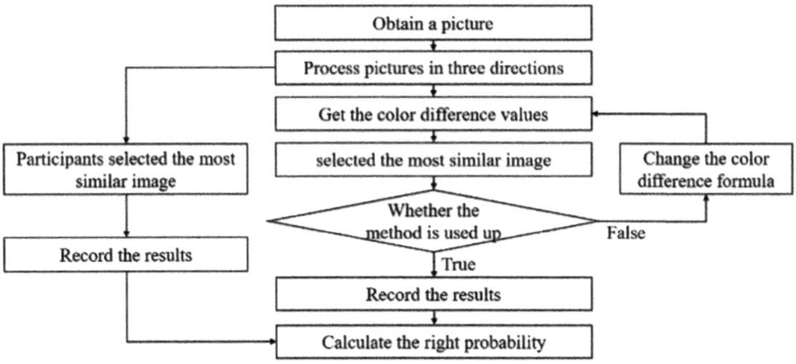

Fig. 5. Experimental procedure.

5.2 Result Analysis

In the experiment, the effective result was taken while the participants can clearly distinguish the sequence number of the photos, which the facial color is the closest to the original image in the image group and the program can effectively identify the facial sampling points to take the color values. After experiments, a total of 1894 cases were obtained to be the effective results. The numbers of effective comparison results with different color difference formulas are as Table 1:

Table 1. Result comparison.

	RGB weighted'	HSV'	CIE Lab'
Match number	1270	1193	1435
Probability	67.053%	62.988%	75.766%

From the experiment, the following conclusions can be obtained:

- The face color comparison method has a certain effect on the face color comparison. When CIELAB formula is used, the matching rate between the result of program and the result of human vision selection is more than 75%, which shows that the method has a certain practical value.
- While using the face color comparison method, CIE LAB color difference formula has achieved a good effect, and this formula should be preferred in use.

6 Conclusion

This paper proposed a facial color comparison method, and verified its practical value through experiments. At the same time, through experiments, the application effect of the color difference formula in several color Spaces in this method was compared, and the CIE LAB color space had a good effect.

However, the method proposed in this paper is not perfect, which may be insufficient sampling points and insufficient sample representativeness, and the accuracy needs to be improved. Therefore, the study of facial color evaluation needs to be further studied.

Acknowledgement. This work is supported in part by the National Natural Science Foundation of China (No. 61472092) and Guangdong College Student Innovation Project (No. S201911078074).

References

1. Yang, Z.Y., Wang, Y., Yang, Z.D., Wang, C.D.: Vector-angular distance color difference formula in RGB color space. Comput. Eng. Appl. **46**(6), 154–156 (2010)
2. Zhang, X., Wandell, B.A.: A spatial extension of CIELAB for digital color image reproduction. In: SID International Symposium Digest of Technical Papers, vol. 27, pp. 731–734 (1996)
3. Hong, G., Luo, M.R.: Perceptually-based color difference for complex images. In: Proceedings Volume 4421, 9th Congress of the International Color Association, pp. 618–621. SPIE, Washington, USA (2002)
4. Jin, X., Zhang, S., Li, Q., Du, L., Zhu, C.: Development of color difference formula and its application in fabric color evaluation. J. Silk **50**(5), 33–38 (2013)
5. King, D.E.: Dlib-ml: A machine learning toolkit. J. Mach. Learn. Res. **10**, 1755–1758 (2009)
6. Yang, C.: Research and design of color difference detection system based on machine vision. Zhejiang Sci-Tech University (2015)
7. Androutsos, D., Plataniotis, K.N., Venetsanopoulos, A.N.: A novel vector-based approach to color image retrieval using a vector angular-based distance measure. Comput. Vis. Image Underst. **75**(1–2), 46–58 (1999)
8. Gao, S.: Color-based image retrieval methods and system implementation. Beijing University of Posts and Telecommunications (2006)
9. Lin, X.Y.: The design of the shading variation measurement system based on machine vision. Huazhong University of Science and Technology (2008)

Transfer Learning Based Motor Imagery of Stroke Patients for Brain-Computer Interface

Yunjing Miao[1], Fangzhou Xu[2(✉)], Jinglu Hu[3], Dongri Shan[2], Yanna Zhao[4], Jian Lian[4], and Yuanjie Zheng[4]

[1] School of Electrical Engineering and Automation, Qilu University of Technology (Shandong Academy of Sciences), Jinan 250353, Shandong, China

[2] School of Electronic and Information Engineering, (Department of Physics), Qilu University of Technology (Shandong Academy of Sciences), Jinan 250353, Shandong, China
xfz@qlu.edu.cn

[3] Department of Physical Medicine & Rehabilitation, Qilu Hospital, Cheeloo College of Medicine, Shandong University, Jinan 250012, Shandong, China

[4] School of Information Science and Engineering, Shandong Normal University, Jinan 250358, Shandong, China

Abstract. As deep learning continues to be a hot research issue, analysis of brain activities based on deep neural networks have been in the rapid development. In this paper, the convolution neural network is introduced to decode the electroencephalogram (EEG) from stroke patients to design an effective motor imagery based brain-computer interface system. We develop a novel convolution neural network architecture combined with transfer learning to recognize motor imagery tasks from stroke patients. We also explore intra-subject and inter-subject transfer learning methods to evaluate the performance of the proposed framework and to avoid time-consuming training process to low algorithmic complexities. The fine-tune technology has been employed to transfer model parameters. The proposed framework is a combination of EEGNet model and fine-tune technology. The proposed framework is evaluated on stroke patients, the average classification accuracy is 65.91%. Experimental results validate that the proposed algorithm can obtain satisfactory performance improvement for EEG-based stroke rehabilitation of brain-computer interface system, and show that our model has potential for effective stroke rehabilitation motor imagery based brain-computer interface systems.

Keywords: Brain-Computer interface · EEG · Stroke rehabilitation · Motor imagery · Transfer learning

Q. Liu et al. (Eds.): CENet 2020, AISC 1274, pp. 113–120, 2021.
https://doi.org/10.1007/978-981-15-8462-6_13

1 Introduction

Brain-computer interface (BCI) is a system that does not rely on the peripheral nerve and muscle tissue of the human body to establish a new correspondence and command technology between brain and external device [1]. Electroencephalography (EEG) is a noninvasive way of BCI [2]. At present, there are several commonly used EEG-based paradigm-related signals used in BCI systems, such as visual evoked potential (VEP), P300 evoked potential, motor imagery (MI) and their combinations [3].

MI is defined as a BCI approach involving movement of a part of the body without any muscle contraction, imagine yourself moving a part of your body. Each MI task causes oscillations in the corresponding sensorimotor cortex area of the brain [4]. The goal of MI-based BCI systems is to identify potential MI tasks.

In recent years, the application of deep learning (DL) in BCI has been widely developed, such as predicting the sex of the user [5], epileptic seizure prediction [6] or MI classification. We know that deep learning can offer better classification performance by increasing the size of training data. In past research, a deep belief network (DBN) model [7] and a convolutional neural networks (CNN) model [8] was applied for MI classification. However, traditional DL methods require large amounts of data to avoid overfitting, since data collection is relatively expensive and time-consuming, its application in the field of neuroscience is challenging [9]. Transfer learning (TL) is a method of machine learning where a pre-trained model is reused for another task. In the context of deep learning, TL can solve the above problems, it can not only allow too little data for deep learning training but also keep the model performance at an acceptable level [10].

In this paper, we propose a CNN based deep learning framework with transfer learning for EEG analysis of stroke patients to improve performance of the stroke rehabilitation based BCI system. The proposed architecture is a combination of EEGNet and fine-tune. We employ EEGNet architecture combined with transfer learning to recognize motor imagery tasks from stroke patients. We also propose intra-subject and inter-subject transfer learning methods to evaluate the performance of the proposed framework and to avoid time-consuming training process. The transfer model parameters can be optimized using the fine-tune technology. The proposed framework is evaluated on dataset of MI-based EEG signals acquiring from the stroke patients and the normal subjects. The propose framework can achieve the average classification accuracy is 65.91%. Moreover, the transfer learning method can avoid time-consuming training process to low algorithmic complexities. The results confirm that the proposed framework can obtain satisfactory performance improvement for EEG-based stroke rehabilitation of BCI system, and show that our framework is simple and efficient.

The rest of this article is organized as follows: Sect. 2 describes the experimental data and the proposed framework, Sect. 3 gives the experimental results, Sect. 4 discusses the conclusions.

2 Method

2.1 Experimental Data

The EEG dataset of 11 stroke patients has been collected in the Deparment of Physical Medicine & Rehabilitation, Qilu hospital, Cheeloo College of medcine, Shandong University. The dataset includes trials of 5 healthy subjects and 6 stroke patients. During the signal acquisition procedure, the subjects have performed imagination of left or right hand movements. In these trials, the 64-channel signal acquisition device has been used. Each trial has been recorded for 3 s, followed by a 4-second interval. Data collection required 60 trials, which include 30 trials of right-hand imagination and 30 trials of left-hand imagination. The whole experiment of one subject takes approximately 9 min. In order to prevent from the visually evoked potentials, each trial has recorded 1 s after the end of the visual cue. EEG signal has been recorded at a sampling rate of 1000 Hz. The timing diagram of single-trial recorded in the whole experiment has been shown in Fig. 1.

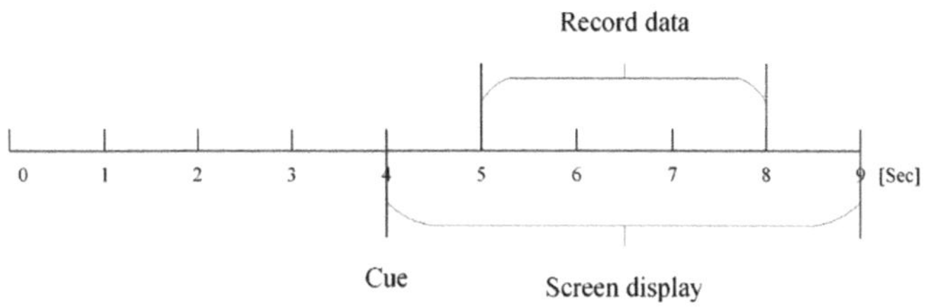

Fig. 1. Timing diagram of single-trial recorded in the whole experiment.

2.2 Transfer Learning

The definition of symbol includes two parts, feature space X and edge allocation $P(X)$ of feature space. $x = \{x_1, x_2, \ldots, x_n\}$ belongs to X, for two different domains, they are different in the feature space or edge probability distribution. The domain is expressed as $D = (X,\ P(X))$. The mission include two parts, label expanse Y and a prediction objective function $f(.)$. A task is expressed as $T = \{Y, f(.)\}$, the prediction objective function cannot be directly and accurately observed, but can be obtained through training. From the perspective of probability theory, the target prediction function $f(.)$ can be expressed as $P(Y|X)$. The task is expressed as $T = \{Y,\ P(Y|X)\}$.

The definition of transfer learning is described as follows. Given the source domain D_S, the source domain learning task T_S, the target domain D_t, and the target domain task T_t. D_S is not equal to D_t and T_S is not equal to T_t. Transfer learning uses the knowledge enhancement in the source domain D_S and T_S to improve or optimize the learning efficiency of the prediction objective function $f_t(.)$ in the target domain D_t.

The transfer learning method used in this article is fine-tune technology. The pre-trained models can be transferred to your own model. The pre-trained network weights are usually used to initialize their own network weights instead of the original random initialization.

The adaptive layer between feature extraction layer and classification layer can be added to make the data distribution of target domain and source domain closer. An adaptation layer is added after the feature layer to calculate the distance between the target domain and the source domain. It is added to the loss of the network for training as $\ell = \ell_c(D_s, y_s) + \lambda\ell_A(D_s, D_t)$.

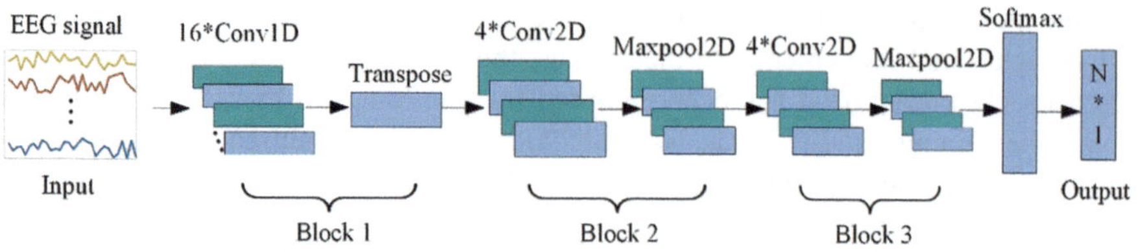

Fig. 2. The structure of our propose EEGNet model. There are 5 layers in total, and the middle 3 layers are convolutional layers.

2.3 EEGNet Architecture

EEGNet network has been successfully applied to MI-based BCI systems, mainly performing automatic feature extraction and classification. As shown in Fig. 2, the EEG signal processing at different stages is described.

EEGNet is a CNN architecture which is compact. It can be applied to different BCI paradigms, such as MI, P300 visual evoked potential, and motor-related cortical potential (MRCP) [11]. One advantage of EEGNet is that it can be trained basing on a limited amount of data obtained in the calibration phase. The structure of the model in the paper can be defined as follows:

1) In block 1, a depthwise convolution of size (1, C) is used, where C denotes the number of channels. In the field of CNN for pattern recognition, the advantage of convolution is to decrease the amount of trainable parameters to be fitted, because they do not connect all the previous feature maps together. The next step is to normalize. Dropout technology is used for normalizing or modeling.
2) In block 2, a depthwise convolution of size (2, 32) is employed. In block 3, a depthwise convolution of size (8, 4) is adopted to perform convolution. The advantage are that the number of parameters for fitting reduced, and the kernel of each feature map can be learned well. Then, the largest pooling layer of a size of (2, 2) is adopted for size reducing.
3) Softmax, which has a category of N, receives the feature output from the previous block, where N is the number of classes for MI tasks. The softmax function is adopted because the EEGNet is a multi-class model [12].

3 Results

3.1 Classification Accuracy

The paper mainly introduced two parts of results. The first part is the results of intra-subject. An experimental of one subject can be divided into training and test subsets. The second part is the results of inter-subject. The dataset of all subjects can be separated into two groups for analysis. One subset consists of the data description from 10 subjects as a training prediction model, and the other subset consists of the remaining data from 1 subject. The role of the other subset is to assess the performance of the proposed model. The whole process is performed 11 times and each subject can be selected for performance evaluation.

We has employed three different band-pass filterings to process EEG data. These bandpass filterings include bandpass filtered between 8 Hz and 24 Hz, bandpass filtered between 8 Hz and 30 Hz, and bandpass filtered between 8 Hz and 40 Hz. Model parameters has been introduced in the definition of model structure mentioned in Sect. 2. The "adam" optimizer is used for model fitting. It creates learning models in the pytorch environment. Table 1 described the paramters of EEGNet architecture, where M is the number of samplings.

In the intra-subject analysis, EEGNet trained for different training times during training period. However, we found that the experimental results of training in different frequencies did not obviously change. The averages classification accuracies under different training times have been described as follows. The average accuracy is $61.00 \pm 5.00\%$ when the training times is 100. The average accuracy is $65.00 \pm 6.00\%$ when the training times is 300. The average accuracy is $70.00 \pm 5.00\%$ when the training times is 500.

In the inter-subject analysis, the accuracy rate has not been improved with the increase of training times. The average test classification accuracies under different training times is as follows. The frequency range is from 8 Hz to 24 Hz, and the dropout is 0.25. The test accuracy is $60.00 \pm 4.00\%$ when the training cycles is 100. The test accuracy is $61.00 \pm 5.00\%$ when the training cycles is 300. The test accuracy is $59.00 \pm 5.00\%$ when the training cycles is 500.

3.2 Comparative Results

We compared EEGNet with another CNN model [13]. The results of the two models were compared according to different subjects and the results obtained are shown in Table 2. We can infer that the proposed model commonly outperform another CNN model. The average accuracy of our framework has been increased about 7.27% compared with another CNN model.

Table 1. The parameters of EEGNet architecture.

Block	Layer	Input	Size	Output
1	Conv1D	C*M	(1, 64)	16*1*M
	BatchNorm	16*1*M		16*1*M
	Transpose	16*1*M		1*16*M
	Dropout	1*16*M	0.25	1*16*M
2	Conv2D	1*16*M	(2, 32)	4*16*M
	BatchNorm	4*16*M		4*16*M
	Maxpool2D	4*16*M	(2, 2)	4*8*(M/2)
	Dropout	4*8*(M/2)	0.25	4*8*(M/2)
3	Conv2D	4*8*(M/2)	(8, 4)	4*8*(M/2)
	BatchNorm	4*8*(M/2)		4*8*(M/2)
	Maxpool2D	4*8*(M/2)	(2, 2)	4*4*(M/4)
	Dropout	4*4*(M/4)	0.25	4*4*(M/4)
Classifier	Softmax	4*4*(M/4)		N

Table 2. The accuracies obtained by using CNN model and EEGNet model.

Subject	Test Acc (%)	Test Acc (%)
Model	CNN	EEGNet
A01	54.00	59.00
A02	57.00	61.00
A03	65.00	64.00
A04	56.00	61.00
A05	61.00	75.00
A06	58.00	70.00
A07	53.00	64.00
A08	67.00	71.00
A09	60.00	65.00
A10	52.00	65.00
A11	62.00	70.00
Mean	58.64	65.91

4 Discussion and Conclusions

Our proposed framework is focus on EEG dataset on stroke patients. The EEG data acquisition from the experiments are difficult to distinguish. It is well known that brain lesions may seriously change the dynamics of EEG signals, so the instability of data distribution is increased. Therefore, the performance of brain activities from stroke patients equivalent to healthy subjects may not be expected. Furthermore, the sizes of datasets from patients are very limited. Thus, the EEGNet model has been employed to solve these problems. The inter-subject prediction performance of EEGNet is also better than other models, and the transfer learning can be used to reduce computational cost.

The purpose of this study is to understand whether the pre-trained model can further improve the performance of deep learning framework on the limited number of dataset. The experimental results have shown that although only the small size of data has been trained, the knowledge of feature learning has been successfully transferred between experiments by fine-tune strategy. It is known that the pre-trained EEGNet model was trained on a large amount of EEG data, which has made our proposed model more robust. The EEGNet combined with the transfer learning can be used to reduce computational cost. Through the experimental performance, it can be proved that the finetune can be employed to improve the performance of our proposed architecture.

It can be inferred that the proposed framework can transfer relative knowledge to recognize different MI tasks. In future work, much larger amounts of EEG data will be utilized to monitor EEG-biomarkers and to evaluate the system performance.

Acknowledgements. This research is supported in part by the National Natural Science Foundation of China under Grant No. 61701270, Grant No. 81472159, Grant No. 81871508, and Grant No. 61773246, in part by the Program for Youth Innovative Research Team in University of Shandong Province, China under Grant No. 2019KJN010, in part by the Key Program for Research and Development of Shandong Province in China (Key Project for Science and Technology Innovation, Department and City Cooperation) under Grant No. 2019TSLH0315, in part by the Jinan Program for Development of Science and Technology, in part by the Jinan Program for Leaders of Science and Technology, in part by the Key Research and Development Plan of Shandong Province Grant No. 2017G006014, in part by the Natural Science Foundation of Shandong Province in China under Grant No. ZR2016FQ06, in part by the Major Program of Natural Science Foundation of Shandong Province in China under Grant No. ZR2019ZD04, Grant No. ZR2018ZB0419, and in part by the Taishan Scholar Program of Shandong Province, China under Grant No. TSHW201502038.

References

1. Graimann, B., Allison, B., Pfurtscheller, G.: Brain–computer interfaces: a gentle introduction. In: Graimann, B., Pfurtscheller, G., Allison, B. (eds.) Brain-Computer Interfaces. The Frontiers Collection, pp. 1–27. Springer, Heidelberg (2010). https://doi.org/10.1007/978-3-642-02091-9_1
2. Hossain, I., Khosravi, A., Hettiarachchi, I., Nahavandi, S.: Multiclass informative instance transfer learning framework for motor imagery-based brain-computer interface. In: Computational Intelligence and Neuroscience, pp. 1–12 (2018)
3. Wolpaw, J.R., Birbaumer, N., McFarland, D.J., Pfurtscheller, G., Vaughan, T.M.: Brain–computer interfaces for communication and control. Clin. Neurophysiol. **113**(6), 767–791 (2002)
4. Tabar, Y.R., Halici, U.: A novel deep learning approach for classification of EEG motor imagery signals. J. Neural Eng. **14**(1), 016003 (2016)
5. Van Putten, M.J., Olbrich, S., Arns, M.: Predicting sex from brain rhythms with deep learning. Sci. Rep. **8**(1), 1–7 (2018)
6. Hosseini, M.P., Soltanian-Zadeh, H., Elisevich, K., Pompili, D.: Cloud-based deep learning of big EEG data for epileptic seizure prediction. In: 2016 IEEE Global Conference on Signal and Information Processing, GlobalSIP, pp. 1151–1155. Washington, D.C. (2016)

7. An, X., Kuang, D., Guo, X., Zhao, Y., He, L.: A deep learning method for classification of eeg data based on motor imagery. In: Huang, D.-S., Han, K., Gromiha, M. (eds.) ICIC 2014. LNCS, vol. 8590, pp. 203–210. Springer, Cham (2014). https://doi.org/10.1007/978-3-319-09330-7_25
8. Yang, H., Sakhavi, S., Ang, K.K., Guan, C.: On the use of convolutional neural networks and augmented CSP features for multi-class motor imagery of EEG signals classification. In: 2015 37th Annual International Conference of the IEEE Engineering in Medicine and Biology Society, pp. 2620–2623. IEEE Xplore, Milan (2015)
9. Uran, A., van Gemeren, C., van Diepen, R., Chavarriaga, R., Millán, J.D.R.: Applying transfer learning to deep learned models for EEG analysis. ArXiv Preprint ArXiv: 1907.01332 (2019)
10. Abdulkader, S.N., Atia, A., Mostafa, M.S.M.: Brain computer interfacing: applications and challenges. Egypt. Informat. J. **16**(2), 213–230 (2015)
11. Raza, H., Chowdhury, A., Bhattacharyya, S.: Deep learning based prediction of EEG motor imagery of stroke patients' for neuro-rehabilitation application. In: International Joint Conference on Neural Networks, pp. 1–8. Glasgow (2020)
12. Lawhern, V.J., Solon, A.J., Waytowich, N.R., Gordon, S.M., Hung, C.P., Lance, B.J.: EEGNet: a compact convolutional neural network for EEG-based brain–computer interfaces. J. Neural Eng. **15**(5), 056013 (2018)
13. GitHub. https://github.com/hauke-d/cnn-eeg

Intelligent Virtual Lipstick Trial Makeup Based on OpenCV and Dlib

Yuanyong Feng, Zhidong Xie, and Fufang Li(✉)

School of Computer Science and Cyber Engineering,
Guangzhou University, Guangzhou 510006, China
liff@gzhu.edu.cn

Abstract. In order to try different lipstick products in a reusable, low-cost, and hygienic manner, an intelligent virtual lipstick trial algorithm is proposed with assistance of Dlib and OpenCV. In the proposed method, we perform face detection through the pre-trained face detection model of dlib.get_frontal_-face_detector in the Dlib library, and then extract feature points from the face in the video according to the above pre-trained model. When lips and other important face parts are recognized, they are then filled with specific color patterns according to lighting environment, highlighting the import characteristics for target lipstick, making it reliably displayed on a smart phone screen. The process is full of simplicity, quickness and convenience.

Keywords: Lipstick makeup · Virtual reality · OpenCV · Dlib

1 Introduction

With the improvement in global people's life, there evidence great increase in the field for cosmetic consumption. In the traditional entity shop sale model, consumers are obligated to face with the problem of hygiene when they are trying makeup products, and imposing business owners to the sale cost expanding. Confronting the predicament of the global epidemic in 2020, many major e-commerce companies have launched new business model of online live streaming with goods, providing consumers an unprecedented shopping experience. Among others, cosmetics section enjoys the great benefits under this model, while their consumers blindly experience the makeup effect vision online. To overcome this dilemma, online virtual makeup reality methods are proposed. Constructed on the maturity of face recognition and image processing technology, such shopping experience will inevitably become hot topics in the field of computer graphics.

As our best knowledge, there are very few researches on online virtual makeup, while most of them are image-based. In 2007, Tong et al. [1] proposed a virtual makeup test method based on images of an identical person before and after makeup as an example. The steps of deformation, segmentation, and repair were used. The makeup was preserved completely but the process was cumbersome. In 2009, Guo et al. [2] proposed improvements, by inputting a single reference makeup image to migration, weaken precondition constraint. However, this method uses the Poisson equation to fuse the highlights and shadows of the image. Due to the lack of continuity

Q. Liu et al. (Eds.): CENet 2020, AISC 1274, pp. 121–127, 2021.
https://doi.org/10.1007/978-981-15-8462-6_14

in the solution of the Poisson equation, the visual effect is still flawed. In 2010, Zhen et al. [3] proposed a digital face makeup technology based on sample pictures with lacks of automation in makeup transferring, and complexity in work mechanism. In 2013, Du et al. [4] proposed a multi-sample makeup transfer technology to transfer different sample makeups to the same face, using image fusion technology to achieve the effect. In 2015, Lin et al. [5] proposed a method to achieve high-fidelity transplantation of digital face image makeup using suppressed lighting and edge adaptive filtering. This method can overcome the stitching effect generated by Poisson equation fusion. In 2018, Liu et al. [6] proposed a virtual makeup test algorithm based on a planar grid model, which realized real-time video-based makeup test.

This paper proposes a real-time virtual lipstick test algorithm based on Dlib library and OpenCV. During each iteration, a camera frame is retrieved first in real time. Then, some specific facial feature points are extracted and the lip contour is recognized. Subsequently, the lip part is mixed with target makeup pattern. Finally, an output face effect image is obtained by merging the processed lips and the original face, within a very short period in real time. The algorithm has the following advantages: 1. Real-time feedback of the processing effect; 2. The simplicity of the algorithm model, which can be achieved by only a simple operation.

2 Lipstick Makeup Migration Method Based on OpenCV and Dlib Library

The lipstick makeup transfer method based on OpenCV and Dlib libraries needs to call the camera in real time to obtain the face photos that need to be processed, by using the Dlib library for face detection and facial feature points extraction, and then does image processing and target lipstick weighted fusion on OpenCV, producing the real-time lipstick makeup test effect. The algorithm flow is shown in Fig. 1.

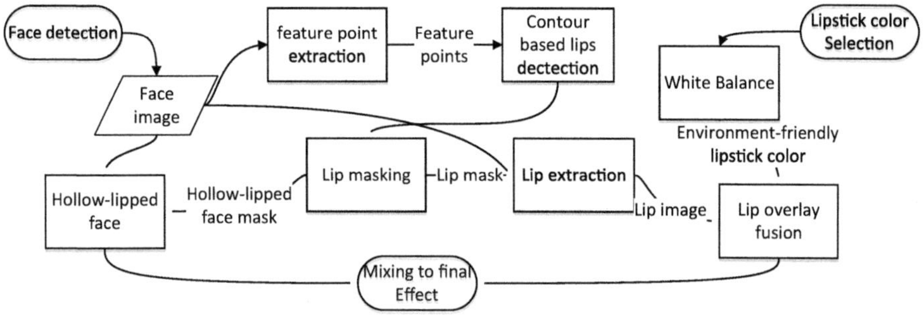

Fig. 1. Lipstick makeup migration flowchart.

2.1 Dlib Face Detection and Face Feature Point Calibration

The model trained by Dlib face detection HOG feature/direction gradient histogram and cascade regression tree algorithm [7] can detect the number and position of faces in the picture and perform calibration. The feature extractor predictor and pre-training model provided in the Dlib library can perform feature point detection and calibration on the calibrated face [8]. This article uses Dlib's face detector and feature extractor predictor for face detection and face feature point calibration (as is shown in Fig. 2).

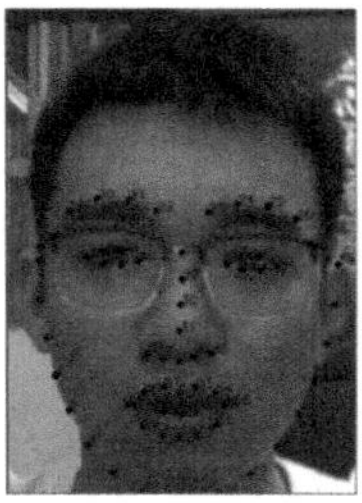

Fig. 2. Face feature points.

2.2 Outline Filling of Feature Points

In order to better process the lip information, this paper performs contour filling based on lip feature points. In order to achieve virtual lipstick test makeup in real application scenarios, the filling is divided into two parts, upper and lower lips. The image obtained at this time is that the upper and lower lips are white, and the rest is black, which is denoted as a mask.

2.3 Face Image with Black Lips

For the subsequent implementation of the algorithm, a face image with black lips needs to be obtained. First, use the OpenCV inversion operation on the mask image to obtain the image, which is recorded as not_mask. Then, the open sequence image img and the not_mask image are combined with OpenCV to obtain the face image with black lips, which is recorded as not_z.

2.4 White Balance Lipstick Color Under Current Lighting Environment

The human visual system has color constancy, and can obtain the invariant characteristics of the color of the surface of the object from certain changing lighting environments and imaging conditions. However, the imaging device does not have this "adjustment" function, and different lighting environments will cause a certain degree of deviation between the color of the acquired image and the true color of the object. This deviation will affect the accuracy and robustness of subsequent image analysis [9]. In order to reduce the influence of the color difference on the selected lipstick color and the device imaging under different lighting environments, this paper uses the gray world algorithm to do the white balance of the selected lipstick color under the current lighting environment. Finally, the lipstick color image after white balance is obtained, which is recorded as lipstick color.

2.5 Face Images that Only Retain Lip Information

For the subsequent implementation of the algorithm, it is necessary to obtain a face image that only retains lip information. The frame sequence image img and mask image are combined and operated with OpenCV to obtain a face image that only retains lip information, and is denoted as z.

2.6 Linear Overlay Image Fusion

In order to achieve the effect of lipstick virtual makeup test, the linear superimposed fusion method is used to linearly superimpose and fuse the face image z obtained above and the lipstick color image lipstick_color after white balance to obtain the merged image mixed.

2.7 Final Effect

Finally, the face image not_z with a black lip and the linearly fused lip image are ORed with OpenCV to obtain the face image after the final lip lipstick has been applied for a virtual makeup test, and is recorded as final_img.

3 Experimental Results and Analysis

3.1 Experiment Settings

This article experiment runs on the glory magicbook laptop. Hardware environment: AMD Ryzen 5 2500U with Radeon Vega Mobile Gfx 2.00 Ghz. Software environment: Windows 10 operating system, Pycharm 2019.1.3, python 3.6, Dlib 19.6.1, OpenCV 2.3.1.

Because there is no objective quantitative evaluation index for lipstick virtual makeup test, it can only be judged by comparing the subjective visual effects and processing details. Next, several other sets of experiments are given to compare with this experimental method. Since several other groups of experiments are overall virtual makeup tests, and this article only focuses on lipstick virtual makeup tests, only the lipstick treatment effect is compared when comparing.

The experiment is based on the sequence of frames entered by the camera in real time. Here, only one frame of the processed image is taken as the experimental explanation, and the other frames are the same. The images needed for the experiment include unprocessed face image video frames and selected lipstick color images, which are denoted as a and b, respectively. In order to avoid the problem of lipstick copyright, no specific lipstick-related information is given here, only the lipstick RGB three-channel values are given as 32, 42, 161. The experimental effect diagram, marked as c. The detailed drawing of the lips is denoted as d. as Fig. 3 shows.

(a) Original Frame (b) Lip Color Pattern (c) Made-up Face (d) Colored Lip Detail

Fig. 3. Makeup fusion example.

Experiment 1. The images needed for Tong's experiment are the images of the same person before and after makeup, which are denoted as a and b, respectively. The image requiring virtual makeup test is denoted as c. The experimental effect chart, marked as d.

Experiment 2. The image needed for Guo's experiment has a reference image after makeup, which is a. The image requiring virtual makeup test is denoted as b. The experimental effect diagram, marked as c.

Experiment 3. The image required for Du's experiment has a reference image after makeup, which is denoted as a. The image requiring virtual makeup test is denoted as b. The experimental effect diagram, marked as c.

Experiment 4. The images used in Liu's experiment are the reference images of eye shadow, eyelashes, blush, and eyebrows, respectively, which are a, b, c, d. The image requiring virtual makeup test is denoted as e. The experimental effect chart is denoted as f.

4 Analysis of the Results

As Table 1 shows, Tong's method requires two reference images, with good detail, but the steps are more complicated. Guo's method is prone to discontinuous problems, and the block effect is more obvious. Du's method has a big difference between the skin color achieved in the effect image and the original skin color in the target image, and the real experience of makeup has been reduced. Lin's method is also more sensitive to skin color differences. Liu's method is not very effective in the fusion of lip lipstick. In summary, the method of this paper does not set a reference image during the lipstick virtual makeup test, and only requires the original face image and the selected lipstick color image, which is in line with the real application scenario when people buy lipstick in reality. Secondly, the method in this paper performs white balance processing on the lipstick difference generated under different lighting environments, so that the effect of the real test makeup effect due to lighting factors is reduced. Thirdly, the method in this paper realizes the virtual makeup test of lipstick based on video, which has better instantaneousness. Fourth, the implementation steps and procedures of the method in this paper have good simplicity. The shortcomings of this article are that the method used for the fusion of lipstick and lip information is a simple linear superposition method, the detailed information of the lips cannot be completely retained, and the processing effect of the edges needs to be improved.

Table 1. Makeup effect comparison.

Method	Reference Image	Original Image	Result Image
Tong's	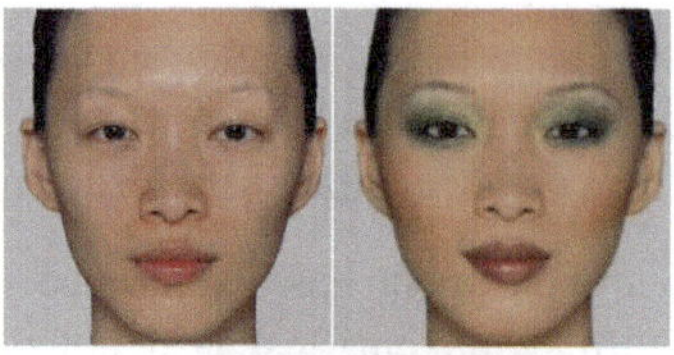	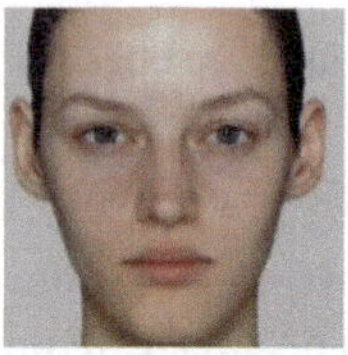	
Guo's	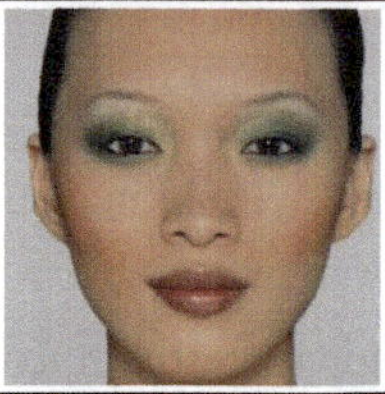	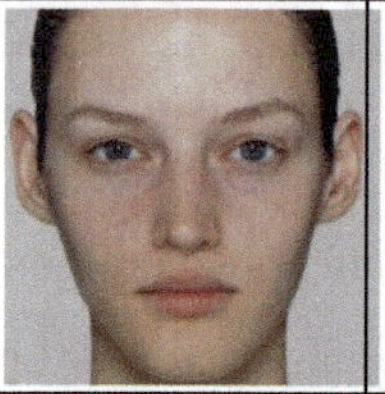	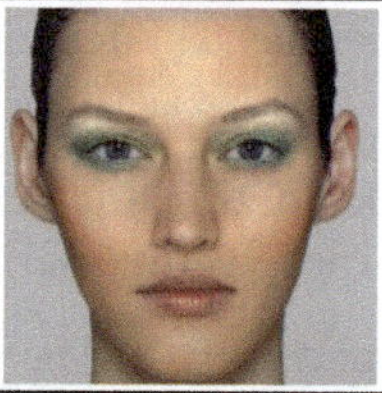
Du's	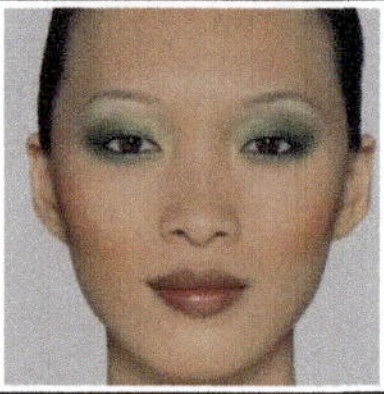	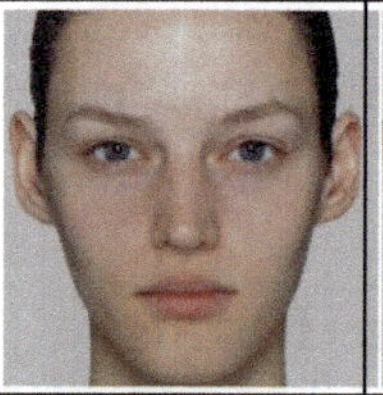	
Liu's	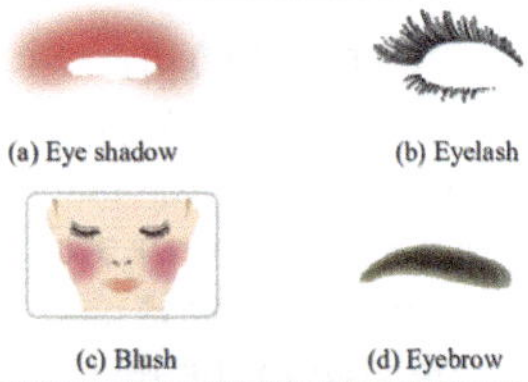	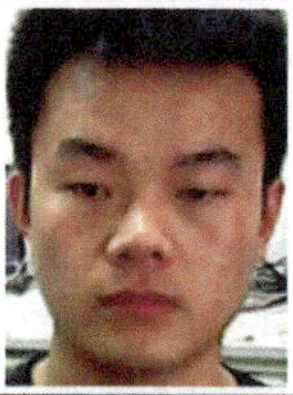	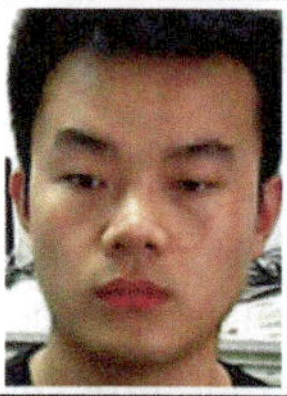
Ours			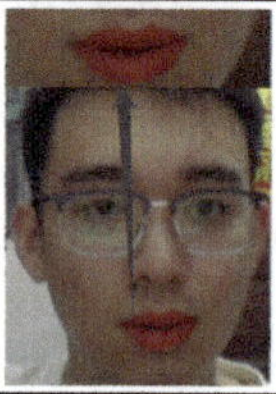

5 Conclusion

This paper uses the face detector and face feature extractor of the Dlib library to extract face feature points from the frame sequence captured by the camera, and uses the gray-scale world white balance method to perform the selected lipstick color under the current lighting environment. The white balance processing uses linear superposition fusion method to perform image fusion processing on the face lips and the processed lipstick color, which achieves the effect of real-time lipstick makeup test. Judging from the process and experimental results, this method has the characteristics of easy implementation, small calculation complexity, and good color difference processing effect. The experimental results also show that there is room for further improvement in edge processing and detail retention. The next step is to apply it to mobile applications and applets.

Acknowledgement. This work is supported in part by the National Natural Science Foundation of China (No. 61472092) and Guangdong College Student Innovation Project (No. S201911078074).

References

1. Tong, W., Tang, C., Brown, M., Xu, Y.: Example-based cosmetic transfer. Cosmet. Toiletries **15**(4), 211–218 (2007)
2. Guo, D., Sim, T.: Digital face makeup by example. In: 2009 IEEE Conference on Computer Vision and Pattern Recognition, pp. 73–79. IEEE, NJ, USA (2009)
3. Zhen, B., Wu, H., Xu, D.: An example based digital face makeup algorithm. J. Yunnan Univ. (Sci. Edn.) **32**(S2), 27–32 (2010)
4. Du, H., Shu, L.: Makeup transfer using multi-example. In: Proceedings of the 2012 International Conference on Information Technology and Software Engineering, pp. 467–474. Springer, Heidelberg (2013). https://doi.org/10.1007/978-3-642-34531-9_49
5. Lin, J., Xiao, Q., Wang, S.: A high-fidelity makeup face image transfer method. Comput. Appl. Softw. **32**(08), 187–189 + 210 (2016)
6. Liu, J., Li, J., He, H.: Application of virtual trial makeup based on video. J. Syst. Simul. **30** (11), 4195–4202 (2018)
7. Kazemi, V., Sullivan, J.: One millisecond face alignment with an ensemble of regression trees. In: IEEE Conference on Computer Vision and Pattern Recognition, pp. 1867–1874. IEEE, NJ, USA (2014)
8. Davis, K.: Dlib-ml: a machine learning toolkit. J. Mach. Learn. Res. **10**(60), 1755–1758 (2009)
9. Xu, X., Cai, Y., Liu, X., Shen, L.: Improved grey world color correction algorithms. Acta Photonica Sinica **39**(03), 559–564 (2010)

A Hybrid Chemical Reaction Optimization Algorithm for N-Queens Problem

Guangyong Zheng[1(✉)] and Yuming Xu[2]

[1] College of Computer Science and Technology,
Hengyang Normal University, Hengyang 421002, China
1224461813@qq.com
[2] College of Computer Science,
Changsha Normal University, Changsha 410199, China

Abstract. The N-queens problem is a classical NP hard problem with many solving methods. In this paper, a hybrid chemical reaction optimization algorithm is proposed to solve the N-queens problem. This algorithm combined with chemical reaction optimization algorithm and greedy algorithm, is presented to solve the problem of mixed chemical molecular structure, and design the appropriate molecular coding and chemical reaction of the four basic reaction process and the objective function, the simulation experimental results show that the design of hybrid optimization algorithm to solve the N-queens problem of the chemical reaction have got improved.

Keywords: N-queens · Hybrid · Chemical reaction optimization · Molecule

1 Introduction

The N-queens problem is a classical NP hard problem which can be solved by many methods such as backtracking method. Because its time complexity is O(n!), (n is queen number). So it is a NP hard problem. In this paper, a Hybrid Chemical Reaction Optimization (HCRO) algorithm is adopted to solve the N-queens problem. In literature [1–3], Chemical Reaction Optimization (CRO) is proposed to solve task scheduling problem in heterogeneous environment. In literature [4, 5], CRO is proposed to solve 0-1 Knapsack problem, etc., while literature [6] theoretically discusses the global convergence of CRO algorithm and proves its good convergence.

The Eight queens is a well-known NP-complete problem proposed by the famous mathematician Gauss in 1850. Using a regular chess board, the challenge is to place eight queens on the board such that no queen is attacking any of the others. In chess, a queen can attack horizontally, vertically, and diagonally, but no queen can attack any other queens on the board. That is to say, any two queens are not in the same line, same column or the same slash direction and the oblique line parallel to the main diagonal. Due to the large amount of experiments and tests are needed to solve the eight queen's placement problem, all the solutions were not obtained at that time. This problem slowly evolved into the N-queens problem [7], that is, to put N-queens on an $N \times N$ international chessboard so that no conflict would occur between any two queens. The N-queens problem is an NP hard problem, and its difficulty increases exponentially with the increase of N [8].

Q. Liu et al. (Eds.): CENet 2020, AISC 1274, pp. 128–137, 2021.
https://doi.org/10.1007/978-981-15-8462-6_15

2 The Related Work

2.1 Chemical Reaction Optimization (CRO) Algorithm

Algorithm Introduction. Meta-heuristic method is a set of concepts used to define heuristic algorithms. Its goal is to solve some common computing problems. Some random search algorithms, such as Evolutionary Algorithm, Ant Colony Algorithm, Particle Swarm Optimization (PSO) and other Intelligent Algorithms with heuristic frameworks are all meta-heuristic algorithms.

The change of chemical reaction products can be marked by the change of potential energy of the reaction system. The whole reaction process is a process in which the potential energy of the reaction gradually decreases. When the reaction process is done, the potential energy of the system reaches the minimum and the state tends to be stable. Thus, it can be seen that chemical reaction is actually a process of releasing energy. At the initial stage of the reaction, the molecular movement is relatively intense with relatively high energy. After a series of reactions, the molecule is stabilized to obtain the lowest energy state and the chemical reaction process is an optimization process to seek the minimum potential energy of the system [9].

Chemical Reaction Optimization (CRO) is one of the meta-heuristic methods, which simulates the natural phenomenon of molecular movement in Chemical reactions and obtain better solutions by capturing molecules with lower Potential Energy (PE) [8]. Chemical reactions are mainly composed of four basic reactions of molecules: monomolecular collision, monomolecular decomposition, intermolecular collision and molecular synthesis. The final result of chemical reactions is the product of chemical reactions.

The molecule must have some properties. The basic ones are:
(a) The molecular structure (b) The potential energy (PE)
(c) kinetic energy (KE) (d) Numhit (Number of hits)
(e) MinHit (the minimum hit number) (f) MinPE (the minimum PE)

Algorithm Solving Process. The solving process of CRO algorithm can be divided into the following three stages:

(1) Initialization stage: It mainly defines solution space and also some algorithm functions, such as target function, decomposition function, synthesis function, etc., and sets control parameters for the execution of the algorithm. Meanwhile, it also sets the molecular group that initially participates in the reaction.

In addition to defining the molecular structure (that is, the solution form of the problem to be solved) and the solution space (the range of the solution to the problem to be solved). We also need to define the relevant functions and parameters and give the initial values [11], as shown in Table 1:

Table 1. The functions and parameters of CRO algorithm.

Category	Symbol	Implication
Process, function:		
	f()	The objective function
	OnwallCollision()	Single molecule collision
	Decomposition()	Monomolecular decomposition
	IntermolecularCollision()	Intermolecular collision
	Synthesis()	Molecular synthesis process
Parameters:		
	PopSize	The initial number of molecules in the container
	KELossRate	The upper limit of the percentage of KE lost in the wall impact
	MoleColl	Determinants of the type of molecular reaction
	Buffer	Molecular kinetic energy cache
	InitialKE	The initial value of the molecule KE
	α	Monomolecular reaction determinants
	β	Multi-molecular reaction determinants
	POP	Molecular collection
	PE	Molecular potential energy
	KE	Kinetic energy of molecules

(2) Iteration phase: When the algorithm is executed, chemical reaction is repetitive through multiple iterations, and each iteration is a basic reaction. The main steps are as follows: First, determine the reaction type according to the parameter value generated randomly. Second is to randomly select the corresponding number of molecules according to the type of reaction and the Third is, depending on the molecular reaction. If you do not meet the stop condition, you go to the first step. If the set stop condition is reached (such as the set number of iterations, etc.), the following program is executed.
(3) Finally determine the optimal solution: After the iterative operation of CRO algorithm, output the optimal solution that meets the conditions in all molecules.

2.2 Greedy Algorithm

The greedy algorithm is simply described as follows: process a group of data, find the optimal value, process again, find the optimal value, and repeat until the optimal value that meets the conditions appears or the problem processing step is completed. That is to say, greedy algorithm is an algorithm that takes the best or optimal choice in the current state in every step of selection, so as to get the best or optimal result. The essence of the greedy algorithm is to construct global solutions with local solutions so as to obtain better solutions as quickly as possible. When steps in a certain algorithm can no longer move forward, the algorithm stops [10].

In the chemical reaction optimization algorithm, the greedy algorithm process can accelerate the convergence speed, can integrate the advantages of the two algorithms and achieve a better result.

3 Algorithm Design

Hybrid Chemical Reaction Optimization (HCRO) algorithm combines the advantages of greedy algorithm and chemical reaction optimization algorithm to accelerate the search speed of the optimal solution, and forms a hybrid chemical reaction optimization algorithm.

3.1 Molecular Coding

For the N-queens problem, some algorithms use traditional binary encoding and some use multi-value encoding. Take $N = 10$ as an example to introduce molecular coding. The multi-value coding method with conflicting statistics is adopted. The molecule of N-queens problem is represented by a two-dimensional vector b.

Define b = {b(c0,0), b (c1, 1), b (c2, 2), b (c3, 3), b (c4, 4), b (c5, 5), b (c6, 6), b (c7, 7), b (c8, 8), b (c9, 9)}, including $b(c_i, i)$ is a natural number, said the queen and other conflicts number i; $c_i \in \{0,1,2,3,4,5,6,7,8,9\}$ and $\forall c_i, c_j, c_i \neq c_j$, namely the value cannot be the same. The two-dimensional vector is expressed and explained as follows (Table 2):

Table 2. Two-dimensional Vector.

Element value	$b(c_0,0)$	$b(c_1,1)$	$b(c_2,2)$	$b(c_3,3)$	$b(c_4,4)$	$b(c_5,5)$	$b(c_6,6)$	$b(c_7,7)$	$b(c_8,8)$	$b(c_9,9)$
Boardline	c0	c1	c2	c3	c4	c5	c6	c7	c8	c9
Board column	0	1	2	3	4	5	6	7	8	9

Calculation method for the number of conflicts of each element: The initial value of each vector element $b(c_i, i)$ is 0, and the conflict between the queen of column 0 and the queen of column 1-9 is compared. For every conflict, the value of $b(c_1,1)$ increases by 1. The value of $b(c_2,2)$ was increased by 1 for every conflict between the queen in column 2 and the queen in column 1, 3–9. And so on, and you get the values of the vector elements (Table 3).

Table 3. The algorithm description of conflict number calculation.

```
Algorithm 1: calculate the number of conflicts (that is, the value of vector elements)
  Input: vector b          // A vector to recalculate the number of collisions
  Output: vector b1        // The right vector for the Number of conflict
for (i=0;i<=9;i++)
    for(j=0;j<=9;j++)
      b(ci,j)=0;
for (i=0; i<=9; i++)
   for (j=0; j<=i-1 ||(j>=i+1 && j<=8); j++)
     { if(j==i)
         continue;
     if (|(ci-cj)/(i-j)|==1) // Indicates conflict
         a(ci,i)++ ;}
return
```

3.2 Objective Function

In the N-queens problem, the queen cannot be in the same row and the same column, which means that the values of ci of each element of the vector cannot be repeated. Also, consider the case where you are not on the same slash, when the absolute value of the ratio of the difference between the number of rows and the number of columns between the two queens is 1($|(c_j - c_i)/(j-i)| = 1$), then the two queens are on the same diagonal (on or parallel to the two main diagonals), indicating conflict.

Therefore, to obtain a solution to the problem of *N* queens, it is necessary to ensure that the total number of conflicts is 0. Therefore, the objective function is designed to f = f(b), for the above two dimensional vector b, $f(b) = \sum_{i=1}^{N} b(ci, i)$, it represents the sum of the conflict Numbers of *N* queens. For each vector b, the algorithm has searched a solution b when f (b) = 0. When there is a feasible solution or iteration is to reach the set value, the algorithm end the solving process.

3.3 Four Kinds of Chemical Reaction Operators

(1) Initial group of reaction
Randomly select *M* different molecules as the initial population of the reaction (*M* can be more than 10 times larger than *N*).

(2) Greedy algorithm for single-molecule collision
In the single-molecule collision reaction, the molecular structure changes little and the main purpose of this reaction is to explore the potential energy surface in a small range, so it can change the molecular solution structure in a small range.

(3) Greedy algorithm for single-molecule decomposition
The chemical reaction operator is to use the existing molecule *m* to produce two new molecules m_1 and m_2. The collision is so strong that the molecule is split, which is a search for a new solution.

(4) Greedy algorithm for collision between molecules
Change two existing molecules m_3 and m_4 in the event of a collision to produce new molecules m_5 and m_6. Referring to the process of single-molecule collision, the single-molecule collision is performed on the vectors m_3 and m_4 respectively, until a better solution appeared or the test vertex reached half of the number of vertices.

Table 4. The pseudocode of HCRO algorithm for solving N-queens problem.

Algorithm 2: HCRO algorithm

Input: For the initial reaction molecule group POP of the N queen problem, set the value of N

Output: The optimal solution to the N-queens problem

```
Establish the initial reaction molecule group POP;
while (When the end of the reaction condition is not satisfied)
 { if ( Monomolecular reaction)
     { pick a random molecule v from POP;
     if ( Monomolecular decomposition reaction)
         Greedy algorithm for single molecule decomposition
         if the newly generated molecules m1 and m2 are covered by figure G,
                 they are added to POP and the original molecule m is removed
                 from it
     else
         Greedy algorithm single molecule collision
         if the newly generated molecule m3 is covered by figure G,
                 it is added to POP and the original molecule is removed from it
       Replace v with a better solution }
 if (Polymolecular reaction)
     { A molecule m1 and m2 were randomly selected from POP;
     if (Multi-molecule synthesis )
         Greedy algorithm molecule synthesis m3, if m3 is a better solution, then
                 add to POP and remove m1 and m2;
     else
         The greedy algorithm produces m4 and m5 in the intermolecular collision
                 reaction if m4 or m5 is a better solution, add it to POP and remove
                 m1 and m2;
     }
   Check for the new minimum solution // Identifies the molecule that currently
                                        has the minimum value of the
                                        target function
 }
return the global optimal solution
```

3.4 HCRO Algorithm Description

Basic idea of HCRO algorithm: In the process of chemical reaction optimization algorithm to search for the global optimal solution, greedy algorithm is used to search the local optimal solution to speed up the accuracy and speed of the solution.

The algorithm can be divided into three phases. The first phase is initialization. The second phase is iteration to search for better solution. The third phase is the optimal solution confirmation and to get solution after the reaction is stopped (Table 4).

4 Simulation Results and Analysis

4.1 Experimental Environment

On the Visual studio 2010 integrated development environment platform, an object-oriented C# language was used to realize the CRO algorithm for solving the N-queens problem.

In CRO, the basic unit of algorithm is molecule, and there are four basic reaction processes, whose operation object is molecule. An object-oriented C# language can be used to define a molecular class and four reaction processes. The definition of a molecule is as follows:

```
class Molecule              // molecular structure
  { public int[] Atoms;       // Task priority queue
   public double PE;         // Molecular potential energy
   public double KE;         // Molecular kinetic energy
   public int NumHit;        // Molecular collision number
   public int MinHit;
   public double MinPE;
                              // The minimum potential energy of the molecule
                              //throughout the reaction
   public int[] MinStruct;
                        // The structural state of a molecule at the minimum potential
                        //energy of the reaction
   public int[] Collision;
  }
```

The initial value of the parameter is set as follows:

Popsize = 1000, KElossrate = 0.2, Molecoll = 0.2, InitialKE=1000, α=500, β=4, Iteration=1000.

4.2 Experimental Results

The optimal solution is obtained after the program was run. However, due to the randomness of the heuristic algorithm, the time required for each run is different, so the running time is averaged for three times. The termination condition of the algorithm is to find the optimal solution or the number of iterations reaches the set value. When N is equal to 10, One result of running the program is {(1,0), (6, 1), (9, 2), (5, 3), (2, 4), (0, 5), (3, 6), (7, 7), (4, 8), (8, 9)}, the Program running results are as follows (Fig. 1):

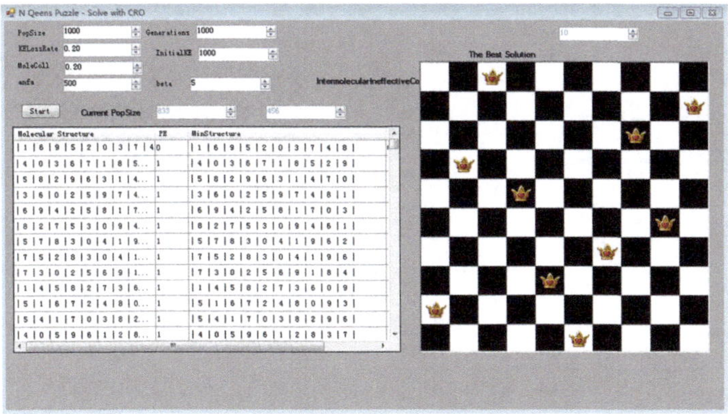

Fig. 1. The results after running (N = 10).

When N is 15, 30, 50, 100, the program operation results are shown as follows (Figs. 2, 3, 4, 5):

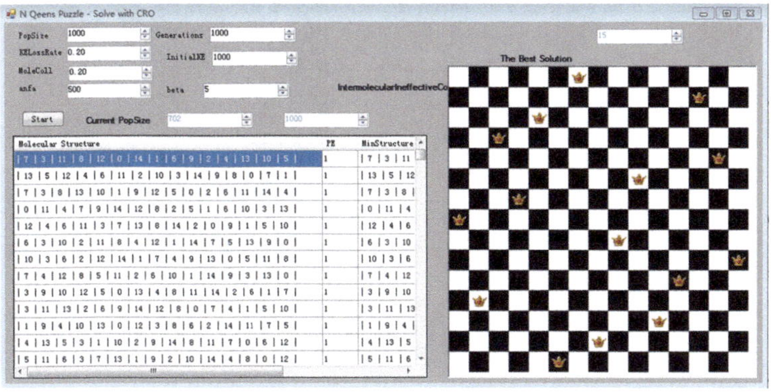

Fig. 2. The results after running (N = 15).

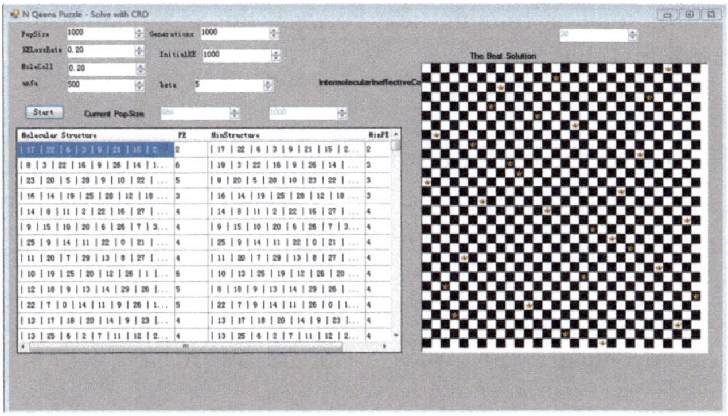

Fig. 3. The results after running (N = 30).

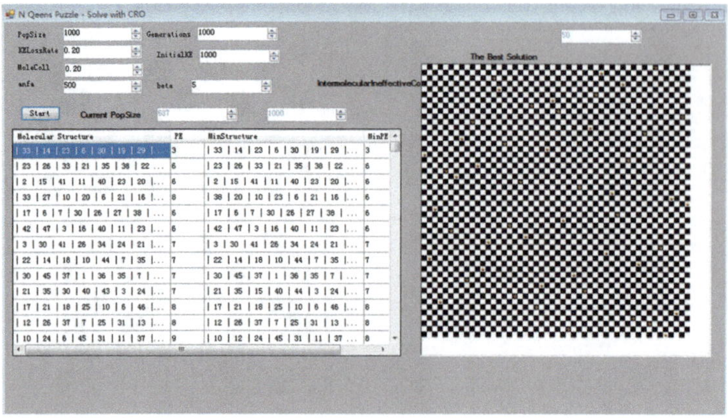

Fig. 4. The results after running (N = 50).

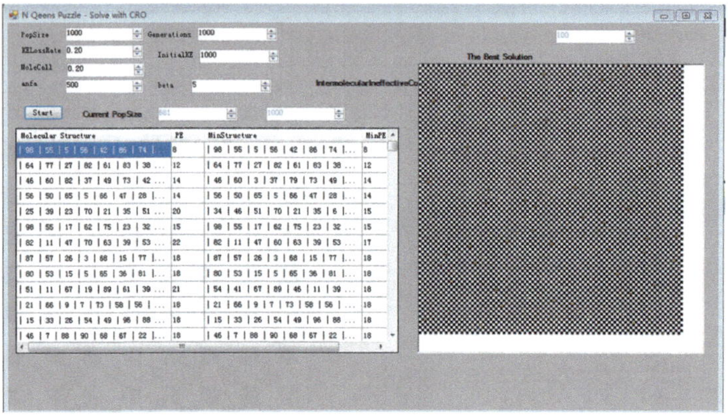

Fig. 5. The results after running (N = 100).

4.3 Analysis of Experimental Results

The comparison of solution running time between this algorithm and other methods [11, 12] is shown in Table 5:

Table 5. Comparison table of running time between HCRO and Other methods (unit: second).

Queen's number	10	15	30	50	100	180	500
HCRO Algorithm	0	0.024	0.03	0.10	0.31	2.12	4.30
Backtracking	0.47	578.04	–	–	–	–	–
Ant Colony Algorithm	0	0	0	0.11	0.31	2.15	4.32
Genetic Algorithm	0	0	215.23	965.76	–	–	–

When the backward method is used to solve more than 30 queens, the solution time is too long, so the solution time cannot be given. As can be seen from the above table, the efficiency of hybrid chemical reaction optimization algorithm for solving N-queens problem is improved to a certain extent compared with the backtracking method. When the number of queens is large, the solving time is better than that of ant colony algorithm, but it is better than that of genetic algorithm.

5 Conclusions

This paper introduces the basic idea and process of solving N-queens problem by HCRO algorithm, which is realized by C# language programming, and good simulation results are obtained. When the mixed chemical reaction optimization algorithm is applied, the experimental results may be slightly different due to the different molecular groups involved in the reaction, but in general, it is improved for solving the N-queens problem.

Acknowledgment. This research was supported by the Scientific Research Project of Human Provincial Department of Education (Grant No. 18A332, Grant No. 17A017).

References

1. Li, K.L., Zhang, Z.M., Xu, Y.M., Gao, B., He, L.G.: Chemical reaction optimization for heterogeneous computing environments. In: IEEE 10th International Symposium on Parallel and Distributed Processing with Applications (ISPA) (2012)
2. Xu, Y.M., Li, K.L., He, L.G., Truong, T.K.: A DAG scheduling scheme on heterogeneous computing systems using double molecular structure-based chemical reaction optimization. J. Parallel Distrib. Comput. **73**(9), 1306–1322 (2013)
3. Xu, J., Lam, A.Y.S., Li, V.O.K.: Chemical reaction optimization for task scheduling in grid computing. IEEE Trans. Parallel Distrib. Syst. **22**(10), 1624–1631 (2011)
4. Truong, T.K., Li, K.L., Xu, Y.M.: Chemical reaction optimization with greedy strategy for the 0-1 knapsack problem. Appl. Soft Comput. **13**(4), 1774–1780 (2013)
5. Pan, B., Lam, A.Y.S., Li, V.O.K.: Network coding optimization based on chemical reaction optimization. In: Global Telecommunications Conference (2011)
6. Lam, A., Li, V., Xu, J.: On the convergence of chemical reaction optimization for combinatorial optimization. IEEE Trans. Evol. Comput. **7**(99), 1–10 (2010)
7. Liu, J., Ouyang, J.Q., Chen, L.J.: The hybrid genetic algorithm was used to solve the N queen problem. Nat. Sci. J. Xiangtan Univ. **06**, 37–41 (2007)
8. Shuai, X.B., Ma, S.: Applying Boolean genetic operator to solve N queen problem. Comput. Eng. Appl. **47**(16), 49–68 (2011)
9. Zhang, Z.M.: Research on grid task scheduling based on chemical reaction optimization. Hunan University (Master Thesis) (2012)
10. Gong, G.S.: Application of greedy algorithm in solving P class problem. Comput. Knowl. Technol. **7**(2), 18–20 (2011)
11. Cheng, X.M.: Principle of ant colony algorithm and its improvement. J. Guangdong Normal Univ. Technol. **04**, 68–70 (2006)
12. Wang, N.: Ant colony algorithm is applied to solve N queen problem. Modern Commun. **05**, 245–246 (2016)

Apple Soluble Solids Content Prediction Based on Genetic Algorithm and Extreme Learning Machine

Xingwei Yan, Tao Shen, Shuhui Bi(✉), and Qinhua Xu

School of Electrical Engineering, University of Jinan, Jinan 250022, China
497754391@qq.com, {cse_st, cse_bish}@ujn.edu.cn

Abstract. Soluble solids contents (SSC) is the general term of monosaccharide, disaccharide and polysaccharide, and is one of the key evaluation indexes of apple taste and nutritional quality. According to the prediction of soluble solid content in Red Fuji apple, a method based on Genetic Algorithm and Extreme Learning Machine was proposed. Genetic Algorithm (GA) was used to select the best near-infrared spectrum of Fuji apple, and the prediction model of Extreme Learning Machine (ELM) was established. The prediction correlation coefficient of the model was 0.9723, and the prediction root mean square error was 0.1855. The experimental results show that the predicted results are in good agreement with the actual data, which plays a good role in the prediction of soluble solid content.

Keywords: Apple · Near infrared spectroscopy · Genetic algorithm · Extreme learning machine

1 Introduction

China's apple production and planting area are in the first place in the world, and apple industry is the pillar industry of China's agriculture [1]. With the improvement of people's living standard, people pay more and more attention to the internal quality of fruit. Although China's apple output ranks first in the world, its export volume is far lower than that of Europe and the United States. The main reason for this situation is that the internal quality standard of China's apple is too low, so it cannot be well classified and sorted for commercialization. Therefore, it is of great significance to improve the internal quality grading level of apple industry in China to improve the international competitiveness of apple industry and meet the needs of different grades of domestic people [2]. Compared with the traditional chemical methods, NIR technology has the advantages of fast, non-destructive, low-cost, no need for sample pretreatment and so on. It has gradually become a new high-speed development technology in recent years [3]. Li Zhicheng et al. obtained the near-infrared spectrum of apple by near-infrared spectroscopy, tested the hardness of Apple quickly, and realized the fast nondestructive testing of apple [4]. Miao Xuexue et al. optimized the characteristic wavelength of rice near-infrared spectrum [5]. The results showed that the optimization method of characteristic wavelength not only improved the prediction

Q. Liu et al. (Eds.): CENet 2020, AISC 1274, pp. 138–145, 2021.
https://doi.org/10.1007/978-981-15-8462-6_16

performance and accuracy of the model, but also effectively saved the modeling time; Guo Wenchuan et al. used near infrared spectroscopy combined with ELM to identify the damaged Kiwifruit during storage, which provided an effective method to reduce the loss of fruit decay and cross infection [6].

In this paper, the near-infrared spectrum of Red Fuji apple is collected by near-infrared spectrum technology, and the optimal characteristic wavelength is selected by GA in the whole spectrum range, which is used as the input variable to establish the prediction model of soluble solids and near-infrared spectrum.

2 Materials and Methods

2.1 Experimental Materials and Data Collection

The Red Fuji apple produced in Qixia, Yantai was used as the sample. 439 apples without defect and damage were selected, 70% of which were used as the correction set and the remaining 30% as the prediction set. Before the experiment, the apple was placed in the laboratory for 12 h to make the overall temperature of the apple sample consistent with the ambient temperature, and keep the laboratory temperature basically unchanged during the experiment.

The near-infrared spectrum acquisition equipment of Red Fuji apple is the near-infrared detector of Antaris II, which adopts InGaAs detector, and the sampling mode adopts integral sphere diffuse reflection. Each apple sample is collected three times, and the acquisition point is at the equator of the apple with equal interval of 120°. The average value of the three spectral data is the original spectrum of the sample. After the spectral acquisition, the content of soluble solids was measured at the spectral acquisition location, and the content of soluble solids was read by dropping the juice from the pulp on the sugar meter. The content of soluble solids at three spectral acquisition locations of each apple was taken as the reference value of the sample.

2.2 Algorithm Introduction

Genetic Algorithm is a random global search and optimization method [7]. It imitates the natural selection mechanism of nature, based on evolution theory and genetic theory, and uses the survival principle of the fittest, so that excellent genes can be transmitted to the next generation. For wavelength optimization, the basic idea is to take the wavelength combination with the best modeling effect in the "parent" as an individual to enter the next generation; the individuals in the next generation generate a batch of new individuals through random selection, crossover and mutation and other genetic operations. With the continuous reproduction of individuals, the individuals of the previous generation are constantly replaced by adaptive individuals. After several iterations, the calculation Finally, the method converges to the best individual, that is, the optimal combination solution of wavelength selection.

The operation flow of GA is shown in Fig. 1:

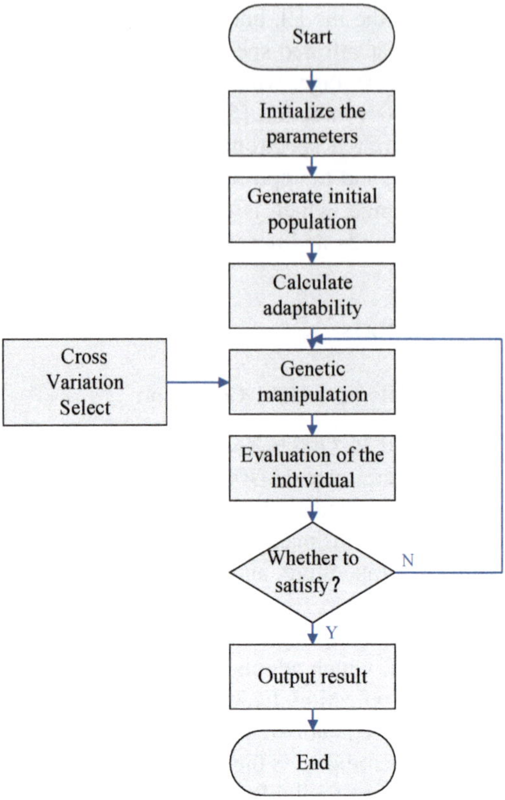

Fig. 1. Flow chart of GA.

Extreme Learning Machine (ELM) is an efficient training algorithm for Single-hidden Layer Feedforward Neural Networks (SLFNs) [8]. ELM is different from the classical neural network, it does not need the tedious iterative process in the gradient descent algorithm to adjust parameters and consumes a lot of time. It randomly generates all weights and hidden layer node thresholds. And it is always the same in the training, only the number of nodes need to be set artificially, and then the inverse matrix is calculated to get the output weight, then the optimal value can be calculated. Compared with the traditional SLFNs, the training speed of elm is significantly improved, the efficiency is much higher than the previous algorithm, and the generalization performance is good.

3 Data Processing and Analysis

3.1 Spectral Pretreatment

The specific steps of the experiment are shown in Fig. 2. Firstly, in the process of spectrum acquisition, it may be affected by the state, background and other factors of the spectral instrument, resulting in noise interference in the spectrum, affecting the accuracy and stability of the modeling. Therefore, it is necessary to preprocess the original spectrum, reduce the impact of noise, and improve the accuracy of the spectrum. In this paper, the effects of noise and baseline drift in the spectrum are eliminated by smoothing and multiple scattering correction. As shown in Figs. 3 and 4, the collected original spectrum and the spectrum after simple pretreatment are given respectively.

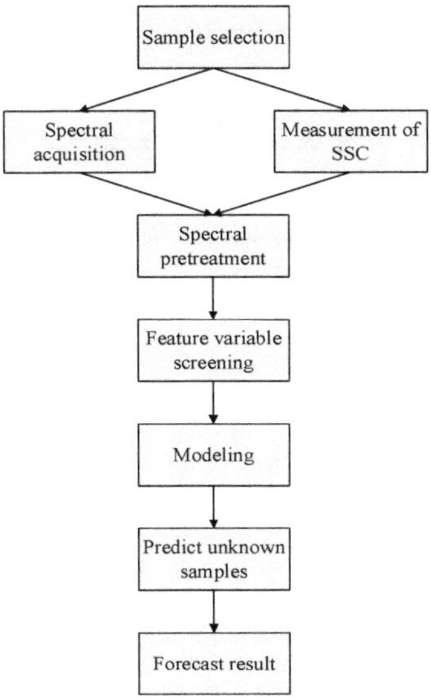

Fig. 2. Experiment process.

3.2 Optimization of Spectral Characteristics

As a practical, efficient and robust optimization technology, GA is an optimization method which imitates the evolution mechanism of natural organisms. In the application of near infrared spectrum detection, GA is used to optimize the whole spectrum range and select the most effective characteristic wavelength. In this paper, combined with the use of elm, the RMS error between the predicted value and the actual value of the dependent variable in elm interactive verification is used as the fitness function of GA. The main parameters of GA are as follows: the initial population is 10, the crossover probability is 0.7, the mutation probability is 0.1, and the number of genetic iterations is 50. The change of the optimal individual fitness value in the optimization process of GA is shown in Fig. 5. From the average fitness and the optimal fitness curve, it can be seen that the convergence of the algorithm is very smooth, and there is no oscillation. The selection of characteristic wavelengths is represented by characters containing 0/1, where 1 and 0 respectively indicate whether the corresponding wavelengths are selected, and Fig. 6 shows the number of selected wavelengths.

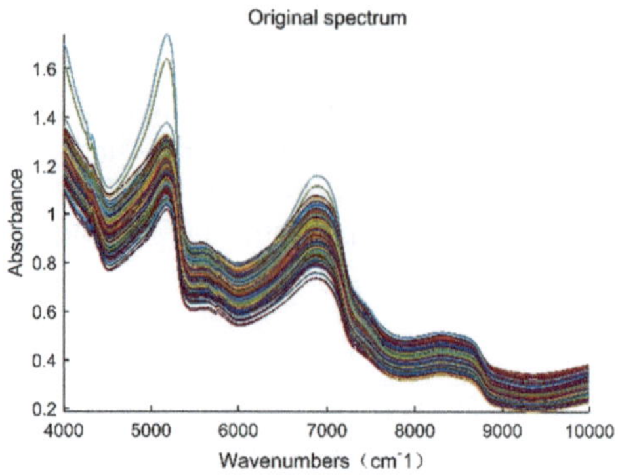

Fig. 3. Original spectrum.

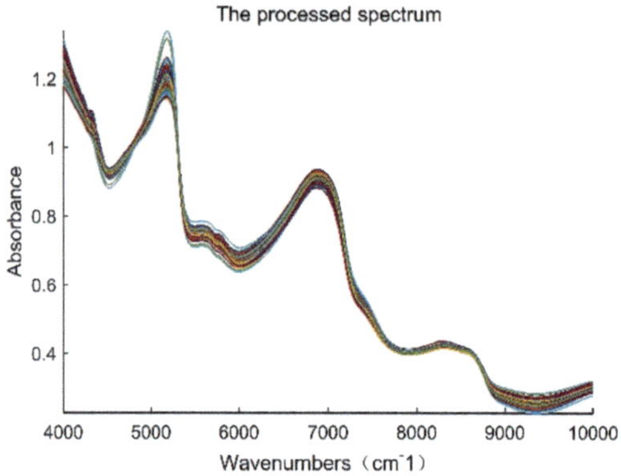

Fig. 4. Spectrogram after processing.

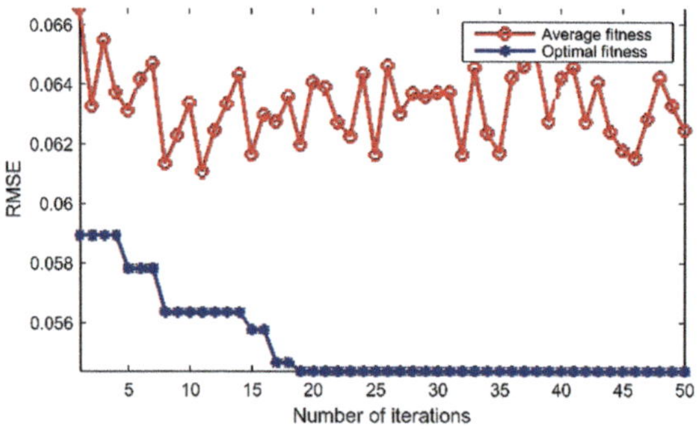

Fig. 5. Optimal individual fitness value.

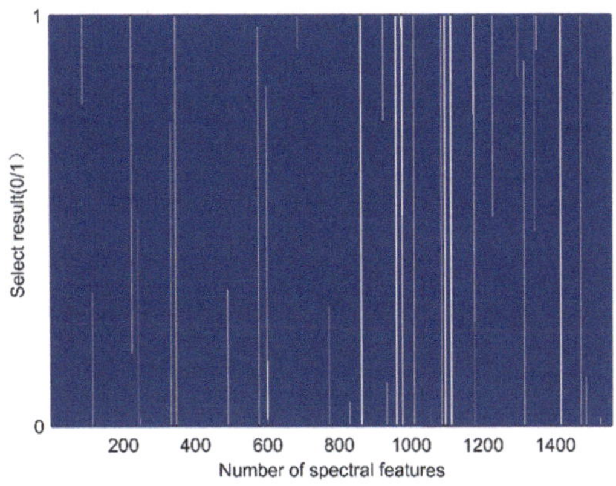

Fig. 6. Screened best characteristic wavelength.

3.3 Model Building

The characteristic wavelength selected by GA is used as the input of modeling. Originally, elm is used for modeling. The learning algorithm of elm mainly includes the following steps:

1) The number of hidden layer neurons was determined, and the connection weight W between input layer and hidden layer and the bias B of hidden layer neurons were set randomly;
2) An infinitely differentiable function is selected as the activation function of the hidden layer neuron, and then the hidden layer output matrix H is calculated;
3) Calculate output weight $\hat{\beta}$: $\hat{\beta} = H^{+}T'$.

The activation functions of neurons in the hidden layer can be 'sig', 'sin' and 'hardlim'. Through experimental comparison, it can be seen from Table 1 that sig has the best effect. In this paper, Sig is selected as the activation function of neurons in the hidden layer, and the number of neurons in the hidden layer is 100.

Table 1. Performance indicators of different activation functions.

Activation function	Correlation coefficient	RMSE
Sig	0.9723	0.1854
Sin	0.9668	0.2028
Hardlim	0.9674	0.2012

3.4 Model Verification and Evaluation

In this paper, the ratio of modeling set to prediction set is 7:3, and prediction correlation coefficient and prediction root mean square error are selected as the evaluation indexes of the model. Generally speaking, the larger the correlation coefficient and the smaller the root mean square error are, the better the model is. The effect diagram of the predicted and actual values of the content of Apple soluble solids in the prediction concentration is shown in Fig. 7. From Fig. 7, it can be seen that the actual and predicted values of the content of Apple soluble solids fit Chengdu well, which can achieve the prediction and evaluation effect.

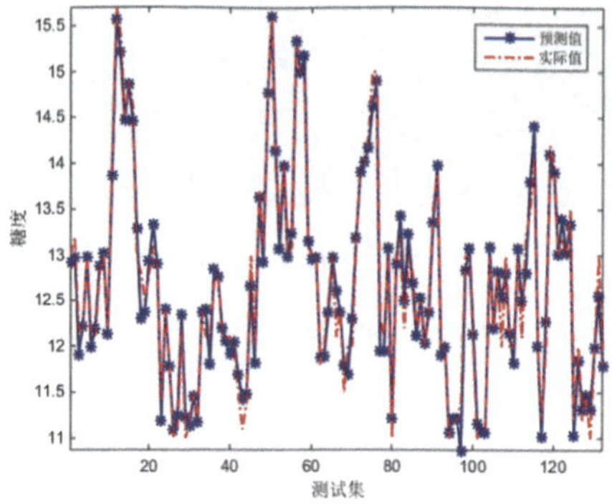

Fig. 7. Effect chart of real and predicted values of sample SSC.

Figure 8 shows the correlation curve between the real value and the predicted value of SSC in the prediction set. It can be seen from Fig. 8 that each point of the sample is distributed near the trend line (45° to the axis), which shows that the predicted value of near-infrared spectrum is basically the same as the actual value. The correlation coefficient R^2 of the prediction set model is 0.9723, and the root mean square error RMSE is 0.1855.

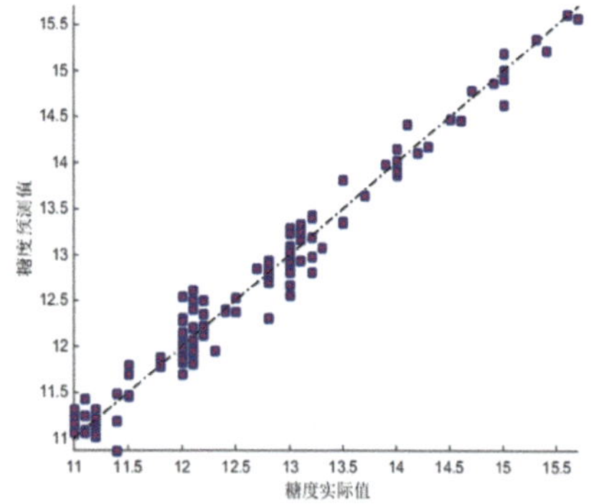

Fig. 8. Correlation between the true value of the sample SSC and the predicted value of the prediction set.

4 Conclusion

The spectral data of apple were collected by near-infrared spectroscopy technology. The spectral data was preprocessed simply and the original spectrum was screened by GA. The model of the screened characteristic spectrum was established, and the ideal model and prediction effect were obtained. The results showed that the coefficient R^2 between the predicted value and the actual value was 0.9723, and the root mean square error (RMSE) was 0.1855. It can be seen that the rapid detection of soluble solid content of Apple by near infrared spectroscopy can quickly and accurately obtain the prediction value of soluble solid content of samples without damage.

References

1. Meng, X.N., Zhang, Z.H., Li, Y., et al.: Research status and progress of apple grading. Decid. Fruits **51**(6), 24–27 (2019)
2. Chen, X.S., Han, M.Y., Su, G.L., et al.: Discussion on today's world apple industry trends and the suggestions on sustainable and efficient development of apple industry in China. J. Fruit Sci. **27**(4), 598–604 (2010)
3. Chu, X.l., Yuan, H.F., Lu, W.Z.: Progress and application of spectral data pretreatment and wavelength selection methods in NIR analytical technique. Progress Chem. (04), 528–542 (2004)
4. Li, Z.C., Wang, W.C.: Rapid and nondestructive detection of apple firmness by near infrared spectroscopy. China Fruit Veget. **38**(11), 29–31 (2018)
5. Miao, X.X., Gong, H.R., Tao, S.H., et al.: Determination of moisture content in rice by near infrared spectroscopy combined with characteristic wavenumber. Food Sci. Technol. **44**(10), 335–341 (2019)
6. Guo, W.C., Wang, M.H., Gu, J.S., et al.: Identification of bruised kiwifruits during storage by near infrared spectroscopy and extreme learning machine. Opt. Precis. Eng. **21**(10), 2720–2727 (2013)
7. Liang, L.H., Yuan, J., Zhang, S.T., et al.: Design a software real-time operation platform for wave piercing catamarans motion control using linear quadratic regulator based genetic algorithm. PLoS ONE **13**(4), 1–23 (2018)
8. Uzair, M., Mian, A.S.: Blind domain adaptation with augmented extreme learning machine features. IEEE Trans. Cybern. **47**(3), 651–660 (2017)

Structural Damage Semantic Segmentation Using Dual-Network Fusion

Lingrui Mei[1], Jiaqi Yin[2], Donghao Zhou[3], Kai Kang[4], and Qiang Zhang[1(✉)]

[1] State Key Laboratory of Vehicle NVH and Safety Technology, China Automotive Engineering Research Institute, Chongqing 401122, China
zhangqiang2@caeri.com.cn

[2] Chongqing University, Chongqing 400000, China

[3] Chongqing Communications Planning Survey and Design Institute, Chongqing 401121, China

[4] The Hong Kong University of Science and Technology, Hong Kong, China

Abstract. Road cracks seriously affect road life and driving safety, and road crack segmentation has been a key task in many research fields and industries. Scenes of existing road crack datasets are relatively homogeneous, most datasets only have local crack images taken from a single perspective, and the images lack complex interference pattern other than the cracks and the road itself. However, real-world images of roads are much more complex, and roads may have other dark patterns that interfere with crack segmentation. Perspective and lighting conditions may also be different between every image in the dataset, which makes many road crack segmentation methods extremely low accuracy and severely affects the generalization ability of the semantic segmentation model, while many applications require higher precision rather than recall. In this paper, two methods are proposed to improve the accuracy: augmenting complex patterns in road images and the fusion of classification networks. The methods greatly improve the performance of road crack segmentation.

1 Introduction

The construction and monitoring of the road network is a top priority for infrastructure development and maintenance, and for a variety of reasons, roads will inevitably have cracks. Road cracks are an important factor directly related to road life and traffic safety [1]. And testing poses a great safety risk. Therefore, how to accurately detect the cracked areas of roads by road images without affecting traffic has become an urgent task. Cracks are a form of road disease that has influence on road performance and threatens road safety [2]. Governments have made tremendous efforts to achieve the goal of building high quality road networks. National governments are more fully aware of the need for road inspection and maintenance. Crack inspection is an important part of the road maintenance system and has attracted increasing participation. In addition, traditional

These authors contributed equally to this work.

Q. Liu et al. (Eds.): CENet 2020, AISC 1274, pp. 146–157, 2021.
https://doi.org/10.1007/978-981-15-8462-6_17

methods of visual inspection and manual pavement crack analysis are costly, time-consuming, dangerous, labor-intensive, and highly subjective. Manual inspection, due to its high variability, inability to provide meaningful quantitative information and other disadvantages, can lead to crack information in different measurements with a high degree of inconsistency. With the development of computer vision technology and image processing technology, visual defect recognition adds a new way to solve this engineering problem. Based on the fact that the crack pixels in the pavement image are darker and continuous than the surrounding colors, a new type of pavement crack based on fuzzy logic is proposed [3]. Detection algorithm. However, this method did not meet the required accuracy requirements and solve practical problems. Subsequently, a research team from the Chinese Academy of Sciences proposed a "crack forest" [4] for crack identification based on a random structural forest. This algorithm has been developed in terms of efficiency and operability compared to other methods, and it has been shown experimentally that the "crack forest" has a higher accuracy than its competing algorithms.

With the development of deep learning research and applications [5, 6], more and more crack detection based on deep learning methods have emerged. Silva et al. present a deep learning based classification method for cracked images [7]. In this article, they first create an aerial view of a road taken using UAV and then label the captured images. However, the images labeled by this method do not involve pixel-level segmentation, but merely a bicategorization of whether the images contain cracks or not. They then used a VGG16 network [8] to train a two-class classifier on the collected images. This method uses a deep learning model for training, which has better recognition accuracy than the original method, but the method only uses the VGG16 network to train the classifier, which only provides ability to determine the presence of cracks in an image and but to perform pixel-level semantic segmentation, and it makes the model not ideal for application in real-life scenarios. Liu et al. proposed a semantic segmentation method to detect structural cracks [9], which first collects images, and then analyzes them. These images were pixel-level annotated and then the model was trained using the network proposed by Xie et al. [10]. The model was initially used for edge detection, but road crack detection is not a simple edge detection problem, and in the face of complex scenarios, the robustness of the algorithmic model using simple correlation such as edge detection is not ideal.

Although U-Net [11] has achieved good results on the segmentation problem, it is still lacking in large-scale applications in real-life scenarios, and may still yield dangerous results under rare circumstances in areas requiring high precision, such as automated driving tests. Identifying road cracks is a very important task in the field of civil engineering and automated driving testing. There are already many approaches that can perform the task of road crack detection and semantic segmentation, but such approaches are not very effective in improving the segmentation performance. While at the same time, it was difficult to solve some of the corner cases, some of them are shown in Fig. 1.

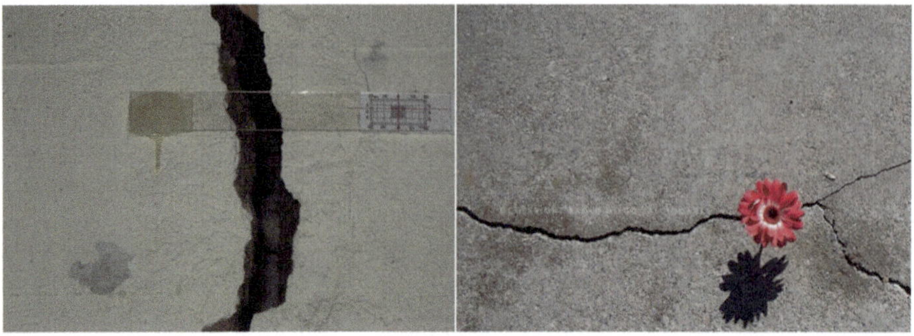

Fig. 1. Two different kinds of bad cases. There may be many other types of objects on the road, such as the ruler in the picture on the left and the flower on the right. The proportion of such objects in the training set is extremely small, which can seriously affect the performance of the segmentation network.

There are many shadows and other dark clutter on the road. Due to the lack of large-scale datasets for the road crack semantic segmentation task and the fact that the existing datasets are specific to a single scene. Lack of background for other types of objects and extremely monotonous road pattern with no other fixed shapes and patterns to interfere with the perspective. Comparison of uniformity. The low diversity of the dataset can cause the neural network to overfit to a single scene, resulting in neural networks trained on existing datasets Unable to complete the task of semantic segmentation of complex scenes. In practice, we may encounter many complex scenes:

- Images containing shadows. such as tree shadows, vehicle shadows, etc.
- Images of roads with other patterns, such as road edges, dark paint, etc.
- Images containing other scenes, such as surrounding soil, trees, vehicles and pedestrians, etc.
- Images taken from different perspectives.

In most cases, the low diversity of the data resulted in widespread misreporting. Many areas in the images where road cracks did not exist were incorrectly assumed to have road cracks. To address the above problem of excessive false alarms encountered in real-world testing, we propose two approaches:

- We present a method for using image generation to enhance the original dataset. The method works by generating reasonable shaded images in a pre-prepared library of clips and adding shadows to the original dataset, and performing data augmentation (e.g., affine transformation), and finally model training.
- Trained a classification network and used it to identify the presence or absence of cracks in subregions. The network is trained on a new dataset that is easier to label, which is divided into two categories: images of roads containing cracks and is the picture of the road without cracks. After completing the training of the semantic segmentation model, we will use the trained classification model to slide the window on the image. Finally we will select and fuse the results of the two models to get the final output.

2 Related Work

2.1 U-Net

U-Net is a fully convolutional network based on FCN improvements, and its structure resembles a U-shape, hence the name U-Net. U-Net was originally designed to solve the medical image segmentation problem, and it proposes a data enhancement strategy to effectively exploit the labeling data, i.e., by applying elastic deformation to the original image to obtain a complementary image, which allows the network to learn elastic deformation invariance. In addition, U-Net employs an overlapping tile strategy, i.e., due to the lack of contextual information for the pixels in the boundary region, by using a tile in the original image The practice of overlapping a circle around the periphery to complement the context. In addition, another U-Net innovation is the increased weighting of the loss of background pixels between sticky objects of the same kind, so that every segmentation contour of the object is clear. Compared with other convolutional neural networks, this network requires fewer training sets and higher segmentation accuracy. The network structure of U-Net includes an auto-encoder and a contracted shrink path which is used to get context information(encoder) and the expand path to the reconstruction location(decoder).

The final feature map of U_Net is computed by a pixel-wise soft-max with the cross entropy loss function. The soft-max is defined as:

$$p_k(x) = \frac{exp(a_k(x))}{\sum_{k'=1}^{K} exp\left(a_k^{'}(x)\right)} \quad (1)$$

where $a_k(x)$ denotes the activation in feature channel k at the pixel position x, and K is the number of classes and $p_k(x)$ is the approximated maximum-function. The cross entropy then calculates the deviation of $p_{l(x)}(x)$ from 1 at each position using:

$$E = \sum_{x \in \Omega} \omega(x) \log\left(p_{l(x)}(x)\right) \quad (2)$$

where l is the ground truth label of each pixel and ω is a weight map that introduced to give some pixels more importance during the training process.

2.2 ResNet [12]

The VGG network tries to find out how deep a deep learning network can be in order to consistently improve classification accuracy. The general impression is that the deeper the network is, the better its performance. However, we found that adding more layers to a deep CNN network after it reaches a certain depth does not lead to better classification performance. Improving, instead, invites the network convergence to become slower and the classification accuracy of the test dataset to become worse. Excluding problems such as model overitting due to small datasets, we find that a deep network still degrades the classification accuracy (as opposed to a shallower network).

When the network is degraded, the shallow network can achieve better training than the deep network, which is when we can achieve better training if we pass the low-level features to the higher layers, then the effect should be at least as good as a shallow network, or if a VGG-100 network is at layer 98 Using exactly the same features as VGG-16 layer 14, then VGG-100 should have the same effect as VGG-16 has the same effect. Therefore, we can add a direct mapping between the 98 and 14 layers of the VGG-100 to achieve this effect.

From an information-theoretic point of view, due to the existence of data processing inequalities, the image information contained in the feature map decreases layer by layer as the number of layers is deepened during forward transmission, and the addition of a direct mapping of the residual network ensures that the network of layers must contain more image information than the layers.

The residual network is composed of a series of residual blocks. A residual block can be represented as:

$$x_{l+1} = x_l + F(x_l, W_l) \tag{3}$$

The residual block is divided into two parts: the direct mapping part and the residual part. The direct mapping x_l, and the residual part x_{l+1}, generally consists of two or three convolutional operations. However, in a convolutional network, the number of features may be different from the number of residuals of the direct mapping. Dimensionality reduction. At this point the residual block is expressed as:

$$x_{l+1} = h(x_l) + F(x_l, W_l) \tag{4}$$

where $h(x_l = W_l'x)$, $W_l'x$ is 1×1 convolution.

Based on this idea of using direct mapping to connect the different layers of the network, the residual network was created.

3 Dataset

In this paper, we trained our models on Deep Crack dataset [9] and Concrete Crack Images dataset [13]. The Deep Crack dataset was developed by Computer Vision and Remote Sensing (CVRS) Lab, School of Remote Sensing and Information Engineering, Wuhan University, Wuhan, Hubei, PR China. The benchmark dataset contains 537 images with manual annotation maps. The authors of this dataset used a combination of extended Fully Convolutional Neural networks (FCNs) and Deep Supervised Nets (DSNs) to create the automated crack detection network known as "Deep Cracks". At that time, the validity of the "Deep Crack" network was verified using this dataset. The method presented in the paper achieved state-of-the-art performance on the chosen dataset. The Crack Images for Classification dataset consists of 2000 labeled "Positive" images and 2000 labeled "Negative". The dataset is composed of the actual image with reduced quality. In their article, the authors use a convolutional neural network produced by the TensorFlow Keras API, a network structure with 3 convolutional layers was created. The first layer is the Flatten layer, the second layer is a Fully Connected layer made of 64 nodes, and the last is the output layer. The author uses GPU for training, Sigmoid activation function for the output layer, and ReLU (Rectified Linear

Unit) for the rest of the layers' activation, and using this dataset, the cracked image classification was achieved on the nice effect.

When we trained the semantic segmentation network, we also added a small dataset of our own images containing complex scenes. Some of these images are from the Internet, and the others are taken by our camera.

4 Proposed Method

4.1 Data Augmentation and Segmentation

Before the semantic segmentation network starts training, we augment the training data using our proposed shadow generation method. In the actual training data, most of the difficulty in segmentation is mainly due to the presence of a large number of complex shadows blocking the cracks in the image, so this paper considers mainly the case of shadows. We prepare a large number of possible objects on the road, which are processed and transformed into shadows to be cast on the road image. The shadow generation is mainly based on the physical objects that may appear near the road surface in the real world, we detect the outline of the physical objects to select the physical foreground, the background Set it to transparent, and then binarize the physical foreground to get the black physical foreground. In order to make the black foreground more shadow-like, we randomly set a reasonable amount of transparency to the shadows when we add them and apply a Gaussian Blur Filter. Figure 2 shows two sample images after data augmentation.

Fig. 2. Image samples of road cracks after data augmentation.

The shaded images generated by the above method can largely avoid the false shadows caused by excessive network segmentation positive while using shading to enhance the dataset, we also used other methods to enhance the dataset, such as Random affine transformation, random cropping, random rotation, random changes in brightness, etc. Using other data enhancement methods can reduce the risk of model overfitting.

Once we had completed the data augmentation, we trained our segmentation model using U-Net architecture as shown in Fig. 3.

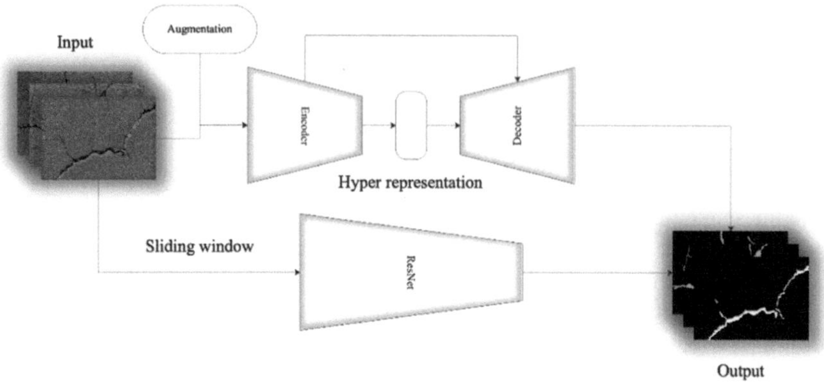

Fig. 3. Our fusion network. The network has two parts: the semantic segmentation part and the classification part.

4.2 Classification Network

The problem of too many false positives may still not be completely eliminated by using the segmentation model alone, since there are of false positive does not come from shadows, but from other, such as more complex patterns which may be difficult to generate through a purely generative model. However, we can use other perceptual models to determine the result splitting as a way to filter out the remaining false positive. Our approach is to train a two-class classification network to determine whether a subregion of the image contains road cracks or not. The network is based on ResNet, and is trained on Concrete Crack Images.

4.3 Dual-Net Fusion

We use ResNet to train the classifier. After training is complete, we enter the images into the classification network via a sliding window. In this way, we can determine whether each subregion of the input image contains a crack or not. In our judgment strategy, we will only output a subregion with segmentation result if both the segmentation network and the classification network think that it contains a crack, otherwise we will only output a full black background color and ignore the segmentation result of this subregion. The vast majority of false alarms can be eliminated by the classifier due to the high accuracy it has on the subregion classification task. This method has a great advantage in the task of offline road crack segmentation which is insensitive to the amount of calculation.

5 Experiments and Results

We used the PyTorch deep learning framework to build models and trained semantic segmentation and classification models on a workstation with AMD Ryzen 3900X and NVIDIA TITAN RTX. We used Adam as our optimizer [14].

First, we train the semantic segmentation network on the modified Deep Crack dataset and record the loss of the network training process, the loss curve is shown in Fig. 4. We find that images that have been augmented will always lose less during training than those that have not been augmented with data. We think that this is due to the numerical differences in the images after data augmentation, but otherwise the two loss functions do not converge speed have no significant difference. We believe that the current data enhancement algorithm does not make the model significantly underfit. Samples of the prediction results of our U-Net are shown in Fig. 5.

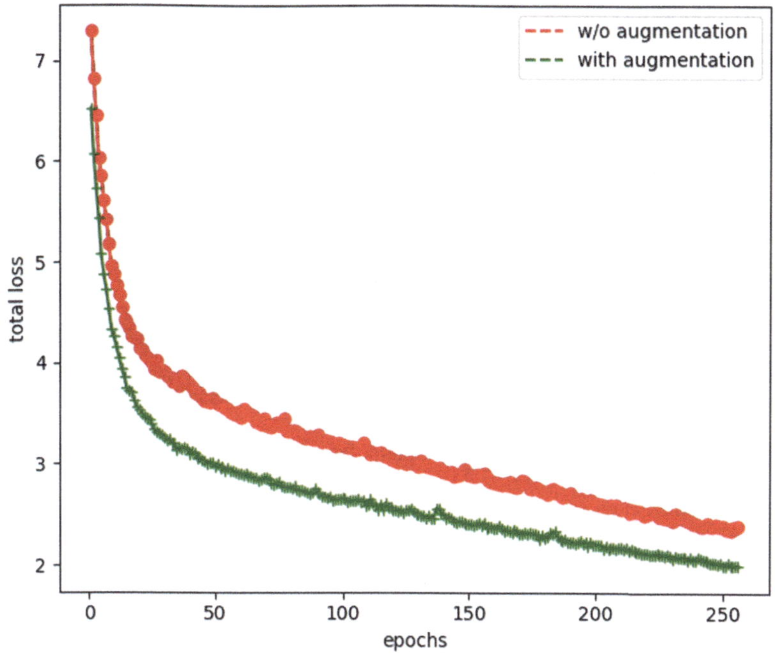

Fig. 4. Loss during semantic segmentation network training process.

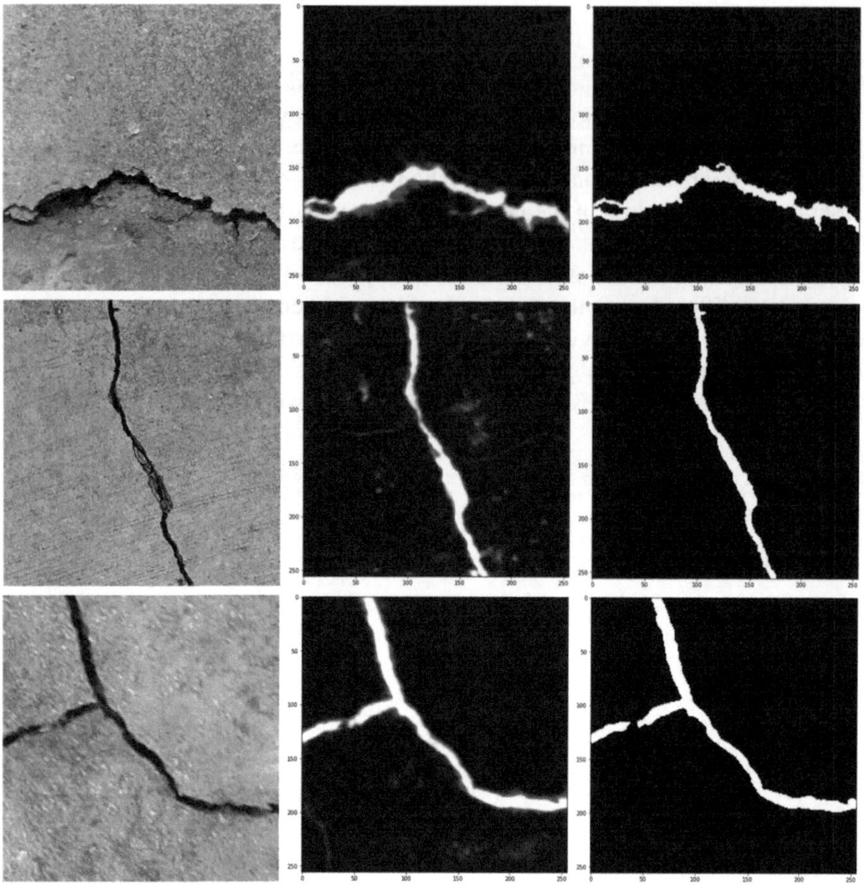

(a) nput (b) Output(soft-max) (c) Ground Truth

Fig. 5. The segmentation result of the semantic segmentation network. The second column of pictures is the result of the U-Net network output through soft-max. It can be seen intuitively that the semantic segmentation model we trained can find the location of the crack in pixel-wise very well.

We evaluated the performance of our proposed method on the images in the test set, comparing our method with several other original U-Net. We set thresholds, binarize the U-Net predicted images, and then calculate Precision(Pr) and Re(Re) for the predicted and labeled images. These metrics are commonly used in evaluating classification tasks. Their precise definitions are as follows:

$$P_r = \frac{TP}{TP + FP} \tag{5}$$

$$R_e = \frac{TP}{TP + FN} \tag{6}$$

where TP is the number of true positives, FP is the number of false positives, and FN is the number of false negatives.

As can be seen in Table 1, our proposed method makes the semantic segmentation performance of road cracks much better.

Table 1. Model comparison.

Model	Precision	Recall
w/o augmentation	0.91041	0.97231
with augmentation	0.98216	0.84643

We also trained the classification network to calibrate the prediction results of the semantic style network. The classification network we trained went through 10 epochs, and the accuracy of the classification network on the test set reached 98%. The prediction results of the fusion of the two models are shown in Fig. 6. It can be seen from the figure that the images processed by the model classification network are somewhat sufficient to eliminate false alarms caused by complex patterns. If the input image contains complex interference patterns, the se mantic segmentation network alone is often unable to output accurate segmentation results, but with the road crack classifier to perform secondary verification on the results of semantic segmentation, both the

(a) Test image (b) Model prediction

Fig. 6. Model prediction using augmented network fusion. We constructed a test dataset with a number of images containing complex patterns and other objects that are not road cracks, and the use of the data without such A network trained on a dataset of objects is difficult to correctly segment a road crack. As shown in Figure (a), a flower grows out of the road crack, whereas direct segmentation through the semantic segmentation network would treat the flower stalk as a road also part of the rift (as shown in Figure (b)). In Figure (b), we used a classification network for a sliding window on the image, and the Negative region is where the classification network thinks that the non Areas containing cracks, Positive is the area that the classification network considers to contain road cracks. Using the classification network was able to exclude a large number of false positive test images.

precision and accuracy of segmentation can be improved as shown in Fig. 6. At the same time it does not have an excessive negative impact on the correct prediction of the semantic segmentation network, as the Re does not decrease. In some cases, the Re can be even higher.

6 Conclusion

In this paper, we make an analysis of the low diversity of existing road crack semantic segmentation datasets and propose that Two solutions are proposed. First, we propose data augmentation by adding patterns to our simple original datasets, allowing the training data to cover a more complex scenarios. After training the semantic segmentation network, we also trained the classification network separately on the classification dataset and on the semantic segmentation network. We demonstrated the data augmentation and model fusion methods in the paper, we trained the model, and showed the prediction results of the two models. The results of the sliding-window calibration were performed to improve the accuracy of the model. Overall, our methods enable the model to cope with more complex road images, and our methods are used in practical scenarios with a higher level of robustness. Our method has achieved excellent results in road segmentation tasks that are not sensitive to the amount of computation but require high accuracy.

Acknowledgments. This research was funded by the Self-Driving Test and Evaluation Equipment Project (Project ID: CPYF-2019-02) and the State Key Laboratory of Vehicle NVH and Safety Technology.

References

1. Shen, G.: Road crack detection based on video image processing. In: 2016 3rd International Conference on Systems and Informatics (ICSAI), pp. 912–917. IEEE, NJ (2016)
2. Oliveira, H., Correia, P.L.: Automatic road crack segmentation using entropy and image dynamic thresholding. In: 2009 17th European Signal Processing Conference, pp. 622–626. IEEE, NJ (2009)
3. Cheng, H., Chen, J.R., Glazier, C., Hu, Y.: Novel approach to pavement cracking detection based on fuzzy set theory. J. Comput. Civil Eng. **13**(4), 270–280 (1999)
4. Shi, Y., Cui, L., Qi, Z., Meng, F., Chen, Z.: Automatic road crack detection using random structured forests. IEEE Trans. Intell. Transp. Syst. **17**(12), 3434–3445 (2016)
5. Krizhevsky, A., Sutskever, I., Hinton, G.E.: ImageNet classification with deep convolutional neural networks. In: NIPS 2012: Proceedings of the 25th International Conference on Neural Information Processing Systems, vol. 1, pp. 1097–1105. Curran Associates Inc., NY (2012)
6. LeCun, Y., Bottou, L., Bengio, Y., Haffner, P.: Gradient-based learning applied to document recognition. Proc. IEEE **86**(11), 2278–2324 (1998)
7. Silva, W.R.L., Lucena, D.S.: Concrete cracks detection based on deep learning image classification. In: Proceedings of the 18th International Conference on Experimental Mechanics, pp. 489–489. MDPI, Basel, (2018)

8. Simonyan, K., Zisserman, A.: Very deep convolutional networks for large-scale image recognition. In: ICLR 2015: The 3rd International Conference on Learning Representations (ICLR2015). arXiv:1409.1556 (2014)
9. Liu, Y., Yao, J., Lu, X., Xie, R., Li, L.: Deepcrack: a deep hierarchical feature learning architecture for crack segmentation. Neurocomputing **338**, 139–153 (2019)
10. Xie, S., Tu, Z.: Holistically-nested edge detection. In: Proceedings of the IEEE International Conference on Computer Vision, pp. 1395–1403. IEEE (2015)
11. Ronneberger, O., Fischer, P., Brox, T.: U-Net: convolutional networks for biomedical image segmentation. In: Navab, N., Hornegger, J., Wells, W.M., Frangi, A.F. (eds.) MICCAI 2015. LNCS, vol. 9351, pp. 234–241. Springer, Cham (2015). https://doi.org/10.1007/978-3-319-24574-4_28
12. He, K., Zhang, X., Ren, S., Sun, J.: Deep residual learning for image recognition. In: Proceedings of the IEEE conference on computer vision and pattern recognition, pp. 770–778. IEEE, NJ (2016)
13. Özgenel, Ç.F. Concrete crack images for classification. Mendeley Data, vl (2017)
14. Kingma, D.P., Ba, J.: Adam: a method for stochastic optimization. In: The 3rd International Conference for Learning Representations. arXiv preprint arXiv:1412.6980 (2014)

Improved UAV Scene Matching Algorithm Based on CenSurE Features and FREAK Descriptor

Chenglong Wang[1(✉)], Tangle Peng[1], Longzhi Hu[2], and Guanjun Liu[1]

[1] College of Computer Science and Technology, Hengyang Normal University, Hengyang 421002, China
wclbaozi@163.com

[2] Zhongxing Telecommunication Equipment Corporation, Shenzhen 518000, China

Abstract. The traditional local invariant features scene matching algorithms have many redundant points, poor real-time and Low anti-geometric transformation. So we propose an unmanned aerial vehicle scene matching algorithm based on CenSurE features and FREAK descriptor. Steps of the algorithm are as follows: use the CenSurE-star operator to extract the feature points, calculate feature vectors by FREAK descriptors, then complete matching process through KNN ratio method, finally use RANSAC location model to get image transform relations, latitude and longitude. Experiments show that: compared with SIFT, SURF, ORB algorithms, this algorithm not only has better robustness for various transformation, but also greatly shorten the processing time of about 86 ms. The positioning error within 0.6 pixel, scale error within 0.02 times, rotation Angle error within 0.04°.

Keywords: Unmanned aerial vehicle scene matching algorithm · CenSurE features · FREAK descriptor

1 Introduction

In the field of UAV assisted navigation. Scene matching technology use the real-time image collected by the image sensor during flight to match the pre-prepared reference image, so as to obtain accurate position navigation information, complete the correction of inertial navigation error, and greatly improve the navigation accuracy [1]. Because of the fast flight speed and complex angle of view transformation of the aircraft, it becomes the key to provide a scene matching and positioning algorithm with high reliability and real-time performance.

Because the traditional scene matching methods based on gray scale and transform domain cannot adapt to the rotation, scale, lighting and other changes between airborne images, and the processing time is too long, it is difficult to apply them to UAV technology. In 2004, Lowe et al. focused on local feature spots of images and proposed SIFT algorithm, marking a new stage of feature detection and descriptor development. SIFT algorithm has excellent robustness, but its solution is complex and time-consuming; In 2008, Bay et al. proposed SURF algorithm on this basis, which cost 1/3

Q. Liu et al. (Eds.): CENet 2020, AISC 1274, pp. 158–167, 2021.
https://doi.org/10.1007/978-981-15-8462-6_18

of SIFT. However, feature vectors extracted from multiple main directions would generate too many matching points, resulting in increased mismatches. In order to improve the computational speed, local binary descriptors have been developed rapidly in recent years. Calander et al. proposed BRIEF algorithm in 2010 and Rublee et al. proposed ORB algorithm in 2011, both of which used binary chain code or operation to accelerate matching. However, BRIEF descriptors are very sensitive to rotation and ORB descriptors do not have scale invariance, which limits the use of spot detectors and binary descriptors in UAV scene matching algorithms [2, 3]. Therefore, this paper proposes a star filter spot feature detection method based on CenSurE (center around extreme value) and a fast robust UAV scene matching algorithm based on FREAK binary feature description method.

2 Scene Matching Algorithm Based on CenSurE Feature and FREAK Descriptor

2.1 CenSurE Characteristics

CenSurE is a new local spot feature proposed by Agrawal in 2008. It has high precision, good stability and high computational efficiency, and has been applied to outdoor robot navigation and virtual and real registration [4, 5]. CenSurE feature extraction mainly consists of three steps:

A) The scale space is constructed with a double layer filter.
CenSurE uses a simple bilayer filter to approximate The Laplace transform of Gauss. Figure 1 shows a square filter, also known as CenSurE-DOb. The kernel size of CenSurE-DOb filter is $(2n + 1) \times (2n + 1)$, and the outer core size is $(4n + 1) \times (4n + 1)$. Let I_n be the kernel weight coefficient and O_n be the outer kernel weight coefficient. In order for the DC response of this filter to be 0, the weight coefficient needs to satisfy the following equation:

$$I_n(2n + 1)^2 = O_n(4n + 1)^2 \tag{1}$$

Normalize the scale:

$$I_n(2n + 1)^2 = I_{n+1}(2(n + 1) + 1)^2 \tag{2}$$

It is obvious that the square filter is not a rotation-invariant kernel function, especially when the plane is rotated by 45°. For UAV scene matching with high rotation scale performance requirements, this paper uses CenSurE improved version-star filter mentioned in literature [6], namely Censure-Star. The star filter is a combination of a square filter and a 45° rotating square filter, as shown in Fig. 2. The response value of the center surround haar wavelet at each pixel point is also superposed from the original integral image and the 45° rotation integral image.

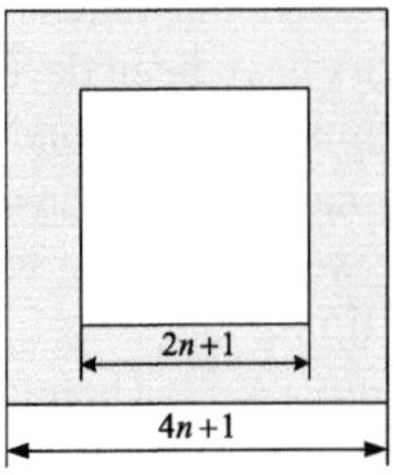

Fig. 1. CenSurE square filter

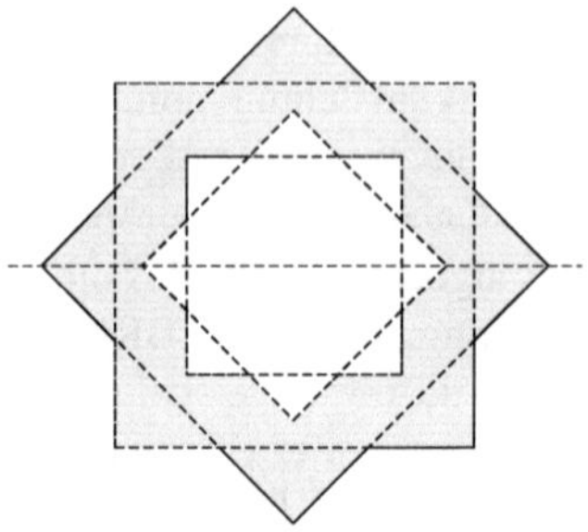

Fig. 2. CenSurE-star filter

B) Detect local extremum points.
The method of non-maximum suppression was used to detect the scale space extremum as candidate feature points. The response values of each pixel in the scale space were compared with 8 neighborhood points in the scale layer and 18 neighborhood points in the upper and lower adjacent scale layer to filter out the extreme value points less than the set threshold.
C) Filter out the unstable points on the edge.
The principal curvature is calculated using the scale-adapted Harris method, which can be expressed as:

$$H_a = \begin{bmatrix} \sum L_x^2 & \sum L_x L_y \\ \sum L_x L_y & \sum L_y^2 \end{bmatrix} \tag{3}$$

L_x and L_y are the partial derivatives of response function L in the directions of x and y. The corresponding principal curvature can be calculated by calculating the trace and determinant of H_a. The principal curvature threshold is 10. The candidate points larger than the threshold are filtered out, that is, the edge response is eliminated. and the remaining candidate points are CenSurE-star feature points containing location and scale information.

2.2 FREAK Descriptor

FREAK descriptor was proposed by Alexandre Alahi et al. in 2012. It is inspired by the human visual system and adopts a topological structure close to the human retina to process information [7]. The sampling model is shown in Fig. 3. Fovea area processes high-precision image information, while Para area processes low-precision image information (Fig. 3).

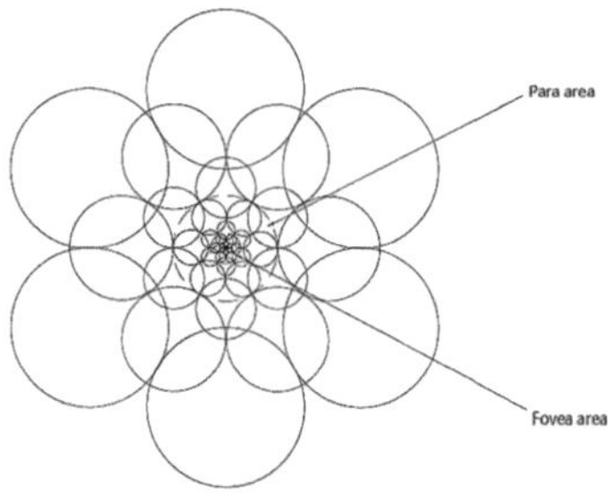

Fig. 3. FREAK descriptor sample model

It can be seen from the figure that sampling points are uniformly distributed on concentric circles centered on feature points. The closer to the feature point, the denser the sampling point. The sensing domain of adjacent sampling points (that is, the circle with the sampling point as the center and the Gaussian core as the radius) is designed as an overlapping structure to obtain more detailed information. During the sampling, each sampling point should be filtered by Gaussian filter, and the size of the Gaussian core of the sampling point is proportional to the concentric circle radius of the current sampling point. After sampling, the FREAK descriptor is built with the sampling information. F is used to represent the binary sequence of the FREAK descriptor, and each bit of F can be obtained through the intensity comparison of the sampling point region:

$$F = \sum_{0 \le \alpha \le N} 2^{\alpha} T(P_{\alpha}) \tag{4}$$

$$T(P_{\alpha}) = \begin{cases} 1 & if(I(P_a^{r1}) - I(P_a^{r2}) > 0) \\ 0 & otherwise \end{cases} \tag{5}$$

where, P_{α} represents the sampling point pair, α represents the binary left shift bit value, N represents the descriptor length, and if the sampling point is S, $N = C_S^2$. $I(P_a^{r1})$ and $I(P_a^{r2})$ are the intensity values of the two sample point regions.

To ensure that the descriptor has rotation invariance and to determine the principal direction of the descriptor, FREAK chooses 45 sampling point pairs with central symmetry to obtain the principal direction. Let M represent the number of sampling point pairs, G represents the set of sampling point pairs, and P_o represents the set of sampling point pairs, then the principal direction O of the characteristic point is:

$$O = \frac{1}{M} \sum_{P_o \in G} (I(P_o^{r1}) - I(P_o^{r2})) \frac{P_o^{r1} - P_o^{r2}}{\left\| P_o^{r1} - P_o^{r2} \right\|} \tag{6}$$

Due to the scene matching just keep the dimension of the large amount of information, so be screened dimensions: in FREAK sampling mode, a total of 43 sampling point area, 903 points on the combination of correlation of the descriptor matrix calculation, the top of 512 is the most relevant, add more does not improve the effect, so before the 512 as the final descriptor [8].

2.3 Knearest Neighbor Ratio Method

After building the FREAK descriptor on the feature points on the baseline and real-time graph respectively, two sets of descriptors are obtained. Next, this paper USES the K-nearest neighbor ratio method to match two sets of descriptors. Firstly, hamming distance is used as similarity measure of feature points. The Hamming distance between two binary strings of equal length is realized by bitwise xor operation, and the number of statistical results of 1 is the distance result. Secondly, the k-nearest neighbor algorithm (K = 2) is used to obtain the candidate matching points. The method is to use violent search to find the two points with the smallest distance between hamming and each feature point of the real-time image, namely the nearest neighbor point and the second neighbor point [9]. Finally, the ratio of nearest and subnearest distance is used for threshold filtering, as shown in Formula 7, to obtain the final matching point pair.

$$\frac{D_{FirMin}}{D_{SecMin}} \leq N \tag{7}$$

D_{FirMin} is the hamming distance of the nearest neighbor point, D_{SecMin} is the hamming distance of the second neighbor point, and N is the threshold value.

2.4 RANSAC Based on Location Model

After matching points are obtained, this paper uses the idea of RANSAC algorithm to establish a scene matching and positioning model. The model can not only obtain the transformation equation between the base graph and the real-time graph, but also remove the wrong point pairs in the matching points and obtain the latitude and longitude of the flight position.

Perspective transformation exists in scene matching datum graph and real-time graph. The transformation equation is as follows:

$$\begin{bmatrix} x' \\ y' \\ 1 \end{bmatrix} = H \begin{bmatrix} x \\ y \\ 1 \end{bmatrix} = \begin{bmatrix} h_{11} & h_{12} & h_{13} \\ h_{21} & h_{22} & h_{23} \\ h_{31} & h_{32} & h_{33} \end{bmatrix} \begin{bmatrix} x \\ y \\ 1 \end{bmatrix} \tag{8}$$

where (x', y') represents the pixel coordinates of the baseline image and (x, y) represents the pixel coordinates of the real-time image. $\begin{bmatrix} h_{11} & h_{12} \\ h_{21} & h_{22} \end{bmatrix}$ represents linear transformation, such as scaling and rotation, $\begin{bmatrix} h_{13} & h_{23} \end{bmatrix}^T$ represents translation transformation,$\begin{bmatrix} h_{31} & h_{32} \end{bmatrix}$ generates perspective transformation, and $h_{31} = h_{32} = 0$ represents affine transformation. You normally normalize the matrix by taking $h_{33} = 1$.

RANSAC is a random sampling test method that can model with a small amount of data and filter all the data. Using the idea of RANSAC algorithm for reference, an optimal homography matrix H is found to make the number of data points satisfying the matrix maximum. To solve the 8 unknown parameters of the H matrix, at least 8 linear equations are needed, that is, at least 4 sets of matching point pairs. The specific steps are as follows:

1) Randomly select 4 groups from the matching point pairs obtained in Sect. 2.3, calculate the transformation matrix H, which is denoted as model M;
2) The projection error of all matching points and model M is calculated. If the error is less than threshold, the matching points are added into the inner point set:

$$\sum_{i=1}^{n}\left(x_i' - \frac{h_{11}x_i + h_{12}y_i + h_{13}}{h_{31}x_i + h_{32}y_i + h_{33}}\right)^2 + \left(y_i' - \frac{h_{21}x_i + h_{22}y_i + h_{23}}{h_{31}x_i + h_{32}y_i + h_{33}}\right)^2 \leq \varepsilon \tag{9}$$

3) Select 4 matching point pairs and repeat steps 1 and 2;
4) If the number of elements in the current interior point set is greater than the optimal interior point set, the current set is updated to be the optimal interior point set until the number of iterations is greater than K and it exits.

At this point, the parameters of the transformation matrix H with the optimal set of internal points are the final transformation parameters, thus the perspective transformation model between the base graph and the real-time graph is obtained. Matching point pairs that do not meet this model are considered as mismatched and eliminated, which further improves the accuracy of scene matching algorithm.

When the aircraft is flying horizontally, it is considered that the center point of the real-time image taken by the airborne sensor is the position of the aircraft, and its pixel is marked as (x_c, y_c). Through the perspective transformation model calculation, the corresponding pixel coordinates of point (x_c', y_c') in the reference diagram can be obtained. The transformation formula of longitude and latitude and pixel coordinates can be used to calculate the flight position of the aircraft as follows:

Longitude:

$$Lon_c = x_c' \times \frac{\max Lon - \min Lon}{l} + \min Lon \tag{10}$$

Latitude:

$$Lat_c = \max Lat - y_c' \times \frac{\max Lat - \min Lat}{h} \tag{11}$$

where, l is the number of pixels in the horizontal direction, h is the number of pixels in the vertical direction, and the geographical coordinate region $(\min Lon, \max Lon, \min Lat, \max Lat)$ of the datum graph is all known parameters.

3 The Experiment

3.1 Match Threshold Setting

In the process of screening the matching points, the algorithm in this paper firstly USES the matching points to conduct threshold screening for the ratio of nearest neighbor distance and second neighbor distance, and then performs threshold screening for the projection error of matching points based on RANSAC algorithm, so as to determine the final matching points. If the ratio threshold N and the error threshold are set too small, it may cause a large number of correct matching points to be filtered out; if the setting is too large, it may lead to the existence of wrong matching points after filtering. Therefore, the reasonable setting of the two determines the accuracy of matching. The correct matching rate is the matching accuracy, which is defined as the ratio of the correct matching number to the total matching number. The formula is as follows:

$$precision = \frac{correct\,matches}{all\,matches} \tag{12}$$

The scale rotation change group and the brightness change group of the standard test image library were used for the test. By adjusting the size of the two thresholds, the relationship between them and the correct matching rate was obtained. The experimental results are shown in Fig. 4 (Tables 1, 2, 3 and 4).

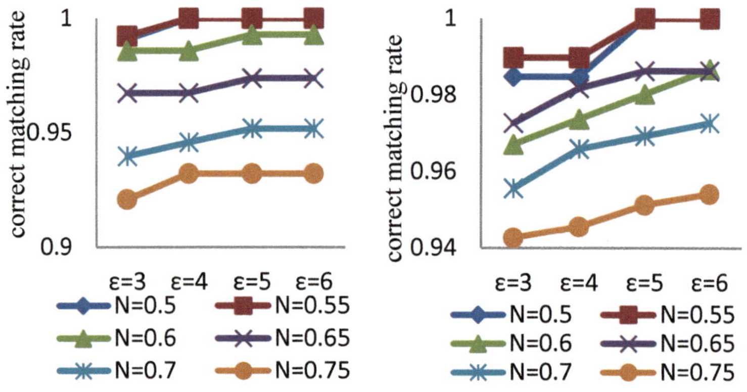

Fig. 4. The correct matching rate varies with the threshold. Scale rotation changes (left). Brightness change (right).

As can be seen from the experimental results, in the case of scale rotation and brightness changes with obvious differences in ssscene matching images, the correct matching rate was higher than other values at the ratio threshold of N = 0.55, and the correct matching rate was significantly increased when the error threshold of ε = 5, and then the improvement was slowed down. Therefore, N = 0.55 was used as the ratio threshold and ε = 5 as the error threshold.

3.2 Robustness Experiment of the Algorithm

The algorithm in this paper adopts the matching algorithm combining Censu-star spot feature detector and FREAK binary feature descriptor. Therefore, SIFT algorithm, SURF algorithm and ORB algorithm, which are typical algorithms for spot feature detection, as well as typical algorithms for binary feature description, are used to conduct several robust performance experiments. The experiment still USES the 4 groups of image changes in the standard test image library, and the experimental results are as follows:

Table 1. Experimental results of scale change group.

	SIFT algorithm	SURF algorithm	ORB algorithm	Algorithm in this paper
Correct matches	423	124	84	8
Error matches	9	4	3	0
Accuracy (%)	97.91	96.87	96.55	100.00
The number of feature points	6268	2478	947	782

Table 2. Experimental results of the brightness change group.

	SIFT algorithm	SURF algorithm	ORB algorithm	Algorithm In this paper
Correct matches	533	334	63	48
Error matches	8	14	2	3
Accuracy (%)	98.52	95.98	96.92	94.12
The number of feature points	1563	1126	500	247

Table 3. Experimental results of noise change group.

	SIFT algorithm	SURF algorithm	ORB algorithm	Algorithm in this paper
Correct matches	260	289	63	15
Error matches	6	8	1	0
Accuracy (%)	97.74	97.31	98.44	100.00
The number of feature points	734	975	520	92

Table 4. Experimental results of visual Angle change group.

	SIFT algorithm	SURF algorithm	ORB algorithm	Algorithm in this paper
Correct matches	782	154	Matching failure	93
Error matches	4	2		0
Accuracy (%)	99.49	98.72		100.00
The number of feature points	3666	2672		942

Experimental results show that the proposed algorithm has excellent adaptability in scale rotation, noise and Angle of view, and even achieves zero mismatching. SIFT and SURF algorithms are inferior to the algorithm in terms of anti-noise performance and adaptability of scale scaling. The ORB algorithm even fails to match the situation when the perspective changes greatly, which is due to the scale-free invariance of its feature detector. However, in terms of brightness change, the algorithm presented in this paper is not as good as other algorithms. In addition, when the correct matching rate is the same, the number of feature points of this algorithm is far less than that of other algorithms, indicating that this algorithm has high efficiency and low redundancy.

3.3 Real-Time Experiment of the Algorithm

In order to show the advantages of the proposed method in operation time, SIFT and SURF methods were used for comparison. The images used in the experiment were captured from Google Earth. The base image size was 320 × 249 and the real-time image size was 240 × 180. The time consumption of the three methods in feature point extraction, feature description, feature matching and RANSAC location was counted, and the results were shown in Table 5.

Table 5. Experimental results of visual Angle change group.

	Feature point extraction Time (ms)	Character description Time (ms)	Feature matching Time (ms)	RANSAC location Time (MS)	The total Time (ms)
Algorithm in this paper	9.67	67.76	6.77	1.54	85.74
SURF algorithm	69.27	80.71	8.07	1.58	155.63
SIFT algorithm	305.16	596.08	19.70	1.60	922.54

It can be seen from Table 5 that the time of feature point extraction, feature vector generation and feature vector matching of this algorithm is much less than that of SIFT and SURF. The whole registration process only takes 85.74 ms, which makes this algorithm have a broad application prospect in aircraft navigation occasions requiring real-time application.

4 Conclusion

Aiming at the requirement of real-time and robustness of UAV scene matching, this paper proposes a scene matching algorithm based on CenSurE feature and FREAK de. The censu-Star feature point detection method adopted in this paper is not only extremely fast, but also able to complete the correct matching with a small number of feature points with low redundancy. The repeatability rate is superior to DOG operator under various changes and Hassian operator under scaling changes. In addition, using the FREAK descriptor to simulate the human retinal imaging mechanism, the binary xOR operation hamming distance calculation method greatly accelerates the matching speed. Experiments show that the overall performance of the algorithm is excellent, and its adaptability in scale rotation, noise and perspective change is superior to SIFT, SURF and ORB algorithms. Moreover, it can complete the matching within 86 ms and provide positioning coordinates. The position error of matching point is within 0.6 pixels, the scale error is within 0.02 times, and the rotation Angle error is within 0.04°. Such high accuracy and high timeliness make it have a broad application prospect in the field of UAV visual aid navigation.

References

1. Liu, X.C.: Research on scene matching navigation technology based on real-time image and satellite image. University of National Defence Science and Technology, Beijing, pp. 3–6 (2008)
2. Xu, Y.X., Jiang, Y.L., Chen, F.: Quick scene matching algorithm based on BRISK in inertial integrated navigation system. J. Optoelectr. Laser **23**(8), 1589–1596 (2012)
3. Wang, C.J., Sun, T., Chen, J.: Rapid scene matching based on FREAK characteristics. J. Electr. Meas. Instrum. **29**(2), 204–210 (2015)
4. Hu, J., Tang, G.M., Sun, Y.F., Liu, J.: Study on virtual reality registration method based on CenSurE characteristics. Comput. Appl. Res. **30**, 1910–1913 (2013)
5. Agrawal, M., Konolige, K., Blas, M.R.: CenSurE: center surround extremas for realtime feature detection and matching. In: 10th European Conference on Computer Vision, pp. 102–109 (2008)
6. Kosaka, N., Ohashi, G.: Vision-based nighttime vehicle detection using CenSurE and SVM. IEEE Trans. Intell. Transp. Syst. **16**(5), 2599–2602 (2015)
7. Li, J.J., Zhao, Y., Wang, A.X.: Real-time augmented reality registration algorithm based on modified FREAK. Minicomput. Comput. Syst. **1**(1), 173–177 (2016)
8. Alahi, A., Ortiz, R., Vandergheynst, P.: FREAK: fast retina keypoint. In: IEEE Conference on Computer Vision and Pattern Recognition (2012)
9. Li, J.S., Zhang, J., Yang, W.: Feature-Based Infrared Image Target Matching and Tracking Technology. Science Press, Beijing (2014)
10. Sun, J.Y., Sun, X.D., Wang, Z.: Forward Infrared Image Matching Technology. Science Press, Beijing (2011)

Analysis of Image Quality Assessment Methods for Aerial Images

Yuliang Zhao, Yuye Zhang(✉), Jikai Han, and Yingying Wang

Qingdao Branch, Navy Aeronautical and Astronautical University, Qingdao 266041, China
zyl_playboy@126.com

Abstract. Due to its special imaging conditions, aerial reconnaissance images are easy to be degraded by noise, atmospheric condition, defocus, light exposure, motion-displacement and so on. And the image quality degradation comes down to noise, low contrast and low definition. In order to choose image automatically from massive image data to use in interpretation, appropriate image quality assessment method fit to aerial images should be chosen. The experiments were designed to compare commonly used methods. And the conclusions were that the Mean square error and Mean gradient methods assess noise more accurately. Edge preserving index method assesses definition more accurately, Standard deviation method assesses contrast more accurately, and Structure Similarity (SSIM) method assesses more accurately in image comprehensive quality.

Keywords: Aerial reconnaissance image · Image quality assessment · Structure similarity

1 Introduction

Aerial camera real-time reconnaissance has stronger information timeliness and accuracy than other remote reconnaissance methods such as satellite remote sensing reconnaissance, and is one of the important sources of information for military intelligence. Due to their special imaging conditions, aerial reconnaissance images are quite different from the visual characteristics of images taken by ground cameras and satellite remote sensing images. The factors that cause image degradation are also different. And it is easily affected by problems such as noise, atmospheric conditions, defocus, exposure, moving image shift, etc., resulting in degraded images. In order to automatically filter available images from massive image data and further use it for intelligence interpretation, it is necessary to study objective, comprehensive and effective image quality assessment methods. In this paper, based on the analysis of the factors that affect the degradation of aerial reconnaissance images, a comparative study of image quality assessment methods is carried out.

Q. Liu et al. (Eds.): CENet 2020, AISC 1274, pp. 168–175, 2021.
https://doi.org/10.1007/978-981-15-8462-6_19

2 Quality Reduction Characteristics of Aerial Reconnaissance Image

It is necessary to understand the many factors that affect the image quality of aerial reconnaissance in order to achieve image quality judgment purposefully. The temperature and pressure conditions of aerial photography are different from those on the ground. The main factors that affect the quality of aerial images include [1]: noise, atmospheric conditions, defocus, exposure time, and image shift.

2.1 Noise

The noise pollution of aerial reconnaissance images can be divided into dark current noise, shot noise, transfer noise, and reset noise. The dark current noise is produced by the CCD imaging device of the aerial camera, which is one of the white noises. The magnitude of dark current noise is related to temperature and light integration time. If you want to reduce the noise generated by dark current, you should give the CCD imaging device a suitable working temperature environment and illumination environment, and the aerial camera works at an altitude of several kilometers, it is inevitable to be affected by dark current noise. In addition, reset noise is also called ktc noise, where k stands for Boltzmann's constant, t stands for temperature, and c stands for floating capacitor value. The aerial camera CCD imaging device uses a floating capacitor to convert the generated electronic signal into a voltage, and uses a floating diffusion capacitor to output. When the floating diffusion capacitor reset switch is turned on, it is easy to have KTC noise added to it.

2.2 Atmospheric Condition

During the aerial reconnaissance process, the following two atmospheric conditions are usually affected:

The Influence of Atmospheric Instability. There are three situations: flicker, pulsation and jitter when taking photos through the atmosphere. The flicker phenomenon refers to the instantaneous change of the color and brightness of the shooting target, which eventually leads to the deterioration of the image contrast; the pulsation phenomenon refers to the instantaneous change of the shape of the shooting target, which ultimately leads to the blurring of the image and the deterioration of the clarity. The jitter phenomenon refers to the instantaneous change of the position of the shooting target, which eventually causes the image to shift.

The Effect of Atmospheric Fog. When the atmosphere is fogged, the image surface seems to be attached with a layer of gray fog, which causes the grayscale of the scene to deteriorate, and loses many edge details and reduces the contrast of the image.

2.3 The Defocusing Phenomenon

It is known from the imaging formula that the object distance, image distance, and focal length are directly related. If the aerial photograph is taken, the focal plane position cannot be changed with the change of the object distance and image distance, and defocusing will occur. The biggest impact of defocusing on the image is to make the image blurred, and secondly, it will cause poor contrast of the image. Only when the focus is correct, the image will be clear and the edges will be clear. The defocusing phenomenon is caused by the relative position distance between the photosensitive device and the CCD receiving surface, that is, the focal length cannot be adjusted in time during aerial photography. The causes usually include changes in atmospheric pressure, temperature, and shooting distance during aerial photography.

2.4 Exposure Time

Exposure time will be affected by many factors. Common factors include the following: the sensitivity of the CCD, the dynamic range of the CCD, the basic parameters and optical characteristics of the camera, the scene illumination and the brightness coefficient.

If the exposure time is too long, the charge stored in the CCD device reaches the maximum peak, which is also called the full well capacity. After exceeding the capacity of the full well, all the incoming image information will be lost, resulting in distortion of the image and a sharp drop in resolution; If the exposure time is too short, the illuminance of the CCD image surface is too low, and the dark areas and edges in the captured image will be less effective. Both will decrease the image resolution and contrast.

2.5 The Image Shift

The image shift of an aerial camera is mainly generated at the moment of exposure because the aircraft movement drives the camera movement. There are six degrees of freedom in the movement of aircraft in the air, which are likely to cause image shift in aerial images.

Image shift will cause the shape and tone of the target in the image to change. The degree of influence is related to the shooting scale. The impact on the large-scale image is relatively small, and the impact on the small-scale image is relatively large. Changes in appearance and color tone will eventually cause the image to become blurred and the sharpness to decrease.

By analyzing the above factors that affect the imaging quality of aerial reconnaissance images, the resulting degradation will be classified into the following three categories: noisy image noise, poor image contrast, and poor image clarity.

3 Common Image Quality Evaluation Methods

At present, popular image quality evaluation methods include subjective evaluation method and objective evaluation method. The objective evaluation method [2] includes full reference and no reference methods according to whether it has original image reference or not. The common full reference quality assessment method includes four methods [3]: mean square error (MSE), peak signal-to-noise ratio (PSNR), edge preserving index (EPI) and structure similarity (SSIM); And no reference method includes three methods: standard deviation assessment, information entropy assessment and average gradient assessment.

3.1 The Mean Square Error (MSE) Method

The mean square error (MSE) method is a full reference evaluation method and the most classical algorithm based on the gray-scale error of image pixels. MSE is defined as [4]:

$$MSE = \frac{1}{MN}\sum_{i=1}^{M}\left\{\sum_{j=1}^{N}[X(i,j) - Y(i,j)]^2\right\} \tag{1}$$

M and N represent the total number of image pixels in the direction of horizontal and vertical axis, the product of which represents the size of the image, and X (i, j) and Y (i, j) respectively represent the value of the original image and the image to be tested on the same pixel point(i, j). By comparing the gray difference of each pixel of the two images before and after processing, we can judge the similarity of the two images. Generally speaking, the smaller the mean square error result, the better the imaging quality.

3.2 The Method of Peak Signal to Noise Ratio

The method of peak signal to noise ratio is the result of further optimization of MSE method, which is defined as:

$$PSNR = 10 * \log_{10}\frac{L^2}{MSE} \tag{2}$$

Where L is the range of gray value in the image. For general 8-bit gray image, l is 255. PSNR is usually used to reflect the distortion degree of the whole image. Generally speaking, the larger the PSNR, the better the image quality.

3.3 Edge Preserving Index (EPI)

Edge preserving index (EPI) is an index to compare and calculate the ability of horizontal and vertical edge retention before and after image processing:

$$EPI = \frac{\sum_{i=1}^{m}\left|G_{R1} - G_{R2}\right|_{滤波后}}{\sum_{i=1}^{m}\left|G_{R1} - G_{R2}\right|_{滤波前}} \tag{3}$$

M is the total number of pixels in the image, and G_{R1}, G_{R2} are the gray values of two adjacent pixels, left and right or up and down. In general, the closer the EPI value of the final calculation to 1, the stronger the edge retention ability of the algorithm, that is, the better the image quality.

3.4 Structure Similarity (SSIM)

Structure similarity (SSIM) method compares the three modules of image structure, brightness and contrast, and then gets the final evaluation conclusion by weighting. Structural similarity evaluation is often used to evaluate the comprehensive quality of images. The calculation method is as follows [5]:

$$SSIM(X,Y) = l(X,Y) \cdot c(X,Y) \cdot s(X,Y) = \frac{(2\mu_X\mu_Y + C_1)(2\sigma_{XY} + C_2)}{(\mu_X^2 + \mu_Y^2 + C_1)(\sigma_X^2 + \sigma_Y^2 + C_2)} \tag{4}$$

Here,$l(X,Y) = \frac{2\mu_X\mu_Y + C_1}{\mu_X^2 + \mu_Y^2 + C_1}$ represent the brightness comparison function; $c(X,Y) = \frac{2\sigma_X\sigma_Y + C_2}{\sigma_X^2 + \sigma_Y^2 + C_2}$ represent the contrast comparison function; $s(X,Y) = \frac{\sigma_{XY} + C_3}{\sigma_X\sigma_Y + C_3}$ represent the structure similarity comparison function.

The original image and the image to be tested are respectively represented by X and Y. μ is the image mean, σ is the image standard deviation, and the covariance of X and Y is $\sigma_X\sigma_Y.l$, c and s are three parameters used to adjust brightness, contrast and structure distortion. C1, C2 and C3 are quoted to avoid denominator zero in the formula. Generally, the closer the result is to 1, the better the image quality is: the closer the result is to 0, the worse the image quality is.

3.5 Standard Deviation Evaluation Method

Standard deviation evaluation method is a non-reference image quality evaluation method based on image histogram features. After making the image to histogram, the gray mean value and gray variance of the image can be obtained by calculating the first-order origin moment E_1 and the second-order center distance M_1, and then the standard deviation M_2 can be obtained by operation.

$$E_1 = \sum_{j=1}^{L} r_j * p(r_j) \tag{5}$$

$$M_1 = \sum_{j=1}^{L} (r_j - E)^2 * p(r_j) \tag{6}$$

$$M_2 = \sqrt{M_1} \tag{7}$$

L is the number of gray levels, $p(r_j)$ is the size of the j gray value, represents the probability of the j gray value, and E is the gray mean value. The mean value indicates whether the image is bright or dark, which can only evaluate the overall brightness of the image, so no in-depth experiments are conducted; while the standard deviation reflects the overall contrast of the image.

3.6 Information Entropy Evaluation

Information entropy evaluation is another common method of histogram statistical characteristics. Entropy is an abstract concept. The formula is defined as:

$$H_r = -\sum_{j=1}^{L} p(r_j) * \log_2 p(r_j) \tag{8}$$

The larger the entropy is, the more information the image contains. This method is used to measure the amount of information, also known as information entropy.

3.7 The Average Gradient Method

The average gradient reflects the contrast of small details and the characteristics of texture changes in the image. It is a common non-reference evaluation method, which is defined as [6]:

$$\nabla g = \sum_{i=0}^{M-1} \sum_{j=0}^{N-1} \frac{\sqrt{(\Delta I_X^2 + \Delta I_Y^2)/2}}{MN} \tag{9}$$

Where ∇g is the average gradient of the image, m and N are the total number of image pixels in the horizontal and vertical directions respectively, and ΔI_X^2 and ΔI_Y^2 are the square of the first-order partial derivative of image I in the X and Y directions respectively.

4 The Experiment Comparison of the Evaluation of the Image Quality

In order to compare these seven methods, we choose the image which has no noise, high definition and moderate contrast. According to the characteristics of human vision, the original image is divided into three categories, including 18 contrast experimental images: six images with different noise levels ranging from basically unrecognizable to seriously affect the quality of the image, six images with different clarity levels ranging from very clear to seriously blurred, and six images with different contrast levels ranging from relatively lowest to the highest visual perception contrast. Seven methods were used to evaluate the quality of three groups of images.

The conclusion can be draw by experiments as follow:

(1) Observe the trend and rate of change of the broken line, evaluate the noise separately, the mean square error method and the average gradient method have the best evaluation effect on the noise image; evaluate the clarity separately, the edge preserving index (EPI) method has the best evaluation effect; evaluate the contrast separately, the image standard deviation method has the most obvious evaluation effect.

(2) For an aerial image used for information interpretation, it's better to get smaller noise, and better definition. But not the greater the contrast, the better the image is. The contrast should be moderate. Therefore, the experimental results of the method of comprehensive evaluation of experimental image quality are usually monotonous with the increase of noise and blur. However, the contrast curve should be from low to moderate to high to get arc curve.
There are three methods, which are mean square error (MSE), peak signal-to-noise ratio (PSNR) and structural similarity (SSIM), which are generally consistent with the trend of noise, sharpness and contrast of experimental images. That is to say, these three methods can comprehensively evaluate the image quality, and these three methods are all full reference evaluation methods.
In terms of the change rate of broken line, SSIM method is the best. Combined with its principle, the structure similarity evaluation method considers three aspects of image structure, brightness and contrast, and gives a certain weighted value for calculation to get the final accurate evaluation results. The structure similarity method is also the most commonly used comprehensive image evaluation method at present.
(3) Image entropy method is not suitable for image quality evaluation, but it can be used for aerial image classification.

5 The Next Step

SSIM is a full reference evaluation method, which needs high-quality clear reference images for comparison, so the application of this method in aerial reconnaissance image evaluation has great limitations. The next step is to design a non-reference comprehensive evaluation method for the aerial reconnaissance image based on the characteristics of the aerial reconnaissance image and the theoretical experience of the structural similarity evaluation method. However, the full reference evaluation method still has the following uses:

(1) The detection ability of the two cameras and the selection of parameters used to adjust the camera to cope with various factors are identified. First, we can use advanced methods to get clear images of the photographing area, then we can change the camera and adjust the camera parameters to get different quality images. Finally, we can use the full reference method to compare with it to get accurate evaluation results and select the best ones;
(2) When choosing an image enhancement method that is more suitable for different situations, we can first use advanced means to obtain clear images of the photographing area, then use the method to be tested to enhance the captured image, and finally use the full reference method to compare the acquired image with the processed image and the clear image, so as to select the best one;
(3) After the design of a new evaluation method, we can use full reference evaluation method which the theory has been very mature to get accurate evaluation conclusions. Then use the new evaluation method to evaluate the image, compare if the two conclusions are consistent, so as to determine whether the new design method is accurate and feasible.

References

1. Zhao, Y.: Principles of Aerial Photography. National Defense Industry Press, Beijing (2017)
2. Miao, Y.: Research on Image Quality Evaluation Algorithm. Kunming University of Technology (2016)
3. Hong, X.: Research on Algorithm of Full Reference Image Quality Evaluation. Xi'an University of Electronic Science and technology (2014)
4. Zhang, Y.: Image Processing. Tsinghua University Press, Beijing (2014)
5. Yang, C., Chen, G., Xie, S.: Research on image quality evaluation method based on gradient information. Acta Electronica Sinica **35**(7), 1313–1317 (2007)
6. Zhu, L., Li, Q., Zhang, T., Wang, G.: Metric of image quality based on structural similarity. Opto-Electron. Eng. **34**(11), 108–113 (2007)

The Applications of AI: The "Shandong Model" of E-commerce Poverty Alleviation Under Technology Enabling Direction

Xinmei Wu[1], Zhiping Zhang[2], and Guozhen Song[3](✉)

[1] School of Accounting, Shandong University of Finance and Economics, Jinan 250014, China
[2] School of Computer Science and Technology, Shandong University of Finance and Economics, Jinan 250014, China
[3] School of Statistics, Shandong University of Finance and Economics, Jinan 250014, China
llpwgh@163.com

Abstract. Based on the three typical patterns of e-commerce poverty alleviation found in the field research in many places in Shandong province—"peasant household + e-commerce platform", "e-commerce park + enterprise + peasant household", "peasant household + cooperative + e-commerce platform", this paper conducts an in-depth analysis of the three models from the two aspects of mode operation mechanism and enabling mechanism, and compares the development stage, applicable conditions, enabling characteristics and poverty alleviation effect among the models, summarizes the advantages of various models, and is committed to providing a certain reference for regions that have not yet been lifted out of poverty.

Keywords: E-commerce poverty alleviation model · Technology enabling · Case analysis

1 Introduction

E-commerce poverty alleviation is a new research topic in recent years. From the existing literature, the research on e-commerce poverty alleviation focuses on two aspects; one is to discuss the definition and mechanism of the e-commerce poverty alleviation concept, the other is to discuss the different e-commerce poverty alleviation models and compare them. However, the comparison between different models and the relevant applicable conditions for specific provincial areas are still rare, and there is a lack of discussion on the mechanism of e-commerce poverty alleviation from the perspective of technical empowerment.

With the trend of diversification of e-commerce poverty alleviation model in Shandong province in recent years, the ways of poverty alleviation in poor regions have increased. However, the problems have also followed: due to the lack of relevant research above, what kind of e-commerce poverty alleviation model should be chosen in poor regions to achieve the best poverty alleviation effect? Choosing not to apply local channels of poverty alleviation may lead to a series of follow-up problems, such

Q. Liu et al. (Eds.): CENet 2020, AISC 1274, pp. 176–182, 2021.
https://doi.org/10.1007/978-981-15-8462-6_20

as the inefficient implementation of e-commerce poverty alleviation, the difficulty of achieving scale, effectiveness, sustainable poverty eradication effect, and so on. It can be seen that in order to promote the smooth implementation of e-commerce poverty alleviation policy and the effective use of poverty alleviation resources, it is necessary to carry out in-depth and systematic research on the typical model of e-commerce poverty alleviation in Shandong province from the perspective of technology enabling at the present stage. Taking this necessity as the starting point, the aim of this paper is to make an in-depth study on three typical e-commerce poverty alleviation models in Shandong province from the two angles of mode operation mechanism and technology enabling. It compares them with each other, finally obtains its advantages, differences and applicable conditions, and devotes itself to providing some reference for poor areas that have not yet carried out e-commerce poverty alleviation or carried out poor results.

2 E-commerce Poverty Alleviation Model

2.1 Summary of Data Collection

This paper is discussed on the basis of the research results in Mengyin County, Linyi City, Shandong Province, Cao County, Heze City, and Xinxian County, Liaocheng City, 2018–2019. In order to verify the typical significance of e-commerce poverty alleviation model, promote value and explore the key factors that e-commerce poverty alleviation has achieved good results, 50 paper questionnaires were distributed and 150 valid questionnaires were collected in the field survey for each of the three poor villages. According to the data analysis, more than 70% of the rural households in the three places maintained a high level of access to e-commerce poverty alleviation, among which the highest income acquisition of e-commerce poverty alleviation, and more than half of the research subjects in the three places have a certain understanding of the concept of “e-commerce poverty alleviation”, which is enough to show that the model of the three places has certain promotion value.

2.2 Three Typical Models of “Shandong Model” for E-commerce Poverty Alleviation

“Farmers + E-commerce Platform” Model. “Farmers + e-commerce platform” in the model of e-commerce poverty alleviation, “people” is the direct enabling object of farmers. The model is a successful case of e-commerce creation by e-commerce leaders, so that most ordinary farmers see the possibility of increasing income, stimulate their enthusiasm, let them gradually understand e-commerce, master the relevant technology, and then trigger a certain degree of fission effect. With the support of government funds and policies and the gradual improvement of the platform, the model can realize further scale effect and agglomeration effect (Fig. 1).

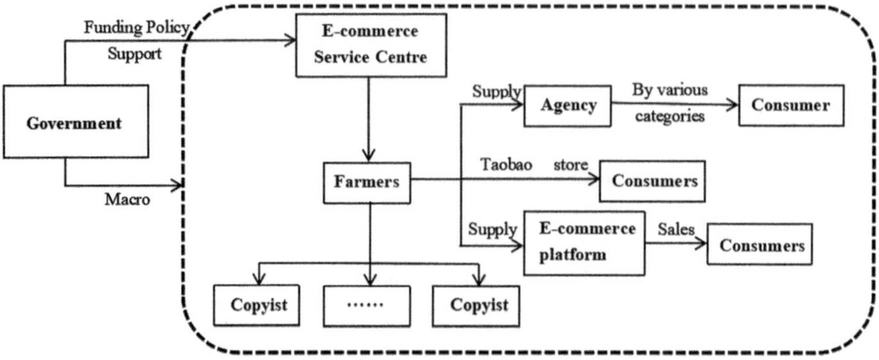

Fig. 1. "E-commerce Park + Enterprise + Farmers" Model.

"E-commerce Park + Enterprise + Farmers" Model. Under the model of "E-commerce Par + enterprises + farmers", "industry" is the direct enabling object of e-commerce. Through the e-commerce platform, the agricultural special industry can connect with the large market, and the expanded market demand forces the supply side to realize self-improvement, improve the quality of products, and realize the energy increase of agricultural products. Industry, as the enabling object of e-commerce, also plays the role of enabling transmission, directly benefiting the poor households in the upper reaches of the industrial chain, so that the poor households can realize the dual increase of income and technology, and take the road of getting rid of poverty and becoming rich. The spillover effect of industrial chain can also drive the development of county economy, attract more enterprises and poor households to participate, and further expand the scale of industry and the effect of poverty alleviation (Fig. 2).

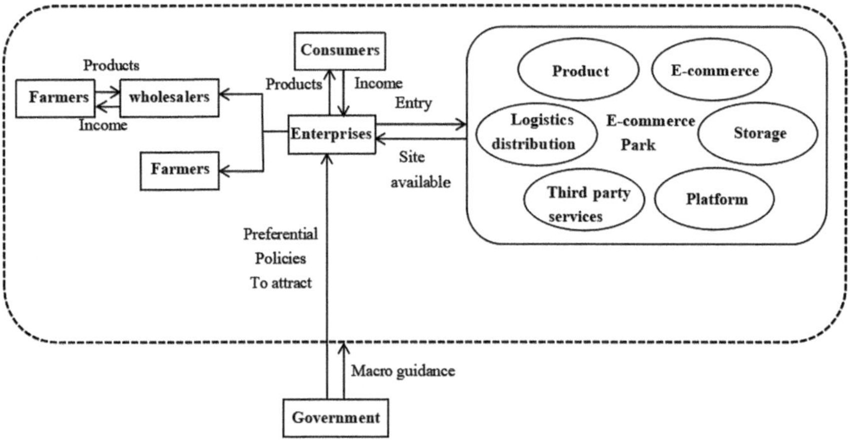

Fig. 2. "Farmers + Cooperative + E-commerce Platform" Model

"Farmers + Cooperatives + E-commerce Platform" Model. The "organization" is direct enabling object for e-commerce in the model of "farmers + cooperatives and e-commerce platform". The e-commerce platform realizes the promotion of characteristic agricultural products in the market and the further creation of characteristic brands. Cooperatives play an enabling role in this model, through the provision of preferential means of production to farmers, technical training and other ways to enable farmers, so that farmers to achieve economic and technological dual energy, the real realization of "people to fish" type of poverty alleviation (Fig. 3).

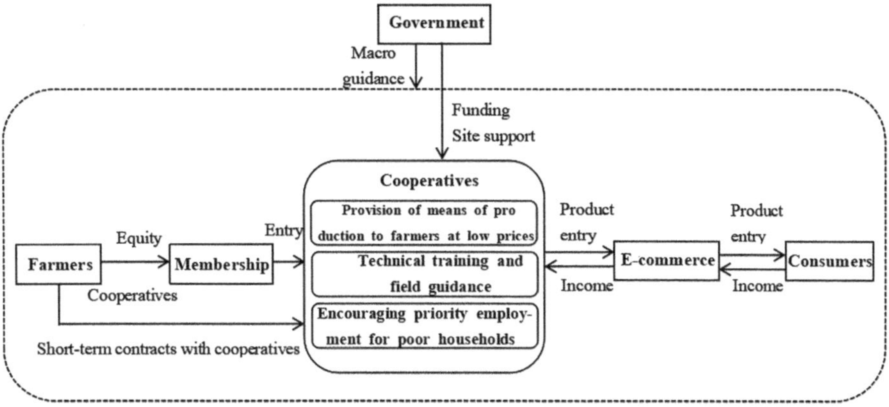

Fig. 3. "Enterprises + Farmers + E-commerce Park" Model

2.3 Model Comparison and Advantage Analysis

Pattern Contrast. Based on the above analysis of the operation mechanism and enabling mechanism of various e-commerce poverty alleviation models, we can compare the above research models through four aspects: development stage, geographical application conditions, enabling characteristics and poverty alleviation effect, in order to explore the differences and unique advantages between the models (Table 1).

Table 1. Comparison of the three models

	Development phase	Font size and style	Enabling characteristics	Poverty alleviation
"Farmers + E-commerce Platform" Model	Family small workshop | Taobao Town/Street | Taobao Industrial Park	1. Entrepreneurial farmers have a certain understanding of the Internet Taobao 2. Product positioning accuracy	1. Enabling farmers 2. To the folk bottom-up e-commerce poverty alleviation practice, supplemented by the top-down guidance of the government	1. E-commerce ecology gradually 2. Entrepreneurship and economic poverty alleviation 3. Traditional enterprise transformation
"Farmers + Cooperative + E-commerce Platform" Model	Government-led | Scale-up | Industrial chain formation	1. Local characteristic agriculture relies on industry 2. Government's strong financial and policy support to e-commerce pro-poor cooperatives	1. Organizational empowerment 2. Led by top-down government	1. Progressive formation of the e-commerce industry chain 2. Technology and economic poverty alleviation by farmers 3. Increased number of cooperatives with significant benefits
"Enterprises + Farmers + E-commerce Park" Model	Government-led Building e-commerce parks Industrial integration	1. Local characteristic agriculture relies on industry 2. Strong Government support and guidance	1. Industrial empowerment 2. Led by top-down government	1. Integration of resources and expansion of production scale 2. To realize industrial integration, promote the development of county economy and realize spillover effect

Model Advantage. According to the comparison between the above models, it can be seen that based on different organizational forms, mode operation mechanism, government support form and development stage planning, the enabling effect, applicable region and poverty alleviation effect of e-commerce poverty alleviation model are different. Therefore, the advantages of various models can be briefly analyzed according to the comparative results below.

"Farmers + e-commerce platform" model is the only one of the three modes of direct empowerment of farmers, poverty alleviation by the people. In contrast, the model is easier to mobilize the enthusiasm of farmers and improve the concept of backward farmers, reducing the possibility of returning to poverty. Thus, based on the good enabling effect, the model can induce the imitation of other farmers through the rural community network, and promote the scale effect and agglomeration effect.

The "e-commerce park + enterprises + farmers" model takes industry as a direct enabling object. Based on the top-down support of the local government to attract leading enterprises to enter, and then make the market demand for agricultural products expand, forcing the industrial chain to achieve optimization, to achieve the quality of supply-side products. This model can drive local industrial agglomeration and the improvement of industrial chain, and its spillover effect can better drive the development of county economy, attract more enterprises to enter and join farmers, and form a virtuous circle.

The model of "farmer + cooperative + e-commerce platform" takes organization as the direct enabling object. Through the multi-cooperation way and the comprehensive empowerment transmission to the farmer, it is helpful to achieve the effect of getting rid of poverty and getting rich more centrally and efficiently within the organization. Top-down government guidance and adequate funding also ensured the successful operation of the model.

3 Conclusions

3.1 E-commerce Development Under Technology Enabling Orientation

From the analysis of the above three models, it can be seen that technology empowerment is one of the essential characteristics of e-commerce. E-commerce has optimized the structure of the industrial chain by enabling technology to the main body of different organizations, realizing its own interests connection with other main bodies, and relying on the local superior industries. Therefore, e-commerce has realized the industrialization and large-scale development of different effects in different regions, in which technology empowerment plays a key role.

3.2 Effect of E-commerce on Poverty Alleviation Under Technology Enabling Direction

E-commerce poverty alleviation refers to the means of e-commerce to achieve the effect of poverty alleviation and enrichment. Through the above research and analysis, it is concluded that technology empowerment is the essential reason that e-commerce, an emerging means, can effectively act on precision poverty alleviation. On the one hand, e-commerce platform through technology to promote agricultural products uplink, so that poor households to achieve income and employment, complete economic poverty relief. On the other hand, with the support of government policy or the practice of poor households, the concept of Internet is integrated into the countryside and the spirit is lifted out of poverty.

3.3 Main Factors Influencing the Effectiveness of E-Commerce in Poverty Alleviation

Through the comparison and research data analysis of different models of e-commerce poverty alleviation above, it is found that the achievement of good effect of e-commerce poverty alleviation is closely related to the local industrial base, e-commerce infrastructure and the possession of the main body of e-commerce business. The three also play a key role in establishing a long-term mechanism for e-commerce to help the poor and prevent the return to poverty. Local transformation of traditional superior industries through e-commerce means, the "new kinetic energy" into traditional industries, promote the transformation of traditional industries, enhance industrial competitiveness, force the improvement of local e-commerce infrastructure and the opening of related industrial chains, achieve the dual enabling of e-commerce operators in material and spiritual aspects, in the process of large-scale development of continuous spillover effects, promote the development of local county economy, forming a virtuous circle.

References

1. Liu, J.Y.: Adhere to precision poverty alleviation precision poverty relief, constantly win a new victory in the work of poverty relief, Heze (2017)
2. Wang, X.D., Liang, C.X.: A New Countryside and Electronic Commerce, pp. 139–151. China Agricultural Science and Technology Press (2014)
3. Tang, J.L.: Comparative study on rural electronic commerce model. SME Manag. Technol. **2**, 271–272 (2013)
4. Wang, X.D., Zhang, C.M.: The enlightenment of the "Sand collection model", China's new thinking of poverty alleviation in rural areas in the Internet age. Inf. Constr. **1**, 6–9 (2011)
5. Lin, G.Y.: A study on the action mechanism of rural e-commerce in poverty alleviation and its promotion mechanism. Graduate School of Chinese Academy of Social Sciences (2016)
6. Menhao: A study on market occurrence mechanism of Taobao village in China—focusing on "Cao Cun". Jilin University (2018)
7. Mazepo: Farmer endowment, regional environment and willingness of e-commerce to participate in poverty alleviation questionnaire survey based on 630 farmers in frontier ethnic areas. China's Circ. Econ. **31**(5), 47–54 (2017)

Emotional Research on Mobile Phone E-Commerce Reviews on LSTM Model

Chunmei Zhang[1(✉)], Mingqing Zhao[2], and Gang Dong[1]

[1] Big Data School, Qingdao Huanghai University, Qingdao 266000, China
158434851@qq.com
[2] College of Mathematics and System Science,
Shandong University of Science and Technology, Qingdao 266000, China

Abstract. Emotional analysis has been omnipresent. Grabbing adequate information from online shopping evaluation and analyzing it can provide a basis to choose for businesses or consumers. Taking "Jingdong Mall" mobile e-commerce review as an example, the authors of this paper use web crawler technology to obtain sufficient data of 12 products under 7 mobile phone brands in the sales ranking. The authors then use Python's Jieba library to segment the reviews and gensim module to train word2vec word vector model and classify the mobile e-commerce reviews by constructing LSTM neural network model. It is concluded that the consumers' views that there exists some advantages and disadvantages of "Apple" and "Huawei" mobile phones.

Keywords: Financial fraud · Listed company · Financial indicators · News sentiment · Logistic model

1 Introduction

With the rapid development of the Internet economy, convenient and fast online shopping has undoubtedly become the people's first choice to go shopping. However, online shopping cannot directly judge the quality of goods through direct contact, so an online evaluation of goods in major e-commerce platforms is particularly critical. It can not only enable consumers to understand the satisfaction of the goods they have purchased, but also enable merchants to grasp the real consumer experience in time, and then provide consumers with more satisfactory quality and service. As the most important smart terminal, the scale of mobile netizens has reached 847 million. Facing such a huge user group, capturing the real consumption experience of consumers, and choosing the products they like when facing the trend of serious homogenization of mobile phones. Therefore, how to make accurate positioning and analysis of products by commodity reviews is worth studying.

Text sentiment analysis is also called opinion grabbing. Combined with product evaluation, it can analyze and summarize consumer comments on a particular product, and then find out the emotional tendency of consumers, which is very helpful to improve the product quality, service attitude, and business scale of businesses. Therefore, the research on emotion analysis technology helps promote the vigorous development of e-commerce.

Q. Liu et al. (Eds.): CENet 2020, AISC 1274, pp. 183–192, 2021.
https://doi.org/10.1007/978-981-15-8462-6_21

In the research of emotion analysis and LSTM model, Ma et al. [2] proposed that in order to display emotional dependence, an overlay layer comprising a target layer and sentence layer was added to the LSTM model realize the integration of explicit information and implicit information. Meanwhile, an extension of LSTM model called Sentic LSTM was proposed, which is a mixture of an LSTM model and a recursive additive network to simulate a new network, and the experimental results show that the performance of solving the target context and emotional polarity has been improved. Sherstinsky et al. [3] provided a detailed description of all equations related to the LSTM system and their constituent entities. Although unconventional, the choice of representation and method for representing the LSTM system was easy to understand.

Because there are significant differences between Chinese text and English text in terms of grammatical structure and semantic expression, Liu et al. [4] proposed to use SVN and RNN to construct a classifier to classify the text emotion and evaluate the classification results to get the emotional keywords in the text. However, the text's emotional features include emotional keywords, negative words, and turning words and determine the critical features of the text emotion. Liang et al. [5] put forward the emotion analysis based on polarity transfer and LSTM, and introduced the emotion polarity transfer model according to the relevance between contexts, which has a good effect on the text emotion analysis.

2 Model Building

2.1 Word2vec Model

The basic function of word2vec model is to express every word in natural language as a word vector with unified meaning, because only by transforming natural language into vector form, can we build relevant algorithms on it. In short, that is by studying the text, and then using the word vector to represent the meaning information of the word. In English text, you only need to divide sentences into words according to blank spaces, while in Chinese text, you can't use a single character to map. Therefore, Chinese sentences need to be divided into words according to semantics before they can be used for mapping.

2.2 LSTM Model

LSTM is the abbreviation of Long Short Term Memory, which can be translated into short and long term memory artificial neural network. It is a variant model of RNN neural network model, which can effectively solve the problem of gradient disappearance or gradient explosion in RNN model. LSTM model optimizes the hidden layer of RNN model and adds a core element of memory cell to control the memory and forgetting of information. Compared with RNN model, LSTM model has better performance in longer sequence.

3 Empirical Analysis

3.1 Sample Selection

There is the function of commodity review in all major e-commerce platforms, but only the "Jingdong Mall" e-commerce platform gives consumers' star value scores for commodities at the same time in the commodity review interface. In this study, comments are bound with star score, which is taken as the score of comment statement. Full score of star value is 5 stars. When the star value score is more than 3 stars, we can treat comments as positive comments; when the star value score is less than 3 stars, we can treat comments as negative comments. Since the comments with a star value score of three stars are ambiguous and cannot be classified, the comments with a star value score of three stars are excluded from the sample. Therefore, according to a mobile phone sales ranking released by the "Jingdong Mall" e-commerce platform at the beginning of 2020, what we select in this study is not only the comments of official mobile phone stores, but also the ones of some high-level private mobile phone stores. At the same time, we need to filter samples and eliminate the relatively small number of mobile phone models on the leaderboard. Finally, 12 products under 7 popular mobile phone brands were obtained, and the specific product models are shown in Table 1.

Table 1. Brands and models selected samples.

Brands	Models
Apple	iPhone 11, iPhone X, iPhone XR
Huawei	Huawei Mate 30, Huawei Mate 20, Huawei P30
Glory	Glory 20S, Glory 9X
Mi	Mi 9
Redmi	Redmi K20
OPPO	OPPO Reno
VIVO	VIVO IQOO

According to Table 1, 12 mobile phones come from different brands. In terms of price, Apple and Huawei have six high-end mobile phones. The six mobile phones with high cost performance are Glory, Mi, Redmi, OPPO and VIVO, accounting for 50% of the total. The price of the mobile phone model selected in this sample basically covers all the price fields, that is to say, the expression of consumer language in all the price fields of mobile phone consumption is covered. From this point of view, this sample can make the model have better generalization.

3.2 Data Preprocessing

Web crawler is a program or script that automatically simulates human operation to obtain information according to certain rules. In this study, we use the strategy of combining selenium library and requests library to carry out anti-crawler. At the same time, we slice the comment data and get 22000 comments.

Firstly, Data Cleaning

The crawled data is cleaned (including removing URL, punctuation, emoticons, pictures, etc.) and the duplicates are eliminated. Finally, there are 20057 comments left. According to the star score of more than 3 stars, the positive tag is 1, and the negative tag is 0. For the convenience of statistical summary, 20000 comments are selected here, including 13479 comments with the tag of 1 and 6521 comments with the tag of 0. At last, 5000 comments are randomly selected from 20000 comments as independent samples to evaluate the model. The corpus set is shown in Table 2.

Table 2. The example of crawling corpus set.

Emotion category	Sample text	Label
Positive	Large screen, full screen cool; powerful configuration, function and photography, and to know its advantages by using	1
	The affordable price and high cost performance. The same configuration is nearly half the price cheaper than others	1
Negative	Fingerprint effect is a little stuck, delayed	0
	The physical examination was poor, and the mobile phone was average. Not working twice is quite disgusting	0

Secondly, Word Segmentation

In Chinese text, there is no fixed meaning before a single Chinese character is composed of words, so it is impossible to train word2vec model with a single Chinese character, and it is necessary to divide comment statements into words. Here, we use Python's jieba library to divide sentences into words in precise patterns. Because there are many special terms in the mobile phone industry, as well as mobile phone brand and model name, in order to more accurately segment words, we have established a dictionary about some special terms in the mobile phone industry, as shown in Table 3.

Table 3. Summary of special terms, brands and models.

Terms, brands and models for mobile phone industry					
Huawei	Huawei	Mi	Xiaomi	Apple	Apple
Glory	Honor	Oppo	Vivo	Redmi	Redmi
Mate30	Mate20	P30	20S	9X	K20
Reno	IQOO	Appearance	Pixel	Endurance	Charging
Sound effect	Performance	Sound quality	Screen	Fluent	Heat dissipation
Photo taking	System	Stuck	Signal	Delicate	Heating
Fingerprint unlocking	Hand feeling	Screen protector	Three photos	Linear motor	Cost performance
AMOLED	OLED	Feeling	Bangs	Standby	Configure
Function	Dark screen	Camera	Scratch	Battery	Double cassette
Curved screen	Crash	Processor	Rollover	Unicorn	Dragon

The positive comment sample segmentation and the negative ones are separated, and the segmentation results are visualized with a word cloud diagram. In the word cloud diagram, the keywords with high frequency of text data words will be displayed in a more prominent way.

Figure 1 is a cloud map of positive emotions in mobile phone reviews. Because there are too many words, we only select words with frequency more than 50. From Fig. 7, we can directly see the positive comment words such as large memory, satisfied color, clear camera, good sound effect, fingerprint unlocking, narrow frame, etc. From these comments, we can distinguish that consumers attach great importance to these functions. If mobile phone manufacturers devote to these aspects of mobile phones, consumers will be generous to give praise.

Fig. 1. Words cloud map of positive mood.

Fig. 2. Words cloud map of negative mood.

In the same way, Fig. 2 is a cloud of words for mobile phone comments on negative emotions. Because there are too many words, we only select words with frequency greater than 50. In Fig. 8, it can be seen that the most intolerable problems when consumers buy mobile phones: stuck, crash, system card, serious heat, blurred photo, poor signal, cut-off and so on.

3.3 Model Building and Training

Text Vector Processing Based on Word2vec Model. According to the above processing, the comment sentences have been converted into words, but words can not directly participate in the training of the model, so it is necessary to train the comment emotion dictionary through word2vec model, and convert the comment sentences into vectors to express. The specific parameter settings based on word2vec model are shown in Table 4.

Table 4. Word2Vec parameter setting.

Parameter	Parameter description	Setting up
cpu_count	Computing core	multiprocessing.cpu_count()
vocab_dim	Eigenvector dimension	200
n_exposures	Minimum threshold for word occurrence	10
n_iterations	Number of iterations	1
window_size	Context window length	5
n_epoch	Number of training rounds	100
maxlen	Word vector matrix length	100
cbow	Use cbow model or not (0 is used.)	0

After inputting the segmented text into the trained model, the model will convert the segmented text into sentence vector according to the word vector, and then the algorithm of LSTM model can be constructed on this basis.

Building LSTM Model. In this paper, the LSTM neural network model built by keras framework is composed of three basic structures, such as input layer, hidden layer and output layer. The parameters of each layer of the model are shown in Table 5.

Table 5. Summary of parameters of each layer of the model.

Layer (type)	Output shape	Param #
embedding_1 (Embedding)	(None, 100, 200)	1685400
lstm_1 (LSTM)	(None, 100)	120400
dropout_1 (Dropout)	(None, 100)	0
dense_1 (Dense)	(None, 1)	101
Total params: 1,805,901 Trainable params: 1,805,901 Non-trainable params: 0		

Training Results of LSTM Neural Network Model. After repeated experiments, when the model loss function is binary_crossentropy, sigmoid for activation function, Adam for optimizer, the model performs best in the verification set, with the highest accuracy of 0.92. The loss function and accuracy curve of training set and verification set with the number of training rounds are as follows.

It can be seen from Fig. 3 that the loss value of training set and verification set decreases with the increase of training rounds, and tends to be stable from 25 rounds.

It can be seen from Fig. 4 that with the increase of training rounds, the accuracy of the training set and the verification set increases continuously, and the accuracy of the training set is almost the same as that of the verification set, indicating that there is non-fitting phenomenon.

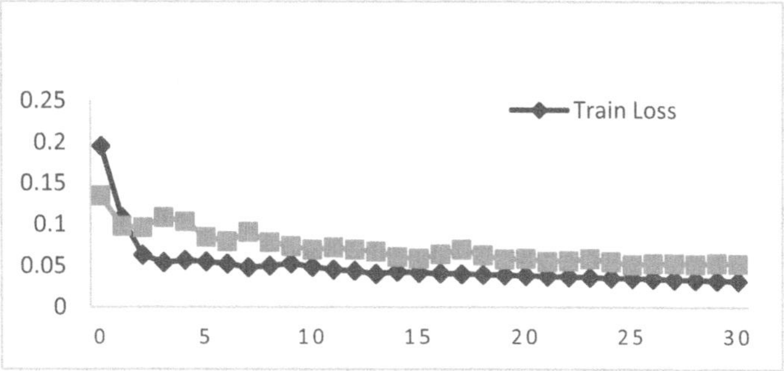

Fig. 3. Loss of training set and verification set with training rounds.

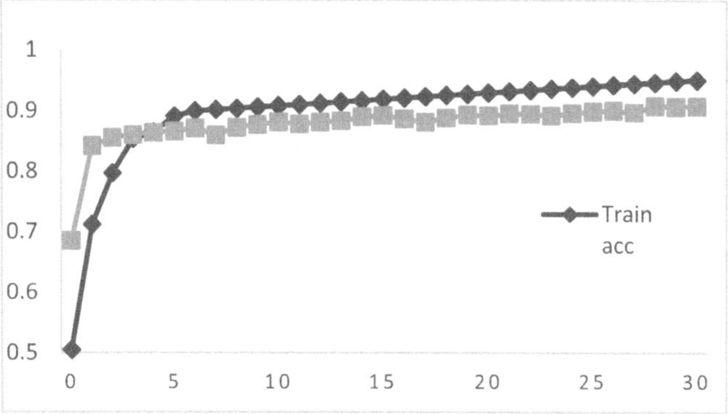

Fig. 4. The accuracy of training set and verification set varies with the number of training rounds.

3.4 Model Test and Emotional Analysis of Mobile E-Commerce Comments

LSTM Model Inspection. Put the independent samples processed before into the trained LSTM neural network model for testing, and the examples are shown in Table 6.

Table 6. The example of independent sample test.

Sample text	Model predicts	Real emotions
Good appearance, fresh and beautiful photos with the front.Aha	1	1
Easy to use. I can't love it any more	1	1
Standby time is too short, and the speed is very slow	0	0
Today, just get the mobile phone and watch the video isn't fluent. Poor quality	0	0

The trained LSTM neural network model performs well in the test of independent samples, and the recognition accuracy of 5000 independent sample review data is 4583, with an accuracy rate of 91.66%. It shows that the model achieves good training effect and can be applied to solve practical problems.

Emotional Analysis of Mobile E-commerce Comments. According to the prediction results of the model, the independent samples are divided into positive samples and negative samples according to the emotion contained in the sample sentences. After the processing, 1355 positive samples and 3745 negative samples are obtained.
Here, we use the “Huawei” and “Apple” mobile phone brands as an example to analyze the model. In order to show the data more intuitively, we choose the first 20 words cloud which is closest to their word vector to do the data visualization processing. The words are as follows:

In Fig. 5, we can see clearly that the system smoothness font of is the largest in the evaluation of Apple mobile phone, indicating that the frequency of system smoothness in the review words is quite high. At the same time, double-sided glass and piano baking varnish appear, and the adjective of external shape shows that the shape of Apple mobile phone is also popular. In addition, words like face ID, dual card and dual standby and so on,also express the consumers’ praise for these functions. In Fig. 6, we can clearly see the words like high color, good feel in the cloud picture of Huawei mobile positive comment words, which shows that the appearance of Huawei mobile phones is more popular with consumers.

Fig. 5. “Apple” positive comments.

Fig. 6. “Huawei” positive comments.

In Fig. 7, we can see that there are defects, carton, black screen, poor hand feel and fever in the cloud picture of Apple Mobile negative comment words. In Fig. 8, we can see clearly that there are scratches, operation cards, poor network and other problems in Huawei mobile negative comment cloud. In this regard, mobile phone manufacturers can update or upgrade mobile phones in combination with the problems shown in the cloud picture. At the same time, consumers can be wary of these problems in the word cloud, so as to avoid buying mobile phones with the above problems.

Fig. 7. "Apple" negative comments.

Fig. 8. "Huawei" negative comments.

With the positive and negative samples, the high-frequency words in the positive and negative samples are counted, and the performance, appearance, photo, screen, sound effect, operation, standby and other seven dimensions are selected from the statistical results for evaluation. as shown in Fig. 9.

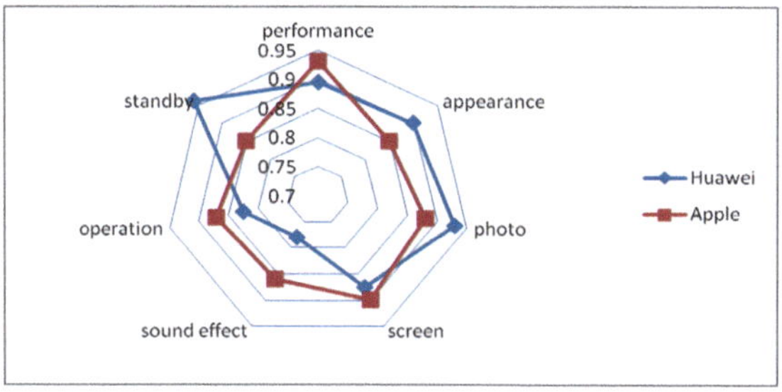

Fig. 9. Radar chart of product dimension.

It can be seen from Fig. 9 that "Huawei" and "Apple" mobile phones have their own advantages and disadvantages in the performance of these seven dimensions, "Apple" mobile phones perform better than "Huawei" in terms of performance, running speed, sound effect, screen, etc., while "Huawei" are better in terms of standby, appearance and photo taking.

For consumers, they can use radar chart to choose between Huawei and Apple mobile phones to meet their own needs. For mobile phone manufacturers, we can combine the radar chart to analyze the advantages and disadvantages of themselves and their friends in the eyes of consumers, and then optimize and adjust mobile phones.

4 Conclusion

The comment text contains a lot of emotional information which can mine the user's emotion and business value. This paper has reference significance for consumers and mobile phone manufacturers. The research results and shortcomings of this paper are reflected in the limitations of e-commerce platform to access comment data, compared with millions of sales volume, the conclusion of research response is limited. This research method is not limited to the mobile phone industry, but also applicable to other business areas with a large number of user comment data.

References

1. Derry, J., Moch, A.B., Arie, A.S.: Word2Vec model analysis for semantic similarities in English words. Procedia Comput. Sci. **157**, 160–167 (2019)
2. Ma, Y.K., Peng, H.Y., Khan, T., Cambria, E., Hussain, A.: Sentic LSTM: a hybrid network for targeted aspect-based sentiment analysis. Cogn. Comput. **10**(4), 639–650 (2018)
3. Sherstinsky, A.: Fundamentals of recurrent neural network (RNN) and long short-term memory (LSTM) network. Physica D **404**, 132306 (2020)
4. Liu, M., Zan, H.Y., Yuan, H.B.: Judgment and extraction of text sentiment key sentences based on SVM and RNN. J. Shandong Univ. (Sci. Edn.) **49**(11), 68–73 (2014)
5. Liang, J., Chai, Y.M., Yuan, H.B., Gao, M.L., Zan, H.Y.: Sentiment analysis based on polarity transfer and LSTM recursive network. J. Chin. Inf. Process. **29**(05), 152–159 (2015)

Research on Intelligent Robot System for Park Inspection

Lianqin Jia(✉), Liangliang Wang, Qing Yan, and Qin Liu

Shandong Institute of Commerce and Technology, Jinan, China
20501121@qq.com

Abstract. With the acceleration of urbanization in China, as an essential part of it, parks play an important role in improving the living environment, to promote city brand, and to enhance citizens' sense of happiness and gain. Currently, the management of the park mainly adopts the manual inspection method that has many disadvantages. This paper proposes a park inspection system based on artificial intelligence. The system has the following main advantages: using multi-source sensing technology and location algorithm, using artificial intelligence, automatic control, not affected by time, place, and other external factors, can improve the work efficiency, enhance the coverage, make the inspection work accurate and effective. It can also reduce maintenance costs and facilitate the construction of intelligent monitoring networks in public places such as parks during the epidemic prevention and control period.

Keywords: Artificial intelligence · Park inspection system · Multi-source sensing technology · Location algorithm

1 Introduction

With the acceleration of urbanization in China, parks play an essential role in improving the living environment, to promote the city brand, and to enhance citizens' sense of happiness and gain. Since opening the park, the operation, and maintenance of each subsystem (such as water supply and drainage system, power supply and distribution system, lighting system, etc.) is relatively heavy work, the traditional management mode as manual inspection, records and report has many disadvantages, such as limited coverage, high cost, reporting information not accurate enough etc. During the epidemic prevention and control period, especially the manual inspection has risks, so it is necessary to develop the park inspection system, especially based on artificial intelligence. Specifically, the system has the following main advantages: automatic control by artificial intelligence, not affected by time, place and other external factors, can not only improve the work efficiency, improve the coverage, accuracy and real-time inspection work; Maintenance costs can also be reduced since the running costs of machines are much lower than Artificial cost; It can also help build intelligent monitoring networks in public places such as parks during the COVID-19 epidemic prevention and control period.

The research of park inspection system mainly includes intelligent robot technology, multi-source sensing technology and the organic integration of the two.

Q. Liu et al. (Eds.): CENet 2020, AISC 1274, pp. 193–199, 2021.
https://doi.org/10.1007/978-981-15-8462-6_22

Abroad for those technologies research originated in the 1980s [1], in 1984, the U.S. military developed first stage ground of autonomous vehicle in the world [2], in 1987, Smith put forward the robot localization at the same time [3]. In 2017, Endres et al. proposed new SLAM, which opened up new applications of RGB-D camera [4–6] and introduced the study of SLAM into a new field of vision.

The related researches started relatively late in China. In 1999, Shandong electric power company started the first study of the inspection robot and developed the prototype in 2004 [1]. On this basis, the company developed a series of substation inspection robots, which realized the robot's intelligent inspection in the substation and promoted its application. Before this, no robot had been used in substation equipment in the world [7, 8]. In 2009, the Harbin Institute of Technology developed is suitable for the coal mine detection robot. The robot uses three joint track [9]. Caterpillar track structure is adopted for the inspection robot, which has better maneuverability and obstacle ability [1]. In 2013, Central South University developed a reconfigurable tracked mobile robot, which can adjust the center of gravity when driving, to increase the stability of the barrier [10]. Besides, under the support of the national policy, the domestic many colleges and universities research institutions are also has been a rapid development of mobile robot research, Chinese academy of sciences, Shenzhen Institute of Advanced Technique to make the installation has a depth camera care robot [1].

The intelligent robot will replace some jobs with high risk, high strength, and low efficiency in the future, especially in the manufacturing and services industries, application advantages are gradually highlighted in these industries. In February 2020, the COVID-19 outbreak has broken out. We also saw the intelligent robot's excellent performance in fighting the outbreak, such as robots catering for staff is suspected of observing the epidemic area and the isolation ward, the auxiliary medical robot that delivers laboratory tests, medicines, and temperature readings for multiple patients in Vulcan Hill Hospital. 5G care robot can be capable of patient care, consultation, and Q&A, epidemic prevention publicity, and video interaction in Shanghai no. 6 People's Hospital. Using intelligent robots effectively reduces the risk of cross-infection and improves the level of quarantine control in the ward.

With the application advantages of intelligent robot gradually prominent, it will be an inevitable trend of science and technology development for intelligent robot to enter into urban public service field. Therefore, it is necessary to study the robot system of intelligent park inspection.

2 The Research Target

This research project integrates the graphic image recognition technology of artificial intelligence with the information acquisition and transmission system of the IOT technology. By using the running and climbing robots equipped with various sensors, the route of inspection is designed through the ranging and positioning modules and the automatic obstacle avoidance to collect all kinds of information. Take pictures of the surrounding environment of the park through the HD camera, use the graphic image recognition technology to identify the qr code "smart" badge, small animals in the grass visitors, garbage and other objects in the park, and realize intelligent inspection with the help of the Internet of things alarm system. In this way, the problems of low efficiency, heavy workload, missed inspection, slow information transmission and high operation and maintenance cost of manual inspection are reduced, and It can realize the standardization, information, data and intelligence of the park inspection.

3 The Research Content

Intelligent park inspection robot system is an intelligent Internet of things system with functions such as security inspection, efficient information collection and foreign body invasion alarm, etc. It needs to complete the construction of robot hardware structure, the design of IOT communication system and the construction of software control system. The whole structure consists of six parts: background management system (hereinafter referred to as BMS), Network communication system (hereinafter referred to as NCS), remote centralized control center (hereinafter referred to as RCCC), environmental monitoring and analysis system (hereinafter referred to as EMAS), inspection robot, robot charging and so on. As the control center of the whole intelligent inspection system, the RCCC is responsible for executing the instructions issued by the park staff and sending them to the robot BMS to control the operation of the whole system. The overall structure is shown in Fig. 1.

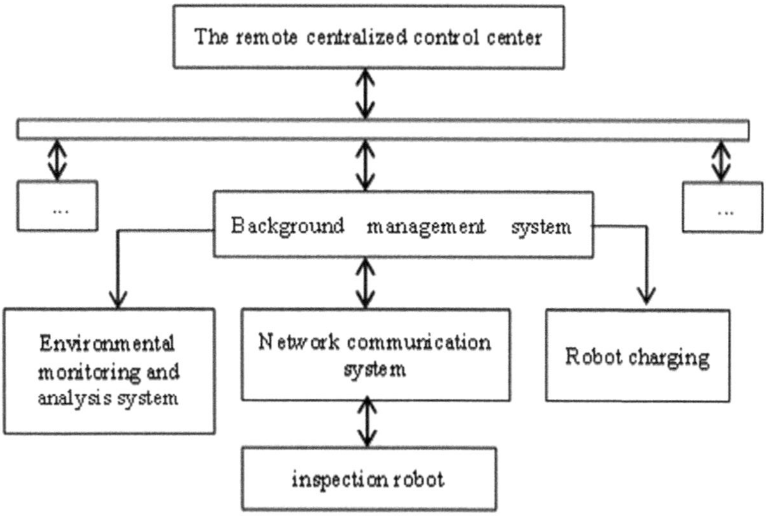

Fig. 1. Structure of park inspection system.

3.1 Robot Physical Framework Design and Construction

The robot physical framework includes shape module, chassis movement module and power management module. The shape module mainly refers to the appearance of the inspection robot, including the head, the holder, the shell, driving mechanism and chassis. Chassis movement module driving ability is strong, and its load capacity is high, which can adapt to different road environment changes and can climb a small slope. Because robots often encounter obstacles when performing patrol tasks, they need to have a certain ability to get over obstacles. The robot chassis movement module is recommended to adopt the four-wheel suspension structure with semi-independent suspension. This form of suspension system, can Prevent the car body left and right appearing big fluctuation in the crossing of small obstacles, can keep the car running smoothly. The battery management module is equivalent to the human heart, and it delivers the required voltage and current to the functional modules of the whole robot system to ensure the normal operation of the inspection robot.

3.2 Carrying of Control Parts and Various Sensors in the Robot Chassis

Inspection robot chassis consists of four-wheel vehicles and six foot mechanical arm, containing mechanical module, control module, perception module, the wifi module, voice module, etc. It can run or climb with the navigation route, and it will collect the relevant data sent to the control center by the wireless network automatically, and it will use image recognition technology to give a verbal alarm. The control module includes industrial computer, PLC, controller, dc motor, etc. Ipc is the core of the entire hardware system, which mainly completes the tasks of positioning and navigation, path planning, releasing patrol inspection, communicating with other devices and unifying and centralizing processing and control of all kinds of data.

The sensor module consists of HD-camera, infrared thermal imager and various sensors. HD-camera, infrared thermal imager and pick-up is mainly responsible for collecting high-definition images, instrument identification, temperature monitoring and sampling site environments such as voice, including the information fusion of odometer, laser radar sensor, vision sensor and other equipment, using for positioning and navigation. The location navigation module can detect and upload the robot's location in real time.

Information fusion refers to the information processing process of using computer technology to analyze and optimize the observation information of several sensors obtained in time sequence under certain criteria to complete the required decision-making and estimation tasks. Multi-sensor information fusion can improve the accuracy and reliability of the environment observation in the course of robot positioning and navigation. The navigation structure, laser sensor, ultrasonic sensor and visual sensor collect data and input them into the sensor fusion module for matching and fusion. In the sensor data fusion algorithm, due to the different processing cycles of different sensors, the system needs to determine a unified time reference. At the same time, the robot's posture and speed are adjusted and integrated into the positioning algorithm through the data obtained by the speedometer, and then a complete route plan is obtained.

3.3 Background Management System

Background management system is the center of inspection robot management and control, which mainly includes system management, data management, input management and output management and other functions. System management includes the use of a variety of function settings and password management, while data management is the key to the inspection of some key information, data and other information management. Input management refers to the automatic import of the key image, audio and other files generated by judgment after detection. Output management is to analyze and diagnose the acquired images and audio materials, to generate the results to be executed and output them to the corresponding instructions to be executed by the inspection robot.

3.4 Communication System Construction

The intelligent park inspection robot realizing the wireless control of the system and the remote return function of the robot to collect information by communicating with the wired LAN and the outdoor wireless network. The background management system is wired to the network communication system, environmental monitoring system and robot charging, so as to receive real-time environmental information and control the charging switch according to the robot status. The background management system is mainly responsible for issuing the task of inspection to the robot, calculating and analyzing the environmental information of the implementation park detected by the robot, and diagnosing the implementation state of the park. In order to facilitate the centralized control of multiple park inspection robots, the background management system will upload the corresponding data information to the RCCC through the private network, and the centralized control center will be responsible for the coordination and control and information collection.

3.5 Environmental Monitoring and Data Analysis System

The basic data obtained by the intelligent inspection robot and the inspection data are stored in the local server as files. The data analysis system recognizes the file names and associates them with the inspection database. The system makes corresponding diagnosis of the files. It will start the alarm if there is any abnormality. Data analysis includes the following parts: acquisition sub-module, storage sub-module, cleaning sub-module, analysis sub-module, decision application sub-module, etc.

The acquisition sub-module mainly collects real-time environmental information on the site of the park to provide the most original analysis data for the whole analysis system. The storage submodule is mainly used to store the collected park environment information data. According to the pre-set database in normal state, the cleaning sub-module checks and tests the collected information to determine the redundant data, wrong data, defective data and other undesirable data values, so as to eliminate these invalid data and realize the optimization of data quality. The analysis sub-module will model and process the cleaned data, resulting in a normal operation model defect. When the park inspection robot is working, the real-time data collected from the park will be compared with the model in real time. If it is within the normal range, the correct result will be generated. If there is an exception, the abnormal result will be produced. The decision application sub-module needs to make judgment and decision according to the analysis results, and finally determine whether the alarm signal should be issued.

4 Key Issues to Address

By optimizing the movement mode and structure of the robot, the obstacle surmounting ability of the inspection robot can be solved. It can not only run quickly on the flat ground, but also crawl smoothly on the uneven section of the lawn. Multi-source sensing technology is adopted to realize multi-sensor information fusion and location algorithm, so as to collect real-time information of each point in the park and upload the background management system through wireless network in a timely manner. It adopts HD camera, and uses artificial intelligence system to solve the problem of graphic image recognition accuracy, and correctly identifies "smart" badge, lawns, gardens, tourists, small animals and garbage. It will use image recognition technology to identify invasive small animals and tourists who mistakenly step on the grass and garden, and trigger the IOT alarm system to automatically voice drive away and persuade away.

5 Conclusion

The inspection robot integrates various control modes of the robot, so that the movement of the robot is not restricted by natural conditions. Artificial intelligence is deeply integrated with technologies such as the IOT and big data. It can collect information in real time through various sensors and HD cameras, and realize intelligent monitoring and intelligent alarm by combining with graphic image recognition technology. The system can facilitate the construction of intelligent monitoring networks in parks and other public places during the prevention and control of the COVID-19. The results of this research can help the development of intelligent robot functions and application fields to some extent.

References

1. Zhang, M.N.: Design and implementation of intelligent inspection robot system based on multi-source perception. Zhejiang University of Technology (2019)
2. Zhu, X.L., Wang, H.G., Fang, L.J., Zhao, M.Y., Zhou, J.P.: Analysis of walking dynamic characteristics and posture of transmission line inspection robot. J. Mech. Eng. **12**, 143–150 (2006)
3. Endres, F., Hess, J., Sturm, J., et al.: 3-D mapping with an RGB-D camera. IEEE Trans. Robot. **30**(1), 177–187 (2014)
4. Huang, A.S., Bachrach, A., Henry, P., et al.: Visual odometry and mapping for autonomous flight using an RGB-D camera. Robot. Res. pp. 235–252 (2017)

5. Fu, M.Y., Lu, X.W., Liu, T., Yang, Y., Li, X.H., Li, Yu.: Real-time SLAM algorithm based on RGB-D data. Robotics **37**(06), 683–692 (2015)
6. Lu, H.Y., Qian, Q., Zhang, B., et al.: Development of a mobile robot for substation equipment inspection. Autom. Electr. Power Syst. (2006)
7. Guo, R., Li, B., Sun, Y., et al.: A patrol robot for electric power substation. In: International Conference on Mechatronics and Automation, pp. 55–59. IEEE, China (2009)
8. China's first mine rescue robot was successfully developed in Tangshan city. China Sci. Technol. Inf. (13) 7 (2009)
9. Luo, Z.R., Shang, J.Z., Zhang, Z.X.: A reconfigurable tracked mobile robot based on four-linkage mechanism. J. Central South Univ. **20**(1), 62–70 (2013)

Intelligent Network Operation and Maintenance Based on Deep Learning Technology

Xidian Wang, Lei Wang, Zihan Jia(✉), Jing Xu, Yanan Wang, Duo Shi, Yang Xue, and Zhenlin Nie

China Mobile Design Institute, Beijing 100080, China
jiazihan@cmdi.chinamobile.com

Abstract. Artificial Intelligence is booming nowadays and has developed tremendous effect in many areas. For Telecommunication, due to the growing network complexity and business diversity, it has been a challenge for operation and maintenance (O&M) of a wireless network. This paper analyzes the requirement for intelligent network O&M, sorts out the selection of artificial intelligence (AI) algorithm, and discusses the application scenarios of relevant algorithms in the field of network O&M. Based on 4 cases of intelligent application, this paper reveals how efficiency and quality of network O&M will be improved by AI technology.

Keywords: Artificial intelligence (AI) · Network operation and maintenance (O&M) · Deep learning · Network intelligence

1 Introduction

Network quality is the lifeline of telecommunication enterprises. Because the network structure of China Mobile is complicated, it is challenging for network O&M to overcome the difficulties, understaffing, and process differences. At present, network maintenance and support are still following traditional patterns, where malfunction handling is passive.

With the evolution of AI and deep learning technology, the massive application of cloud computing and GPU, the development of the Internet of Things (IoT), the enrichment of big data and its maturing processing technology [1], AI has exceeded human in many areas [2]. The combination of AI and industries has accelerated the industrial transformation, which means intelligent recognition and data analysis could partially or even wholly substitute for human work. How to use AI technology to promote the intelligence of the telecommunication industry itself is exceptionally urgent, directly affecting the implementation quality of national and enterprise strategies [3].

Based on the perspective of network O&M in the telecommunication area, this paper studies network intelligent O&M technology, replacing passive handling with active precautions, aiming at the front-line O&M problems. In this way, labor-intensive

Q. Liu et al. (Eds.): CENet 2020, AISC 1274, pp. 200–207, 2021.
https://doi.org/10.1007/978-981-15-8462-6_23

O&M will be substituted, which leads to O&M methods innovation, efficiency improvement and better resource utilization.

2 Requirement Analysis and Technology Selection for Intelligent Network O&M

More than 300 kinds of AI algorithms, among which over 40 are based on neural networks. The requirement of the AI algorithm for Intelligent Network O&M can be concluded as follows.

First, the algorithms should be applied for modeling and prediction using critical data of network malfunction, performance, and power environment [4]. Meanwhile, they should be able to cover the deep mining, learning, and decision task based on the network O&M experiences, cases and handling orders. Furthermore, they should be able to achieve dimension reduction of high-dimensional network structure and features, the portrait of scenario feature and label modeling. Besides, they should be able to analyze the anomaly, determine the reason and locate the problems for network hazard, alert and complaint problems.

2.1 Requirement Analysis of Intelligent Network O&M

Traditional network O&M requires heavily human work. Problems like inefficiency of certain processes and delay of resource allocation also exist. For example, difficulties for hazard detection, lack of pertinence for inspection, troubles for root cause locating, inefficiency for human analysis of orders. Meanwhile, the value of abundant experience gathered during the long-term works of malfunction handling, alert analysis and inspection has not been fully exploited [5].

AI technology possesses natural advantages solving problems above. Relying on machine learning and deep learning algorithms, new methodology of network O&M transforms passive handling to active precaution with the early discovery of network problems. Thus the resource utilization and efficiency of O&M are elevated, and the cost is reduced.

AI technology have superiority in fields such as malfunction prediction, base station anomaly detection, intelligent inspection, malfunction root cause classification, intelligent order assignment, performance and alert associate analysis, etc.

Hazard Prediction. Adopting LSTM network, this module, trained with history data of alert and performance, realizes important alert prediction. It predict the probability of whether base station malfunction will happen in n hours, when certain performance fluctuation or minor alert anomaly feature was detected.

Base-station Anomaly Detection and Intelligent Inspection. Adopting VAE algorithm, this module, trained with history data of inspection, alert, performance, malfunction handling orders, realizes base-station anomaly operating situation detection. Combined with current network data of alert and performance, it plans inspection route, generates inspection order, guiding the intelligent inspection by maintenance crew.

Malfunction Root Cause Classification and Intelligent Order Allocation. Adopting DBN algorithm, this module, trained with history data of malfunction, malfunction handling orders, performance and power environment, realizes malfunction root cause classification. It can provide information for intelligent decision, directing malfunction handling by maintenance crew.

2.2 Deep Learning Technology Selection in Network O&M

Based on the characteristics and various demand of Network O&M business, certain AI technologies are selected and researched, including prediction, anomaly detection, classification, associated analysis, etc. Specifically, algorithms for each technology is selected as below:

Prediction-Long-Short Term Memory Neural Network (LSTM). Time series prediction predicts statistic values of some event's features in a future period based on its feature in a past period. The most powerful tool of time series prediction is recurrent neural network (RNN). The result computed by each hidden layer is related with both current input and result from its former hidden layer. In this way, RNN's output is equipped with information of its earlier few outputs. LSTM is designed to solve the problem of long-term dependency of regular RNN, which makes it very suitable for time series prediction problems.

Anomaly Detection-Variational Auto Encoder (VAE). Anomaly Detection is also called 'outlier detection', which is ubiquitous needed and has practical value in machine learning area. Auto Encoder is an artificial neural network that effectively learns the representation of input data by unsupervised learning. Anomaly detection using auto encoder is based on deviation, which adopts reconstruction error as the anomaly indicator. Data points with high reconstruction error are considered anomaly points. VAE uses probability measurement, which is known as reconstruction probability, instead of reconstruction error as the anomaly indicator. The advantage is that reconstruction probability is more objective and general compared to the reconstruction error, and model threshold is not needed for the anomaly judgement.

Classification-Deep Belief Network (DBN). DBN is a type of neural network that composed with multiple hidden layers. Compared to traditional shallow model, it can avoid problems like overfitting and insufficient capacity. DBN is composed with a few layers of Restricted Boltzmann Machine (DBN) and a layer of BP neural network, which possesses not only effective feature extraction ability but also strong classification result. DBN is trained layers by layers, resolving the optimization problems for deep neural network, which gives a better-adjusted initial value for the whole network. Thus the optimal solution will be achieved with a few times of fine tuning.

3 Typical AI Applications in Network O&M

3.1 Hazard Prediction Technology Based on LSTM

Currently, when malfunction occurs during operation, the base station will be sending alerts. Some of them are major alerts (hazard), which generate malfunction orders. Maintenance crew who gets the order will be handling the malfunction. Major alerts means that the base station is out of service (OOS) and users won't be able to connect to the network. Therefore, the communication is affected. At present, O&M for 4/5 G base station is passive. Only when the major alert (hazard) is happened will the O&M orders be sent. Normally, before the major alert (hazard) occurs, there also exists some minor alerts, performance fluctuation and power environment data anomaly. Based on this phenomenon, it is possible for AI prediction model to predict probability of major alert (hazard) in a future period using the history data of minor alert, performance and power environment. For those base stations that predicted to be highly possible to break down, O&M crew can be assigned of handling in advance, by which the OOS is evitable. The implementation basis is shown in Fig. 1.

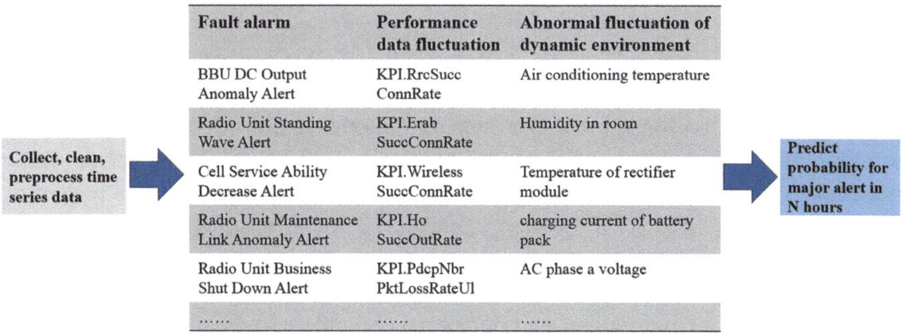

Fig. 1. Implementation basis of hazard prediction.

Utilizing time series prediction ability of LSTM, the model is trained with history data of alert, performance and power environment, which learns the implicit variation trend within the data and capable of inference the probability of whether major alert will happen in the future.

The implementation is obtained as follows:

1) Collect network feature data: collect, clean and preprocess history data of network alert, performance, power environment.
2) Build the training dataset: associate samples with their base station (cell) ID and timetable to make a time series dataset.
3) Label the training dataset: label samples that generate a major alert in a future period with '1', as the positive samples. Otherwise label it with '0', as the negative samples.
4) Train the Model and Tune: train the LSTM model with training samples, and tune the parameter to obtain optimal model.

5) Real-time prediction: associate with the real-time network data of alert, performance and power environment, output the probability of major alert happening in N hours using the well-trained model.

Currently, the accuracy of the model is over 78% predicting the major alert in 72 h, providing preliminary foundations for active O&M. As the data dimension, quantity and quality improves, the model can be further optimized, and the accuracy will increase as well.

This application is implemented during the alert statistic analyzing and predicting section. It predicts the future hazard situation of base station, guiding the optimization of O&M resource. In this way, O&M crew would be able to solve potential malfunction ahead, avoiding base station OOS. The base station inspection can also be arranged according to the hazard priority to reduce the cost, so that the limited inspection resources can be distributed to those sub-health base stations. In addition, this application can help build a library for rare alert situations and providing experience for newly happened alert.

3.2 Base Station Anomaly Detection and Intelligent Inspection Based on VAE

Currently, guiding methods is shorted for allocation of inspection resources, the inspection work stays in the state of wide casting and low efficiency for a long time. Traditional inspection assignment is arranged rotational as planed, one area is inspected each time. The health level and seasonal potential malfunction is neglected. Aimless inspection is not associated with the potential alerts, which hasn't caused a breakdown, and the precaution effect of inspection is unclear. Therefore, how to properly allocate and centralize inspection resource to maximize its value is critical for network O&M.

Based on the VAE algorithm, multiple dimensional anomaly detection of data from malfunction, power environment and performance gathered across the whole network is obtainable. With unified assessment, prediction and sorting, the hazard priority, precise inspection list and key inspection items can be outputted for every base station. As the limited resources can be precisely allocated to the sub-health base station, inspection cost will be significantly reduced. The principle of anomaly detection is shown in Fig. 2.

The implementation is obtained as follows:

1) Collect network feature data: collect, clean and preprocess history data of network malfunction alert, performance alert, power environment alert.
2) Build the training dataset: associate data mentioned before with their base station (cell) ID and timetable to make a dataset. Note that unsupervised learning doesn't request labeling.
3) Train the Model and Tune: train the VAE model with training samples, and tune the parameter to obtain optimal model.
4) Real-time Anomaly Detection: collect network alert data periodically and generate samples, output the anomaly level of each base station with the well-trained model. Based on the VAE network model, anomaly labels will be projected into different

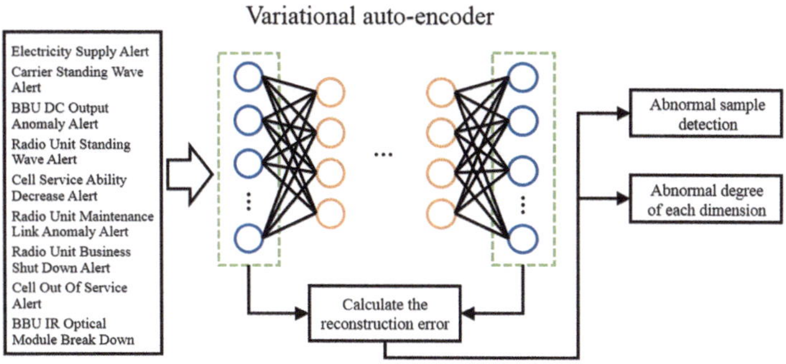

Fig. 2. The principle of anomaly detection.

categories. After the dimension reduction, abnormal base stations will be visualized as the outliers.

5) Generate Inspection Plan: Combine the anomaly level outputted by the model with weights, including time-wise weights and apparatus-room-wise weights. Set alerts with high anomaly level as critical items and output inspection schedule.

Application Effect: Compared to rotational method, the number of inspection orders and cost of O&M is reduced, frequency and duration of OOS is reduced, as well as the number of daily malfunction. The number of problems exposed by precise inspection is increased, which eliminate the potential malfunction. Inspection progress is simplified as the critical items are pointed out, which evades the problems of unbalanced level for the O&M crew.

With the intelligent inspection planning, the condition of base stations located in the important area or complicated area with high frequency of malfunction, poor performance and deteriorated environment will be monitored. And the aimless wide casting and low efficiency state will be revised. Concentrated to the base station with frequent malfunction, recurrence of historical breakdown is prevented. At last, waste of O&M resources and cost for single base station O&M is cut down, so the efficient management of O&M inspection is achieved.

3.3 Root Cause Classification and Intelligent Order Allocation Based on DBN

Currently, when network malfunction occurs, local telecom operator will receive a mass of alert messages. According to the level of the messages, handling orders will be allocated selectively. After the O&M crew solves the malfunction at the scene, they will fill in the reason of this malfunction. Due to some of these huge amount of alerts are just caused by the black out, and after the electricity supply recovers, the network will naturally back to normal, orders for these condition is unnecessary. In other situations, O&M crew can only propose solutions after they arrive the scene and inspect the problems, which dispensable extends the duration of the breakdown.

Therefore, aimed at these situation that has no effective classification and allocation scheme for the massive orders, A DBN classification model is trained with history data of alert and reasons in the malfunction orders. Apply this model in the current network, based on the newly generated alert data, the reason of malfunction can be classified with different reasons such as black out, software fault, false alarm, transmission fault, etc. Thus, the O&M crew will be aware of the type of malfunction in advance. In this way, the strategy of order allocation can be optimized, the number of orders will be reduced and the handling time will be decreased.

The implementation is obtained as follows:

1) Collect network feature data: collect, clean and preprocess history data of network alert, malfunction order.
2) Build the training dataset: associate each malfunction order with its former and later alert data to make a dataset.
3) Label the training dataset: based on the reasons in malfunction orders, label major alert, which also means a malfunction, with their root cause, such as black out, software fault, false alarm, transmission fault, etc., as samples.
4) Train the Model and Tune: train the DBN model with training samples, and tune the parameter to obtain optimal model.
5) Real-Time Classification: associate with the real-time network data of alert. When major alert occurs, use the data to generate samples and output the reason using the well-trained model.
6) Optimize order allocating scheme: O&M crew can adopt different order allocating schemes with respect to different malfunction causes. For example, as studied with history data, black out takes up nearly 60% of the malfunction reasons, among which 25% will be recover automatically in 20 min. Therefore, the allocating of orders with black out reason can be postponed for 20 min, which cut down 15% total order number.

In conclusion, this application utilizes AI algorithm to achieve effective malfunction root cause classification, provides intelligent decision making for handling, guides the maintaining process for O&M crew, realizes precisely order allocation. By smart decision-making schemes, the number and quality of orders is reduced and increased respectively, the duration of malfunction is shortened, and the high efficiency of front-line O&M crew is guaranteed.

4 Conclusion

For telecom operators, AI is both a challenge and an opportunity. The enabling technology provided by AI can be applied to various aspects of telecom networks and related vertical businesses, and will become an important enabling technology for the next generation of telecom networks. The introduction of AI technology will bring a new state to network operation. Research on the application of AI in network O&M will tremendously improve the level of network intelligent operation and maintenance, and help network operation reduce cost and increase efficiency.

References

1. Anandakumar, H., Arulmurugan, R.: Artificial intelligence for cognitive telecommunication network. In: Third International Conference on I-SMAC (IoT in Social, Mobile, Analytics and Cloud) (I-SMAC), Palladam, India, pp. 290–294 (2019)
2. Tanaka, T., Kuwabara, S., Oda, T., Kitamura, K., Inuzuka, F., Inui, T.: Advances toward ai-assisted autonomous network diagnosis. In: Asia Communications and Photonics Conference (ACP), Chengdu, China, pp. 1–3 (2019)
3. Liu, Y., Tu, Y., Hsu, C., Chao, H.: Predicting malfunction of mobile network base station using machine learning approach. In: 20th Asia-Pacific Network Operations and Management Symposium (APNOMS), Matsue, Japan, pp. 1–4 (2019)
4. Kibria, M.G., Nguyen, K., Villardi, G.P., Zhao, O., Ishizu, K., Kojima, F.: Big data analytics, machine learning, and artificial intelligence in next-generation wireless networks. IEEE Access **6**, 32328–32338 (2018)
5. Cayamcela, M.E.M., Lim, W.: Artificial intelligence in 5G technology: a survey. In: 2018 International Conference on Information and Communication Technology Convergence (ICTC), Jeju, Korea, pp. 860–865 (2018)

Research on Autonomous Driving Perception Test Based on Adversarial Examples

Yisi Wang, Wei Sha, Zhongming Pan, and Yigui Luo(✉)

Tongji University, Shanghai 201804, China
{wangyisi,shawei,panzhongming,luoyigui}@tongji.edu.cn

Abstract. Recently some accidents brought the safety of autonomous vehicles to the public's attention. Autonomous vehicles are not only hardware products, but also a system of multi-sensor perception modules and Deep Neural Networks (DNNs) software. There are many methods to test traditional software. However, finding a good enough software testing method for vehicles on-board DNNs models is still an open problem. In this paper, we provide a testing method based on boundary value testing. For DNNs, many studies have shown that adversarial examples can disable DNNs, and they are often closed to the decision boundary. Therefore, adversarial examples can be used as test cases. To accomplish this task, we use two novel algorithms to design boundary value test cases: a white-box method based on gradient information and a black-box method using Deep Reinforcement Learning. Further, we propose a whole testing procedure, including a computer simulation test, sensor unit test and real vehicle test. The experimental results show that it is feasible to use adversarial examples for testing and our method could increase efficiency. Our work may have inspirations for automobile manufacturers and researchers.

Keywords: Autonomous driving · Deep learning · Software testing · Adversarial examples

1 Introduction

Many DNNs models are deployed on the perception system of autonomous vehicles. DNNs models play an important role in classification, object detection, localization, and other tasks [1]. However, studies have shown that DNNs are often not robust to some input data with small perturbations [2, 3]. This kind of imperceptible perturbations can often disable the state-of-art DNNs, which will lead to the failure of the perception module and seriously lead to accidents. In order to detect possible vulnerabilities, testing is significant. Our interest here is mainly focused on searching for an effective software testing method to test the on-board DNNs models.

One of the efficient methods in software testing is boundary value testing [4]. In practice, many software bugs are concentrated on the boundary of its equivalent class. Good enough boundary value test cases will greatly help save the cost of testing and increase the testing efficiency.

In order to obtain boundary value test cases, it is a good idea to design test cases by adversarial examples. Adversarial examples are the samples that add small

Q. Liu et al. (Eds.): CENet 2020, AISC 1274, pp. 208–218, 2021.
https://doi.org/10.1007/978-981-15-8462-6_24

perturbations to the original inputs and could disable the DNNs models. Although adversarial attacks are not designed to find decision boundaries, adversarial examples are often closed to the DNNs decision boundaries. We design two novel algorithms to generate adversarial examples of test cases under a white-box and black-box test. Then we design a testing procedure based on the test cases. We also optimize the transformation method of the digital cases to the physical cases for a real-world test.

Experiments verify the proposed method. We show the improvement of our two test case generation algorithms compared with FGSM, Deepfool, GA, and Gan algorithms. After that, we show the result of the testing procedure of the computer simulation test and sensor unit test. Finally, we summarize the conclusion according to the experimental data.

2 Background

2.1 Autonomous Vehicles Perception

Many reports show that autonomous vehicles have achieved high performance in millions of miles testing on the road. The most basic and crucial part of autonomous vehicles is the perception system, including camera, millimeter-wave radar, laser radar (LiDAR), IR sensor and GPS, which can simulate human drivers sensing the information from the environment. DNNs are often used to extract information features from inputs [5]. The latest research [6] also focuses on applying deep learning, such as 3D object detection framework based on a single RGB camera. Figure 1 shows the perception modules of autonomous vehicles.

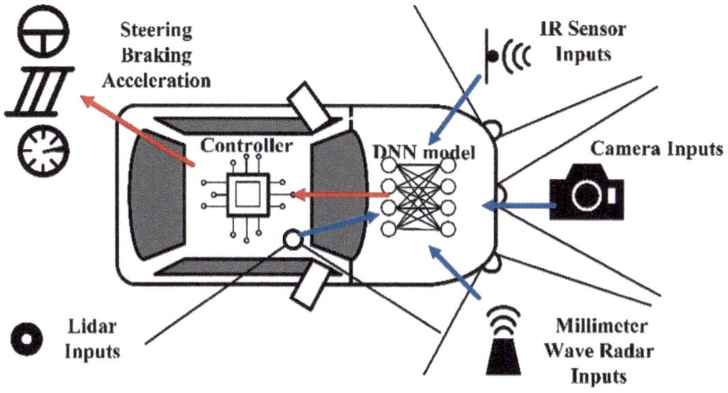

Fig. 1. A basic autonomous vehicle system structure. DNNs models are deployed on multiple perception sensors such as cameras, millimeter-wave radars, lidars and IR sensors. The outputs of DNNs are entered into the controller and the controller makes steering, braking and acceleration decisions to control the vehicle.

2.2 DNNs and Adversarial Examples

A typical DNNs model is composed of multiple processing layers which can extract different features of the inputs [5]. In most cases, DNNs models can achieve excellent performance. For example, AlexNet [7] won the champion of ILSVRC2012. YOLO-v3 [8] object detection model is much efficient and rapid, deployed on a variety of experimental autonomous vehicles and robots.

However, some studies found that DNNs models may be vulnerable to adversarial examples made of well-designed perturbations added to the inputs [2]. J. Goodfellow et al. [3] proposed a gradient-based attack method called FGSM, which directly adds the perturbations based on the gradient sign to the inputs. Kurakin et al. [9] improved FGSM for smaller perturbations. Moosavi-Dezfooli et al. [10] proposed Deepfool. This method iterates along the gradient direction perpendicular to the decision boundary to find the minimum perturbation. Heo et al. [11] used the improved Deepfool algorithm to find boundary supporting sample in the knowledge distillation task. For the black-box testing task, Hu et al. [12] trained a Generative Adversarial Network (GAN) to generate adversarial examples. Su et al. [13] used the differential Evolution Algorithm (EA) to generate one-pixel adversarial examples. Our study in the black-box case uses the state-of-art Deep Learning method called Reinforcement Learning (RL). RL [14] is a method of searching the optimal solution through continuous trial-and-error interaction with the environment.

2.3 Software Testing on Autonomous Driving Vehicles

Some vechicle manufacturers abroad including Tesla, GM, BMW and some domestic including SAIC, NIO, BYD are building testing system for new autonomous vehicles. Several real-world accidents show that autonomous vehicles are not as safe as people expected and many faults can be attributed to the incorrect or unexpected corner-case faults of the DNNs-based software on board [5].

Boundary value testing [4] is a good way to test the corner-case faults. Adversarial examples are often closed to the DNNs decision boundary [15]. When we finish the simulation test, the test need be done in the real world. Adversarial examples work not only in digital images, but also in the physical world [9]. Eykholt et al. [16] proposed the PR2 method to transform digital samples into physical world.

Also, a variety of types of DNNs in the perception modules can be tested. Xie et al. [17] successfully attacked the semantic segmentation network of FCN and the object detection network of Faster RCNN. Moreover, our work takes the camera inputs as example, and the methods to test other sensors are similar.

3 Method to Design Test Cases

3.1 Iterative Method for White-Box Test

Let us denote the inputs of the DNNs model to be x, the model under test to be $f(x, \theta)$, and the adversarial perturbation that disables the model to be r, θ is the parameters of $f(x, \theta)$, such as weights and bias. Our goal is to find a series of adversarial examples closed to the decision boundary as boundary value test cases (BVTC). In our work, iterative method for white-box test could be regarded as a kind of constrained optimization problem. It can be formulated as follows:

$$\max f(x + r, \theta) \neq f(x, \theta) \tag{1}$$

$$\min r, \| r \|_p < \in \tag{2}$$

$$\min r, S(r) > \varphi \tag{3}$$

where $||r||_p$ denotes the L_p norm of a vector, $\in$ constrains the perturbation magnitude. S is the function of image similarity, φ ensures that the adversarial example is similar to original input in human vision.

$$r_i^k = -\frac{f_k(x_i) - f_b(x_i)}{||\nabla f_k(x_i) - \nabla f_b(x_0)||_2^2}(\nabla f_k(x_i) - \nabla f_b(x_0)) \tag{4}$$

$$x_{i+1} = x_i - \lambda_{constrain}\left(\min_k r_i^k\right) \tag{5}$$

$$f_{fc}(x_\lambda) < f_{fc}(x_i)\, and f(x_\lambda) = f(x_i) \tag{6}$$

To find the BVTC, we move the sample to the direction perpendicular to the decision boundary of the nearest class. The perturbation that each step added to the sample can be represented in (4), where $\nabla f_k(x_i)$ denotes the gradient of the k-class and $\nabla f_b(x_0)$ denotes the gradient of the original class. It is a kind of distance formula and it was firstly proposed in Deepfool. However, sometimes it is abnormally large due to a small gradient. And similar to a kind of Newton method solving nonlinear equations, it has a high requirement on the initial value, which is based on the assumption that the neural networks have high linearization characteristic. But sometimes the decision boundary has a nonlinear feature locally such as Relu neural networks. It is possible to miss the nearest adversarial example in the approximation process and get a further one. Figure 2 explains this phenomenon in the one-dimensional case.

Moreover, it operates in a greedy way and is not guaranteed to converge to the optimal perturbation. To improve this situation, we introduce a punishment parameter λ to control the step size. λ operates on the gradient, starts with a higher multiple of the slope, stops under the constraint in (6). f_{fc} denotes the outputs before softmax. Choose proper λ that makes the f_{fc} value go down and dose not make the label change.

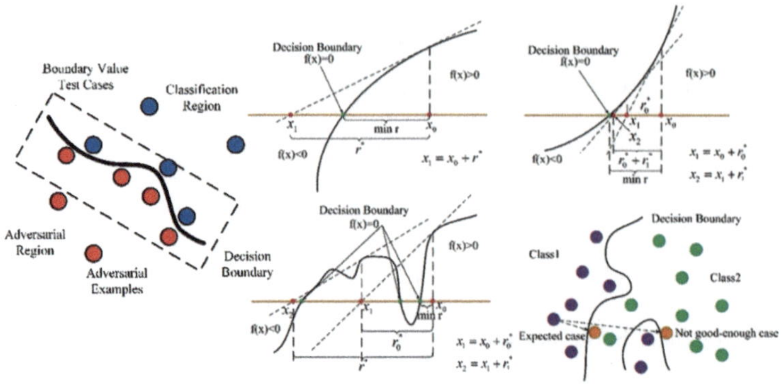

Fig. 2. Use adversarial example as BVTC and results of linear approximation in 3 kinds of nonlinear conditions in the one-dimensional case. It shows nonlinear conditions may miss the nearest case sometimes.

Algorithm 1 white-box Iterative Method to generate test cases

Input: Original image x, model under test f.
Output: Test cases $\hat{x}$.
Initialize: $i \leftarrow 0$, max_ite, b: basic class, k, j: loop control factor, $r \leftarrow 0$, classes.
while $f(x_i) = f(x_0)$ **and** $i < max\_ite$ **do**:
  for k **in** classes:
    $r_i \leftarrow -\frac{f_k(x_i)-f_b(x_i)}{||\nabla f_k(x_i)-\nabla f_b(x_0)||_2^2}(\nabla f_k(x_i) - \nabla f_b(x_0))$; record r_i;
  $r_{i\_min} \leftarrow \min(r_i)$; //get the minimal perturbation direction class
  end for
  for λ **in** Slope: //Slope←[8,4,2,1]
    $j \leftarrow 0$; $r_j' \leftarrow \lambda r_{i\_min}$; $x_j \leftarrow x_i + r_j'$;
    if $f(x_j) \neq f(x_i)$ **and** $f_{fc}(x_j) < f_{fc}(x_i)$: $r_i \leftarrow r_{j-1}'$; break;
    $j \leftarrow j + 1$; $r_i \leftarrow r_j'$;
  end for
  $i \leftarrow i + 1$;
end while
$\hat{x} = \sum_{i=0} r_i + x_0$;
return $\hat{x}$

3.2 Reinforcement Learning Method for Black-Box Test

Reinforcement learning (RL) is a machine learning method in trial-and-error way. The goal of RL algorithm is training the agent to learn a policy to solve a complex optimization problem. For the black-box test, we can only observe from the inputs and the outputs. This is very similar to the model-free RL task, where agent interacts with the environment without guidance to get the maximum reward. Therefore, using RL method to explore the design of BVTC in black-box test is a novel and interesting work.

Our work is to train a Deep Q-learning (DQN) network [18] to generate BVTC. The state at time t is represented as $s_t \in (x_t, p_t)$, where x_t denotes the input sample and p_t denotes the probability outputs of the model under test. The action $a_t \in A$, where a_t is the actions of modifying the pixels of inputs. We can set A by original rules such as modify 10% pixels, or we can set A from a pretraining set of Genetic Algorithm. The agent takes actions to modify the inputs in a Markov process. In the training, agent takes actions in ϵ*–greedy* way, which means it takes random actions firstly, reduces random actions and increases the network prediction as the training progressed. If the agent succeeds in changing the label, it receives a high positive reward. If it fails, it receives a negative reward, and for each action, it receives −0.1 reward for punishment. The goal is to find a policy π that maximizes the Q value function following the Bellman equation in (7).

$$Q(s_t, a_t) = r_t + \gamma \times maxQ(s_{t+1}, a_{t+1}, \theta) \tag{7}$$

where γ is a factor and θ are the parameters of the DNNs. We use DQN network to fit the optimal Q value function. Figure 3 shows the structure of RL method:

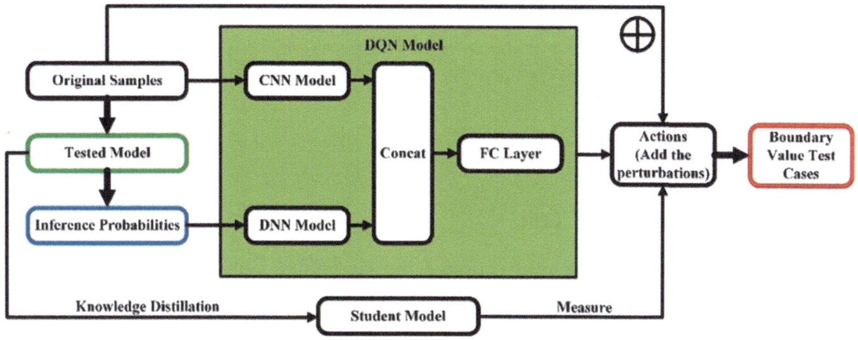

Fig. 3. Details of RL algorithms to design the BVTC.

Original Samples are represented by a CNN and the inference probabilities of the model under test are represented by a DNN. The concat of the outputs of CNN and DNN is entered to a FC layer. Because we don't know the parameters in black-box test, we use knowledge distillation [19] to train a 10-layer student model from the model under test, and we use the parameters of the student model to evaluate the quality of the BVTC.

Algorithm 2 black-box Reinforcement Learning Method to generate test cases

```
Input: Data sets x, model under test f, Deep Q-learning Network m.
Output: Test cases x̂, well-trained DQN m̂.
Initialize: i ← 0, exp_pool: a list to store exp, Q_sa: a matrix to store Q value.
for x_i in x do:
    use ∈ −greed policy to choose random action or use m(x_i) as action;
    x̂ ← x_i + action;
    if f(x_i) ≠ f(x̂):
        reward = 10; add(exp)to(exp_pool); record(x̂); break;
    else:
        reward = −0.1; add(exp)to(exp_pool);
    reward = -1.0; add(exp)to(exp_pool);
    if exp_pool is full:                    //use the exps to update DQN
      for exp in exp_pool:
          if exp.reward = 10 or -1.0:   Q_sa[exp.action] ← exp.reward;
          else:   Q_sa[exp.action] ← exp.reward + γ × max(m(exp.image_next));
      end for
      use back propagation to train m to m̂ on the Q_sa dataset to fit Q value
end for
return m̂ and test case sets of x̂
```

4 Test Procedure

4.1 Computer Simulation Test

Before the real-world test, we first propose the computer simulation test, which conforms to the engineering design process and can also save a lot of testing cost. There are some mature simulation software for researchers to use, such as CARLA, AirSim, PreScan based on Unity3D and other 3D engines.

We first use the algorithms in chapter 3 to generate the test cases according to the test requirements, and then deploy them in the simulation environment by modifying the physical asserts of the simulator. We use CARLA [20] as simulator and we write a script to use the vehicle camera in the simulator as the input of DNNs models for traffic sign classification and detection. At the same time, we draw a virtual map based on some roads in Jiading campus of Tongji University to ensure that the simulation environment is closed to the real environment.

4.2 Sensor Unit Test

After passing computer simulation test, we build the perception module under test in lab environment, which generally includes the on-board computer, sensors and power supply. We transplant the DNNs from PC into the on-board computer.

The we need to deploy the test cases in 4.1 in physical environment. Due to a variety of factors such as printers, lighting and angles, digital test cases maybe fail in the real world. Therefore, inspired by Eykholt et al. [16], we set the optimization objective of the two algorithms in chapter 3 as follows for real-world deployment:

$$\arg\min_r \lambda d(M_x \cdot r) + \sum\nolimits_{x_i \sim x} J(f_\theta(x_i + T_i(M_x \cdot r), y^*) + NPS + CDS \tag{8}$$

$$NPS = \sum\nolimits_{p_{digit} \in P} \min_{c_{print} \in C} |p_{digit} - c_{print}| \tag{9}$$

$$CDS = \sum\nolimits_{u_{train\_sample} \in U} \min_{v_{test\_camera} \in V} |u_{train\_sample} - v_{test\_camera}| \tag{10}$$

where M_x denotes the mask matrix to constrain the perturbations generating region, which is the part of the main object. r are the perturbations. T_i denotes a variety of transformation functions for ensuring the physical cases valid in many conditions, including angle, distance, rotation, and so on. d denotes the modification magnitude. f_θ denotes the DNNs model under test. J represents a loss function to measure the difference between the output label of the DNNs model and the real label. The newly added item NPS (non-printability score) denotes the color difference between the pixels of the digital image and the printer image. CDS (camera-difference score) represents the color difference between the training samples and the camera on the autonomous vehicle.

4.3 Real Vehicle Test

Finally, we can test a real vehicle. We can make the vehicle run automatically on the road, turn on the sensors under test, and make it approach the physical test case at different speeds and angles. Finally, we record the data each time, and form the test report.

5 Experiment

5.1 Performance on White-Box Iterative Algorithm

First, we experimented with the Iterative Algorithm on CIFAR-10 datasets (LeNet) and compared with two other algorithms including FGSM, Deepfool. We use SSIM [21], FSIM [22] to evaluate the structure similarity. And we use L_2 norm to measure the distance. Table 1 shows the result of our experiment. It shows that our Iterative Algorithm could generate closer adversarial examples and the structure similarity is higher. However, our method cost more time for each sample, the real-time performance of testing task is not high, so the cost is acceptable.

Table 1. Comparison results of three white-box algorithms.

Algorithm	Success rate	Distance	SSIM	FSIM	Time consumption
FGSM	96.4%	0.403	0.995	0.992	0.02 s
Deepfool	93.7%	0.272	0.996	0.994	0.05 s
Iterative algorithm	94.2%	0.254	0.997	0.995	0.08 s

5.2 Performance on Black-Box Reinforcement Learning Algorithm

Also, we experimented with the RL Algorithm on CIFAR-10 datasets and compared with two other algorithms including GA, GAN. We take the action set A from the pretrained set of GA in 100-generation and the action space is to modify 6 pixels. Table 2 shows that RL Algorithm could also generate better adversarial examples, and the time consumption is higher.

Table 2. Comparison results of three black-box algorithms.

Algorithm	Success rate	Distance	SSIM	FSIM	Time consumption
GA	36.5%	1.217	0.989	0.986	2.1 s
GAN	22.7%	1.468	0.975	0.973	>5 s
RL algorithm	41.3%	1.109	0.991	0.987	>10 s

5.3 Computer Simulation Test

We trained a classification DNN based on VGG-16 and an object detection DNN based on YOlO-v3 on our traffic signs datasets to classify and detect "No-left" sign. Then we generate 10 test cases using Iterative Algorithm and other 10 cases using RL Algorithm. We choose the 2 best performance test cases and deploy them into the simulator. And we make the agent vehicle drive automatically. Figure 4 shows the result of our test. It shows we choose 3 different visual angles of 30° left, 30° right and the front to test the classification network. We test for 100 times at each visual angles, and the failure rate of the camera is 37%, 28%, 56%. Also, it disables the detection network sometimes, the failure rate is 8%. This is because of the generalization of adversarial examples.

5.4 Sensor Unit Test

We use NVIDIA Jetson TX2 embedded computer which is popular in autonomous vehicles and a USB camera to build the sensor unit test platform. TX2 runs a custom Linux and we transplant the DNNs models under test from PC into it. We further optimize the test cases in 5.3 using the constraints mentioned in 4.2, and print them into

Fig. 4. Computer simulation test of no-left traffic sign.

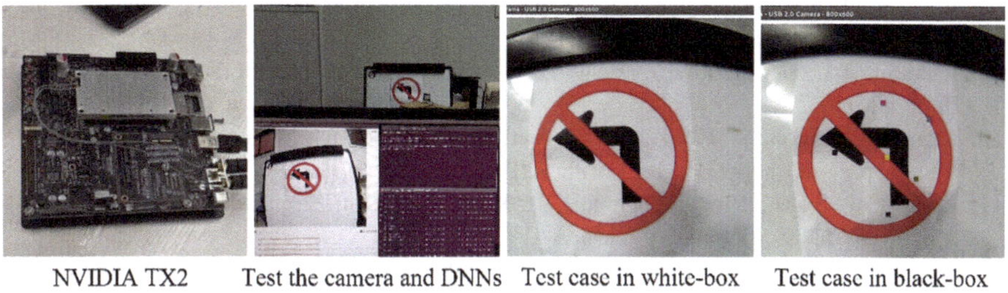

Fig. 5. Sensor unit test of no-left traffic sign in physical world.

asserts in the lab. We also test for 100 times at each visual angle, and the failure rate of the camera is 26%, 21%, 36%. As for detection network, the failure rate is 5%. It seems that the testing success rate of physical cases is lower than that of the digital cases (Fig. 5).

6 Conclusion

In this paper, we propose to use adversarial examples closed to the decision boundary of DNNs models to be the test cases of boundary value test. We propose an algorithm improved from the iterative method in white-box task and an algorithm, novel and efficient, comes from the state-of-art RL in black-box task to design test cases. Based on the test cases, we propose an entire software testing procedure including computer simulation test, sensor unit test and real vehicle test. Moreover, for real-world test, we improve the digital test cases and deploy them into the physical world. Experiments have shown that two test cases generating algorithms could generate test cases that meet the testing requirements. Also, it has shown that the test procedure has good result and can detect some bugs of the autonomous driving perception system. Therefore, our work can be useful for researchers and automotive manufacturers on vehicle development.

References

1. Baidu Apollo Homepage. https://github.com/ApolloAuto/apollo. Accessed Apr 2020
2. Szegedy, C., Zaremba, W., Sutskever, I., Bruna, J., Erhan, D., Goodfellow, I., Fergus, R.: Intriguing properties of neural networks. In: ICLR (2014)
3. Goodfellow, J., Jonathon Shlens, I., Szegedy, C.: Explaining and harnessing adversarial examples. In: ICLR (2015)
4. Patton, R.: Software testing, pp. 46–52. China Machine Press, Beijing (2006)
5. Tian, Y.C., Pei, K.X., Jana, S., Ray, B.: DeepTest: automated testing of deep-neural-network-driven autonomous cars. Arxiv preprint. arxiv:1708.08559 (2017)
6. Li, B.Y., Ouyang, W.L., Sheng, L., Zeng, X.Y., Wang, X.G.: GS3D: An efficient 3D object detection framework for autonomous driving. In: CVPR (2019)
7. Krizhevsky, A., Sutskever, I., Hinton, G.: ImageNet classification with deep convolutional neural networks. In: Advances in Neural Information Processing Systems (2012)
8. Redmon, J., Farhadi, A.: YOLOv3: an incremental improvement. Arxiv preprint. arxiv: 1804.02767 (2018)
9. Kurakin, A.J., Goodfellow, I., Bengio, S.: Adversarial examples in the physical world. In: ICLR (2017)
10. Moosavi-Dezfooli, S., Fawzi, A., Frossard, P.: Deepfool: a simple and accurate method to fool deep neural networks. In: CVPR (2016)
11. Heo, B., Lee, M., Yun, S., Choi, J.Y.: Knowledge distillation with adversarial samples supporting decision boundary. In: AAAI (2019)
12. Hu, W.W., Tan, Y.: Generating adversarial malware examples for black-box attacks based on GAN. Arxiv preprint. arxiv: 1702.05983 (2017)
13. Su, J.W., Vargas, D.V., Sakurai, K.: One pixel attack for fooling deep neural networks. In: CVPR (2017)
14. Kaelbling, L.P., Littman, M.L., Moore, A.W.: Reinforcement learning: a survey. J. Artif. Intell. Res. **4**, 237–285 (1996)
15. Cao, X., Gong, N.Z.: Mitigating evasion attacks to deep neural networks via region-based classification. Arxiv preprint. arxiv:1709.05583 (2017)
16. Eykholt, K., Evtimov, I., Fernandes, E., Li, B., Rahmati, A., Xiao, C.W., Prakash, A., Kohno, T., Song, D.: Robust physical-world attacks on deep learning visual classification. In: CVPR (2018)
17. Xie, C.H., Wang, J.Y., Zhang, Z.H., Zhou, Y.Y., Xie, L.G., Yuille, A.: Adversarial examples for semantic segmentation and object detection. In: ICCV (2017)
18. Mnih, V., Kavukcuoglu, K., Silver, D., Graves, A., Antonoglou, I., Wierstra, D., Riedmiller, M.: Playing atari with deep reinforcement learning. Arxiv preprint. arxiv:1312.5602 (2013)
19. Hinto, G., Vinyals, O., Dean, J.: Distilling the knowledge in a neural network. In: NIPS (2015)
20. Dosovitskiy, A., Ros, G., Codevilla, F., Lopez, A., Koltum, V.: CARLA: an open urban driving simulator. https://carla.org. (2017)
21. Wang, Z., Bovik, A.C., Sheikh, H.R., Simoncelli, E.P.: Image quality assessment: from error visibility to structural similarity. IEEE Trans. Image Process. **13**(4), 600–612 (2004)
22. Zhang, L., Zhang, L., Mou, X., Zhang, D.: FSIM: a feature similarity index for image quality assessment. IEEE Trans. Image Process. **20**(8), 2378–2386 (2011)

Application Research of Artificial Intelligence Technology in Intelligent Agriculture

Lianqin Jia(✉), Jun Wang, Qin Liu, and Qing Yan

Shandong Institute of Commerce and Technology, Jinan, China
20501121@qq.com

Abstract. With the rapid development of Internet technology and artificial intelligence technology, traditional agriculture has ushered in new development opportunities. Integrating artificial intelligence technology into the construction of intelligent agriculture and organically integrating with IoT technology can promote the vigorous development of intelligent agriculture in China. Through the research and analysis of some major problems existing in the current application of artificial intelligence technology in agriculture in China, this paper puts forward Suggestions that artificial intelligence technology can be applied in the implementation of technological innovation in the construction of intelligent agriculture. The example is the construction of fruit and vegetable epidemic disease accurate diagnosis information system based on artificial intelligence technology. This paper discusses the specific application of artificial intelligence technology to modern agriculture and discusses the positive significance of artificial intelligence technology to the development of intelligent agriculture.

Keywords: Artificial intelligence · Intelligent agriculture · The IOT · The internet

1 Introduction

Artificial Intelligence (AI) is to simulate the conscious thinking rules and processes (such as learning, reasoning, thinking, etc.) of human beings, to build Artificial systems with individual intelligence, which has become the cornerstone of the new round of industrial revolution and has been widely applied to many fields of science. With the double growth of world population and economy, the contradiction between the improvement of diet structure and food demand is increasingly prominent, and the challenges in the field of agriculture are more intense than those in other fields. Intelligent agriculture based on artificial intelligence is an effective way to solve this challenge, which has an excellent application potentially [1, 2].

At present, the research on the combination of artificial intelligence and agricultural technology is being carried out at home and abroad., mainly concentrated in the special agricultural sensors, livestock biological knowledge, big agricultural data, mechanized farm network, enclosed ecosystems, intelligent plant production factories, agriculture unmanned aerial vehicle (UAV), precision agriculture technology and the research of intelligent equipment, etc. [3, 4]. It has become the trend to improve agricultural

Q. Liu et al. (Eds.): CENet 2020, AISC 1274, pp. 219–225, 2021.
https://doi.org/10.1007/978-981-15-8462-6_25

productivity greatly through artificial intelligence technology and reform traditional agricultural production and management.

2 Problems in the Application of Artificial Intelligence in Agriculture

With the continuous optimization of China's agricultural structure and the continuous improvement of China's residents' consumption ability, people are paying more attention to the quality and safety of agricultural products. It is urgent to introduce AI technology to create intelligent agriculture. However, the development of intelligent agriculture in most parts of China is still in the early stage, and the specialized, modernized, whole-process, and large-scale intelligent agriculture system has not been fully formed. Compared with China's status as a major agricultural country, there is still a big gap for accelerating the quality efficiency and transformation of agriculture, especially, which mainly focuses on the following aspects [5–8].

2.1 Intelligent Agricultural Infrastructure is Still Inadequate

Due to the lack of infrastructure, some agricultural labor in many places is still largely manual. With the increase of the aging of our population, the rural labor force is gradually transferred to the city, which leads to the lack of manpower, and it makes agricultural products not be picked in time to market during the busy agricultural period. At the same time, the lack of modernization of machinery leads to a great increase in labor intensity. It will completely replace the complex human labor of selective harvesting by promoting the widespread adoption of picking robots.

2.2 The Technology Popularization of Intelligent Agriculture Lags Behind

Standards for facilities, equipment, temperature control, and operation specifications in all aspects of agriculture are not uniform. Agricultural information and communication facilities are severely lacking. The quality of agricultural users needs to be improved. Their ability to apply and accept modern computer technology is also weak. All these are not conducive to the popularization of intelligent agriculture.

2.3 Backward Informatization Level

Agricultural information and communication facilities are severely lacking. Without information and communication facilities, modern computer technology cannot be used. However, most rural areas in China lack of investment in the construction of communication facilities, so that the level of agricultural digitalization cannot be developed, and it is also unable to improve the timeliness and accuracy of agricultural information. Simultaneously, the lack of unified technical standards for the IoT will also hinder the further development of intelligent agriculture, resulting in the unmet demand for resources in standardized agricultural production. It also restrains the scientific research team from collecting and feedbacking agricultural information widely and timely and restricts the development of intelligent agriculture.

3 The Innovative Application of Artificial Intelligence Technology in Agriculture [9–12]

3.1 Greenhouse Control Technology

The greenhouse control technology is a typical application of artificial intelligence technology and IoT technology in agriculture. Relevant parameters in the greenhouse, such as air temperature and humidity, carbon dioxide concentration, light intensity, soil moisture and humidity, and soil temperature, which can be obtained through the remote monitoring system of the IoT. It can even get some real-time video images. Through the analysis of the standard model parameters preset, it can also judge by artificial intelligence technology of related equipment, such as temperature and humidity control machine, sunshade control system, sprinkler irrigation system, cooling system, fill light equipment for automatic control, and automatically adjust the related parameters in the barn, reduce labor and increase the speed of the growth of crops. Of course, this system supports not only computers but also mobile phones, tablets and other device terminals, which can feedback real-time monitoring information and alarm information to users.

3.2 Outdoor Crop Control Technology

The technology of outdoor crop control is to realize the real-time collection of crop growth information by installing information acquisition equipment such as high-definition cameras in the appropriate place of the crop base. At the same time, the agricultural environment monitoring system can be introduced to the base, and the information acquisition equipment can be connected to form the base monitoring and control system. It can monitor wind speed, temperature, and humidity, ultraviolet ray, soil, etc. in real-time. It will also monitor information such as solar illumination and rainfall, and send information to users in real-time. Simultaneously, the system can compare the collected information according to the preset healthy growth model of crops. In case of any abnormality, real-time warning information can be sent to users. It is convenient for users to prevent disasters timely, reduce risks caused by objective factors such as weather, and improve crop yield.

3.3 Intelligent Agricultural Chip Technology

There are many research fields of intelligent agricultural chip technology, and the development space is massive. It shall include agricultural sensor sensing and front-end processing of weak signal simulation, ultra-low power IoT communication, intelligent sensor standard protocol interface and IP research and development of integrated circuit corresponding to special optical perception; the research on agricultural special environment chip materials and packaging technology, sensor information fusion, intelligent control algorithm, agricultural decision support system intelligent chip architecture and packaging technology; to develop intelligent agricultural chip development platform with high integration and multi-layer structure to meet the needs of different agricultural application scenarios, reduce application and energy consumption, and improve the reliability and stability of target products; to create a multi-source anti-lightning diagnostic information acquisition and sensing chip with integrated 3D, Ultraviolet, near-infrared, temperature and other functions; to develop end-to-end solutions that meet the requirements of platform-level services such as equipment connectivity, protocol adaptation, data storage, data security and big data analysis to improve the chip ecosystem and promote the industrialization of agricultural chips.

3.4 Intelligent Agricultural Electronic Tagging Technology

Intelligent agricultural electronic tag technology mainly involves the research of electronic tag chips' key technologies such as low-power circuits, security algorithms, and data automatic modulation and storage. It will promote intelligent electronic tags' technical upgrade, focusing on developing intelligent electronic tags for agricultural products with sensing functions such as temperature, humidity and THREE-DIMENSIONAL acceleration on realizing intelligent perception and transmission of agricultural products information. The design method and technical system of an intelligent electronic tag are established and verified.

3.5 Intelligent Agricultural Equipment Transmission Technology

Research on intelligent agricultural professional equipment transmission technology mainly includes sensing methods of soil moisture, nitrogen, organic matter, trace elements, heavy metals, and other components based on new principles such as optics and electromagnetism. The sensor's physical mechanism and action rule are clarified, and a series of sensors and detection systems are developed. The optical sensing equipment and instruments for crop growth detection will be studied, and the stable quantitative regression model is established to obtain the information of crop populations and individual traits at different scales. Study the sensing method of harmful gas, dust, and water quality in the breeding environment, develop the corresponding sensor, and realize the real-time monitoring of livestock and poultry houses and aquaculture environments. A new sensing method for animal nutrition status, epidemic disease, weight, fat content, and other physiological indicators will be developed. It will improve the sensor accuracy and develop miniaturized, wearable and invasive sensors. The non-destructive and rapid sensing mechanism of agricultural products components and quality is studied, and corresponding sensors and rapid testing equipment are developed.

3.6 Agricultural UAV Information Acquisition Technology

UAV has more and more application value in agriculture. Research on information acquisition technology of agricultural UAV mainly includes the following aspects: Select and optimize the UAV platform with long navigation time, high stability and reliability, and construct the UAV load high-precision self-stabilizing cradle head and stabilizing control technology; For complex farmland operating environment, research agricultural UAV high-throughput, real-time measurement and control communication technology, to achieve automatic obstacle avoidance, path planning and intelligent flight navigation control; For the unstable platform of UAV, we have innovatively developed micro hyperspectral resolution gaze imaging spectrometer and micro high spatial resolution multi-spectrometer, and carried out the integration of Li DAR, optics, radar and other multi-load sensors to realize the collaborative control and synchronous data acquisition of multi-load sensors; The analytical model of agricultural key parameters, such as geometry of remote sensing data, radiation correction, physico-chemical ecology and structure of crops, will be studied, and the UAV data processing software system will be built. Carry out ground experiments to evaluate UAV remote sensing platform, load and processing model accuracy.

3.7 Intelligent High-Resolution Remote Sensing Data Analysis Technology

Agricultural high-resolution remote sensing data product classification and database specifications, data sharing technical service specifications, to achieve multi-source mass high-resolution remote sensing data-efficient management; The geometric and radiometric calibration techniques of high-resolution remote sensing data in a large area are established to realize accurate preprocessing of remote sensing data under complex terrain and meteorological conditions. The multi-load high-resolution satellite remote sensing information fusion analysis technology will be established to realize the quantitative analysis of crop physicochemical, ecological parameters and farmland environmental factors. The ground reliability detection system of high-resolution satellite remote sensing in typical agricultural production areas is established to realize the comprehensive test of the loading performance, geometric, radiometric correction and the precision of analytical parameters of high-resolution satellite remote sensing. The establishment of high resolution remote sensing big data cloud service platform and system for agriculture can realize efficient processing and service of multi-scale high resolution remote sensing data.

We have discussed the innovative application of AI technology in agriculture above. Next, we will take a precise diagnosis of fruit and vegetable epidemic disease as an example to explore the accurate diagnosis information system of fruit and vegetable epidemic disease based on artificial intelligence technology.

4 Accurate Diagnosis Information System of Fruit and Vegetable Epidemic Disease Based on AI Technology

Based on the existing fruit and vegetable truck system, based on AI technology, the quarantine system, the analysis system of epidemic areas, the reporting system of an epidemic situation will be added, and a new set of intelligent quarantine equipment will be designed. The quarantine results of common diseases and insect pests fed back by the logistics vehicles in the process of logistics transportation can be used to prohibit the logistics vehicles carrying epidemic diseases from warehousing. In addition, big data analysis based on Ali cloud and accurate artificial intelligence monitoring will be combined to analyze the current epidemic situation and report it to relevant departments for emergency decisions. The research contents are as follows:

4.1 Design of Inspection Equipment for Various Diseases and Insect Pests in Fruits and Vegetables

Firstly, a variety of diseases and insect pests detection equipment using a vehicle-mounted power supply will be designed. It is planned to be a box type. After fully stirring and mixing the 50 kg fruits and vegetables sampled for a period of time, PCR amplification will be conducted for multiple rounds. Then the electrophoretic PCR results will be sequenced for DNA to detect the presence of associated infection.

4.2 Upgrade and Improve the Existing Fruit and Vegetable Logistics Traceability System

There are real-time tracing functions in the existing fruit and vegetable tracing system, but there is no big data monitoring and precise positioning and reporting of epidemic areas. To save investment, eliminate information-island, and achieve multi-department data sharing, we will add epidemic monitoring, epidemic area positioning and real-time reporting functions to the existing fruit and vegetable tracing system.

4.3 Big Data Analysis and Artificial Intelligence Accurate Monitoring Based on Ali Cloud

At present, there are thousands of logistics vehicles and corresponding warehouses, all of which need to analyze the epidemic situation of fruits and vegetables. A cloud big data analysis system needs to be built to analyze and summarize the epidemic situation results. We can use Ali cloud as a one-stop big data solution. Ali plus provides convenient enterprise-level data warehouse, business intelligence, machine learning, data visualization and other practical functions, which can reduce the development difficulty, quickly generate logistics thermal maps, quarantine thermal maps and precise positioning of epidemic areas, and reduce the overall cost of ownership.

4.4 Interface with the Government Epidemic System

The system interfaces with the government epidemic system, allows rapid detection of endemic diseases and pests during the incubation period. The rapid positioning and reporting of fruit and vegetable outbreaks can avoid the economic losses of users' warehouse logistics and enable government agencies to respond quickly before the outbreak of epidemics. It is of great positive significance to the national economy and people's livelihood to cull the epidemic in the bud.

5 Conclusion

The development of artificial intelligence and IoT brings new opportunities and challenges for the development of new intelligent agriculture. New technologies have gradually subverted the traditional agricultural model, and machines have replaced many traditional human jobs. Machine learning will become more and more important in intelligent agriculture. Using artificial intelligence and other new technologies to transform traditional agriculture and carrying out precise management, intelligent control, scientific planting, and production-oriented development of agricultural production factors will liberate human resources and bring significant improvement in productivity and continuous improvement in the quality of agricultural products. This is bound to be the trend of agricultural development in the future.

References

1. Sun, G., Fang, Y., et al.: Application of artificial intelligence in intelligent agriculture. J. Jilin Normal Univ. Eng. Technol. **35**, 93–95 (2019)
2. Chen, G.F., Li, J., Chen, H., et al.: Research progress of artificial intelligence technology in agriculture in the era of big data. J. Jilin Agric. Univ. **4**, 502–510 (2018)
3. Ling, Y.: Application of artificial intelligence technology in agriculture. Comput. Knowl. Technol. (29), 181–183 202 (2017)
4. Chen, T.J., Zeng, J., Xie, C.J., et al.: Intelligent pest and disease identification system based on deep learning. China Plant Protect. Guide (4) 6 (2019)
5. Zhao, C.J.: Artificial intelligence leads agriculture into a new era. China Rural Sci. Technol. (1), 29–31(2018)
6. Chen, G.F., Li, J., Chen, H., et al.: Research progress of artificial intelligence technology in the field of agriculture in the era of big data. J. Jilin Agric. Univ. **40**(4), 502–510 (2015)
7. Deng, W.T., Zhao, Y.C., Yuan, S., et al.: Development and application of intelligent agriculture based on IOT. Agric. Netw. Inf. **29**(12), 9–12 (2014)
8. Liu, X., Zheng, H.Y., Shi, N.Q., et al.: Progress in the application of artificial intelligence in agricultural production. J. Fujian Agric. **28**(6), 609–614 (2013)
9. Li, F.: Agricultural technology innovation based on artificial intelligence. J. Agric. Netw. Inf. **26**(11), 20–23 (2017)
10. Mao, L., Cheng, W.L.: Application of artificial intelligence technology in modern agricultural production. Agric. Netw. Inf. **5**, 14–18 (2018)
11. Wang, Y., Zheng, Z.H.: Application direction and development path of artificial intelligence technology in agriculture. Inf. Commun. Technol. Policy **6**, 29–31 (2019)
12. Shi, Y.L.: Application of artificial intelligence technology in modern agricultural production. South. Agric. Mach. **50**(14), 73 (2019)

Facial Expression Extraction Based on Wavelet Transform and FLD

Lichun Yu[1(✉)] and Jinqing Liu[2]

[1] Fuzhou University of International Studies and Trade, Fuzhou 350202, China
84246706@qq.com
[2] Fujian Normal University, Fuzhou 350007, China

Abstract. In order to improve the recognition rate of face recognition, a face expression extraction method is designed. Firstly, wavelet transform was used to extract low-frequency components, then PCA was used to map low-frequency images to low-dimensional space, and finally, FDA method was used to extract face features in low-dimensional space. Experimental results show that this method can effectively overcome the influence of facial expression and gesture changes on face recognition, and not only improve the face recognition rate, but also improve the recognition speed.

Keywords: Wavelet transform · Principal component analysis · Fisher linear discriminant analysis · Eigen face

1 Introduction

Face recognition is a challenging research topic in the field of pattern recognition and artificial intelligence. Face recognition is an important biometrics technology which uses computer to analyze face image and extract the effective identification information for personal identity authentication. Face recognition involves artificial intelligence, pattern recognition, computer graphics, computer vision and other fields, and has a very wide range of applications in public security, commercial enterprises, places access and other fields, such as public security criminal investigation and recovery, access control, identification, video monitoring, etc. [1]. In recent years, face recognition technology has developed rapidly and achieved a lot of research results, but there are still many challenging problems. Such as illumination change, size change, facial expression change and other reasons lead to a sharp decline in face recognition rate. It is relatively simple to overcome the influence of light change and size, and significant effects can be achieved by preprocessing the face image before recognition [2]. It is difficult to solve the problem of recognition rate decline caused by facial expression change, so the research on facial expression recognition is a challenging work [3]. In order to solve this problem, a face recognition method based on weighted wavelet transform and FLD is proposed.

Q. Liu et al. (Eds.): CENet 2020, AISC 1274, pp. 226–232, 2021.
https://doi.org/10.1007/978-981-15-8462-6_26

2 Weighted Wavelet Transform of Face Images

Wavelet analysis is a new time-frequency analysis method, which is the development and extension of Fourier analysis theory. Its essence is to filter signals with a set of multi-scale high-pass and low-pass filters, decompose the low-frequency and high-frequency parts of signals into different frequency bands, and then do further analysis and processing [4]. The advantage of wavelet transform is that it has multi-resolution analysis and multi-scale decomposition, and can characterize the local characteristics of signals in both time and frequency domains. In the process of image processing, a picture is regarded as a two-dimensional matrix, and the face image is decomposed by wavelet transform in frequency domain. After a single wavelet transform, the image is decomposed into four sub-band images, whose size is 1/4 of the original image. The low frequency sub band LL1, the high frequency sub-band horizontal component, vertical component and diagonal component are LH1, HL1, HH1, respectively. Low-frequency sub band LL1 can be further decomposed into four sub bands LL2, LH2, HL2 and HH2, whose size is 1/16 of the original image. One-dimensional and two-dimensional decomposition of face images are shown in Fig. 1. Research shows that facial expression changes only affect the high-frequency part, while the low-frequency part is not sensitive to facial expression changes. Therefore, only the low-frequency part is retained and the high-frequency part is removed, the structural feature information of the face can be retained. Moreover, with the decomposition of wavelet, the size of low-frequency sub band image decreases gradually, which effectively reduces the storage space and computational complexity, which is more important to improve the recognition speed.

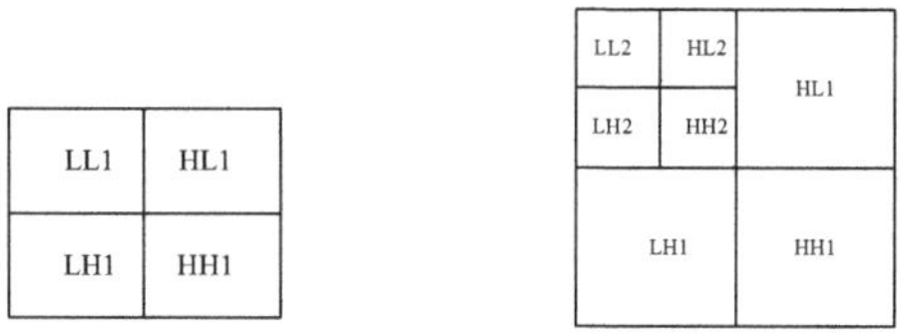

a. one-dimensional Decomposition b. Two-dimensional Decomposition

Fig. 1. Wavelet decomposition.

a. original image b. one-dimensional decomposition

c. two-dimensional decomposition d. LL2 Low frequency component

Fig. 2. Wavelet two-dimensional decomposition.

Figure 2 is a face image from Yale face database, which shows the two-dimensional decomposition process of face wavelet. LL2 is a two-dimensional decomposition of low-frequency components, the size of which is 1/16 of the original image. From the picture, we can see that the main expression of facial expression is eyes and mouth. In horizontal sub-band HL, eye and mouth features are more obvious, while in vertical subband LH, vertical features are more prominent, and the features of face contour and nose are more obvious. The information of diagonal sub-bands represents the overall structural features, which is more important in representing the features of rigid objects, but the face is a non-rigid structure, so the diagonal sub-bands are vulnerable to the influence of expression and posture, and the stability is poor.

Most of the information of the face image is concentrated in the low-frequency part, which reflects the overall profile of the face. The stability of low frequency sub-band is better for the change of facial pose and expression, which can overcome the influence of the change of facial pose and expression. Horizontal subband and vertical subband contain facial expression information and edge contour information respectively, while diagonal subband contains overall information. Four wavelet sub-bands are weighted and fused to extract the facial feature information which is beneficial to face recognition. The feature extraction formulas are as follows:

$$X = w_1 \times LL2 + w_2 \times LH2 + w_3 \times HL2 + w_4 \times HH2 \quad (1)$$

w_1, w_2, w_3 and w_4 represent different weights.

3 EigenFace

EigenFace's idea is to transform face from pixel space to another subspace, and then do similarity calculation in subspace [5]. The spatial change method used by EigenFace is principal component analysis [6] (PCA). EigenFace maps face image from high-dimensional image space to low-dimensional feature space by principal component analysis, and then classifies and recognizes the reduced image data. Its basic principle is to use K-L transform to express face image and generate a set of orthogonal base plane images. This method considers that different expressions can be separated in low-dimensional face images. The K-L transform feature matrix is arranged in descending order, and the corresponding feature is U = [u1, u2,..., un]. The face image is decomposed into several feature components. Because the feature components of the face image are very similar to those of the face, they are called "EigenFace" and can reconstruct the original face image from the partial weighted coefficients of the feature faces. Finally, these feature vectors are used as discriminant features for face classification and recognition. M main eigenvectors (m < n) are selected as orthogonal bases. The training samples and recognition images are projected into the new m-dimensional space respectively, and the projection coefficients are obtained to represent the determined faces. At this time, the problem of face recognition is transformed into the classification of coordinate coefficients. There are two main problems in EigenFace method. Firstly, PCA trains the total scatter matrix of samples, so it cannot train the class information of samples. It means the feature extraction method adopted by PCA is unsupervised. Secondly, the image matrix can be transformed into image vector, which can simplify the computation, but with the increase of the number of training images, the whole feature extraction process has a large amount of calculation. In the process of training and recognition, there are still difficulties in high-dimensional vector and matrix operations.

4 Linear Discriminant Analysis

Linear Discriminant Analysis (FLD), also known as Fisher Linear Discriminant (FLD), is an improvement based on eigenface method. Its purpose is to extract the most discriminant low-dimensional features from the high-dimensional features, and these low-dimensional features can make the same category of samples as concentrated as possible, and different categories of samples as separate as possible, so as to minimize intra-class distribution and maximize out-class distribution.

Suppose there are N facial expression images, $\{P_i\}$, i = 1, 2,…, N. Each image is represented by n-dimensional vectors. Then the inter-class scatter matrix S_b and the intra-class scatter matrix S_w are respectively

$$S_b = \frac{1}{N}\sum_{i=1}^{c} N_i(m_i - m)(m_i - m)^T \tag{2}$$

$$S_w = \frac{1}{N}\sum_{i=1}^{c} N_i S_i \tag{3}$$

N is C_i (i = 1, 2,… (c) the number of training samples of class; m_i is the mean vector of class C_i.

$$m_i = \frac{1}{N_i}\sum_{x \in c_i} x_i \tag{4}$$

m is the mean of all samples,

$$m = \frac{1}{N}\sum_{x} x \tag{5}$$

The goal of LDA is to find an optimal projection direction, W_{opt}, to maximize the ratio of inter-class dispersion Sb to intra-class dispersion S_w. Therefore, the Fisher criterion can be defined as:

$$J(W_{opt}) = \arg\max \frac{|W^T S_b W|}{|W^T S_w W|} \tag{6}$$

The selected features make the ratio of inter-class dispersion Sb and intra-class dispersion Sw maximum, which can better represent the differences between classes and facilitate classification. PCA method can find the similarity between the two images very well. LDA method focuses more on the details of the image. Because the details are the key to distinguish facial expressions, LDA has more advantages in facial expression recognition. In practical applications, it is used to extract facial features directly. This is because the number of LDA training samples is much smaller than the dimension of image vector, which will lead to the singularity of the intra-class scatter matrix SW.

5 Face Recognition Based on Weighted Wavelet Transform and FLD

Although the feature face method can achieve the goal of dimensionality reduction, it has a large amount of computation, and does not consider the classification information of training samples. Therefore, it can first reduce the amount of data by weighted two-dimensional wavelet transform. LDA method can distinguish the information of training samples, but it is difficult to solve the problem. Therefore, an effective improved method is the combination of PCA method and FLD method, namely Fisher linear discriminant method. This paper presents a facial expression recognition method based on wavelet transform and Fisher linear discriminant. The recognition process of the algorithm is divided into two stages: training stage and recognition stage. The system identification process is shown in Fig. 3. In the training stage, all samples are trained by two-dimensional discrete wavelet transform. The low-frequency sub-image is obtained instead of the original image. Low-frequency sub-image engraving is a face invariant feature, which has good stability and reduces the amount of computation. Then PCA transform training samples of each low-frequency sub-image are used to construct subspace. On this basis, the feature subspace of LDA method is obtained and the eigenvector library is calculated.

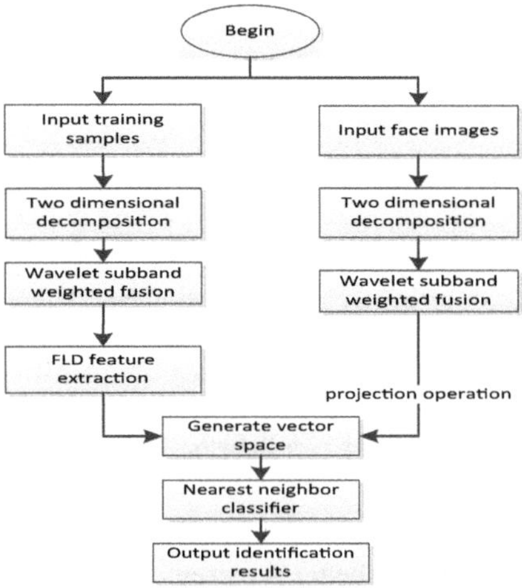

Fig. 3. Face recognition process.

For the recognition stage, firstly, two-dimensional wavelet transform is applied to the face image to get the low-frequency features projected into the corresponding feature subspace. The similarity matrix is obtained, and then face recognition is carried out by using the nearest neighbor classification (KNN), i.e. calculating the Euclidean distance between the feature vector space extracted from the face image of the test sample and the feature vector of each training sample set, and taking the class of the training sample belonging to the nearest distance of the face image of the test sample as the class of the training sample. Classification of Test Sample Images.

6 Experiment

The experiment uses Yale Face Database [7], which contains 15 people, 11 photos per person and 165 photos, mainly including changes in lighting conditions and facial expressions. In the experiment, the first six pieces of each person are taken as training samples, the last five as testing samples. The total training samples are 90 pieces and the test samples are 75 pieces. Four algorithms, PCA, FLD, 2DWT + PCA and 2DWT + FLD, are used to carry out the experiment. The training time, recognition rate and recognition time are shown in Table 1.

From Table 1, we can see that FLD algorithm is superior to PCA algorithm in expression recognition from the recognition rate. The recognition rate of 2DWT + FLD algorithm is obviously better than the other three algorithms. Compared with training time and recognition speed, 2DWT + FLD training time is the shortest and recognition speed is the fastest. Therefore, it can be seen that the combination of two-dimensional wavelet transform and FLD face expression recognition method not only improves the recognition rate, but also improves the recognition speed.

Table 1. Face recognition algorithm comparison.

Method	Training (sec)	Recognition rate (%)	Recognize (sec)
PCA	75.51	88.7	2.87
FLD	66.34	91.23	2.31
2DWT + PCA	24.29	94.12	1.92
2DWT + FLD	19.7	95.8	1.36

7 Conclusion

A facial expression recognition method based on facial expression recognition is proposed for facial expression transformation. In the aspect of wavelet features, firstly, low frequency coefficients are selected as features by two-dimensional wavelet transform, and then Fisher linear method is used to further extract the features. Wavelet decomposition is used as the result of image scaling. Generally, the noise is reduced and the recognition rate is improved. The results show that FLD linear discriminant method is more suitable for facial feature extraction. Wavelet function combined with

FLD can further improve the final recognition rate and the execution efficiency. In feature selection, this paper only refers to principal component analysis. Later, we will try to use two-dimensional principal component analysis (2D-PCA) and two-dimensional principal component analysis (BD-PCA) to extract features. At the same time, in the face database, we only use the Yale facial database experiment. In the future, we will use other facial expression databases.

References

1. Xiao, J.F.: Research on the application of face recognition technology in the field of public security. computer science (S2) (2016)
2. Zhang, R.F., Wei, F.S.: Image scene classification combined with fisher discriminant analysis and sparse coding, J. Comput. Aid. Des. Graph. (05) (2015)
3. Wang, D.F., Zhou, J.S., Mei, H.T.: A review of facial expression recognition. Comput. Eng. Appl. (20) (2014)
4. Gu, S.F.: Face preprocessing algorithm under complex light conditions. Comput. Eng. Appl. (19) (2014)
5. Zuo, T.F.: Overview of face recognition technology. Softw. Guide **16**(02), 182–185 (2017)
6. Zhang, N.F., Wu, X.S., Ma, Y.T.: Research on the application of DCT coefficient recombination in eigenfaces. Comput. Eng. Appl. (19) (2011)
7. YALE Homepage. http://vision.ucsd.edu/datasets/yale_face_dataset_original/yalefaces.zip. Accessed 21 Apr 2020

Research on Content-Based Reciprocal Recommendation Algorithm for Student Resume

Jianfeng Tang(✉) and Jie Huang

School of Software Engineering, Tongji University, Shanghai 201804, China
{billtangjf,huangjie}@tongji.edu.cn

Abstract. At present, the emergence of this kind of recruitment website promotes the recommendation calculation method's rapid progress. However, the old recommendation algorithm is based on the preferences or requirements of individual participants. The current job-hunting situation should give not only excellent employees but also give suitable jobs to job seekers. So there is this kind of Reciprocal Recommendation Method. The purpose of such a way is to recommend customers one-to-one. There is a similarity between the two, and use this similarity to ensure the suitability and stability of the job. In this system, enterprises can confirm the keywords needed within the range of resume selection setting, and the system will select students' resumes according to the keywords set. In this paper, we combine the characteristics of Chinese resume, analyze the two-way reciprocal recommendation mode of school enterprises and students, and finally use the simulation test to test the accuracy and effect of the content-based recommendation algorithm.

Keywords: Reciprocal recommendation · Preference · Accuracy rate · Recall rate

1 Content-Based Reciprocal Recommendation of Student Resume

In this paper, we will mainly analyze the content-based resume reciprocal recommendation process. We first discuss the significance of reciprocal recommendation of resumes and then extract the characteristics of Chinese resumes based on customers' requirements, then push to the customers.

1.1 Reciprocal Recommendation Method

At present, the emergence of this kind of recruitment website promotes the rapid progress of the recommendation calculation method [1]. Nevertheless, the old recommendation algorithm is based on the preferences or requirements of individual participants. The current situation of job-hunting should give not only excellent employees but also give suitable jobs to job seekers. So there is this kind of Reciprocal Recommendation Method [3]. The purpose of such a way is to recommend customers

Q. Liu et al. (Eds.): CENet 2020, AISC 1274, pp. 233–239, 2021.
https://doi.org/10.1007/978-981-15-8462-6_27

one-to-one [4]. There is a similarity between the two, and use this similarity to ensure the suitability and stability of the job. The following describes the detailed process.

Let be the recommended method, customer user selected and Object is the recommended project as a whole. It is judged that the customer's preference for project j is lower than that for a project i. Formula 1 is the recommended formula.

$$Recommend1(user) = \{i : prefer1(user, i) > prefer1(user, j)\} \quad (1)$$

Here, the degree of preference *prefer1* indicates the customer's preference for a project.

Reciprocal recommendation should also consider the project, so a second recommendation should be established. Formula 2 is the detailed recommended formula.

$$Recommend2(i) = \{user : prefer2(i,\ user) > prefer2(i,\ userv)\} \quad (2)$$

Here, the symbol that indicates the customer in another collection is userv, and the symbol that indicates the project's preference for customers is *Recommend2(i)*.

Therefore, the recommended method should meet the requirements of the above two formulas. Formula 3 is the detailed formula.

$$RR(user) = \{i :\ i \in Recommend1(user)\ and\, user\ \in Recommend2(i)\} \quad (3)$$

In terms of reciprocal recommendation, it will produce a unique final value, and the final value can also be presented in the form of a list. It's best to use the form of a list, which can achieve the degree of reciprocity, that is, the evaluation score requirements of perfer1 and 2. Then, arrange the values from high to low, and formula 4 is the detailed formula of reciprocity degree.

$$preferRR(user, i) = w_1\, prefer1(u, i) + w_2 prefer2(i, u) \quad (4)$$

Here, the weight factor is represented by w_1 and w_2,and the sum of the two is 1.

1.2 Feature Extraction of Chinese Resume

Compared with the general text, this Chinese resume is easy to obtain [2]. This is because the resume has a relatively fixed form, and the matching operation of the expected content of the customer and the company is relatively simple.

Table 1 is the corresponding relationship table of the corresponding association between the attributes of job seeking and recruitment enterprises.

Detailed discretization of various resume categories should be carried out here. For example, the salary is divided into four levels according to the requirements of the company, which includes 10 k–15 k, 8 k–10 k, 4 k–8 k, and detailed salary should be discussed face to face. The educational background is divided into four levels,

Table 1. Corresponding relationship between the attributes of job seeking and recruitment Enterprises

Category in	Complex categories	Basic categories
Resume content	Previous work experience, Education background, Professional skills, Expected responsible position	Name, Age, Diploma
Job seeker expectations	Expected salary, Vacation mode, Overtime or not, etc.	Expected salary and benefits, Expected future workplace, Position applied for
Company requirements	Work or practice experience, Education background, Major, Skill	The salary that can be given, The location of the work, The nature of the work, The position offered

including doctor, master, undergraduate, junior college, etc. In this way, the resume information can be operated exactly according to the category. Formula 5 uses a matching formula to represent a detailed association.

$$compare\left(Q_{company}, P_{user}\right) \tag{5}$$

Here, the company's expectations are represented by the symbol $Q_{Company}$, and customer preferences are represented by the symbol P_{user}.

1.3 Content Based Reciprocal Recommendation Algorithm

Because students are the object of this comprehensive management system of school enterprise cooperation, they should be converted according to the real requirements according to the attribute categories mentioned above. In this system, enterprises can confirm the keywords needed within the range of resume selection setting, and the system will select students' resumes according to the keywords set. In this process, students and enterprises are recommended customers and served customers respectively, which can improve the accuracy of recommendation.

The reciprocal recommendation method of resume can meet the needs of students and enterprises. Using the established keyword search can improve the success rate of recruitment for internships. According to the variables set forth above, formula 6 below is the formula for students' basic data to meet the requirements of the enterprise.

$$compare\left(Q_{stu}, P_{job}\right) \tag{6}$$

The following formula 7 is the formula for enterprises to meet the requirements of students.

$$compare\left(Q_{job}, P_{stu}\right) \tag{7}$$

The following formula 8 shows the fit degree between the actual experience of the students and the positions they apply for.

$$compare\left(Stu_{resumw}, Company_{job}\right) \tag{8}$$

According to the calculated average of the two, we get the reciprocal score between the applied position and the reward. Rank according to the scores and push them to the enterprise top N.

(1) The statistics of students' basic data to reach the school enterprise preference are shown in the following formula 9.

$$compare\left(Q_{stu}, P_{job}\right) = \frac{\sum_{j=1}^{3}\left(w_j \times Q_{\text{stuj}} \times prefer_{jobj}\right)}{\sqrt{\left(\sum_{j=1}^{3} Q_{stuj}^2\right) \times \left(\sum_{j=1}^{3} prefer_{jobj}^2\right)}} \quad and\ w_j = \frac{\sum_{k=1}^{M} a_{jk}}{\sum_{j=1}^{4}\sum_{k=1}^{M} a_{jk}} \tag{9}$$

Here, the weight of each preference value of the enterprise is w_j. In most cases, a position will receive a lot of resumes before a real position is determined. M is used to represent the number of resumes received. Before the job of this position determines the personnel, if the j-th characteristic of this position obtains the k-th resume, the enterprise considers that the k-th resume meets the conditions. Then 1 is assigned to a_{jk}, otherwise 0 is assigned to a_{jk}.

(2) The satisfaction of the students with the basic data of the enterprise is shown in the following formula 10.

$$compare(Qjob, P_{stu}) = \frac{\sum_{j=1}^{4}\left(w_i \times Q_{\text{jobi}} \times prefer_{stuj}\right)}{\sqrt{\left(\sum_{i=1}^{4} Q_{jobi}^2\right) \times \left(\sum_{i=1}^{4} prefer_{stuj}^2\right)}} \tag{10}$$

Here, students' preferences are represented by symbols w_i. In the real process of putting in resumes, stu represents students, and N represents the number of positions put in resumes. If the i-th attribute of the student's satisfaction matches the demand in the k-th position, then assign 1 to a_{ik}, otherwise, assign 0 to a_{ik}.

(3) For job seekers, their work experience should be in line with the needs of the recruitment company. However, because the target of this system is students, if the job-hunting content includes work experience, the statistics of this content can

not be counted in the total number. If you want to count in the total, you can use cosine similarity to achieve. Formula 11 is the detailed calculation formula.

$$compare\left(Stu_{resume},\ Company_{job}\right) = \frac{Stu_{resume} \bullet Company_{job}}{\left|Stu_{resume}\right| \times \left|Company_{job}\right|} \tag{11}$$

(4) Using harmonic mean to represent the relationship between enterprises and students when their preference degree is the maximum, as shown in formula 12 below.

$$r(Stu, Job) = \frac{3}{\left(c\left(Q_{stu}, prefer_{job}\right)^{-1}\right) + \left(c\left(Q_{stu}, prefer_{job}\right)\right)^{-1} + \left(c\left(Stu_{resume}, Company_{job}\right)\right)^{-1}} \tag{12}$$

(5) Arrange the reciprocal scores obtained in the previous link in order, select top 5 or top 10 according to the order, and then push them to the enterprise.

2 Simulation Experiment

In order to ensure the effectiveness of the established recommendation method based on content recommendation and information extraction, we can take simulation experiments to test its effect and ensure that this method can be used in real situations.

2.1 Experimental Design

In the process of establishing simulation experiment, we should design simulation test data set first. Here, we select the real recruitment information in the computer recruitment industry from the network as the data set of this simulation experiment.

For example, choose computer industry positions, which cover 80 positions, such as front-end UI designer, senior software engineer and software programmer, etc. The number of resumes submitted for these 80 positions is 2700. From the past information analysis, 14000 resumes were put in, and 12800 were job information. Here, 30% of the information will be selected as the basic data to be measured.

According to the experimental information described above, use the rule-based resume extraction method to extract the basic information such as work place, education background, number and major. This paper uses the method of resume data extraction based on machine learning and statistics to extract the more complex work nature, work scope, work experience and work content. Then it analyzes the extracted information, and the content-based reciprocal recommendation method can meet the needs of school enterprises for recruitment, as well as the needs of students for positions and wages.

Here, we should also develop experimental standards, which can make use of the standards commonly used in the recommendation system, including F1 value, recall rate and accuracy rate. Formula 13, 14 and 15 are the detailed formulas.

$$\text{Accuracy Rate} = \frac{\sum_{job \in J} R(job) \cap T(job)}{\sum_{job \in J} T(job)} \tag{13}$$

$$\text{Recall Rate} = \frac{\sum_{job \in J} R(job) \cap T(job)}{\sum_{job \in J} R(job)} \tag{14}$$

$$F_1 = \frac{2 \times \text{Accuracy Rate} \times \text{Recall Rate}}{\text{Accuracy Rate} + \text{Recall Rate}} \tag{15}$$

Here, the list of job application forms pushed is *R(job)* and the job application form list for recruitment is *T(job)*. Using the F1 value, recall rate and accuracy rate and other statistical results described above, the operation results of the content based reciprocal recommendation algorithm are discriminated to determine the feasibility, accuracy and effectiveness of the results.

2.2 Result Analysis

At present, collaborative filtering algorithm is the most recommended algorithm, so this paper compares the old collaborative filtering algorithm with the recommendation algorithm used in this paper, as shown in Fig. 1.

We can draw a conclusion that the recommendation method used in this paper is more perfect. For example, the F1 of collaborative filtering recommendation method is 26%, while the F1 used in this paper is 36%. The recommendation method of this paper refines the main information and classifies every module in the resume. These are used to evaluate the categories of resumes, and the standards are also divided in detail.

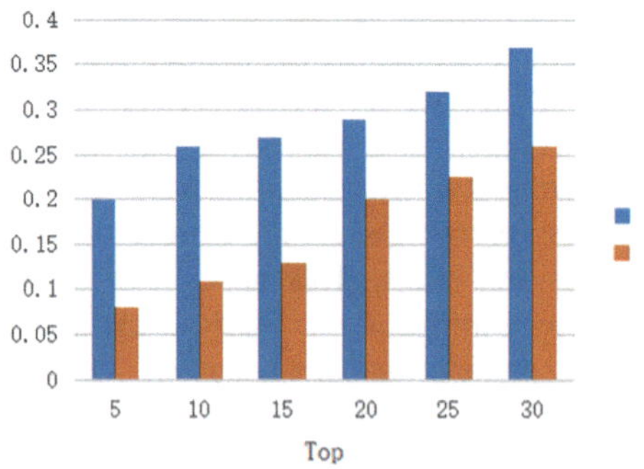

Fig. 1. The comparison between the recommended algorithm and the traditional collaborative filtering algorithm.

We divide education into four categories. Therefore, we can classify the detailed recommendation data according to the requirements of real students and enterprises to obtain more perfect results.

F1 Value the recommendation algorithm in this paper the collaborative filtering algorithm.

In general, using the comparison results in Fig. 1, we can see that the F1 value of the collaborative recommendation algorithm is smaller than that of the reciprocal recommendation algorithm used in this paper. Therefore, the mutual recommendation method adopted in this paper is more perfect than the traditional collaborative filtering algorithm.

3 Summary

Based on the characteristics of Chinese resume, this paper analyzes the algorithm of student resume based on content recommendation. Then it analyzes the two-way reciprocal recommendation mode of school enterprise and students. Finally, the simulation test is used to test the accuracy and effect of the content recommendation based resume algorithm.

References

1. Xia, P., Liu, B.Y., Sun, Y.Z., Chen, C.: Reciprocal recommendation system for online dating. In: Proceedings of the 2015 IEEE/ACM International Conference on Advances in Social Networks Analysis and Mining, pp. 234–241. Association for Computing Machinery Inc., Paris, France (2015)
2. Gu, N.N., Feng, J., Sun, X., Zhao, Y., Zhang, L.: resume information automatic extraction and recommendation algorithm. Comput. Eng. Appl. **53**(18), 141–148 (2017)
3. Xia, P., Zhai, S.F., Liu, B.Y., Sun, Y.Z., Chen, C.: Design of reciprocal recommendation systems for online dating. Soc. Netw. Anal. Min. **6**(1), 106 (2016)
4. Koprinska, I., Yacef, K.: People-to-people reciprocal recommenders. In: Recommender Systems Handbook, 2rd Edition, pp. 545–567 (2015)

Research on Personalized Recommendation Based on Big Data Technology

Jinhai Li(✉), Shichao Qi, Longfeng Chen, and Hui Yan

School of Computer Science and Technology,
Taizhou University, Taizhou 225300, China
ljh-hk@163.com

Abstract. Aiming at the sparsity problem existing in the traditional collaborative filtering algorithm, this paper proposed an improved similarity computing method that integrated user rating behavior and item attributes. The sparse matrix is evaluated and predicted by the similarity calculation method, then the prediction rating was filled in the sparse matrix. At the same time, in the context of big data, the data scale was too large to affect the execution efficiency of the recommendation system. Hadoop platform was adopted to implement collaborative filtering recommendation algorithm based on the improved similarity model. Based on large-scale data segmentation, the distributed parallel processing was carried out. The proposed improved algorithm is verified by Movielens which was an internationally standard data set. The verification results show that the personalized recommendation system based on Hadoop platform and improved recommendation algorithm has better recommendation performance.

Keywords: Big data · Personalized recommendation · Similarity algorithm · Data sparsity · Distributed parallel processing

1 Introduction

Personalized recommendation is a new development opportunity, according to the big data to carry on the design and optimization, improve the efficiency of the data processing of personalized system, take the initiative to filter the useless information, and recommend suitable products to improve the user experience is a realistic problem [1, 2].

At present, the most widely used and successful technology in recommendation field is collaborative filtering recommendation algorithm. The collaborative filtering algorithm is based on the basic idea of finding the user set or item set similar to the target users and the target items [3]. The user-item scoring matrix becomes huge as the number of users and items grows rapidly. Also because of differences in personality characteristics such as personal interests and hobbies, users usually use only a small number of items. This results in insufficient data for users to comment less on the project. This presents the problem of sparse user-project rating matrix in collaborative filtering recommendation system [4]. In addition, due to the increasing amount of information, the collaborative filtering recommendation system also faces the problems of storage space and scalability of analysis and calculation. Nowadays, most of the research of recommendation

Q. Liu et al. (Eds.): CENet 2020, AISC 1274, pp. 240–247, 2021.
https://doi.org/10.1007/978-981-15-8462-6_28

algorithms focuses on improving the efficiency of the algorithm itself, but because the number of users and goods in the Internet environment is increasing rapidly, the application value of the improved algorithm to speed up the calculation is very limited, and the promotion space of the algorithm is gradually becoming smaller [5]. In addition, the current operation mode of single machine large server is often constrained by hardware performance, only improving the computing power of large server and expanding storage space cannot solve these problems perfectly [6].

Therefore, while improving the recommendation algorithm of collaborative filtering to improve the recommendation accuracy, this paper introduces big data technology, based on the distributed parallel processing characteristics of Hadoop platform, improves the execution efficiency of the recommendation algorithm and shortens the recommendation time that users need to wait for.

2 Big Data Distributed Parallel Processing Platform Construction

The construction of big data distributed parallel processing platform mainly involves HDFS distributed file system and MapReduce distributed data processing model.

2.1 HDFS Distributed File System

For HDFS distributed file systems, there are mainly two different types of nodes, one of which is the Namenode node and the other is the Datanode node [7]. These nodes are run in a manager-worker mode. The Namenode node is the manager of the whole cluster, whose responsibility is to manage the namespace of the whole distributed file system and maintain Metadata of the file system. Datanode is the worker in the entire distributed cluster, mainly completing the storage of data. where the data is split into multiple file blocks of the same size that are distributed to multiple nodes in the cluster and need to send their own required storage information to the Namenode node periodically.

Based on the HDFS distributed file system, the user data and the project data in the recommendation system are stored to realize the distributed storage of the massive recommendation data.

2.2 MapReduce Distributed Data Processing Model

MapReduce is a model used for computing in the entire Hadoop ecosystem and a distributed computing method for computing massive amounts of data [8]. This structure makes it easier for developers to process data in large distributed systems. The calculation of specific business logic is as follows: the first is to big file for decentralized operation processing, split into multiple small files with the same size, these small files are processed in parallel by multiple map processes, and the processed results are directly transferred to the reduce process, which will directly summarize and calculate the map results, and the final calculation results will be directly output to HDFS.

3 Improved Collaborative Filtering Algorithm Based on User Behavior Model

3.1 Traditional Similarity Algorithm Analysis

Similarity computation is the most critical step in the collaborative filtering recommendation algorithm. The accuracy of the similarity calculation is related to the recommendation accuracy of the recommendation system. For traditional similarity measurement methods, the more typical calculation method is the improve cosine similarity calculation method [9].

$$sim(i,j) = \frac{\sum_{u \in U} (R_{u,i} - \overline{R}_u)(R_{u,j} - \overline{R}_u)}{\sqrt{\sum_{u \in U} (R_{u,i} - \overline{R}_u)^2 \sum_{u \in U} (R_{u,j} - \overline{R}_u)^2}} \tag{1}$$

Where $\overline{R}_u$ represents the average of all user ratings. By calculating the similarity between the items that have been calculated, and then making an effective prediction of the items that have not been graded, a weighted summation is mainly used in the process of prediction. The specific formulas are as follows:

$$p_{u,i} = \frac{\sum_{all\ similar\ items,N} (s_{i,N} * R_{u,N})}{\sum_{all\ similar\ items,N} (|s_{i,N}|)} \tag{2}$$

In formula (2), $S_{i,N}$ represents the similarity between the current item and the other items N, and $R_{u,N}$ represents the current user's rating of the formation of the item N.

3.2 Problems in Traditional Similarity Algorithms

For the traditional similarity calculation process, generally based on the two users to do the common scoring vector operation processing, but in general, each user's own energy is relatively limited. It's impossible for each user to rate all participating projects, and many users simply rate fewer projects. Therefore, in the process of recommending system design, the actual number of user rating items may be far less than the total number of user rating information. In this case, the problem of sparsity may exist in the later stage of computation.

In order to solve the problems caused by the problem of sparse scoring data, many scholars and related experts have proposed this model of rating matrix filling between users-items [3]. The simplest way is to set a fixed default for missing items. Generally, it is set to the average information in the project.

Aiming at the above problems, this paper proposes an improved similarity calculation method that integrates user rating behavior and item attributes. The sparse matrix is predicted by the similarity calculation method, and then the sparse matrix is filled with the prediction score. Finally, the similarity calculation of the filled matrix is recommended.

3.3 Improved Collaborative Filtering Algorithm Based on User Behavior Model

In the actual behavior of user rating, the behavior of user rating can show the similarity of user's interest. At the same time, the more users score project i and project j, and the closer the score is, the better the similarity between the two projects can be explained. In order to measure the similarity between the two items, two factors of user rating behavior were considered: the number of users with similar ratings and the number of users without similar ratings. The paper uses $\left(U_{\mathrm{i}} \cap U_{\mathrm{j}}\right)$ to represent the set of users who jointly rate item i and item j and have a difference of less than or equal to 2. The formula is defined as follows:

$$\left(U_{\mathrm{i}} \cap U_{\mathrm{j}}\right) = \left\{u \middle| \left(R_{u,i} \neq 0\right) \cap \left(R_{u,j} \neq 0\right) \cap \left|R_{u,i} - R_{u,j}\right| \leq 2\right\} \tag{3}$$

n = $\left|\left(U_{\mathrm{i}} \cap U_{\mathrm{j}}\right)\right|$ represents the number of people who are jointly rated for item i and item j and rated very close. The following formula for calculating the improved similarity is obtained for the comprehensive scoring behavior:

$$sim_x(i,j) = \frac{min(n,r)}{r} sim(i,j) * (1 - \frac{\alpha * m}{m+n}) \tag{4}$$

where r denotes the user number threshold for scoring, such as 300, sim(i,j) through the improved cosine similarity formula (1) calculation, m denotes the number of people who only rate one for item i and item j, and α is the impact factor, indicating the degree of influence of the number of non-common scoring users on the overall similarity calculation during the similarity calculation.

The traditional collaborative filtering algorithm is used to calculate the item similarity through the scoring information table, and the attribute characteristics of the item can also provide the important auxiliary factors for the similarity calculation. Especially when the rating information is not enough to calculate the similarity of the item, the attribute information of the item is more useful. The item attribute similarity is calculated by the item attribute matrix [10].

Throughout the project calculation process, the basic attribute information represented by item i and item j is presented using set Ci and set Cj, and the similarity is calculated by the following formula:

$$sim_y(i,j) = \frac{\left|c_i \cap c_j\right|}{\left|c_i \cup c_j\right|} + \theta \tag{5}$$

In formula (5), $\left|\left(C_{\mathrm{i}} \cap C_{\mathrm{j}}\right)\right|$ represents the number of attributes common to all of these properties of the entire item i and j. Similarly, $\left|\left(C_{\mathrm{i}} \cup C_{\mathrm{j}}\right)\right|$ represents the number of all non-repeated attribute elements in the entire project, θ is a modified constant term that prevents most items from having a similarity of 0 because they do not have common attributes, in this study $\theta = 0.01$.

By combining the common influence of user scoring behavior and project attributes, the weighting method is used to combine the two similarity calculation methods. Meanwhile, to improve the accuracy and stability of this similarity calculation, two weights βx and βy are introduced. The final formula for calculating the similarity after weighting is as follows:

$$sim' = \beta_x sim_x(i,j) + \beta_y sim_y(i,j) \tag{6}$$

Among them, $\beta_x = \frac{sim_x^2(i,j)}{sim_x^2(i,j) + sim_y^2(i,j)}$, $\beta_y = \frac{sim_y^2(i,j)}{sim_x^2(i,j) + sim_y^2(i,j)}$.

The traditional collaborative filtering algorithm mainly calculates the similarity between items by using the pattern of scoring, but the attribute characteristics of the whole item also provide a very important auxiliary factor in the process of calculating the similarity directly. Especially when the rating information is not enough to directly participate in the similarity calculation in the project, so that the attribute information of the project can play a basic role. This study is mainly aimed at these problems by adopting comprehensive user scoring behavior and its overall project attributes as a common factor, that is, using a weighted method to calculate the similarity between items. By using this similarity calculation method to fill the user's scoring matrix effectively, then the scoring operation is targeted, and the non-rating items are processed by the results of the prediction. Therefore, because there are more items in the whole user's scoring, this method can solve the problem of the traditional data sparsity, and can find the nearest neighbor between the items and improve the recommended performance of the collaborative filtering algorithm.

4 Improved Collaborative Filtering Recommendation Algorithm Based on Hadoop

The improved collaborative filtering recommendation algorithm is applied to the platform of big data, and the improved collaborative filtering algorithm based on user behavior model is realized by using Hadoop platform. First of all, it is necessary to carry out the related offline operation, and save the results of the offline operation directly to the corresponding relational database. On-line operations are performed when a user needs a recommendation. In the whole process of off-line and online calculation, the content of the calculation includes four items, namely, the average score information in the project, the similarity information existing between the items, the rating prediction information and the common score user number information in the project and the number of users formed by the non-common score, the similarity between the mutual attributes of the items and the non-rating items. In the process of using the similarity model to calculate the predictive score information of the non-grade item, the score is directly added to the data set of the original user score in the format of <user, item, score> . After the later filling is completed, MapReduce can be used to implement the improved collaborative filtering recommendation system based on user behavior.

4.1 Algorithm Flow of Recommendation Module

The specific algorithm steps are as follows:

Step 1: Get the rating dataset and the project attribute dataset and construct the user rating matrix and the project attribute set matrix.

Step 2: a. Find items that the current user does not score in the user scoring matrix (the score is NULL).

b. Calculate user rating similarity for each ungraded item (formula 4) and item attribute similarity (formula 5).

c. The final weighted similarity of the project is calculated by combining the influence of user rating similarity and project attribute similarity (formula 6).

Step 3: Predict the final weighted similarity of the scoring items according to the need, give the predictive score (formula 2), fill in the user scoring matrix.

Step 4: The similarity between users (formula 7) was calculated by filling in the scoring matrix to find the k nearest neighbor users of the current users.

$$sim(u,v) = \frac{\sum_{a \in I_{uv}} \left(R_{ua} - \overline{R_u}\right)\left(R_{va} - \overline{R_v}\right)}{\sqrt{\sum_{a \in I_{uv}} \left(R_{ua} - \overline{R_u}\right)^2}\sqrt{\sum_{a \in I_{uv}} \left(R_{va} - \overline{R_v}\right)^2}} \tag{7}$$

where Iuv is a combination of items jointly evaluated by users u and v. $\overline{R}_u$ and $\overline{R}_v$ represent user u and v average ratings for all items, respectively. The greater the Sim(u,v) $\in [0, 1]$, the greater the similarity between user u and v.

Step 5: Calculate the average of the k nearest neighbor users for the item as the current user rating for the item

Step 6: Rank all ungraded items in Step 2 according to the score in Step 5, and recommend the first n items to the current user

4.2 Experimental Results and Analysis

Using the international standard public data set Moviens as the test data set, this data set includes more than 100,000 ratings recorded by 943 different users for 1682 different films, each of which is able to score at least 20 films. The data set includes 19 different categories of films, at least one of which for each film, and some may also belong to multiple different categories at the same time.

In the course of the experiment, the influence factor α is mainly studied, and the user number threshold r of the actual score is set to the maximum value of 328. First, the user scoring data set is obtained and constitutes the user scoring matrix. Secondly, the value of the influence factor α is specified, and the similarity of the user scoring behavior is calculated according to the model and sorted to form the set N(i), then specify the value of the nearest neighbor number k, calculate the predicted score value of the target user for the ungraded item according to the model, and finally compare with the standard set to calculate the average absolute deviation (MAE), adjust the value of the influence factor α for the next calculation.

If there are n user scores in the test set, the actual user score set is A and the predicted user score set is P, then the formula of MAE is:

$$MAE = \frac{\sum_{i=1}^{n} |A_i - P_i|}{n} \tag{8}$$

The impact factor α validation results are shown in Fig. 1:

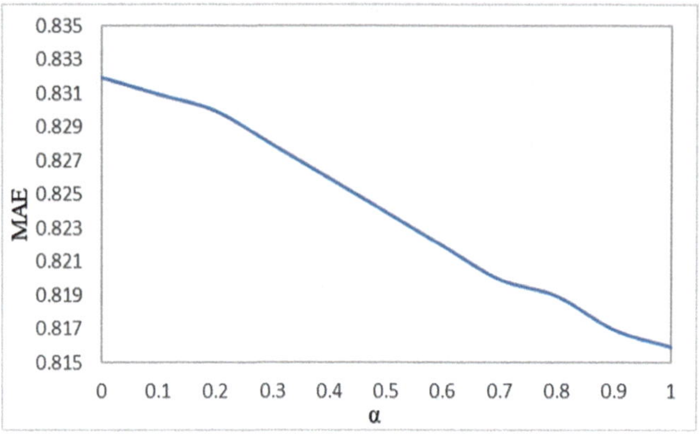

Fig. 1. Impact factor α validation results.

It can be seen from the figure that as the impact factor α increases from 0 to 1, the MAE is decreasing rapidly. When the impact factor takes 1.0 of the maximum value, the MAE is minimum. Therefore, when the impact factor α takes bigger, the influence degree of non-common scoring users plays a more important role in the calculation of the similarity, that is, simx(i,j) can achieve a better effect.

5 Conclusion

This paper calculates the missing value of filling item score by combining the user's behavior and the improved similarity of item attribute, which makes up for the deficiency of traditional fixed default value or average value. Based on the similarity between users, the accuracy of finding the nearest neighbor is guaranteed. Aiming at the problem that the data scale too large affects the execution efficiency of the recommendation system in the big data environment, the Hadoop platform is used to implement the collaborative filtering recommendation algorithm based on the similarity model, which is distributed parallel processing on the basis of the large-scale data segmentation, which improves the execution efficiency of the recommendation algorithm, especially when the large data volume is calculated, the execution time of the algorithm remains stable.

This paper selects the international standard public data set Movierens as the test data set, which contains more than 100,000 scoring records relative to the large data, still the data scale is too small. In order to further verify the validity of this algorithm, we need to test based on the larger data set, which is also a further prospect of later research.

Acknowledgment. This study was funded by the Humanity and Social Science Youth foundation of Ministry of Education of China (18YJCZH077), the Natural Science Foundation of the Jiangsu Higher Education Institutions of China (17KJB520038), Social Science Foundation of Jiangsu Province (16TQB009), and the Scientific Research Foundation of Taizhou University (QD2016036).

References

1. Yao, K., Tu, P., Chen, Y.X., Su, M.: Research on the effectiveness of personalized recommender system based on multi-source big data. J. Manage. Sci. **31**(05), 3–15 (2018)
2. Liu, H.O., Chen, J., Sun, J.J., Zhang, Y.M.: Research on mobile digital library situational recommendation system oriented to big data. Lib. Work Study **1**(9), 58–64 (2018)
3. Bao, Z.Q., Song, J.X.: Improved collaborative filtering recommendation algorithm based on association rules filling. Mod. Electron. Tech. 42(03), 78–81 + 86 (2019)
4. Huang, X.Y., Long, S.Y., Xie, J.: Collaborative filtering recommendation algorithm combined with user interest degree clustering. Appl. Res. Comput. 36(09), 2609–2612 + 2617 (2019)
5. He, R., Gonzalez, H.: Numerical synthesis of pontryagin optimal control minimizers using sampling-based methods. In: 2017 IEEE 56th Annual Conference on Decision and Control (CDC), pp. 733–738. IEEE, NJ, USA (2017)
6. Leng, Y.J., Lu, Q., Liang, C.Y.: Survey of recommendation based on collaborative filtering. Pattern Recog. Artif. Intell. **27**(8), 720–734 (2014)
7. Liu, J., Leng, F.L., Li, S.Q., Bao, Y.B.: A distributed file system based on HDFS. J. Northeast. Univ. (Nat. Sci.) **40**(6), 795–800 (2019)
8. Xu, X., Tang, M.: A new approach to the cloud-based heterogeneous mapreduce placement problem. IEEE Trans. Serv. Comput. **9**(6), 862–871 (2017)
9. Karabadji, N.E.I., Beldjoudi, S., Seridi, H., Aridhi, S., Dhifli, W.: Improving memory-based user collaborative filtering with evolutionary multi-objective optimization. Expert Syst. Appl. **98**, 153–165 (2018)
10. Sun, J.G., Ai, L.R.: Collaborative filtering recommendation algorithm based on item attribute and cloud model filling. J. Comput. Appl. **32**(03), 658–660 (2012)

Agricultural Product Quality Traceability System Based on the Hybrid Mode

Jun Wang[1,2](✉), Lianqin Jia[1,2], and Qian Zhou[1,2]

[1] Shandong Institute of Commerce and Technology, Jinan, China
wangjun26866@163.com
[2] The National Engineering Research Center for Agricultural Products Logistics, Jinan, China

Abstract. With the consumer's request to agricultural products is more and more strict, a traceability system of agricultural products quality based on hybrid mode is designed and developed in this paper. The system connects the various links of crop planting, acquisition, processing, distribution and sales, and obtains the real-time information from crop seed cost to agricultural product sales. It can make farmland scientifically be managed according to the information of crop growth environment, improve the use efficiency of agricultural resources, and is conducive to the sustainable development of agriculture. In addition, information processing and transmission, and sales of the crops from the seeding to the harvest and to the agricultural products are recorded in real time in the database so that the consumers can understand real-time information on produce. When unqualified agricultural products are detected, the production process of such products can be immediately controlled from the source, and the circulation of unqualified agricultural products can be more effectively controlled. In this way, the rights in the interest of consumers can be effectively guaranteed.

Keywords: Traceability system · Intelligent agriculture · Quality and safety

1 Introduction

Intelligent agriculture integrated with the IOT technology is a heart-shaped agricultural development model. It appears when Modern Agricultural Technology has grown to a certain phase [1]. Smart agriculture integrates information technologies such as Internet of Things (referred as IOT), Internet and wireless communication. Various sensor nodes are deployed in the detection area, maijorly collect the valuess of air humiture, illumination, soil humiture, and other relevant environmental information parameters of such heavy metals as nitrogen, phosphorus, potassium, cadmium, mercury and other components contained in the soil. At present, in order to promote the sustainable development of agriculture, it is necessary to greatly reduce the cost and corresponding consumption of agriculture. Therefore, it is necessary to adjust the equipment that implements automatic control in a timely manner, such as visual remote monitoring and operation, catastrophic warning, and agricultural precision planting. On this basis, the effective use of agricultural resources can promote the realization of intelligent agriculture [2].

Q. Liu et al. (Eds.): CENet 2020, AISC 1274, pp. 248–253, 2021.
https://doi.org/10.1007/978-981-15-8462-6_29

With the improvement of living standard, people more care about both the safety and the quality of food. Food safety is a significant matter in China, advancing a request to the government to improve the traceability Security System of Agricultural Product Quality. The process from agricultural production to marketing is monitored to ensure that residents get healthy and nutritious agricultural products.

The traceability system related to agricultural product quality has many composite sensors, wireless data transmission, intelligent gateway [3–5], Web Service [6], two-dimensional code (namely, QR code) [7] and mobile terminal technology, collecting the relevant information such as farming of agricultural product, fertilization, irrigation, weeding, pest control, harvesting. In addition, this mixed model also collects and records some environmental information in real time, covering temperature, humidity and light intensity during the growth of crops, as well as soil heavy metal detection, agricultural product processing, transportation and sales. They are included in the database in a timely manner, allowing government quality inspection departments and consumer groups to track the process of crops from growing to final processing as agricultural products without lagging behind. If the crops are determined to be unqualified by the government quality inspection department, the source of the quality problems will be quickly found, so as to effectively respond to such unsafe problems and prevent the circulation of unqualified agricultural products. Consumers can trace quality safety information while purchasing agricultural products in anytime, to safeguard their legal interest.

2 Methodology

The traceability system goes from crop planting to terminal distribution. The agricultural product quality and safety traceability system based on this model involves giving each field a unique ID logo, which involves the whole process of crop cultivation to final distribution. The ID becomes an important indicator for recording the aforementioned actions, such as sowing, fertilizing, crop healing, harvesting, and processing, transportation, as well as the information related to sales of agricultural products. Moreover, consumers can also track the relative information.

3 System Design

The system consists of hardware and software. The former is mainly responsible for collecting the real-time data of agricultural product planting, processing, circulation, distribution and sales, and loading it into the database. The latter is divided into Web services and Android clients, but Android clients cannot directly interact with the database. This article therefore mainly uses Web services to obtain data, and the Android clients implement data acquisition through the SOAP protocol, so that users can check and trace information.

3.1 Design of Hardware Environment

The main component of the hardware is various modules related to power supply, sensor, ZigBee and gateway, among which the power supply module is responsible for the power supply for other modules, sensor modules for the real-time collection of Environmental Data, and then the transfer to the gateway with the help of ZigBee module. The gateway module uploads thereafter the data to the database via a specific serial communication program.

3.2 Software Design

On the basis of the effective content read from the database according to the Web service, the hardware gateway module allows the Android client to complete the acquisition through the SOAP protocol. The end user finally contacts the database through a Web service to implement the software architecture of the data query function shown in Fig. 1.

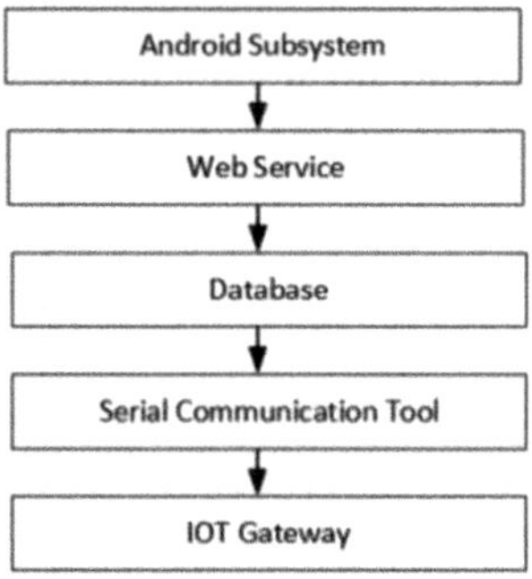

Fig. 1. Software architecture.

4 System Implementation

As mentioned above, the software system includes Web service and Android client. The mobile terminal is the main component of the system with the adoption of Android so that administrators and users can quickly query agricultural product information. While the back-end database selects SQL Server as a tool for managing data. The information completely from the SQL Server database is finally displayed on the Android side. The system uses the Web service as a bridge in the development process to realize the direct transmission of data, and communicates with the Web service through the mobile terminal to ensure indirect communication with the SQLSERVER database.

4.1 Web Service

For the query of the necessary data, the software structure of the Web service can be retrieved from the SQL Server database. Figure 2 exhibits the process of invoking data through Web services, including the information of each link of the crop planting process, as well as other information on pesticide residues, the environment collected by sensors, soil heavy metal detection and agricultural product quality tracking. To obtain the required data, click the corresponding hyperlink. For example, click "selectAllCargoInfor30" to call content related to the quality of agricultural products.

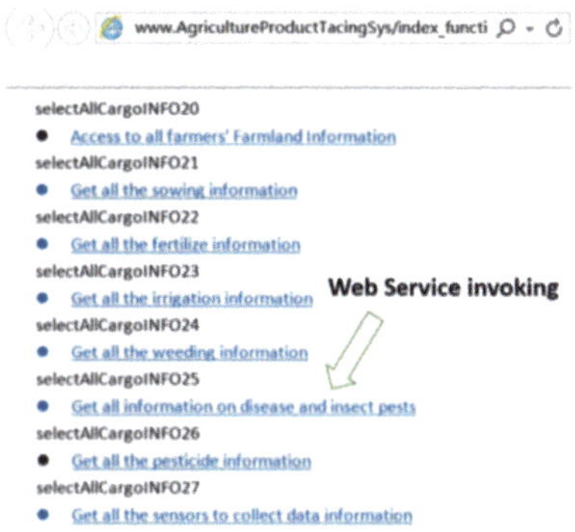

Fig. 2. Web service invocation method.

4.2 Android Client

The Android program includes four modules, which are responsible for farmland cultivation, crop growth, real-time monitoring of the growth environment and quality tracking of agricultural products. For the first two aspects, see Fig. 3.

Fig. 3. Real-time mobile monitoring of growth and the traceabilitymodule of agricultural products.

The entire system is composed of three subsystems, namely production management, logistics unit and traceability information query, as detailed below.

Traceability Information Management Centre, as the Core of the System, Manages System Users and Their Permissions. In a shared database, it is mainly used to store index information that can be traced throughout the agricultural product supply chain, as well as related information provided by government regulatory departments and inspection and quarantine departments such as agriculture and quality supervision, and after the information is entered, analyze and output whole system.

The Subsystem of Manufacture and Logistics Parts can not sorely run independently as the identity of the unit's information management system, but obtain access rights after logging in to the system, trace, enter and modify the index information of each link of the agricultural product supply chain.

Traceability Information Query Subsystem allows consumers to trace information through various channels such as trace websites, self-service terminals, Android mobile systems, telephone customer service, etc., and promote, recall and complain about problematic products.

5 Hybrid Traceability Quality and Safety Query for Agricultural Products

Under the background of fast growth of the Internet, acceleration of mobile networks and popularization of smart phones, the system should gradually be widely adapted, mainly including:

PC Web Site Query. You can log in "agricultural product quality safety to trace a system" website, t, carry on the inquiry, and search the traceability information by using any Networked computer.

Self-service Terminal Query. Wholesale centers, supermarkets, and the markets of farm produce set up special business terminals to facilitate consumers to trace agricultural product information.

QR-tag Scanning Query. Scanning of QR codes with smart phones can open the quality traceability website in an automatic way, or directly decode the information traced of the agricultural products.

Android Client Query. Consumers can search source information whenever and wherever through We Chat or brief messages.

In the quality and safety traceability system, these methods are combined with each other to form a mixed query model. This is a typical performance that fully utilizes the mainstream development technology that currently appears in the field of information technology, providing consumers with more convenient methods for querying traceability information as shown in Fig. 4.

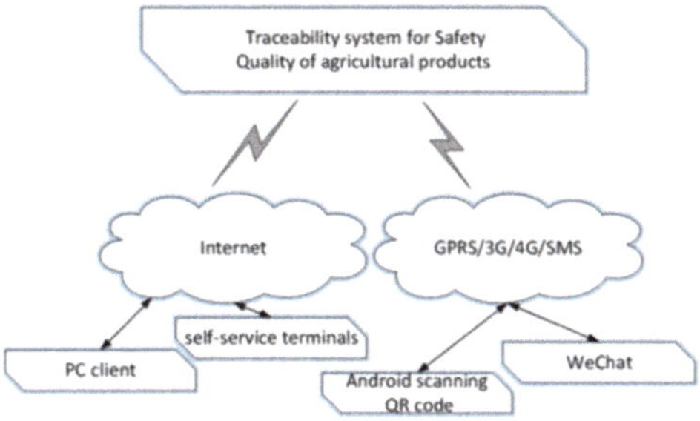

Fig. 4. Hybrid query mode for traceability information.

6 Conclusion

The system successfully links the entire process related to crops, including planting, management, procurement, processing, transportation and distribution. This aspect helps farmers manage agricultural production scientifically based on the external growth environment of crops, and improves the utilization rate of agricultural resources on the basis of improving the transparency of logistics; on the other hand, the detailed content of each link is recorded in the database in real time. In this way, consumers can more easily understand and grasp the traceability information of agricultural products in order to solve any safety management problems that may occur.

References

1. Pivoto, D., Waquil, P.D., Talamini, E., Corte, V.F.D., Mores, G.D.V.: Scientific development of intelligent farming technologies and application in Brazil, Information Processing in Agriculture. J. (2018)
2. Ibrahim, A., Muhammad, R., Alshitawi, M.M.: Intelligent green house application based remote monitoring for precision agricultural strategies: a survey. J. Appl. Sci. **7**, 947–952 (2017)
3. Yang, Y., Wei, K.: The key technology research of quality and safety traceability systems of agricultural products. In: International Conference on Intelligent Transportation, pp. 880–882 (2018)
4. He, C.L., Shen, M.X.: Design and realization of a greenhouse temperature intelligent system based on NB-IoT. J. South Chin. Agric. Univ. **39**(2), 117–124 (2018)
5. Zheng, Q., Peng, L.: The design of remote greenhouse monitoring system based on the embedded web server. J. Agric. Mech. Res. (2013)
6. Chen, N., Zhang, X., Wang, C.: Integrated open geospatial web service enabled cyber-physical information infrastructure for precision agriculture monitoring. Comput. Electron. Agric. **111**, 78–91 (2015)
7. Tarjan, L., Šenk, I., Tegeltija, S., Stankovski, S.: A readability analysis for QR code application in a traceability system. Comput. Electron. Agric. **109**, 1–11 (2016)

The Research of Intelligent Bandwidth Adjustment System

Yong Cui(✉), Yili Qin, Qirong Zhou, and Bing Li

Satellite Maritime Tracking and Control Department, Jiangyin 214431, China
tougaoqin@163.com

Abstract. An automatic intelligent implementation scheme is proposed based on the analysis of the characteristics of the information network transmission services of surveying ship by corresponding service strategies designation, which can give high quality services. The key of this scheme is an intelligent bandwidth allocation system that is designed for this scheme. This bandwidth allocation method effectively accomplish the intelligent redistribution of the limited bandwidth and improve the efficiency and reliability of the whole public information network of the surveying ship through the mirror data grabbing and analyzing and accomplishing the remote controlling of the encoder and the switch by remote bandwidth adjustment instructions.

Keywords: Intelligent bandwidth · Python · Tcpdump · Telnet

1 Introduction

Ships can only communicate with the shore through the satellite link during the voyage. With the increasing of the traffic volume of the ship-to-shore communication, the limited bandwidth resource of the satellite becoming more strained. Formulating corresponding service strategies within limited bandwidth and providing reliable communication services for related businesses become a serious problem for people in related positions to overcome. For example, in this article, an important Service “a” is newly added to the ship-to-shore communication. Without any change in the original bandwidth and services, the new Service “a” will improve the bandwidth utilization between the shore and ship and guarantee the quality of all business services at the same time.

2 Business Analysis

The bandwidth of Satellite between the shore and ship during the voyage remains unchanged at x kbit/s. Bandwidth distribution is shown in Fig. 1: Telephone service y kbit/s, Image service z kbit/s, for a total of $(y + z)$ kbit/s. When $(y + z) < x$, the bandwidth usage does not reach the speed limit value, the services would not affect each other. Under the condition that the total bandwidth is unchanged between the shore and the ship, if the added service “a” exceeds the $(x-y-z)$ kbit/s, if there is no service policy adjustment be made in advance, when all business data is transmitted at

Q. Liu et al. (Eds.): CENet 2020, AISC 1274, pp. 254–260, 2021.
https://doi.org/10.1007/978-981-15-8462-6_30

the same time, the total traffic will exceed the total bandwidth and cause business interaction or communication interruption.

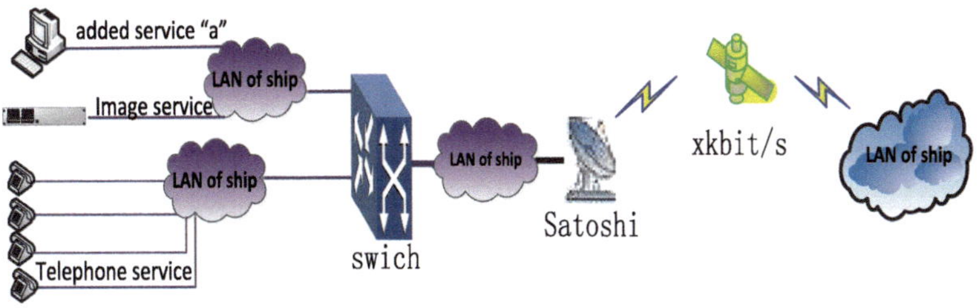

Fig. 1. Network structure between the ship and shore.

In the following, packet capture analysis is performed on important services "a", Telephone service and Image services between the shore and ship network transmission. The Image service is a full-time service and needs to be sent to the shore without interruption. The Image service data and the important service "a" data has been captured at the same time, as shown in Fig. 2 (the blue curve represents the Image service data, and the red curve is the important service "a" data rate). It can be known from Fig. 2, since the Image service needs to be transmitted throughout the time period, during the data transmission of the important service "a" the bandwidth occupied by the service "a" and the Image service is about (m + z) kbit/s.

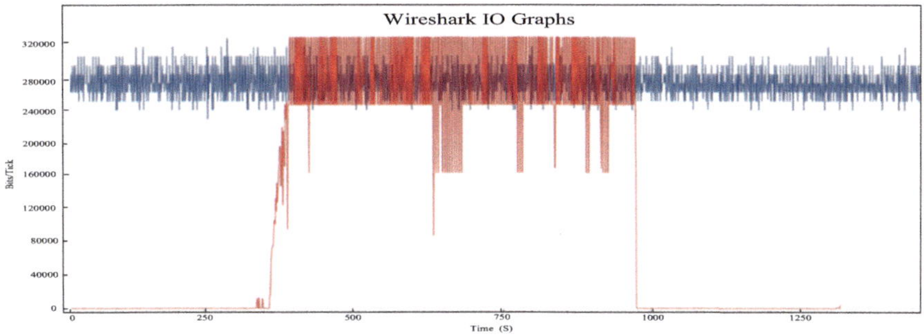

Fig. 2. I/O diagram of service "a" and the Image service data.

The Image data transmitted from the ship to the shore is specified as the image of the ship's front left direction. The picture does not change much between sailing or berthing. It is found through experiments that the fixed picture does not require much image transmission rate. If video encoder encoding rate reduced from the specified b kbit/s to c kbit/s, when the image is same or changes little it will not affect the new services. Therefore, when traffic is congested, the encoding rate can be appropriately reduced.

In order to avoid traffic congestion and packet loss at the router exit and affect the quality of business services, it is necessary to formulate relevant business service strategies according to the different needs of each business. Business requirements are shown in Table 1.

Table 1. Business demand statistics.

Business type	Priority	Usage	Rate (kbit/s)	Delay/Packet loss	Service quality
Service "a"	High	Base on needs	Highest m	High	Reliable, Complete transmission
Telephone service	High	High	Highest y	High	Real-time, Complete transmission
Ship image	General	Full time	Adjustable, Highest z	General	Real-time transmission

3 Solution and Key Technology

Because the three types of services have different characteristics and different service quality requirements, corresponding service strategies can be formulated to achieve continuous transmission of business data that meets the required service quality within a limited bandwidth.

Design the software on a server with dual network cards. The first network card is connected to the firewall mirror port for real-time packet capture, and the second network card is connected to the access switch for network connection.

As shown in Fig. 3: The firewall conFig.s mirror port which connect the plug-in server network card 1 for real-time packet capture, and analysis and judge the data packet based on the judgment result, and the network card 2 is used to log into the aggregation switch to perform the corresponding operations.

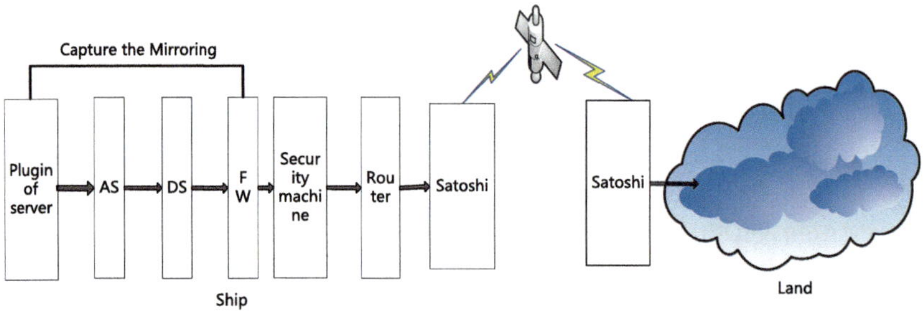

Fig. 3. Server distribution topography.

We consider the state of data transmission in the ship's intranet as the without the data transmission of Service "a" and with the data transmission of Service "a", which is the initial state and Service "a" transmission status. The judgment is based on whether Service "a" data is detected in the network.

Manually operating bandwidth redistribution requires a technician to observe the data transmitted in the intranet or receive related notifications to switch the data transmission state. This method requires a technician to log in to the CODEC and Switch manually set the relevant parameters. And it needs to restore the initial state in the end. Although this method is simple and easy to implement, but it's time-consuming and inefficient, and the data transmission service quality could be affected by misoperation which cause this method lacks of reliability and real-time performance. Intelligent bandwidth redistribution can detect the Service "a" data, it can operate the encoder and switch, automatically determine after the data transmission is completed, thereby restoring the initial state without the need for personnel supervision and review. This method is responsive, accurate, and time-efficient, while reducing the cost of communication between different departments.

4 Solution Realization

4.1 Web Page Automatic Login Operation

When sending Service "a" file, the first step is to log in to the CODEC web page, reduced encoding rate from y kbit/s to y1 kbit/s. The open () function of the webbrowser module is used to start a new browser and open the specified URL.

```
import webbrowser
webbrowser.open('http://x.x.x.x/') #open the Web page
```

In the "GUI Automation" module, which is the pyautogui module, it contains functions that can simulate mouse movements, keystrokes, and mouse wheel scrolling.

```
import pyautogui
```

pyautogui.typewrite() sends virtual keystrokes order to the computer, and simulates the keyboard to type a given character string to execute the browser login and fill in encoding code rate.

pyautogui.position() recognizes the current screen coordinate of the mouse by scanning the screen in real time, and locates the exact positions of the buttons such as "basic configuration", "bit rate", "save" and "close browser".

Send a virtual order of mouse click to the computer, call the pyautogui.click() function, and by default use the left button to click on the given coordinates.

During the running of the program, because the browser has a certain response time, the time.sleep() function is introduced to allow the program to wait for a few seconds to continue running.

4.2 Automatic Login and Operating Switch

When sending Service “a” file, both of the second and third steps need to log in and accessing the switch of the CODEC and the switch of the Service “a” device to perform speed limit adjustments or cancel the speed limit.

The telnetlib module which used in this article has related telnet remote login function. By using this function, server can remotely log into the aggregation switch and enter the corresponding configuration and other operation orders.

4.3 Data Capture, Analysis and Judgment

Use the tcpdump packet capture tool to capture the data with network card 1. Firstly, analyze the data packet when sending Service “a” file, and design an effective and convenient method to determine the sending status of Service “a” file. Through analysis, it’s found that the protocol of the data packet is ESP when the “a” device which address is B.B.B.B server transfer file, at this point, the system will consider that there is Service “a” data being transmitted. In the network card 1 enp7sof1, define IPfind () as “a” device that uses the IP address of a device (B.B.B.B) as the packet-capturing host to filter. Every 200 packets as one circulated, and find the ‘B.B.B. B: ESP’ characteristic value in each cycle.

Def IPfind ()

Text1 = subprocess. check_output ([‘tcpdump’, ‘-i’, ‘enp7sof1’, ‘-c’, ‘200’, ‘host’, ‘A.A.A. A’]) #A.A.A.A is Service “a” IP address

Flag1 = (text1.find(‘A.A.A.A: ESP’)! = -1)

Return (Flag1)

4.4 Circular Judgment

In the previous section, the packet was captured on the mirror interface by tcpdump module. By looking for the ESP protocol in the variable data, it is determined whether the “a” device is uploading a file or not. After repeated testing, discussion and summary, this article has refined the method of cyclic judgment. The process is shown in Fig. 4.

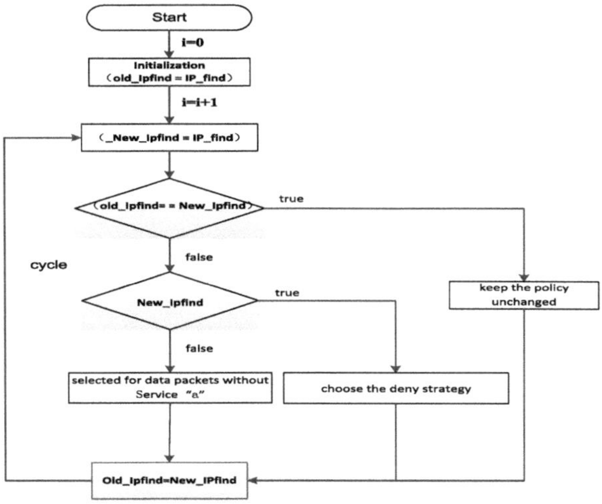

Fig. 4. Loop judgment logic diagram.

As shown, the process is mainly divided into five steps:

Definition OLD_ipfind whether Service "a" package was caught at last time; IP_find represents the judgment of whether a service package was caught; NEW_ipfind represents the judgment of whether Service "a" package was caught currently.

Step 1: Start initializing the defined IP_find assignment to OLD_ipfind; (i = 0)

Step 2: Copy IP_find to NEW_ipfind (indicating whether the packet of Service "a" is currently caught) (i = i + 1)

Step 3: When OLD_ipfind is equal to NEW_ipfind, it indicates that the last time packet capture check is consistent with the current packet capture, so no operation is required to keep the policy unchanged.

Step 4: Determine whether the current NEW_ipfind is true; if it's true, find the packet of Service "a" and choose the deny strategy. If it's false, it means that the permit policy is selected for data packets without Service "a".

Step 5: Update NEW_ipfind to OLD_ipfind

Cycle through steps 2 to 5.

5 Effect Verification

Test the system during the evening when there are many Telephone service. As shown in Fig. 5, the Image rate is rapidly reduced from x kbit/s to y1 kbit/s when sending Service "a". The rate of "a" file is paused at y1 kbit/s during the initial transmission and it reached the highest value m kbit/s quickly after three second, and the image quickly recovered to x kbit/s after the Service "a" was transmitted. The amount of three type's service flows did not exceed the total bandwidth value x kbit/s during the whole process. We achieved the desired effect, the phone was unblocked, "a" file has transferred fast, the ship image transmitted normally.

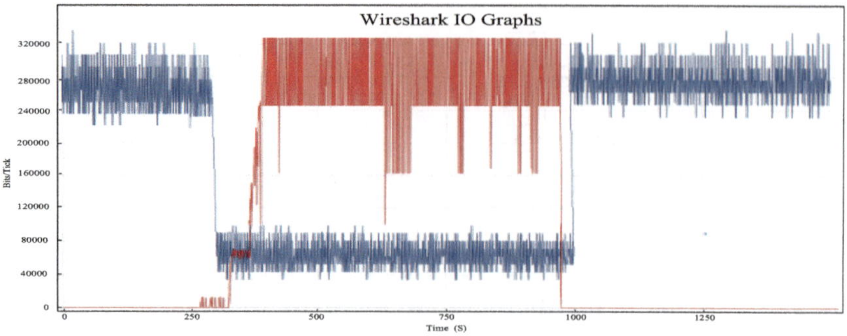

Fig. 5. Final renderings.

6 Conclusion

Under of the limited bandwidth resources situation between the shore and ship, this paper has carried out a detailed analysis of the transmission services between the ship and the shore. Based on the transmission needs of each business, a corresponding service strategy was formulated, and an intelligent solution was designed and implemented through analysis. This intelligent solution uses tcpdump packet capture tool and self-developed analysis software to complete the capture, analysis and judgment of mirrored data, complete the remote sending of related instructions, and realize the intelligent allocation of WAN network bandwidth and intelligent switching status. The intelligent switch provides a feasible solution for the effective use of the limited satellite communication bandwidth, and improves the efficiency and reliability of ship-to-shore communication. The intelligent bandwidth allocation system has a certain degree of openness, which provides a way of thinking for other bandwidth optimization problems in the future.

References

1. Dusty, P.F.: Python 3 Object-Oriented Programming, 2nd edn. Electronic Industry Press, Beijing (2018)
2. Mark, G., Sobell, F., Matthew Helmk, S.: Linux Command Line and Shell Programming Combat, 4th edn. Tsinghua University Press, Beijing (2018)
3. Zhao, H.F., Bao, G.S., Ma, D.T.: Python Network Programming. Tsinghua University Press, Beijing (2018)
4. Yoram, O.F.: Network Analysis in Action. People Post Press, Beijing (2016)
5. Chen, L.F., Yuan, J.S., Zhan, Y.Q.: Method and system for automatically adjusting network bandwidth. CN107094122, Wuhan (2017)
6. Chen, H.F.: Research on control scheduling methods of resource-constrained network control systems. J. Autom. 71–73 (2017)
7. Wang, H.F., Lian, J.S., Xia, H.T.: Non-uniformly sampled switched control for communication resource con-strained network systems. Acta Elect R Onica Sinica. **11**, 2597–2603 (2018)
8. Wang, J.F.: Scheduling of networked control systems under resource constraints. Control Decis. **23**(5), 551–559 (2008)

Research on Tibetan Phrase Classification Method for Language Information Processing

Zangtai Cai[1,2,3(✉)], Nancairang Suo[1,2,3], and Rangjia Cai[1,2,3]

[1] Computer College of Qinghai Normal University, Qinghai Xining 810016, China
1476082137@qq.com

[2] Key Laboratory of Tibetan Information Processing, Ministry of Education, Qinghai Xining 810008, China

[3] Tibetan Intelligent Information Processing and Machine Translation Key Laboratory, Qinghai Xining 810008, China

Abstract. Phrases as a level of linguistic analysis, occupying a very important position in this field. Effective phrase analysis is crucial to reduce difficulty of subsequent syntactic analysis, and reduce search space of syntactic analyzer, as well as improving accuracy of machine translation. At present, research in Tibetan phrases for information processing has just started; it needs to be further developed. Based on previous studies on boundary between Tibetan phrases and Tibetan sentences, with characteristics and requirements of Tibetan information processing, in accordance with grammatical function and the principle of automatic analysis to classification of phrases processing, and specify markup codes for Tibetan phrase units in information processing are discussed in this paper.

Keywords: Tibetan · Phrase · Classification · Mark

1 Preface

It has been more than 1400 years since the creation of Tibetan written language by Tonmi. During this period, many of Tibetan scholars annotated and expand, forming a traditional Tibetan grammar system with explicit structural features centered on function words and verbs based on "Thirty-Year" and "Sound Theory", which both were composed by Tonmi as the blueprint. Looking at the study of Tibetan grammar, the books and research results of Tibetan grammar in the past dynasties are full of enthusiasm. The study of Tibetan traditional grammar has made a detailed description of composition of Tibetan characters, temporal changes of words, association methods of function words, and the structure of Tibetan sentences. However, there is no in-depth study on Tibetan phrases in terms of information processing. With the continuous development and improvement of natural language information processing, the original word-based grammatical analysis method is transformed into the idea that the phrase is the grammatical analysis of the processing unit [1]. Especially in the fields of syntactic analysis, machine translation, text retrieval and other fields, the application of phrases is relatively mature.

Q. Liu et al. (Eds.): CENet 2020, AISC 1274, pp. 261–267, 2021.
https://doi.org/10.1007/978-981-15-8462-6_31

Because of the reasons above, my research crew have made in-depth research on structure of Tibetan phrases, definition of Tibetan phrases and Tibetan sentences, classification of Tibetan phrases, and marking methods of Tibetan phrases in dictionary. It lays a solid foundation for the study of Tibetan syntactic analysis, statistical machine translation, and Tibetan text retrieval based on Tibetan phrases.

2 Classification of Tibetan Phrases

For the classification and labeling of English phrases, the larger research projects are Lancaster Treebank of UCREL, Lancaster University, UK [8]. And Penn Treebank of Pennsylvania University in the United States [9] The former has a large set of tokens, and the detailed phrase syntax information is described by organizing into different levels. The latter's markup set is more concise, with only 14 syntactic markers, but it is characterized by the addition of four markers indicating different null elements (Null Elements). In China, the division and labeling of Chinese phrases mainly includes the determination of the set of Chinese phrase labeling marks completed by the Institute of Computational Linguistics of Peking University [10]. This paper proposes a set of syntactic markers for Chinese phrase segmentation and annotation, which provides a unified specification for automatic processing and manual proofreading of Chinese phrase annotation.

In the process of studying Tibetan phrases, it mainly draws on the content of the literature [10], because both Chinese and Tibetan belong to the same language family, and have common features in research ideas, language functions and structures. In [10], Chinese phrases are divided into:

1) Noun phrases, such as: beautiful hats;
2) Nominal quasi-phrases, such as: workers;
3) Verb phrases, such as: watching movies;
4) Verb quasi-phrases, such as: take a look;
5) Adjective phrases such as: particularly quiet;
6) Descriptive quasi-phrase, such as: red;
7) Adverbial phrases, such as: open-minded;
8) Prepositional phrases such as: in Beijing;
9) Distinguish part-of-speech phrases, such as: this one;
10) Time part-of-speech phrases, such as: early war;
11) Place word phrase, such as: in the village;
12) Quantitative part-of-speech phrases, such as: this group;
13) Single sentence pattern, such as: then;
14) A complex sentence, such as: if she is willing;
15) Whole sentence, such as: Can you go or not?

After seeing the classification of the above Chinese phrases, it was found that the categories of Chinese phrases and their mark sets cannot be directly borrowed into the study of Tibetan processing. Although Chinese and Tibetan are of the same language family, they are two completely different languages. From the analysis of the characteristics of the grammatical structures of these two languages, there are three main differences:

(1) The word order is different. The Chinese word order is usually SVO (S is the subject, V is the predicate, O is the object), and the Tibetan word order is SOV. For example: I am a teacher, S + V + O (Chinese structure), ང་དགེ་རྒན་ཡིན། S + O + V (Tibetan structure). Another manifestation of different word order is that when adjectives, numerals, and pronouns are used as attributives, they are all in front of the central words in Chinese, while in Tibetan they are usually behind the central words, such as: recruits/that person, adjectives + central words (Chinese structure), དམག་མི་གསར་པ། མི་གན།Center word + adjective (Tibetan structure). As can be seen from the above examples, the position of the central word in the phrase changes when the language is different. When the position of a word changes in a phrase, the categorization of the phrase also changes.
(2) The morphological changes are different. Chinese lacks morphological changes, while Tibetans have morphological changes. The tense form of verbs is a prominent expression [11]. Tibetan verbs in the form of tense, such as "eat" ཟ་བ(present tense), བཟས་པ(past style) and བཟའ་བ(Future style) three tense forms; while Chinese verbs have no tense form, such as "eat", there is only one form in Chinese, the tense is represented by a specific vocabulary, and if the past tense is added "to", it means the future Add "will". For example, "eat" is a phrase in Chinese and is classified into verb-based quasi-phrases. And in Tibetan བཟས་པ"Not a phrase, but still at the level of the word.
(3) Different expressions. Chinese is mainly expressed by word order, while Tibetan is mainly expressed by means of case markers. The addition of case markers in Tibetan is directly related to the correctness of the meaning expressed. In addition, there is a fixed collocation phrase in Tibetan, whose structure is similar to idioms (idioms are regarded as words), but not idioms. In general, the formation of idioms has a background story, and it does not. E.g:རྒྱ་མ་བོད། ར་མ་ལུག Such fixed collocation phrases do not appear in Chinese grammar, so when categorizing Tibetan phrases, such phrases are separately classified into "fixed phrases." Another example is the phrase "red" (དམར་པོར་གྱུར།The central word in") is "red", and "red" is an adjective, so it is classified into the adjective quasi-phrase in Chinese phrases. In Chinese, "red" is expressed in Tibetan as "reddened", the central word in "reddened" becomes "change", and "change" is a verb, "redder" Phrases should be categorized into verb phrases. It shows that the Tibetan language pays attention to the "red" process, while the Chinese emphasizes the result of "red".

2.1 Classification Principles of Tibetan Phrases

The classification system of Tibetan phrases for language information processing is mainly based on the Tibetan traditional grammar classification system. According to the characteristics and requirements of Tibetan information processing, and referring to the research results of Chinese phrase classification, according to the grammatical function and the automatic analysis and processing of computers. The principle divides the phrase. Intersections are not allowed between large classes, including the annotation formal layer and the semantic layer; the intersection of the annotation formal layers is allowed between the small classes, but the intersection of the semantic layers is not allowed.

2.2 Classification of Tibetan Phrases

The classification system divides Tibetan phrases into eight broad categories, namely noun phrases, verb phrases, adjective phrases, quantifier phrases, judgment phrases, fixed phrases, time phrases, and pronoun phrases. among them:

a) Noun phrases include noun phrases of noun structure, noun phrases of suffix structure, noun phrases of partial structure, noun phrases of parallel structure, noun phrases of position and structure, and noun phrases.
b) The verb phrase includes the verb phrase of the object structure, the verb phrase of the parallel structure, the verb phrase of the supplementary structure, the verb phrase of the dynamic complement structure, the verb phrase of the applied structure, the verb phrase of the biased structure, the orientation and the structure of the premises. 8 subcategories such as verb phrases and verb phrases of subject-predicate structure;
c) adjective phrases include two subclasses: adjective phrases in a side-by-side structure and adjective phrases in a biased structure;
d) Quantifier phrases, judgment phrases, fixed phrases, time phrases, and pronoun phrases are not subcategories.

3 Tag Code for Tibetan Phrases

Principle of marking code, according to the international common practice, the mark code mainly uses letters of English terms. For example, "noun phrase", using NP; "numerical phrase", using MP as the tag code. It is unique in Tibetan language or inconvenient to use English terminology in Tibetan. According to the usual practice in China, the mark code uses Chinese pinyin letters. For example, "judgment phrase" uses the initials of Chinese Pinyin as the mark code PP. The details of the tag code, sub-category information, phrase structure and Tibetan examples of Tibetan phrases are shown in the table.

The leftmost column in the table represents the tag codes for the eight broad categories in Tibetan phrases. NP, VP, AP, MP, PP, GP, TP, and RP represent noun phrases, verb phrases, adjective phrases, quantifier phrases, judgment phrases, fixed phrases, time phrases, and pronoun phrases in Tibetan phrases, respectively. The second column is the name of the Tibetan phrase. The third column represents the sub-category tag codes classified in the noun phrase, the verb phrase and the adjective phrase of the Tibetan phrase, and the remaining quantified word phrases, judgment phrases, fixed phrases, time phrases and pronoun phrases are not divided into sub-categories. Therefore, the contents of the table here are empty. The fourth column of the table is the subclass name. The fifth column represents the structure of Tibetan phrases, in which lowercase English letters represent part of speech; uppercase English letters represent phrases; "+" means connection; "/" means or; "[]" means dispensable; "…." indicates that the preceding words are repeatedly connected; the Tibetan text appearing in the phrase structure indicates a fixed collocation of the Tibetan vocabulary words; ";" is a different phrase structure before and after. For example, in the noun

Table 1. Classification of tibetan phrases and their tag codes.

mark	Phrase name	Subclass tag	Subclass name	Phrase structure	Tibetan example
NP	Noun phrase	NPM	Name complement structure	n + a; n + f; n + ub	ཁང་པ་ཆུང་ཆུང་། ཟེ་ལིང་རྩུབ་ཚད།གླང་ཆེན་སྣ་ཐུ།
		NPH	Suffix structure	n/v+h	ཡོན་ཏན་ཅན།ཁཤད་སྟོལ།འགྲོ་སྣང་ས།སྣ་ལུགས།
		NPP	Partial structure	n+[gz]+n; a/r+gz+n; t+[gz]+n;	ཁ་སྐད་བཀྲ་ཤིས།ངའི་དཔེ་ཆ། དེ་རིང་གི་ཉེ་མཁྲིས་པའི་ཐང་ཀ།
		NPB	Parallel structure	n/r+c+n/r; n+n+…	དཔལ་འབྱོར་དང་སྤྱི་ཚོགས།རབས་དང་རིམ་པ།
		NPS	Location	n+gz+f;n+[gz]+s; f+[gz]+n	ཁང་པའི་ཤར་ཕྱོགས།སློབ་ཁང་ནང་། ལྷོ་ཕྱོགས་ས་ཁུལ།
		NPF	Compound finger	n+rz;n+n	མི་རིགས་སོ་སོ།ཕོ་བྲང་པོ་ཏ་ལ།མཚོ་སྔོན་ཁྲི་ཤོར་རྒྱལ་མོ།
VP	Verb phrase	VPB	Guest structure	n+vt/; n / r + vi; n / s / s / f + G L / c + v	ཡི་གེ་བྲིས།ཆར་པ་བབས། ལོ་གསར་ལ་ཐོན།བཀོ་བཞེངས་མཛད།
		VPL	Parallel structure	v + v … ..; v + c + v	འཆད་རྩོད་རྩོམ།འཁྲུག་པ་དང་བཅོམ་པ།བཤད་པའམ་བྲིས་པ།
		VPD	Replenishment structure	v + c + v	སོང་སྟེ་ཐོན།ལྡུག་སྟེ་ངུ།དགའ་སྟེ་རྒྱུགས།
		VPB	Dynamic complement structure	v+us; v + ux;and the +;v + [gl] + v	འགྲོ་བཞིན། ཤེས་དཀའ། འགྱུར་སླ།བརྗོད་འདོད།
		VPS	Structure	v + gl + v	ཟ་རུ་བཅུག
		VPP	Partial structure	d + [gl] / [gx] + h	ཤིན་ཏུ་གུས་པ།
		VPF	Location	n/r/s/f+c（ནས/ལས）+v	རི་ལས་བབས།
		VPZ	Subject-predicate structure	n / r + gx + VT; n / s + we	སློབ་མས་ཀློག།ཆར་པ་བབས།རླུང་གཡུག་ཡོ་སོང་།རིན་ཐང་བྲལ་བ།
		APB	Parallel structure	a+c+a	སྙན་ཞིང་འཇེབས།

(Continued)

Table 1. (*Continued*)

AP	Adjective phrase	APP	Partial structure	d + a; n / r / m / MP + c (ལས) + a; n + c + a	ཤིན་ཏུ་མཛེས། ཐེས་འཁོར་ལས་བཟང་། རྟ་བས་མགྱོགས། དེ་བས་བཟང་།
MP	Quantitative word phrase			q+m; m+q; n/r+m; mj+mg	ཕྱུ་གཅིག།༧༦༦༤ལོ། སློབ་གྲོགས་ཚོ། ཆིག་སྟོང་ཚམ། དཀོན་མཆོག་གསུམ། ལུག་གཅིག
PP	Judgment phrase			n + etc. / up / to /	འབྲོག་པ་མིན། ཁྲུམ་པའོ།
GP	Fixed phrase				ར་མ་ལུག རྒྱ་མ་བོད།
City	Time phrase			v/r/t+[gz]+t; t+gl; v/t/vp+t	སློབ་པའི་ཚོ། སྐབས་སུད་ནས་བཟུང་། འགྲོ་སྐབས།
Rp	Pronoun phrase			n/r/v+ladon/gz/gx; r+སྐད	རི་སྟོད་དུ། ནམ་ཞིག ཅེས་ཟེར། ཞེས་བདགས། ཞེས་བརྗོད།

phrase (NPM) of the noun structure, the noun is the central word. Adjectives, orientation words, or auxiliary words after the noun are used to modify the preceding noun. For example, in the phrase of a biased structure, the overall function is the same as the function of one of the components, and this component is the central word. The central word of a noun phrase with a positive relationship is a noun, the former part is the attributive of the modified central word, and the attributive part is composed of a noun or a noun phrase, a verb or a verb phrase, an adjective (other than an adjective of a monosyllabic), or an adjective phrase, a pronoun. Attributes are used between the attributive and the central word. Sometimes the genitive can be omitted, but its meaning and function are unchanged. In order to control the length, do not list them one by one. The last column is a Tibetan example Table 1.

4 Conclusion

With the continuous development and improvement of natural language information processing, the original word-unit grammar analysis method is converted into a phrase for the grammatical analysis of the processing unit. With such a background, Tibetan information processing is no exception. Tibetan phrases such as Tibetan syntactic analysis, Tibetan-Chinese machine translation, and Tibetan text retrieval are all used in Tibetan language. Therefore, definition of Tibetan phrases, the differences between Tibetan phrases and Tibetan sentences are studied in this paper. Based on this, the Tibetan phrases are classified and the tag codes of the Tibetan phrase category units in the information processing are specified. Next, we will build a large-scale Tibetan corpus based on the automatic extraction of Tibetan phrases, and then build a Tibetan phrase information database for Tibetan sentence analysis based on Tibetan phrases, statistical machine translation, and Tibetan text retrieval hoping to lay a solid foundation for these field of researches.

References

1. Wan, F.C., Yu, H.Z.: Research on tibetan phrase syntax for machine translation. Comput. Eng. Appl. **51**(13), P211–216 (2015)
2. Yu, M., Zhang, G.Z., Wang, X.H.: Research on the classification of chinese phrase boundaries based on rules. Suzhou Univ. Acad. J. (Nat. Sci.) **10**(3), P226–232 (1994)
3. Ji, T.J.: General Introduction to Modern Tibetan Grammar. Gansu Nationalities Publishing House, Gansu (2000)
4. More Knowledge. Tibetan Grammar Notes. Gansu: Gansu Nationalities Publishing House 5, 28 (1987)
5. Ji, T.J.: Tibetan Sentence Method. Beijing: China Tibetology Press 7, 43 (2013)
6. Cai, Z.T.: Research on automatic recognition method of tibetan sentence boundary based on maximum entropy classifier. Comput. Sci. Eng. **34**(6), 130–133 (2012)
7. Cai, Z.T.: The construction of Tibetan corpus and its deep processing methods. Comput. Eng. Appl. **48**(66), 127–131 (2012)
8. Garside, R., Leech, G., Sampson, G.: The computational aualysis of English, longman (1987)
9. Mitchell, P.M.: Mary annmarcinkiewicz, and beatricesantorini, buiding a large annotated corpus of English: the Penn Treebank. Comput. Linguist. **19**(2), 313–330 (1993)
10. Zhou, Q., Yu, S.W.: Determination of Chinese phrase marker set. Chin. J. Inf. (4) (1996)
11. Jiang, W.: The syntactic and semantic classification of modern Tibetan verbs and related grammatical sentences. Chin. J. Inf. **20**(1), 111–115 (2006)

A Garbage Image Classification Framework Based on Deep Learning

Chengchuang Lin[1,2,3](✉), Gansen Zhao[1,2,3], Lei Zhao[1,2,3], and Bingchuan Chen[4]

[1] School of Computer Science, South China Normal University, Guangzhou, China
{chengchuang.lin, gzhao, zhaolei}@m.scnu.edu.cn
[2] Key Lab on Cloud Security and Assessment Technology of Guangzhou, Guangzhou, China
[3] SCNU & VeChina Joint Lab on BlockChain Technology and Application, Guangzhou 510631, China
[4] Guangdong University of Finance and Economics, Guangzhou 510320, China
chbingch@mail2.sysu.edu.cn

Abstract. In this paper, we present a garbage image classification framework to tackle the waste sorting problem which besets residents around the world every day. The proposed framework consists of two modules, a convolutional neural network backbone transferred from the ImageNet classification task and a customized network header designed for the garbage image classification task. To friendly deploy into mobile devices, the proposed method makes an artful tradeoff between classification accuracy and running efficiency. The proposed framework yields 95.62% online classification accuracy on the test dataset provided by Huawei cloud with 95 ms per image inference time occupying 897 Megabytes GPU memory.

Keywords: Transfer learning · Garbage classification · Weakly supervised learning

1 Introduction

Garbage classification and reduction are imperative in many developing countries, where the imbalance between the rapid growth of solid waste and insufficient disposal capacity can be challenging [1]. Taking Shanghai (One of the biggest cities in China) as an example, the daily output amount of recyclable garbage in 2019 reached 5,605 tons, which has been increased by 5 times as against 2018 [2]. Since July 1, 2019, the Regulations on the management of domestic garbage in Shanghai was officially implemented, garbage classification is a nationally-involved residents action in China. According to Shanghai garbage classification regulation, all solid domestic waste is divided into four categories (recyclable garbage, hazardous garbage, wet garbage, and dry garbage), a total of forty-three sub-categories. How to correctly classify garbage has become a difficult problem in residents' life. Current deep learning based computer vision techiniques can assist residents to classify solid waste correctly.

Q. Liu et al. (Eds.): CENet 2020, AISC 1274, pp. 268–275, 2021.
https://doi.org/10.1007/978-981-15-8462-6_32

Since Alex et al. [3] first proposed a deep convolutional neural network-based method to image classification task and obtained much better result than traditional methods done, a large number of models [4, 5] based Deep Learning (DL) established the stated of the art in many areas of computer vision. Victoria et al. [6] presented an automatic image-based waste classification method utilizing Inception-ResNet [7] architecture for the TrashNet [8] dataset and obtained 88.6% classification accuracy. Deep learning-based computer vision techniques begin to assist residents to classify solid waste correctly and expediently.

However, to solve garbage image classification using machine learning techniques based on deep learning still exist the following challenges. Firstly, training a deep learning-based model from scratch to fine requires a large volume of labeled data where acquiring such data is expensive and time-consuming [9]. Besides, though high classification accuracy is significant, the less GPU memory occupying and shorter average inference time of the model is also critical because such models for garbage image classification is usually running on mobile devices.

To combat the challenges mentioned above, we proposed a garbage image classification framework based on weakly-supervised (WSL) transfer learning. The proposed framework consists of a backbone network transferred from ResNeXt [5] and a customized network header designed for garbage classification task. We load WSL weights [10] pre-trained by millions of social media images as the initial weights of the backbone. Only one hour is required to finish the training of the proposed framework from coarse-to-fine on a Nivida GTX-2080Ti GPU. The online test result with 95.62% accuracy, 897 M occupying GPU memory, and 95 ms average inference time per image demonstrates the effectiveness tradeoff of the proposed model on accuracy and efficiency.

The contribution of this paper is multi-folds. First, we address the problem in a formalization manner and propose the corresponding framework to solve it. After that, we proposed a training method, which can complete the training and fine-tuning of the model in about one hour. Last but not least, we conduct a series of rigorous experiments to demonstrate the effectiveness of the proposed method.

2 Method

2.1 Problem Formalization

We use $I, I \in \mathbf{R}$ to denote an garbage image where $\mathbf{R}$ is the three-dimensional tensor with height, width, the color channels of the I respectively. We use a one-hot vector y to depict the label of the garbage image I. Therefore, we can use symbol $X = \{I_j\}_{j=0}^{N-1}$, and $Y = \{y_j\}_{j=0}^{N-1}$ to represent the training image set and the corresponding label set where N denotes the total number of the training images. After that, we use symbol T to represent the online image test set, and $\widehat{Y}$ to the ground-truth label set of T.

We use the symbol Φ to represent the implicit expression of a model learned from X, Y. Furthermore, we use $\widehat{Y}', \widehat{Y}' = \Phi(T)$ to denote the predict output set on the online test set T of a model. After that, we can use the function $\mathrm{L}\left(\widehat{Y},\widehat{Y}'\right)$ to represent the deviation between $\widehat{Y}$ and $\widehat{Y}'$. The smaller L represents higher classification accuracy. Lastly, we use ζ to represent the GPU memory occupation and τ to average inference time per image of the model.

Consequently, the garbage image classification problem which requires classification accuracy and efficiency (GPU Memory and Average Inference Time) as higher as possible can be formalized as $\prod_{\min}(L, \zeta, \tau)$. However, models with higher accuracy regularly require more resources. Accordingly, when solving this problem, we need to design the model very carefully to achieve the best trade-off between performance and efficiency.

2.2 The Framework Architecture

Transfer learning is one of the most effective solutions to find a tradeoff between garbage classification accuracy and running efficiency on the condition of the limited training dataset. Hence, the proposed framework consists of two modules, a feature encoding backbone network transferred from ResNeXt-101-32 × 8d [5], and a customized network header for improving the generation performance of the model.

The ResNeXt backbone is an orderly stack of regular neural network modules called building block as Fig. 1 depicted. A general building block is assembled by multiple paths of convolutional neurons. The total quantity of assembled paths is termed by *cardinality*. The channel amount of the convolutional neurons in the building block is the width of the block, termed by the symbol d. We use F to denote the input and output tensor dimensions of the building block. In the backbone network, we apply ResNeXt101-32 × 8d as the concrete deep neural network, which means the *cardinality* of the building block is 32 and the *width* is 8.

To improve the applicability of the model in the garbage image classification application, we design 9 layers of neurons as the customized header adding to the backbone network as the output of the classifier. The first layer of the network header is a mixed *Pooling* layer, where the *AveragePooling* layer is designed to take into account the global features of garbage images, and the *AveragePooling* layer is designed to differentiate features encoded by the backbone module. The second layer is a *Flatten* layer that makes an n-dimensional feature into a flatten vector. The third and seventh layers are *Regularization* layers that improve the generalization ability of the header. The fourth and eighth layers are *Dropout* layers, whose function is to prevent overfitting of the header. The other layers are *Linear* layers that act as activation functions.

Table 1 summarizes the detail of the proposed garbage image classification model. The backbone model consists of five blocks termed by *conv1* to *conv5*. For convenience, we use *block* () to depict the building block, and symbol c to denote the *cardinality* parameter. We use *block* () to depict the building block, and symbol c to denote the *cardinality* parameter.

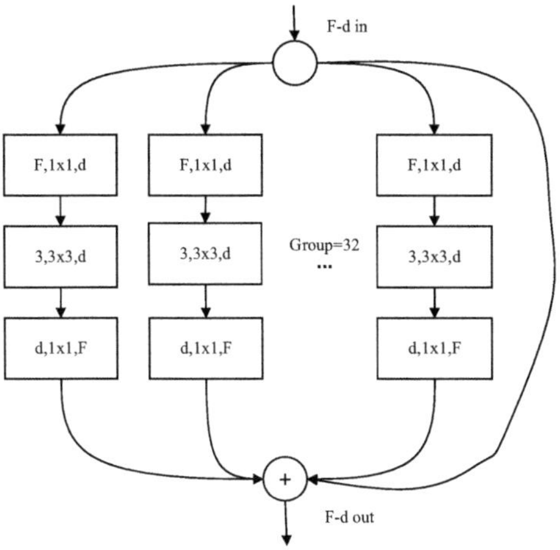

Fig. 1. This figure describes a standard building block assembled by multiple paths of convolutional neurons.

Table 1. The detail of the proposed method.

Block name	Detail	Output shape
conv1	conv2d (7 × 7, 64, stride = 2)	112 × 112 × 3
conv2	max_pooling (3 × 3, stride = 2) block (F = 128, d = 8, c = 32) × 3	56 × 56 × 128
conv3	Block (F = 256, d = 8, c = 32) × 4	28 × 28 × 256
conv4	Block (F = 512, d = 8, c = 32) × 23	14 × 14 × 512
conv1	Block (F = 1024, d = 8, c = 32) × 3	7 × 7 × 1024
customized header	Pooling2d () Flatten () BatchNormal1d (4096) Dropout (0.25) Linear (in = 4096, out = 512) Relu () BatchNormal1d (512) Dropout (0.5) Linear (in = 512, out = 43)	43 × 1

2.3 Training Methodology

Although deep learning is promising in computer vision tasks, the hyper-parameters setting is still an art task that requires extensive experience [10]. Consequently, to promote convergence of the proposed model, we introduce the following the training method employed in our training of the proposed model.

Firstly, we load the WSL weights [10] into the backbone as its initial weights, which means that our backbone network has coarse initial weights to avoid optimizing from random weights directly. Secondly, as neural layers of the header network have stochastic weights. We freeze the backbone network and then quickly adjust the weights of the header utilizing one cycle learning [11] with five *epochs* and the discriminative learning rate [12] with *slice (1e-5, 4e-5)*. After the above stage, we unfreeze the backbone to fine-tune the whole network from coarse-to-fine utilizing one cycle learning with ten *epochs* at a quite small learning rate with *slice(1e-6, 4e-6)*. Finally, we restore the best weights when finishing all training epochs.

The most significant benefit of our training approach is that it can make the proposed model convergence within 15 epochs. It means that we can finish the training of the model using a GTX-2080Ti GPU in less than one hour (about 3 min per epoch).

3 Experiments

3.1 Dataset Description

The training dataset used for our training is publicly available from Huawei cloud[1]. This dataset contains a total of 14,802 image samples in 43 categories. The dataset has the following characteristics.

- Although the image samples in this dataset do not have a uniform resolution, the resolution of most image samples exceeds 200 × 200 pixels.
- The distribution of categories in the dataset is unbalanced. For example, the total number of image samples for *abandoned electrical wire* is eight times the total number of the *toothpick*.
- It is a common phenomenon that a given garbage image sample contains several categories of garbage contents.

During the training process, we divide 80 percent of the samples as the training set and the rest as validation set utilizing stratified sampling. We resize all image samples into 224 × 224 pixels. Finally, we test and evaluate our model on the online test dataset provided by Huawei cloud.

3.2 Experiments Settings

We implement our proposed method using PyTorch and FastAI toolkits. The experiments are accelerated by Nvidia GeForce RTX 2080Ti GPU with 11019MiB GPU memory.

In our experiments, we use label smoothing cross-entropy loss as the loss function, and set its hyper-parameter α to 0.1. The batch size parameter *bs* is set to 16 to ensure our proposed method can successfully run on a single GTX-2080Ti GPU. We regularize each batch of data using the ImageNet distribution. We set the discriminative learning rate *lr* to *slice (1e-5, 4e-5)* and the *epochs* parameter of one cycle learning to 5 during the training of the network header. The discriminative learning rate *lr* is set to *slice (1e-6, 4e-6)* and the *epochs* parameter of one cycle learning to 10 in fine-tune stage.

[1] https://modelarts-competitions.obs.cn-north-1.myhuaweicloud.com/garbage_classify/dataset/garbage_classify.zip.

3.3 Experimental Results

To evaluate the general performance of the model on precise and efficiency on the online test set, we adopt accuracy (*acc*), GPU memory occupation (ζ) and average inference time (τ) metrics to evaluate our proposed method.

The *acc* metric is calculated by dividing the total number of accurately classified samples by the total number of samples in the test set. The ζ metric is calculated by subtracting the remaining GPU memory after loading the inference model from the total GPU memory. When calculating the τ metric, we first summarize the inference time of all test set samples and then divide by the number of samples in the test set. Note that the inference batch size is set to 1 when evaluating the average inference time.

Table 2 summarizes the overall evaluation results with the ablation study on the online test dataset. First of all, we build our baseline model by ResNet-50 [4]. The baseline model yields 78.21% online classification accuracy which is an undesirable result. However, the most significant characteristic of the ResNet-50 is high running efficiently with low GPU memory occupation and fast inference time.

After that, we rebuild the model transferred from ResNeXt-10132 $\times$ 8 [5] as backbone and evaluate it on the online test dataset. Compared to the baseline model, the evaluation result yields 91.17% classification accuracy, which obtains 12.96% accuracy boosting. However, it requires 879 M GPU running memory and 90 ms average inference time per image sample.

To evaluate the effectiveness of the proposed customized header, when we add the customized header to the backbone, the proposed model yields 1.52% classification accuracy improvement. Meanwhile, the GPU memory occupation of the model only requires 897 M, and the average inference time per image sample only requires 95 ms. Without adding more GPU occupation and inference time, we train the model applying our proposed training method, which earns 2.93% online classification accuracy improvement. The result demonstrates the effectiveness of the proposed training method.

To further improve the classification accuracy, we evaluate a more complicated backbone such as ResNeXt-101 32x16. The online test result shows that a more complicated backbone can further improve little accuracy performance. However, the GPU memory occupation and average inference time grow much faster than accuracy, which is inhospitable to deploy into mobile devices.

The above experimental results shown in Table 2 demonstrate that the proposed model is an effective and promising approach to tackle the garbage classification problem.

Table 2. Summary of the experimental results.

	Accuracy	Improvement	GPU Mem	Inference time
ResNet-50 [4]	78.21%	baseline	797 M	50 ms
ResNeXt-101 328 [5]	91.17%	12.96%	879 M	90 ms
Proposed method	92.69%	1.52%	897 M	95 ms
Proposed method +training method	**95.62%**	**2.93%**	**897 M**	**95 ms**
ResNeXt-101 32 × 16 [5] +proposed model +training method	96.08%	0.46%	1143 M	120 ms

4 Conclusion

In this paper, we propose a novel garbage image classification approach to assist residents classifying their domestic garbage accurately and conveniently. Our proposed method makes a tradeoff between classification accuracy and running efficiency to friendly deploy into mobile devices. Firstly, we transferred ResNeXt-101 32 × 8 to build the backbone of the proposed method. After that, to promote the model generalization, we design a customized header. Meanwhile, to improve the model's convergence, we introduce an effective training method. Finally, we demonstrate the effectiveness of the proposed model and corresponding training method though the online test set. According to the experimental results, we can draw the follow conclusions:

- We present a novel garbage image classification method to address the domestic garbage image classification problem on the condition of the limited training dataset.
- For the convenience of residents, we make an artful trade-off between high accuracy and running efficiently, which makes the proposed model friendly to deploy into mobile devices.
- The experimental result with 95.62% online classification accuracy on the test dataset, 95 ms per image inference time, and 897 Megabytes GPU memory demonstrate the effective and efficiency of the proposed method.

Acknowledgements. Supported by National Key R&D Program of China (2018YFB1404402) Guangdong Science & Technology Fund (No.2019B010137003, No. 2016B030305006, No. 2018A07071702, and No. 201804010314), Guangzhou Science & Technology Fund (No. 201804010314, No. 2012224-12), VeChain Foundation (No. SCNU-2018-01), and Guangdong Provincial Department of education characteristic innovation project (Natural Science, No. 2017ktscx074).

References

1. Li, X.R., et al.: Garbage source classification performance, impact factor, and management strategy in rural areas of China: a case study in Hangzhou. Waste Manag **89**, 313–321 (2019)
2. Tong, Y.Q., Liu, J.F., Liu, S.Z.: China is implementing garbage classification action. Environ. Pollut. **259**, 113707 (2020)
3. Krizhevsky, A., Ilya, S., Geoffrey, E.H.: Imagenet classification with deep convolutional neural networks. In: Advances in Neural Information Processing Systems (2012)
4. He, K.M., et al.: Deep residual learning for image recognition. In: Proceedings of the IEEE Conference on Computer Vision and Pattern Recognition (2016)
5. Xie, S.N., et al.: Aggregated residual transformations for deep neural networks. In: Proceedings of the IEEE conference on computer vision and pattern recognition (2017)
6. Ruiz, V., et al.: Automatic image-based waste classification. In: International Work-Conference on the Interplay Between Natural and Artificial Computation. Springer, Cham (2019)
7. Szegedy, C., et al.: Inception-v4, inception-resnet and the impact of residual connections on learning. In: Thirty-first AAAI Conference on Artificial Intelligence (2017)
8. Yang, M., Gary, T.: Classification of trash for recyclability status. CS229 Proj. Rep. **2016** (2016)
9. Nalepa, J., et al.: Data augmentation via image registration. In: 2019 IEEE International Conference on Image Processing (ICIP). IEEE (2019)
10. Mahajan, D., et al.: Exploring the limits of weakly supervised pretraining. In: Proceedings of the European Conference on Computer Vision (ECCV) (2018)
11. Smith, L. N.: A disciplined approach to neural network hyper-parameters: Part 1–learning rate, batch size, momentum, and weight decay. arXiv preprint arXiv:1803.09820 (2018)
12. You, Y., Igor, G., Boris, G.: Large batch training of convolutional networks. arXiv preprint arXiv:1708.03888 (2017)

Ancient Ceramics Classification Method Based on Neural Network Optimized by Improved Ant Colony Algorithm

Yanzhe Liu and Bingxiang Liu(✉)

Jingdezhen Ceramic Institute, Jingdezhen 333403, Jiangxi, China
lbx1966@163.com

Abstract. Aiming at the problem of ancient ceramics classification, this paper proposes a method of ancient ceramics classification based on BP neural network. Making appropriate improvements to the shortcomings of the model: using ant colony algorithm to solve the problem that the network depends on the initial value; using the method of dynamically adjusting the learning rate to solve the shortcomings of the network which is easy to overfitting; optimizing the ant colony algorithm by improving the path selection strategy and the pheromone update strategy to solve the problem that the ant colony algorithm is easy to fall into the local optimum. Finally, the ancient ceramic classification model based on the BP neural network optimized by the improved ant colony algorithm is constructed. The trained model is evaluated, and the results show that the model has a good effect on the classification of ancient ceramics.

Keywords: Classification of ancient ceramics · BP neural network · Ant colony optimization

1 Introduction

At present, there are two main methods to judge the age of ancient ceramics: C14 dating and thermoluminescence dating [1–3]. But both methods have flaws. C14 dating technology is complicated and expensive, which is often used in measuring ancient porcelain with a long history. The thermoluminescence dating method will produce a larger error if the ceramic undergoes high temperature again after firing, so a new method is needed to judge the age of ancient ceramics. According to the findings of the researchers, certain elements in ceramics can be used as characteristic elements representing the characteristics of this ceramic. By analyzing the content of characteristic elements, the manufacturing place and manufacturing time of ceramics [4] could be judged. This discovery provides a new idea for the classification of ancient ceramics.

This paper takes the chemical elements of ancient ceramics as eigenvalues, proposing a classification method of ancient ceramics based on BP neural network to solve the shortcomings of neural networks. This paper made some improvements. The ant colony algorithm is used to assign initial values to the neural network to solve the problem that the network is dependent on the initial values too much. The learning rate of the neural network is adjusted dynamically to solve the shortcomings of the neural

Q. Liu et al. (Eds.): CENet 2020, AISC 1274, pp. 276–282, 2021.
https://doi.org/10.1007/978-981-15-8462-6_33

network easy to fall into overfitting. To solve the problem that the ant colony algorithm is easy to fall into the local optimal solution, the ant colony algorithm is optimized by improving the path selection strategy and the pheromone update strategy. Finally, the ancient ceramic classification model based on the BP neural network optimized by the improved ant colony algorithm is constructed. After the model is trained, the performance of the optimized model is evaluated through the confusion matrix and ROC model evaluation indicators.

2 Model Theory

2.1 BP Neural Network

BP (back propagation) network is a multi-layer feed-forward neural network trained by inverse error propagation algorithm [5]. This algorithm has many flaws, for example, the BP neural network learns each data in the sample set one by one, so the network training process convergence is slow. At the same time, the algorithm takes the sum of squared errors as the objective function, and using the gradient method to find its minimum value. This method leads to the fact that it will inevitably produce local minima unless the squared error function is positive defined. In addition, the learning process of the BP neural network has a great relationship with the setting of the initial learning rate. If the learning rate is too large or too small, it is easy to produce overfitting or underfitting [6]. These defects need to be optimized to get better results.

2.2 Ant Colony Algorithm

Ant Colony Optimization Algorithm (ACO) is an intelligent optimization algorithm proposed that imitates the behavior of ant colony. And the design of ant colony algorithm mainly refers to three important strategies of ants in the foraging process: selection strategy, update strategy and collaborative strategy.

In the ant colony algorithm model, the probability p of ants from i to j at time t is:

$$p_{ij}^{k} = \begin{cases} \frac{\tau_{ij}^{\alpha}(t)\times\eta_{ij}^{\beta}(t)}{\sum_{s\in allowed}\tau_{is}^{\alpha}(t)\times\eta_{is}^{\beta}(t)}, \text{if } j \in allowed \\ 0, \text{otherwise} \end{cases} \quad (1)$$

The ant colony algorithm also has its shortcomings: the amount of information initially set by the algorithm is the same, thus the ants may not necessarily be the optimal path direction when creating the first path. This situation can easily mislead subsequent selection. The algorithm uses the positive feedback mechanism to strengthen the current better solution, but it will cause the algorithm to hover around the current optimal solution, thus easily making the algorithm stagnate prematurely and finally falling into the local optimal solution. Therefore, it is necessary to optimize the ant colony algorithm to get a good solution.

2.3 Algorithm Optimization

Facing the problem that the neural network is very dependent on the initial value, this article changes the setting of the learning rate from a fixed value to a dynamic adjustment. For example, an improvement can set the initial value of the learning rate to a larger number, so that the neural network can converge at a faster rate in the early stage of training. At the same time, the learning rate will be gradually reduced during the training to avoid overfitting. There are many ways to dynamically adjust the learning rate, in which the commonly used methods are the piecewise function, exponential function and polynomial function. Combined with the classification problem of ancient ceramics, after many experiments, this article sets the method of dynamically adjusting the learning rate in the classification model as a piecewise function, and the specific update formula is as follows:

$$\eta(t+1) = \frac{\eta(t)}{2^n} \quad \text{if} \ t_i \leq t \leq t_{i+1} \tag{2}$$

To solve the problem that the ant colony algorithm is easy to fall into the local optimal solution, this paper optimizes the path selection strategy and the pheromone update strategy respectively. In the path selection strategy, this paper uses a combination of deterministic selection and random selection. At the same time, the probability of deterministic selection is dynamically adjusted during the search process, so that after the algorithm has basically determined the evolution direction, it can narrow the gap between the best path and the worst path pheromone concentration, to make algorithm search more comprehensive. This algorithm determines the next city s to which ant k is transferred from i according to the following formula [7].

$$s = \begin{cases} max_{u \in allowed_k}\left\{\tau_{iu}^{\alpha}\eta_{iu}^{\beta}(t)\right\}, \text{if } r \leq p_0 \\ \text{choose according to the probability in}(1), \ else \end{cases} \tag{3}$$

In the pheromone update strategy, a collaborative update strategy of pheromone global update and local update is used [8]. The optimized local update formula of pheromone is as follows:

$$\tau_{ij}(t+1) = (1-\rho) \times \tau_{ij} + \rho \times \min(\tau_{ij} | 1 \leq r \leq k_r) \tag{4}$$

The pheromone global update formula is:

$$\tau_{ij}(t+1) = \mu(t)(1-\rho) \times \tau_{ij}(t) + \Delta\tau_{ij} \tag{5}$$

$$\mu(t) = \frac{1}{e^{\frac{t}{20}}}, \Delta\tau_{ij} = \frac{Q}{e\sqrt{\frac{E_t}{M \times N}}} \tag{6}$$

2.4 Optimized BP Neural Network with Improved Ant Colony Algorithm

In BP neural networks, the initial values of weights and thresholds have a great influence on the learning effect of the network. Since the value ranges of the weight and the threshold are both *[0, 1]*, the weight and the threshold can be equally divided to a limited number of value intervals. Using the improved ant colony algorithm to optimize the BP neural network is to use the global optimization ability of the ant colony algorithm to initialize the weights and thresholds of the BP neural network and narrow the range of the weights and thresholds.

Assuming that the BP neural network has p input nodes, q output nodes and h hidden nodes. In the ant colony algorithm, there will be $(p + q + 1) * h + q$ vertices when the weight and threshold are regarded as a vertex. There are e routes from the i-th vertex to the j-th vertex, which represents that the i-th vertex has e selectable value ranges. Let the ants start from a certain vertex, selecting a certain path according to the formula to reach the next vertex, and so on, until all vertices are walked. Each time when the ant completes a cycle, it will generates a solution. After n cycles of m ants, the best solution among all the solutions will be selected as the initial value of BP neural network weight and threshold.

3 Prediction Experiment

3.1 Data Processing

A total of 222 sets of ceramic sample data were collected in this paper, including ceramic raw material chemical components Na_2O, MgO, Al_2O_3, SiO_2, K_2O, CaO, TiO_2, Fe_2O_3, CoO, MnO, CuO, ZnO, PbO_2, Rb_2O, SrO, Y_2O_3, ZrO_2, P_2O_5, a total of 18 chemical elements. The labels were divided into Yuan, Ming and Qing. The "hold-out" and "stratified random sampling" were used to divide the training sets and the test sets, and then the 222 samples were divided into 162 training sets and 60 test sets, in which the training sets and test sets accounted for 72.97% and 27.03%. The number of tags labeled Yuan, Ming and Qing in the total data were 114, 68 and 40, which respectively accounted for 51.4%, 30.6% and 18.0%; the numbers in the training sets were 83, 50 and 29, accounting for 51.2% and 30.9% and 17.9%; the number of test sets were 31, 18 and 11, accounting for 51.7%, 30.0% and 18.3%. Some elements are missing from the original data, so the missing value is set to 0 in the preprocessing from avoiding the missing value from affecting the experiment. At the same time, the three-digit 01 code are used to indicate three kinds of tags, namely, 001 for the Yuan Dynasty, 010 for the Ming Dynasty, and 100 for the Qing Dynasty. Part of the data is shown in Table 1.

Before conducting the experiments, it is usually necessary to process the acquired data samples before input them into the network for calculation. Under normal circumstances, the data will be quantified as a real number between *[0, 1]*, which means that the input samples are normalized. The normalization formula is:

Table 1. The main chemical composition of porcelain.

Serial	Na_2O	MgO	Al_2O_3	SiO_2	CaO	MnO	CuO	…	Classification
1	2.07	0.31	23.01	69.32	0.32	530	10	…	001
2	2.68	0.05	12.95	70.10	9.02	579	60	…	001
3	2.79	0.16	12.77	66.75	10.55	690	30	…	001
4	0.88	0.34	14.33	73.16	2.46	45350	40	…	010
5	0.67	1.38	14.14	70.56	5.22	640	90	…	010
6	0.03	0.35	22.97	69.06	7.18	620	20	…	010
…	…	…	…	…				…	…

$$X_i' = a \times \frac{X_i - \min(X_i)}{\max(X_i) - \min(X_i)} + b \tag{7}$$

3.2 Experiment and Result Analysis

The number of nodes are set to 18, 37 [9] and 3 in the input layer, hidden layer and output layer of the neural network, the maximum number of iterations is 450, and the initial learning rate is 0.4. According to the theory described above, the optimized BP neural network is used to establish the ancient ceramic classification model, the training set is used to train the network, and then the test set is used to test the generalization ability of the model. Finally, the accuracy of the model before and after the optimization of the algorithm will be obtained. As shown in Table 2, the accuracy of the optimized model has been greatly improved. In order to show the effect of the optimization algorithm in more detail, this paper introduces confusion matrix and ROC curve to evaluate the model.

Table 2. Experimental results.

	Accuracy rate	Qing	Ming	Yuan
Before optimized	63.33%	36.36%	50.00%	80.64%
After optimized	78.33%	63.64%	72.22%	87.10%

Figure 1 is a diagram of the confusion matrix before and after the model improvement. From the Fig., it can be found that the samples labeled as Ming and Qing have a higher probability of being misjudged as Yuan when the algorithm is not improved, so the accuracy of the label as Yuan is higher than that of Ming and Qing a lot. The improved model corrects this deviation, greatly reducing the probability of being misjudged as the Yuan, so that the accuracy of each label has been improved, especially the accuracy rate of the Qing has been greatly improved from 36.36% to 63.64%.

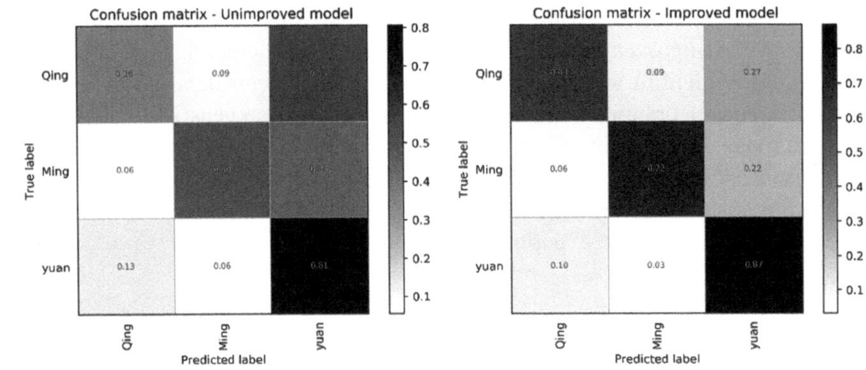

Fig. 1. The confusion matrix of the model before and after improvement.

Figure 2 is the ROC curve before and after the model improvement. The performance of the model can be compared through the area under the ROC curve – the AUC value. The AUC values labeled Yuan, Ming and Qing before the improvement were respectively 0.73, 0.92 and 0.79, which were increased to 0.76, 0.94 and 0.85 after the improvement. The AUC values of the three types of labels have been improved. It can be clearly seen that the classification effect of the optimized model is better than the model before optimization.

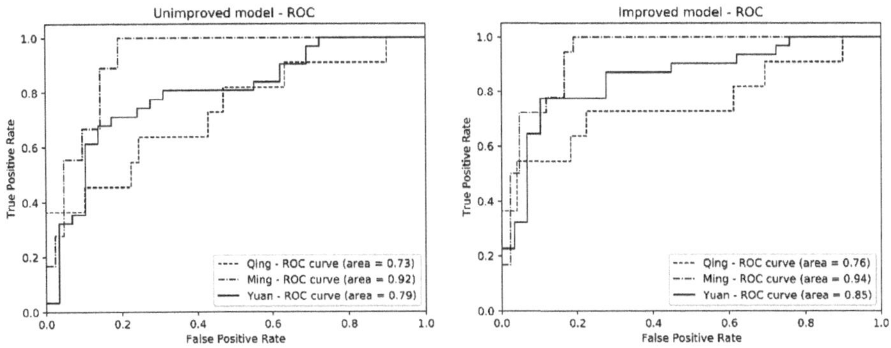

Fig. 2. ROC curve of the model before and after improvement.

4 Conclusion

In this paper, the optimized BP neural network is applied to the classification of ancient ceramics. A classification model is constructed and the model is optimized by using ant colony algorithm. The final experimental results showed that the overall accuracy of the unimproved model reaches 63.33%, while the overall accuracy of the optimized model increased to 78.33%. The results prove that the model has a good classification effect in the ancient ceramic classification problem, and the improved ant colony algorithm has a good optimization effect on the network model. In summary, the neural network optimized by the improved ant colony algorithm has a good classification effect on the classification problem of ancient ceramics, which a certain practical value. However, the network model optimization in this paper is not comprehensive enough. In the future, the model optimization can be discussed in a more detailed and comprehensive way to further improve the optimization effect. Meanwhile, the lack of ancient ceramic data also has a certain impact on the training of the model. More data can be collected in the future to improve the experiment.

References

1. Siiriäinen, A.: Archaeological background of ancient lake Päijänne and geological dating of the meso-neolithic boundary in Finland. Bull. Geol. Soc. Finl. **42**, 119–127 (1970)
2. Laura, M., Silvano, M., Caterina, D.V., Aida, M.C.: Evaluation of a FTIR data pretreatment method for principal component analysis applied to archaeological ceramics. Microchem. J. **125**, 224–229 (2016)
3. Fleming, S.J.: Thermoluminescence dating of ancient ceramics and bronzes. Med. Sci. Law **14**(1), 11–14 (1974)
4. Li, W.D., Luo, H.J., Li, J.N., Li, J.Z., Guo, J.K.: Studies on the microstructure of the black-glazed bowl sherds excavated from the Jian kiln site of ancient China. Ceram. Int. **34**(6), 1473–1480 (2007)
5. Chu, J.L., Li, H.Y., Chen, X.J.: Research on improved BP learning algorithm of BP neural network. Adv. Mater. Res. **765–767**, 1644–1647 (2013)
6. Wang, L., Wang, D., Ding, N.: Research on BP neural network optimal method based on improved ant colony algorithm. In: Second International Conference on Computer Engineering & Applications. IEEE Computer Society (2010)
7. Dreo, J., Siarry, P.: Continuous interacting ant colony algorithm based on dense heterarchy. Future Gen. Comput. Syst. **20**(5), 841–856 (2004)
8. Zhang, Y.R.: Research and application of BP neural network optimization based on improved ant colony algorithm. Northeastern University (2013) (in Chinese)
9. Gikhman, I.I.: (Iosif Il′ich). Introduction to the Theory of Random Processes, pp. 15–24. American Mathematical Society, Providence (2002)

Design of ISAR Image Annotation System Based on Deep Learning

Bingning Li(✉), Chi Zhang, Wei Pei, and Liang Shen

Xi'an Satellite Control Center, Xi'an 710000, China
lbnvivian@126.com

Abstract. Imaging radar can obtain excellent images of non-cooperative moving targets, which is an essential means of target detection. According to the scattering point and time-frequency characteristics of the ISAR image, it is possible to determine the physical properties such as the target's component structure. Combining with other radar electromagnetic scattering characteristics, it is the primary technical approach to carry out the full recognition of the target.

Manually captioning of the target components and the main structure is a traditional and straightforward method that lacks a quantitative standard, requires heavy workload, and the captioning result may be affected by the subjective judgment. With the development and application of deep learning technology in recent years, remarkable achievements have been made in the fields of image classification, image detection, and machine translation, and the performance of automatic image captioning has been greatly improved. Through intelligent segmentation of ISAR images and feature extraction of segmented regions, we build a database of image features and generate the corresponding text description according to the characteristics of different parts of the image applying the text generation network. Based on the above theory, this paper brings out an ISAR image annotation system design scheme based on deep learning.

Keywords: First ISAR image · Deep learning · Automatic image annotation · Convolutional neural network · Multi-area

1 System Design

1.1 Demand Analysis

When processing the echo data of imaging radar, a large number of real-time ISAR images and post images are generated. Selecting high-quality images from a large number of images to caption the physical attributes such as the structure of its components is based on the subsequent target recognition work. At present, the method of manual image judgment is of heavy workload and low efficiency and can be easily affected by subjective assumptions due to the lack of the unified standard. Through the establishment of the annotation model [1], we can conduct automatic ISAR image analysis, feature extraction, and classification recognition effectively, and the recognition results can be quickly obtained. Compared with manual image annotation, the quantitative index of the system is precise, and the conclusion is objective, which can

Q. Liu et al. (Eds.): CENet 2020, AISC 1274, pp. 283–288, 2021.
https://doi.org/10.1007/978-981-15-8462-6_34

ensure the rapidity, accuracy, and real-times of recognition. The overall framework of the system is shown in Fig. 1.

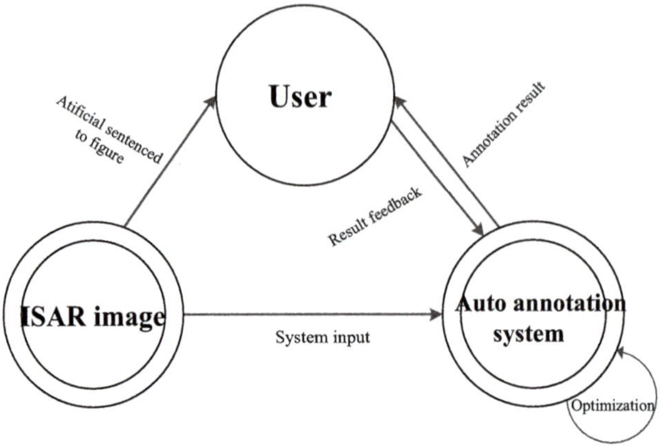

Fig. 1. System architecture diagram.

1.2 Functional Module Design

According to the related content of demand analysis, the system is divided into four functional modules, namely image evaluation, feature extraction module, associated annotation module, and ISAR image feature library. The central function module is shown in Fig. 2.

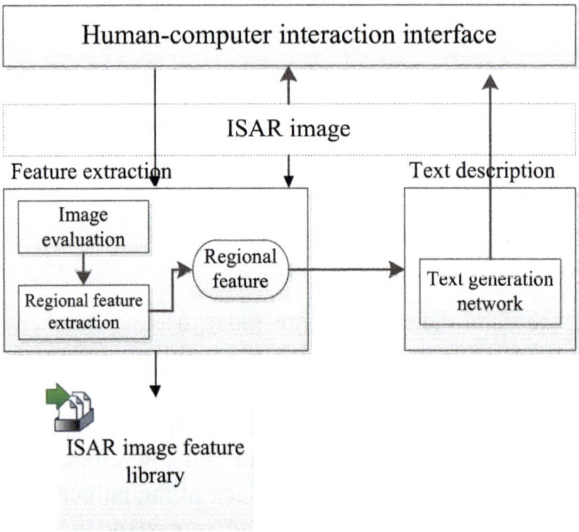

Fig. 2. Central function module.

Image Quality Evaluation. That is data selection and preprocessing, forming a training data set. To ensure the recognition effect of the subsequent model, it is necessary to filter the massive ISAR images, remove the image data with poor quality and unclear structure. The image enhancement will reinforce the target part and depress the background noise.

Feature Extraction Module. It is the basis of the follow-up work. The accuracy of image region feature extraction directly affects the result of text annotation. Convolution neural network is used to select features after region segmentation and calculate feature parameters.

Associate the Annotation Module. That is to select the annotation model and form the corresponding text description according to the characteristics of different regions.

Build an ISAR Image Feature Library. According to the region segmentation results of the above two models, extract the region feature value and then get the ISAR image feature library [2]. The features extracted in this paper can be divided into panel size, contour, components, primary structure, etc.

1.3 System Architecture Design

Because deep learning requires high hardware configuration, large network model, long training time, and running time, we consider building a deep learning server to solve the problem of client machine configuration [3], using a convolutional neural network (CNN) to obtain image region features for automatic image annotation [4]. At present, the mainstream deep learning frameworks include TensorFlow, Caffe, Torch, etc. In this paper, the TensorFlow learning framework is adopted. If other mainstream deep learning frameworks are added in the later stage, only one-time modification or creation of the corresponding image is needed. A similar image is imported into the server, without tedious operations of compilation and deployment for dependency libraries. In addition to meeting the relevant functional requirements, to ensure the smooth and reliable operation of the system, the following criteria shall also be considered:

Stability. The input data flow of the system is a large short-term amount of data and a small amount of picture data. Performance optimization and redundancy design should be considered to avoid jamming in the process.

Friendliness. The human-computer interface shall be intuitive and straightforward, clearly marked, and easy to operate.

Scalability. Considering the continuous expansion of the characteristics of different target structural components and the input of optical pictures, the system design should be scalable in function and data storage.

2 System Implementation

2.1 Model Construction

The hierarchical structure of image understanding includes three layers. The bottom layer understanding is based on the image data input and outputs the visual features [5]; the middle layer understanding forms the feature representations matching the target; the top layer understanding is based on the feature data of the bottom layer and the middle layer as the input to realize the image semantic interpretation. Without relying on the characteristics of artificial design, the convolutional neural network is constructed to transform the underlying data into high-level understanding [6]. In this paper, the multi-region based image automatic annotation method is used to generate text description for different regions in the image, and the constructed model is shown in Fig. 3.

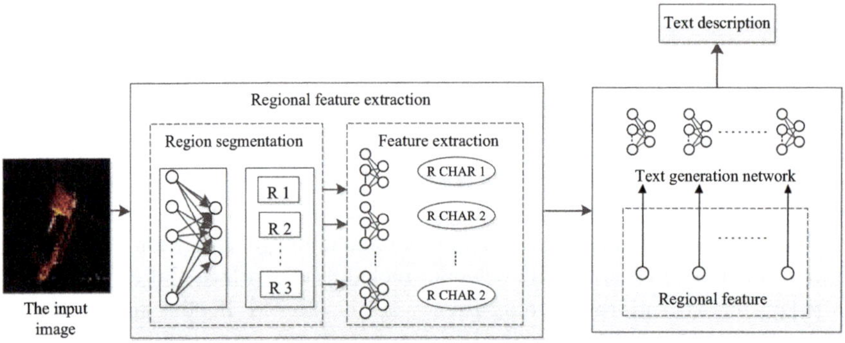

Fig. 3. System module.

In this model, the region feature extraction network monitors the different object regions in the image, extracts the visual features, and the text generation network generates the corresponding text description according to the modified object region features. Combining the features and text output by the two networks, the image annotation is realized [7].

2.2 Automatic Annotation Method

The annotation method process adopted in this paper is shown in the following figure. Firstly, based on the image quality evaluation of the surface target model, the image is divided into three grades: "excellent", "medium" and "poor". The images rated as "excellent" are selected for PCA whitening preprocessing [8]. In the next step, 1500 of the 2000 original images are selected as the training set, 500 as the test set, and 100 copies of the training set are manually captioned to train the initial model. Subsequently, import the uncaptioned image through iteration, extract the visual features and text description, and finally generate the annotation results. The iterative training keeps updating the algorithm model [9]. Simultaneously, the method of confidence threshold filtering is used to obtain the annotation results of high confidence. At last, the results were verified artificially, and training intervention was carried out [10].

2.3 Method Validation

With 20 categories of targets, 1500 pictures of 75 pictures for each category were selected as the training set, and 25 pictures for each category that is a total of 500 pictures as the test set. After the training iterated 11 times, the test accuracy reached 93.6%. Table 1 displays the detailed process.

Table 1. Method testing procedure.

Iterations	Iterations time (min)	Number of test sets	Test set accuracy
1	3	500	73.8%
2	4.7	729	77.2%
3	5.1	876	80.8%
4	6.9	904	83.1%
5	8.3	1056	85.5%
6	9.9	1188	88.3%
7	10.7	1312	89.7%
8	11.3	1407	90.1%
9	11.9	1510	91.6%
10	12.4	1620	92.7%
11	13.6	1809	93.6%

3 Conclusion

The most fundamental problem of automatic image annotation is to obtain robust and efficient features. Image annotation is to process visual and knowledge information and analyze image content comprehensively through deep learning methods. The experimental results verified the robust and effectivity of the image annotation method based on deep learning proposed in this paper, which combines the features obtained from image vision and text description. In conclusion, the ISAR image annotation system based on deep learning can improve the efficiency of manual annotation and reduce the error rate. This system demonstrates an advantage in high application value, which can not only be used for fast training of convolutional neural network model, but also be widely used in other network models of supervised learning, not only be applied to process ISAR image data, but also other images such as optical, remote sensing, etc.

References

1. Datta, R., Joshi, D., Li, J., Wang, J.Z.: Image retrieval: ideas, influences, and trends of the new age. ACM Comput. Surv. **40**(2), 1–60 (2008)
2. Chen, F., Ji, R., Sun, X., Wu, Y., Su, J.: Group cap: group-based image captioning with structured relevance and diversity constraints. In: The IEEE Conference on Computer Vision and Pattern Recognition (CVPR), NJ, USA, pp. 1345–1353. IEEE (2018)
3. Tian, X., Wang, L., Ding, Q.: Review of image segmentation based on deep learning. J. Softw. **30**(2), 440–468 (2019)
4. Goodfellow, I.J., Pouget-Abadie, J., Mirza, M., Xu, B., Warde-Farley, D., Ozair, S., Courville, A., Bengio, Y.: Generative adversarial networks. In: Advances in Neural Information Processing Systems, NY, USA, vol. 27, pp. 2672–2680. ACM (2014)
5. Wei, Y.C., Zhao, Y.: A review on image semantic segmentation based on DCNN. J. Beijing Jiaotong Univ. **40**(4), 82–91 (2016)
6. Girshick, R., Donahue, J., Darrell, T., Malik, J.: Rich feature hierarchies for accurate object detection and semantic segmentation. In: Proceedings of the IEEE Conference on Computer Vision and Pattern Recognition, NJ, USA, pp. 580–587. IEEE (2014)
7. Li, F.F., Iyer, A., Koch, C., Perona, P.: What do we perceive in a glance of a real-world scene. J. Vis. **7**(1), 10 (2007)
8. Gao, L., Fan, K., Song, J., Liu, X., Xu, X., Shen, H.T.: Deliberate attention networks for image captioning. In: Proceedings of the AAAI Conference on Artificial Intelligence, CA, USA, pp. 8320–8327. AAAI Press (2019)
9. Fu, K., Li, J., Jin, J., Zhang, C.: Image-text surgery: efficient concept learning in image captioning by generating pseudopairs. IEEE Trans. Neural Netw. Learn. Syst. **29**(12), 5910–5921 (2018)
10. Liu, S., Qi, X., Shi, J., Zhang, H., Jia, J.: Multi-scale patch aggregation (MPA) for simultaneous detection and segmentation. In: Proceedings of the IEEE Conference on Computer Vision and Pattern Recognition, NJ, USA, pp. 3141–3149. IEEE (2016)

Research on Public Sentiment Data Center Based on Key Technology of Big Data

Zhou Qi[1], Yin Jian[2(✉)], and Liangjun Zhang[3]

[1] Open University of Guangdongy, Guangzhou 510630, China
zhoudake77@163.com
[2] Sun Yat-Sen University, Guangzhou 510630, China
issjyin@mail.sysu.edu.cn
[3] Guangzhou Teddy Intelligent Technology Co., Ltd., Guangzhou 510630, China

Abstract. Through the construction of public opinion data center as the research object, a series of data acquisition, monitoring and analysis and mining tasks based on public opinion information can provide natural language processing functions such as web data crawling and monitoring, web data parsing, text data preprocessing, text analysis and mining. Research shows that automatic summarization technology, event recognition technology, event context combing and emotion analysis technology can be based on text mining algorithm to achieve automatic keyword extraction, content simplification, similarity calculation, tracking and monitoring, and subjective text with emotional color analysis has a very good application value and efficiency.

Keywords: Public opinion information · Automatic promotion · Event recognition · Event context · Sentiment analysis

1 Introduction

This platform is mainly composed of two modules: (1) general module; and (2) business theme module. The general module mainly provides the function point key of the whole process of public opinion analysis (including data acquisition and monitoring, data analysis, data preprocessing, mining modeling, result display, etc.), which is used to support the background business function realization and the front-end custom implementation. Collection, monitoring and analysis of mining tasks.

2 General Module Design for Text Collection and Analysis and Mining

This module is mainly embedded in the public opinion analysis tool developed by Teddy Technology. It can provide natural language processing functions, such as web data crawling and monitoring, web data parsing, text data preprocessing, text analysis and mining. Based on this module, on the one hand, the front-end users can operate by dragging and dropping, and connect the function keys in a flow way to achieve the purpose of customizing and realizing the requirements of text mining. On the other hand,

Q. Liu et al. (Eds.): CENet 2020, AISC 1274, pp. 289–301, 2021.
https://doi.org/10.1007/978-981-15-8462-6_35

the above-mentioned function keys support the development of back-end business functions and save time. And cost. The function of this module is shown in Table 1.

Table 1. General module function of text collection and analysis mining.

Modular	Subfunction	Describe
Data acquisition and monitoring	Web crawler	Automatically fetching web page information according to certain rules
	Data monitoring	Monitor the parameters of web crawler process, such as page count, time consumption and so on
	Timed task	Setting timed crawling tasks for web crawlers
Web page parsing	Web page parsing	Extract the content from the crawled web page information
Data preprocessing	Clause	The long text is divided into short text by sentence by a character
	Participle and part of speech tagging	The text is segmented into words, and each part of speech is tagged, such as place adverbs, time adverbs, etc.
	Named entity recognition	Identifying entities with specific meanings in text, including person names, place names, organization names, proper nouns, etc.
	Discontinuation of words	Filtering some words or words in the result of word segmentation based on the word list
	Word vector	Methods to convert text into numeric vectors, including Skip-gram, CBOW, word2wec, etc.
	TF-IDF	Calculate TF-IDF values of words
	Similarity computation	Realization based on cosine theorem
Mining and modeling	Text clustering	Contains common algorithms such as LDA, K-means and DBSCAN
	Text classification	Including common algorithms such as SVM, Na, ve Bayes, KNN and so on
	Emotional analysis	Analysis of subjective texts with emotional color
Results show	Text specific display	Such as word cloud chart, word meaning association diagram, propagation path map, etc.
	Other	Traditional bar charts, pie charts, broken lines, etc.

This paper mainly analyzes and studies the key technologies of automatic summarization technology, event recognition technology, event context combing and emotion analysis technology.

3 Automatic Summarization Technology Design

Automatic summarization technology refers to the automatic extraction of simple and coherent text from the original text to reflect the central content. According to the requirement of construction, this technology is applied to the media report text to generate the document summary which reflects the purpose of the report. On the one hand, it can be used to identify the media news event, that is, the media report of the same event can be merged by combining the report summary and the basic information of the report with the clustering technology; on the other hand, it is helpful to the news event. The context is organized to fully grasp the development process of events.

It is generally believed that a document is organized by multiple topics. Therefore, automatic summarization technology uses clustering method to gather sentences under the same topic, and then selects the central statements of each topic to generate an appropriate summary. Thus, the abstraction of text summaries in media reports is mainly like the three processes in Fig. 1:

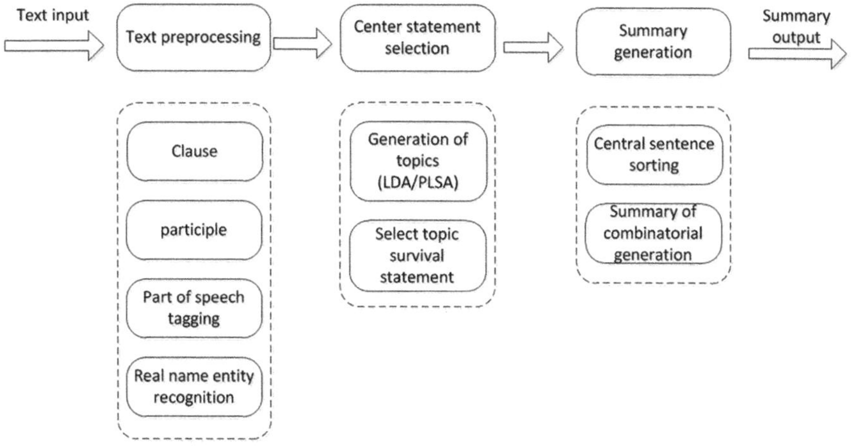

Fig. 1. Flow chart of automatic text summarization.

3.1 Text Preprocessing Design

For each media report document, the text processing method is used to split the sentences, and then the Chinese word segmentation technology is used to segment the sentences, and the corresponding parts of speech (such as verbs, adjectives, time words, place names, etc.) of the words after segmentation are marked. Named entity recognition based on part of speech is to identify the specific meaning of the entity in the sentence (such as person name, place name, time, proper noun, etc.), so as to filter out the main information of the sentence.

Generally speaking, the corpus-based segmentation method is more scalable and accurate than the sequential annotation-based segmentation method for specific scene, industry and domain text, but the premise is that the corpus is complete and complete.

As the media reports cover all walks of life, the corpus needs to be more complete and adaptable, so it is particularly important to collate the reporting corpus. According to the reported data, a total of 18 word sets, 7 word sets according to emotional polarity, and a public lexicon (such as modern Chinese dictionary) are preliminarily sorted out, as shown in Table 2.

Table 2. Collection of Thesaurus.

Lexicon	Contained word set
Industry classification Thesaurus	Agriculture, forestry, animal husbandry, fishery, medicine, health, building materials, metallurgy, minerals, petroleum, petrochemical, water conservancy, hydropower, transportation, information industry, machinery, electromechanical, light industry, food, clothing, textile, professional services, safety protection, environmental protection, tourism, leisure, office culture, education, electronics, toys, furniture and gifts
Emotional lexicon	Positive emotions/negative emotions/degree modifiers/adjective modifiers/negative words/threatening words/extreme words
Named entity Thesaurus	Name/city information/government agencies/laws and regulations
Modern Chinese lexicon	Modern Chinese dictionaries/idioms/dictionaries/network language dictionaries
Custom Dictionary	Reporting custom words

3.2 Central Sentence Selection

Taking the sentences pretreated above as granularity, the topic model based on semantics (such as LDA) is used to generate some topics of the report document, and the importance scores of each sentence in each topic are calculated considering the factors such as the position of the sentence in the media report document and the number of keywords involved, so as to select the topic headings. Sentence. Taking the content of a report as an example, the results are as follows:

Report content
The total import and export value of China and Russia in the first seven months of this year was 46.822 billion US dollars, up 21.8% year-on-year, according to data released by the General Administration of Customs of China. Experts from well-known Russian institutions said that China-Russia economic and trade cooperation has a solid foundation and strong willingness to cooperate. The bilateral trade volume is expected to reach 80 billion US dollars in the whole year. Russian Research Institute Carnegie Moscow Center research project director Gabyev told

Xinhua News Agency that in recent months, the international market metal prices have risen significantly, as raw materials and commodities account for a large proportion of Russian exports, if metal prices continue to maintain high, and oil prices remain around $50 a barrel, plus the second half of the Russian Central Committee The bilateral trade volume is expected to reach US $80 billion this year. Gabyev said that Russia's exports of non-raw materials to China are increasing. Factors contributing to this trend include the low exchange rate of the ruble, the efforts of the two governments to reduce non-tariff barriers and the facilitation of bilateral food trade by building transport infrastructure. The Russian export center is the foreign trade promotion agency of the Russian Foreign Economic bank. Ma Monov, the managing director of the center responsible for promoting international trade, told Xinhua that Russia's exports to China are mainly energy products, timber, forestry processing products and machinery and equipment, of which energy products account for more than half. At the same time, Russian exports of nuclear reactor equipment, turbojet aeroengines and special motors to China have increased significantly. ...

Centers
Tatement
ection

Result of topic centered sentence selection	
Topic sentence 1	In the first 7 months of this year, the total value of imports and exports of China and Russia was US $46 billion 822 million, an increase of 21.8% over the same period last year.
Topic sentence 2	China-Russia economic and trade cooperation has a solid foundation and strong willingness to cooperate. This year's bilateral trade volume is expected to reach 80 billion US dollars.

3.3 Summary Generation

After choosing the sentence as the center sentence, the narrative of a complete event usually covers the information of time, place, person, cause, course and result. Combining the central statements according to the time and the order of the original documents, a coherent and concise summary of media reports is generated. If the above report is selected, the final summary will be as follows:

Result of topic centered sentence selection	
Topic sentence 1	In the first 7 months of this year, the total value of imports and exports of China and Russia was US $46 billion 822 million, an increase of 21.8% over the same period last year.
Topic sentence 2	China-Russia economic and trade cooperation has a solid foundation and strong willingness to cooperate. This year's bilateral trade volume is expected to reach 80 billion US dollars.

Summary generation

Report summary
In the first seven months of this year, the total import and export value of China and Russia was 46.822 billion US dollars, up 21.8% over the same period of last year. China-Russia economic and trade cooperation has a solid foundation and strong willingness to cooperate. This year's bilateral trade volume is expected to reach 80 billion US dollars.

4 Event Recognition Technology Design

Different countries and media organizations can report the same event, that is, there may be duplicate reports on the same thing. Therefore, how to distinguish whether a new report document is a new report event or is divided into the same media event with the historical report is to further achieve relevance analysis, emotional analysis and trend prediction. The key. One of the simplest and traditional principles is to identify by similarity of text.

The common algorithm of text similarity calculation is based on space vector model (VSM) and cosine distance expansion. It simplifies the processing of text content into vector operation in vector space, and it expresses semantic similarity with spatial similarity, which is intuitive and easy to understand. It has been widely used in short text similarity calculation and detection. Cable. But in fact, a network media report data is often a long text, there are many redundant information describing the event, which affects the performance of space vector model based on TF-IDF. Considering this factor, the foregoing text for the media reports automatically extract the key information of the events described by the media, condense the long text into short text, improve the efficiency and accuracy of event recognition.

The recognition of media events includes two parts: the classification of the same event in historical reports and the detection of new and old events in new data streams. However, the former can be classified as the special case of the latter. Media event detection is to distinguish whether the reported data stream belongs to a known media event, and to extend and extend the known media events. The basic implementation process is shown in Fig. 2 below.

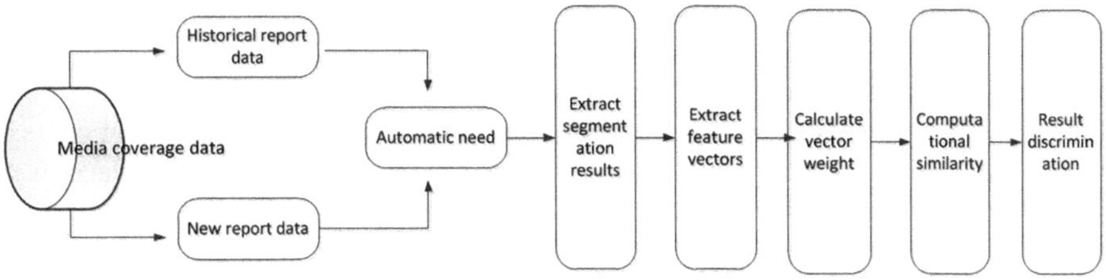

Fig. 2. Basic process of event recognition.

(1) Automatic abstraction of historical reports and new reports is realized by clauses, word segmentation and topic extraction, and the segmentation results are saved.
(2) Extract the corresponding parts of summary sentences from the segmentation results for further construction of space vector model.
(3) Remove the stop words from the segmentation results and use the remaining words or phrases as the feature vectors of the spatial model.

Since the abstract is the main content of the event, it is not necessary to select text features by chi-square detection, information gain and other methods like other space vector models.

(4) The TF-IDF values of computed features are standardized as vector weights to construct a space vector model.

TF: Term frequency is the frequency of keywords in an article. For example, if there are N keywords in an M-word article, then TF = N/M.

IDF: Inverse document frequency refers to the reverse text frequency, which is an index used to measure the weight of keywords. It is calculated by formula $\mathrm{IDF} = \log\left(\frac{DW}{\mathrm{D}}\right)$, where D is the total number of articles, and Dw is the number of articles with keywords.

(5) Calculate the similarity between new document and any old document based on cosine distance.
(6) Comparison method is used to distinguish the results: if the similarity value does not exceed the preset threshold, the new reported data becomes a new event independently; if the similarity exceeds a certain threshold, the reported event is classified as the event with the highest similarity as belonging to the same event class, and the event is classified as recognition.

Based on the above methods of event detection, we can update the reports in real time and grasp the latest trends of media events. But in fact, the process of similarity calculation is based on word frequency information. The similarity of two documents depends on the number of common words. It is impossible to distinguish the semantic fuzziness of natural language and the correlation between words. Process.

5 Incident Context Combing Design

Users want to understand a hot event, need to browse a large number of reports about the media event, repeated reading, time-consuming and laborious. The “event context” technology can extract the basic elements of each important development stage in the event life cycle, and take time as the main line to show. It can improve the efficiency of user access to information, help users easily understand the development of the whole event, and guide users to read in depth.

The goal of event sketching is to give a list of important news that covers important information at all important stages of the news event.

From a developmental point of view, when events are at an important stage of progress, there will be some high-quality reports, and the more reports, the more concentrated. From this point of view, we use time slice clustering method to mine the main stages of events. Of course, the better the quality of media events, the longer the coverage time window, the more beneficial for mining context. The process of event context mining based on time slice clustering is shown in Fig. 3.

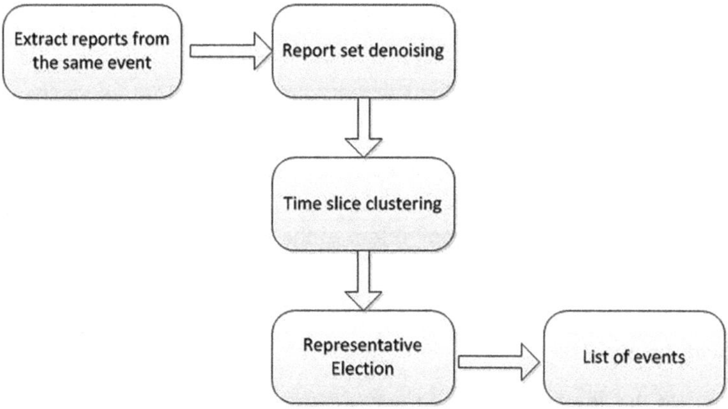

Fig. 3. Time series clustering process based on time slice clustering.

After event recognition, the data stored and managed in the database can be directly used for event skeleton combing and mining, and only need to retrieve the event reports that need to be mined.

5.1 Report Set Denoising

The media event report has the timeliness, generally speaking, is more timeliness than the authoritative media website to the event report, but the general small media website will have certain time lag. For example, the same content of media reports, Xinhua launched at 23:10 on the day of the incident, and some small sites are likely to have to wait until the next day 10:00 to release. This will lead to the same description and belong to the same event stage of the report, is divided into different stages, thus affecting the time slice clustering stage division effect.

The above noise reports can still be weighed by the method of text similarity measurement. If two media reports belong to the same news information point or the relationship between reprint and reprint, then the similarity between the two documents will be very high. Therefore, the similarity between the two documents is calculated according to the report set corresponding to the media event. If the similarity between the two documents is very high, the two documents are merged and the earliest report document is taken as the representative to participate in time slice clustering.

5.2 Time Slice Clustering

For each media report, extract its release time, and then according to the release time of the report, a media event set of reports are arranged. Projecting the report to the timeline according to the release time, if the event has a clear stage of development, according to the previous assumption, you can see some news-intensive areas on the timeline, these areas are the various stages of the event we want to obtain.

There are several common clustering algorithms, such as *K*-Means and *K*-mediods, for computing dense time segments. However, these clustering algorithms need to determine the number of event stages beforehand, and it is really difficult to determine this result in advance. Therefore, this paper chooses the agglomeration hierarchical clustering algorithm. The specific approach is: first, each report is regarded as a time segment (point), and then each time the nearest two fragments are merged until the distance between any two fragments is greater than the predetermined threshold.

5.3 Selection of Representative Reports

After getting some progress of the event through time slice clustering method, we should extract a representative report from each progress report set and summarize the content in the summary of the report.

As to how to extract the representative report of the event stage, the following factors are generally considered: whether the website of the report source is authoritative, when the report is published (in the same clip), whether the report is reproduced more, and so on.

After the above steps, we can get the summary list of the core reports in each stage of the event, further extract the publishing time of the core reports, and sort them in order of time, so we can get the main context of the media event. The key technology of this method is time segment clustering, which is very important for measuring the distance between reports. The conventional "natural time distance" often ignores the difference of the number of reports in 24 h a day. Therefore, in practice, the measurement of "news time distance" can be combined with the frequency of different time periods to give a comprehensive evaluation.

6 Emotion Analysis Technology Design

The emotional analysis of netizens is mainly based on the media events classified in the preceding section. By understanding the emotions expressed in the netizens'sentences and words, it reflects the netizens' positive or negative views on media events and state policies.

The analysis of emotional tendency is based on two aspects: one is the direction of emotional orientation, the other is emotional orientation.

Emotional orientation, also known as emotional polarity, can be understood as the attitude of netizens to express their views on current media events is positive, negative and neutral. For example, "praise" and "support" are both commendatory words to express positive emotions, while "opposition" and "condemnation" are derogatory words to express negative emotions. In addition, positive emotional words are also influenced by emotional modifiers, which change the direction of emotional orientation, such as: "Clamoring to mess up the" one belt road "is obviously irrational behavior." reason "is originally a positive emotion word, and it expresses negative emotions under the modification of "no" .

Emotional tendencies refer to the extent to which netizens express positive or negative emotions toward media events. Different emotional states are often reflected by different emotional words or modifiers. For example, "advocacy" and "support" are both positive emotions and commendatory words. But "advocacy" is far more intense than "support" in emotional expression. Similarly, the intensity of "condemnation" will be further enhanced under the "strong" modification.

In order to judge the emotional state of netizens, it is necessary to define the polarity of emotional words (positive, negative and neutral), and then give different weights to each emotion according to modifiers. For example, in a media report, it is pointed out that "one belt and one road" is extremely beneficial to the long-term development of foreign trade. "Favorable" expresses positive emotions, while the word "extreme" emphasizes the high inclination of positive emotions.

According to the report data, we can mine the netizens'emotional viewpoint or analyze their emotional tendency. The brief process is shown in Fig. 4.

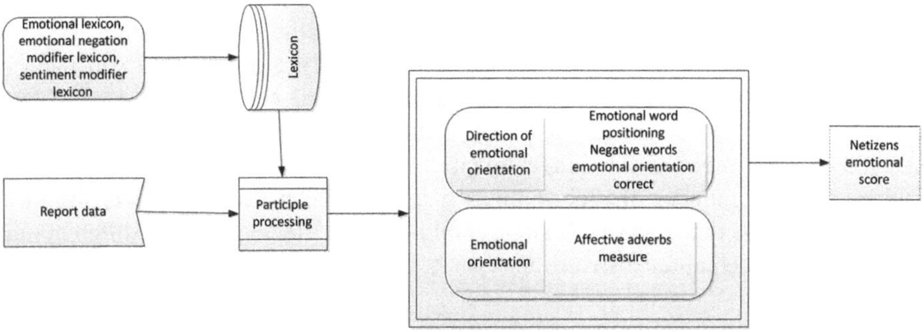

Fig. 4. Basic process of emotional score acquisition.

6.1 Participle Processing

The key point of this part of text segmentation is to have enough perfect emotional lexicon, including emotional lexicon, emotional negative modifier lexicon, emotional degree modifier lexicon.

Emotional lexicon: Customize the lexicon, determine the direction of emotional orientation, have a dictionary to express positive emotions, express negative emotions dictionary; merge positive and negative emotions dictionary.

Affective Negative Modifier Lexicon: Customize the commonly used negative lexicon (no, no, no, no, no, no, no, no, no, no, no, no, no, no, no, no, no, no, no, no, no, no, no, no, no, no, no, no, no, no, no, no, no, etc.), which is used to determine the direction of emotional orientation.

Affective Degree Modifiers: Customize Affective Degree Modifiers, and classify adverbs of degree into six levels to distinguish the differences of emotional enhancement. The higher the level, the greater the degree of emotional enhancement, as shown in Table 2.

Tab. 2. Degree adverb scale.

Adverb of degree	Grade
Over	5
Very	4
Extreme/most	3
More	2
Insufficiently	1

6.2 Emotional Positioning

Emotional orientation is based on the result of word segmentation for combination judgment. The first step is to accurately locate the emotional orientation of Internet users. Affective words should be judged firstly, and then retrieved whether they are modified by negative determinants to locate them accurately.

This step mainly relies on the measurement of emotional degree modifiers. The detailed process is as follows:

Step one: orientation of netizens' emotional orientation.

Preliminary positioning of emotional words: matching the result of segmentation with the emotional lexicon, matching the positive emotional words, its emotional assignment is a positive value; matching the negative emotional words, its emotional assignment is a negative value; not matching the emotional value, positioning as neutral, the emotional value is 0.

Negative modifiers eventually assess the direction of emotional orientation: there are multiple negative phenomena in Chinese, that is, when the negative word appears odd, it means negative; when the negative word appears even, it means positive. According to Chinese habits, if there are odd negative words in front of each affective word, it will be adjusted to the opposite direction of emotional orientation, and the polarity of emotional value is opposite, that is, the positive value is adjusted to negative value, and the negative value is adjusted to positive value.

The second step: Internet users' sentiment inclination measurement.

After the first step, the direction of emotional orientation of the netizens has been determined. When the emotions are positive or negative, different netizens may have different emotional strength, that is, emotional orientation. The important factor that affects people's emotional orientation comes from the degree adverbs modified before their emotional words. Whether degree adverbs appear before searching for emotional words, if degree adverbs appear, the emotional value can be adjusted according to the level of the corresponding degree adverbs (the granularity or distance between levels can be adjusted according to specific circumstances). An example of emotional score calculation is shown in Fig. 5.

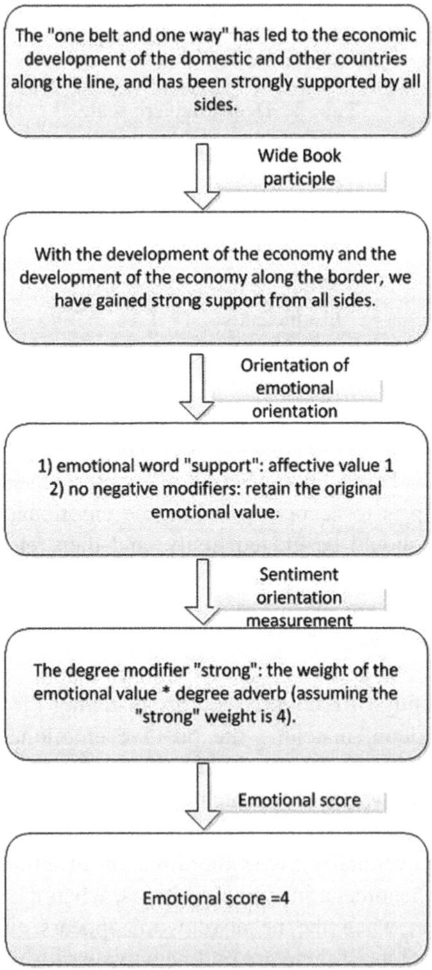

Fig. 5. Example of emotional score calculation.

6.3 Emotional Aggregation

After these processes, we can get emotional scores of netizens. According to the result of emotional score, we can classify the overall emotions of the netizens. When the emotional value is 0, it means that the netizens are emotionally neutral; when the emotional value is positive, the netizens are positively polarized; when the emotional value is negative, the netizens are passive. In view of the negative crowd can be in-depth analysis, analysis of the real reasons for the negative opposition of netizens, is conducive to understanding the events and policies behind the more real benefit or damage groups.

7 Conclusion

Keyword automatic extraction. Based on text mining algorithm, text keywords are automatically extracted, which lays a foundation for further keyword-based retrieval.

Abstract automatic extraction. Based on text mining algorithm, text summary is extracted automatically, and text content is refined to facilitate users to preview text content quickly.

Event induction and recognition based on similarity. Based on the results of document similarity calculation, the induction and recognition of the same event in multi-language, multi-channel and long time is realized, that is, the information with repeatability or approximation is automatically classified.

Hotspot recognition. Based on the evaluation model, hot articles or events can be identified by synthesizing the multi-dimensional indexes of netizens'hotspot, communication power and influence in different channels and regions.

Sort out and track the incident. Combing the development context of the same event/subject, realizing the long-term tracking and monitoring of a public opinion event, so as to grasp the development trend of the event at any time.

The way of communication is analyzed. This paper analyzes the participation and reprinting of events among the mainstream media, and analyzes the influence of events from the perspective of the depth and breadth of events.

Public opinion monitoring and early warning. Focus on positive public opinion (enhanced cooperation and exchanges) and extreme negative public opinion (seeking more solutions and cooperation programs) and set early warning conditions. The public opinion to achieve early warning conditions, support stations, SMS, e-mail and other means of early warning.

Acknowledgement. Foundation project: Major scientific research project of key platform in Guangdong (2016KTSCX192).

References

1. Meng, X.F., Ci, X., et al.: Big data management: concept, technology and challenge. Comput. Res. Dev. **25**(1), 146–169 (2013)
2. Cheng, X.Q., Jin, X.L., Wang, Y.Z., et al.: Summary of big data systems and analysis techniques. J. Softw. **25**(9), 1889–1908 (2014)

Fire Detection from Images Based on Single Shot MultiBox Detector

Zechen Wan[1], Yi Zhuo[2], Huihong Jiang[2], Junfeng Tao[2], Huimin Qian[1(✉)], Wenbo Xiang[3], and Ying Qian[4]

[1] College of Energy and Electrical Engineering, HoHai University, Nanjing 211100, China
amandaqian@hhu.edu.cn
[2] Zhejiang Huayun Information Technology Co., LTD., Hangzhou 310008, China
[3] College of Automation, Nanjing University of Science and Technology, Nanjing 210008, China
[4] School of Computer Engineering, Nanjing Institute of Technology, Nanjing 211167, China

Abstract. Fire detection is quite necessary because fire is very harmful to both lives and properties of humans. However, the causes of fire are various and complicated, which leads to the difficulty of fire prevention. As the increase of video surveillance systems, fire detection from images has become a research hotspot. Traditional algorithms focused on the contour, color and movement features of fire, which are only efficient in particular scenes and cannot be generalized. With the development of deep learning, convolution neural networks with good generalization performance have been broadly used in object detection. In this paper, a one-stage object detector based on deep neural network, namely Single Shot MultiBox Detector (SSD), is introduced to fire detection. In order to improve the generalization performance of detector and tackle the training difficulty of the lack of training samples, a dataset with a large number of flame and fire images, namely HHFire are constructed. The format of annotation is the same as that of Pascal Visual Object Classes. Images from the dataset covers lots of real scenes, including forest, buildings, fields, indoors, etc. Experimental results have demonstrated the efficiency of SSD on HHFire.

Keywords: Fire detection · Deep learning · Object detection · SSD

1 Introduction

Fire often results in disaster in forests, factories, buildings and other public places. The lives and properties of humans are always damaged by fires. Therefore, detecting the fire and preventing it timely are quite important. However, the causes of fire are various and complicated, which leads to the difficulty of fire prevention.

Traditional fire detection is based on temperature, smoke or light sensors. For example, Xu [1] set up an intelligent automatic fire alarm system, which can detect fire by the changes of air temperature and pollution captured by the fire alarm controller, temperature and smoke detector, and other physical devices. Hu [2] introduced the

Q. Liu et al. (Eds.): CENet 2020, AISC 1274, pp. 302–313, 2021.
https://doi.org/10.1007/978-981-15-8462-6_36

multi-band infrared flame detector to flame detection in view of the shortcomings, like short detection distance and slow response, of conventional physical devices such as thermal smoke detector. This kind of method strongly depends on the proper environments, can only cover a small area, and has a high cost.

Currently, most of the public places are equipped with video surveillance systems, which still act as a recorder for finding the cause of fire after the disaster. It is urgent to develop the function of video surveillance system to give a timely alarm at the beginning of fire. Therefore, detecting fire from images has become the hotspot of fire detection. Classic image-based fire detectors aim to extract the features of fire motion, color model and contour. Chen et al. [3] studied the dynamic behavior and irregularity detection of flames in RGB and HSI color Spaces. Celik and Demirel [4] designed the classification rules according to the separability of chromaticity and brightness in YCbCr space. The accuracy of this algorithm is high only when the detection distance is short and fire is serious. FOGGIA et al. [5] set up a multi-expert framework for real-time flame detection considering the combination of shape, color and motion attributes. It is the commonly-used fire detection method in real systems currently. However, the false alarm rate is still high.

With the development of deep neural networks, such as LeNet [6], AlexNet [7], VGG [8], ResNet [9], the performance of object detector guarantees the application in real systems. Fire detection is a task of specific object detection, and the introduction of deep neural networks will bring a new era for fire detection. Hui et al. [10] introduced Faster R-CNN into fire detection, revised the network and constructed a dataset to improve detection accuracy. Hong et al. [11], presented fire detector based on R-FCN with ResNet as base network. These papers have announced high fire detection accuracy on the specified dataset. Nevertheless, further analysis reveals that the type of fire and the scene are incomplete. And they cannot obtain satisfying detection results in other applications.

According to the circumstances, fire images with various fire types and under various real scenes are collected through internet to construct a big fire-image dataset. Furthermore, considering the variety of flame/fire scale, single shot multibox detector, abbreviated as SSD, has been employed in fire detection. The fusion of features extracted by lower network layers and deeper network layers devotes to the high accuracy of fire detection.

2 Illustrations of SSD

2.1 Object Detectors Based on Convolutional Neural Network

Convolutional Neural network (CNN), is a deep network inspired by biological vision [12], which has been broadly used in computer vision. LeNet, which obtained success in handwritten number recognition is the pioneer of CNN. Successively, CNNs including AlexNet, VGG, GoogleNet have been broadly applied in image classification, object positioning, behavior recognition, and object detection. Typical CNN structure is shown in Fig. 1.

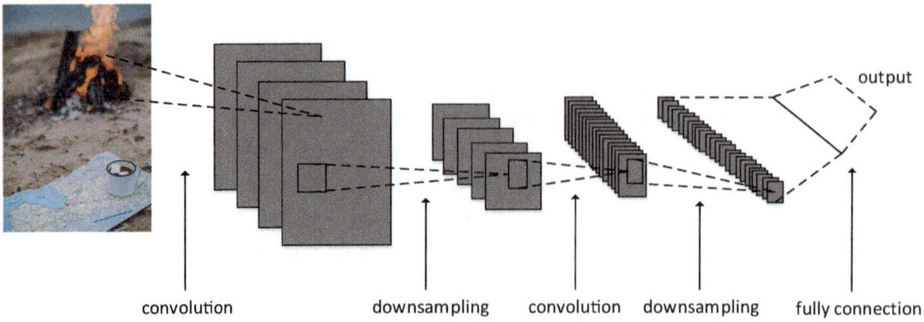

Fig. 1. Typical CNN structure diagram

CNN has found great applications in object detection. And the commonly-used object detectors can be divided into two types: two-stage object detectors and one-stage object detectors. For two-stage object detectors, the first stage is to extract region proposals through sliding windows [13] or Selective Search [14] or RPN network [15], and the second stage is to achieve object classification and position regression. Typical two-stage object detectors include R-CNN [16], Fast R-CNN [17], Faster R-CNN [18] and other extensions. The one-stage object detectors complete position regression and object classification from the original images without an advanced region proposals extraction. Thus, the detection speed of one-stage object detector is faster than that of two-stage object detector. SSD is a typical detector of one-stage object detector.

2.2 SSD

Single Shot MultiBox Detector (SSD) [19], is a multiple scale object detector. Commonly-used traditional methods of acquiring scale invariance are constructing the pyramid of original images and combining the detection results at all scales. Considering that deep neural networks extract features layer by layer and from coarse to fine, SSD proposed to combine multiple feature maps of the out-puts of multiple convolution layers. SSD has the next three characteristics com-pared to other detectors.

Feature Maps From Multiple Convolution Layers are Combined. Objects in the images have different scales. More detail information can be retained in the low-level feature images and more semantic information can be abstracted in the high-level feature image. Traditional object detection algorithm only performs detection on the last layer of feature image, which results in the inefficiency in detecting objects with scale variance. SSD proposed to combine feature maps of low-level and high-level convolution layers. More Specific, SSD revised the network structure of VGG16 by alerting the last two full connection layers of VGG16 into convolution layers, and augmenting 4 convolution layers. Detailed network structure is listed in Fig. 2. The outputs of five convolution layers, which are marked in blue, are combined to achieve classification and position regression.

Establishing Default Boxes to Achieve the Same Function of Region Proposals. Default boxes refers the concept of anchor in PRN proposed in Faster R-CNN. In

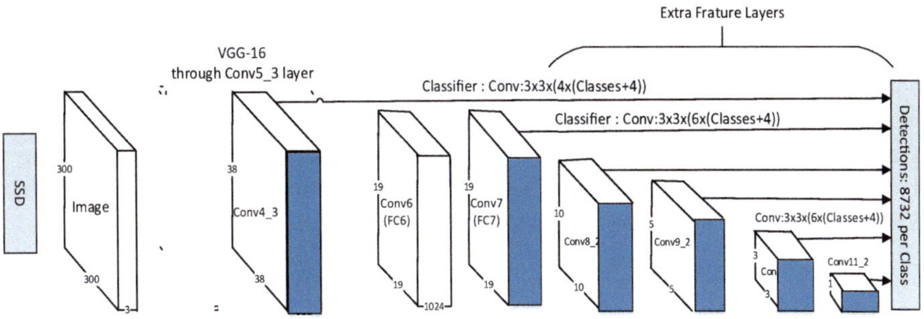

Fig. 2. Structure of SSD [19]

detail, the feature map is firstly divided into some small grids, and then a series of default boxes with different length-width ratios are manually set in each grid. As shown in Fig. 3, for an 8 × 8 feature image, four candidate boxes with different aspect ratios are set in each cell, and the total number of default boxes is 256.

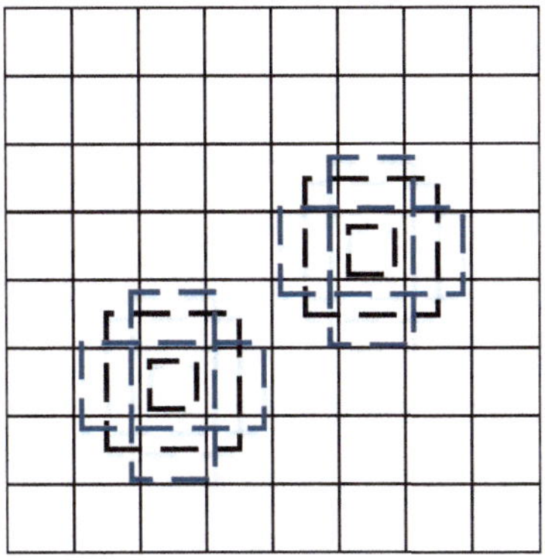

Fig. 3. Default box generation in SSD [19]

Default box in SSD is different from anchor in RPN. SSD employs multi-scale feature maps, that means default boxes should be set on feature map under each scale. The scale S_k of standard default box in each feature map can be computed by Eq. (1), where m represents the number of feature images, k represents the number of current feature image, S_{min} is set as 0.2, and S_{max} is set as 0.9, indicating the scale of lowest and highest layers employed in detection. the values of S_{min} and S_{max} are chosen by experience.

$$s_k = s_{\min} + \frac{s_{\max} - s_{\min}}{m-1}(k-1), k \in [1, m] \tag{1}$$

When the scale of standard default box for each feature map has been set, the sizes of all the default boxes, which are acquired by setting different length-width ratios under the standard default box, can be determined by Eq. (2). In which, a_r is the height-width ratio, and w_k^a and h_k^a are the width and height of each default box, respectively. It should be noted that there are 6 default boxes totally including the standard default box and the expanded default boxes under various width-length ratios. The scale diversity of default boxes on each considered feature maps helps the detector to adapting to objects with various shapes and sizes in the images.

$$\begin{aligned} w_k^a &= s_k\sqrt{a_r} \\ h_k^a &= \frac{s_k}{\sqrt{a_r}} \\ a_r &\in \left\{1, 2, 3, \frac{1}{2}, \frac{1}{3}\right\} \end{aligned} \tag{2}$$

Convolution Prediction is Employed to Reduce the Number of Parameters. The fully connected layers in VGG16 are substituted by full convolution layers with 3 × 3 kernel size, which helps to reduce the parameter number dramatically. Moreover, two kinds of different 3 × 3 convolution kernels are applied to the same feature map to support box regression and object classification respectively. Figure 4 is a schematic diagram of SSD convolution prediction.

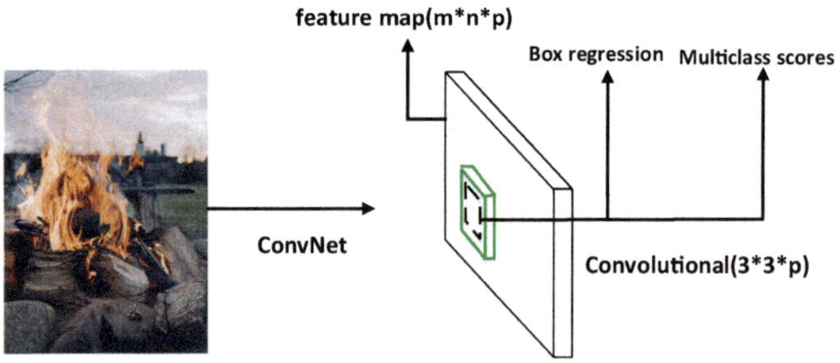

Fig. 4. Structure diagram of long-term recurrent convolutional network

In the process of position regression, four offsets, (x, y, w, h), for each default box will be generated to represent the position of the predicted bounding box. These four offsets respectively represent the coordinate offsets of the bounding box center according to the corresponding default box as well as the width and height. And at the same time, the part of object classification will output m classification values to represent the probabilities of the object belonging to the specified classes.

2.3 Loss Function of SSD

In the training of a classifier, a loss function should be defined to estimate the distance between the predicted value and the real value. Loss function is always a non-negative real value function. The process of training the classifier can be seen as finding the minimum value of loss function. The loss function employed in SSD is listed in Eq. (3).

$$L(x, c, l, g) = \frac{1}{N}\left(L_{conf}(x, c) + \alpha L_{loc}(x, l, g)\right) \quad (3)$$

Where, L is the loss of prediction bounding box, and L is the weighted sum with weight coefficient α of the prediction category loss L_{conf} and the position offset loss L_{loc}. N is the number of matched default boxes, x is a binary value with 0 or 1, denoting whether the matched box belongs to a certain category or not, c is the confidence of class prediction, l is the prediction position, and g is the true position.

3 Fire-Image Data Set

Commonly used public data sets for object detection include PASCAL VOC (visual object classification) [20], ImageNet [21], MS COCO [22] and others. PASCAL VOC was constructed originally in 2005 and it is set up for the PASCAL VOC challenge. The PASCAL VOC data set contains 20 types of objects, including people, animals (such as cats, dogs, islands), vehicles (such as cars, boats, planes), furniture (such as chairs, tables, sofas), and etc. The total number of images is about 11,000 and that of objects is over 27,000. Microsoft Common Objects in Context (MS COCO), originated in 2014 and is built by Microsoft. COCO dataset can be used for object detection, segmentation and caption. It is mainly captured from complex everyday-life scenes, including 91 types of objects, 328,000 images and 2,500,000 tags. ImageNet is set up for image under-standing, and it contains more than 20,000 categories objects. The dataset contains more than 14 million manually annotated images to indicate the target in the image, and at least a million of images have been labeled the bounding boxes.

However, it is found that neither of PASCAL VOC, ImageNet or MS COCO data sets have the category of fire or flame. And it should also be acknowledged that some fire datasets can be found, like Bilkent university flame video library [23], dataset of National Institute of Standards and Technology [24], dataset of Computer Vision and Pattern Recognition Laboratory of Keimyung University [25] Unfortunately, either the number of samples or the diversity of fires or scenes are not enough for the purpose of this paper. Therefore, in order to meet the requirements of our detection task and to improve the performance of fire detector, a home-made dataset, namely HHFire is constructed.

The creating process of HHFire data set can be divided into three steps.

Image Collection. We collected fire images and videos from internet sites like Baidu, Google, Unsplash, YouTube and others. And then only a small part of fire images are selected out from fire videos and considered in HHFire. Fire images of the existed fire dataset are also included in HHFire. Finally, about 3100 fire images covering variety of backgrounds, fire types are collected. Some samples are shown in Fig. 5. From which, it can be seen that the commonly occurred candle-stick fire, building fire, field fire,

Fig. 5. Some examples of fire images

forest fire, fire in indoors, flames, artillery fire, and others, are all included in HHFire. And some of images contains multiple fire objects. The diversity of fire types and occurrence scenarios can help to improve the detection accuracy of detectors trained on this dataset.

Data Annotation. In this paper, all the fires in the images are annotated ac-cording to the format of PASCAL VOC. Annotation software, namely La-belImg [25], is used in this paper. After annotations, an XML file is corresponding to each image, which indicates the right positions of fire objects in the image.

Allocating of Training Set and Test Set. In this paper, all the collected images are divided into training set and test set. The training set is used to train the SSD and the test set is applied to test the performance of trained detector. In this paper, we randomly selected 90% of the fire images in HHFire as training samples, and the left 10% as test samples.

To this end, training set and testing set have been constructed for experiments.

4 Experimental Results

4.1 Experimental Platform

In this paper, the experimental platform is composed of Ubuntu16.04 (Linux), Ge Force GTX1080Ti with 8 GB memory and deep learning Tensorflow framework. Tensorflow is a deep learning framework produced by Google. It supports Java, Python and C++ scheduling, and can achieve the implementation and revision of various object detection algorithms easily.

4.2 Experimental Process

The experimental process of this paper can be divided into three steps: data preparation, model training, and model test.

Data Preparation. The annotations of images are firstly converted into tfrecords formats to adapt to the Tensorflow framework used in this experiment.

Model Training. The original Tensorflow framework of SSD was built on VOC data set with 20 categories of objects. In this paper, it is employed to detect the fire. Therefore, the prediction categories are two, fire or not.

Batch size, learning rate and maximum iteration steps all have influence on the speed of model training and the accuracy of experimental results. need to be adjusted. Batch size is the number of samples that are sent to the network for training each time. A large batch size makes the training procedure require large memory space and might lead to the local minimum value. But a small batch size results in time-consuming training procedure and the values of loss function will shaking. Adjusting the learning rate can control the updating speed of parameters. Reducing learning rate can decrease convergence speed but increasing training time. While, enlarging learning rate leads to the oscillation of parameters near the optimal solution. Maximum iteration steps should be adjusted according to the number of samples in the data set to avoid repetitive and useless work. In this paper, the batch size, learning rate and maximum iteration steps are set to 8, 0.001, and 50000 respectively.

Model Testing. There are about 300 fire images in the test set. All the images are used to test the performance of SSD. In addition, to test the false alarm rate of the proposed algorithm, another 100 images containing the non-fire, but fire-like objects were downloaded from the Internet. And another test experiment is con-ducted in those images.

4.3 Results Analysis

Two types of test experiments are conducted in this paper. Firstly, the images in test set are detected by the trained SSD. Figure 6 has shown some detection examples. All the fires in those images are correctly detected as fire with high confidence.

In this paper, we employ four parameters, namely True Positive Rate (TPR), False Negative Rate (FNR), True Negative Rate (TNR) and False Positive Rate (FPR), to illustrate the performance of fire detection algorithm. Among them, TPR indicates those fire images which are correctly detected as fire-image. FPR indi-cates those non-fire images which are incorrectly detected as fire-image. TNR indicates those non-fire images which are correctly detected as non-fire-image. And FNR indicates those fire images which are incorrectly detected as non-fire image. In the first experiment, we calculated TPR and FNR, which are listed in Table 1. More specific, 241 images of the total 300 images are correctly detected as fire-image, and 59 images are incorrectly detected as non-fire-image. Some samples are listed in Fig. 7. After analysis of the incorrectly detected images, it can be seen that SSD is still difficult in finding objects with small size. The rea-son is that SSD has not been trained for small-size objects.

Fig. 6. Image samples of test results

Table 1. Test results on fire-image

Testing set information		Detection results of SSD	
Total number	Contain fire	TPR	FNR
300	Yes	80.3%	19.7%

Fig. 7. Image samples of test results

It results from the control principle of positive-false-sample-ratio. Under this principle, positive samples with small-size objects will be discarded in the training process.

Secondly, another 100 images that contains non-fire but fire-like objects are tested by our SSD. There are 86 non-fire images in the set which are correctly detected as non-fire image. And another 14 images are incorrectly detected as fire-image. Some examples have been listed in Fig. 8, which are listed in Table 1. It can be seen that most of the incorrectly detected objects are strong headlights or street lights in the evening, which are very similar with fire in shapes or colors.

Fig. 8. Samples of non-fire-image with false detection

Table 2. Test results on non-fire-image

Testing set information		SSD	
Total number	Contain fire	TPR	FNR
100	No	85.5%	14.5%

In conclusion, SSD algorithm trained on the proposed HHFire can reach a detection rate of over 80%. And it should be improved to deal with fires with small size and bright lights at night.

5 Summary and Prospect

Fire detection is an urgent but unresolved problem in real applications. The existed fire detection algorithm from images cannot acquire satisfying performance. Deep neural networks have acquired lots of successes in general object detection. However, the performance of fire detection should be improved due to the lack of specialized dataset. In this paper, a fire dataset named HHFire containing al-most 3100 fire images is constructed. The diversity of fire types and backgrounds can help to develop the accuracy of detectors. Single shot multibox detector is trained in this paper to detect fire images. Experimental results demonstrated the efficiecy of the proposed algorithm in most of the real scenes.

However, the detection accuracy is still unsatisfying due to the deficiency of SSD in recognizing objects with small size, and the excessive similarity between fires and some objects including headlights or street lights at night. One reason of small-size fire cannot be detected is that deeper feature map with semantic in-formation used in detection cannot cover the information of original image region of small-size fire due to multiple pooling operations. The other reason is related with IOU of prior box against ground-truth box. Objects with small size have much less positive ROI samples in training, which results in the incomplete learning of small size objects. Follow-up work includes improving SSD by revising the control principle of positive-false-sample-ratio in training, employing other detectors based on CNN, and enlarging the fire dataset.

References

1. Xu, Y.: The Structure of Automatic Fire Alarm System Based on Virtual Instrument. Tianjin University of Technology, Tianjin (2013)
2. Hu, X.: Research and Product Development of Mir Flame Detector System. Zhejiang University, Hang-zhou (2013)
3. Guo, X., Wu, W., Chen, Z.: Multiple complex-coefficient-filter based phase-locked loop and synchronization technique for three-phase grid-interfaced converters in distributed utility networks. IEEE Trans. Industr. Electron. **58**(4), 1194–1204 (2011)
4. Celik, T., Demirel, H.: Fire detection in video sequences using a generic color model. Fire Safety **44**(2), 147–158 (2009)
5. Foggia, P., Saggese, A., Vento, M.: Real-time fire detection for video-surveillance applications using a combination of experts based on color, shape, and motion. IEEE Trans. Circ. Syst. Video Technol. **25**(9), 1545–1556 (2015)
6. Lecun, Y., Bottou, L., Bengio, Y., et al.: Gradient-based learning applied to document recognition. Proc. IEEE **86**(11), 2278–2324 (1998)
7. Krizhevsky, A., Sutskever, I., Hinton, G.: ImageNet classification with deep convolutional neural networks. In: Proceedings of Conference and Workshop on Neural Information Processing Systems, pp. 1105–1907 (2012)
8. Simonyan, K., Zisserman, A.: Very deep convolutional networks for large-scale image recognition [EB/OL] (2014). arXiv:1409.1556 [cs. CV]
9. He, K., Zhang, X., Ren, S., et al. Deep residual learning for image recognition. In: IEEE Conference on Computer Vision and Pattern Recognition (CVPR), pp. 770–778. USA: IEEE Las Vegas, NV (2016)
10. Hui, T., Halidan, A., Du, H.: Multi-type flame detection combined with Faster R-CNN. J. Image Graph. **24**(01), 0073–0083 (2019)
11. Hong, W., Li, C.: Flame detection based on region full convolution network combined with residual network. Adv. Laser Optoelectron. **627**(4), 168–173 (2018)
12. Szegedy, C., Liu, W., Jia, Y., et al.: Going deeper with convolutions [EB/OL] (2014). arXiv: 1409.4842 [cs. CV]
13. Dalal, N., Triggs, B.: Histograms of oriented gradients for human detection. In: Proceedings of the 2005 IEEE Computer Society Conference on Computer Vision and Pattern Recognition (CVPR), pp. 886–893. USA: IEEE, Santiago, CA (2005)
14. Uijlings, J., Gevers, T., et al.: Selective search for object recognition. Int. J. Comput. Vis. **104**(2), 154–171 (2013)
15. Ren, S., He, K., Girshick, R., Sun, J.: Faster R-CNN: towards real-time object detection with region proposal networks. IEEE Trans. Pattern Anal. Mach. Intell. **39**(6), 1137–1149 (2017)
16. Girshick, R., Donahue, J., Darrell, T., et al.: Rich feature hierarchies for accurate object detection and semantic segmentation. In: IEEE Conference on Computer Vision and Pattern Recognition, pp. 580–587. USA: IEEE, Columbus, OH (2014)
17. Girshick, R.: Fast R-CNN. In: IEEE International Conference on Computer Vision (ICCV), pp. 1440–1448. USA: IEEE, Santiago (2015)
18. Liu, W., Anguelov, D., et al.: SSD: Single Shot MultiBox detector. In: Proceedings of European Conference on Computer Vision 2016 (ECCV), pp. 21–37. Switzerland: Springer, Zurich (2016)

19. Everingham, M., Winn, J., Zisserman, A., et al.: The PASCAL visual object classes (VOC) challenge. Int. J. Comput. Vision **88**(2), 303–338 (2010)
20. Russakovsky, O., Deng, J., Su, H., et al.: ImageNet large scale visual recognition challenge [J]. Int. J. Comput. Vision **115**(3), 211–252 (2015)
21. Lin, T., Maire, M., Belongie, S., et al.: Microsoft COCO: common objects in context. In: Proceedings of the 13th European Conference on Computer Vision (ECCV), pp. 740–755. Switzerland: Springer, Zurich (2014)
22. http://signal.ee.bilkent.edu.tr/VisiFire/
23. https://www.nist.gov/video-category/fire
24. https://cvpr.kmu.ac.kr/Dataset/Dataset.html
25. https://blog.csdn.net/jesse_mx/article/details/53606897

Human Activities Recognition from Videos Based on Compound Deep Neural Network

Zhijian Liu[1], Yi Han[2], Zhiwei Chen[2], Yuelong Fang[2], Huimin Qian[1(✉)], and Jun Zhou[1]

[1] College of Energy and Electrical Engineering, HoHai University, Nanjing, Jiangsu 211100, China
amandaqian@hhu.edu.cn

[2] Zhejiang Huayun Information Technology Corporation, Hangzhou, Zhejiang 310012, China

Abstract. The two-stream deep neural networks learn contour evolution information of human activities from image frames and motion information from information evolution between image frames. Optical flow image is one of the commonly-used methods to depict the motion information between frames. The existing optical flow images are usually obtained by off-line calculations. In this paper, a deep neural network named FlowNet2.0 is used to generate optical flow images, and based on which a compound deep neural network is proposed to merge spatial and temporal information to complete the recognition of human activities from videos. Specifically, the proposed compound deep neural network is composed of two sub-networks, static-data-stream learning network and dynamic-data-stream learning network. The former is applied to extract spatial evolution information from RGB images, and the latter generates optical flow prediction images from RGB image sequence first, and extracts temporal evolution information from optical flow images. Finally, the results of two sub-networks are combined under the fusion algorithm to achieve human activities recognition. Experimental results show that the compound deep neural network proposed in this paper can effectively identify human activities from video sequences, and the average classification accuracy on data set UCF101 can reach 83.35%.

Keywords: Human activities recognition · Convolution neural network · Recurrent neural network · Compound deep neural network · Model fusion

1 Introduction

The recognition of human activities from videos has broad applications in the fields of intelligent video surveillance, smart home, human-computer interaction system, and content-based video retrieval. Human activities recognition from videos aims to analyze and understand human actions or activities in video sequence, so as to achieve the understanding of video data. With the development of deep neural network in computer vision technology, human activities recognition technology based on deep neural network has attracted much attention [1–9]. Currently, deep neural networks commonly-used in human activities recognition include spatio-temporal convolutional

Q. Liu et al. (Eds.): CENet 2020, AISC 1274, pp. 314–326, 2021.
https://doi.org/10.1007/978-981-15-8462-6_37

neural network [1–5], convolutional neural network based on dual-flow structure [2, 7, 14], deep neural network based on generative model (such as Autoencoder) [3], Restricted Boltzmann Machine (Restricted Boltzmann Machine, RBM) [4], Generative Adversarial Networks, GANs [5]), time consistency network [6], etc.

Video sequence is three-dimensional data (or spatio-temporal data), which includes the static data stream describing the spatial contour evolution of human activities in the two-dimensional image planes, and the dynamic data stream describing the motion evolution of human activities along the temporal axis. Static data stream is composed by RGB image frames, while dynamic data stream is exhibited as motion information along temporal sequence, which is usually depicted by optical flow image. The two-stream convolutional neural network is actually designed to learn the spatial contour evolution information of human activities and temporal motion evolution by different deep neural networks. For example, the two-stream Convolutional Networks proposed by Karen et al. [2] were designed as follows. Firstly, the optical flow images describing human motion information were generated according to the original video sequence; then, RGB images and optical flow images were fed into the corresponding convolution neural networks respectively to generate corresponding feature vectors, and then the resulted feature vectors were combined. Finally, human activities were classified by trained SVM classifiers. Feichtenhofer et al. [7] discussed the influence on model performance when temporal information and spatial information were combined at different network stages and in different ways. According to which, a fusion network with better performance and fewer parameters was proposed. Donahue et al. [14] proposed Long-term Recurrent Convolutional Network (LRCN) by combining Convolutional Neural Networks (CNN) and Long Short-term Memory Recurrent Neural Networks (LSTM-RNN).

The traditional method of optical flow calculation is proposed by Horn-Schunck [16], which is complicated and time-consuming. And it is easy to be disturbed by noise. In recent years, neural networks have been employed in optical flow calculations, like FlowNet 2.0 presented by Eddy Ilg [11]. In this paper, a novel compound neural network with two-stream structure will be proposed. The compound neural network is composed of two sub-networks, one of which is a static-data-stream learning network, and the other is a dynamic-data-stream learning network. The former model learns spatial representations of human activities, and the latter model learns temporal representations. Both are combined to identify human activities.

2 Compound Deep Neural Network

In this paper a compound deep neural network (CDNN), composed of static-data-stream learning network and dynamic-data-stream learning network, is presented. The RGB images from video are fed into the static-data-stream learning network to generate spatial representations and predict the types of human activities, and the RGB image sequence from video is fed into dynamic-data-stream learning network to generate the corresponding optical flow images and furthermore to generate temporal representations and predict the types of human activities.

In specific, dynamic-data-stream learning network can be seen as the combination of an optical flow image prediction network and an optical flow image learning network. While the static-data-stream learning network is an RGB image learning network. In this paper, the same neural network is used in optical flow image learning network and RGB image learning network. And we will discuss the performance of compound deep neural network when choosing data-stream learning network as single-frame neural network or LRCN.

Next, the above-mentioned network models including optical flow image prediction network, single-frame neural network model and LRCN will be illustrated.

2.1 Optical Flow Image Prediction Network

The optical flow image prediction network is composed of neural network named FlowNet [10]. Where, FlowNet can be divided into two network structures: FlowNetSimple (FlowNetS) and FlowNetCorrelation (FlowNetC). More specific, FlowNet includes compression module composed by 9 convolution layers and extension module composed by 4 deconvolution layers. The features are extracted from consecutive video frames by compression module. However, the multiple pooling operations after convolution operations lead to the low-resolution feature image, in which much information will be lost. Therefore, the deconvolution module is set behind the convolution module. FlowNet is efficient in generating the optical flow image when the objects in sequential images have large displacement. Moreover, the generation process is also time-consuming. According to this, FlowNet2.0 [11] is presented. The structure of FlowNet2.0 network is shown in the following figure:

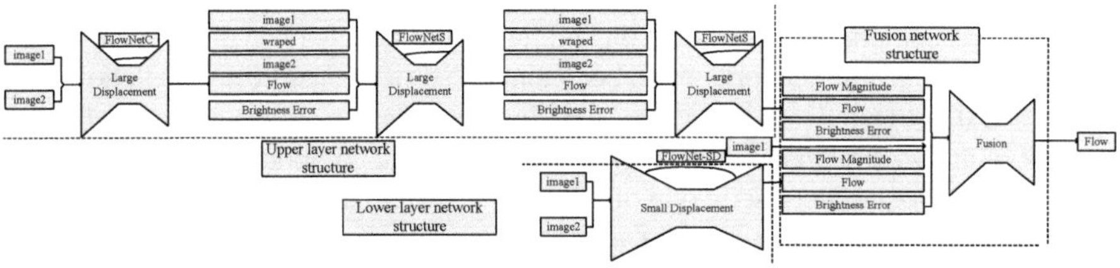

Fig. 1. Structure of FlowNet2.0

As we can see FlowNet 2.0 is composed of an upper network, an lower network and a fusion network. The upper network denoted as FlowNet-CSS, is stacked by a FlowNetC network and two FlowNetS networks. FlowNet-CSS is used to predict the optical flow of objects with large displacement between image frames. The lower network is an improved FlowNetC network called FlowNet-SD, which is developed to predict the optical flow of objects with small displacement between image frames. the prediction results of FlowNet-CSS and FlowNet-SD are combined by the fusion network to generate the final optical flow image.

The inputs of compound deep neural network are RGB image sequence extracted from the original video. Specifically, one frame is taken from every 30 frames of the original video and added to the RGB image sequence. The generation process of optical flow image can be depicted by Fig. 1 and Fig. 2 show the created optical flow images of human behavior.

(a) Apply Eye Makeup (b) Archery

(c) BaseballPitch

Fig. 2. Partial optical flow image frames of some human activities

2.2 RGB/Optical Flow Image Learning Network

In our compound deep neural network, the same neural network will be employed in optical flow image learning network and RGB image learning network. To explore the best performance of the compound deep neural network, two types of neural network are applied, which are single-frame network and long-term recurrent convolutional network.

Single-Frame Network

The single-frame network adopts ZFNet convolutional neural network [12] as the basic network. The structure of the single-frame network is shown in Fig. 3. Firstly, CNNs and fully connected layer are used to extract features from the input images. And then the feature vectors are classified by Softmax classifier [13] to acquire the predictions. In Fig. 3, the prediction vector is denoted as $\mathbf{y} = \left[y^{(1)}, y^{(2)}, \ldots, y^{(n)}\right]$ where $y^{(i)}$ represents the predictive probability vector of current image frame.

Long-Term Recurrent Convolutional Network

The long-term Recurrent Convolutional Network (LRCN) [14] is capable of learning evolution information in time sequence. And LRCN can be obtained by adding LSTMs after each CNN in single-frame network, which is described in Fig. 4. The additional LSTMs can learn temporal information from the inputs.

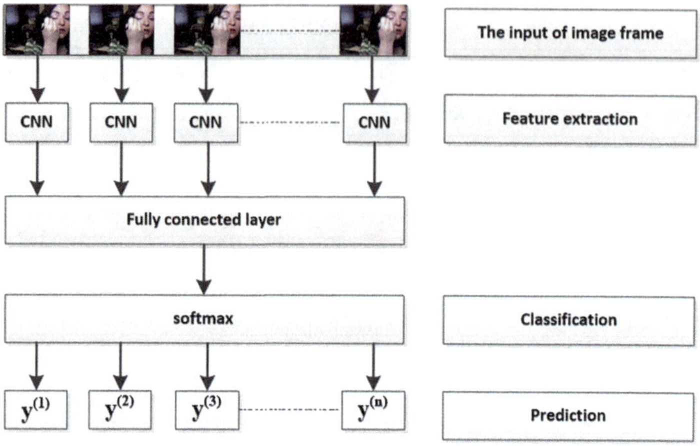

Fig. 3. Structure diagram of single-frame network

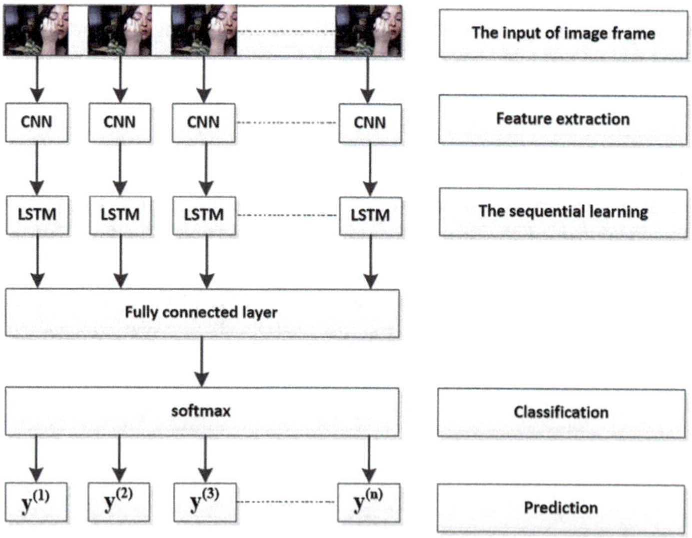

Fig. 4. Structure diagram of long-term recurrent convolutional network

2.3 Prediction Fusion

In the proposed compound deep neural network, the predictions of static-data-stream learning network and those of dynamic-data-stream learning network will be combined to give an overall class prediction of human activities from videos. Specifically, the results of static-data-stream learning network, creating predictions through single-frame network or LRCN by taking RGB images as inputs, and the results of dynamic-data-stream network, creating predictions through single-frame network or LRCN by taking optical flow images as inputs, are combined by weighted fusion algorithm [15]. The weighted fusion algorithm can be described as follows.

As above-mentioned, the output of single-frame network or LRCN is prediction vector $\mathbf{y} = \left[y^{(1)}, y^{(2)}, \ldots, y^{(n)}\right]$ Each $y^{(k)}$ represents the probability that the human activity of input video segment belongs to the k-type activity. Each $y^{(k)}$ from prediction vector of static-data-stream learning network and that of dynamic-data-stream learning network are added with weights When model fusion is carried out, the probability vectors output from the model based on RGB image frames and the model based on optical flow image frames are added up according to different weights, and the term with the highest probability obtained is taken as the final prediction result of the sample. The model fusion method is shown as follows:

$$V = \lambda \frac{\sum_n \mathrm{y}^{(i)}_{\text{optical flow frame}}}{n} + (1 - \lambda) \frac{\sum_n \mathrm{y}^{(i)}_{\text{RGB frame}}}{n} \tag{1}$$

In the above equation, m is a variable between (0,1), representing the weight part of the fusion model; $y^{(i)}_{optical\,flow\,frame}$ represents the recognition probability vector of human behavior in reading the current prediction optical flow frame in the video segment. $y^{i}{}_{RGB\,frame}$ represents reading from video segment and extracting the probability vector of human behavior recognition of RGB image frame. N represents the number of video frames intercepted in the video section read; V represents the resulting behavior prediction probability vector. During the experiment, by comparing the final prediction probability vector, the behavior with the highest probability vector is regarded as the final classification result.

3 Training of Compound Deep Neural Network

The training of compound deep neural network includes the training of FlowNet2.0, the training of single-frame network or the training of LRCN.

3.1 Training of Optical Flow Image Prediction Network

The training of optical flow image prediction network includes the following steps.

Step 1: Randomly initialize parameters of optical flow prediction model based on convolutional neural network (set network weight and bias).
Step 2: Pre-train the optical flow prediction model on the Flychairs and Fly-Things3D datasets.

Step 3: The training set of optical flow prediction model was extracted based on video frame extraction algorithm.
Step 4: Read the first frame of the image, and set the current number of iterations $p = 1$, input the total number of video frames P, the total number of iterations N, learning rate $k = 0.0001$.
Step 5: Determine whether $p + 30$ is less than P, if less than P, Step 6 is executed, Otherwise end the current loop.
Step 6: The p frame and the RGB image frame of p + 30 frame were input as the optical flow prediction model to generate the predicted optical flow frame.

3.2 Training of Single-Frame Network

The training of single-frame network includes the following steps.

Step 1: Randomly initialize the parameters of ZFNet network (All the weights are supposed to subject to a normal distribution with a standard deviation of 0.1, and all the bias values are initialized as 0);
Step 2: Pretraining ZFNet on the subset of ImageNet;
Step 3: Importing the image frames of training samples, and set original iteration number $i = 1$, total iteration number N, initial learning rate $\alpha = 0.001$, learning rate decay $k = 1$, and the decay interval n;
Step 4: If current iteration number i is less than or equal to the total iteration number N, go to Step 5; otherwise, finish the training process;
Step 5: If the current iteration number i is equal to the product of n iterations and the learning rate decay number k, go to Step 6; otherwise, go to Step 7.
Step 6: After n iterations, the learning rate α decreases to 10% of the original learning rate, and the learning rate attenuation times are increased by 1.
Step 7: Calculate the loss value and update the weight value and offset;
Step8: The number of iterations $i + 1$, and go to Step 4;

The definition of loss function in the training process is shown in Eq. (2) [20].

$$L = -\frac{1}{N}\sum_{i} \log[\text{Softmax}(x_i)] \tag{2}$$

Where, $i = 1, 2, \cdots, N$, x_i is the sample, N is the number of input samples for one iteration, and $\text{Softmax}(x_i)$ is the probability vector output of Softmax layer. The specific calculation process of $\text{Softmax}(x_i)$ is shown in Eq. (3):

$$\text{Softmax}(x_i) = \frac{e^{x_i}}{\sum_{j=1}^{K} e^{x_j}} \tag{3}$$

Where, $i = 1, 2, \cdots, K$, x_i and x_j are samples, and K is the number of categories.

3.3 Training of LRCN

The training of LRCN can be achieved on the basis of the training of single-frame network. The initialization parameters of LRCN, either for RGB images or optical flow images, are chosen as parameters of the corresponding single-frame network. And the parameters of newly added LSTM layers are initialized randomly. The loss function in the training process is defined as the same of single-frame network. In detail, the steps of training LRCN are listed in the next.

Step 1: initialize the layer corresponding to the LRCN model with the parameters of the single-frame network;
Step 2: randomly initialize LSTM layer;
Step 3: read the image frames of the training samples, and set the number of iterations $i = 1$, initial learning rate $\alpha = 0.001$, and learning rate decay $k = 1$.
Step 4: If the current iteration number i is less than or equal to the total iteration number N, go to Step 5; otherwise, the current training will be terminated.
Step 5: If the current iteration number i is equal to the product of n iterations and the learning rate decay number k, go to Step 6; otherwise, go to Step 7.
Step 6: after n iterations, the learning rate a decreased to 10% of the original learning rate, and the learning rate attenuation times were increased by 1.
Step 7: calculate the loss value and update the weight and offset;
Step 8: the number of iterations $i + 1$, and go to Step 4;

It should be illustrated that some of the training parameters are different for RGB images and optical flow images. In the single frame network model, a total of 5000 iterations were carried out for RGB images, and the learning rate was reduced to 10% of the former per iteration for 3000 times. The weight attenuation was set as 0.005, and the momentum was set as 0.9. At the same time, a total of 50000 iterations were carried out in total for optical flow images, and the learning rate was reduced to 10% of the former per iteration of 20,000 times. The weight attenuation was set as 0.005, and the momentum was set as 0.9. In the LRCN network model, a total of 30000 iterations were carried out for RGB images, and the learning rate was reduced to 10% of the former per iteration for 10000 times. The weight attenuation was set as 0.005, and the momentum was set as 0.9. At the same time, a total of 70000 iterations were carried out in total for optical flow images, and the learning rate was reduced to 10% of the former per iteration of 20,000 times. The weight attenuation was set as 0.005, and the momentum was set as 0.9.

4 Experimental Results and Analysis

4.1 Experimental Platform and Data Set

The deep learning software platform used in this experiment is the open source Caffe environment [21], and the hardware platform is the GPU of NVIDIA RTX2080Ti. The data set of the experiment in this paper is UCF101 public database of human activities recognition [19]. There are 101 types of activities and 13,320 video fragments.

Fig. 5. Examples of partial human activities in UCF101

Each type of activity is completed by 25 people, and each person acts the same activity for 4–7 times. The resolution of video frame is 320 × 240. Some examples of activities are shown in Fig. 5:

In order to facilitate the training and testing of ZFNet single-frame network and LRCN, the UCF101 data set is divided into two parts, namely, the training set and the test set. Among them, the training set contains 9,537 video samples and the test set contains 3,783 video samples. Then the video sample was preprocessed according to Sect. 3 to obtain RGB image frames and optical stream image frames. The corresponding RGB single-frame network and optical flow single-frame network were obtained by training the RGB single-frame network and optical flow single-frame network respectively. Similarly, the LRCN was trained on the RGB image frames and optical flow image frames respectively to obtain the corresponding LRCN based on the RGB image frames and the LRCN based on the optical flow image frames.

4.2 Experimental Results and Analysis

The proposed compound deep neural network is composed of static-data-stream learning network and dynamic-data-stream learning network. And two types of data-stream learning network, namely single-frame network and LRCN, are discussed in this paper. Dynamic-data-stream learning network includes optical flow image prediction network and optical flow image learning network. For the optical flow image prediction network, the generation method of optical flow images is improved by employing FlowNet2.0. Therefore, three kinds of comparison experiments are conducted in this paper.

1) Comparison between single-frame network and LRCN for data-stream learning network

   The next experiments are conducted. Firstly, the performances of static-data-stream learning network and dynamic-data-stream learning network are separately tested. And the performance of dynamic-data-stream learning network is effected by the generation methods of optical flow images. Then, two-stream deep neural network in [14] and the proposed compound deep neural network are compared. For all the above comparison experiments, single-frame network and LRCN are respectively used in data-stream learning network.

Table 1. Comparison of the classification performance of single frame and LRCN

Input accuracy (%) model	SDSLN	DDSLN	DDSLN-H	(1/2) SDSLN + (1/2) DDSLN-H	(1/2) SDSLN + (1/2) DDSLN–F
Single-frame network	67.32	74.26	76.47	75.25	78.10
LRCN	68.33	77.23	80.2	80.20	81.72

In Table 1, Static-data-stream learning network (SDSLN) is used to implement human behavior recognition based on RGB image frames. Dynamic-data-stream learning network (DDSLN) is used to implement human behavior recognition based on optical flow frame. Dynamic-data-stream learning network with Horn-Schunk optical flow images (DDSLN-H) is used to implement based on the traditional optical flow frame of human behavior recognition. Dynamic-data-stream learning network with the proposed optical flow prediction images (DDSLN-F) is used to implement recognition based on improved optical flow frame of human behavior. At the same time, we can get the next conclusions: first, the performance of networks when LRCN is used for data-stream learning network is better than that when single-frame network is used; second, dynamic-data-stream is more efficient than static-data stream for human activities recognition; third, the proposed optical flow prediction network can improve the performance of compound deep neural network.

2) Comparison of static-data-stream and dynamic-data-stream fusion methods
   Static-data-stream learning network is based on RGB images from videos to learn the spatial contour evolution of human activities, and dynamic-data-stream learning network is based on optical flow images to learn the temporal evolution of human activities. Both are essential to the efficient recognition of human activities. Therefore, the prediction results of both networks will be combined by the method described in Sect. 2.3. In this part, we chose the network weight λ in Eq. (1) as 0, $\frac{1}{2}$, $\frac{2}{3}$, and 1. Also, the single-frame network and LRCN are employed in the data-stream learning network. The comparison results are shown in Table 2. For single-frame network and LRCN, the same conclusion is acquired as that in the above experiments, that is, the recognition accuracy is higher when LRCN is employed than that when single-frame network is used. And it can be seen that the recognition accuracy of compound deep neural network is better than static-data-stream learning network and dynamic-data-stream learning network, and increasing the weight of dynamic-data-stream learning network, the recognition accuracy can also be increased.

Table 2. Comparison of classification performance of models with different weights

Fusion weight λ	Classification accuracy(%)	
	Single frame network	LRCN
$\lambda = 0$	67.32	68.33
$\lambda = \frac{1}{2}$	78.10	81.72
$\lambda = \frac{2}{3}$	80.43	83.35
$\lambda = 1$	76.47	80.20

3) Comparison between the proposed method and the existing method
The classification accuracies of some existing methods are compared with the proposed compound deep neural network, which is shown in Table 3. As it can be seen, the proposed compound deep neural network(CDNN) has achieved a higher accuracy.

Table 3. Comparison of the classification performance of different methods

Recognition method	Classification accuracy(%)
Bi-LSTM [22]	64.8
Slow Fusion CNN [23]	65.4
Composite LSTM Model [5]	75.8
LMDR [24]	68.7
LRCN [14]	81.14
CDNN	83.35

5 Conclusion

Two-stream structure network is efficient for human activities recognition from videos since the spatial and temporal information are both important in recognizing human activities. In this paper, a compound deep neural network, composed of static-data-stream learning network and dynamic-data-stream learning network, is proposed. Furthermore, an optical flow images prediction network is proposed to obtain the temporal evolution representation of video. Experimental results demonstrated the efficiency of the proposed compound deep neural network. In the future, the compound deep neural network will be improved by applying deeper base-network, like VGGNet, GoogleNet, etc., and unify the optical flow image prediction network and learning network as an unified network so as to reduce the complexity of network and to achieve the best optimization.

References

1. He, K., Zhang, X., Ren, S., et al.: Deep residual learning for image recognition. In: IEEE Conference on Computer Vision and Pattern Recognition, pp. 770–778 (2015)
2. Ji, S., Xu, W., Yang, M., et al.: 3D convolutional neural networks for human action recognition. Pattern Anal. Mach. Intell. IEEE Trans. **35**(1), 221–231 (2013)
3. Simonyan, K., Zisserman, A.: Two-stream convolutional networks for action recognition in videos. In: Advances in Neural Information Processing Systems, pp. 568–576 (2014)
4. Yan, X., Chang, H., Shan, S., et al.: Modeling video dynamics with deep dynencoder. In: European Conference on Computer Vision, pp. 215–230 (2014)
5. Chen, B.: Deep Learning of Invariant Spatio-Temporal Features from Video. University of British Columbia, Columbia (2010)
6. Srivastava, N., Mansimov, E., Salakhutdinov, R.: Unsupervised learning of video representations using LSTMs. Proc. Mach. Learn. Res. **37**, 843–852 (2016)
7. Misra, I., Zitnick, CL., Hebert, M.: Unsupervised learning using sequential verification for action recognition. In: European Conference on Computer Vision, pp. 873–896 (2016)
8. Feichtenhofer, C., Pinz, A., Zisserman, A.: Convolutional two-stream network fusion for video action recognition. In: Proceedings of IEEE Conference on Computer Vision and Pattern Recognition, pp. 1933–1941 (2016)
9. Sathasivam, S.: Logic learning in hopfield networks. Mod. Appl. Sci. **2**(3), 57–63 (2008)
10. Han, J., Tao, J., Wang C.: Flownet: a deep learning framework for clustering and selection of streamlines and stream surfaces. IEEE Trans. Vis. Comput. Graph. (2018)
11. Ilg, E., Mayer, N., Saikia, T.: FlowNet 2.0: evolution of optical flow estimation with deep networks. IEEE Trans. Vis. Comput. Graph. (2018)
12. Hochreiter, S., Schmidhuber, J.: Long Short-term memory. Neural Comput. **9**(8), 1735–1780 (1997)
13. Zeiler, M.D., Fergus, R.: Visualizing and understanding convolutional networks. In: European Conference on Computer Vision, pp. 818–833 (2014)
14. Simon, H.: Neural networks and machine learning. In: Shen F., Xu, Y., Zheng, J., Chao, Jing.: Translated. Beijing: Machinery industry press (2011)
15. Donahue, J., Hendricks, L.A., Guadarrama, S., et al.: Long-term recurrent convolutional networks for visual recognition and description. In: Proceedings of IEEE Conference on Computer Vision and Pattern Recognition, pp. 2625–2634 (2015)
16. Simonyan, K., Zisserman, A.: Very deep convolutional networks for large-scale image recognition. In: International Conference on Learning Representations. arXiv:1409.1556 (2015)
17. Horn, B.K.P., Schunck, B.G.: Determining Optical Flow. Massachusetts Institute of Technology, Cambridge (1980)
18. Deng, J., Dong, W., Socher, R., et al.: ImageNet: a large-scale hierarchical image database. In: Proceedings of IEEE Computer Society Conference on Computer Vision and Pattern Recognition, 20-25 June 2009, Miami, Florida, pp. 20–25 (2019)
19. [EB/OL]. http://www.imagenet.org/challenges/LSVRC/
20. Soomro, K., Zamir, A.R., Shah, M.: UCF101: a Dataset of 101 human actions classes from videos in the wild. In: IEEE Conference on Computer Vision and Pattern Recognition. arXiv:1212.0402 (2012)
21. Jia, Y., Shelhamer, E., Donahue, J., et al.: Caffe: convolutional architecture for fast feature embedding. In: ACM International Conference. Multimedia, pp. 675–678. ACM (2014)

22. Rui, G., Xin, X., Yi, J., et al.: Action recognition with hierarchical convolutional neural networks features and bi-directional long short-term memory model. Control Theory Appl. **34**(6), 790–796 (2017)
23. Karpathy, A., Toderici, G., Shetty, S, et al.: Large-scale Video Classification with Convolutional Neural Networks. In: IEEE Computer Society Conference on Computer Vision and Pattern Recognition, pp. 1725–1732 (2014)
24. Peng, X., Qiao, Y., Peng, Q., et al.: Large margin dimensionality reduction for action similarity labeling. IEEE Sig. Process. Lett. **21**(8), 1022–1025 (2014)

Deep Q-Learning Based Collaborative Caching in Mobile Edge Network

Ruichao Wang[1(✉)], Jizhao Lu[2], Yongjie Li[2], Xingyu Chen[1], and Shaoyong Guo[1]

[1] State Key Laboratory of Networking and Switching Technology, Beijing University of Posts and Telecommunications, Beijing 100876, China
rcwang@bupt.edu.cn

[2] Information and Communication Company of State Grid Henan Electric Power Co., Ltd., Henan 450000, China

Abstract. Mobile edge computing technology which includes computing, storage, and communication enhanced the service capability of mobile network. With the development of mobile edge computing technology, caching content in the edge of network has been paid more and more attention. Considering the limitation of caching space, it is very important to design cache replacement algorithm. In this paper, we considered the device-to-device (D2D) communication aided edge cache. And proposed a deep Q-learning based cache replacement algorithm to reduce the delay of user request. Simulation results show that the proposed algorithm can reduce the delay compared to traditional replacement algorithm.

Keywords: Mobile edge computing · Cache replacement · D2D communication

1 Introduction

With the rapid development of intelligent terminal equipment and cellular network, mobile video has increasingly become the main application of mobile users, which brings huge network traffic to mobile edge network. The huge traffic will become heavy burden to the core network, leading to the decline of the network performance and the deterioration of user's quality of experience. Solving the above problems not only requires better network infrastructure, but also needs to optimize the mobile edge network from the perspective of traffic management.

In recent years, there are many researches on mobile edge network caching. The MEC servers can provide users with low-latency services. J. Gao et al. in [1] proposed a probabilistic caching algorithm. The model they proposed considers the overlapping coverage of mobile access network and calculates the probability of cache contents according to the spatial coupling between base stations, which can reduce the occupation of cache space by repeated contents. A. V. Ribeiro et al. in [2] proposed a user clustering method based on the degree of affinity, and the base station selected the main cache contents according to the maximum common interest of its user group, obtaining a higher cache hit rate. Zhang, F., Dai, J., Zhang, Y., Liu, X. et al. [3–6]

Q. Liu et al. (Eds.): CENet 2020, AISC 1274, pp. 327–334, 2021.
https://doi.org/10.1007/978-981-15-8462-6_38

considered the impact of mobility of end devices on cache placement. Zhang, F. et al. [3] and Dai, J. et al. [4] respectively proposed active caching methods based on mobility perception. Zhang, Y. et al. [5] and Liu, X. et al. [6] studied service migration algorithm across base stations.

The rapid development of smart end devices not only brings new mobile applications, but also has an important impact on the structure of mobile network. A large amount of idle storage space on the user side can provide content caching service for other users through D2D communication. There is a mass of user-side resources in the mobile network unused. Li, B., Im, H. Y. R., Jiang, C. L., Wang, C. et al. [7–10] studied joint resource scheduling scheme and proved the potential of user-side resources. Based on the above analysis, we modeled the mobile edge network and proposed a cache replacement algorithm based on deep Q-learning for minimizing the delay.

2 System Model

In this paper, a mobile edge network model that includes several macro base stations (MBS) and smart end devices (SE) is considered, as shown in Fig. 1. Denote $\mathcal{M} = \{1, 2, \ldots, M\}$ as set of MBSs and $\mathcal{N} = \{1, 2, \ldots, N\}$ as set of SEs. The smart end belongs to one home base station. It can communicate with that base station through the cellular link or communicate with other neighboring SEs through the D2D link. Without loss of generality, interference between the D2D links and cellular links is ignored considering wireless spectrum does not overlap. MBSs are fully connected via dedicated wired links to quickly share their own cached contents.

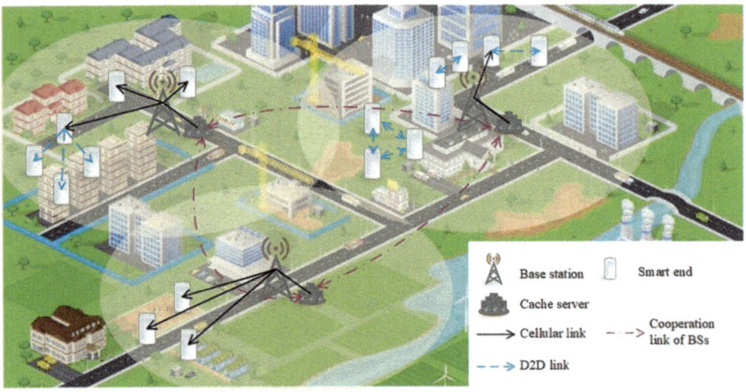

Fig. 1. System model of mobile edge network.

For the cellular links from the MBS to SEs, considering that all users within the service range of the MBS access the service equitably, which means the link rates of all user are roughly. The rate of the cellular link is denoted as R_F. For D2D link, we assume that only two users stay within a certain distance can establish D2D communication. The rate of D2D link denote as R_D. And the rate of the dedicated wired link between the MBSs and the backhaul link between the MBS and the cloud is also a constant value, denote as R_A and R_C respectively.

Denote $\mathcal{K} = \{1, 2..., K\}$ as the video contents requested by users. For each video file k, the size of the video file is denoted by V_k. The popularity ranking of the video file is denoted by $f_k \in \{1, 2..., K\}$, where $f_k = 1$ means the highest popularity and $f_k = K$ means the lowest popularity. The popularity of each video file follows the Zipf distribution of parameter α

$$p_k = \frac{f_k^{-\alpha}}{\sum_{k=1}^{K} f_k^{-\alpha}} \tag{1}$$

where α represents the density of Zipf distribution, and the general value range of α is $\alpha \in [0, 2]$. Larger α means that user requests are more focused on the few most popular files, smaller α means diverse requests on the contrary. Specially, $\alpha = 0$ means that the popularity of all video files is equal, thus the probabilities of files being requested are equal. Assume that the number of total requests q follows the Poisson process with parameter λ, which represents the intensity of arrival. The probability of each user requesting content is equal, denoted as $p_q^{(i)} = q/N$. According to the popularity and request probability of video content, the probability of user i requesting a video file k can be calculated as $p_k^{(i)} = p_k p_q^{(i)}$.

In our proposed caching model, SEs can provide part of the cache space when idle to meet the needs of other users. At the level of user, denote C_i as the cache space provided by SE i. At the level of base station, similarly, denote W_j as the cache space provided by MBS j. SEs and MBSs should cache as much content as possible within the allowed cache capacity to improve the service capability of mobile edge network. Using binary variables X and Y represent the cache decisions of SEs and MBSs respectively.

3 Deep Q-Learning Based Caching Replacement Algorithm

3.1 Problem Formulation

Our goal is to meet user's needs as close to the user as possible, so that the total delay to meet the needs of all users within the network is minimal. Here, assume that the order in which the user sends the request is D2D neighbors, home base station, other base station and cloud. Binary variable S_D, S_B, S_C and S_A indicates where the request is satisfied, and T_D, T_B, T_C and T_A indicates the delay accordingly. It should be noted that streaming video does not need to wait until all video clips are downloaded to start playing when the data rate reaches the requirement of smooth playback. Therefore, on the premise of guaranteeing the quality of user's experience, T_{req} is used to represent the delay that the user needs to wait.

$$T_{req} = \rho \sum_{i=1}^{N} \sum_{k=1}^{K} p_k^{(i)} * [S_D T_D + S_B T_B + S_A T_A + S_C T_C] \tag{2}$$

where ρ represents the minimum buffer ratio that the video fragment can play smoothly.

In fact, given the stable and bounteous communication links between base stations, the delay increase caused by the collaborate between the base stations to satisfy the request is relatively small and therefore acceptable. It is better to cache as many different contents as possible to satisfy the user's request. When the bandwidth of collaboration link between base stations is large enough, the extra delay introduced by collaboration can be ignored, thus (2) can be can be simplified to

$$T_{req} = \rho \sum_{i=1}^{N} \sum_{k=1}^{K} p_k^{(i)} \left\{ \max_{j \in \mathcal{NS}(i)} \left\{ x_k^{(j)} \right\} \frac{V_k}{R_D} + \left(1 - \max_{j \in \mathcal{NS}(i)} \left\{ x_k^{(j)} \right\} \right) \left[\frac{V_k}{R_B} + \left(1 - \max_{g \in \mathcal{M}} \right) \left\{ y_k^{(g)} \right\} \left(\frac{V_k}{R_C} + T_r \right) \right] \right\} \tag{3}$$

where *NS(i)* represents the neighborhood set of the user *i*. $\max_{j \in \mathcal{NS}(i)} \left\{ x_k^{(j)} \right\} = 1$ indicates that when any user *j* in the neighbor of user *i* caches content *k*, user *i* will download the content directly from the D2D link. Otherwise $\max_{j \in \mathcal{NS}(i)} \left\{ x_k^{(j)} \right\} = 0$ indicates that when all the neighbors of user *i* are unable to satisfy its request, it will request resources from base station. $\max_{g \in \mathcal{M}} \left\{ y_k^{(g)} \right\} = 1$ indicates that content *k* is cached in the base station. Or the request will be forwarded to cloud.

Therefore, the delay optimization problem is described as

$$\min_{X,Y} T_{req} \tag{4}$$

$$s.t. \quad C_1 : x_k^{(i)}, y_k^{(g)} \in \{0, 1\} \tag{5}$$

$$C_2 : \sum_{k=1}^{K} x_k^{(i)} V_k \leq C_i , \ \forall i \in \mathcal{N} \tag{6}$$

$$C_3 : \sum_{k=1}^{K} y_k^{(j)} V_k \leq W_j , \ \forall j \in \mathcal{M} \tag{7}$$

$$C_4 : \sum_{m}^{M} y_k^{(m)} \leq 1 , \ \forall k \in \mathcal{K} \tag{8}$$

where (5) indicate that the cache decision variable is binary. (6) and (7) indicate that the cached contents in smart end and base station cannot exceed their capacity limit respectively. (8) indicate that for the same content *k*, only one base station will cache it at the same time.

The optimization of delay is NP-complete problem. When the number of smart ends and base stations in the network is large, it is difficult to solve the problem. Therefore, we proposed a cache replacement algorithm based on deep Q-learning.

3.2 Caching Replacement Algorithm

We formulate the delay of collaborative mobile edge cache and obtained a complex NP complete problem. In this section, we solve the problem with the deep Q-learning method and propose a caching replacement algorithm.

Using a tuple $\{\mathcal{S}, \mathcal{A}, \mathcal{R}\}$ to describe the mobile edge network, which represent the caching state, replacement action and reward respectively.

$\mathcal{S}$: Cache state. In the proposed network model, there are N SEs and M MBSs, and the contents to be cached are a set of K video files. The cache state of the network can be described by a vector with dimension of (N + M), which is defined as the state vector S_{N+M}. $S_i = k$ indicates device i caches content k.

$\mathcal{A}$: Replacement action. Using a vector with dimension of (N + M) represents the replacement of cached contents. $A_i = k$ indicates that device i discards content k from the cache buffer and caches the new arrival content k'.

$\mathcal{R}$: Reward function. Define $R(s, a)$ to represent the benefit of selecting action a under state s. The reward function is defined as follows:

$$R(s,a) = T^{s'}_{req} - T^{s}_{req} \tag{9}$$

the meaning of (9) is the reduction of the delay mentioned in when the cache state of edge network is in s, and then action a is selected to enter the state s'.

Policy $\pi : S \rightarrow A$ represents the mapping of the state space to the action space. Different π corresponds to different action value function $Q^{\pi}(s, a)$ which represent the reward of choosing a strategy π after being in state s and performing an action a

$$Q^{\pi}(s,a) = \mathbb{E}_{\pi}\left\{\sum_{i=0}^{\infty} \gamma^{i} R[s_i, a_i] | S_t = s, A_t = a\right\} \tag{10}$$

The goal of this paper is to find the optimal strategy π^*, which is to replace cache content based on network state to reduce latency. When the strategy is optimal, the corresponding action value function $Q^*(s, a)$ simultaneously get a maximum value, which satisfies Bellman optimal equation. Therefore the solution of the optimal strategy can be transformed into the solution of the action value function, which can be realized by iteration

$$Q^{t+1}(s,a) = (1-\alpha)Q^{t}(s,a) + \alpha\left[R^{t}(s,a) + \gamma \max_{a'} Q^{t}(s',a')\right] \tag{11}$$

where α represents the learning rate, s' represents the state entered next step with current state-action pair, and a' represents the action selected in the next state. However, with the increasing size of the network, it becomes impossible to save a table about all Q value. Therefore, the deep Q-network is used to save the value of the $Q^{\pi}(s, a)$. The deep Q-learning based cache replacement algorithm is described as follows.

First, train the Q-network with history requests. In each episode, update the cache according to the given policy. When the request can be satisfied by itself, the cache state of the network remains unchanged. When the request is satisfied by its neighbor or by its home or other base station, the user caches the content for his own usage. When the requested content is not cached by the edge network, the cache replacement in base station layer is carried out according to the result of Q-learning, ensuring that the replacement process improves the cache benefit of the network. After every few episodes, the parameters of Q network are updated.

4 Numerical Simulation

A numerical simulation of the proposed cache replacement algorithm is performed and compared with the traditional algorithm, that is LRU and LFU.

The network model is assumed to be a mobile network consisting of 2 MBSs and several SEs. The users are randomly distributed in a rectangular area of size 200 m × 200 m. Time for transmitting a video clip from base station to a user is set to 500 ms. The threshold of D2D communication distance is 20 m, and the average delay of D2D-satisified is 350 ms. The total number of video clips is 2000, and the parameter of zipf distribution is 0.8. Suppose the cache capacity of the SE is 1 clip, and the cache capacity of the base station is 50 clips. The delay of the collaboration between base stations is ignored. And the additional delay of retrieving contents from the cloud is set to 1000 ms. Assume that the arrival rate at each step is 30.

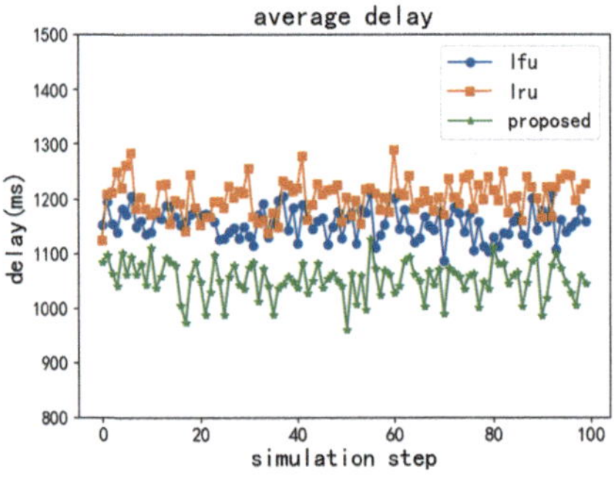

Fig. 2. Average delay with 200 users.

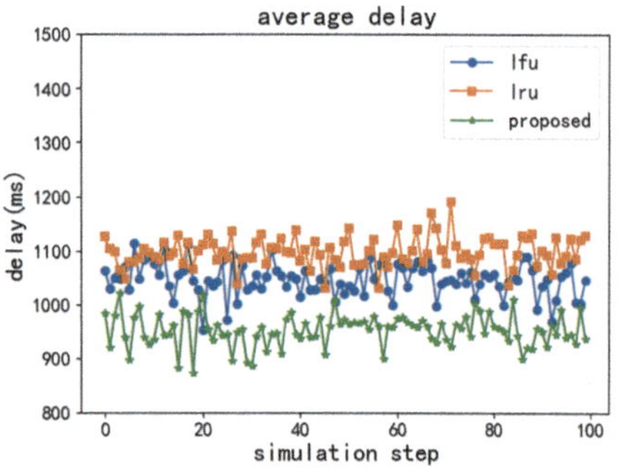

Fig. 3. Average delay with 100 users.

First, we set the number of end nodes of users in the network as 100 and simulate the delay of the proposed algorithm. A total of 100 steps of simulation is carried out, in each step, the user randomly generated requests. The result is shown in. Figure 2.

In each step, the range of average delay of the proposed algorithm is about 1000–1100 ms, which is lower than its of LRU and LFU algorithms. Analyzing all the requests in the whole simulation process, the average delay of the proposed algorithm is 1052 ms, which is lower than the average delay of LFU algorithm (1155 ms) and the average delay of LRU algorithm (1203 ms), reducing the delay by about 9% or 12% than LFU or LRU respectively. Compared with LRU and LFU, the proposed deep Q-learning based cache replacement algorithm has a lower average delay.

Furthermore, we study the performance of the proposed algorithm when the number of users increases. The number of users is set to 200, at which point the user provides more cache capacity than the base station provides. With other parameters unchanged, 100-step simulation is carried out for delay and the results are shown in Fig. 3. When the number of users in the region increases, the average delay of the proposed algorithm drops to 952 ms, while the average delay of LRU algorithm and LFU algorithm is 1099 ms and 1047 ms respectively. The proposed algorithm still has better performance.

5 Conclusion

In this paper, we study the cache replacement of mobile edge network. We consider D2D communication between smart ends of users and make full use of cache resources from users to provide services to other users in the network through hierarchical collaboration. The main goal of this paper is to minimize the delay. However, the optimization problem is proved a NP-complete problem. It can be too complicated to solve in real time when the scale is large. To solve the problem, we propose a cache replacement algorithm based on deep Q-learning. Simulation results show that compared with traditional replacement algorithms, such as LFU and LRU, our proposed algorithm performs better in terms of delay.

Acknowledgement. This work has been supported by State Grid Corporation of China science and technology project "Key technology and application of new multi-mode intelligent network for State Grid" (5700-202024176A-0-0-00).

References

1. Gao, J., Zhao, L., Sun, L.: Probabilistic caching as mixed strategies in spatially-coupled edge caching. In: 2018 29th Biennial Symposium on Communications (BSC), Toronto, ON, pp. 1–5 (2018)
2. Ribeiro, A.V., Sampaio, L.N., Ziviani, A.: Affinity-based user clustering for efficient edge caching in content-centric cellular networks. In: 2018 IEEE Symposium on Computers and Communications (ISCC), Natal, pp. 00474–00479 (2018)
3. Zhang, F., et al.: EdgeBuffer: caching and prefetching content at the edge in the mobilityfirst future internet architecture. In: 2015 IEEE 16th International Symposium on a World of Wireless, Mobile and Multimedia Networks (WoWMoM), Boston, MA, pp. 1–9 (2015)

4. Dai, J., Liu, D.: MAPCaching: a novel mobility aware proactive caching over C-RAN. In: 2017 IEEE 28th Annual International Symposium on Personal, Indoor, and Mobile Radio Communications (PIMRC), Montreal, QC, pp. 1–6 (2017)
5. Zhang, Y., Li, C., Luan, T.H., Fu, Y., Shi, W., Zhu, L.: A mobility-aware vehicular caching scheme in content centric networks: model and optimization. IEEE Trans. Veh. Technol. **68** (4), 3100–3112 (2019)
6. Liu, X., Zhang, J., Zhang, X., Wang, W.: Mobility-aware coded probabilistic caching scheme for mec-enabled small cell networks. IEEE Access **5**, 17824–17833 (2017)
7. Li, B., Gao, Z., Shan, X., Zhou, W., Ferrara, E.: SoReC: a social-relation based centrality measure in mobile social networks. In: 2019 26th International Conference on Telecommunications (ICT), Hanoi, Vietnam, pp. 485–489 (2019)
8. Im, H.Y.R., Thar, K., Hong, C.S.: Q-learning based social community-aware energy efficient cooperative caching in 5G networks. In: 2019 Eleventh International Conference on Ubiquitous and Future Networks (ICUFN), Zagreb, Croatia, pp. 500–503 (2019)
9. Jiang, C., Gao, L., Luo, J., Gong, S.: Crowdsourcing for mobile edge caching: a game-theoretic analysis. In: ICC 2019–2019 IEEE International Conference on Communications (ICC), Shanghai, China, pp. 1–6 (2019)
10. Wang, C., Wang, S., Li, D., Wang, X., Li, X., Leung, V.C.M.: Q-learning based edge caching optimization for d2d enabled hierarchical wireless networks. In: 2018 IEEE 15th International Conference on Mobile Ad Hoc and Sensor Systems (MASS), Chengdu, pp. 55–63 (2018)

VNF Instance Dynamic Scaling Strategy Based on LSTM

Hongwu Ge[1], Yonghua Huo[1], Zhihao Wang[1], Ping Xie[1], and Tongyan Wei[2](✉)

[1] The 54th Research Institute of CETC, Shijiazhuang 050002, China
[2] State Key Laboratory of Networking and Switching Technology, Beijing University of Posts and Telecommunications, Beijing 100876, China
wtybupt2016@163.com

Abstract. In order to solve the high latency problem caused by traffic overloading on the control plane, there is the new generation architecture with the support of the network function virtualization in 5G core network. In this architecture, the new network function AMF interacts directly with UE and is only responsible for processing access requests, so as to shorten latency of receiving requests. However, the 5G architecture still lacks a strategy to control the deployment scale of AMF instances in order to timely complete the reasonable scheduling of resources under the premise of ensuring the quality of service. We put forward a dynamic scaling strategy for 5G core network based on LSTM model with the instance scalability. First, we predict the total volume of the traffic loading in the future through the LSTM model of recurrent neural network. Then, the deployment of instances would be controlled by dynamic threshold algorithm, the increase or decrease of instances deployment number is controlled by the dynamic threshold with upper and lower limits, which would be changed according to the rate of traffic change, so as to complete the deployment of instances in the shortest time. After that, we make the simulation test using by the actual telecommunication data set, and compare our strategy with the deep neural network and the static threshold algorithm, to demonstrate the feasibility and excellent performance of our strategy.

Keywords: 5G architecture · AMF scaling · LSTM · Dynamic threshold

1 Introduction

With the successful commercial use of 4G mobile network, the amount of user equipment accessing to the network is increasing rapidly. As of the end of the second quarter of 2018, there were 3.6 billion LTE users worldwide, accounting for 42% of global mobile users. By the beginning of November 2018, 270 commercial LTE-A networks were identified in 128 countries or regions, and 310 operators have invested LTE-A in 130 countries or regions [1].

As the increase of the equipment amount, the traffic generated by the data plane and control plane in the network has been growing rapidly, so a new generation mobile communication technology has emerged [2]. The architecture is applied for meeting the

Q. Liu et al. (Eds.): CENet 2020, AISC 1274, pp. 335–343, 2021.
https://doi.org/10.1007/978-981-15-8462-6_39

new scenarios such as autonomous vehicles, internet medical, HD live broadcast. In view of these new scenarios, how to achieve a real millisecond level instance scaling response, so that users can obtain better quality of service, has become an urgent problem [3].

In 2018, 3GPP proposed a reference architecture for the new generation mobile networks [4]. With the support of NFV, the functional components in 5G core network architecture can be modularized and virtualized at the same time. The original network function MME of 4G network is divided into new network function AMF and SMF. Among them, AMF is responsible for receiving requests for new session generated by user equipment and processing all tasks related to the access and mobility management, and exists in the virtual machine or containers as the form of instance.

However, the simple function splitting only improves the speed of processing user access requests. If a large amount of sudden accesses is encountered, the mobile network under this architecture will still be difficult to process all requests in time, finally resulting in the latency or even access rejection. Therefore, a dynamic and scalable network function arrangement strategy is essential. It will help the mobile network choose to deploy or shut off the instances related to the access management at the appropriate time, so as to realize the reasonable allocation of resources.

Main contributions of this paper are following: (1) The dynamic instance scaling method based on LSTM prediction model is designed. With the attribute of RNN that it could predict time sequence data more accurately, LSTM model predicts traffic data in a specific future period. And dynamic scaling algorithm makes predicting results as the input, controls the instance scaling and exports instance amount during the specific period. (2) The simulation test with the actual traffic data set is carried out to verify the feasibility of our method. (3) Our method is compared with DNN model in the preliminary predicting stage and static threshold algorithm in instance scaling stage to show its performance.

The remainder of this paper is organized as follows. Section 2 mainly introduces the related work of dynamic scalable strategy in 5G network. Section 3 clarifies the problem statement and puts forward the detailed design of the whole algorithm. In Sect. 4, the feasibility and performance of the algorithm are proved by simulation and contrast test. Section 5 is the summary and the future work focus.

2 Related Work

In [5] and [6], the author has introduced the technical composition and structure of NFV, which provides us some ideas. And from the 3GPP standards of 5G, it has adopted the new telecom structure based on NFV.

In [7], authors propose an adaptive scaling mechanism based on Q-Learning and Gaussian process (GPs), which can handle the change of traffic load and make more reasonable scaling decision. The addition of Q-learning makes the mechanism itself more flexible and adaptable to the actual traffic change trend. However, every learning process will consume a lot of time, which is not fully applicable to 5G practical application scenarios.

In [8], authors propose a 5G core network service management architecture based on NFV/SDN, constructs a more efficient management deployment framework, and realize the flexible arrangement about NFs and workload optimization. This method uses the technical characteristics of NFV/SDN to build 5G architecture, and focuses on an optimal workload allocation algorithm. Although it can meet the requirement of load balancing to a certain extent, there is no scheduling strategy for the NF which handle access requests.

According to [9], the dynamic auto scaling algorithm (DASA) achieves the balance between cost and performance by determining the most suitable total amount of VNF in 5G architecture in each time period. Because AMF belongs to VNF, this algorithm provides some inspiration for us, that is, to improve the QoS of mobile network by dynamically controlling the VNF scale. However, the control object of this algorithm is vEPC, which has a large amount and will make the control process more complicated.

3 Design of Dynamic Threshold Algorithm Based on LSTM

3.1 Problem Statement

In order to meet the requirement of low latency in 5G vision, the new 5G core network architecture needs to shorten the response time. AMF is the core of management access and mobile tasks in the architecture. With NFV support, AMF can dynamically scale on the control plane according to the traffic change trend generated by telecom activities through instantiation to ensure timely processing of access requests. Therefore, we need to predict traffic load in one period in the future, and then determine the deployment or shutting-off of the AMF instance according to the prediction results. The specific description is as follows: firstly, long short-term memory model in RNN is used to predict the traffic in a fixed area in a certain period in the future, and then the number of AMF instances is controlled by dynamic threshold algorithm according to the prediction results to avoid overloading and energy saving. This method is added to 5G architecture as the AMF scaling strategy.

3.2 LSTM Prediction Model

In the first step, LSTM is the traffic prediction model in the early stage. Long short-term memory network is a time recurrent neural network designed to solve the long-term dependence problem in traditional RNN. The detailed structure of LSTM unit is in Fig. 1. The main structure includes input, control gates and output. The control gates are composed of input gate, forget gate and output gate, and the output is the cell state at this time and final output result. Cell state is responsible for memory storage [10, 11].

The generate formula of control gates is formula (1), where x is the input of control gates, that is vector $[h_{t-1}, x_t]$. W represents for the weight matrix of control gates, to measure the importance of inputs from different gates. b is the bias term which is as the intercept in the linear equation to make the model more suitable for solving actual problems.

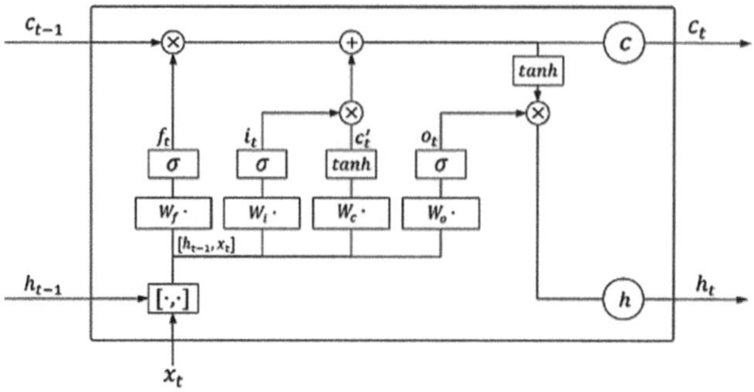

Fig. 1. LSTM unit structure.

$$g(\boldsymbol{x}) = \sigma(W\boldsymbol{x} + \boldsymbol{b}) \tag{1}$$

The cell state c_t at the current time (formula (2)) is composed of the dot product of forgetting gate and the cell state at the previous time plus the dot product of input gate and the cell state at this time.

$$\boldsymbol{c}_t = \boldsymbol{f}_t \circ \boldsymbol{c}_{t-1} + \boldsymbol{i}_t \circ \boldsymbol{c'}_t \tag{2}$$

It also shows that the cell state of every moment is affected by the previous state and existing input, that is, it has long-term memory while also constantly adds new information. The final output h_t (formula (3)) depends on the output gate and the current cell state.

$$\boldsymbol{h}_t = \boldsymbol{o}_t \circ \tanh(\boldsymbol{c}_t) \tag{3}$$

3.3 Dynamic Threshold Algorithm

In the second step, the number of AMF instances is controlled by dynamic threshold algorithm based on the prediction results. The dynamic threshold algorithm is divided into two parts: scaling-in part and scaling-out part. First, set a threshold according to the traffic load capacity of AMF, and the threshold has the upper and lower limits. Then determine the AMF scale according to the traffic in the next week predicted by LSTM before. When the traffic load to be processed reaches the upper limit, the deploying (scaling-out) process will be carried out. When the lower limit is reached, the shutting off (scaling-in) process will be carried out. At the same time, set a monitoring period and control the upper and lower limit of the threshold regularly, so that the number of instances scaling changes more reasonable.

The automatic scaling-out algorithm with dynamic threshold is Algorithm 1.

Algorithm 1: Scaling-out

---

while AMF is processing the UE requests **do**
  if $U(i+1) \geq U(i)$ **then**
    if $M^{+} > 50\%C$ **then**
      $M^{+} = M^{+} - \alpha M^{+}$
      Scaling-out starts, the increasing number of AMF is $\lceil \frac{U - N \times \lambda C}{\lambda C} \rceil$
    else
      the number of AMF should not be changed
    end if
    Add up the number of AMF
  end if
  if there is no scaling-out operation in $2n$ periods
    $M^{+} = \min(75\%C, M^{+} + \alpha M^{+})$
  end if
end while

Suppose that max load that one AMF can support is C, the total number of AMF instances needed in the mobile network is N, and the total amount of traffic that the network needs to handle at the time i is $U(i)$. We set the load threshold value as M. in order to enable AMF to process all requests normally, the initial value of the upper threshold value is set to $75\%C$, the minimum value is $50\%C$, while the initial value of the lower threshold value is set to $5\%C$, and the maximum value is $30\%C$. When the traffic in the next time period is greater than that in the current time, the scaling-out operation is started.

Where λ is the balance coefficient in the range of 0 to 1 to ensure that AMF will not be under full load. At the same time, if the upper limit of the threshold is still greater than its lowest value, it will decrease in a proportion of $\alpha\%$, which makes the scaling-out process start to respond more quickly. After the scaling-out process is completed, the load of each AMF is reduced accordingly. If the demand of traffic load at this time is met, the scaling-out operation will not be carried out. Set the monitoring period as n. after two monitoring periods, if the scaling-out operation does not occur, then the upper limit of the threshold should be increased automatically to avoid rapid scaling-out.

The automatic scaling-in algorithm with dynamic threshold is Algorithm 2.

Algorithm 2: Scaling-in

while AMF is processing the UE requests **do**
  if $U(i+1) \leq U(i)$ **then**
    if $M^- < 30\%C$ **then**
      $M^- = M^- + \alpha M^-$
      Scaling-in starts, the decreasing number of AMF is $\lfloor \frac{N \times \lambda C - U}{\lambda C} \rfloor$
    else
      the number of AMF should not be changed
    end if
    Add up the number of AMF
  end if
  if there is no scaling-in operation in $2n$ periods
    $M^- = \max(5\%C, M^- - \alpha M^-)$
  end if
end while

Similarly, when the traffic in the next time period is less than that in the current time, the scaling-in operation is started. If the upper limit of the threshold is still less than its maximum value, it will decrease in the proportion of $\alpha\%$. After the scaling-in is completed, the load of each AMF is increased accordingly. Set the monitoring period as n. After two monitoring periods, if the scaling-in operation does not occur, the lower threshold should be automatically reduced.

4 Feasibility and Performance

4.1 Environment Configuration

Data set of telecom activities provided by 'big data challenge of Telecom Italy' in 2014 is selected as the original input of simulation test [12]. Basic environment depends on Anaconda 3 and the additional libraries–tensorflow and keras.

4.2 Feasibility Assessment

After the completion of the building and training of LSTM model, we predicted traffic data in the corresponding time of the test set, and obtained the comparison chart between predicted results and test set. The trend of prediction results is similar as the actual situation, and the deviation between those is small. The accuracy of LSTM model is 98.4%. At the same time, we could summarize the change rule of the traffic changing. From 11:00 to 13:00, three kinds of traffic are at the peak, while from 1:00 to 5:00, three kinds of traffic are at the trough. And the traffic on weekends is obviously less than that those on weekdays. This is consistent with the actual activities, which shows that our model is feasible.

The network traffic prediction data of a certain day in the test set time was selected as the input of dynamic threshold algorithm simulation test. The initial value of AMF instance deployment is 10. Initial value of the instance number is reduced to 5 under the effect of scaling-in algorithm. At 1:00, it decreases to the lowest value 4. From 11:00 to 13:00 a.m. and from 16:00 to 17:00 a.m., it increases to the maximum value 12. After 18:00, it gradually decreased to 6. Therefore, the dynamic threshold algorithm is feasible.

4.3 Performance Assessment

DNN model can also be used for prediction. In order to show the performance of LSTM. Taking the network traffic of a week in the test set as an example, we put the actual data, LSTM prediction results and DNN prediction results for comparison. The results are illustrated in Fig. 2. We can see that although the overall trend predicted by the two models is consistent with the actual value, it is obvious that LSTM model has better fitting effect on the actual situation, and DNN prediction results are far from the actual value. At the same time, we further measure the performance of different models by predicting deviation rate. The deviation rate of LSTM is 1.534%, smaller than that of DNN is 8.202%. LSTM model exhibits a better performance, which is conducive to the dynamic scaling of AMF instances.

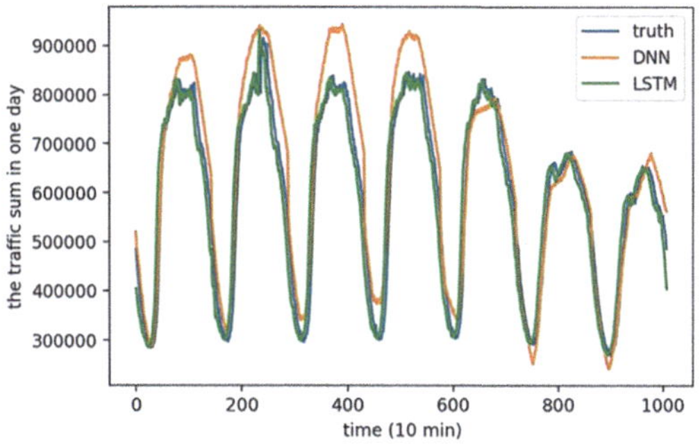

Fig. 2. Comparison of LSTM prediction result, DNN result and actual data.

In order to verify that dynamic threshold algorithm can get the optimal solution in the deployment of AMF instances, we took the static threshold method as its control group to carry out the controlled test. The logic of static threshold method is basically same as that of dynamic threshold method. Only the dynamic change part of the threshold is deleted. The number of AMF instance deployments controlled by different methods is shown in Fig. 3.

Set the initial value of the deployed AMF instance to 10. When telecom activities are in the trough during 1:00 to 6:00, dynamic threshold algorithm can start scaling-in operation more quickly by increasing the lower limit of the threshold, and the number of AMF under its control is far lower than that deployed under the static threshold. Its minimum deployment duration is more than twice that of static threshold, which

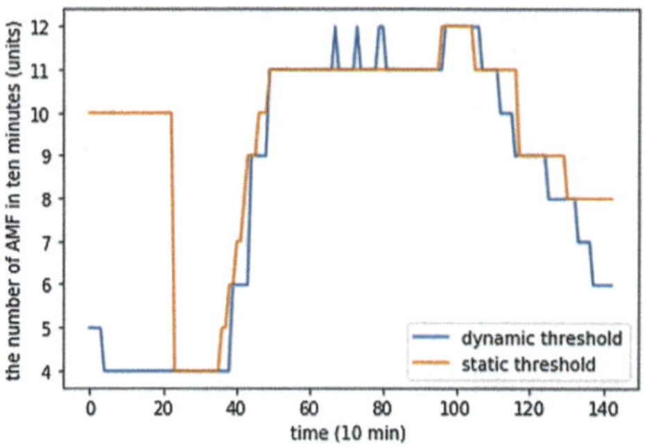

Fig. 3. AMF deployment comparison.

effectively solves the resource waste at night. From 12:00 to 13:00 at noon, there are three small fluctuations controlled by dynamic threshold, indicating that the dynamic threshold method is more sensitive. The traffic began to decrease at around 18:00, final number of AMF deployed was 6, which was less than that under the control of the static threshold. So dynamic threshold algorithm has better performance in instance deployment.

5 Conclusion

In this paper, an AMF instance scaling algorithm based on LSTM and dynamic threshold is designed. By controlling the deployment scale of AMF instances in 5G core network, the algorithm predicts the traffic of telecom activities to be generated in a certain period in the future and takes the prediction results as input to adjust the number of AMF instances scaling in mobile network through a dynamic threshold. We have shown the feasibility and performance of our algorithm through actual telecommunication activity data set. In the future, we will continue to collect records generated by AMF, for example, the instance CPU utilization of request dealing, the delay loading new resources and other needed parameters of AMF during equipment accessing. Dynamic threshold algorithm will be optimized to be more suitable for the actual scene, such as set more benchmarks.

References

1. GSA: gigabit LTE: global status december 2019. https://www.gsacom.com
2. OCLC: NextSpace: the OCLC newsletter No. 24. OCLC Library. https://library.oclc.org/digital/collection/p15003coll11/id/23
3. Huawei: 5G wireless network plan solution white paper (2018). https://www.huawei.com/cn/industry-insights/outlook/mobile-broadband/insights-reports/5g-wireless-network-plan-solution-whitepaper
4. 3rd Generation Partnership Project (3GPP): System architecture for the 5G system. https://tinyurl.com/y7monxmh

5. Lao, B., Deng, J.Y.: Key technologies and advantages of NFV in core network. Chin. New Telecommun. **20**(17), 56–57 (2018)
6. Schaffrath, G.: Network virtualization architecture: proposal and initial prototype. In: ACM SIGCOMM Workshop on Virtualized Infrastructure Systems & Architectures DBLP (2009)
7. Arteaga, C.H.T., Rissoi, F., Rendon, O.M.C.: An adaptive scaling mechanism for managing performance variations in network functions virtualization: a case study in an NFV-based EPC. In: CNSM, pp. 1–7. IEEE (2017)
8. Lu, M., Wen, X., et al.: An SDN/NFV based framework for management and deployment of service based 5G core network. China Telecommun. (2018)
9. Ren, Y., Phung-Duc, T., Yu, Z.W., et al.: Dynamic auto scaling algorithm (DASA) for 5G mobile networks. IEEE Globecom, IEEE (2016)
10. Colah's blog: Understanding LSTM networks. http://colah.github.io/posts/2015-08-Understanding-LSTMs/
11. Qiu, X.P.: Neural network and deep learning: "Chapter 6-LSTM", Mechanical Industry Press (2020)
12. Italia, T.: Telecom italia big data challenge. http://www.telecomitalia.com/tit/en/bigdatachallenge/contest.html

A Flow Control Method Based on Deep Deterministic Policy Gradient

Junli Mao[1], Donghong Wei[1], Qicai Wang[1], Yining Feng[2(✉)], and Tongyan Wei[2]

[1] The 54th Research Institute of CETC, Shijiazhuang 050000, China
[2] State Key Laboratory of Networking and Switching Technology, Beijing University of Posts and Telecommunications, Beijing 100876, China
fengyining@bupt.edu.cn

Abstract. With the rapid development of modern network technology, business traffic in the network is exploding, and network traffic control technology is facing greater challenges. Traditional network flow control methods often cannot calmly deal with complex and dynamic network states, and it is difficult to avoid network congestion. Therefore, we need an intelligent method that can update flow control decisions according to the current network state.

Software defined network (SDN) has brought new ideas for controlling network traffic. In the SDN network architecture, the controller of the control plane can send the network routing policy to manage the traffic forwarding in the data plane layer, so as to implement network management and resource scheduling. The programmability of SDN also provides the possibility of implementing flow control algorithms. By combining SDN and reinforcement learning (RL), a good traffic forwarding policy can be obtained, and the problem of traffic control can be solved effectively. In order to achieve intelligent control of network traffic, this paper introduced the deep deterministic policy gradient (DDPG) algorithm in RL. With the goal of maximizing network performance, we designed and implemented a network flow control algorithm based on DDPG. We trained the model, and finally verified the effectiveness and superiority of the flow control method based on DDPG through simulation experiments.

Keywords: Flow control · SDN · Reinforcement learning · DDPG

1 Introduction

In recent years, with the advancement of network infrastructure, the development of Internet communication technology and the growth of the number of users, the business traffic in the network has also shown a trend of rapid growth. How to balance the traffic load through a reasonable control method has become an important research topic in the field of computer networks. Traditional flow control methods include the shortest path method [1] and the multipath routing method [2]. The shortest path algorithm selects the path with the smallest sum of link weights for the communication session to forward traffic. It can find the optimal path for a single traffic. ECMP [3] is a multi-path routing method that can balance all equal-cost paths. However, these two methods can

Q. Liu et al. (Eds.): CENet 2020, AISC 1274, pp. 344–351, 2021.
https://doi.org/10.1007/978-981-15-8462-6_40

only be aimed at specific network and traffic scenarios, but cannot effectively deal with complex and changing network conditions, which is easy to cause partial network congestion. Therefore, an intelligent network flow control method that can adapt to the changes of the network environment is required.

SDN is a new type of network architecture [4]. Its core idea is to separate the control plane from the data plane. The control plane sends the traffic processing policy, and the data plane executes the policy. SDN controller has programmability [5], which is convenient for developers to flexibly manage the network. In this paper, we will use this feature of SDN to implement flow control. To this end, we need to develop the flow control policy to implement the traffic forwarding in data plane. In this paper, we will introduce DDPG algorithm [6] in RL to analyze and make decisions. DDPG has an Actor-Critic dual network structure. The two networks are respectively responsible for generating and evaluating policy. The interaction of the two networks can better iterative and optimize, so as to help output the optimal policy faster.

In this paper, we propose a network traffic control algorithm based on DDPG under the SDN architecture, in order to rationally distribute traffic in random network topology, as well as avoiding network congestion and improving network service quality. The main contributions of this paper are as follows: First, we establish the model of the flow control problem, and design the objective function. In the process of designing DDPG-based network traffic control algorithms, we use rate and delay as parameters to measure network performance. We set input, output and reward function for the algorithm, and clarify the algorithm flow. Second, we design the system in the SDN network architecture. We implement the network flow control algorithm based on the DDPG algorithm in the network simulation platform Mininet. We adjust the parameters and train the model. Third, we test the performance of the algorithm. We use two traditional network flow control algorithms for comparative evaluation. For each algorithm, the network rate and delay under different network traffic demands are output respectively. The evaluation results show that the performance of the DDPG algorithm is better than the baseline algorithms, which verifies the rationality and effectiveness of the DDPG algorithm.

The rest of this paper is organized as follows. Section 2 introduces related work. Section 3 introduces the design process of the network flow control algorithm based on DDPG. Section 4 gives the simulation and evaluation results. Section 5 makes a summary.

2 Related Work

The flow control method based on SDN and RL has achieved certain research results. SDN network architecture provides new ideas for flow control.

Ke proposed the SDN-oriented multipath ant colony routing algorithm [7]. The algorithm used multi-path routing. Considering that traditional routing can only select one path with the smallest number of hops, while the actual network often has multiple shortest paths with the same number of hops, multi-path routing can provide multiple alternative paths for data packets. At the same time, by collecting link state information such as remaining bandwidth, delay and rate of packet loss, and dynamically changing the parameters used in the calculation, the link congestion problem was better solved and the link utilization was improved. Gao et al. proposed a highly deterministic flow control method for SDN-based cloud architecture networks [8]. This method used the characteristics of the SDN controller to have the entire network view to obtain the global network topology and flow information. On this basis, the optimal path of the business flow was planned, and reasonable queue scheduling parameters were assigned to the switch. In addition, data center TCP (DCTCP) was also used to reduce network congestion and ensure bandwidth and delay of various services.

RL has also been used to control network traffic. Ruhul et al. proposed a route optimization technique based on Q-learning algorithm [9], which used a Q value table based on link delay to replace the routing table in the distance vector algorithm to optimize the routing for data forwarding.

Xu, Z. Y. et al. proposed an experience-driven network traffic scheduling method [10], which used deep reinforcement learning to achieve model-free control in communication networks. The algorithm input global network state information and output all links' split ratio. Simulation experiments showed that the algorithm can effectively improve the throughout of the network, reduce the delay, and improve the network utilization.

In this paper, we will use the DDPG algorithm in RL to complete the decision process, and use the SDN network architecture to control the global network. In addition, we will set different priorities for different communication sessions to ensure the reliability of the network.

3 Algorithm Design

3.1 Problem Model

In this paper, we have a random network topology as shown in Fig. 1.

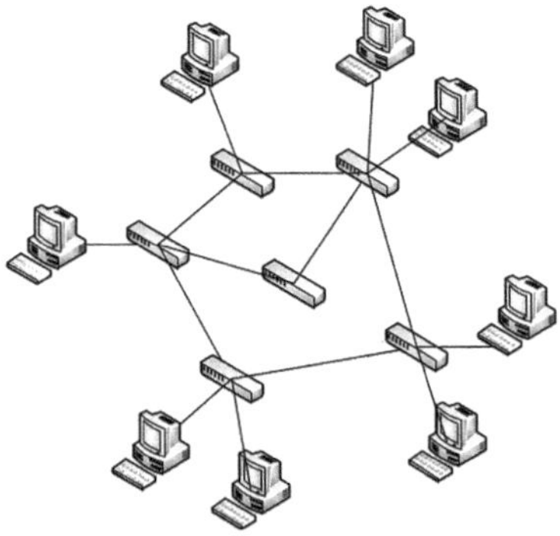

Fig. 1. Random network topology.

There are several hosts and several links in the network. Traffic flow is initiated by the host. There are multiple communication sessions in the network at the same time. All the data packets with the same source and destination node are grouped into one flow. As the transmission rate requirement of each flow increases, a single path cannot meet the flow demand. In order to avoid the occurrence of network congestion, the SDN controller should deliver the traffic distribution policy to the switch. The switch forwards the data packets of each flow to multiple outgoing ports, so that each flow can be transmitted through multiple paths. Different paths will transmit different amount of flow due to different weights.

The split ratio of each flow transmitted in multiple paths will affect the global network state. So, before determining the split ratio, we need to use RL to train the SDN controller [11]. During the training process, the quality of the network state is used as a reward signal. For the purpose of obtaining greater rewards, the controller updates its distribution strategy iteratively to find the optimal split ratio for each flow. After the training, the controller will get the optimal split strategy.

In the business scenario of this article, we control the flow by assigning each business flow a different weight to multiple paths. Therefore, the flow control in this paper focuses on solving the problem of load balancing of flow.

We design the objective function for the flow control problem based on RL under the SDN architecture. The relevant definitions are as follows:

Define that there are K communication sessions in the network at the same time. The rate of the kth communication session is x_k, the delay is y_k, the priority is p_k, the set of available paths is N_k. The traffic allocated to the j th available path of the k th communication session is $f_{K,j}$, and its proportion is $w_{K,j} = f_{k,j} \big/ \sum_j f_{k,j}$

According to the α-fireness model based on weight [12], we set the objective function as:

$$max \sum_{k=1}^{K} U(x_k, y_k), \tag{1}$$

$$\text{with } U(x_k, y_k) = U_{\alpha_1}(x_k) - \sigma\, U_{\alpha_2}(y_k), \tag{2}$$

$$U_\alpha(x_k) = x^{1-\alpha} \big/ (1 - \alpha) \tag{3}$$

For the above formula, the process of solving the priority p_k in this paper is as follows:

First, obtain global routing information, get all available paths for each communication session, and the number of hops for each path.

Second, get the average number of hops of the path in the kth communication session:

$$t_k = Sum\ of\ links\ across\ all\ paths / Number\ of\ paths \tag{4}$$

Third, get the average number of hops for the effective paths of all K communication sessions in the network topology:

$$t = \sum\nolimits_{j=1}^{K} t_j / K \tag{5}$$

If $t_k > t$, set the priority of the kth communication session $p_k = 2$, else, $p_k = 1$.

We set higher priorities for communication sessions with longer average paths and longer distances, in order to prioritize resource allocation and improve reliability.

3.2 Flow Control Algorithm Based on DDPG

We set the state space, action space and reward function for the algorithm. The network state will be expressed by the rate and delay of the communication session in the network. We set the state vector as:

$$s = [(x_1, y_1), \ldots, (x_k, y_k), \ldots, (x_k, y_k)] \tag{6}$$

The action is the forwarding of the network traffic, we use the split ratio to represent the action. The action vector is defined as:

$$a = \left[w_{1,1}, \ldots w_{k,j}, \ldots, w_{k,|N_k|}\right], \tag{7}$$

$$\text{with} \sum\nolimits_{j=1}^{|N_k|} w_{k,j} = 1 \tag{8}$$

The goal of the RL algorithm is to maximize the reward function, so we use the objective function to define the reward:

$$r = \sum\nolimits_{k=1}^{K} U(x_k, y_k) \tag{9}$$

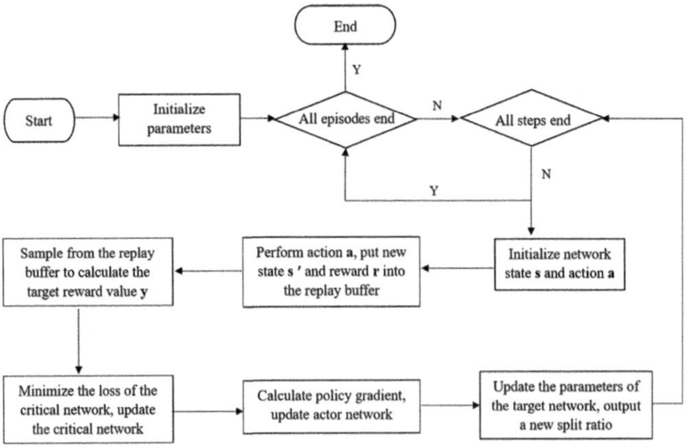

Fig. 2. Flow control algorithm based on DDPG.

After completing the variable design, we define the flow control algorithm flow based on DDPG as shown in Fig. 2:

4 Simulation

4.1 System Model

In order to realize the flow control algorithm based on DDPG, we construct a system model based on SDN architecture as shown in Fig. 3. The control layer has three modules. The link state collection module is used to collect network topology information and network state information. It is responsible for sending the information to the DDPG training module. The DDPG training module is used to train the controller, as well as updating the model and policy. The decision delivery module uses the trained model to deliver the distribution policy to the forwarding layer, and instructs the switch to forward the data stream according to the optimized distribution policy.

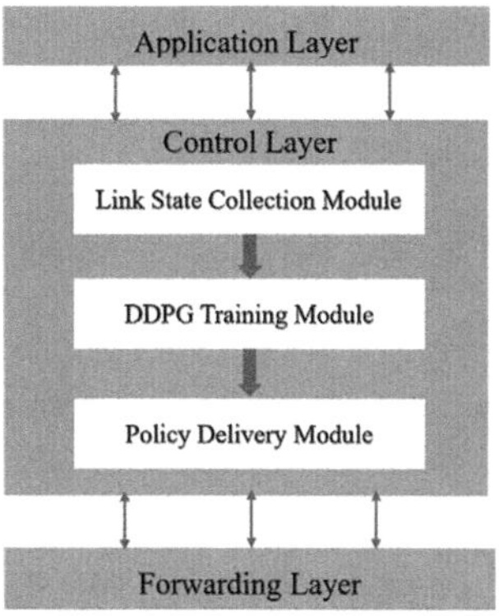

Fig. 3. System model.

Table 1. Parameters of training process.

Parameter	Value
Reward discount	0.99
Episodes	5000
Update rate	0.01
Size of reply buffer	1600
Size of mini-batch	32
Critic network parameter	0.01
Actor network parameter	0.001

We build the SDN network architecture on the Mininet platform [13] in the virtual machine. We write a Ryu controller [14], which is Python-based, to implement the system model and control network traffic. When implementing the algorithm, we set parameters as shown in Table 1:

4.2 Simulation Results

In the simulation experiment, in order to verify the effectiveness of the flow control algorithm based on DDPG, we choose Dijkstra [14] and ECMP as the baseline algorithms, which are two traditional network flow control algorithms. Dijkstra is an algorithm that calculates the shortest path based on the number of hops. ECMP is a load balancing algorithm that balances all equal-cost paths.

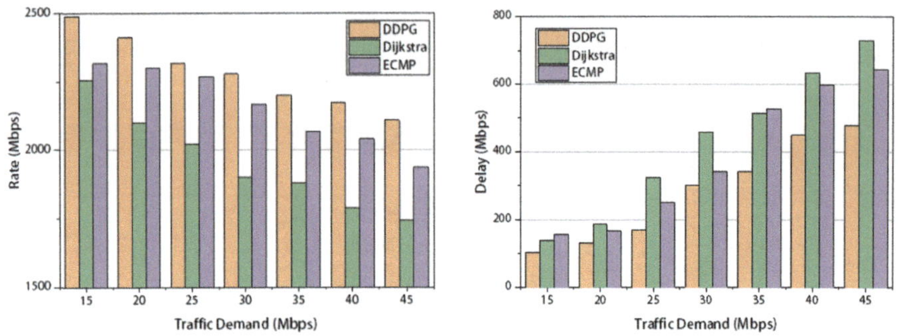

Fig. 4. Performance of algorithms in random network topology.

To make the results more convincing, we compare the two dimensions of rate and delay. The results show that, compared with Dijkstra algorithm and ECMP algorithm, the DDPG algorithm relatively improves the average network rate by 16.8% and 6.1% and relatively reduces the average network by 33.7% and 26.3% Fig. 4.

According to the state of the global network, the DDPG algorithm delivers a reasonable split ratio for each communication session in a more targeted manner, so it has better control effect on network traffic.

5 Conclusion

In this article, we studied the problem of network flow control, and proposed a method of introducing DDPG algorithm in RL to control network traffic. Using the programmability of the SDN architecture network, we implemented an algorithm to control network traffic and completed simulation experiments. The results showed that the performance of network flow control algorithm based on DDPG worked better than traditional network flow control algorithm, and it reasonably allocated network traffic according to the network status and improved the overall performance of the network.

References

1. Nelson, L.S., Da, F., Raouf, B.: Openflow and SDN for clouds. In: Cloud Services, Networking, and Management, pp. 129–152. IEEE (2015)
2. Mpho, N., Albert, A.L., Sabelo, D.: Multi-path Load balancing for SDN data plane. In: The International Conference on Intelligent and Innovative Computing Applications (ICONIC 2018) (2018)
3. Open Shortest Path First (OSPF). https://en.wikipedia.org/wiki/Open Shortest Path First
4. Foundation, O.N.: Software-defined networking: the new norm for networks (2012)
5. Zuo, Q.Y., Chen, M., Zhao, G.S., Xing, C.Y., Zhang, G.M., Jiang, P.C.: Research on SDN technology based on openflow. J. Softw. **24**(05), 1078–1097 (2013)
6. Lillicrap, T.P., Hunt, J.J., Pritzel, A., Heess, N., Erez, T., Tassa, Y., Silver, D., Wierstra, D.: Continuous control with deep reinforcement learning, ICLR 2016
7. Ethier, S.N., Kurtz, T.G.: Markov processes: characterization and convergence (2008)
8. Csaba, S.: Algorithms for reinforcement learning. Algorithms for Reinforcement Learning, Morgan & Claypool (2010)
9. Chen, R., Liang, Q., Jiao, X.: Method and device for processing openflow-based group table and group table configuration unit (2016)
10. Fang, B.Y., Zhang, G., Zhang, Y.Y., et al.: Research on the development status of open source SDN controller. Post Telecommun. Des. Technol. **07**, 39–46 (2014)
11. Jawaharan, R., Mohan, P.M., Das, T., Gurusamy, M.: Empirical evaluation of SDN controllers using mininet/wireshark and comparison with cbench. In: 2018 27th International Conference on Computer Communication and Networks (ICCCN), pp. 1–2 (2018)
12. Kanthi, N., Bharadia, D., Mao, H.Z., Chinchali, S., Alizadeh, M., Katti, S.: NUMFabric. In: Proceedings of the 2016 Conference on ACM SIGCOMM 2016 Conference, SIGCOMM 2016 (2016)
13. Kaur, K., Singh, J., Ghumman, N.S.: Mininet as software defined networking testing platform. In: International Conference on Communication, pp. 139–142 (2014)
14. Yang, Y.: An efficient implementation of shortest path algorithm based on Dijkstra algorithm. J. Wuhan Tech. Univ. Surv. Mapp. (WTUSM) (1999)

Design of Forest Fire Warning System Based on Machine Vision

Jiansheng Peng[1,2(✉)], Hanxiao Zhang[2], Huojiao Wu[1], and Qingjin Wei[1]

[1] School of Physics and Mechanical and Electronic Engineering, Hechi University, Yizhou, Hechi 546300, China
sheng120410@163.com
[2] School of Electrical and Information Engineering, Guangxi University of Science and Technology, Liuzhou 545000, China

Abstract. Forests are one of the most important natural resources in the world. It brings great economic value to people's production and life. However, the occurrence of forest fires will burn plants and kill animals. Forest fires have even caused incalculable losses. The article aims to study a forest fire prevention early warning image processing system based on machine vision. First, the foreground detection of moving targets is performed. Then use the mixed Gaussian model algorithm to determine whether there are moving objects in the video. Then track and identify the flames in the video. Then determine whether it is flame. Draw its outline and give an alarm to the supervision office when the result is fire. Then the early warning system waits for the relevant staff to deal with it. Finally, it summarizes the shortcomings of the design of a machine vision forest fire prevention early warning system in this paper and looks forward to the research prospects.

Keywords: Machine vision · Early warning · Flame tracking recognition

1 Introduction

Forests are one of the most important natural resources in the world. Many plants will be burned and animals will be burned to death when a fire breaks out in the forest. Furthermore, forest fires will cause huge economic losses. Our territory is vast and contains many forest resources. Tens of thousands of fires occur each year, resulting in the loss of forest area accounting for about 5–8% of the total global burnout. The forest fire prevention early warning system is an intelligent system that uses modern advanced technology for forest monitoring and management. The early warning system can accurately know where the fire occurred and the size of the disaster at the initial stage of the fire. Then the early warning system will remind the relevant staff to carry out the relevant processing work. With the continuous development of image processing and machine vision, it has brought new directions to the research of fire detection systems. Video surveillance can detect a wider range and can obtain more intuitive information, making fire detection algorithms based on machine vision a hot pot.

Q. Liu et al. (Eds.): CENet 2020, AISC 1274, pp. 352–363, 2021.
https://doi.org/10.1007/978-981-15-8462-6_41

The forest fire prevention platform was first used in Western countries. For example, Canada developed a grading system for fire detection in 1987. The United States developed a national fire risk prediction system in 1972. China Forestry University and China University of Science and Technology have successfully developed a fire warning system. They have also begun trials in various forest areas in my country. China's per capita forest area is not large, resource strength is not high and regional distribution is uneven. With the improvement of national legislation to protect forest resources and the enhancement of protection, forest fires still occur frequently. There are some environmental problems in our country at present. Problems such as river water pollution are hazards caused by forest fires. How to successfully prevent and detect fires in time is an important research direction for us to explore and give effective solutions [1]. Now the internationally well-known forest fire alarm technology is the German Fire-Watch forest fire automatic warning system. The monitoring radius of the system can cover a range of 10 km, but its installation cost also becomes high. The "Earth" satellite of the United States is specifically responsible for the forest fire prevention of American land. The satellite orbits about 705 km above the ground and runs around the earth, which can effectively detect the temperature distribution on the ground. The satellite alerts the areas with higher temperatures. In addition, the United States also uses drone technology to monitor forests in real time. These results show that some developed countries are much more advanced than the technology currently used by China. The human capital they invest is not high, but the technical requirements and costs are very high [2].

In China, ground patrol-type forest fire prevention, observation deck monitoring and air patrol are widely used [3]. In comparison, our technical work still has a lot of room for improvement. And China's terrain is complex and the climate is changing. Moreover, the economic development levels of various regions in my country are inconsistent. Therefore, it is difficult to unify forest protection. Different measures should be worked out for different forest distributions and terrain environments [4]. These problems require us to find a better solution.

The second Section of this paper introduces the research status of forest fire prevention. Section 3 introduces the foreground detection of moving targets and extracts the foreground of flame. The fourth Section introduces the flame tracking and identification. The target is tracked and identified by the color tracking identification principle to determine whether it is flame. Section 4 introduces the flame contour extraction process. Finally, it further summarizes the shortcomings of the existing methods and prospects.

2 Moving Object Foreground Detection

2.1 Collection End

The cameras deployed in the forest area use remote color-to-black and white 24 h all-weather cameras with waterproof and anti-lightning functions. The definition of the camera should choose a high-definition camera, in order to eliminate unnecessary interference and facilitate accurate judgment. The monitoring range of the camera is

12 km^2, covering a forest area within 3 km. The gimbal adjusts the position. Install the camera to choose to install in a place with wide field of view and wide viewing range. Try to choose a location with a fixed power supply to ensure the normal operation of the camera. Taking 200,000 km^2 of forest area as an example, it is necessary to install 200 cameras to basically cover all corners of the forest area. Considering that the camera shake will affect the judgment of the movement prospect, it is necessary to ensure that the installation is reliable when installing.

2.2 Target Foreground Detection

Target foreground detection means when the video at the camera end is found to have a frame difference, the system starts to detect the image in the video. The image of the current frame is set as the foreground image and the frame immediately before the current frame is set as the background image. After background modeling and constantly updating the foreground to determine whether there are moving things in the video [5]. The flame is not stationary when it is burning, it will make irregular movements with the blowing of the wind or its own combustion. Therefore, the foreground detection of the target is used to capture the moving flame in the video. Moving target foreground detection is the basis of subsequent flame image analysis. Therefore, reasonable detection is very important for flame analysis. It is difficult to detect due to disturbances in the forest due to changes in light, weather and other factors. The camera in the forest area is generally divided into static and dynamic. However, the current research is based on the situation where the camera is stationary, so the analysis is based on the target with a static background.

2.3 Hybrid Gaussian Background Modeling Algorithm

The mixed Gaussian background model is a further optimized algorithm on the single Gaussian model. The mixed Gaussian background model fits the background by establishing multiple Gaussian distributions. Therefore, the description of the multimodal distribution background is realized and some shortcomings of the single Gaussian background model are optimized [6]. The mixed Gaussian background modeling algorithm uses Gaussian probability density function as the calculation basis. This algorithm decomposes a thing into a model formed by several functions as needed. The change in pixel value is determined by the function value on the pixel.

$$P(X_t) = \sum\nolimits_{i=1}^{k} W_{i,t} x \varepsilon(x_t, \mu_{i,t}, \tau_{i,t}) \tag{1}$$

$$\varepsilon(x_t, \mu_{i,t}, \tau_{i,t}) = \frac{1}{|\tau_{i,t}|^{1/2}} e^{-1/2(x_t, \mu_{i,t}) T_{i,t}^{-1} (x_t - \mu_{i,t})} \tag{2}$$

$$\tau_{i,t} = \delta_{i,t}^2 I \tag{3}$$

where k is the total number of distribution modes. $\varepsilon(x_t, \mu_{i,t}, \tau_{i,t})$ is the i-th Gaussian distribution at time t. $\mu_{i,t}$ is its mean value. $\tau_{i,t}$ is its covariance matrix. $\delta_{i,t}$ is the variance. I is the three-dimensional identity matrix. $W_{i,t}$ is the weight of the Ith Gaussian distribution at time t. The approximate algorithm flow is as follows:

(1) Each current pixel model is calculated and compared with its previous model by $\left|x_t - \varepsilon_{i,t=1}\right| \leq 2.5\sigma_{i,t=1}$ formula. Until a new pixel appears, it means that the deviation should be guaranteed within 2.5σ.
(2) If the matching result meets the background condition, the pixel belongs to the background.
(3) The mean μ and standard deviation σ of pixels without updating the background remain unchanged. Update according to formulas (4), (5), (6):

$$P = \alpha * \varepsilon(x_t|\mu_k, \sigma_k) \tag{4}$$

$$\mu_t = (1 - \rho) * \mu_{t-1} + \rho * x_t \tag{5}$$

$$\sigma_t^2 = (1 - \rho) * \sigma_{t-1}^2 + \rho * (x_t - \mu_t)^T \tag{6}$$

(4) If (2) is not carried out, the pixels will be arranged in descending order of weight.

The hybrid Gaussian modeling algorithm improves the deficiencies of the single Gaussian modeling algorithm. The hybrid Gaussian modeling algorithm has a good performance in extracting the target in the video, and can be applied in many different places. Fig. 1 and Fig. 2 show the process of mixed Gaussian modeling. Figure 1(a) is the frame image when the moving target has not appeared. Figure 1(b) is the input frame of Fig. 1(a). Figure 2 is the 25th frame image captured after the appearance of a moving target. Figure 2(a) is the original input image. Figure 2(b) is the modeled output image.

(a) Output image corresponding to no frame difference.

(b) Picture without frame difference.

Fig. 1. Screen when no moving objects appear.

(a) Output image after mixed Gaussian Modeling. (b) Input image of the 25th frame.

Fig. 2. Screen when moving objects appear.

When the system works, no moving objects appear. The system screen has not changed, as shown in Fig. 1. As shown in Fig. 2, when the system detects that the frame difference of pixels changes, the background image is updated. Take the current frame as the foreground, and the previous frame of the current frame as the background image. Moving objects are constantly moving and the system is constantly updated. The system preprocesses the image to determine whether it is a fire.

2.4 Target Image Preprocessing

The preprocessing of the target image is a preparatory work to extract the target. The purpose of pre-processing is to remove interference factors and make the target image clearer to obtain more accurate results. The entire process has the following steps:

(1) Calculate the average value of the three channels R, G, B, and then attach the average value to the three components. That is, the color image is converted into a grayscale image, as shown in Fig. 3. The purpose of the grayscale color image is to reduce the amount of system calculation. Image graying can minimize the amount of calculation when processing pictures.

(a) Original image of fire occurrence. (b) Fire gray image.

Fig. 3. Image before and after graying.

(2) Binary processing is performed on the grayscale picture in Fig. 3. Set a threshold, when the pixel value is greater than this value, it is marked as 1, which is black, otherwise it is 0, which is white. After binarizing the grayscale image, only two colors of black and white are left in the image. In this way, the outline of the image is clearer, which is convenient for further analysis of the outline of the figure [7]. Although the image has been binarized in grayscale, it has not changed the outline of the image and has pixels. The originality of the image is maintained, as shown in Fig. 4.

Fig. 4. Image after further binarization.

(3) Figure 5 is obtained after performing the morphological gradient operation in Fig. 4. The basic morphological gradient is the difference image obtained by subtracting the eroded image from the expanded image. When transmitting video images, there will always be noise interference, such as pepper and salt noise. The morphological operation of the image is to remove noise interference and obtain a more accurate image. The image after morphological operation will also remove the interference caused by some environmental factors, such as wind. After the operation, the color outline of the image is more clear and obvious, which reduces the amount of calculation for the channel separation performed later.

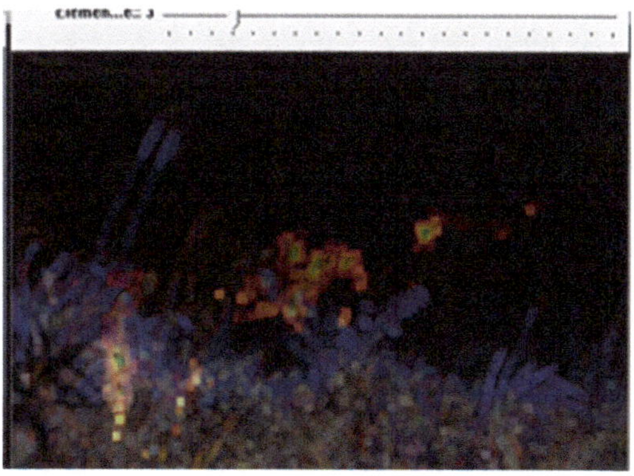

Fig. 5. Image after morphological manipulation.

3 Flame Tracking and Identification

Flame tracking and recognition methods include: color recognition, reliability model recognition, etc. [8]. The color recognition algorithm is to separate the three RGB channels and extract the R channel. RGB is the basic color that constitutes color. If the RGB criterion is used to extract the outline of the flame, the outlines of objects with similar colors will be recognized together and the result is not accurate. Therefore, it is necessary to convert RGB into a threshold image calculated by the HIS model to judge whether the detected image is a fire image. If it is a fire image, follow it.

3.1 RGB to HSI

For the flame, the red and green components will be larger than the blue component, as shown in formula (1). RGB color space is inseparable, but HIS can be separated. HIS is based on color hue (Hue), color saturation (Saturation or Chroma) and brightness (Intensity or Brightness) to describe the color [9]. Turn the RGB color space to HIS to calculate, the calculation formula is shown in (9). The main pixels of the flame are the chroma and saturation of the red component. The basic operation is to separate the three RGB channels first. According to the threshold value of formula (8), the R channel is obtained. Then calculate the value of S from formula (9). From this result, it is determined whether the position is a flame pixel. If it is a flame pixel, the point is displayed as white, if it is not a flame pixel, it is displayed as black. Since only the S component in HSI is used in (9), it is sufficient to directly calculate the S component without using the color model conversion function.

$$R > G > B \tag{7}$$

$$R > R_T \tag{8}$$

$$S \geq ((255 - R) * S_T / R_T \tag{9}$$

Among them, R_T is the threshold of R, which is verified by the color histogram. When the threshold of R is 49 and the threshold S_T of S is 7, the extracted flame profile is satisfactory.

3.2 Histogram Calculation of RGB

First, use the image pixel histogram to count the frequency of pixel values (0–255) of the entire image in the gray range, as shown in Fig. 6(a). Among them, the red line represents the frequency of R pixels. Green represents G pixels. Blue represents B pixels. The threshold size of the R channel can be estimated from Fig. 6(b).

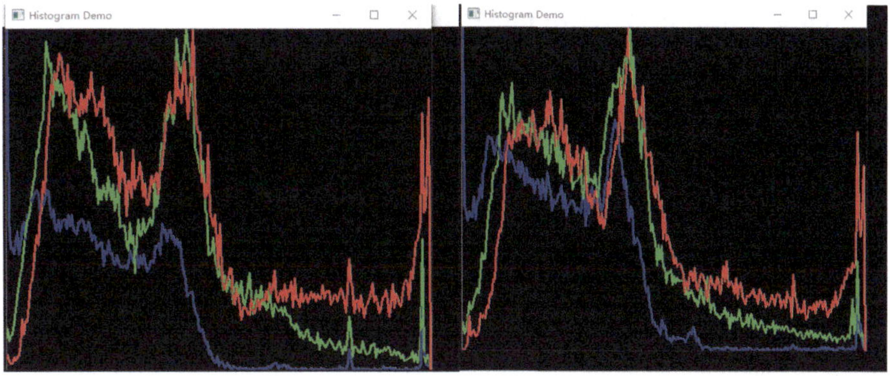

(a) Statistics of pixel values in the gray Range. (b) RGB pixel distribution map.

Fig. 6. Histogram calculation result graph.

3.3 Flame Tracking

When the color recognition algorithm recognizes the presence of flames in the video image, a positioning query algorithm that draws a rectangular frame is used. Perform pixel location tracking on the flame area in the video. Mark the fire zone with a rectangular frame. No matter what happens to the position of the flame, it can always track and locate. As shown in Fig. 7.

Fig. 7. The system locates the fire.

When the flame is tracked, the system starts to search and draw the outline of the fire. The system uses the method of searching contour and point topology to search the maximum edge of the flame. Then draw the outline of the largest edge. When the moving object recognized by the system is not a pyrotechnic image, the system returns to the motion foreground detection state.

4 Flame Contour Extraction Process

The system first converts the picture to a grayscale picture. The grayscale image can reduce the amount of calculation and increase the calculation speed [10]. Binarize the grayscale image. The threshold is estimated in the histogram with images. Pixels larger than the threshold range are marked with 1, which is white. Images smaller than the threshold are 0, that is, black. Then perform morphological gradient processing on the binary image. This makes the red component of the image more visible. It is convenient for the separation of RGB three channels to make the outline of the image stand out. Next, draw the extracted outline. Use the outline drawing API in opencv. When the code is running, locate and track the flame contour. Then draw the outline of the flame. Finally, an alarm is sent to the monitoring center to deal with the fire. The flow chart is shown in Fig. 8.

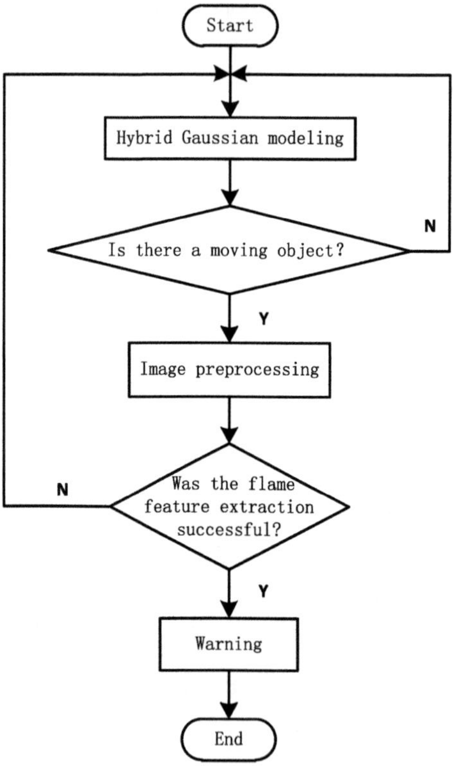

Fig. 8. System design flow chart.

Figure 9 is the flame extraction at the 50th frame in the video. It can be seen from the figure that the outline of the flame has been basically calibrated and extracted. Figure 10 is the flame shape extraction in the 53rd frame of the video. Flame pictures from 50 to 53 frames can be seen. When the frame keeps getting bigger, the system can track the movement of the flame.

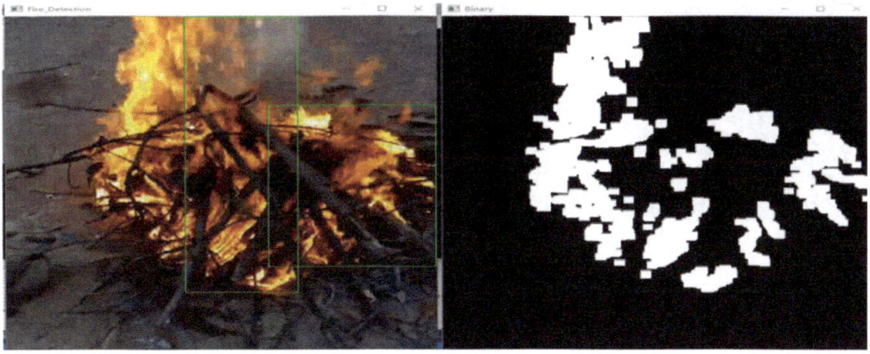

(a) 50th frame image in the video. (b) flame contour extraction map.

Fig. 9. Identify the 50th frame of wood burning image.

Figure 11 is the tracking identification when the burning material is hay. It can be seen from the picture that the flame outline can basically be completely recognized. When the system recognizes that the image in the video is a fire, the system sends an alarm to the supervision center. Then the system reminds the staff to deal with it accordingly. After repeated experiments, the fire recognition rate of the forest fire prevention and warning system based on machine vision can reach 94%. As shown in Table 1. It can be seen from the table that the system can successfully track and identify flames regardless of whether the burning material is hay or wood without changing the threshold. Because of some interference, the outline drawing area of the flame did not reach 100%. But the area of the main local part can be well drawn.

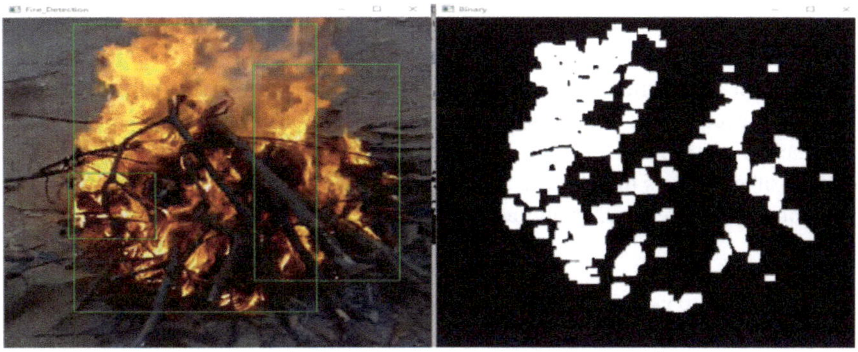

(a) The 53rd frame image in the video. (b) The flame outline drawn corresponding to the 53rd frame.

Fig. 10. Intercept frame 53 to identify the flame image when wood is burning.

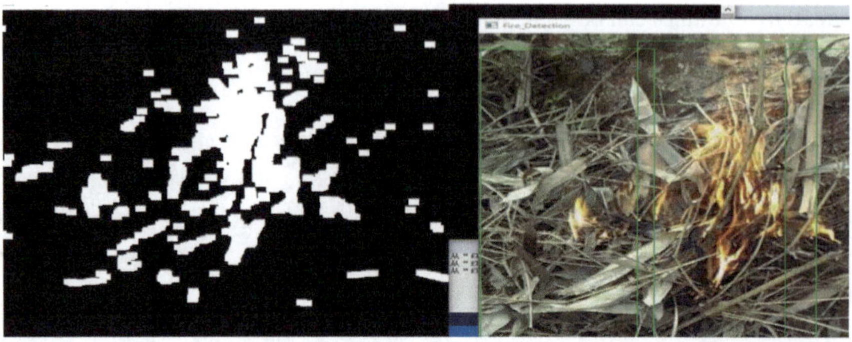

(a) Fire outline at frame 60 (b) Input image at frame 60

Fig. 11. Recognition image of hay burning.

Table 1. Four selected video flame recognition situation table.

Video sample	Combustion materials	Whether to track and recognize the flame in the video	Flame outline drawing area
Video one	Timber	Yes	95%
Video two	Timber	Yes	94%
Video three	Timber	Yes	94%
Video four	Hay	Yes	92%

5 Conclusion

This paper's machine vision forest fire prevention early warning system has made great progress. Using machine vision to detect the flame area not only greatly reduces the computational complexity, but also improves the accuracy of detection. However, real-time, robustness and scalability have not met the requirements of people's production and life. This article is to test the flame. The detection accuracy of the system needs to be improved. Because the detection object is only flame, it is difficult to meet the detection requirements of real life.

The forest fire warning system of machine vision can also use other schemes to improve the accuracy of the forest fire warning system of machine vision. Fire is often accompanied by smoke. Detecting smoke can effectively prevent fires. Analyze the fire source in other aspects to get more accurate results as possible. More methods can also be combined to more quickly and accurately identify the pre-disaster protection of the fire.

Acknowledgement. The authors are highly thankful to the Research Project for Young and Middle-aged Teachers in Guangxi Universities (ID: 2019KY0621), to the Natural Science Foundation of Guangxi Province (No. 2018GXNSFAA281164). This research was financially supported by the project of outstanding thousand young teachers' training in higher education institutions of Guangxi, Guangxi Colleges and Universities Key Laboratory Breeding Base of System Control and Information Processing.

References

1. Zhao, T.: Design and implementation of forest fire prevention platform. Fudan University (2012)
2. Huang, R.L.: Research on forest fire smoke recognition method based on video images. Beijing Forestry University (2012)
3. Chen, S.Y., Liu, S.: Based on the Realization of Computer Vision Technology, pp. 35–75. Science Press, Beijing (2008)
4. Bai, F., Zhang, L.P., Zhou, D.Y.: World forest fire prevention and monitoring technology overview. For. Labor Saf. **12**(3), 20–26 (2008)
5. Chen, X.J., Chai, X.D.: An infrared image segmentation method based on simplified PCNN. J. Anhui Univ. (Nat. Sci. Edn.) **34**(1), 76–77 (2010)
6. Guan, Y.D., Han, T.: Design of forest fire prevention system based on image recognition algorithm. Sci. Technol. Herald **26**(5), 77–80 (2008)
7. Huang, Z.H., Li, Z.M.: Image-based wireless remote monitoring system for forest fires. Sensors Microsyst. **28**(3), 86–88 (2009)
8. He, P., Li, C.L.: The design of thermal imaging forest fire alarm system and its image processing method. Microcomput. Inf. **22**(1–1), 98–99 (2006)
9. Jia, X.J., Zhang, Z.J.: Automatic segmentation method of structured light image based on simplified PCNN model. Optoelectron. Laser **22**(3), 457–458 (2011)
10. Jia, Z.C., Hu, T.X., Liu, C.J.: Preliminary research on forest fire monitoring technology based on machine vision. For. Mach. Woodwork. Equip. **36**(10), 21–23 (2008)

Design of an Intelligent Glass-Wiping Robot

Jiansheng Peng[1,2(✉)], Hemin Ye[2], Xin Lan[1], Qingjin Wei[1], and Qiwen He[1]

[1] School of Physics and Mechanical and Electronic Engineering, Hechi University, Yizhou, Hechi 546300, China
sheng120410@163.com

[2] School of Electronic Engineering, Guangxi Normal University, Guilin 541004, China

Abstract. In order to solve the heavy and dangerous problems of people's daily glass cleaning, an intelligent glass cleaning robot is designed using the Arduino MEGA 2560 control board combined with a vacuum compressor, a steering gear, a two-position two-way solenoid valve, a vacuum generator and a vacuum suction cup. The intelligent glass-wiping robot uses the Arduino programming software as the control center, and controls the running angle of 8 servos through the Arduino MEGA 2560 expansion board, realizing the gait of 4 servos bionic legs on the glass. Experiments have shown that during the cleaning process of the intelligent glass-wiping robot, the safety rope at the front end can effectively prevent the robot from falling. The intelligent glass-wiping robot uses four bionic legs of the steering gear to move, which can realize the basic functions of movement, adsorption, obstacle avoidance and cleaning.

Keywords: Glass-wiping robot · Vacuum suction · Arduino programming

1 Introduction

The intelligent glass-wiping robot is classified as a wall-climbing robot and is one of the extreme operating robots. It helps humans get rid of the danger of high-altitude cleaning operations. The living standards of our people are constantly improving, and the living environment has also been greatly improved. There are now more and more buildings decorated with glass curtains. Although the glass curtain is beautiful and the cleaning of the glass curtain is very troublesome. Therefore, it is of great significance to design an intelligent glass-wiping robot whose price is easily accepted by the public.

In the 1860s, Japan began research on robots and developed the world's first wall-climbing robot. Then Germany, South Korea, Britain and other countries have also joined the ranks of the research [1]. South Bank University is an early university in the UK studying wall robots. The Robug III wall mobile robot is a multi-foot vacuum adsorption wall mobile robot developed by South Bank University [2]. At present, the common foreign sales products on the market are German Bo Niu series window cleaning robots [3].

Chinese first window cleaner for exterior walls was designed and manufactured by the Institute of Construction Mechanization, China Academy of Building Research.

Q. Liu et al. (Eds.): CENet 2020, AISC 1274, pp. 364–375, 2021.
https://doi.org/10.1007/978-981-15-8462-6_42

Its appearance fills the gap in our country in this field [4]. At this stage, my country has also made great achievements in the field of window cleaning robots. Domestic home robot manufacturer Cobos has developed a number of “window treasure” series products [5]. According to the research results of scholars at home and abroad, the main moving methods of window cleaning robots include wheel type, foot type, frame type and crawler type [6].

On the basis of previous studies, the system design of the intelligent glass-wiping robot has continuously developed design analysis and simulation experiments, and finally formulated the design plan of the intelligent glass-wiping robot. The main research and design contents are as follows: (1) The design of the movement mode and gait of the glass-wiping robot, the selection of the overall hardware material and the design of the model, and the hardware construction of each module. (2) Commissioning when we use ultrasound. During the cleaning operation of the smart glass cleaning robot, the Arduino MEGA 2560 control board executes the steering obstacle avoidance program after receiving the ultrasonic signal. (3) We use Bluetooth for remote control. The robot may encounter dead corners during the cleaning process, which makes it impossible to clean some places. At this time, the user can remotely control the smart glass cleaning robot through Bluetooth to change the position and posture of the smart glass cleaning robot. (4) The time plan for the relay to control the power on and off of the solenoid valve. The solenoid valve power-on and power-off time determines the suction time of the suction cup, and then adjusts the movement time of the intelligent glass-wiping robot according to the suction time of the four suction cups.

2 Hardware System Design of Intelligent Glass-Cleaning Robot

2.1 System Composition of Intelligent Glass-Wiping Robot

Figure 1 is a framework diagram of the intelligent glass-wiping robot system. The hardware part of the intelligent glass-wiping robot is composed of six parts: a mobile module, an obstacle avoidance module, an adsorption module, a cleaning module, a communication module, and a safe battery life module.

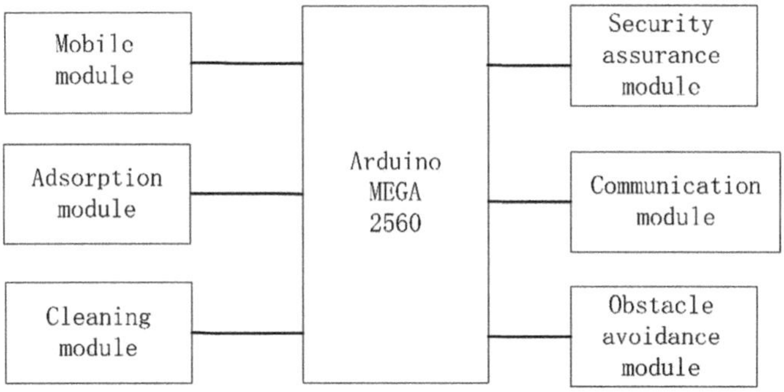

Fig. 1. Robot system framework.

2.2 Mobile Module Design

The mobile module of the intelligent glass-wiping robot is mainly composed of four bionic legs built by the steering gear and the servo head and the copper plate with a size of 300 mm × 200 mm × 1 mm. Each bionic leg is constructed by two types of MG996 steering gear, one U-shaped steering gear bracket, and two multi-functional steering gear brackets. Each bionic leg can achieve two degrees of freedom. The bottom servo controls the horizontal movement of the bionic leg, and the top servo controls the pitch movement of the bionic leg. The initial state of the intelligent glass-wiping robot is that the four legs are placed symmetrically on both sides of the robot. Figure 2 is a block diagram of the mobile module. The initial action is left front and right back, and the pitch motion is performed on the short U-shaped steering gear bracket of the top steering gear. The two bottom steering gears move horizontally and stop turning after reaching the preset angle. Subsequently, the short U-shaped steering gear bracket mounted on the top steering gear performs a pitching motion again, thereby driving the suction cup connected to the short U-shaped steering gear bracket to move downward. After the left front and right rear servos have completed the action instructions, the left rear and right front servos perform the same actions as the left front and right rear servos. Subsequently, the four bottom steering gears moved forward at a certain angle (each leg moved forward with respect to the robot, the same applies below), and returned to the initial state. The intelligent glass-wiping robot completed a cycle of movement.

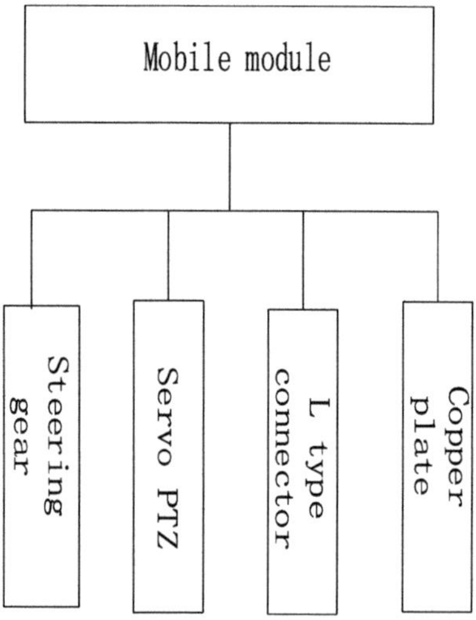

Fig. 2. Mobile block diagram.

The mobile module adopts MG996 type servo, the working voltage is 3.0–7.2 V, the working dead zone is 4 subtle, and the working torque is 13/cm. The control principle of the steering gear is to control by changing the pulse width (pwm) signal.

PWM takes 20 ms as a cycle. To control the change of the servo, it is necessary to adjust the ratio of high and low potential in a cycle.

In the mobile module, four L-shaped connectors are designed with the help of aluminum sheets. The L-shaped connectors are shown in Fig. 3. The horizontal side of the L-shaped connector is connected to the suction cup. The vertical side is fixed to the short u-shaped steering gear bracket by 4 screws with a diameter of 1 mm. Each L-shaped connector is 65 mm long and 3 mm wide, and 1 mm thick. The hole diameter of the L-shaped connector is 13 mm, and its hole diameter is made according to the size of the metal connecting rod of the suction cup, so as to better show the effect of the servo driving the suction cup to move.

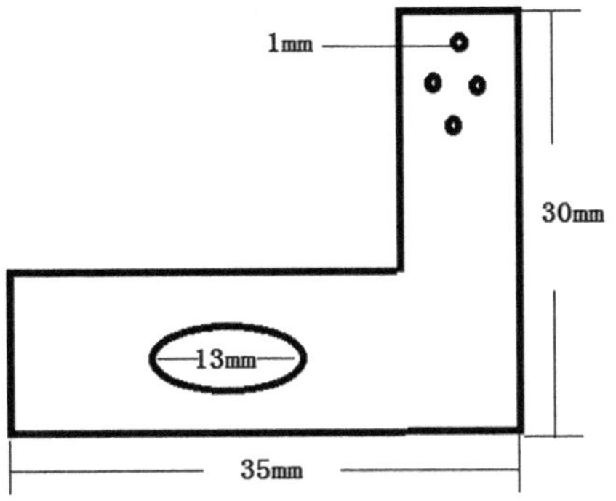

Fig. 3. L type connector.

2.3 Adsorption Module Design

So far, there are many adsorption methods for glass-wiping robots, including those that rely on magnetic force to adsorb glass, those that rely on thrust to adsorb on glass, and the use of vacuum generators to generate negative pressure adsorption [7]. The adsorption method selected by the intelligent glass-wiping robot is vacuum adsorption, which uses a vacuum generator to generate negative pressure and is adsorbed on the glass. The main equipments used in the suction module of the intelligent glass-wiping robot are air compressor, trachea, 5 V relay, 12 V solenoid valve, vacuum generator and vacuum suction cup. After the air flows out of the air compressor, it flows into the solenoid valve through the air pipe. When the solenoid valve A is energized, compressed air flows into the vacuum generator A from the air outlet of the solenoid valve A. When the solenoid valve B is energized, compressed air flows into the vacuum generator B from the outlet of the solenoid valve B. After the compressed air flows into the vacuum generator, a suction motion is generated around the outlet of the jet tube. As the air around the nozzle is gradually drawn away, the pressure in the adsorption chamber is lower than atmospheric pressure. The negative pressure generated by the vacuum generator is transmitted to the suction cup through the trachea, and there is a relative degree of vacuum between the air pressure in the suction cup and the atmospheric pressure. When a closed space is formed in the suction cup, the suction cup will be adsorbed on the glass. Figure 4 is a frame diagram of an adsorption module.

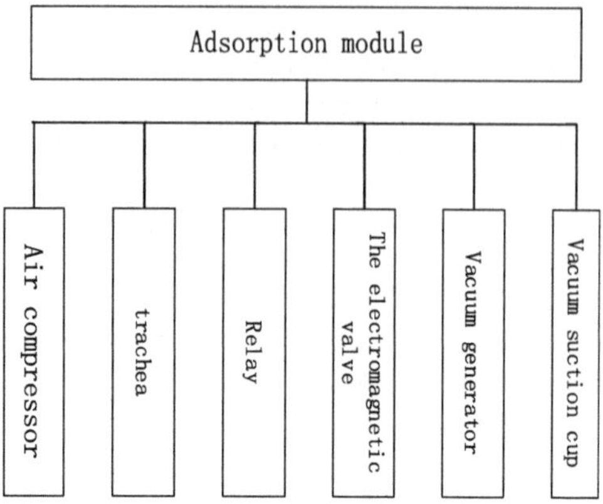

Fig. 4. Frame diagram of adsorption module.

The relay is an indispensable part of the adsorption module, which controls the state of the solenoid valve being energized and de-energized, and thus controls the adsorption and disengagement of four suction cups. The circuit diagram of the relay is shown in Fig. 5. The optocoupler 4N25 performs signal isolation and uses low voltage to control high voltage. Transistor NPN amplifies the current and amplifies the weak signal into an electrical signal with a larger amplitude. The fuse RL is a current protector. When the current exceeds the specified value, the circuit is disconnected. The relay uses an electromagnet to control the power on and off of the working circuit, and its working principle is electromagnetic effect.

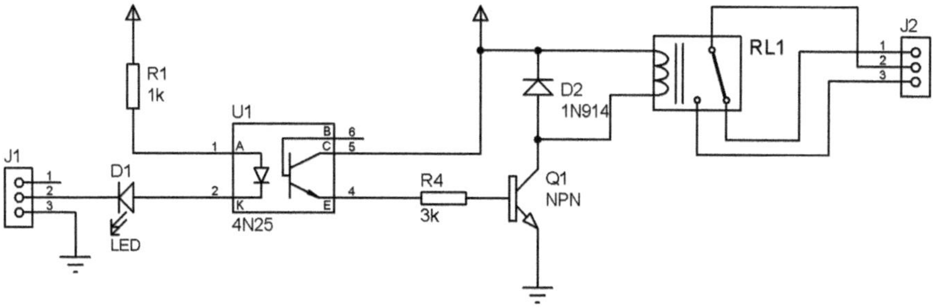

Fig. 5. Relay circuit diagram.

2.4 Design Analysis of Obstacle Avoidance Module

The obstacle avoidance module uses ultrasonic waves to automatically avoid obstacles. Figures 2, 3, 4, 5 and 6 is the ultrasonic wiring diagram used by the intelligent glass-wiping robot. Ultrasonic adopts model HC-SR04-P for distance measurement. When the distance detected by the ultrasonic wave is less than the set value, the Arduino MEGA 2560 control board will receive the signal, and then execute the steering procedure of the left and right turning of the servo, thereby achieving obstacle avoidance.

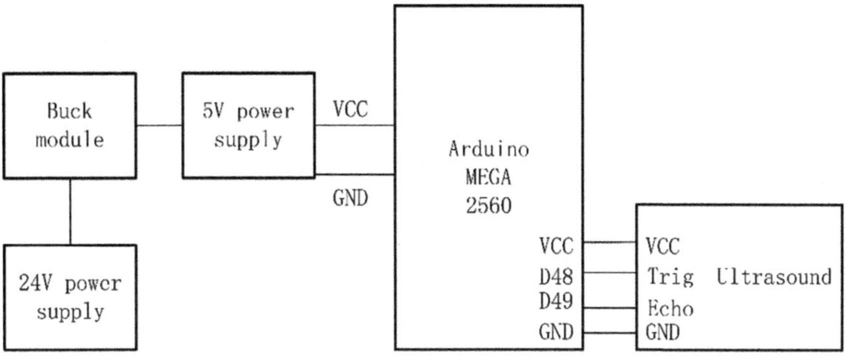

Fig. 6. Ultrasonic wiring diagram.

2.5 Communication Module Design

The communication module of the intelligent glass-wiping robot adopts Bluetooth remote control, and the Bluetooth serial communication uses HC-06. When the manual mode is selected, and the movement of the robot is controlled by the Bluetooth module. To manually control the robot, you need a mobile phone to download the Bluetooth APP. The intelligent glass-wiping robot uses a Bluetooth serial assistant APP to control the gait of the robot. After downloading, we customize the settings of the interface of the Bluetooth assistant. The customized commands include forward, backward, left turn, right turn and stop. The smart glass cleaning robot Fig. 7 is the wiring diagram of the Bluetooth module.

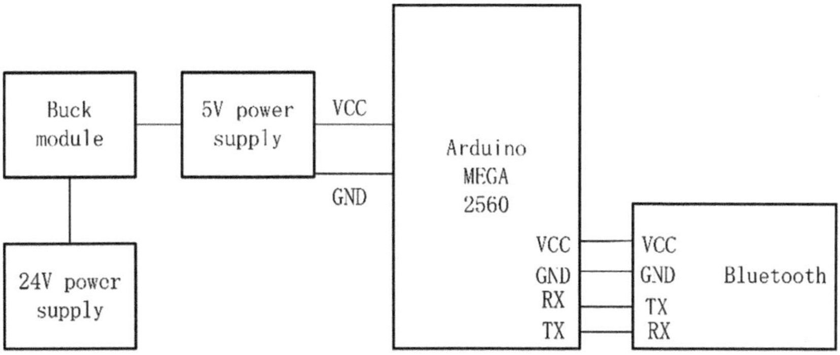

Fig. 7. Bluetooth module wiring diagram.

2.6 Clean Module Design

The 5 V relay used in the cleaning module controls the water spraying time by controlling the power on and off of the water pump. The cleaning device is composed of a micro water pump, a water tank, a nozzle, a water pipe and a cleaning cotton. The voltage of the miniature water pump is controlled by horizontal 5 V voltage. The water tank is modified from a sealed candy box. The nozzle adopts a 2 mm two-section, quick-insertion fine atomizing nozzle with a water spray aperture of 0.2 mm. The cleaning part of the bottom of the copper plate in the intelligent glass-wiping robot is composed of 9 nano sponge blocks with a size of 10 cm × 6 cm × 2 cm. Figure 8 shows the frame diagram of the robot cleaning module. The nozzle installed at the front of the robot is connected to the water pump in the water tank through a water pipe. When cleaning, the water pump sprays a certain amount of water onto the glass at regular intervals. The cleaning cotton installed under the copper plate of the robot continuously scrubs the glass during the movement of the robot. The 9 cleaning cotton has a large scrubbing area, which can improve the cleanliness of the intelligent glass cleaning robot.

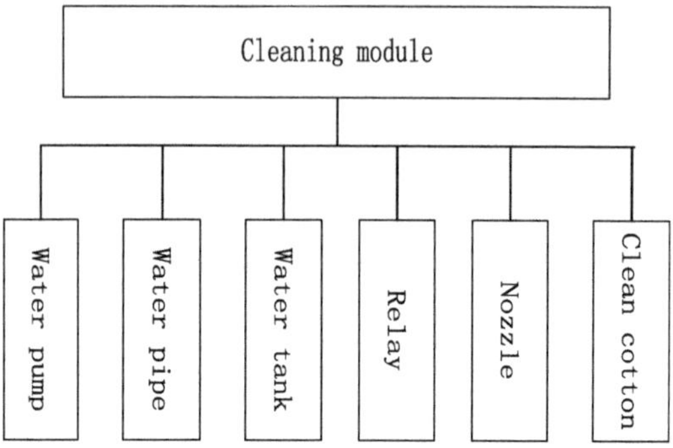

Fig. 8. Clean module frame.

3 Software System Design of Intelligent Glass-Wiping Robot

3.1 Design of the Main Program of the Intelligent Glass-Wiping Robot

Before the intelligent glass-wiping robot starts, it must first choose between automatic operation or manual operation. If the automatic operation mode is selected, and the intelligent glass-wiping robot is controlled by ultrasound in real time. When the ultrasonic wave detects that the distance between the robot and the window is less than 6 cm. The robot will execute the ultrasonic obstacle avoidance procedure, thereby changing its running direction and performing normal cleaning operations. When manual operation is selected, the smart glass-wiping robot is manually controlled by the Bluetooth module. After selecting the manual mode, the Bluetooth device initializes and detects whether the connection is incorrect. After the connection is correct, the robot is controlled on the Bluetooth serial port APP of the mobile phone to realize the normal cleaning operation of the robot. The main program flow chart of the intelligent glass-wiping robot is composed of 5 sub-program flow charts, which are the flow chart of the ultrasonic obstacle avoidance module, Bluetooth remote control, mobile module, adsorption module and cleaning module. Figure 9 is a flowchart of the main program of the intelligent glass-wiping robot.

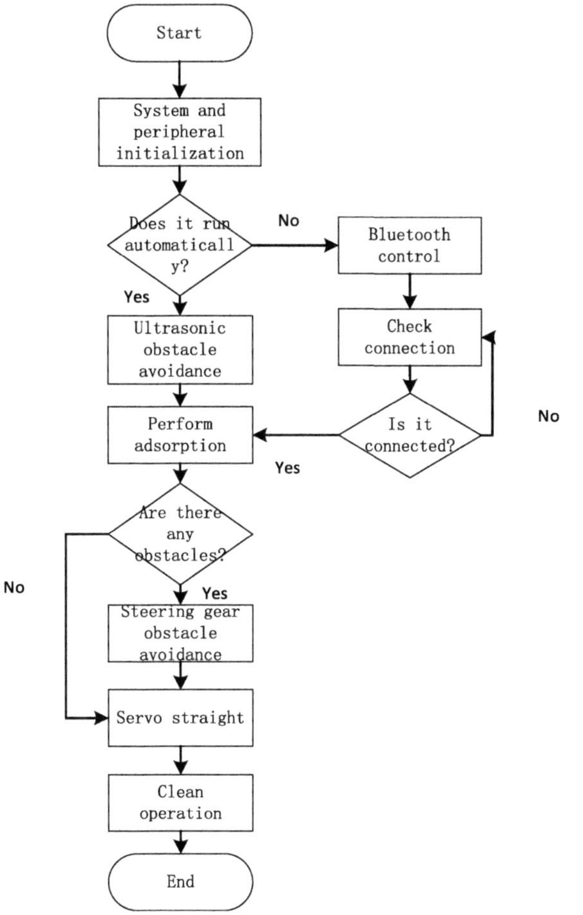

Fig. 9. Flow chart of main program of intelligent glass-wiping robot.

3.2 Subprogram Design of Intelligent Glass-Wiping Robot

Ultrasonic testing is the first subroutine of the intelligent glass-wiping robot. If the ultrasonic wave detects the window edge, and it sends a signal to the Arduino MEGA 2560 control board. The control board receives signals to control the operation of the steering gear, and the steering gear executes the obstacle avoidance procedure. Figure 10 is a flowchart of the ultrasonic detection window.

The second subroutine is the subroutine to move the servo in the mobile module. If the window is detected by the ultrasonic wave, and the servo executes the straight-through procedure. If the window is not detected by the ultrasound, and the left, right, forward, and backward commands are executed. Figure 11 is a flow chart executed by the steering gear.

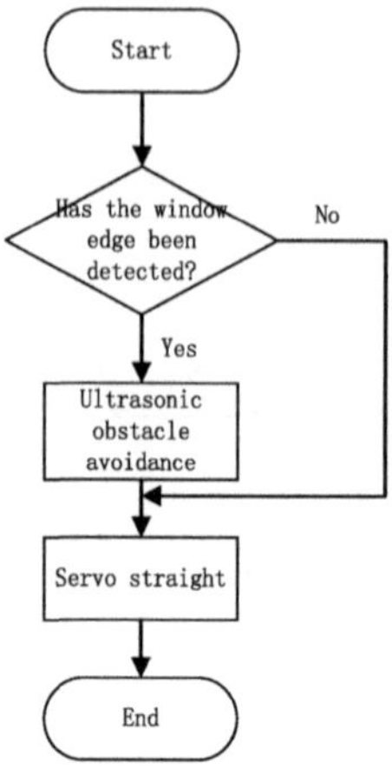

Fig. 10. Flow chart of ultrasonic detection window.

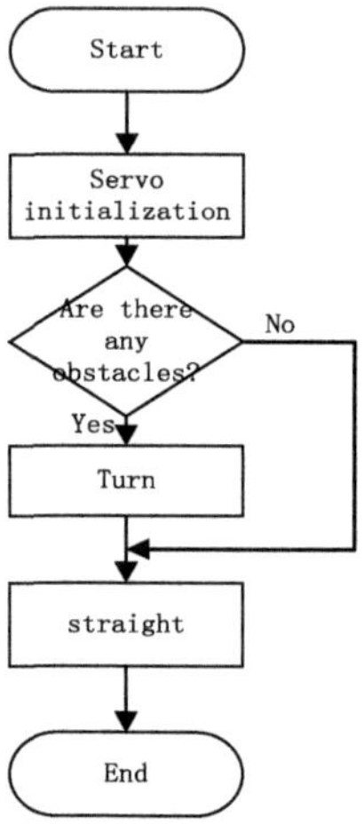

Fig. 11. Flow chart executed by the steering gear.

The third subroutine of the smart glass-wiping robot is the Bluetooth remote control program. We use Bluetooth wireless transmission for manual remote control of smart glass-wiping robots. If Bluetooth is connected. Then it sends forward, backward, left, and right commands to remotely control the steering gear to change the gait of the smart glass-wiping robot, so as to achieve the purpose of controlling the movement of the smart glass-wiping robot. Figure 12 is a flowchart of the Bluetooth program.

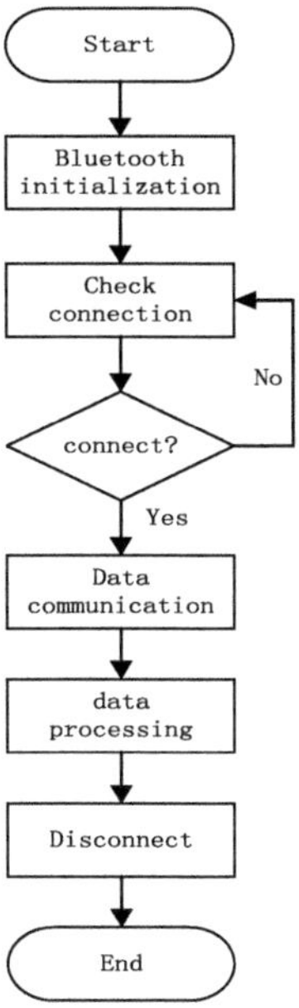

Fig. 12. Bluetooth program flow chart.

The fourth subroutine is the procedure of the intelligent glass-wiping robot adsorption module, as shown in Fig. 13. When the air compressor outputs compressed air, the air inlet is selected to the vacuum generator through the solenoid valve. The solenoid valve uses the high and low levels of the relay to control its power-on and power-off states. If the solenoid valve A is energized, the solenoid valve A port outputs compressed air. The compressed air flows through the No. 1 vacuum generator to produce negative pressure. The suction cups No. 1 and No. 4 form a closed space and generate an air pressure difference with the atmospheric pressure. The suction cups No. 1 and No. 4 are adsorbed on the glass. When the solenoid valve B is energized, the solenoid valve B port outputs compressed air. Compressed air flows through No. 2 vacuum generator to produce negative pressure. No. 2 and No. 3 suction cups form a closed space and the atmospheric pressure creates a pressure difference, No. 2 and No. 3 are adsorbed on the glass.

In the intelligent glass cleaning robot cleaning module, the relay is used to control the power supply of the water pump. Figure 14 is a flow chart of the cleaning module. When the relay is energized, the water pump draws water and sprays the water on the glass through the spray nozzle. If the pump stops pumping when the relay is de-energized, the water in the water tank cannot be ejected.

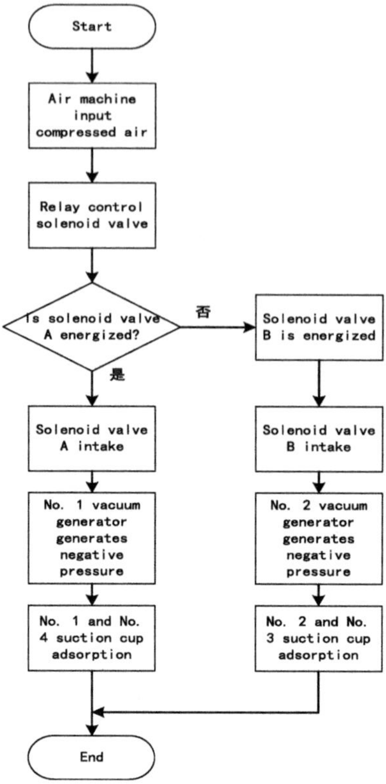

Fig. 13. Flow chart of adsorption module.

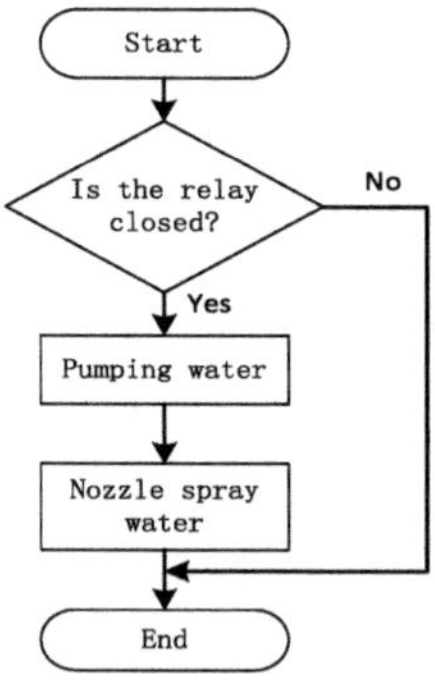

Fig. 14. Flow chart of the cleaning module.

4 Conclusion

The special plane movement of the intelligent glass-wiping robot needs to overcome the influence of gravity, friction and other factors. The main difficulty is the design of the gait of the robot on the glass. The intelligent glass-wiping robot has extremely high requirements on the height of the suction cup during the movement. When the servo drives the suction cup to fall, the suction cup must be close to the glass to achieve the effect of adsorption. Based on this, we have created an intelligent glass-wiping robot controlled by Arduino MEGA 2560.

After a lot of practice and debugging, the intelligent glass-wiping robot has realized that the four bionic legs in the mobile module of the intelligent glass-wiping robot can run smoothly. The intelligent glass-wiping robot can be safely adsorbed on the glass. When the power is insufficient, the safety protection module of the intelligent glass-wiping robot can issue an alarm. Its cleaning module can spray a small amount of water on the glass at regular intervals. Due to gravity and the weight of the smart glass-wiping robot, the robot cannot currently run smoothly on a 90-degree vertical glass plane. But it can run smoothly on a glass plane with an inclination angle of 60°. Therefore, the system looks forward to further improvement.

Acknowledgement. The authors are highly thankful to the Research Project for Young and Middle-aged Teachers in Guangxi Universities (ID: 2019KY0621), to the Natural Science Foundation of Guangxi Province (No. 2018GXNSFAA281164). This research was financially supported by the project of outstanding thousand young teachers' training in higher education institutions of Guangxi, Guangxi Colleges and Universities Key Laboratory Breeding Base of System Control and Information Processing.

References

1. Fu, Y.L., Li, Z.H.: Research progress of wall-climbing robots. Mech. Des. **04**, 1–5 (2008)
2. Chen, Z.: Research on wall crawling robot design and path planning. Shenyang University of Technology (2006)
3. He, J.H.: Design and performance analysis of virtual prototype of window cleaning robot. Northern Polytechnical University (2018)
4. Zhang, H., Xue, B.X., Xiao, X.F., He, L.L.: Development status and trend of window cleaners for high-rise buildings in my country. Archit. Sci. **05**, 31–33 (1998)
5. Xu, L.J.: Structural design and key technology research of window cleaning robot. Hebei University of Technology (2017)
6. Luo, X.: Structural design and reliability study of walking pneumatic wall-climbing robot. Xiangtan University (2016)
7. Wang, J.B., Li, L.R., Chen, Y.H., Chen, Y., Chen, G., Chen, J.J.: The research status and development of the adsorption method of wall mobile robots. Machinery **39**(01), 1–5 (2012)

Design and Implementation of ROS-Based Rapid Identification Robot System

Qingjin Wei[1], Jiansheng Peng[1,2(✉)], Hanxiao Zhang[2], Hongyi Mo[1], and Yong Qin[1]

[1] School of Physics and Mechanical and Electronic Engineering, Hechi University, Yizhou, Hechi 546300, China
sheng120410@163.com
[2] School of Electrical and Information Engineering, Guangxi University of Science and Technology, Liuzhou 545000, China

Abstract. Weeds in the corn field are extremely harmful to the growth of corn seedlings and should be controlled in time. The main domestic weeding method is still based on large-scale spraying of chemical agents. This method not only causes waste of resources, but also pollutes the environment. In order to accurately identify corn seedlings and weeds in the corn field, we should do selective spraying of chemicals. This article aims to study a ROS-based rapid identification robot system. First, this article introduces the design of the robot system. Then this article introduces the model generation subsystem. The system can identify the collected images and classify and locate the corn and weeds in the images. Next, this article introduces the communication subsystem. The communication subsystem realizes the remote control of the Spark robot, and can transfer the collected images to the model for detection. Moreover, this article introduces the experimental process of the system. Finally, the shortcomings of the design of the ROS-based seedling rapid recognition robot system are summarized and the research prospect is prospected.

Keywords: Seedling identification · ROS · Spark

1 Introduction

In the field of agricultural production, corn has always been my country's main food crop. It is very important to remove weeds efficiently and accurately in the early stages of corn seedling growth. So far, people have mostly used weed removal and chemical spraying to remove weeds [1]. Manual weeding has increased labor costs. At present, the spraying of chemical agents generally adopts extensive large area spraying, and spraying chemical agents for weeding has quick results. Weed distribution has the characteristics of uneven distribution. Therefore, such excessive spraying of chemical agents will not only cause environmental pollution, but also increase the cost of weed removal [2]. Therefore, selective spraying of chemicals, that is, variable spraying, is very important. This is now the popular intelligent weeding. The premise of intelligent weeding is the application of rapid grass identification technology, which can remove weeds around the crop without harming the crop. Intelligent weeding can reduce the

Q. Liu et al. (Eds.): CENet 2020, AISC 1274, pp. 376–387, 2021.
https://doi.org/10.1007/978-981-15-8462-6_43

release of pesticides, reduce environmental pollution, increase the level of intelligence, and improve operating efficiency.

In 2008, Tellaeche et al. proposed an automatic computer vision-based weed identification method through Bayesian decision theory to solve the detection and variable spraying of weeds in corn fields. This method is suitable for the use of herbicides when weeds and corn exhibit similar spectral characteristics and irregular weed distribution [3]. In 2009, Gonzalez Andujar developed an expert system for pest identification of olive crops. The system uses 9 kinds of weeds, 14 kinds of insects and 14 kinds of diseases as the database. According to the verification results, the system can be used as a recognition tool for farmers and technicians [4]. In 2011, Kishore C. Swian et al. proposed the concept of active shape model and developed an automatic active shape matching system. According to the morphology of weeds and plants, the model effectively identified nightshade plants with a recognition rate of approximately 90% [5]. In 2013, Montalvo et al. designed a new automated expert system for image segmentation in corn fields. Its main idea is to automatically and continuously perform image processing tasks by mapping expert knowledge. The expert system can solve the problem of artificial irrigation and natural rainfall making the green of weeds and crops polluted by soil materials in reality. The system can effectively identify weeds and crops with considerable effects [6]. In 2009, Wu Lanlan et al. studied the color image features of weeds, soil and corn, and proposed a new gray-scale image technology. The technology uses shape features as input vectors of support vector machines to achieve weed identification in corn fields. Under the selection of a suitable kernel function, the recognition rate can reach 98.3% [7]. In 2014, Wang Hongyan et al. optimized the kernel function of the support vector machine by screening the texture characteristics of the pretreated cornfield weed leaves. Then they sorted the weeds. Experiments show that when the weights of the polynomial kernel function (second order) and the radial basis function in the combined function are 0.8 and 0.2, respectively, the recognition rate is as high as 86% [8]. With the development and development of deep learning, target detection under the framework of deep learning has become a popular area in agricultural research.

The second chapter of this article introduces the subsystem of model generation. The subsystem generated by the model is used to identify the acquired images. Section 3 introduces the communication subsystem. The communication subsystem can realize the remote control of the Spark robot and complete the real-time detection of seedlings. Section 4 introduces the experimental process. Finally, it further summarizes the shortcomings of the existing methods and prospects.

2 Model Generation Subsystem Analysis

The design of the model generation subsystem is a prerequisite for rapid identification based on ROS seedlings. First, we put the data set of corn and weeds into the built network model. Next, the model refines its features by extracting the features of corn and weeds. Besides, the model classifies and locates the corn and weeds in the image. We use a lot of training to generate a model that recognizes the collected images.

2.1 Make a Dataset

In the field of image recognition, the quality of the data set is very important. In the field of target detection, having a high-quality data set is a prerequisite for target detection to be correctly identified [10]. First you need to make a data set. The flow chart of data set creation is shown in Fig. 1 below. This design uses a mobile phone high-definition camera to collect about 1,000 pictures containing corn seedlings and weeds in the corn field. There are about 500 pictures in each category. For the environment around corn seedlings and weeds, many situations including light intensity and the like are considered. Combined with the actual situation, the collection is carried out in the two time periods of morning and afternoon.

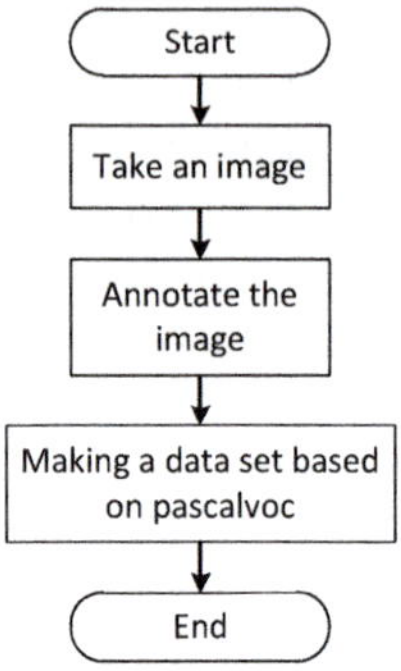

Fig. 1. Data set production process.

As we all know, most deep learning frameworks are experimented with the PASCAL VOC data set as the standard. In order to facilitate the training of the model, we made the collected pictures of corn and weeds according to the PASCAL VOC data set format. First we label the pictures of corn and weeds. This design divides the sample into two categories according to the actual situation. Among them, corn is a kind, weed is a kind. More than 1,000 pictures including the data set are manually completed using the labeling tool Labellmg. An example of the process of labeling corn seedlings and weeds is shown in Fig. 2.

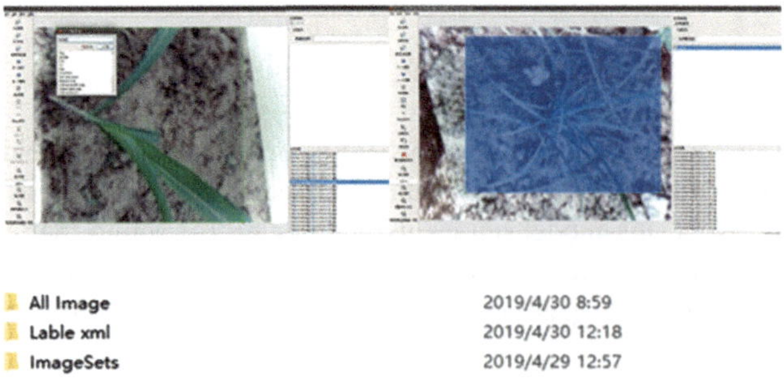

Fig. 2. An example of the process of marking corn seedlings and weeds.

2.2 Corn and Weed Target Detection Based on Optimized RetinaNet Depth Model

Basic Principles of RetinaNet Deep Model. The RetinaNet model is composed of three modules, namely Deep Residual Network (ResNet), Feature Pyramid Network (FPN), Classification Subnet and Positioning Subnet. Its network structure is shown in Fig. 3. Figure 3(a) ResNet shows that it is a deep residual network. Its role is to complete the initial extraction of image features. Figure 3(b) shows that FPN (Feature Pyramid Networks) is a feature pyramid network, and its function is to recombine the feature maps generated in Fig. 3(a). Next, the network completes the refined extraction of image features, which is convenient for better expression of image information. Figures 3(c) and (d) shows the sub-network for classification and positioning.

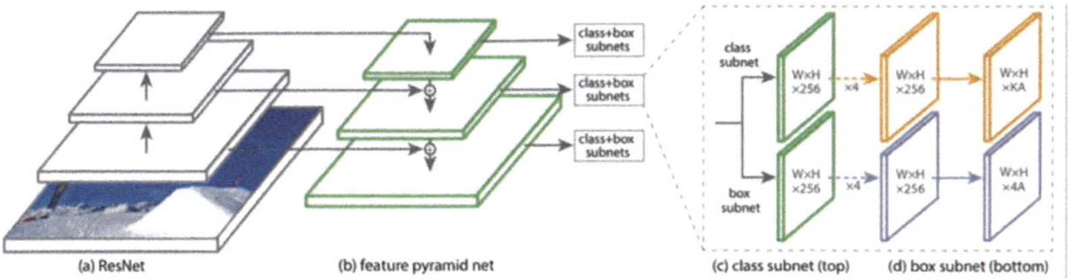

Fig. 3. RetinaNet network structure.

Feature Pyramid Network (FPN). The low-level features of image samples have less semantic information and the target position is accurate. High-level features have many semantic information and rough target positions. Therefore, someone proposed the FPN method. FPN merges the feature maps of different layers through bottom-up connections, top-down connections, and horizontal connections. It mainly solves the multi-scale problem in target detection. Through simple network connection changes, the model can greatly improve the performance of object detection.

From bottom to top is the CNN forward propagation process. In the forward process, ResNet has five feature extraction processes. The size of the feature map generated by it does not change, and it is named C1, C2, C3, C4, and C5, respectively. After the convolution kernel of C5 is 3×3 and the step size is 1, the first layer network structure P5 of FPN is obtained. The top-down process is performed by upsampling. The horizontal connection is to add the results of the upsampling and the feature maps of the same size generated from bottom to top. The convolution of each addition operation will use a 3 * 3 convolution test to eliminate the aliasing effect of upsampling. That is, the network performs a convolution kernel on C4 of 1×1, and the convolution operation with a step size of 1 adds the upsampling result of P5. The network performs a convolution kernel with a convolution kernel of 3×3 and a step size of 1 to obtain the second layer network structure P4 of FPN. By analogy, the network generates P3, P2, P1. The structure of FPN can effectively construct multi-scale feature maps from a single image, so that each layer of the pyramid can be used for target detection of different sizes.

RetinaNet Depth Model Optimization. In the field of deep learning, straight-line CNN is generally used to extract image features autonomously. If this kind of network wants to further improve the accuracy, it is necessary to increase the number of network layers as high as possible. This linear network will also enhance the ability to describe pictures. But at the same time, a lot of errors will also occur, resulting in a rapid decline in training accuracy and test accuracy. The emergence of ResNet changed the shortcomings of CNN. ResNet transmits part of the original input data directly to the next layer without matrix multiplication and nonlinear transformation. The network solves the problems of loss and loss existing in the traditional standard convolutional layer when transferring data. The network guarantees the integrity of information to a certain extent. At present, the highest layer of ResNet is 152 layers. After comprehensively considering the purpose and practicality of this design, the structure of building ResNet network with 50 layers is used to extract the features of the image. The convolutions of the three-layer networks 1 * 1, 3 * 3, and 1 * 1 are connected in series as a residual module, as shown in Fig. 4.

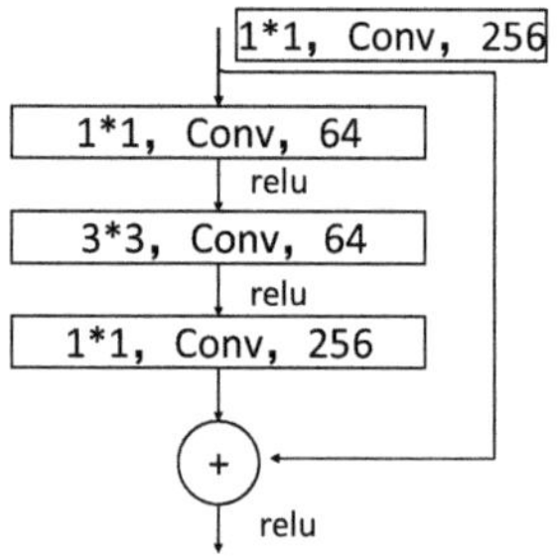

Fig. 4. Three-layer ResNet residual learning module.

The deep neural network learning RetinaNet model used in this design has the disadvantages of long training time and large model. The large size of the model makes it face the problems of insufficient memory and slow detection speed. The speed of corn and weed detection in this design should be as fast as possible. This design uses the idea of MobileNet to accelerate and compress ReNet50, which belongs to the first direction. MobileNet uses deep separable convolution. Its core idea is to decompose the original standard convolution operation into a deep convolution (Depthwise convolution) and a point convolution (Pointwise convolution). The network reduces the redundant expression of the convolution kernel, and significantly reduces the amount of calculation and the parameter of the model. This paper divides the standard 3×3 convolution operation in ResNet50 into a 3×3 depth convolution module and a 1×1

point convolution module. This paper builds the network through BN (Batch Normalization) and ReLU activation function. The network structure is shown in Fig. 5.

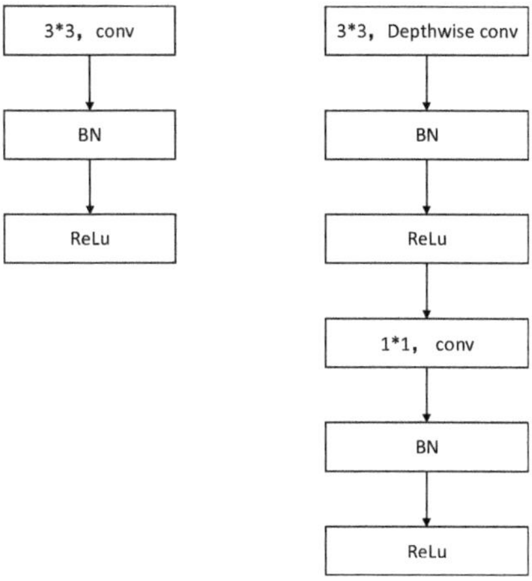

Fig. 5. MobileNet structure diagram.

2.3 Model Training

This design uses transfer learning to train corn and weed target detection models. The model changes the number of categories in the input_data layer and the number of outputs in the cls_score layer to 3, that is, corn, weeds, and background. The hyper-parameter lr of the model is initialized to le−5. The training layer number epoch is initialized to 50. The number of training steps per layer is 1000. The batch-size is initialized to 1. The whole training process iterated 50,000 times. The specific training steps are as follows: (1) The model uses the corn and weed data set, and specifies the proportion of the training data set and the test data set in the data set. (2) The model uses software Pycharm, language python3.5, and deep learning framework Keras to build the network. The network inputs the processed data set into the ResNet50 network structure joined with the MobileNet idea to obtain the five-layer feature map (feature_map) of the original picture. The first layer feature_map_64 and the second layer feature_map_256 learn low-level features such as edges. The third layer feature_map_512 learns complex features such as corn and weed network textures. The fourth layer feature_map_1024 learns the distinctive features such as corn and weeds. The fifth layer feature_map_2048 learns the key features. (3) The model uses the five-layer feature map obtained by ResNet50 to construct an FPN network. Select a feature map of any number of channels in the FPN network to generate candidate regions. (4) The model inputs the candidate regions generated in (3) to the classification network and position regression sub-network for training. Until the final network convergence, the construction of corn and weed detection model is completed.

3 Communication Subsystem

The design of the communication subsystem is to meet the following two functions: (1) The computer establishes communication with the Spark robot through the ROS platform, which can realize the remote control of the Spark robot. (2) The image collected by the upper camera during the movement of the Spark robot can be transferred to the model for detection.

3.1 Spark Robot Control

Spark robot is a teaching platform for ROS [9]. It has high flexibility and strong expansibility. We publish the topic through the computer, and the Spark robot subscribes to the topic to establish communication between the computer and the Spark robot. The computer issues a remote control node to the Spark robot. So we can use the keyboard to control the Spark robot. The keyboard control is shown in Table 1.

Table 1. Keyboard control.

Button	Spark robot
W	Go ahead
S	Back
A	Turn left
D	Turn right

3.2 Image Transmission

In order to ensure that the speed of image transmission to the computer can complete real-time detection, the camera is directly connected to the computer's USB through a serial line. We call the opencv library and open the camera connected to the USB serial port. The system transmits the image to the computer in real time and displays it.

4 Experiment

4.1 Model

In this paper, the corresponding experiments are carried out under the system where the computer configuration is Ubuntu16.04, CUDA8.0, cu DNN5.1, NVIDIA GTX1070TI. This design uses corn (maize) average accuracy (AP), weed (weed) average accuracy (AP), and MAP (average AP for each class) as the model's performance evaluation indicators. The experiment can be divided into the following two parts:

(1) Comparison of detection effects of different models on the data set of this design. First specify the training set as 700 pictures. In the picture, there are 350 corns and weeds respectively, which were put into the original RetinaNet model and the

optimized RetinaNet model for training. The model generated after training detects 300 pictures in the test set and finds the average detection speed. The experimental results are shown in Table 2. By comparing the experimental results of the original RetinaNet and this design method, it can be found that the average detection speed of the RetinaNet model optimized by this design is about 1.1 s higher than that of the original RetinaNet model. The MAP value of the model is only reduced by about 0.5% points. The RetinaNet model optimized by considering this design is an effective model for identifying corn and weeds, which basically meets the real-time requirements.

Table 2. Comparison of detection effects of different models on the data set of this design.

Model name	AP		MAP	Average detection speed (s)
	Maize	Weed		
RetinaNet	0.9368	0.9548	0.9458	2.252
Optimized Retinanet model	0.9303	0.9505	0.9404	1.122

(2) The influence of the proportion of the training set and the test set on optimizing the accuracy rate (MAP) of the RetinaNet model. The corn and weed data set produced by this design should keep the same format as the PASCAL VOC data set. An important step is to specify the proportion of the training and test sets in the corn and weed dataset. In this part of the experiment, we explore the effect of the proportion of the training set and the test set on the identification accuracy of corn and weeds. The purpose is to find the ratio of the training set and the test set in the data set when the accuracy of the corn and weed identification is the highest. This part of the experiment begins with the training set and the test set each accounting for 50% of the data set. Slowly increase the proportion of the training set to the data set, and decrease the proportion of the test set to the data set. Observe the resulting changes in the accuracy of corn and weed identification. The influence of the proportion of the training set and the test set on the accuracy of the optimized RetinaNet model is shown in Table 3.

Table 3. Influence of the proportion of training set and test set on the accuracy of RetinaNet model optimization.

Training set ratio (%)	Test set proportion (%)	Number of iterations	MAP
60	40	50000	0.9204
70	30	50000	0.9404
80	20	50000	0.9106
90	10	50000	0.8924

The experimental results show that with other conditions unchanged, using 70% of the training set and 30% of the data set, the model has the highest recognition accuracy for corn and weeds. Finally, this design visualizes the test results. The results of corn and weed identification in some test sets are shown in Fig. 6.

Fig. 6. Recognition results of corn and weeds in some tests.

4.2 Miaocao Rapid Identification Robot System Actual Debugging

Simulated Venue. The Miaocao rapid recognition robot system is composed of Spark robot, camera, computer and so on. In order to detect the success rate of the entire system identification, it is necessary to simulate the corn field for testing. Due to the limitation of conditions, we finally chose to spray-paint some pictures of the test set as posters, which were used as the test site. As shown in Fig. 7, there were 9 corn seedlings and 9 weeds on the test site. A total of 18 pictures are randomly distributed on the 1 m * 1 m field. The computer publishes the theme through the ROS platform. The Spark robot subscribes to this topic to complete the control of the Spark robot. Control the movement of the Spark robot on the simulated field, as shown in Fig. 8. The pictures collected by the camera on the Spark robot are transmitted to the model, thus completing the real-time recognition of corn and weeds.

Test Results and Analysis. When the Spark robot simulates the movement of the venue, it turns on the camera and transmits the collected pictures to the model for verification. The detection results are shown in Fig. 9. The result of the recognition is not satisfactory. Although the model can detect the location of corn and weeds, it will identify corn seedlings as weeds. The results did not meet the expectations of this design.

Fig. 7. Test site.

Fig. 8. Spark moves on a simulated field.

Fig. 9. Test results.

After investigation, it was found that the camera used to collect pictures has lower pixels, as shown in Fig. 10. In the process of making posters, the picture will be degraded, which is far from the original picture, as shown in Fig. 11. Through the comparison between Fig. 10 and Fig. 11, it can be seen that due to the accuracy of the camera and the site, the detection effect of the system during operation is very poor. The model can only identify weeds and not corn seedlings.

Fig. 10. Pictures taken by mobile phone.

Fig. 11. Pictures collected by the camera.

In order to solve the above problems, we produce enough data to allow the model to constantly modify itself to obtain satisfactory results. Therefore, we need to process and collect the pictures again: (1) First, the original data set is expanded by reducing the picture pixels on the image of the original data set, so that the training sample covers as many cases as possible. (2) Use the camera on the Spark robot platform to collect pictures of the test site, and re-mark the collected pictures.

5 Conclusion

This design introduces deep learning models into the detection of corn seedlings and weeds in complex farmland scenes. The design optimizes the commonly used first-order target detection model RetinaNet, and uses the RetinaNet network in deep learning to optimize the model. The recognition rate of the model is fast, which can meet the real-time requirements, and can realize the rapid recognition of corn and weeds. Spark robot is a teaching robot development platform based on ROS, which saves a lot of repetitive work. This design not only meets the speed and accuracy of the identification of corn and weeds, but also it has a high expansion compatibility.

There are many types of weeds. The color and shape of weeds change as weeds grow and light changes. Therefore, it is a further research goal to study the distinctive characteristics of different crops and weeds, establish a database of crop characteristics, and develop a recognition system specifically for crops and weeds. In addition to the research and optimization of algorithms and the construction of models, we can also improve the quality of hardware facilities to increase processing speed.

Acknowledgement. The authors are highly thankful to the Research Project for Young and Middle-aged Teachers in Guangxi Universities (ID: 2019KY0621), to the Natural Science Foundation of Guangxi Province (No. 2018GXNSFAA281164). This research was financially supported by the project of outstanding thousand young teachers' training in higher education institutions of Guangxi, Guangxi Colleges and Universities Key Laboratory Breeding Base of System Control and Information Processing.

References

1. Yan, B.: Application of machine vision technology in weed identification at corn seedling stage. Agric. Mech. Res. **40**(03), 212–216 (2018)
2. Li, X., Zhu, W., Ji, B.: Cotton field weeds based on feature optimization and LS-SVM. J. Agric. Mach. **41**(11), 168–172 (2010)
3. Ji, S., Wang, R., Chen, J., Zhao, X.: Research on the application of computer image processing technology to identify weeds in corn seedling field. Trans. Chin. Soc. Agric. Eng. **17**(02), 154–156 (2001)
4. Tong, X., Zhang, J.: Target detection of ROS platform based on SSD_MobileNet model. Comput. Syst. Appl. **28**(1), 94–99 (2019)
5. Howard, A.: MobileNets: Efficient Convolutional Neural Networks for Mobile Vision Applications (2017). https://blog.csdn.net/qq_31914683/article/details/79330343
6. Gong, L.: Crop image recognition based on ComVI and double threshold OTSU algorithm. J. Drain. Irrig. Mach. Eng. **32**(04), 363–368 (2014)
7. Li, B., Yi, J., Huang, F., Yan, P., Zhang, Y., Wu, J.: Research on license plate recognition algorithm based on TensorFlow deep learning. J. Chengdu Polytech. Univ. **21**(04), 28–33 (2018)
8. Ju, Y., Wang, X., Chen, X., Zheng, W.: Research on OMR recognition of convolutional neural network based on TensorFlow platform. Wirel. Internet Technol. **16**(04), 54–57 (2019)
9. Jia, H.: Design of robot target recognition system based on ROS platform. Electron. Test (02), 49–50+59 (2018)

Lane Recognition System for Machine Vision

Yong Qin[1], Jiansheng Peng[1,2(✉)], Hanxiao Zhang[2], and Jianhua Nong[1]

[1] School of Physics and Mechanical and Electronic Engineering, Hechi University, Yizhou, Hechi 546300, China
sheng120410@163.com

[2] School of Electrical and Information Engineering, Guangxi University of Science and Technology, Liuzhou 545000, China

Abstract. With the rapid increase in the number of cars, driving safety has been paid more and more attention. The lane line recognition algorithm has also become a hot research area. In order to realize the unmanned driving of smart cars, this article aims to study a lane recognition system for machine vision. First, the realization of the host computer is introduced. The host computer is convenient to control the trolley. The host computer can also change the model, making the car adapt to driving in different environments. Next, this article introduces lane recognition. In this paper, a fully convolutional network is used to semantically segment the image to obtain the feasible region of the road. Then this article introduces the communication and control part of the system. The emphasis is on the serial communication and control implementation of the system. The system is controlled by computer and Arduino is used as the control board of the car. In addition, this article introduced the experimental process. Finally, the shortcomings of the design and implementation of a machine vision lane recognition system in this paper are summarized and the research prospect is prospected.

Keywords: Machine vision · Lane recognition · Convolutional neural network · Unmanned driving

1 Introduction

With the rapid development of automotive technology and artificial intelligence, driverless driving has become more and more popular. People will be able to see intelligent driverless smart cars in reality in the near future. Unmanned intelligent vehicles, also known as unmanned ground vehicles (UGV). UGV is a typical outdoor robot [1]. In view of the current technological development, it is more advantageous to identify the feasible area or road sign of the road through the camera. A lane recognition system that studies machine vision can detect road information in real time. The system provides the basis for intelligent robot navigation. And the system is of great significance to the navigation of driverless smart cars [2].

In the 1960s, one of the earliest prototypes of autonomous driving, the "Stanford" [3], made great progress with the efforts of Hans Moravik. He uses remote images to control the movement of the "Stanford". But in the end it went directly on other roads

Q. Liu et al. (Eds.): CENet 2020, AISC 1274, pp. 388–398, 2021.
https://doi.org/10.1007/978-981-15-8462-6_44

without control. In recent years, unmanned driving technology has developed rapidly in various countries around the world. This benefited from the leapfrog development of advanced technologies such as electronic communication technology, intelligent control technology and intelligent computing technology [4]. At present, researches on the feasibility and practicability of developed countries such as Europe and the United States include military uses and urban environments [5]. Driverless cars have gradually matured from the initial concept research to the small batch production realized on the ground. Since the 21st century, countries have successively formulated plans for the development of autonomous driving, which shows that the future trend of the popularity of driverless cars is inevitable. The tremendous progress made in the field of machine learning in recent years has made autonomous vehicles have made great progress in environmental cognition. The ability of autonomous vehicles to perform unmanned driving in actual traffic environments is one of its main achievements [6]. The first unmanned vehicle highway test in my country developed by the Military Transportation Academy has been successfully tested [7]. The unmanned vehicle also uses a camera to recognize the road.

The second chapter of this paper introduces the host computer implementation of lane recognition system. The host computer is mainly for convenient control of the driving and speed of the car, and can also be adapted to driving in different environments. Section 3 introduces the system's lane recognition. Identify the target as a feasible area of the road. Section 4 introduces the communication and control part of the system. In this article, the machine vision lane recognition system uses a computer as the main control, and the Arduino is the control board of the car. Section 5 introduces the platform construction and experiment process of the system. Finally, it further summarizes the shortcomings of the existing methods and prospects.

2 The Realization of the Host Computer

The upper computer is part of the front desk. It is mainly for the convenience of controlling the running and speed adjustment of the car. The predicted model can also be directly replaced by the host computer to adapt to driving in different environments. The realization of the host computer is mainly divided into two modules, namely automatic driving module and manual driving module. Only one of the two modes can be selected at the same time. The upper computer uses Python language, which is realized through the Pyqt5 library. First, initialize the interface and serial port. Determine whether there is a serial port that can be opened, and then open the serial port to send the cart control check code. Before carrying out the control of the car, it is necessary to verify the information and complete the initialization of the car control. Finally, monitor whether the button is triggered by the slot function.

3 Lane Recognition

Lane recognition is an important part of the entire system. The system is mainly to identify the feasible area of the road. This design uses a fully convolutional network to semantically segment the image to obtain a feasible area of the road. The flow chart for lane recognition is shown in Fig. 1.

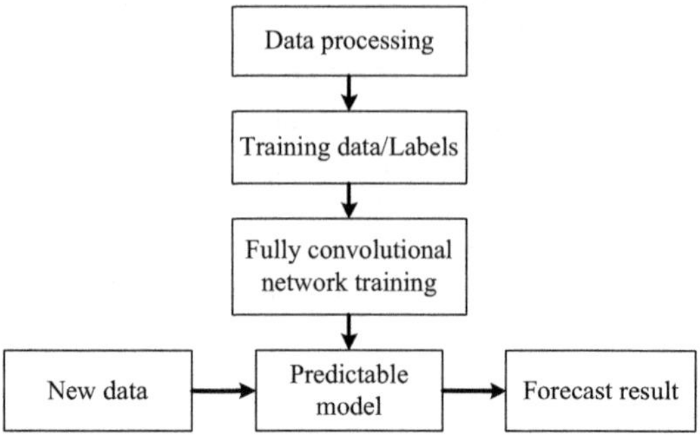

Fig. 1. Lane recognition flow chart.

The main process of lane recognition is to obtain road information picture data. Next we process the road picture data and label the data. We then make the training data to complete the production of data training data. Finally, a model can be used to predict the feasible area of the road through full convolutional network training. When there is input road data, the feasible area of the road can be predicted.

3.1 Data Processing

The data processing part includes data collection, data segmentation, data selection, data labeling, data dimensionality reduction, and data packaging. This part of data processing directly affects the accuracy of road prediction.

Data Collection. The main work of this part is to collect data and segment and select the video file data. The source of the data is divided into two parts. First get the road information through the camera and record the road information into video first. Then the system divides the video into pictures. Because the training data of the neural network requires a large number of labeled pictures, the information of the video should be divided into the information of the pictures. The process of dividing the video into pictures is to load the data frame of the video file into a variable through Python and OpenCV. Next, the system save the data of each frame as a picture to complete the video to picture processing process. In the process of video to picture processing, in order to maintain the diversity of data, we save a picture every 20 frames of data. Moreover, in the same road section, the data of cloudy and sunny days are taken and processed separately to ensure the diversification of data. For this part, the data picture obtained is $720 \times 1280 \times 3$ color picture. The other comes from the online dataset and uses the road dataset from KITTI. Collect the data set by the above two methods to complete the data collection.

Training Data Production. The production process of training data can be divided into labeling data, dimensionality reduction of image data and data packaging. The effect of making road data labels is shown in Fig. 2.

Fig. 2. Road label illustration.

The G-channel data is used for the road labels. The value of the feasible region of the G channel is set to 255, and the other background values are all set to 0. We use Photoshop to complete the label production. The G channel refers to a color image with three channels, which are RGB channels. Here we only use the G channel. But in practice, any channel can be used or all three channels can be used. But in order to speed up the training, only one channel can be selected as the label. After the labeling of the picture is completed, it cannot be directly used as training data. In order to save the calculation time, first compress the picture size data 720 × 1280 × 3 obtained by the camera to 80 × 160 × 3 size pictures. The label part is also compressed to the same size, and only the G channel data of the label data is obtained. The final training image size is 80 × 160 × 3 color images, and the label image size is 80 × 160 × 1. At this point, the training data is completed. After completing the above operations, although the data are all labeled, the training model only receives matrix-type data. Therefore, it is necessary to save the above-labeled picture data in a PKL file in the form of a matrix, so that it can be easily called when the model is trained. After the data collection and label production are completed, the amount of data is relatively small. Only more than 1,000 pieces of road picture data have been created. For convolutional neural networks, it is difficult to train a good model with a small amount of data. In order to increase the amount of data, we upsample the pictures on the original data set. After upsampling the data several times, the data is raised to more than 10,000 pieces of training data. But in this process, the diversity of samples should be maintained.

3.2 Model Optimization

The training model is built using Python language and Keras library. In order to get a better prediction effect, the final model is obtained after multiple adjustments. There are many models in the middle but the effect is not good. Therefore, in this article, only three cases are selected for analysis. They are under-fitting, over-fitting and optimal conditions.

Underfitting. In the experiment, start the experiment from a small number of hierarchical structures. At the beginning of the experiment of model selection, 3 convolution layers, 2 dropout layers, 2 pooling layers, 3 deconvolution layers, and 2 upsampling were used. The main function of the convolutional layer is to extract the features of the image. After sampling the convolutional layer and a series of operations, the pixel classification is completed. Before training, we must first choose the optimizer and loss function. The optimizer used in the model is the Adam optimizer. The loss change obtained through training is shown in Fig. 3.

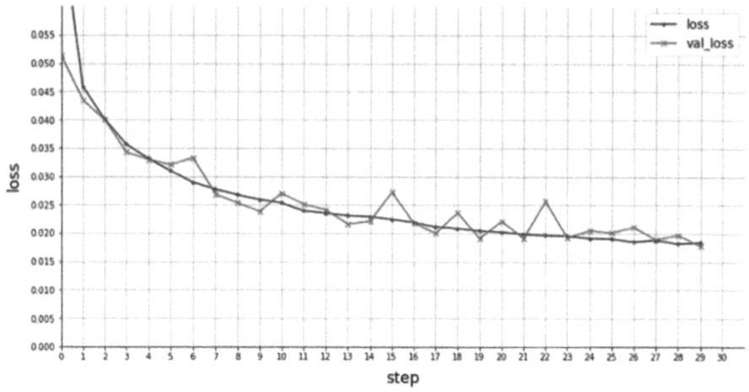

Fig. 3. Underfit model loss graph.

The training data is divided into training set and test set. There are two types of training set: validation set and training set. Loss represents the loss of the verification set, and val_loss represents the loss of the test set. It can be seen from the changes in the above graph that after 30 trainings, the minimum loss loss is about 0.018.

Overfitting. This paper uses 12-layer convolutional layer, 20-layer Dropout layer, 3-layer pooling layer, 12-layer deconvolution layer, 3-layer upsampling. The loss change obtained through training is shown in Fig. 4. After 30 rounds of iterative training, the total loss is about 0.006.

Final Model. In this process, it is mainly to increase neurons or increase the level of the network. After many adjustments, the difference from the original model is that the number of neurons is increased in some layers, or the convolutional layer of the network is increased. The final model structure is shown in Fig. 5. There are a total of 8 convolutional layers, 3 pooling layers, 12 dropouts, 8 deconvolution layers, and 3 upsampling layers. The construction of the model is realized using the Kears library.

The loss function uses the mean square error loss. The optimizer it uses is also the Adam optimizer. The loss changes obtained through training are shown in Fig. 6.

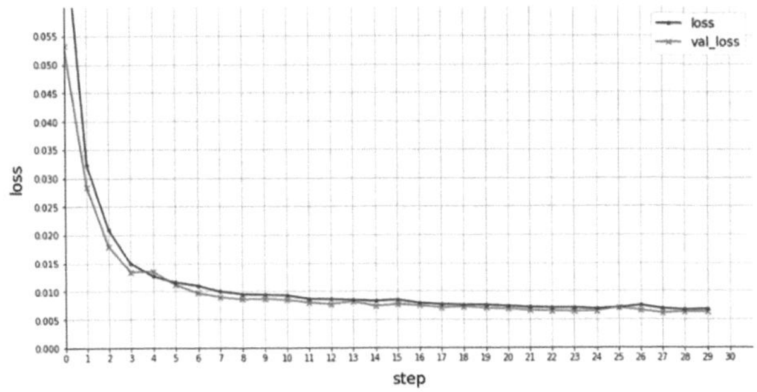

Fig. 4. Overfit model loss graph.

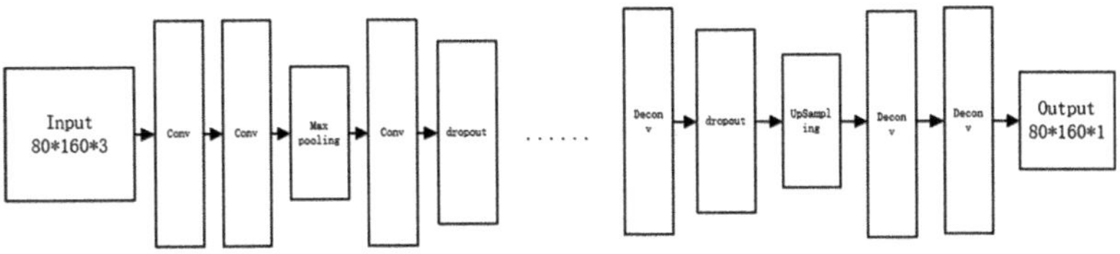

Fig. 5. Model structure diagram.

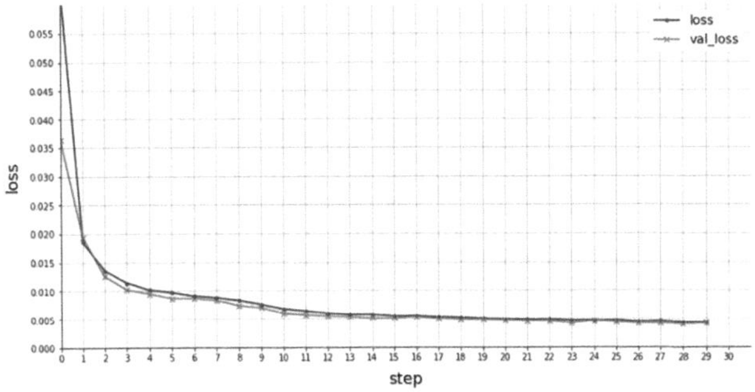

Fig. 6. Final model loss graph.

It can be seen from Fig. 6 that when the model is trained to the 10th time, the loss of the test set has fallen below 0.01. After 30 rounds of training, the loss fell below 0.005. Compared with the previous model, the loss has decreased. By comparison, this model is finally selected as the final model. Through the above training, a model that can be used for road feasible area prediction is finally obtained.

4 Communication and Control

4.1 Serial Communication

The computer is the main control, and the Arduino is the control board of the car. In order to realize the control of the car by the computer, the text adopts the way of serial port transmission. After the computer recognizes the lane, it sends data to the control board of the car through the serial port to control the car to execute. The realization process of the communication between the computer and the car control board is shown in Fig. 7.

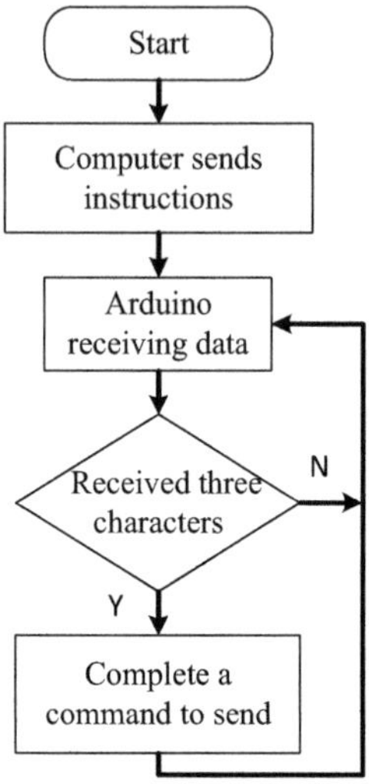

Fig. 7. Serial port sending command diagram.

The serial port of the computer is sent through the Python language. In order to better control the speed and convert the sent characters into numbers, this system uses 3 characters as an instruction. The three-character number selection range is 000 to 999. When the car control panel receives three characters, it represents an instruction. Then we parse the characters and convert them to numbers. Finally, the analysis of data transmission and reception is completed.

4.2 Control Realization

The image processing part is realized on the computer. The control part of the car mainly includes receiving status commands sent by the computer. Operate forward or left after receiving the command to control the development complexity and hardware cost. In the control part of the car, the Arduino development board was selected as the control board of the car. The process of the control board receiving instructions and executing is shown in Fig. 8 below.

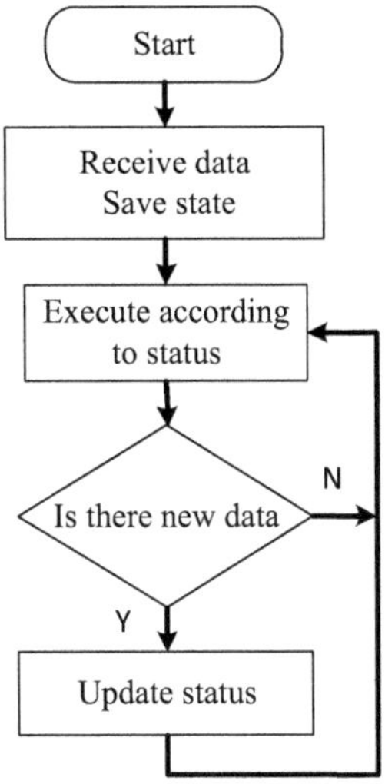

Fig. 8. Car control flow chart.

The car receives the status command sent by the computer. The received command range can be divided into four states, namely forward, left, right and back. After receiving the data sent by the serial port, control the car to drive. The speed in this system is to control the pulse width of PWM. The control part of the car is mainly to identify the road first. As shown in Fig. 9, the predicted feasible region is divided into three states. When the predicted feasible region is as shown in Fig. 9(a), it can be judged as the forward state. When the predicted feasible area is shown in Fig. 9(b), it can be judged that the predicted feasible area is on the left, and the car needs to turn left at this time. Similarly, when the predicted feasible region is shown in Fig. 9(c), a right turn is required. However, there is a range in the state of going straight, and it is necessary to turn only after exceeding this range.

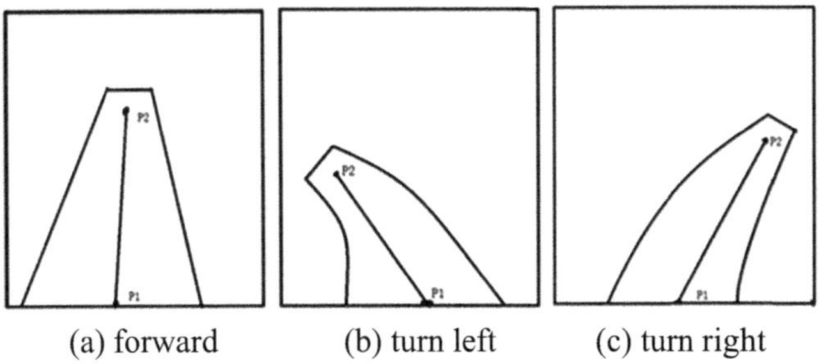

Fig. 9. Feasible area map.

5 Experiment and Analysis

5.1 Platform Construction

The car uses Neptune 4WD DZ independent all-wheel drive wheeled robot. The camera chooses SJ7000. The complete hardware platform is shown in Fig. 10. The hardware consists of car, computer, camera and Arduino development board.

Fig. 10. Feasible area map.

5.2 Experiment

This design is mainly to study the machine vision to identify the feasible area of the lane, and let the car drive along the feasible area. The main content is two parts, the front and back. The front desk is mainly the realization of the host computer. The background is mainly road identification and car control. The framework of the entire system is shown in Fig. 11.

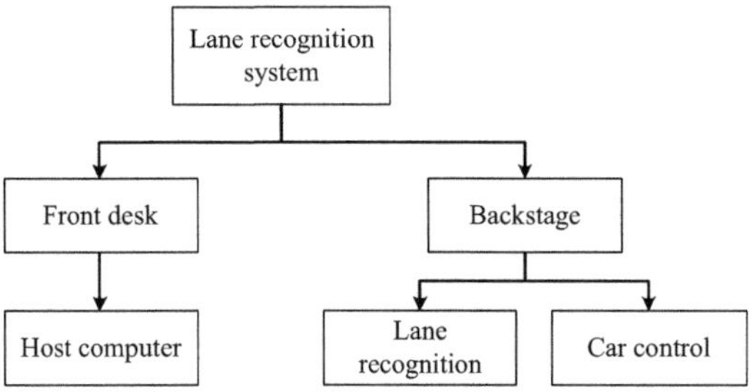

Fig. 11. System framework diagram.

The experiment of the host computer is mainly divided into automatic mode test and manual mode test. When it is successfully opened, a prompt to complete the initialization will be displayed in the text box. After the host computer test, make feasible area prediction. After the data acquired through the camera, first zoom the acquired image data to an image of $80 \times 160 \times 3$ size. Two road sections under different road conditions are selected for prediction. One is the road with a clear edge, as shown in Fig. 12. The other is the prediction of unclear edge roads, as shown in Fig. 13.

Fig. 12. Clear edge road prediction map.

Fig. 13. Unclear edge road prediction map.

It can be seen from the above predictions that the prediction of roads with clear edges is better. However, the effect of the system on road predictions with unclear edges is poor. Next, carry out the car control experiment. First of all, the data of the track and field are collected to make training data. Then we train on the lane data. After obtaining the model through training, test the recognition of the car on the runway. Figure 14 shows the car driving test under sunny conditions.

Fig. 14. Car driving test chart.

The effect of lane recognition depends largely on the part of the data. When the training data in the model is the same as the driving environment of the car, the feasible area predicted by the car is more accurate. When the training data in the model is different from the driving environment of the car, the prediction effect is not good.

6 Conclusion

Traditional lane detection has the problems of complex recognition algorithm, slow processing speed and lack of robustness. The machine vision lane recognition system proposed in this paper has made great progress on these issues. The pre-stage system obtains video data through the camera. The system divides the video data into picture data at certain intervals and tags the picture data of these roads. The system identifies the feasible area of the lane through the fully convolutional neural network, and allows the car to move along the lane. However, in the case of unclear road edges and interference with the road surface, the accuracy of this method of lane line recognition needs to be improved. This system uses only one camera sensor. In the follow-up, more sensors need to be added to assist in the identification of the road, the steering of the car, and the convenience of obtaining more information around.

Acknowledgement. The authors are highly thankful to the Research Project for Young and Middle-aged Teachers in Guangxi Universities (ID: 2019KY0621), to the Natural Science Foundation of Guangxi Province (NO. 2018GXNSFAA281164), This research was financially supported by the project of outstanding thousand young teachers' training in higher education institutions of Guangxi, Guangxi Colleges and Universities Key Laboratory Breeding Base of System Control and Information Processing.

References

1. Liu, S.: The First Self-Driving Technical Book. Electronic Industry Press, Beijing (2017)
2. Su, B.: Advanced technology and development of driverless cars. Internal Combust. Engine Parts (09), 236–238 (2020)
3. Maurer, H.: The standford cart and the CMU rover. Process. IEEE **71**(7), 872–884 (1983)
4. Chen, H., Xiong, G., Gong, J., Jiang, Y.: Introduction to Driverless Cars. Beijing Institute of Technology Press, Beijing (2014)
5. Qiao, W., Xu, X.: Development status and direction of driverless cars. Shanghai Autom. **33**(07), 40–43 (2007)
6. Li, D.: Autonomous driving grade division and technical route analysis. Autom. Pract. Technol. **06**, 22–25 (2019)
7. Duanmu, Q., Ruan, J., Ma, J.: Advanced technology and development of driverless cars. Agric. Equip. Veh. Eng. **52**(03), 30–33 (2014)

Review on Deep Learning in Intelligent Transportation Systems

Yiwei Liu, Yizhuo Zhang, and Chi-Hua Chen(✉)

Fuzhou University, Fuzhou 350100, China
chihua0826@gmail.com

Abstract. Intelligent transportation system (ITS) contributes to allocate transportation resources, from citywide ones to nationwide, more efficiently with the help of algorithms. Due to the fact that the ITS is a quite comprehensive field, it is necessary for researchers to have a better understanding of dominant methods and which one is proper for the targeted subjects of ITS. More importantly, with a knowledge of the remained challenges in developing ITS and relevant techniques, researchers may have a clearer direction to work on. To provide researchers with dedicated information on specific machine learning (ML) techniques used in object recognition and traffic prediction, two essential study subjects in ITS, this paper mainly focuses on deep learning and neural network (NN), one of widely-used ML algorithms, and aims to conduct a brief review on its recent applications in ITS, as well as to mine its potential usage. As a result, this review introduces some popular NN, convolutional neural network (CNN), long short-term memory (LSTM) network, gated recurrent unit (GRU) network, and their hybrid mechanism, first. Then their applications and performance in ITS are described. Finally, this paper discusses constraints on some of them and suggests some promising research directions.

Keywords: Deep learning · Convolutional neural network · Long Short-Term memory · Gated recurrent unit · Intelligent transportation system

1 Introduction

As the traffic demand boosts with the increasing urbanization level, constructing an intelligent transportation system (ITS) is a necessity, not only would it bring more convenience to people, but also it makes the management of citywide or nationwide transportation more efficient. Machine learning (ML), progressed rapidly on the basis of artificial intelligence (AI) in the last decades, has achieved a lot in classification tasks and regression problems. Since there is also a strong need of recognizing objects and estimating traffic in ITS, such as vehicle detection in autonomous driving system or travel time prediction in routing planning, many ML techniques have been adopted in various subjects of ITS consequently, for example, k-means in speed prediction [1] and Bayes Network in car detection [2]. However, most of these traditional ML techniques failed to learn the complex features in a large-scaled transportation system. Hence, recent mainstream researches pay attention to the usage of neural networks (NNs) in ITS, since NN performs well in capturing features and obtains higher accuracy

Q. Liu et al. (Eds.): CENet 2020, AISC 1274, pp. 399–408, 2021.
https://doi.org/10.1007/978-981-15-8462-6_45

compared to those ML algorithms. Considered these facts, this study will introduce three popular NN structures, which are convolutional neural network (CNN), long short-term memory (LSTM) network, and gated recurrent unit (GRU) network. The ITS applications based on these NNs will presented and discussed in this study.

The structure of this study is organized as follows. To make these methodologies better understood, some theoretical background will be described in Sect. 2. The detailed usage of each NN is presented in Sect. 3. The conclusions of this study are presented in Sect. 4.

2 Theoretical Background

This section presents three neural network models which include CNN, LSTM network, and GRU network.

2.1 CNN

CNN consists of an input layer, convolutional layers, pooling layers, and an output layer, all of which are shown in Fig. 1. The core part of CNN is the filters in the convolutional layer, which are capable of extracting features from a specific input matrix. If multiple convolutional layers are implemented in CNN, multi-dimensional features could be mined accordingly. Besides, the pooling layer here helps to decrease computation by reducing the dimension of outputs from convolutional layers. Usually, a softmax activation function is used in the last layer to generate the final results of classification [3].

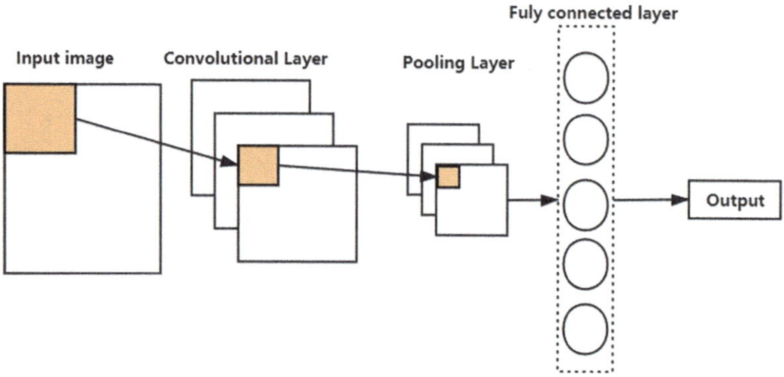

Fig. 1. The structure of CNN [3].

2.2 LSTM Network

Long short-term memory (LSTM) has proven to alleviate the problems of exploding and vanishing gradient when input with a long sequence. This feature comes from the complex structure of LSTM unit in Fig. 2. At time step *t*, input gate and forget gate control the maintenance of information in memory cell c_t together. Input $\mathbf{x}_t$, concatenated with hidden state $\mathbf{h}_{t\text{-}1}$ from last time step, generates gates i_t, f_t, and o_t, respectively [4]. The complete formulation for each variable is described in Eqs. (1)–(6) where • means Hadamard product and stands for sigmoid activation function.

$$i_t = \sigma\left(\mathbf{W}_i[\mathbf{h}_{t-1}, \mathbf{x}_t]^T + \mathbf{b}_i\right). \tag{1}$$

$$f_t = \sigma\left(\mathbf{W}_f[\mathbf{h}_{t-1}, \mathbf{x}_t]^T + \mathbf{b}_f\right). \tag{2}$$

$$o_t = \sigma\left(\mathbf{W}_o[\mathbf{h}_{t-1}, \mathbf{x}_t]^T + \mathbf{b}_o\right). \tag{3}$$

$$c_t' = \tanh\left(\mathbf{W}_c[\mathbf{h}_{t-1}, \mathbf{x}_t]^T + \mathbf{b}_c\right). \tag{4}$$

$$c_t = f_t \bullet c_{t-1} + i_t \bullet c_t'. \tag{5}$$

$$h_t = o_t \bullet \tanh(c_t). \tag{6}$$

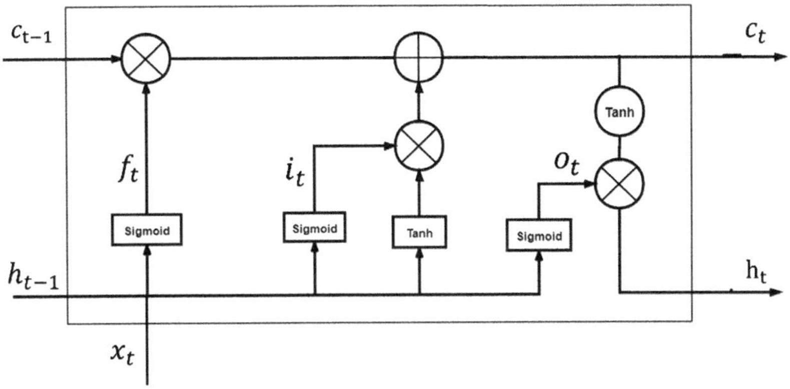

Fig. 2. The structure of LSTM unit [4].

2.3 GRU Network

Gated recurrent unit (GRU) could be regarded as a simplified LSTM unit in Fig. 3 [5]. It only has two gates, reset gate and update gate, and remove the memory cell. The formulized definitions are described in Eqs. (7)–(10). The update gate, z_t, is responsible to decide how much information from last time step $t-1$, could be retained. The bigger value z_t holds, the more information at $t-1$ be reserved.

$$r_t = \sigma\left(\mathbf{W}_r[\mathbf{h}_{t-1}, \mathbf{x}_t]^T\right). \tag{7}$$

$$z_t = \sigma\left(\mathbf{W}_z[\mathbf{h}_{t-1}, \mathbf{x}_t]^T\right). \tag{8}$$

$$h_t' = \tanh\left(\mathbf{W}_h[\mathbf{h}_{t-1} \bullet r_t, \mathbf{x}_t]^T\right). \tag{9}$$

$$h_t = z_t \bullet h_{t-1} + (1 - z_t) \bullet h_t'. \tag{10}$$

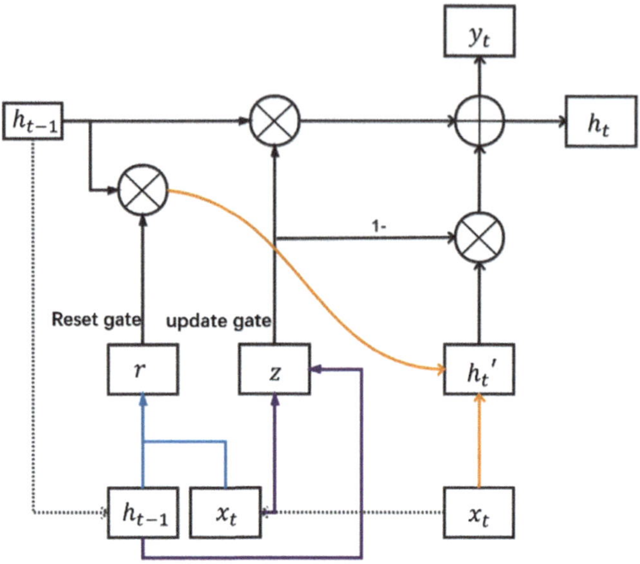

Fig. 3. The structure of GRU unit [5].

2.4 Summary

CNN utilizes convolutional filters in layers to capture features. LSTM network and its variation GRU are capable of storing information of recent inputs because of theirs well-designed cells, which evolved from recurrent neural network (RNN), an original model on short-term memory [6].

3 Deep Learning in Intelligent Transportation Systems

Three topics which include (1) CNNs in ITSs, (2) LSTM networks and GRU networks in ITSs, and (3) hybrid of CNNs and LSTM networks in ITSs are described in the following subsections.

Articles reviewed in this study come from online academic searching engines (e.g., Baidu Xueshu and Google Scholar) and an academic database (i.e., IEEE Xplore), with relevant keywords, such as CNN, LSTM, transportation, vehicles, etc. Most of the discussed papers hold a certain amount of citations and thus their experimental outcomes are quite convincing.

3.1 CNNs in ITSs

CNN has notable achievements in object classification and recognition from images and videos [7, 8]. Hence, many studies employed CNN in this field. Firstly, traffic signs recognition techniques have developed rapidly with the help of CNN. In fact, researchers modified part of traditional CNN layers to increase the accuracy of signs recognition. Authors in [9, 10] both altered CNN to feed outputs of two stages to classifier separately, as shown in Fig. 4, providing the classifier with variable-scales features. Meanwhile, to speed up detection, in [10] a layer was implemented with fixed filters to locate specific regions in images that most possibly contain traffic signs. Instead of using cross-entropy as the measurement of model performance, the design of CNN model [11] borrowed the ideas of hinge loss from support vector machine to propose a new cost function, driving CNN to make better use of its capacity to focus on misclassified training samples. Another improvement in traffic signs recognition was obtained by a hierarchical CNN [12], by partitioning a tricky problem to subsets. Also, owing to the characteristic of clustering algorithm, the whole training process could be done in an unsupervised way, which made it possible to avoid the error caused by manual labelling beforehand.

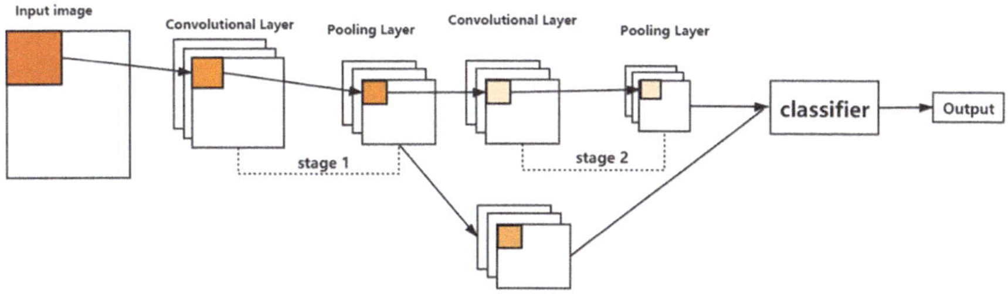

Fig. 4. The framework of CNN, capturing multi-scale features [9].

Secondly, CNN works well in detecting mobile objects in ITS. Liu et al. adopted CNN to detect passengers in preprocessed video images for predicting crowd flow [13]. A fully convolutional neural network was employed to undertake detection tasks on 3D range scan data [14] and was experimented in an autonomous driving system. With the same aim of extracting multi-scale feature though, Chen et al. presented a different approach by modifying deep convolutional neural network (DCNN) [15]. The last layer in DCNN was divided into blocks with various-sized convolution filters and pooling fields. A simple example is shown in Fig. 5. The convolutional layer and pooling layer in each block shown here are set to hold filters, whose size is adjusted according to the desired scales of extracted features. Additionally, CNN made a contribution to identifying transportation mode on a large scale [16], either.

In terms of traffic regression problems, researchers managed to adapt CNN to capture spatial and temporal features in the road network simultaneously. [17, 18] generated images or traffic flow matrix from a transportation network. The extraction of traffic features is slightly different from that in image classification. Convolution a pooling operation was carried out in two dimensions, which is capable of mining spatio-temporal relations of road segments.

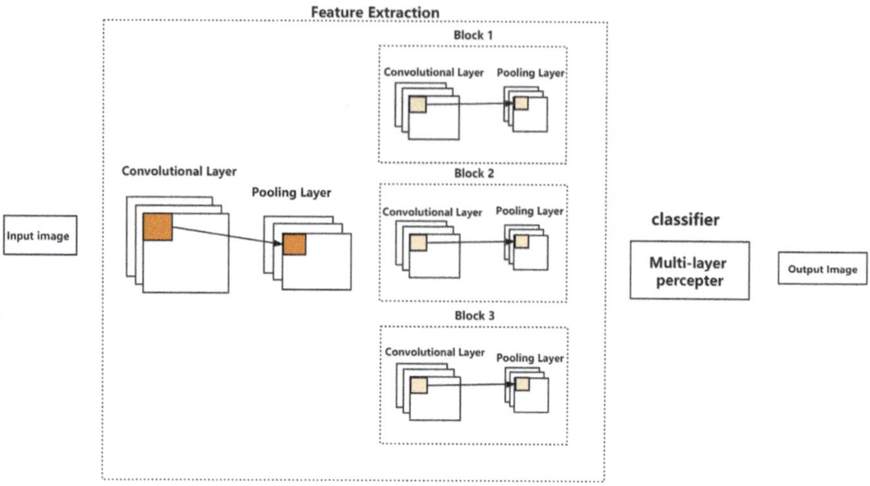

Fig. 5. The architecture of DCNN [15].

3.2 LSTM Networks and GRU Networks in ITSs

LSTM has been popular for predicting time series for its outstanding performance in capturing long-term temporal features [19, 20]. GRU also turns out to be suitable in this aspect [21, 22] and it even outstrips LSTM in traffic flow prediction, according to [23]. However, only temporal factors are not enough to ensure the accuracy of traffic regression, because the traffic condition of a segment is also influenced by its spatial features. Hence, incorporating those factors when designing a prediction model is necessary. In [24], researchers proposed a stacked bidirectional LSTM (Bi-LSTM) network, due to the fact that targeted segments are usually affected by the adjacent ones, both upstream and downstream. For Bi-LSTM network in Fig. 6, the forward layer output is denoted as $\left(h_1^+, h_2^+, h_3^+\right)$, while the counterpart is $\left(h_1^-, h_2^-, h_3^-\right)$. Then, such outputs are concatenated to produce the final result $\left[h^+, h^-\right]$ denoted as h. As for the proposed stacked Bi-LSTM model shown in Fig. 7, inputs to the horizontal Bi-LSTM units are data of segments along a path, while the vertical ones correspond to the state of path at each time step. Another research [25] came up with a novel coefficient matrix to handle spatial correlations among traffic observation points and then merge these matrices into the input of LSTM unit in time series.

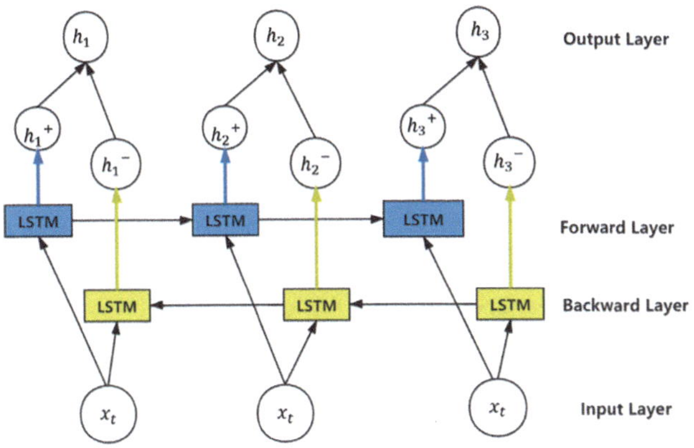

Fig. 6. The structure of Bi-LSTM layer [24].

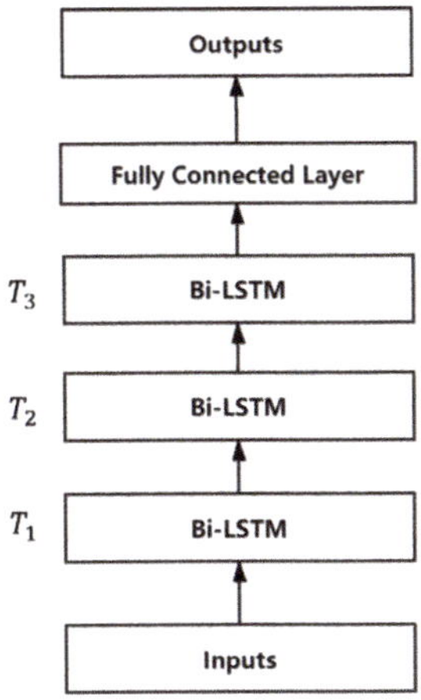

Fig. 7. The framework of stacked Bi-LSTM model [24].

3.3 Hybrid of CNNs and LSTM Networks in ITSs

To deal with spatiotemporal issues in traffic prediction, researches also employed convolutional LSTM (ConvLSTM), a variation of LSTM unit first introduced in [26]. What distinguished it from LSTM unit is that at a time step, all of its variables are 3D sensors, containing spatial structure in the last two dimensions. Then states and gates are calculated by convolution operations shown in Eqs. (11)–(16), where $\circ$ denotes convolution operation. According to the result of [27], ConvLSTM outperformed LSTM in bus travel time prediction. Researchers aggregated spatiotemporal features and external factors, such as weather impacts, and then fed them into the ConvLSTM network in order to estimate passenger demand [28]. To cope with spatial heterogeneous challenge, an ensemble mechanism was added on top of initial ConvLSTM [29].

$$i_t = \sigma\left(\mathbf{W}_i[\mathbf{h}_{t-1}, \mathbf{x}_t]^T + \mathbf{W}_{i,c} \circ c_{t-1} + \mathbf{b}_i\right). \tag{11}$$

$$f_t = \sigma\left(\mathbf{W}_f[\mathbf{h}_{t-1}, \mathbf{x}_t]^T + \mathbf{W}_{f,c} \circ c_{t-1} + \mathbf{b}_f\right). \tag{12}$$

$$c_t' = \tanh\left(\mathbf{W}_c[\mathbf{h}_{t-1}, \mathbf{x}_t]^T + \mathbf{b}_c\right). \tag{13}$$

$$c_t = f_t \circ c_{t-1} + i_t \circ c_t'. \tag{14}$$

$$o_t = \sigma\left(\mathbf{W}_o[\mathbf{h}_{t-1}, \mathbf{x}_t]^T + \mathbf{W}_{o,c} \circ c_t + \mathbf{b}_o\right). \tag{15}$$

$$h_t = o_t \circ \tanh(c_t). \tag{16}$$

4 Conclusions and Future Work

From these representative previous works, NN turned to be useful in increasing the accuracy of recognizing objects and regression problem in the background of transportation. CNN, LSTM and GRU are adapted according to the specific tasks in ITS. Meanwhile, most of studies in this paper used various methods to preprocess data before feeding it to NN. In [30], temporal hierarchical clustering assisted an attention-based Bi-LSTM model to excel in traffic speed prediction. Thus, it might be a good choice for researchers to prepare data with relevant technique. However, researches about object recognition mostly failed to do real-time detection. Thus, how to reduce the delay of detection would be an interesting area to explore. Besides, ConvLSTM is computationally expensive, so how to make a balance between performance and computation cost is a concern. In the future, it would be also promising to take advantage of more advanced NN, such as ResNet and DenseNet, to improve model performance and bring about more applications for ITS.

Acknowledgements. This work was partially supported by the National Natural Science Foundation of China (Nos. 61906043, 61877010, 11501114 and 11901100), Science Foundation of the Fujian Province, China (No. 2019J01243), Funds of Education Department of Fujian Province (No. JAT190026), and Fuzhou University (Nos. 510730/XRC-18075, 510809/GXRC-19037, 510649/XRC-18049 and 510650/XRC-18050).

References

1. Nath, R.P.D., Lee, H.J., Chowdhury, N.K., Chang, J.W.: Modified k-means clustering for travel time prediction based on historical traffic data. Knowl.-Based Intell. Inf. Eng. Syst. **6276**, 511–521 (2010)
2. Zhao, T., Nevatia, R.: Car detection in low resolution aerial images. Image Vis. Comput. **21** (8), 693–703 (2003)
3. Lecun, Y., Bottou, L., Bengio, Y., Haffner, P.: Gradient-based learning applied to document recognition. Proc. IEEE **86**(11), 2278–2324 (1998)

4. Hochreiter, S., Schmidhuber, J.: Long short-term memory. Neural Comput. **9**(8), 1735–1780 (1997)
5. Cho, K., van Merrienboer, B., Gulcehre, C., Bahdanau, D., Bougares, F., Schwenk, H., Bengio, Y.: Learning phrase representations using RNN encoder-decoder for statistical machine translation. In: Proceedings of the 2014 Conference on Empirical Methods in Natural Language Processing (EMNLP). ACL, PA, USA, pp. 1724–1734 (2014)
6. Zachary, C.L., Berkowitz, J., Elkan, C.: A critical review of recurrent neural networks for sequence learning (2015). arXiv: 1506.00019
7. Krizhevsky, A., Sutskever, I., Hinton, G.E.: ImageNet classification with deep convolutional neural networks. Commun. ACM **60**(6), 84–90 (2017)
8. Simonyan, K., Zisserman, A.: Two-stream convolutional networks for action recognition in videos. In: NIPS'14: Proceedings of the 27th International Conference on Neural Information Processing Systems. ACM, NY, USA, vol. 1, pp. 568–576 (2014)
9. Sermanet, P., LeCun, Y.: Traffic sign recognition with multiscale convolutional networks. In: Proceedings of the 2011 International Joint Conference on Neural Networks (IJCNN), NJ, USA, pp. 2809–2813. IEEE (2011)
10. Wu, Y., Liu, Y., Li, J., Liu, H., Hu, X.: Traffic sign detection based on convolutional neural networks. In: Proceedings of the 2013 International Joint Conference on Neural Networks (IJCNN), NJ, USA, pp. 2809–2813. IEEE (2013)
11. Jin, J., Fu, K., Zhang, C.: Traffic sign recognition with hinge loss trained convolutional neural networks. IEEE Trans. Intell. Transp. Syst. **15**(5), 1991–2000 (2014)
12. Mao, X., Hijazi, S., Casas, R., Kaul, P., Kumar, R., Rowen, C.: Hierarchical CNN for traffic sign recognition. In: Proceedings of the 2016 IEEE Intelligent Vehicles Symposium (IV), NJ, USA, pp. 130–135. IEEE (2016)
13. Liu, G., Yin, Z., Jia, Y., Xie, Y.: Passenger flow estimation based on convolutional neural network in public transportation system. Knowl.-Based Syst. **123**, 102–115 (2017)
14. Li, B., Zhang, T., Xia, T.: Vehicle detection from 3D Lidar using fully convolutional network (2016). arXiv: 1608.07916
15. Chen, X., Xiang, S., Liu, C., Pan, C.H.: Vehicle detection in satellite images by hybrid deep convolutional neural networks. IEEE Geosci. Remote Sens. Lett. **11**(10), 1797–1801 (2014)
16. Dabiri, S., Heaslip, K.: Inferring transportation modes from GPS trajectories using a convolutional neural network. Transp. Res. Part C: Emerg. Technol. **86**, 360–371 (2018)
17. Zhang, W., Yu, Y., Qi, Y., Shu, F., Wang, Y.: Short-term traffic flow prediction based on spatio-temporal analysis and CNN deep learning. Transportmetrica A: Transp. Sci. **15**(2), 1688–1711 (2019)
18. Ma, X., Dai, Z., He, Z., Ma, J., Wang, Y., Wang, Y.: Learning traffic as images: A deep convolutional neural network for large-scale transportation network speed prediction. Sensors **17**(4), 818 (2017)
19. Duan, Y., Lv, Y., Wang, F.Y.: Travel time prediction with LSTM neural network. In: Proceedings of the 2016 IEEE 19th International Conference on Intelligent Transportation Systems (ITSC), NJ, USA, pp. 1053–1058. IEEE (2016)
20. He, P., Jiang, G., Lam, S.K., Tang, D.: Travel-time prediction of bus journey with multiple bus trips. IEEE Trans. Intell. Transp. Syst. **20**(11), 4192–4205 (2019)
21. Zhang, D., Kabuka, M.R.: Combining weather condition data to predict traffic flow: a GRU-based deep learning approach. IET Intell. Transp. Syst. **12**(7), 578–585 (2018)
22. Zhao, J., Gao, Y., Qu, Y., Yin, H., Liu, Y., Sun, H.: Travel time prediction: Based on gated recurrent unit method and data fusion. IEEE Access **6**, 70463–70472 (2018)
23. Fu, R., Zhang, Z., Li, L.: Using LSTM and GRU neural network methods for traffic flow prediction. In: Proceedings of the 2016 31st Youth Academic Annual Conference of Chinese Association of Automation (YAC), NJ, USA, pp. 324–328. IEEE (2016)

24. Wang, J.W., Chen, R., He, Z.: Traffic speed prediction for urban transportation network: A path based deep learning approach. Transp. Res. Part C: Emerg. Technol. **100**, 372–385 (2019)
25. Zhao, Z., Chena, W., Wu, X., Chen, P.C.Y., Liu, J.: LSTM network: a deep learning approach for short-term traffic forecast. IET Intell. Transp. Syst. **11**(2), 68–75 (2017)
26. Shi, X., Chen, Z., Wang, H., Yeung, D.Y., Wong, W.K., Woo, W.C.: Convolutional LSTM network: a machine learning approach for precipitation nowcasting (2015). arXiv: 1506.04214
27. Petersen, N.C., Rodrigues, F., Pereira, F.C.: Multi-output bus travel time prediction with convolutional LSTM neural network. Expert Syst. Appl. **120**, 426–435 (2019)
28. Ke, J., Zheng, H., Yang, H., Chen, X.: Short-term forecasting of passenger demand under on-demand ride services: a spatio-temporal deep learning approach. Transp. Res. Part C: Emerg. Technol. **85**, 591–608 (2017)
29. Yuan, Z., Zhou, X., Yang, T.: Hetero-ConvLSTM: a deep learning Approach to traffic accident prediction on heterogeneous spatio-temporal data. In: KDD' 2018: Proceedings of the 24th ACM SIGKDD International Conference on Knowledge Discovery & Data Mining. ACM, NY, USA (2018)
30. Liu, D., Tang, L., Shen, G., Han, X.: Traffic speed prediction: An attention-based method. Sensors **19**(18), 3836 (2019)

Glacier Area Monitoring Based on Deep Learning and Multi-sources Data

Guang Wang[1], Yue Liu[1], Huifang Shen[2](✉), Shudong Zhou[1], Jinzhou Liu[1], Hegao Sun[2], and Yan Tao[1]

[1] Liaoning Technical University, Huludao 125105, China
[2] Quanzhou Institute of Equipment Manufacturing Haixi Institutes, Chinese Academy of Sciences, Quanzhou 362216, China
shenhf@fjirsm.ac.cn

Abstract. Glaciers are the source of fresh water that have obvious response on climate change. It is crucial and meaningful to monitor glacier area. In this paper, a glacier detection method based on features derived from low-resolution optical data, thermal data and TerraSAR-X images is proposed. The Land Surface Temperature (LST) was firstly obtained by Convolutional Neural Network (CNN). Combined with the velocity information, a low-resolution binary mask was derived for the supervised classification of SAR imagery. Afterwards, a set of suitable features was derived from the SAR intensity image, such as texture information generated based on the gray level co-occurrence matrix (GLCM), and intensity values. With these features above, the glaciers were classified by Random Forest (RF) to distinguish the glacier from the non-glacier areas. Compared to the unsupervised classification only using SAR data, the glacier detection method proposed in this paper achieved a better performance with the overall classification accuracy of 90.88%.

Keywords: Glacier area · CNN · LST · Optical data · Thermal data · SAR

1 Introduction

The study of glacier contributes to the understanding of the law of glacier change [1], which is one of the important research contents in the field of glacier change and global climate change [2]. LST can be used as a feature to distinguish the glacier from the surrounding environment, especially in summer. The traditional method assumes a simple linear relationship between surface temperature and brightness and emissivity, and then using the relationship to obtain LST retrieval results [3]. However, these methods simplify many preconditions, such as atmospheric constant correction, absorption by surface vegetation, and ignoring atmospheric scattering effects. These assumptions affect the accuracy of retrieval to some extent.

The neural network algorithm has less dependence on physical equation of radiation transmission, which can solve the problem of ill-condition retrieval of physical model [4]. Neural network does not need to deduce the physical process for specific problems, only needs to input a representative data set containing multiple samples to train the neural network, and the trained neural network model is used for the retrieval

Q. Liu et al. (Eds.): CENet 2020, AISC 1274, pp. 409–418, 2021.
https://doi.org/10.1007/978-981-15-8462-6_46

[5]. In this paper, CNN is used to achieve the goal of LST retrieval. It is meaningful to introduce CNN to construct the relationship between Brightness Temperature (TB) and Land Surface Temperature (LST). In addition, Synthetic Aperture Radar (SAR) is widely applied to map glaciers because SAR is not affected by weather conditions, and can work all day and all night [6].

In this paper, the LST map is obtained by combining the optical data and thermal data with the CNN. Combining with velocity information, the classification mask is deduced for the automatic training of classification model. With texture information and intensity values extracted from high resolution TerraSAR-X images, glaciers were classified by RF. The main contributions which are specific to mapping the glacier area can be stated as:

- The Land Surface Temperature (LST) map was firstly obtained by Convolutional Neural Network (CNN).
- Multi-sources data were used for the classification of glacier area, which include optical data, thermal data and SAR imagery.
- The feasibility and superiority of the method proposed in this paper was demonstrated compared to the unsupervised classification by only using SAR data.

The paper is organized as follows: Sect. 2 introduces the study area and datasets. Section 3 describes the proposed methodology for LST retrieval, and the method of glacier classification. In Sect. 4, the experimental results are presented and analyzed. Finally, conclusions and recommendations are described in Sect. 5.

2 Study Area and Datasets

The Taku glacier is a tidewater glacier of a small scale low-latitude glacial system, which located in Juneau Icefield in the southeastern coast mountains of Alaska, USA. It is considered to be the deepest and thickest glacier known in the world [7]. In Fig. 1, an illustration of the Taku glacier and the Juneau Icefield is given. In this paper, TerraSAR-X images obtained in the summer of 2009 to classify glaciers. In addition, Digital Elevation Model (DEM) is used to provide topographic information. Moreover, the existing DEM data are used to carry out orthogonal correction and geocoding for TerraSAR-X data. The enhanced data had a ground resolution of 3 m (much better than the optical and thermal images used in the experiments) with a pixel spacing of 1.5 m in both azimuth and distance directions.

Meanwhile, TB data observed by Aster (with a resolution of 90 m) was used as the independent variable of the LST retrieval algorithm. Due to the lack of synchronous ground measurement data, an almost simultaneous acquisition of the Modis LST product was used as a dependent variable for CNN input. The MODIS and ASTER sensors are aboard the Terra satellite.

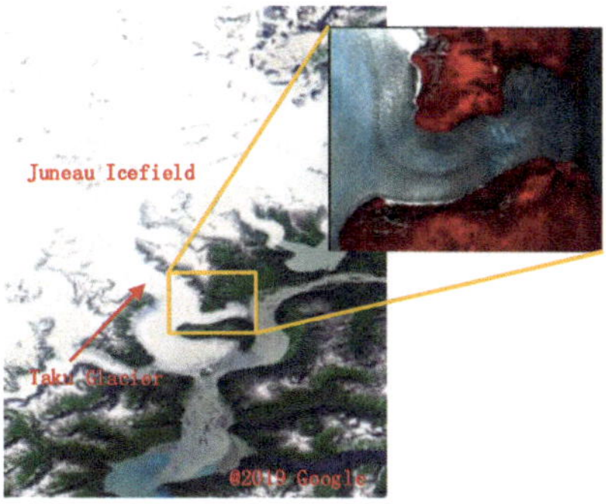

Fig. 1. Illustration of Taku glacier.

3 Methodology

The work proposed in this paper is to classify glaciers and non-glaciers. For this reason, CNN was firstly used to obtain LST, in order to derive a low-resolution binary mask with the resolution of 30 m. Then, the texture information and intensity values of SAR imagery were used. Afterwards, the glaciers were classified by RF. In addition, in order to compare with the glacier classification method proposed in this paper, the unsupervised classification using only SAR data was also carried out in this paper. The overall workflow for the proposed glacier classification approach is presented in this paper are shown in Fig. 2.

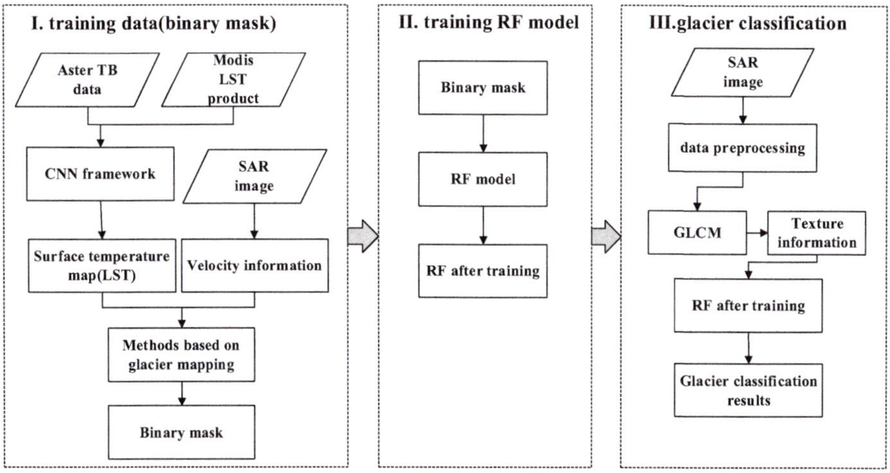

Fig. 2. Overall workflow for the proposed glacier classification approach.

3.1 LST Retrieval of CNN

In this paper, the nonlinear relationship between TB data and LST constructed by CNN is introduced. TB data observed by Aster and Modis LST product were used as reference data, which were randomly divided into training data set and test data set. A typical CNN is shown in Fig. 3. CNN is mainly composed of input layer, convolution layer, pooling layer, full connection layer and output layer [8].

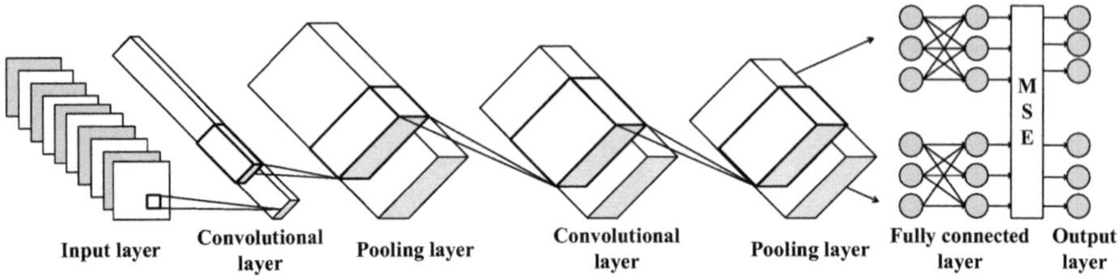

Fig. 3. CNN architecture.

The size input patch is 64 × 64, and the convolution kernel size is 3 × 3, all the convolution stride is fixed to 1 pixel [9]. The convolution operation of CNN is calculated by Eq. (1):

$$x_j^l = f\left(\sum_{i \in M_j} x_i^{l-1} \cdot k_{ij}^l + b_j^l\right) \tag{1}$$

In Eq. (1), l is the number of layers where the convolution layer is located, k is the convolution kernel, b is the bias term, f is the activation function, and M_j is the input characteristic data of the previous layer. ReLU function is used as the activation function, and the calculation Equation is as follows:

$$f(x) = \max(x, 0) \tag{2}$$

For ReLU, if the input is greater than 0, the output is equal to the input; Otherwise, the output is 0. The pooling layer adopts the max pooling method, and the Equation is as follows:

$$x_j^l = f\left(\beta_j^l \max\left(x_j^{l-1}\right) + b_j^l\right) \tag{3}$$

In Eq. (3), l represents the number of layers, and β represents the maximum pool coefficient, while b is the bias term and f is the activation function [10]. The pooling layer uses ReLU as the activation function. The pooling kernel is set to a matrix size of 2. The CNN output layer uses the mean square error (MSE) as the loss function. The loss function Equation of each layer is as follows:

$$MSE(y, y') = \frac{\sum_{i-1}^{n} \left(y_i - y_i'\right)^2}{n} \tag{4}$$

In Eq. (4), y_i is the target value of the i th data in a training batch, and y_i' is the retrieval value given by CNN. The optimization process of CNN includes two stages: forward propagation stage and backward propagation stage. In the forward propagation stage, the training samples are input and then transmitted to the output layer layer-by-layer through the convolution kernel pool [11]. The predicted value of the forward propagation phase is calculated and compared with the target value to obtain the error between the two. The back-propagation stage is used to calculate the gradient of each parameter of the loss function. Then, the gradient descent algorithm of each parameter is updated according to the gradient and learning rate, so that the loss function of the training stage is as small as possible [12]. To build CNN, the paper uses the Keras interface provided by python and the TensorFlow deep learning framework.

3.2 Features from SAR Data

Velocity Information. The main components of glacier are ice, snow and water. Due to the alternations of seasons and the gravity and pressure, the glacier ice moves along the direction of movement with different motion amplitudes, with summer fast and winter slow. Moreover, the glacier is mainly surrounded by rocks or others. Therefore, the existence of the movement state of the glacier and the static state of the surrounding environment, which contributes to the classification of the glacier [13]. In this paper, the approaches for estimating glacier surface motion using repeat SAR intensity images via point-like features (PLF) and the robust phase correlation (PC) algorithm [14].

Texture Information. Texture is one of the most important features for image classification. The second-order gray-level co-occurrence matrix (GLCM) is an effective method for image texture feature analysis, which statistics the distribution law of Gray value between pixels in Gray image. The GLCM approach for texture analysis is based on the assumption: texture information in the image is contained in the overall or average spatial relationship between image grays [15].

In this paper, the GLCM of each pixel is generated in four directions and compared with the eight neighborhoods around the central pixel. Afterwards, four features of GLCM were calculated: contrast, correlation, energy and homogeneity. Then calculate the mean values of the four features in four different directions, and finally define four texture features for the experiment. Most areas of the glacier surface have unique texture characteristics, which makes texture a useful classification information.

3.3 Classification Method

In this paper, the research area is divided into glacial area and non-glacial area, and the Random Forest (RF) is used for supervised classification [16]. RF outputs classification results based on lots of decision trees during training, that is, the pattern of classes provided by a single tree. Each tree in the forest classifies the samples based on an independent set of sampled random variables, which are used for optimal segmentation of nodes under certain conditions. When each tree grows to the size of the smallest node, the forest chooses the category with the most votes.

4 Experiment Results

4.1 Feature Set

Land Surface Temperature. The method of LST retrieval described above was used to estimate the LST map of the study area (Fig. 4). Moreover, differences in surface temperature values between the glaciers and the surrounding areas contribute to classification.

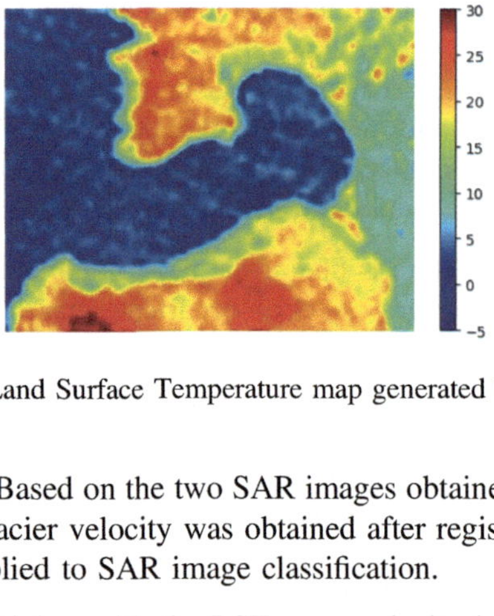

Fig. 4. Land Surface Temperature map generated by CNN.

Velocity Information. Based on the two SAR images obtained from the Taku glacier at different times, the glacier velocity was obtained after registration (Fig. 5). Finally, velocity feature was applied to SAR image classification.

Binary Mask. By combining with the LST map, velocity information and adopting the pixel-based glacier mapping method, a binary mask with a resolution of 30 m was obtained, as shown in Fig. 6.

Texture Information. Figure 7 shows the mean values calculated from GLCM. In contrast, the water region is considered a dark region and a bright region in energy and homogeneity. The water region does not show signal information in SAR images due to specular backscattering. This contributes to distinguish between glaciers and water.

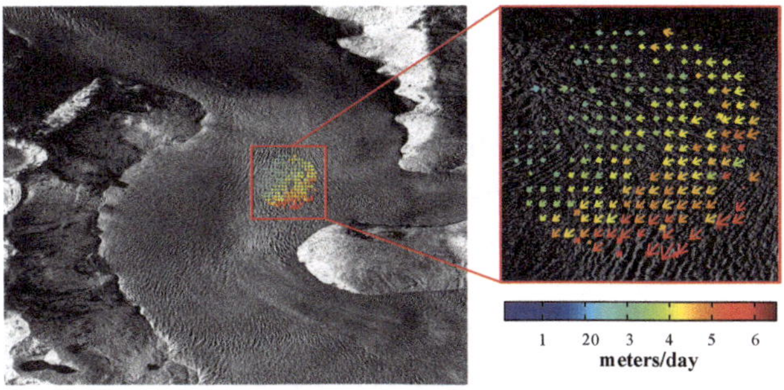

Fig. 5. The velocity feature of glacier.

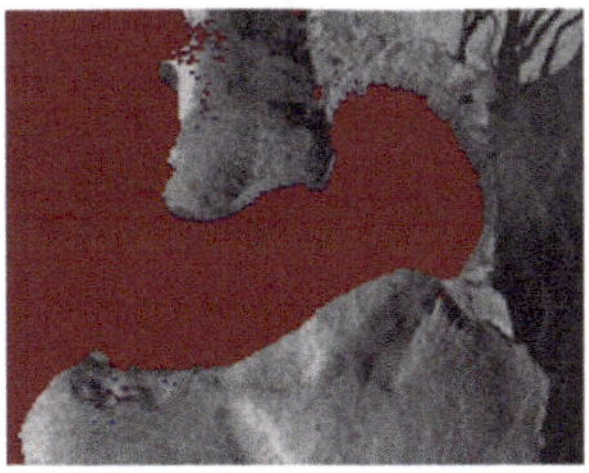

Fig. 6. Glacier surface classification results (red) based on optical and thermal data were plotted on a geologically coded grayscale SAR scene with 3 m resolution.

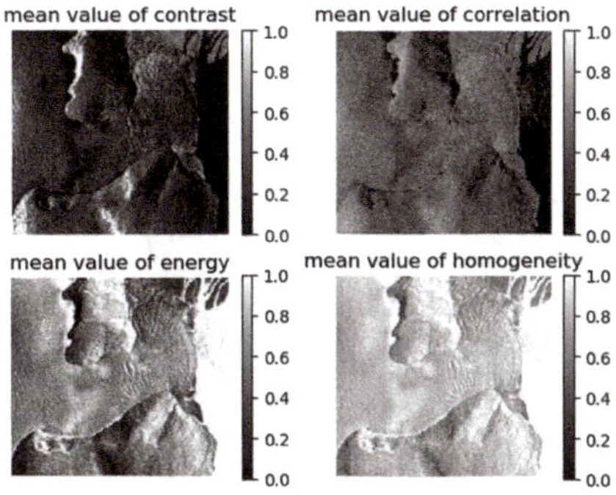

Fig. 7. Upper left: mean value of contrast. Upper right: mean value of correlation. Bottom left: mean value of energy. Bottom right: mean value of homogeneity.

4.2 Glacier Classification Results

Afterwards, the binary mask was applied to RF's supervised classification process based on TerraSAR-X imagery, aiming to distinguish the glacier from the surrounding environment. Finally, the glacier classification method proposed in this paper was used, the overall classification accuracy reached 90.88% (Table 1 and Fig. 8) Table 2.

Table 1. Confusion matrix of the glacier classification method proposed in this paper.

Category	Glacier (predicted)	Non-glacier (predicted)
Glacier (Actual)	90.53%	9.47%
Non-glacier (Actual)	8.77%	91.23%

Overall accuracy = 90.88%.

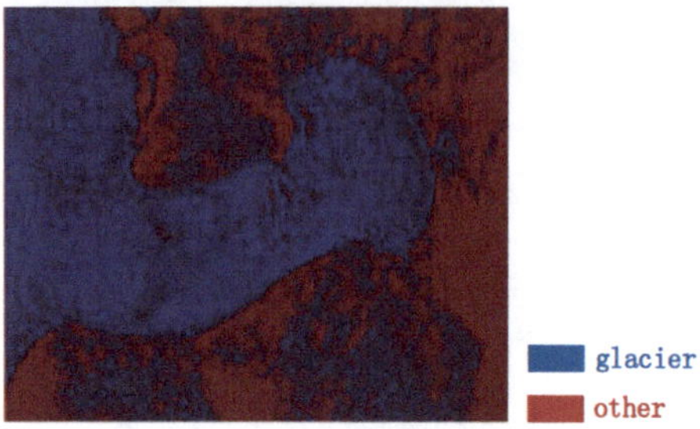

Fig. 8. Results of the glacier classification method proposed in this paper. Blue: glacier area; Red: non-glacial area.

Table 2. Confusion matrix of the glacier classification result only using SAR data.

Category	Glacier (predicted)	Non-glacier (predicted)
Glacier (Actual)	51.32%	48.68%
Non-glacier (Actual)	20.14%	79.86%

Overall accuracy = 65.59%.

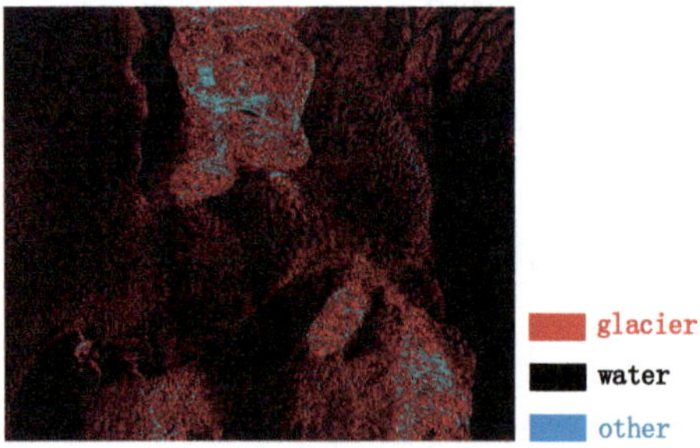

Fig. 9. Results of the glacier classification only using SAR data. Blue: glacier area; Red: non-glacial area.

The accuracy of unsupervised classification results only using SAR data is 65.59%, which is far lower than the glacier classification method proposed in this paper. In Fig. 9, we found that many non-glacier pixels were divided into glaciers, and some glacier pixels were divided into water labels, resulting in inaccurate classification results. In sum, the supervised classification by using multi-sources data helped to improve the classification accuracy.

5 Conclusion and Recommendation

This paper proposes a framework for glacier monitoring based on CNN and SAR intensity image. First, LST map was generated from optical and thermal data by CNN. Second, unique texture information was obtained from single-channel SAR image. By the method of LST retrieval of CNN, the radiation transmission process is simulated. Above all, the nonlinear relationship between BT and LST is solved. In the process of single-channel SAR classification, the feature of velocity is used to enrich the classification characteristics, and enhance the classification accuracy. The method of glacier classification proposed in this paper has been proved to be effective.

However, there are still improvements need future research. For example, other high-level features should be used instead of GLCM texture information (the other glacier physical information).

In future work, the accurate understanding of the glacier region will contribute to the mapping of the two-dimensional glacier changes. And it will become an important index for the estimation of the glacier volume in the geophysical model, laying the foundation stone for the follow-up study of glaciers.

References

1. Houghton, J.: Mass Balance of the Cryosphere: Observations and Modelling of Contemporary and Future Changes. Cambridge University Press, Cambridge (2004)
2. Sun, X., Zhang, R., Huang, W.: The response between glacier evolution and eco-geological environment on the Qinghai-Tibet Plateau. China Geol. **2**(1), 1–7 (2019)
3. Duan, J.K.: Comparative analysis of land surface temperature retrieval algorithm based on landsat8 data——A case study in Beijing city. Anhui Agric. Sci. Bull. **2019**(17), 58 (2019)
4. Denli, H., Subrahmanya, N.: Geophysical inversion with convolutional neural networks. Pat. Appl. **16**(57), 356 (2019)
5. LeCun, Y., Bengio, Y., Hinton, G.: Deep learning. Nature **521**(7553), 436–444 (2015)
6. Shimoni, M., Borghys, D., Heremans, R.: Fusion of PolSAR and PolInSAR data for land cover classification. Int. J. Appl. Earth Obs. Geoinf. **11**(3), 169–180 (2009)
7. Pelto, M.: Utility of late summer transient snowline migration rate on Taku Glacier Alaska. Cryosphere **5**(4), 1127 (2011)
8. Deng, L., Yu, D.: Deep learning: methods and applications. Found. Trends Sig. Process. **7**, 197–387 (2014)
9. Krizhevsky, A., Sutskever, I., Hinton, G.: ImageNet classification with deep convolutional neural networks. In: Proceedings of the Advances in Neural Information Processing Systems, pp. 1097–1105 (2012)

10. Oquab, M., Bottou, L., Laptev, I., Sivic, J.: Learning and transferring mid-level image representations using convolutional neural networks. In: Proceedings of the 2014 IEEE Conference on Computer Vision and Pattern Recognition, pp. 23–28 (2014)
11. Amato, F., López, A., Peña-Méndez, E.: Artificial neural networks in medical diagnosis, vol. 11(2), pp. 47–58 (2013)
12. Daxini, S.D., Prajapati, J.: Numerical shape optimization based on meshless method and stochastic optimization technique, pp. 1–22 (2019)
13. Maksymiuk, O., Mayer, C., Stilla, U.: Velocity estimation of glaciers with physically-based spatial regularization-experiments using satellite SAR intensity images. Remote Sens. **172**(51–52), 190–204 (2016)
14. Fang, L., Xu, Y., Yao, W.: Estimation of glacier surface motion by robust phase correlation and point like features of SAR intensity images. ISPRS J. Photogrammetry Remote Sens. **121**, 92–112 (2016)
15. Mahmoudi, F., Arabsaeedi, A., Alavipanah, S.: Feature-level fusion of landsat 8 data and SAR texture images for urban land cover classification. J. Ind. Soc. Remote Sens. **47**(3), 479–485 (2019)
16. Yu, Y., Li, M., Fu, Yu.: Forest type identification by random forest classification combined with SPOT and multitemporal SAR data. J. For. Res. **29**(5), 1407–1414 (2017). https://doi.org/10.1007/s11676-017-0530-4

3D Convolutional Neural Networks with Image Fusion for Hyperspectral Image Classification

Cheng Shi(✉), Jie Zhang, Zhenzhen You, and Zhiyong Lv

School of Computer Science and Engineering, Xi'an University of Technology, Xi'an, China
Chengc_s@163.com

Abstract. Image fusion can extract redundant information of multiple images into one image, and the goal of image fusion is to better apply to classification tasks. Convolutional neural networks have proved to be an effective way for accuracy of classification. However, fusion and classification usually considered separately. In this paper, we design a 'fusion-classification networks', and introduce image fusion technology and 3D convolutional neural networks (3D CNNs) into HSI classification. In the proposed method, the fusion process is guided by the classification result, and the classification accuracy is improved by the fusion process. Image fusion technology is performed on spectral bands to exploit the redundancy information of HSI, and 3D CNNs are applied on the fused image to extract more robust spectral-spatial features. The proposed method is tested on two datasets. Its outstanding performance is validated in comparison with other state-of-the-art approaches.

Keywords: Hyperspectral image classification · 3D convolutional neural networks · Image fusion

1 Introduction

Hyperspectral image (HSI) has hundreds of spectral bands for each image pixel. This rich information can provide useful information for image classification and recognition [1]. Furthermore, the increasing spatial resolution makes it more vital to analyze the spatial information. HSI classification has developed a lot in remote sensing image analysis. How to learn the effective features from the rich spatial and spectral information is a challenging task.

The typical feature extraction method is subspace-projection method, such as principal component analysis (PCA) [2], independent component analysis (ICA) [3], and linear discriminant analysis [4]. Moreover, wavelet transform has been investigated into the hyperspectral image classification [5, 6]. The limitation of these methods is that they can extract the spectral information only. But the spatial information is neglected, which also plays an important part in HSI classification. Through the combination of the spatial and spectral information, some spatial-spectral classification methods were proposed, such as Extended Multi-Attribute Profiles (EMAPs) [7], active learning [8], decision fusion [9, 10], edge-preserving filtering [11, 12].

Q. Liu et al. (Eds.): CENet 2020, AISC 1274, pp. 419–428, 2021.
https://doi.org/10.1007/978-981-15-8462-6_47

Recently, deep neural networks are demonstrated a useful tool for image feature extraction. It can learn more complex features from the image than the traditional neural networks (less than three layers). Now the commonly used deep neural networks include stacked autoencoder networks (SAEs) [13], Deep Belief Networks (DBNs) [14], deep Boltzmann machines (DBMs) [15], and deep convolutional neural networks (CNNs) [16–18]. In [19, 20], SAEs and DBNs were firstly introduced into the HSI classification problem, and achieved ideal results. Among all of these networks, CNNs are the two-dimension (2D) neural networks, which make it more suitable for exploiting the deep structure information of the image. Because of the superior performance on 2D image, CNNs have been introduced into handwritten characters recognition [17], document image classification [16, 18], and HSI classification [21, 22]. In [21], CNNs were introduced into the HSI classification in spectral domain. The convolutional and subsampling processes were considered as the one-dimension operator. In [22], the spatial and spectral information was extracted separately to form some two-dimension (2D) feature maps. And then the CNNs were trained based on each feature map to get useful high-level features. Due to the high spectral dimension of HSI, [23] and [24] proposed HSI classification method by combining dimension reduction method and CNNs.

Traditional CNNs are designed to process the 2D data. So, dealing with the 3D data, we should convert it to the 2D model firstly. HSI is a 3D cube, the spatial and spectral information are all useful for classification. And it is believed that the spectral and spatial information should be considered simultaneously to further improve the classification accuracy [25]. To capture the effective space features, some 3D feature learning method was proposed, such as 3D Gabor wavelet [25] and 3D DWT [26] based HSI classification methods. These methods have proved to obtain higher classification accuracy, but they have a problem that 3D features extraction method need to determine the number of features, and it will bring large number features which need to be selected. The feature extraction process and the feature selection process have great influence on the classification accuracy. Deep neural networks can adaptively learn the features of the image, and therefore, some researchers extended the 2D CNNs into 3D mode and applied on video data, 3-D object recognition, and HSI classification. 3D convolutional neural networks (3D CNNs) can automatically learn the features from low-level to high-level, and the feature extraction and the feature selection are fused to learn the appropriate features. Hence 3D CNNs can learn more discriminative features.

To inherit the advantages of 3D CNNs, we propose 3D convolutional neural networks with image fusion (3D CNNs_IF) model for HSI classification in this paper. 3D CNNs_IF consists of three parts: band fusion, 3D feature extraction, and classification. Hence, 3D CNNs_IF can be considered as a 'fusion-classification networks'. 3D CNNs require high spatial and time complexity, so it is necessary to reduce dimension of the HSI first. The reduced image affects the performance of the following feature extraction. Image fusion is to retain the most useful information of multiple images into one image. According to the idea of image fusion, we design a band fusion layer to reduce the dimension of the HSI, and maintain the important spectral and spatial information in the image after dimension reduction. Band fusion layer is to fusion the adjacent several bands into one, which can remove the noisy pixels and reduce the redundant information of HSI. We note in [11], the combination of band fusion and recursive filtering were proposed for HSI classification. Compared with [11] and other traditional dimension reduction methods, the band fusion layer is connected with the 3D feature extraction layer, and the reduced image can not only reduce the computational complexity of 3D CNNs, but also can extract more effective spectral and spatial information under the guidance of the 3D deep features. Based on the proposed 3D CNNs_IF model, the proposed HSI classification method can capture more spectral and spatial features with less loss of information.

The proposed method is tested on two well-known datasets: Pavia University and Salinas images. Experimental results show that the proposed 3D CNNs_IF method can improve the classification accuracy obviously. This paper is organized in four sections. Section 2 describes the structure of 3D convolutional neural networks with image fusion. Section 3 consists of the experimental results and analysis, and Sect. 4 presents the concluding remarks.

2 The Structure of 3D Convolutional Neural Networks with Image Fusion

In this section, we propose 3D convolutional neural networks with band fusion method for hyperspectral image classification. We first construct the 3D cube sample by taking a 3D neighborhood window on the HSI, and then feed each 3D sample into 3D CNNs_BF model for training. The proposed 3D CNNs_BF model consists of three parts: band fusion, 3D feature extraction, and classification, which is shown in Fig. 1.

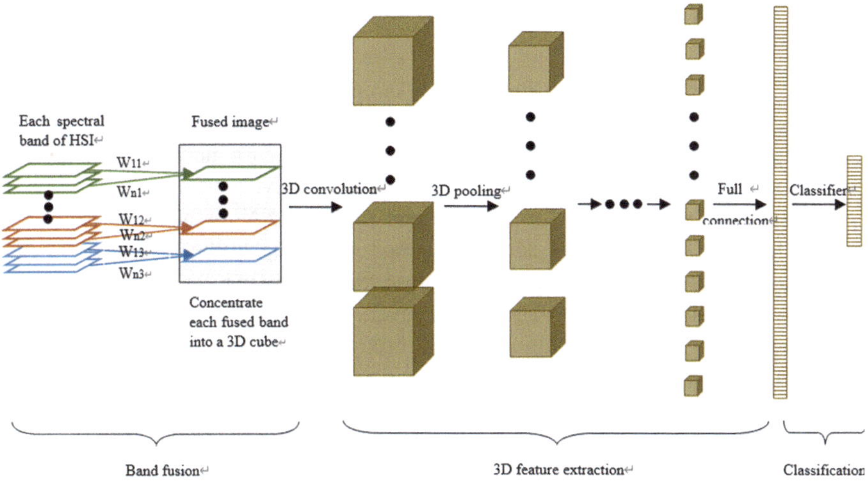

Fig. 1. Framework of 3D convolutional neural networks with image fusion.

2.1 Image Fusion

HSI usually consists hundreds of spectral bands, which contain a lot of redundant information. In order to effectively use the redundant information and reduce the complexity of the subsequent feature extraction, we first generate a band fusion layer to extract valid information.

Firstly, we divided the spectral bands into M groups, and each group contains N spectral bands $(X_1, \ldots, X_N)$. The spectral bands of each group are fused into one band by weighted fusion rule. And then pass the fused band into a nonlinear activation function.

$$\mathbf{F}_j = s\left(\sum_{i=1}^{N} \mathbf{W}_{ij} X_i\right) \quad j = 1, \ldots, M \tag{1}$$

where $\mathbf{W}_{ij}$ is the fusion weight connecting $i-th$ input and $j-th$ output, $\mathbf{F}_j$ means the $j-th$ output band, and $s(\bullet)$ is the sigmoid function.

According to Eq. (1), the valid information of multiple bands can be fused into a band through the fusion process. Different from traditional fusion method, the weight $\mathbf{W}_{ij}$ is not a pre-fixed value, but is updated by the supervised back propagation algorithm. Hence the fused image can be more effective on the extraction of redundant spectral information.

Through Eq. (1), we can get M fused spectral bands. And then the obtained bands $\mathbf{F}_j, j = 1, \ldots, M$ are concentrated into a 3D cube for the following process of feature extraction.

2.2 3D Feature Extraction

In Sect. 2.1, the spectral dimension of each sample is reduced and these samples are concentrated into a 3D cube for spectral-spatial feature learning. To extract more discriminative spectral-spatial feature, 3D convolutional neural networks are applied to extract the spectral and spatial information simultaneously.

Similar to 2D CNNs, the forward process for 3D CNNs is also a layer-by-layer based calculation. The features can be obtained according to the feature maps in the previous layer. The calculation process is described following.

Convolution layer: At the convolution layer, the input 3D feature maps are convolved with the 3-D filters and then passed through a nonlinear function to obtain the 3D output feature maps. The process is formulated as Eq. (2).

$$\mathbf{F}_j^l = s\left(\sum_{i \in M} \mathbf{F}_i^{l-1} * \mathbf{K}_{ij}^l + b_j^l\right) \quad i = 1, \ldots, p, j = 1, \ldots, q \tag{2}$$

where $\mathbf{F}_j^l$ means the output 3-D feature map in l layer, q is the number of the output feature maps. $\mathbf{F}_i^{l-1}$ is the input 3-D feature map in $l-1$ layer, and p is the number of the input feature maps. $\mathbf{K}_{ij}^l$ is a 3-D filter, and b_j^l is the bias for $j-th$ output feature map.

Pooling layer: The max-pooling method is adopted in this paper. The purpose of the pooling layer is to sub-sample the previous feature map and make it size reduced. The formula is shown in Eq. (3).

$$\mathbf{F}_j^l = s(g(\mathbf{F}_i^{l-1}) + b_j^l) \quad i = 1, \ldots, p, j = 1, \ldots, q \tag{3}$$

where $g(x)$ represents a 3-D sub-sampling operation. The previous feature map F_i^{l-1} is divided into several small 3-D cubes. For each cube, map its maximum value into the output feature map.

Full connection layer: If each feature map in previse layer is a 3-D cube, it should be flatted into a vector first. And then each value of the vector is connected to all the units of full connection layer. The formula is shown in Eq. (4).

$$f^l = s(\mathbf{W}^l g^{l-1} + b^l) \tag{4}$$

where g^{l-1} is the flatted vector in the previous layer, and $\mathbf{W}^l$ is the connection weight matrix.

2.3 Classification

Softmax classifier is used for classification in 3-D convolutional neural networks. The softmax model is shown in Eq. (5).

$$y^L = \frac{1}{\sum_{j=1}^{numclass} e^{w_j^T f^l}} \begin{bmatrix} e^{w_1^T f^l} \\ e^{w_2^T f^l} \\ \vdots \\ e^{w_{numclass}^T f^l} \end{bmatrix} \tag{5}$$

where *numclass* means the class number in the classification problem. The output $\mathbf{y}^L$ is a vector for each sample. The value in vector $\mathbf{y}^L$ represents the probability of all the classes. The position where it has a maximum value in vector $\mathbf{y}^L$ is the label for the sample. The mini-batch gradient descent method is used for training the proposed model.

3 Experimental Results

3.1 Datasets

Two different datasets: The Pavia University and Salinas images are used in the following experiments to verify the effectiveness of the proposed 3D_CNN_IF method.

The Pavia University image was recorded by Reflective Optics System Imaging Spectrometer. It has 115 bands with size of 610 × 340. Only 103 bands are used for experiments due to water absorption. Its spectral range is from 0.43 to 0.86 μm. 42776 labeled pixels with 9 classes are contained in the ground truth map.

The Salinas image was recorded by the AVIRIS sensor over Salinas Valley, CA, USA. It has 224 bands with size 512 × 217. Remove water absorption, 204 bands are used for the experiments. Its spectral range is from 0.4 to 2.5 μm. 54156 labeled pixels with 16 classes are from the ground truth map.

In our experiments, the parameter setting for the proposed method is shown as follows: The spatial size of the input sample is 17 × 17, the number of group is set as 20, the first convolution layer has 12 filters with size 4 × 4 × 5, the second convolution layer has 36 filter with size 3 × 3 × 3. Max-pooling is used after the first convolution layer. The full connection layer has 100 units. And the end of the 3-D CNNs is the softmax classifier. The iteration number is set as 500 epochs for training, the learning rate is set as 0.05. 10 independent replications of all the experiments are conducted. The average values of overall accuracy (OA), average accuracy (AA), and kappa coefficient are used to evaluate the classification accuracy. For each replication, 10% of the labeled samples are randomly selected to generate the training set, and the rest samples for testing. The false color image of two datasets are shown in Fig. 2.

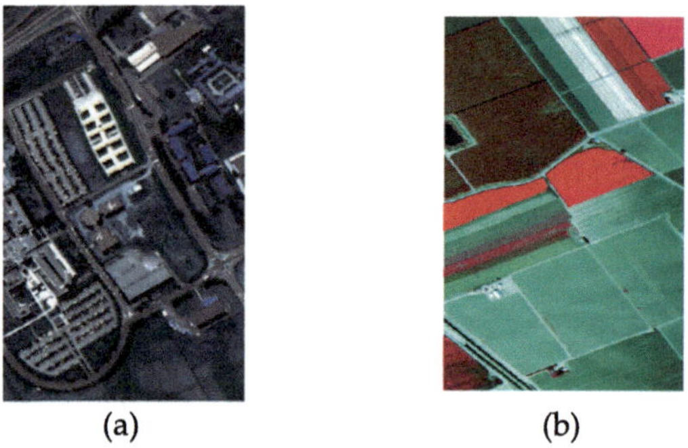

Fig. 2. Experiment datasets (a) False color image of Pavia University; (b) False color image of Salinas.

3.2 Comparisons to Other Methods

In this section, PCA, EMAPs [7], 1-D CNNs [21], SAEs [19], and 3D DWT [26] are used to make a comparison. For EMAPs method, the threshold value of standard deviation attribute is [2.5%, 5%, 7.5%, 10%], and the threshold value of area attribute is set as 200, 500, and 1000. For 3D DWT method, haar wavelet is utilized for feature extraction. The other parameters are set as the fault in their papers.

Figures 3, 4 show the RGB composition map for classification results and Tables. 1–2 show the classification accuries. PCA and 1D CNNs methods only use the spectral information for classification, and EMAPs and SAEs are the spectral-spatial based classification methods. 3D DWT and the proposed method are the 3D based methods, which can capture the spectral and spatial information simultaneously. We can observe from Figs. 3 and 4 that PCA method is sensitive to noise, and the classification maps present the noisy classification results. Although 1D CNNs belong to the spectral-based classification method, but it can capture the high-level features, and suppress the noise in classification map. Obviously, the classification results of spectral-spatial methods are better than the spectral-based method. Some misclassifications of the mixed spectral pixels are reduced in classification map, especially for class 3 of Pavia University and class 8 of Salinas. Of all the compared methods, 3D based methods shows a competitive classification result. Since 3D DWT cannot represent the image adaptively, which affects its performance. The proposed method can capture the high-level features of the HSI, and obtains much higher classification accuracy than other methods do Tables 1 and 2.

Table 1. Classification accuracies of Pavia University.

Classes	PCA	EMAPs	1D CNNs	SAEs	3D DWT	3D CNNs_IF
OA/total	82.06	96.36	93.28	95.18	95.99	97.77
AA	76.31	95.10	90.38	93.18	94.28	97.13
Kappa	74.96	95.17	91.04	93.59	94.67	97.04

Table 2. Classification accuracies of Salinas.

Classes	PCA	EMAPs	1D CNNs	SAEs	3D DWT	3D CNNs_IF
OA/total	96.95	97.45	91.73	92.01	94.25	97.97
AA	70.86	97.52	94.68	96.56	97.54	98.15
Kappa	74.19	97.16	90.77	91.10	93.60	97.75

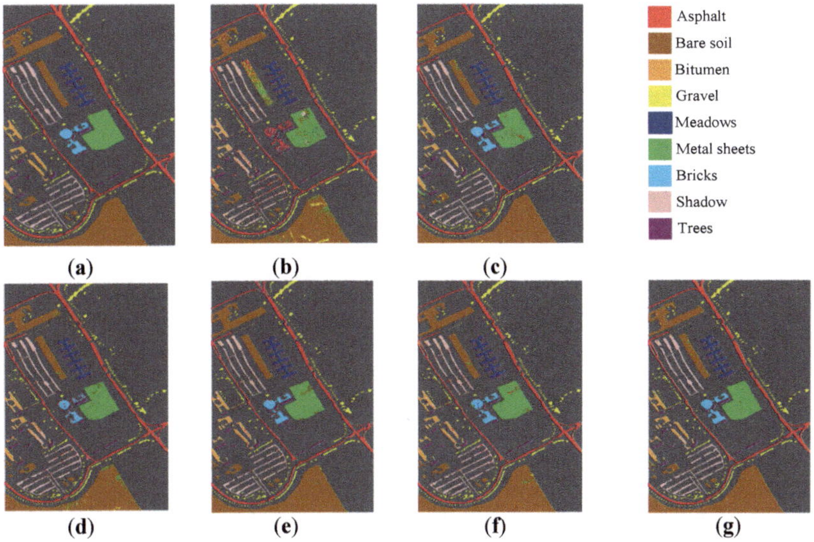

Fig. 3. Classification maps for Pavia University dataset; (a) Ground truth map; (b) PCA; (c) EMAPs [7]; (d) 1D CNNs [21]; (e) SAEs [19]; (f) 3D DWT [26]; (g) 3D CNNs_IF

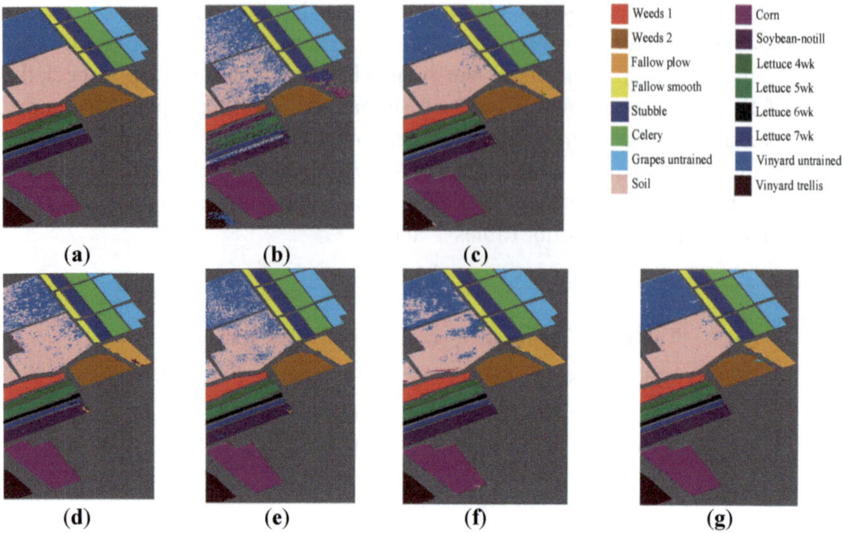

Fig. 4. Classification maps for Pavia University dataset; (a) Ground truth map; (b) PCA; (c) EMAPs [7]; (d) 1D CNNs [21]; (e) SAEs [19]; (f) 3D DWT [26]; (g) 3D CNNs_IF.

4 Conclusions

In this paper, a novel framework for HSI classification via 3D CNNs_IF is proposed. Image fusion technology is applied on spectral band of HSI to exploit the effective information, and 3D CNNs are used for exploiting the deep features of HSI. In our method, the "fusion-classification" networks can obtain more robustness and discrimination features, and the classification accuracy is improved obviously. For hyperspectral images, reducing spectral redundancy is very important for classification. In our further work, we will study more effective "fusion-classification" networks to further improve the classification accuracy.

Acknowledgements. This work was supported by the National Natural Science Foundation of China (61902313, 61701396, 61702409), Natural Science Foundation of Shaanxi Province [2020JQ-652], Research Project of Shaanxi Provincial Education Department.

References

1. Fauvel, M., Tarabalka, Y., Benediktsson, J.A., Chanussot, J., Tilton, J.C.: Advances in spectral-spatial classification of hyperspectral images. Proc. IEEE **101**, 652–675 (2013)
2. Prasad, S., Mann Bruce, L.: Limitations of principal components analysis for hyperspectral target recognition. IEEE Geosci. Remote Sens. Lett. **5**, 625–629 (2008)
3. Villa, A., Benediktsson, J.A., Chanussot, J.E.F.: Hyperspectral image classification with independent component discriminant analysis. IEEE Trans. Geosci. Remote Sens. **49**, 4865–4876 (2011)

4. Li, W., Prasad, S., Fowler, J.E.E.F.: Locality-preserving dimensionality reduction and classification for hyperspectral image analysis. IEEE Trans. Geosci. Remote Sens. **50**, 1185–1198 (2012)
5. Schmidt, F., Doute, S., Schmitt, B.: Wavanglet: an efficient supervised classifier for hyperspectral images. IEEE Trans. Geosci. Remote Sens. **45**, 1374–1385 (2007)
6. Daamouche, A., Melgani, F.: Swarm intelligence approach to wavelet design for hyperspectral image classification. IEEE Geosci. Remote Sens. Lett. **6**, 825–829 (2009)
7. Dalla Mura, M., Benediktsson, J.A., Waske, B.E.F.: Morphological attribute profiles for the analysis of very high resolution images. IEEE Trans. Geosci. Remote Sens. **48**, 3747–3762 (2010)
8. Jun, L., Bioucas-Dias, J.M., Plaza, A.: Spectral-spatial classification of hyperspectral data using loopy belief propagation and active learning. IEEE Trans. Geosci. Remote Sens. **51**, 844–856 (2013)
9. Li, J., Huang, X., Gamba, P.E.F.: Multiple feature learning for hyperspectral image classification. IEEE Trans. Geosci. Remote Sens. **53**, 1592–1606 (2015)
10. Tuia, D., Flamary, R., Courty, D.: Multiclass feature learning for hyperspectral image classification: Sparse and hierarchical solutions. J. Photogrammetry Remote Sens. **105**, 272–285 (2015)
11. Kang, X.D., Li, S.T., Benediktsson, J.A.: Feature extraction of hyperspectral images with image fusion and recursive filtering. IEEE Trans. Geosci. Remote Sens. **52**, 3742–3752 (2014)
12. Kang, X.D., Li, S.T., Benediktsson, J.A.: Spectral-spatial hyperspectral image classification with edge-preserving filtering. IEEE Trans. Geosci. Remote Sens. **52**, 2666–2677 (2014)
13. Bengio, Y., Lamblin, P., Popovici, D.E.F.: Greedy layer-wise training of deep networks. In: Proceedings of the Twentieth Conference on Neural Information Processing Systems. Advances in Neural Information Processing Systems 19, Vancouver, British Columbia, Canada. DBLP, pp. 153–160, December 2007
14. Hinton, G.E., Osindero, S., Teh, Y.: A fast learning algorithm for deep belief nets. Neural Comput. **18**, 1527–1554 (2006)
15. Salakhutdinov, R., Hinton, G.E.: Deep Boltzmann machines. J. Mach. Learn. Res. **5**, 1967–2006 (2009)
16. Lecun, Y.L., Bottou, L., Bengio, Y.E.F.: Gradient-based learning applied to document recognition. Proc. IEEE **86**, 2278–2324 (1998)
17. Lecun, Y.L., Kavukcuoglu, K., Farabet, C.: Convolutional networks and applications in vision. In: Proceedings of 2010 IEEE International Symposium on Circuits and Systems (ISCAS), pp. 253–256 (2010)
18. Kang, L., Kumar, J., Ye, P.E.F.: Convolutional neural networks for document image classification. In: IEEE 22nd International Conference on Pattern Recognition (ICPR), pp. 3168–3172 (2014)
19. Chen, Y., Lin, Z., Zhao, X.: Deep learning-based classification of hyperspectral data. IEEE J. Sel. Top. Appl. Earth Observations Remote Sens. **7**, 2094–2107 (2014)
20. Chen, Y., Zhao, X., Jia, X.: Spectral-spatial classification of hyperspectral data based on deep belief network. IEEE J. Sel. Top. Appl. Earth Observations Remote Sens. **8**, 1–12 (2015)
21. Hu, W., Huang, Y.Y., Wei, L.E.F.: Deep convolutional neural networks for hyperspectral image classification. J. Sens. **6**, 468–477 (2015)
22. Yue, J., Zhao, W.Z., Mao, S.J.E.F.: Spectral-spatial classification of hyperspectral images using deep convolutional neural networks. Remote Sens. Lett. **6**, 468–477 (2015)

23. Zhao, W.Z., Du, S.H.: Spectral-spatial feature extraction for hyperspectral image classification: A dimension reduction and deep learning approach. IEEE Trans. Geosci. Remote Sens. **4**, 4544–4554 (2016)
24. Zhao, W.Z., Du, S.H.: Learning multiscale and deep representations for classifying remotely sensed imagery. J. Photogrammetry Remote Sens. **113**, 155–165 (2016)
25. Shen, L., Zhu, Z.: Discriminative Gabor feature selection for hyperspectral image classification. IEEE Geosci. Remote Sens. Lett. **10**, 29–33 (2013)
26. Qian, Y., Ye, M.: Hyperspectral image classification based on structured sparse logistic regression and three-dimensional wavelet texture features. IEEE Trans. Geosci. Remote Sens. **51**, 2276–2291 (2013)

Traffic Flow Prediction Model and Performance Analysis Based on Recurrent Neural Network

Haozheng Wu[1], Yu Tang[1], Rong Fei[1(✉)], Chong Wang[2], and Yufan Guo[1]

[1] Xi'an University of Technology, Xi'an 710048, China
annyfei@xaut.edu.cn
[2] University of Electronic Science and Technology of China, Chengdu, China

Abstract. Traffic flow has the characteristics of randomness, nonlinearity and periodicity. Using deep learning as the method of traffic flow prediction can make full use of traffic flow characteristics for prediction. In this paper, a recurrent neural network-LSTM (Long Short-Term Memory) model and its variant GRU (Gated Recurrent Units) are used to solve the traffic flow prediction problem. According to the road traffic information collected by the automatic measurement station (LAM) of the Dutch traffic management department, the vehicle traffic flow data has been predicted every five minutes. It's verified that the GRU traffic flow prediction model performs well by comparing the prediction accuracy of LSTM and GRU.

Keywords: LSTM · GRU · Traffic flow

1 Introduction

Traffic flow refers to the vehicle flow formed by motor vehicles driving on the urban roads. It involves three parameters such as the number of vehicles currently driving on the road, speed of vehicle and road density [1]. The change laws of the can be studied to discover various characteristics of traffic flow, which are currently the first choice for many scholars to study traffic flow problems, and which are also an important basis for using intelligent transportation systems for traffic control and guidance [2].

At present, as there are few public datasets in China, and the traffic databases of most cities are still under construction. A large amount of high-quality data about traffic flow can't be obtained. Therefore, in this paper, the data used in the experiment comes from the Finnish Public Transport Authority, using the automatic measurement stations (LAM) to collect road traffic information. There are currently about 500 active LAM sites in Finland. Information is shared between the original form and the completed report.

This paper mainly analyzes and studies the LSTM and GRU models to effectively predict the short-term traffic flow, and compares their performance. First, the characteristics of the short-term traffic flow data in LAM points: influencing factors, processes and other characteristics are analyzed. In view of the characteristics such as

Q. Liu et al. (Eds.): CENet 2020, AISC 1274, pp. 429–438, 2021.
https://doi.org/10.1007/978-981-15-8462-6_48

dependence, timing and nonlinearity of flow data, a short-term traffic flow prediction model based on recurrent neural network LSTM and GRU was designed, and experiments have verified that performance of GRU is better.

2 Related Theories

LSTM (Long short term memory) is a neural network improved the defects of recurrent neural network, which adds a “processor” to the RNN neurons to determine whether the current information is to be forgotten [3]. The structure of this processor is also named cell, thus making the RNN network to have a long-term memory function and solving the problem of handling long-distance dependence of RNN [9], which is proposed by Hochreiter and Schmidhuber (1997). There are four controllable gates (the forgotten gate, input gate, candidate gate, and output gate) in each memory cell of the hidden layer, and the sequential data are interacted and transmitted in a unique way between them. As shown in Fig. 1.

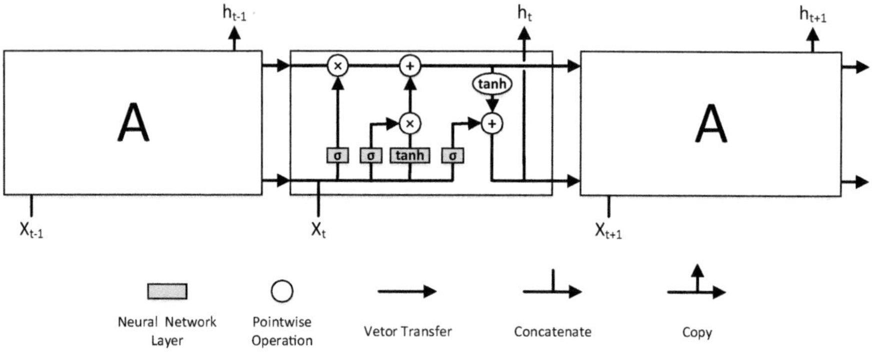

Fig. 1. The sketch of LSTM.

In the Fig. 1, vectors means to transfer, concatenate, and copy the data, as well as input and output. The circle represents the addition, multiplication and other operations of the corresponding vector, and the matrix A is the learned neural network layer.

The most important feature of LSTM is the existence of cell state. Combining the characteristics of the cycle, the cycle of the cell state is like a chain on which the information runs. There are only a few interactions, and the information can be continuously transferred. The controllable gate allows information to pass through selectively, which contains a sigmoid function and a vector multiplication operation. The Sigmoid function can receive the input value and then output a value between 0 and 1, which describes how much the current input value can pass through the gate. 0 means “not allow any quantity to pass” and 1 means “allow all quantities to pass” [4–6].

In the process of LSTM training, the first step of inputting information is to pass the forgotten gate, and then decide the information to forget from the cell state, that is, read h_{t-1} and x_t, and then output a number between [0, 1] to each cell state C_{t-1}, among which, 1 means "completely reserved" and 0 means "completely abandoned". The structure of the forgotten gate is shown in Fig. 2 below.

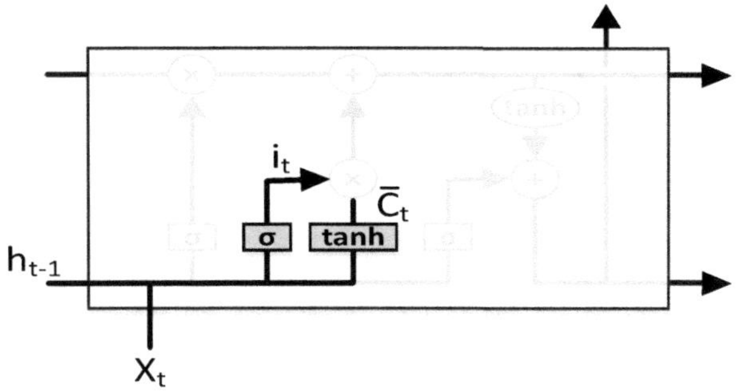

Fig. 2. The substructure of forgotten gate and function.

$$i_t = \sigma(W_i \cdot [h_{t-1}, x_t] + b_i) \tag{1}$$

$$\widetilde{C}_t = \tanh(W_C \cdot [h_{t-1}, x_t] + b_C) \tag{2}$$

where h_{t-1} indicates historical information, and x_t indicates new information currently flowing into the cell, and decides which historical information to forget.

The second Step of the training process is to prepare for updating the cell state, divided into two parts. The first part is the input gate where input data is saved into the cell unit. The second part is to create a new candidate vector and store the output value of the previous cell state, which can be compared to memorizing information of the previous cell state. Just like $\tilde{C}_{t-1}$ will be added to the state, combining these two parts to create a new memory, which is the next step, using two pieces of information to update the cell state. The structure of input gate is shown in the Fig. 3.

$$f_t = \sigma\left(W_f \cdot [h_{t-1}, x_t] + b_f\right) \tag{3}$$

$$C_t = f_t * C_{t-1} + i_t * \widetilde{C}_t \tag{4}$$

The third step is to update the old cell status, that is, update C_{t-1} to C_t. By multiplying C_{t-1} and f_t, the forgotten information is selected and forgotten. Adding $i_t * \tilde{C}_t$, this is the way to obtain C_t, and then update C_{t-1}.

The last Step is to pass the output gate. The information is filtered to determine the information needs to be output, by running a sigmoid function. Then, after processing the cell state through the tanh function, a value between [−1, 1] can be obtained, which after being multiplied with the output of the sigmoid function can determine and output the information we need (Fig. 4).

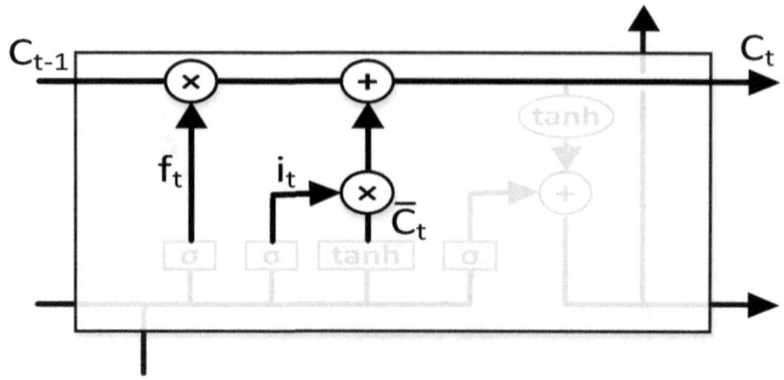

Fig. 3. Update old cell status and functions.

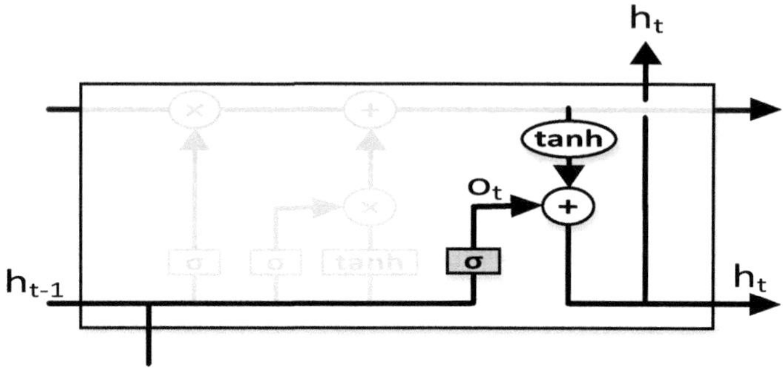

Fig. 4. The output process and function.

$$o_t = \sigma\left(W_o[h_{t-1}, x_t] + b_o\right) \tag{5}$$

$$h_t = o_t * \tanh(C_t) \tag{6}$$

A summary is as follow. The forget gate can determine the information to be abandoned in the memory cell, that is, how much information in the memory cell C_{t-1} at the previous moment can be transferred to the memory cell C_t at the current moment. The main function of the input gate is to control the input information to enter the memory cell C_t at the current moment selectively. The main function of the output gate is to determine how much information in the current memory cell C_t can enter the hidden layer state h_t. Use the forgotten gate and the input gate to update the memory cell, and use the tanh function to generate the candidate memory cell, that is, the information about to enter the output gate.

LSTM controls the current memory cell to forget, memory, update and output through the interaction of the four gates and also realize the continuous accumulation of historical information, that is, memory, through continuous cyclic update.

The training algorithm of LSTM is still BPTT, and the output value of each nerve is calculated forward, that is, the values of the five vectors f_t, i_t, C_t, O_t, S_t. The parameters that LSTM needs to learn and update are the weight matrix and the bias term of the four gates.

The state calculation formula of RNN is $S_t = f(W * S_{t-1} + U * x_t)$. According to the chain rule, the gradient will change to a multiplicative form, and the function value of sigmoid is less than 1, and the multiplicative form will lead to the situation that eventually approaches 0. In order to solve this problem, scientists used the accumulation form $S_t = \sum_{\tau=t}^{t} \nabla S_\tau$, whose derivative is also accumulation, so as to avoid the disappearance of the gradient. LSTM is used the accumulation form which can be seen from $C_t = f_t * C_{t-1} + i_t * \tilde{C}_t$.

The above is the basic principle of the LSTM model, and GRU is a variant of LSTM, proposed by Cho et al. [7, 8] in 2014. GRU only contains update and reset gates, which is z_t and r_t in the figure below. The role of the update gate is to determine how much information at the previous moment can enter the current cell state. The larger the value of the update gate is, the more information that enters the current cell state will be. The role of the reset gate is to select how much information to forgot at the previous moment. The larger the value of the reset gate is, the less information will be forgotten.

Through the Fig. 5, analogy with the LSTM model, the process and formula of its forward propagation can be obtained:

$$r_t = \sigma(W_t \cdot [h_{t-1}, x_t]) \tag{7}$$

$$z_t = \sigma(W_z \cdot [h_{t-1}, x_t]) \tag{8}$$

$$\tilde{h}_t = \tanh(W_{\tilde{h}}[r_t * h_{t-1}, x_t]) \tag{9}$$

$$h_t = (1 - z_t) * h_{t-1} + z_t * \tilde{h}_t \tag{10}$$

$$y_t = \sigma(W_o \cdot h_t) \tag{11}$$

z_t is the update gate, the logic gate when updating activation. r_t represents reset gate, and decides whether to abandon the previous activation h_t when candidate activation, $\tilde{h}_t$ represents candidate activation to receive $[x_t, h_{t-1}]$, and h_t represents activation the hidden layer of GRU, to receive $[h_{t-1}, \tilde{h}_t]$.

The update gate mainly performs linear transformation on the information at the previous time and the current time, that is, multiplies the weight matrix by the right, and then enter the added data into the update gate, that is, multiplies the sigmoid function, and the value obtained is between [0, 1].

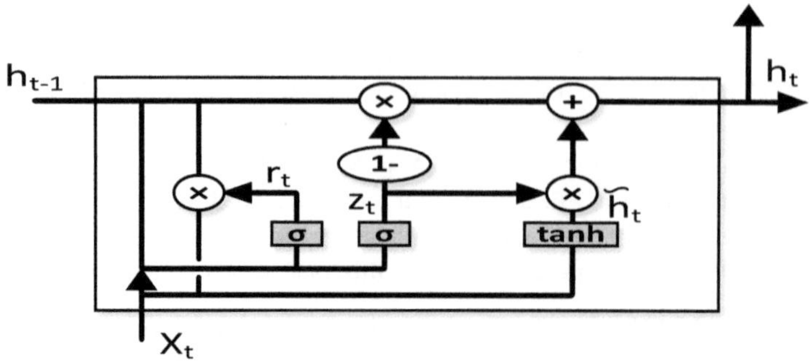

Fig. 5. The output process and function.

The main function of the reset door is to determine how much historical information cannot be passed on to the next moment. Similar to the data processing of the update gate, the information at the previous time and the current time is linearly transformed, that is, the weight matrix is respectively multiplied by the right, and then the added data is sent to the reset gate, which is multiplied by the sigmoid function and obtain the resulting value between [0, 1]. The difference lies in the value and use of these two weight matrix.

GRU no longer uses separate memory cells to store memory information, but directly uses hidden units to record historical states, using the reset gate to control the data volume of current information and memory information and generating new memory information to continue to forward. Because the output of the reset gate is between [0, 1], the amount of data that can continue to be transferred forward by using the reset gate to control the memory information. When the reset gate is 0, that means all the memory information is cleared, otherwise the reset gate is 1, it means all the memory information has passed.

The update gate is used to calculate the output of the hidden state at the current moment. The output information of the hidden state includes the hidden state information h_{t-1} at the previous moment and the hidden state output h_t at the current moment. The update gate is used to control the amount of data to pass on. The training method of GRU is similar to LSTM.

3 Model Design

3.1 Experimental Environment and Dataset

Operating System: Windows 10. Hardware platform: a universal PC. Development tools: Python3, Tensorflow, Keras.

Dataset: The one-month data from January 1, 2019 in Finland is used as the training set, and the data of the following week is used as the test set [9].

3.2 Experimental Environment and Dataset

In this paper, in order to compare the performance differences between the two models, the same construction method is used for both models, that is, they both involve a neural network including two hidden layers. Python is used for development, with Keras, a high-rise neural network API based on Tensorflow, Theano and CNTK backends. Using Keras framework, the speed we build models has been accelerated. This paper first generates a standardized object scaler from the processed training data and test data, using the object to standardize the dataset. Because the prediction task of sequential data needs to use historical data to predict the future data, the training data needs to be divided. A time-lagged variable is set up that lags = 12, the historical data of the previous one hour is used to predict the data of the next time node, and finally the training data set becomes the form to be (samples, lags).

The data set divided still has the characteristics of time sequence in order. Even though keras can choose to shuffle the data during training, the execution order is to sample the VAL data before shuffling, and the sampling process is still in order. Therefore, NP.random.shuffle is used to shuffle the data in advance to disrupt the order of the data.

Next is the construction of the model. Each hidden layer of the model contains 64 neurons. A dropout layer is added between the hidden layer and the output layer to reduce interdependence between nodes and realize the regularization of the neural network. Finally, the output layer contains 1 neural unit, that is, a number should be output.

There have been set 2 hidden layer LSTM2 and 2 hidden layer GRU networks.

LSTM and GRU are trained according to the normal RNN network, using the train_model () function to train ans RMSprop (lr = 0.001, rho = 0.9, epsilon = 1e−06) as the optimizer, with the batch_szie to be 256, epochs to be a total of 600, and lags to be 12 (scilicet time-lagged length is one hour).

Model regression prediction results are evaluated using several indicators as MAE, MSE, RMSE, MAPE, R2, explained_variance_score.

After the model training is completed, the processed test set is input into the model for prediction to obtain prediction accuracy, and the prediction result is shown in the Table 1. The results of evaluation indicators are as follows:

Table 1. Table captions should be placed above the tables.

	LSTM	GRU
MAPE	31.003164%	24.585686%
MAE	6.691722	6.703402
MSE	83.585697	85.958260
RMSE	9.142521	9.271368
R2	0.932984	0.931082
Explained_variance_score	0.933180	0.931228

In order to obtain the general rule, the parameters of the two models were adjusted, and the same dataset is used to train and test again. The process is the same as above. In this experiment, the optimizer was adjusted and adopted adam optimization (Lr = 0.001, beta_1 = 0.9, beta_2 = 0.999, epsilon = 1e−08) to train the model. The results of evaluation indicators are as follows (Table 2):

Table 2. The result of evaluation indicators.

	LSTM	GRU
MAPE	26.143534%	27.275156%
MAE	6.561796	6.570125
MSE	81.740679	81.411213
RMSE	9.041055	9.022816
R2	0.934463	0.934727
Explained_variance_score	0.934714	0.934815

At the same time, a three-layer LSTM and a GRU neural network model were also chosen to be built to compare the performance differences between the two models. Inputting the same dataset for training and testing, this experiment still used the RMSprop optimizer, compared with the results of the two-layer neural network:

The results of evaluation indicators are as follows (Table 3):

Table 3. The result of evaluation indicators.

	LSTM	GRU
MAPE	24.241568%	25.885307%
MAE	6.806903	7.819329
MSE	89.589734	125.560051
RMSE	9.465185	11.205358
R2	0.928170	0.899330
Explained_variance_score	0.928381	0.900794

Then change the optimizer used for training, the adma optimizer was used to train and test again. The following results were obtained:

The results of evaluation indicators are as follows (Table 4):

Table 4. The result of evaluation indicators.

	LSTM	GRU
MAPE	24.209921%	24.256364%
MAE	6.618004	6.649293
MSE	82.718285	86.228644
RMSE	9.094959	9.285938
R2	0.933679	0.930865
Explained_variance_score	0.933702	0.931027

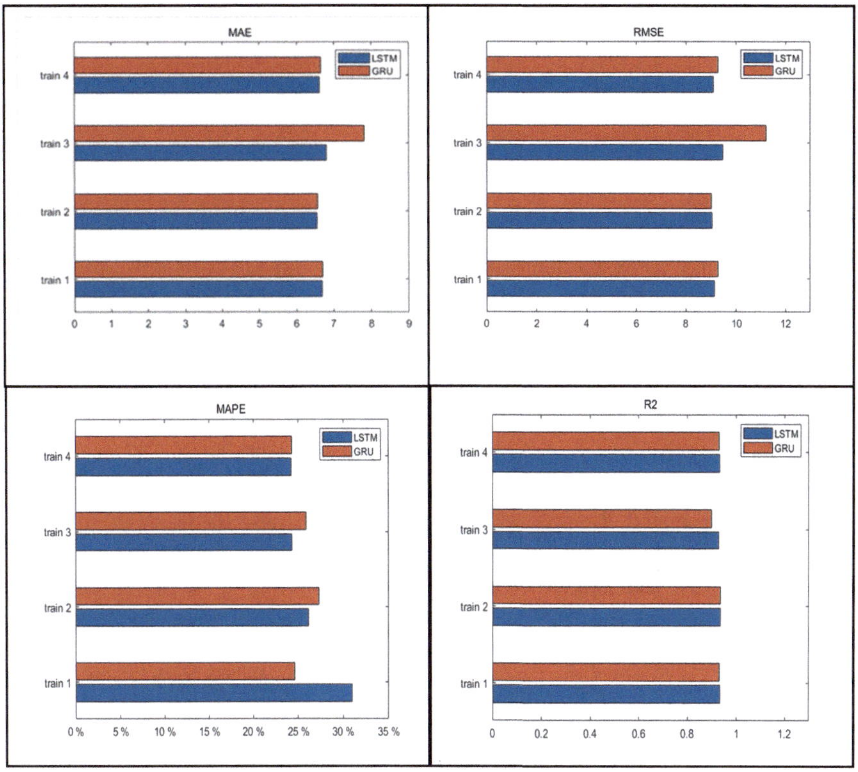

Fig. 6. The results of MAE, RMSE, MPE and R2.

4 Results Analysis

From Fig. 6, it can be seen from the above data that MAE, MSE and RMSE are the descriptions of the model prediction results and the error of the true value. It's knowen that LSTM and GRU perform poorly when using a three-layer neural network and using the RMSprop optimizer. Not only the error is relatively large, but the fitting effect is not good, and when the adma optimizer is used, as shown in training 4, the performance of the two models is almost the same, and the advantages and disadvantages cannot be distinguished. When using a two-layer neural network, the prediction error is significantly smaller than that of the three-layer neural network, and the fitting effect described by R2 is also much better than that of the three-layer. Therefore, two-layer neural network is taken as the analysis object to compare the two models. The MAPE chart should be focused to analyze, since the value of MAPE is a relative value, which is the average absolute percentage error. It not only take the error between the predicted value and the true value into consideration, but also considers the ratio of error to true value, used to evaluate the same set of data of different models. When using the RMSprop optimizer, the MAPE value of GRU is 5 percentage points points smaller, indicating that GRU is more accurate than LSTM in the longer-term prediction is the result. While using the Adam optimizer, the difference between LSTM and GRU is within one percentage point. The conclusion can be obtained that when predicting traffic flow data, the two-layer neural network performs better, and the performance of the GRU model is slightly better than LSTM.

5 Conclusion

This paper mainly analyzes and studies the LSTM and GRU models to effectively predict short-term traffic flow, and compares their performance. First, the research background of short-term traffic flow and the current research status based on this problem are introduced. Then, a systematic analysis of the problem is carried out. Taking traffic flow as the research object, the characteristics of traffic flow, influencing factors, and forecasting process are analyzed.

After analyzing and processing the data of the Dutch Transport Agency, the corresponding data sets were established. The mean method used to fill in missing data and the data normalized, the processed data is converted into the model's input matrix format for the training. In view of the characteristics such as dependence, timing and nonlinearity of flow data, a short-term traffic flow prediction model based on recurrent neural network LSTM and GRU is adopted, and the advantages and principles are understood and analyzed. Model established, trained and tested, from the result analysis GRU model possessed the better performance. In future research, the accuracy of the final prediction structure needs to be further improved.

References

1. Lv, Y., Duan, Y.: Traffic flow prediction with big data: a deep learning approach. IEEE Trans. Intell. Transp. Syst. **16**(2), 865–873 (2014)
2. Ahmed, M.S., Cook, A.R.: Analysis of Freeway Traffic Time Series Data by Using Box. Jenkins Techniques, vol. 722, pp. 1–9. Transportation Research Board, Washington, D.C. (1979)
3. Okutani, I., Stephanedes, Y.J.: Dynamic prediction of traffic volume through Kalman filtering theory. Transp. Res. Part B **18**(1), 1–11 (1984)
4. Castro-Neto, M., Jeong, Y.S., Jeong, M.K., et al.: Online-SVR for short-term traffic flow prediction under typical and a typical traffic conditions. Expert Syst. Appl. Int. J. **36**(3), 6164–6173 (2009)
5. Jeong, Y.S., Byon, Y.J., Castro-Neto, M.M., et al.: Supervised weighting-online learning algorithm for short-term traffic flow prediction. IEEE Trans. Intell. Transp. Syst. **14**(4), 1700–1707 (2013)
6. Smith, B.L., Demetsky, M.J.: Short-term traffic flow prediction: neural network approach. Transp. Res. Rec. (1453) (1994)
7. Cho, K.B., van Merrienboer, C., Gulcehre, F., Bougares, H., Schwenk, D., Bahdanau, Y., Bengio.: Learning phrase representations using RNN encoder-decoder for statistical machine translation, arXiv preprint arXiv:1406.1078 (2014)
8. Kuremoto, T., Kimura, S., Kobayashi, K., et al.: Time seried forecasting using a deep belief network with restricted Boltamann machines. Neural Comput. **137**, 47–56 (2014)
9. Hochreiter, S., Schmidhuber, J.: Long short-term memory. Neural Comput. **9**, 1735–1780 (1997)

Review on Deep Learning in Feature Selection

Yizhuo Zhang, Yiwei Liu, and Chi-Hua Chen(✉)

Fuzhou University, Fuzhou 350100, China
chihua0826@gmail.com

Abstract. Feature selection (FS) plays an important role in the machine learning (ML) field. Since FS solves the problem of dimensional explosion in ML very well, more and more people are paying attention to FS. Not only that, but this technique also takes advantage of the computational complexities and time reductions. Inspired by the points mentioned above, more and more FS algorithms solved by deep learning framework are appearing. Due to the importance of FS, it is necessary to conduct further research. However, FS is wide coverage, and the algorithms involved are numerous, which makes researchers need to spend a lot of time searching and reading the literature. In order to provide researchers with dedicated information and enable them to quickly have an overall understanding of the FS field, this article will from three aspects, including the main functions and framework of FS, search strategies of FS, and the evaluation strategy and algorithms in related fields to introduce FS from whole to part. Finally, this article discusses some existing problems and points out some promising research directions.

Keywords: Feature selection · Dimension reduction · Deep learning · Big data

1 Introduction

Nowadays, the Internet of things (IoT) has been developed rapidly, a large amount of data is collected by the enterprises and industries [1]. In the information age, people's daily lives and devices on the Internet also produce a large amount of data. These data meet the needs of ML, but the amount of data is too large to mark every piece of data, and not all features are useful, some data even have errors [2]. Another signification problem is that meaningless data need to be removed which may cause negative impacts on learning tasks. Therefore, there is a need for a method that can extract useful subsets from the original data and reduce the impact of redundant data to improve ML accuracy and stability. FS aims to reduce the data dimensions by removing some irrelevant features, so it is very suitable for solving these problems. Because deep neural networks proved working well in processing massive data, FS based on deep neural networks had been proposed to deal with these problems [3–5]. In recent years, there are many constructive methods integrated with ML top research theories have been proposed, such as Knockoff generative adversarial network (GAN), FS guided auto-encoder.

In order to have a better understanding of the FS, this study will raise three general questions (GQ), which are focus on the overall FS and domain-specific methods.

Q. Liu et al. (Eds.): CENet 2020, AISC 1274, pp. 439–447, 2021.
https://doi.org/10.1007/978-981-15-8462-6_49

Table 1 describes the general questions. Answering these questions can give a general understanding of the FS.

Table 1. General questions.

Identifier	General Questions (GQ)
GQ1	What is the main purpose of the FS?
GQ2	What are the search strategies for FS?
GQ3	What are the evaluation methods for FS?

The remainder of the paper is organized as follows. Section 2 presents search strategies and sources. In Sect. 3, review results are discussed for these general questions. The conclusions and future work are given in Sect. 4.

2 Sources

The articles involved in this study are from journals and conferences by searching in online academic searching engines (e.g., Web of Science). The relevant keywords used for searching such as FS, search strategy, filter, wrapper, evaluation methods. When selecting papers, the quality, amount of citations and contribution of algorithms are considered.

3 Results and Discussions

In this section, the relevant papers have been reviewed and discussed to answer the three general questions in Table 1.

3.1 GQ1–What Is the Main Purpose of the FS?

Due to the rapid development of the IoT and information technology, the data from various collection devices usually contain a large amount of redundant information and error information. This kind of data may cause an adverse impact on big data analysis. To solve this problem, the data is usually pre-processed which is dimension reduction. There are two popular dimension reduction methods which are feature extraction (FE) and FS. The result of FS is a subset of the original data, and the result of FE is a mapping of the original data. Figure. 1 shows the process of FS. FS aims to approximate the original data by selecting a set of essential features. So, FS method can make a model more reliable and be furnished with better interpretability. In FS, search strategies and evaluation methods are critical to the quality of selection results.

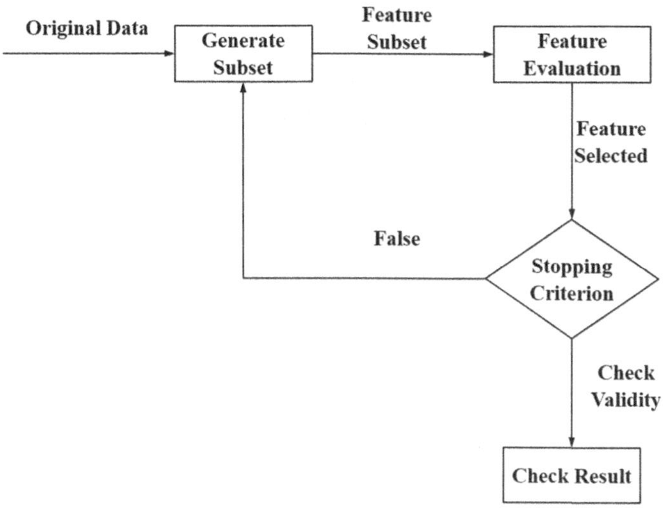

Fig. 1. The process of FS.

3.2 GQ2-What Are the Search Strategies for FS?

The current research generally divides the search strategy of FS into three parts: (1) global search strategy, (2) random search strategy, and (3) heuristic search strategy [6].

The global search strategy can analyze and evaluate each subset of data to decide whether to select this feature. Generally, there are enumeration method and branch-and-bound method [7]. The enumeration method is easy to achieve but they will consume a lot of time. The branch-and-bound method is preferred if the amount of data is large. The branch-and-bound method will define an evaluation function and a number of features that would be selected before the algorithm runs [8].

The random search strategy can weight each piece of data according to the validity of its feature, and select depend on a defined threshold. The random search strategy usually combines FS problems with other artificial intelligence algorithms such as simulated annealing algorithm (SAA) and genetic algorithm (GA) [9]. Therefore, if the number of epochs in these algorithms is too small, the accuracy of the results may be unstable. When the number of epochs is large enough and the parameter settings are appropriate, the random search strategy methods would show a better result.

The heuristic search strategy can create a sequence for the original data. In this sequence, a forward selection method and a backward selection method are used to add and delete a feature, until the optimal feature subset is reached [10]. Four types of methods for the heuristic search strategy include sequence forward selection method (SFS), generalized sequence forward selection method (GSFS), sequence backward selection method (SBS), generalized sequence backward selection method (GSBS).

3.3 GQ3-What Are the Evaluation Methods for FS?

The evaluation methods of FS are generally divided into filter methods, wrapper methods, and embedded methods [11, 12].

Filter Methods. The filter method generally uses the statistical performance of all training data directly to evaluate features, so it is independent of the subsequence learning algorithm. It can quickly remove redundant features in large data sets [13]. A variance threshold method (VTM) and a mutual information method (MIM) will be introduced in this subsection.

Variance Threshold Method. The VTM can calculate the variance of each feature and select the features whose variance is greater than the defined threshold [14].

The variance of feature is defined as Eq. (1).

$$V(t) = \frac{1}{m}\sum_{i=1}^{m}\left(f_i(t) - \bar{f}(t)\right)^2 \tag{1}$$

In Eq. (1), m is the number of categories, and the $f_i(t)$ means the frequency of feature t in the i-th category. $\bar{f}(t)$ notes the average of the frequency of feature t in Eq. (2).

$$\bar{f}(t) = \frac{1}{m}\sum_{i=1}^{m} f_i(t) \tag{2}$$

The variance of feature $V(t)$ denotes the discrete trend of feature t. The distribution of feature t in m categories is more concentrated with a larger variance $V(t)$. Equation. (3) could be used to normalize variances $V(t)$ for avoiding the suppression of low frequencies by high frequencies.

$$N(t) = \frac{\frac{1}{m}\sum_{i=1}^{m}\left(f_i(t) - \bar{f}(t)\right)^2}{\frac{1}{m}\sum_{i=1}^{m}\left(f_i(t)\right)^2} = \frac{V(t)}{\frac{1}{m}\sum_{i=1}^{m}\left(f_i(t)\right)^2} \tag{3}$$

The distribution probability of feature t in category $c_j$is defined as Eq. (4).

$$\theta_j = \frac{f_{c_j}(t)}{|c_j|} \tag{4}$$

In Eq. (4), $f_{c_j}(t)$ denotes the frequency of feature t in category c_j, and $|c_j|$denotes the sum number of category c_j occurred in data. Therefore, the feature evaluation function is defined as Eq. (5).

$$T(t) = V(t) \times \max_{1 \le j \le m} \theta_j \tag{5}$$

After calculating the all features evaluation score $T(t)$, the features whose score is greater than the defined threshold can be selected.

Mutual Information Method. The MIM is a way of representing the correlation between features. The information entropy of **X** can be estimated by Eq. (6), and the conditional information entropy of **X** after the known variable **Y** is calculated by Eq. (7) [15].

$$H(\mathbf{X}) = -\sum_{x \in \mathbf{X}} \Pr(x) \times \log \Pr(x) \tag{6}$$

$$H(\mathbf{X}|\mathbf{Y}) = -\sum_{y \in \mathbf{Y}} \Pr(y) \times \sum_{x \in \mathbf{X}} \Pr(x|y) \times \log \Pr(x|y) \tag{7}$$

In Eqs. (6) and (7), $\Pr(x)$ means the probability that feature **X** takes x, and $\Pr(x|y)$notes the probability that feature **X** takes x when feature **Y** takes y. The mutual information $I(\mathbf{X};\mathbf{Y})$ between feature **X** and feature **Y** can be calculated by Eq. (8).

$$\begin{aligned} I(\mathbf{X};\mathbf{Y}) &= H(\mathbf{X}) - H(\mathbf{X}|\mathbf{Y}) = H(\mathbf{Y}) - H(\mathbf{Y}|\mathbf{X}) \\ &= \sum_{y \in \mathbf{Y}} \sum_{x \in \mathbf{X}} \Pr(x,y) \times \log \frac{\Pr(x,y)}{\Pr(x)\Pr(y)} \end{aligned} \tag{8}$$

Furthermore, the correlation between feature **X** and feature **Y** can be calculated by Eq. (9).

$$R(\mathbf{X};\mathbf{Y}) = \frac{I(\mathbf{X};\mathbf{Y})}{H(\mathbf{X}) + H(\mathbf{Y})} \tag{9}$$

The value range of $Sim(x,y)$ is between [0, 1]. If the value of $R(\mathbf{X};\mathbf{Y})$ is 0, it means that two features are not related; if the value of $R(\mathbf{X};\mathbf{Y})$ is 1, it means that two features are completely related.

Wrapper Methods. Because the evaluation of the wrapper methods depends on the training accuracy of the subsequent learning algorithm, they have fewer errors than the filter methods. However, the computation cost of wrapper methods is large, it is not suitable for large data set operations.

In related research, Guyon et al. used the support vector machine (SVM) classification algorithm to measure the importance levels of features, and constructed a classifier with better classification performance [16]. Gui et al. proposed a method based on attention mechanism with high accuracy and stability for noisy and small data sets [17]. Ye et al. proposed a fast wrapper FS method called fast feature subset ranking (FFSR) [18].

Embedded Methods. The embedded methods which are different from the filter methods and the wrapper methods combine the variable selection and training process with the advantages of high efficiency and better integration ability. The embedded methods also consider the relationship between each two-features and usually incorporate the relationship into the learning phase of another algorithm.

Several embedded unsupervised FS methods use regularization for selecting discrete features. However, by using regularization, the embedded methods usually require higher computation cost to get the final results. An embedded end-to-end FS method brings a good solution to this problem. The proposed method introduces a concrete auto-encoder for differentiable FS and reconstruction [19]. And the concrete auto-encoder contains a concrete selector layer encoder and a standard neural network decoder.

Han et al. proposed an auto-encoder feature selector (AEFS) method which combines group lasso tasks and auto-encoder [20]. By excavating the linear and nonlinear information among features, the AEFS can select the optimal features. Most traditional embedded based methods (e.g., least absolute shrinkage and selection operator (LASSO) [21]) only work in the linear relationship and ignore the nonlinear relationship between each two-features. Therefore, the neural networks are used to learn the flexible nonlinear relationship between features, which can flexibly handle various tasks (e.g., Image segmentation [22] and image recognition [23, 24]). The studies used an auto-encoder to improve the effect on unsupervised learning and selected features with high recognition.

Auto-Encoder Feature Selector. The goal of AEFS is to select t optimal features from $\mathbf{X}$. The matrix $\mathbf{X} = [x_1, x_2, \ldots, x_m]^T$ is the unlabeled sample matrix from data; d is the sample dimension, and m is the number of features.

The auto-encoder in AEFS has two fully connected layers and uses a typical h-dimensional hidden layer including a coding function $g(\mathbf{X})$ and a decoding function $h(g(X))$ as Eqs. (10) and (11).

$$f(\mathbf{X}) = \sigma_1\left(\mathbf{X}\mathbf{W}^{(1)}\right) \tag{10}$$

$$\hat{\mathbf{X}} = h(g(\mathbf{X})) = \sigma_2(g(\mathbf{X})\mathbf{W}^{(2)}) \tag{11}$$

In Eqs. (10) and (11), σ_1 and σ_2 represent the activation functions in the hidden layer and the output layer; $\Theta = \{\mathrm{w}^1, \mathrm{w}^2\}$ is the weight parameters, and $h(g(\mathbf{X}))$ is the overall function of the auto-encoder. In the learning process, the article uses the least square method to describe the loss function of the auto-encoder. The loss function is expressed as Eq. (12).

$$\Im(\Theta) = \frac{1}{2m}\|\mathbf{X} - h(g(\mathbf{X}))\|_F^2 \tag{12}$$

The goal is optimizing loss function. After learning process, the auto-encoder can reduce the dimension of the matrix $\mathbf{X}$ and output the decoding matrix as $\hat{\mathbf{X}} = h(g(\mathbf{X}))$.

In the FS stage, by comparing the reconstructed data result $\hat{X}$ and original data $\mathbf{X}$ after discarding some features; if the features of these two matrices are similar, redundant features can be removed. The matrix $\mathrm{W}^{(1)} = [w_1, w_2, \ldots w_d]^T$ is used to represent the weights among the neurons in the input layer and the neurons in the hidden layer. If $\|w_i\|_2 \approx 0$, the i-th feature is very sparsely connected; otherwise, the connection is very closed. In order to be able to select the more important feature from $\mathbf{X}$, the row-sparse regularization is used for $\mathbf{W}^{(1)}$. The row-sparse regularization can be calculated by Eq. (13) which is used to find the L2 norm and the L1 norm for $\mathbf{W}^{(1)}$ in turn ($\|\mathbf{W}^{(1)}\|_{2,1}$). In Eq. (13), h is the dimension of hidden layer; d is the sample dimension; d represents the weight parameters of the i-th neuron in the l-th layer to the j-th neuron in the (l+1)-th layer.

$$\|\mathbf{W}^{(1)}\|_{2,1} = \sum_{i}^{d} \sqrt{\sum_{j}^{h} (w_{ij}^{(1)})^2} \tag{13}$$

Therefore, the Eq. (12) can be reformulated as Eq. (14).

$$\Im(\Theta) = \frac{1}{2m} \|\mathbf{X} - h(g(\mathbf{X}))\|_F^2 + \alpha \|\mathbf{W}^{(1)}\|_{2,1} \tag{14}$$

In the training process, in order to avoid the problem of overfitting, a weight decay term is necessary. The final loss function is defined as Eq. (15).

$$\Im(\Theta) = \frac{1}{2m} \|\mathbf{X} - h(g(\mathbf{X}))\|_F^2 + \alpha \|\mathbf{W}^{(1)}\|_{2,1} + \frac{\beta}{2} \sum_{i=1}^{2} \|\mathbf{W}^{(i)}\|_F^2 \tag{15}$$

In Eq. (15), β is a penalty parameter. The AEFS [19] can be used to optimize Eq. (15).

Summary and Discussions. In summary, there are three kinds of evaluation methods in FS, which are filter method, wrapper method and embedded method. For the filter method, it mainly focuses on the correlation between the features in the original data. It is responsible for selecting optimal features and sending these features to the subsequent learning algorithm. Therefore, the filter method is more versatile and wound not be affected by the subsequent learning process. If the amount of original data set is huge, the filter method is a good solution, and it can quickly remove a large number of redundant features. However, because it is independent of the learning algorithm, the accuracy of FS of the filter method is usually lower than the wrapper method. For the wrapper method, it relies on the learning effect of the subsequent learning algorithm to evaluate the feature subset. Therefore, if you want to perform FS and evaluate FS performance on an original data set, it is necessary to train a classifier and evaluate the feature subset based on the performance of the classifier. The performance of the wrapper method is better than the filter method, and the wrapper method is more accurate and reliable. The disadvantages of the wrapper method are that it requires training and evaluate each generated subset, so it is not suitable for large-scale data set operations. The embedded method is similar to the wrapper method, but the difference from the first two methods is that the embedded method uses the FS algorithm itself as a part of the learning algorithm and selects features during each learning process. Embedded needs to redesign the learning algorithm to combine FS, so the complexity and difficulty of implementation will increase. But the embedded method would save a lot of time in the model training stage. Embedded as a hotspot of current research, it has the advantages of high efficiency and integration with machine learning.

4 Conclusions and Future Work

This paper first analyzes the main goals and general architecture of FS, then classifies FS according to search strategies and evaluation strategies. To solve high-dimensional problems, FS can remove redundant features according to global search strategy, random search strategy, or heuristic search strategy. It can choose the filter method to deal with massive data or the wrapper method to improve the quality of generated subsets. It also can use the embedded method and combine other ML algorithms to make better performance. However, there are still some shortcomings such as a single algorithm that cannot support both high-dimensional and low-dimensional data. In the future, the development direction of FS would focus on combining with other ML algorithms, such as adversarial neural networks, convolutional neural networks, etc., to be more suitable for practical applications. And the multi-view FS would also be a hot topic.

Acknowledgements. This work was partially supported by the National Natural Science Foundation of China (Nos. 61906043, 61877010, 11501114 and 11901100), Science Foundation of the Fujian Province, China (No. 2019J01243), Funds of Education Department of Fujian Province (No. JAT190026), and Fuzhou University (Nos. 510730/XRC-18075, 510809/GXRC-19037, 510649/XRC-18049, and 510650/XRC-18050).

References

1. Yin, S., Ding, S.X., Xie, X., Luo, H.: A review on basic data-driven approaches for industrial process monitoring. IEEE Trans. Ind. Electron. **61**(11), 6418–6428 (2014)
2. Guyon, I., Elisseeff, A.: An introduction to variable and feature selection. J. Mach. Learn. Res. **3**, 1157–1182 (2003)
3. Wang, Q., Zhang, J., Song, S., Zhang, Z.: Attentional neural network: feature selection using cognitive feedback. In: NIPS 2014: Proceedings of the 27th International Conference on Neural Information Processing Systems, vol. 2, pp. 2033–2041. ACM, NY, USA (2014)
4. Li, Y., Chen, C.Y., Wasserman, W.W.: Deep feature selection: theory and application to identify enhancers and promoters. Lecture Notes in Computer Science, vol. 9029, pp. 205–217 (2015)
5. Roy, D., Murty, K.S.R., Mohan, C.K.: Feature selection using deep neural networks. In: Proceedings of the 2015 International Joint Conference on Neural Networks, pp. 1–6. IEEE, NJ, USA (2015)
6. Belkin, M.: Problems of Learning on Manifolds. The University of Chicago, IL, USA (2003)
7. Narendra, P.M., Fukunaga, K.: A branch and bound algorithm for feature subset selection. IEEE Trans. Comput. **26**(9), 917–922 (1977)

8. Li, M., Kamil, M.: Research on feature selection methods and algorithms. Comput. Technol. Dev. **2013**(12), 16–21 (2013)
9. Wu, B., Abbott, T., Fishman, D., McMurray, W., Mor, G., Stone, K., Ward, D., Williams, K., Zhao, H.: Comparison of statistical methods for classification of ovarian cancer using mass spectrometry data. Bioinformatics **19**(13), 1636–1643 (2003)
10. Pudil, P., Novovicova, J., Kittler, J.: Floating search methods in feature selection. Pattern Recogn. Lett. **15**(11), 1119–1125 (1994)
11. Li, J., Cheng, K., Wang, S., Morstatter, F., Trevino, R.P., Tang, J., Liu, H.: Feature selection: a data perspective. arXiv: 1601.07996 (2016)
12. Wang, Y., Xu, C., Xu, C., Tao, D.: Beyond RPCA: flattening complex noise in the frequency domain. In: NIPS 2014: Proceedings of AAAI 2017: Proceedings of the Thirty-First AAAI Conference on Artificial Intelligence, pp. 2761–2766. ACM, NY, USA (2017)
13. Yao, X., Wang, X., Zhang, Y., Quan, W.: Summary of feature selection algorithms. Control Decis. **27**(2), 161–313 (2012)
14. Yuan, Y., Wang, X.: Text feature selection algorithm based on variance. Comput. Eng. **2012** (12), 155–157 (2012)
15. Jiang, S., Wang, L.: Feature selection based on feature similarity measure. Comput. Eng. Appl. **46**(20), 153–156 (2010)
16. Guyon, I., Weston, J., Barnhill, S., Vapnik, V.: Gene selection for cancer classification using support vector machines. Mach. Learn. **46**, 389–422 (2002)
17. Gui, N., Ge, D., Hu, Z.: AFS: An attention-based mechanism for supervised feature selection. In: Proceedings of AAAI 2019: Proceedings of the Thirty-Third AAAI Conference on Artificial Intelligence, pp. 3705–3713. AAAI, CA, USA, (2019)
18. Ye, J., Gong, X.: A novel fast wrapper for feature subset selection. J. Changsha Univ. Sci. Technol. (Nat. Sci.) **2010**(4), 69–73 (2010)
19. Balin, M.F., Abid, A., Zou, J.Y.: Concrete autoencoders: differentiable feature selection and reconstruction. In: Proceedings of ICML 2019: Proceedings of the 36th International Conference on Machine Learning, PMLR 97:, pp. 444–453. JMLR, Inc. and Microtome Publishing, USA (2019)
20. Han, K., Wang, Y., Zhang, C., Li, C., Xu, C.: Autoencoder inspired unsupervised feature selection. In: Proceedings of the 2018 IEEE International Conference on Acoustics, Speech and Signal Processing (ICASSP), pp. 2941–2945. IEEE, NJ, USA (2018)
21. Tibshirani, R.: Regression shrinkage and selection via the lasso. J. R. Stat. Soc. **58**(1), 267–288 (1996)
22. Long, J., Shelhamer, E., Darrell, T.: Fully convolutional networks for semantic segmentation. In: Proceedings of the 2015 IEEE Conference on Computer Vision and Pattern Recognition (CVPR), pp. 3431–3440. IEEE, NJ, USA (2015)
23. He, K., Zhang, X., Ren, S., Sun, J.: Deep residual learning for image recognition. In: Proceedings of the 2016 IEEE Conference on Computer Vision and Pattern Recognition (CVPR), pp. 770–778. IEEE, NJ, USA (2016)
24. Wang, Y., Xu, C., You, S., Tao, D., Xu, C.: Cnnpack: packing convolutional neural networks in the frequency domain. In: Advances in Neural Information Processing Systems 29 (NIPS 2016), pp. 253–261 (2016)

Design and Implementation of Handwritten Digit Recognition Based on K-Nearest Neighbor Algorithm

Ying Wang[1,2], Qingyun Liu[1,2(✉)], Yaqi Sun[1,2], Feng Zhang[1,2], and Yining Zhu[1]

[1] College of Computer Science and Technology, Hengyang Normal University, Hengyang 421002, China
liuqy@139.com

[2] Hunan Provincial Key Laboratory of Intelligent Information Processing and Application, Hengyang 421002, China

Abstract. In this paper, a handwritten digit recognition system based on K-nearest neighbor algorithm is designed and implemented. By analyzing the application principle and steps of k-nearest neighbor algorithm in machine learning classifier, we design handwritten digit recognition in Python language. In order to improve the accuracy and accuracy of handwritten character recognition, the method of smooth denoising and character segmentation methods are used to preprocess the data, extract the structure and statistical features. The experiments had shown that the accuracy of the algorithm for handwritten character recognition can reach over 97%.

Keywords: K-nearest neighbor algorithm · Handwritten digit recognition · Image preprocessing

1 Introduction

At present, there are still a large number of handwritten digital documents that need to be input into the computer for management, such as handwritten express documents in the logistics industry, handwritten checks and remittance documents in the banking industry, company account books, etc. [1]. Due to the randomness of handwritten numbers, the accuracy of recognition is easily affected by font size, stroke thickness and tilt angle, so the design of handwritten number recognition method and system has important theoretical value and practical significance. The K-nearest neighbor algorithm is an important artificial intelligence algorithm based on statistical characteristics [2]. It is easy to understand and has low requirements for the system. Python is an interpreted, object-oriented and dynamic high-level scripting language. It can realize fast machine learning and artificial intelligence algorithm by the compiler. In this paper, Python is used as the programming language to develop a small handwritten number recognition system.

Q. Liu et al. (Eds.): CENet 2020, AISC 1274, pp. 448–455, 2021.
https://doi.org/10.1007/978-981-15-8462-6_50

2 KNN Algorithm Principle

KNN algorithm is a simple and classic machine learning classification algorithm [3]. It calculates the distance between the samples to be classified X and each training sample according to the distance function, and selects the K samples with the smallest distance from the samples to be classified as the k nearest neighbors of X. Finally it takes the category of most of the k nearest neighbors of X as the category of X. The classification sample can be used for handwritten numeral recognition by replacing it with handwritten numeral sample.

KNN algorithm consists of the following three steps.

(1) Calculate distance: given a test object, calculate its distance from each object in the training set.
(2) Find neighbor: circle the nearest K training objects as the nearest neighbor of the test object.
(3) Classify: classify the test objects according to the main categories of the k nearest neighbors.

Therefore, the most critical is the calculation of distance. Generally speaking, to define a distance function d (x, y), we need to meet the following criteria [4].

$$d(x,y) = 0$$

$$d(x,y) \geq 0$$

$$d(x,y) = d(y,x)$$

$$d(x,k) + d(k,y) \geq d(x,y)$$

In the calculation of distance, Euclidean measure (also known as Euclidean distance) is generally used. Euclidean distance in two-dimensional and three-dimensional space is the actual distance between two points, and its calculation formula is as follows:

$$\mathrm{d(x,y)} = \sqrt{\sum_{i=1}^{n}(x_i - y_i)^2} \tag{1}$$

3 The Implementation of KNN Algorithm for Handwritten Digit Recognition

Handwritten digit recognition mainly includes image preprocessing module, feature extraction module, classifier design module and digit recognition module [5]. The flow is shown in Fig. 1.

3.1 Selection of Handwriting Data Set

The research of handwritten numeral recognition algorithm needs a good handwritten numeral set (i.e. classification set) for training and testing. MNIST handwritten data set is a collection of handwritten numbers of different age groups organized by Google Labs, New York University and Microsoft Research Institute, including 60000 training sets and 10000 test sets [6]. Therefore, we choose data sets of diverse data types and large amount of MNIST data as data sets for testing and training in this research. Some handwritten digital pictures in MNIST dataset are shown in Fig. 2.

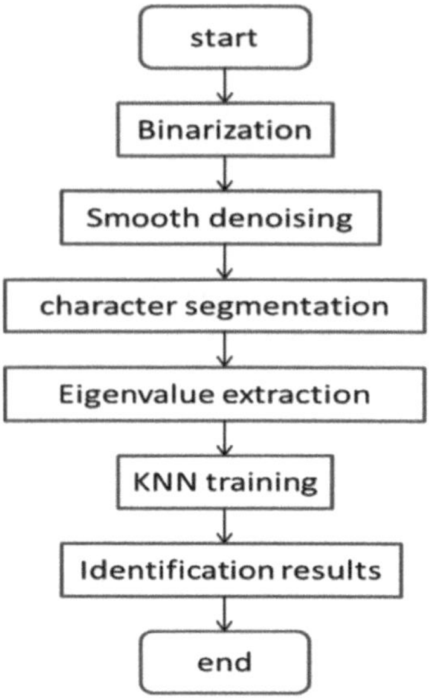

Fig. 1. Flow chart of handwritten digit recognition

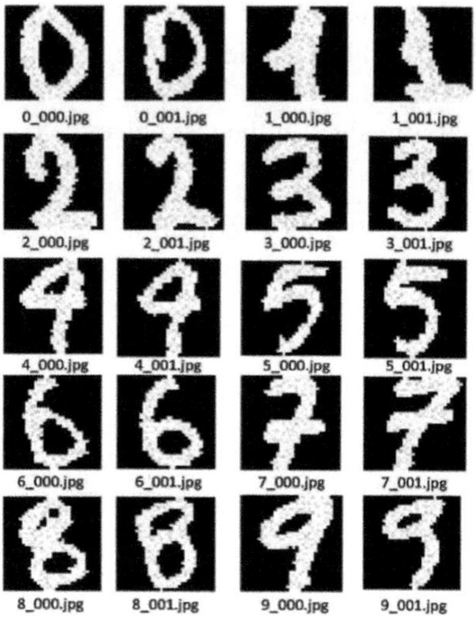

Fig. 2. Some samples of handwritten figures.

3.2 Image Preprocessing

The image acquisition is greatly affected by the external environment, resulting in a great deal of interference in the image itself, so the direct image cannot be directly used, and needs to be preprocessed. The purpose of preprocessing is to remove the noise in the image on the premise of preserving the original image information as much as possible, so that the image can adapt to the needs of the model. Therefore, generally speaking, for different models, the way of preprocessing is different. At present, the commonly used preprocessing methods include binarization, smoothing and denoising.

(1) Binarization

Setting the gray value of the pixels on the image to 0 or 255 is the process of rendering the whole image with obvious black-and-white effect. The original digital image contains gray-scale information which is not needed by neural network recognition.

This gray-scale information will seriously affect the processing speed of the program. The purpose of binarization is to remove this unnecessary image information, separate the objects in the image from the background, and speed up the processing speed of the program.

There are many methods of binarization. In this system, the global threshold binarization method is to determine a threshold value according to the histogram or the spatial distribution of the gray level of the image, and to transform the gray level image to the binary image according to the threshold value [7].

(2) Smooth denoising

In the process of image capturing, because the occlusion and other objects in the air will make the image captured by the camera noisy, and after binarization processing, the image will increase some noise points, which needs to be filtered through image denoising, so as to reduce the non-target information in the image recognition process, and the processed image brightness and contrast will be adjusted appropriately, The image looks smoother. The system uses wavelet threshold to denoise.

3.3 Eigenvalue Extraction

In the task of image classification, feature extraction is a very important step. The result of feature extraction can directly affect the final result. Although people are very proficient in the recognition of numbers, but because of the lack of understanding of the recognition mechanism of the brain, it is not clear what features people pay attention to when they recognize a number. Generally, people choose features based on experience, which are verified by experiments. In generally, feature extraction is particularly important in traditional pattern recognition and in neural network pattern recognition, and selecting suitable features affect the final recognition results directly. Therefore, the results of feature extraction are affected by different methods and features. In this algorithm, the image is grayed and converted to black-and-white image, then the image is zoomed to 32×64 size, and then each pixel is processed into binary data.

3.4 Implementation of KNN Algorithm

Because the data set is a binary TXT text format, it is necessary to convert the 32×64 test set and training set into 1×1024 one-dimensional vector, and calculate the Euclidean distance between each sample in the test set and each sample in the training respectively. The calculation results are classified and extracted from each sample label according to KNN principle. KNN core algorithm code is as follows:

```
class KnnClassifier (object):
def __init__(self, labels, samples):
"""Initialize classifier with training data."""
# Initialize classifier with training data
self.labels = labels# Corresponding mark
self.samples = samples# Sample set: it can be a number or a string, etc
def classify (self, point, k = 3):
```

```
"""Classify a point against k nearest
in the training data, return label."""
# K-nearest neighbor classification is used in training data, and marks are returned
dist = array ([L2dist(point, s) for s in self.samples])
ndx = dist.argsort()#Sort them out
votes = {}# Using dictionary to store k-nearest neighbor
for i in range(k):
label = self.labels[ndx[i]]
votes.setdefault (label, 0)
# If the dictionary contains a given key, the value corresponding to the key is returned; otherwise, the value set for the key is returned.
votes[label] += 1
return max (votes, key=lambda x: votes.get (x))
def L2dist (p1, p2):
return sqrt (sum((p1-p2) **2))
def L1dist (v1, v2):
return sum (abs (v1-v2))
```

4 Experimental Results and Analysis

The main interface of the system mainly includes sample library recognition and tablet recognition, as shown in Fig. 3, 4 and 5.

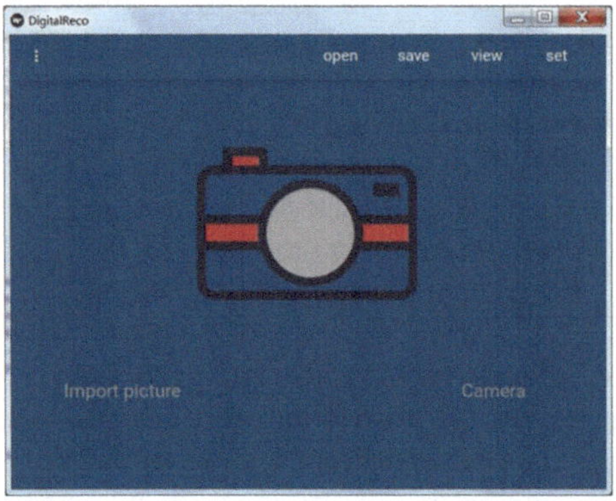

Fig. 3. System interface.

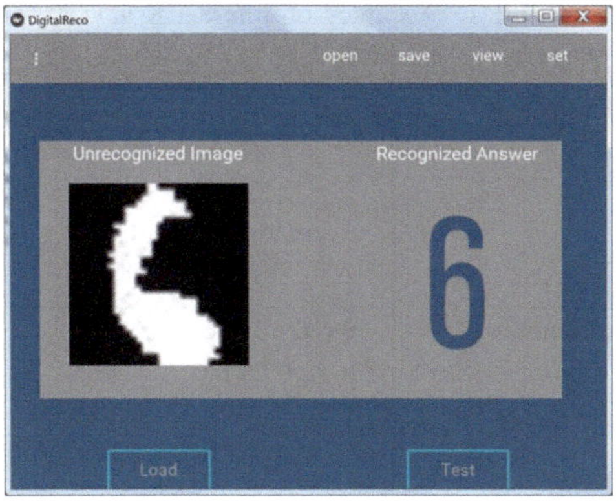

Fig. 4. Recognition result.

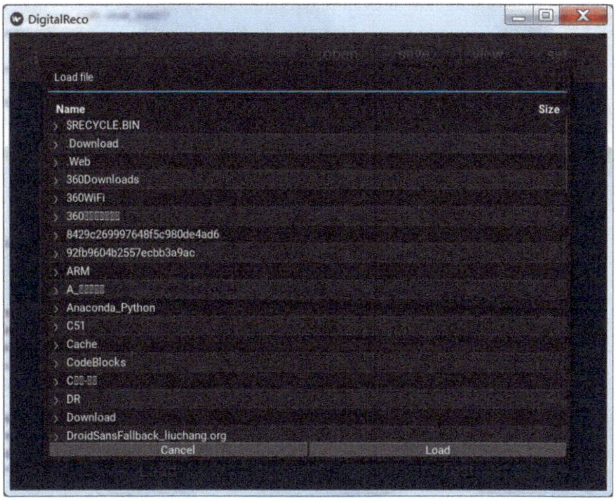

Fig. 5. Choose picture interface.

As a recognition system, handwritten numeral recognition needs to use some indexes to evaluate its performance. The common evaluation indexes are correct. The formula is defined as follows:

$$\text{the correct rate} : A = \frac{\text{Correctly identify the number of samples}}{\text{Total samples}} * 100\%$$

In the system, test samples are input, pyqt5 is used as the display interface, KNN operation is performed with all samples in the training set, and the comparison between predicted numbers and real numbers is completed respectively. When k is taken as 3, 5, 7 and 9 respectively, the accuracy of the operation is shown in Table 1, 2, 3, 4.

Table 1. Accuracy of different K values.

	K = 3	K = 5	K = 7	K = 9
Accuracy	98.94%	98.20%	97.78%	97.78%

Table 2. Accuracy at k = 1.

Digit	0	1	2	3	4
Accuracy	1	98.97	1	96.47%	1
Digit	5	6	7	8	9
Accuracy	97.22%	1	1	94.51%	98.88%

Table 3. Accuracy at k = 3.

Digit	0	1	2	3	4
Accuracy	1	98.97	1	98.82%	1
Digit	5	6	7	8	9
Accuracy	98.15%	1	1	95.60%	97.75%

Table 4. Accuracy at k = 9.

Digit	0	1	2	3	4
Accuracy	1	98.97	97.83%	96.47%	98.25%
Digit	5	6	7	8	9
Accuracy	97.22%	1	1	91.21%	97.75%

From the above table, it can be seen that when k = 3, a better classification effect can be obtained, with the highest accuracy of 98.94%. The larger the value of K is, the lower the accuracy is.

5 Conclusion

In general, KNN algorithm has advantages and disadvantages. The advantages lie in its simple operation and easy understanding. The disadvantage is that it takes a relatively long time to calculate the result when the sample is too large. It can be applied to some places with small computation or high fault tolerance. In engineering applications, people hope that the recognition ability is both efficient and accurate. Technicians can try other algorithm models to improve the recognition effect.

Acknowledgments. This work was supported by National Natural Science Foundation of China (61503128, 61772179), Hunan Provincial Natural Science Foundation of China (2020JJ4152), Science and Technology Plan Project of Hunan Province (2016TP1020), Scientific Research Fund of Hunan Provincial Education Department (18A333), Double First-Class University Project of Hunan Province (Xiangjiaotong [2018] 469), Postgraduate Scientific Research

Innovation Project of Hunan Province (CX20190998), Degree & Postgraduate Education Reform Project of Hunan Province (2019JGYB266), Postgraduate Teaching Platform Project of Hunan Province (Xiangjiaotong [2019] 370–321).

References

1. Singh, P.K., Sarkar, R., Nasipuri, M.: A study of moment based features on handwritten digit recognition. Appl. Comput. Intell. Soft Comput. **2016**, 111–127 (2016)
2. Pang, G., Jin, H., Jiang, S.: CenKNN: A scalable and effective text classifier. Data Mining Knowl. Discov. **29**(3), 593–625 (2015)
3. Nguyen, B., Morell, C., Baets, B.D.: Large-scale distance metric learning for k-nearest neighbors regression. Neurocomputing **214**, 805–814 (2016)
4. Yigit, H.: ABC-based distance-weighted kNN algorithm. J. Exp. Theoretical Artif. Intell. Artif. Intell. **27**(2), 189–198 (2015)
5. Majumder, S.: Handwritten digit recognition by elastic matching. J. Comput. 1067–1074 (2018)
6. Babu, U.R., Venkateswarlu, Y., Chintha, A.K.: Handwritten digit recognition using k-nearest neighbour classifier. In: World Congress on Computing & Communication Technologies. IEEE, 60–65 (2014)
7. Zhao, H., Rosin, P.L., Lai, Y.K.: Automatic semantic style transfer using deep convolutional neural networks and soft masks. Vis. Comput. **36**(7), 1307–1324 (2019)

Gender Identification for Coloring Black and White Portrait with cGan

Qingyun Liu[1,2(✉)], Mugang Lin[1,2], Yaqi Sun[1,2], and Feng Zhang[1,2]

[1] College of Computer Science and Technology, Hengyang Normal University, Hengyang 421002, China
liuqy@139.com
[2] Hunan Provincial Key Laboratory of Intelligent Information Processing and Application, Hengyang 421002, China

Abstract. This paper proposes a method to color black & white portrait images with gender recognition model and condition generation network (cGan). As a cGan model essentially, a method named pix2pix interprets image style transformation as a translation process from input pixels to target pixels. Lots of input to target image pairs should be prepared for model training for different application scenarios. When pix2pix is applied to color black & white portraits, there are obvious differences in color and saturation between portraits of different genders while the ability of gender recognition isn't owned by pix2pix in essence. Therefore, a gender recognition model is added to the network to produce more realistic and gender specific color portraits. The experiments show that our work has achieved better results than some other common methods.

Keywords: Gender identification · GANs · Loss function · Portrait style transfer

1 Introduction

Since 2016, Deep Neural Networks have been applied in a new field, with the purpose of interpreting images [1–3], voice and text [4, 5] by simulating the mechanism of human brain. It is a task with wide application value and skill to color black & white portraits. In particular, manually coloring old black & white portraits requires the artist's keen observation and understanding of the objects in black & white photos [6]. Real features and actual scenes of black & white portraits can be presented after reasonable colors are added, and these improvements can leave a deeper impression on the audience. In experiments of coloring black & white portraits with pix2pix [7], the results show that the model can automatically understand the facial features of the portrait through a large number of conversion training. The network can add corresponding colors to black & white portraits in different facial areas after training. However, the difference between male and female in facial features is not obvious. If the training set contains a large number of portraits of different genders, the experimental results show that the network can't identify the gender of the portrait when coloring. The network is likely to add feminine colors to male portraits and masculine colors to female portraits. In fact, there are obvious differences in color distribution

Q. Liu et al. (Eds.): CENet 2020, AISC 1274, pp. 456–464, 2021.
https://doi.org/10.1007/978-981-15-8462-6_51

between male and female portraits, for example, female portraits are more gentle and warm, and lips are brighter while male portraits tend to be cold in color. Therefore, we classify and train different gender portraits to get generators for different gender characteristics. A model for gender identification is designed and added in our work to detect and classify the portraits in test set. Later, gender specific portraits will be applied to different gender generator which has been trained before. And each portrait will be colored with gender specific colors automatically.

2 Related Work

2.1 Coloring Black and White Portraits by Pix2pix

The case for coloring black & white portraits can better explain the design principle of pix2pix. Many problems in image conversion can be posed as "translating" an input image into a corresponding output image in pix2pix. Not only the mapping relationship between input and output images can be learned automatically, but also the loss functions used to guide model training can be obtained after training. In the black & white portrait coloring experiment, the input image and the target image have the same image structure. And the main difference lies in the color of filling. Through a lot of training, the model can well understand the contour features of the facial features of the portrait and add realistic colors to them, as shown in Fig. 1.

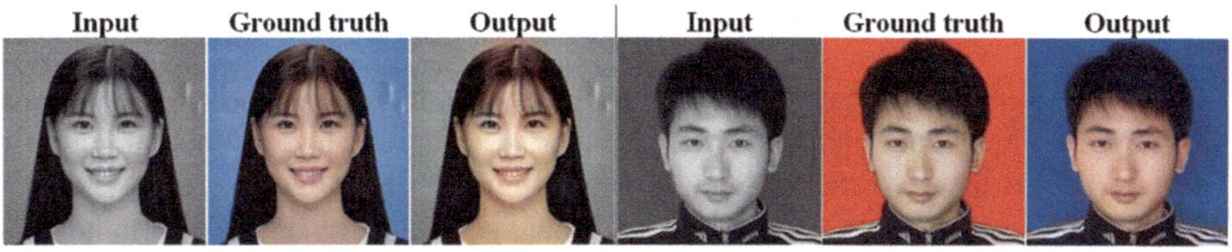

Fig. 1. Some results of coloring black & white portraits by pix2pix.

However, the gender of portraits hasn't been distinguished properly when coloring. After trained with male-female data set, the generator may add warm colors for male portraits that are more suitable for female portraits, and cold colors for female portraits that are suitable for male portraits, as shown in Fig. 2.

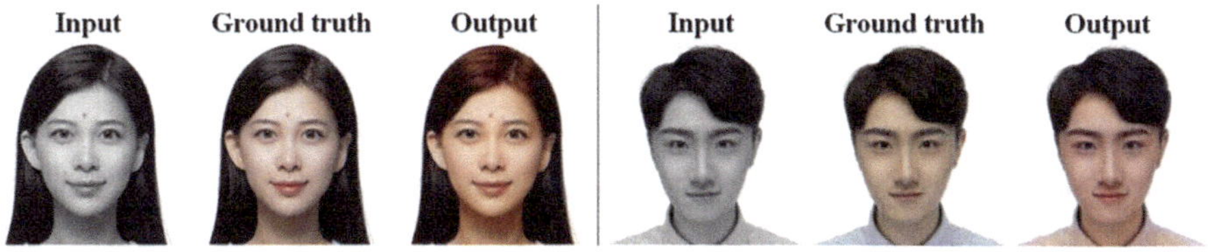

Fig. 2. Some unsatisfied results with colors that don't quite match their gender characteristics.

In addition, we tried to improve the performance of pix2pix model by constraining the training set or modifying the network model, so that it cannot only recognize the gender of the portrait, but also have the realistic coloring ability. As it turns out, the task is arduous. Therefore, we designed an additional gender detection system to classify the black & white portraits to be input into the generator of pix2pix.

2.2 Gender Identification Model

In order to classify the gender of black & white portraits to be colored, a simple BP neural network [8] was designed in our work. The gradient descent method and sigmoid function are used to train the network. By inputting a large number of marked male and female portraits for model training, a neural network which can effectively identify gender features is finally obtained. The BP network model for gender identification is shown in Fig. 3.

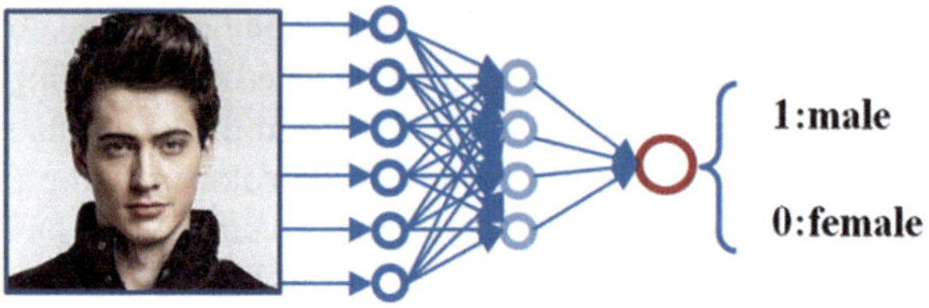

Fig. 3. Some unsatisfied results with colors that don't quite match their gender characteristics.

The neural network consists of an input layer, a hidden layer and an output layer. All the portraits in train and test set are converted to grayscale with the resolution of 28×28 in advance. So the number of neurons in the input layer is 784, while the number of neurons in the hidden layer is X which can be changed before training. The number of output neurons is 1, which is used to represent the gender of the portrait in test. When the value of output is greater than 0.5, it indicates that the portrait is male, otherwise it is a female portrait. We test the model's ability to recognize gender by modifying the epoch of iterations and the number of neurons of the hidden layer. Figure 4 and Fig. 5 respectively show the influence of different iterations and the number of neurons in the hidden layer on the network's ability to recognize gender.

Through a large number of training and test, we find that when the number of iterations is 2000 and the number of neurons in the hidden layer is 30, a satisfactory recognition result with the accuracy of more than 95% can be obtained by the model.

In this paper, on the basis of pix2pix model, we propose a method for coloring black & white portraits with a gender recognition network. The main contributions of this paper are summarized as follows:

- We find that the gender of black & white portraits aren't distinguished when coloring in pix2pix network. The black & white portraits may be colored with different gender characteristics.

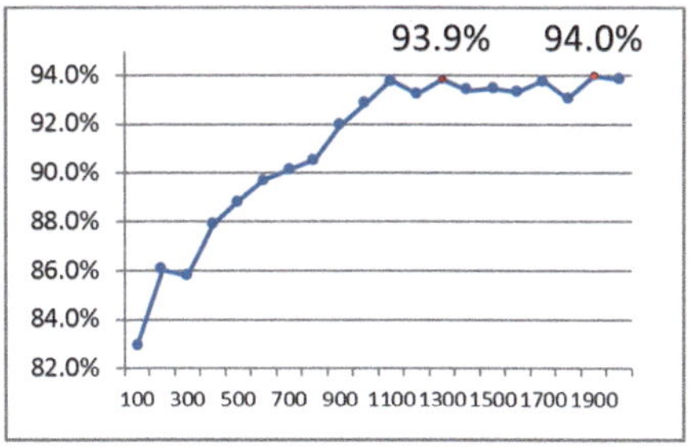

Fig. 4. The accuracy of gender recognition under different iterations.

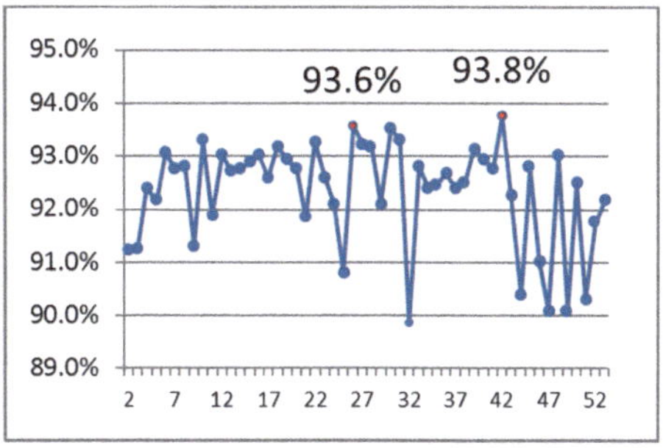

Fig. 5. The accuracy of gender recognition under different number of neurons in hidden layer.

- We designed a simple BP neural network. And through a large number of training and testing, the accuracy of gender recognition can reach 95%.
- We combined the gender recognition network and pix2pix network. All black & white portraits are first input into the gender recognition network for gender classification, and then input into the corresponding G network for coloring with corresponding gender characteristics according to the classification results.

3 Method and Design

3.1 Pix2pix Training Model

Pix2pix network is picked for coloring in our work. In pix2pix, G network adopts u-net [9] structure with skips. The input black & white portrait and the output color portrait have the same characteristics. The skips can transfer some low-level information between layer i and layer n-i in G. A discriminator of Markov [10] is designed for D network in pix2pix. After continuous iteration and confrontation training, G has stronger generation ability, while the identification ability of D has been improved. Finally, G and D achieve a balance. Finally, G can automatically learn the mapping relationship from input image to target image after training. Such as, the ability to color black & white portraits in our work. The training model is shown in Fig. 6.

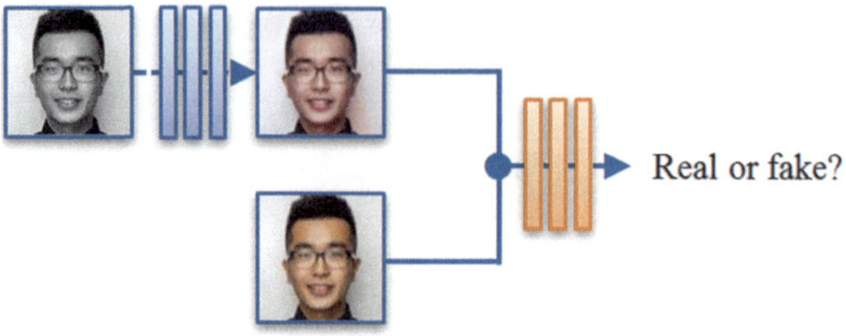

Fig. 6. The pix2pix training model for coloring black & white portraits.

The common conditional GAN [11, 12] can be expressed as:

$$\min_G \max_D V(D,G) = E_{x\sim P_{data}}[logD(x|y)] + E_{z\sim P_z(z)}[\log(1 - D(G(z|y)))] \quad (1)$$

where G tries to minimize this objective while D tries to maximize it conversely. In a large number of experiments, we found that pix2pix can better understand the facial contour of the portrait and add corresponding colors to it, but it does not distinguish the gender of the portrait. Therefore, we build different data sets for different genders in training, so as to obtain serveral G for coloring different gender portraits.

3.2 Generation Model with Gender Recognition

In order to equip the whole model with the ability of gender recognition and coloring for different gender portraits, we designed different data sets for different genders, and designed and trained a BP neural network to identify the gender of portrait. In the test of coloring black & white portraits, the portraits that need to be colored are first input into the gender recognition network for gender classification, and then input into the corresponding G network for coloring, so as to get the coloring results which meet the

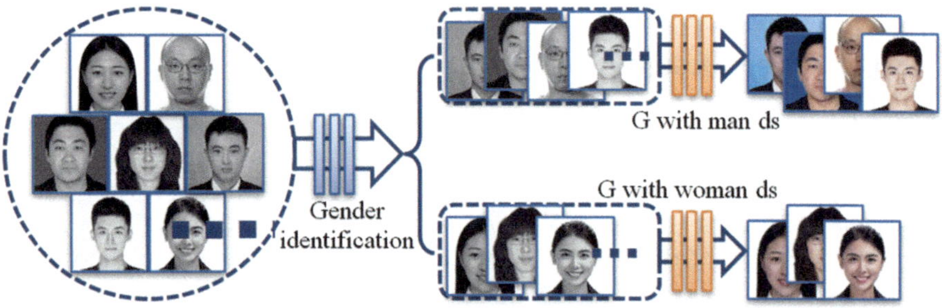

Fig. 7. Generation model with gender recognition.

gender characteristics. The overall framework of the generation model with gender recognition is shown in Fig. 7.

4 Experiments

In the experiment of training and test, all portraits are 1–3 channel images with the resolution of 256 × 256. Our facade training set consists of 150 images from black & white portrait to colored portrait for both male and female. The resolution of the portraits to be input into the gender recognition network for gender classification is also 256 × 256. However, the gender recognition network will resize them at 28 × 28

Fig. 8. Some results by pix2pix under the separate and mixed data sets of gender.

resolution. The pix2pix is trained and tested on a single NVIDIA GPU (GeForce GTX 1080) while the gender recognition network is on a single CPU (Core i7).

Firstly, we tested the coloring effect of different gender portraits with G-Network after training under the separate and mixed data sets of gender. As shown in Fig. 8.

The results in Fig. 8 show that the model can add satisfactory colors to the black & white portraits of corresponding genders under the separate training set of male and female genders. In the mixed gender training set, G does not understand the gender of the portrait well, so it is possible to add colors with no obvious gender characteristics.

In order to evaluate the effectiveness and performance of our model, we use the same dataset and test portraits. As an extensive research field, portrait transformation has been widely studied and applied, we compared the results of three other generation methods for coloring, the comparison results are shown in Fig. 9.

The test results in Fig. 9 show the effect of different models in coloring black & white portraits. From the test results, both Gatys [13] and CNN-MRF [14] can add realistic colors to the input portraits under the guidance of the ground truth portraits. However, the coloring ability of these two models will disappear without the guidance of the target portraits. Unfortunately, MGAN [15] model basically does not show the

Fig. 9. Comparison of black & white portrait coloring effects in different networks.

Fig. 10. Some results coloring for old black & white portraits.

ability of coloring. Moreover, the results also show that the gender of portraits are well understood, and the added color is realistic and matches the gender characteristics of the portrait in our work.

In addition, we also tested some old black & white portraits to pass gender detection and add color. The test results are shown in Fig. 10.

The test results in Fig. 10 show the coloring effect of the old black & white portrait. The experiment shows that the model we designed can also show satisfactory coloring ability on the old black & white portrait. We believe that our work can provide some reference and help for the research of the coloring direction of the black & white photos.

5 Discussion and Conclusion

The experimental results show that the pix2pix generation network with gender identification can add more gender specific colors to black & white portraits of different genders. Similarly, we infer that portraits of different ages, skin colors and expressions will also have different color characteristics. If different types of portraits are put in the same data set to be trained, pix2pix can't distinguish different types of portraits and add colors suitable for their classification. Experiments show that different portrait features have some mutual interference. If we can classify black & white portraits according to age, skin color and expression before coloring, and design different training sets for them, we believe that these changes can also have a positive impact on the coloring effect.

Acknowledgements. This work is supported by the National Natural Science Foundation of China (Nos. 61503128, 61772179), Hunan Provincial Natural Science Foundation of China (Nos. 2020JJ4152, 2019JJ40005), the Science and Technology Plan Project of Hunan Province (No. 2016TP1020), the General Scientific Research Fund of Hunan Provincial Education Department (Nos. 17C0223, 18A333), the Double First-Class University Project of Hunan Province (No. Xiangjiaotong [2018] 469), Postgraduate Research and Innovation Projects of Hunan Province (No. Xiangjiaotong [2019] 248–998), and Hengyang guided science and technology projects and Application-oriented Special Disciplines (No. Hengkefa [2018] 60-31).

References

1. Zhao, H.H., Rosin, P., Lai, Y.K., Zheng, J.H., Wang, Y.N.: Adaptive gradient-based block compressive sensing with sparsity for noisy images. Multimed. Tools Appl. (2019)
2. Kazemi, H., Taherkhani, F., Nasrabadi, N.M.: Unsupervised facial geometry learning for sketch to photo synthesis (2018)
3. Wang, T.C., Liu, M.Y., Zhu, J.Y., Tao, A., Kautz, J.: Bryan Catanzaro high-resolution image synthesis and semantic manipulation with conditional Gans. In: Computer Vision and Pattern Recognition (2017)
4. Zhao, H.H., Rosin, P., Lai, Y.K.: Image neural network style transfer with global and local optimization fusion. IEEE Access (2019)

5. Zhao, H., Rosin, P.L., Lai, Y.K.: Automatic semantic style transfer using deep convolutional neural networks and soft masks. Vis. Comput. (2019)
6. Reinhard, E., Ashikhmin, M., Gooch, B., Shirley, P.: Color transfer between images. IEEE Comput. Graphics Appl. **21**, 34–41 (2001)
7. Isola, P., Zhu, J.Y., Zhou, T., Efros, A.A.: Image to-image translation with conditional adversarial networks. In: IEEE Conference on Computer Vision and Pattern Recognition, vol. 2–3, pp. 5–8, 13 (2017)
8. Krizhevsky, A., Sutskever, I., Hinton, G.: ImageNet classification with deep convolutional neural networks. NIPS. Curran Associates Inc. (2012)
9. Ronneberger, O., Fischer, P., Brox, T.: U-net: convolutional networks for biomedical image segmentation. In: MICCAI, pp. 234–241 (2015)
10. Wang, T.C., Liu, M.Y., Zhu, J.Y., et al.: High-resolution image synthesis and semantic manipulation with conditional GANs (2017)
11. Mirza, M., Osindero, S.: Conditional generative adversarial nets. arXiv preprint arXiv (2014)
12. Ronneberger, O., Fischer, P., Brox, T.: U-Net: Convolutional networks for biomedical image segmentation (2015)
13. Gatys, L.A., Ecker, A.S., Bethge, M.: Image style transfer using convolutional neural networks. In: IEEE Conference on Computer Vision & Pattern Recognition. IEEE (2016)
14. Li, C., Wand, M.: Combining markov random fields and convolutional neural networks for image synthesis. arXiv:1601.04589 (2016)
15. Li, C., Wand, M.: Precomputed real-time texture synthesis with Markovian generative adversarial networks. In: ECCV (2016)

Communication System Detection, Analysis and Application

Considerations on the Telemetry System in Fight Test

Tenghuan Ding(✉), Tielin Li, and Wenyu Yan

Chinese Flight Test Establishment, Xi'an 710089, China
dth1991@163.com

Abstract. At first, this paper reviews the present situation and some developing trends of the telemetry system in flight test. Then the paper introduces the new requirements of the telemetry system brought by the development of flight test assessment. At last, this paper shows some considerations in developing the telemetry system, including those on the architecture of network telemetry system, radio frequency resource, the control of telemetry data stream, band efficiency and coding and error control, and RF environment and classification.

Keywords: Telemetry · Flight test · Network telemetry system

1 Brief Introduction of the Telemetry System in Flight Test

The telemetry system in flight test is a system (or systems) to meet the remote measuring demands in flight test. The systems often include both the airborne testing system (data acquire units) and the remote testing system (such as the radar, the electronic theodolite, and the tacheometry), which are to measure, acquire and record the necessary data of a testing object (aircrafts, missiles) and to build a communication between the testing object and the grounded station.

During the past few decades, FTT (flight test technology) has made a great progress. As a result, there was a change on the concept of the telemetry system. The telemetry system covers a very big range now. It includes not only the data transmission link (wired or wireless), but also the data acquiring systems (on board or remote test systems), the receiving station on ground and the interface to the ground network (gNET). The data acquiring systems firstly acquire the data, and then transmit the data to the receiving station on ground by the wired or wireless way. The receiving station will translate the data, and finally provides the data to gNET for further use through the interface.

The telemetry system is to serve the flight test assessment. In recent years, the object of the assessment is developing gradually from the single platform (e.g. a single aircraft or a single weapon) to systems (e.g. a combat system including several aircrafts, some radars and intelligence systems). The telemetry system is now given some new duties as a result. It demands more for the telemetry system.

Q. Liu et al. (Eds.): CENet 2020, AISC 1274, pp. 467–473, 2021.
https://doi.org/10.1007/978-981-15-8462-6_52

2 The Current Situation and Trend of the Telemetry System in Flight Test

2.1 The Current Situation of the Telemetry System in Flight Test

On the beginning of flight test, the assessment is focused on the single platform. So, the telemetry is designed to serve just the single platform. The IRIG106 telemetry standard was initialized to fitted the single platform assessment very well by the IRIG (Inter-Range Instrumentation Group) of RCC (Range Commanders Council). The early version of IRIG106 only provides a single serial transmission (SST) link standard based on the PCM. The PCM telemetry system is a very mature and stable system which has been being applicated in flight test for decades. It has been playing a very important role in flight test by providing a stable real-time data transmission to make the security monitoring and real-time assessment possible.

2.2 The Trend of the Telemetry System in Flight Test

In 21st century, the combat system has made a transition from PCW (Platform Centric Warfare) to NCW (Network Centric Warfare) gradually. There appear some big weapon systems in which the network is center, such as C4ISR, Air-Ground Integration, Sea-Ground Integration and GIG (Global Information Grid). According to the new demands "the test means the real combat and training", the flight test assessment will be focused on the whole performance of the SoS (system of system) instead of just a single object or a single aspect of the object [1]. To meet the new demands of flight test assessment, the flight test system will be synchronously changed into a network-based system, and flight test will be conducted in a complex electromagnetic environment, even in the cyberspace.

US has been leading the trend of the telemetry system in flight test. In 2004, the CTEIP (Central Test and Evaluation Investment Program) has lunched the iNET (integrated network enhanced telemetry). The iNET is aiming at a telemetry network covering the sea, the land, the sky and the space, which offers the wide-band transmission link between all members. FTE (Flight Test Engineer) in China brought up the concept of agNET (air-ground NET) [2] which is based on iNET but concludes more (such as TENA, cyberspace etc.).

3 Considerations on the Flight Test Telemetry System

3.1 The New Requirements on the Flight Test Telemetry System

With the change from PCW to NCW, the flight test assessment is more about more about the multi-aircraft cooperative test than a single aircraft test, and the effectiveness of a whole system. The "GlobalReach" combat means the test range is explosive. A variety of radio communication facilities and electronic warfare equipment are got used in the modern warfare and test, which has made the RF environment even worse. All above has required the flight test telemetry to develop in some aspects:

The Explosive Data Bandwidth in Flight Test. In recent years, aviation industry and flight test has improved very much, and much more data is needed in the assessment in flight test. To take the US aviation for example, the data width in flight test of F-16A and F-18A are approximately 100 kbps, and as for F-15E is above 1 Mbps. It is even up close to 10 Mbps in that of F-22. However, the flight test of F-22 was happened in about 1990s. It is predictable that the data bandwidth will be getting wider. From other perspective, all flight test instruments manufacturers around the world are all focus on the Gigabit Ethernet devices, which implies that the data width will increase rapidly.

The increase of the data bandwidth challenges the data acquiring system and data recording system, while it is a much more tough challenge to accomplish a big bandwidth of telemetry.

The Increasing Assessment on the Whole System. Flight test is more focusing on the multi-aircraft in formation and in cooperative. With the combat system getting bigger, the flight test assessment will cover the whole combat system including more and more aircrafts, weapons, radars and other subsystems.

For example, in one battle, the E-2D airborne early warning aircraft (AEWAC) cooperates with the command posts, naval vessels, other AEWACs, fighter aircrafts, unmanned aircrafts and so on. So, when you want to assess the performance and effectiveness of the battle, you should assess all the members in the battle together and then you can get the comprehensive performance and effectiveness.

But now, flight test system on one aircraft is independent from other systems on other aircrafts, one cannot communicate with others. By the telemetry link, several aircrafts in a limit range can communicate with the grounded station. When you want to combine or compare data from these aircrafts, you can only stay in front of a computer in the grounded station, however the pilots on the aircrafts will not able to do so.

All of these forced the flight test system into a network test system, which can build the links between all members to share, exchange data.

The Expansion of Flight Test Field. With the concept of "WorldReach" battle, the flying range of modern aircrafts is getting much longer. The flight test field is also expanding. So, it is required for the flight test telemetry to cover a larger range of flight test field.

Since the current telemetry system is a star topology in which the grounded station is the center, the largest distance in which a single aircraft can communicate with the grounded station confines the range the telemetry can cover. So, to cover a bigger range is another challenge to the telemetry in flight test.

The More complex RF Environment. There is a growing on the radio utility that is to detect, to communicate, to disturb other radio. All of these make the RF environment worse. In a certain band (such as S band), there are lot of devices which share the same band. Of course, these devices will interfere with each other. So how to guarantee the stability, reliability of the telemetry in flight test in the more complex RF environment is to be solved. The classification of the transmitting data from being captured in this RF environment is another problem.

The 4 aspects all require the telemetry system in flight test to be improved to a network which can supply a big communication bandwidth, cover a bigger range, tolerate the disturbance, reject to be grabbed and so on.

3.2 Consideration on the Telemetry System in Flight Test

Self-organizing Network Telemetry with Big Bandwidth and High Precise of Synchronization. In current situation, telemetry system is a star topology in which the grounded station is the center, and the transmission stream is PCM stream in a single direction (from an aircraft to the grounded station). There will be several disadvantages:

Limited Bandwidth. The PCM stream in data acquiring and recording is very stable, but the bandwidth of the wired transmission in airborne test system is limited. That makes the bandwidth of telemetry is limited, too. Now, the bandwidth of PCM stream is less than only 20 Mbps.

Single Direction. The transmission is in one single direction from an aircraft to the grounded station. The airborne test system is out of control now. However, in some cases, it is needed to control the airborne test system.

Limited Range. In the current telemetry system, which is a star topology, every aircraft communicates with the grounded station. The limited distance of the valid telemetry link confines the range of telemetry in flight test. Without a network or relay satellite, if you want to cover more range, there seems to be the only way to make a telemetry relay by arranging many distributed grounded stations.

Unable to Share Test Data Between Aircrafts. The topology of the current telemetry system makes it impossible to communicate between the aircrafts in the air. When it comes to the cooperative subject, the pilot or engineer in one aircraft is not able to get the information of other aircraft in real time.

Thus, the telemetry system will turn into a network telemetry system (known as TmNS). In the 2017 and later version of IRIG106 has also described the standards on the TmNS. The TmNS will use network package instead of PCM frames to gain a wide bandwidth. And the TmNS will just like a network to connect all the member in bi-direction, which will make the data of every aircraft available to all the members.

By the network telemetry, it will extend the communication bandwidth, expand the telemetry range, satisfy the cooperative assessment. Meanwhile the bi-direction link will make it possible to upload some control commands, such as to power on/off the airborne system, to change the telemetry frequency and to ask for the particular data (just like the need in next section).

But there are also lots of problems to be solved:

Because the network is much more flexible, and the member may fly so fast. Consequently, on the network layer, there should be a self-organizing network which can automatically select an appropriate point as the relay. On the Transport Layer, maybe the TCP protocol will be used to get a reliable transmission instead of UDP which is often used in the airborne test system to get a high speed and to reduce the burden of the networks (most of them are wired). And on the physical layer, it is concerned that the Doppler effect caused by the high speed of aircrafts. Finally, you will get the reliable link among each other.

Meanwhile, since the assessment in flight test is very sensitive to time, so the wireless TmNS must be in a precise time synchronization by some technologies. Now the airborne test system obtains the GPS/BDS signals as the outer time reference, and then inside the test system, PTP and sync pluses are applied to gain a high precise synchronization. Since test systems in different areas will obtain different time because the GPS/BDS signals is disturbed by the cloud, solar motions and so on, it is difficult to synchronize all the test systems in the same time reference.

The Use of Higher and Wider Bandwidth in the Spectrum of the Radio. Before the 21st century, in China, there was not that many aircrafts which are in flight test at one time and there was little data to acquire. The Radio Commission of China distributed a 200 Mhz bandwidth in S band to the flight test. At that time, the 200 MHz bandwidth is enough. But now, on one hand, there are more and more aircrafts in fight test at the same time, and the data needed is getting bigger. So, the bandwidth needed is far beyond the 200 MHz. On the other hand, more and more telecom companies used the S band now. For example, 4G communication of the China Mobile, the China Telecom and the China Unicom are all using a part of S band. That will disturb the telemetry link of flight test more or less.

Consequently, it is necessary to get the higher and wider band radio used in the telemetry in flight test (which is an easier and direct way). In the later version of IRIG106, the C band is planned to be used in telemetry of flight test. Meanwhile many FTI manufacturers are on the way to the C band telemetry devices (such as the transmitter, amplifier, receiver). So, the frequency of telemetry in future is getting higher, even to the terahertz. But in flight test, it is concerned that the higher frequency will cause a higher loss of power and a shorter valid distance of communication, and at the same time, the stability and costs of higher frequency radio must be taken into consideration.

The Control of Telemetry Data Stream. In the test system based on PCM, all the devices use PCM to code the data and then composite the minor/major PCM frames. The recording and telemetry of the data is all in the PCM frames now. The size and location of data in the PCM frames are all fixed (so we can find the correct data by the location definition). In the 2015 and later version of the IRIG106, the chapter 7 describes the standards to transmit and translate the network packages in PCM frames. It is more flexible to put different formats of data in PCM.

Anyway, the data and the bandwidth of the PCM frames in telemetry are fixed. Since the bandwidth of the PCM streams is limited (described above), you cannot transmit all data, so you can only select a part you need to transmit before the aircrafts take off. And the PCM frames will occupies a fixed bandwidth of telemetry.

The TmNS will benefit on the control of telemetry data stream. Since there are lots of mature protocols which can be applied in the TmNS. On the application layer there is the Request-Response protocol. By this, the telemetry data stream is to be controlled by your requests. That will make it much more flexible to get what you want. And at the same time, the dynamic data stream means the flexibility to allocate the bandwidth of telemetry [3].

The Higher Bandwidth Efficiency and Error Control Coding. In current situation, FM (frequency modulation) is still in a large scale of use, because FM is very mature and stable, costs little on the device, and with a high SNR (signal to noise ratio). But the modulation index and band efficiency of FM is low, which costs too much bandwidth to transmit little data. The band efficiency should be improved by the new modulation way such as SOQPSK/16QAM to save the finite bandwidth.

At the same time, the improvement of the band efficiency will increase the BER (bit error rate) with the same SNR. So, if you want to lower the BER, you can only to improve your transmitter and amplifier to increase the SNR, that will cost a lot.

The error control coding will do benefit, too. In the bi-direction telemetry system, it is easier to control the error by ARQ (automatic repeat request) than that in a single direction system.

RF Environment and Classification. The test will approach to the real combat in the future. So, the RF environment in flight test will be considered. Now there are some artificial ways to avoid to be interfered by the bad FR environment. For example, in some cases, there are many restrictions on use of radio around the test field. When the test field is getting bigger and bigger, it is impossible to control the FR environment by man. So, some technology should be used in telemetry system.

Besides, many flight tests are classified on military or commercial thoughts, which means that the telemetry system should be classified to get rid of some malicious capture and monitoring.

The Use of Intelligent Technology. Now the use of the intelligent technology is very little in the telemetry system. There are great demands of the intelligent technologies in flight test. For example, the control of telemetry data stream according to the certain test tasks should be in an intelligent way, and with the telemetry data, some AI surveillance or monitoring could be done to do some warning or security monitoring.

4 Conclusion

The trend of flight test system is led by the development of flight test. The flight test is now in a transition from platform-centric test to network-centric test. The flight test system will be surely developed into a network system which covers the sky, the land, the sea, and the space. The telemetry system in flight test will be developed also into a TmNS which covers a very wide range of the test field, with a wide bandwidth and a low latency communication in bi-directions. In the transmission, the band efficiency should be increased, and the error control coding will be in sue. The bad FR environment resistance is to be improved. The classification should also be in consideration.

References

1. Yang, T., Zhang, Z.: Development of integrated air-ground test network. Comput. Meas. Control **24**(5), 1–4 (2016)
2. Yang, T.: Development and thinking of integrated test network architecture in 5-D space. Flight Dyn. **34**(151), 1–6 (2016)
3. Reinwald, C.: Telemetry data on demand: the key to understanding the telemetry network revolution. In: International Telemetering Conference Proceedings, vol. 54. International Foundation for Telemetering, CA, USA (2018)

Design and Implementation of Remote Visitor System Based on NFC Technology

Zhiqiang Wang[1,2,3](✉), Shichun Gao[2], Meng Xue[2], Xinyu Ju[2], Xinhao Wang[2], Xin Lv[1], Yang Li[1], and Tao Yang[3]

[1] State Information Center, Beijing 100045, China
wangzq@besti.edu.cn
[2] Beijing Electronic Science and Technology Institute, Beijing 100070, China
[3] Key Lab of Information Network Security, Ministry of Public Security, Shanghai 200000, China

Abstract. Most visitor systems' information management is at a low level, and the processing method is still semi-manual registration. To improve the level of automation of the visitor system, this paper designs a remote visitor system based on NFC technology deployed on the Android smart-phone platform. It reads and emulates card data through NFC technology and adopts SM series algorithm to encrypt and decrypt the transmitted data. The system effectively solves the problems about security and automation in the traditional guest system.

Keywords: Remote visitor system · Android system · NFC technology · SM series cryptographic algorithm

1 Introduction

With the rapid development of modern information technology, convenient, efficient, and fast digital management and network technologies significantly change our lives, but at present, visitor management in some companies and apartment is still in the stage of manual registration [1]. The shortcomings of traditional management are increasingly prominent, such as the lack of permission from the interviewee, the visitors' false identity, and volatile and statistically inconvenient access data. To solve the problems, this paper designs a remote visitor system based on NFC technology, which is based on Android smart phone [2]. It sends personal information and other related information to the client deployed in the interviewee's Android smart-phone. With the help of this system, the interviewee has a more clearly understanding of the visitor's identity and the reason for the visit and handles related access requests. The visitor can also see the status of this visit in real time [3]. The design and implementation of this system provides a new solution to improve the visitor system's availability and security in the current environment.

The paper is organized as follows. We introduce related technologies and algorithms in Sect. 2. The system design is given in Sect. 3. Section 4 describes the realization and environment configuration of the system. The simulation test is in Sect. 5. Section 6 concludes the paper.

Q. Liu et al. (Eds.): CENet 2020, AISC 1274, pp. 474–482, 2021.
https://doi.org/10.1007/978-981-15-8462-6_53

2 Related Technologies and Algorithms

2.1 NFC Technology

Near-field communication (NFC) is a short-range wireless technology that makes your smartphone, tablet, wearables, payment cards, and other devices even smarter. With NFC, you can transfer information between devices quickly and easily with a single touch [4]. With Android system updated to 2.3 and API Level updated to 9, the Android introduced the APIs of NFC, marking the official landing of NFC technology on the Android system. With the further update of Android system, NFC technology has been fully implemented and applied on Android. NFC has three working modes: read-write mode, point-to-point mode and card emulation mode.

(1) Read-Write mode. NFC device can be used as a contact-less reading tool to read the information of the contact-less smart card directly. The contact-less smart card protocols include the ISO14443 Type-A, Type-B and FELICA standards. This contact-less smart card is commonly called IC card, bus card, access card or all-in-one card.

(2) Point to Point mode. Two NFC devices establish wireless connection and exchanges data. This mode follows the ISO18092 standard, and the data transmission format defined as NDEF (NFC Data Exchange Format) is very strict.

(3) Card emulation mode. NFC device (such as Android phone) is simulated as a traditional contact-less smart IC card, which can write data to the NFC's confidential unit, and can be recognized and read by the card reader. Therefore, the access control system can be realized without additional cost.

2.2 SM Algorithms

In the solution proposed in this paper, SM2 and SM4 algorithms are used.

First introduce the SM2 algorithm [5]. Asymmetric encryption algorithm SM2 was released by the Chinese cryptography administration in 2010. The SM2 algorithm is based on the point-group discrete logarithm problem in the elliptic curve algorithm. The following cubic equation:

$$y^2 = x^3 + a * x + b$$

is used to represent the elliptic curve. By specifying a and b coefficients, a unique standard curve was determined. Compared with RSA algorithm, SM2 algorithm is a more advanced and secure public key cryptography algorithm, which is often used as a substitute for RSA algorithm in commercial cryptography applications.

Secondly introduce the algorithm SM4 [6]. The SM4 symmetric cryptographic algorithm is used to encrypt and decrypt data, effectively ensuring the confidentiality of data and information. The packet length of the algorithm is 128 bits and the key length is 128 bits. Both the encryption algorithm and the key expansion algorithm adopt a 32-round nonlinear iterative structure [7]. The structure of the decryption algorithm is the same as that of the encryption algorithm, except that the use order of the round keys are reversed.

3 System Architecture

The system includes the Android client, the server and the database deployed on Tencent Cloud. The architecture is shown in Fig. 1. The Android client is responsible for the interaction with the user and the reading of the data of the access control card and the emulation of the access control card. The server responds to the user's operation and the database stores the corresponding information. There are five steps: ①Visitor B is registered for authentication. ②Visitor B submits an access request to resident A. ③Resident A receives an access request from visitor B after A refresh. ④Resident A responds to the access request of visitor B. If resident A agrees to the access request, resident A select an uploaded access card to authorize visitor B. ⑤Visitor B simulates the data of resident A's access control card through the NFC function of the Android phone, and achieves access through the access control.

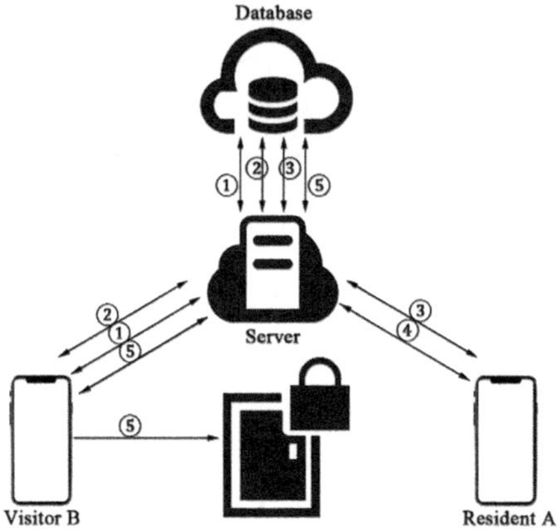

Fig. 1. System architecture diagram.

4 System Implementations

4.1 Android Client

The Android Client mainly includes 3 modules: transmission of SM4 symmetric keys, NFC read card data, NFC card Emulation. The following will describe the implementation the modules.

Transmission of the SM4 Symmetric Keys. As shown in Fig. 2, the Android Client firstly receives the public keys of SM2 from the web server. Secondly generates the SM4 symmetric keys and SM2 key pair. Thirdly encrypts the SM4 symmetric keys by the received SM2 public keys. Fourthly obtains the digest of the client identifier by SM3 hash algorithm. Fifthly signs the digest by the generated SM2 private keys. Finally sends the encrypted keys, the digest and digital signature to the web server.

The above steps complete the safe transmission of the SM4 symmetric keys. In the subsequent network communication, web server and Android client will use the SM4 symmetric keys for encryption and decryption. Compared with using the SM2 algorithm for encryption and decryption alone, this method consumes less time and memory resources and more efficient.

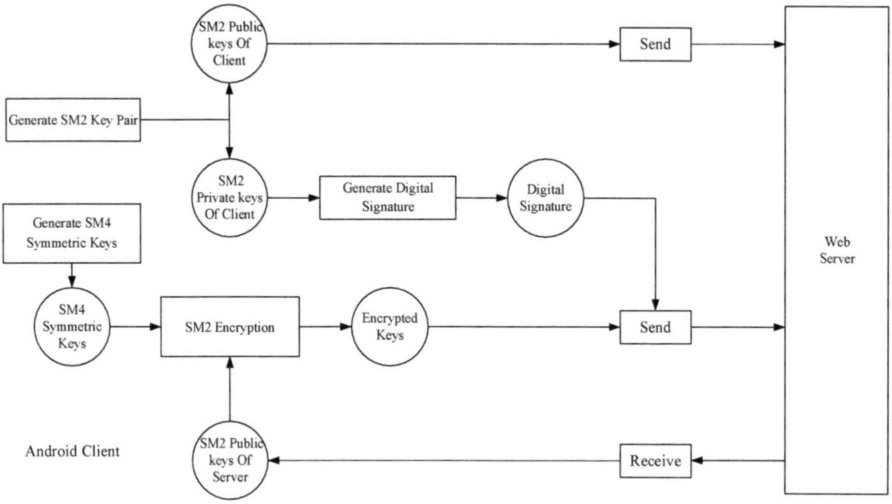

Fig. 2. Transmission of the SM4 symmetric keys.

Read Card Data Through NFC. The purpose of this module is to read the data from access cards, whose flowchart is shown in Fig. 3. The data in these cards will help the visitors to pass the gates and achieve visits.

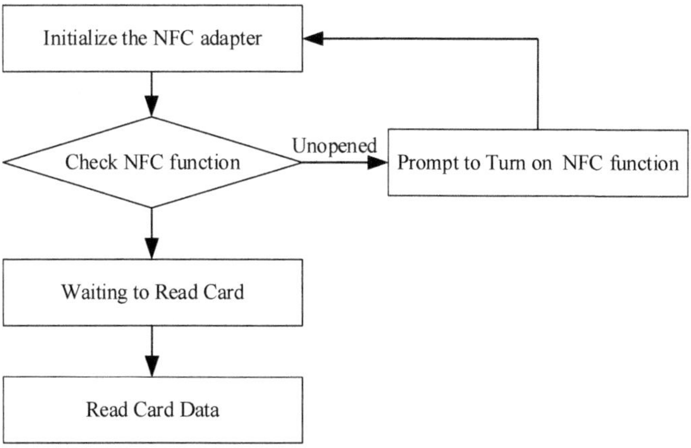

Fig. 3. Read card data through NFC.

After jumping to the page of NFC read card data, the system will first initialize the NFC adapter, then check whether the NFC function of the Android phone has been started, if the function is ready, then initialize the settings of NFC and wait for cards. Android phones will read the data of card when the NFC device detects card is approached.

Emulate Card Through NFC (Fig. 4). First get the card data from the web server, then create a configuration file and write the data to the file. After the data is successfully written to the file, NFC devices apply this configuration file. The emulation of card is realized in this module.

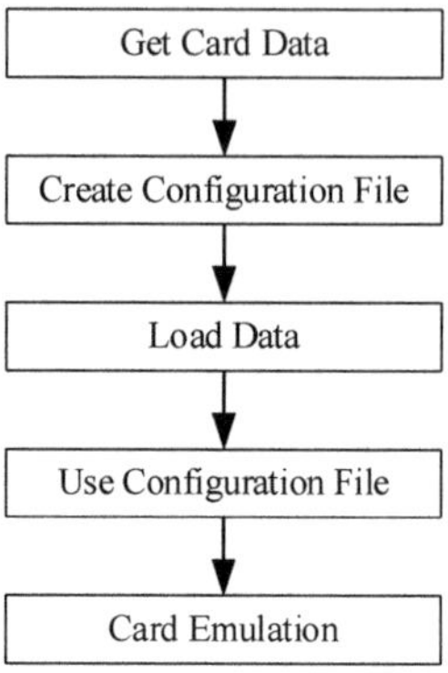

Fig. 4. NFC card emulation.

4.2 Web Server

Web server is responsible for responding to the network request of the Android client. The server judges the URL request of the Android client and processes the received data differently. After processing, the processed data will return to the Android client.

Reception of Symmetric Keys (Fig. 5). When Web server detects the network connection of the Android client, Web server will generate SM2 key pair containing public keys and private keys. Then Web server sends the SM2 public key to the Android client after encoding, meanwhile generate SessionID to identify this session and save the private keys in the memory space allocated by the SessionID. After the server receives the encrypted keys, client SM2 public key and the digital signature sent by the Android client, the server will first verify the digital signature through client SM2 public keys, if the result is correct, the server will decrypt the encrypted keys by the SM2 private keys which is just generated. In this way, the server obtains the SM4 symmetric keys generated by the Android client.

User Registration (Fig. 6). After receiving the registration information from the Android client, the server queries the user table in the MySQL database to check whether the user information has been registered. If the phone number has been registered, the failed status will be returned. If it has not been registered, the user table in the MySQL database will perform the registration operation, and the successful status will be returned.

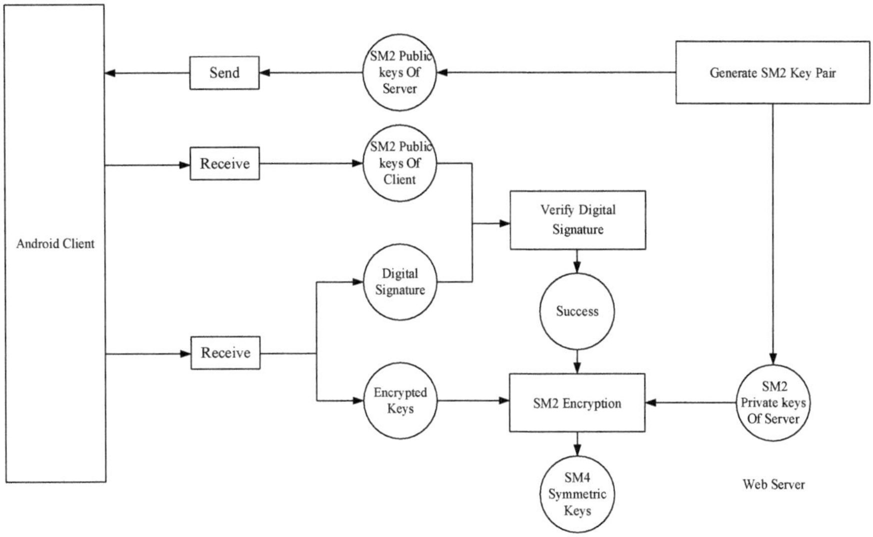

Fig. 5. Reception of symmetric keys.

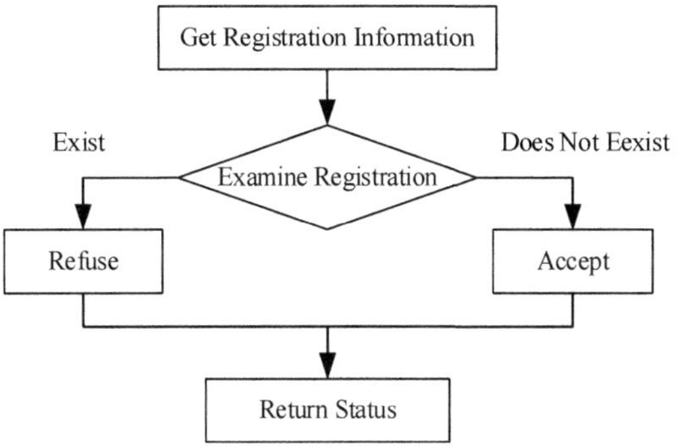

Fig. 6. Process the registration request on the Server.

4.3 Environment Settings

All operations on the client are based on the Android system under MI 8. The network middleware Apache Tomcat is deployed on Tencent Cloud (Ubuntu Server 18.04.1) and processes client network requests. We use the Windows 10 operating system for project implementation and remote access and control of the server. The experimental environment is shown in Table 1.

Table 1. The experimental environment.

Platforms	Tencent Cloud
Hardware dependencies	Mi 8
Software dependencies	Android 9
	Ubuntu Server 18.04.1 LTS
	Apache Tomcat 8.5.43
	MySQL 5.7.29

5 Experiments

After registering, the visitor initiates an application for visit, as shown in Fig. 7.

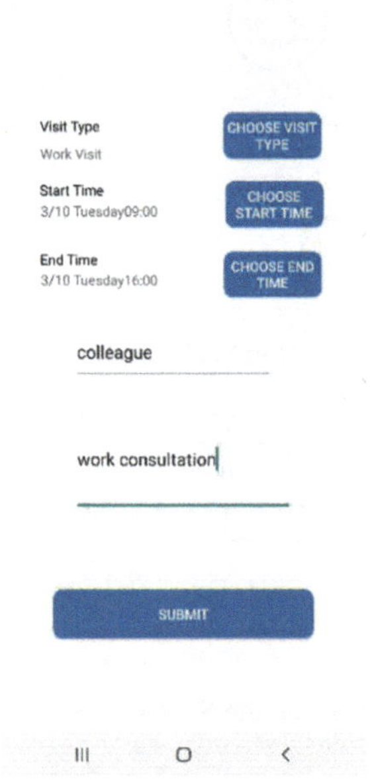

Fig. 7. Initialization.

Fig. 8. Refuse or accept.

Resident will process the visit. After clicking accept button (Fig. 8), program will ask residents to select an access card which is already uploaded, as shown in Fig. 9. After the visitors refresh the request page, they can emulate the card through NFC by clicking the button, as shown in Fig. 10.

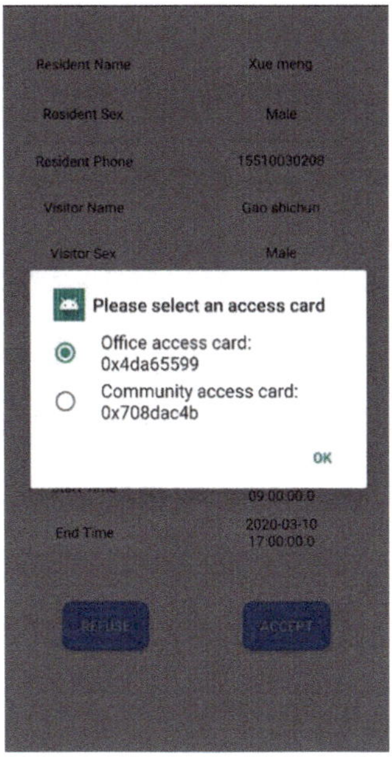

Fig. 9. Select a access card.

Fig. 10. NFC emulation.

6 Conclusion

With the development and application of intelligent visitor system, the internal security management will be strengthened greatly. In the paper, based on the NFC technology, a remote visitor system is designed. The SM series encryption algorithms were used to encrypt and decrypt data transmitted over the network. The system can ensure the security of data transmission and improve the level of informatization greatly.

In the future research, we will optimize the implementation details of the solution to provide convenient and secure access for users.

Acknowledgments. This research was financially supported by the National Key Research and Development Plan (2018YFB0803401), Key Lab of Information Network Security, Ministry of Public Security (C19614), Special fund on education and teaching reform of Besti (jy201805), the Fundamental Research Funds for the Central Universities (328201910), China Postdoctoral Science Foundation (2019M650606), 2019 Beijing Common Construction Project-Teaching Reform and Innovation Project for Universities in Beijing, key laboratory of network assessment technology of Institute of Information Engineering, Chinese Academy of Sciences. The authors grateful acknowledge the anonymous reviewers for their valuable comments.

References

1. Zhao, Z.H.: Development of real-time visitor safety management platform based on RFID. Jilin University (2015)
2. Yao, Y.M., Liu, W.G.: Study of android's architecture and its application development. Comput. Syst. Appl. **11**, 110–112+24 (2008)
3. Geng, D.J., Suo, Y., Chen, Y., Wen, J.L., Yong, Q.: Design and implementation of android phone based access and control in smart space. J. Comput. Appl. **31**(02), 559–561+571 (2011)
4. Sun, H.: Design and implementation of smartphone campus card based on NFC technology and cloud services. Res. Explor. Lab. **35**(07), 120–126 (2016)
5. Wang, C.H., Zhang, Z.F.: Overview on public key cryptographic algorithm SM2 based on elliptic curves. J. Inf. Secur. Res. **2**(11), 972–982 (2016)
6. Lv, S.W., Su, P.Z., Wang, P., Mao, Y.Y., Huo, L.L.: Overview on SM4 algorithm. J. Inf. Secur. Res. **2**(11), 995–1007 (2016)
7. Yan, H.B.: Research on critical technology of network key management. Comput. Appl. Softw. **28**(12), 68–70 (2011)

Bit Slotted ALOHA Algorithm Based on a Polling Mechanism

Hongwei Deng[1,2(✉)], Wenli Fu[1], Ming Yao[1], Yuxiang Zhou[1], and Songling Xia[1]

[1] Hengyang Normal University, Hengyang 421002, China
dhwwhd@163.com
[2] Hunan Provincial Key Laboratory of Intelligent Information Processing and Application, Hengyang 421002, China

Abstract. In this study, we propose a bit slotted ALOHA algorithm based on a polling mechanism (PBSA) to address the tag collision problem of dynamic radio frequency identification (RFID) systems. This algorithm is based on a combination of the dynamic framed slotted ALOHA (DFSA) algorithm and the free framed slotted ALOHA (FSA) algorithm. The polling mechanism is introduced to change the tag removal protocol. While removing the idle slots, the collision slot is converted into a single slot, which improves the tag recognition efficiency per unit frame. Simulation analysis shows that compared with the dynamic framed slotted ALOHA (DFSA) algorithm, the grouped dynamic framed slotted ALOHA (GDFSA) algorithm, and the cancel idle framed slotted ALOHA (CIFSA) algorithm, the PBSA algorithm exhibits lower slot overhead, higher tag throughput, and lower lag loss rate, which improves the efficiency of tag identification in dynamic RFID systems.

Keywords: Polling mechanism · Bit slots · Grouping dynamic frame slots · Removal of idle slots

1 Introduction

Anti-collision algorithms for radio frequency identification (RFID) systems can be classified into ALOHA algorithms and binary tree algorithms [1]. In mobile scenarios such as production line conveyors and highway toll stations, mobile tags are affected by the reader's reading range, tag movement speed, tag density, and tag arrival rate, which can cause random post-recognition and missed reading of tags, thereby severely affecting the identification efficiency of the system. Therefore, the exploration of tag anti-collision algorithms for dynamic RFID systems is imperative for improving the tag identification efficiency.

In recent years, several studies [2–8] have focused on anti-collision algorithms for mobile RFID systems. Among them, a frame slotted ALOHA algorithm was proposed based on continuous slot prediction [9]. This algorithm predicted the slot status by intra-frame, skipped invalid slots to identify valid slots and used a hashing mechanism to reduce the number of additional intra-frame collision tags, leading to a stable throughput of approximately 65%. However, owing to random nature of the hash

Q. Liu et al. (Eds.): CENet 2020, AISC 1274, pp. 483–492, 2021.
https://doi.org/10.1007/978-981-15-8462-6_54

mechanism, collision slots and idle slots could not be completely eliminated. To avoid the impact of idle slots on the system identification efficiency, a modified ALOHA algorithm for canceling idle frame slots (CIFSA) was proposed [10]. This algorithm adopted a reservation mechanism, where the tags were randomly modified based on the predetermined time slot sent by the reader, and then the idle slots of each frame were systematically shielded. This reduced the frame identification cycle and effectively improved the system throughput. However, the collision time slot was not processed in this algorithm. Therefore, the identification efficiency of the system can be further improved.

In this study, we propose a bit slotted ALOHA (PBSA) algorithm based on CIFSA and grouped dynamic frame slotted ALOHA (GDFSA) algorithms [11]. The PBSA algorithm introduces a polling mechanism [12] to change the tag removal strategy. After removing the idle slots, the slot counter counts the number of tags in the same slot, which is recorded as K. Among them, K−1 tags are randomly silenced and moved to the next frame, and the collision slot is converted into a single slot. Thus, the tag recognition efficiency of a unit frame is effectively improved.

2 Description of the PBSA Algorithm

2.1 Algorithm for Removing the Idle Frame Slots

The specific process of the ALOHA algorithm for removing idle frame slots is shown in Fig. 1. If the length of the frame slot is L, and the tag entering the identification area randomly selects its own communication slot m from 0 to L−1 and then generates a reservation sequence R consisting of L bits, the mth bit corresponding to the reservation sequence R is 1, while others are 0. According to the received sequence, the reader generates a decoding signal consisting of 0 and X. Each tag compares its own reserved sequence with the decoding sequence. If the reserved bit has i zeros (idle slots) before the corresponding bit in decoding sequence, then the reserved bits of the tag move forward by i positions to obtain the post-idle sequence.

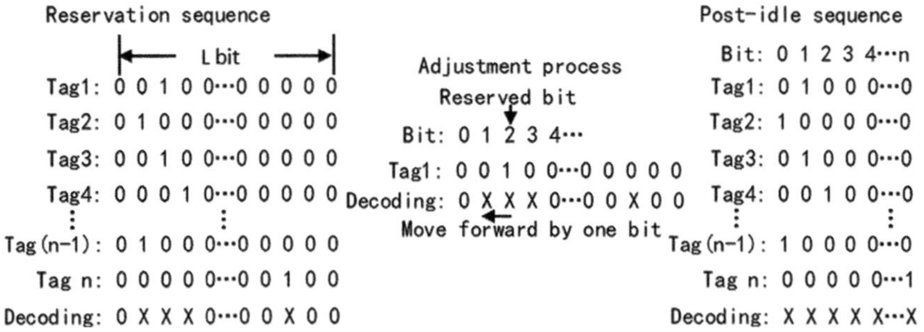

Fig. 1. Procedure for adjusting the idle time slot.

2.2 Polling Mechanism

The polling mechanism is divided into a pre-frame preparation phase and an intra-frame identification phase. In the pre-frame preparation phase, the tag is initially grouped and serialized to generate a reservation sequence, and the idle slots are removed. Subsequently, the reader counts the number of tags K in the same collision time slot through the time slot counter, and K−1 tags are randomly silenced. The collision slot is then converted into a single slot. In the intra-frame identification phase, the single slots are initially identified, and the silent tags are woken up to generate a new reservation sequence. Then, we are ready to start the identification of the next frame.

The principle and example of tag removal are shown in Figs. 2 and 3, respectively. Assuming that there are K tags (K > 0) in a slot, K−1 = 0 indicates that the slot is a single slot and the tag can be identified. Therefore, there is no need to remove it. However, K−1 $\neq$ 0 indicates that the time slot is a collision slot. In this case, K−1 tags are randomly silenced and moved to the next frame.

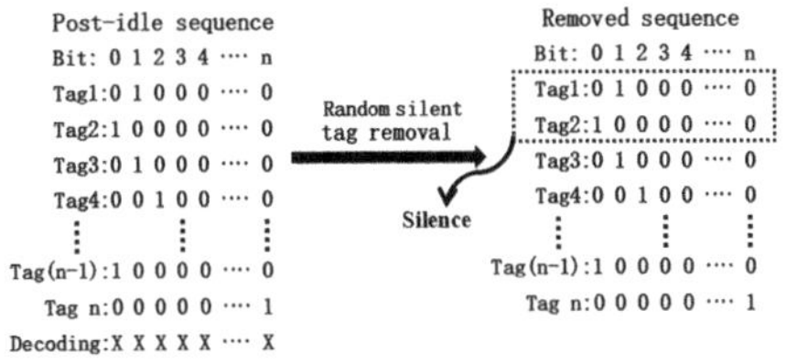

Fig. 2. Principle of tag removal.

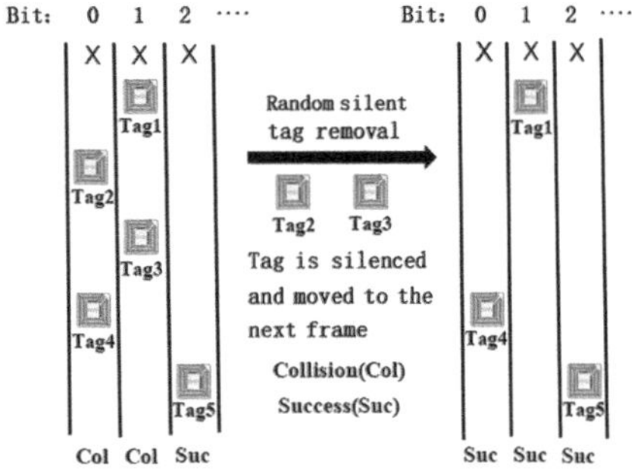

Fig. 3. Example of tag removal.

3 Algorithm Analysis

3.1 Data Analysis

Assuming that the frame length is L, the total number of tags to be identified is n, and the number of tags to be identified in a slot is r, then the probability of every slot to be selected is 1/L, which can be inferred as follows:

The probability of tag response in a time slot is:

$$P = C_n^r \left(\frac{1}{L}\right)^r \left(1 - \frac{1}{L}\right)^{n-r} \tag{1}$$

According to Eq. (1):

When r = 0, the probability of idle time slot:

$$P_i = \left(1 - \frac{1}{L}\right)^n \tag{2}$$

When r = 1, the probability of successful time slot:

$$\mathbf{P_s} = \frac{\mathbf{n}}{\mathbf{L}} * \left(\mathbf{1} - \frac{\mathbf{1}}{\mathbf{L}}\right)^{\mathbf{n-1}} \tag{3}$$

When r = 2, the probability of collision time slot:

$$\mathbf{P_c} = \mathbf{1} - \mathbf{P_i} - \mathbf{P_s} = \mathbf{C_n^r}\left(\frac{\mathbf{1}}{\mathbf{L}}\right)^{\mathbf{r}}\left(\mathbf{1} - \frac{\mathbf{1}}{\mathbf{L}}\right)^{\mathbf{n-r}} \tag{4}$$

Idle slot expectation:

$$\mathbf{E_i} = \mathbf{L} * \left(\mathbf{1} - \frac{\mathbf{1}}{\mathbf{L}}\right)^{\mathbf{n}} \tag{5}$$

Success slot expectation:

$$\mathbf{E_s} = \mathbf{n} * \left(\mathbf{1} - \frac{\mathbf{1}}{\mathbf{L}}\right)^{\mathbf{n-1}} \tag{6}$$

Collision time slot expectation:

$$\mathbf{E_c} = \mathbf{L} - \mathbf{E_i} - \mathbf{E_s} = \mathbf{C_n^r}\left(\frac{\mathbf{1}}{\mathbf{L}}\right)^{\mathbf{r-1}}\left(\mathbf{1} - \frac{\mathbf{1}}{\mathbf{L}}\right)^{\mathbf{n-r}} \tag{7}$$

According to formula (7), the throughput of this frame can be inferred:

$$\mathrm{s} = \frac{\text{Number of tag successfully identified in the frame}}{\text{The frame length}} \tag{8}$$

According to the above analysis, if the number of removed idle slots is L_0, the frame length is $L-L_0$. Tag removal protocol is introduced to change the collision slot into a single slot, and all the slots ultimately become successful slots. Therefore, the frame length is equal to the number of slots $(L-L_0)$, and the throughput S is

$$\mathbf{s} = \frac{\mathbf{L} - \mathbf{L_0}}{\mathbf{L} - \mathbf{L_0}} = \mathbf{1} \tag{9}$$

In the tag identification process of dynamic RFID systems, the tag may be lost when it leaves the signal area because the reader has limited time to identify the tags passing through this area. In dynamic RFID systems, the tag loss rate (TLR) is an important performance indicator, which is defined as follows:

$$\mathbf{TLR} = \frac{\text{Number of unrecognized tags leaving the signal area}}{\text{Total number of tags to identified by the system}} \quad (10)$$

Here, we adopt an earlier reported idea [15] and analyze the TLR of each algorithm through simulation using the simplified dynamic RFID model instead of a complicated mathematical framework.

3.2 Flowchart and Specific Steps

As shown in Fig. 4, the specific implementation steps of the PBSA algorithm are as follows:

(1) Estimate the total number of tags and group them.

In the pre-frame preparation phase, the Vogt algorithm [16] is used to estimate the total number of tags N. The tags are grouped to obtain the number of groups M and to initialize the variable flag. According to the time of entering the signal area, the reader sends a serialization command (SC) [17] to obtain the time serial number (SN) of each group of tags.

(2) Determine whether the earliest time group exists.

After receiving the start command, the tag determines the earliest time group. (i) If the earliest time group tag does not exist in the signal area, the reader recognizes the next set of tags, and the head pointer moves forward by one, i.e., flag = flag+1. (ii) If the earliest time group tag exits in the signal area, it indicates that there are unidentified tags and an acknowledgment code (ACK) is sent.

(3) Generate reservation sequence.

The reader receives the ACK signal, broadcasts the frame start command, and selects any bit from 0 to L−1 to receive a start command to make a reservation for generating a reservation sequence R.

(4) Generate the decoding sequence.

The reader detects all the received reservation sequences and compares them bit by bit to generate a decoding sequence.

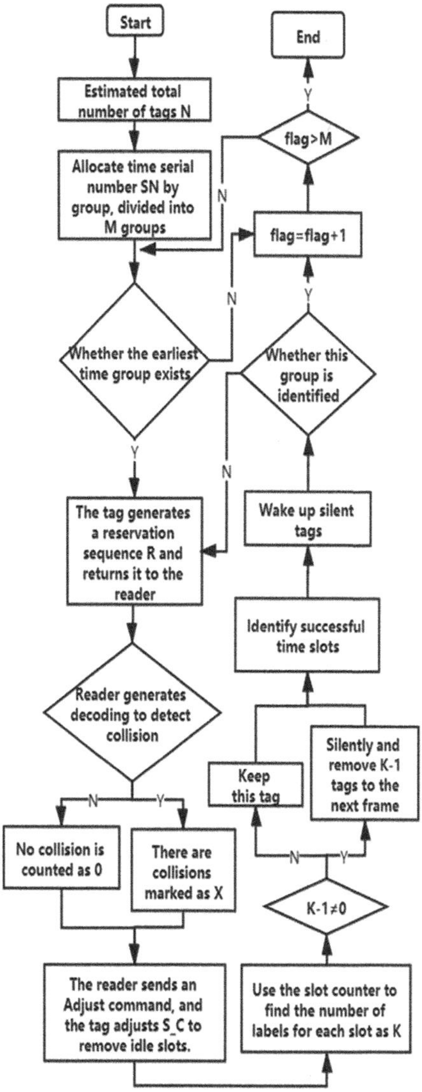

Fig. 4. Flowchart of PBSA.

The collision bits are recorded as X, and the remaining bits are recorded as 0.

(5) Remove idle slots

The reader compares the reserved bit of the tag with the decoding sequence and sends an Adjust command to adjust SC. If the reserved bit has i zeros (idle slots) before the corresponding slot in the decoding sequence, the reserved bit of the tag moves i positions to obtain the new serial number.

(6) Remove collision bit tags.

The number of tags K in the same time slot is counted by the time slot counter. If K $-1 = 0$, the slot is a single slot. If $K-1 \neq 0$, the slot is a collision slot, and $K-1$ tags are selected and silenced randomly. These tags are shifted to the next frame, converting the collision slot into a single slot.

(7) Identify the intra-frame.

The single slot is first identified, and then the silent tags in the pre-frame preparation phase are woken up to generate a new random sequence. Then, step (3) is repeated until a set of tags are identified. Subsequently, step (2) is implemented.

(8) Test whether the procedure has ended.

If flag < M, it indicates that there is an unidentified tag group. Then, step (2) is repeated to complete the identification of the tag group.

If flag $\geq$ M, it indicates that all the tag groups have been identified, and the algorithm ends.

4 Results and Discussion

To verify the effectiveness of the proposed algorithm, we compared the performances of dynamic frame slotted ALOHA (DFSA) [10], GDFSA [11], CIFSA, and PBSA algorithms. The simulation experiments were performed in Windows 10 environment with 4 GB DDR4 RAM, Core i5-7500 CPU, using MATLAB 2018b. We assumed that the tag space was evenly distributed, and the initial value of the number of tags to be identified was 50. In combination with the actual conditions, the upper limit of the number of tags was not more than 1600. For more accurate analysis and verification, the simulation results were obtained as the average value of 100 experimental data under the same environment, and the system throughput, slot overhead, and TLR of the four algorithms were compared.

4.1 Throughput Analysis

The simulated throughput curves for the four algorithms (DFSA, GDFSA, CIFSA, and PBSA) are shown in Fig. 5. The PBSA algorithm avoids the rapid decline in the throughput observed in the DFSA algorithm with the increase in the number of tags by implementing the Vogt tag estimation algorithm, tag grouping, first-come-first-serve policy, removal of idle frame slot based on reservation mechanism, and tag removal mechanism. Further, compared with GDFSA and CIFSA algorithms, PBSA algorithm exhibits superior throughput and stability. When the number of tags $N \leq 354$, the throughputs of DFSA and GDFSA algorithms are basically identical. When $N > 354$, the throughput of DFSA algorithm decreases sharply. When $N = 1600$, the throughput of DFSA algorithm is extremely low (0.076). GDFSA algorithm avoids this deterioration by using tag grouping and first-come-first-serve strategy, and its throughput is stabilized at ~0.36. The CIFSA algorithm uses tag grouping, first-come-first-serve strategy, and reservation mechanism to eliminate the impact of idle slots on the throughput, and it achieves a stable throughput of ~0.75. The PBSA algorithm introduces a polling mechanism to change the tag removal strategy and converts the collision slots into successful slots while removing the idle slots, which significantly improves the tag recognition efficiency of a unit frame, facilitating a high throughput of ~1.00. Compared with DFSA, GDFSA, and CIFSA algorithms, the PBSA algorithm exhibits the highest efficiency, which becomes more evident when the number of tags is large.

4.2 Analysis of Total Time Slot Consumption

In RFID systems, the smaller the total number of time slots needed for tag identification, the better the system performance. The simulation results for the total number of slots consumed by the four algorithms are shown in Fig. 6. It is clear that when the number of tags increases to 1000, the total number of slots of the DFSA algorithm increases exponentially, while that of the CIFSA, GDFSA, and PBSA algorithms increases linearly. Further, the growth rate of the total number of slots of PBSA algorithm is the smallest. This validates the superiority of the PBSA algorithm over the other three algorithms.

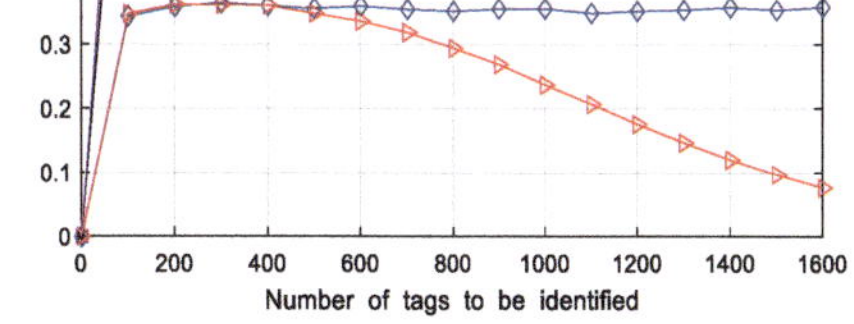

Fig. 5. Throughput analysis.

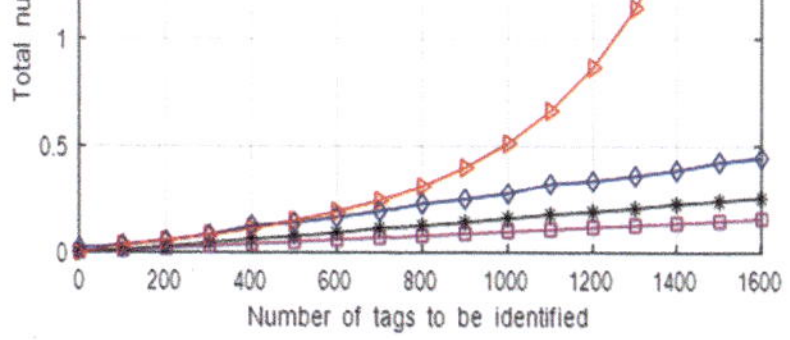

Fig. 6. Analysis of total number of slots.

4.3 TLR Analysis

The simulation model does not consider the time consumed by the reader to send instructions to the tag and only considers the time consumed by the tag to send information to the reader (represented by slots). Therefore, the major time loss of RFID communication system occurs when the tag sends information to the reader. By establishing a simplified dynamic RFID model for tag identification as shown in Fig. 7, we can find that the time period of tag stay is T = 500 (slots), and the initial density of tags in the signal area is 100. Here, we change the tag density by adjusting the number of tags in the signal area and analyze the impact of these changes on the performance indicators of mobile RFID systems.

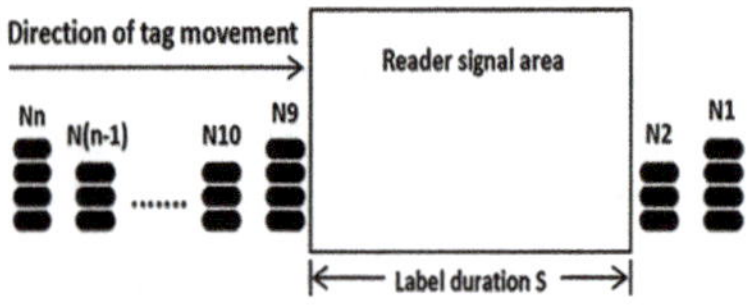

Fig. 7. Simplified RFID model.

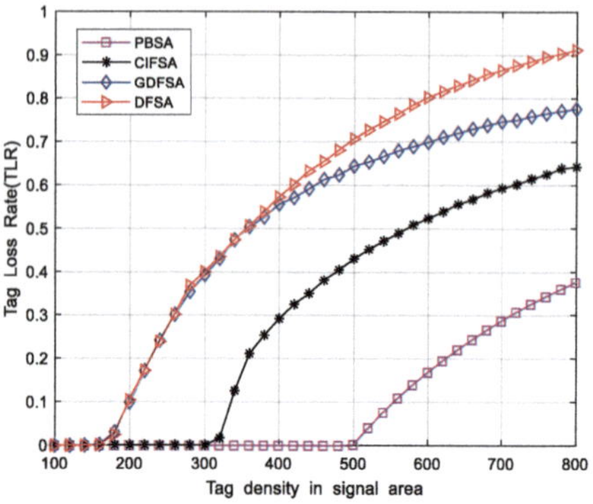

Fig. 8. Loss rate analysis.

The following conclusions can be drawn from Fig. 8. (1) Under the same conditions, when the number of tags $N \leq 354$, the TLR curves of DFSA and GDFSA algorithms almost coincide. When $N > 354$, the TLR of the DFSA algorithm exhibits an increasing trend with an increasing rate higher than that of the GDFSA algorithm. This is because the GDFAS algorithm groups and serializes the tags, which improves the tag recognition efficiency per unit frame and reduces the TLR. (2) Under the same conditions, the TLR of the PBSA algorithm is much lower than that of the DFSA and the GDFSA algorithms. The is because the PBSA algorithm eliminates the impact of idle slots. (3) When $N < 300$, the TLR of the CIFSA and PBSA algorithms are basically identical. For $N > 300$, the TLR of the CIFSA algorithm shows a steep increasing trend, while that of the PBSA algorithm exhibits an increasing trend for $N = 500$. This is because after completely removing the idle slots, the PBSA algorithm converts the collision slot into a single slot through the tag removal mechanism, which significantly improves the tag identification efficiency of the system and reduces the overall tag loss. (4) When the number of tags or tag density in the signal area is very large, the increase of the TLR of all the algorithms is evident.

5 Conclusion

We proposed a bit slotted ALOHA algorithm based on a polling mechanism. In this algorithm, the number of tags was initially estimated. When the number of tags was greater than 354, the tags were grouped and serialized. After utilizing first-come-first-serve strategy and reservation mechanism to remove the idle slots, the tag removal mechanism was employed for converting the collision slots into single slots to realize fast identification of tags. Simulation experiments showed that compared to the other algorithms, the throughput of PBSA algorithm increased much rapidly with the number of tags, especially when the number was more than 1000. Further, the PBSA algorithm exhibited smaller slot overhead, higher tag throughput, better stability, and enhanced recognition efficiency than the other algorithms. The simulation of the dynamic tag recognition model revealed extremely low TLR of the PBSA algorithm. It may be noted that the proposed algorithm has a maximum frame length of 256. The tag requires only 8-bit binary circuits, and the required storage space is 8 bits. The complexity of the tag circuit is low, and the production cost of the tag is significantly reduced. Therefore, the proposed algorithm exhibits immense potential for applications requiring fast and rapid tag identification.

Acknowledgment. This research was supported by the Scientific Research Project of Hunan Provincial Department of Education (18A332), the Science and Technology Plan Project of Hunan Province (2016TP1020), and the Application-Oriented Special Disciplines, Double First-Class University Project of Hunan Province (Xiangjiaotong [2018] 469).

References

1. Zhang, X., Zhou, W.: Research on dynamic framed binary tree anti-collision algorithm for RFID system. J. Syst. Simul. **30**(3), 1063–1073 (2018)
2. Joo, Y.I., Seo, D.H., Kim, J.W.: An efficient anti-collision protocol for fast identification of RFID tags. Wireless Pers. Commun. **77**(1), 776–777 (2014)
3. Chen, M., Chen, S.: An efficient anonymous authentication protocol for RFID systems using dynamic tokens. In: IEEE 35th International Conference on Distributed Computing Systems, Columbus, OH, pp. 756–757 (2015)
4. Mu, Y., Zhang, X.: Adaptive tree grouping and blind separation anti-collision algorithm for radio frequency identification system. J. Comput. Appl. Math. **35**(1), 19–22 (2015)
5. Wang, Y.: Research on identification efficiency of RFID system in large-scale tag environment. Yunnan Minzu University (2016)
6. Al-Zewairi, M., Suleiman, D., Shaout, A.: Multilevel fuzzy inference system for risk adaptive hybrid RFID access control system. In: Cybersecurity and Cyberforensics Conference (CCC), Amman, Jordan, pp. 1–7 (2016)
7. Ding, Z., Ding, L., Tang, H.: Anti-collision algorithm based on dynamic reservation mechanism. Comput. Eng. **43**(2), 317–321 (2017)
8. Song, Y., Li, X.: Method for evaluating performance of single reader mobile RFID system. Comput. Eng. Appl. **55**(23), 228–234 (2019)
9. Zhang, X., Hu, Y.: Research on RFID anti-collision algorithm for adaptively allocating time slots in packets. Acta Electronica Sinica **44**(06), 1328–1335 (2016)

10. Yang, L.: Research on anti-collision of mobile tag in dynamic RFID system. Southwest Petroleum University (2016)
11. Dong, H., Sheng, K., Ma, J., Yao, H.: Aloha anti-collision algorithm for adaptive frame slot based on tag packet first come first served. J. Guangxi Univ. Sci. Technol. **29**(01), 61–68 (2018)
12. Lei, X., Sanglu, L.: Radio Frequency Identification Technology-Principle, Protocol and System Design, 2nd edn., pp. 205–210. Beijing: Science Press, Beijing (2016)
13. Yu, S.: Anti-collision protocol in large-scale RFID system. Nanchang University (2016)
14. Mo, L., Chen, W., Ren, J.: Anti-collision algorithm for RFID based on counter and bi-slot. J. Comput. Appl. **37**(08), 2168–2172 (2017)
15. Li, X.W., Feng, Q.Y., Chen, Y.H.: System performance evaluation method of anti-collision protocol in mobile RFID system. J. China Railway Soc. **36**(4), 42–47 (2014)
16. Zhou, W.: Research on the anti-collision algorithm of multi-tag adaptive identification of RFID system and its performance analysis. Jiangxi University of Technology (2017)
17. Li, X.; Research on tag identification technology in mobile RFID systems. Southwest Jiaotong University (2015)
18. Chen, H., Gao, Q., An, X.: Research on anti-collision algorithm based on dynamic frame slotted Aloha. Wireless Internet Technol. **17**, 107–109 (2017)
19. Yuan, L., Du, Y., He, Y., Lü, M., Cheng, Z.: Grouped dynamic frame slotted Aloha tag anti-collision algorithm based on parallelizable identification. J. Electr. Eng. Technol. **40**(4), 944–950 (2018)

A Mobile E-Commerce Recommendation Algorithm Based on Data Analysis

Jianxi Zhang[1] and Changfeng Zhang[2(✉)]

[1] Computational Sciences, Shandong Institute of Commerce and Technology, Ji'nan 250000, Shandong, China
[2] Food Sciences, Shandong Province Key Laboratory of Storage and Transportation Technology of Agricultural Products, Ji'nan 250000, Shandong, China
zcf202@163.com

Abstract. With the development of the Internet and the change of consumption concept, e-commerce platform has become the primary way of shopping. In this case, personalized recommendation system has become the main way to find favorite businesses, and also become the main means to sell goods on the platform. n the recommendation system of e-commerce platform, collaborative filtering recommendation algorithm is the most widely used. The main idea of the algorithm is to find the similar neighbor set of the target user, and recommend the user preference of the neighbor set to the target user as a recommendation.

In this paper, a collaborative filtering recommendation algorithm based on context information is proposed, which focuses on the influence of user context information on the similarity calculation between users in the recommendation algorithm in the mobile Internet environment. The algorithm takes the user location information and user trust relationship as the key factors. First, the algorithm preprocesses the original score data through SVD to alleviate the scarcity of data, then generates a list of merchant candidate recommendations based on the score similarity of users Using collaborative filtering. And then calculates the travel cost of each merchant in the candidate recommendation list to the target user according to the location context, and adjusts the recommendation weight of the merchant in the list, Finally, the final Top-N merchants are generated to recommend to the target users.

In the end of this paper, through the experimental simulation of the proposed algorithm, and compared with the traditional collaborative filtering recommendation algorithm, it is proved that the algorithm further improves the effectiveness and accuracy of the recommendation algorithm.

Keywords: Collaborative filtering · Recommendation algorithm · Context information

Q. Liu et al. (Eds.): CENet 2020, AISC 1274, pp. 493–501, 2021.
https://doi.org/10.1007/978-981-15-8462-6_55

1 Introduction

In the environment of mobile internet, e-commerce platform can obtain more information about users, such as location, time, service, environment, etc. In these information, location is usually the primary concern of users, because in real life, people's activities show the characteristics of localization [1]. As the distance between merchants increases, so will people's interest in them. Therefore, in the process of business recommendation, it is not only necessary to meet the personalized needs of users to the business, but also to explore the relationship between the recommended business and the user's location context. For example, we can't recommend merchants too far to users. However, the traditional business recommendation algorithm based on collaborative filtering [2] only calculates the similarity between users or items through the "user-item" two-dimensional scoring matrix, and recommends based on similar users or items, ignoring the impact of context information.

In view of the above problems, this paper singular value decomposition (SVD) [3] is introduced to solve the problem of sparsity of scoring data and introduces the context of location and user rating into the recommendation process, and proposes a business recommendation algorithm based on location and trust relationship (BR-LaT).

2 Literature Review

Context aware recommendation [4] is an application of context aware computing theory in recommendation field. Traditional recommendation techniques only focus on the "user-project" relationships, but ignore their context. Adomavicius et al. [5] first proposed to integrate contextual information into the recommendation system to improve the accuracy of recommendation. According to the application phase of context in recommendation, context-aware recommendation is divided into three categories: context prefiltering, post-context filtering and context modeling.

As an important contextual information, location has been widely used in the recommendation field. In literature [1], a location-aware recommendation system was constructed based on the pyramid model. In literature [6], a location-based recommendation system is introduced. It builds a user preference model by mining the GPS location track data, and then makes business recommendations to users based on the model. In the literature [7], the user's activity range is first established by analyzing the user's historical location data, and then merchants are recommended to the user within the range. But, when the user moves out of that range, the results are terrible. Literature [8] has improved this defect. It establishes the user's food taste preference model by clustering, and obtains better recommendation results. In literature [9], the location preference of users is modeled by bayesian model, and on top of that, a personalized business recommendation system based on location is constructed. In literature [10], not only the location context, but also the weather, traffic conditions and other contextual factors are taken into account when personalized merchant recommendation is made. With the rapid development of social networks and the popularity of mobile location-based devices, Some researchers have proposed location-based social networks by combining social networking with location context [11, 12], users' location data can be obtained through their check-in behavior, and then combined with the content of friends' relationship and time in the social network to make recommendations.

Based on the above research, the location and user evaluation context are further combined to improve the traditional merchant recommendation algorithm based on collaborative filtering [2] so as to improve the accuracy of the recommendation.

3 Description of the BR-LaT

3.1 Data Model

A data model is established based on user information, merchant information and user evaluation information of merchants in the data set. It mainly consists of four data sets:

(1) User collection: $\boldsymbol{U} = \{\boldsymbol{u}_i | \boldsymbol{i} \in [1, \boldsymbol{M}]\}$, Where $\boldsymbol{M}$ is the number of users;
(2) Business Collection: $\boldsymbol{B} = \{\boldsymbol{b}_j | \boldsymbol{j} \in [1, \boldsymbol{N}]\}$, Where $\boldsymbol{N}$ is the number of businesses;
(3) Business location collection: $\boldsymbol{L} = \{\boldsymbol{l}_k | \boldsymbol{k} \in [1, \boldsymbol{K}]\}$, Where $\boldsymbol{K} = \boldsymbol{N}$, Position is represented by latitude and longitude coordinates;
(4) "User–Business" rating matrix: $\boldsymbol{R} = \{\boldsymbol{r}(\boldsymbol{u}_i, \boldsymbol{b}_j) | \boldsymbol{u}_i \in \boldsymbol{U}, \boldsymbol{b}_j \in \boldsymbol{B}\}$, Every row in the matrix $\boldsymbol{r}(\boldsymbol{u}_i, \boldsymbol{b}_j)$ represents the score by user $\boldsymbol{u}_i$ to business $\boldsymbol{b}_j$, Where $\boldsymbol{r}(\boldsymbol{u}_i, \boldsymbol{b}_j) \in [0, 5]$, the higher the score, the higher the user's preference for the business.

3.2 SVD Alleviates Data Scarcity

In the actual recommendation system, there are a large number of users and merchants, but only some users rate a small number of merchants. As a result, the "User–Business" two-dimensional scoring matrix is often sparse, and sparsity is one of the important factors that affect the accuracy of recommendation [4]. In order to solve this problem, the algorithm in this paper uses ***SVD*** to reverse fill the original matrix to alleviate the sparsity of data [13].

Procedure is as follows:

(1) Pre filling of "User–Business" rating matrix. According to the "User–Business" rating matrix $\boldsymbol{R}$. in the data model, the average score of each user to the business is calculated to fill in the vacancy of the corresponding row element in $\boldsymbol{R}$, forming the matrix $\boldsymbol{R}'$.

(2) ***SVD*** decomposition matrix $\boldsymbol{R}'$. The high-dimensional matrix $\boldsymbol{R}'$ is reduced to three low-dimensional matrices $\boldsymbol{U}$, $\boldsymbol{S}$ and $\boldsymbol{V}$ by ***SVD***. For matrix $\boldsymbol{S}$, select top $\boldsymbol{k}(\boldsymbol{k} \leq \boldsymbol{rank}(\boldsymbol{R}')$ largest singular values, set remaining values to $\boldsymbol{0}$. And then delete all rows and columns with the value of $\boldsymbol{0}$ to get the simplified matrix $\boldsymbol{S}_k$. Simplified matrix $\boldsymbol{U}_k$ and $\boldsymbol{V}_k$ is obtained in the same way.

(3) Reverse Filling of Raw Matrix $\boldsymbol{R}$. The square root of matrix $\boldsymbol{S}_k$ is denoted as $\boldsymbol{S}_k^{\frac{1}{2}}$, the forecast score of User $\boldsymbol{u}_i$ for business $\boldsymbol{b}_j$ is: $\boldsymbol{P}(\boldsymbol{u}_i, \boldsymbol{b}_j) = \overline{\boldsymbol{r}_{u_i}} + \boldsymbol{U}_k\boldsymbol{S}_k^{\frac{1}{2}}(\boldsymbol{i}) \cdot \boldsymbol{S}_k^{\frac{1}{2}}\boldsymbol{V}_k^T(\boldsymbol{j})$, Where $\overline{\boldsymbol{r}_{u_i}}$ is the average user rating of the business, $\boldsymbol{U}_k\boldsymbol{S}_k^{\frac{1}{2}}(\boldsymbol{i})$ represents the i-th row of

the matrix, $S_k^{\frac{1}{2}}V_k^T(j)$ j-th column of the matrix. A dense matrix $\boldsymbol{R}''$ is formed by reverse filling the corresponding vacant positions of the original matrix $\boldsymbol{R}$ according to the predicted $\boldsymbol{P}(\boldsymbol{u}_i, \boldsymbol{b}_j)$.

3.3 Similarity Calculation

After the original score matrix is backfilled by ***SVD***, the similarity between u, r, s is calculated by the filled matrix $\boldsymbol{R}''$. The similarity calculation methods in the recommendation system mainly include: (1) Cosine Similarity; (2) Improved Cosine Similarity; (3) Pearson Correlation Coefficient [14] Etc. Pearson correlation coefficient is most widely used based on users' common score set. Therefore, Pearson correlation coefficient is adopted to calculate the similarity in this paper. The formula is as follows:

$$Sim(u_x, u_y) = \frac{\sum_{b_k \in B(u_x,u_y)} (r(u_x, b_k) - \overline{r_{u_x}}) \times (r(u_y, b_k) - \overline{r_{u_y}})}{\sqrt{\sum_{b_k \in B(u_x,u_y)} (r(u_x, b_k) - \overline{r_{u_x}})^2 \times \sum_{b_k \in B(u_x,u_y)} (r(u_y, b_k) - \overline{r_{u_y}})^2}} \tag{1}$$

In formula (1), $\boldsymbol{B}(\boldsymbol{u}_x, \boldsymbol{u}_y)$ represents the common scoring business collection of the user $\boldsymbol{u}_x$ and $\boldsymbol{u}_y$. $\boldsymbol{r}(\boldsymbol{u}_x, \boldsymbol{b}_k)$ and $\boldsymbol{r}(\boldsymbol{u}_y, \boldsymbol{b}_k)$ represents the ratings of business $\boldsymbol{b}_k$ in the collection by the user $\boldsymbol{u}_x$ and $\boldsymbol{u}_y$. $\overline{\mathbf{r}_{\mathbf{u}_x}}$ and $\overline{\mathbf{r}_{\mathbf{u}_y}}$ represents the average ratings of business by the user $\boldsymbol{u}_x$ and $\boldsymbol{u}_y$.

3.4 Generate a List of Recommended Candidates

After getting the similarity between users, for the target user $\mathrm{u_x}$, select some users with the highest similarity to Top-N nearest neighbor selection policy as neighbors. Then the score prediction is made based on the neighboring users. Traditional score predictions are generally calculated using the formula (2) [14]:

$$r(u_x, b_j) = \overline{r_{u_x}} + \frac{\sum_{u_y \in N_{u_x}} (Sim(u_x, u_y) \times (r(u_y, b_j) - \overline{r_{u_y}}))}{\sum_{u_y \in N_{u_x}} Sim(u_x, u_y)} \tag{2}$$

In the formula, $\boldsymbol{N}_{\boldsymbol{u}_x}$ represents the set of neighbors of the target user $\mathrm{u_x}$. $\boldsymbol{Sim}(\boldsymbol{u}_x, \boldsymbol{u}_y)$ represents the similarity of target user $\boldsymbol{u}_x$ to user $\boldsymbol{u}_y$ in the neighbor set $\boldsymbol{r}(\boldsymbol{u}_y, \boldsymbol{b}_j)$ represents the score of the user $\boldsymbol{u}_y$ to user $\boldsymbol{b}_j$.

In formula (2), the user's score in the neighbor set is regarded as equally important in scoring prediction. But in real life, these scores from different users usually have different credibility [15]. For example, a generally accepted business generally has a high rating confidence. Therefore, in this paper, different prediction weights are given to users based on their rating confidence. By analyzing the characteristics of the data set in the experiment, we find that: One user can label the comment text as "useful", "funny" or "cool" for another user, and the corresponding number of "useful", "funny" or "cool" for the commented user will be increased. Based on this feature, the more "useful", "funny" and "cool" a user gets, the more recognized that user is. And then the user has a higher rating confidence and prediction weight. This weight calculation method is defined as a formula (3).

$$w_{u_x} = \frac{N_{useful}(u_x) + N_{funny}(u_x) + N_{cool}(u_x)}{Max(N_{useful} + N_{funny} + N_{cool})} \tag{3}$$

In formula (3), The molecule represents the sum of the number of "useful", "funny" and "cool" the users $\boldsymbol{u_x}$ gets. Denominator represents the maximum of the sum of the three components in a data set. Based on the formula (3), the formula (2) is improved to predict the score formula (4). By $\boldsymbol{r'(u_x, b_j)}$, generate the recommended candidate list $\boldsymbol{S}$ of the businesses By ranking the unrated businesses in descending order.

$$r'(u_x, b_j) = \overline{r_{u_x}} + \frac{\sum_{u_y \in N_{u_x}} (w_{u_y} \times Sim(u_x, u_y) \times (r(u_y, b_j) - \overline{r_{u_y}}))}{\sum_{u_y \in N_{u_x}} w_{u_y} \times Sim(u_x, u_y)} \tag{4}$$

3.5 Calculate Travel Penalty

User activities are localized. Therefore, when recommending merchants to users, we should not only consider the score similarity between users, but also explore the relationship between the location of the recommended merchants and the activity range of users. This paper uses the "Travel Penalty" proposed in literature [1] to measure this relationship, which represents the cost of the location of merchant $\boldsymbol{b_j}$ to user $\boldsymbol{u_x}$. The basic idea of calculation is to calculate the average distance between the location of the merchant $\boldsymbol{b_j}$ and the location of the merchant with user $\boldsymbol{u_x}$'s historical score. The formula can be expressed as:

$$TravelPenalty(u_x, b_j) = \frac{\sum_{b_k \in B_{u_x}} d(b_j, b_k)}{|B_{u_x}|} \tag{5}$$

In formula (5), $\boldsymbol{B_{u_x}}$ represents the history score merchant collection for user $\boldsymbol{u_x}$. $|\boldsymbol{B_{u_x}}|$ represents the number of sets. $\boldsymbol{d(b_j, b_k)}$ represents the spatial distance between a business and $\boldsymbol{b_k}$ in a candidate recommendation list. Spatial distance is calculated using a Euclidean distance-based method [16]. The travel cost usually calculated from the formula (5) is a larger value, so it needs to be normalized into the scoring range. According to the characteristics of the dataset in this experiment, it is normalized to [0, 5] using arctangent function. The formula can be expressed as:

$$TravelPenalty'(u_x, b_j) = \frac{10}{\pi} arctan(TravelPenalty(u_x, b_j)) \tag{6}$$

3.6 Generate Final Top-N Recommendations

In the previous sections, the travel cost has been measured for each business in Business Recommendation Candidate List $\boldsymbol{S}$ for target user $\boldsymbol{u_x}$. Considering the Score $\boldsymbol{u_x}$ for $\boldsymbol{b_j}$ and the cost of $\boldsymbol{b_j}$, the recommended weight of final merchant $\boldsymbol{b_j}$ for target user $\boldsymbol{u_x}$ is expressed as:

$$RecScore(u_x, b_j) = r'(u_x, b_j) - TravelPenalty'(u_x, b_j) \tag{7}$$

The final ***Top-N*** is the highest N merchants in $\boldsymbol{RecScore(u_x, b_j)}$ for the user $\boldsymbol{u_x}$.

4 Experiment and Result Analysis

4.1 Experimental Datasets

The data set used in this experiment is Yelp Dataset, a public data set containing business information [17]. Preprocessing the original data, only business-related parts were selected from the data set, and users with more than 5 ratings were selected. The final experimental data included more than 80,000 comments from more than 5,000 users on more than 4,000 businesses and more than 60,000 friendships. Comments are scored as integer values between [0, 5]

In order to test the performance of the recommendation algorithm, the data set was divided into a training set and a test set. This experiment randomly selected 80% as the training set and the other 20% as the test set.

4.2 Experimental Evaluation Index

Experimental use ***P@N*** Index [14] is used to evaluate the recommendation results to verify the performance of the algorithm. The formula can be expressed as:

$$P@N = \frac{\#relevant\ items\ in\ the\ Top - N\ items}{N} \tag{8}$$

The higher the ***P@N*** value is, the higher the accuracy of the recommendation will be; conversely, the lower the accuracy of the recommendation will be.

4.3 Experimental Description and Result Analysis

In order to verify the effectiveness and accuracy of the proposed algorithm, the following three algorithms are compared in this section:

Algorithm 1: **UserCF** algorithm proposed in reference [2];
Algorithm 2: **LocationCF** algorithm proposed in reference [1];
Algorithm 3: **BR-LaT** algorithm proposed in this paper.

According to the top-n selection strategy, the number of similar neighbors is 5, 10, 15, 20.5 and 10 businesses are recommended to the target users. The experimental results are shown in Fig. 1 and Fig. 2.

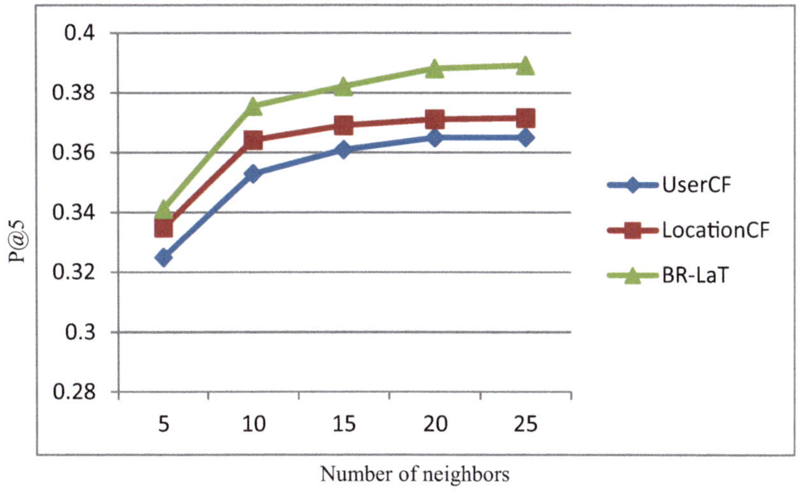

Fig. 1. P@5 comparison of different algorithms.

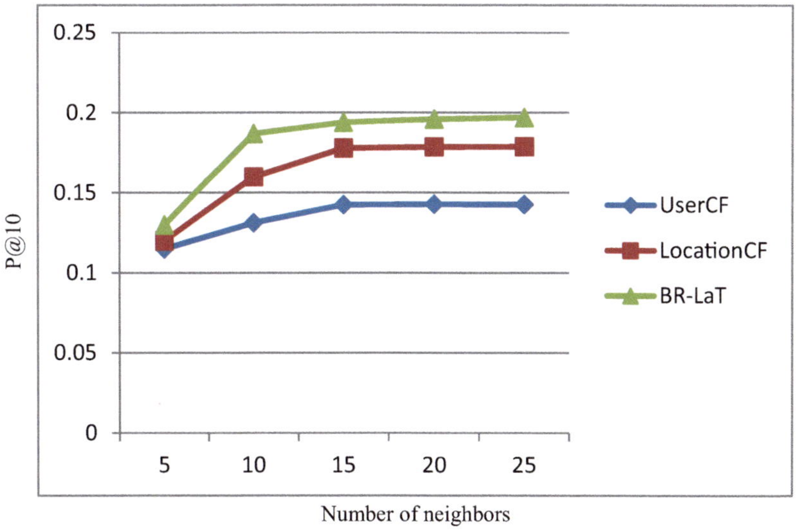

Fig. 2. P@10 comparison of different algorithms.

According to the experimental results in Fig. 2-1 and Fig. 2-2, it can be shown that: compared with the UserCF algorithm, the LocationCF algorithm has higher P@5 and P@10. It shows that the accuracy of recommendation can be improved by introducing location context on the basis of traditional collaborative filtering recommendation algorithm. compared with the LocationCF algorithm, the BR-LaT algorithm has higher P@5 and P@10. It shows that the BR-LaT algorithm has higher recommendation accuracy than LocationCF algorithm. Because BR-LaT algorithm is improved on the basis of LocationCF algorithm. Firstly, SVD is used to preprocess the original data, which can alleviate the sparsity of the data to a certain extent. It introduces the rating credibility and gives different forecasting weights according to the rating credibility of users, which is more practical.

5 Summary

Compared with the traditional business recommendation algorithm based on collaborative filtering, this paper focuses on the influence of user activity localization and user evaluation on the recommendation effect, and proposes a business recommendation algorithm based on location and user trust context. ***SVD*** technology is introduced to preprocess data matrix, and the final ***Top-N*** merchants are recommended to target users. The experimental results show that the proposed algorithm can improve the accuracy of recommendation.

Acknowledgements. This work was supported by Key Science and Technology Project of Ningxia government in China.

References

1. Levandoski, J.J., Sarwat, M., Eldawy, A., et al.: Lars: a location-aware recommender system. In: 2012 IEEE 28th International Conference onData Engineering (ICDE), pp, 450–461. IEEE (2012)
2. Adomavicius, G., Tuzhilin, A.: Toward the next generation of recommender systems: a survey of the state-of-the-art and possible extensions. IEEE Trans. Knowl. Data Eng. **17**(6), 734–749 (2005)
3. Sarwar, B., Karypis, G., Konstan, J., et al.: Application of dimensionality reduction in recommender systems. In: ACM Webkdd Workshop (2000)
4. Saini, B., Singh, V., Kumar, S.: Information retrieval models and searching methodologies: Survey. Inf. Retrieval **1**(2), 20 (2014)
5. Adomavicius, G., Sankaranarayanan, R., Sen, S., et al.: Incorporating contextual information in recommender systems using a multidimensional approach. ACM Trans. Inf. Syst. (TOIS) **23**(1), 103–145 (2005)
6. Takeuchi, Y., Sugimoto, M.: CityVoyager: An outdoor recommendation system based on user location history. In: Ma, J., Jin, H., Yang, L.T., T, J.J.-P. (eds.) UIC 2006. LNCS, vol. 4159, pp. 625–636. Springer, Heidelberg (2006). https://doi.org/10.1007/11833529_64

7. Gupta, G., Lee, W.C.: Collaborative spatial object recommendation in location based services. In: 2010 39th International Conference on Parallel Processing Workshops (ICPPW), pp. 24–33. IEEE (2010)
8. Hasegawa, T., Hayashi, T.: Collaborative filtering based spot recommendation seamlessly available in home and away areas. In: 2013 IEEE/ACIS 12th International Conference on Computer and Information Science (ICIS), pp. 547–548. IEEE (2013)
9. Park, M.-H., Hong, J.-H., Cho, S.-B.: Location-based recommendation system using bayesian user's preference model in mobile devices. In: Indulska, J., Ma, J., Yang, L.T., Ungerer, T., Cao, J. (eds.) UIC 2007. LNCS, vol. 4611, pp. 1130–1139. Springer, Heidelberg (2007). https://doi.org/10.1007/978-3-540-73549-6_110
10. Lee, B.-H., Kim, H.-N., Jung, J.-G., Jo, G.-S.: Location-Based service with context data for a restaurant recommendation. In: Bressan, S., Küng, J., Wagner, R. (eds.) DEXA 2006. LNCS, vol. 4080, pp. 430–438. Springer, Heidelberg (2006). https://doi.org/10.1007/11827405_42
11. Jiang, D., Guo, X., Gao, Y., et al.: Locations recommendation based on check-in data from location-based social network. In: 2014 22nd International Conference on Geoinformatics (GeoInformatics), pp. 1–4. IEEE (2014)
12. Wei, W., Zhu, X., Li, Q.: LBSNSim: Analyzingand modeling location-based social networks. In: 2014 Proceedings IEEE INFOCOM, pp. 1680–1688. IEEE (2014)
13. Koren, Y., Bell, R., Volinsky, C.: Matrix factorization techniques for recommender systems. Computer **8**, 30–37 (2009)
14. Shah, N.J., Bhanderi, S.D.: A survey on recommendation system approaches. Data Min. Knowl. Eng. **6**(4), 151–156 (2014)
15. Luo, C., Luo, X.R., Schatzberg, L., et al.: Impact of informational factors on online recommendation credibility: the moderating role of source credibility. Decis. Support Syst. **56**, 92–102 (2013)
16. Hjaltason, G.R., Samet, H.: Distance browsing in spatial databases. ACM Trans. Database Syst. (TODS) **24**(2), 265–318 (1999)
17. Yelp Dataset, https://www.yelp.corn/academicdataset

Heat Dissipation Optimization for the Electronic Device in Enclosure Condition

Bibek Regmi[1,2(✉)] and Bishwa Raj Poudel[3]

[1] College of Mechanical and Electrical Engineering, Hohai University, Changzhou, China
regmi_99@yahoo.com
[2] Suzhou Fanglin Technology Company Ltd., Suzhou, China
[3] Institute of Engineering, Tribhuvan University, Kirtipur, Kathmandu, Nepal
bishwapoudel7@gmail.com

Abstract. The development of optimized miniaturize and powerful electronic components keep the strain on engineers to produce optimal cooling designs for better results. One method for cooling these electronic components is with heat sinks which effectively increase the surface area available for extracting the heat from the electronic components. In this paper, a methodology is developed where plane fitting methods are used to make boundary planes that make the boundary for the design of heat sink. This paper helps engineers and researchers with less knowledge of thermal analysis to design a proper heat sink.

Keywords: Heat sink · Heat sink optimization · Pin fin · Optimization · ICEPAK · Plane fitting

1 Introduction

Information technology is prolonged in the tracking of miniaturized and powerful electronic devices which results in an increase in power densities in electronic products. When power densities increase temperature also increases [1]. Maximum temperature in the device circumscribes the longevity of these devices. To dissipate heat from the heat sink needs more heat transfer surface area that decreases the thermal resistance of the package. However, heat sink increases the sizes, cost and weight of the components. These reasons create the essence of an optimal heat sink with an effective tool and design.

Generally, Compact modelling and Computational Fluid Dynamic (CFD) are two procedures used on cooling packages in thermal design. In the Compact method [2] for the entire package, the single heat transfer coefficient is estimated. It is the most common method used, computationally inexpensive, unsophisticated optimization but results are very crude. On the other hand, CFD is computationally expensive and optimization is rarely estimated. Majority of heat sinks problems are solved with geometry, selected properties and lowest thermal resistance [3]. Influence of parameter on thermal resistance can be undaunted through the chart that shows characteristics of commercially available heat sinks [4]. Performance of design is completely based on

Q. Liu et al. (Eds.): CENet 2020, AISC 1274, pp. 502–513, 2021.
https://doi.org/10.1007/978-981-15-8462-6_56

expertise and knowledge of a designer, not on optimal design. Combine method [5], uses both methods, is estimated in using flow paths from geometry and estimated the flow using network solver. Heat transfer coefficient is resolute through derived equations or curve fits and temperature distribution is numerically modelled with a curvilinear form of the discredited diffusion equation.

ICEPAK, the heat sink design tool [6], uses this method to solve the temperature distribution in heat transfer and the fluid flow. ICEPAK consists of built-in fan and heat sink models due to which cooling model can be designed and solved. It also contains the options of optimizing the design for any specified objectives functions by selecting the specified parameter as a variable in optimization procedure. It is simple and can be designed quickly with accurate output. It reduces the computational and actual cost of the design process.

The number of experiments needed to get adequate results for testing and generating the correlation is extremely high because of the number of variables that has an influence on the contraction heat transfer coefficient. Therefore, CFD is used as an analysis tool to rapidly generate sufficient data for the testing and generating of the correlation rather than doing an experiment and gives more detailed results than experiments.

2 Theoretical Approach

2.1 Basic Principle of HEAT SINK

The frequently used fluid medium by heat sink is air, whereas it also can be water or refrigerants and oil in the case of heat exchangers. According to Fourier's law heat conduction, in the presence of temperature gradient, Heat will transfer from a higher temperature to a low-temperature region in the body. This transfer of heat can be achieved in three different ways, which are convection, radiation and conduction.

The heat transfer rate by conduction, q_k, is proportional to the product of the cross-sectional area through which heat is transferred and the temperature gradient.

$$q_k = -KA.\frac{dT}{dx}$$

In the above Fig. 1, the air flows through the duct where the base of the heat sink is higher in temperature than that of air. Using the conservation of energy for steady-state condition and applying the Newton's Law of cooling for the temperature nodes, Fig. 1 gives the following equations,

$$\dot{Q} = \dot{m}C_P\left(T_{air,out} - T_{air,in}\right)$$

$$\dot{Q} = \frac{T_{hs} - T_{air,av}}{R_{hs}}$$

Where, $T_{air,av} = \frac{T_{air,out} - T_{air,in}}{2}$

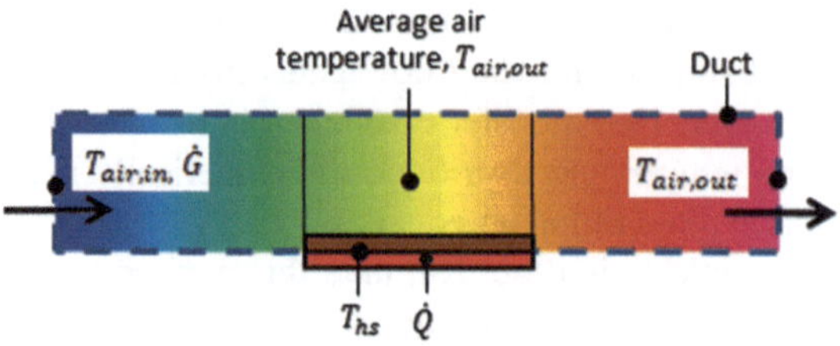

Fig. 1. Heat sink in a duct

The above equations illustrate that the average air temperature increases as the airflow through the heat sink decreases due to which the base temperature of heat sink as well as thermal resistance of the heat sink also increases.

2.2 The Pipeline of the Process

Design of heat sink and its optimization is a very complicated and tedious procedure. In this research, we will analyze the thermal performance of an electronic component with ANSYS ICEPAK. First of all, the enclosure's thermal analysis will be done to see the heat flow inside it and a point cloud will be adjusted in the heat flow direction. These point cloud will be extracted to external software (GOM) for the optimization process. Gaussian best-fit method and Chebyshev best fit method will be used to create the heat sink boundary. The optimized planes will further be taken to 3D software for further design. At last, heat sink with different shapes and materials will be analyzed to get the best heat sink. The detailed algorithm is described in the section below Fig. 2.

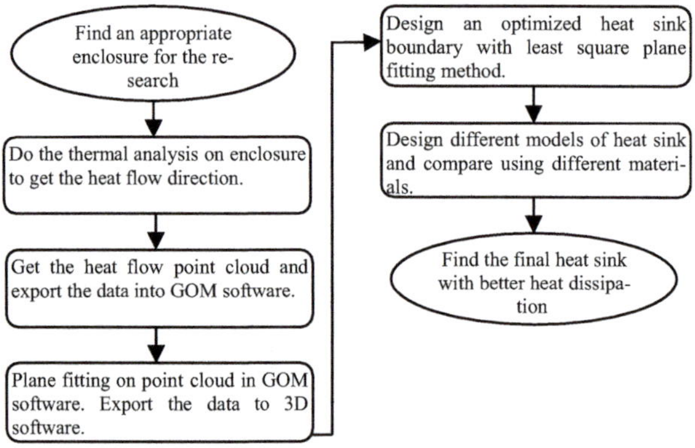

Fig. 2. The flowchart of the optimization process

2.3 Boundary Plane Fitting Method

This paper uses Gaussian best-fit method and Chebyshev best fit method for the optimization of heat sink boundary planes.

Gaussian best-fit method is one of the most popular methods to get the best-fit plane. At first, for fitting any plane we need to find out an average point on the plane and the normal of the plane. Here, we mainly adopted least square plane using Eigen-value method which meets the condition [7],

$$a^2 + b^2 + c^2 = 1$$

Under this above equation condition, according to the plane equation,

$$ax + by + cz = d$$

Above equation gives the plane parameters where a, b, c are the unit normal vectors of plane and d is the distance from the plane to the origin of coordinates. Therefore, these 4 parameters a,b,c,d can determine the planar characteristics. Now, use the above equation for measurement of a plane with n number of points.

After we get a points set $P_i\{i = [1, 2, 3. \ldots n]\}$ with n number $\left(\{(x_i, y_i, z_i)\}_{i=1}^{n_i}\right)$ of the plane, then use the above equation and calculate the distance from each point to the plane:

$$d_i = |ax_i + by_i + cz_i - d|$$

If we want to get the best fitting result, we have to fit the equation:

$$\sum_i d_i^2 = \sum_i (ax_i + by_i + cz_i - d)^2 \rightarrow \min$$

Using the Lagrange multiplier method, finding out the Extreme value of the below function:

$$f = \sum_i d_i^2 - \lambda\left(a^2 + b^2 + c^2 - 1\right)$$

Using the above equation, seek for the partial derivative of a, b, c, d

$$\begin{bmatrix} \sum_i \Delta x_i \Delta x_i & \sum_i \Delta x_i \Delta y_i & \sum_i \Delta x_i \Delta z_i \\ \sum_i \Delta y_i \Delta x_i & \sum_i \Delta y_i \Delta y_i & \sum_i \Delta y_i \Delta z_i \\ \sum_i \Delta z_i \Delta x_i & \sum_i \Delta z_i \Delta y_i & \sum_i \Delta z_i \Delta z_i \end{bmatrix} \begin{bmatrix} a \\ b \\ c \end{bmatrix} = \lambda \begin{bmatrix} a \\ b \\ c \end{bmatrix}$$

where,

$$\Delta x_i = x_i - \frac{\sum_i x_i}{n}, \Delta y_i = y_i - \frac{\sum_i y_i}{n}, \Delta z_i = z_i - \frac{\sum_i z_i}{n}$$

From the above, we can get the co-ordinates for the center point,

$$x_0 = \frac{\sum_i x_i}{n}, y_0 = \frac{\sum_i y_i}{n}, z_0 = \frac{\sum_i z_i}{n}$$

Now we can get,

$$\mathrm{d} = \mathrm{ax_0} + \mathrm{by_0} + \mathrm{cz_0}$$

This above solution provides the least square plane $\mathrm{d} = \mathrm{ax_0} + \mathrm{by_0} + \mathrm{cz_0}$. After getting the equation of plane, we need to determine the boundary for plane that contains all the point cloud data. Here, we calculated the boundary in two directions. As we already know the center point $(\mathrm{x_0}, \mathrm{y_0}, \mathrm{z_0})$ and the unit normal vector $(\mathrm{a}, \mathrm{b}, \mathrm{c})$ of the plane, we can use it to get the boundary. Let's assume the plane's unit normal vector is in z-axis direction or the plane is in XOY plane. Now project all the points on the least square plane and calculate the distance from the center point to the projected points. The orthogonal distance from the least-squares plane to the given sample point can be taken from,

$$D_i = \frac{d_i - (ax_i + by_i + cz_i)}{\sqrt{(1 + a^2 + b^2 + c^2)}}$$

Then we can get maximum value and minimum value in OX and OY direction as Max(x), Max(y), Min(x) and Min(y) which works as the boundary distance. The maximum value is the boundary distance in positive direction whereas; Minimum value is the boundary distance in the negative direction from the center point.

As for Chebyshev best fit method, Chebyshev polynomials are defined over $[-1, 1]$, A Chebyshev polynomial of order n can be defined by the following closed-form:

$$\mathrm{T_n(x)} = \cos\left(\mathrm{n}\cos^{-1}\mathrm{x}\right)$$

However, the definition is not friendly to compute. There is a recursive formulation as well, which enables a simple calculation of higher-order Chebyshev polynomials [8]:

$$\mathrm{T_{n+1}(x)} = 2\mathrm{xT_n(x)} - \mathrm{T_{n-1}(x)} \mathrm{forn} > 1$$

3 Test Setup and Procedure

In this section, the first part gives the detail on ICEPAK and the parameters for analysis whereas the second part says about the design and optimization of the new heat sink.

3.1 Enclosure Module

In this paper, we use the existing electronic device for the experiment. The housing is 110 mm × 100 mm × 40 mm in dimensions. The material used for the outside wall of enclosure is ABS + PC. This enclosure has opening only at the bottom and the top part.

a. b.

Fig. 3. Sample enclosure for the analysis

Above Fig. 3 shows two PCBs inside of enclosure which contains many other smaller electronic components on it. This device uses natural airflow as a cooling method. The main component that produces heat inside the device is NETX52 (Fig. 3b). The power given to NETX52 is 1.5 W.

3.2 CFD Simulation Approach

For the optimized design of heat sink, we first have to import the 3D design of this enclosure into the ANSYS ICEPAK or design enclosure geometry where the assembly is made like actual product. Necessary electric components are contained in the enclosure. Since we are more focused on minimizing the maximum heat around the NETX52 component, we only kept some parts that are near to the NETX52 and deleted other parts that are not much affected by heat. We also considered the height of the semiconductors so that they won't come in contact with optimized heat during the installation.

For the enclosure walls, ABS + PC material is chosen. The specific heat capacity of ABS + PC material is 1900 J/Kg.K with thermal conductivity of 0.26 W/m.K. As for the Thermal Specification of the walls, External Conditions is kept as Temperature in ICEPAK. The next part is the meshing of the geometry. For the meshing type, Mesher-HD is selected where the Mesh parameter is kept as Normal. The unit for the mesh is chosen as mm. Mesh assemblies are drawn separately.

In the "General Setup" tab, "Radiation" is kept On with "Ray tracing radiation" model. Radiation heat transfer becomes significant at high temperatures and is typically more important for natural convection problems as compared to forced convection problems in electronics cooling applications. In this part, first, we do thermal analysis on the enclosure to see the heat flow direction. In order to get the proper heat flow direction, Zero Equation of Turbulent Flow regime is kept.

The Number of iterations in Basic settings is kept 20 whereas, convergence criteria is kept default values.

Below Fig. 4a. shows the Heat Flow with Zero Equation Turbulent Flow Regime which is further imported into ANSYS CFD-Post Result to get the heat flow data. Streamline is drawn to the wireframe to see the heat flow direction. Here, the ambient temperature is kept 26 °C and Volume and Isosurface are selected to see the solid model of heat flow where the temperature is more than 41 °C. Also Fig. 4c. and 4d. show the plot of volume and Isosurface on the direction of heat flow. We can see that hot air is flown towards the upper opening. It also shows the Points cloud on the volume solid and Isosurface. These points are further extracted to csv file which is used for the optimization to make a suitable heat sink.

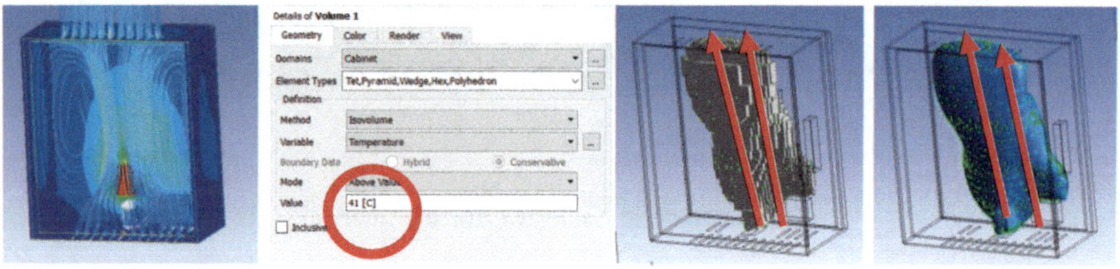

a. Heat Flow with Zero Equation Turbulent Flow Regime

b. Volume Edit Box

c. Volume

d. Isosurface on heat flow direction

Fig. 4. Hot airflow

The points that are extracted to csv files are then imported to external software. GOM INSPECT software is used to open the csv file. This software further analyzes the point cloud for the optimization process.

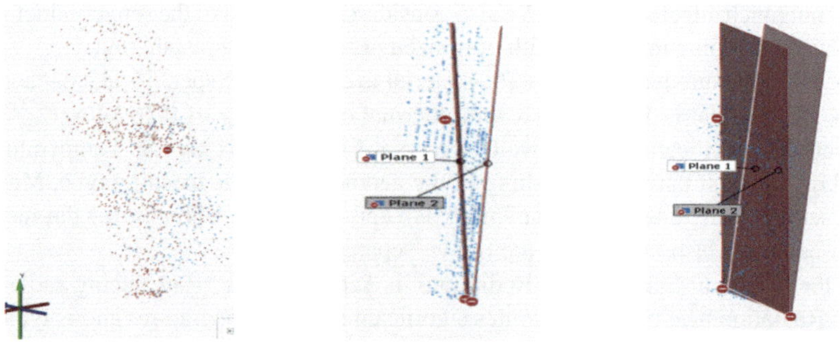

a. Point Cloud in GOM Inspect Software

b. Plane 1 - Gaussian best-fit method Plane
Plane 2 - Chebyshev best fit method Plane

Fig. 5. Simulation in GOM inspect

First the points on point cloud shown in Fig. 5a. have to be selected for plane fitting. Then, Gaussian best-fit method and Chebyshev best fit method is used for the plane fitting which is shown in Fig. 5b.

Now, these two boundary planes are imported into 3D software for the further designing process. We use PTC CREO for 3D design and modelling. The fitting planes are exported using.iges file from GOM Inspect software. The below Fig. 6a. shows the fitting planes in PTC CREO and Fig. 6b. shows the heat sink inside the boundary of 2 planes.

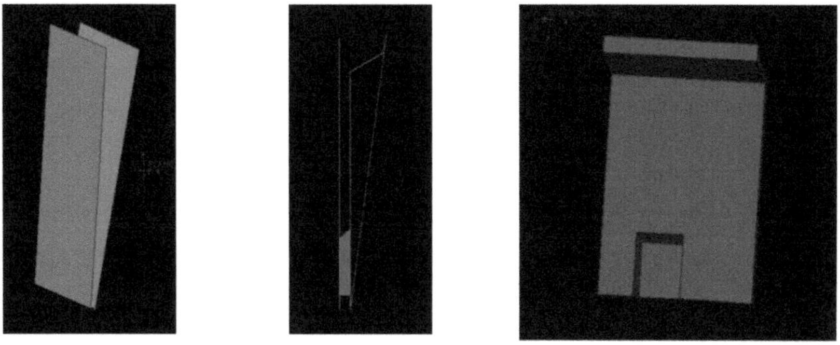

a. fitting planes in PTC CREO b. the heat sink boundary c. New optimized heat sink

Fig. 6. Fitting plane

The Above Fig. 6c. shows the optimized heat sink which can be exported to ANSYS ICEPAK for further comparison using different shapes and materials.

4 Heat Sink Comparison

This section contains a series of methods to evaluate the heat sink. For this comparison, we take aluminum as materials for testing. Even though copper has excellent heat sink properties for thermal conductivity, corrosion resistance, biofouling resistance, and antimicrobial resistance but copper is also three times as dense [6] and more expensive compared to aluminum. Also, aluminum heat sinks can be extruded but not possible for copper because aluminum is more ductile compared to copper.

A heat sink design must fulfil both its thermal as well as its mechanical requirements. Concerning the latter, the component must remain in thermal contact with its heat sink with reasonableshock and vibration. We compared thermally conductive tape and silicone adhesive (Dow Corning SE 9184 White RTV) as heat sink attachment methods. The thermally conductive tape is one of the most cost-effective heat sink attachment materials. It is suitable for low-mass heat sinks and for components with low power dissipation. Whereas, epoxy is more expensive than tape, but provides a greater mechanical bond between the heat sink and component, as well as improved thermal conductivity, due to which, we opted to use Epoxy for our further analysis.

In this paper, two types of the heat sink are compared which are:

1. Optimized Heat Sink
2. Pin Fin Heat Sink.

4.1 Optimized Heat Sink

In this part, we show the thermal analysis of optimized heat sink with a further modified design. Three different shapes are made: optimized, optimized with holes and plane with fins and holes. Figure 7a. is the normal optimized heat sink and Fig. 7b. is modified heat sink with opening holes at the upper side of the heat sink. Whereas Fig. 7c. is also modified heat sink but with fins on the surface and holes at the upper side of the heat sink. Further, we do the thermal analysis using these heat sinks in order to see the final result.

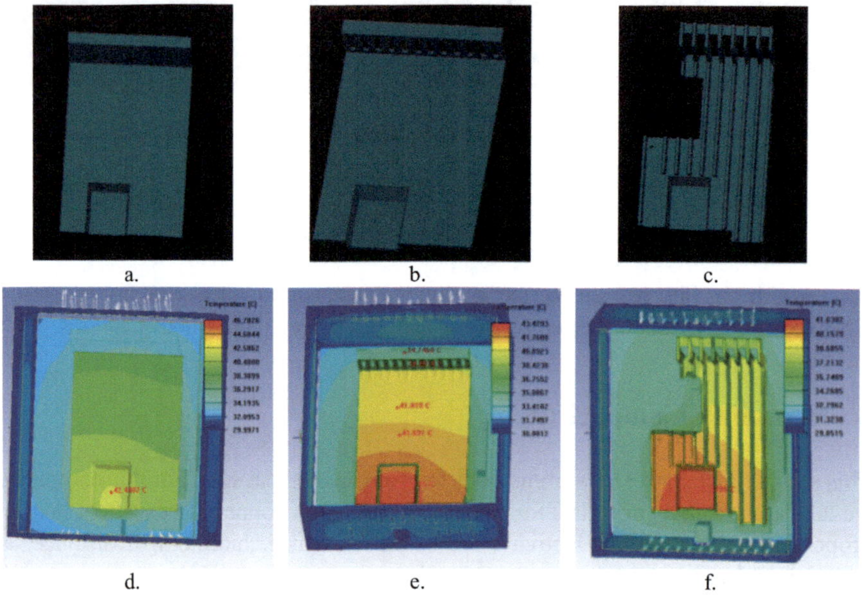

Fig. 7. Thermal analysis on optimized heat sinks

From the below Table 1, we can see that Optimized heat sink with Fins and Holes dissipate more heat from the electric semiconductor in compared to other optimized heat sinks. Due to which, this new optimized heat sink design is used for further analysis.

Table 1. Temperature on optimized heat sinks

No.	Shape	Overall temp.	Heat sink temp.
1.	Normal optimized	46.7826 °C	42.4402 °C
2.	Optimized with holes	43.4293 °C	43.2725 °C
3.	Optimized with fins and holes	41.6302 °C	41.5286 °C

4.2 Pin Fin Heat Sink

Pin Fin Heat Sink, which is often used in experimental research, is chosen [9]. Pin fin heat sinks for surface mount devices are available in a variety of configurations, sizes and materials.

A fin of a heat sink may be considered to be a flat plate with heat flowing in one end and being dissipated into the surrounding fluid as it travels to the other [6]. As heat flows through the fin, the combination of the thermal resistance of the heat sink impeding the flow and the heat lost due to convection, the temperature of the fin and, therefore, the heat transfer to the fluid, will decrease from the base to the end of the fin.

The Pin Fin heat sink size we use here is 14 × 14 × 10 mm. The unique pin fin design generates significant cooling power and is highly suitable for "hot" devices and applications that have limited space for cooling. For the testing process, we take four kinds of fin heat sink, In-line Pin Fin, Staggered Pin Fin, Extruded Fin and Cross-cut Extrusion. Below Fig. 8 shows a different kind of chosen heat sink. The first two heat sinks have 4 × 4 Fins with round and cross-cut shapes. The first In-line Pin Fin heat sink has equaled the number of gaps between pins and is in line to each other whereas staggered pin fin has fins arranged in Zig-Zac order. Extruded Fin heat sink has four extruded fins which help airflow in one particular direction. The last cross-cut extrusion's fins are square in shape and are arranged in 4 × 5 orders, with more fins in the airflow direction.

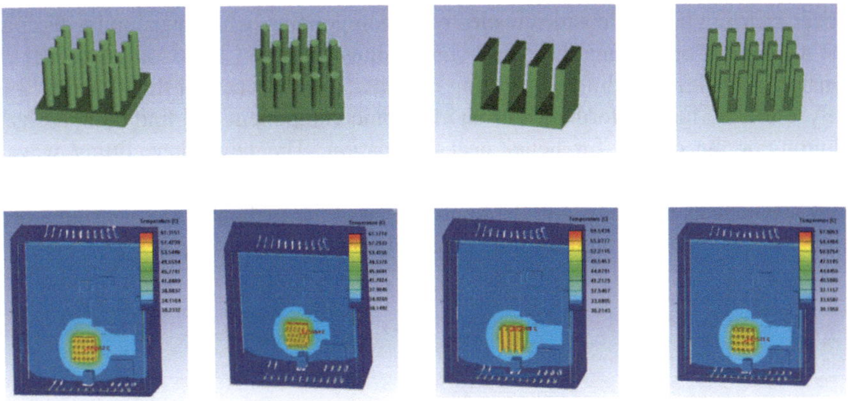

a) In-line Pin Fin 57.6957 ºC

b) Staggered Pin Fin57.5654 ºC

c) Extruded Fin55.7708 ºC

d) Cross-cut Extrusion54.1571 ºC

Fig. 8. Heat dissipation in pin fin heat sinks

From the above Fig. 8, we can see that Cross-cut Extrusion arrangement style dissipate more temperature from the electronic component. Although all the results are somewhat similar with a temperature difference of only about 2–3 °C, cross-cut extrusion heat sink dissipates more heat.

4.3 Comparison Between Cross-Cut Extrusion and Optimized Heat Sink

The below Table 2 shows the best heat sink of its kind. The better Pin Fin Heat Sink (Cross-cut Extrusion Heat sink) is compared with the Optimized Heat Sink to show the heat dissipation result. Ambient temperatures for both of these heat sinks are kept 26 ° C. After ICEPAK simulation in both heat sinks, we see that overall temperature for Cross-Cut Extrusion Heat Sink is 57.90 °C whereas for optimized heat sink, it is 41.63 °C. This shows that an increase in temperature in the first heat sink is 16.27 °C more than in second heat sink. Also, the temperature in the first heat sink is 54.15 °C with later giving 41.52 °C. These results show that Optimized heat sink dissipates more heat than the normal Pin Fin heat sink used.

Table 2. Heat dissipation on heat sinks

No.	Heat sink	Ambient temp.	Overall temp.	Heat sink temp	Increase in temp
1.	Cross-cut extrusion heat sink	26 °C	57.90 °C	54.15 °C	31.9 °C
2.	Optimized heat sink	26 °C	41.63 °C	41.52 °C	15.63 °C
Temperature difference					16.27 °C

5 Conclusion

The packaging industry is continually shrinking electronic package sizes. This results in ever-growing power densities in electronic components, which need to be dissipated. One method for optimizing these heat sink designs is to use CFD (Computational Fluid Dynamics) to simulate different cooling designs. The objective of this research was to analyze the thermal performance of an electronic component and find an optimization solution for the design for better heat dissipation. The new plane fitting ways for optimization of heat sink give boundary for designing of a better heat sink. This paper helps mechanical engineers with less knowledge of thermal analysis to design a proper heat sink. This method reduces time consumption while selecting a heat sink.

References

1. Khan, W., Culham, J., Yovanovich, M.: Performance of shrouded pin-fin heat sinks for electronic cooling. J. Thermophys. Heat Transfer-J. Thermophys. Heat Transfer **20**, 408–414 (2006). https://doi.org/10.2514/1.17713
2. Brucker, K.A., Majdalani, J.: Equivalent thermal conductivity for compact heat sink models based on the churchill and chu correlation. IEEE Trans. Compon. Packag. Technol. **26**(1), 158–164 (2003). https://doi.org/10.1109/TCAPT.2002.806173
3. Shi, Z.F., Lu, A.C.W., Tan, Y.M., K.H, Tan, R., Tan, E.: Heat sink design optimization for optical transponders. In: International Symposium on Microelectronics, vol. 4931, pp. 330–335 (2002)
4. Lee, S.: How to Select a Heat Sink, Advanced Thermal Engineering Aavid Thermal Technologies Inc
5. Radmehr, A., Kellar, K.M., Kelly, P., Patankar, S.V., Kang, S.S.: Analysis of the effect of bypass on the performance of heat sinks using flow network modeling (FNM). In: Fifteenth Annual IEEE Semiconductor Thermal Measurement and Management Symposium, pp. 42–47 (1999)
6. Sergent, J., Krum, A.: Thermal Management Handbook for Electronic Assemblies, 1st edn. McGraw-Hill, New York (1998)
7. Regmi, B., Liu, S., Wen, Z.: An actual mating envelope plane searching algorithm for minimum zone flatness evaluation. In: 2014 International Academic Conference of Postgraduates, NUAA, vol. 2 (2014)
8. Sajikumar, S., Anilkumar, A.K.: Image compression using Chebyshev polynomial surface fit. Int. J. Pure Appl Math. Sci. **10**(0972–9828), 15–27 (2017)
9. Anon, Unknown, Heat sink selection, Mechanical engineering department, San Jose State University, 27 January 2010

Transformer Fault Diagnosis Based on Stacked Contractive Auto-Encoder Net

Yang Zhong(✉), Chuye Hu, Yiqi Lu, and Shaorong Wang

State Key Laboratory of Advanced Electromagnetic Engineering and Technology, Huazhong University of Science and Technology, Wuhan, Hubei Province 430074, China
455541748@qq.com

Abstract. Dissolved gas analysis (DGA) is an effective method for oil-immersed transformer fault diagnosis. This paper proposes a transformer fault diagnosis method based on Stacked Contractive Auto-Encoder Network (SCAEN), which can detect the transformer's internal fault by using DGA data, including H2, CH4, C2H2, C2H4, C2H6. The network consists of a three-layer stacked contractive auto-encoder (SCAE) and a backpropagation neural network (BPNN) with three hidden layers. A large amount of unlabeled data is used to train to obtain initialization parameters, and then a limited labeled dataset is used to fine-tune and classify the faults of trans-formers. The proposed method is suitable for transformer fault diagnosis scenarios, which contains very limited labeled data. when tested on real DGA dataset, the fault diagnosis accuracy is up to 95.31% by SCAEN, which performs better than other commonly used models such as support vector machine (SVM), BPNN, auto-encoder (AE), contractive auto-encoder (CAE) and SCAE.

Keywords: Oil-immersed transformer · Transformer fault diagnosis · Auto-encoder · Dissolved gas analysis · Regularization

1 Introduction

The power transformer is the leading equipment of the power grid, which plays the role of voltage transformation and power transmission. During long-term operation, under the influence of overheat and discharge, the transformer oil and organic insulating materials will gradually age and decompose, producing a small amount of low molecular hydrocarbons and other gases [1, 2].

DGA is an effective method for transformer fault diagnosis. It mainly detects the dissolved gas composition in transformer oil, including H2, CH4, C2H2, C2H4, C2H6. Since the composition and content of dissolved gas in oil can reflect the transformer's internal overheat and discharge faults, many DGA data-driven fault diagnosis methods such as SVM and BPNN have been widely used [3–5]. However, these methods have the problems that standard features are challenging to transfer, and the accuracy is not high enough.

This paper proposes a transformer fault diagnosis method based on the Stacked Contractive Auto-Encoder Network (SCAEN), which can analyze DGA data to detect

Q. Liu et al. (Eds.): CENet 2020, AISC 1274, pp. 514–522, 2021.
https://doi.org/10.1007/978-981-15-8462-6_57

internal faults in the transformer. The model consists of a three-layers SCAE and a BPNN with three hidden layers, and the number of neurons in each BPNN hidden layer is the same as SCAE. The SCAE uses a large amount of unlabeled data in unsupervised training to obtain the optimized parameters, which as the initial parameters of BPNN. Then the BPNN uses a limited labeled dataset to fine-tune and classify the faults of transformers.

The rest of the paper is organized as follows: The AE and the corresponding variants is described in Sect. 2. The experimental setups and comparison strategy are described in Sect. 3. The experimental results and discussion are presented in Sect. 4. The conclusion is presented in Sect. 5.

2 Auto-Encoders Variants

2.1 Auto-Encoder (AE)

AE [6] is a neural network including input layer, hidden layer and output layer, as shown in Fig. 1.

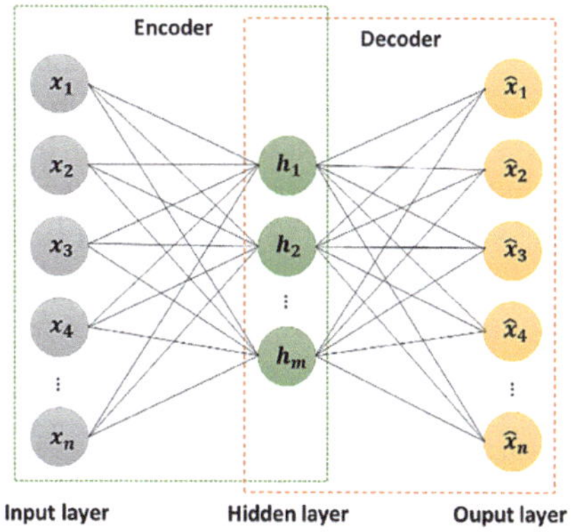

Fig. 1. Auto-encoder structure.

AE includes two parts: encoder and decoder. For most AEs, the input signals are encoded into multiple feature representations by the encoder, and then these feature representations are restored to the original form of the signals by the decoder. The only difference between AE and the forward propagation neural network is that the number of neurons in the output layer of AE is the same as the number of neurons in the input layer. In other words, the learning process of AE is reconstructing the input signal. More specifically, the role of the encoder is to map the original space of the input signals to the feature space. The behavior can be characterized by the following function:

$$h = f(\mathrm{x}) = s_f\left(W^1 x + b_h\right) \tag{1}$$

The role of the decoder is to recover the original signals from the feature space. The behavior can be characterized by the following function:

$$y = g(\mathrm{h}) = s_g\left(W^2 h + b_y\right) \tag{2}$$

Where s_f and s_g represent activation functions, W^1 and W^2 represent weight matrices, and b_h and b_y represent biases. the parameter $\theta = \left\{W^1, W^2, b_h, b_y\right\}$ to minimize the reconstruction error between the output layer and input layer signals. The reconstruction error is characterized by the following equation:

$$J_{AE}(\theta) = \sum\nolimits_{x \in \chi n} L(x, g(f(\mathrm{X}))) \tag{3}$$

Where L represents the reconstruction error. The squared error $L(x, y) = \|x - y\|^2$ and the cross-entropy loss $L(x, y) = \sum_{i=1}^{n} x_i \log(y_i) + (1 - x_i)\log(1 - y_i)$ are two typical reconstruction error functions.

2.2 Stacked Auto-Encoder (SAE)

SAE [7] is a deep auto-encoder formed by stacking multiple AEs. SAE optimized neural network models by mining deeper features in input signals.

SAE adopts a layer-by-layer greedy training method from bottom to top to train each layer of AE to obtain optimized parameters, which as the initial parameters of BPNN. Specifically, the input signals are sent to the first layer of AE, and obtains the first-order feature representation $x_1(i)$; then the features of the first layer are sent to the second layer, and obtains the second-order feature representation $x_2(i)$; the n-th-order features of the original signals output by the last layer of AE and represent by $x_n(i)$. When the pre-training completed, supervised fine-tuning begins. The fine-tuning process is to derive each parameter of the overall loss function, and then uses the gradient descent method to obtain the optimized parameters.

2.3 Contractive Auto-Encoder (CAE)

AE can automatically learn the feature representation from input signals by unsupervised learning. But experiments show that when the number of neurons in the hidden layer is more than the dimension of the input signals, AE will tend to learn an identity function. In other words, the input signals are directly mapped to the output signals without any conversion. In this case, we cannot get any useful information from AE. To solve this problem, we can limit the number of neurons in the hidden layer so that they are smaller than the dimensions of the input signals. In addition, there is another effective method that limiting the loss function of the AE. Experiments show that by this method, even if the number of neurons in the hidden layer is greater than the dimension of the input signals, AE can still learn useful features.

CAE was proposed by Rifai [8] et al. This auto-encoder variant obtains more features by adding regularization term to the loss function. Rifai et al. Proved that by adding regularization term in the form of a Jacobian matrix, the feature mapping is more strongly contracting at the training signals. Specifically, the loss function of CAE is as follows:

$$J_{\mathrm{CAE}}(\theta) = \sum\nolimits_{x \in \chi n} L\left(x, g(f(x)) + \lambda \|J_f(x)\|_F^2\right) \tag{4}$$

$$\lambda \|J_f(x)\|_F^2 = \sum_{ij} \left(\frac{\partial y_i(x)}{\partial x_i}\right) \tag{5}$$

Where λ represents the contractive rate, and the regularization term represents the sum of squares of the partial differential of the hidden layer to the input signals, that is, the Jacobian matrix of the hidden layer to the input signals. By adding the regularization term to the loss function, the disturbance of the input in all directions can be reduced to achieve the purpose of anti-noise, thereby improving the robustness of the extracted features.

The SCAEN proposed in this paper consists of a three-layers SCAE and a BPNN with three hidden layers. The SCAE is obtained by stacking three layers of CAE, and the number of neurons in each BPNN hidden layer is the same as SCAE.

3 Experiments Setting

3.1 Input Layer

Setting the main components of dissolved gas in the oil as the input of the network model, including H_2, CH_4, C_2H_2, C_2H_4, C_2H_6. Therefore, the dimension of the input vector is 5. Before experiments, Z-score standardization [9] is used for data normalization, and the conversion formula is:

$$x^* = \frac{x - \bar{x}}{\sigma} \tag{6}$$

Where $\bar{x}$ is the average value of the input data and σ is the standard deviation of the input data. After normalization, the average value of the input data is 0, and the standard deviation is 1.

3.2 Output Layer

The six states of transformer [10]: Partial discharges (PD), discharges of low energy (D1), discharges of high energy (D2), thermal faults <700 °C (T1 & T2), thermal faults >700 °C (T3) and normal state (Normal). Therefore, the dimension of the output vector is 6.

3.3 Data Preprocessing

This paper collects 500 sets of labeled DGA data and 3000 sets of unlabeled DGA data from IEC 60599, ICT TC 10 database [10] and IEEE DataPort [11]. 3000 sets of unlabeled data are used for unsupervised training and 500 sets of labeled data is divided into a training set and a test set in a 4:1 ratio for supervised fine-tuning. Some training data and test data are shown in Table 1.

Table 1. DGA data examples.

H_2	CH_4	C_2H_2	C_2H_4	C_2H_6	Types
36036	4704	10	5	554	PD
1230	163	692	233	27	D1
7150	1440	1760	1210	97	D2
3420	7870	33	6990	1500	T1&T2
6709	10500	750	17700	1400	T3
105	125	10	166	71	Normal

3.4 Model Structure

The SCAEN consists of a three-layers SCAE and a BPNN with three hidden layers. The number of neurons in the input layer is 5. The middle three layers are hidden layers, the number of neurons is 12, 10, and 8. The number of neurons in the final output layer is 6, and the extracted features are classified by the softmax classifier. The SCAE structure is shown in Fig. 2. The BPNN structure is shown in Fig. 3.

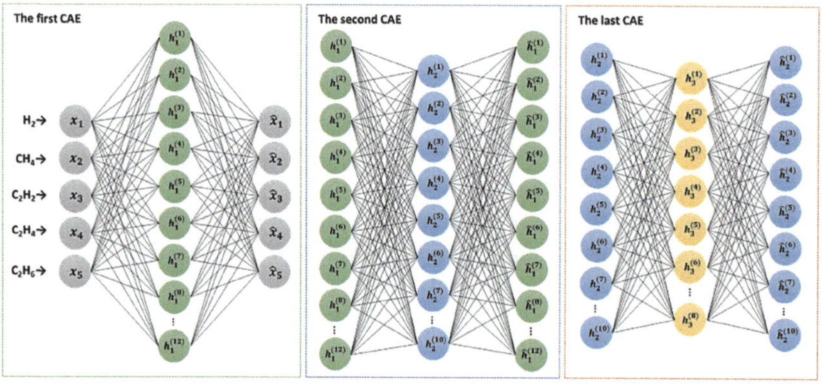

Fig. 2. SCAE structure.

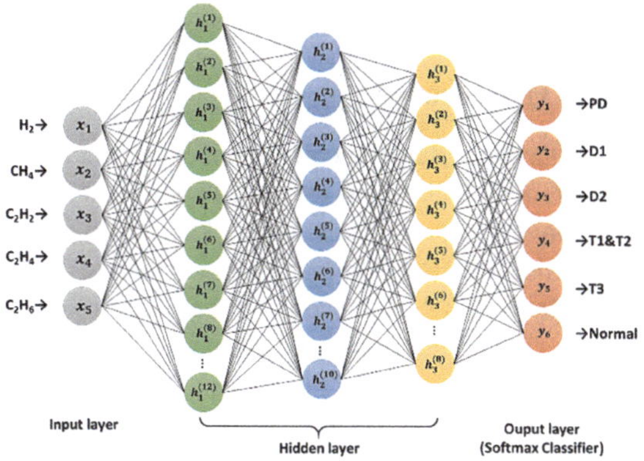

Fig. 3. BPNN structure.

3.5 Experiment Steps

Step 1: Normalizing the DGA data to obtain the input of the model;
Step 2: Using a large amount of unlabeled training data and adopting the greedy training method from bottom to top to complete the pre-training (as showing in Fig. 4). Then obtaining the initialization parameters of BPNN;
Step 3: Using the back-propagation algorithm and gradient descent algorithm for BPNN with the labeled training dataset, and fine-tuning from top to bottom to obtain the optimal network parameters, and output the classification type of the transformer by the softmax classifier;
Step 4: Input the labeled test dataset to the trained BPNN to get the accuracy of the model.

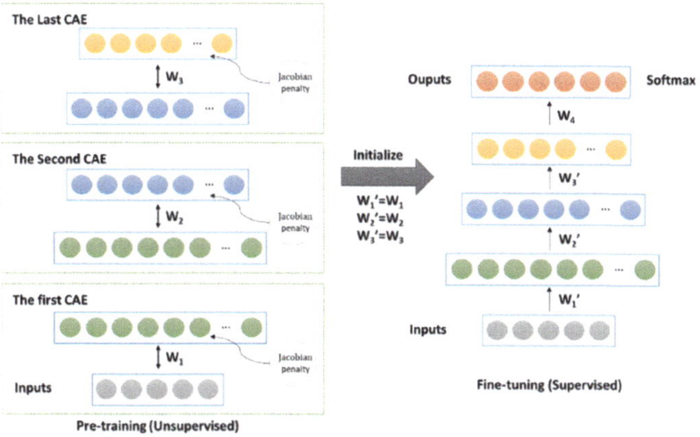

Fig. 4. Schematic diagram of the training process.

4 Experiments Result

4.1 Hyperparameter Optimization

When setting different learning rates, the convergence speed and convergence value of the training loss are different. It is found through experiments that when LR = 0.005 training loss quickly converges to a lower value. Therefore, we set the model LR = 0.005. Figure 5 shows the effect of different contractive rates λ on the classification accuracy on the testing dataset. It can be seen that the accuracy is the highest when the $\lambda = 0.01$. Therefore, we choose the optimized contractive rate $\lambda = 0.01$.

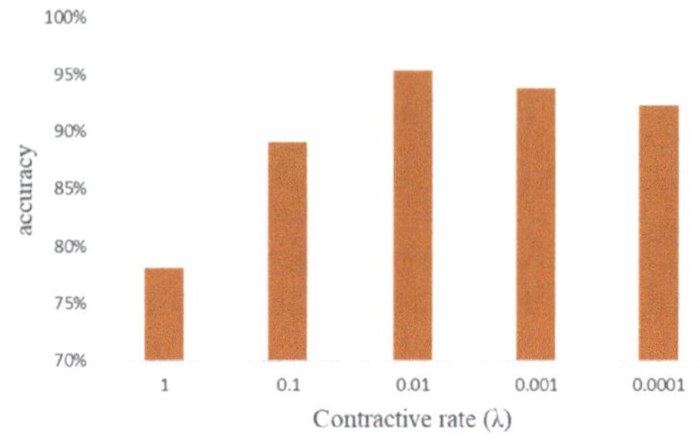

Fig. 5. The classification accuracy with different contractive rates λ.

4.2 Models Performance Analysis

In order to verify the effectiveness of this method, this paper also conducted simulation tests on several other commonly used fault diagnosis methods. The methods selected for comparison with the model in this paper are:

SVM: Support vector machine with radial basis function (RBF) core;

BPNN: Back propagation neural network with three hidden layers, and the number of neurons in each layer is 5, 12, 10, 8, 6, which is the same as the BPNN model in this paper;

AE: There is only one hidden layer of auto-encoder, and the number of hidden neurons is 12;

CAE: There is only one hidden layer of contractive auto-encoder, and the number of neurons in the hidden layer is 12;

SAE: A three-layers stacked auto-encoder, the number of neurons in each hidden layer is 12, 10, 8, which is the same as the SCAE model in this paper.

All experimental results are listed in Table 2. It can be seen that the SCAEN proposed in this paper has the highest test accuracy rate, reaching 95.31%. In addition, with the deepening of the network structure, the performance of the model can generally increase. The specific performance is that the test accuracy of deep networks such as DCAE, SAE and BPNN is higher than that of shallow networks CAE, AE and

SVM. Besides, under the same network structure, the accuracy of the pre-trained SAE is 92.19%, which is better than the BPNN's 87.50%, which illustrates the pre-training method can help deep learning to get better performance. Finally, the method in this paper is superior to the performance of SAE, and the method of CAE is superior to the performance of AE, indicating that the limitation of the contraction term can help the auto-encoder to mine more representative and distinguishing features, thereby improving the recognition accuracy of the model. In general, the proposed method SCAEN has the best performance.

Table 2. Test results of various models.

Model	Classification accuracy
SVM (with RBF kernel)	78.13%
BPNN	87.50%
AE	81.25%
CAE	84.38%
SAE	92.19%
SCAEN	95.31%

5 Conclusions

This paper proposes a transformer internal fault diagnosis method based on SCAEN, this method uses SCAE for unsupervised pre-training, which is very suitable for the scenario where transformer equipment has a lot of unlabeled data, and the use of contraction penalty terms can make feature learning ability stronger and more robust. Supervised fine-tuning on BPNN can train the model to classify the states of transformer. Through the experiments we can see the performance superiority of SCAEN.

This paper can achieve automatic feature extraction, but the network parameters such as the number of layers of the network, the number of neurons in each layer, and the learning rate still need to be manually adjusted. Besides, except the DGA data, the operating parameters of the transformer also have some indicators such as furfural content in oil and winding insulation resistance. In future work, we will continue to study how to use multi-dimensional information fusion in SCAEN, to achieve a higher recognition accuracy and more robust fault diagnosis ability.

References

1. Mirowski, P., LeCun, Y.: Statistical machine learning and dissolved gas analysis: a review. IEEE Trans. Power Deliv. **27**, 1791–1799 (2012)
2. Zahriah, B.S., Rubiyah, B.Y.: Support vector machine-based fault diagnosis of power transformer using k nearest-neighbor imputed DGA dataset. J. Comput. Commun. **02**, 22–31 (2018)
3. Wang, Z.Y., Liu, Y.L., Griffin, P.J.: A combined ANN and expert system tool for transformer fault diagnosis. IEEE Trans. Power Deliv. **13**, 1224–1229 (1998)

4. Yang, H.T., Liao, C.C., Chou, J.H.: Fuzzy learning vector quantization networks for power transformer condition assessment. Dielectr. Electr. Insul. IEEE Trans. **8**, 143–149 (2001)
5. Seifeddine, S., Khmais, B., Abdelkader, C.: SVM-based decision for power transformers fault diagnosis using Rogers and Doernenburg ratios DGA. In: International Multi-conference on Systems, pp. 1–6 (2013)
6. Hinton, G.E.: Reducing the dimensionality of data with neural networks. Science **313**, 504–507 (2006)
7. Suck, B.R., Sung, K.O.: Design of fuzzy k-nearest neighbors classifiers based on feature extraction by using stacked autoencoder. Trans. Korean Inst. Electr. Eng. **64**, 113–120 (2015)
8. Salah, R., Pascal, V., Xavier, M. et al.: Contractive auto-encoders: explicit invariance during feature extraction. In: ICML (2011)
9. Hadley, B.C., et al.: Mapping inundation uncertainty with a standard score (Z-score) technique, in Agu Fall Meeting (2010)
10. Duval, M., Depabla, A.: Interpretation of gas-in-oil analysis using new IEC publication 60599 and IEC TC 10 databases. Electr. Insul. Mag. IEEE **17**(31), 31–41 (2001)
11. Ali, A.: Dynamic assessment and prediction of equipment status in 220 kV substations by big data and AI technology, http://dx.doi.org/10.21227/jve3-nh57. Accessed 26 May 26

Control Strategy for Vehicle-Based Hybrid Energy Storage System Based on PI Control

Chongzhuo Tan, Zhangyu Lu, Zeyu Wang, and Xizheng Zhang(✉)

Hunan Institute of Engineering, Xiangtan 411104, Hunan, China
z_x_z2000@163.com

Abstract. At present, electric vehicles have a general problem of short battery life and short travel distance. Therefore, introducing supercapacitors and DC/DC converters to form a hybrid energy storage system (HESS) with the battery to make up for the lack of pure battery energy. In this paper, a control strategy for an on-board hybrid energy storage system based on PI control is proposed. First of all, This strategy is through energy management strategy to distribute the required power, and then by changing the duty cycle of the IGBT switch in the DC-DC converter, to control the battery and supercapacitor output currents and to keep the DC bus voltage stable. The simulation results on the build simulation model, Simulink, shows that the proposed control method can accurately track the reference values of battery current and supercapacitor current and stabilize the DC bus voltage, which proves the effectiveness of the control strategy.

Keywords: Hybrid energy storage system · DC/DC converter · PI control · Simulink

1 Introduction

With the increasing problems of energy crisis and environmental pollution, electric vehicles have become the future development direction of the automotive industry. However, the short driving distance of electric vehicles is one of the important reasons why electric vehicles are so difficult to be widely popularized. Batteries have a high energy density, but their power density is low, and their cycle life is short. SC has high power density and long cycle life. Therefore, the SC and battery complement each other and hybrid with DC/DC converter to form HESS, which can effectively improve the driving mileage of electric vehicles [1].

At present, how to control the output current of the battery and supercapacitor is the key focus of the control research of HESS. Reference [2] has proposed an advanced hybrid energy storage system topology, which connects a bi-directional DC/DC converter in series with each battery and supercapacitor to the DC bus, but this structure has the disadvantage of high circuit loss. The reference [3] proposes a control strategy for hybrid energy storage systems based on sliding mode control, but its control method is too complicated. The reference [4] linearizes the demand power, the charge state of the battery and supercapacitor, to achieve the control of the hybrid energy storage system, however, HESS is a non-linear system and the control stability is not strong by using linear control strategy.

Q. Liu et al. (Eds.): CENet 2020, AISC 1274, pp. 523–529, 2021.
https://doi.org/10.1007/978-981-15-8462-6_58

Based on the above research on HESS, this paper analyzes the HESS topology and proposes a control strategy for a vehicle-mounted hybrid energy storage system based on PI control. The control strategy first uses a rule-based energy management strategy to distribute the load-demanding power, then generating the battery current and supercapacitor current reference values. Then use two PI controllers to make the battery current and supercapacitor current follow the reference value stably, while a voltage stabilization term is added to the supercapacitor reference value to keep the DC bus voltage stable. The effectiveness of the proposed control strategy was demonstrated through simulation.

2 HESS System Architecture

The structure of the hybrid energy storage system is shown in Fig. 1, where the battery is connected to the load through the boost converter to provide average power to the load, and the supercapacitor is connected to the load via a Buck-Boost converter as an auxiliary power source to provide peak power to the load. The Boost converter consists of the switching device IGBTS_1, the diode D, and the high frequency inductor L_{bat}. The Buck-Boost converter consists of the switching devices IGBTS_2 and S_3 and the high frequency inductor L_{sc}.V_{bat}, V_{sc}, and V_{dc} correspond to the voltage of the filter capacitors C_{bat}, C_{sc}, and C_{dc}. R_{bat}, R_{sc} are the internal resistance of the battery and supercapacitor, respectively. i_{bat} is the battery current, i_{sc} is the supercapacitor current and i_{load} is the current flowing through the load. The structure uses the Boost converter and the Buck-Boost converter as components to control the output power of the battery and supercapacitor and controls the power transfer between circuits by adjusting the duty ratio of the IGBTs switch. From this, the equations of state for battery current and supercapacitor current is

$$\dot{i}_{\text{bat}} = -(1 - D_1)\frac{V_{\text{dc}}}{L_{\text{bat}}} - i_{bat}\frac{R_{bat}}{L_{bat}} + \frac{V_{bat}}{L_{bat}} \tag{1}$$

$$\dot{i}_{\text{sc}} = -(1 - D_2)\frac{V_{\text{dc}}}{L_{\text{sc}}} - i_{\text{sc}}\frac{R_{\text{sc}}}{L_{\text{sc}}} + \frac{V_{\text{sc}}}{L_{\text{sc}}} \tag{2}$$

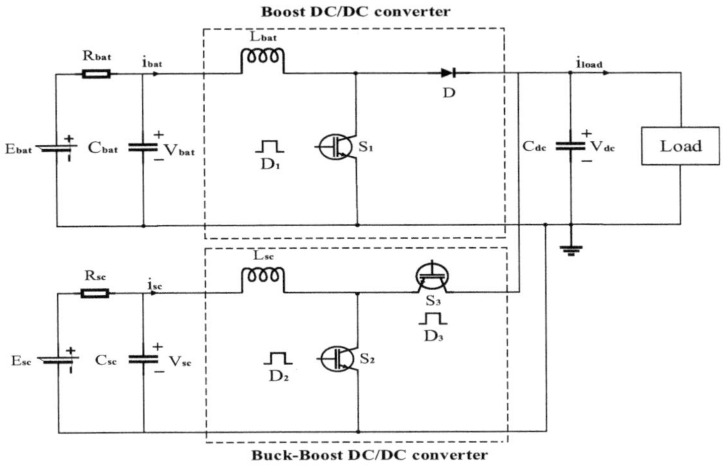

Fig. 1. HESS circuit model.

3 Control Strategy Design

As shown in Fig. 2, this article first adopts a rule-based energy management strategy to allocate the load-demanding power P_{demand} to generate the battery current and supercapacitor current reference values. Then, the battery current and supercapacitor current are stabilized by two PI controllers to follow the reference value, while a voltage stabilization term is added to the supercapacitor reference current to keep the DC bus voltage stable. The specific design is as follows.

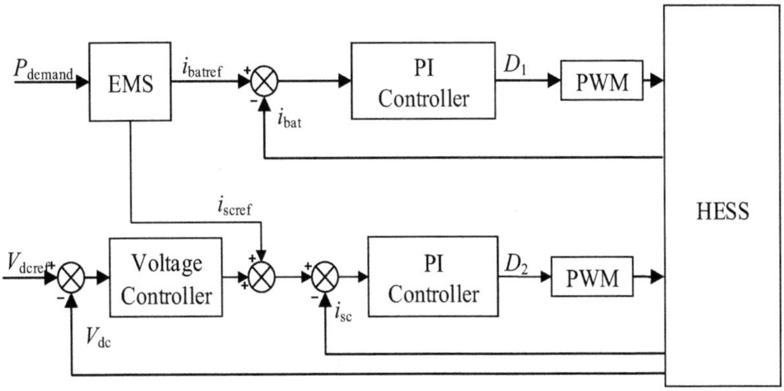

Fig. 2. Overall block diagram of the control strategy.

3.1 Energy Management Strategy Design

In this paper, the energy management of the hybrid energy storage system is selected based on the regular energy management strategy, the specific design is shown in Fig. 3, where P_{bat} is the battery power, P_{sc} is the supercapacitor power, V_{sc} is the supercapacitor voltage, $V_{sc,max}$ is the supercapacitor rated voltage, and P_{min} is the power threshold value.

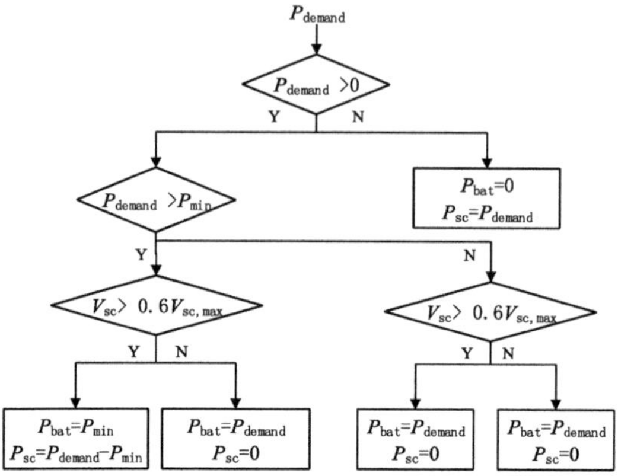

Fig. 3. Energy management strategies.

3.2 PI Controller Design

Using typical PI control methods, the current of the battery and supercapacitor can be accurately controlled to follow the reference value to ensure that the current will not fluctuate greatly, giving full play to the advantages of hybrid energy storage systems. The difference between the reference and actual values of the battery and supercapacitor as the input to the PI controller is

$$e_2 = i_{batref} - i_{bat} \tag{3}$$

$$e_3 = i_{scref} - i_{sc} \tag{4}$$

where i_{batref} is the battery current reference value and i_{scref} is the supercapacitor current reference value, Can be obtained from energy management strategy:

$$i_{batref} = \frac{P_{bat}}{V_{bat}}, \quad i_{scref} = \frac{P_{sc}}{V_{sc}} \tag{5}$$

To keep the DC bus voltage stable, the voltage stabilization term is added to the i_{scref} to introduce the error variable is

$$e_1 = V_{\mathrm{dcref}} - V_{\mathrm{dc}} \tag{6}$$

where V_{dcref} is the expected value of the DC bus voltage, and in order to stabilize the DC bus voltage at all times at the expected value, the control rules are

$$\mathrm{u}_{\mathrm{dc}} = K_{\mathrm{P,dc}} e_1 + K_{\mathrm{I,dc}} \int e_1 dt = K_{\mathrm{P,dc}}(V_{\mathrm{dcref}} - V_{\mathrm{dc}}) + K_{\mathrm{I,dc}} \int (V_{\mathrm{dcref}} - V_{\mathrm{dc}}) dt \tag{7}$$

where $K_{\mathrm{P,dc}}$ is the scale factor and $K_{\mathrm{I,dc}}$ is the integration factor. From this one can derive

$$i_{\mathrm{scref}} = \frac{P_{\mathrm{sc}}}{V_{\mathrm{sc}}} + \mathrm{u}_{\mathrm{dc}} V_{\mathrm{dc}} \tag{8}$$

Therefore, the control rules for the battery-side PI controller and the supercapacitor-side PI controller are

$$\mathrm{u}_{\mathrm{bat}} = K_{\mathrm{P,bat}} e_2 + K_{\mathrm{I,bat}} \int e_2 \mathrm{d}t = K_{\mathrm{P,bat}}(i_{\mathrm{batref}} - i_{\mathrm{bat}}) + K_{\mathrm{I,bat}} \int (i_{\mathrm{batref}} - i_{\mathrm{bat}}) dt \tag{9}$$

$$\mathrm{u}_{\mathrm{sc}} = K_{\mathrm{P,sc}} e_3 + K_{\mathrm{I,sc}} \int e_3 \mathrm{d}t = K_{\mathrm{P,sc}}(i_{\mathrm{scref}} - i_{\mathrm{sc}}) + K_{\mathrm{P,sc}} \int (i_{\mathrm{sdcref}} - i_{\mathrm{sc}}) dt \tag{10}$$

where $K_{\mathrm{P,bat}}$ is the battery side proportional factor, $K_{\mathrm{I,bat}}$ is the battery side integration factor, $K_{\mathrm{P,sc}}$ is the super capacitor side proportional factor, $K_{\mathrm{I,sc}}$ is the super capacitor side integration factor. The PI controller output is calculated and the duty cycle of the PWM signal adjustment converter IGBT is generated to control the battery current and supercapacitor current so that it follows the reference value steadily.

4 Simulation Results and Analysis

To verify the effectiveness of the control strategy proposed in this paper, a hybrid energy storage system model was built in Simulink, where the control current source was used instead of the load, the DC bus voltage reference value was set to 200 V, the initial *SOC* value of the battery was set to 100%, the initial voltage was set to 180 V, the initial voltage of the supercapacitor was set to 150 V, and the P_{min} was set to 2000 W. The model parameters were selected as shown in Table 1, and the proportional and integral coefficients of controller were selected as shown in Table 2. The load demand power is shown in Fig. 4(a). The simulation results are shown in Figs. 4 (b)–(d). It can be seen that when the load power changes, the battery current and supercapacitor current can track the reference value, and the DC bus voltage is also stable at the reference value of 200 V.

Table 1. Parameter of model.

Parameter	Value
L_{bat}, Battery side inductance (μH)	260
L_{sc}, SC side inductance(μH)	200
R_{bat}, Inductor series resistance (Ω)	0.25
R_{sc}, Inductor series resistance (Ω)	0.16
C_{dc}, Load side capacitance (mF)	15
C_{bat}, Battery side filter capacitance (mF)	70
C_{sc}, SC side filter capacitance (μF)	70

Table 2. Proportional and integral coefficients.

Parameter	$K_{P,dc}$	$K_{I,dc}$	$K_{P,bat}$	$K_{I,bat}$	$K_{P,sc}$	$K_{I,sc}$
Value	6	1	10	0.5	10	0.5

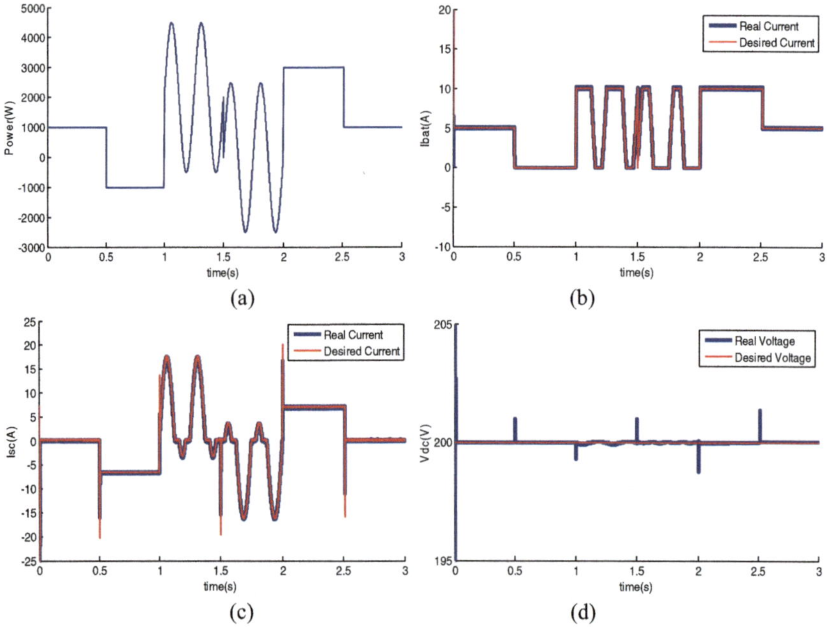

Fig. 4. (a) Custom load demand power. (b) Battery current change curve. (c) Supercapacitor current change curve. (d) DC bus voltage change curve.

5 Conclusion

In this paper, a control strategy for an on-board hybrid energy storage system based on PI control is proposed. This strategy first distributes the required power through an energy management strategy, and then controls the battery and supercapacitor output currents by changing the IGBT switching duty cycle and keeping the DC bus voltage stable. The simulation results show that the effectiveness of the strategy.

Acknowledgment. This work was supported by the National Natural Science Foundation of China (61673164), the Natural Science Foundation of Hunan Province (2020JJ6024) and the Scientific Research Fund of Hunan Provincal Education Department (17A048, 19K025).

References

1. Zhang, Q., Li, G.: Experimental study on a semi-active battery-supercapacitor hybrid energy storage system for electric vehicle application. IEEE Trans. Power Electron. **35**(1), 1014–1021 (2020)
2. Jung, H., Wang, H., Hu, T.: Control design for robust tracking and smooth transition in power systems with battery/supercapacitor hybrid energy storage devices. J. Power Sour. **267**, 566–575 (2014)
3. Song, Z.Y., Hou, J., Hofmann, H., Li, J.Q., Ouyang, M.G.: Sliding-mode and Lyapunov function-based control for battery/super-capacitor hybrid energy storage system used in electric vehicles. Energy **122**, 601–612 (2017)
4. Rotenberg, D., Vahidi, A., Kolmanovsky, I.: Ultracapacitor assisted powertrains: modeling, control, sizing, and the impact on fuel economy. IEEE Trans. Control Syst. Technol. **19**(3), 576–589 (2011)

Research on Task Scheduling in Distributed Vulnerability Scanning System

Jie Jiang[1,2(✉)] and Sixin Tang[1,2]

[1] College of Computer Science and Technology, Hengyang Normal University, Hengyang 421002, China
850369627@qq.com
[2] Hunan Provincial Key Laboratory of Intelligent Information Processing and Application, Hengyang 421002, China

Abstract. Network threats caused by system vulnerabilities is increasing gradually, the distributed vulnerability scanning system can scan large-scale and complex networks and report the vulnerability information. Task scheduling is one of the core components in a distributed system. In this paper, we use dynamic optimization algorithms to improve the task scheduling efficiency of distributed vulnerability scanning system, we propose a PSO-based task scheduling scheme and improves the search ability of particles by adjusting algorithm parameters. We compared the time consume when using existing 'Resource Aware Scheduling algorithm' (RASA) with the basic particle swarm optimization (PSO) algorithm and the improved particle swarm optimization (IPSO) algorithm. Our results show that IPSO has better performance than other scheduling methods.

Keywords: Vulnerability scanning · Distributed · Task scheduling · PSO algorithm

1 Introduction

With the development of computer network and Internet widely available, the number of network security incidents and the losses caused by them are also increasing gradually. Vulnerability scanning can find out the vulnerabilities existing in network system by conducting vulnerability detection and result analysis on the hosts and network devices. Vulnerability scanning is an important way to strengthen the security of the system and ensure the robustness of the system [1]. Through vulnerability scanning, users can search for potential insecure factors in the system and analyze them in advance to detect and identify known or potential security risks and hidden dangers.

Vulnerability scanning has two common scanning methods, one is a host-based scanning method [2]. This method checks the settings of security threats in the system by running a scanner on the host, One is a network-based scanning method, which detects vulnerabilities of detected objects through the network according to the known vulnerability set and records the response results of the detected objects [3].

Distributed vulnerability scanning [4] is a method of arranging scanning engines on hosts in different network regions through network-based vulnerability scanning to

Q. Liu et al. (Eds.): CENet 2020, AISC 1274, pp. 530–536, 2021.
https://doi.org/10.1007/978-981-15-8462-6_59

complete scanning tasks in parallel. Through distributed cluster control, this system can scan networks in different areas at the same time, which has good results for large-scale and complex networks, and by using appropriate task scheduling methods for each node in distributed vulnerability scanning, it will improve the performance of system.

We adopted a meta-heuristics method called Particle Swarm Optimization (PSO) to effectively distribute the scanning tasks to each host. PSO is a global search adaptive optimization method [5]. The algorithm is similar to other dynamic optimization algorithm like Genetic algorithms but, it considers the social behavior of the particles. Each particle adjusts its trajectory based on its best position (local best) and the position of the best particle (global best) of the entire population in every generation. This concept increases the stochastic nature of the particle and converges quickly to a global minimum with a reasonable good solution.

The remaining parts of this paper are organized as follows: Sect. 2 presents some related works, we describe a PSO-based task scheduling scheme and improves the search ability of particles by adjusting algorithm parameters. In Sect. 3 we compare three task scheduling schemes. Section 4 concludes the paper and presents future work.

2 Task Scheduling in Distributed Vulnerability Scanning

The main work of network-based vulnerability scanning is to construct data packets according to the scanning rules, send it to multiple devices on the network. Due to the complex scanning rules and the variety of network devices, traditional vulnerability scanning has the problems of time consumption and low efficiency [6]. The task scheduling of distributed vulnerability scanning refers to the process of formulating tasks according to the active detection target according to the distributed scanning system and dividing system resources for it according to a certain strategy. It is necessary to reduce the total completion time of the task and the system under the conditions of meeting user needs. Execution cost and energy consumption, while making full use of the system's resources, and ultimately obtain an orderly, high-quality scheduling program. Need to minimize task execution costs and maximize use of system resources.

2.1 Related Work

Task scheduling in a distributed system is known as an NP-hard problem for general cases. Common scheduling methods are classified into static scheduling [7] and dynamic scheduling [8]. Common methods of static scheduling are: Min-Min and Max-Min algorithms and First Come First Serve algorithm. In [9], a Resource Aware Scheduling Algorithm (RASA) algorithm was proposed to combining Min-Min Max-Min advantage for task allocation and processing. In [10], a FCFS schedules task was introduced to optimize task scheduling. This kind of method has advantages for scheduling larger tasks or resource-intensive tasks. Common methods of dynamic scheduling include Simulated Annealing Algorithm (Simulated Annealing, SA) [11], Genetic Algorithm [12], and PSO [13]. These dynamic scheduling methods are good at certain fields. According to [14], PSO has advantages in distributed resource scheduling, in this article we intend to use this method for task scheduling. To evaluate effect of the algorithm, we usually use the following indicators: times consume, service quality, load balancing, and cost performance. Time consume refer to the total response time consumed by the system to complete all task requests. Increasing time efficiency can effectively improve system performance, so this paper compares resource scheduling methods based on time indicators.

2.2 Task Scheduling Based on Particle Swarm Optimization

In the task scheduling problem, encode the particles and a fitness function was needed. Considering there are n scanning tasks to be executed in m scanners, then a number in the particle population composed of M particles is The particle position code of i is: $Xi = \{x_{i1}, x_{i2}, \ldots, x_{jk}, \ldots x_{in}\}$, where $i \in [1, M]$, j represents the task number, $j \in [1, n]$, and the value of x_{ij} represents being assigned The scanner number of the task, $x_{ij} \in [1, m]$.

We use the total task completion time to measure the particles. The n scanning tasks are executed in m scanners. The running time can be expressed as an $n \times m$ matrix ETC. The running time of the ith task in the jth scanner can be expressed by the value of the matrix ETC (i, j), then the total task completion time of the jth scanner $T(\mathrm{j}) = \sum_{i=1}^{n} \mathrm{ETC(i,j)}$, where $i \in [1, n], j \in [1, m]$. The completion time of all tasks is $TT = \max(T(j))$, where $\mathrm{j} \in [1, m]$. Particle swarm optimization usually takes the larger value of the moderation function as the better solution, so the fitness function defined in this paper is:

Fitness $= \frac{1}{\mathrm{TT}}$

Through the analysis of task scheduling in distributed vulnerability scanning, the basic PSO formula is obtained:

$$v_{ij}^{k+1} = w\, v_{ij}^{k} + c_1 r_1 \left(pbest_{ij}^{k} - x_{ij}^{k}\right) + c_2 r_2 \left(gbest_{ij}^{k} - x_{ij}^{k}\right) \tag{2.1}$$

$$x_{ij}^{k+1} = x_{ij}^{k} + v_{ij}^{k+1} \tag{2.2}$$

where:

w inertia weight

v_{ij}^{k} velocity of particle i in j dimension at iteration k

v_{ij}^{k+1} velocity of particle i in j dimension at iteration $k + 1$

c_l acceleration coefficient; $l = 1, 2$

x_{ij}^{k} current position of particle i in j dimension at iteration k

x_{ij}^{k+1} current position of particle i in j dimension at iteration $k + 1$

r_l random number between 0 and 1; $l = 1, 2$

$pbest_{ij}^{k}$ position of particle i at iteration k

gbest_{ij}^{k} position of best particle in a population

Details of baisc PSO algorithm in the task scheduling operation is as follows:

Step 1 Initialize m particles according to the coding method, set the speed and position of these m particles

Step 2 Find the ETC value by analyzing the distribution of each particle to the scanner

Step 3 Use the ETC matrix calculation to obtain a moderate function value

Step 4 Update the global optimal value and individual optimal value by iteratively updating the moderate function value

Step 5 When the iteration reaches the set maximum number of iterations, go to step 7

Step 6 According to the formula, the basic formula of particle swarm, update the speed and position of each generation of particles, execute step 2

Step 7 end progress

2.3 Improved Particle Swarm Optimization

The task scheduling of distributed vulnerability scanning is a discrete problem, particle swarm optimization is easy to fall into local optimum and slow convergence. Due to this reason, it is usually necessary to adjust the parameters according to the characteristics of the task to reduce the blindness of the search and improve the probability of optimization [15].

In the PSO search, we can judge the operation of the algorithm based on the fitness function value. if we can find a value for comparison, that is said if current fitness function value is larger than the value, we think it is a better solution, on the contrary, if it is smaller than the value, it is a worse solution, then we can count the number between better and worse solutions, and then judge the next algorithm search solution. If the proportion of better schemes is higher, we take a smaller weight value to control the algorithm to explore in the current solution space finely. If the proportion of poor schemes is higher, we take a larger weight value to expand the exploration space. In order to solve this problem we introduce the concept of distance, in each iteration, the difference between the value of the current fitness function and the minimum fitness function value or the maximum fitness function value in this iteration is called as low-distance or high-distance, So we can count the ratio of low-distance and high-distance for calculation the number of better or worse solution, if the ratio value great then 1, then the number of current better statistical schemes is increased by one. On the other wise, the number of current worse statistical schemes is increased by one. After counting the current fitness function values of all particles, different weights are taken according to the number of smaller and larger values, and if the numbers of smaller and larger values are equal, the median weight is taken. We call this kind of PSO is improved dynamic weights PSO (IPSO) to distinguish the previous basic PSO.

3 Simulation Results

Our experiments environment includes multiple servers in different network segments, of which one Master node is used for task distribution and collection. Run vulnerability scanners in other scan node cluster to scan devices in different network segments, performance comparison experiment topology as follow (Fig. 1).

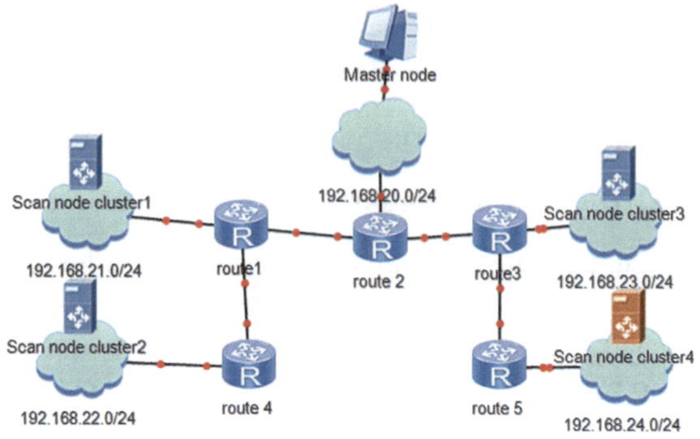

Fig. 1. Performance comparison experiment topology.

We estimate the task execution time by counted task running time, task complexity and device performance. We set up a detection scheme according to the security level of the scanned object in advance, such as comprehensive detection of high-risk targets such as routing and bastion hosts, especially for specific targets. Applications, such as databases and websites, are enabled suspicious vulnerabilities simulated interactive scanning; scanning of common target hosts is mainly for common system vulnerabilities such as buffer overflow vulnerabilities, default security configuration vulnerabilities, etc. The host performs basic detection, such as port detection and weak password detection tasks; by calculating the resource consumption of various tasks, this is used as the statistical basis for task scheduling. Then we compared the operating efficiency of three task scheduling schemes on the container cluster. First, we test Resource Aware Scheduling Algorithm (RASA). Second, we test basic PSO task scheduling scheme. Third we test the improved dynamic weights PSO was introduced in this paper (IPSO), because task combination and container allocation have different completion times under the same scheduling scheme, we recorded 50 tasks scheduled on 20 scanners, 100 tasks scheduled on 40 scanners, and 150 tasks scheduled on 60 scanners The time consumption of scheduling, using the same set of tasks and the same program to run on the cluster 20 times to take the average to evaluate the effect of the program, the experimental results are as follows (Fig. 2).

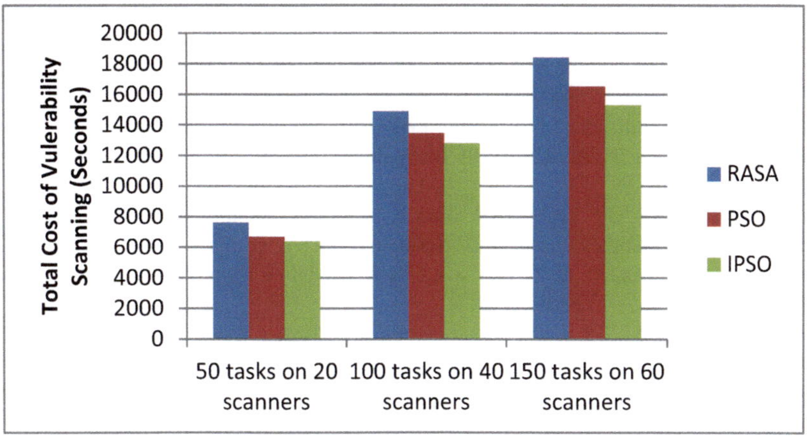

Fig. 2. Comparison of execution time of RASA, PSO and IPSO.

4 Conclusions and Future Work

This paper describes a task scheduling strategy based on particle swarm optimization (PSO) in a distributed vulnerability scanning system. We propose a PSO-based task scheduling scheme and improves the search ability of particles by adjusting algorithm parameters. Resource Aware Scheduling Algorithm (RASA) and Particle Swarm Optimization (PSO) and the Improved Particle Swarm Optimization proposed in this paper, we found that by dynamically adjusting the PSO parameters for task scheduling, the task time consumption is 5–15% of the other two solutions Advantage.

In the future, we plan to design multi-objective optimization schemes for the system based on more optimization indicators such as system service quality and energy consumption.

References

1. Yoshimoto, M., Bista, B.B., Takata, T.: Development of security scanner with high portability and usability. In: International Conference on Advanced Information Networking & Applications (2005)
2. Taylor, P., Mewett, S., Brass, P.C., et al.: Vulnerability assessment and authentication of a computer by a local scanner.: US, 7178116B1 (2007)
3. Thacker, B.H., Riha, D.S., Fitch, S.H.K., et al.: Probabilistic engineering analysis using the NESSUS software. IEEE Int. Conf. Neural Netw. **28**(1–2), 83–107 (2006)
4. Zhang, W., Teng, S.H., Fu, X.F.: Scan attack detection based on distributed cooperative model. In: International Conference on Computer Supported Cooperative Work in Design (2008)
5. Kennedy, J., Eberhart, R.: Particle swarm optimization. IEEE Int. Conf. Neural Netw. **4**, 1942–1948 (1995)

6. Kim, Y., Baek, S.Y., Lee, G.: Intelligent tool for enterprise vulnerability assessment on a distributed network environment using nessus and oval. In: Khosla, R., Howlett, R.J., Jain, L.C. (eds.) KES 2005. LNCS (LNAI), vol. 3682, pp. 1056–1061. Springer, Heidelberg (2005). https://doi.org/10.1007/11552451_146
7. Etminani, K., Naghibzadeh, M.: A min-min max-min selective algorithm for grid task scheduling. In: The Third IEEE/IFIP International Conference on Internet, Uzbekistan (2007)
8. Clerc, M., Kennedy, J.: The particle swarm-explosion, stability, and convergence in a multidimensional complex space. IEEE Trans. Evol. Comput. **6**(1), 58–73 (2002)
9. Parsa, S., Entezari-Maleki, R.: RASA: a new task scheduling algorithm in grid environment. World Appl. Sci. J. **7**, 152–160 (2009)
10. Agarwal, A., Jain, S.: Efficient optimal algorithm of task scheduling in cloud computing, environment. International Journal of Computer Trends & Technology (2014)
11. Shu, W., Zheng, S., Gao, L., et al.: An improved genetic simulated annealing algorithm applied to task scheduling in grid computing. In: International Conference on Complex Systems & Applications (2006)
12. Woo, S-H., Yang, S.B., Kim, S.D.: Task scheduling in distributed computing systems with a genetic algorithm. In: High Performance Computing on the Information Superhighway, Hpc Asia. IEEE, (1997)
13. Tao, Q., Chang, H.Y., Yi, Y., et al.: A rotary chaotic PSO algorithm for trustworthy scheduling of a grid workflow. Comput. Oper. Res. **38**(5), 824–836 (2011)
14. Tasgetiren, M.F., Liang, Y.C., Sevkli, M., Gencyilmaz, G.: A particle swarm optimization algorithm for makespan and total flowtime minimization in the permutation flowshop sequencing problem. Eur. J. Oper. Res. **177**(3), 1930–1947 (2007)
15. Yin, P.Y., Yu, S.S., Wang, Y.T.: A hybrid particle swarm optimisation algorithm for optimal task assignment in distributed systems. Comput. Stand. Interfaces **28**(4), 441–450 (2006)

Delta Omnidirectional Wheeled Table Tennis Automatic Pickup Robot Based on Vision Servo

Ming Lu(✉), Cheng Wang, Jinyu Wang, Hao Duan, Yongteng Sun, and Zuguo Chen

School of Information and Electrical Engineering, Hunan University of Science and Technology, Xiangtan 411201, China
mlu@hnust.edu.cn

Abstract. The survey found that during table tennis training there are often many scattered balls that need to be picked up manually, affecting the efficiency of the players' training. Currently, table tennis is picked up manually using a table tennis picker. In this paper, delta omnidirectional wheeled table tennis automatic pickup robot based on the vision servo achieves fully automatic table tennis pickup by combining the characteristics and requirements of table tennis sports. In this paper, we design a vision algorithm for a table tennis automatic pickup robot system, give a kinematic inverse solution for the delta omnidirectional wheel, complete the design of the robot control system, and build the platform for the experiment. A pickup experiment is conducted on scattered ping-pong balls of different colors. Experiments have shown that the robot can safely clean up scattered table tennis balls on the table tennis field with high positioning accuracy and top pick up rate. It is proved that the robot designed in this paper has extensive application value and prospects.

Keywords: Pitch cleanup · Image processing · Motion control · Omnidirectional wheeled moving platform

1 Introduction

During daily training, table tennis players often have to use lots of tables tennis balls for repeated training, which results in a large number of scattered table tennis balls on the field. Cleaning up manually these ping pong balls is a tedious and time-consuming task. In recent years, scholars have contributed to the research on table-tennis sports such as table-tennis teeing machines [1], table-tennis hitting robots [2], table-tennis trajectory tracking and estimation methods [3, 4]. However, there is little research on court-cleaning robots, and the widely used table-tennis ball-pickers which assist in artificial ball-picking can reduce the workload of staff, but they are still not separated from manual labor. Therefore, introducing smart stadium cleaning robots is an imperative and has a wide scope for development. There have been some attempts at ball-picking robots, such as Chen X et al. proposed a tennis ball collecting robot based on deep learning [5]. There are numerous randomly distributed balls scattered during

Q. Liu et al. (Eds.): CENet 2020, AISC 1274, pp. 537–542, 2021.
https://doi.org/10.1007/978-981-15-8462-6_60

table tennis training, so it is important to accurately identify and locate multiple balls. Due to the robot has to avoid indoor objects such as training equipment, athletes, and walls. This requires robots that can walk indoors at any angle and in any direction without collision.

2 Robotic Software Systems

This paper get target coordinate information by machine vision, bring the coordinate information into the established kinematic model of the delta omnidirectional wheel and get the control information of the robot by inverse solution. The control information is imported into the controller to complete the robot's motion control. The design ideas are shown in the Fig. 1.

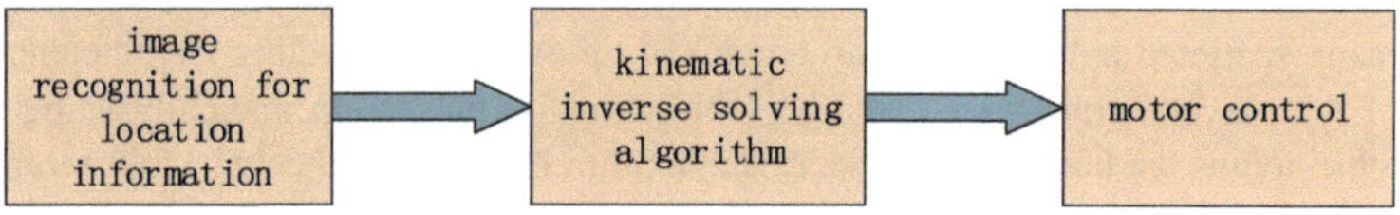

Fig. 1. Software system architecture diagram.

2.1 Image Recognition

Paul Hough proposed an algorithm for detecting circles in the mid-20th century which is called the Hough transform detected specific curves. Through the duality of points and lines, a particular curve of the original image space is transformed into a point in the parameter space by certain rules, and the points in the parameter space are accumulated to form an accumulator, and the peak point of the accumulator is the final information obtained. The classical Hough circle detection principle has the advantages of high accuracy and strong noise suppression, but the calculation volume is very large in three-dimensional space, so the classical Hough circle detection is not suitable for practical applications [6].

This paper used the more convenient and accurate Hough gradient method for the round detection [7–9]. The center of the circle is firstly found according to the modal vector of each point, so that the three-dimensional cumulative plane is transformed into a two-dimensional cumulative plane; then the radius is determined according to support of non-zero pixels at the edge of all candidate centers, completing the identification of the circle.

After pre-processing the high-definition images from the camera, the gradient-Hoff transform is used to identify the table tennis balls and get the location information [10]. The identification of the table tennis program using the gradient-Hoff transform was run as shown in Fig. 2.

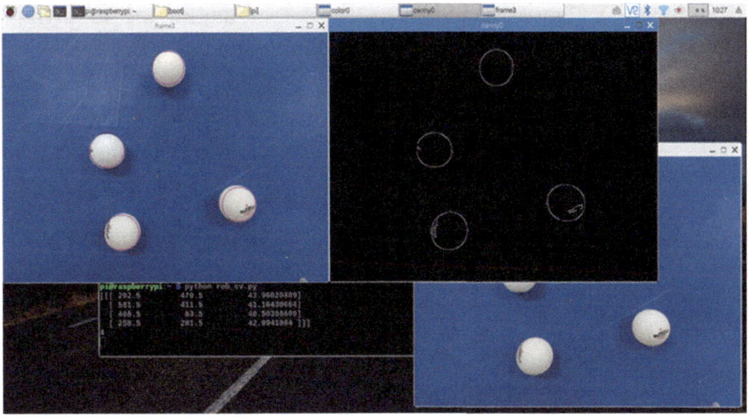

Fig. 2. Hough circle detection program running.

2.2 Kinematic Inverse Solving Algorithm

Establish the coordinate system of the robot, as shown in Fig. 3.

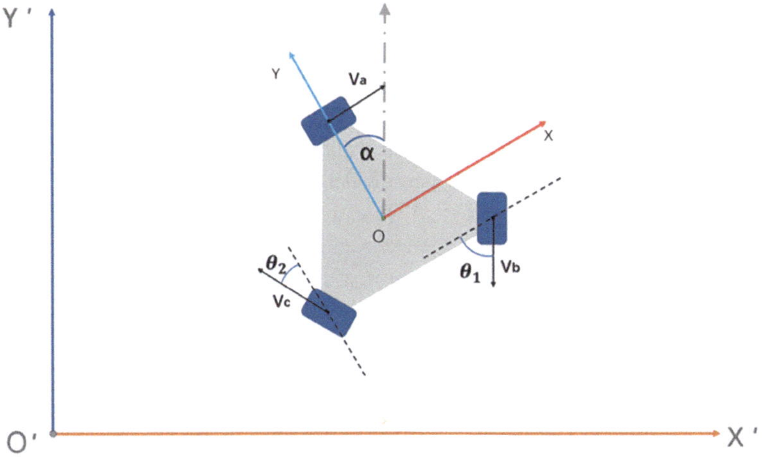

Fig. 3. DELTA omnidirectional wheel kinematics modeling.

$X'O'Y'$ is the world coordinate system and XOY is the robot coordinate system. Where the distance from each omnidirectional wheel to the chassis center is L, the angular speed of the robot rotation is ω, and the clockwise direction is set as the positive direction of the angular speed, and the speed of the three omnidirectional wheels is V_a, V_b, V_c. In the robot coordinate system, it decomposes the speed of motion as V_x, V_y. The clamping angle $\theta_1 = 60°$, $\theta_2 = 30°$, and α is the clamping angle of two unique coordinate systems. Thus, it is possible to establish the following velocity relation Equation for the robot coordinate system, as shown in Eq. 1 [11, 12].

$$\begin{bmatrix} V_a \\ V_b \\ V_c \end{bmatrix} = \begin{bmatrix} 1 & 0 & L \\ -cos\theta_1 & -sin\theta_1 & L \\ -sin\theta_{12} & cos\theta_2 & L \end{bmatrix} \begin{bmatrix} V_X \\ V_Y \\ \omega \end{bmatrix} \tag{1}$$

Transform the velocity relation Equation to obtain the following velocity relation Equation for the world coordinate system, as shown in Eq. 2.

$$\begin{bmatrix} V_a \\ V_b \\ V_c \end{bmatrix} = \begin{bmatrix} \cos\alpha & \sin\alpha & L \\ -\cos\theta_1\cos\alpha + \sin\theta_1\sin\alpha & -\cos\theta_1\sin\alpha - \sin\theta_1\cos\alpha & L \\ -\sin\theta_2\cos\alpha - \cos\theta_2\sin\alpha & -\sin\theta_2\sin\alpha + \cos\theta_2\cos\alpha & L \end{bmatrix} \begin{bmatrix} V_{X'} \\ V_{Y'} \\ W \end{bmatrix} \tag{2}$$

2.3 Motor Control

PID algorithms are extremely widely used in industrial control and are not dependent on the specific model of the control object, but are also easy to understand and refer to. The PID controller has many derivative PID algorithms in digital control, such as incremental PID, positional PID, feed-forward PID, fuzzy PID, etc. [13, 14]. Since the value of Δu(k) in the incremental PID is only related to the last three samples, there is no accumulation of errors, the system is more stable, and the amount of calculation is greatly reduced, which is suitable for control objects with integral terms. Therefore, we use an incremental PID algorithm to control the speed of the motor [15].

When the parameters of the PID controller are $K_p = 4.2$, $K_i = 0.1$, it shows the control effect of the PID in the Fig. 4. It can be seen that the system is less over-regulated, more responsive and has better tracking performance.

Finally, the parameters of the simulation are imported into the controller, and then minor adjustments are made for the actual situation, resulting in a stable control effect.

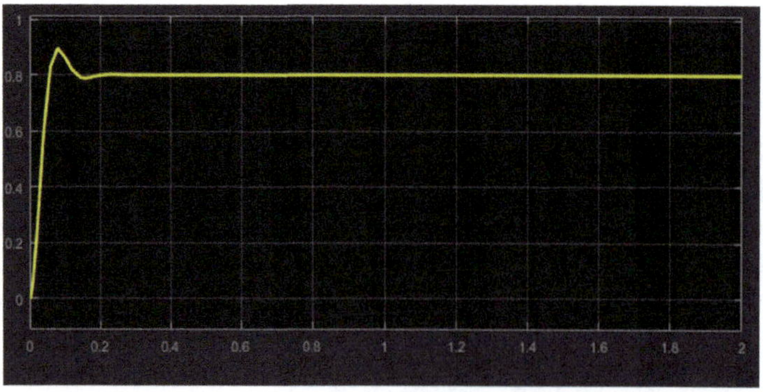

Fig. 4. Schematic of the simulation results.

3 Experimental Validation

In this paper, it builds the experimental platform using the Raspberry Pie 3B + processor, high-definition color camera, millimeter-level LiDAR, and motion actuator as the core hardware foundation. The experimental platform is shown in Fig. 5. Pickup experiments were performed on 50 white balls, 50 orange balls and 25 each of white and orange balls mixed, and the experimental data are as follows (Table 1).

Table 1. Experimental data.

Ball color	Total	Number of picks	Pick-up rate
orange	50	49	98.0%
white	50	50	100%
mixed	50	48	96.0%

After test and statistics, the robot needs about 2S for each ping pong ball it draws, and the internal storage box can hold 50 ping pong balls. A 98.0% success rate for table tennis collection is obtained by three repeated experiments. Experiments have shown that the robot has a higher pickup rate, faster pickup speed and can meet the demand better.

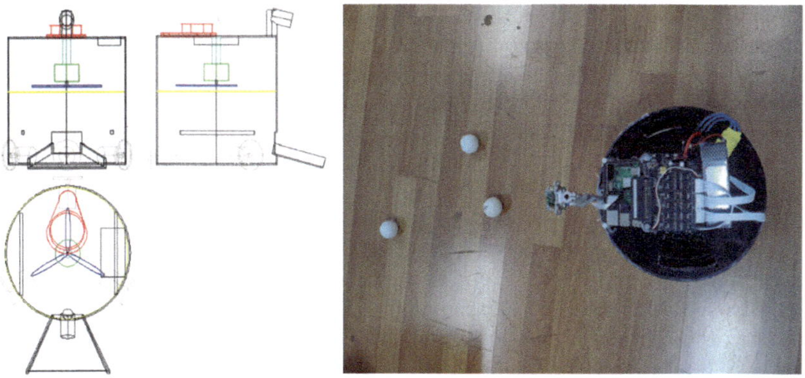

(a) (b)

Fig. 5. (a) Robots three view. (b) Table tennis court cleaning robot in kind.

4 Conclusion

To overcome the difficulty of traditional table tennis picking methods that require manual execution and are inefficient, this paper designs a vision servo-based delta omnidirectional wheel table tennis picking robot that combines radar information with kinematic inverse solving algorithms to control the delta omnidirectional wheel so that the robot can avoid obstacles and has high flexibility. The test results show that the machine vision-based approach has a beneficial recognition effect and provides an important reference for robot-assisted motion.

Acknowledgement. This research was funded by the National Natural Science Foundation of China (grant number 61672226, 61903137). This research was funded by the Natural Science Foundation of Hunan Province (grant number 2020JJ4316).

Refrences

1. Hayakawa, Y., Nakashima, A., Itoh, S., et al.: Ball trajectory planning in serving task for table tennis robot. SICE J. Control Meas. Syst. Int. **9**(2), 50–59 (2016)
2. Zhang, K., Cao, Z., Liu, J., et al.: Real-time visual measurement with opponent hitting behavior for table tennis robot. IEEE Trans. Instrum. Meas. **67**(4), 811–820 (2018)
3. Liu, Y., Liu, L.: Accurate real-time ball trajectory estimation with onboard stereo camera system for humanoid ping-pong robot. Robot. Auton. Syst **101**, 34–44 (2018)
4. Lin, H.I., Huang, Y.C.: Ball trajectory tracking and prediction for a ping-pong robot. In: 2019 9th International Conference on Information Science and Technology. pp. 222–227, IEEE (2019)
5. Gu, S., Chen, X., Zeng, W., et al.: A deep learning tennis ball collection robot and the implementation on nvidia jetson tx1 board. In: 2018 IEEE/ASME International Conference on Advanced Intelligent Mechatronics. pp. 170–175,. IEEE (2018)
6. Pinto, R.M., Becker, M., Tronco, M.L.: Sweet citrus fruit detection in thermal images using fuzzy image processing. In: Applications of Computational Intelligence: Second IEEE Colombian Conference, ColCACI 2019, Barranquilla, Colombia, June 5–7, Revised Selected Papers. Springer Nature (2019)
7. Wang, Y., Cheng, G.: Application of gradient-based Hough transform to the detection of corrosion pits in optical images. Appl. Surface Sci. **366**, 9–18 (2016)
8. He, K., Hu, Y.P.: An improved method of hough transform for circle recognition based on optimized gradient. In: 2017 5th International Conference on Frontiers of Manufacturing Science and Measuring Technology. Atlantis Press (2017)
9. Calatroni, L., van Gennip, Y., Schönlieb, C.B., et al.: Graph clustering, variational image segmentation methods and Hough transform scale detection for object measurement in images. J. Math. Imaging Vis. **57**(2), 269–291 (2017)
10. Solsona, S.P., Maeder, M., Tauler, R., et al.: A new matching image preprocessing for image data fusion. Chemom. Intell. Lab. Syst. **164**, 32–42 (2017)
11. Ayyıldız, M., Çetinkaya, K.: Comparison of four different heuristic optimization algorithms for the inverse kinematics solution of a real 4-DOF serial robot manipulator. Neural Comput. Appl. **27**(4), 825–836 (2016)
12. Zhang, Y., Huang, H., Yan, X., et al.: Inverse-free solution to inverse kinematics of two-wheeled mobile robot system using gradient dynamics method. In: 2016 3rd International Conference on Systems and Informatics (ICSAI). pp. 126–132, IEEE (2016)
13. Zhang, J., Yang, S.A.: Novel PSO algorithm based on an incremental-PID-controlled search strategy. Soft. Comput. **20**(3), 991–1005 (2016)
14. Eltag, K., Aslamx, M.S., Ullah, R.: Dynamic stability enhancement using fuzzy PID control technology for power system. Int. J. Control Autom. Syst. **17**(1), 234–242 (2019)
15. El-Samahy, A.A., Shamseldin, M.A.: Brushless DC motor tracking control using self-tuning fuzzy PID control and model reference adaptive control. Ain Shams Eng. J. **9**(3), 341–352 (2018)

RTP-GRU: Radiosonde Trajectory Prediction Model Based on GRU

Yinfeng Liu[1], Yaoyao Zhou[2], Jianping Du[1], Dong Liu[1], Jie Ren[1], Yuhan Chen[3], Fan Zhang[2], and Jinpeng Chen[2(✉)]

[1] Beijing HY Orient Detection Technology Co., Ltd., Beijing 102206, China
[2] Beijing University of Posts and Telecommunications, Beijing 100786, China
jpchen@bupt.edu.cn
[3] Beijing University of Technology, Beijing 100083, China

Abstract. Radiosonde has always played a very important role in meteorological detection, so how to properly schedule the radiosonde to reach the sensitive region is an urgent problem. In this paper, deep learning is applied to this field for the first time to provide a basis for the reasonable scheduling of radiosonde by predicting the motion trajectory of radiosonde. Based on the radiosonde data from February 2019 to October 2019, this paper uses the radiosonde trajectory prediction model based on GRU (RTP-GRU) to predict the radiosonde trajectory in a period of time in the future. The experimental results show that this model has better performance than baseline methods such as RNN and LSTM. The results show that it is feasible and valuable to explore this field with deep learning method.

Keywords: Radiosonde · RTP-GRU · Trajectory prediction

1 Introduction

With the development of science and technology, the scope of human production and living has been further expanded, and meteorological information plays an increasingly important role in human activities. Therefore, people have created a variety of methods for meteorological detection, such as meteorological radar, meteorological satellites, radiosonde etc. The radiosonde has the advantages of low investment, low cost, fast effect, long flight time, and high accuracy of observation data, so it is widely used.

The radiosonde uses double-layer balloons to reach high altitude to measure meteorological elements such as temperature, atmospheric pressure, humidity, wind speed and direction. The meteorological detection process is divided into three stages: firstly, it enters the ascending stage due to buoyancy; then the outer balloon explodes at about 30 km, and the inner balloon carries the radiosonde to move horizontally in the stratosphere; finally, the ground sends a "fuse" command to separate the radiosonde from the sounding balloon, and the radiosonde falls. At each stage, the radiosonde collects meteorological information.

Although the radiosonde has many advantages, it has a major disadvantage in that it has no power system. It is only affected by wind and buoyancy, and its flight trajectory is not controlled. It is difficult to predict its flight trajectory and detection range. In

Q. Liu et al. (Eds.): CENet 2020, AISC 1274, pp. 543–550, 2021.
https://doi.org/10.1007/978-981-15-8462-6_61

order to solve this problem, Liu et al. [1] proposed a method of combining a drone with a sounding balloon. However, this method is costly and difficult to adopt in practical applications. This paper uses the method of predicting the radiosonde trajectory to solve this problem.

The application scenarios faced by this paper are: a meteorologically sensitive area needs to be meteorologically detected to determine whether the radiosonde in the air can reach the sensitive area in the future. This paper proposes a GRU-based radiosonde trajectory prediction method (RTP-GRU) to achieve the radiosonde trajectory prediction.

The main contributions of this article are as follows:

1) This paper uses neural network algorithm to predict the trajectory of the radiosonde for the first time.

2) This paper proposes a novel radiosonde trajectory prediction model based on GRU (RTP-GRU).

3) In this paper, the experiment of RTP-GRU method is carried out on real radiosonde historical data set, and the experimental results show that the method proposed in this paper is effective.

2 Related Work

This part mainly introduces two aspects: the existing solutions to the uncertainty of the trajectory of the sounding balloon and the related algorithms for trajectory prediction.

2.1 Air Balloon

In order to solve the problem of uncertainty in the trajectory of the sounding balloon, Liu et al. [1] proposed a method of combining the sounding balloon with an unmanned aerial vehicle (UAV). This method makes full use of the flying height of the sounding balloon and the advantage that the drone can be controlled, so that the combination of the sounding balloon and the drone can not only realize the high-altitude detection task, but also achieve the purposeful flight to reach the designated position. The advantage of this research work is that it can realize the artificial control of detecting the balloon trajectory, but the disadvantage is that it needs to use a drone, and the cost is high.

2.2 Trajectory Prediction

Neural network algorithm is the most popular method applied to trajectory prediction recently. Lee et al. [2] proposed that RNN and context information can be used to generate robot trajectories. Choi et al. [3] proposed that RNN and Maximum Margin Inverse Reinforcement Learning (IRL) can be used to predict the trajectory of objects in dynamic scenes. Shi et al. [4] proposed an LSTM-based aviation flight trajectory prediction model. Syed et al. [5] proposed that the SSeg-LSTM method uses semantic segmentation to merge scene information to solve the problem of human motion in crowded environments. Zhang et al. [6] proposed that the trajectory of pedestrians has a great relationship with the intentions of neighboring passers-by. Xue et al. [7] propossd the Location-Velocity-Temporal Attention (LVTA) LSTM model for predicting pedestrian trajectories in 2020. This paper uses GRU-based radiosonde trajectory prediction method for trajectory prediction, and then introduces RTP-GRU in detail.

3 Trajectory Prediction Framework of Radiosonde

3.1 Problem Setting

Considering that the position information of the radiosonde can be expressed by longitude, latitude and altitude, so if we want to predict the movement track of the radiosonde in the future, we only need to use the historical information such as temperature, pressure, humidity, the speed of the radiosonde and the position of the radiosonde to predict the longitude, latitude and altitude of the radiosonde at each time in this time period. They can be represented as a tuple $(L_t,\ W_t,\ A_t)$ where L_t represents the longitude at the time t of the radiosonde, W_t represents the latitude at the time t of the radiosonde, and A_t represents the altitude at the time time t of the radiosonde.

3.2 Overview of RTP-GRU

In the section, we introduce our approach, which base on the GRU algorithm and is divided into two steps. Firstly, we introduce the Input Layer. Secondly, we describe in detail the RTP-GRU model. We explain these steps as follows.

Input Layer. First, we need to use a series to supervised function, which transforms a time series dataset into a supervised learning dataset. This function mainly uses the Pandas shift () function. After the transformation, the historical data of the radiosonde turns into the form as shown in Table 1.

Table 1. The form of transformation.

Features				Target						
P1 (t)	P2 (t)	…	P9 (t)	Q1 (t + 1)	Q2 (t + 1)	Q3 (t + 1)	…	Q1 (t + x)	Q2 (t + x)	Q3 (t + x)

P1(t) to P9(t) represent 9 features of the input at time step t, that are the input of TPR-GRU model, Q1(t + x), Q2(t + x), Q3(t + x) represent the prediction of the longitude, latitude, altitude of the x-minute radiosonde.

Finally, in order to make the model to have better performance, the data goes through a normalization process using the MinMaxScaler function in the sklearn package, and is unified to [0, 1]. The formula follows:

$$x0 = \frac{x - \min(x)}{\max(x) - \min(x)} \tag{1}$$

RTP-GRU. RTP-GRU has great advantages in processing and predicting time series data. It is a special form of RNN. Both RTP-GRU and RNN have a network module with chain structure. In RNN, the module consists of a single neuron structure. But in RTP-GRU, this module consists of cells with two gates. The cell relies on two gates for feature selection, including the update and gate the reset gate. These two gating mechanisms are special in that they are capable of preserving information in long-term sequences and are not removed over time or removed without being related to predictions. Apart from these two gates, the RTP-GRU consists of three nodes: the input node x_t, the hidden node h_t and the output node y_t. The loop body of RTP-GRU is shown in Fig. 1.

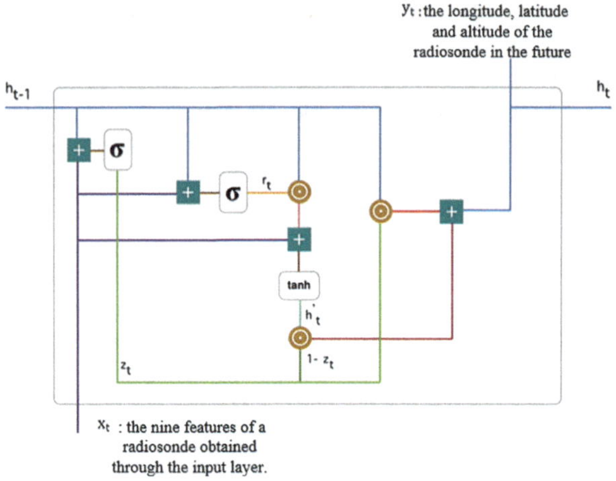

Fig. 1. The loop body of RTP-GRU.

In the RTP-GRU approach, we take advantage of the nine features of a radiosonde obtained through the input layer as the input x_t. Then the calculation methods for these two types of gates are illustrated by the following:

$$z_t = \sigma(W_z * [h_{t-1}, x_t]) \tag{2}$$

Equation (2) describes the calculation process of the update gate in the cell. This update gate at the time step t for the radiosonde determines how much the historical information of the radiosonde to be discarded and the historical information of the radiosonde to be added. Where $\sigma$is the sigmoid function, W_z is the weights of the update gate, h_{t-1} is the output at time step $t-1$, and [] indicates that two vectors are connected.

$$r_t = \sigma(W_r * [h_{t-1}, x_t]) \tag{3}$$

Equation (3) describes the calculation process of the reset gate in the cell. This reset gate at the time step t for the radiosonde determines how much historical information of the radiosonde is forgotten. Where W_r is the weights of the reset gate.

$$\tilde{h}_t = \tanh\left(W_{\tilde{h}_t} * [r_t * h_{t-1}, x_t]\right) \quad (4)$$

The new memory content will use the reset gate to store the related information of the radiosonde. Its calculation expression is shown in Eq. (4). Where $W_{\tilde{h}}$ is the weights.

$$h_t = (1 - z_t) * h_{t-1} + z_t * \tilde{h}_t \quad (5)$$

Equation (5) describes the final memory of the current time step. The update gate is used in this process, which determines the current memory content $\tilde{h}_t$ of the radiosonde and what information of the radiosonde needs to be collected in the previous time step h_{t-1}.

$$y_t = \sigma(W_o * h_t) \quad (6)$$

Equation (6) describes the output of the current time step, which mainly includes the longitude, latitude and altitude of the radiosonde in the future. Where W_o is the σ is the sigmoid function.

4 Experiments

4.1 Dataset

The dataset used in this experiment is the historical data of the ten months from January 2019 to October 2019, which provided by Beijing HY Orient Detection Technology Co., Ltd. The data is collected once a second. There are in total 3,738,060 pieces of data. Each record mainly contains the pressure, temperature, humidity of the current position of the radiosonde, and the northward speed, eastward speed, ascent speed, longitude, latitude, altitude of the current time of the radiosonde.

4.2 Data Pre-processing

Since the radiosonde is affected by the pendulum effect in the air, the speed change will fluctuate. Therefore, we first need to fit the data of the radiosonde's velocity column. In the experiment, we used the least squares method to fit every three hundred data. At the same time, the time interval between the two records of the original data set is one second, and the position of the radiosonde changes little, so the average method is used to take an average of every 60 records as a new record. The processed data is shown in Table 2.

Where P is the pressure, which measured in hPa. T is the temperature, which measured in °C. H is the humidity, which measured in %. NS is the northward speed of the radiosonde, which measured in m/s. ES is the eastward speed, which measured in m/s. AS is the ascent speed, which measured in m/s. L is the longitude, which measured in °. W is the latitude, which measured in °. A is the altitude, which measured in m.

In the experiment, we used the data from January to September 2019 as the training dataset, and performed the 5-fold cross-validation on the training set, then used the October 2019 data as the test dataset. Then, we predicted the trajectory of the

Table 2. The processed data.

Time	P	T	H	NS	ES	VS	L	W	A
10-21 07:22	988.50	16.96	84.85	−0.057	−0.005	−0.009	111.36	30.73	251.98
10-21 07:23	988.51	16.92	84.48	−0.081	−0.007	−0.013	111.36	30.73	252.45
10-21 07:24	988.56	17.03	84.46	−0.038	−0.004	−0.005	111.35	30.73	249.85

radiosonde in the next 15 min, the next 30 min, the next 60 min, the next 90 min, and the next 120 min.

4.3 Baseline Methods

To evaluate the performance of the proposed model RTP-GRU, another two models are chosen to be baseline methods.

RNN. First developed in 1980s, RNN obtained its specialness due to its structure: the neurons are connected with each other and self-looped, thus the structure is able to display dynamic temporal behaviors and remember the information from last process.

LSTM. LSTM, a type of RNN, which is similar to GRU and also used memory cells and gates to control the long-term information saved in the network. Compared with GRU, LSTM has one more gate, which increases some matrix multiplication.

4.4 Experimental Setup

The RTP-GRU model, RNN model and LSTM model used in the experiment are based on keras library and the python deep learning library. The amount of hidden layer nodes determines the accuracy of the model and the training time of the model. By experiments, all the models are trained for 200 epochs. Batch size affects the amount of data processed at a time, the training is carried out in mini-batches with the batch size of 72. Adam, an optimization algorithm used to replace stochastic gradient descent in deep learning models, is chosen as the optimizer of these models, as it combines the best performance of AdaGrad and RMSProp algorithms and can provide optimization methods for solving sparse gradient and noise problems.

4.5 Evaluation Metric

In our experiments, we adopt Mean Absolute Error (MAE) defined as follow as the evaluation metrics.

$$MAE = \frac{1}{m}\sum_{i=1}^{M}|y_i - \hat{y}_i| \tag{7}$$

Where y_i is the observed i^{th} data point, $\hat{y}_i$ is the predicted and m denotes the total number of data points in the test set.

4.6 Results and Analysis

Table 3 shows the results from our experiments. From Table 3, we observe that RTP-GRU statistically obtains significant improvement over two baseline methods. Thus, the RTP-GRU model has the best prediction accuracy. Next, we note that LSTM and RTP-GRU consistently outperform RNN, and the prediction results of RTP-GRU and LSTM are less different. We believe that this can be attributed to the fact that RNN can't learn long-term dependent information. In other words, RNN can't do long-term memory, only consider the latest state, RTP-GRU and LSTM can remember the long-term state, and use gating to decide which state can be retained and which state should be forgotten. RTP-GRU is a variant of LSTM. The final model is simpler than the standard LSTM model, and its parameters are less than LSTM. In most cases, the effect is similar to LSTM. In this experiment, the prediction result of RTP-GRU is slightly better than that of LSTM. At the same time, we find that with the increase of prediction time, the MAE value increases gradually, that is to say, the prediction accuracy of the model will decrease correspondingly.

Table 3. Comparison with baseline methods in terms of the Mean Absolute Error.

Forecast duration	RNN	LSTM	RTP-GRU
15	0.0160	0.0130	0.0115
30	0.0265	0.0231	0.0221
60	0.0532	0.0425	0.0416
90	0.0724	0.0612	0.0598
120	0.0867	0.0755	0.0745

5 Conclusion

The prediction of the position of the radiosonde can help us to better dispatch the radiosonde. This paper proposes a method of the predicting position of radiosonde using RTP-GRU network which is a kind of deep learning neural network derived from RNN. Experimental results have shown that RTP-GRU statistically obtains significant improvement over RNN and LSTM.

6 Acknowledgment

This work is supported by National Key R&D Program of China under Grant No. 2018YFC1506204 (Project No. 2018YFC1506200).

References

1. Liu, P.Y., Huang, Y.W., Kuo, C., Chou, Y.H.: The effective retrieval of the sounding balloon combined with unmanned aerial vehicle. 2015 IEEE International Conference on Systems. Man, and Cybernetics, pp. 26–31. IEEE, NJ, USA (2015)
2. Lee, Y.M., Kim, J.H.: Trajectory generation using RNN with context information for mobile robots. In: Kim, J.H., Karray, F., Jo, J., Sincak, P., Myung, H. (eds.) Robot Intelligence Technology and Applications 4. Advances in Intelligent Systems and Computing, vol. 447, pp. 21–29. Springer, Cham (2016)
3. Choi, D., An, T.H., Ahn, K., Choi, J.: Future trajectory prediction via RNN and maximum margin inverse reinforcement learning. In: 2018 17th IEEE International Conference on Machine Learning and Applications (ICMLA), pp. 125–130. IEEE, NJ, USA (2018)
4. Shi, Z., Xu, M., Pan, Q., Yan, B., Zhang, H.: LSTM-based flight trajectory prediction. In: 2018 International Joint Conference on Neural Networks (IJCNN), pp. 1–8. IEEE, NJ, USA (2018)
5. Syed, A., Morris, B.T.: SSeg-LSTM: semantic scene segmentation for trajectory prediction. In: 2019 IEEE Intelligent Vehicles Symposium (IV), pp. 2504–2509. IEEE, NJ, USA (2019)
6. Zhang, P., Ouyang, W., Zhang, P., Xue, J., Zheng, N.: SR-LSTM: state refinement for LSTM towards pedestrian trajectory prediction. In: 2019 IEEE/CVF Conference on Computer Vision and Pattern Recognition (CVPR), pp. 12077–12086. IEEE, NJ, USA (2019)
7. Xue, H., Huynh, D.Q., Reynolds, M.: A location-velocity-temporal attention LSTM model for pedestrian trajectory prediction. IEEE Access **8**, 44576–44589 (2020)

Life Prediction Method of Hydrogen Energy Battery Based on MLP and LOESS

Zhanwen Dai(✉), Yumei Wang, and Yafei Wu

Institute 15 of China Electronics Technology Group Corporation, Beijing 100089, China
1466671891@qq.com

Abstract. Proton exchange membrane fuel cell (PEMFC) has the advantages of stability and high efficiency, but it has little life problems. Therefore, through life prediction and other work, we can know the fuel cell's health status in time and promote its development. In this paper, aiming at the problem of life prediction, a method based on MLP and locally weighted scatterplot smoothing are proposed, which can reduce the data and improve the prediction efficiency by reconstructing the data, can not only retain the original trend of the data but also remove the noise and peak. Finally, the processed data is applied to the MLP model for life prediction. Experiments show that accuracy can reach more than 96%.

Keywords: Hydrogen energy battery · Life prediction · Smoothing · Sequential data

1 Introduction

For the life prediction of fuel cells, Ma et al. developed the PEMFC power generation system's life prediction method using gray theory and extendable theory. The results prove that the gray theory algorithm is sufficient for the complex state monitor-ing and life prediction problems of fuel cells [1]. Shen et al. proposed a prediction method based on the correlation vector machine, predicting the downward trend of the output voltage. The results show that the error between the predicted value and the experimental result is small [2]. Morando et al. proposed the PEMFC aging prediction algorithm based on the echo state network. The voltage data preprocessed by the short-time Fourier transform method is input into the neural network, and the neural network is used to replace the neural network of the original hidden layer. The exist-ing voltage degradation data of PEMFC is used to train the network parameters, and the iterative structure is used to predict the PEMFC. The voltage of the stack, thereby estimating the remaining service life of PEMFC [3–5].

The emergence of prediction and health management technology (PHM) provides a new way to study the durability of PEMFC. PHM technology's core idea is to select the relevant indicators of PEMFC's battery health status by analyzing the real monitoring data and using the indicators to monitor the PEMFC's battery health status and predict the remaining service life of PEMFC. Adjust the system working conditions in time according to the prediction results to improve the durability and safety of PEMFC.

Q. Liu et al. (Eds.): CENet 2020, AISC 1274, pp. 551–562, 2021.
https://doi.org/10.1007/978-981-15-8462-6_62

Combined with proton exchange membrane fuel cells' characteristics, the original hydrogen energy battery data is not right, through data reconstruction and smoothing methods to improve data quality and extract representative data. The extracted data is used to predict the proton exchange membrane fuel cell's remaining life through the life prediction algorithm model.

2 Hydrogen Battery Life Prediction Algorithm

2.1 LightGBM Algorithm

LightGBM is a framework that implements the GBDT algorithm and supports efficient parallel training. There are two main algorithms.

One is gradient-based one-sided sampling (from a reduced sampling angle). It excludes most samples with small gradients, and only uses the remaining samples to calculate the information gain. Although the GBDT algorithm does not have data weights, according to the definition of information gain, each data instance has a different gradient. Large gradient examples have a greater impact on information gain. Therefore, a large gradient of the sample should be maintained during the sampling process. The specific method is to set a threshold in advance, or randomly remove samples with a small gradient between the highest percentiles. Experiments show that under the same sampling rate, the measurement results of this method are more accurate than the results of random sampling, especially in the case of a large information gain range.

The other is exclusive feature bundling. In the sparse feature space, many features are almost mutually exclusive, for example, many features rarely use non-zero values at the same time. Usually in applications, although there are many feature quantities, but because the feature space is very sparse, design a lossless method to reduce the effective features, especially in the sparse feature space, many features are almost mutually exclusive, we can Features are bound to a single feature. Then, the bundling problem is reduced to the coloring problem of the graph, and the approximate solution is obtained by the greedy algorithm.

2.2 Decision Tree Algorithm

Decision tree is a non-parametric method for analyzing hierarchical relationships. Trees can identify and express non-linear and non-additive relationships in a simple form. The idea of the decision tree is to use discriminative variables to recursively subdivide the training sample set into homogeneous groups. The variable selection criterion is based on the entropy measure in information theory. Therefore, a variable is selected so that the data set has the best discrimination in a given class. Repeat this process for each new subset of the data set until all data in the subset belongs to the pure subset as much as possible.

The basic concept of the decision tree: the decision tree is composed of the upper and lower divide-and-conquer mode; all attributes are classified, and the attributes of continuous values must be discretized; all the rule samples are at the beginning; the attribute-based determination of the partition recursively The nodes are sampled; the choice of attributes is based on heuristics or statistical measurements. Stop division conditions: all samples from a given node belong to the same category; characters that do not need to be further divided—classify leaf nodes with a majority vote; there are no samples on a given node.

2.3 MLP Algorithm

The entire data processing process can be roughly divided into two stages.

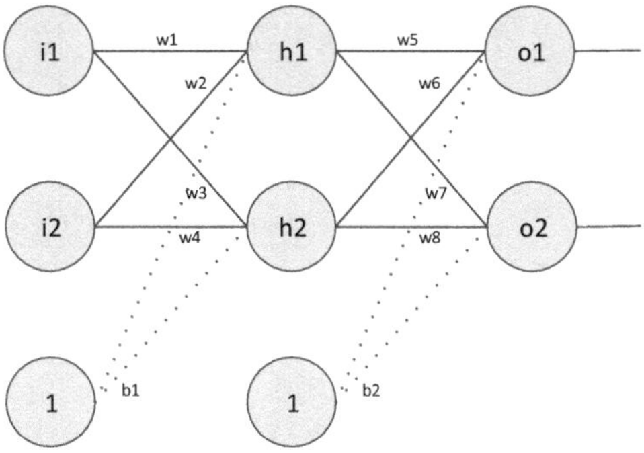

Fig. 1. MLP neural network.

1) Forward propagation of MLP

Forward propagation is the prediction process, which is the process from input to output. The calculation process is the original input, weighted calculation, to the intermediate node of the hidden layer, using the activation function, to the output node.

2) Back propagation of MLP

The back propagation process is the training process for the neural network, that is, training and updating the weights on the above sides. Back propagation includes three processes, the first process is to calculate the error between the correct value of the sample and the predicted output, and the second process is the update of the output layer parameters. The third process is the update of hidden layer parameters.

a. Error calculation process:

$$E_{total} = \sum \frac{1}{2}(target - output)^2 \tag{1}$$

target is the original correct value of the sample, and output is the value predicted for the sample.

b. Output layer parameter update process:

For a certain weight, use the overall error to calculate the partial derivative according to the chain derivative rule.

$$\frac{\partial E_{total}}{\partial w_o} = \frac{\partial E_{total}}{\partial out_o} * \frac{\partial out_o}{\partial net_o} * \frac{\partial net_o}{\partial w} \tag{2}$$

c. Hidden layer parameter update process:

Similar to the parameter update of the output layer, a partial derivative of the hidden layer weights is chained with errors.

$$\frac{\partial E_{total}}{\partial w_h} = \frac{\partial E_{total}}{\partial out_h} * \frac{\partial out_h}{\partial net_h} * \frac{\partial net_h}{\partial w} \tag{3}$$

According to the above formula, the weight is updated to:

$$w' = w - \eta * \frac{\partial E_{total}}{\partial w} \tag{4}$$

Among them, η is the learning rate.

2.4 XGBoost Algorithm

XGBoost is short for "Extreme Gradient Enhancement" and is suitable for classification and regression. The core idea is to continuously add trees, continuously split the features to grow a tree, and add one tree at a time, in fact, learn a new function f(x) to fit the last predicted residual. When we finish training to get k trees, we have to predict the score of a sample. In fact, according to the characteristics of this sample, a corresponding leaf node will fall in each tree, and each leaf node corresponds to a score. Finally, it is only necessary to add up the scores corresponding to each tree to be the predicted value of the sample:

$$y_i^{(t)} = \sum\nolimits_{k=1}^{t} f_k(x_i) \tag{5}$$

3 Introduction of Hydrogen Battery Life Prediction Data

This experiment uses data provided by the FCLAB Research Federation in the 2014 IEEE competition [6].

The stack test platform is suitable for fuel cells with power up to 1 kw (electric power). The 1 kW fuel cell stack used in this article was assembled in FCLAB test. The stack is composed of 5 single cells. The activation area of each single cell is 100 cm^2. The nominal current density of the battery is 0.70A/cm^2, and the maximum current density is $1A/cm^2$. In order to grasp the working conditions of the fuel cell as accurately as possible, many physical parameters involved in the reactor can be measured and controlled, a total of 24 columns (Table 1).

Table 1. Hydrogen battery physical parameters.

Parameters	Physical meaning
TinH2; ToutH2	Hydrogen inlet and outlet temperature (°C)
TinAIR; ToutAIR	Air inlet and outlet temperature (°C)
PinAIR; PoutAIR	Air inlet and outlet pressure (mbara)
DinH2; DoutH2	Hydrogen inlet and outlet flow rate (l/mn)
DinAIR; DoutAIR	Air inlet and outlet flow rate (l/mn)
DWAT	Cooling water flow rate (l/mn)
HrAIRFC	Humidity of air inlet (%)
TinWAT; ToutWAT	Cooling water inlet and outlet temperature (°C)
PinH2; PoutH2	Hydrogen inlet and outlet pressure (l/mn)
I	Current (A)
J	Current density (A/cm^2)
U1-U5	Single battery (V)
Utot	Stack voltage (V)

4 Hydrogen Battery Origin Data Analysis

Figure 2 shows the comparison between the original data of the hydrogen battery and the reconstructed data, showing the relationship between time and other physical characteristics. It can be seen that with the increase of the operating time of the fuel cell, the voltage of each unit cell of the stack has a downward trend, but the relationship between the remaining features and time is not obvious, and there are a lot of noise and spikes. Observe by reconstructing the data and removing noise. The middle line is the reconstructed data. It can be seen that the reconstructed data removes a lot of noise spikes, which makes the data look smoother and the trend remains better. As far as the trend of each curve is concerned, the voltage of the stack shows a certain downward trend, and other physical characteristics do not show a significant change trend. The voltage of each unit cell of the stack has certain reference value as the characteristic of life prediction. The curves of other attributes almost all show a horizontal or fluctuating trend, and there is no obvious relationship with time. Therefore, the relationship between the life time of the hydrogen battery and other characteristics is relatively small. Therefore, this study decided to use the stack voltage as a health indicator of PEMFC through the diagram presented in Fig. 1 and the reading of external literature.

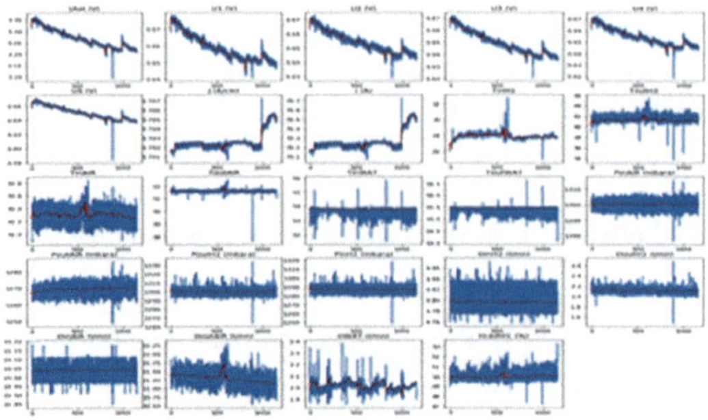

Fig. 2. Relationship between time and other dimension features in original data and reconstructed data.

Through the analysis of the original hydrogen battery data, the algorithm flow chart for this scenario is proposed:

The flow of PEMFC remaining service life prediction based on the smoothed model is shown in Fig. 3, the specific process is as follows:

1). Obtain the actual original aging data set through the PEMFC system;

2). Reconstruct the data according to the method of extracting a set of data points at an interval of one hour;

3). Use the local weighted regression scatter smoothing method to smooth the voltage data points after the reconstruction, and remove the abnormal data such as noise and peaks in the data;

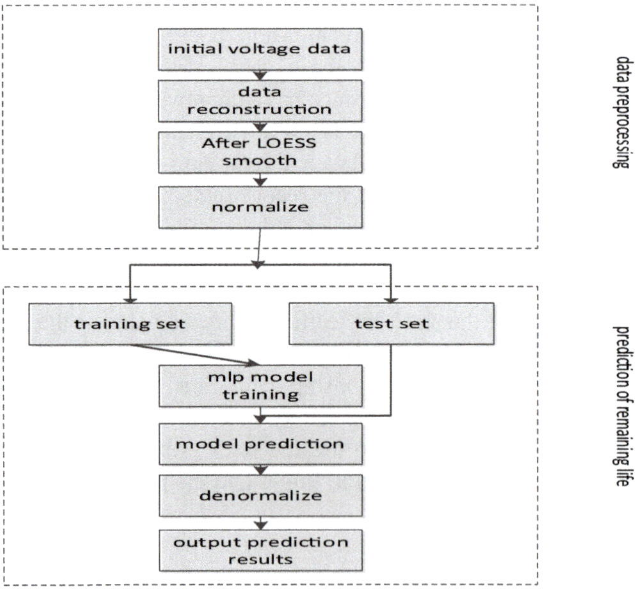

Fig. 3. Algorithm flowchart.

4). Use normalized sample data to reduce model training time;

5). By selecting [0 h, 550 h] samples as the training set, extracting [551 h, 1154 h] samples as the test set;

6). Send the training set data into the model, and randomly initialize the hyper-parameters, determine the parameters through multiple experiments, and get the prediction model. Send the test set into the prediction model for testing;

7). Denormalize the prediction result and write the voltage data into the csv;

8). By reading the csv file, the predicted time is output.

5 Data Reconstruction

By observing the curve in Fig. 1, it can be seen that the 143862 group of raw data contains a lot of noise and spikes, which will cause a lot of time in calculation. Therefore, data preprocessing is required for the raw voltage data. By sampling the sample data set at 1 h intervals, 1154 sets of information data are selected to represent the original data.

The reconstructed data has better improvement on the spike, but it is not ideal for noise and other processing, so the data needs to be further deal with. In this scenario, the voltage gradually declines with time. When encountering this trending data, we cannot use a simple mean plus or minus 3 times the standard deviation to remove outliers. We need to consider the trend and other conditions. Using local weighted regression, you can fit a trend line, using this line as the baseline, and those far away from the baseline are true outlier points. Therefore, consider smoothing the data before proceeding with further model training.

6 Local Weighted Regression Scattering Smoothing Algorithm

For the study of the life prediction of hydrogen energy batteries, it is not necessary to pay too much attention to every detail of the curve, but to grasp the overall trend of the data. LOESS can achieve this purpose, smooth the data, remove noise and spikes, and retain the overall trend. LOESS is a non-parametric method used for local regression analysis. It mainly divides the sample into cells, performs polynomial fitting on the samples in the interval, and repeats this process continuously to obtain weighted regression curves in different intervals. Then connect the centers of these regression curves together to form a complete regression curve. The specific process is as follows:

1). Decide the number and position of fitting points

2). With the fitting point as the center, determine the k closest points

3). Calculate the weight of these k points by the weight function

4). Polynomial fitting (first or second order) by weighted linear regression

5). Repeat the above steps for all fitting points

Regarding the determination of the weight, first of all, it needs to determine the distance between the point in the interval and the fitted point. This distance refers to the distance on the x-axis. We also need to find the maximum distance in the interval, and then normalize the other distances:

$$w_i(x_0) = W\left(\frac{|x_0 - x_i|}{\Delta(x_0)}\right) \tag{6}$$

This weight is smaller as the distance from the fitting point is smaller, so we need to make a transformation:

$$\mathrm{W}(\mu) = \left(1 - \mu^3\right)^3 \tag{7}$$

The exponent can be selected to be quadratic (B function) or cubic (W function). The cubic decreases the surrounding weight faster and has a smoother effect. It is more suitable for most distributions, but it increases the variance of the residual. That is to say, the W function is used more in the first iteration, and the B function is selected in the second iteration.

Regarding the fitting of the weighted linear regression on the scattered points in the interval, the reason why we use weighted linear regression instead of ordinary linear regression here is because considering the fitting point, the value of the points near it fits the fitting line The effect of should be greater, and the effect of points farther away should be smaller, so when we want to define the loss function, we should give priority to reducing the error between the nearby points and the fitted line. This is what we have to add to the ordinary least squares method. The reason for the weight is actually weighted least squares:

$$\mathrm{J}(\mathrm{a}, \mathrm{b}) = \frac{1}{\mathrm{N}}\sum\nolimits_{i=1}^{N} w_i(y_i - ax_i - b)^2 \tag{8}$$

It can be seen that after adding the loss function to the weight, when we minimize the loss function, we will consider more points with greater weight, hoping that they will be better. The result of this fitting will naturally be more biased to the point with greater weight. That is to say, the scattered points closer to the fitting point have a greater impact on the fitting line.

Assuming that the data to be smoothed is determined by x and y, the smoothed value of (x, y) can be obtained by specifying the data range adjacent to x and performing weighted regression. The regression coefficient is defined as follows:

$$w_i = \begin{cases} \left(1 - \left|\frac{x - x_i}{d(x)}\right|^3\right)^3, & \left|\frac{x - x_i}{d(x)}\right| \leq 1 \\ 0, & \left|\frac{x - x_i}{d(x)}\right| \geq 1 \end{cases} \tag{9}$$

The above process is repeated for each point, and finally a set of smooth points (xi, yi) will be obtained. Connect these smooth points with short straight lines to get the LOESS curve.

In the corresponding smooth fitting equation, the gradient a and the constant b are defined as:

$$\mathrm{a} = \frac{\sum w_i^2 (x - \bar{x})(y - \bar{y})}{\sum w_i^2 (x - \bar{x})^2} \tag{10}$$

$$\mathrm{b} = \bar{y} - a\bar{x} \tag{11}$$

$\bar{x}$ and $\bar{y}$ are the weighted average of x and y, respectively.

Through this method, the reconstructed data is smoothed to improve the quality of the data. It can be seen from Fig. 4 that the spikes are the original data, the spikes are the reconstructed data, and the spikes are the smoothed data. The smoothed data can be retained the main trend of the original data, and can effectively remove noise and spikes.

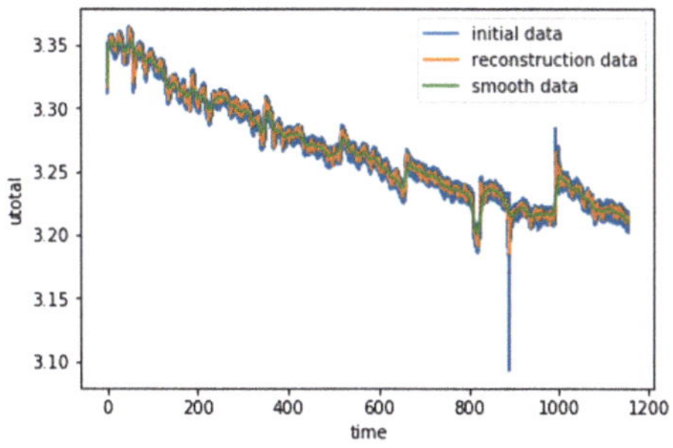

Fig. 4. Data vision.

Considering that the dimensional difference between the parameters is large, it is easy to cause data distortion and it is difficult to reflect the true role relationship of related variables. Therefore, it is necessary to normalize the smoothed data and map the smoothed data to [−1, 1] to reduce the impact of large variable differences on model performance and better achieve life prediction.

After normalization, the range of certain attribute values will not be too large, which will lead to a greater correlation with the life of the hydrogen battery to be predicted, thus weakening the correlation of certain essential attributes., The normalized voltage and time relationship is still very reliable.

7 Algorithm Evaluation Index

In this paper, the decision coefficient: R2 (R-Square) is used to evaluate the fitting degree and error rate of the algorithm to judge the quality of the model.

The R2 (R-Square) evaluation formula is:

$$R^2 = 1 - \frac{\sum_i (\widehat{y_i} - y_i)^2}{\sum_i (\overline{y_i} - y_i)^2} \tag{12}$$

Among them, the numerator part in the fraction of the formula represents the sum of the square of the difference between the predicted value and the true value, and the denominator part represents the sum of the square of the difference between the mean and the true value.

According to the results of R-Squared, to judge the quality of the life prediction model, its value range is [0, 1]. If the result is 0, the result of the prediction model is almost different from the original result, and the fitting effect is very poor; if the result is 1, the difference between the prediction result of the prediction model and the real result is very small, and the fitting effect of the model is very good. In general, the larger the R-Squared, the better the model fitting effect.

8 Result Analysis

Observation of Table 2 shows that for the LightGBM algorithm, the score before smoothing is 0.778, the score after smoothing is 0.785, which is an increase of 0.007; for the decision tree algorithm, the score before smoothing is 0.832, the score after smoothing is 0.863, which is an increase of 0.031; For the MLP algorithm, the score before smoothing was 0.889, the score after smoothing was 0.962, up 0.073; for the XGBoost algorithm, the score before smoothing was 0.857, the score after smoothing was 0.870, up 0.013. It can be seen from the results that, for time series data, the use of neural network models can have a better fitting trend than traditional machine learning methods, and the prediction results are more accurate. Although the accuracy of the XGBoost algorithm is not very high, it still has a large advantage compared with other machine learning algorithms.

Table 2. Comparison of results of different algorithms based on smoothed data.

Algorithm	The R-Square before smoothing	The R-Square after smoothing
LightGBM	0.778	0.785
decisiontree	0.832	0.863
MLP	0.889	0.962
XGBoost	0.857	0.870

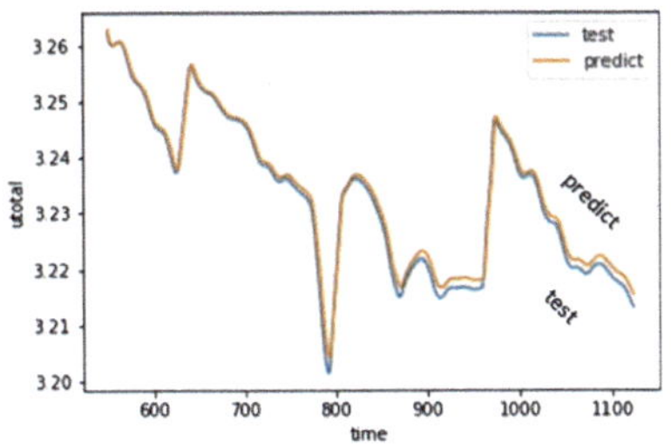

Fig. 5. Algorithm flowchart.

It can be seen from Fig. 5 above that the upper curve in the figure is the prediction result curve based on the MLP and LOESS-based hydrogen battery life prediction method, and the lower curve in the figure is the curve after the original data has been smoothed. The accuracy is obvious.

9 Conclusion

Through the analysis in this article, it can be obtained that the physical parameters of the voltage of the hydrogen battery are better than other physical parameters, which has better representativeness, can roughly estimate the life of the battery, and has better practicability.

Reconstitute the original data by sampling at equal intervals, and use the local weighted regression scatter smoothing method to smooth the reconstructed data. Not only does it retain the original data trend, but also removes noise and spikes, reducing calculation costs and making predictions. Efficiency is greatly improved.

The R2-Score index is used to evaluate the prediction results, ensuring the validity and correctness of the results.

Through the analysis of the prediction results, after smoothing, in these several algorithms, the highest score increased to 0.962.

Through the above analysis, this method has a certain generality and can be applied to the life prediction of hydrogen battery in life.

References

1. Wang, M., Tsai, H.H.: Fuel cell fault forecasting system using grey and extension theories. IET Renew. Power Gener. **6**(6), 373–380 (2012)
2. Yin, S., Xie, X.C., Lam, J., et al.: An improved incremental learning approach for KPI prognosis of dynamic fuel cell system. IEEE Trans. Cybern. **46**(12), 3135–3144 (2016)
3. Morando, S., Jemei, S., Gouriveau, R., et al.: Fuel cells prognostics using echo state network. In: Industrial Electronics Society. pp. 1632–1637, Vienna: IEEE (2013)
4. Morando, S., Jemei, S., Gouriveau, R., et al.: Fuel cells remaining useful lifetime forecasting using echo state network. In: 2014 IEEE Vehicle Power and Propulsion Conference. pp. 1–6, Coimbra: IEEE (2014)

5. Morando, S., Jemei, S., Hissel, D., et al.: A novel method applied to proton exchange membrane fuel cell ageing forecasting using an echo state network. Math. Comput. Simulat. **131**, 283–294 (2017)
6. Fclab Research. IEEE PHM 2014 Data Challenge. http://eng.fclab.fr/ieee-phm-2014-data-challenge/ (2014)
7. Zaadnoordijk, W.J., Bus Stefanie, A.R., Lourens, A., Berendrecht, W.L.: Automated time series modeling for piezometers in the national database of the Netherlands. Ground Water **57**(6), 834–843 (2018)
8. Behzad, A., Doug, W., Robert, M.: Building a telescope engineering data system with Redis, InfluxDB and Grafana. Astronomical Telescopes + Instrumentation (2018)
9. Pelkonen, T., Franklin, S., Teller, J., et al.: Gorilla: a fast, scalable, in-memory time series databas. Proc. VIDB Endow. **8**(12), 1816–1827 (2015)
10. Li, Q., Wang, T.H., Dai, C.H., et al.: Power management strategy based on adaptive droop control for a fuel cell-battery-supercapacitor hybrid tramway. IEEE Trans. Veh. Technol. **67**(7), 5658–5670 (2018)
11. Liu, J.X., Gao, Y.B., Su, X.J., et al.: Disturbance observer-based control for air management of PEM fuel cell systems via sliding mode technique. IEEE Trans. Control Syst. Technol. **27**(3), 1129–1138 (2018). https://doi.org/10.1109/TCST.2018.2802467

Application of DS Evidence Theory in Apple's Internal Quality Classification

Xingwei Yan, Liyao Ma, Shuhui Bi(✉), and Tao Shen

School of Electrical Engineering, University of Jinan, Jinan 250022, China
cse_bish@ujn.edu.cn

Abstract. Aiming at the classification of Red Fuji apples based on the content of soluble solids, a multi-classifier model fusion method based on Dempster and Shagfer (DS) evidence theory is proposed. Multi-model fusion is carried out by integrating different modeling methods such as Partial Least Squares (PLS) and Extreme Learning Machine (ELM). This method uses the prediction results of each model to construct the DS probability distribution function, and outputs the final fusion result through the DS fusion rule. Experimental results show that, compared with a single model, the designed multi-model fusion can improve the accuracy of apple classification to a certain extent, which is conducive to improving the efficiency of apple classification.

Keywords: Apple · D-S evidence theory · Soluble solids · Grade classification

1 Introduction

China is rich in fruit varieties and has a long cultivation history, making my country's fruit industry second only to vegetables and food. Compared with other fruits, my country's apple production and consumption are both in the first place in the world. Although my country's domestic sales are all ranked first in the world, and its output is also on the rise, my country is not an apple export trade powerhouse. Not only can it not be compared with other countries such as Europe and the United States, but also far below the world average. The main reason for this situation is that the internal quality standards of apples in my country are too low, and they cannot be well classified for commercial processing such as grading and sorting, and the quality and safety standards are too low. Therefore, grading the internal quality of apples is extremely important for promoting the development of my country's apple industry [1]. Compared with traditional chemical physical and chemical analysis methods, visible/near infrared spectroscopy non-destructive testing technology has the advantages of non-destructive, fast, green, efficient, low cost, does not require sample pretreatment, etc., has gradually become a high-speed in recent years Emerging technologies for development [2, 3].

This paper uses near infrared spectroscopy technology to collect the near infrared spectroscopy of Red Fuji Apple, and uses genetic algorithm to optimize the characteristic wavelength of the entire spectrum interval, determine the best characteristic wavelength, and use it as an input variable to establish soluble solids and near infrared spectroscopy. Corresponding prediction model. It further proposes the use of DS

Q. Liu et al. (Eds.): CENet 2020, AISC 1274, pp. 563–571, 2021.
https://doi.org/10.1007/978-981-15-8462-6_63

evidence theory for the fusion of multiple modeling methods, the prediction results of the two modeling methods of PLS [4] and ELM [5] are used as the probability distribution function of DS, and then the final classification results are output by DS fusion.

2 Materials and Methods

2.1 Experimental Materials and Data Collection

In this experiment, the red Fuji variety was selected, and its origin is Qixia City, Shandong Province. This type of apple has high quality. In order to enrich the representativeness of the samples, 439 apple samples with no obvious defects on the surface and uniform color were selected, of which 70% were used as the correction set and the remaining 30% were used as the prediction set. The experimental near-infrared spectrum acquisition equipment is Antaris II's near-infrared detector, which uses InGaAs detector, and the sampling mode adopts integrating sphere diffuse reflection. During the spectrum collection process, an average of three points is collected at the equatorial position of the apple, and the average value of the spectral data of the three points is used as the original spectrum of the sample. The spectrum diagram is shown in Fig. 1. After the spectrum collection is completed, measure the content of soluble solids at the location of the spectrum collection, drip the juice from the pulp on the sugar meter to read the content of soluble solids, and the content of soluble solids at the 3 spectrum collection locations per apple. The reference value for this sample [6].

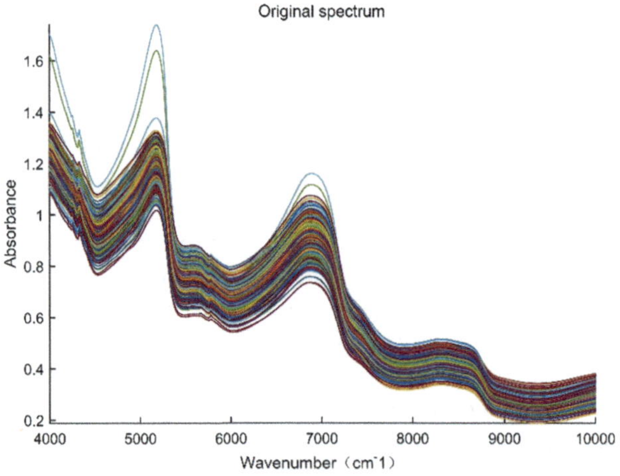

Fig. 1. Original spectrum.

2.2 Algorithm Introduction

DS evidence theory [7, 8] is a set of mathematical theories established by Dempster and Shafer, which provides an effective calculation method for multi-information fusion. The DS evidence theory is built on a recognition framework U, and all the elements in U are incompatible with each other, then the function $m : 2^U \rightarrow [0, 1]$ meets the following conditions:

$$m(\emptyset) = 0$$

$$\sum_{A \subset U} m(A) = 1$$

When, called $m(A)$ is A basic probability assignment (BPA).

If a subset of the frame of discernment U is A, with $m(A) > 0$, then A is called the focal element of the belief function (BEL). The combination rule of DS evidence theory provides the rule of combining 2 evidences. Suppose BEL_1 and BEL_2 two trust functions on the same recognition frame U, m_1 and m_2 are their corresponding basic probability assignments, and the focal elements are $A_1, \cdots, A_k$ and $B_1, \cdots, B_r$ respectively. Assume

$$K_1 = \sum_{A_i \cap B_j = \emptyset} m_1(A_i) m_2(B_j) < 1$$

Then,

$$m(C) = \begin{cases} \frac{\sum_{A_i \cap B_j = C} m_1(A_i) m_2(B_j) < 1}{1 - K_1} & \forall C \subset UC \neq \emptyset \\ 0 & C = \emptyset \end{cases}$$

3 Experimental Design and Analysis

3.1 Experimental Method Design

According to the classification results of the prediction model, calculate the classification accuracy rates of different types of apples, and set the classification accuracy rates of the first, second, and third types of fruits as P_1, P_2 and P_3, respectively, and the corresponding classification error rate as the quality function of its uncertain category. Let A represent first-class fruit, B represent second-class fruit, and C represent third-class fruit.

When the predicted soluble solids content y_{pre} is in the interval (13, 16], the recognition framework is $U = \{A, B, C, AB, AC, BC, ABC\}$. Define the soluble solids content at both ends of the interval $m(A)$ are 0.6 and 0.99 respectively, indicating that the closer to 16, the higher the support rate for the first-class fruit, and the closer to 13, the lower the support rate for the first-class fruit, which is the support for the second-class fruit The rate increases.

$$m_{y_{pre}}(A) = P_1\left[m_{maxA} - \frac{(y_{\max A} - y_{pre}) \times (m_{maxA} - m_{minA})}{y_{\max A} - y_{\min A}}\right];$$

$$m_{y_{pre}}(B) = P_1\left[m_{minB} + \frac{(y_{\max A} - y_{pre}) \times (m_{maxA} - m_{minA})}{y_{\max A} - y_{\min A}}\right];$$

$$m_{y_{pre}}(AB) = 1 - P_1.$$

where P_1 is the classification correct rate of the interval, m_{maxA} is the maximum probability assignment of the interval A, m_{minA} is the minimum probability assignment of the interval A, m_{minB} is the minimum probability assignment of the interval B, and $y_{\max A}$ is the maximum soluble solid of the interval Content value, $y_{\min A}$ is the minimum soluble solid content value in the interval, and y_{pre} is the predicted soluble solid content value.

When the predicted soluble solid content y_{pre} is within the interval [11, 13], the recognition frame is $U = \{A, B, AB, AC, BC, ABC\}$. The $m(B)$ of the soluble solids content at the two ends of the interval and the middle of the interval are 0.6 and 0.9, respectively, indicating that when y_{pre} is closer to the middle of the interval, we think that it is the most likely second-class fruit. Has the highest support rate; the further away from the middle position, the closer to the two ends of the interval, it is considered that the support rate of the second-class fruit is reduced, and the support rate of the first-class fruit or the third-class fruit is increased.

When the content of soluble solids is predicted to be in the interval [11, 12]:

$$m_{y_{pre}}(A) = T - T \times \frac{(y_{pre} - y_{\min B}) \times 0.5}{y_{midB} - y_{\min B}};$$

$$m_{y_{pre}}(B) = P_2\left[m_{minB} + \frac{(y_{pre} - y_{\min B}) \times (m_{maxB} - m_{minB})}{y_{midB} - y_{\min B}}\right];$$

$$m_{y_{pre}}(C) = T + T \times \frac{(y_{pre} - y_{\min B}) \times 0.5}{y_{midB} - y_{\min B}};$$

among them,

$$T = \frac{1}{2}P_2\left[1 - \left(m_{minB} + \frac{(y_{pre} - y_{\min B}) \times (m_{maxB} - m_{minB})}{y_{midB} - y_{\min B}}\right)\right];$$

$$m_{y_{pre}}(ABC) = 1 - P_2.$$

When the content of soluble solids is predicted to be in the interval [12, 13]:

$$m_{y_{pre}}(A) = T + T \times \frac{\left(y_{maxB} - y_{pre}\right) \times 0.5}{y_{\max B} - y_{midB}};$$

$$m_{y_{pre}}(B) = P_2\left[m_{minB} + \frac{\left(y_{maxB} - y_{pre}\right) \times (m_{maxB} - m_{minB})}{y_{\max B} - y_{midB}}\right];$$

$$m_{y_{pre}}(C) = T - T \times \frac{\left(y_{maxB} - y_{pre}\right) \times 0.5}{y_{\max B} - y_{midB}},$$

among them,

$$T = \frac{1}{2}P_2\left[1 - \left(m_{minB} + \frac{\left(y_{maxB} - y_{pre}\right) \times (m_{maxB} - m_{minB})}{y_{\max B} - y_{midB}}\right)\right];$$

$$m_{y_{pre}}(ABC) = 1 - P_2.$$

Among them, P_2 is the classification correct rate of the interval, m_{maxB} is the maximum probability assignment of the interval B, m_{minB} is the minimum probability assignment of the interval B, $y_{\max B}$ is the maximum soluble solid content value of the interval, $y_{\min B}$ is the interval The minimum soluble solid content value of y_{midB} is the intermediate value of the soluble solid content value in the interval, and y_{pre} is the predicted soluble solid content value.

When the predicted soluble solids content y_{pre} is in the interval (8, 11], the recognition framework is $U = \{U = \{A,\ B,\ C, AB, AC, BC, ABC\}\}$. The $m(C)$ of the soluble solids content at the two ends of the interval are 0.99 and 0.6, respectively. The closer to 8, the higher the support rate for the third-class fruit, and the closer to 11, the support for the third-class fruit. The rate decreases, which is an increase in the support rate of the second-class fruit.

$$m_{y_{pre}}(B) = P_3\left[m_{minB} + \frac{\left(y_{pre} - y_{\min C}\right) \times (m_{maxC} - m_{minC})}{y_{\max C} - y_{\min C}}\right];$$

$$m_{y_{pre}}(C) = P_3\left[m_{maxC} - \frac{\left(y_{pre} - y_{\min C}\right) \times (m_{maxC} - m_{minC})}{y_{\max C} - y_{\min C}}\right];$$

$$m_{y_{pre}}(BC) = 1 - P_3.$$

Among them, P_3 is the classification correct rate of the interval, m_{maxC} is the maximum probability assignment of the interval C, m_{minC} is the minimum probability assignment of the interval C, m_{minB} is the minimum probability assignment of the interval B, and $y_{\max C}$ is the maximum soluble solid of the interval Content value, $y_{\min C}$ is the minimum soluble solid content value in the interval, and y_{pre} is the predicted soluble solid content value.

3.2 Analysis of Results

Using the soluble solids content predicted by ELM and PLS to classify apples, the classification accuracy is shown in Fig. 2 and Fig. 3:

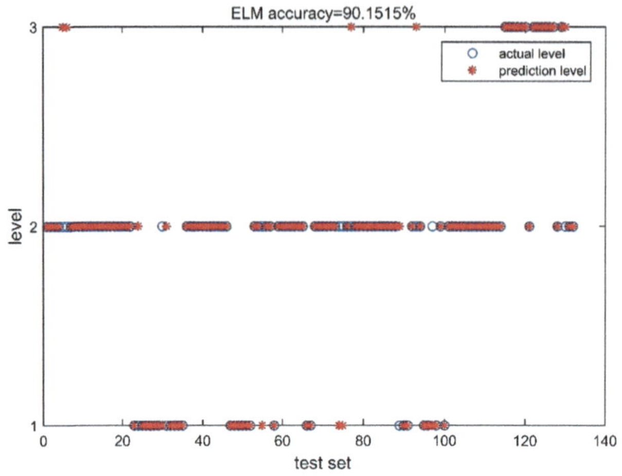

Fig. 2. Classification accuracy of ELM test set.

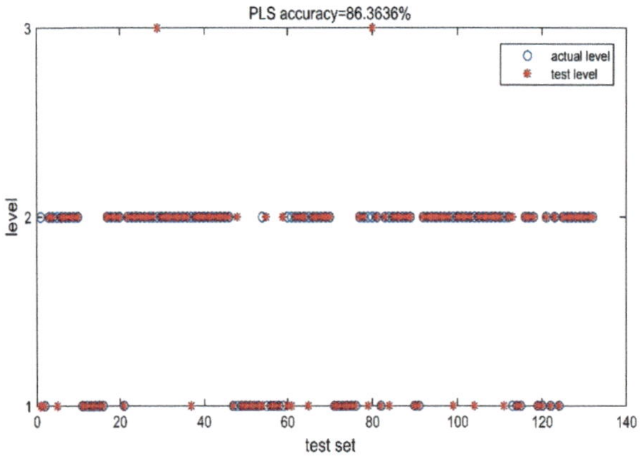

Fig. 3. Classification accuracy of PLS test set.

Among them, the test set has a total of 132 sets of data, accurately identified as 119 sets, and the classification accuracy rate is 90.15%.

Among them, the test set has a total of 132 sets of data, which are accurately identified as 112 sets, and the classification accuracy rate is 86.36%.

Taking one set of prediction results as an example, the actual measured soluble solids content is 11.2, the ELM method predicts the soluble solids content is 10.83, and the PLS method predicts the soluble solids content is 11.69.

First, we construct the recognition framework:

$$U = \{A, B, C, AB, AC, BC, ABC\}$$

According to the predicted soluble solids content, the basic probability assignments of the ELM prediction model and the PLS prediction model are obtained by entering the formula, as shown in the following Table 1:

Table 1. Basic probability assignment table.

Prediction model	A	B	C	AB	AC	BC	ABC
ELM model m_1	0.59	0.31	0	0.10	0	0	0
PLS model m_2	0.05	0.69	0.11	0	0	0	0.15

We combined the theory of DS evidence with the ELM soluble solids content prediction model and the PLS soluble solids content prediction model.

$$m_1(A) = 0.59.\, m_1(B) = 0.31.\, m_1(C) = 0.\, m_1(AB) = 0.10.\, m_1(AC) = 0.$$

$$m_1(BC) = 0.\, m_1(ABC) = 0.$$

$$m_2(A) = 0.05.\, m_2(B) = 0.69.\, m_2(C) = 0.11.\, m_2(AB) = 0.\, m_2(AC) = 0.$$

$$m_2(BC) = 0.m_2(ABC) = 0.15.$$

Solve $m_1 \oplus m_2$.

According to the formula of the synthesis rule:

$$m(C) = \begin{cases} \frac{\sum_{A_i \cap B_j = C} m_1(A_i) m_2(B_j) < 1}{1 - K_1} & \forall C \subset UC \neq \emptyset \\ 0 & C = \emptyset \end{cases}$$

Inferred:

$$(m_1 \oplus m_2)_A = 0.3571.\, (m_1 \oplus m_2)_B = 0.5993.\, (m_1 \oplus m_2)_C = 0.$$

$$(m_1 \oplus m_2)_{AB} = 0.0436.\, (m_1 \oplus m_2)_{AC} = 0.\, (m_1 \oplus m_2)_{BC} = 0.$$

$(m_1 \oplus m_2)_{ABC} = 0$. In the end, it was designated as a second-class fruit.

The prediction results of different methods of all prediction sets are integrated with DS evidence theory. The categorized results after fusion are shown in Fig. 4:

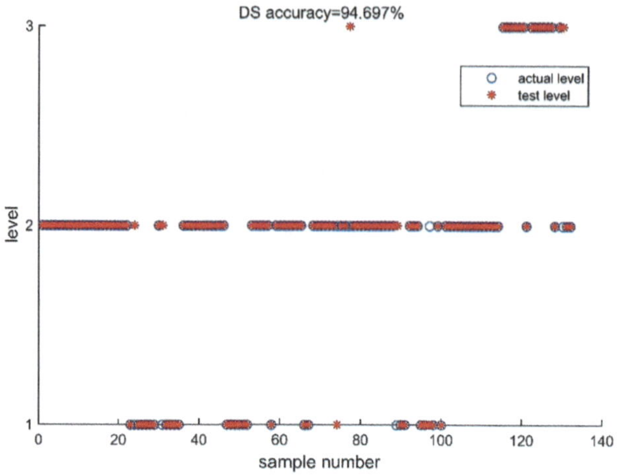

Fig. 4. Accuracy of classification level of test set after DS fusion.

4 Conclusion

It can be seen from the experimental results that the extreme learning machine and partial least squares are selected as the single classifier. In view of the limitations of the single classifier, a multi-classifier integration method based on DS evidence theory fusion is proposed. The advantage is to improve the classification performance of the classifier. The experimental results show that the recognition algorithm based on the DS evidence fusion of the classifier has significantly improved the classification accuracy rate compared with the single classifier. The classification accuracy of the proposed method reaches more than 94.697%, which is much higher than that of the single modeling method.

Acknowledgment. This work is supported by Key R&D projects of Shandong Province under grant 2019GNC106093, Shandong Agricultural machinery equipment R&D innovation plan project under grant 2018YF011, and Shandong Provincial Natural Science Foundation ZR2018PF009.

References

1. Meng, X.N., Zhang, Z.H., Li, Y., et al.: Research status and progress of apple grading. Deciduous Fruits **51**(6), 24–27 (2019)
2. Liu, Q.L., Tan, B.H.: Nondestructive testing study of apple on Near-infrared spectroscopy. J. Hubei Univ. Technol. **32**(4), 26–28 + 38 (2017)
3. Chu, X.L., Yuan, H.F., Lu, W.Z.: Progress and application of spectral data pretreatment and wavelength selection methods in NIR analytical technique. Prog. Chem. **04**, 528–542 (2004)
4. Zou, X.H., Hao, Z.Q., Yi, R.X., et al.: Quantitative analysis of soil by laser-induced breakdown spectroscopy using genetic algorithm-partial squares. Chin. J. Anal. Chem. **43** (02), 181–186 (2015)
5. Liu, X.L., Liu, L.S., Wang, L.L., et al.: Performance sensing data prediction for an aircraft auxiliary power unit using the optimized extreme learning machine. Sensors **19**(18), 1–21 (2019)
6. de Araújo Gomes, A., Galvão, R.K.H, de Araújo, M.C.U., et al.: The successive projections algorithm for interval selection in PLS. Microchem. J. **811**, 13–22 (2014)
7. Shafer, G.: A Mathematical Theory of Evidence. Princeton University Press, Princeton (1976)
8. Dempster, A.P.: Upper and lower probabilities induced by a multivalued mapping. Math. Stat. **38**(2), 325–339 (1967)

Scheduling Algorithm for Online Car-Hailing Considering Both Benefit and Fairness

Rongyuan Chen[1,2], Yuanxing Shi[1,2], Lizhi Shen[1,2], Xiancheng Zhou[1,2(✉)], and Geng Qian[1,2]

[1] Key Laboratory of Hunan Province for Statistical Learning and Intelligent Computation, Hunan University of Technology and Business, Changsha 410205, China
zxc6501@126.com
[2] Mobile E-Business Collaborative Innovation Center of Hunan Province, Hunan University of Commerce, Changsha 410205, China

Abstract. Few existing online car-hailing scheduling algorithms simultaneously take into consideration benefits and fairness. In this paper, a weighted sum of platform profit, driver income and passenger waiting time was used as the objective function of the assignment model. The improved Hungarian algorithm was employed to solve the problem. The imbalanced assignment problem was converted into a standard assignment problem by a virtual person (car) method. The interests of the platform, driver, and passenger were balanced through weight coefficients. Three groups of experiments have verified that compared with the global search algorithm and the greedy algorithm, the performance of the results obtained by the presented algorithm has a great improvement, and the solution time complexity is lower than the global search algorithm.

Keywords: Online car-hailing scheduling · Assignment model · Hungarian algorithm · Multi-objective optimization

1 Introduction

Online car-hailing enhances the efficiency of the use of resources, solves part of the employment problem, and alleviates the urban traffic tension. The 2019 online car-hailing market report is shown that the overall market transaction size of mobile travel in 2018 reached 311.3 billion yuan, of which the express train sector accounted for 71.48%. From 2015 to 2018, the overall market maintained rapid growth, with an average compound annual growth rate of 50.01%. The popularity of online travel has catered to people's demand for the optimization of travel mode.

The online car-hailing scheduling strategy is directly related to platform revenue, driver income and passenger experience. Many scholars try to use operations research and machine learning methods to solve vehicle-scheduling problems. Zhu [1] proposed a nonlinear mixed integer model to solve the problem of car-hailing scheduling based on maximizing the profit of the car-hailing platform, the nonlinear problem is transformed into a linear problem by means of linearization reconstruction and solved by variable neighborhood search algorithm, although the solution is fast, it usually

Q. Liu et al. (Eds.): CENet 2020, AISC 1274, pp. 572–582, 2021.
https://doi.org/10.1007/978-981-15-8462-6_64

converges to the local optimal solution; Meghjani et al. [2] proposed a hybrid informant vehicle matching method combining greed and dichotomies with the goal of minimizing the total stroke, this method first quickly matches the taxi and passenger according to the Euclidean distance, and then iterates according to the real distance, but this method is suitable for small-scale human-vehicle matching problems; Dai [3] proposed a human-vehicle matching algorithm based on RRA-LSP vehicle recommendation, the algorithm selects a suitable recommendation to passengers from vehicles near the passengers, however, this method gives priority to passengers' experience, and does not fully consider the platform and social benefits, which is not suitable for passenger-intensive situations; Liu [4] proposed a taxi dispatching framework based on two-stage control, first calculate the priority of the person-vehicle matching degree, then use the priority as the weight of the bipartite graph, and then use the Kuhn-Munkres method to solve the bipartite graph, the framework has good flexibility, taking into account passenger experience and company profitability to a certain extent, but the parameter setting depends on experience; Xie et al. [5] used the idea of swarm intelligence to propose a taxi intelligent scheduling algorithm based on artificial fish swarm algorithm, which has a good global optimization ability; He [6] proposed a scheduling method for car-hailing based on reinforcement learning, which first meshed the city and then planned the route, this method can effectively handle random travel requirements; Su et al. [7] converted the car-hailing scheduling problem into an assignment problem and applied the difference method to solve it, this method is faster, but the final solution depends greatly on the difference; Fang et al. [8] based on the idea that the cost of each sub-program is the smallest, then the cost of the whole program is the smallest, using the greedy algorithm to perform a global search algorithm for the unbalanced assignment model, this method is faster, but it often converges to the local optimal solution for large-scale problems; Chen et al. [9] considered the influence of the allocation order on the optimization goal, and applied a genetic algorithm with improved genetic coding to solve it. When the number of iterations is large, it converges to the optimal solution with a high probability.

Most of the above methods do not take into account the concerns of the platform, driver and passenger at the same time. This paper proposes a method of car-hailing scheduling that takes into account efficiency and fairness. This method expresses the benefits of the driver and platform and the fairness of waiting time through empty driving distance, In order to avoid the long waiting time of passengers, the distance cost between the car and the passenger is dynamically adjusted through the time function, and the weighted sum of the empty driving distance and the waiting time is taken as the goal, and then the human and vehicle configuration problem is modeled as an assignment model, using the improved Algorithmic solution; In order to increase the enthusiasm of drivers, design a dynamic profit distribution plan, Experiments show that this method has a fast solution speed and can effectively balance the interests of platform, driver and passenger.

2 Problem Description and Model Construction

In essence, the network scheduling problem is the matching problem of people and cars. That is to say. For M cars and N people, a matching scheme should be found within a limited time to make the total waiting time of passengers as short as possible and the total distance of empty driving of vehicles as small as possible. This paper considers three situations: 1) more people and fewer cars; 2) more cars and fewer people; 3) as many cars as people. For the first two cases, this paper converts the imbalance problem into a balance problem through virtual cars and people, and then models the car-hailing scheduling problem as an assignment model.

For ease of description, the following simplifications are made: 1) All vehicles carry the same average speed regardless of whether they carry passengers; 2) Through the adjustment of the distribution plan, the driver highly cooperates with the scheduling of the platform and unconditionally executes the scheduling instructions of the platform; 3) Only one car is required for each ride; 4) Passenger experience is only related to waiting time. x_{ij} variable 0–1 is used to indicate whether vehicle i is associated with passenger j, 1 indicates that vehicle is associated with passenger, and vice versa is 0.

$$x_{ij} = \begin{cases} 1, & \text{Car i received j} \\ 0, & \text{Otherwise} \end{cases} \tag{1}$$

The passenger's waiting time when vehicle i receives passenger j, t is the time the passenger waits when the driver picks up the passenger, s is the distance from the driver's current location to the passenger's departure point, v is the average vehicle speed, q is the distance from the driver's current location to the passenger's departure point:

$$t_{ij} = \frac{s_{ij}}{v} + q_{ij} \tag{2}$$

Passengers want the objective function value f_1 to be as small as possible, i.e. the total waiting time for all waiting passengers should be as small as possible:

$$\min f_1 = \sum\nolimits_i \sum\nolimits_j t_{ij} x_{ij} \tag{3}$$

In the subsequent discussion, it is assumed that there are multiple batches of vehicles matching multiple batches of passengers, and there is a situation where passenger j waits for the vehicle for a long time. Considering the fairness of passengers, to avoid the long-term waiting of passenger j to no avail, use the exponential function e to adjust the value of this function increases the priority of passengers waiting for a long time, and reduces the situation that the same passenger cannot wait for the car many times, b is the batch waiting for passengers when received by the driver:

$$\min f_1 = \sum_i \sum_j e_{bj} t_{ij} x_{ij} \tag{4}$$

Drivers want the value of the objective function to be as small as possible, that is, the total empty driving distance of all drivers during the process of picking up passengers is as small as possible:

$$\min f_2 = \sum_i \sum_j s_{ij} x_{ij} \tag{5}$$

Drivers want the objective function value f3 to be as large as possible, because the platform draws from the driver's revenue, and this is also what the platform expects, that is, all drivers will send passengers as far as possible from the departure to the destination, d is the distance from passenger's departure point to destination:

$$\max f_3 = \sum_i \sum_j d_{ij} \tag{6}$$

Considering passengers, drivers, platforms and other aspects, the goal is to minimize costs in all aspects. After standardizing the objective function, the optimal scheme is solved by introducing weights of α and β to adjust the relationship between them:

$$\min f_4 = \alpha \frac{f_1}{f_{min1}} + \beta \left(\frac{f_2}{f_{min2}} - \frac{f_3}{f_{min3}} \right) \tag{7}$$

In the formula, $0 \leq \alpha \leq 1, 0 \leq \beta \leq 1$, which are adjustable coefficients, it can be seen that when $0 \leq \beta \leq \alpha \leq 1$, the optimal solution is more biased towards $\min(\sum_i \sum_j e_{bj} t_{ij} x_{ij})$, When $0 \leq \alpha \leq \beta \leq 1$, the optimal solution is more biased towards $\min(\sum_i \sum_j (s_{ij} - d_{ij}) x_{ij})$. Therefore, the driver and passenger matching efficiency is:

$$c_{ij} = \sum_i \sum_j \alpha e_{bj} t_{ij} + \beta (s_{ij} - d_{ij}) \tag{8}$$

Then the standard form of the assignment problem is:

$$\min z = \sum_{i=1}^{m} \sum_{j=1}^{n} c_{ij} x_{ij} \tag{9}$$

Constraint C1 means that for any vehicle, at most one order can be taken:

$$C_1 : \sum_i x_{ij} \leq 1 \tag{10}$$

Constraint C2 means that for any passenger, at most one car:

$$C_2 : \sum_i x_{ij} \leq 1 \tag{11}$$

3 Optimization Algorithm

For the assignment model, commonly used solving methods are particle swarm optimization, genetic algorithms, ant colony algorithms, simulated annealing algorithm, and artificial bee colony. Yin [10] constructed the ant path graph to solve the assignment problem with the cost matrix elements as nodes, trying to reduce the number of calculations for the ants to find the path each time, ant colony algorithm is very robust, and the solution efficiency is not high; Gao [11] uses particle swarm algorithm combined with cross operations to solve the assignment problem, and crosses the current solution with the individual optimal solution and the global optimal solution respectively, and uses the resulting solution as the new position solution, the algorithm needs to adjust fewer parameters, but the global convergence is not good; Zhao [12] used a combination of simulated annealing and genetic algorithm to solve the assignment problem, the algorithm converges faster and the search efficiency is higher, but sometimes it converges to the local optimal solution.

In the assignment model, the coefficients of the objective function can be written as a matrix of n $\times$ n dimensions, and the efficiency matrix E is used to represent this set of parameters. The optimal solution of the assignment problem can be obtained by processing the efficiency matrix. The goal of the algorithm is to make a set(n) of zero elements in different rows and different columns appear in the matrix. (The optimal solution is to make $x_{ij} = 1$ corresponding to these zero element positions and $x_{ij} = 0$ at other positions).

The specific steps of the Hungarian algorithm are as follows [13–16]:

Step 1: Subtract the minimum value of this row from each row of the association matrix

Step 2: Subtract the minimum value of this column from each column of the new matrix.

Step 3: Use the fewest row and column lines to go through all the zeros in the new matrix. If the row and column lines do not wear all the elements of the matrix, go to the next step, otherwise go to Step 5.

Step 4: Find the smallest element among the elements that are not worn by the row and column lines, and subtract the smallest element of the remaining elements. The element corresponding to the intersection of the row and column lines plus the smallest element.

Step 5: Find the 0 element corresponding to each row and the 0 element corresponding to the column, and find the optimal allocation based on the 0 element.

4 Experiment and Analysis

In order to test the effectiveness of the algorithm in this paper, the greedy algorithm [17], the global search algorithm [8] and the method of this paper were used to conduct experiments on the two sets of vehicle height data, and compared qualitatively and quantitatively.

4.1 Experiment 1

In this experiment, 2019 Shanghai Didi taxi data are used to simplify the latitude and longitude into a rectangular coordinate system, with km as the unit. The data gives the taxi demand data for a certain day in a city within 3 min and the empty car position information at the same time, including the coordinates of 667 empty cars, the boarding coordinates of 1001 passengers, and the destination coordinates, as shown in Fig. 1:

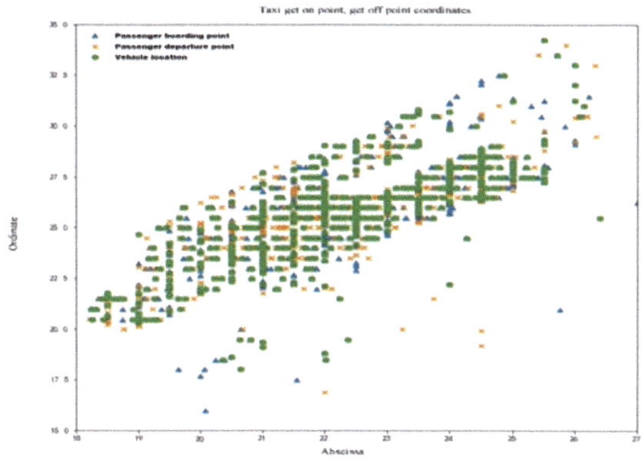

Fig. 1. Passengers' boarding point, getting off point, taxi location data.

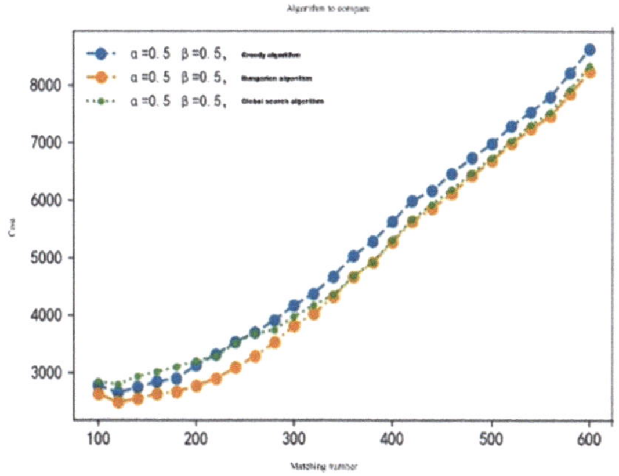

Fig. 2. Scheduling effect of three algorithms.

The performance of the greedy algorithm, the global search algorithm and the algorithm in this paper is given in Fig. 2. With the increase of the number of match multipliers, it can be seen that the cost of the online car-hailing platform shows an upward trend, for small-scale vehicle scheduling, the performance of the three algorithms scheduling is very close, the cost is basically the same, but as the scale of vehicle scheduling gradually increases, the performance of the three algorithm scheduling gradually stabilizes. Experiments demonstrate that compared with the other two algorithms, this algorithm can significantly reduce the total cost.

4.2 Experiment 2

The data was highlighted at this experiment include: taxi ID, time, longitude, latitude, angle, taxi's instantaneous speed and taxi's passenger carrying status. Nearly 100,000 pieces of data, after preliminary data cleaning, the abnormal data is removed, the latitude and longitude coordinates are simplified to rectangular coordinates, and the two times periods 7:30–11:30 and 18:30–23:30 are pressed every hour Divide into 11 batches of data, the scheduling results of the three algorithms are shown in Fig. 3. As can be observed in Fig. 3, the cost of the algorithm in each time period of this paper is less than the other two algorithms, and the passenger waiting time is greatly reduced.

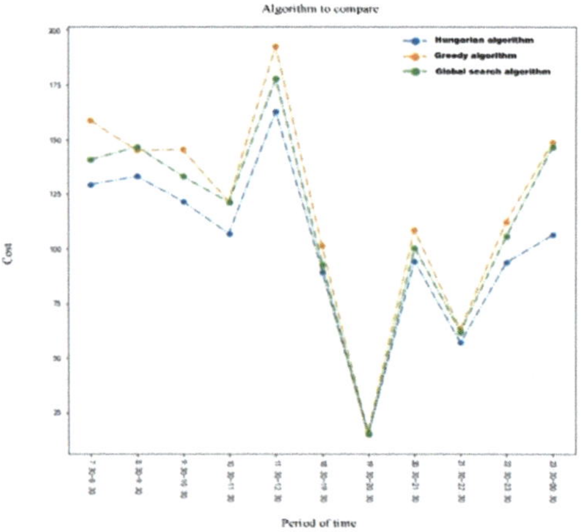

Fig. 3. Performance of different scheduling algorithms.

4.3 Experiment 3

This experiment analyzes the influence of the weight setting in the algorithm of this paper on the car-hailing scheduling. In order to balance the fairness of passengers' sense of experience and the benefit of drivers, the weighted Ben and sun are introduced to adjust the relationship between them. When $0 \leq \alpha \leq \beta \leq 1$, the weight of passengers is greater than that of drivers, that is, online car-hailing platforms pay more attention to the fairness and experience of passengers, When $0 \leq \beta \leq \alpha \leq 1$, the weight of passengers is less than the weight of the driver, that is, the online car-hailing platform pays more attention to the benefits for the driver. Randomly select 70 cars and 95 passengers, under different α and β weights, the vehicle-passenger association relationship paired by this algorithm is shown in Fig. 4. Figure 4 simulates the situation of the ride-hailing platform that focuses more on passenger experience and fairness, and gives the match-match relationship of $\alpha = 0.3$ and $\beta = 0.7$, and the loss cost is 897.73; Fig. 4 shows the multiplier matching relationship of $\alpha = 0.5$ and $\beta = 0.5$, and the loss cost is 903.17; Fig. 4 shows the situation where the simulated car-hailing platform focuses more on the driver's efficiency, and gives the multiplier matching relationship with $\alpha = 0.8$ and $\beta = 0.2$, and the loss cost is 1048.59.

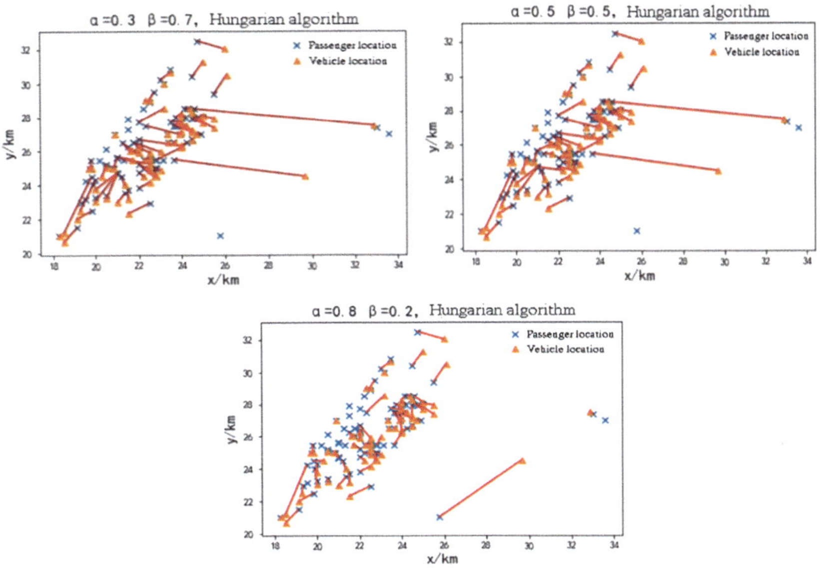

Fig. 4. Correlation diagram of the multiplier matching under different weights

Since the profit of the online car-hailing platform is basically directly linked to the driver's income, after improving the fairness and experience of passengers and reducing the effectiveness of the driver, you will find that the online car-hailing platform's performance for vehicle scheduling varies with the number of drivers Got worse. Therefore, in the process of vehicle scheduling, the algorithm compares the cost of different weight coefficients one by one. By testing six different sizes of vehicle scheduling, the experimental results are shown in Tables 1 and 2.

Table 1 shows the time cost of passengers with different weights and matching numbers under different weights. With the increase of α weight and the decrease of β weight, the waiting time of passengers also becomes longer, resulting in the experience of passengers riding on the online car Get worse.

Table 1. Waiting time for passengers with different weights and matching numbers under different weights.

Weights	Passenger waiting time (Second)					
	Matching number = 100	Matching number = 200	Matching number = 300	Matching number = 400	Matching number = 500	Matching number = 600
$\alpha = 0.2$ $\beta = 0.8$	246.191	519.429	817.954	1147.756	1387.003	1600.784
$\alpha = 0.35$ $\beta = 0.65$	406.000	836.772	1361.087	1868.524	2266.630	2594.130
$\alpha = 0.5$ $\beta = 0.5$	561.586	1126.548	1799.515	2473.781	3039.660	3509.515
$\alpha = 0.65$ $\beta = 0.35$	662.498	1340.659	2140.307	3027.516	3742.823	4382.436
$\alpha = 0.8$ $\beta = 0.2$	693.397	1456.758	2381.272	3503.700	4424.334	5222.524

Table 2. Driver benefits under different weights and multiplier matching numbers.

Weights	Driver benefits (Yuan)					
	Matching number = 100	Matching number = 200	Matching number = 300	Matching number = 400	Matching number = 500	Matching number = 600
$\alpha = 0.2$ $\beta = 0.8$	1129.628	1648.394	2312.263	3220.286	3696.936	3504.216
$\alpha = 0.35$ $\beta = 0.65$	936.325	1390.109	1924.696	2711.397	3336.196	2989.642
$\alpha = 0.5$ $\beta = 0.5$	732.726	1121.548	1586.786	2239.607	2724.105	2458.114
$\alpha = 0.65$ $\beta = 0.35$	565.102	875.053	1267.470	1697.157	2067.844	1846.463
$\alpha = 0.8$ $\beta = 0.2$	415.857	627.600	904.177	1107.588	1302.206	1163.035

Table 2 shows the driver benefits of different match multipliers under different weights. As the α weight increases and the β weight decreases, the driver's income becomes lower and lower. The gap is more obvious.

Figure 5 shows in the performance of vehicle scheduling shown by the online car using different weights under different vehicles and passenger numbers. Take $\alpha = 0.35$, $\beta = 0.65$ and $\alpha = 0.65$, $\beta = 0.35$ to compare the loss cost when the multiplier matching number is 100, 300, 600, through simple estimation, when the number of division multiplication matches is 100, the cost difference is 115, when the number of division multiplication matches is 300, 122, and when the number of division multiplication matches is 600, the cost difference is 645. It is shown that the larger the α-weight, the worse the performance of online car-scheduling. As the number of multipliers and multipliers increases, the difference before the different α-weight becomes larger and larger. Therefore, it is more friendly to appropriately reduce the weight of passengers' fairness and experience, and increase the driver's efficiency weight for vehicle scheduling on the wired car-hailing platform.

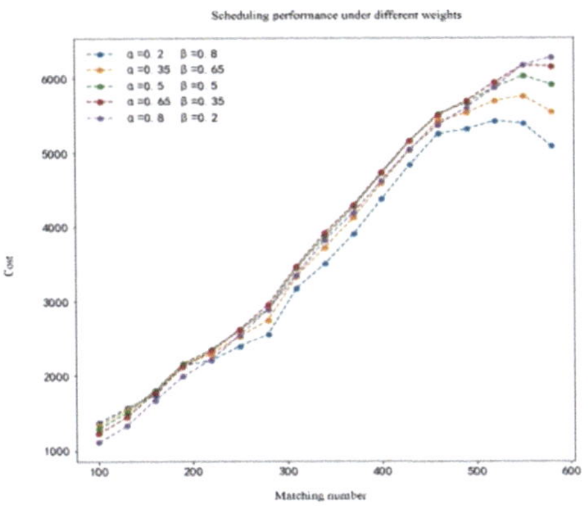

Fig. 5. Performance of the multiplier match number under different weights.

5 Conclusion

Based on comprehensive consideration of passenger experience, driver revenue and platform profit, this paper models the car-hailing scheduling problem as an assignment model and uses the Hungarian algorithm to solve it, which improves the efficiency of car-hailing scheduling and reduces the car-hailing scheduling the waste of manpower and material resources caused by unreasonable vehicle scheduling in China. Discussed the car-to-car scheduling schemes in the three situations of people and vehicles balance, more people than cars and more cars than people. It not only examines the distance between passengers and empty cars, but also allocates them optimally according to the passenger's destination distance. Compared with the greedy algorithm and the global search algorithm, the Hungarian algorithm in this paper has achieved better scheduling results, which not only save manpower and resources, but also take into account the customer experience. This assignment method is clearly superior to the case where only the distance between the void driver and the passenger is considered. After filling out the order, the driver can join the operation faster and improve the overall efficiency.

Fund Project. Hunan province key research and development plan project (2018GK2058), Hunan Provincial Natural Science Foundation of China (2020JJ4248, 2020JJ4251, 2018JJ3264), Research project of degree and graduate education reform in Hunan Province (2019JGYB242).

References

1. Zhu, R.: Ride-haiing scheduling study based on Platform Economic. Shandong University (2019)
2. Meghjani, M., Marczuk, K.: A hybrid approach to matching taxis and customers. In: IEEE Region 10 Conference (2017)
3. Dai, G.: Research and implementation of online balanced assignment algorithms for online car-hailing services. Xi'an Electronic University of Science and Technology (2018)

4. Liu, Y.W.: Optimization of urban taxi dispatch using distributed computation intelligence. South China University of Technology (2019)
5. Xie, R., Pan, W.: Intelligent taxi dispatching based on artificial fish swarm algorithm. Syst. Eng-Theory Pract. **37**(11), 2938–2947 (2017)
6. He, S.X.: Grid-based taxi dispatching method based on reinforcement learning. Appl. Res. Comput. **36**(03), 762–766 (2019)
7. Su, X.D., Sun, T., Ma, L.: Application of difference method in unequally assignment problem. Comput. Eng. **2005**(22), 188–190 (2015)
8. Fang, M.Y.: Research on unbalanced assignment problem based on global search algorithm. Math. Pract. Theory **43**(03), 184–188 (2013)
9. Chen, Z.W.: Modeling and genetic algorithm solution of UAV collaborative target assignment considering assignment order. Control Theory Appl. **36**(07), 1072–1082 (2019)
10. Ying, R.K.: Research and application of the ant colony algorithm in the assignment problem. Comput. Eng. Sci. (2008)
11. Xue, F.: Solving 0-1 integer programming problem by hybrid particle swarm optimization algorithm. Comput. Technol. Autom. **30**(01), 86–89 (2011)
12. Zhao, L.: A study of assignment problem based on mixed algorithm of simulated annealing algorithm and genetic algorithm. Logistics Sci-Tech **34**(12), 85–88 (2011)
13. Li, T., Li, Y., Qian, Y.: Improved Hungarian algorithm for assignment problems of serial-parallel systems. J. Syst. Eng. Electron. **27**(04), 858–870 (2016)
14. Laha, D., Gupta, J.N.D.: A Hungarian penalty-based construction algorithm to minimize makespan and total flow time in no-wait flow shops. Comput. Ind. Eng. **98**, 373–383 (2016)
15. Hu, M.H.: Optimization of airport slot based on improved Hungarian algorithm. Appl. Res. Comput. **36**(07), 1–7 (2019)
16. Rabbani, Q., Khan, A., Quddoos, A.: Modified Hungarian method for unbalanced assignment problem with multiple jobs. Appl. Math. Comput. **361**, 493–498 (2019)
17. Shen, J.: Appointment simulation of unattended laboratory instruments based on greedy algorithm. Comput. Simul. **36**(12), 163–167 (2019)

The Shortest Time Assignment Problem and Its Improved Algorithm

Yuncheng Wang(✉), Chunhua Zhou, and Zhenyu Zhou

PLA Strategic Support Force Information Engineering University, Zhengzhou 450000, China
wrtpzcd@sina.com

Abstract. In this paper, a minimum adjustment method is proposed to solve the shortest time assignment problem. The algorithm can meet practical needs, but the complexity is high. The paper improves the minimum adjustment algorithm, the new algorithm can take into account the requirements of the shortest time and the shortest total time simultaneously, and its solving process is efficient and straightforward. An example is used to verify the effectiveness of the algorithm.

Keywords: Operational research · Assignment problem · The shortest time limit · Minimum adjustment method

1 Introduction

The classic assignment problem is a significant optimization problem in operations research, which can be described as there are n tasks, requiring n people to complete one of them, but due to the nature of the task and the different expertise of each person, the efficiency (or cost) of each person to complete different tasks is different. The problem arises as to which person should be assigned to which task so that the total efficiency (or time required) to complete n tasks is highest. The classic assignment problem can be realized by using the Hungarian method. However, we often meet with the deadline for holding large-scale conferences and trade fairs in real life. Transfer of disaster relief supplies, emergency supplies, personnel, and first-aid medical supplies, as well as some items with a short shelf-life; Railway, mine accident rescue emergency and so on, need in the shortest possible time to realize, in order to avoid causing a more significant loss, then people often need to consider the whole process is not the least total time. However, in the shortest, it conforms to the actual needs in many cases. Therefore, the shortest time limit algorithm has more practical meaning.

The problem of the shortest time assignment was first proposed by Gross, which is to solve the problem of the shortest time for the whole process instead of the sum of time. Many scholars in China have studied the problem of the shortest assignment. The main algorithms for solving the problem of shortest time assignment include significant M method [1], stepwise optimization algorithm [2], growth tree method and label meth-od [3, 4], the method using the binary graph matching idea [5], the shortest time approximation method [6] and the method based on the idea of minimum adjustment method [7], among which the literature [8, 9] is representative. The method in literature [8] does

Q. Liu et al. (Eds.): CENet 2020, AISC 1274, pp. 583–588, 2021.
https://doi.org/10.1007/978-981-15-8462-6_65

not have the foresight for the independent circle element, so it still needs to conduct a trial assignment for the problem first. If the trial assignment does not meet the row balance and column balance, it needs to carry out correction steps.

Moreover, when the problem's size increases, more steps are needed, which is not conducive to the computer solution. Literature [9] is an algorithm for solving the shortest time assignment problem based on the Hungarian algorithm. It requires steps such as transformation matrix, trial assignment and drawing the minimum zero element coverage line, etc. The process is complicated and not conducive to computer solutions. In this paper, the optimal solution of the shortest time assignment problem is obtained by calculating each column's minimum value and drawing a circle with the methods of direct adjustment and indirect adjustment.

2 Build the Shortest Time Assignment Problem Model

Suppose n people are assigned to do n tasks. Each person is assigned to do one task. Each task is assigned to one person. Given the efficiency (such as the time required) of each person to complete each task, it is given by the matrix $c = (c_{ij})_{n \times n}$, which is called the relational matrix or the time rate matrix. Ask how to assign tasks so that they can be completed in the shortest possible time. It is easy to see that the above problems can be summarized into the following planning model:

$$\begin{cases} \min f = \max\{c_j x_{ij},\ 1 \le i \le n, 1 \le j \le n\} \\ \text{st. } \sum_{i=1}^{n} x_{ij} = 1 \quad (j = 1, 2, \ldots, n) & (1) \\ \sum_{j=1}^{n} x_{ij} = 1 \quad (i = 1, 2, \ldots, n) & (2) \\ x_{ij} = 0 \ \ or \ \ 1 & (3) \end{cases}$$

$$x_{ij} = \begin{cases} 1, & \text{when the ith person is assigned to do the jth job} \\ 0, & \text{when the ith person is not assigned to do the jth job} \end{cases}$$

3 Model Solution and Improved Algorithm

3.1 Theorem [4]

Any subset of the optimal assignment set is the optimal assignment of their intersecting submatrices.

Proof: By means of contradiction, the original optimal assignment set is not the optimal assignment of the problem if a subset of the optimal assignment set is not the optimal assignment of their intersection matrix. Discuss in two cases: (1) If the subset contains all C_0, but there is an assignment in the intersection matrix, and its maximum element is less than C_0, then this assignment can be used to replace the original assignment, and a better solution of the objective function formula can be obtained.

Therefore, the original optimal assignment is not the optimal assignment. (2) If there is an assignment in the intersecting submatrix, and its sum is less than the sum of the original assignment, then we can get a better solution of the objective function by replacing the original assignment with this assignment. Therefore, the original optimal assignment is not the optimal assignment.

3.2 Solution Model

The method of minimum adjustment is used to solve the problem of shortest time assignment. The determination of the minimum scheme and the idea of adjustment according to the principle of minimum increment remain unchanged. However, the calculation of function value and increment is different due to the difference of objective function. The specific steps are as follows:

Step 1: take the minimum value of each column and draw a circle, and assign work according to this circle element, which is called the minimum scheme, denoted as X_0, that is, the variable $X_{kj}^{(0)} = 1$ corresponding to the C_{ij} of the circle element, and the remaining $X_{ij}^{(0)} = 0$.

Step 2: check the feasibility. The minimum solution X_0 clearly meets the column equilibrium condition (1) and becomes the optimal assignment scheme if the row equilibrium condition (2) is also met. At this point, the optimal objective function value of the model is:

$$\min f = \max\{c_{ij}x_{ij}, 1 \le i \le n, 1 \le j \le n\} = \max\{c_{kj}\} \tag{4}$$

Otherwise, it is an unfeasible scheme, which should be adjusted to meet the line balance condition (2). If the number of circle elements is greater than 1, it is called out line. If it's equal to 1, it's called the equilibrium row; If it is equal to 0, it is called a call-in line, and step 3 is performed for the minimum plan X_0.

Step 3: begin to adjust, remove the line out of a circle, then add a circle in the same column elsewhere (always maintain equilibrium conditions (1) set up), if the increase of the circle line in the balance, the balance line of the original circle should be removed and change elsewhere within the same column, line until a call in one element in the painting circle, is an adjustment. Do so until the uncalled action is complete. Each adjustment is based on the principle that the increment of the objective function (whose value is the largest of all looping elements) is the minimum, which is called the adjustment difference. The details are as follows:

First, confirm the objective function value of the minimum scheme X_0, namely the one with the largest value among all drawing elements, and write it as $C_{i_0j_0} = \max\{C_{kj}\}$ Check whether the minimum scheme X_0 has balance lines. If not, go to (1).

Direct Adjustment. Calculate the minimum direct difference $h_1^{(i)}$ and set $h_{sj} = \max\{C_{sj} - C_{i_0j_0}, 0\}$, where S is the call-in row and j is the column of C_{kj}, then $h_1^{(i)} = \min\{h_{sj}\}$. If $h_{(1i)} = h_{s_1j_1}$, it means that the adjustment is to transfer the circle of the drawing element C_{kj} of the calling out line to the element C_s at the column j_1 of the calling in line S_1. Let the scheme after adjustment be X_1 to (2).

Indirect Adjustment. If have balance, besides considering the minimum differential directly $h_{(1i)}$, calculate indirectly $h_{(1ii)}$, and bring up the lines of a circle to balance line of certain elements within the same column, then the balance of the circle turn out line to keep balance, finally until line to transfer into a particular element, thus the so-called indirect adjustment routes and corresponding indirectly is poor, with minimizing the objective function value. If the route is adjusted, it can be expressed as:

$$c_{k_0j_0} \rightarrow c_{s_1j_0} \rightarrow c_{k_1j_1} \rightarrow \ldots \rightarrow c_{k_tj_t} \rightarrow c_{s_{t+1}j_t} \tag{5}$$

The corresponding indirect adjustment is:

$$h_{s_{t+1}j_t} = \max\{c_{s_{p+1}j_p} - c_{i_0j_0}, 0;\ p = 0, 1, \ldots, t\} \tag{6}$$

The corresponding minimum indirect difference:

$$h_{(1ii)} = \min\{h_{s_{t+1}j_t}\} \tag{7}$$

Where $c_{i_0j_0}$ is the maximum value of the circle element in the minimum scheme X_0, k_1, k_2 ..., K_t is the equilibrium line, $c_{k_pj_p}$ (p = 1, 2 ..., t) is the circle element of each balance line, and St + 1 is the call-in line. Compare $h_{(1i)}$ and $h_{(1ii)}$ and take the smaller one as adjustment difference h1, and adjust it according to the corresponding adjustment route.

Step 4: let the scheme be X_1 after adjustment, and let $c_{i_1j_1}$ be the one who draws the maximum value of circle element in X1 after one-step adjustment, and $c_{i_1j_1}$ be the one who draws the maximum value of circle element in Xt after t-step adjustment. Return 2, repeat the adjustment process of the minimum scheme X0, then the feasible scheme is the optimal scheme.

3.3 Improvement Algorithm

To solve the above problems, the algorithm is improved with the following idea: first, find the shortest time limit assigned, that is, the optimal solution f of the target, then there must be n independent elements in the matrix C that are not greater than f, so as long as their sum is minimum. The steps are as follows:

Step 1: find the shortest time, find the minimum values of each row and each column in the time matrix, and record the maximum value as F_1.
Step 2: the elements in the time matrix that are not greater than F_1 are denoted as 0 and denoted as the matrix $C(F_1)$.
Step 3: if the zero element in a row (column) in the matrix $C(F_1)$ is both row and column independent, select it and cross out the row and column it is in. If there are multiple independent zeros in a column (row) in $C(F_1)$, find the subminor elements in their rows (columns), select the independent zeros in the row (column) with the largest subminor value, and cross out the row and column in which it is located.

In more than one of these cases, the independent zeros in the row or column with the largest subminor element are assigned first. If all the rows (columns) in $C(F_1)$ have no independent zero element, one of the rows (columns) with the least zero element is selected arbitrarily (most comity few).

Step 4: if n independent zero elements are selected, their corresponding variables are set to 1 and the other variables are set to 0, thus obtaining the optimal assignment solution. Otherwise, go to step 5.

Step 5: take $F_2 = \min\{c_{ij} | c_{ij} > F_1, \text{i,j} = 1, 2, \ldots, n\}$, use F_2 instead of F_1, and go to step 2.

4 Example of Improved Algorithm

Given that the time matrix of a multi-objective shortest time assignment problem is:

$$C = \begin{pmatrix} 19 & 26 & 13 & 11 \\ 10 & 8 & 17 & 14 \\ 28 & 18 & 15 & 11 \\ 14 & 11 & 10 & 8 \end{pmatrix} \tag{8}$$

$F_1 = 11$ is obtained from step 1, and C is obtained by changing the element no larger than $C(F_1)$ in C to 0.

$$C(F_1) = \begin{bmatrix} 19 & 26 & 13 & 0 \\ 0 & 0 & 17 & 14 \\ 28 & 18 & 15 & 0 \\ 14 & 0 & 0 & 0 \end{bmatrix} \tag{9}$$

Select 0 according to the requirements in step 3, and the results are as follows:

$$C(F_1) = \begin{bmatrix} 19 & 26 & 13 & 0 \\ 0 & 0 & 17 & 14 \\ 28 & 18 & 15 & 0 \\ 14 & 0 & 0 & 0 \end{bmatrix} \tag{10}$$

Since the number of 0 selected is 3, which does not meet the requirements in step 4, $F_2 = 13$ is selected again.

$$C(F_2) = \begin{bmatrix} 19 & 26 & 0 & 0 \\ 0 & 0 & 17 & 14 \\ 28 & 18 & 15 & 0 \\ 14 & 0 & 0 & 0 \end{bmatrix} \rightarrow C(F_2) = \begin{bmatrix} 19 & 26 & 0 & 0 \\ 0 & 0 & 17 & 14 \\ 28 & 18 & 15 & 0 \\ 14 & 0 & 0 & 0 \end{bmatrix} \rightarrow C(F_2) = \begin{bmatrix} 0 & 0 & 1 & 0 \\ 1 & 0 & 0 & 0 \\ 0 & 0 & 0 & 1 \\ 0 & 1 & 0 & 0 \end{bmatrix} \tag{11}$$

It can be seen that the above method finds the optimal solution to the shortest time assignment problem. Compared with the traditional method, this algorithm has relatively simple steps to solve the problem, faster speed and less computation. To obtain the shortest time and at the same time require the minimum total time, it only needs to use the Hungarian method learned in operations research to continue to solve the problem after obtaining the matrix $C(F_1)$, which has good adaptability.

5 Summary

Assignment problem is an important knowledge point in the chapter of integer programming in operations research. Based on the textbook content, this paper proposes the assignment problem of the shortest time, which is more general for solving practical problems. Secondly, the traditional method of solving the problem of shortest time assignment is improved, and a more efficient and succinct solution is put forward. Besides, it can meet the requirements of the shortest total time at the same time, and has good adaptability and generality.

References

1. Gross, O.: The bottleneck assignment problem. The Rand Corporation, Santa Monica (1959)
2. Shaugang, X.F.: The shortest time assignment problem is solved based on the minimum adjustment method. Pract. Underst. Math. **39**(17), 179–187 (2009)
3. Xu, C.: Assignment problem with time factor. J. Qingdao Univ. (Nat. Sci. Edn.) **13**(2) (2000)
4. Huang, Z., Ding, G.H., Guo, D.W.: A stepwise optimization algorithm for the assignment problem of the shortest time limit. China Sci. Technol. Paper **11**(5) (2016)
5. Li, Z.P., Wang, L.: A solution to the problem of shortest time default assignment. Oper. Manage. **9**(2) (2000)
6. Ren, D., Lu, G.Z.: A method of solving the problem of shortest time and least cost assignment. Autom. Instr. **20**(3) (2005)
7. Han, Y.N., Guo, D.W.: Optimization algorithm for multi-objective linear non-differentiable assignment problem. J. Inner Mongolia Norm. Univ. (Chin. Version Nat. Sci.) **46**(3), 325–328, 333 (2017)
8. Ahuja, R.K., Magnanti, T.L., Orlin, J.B.: Network Flows. Prentice Hall Press, New Jersy (1993)
9. Xiong, G., Hu, Y.W., Shi, J.M.: A new algorithm for solving the minimum cost flow based on dual principle. J. Shandong Univ. (Sci. Edn.) **46**(6) (2011)

Injection Simulation Equipment Interface Transforming from PCIe to SFP Based on FPGA

Jing-ying Hu(✉) and Dong-cheng Chen

Jiangsu Automation Research Institute, Lianyungang 222006, China
hujingying3@163.com

Abstract. To resolve the difficulty of sampling real-time data when debugging algorithm on digital signal processing equipment, a board based on FPGA which can transmit data from PCIe to fiber is designed. Computer sends data to DDR3 on the designed board through PCIe. The ARM in FPGA sends data to the PL of the chip through DMA. The PL sends data to digital signal processing equipment through fiber after ping-pang operation in the PL. The digital signal processing equipment exchanges data through fiber with the developed board. The digital signal processing equipment can use the data receiving from the board developing the processing algorithm. The experiment shows that the designed board can meet the simulation requirement when developing the digital signal processing algorithm and can improve the developing efficiency.

Keywords: ZYNQ · PCIe · ARM · SFP

1 Introduction

Real-time signal processing is usually used on equipment which has special sensors like radar and sonar. It is impossible for the signal processing equipment to sample data using the sensor every time, for it will cost very much if sampling data using real sonar or radar. In order to solve the problem, situation simulation is always used. But in some situation simulation, expensive equipment is needed, and the simulation data can't reflect the equipment working situation. Then the algorithm can't be proved well. Digital signal data injection simulation is a very effective solution for the digital signal processing equipment simulation. The equipment injects data into the simulated digital signal processing equipment through some special channel. The digital signal processing equipment can process the working state data in the lab. And the simulation can be repeated as will, which can prove the processing algorithm stability.

As with the development of the communication technology, high speed differential signal bus is the developing direction. The PCIe (Peripheral Component Interconnect express) was first released in 2001. It is developed by Intel based on the third IO bus standard. PCIe bus is a serial bus technology. It can send data by very high frequency, so it can ensure the integrity and the high band width of the communication. The first generation PCIe can transmitting data at 2.5 Gbps us single lane. The second generation can reach 5.0 Gbps per lane, and the third generation can reach 8.0 Gbps per lane.

Q. Liu et al. (Eds.): CENet 2020, AISC 1274, pp. 589–595, 2021.
https://doi.org/10.1007/978-981-15-8462-6_66

The fiber is also used in high speed differential signal communication. The send data rate can reach 10 Gbps per lane using the FPGA. And the fiber communication is a high speed, long distance and high safety communication form and widely used in many fields.

In this paper, an injection simulation equipment by transmitting PCIe data to fiber based on FPGA is designed. The computer writes data to the DDR3 memory on the FPGA board through PCIe. The FPGA move the data from DDR3 to the GTP, then send to the digital signal processing equipment via fiber. The GTP send data rate sets to 4.0 Gbps, then the effective data rate will be 400 MB/s. It can meet bandwidth requirement of digital signal processing equipment.

2 System Design

There two ways to make the system design of PCIe interface. First is using bridge chip, convert the PCIe bus to local bus, just like most commercial computer. Because of the low speed local bus, this converting type has relatively low sending speed usually. The other type is making it using FPGA. Many FPGA integrate hard PCIe IP core in it. Which can be used to develop PCIe based equipment quickly.

In this paper, the designed device based on FPGA makes of central processing FPGA, the DDR3 data buffer unit, fiber optical module, PCIe interface, data transmitting computer and digital signal processing equipment side FPGA. The system designed is shown in Fig. 1.

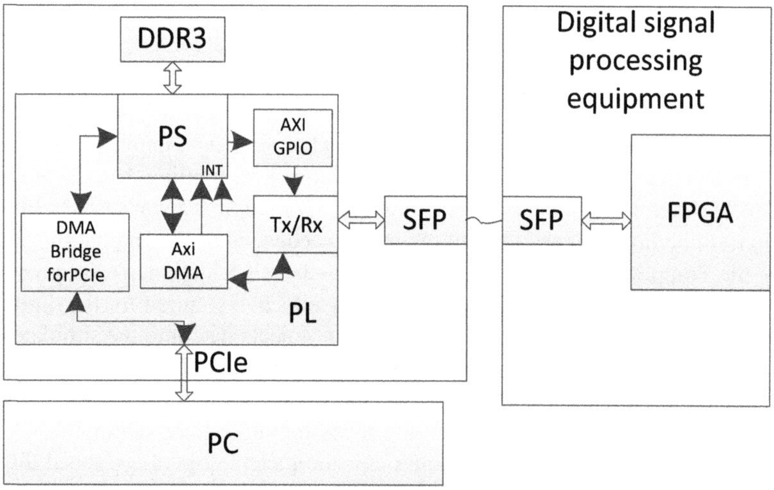

Fig. 1. System function.

The central processing FPGA is the XC7Z015-CLG485 chip, which is a ZYNQ7000 series SoC FPGA by Xilinx. ARM + FPGA SOC technology is used by the chip. The FPGA logic and a dual core ARM Cortex-A9 are integrated on a single chip. AMBA® interconnect, internal memory, memory interface, and external device are all integrated on the chip. The chip is made of PS and PL. PS is a dual core ARM Cortex-A9 processor, which is ARM-v7 architecture, can run at 766 MHz. Each CPU core has 32 KB 1st level instruction and data buffer and 512 KB 2nd level CPU shared buffer. The external memory interface support 16/32 bit DDR2 and DDR3. The PL has lots of logic resources. There are 4 high speed GTP transceivers, which can at most support PCIe Gen2 X4. For we use PCIe and SFP fiber in the design, the PCIe Gen2 X2 using 2 GTP high speed transceivers and 1 GTP high speed transceiver as fiber is deigned.

The DDR3 memory is attached on the PS side of the ZYNQ chip. It is accessed by the ARM core directly. In this system, the computer write the data to DDR3 memory through PCIe, the ARM core access the data and send it to PL side through DMA. Two DDR3 chips are attached to the chip, the data width is 32, the clock rate is 533 MHz, then the data accessing rate can reach 1066 MHz. If the accessing efficiency of the DDR3 is 70%, the effective data accessing band width is 2.9 GB/s. It can meet the requirement of the digital signal processing device injection simulation system.

As there is only one GTP bank with two sets of GTP bank on the adopted ZYNQ7000. In this design, the PCIe and SFP fiber interface must share the only GTP bank. The bank has two differential reference clock input ports, the PCIe need 100 MHz reference clock and SFP needs 125 MHz reference clock. In this design, for there is only one GT_COMMON module in the GTP bank, and the GT_COMMON module has two QPLL output, so the PCIe and SFP must share the GT_COMMON module. The PLL0OUT must connect to the PLL0 of the GT_CHANNEL, the PLL1OUT must connect to the PLL1 of the GT_CHANNEL, then the PCIe and the SFP can share the same GTP bank.

As the central processing unit, ZYNQ do all the data processing task of the system. The computer sends data through PCIe interface using DMA to the PS memory. Write data to the DDR3 of the ZYNQ. The PCIe core is a hard core on the chip, the logic used is XDMA core, which includes PCIe interface core and DMA core. It is used for sending and receiving dada between the computer and the board. The data sending engine organizes data to TLP data in PCIe protocol type, and sends to the PCIe core. The sending DMA module transmits data reading from DDR3 to TLP type and sends data to the sending engine. Receiving engine receives different tyoe of TLP data from PCIe core. The receiving DMA module is used to implement sending flow control for memory request packet and receiving data width transmitting. The DMA state control register module is mainly used for analyzing the register command and setting from the computer. And announce the sending engine, receiving engine and other modules to response to the notice.

The PS read data from DDR3 to PL using AXI_DMA module. The computer writes the data to a given address of the DDR3. When the PL interrupts the PS for data request, the PS starts the DMA module to send data. The AXI_DMA module reads data from DDR3 and sends to PL side. The PL side analyzes data using axi protocol and writes to FIFO waiting for writing to the next data buffer. When the AXI_DMA module finished sending data the core interrupts the PS.

The AXI_GPIO is used for sending state from PS to PL. When the computer writes data to the DDR3 on board for resetting or the PS writes data after finishing DMA to the PL, the PS writes resetting state or finishing DMA to PL through AXI_GPIO core. The PL will do some operation when receiving the GPIO state. And the PL can also writes state to the PS side.

Two AXI high performance AXI port with 64 bit data width is enabled. The port is used by the AXI_DMA core and XDMA core. In this system, the data clock rate is 200 MHz, then the data bandwidth can reach 1600 MB/s. The high performance AXI port provides data channel for PL to access DDR3 and OCM (on chip memory). Each port includes two FIFO buffers, which are used for reading and writing data stream. A general purpose AXI master port is enabled for PS to write setting data to AXI_GPIO core, AXI_DMA core and other AXI cores. Three PS external interrupts are enabled, one for AXI_DMA finish data transmitting and the other two for data transmitting beginning and data requesting. Data transmitting beginning and data requesting interrupt the PS using a rising edge, when receiving the interrupt from PL, the PS starts DMA to write data to PL side. The clock of every AXI module is 200 MHz provided by the PS.

The SFP module at the PL side sends data using two buffer ping pong operation. In this system, the data rate is 104 KB/5.12 ms, so two 128 KB dual ports ram is used. The input side of dual port ram is connected to the AXI data receiving FIFO output side. Every 104 KB data will write to one of the two dual port ram. At the data transmitting side, the program reads data from the dual port ram and sends to the digital signal processing system at the required rhythm. The data transmitting module and the digital signal processing equipment exchange data through the fiber cable. The digital signal processing equipment sends data requirement when need data, this system starts the data transmitting state machine. Packet head is added to every data packet, which is used by the digital signal processing equipment for data analyzing and arraying. The data transmitting state machine is shown in Fig. 2.

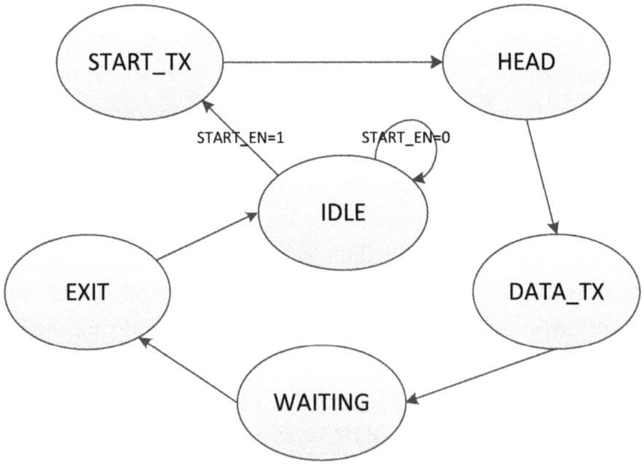

Fig. 2. Data transmitting state machine.

3 Software Design

In this paper, the injection simulation equipment key technic is the design of hardware logic and the design of software in the ARM core. The hardware logic was introduced in system design. The central control and data exchange is implemented by the ARM core. The ARM on the chip is a dual core Cortex-A9 processor. The main working frequency is 667 MHz. In this system, only one processing core is used. The core dose all the initialization, data moving control and respond to all the interrupts. After all the initialization, the program starts a DMA to move data from DDR3 to PL, then the program waits for interrupts from the PL side. In the processing flow, the first interrupt will be the first 104 KB data finishing interrupt from the AXI_DMA core. Then the ARM will wait for data starting or data requesting interrupt. When the data starting or data requesting interrupt comes, the ARM starts another data DMA. For example, in this design, 104 KB data will be transmitted by the DMA. The DMA moves data through multi-descriptor. Each descriptor has as more as 8233607 bytes data buffer, but for the logic resources is limited, the AXI data analyzing module has only 4 KB buffer FIFO, and 104 KB data will be move for a DMA. So in the program, the 104 KB data is divided into 52 multi-descriptors. Each descriptor move 2 KB data to the PL side. The data buffer in the PL side will not be full, and meet the bandwidth requirement of the simulation system.

After all the initialization, the program will wait for the computer operation and PL interrupts. When the initialization button is pushed on the computer, the computer writes data mark to the restricted address in the DDR3 through PCIe. Then the computer choose the transmitted data file, after choosing file the arm starts a DMA to transmit data to one of the two dual port rams on the PL side. When the start button pushed, the ARM will wait for the DDR3 in some restricted address being set to special value. The program flow is shown in Fig. 3.

4 Experiment and Discuss

The designed the injection simulation equipment communicates with the computer though PCIe and communicates with the digital signal processing equipment through fiber cable. The main function is implemented in the FPGA. The result verifying will influence the system designing efficiency. The Vivado provides very good solution for hardware logic design progress. The integrated logic analyzer IP core can view the wave of any signal. The sampling depth can be set as will. It is a very good tool for hardware logic designing.

In the system design, the PS control the DMA to read data from DDR3 to PL side, and the PL side communicate with digital signal processing equipment through fiber cable is the key data needed monitored. The Fig. 4 shows the DMA writes data to dual port ram in the PL side. When the data valid signal is high, the data is wrote to the ram. The clock frequency of DMA module writes data to AXI data analyzing module is 200 MHz, the module buffers data in the FIFO, at the same time, the PL read data from FIFO to dual port ram for ping pong operation. When a batch data wrote to the ram, after a batch data finishing transmitting through fiber cable, the ping pong mark will reverse and the PL will interrupt the ARM for data requesting. The new data wrote to the PL will be wrote to the other dual port ram.

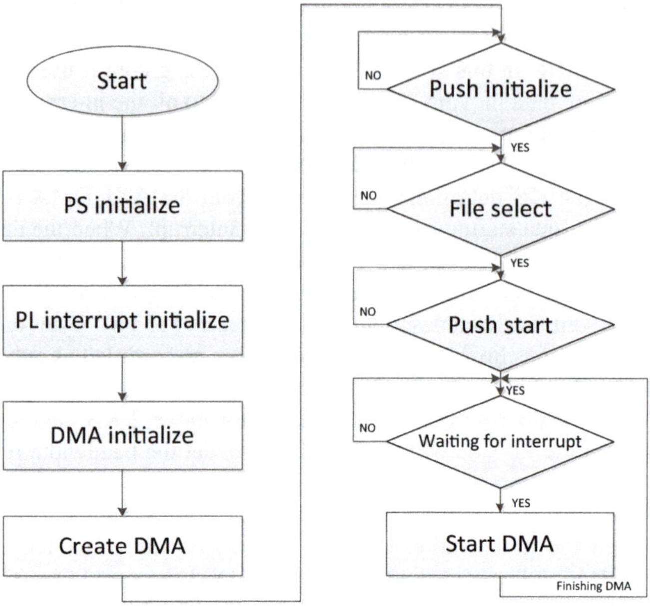

Fig. 3. Software flow.

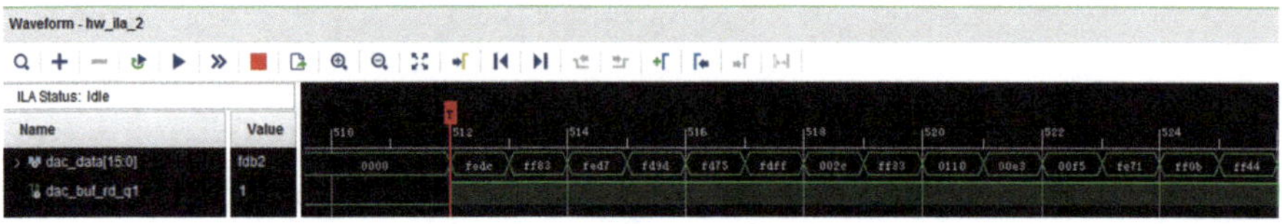

Fig. 4. Waveform of DMA writing data to PL side.

Figure 5 shows the fiber module transmitting and receiving data waveform. In the figure, when receiving data transmitting enable signal from the digital signal processing equipment, the data_request_i will be set to high. When the PS finishes transmitting a batch data to PL side, the PS writes a state to the PL through GPIO core. It is the Start_Tx in this design. When the GPIO revered, the Start_Tx will be set high. Only the Start_Tx signal setting high can trigger the data transmitting state machine starting and reading data from the dual port ram.

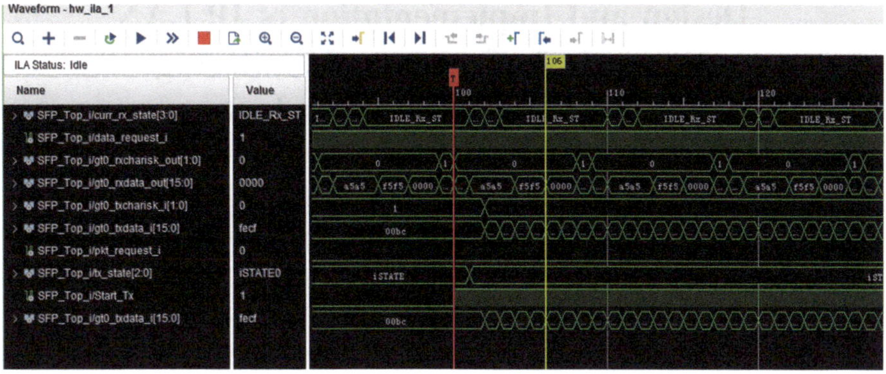

Fig. 5. Communication between the PL and digital signal processing equipment.

5 Conclusion

An injection simulation equipment transforming data from PCIe to SFP based on FPGA is designed. The ZYNQ 7000 FPGA is used as the central processing unit. The computer writes data to the DDR3 memory of the FPGA board. The PS side on the FPGA transmits data to the PL side using DMA. Ping pong operation is implemented using two dual port ram. The board communicates data with the digital signal processing equipment using the fiber cable at the required rhythm. Then the digital signal processing equipment can work at the real situation. Both data and rhythm is the same as reality. It is very meaningful for the digital signal processing equipment flow and algorithm debugging. The digital signal processing equipment can do the real job in the lab, which can reduce the experiment cost and equipment developing cycle.

References

1. Fan, B., Wang, Y.J., Sun, H.: High speed image injected simulator based on PCIe bus. Comput. Eng. Des. **035**(003), 1056–1060+1099 (2014)
2. Chen, D., Hu, J., Lv, W., Zeng, F.: Design and implementation of muti-interface LCD controller based on IP cores. Chin. J. Liquid Crystals Displays **23**(2), 41–47 (2017)
3. Fan, B.: Study on the Key Technology of High-Speed Image Acquisition and Processing Based on FPGA. University of Chinese Academy of Sciences, Beijing (2013)
4. Crockett, L.H., Elliot, R.A., Enderwitz, M.A., Stewart, R.W.: The Zynq Book: Embedded Processing with the Arm Cortex-A9 on the Xilinx Zynq-7000 All Programmable SoC. Strathclyde Academic Media, Glasgow (2014)

Design and Implementation of IP LAN Network Performance Monitoring and Comprehensive Analysis System

Yili Qin[1(✉)], Qing Xie[2], and Yong Cui[1]

[1] China Satellite Maritime Tracking and Control Department, Jiangyin 214400, Jiangsu, China
tougaoqin@163.com
[2] Space Engineering University, Huairou, Beijing 101400, China

Abstract. This paper designs and implements a LAN-based IP network performance monitoring and comprehensive analysis system. Aiming at the specific network architecture and environment of LAN, the system has three functional modules: network environment monitoring, service flow data capture analysis, and network failure analysis. When network failure occurs, the system can analyze the network status comprehensively to implement the initial judgment and positioning of the fault, which provides a convenient network management method for network maintenance and management personnel.

Keywords: IP · LAN · Performance monitoring · Comprehensive analysis

1 Introduction

With the development of IP-based technology, the application of network technology in aerospace measurement experiments is deepening. Especially in the sea, the real-time data transmission of measurement vessels performing related telemetering and control services through satellite links and ground stations has multiplied rate and increased amount. The network monitoring requirements for network environment monitoring, service flow performance analysis and network fault location is becoming increasingly urgent and important. This paper designs and implements a LAN-based IP network performance monitoring and comprehensive analysis system. According to the specific network architecture and environment of LAN, the real-time monitoring of the business data can be monitored, the network failure can be judged and located, which provides a very convenient network management means for network maintenance managers [1, 2].

2 Requirements Analysis

2.1 Situation Analysis

There are three types of network monitoring methods currently used in local area networks:

Q. Liu et al. (Eds.): CENet 2020, AISC 1274, pp. 596–603, 2021.
https://doi.org/10.1007/978-981-15-8462-6_67

The first is Ping. Ping commands are used to perform connectivity tests on the access devices in the LAN to obtain information such as network connectivity status, transmission delay, and packet loss rate between devices.

The second is the network topology status query system. Generally, this kind of system uses SNMP protocol to obtain the interface status information of network devices, thereby determining the network topology status. However, in order to improve compatibility, most systems use polling to obtain MIB information of SNMP protocol, resulting in slow real-time update of network topology status and a lack of timeliness.

The third is packet capture software. Such as Wireshark and Ominpeak, the more mature packet capture software on the market. The advantage of such software is that it can capture and display the data of network monitoring points in real time, comprehensively and quickly. The Disadvantage is that it cannot directly parse the data of the dedicated application protocol. It cannot analyze, process and judge the captured information for the service flow data, and cannot directly filter and display the information required by network maintenance managers.

2.2 Functional Requirements

In addition to the above three types of monitoring functions, the IP network performance monitoring and comprehensive analysis system needs to implement the following functions to meet the network management needs of LAN.

Network Environment Monitoring. The system has the function of adding the IP address of the device, and obtains the network connectivity status of the device through Ping commands; it has the ability to log in to the network switching device through Telnet, and use the switch configuration commands to obtain the consistent result of device configuration parameter comparison and the network performance status of the interconnected interface, such as optical power loss.

Monitoring and Analysis of Service Flow. The network structure used in LAN is a "routing-convergence-access-terminal" four-layer model, as shown in Fig. 1. The system uses distributed deployment background packet capture software which captures packets at the monitoring points of each layer and stores the acquired special protocol packet information in the system database after parsing. The database uses the parsed packet field information to perform quantity statistics, rate calculation, packet loss and disorder judgment, and displays the results on the service flow data monitoring interface.

Network Fault Analysis. Based on network environment monitoring and service flow monitoring analysis, a procedure for fault judgment and location is designed, and a preliminary analysis of network faults in the LAN is performed.

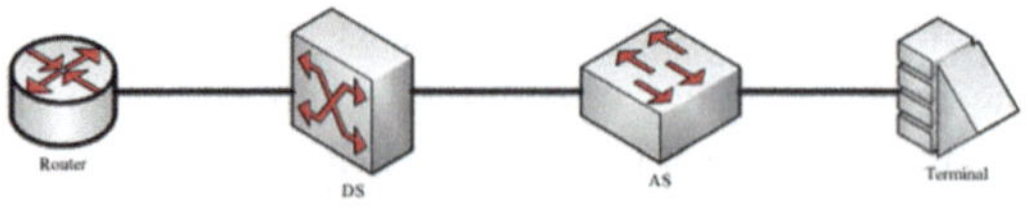

Fig. 1. LAN network structure mode.

3 Design and Implementation of Functional Modules

3.1 Overall Architecture

The system development uses Visual Studio 2010 and SQL 2008 database. Integrating ASP .NET, AJAX 1.0, etc., VS [3] can also efficiently develop Web applications. SQL Server 2008 was released on the Microsoft Data Platform and can store data for structured, semi-structured, and unstructured documents directly into the database [4].

The overall architecture of the system is divided into three major parts and four functional modules as shown in Fig. 2.

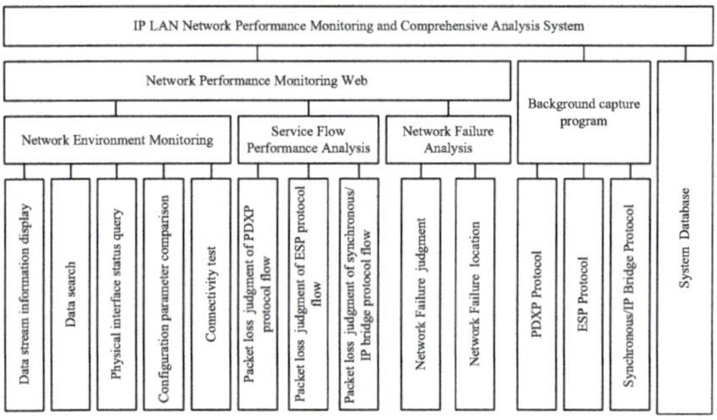

Fig. 2. Overall system architecture.

(1) The part of background packet capture program is deployed in each monitoring node within the LAN and can parse the data protocol of different service flow.
(2) The part of network performance monitoring is used to display the various performance status of LAN, as well as information statistics of packet data, packet loss judgment, and network failure judgment and positioning.
(3) The part of System database completes data collection and storage.

3.2 Network Monitoring Module

Data Stream Information Display. Establish communication with the database, displays the background packet capture information stored in the database to the front-end Web interface in real time, and set the time for real time refresh. According to the same source IP address, destination IP address, and related data identification fields, the home page displays the total packet information of the three protocol data streams. The pages of different capture monitoring nodes display the full state information of the data.

Data Search. The system provides a search function for packet capture data information. It can distinguish data protocols, source IP addresses, destination IP addresses, packet sequence numbers, and packet capture times for classified searches. For PDXP protocol data, the relevant data identification fields can be distinguished for searching.

Physical Interface Status Query. Write a Telnet-type program, define the Telnet communication protocol, and use sockets to establish Telnet communication with remote devices. The optical power loss value of the interface of the network switching device is obtained and whether it meets the index requirements can be judged by entering the "display transceiver interface" command in the program.

Configuration Parameter Comparison. Remote login to the network switching device added to the system database through the Telnet category. Enter the "compare current-configuration" command in the program to get the comparison result of the network switching device configuration.

Connectivity Test. Call the Ping category to implement a Ping test on the terminal device added to the system database, and display the number of successful, failed as well as loopback delay results of the Ping test in real time.

3.3 Service Flow Performance Analysis Module

The main function of the service flow performance analysis module is to use the packet sequence number fields in the packet headers of the three data protocols to perform packet loss and disorder judgments.

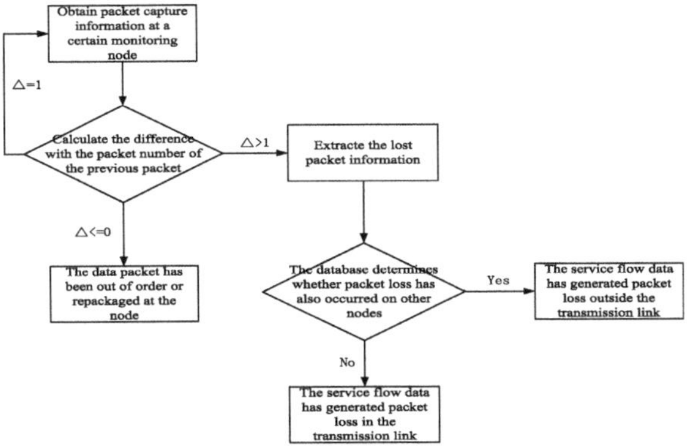

Fig. 3. Service flow performance analysis process.

Use the background packet capture program to obtain packet capture information at a monitoring node. For the same service flow data, each time a packet is captured, the difference between the packet sequence number of the previous packet and the previous packet is calculated. If the difference $\Delta = 1$, it means that no packet loss has occurred; if $\Delta < 0$, it means that the data packet has been out of order at this node; if $\Delta = 0$, it means that the data packet has repacked at this node; if $\Delta > 1$, the relevant packet header fields of the packet loss are extracted from the database, including the data identifier, packet sequence number, etc. Based on this information, search the packet capture information of other monitoring nodes in the database to determine whether packet loss has also occurred on other nodes. If yes, it means that the service flow data has lost packets outside the transmission link. If not, it indicates that packet loss has occurred inside the transmission link. The specific analysis and judgment process are shown in Fig. 3.

3.4 Network Failure Analysis Module

The direct manifestation of network failure is packet loss and disorder in the service data transmission. According to the cause, network failures can be divided into: device connectivity failure caused by abnormal physical interface status; network link failure caused by incorrect routing configuration of network switching equipment; network congestion failure caused by sudden service data traffic. If all status checks of the network are normal, and there are still abnormalities such as packet loss of service data, it can be determined that the data sending source or receiving end is faulty.

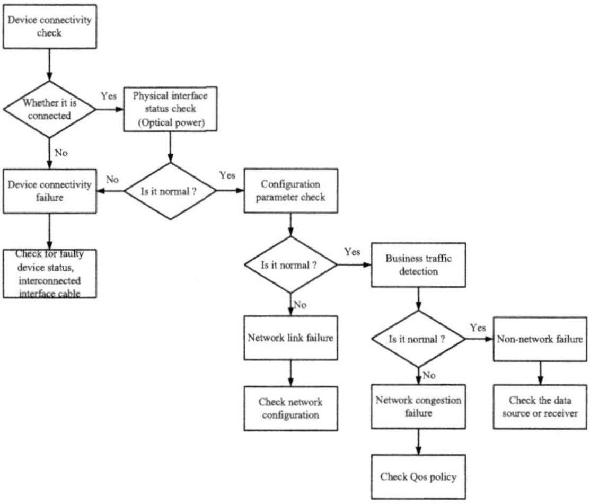

Fig. 4. Network fault judgment and location process.

The system detects service data packet loss through the service flow performance analysis module, and compares the packet loss information in the database to determine the initial location of the packet loss. Use the network environment monitoring module to conduct a performance test on the device/terminal at the initial location of the packet loss to determine the fault type and locate the fault location initially. The specific judgment process is shown in Fig. 4.

3.5 Background Capture Program Modules

Capturing Function. Based on Socket communication technology, the background packet capture program is designed. Applications usually send requests to a network or respond to network requests through "sockets". Socket is used to establish a network connection. When the connection is successful, a socket instance [6] will be generated at both ends of the application. The required functions can be completed by operating this instance.

A socket instance is created in the background packet capture program, and the network adapter address of the terminal where the program is located is obtained through the ManagementObjectSearcher collection. The SqlConnection connection string is used to establish communication with the system database, design the DataArrived function, and store the packet information on the specified network adapter to the database [7, 8].

Data Protocol Analysis Function. At present, there are mainly three kinds of service flow data protocols in LAN, which are PDXP protocol, ESP protocol and Synchronous/IP bridge protocol. According to different transmission ports adopted by the data application protocol packet at the transport layer, the three protocols are distinguished in the packet capture program. At the same time, according to the header field information of the three protocols, the buffer interval and format of the captured data are designed to realize the parse of data packets.

4 Deployment and Application

4.1 System Deployment

The system is divided into three parts to deploy:

(1) Background packet capture program. Compile by the Visual Studio 2010 development platform to generate an executable installer. The system needs a background packet capture program to capture IP data flowing through routers, aggregation switches, access switches, and service flow terminals. Therefore, a client that installs a background packet capture program needs to be accessed under the router, the aggregation switch, and the access switch. After the installation is complete, the node name, system database IP address, username and password should be set up.
(2) System database. The installation is deployed on a database server. The data table contained in the database is shown in Table 1.
(3) System Web site. The installation is deployed on the web server and published in the LAN through IIS.

Table 1. Database table design.

Table name	Table contents
PdxpHead	PDXP datagram header information
PdxpPacketSta	PDXP data stream information statistics
PdxpPacketSave	PDXP packet information statistics
EspHead	ESP datagram header information
EspPacketSta	ESP data stream information statistics
EspPacketSave	ESP packet information statistics
BrigeHead	Synchronous/IP bridge datagram header information
BrigePacketSta	Synchronous/IP bridge data flow information statistics
BrigePacketSave	Synchronous/IP bridge packet information statistics
PingTerminal	Ping terminal information statistics
TelnetTerminal	Telnet terminal information statistics

4.2 System Application

The main web page of the system mainly includes the statistics of the total number of service flow packets and the display of packet loss information of the four monitoring nodes. Click the picture of each node in the topology map to get the full state information display of all the packets captured by each node. The configuration parameter comparison page displays the configuration comparison result of the network switching device. The Ping test page displays the network terminal connectivity status results.

5 Conclusion

Based on the actual LAN environment, an IP network performance monitoring and comprehensive analysis system is designed. The system has functions such as network environment monitoring, service flow data capture analysis, network failure analysis, etc., which can monitor the performance of the LAN. When a network failure occurs, the system can comprehensively analyze the network status to achieve preliminary judgment and location of the failure. A serious of space test tasks shows that the system can help network administrators to monitor network status and analyze faults in real time and quickly. Subsequent research will focuses on how to improve the scalability of the system.

References

1. Das, R.F., Tuna, G.S.: Packet tracing and analysis of network cameras with Wireshark. In: International Symposium on Digital Forensic and Security. IEEE, Tirgu Mures (2017)
2. Tang, M.F., Wang, C.S.: Developing LUA-based PDXP protocol data analysis plug-in. Comput. Appl. Softw. **32**(1), 121–135 (2015)
3. Anderson, F.: ASP NET Advanced Programming. Tsinghua University Press, Beijing (2002)

4. Jeffer, R.S.: SQL Server 2005 Reference Encyclopedia. Tsinghua University Press, Beijing (2013). translated by Zhou, Z. and Huang, M
5. Niu, Y.F.: Design and implementation of ATC automatic integrated track display system based on B/S. J. Civ. Aviat. Flight Univ. China **1**, 68–73 (2020)
6. Liu, F.F., Liu, G.S., Wang, R.T.: C # Programming Tutorial, 2 edn. Electronic Industry Press, Beijing (2013)
7. Yin, G.F., Wei, G.S.: Application design of data packet capturing based on sharppcap. In: Fourth International Joint Conference on Computational Sciences and Optimization, pp. 861–864. IEEE (2011)
8. Guo, S.F.: C# .NET programming Tutorial. Tsinghua University Press, Beijing (2012)

Research on RP Point Configuration Optimizing of the Communication Private Network

Yili Qin[1(✉)], Yong Cui[1], and Qing Xie[2]

[1] China Satellite Maritime Tracking and Control Department, Jiangyin, Jiangsu, China
tougaoqin@163.com
[2] Space Engineering University, Huairou, Beijing 101400, China

Abstract. Compared with unicast and broadcast, multicast communication technology can save network bandwidth under specific communication conditions. But there are some problems appear with the growing application, the analysis of multicast RP election and packet loss during RPT switching to SPT process are introduced, three solutions for this kind of problem are proposed. Because of the special network environment and transmission mode of communication private network, manual configuration method that disables SPT switching has failed, and the introduction method for static multicast group has disadvantages, finally, the RP point is selected from the dynamic pointing converging switch to the dynamic pointing the core switch. It is proved by experiment that the method is effective.

Keywords: Multicast · RP · RPT · SPT

1 Introduction

The transmission data of the communication private network mainly includes real-time voice, image services, PDXP data, etc., which require high reliability and uninterrupted transmission of services. Multicast technology realizes efficient data transmission in IP networks, which can effectively save bandwidth, control network traffic, and reduce network load. Therefore, it is widely used in business-critical transmission. Due to the expanding business needs, the communication private network has been upgraded. The data intranet is connected to the communication private network in the form of subnet. Among them, the core access switch of data intranet is connected to the communication private network aggregation switch, and still maintains the RP point strategy before the transformation (the aggregation switch is dynamically elected as the RP point). During the service test, it was found that when any source multicast data is transmitted in the data intranet, when the multicast shared tree (RPT) is switched to the shortest path number (SPT), initial random multicast data packet loss occurs. How to better solve such problems is the content of this article.

Q. Liu et al. (Eds.): CENet 2020, AISC 1274, pp. 604–611, 2021.
https://doi.org/10.1007/978-981-15-8462-6_68

2 Multicast Data Transmission Process from Any Source in a Communication Private Network

In the communication private network, the intranet service data is mainly transmitted by random source multicast. Among them, secondary switches, core switches, communication private network aggregation switches, etc. all enable sparse mode PIM SM multicast routing protocol. Dynamic aggregation is used to elect the private network aggregation switch as RP [1–3].

2.1 The Working Process of PIM-SM

Protocol Independent Multicast-Sparse Mode (PIM-SM) is a protocol-independent multicast sparse mode. It uses the unicast routing table established by the unicast routing protocol to complete the RPF [4] check. It does not need to maintain a multicast routing table to implement multicast forwarding. Therefore, it does not need to send or receive multicast routing updates like other protocols, so the overhead of PIM is much lower. The operation of PIM-SM revolves around a one-way shared tree from source to receiver. On the shared tree, there is a root node (RP). The multicast data flow on the shared tree depends on the RP to forward down. Therefore, the shared tree is also called the RP tree, often referred to as RPT.

IP multicast is defined in the IP layer network and sends data packets to a specific set of nodes in the network in a best-effort manner. This set is called a multicast group. The basic idea involved is that only one copy of the same data needs to be sent from the source host, and all hosts in the same multicast group can receive the same data. Therefore, the IP multicast technology can effectively meet the application requirements of "single point transmission and multiple point reception".

According to different multicast sources, multicast can be divided into designated source multicast and random source multicast. The multicast source of the specified source multicast is a certain host, and group members are notified of the IP address of the multicast source in advance. The multicast source of an random source multicast is an uncertain host, and group members are not informed of the IP address of the multicast source in advance.

2.2 The Transmission Process of Any Source Multicast Data in the Communication Private Network

In the communication private network, when server A sends random source multicast data to server B, a small amount of packet loss occurs. In order to analyze the transmission process of random source multicast data packet and the reason for packet loss, a simulation network model is set up for experiments.

The entire random source multicast data is transmitted through the shared tree [4]. The process can be divided into two parts, one is from the receiver to the RP, and the other is the source to the RP. The receiver-RP shared tree is established by the receiver sending a join message to the first-hop router, and then the first-hop router sends a join message toward the RP. At the same time, the corresponding (*, G) routing entries are established on all the routers on the way, and the corresponding outbound interfaces are added (Fig. 1).

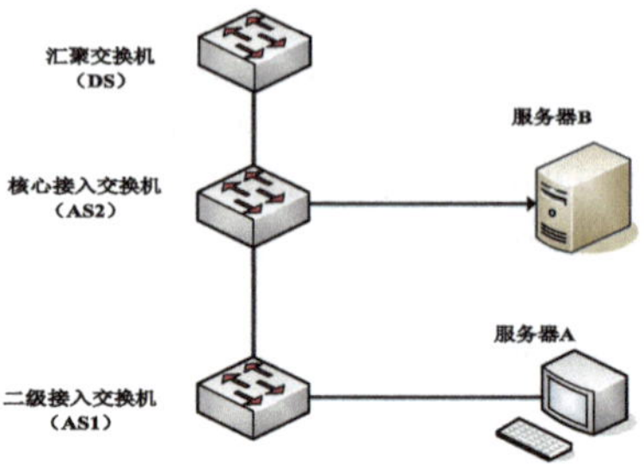

Fig. 1. Network simulation experiment environment.

The establishment of the source-to-RP shared tree is a unicast PIM registration message to the RP through the sender DR. The RP decapsulates the message, checks the corresponding route, forwardings the data, and sends a PIM reservation message to the DR to terminate the DR registration and receive the unencapsulated data stream. At this point, a multicast data stream flows from the source to the receiver through the RP.

3 Packet Loss Rule Analysis of RPT Switching SPT

3.1 RPT Switch SPT Process

In the actual multicast data forwarding process, the shortest path between the multicast source and the receiver is often not the RPT path. At this time, the PIM SM multicast routing protocol will switch this path to the shortest path of the SPT. This switching process is called RPT to SPT handover. After switching, the data packets still being forwarded in RPT will continue to be forwarded to DR. DR queries the source direction of these packets. If the path direction of the packets is not consistent with the SPT, the DR will not be received, resulting in packet loss of multicast data [5].

By default, DR joins the shortest path tree immediately after detecting the multicast source, and starts to switch from RPT to SPT. In the network simulation test, the switch is to send the SPT registration message to the multicast source from the last hop router (DR) along the shortest path. When the multicast source receives the registration message, it starts to forward the multicast data along the SPT.

When AS1 receives the SPT join message, it adds the interface to AS2 to the output interface table of the existing (S, G) entry. In this way, the data stream of server A no longer passes through DS1, but is directly transmitted to server B through AS2.

In the network simulation model of this paper, after the two LANs are connected, the aggregation switch is elected as the RP. When server A sends multicast data to server B, it needs to go through RP (DS1) and forward the path: server A-AS1-AS2-DS1-AS2-server B. When server B receives the first packet of data, it starts to switch from RPT to SPT and changes the forwarding path to server A-AS1-AS2-server B.

After the switchover is complete, the data packets still being forwarded in the path server A-AS1-AS2-DS1-AS2 will arrive from port G0/0/1 of AS2, and the port in the SPT direction is G0/0/2, so all these packets are dropped.

3.2 Analysis of Packet Loss Rules

In the simulation environment, the multicast test software is used to send multicast data from the simulation server A to the server B at different average packet intervals (in the network no-load mode). The test results are shown in Table 1.

Table 1. Multicast packet test results at different intervals.

Source IP address port number	Destination IP address port number	Multicast packet length (byte)	Average packet interval (ms)	Total number of packets received	Number of lost packets
x.x.x.75 33333	225.0.x.x 33333	200	2.45	86325	0
		200	2.33	65324	1
		200	2.21	75426	1
		200	2.09	65342	1
		200	1.97	53246	2
		200	1.85	86245	2
		200	1.73	95324	3
		200	1.61	78421	4
		200	1.49	65334	5
		200	1.37	59654	6
		200	1.25	58624	7
		200	1.13	57891	9
		200	1.01	123561	10
		200	0.89	124653	12
		200	0.77	156235	15
		200	0.65	132654	21
		200	0.53	122465	26

As can be seen from Table 1, no packet loss occurs when the average packet interval is set to be greater than 2.45 ms, and packet loss is started when the average packet interval is set to 2.33 ms, and the number of packet loss and disorder occurs when the average packet interval is set to less than 2.33 ms.

According to the experimental data, the relationship between the average packet interval and the number of lost packets is shown in Fig. 2. It can be seen that when the server B receives the first packet and starts to switch STP, the faster the server A sends the packet to the server B, the more the number of multicast packets staying at the RPT after the path switch, resulting in multicast data packet loss.

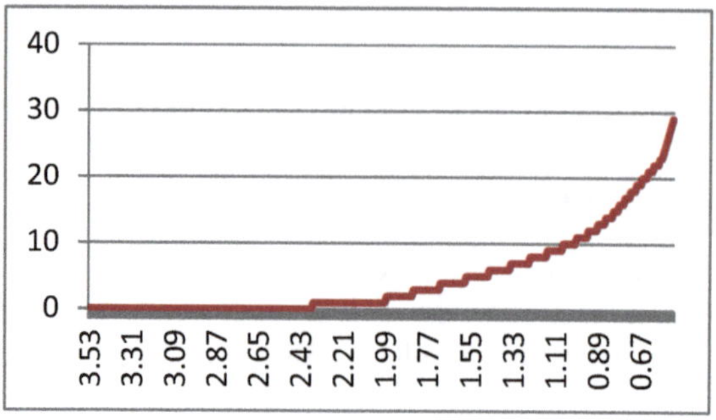

Fig. 2. Relationship between average packet interval and number of packet loss

4 Solution of Packet Loss

Three methods are generally used to solve the data loss caused by RPT switching SPT. The first method is to set the threshold SPT-Switchover on the receiver DR to directly prohibit switching to SPT. The second method is to manually conFig. the multicast static import on the receiver DR so that the receiver DR generates the SPT directly, and it would not be forwarded to the receiver after aggregated at the RP point. The third method is to set the RP point to a switch on the SPT link to make the RPT consistent with the SPT [6–8].

4.1 Disable SPT Switching Mode

Enter the PIM mode on random source multicast receiver DR to conFig. the spt switch threshold infinity. The receiver DR will not initiate an RPT-to-SPT switchover application "SPT-join" to the source DR. In this way, data packets are often not transmitted along the optimal path, but the private network multicast data traffic is small, so it will not affect the private network RP points (aggregation switches).

In the network simulation environment, conFig. "spt-switch-threshold infinity" on the receiving end DR (AS2) to prohibit AS2 from sending a switch request to the multicast source. However, after experimental verification, although AS2 did not send an SPT-join message, the multicast data transmission link still switched from RPT to "SPT" and packet loss occurred.

As mentioned above, after receiving the registration message, the RP forwards (S, G) the joining information to the multicast source, and the information is transmitted to the multicast source DR hop by hop. Although the receiver AS2 is forbidden to send a switch request to the source, since the RP must pass through AS2 in the process of sending (S, G) join information to the source, the receiver AS2 has formed (S, G) entries. As a result, the multicast stream still "switching" to SPT in AS2.

4.2 Multicast Static Introduction

For hosts that receive data from a multicast group stably for a long time, Manual and static mode to join the multicast group can be used to avoid this packet loss problem. After configuring IGMP static group, DR will directly generate SPT, which will not be forwarded to the receiver after aggregating at the RP point. After statically joining a group of multicasts, the switch will continuously maintain the (S, G) entries of the group, regardless of whether there are members of the group on the network segment, and always maintain the shortest tree multicast path.

But such methods require a lot of configuration work. ConFig. each source in the access switch where the receiver resides, and reconFig. when adding a new multicast group. Server A currently has eight different IP multicasts, one of which is sent to one of the four multicast groups. If the multicast static import mode is adopted, at least 32 static import routes must be conFig. d on the core access switch. If the IP address need to be added or changed in the future, the corresponding configuration on the core switch must be added or modified, which brings great inconvenience to the management staff.

4.3 Modify RP Point

The internal network core access switch model is S7706, and the communication private network aggregation switch model is S9303. The processing capability of S7706 is equivalent to that of S9303 [9, 10]. Therefore, changing the RP point on both devices does not affect the transmission performance of any source multicast data. Delete the aggregation switch 1 (DS1) dynamic election RP configuration, and conFig. the core access switch 1 dynamic election as the preferred RP point in the network so that the RPT and SPT paths of the multicast data from server A to server B are the same. This prevents packet loss caused by the RPT switching SPT process. This kind of packet loss can be solved by modifying RP point to core access switch to solve.

5 Experimental Verification

In order to verify whether the modification of RP point as the core access switch will cause influence on the transceiver of random source multicast data stream, random source multicast group transceiver node in the communication private network are simulated, and simulation experiments of full-state transmission and reception are performed. The experiment results show that when the RP point is modified as the core switch, packet loss and disorder will not be caused to random source multicast data reception in the communication private network (Table 2).

Table 2. Communication private network multicast list.

Name	Source node	User node	Multicast form	Path switch	Loss & disorder case	Strategy
Source 1	Secondary access switch	Core access switch 1	Arbitrary source	None	None	RPT and SPT same
Source 2	Secondary access switch 2	Core access switch 1	Arbitrary source	None	None	RPT and SPT same
Source 3	Secondary access switch 3	Core access switch 1, Primary access switch 1	Arbitrary source	No	No	RPT and SPT same
Source 4	Primary access switch 1	Primary access switch (3, 4, 5)	Specify source	None	None	Multicast static introduced
Source 5	Primary access switch 2	Primary access switch (3, 5)	Specify source	None	None	Multicast static introduced
Source 6	Primary access switch 1	Primary access switch 6	Specify source	None	None	Multicast static introduced

6 Concluding Remarks

The transmission process of random source multicast data packet in the private network data internal network is introduced. For the establishment of the multicast route, the switching process from RPT to SPT and the causes of packet loss are analyzed, and the corresponding rules are found. For multicast packet loss failure, three solutions are proposed, analyzed and compared. Practice has proved that the solution of changing the original RP strategy that dynamically points to the aggregation switch to the dynamic RP that points to the core switch is the most reliable, simple, and convenient.

References

1. Internet Group Management Protocol, Version 2. IETF RFC2236 (1997)
2. Internet Group Management Protocol, Version 3. IETF RFC3376 (2002)
3. Protocol Independent Multicast–Sparse Mode (PIM-SM): Protocol Specification. Draft-ietf-pim-sm-v2-new-11 (2004)
4. Source-Specific Multicast for IP. Draft-ietf-ssm-arch-03 (2003)
5. Cain, B.F., Deering, S.S., Kouvelas, I.T.: Internet group management protocol (version 3) [EB/OL] (2002). http://www.ietf.org/internet-drafts/draftietf-idmr-igmp-v3-11.txt

6. Li, X.F., Han, G.S., Liu, H.T.: Research on packet loss during RPT to SPT switch. Comput. Eng. **7**(33), 107–108 (2007)
7. Li, K.F., Chen, X.S., Zhao, Q,T.: Methods to solve familiar multicast faults of spaceflight communication IP network. J. Telemetry, Track. Command **2**(33), 58–61 (2012)
8. Zhou, Y.F., Wei, W.S.: Research on packet loss during RPT to SPT switch in PIM sparse mode. Electron. Des. Eng. **18** (25) (2017)
9. Huawei Technologies Co., Ltd. Quidway S9300 Terabit Routing Switch V100R003C00 Configuration Guide-MPLS (2010)
10. Huawei Technologies Co., Ltd. Quidway NetEngine20/20E Series Router V200R005 (2008)

Cross Polarization Jamming and ECM Performance of Polarimetric Fusion Monopulse Radars

Huanyao Dai(✉), Jianghui Yin, Zhihao Liu, Haijun Wang, Jianlu Wang, and Leigang Wang

State Key Laboratory of Complex Electromagnetic Environment Effects on Electronics and Information System, Luoyang 471000, China
30477205@qq.com

Abstract. The cross polarization jamming is an novel jamming technology with good effect on conventional monopulse radars. By theoretical analysis, the angular measurement error can reach the half-beam width. However, the polarimetric array fusion monopulse radars can obtain the angular measurements through two polarization arrays, work out the maximum likelihood estimation of the target angle to achieve the final angular. In this way, the radar can have its anti-jamming capability improved. In this paper, a novel model has been proposed for polarization domain countermeasures. The countermeasure performance of cross polarization jamming and polarization fusion monopulse radars has been analyzed based on computer simulation. The results show that the polarization monopulse radar is insensitive to cross polarization jamming and the jamming effect is not ideal, so it is necessary to design pertinent jamming.

Keywords: Monopulse radar · Polarization fusion · Jamming · Cross polarization

1 Introduction

Electronic Countermeasures (ECM) [1, 2] is a special combat mode and an indispensable combat force for modern warfare. It can be used in various ways in strategic, operational and tactical operations such as strategic deterrence, combat support, weapon platform self-defence, position protection, and anti-terrorism stabilization. The Polarimetric Fusion Monopulse Radar (PFMR) using the polarized integration monopulse angle measurement method can accurately measure the target angle [3] and plays an important role in the ECM. PFMR is obviously superior to traditional single-polarized array radar in the fields of anti-jamming, target recognition, imaging, etc. It is an important new system radar [4]. PFMR makes full use of the advantages of strong anti-jamming capability and high multi-target resolution of the array radar, and can use the polarization information [5] of the target to further improve the accuracy of the target angle measurement. The Polarimetric Fusion [6] mono-pulse angle measurement method is an improvement over the mono-pulse angle measurement method [7]. The measurement results of the H polarization channel and the V polarization channel are

Q. Liu et al. (Eds.): CENet 2020, AISC 1274, pp. 612–622, 2021.
https://doi.org/10.1007/978-981-15-8462-6_69

weighted and fused, then the final measurement angle is obtained. The performance of this method is better than traditional mono-pulse angle measurement method. Because the polarized integration mono-pulse array radar adopts array antennas and polarization integration technology, it has strong anti-interference ability to the traditional single-polarized suppression interference and angle deception interference [8]. In view of the importance of angle measurement in target tracking, positioning and guidance, seeking an effective angle deception jamming technique has been a hot issue in radar countermeasure research. According to the inconsistency between the radar antenna main polarization and cross-polarization receiving vectors, cross-polarization interference [9] emits electromagnetic waves which have same frequencies and orthogonal polarization to the main polarization of the radar antenna to illuminate the radar to achieve the purpose of angle deception. Because cross-polarization interference does not require spatially separated multiple interference sources, it has great application potential for important target protection or missile penetration, and it is widely considered effective technical means to deal with mono-pulse angle measurement radars. In this paper, the feasibility of cross-polarization interference to the PFMR is analysed. It is proved that cross-polarization can effectively interfere with the angle measurement performance of the polarized integration mono-pulse array radar.

2 Processing Flow of Polarimetric Monopulse Radars

This section has presented the processing flow of the monopulse angular measurement based on polarimetric fusion, as shown in Fig. 1. Process the data received from horizontal and vertical polarization channels separately: According to the principle of array radar monopulse, obtain the angular measurements of two polarization arrays respectively; work out the maximum likelihood estimation of the echo amplitude on the basis of the angular measurements achieved. Finally, carry out weight fusion of the target angle information to get the final angular measurement.

Based on the angular measurements $\hat{\theta}_{\mathrm{H}}, \hat{\theta}_{\mathrm{V}}$, use the maximum likelihood estimation to get the complex amplitudes of H and V polarization directions. Take the horizontal polarization as an example and the received signal model can be rewritten as:

$$\mathbf{x}_{\mathrm{H}} = E_{\mathrm{H}}\mathbf{s}(\hat{\theta}_{\mathrm{H}}) + \mathbf{n}_{\mathrm{H}} \tag{1}$$

The probability density function (PDF) is:

$$p(\mathbf{x}_{\mathrm{H}}; E_{\mathrm{H}}) = \frac{1}{(2\pi\sigma^2)^{\frac{N}{2}}} \exp\left[-\frac{1}{2\sigma^2}\left(\mathbf{x}_{\mathrm{H}} - E_{\mathrm{H}}\mathbf{s}\left(\hat{\theta}_{\mathrm{H}}\right)\right)^{\mathrm{H}}\left(\mathbf{x}_{\mathrm{H}} - E_{\mathrm{H}}\mathbf{s}\left(\hat{\theta}_{\mathrm{H}}\right)\right)\right] \tag{2}$$

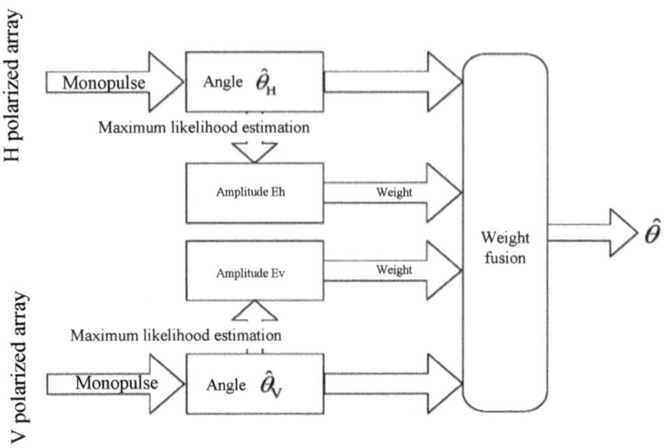

Fig. 1. Processing flow of the monopulse angular measurement based on polarimetric fusion.

The data $\mathbf{x}_H$ is fixed and the probability density function is a function of E_H, that is, likelihood functions. Based on the theory of derivation of real functions from complex numbers, work out the derivative of the logarithmic likelihood function with respect to E_H, make it equal to zero and get:

$$\frac{\partial \ln p(\mathbf{x}_H; E_H)}{\partial E_H} = \mathbf{s}^H\left(\hat{\theta}_H\right)\mathbf{x}_H - \mathbf{s}^H\left(\hat{\theta}_H\right)\mathbf{E}_H\mathbf{s}\left(\hat{\theta}_H\right) = 0 \tag{3}$$

Obtain the maximum likelihood estimator of the amplitude of the horizontal polarization component:

$$\hat{E}_H = \mathbf{s}^H\left(\hat{\theta}_H\right)\mathbf{x}_H / N \tag{4}$$

Similarly, obtain the maximum likelihood estimator of the amplitude of the vertical polarization component:

$$\hat{E}_V = \mathbf{s}^H\left(\hat{\theta}_V\right)\mathbf{x}_V / N \tag{5}$$

At the same time, also obtain the polarization state of the target echoes from the above Eqs. (4) and (5).

Carry out fusion processing of the angular measurements $\hat{\theta}_H, \hat{\theta}_V$ of the H and V polarization arrays:

$$\hat{\theta} = \alpha_1\hat{\theta}_H + \alpha_2\hat{\theta}_V \tag{6}$$

Optimize the variance of the weighted coefficients α1, α2 to minimize the variance σ2 of $\hat{\theta}$ and solve as follows by means of the Lagrange multiplier method:

$$\begin{cases} a_1 = \dfrac{\sigma_V^2}{\sigma_H^2 + \sigma_V^2} \\ a2 = \dfrac{\sigma_H^2}{\sigma_H^2 + \sigma_V^2} \end{cases} \tag{7}$$

Where, σ_H^2, σ_V^2 are variances of $\hat{\theta}_H, \hat{\theta}_V$ respectively. In practical applications, σ_H^2, σ_V^2 are unknown, but it can be proved that they are inversely proportional to the ratio, that is:

$$\sigma_H^2 = \frac{k_2 \sigma_n^2 \theta_{3dB}^2}{\left|\hat{E}_H\right|^2}, \sigma_V^2 = \frac{k_2 \sigma_n^2 \theta_{3dB}^2}{\left|\hat{E}_V\right|^2} \tag{8}$$

Where, k2 ≈ 0.19 N/(N2 − 1) is a constant and θ3 dB is 3 dB beamwidth. Obtain the final angle estimate as follows:

$$\hat{\theta} = \frac{\hat{\theta}_H \left|\hat{E}_H\right|^2 + \hat{\theta}_V \left|\hat{E}_V\right|^2}{\left|\hat{E}_H\right|^2 + \left|\hat{E}_V\right|^2} \tag{9}$$

3 Mathematical Modeling of Cross Polarization Jamming Mechanism

It is assumed that the polarization array radar adopts a dual polarization mode for working to transmit horizontal polarization waves, receive horizontal and vertical polarization waves at the same time and then use the polarization fusion method for angular measurement. If vertical polarization waves are transmitted from the target to interfere with the polarization array antenna, the signal-to-noise ratio of the vertical polarization channel is enhanced and the angular measurement is more accurate, but, due to the existence of cross polarization jamming, the error of angular measurement will increase for the horizontal polarization channel. After the fusion of the two channels, the angular measurement error is likely to increase, which makes the polarization fusion algorithm invalid.

In the simulation, the sinc function is adopted to simulate the antenna main polarization patterns of Beam 1 and Beam 2:

$$G_m = [\sin(kx)/(kx)]^2 \tag{12}$$

The following function is adopted to simulate the pattern of the cross polarization amplitude of Beam 1:

$$G_{c1} = L_1 \cdot \frac{2k_1^2 x \sin(k_1 x) \cdot \cos(k_1 x) - 2k_1 \sin^2(k_1 x)}{(k_1 x)^3} \tag{13}$$

The following function is adopted to simulate the pattern of the cross polarization amplitude of Beam 2:

$$G_{c2} = L_2 \cdot \frac{2k_2^2 x \sin(k_2 x) \cdot \cos(k_2 x) - 2k_2 \sin^2(k_2 x)}{(k_2 x)^3} \tag{14}$$

Where, the values of k_1 and k_2 are determined by the number of the sidelobes in $[0, \pi]$, L_1 and L_2 as attenuation regulators, can be set as required, and x indicates the corresponding angle value in radians.

So far, there is no effective mathematical model to simulate the phase pattern of cross polarization and the randomness is huge. In this paper, a group of fixed random numbers is used to simulate the phase pattern of cross polarization. In order to be consistent with the actual situation, this random array remains unchanged in the simulation analysis. $P_{c1}(\theta)$ and $P_{c2}(\theta)$ indicate the cross-polarization phase patterns of Beam 1 and Beam 2 respectively. The co-polarized phase patterns can be considered consistent; if set as a constant $P_m(\theta) = 0°$ in this paper, the echoes received by the two beams do not have the phase difference caused by the antenna.

Then, the difference signal obtained by the horizontal polarization antenna array is as follows:

$$\begin{aligned} \Delta_h &= \Delta_{mh} + \Delta_{ch} \\ &= G_m \cdot E_h\left(1 - e^{-j\Delta\varphi}\right) + G_{c1} \cdot P_{c1}(\theta) \cdot (E_v + E_{iv}) \\ &- G_{c2} \cdot P_{c2}(\theta) \cdot (E_v + E_{iv}) \cdot e^{-j\Delta\varphi} \end{aligned} \tag{15}$$

The sum signal is as follows:

$$\begin{aligned} \Sigma_h &= \Sigma_{mh} + \Sigma_{ch} \\ &= G_m \cdot E_h\left(1 + e^{-j\Delta\varphi}\right) + G_{c1} \cdot P_{c1}(\theta) \cdot (E_v + E_{iv}) \\ &+ G_{c2} \cdot P_{c2}(\theta) \cdot (E_v + E_{iv}) \cdot e^{-j\Delta\varphi} \end{aligned} \tag{16}$$

Where, E_h refers to the echo of the horizontal polarization wave transmitted by the polarization radar and reflected by the target, E_{iv} indicates the vertical polarization wave transmitted by jamming and is usually about 20 dB larger than the target signal; $\Delta\varphi = \frac{2\pi D}{\lambda}(\sin\theta - \sin\theta_0)$ indicates the phase difference. Then the angle can be estimated according to the formula $\hat{u} = u_0 + k_1 \mathrm{Im}\left[\frac{\Delta}{\Sigma}\right]$.

The difference signal obtained from the vertical polarization array antenna is as follows:

$$\begin{aligned} \Delta_v &= \Delta_{mv} + \Delta_{cv} = G_m \cdot (E_v + E_{iv}) \cdot \left(1 - e^{-j\Delta\varphi}\right) \\ &+ G_{c1} \cdot P_{c1}(\theta) \cdot E_h - G_{c2} \cdot P_{c2}(\theta) \cdot E_h \cdot e^{-j\Delta\varphi} \end{aligned} \tag{17}$$

The sum signal is as follows:

$$\begin{aligned}\Sigma_v &= \Sigma_{mv} + \Sigma_{cv} = G_m \cdot (E_v + E_{iv}) \cdot \left(1 + e^{-j\Delta\varphi}\right) \\ &+ G_{c1} \cdot P_{c1}(\theta) \cdot E_h + G_{c2} \cdot P_{c2}(\theta) \cdot E_h \cdot e^{-j\Delta\varphi}\end{aligned} \tag{18}$$

Where, E_v is the vertical polarized echo of the horizontal polarized wave transmitted by the polarimetric radar and reflected by the target.

For horizontally polarized receiving arrays, the received signal is as follows:

$$\mathbf{x}_h = G_m \cdot E_h \cdot \mathbf{s}(\hat{\theta}_h) + G_c \cdot (E_v + E_{iv}) \cdot \mathbf{s}(\hat{\theta}_h) + \mathbf{n}_h \tag{19}$$

Where, $G_c \cdot E_{iv} \cdot \mathbf{s}(\hat{\theta}_h)$ indicates the changes of the received signals from cross polarization components. According to the polarization fusion algorithm, also use the formula $\hat{E}_h = \mathbf{s}^{\mathrm{H}}\left(\hat{\theta}_h\right)\mathbf{x}_h/N$ to estimate the amplitude of the horizontal polarization component, which will bring about a comparatively large estimation error.

The vertical polarization receiving array can receive such signals as follows:

$$\mathbf{x}_v = G_m \cdot [E_v + E_{iv}] \cdot \mathbf{s}(\hat{\theta}_v) + G_c \cdot E_h \cdot \mathbf{s}(\hat{\theta}_v) + \mathbf{n}_v \tag{20}$$

Equivalently, the amplitude of the received signal has increased.

Due to the existence of vertical polarization jamming components, the polarization state of the echo obtained has changed greatly compared with that without jamming. After obtaining the angle and amplitude estimations of the horizontal and vertical polarization components, the final angle estimation can be obtained according to the formula (9).

Simulation settings: Number of array elements N = 16, the array element spacing is set to be half wavelength, the wavelength is 6 cm, the beam width is θ_{3dB} = 6°, the beam pointing is θ_0 = 0°, the target local angle is 2°, the Monte Carlo times are set to be M = 1000 and the polarization parameter of the target echo is set to be $\eta = 0$. In the simulation, set the cross-polarization component 20 dB smaller than the co-polarization component and the cross-polarization jamming component 20 dB larger than the co-polarization component. Use the angle estimation root-mean-square error RMSE = $\sqrt{E\left[\left(\hat{\theta} - \theta\right)^2\right]}$ to estimate the angular performance.

The main polarization pattern of the sub-array is simulated by a sinc function.

$$G_m(\theta) = [\sin(k_m\theta)/(k_m\theta)]^2 \tag{21}$$

The amplitude pattern of the sub-array cross polarization is simulated by the variation function of the first derivative of the sinc function. It has many control parameters and can flexibly simulate the cross polarization pattern.

$$G_c = L \cdot \frac{2k_c^2\theta \sin(k_c\theta) \cdot \cos(k_c\theta) - 2k_c \sin^{\alpha}(k_c\theta)}{(k_c\theta)^{\beta} + 1} \tag{22}$$

Where, the k value is determined by the number of the sidelobes in $[0, \pi]$, L, as an attenuation regulator, can be set as required, α and β are beam shape parameters and θ indicates the corresponding angle value in radians. In Formula (21), if $k_m = 40$ is adopted, the beam width should be $\theta_{3dB} = 4°$. Set the cross polarization amplitude patterns of the two sub-arrays of the horizontal polarized array antenna as $k_{\mathrm{H}1c} = 39$, $L_{\mathrm{H}1c} = 32$, $\alpha_{\mathrm{H}1c} = 1.8$, $\beta_{\mathrm{H}1c} = 1.5$, $k_{\mathrm{H}2c} = 40$, $L_{\mathrm{H}2c} = 32$, $\alpha_{\mathrm{H}2c} = 1.4$ and $\beta_{\mathrm{H}2c} = 2$; set the cross polarization amplitude patterns of the two sub-arrays of the vertical polarized array antenna as $k_{\mathrm{V}1c} = 35$, $L_{\mathrm{V}1c} = 33$, $\alpha_{\mathrm{V}1c} = 1.2$, $\beta_{\mathrm{V}1c} = 2$, $k_{\mathrm{V}2c} = 30$, $L_{\mathrm{V}2c} = 30$, $\alpha_{\mathrm{V}2c} = 2$ and $\beta_{\mathrm{V}2c} = 2$.

So far, there has been no effective mathematical model to simulate the phase pattern of cross polarization and the randomness is huge. In this paper, a group of fixed random numbers are used to simulate the phase pattern of the sub-array cross polarization. In the simulation analysis, a set of random numbers corresponding to each sub-array remain unchanged.

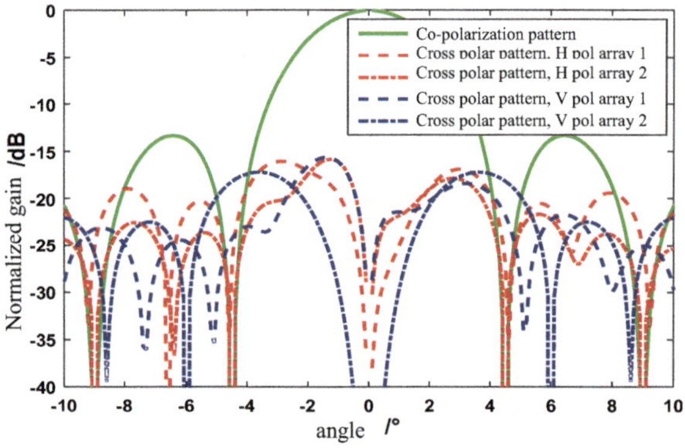

Fig. 2. Principal and cross polarization amplitude patterns of each sub-array.

Figure 2 gives the amplitude patterns of the main and cross polarization for each sub-array. As is shown in the Fig., the amplitude of the cross polarization in the main lobe is obviously lower than that of the main polarization, about 20 dB lower, which is consistent with the measured data of the antenna. The amplitude patterns are also quite different for cross polarization of each sub-array, which provides the physical conditions for the effective realization of cross polarization jamming.

4 Performance Simulation of Polarimetric Monopulse Radar

4.1 Experimental Datasets

Relationship between angular measurement accuracy and SNR.

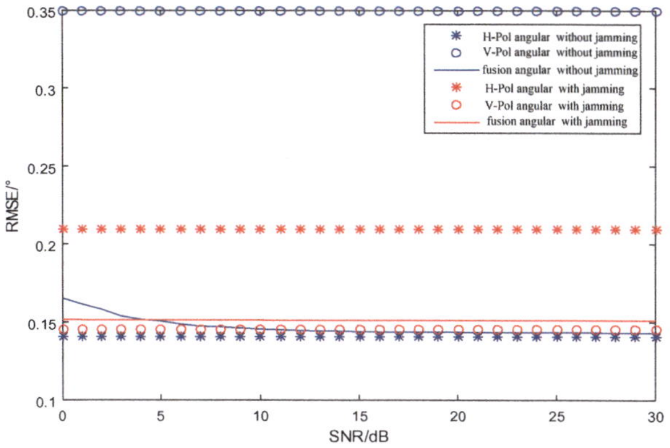

Fig. 3. Relationship between angular measurement accuracy and SNR with or without cross polarization jamming.

As can be seen from Fig. 3, the angular measurement performance is the worst for vertically polarized arrays without cross polarization jamming because the vertically polarized echo received by the array antenna is 20 dB smaller than the horizontally polarized echo and then the horizontally polarized echo is cross polarization jamming for vertically polarized receiving antennas, whose angular measurement error is largest accordingly. The vertically polarized echo is very small, so it almost has no effect on the angular measurement performance of horizontally polarized receiving antennas. However, vertical polarization antennas are characterized by large angular measurement error, so the angular measurement performance subject to polarization fusion is worse than that of horizontal polarization antennas after polarization fusion. When cross polarization jamming exists, the vertical polarization echo is enhanced equivalently, which improves the angular measurement performance of vertical polarization antennas. However, horizontal polarization antennas are subject to cross polarization jamming, so that the performance of angular measurement is significantly reduced. After polarization fusion, the performance of angular measurement is better than that of horizontal polarization antennas, but worse than that of vertical polarization.

4.2 Relationship Between Angular Accuracy and Cross Polarization Jamming Intensity

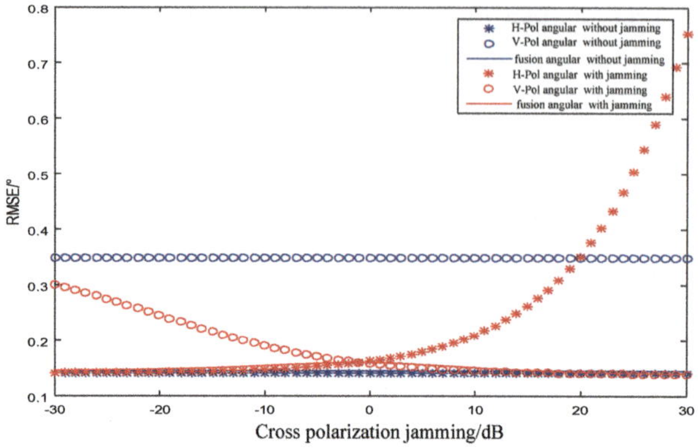

Fig. 4. Relationship between angular accuracy and cross polarization jamming intensity with or without cross polarization jamming.

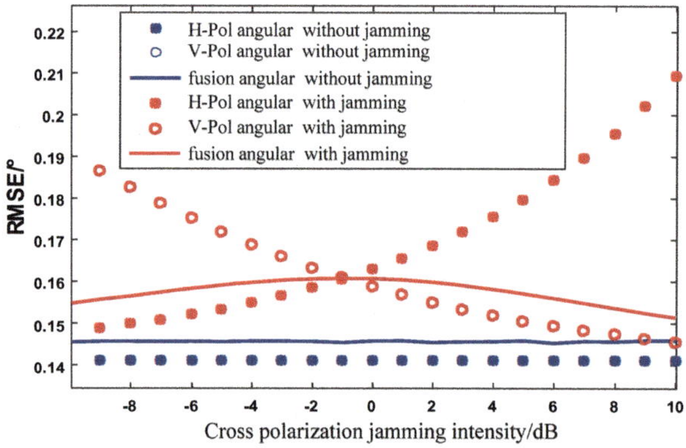

Fig. 5. Relationship between angular measurement accuracy and cross polarization jamming intensity.

In simulation (Fig. 4 and Fig. 5), SNR = 10 dB. When cross polarization jamming exists, the angular measurement performance of horizontal polarization antennas will decrease gradually with the rise of jamming intensity while the angular measurement performance of vertical polarization antennas will increase gradually. When the cross polarization jamming intensity is comparatively small or large, the horizontal and vertical polarization antennas will have their angular measurement performance different obviously, so it is close to the high-performance side after polarization fusion. When the cross polarization jamming is about 0 dB, the difference is very small between the horizontal and vertical polarization antennas and after polarization fusion;

the angular measurement performance is not yet improved. Figure 5 shows the main local amplification of the relationship between the angular measurement accuracy of polarized array radars and the cross polarization jamming intensity and compares it with the jamming-free condition; when the cross polarization jamming intensity is in the range of −10 dB–10 dB, the angular measurement accuracy subject to jamming should be worse than that subject to no jamming. Figure 5 has not given the curve for angular measurement accuracy of the vertical polarization array subject to no jamming; the vertical polarization component of the target echo is very small and the horizontal polarization component of the target echo causes cross polarization jamming against it, so that the angular measurement has lost its significance for comparison due to greater deviation. When the cross polarization jamming is about 0 dB, that is, when the cross jamming intensity is close to the horizontal polarization component intensity of the target echo, the angular measurement performance of horizontal polarization array antennas is degraded due to jamming while the angular measurement performance of vertical polarization array antennas is improved slightly; after polarization fusion, the angular performance for polarization fusion of polarization array antennas has dropped by the maximum of 10.2%.

5 Conclusion

In this paper, we have analyzed the jamming effect of cross polarization jamming on monopulse radars with polarization array fusion. Based on the polarization fusion algorithm, we can obtain the angular measurements through two polarization arrays, carry out weight fusion of the target angle to achieve the final angular measurement, improve the anti-jamming capability of the radar and make the polarization monopulse radar insensitive to cross polarization jamming; after polarization fusion, the angular measurement performance of polarization array radars will decrease greatly up to 10.2% and the decrease of the angular measurement performance has not yet reached the jamming effect actually required, so, as the jammer, it needs a special jamming technology to be designed, for example, alternately changing the jamming polarization, so that the fusion algorithm does not converge to destroy the radar angular measurement.

References

1. Wang, H.J., Li, J.B., Wang, T., Wang X.S.: Calculate the polarized scattering properties of the canted rain media with M-P DSD using extended mie scattering method. In: International Conference on 2015 IET, Hangzhou, China, October, pp. 1–5 (2016)
2. Wang, H.J., Zheng, G.Y., Zeng, Y.H., Wang, L.D.: Experimental analysis on the passive bistatic radar based on dual-polarized integration technology. Moder Radar **39**(10), 11–20 (2017)
3. Nichel, U.: Overview of generalized monopulse estimation. IEEE Aerosp. Electron. Syst. Mag. **21**(6), 27–56 (2006)
4. Li, M.Q., Li, Y.Z., Dong, J., Wang, X.S.: Polarization characteristic analysis of crossed dipoles. Chin. J. Radio Sci. **27**(2), 396–401 (2012)

5. Zhu, A.S., Chen, Z.L., Liu, X.Q., Zhen, Y.H., Yang, Y.: Numerical calculation of radiation characteristic of monopole and dipole antenna. J. Microwaves **31**(2), 33–38 (2015)
6. David, L.A.: EW104: EW Against a New Generation of Threats. Artech House Press (2015)
7. Kalinbacak, I., Pehlivan, M.: Cross polarization monopulse jammer located on UAV. In: Proceedings of 8th International Conference on Recent Advances in Space Technologies, RAST 2017, pp 337–341 (2017)
8. Ma, J.Z., Shi, L.F., Cui, G., Xiao, S.P.: Further analysis of retrodirective cross-eye jamming, polarization considerations. In: 2018 IET Conference Publications CP (2018)
9. Ma, J.Z., Shi, L.F., Xiao, S.P., Wang, X.S.: Mitigation of cross-eye jamming using a dual-polarization array. J. Syst. Eng. Electron. **29**(3), 491–498 (2018)

The Customization and Implementation of Computer Teaching Website Based on Moodle

Zhiping Zhang[1(✉)] and Mei Sun[2]

[1] School of Computer Science and Technology, Shandong University of Finance and Economics, Jinan 240014, China
zhipingzhang@sdufe.edu.cn
[2] School of Public Finance and Taxation Heidelberg, Shandong University of Finance and Economics, Jinan 240014, China

Abstract. Moodle is based on the theory of constructivism, it is used for construction of internet-based courses and the package of web site's open source code. Moodle supports diversified ways of teaching and learning. It can create a good network learning environment for learners' autonomous learning, collaborative learning and personalized learning. It also can bring convenience for teachers and school teaching. In view of the advantages of Moodle system, we select it as a platform for the construction, development and application of the college of computer science and technology subject. At first, this paper used the literature research to understand the development idea and the function characteristics of Moodle, and analyzes the architecture of Moodle and related technologies, especially the required environment, conventions and ways of the second exploitation of Moodle theme. It shows the difficulty of the development of each way. Then redefining professional subdirectory according to different requirements of activities module and function module.

We designed a typical of college computer science and technology of Moodle theme taking into account from the overall style, color system, font and layout, fresh graduates courses and graduate thesis submit. Finally summarizes the main work of development theme and the problems still needing to improve, equally looks forward to the future work.

Keywords: Moodle · Theme · Secondary development

1 Background

1.1 What is Moodle

Moodle is a management system co-developed by Dr. Martin dougiamas [1] of Australia based on Constructivism education theory. It has three functions: website management system, course management system and learning management system. It is a free open source software, which has been widely used in various countries. Moodle is the abbreviation of Modular Object-Oriented Dynamic Learning Environment. It is a software package for building Internet based courses and websites. The Moodle

Q. Liu et al. (Eds.): CENet 2020, AISC 1274, pp. 623–627, 2021.
https://doi.org/10.1007/978-981-15-8462-6_70

platform is based on the social constructivism teaching idea, that is, educators (teachers) and learners (students) are equal subjects. In teaching activities, they cooperate with each other and build knowledge together according to their own experience.

On the Moodle platform, users can adjust the interface and increase or decrease content at any time according to their own needs. The course list shows the description of each course on the server, including the information of each course and the access rights of the students. Courses can also be classified and searched. A Moodle website can build thousands of courses. Visitors can search courses by category and study courses according to their own needs.

The Moodle platform is also compatible and easy to use, easy to install, and can be installed on any web server that supports PHP. It only needs one database and can be shared. All form transfers must be confirmed, data verified, cookies encrypted, and so on. When users register, they log in for the first time through email, and the same email address can't be registered repeatedly in the same course, all of which makes Moodle's security enhanced. At present, some primary and secondary school teachers in China have begun to use Moodle to manage their own teaching activities, while the Moodle project is still in constant development and improvement.

1.2 Research Status at Home and Abroad

After entering the 21st century, various countries in the world have adopted a variety of curriculum management systems, learning management systems or learning activity management systems to organize teaching in the network environment, so e-learning, online learning, virtual learning and network-based learning have gradually emerged.

University of Phoenix started to provide online learning mode in 1989. Today, there are 110 campuses and learning centers in 21 states in the United States, Puerto Rico and Canada. It is the largest private university in the United States. At present, University of Phoenix has nearly 200 thousand students and nearly 10 thousand online teachers. The Open University, Dublin City University Moodle system, Moodle curriculum school built in the United States, Moodle curriculum school built in Thailand, etcetera, have introduced Moodle online learning platform in an all-round way. Because the open-source software Moodle is free and has powerful functions, so far, more than 2000 institutions in nearly 100 countries have adopted Moodle for online education.

Moodle is the most commonly used learning curriculum development management system in UK universities [4], and The Open Universities have also won the MACT Award for Moodle curriculum management system (This award is awarded by information technology research, one of the six core projects of the famous Andrew Mellon Foundation. It is awarded to the leader in collaborative development of open source software tools in higher education and non-profit activities.), The Open University has invested thousands of hours of programming work in Moodle and its community, becoming the leader of Moodle community [4].

2 Second Development of Theme Based on Moodle Platform

2.1 Construction of Development Environment

Moodle is developed with Apache, MySQL and PHP. It was originally built on Linux. Now it can be downloaded to the complete installation package supporting windows and other operating systems on the official website. Users familiar with web servers and databases can download the operation package, and then configure the server themselves, which is helpful to understand PHP and MySQL servers. Users who download the complete installation package can decompress the installation package after downloading, and then start the server to build the development environment.

The following is the choice of development tools, because PHP is a script language that does not need to be compiled for interpretation and execution, so we can use the text editor of windows system to directly modify the source file, and then we can develop Moodle. However, we know that because the simple use of text editor can not give error prompt to PHP syntax and keywords and other reasons, the Moodle community also recommends the open source visual development tool eclipse.

2.2 The Form of Secondary Development of Moodle

As a third-party secondary development, Moodle focuses on the design and development of its external functional modules to meet the specific needs, that is, the module development around Moodle's core code, rather than the modification of Moodle's core code and modules. Table 1 shows the secondary development form of Moodle [12].

Table 1. The form of secondary development of Moodle

Form	Suitable for users	Concrete content	Difficulty level
Language pack development	Understand Moodle's functions and online courses, and have the ability of language translation	Provide language package for Moodle's interface. In addition to the interface language package, there are many documents that need to be translated	Simple
Theme style	Understand the function of Moodle and have some knowledge of CSS, XML and HTML	Provide different appearance for Moodle, mainly including the modification of layout, picture, font, color and other interface appearance	Commonly
Functional modules and plug-ins	Understand the functions of Moodle and network courses, have a certain knowledge of relational database and PHP, and have a certain knowledge of web design	Expand specific requirements	Difficult

3 Customization and Implementation of Finance and Taxation Teaching Website

3.1 Registration and Installation of Moodle

The software we use to implement our teaching website is the latest version of Moodle. First, install xampp (Apache + MySQL + PHP + PERL). Secondly, download the installation package on the official website of Moodle and unzip it to the directory of xampp/htdocs.

3.2 Subject Changes

For the finance and taxation teaching website, we first change the theme interface and layout. Before the change, the first thing we need to do is create the files and directories that will be used and build the related theme files. Secondly, configure the theme.

3.3 Curriculum and Specialty Setting

We classify them according to the specialty setting and discipline setting. At the same time, it classifies the graduation class thesis submission and the instructor, which is convenient for the teacher to review and the student to submit.

For the course, we can change the name of the unit, write down the key points of the unit, the time plan of the unit, and then start to plan what the unit needs to teach, so that students can learn, there must be content, which used to be lectures, now is text, pictures, and video. After the consolidation of self-study knowledge, students can do homework to consolidate. The teacher can observe whether the students log in and whether the students' test scores already exist in the background.

Courses can be selected after the course is set up. Users (students) must have an account to learn. Click to join the user. There is no user in the account by default. So if you want to select courses, you need to add a small number of users under the account in website management. Select batch processing to upload a large number of users with Excel tools, and then add these users to our courses.

4 Conclusion

This paper is mainly based on Moodle to customize and implement a finance and tax teaching website, making full use of Moodle's functional modules, combining with the characteristics of the college and professional settings, making a teaching website.

In the next step, we will further modify and improve our teaching platform in order to better carry out the teaching work of our teachers.

References

1. Wang, X.Y.: Research and application of secondary development of Moodle. Chengdu Univ. Technol. **5**, 12 (2010)
2. Jason, C., Helen, F.: Using Moodle, 2nd edn., pp. 4–5. O'Reilly Community Press, Sebastopol (2007)
3. Li, J.H.: Information course design–the creation of Moodle information learning environment, vol. 1. East China Normal University Press, Shanghai (2007)
4. Wang, D.: Research of network course based on Moodle platform. Sichuan Normal University (2008)
5. Qu, Y.L., Qu, H.Y.: A new structure of the application based B/S schema and its implementation. Appl. Res. Comput. **17**(6), 56–58 (2000)
6. Baidu Encyclopedia Apache server definition. https://baike.baidu.com/view/28283.htm. Accessed 5 2015
7. Baidu Encyclopedia MySQL database definition. https://baike.baidu.corn/view/24816.htm. Accessed 5 2015
8. Baidu Encyclopedia PHP language definition. https://baike.baidu.com/view/99.htm. Accessed 5 2015
9. Baidu Encyclopedia Introduction to W3C. https://baike.baidu.com/view/7913.htm. Accessed 5 2015.
10. Baidu Encyclopedia Introduction to XML. https://baike.baidu.com/view/63.htm. .Accessed 5 2015
11. Baidu Encyclopedia Introduction to XHTML. https://baike.baidu.com/view/15906.htm. Accessed 5 2015
12. Ye, H.S., Ji, J.: Second developing and design for Moodle. e-Education Res. **25**(4), 52 (2007)

Recommendation of Community Correction Measures Based on FFM-FTRL

Fangyi Wang(✉), Xiaoxia Jia, and Xin Shi

North China Institute of Computing Technology, Beijing 100083, China
wfy0917@outlook.com

Abstract. Community correction refers to the correction on the five types of objects. Community correction work is to correct the community correction objects' psychology and behavioral habits by taking proper correction measures for corresponding correction objects, promote them to return to society. Selection of correction measures is a key component in community correction work. However, this work is still at the manual stage in China, which consumes much manpower and resources. To this end, we propose an algorithm namely FFM-FTRL which is the first time that machine learning is applied to addressing the recommendation of community correction measures, improving the efficiency of community correction work. Our method is specifically designed to solve sparse data, and reduce model training time. To evaluate its effectiveness, we select FM and FFM-SGD as our baselines. Experimental results show that F1 score is improved by 4% and 0.5%, and the training time is shortened.

Keywords: Community correction · FFM-FTRL · Correction measures recommendation · Feature engineering

1 Introduction

Community correction is a kind of non-jail punishment. Convicts who are sentenced to supervision without incarceration, granted probation or parole, or permitted to temporarily serve a sentence outside an incarceration facility shall be subject to community correction according to the law. Community correction work aims to correct their criminal psychology, bad behavior and to promote them return to society. Selection of correction measures, one of the most important part of community correction work [1], stays at low-automation level and lack of effective recommendation methods [2].

As for corrective measures, the recommendation process should always be based on evidence-based thinking which is widely recognized [3]. We take evidence-based thinking into account, present an online recommendation method, apply machine learning algorithm to the problem of community corrective measures recommendation for the first time.

In our method, the data processing layer constructs a correction object-measure dataset, and processes the data by feature engineering. The algorithm layer trains the data by field-aware factorization machine (FFM) and optimize by Follow-The-Regulated D-Leader (FTRL) as our optimizer.

Q. Liu et al. (Eds.): CENet 2020, AISC 1274, pp. 628–634, 2021.
https://doi.org/10.1007/978-981-15-8462-6_71

2 Related Work

The complete recommendation system includes user set and item set, which contain different attributes. According to user attributes and preference records, personalized preference lists are recommended for different users. The most classic and most widely used is collaborative filtering algorithm [4], but the collaborative filtering algorithm cannot solve the problem of cold-start, and reflect side feature. The community correction data in our paper are unbalanced, sparse, have field feature. The CTR-prediction method is proved that can address the problems of unbalanced samples, sparse data, high dimension and multi-field features intersection in the data [5]. Therefore, this paper abstracts the recommendation problem into CTR-prediction. For research of CTR, common solutions are Logistic regression (LR), degree-2 poly-nomial mappings (Ploy2) [6], Factorization Machine (FM) [7] and Field-aware Factorization Machine (FFM) [8]. Compared with LR, FFM has better performance of nonlinear fitting [9]. As for FM, each feature has only one hidden vector to learn the potential impact of other features [10].

Therefore, on the issue of community correction measures recommendation, our method is based on FFM. As for optimizer, FTRL optimizer has shown to be more effective at producing sparsity and better performance [11] while optimizer based on Stochastic (online) Gradient Descent (SGD) are not sparse enough [12]. A research that incorporates FM + FTRL outperforms FM + SGD and has a faster rate of convergence [13]. So we establish FFM algorithm by introducing FTRL optimization algorithm, realizing the recommendation of correction measures needed by community correction objects.

3 Methodology

Due to that correction measures and objects data are sparse, FFM algorithm and FTRL optimization algorithm is introduced to enhance the sparseness of the algorithm and shorten the calculation time. The model structure consists of two parts, the data processing layer and the algorithm layer from left to right (see Fig. 1).

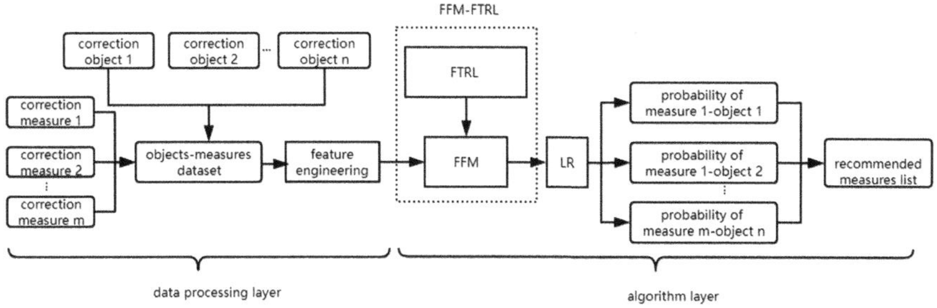

Fig. 1. Model structure.

The data processing layer combines the correction objects data and corrective measure data, encodes the correction object data and corrective measure, constructs the corrective objects-measures dataset, and divides its domain features in the feature engineering stage. The preprocessed data are input to the algorithm layer, which uses FFM-FTRL algorithm and FTRL optimization algorithm to optimize the calculation process of the characteristic parameters of FFM algorithm. The characteristic data pass through a logical regression function to predict the recommended probability of corrective measures and generate the final recommended correction measures list according to thresholds.

3.1 Data Processing Layer

In this paper, the data of correction objects and their correction plans are collected from the database of the community correction system. There are some problems in the original community correction objects data, such as missing value, invalid value. Low-quality data usually lead to serious deviation of experimental results. Therefore, it is necessary to preprocess the collected data to build a high-quality community correction objects dataset. We defined 27 attributes of original objects data.

In the feature division part, this paper divides the correction object-measure dataset into five feature fields, so as to facilitate the subsequent improvement algorithm to extract cross information between features and potential information between different fields. The data format after data preprocessing and feature engineering processing is as follows:

$$label\ field_1 : feat_1 : val_1\ field_1 : feat_2 : val_2\ \cdots \tag{1}$$

where *field* is the feature field number, *feat* is the feature number under one field, *val* indicates the value of feature. If the feature value equals 0, the data will be omitted directly to reduce the use of memory and improve the calculation efficiency of the algorithm. We finally defined the dataset as $\{(x^i, y^i)|i = (1, \cdots, L)\}$, and x^i denotes the input feature data, $y^i \in \{0, 1\}$ indicates whether this corrective measure is suitable for correction object.

3.2 Algorithm Layer

FFM algorithm introduces the concept of Field-aware to attribute features with the same properties to the same Field, in which the relationship between the fields and features is one-to-many (see Fig. 2).

The formula of FFM algorithm model equation is as follows:

$$y(\boldsymbol{x}) = w_0 + \sum_{i=1}^{n} w_i x_i + \sum_{i=1}^{n} \sum_{j=i+1}^{n} (\boldsymbol{w}_{i,f_j} \cdot \boldsymbol{w}_{j,f_i}) x_i x_j \tag{2}$$

where w_0 is the initial weight in the linear expression, w_i denotes the weight corresponding to the first-order feature of each feature item x_i, $\boldsymbol{w}_{i,f_j}$ is the weight

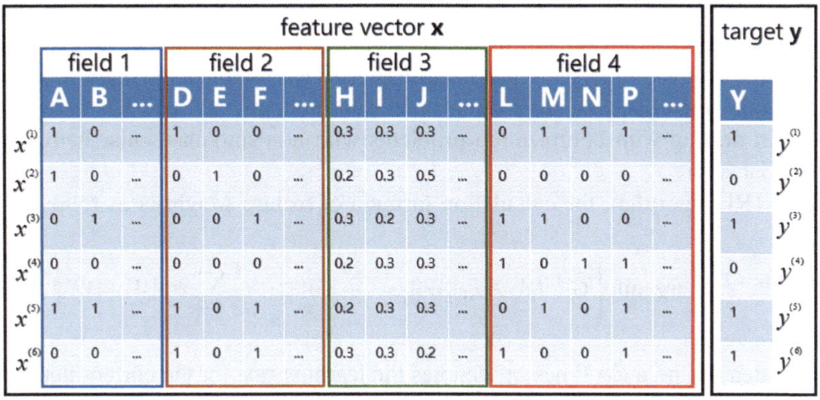

Fig. 2. Flied feature diagram.

corresponding to the second-order cross feature of different feature fields. The linear item is as follows, easy to calculate in the training process.

$$y(x) = w_0 + \sum_{i=1}^{n} w_i x_i \tag{3}$$

We give priority to the calculation of its cross-term features. The cross-term of FFM algorithm is expressed as:

$$\phi_{FFM}(\boldsymbol{w}, \boldsymbol{x}) = \sum_{i=1}^{n} \sum_{j=i+1}^{n} (\boldsymbol{w}_{i,f_j} \cdot \boldsymbol{w}_{j,f_i}) x_i x_j \tag{4}$$

In this paper, the logistic regression function is used to predict the matching probability of different corrective measures and correction objects. The formula is as follows:

$$h(x) = \frac{1}{1 + \exp(\phi_{FFM}(\boldsymbol{w}, \boldsymbol{x}))} \tag{5}$$

In order to quantify the accuracy of the results, we select the Logarithmic Loss Function as the loss function of the improved algorithm, and the formula is as follows:

$$l(\boldsymbol{w}) = -y \log(\phi_{FFM}(\boldsymbol{w}, \boldsymbol{x})) - (1 - y) \log(1 - \phi_{FFM}(\boldsymbol{w}, \boldsymbol{x})) \tag{6}$$

Due to the introduction of feature filed dimension in FFM algorithm, the cross term of the algorithm contains cross term features of feature field, and the cross term part cannot be simplified and optimized at the mathematical level. In this paper, we introduce FTRL optimization learning algorithm to optimize the weight calculation process of each feature term in FFM algorithm, so as to ensure the accuracy, improve the training time of the algorithm and enhance the sparsity of the model.

FTRL, as an online learning optimization algorithm, uses different learning rates to update the training for different dimensions of parameter weights, which can improve the accuracy of recommendation estimation. For FTRL optimization algorithm, it takes into account the characteristics of high precision and sparsity, and has better performance in dealing with optimization problems with non-smooth regular items such as L1 regular item.

In FTRL algorithm, the calculation formula of feature weight is as follows:

$$W^{t+1} = \arg\min_{W}\left\{ G^{(1:t)}W + \lambda_1 \|W\|_1 + \frac{\lambda_2}{2}\|W\|_2^2 + \frac{1}{2}\sum_{s=1}^{t}\sigma^s \|W - W^s\|_2^2 \right\} \tag{7}$$

where t denotes iteration times; η_s denotes the learning rate for the current iteration, G^s is the gradient value of the loss function for the current iteration; $G^{(1:t)} = \sum_{s=1}^{t} G^s$ is the cumulative loss function gradient iterations; $\sigma^s = \frac{1}{\eta^s} - \frac{1}{\eta^{s-1}}$ denotes the difference between the reciprocal of the current iterative learning rate and the reciprocal of the previous iterative learning rate; $\lambda_1 > 0$ is the L1 regularization coefficient and $\lambda_2 > 0$ is the L2 regularization coefficient.

For the weight calculation formula (7) in FTRL algorithm, the quadratic term is expanded, and eliminate its constant term. After calculation, the closed solution in the form of piecewise function is as follows:

$$w_i^{t+1} = \begin{cases} 0, & |z_i^t| \le \lambda_1 \\ -(\eta_i^t + \lambda_2)\left(z_i^t - \operatorname{sgn}(z_i^t)\lambda_1\right), & |z_i^t| > \lambda_1 \end{cases} \tag{8}$$

Note that $Z^t = G^{(1:t)} - \sum_{s=1}^{t} \sigma^s W^s$, FTRL algorithm adopts Per-Coordinate Learning Rates for learning rates, and the learning rates for dimension i are expressed as follows:

$$\eta_i^t = \frac{\alpha}{\beta + \sqrt{\sum_{s=1}^{t}(g_i^s)^2}} \tag{9}$$

where $\alpha > 0$ to ensure that the learning rate is always positive; $\beta > 0$ to ensure denominator is not 0; g_i^s indicates the gradient value of the loss function for the dimension i when the number of iterations is s.

After the training by FFM-FTRL algorithm, the logical regression function is used to predict the corrective object-measure adaptation probability, and reasonable thresholds will be set through experiments to generate the final correction measures list.

4 Experiments

4.1 Datasets

The correction object-measure dataset in our paper contains 130699 rows, includes 2 categories of corrective measures and 27 attributes of correction objects data. In the experiment, the training set and the validation set were separated by the ratio of 9:1.

4.2 Implement

Firstly, we select proper probability thresholds so that the log loss function value used in our model can be minimized. Dichotomy thinking is adopted to search the best parameter of the threshold. The threshold range is delineated in the initial interval, and then the parameter adjustment granularity is decreased to 0.01. When the threshold equals 0.52, the log loss function value is the lowest, and the recommendation prediction effect is best at this time. In order to better reveal the concrete effect of our method, we select FM and FFM-SGD as our baselines under the same dataset, and we compare the F1 score and training time.

4.3 Evaluation

The left figure of Fig. 3 shows the F1 score of different algorithms in 5 times independent experiments. Our method has a larger improvement in F1 score compared with FM, has a smaller improvement compared with FFM-SGD, an average improvement of 4% and 0.5% respectively.

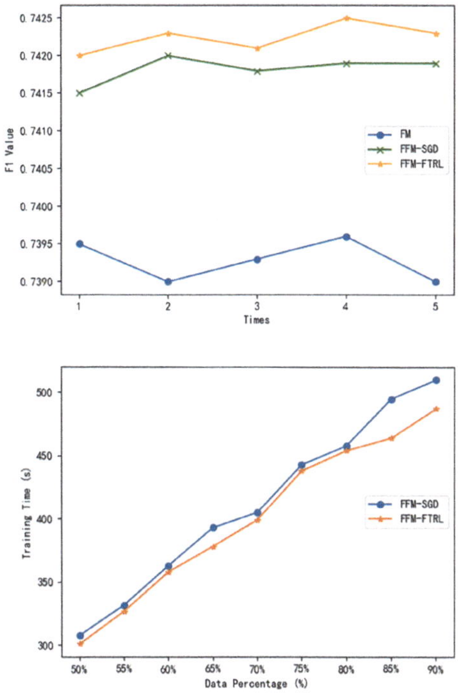

Fig. 3. Comparison of F1 score (Left) and training time (Right).

In the right chart, it can be seen intuitively that the training time of our method is similar to that of FFM-SGD with 50%-80% of input data, and it is slightly reduced.

When the input data is up to 85%, the training time is greatly reduced compared with FFM-SGD.

5 Conclusion

In this work, we proposed a novel method FFM-FTRL to address the recommendation problem of correction measures for community correction objects. Our method consists of a data processing layer and an algorithm layer in which FTRL optimization algorithm is introduced to optimize FFM. Our method overcomes drawbacks in baselines, and the algorithm training time is also shortened. In the future, we will decrease the granularity of recommendation results to compose more definite correction plan, discuss whether recommended measures are suitable in actual work.

References

1. Li, E.: China's community corrections: an actuarial model of punishment. Crime Law Soc. Change **64**(1), 1–22 (2015)
2. Jiang, S.H., et al.: Community corrections in China: development and challenges. Prison J. **94**(1), 75–96 (2014)
3. Taxman, F.S., Steven, B.: Implementing Evidence-Based Practices in Community Corrections and Addiction Treatment. Springer, Heidelberg (2011)
4. Yao, Y., et al.: Overview of collaborative filtering techniques. Zhuangbei Zhihui Jishu Xueyuan Xuebao (J. Acad. Equip. Command Technol.) **22**(5), 81–88 (2011)
5. Richardson, M., Ewa, D., Robert R.: Predicting clicks: estimating the click-through rate for new ads. In: Proceedings of the 16th International Conference on World Wide Web (2007)
6. Chang, Y.W., et al.: Training and testing low-degree polynomial data mappings via linear SVM. J. Machi. Learn. Res. **11**, 1471–1490 (2011)
7. Rendle, S.: Factorization machines. In: 2010 IEEE International Conference on Data Mining. IEEE (2010)
8. Juan, Y.C., et al.: Field-aware factorization machines for CTR prediction. In: Proceedings of the 10th ACM Conference on Recommender Systems (2016)
9. Ren, J., Jian, Z., Jing, L.: Feature engineering of click-through-rate prediction for advertising. In: International Conference in Communications, Signal Processing, and Systems. Springer, Singapore (2018)
10. Rendle, S.: Factorization machines with libfm. ACM Trans. Intell. Syst. Technol. (TIST) **3** (3), 1–22 (2011)
11. McMahan, H.B., et al.: Ad click prediction: a view from the trenches. In: Proceedings of the 19th ACM SIGKDD International Conference on Knowledge Discovery and Data Mining (2013)
12. Bottou, L.: Online algorithms and stochastic approxima-P tions. Online learning and neural networks (1998)
13. Ta, A.P.: Factorization machines with follow-the-regularized-leader for CTR prediction in display advertising. In: 2015 IEEE International Conference on Big Data (Big Data). IEEE (2015)

The Current Situation and Improvement Suggestions of Information Management and Service in Colleges and Universities—Investigation of Xianyang City, Western China

Yang Sun(✉)

School of Education Science, Xianyang Normal University, Xianyang 712000, China
sunyangdjzxx@163.com

Abstract. In recent years, with the rapid development of information technology, artificial intelligence and Internet of things technology, the information construction of colleges and universities in China is also accelerating. This study investigates the current situation of the information management and service of nine colleges and universities in Xianyang City, Western China. From the three aspects of information resource construction, management and service, it understands the construction of campus network in the third-tier city in Western China, the construction and application of multimedia network classroom, network system management, application platform management, teaching system management, user service management, learning resources and basic services etc. In view of the shortcomings in management and service, some suggestions for improvement are put forward.

Keywords: Colleges and universities · Information management · Information service · Suggestion

1 Introduction

The continuous innovation of modern information technology is leading the development of global economy and society. Human society is moving from industry and information society to intelligent and internet of things society. Intellectualization and IOT have become the strategic commanding point for the development of world powers. With the continuous development of information management and service construction for the purpose of "Digital Campus", digital management and one-stop service in Colleges and universities are promoted to be actively applied.

At present, almost all colleges and universities use information-based means to manage students, teachers and other users. By configuring hardware equipment and purchasing management system, the information-based means are applied to the management of university affairs. By means of innovative management and optimized service, the information management and service in colleges and universities can meet the needs of users more. Service-oriented is the principle of management and service in colleges and universities in the information age. The goal of information management

Q. Liu et al. (Eds.): CENet 2020, AISC 1274, pp. 635–642, 2021.
https://doi.org/10.1007/978-981-15-8462-6_72

in colleges and universities is to solve the specific problems of users in the process of using information, optimize the experience effect of service, improve the efficiency of college work, and basically satisfied the needs and feedback of most users [1].

2 Research Purpose and Method

Based on the investigation of the information management and service experience of 9 universities in Xianyang City, this study finds out the problems existing in the management and service of the information management system from the perspective of infrastructure construction and user experience, and puts forward constructive solutions. This study can provide the basis for the information management of colleges and universities, and has a certain significance to promote the improvement of the information service level.

The research mainly uses literature research method to collect data about campus information management and service through CNKI, Google academic search and other tools, analyzes and defines the concept and field of campus information management and service, and develops a questionnaire based on the relevant content learned, which is distributed and recycled in 9 universities in Xianyang City, using SPSS, excel and other tools analyze the survey data.

3 Research Design and Implementation

3.1 Design of Questionnaire

According to the specific situation of the university, the construction of the campus infrastructure is divided into campus network construction and multimedia network classroom construction; the information management is divided into four aspects: network system management, application platform management, teaching system management and user service management; university information service is divided into learning resources construction and specific service categories [2].

3.2 Research Object Selection and Questionnaire Distribution

In October 2019, the researcher asked the students of grades 1–4 from nine universities in Xianyang, Shaanxi University of Chinese Medicine, Xianyang Normal University, Xizang Minzu University, HAO JING College of Shaanxi University of Science & Technology, Shaanxi Polytechnic Institute, Xianyang Vocational Technical College, Shaanxi Technical College of Finance & Economics, Shaanxi Post and Telecommunications College, and Shaanxi Energy college. A total of 270 questionnaires were distributed and 257 questionnaires were actually recovered. In the survey of information management and network resources in colleges and universities, the author collected data by visiting information construction centers and network resources and management websites, and distributed questionnaires to the managers of information construction management offices in colleges and universities, nine copies in total and nine copies effectively recovered.

4 Data Analysis

4.1 Infrastructure Construction

Campus Network Construction. The campus network construction is the foundation of university operation and the basic premise of information management. From the survey, we can know that nine universities in Xianyang city have access to China Education and Research Network (CERNET), and 66.7% of the schools have access to China Unicom, 88.9% of the schools have access to China Telecom, 44.4% of the schools have access to China Mobile as a supplement to the bandwidth.

66.7% of the universities basically cover wireless network, 11.1% of the universities cover 40%, 60% and 80% school region, and 77.8% cover more than 80% school region. The author also investigates the backbone broadband of the campus network of each university, and the results show that more than 90% of the universities have access to gigabytes and megabytes networks. The network construction of the nine universities in Xianyang city has achieved obvious results, which has laid a solid foundation for further campus information management and service, and is a good start for the construction of digital campus.

Multimedia Network Classroom. The construction of multimedia network classroom is the basic hardware construction of the development of colleges and universities, and it is a standard to measure the construction of school hardware equipment. The number of classrooms in three schools is 400–500, among which one school's multimedia network classroom accounts for 78%, one for 56%, and one for 33%, The number of classrooms in three schools is 300–400, among which one school's multimedia network classroom accounts for 100%, the other two accounts for 43%, the number of classrooms in two schools is 200–300, the two schools' multimedia network classroom accounts for 60%, the number of classrooms in another school is 100–200, and all the teaching rooms are multimedia network classrooms.

In general, the number of multimedia classrooms in all schools is 100–200, and there are two schools with 300–400 multimedia classrooms, which basically meet the needs of teaching infrastructure in each school.

4.2 Information Management

Network System Management. Network system management provides important business services such as software and hardware configuration, network security management, information filtering system configuration, network management regulations, etc. Through investigation, we know that 9 universities have built network security system, including information filtering system, network anti-virus system, network operation fault monitoring system.

Six of the schools surveyed have launched IPv6 network services, and three have not. The use of IPv6 can not only solve the problem of the number of network address resources, but also solve the barrier of multiple access devices connecting to the Internet [3]. But not all of the nine colleges and universities in Xianyang are popular.

Application Platform and System Management. According to the survey, 55.5% of colleges and universities have completed the construction of digital campus, and these schools have applied all application management systems to the campus network platform; 44.4% of colleges and universities are in the stage of business re integration, and are in the process of integrating the campus system platform. The data sharing information system has been improved. The systems that can realize the efficient data sharing include the teaching school information system, the library information system, the office automation system, the scientific research information system, the personnel management information system, the student management information system, the graduation information system, the file management information system and the school enterprise sharing information system.

Teaching System Management. The content of educational administration network management in colleges and universities mainly includes user login and query, teaching arrangement, file download and management regulations. Among them, user login, query and teaching arrangement account for 27.3%, file download and management regulations account for 12.1% and 21.2% respectively. In some universities, it also includes other educational administration network management contents, such as school calendar, discipline construction, management personnel, teaching arrangement and other items.

According to 257 questionnaires surveyed by students, the educational administration system in colleges and universities is relatively perfect in information inquiry and examination registration, but students are less satisfied with the course selection management system. It is believed that the course selection system often presents problems such as jam and crowding at the beginning of the semester, which is related to a large number of students' course selection operation at the same time. It also shows that for the course selection management party. The educational administration should make some plans to solve these problems.

User Service Management. The user management service platform of these universities generally includes account login, data improvement, user payment, information change, internet fee service, user self-service, account opening and other specific functions. The information management part is divided into four modules, including data management, user authority management, account password and security strategy. These are the basic parts of University user information management. Data management includes identity identification of all users in University, management of basic data, user authority provides basic service authority for campus users, and account password is the basic measure to ensure user information security. The security policy is to solve the security vulnerabilities of all users and to repair them.

4.3 Information Service Status

Information service can be divided into two categories: digital learning resource construction and basic information service. The subprojects of learning resources include network teaching platform construction, campus library construction, public network resources construction and school-based curriculum construction; basic information services are based on information management platform, including campus portal information, campus service platform, campus all-in-one card and student employment information platform, which can be summarized as teaching management information and data Resource sharing informatization and service project informatization. The research shows three remarkable characteristics of information service in universities in Xianyang City: first, to meet the needs of users for multi-dimensional services, the service is convenient and mobile, breaking the shackles of traditional ideas. Second, the data sharing system is huge. The three is online examination, daily attendance card, network payment and other multi-functional services.

Construction of Learning Resources. Learning resources construction is a platform for teachers and students to provide learning resources, improve teachers and students more convenient access to, use of knowledge, increase the storage of knowledge, improve their knowledge literacy. 71.21% of colleges have digital libraries, 80.93% of colleges provide online learning resources, and 63.8% of colleges have school-based courses. The average number of e-books in most colleges is more than 1 million.

Through the survey on the satisfaction of college students with learning resources, 70% of the students are very satisfied with the construction and function of campus network; 44% of the students are very satisfied with the category and quality of books; 50% of the students think that the effect of learning with network learning resources is satisfied.

Basic Services. University in Xianyang has realized all-in-one card services including identity recognition, book borrowing, medical treatment, dining, bathing, access control, payment and water purchase. All schools have realized the three basic functions of identity recognition, book borrowing, dining. Different colleges all-in-one card contains different services, some of which have more functions and some of which have less functions. The information service and construction level of some schools need to be further improved.

The campus information release platform can provide all teachers and students with the latest campus information, policies and services. According to the results of the survey, most universities in Xianyang have opened information publishing platforms, including school portal, official Wechat, official microblog, and official QQ campus service platforms, accounting for 24%, 27%, 19%, and 19.5% respectively. Colleges and universities in Xianyang city have made remarkable achievements in the construction of information release platform, which brings convenience to the acquisition of information for teachers and students.

According to the survey on the satisfaction of the information services provided by colleges and universities, 15.2% of the people think that they are very satisfied, 35.4% think that they are satisfied, 42% think that they are average, and 7.4% of the students are not satisfied. There is still a lot of space to improve the information service provided by colleges and universities. It is necessary to improve the exact scheme according to the specific problems, and strive to obtain more satisfactory service.

5 Research Results and Problems

5.1 Infrastructure Construction Has Been Preliminarily Completed, and Information Management Projects Are Becoming More and More Perfect

Xianyang university campus network entrance, network speed and wireless network coverage have reached the requirements of the 13th five-year plan, and the construction of hardware resources has also reached the requirements of development.

The information service provided by colleges and universities is very convenient t. from the aspects of students' study, life and work, colleges should strengthen the construction of digital library, network resources, comprehensively popularize the service of teachers and students' all-in-one card, provide the service platform of campus life, and provide the service platform of students' work.

University information management project includes the comprehensive management of network system, vulnerability detection and fault monitoring; the integration of application platform, the construction of data sharing platform, the most important step for the construction of digital campus; the teaching system carries out detailed management of student work, teacher work and curriculum arrangement; user service provides basic services and ensure the security of the user's account. Xianyang university has achieved comprehensive functional coverage in the above information management projects.

5.2 Information Service Embodies Users as the Center

Most teachers and students are satisfied with the all-in-one card service, the effect of network learning resources, the construction of digital library and school-based curriculum. It reflects that these colleges and universities pay attention to the user centered information service experience, which is an important basis for management evaluation.

In general, the information management projects in Xianyang universities are becoming more and more perfect. New management projects are becoming more and more diversified and integrated. It not only makes the whole university's teaching work more convenient, but also makes it easier for managers to be systematic, more simple and efficient.

5.3 The Problems of Information Management and Service in Colleges

Some colleges are not strong enough in the construction of campus informatization, with less investment in equipment, lack of network resources, and poor use effect. In the nine colleges and universities surveyed, some leaders do not have a good understanding of information management and pay enough attention to it, which directly affects the development of information management and service. Each university has its own information or network management center, but the degree and quality of its construction are different. No matter the professional team hired or the university own team has not reached the ideal management goal [4].

The sharing degree of information system in Colleges and universities is not enough, and there is the problem of repeated construction. In the process of database system construction and development, colleges lack of overall thinking and unified planning, resulting in scattered information among all departments of colleges. From the technical point of view, university information managers should integrate technology and resources, so that all departments can make unified planning in data sharing, management, software and hardware operation and maintenance and teaching.

6 The Promotion Strategy of University Information Management and Service

6.1 Information Data Should Be Fully Coordinated Among All Departments in the University

Information sharing not only benefits all teachers and students, but also provides scientific research services, talents and information services. To promote the collaborative development of university's business, service and demand, it requires the full front-end analysis of university's information management project evaluation, including analysis of the objects of university management and service, comprehensive statistics of department's needs, and preliminary opinion collection of the information service to be implemented by University. These are important factors to promote the coordinated development of information management and service in Colleges and universities in Xianyang city.

6.2 Optimize the Allocation of Resources, Pay Attention to the Operation and Maintenance of Software and the Protection of Hardware

Information management is a complex project, the design of information management system is also a system engineering, because it includes not only many program background, software system, but also hardware resources, equipment, network system, which are necessary factors to form a complete information management system. Information management should optimize the allocation of resources, and university information management and maintenance personnel should regularly test the software and hardware to extend the service life of the system.

6.3 Attach Importance to the Overall Planning of Information System and Improve the Ability of Professional Managers

Information data is scattered and sharing degree is not high, which will lead to repeated data entry and affect the quality of service. College managers should innovate the concept of information technology, strengthen the unified construction planning, and make the data fully shared. At the same time, increase the training and learning of management personnel, constantly enrich their experience and technology, and improve the level of information management personnel.

Acknowledgement. This paper is the research results of the special scientific research plan project of Shaanxi Provincial Department of education "The research on the current situation of informatization course construction and students' informatization learning ability of primary and secondary schools in Shaanxi Province" (No. 18JK0815); Research project of education and teaching reform of Xianyang Normal University in 2019 (No. 2019Y019); The project of scientific research plan project of Xianyang Normal University in 2019 (No. XSYK19035).

References

1. Ministry of Education.: Outline of national medium and long term education reform and development plan (2010–2020) [EB/OL] (2010). https://www.moe.gov.cn/srcsite/A01/s7048/201007/t20100729_171904.html?gs_ws=tqq_635879677144434007
2. Xie, Y.R., Li, K.D.: Fundamentals of research methods in educational technology, vol. 12, pp. 56–57. Higher Education Association, Beijing (2006)
3. Jiang, T., Li, Z.G., Dai, S.X.: Analysis of the problems and countermeasures in the information management of higher education. China Audio Vis. Educ. **11**, 33–35 (2008)
4. Sun, Y.: Research on the current situation and countermeasures of e-book package based learning mode. Shaanxi Normal University (2014)

Research on Location Method of Network Key Nodes Based on Position Weight Matrix

Shihong Chen(✉)

Department of Information Engineering, Guangdong Eco-Engineering Polytechnic, Guangzhou 510520, China
48131708@qq.com

Abstract. Aiming at the problem that the current network key node positioning effect is not good, the network key node positioning method is optimized and researched by combining the position weight matrix, the characteristic values of the network key node positioning are collected and analyzed by combining the characteristic collection principle, and the distribution area of the network nodes is selected and accurately positioned according to the characteristic collection result. Finally, the experiment proves that the network key node location method based on the position weight matrix has high accuracy and stability, and fully meets the research requirements.

Keywords: Location · Weight matrix · Network · Node location

1 Introduction

As an important way to locate key nodes in the network, dynamic sensor networks have been widely used in many fields in recent years, such as environmental monitoring, animal tracking, medical and health care, underground personnel search and rescue, etc. all depend on the mobility of sensor nodes. Mobile nodes not only improve the flexibility of event monitoring, but also enhance the self-organization and adaptability of wireless sensor networks. For dynamic sensor networks, the location information of nodes is the premise of their various applications. The mobility of nodes makes the node localization of dynamic sensor networks more complicated than that of static sensor networks [1]. How to locate nodes with low power consumption, low complexity and high accuracy under dynamic conditions is one of the challenges faced by wireless sensor networks. This paper studies the localization problem of dynamic sensor network model with unknown nodes stationary and anchor nodes moving, and proposes a mobile anchor node localization algorithm based on virtual beacon selection. This algorithm studies how to reduce the error accumulation and ensure the minimum positioning error in the presence of ranging error from a geometric perspective [2]. It quantitatively analyzes the errors introduced by the relative positions between nodes, and proves the theorem of minimum positioning error when the positions of the three virtual beacons participating in positioning are equilateral triangles through mathematics. According to this principle, the unknown nodes select the virtual beacons appropriately in the localization process, and obtain more accurate location information.

Q. Liu et al. (Eds.): CENet 2020, AISC 1274, pp. 643–654, 2021.
https://doi.org/10.1007/978-981-15-8462-6_73

2 Network Key Node Positioning Method

2.1 Network Key Node Location Feature Collection

Sensor node is the basic component unit of network key node positioning. The energy required by sensor node operation is provided by the positioning node module. The network node control module is the core part of sensor nodes and plays an important role in the storage, communication and sensing functions of regional information. The sensor nodes work under the unified command and coordination of the network key node positioning control module. Information collection and data conversion in the positioning area are mainly completed by the sensing module, and various messages, events, tasks and information in the working process are stored in the storage module [3]. If you want to communicate with other nodes wirelessly, or exchange control messages, send and receive data, etc., you need the support of the communication module. Communication in a complex network environment is prone to locating faults due to the large scale of transmitted data and strong disturbance [4]. Therefore, fault data can be accurately mined to realize rapid fault location. With the enhancement of data coupling, the fault characteristics of related data cannot reflect the communication condition information under the network environment, resulting in low accuracy of data mining. To solve this problem, a fault data mining method based on dependency graph is proposed to analyze fault data in time domain. A set of relatively stable fiber network communication fault data signal analysis models are established by time domain analysis method, as shown in the Fig. 1:

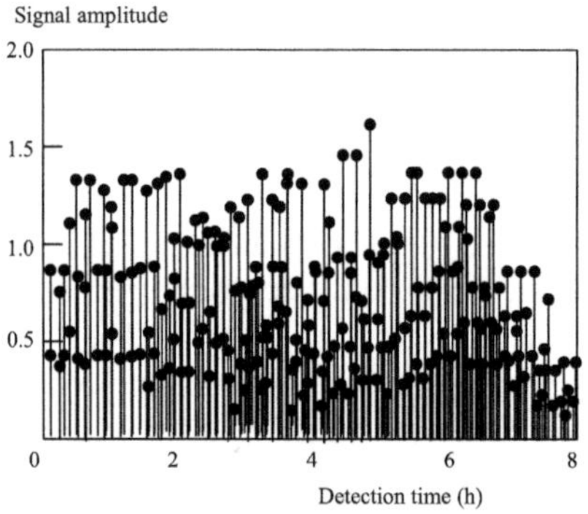

Fig. 1. Network positioning data signal acquisition.

In the relatively stable fault data signal model shown in the above figure, time domain analysis shows that the scalar time series of optical fiber communication data is as follows: $a(t)$, $t = 0, 1, \ldots, n-1$, the basic characteristic signal function can be described as:

$$x = [x_1, x_2, \ldots, x_N] \in Q^{mN} \tag{1}$$

The time domain feature analysis method is adopted to collect the horizontal and vertical feature coordinate gradient values of the network key node positioning area coordinates and decompose the feature coordinate gradient values so as to obtain the maximum gradient feature difference value of data mining. The specific algorithm can be recorded as follows:

$$f = \frac{1}{i \times j} \sum_{x=1}^{j} \sum_{y=1}^{i} |f_x(x, y)| \tag{2}$$

In the above formula, x and y can be recorded as feature mining correlation coefficients of network node feature data, and the correlation directivity values of feature data are further calculated [5]. The specific algorithm is as follows:

$$\lambda = \frac{\|Q - f\| - 1}{\sqrt{H(y) - L(x)}} \tag{3}$$

In the above formula: H(y) and L(x) are the energy mean values respectively. when the network node location area sends a narrow pulse signal, the signal thus transmitted will reach the carrier frequency component of the receiver along each path as follows:

$$y(n) = \frac{n(h) + a}{2\lambda H(y) + fL(x)} \tag{4}$$

$$x(n) = \frac{2[n(h) + a - 1]}{fH(y) + \lambda L(x)} \tag{5}$$

In the above formula: n(h) is the noise component, and a is the transmission delay of the I-th communication channel, and the time domain distribution characteristics of the node characteristic data thus obtained are:

$$K(t) = \sum_{i=1}^{n} \mu p_i [\sin y(n) + \cos x(n)]^{n-i} \tag{6}$$

In the above formula: μ Indicates the broad band of the data set; p_i Indicates the fault data information after removal processing [6]. Recursive characteristic value of n data. Further assuming the continuous frequency domain characteristic values of network data are calculated, the specific algorithm is as follows:

$$a_n = K(t)[p_i(i + n\bar{t})] + \vartheta \tag{7}$$

In the above formula: $\bar{t}$ is the time difference. ϑ is the measurement error. According to the analysis results of network node data time domain characteristics through the above algorithm, an accurate data foundation is provided for data mining, the smallest subset in the entity set is found in a certain stage, which indicates that the probability of basic fault sources included in all early warnings is the largest, the initialization result set is an empty set, the initial binary set is input, and early warning information is received [7]. According to the division stage and selection stage, the design fault location workflow is shown in the Fig. 2.

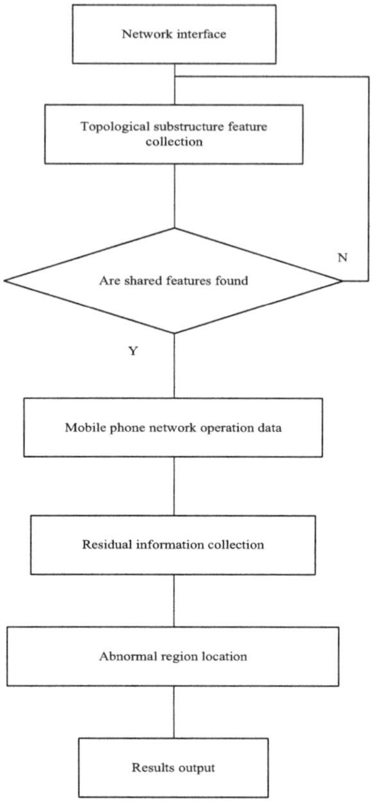

Fig. 2. Fault location workflow.

The purpose of the fault location algorithm is to find possible faults in all locations and ensure the normal operation of network communication [8]. Based on the flow shown in the above figure, the steps are analyzed in detail as follows:

(1) acquiring two early warning information in any early warning set;
(2) finding a node which can explain two early warning information at the same time;
(3) using these nodes to replace the original early warning information;
(4) repeatedly obtaining early warning information until there is no intersection explanation between the two early warning information [9].

According to the execution results of the above steps, the best explanation of all the early warning data is obtained, namely the fault node [10]. In order to obtain each alert explanation, it is necessary to find the upper node of the alert node. According to the dependency graph model, cross chain storage is used between each node, which is convenient to find the upper nodes of each node [11].

2.2 Selection of Abnormal Region of Node Features

According to the above algorithm, the abnormal region of the network key node is further judged, and the cluster channel value of the designed network key node shows that the attacked node has dynamic attribute, so it is easier to locate it [12]. The main parameter set algorithm for setting the anomaly value of vulnerable mobile nodes in the network is as follows:

$$f = \{(x_1, y_1, z_1), (x_2, y_2, z_2), \cdots, (x_n, y_n, z_n)\} \tag{9}$$

In the above algorithm: (x_n, y_n, z_n) Attacks the coordinates of common nodes for the nth malicious act. The distance between malicious behavior and common nodes can be shortened, and then the general position of the attacked node can be displayed until the attack behavior is completed. When the spatial position of the vulnerable area of the network key node changes, malicious behavior 3 will intercept the data packet and release the interference, resulting in the loss of the data packet [13]. At this time, the channel tracking fusion technology is used to determine the position of the attacked node, as shown in the figure.

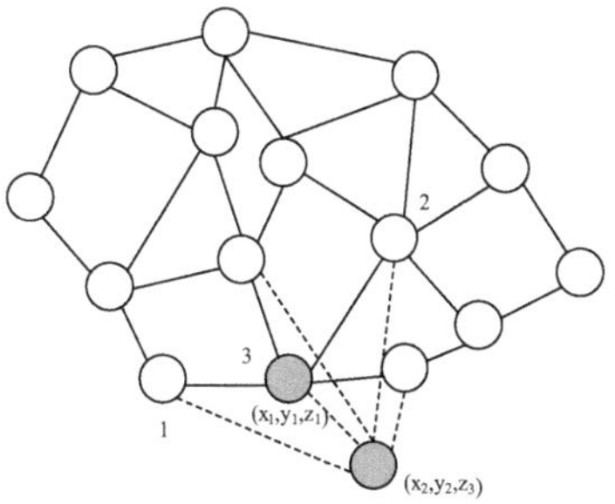

Fig. 3. Network node abnormal region judgment.

As can be seen from Fig. 3, the spatial coordinates of malicious act 3 are (x_1, y_1, z_1) However, the location result is affected by the false location, which leads to the reduction of the location accuracy of vulnerable nodes in network movement [14]. Assuming that the proportion of nodes attacked by malicious behaviors is P and the total number of network nodes is N, then I nodes selected from N common nodes are combined as follows: $A = C_N^i$ Of the A selection combinations, the probability that at least one combination does not include the attacked node is:

$$P' = 1 - \left[1 - (1 - P)^{i}\right]^{A} \tag{10}$$

According to the above algorithm, the value size is further determined, and the node combination scheme is selected according to the determination result. Use coordinate fusion method to calculate the real position of nodes:

$$E = \frac{\left(z_n - \sqrt{\left(x_j - x_1\right)^2 + \left(y_j - y_1\right)^2 + \left(z_j - z_1\right)^2}\right)}{i}, j = 1, 2, \cdots, A \tag{11}$$

In the above formula: $\left(x_j, y_j, z_j\right)$ In order to select the mobile vulnerable node coordinates of the J scheme. According to the above process, the position of the mobile vulnerable node can be determined, and the channel tracking fusion technology can effectively improve the positioning accuracy [15]. By analyzing the status of feeder terminal units in distribution automation, it is determined whether the switches in each area can be started normally, which reflects the relationship between each element of the distribution network and the switches. For a switch function that can be defined as:

$$F(\mathrm{x}) = \left[1 - \sum_{m=1}^{m \neq n} x(m)\right] \|x(m+1)\| \ldots \|x(m+G-1)\| \tag{12}$$

In the formula: $F(\mathrm{x})$ represents a switching function; $x(m)$ Indicates the state value of the M th component downstream of the switch; G represents the number of components downstream of the switch. Further check and calculate the fitness function of the key nodes of the network as follows:

$$k(\mathrm{x}) = \sum_{j=1}^{h} |\mathrm{f}(\mathrm{x}) - F(\mathrm{x})| \tag{13}$$

In the above algorithm $k(\mathrm{x})$ Is the fitness value in space; $\mathrm{f}(\mathrm{x})$ is the true state of the switch. Fault location is calculated by searching the network space for the minimum difference between the actual state value and the derived state value, which is the minimum solution in the formula. According to its minimum solution, the node region can be located accurately.

2.3 Implementation of Network Key Node Localization

Based on the above algorithm, the positioning of key nodes in the network is further realized, and the self-positioning operation algorithm of the wireless sensor network is further optimized. Through positioning the operation direction of the mobile node, the movement rate value of the network key node is judged, so that the movement state of the target position in the working equipment can be accurately obtained, the movement trend of the position can also be accurately judged, and the movement data at different times are recorded and stored in an upper system database, so that the specific positioning information and time information of the network node are judged, and for convenience of recording, the initial point of the positioning target is recorded as follows $x(t_1, a_1, b_1)$, $y(t_2, a_2, b_2)$. Where t_2 is greater than t_1. According to the characteristics of continuous operation of the network, assuming that the mobile device has been moving linearly for a certain period of time, the numerical algorithm for predicting the movement rate and movement length of the mobile device is as follows:

$$V_{\mathrm{a}} = \frac{a_2 - a_1}{t_2 - t_1} \tag{14}$$

$$V_{\mathrm{b}} = \frac{\mathrm{b}_2 - b_1}{t_2 - t_1} \tag{15}$$

Further estimating the stay positions of key nodes of the positioning target for screening and calculation, the specific algorithm is as follows:

$$S = (b_1 + V_{\mathrm{a}}(a_2 - t_2), b_2 + V_b(a_1 - t_2)) \tag{16}$$

The positioning sensor node is redesigned according to the movement speed and the staying position of the mobile equipment, and is calculated by combining the principle of minimum point jump number. The characteristic information exchange value of anchor node data is obtained by using the energy long-distance exchange theorem for analysis, and the corresponding node positioning characteristic data information is obtained. The known data information N is output as the data identification of anchor node in the positioning area for calculation, and finally the coordinates of anchor node are obtained (x_i, y_i) Record the number of completed steps of this point in different networks, and all points in the future can adopt compatible combination of information and data to obtain its number of point hops, which is recorded as $\left(x_j, y_j\right)$. However, if there is a situation in the middle where the node has multiple moving paths to correspond to various different number of point hops, the path U with the smallest number of point hops is selected, because the specific algorithm for obtaining the minimum number of point hops at this time is:

$$HopSize\,\lambda = SUN \sum \frac{\lim\limits_{x \to \infty} \sqrt{(x_i - x_j)^2 + (y_i - y_j)^2}}{2\sum\limits_{i \neq j} H_{ij}} \tag{17}$$

If there are B anchor nodes in the positioning system and the number of hops between the two points is E, the average hop distance L of the anchor nodes can be obtained through calculation, and then the positioning information is checked by the collected data. Since an anchor node has only the first area check information, in order to ensure the accuracy of the obtained check data, other data are usually deleted after the first check data of the anchor node is not known. In order to ensure the accuracy of the deleted data, redundant data characteristics are calculated, specifically as follows:

$$d_{ij} = HopSize\,\lambda * \sum\sum (SH_{ij} - B)E/2L \tag{18}$$

According to the MEMS error data algorithm, the influence degree on the accuracy of the positioning value is calculated. If the inertial tracking value is restricted to a fixed bit P, most of the noise reduction measures adopted are to reduce the noise of the small acoustic wave of the positioning node, and the data signal F is decomposed into lower frequency noise A and higher frequency noise D. After m times of decomposition, it can be obtained:

$$C_{ij} = M + \sum \frac{\log{:}d_{ij} * F(P + A)}{\lim_{0 \to \infty} 2(D - S)} \tag{19}$$

According to the different characteristics of data signal and noise in different scales, a reasonable threshold standard is selected to analyze and process the small acoustic wave coefficient. Finally, the data signal is reconstructed to realize the isolation of data signal and noise. So as to modify and locate the target area. The following are the basic principles of target location (Fig. 4):

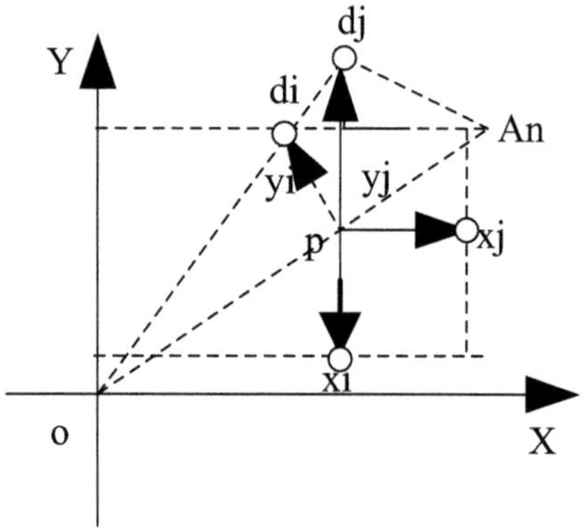

Fig. 4. Principle of network target location.

In the above figure, p represents the actual positioning of the target, o represents the origin position and is set in the center of the photo-sensitive positioning image. the coordinate positioning of the specific pixel is obtained by realizing the coaxial data image. Once there is a light weight and small volume unmanned aerial vehicle (UAV) system error occurs, the target positioning is modified twice to ensure the modified data is scientific and reasonable, thus realizing accurate calculation of the positioning area of the wireless sensor network itself. Once the lower foot completely contacts the ground during the positioning operation period, it is the zero speed interval. By analyzing the distance change of the moving speed, the zero speed interval of the moving distance of the recognition target is analyzed, and the real-time updating of a certain speed value in the region is carried out. According to a plurality of

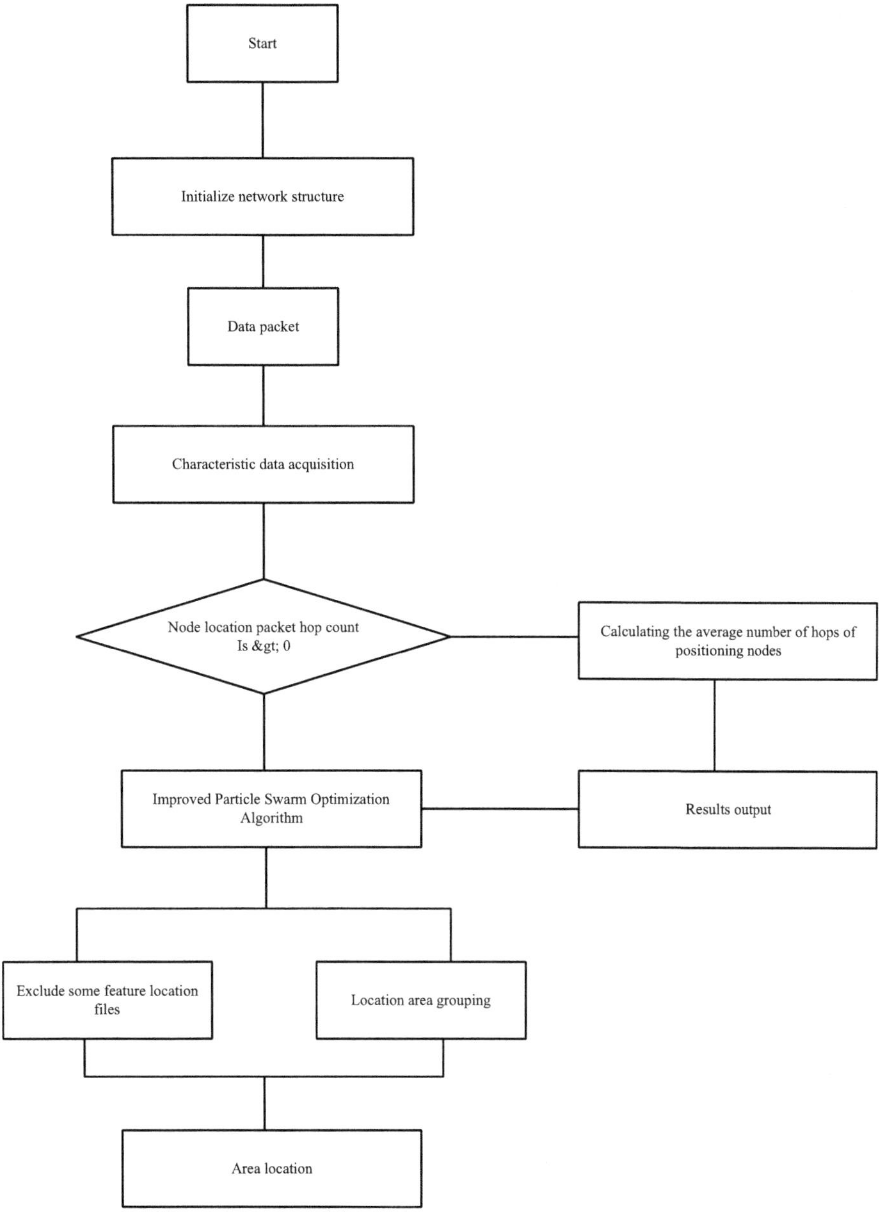

Fig. 5. Wireless sensor network positioning process.

conditional threshold values, the algorithm is judged to judge the positioning detection interval again, and the collected moving filter change is judged by combining the minimum two-layer value, so that the noise value in the positioning process is eliminated, the estimation data of the previous time and the current real-time observation data are analyzed to complete the update estimation of the current state, and if the deviation value of the moving positioning is n, the specific algorithm of the estimated positioning deviation value is as follows:

$$G_i(x_n, y_n) = \sum \lim_{0 \to \infty} L_i(m+1) + d_{ij} - n \tag{20}$$

If the above calculation result is minimized, the error value between the anchor node and its actual position is obtained to be minimum, and the minimum value of coordinates calculated at this time is the optimal solution. The positioning process is optimized based on the above algorithm. In order to solve the problem of non-linear optimization, the applicability function of the anchor node group is further modified, the algorithm of the anchor node group is further modified, the data information communication calculation ability of multiple anchor node groups is enhanced, and the optimal answer is found. The detailed calculation process of the above algorithm is as follows (Fig. 5):

Based on the above steps, the relocation of network key nodes can be effectively realized, the accuracy and effectiveness of network node location can be ensured, and the safety effect of network operation can be improved.

3 Analysis of Experimental Results

In order to verify the practical application effect of the network key node location method based on the position weight matrix, comparative testing is carried out. In order to ensure the accuracy and effectiveness of the experimental results, the experimental environment and experimental parameters are set in a unified standard. Assuming that a large number of network communication nodes are distributed in a uniform array area of 2,000 m × 2,000 m, the experimental parameter settings are shown in Table 1.

Table 1. Setting of experimental parameters.

Experimental parameters	Experimental value
Network frequency band	4 kHz–6 kHz
Carrier frequency time width	5 ms
normalized frequency	0.25 Hz
Number of sampling points	226
Signal-to-noise ratio variation range	−10 dB–10 dB

In the above experimental environment, further record and analyze the positioning accuracy of the traditional positioning method and the detection results of the network key node positioning effect based on the position weight matrix proposed in this paper, and judge the specific detection results of the two methods in the positioning process as shown in the following Fig. 6:

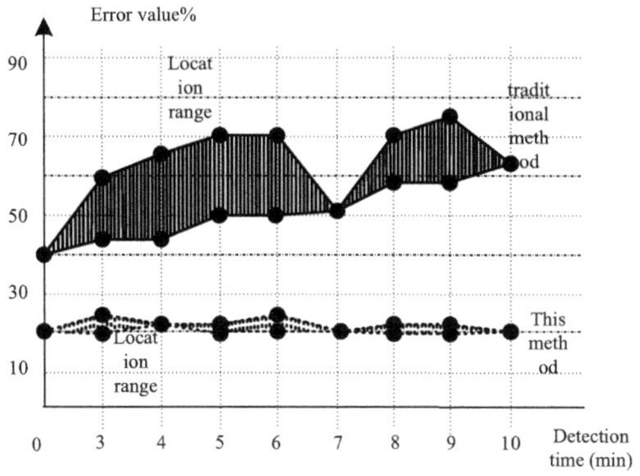

Fig. 6. Comparison test results.

Observation of the above detection results shows that, compared with the traditional detection methods, the network key node positioning method based on the position weight matrix proposed in this paper has relatively low error value and small positioning range in the detection process, thus proving that the network key node positioning effect based on the position weight matrix is relatively better, which can better meet the rapid and accurate positioning of network key nodes and ensure the accuracy of positioning results.

4 Conclusion

In static wireless sensor networks, beacon nodes cannot completely cover the entire network. In the above network structure, uncovered nodes can only approximately estimate their specific locations. In order to better solve the above problems, the blind areas in the mobile network that are not covered by nodes are improved to further provide more accurate positioning coordinates. In static wireless sensor networks, the combination of hot spot effect will lead to network disconnection and affect data transmission and collection. The location method of key nodes based on position weight matrix has high accuracy, effectively solves the above problems, can transfer high energy consumption areas, balance the energy consumption of nodes, can communicate with nodes directly, and improve data collection performance.

References

1. Yan, H., Gao, D.Y., Su, W.: K-medoids based intra-cluster hash routing for named data networking. Tien Tzu Hsueh Pao/acta Electronica Sinica **45**(10), 2313–2322 (2017)
2. Lin, X.Q., Bergman, J., Gunnarsson, F., et al.: Positioning for the Internet of Things: a 3GPP perspective. IEEE Commun. Mag. **55**(12), 179–185 (2017)
3. Maria, G.D., Elena, M., Antonino, F., et al.: Optical trapping and optical force positioning of two-dimensional materials. Nanoscale **10**(3), 12–45 (2017)

4. Emil, A., Jonathas, C., Cláudio, S., et al.: Dynamic scene graph: enabling scaling, positioning, and navigation in the universe. Comput. Graph. Forum **36**(3), 459–468 (2017)
5. Shieh, W.Y., Hsu, C.C.J., Wang, T.H.: Vehicle positioning and trajectory tracking by infrared signal-direction discrimination for short-range vehicle-to-infrastructure communication systems. IEEE Trans. Intell. Transp. Syst. **19**(2), 368–379 (2018)
6. Dilusha, W., Malin, P., Sarath, D.G., et al.: Controlling resonance energy transfer in nanostructure emitters by positioning near a mirror. J. Chem. Phys. **147**(147), 74–117 (2017)
7. Gloy, Y.S., Cloppenburg, F., Gries. T.: Integration of the vertical warp stop motion positioning in the model-based self-optimization of the weaving process. Int. J. Adv. Manuf. Technol. **90**(9–12), 1–14 (2017)
8. Wu, D.F., Liu, X.J., Ren, F.K., et al.: An improved thrust allocation method for marine dynamic positioning system. Naval Eng. J. **129**(3), 89–98 (2017)
9. Huang, H.Q., Yang, A.Y., Feng, L.H., et al.: Artificial neural-network-based visible light positioning algorithm with a diffuse optical channel. Chin. Opt. Lett. **15**(5), 50601–50605 (2017)
10. Pyung, S.K., Eung, H.L., Mun, S.J., et al.: A finite memory structure filtering for indoor positioning in wireless sensor networks with measurement delay. Int. J. Distrib. Sens. Netw. **13**(1), 1550147–1668541 (2017)
11. Zain, B.T., Dost, M.C., Muhammad, Z.K., et al.: Non-GPS positioning systems: a survey. ACM Comput. Surv. **50**(4), 1–34 (2017)
12. Cui, D., Zhang, Q.: The RFID data clustering algorithm for improving indoor network positioning based on LANDMARC technology. Cluster Comput. **24**(5), 1–8 (2017)
13. Lu, C., Wang, Y.: Positioning variation analysis for the sheet metal workpiece with N-2-1 locating scheme. Int. J. Adv. Manuf. Technol. **89**(9–12), 1–15 (2017)
14. Dong, X., Zhang, Z., Liu, D., et al.: Numerical study on boiling critical characteristics of rod beam channel positioning lattice. Hedongli Gongcheng Nucl. Power Eng. **39**(6), 5–10 (2018)
15. Paweł, P., Mieczysław, B., Roman, G.: The integrated use of GPS/GLONASS observations in network code differential positioning. GPS Solutions **21**(2), 627–638 (2017)

Design and Implementation of E-Note Software Based on Android

Zhenhua Li(✉)

School of Computer Science and Technology, Shandong University of Finance and Economics, Jinan 250014, China
82911101@163.com

Abstract. Nowadays, people are increasingly using mobile terminals. In many cases, they are inclined to use E-notes software in their mobile phones to record some simply information. Thus, I developed this E-Note software which named A-Note to suit their requirements. This software was developed by Android Studio 2.3 from google in Java.

Also, I use My Eclipse to manage data in the server. In this version, I just implement a simple UI and two basic ways for note inputting. All what I do on A-Note just to form a simple method for recoding online and to make few contribution for the developing of E-Notes software.

Keywords: Android · Note software · Android studio · Mobile development

1 Introduction

Note-taking is a kind of study method and its carrier that people often use in study. With the development of network technology, note-taking has given birth to a new form with the rise of Internet. But traditional handwritten notes have not disappeared from people's habits, which is not only due to its their extremely low base usage costs and high cost performance, but also their convenience and freedom compared to the current network of cloud notes. In particular, when users study the systematic textbooks designated by the government, the handwritten notes or even the notes directly recorded in the textbooks still have their advantages that can not be ignored.

Today's mobile phones have transformed from simple communication tools to versatile personal portable data processing terminals, which have greatly changed people's lifestyles and pace of life. A more convenient way of communication leads to a different life. The free way of taking notes can also give life more possibilities and bring a more convenient and efficient life.

2 Related Work

So far, most notebook systems have complete document management, cloud services, online sharing and co-editing functions. Existing commercial products such as Evernote, Youdao Cloud Notes, Learning Trails, and Graphite Documents are only limited

Q. Liu et al. (Eds.): CENet 2020, AISC 1274, pp. 655–660, 2021.
https://doi.org/10.1007/978-981-15-8462-6_74

to document editors with cloud backup functions, which have their own advantages and disadvantages compared to handwritten notes in real life.

For example, compared to handwritten notes, note-taking software can store most multimedia files but still cannot provide the same experience as textbook notes. Compared to software, handwritten notes still can't be managed quickly and efficiently on a large scale, edited by multiple people, or edited by multiple people. However, note software does not have such shortcomings as the need to get used to the operation methods and techniques one by one according to the characteristics of the writing tool itself. This is the innate advantage of the note-taking software itself with birth. In terms of academic research, the existing papers based on existing theoretical research are more based on user analysis of note-taking software based on expert systems rather than focusing on the radical improvement and implementation of traditional notes in terms of functionality.

But obviously, the notebook software should not stop there. The popularity of electronic devices is predictable. As people take notes from the pen back to the hand, or from the pen to the electronic pen, it will inevitably require the evolution of note-taking software, But this evolution only exits in some foreign large enterprise products, such as Microsoft's OneNote; some ppt presentation software at home and abroad, such as WPS, office, etc., have to say that it is still a pity.

3 Requirement Analysis

This electronic note software development pursues a simple and easy-to-use interface, and strives to meet the Google Material Design mobile software design recommendations, "simple and easy to use, lightweight and convenient". The name of the software as "A Note" also hopes that users who use this electronic note software can use it more efficiently and conveniently.

The software is divided into three simple pages. After opening the software, the user enters the login page firstly. The user enters the account and password here to log in. After successful login, the user enters the main interface of the program to browse the interface. The main interface has three tab pages that can be swiped horizontally. They are a normal note page that displays a list of ordinary notes, a handwritten note page that displays a list of handwritten notes, and other pages that display software-related information.

After the user clicks on a note item in the normal or handwritten note list, a new window will pop up to display the normal note edit page or the handwritten note edit page. When you long press a certain item in the normal or handwritten note list, the function of deleting this note will pop up, and the user can use this function to delete the note.

The overall top-level use case diagram is shown in Fig. 1.

The following is a demand analysis for each module based on the overall functional requirements:

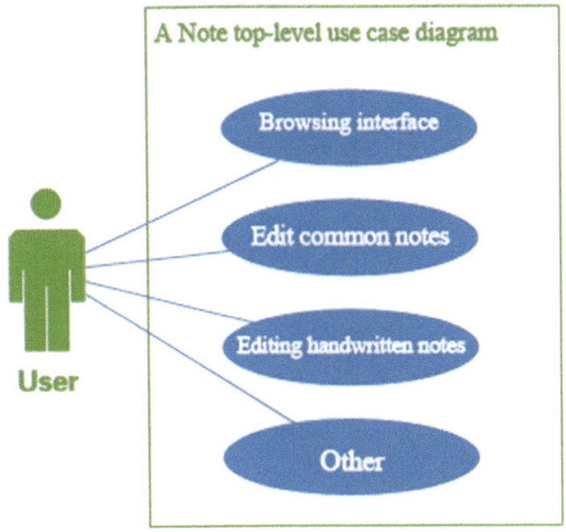

Fig. 1. A Note top-level use case diagram.

3.1 Browse the Interface Module

The browse page is used to display a list of notes and to Manage Normal or handwritten notes. The main content includes three sub pages and two buttons for adding different notes. The display contents are the list of two notes, other information related to software and the buttons for opening and editing two note pages continuously displayed.

The browse interface use case diagram is shown in Fig. 2.

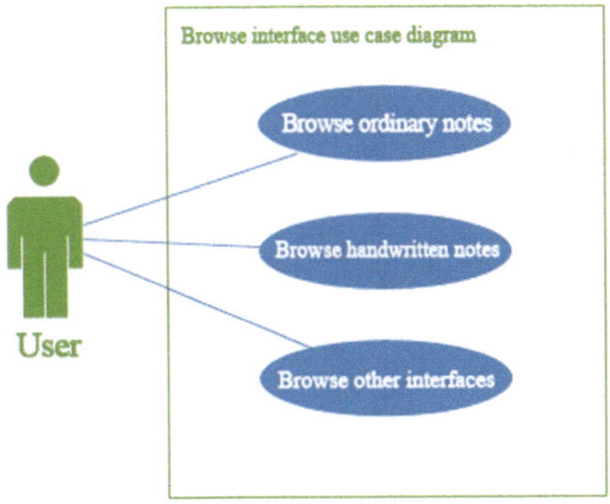

Fig. 2. Browse interface use case diagram.

3.2 Edit General Notes Module

This module is a page for users to edit ordinary notes. This page users can return to the main interface by the back button in the upper left corner to manage their own notes, add notes, delete notes and access to the general notes edit page.

Edit the normal note page use case Diagram as shown in Fig. 3.

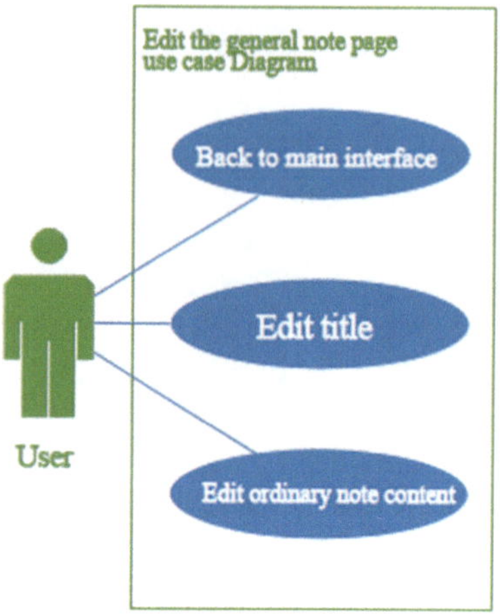

Fig. 3. Edit the general note use case diagram.

3.3 Edit the Handwritten Note Module

This module for Users to edit personal handwritten notes page, users, when they switch to this page, are able to edit handwritten notes, and through the upper left-hand corner back button to return to the main interface.

Edit the use case diagram for the handwritten note page as shown in Fig. 4.

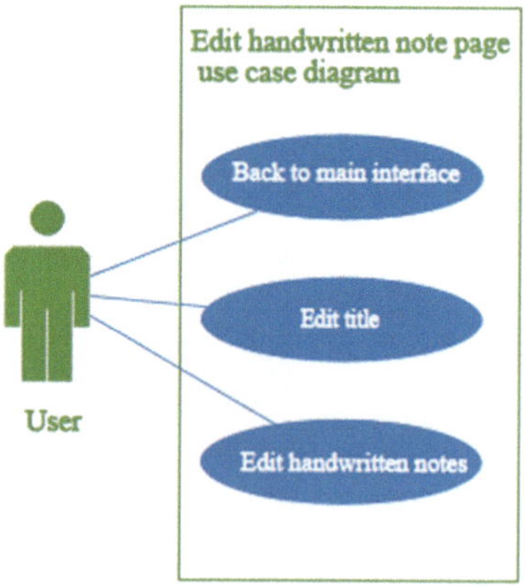

Fig. 4. Edit the use case diagram for the handwritten note page.

3.4 Connect Back End Service Module

Through this module, the software interacts with the web service function published by the back-end server to obtain data returned by the server, send various request information to the server, and upload local pictures to the server for other activities in the software.

The use case diagram of the connection back-end service module is shown in Fig. 5.

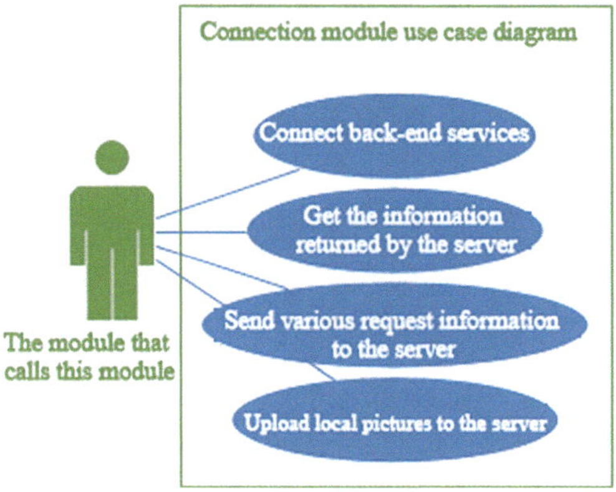

Fig. 5. Use case diagram of the connection module.

3.5 Other Modules

When using the A Note software, users can learn the basic information and feedback ways of the software here, and log out.

4 Conclusion

This paper presents an android-based electronic note-taking software, based on the design idea of "easy to use, lightweight and convenient", by analyzing the advantages and disadvantages of traditional handwritten note-taking and comparing the relevant software in the current market. This software has realized the function of adding and deleting ordinary notes, handwriting notes, picture uploading and so on. However, compared with other mature software in the market, there are still some shortcomings such as relatively few functions. The related functions in the a Note software will be improved continuously in the future, integrating future market trends to improve user experience.

References

1. Zhou, Y.: Research and implementation of cloud note system based on Android, pp. 1–95. Xi'an Jiaotong University (2015)
2. Shi, B.B.: Research on the design of mobile application software for college students' class notes based on mental model, pp. 1–74. Zhejiang University of Technology (2015)
3. Yin, W.J.: Design and implementation of teaching-oriented note-taking wiki system, pp. 1–95. Dalian University of Technology (2014)
4. Zhang, J., Deng, X.X.: Design and development of personalized web-based learning note system. J. Hunan Univ. Sci. Eng. **31**(8), 76–77 (2010)
5. Wang, Y.M.: Design and implementation of multi-person collaborative note-taking system, pp. 1–79. Beijing University of Posts and Telecommunications (2014)
6. Pan, H.Y., Tao, L.M., Xu, G.Y.: A survey of electronic note-taking system based on natural interface. Comput. Inf. Technol. **2009**(Z2), 36–38 (2009)

Research and Design of Context UX Data Analysis System

Xiaoyan Fu and Zhengjie Liu(✉)

Dalian Maritime University, Dalian 116026, China
18203555731@163.com, liuzhj@dlmu.edu.cn

Abstract. At present, user researchers have problems in UX (User experience) data analysis, such as low efficiency, inaccurate context data identification, and low satisfaction of analysis process. Therefore, in order to solve these problems, this paper proposes a design context UX data analysis system to compensate for the shortcomings in the data analysis process. This paper takes the analysis of UX data collected by the CAUX (Context-Aware User Experience) tool as an example, using the relevant methods in Cognitive Task Analysis (CTA), and on the basis of sensemaking loop model, explore the data analysis process of UX researchers through experiments. And carry out demand research for each stage of the analysis process, design a context UX data analysis system according to the requirements. This thesis summarizes the model of UX data analysis process, completes the design of context UX data analysis system, and evaluation experiment proves that the system can effectively solve the problems in the UX data analysis process, and provides a new idea for the UX research practice in the mobile Internet environment.

Keywords: Context aware · Cognitive task analysis · Data analysis

1 Introduction

With the rapid development of mobile Internet, user experience has become an important factor in determining the success or failure of information technology products. Therefore, in order to improve the user experience, user researchers will collect various "data" through data collection tools to analyze user behavior. The collected UX data will be analyzed by data analysis tools. Currently, common data analysis tools are mostly independent third-party analysis tools. PA Nogueira [1] designed and developed an analysis tool for digital game UX data to ensure that user researchers can perform objective data analysis without disturbing the game player experience; Miller A [2] established automatic data analysis and visualization tool evaluation standards; S Garg [3] developed a visual analysis tool based on machine learning and deductive reasoning techniques to help users easily analyze relationships between complex data sets. However, these tools still have many shortcomings in the actual case study. For example, UX researchers need to manually sort and filter data in the data analysis process, and the data that is closely coupled with the context cannot be effectively analyzed and presented, lead to low efficiency and low satisfaction in the

Q. Liu et al. (Eds.): CENet 2020, AISC 1274, pp. 661–669, 2021.
https://doi.org/10.1007/978-981-15-8462-6_75

data analysis process. Therefore, this paper proposes a design context UX data analysis system to compensate for the shortcomings in the data analysis process.

In order to solve these problems, we need to understand the process of researchers analyzing UX data. This paper uses the CTA method to understand this process. CTA uses a variety of interviews, observations and other methods to acquire the knowledge of experts to complete complex tasks. It not only determines the complex cognitive process and required knowledge of workers when they complete the task level, but also further determining what reasoning and decision-making the worker used in completing the task [4]. The method can effectively obtain the data analysis process of the researcher.

2 Overview

This paper takes the UX data collected by the CAUX tool as an example, uses CTA method to establish model of researcher data analysis through iterative experiment, extracts the demand according to the model, then designs the UX data analysis system, and finally the system is evaluated using the task walkthrough and system usability scale.

CAUX is a remote user experience data collection tool based on context awareness. It combines context-aware model with remote UX research technology, through sensors, system status information, etc. to perceive the context of smartphone users (such as: time, location, system operation Status, user behavior, etc.), depending on the context and collection conditions, whether to trigger the data acquisition instruction. After triggering the instruction, collect the main and objective data related to the user experience by means of sensors, questionnaires, logs, recordings, etc. Researchers analyze the data and summarize the user experience problems and user needs to support user experience research and product design and development.

Cognitive task analysis is an extension of traditional task analysis techniques. It is used to generate knowledge target structure and thinking process under observable task behavior. It is used to capture explicit observable behavior and implicit cognitive function to form a unified whole. Cognitive tasks analysis is mainly in the following five stages [5]:

1. Collecting preparatory knowledge
2. Clear knowledge representation
3. Key knowledge leads
4. The result is transformed into a model
5. Analyze and verify the data obtained and guide the application

3 Exploring UX Data Analysis Process Based on CTA

Firstly, the steps of cognitive task analysis were planned, and the decision was made to gradually study according to the three stages of collecting preparatory knowledge, clear knowledge representation, and transforming the results into models. Before the official study began, we confirmed that the screening criteria for the study participants were: UX researchers with certain project practical experience, knowledge of CAUX data collection tools, and using their real UX case studies, and finally confirmed a total of 5 participants participated in the study, including 3 males and 2 females. Table 1 is user information table.

Table 1. User information

Number	Gender	Experience	Research case	Case goal
1	Male	Three years	Mobile Shopping Software User Experience Evaluation Research	Explore the ability of CAUX to evaluate remote user experience and summarize the methodology of remote evaluate
2	Female	Three years	Research on user behavior across APP	Discover user experience issues to enhance the user experience
3	Male	Four year	Research on user experience evaluation method of mobile website	Explore the advantages of CAUX mobile website evaluation and summarize usage methodology
4	Female	Two year	The establishment of mobile phone game characters	Research CAUX's ability to build personas
5	Male	One year	Research the behavior of users' WeChat reading	Discover user experience issues

3.1 Collecting Preparatory Knowledge

Before starting the experiment, must first make data collection instruction to collect user data of interest to UX researchers. The instruction is semantically the structure of "IF context trigger condition THEN data acquisition operation". For example, if "user turns on application X", the "current network status" information is collected. In each instruction, the context trigger condition corresponds to a trigger-related value, and the data collection operation corresponds to an operation-related value. According to the needs of the five research cases, the CAUX acquisition instructions are as shown in Table 2.

3.2 Clear Knowledge Representation

The experiment lasted 7 days, 5 researchers conducted a total of 4 rounds of experiments, collected 65,914 UX data through CAUX tools, this stage using observation and unstructured interviews to conduct research. Let researchers analyze data based on case objectives in their own workplace, observes the researcher's data analysis process and

Table 2. Rules setting

IF	THEN
CAUX, Screen open	Time, location, battery, network, URL, illumination
CAUX, Screen close	Time, location, battery, network, URL, illumination
Shopping APP, Read app, Game app, CAUX browser open	Time, location, battery, network, URL, recording, screenshot, Write log
Shopping APP, Read app, Game app, CAUX browser open	Time, location, battery, network, URL, recording, screenshot, Write log

records the problem, and then interviews based on the problem. The interview questions mainly involve the following aspects:

(1) What is the progress of the case study and what are the findings for the user?
(2) How do you analyze UX data collected by CAUX?
(3) Have you encountered any problems during the analysis, how to solve them?
(4) More ideas in the analysis.

Through this phase of research, following overall process of data analysis was obtained:

1. First of all, researchers use pre-experimental methods to exploratory obtain as much data as possible from the user's research objectives, and to further understand other data related to the target data. At this time, a preliminary analysis will be performed on the collected data to determine whether useful results can be obtained from it. If the data is invalid, the data acquisition instructions are modified. In most cases, it is impossible to draw conclusions based solely on the analysis of UX data. It is necessary to obtain supplementary data through other methods such as recording or log method, and then analyze the supplementary data together with the UX data.
2. Then, began to look for user data characteristics, and come up with some scenario assumptions, and then asked the user some questions to confirm. At this point, the researchers have basically mastered the user's regular activities and behavioral habits, so they pay more attention to the specific details of the user data. The researchers' assumptions at this time are more based on the user's scenario, and the user will be at a key point in the process of discovering the problem, dig into complete context information to find out where the user's operation is problematic.
3. At last, begin to confirm data characteristics observed before and adjust the assumptions. At this time, the data is gradually finished, summarize some user experience issues or find some singular data and restore user scenarios, but in the actual observation, it is found that the UX researchers are limited by the efficiency of the analysis tool. The original data is often directly described as the context. Therefore, there are problems such as repeated description of the context or inaccurate contexts caused by data inconsistency.

3.3 The Result is Transformed into a Model

Through the previous stages of research, we obtained the UX data analysis process of researchers and identified the key steps. In order to better extract the requirement, we need a more complete description of the analysis process. Therefore, the sensemaking loop model is added to the research, which not only can better express the cognitive process of the analyst, but also can be based on it to explore effective data analysis strategies [6]. We sort out the task sequences obtained in the experimental results, and obtain the sensemaking loop model as shown in Fig. 1.

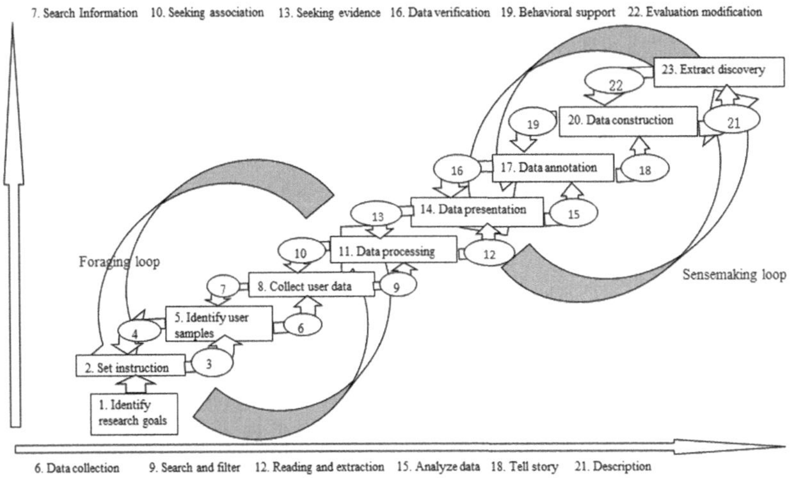

Fig. 1. Sensemaking loop model

The figure above shows the process of UX researchers' data analysis. The boxes in the figure represent the products of the phase, and the process on the arrow represents the related operations of the phase products. The whole analysis process is divided into two main cycles: (1) Foraging loop includes collecting and querying UX data, and extracting user information from it to form a set of behavior and viewpoint information related to the research target. (2) Sensemaking loop includes data presentation, data annotation, data construction, etc., according to the key behavior patterns and related scenario elements, analyze the data and extract the findings.

4 Context UX Data Analysis System Design and Evaluation

4.1 Requirement Analysis

The previous research focused on exploring the process of data analysis, and the requirements of some stages are not well captured. Therefore, in this stage of research, we supplemented the use of user test methods to obtain the requirements of UX researchers in the data analysis process. First, determine the task of the user test as follows, select the real user data in the case study conducted by the participants in the previous stage, and let the user use the existing analysis means to perform the actual analysis operation on the data. In the process, the "Think Aloud" method [7] is used to describe the problems found and related ideas, and finally the five categories of 14 demand points were summarized in Table 3.

Table 3. Requirement description

Requirement	
Data query	Unified data query; Multi-user multi-type query; Search Query
Data presentation	Data presentation type selection; Data hiding and display Data linkage; Data is presented on the timeline
Data annotation	Labeling method; Labeling operation; Recording next to the data
Data construction	Scenario Hypothesis; Personas; Behavior Mode
Extraction found	Classification List Discovery

4.2 Functional Design

According to the requirements of UX researchers in data analysis process, function and structure of data analysis system were designed, as shown in Fig. 2. Here, We divide the data analysis system into two functional modules: data query and data presentation. Data query unit mainly includes objective data query and context data query, wherein the data query can extract data from the database and filter and display all the data; Data presentation unit includes data visualization, data tag, data construction; data visualization is a graphical display of various types of information in a sub-view, which can switch between different users or different data sources, and ensure the correlation between data, and construct different user scenarios through data analysis. Data tag function is provided; data construction is based on the mental model of the researcher data analysis, and the corresponding behavior pattern of the user data is obtained, and finally the summary analysis results are extracted discovery.

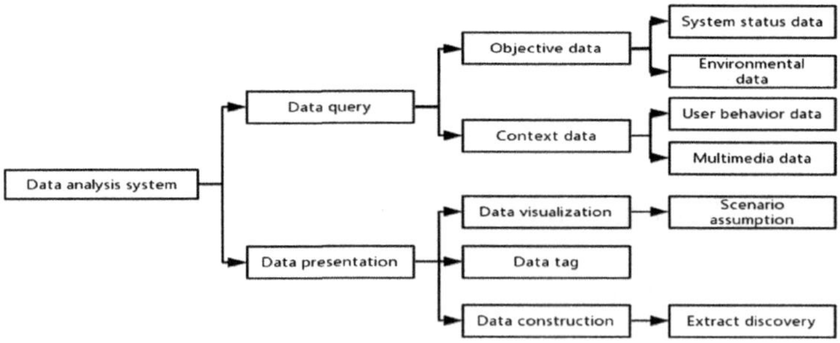

Fig. 2. Functional structure diagram

4.3 System Implementation

At this stage, we implemented the page layout of the UX data analysis system based on HTML, and added the rendering and interaction functions of the real data through the relevant framework of JavaScript. Data exploration module is designed to make elements in the page interactive, we use the Echarts [8] framework to achieve the display of most of the charts. The data construction module is based on the open source JavaScript class library JsMind, which displays and edits the charts. Figure 3 is main interface of the system.

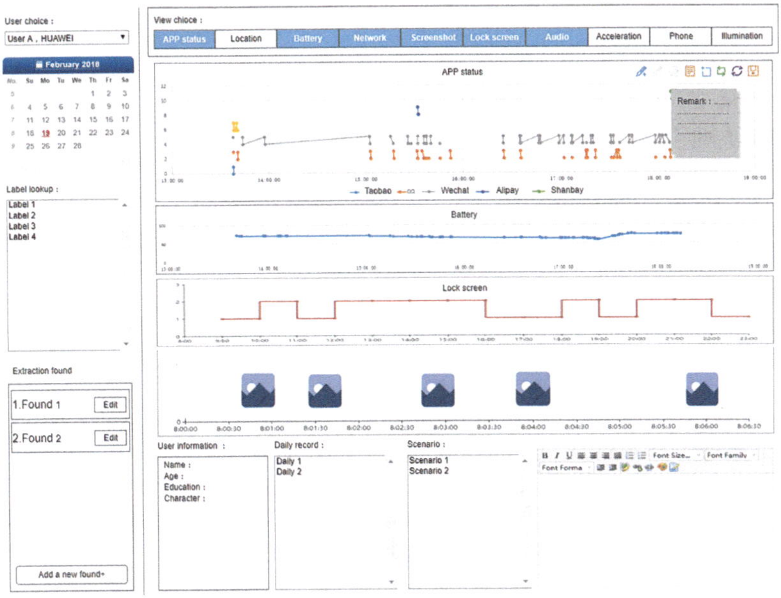

Fig. 3. Main interface of the system

4.4 System Evaluation

In this paper, the task walk-through method is used to evaluate the system, and the SUS scale is used to understand the test user's recognition of the system function. We recruited 20 people as participants in this test evaluation. Twelve of them used CAUX to conduct a preliminary UX study. 8 people did not use CAUX for research, but had a deeper understanding of CAUX functions, and 20 testers were using data analysis systems for the first time. The test task is: using the data analysis system to analyze and discover the data of three users for three days in the context of the user's cross-APP behavior research case. After each test task is completed, the system satisfaction is scored using the SUS scale. The positive problem conversion score is $x - 1$, and the negative problem is $5 - x$. After all the questions are scored, multiplied by 2.5 to get the total score. In the following stages, "0–50" is F, "50–60" is E… "90–100" is A [9], as shown in Fig. 4, the user satisfaction is calculated according to the score of the subject: 82.3, at level B, indicates that the tester is satisfied with the system but there are also areas where improvement is needed.

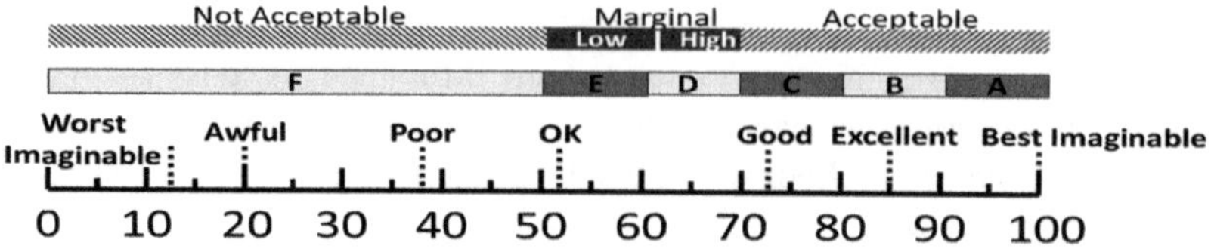

Fig. 4. SUS standard scale

5 Conclusion

The paper successfully acquires the sensemaking loop model of UX researchers in the process of data analysis through cognitive task analysis method. This model combines with the "Think Aloud" method to extract the 5 categories of 14 demand points in the data analysis process. According to these demand points, the functional structure of the context UX data analysis system is designed, and the development of the system is realized by using Web related technology. Finally, the evaluation verifies the effectiveness and availability of the system, which provides good help for future UX researchers data analysis. The design of the UX data analysis system has reached the expected goal, but due to the limitations of time and resources, some of the research content of this paper still needs to be improved. Because the UX researchers required the experience of using CAUX tools in the screening conditions of this study, it is difficult to select more subjects in a wider range, so there are certain limitations in the selection of samples. In the future, user samples can be enriched and UX researchers using other data collection tools can be added to improve the universality of research.

References

1. Nogueria, P.A., Torres, V., Rodrigues, R.: Automatic emotional reactions identification: a software tool for offline user experience research. Lecture Notes in Computer Science, vol. 8215, pp. 164–167 (2013)
2. Miller, A., Lekar, D.: Evaluation of Analysis and Visualization Tools for Performance Data. Uni Stuttgart - Universitätsbibliothek (2014)
3. Sachs, G., Morgan, C., Db, A.G.: Tools to enrich user experience during "visual analysis", pp. 336–340. State University of New York at Stony Brook (2010)
4. Schraagen, J.M., Chipman, S.F., Shalin, V.: Cognitive Task Analysis. L. Erlbaum Associates (2000)
5. Hoffman, R.R., Militello, L.G., et al.: Perspectives on cognitive task analysis: historical origins and modern communities of practice (expe. Portuguese Studies) **31**(1), 94–106 (2015)
6. Pirolli, P., Card, S.: The sensemaking process and leverage points for analyst technology as identified through cognitive task analysis. In: International Conference on Intelligence Analysis (2005)

7. Padilla, J.L., Leighton, J.P.: Cognitive interviewing and think aloud methods. In: Understanding and Investigating Response Processes in Validation Research. Springer (2017)
8. Cong-Jing, R., Xiao-Fei, X.U.: Research on visualization method of patent citation relationship based on Node JS+ ECharts. Inf. Sci. (2018)
9. Bruun, A., Stage, J.: The effect of task assignments and instruction types on remote asynchronous usability testing. In: SIGCHI Conference on Human Factors in Computing Systems. ACM (2012)

Trust Evaluation of User Behavior Based on Entropy Weight Method

Yonghua Gong(✉) and Lei Chen

Nanjing University of Posts and Telecommunications, Nanjing, China
gongyh2008@163.com

Abstract. Untrustworthy behaviors such as false redundant information flooding and malicious recommendation hinder the healthy development of social e-commerce, and predicting and evaluating the credibility of users is of great significance to social e-commerce marketing promotion. Giving objective and appropriate evaluation weights to decision attributes is the first step to predict the credibility of users. This paper proposes a dynamic trust quantification model based on entropy weight method, which can realize adaptive weight setting. Experiments show that compared with similar models, the model is more adaptable and has a higher transaction success rate.

Keywords: Trust model · Entropy weight method · Social commerce

1 Introduction

In social commerce, improving the credibility of the network environment and adopting appropriate access control methods are important ways to ensure transaction security. Among them, the trust management of buyers and sellers is introduced into social commerce to further adapt to the security requirements of the transaction system, highlighting the dynamic management of the transaction behavior of the entity nodes, and providing support for establishing an efficient and secure network access environment and enhancing the trust perception of the entity nodes. Trust quantification and evaluation is an important part of trust management, which determines whether the trust evaluation mechanism can fully and accurately reflect the entity trust value. However, because trust is an abstract subjective concept which has the characteristics of asymmetry, selectivity, and transitivity, it is difficult to objectively reflect and measure. From the perspective of psychology, trust is an abstract psychological cognition that is difficult to quantify. The relationship of trust is ambiguous and changes dynamically with the context of behavior between entities. It often involves knowledge, ability, goodwill, reputation and other factors which makes it difficult to establish a relatively complete and dynamically described trust evaluation model. Therefore, constructing the trust model by extracting the trust transfer mechanism between entities which are under specific situation, and calculating the trust value of each entity in the trust transfer over time is of great significance for solving the network security problem and realizing the trusted interaction of the entity. The trust model also has broad application prospects in the fields of mobile commerce and Internet of Things.

Q. Liu et al. (Eds.): CENet 2020, AISC 1274, pp. 670–675, 2021.
https://doi.org/10.1007/978-981-15-8462-6_76

In the corresponding network environment, it is comprehensively considered the subjective factors affecting trust computing and the changes of objective evidence in trust model. Based on the user behavior assessment, the model is dynamically adjusted, and the trustworthiness of the object is calculated and quantified to help subjects choose authentic and reliable services and avoid malicious transactions according to the requirements of the subject. Generally, the trust model monitors, acquires, and analyzes contextual information through interactions such as scoring, evaluating, and recommending between entities, and uses the results of the analysis as a basis for trust evaluation. Reference [1] proposes the RulerRep model which is based on the direct transaction between the entity node and the service provider that measures the accuracy of the evaluation information of other nodes, and calculates the deviation degree of other nodes from their evaluation information to the service provider. In this model, the weight of the evaluation with a large degree of deviation is small, while the weight of the evaluation with a small degree of deviation is significant. Reference [2] proposed a SVD signs based clustering algorithm to process the trust and distrust relationship matrix in order to discover the trust communities. Besides, a sparse rating complement algorithm is proposed to generate dense user rating profiles. The experimental results show that the proposed model can alleviate sparsity and cold start problems to a large extent. Reference [3] proposed a method that can predict rating for personalized recommender systems based on similarity, centrality and social relationships. Considering that the social trust values are taken into the proposed mechanism, the model is better than traditional collaborative filtering approaches. Moreover, the proposed model can predict user rating for products based on user-item rating matrix with probabilistic matrix factorization method. Reference [4] proposed to design pro-actively access control so that access control systems can respond to malicious access based on the analysis of the shortcomings of existing access control models. And the authorization strategy based on game theory was designed to regulate the access behavior of malicious entities. Reference [5] proposed two combination strategies for two factors to find the top-k most trusted neighbors, one factor interest topic which was used to measure the semantic similarity between users, while the other factor topology feature was calculated from trust propagation ability of users. Besides, the traversal depth was constrained according to "small world" theory during trust inference. Reference [6] proposed a trust management model to achieve enhanced quality of service parameters by solving challenges such as the presence of attack-prone nodes and extreme resource limitations. The trust values were calculated by trust management model according to P2P and link evaluations. Reference [7] proposed SNTrust model to find the trust of nodes in a network using a local and global and investigated trust, influence and their relationship in social network communities. Different influence evaluation approaches are used in the process of finding influential nodes. The result of reference showed that a strong linear relationship existed between the influence of nodes in social network and community.

2 Proposed Trust Model

In e-commerce or a recommendation system, an entity node usually performs a service evaluation on a certain evaluation index or purchased product to form a product evaluation system, and then recommends the product to a nearby friend. For an entity service evaluation network of size n with an evaluation index of m, the structural evaluation matrix T is:

$$T = \begin{bmatrix} t_{11} & t_{12} & \cdots & t_{1n} \\ t_{21} & t_{22} & \cdots & t_{2n} \\ \cdots & \cdots & \cdots & \cdots \\ t_{m1} & t_{m2} & \cdots & t_{mn} \end{bmatrix}$$

Among them, t_{ij} represents the recommended trust level of entity node i with respect to the decision attribute j$(1 \leq i \leq m, 1 \leq j \leq n)$. Then calculating the sum of the recommended trust evaluations of the entity node i of the N decision attributes in the recommended trust evaluation matrix, specific as follows: $\sum_{j=1}^{N} t_{ij}$. Based on this, the recommended trust entropy value of the entity node i is calculated as: $e_i = -K \sum_{j=1}^{n} p_{ij} \ln p_{ij}, K = 1/\ln n$. Thus, the recommended entropy weight is: $w_i = \frac{1-e_i}{m-\sum_{i=1}^{m} e_i}$.

However, since the trust relationship is a subjective concept that is difficult to objectively express and quantify, in the evaluation of the recommendation trust, the potential evaluation bias and the uncertainty of the traceability data are often generated due to the noise of the data. So we use the membership function method for classification and evaluation to reduce the data noise and decrease the uncertainty. We use the membership function method for classification and evaluation. Defining the membership matrix as $B_{m \times n}$, the element b_{ij} in the matrix is represented as the membership of the recommended trust value, and b_{ij} is constrained to $\sum_{j=1}^{n} b_{ij} = 1, b_{ij} \geq 0$. The entity node regards the discourse right of the decision attribute as the same, which calculates the degree of consistency of the recommendation of the m entity nodes for the n decision attributes, and defines $t_j = \sum_{i=1}^{m} t_{ij}/m$ as the average recommendation degree. Defining the uncertainty of the cognition bias of the entity node for the decision attribute and the difference from the recommendation expectation is the recommendation blindness Q_j, and Q_j is constrained to

$$Q_j = \sum_{i=1}^{m} b_{ij} \left| \left\{ \begin{array}{c} [\max(t_{1j}, t_{2j}, \cdots, t_{mj}) - t_j] \\ + [\min(t_{1j}, t_{2j}, \cdots, t_{mj}) - t_j] \end{array} \right\} /2 \right| \quad (1)$$

The goal of this paper is to set the overall recommendation blindness to a minimum, that is:

$$\min Q = \sum_{j=1}^{n}\sum_{i=1}^{m} b_{ij}\left|\left\{\begin{array}{l}[\max(t_{1j}, t_{2j}, \cdots, t_{mj}) - t_j] \\ +[\min(t_{1j}, t_{2j}, \cdots, t_{mj}) - t_j]\end{array}\right\}/2\right| \tag{2}$$

The entropy weight method is an effective method for objectively sorting various influencing factors, that is, the method of calculating the weights of multiple decision attributes by using information entropy. Information entropy can reflect the size of the information source containing information. The larger the information entropy, the larger the amount of information it contains, and the smaller the uncertainty. In this context, the greater the information entropy means higher credibility. Overall information entropy of membership matrix $B_{m\times n}$ can be measured by $H = -\sum_{j=1}^{n}\sum_{i=1}^{m} b_{ij}\log b_{ij}$. To minimize the overall recommendation blindness and increase the overall information entropy, set the equation

$$Z = \min\left\{\begin{array}{l}\sum_{j=1}^{n}\sum_{i=1}^{m} b_{ij}\log b_{ij} - \\ \sum_{j=1}^{n}\sum_{i=1}^{m} b_{ij}\left|\left\{\begin{array}{l}[\max(t_{1j}, t_{2j}, \cdots, t_{mj}) - t_j] \\ +[\min(t_{1j}, t_{2j}, \cdots, t_{mj}) - t_j]\end{array}\right\}/2\right|\end{array}\right\} \tag{3}$$

Constructing Lagrangian function according to formula and Lagrangian multiplier method:

$$\begin{array}{l}F(b_{ij}, t_j, \lambda) = \sum_{j=1}^{n}\sum_{i=1}^{m} b_{ij}\log b_{ij} - \\ \sum_{j=1}^{n}\sum_{i=1}^{m} b_{ij}\left|\left\{\begin{array}{l}[\max(t_{1j}, t_{2j}, \cdots, t_{mj}) - t_j] \\ +[\min(t_{1j}, t_{2j}, \cdots, t_{mj}) - t_j]\end{array}\right\}/2\right| + \lambda(\sum_{j=1}^{n} b_{ij} - 1)\end{array} \tag{4}$$

Among them λ is a Lagrangian factor, and make $\frac{\partial F}{\partial b_{ij}} = 0$, $\frac{\partial F}{\partial t_j} = 0$ and $\frac{\partial F}{\partial \lambda} = 0$. Then we can get:

$$b_{ij} = \exp(\sum_{j=1}^{n}\left|\left\{\begin{array}{l}[\max(t_{1j}, t_{2j}, \cdots, t_{mj}) - t_j] \\ +[\min(t_{1j}, t_{2j}, \cdots, t_{mj}) - t_j]\end{array}\right\}/2\right| - mn/\ln 2) \tag{5}$$

3 Analysis of Experiment

The paper uses peersim to build a simulation experiment environment, which supports structured and unstructured P2P network simulation, the total number of users is set to 1,000, the normal interaction threshold is 0.6, the malicious interaction threshold is 0.3, and the satisfaction default evaluation value is 0.5. The experimental investigation index is the satisfaction rate of user interaction, that is, the ratio of the number of nodes that are satisfactory to the interaction evaluation to the number of nodes in a specific experimental simulation period. The experimental comparison models are Extended ABAC model [8] and UnCM model [9].

Figure 1 shows the comparison of the interaction satisfaction rate under the non-malicious node which all nodes are honest nodes. Under this condition, regardless of the simulation experiment period, the interactive satisfaction rate curve has a stable trend, and the curve is smooth and close to a straight line. However, the interactive satisfaction rate of the proposed model is higher than that of Extended ABAC model and UnCM model. The reason is that this paper adopts the entropy weight method, and gives appropriate emphasis on the decision attributes with great influence when evaluating the weights, thus improving the overall interaction satisfaction degree.

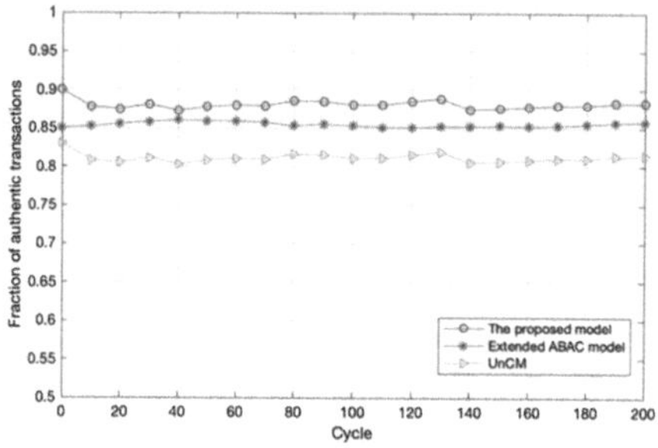

Fig. 1. Interaction satisfaction under non-malicious nodes.

As in Fig. 2 and Fig. 3, When the proportion of malicious nodes increases, the interaction satisfaction rate will decrease to a certain extent, and the degree of decline increases with the percentage of malicious nodes. In addition, the degree of decline of UnCM model and graph fluctuation are more obvious in the three models. The model proposed in this paper can weaken the influence of malicious nodes on trust evaluation, and effectively block malicious nodes to a certain extent, thus improving the effectiveness and objective of trust evaluation decision.

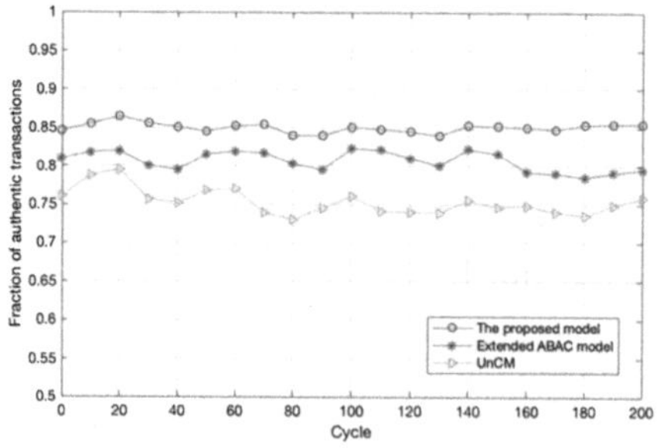

Fig. 2. Interaction satisfaction with 20% malicious nodes.

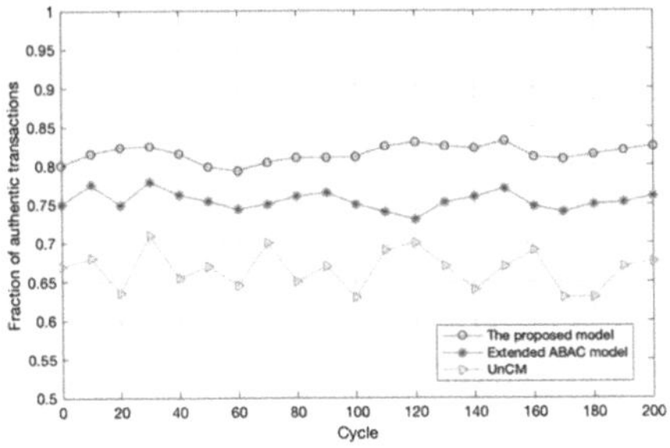

Fig. 3. Interaction satisfaction with 40% malicious nodes.

4 Conclusion

In view of the current inaccuracy and lack of validity of trust estimation, this paper proposes a dynamic trust quantification mechanism based on entropy weight method. The model can dynamically adapt to decision attributes to achieve weight distribution. This paper provides a new idea for the establishment of trust model in the area of social e-commerce and Internet of Things. It should be noted that since trust is a complex concept, the trust model of this paper is incomplete or even improperly considered in some aspects. For example, the trust model in this paper does not take into account the multiple influencing factors of trust and the time sensitivity or asymmetry of trust, which can be seen as a basis for building a trust model in future research.

Acknowledgement. Supported by Natural Science Foundation of NJUPT (No. NY220044).

References

1. Shan, M.H., Gong, J.W., Niu, E.L.: RulerRep: filtering out inaccurate ratings in reputation systems based on departure degree. J. Comput. **7**(33), 1226–1235 (2010)
2. Ma, X., Lu, H.W., Gan, Z.B.: An explicit trust and distrust clustering based collaborative filtering recommendation approach. Electron. Commer. Res. Appl. **25**, 29–39 (2017)
3. Davoudi, A., Chatterjee, M.: Social trust for rating prediction in recommender systems: Effects of similarity, centrality, and social ties. Online Soc. Netw. Media **7**, 1–11 (2018)
4. Zhang, Y.X., He, J.S., Zhao, B.: Towards more pro-active access control in computer systems and networks. Comput. Secur. **49**, 132–146 (2015)
5. Mao, C.Y., Xu, C.F., He, Q.: A cost-effective algorithm for inferring the trust between two individuals in social networks. Knowl.-Based Syst. **164**, 122–138 (2019)
6. Shabut, A.M., Kaiser, M.S., Dahal, K.P.: A multidimensional trust evaluation model for MANETs. J. Netw. Comput. Appl. **123**, 32–41 (2018)
7. Asim, Y., Malik, A.K., Raza, B.: A trust model for analysis of trust, influence and their relationship in social network communities. Telemat. Inf. **36**, 94–116 (2019)
8. Smari, W.W., Clemente, P., Lalande, J.F.: An extended attribute based access control model with trust and privacy: application to a collaborative crisis management systems. Future Gener. Comput. Syst. **31**, 147–168 (2014)
9. Piunti, M., Venanzi, M., Falcone, R.: Multimodal trust formation with uninformed cognitive maps (UnCM). In: Proceedings of the 11th International Conference on Autonomous Agents and Multiagent Systems, vol. 3 (2012)

Byzantine Fault-Tolerant Consensus Algorithm Based on the Scoring Mechanism

Cheng Zhong[1(✉)], Zhengwen Zhang[1], Peng Lin[2], and Shujuan Sun[1]

[1] State Grid Xiongan New Area Power Supply Company, Heibei 071000, China
zhongc@126.com
[2] Beijing Vectinfo Technologies Co., Ltd., Beijing 100088, China

Abstract. In order to fetch up the shortages of default consensus mechanism of Ethereum, this paper proposes a practical Byzantine fault-tolerant consensus algorithm based on the scoring mechanism. In view of the above two defects of PBFT, this paper proposes a Scoring algorithm based on PBFT (SPBFT). The proposed algorithm takes the advantages of high expansion and high reliability on the basis of solving the Byzantine general problem. Experiments show that our proposed improvement algorithm can not only meet the dynamic scalability of nodes, but also ensure that elected master node in system is more reliable.

Keywords: Ethereum · Byzantine fault-tolerant consensus algorithm · Scoring mechanism

1 Introduction

The consensus mechanism adopted by Ethereum is PoW (Proof of Work). The core idea of PoW is to obtain accounting rights through competing computing power. Nodes with larger workloads have higher competitiveness and are more likely to obtain accounting rights. Although PoW can safely complete the consensus process, each consensus requires a large amount of computing power, causing a lot of waste of resources [1]. Based on the above analysis, this paper proposes an improved algorithm for the practical Byzantine fault tolerance algorithm based on the scoring mechanism. This improved algorithm is applied to the consensus process of Ethereum, so that the consensus of Ethereum can be completed without competing computing power, and then complete the transaction data on-chain.

The Byzantine generals problem is a problem that must consider in blockchains consensus algorithms [2]. The Byzantine fault tolerance problem can be roughly described as: how to make a decision in a network with inactive nodes or malicious nodes to ensure the correct operation of the system and the authenticity of transaction data. The PBFT algorithm is used to solve the Byzantine general problem, but this algorithm has certain defects [3]. First, the number of nodes in the entire consensus network is fixed. Once the nodes join or exit, the system cannot perceive them and does not have dynamic scalability. Second, the selection method of the master node is to select the master node in turn according to the number of the bookkeeping node. This method of selecting the master node is unreasonable because there is no reliability verification of the next master node, the next master node is likely to be a malicious

Q. Liu et al. (Eds.): CENet 2020, AISC 1274, pp. 676–684, 2021.
https://doi.org/10.1007/978-981-15-8462-6_77

node [4]. Aiming at the above-mentioned two shortcomings of PBFT, this paper proposes a scoring-based algorithm SPBFT (PBFT based on Scoring, SPBFT). This algorithm has the advantages of high scalability and high reliability on the basis of solving the Byzantine general problem. The SPBFT algorithm will be introduced in detail from three aspects, including system model, consensus process and simulation analysis.

2 System Model

2.1 Network Model

In order to solve the problem that the PBFT algorithm does not support the consensus nodes to dynamically join or leave the network, in the SPBFT algorithm, the nodes in the blockchain network are divided into five categories: ordinary nodes, alternative nodes, accounting nodes, master nodes, and abolition nodes. The purpose of each node is as follows:

1) The ordinary node set *NP* refers to the set of independent nodes that exist in the network. This type of node has not yet been added to the blockchain network. Its number can be set to N_P, and its number is $\{0, 1, \ldots, N_{P-1}\}$.
2) The candidate node set *NA* is a set of nodes in the blockchain. This type of node is added to the blockchain network by ordinary nodes performing real-name authentication. Candidate nodes are a type of nodes divided to satisfy the common nodes' dynamic joining into the network. The number of candidate nodes can be set to N_a, and numbered $\{0, 1, \ldots, N_{a-1}\}$.
3) The accounting nodes set *NB* are equivalent to the set of slave nodes in the PBFT. This type of node cooperates with the master node to complete the consensus process. The billing nodes are stored in a first-in-first-out queue. Before the next consensus starts, candidate nodes are dynamically added to the end of the queue of accounting nodes to achieve the dynamic scalability of the nodes in the network. The number of nodes can be set to N_b, numbered $\{0, 1, \ldots, N_{b-1}\}$.
4) The master node *P* is used to package the transactions sent by the client and complete the final transaction data on-chain. The formula for selecting the master node is given below.
5) The abolition nodes set *NR* is a collection of downtime nodes and malicious nodes in the blockchain network. The role of this type of node is to prevent malicious nodes and inactive nodes from being elected as the master node, making the elected master node more honest and reliable, and further improve the consensus efficiency, and its number can be set to N_r, numbered $\{0, 1, \ldots, N_{r-1}\}$.

From the introduction above, we can get the conversion relationship between the five types of nodes, as shown in Fig. 1.

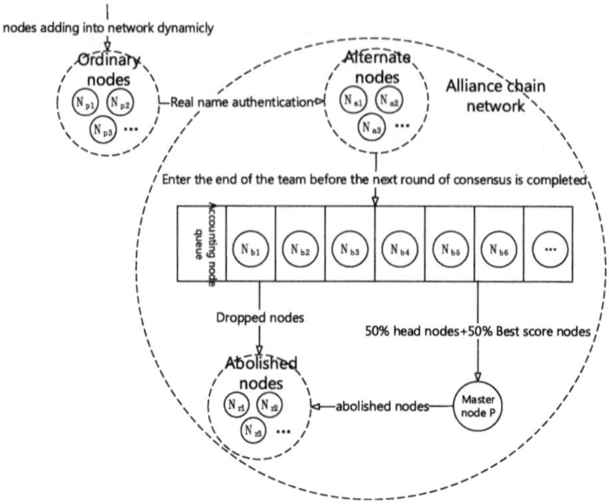

Fig. 1. Conversion relationship between the five types of nodes.

2.2 Network Algorithm

Candidate Nodes Join the Alliance Chain. Before the next round of consensus, the elements in the candidate node set are added to the tail of the accounting node queue, so as to achieve the purpose of the candidate nodes dynamically joining the alliance chain network. Therefore, the set of accounting nodes can be described by the following formula:

$$N_b = \left\{N_{b_1}, N_{b_2}, N_{b_3}, \ldots, N_{b_{n_b}}, N_{a_1}, N_{a_2}, N_{a_3}, \ldots, N_{a_{n_a}}\right\} \tag{1}$$

Among them, N_{b_i} is the i-th element in the set N_B, N_{a_i} is the i-th element in the set N_A.

Scoring Mechanism. In order to record how each consensus node performs in the consensus process, a scoring mechanism is introduced. The scoring mechanism is a mechanism for scoring the participating nodes based on their performance. The rules of the mechanism are as follows:

If the master node M does not complete the on-chain work of the block on time, the master node M will be deducted 2 points, that is $S(M) - = 2$.

If the master node M is a malicious node, the master node M will be scored as $S(M) = -1$ and added to the abolition node set. There will never be a chance to be elected as the master node.

If the master node M completes the on-chain work of the block on time, the master node M will be awarded 2 points, that is, $S(M) + = 2$.

When switching views, if the accounting node N_{b_i} positively judges the master node, it will be awarded 1 point, that is, $S(N_{b_i})+=1$; if the accounting node N_{b_i} misjudges the master node, 1 point will be deducted, that is, $S(N_{b_i})-=1$.

In summary, the scoring mechanism is as follows:

$$S(N_{a_{ki}}) = \begin{cases} S(N_{a_{ki}}) + 2, \textit{Successful completion of consensus} \\ S(N_{a_{ki}}) - 2, \textit{Timeout did not complete consensus} \\ S(N_{a_{ki}}) + 1, \textit{Positive judgment master node} \\ S(N_{a_{ki}}) - 1, \textit{Misjudging the master node} \\ -1, \textit{Failed to complete consensus} \end{cases} \tag{2}$$

The system's main node scoring process is shown in Fig. 2.

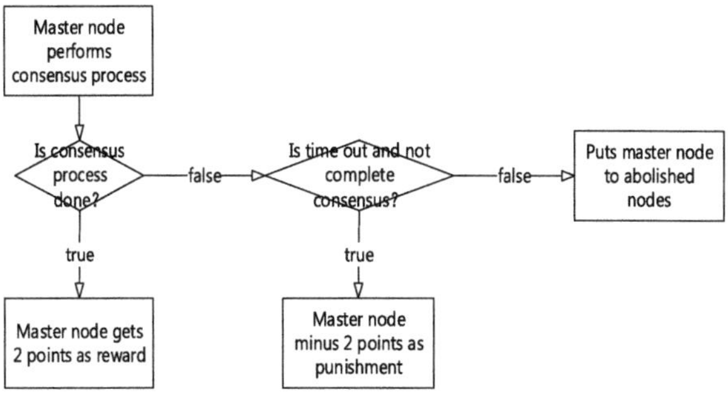

Fig. 2. Master node scoring flowchart.

The process of scoring the accounting nodes by the system is shown in Fig. 3.

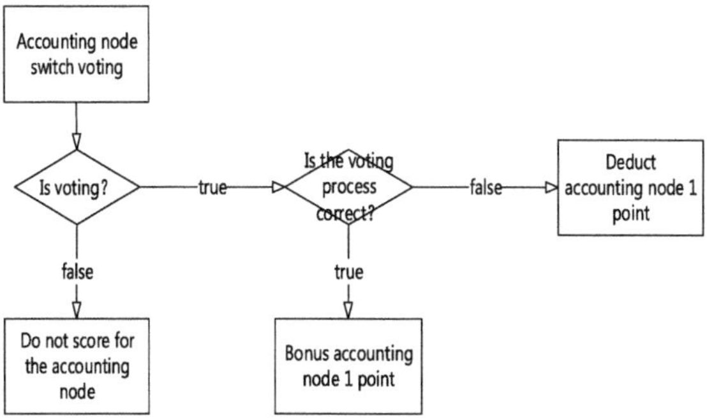

Fig. 3. Flow chart of bookkeeping node scoring.

The score of each node represents the accuracy of the consensus behavior of the node. The higher the node's score, the more honest and reliable the node's performance in the consensus process is; On the contrary, the lower the node's score, the more dishonest and unreliable the node is.

Selection of the Master Node. The selection of the master node is relatively innovative, in order to take into account the scoring and fairness. We select the head element of the accounting node queue with the probability of coefficient a + the element with the highest score selected with the probability of coefficient (1 − a) as the master node. Let's consider the master node as M, then P satisfies the following formula:

$$M = a * first(N_{b_i}) + (1 - a) * \frac{S(N_{b_i})}{\sum_{i=0}^{i=N_b} S(N_{b_i})}, 0 < a < 1 \quad (3)$$

The first function is used to determine whether N_{b_i} is the head node of the accounting node queue, if N_{b_i} is the head node, then $first(N_{b_i}) = 1$; if N_{b_i} is not the head node, then $first(N_{b_i}) = 0$. $\frac{S(N_{b_i})}{\sum_{i=0}^{i=N_b} S(N_{b_i})}$ calculates the probability that the accounting node is selected as the master node according to the points of the accounting node. Suppose $a = 0.5, \sum_{i=0}^{i=N_b} S(N_{b_i}) = 100$, Then the relationship among the probability of head node of the accounting queue, non-head node of an accounting queue with scores, and remaining accounting node being selected as the master node is shown in Fig. 4.

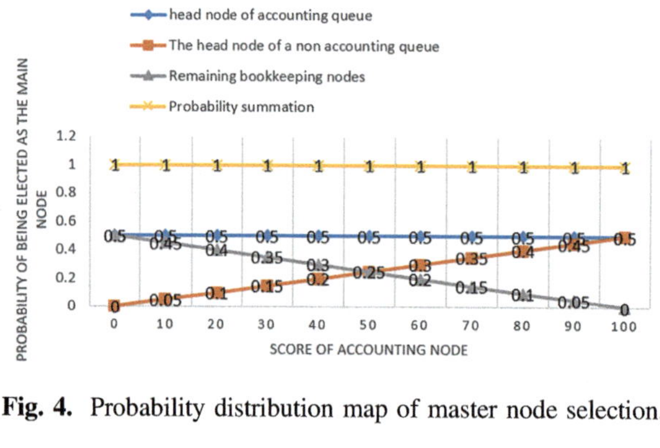

Fig. 4. Probability distribution map of master node selection.

Suppose $P(N_{b\_all})$ represents the sum of the probabilities of all nodes in the accounting queue being elected as the master node, $P(N_{b\_first})$ represents the probability that the head node of the accounting queue is elected as the master node, $P(N_{b\_someone\_not\_first})$ represents the probability that the non-head node of a certain accounting node queue is elected as the master node, $P(N_{b\_other})$ represents the probability that the remaining accounting node is elected as the master node, then the relationship among $P(N_{b\_all})$, N_{b_first}, $P(N_{b_{\mathrm{someone}_{not_{first}}}})$ and $P(N_{b\_other})$ is:

$$P(N_{b\_all}) = P(N_{b\_first}) + P(N_{b\_someone\_not\_first}) + P(N_{b\_other}) = 1 \tag{4}$$

When the master node completes a round of consensus, it is either abolished by reporting or the task is completed and placed at the end of the accounting node queue. The accounting queue elements are represented as follows:

$$NB = \{N_{a_{m1}}, N_{a_{m2}}, N_{a_{m3}}, \ldots, N_{a_{mn}}\}, m_i \neq m_j \, ifi \neq j \tag{5}$$

$(m1, m2, m3, \ldots, mn\}$ is the subsequence of $(k2, k3, k4, \ldots, kt, k1)$, the missing part is the abolished bookkeeping nodes. The selection scheme of the master node can not only ensure that each node has a small probability of being selected as the master node, but also ensure that the node with a high score will be selected as the master node, thereby making the selection of the master node more reasonable and accurate.

3 Consensus Process

This chapter mainly introduces how to achieve consensus on transaction information across the network between the master node and the bookkeeping node. The general process of a round of consensus is as follows (Fig. 5):

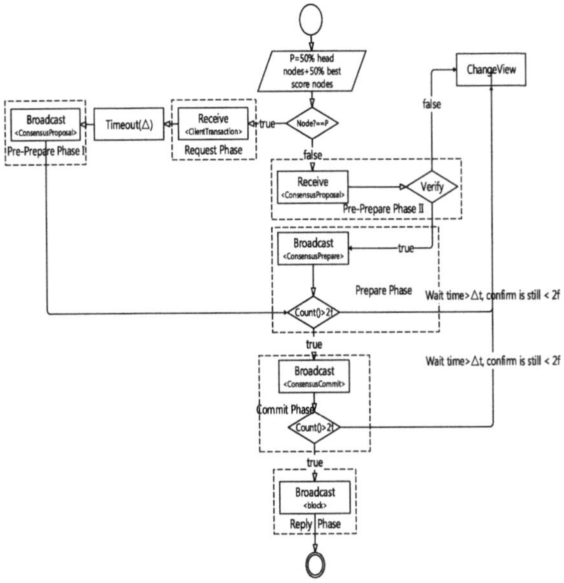

Fig. 5. Consensus flowchart.

4 Simulation Analysis

4.1 Scalability

In order to verify the scalability of the SPBFT algorithm, two sets of experiments were performed. The specific process is as follows:

The change in the number of nodes participating in consensus in the PBFT algorithm and SPBFT algorithm is shown in Fig. 6.

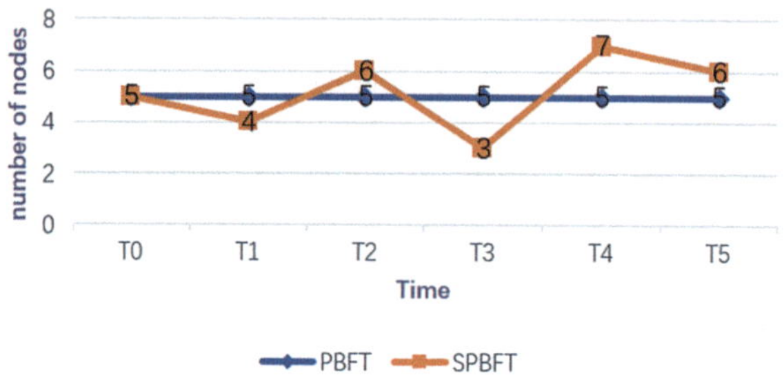

Fig. 6. Change graph of the number of nodes in the consensus algorithm.

Analysis: At time T0, the number of nodes in the consensus network is 5. At this time, a node exits the consensus network, then the signal of node withdrawal will be detected by the system before the next round of consensus, so the number of nodes in the consensus network monitored at time T1 is 4. Similarly, at time T1, the number of nodes in the consensus network is four. At this time, two nodes join the consensus network, then the signal added by the nodes will be discovered by the system before the next round of consensus, so the consensus network is monitored at time T2. The number of nodes is 6.

Conclusion: It can be known from the experimental results that the PBFT algorithm does not support the joining and exiting of consensus nodes from the network, while the SPBFT algorithm supports the joining and exiting of consensus nodes from the network.

4.2 Reliability

The master node of the SPBFT algorithm is selected by the following formula (6). From this formula, it can be known that when the coefficient a is constant, the higher the score of the consensus node is, the greater the probability that the node will be elected as the master node is, and the higher the reliability of the master node is.

$$M = \begin{cases} a * N_{b\_first}, \\ (1-a) * \frac{S(N_{b_i})}{\sum_{i=0}^{i=N_b} S(N_{b_i})}, 0<a<1 \\ N_{b_{first}}, S(N_{b_i}) = 0 \end{cases} \tag{6}$$

the situation where the $N_{b\_not\_first}$ node is elected as the master node in the SPBFT algorithm is shown in Fig. 7.

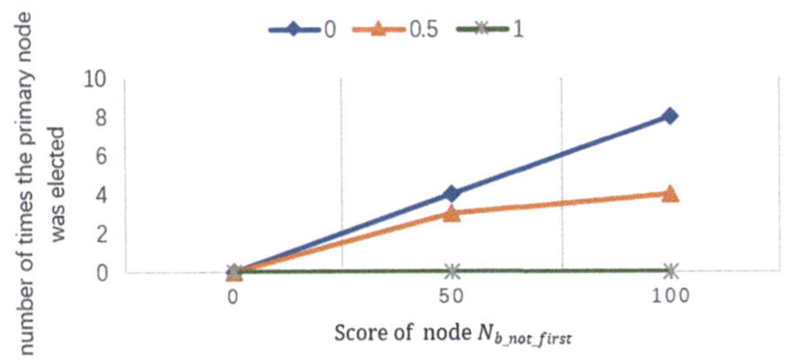

Fig. 7. Graph of the times that node $N_{b\_not\_first}$ is elected as the primary node.

The analysis of the experimental results shows that in the SPBFT algorithm, the smaller the a is, the greater the probability that $N_{b\_not\_first}$ is elected as the master node is; when the score is larger, the probability that $N_{b\_not\_first}$ is elected as the master node is greater. The master node has the right to book, and each time the book is successful, it will get a certain salary reward. In order to get more salary rewards, the master node must work honestly, so as to get more scores, so that there will be a greater probability of being elected as the master node, and thus get more mining salary. Over time, the consensus of the SPBFT algorithm will become more and more reliable.

5 Conclusion

Based on the defect that Ethereum's default consensus mechanism PoW requires a lot of resources, this paper presents a practical Byzantine fault-tolerant consensus algorithm (SPBFT) based on a scoring mechanism. Based on the PBFT algorithm, this algorithm introduces a node classification and scoring mechanism, which solves the problems that the PBFT algorithm does not support dynamic network scalability and unreasonable master node selection. Applying the SPBFT algorithm to Ethereum can not only avoid a large amount of computing power to complete consensus, but also make the consensus process have the advantages of high scalability and high reliability.

Acknowledgments. This work is supported by State Grid Technical Project "Research on key technologies of secure and reliable slice access for Energy Internet services" (5204XQ190001).

References

1. Zheng, K., Liu, Y., Dai, C., Duan, Y., Huang, X.: Model checking PBFT consensus mechanism in healthcare blockchain network. In: Proceedings of the 9th International Conference on Information Technology in Medicine and Education, pp. 877–881 (2018). https://doi.org/10.1109/ITME.2018.00196
2. Platania, M., Obenshain, D., Tantillo, T., Amir, Y., Suri, N.: On choosing server-side or client-side solutions for BFT. ACM Comput. Surv. **48**(4), 30 (2016). https://doi.org/10.1145/2886780
3. Sukhwani, H., Martínez, J.M., Chang, X., Trivedi, K.S., Rindos, A.: Performance modeling of PBFT consensus process for permissioned blockchain network (hyperledger fabric). In: Proceedings of 2017 IEEE 36th Symposium on Reliable Distributed Systems, pp. 253–255. IEEE (2017). https://doi.org/10.1109/SRDS.2017.36
4. Castro, M., Liskov, B.: Practical byzantine fault tolerance and proactive recovery. ACM Trans. Comput. Syst. **20**(4), 398–461 (2002). https://doi.org/10.1145/571637.571640
5. Zheng, Z., Xie, S., Dai, H., Chen X., Wang, H.: An overview of blockchain technology: architecture, consensus, and future trends. In: Proceedings of 6th IEEE International Congress on Big Data, pp. 557–564. IEEE (2017). https://doi.org/10.1109/BigDataCongress.2017.85
6. Gramoli, V.: From blockchain consensus back to Byzantine consensus. Future Gener. Comput. Syst. **107**, 760–769 (2017). https://doi.org/10.1016/j.future.2017.09.023
7. Tsai, W., Bai, X., Yu, L.: Design issues in permissioned blockchains for trusted computing. In: Proceedings of 2017 IEEE Symposium on Service-Oriented System Engineering, pp. 153–159. IEEE (2017). https://doi.org/10.1109/SOSE.2017.32

Network Topology-Aware Link Fault Recovery Algorithm for Power Communication Network

Huaxu Zhou[1], Meng Ye[1], Yaodong Ju[1], Guanjin Huang[1], Shangquan Chen[1], Xuhui Zhang[1], and Linna Ruan[2(✉)]

[1] CSG Power Generation Co., Ltd., Beijing 100070, China
[2] Beijing University of Posts and Telecommunications, Beijing 100876, China
29200011@qq.com

Abstract. In power communication networks, in order to solve the problem of low user satisfaction caused by network link failures, this paper proposes a network topology-aware power communication network link failure recovery algorithm. The algorithm includes three parts: solving the number of services affected by the faulty link, calculating the recovery value of the faulty link, and performing fault recovery based on the recovery value of the faulty link. In terms of the number of services affected by the faulty link, the shortest route used by the power service is used as a criterion to calculate the number of services related to each faulty link. In terms of calculating the recovery value of the failed link, the value of the recovery of the failed link is calculated from four aspects: the value of the traffic and the value of the service traffic brought by the recovery of the failed link, the centrality of the failed link, and the resource overhead of the recovery of the failed link. In the simulation experiment part, the algorithm in this paper is compared with the traditional algorithm, which verifies that the algorithm in this paper has better performance in terms of user satisfaction.

Keywords: Power communication network · Fault recovery · Network topology · User satisfaction

1 Introduction

With the rapid development and application of smart grid services, the number and types of power services carried on power communication networks are increasing. However, due to natural or man-made causes, large-scale power communications network failures will still occur. Therefore, when the power communication network fails, how to quickly restore the damaged power business has become an urgent problem to be solved. In the existing fault recovery research, it can be divided into three types: fault recovery based on mathematical model ideas, fault recovery based on network characteristics, and fault recovery based on SDN new technology.

In the study of fault recovery based on mathematical model ideas, literature [1, 2] uses integer programming techniques to model the problem, and classifies the recovery flow problem into single-stage recovery and multi-stage recovery. Reference [3] divides the network to be recovered into multiple recovery areas for the problem of

Q. Liu et al. (Eds.): CENet 2020, AISC 1274, pp. 685–693, 2021.
https://doi.org/10.1007/978-981-15-8462-6_78

large-scale network failure recovery, and selects a highly centralized network resource for recovery in each area, thereby minimizing the network resource overhead. Reference [4] modeled the recovery failure problem as a linear programming problem, and proposed a heuristic failure recovery algorithm to solve the optimal solution for traffic recovery. Reference [5] adopts the idea of grouping and merging, and proposes a fault recovery algorithm based on group-based multi-stage merging. This document first divides network faults into multiple groups, and then combines the fault recovery tasks of multiple groups into a single fault recovery task based on the relationship between the groups to achieve global optimization of the problem solving.

In terms of fault recovery research based on network characteristics, literature [6] uses a combination of multicast and anycast strategies based on the service operation time and bandwidth requirements that need to be recovered to unblock the damaged traffic, thereby achieving faults recovery with minimal link overhead. Reference [7] uses the value of the faulty link to the entire network to evaluate the faulty link, giving priority to recovering the faulty link that has the greatest impact on the entire network environment. Literature [8] uses network connectivity to measure the impact of faulty links on network services, and finds critical faulty links, thereby minimizing recovery resource overhead. Reference [9] adopts the method of identifying key nodes, and proposes a resource backup strategy for network nodes to effectively solve the problem of imbalance in traffic recovery after network failure. In terms of fault recovery based on the new technology of SDN, literature [10] takes the SDN network as an experimental scenario for the problem of low fault recovery ability of existing research, and judges the faulty link that needs to be recovered based on the fault threshold of the customer information link, thus effectively improve the accuracy of fault recovery. Reference [11] adopts Openflow architecture model and proposes a link failure reporting mechanism, which has rapid fault discovery and recovery capabilities.

Existing studies have achieved more results in minimizing restoration resources and maximizing restoration business. However, existing studies have focused on maximizing the recovery traffic and maximizing the number of recovery services, mainly from the perspective of the power network construction unit, without considering user satisfaction. Generally speaking, the network traffic used by important users is large, but the number of general users is large. Therefore, in order to balance the satisfaction of users, it is necessary to simultaneously consider the restoration of the number of services and the restoration of business traffic. In order to solve this problem, based on the analysis of network topology characteristics, this paper proposes a network topology-aware power network link fault recovery algorithm with the goal of balancing the number of recovered services and the characteristics of business flows. And through experiments, it is verified that the algorithm in this paper has achieved good results in improving customer satisfaction.

2 Problem Description

The power communication network is modeled as an undirected graph $G(N, E)$, which includes a network node $n_i \in N$ and a network link $e_{ij} \in E$. The bandwidth load capacity of the network link is represented by $C_{e_{ij}}$. After the network link fails, Θ is used to denote the damaged power service set, and the number of power services $\theta \in \Theta$ contained in the set is expressed by $|\Theta|$. The power service is denoted by θ, and the bandwidth requirement of the network link of the power service is denoted by b^{θ}. This paper uses a path from the source node $s(\theta)$ to the destination node $t(\theta)$ to represent a power service. When link $e_{ij} \in E$ is included in the path of the power service, $x_{e_{ij}}^{\theta} = 1$ is used. If a network link fails, the bandwidth load capacity of the network link will be affected, reducing from $C_{e_{ij}}$ to $s_{e_{ij}}(0 \leq s_{e_{ij}} \leq C_{e_{ij}})$.

In order to quickly recover from network failures, power companies generally have emergency resources. Use R to indicate the amount of recovery resources that can be used to recover from network failures. Use $Num_{e_{ij}}$ to indicate the number of power services that can be recovered. If the required amount of resources $r_{e_{ij}}$ is provided for a damaged link $e_{ij} \in E$, the damaged link $e_{ij} \in E$ will be recovered successfully, which is indicated by $x_{ij} = 1$. When all the faulty links passed by the power service θ are restored, the power service will be successfully restored, indicated by $f_{\theta} = 1$.

3 Recovery Link Value Evaluation Model

In order to find the network link that needs to be recovered from the damaged network link, this paper proposes a recovery link value evaluation model.

3.1 Number of Services to Restore Links

Use $Num_{s(\theta)t(\theta)}$ to indicate the number of network links included in the path that power service $\theta \in \Theta$ traverses from the source point to the destination node. Use $Num_{e_{ij}}$ to indicate the number of power services that can be recovered after recovering the damaged link $e_{ij} \in E$. Considering that the power service will pass through multiple network links, this paper takes the number of power services in the damaged power service set Θ passing through the current damaged link $e_{ij} \in E$ as the value of $Num_{e_{ij}}$. The value of the number of services recovered from a faulty link is denoted by b_{ij} and calculated using formula (1), where $\sum_{\theta \in \Theta} Num_{s(\theta)t(\theta)}$ represents the sum of the number of links contained in the set Θ of all damaged services.

$$b_{ij} = \frac{Num_{e_{ij}}}{\sum_{\theta \in \Theta} Num_{s(\theta)t(\theta)}} \tag{1}$$

3.2 Service Traffic of the Restored Link

Use $Flow_{s(\theta)t(\theta)}$ to denote the flow of electricity business θ in the impaired business set Θ. Using $Flow_{e_{ij}}$ means to restore the traffic of the service recovered by each damaged link $e_{ij} \in E$. Considering that each power business will pass through multiple network links. In this paper, the sum of all the damaged power traffic passing through link $e_{ij} \in E$ is used to represent the value of $Flow_{e_{ij}}$.

The value of the service traffic recovered from the faulty link is denoted by f_{ij} and calculated using formula (2), where $\sum_{\theta \in \Theta} Flow_{s(\theta)t(\theta)}$ represents the sum of the traffic included in all damaged services.

$$f_{ij} = \frac{Flow_{e_{ij}}}{\sum_{\theta \in \Theta} Flow_{s(\theta)t(\theta)}} \tag{2}$$

3.3 Centrality of the Restored Link

Because recovery resources are scarce resources, it is of great significance to maximize the use of recovery resources. This paper analyzes the value of resources in the network from the perspective of distance, so as to maximize the utilization of recovered resources. Because the power business generally uses the shortest path for routing, the nodes at the center of the network pass through a large number of power services. The links connected by these central nodes will also pass more power services.

Use p_{ij} to indicate the centrality of network link $e_{ij} \in E$, and use formula (3) to calculate. Among them, $n_k \in \varphi(n_i)$ represents other network nodes except node n_i in the network. d_{ik} represents the number of links in the shortest path from the starting node $e_{ij} \in E$ of the network link $e_{ij} \in E$ to any other node n_k in the network. $n_k \in \varphi(n_j)$ represents other network nodes except node n_j in the network. d_{jk} represents the number of links in the shortest path from the destination node n_j of the network link $e_{ij} \in E$ to any other node n_k in the network.

$$p_{ij} = \frac{1}{\sum_{n_k \in \varphi(n_i)} d_{ik}} + \frac{1}{\sum_{n_k \in \varphi(n_j)} d_{ik}} \tag{3}$$

3.4 The Value of the Restored Link

Based on the above analysis, for each failed link that needs to be restored, this article analyzes the link that needs to be restored from the four aspects of service quantity, service flow, centrality, and resource volume. In terms of the amount of resources, use r_{ij} to indicate the resource volume needed to recover the damaged link $e_{ij} \in E$. The smaller the r_{ij} resource, the greater the recovery value of the faulty link.

In order to balance the four indicators of business number, business flow, centrality and resource volume, this paper uses the entropy weight method to solve the weight of each indicator. Among them, the w_b represents the weight of the recovered business quantity b_{ij}, the w_f represents the weight of the restored business flow f_{ij}, the w_p represents the weight of the restored resource centrality p_{ij}, and the w_r represents the weight of the resource volume of r_{ij}. Taking the calculation of the weight w_b of b_{ij} as an example, f_{ij}, p_{ij}, and r_{ij} are similar.

The calculation method of weight w_b of b_{ij} is as formula (4). Among them, e_b represents the entropy value of the recovered service quantity b_{ij} (the calculation method is shown in formula (5)), and m represents the index dimension used in the evaluation of the value of the recovered link. This refers to the four dimensions of business number, business flow, centrality, and resource volume. In the calculation formula (5) of e_b, r_{ij-z} represents an evaluation matrix element constructed by m evaluation indicators of N failed links, that is, represents the value of the z-th indicator of link $e_{ij} \in E$.

$$w_b = \frac{1 - e_b}{\sum_{b=1}^{m} (1 - e_b)} \tag{4}$$

$$e_b = -\frac{1}{\ln N} \times \ln \frac{r_{ij-z}}{\sum_{z=1}^{N} r_{ij-z}} \times \sum_{z=1}^{N} \frac{r_{ij-z}}{\sum_{z=1}^{N} r_{ij-z}} \tag{5}$$

Because the ranges of the four indicators of the restoration link, such as the number of services, service flow, centrality, and resource volume, are different, in order to balance the value of each indicator, this article uses the co-chemotactic function $u(x) = \frac{x}{\sqrt{\sum x^2}}$ to balance the comparison indicators. Based on this, the value of each restored link $e_{ij} \in E$ can be calculated using formula (6).

$$\begin{aligned} P_{(ij)} = {} & w_b \frac{b_{ij}}{\sqrt{\sum_{z=1}^{N} (b_{ij-z})^2}} + w_f \frac{f_{ij}}{\sqrt{\sum_{z=1}^{N} (f_{ij-z})^2}} \\ & + w_p \frac{p_{ij}}{\sqrt{\sum_{z=1}^{N} (p_{ij-z})^2}} + w_r \frac{r_{ij}}{\sqrt{\sum_{z=1}^{N} (r_{ij-z})^2}} \end{aligned} \tag{6}$$

4 Algorithm

The network topology-aware link fault recovery algorithm for power communication network (NTLFRA) proposed in this paper is shown in Table 1. The algorithm includes solving the number of services affected by the faulty link, calculating the recovery value of the faulty link, and performing fault recovery based on the recovery value of the faulty link. In terms of the number of services affected by the faulty link, the shortest route for the power service is used as the criterion to calculate the number of services related to each faulty link. In terms of calculating the recovery value of a faulty link, the recovery value of the faulty link is calculated from four aspects: the value of the business quantity and the value of the service flow brought by the recovery of the faulty link, the centrality of the faulty link, and the resource cost of the faulty link recovery. In terms of fault recovery based on the recovery value of the faulty link, the recovery resource is used as the constraint to recover the faulty link with a higher recovery value until the recovery resources are used up.

Table 1. Algorithm NTLFRA.

Input: power communication network $G(N,E)$, damaged power communication network $G'(N',E')$, business collection damaged after failure Θ, total amount of recovered resources R
Output: Recovery strategy
1. Find the number of services associated with each faulty link $e_{ij} \in E'$ in the faulty link set E':
1) Use the shortest path method to analyze the links used by each service θ in the damaged service set Θ;
2) Count the number of services associated with each failed link $e_{ij} \in E'$;
2. Calculate the recovery value of each failed link $e_{ij} \in E'$
1) Use formula (1) to calculate the value b_{ij} of the number of services recovered from a failed link
2) Use formula (2) to calculate the value of the service traffic recovered from the failed link f_{ij}
3) Use formula (3) to calculate the centrality p_{ij} of the faulty link
4) Based on the entropy weight method, use formula (6) to calculate the recovery value $P_{(ij)}$ of the faulty link
5) Sort each failed link $e_{ij} \in E'$ in the failed link set E' based on $P_{(ij)}$ in descending order to obtain the set E'_{des}
3. In set E'_{des}, select the faulty links with the highest recovery link value one by one to recover until resource R is used up and output the recovery strategy.

5 Performance Analysis

To verify the performance of the algorithm, the GT-ITM [12] tool was used to generate a power communication network simulation environment with 50 and 100 network nodes. The bandwidth resources of each network link are subject to an even distribution between (1, 30). Randomly select network nodes from the network and use the shortest path algorithm to simulate power business. The bandwidth resource requirements of each power service are subject to an even distribution between (1, 5). The number of power businesses increased from 50 to 150. In terms of network link failure simulation, a link is randomly selected from the network links as a failed link with a probability of 0.05. After a link fails, its remaining link capacity is subject to an even distribution of (0, 10).

In terms of algorithm comparison, the algorithm NTLFRA in this paper is compared with the traditional algorithms of maximizing recovery services (MRSA) and maximizing recovery flow (MRFA). In terms of comparative indicators, user satisfaction u is used for averaging, and it is calculated using formula (7). Among them,

α and β are used to adjust the weighting factors of the recovery service flow and quantity. *num* represents the number of services whose recovery traffic is $flow_\theta$.

$$u = \alpha \sum\nolimits_{\theta \in \Theta} num * flow_\theta + \beta \sum\nolimits_{\theta \in \Theta} f_\theta \tag{7}$$

In terms of comparison indicators, the algorithm is compared in terms of the impact of the amount of recovered resources on user satisfaction and the number of deployed services on user satisfaction. In terms of the comparison of the impact of the amount of recovered resources on user satisfaction, the algorithm performance of 50 network nodes at the network scale was compared. Deploying 100 services, the comparison results of 50 network nodes are shown in Fig. 1. The X axis represents the total amount of recovered resources from 50 to 100, and the Y axis represents user satisfaction. As can be seen from the figure, as the total amount of recovered resources increases, all three algorithms tend to stabilize. It shows that the amount of resources gradually meets the demand for restoring resources. The result of MRFA is the worst, because there are relatively few high-traffic services, which have little impact on user satisfaction. The algorithm NTLFRA in this paper is optimal, which shows that the algorithm in this paper can fully balance the performance of the number of services and the traffic flow under the condition of different total restoration resources.

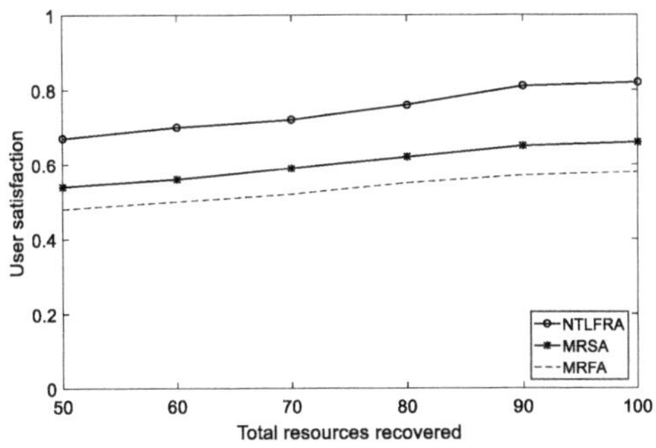

Fig. 1. The impact of recovery resources on user satisfaction.

In terms of the impact of the number of deployed services on user satisfaction, the algorithm performance under 50 network nodes is compared. Among them, the total amount of recovered resources under the environment of 50 network nodes is 50. The comparison results are shown in Fig. 2. The X axis represents the deployment business volume changes from 50 to 100, and the Y axis represents user satisfaction. As the number of deployed services in the network increases, all three algorithms tend to stabilize, indicating that the resource capacity is gradually used up. The results of the algorithm NTLFRA in this paper are optimal, which shows that the algorithm of this paper can fully balance the performance of the number of services and the traffic flow under the conditions of the total amount of deployed services.

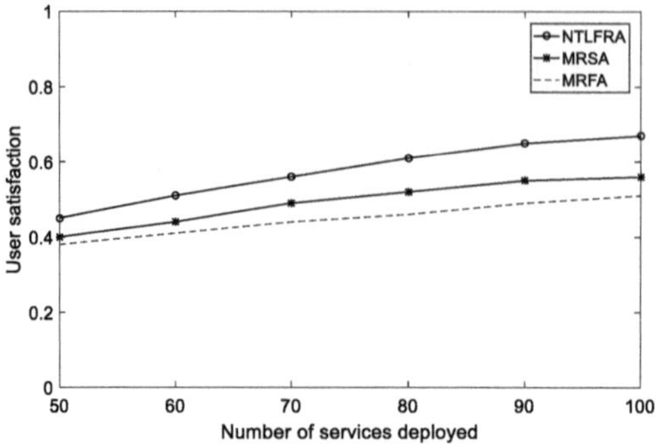

Fig. 2. The impact of the number of deployed services on user satisfaction.

6 Conclusion

In the context of the rapid development of smart grids, the number and types of power services carried on power communication networks are increasing. In order to solve the problem of low user satisfaction caused by network link failures, the network topology-aware power network link failure recovery algorithm proposed in this paper includes solving the number of services affected by the faulty link, calculating the recovery value of the faulty link, and fault recovery based on the value of faulty link recovery. Through simulation experiments, it is verified that this algorithm has achieved good results in terms of user satisfaction. With the construction and application of 5G networks, the power business based on 5G networks in power communication networks has also gradually increased. In the next work, based on the research results of this paper, we will study the fault recovery algorithm under 5G network, so as to improve user satisfaction in the 5G network environment.

Acknowledgments. This work is supported by China Southern Power Grid Technology Project "Research on application of data network simulation platform" (STKJXM20180052).

References

1. Yu, H., Yang, C.: Partial network recovery to maximize traffic demand. IEEE Commun. Lett. **15**(12), 1388–1390 (2011)
2. Kaptchouang, S., Ouédraogo, I.A., Oki, E.: Preventive start-time optimization of link weights with link reinforcement. IEEE Commun. Lett. **18**(7), 1179–1182 (2014)
3. Bartolini, N., Ciavarella, S., La Porta, T.F., et al.: Network recovery after massive failures. In: 2016 46th Annual IEEE/IFIP International Conference on Dependable Systems and Networks (DSN), pp. 97–108. IEEE (2016)
4. Liu, K.: Large-scale network fault recovery mechanism based on linear programming. Comput. Eng. **42**(7), 104–108 (2016)
5. Genda, K., Kamamura, S.: Multi-stage network recovery considering traffic demand after a large-scale failure. In: 2016 IEEE International Conference on Communications (ICC), pp. 1–6. IEEE (2016)

6. Bao, N.H., Yuan, Y., Liu, Z.Q., et al.: Optical data center network service recovery scheme based on link lifetime. J. Commun. **39**(8), 125–132 (2018)
7. Wang X.L.: Network recovery and augmentation under geographically correlated region failures. In: 2011 IEEE Global Telecommunications Conference-GLOBECOM 2011, pp. 1–5. IEEE (2011)
8. Kamrul, I.M., Oki, E.: Optimization of OSPF link weight to minimize worst-case network congestion against single-link failure. In: 2011 IEEE International Conference on Communications (ICC), pp. 1–5. IEEE (2011)
9. Zhang, L.M., Chen, X.C.: IP fast recovery traffic balancing method based on multi-route configuration. Comput. Technol. Dev. **06**, 90–94 (2019)
10. Pan, Z.A., Liu, Q.J., Wang, X.Y.: Software-defined network customer information link failure recovery simulation. Comput. Simul. **05**, 241–244 (2018)
11. Gyllstrom, D., Braga, N., Kurose, J.: Recovery from link failures in a smart grid communication network using openflow. In: 2014 IEEE International Conference on Smart Grid Communications (SmartGridComm), pp. 254–259. IEEE (2014)
12. Zegura, E.W., Calvert, K.L., Bhattacharjee, S.: How to model an internetwork. In: Proceedings of IEEE INFOCOM 1996. Conference on Computer Communications, vol. 2, pp. 594–602. IEEE (1996)

An Orchestration Algorithm for 5G Network Slicing Based on GA-PSO Optimization

Wenge Wang[1(✉)], Jing Shen[1], Yujing Zhao[1], Qi Wang[2], Shaoyong Guo[3], and Lei Feng[3]

[1] State Grid Henan Electric Power Company Information and Communication Company, Zhengzhou 450052, China
sg_wangwenge@163.com

[2] State Grid Henan Electric Power Company, Zhengzhou 450052, China

[3] State Key Laboratory of Networking and Switching Technology, Beijing University of Posts and Telecommunications, Beijing 100876, China

Abstract. Network slicing is an important technology for implementing on-demand networking based on the 5G network architecture of SDN/NFV. By analyzing the main scenarios of 5G, a network slice orchestration algorithm based on GA-PSO optimization under the SDN/NFV architecture is proposed. This algorithm uses the particle swarm optimization algorithm to quickly converge on the characteristics of the global optimal solution, and design the evaluation function of network slice performance. Moreover, the ability of genetic algorithm to quickly search randomly is used to update and optimize the network slice, and the particle swarm is used to chase the local optimal solution and the global optimal solution to obtain the optimal network slice. Simulation experiment results show that the algorithm can realize the personalized creation of network slices in multi-service scenarios, give full play to the advantages of SDN's centralized control mode, and reduce network energy consumption while improving network resource utilization.

Keywords: 5G · Network slicing · Genetic algorithm-particle swarm optimization · Orchestration

1 Introduction

With the increase in the number of user terminals, the increase in the scale of traffic, and the diversification of user needs, the current core network EPC (Packet core) gradually became unable to cope with the diversified service requirements. In the 5G era, Internet service objects and application scenarios have become diversified. It is neither realistic nor efficient to construct a dedicated physical network for each service the network slicing (NS) technology can realize the transition from "one size fits all" to "one size per service", providing a brand-new solution for the existing network [1].

The prerequisite for network slicing is that various network elements and centralized control of SDN can be virtualized [2]. With the maturity of network function virtualization (NFV) technology, software and hardware decoupling and sharing foundations are realized Facility resources and on-demand scheduling [3], while

Q. Liu et al. (Eds.): CENet 2020, AISC 1274, pp. 694–700, 2021.
https://doi.org/10.1007/978-981-15-8462-6_79

software-defined networking (SDN) decouples the data plane from the control plane, simplifies network management and flexibly configures routing [4]. Therefore, in NFV and SDN, Under the network architecture, the orchestration and deployment of network slices becomes feasible [5].

With different requirements and performance requirements, in the NFV and SDN network architecture, the arrangement of slices will directly affect the network load, resource utilization, and energy consumption. Based on SDN technology, researchers have done a lot of research in optimizing network slice orchestration and improving resource utilization. These research methods are all based on the optimization of network resources in a data center with a relatively simple network status. They do not consider the complex requirements of application services in terms of bandwidth, delay, and reliability. Most of them are only aimed at network resource utilization or QoS.

In order to solve the above problems, a network slicing algorithm based on GA-PSO was proposed. Specifically, this paper converts the optimization of the network into the orchestration of network slices. Through statistical analysis of user traffic, we know the distribution characteristics of the entire network, pre-construct the basic slice, and then analyze the load and demand for real-time traffic analysis. Slicing and deploying the constructed result on the switching node in the form of OpenFlow protocol flow table.

2 GA-PSO Slicing Generation Algorithm

In the routing model based on network slice division, the quality of slice division directly determines the load and resource utilization of the network. Therefore, the generation of slices is very important for the network slice architecture system based on NFV/SDN. In addition, the existing slicing algorithms generally adopt greedy strategy to divide resources and select routes one by one according to the needs of the network. The lack of global optimization does not give full play to SDN's advantages of mastering global information and centralized control. And when the network load is very large, the time complexity of dividing one by one is too large to meet the real-time demand.

2.1 Particle Swarm Optimization

Particle swarm optimization (PSO) is a swarm intelligence algorithm developed by Kennedy J and Eberhart R C et al. [7] in 1995. It is a simulation of the migration and aggregation of the bird's foraging process. The community of simple individuals and the interactive behavior between individuals simulate the search for the global optimal solution. The basic PSO algorithm defines two very important parameters: in a certain generation of population, the particle with the highest fitness is called P_{BEST}; and the global optimal solution found by the particles of all populations so far is called G_{BEST}. And save the position it finds, used to guide and update the position and speed of particles. The position $X_{ij}(t + 1)$ and speed update equations $V_{ij}(t + 1)$ are as follows:

$$V_{ij}(t+1) = w \cdot V_{ij}(t) + C_1 random(0,1)[p_{ij} - X_{ij}(t)] + C_2 random(0,1)[g_{ij} - X_{ij}(t)] \tag{1}$$

$$X_{ij}(t+1) = X_{ij}(t) + V_{ij}(t+1),\ j = 1, 2, \ldots, d \tag{2}$$

Where w is the inertial weight, C1 and C2 are positive learning factors, and random (0, 1) is the random number distributed uniformly from 0 to 1.

2.2 GA-PSO

The Basic Idea of the Algorithm. This paper draws on the ideas of hybridization and mutation in GA (genetic algorithm) [8] and applies these two ideas to slice optimization. A slicing as a subgraph represents a feasible solution, so the evolution of the population and the migration of particles are transformed into a process of subgraph hybridization.

The basic idea of the algorithm is to obtain two types of primitive particles according to the 3 main application scenarios proposed by 3GPP, and then the two types of primitive basic particles are further initialized to obtain a certain number of primitive particles to form an initial population. Each particle represents a topological subgraph, evaluate the fitness of each particle, store the current individual optimal particle and global optimal particle, update and optimize the subgraph according to the idea of hybridization and mutation, and generate a new topological subgraph. Particle swarm follows the current optimal particle and searches in the solution space, that is, iteratively finds the optimal subgraph as the final routing scheme.

Basic Particle Fitness Evaluation Function. In the "5G White Paper" [6], the main application scenarios of the future mobile Internet and the Internet of Things are divided into uRLLC, mMTC, and eMBB. The key capability indicators required for each scenario are high bandwidth experience rate, ultra-high traffic density, and ultra-high Connection density, low latency and high reliability.

Combining these four key capability indicators, this paper selects the two parameters of delay and bandwidth to characterize the performance indicators of future 5G application scenarios.

The performance of network slicing is characterized by different performance parameters. However, different parameters have different value ranges and units, so quantitative comparison and analysis cannot be performed. In this paper, the zero mean normalization method is used to normalize different transmission parameters, and the normalization calculation formula is:

$$P^{nor} = \frac{p - \mu}{\sigma} \tag{3}$$

Among them, P^{nor} is the normalized value of the performance parameter; p is the performance parameter; μ is the average value of the performance parameter; σ is the variance of the performance parameter.

The evaluation of the fitness of a particle is made according to the transmission parameters of the slice it represents, that is to say the fitness of the particle is the fitness of NS. This paper designs an evaluation of particle fitness based on an exponential function the function is as follows:

$$Fitness(\alpha, \beta, D, B) = -\alpha e^{D} + \beta e^{B} \quad (4)$$

Among them, D is the path delay value with the largest delay in a normalized sub-graph; B is the minimum bandwidth of the link in a normalized sub-graph; α is the proportion of low-delay requirement slices in all slices; β is the proportion of high-bandwidth demanded slices in all slices.

Algorithm Implementation Steps. The specific steps of the network slice orchestration algorithm based on GA-PSO optimization are as follows:

Step 1 The transmission parameters that characterize the performance of the network slice are normalized. The normalization method is based on Eq. (3).

Step 2 low-latency slices, high-bandwidth slices, and high-reliability slices to form a hybridization pool. The slices in the pool are randomly hybridized and initialized according to the hybridization and mutation algorithms in Sect. 3.2.4, and initialized 2n particles, each particle is an NS, representing a topological subgraph G.

Step 3 For each type of slice, select the appropriate parameters, calculate the fitness of the population particles according to Fitness (α, β, D, B), and sort them, and select the n particles with the highest fitness to form the initial population.

Step 4 Record the slice with the highest fitness in Step 3, which is the local optimal particle Gpb and the global optimal particle Ggb.

Step 5 Set the number of iterations m and the optimal solution control threshold τ describing the stability of the optimal solution.

Step 6 According to the algorithm in Sect. 3.2.4, according to the local optimal NS and the global optimal NS, all particles in the particle swarm are updated by means of hybridization and mutation.

Step 7 Calculate the fitness of the population particles according to Fitness (α, β, D, B), update the local optimal particle Gpb, if the current local optimal particle Gpb fitness value is higher than the current global optimal particle Ggb, then update the global optimal particle Optimal particles; otherwise the optimal solution control counter is incremented by 1.

Step 8 Check the iteration termination condition, if the number of iterations reaches m times or the value of the optimal solution control counter is greater than the optimal solution control threshold τ, then go to Step 9, otherwise go to Step 6.

Step 9 Output the optimal subgraph Ggb

3 Experimental Analysis

3.1 Experimental Setup

In order to verify the performance of the network slice orchestration algorithm based on GA-PSO optimization proposed in this paper, an experimental environment is designed, and the topology is shown in Fig. 1. In the network environment, the source (O) nodes O1, O2, …, On are the nodes that accept user traffic and are responsible for receiving user traffic, and D1, D2, …, Dn are the destination (D) nodes; S1, S2, S3, …, Sm is a switch running the OpenFlow protocol; the controller is the controller of the entire SDN.

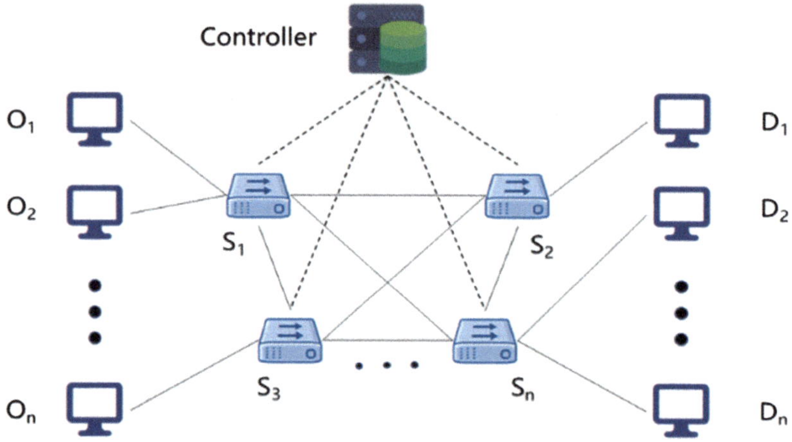

Fig. 1. Topological schematic diagram of experimental environment.

3.2 Experimental Design and Result Analysis

This paper chooses two algorithms to compare with the proposed GA-PSO. Method one is the OSPF algorithm. Another method is the greedy strategy [9].

In the following experimental analysis process, keeping the traffic demand, that is, the total amount of slices unchanged, by adding the nodes of the network (the network size changes) to compare the time taken by the three algorithms to generate routing strategies and the energy consumption after routing deployment.

In the case of the same number of traffic requirements (the number of traffic in this article is 30), this article compares the performance of the three methods under different network scales.

(1) Time complexity: The algorithm's Time complexity is a very important indicator, which indicates whether the algorithm can achieve the expected effect in the actual deployment environment. When the network scale is small, the algorithm using the greedy strategy is similar to or even better than GA-PSO in time complexity. However, the time used by the network scale also increases rapidly

and the performance deteriorates. Although the time used by GA-PSO algorithm is not very small when the network size is small, with the increase of the network size, the time used by GA-PSO algorithm grows slowly and has good time stability for different networks, which is obviously better than the queuing algorithm based on greedy strategy.

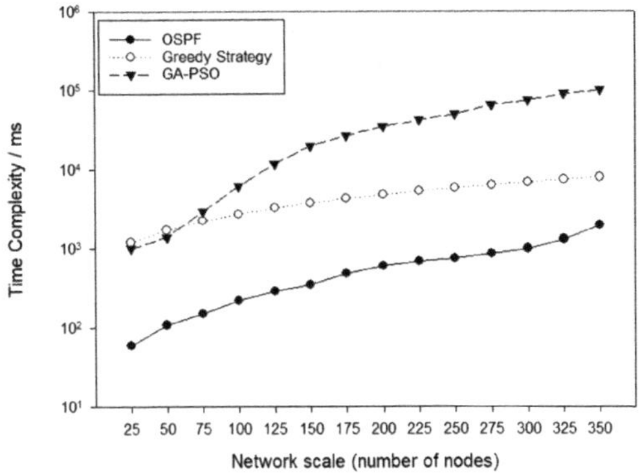

Fig. 2. Time complexity of three algorithms under different network scales.

(2) Energy consumption: In the future, energy saving becomes an unavoidable and important issue. As can be seen from Fig. 3, the larger the network size, the GA-PSO algorithm has energy consumption advantages over the other two algorithms. The more obvious, at the same time, as the network scale increases, the network energy consumption of the GA-PSO algorithm grows slowly, showing good balance ability and energy consumption stability. Although it can be seen from Fig. 2, when the network scale is 350, GA-PSO takes about 14% more time than OSPF, but the energy cost decreasing in Fig. 3 is about 32%.

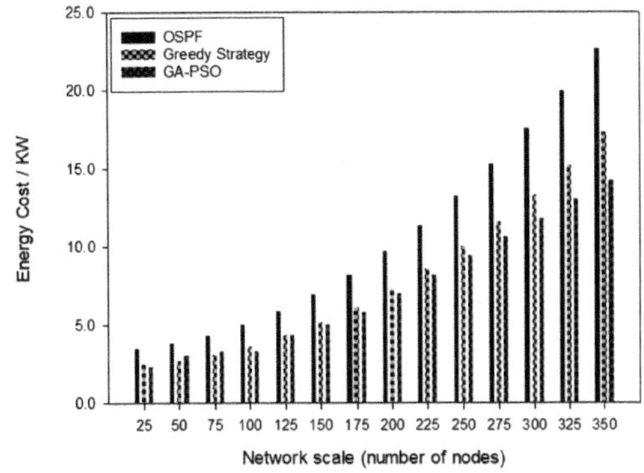

Fig. 3. Energy consumption of three algorithms under different network scales.

4 Conclusion

In this paper, through the analysis of the current main application scenarios and the research on the algorithm of network slicing, a network slicing algorithm based on GA-PSO optimization is proposed. The characteristics of this algorithm are: drawing on the idea of hybridization and mutation in the GA algorithm, the traditional PSO algorithm Improvements were made, and the PSO's group intelligence feature was applied to the optimization problem of network subgraphs, which made the algorithm's global search performance substantially improved, and gave full play to the superiority of the centralized control of the SDN architecture. Experimental data shows that the algorithm has a good effect on the optimization of high-load traffic in large-scale networks.

Acknowledgement. This work has been supported by the State Grid Henan Science and Technology project "Research on key technologies of service terminal ubiquitous access and slicing bearing in power IoT based on 5G network".

References

1. Sama, M. R., An, X., Wei, Q., Beker, S.: Reshaping the mobile core network via function decomposition and network slicing for the 5G Era. In: 2016 IEEE Wireless Communications and Networking Conference, pp. 1–7. IEEE, NJ (2016)
2. Pries, R., Morper, H., Galambosi, N., Jarschel, M.: Network as a service - a demo on 5G network slicing. In: 2016 28th International Tele traffic Congress (ITC 28), pp. 209–211. IEEE, NJ (2016)
3. Giannoulakis, I., Kafetzakis, E., Xylouris, G., Gardikis, G., Kourtis, A.: On the applications of efficient NFV management towards 5G networking. In: 1st International Conference on 5G for Ubiquitous Connectivity, pp. 1–5. IEEE, NJ (2014)
4. Ksentini, A., Bagaa, M., Taleb, T.: On using SDN in 5G: the controller placement problem. In: 2016 IEEE Global Communications Conference (GLOBECOM), pp. 1–6. IEEE, NJ (2016)
5. Zhang, J., Xie, W., Yang, F.: An architecture for 5G mobile network based on SDN and NFV. In: 6th International Conference on Wireless, Mobile and Multi-Media (ICWMMN 2015), pp. 87–92. IEEE, NJ (2015)
6. NGMN Alliance. Next generation mobile networks 5G white paper. NGMN 5G Initiative, Main, Germany (2015)
7. Kennedy, J.: Particle swarm optimization. In: Encyclopedia of Machine Learning, pp. 760–766. Springer, Boston (2010)
8. Deb, K., Pratap, A., Agarwal, S., Meyarivan, T.: A fast and elitist multi-objective genetic algorithm: NSGA-II. IEEE Trans. Evol. Comput. **6**(2), 182–197 (2002)
9. Wu, Y.N.: Research and implementation on resource control mechanism based on network slicing. Southeast University (2016)

A Path Allocation Method for Opportunistic Networks with Limited Delay

Zhiyuan An[1(✉)], Yan Liu[1], Lei Wang[1], Ningning Zhang[1], Kaili Dong[1], Xin Liu[2], and Kun Xiao[2]

[1] Communication and Dispatching Center State Grid Henan Information and Telecommunication Company, Zhengzhou 450000, China
19801116781@163.com
[2] State Key Laboratory of Networking and Switching Technology, Beijing University of Posts and Telecommunications, Beijing 100876, China

Abstract. In the future heterogeneous integrated wireless network environment, delay limited network has the characteristics of distributed self-organization, multi hop transmission, delay tolerance and intermittent link connection. Although it has great application value in many scenarios, it still needs to provide certain QoS guarantee according to the business requirements. Therefore, how to effectively transmit information to meet the needs of different services in terms of transmission rate and delay is one of the key issues in the research of path allocation in delay constrained networks. In this paper, a path allocation method based on service priority is proposed. Network nodes use local information to evaluate the network state and calculate the initial service priority, and then select relay nodes according to congestion degree and encounter probability. Simulation results show that the algorithm improves the quality of service of high priority services in congestion, and provides high performance services for low priority services when the network resources are sufficient.

Keywords: Opportunistic network · Service priority · Path allocation

1 Introduction

In view of how to efficiently transmit information in the network, meet the delivery rate and delivery delay requirements of different business information, some studies have proposed such as selection methods based on arrival probability or reference degree [1], but these methods have certain defects, such as low priority business may still not get transmission opportunities when resources are sufficient, and may not be able to when resources are insufficient Guarantee the service quality of high priority business. The QoS policy constraint policy proposed in reference [2] solves the above problems, but requires information such as the number of global nodes, and takes up too much space resources.

Q. Liu et al. (Eds.): CENet 2020, AISC 1274, pp. 701–706, 2021.
https://doi.org/10.1007/978-981-15-8462-6_80

To solve these problems, this paper proposes a path allocation method based on service priority. This algorithm can use local information to improve the quality of service of high priority service when the network is congested, and provide high performance service for low priority service when the network resources are sufficient.

2 Related Work

The performance of wireless network will be directly affected by the mobile model itself and its diversity [3]. The results of the same algorithm running in different mobile models are very different. The movement model of nodes is used to describe the physical movement rules of nodes, such as the location of nodes, the range of activities, the movement speed, the direction of advance and the dwell time. It is a mathematical model of the movement mode of nodes in the actual scene [4]. M. S. Rahim et al. Compared the effects of random walk (RW), random walk direction (RD) and shortest path graph (SPMB) on the performance of a variety of routing protocols through three indicators: transfer probability, average delay time and overhead ratio [5]. The results show that the mobile model is closely related to the dynamic change of network topology, and it is an important basis for the design, improvement and perfection of routing planning method, congestion control mechanism and power control algorithm in mobile environment.

3 Problem Model

The network model of the system is defined as follows:

(1) The entire network is a set of nodes, set to N, there are N nodes in the set, and each node follows the random walk model to move in the network. Any node in the network can be summarized into the set of Eq. 1:

$$N = \{n_j \leq i \leq n\} \tag{1}$$

(2) The set S_i is defined as all nodes contacted by node i in the past period T, which is called the social group of node i. Each node can evaluate the current network cache resource pressure through the average congestion degree of the social group.

4 Algorithm Design

4.1 Service Initial Priority Calculation Method

Network Status Evaluation. Consider using local information to assess the congestion of the network. Let br be the cache space occupancy rate, where $br \in [0, 1]$, the set S_i be the set of all nodes that node i has met in the past period T, which is called the social group of node i. Each node evaluates the current network cache status by the average congestion of the social group. Define social group congestion SBR_i as shown in Eq. 2.

$$SBR_i = \frac{1}{|S_{i,j}|}\sum_{k \in S_{i,j}} br_k \tag{2}$$

Service Priority Mapping. In a delay limited network, when providing network services for a service, it is necessary to evaluate in advance whether the network can provide the required quality of service for the service to ensure the normal operation of the network [6].

Assume the communication area of the node as d, each node follows a random movement model, and the area length of the network is Len.
The average delay of the network:

$$ED_{avg} = 0.5Len(0.34\log Len - \frac{2^{d+1} - d - 2}{2^d - 1}) \tag{3}$$

In the best case:

$$ED_{\exp} = \frac{H_{N-1}}{N-1} ED_{avg} \tag{4}$$

Where: $H_n = \sum_{i=1}^{n} \frac{1}{i}$

The mapping method of the initial priority pri of the service is shown in Eq. 5:

$$pri(ED_{hope}) = \begin{cases} 2, & ED_{hope} \in [ED_{\exp}, e^{\frac{1}{OL}} ED_{\exp}] \\ 1, & ED_{hope} \in [e^{\frac{1}{OL}} ED_{opt}, e^{\frac{OH}{OL}} ED_{avg}] \\ 0, & ED_{hope} \in [e^{\frac{OH}{OL}} ED_{avg}, \infty] \end{cases} \tag{5}$$

For priority as high priority, the level of 2, if the target delay in $[e^{\frac{OH}{OL}} ED_{opt}, e^{\frac{1}{OL}} ED_{opt}]$ interval, the downgrade flag set to 1, otherwise 0. For priority as the priority, i.e., level 1, if the target delay in $[ED_{avg}, e^{\frac{OH}{OL}} ED_{avg}]$ interval, the downgrade flag set to 1, otherwise 0.

4.2 Relay Node Selection Method Based on Congestion Degree and Delivery Prediction

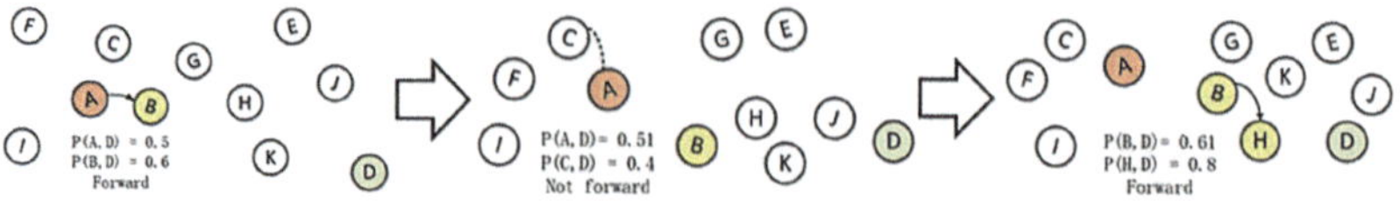

Fig. 1. Message forwarding process based on delivery probability.

There are five nodes in the figure, A, B, C, H and D. Where A is the source node of the message, B, C and H are other mobile nodes in the network, and D is the destination node of the message. $P(i,j)$ represents the predicted contact probability between node i and node j. It can be seen that when node A encounters node B, because the probability of node B to the destination node is higher, it chooses to forward the message to node B, and node B carries the message to move in the network. Node A encounters node C, because the probability of node C to the destination node is lower, node A will not transmit messages to C. When node B encounters node H, select node H after message passing (Fig. 1).

The congestion degree of node i is br_i, and the congestion degree of the node social group is SBR_i. The calculation formula of the measure BP of the relay node is as follows.

$$BP_{node(i,j)} = \sin((\pi \times br)/2) \times SBR_{node} \times P_{(i,j)} \tag{6}$$

For the message m, when the node holding the message encounters other nodes, a measure based on the degree of congestion and contact probability is calculated according to the above formula to make a decision whether to delivery the message to the meeting node.

The prediction of the contact probability between two nodes is mainly divided into three parts: (1) Node update. (2) Node contact attenuation. (3) Node transfer.

Update strategy: When two nodes are in contact, the contact prediction value will increase, and the specific increase factor is determined by a weight value.

$$P_{(i,j)} = P_{(i,j)pre} + (1 - P_{(i,j)pre})\lambda_p P_{ini} \tag{7}$$

where P_{ini} represents an initial constant.

Attenuation strategy: If two nodes do not contact within a time interval, the future contact probability of these two nodes will be smaller, so the contact probability will be attenuated with reference to the coefficient.

$$P_{(i,j)} = P_{(i,j)pre} \times \gamma^{\kappa} \tag{8}$$

where $\gamma \in [0, 1)$ is the attenuation coefficient, κ is the number of time interval blocks that have been touched in order from the most recent to the current time.

Delivery strategy: The algorithm believes that if two nodes are in frequent contact with a third node, the probability of successful message delivery in these two nodes will be higher, thereby increasing the contact probability of these two nodes.

$$P_{(i,j)} = P_{(i,j)pre} + (1 - P_{(i,j)pre}) \times P_{(i,h)} \times P_{(j,h)} \times \beta \tag{9}$$

Where $\beta \in [0, 1)$ is the delivery coefficient.

5 Simulation

Based on the simulation platform ONE, the experimental simulation of the network is established, and the node's movement model is a random walk model. There are 100 nodes, the mobile speed of the nodes is between 0.5–1 m/s, the effective communication range of the nodes is 20 m, and the minimum number of copies is 8, P_{ini} is 0.75, γ is 0.97 and β is 0.25, the size of message is between 100–300 kb.

In the experiment, according to the different requirements of different services for network performance, three types of services are defined, they are defined as H-class with a target delay of 800–1200 s; M-class with a target delay of 1200–1800 s; L-class with a target delay of 1800–3000 s. In network simulation, the ratio of the number of messages generated by these three types of system is 1: 3: 6. The algorithm mentioned in this paper is abbreviated as DCNPA. The experimental comparison algorithm includes priority-independent congestion-controlled Spray and Wait routing (CSAW), and priority-based routing QoS Policy 1, which use the infection strategy.

The simulation experiment mainly compares the routing algorithm performance by comparing the routing conditions in different buffer spaces (Fig. 2).

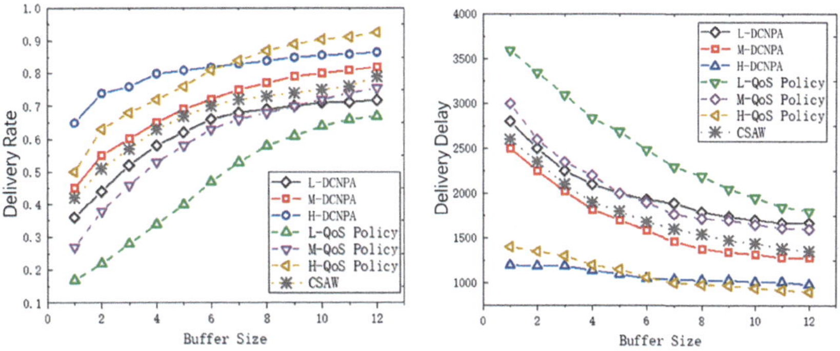

Fig. 2. Relationship between buffer size and delivery rate, delivery delay.

When the available resources cannot achieve the desired performance for all messages, the first is to satisfy H-class, then M-class, and then L-class.

When network resources are insufficient, even if a large amount of network resources are allocated for high-priority services, the delivery rate and network delay are still not as good as the DCNPA algorithm. This is because the QoS Policy algorithm uses infectious disease routing, which has a greater demand on the network buffer area. As the node cache area increases, the growth of class H tends to be flat, and the performance of M-class and L-class gradually improves. The performance of M-class is steadily higher than that of class L. In the case of sufficient network resources, although the H-class delivery rate of the DCNPA algorithm is slightly lower than the H-class delivery rate of the QoS Policy algorithm, the M-class and L-class delivery rates of the DCNPA algorithm are much higher than the same category of the QoS Policy Message delivery rate. It shows that DCNPA has higher adaptability and flexibility to network resources.

6 Conclusion

In this paper, a path allocation algorithm based on service priority is proposed. The algorithm can use node information of local network to improve service quality guarantee for high-priority services when the network is congested, and improve high-performance services for low-priority services when network resources are sufficient. Simulation results show that compared with other routing protocols, the path allocation algorithm based on service priority can provide higher-performance network services, and is better adapted to the dynamic changes of delay-limited networks.

Acknowledgment. This work was supported by the Science and Technology Project of State Grid Henan Electric Power Co., Ltd. (5217Q018003C).

References

1. Patel, A., Kamboj, P.: Survey on positioning and tracking methods for disruption tolerant network. In: 2017 International Conference on Innovative Mechanisms for Industry Applications (ICIMIA), pp. 215–218. IEEE Press, Bangalore (2017)
2. Matzakos, P., Spyropoulos, T., Bonnet, C.: Joint scheduling and buffer management policies for dtn applications of different traffic classes. IEEE Trans. Mob. Comput. **17**(12), 2818–2834 (2018)
3. Chen, D. L., He, X., Liu, T.X.: The overview of mobile model in opportunistic network. Inf. Secur. Technol. (5), 42–45 (2015)
4. Spyropoulos, T., Psounis, K., Raghavendra, C.S.: Spray and wait: an efficient routing schemefor intermittently connected mobile networks. In: Proceedings of the 2005 ACM SIGCOMM Workshop on Delay-Tolerant Networking, pp. 252–259. ACM Press, Philadelphia (2005)
5. Hossen, S., Rahim, M.S.: Impact of mobile nodes for few mobility models on delay-tolerant network routing protocols. In: 2016 International Conference on Networking Systems and Security (NSysS), pp. 1–6. IEEE Press, Dhaka (2016)
6. Tajima, S., Asaka, T., Takahashi, T.: Priority control using multi-buffer for DTN. In: The 16th Asia-Pacific Network Operations and Management Symposium, pp. 1–6. IEEE Press, Hsinchu (2014)

The Fault Prediction Method Based on Weighted Causal Dependence Graph

Yonghua Huo[1], Jing Dong[2], Zhihao Wang[1], Yu Yan[2], Ping Xie[1], and Yang Yang[2(✉)]

[1] The 54th Research Institute of CETC, Shijiazhuang 050000, China
[2] State Key Laboratory of Networking and Switching Technology, Beijing University of Posts Telecommunications, Beijing 100876, China
yyang@bupt.edu.cn

Abstract. The trunked system uses many computer nodes in close collaboration to carry a large number of high-performance computing applications, and guarantees strict QoS (Quality of Service) requirements through distribution and backup mechanism. Because of the cooperative and interactive nature of the trunked system, the failure of one node may lead to the failure of other nodes. The system log of the cluster is a very important failure prediction resource, which can use the detailed information of the system running condition recorded in the log to track the system behavior for a long time. At present, three steps mainly affect the performance of the trunked system failure prediction method: event filtering, event mining and event prediction. In the event prediction stage, the mined causal relationship is used for reasonable reasoning to achieve the purpose of fault prediction. However, the prediction method based on rule reasoning needs to scan the database for many times, which costs a lot of time in the pattern matching stage, and the reasoning efficiency is low. Aiming at the above problems, an abstract Weighted Causal Dependency Graph (WCDG) is designed to represent the event rules, and the forward uncertainty reasoning based on the Weighted Causal Graph can achieve the purpose of fault prediction. Compared with other prediction models, this method can store and update event rules more easily and meet the real-time requirement of the system.

Keywords: Trunked system · Failure prediction · Causal dependency graph

1 Introduction

Traditional fault management mechanisms include fault detection, fault diagnosis and location, fault repair and other technologies. However, these are passive fault management methods, which usually bring huge extra costs when dealing with errors or component faults, and cannot meet the new requirements of the dynamic operating environment of the system. Fault prediction is based on the analysis of the historical status and current behavior of the system, and predicts whether the system is about to generate a fault and the trend of the fault. It can avoid or reduce the extra costs caused by faults and fault repairs to the maximum extent, and based on the prediction results, timely and efficient fault isolation and resource reconfiguration, to ensure the failure

Q. Liu et al. (Eds.): CENet 2020, AISC 1274, pp. 707–714, 2021.
https://doi.org/10.1007/978-981-15-8462-6_81

elasticity of the system and efficient and stable operation of services [1]. Therefore, it is of great significance to study the active fault prediction mechanism to improve the reliability of the system.

To sum up, considering the complexity, dynamics and fault sensitivity of the trunked system, this paper focuses on the fault prediction method of the trunked system. This paper proposes a fault prediction method based on weighted causal dependency graph for system faults, uses the summary information to conduct fault correlation analysis at a higher level, predicts the propagation path and influence range of node faults, predicts the possibility of large-scale system faults, and then guides the subsequent fault isolation and task redistribution.

In the paper, a weighted causal dependence graph (WCDG) is designed to represent event rules, and a fine-grained event prediction algorithm based on WCDG is proposed. Compared with other prediction models, the event rules can be stored and updated more easily to meet the real-time requirements of the system. Compared with the existing methods, this method can provide higher prediction accuracy and higher execution efficiency, which is of great significance for realizing intelligent fault management and effectively ensuring the availability of the trunked system.

The rest of this article is organized as follows. Section 2 reviews some related work. Section 3 describes our failure prediction method WCDG. Simulation results and corresponding discussions are presented in Sect. 4. Finally, Sect. 5 summarizes this paper.

2 Related Work

The early research mainly used the classical reliability theory for fault prediction, that is, the TBF of the system satisfies certain stable distribution, such as exponential distribution, Weibull distribution, etc. [2], but this kind of model is only applicable to components or systems with constant failure rate. Some researchers believe that this hypothesis is not valid, because when the system is in a fault process, the probability distribution of TBF is constantly changing, so statistical prediction methods such as regression analysis and time series analysis can be used for modeling and fitting. Rui Ying, L. et al. [3] used the autoregressive moving average model (ARMA) to fit and predict the failure rate. Rocco, S. et al. [4] used SSA model to decompose fault behavior and predicted fault behavior by using time series related to faulty data.

By collecting various performance data of Apache Web server, Li, L. et al. [5] established an auto-repressible system model with the help of auxiliary means (ARX) to predict the future changes of performance indicators. Hoffmann, G. et al. [6] proposed a UBF (Universal Basis Function), a fault prediction model suitable for network and telecommunication environments. The model also predicts based on the time when server resources are exhausted. Lan, Z. et al. [7] proposed a dynamic meta-learning fault prediction method. The innovation of this method lies in three aspects: first, the meta-learning integrates several basic prediction methods to obtain more fault modes and fault derived rules; the second is to revise the stored rules by comparing the predicted results with the correct results in the training process.

3 Weighted Causal Dependence Graph (WCGD)

We draw on the idea of the literature [3] to propose a new weighted causal dependency graph (WCDGs) to represent the event rules. In the whole life cycle of the trunked system, we will update WCDGs automatically. And then we use the causal map to achieve fault prediction.

Construction and Updating of Weighted Dependency Graphs. The weighted causal dependency graph is a directed dependency graph (V, G). A vertex in WCDG represents an event whose sub-vertex is its posterior event in the event rule. The edges of WCDG represent the causal correlation of two events connected by edges and the pointer represents the chronological order. There are four attributes per edge: head vertex, tail vertex, support count and confidence. For example, for the 2-element event rule (A, B), the head vertex is B, and the tail vertex is A; the support count and confidence of the edge are the support and confidence of the rule event (A, B) respectively. The confidence can be used to indicate the strength of the association between the two events. The construction of WCDGs consists of three main steps:

Step 1: Construct a set of WCDGs based on the event rules generated by Apriori algorithm. The WCDG represent related situation of an event tag.

Step 2: Deletion of WCDGs. Introduce loops into the diagram. For example, AB and BA may appear at the same time, but this violates the theory of causality. To trim these edges, we compare their support counts. A lower support count indicates a weak statistical dependency between events, and their simultaneous appearance may be just a coincidence. Therefore, we remove the less supported edges from the graph.

Step 3: Create a WCDGs index and save the location of the events in WCDGs. [WCDG ID, WCDG entry vertex] is the index of WCDG. We use these indexes to locate events. After these three steps, a series describing event correlations of WCDGs is constructed. Figure 1 shows an event WCDG from the Blue Gene/L log.

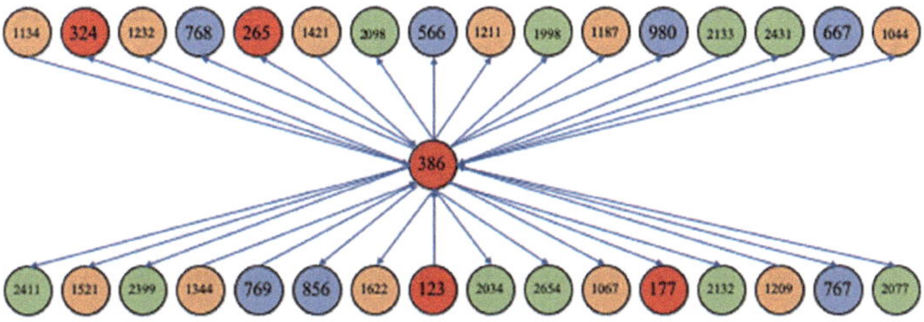

Fig. 1. Part of an event WCDG from the Blue Gene/L log.

In addition, WCDGs show a small world phenomenon in which several vertices are connected to many other vertices. What is more, we can find that these server nodes fail more frequently than others by checking the node locations of these events. This may indicate that these nodes are the core of failure propagation and therefore worthy of attention in fault analysis.

Failure Prediction Based on WCDGs. Based on saved WCDGs we can predict the probability of subsequent events occurring when an event occurs. We believe that the probability of occurrence of a fault event may be affected by two factors: the intensity of the influence of the pre-order event, and the degree of association between the events. Expressed as (1):

$$\text{Probability}(e_j) = \sum_{e_i \to e_j} power(e_i) \times \text{corr}(e_i \to e_j) \quad (1)$$

Power(e$_i$): It represents the intensity of the impact of event e_i, which is influenced by two factors, one is the weight of the event, which represents the importance of the event and the amount of information contained. The second is the position of the event in the causal map. It is known from the characteristics of the small world that the point with more branches has a greater influence.

$$power(e_i) = w(e_i) \times \lg d \quad (2)$$

In the formula, $w(e_i)$ is the weight of the event, and d is the degree of the node (that is, the number of edges directly connected to this node). It tests that using logarithm of d can achieve better results.

Corr(e$_i$ → e$_j$): It represents the degree of causality between the two events, and the calculation formula is as follows. confidence($e_i \to e_j$) is the confidence of the edge, which represents the strength of the correlation between events, and δ is the time-fading factor, so that the credibility of reasoning will be weakened with the increase of inference step. In this paper, δ = 0.7

$$\text{corr}(e_i \to e_j) = \delta \times \text{confidence}(e_i \to e_j) \quad (3)$$

For failure event prediction, two important periods need to be considered: The observation time window Δt_d and the prediction time window Δt_p. For real-time failure prediction, $\Delta t_d = 1$, this paper is set to 10 s. The principle of failure prediction is to predict the type of event that may occur within Δt_p by the observed event in Δt_d and its probability of occurrence. Once the predefined predictive probability threshold Pth is exceeded, a warning is issued and fine-grained failure prediction information (including event occurrence, location, event level, and event type and event description) will be given. The event prediction algorithm based on WCDGs is described as follows:

Step 1: Define the predicted probability threshold Pth and the predicted time window Δt_p.

Step 2: When an event occurs, we will search the index of the event and find the matching WCDG. The probability of the vertex of the head is calculated according to (1). If a head vertex is connected to a marked vertex of two different sides, we will calculate two probabilities and choose the largest one as the probability of the head vertex.

Step 3: If the probability of a head point is higher than the predicted probability threshold Pth, then the head point is a predicted event and is marked.

Step 4: Loop through step 2 until the probability of no event appearing greater than Pth, then output all events predicted.

The working flow chart of this method is shown in Fig. 2.

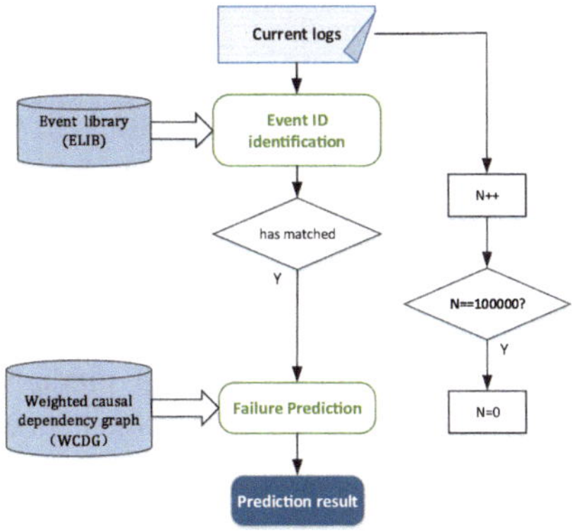

Fig. 2. Method framework.

4 Experiment

There are two data sets used in this paper: LANL, Blue Gene/L (Table 1). The above data set can be downloaded from CFDR (CFDR, Computer Failure Data Repository) [8] and Argonne Leadership Computing Facility (ALCF) [9]. We compare the fault method proposed in this paper with other methods using the same log. There are three comparison algorithms: Dynamic [7], LogMaster [10], CDG-Based [11], and SUCEG [12]. All experimental results given in this section are based on 10-fold cross-validation.

Table 1. The information about the logs of LANL, Blue Gene/L.

Prophetic	LANL	Blue Gene/L
Size	31.5 MB	708.8 MB
Record Count	433,490	4,399,503
Start Date	2005-07-31	2005-06-03
End Date	2006-04-04	2006-01-03

The calculation results of this section are based on the following parameter Settings: Pth = 0.5, $\Delta t_p = 60$min, $\Delta t_d = 60$min, slide time window length $W = 60$min, $minSup = 0.20$, $minCof = 0.25$.

Firstly, the time performance of the fault prediction method is analyzed. Table 2 shows the comparison of the average prediction time (unit: minutes) between the proposed WCGD algorithm and the LogMaster algorithm under different rule Numbers (knowledge base capacity). (note: the rule reasoning part of the rest of the algorithm works on the same principle as the LogMaster).

Table 2. Average forecast time comparison.

Rule number	Blue Gene/L		LANL	
	WCGD	LogMaster	WCGD	LogMaster
10000	1.22	2.96	1.07	2.57
20000	2.42	4.54	2.78	4.50
30000	5.66	8.92	5.50	9.99

It can be found from Table 2 that the average prediction time of all methods increases with the increase of the number of rules, which is because the increase of the number of rules leads to the increase of the number of rule matches and the increase of the number of steps of reasoning.

Secondly, we use Precision and Recall, F-1 values to analyze the failure prediction accuracy, as shown in Figs. 3, 4 and 5.

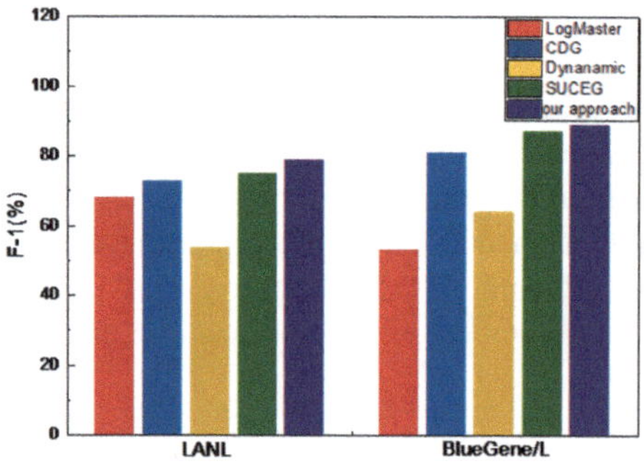

Fig. 3. Prediction of F-1 values by different algorithms.

Figure 3 shows the comparison of F-1 index between the proposed algorithm and the comparison algorithm on two public data sets. On data set LANL, the F-1 indexes of LogMaster, CDG, Dynamic, SUCEG and the algorithm in this paper are 68%, 73%, 54%, 75% and 79%, respectively. On Blue Gene/L data set, the F-1 indexes of Log-Master, CDG, Dynamic, SUCEG and the algorithm in this paper are 53%, 81%, 64%, 87% and 89%, respectively.

Figure 4 shows the comparison of the Precision index between the proposed algorithm and the comparison algorithm on two public data sets. On data set LANL, the Precision indexes of LogMaster, CDG, Dynamic, SUCEG and the algorithm in this paper are 81%, 77%, 79%, 78% and 83%, respectively. On Blue Gene/L data set, the Precision indexes of LogMaster, CDG, Dynamic, SUCEG and the algorithm in this paper are 89%, 88%, 90%, 89% and 92%.

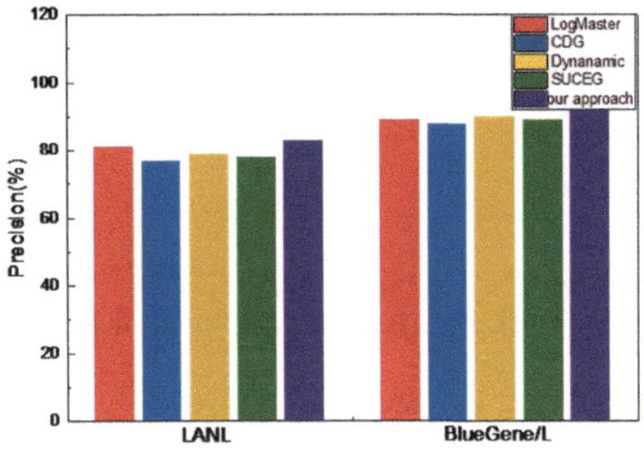

Fig. 4. Prediction Precision values of different algorithms.

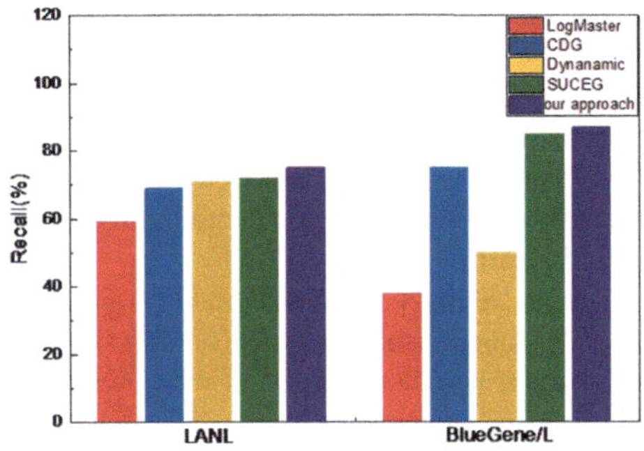

Fig. 5. The predicted Recall value of different algorithms.

Figure 5 shows the comparison of the Recall index between the proposed algorithm and the comparison algorithm on three public data sets. On the data set LANL, the Recall index of LogMaster, CDG, Dynamic, SUCEG and the algorithm in this paper are 59%, 69%, 71%, 72% and 75%, respectively. On Blue Gene/L data set, the Recall indexes of LogMaster, CDG, Dynamic, SUCEG and the algorithm in this paper are 38%, 75%, 50%, 85% and 87%, respectively.

5 Conclusion

For system faults, this paper proposes a fault prediction method based on weighted causality dependence graph. In the fault prediction stage, a causal association graph structure is designed to store and update event rules, and a fault prediction method based on causal association graph reasoning is proposed, which has strong intuitiveness, flexibility and timeliness. It can automatically complete the rule update and fault prediction during the entire life cycle of the cluster system. The experimental results of the two real log data sets LANL and Blue Gene/L also verify that the method has higher accuracy under the condition of ensuring prediction accuracy.

Acknowledgment. This work is supported by Open Subject Funds of Science and Technology on Information Transmission and Dissemination in Communication Networks Laboratory (SKX182010049).

References

1. Salfner, F., Lenk, M., Malek, M.: A survey of online failure prediction methods. ACM Comput. Surv. **42**(3), 1–42 (2010)
2. Zhong, J., Wang, Z., Su, L.: Study on adaptive failure prediction algorithm for supercomputer. J. Inf. Comput. Sci. (JICS) **12**(9), 3697–3704 (2015)
3. Li, R.Y., Rui, K.: Research on failure rate forecasting method based on ARMA model. Syst. Eng. Electron. **30**(8), 1588–1591(2008)
4. Rocco, S., Claudio, M.: Singular spectrum analysis and forecasting of failure time series. Reliabil. Eng. Syst. Saf. **114**, 126–136 (2013)
5. Li, L., Nathan, K.V., Trivedi, K.S.: An approach for estimation of software aging in a web server. In: Proceedings International Symposium on Empirical Software Engineering, pp. 91–100. IEEE, Piscataway (2002)
6. Hoffman, G., Malek, M.: Call availability prediction in a telecommunication system: a data driven empirical approach. In: 2006 25th IEEE Symposium on Reliable Distributed Systems (SRDS 2006), pp. 83–95. IEEE, Piscataway (2006)
7. Lan, Z., Gujrati, P., Sun, X.H.: Fault-aware runtime strategies for high-performance computing. IEEE Trans. Parallel Distrib. Syst. **20**(4), 460–473 (2009)
8. CFDR Data [EB/OL]. https://www.usenix.org/cfdr-data. Accessed 22 Apr 2020
9. Cluster-trace-v2018 [EB/OL]. https://github.com/alibaba/clusterdata/tree/v2018. Accessed 22 Apr 2020
10. Fu, X., Ren, R., Zhan, J., et al.: LogMaster: mining event correlations in logs of large-scale trunked systems. In: 2012 IEEE 31st Symposium on Reliable Distributed Systems, pp. 71–80. IEEE, Piscataway (2012)
11. Fu, X., Ren, R., McKee, S.A., et al.: Digging deeper into trunked system logs for failure prediction and root cause diagnosis. In: 2014 IEEE International Conference on Cluster Computing (CLUSTER), pp. 103–112. IEEE, Piscataway (2014)
12. Yu, Y., Chen, H.: An approach to failure prediction in cluster by self-updating cause-and-effect graph. In: International Conference on Cloud Computing, pp. 114–129. Springer, Berlin (2019)

Network Traffic Anomaly Detection Based on Optimized Transfer Learning

Yonghua Huo[1], Libin Jiao[1], Ping Xie[1], Zhiming Fu[2], Zhuo Tao[2], and Yang Yang[2(✉)]

[1] The 54th Research Institute of CETC, Shijiazhuang 050000, China
[2] State Key Laboratory of Networking and Switching Technology, Beijing University of Posts, Beijing, China
yyang@bupt.edu.cn

Abstract. With wide use of internet, the problem of anomaly detection in network security has become an import issue of concern. In order to solve the problem, this paper proposes a new method FEN-TLSAE which uses transfer learning to detect network traffic when plenty of labeled data are unavailable. First, we use feature extraction to get two feature subsets and use two different networks to train them. Then we add BN and dropout to the improved autoencoder for regularization. Finally, we use joint inter-class and inter-domain distributional adaptation (JCDA) in the process of transfer learning. Minimizing the marginal and conditional distribution distance between the source and target domains while maximizing the distribution distance of samples between different classes in the source domain. The experiments on the NSL-KDD data set indicate that the proposed method is more efficient than that based on TLSAE. Also, the method has higher detection accuracy, recall and precision.

Keywords: Feature extraction · Transfer learning · Anomaly detection

1 Introduction

Recently, with wide use of internet, while the network brings convenience to people, the attack behavior in the network is also constantly developing. In the current technology related to network security issues, intrusion detection has become an important protective method, aiming to detect anomalies in time and accurately for the system to make subsequent responses. As an important part of intrusion detection, network traffic anomaly detection can quickly detect network traffic anomalies that deviate from normal network traffic behavior from the traffic data.

Transfer learning is an emerging machine learning method. Its content is to reuse models developed for specific tasks on other similar tasks. This technology has the advantages of transfer data annotation, model transfer, and adaptive learning. Take advantage of transfer Learning techniques can reduce the dependence of deep learning on strong computing power, and can also reduce the dependence on large amounts of labeled data in traditional machine learning models.

First, for different feature subsets, this paper performs different data processing. For efficient feature subsets, this paper uses multi-layer perceptrons to process and analyze

Q. Liu et al. (Eds.): CENet 2020, AISC 1274, pp. 715–722, 2021.
https://doi.org/10.1007/978-981-15-8462-6_82

the deep data information. For the remaining feature subsets, only one fully connected layer is used to supplement the former data information. Then the AutoEncoder is improved by combining regularization methods such as Batch Normalization and dropout. Finally, Establishing the objective function by minimizing the marginal distribution and conditional distribution distance between the source domain and the target domain, while maximizing the distribution distance of samples between different categories in the source domain.

2 Related Works

Li, X. K. et al. [1] propose an effective deep learning method, namely AE-IDS (Auto-Encoder Intrusion Detection System) based on random forest algorithm. This method constructs the training set with feature selection and feature grouping. After training, the model can predict the results with auto-encoder, which greatly reduces the detection time and effectively improves the prediction accuracy. Alazzam, H. et al. [2] also propose wrapper feature selection algorithm for IDS. This algorithm uses the pigeon inspired optimizer to utilize the selection process. A new method to binarize a continuous pigeon inspired optimizer is proposed and compared to the traditional way for binarizing continuous swarm intelligent algorithms.

Qureshi, A. S. et al. [3] exploit the concept of self-taught learning to train deep neural networks for reliable network intrusion detection. Self-taught learning-based extracted features, when concatenated with the original features of NSL-KDD dataset, enhance the performance of the sparse auto-encoder. Long, Wen. et al. [4] propose a new DTL method. It uses a three-layer sparse auto-encoder to extract the features of raw data, and applies the maximum mean discrepancy term to minimizing the discrepancy penalty between the features from training data and testing data.

Wang, J. et al. [5] propose a novel transfer learning approach, named as Balanced Distribution Adaptation (BDA), which can adaptively leverage the importance of the marginal and conditional distribution discrepancies, and several existing methods can be treated as special cases of BDA. Based on BDA, they also propose a novel Weighted Balanced Distribution Adaptation (W-BDA) algorithm to tackle the class imbalance issue in transfer learning.

3 Traffic Anomaly Detection Model of Deep Autoencoder Based on Transfer Learning

Regardless of the method based on statistical theory or the traditional machine learning method, a large amount of labeled data is required to train the model, but in the real environment, the label data is insufficient. In order to solve the above problems, this paper proposes an improved deep autoencoder model FEN-TLSAE. The basic process is in the following steps:

Step 1: Preprocess the source domain data and target domain data. Efficient feature subsets and other feature subsets are obtained through feature extraction.
Step 2: Use two small neural networks to perform feature learning on two feature subsets.
Step 3: Input features into AutoEncoder with BN and dropout to get weight parameters.
Step 4: Establish the objective function by minimizing the marginal distribution and conditional distribution distance between source domain and target domain, while maximizing the distribution distance of samples between different categories in source domain. And realize transfer learning between source domain and target domain.
Step 5: Add a classifier on top of the model and fine-tune the model using target domain data.
Step 6: Use the trained model to determine whether the network traffic is abnormal.

The flow chart of FEN-TLSAE is as follows (Fig. 1).

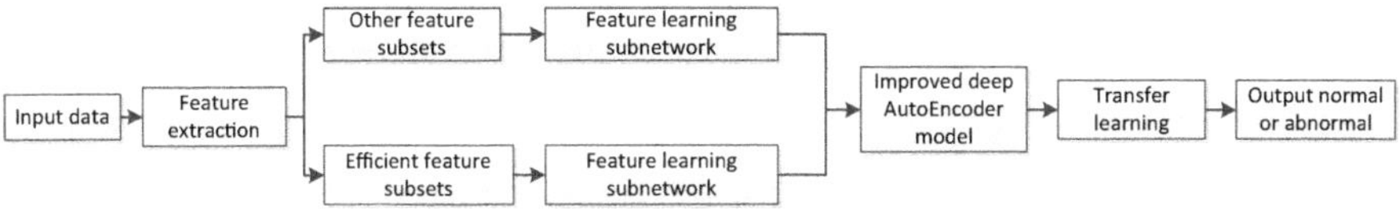

Fig. 1. The flowchart of FEN-TLSAE.

3.1 Diverse Feature Extraction

For different feature subsets, this paper performs different data processing. For efficient feature subsets, this paper uses multi-layer perceptrons to process and analyze the deep data information. For the remaining feature subsets, only one fully connected layer is used to supplement the former data information.

We can get the multi-dimensional input X_1 from the efficient feature subsets and the multi-dimensional input X_2 from the remaining feature subsets. Through feature extraction, we can get the final input data to the autoencoder.

$$Y_1 = f(W_2 f(W_1 X_1 + B_1) + B_2) \tag{1}$$

$$Y_2 = f(W_3 X_2 + B_3) \tag{2}$$

Where W_1, and W_3 are feature transform weight matrices. B_1, B_2 and B_3 are bias matrices. The ReLU function is used as the activation $f(\cdot)$. Y_1 and Y_2 are the high-dimensional features of the final output (Fig. 2).

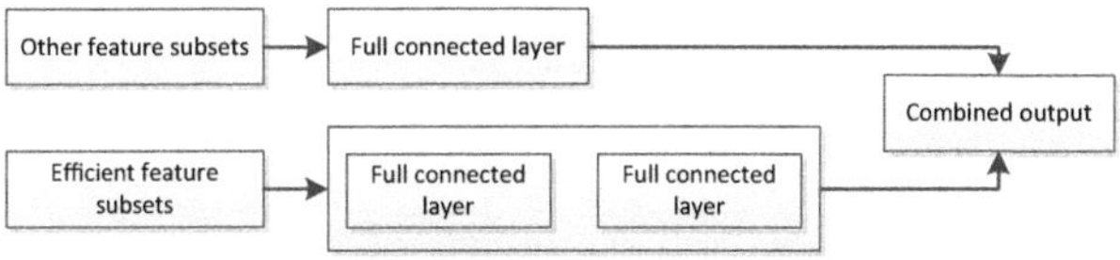

Fig. 2. The flowchart of feature extraction.

3.2 AutoEncoder with Regularization

The AutoEncoder is improved by combining regularization methods such as Batch Normalization and dropout.

In the improved AutoEncoder model, we add a layer of BN to batch normalize the input data.

Calculating the mean and standard deviation:

$$\mu_{\beta} = \frac{1}{m}\sum\nolimits_{i=1}^{m} x_i \tag{3}$$

$$\sigma_{\beta}^{2} = \frac{1}{m}\sum\nolimits_{i=1}^{m} \left(x_i - \mu_{\beta}\right)^2 \tag{4}$$

Normalizing each feature:

$$\hat{x}_i = \frac{x_i - \mu_{\beta}}{\sqrt{\left(\sigma_{\beta}^{2} + \varepsilon\right)}} \tag{5}$$

Panning and zoom again:

$$y_i = \gamma \hat{x}_i + \beta \tag{6}$$

Adding a BN layer between layers to normalize parameters so that the distribution between layers is relatively independent from other layers. Since the BN layer normalizes the feature distribution and reconstructs it, the BN layer can also have certain regularization capabilities.

Although dropout has a problem that the model structure is very unstable due to gradient oscillations, it cannot be denied that dropout has unparalleled advantages in regularization to improve the network generalization ability.

3.3 Transfer Learning Method for Joint Adaptation Between Inter-class and Inter-domain Distribution

When the transfer learning method of inter-domain distribution adaptation minimizes the distribution difference between the source domain and the target domain, it usually loses or even weakens some important properties of each domain itself. Therefore, in the process of adapting the distribution between domains, it is also important to consider the data nature of the domain itself.

Based on this situation, this paper proposes a transfer learning method with Joint Inter-class and Inter-domain Distributional Adaptation (JCDA).

Build an Optimization Model. The distance of the same sample distribution in the source domain and the target domain is

$$Dist\left(P_s^{(c)}, P_t^{(c)}\right) = \sum_{c=1}^{C} \left\| \frac{1}{n_s^{(c)}} \sum_{i=1}^{n_s^{(c)}} \emptyset\left(x_{s_i}^{(c)}\right) - \frac{1}{n_t^{(c)}} \sum_{j=1}^{n_t^{(c)}} \emptyset\left(x_{t_j}^{(c)}\right)^2 \right\| \\ = \sum_{c=1}^{C} tr(A^T K M_c K^T A) \tag{7}$$

$$(M_c)_{i,j} = \begin{cases} \frac{1}{n_s^{(c)} n_s^{(c)}}, x_i \in X_s^{(c)}, x_j \in X_s^{(c)} \\ \frac{1}{n_t^{(c)} n_t^{(c)}}, x_i \in X_t^{(c)}, x_j \in X_t^{(c)} \\ \frac{-1}{n_s^{(c)} n_t^{(c)}}, x_i \in X_s^{(c)}, x_j \in X_t^{(c)}\ or\ x_i \in X_t^{(c)}, x_j \in X_s^{(c)} \\ 0,\ \ other \end{cases} \tag{8}$$

Where c = 1, 2, …, C is the sample category number. $n_s^{(c)}$ is the total number of samples of category c in the source domain. $n_t^{(c)}$ is the total number of samples of category c in the target domain. $X_s^{(c)}$ is a set of samples of category c in the source domain. $X_t^{(c)}$ is a set of samples of category c in the target domain. K is the kernel matrix of the source and target domain samples. A is the feature transform matrix.

The distance between different types of samples in the source domain is

$$Dist\left(P_s^{(c)}, P_t^{(\bar{c})}\right) = \sum_{c=1}^{C} \left\| \frac{1}{n_s^{(c)}} \sum_{i=1}^{n_s^{(c)}} \emptyset\left(x_{s_i}^{(c)}\right) - \frac{1}{n_t^{(c)}} \sum_{j=1}^{n_t^{(\bar{c})}} \emptyset\left(x_{t_j}^{(\bar{c})}\right)^2 \right\| \\ = \sum_{c=1}^{C} tr(A^T K \bar{M}_c K^T A) \tag{9}$$

$$(\bar{M}_c)_{i,j} = \begin{cases} \frac{1}{n_s^{(c)} n_s^{(c)}}, x_i \in X_s^{(c)}, x_j \in X_s^{(c)} \\ \frac{1}{n_s^{(\bar{c})} n_s^{(\bar{c})}}, x_i \in X_s^{(\bar{c})}, x_j \in X_s^{(\bar{c})} \\ \frac{-1}{n_s^{(c)} n_s^{(\bar{c})}}, x_i \in X_s^{(c)}, x_j \in X_s^{(\bar{c})}\ or\ x_i \in X_s^{(\bar{c})}, x_j \in X_s^{(c)} \\ 0,\ \ other \end{cases} \tag{10}$$

Where $\bar{c}$ is a category number other than c. $n_s^{(\bar{c})}$ is the total number of samples in the source domain whose category is not c. $X_s^{(\bar{c})}$ is a set of samples in the source domain whose category is not c.

By minimizing the inter-domain distribution distance, maximizing the inter-class distribution distance in the source domain, using the weight adjustment coefficient β to weigh the importance between the two, the objective function is established, and the sample variance is used as a constraint condition to obtain the optimization problem:

$$min\, tr(A^T K M_0 K^T A) + \sum_{c=1}^{C} tr(A^T K M_c K^T A) - \beta \sum_{c=1}^{C} tr(A^T K \bar{M}_c K^T A) + \lambda tr(A^T A) \\ s.t. A^T K H K^T A = I \tag{11}$$

Solve the Optimization Model. For optimization problems with constraints, this paper uses Lagrange function to solve. Let K_s be the matrix of the first n_s columns of K, and K_t be the matrix of the last n_t columns of K. $A^T K_s$ and $A^T K_t$ are new feature representation of source domain and target domain.

Algorithm Description. JCDA learns the new feature representations of source and target domain samples by combining inter-class distribution adaptation and inter-domain distribution adaptation.

Step 1. Initialize the dimension k of the new feature space, the regularization coefficient λ, the weight adjustment coefficient β and the number of iterations T. Standardize the samples in the source and target domains, and select the kernel function.
Step 2. Use supervised learning training (X_s, Y_s) and get a classifier, use the classifier to learn the label Y_t of X_t, calculate the kernel matrix K and H.
Step 3. Calculate M_0 and $\bar{M}_c$ according to (X_s, Y_s) and (X_t, Y_t).
Step 4. Select the eigenvector corresponding to the first k smallest eigenvalues to form the transform matrix A. Solve $Z_s = A^T K_s, Z_t = A^T K_t$.
Step 5. Use supervised learning training (Z_s, Y_s) and get a classifier, use the classifier to learn the label Y_t of Z_t.
Step 6. If $iter < T$, let $iter = iter + 1$, go to step 3; otherwise, the algorithm ends.

4 Simulation

In this section, we will analyze the simulation results of FEN-TLSAE. We use NSL-KDD as the data set [6] and evaluation indicators include accuracy, recall and precision.

4.1 Data Preprocessing

In this paper, the character-type features are encoded by means of dummy variable coding. Since each feature dimension in the feature vector is different, the range of values is different, so standardization is required.

The specific steps of data preprocessing are: first use dummy coding for character attribute mapping, then normalize the data attributes processed, finally Use feature extraction.

4.2 Simulation Results

In this paper, we choose 50, 100, 150 and 200 as different numbers of epochs. We compare FEN-TLSAE with TLSAE [7].

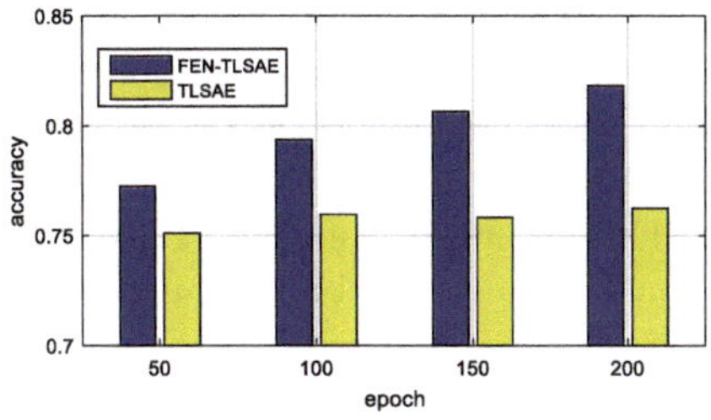

Fig. 3. Accuracy comparison between FEN-TLSAE and TLSAE

From Fig. 3, we can see that the accuracy has reached a certain level. When the number of epochs is 200, the accuracy of FEN-TLSAE is 81.82% and that of TLSAE is 76.23%. And the accuracy of FEN-TLSAE proposed in this paper is higher than that of TLSAE on different sizes of epoch.

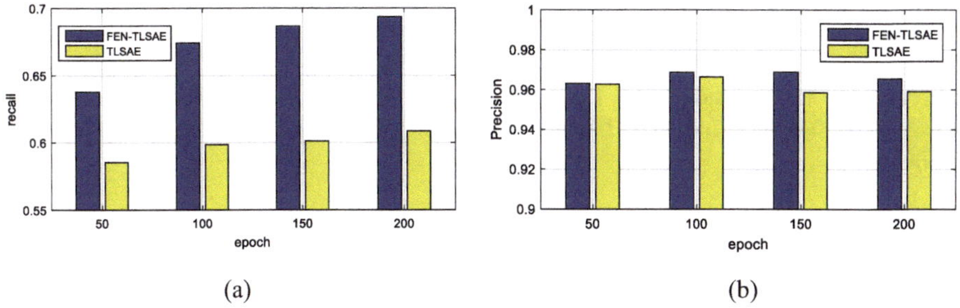

Fig. 4. Recall and precision comparison between FEN-TLSAE and TLSAE.

From Fig. 4(a), we can see that when the number of epochs is 200, the recall of FEN-TLSAE is 69.35% and that of TLSAE is 60.85%. Compared with TLSAE, FEN-TLSAE is improved by 8.5%. From Fig. 4(b), we can see that when the number of epochs is 50, the precision of FEN-TLSAE has reached 97.12%. when number of epochs changes, the precision of FEN-TLSAE is always higher than that of TLSAE.

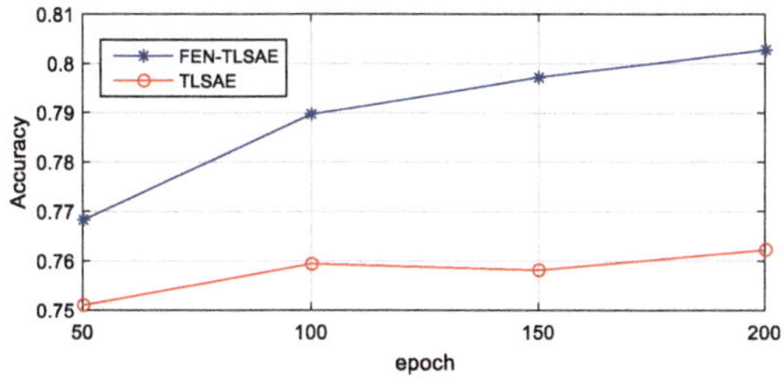

Fig. 5. Accuracy changes at different epochs.

From Fig. 5 we can see that as the epoch size increases, the accuracy of FEN-TLSAE and TLSAE increases. As the number of epochs increases, the number of weight update iterations in the neural network increases, the model gradually enters the fitted state. We compare FEN-TLSAE with TLSAE on accuracy, recall and precision. From the simulation results, we can see that FEN-TLSAE has better performance than TLSAE. FEN-TLSAE can help to detect abnormal traffic well.

5 Conclusion

In this paper, we have proposed FEN-TLSAE which uses transfer learning to detect network traffic when plenty of labeled data are unavailable. We use feature extraction, regularization and transfer learning to detect abnormal traffic well. During transfer learning, we consider joint inter-class and inter-domain distributional adaptation. The experiments on the NSL-KDD data set indicate that FEN-TLSAE is more efficient than that based on TLSAE. Also, the method has higher accuracy, recall and precision.

Acknowledgment. This work is supported by National Key Research and Development Program of China (2019YFB2103200), the Fundamental Research Funds for the Central Universities (500419319 2019PTB-019), Open Subject Funds of Science and Technology on Information Transmission and Dissemination in Communication Networks Laboratory (SKX182010049), the Industrial Internet Innovation and Development Project 2018&2019 of China.

References

1. Li, X.K., Chen, W., Zhang, Q., et al.: Building auto-encoder intrusion detection system based on random forest feature selection. Comput. Secur. **95**, 101851 (2020)
2. Alazzam, H., Sharieh, A., Sabri, K.E.: A feature selection algorithm for intrusion detection system based on pigeon inspired optimizer. Expert Syst. Appl. **148**, 113249 (2020)
3. Qureshi, A.S., Khan, A., Shamim, N., et al.: Intrusion detection using deep sparse auto-encoder and self-taught learning. Neural Comput. Appl. **32**, 3135–3147 (2020)
4. Wen, L., Gao, L., Li, X.: A new deep transfer learning based on sparse auto-encoder for fault diagnosis. IEEE Trans. Syst. Man Cybern. Syst. **49**(1), 136–144 (2017)
5. Wang, J., Chen, Y., Hao, S., et al.: Balanced distribution adaptation for transfer learning. In: 2017 IEEE International Conference on Data Mining (ICDM), pp. 1129–1134. IEEE (2017)
6. Tavallaee, M., Bagheri, E., Lu, W., et al.: A detailed analysis of the KDD CUP 99 data set. In: 2009 IEEE Symposium on Computational Intelligence for Security and Defense Applications, pp. 1–6. IEEE (2009)
7. Shone, N., Ngoc, T.N., Phai, V.D., et al.: A deep learning approach to network intrusion detection. IEEE Trans. Emerg. Top. Comput. Intell. **2**(1), 41–50 (2018)

Constant-Weight Group Coded Bloom Filter for Multiple-Set Membership Queries

Xiaomei Tian[1,2(✉)] and Huihuang Zhao[1,2]

[1] College of Computer Science and Technology, Hengyang Normal University, Hengyang 421002, China
tianxm@hynu.edu.cn
[2] Hunan Provincial Key Laboratory of Intelligent Information Processing and Application, Hengyang 421002, China

Abstract. Multiple-set membership query problem plays a very important role in many network systems, including packet forwarding architectures in big data centers, routing switching devices and internet firewall systems. Many solutions to multiple-set membership query problem are based on Bloom filters. And each of these solutions has its own merits. Some have high query speed and some have good accuracy. But generally, they cannot have both at the same time. In this paper, we propose a probabilistic data structure named constant-weight group coded Bloom Filter (CWGCBF) for fast and accurate multi-set membership query. The key technique in CWGCBF is encoding each element's set ID to constant weight code, and directly recording this constant-weight code in the bloom filter vector. Experimental results show that in terms of false positive ratio, CWGCBF, Shifting Bloom Filter and Combinatorial Bloom Filter have great advantages to the other compared Bloom filters, and in terms of time efficiency, CWGCBF and ID Bloom Filter are several times faster than the others.

Keywords: Multiple-set membership query · Constant-weight group coded bloom filter · Data center

1 Introduction

Given s disjoint sets, the problem of multiple-set membership queries is to find out whether an item x belongs to one of these multiple sets, and if so, answer which particular set x belongs to. Answering membership query for multiple sets without intersection is a fundamental work in various network applications, such as packet forwarding in a layer-2 switch [1], IP lookup in a router [2], distributed web caching [3], deep packet inspection [4], packet classification [5], network traffic measurement [6].

For example, in data centers for cloud computing, Layer-2 switches hold MAC address tables which consist of a large number of destination MAC addresses and theirs corresponding outgoing ports. When a packet arrives, the data center switch finds out the forwarding port after looking up its MAC address table according to the packet's destination MAC address. In order to speed up packet forwarding, these switches used to utilize hash tables to store MAC address table and query the hash tables for the

Q. Liu et al. (Eds.): CENet 2020, AISC 1274, pp. 723–730, 2021.
https://doi.org/10.1007/978-981-15-8462-6_83

forwarding port. However, due to the large number of memory space used by enormous entries in MAC address tables, hash tables are stored in slower DRAMs, not in faster SRAMs. Moreover, hash tables often have too low query performances in the worst cases. Therefore, many recent works, such as [7] and [8], are devoted to exploring more efficient mechanisms for packet forwarding. In order to match line speed to the greatest extent, these works utilize bloom filters and multiple-set query mechanism to forward a large number of packets.

The rest of this paper is organized as follows. Section 2 discusses the related work. Constant-weight Group Coded Bloom filter is proposed in Sect. 3. Section 4 gives experimental results of our algorithm and the comparative algorithms. Then, we conclude the paper in Sect. 5.

2 Related Work

There are many works which are devoted to multiple-set membership query. Most of these works are based on bloom filters, for example, Magic Cube Bloom Filter [1], Bloomier filter [5], Shifting bloom filter [6], Additional bloom filter [7], Difference Bloom Filter [8], Buffalo [9], ID Bloom Filter [10], Coded Bloom Filter [11], Combinatorial Bloom Filter [12], kBF [13] and iSet [14]. These variants of standard bloom filter are usually composed of multiple bloom filter vectors. At the beginning, standard bloom filters were only used to solve single-set membership query problem. Gradually, various variants of standard bloom filters were established to solve various network problems. This paper only addresses multiple-set membership query problem.

Multiple-set membership query problem is different from single-set membership query problem. To solve single-set membership query problem, data structures only need to record attribution information of the elements, while in multiple-set membership query problem, data structures still need to record set IDs of the elements explicitly or implicitly. The recent common structures for multiple-set membership query are discussed briefly as follows. For ease of understanding and presentation, the notations used for subsequent discussions are given, as shown in Table 1.

Usually one Bloom filter vector encodes one single set. To encode s sets, a straightforward approach is to use s bloom filter vectors to represent s sets, i.e. one bloom filter vector for each set. The data structure is composed of s bloom filter vectors (B_1, B_2, …, B_s) and a certain number of hash functions (h_1, h_2, …, h_k). When inserting an element e from set S_i, e is hashed to the i-th bloom filter vector B_i. If querying for x's membership, all bloom filter vectors are examined. This leads to very low memory access efficiency [14]. Therefore, many works have done in two aspects: query exactness and memory access efficiency. Some works aimed to promote the exactness of membership query, some works aimed to better the memory access efficiency, and there are still some works tried to do better in both query exactness and memory access efficiency.

Bloomier filter [5] uses cascaded bloom-filter-pairs (A_i, B_i) to encode an element e from one set. If the i-th binary bit of e's set ID is 1, e is inserted into A_i, otherwise it is inserted into B_i. So, the item e is inserted into $\lceil log_2 s \rceil$ bloom filter vectors. Suppose e's set ID is 1011. The item e will be inserted into A_1, B_2, A_3 and A_4. When looking up elements, it checks the bloom-filter-pairs one by one. If the first bloom-filter-pairs say

Table 1. Notations and their meanings.

Notations	Meanings
e	the inserted element
x	the queried item
s	the number of different sets
k	the number of hash functions
B_i	the i-th bloom filter vector
h_i	the i-th hash function
(A_i, B_i)	the i-th bloom-filter pair
g	the number of groups of hash functions
$h_i(x)$	the i-th hash value of x
$B[h_i(x)]$	the $h_i(x)$-th bit of bloom filter vector B
p	a false positive
l	the length of constant weight code for sets
n	the count of all elements in all sets
m	the size of bloom filter
$m{:}n$	the proportion of bit space to element

yes, then check the second bloom-filter-pairs, and so on; if say no, stop checking the subsequent bloom-filter-pairs.

Coded Bloom filter [11] has $\lceil log_2 s \rceil$ bloom filter vectors. It inserts an element e based on e's set ID. If the i-th binary bit of e's set ID is 1, e is inserted into the i-th bloom filter vector B_i. So, the element e is inserted into the first bloom filter vector B_1, the third bloom filter vector B_3, the fourth bloom filter vector B_4. Coded Bloom filter queries items by checking all $\lceil log_2 s \rceil$ bloom filter vectors.

Another solution is Combinatorial Bloom Filter [12] which is proposed by Hao et al. It has only one Bloom filter vector, but uses g groups of hash functions, and g is bigger than $\lceil log_2 s \rceil$. It selects the i-th group of hash functions to map e to the bloom filter vector if the i-th binary code of e's set ID is 1. It should be noted that Combinatorial Bloom Filter uses constant-weight error-correcting code to encode the set id.

Yang et al. [6] proposed Shifting Bloom Filter to solve multiple-set query question. Shifting Bloom Filter constructs only *one* bloom filter vector as same as Combinatorial Bloom Filter, but Shifting Bloom Filter needs only *one* group of hash functions. To keep record of e's set ID, it inserts e into filter not only relying on e's hash values, but also relying on an offset which represents e's set ID. That is to say, if e's hash value is $h_i(e)$ and e's set ID is id, then $B[h_i(e) + id]$ is set to 1. When querying x, it examines the *bitwise-AND* binary results of $B[h_1(x) + j]$, $B[h_2(x) + j]$, …, $B[h_k(x) + j]$, where j ranges from 1 to s. If all binary bits of the j-th result are 1, then x is considered to be in the j-th set.

Liu et al. presented ID Bloom Filter in the work [10]. The key technique of ID Bloom Filter is directly recording e's set ID at its mapping positions in one bloom filter vector. More concretely, if e's hash value is $h_i(e)$ and the binary string of e's set ID is 1011, then $B[h_i(e)]$, $B[h_i(e) + 1]$, $B[h_i(e) + 2]$, $B[h_i(e) + 3]$ are *bitwise-OR*ed with the binary string 1011. When querying x, it retrieves k binary strings, $B[h_i(x)]B[h_i(x)+1]$ $B[h_i(x)+2]B[h_i(x)+3]$, where i ranges from 1 to k. Then it calculates the *bitwise-AND* results of these k binary strings as the estimated set ID for x.

The most recent work is magic cube bloom filter which is presented in [1]. In [1], the data structure is recognized as a magic cube, which is comprised of d bloom filter vectors, while the length of each bloom filter vector is w bits. When an element from set i is inserted to magic cube bloom filter, it is mapped randomly to k bits in a $d \cdot w$ bit-array. Thus, the items from the same set are distributed in different w-bit-arrays. Magic cube bloom filter improves lookup accuracy and query speed through utilizing spatial locality and redistributing items.

3 Constant-Weight Group Coded Bloom Filter

In network systems, Bloom filters [15] are often utilized to judge whether an element e is the member of set S or not. The core of standard Bloom filter [15] is an m-bit vector B and a group of hash functions $h1, h2, \ldots, hk$. The m-bit vector is set to 0s at the beginning. Each element in S is mapped into k bits in B with k hash functions, and these k bits ($B[h_1(e)], B[h_2(e)], \ldots, B[h_k(e)]$) are thereby set to 1. In order to judge whether x is in S or not, $h_1(x), h_2(x), \ldots, h_k(x)$ are calculated at first, then the bit vector B is checked whether $B[h_1(x)], B[h_2(x)], \ldots, B[h_k(x)]$ are all 1s or not. If all 1s, then x is determined to belong to S; otherwise, as long as any of $B[h_1(x)], B[h_2(x)], \ldots, B[h_k(x)]$ is 0, x must not exist in S. It is possible to misjudge item p which is not in S as being in S. The item p is called a false positive.

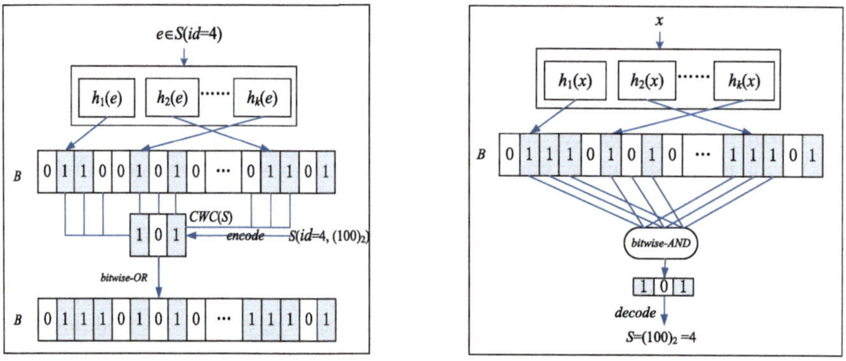

(a) Inserting an element e from set S (b) Querying an item x

Fig. 1. Constant-weight group coded Bloom filter.

Figure 1 shows the inserting and querying procedures of Constant-weight group coded Bloom filter. For the convenience of description, this algorithm is abbreviated as **CWGCBF**. The algorithm is described as follows.

When inserting an element e from set S into the bit vector B, firstly we compute k hash values using MurmurHash3 algorithm [16] and find these k positions in the filter vector B. Before performing *bitwise-OR* operation, e's group ID S is encoded to a constant-weight code, then this code is *bitwise-OR*ed with the l-bit binary string starting from each position in the filter vector B. When querying an item x for its membership, after finishing k hash computations in the same way, the algorithm retrieves k l-bit binary strings, then *bitwise-ANDs* these l-bit strings to get the estimated set *id* for x. Finally, *id* is decoded to the real set ID of x.

4 Simulation Experiments

In this section, simulation experiments are carried out to verify the effects of our algorithm and the comparative algorithms. During the experiments, our method was compared with Shifting Bloom Filter [6], ID Bloom Filter [10], Coded Bloom Filter [11] and Combinatorial Bloom Filter [12].

4.1 Experimental Setting

For Shifting Bloom Filter (SHIFTBF), we implemented the data structure like [6]. For ID Bloom Filter (IDBF), the bit array is constructed according to [10]. Coded Bloom Filter (CODEDBF) is built using the algorithm in [11] and Combinatorial Bloom Filter (COMBBF) is set up according to [12].

In our experiments, we use MurmurHash3 [16] to compute hash values. Murmur hash is a kind of non-encrypted hash function, which is very suitable for general hash based search, and it has many varieties. We compute 4 hash values using MurmurHash3 for each bloom filter. During these algorithms, 1000 elements are randomly generated, and they are randomly and evenly distributed in 255 disjoint sets. In CWGCBF, every set ID is encoded to an 11-bit binary string whose weight is 5 constantly. All of the bit vectors are 48000 or 80000 bits. For each configuration, we report the average results of 100 independent runs in Figs. 2, 3, 4 and 5.

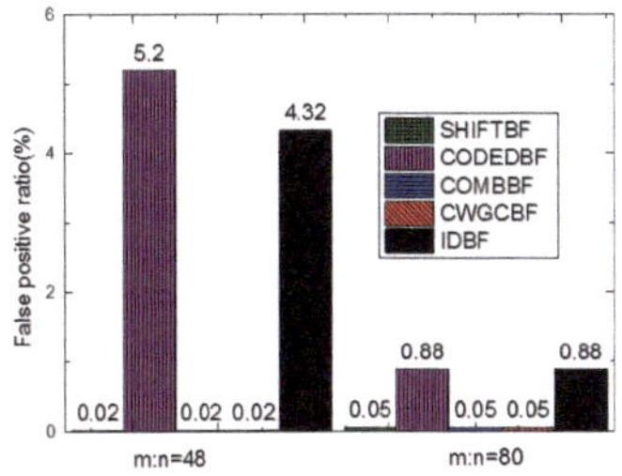

Fig. 2. False positive ratios comparison.

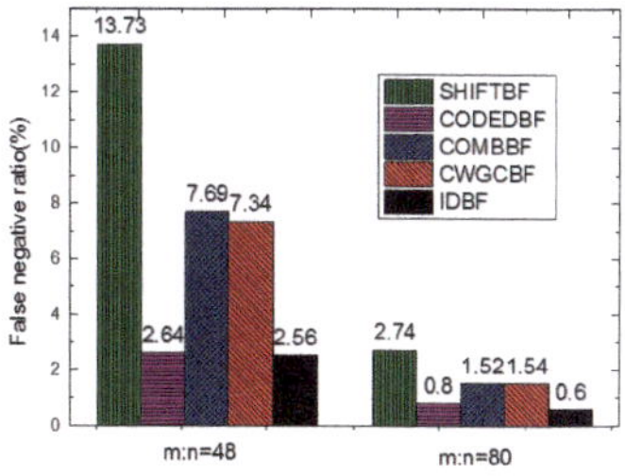

Fig. 3. False negative ratios comparison.

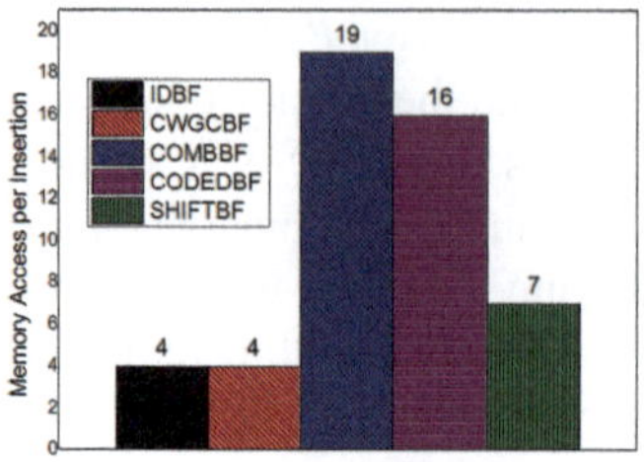

Fig. 4. Memory access times per insertion.

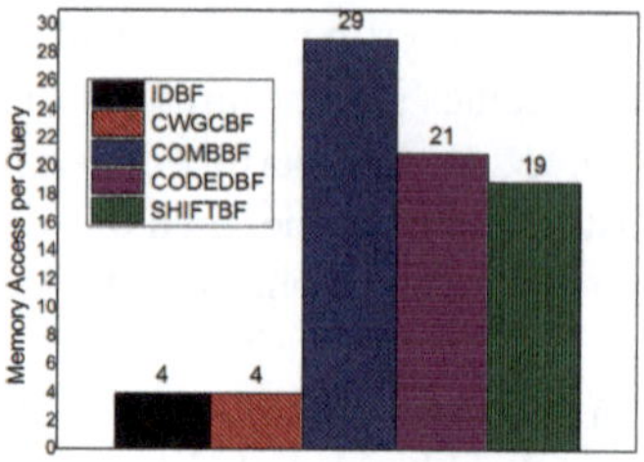

Fig. 5. Memory access times per query.

4.2 Results and Discussion

From Fig. 2, it can be seen that the false positive ratios of CWGCBF, SHIFTBF and COMBBF are always significantly less than those of CODEDBF and IDBF, and they are basically the same. That is to say, CWGCBF is one of the best algorithms among these bloom filters as for false positive ratio.

As for false negative ratio, it can be seen from Fig. 3 that CWGCBF has roughly the same ratio as COMBBF, and it is better than SHIFTBF, but worse than IDBF and CODEDBF.

Although Fig. 2 and Fig. 3 only show false positive ratios or false negative ratios when m:n is 48 or 80, in fact when m:n takes other values, the situation is still similar. CWGCBF have the best false positive ratio in all algorithms. At the same time, its false negative ratio is in the middle of all algorithms in the experiments.

Figure 4 and Fig. 5 show memory access times for all Bloom filters. From these two figures, it can be seen that CWGCBF has the same memory access performance as IDBF, and has absolute advantage over COMBBF, CODEDBF and SHIFTBF. Memory access times per insertion and memory access times per query are independent of m:n.

Among the above typical multiple-set membership query algorithms, our method CWGCBF shows very good performances in false positive ratio and memory access times. From Figs. 2, 3, 4 and 5, we can conclude that CWGCBF and IDBF have the best comprehensive performances. Because CWGCBF uses constant-weight code to encode items' set ID, and this code usually has more 1s than set ID, when other experimental parameters are same, it needs larger m:n to get almost false negatives ratio than IDBF. Therefore in the same experimental configuration, CWGCBF is a little weaker than IDBF in false negative ratio, but CWGCBF has better performances than IDBF in false positive ratio, and our method has the same memory access efficiency as IDBF.

5 Conclusion

In this paper, we proposed constant-weight group coded bloom filter. This algorithm can be used to solve multiple-set membership query problem efficiently. It has the characteristics of easy realization, high query precision and high time efficiency, so it can be widely used in packet forwarding in big data center, IP lookup in routers, distributed web caching, and so on.

Acknowledgment. This work was partly supported by the National Science Foundation of China under (Nos. 61503128, 61772182, 61772178), Hunan Provincial Natural Science Foundation of China (No. 2020JJ4152), Double First-Class University Project of Hunan Province (Xiangjiaotong [2018] 469), the Science and Technology Plan Project of Hunan Province (No. 2016TP1020), the General Scientific Research Fund of Hunan Provincial Education Department (No. 18A333), Postgraduate Research and Innovation Projects of Hunan Province (No. Xiangjiaotong [2019]248-998). Hengyang guided science and technology projects and Application-oriented Special Disciplines (No. Hengkefa [2018]60-31) and Postgraduate Scientific Research Innovation Project of Hunan Province (No. CX20190998).

References

1. Sun, Z., Gao, S., Liu, B., Wang, Y., Yang, T., Cui, B.: Magic cube bloom filter: answering membership queries for multiple sets. In: Proceedings of IEEE International Conference on Big Data and Smart Computing (BigComp). Kyoto, Japan, pp. 1–8 (2019)
2. Yang, T., Xie, G.G., Liu, A.X., et al.: Guarantee IP lookup performance with fib explosion. Comput. Commun. Rev. **44**(4), 39–50 (2014)
3. Fan, L., Cao, P., Almeida, J., et al.: Summary cache: a scalable wide-area web cache sharing protocol. IEEE/ACM Trans. Networking (TON) **8**(3), 281–293 (2000)
4. Dharmapurikar, S., Krishnamurthy, P., Sproull, T.S., Lockwood, J.W.: Deep packet inspection using parallel bloom filters. IEEE Micro. **24**(1), 52–61 (2004)
5. Chazelle, B., Kilian, J., Rubinfeld, R., et al.: The Bloomier filter: an efficient data structure for static support lookup tables. In: Proceedings of the Fifteenth Annual ACM-SIAM Symposium on Discrete Algorithms, pp. 30–39 (2004)
6. Yang, T., Liu, A.X., Shahzad, M., Yang, D.S., Fu, Q.B., Xie, G.G., Li, X.M.: A shifting framework for set queries. IEEE/ACM Trans. Networking **25**(5), 3116–3131 (2017)
7. Lee, H., Nakao, A.: Improving bloom filter forwarding architectures. IEEE Commun. Lett. **18**(10), 1715–1718 (2014)
8. Yang, D., Tian, D., Gong, J., Gao, S., Yang, T., Li, X.: Difference bloom filter: a probabilistic structure for multiple-set membership query. In: Proceedings of 2017 IEEE International Conference on Communications, pp. 1–6. Paris (2017)
9. Yu, M., Fabrikant, A., Rexford, J.: Buffalo: bloom filter forwarding architecture for large organizations. In: Proceedings of ACM Conference on Emerging NETWORKING Experiments and Technology, Rome, Italy, pp. 313–324 (2009)
10. Liu, P., Wang, H., Gao, S.: ID bloom filter: achieving faster multi-set membership query in network applications. In: IEEE 2018 IEEE International Conference on Communications, Kansas City, MO, USA, pp. 1–6 (2018)
11. Chang, F., Feng, W.C., Li, K.: Approximate caches for packet classification. In: Proceedings of the 23rd Annual Joint Conference of the IEEE Computer and Communications Societies (INFOCOM), Hong Kong, China, pp. 2196–2207 (2004)
12. Hao, F., Kodialam, M., Lakshman, T.V., Song, H.: Fast multiset membership testing using combinatorial bloom filters. IEEE/ACM Trans. Networking **20**(1), 295–304 (2012)
13. Xiong, S., Yao, Y., Cao, Q., He, T.: KBF: a bloom filter for key-value storage with an application on approximate state machines. In: International Conference on Computer Communications, pp. 1150–1158 (2014)

14. Qiao, Y., Chen, S., Mo, Z., Yoon, M.: When bloom filters are no longer compact: multi-set membership lookup for network applications. IEEE/ACM Trans. Networking **24**(6), 1–14 (2016)
15. Bloom, B.H.: Space/time trade-offs in hash coding with allowable errors. Commun. ACM **13**(7), 422–426 (1970)
16. Source code of the murmurhash3. https://chromium.googlesource.com/external/smhasher/+/c2b49e0d2168979b648edcc449a36292449dd5f5/MurmurHash3.cpp

Information Security and Cybersecurity

A Image Adaptive Steganography Algorithm Combining Chaotic Encryption and Minimum Distortion Function

Ge Jiao[1,2](✉), Jiahao Liu[1], Sheng Zhou[1], and Ning Luo[1]

[1] College of Computer Science and Technology, Hengyang Normal University, Hengyang 421002, China
jiaoge@126.com

[2] Hunan Provincial Key Laboratory of Intelligent Information Processing and Application, Hengyang 421002, China

Abstract. Digital image has the characteristics of high redundancy and convenient processing, which is the ideal carrier of information hiding application. Therefore, digital image steganography has become a hot research direction in the field of information security. In order to ensure the concealment, reliability and security of image steganographic information, an image adaptive steganographic algorithm combining chaotic encryption and minimum distortion function is designed. The algorithm generates random keys, encrypt the secret information using Logistic and ChebyShev mapping, and then embeds the key and the encrypted secret information into the carrier image using the HILL steganographic algorithm. The algorithm stores the embedded key and the decryption key separately, so the attacker cannot get the correct secret information even if he gets the embedded key, which improves the security of the key. Experiments show that the pixel change rate of the algorithm is 6.76%, the peak signal-to-noise ratio is 59.51 db, and it has a very good anti-steganographic analysis ability.

Keywords: Image adaptive steganography · Chaotic encryption · Cost · PSNR

1 Introduction

Steganography is a technique and science of information hiding, which means that no one other than the intended recipient is aware of the event or content of the transmission of information. Image steganography is a kind of covert communication technology in which secret messages are embedded in image carriers for information transmission. The early non-adaptive image steganographic methods mainly include LSB [1], F5 [2], OutGuess [3], MB [4], nsF5 [5], etc. These algorithms have the advantages of simple design and easy operation, but poor security.

In recent years, with the development of the steganographic techniques, adaptive steganography has gradually become a hot research direction in the field of current image steganography. Combined with the structural characteristics of the image itself, the algorithm adaptively selects the regions in the image that are relatively difficult to detect and insensitive to embed the message, which preserves the more complex image

Q. Liu et al. (Eds.): CENet 2020, AISC 1274, pp. 733–740, 2021.
https://doi.org/10.1007/978-981-15-8462-6_84

statistical characteristics and greatly improves the security of steganography. The algorithm is combined with the structure of the image features itself, adaptive selection is relatively difficult to detect, not sensitive areas of the image for message embedding, retain the more complex the statistical properties of the image greatly improves the security of steganography. At present, several mainstream adaptive stegographic algorithms, such as HUGO [6], WOW [7], S-UNIWARD [8], HILL [9], MiPOD [10], are mainly based on the minimum distortion cost function. Message embedding is realized in the form of encoding through the embedded generation value of different pixels, during which the total distortion value of all pixels should be kept to the minimum, which can effectively improve the anti-detection ability and maximize the original image characteristics. Finally, the corresponding steganographic image is obtained by the adaptive steganographic coding method STC [11]. Therefore, better definition of distortion cost function can effectively improve the security of adaptive steganographic algorithm.

2 Preliminaries on Chaotic Systems and Image Steganography

2.1 Logic Map

Logistic map is a typical non-linear chaotic equation, which can generate complex chaotic behavior [12, 13]. It generated chaotic sequence has better randomness, the x_{n+1} are all distributed on (0, 1) when the $\mu \in (3.5699456,4]$.

$$x_{n+1} = \mu x_n(1 - x_n),\ \ x_n \in (0, 1) \tag{1}$$

2.2 ChebyShev Map

ChebyShev map has good initial sensitivity and long-term unpredictability of chaotic sequences, which is in chaos when $k \geq 2$ [14, 15].

$$x_{n+1} = \cos(k\arccos(x_n)),\ \ x_n \in [-1, 1] \tag{2}$$

2.3 HILL-Based Minimized Distortion Function

HILL algorithm is a steganographic algorithm with excellent detection resistance and computational speed under minimum embedded distortion system. Compared with WOW algorithm, HILL algorithm uses a smaller and more concise filter. The HILL algorithm uses a high pass filter to determine the texture area and two low pass filters to aggregate the pixels with lower modification costs. The formula for calculating the modification cost for the HILL algorithm is shown below.

$$D(x, y) = \sum\nolimits_{i=1}^{n} \rho_i(x_i, y_i) \quad (3)$$

where, $x = (x_1, x_2, \ldots, x_n)$ is the carrier image to be embedded, $y = (y_1, y_2, \ldots, y_n)$ is the embedded secret image, and $p(x_i, y_i)$ is the distortion cost of modifying the ith carrier pixel x_i to y_i.

2.4 Syndrome-Trellis Codes (STC)

STC code is binary steganographic code, and the calculation formula is as follows:

$$Emb(c, m) = \arg \min d(c, s) \quad (4)$$

The encoding process is the process of finding the code word s with the minimum hamming distance from carrier *c* in the cosset of the secret message *m*. After receiving *s*, the receiver can multiply *H* to obtain the secret message *m*. Where, *d(c, s)* is the hamming distance between *c* and *s*, and *H* is the check matrix of parameters Shared by both receiving and receiving parties.

3 Design and Implementation of Image Adaptive Steganography Algorithm

Our algorithm uses the Logistic and ChebyShev maps to the secret information is encrypted, according to the image noise and texture complexity, HILL cost function is used to calculate the corresponding embedding cost, and then the key and the encrypted secret information is embedded into the carrier image, statistics the total image distortion, to use on STC embedded coding minimizes the distortion in order to get the secret image. The algorithm process is as follows [16–18]: (see Fig. 1).

(1) Convert secret information into ASCII code *A[M*N]* by character;
(2) Convert *A[M*N]* into binary sequence *B[M*N]*;
(3) Randomly generated two keys are key_L ($key_L \in (0,1)$) and key_C ($key_C \in [-1,1]$), where key_L is the initial key for Logistic mapping and key_C is the initial key for ChebyShev mapping.
(4) Take key_L as the initial key, use Logistic mapping to iterate for 100 times (eliminate the influence of transient), and use ChebyShev to iterate for 1 more time, and save the result in C_x.
(5) Take key_C as the initial key, iterate 100 times with ChebyShev mapping (eliminate the influence of transient), take the absolute value of the result, use Logistic iteration for 1 time, and save it in L_x.
(6) Extract elements from *B[M*N]*. If the position number of this element is odd, then use ChebyShev chaotic map to iterate with C_x as the initial key, and record the result of each iteration as the next odd element point iteration encryption; If the element's position number is even, Logistic chaos mapping is used to iterate with L_x as the initial key, and the result of each iteration is recorded as the element point corresponding to the next even number point.

(7) Take the element point being encrypted and the element point at the previous position for xor operation;
(8) Repeat steps (6) and (7), and finally output ciphertext *A'*;
(9) Obtain the carrier image and extract the pixel matrix *I*;
(10) Use the HILL cost function to embed the key and encrypted secret information into the carrier image *I' = HILL(I, A')*;
(11) The corresponding steganographic image *SI = STC(I')* was obtained by the adaptive steganographic coding method STC.
(12) Obtain the carrier image and extract the pixel matrix with the embedded key;
(13) Extract key and secret information;
(14) Use the key to decrypt the secret information through the decryption algorithm;
(15) Convert binary secret information into a string, that is, decrypted information.

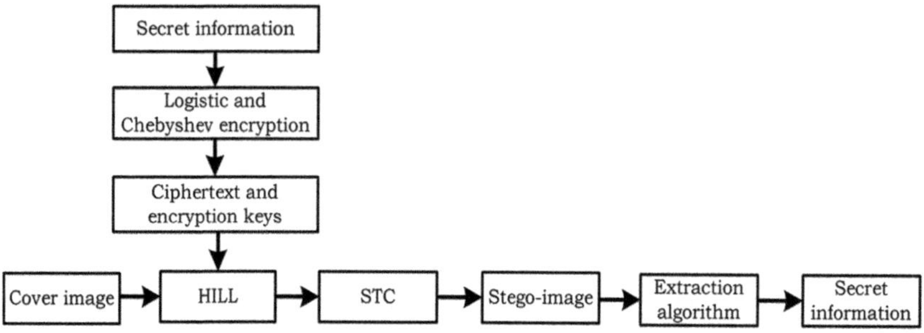

Fig. 1. Image steganography and steganalysis algorithm flow chart.

4 Experimental Results and Analysis

4.1 Imperceptibility Analysis

Imperceptibility refers to the comparison between the densified image and the original image to see whether it can achieve an indistinguishable effect, which is the simplest evaluation method for the image steganographic algorithm. Figure 2 shows the carrier image and the corresponding steganographic image, which show no difference from the naked eye.

4.2 Pixels Change Rate

The pixel change rate is one of the indicators to judge the steganographic algorithm of images. By comparing the pixel change rate between the original image and the densified image, we can see the change size of the whole image and the original image before and after embedding. The larger the pixel change rate is, the more the image changes and is more vulnerable to hackers and viruses. It can be seen from Table 1 that the pixel change rate of the proposed adaptive steganography algorithm is lower than that of the classic LSB image steganography algorithm, and it has better anti-detection ability.

(a) (b)

Fig. 2. (a) Cover image and (b) corresponding stego image.

Table 1. Comparison of pixel change rates based on different steganographic algorithms.

Existing methods	Pixels change rate (%)
Classic LSB [19]	52.42
Ours	6.76

4.3 PSNR

The PSNR (Peak signal-to-noise ratio) is an objective standard for image evaluation. When PSNR is greater than 38 dB, the image visual quality requirements are met.

$$MSE = \frac{1}{M \times N} \sum\nolimits_{i=1}^{m} \sum\nolimits_{j=1}^{n} [S(i,j) - C(i,j)]^2 \tag{5}$$

$$PSNR = 10 \times \lg \left[\frac{(2^r - 1)^2}{MSE} \right] \tag{6}$$

where S is the densified image, C is the original image, m and n are the height and width of the image, r is the number of sampling bits of each pixel, and MSE is the mean square deviation of the image. The larger the PSNR value of the image, the less distortion of the image. Table 2 shows that the PSNR of our algorithm is higher than that of reference [19–22], and slightly lower than that of reference [23]. This indicates that steganographic image distortion using our algorithm is less.

Table 2. Analysis of different image steganographic algorithms based on PSNR.

No	Existing methods	PSNR values (dB)
1	Classic LSB [19]	52.2416
2	Karim [20]	52.2172
3	Channali et al. [21]	51.9764
4	SCC [22]	52.2023
5	Ours	59.5053
6	Khan et al. [23]	63.0034

4.4 Histogram Analysis

Histogram describes the changes in the image and is a common method to evaluate the image processing. Figure 3 depicts the change of histogram before and after steganography. It is difficult to detect histogram changes in images after steganography with our algorithm, so there is little change in images after steganography, indicating that the algorithm has a good effect after steganography and can effectively resist attacks from statistical methods.

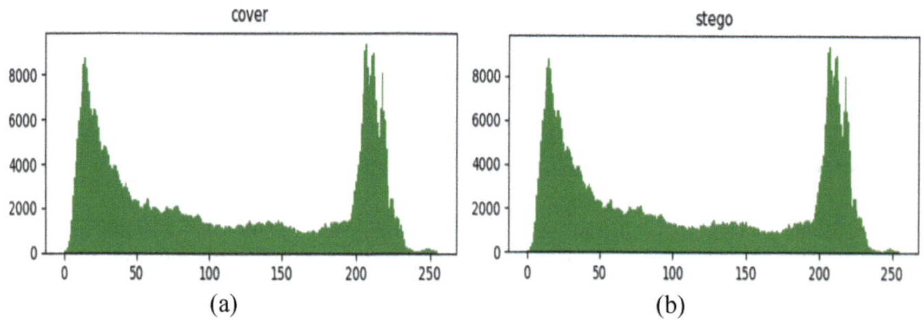

Fig. 3. (a) Histogram of the cover image and (b) histogram of the stego image.

4.5 Embedded Location Analysis of Secret Information

Different images have different textures, and embedding secret information in places with complex textures makes it harder to detect, in order to better deal with non-statistical attacks. The image steganographic analysis method based on convolutional neural network can extract the image features to analyze whether the image contains secret information. By comparing the embedding position of the original image and the secret information in the steganographic image, the ability of the image to resist the feature attack can be determined. Figure 4 (a) the texture of the mountains is more complex than that of the sky. The pixel value of the sky is single, while the pixel value of the mountains is rich with more changes. Our algorithm is used to embed the secret information in the mountains (see Fig. 4 (b)), which is better than the traditional LSB steganographic algorithm (see Fig. 4 (c)), so our algorithm can better resist the feature attack.

(a) (b) (c)

Fig. 4. (a) Cover image, (b) embedded pixel location based on HILL, (c) embedded pixel location based on LSB.

5 Conclusion

Combining chaos theory and HILL distortion function to design steganographic algorithm has the advantages of fast processing speed and high security. The algorithm uses randomly generated keys to encrypt secret information, and even the user does not know the encrypted keys, which increases the security of the algorithm. The encryption key is embedded into the densive-carrying image, and the densive-carrying image can be transmitted through the public channel due to its concealment. Even if the attacker steals the embedded key, the secret information cannot be obtained.

Acknowledgement. This work is supported by the Scientific Research Fund of Hunan Provincial Education Department (19B082), the Science and Technology Development Center of the Ministry of Education-New Generation Information Technology Innovation Project (2018A02020), the research supported by Science Foundation of Hengyang Normal University (19QD12), the Science and Technology Plan Project of Hunan Province (2016TP1020), the Application-oriented Special Disciplines, Double First-Class University Project of Hunan Province (Xiangjiaotong [2018] 469), the Hunan Province Special Funds of Central Government for Guiding Local Science and Technology Development (2018CT5001), the Subject Group Construction Project of Hengyang Normal University (18XKQ02).

References

1. Petitcolas, F.A., Anderson, R.J., Kuhn, M.G.: Information hiding—a survey. Proc. IEEE **87** (7), 1062–1078 (1999)
2. Westfeld, A.: F5-a steganographic algorithm. In: International Workshop on Information Hiding, pp. 289–302. Springer, Berlin, Heidelberg (2001)
3. Provos, N.: Defending against statistical steganalysis. In: Usenix Security Symposium, vol. 10, pp. 323–336 (2001)
4. Sallee, P.: Model-based steganography. In: International Workshop on Digital Watermarking, pp. 154–167. Springer, Berlin, Heidelberg (2003)
5. Fridrich, J., Pevný, T., Kodovský, J.: Statistically undetectable jpeg steganography: dead ends challenges, and opportunities. In: Proceedings of the 9th Workshop on Multimedia & Security, pp. 3–14 (2001)

6. Pevný, T., Filler, T., Bas, P.: Using high-dimensional image models to perform highly undetectable steganography. In: International Workshop on Information Hiding, pp. 161–177. Springer, Berlin, Heidelberg (2010)
7. Holub, V., Fridrich, J.: Designing steganographic distortion using directional filters. In: 2012 IEEE International Workshop on Information Forensics and Security (WIFS), pp. 234–239. IEEE (2012)
8. Holub, V., Fridrich, J., Denemark, T.: Universal distortion function for steganography in an arbitrary domain. EURASIP J. Inf. Secur. **2014**(1), 1–13 (2014). https://doi.org/10.1186/1687-417X-2014-1
9. Li, B., Wang, M., Huang, J., Li, X.: A new cost function for spatial image steganography. In: 2014 IEEE International Conference on Image Processing (ICIP), pp. 4206–4210. IEEE (2014)
10. Sedighi, V., Cogranne, R., Fridrich, J.: Content-adaptive steganography by minimizing statistical detectability. IEEE Trans. Inf. Forensics Secur. **11**(2), 221–234 (2015)
11. Filler, T., Judas, J., Fridrich, J.: Minimizing additive distortion in steganography using syndrome-trellis codes. IEEE Trans. Inf. Forensics Secur. **6**(3), 920–935 (2011)
12. Yu, S.S., Zhou, N.R., Gong, L.H., Nie, Z.: Optical image encryption algorithm based on phase-truncated short-time fractional Fourier transform and hyper-chaotic system. Opt. Lasers Eng. **124**, 105816 (2020)
13. Zhou, N., Jiang, H., Gong, L., Xie, X.: Double-image compression and encryption algorithm based on co-sparse representation and random pixel exchanging. Opt. Lasers Eng. **110**, 72–79 (2018)
14. Zhou, N., Yan, X., Liang, H., Tao, X., Li, G.: Multi-image encryption scheme based on quantum 3D Arnold transform and scaled Zhongtang chaotic system. Quantum Inf. Process. **17**(12), 1–36 (2018). https://doi.org/10.1007/s11128-018-2104-6
15. Wen, W., Wei, K., Zhang, Y., Fang, Y., Li, M.: Colour light field image encryption based on DNA sequences and chaotic systems. Nonlinear Dyn. **99**(2), 1587–1600 (2019). https://doi.org/10.1007/s11071-019-05378-8
16. Jiao, G., Peng, X., Duan, K.: Image encryption with the cross diffusion of two chaotic maps. TIIS **13**(2), 1064–1079 (2019)
17. Jiao, G., Zhou, S., Li, L., Zou, Y.: Hybrid chaotic encryption algorithm for securing dicom systems. Int. J. Perform. Eng. **15**(5), 1436–1444 (2019)
18. Jiao, G., Li, L., Zou, Y.: Improved security for android system based on multi-chaotic maps using a novel image encryption algorithm. Int. J. Perform. Eng. **15**(6), 1692–1701 (2019)
19. Sharp, A., Qi, Q., Yang, Y., Peng, D., Sharif, H.: A video steganography attack using multi-dimensional discrete spring transform. In: 2013 IEEE International Conference on Signal and Image Processing Applications, pp. 182–186. IEEE (2013)
20. Yang, H., Sun, X., Sun, G.: A high-capacity image data hiding scheme using adaptive LSB substitution. Radio Eng. **18**(4), 509–516 (2009)
21. Channalli, S., Jadhav, A.: Steganography an art of hiding data, vol. 1, (3), pp. 137–141 (2009). arXiv preprint arXiv:0912.2319
22. Joo, J.C., Lee, H.Y., Lee, H.K.: Improved steganographic method preserving pixel-value differencing histogram with modulus function. EURASIP J. Adv. Sig. Process. **1**, 249826 (2010)
23. Muhammad, K., Sajjad, M., Mehmood, I., Rho, S., Baik, S.W.: A novel magic LSB substitution method (M-LSB-SM) using multi-level encryption and achromatic component of an image. Multimedia Tools Appl. **75**(22), 14867–14893 (2015). https://doi.org/10.1007/s11042-015-2671-9

Research on Quantitative Evaluation Method of Network Security in Substation Power Monitoring System

Liqiang Yang[1], Huixun Li[2(✉)], Yingfu Wangyang[1], Ye Liang[2], Wei Xu[1], Lisong Shao[2], and Hongbin Qian[1]

[1] State Grid Huzhou Power Supply Company, Huzhou, China
[2] NARI Group Corporation, Beijing, China
lihuixun123@163.com

Abstract. In this paper, the assets, threats, and vulnerabilities of substation power monitoring systems are identified and assigned according to the relevant requirements of national information security risk assessment, and the assignment is optimized in combination with threat and vulnerability. Finally, a calculation method of asset loss is proposed.

Keywords: Substation · Threat identification · Vulnerability identification · Risk loss calculation · Quantitative assessment

1 Introduction

As an essential part of the smart grid dispatching control system, substation security protection is very important to the whole smart grid dispatching control system cooperative security protection. Based on the network security management platform, according to the relevant requirements of national information security risk assessment, through the identification of assets, threats, and vulnerabilities of substation power monitoring system, substation risk analysis, A quantitative evaluation method of safety protection which accords with the characteristics of substation power monitoring system is proposed to improve the level of substation safety protection.

2 Identification of Assets, Threats and Vulnerabilities of Substation Power Monitoring System

Based on GB/T20984-2007 *Information Security Technology-information Security Risk Assessment Specification*, the identification of assets, threats and vulnerabilities of substation power monitoring system is mainly carried out around the basic elements of assets, risks, vulnerabilities and so on. Each element is defined according to the asset attributes of substation power monitoring system. The asset value of substation power monitoring system is determined by the degree of achievement of the three security attributes, namely, confidentiality, integrity and availability, or the degree of influence

Q. Liu et al. (Eds.): CENet 2020, AISC 1274, pp. 741–748, 2021.
https://doi.org/10.1007/978-981-15-8462-6_85

caused by the failure of its security attributes. The attribute of the threat is the threat subject, object, frequency and motive of substation power monitoring system.

2.1 Identification of Assets

According to the business function of substation power monitoring system, asset identification and importance assignment are carried out to determine its importance in the process of power production. Asset identification is the process of classifying and marking the system components such as equipment, data and personnel of substation power monitoring system. After determining the scope of the evaluation, the value of the assets in it is analyzed. Asset identification is to clarify the purpose, mission and role of assets, and then determine the value of assets preparation.

According to the substation power monitoring system assets form and content are different, generally can be classified into data, software, hardware, personnel four categories. Data refers to all kinds of data materials stored in the information media, including source code, database data, system documents, operation management procedures, plans, reports, user manuals, all kinds of paper documents, etc. Software refers to system software, application software, source program, etc. Hardware refers to network equipment, computer equipment, storage equipment, transmission lines, safeguard equipment, security equipment, etc. Personnel refers to people who have important information and core business.

According to the asset assignment method in the information security risk assessment specification, the importance of substation power monitoring system is classified and defined.

Level 4: very important, the safety in the production control area basically belongs to level 4, has the control function and its safety attribute damage may cause the very serious loss to the organization.

Level 3: important, the safety level in the production control area belongs to level 3, which does not have the control function and may cause serious losses to the organization after its safety attribute is destroyed.

Level 2: more important, the safety in the production control area basically belongs to the level 2, its safety attribute damage may cause the moderate degree loss to the organization.

Level 1: not important, the safety level in the production control area is lower than the second level, and the damage of its safety attribute may cause lower loss to the organization.

Level 0: unimportance, it will not cause damage to power system after its safety property is destroyed.

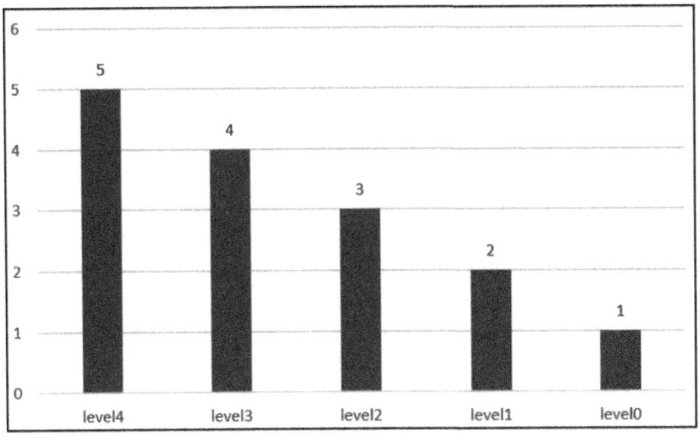

Fig. 1. Classification of asset security attributes.

Asset assignment according to the security level and importance of substation power monitoring system assets, as shown in Table 1.

Table 1. Asset assignment of substation power monitoring system

Asset	Asset assignment	Valuation of assets
Substation Monitoring System	3	200
Five Defense Systems	3	10
Wide area phasor measuring device	2	12
relay protection	4	5
Automatic safety control	4	10
Fire alarm	3	10
Electrical Energy Acquisition Device	2	5
Fault recording	2	10
primary equipment online monitoring	3	5
Auxiliary equipment monitoring	2	5
PM	2	2

2.2 Threat Identification Analysis

Threat identification analysis refers to the determination of the threat faced by substation power monitoring system through technical measures, statistical information and empirical judgment, including the security infringement from internal and external to substation power monitoring system. Threat identification mainly includes two aspects, one is to classify the threats faced by the substation power monitoring system, and to determine the source of the threat, and to complete the identification of the threat; The other is to analyze the possibility of the threat and assign the value of the threat through statistics of the frequency of the threat.

The main threats to substation power monitoring system protection include software and hardware failures, natural disasters and other non-human threats, as well as hackers, viruses, network attacks and other human threats.

The threat assignment of substation power monitoring system is the possibility of occurring in the environment of substation power monitoring system through threat, and the frequency of statistical threat occurring in the past years. Combined with the probability of threat occurrence and the frequency of occurrence, the higher the value, the greater the probability of threat occurrence.

Figure 2 lists the common threat manifestations faced by substation power monitoring systems.

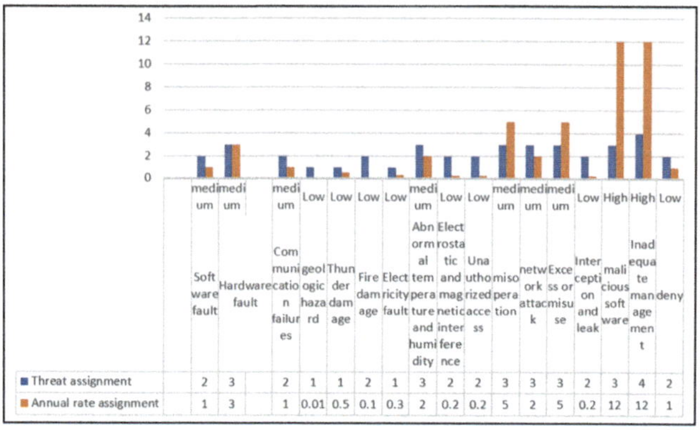

Fig. 2. Common threat manifestations and assignments.

2.3 Vulnerability Identification Analysis

Vulnerability identification is to find the power monitoring system assets and protection measures in substation network security deficiencies. vulnerability may be threatened to be exploited and damage substation power monitoring system assets.

Through two steps of vulnerability identification and assignment, the vulnerability identification of substation power monitoring system is found and analyzed.

The vulnerability identification of substation power monitoring system is aimed at substation power monitoring system assets, first identify the loopholes of the assets themselves, then analyze and find the defects in management, and finally comprehensively evaluate the vulnerability of the assets. The vulnerability identification of substation power monitoring system is mainly analyzed by technical means and management means, which mainly includes network security audit system or audit tools, and management means mainly include interviews and questionnaires.

Vulnerability assignment is based on the analysis of the defects after vulnerability identification, according to the vulnerability is threatened to use, resulting in the size of the impact and ease of assignment.

3 Calculation Method of Risk Quantification in Substation Power Monitoring System

3.1 Vulnerability Assignment Optimization

vulnerability assignment is assigned according to the impact of vulnerability and the degree of difficulty after analyzing the vulnerability of the system, which is subjective. in order to quantify more accurately, it is necessary to optimize the vulnerability assignment. It is necessary to combine the relationship between assets, vulnerability, threat, and the protective effect of existing security measures on vulnerability in the system, carry out vulnerability assignment analysis, select appropriate vulnerability assignment, and ensure the accuracy of vulnerability assignment.

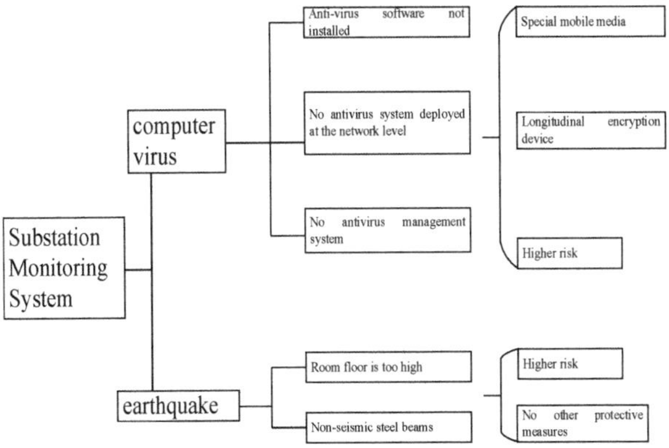

Fig. 3. Examples of vulnerability assignment optimization.

For example, the substation monitoring optimizes the vulnerability assignment, so the vulnerability assignment is 0.3 and 0.1.

3.2 Asset Impact Assignment

The impact degree of assets refers to the impact of the threat on the assets after the risk action against the detected vulnerability. The impact degree of assets includes the scope and degree of impact of assets, and the assignment interval is 0-1.

Calculation formula: asset impact degree = asset impact range assignment × loss degree assignment

Table 2. Asset impact statement.

Threat	Vulnerability	Impact	Loss	Impact degree
Software fault	1) Improper control of resources leads to decreased availability 2) No regular backup of data and backup recovery 3) No restrictions on data format and file type	10–100%	10%	0.01-0.1
Hardware fault	Single point failure of the system	10–100%	5–10%	0.005–0.01
Communication failures	No alternate communication links	100%	10–20%	0.1–0.2
Geologic hazard	No remote disaster response	100%	100%	1.0
Thunder damage	No lightning protection	20%	50%	0.1
Fire damage	No Fire Warning System (Fire Sense) Non-automatic gas extinguishing devices No fire proof material in machine room Fire-proof access to engine room is not clean Other unrelated items in the machine room	100%	20%	0.2
Electricity fault	Dual Power Supply No UPS No backup power supply system	100%	20%	0.2
Abnormal temperature and humidity	No precision air conditioning Temperature and humidity exceeding standard in machine room without alarm device	5–10%	50%	0.025–0.05
Electrostatic and magnetic interference	No protection against static electricity No electromagnetic shielding cabinet	5%	50%	0.025
Unauthorized access Mis operation	Illegal Intranet Access-free system No video surveillance system	10% 10–20%	10% 20%	0.01 0.02–0.04
Network attack Unauthorized access Mis operation	No topology or actual inconsistency No terminal login timeout set Equipment not identified	10% 20% 5%	20% 20% 20%	0.02 0.04 0.01
Network attack	Password strength and complexity do not meet the requirements Inadequate security policy Improper access control policy Minimum service principle not applied High-risk system patches No border protection equipment deployed	5–20% 10–50% 10–100%	10% 5% 10%	0.005–0.02 0.005–0.02 0.01–0.1
Excess or misuse	Undeployed dual-factor authentication Permissions not separated No sensitive mark No restrictions on management addresses No restrictions on terminal access mode or network address range Unopened audit Illegal outreach	5–10% 5–50%	10% 10%	0.005–0.01 0.005–0.05
Interception and leak	Key data not encrypted No residual information protection measures	20% 10%	10% 10%	0.02 0.01
Malicious software	No malicious code protection Malicious code feature library was not updated in time Virus not regularly detected	5–100%	10%	0.005–0.1
Inadequate management deny	Management security issues No security audit opened No protection of audit processes or records Inadequate audit strategy	10–50% 5–50%	10% 5%	0.01–0.25 0.0025–0.025

3.3 Risk Loss Calculation

After the assets identification, threat identification and vulnerability identification of substation power monitoring system are completed, the value of assets, the vulnerability caused by threat are assigned and optimized, the influence degree of assets is calculated, and then the expected loss of assets is analyzed and calculated, so as to achieve the purpose of quantifying the network security risk of substation power monitoring system.

The expected loss formula of substation power monitoring system is as follows:

$$\mathrm{L_{bdz}} = \sum_{i=1}^{i_{\max}} \left(P_i \cdot \sum_{k=1}^{k_{\max}} \left(Y_{i.k} \cdot T_k \cdot V_k \right) \right)$$

The L_{bdz} represents the expected loss of the asset; the P_i represents the i asset value; the T_k represents the annual incidence of the threat corresponding to the k vulnerability; the V_k represents the vulnerability assignment after the optimization of the first vulnerability; and the $Y_{i.k}$ represents the degree of impact on the asset P_i of the security event that the second threat and vulnerability may cause.

After the above identification and optimization, the risk loss of substation power monitoring system can be calculated by applying the risk loss formula, so that the network security risk of substation power monitoring system can be quantified. The following table illustrates the risk loss analysis results of a substation power monitoring system.

The asset value of a substation monitoring system is 2 million yuan according to the asset assignment table of substation power monitoring system. According to the common threat manifestation in Fig. 3 and the threat involved in the assignment, the virus risk loss of substation monitoring system can be calculated as 1800 yuan according to the product of vulnerability assignment and impact assignment in Fig. 2 and Table 2. Finally, the risk loss calculated by all its vulnerabilities is summed up as the risk loss faced by its system. Table 3 below is a comparative table of risk loss calculation for substation monitoring system.

Table 3. Comparison of risk loss calculation for substation monitoring system.

Assets	Value of assets	Threat sources	Threat probability	Vulnerability assignment	Impact assignment	Risk loss
Substation monitoring system	200	Computer virus	3	0.3	0.001	0.18
		Earthquake	0.1	0.01	1	2
		Fire damage	0.1	0.01	0.5	1
		Access control	3	0.3	0.25	45

4 Conclusion

Based on the substation power monitoring system, according to the information security assessment specification, combined with the power system information security inspection specification, through the process of identification, analysis and calculation, the assignment is optimized by the method of asset, threat and vulnerability association analysis. Finally, the asset loss formula is used to calculate the asset loss, forming a quantitative method of information security risk assessment for substation power monitoring system.

References

1. Xi, R.R.: An improved network security situation quantitative assessment method. J. Comput. Sci. **38**(04), 749–758 (2015)
2. Tian, W.J.: Quantitative assessment method of multi-node network security situation based on threat propagation. Comput. Res. Dev. **54**(04), 731–741 (2017)
3. Zhang, S.W.: Research on quantitative assessment method of network security risk based on random model. PLA Information Engineering University (2014)
4. Liu, F.: Research on information system security assessment theory and its key technologies. University of Defense Science and Technology (2005)
5. Zhang X.Y.: Research on network security situation quantitative assessment and prediction method. Chongqing University of Posts and Telecommunications (2016)
6. Lu, P.: Research and application of network security situation quantitative assessment method. University of Electronic Science and Technology of China (2019)
7. Zhang, J.Q.: A study on quantification model of power information security based on GHPN. North China Electric Power University (2016)

Research on FTP Vulnerability Mining Based on Fuzzing Technology

Zhiqiang Wang[1,2(✉)], Haoran Zhang[2], Wenqi Fan[2], Yajie Zhou[2], Caiming Tang[2], and Duanyun Zhang[2]

[1] State Information Center, Beijing 100045, China
wangzq@besti.edu.cn
[2] Beijing Institute of Electronic Science and Technology, Beijing 100070, China

Abstract. With the development of technology, FTP protocol has been widely used but also brought many threats and hidden dangers, including remote attacks, denial of service and so on. Aiming at the above problems, this study develops a vulnerability mining and evaluation technique for FTP protocol based on Fuzzing. This project use Python programming to implement Fuzzing. Using freefloatftp software to build a simple FTP server in Windows XP SP3 operating system. Ftpfuzzer.py is used to attack the server. Besides, Immunity Debugger is used to preliminarily analyze the vulnerabilities mined, so as to provide direction for resisting attacks.

Keywords: Fuzzing test · FTP · Windows XP SP3 · Vulnerability mining · Python

1 Foreword

Now, fuzzing is very popular in the field of vulnerability mining, and FTP protocol is the main protocol of file transmission. The application level of fuzzing technology is very broad. The test of FTP protocol is only a small part of its use. The ability of mining various loopholes by fuzzing technology is difficult to be realized by the previous technology. We can use fuzzing technology to test loopholes and improve various protocols and software to improve the user experience [1].

This paper will study how to use Python programming to implement fuzzing test to mine the vulnerability of FTP server, and make a preliminary analysis of the vulnerability.

2 Background Introduction

The background of this project includes two parts: FTP protocol and Fuzzing.

2.1 FTP Protocol

FTP is the protocol that two computers transmit files on TCP/IP network. FTP is one of the earliest protocols used on TCP/IP network and Internet [2]. It belongs to the

Q. Liu et al. (Eds.): CENet 2020, AISC 1274, pp. 749–754, 2021.
https://doi.org/10.1007/978-981-15-8462-6_86

application layer of network protocol group. The FTP client can send commands to the server to download files, upload files, and create or change directories on the server [3]. FTP uses client server mode. A FTP server process can serve multiple client processes at the same time. The server process of FTP consists of two parts: a main process, which is responsible for receiving new requests; and several subordinate processes, which are responsible for processing a single request.

2.2 Fuzzing

Fuzzing testing is a kind of software vulnerability detection technology, which is an effective test method to find security vulnerabilities [4]. It detects software vulnerabilities by providing unexpected, random or wrong data as input to observe whether the program can tolerate random input. Monitoring the abnormal results to find out the potential vulnerabilities when the software crashes [5]. Fuzzing testing is illogical and just produces a messy data attack program. Using fuzz testing to attack applications can find other security vulnerabilities that are difficult to find by using logical thinking to test [6].

The choice of fuzzing testing method depends on different factors and may vary greatly. There is no absolutely correct fuzzing testing method [7]. The choice of fuzzing testing method depends entirely on the target application, the skills of researchers, and the format of the data to be tested. However, there are several basic stages of fuzzing testing (see Fig. 1):

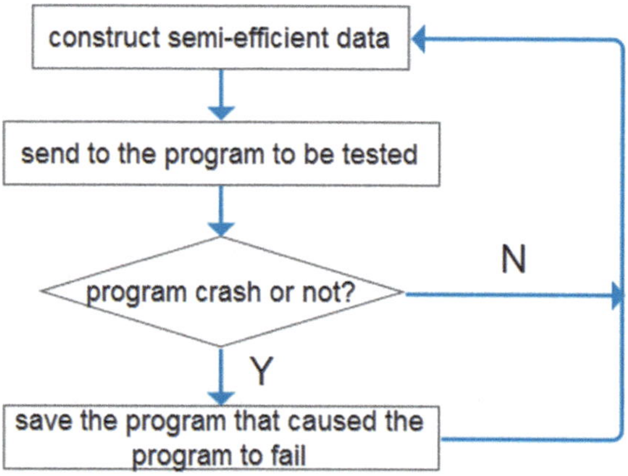

Fig. 1. Flow chart of basic stage of fuzzing test.

3 Architecture

The architecture of this project includes two modules: preliminary vulnerability detection module and memory error detection module (see Fig. 2).

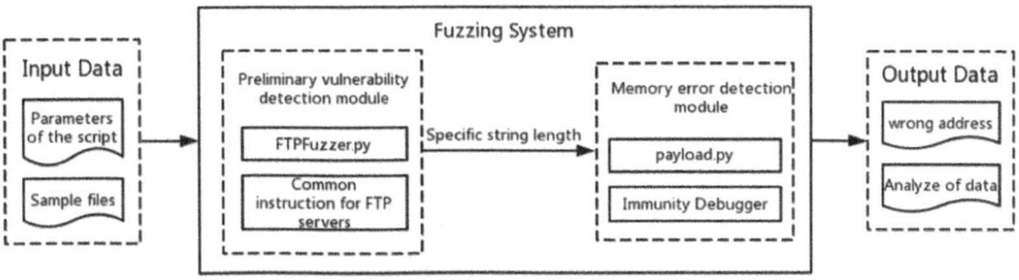

Fig. 2. Architecture diagram.

The preliminary vulnerability detection module is composed of ftpfuzzier.py file, which is mainly used for the fuzzing testing of FTP server [8]. It can be found that which FTP protocol is used by the instructions of the long characters caused the FTP server to crash. The string length of specific instructions can be viewed after the FTP server crashes. This module should contain the common instructions of FTP server. The minimum length of string sent to the server is 20. The incremental size is 100, and the maximum is 2920.

The memory error detection module is composed of payload.py file and Immunity Debugger tool. After the last module gets the length of the concrete string, it can input the corresponding length random characters in the payload.py file and send it to the server. The immunity debugger tool displays the memory address of the specific error when the server crashes [9].

4 Experiment and Evaluation

4.1 Experiment

Implement the Main Code. During the implementation, common instructions from the FTP server are inserted when ftpfuzzer.py is written, and a relatively appropriate threshold is set to increase the length from small. Payload.py imports the 'socket' packet, sets the server's IP address and port to send strings to the correct FTP server for experimentation, and limits the length of characters the server returns to ensure they work properly.

Get Crash String Length. Running FTPFuzzer.py, it can be found that when running to the 'user' instruction and the character length is 420. The freefloatftpserver server running in the Immunology Debugger crashes, and the string length that causes it to crash is 420 (see Fig. 3).

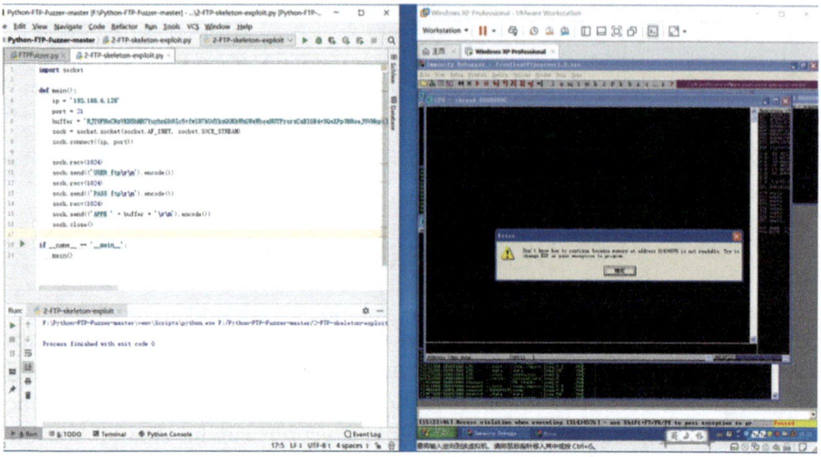

Fig. 3. Server crash and script diagram.

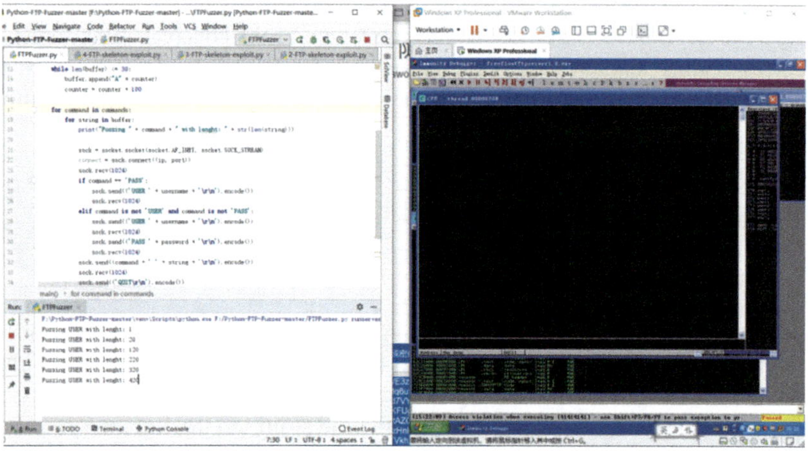

Fig. 4. Error memory address map.

Get a Random String with a Length of 420. Find a string generator in the Internet and produce a random string with a length of 420:

’RJT0FHoCRzVKN5hMH7YuyhnGh9Zc5vfwlN7kUdYkzQUKbWhGWeWboaNUT
FrorxCaBIGB4v5QeXPp3BRoeJ5VNkpxJj4xQRuFVMzqck9MMjhM6phMOrkfm4
Op0Ynhkan68NFLPUxczb4OKkx2yCiKYpFdHXleurgHAHM2wbazclBlnAq6amJG
hEvFhkOI0iT50U9faibZUugOabXy5N3s4OUjWj0U6XWTtmOMGD6I73rlAWgRqa
kK1OvEC14Cebid229wSZe3D98PncxJZlcIBx3c3lbUiSZJ8rxMgj3yCNEyr6VpieUf
GBTwueCJs6osjCjAxx1r0QOM3zVFTDIYBqcmRTptFD7UUnc3f5n00fJ78laWpLO
g0yValUopUMs228uVyTpKVLwrFBrAFXxE0trKsLLofsITwqtZ’

Run Script to Get Memory Error Address. Reopen the freeloadftp server in the Immunology Debugger and run it. Put the 420 length random string into the script payload.py and run it to get the memory error address of 31434576 (see Fig. 4).

4.2 Evaluation

The following Table 1 gives a summary of the project evaluation.

Table 1. Evaluation of experimental results.

Data	Evaluation
Crash string evaluation	In this study, the string causing the FTP server to crash was a 420-length 'USER' instruction. However, for the FTP server, this crash is not unique. Any possible length in the subsequent instructions may cause the server to crash. This study was not in-depth studied for time reasons
Random string evaluation	The random string used in this study is generated by a random string generator on the Internet, and it is not unique. Different strings cause different memory error addresses for server crashes
Memory error address evaluation	The 420-length random string used in this study caused the FTP server to crash at a memory error address of 31434576. But as mentioned in the previous section, if you use a different random string, the memory error address that ultimately causes the server to crash is also different. Any memory address that causes the server to crash can be exploited and eventually attacked

5 Conclusion and Further Research Direction

In this study, two functional modules are implemented to find out the vulnerability of FTP protocol, and find out the wrong memory address, which provides the direction for vulnerability prevention, and also completes the exploration of fuzzing technology. FTP protocol as a very common file transfer protocol, we need to pay more attention to fix the vulnerability to protect file transfer and user security.

The next research can find the specific error byte according to the memory address, and then use the knowledge of reverse engineering to attack the FTP server, and finally realize the invasion of the FTP server.

6 Special Thanks and Author Contributions

6.1 Special Thanks

This research was financially supported by the Key Research and Development Plan (2018YFB1004101), Key Lab of Information Network Security, Ministry of Public

Security (C19614), Special fund on education and teaching reform of Besti (jy201805), the Fundamental Research Funds for the Central Universities (328201910), China Postdoctoral Science Foundation (2019M650606), 2019 Beijing Common Construction Project-Teaching Reform and Innovation Project for Universities in Beijing, key laboratory of network assessment technology of Institute of Information Engineering, Chinese Academy of Sciences. The authors gratefully acknowledge the advisor for his guidance and support in the research process.

Author Contributions. Wang Zhiqiang and Zhang Haoran conceived and designed the framework and the algorithm; Fan Wenqi and Zhang Haoran performed the experiments; Zhou Yajie, Tang Caiming and Zhang Duanyun analyzed the data; Fan Wenqi and Zhang Haoran wrote the paper. Conflicts of Interest: The authors declare no conflict of interest.

References

1. Klees, G., Ruef, A., Cooper, B., Wei, S., Hicks, M.: Evaluating fuzz testing. In: Proceedings of the 2018 ACM SIGSAC Conference on Computer and Communications, NY, USA, pp. 2123–2138. ACM (2018)
2. Cai, Y.: The FTP server technique study and realization. University of Electronic Science and Technology of China (2005)
3. Hao, H.: Principle analysis of FTP. Comput. Netw. **42**(14), 40–41 (2016)
4. Qian, L., Yang, M., Wang, L.: Research and design of software vulnerability mining platform based on directional fuzzy testing. Netw. Secur. Technol. Appl. **01**, 33–34 (2019)
5. Duncan, S.: Fuzzing for software security testing and quality assurance. Softw. Qual. Profess. **11**(3), 47–48 (2009)
6. Su, P., Huang, H., Yu, Y., Zhang, T.: Research review on automatic exploitation of software vulnerabilities. J. Guangzhou Univ. (Nat. Sci. Edn.) **18**(03), 52–58 (2019)
7. Li, H., Qi, J., Liu, F., Yang, F.N.: The research progress of fuzz testing technology. Sci. China Inf. Sci. **44**(10), 1305–1322 (2014)
8. Zhang, X., Li, Z.: Survey of fuzz testing technology. Comput. Sci. **43**(5), 1–8 (2016)
9. Zheng, L., Xie, X.: Analysis of fuzzing vulnerability mining technology. In: Proceedings of the 24th National Symposium on Computer Security, pp. 391–395. Netinfo Security, China (2009)

Research on Integrated Detection of SQL Injection Behavior Based on Text Features and Traffic Features

Ming Li[1(✉)], Bo Liu[2], Guangsheng Xing[2], Xiaodong Wang[2], and Zhihui Wang[2]

[1] College of Intelligence Science and Technology, National University of Defence Technology, Changsha 410073, China
1024587662@qq.com

[2] National Key Laboratory of Parallel and Distributed Processing, College of Computer Science and Technology, National University of Defence Technology, Changsha 410073, China

Abstract. With the rapid development of Internet technology, various network attack methods come out one after the other. SQL injection has become one of the most severe threats to Web applications and seriously threatens various Web application services and users' data security. There are both traditional detection methods and emerging methods based on deep learning technology with higher detection accuracy for the detection of SQL injection. However, they are all for detecting a single statement and cannot determine the stage of the attack. To further improve the effect of SQL injection detection, this paper proposes an integrated detection framework for SQL injection behavior based on both text features and traffic features. We propose a SQL-LSTM model based on deep learning technology as the detection model at the text features level. Meanwhile, the features of the data traffic are merged. By this integrated method, the detection effect of SQL injection is further improved.

Keywords: SQL injection · Integrated method · SQL-LSTM · Traffic features

1 Introduction

With the rapid development of Internet technology, various network attack methods come out one after the other. As the most common attack method in network security, SQL injection has a history of more than ten years. According to the report provided by OWASP, Open Web Application Security Project, SQL injection has become one of the most severe attacks on Web business systems [1]. Through carefully crafted user input, attackers can use Web application vulnerabilities to gain control of the back-end database server and then perform malicious operations, such as tampering with web-page information and stealing data. SQL injection's primary way is to construct a malicious SQL syntax combination as input parameters of the Web application. When the application executes the SQL statement, it will execute the malicious operation synchronously to implement the SQL injection attack. SQL injection can also occur if an application stores or passes a malicious string containing an attacker's input during

Q. Liu et al. (Eds.): CENet 2020, AISC 1274, pp. 755–771, 2021.
https://doi.org/10.1007/978-981-15-8462-6_87

code execution. According to the types of construction parameters, SQL injection is mainly divided into getting an injection, POST injection, Cookie injection, wide byte injection, time-based injection, Boolean injection, error injection, joint query injection, multi-state query injection [2]. Although there are many types of SQL injection, all SQL injection statements must follow specific rules. Therefore, the scripts have certain features in terms of text. Meanwhile, as an attack behavior, SQL injection also has corresponding features in terms of traffics. The existence of these two kinds of features provides the basis for SQL injection detection.

Using traffic features to detect malicious behavior is a fundamental idea commonly used in cyber security. When an attack behavior occurs, data traffic will produce apparent abnormal behavior. SQL injection attacks are no exception, such as generating many requests from the same source IP address from in a short time for the destination IP address. However, many traffic features are not unique to SQL injection attacks. Therefore, using traffic features to detect SQL injection behavior needs to be matched with another detection method.

For the detection method using text features, the traditional methods mainly use such as keyword filtering, black and white list mechanism, pattern matching, and machine learning methods based on Bayesian algorithm to detect SQL injection behavior. For example, the current real-time online detection devices are based on basic string filtering and regular expression. These models are usually deployed on a hardware platform based on Deep Packet Inspection (DPI) technology. They can perform online detection of real-time large-flow data by batching loading rules, but the false detection rate is high.

In recent years, artificial intelligence technology has developed rapidly. In image and text processing, related technologies based on deep learning have been applied effectively. The SQL statement in the real-time network traffic has the dual characteristics of network data and scripting language. Based on this feature, it can learn from the deep learning technology processing ideas in image and text processing.

In order to improve the effect of SQL injection detection, this paper proposes an integrated detection framework for SQL injection behavior based on both text features and traffic features. At the text features level, we propose a SQL-LSTM model based on Long Short-Term Network (LSTM) as the detection model. Meanwhile, the features of the data traffic are merged. By this integrated method, the detection effect of SQL injection is further improved.

The main contributions of our work can be summarized as follows:

- We propose an integrated detection framework for SQL injection behavior and improve the detection effect of SQL injection.
- We use deep learning technology to detect SQL injection at the text features level and build an LSTM-based detection model - SQL-LSTM.
- At the traffic features level, we combine the traffic features with the detection results of the SQL-LSTM model to detect the attack stage of SQL injection.
- We integrate existing data to construct a data set for SQL injection detection and design a data preprocessing method for the experimental data.

In the remaining parts, Sect. 2 presents related work. Section 3 describes the approach. Section 5 reports experimental results. Section 6 concludes this work.

2 Related Work

For the detection and prevention of SQL injection attacks, a lot of research has been done at home and abroad. Sharma et al. [3] introduces the classification of SQL injection, summarizes descriptions and examples of various types of attacks, and expounds detection and prevention techniques for SQL injection attacks. According to the research direction, the research is mainly about the front-end related applications and the network data including SQL injection statements. The main focus of this paper is on data-level research.

2.1 Traditional Detection Methods

For traditional detection methods, there are mainly two methods for SQL injection detection. One is to build corresponding feature expressions or models based on key strings, and the other is to build models based on SQL semantic grammar rules.

Based on the key string aspect, Halfond et al. [4] uses a static analysis method to build a model of valid SQL statements, and dynamically analyses whether the SQL statement is consistent with the established static model. However, this method is only applicable to related specific applications to establish a blacklist-like mechanism, and detection capability is limited. Wan et al. [5] and Wang et al. [6] validate the data content through regular expressions, and filter sensitive words of SQL injection attacks. This method is relatively easy to implement, but the false detection rate is too high.

Based on rules of SQL semantic grammar, Patel et al. [7] and Prabakar et al. [8] propose a new idea of SQL injection detection based o pattern matching and sequence comparison. This idea uses pattern matching algorithm to detect and prevent SQL injection attack. Pattern matching algorithm achieves the detection purpose by analysing whether the injection command string and normal command string can be completely matched. This method can effectively reduce the complexity of time and space, but it is only suitable for SQL injection detection of specific applications and is not widely representative. Han et al. [9] uses syntax tree features matching method to detect the SQL injection. The user input data is input into the specific statement by the numeric type and character type, respectively. Then, the two generated SQL statements are parsed to generate a syntax tree to judge whether there is an injection behavior. This method has a good detection effect, but it requires researchers to spend a lot of time building feature model, and the model is only for the statements of the specific application type.

2.2 Methods Based on Machine Learning

In the aspect of machine-learning-based SQL injection detection, there are many related researches at home and abroad, and it is necessary to extract the original data packet features or data flow features. Joshi et al. [10] uses Bayesian algorithm to model and identify normal SQL statements and malicious SQL statements. Ladole et al. [11] proposes a system to detect the SQL injection attack by using Support Vector Machine (SVM) classification and Fisher Score. The system can also classify the users into normal users or attackers according to the query by the submitted them. Kim et al. [12]

uses the SVM algorithm to model the feature vectors extracted from the internal query tree in the database log, and detects the SQL injection behavior that has occurred, not for real-time online detection. Rawat et al. [13] also uses SVM for classification and prediction of SQL injection attack. The above methods are based on the original data for feature extraction, and the detection accuracy has been improved compared with the traditional methods.

2.3 Methods Based on Deep Learning

Compared with machine learning, deep learning neural networks are deeper. As the size of training data increases, high-level features can be learned directly from the data, and no manual summary extraction is required. At present, there are many preliminary application in the recognition and processing of network scripts and network data. Although there is a few studies on the detection methods of SQL injection behavior based on deep learning in the open literature, there are already many open source code on Github. In the aspect of network script abnormal behavior detection, Yang et al. [14] proposes a keyword-based convolutional Gated Recurrent Unit (GRU) network for detecting malicious URLs. Fu et al. [15] proposes a model based on multi-class feature extraction and Probabilistic Neural Network (PNN) to detect JavaScript malicious scripts. In terms of network data recognition and processing, deep learning technology has also been widely used. Many domestic and foreign researches [16–20] use the similarity of traffic data and image, taking raw data as input and using the deep neural network to complete the data representation learning to realize the network behavior anomaly detection. The above researches use the deep learning technology to construct the corresponding detection and recognition models, and all of the models can achieve high accuracy. It shows that the deep learning technology can also build efficient detection models for SQL scripts carried in network data. Currently, related applications and codes based on deep learning technology for SQL injection recognition and detection have appeared in the Github. According to our research, the existing open source researches are based on the construction of word vectors to generate training data. These methods require the construction and training of two kinds of models - the word vectors models and the detection models, and the process is rather cumbersome.

3 Methodology

This Section describes our methods. First, in the overview, we introduce integrated detection framework for SQL injection behavior based on both text features and traffic features. Then, we introduce the details of data set. Then, details of the dataset and the various parts of the framework are introduced in the following section.

3.1 Integrated Detection Framework for SQL Injection Behavior Based on Both Text Features and Traffic Features

For network scripts recognition and detection, deep learning technology can achieve high accuracy when detecting the same type of data scripts as the training set.

However, deep learning technology can only detect a single SQL statement and this method can not detect the attack stage. The detection of attack stage must rely on the features of data traffic. Therefore, other auxiliary methods must be used for detection. In order to solve this problem and improve the effect of SQL injection detection, we propose an integrated detection framework, and it utilizes both the textual features and traffic features of network data. The framework mainly includes a data down-flow module, data preprocessing module, deep learning detection module based on text features, traffic information extraction module and comprehensive detection module, as shown in Fig. 1.

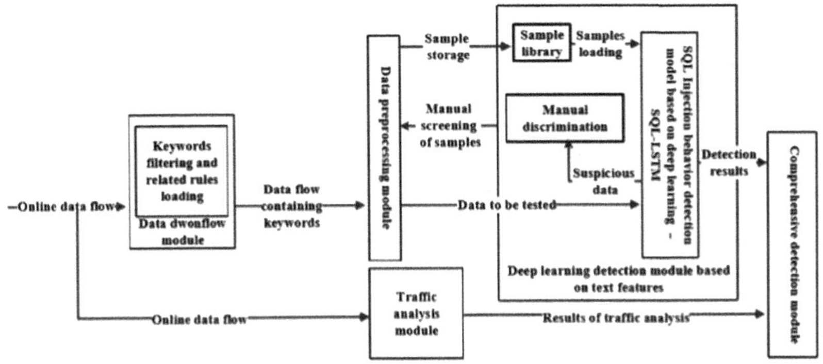

Fig. 1. Integrated detection framework for SQL injection behavior based on both text features and traffic features.

For integrated detection of SQL injection, we design the data detection process. When the online data flow reaches the detection framework, it flows to both the text detection and the traffic detection. For the detection based on text features, in order not to cause data congestion, a data down-flow module needs to be introduced before the SQL statements are detected by using text features. Since the deep learning detection model is more complex and computationally intensive than the traditional detection model, if it is directly loaded onto a high-speed network link for data detection, data congestion will inevitably occur during the detection process. For this module, the current main implementation method is to utilize a high-speed deep packet inspection device. The device can implement data down-flowing through keyword filtering and loading rules, and greatly reduce the amount of data detected by subsequent SQL-LSTM model. After filtering and down-flowing, the data enters the data preprocessing module. The function of the module is to preprocess real-time data containing SQL keywords to better represent the data features. Then, the preprocessed data enters the SQL-LSTM detection model, to detect whether the data belongs to the SQL injection statements. Meanwhile, as various network data scripting protocols are constantly being revised and revised, and the data types and data payloads in different network lines are also different. In order to ensure the detection effect of the model in various scenarios, an adaptive training process is added in this part.

For traffic feature-based detection, the same data flow first passes through the traffic information extraction module. The function of this module is to extract the relevant information of the traffic features. Then, the traffic information and the detection result based on text features are finally merged into the comprehensive detection module to detect the attack stage. The process is shown in Fig. 2.

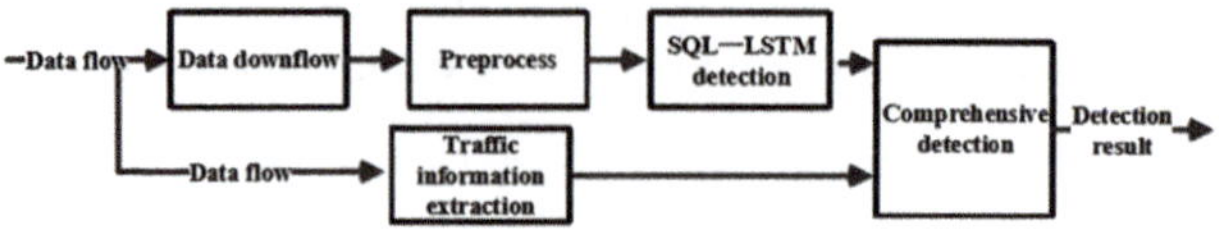

Fig. 2. Data detection process.

3.2 Source and Preprocess of the Data

This Section introduces our data set. First, we introduce the source of the data set. Then, the process of data feature extraction is elaborated.

Source of the Data Set. The original training data source mainly includes three parts. The first part comes from Github. The keyword "SQL injection" is used to search for open source project and obtain data samples. The second part comes from the commonly used injection attack tool - Ming Xiaozi 4.3 and SQLmap 1.3.5 [21]. In the execution of the injection operation, the packets are captured in real time to analyse and extract the data samples. The third part comes from other network data samples of non-SQL injection scripts. It can be seen from the original data samples that the SQL injection samples have feature words of high frequency and corresponding context semantic environments. We calculate the feature words of high frequency in the SQL injection sample. There are 95602 scripts in total and the keywords in the statistics are not case-sensitive. The statistical results are shown in Fig. 3.

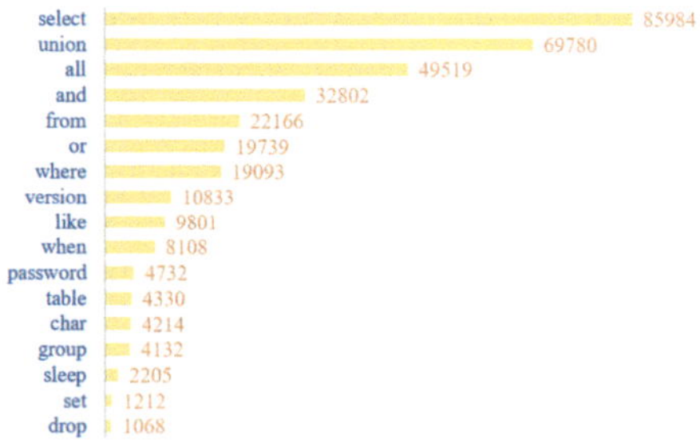

Fig. 3. SQL injection samples keyword statistics.

When the detection model is actually running on the line, in order to effectively reduce the flow rate to lighten the model calculation load, the packet containing the keywords is imported for identification and detection. In order to facilitate the comparison of positive and negative samples in the model training process, we delete the normal scripts without keywords to get the original normal samples, with a total of 57624.

In order to test the generalization detection ability of the model, that is, to test the ability of the model to identify suspicious SQL injection scripts with non-training centralized data features, we collect other types of SQL injection scripts and network scripts with similar structures to SQL keyword syntax to build the generalized test set.

Process of Data Feature Extraction. In this paper, based on the characteristics of the data, we combine the existing network script feature extraction methods to achieve the extraction of data features. Based on the existing research, common network script feature extraction methods are mainly divided into four categories: The first category is based on the characteristics of the data packets and data flow which carry the network script, such as the communication time of the data packet, the number of connections, etc. For the third category, the network script has the feature similar to natural language, so it can be transformed into word vectors. The method mainly includes one-hot encoding [22] and Word2Vec [23], etc. The fourth category is the hybrid use of the first two methods to extract mixed features for processing. On the whole, the first method needs to accumulate network real-time data for feature extraction analysis. Under the end-to-end complex network environment, the sources of data entities are various, the data link is easy to change. Therefore, the requirements for the feature extraction time window are set higher, and the researchers' ability to interpret the original data and related data scripts is tested. The third method is based on the analysis of data features. This method can achieve high accuracy in related research, but the data volume will expand exponentially in the process of segmentation and word vector transformation. In the network with large traffic, the computational ability and real-time detection capability of the model are challenged. Meanwhile, many updates are required when updating the iteration. In this paper, the feature extraction method is mainly to transform the original data into character encoding as the input feature. For the features in the script, the scripts need to perform URL encoding transformation, digital deduplication and replacement, script deduplication, and length uniformization. The difference between our feature extraction method and word vectors feature extraction method can be seen in the model test results.

In this paper, the characteristics of SQL injection samples and normal samples are studied. Each sample is extracted or filled with a length of 400 bytes, and converted into a 20 * 20 gray scale image. Then, the k-Nearest Neighbor algorithm is used to amplify the gray scale image to obtain a series of sample data's gray scale image. The gray scale images of normal samples and SQL injection samples are shown in Fig. 4 and Fig. 5.

From the gray scale image, there are some differences in the general characteristic of the two types of samples. However, since not all scripts in the original samples utilize URL encoding, and there are digital expressions in the scripts. This situation increases the number of useless features in the samples. Therefore, in order to strengthen the features of the original samples, it needs to be further processed for the original samples, mainly including URL encoding transformation, digital deduplication and replacement, script deduplication, and length uniformization. The data processing flow is shown in Fig. 6.

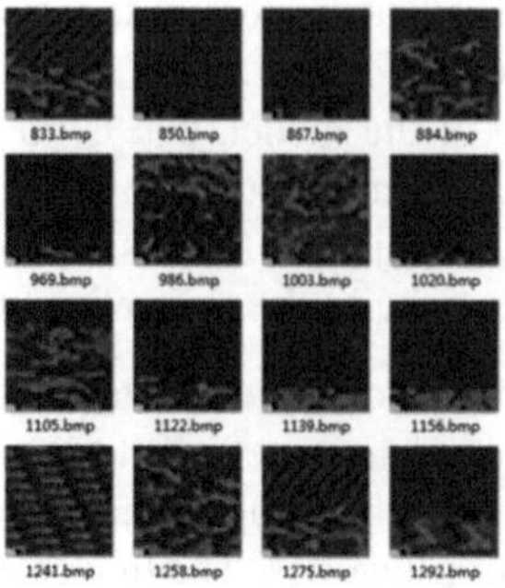

Fig. 4. Normal samples gray scale image.

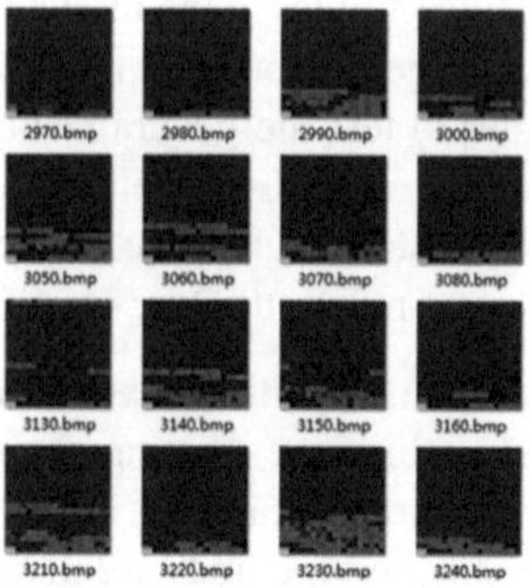

Fig. 5. SQL injection samples gray scale image.

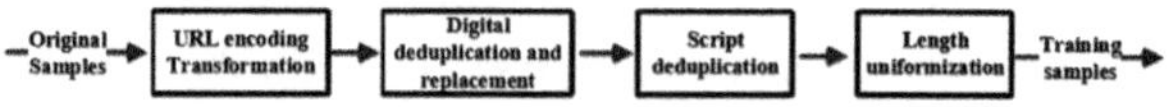

Fig. 6. Original samples processing flow.

URL Encoding Transformation. In the network transmission, some symbols can not be directly transmitted in the URL script, and need to be encoded. The format of the encoding is: % plus character ASCII value hexadecimal form, for example, the encoded value of the space is "%20". Due to the existence of such encodings, the scripts with the same meaning in the original samples have different feature expressions. This redundant feature expression increases the model training cost, and a large number of such encoding methods are mixed in the original samples. For example,"%22%29%29%29%20UNION%20ALL%20SELECT%206781%2C6781%2C6781%2C6781%2C6781%2C6781%2C6781%2C6781%23". In training, such encoding will affect the automatic extraction of features of the model, so it is necessary to decode the encoding format of this class to reduce the number of sample features.

Digital Deduplication and Replacement. There are a large number of decimal or hexadecimal numbers in the samples. For example, in the case of a permanent injection structure such as "1 = 1", it can be replaced with an arbitrary number and the injection effect can also be achieved. Therefore, in order to better express the sample features, it is necessary to replace the numbers in such constructs with specific character encodings. For example, "union select 1, 2, 3, 4, 5, 6, 7, 8, 9, table_name, 11, 12, 13 from information_schema. tables" "(0x7179647371, 0x6141534f415555665645, 0x717a687371), NULL, NULL, NULL, NULL, NULL". The decimal part of the script fragment is replaced with "DD" and the hexadecimal part is replaced with "FF". After the above script fragment is replaced, it will become the form of "union select DD, DD, DD, DD, DD, DD, DD, DD, DD, table_name, DD, DD, DD from information_schema. tables" "(DDxFF, DDxFF, DDxFF), NULL, NULL, NULL, NULL, NULL".

Script Deduplication. After the above two steps, there will be script duplication in the original samples, and the scripts needs to be deduplicated. After the deduplication, the SQL injection samples has 76666 scripts, and the normal samples has 38471 scripts.

Length Uniformization. The original input features of the deep learning model must be fixed in length, so the length of the script in the samples needs to be unified. We calculate the script length distribution in the SQL injection sample and the result is shown in Fig. 7.

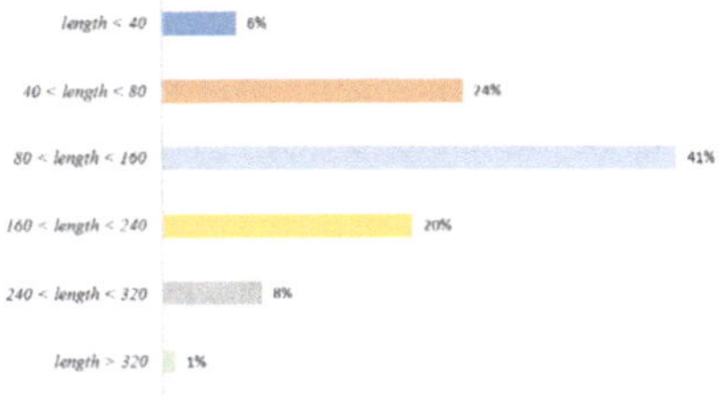

Fig. 7. SQL injection script length statistics.

Based on the statistical results, the lengths of 80, 240, and 400 are selected as the standardization of the samples length, and the corresponding samples are generated separately for comparison experiments. In the process of unification, zero-fill operation is performed on samples of insufficient length. For long samples, interception operation is performed, script data is placed in different positions in the form of sliding windows, and scripts containing no keywords are deleted. For example, "union select DD, DD" contains 18 characters (space counts 1 character). After the unified operation of length 14, the new scripts are: "union select D", "union select DD", "ion select DD,", "on Select DD, D" "n select DD, DD". Since all scripts contain the keyword select, there is no need to delete scripts without keywords.

After the length uniformization, the SQL injection samples and normal samples in the training set are annotated. The SQL injection scripts are marked as "0" and the normal scripts are marked as "1". Then, the two types of samples are divided into a training set and a test set according to a 10:1 ratio. The scripts in the training set are converted into a tfrecord format so tensorflow can read them effectively. The two types of samples in the training set are mainly used for model training and testing. The same pre-processing as above is performed for the generalized test set, but the generalized test set is not need to be converted into the tfrecord format.

3.3 Deep Learning Detection Module Based on Text Features

SQL-LSTM model. The SQL-LSTM model is a three-layer LSTM-based model. Compared to CNN, LSTM is often more advantageous in text processing. The main idea of the model is similar to classification problem of two categories in text processing. In this model, the classification problem is to judge whether the input sample is a SQL injection script. The specific structure of the model is shown in Fig. 8.

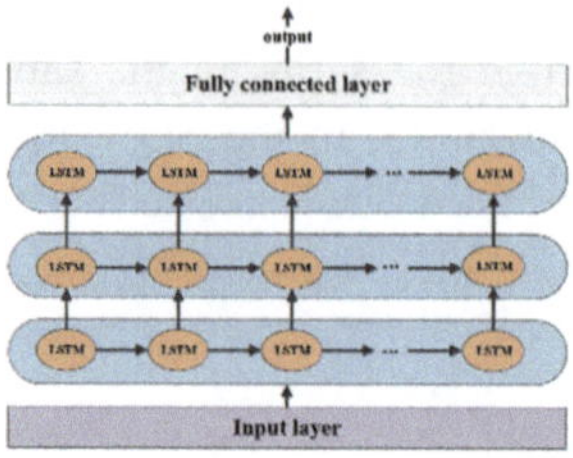

Fig. 8. Structure of SQL-LSTM model.

The first part of the model is the input layer. The role of this layer is mainly to transformer the preprocessed raw data into a vector and then input it into the multi-layer LSTM structure. Next is the three-layer LSTM structure. LSTM can acquire long-distance dependency information and capture certain context information, so it is more suitable for text processing. The function of the three-layer LSTM is mainly to extract script features. The number of hidden units is set as 256, and the multi-layer structure can better acquire the script features. Then, the output of the LSTM is used as the input of the fully connected layer, and the features are further extracted. Finally, the Softmax is used for classification. The input is assumed to be [x1, x2], and after passing through the Softmax function, the output becomes [y1, y2], and y1 + y2 = 1. The threshold is set to 0.5. When y1 > 0.5, the original sample is judged to be SQL injection sample, otherwise it is regarded as normal sample.

In the training, the batch size of each iteration is 128, and the optimization algorithm uses the Adam algorithm. In order to prevent the model from over-fitting, we adopt the dropout strategy with a parameter of 0.5. This strategy will make half of the neural network units invalid in each round of training.

Model Adaptive Training. Because various network data scripting protocols are constantly being revised and revised, and the data types and data payloads in different network lines are also different. Therefore, in order to ensure the detection effect of the model in various scenarios, an adaptive training process is added.

After the pre-processed data is detected by the SQL-LSTM model, the data determined by the model as the SQL injection behavior enter the adaptive training process. Then, the data is discriminated and annotated manually, and the manually filtered data is passed through the data preprocessing module to generate the new training samples. The new training samples are added to the sample library for training by the SQL-LSTM model to improve the adaptability of the model to different application scenarios. The adaptive training process is shown in the Fig. 9.

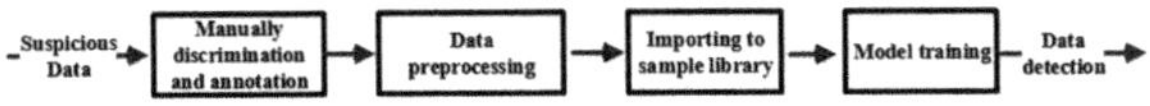

Fig. 9. Adaptive training process.

Through the adaptive training process, the sample library can be continuously enriched. When the data protocol and source change, as long as the data still has the basic features of the SQL statements, the SQL-LSTM model can adapt to the detection of the new data after a period of training. Meanwhile, the detection effect of the original data will not be affected. In order to achieve the above objective, the SQL-LSTM model must be highly sensitive to the key strings of the SQL statements. In this way, the SQL-LSTM model can detect as much data as possible when encountering data with new features to provide complete data for manual discrimination and annotation.

3.4 Traffic Information Extraction Module

The main function of this module is to analyse the features of data traffic and extract the information needed for comprehensive detection. The implement of this module mainly relies on the existing traffic probe technology, and NetFlow v5 is used for features extraction.

The first type of information to be extracted is the IP address, and it is used to determine the parties to the communication in the network. In the detection of SQL injection, the detection based on text features must rely on the SQL statement. However, in the communication between the two parties, the SQL statement is only present in the single direction. For the detection based on traffic features, the traffic information can not be fully utilized if only one direction of the data packets is analysed. Therefore, when extracting the IP address information, if the data packets between the two communication parties contains the SQL scripts, the traffic information extraction module needs to save the information of the two-way data packets of the communication parties together. In addition, the communication port information is also very important in the SQL injection attack, so the port information of both parties should also be included. The format of the IP address information is shown in Table 1.

Table 1. Information of IP address and port.

Source IP address	Source Port
Destination IP address	Destination Port

The second type of information needs to be extracted in the traffic information extraction module is the duration of communication between the two communication parties, the number of data packets and the size of the data packets. When performing SQL injection attacks, a large number of communication packets are generated between the two parties in a short time, or data packets containing a large amount of information are generated, and these are obvious abnormal behaviors. Therefore, the information of the traffic packets is of great significance for the detection of the attack stage. The format of the data packets information is shown in Table 2.

Table 2. Information of data packets.

Start time	End time	Duration
Number of packets	Max size of packets	Total flow

Finally, the two pieces of information are combined to form the complete traffic features information, and the format is shown in Table 3.

Table 3. Information extracted by the traffic information extraction module.

Source IP address		Source port
Start time	End time	Duration
Number of packets	Max size of packets	Total flow

3.5 Comprehensive Detection Module

The detection results of the SQL-LSTM model and the results of the traffic information extraction are finally summarized into the comprehensive detection module for further analysis. The detection results passed to this module by the SQL-LSTM model includes not only the label of the SQL statement, but also the information of the five-element array {source IP address, source port, destination IP address, destination port, protocol} of the data packet corresponding to the SQL statement. The comprehensive detection module first uses these information to match the SQL statement with the information obtained by the corresponding traffic information extraction module. Then, this module uses the fused information for comprehensive detection.

After the information fusion is completed, we analyse the attack stage based on the data traffic features for the SQL injection statement. The analysis is mainly from two aspects. The first is to analyse the frequency of interaction between the two parties. For the database field guessing at the beginning of SQL injection, brute force methods are often used. At this time, a large number of packets sent from the attacker to the sever will be generated in a short time. However, the response information from the sever is usually less. The second is to analyse the size of the data packets and the overall traffic. In the final stage of SQL injection, the attacker often gets a large amount of data from the target database. At this time, the size of the packet returned by the server to the server attacker will be significantly large, and the size of the overall traffic will also be large, indicating that the attack has occurred. In addition, some attacks also need to be combined with the server to analyse the return content of the server for analysis. The detection on this aspect will be expanded in further approaches.

4 Experiments

4.1 Experiments Environment

This section focus on the settings of the experimental environment. In this paper, we use four virtual machine to mimic the real network environment. Three virtual machines represent three different clients, and the IP address of them are 202.200.1.211, 202.200.1.212, 202.200.1.213. The three virtual machines correspond to three different types of traffic, namely, SQL injection traffic, normal traffic to the database, and traffic frequently accessed web pages to simulate the Denial of Service (DoS) attack.

For SQL injection traffic, this type of traffic has SQL injection statements in terms of text features to be used for SQL-LSTM model detection, and also has SQL injection traffic features. The traffic features are used to detect the attack stage by the comprehensive detection module. For normal traffic to the database, it has the normal SQL statements in terms of text features to be used for SQL-LSTM model detection. But there are no obvious traffic features. For the traffic frequently accessed web pages to simulate the DoS attack, it does not have SQL statements, and only has some keywords of SQL statement in the scripts, such as "select", "union", and et al. But it has the same large amount of traffic packets feature as the database field guessing in the SQL injection attack.

Another virtual machine is used to represent the server, and the corresponding IP address is 202.200.1.210. On this virtual machine, we use the open source project of Github to build the SQL injection practise platform - Sqli-labs. The purpose of using this platform is to initiate the real SQL injection attack and achieve a complete attack phase without generating destructive behavior. Meanwhile, the other two types of traffic will also not cause damage. For the injection tool, we choose the SQLmap. For the choice of the traffic packets capture tool, we use the wireshark. The specific framework of the experiment is shown in Fig. 10.

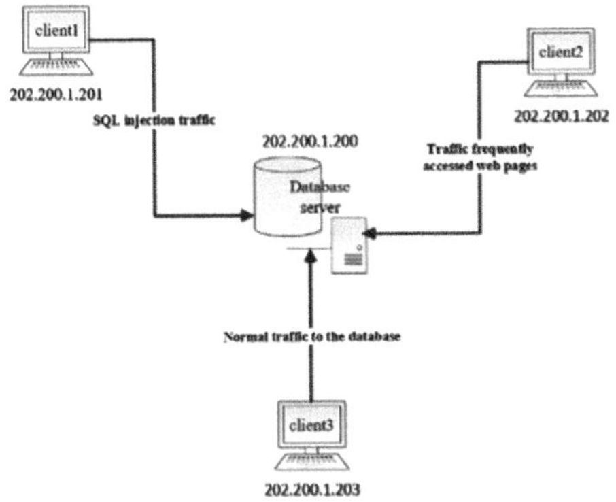

Fig. 10. Framework of experiment.

4.2 Detection Module Based on Text Features

Evaluation Index. A valid anomalous behavioral data detection model should have extremely high accuracy in the detection data of a known range. In the unknown generalized data set, it should be as high as possible. Since new data features exist in the generalized data set, the model may not identify the positive and negative examples well. in this case, all the suspicious data should be extracted to ensure a high recall rate. After the suspicious data is extracted, it is manually annotated, and then the data with new features is added to the sample library. And the enriched training samples are used for the model adaptive training to continuously improve the detection effect of the model. In general, the evaluation of the model should follow the principle of criteria should follow the principle of keeping the balance between accuracy and recall rate, and the recall rate is of more significance.

In this paper, the accuracy rate A, recall rate R, F1 value are used to test and evaluate the SQL-LSTM model. The accuracy and recall rate are the recognition results of the model, and F1 value is the comprehensive evaluation index calculated based on the accuracy rate A and the recall rate R.

Comparative Experiments at Different Lengths. Through the analysis of the length of the scripts, the length of the SQL statements has a large range. In order to have a better detection effect of the SQL-LSTM model, it is necessary to test the performance of the detection model under different input data lengths to determine the most suitable length. The input data of lengths of 80, 240 and 400 are respectively tested using the test set and the generalized test set. The results of the accuracy rate A_1 under the test set, the accuracy rate A_2 and the recall rate R under the generalized test set are shown in Table 4.

Table 4. SQL injection script test results at different lengths.

Length	A_1	A_2	R
80	97.16%	49.87%	79.13%
240	99.89%	46.44%	99.25%
400	99.74%	45.97%	97.29%

Under the test set, the accuracy is the highest when the length is 240, reaching 99.89%. Because the accuracy is not high under the generalized test set, but the recall rate is very high at the length of 240, reaching 99.27%, so 240 is selected as the script input length.

By observing the statistics of script length, scripts with length less than 80 account for about 30%, while scripts with length less than 240 account for more than 90%, and scripts with lengths between 80–240 account for more than 60%. When the length is set to 80, about 70% of the scripts will lose a lot of feature information due to the truncation operation of the scripts. And when the length is set to 400, most of the scripts are not long enough, so padding operation is performed. However, the padding operation will result in a large amount of redundant information in the sample features. Setting the length to 240 can not only contain the complete information of most scripts, but also has less redundant information for detection, so when the length is set to 240 can have the best effect.

Comparative Experiments under Different Feature Extraction Method. In order to compare the data feature extraction processing and the validity of the model construction, we use the Github open source project to build a method based on Word2Vec to extract data features. Base on this method, CNN, LSTM, GRU, Multi-Layer Perception (MLP), SVM models are constructed and compared with SQL-LSTM. By tuning the different model layers and model parameters, the best test results of the corresponding type models are obtained. Comparing the SQL-LSTM model with the above model, the results of the accuracy A_1 under the test set, the accuracy rate A_2, the recall rate R and the F1 value under the generalized test set are shown in Table 5.

Table 5. Model comparison test results.

Model	A_1	A_2	R	F1
SVM	85.21%	84.34%	60.00%	69.77%
MLP	98.96%	73.91%	68.00%	70.83%
GRU	99.98%	83.33%	60.00%	72.73%
LSTM	99.85%	84.21%	64.00%	72.73%
CNN	99.98%	76.19%	64.00%	69.57%
SQL-LSTM	99.89%	46.44%	99.25%	63.36%

Under the test set, the accuracy of the SQL-LSTM model and CNN, LSTM and GRU models based on word vector feature extraction are very high, exceeding 99.5%. Under the generalized test set, SQL-LSTM model has lower accuracy than other models, but the recall rate is very high, reaching 99.25%, and the F1 values between the models are not much different. Although the SQL-LSTM has a higher false detection rate than the other models on the generalized dataset, it can ensure that the detection system does not miss the suspicious SQL injection behavior at runtime, and the other models will have obvious missed detection conditions. Therefore, the SQL-LSTM model can be better applied to our integrated detection framework.

4.3 Experiments of Comprehensive Detection

For SQL injection, this paper mainly divides its attack process into three stages. The first stage is the database field guessing and the detection of injection points. For this stage, most of attackers use brute force methods. A large number of packets sent from the attacker to the server are generated in a short period of time, and the response information of the server is usually small. The traffic features to be analysed at this time are mainly the communication frequencies between the two communicating parties. The second stage is the attacker's elevated privileges and data injection phase. In this stage, there is no obvious feature in terms of the traffic aspect, and the attack traffic is similar to the normal SQL access traffic. The third stage is the acquisition of data after successful injection. In this stage, the attacker often gets a large amount of data from the target database. At this time, the size of the data packet returned by the server to the attacker will be significantly large, and the size of the overall traffic will also be large. The obvious traffic features at this time are the size of the packet and the size of the overall traffic, and this phenomenon indicates that a successful attack has occurred.

For the experimental environment set up in this paper, we use three virtual machines to access the database server and capture the corresponding data packets. After the SQL-LSTM model is detected, the detection results are matched with the corresponding data traffic information. Through this integrated detection method, we have achieved more than 95% accuracy in the judgement of the SQL injection attack stage. Therefore, the detection framework proposed in this paper can achieve a quite good effect on the detection of SQL injection. In practical applications, when the framework detects the first stage of attack behavior, precautions against the source of the data packet will be taken to prevent more serious damage. Usually, the first phase of the SQL injection attack takes a long time, and the SQL-LSTM model of our proposed framework can detect the SQL injection statement in time after the down-flow measures. Meanwhile, the extraction technology of the traffic feature information is also very mature. Therefore, our methods can achieve quick and accurate detection and effectively defend against SQL injection attacks.

5 Conclusion

This paper proposes an integrated detection framework for SQL injection behavior based on both text features and traffic features. At the text features level, we use deep learning technology to build a detection model - SQL-LSTM. This model utilizes a multi-layer LSTM structure and effectively improves the detection effect of the SQL injection scripts. Meanwhile, in order to further improve the effect of the SQL injection detection, we merge the output of the SQL-LSTM with the data traffic features. In the future, our work has three main directions. The first is to optimize the SQL-LSTM model to further improve the accuracy, the second is to introduce the content analysis of the data return by the server into our framework, the third is to extend our framework to other network script type attack behavior detection.

References

1. OWASP Top 10.-The ten most critical web application security risks (2017). https://www.owasp.org/images/7/72/OWASP_Top_10-2017_(en).pdf.pdf
2. Wang, D., Zhao, W., Ding, Z.: Review of detection for injection vulnerability of web applications. J. Beijing Univ. Technol. **42**, 1822–1832 (2016)
3. Sharma, C., Jain, S.: Analysis and classification of SQL injection vulnerabilities and attacks on web applications. In: International Conference on Advances in Engineering and Technology Research. IEEE (2015)

4. Halfond, W., Orso, A.: AMNESIA: analysis and monitoring for Neutralizing SQL-injection attacks. In: IEEE/ACM International Conference on Automated Software Engineering, Long beach, California, USA, pp. 174–183 (2005)
5. Wan, M., Liu, K.: An improved eliminating SQL injection attacks based regular expressions matching. In: International Conference on Control Engineering and Communication Technology, IEEE Computer Society (2012)
6. Wang, W., Li, C., Duan, G.: Design of SQL injection filtering module based on regular expression. Comput. Eng. **37**(5), 158–160 (2011)
7. Patel, N., Shekokar, N.: Implementation of pattern matching algorithm to defend SQLIA. Procedia Comput. Sci. **45**, 453–459 (2015)
8. Prabakar, M., Karthikeyan, M., Marimuthu, K.: An efficient technique for preventing SQL injection attack using pattern matching algorithm. In: International Conference on Emerging Trends in Computing. IEEE (2013)
9. Han, C., Lin, H., Huang, C.: Research on the SQL injection filtering based on SQL syntax tree. Chinese J. Network Inf. Secur. (2016)
10. Joshi, A., Geetha, V.: SQL Injection detection using machine learning. In: International Conference on Control (2014)
11. Ladole, A., Phalke, M.: SQL injection attack and user behavior detection by using query tree, Fisher Score SVM Classification. IRJET **3**, 1505–1509 (2016)
12. Kim, M., Lee, D.: Data-mining based SQL injection attack detection using internal query trees. Expert Syst. Appl. **41**(11), 5416–5430 (2014)
13. Rawat, R., Shrivastav, S.: SQL injection attack detection using SVM. IJCA **42**(13), 1–4 (2012)
14. Yang, W., Zuo, W., Cui, B.: Detecting malicious URLs via a keyword-based convolutional gated-recurrent-unit neural network. IEEE Access, p. 1 (2019)
15. Fu, L., Zhang, H., Huo, L.: JavaScript malicious script detection algorithm based on multi-class features. PR AI **28**(12), 1110–1118 (2015)
16. Torres, P., Catania, C., Garcia, S., et al.: An analysis of recurrent neural networks for botnet detection behavior. Biennial Congress of Argentina, IEEE (2016)
17. Wang, Z.: The applications of deep learning on traffic identification. BlackHat USA (2015)
18. Wang, W., Zhu, M., Zeng, X., et al.: Malware traffic classification using convolutional neural network for representation learning. ICOIN, IEEE (2017)
19. Tang, P., Qiu, W., Huang, Z., et al.: SQL injection behavior mining based deep learning. In: International Conference on Advanced Data Mining and Applications. Springer, Cham, Nanjing, pp. 445–454 (2018)
20. Zhang, L., Liao, P., Zhao, J.: A method of unknown protocol recognition based on convolution neural network. Microelectr. Comput. **35**, 112–114 (2018)
21. SQLmap-Automatic SQL injection and databasetakeover tool. http://SQLmap.org
22. Pan, S., Xue, Z., Shi, Y.: Malicious URL detection based on convolution neural network. Commun. Technol. **8**, 1918–1923 (2018)
23. Mikolov, T., Sutskever, I., Chen, K., et al.: Distributed Representations of Words and Phrases and their Compositionality. NIPS 2013, Lake Tahoe, USA, pp. 3111–3119 (2013)

Android Secure Cloud Storage System Based on SM Algorithms

Zhiqiang Wang[1,2,3(✉)], Kunpeng Yu[2], Wenbin Wang[2], Xinyue Yu[2], Haoyue Kang[2], Xin Lv[1], Yang Li[1], and Tao Yang[3]

[1] State Information Center, Beijing, China
wangzq@besti.edu.cn
[2] Beijing Electronic Science and Technology Institute, Beijing, China
[3] Key Lab of Information Network Security, Ministry of Public Security, Shanghai, China

Abstract. With the rapid development of mobile Internet, the methods of data storage have been greatly enriched. However, many problems exist in mobile internets' current information storage, such as plain-text storage, weak and nondomestic encryption algorithms. This paper designs a secure cloud storage platform with SM algorithms based on the Android platform. The platform uses a combination of encryption and authentication methods that use multiple SM algorithms (SM2, SM3, SM4), which realizes hybrid encryption and authentication of information storage. Finally, lots of experiments are done, and the results show the security and reliability of our platform, the time about regular operations of our platform is in the range of acceptance given performance and security.

Keywords: Mobile internet · SM algorithms · Data storage · Cloud storage platform

1 Introduction

With the reduction of computer hardware costs and improved performance, technologies such as big data, cloud computing, and mobile Internet are more and more widely used in people's daily lives. Cloud computing and storage are gradually affecting the way of people's life and work [1]. Compared with the traditional local storage mode, cloud storage has an ample storage space, multiple device sharing, secure sharing, fast upload, and download speed, etc. Cloud storage provides users with cloud storage services and uploads data to the data center for maintenance management [2]. Cloud storage systems are flexible for users who can customize the space's corresponding size according to their requirements [3]. Users can save numerous software and hardware costs by using cloud storage.

While cloud storage provides convenience to users, uploaded data is challenged by more and more security threats. In 2016, more than 60 million accounts of Dropbox were stolen by hackers. In 2019, Cloud storage service company, MEGA's data, containing 770 million mailboxes, was leaked. iCloud users are frequently exposed to

Q. Liu et al. (Eds.): CENet 2020, AISC 1274, pp. 772–779, 2021.
https://doi.org/10.1007/978-981-15-8462-6_88

data breaches. Protecting the security of users' data in cloud storage is becoming an increasingly important issue [4].

The smart phone is one of the essential ways to acquire information devices in people's daily life [5]. People can significantly reduce data loss risk by backing up their data to the cloud [6]. Traditional cloud storage on Android is implemented at the application layer, and cloud storage applications run on the Android virtual machine. Due to the Android virtual machine's independence, it is very difficult for other applications to access the data stored in the cloud, which needs require additional processing for cross-process communication. Simultaneously, the traditional cloud storage application is not based on the underlying file system, but does similar processing on the application layer, resulting in the data being disjointed from the file system. This has certain limitations in performance and sharing [7].

This paper mainly adopts the hybrid encryption method in cloud storage and designs and realizes an android secure cloud storage platform. The innovations and contributions of this paper are as follows:

(1) Multiple SM algorithms (SM2, SM3, SM4) are adopted in the designed system for encryption and authentication, ensuring the integrity and security of the data uploaded into the Android cloud platform.
(2) Multifunctional cloud storage platform. The designed system can offer multiple functions, such as personal file storage management, file sharing, transmission, etc. Users can store files that need to be kept confidential on the cloud platform according to their personal needs.

2 System Design and Analysis

2.1 System Architecture

The architecture of our cloud storage system is shown in Fig. 1. The system adopts open system architecture and modular design. By reorganizing and filtering the data stream, it can realize strong encryption and authentication to ensure high reliability and stability. The system is divided into three levels: presentation layer, security encryption layer and data processing layer.

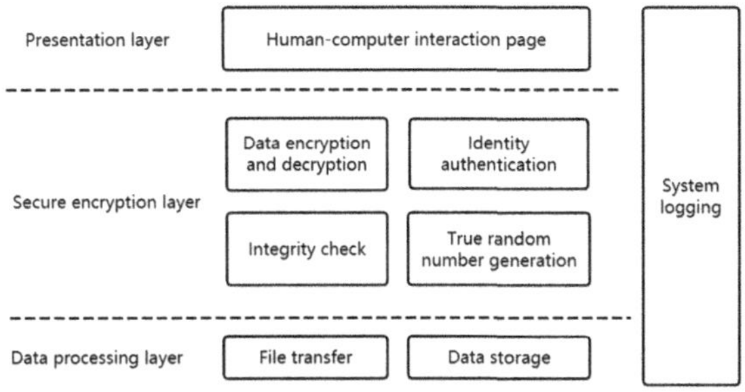

Fig. 1. The system architecture.

In the presentation layer, a simple and easy-to-use human-computer interaction page is designed, which can be used to log in, upload files, load files and share files.

In the security encryption layer, the system uses a variety of encryption algorithms for encryption and authentication. On the one hand, it selects the symmetric encryption algorithm to encrypt and decrypt data, which ensures not only the security of the data, but also the efficiency of encryption and decryption; the asymmetric algorithm is selected for signature verification. On the other hand, it selects the asymmetric algorithm to ensure the security of the key in the process of transmission. In order to ensure the uniqueness and security of the user's identity, the digest algorithm is used to process the id and password, then using the digest value for authentication. The true random number generator is selected to ensure the randomness of the generated key.

In the data processing layer, file transfer and storage functions are designed to transfer and store files. Files can be uploaded to servers and stored in the servers' database by using the protocol HTTP. The data processing interface shields the differences of each mobile operating system and can be used in any mobile operating system that supports HTTP, which enhances the portability of the system [8]. In addition, users can also share files with other users through the system.

2.2 Storage Design

This system adopts object-oriented storage, which is similar to the traditional data access methods based on block (parallel SCSI, SAS, FCP, ATA, SATA) and file-based (NFS, CIFS). Object-based data access model is a kind of storage technology. Unlike traditional block-oriented devices, object storage protected storage units which are similar to logical units, allowing access to data through access to storage objects [9].

Object-oriented storage stores objects in a bucket container and accesses them through a unique identity. It has lower hierarchy than the traditional hierarchical data access model, which saves the node space when data increasing. At the same time, object storage also solves the problem of poor sharing between heterogeneous block storage. The object-oriented storage unit consists of four parts: OID, Data, Metadata and Attributes. OID is the unique identity of users, and each object has only one OID according to which developers can operate the objects in bucket. Data is the content of an object. Metadata holds the metadata of objects, including creation time, ownership, file size, and so on. Attributes stores the properties of objects, including access mode, access type, Qos, and so on. Developers can extend the attributes they need in this part. Some object storage combines Metadata and Attributes, collectively referred to as Metadata.

2.3 Encryption Design

The flowchart of data encryption is shown in Fig. 2. During the encryption process, keys and files are encrypted and transmitted to the server in the form of ciphertext and stored in the server's database, which ensures the security of the data in the process of transmission. In order to ensure these keys completely controlled by users, file keys, master keys and public-private key pairs are used for encryption and they all generated

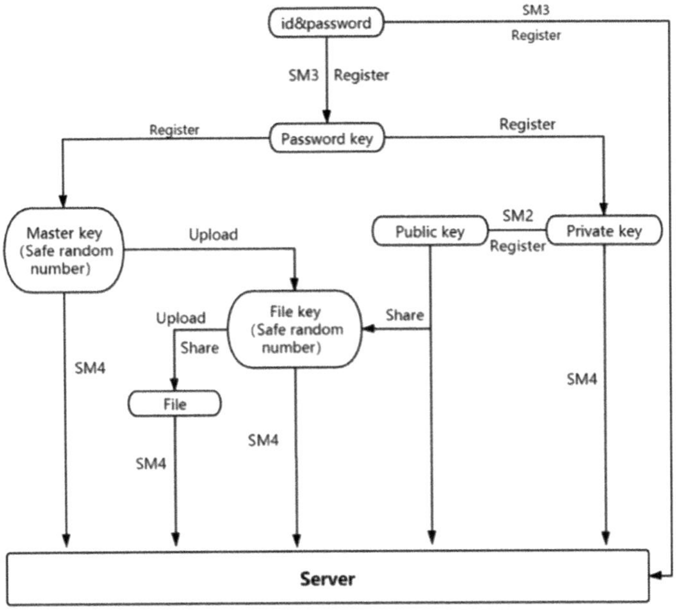

Fig. 2. The flowchart of data encryption.

on the client side. The user has absolutly control over the key, so it will ensure the security and controllability of user data [2].

3 System Implementation

3.1 Registration Module

Users need to input an id, a password and the confirm password when registering. To get encrypted id and password, they are transformed by SM3 respectively. Then the system judges whether the input information is complete. If complete, it will judge whether the password and the confirm password are consistent. If both the passwords are consistent, it will judge whether the encrypted id exists. When the encrypted id doesn't exist, the registration can be continued. After finishing these check process, the user password key is generated by SM3 based on the id and the password. Then, encrypted id and encrypted password are uploaded to the server for subsequent login. Finally, a random security number will be generated as the master key, and a public-private key pair will be generated randomly by SM2. The private key and the master key are encrypted with the SM4 using the password key, and the encrypted private key and the encrypted master key are uploaded to the server database.

3.2 Login Module

Users need to input an id and a password when logging in. The system encrypts the entered id and password with SM3 respectively. The system compares them with the data stored in the server database. If consistent, the system will download the encrypted master key and the encrypted private key from the server. Then the entered id and password together are transformed by SM3 to get the password key of the user. The system uses the password key to decrypt the encrypted master key and the encrypted private key respectively with SM4 to obtain the master key and the private key.

3.3 File Module

When users upload files, they use a random security number as a file key to encrypt the file's content in CBC mode. In addition, the file key is encrypted with the master key obtained at login, and the encrypted file and file key are uploaded to the server subsequently.

Downloading files is the reverse process of uploading files. Users select the files they want to download from the server, download the ciphertext of the file and its corresponding file key to the local, use the master key to decrypt the file key ciphertext to get the file key, and then use the file key to decrypt the ciphertext of the file to get the plaintext.

3.4 Sharing Module

When users share files, they need to enter the id of the target user. The system will perform SM3 operation on the entered id to get the digest value. Find the public key of the target user from the server according to the digest value. Then the user selects the file to be shared from the local computer, generates a random security number as the file key for encryption, and then encrypt the file key with the public key of the target user. In the process of sharing, the system needs to pass the encrypted file information and the encrypted file key to the server. After received by the target user, the encrypted file key is decrypted with the private key first, and then the encrypted file is decrypted with the file key to get the plaintext.

4 Experiment

The deployment diagram of test environment is shown in Fig. 3. The figure describes the test process in detail. Our cloud storage platform is suitable for Android 6.0 or above versions. The mobile applications develop environment is AndroidStudio 3.2 (compileSdkVersion 29, minSdkVersion 21) with development language Java (JDK version 1.8). The cloud server is built based on the SpringBoot technology in JavaWeb. It is a Tomcat server that can interact with the user terminal. The end user can communicate and interact with the server in a good network environment, including the user's login, registration, file upload, download and share functions.

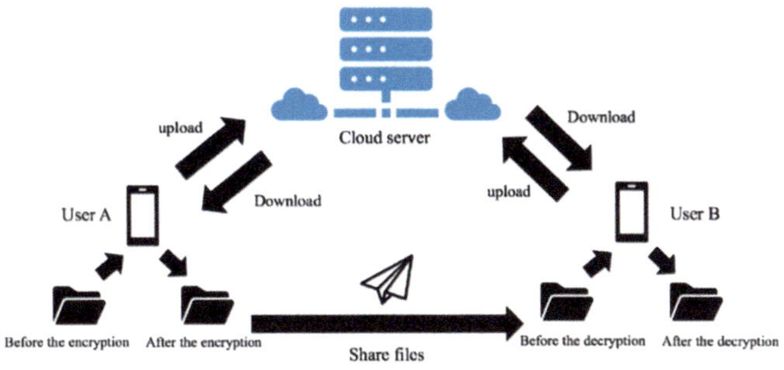

Fig. 3. The deployment diagram of test environment.

5 Evaluation

5.1 Register Module

In the user registration stage, SM3 digest algorithm is mainly used, which mainly tests the time required for different username and password length and complexity.

Table 1. The evaluation of register module.

No	User	Username	Username length	Password	Password length	Time
1	User A	lalalalaouye	12	qaz	3	3 ms
2	User B	QW12qweQWEd	12	qaz	3	4 ms
3	User C	lalalalaouye	12	QHuhu	6	4 ms
4	User D	QW12qweQWEd	12	QHuhu	6	3 ms
5	User E	qwer	4	qaz	3	3 ms
6	User F	qwer	4	QHuhu	6	3 ms

As can be seen from Table 1, the complexity of username and password has little impact on time cost.

5.2 File Upload Module

The combination of SM2 and SM4 encryption algorithms is mainly used in the file upload phase, which mainly tests the time required for different file sizes. In the Table 2, the Time1 represents the upload time of unencrypted files and the Time2 represents the upload time of encrypted files.

Table 2. The evaluation of file upload module.

No	Filename	File size	File type	Time1	Time2
1	File A	1.03 MB	Picture	456 ms	658 ms
2	File B	215 KB	Picture	89 ms	233 ms
3	File C	1.5 MB	Picture	509 ms	858 ms
4	File D	2 MB	Picture	691 ms	1069 ms
5	File E	24.6 MB	PPT	7018 ms	11982 ms
6	File F	9.6 MB	Video	2498 ms	4689 ms

As can be seen from Table 2, the size of the file has a great impact on time cost. The larger the file size, the larger the time cost.

5.3 File Download Module

In the file download stage, SM2 and SM4 encryption algorithms are also used together. In this stage, the time required for different file sizes is mainly tested. As can be seen from Table 3, time cost of file download is far less than it of file upload. In the Table 3, the Time1 represents the download time of unencrypted files and the Time2 represents the download time of encrypted files.

Table 3. The evaluation of file download module.

No	Filename	File size	File type	Time1	Time2
1	File A	970 KB	Picture	984 ms	2698 ms
2	File B	215 KB	Picture	314 ms	476 ms
3	File C	1.5 MB	Picture	1394 ms	2776 ms
4	File D	2 MB	Picture	1895 ms	2898 ms
5	File E	24.6 MB	PPT	13894 ms	27967 ms
6	File F	9.6 MB	Video	6598 ms	7868 ms

6 Conclusion

With the continuous development of information network era, cloud storage technology is becoming more and more important. Many companies at home and abroad, such as Microsoft, Google, Baidu, Tencent, Alibaba, have put forward their own cloud storage security solutions. At present, cloud storage technology has penetrated into all areas of the society. Due to the characteristics of security, flexibility and convenience, the use of mobile terminals to complete various operations has become mainstream. This work designs a secure cloud storage platform based on SM algorithms and Android, adopts the open module design, and uses a variety of algorithms to realize encryption and authentication, which can guarantee the confidentiality, integrity and availability of user files to a certain extent.

Acknowledgments. This research was financially supported by the National Key Research and Development Plan (2018YFB0803401), Key Lab of Information Network Security, Ministry of Public Security (C19614), Special fund on education and teaching reform of Besti (jy201805), the Fundamental Research Funds for the Central Universities (328201910), China Postdoctoral Science Foundation (2019M650606), 2019 Beijing Common Construction Project-Teaching Reform and Innovation Project for Universities in Beijing, key laboratory of network assessment technology of Institute of Information Engineering, Chinese Academy of Sciences. The authors greatefully acknowledge the anonymous reviewers for their valuable comments.

References

1. Zhou, K., Wang, H., Li, C.H.: Cloud storage technology and its applications. ZTE Technol. **16**(4), 24–27 (2010)
2. Fu, Y.X., Luo, S.M., Shu, J.W.: Overview of secure cloud storage systems and key technologies. Comput. Res. Dev. **50**(1), 136–145 (2013)
3. Wu, Z.H.: Cloud Computing Core Technology Analysis. People's Posts and Telecommunications Press, Beijing (2011)
4. Report on internet development in China. Beijing: Internet Society of China (2019)
5. Zhang, N.: Research and application of Android System Architecture. Xi'an University of Science and Technology (2013)
6. Rajaraman, A., Ullman, J.D., Wang, B.Y.: Big Data: Internet Large-Scale Data Mining and Distributed Processing. People's Posts and Telecommunications Press, Beijing (2012)
7. Tian, M. D.: Design and Implementation of Android Cloud Storage File System. South China University of Technology, Guangzhou (2018)
8. Larman, C.: Applying UML and Patterns: An Introduction to Object-Oriented Analysis and Design and Iterative Development. Beijing Machinery Industry Press, Beijing (2006)
9. Min, X.C., Huang, L.C.: Research on Web Service technology based on Android platform. Ind. Control Comput. **24**(4), 92–94 (2011)

Identification System Based on Fingerprint and Finger Vein

Zhiqiang Wang[1,2,3](✉), Zeyang Hou[2], Zhiwei Wang[2], Xinyu Li[2], Bingyan Wei[2], Xin Lv[1], and Tao Yang[3]

[1] State Information Center, Beijing 100045, China
wangzq@besti.edu.cn
[2] Beijing Electronic Science and Technology Institute, Beijing 100070, China
[3] Key Lab of Information Network Security, Ministry of Public Security, Shanghai 200000, China

Abstract. In recent years, with the rapid development of information technology, more and more attention has been paid to information security. Biometrics is an important way to ensure information security, and digital vein recognition is an important branch of biometrics, which has an irreplaceable position in the field of information security. However, the traditional finger vein recognition is vulnerable to attacks such as replay, template tampering, and forging original features, which leads to the destruction of the authentication system. This paper proposes a solution to encrypt the feature points of finger vein with the key generated by the relative difference of fingerprint detail points based on the SM4 algorithm to realize the double encryption authentication of user information, so as to improve the security and accuracy of user identity identification and meet the increasing requirements of high security.

Keywords: Fingerprint · Finger vein · Multi-Biometric feature fusion · Identity authentication · SM4 algorithm

1 Introduction

Currently, in the field of biometrics encryption, the encryption technology of single biometrics is limited to a certain extent in some specific applications, and is vulnerable to attacks such as replay, template tampering, and forging original features [1].

Currently single fingerprint identification or finger vein identification has been widely used in the market, however, the combination of both is still a blank field. Aimed at the above problems, we propose a digital vein identification system based on fingerprint feature key. The system firstly generated a characteristic key based on a specific fingerprint, encrypt the input digital vein feature sample and then store the encrypted digital vein image in a database, which will encrypt twice users' identity information to reduce the chance of the system being cracked [2]. When a user needs to verify his identity information, the corresponding key is generated by entering the fingerprint again, which is used to decrypt the encrypted digital vein image in the

Q. Liu et al. (Eds.): CENet 2020, AISC 1274, pp. 780–788, 2021.
https://doi.org/10.1007/978-981-15-8462-6_89

database and generate the corresponding image. Finally, the user's identity can be determined by comparing the corresponding image with the digital vein image entered by the user.

2 System Design

2.1 Architecture

The system architecture is shown in Fig. 1. It consists of three modules: image acquisition module, the module for processing images, extracting features and encryption, and the matching and identification module.

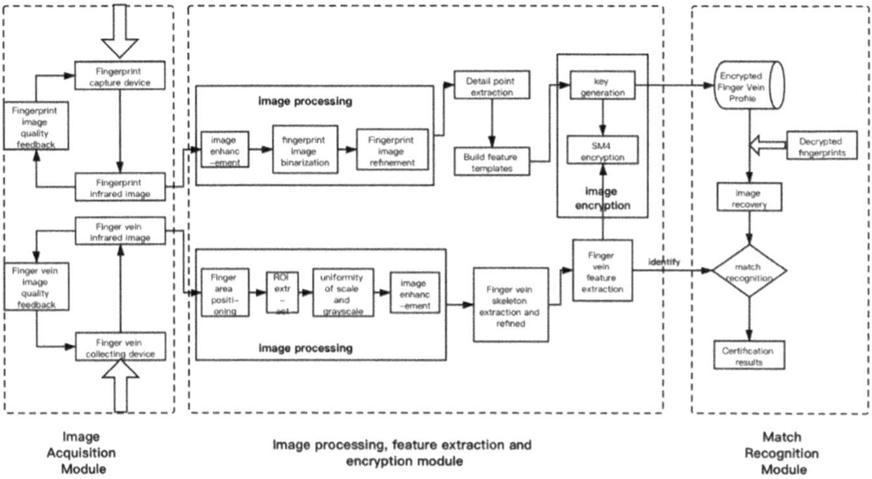

Fig. 1. System architecture.

Firstly, the system obtains the fingerprint image through a fingerprint acquisition instrument, and then performs image enhancement, binarization, and refinement on the fingerprint image. After completing the extraction of the fingerprint details, generate the fingerprint feature key. Secondly, obtain the user's finger vein image by the finger vein acquisition device, and then extract the finger vein image feature information through the steps of ROI extraction, image enhancement, and vein segmentation [3]. Using the SM4 algorithm, the fingerprint feature key is used as the key to encrypt the image of the finger vein feature image, and then store it in a database. When a user is asked to be authenticated, the fingerprint and finger vein need to be collected again, and the fingerprint feature information is extracted to obtain the fingerprint feature key. At this time, the key is used to decrypt the encrypted image in the finger vein feature library, and then a specific matching algorithm is used to match the finger vein image for authentication.

2.2 Fingerprint Key Generation Algorithm

Fingerprint Image Preprocessing. Fingerprint image preprocessing mainly includes steps such as fingerprint image field calculation, fingerprint image segmentation, fingerprint image convergence, fingerprint image enhancement, and binarization and refinement of fingerprint images.

Fingerprint Minutiae Extraction. After preprocessing the fingerprint image, we extract fingerprint minutiae. Fingerprint minutiae are divided into two types, endpoint and bifurcation point. Endpoints refer to the ends of fingerprint lines, and bifurcation points refer to the intersection of two fingerprint lines. The two minutiae feature and pixel models are shown in Fig. 2. Based on the models, we can extract and classify the fingerprint minutiae [4].

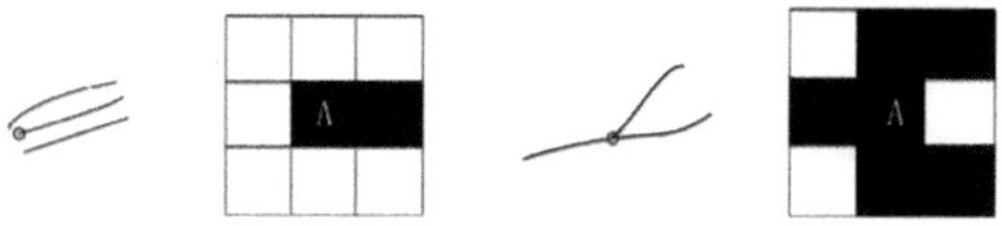

(a) endpoint feature. (b) bifurcation point feature.

Fig. 2. Classification model.

Key Generation Algorithm. Suppose there is a fingerprint image, after the preprocessing, we will get a fingerprint minutiae feature set F:

$$F = \{(x_i, y_i, \theta_i, t_i), i = 1, 2, \ldots, n\} \quad (1)$$

Where $x_i, y_i, \theta_{i,} t_i$ respectively represent the abscissa, ordinate, direction field and minutiae point type of the i-th minutiae point feature, and n is the number of fingerprint minutiae point features.

The directional field describes the general trend of fingerprint lines. In the fingerprint image, as can be seen from Fig. 3, the size of the directional field is the value of the orthogonal decomposition parameter θ of the gradient field.

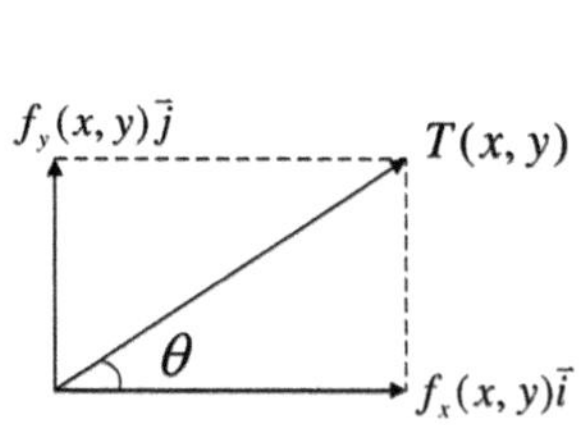

Fig. 3. The direction field is decomposed into coordinates.

According to the tangent value of the angle of the direction field, the angle of the direction field θ (x, y) of the pixel point (x, y) can be calculated.

$$\theta(x, y) = \text{atan}\left(\frac{y}{x}\right) \tag{2}$$

The relative distance d_{ij}, the direction field difference $\Delta\theta_{ij}$ and the type difference Δt_{ij} between the minutiae point m_i and the minutiae point $m_j(m_i \in \{M - m_i\})$ are respectively obtained according to the following formulas:

$$d_{ij} = \sqrt{(x_i - x_j)^2 + (y_i - y_j)^2} \tag{3}$$

$$\Delta\theta_{ij} = |\theta_i - \theta_j| \tag{4}$$

$$\Delta t_{ij} = |t_i - t_j| \tag{5}$$

Then calculate d_{ij}, $\Delta\theta_{ij}$ and Δt_{ij}, $(d_{ij}, \Delta\theta_{ij}, \Delta t_{ij})$ defined as a new feature. There will be *s* new feature vectors, ***s = n*(n-1)/2***. Then we can obtain a new feature template set by transforming the fingerprint minutiae point features, denoted as $F = \{(d_i, \Delta\theta_i, \Delta t_i), i = 1, 2, \ldots, s\}$ [4].

Due to the fuzziness of fingerprint features, they are always within a certain error range. Therefore, before generating the key, we first define a 3d array composed of many small cuboids. Figure 4. is a schematic diagram of the 3d array.

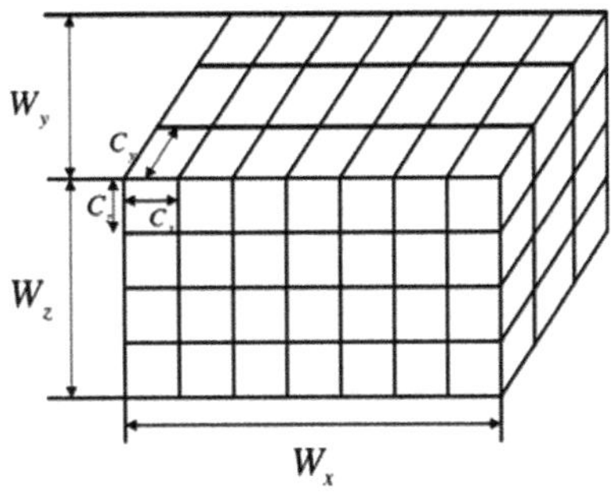

Fig. 4. The 3d array.

Suppose that W_x, W_y and W_z are the length, width and height of the 3d array respectively. In this paper, the height in the 3d array corresponds to the type difference of the feature. For small cuboids, C_x, C_y, and C_z are the length, width, and height of the small cuboid, respectively. So the total number of small cuboids for $a * b * c$, in which $a = \lfloor W_x/C_x \rfloor, b = [W_y/C_y], c = 2$, "$\lfloor\rfloor$" is a return that less than the largest integer of the expression in brackets [4].

In order to ensure that the number of elements in the feature set F is the same in the key generation stage and the key recovery stage, we first define a number of elements s', and sort the elements of the feature set F according to the relative distance d, then take the first 500 points, process the difference in the direction field of each point, and approximate it to normalization by the square root method. Then, the first s' elements in the feature set F, denoted by F':

$$F' = \{(d_i, \Delta\theta_i, \Delta t_i), i = 1, 2, \ldots, s'\} \quad (6)$$

Among them, the i-th and j-th elements in the feature set F' satisfy: when i $\geq$ j, $d_i \geq d_j$ $(i, j = 1, 2, \ldots, s')$. The elements in the feature set F' are divided into two layers according to the differences in their feature types, namely layer 0 and layer 1. Then place each element in its corresponding position on both levels.

If the number of elements in the small rectangle exceeds 1, $b_i = 1$; otherwise, $b_i = 0$. The formula is as follows [4]:

$$b_i = \begin{cases} 1 & ifsum(i) \geq 1 \\ 0 & otherwise \end{cases} \quad i = 1, 2, \ldots, T \quad (7)$$

Where sum(i) is the number of the i-th small rectangle element in the 3d array, and T is the total number of small rectangles in the 3d array. Then generate a fingerprint feature key k with a length T in the order of the small rectangle, denoted as $k = (b_1 b_2 \ldots b_T)$ which will be used as the key.

2.3 Finger Vein Image Preprocessing

Finger vein images contain background information and finger area information, as well as individual difference information. Before extracting the texture of finger veins, it is necessary to preprocess the original images. Image preprocessing generally includes three steps: positioning the finger area, ROI extraction and normalization. In order to obtain a clearer finger vein image, image segmentation is required. The basic idea is: according to the linear features of vein lines, use the direction valley detection to obtain a binary image containing vein lines. The processed image will still have noise and holes, so it is necessary to perform subsequent processing on the image to obtain a clear finger vein skeleton [5].

2.4 Encrypt Finger Vein Image

This paper proposes a finger vein image encryption method based on fingerprint feature key and SM4 algorithm. This method uses the key generated by the key generation algorithm based on the relative difference of fingerprint minutiae points as the initial key of SM4 algorithm, which not only improves the randomness of key generation but also facilitates the generation management.

2.5 Decryption and Authentication

Decryption. In this paper, a hybrid encryption algorithm based on fingerprint feature key and SM4 algorithm was selected to complete image encryption and decryption.

The SM4 algorithm is a dual operation, so the decryption algorithm has the same structure as the encryption algorithm, except that the order of the round keys is reversed. The decryption round key is the reverse order of the encryption round key. Therefore, when the user's fingerprint is obtained again, use the reverse sequence of the key to decrypt the finger vein images stored in the database [6].

The effect of image encryption and decryption is shown in Fig. 5.

Fig. 5. Encryption and decryption.

Finger Vein Image Feature Extraction and Authentication.
(1) Feature extractionThe test picture has a total of 30 feature points. The first column is the abscissa of the feature points, and the second column is the ordinate of the feature points.

(2) MatchingWe design a pixel-based matching algorithm, and then match the feature points of two finger vein feature images. Through feature extraction, the two-dimensional feature matrix of two images can be obtained.

The pixel points P1(i, j) and P2(i, j) are the feature points of the two images. If these two points satisfy the formula below, then the two pixels are considered to be the same feature point. When the pixel coincidence rate of the two feature matrices is bigger than a certain threshold, the two images are considered to be successfully matched [7].

$$\begin{cases} |P_{1x} - P_{2x}| \leq 2 \\ |P_{1y} - P_{2y}| \leq 2 \end{cases} \tag{8}$$

The schematic diagram of pixel coincidence is shown in Fig. 6.

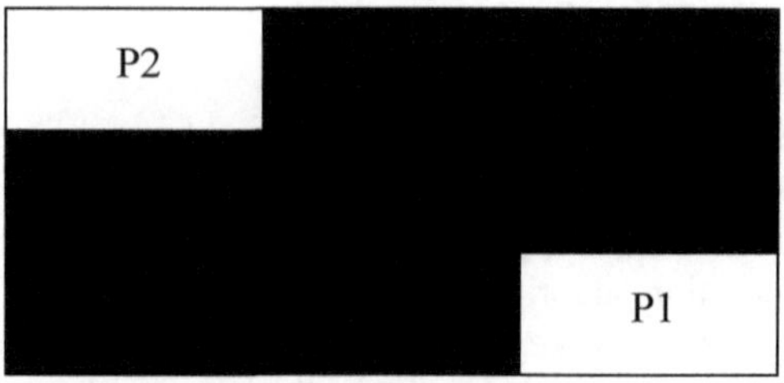

Fig. 6. Schematic diagram of pixels.

The detail points of finger veins are less informative than feature images, so the feature points of the test pictures we choose are generally around 20-30. We need to count the pixel coincidence rate between the template image and the image to be matched. If this coincidence rate is greater than 25%, and the global feature difference between the two images does not exceed 0.0055, we have reason to believe that the two images originate from the same user, Otherwise the match fails [8].

The most important thing in the matching algorithm is to set the threshold. In the research process, according to a large number of test results, we need to constantly modify the threshold of the matching algorithm.

3 Result Analysis

A total of 12 test pictures were derived from 4 different fingers, and 3 pictures were collected from each finger. A total of 12 * 12 = 144 matching tests have been done Table 1.

The algorithm of statistical pixel coincidence number will also affect the statistical coincidence number. Initially, the corresponding coincidence pixels in the characteristic matrix of finger veins to be matched were found based on the characteristics of

Table 1. Matching result record table.

	1–1	1–2	1–3	2–1	2–2	2–3	3–1	3–2	3–3	4–1	4–2	4–3
1–1	√	√	√	×	×	×	×	×	×	×	×	×
1–2	√	√	√	×	×	×	×	×	×	×	×	×
1–3	√	√	√	×	×	×	×	×	×	×	×	×
2–1	×	×	×	√	√	√	×	×	×	×	×	×
2–2	×	×	×	√	√	√	×	×	×	×	×	×
2–3	×	×	×	√	√	√	×	×	×	×	×	×
3–1	×	×	×	×	×	×	√	√	√	×	×	×
3–2	×	×	×	×	×	×	√	√	×	×	×	×
3–3	×	×	×	×	×	×	√	×	√	×	×	×
4–1	×	×	×	×	×	×	×	×	×	√	√	√
4–2	×	×	×	×	×	×	×	×	×	√	√	√
4–3	×	×	×	×	×	×	×	×	×	√	√	√

finger veins stored in the database. We find out that the matching results of A and B images may be not the same as the matching results of B and A images. Therefore, we improve the algorithm of counting the number of overlapping pixels, and obtain the overlapping intersection of feature points in the matching image, so as to eliminate the possibility of obtaining multiple overlapping points for the same feature point.

Then we use false acceptance rate (FAR) and false rejection rate (FRR) commonly used in biometrics to evaluate the performance index of the digital vein recognition algorithm. The calculation formulas of false acceptance rate (FAR) and false rejection rate (FRR) are as follows:

$$\mathrm{FAR} = \frac{NFA}{NIRA} \times 100\% \tag{9}$$

$$\mathrm{FRR} = \frac{NFR}{NGRA} \times 100\% \tag{10}$$

NFA is the number of false acceptances, NIRA is the total number of inter-class tests, and FAR is the proportion of finger veins that shouldn't be matched.

NFR is the number of false rejections, NGRA is the total number of in-class tests, and FRR is the proportion of finger veins that should be matched successfully as mismatched.

Before improving the algorithm, the FAR test result was 1.8520%. The FRR test result was 2.7787%. After improving the algorithm, the FAR test result is 0.0001%. The FRR test result was 5.5556%. The FAR decreases while FRR increases. However, for the entire system, the FAR is more important. If the false acceptance rate of the recognition system is high, it will give the attacker a chance to forge the characteristics of the finger vein, thus providing the opportunity to obtain the important information of the system. Therefore, although the improved algorithm has a higher false rejection rate, it's better to secure the identification system.

4 Conclusion

This paper proposes a hybrid encryption solution based on fingerprint minutiae and SM4 algorithm, which implements an identity authentication system with good security performance and high encryption level. Firstly, fingerprint images are obtained and processed, and generate a fingerprint feature key. Secondly, obtain and extract the finger vein image feature information. Then, encrypt the image with SM4 algorithm, of which the fingerprint feature key is used as the key. When authenticated, a user' fingerprint and finger vein will be collected again and the fingerprint feature key will be generated to decrypt the encrypted image in the database. The specific matching algorithm is used to match the finger vein image for authentication. The security test results show that the key space of this system is large, and the corresponding encryption and decryption algorithms have strong key sensitivity and good statistical characteristics, which can significantly solve the data security issues.

Acknowledgment. This research was financially supported by the Key Research and Development Plan (2018YFB1004101), Key Lab of Information Network Security, Ministry of Public Security(C19614), Special fund on education and teaching reform of Besti (jy201805), the Fundamental Research Funds for the Central Universities (328201910), China Postdoctoral Science Foundation (2019M650606), 2019 Beijing Common Construction Project-Teaching Reform and Innovation Project for Universities in Beijing, key laboratory of network assessment technology of Institute of Information Engineering, Chinese Academy of Sciences. The authors gratefully acknowledge the anonymous reviewers for their valuable comments.

Author Contributions. Zhiqiang Wang and Xin Lv conceived and designed the framework and the algorithm; Zeyang Hou and Zhiwei Wang performed the experiments; Xinyu Li analyzed the data; Bingyan Wei and Tao Yang wrote the paper.

Conflicts of Interest. The authors declare no conflict of interest.

References

1. Li, Y.: Research on Fingerprint encryption Algorithm and its Application. Xidian University (2017)
2. Liu, W.: A new fingerprint-based key protection scheme. China Informatization **08**, 68–69 (2017)
3. Wan, T.: Fusion fingerprint and finger vein image acquisition system. Hangzhou University of Electronic Science and Technology (2017)
4. Zhan, M.: Research on key generation algorithm based on fingerprint characteristics. Hangzhou University of Electronic Science and Technology (2017)
5. Sun, X.: Study on extraction Algorithm of Digital vein image Pattern. Jilin University (2012)
6. Meng, G.: Research on the algorithm of vein encryption and classification recognition. Beijing University of Posts and Telecommunications (2018)
7. Chen, H.: Overview of the principle of finger vein recognition system. Electr. Technol. Softw. Eng. **22**, 81–89 (2016)
8. Sapkale, M., Rajbhoj S.M.: A finger vein recognition system. In: 2016 Conference on Advances in Signal Processing (CASP), NJ, USA, pp. 306–310. IEEE (2016)

Analysis and Design of Image Encryption Algorithms Based on Interlaced Chaos

Kangman Li[1,2](✉) and Qiuping Li[1,2]

[1] College of Computer Science and Technology, Hengyang Normal University, Hengyang 421002, China
lkm@hynu.edu.cn

[2] Hunan Provincial Key Laboratory of Intelligent Information Processing and Application, Hengyang 421002, China

Abstract. For chaotic encryption algorithm, the algorithm of generating chaotic sequence is complex, and the data is float, which directly affects the speed of encryption. In this paper, an interlaced chaotic encryption algorithm is proposed, In this algorithm, the pixel interval is encrypted by chaotic sequence, and the chaotic sequences of other pixels are obtained by XOR the chaotic values of the front and back points, Experiments show that the algorithm can encrypt and decrypt images quickly and has good effect.

Keywords: Interlaced chaotic sequence · XOR · Encryption speed · Encryption effect

1 Introduction

The rapid development of network technology has promoted the application of multimedia in the network. Image and video have gradually become an important way for people to obtain network information [1]. When the image data is transmitted in the network, there is a big hidden danger of information security, which may be attacked illegally, resulting in the destruction or theft of the key data [2]. Therefore, it is urgent to encrypt and protect the image data. Digital image has the characteristics of large amount of information and high redundancy, Using traditional encryption technology to encrypt digital image, which has a long encryption time, low security and poor effect, it has great limitations and is not suitable for digital image encryption [3, 4].

Scholars have put forward many new image encryption algorithms. Chaos theory is widely used in the field of digital image encryption because of its pseudo-random, extremely sensitive to initial value and control parameters, nonlinear and other characteristics [4, 5]. But the encryption time of chaos algorithm is too long Because of the large amount of computation. In this paper, an interlaced chaos algorithm is proposed, which can reduce the computation and improve the encryption speed on the basis of ensuring the basic performance of encryption.

Q. Liu et al. (Eds.): CENet 2020, AISC 1274, pp. 789–795, 2021.
https://doi.org/10.1007/978-981-15-8462-6_90

2 Interlaced Chaotic Encryption Algorithm

2.1 The Principle of Logistic Chaotic Mapping Algorithm

Logistic map is a simple one-dimensional discrete chaotic system. The equation is:

$$x_{n+1} = \mu x_n(1 - x_n) \tag{1}$$

When 3.569945627 < $\mu \leq 4$, logistic map shows chaotic characteristics, and the value of μ is closer to 4, the chaotic characteristics of system are better [6, 7].

The encryption process is as follows:

(1) Import the image file to get the width and height of the image (w × h);
(2) Input the initial value of the sequence (x0), generate the sequence x with the length of n = w × H by the formula (1), and convert the sequence to an integer of 0–255.
(3) The encrypted values are got by XOR the pixel value and the corresponding sequence Xi.

The decryption process is the same as the encryption process.

2.2 Interlaced Chaotic Encryption Algorithm

The time of chaos algorithm is mainly spent on the generation of chaos sequence. Interlaced chaos algorithm uses the method of generating chaos sequence points by every other pixel, which reduce half of the generation calculation of chaos points, so as to improve the encryption speed. The encryption process is as follows:

(1) First, import the image and calculate the pixel number n = w × H.
2) Input the initial value x0 and coefficient μ of formula (1), generate one-dimensional chaotic sequence x with length of N/2, and use formula (2) to generate the key sequence:

$$SX_i = (x_i \times 1000) \bmod 256 \tag{2}$$

3) XOR the every other data sequence point of the original image and the chaotic key sequence, as shown in Fig. 1, and the encryption formula is as follows:

$$P_i' = P_i \oplus SX_n \tag{3}$$

The key sequence values of other pixels X_m are obtained from the XOR the key sequence values SX_{m-1} and SX_{m+1} of the front and back points. The encryption formula for those points is as follows:

$$P_j' = P_j \oplus (SX_{m-1} \oplus SX_{m-1}) \tag{4}$$

In the aspect of program implementation, the following methods can be adopted: The rows and columns number of pixels are divided by 2 to obtain modulus

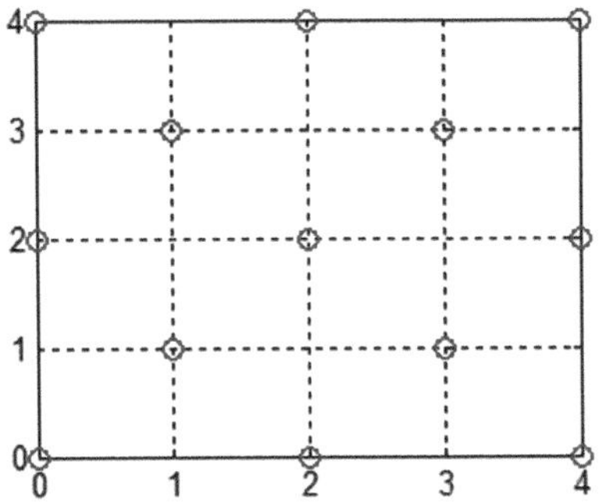

Fig. 1. Interval encryption point.

respectively, If the results are equal, the image is encrypted by formula (3), otherwise, it is encrypted by formula (4).

It can be seen from the algorithm that in fact, every pixel has be encrypted by XOR its corresponding sequence key, the difference from the logistic chaotic mapping algorithm is that half of the sequence key are not generated directly by formulas (1) and (2), but obtained by XOR the sequence key of the two points before and after. In the computer, the XOR operation of integer is obviously faster than floating-point operation. Therefore, the encryption speed of the algorithm has been significantly improved.

The decryption process is the same as the encryption process.

3 Experiment Simulation and Analysis

The experiment is realized by MATLAB, the image encryption speed is compared with different size images, and take 256 × 256 gray Lena image as the experimental object to analyze the performance of interlaced chaotic encryption.

3.1 Encryption Speed

The two methods encrypt several images with different size respectively, and the encrypted time is shown in Table 1.

Table 1. Encryption time of two methods in different resolution.

Size	Logistic chaotic map	Interlaced chaotic
128 × 128	0.1823	0.1534
256 × 256	1.8763	0.7169
512 × 512	135.59	30.36
1024 × 1024	2803.7	581.11

The data in Table 1 show that the improved interlaced chaotic encryption speed is significantly faster, and the larger the image is, the more obvious the difference will be.

3.2 Encryption Effect and Key Sensitivity

Set the initial key value (i.e., x0) to 0.3125, encrypt the Lena image (Fig. 2 (a)) to obtain the encrypted image (Fig. 2 (b)). It can be seen that the encrypted image is disordered, and the traces of the original image can no longer be seen, achieved the encryption effect. And decrypt with the correct key 0.3125 to get the correct decrypted image (Fig. 2 (c)). If we slightly change the initial value of the key to 0.3126, the decrypted image is wrong (Fig. 2 (d)), it can't get any information about the original image. It shows that the algorithm is sensitive to the key. This algorithm is a simplification of chaos algorithm, so its key space is the parameter and initial value space of chaos system, which is large, the difficulty of key attack is great, and the security of the algorithm is high.

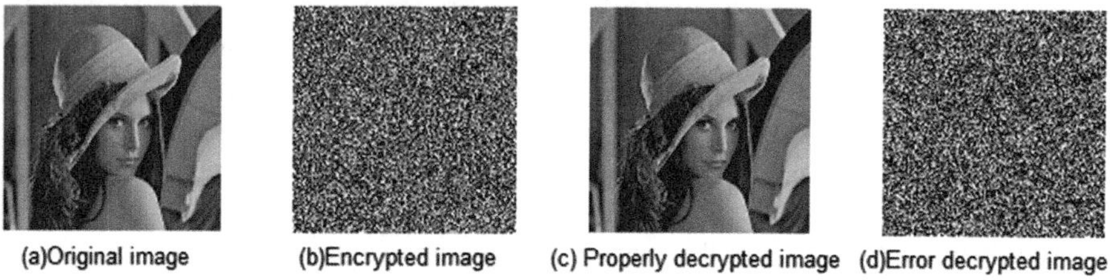

Fig. 2. Results of image encryption and decryption.

3.3 Image Similarity

Image similarity calculation is to score the similarity degree of two images, and judge the similarity degree of image content according to the score. The calculation formula is as follows (5) [8]:

$$\mathrm{XSD}(\mathrm{G}, \mathrm{C}, \alpha, \beta) = 1 - \frac{\sum_{i=1}^{M} \sum_{j=1}^{N} \left[c_{i+\alpha j+\beta} - g_{ij}\right]^2}{\sum_{i=1}^{M} \sum_{j=1}^{N} g_{ij}^2} \tag{5}$$

In the formula, G (G_{ij}) and C (C_{ij}) are two $M \times N$ images, α and β are integers and $0 \leq \alpha < M-1$, $0 \leq \beta < n-1$.

The value range of XSD is [0, 1]. When XSD = 1, it means the two images are the same. In the encryption algorithm, the smaller the similarity between the original image and the encrypted image is, the more successful the encryption is, and the higher the security is [8, 9]. In the experiment, the similarity between the original image and the encrypted image is 0.01219, and the similarity between the encrypted image and the successfully decrypted image is 1, it shown that the encryption and decryption of the algorithm is successful.

3.4 Average Change Value of Gray Scale

The average change of gray level is used to measure the change of gray level in encrypted image. The calculation formula is as follows:

$$\mathrm{GAVE(G,C)} = \frac{\sum_{i=1}^{M}\sum_{j=1}^{N}\left|g_{ij} - c_{ij}\right|}{MN} \tag{6}$$

When the gray level of two images changes evenly (gave is half of the gray level L), the scrambling security of the image is the highest. In this algorithm, the gray level of the image is 256, and the GAVE is 127.33 (about equal to 128), which is close to half of the gray level, and the effect is very good.

3.5 Image Information Entropy

Image information entropy is a kind of statistical form of features, which reflects the average amount of information in the image. The one-dimensional entropy of image represents the amount of information contained in the clustering feature of gray-scale distribution. If P_i is the proportion of pixels whose gray value is i, then the one-dimensional gray-scale entropy of gray image is defined as:

$$\mathrm{H} = \sum_{0}^{255} P_i \log_2 P_i \tag{7}$$

It shown that the larger the uncertainty of the proportion of pixels gray value in a image, the higher the information entropy. When all gray-scale values have the same probability, the information entropy is the maximum. The maximum value of 8-bit gray-scale image should be $H_{max} = \log_2 256 = 8$. The image information entropy obtained in the experiment is 7.9951, which is close to the maximum information entropy.

3.6 Histogram Comparison

Histogram is an image representation of information entropy, which can more intuitively reflect the distribution of gray value of image. The histogram of the original image and the encrypted image is shown in Fig. 3. It can be seen that the histogram of the encrypted image is very uniform, which is very different from the original image, the possibility of decoding from histogram analysis is reduced.

3.7 Anti-interference Analysis

The anti-interference includes anti-cutting and anti-noise analysis.

During the process of image transmission, data may be lost for various reasons. Cut (lose) the encrypted image to 24 × 24 and 128 × 128 respectively, as shown in Figs. 4 (a) and (c), then decrypt the image to get the image as shown in Figs. 4 (b) and (d) respectively. It can be seen that except for the lost part, other pixels can be decrypted normally without being affected by the lost part, which shows that the algorithm has strong anti-cutting performance.

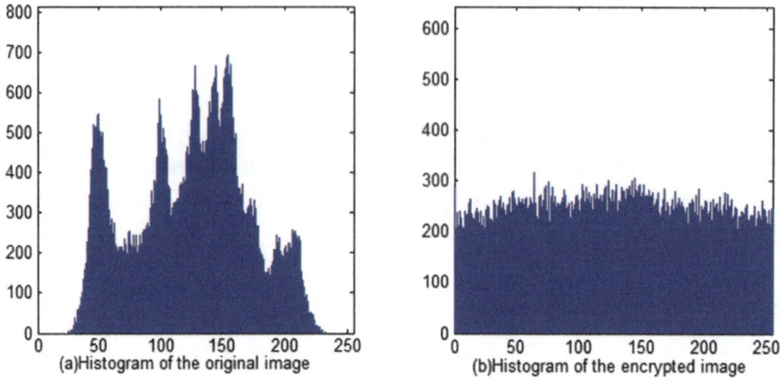

Fig. 3. Histogram of original image and encrypted image.

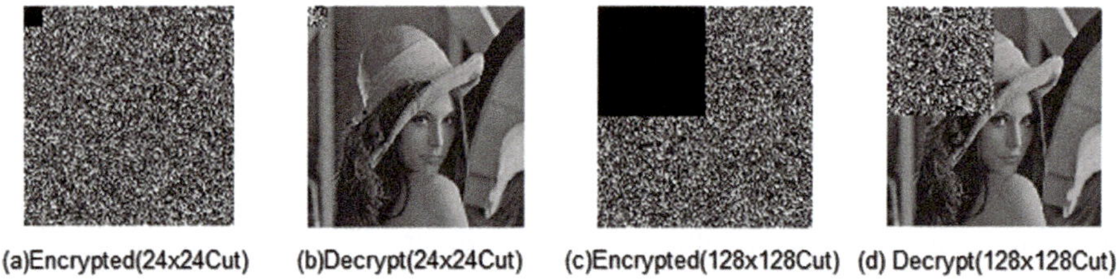

Fig. 4. Anti-cutting effect.

The image will also be interfered by noise in the actual transmission. Next, salt and pepper noise, Gaussian noise and random noise will be added to the encrypted image respectively, and then the image will be decrypted respectively, as shown in Fig. 5.

Fig. 5. Anti-noise effect.

From the perspective of decrypted image, in addition to the distortion caused by noise, the image can be decrypted normally, which shows that the algorithm has good anti-noise performance.

4 Conclusion

With the development of computer technology, image information is more and more widely used, and it is more and more important to ensure the security of image information. In this paper, chaos encryption algorithm is analyzed, and a simplified interval chaos encryption method is proposed to improve the encryption speed on the basis of ensuring the encryption performance. Experiments show that the algorithm has a significant improvement in speed compared with the traditional chaotic encryption algorithm, and all the performance can meet the encryption requirements.

This paper focuses on how to simplify the process and reduce the amount of calculation. Because the algorithm is based on the traditional chaos algorithm, so in the practical application, we can improve the generation algorithm of the password sequence proposed by other literature, and further strengthen the security of encryption.

Acknowledgements. This work was supported by Application-oriented Special Disciplines, Double First-Class University Project of Hunan Province (Xiangjiaotong [2018] 469), The Science and Technology Plan Project of Hunan Province No. 2016TP1020, Science Foundation Project of Hengyang Normal University No. 18D23.

References

1. Zhao, Y.M.: Research on new algorithm of pseudo-random sequence and video security protection technology. Harbin Industrial University (2017)
2. Xue, J., Liu, P.: Multi image optical synchronous encryption algorithm based on chaotic gyrator transform and discrete wavelet transform. J. Electron. Measure. Instrum. **33**(11), 136–146 (2019)
3. Huang, Y.J., Du, Y.X., Shi, W.: Image encryption algorithm based on a novel combinatorial chaotic mapping. Microelectron. Comput. **36**(05), 47–52 (2019)
4. Xu, B., Yuan, L.: Research on image encryption algorithm logistic chaotic based on an improved digital mapping. Comput. Measure. Control **22**(07), 2157–2159 (2014)
5. Zhu, H.G., Lu, X.J., Zhang, X.D., Tang, Q.S.: A novel image encryption scheme with 2D-logistic map and quadratic residue. J. Northeast. Univ. (Nat. Sci.) **35**(01), 20–23 (2014)
6. Liu, J.Y.: Research on digital image encryption algorithm based on chaos system. Hubei Minzu University (2019)
7. Cheng, L.L.: Research on digital image encryption algorithm based on chaos theory. Anhui University (2018)
8. Ma, J.: Image encryption algorithm improvement and performance analysis. Shandong University (2010)
9. Zhang, J.T.: Medical image encryption algorithm based on memristive chaotic system and compressive sensing. Henan University (2019)
10. Huang, S.: A color image encryption algorithm based on improved logistic chaotic map. J. Henan Inst. Eng. (Nat. Sci. Ed.) **27**(02), 63–67 (2015)

Verifiable Multi-use Multi-secret Sharing Scheme on Monotone Span Program

Ningning Wang and Yun Song(✉)

School of Computer Science, Shaanxi Normal University, Xi'an 710062, China
songyun09@snnu.edu.cn

Abstract. Based on the monotone span program, this paper proposes a verifiable multi-secret sharing scheme on the general access structure. Each participant can choose their own secret share and use the RSA cryptosystem to send the share to the dealer through open channels. In the secret recovery phase, the scheme uses the Hash Function to achieve multi-use property and verifiability. Analysis shows that each participant can recover multiple secrets by keeping only one secret share, and shares do not need to be changed when the multi-secrets are renewed. Furthermore, the distribution of secret shares in the scheme does not require a secure channel, which effectively reduces the computational cost of the system. Compared with the (t, n) threshold secret sharing, the scheme has more vivid access structures.

Keywords: Multi-secret sharing · Monotone span program · Hash function · RSA cryptosystem · Verifiability

1 Introduction

Secret sharing [1] is a method to share secrets among a group of participants. It is an important research topic in current cryptography and information security. It is mainly used in many fields such as data security, missile launch and e-commerce. In 1979, Blakley and Shamir proposed a (t, n) threshold secret sharing scheme based on projective geometry theory and Lagrange polynomial interpolation formula [2]. In 1989, a secret sharing scheme based on general access structure is proposed [3]. After many secret sharing schemes for general access structures are put forward [4, 5]. The basic idea is to divide secrets into several shares which are distributed among a group of participants, where a subset of authorized participants can reconstruct the secret, while a subset of non-authorized participants cannot get any information about the secret. But the above schemes are all single secret sharing schemes.

In practical applications, when multiple secrets need to be shared, participants need to save the corresponding shares of each secret, which will reduce the efficiency of the entire scheme. To solve this problem, several multi-secret sharing schemes have been proposed [6–13]. Multi-secret sharing means that each participant only needs to keep one secret share to share multiple secrets. A dynamic multi-use multi-secret threshold sharing scheme based on Lagrange polynomial interpolation was proposed [7]. Shyamalendu et al. [8] proposed a verifiable multi-secret threshold sharing scheme using one-way hash function. Meanwhile, many multi-secret sharing schemes have

Q. Liu et al. (Eds.): CENet 2020, AISC 1274, pp. 796–802, 2021.
https://doi.org/10.1007/978-981-15-8462-6_91

been studied by authors [9–11], but these schemes all need to use a secure channel, and the share of each participant is generated by the dealer. In addition, most of the existing multi-secret sharing schemes are threshold. Due to the particularity of the threshold access structure, each participant has a completely equal status. At the same time, it reduces the flexibility and universality of the scheme, making it difficult to meet the actual operational requirements. Therefore, it is of great significance to study the multi-secret sharing scheme [12, 13] on the general access structure with wider application.

Based on this, this paper proposes a multi-secret sharing scheme based on the general access structure of the monotone span program, which implements verifiability and versatility. In this scheme, participants' secret shares are selected by themselves and each participant has only one secret share. Participants can recover multiple secrets through their own secret shares, and the secure channel is not needed.

2 Preliminaries

2.1 Access Structure

Let the set of participants in the secret sharing scheme be $P = \{P_1, P_2, \cdots, P_n\}$. Γ is an access structure on P which is monotonically increasing

$$(A \in \Gamma,\ A \subseteq C \subseteq P) \Rightarrow C \in \Gamma$$

The participant subset $A \in \Gamma$ is called a minimum authorized subset of P if it satisfies $A \in \Gamma$, $B \subseteq A \Rightarrow B \neq A$, and the set of all minimum authorized subsets in Γ is denoted by $\Gamma_{\min}$. Correspondingly, the adversary structure is monotonically decreasing and the adversary structure is written as $\Delta_{\max}$.

2.2 Monotone Span Program

MSP is a computation model, introduced by Karchmer and Wigderson [14] in 1993.

Definition 1: For a monotone span program (MSP) of a computable Boolean function, it can be written as: $\mathrm{M}(\kappa, \mathrm{M}, \Psi)$,where κ is a finite field, M is a $d \times l$-order matrix on κ, $P = \{P_1, \cdots, P_n\}$ is a set of participants, and $\Psi : \{\text{matrix row label}\} \rightarrow \{P_1, \cdots, P_n\}$ is a surjective, that is, each line in M corresponds to the participant in P.

Definition 2: $\mathrm{M}(\kappa, \mathrm{M}, \Psi)$ is an MSP for the access structure Γ with respect to the target vector $\mathbf{v} \in \kappa$ if and only if the following holds.

$$A \in \Gamma \text{ if and only if } \mathbf{v} \in span\{\mathrm{M}_A\}$$

The target vector $\mathbf{v} \in span\{\mathrm{M}_A\}$ refers to the existence $\mathbf{w}_A$ vector such that $\mathbf{v} = \mathbf{w}_A \cdot \mathbf{M}_A$, where $\mathbf{M}_A$ represents a matrix formed by the rows i of M with $\Psi(i) \in A$. Thus, $\mathbf{v} \in \bigcap_{A \in \Gamma_{\min}} \sum_{i \in A} V_i - \bigcup_{B \in \Delta_{\max}} \sum_{i \in B} V_i$, where V_i is the linear space spanned by the row vector of M, for $1 \leq i \leq n$.

2.3 Linear Multi-secret Sharing Scheme

Based on the above theory, the following shows how to use MSP to implement the LMSS scheme on access structure Γ [14].

Distribution stage: The dealer owns the secret S_O. The dealer constructs the vector **r** so that $S_0 = (\mathbf{v}, \mathbf{r})$, and the dealer distributes the value of $M_i \cdot \mathbf{r}^T$ to P_i.

Reconstruction stage: For each authorized set $A \in \Gamma$, there exists a vector $\mathbf{w}_A$ such that $\mathbf{v} = \mathbf{w}_A \cdot \mathbf{M}_A$, and then $S_0 = (\mathbf{v} \cdot \mathbf{r}) = \mathbf{w}_A \cdot M_A \cdot \mathbf{r}^T = \mathbf{w}_A \cdot (M_A \cdot \mathbf{r}^T)$. Thus, the secret S_O can be calculated linearly by the share of the participants in A.

2.4 Hash Function

Hash function is to transform an arbitrary variable-length input into a fixed-length output after a single-term hash function operation. This output is called a hash value.

The hash function needs to satisfy the following properties:

(1) Given a message M, it is easy to calculate the hash value h.
(2) Given a hash value h, it is computationally infeasible to obtain the message M according to H(M) = h.
(3) Given a message M, it is computationally infeasible to find another message M' such that H(M) = H(M′).
(4) Anti-collision, it is computationally infeasible to find two messages M and M', such that H(M) = H(M′).

The hash function is completely public, and its security depends on its one-way nature.

3 Construction of Our Scheme

3.1 Initialization Phase

Let $\Gamma = \{P_1, \cdots, P_n\}$ denote access structure and $S = \{S_1, \cdots, S_m\}$ denote the secret set to be shared, DC is an honest secret generator. This solution requires a bulletin board. Only the dealer can upload and update the information on the bulletin board and other participants can only download. The Initialization phase is divided into the following steps:

(1) The dealer randomly selects two safe large prime sums p' and q' such that $p = 2p' + 1$ and $q = 2q' + 1$. Dealer computes $N = pq$ and selects key pair (e,d), such that $ed = 1 \bmod (\phi(N))$. Then, the dealer upload (N,e) to the bulletin board and keep d confidentially.
(2) For $1 \le i \le n$, each participant P_i randomly selects an integer sh_i from $[2,N]$, uses (N,e) on the bulletin board to calculate $R_i = sh_i^e \bmod N$, and sends R_i to the dealer through an open channel, while keeping sh_i confidentially.
(3) After the dealer receiving the participant's information $R_{i,}$ ensures $R_i \neq R_k$; otherwise, the dealer informs the corresponding participant to re-select until $R_i \neq R_k (1 \le i \neq k \le n)$. Then, the dealer calculates $sh_i = R_i^d \bmod N (1 \le i \le n)$.

(4) The dealer constructs a monotone span program $\mathrm{M}(\kappa, \mathrm{M}, \Psi)$ that can calculate a $f(\Gamma)$ relative to the target vector $\mathbf{v}$ and $\Psi(i) = P_i$ for $1 \leq i \leq n$.
(5) The dealer chooses a safe Hash function, record it as $\mathrm{H}(\cdot)$, and upload it, $||$ represents the bit string link.

3.2 Distribution Stage

The distribution stage is divided into the following steps:

(1) Let S be the key space based on monotone span program $\mathrm{M}(\kappa, \mathrm{M}, \Psi)$. The dealer randomly selects a vector $\mathbf{r}$, such that $S_0 = (\mathbf{v}, \mathbf{r})$, where S_0 is chosen in S. Then the dealer computes $\mathrm{M}_i \cdot \mathbf{r}^T$.
(2) The dealer randomly selects 2n integers $ID_1, \ldots, ID_n$, $t_1, \ldots, t_n$, where ID_i is the unique identity of each participant. Then, the dealer uploads ID_i and t_i to the bulletin board, for $1 \leq i \leq n$.
(3) For $1 \leq i \leq n$, the dealer computes $I_i = \mathrm{H}(ID_i||sh_i||t_i)$, $L_i = \mathrm{H}(ID_i||I_i||t_i)$, $B_i = \mathrm{M}_i \cdot \mathbf{r}^T \oplus I_i$ and uploads L_i and B_i to the bulletin board.
(4) The dealer computes the information $S'_j = S_0 \oplus S_j (1 \leq j \leq m)$ and uploads S'_j to the bulletin board.
(5) The dealer randomly selects an integer t_0, computes $C_j = \mathrm{H}(t_0||S_j)$, and uploads t_0 and C_j to the bulletin board.

3.3 Reconstruction Stage

Assume that participants in $A = \{P_{i_1}, P_{i_2}, \ldots, P_{i_c}\} \in \Gamma$ want to reconstruct secrets. The reconstruction stage is divided into the following steps:

(1) Each participant P_{i_v} $(1 \leq v \leq c)$ computes $I_{i_v} = \mathrm{H}(ID_{i_v}||sh_{i_v}||t_{i_v})$ by information on the bulletin board and sends I_{i_v} to the DC. Then DC computes $L^*_{i_v} = \mathrm{H}(ID_{i_v}||I_{i_v}||t_{i_v})$ and compares it with L_{i_v} on the bulletin board. If $L^*_{i_v} = L_{i_v}$, it means that the share provided by the participant is correct.
(2) DC gets $\mathrm{M}_{i_v} \cdot \mathbf{r}^T$ by computing $B_{i_v} \oplus I_{i_v} = M_{i_v} r^{\mathrm{T}} \oplus I_{i_v} \oplus I_{i_v}$ and recovers S_0 according to $S_0 = (\mathbf{v} \cdot \mathbf{r}) = (\mathbf{w} \cdot \mathrm{M}_A) \cdot \mathbf{r}^T = \sum_{v=1}^{c} \mathbf{w} \cdot (\mathrm{M}_{i_v} \cdot \mathbf{r}^T)$. Finally, DC downloads S'_j on the bulletin board and computes the shared secrets $S_j = S_0 \oplus S'_j (1 \leq j \leq m)$.

4 Scheme Analysis

4.1 Security Analysis

The security of the scheme can be analyzed by the following three attacks:

Attack 1: Based on the nature of the Hash function, except for the dealer and participants, no adversary can calculate the real share sh_i $(1 \leq i \leq n)$ of participants

through public information ID_i and L_i $(1 \leq i \leq n)$. Therefore, the public information L_i will not reveal the real shares of the participants in the secret distribution stage.
Attack 2: Like Attack 1, based on the nature of the Hash function, no adversary can calculate the real share sh_i of the participant through the public information ID_i and I_i $(1 \leq i \leq n)$. Therefore, when participant sends his share to DC in the secret reconstruction phase, the real share of the participant will not be leaked.
Attack 3: A secret sharing project must meet two basic requirements:

(1) Participants in the authorized subset can easily calculate the shared secret through cooperation;
(2) Participants in unauthorized subset cannot get any information about the secret even through cooperation.

It can be seen from the scheme in this article that if participants in the authorized subset want to recover the secret correctly, each of them needs to take out their own real shares; if the participant or adversary in the unauthorized subset wants to recover the secret, they need to find shares of participants in the authorized subset, according to attack 1 and attack 2. This situation is impossible.

4.2 Verifiability Analysis

The verifiability of this program is reflected in three aspects:

(1) When the DC receives the participant P_i's share I_i, the DC computes $L_i^* = \mathrm{H}(ID_i||I_i||t_i)$ and compares it with L_i on the bulletin board. If $L_i^* = L_i(1 \leq i \leq n)$, it means that the share provided by the participant is correct.
(2) After each participant computes I_i, each of n participants continues to use public information to compute $L_i' = \mathrm{H}(ID_i||I_i||t_i)$ and compares it with L_i on the bulletin board. If $L_i' = L_i(1 \leq i \leq n)$, it means that the values L_i given by the dealer is correct.
(3) After the secret recovery phase, the participant calculates $C_j^* = \mathrm{H}(t_0||S_j)$ and compares it with C_j on the bulletin board. If $C_i^* = C_j(1 \leq j \leq m)$, it means that the DC did not cheat.

4.3 Multi-use Analysis

In the secret reconstruction phase, we see that the participants sent to the DC their own pseudo-shares. According to the security of the hash function, the participants' real shares are not known to the DC and other participants. So, in the next reconstruction phases, instead of revealing their real shares, participants only need to regenerate their own pseudo-shares, which reflecting the property of multi-use.

4.4 Dynamic Analysis

When the shared secret set $S = \{S_1, \ldots, S_m\}$ needs to be updated, participants' shares do not need to be changed. The dealer only needs to update the relevant public information S_i' on the bulletin board.

4.5 Performance Analysis

We can find that this scheme only uses simple XOR operations and Hash functions. For $1 \leq i \leq n$, $1 \leq j \leq m$, the calculation amount of the hash function is as follows:

(1) The dealer aspect: The dealer calculates n times $I_i = \mathrm{H}(ID_i||sh_i||t_i)$, n times $L_i = \mathrm{H}(ID_i||I_i||t_i)$, and m times $C_j = \mathrm{H}(t_0||S_j)$, a total of 2n + m times.
(2) Participants: Each participant calculates a pseudo-share $I_i = \mathrm{H}(ID_i||sh_i||t_i)$ during the reconstruction phase, a total of n times.
(3) DC aspect: DC verifies the participant's share by calculating $L_i^* = \mathrm{H}(ID_i||I_i||t_i)$, a total of n times.
(4) After the secret recovery phase, each participant verifies S_j by computing $C_j^* = \mathrm{H}(t_0||S_j)$ m times, a total of m times.

It can be seen that the function operation in this solution is 4n + 2 m times.

The performance comparison between the existing scheme and this scheme is shown in Table 1 below.

Table 1. Comprehensive performance analysis of the scheme

Performance	Pang.et al. scheme [2]	Liu.et al. scheme [7]	Rong.et al. scheme [10]	Proposed scheme
Multi-secret	No	Yes	Yes	Yes
Access structure	Threshold	Threshold	Threshold	General
Verifiability	Yes	Yes	Yes	Yes
Secure channel	No need	No need	need	No need
Mathematical model	Polynomial interpolation	Polynomial interpolation	Polynomial interpolation	Hash function

It can be seen that the scheme can achieve multi-secret sharing for general access structures, and the scheme does not require a secure channel. The Hash function is used to achieve secret versatility and verifiability of the scheme. The calculation complexity of the proposed scheme is low.

5 Conclusion

This paper proposes a verifiable multi-use multi-secret sharing scheme based on the monotone span program. In this scheme, participants' secret shares can be selected by themselves and each participant can recover multiple secrets by keeping only one secret share. The scheme uses the RSA cryptosystem and Hash function. Hash function achieves verifiability and versatility of shares. The scheme supports dynamic update of shared secret sets, and it does not require a secure channel.

References

1. Shamir, A.: How to share a secret. Commun. ACM **22**(11), 612–613 (1979)
2. Pang, L.J.: Dynamic threshold multiple secret sharing scheme. Comput. Eng. **34**(15), 164–165 (2008). Chinese
3. Itoh, M.: Secret sharing scheme realizing general access structure. Electron. Commun. **72**(9), 56–64 (1989)
4. Mashhadi, S.: Secure publicly verifiable and proactive secret sharing schemes with general access structure. Inf. Sci. **378**, 99–108 (2017)
5. Iftene, S.: General secret sharing based on the Chinese remainder theorem with applications in e-voting. Electron. Not. Theor. Comput. Sci. **186**, 67–84 (2007)
6. Blundo, C., Santis, A.D.: Multi-secret sharing schemes. In: 14th Annual International Cryptology Conference, LNCS, vol. 839, pp. 150–163. Springer, Heidelberg (1993)
7. Liu, Y.: A dynamic multi-secret sharing scheme. Comput. Eng. **35**(23), 120–121 (2009). Chinese
8. Shyamalendu, K.: A verifiable secret sharing scheme with combiner verification and cheater identification. Inf. Secur. Appl. **51** (2020)
9. Wang, C.F.: A dynamically updated password authorization multi-secret sharing scheme. Comput. Eng. Sci. **041**(009), 1597–1602 (2019)
10. Rong, H.G.: Key distribution and recovery algorithm based on Shamir secret sharing. J. Commun. **36**, 60–69 (2015). Chinese
11. Meng, K.J.: Tightly coupled multi-group threshold secret sharing based on Chinese remainder theorem. Discrete Appl. Math. **268**, 152–163 (2019)
12. Binu, V.: Secure and efficient secret sharing scheme with general access structures based on elliptic curve and pairing. Wirel. Pers. Commun. **92**(4), 1531–1543 (2017)
13. Song, Y.: A new multi-use multi-secret sharing scheme based on the duals of minimal linear codes. Secur. Commun. Netw. **8**(2), 202–211 (2015)
14. Karchmer, M.: 'On span programs'. In: Proceedings of the Eighth Annual Structure in Complexity Theory Conference 1993, pp. 102–111. IEEE Computer Society (1993)

Design and Implementation of a Modular Multiplier for Public-Key Cryptosystems Based on Barrett Reduction

Yun Zhao, Chao Cui(✉), Yong Xiao(✉), Weibin Lin, and Ziwen Cai

Electric Power Research Institute of CSG, Guangzhou 510663, China
cuichao2020@163.com

Abstract. This paper refers to a hardware implementation for executing modular multiplication in public-key cryptosystems using the Barrett reduction, which is a method of reducing a number modulo another number without the use of any division. Considering the flexibility of hardware, the modular multiplier we proposed is able to work over 3 prime fields $\mathrm{GF}(p)$, standardized by NIST for use in Elliptic Curve Cryptography (ECC), where the size of primes p are 256, 384, and 521 bits. We designed two methods to optimize the modular multiplier: Firstly, the circuit departed 257×257 multiplier into two 257×129 multipliers and an adder, three parts to optimize for clock frequency. Secondly, we proposed a parallel computing architecture to improve the utilization of multiplier and achieve high throughput. This modular multiplier runs at the clock rate of 300 MHz on 40 nm CMOS and performs a 256-bit modular multiplication in 3 cycles, while 384-bit modular multiplication costs 10 cycles and 521-bit modular multiplication costs 25 cycles. The architecture is very suitable for situations requiring high computing speed, such as online ECC signature verification.

Keywords: Public-Key cryptosystems · Modular multiplication · Barrett reduction · Hardware implementation

1 Introduction

In information system security, especially in digital signature, authentication and key distribution and management, public-key cryptography plays an essential role. Modular multiplication is the basic operation of most public key cryptography, such as RSA, ElGamal, Elliptic Curve Cryptography (ECC) [1], etc.

Modular multiplication [2, 3] is generally expressed as

$$c = (a \times b) \bmod p, 0 \leq a, b \leq p \tag{1}$$

where a and b are k-bit binary large integers. In the design of this paper, we first calculate $z = (a \times b)$, and then calculate the modulus of z to p.

In 1986, Paul Barrett proposed a modular algorithm suitable for hardware implementation [4], which avoids the division operation in modular multiplication and reduces the complexity and hardware cost of modular multiplier. In recent years, fast implementation of modular multiplication algorithm is widely studied [5–10].

Q. Liu et al. (Eds.): CENet 2020, AISC 1274, pp. 803–809, 2021.
https://doi.org/10.1007/978-981-15-8462-6_92

In this paper, we first discuss the Barrett reduction algorithm, and design and optimize the 256, 384, 521 bits width modular multiplier hardware implementation circuit. Finally, we analyze the modular multiplication performance and compare the performance between references and this paper.

2 Preliminaries of Barrett Reduction

Barrett reduction [4] is used to calculate $z \bmod p$, where z and p are positive integers, and p is not a special module. The $c = z \bmod p$ means there exists q such that

$$z = q \times p + c, c < p. \tag{2}$$

The idea of Barrett reduction is to replace q in formula (2) with $\hat{q}$, so as to avoid division operation. $\hat{q}$ is an approximation of q, and $0 \le \hat{q} \le q \le \hat{q} + 2$. So, we have

$$0 \le z - \hat{q}\,p \le z - (q - 2)p < 3p. \tag{3}$$

At last, we need two subtractions at most to get $c = z \bmod p$. Barrett reduction algorithm is shown in Table 1.

Table 1. Barrett reduction steps.

Algorithm 1. Barrett reduction
Input: integers $p, b \ge 3$, $k = \lfloor \log_b p \rfloor + 1, 0 \le z \le b^{2k}, \mu = \lfloor b^{2k}/p \rfloor$.
Output: $z\,mod\,p$.
1. $\hat{q} = \lfloor \lfloor z/b^{k-1} \rfloor \times \mu / b^{k+1} \rfloor$
2. $r = (z \bmod b^{K+1}) - (\hat{q} \times p \bmod b^{k+1})$
3. if $r < 0$, then $r = r + b^{k+1}$
4. repeat $r = r - p$ when $r \ge p$
5. return r.

3 Hardware Implementation Scheme of Barrett Reduction

3.1 Hardware Implementation Scheme of Multiplier

The multiplier is very important in the modular multiplication algorithm. It will be used in the calculation of product and Barrett reduction, and the multiplier is the largest part of the modular multiplication. Because $\mu = \lfloor b^{2k}/p \rfloor$ in Barrett reduction is $k+1$ bits, the multiplier should be able to work with 257 × 257 bits, 385 × 385 bits and 522 × 522 bits. And the modular multiplier can be used in the prime finite fields GF(p), where

$$\lfloor \log_2 p \rfloor \in \{256, 384, 521\}. \tag{4}$$

In this paper, 257 × 257 bits, 385 × 385 bits and 522 × 522 bits multiplier is realized by improved binary multiplication. The binary multiplication's steps are shown as Fig. 1. The multiplier is divided into two parts, which are multiplied by the multiplicand, and the result of partial product is shifted and added to get the result of multiplication.

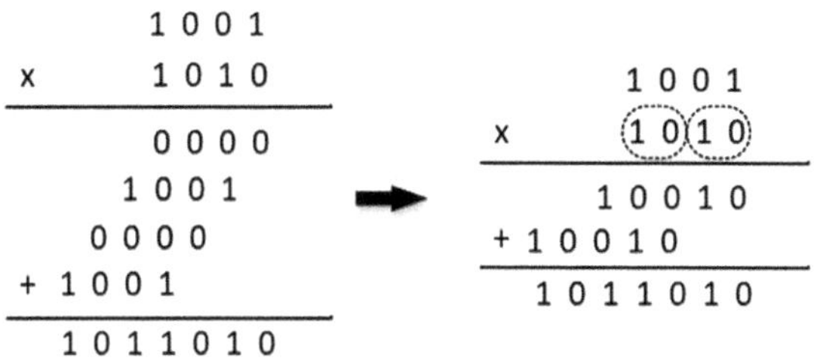

Fig. 1. Modified binary multiplication.

The circuit structure of multiplier is shown in the Fig. 2. The basic unit of multiplier is 257 × 129 bits multiplier. There are three reasons why we choose a 257 × 129 bits multiplier:

1. 1. $\mu = \lfloor b^{2k}/p \rfloor$ in Barrett reduction is $k + 1$ bits.
2. Working frequency of the 257 × 129 bits multiplier is faster than that of a 256 × 256 bits multiplier.
3. Under the working frequency of 300 MHz, the 257 × 129 bits multiplier can get more partial products.

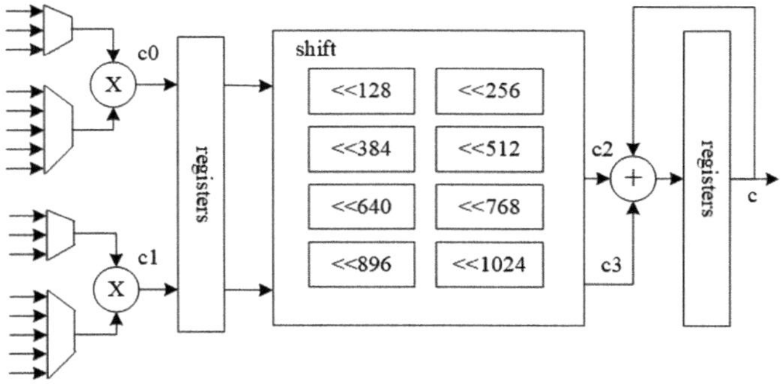

Fig. 2. The circuit structure of pipelined multiplier.

When calculating $c = a[256:0] \times b[256:0]$, the first cycle calculates multiplications $c0 = a[256:0] \times b[127:0]$ and $c1 = a[256:0] \times b[256:128]$, and the second cycle calculates shift and add $c = c0 + c1 << 128$. In this way, we can get a 257 × 257 bits multiplication in 2 cycles. Moreover, it can be used as a pipelined structure in calculate 257 × 257 bits multiplications, with this method, multiplier utilization will be increased to 100%.

385 × 385 bits and 522 × 522 bits multiplication is similar to 257 × 257 bits multiplication. Table 2 shows the specific implementation steps of 385 × 385 bits multiplication.

Table 2. 385 × 385 bits multiplication calculates steps.

Cycle	Multiplication	Shift	Add
1	$c0 = a[255:0] * b[127:0]$,	–	–
2	$c0 = a[255:0] * b[255:128]$,	$c2 = c0 << 0$	$c = c2 + c3$
3	$c0 = a[255:0] * b[384:256]$,	$c2 = c0 << 128$	$c = c + c2 + c3$
4	–	$c2 = c0 << 256$	$c = c + c2 + c3$

As shown in the table above, 4 clock cycles are required to calculate one 385 × 385 multiplication. Besides, 9 clock cycles are required to calculate one 522 × 522 multiplication.

Since the multiplier is the largest part of the modular multiplication algorithm, we should strive to improve the utilization rate of the multiplier.

In 256-bit modular multiplication, we use two-stage pipeline multiplier to calculate modular multiplication.

In the 384-bit modular multiplication calculation, three clock cycles are needed to calculate the multiplication and one clock cycle to calculate the addition for each 385 × 385 bits multiplication calculation, so we input the operands of the next modular multiplication in the fourth clock cycle to improve the utilization of the multiplier.

Similarly, in 521-bit modular multiplication, 8 clock cycles and 1 clock cycle are needed for each 522 × 522 bits multiplication, so we input the operands of the next modular multiplication in the 9th clock cycle to improve the utilization of the multiplier.

3.2 Hardware Implementation Scheme of Barrett Reduction

When calculating the 256-bit modular multiplication algorithm, we choose

$$b = 2^{256}. \tag{5}$$

So, the division and modulus calculations in Barrett reduction turn into shift. Because of

$$\lfloor \log_2 p \rfloor = 256. \tag{6}$$

So, we have k =1. In calculate Barrett reduction, we need to calculate $\mu = \lfloor b^{2k}/p \rfloor$ in advance.

Similarly, when calculating 384-bit modular multiplication and 521-bit modular multiplication, select $b = 2^{384}$ and 2^{521} respectively, $k = 1$. And we need to calculate $\mu = \lfloor b^{2k}/p \rfloor$ in advance as well.

To calculate one reduction, we need to calculate two times of multiplication and one time of subtraction. The two multipliers are z/b^{k-1} times μ and $\hat{q}$ times p, respectively. Data path of Barrett reduction algorithm is shown as Fig. 3.

In order to improve the utilization of multiplier, when calculating 256-bit modular multiplication, the operands of the next cycle multiplication should be prepared for the first multiplication. However, the data of multiplication operands in Barrett reduction is correlated, so calculating two modular multiplication at the same time can maximize the utilization of multiplier.

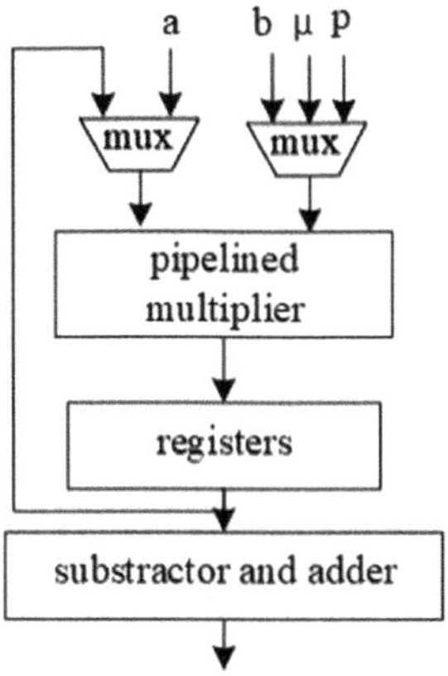

Fig. 3. Data path of Barrett reduction algorithm.

4 VLSI Implementation Results and Performance Comparisons

According to the above design idea, the realization of modular multiplication algorithm is completed. We use the modular multiplication algorithm in the ECC algorithm to verify the correctness of the function of the modular multiplication.

The implementation result is based on 256-bit, 384-bit and 521-bit modular multiplication. In this paper, primes and μ in 256-bit modular multiplication are as follows: p is 0xFFFFFFFE FFFFFFFF FFFFFFFF FFFFFFFF FFFFFFFF 00000000 FFFFFFFF FFFFFFFF, μ is 0x1 00000001 00000001 00000001 00000001 00000002 00000002 00000002 00000003. The 256-bit modular multiplication simulation results are shown in Fig. 4.

After compiling and simulating, we can see that performs a 256-bit modular multiplication in 3 cycles, while 384-bit modular multiplication costs 10 cycles and 521-bit modular multiplication costs 25 cycles when continuously input operands. According to the formula throughput = maximum frequency (main frequency) × data width/total running clock (bit/s), the running speed is calculated as $\frac{300\text{MHz}}{3} \times 256\text{bit} = 25.6\text{Gbit/s}$. The synthesize results under SMIC 40 nm process and performance comparison are shown in Table 3.

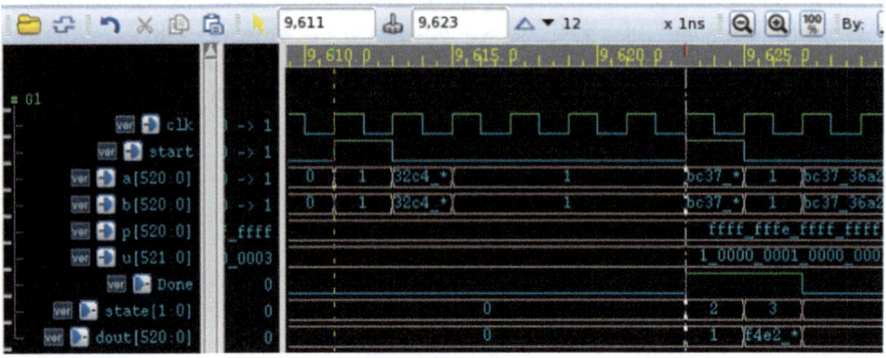

Fig. 4. The simulation result of 256-bit modular multiplication.

Table 3. Performance comparison.

Design	Technology	Area	Frequency	Operands size	Cycles	Throughput
[11]	0.5 μm CMOS	14964 NOR gates	80 MHz	256	208	98.46 Mbit/s
[12]	0.6 μm CMOS	1.33×0.93 mm^2 4100 transistors	66.6 MHz	12	36	22.2 Mbit/s
This paper	40 nm CMOS	436703.4 um^2 608052.6 transistors	300 MHz	256 384 521	3 10 25	25.6 Gbit/s 11.52 Gbit/s 6.252 Gbit/s

Compared with the design in other references, in the design of this paper, the multiplier with large bit width occupies the main area. However, the design of this paper improves the utilization ratio of multiplier to the greatest extent. Compared with other designs, large area brings higher computing performance.

5 Conclusions

The modular multiplication based on Barrett reduction in this paper uses multiplication and subtraction instead of division, which greatly improves the calculation speed and reduces the use area. It supports 256, 384 and 521 bits width modular multiplication which makes it quite flexible. In this paper, we use two methods to improve the performance, one is to shorten the critical path to improve the frequency of work, the other is to improve the utilization of modules that occupy more resources. The above two methods can also be used for reference in the hardware implementation of other circuits.

Compared with other modular multiplication methods, the structure proposed in this paper is more suitable for scenarios with high performance requirements, especially for the implementation of modular multiplication of high-speed ECC encryption.

Acknowledgements. This paper is supported by the independent high safe security chip research for metering and electricity program under Grant No. ZBKJXM20180014/SEPRI-K185011. The authors would like to thank the independent high safe security chip research for metering and electricity program for the support.

References

1. Certicom, A.: ECC the elliptic curve cryptosystem. computer applications & software (4), 1024–1028 (2007)
2. Brickell, E.F.: A fast modular multiplication algorithm with application to two key cryptography. In: Advances in Cryptology: Crypto 82, Santa Barbara, California, USA, August. DBLP (1982)
3. Blakely, G.R.A.: Computer algorithm for calculating the product AB modulo M. IEEE Trans. Comput. **32**(5), 497–500 (1983)
4. Barrett, P.: Implementing the Rivest Shamir and Adleman public key encryption algorithm on a standard digital signal processor (1986)
5. Rao, G.A.V.R., Lakshmi, P.V.: A novel modular multiplication technique for public key cryptosystems (2019)
6. Islam, M.M., Hossain, M.S., Shahjalal, M., et al.: Area-time efficient hardware implementation of modular multiplication for elliptic curve cryptography. IEEE Access **99**, 1–1 (2020)
7. Che, W, Gao, X.: FPGA design and implementation of the carry-save Barrett modular multiplication. Electronic Design Engineering (2016)
8. Liu, C., Ni, J., Liu, W. et al.: Design and optimization of modular multiplication for SIDH. In: 2018 IEEE International Symposium on Circuits and Systems (ISCAS). IEEE (2018)
9. Nedjah, N., Mourelle, L.M.: Hardware architecture for booth-barrett's modular multiplication. Int. J. Model. Simul. (2015)
10. Nadia, N., Mourelle, L.D.M.: A review of modular multiplication methods and respective hardware implementation. Informatica **30**(1), 111 (2006)
11. Tenca, A.F., Todorov, G., Koc, C.K.: High-radix design of a scalable modular multiplier. In: Cryptography Hardware And Embedded System. Springer Verlag, Berlin, Germany, pp. 189–205 (2001)
12. Bernal, A., Guyot, A.: Design of a modular multiplier based on Montgomery's algorithm. In: XIII Conference on Design of Circuits & Integrated Systems (1998)

Near and Far Collision Attack on Masked AES

Xiaoya Yang[1], Yongchuan Niu[2(✉)], Qingping Tang[3], Jiawei Zhang[4], Yaoling Ding[1], and An Wang[1(✉)]

[1] Beijing Institute of Technology, Beijing 100081, China
wanganl@bit.edu.cn
[2] Ant Financial Services Group, Beijing, China
yongchuan.nyc@alibaba-inc.com
[3] Ningbo Digital Co., Ltd., Beijing, China
[4] Data Communication Science and Technology Research Institute, Beijing 100191, China

Abstract. Collision attack is an effective method in the field of side-channel analysis to crack cryptographic algorithms, and masking can be used as a countermeasure. Most collision attacks only utilize the traces that will collide. In this paper, we propose a collision attack method that exploits not only traces tending to collide, but also non-colliding traces. It can bring higher efficiency and reduce the number of needed traces significantly. In addition, our method is a random-plaintext collision attack method instead of a chosen-plaintext attack. The experimental results show that our proposed approach is better than the existing collision-correlation attack proposed by Clavier et al. at CHES 2011 [11]. To achieve a high key recovery success rate at 80%, we use at least 60% less traces than collision-correlation attack.

Keywords: Cryptography · Power analysis attack · Collision attack · Masking

1 Introduction

Side-channel analysis was proposed by Kocher et al. in 1998 [1], since then, a new field of research has exploded in applied cryptography. Cryptanalysts have proposed a large number of techniques in this field such as differential power analysis [2], collision attack [3], correlation power analysis [4], template attack [5], etc.

In this paper, we focus on improving the effectiveness of collision attacks when analyzing masked AES. The collision attack was first proposed in 2003 by Schramm *et al.* [3]. In 2007, Bogdanov proposed a linear collision attack on AES in [6] and some improvements of the collision attack in [7, 8]. In 2015, Ren *et al.* [9] proposed a double sieve collision attack based on bitwise collision detection. All the methods are implemented for primitives without any countermeasures.

In practice, various countermeasures are often employed to protect encrypted devices from side-channel attacks. Masking is believed to be an effective method to counter first-order power analysis. At CHES 2010, Moradi *et al.* [10] mounted a collision attack on a hardware implementation of AES based on Pearson correlation coefficient. However, this success of first-order attacks is due to imperfect masking. At CHES 2011, Clavier *et al.* [11] improved this attack with the collision-correlation

Q. Liu et al. (Eds.): CENet 2020, AISC 1274, pp. 810–817, 2021.
https://doi.org/10.1007/978-981-15-8462-6_93

method. And in 2018, Wang *et al.* [12] used the correctness of the collision rate as a differentiator to detect collisions.

So far, the proposed collision attack methods have been limited to analyzing the information brought about by the collision and ignoring the information that the collision does not occur, which loses a lot of information in the analysis process.

In this paper, we present a near and far collision attacks on software AES implementations protected against first-order power analysis using masked S-Boxes and practical results on simulated curves. Contributions of this paper are as follows:

(1) Our method is a random-plaintext collision attack method, not a chosen-plaintext attack which extremely reduced conditions for the experiment.
(2) We exploit not only traces that tend to collide, but also non-colliding traces. So, the number of needed traces can be decreased significantly.
(3) Our method is much more efficient and generic compared to collision-correlation attack proposed by Clavier et al. at CHES 2011 [11].

2 Preliminaries

2.1 Collision Attack on AES Algorithm

AES is a block cipher algorithm based on Substitution-Permutation Network (SPN). To illustrate, we take the AES-128 for example. After the 128-bit plaintext $P = p_1 \| p_2 \| \cdots \| p_{16}$ is input, it is XOR operation with the whitening key $K = k_1 \| k_2 \| \cdots \| k_{16}$ by byte, and then the first round of conversion is started. Let $S_1, S_2, \cdots, S_{16}$ stand for the 16 S-box operations in the first round and denote their inputs by $x_i = p_i \oplus k_i$, output by $y_i = S_i(x_i)$ for $1 \leq i \leq 16$.

In 2003, Schramm *et al.* [3] proposed the basic concept of collisional attack, which exploits the same intermediate value in some explicit encryption process to detect a linear relationship between key bytes, and then recover the entire key. For example, if x_1 equals x_2 in AES, i.e., $p_1 \oplus k_1 = p_2 \oplus k_2$, we then deduce that $k_2 = k_1 \oplus p_1 \oplus p_2$. In this way, the candidate key space is reduced by 2^8. When all the equations between k_1 and other bytes of the key are established, the space of the candidate key is reduced to 2^8, and the entire key can be recovered by searching for all values of k_1.

The power consumption of operating on two identical intermediate values should be the same. Therefore, the similarity between the power consumption corresponding to the two intermediate values can be used to detect collisions. After the hacker refit the chip and connects it to the probe, he can use an oscilloscope to capture the power/EM signals generated during the operation of the chip. After sending these signals to the computer, the key can be recovered by a collision attack on it.

2.2 An Implementation of Masked AES Algorithm

Herbst *et al.* [13] suggested a smart card implementation of masked AES algorithm, which can also be applied to a microcontroller. Each intermediate value is masked in bytes by an XOR operation with a random mask. The masks are the same in each round during the same encryption and are different in each encryption. There are two masks involved in the S-box operation, m and m', where m is the input mask of the S-box and m' is the output mask of the S-box.

The masked S-box S' can be pre-calculated by the following formula, and it can be reused in each round of AES.

$$y \oplus m' = S'(x \oplus m) \tag{1}$$

3 Near and Far Collision Attack

3.1 Distinguisher Model

For sake of simplicity, we use Hamming weight model [2] in our distinguisher model. In other words, if a, b denote two constants, the power consumption T at the time of the operation about y can be represented as

$$T = aHW(y \oplus m) + b. \tag{2}$$

For N times of masked AES operations corresponding to different random plaintexts, there are N traces. In each trace, a pair of segments corresponding to S_1' and S_2' respectively can be extracted, and each segment contains l reference points. From now on, we will represent the true value by putting a horizontal line over the variable. For example, the first byte of the real key is represented as $\bar{k}_1$.

For each pair of segments, y_1 and y_2 can be calculated by the guessed key byte k_1 and k_2. If $k_1 = \bar{k}_1$ and $k_2 = \bar{k}_2$, then $\mathrm{HW}(y_1 \oplus y_2) = \mathrm{HW}(\bar{y}_1' \oplus \bar{y}_2')$. $\mathrm{HW}(y_1)$ and $\mathrm{HW}(y_2)$ are positively correlated when $\mathrm{HW}(y_1 \oplus y_2) = 0$, while negatively correlated when $\mathrm{HW}(y_1 \oplus y_2) = 8$. We define *near collision* as $\mathrm{HW}(y_1 \oplus y_2) \leq thr (0 \leq thr \leq 4)$ and *far collision* as $\mathrm{HW}(y_1 \oplus y_2) \geq 8 - thr (0 \leq thr \leq 4)$. $Group_1$ consists of segments which are *near collision*, while $Group_2$ consists of segments which are *far collision*. We define $T_j = \{t_{1,j}, t_{2,j}, \cdots, t_{N,j}\}$ as the set of the vertical coordinates of the j-th reference point in each segment corresponding to S_1', and $T_j' = \left\{t_{1,j}', t_{2,j}', \cdots, t_{N,j}'\right\}$ as the set of coordinates corresponding to S_2'. Take

$$\rho\left(T_j, T_j'\right) = \frac{\mathrm{cov}\left(T_j, T_j'\right)}{\sigma_{T_j}\sigma_{T_j'}} = \frac{E\left[\left(T_j - \bar{T}_j\right)\left(T_j' - \bar{T}_j'\right)\right]}{\sigma_{T_j}\sigma_{T_j'}} \tag{3}$$

as the correlation coefficient of T_j and T_j'. Then the distinguisher can be defined as

$$\varepsilon = \max\left\{\rho\left(T_j, T_j'\right)_{Group_1} - \rho\left(T_j, T_j'\right)_{Group_2} \middle| 0 \leq j < l\right\}. \tag{4}$$

T_j and T_j' in $Group_1$ would be positively correlated and T_j and T_j' in $Group_2$ would be negatively correlated when $k_1 = \bar{k}_1$ and $k_2 = \bar{k}_2$. In this case the ε will tend to be larger. On the contrary, the ε will tend to be zero.

3.2 Collision Detection Process

To conduct the near and far collision attack and recover the source key, we perform masked AES operations N times with N plaintexts $P = \{p_{i,1} \| p_{i,2} \| \ldots \| p_{i,16} | 0 \leq i < N\}$ and obtain N traces. We take $Segment_{i,1}$ and $Segment_{i,2} (0 \leq i < N)$ as the segments corresponding to S_1' and S_2' in the i-th trace respectively.

Algorithm 1 describes this process. Here, **ExtractPoint**$(Group, j)$ means extracting two sets. The first set consists of the vertical coordinates of the j-th reference point in $Segment_{i,1}$ in $Group$, while the second set consists of the corresponding vertical coordinates in $Segment_{i,2}$.

Algorithm 1 Near and Far Collision Attack

Input: $P, \{(Segment_{i,1}, Segment_{i,2}) | 0 \leq i < N\}, N, l, thr.$
Output: $\bar{k}_1 \| \bar{k}_2$.

for $k_1 = 0$ **to** 255
  for $k_2 = 0$ **to** 255
    for $i = 0$ **to** $N - 1$
      $\mathrm{HW}(y_{i,1} \oplus y_{i,2}) = \mathrm{HW}(S_1(p_{i,1} \oplus k_1) \oplus S_2(p_{i,2} \oplus k_2))$
      if $\mathrm{HW}(y_{i,1} \oplus y_{i,2}) \leq thr$
        add $(Segment_{i,1}, Segment_{i,2})$ **to** $Group_1$
      if $\mathrm{HW}(y_{1,i} \oplus y_{2,i}) \geq 8 - thr$
        add $(Segment_{i,1}, Segment_{i,2})$ **to** $Group_2$
    end for
    for $j = 0$ to $l - 1$
      $(T_j, T_j') =$ **ExtractPoint**$(Group_1, j)$
      $\rho_1 = \rho(T_j, T_j')$
      $(T_j, T_j') =$ **ExtractPoint**$(Group_2, j)$
      $\rho_2 = \rho(T_j, T_j')$
      calculate $\varepsilon_j = \rho_1 - \rho_2$
    end for
    $\varepsilon_{k_1,k_2} = \max\{\varepsilon_j | 0 \leq j < l\}$
  end for
end for
$\bar{k}_1 \| \bar{k}_2 = \arg\max_{k_1,k_2 \in \{0,255\}} \varepsilon_{k_1,k_2}$
return $\bar{k}_1 \| \bar{k}_2$

4 Experiments and Efficiency Comparison

4.1 Experiments Setup

Here, we present a near and far collision attacks on software AES implementations protected against first-order power analysis using masked S-Boxes, and practical results on simulated traces. For the simulated power traces, we fixed the key, and randomly generated N pairs of plaintexts and masks. Then, we separately set the standard deviation of the noise σ to 1.0 and 3.0, and the adjacent Hamming weights to 1.0. Based on $y_1' = S_1'(p_1 \oplus m \oplus k_1)$, and $y_2' = S_2'(p_2 \oplus m \oplus k_2)$, we generated $2 \times N$ segments of simulated curves where each segment contains 5 reference points.

4.2 Correctness Verification of Near and Far Collision Attacks

To validate our collision distinguisher, we first fixed $\bar{k}_1 = 0x83$, $\bar{k}_2 = 0x56$ and set σ to 1.0. Then we generated 500 traces obtaining 1000 segments and presented our attack at $thr = 3$. Figure 1 shows the relationship between the key we guessed and the distinguisher ε. On this figure, the coordinate of the horizontal axis equals $256 \times k_1 + k_2$. We can find the vertical coordinate with the largest value is 0.6638, and the horizontal coordinate of the point that corresponds to is 65539. Based on the relationship between the horizontal coordinates and the guessed key, we can conclude that the attack result is equal to the correct key, i.e., $\bar{k}_1 = 0x83$ and $\bar{k}_2 = 0x56$.

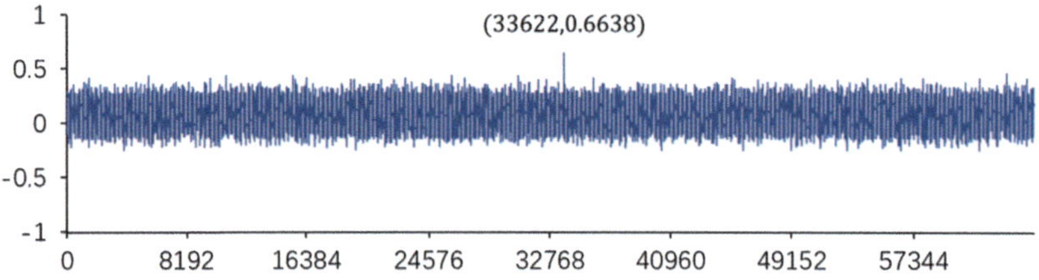

Fig. 1. The relationship between the key we guess and the distinguisher ε.

4.3 Parameter Selection Experiments

To find the best value of the parameter *thr*, we set different values of *thr* and performed several experiments at $\sigma = 1$ and $\sigma = 3$ respectively, and obtain correct rates of key recovery at different number of traces. Since there will be very few traces in $Group_1$ and $Group_2$ when $thr = 0$, so we can ignore this case. From the experimental results shown in Figs. 2 and 3, we can conclude that our method is the most efficient at $thr = 3$.

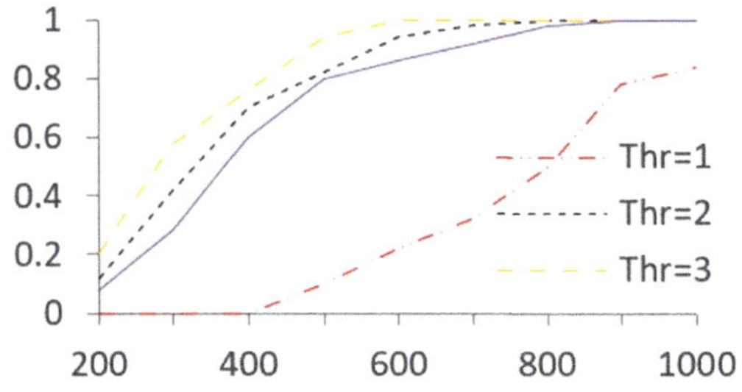

Fig. 2. The relationship between success rate and the number of used traces for $\sigma = 1$.

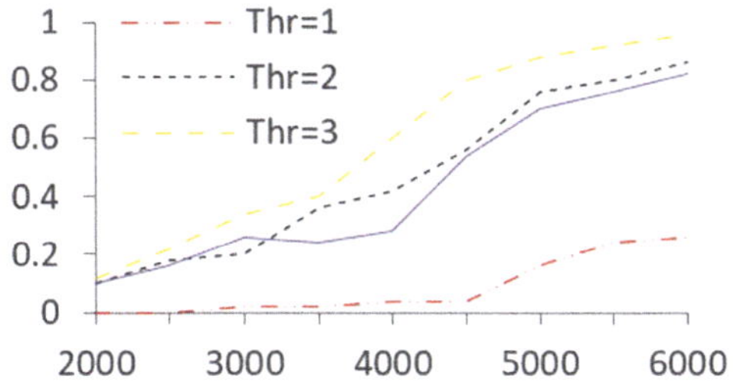

Fig. 3. The relationship between success rate and the number of used traces for $\sigma = 3$.

4.4 Comparison with Improved Collision-Correlation Power Analysis

We did the following two simulation experiments to evaluate the efficiency. We separately set the standard deviation of the noise at 1.0 and 3.0, and the adjacent Hamming weights at 1.0. We fixed the number of reference points at 5 and used our near and far collision attack and collision-correlation attack in CHES 2011. Since the latter method requires collecting 256 traces for a certain defined plaintext, we can conclude from its paper that its 27.5 traces are equivalent to our 1 trace. Therefore, the number of traces corresponding to the collision-correlation attack has been recalculated in the results shown below. The relationship between the number of traces used and the success rate has been shown in Fig. 4 and Fig. 5. It is clear that our near and far collision attack is more efficient than collision-correlation attack. Take success rate of 80% when $\sigma = 1$, for example, approximately 400 encryptions are required to recover the key using our near and far collision attack method, while the collision-correlation method requires approximately 1400 encryptions, meaning that we use at least 60% less traces than collision-correlation attack.

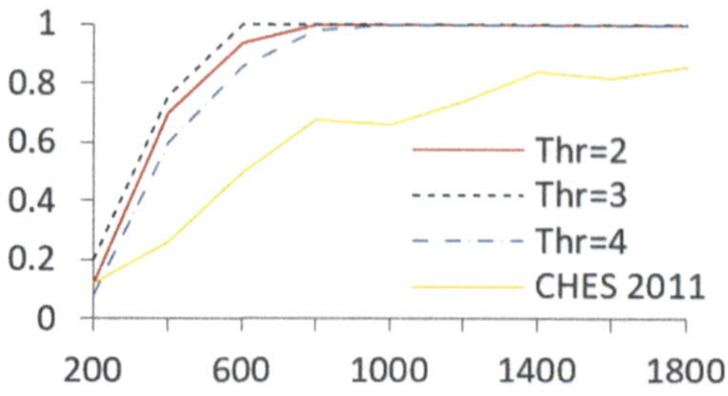

Fig. 4. The relationship between success rate and the number of used traces for $\sigma = 1$.

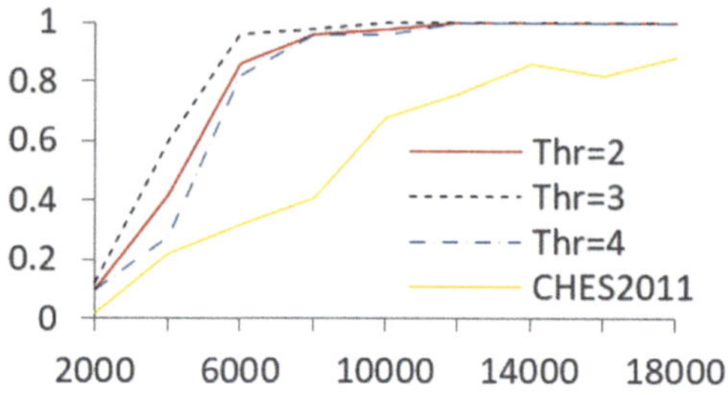

Fig. 5. The relationship between success rate and the number of used traces for $\sigma = 3$.

5 Conclusion

In this paper, we propose a new collision detection method for the masked intermediate values, which not only extremely reduces conditions for the experiment, but also utilizes traces to the maximum extent possible. This gives us a huge advantage in cases which only a few explicit pair of plaintexts and traces are known. In addition, the calculation of correlations can be replaced by the calculation of distance in our method, which greatly enhances its scalability. Research could also continue in this direction thereafter.

Acknowledgement. This work is supported by Beijing Natural Science Foundation (No. 4202070), National Natural Science Foundation of China (Nos. 61872040, U1836101, 61871037), National Cryptography Development Fund (No. MMJJ20170201), Henan Key Laboratory of Network Cryptography Technology (No. LNCT2019-A02).

References

1. Kocher, P.C., Jaffe, J.M., June, B.C.: DES and other cryptographic processes with leak minimization for smartcards and other CryptoSystems, Journal = US Patent 6,278,783 (1998)
2. Kocher, P.C., Jaffe, J., Jun, B.: Differential power analysis. In: Wiener, M. (ed.) CRYPTO 1999. LNCS, vol. 1666, pp. 388–397. Springer, Heidelberg (1999)

3. Schramm, K., Wollinger, T., Paar, C.: A new class of collision attacks and its application to DES. In: Proceedings of 10th International Workshop on Fast Software Encryption, Lund, pp. 206–222 (2003)
4. Brier, E., Clavier, C., Olivier, F.: Correlation power analysis with a leakage model. In: International Workshop Cryptographic Hardware Embedded System, pp. 16–29. Springer, Berlin (2004)
5. Chari, S., Rao, J., Rohatgi, P.: Template attacks. In: Kaliski Jr., B.S., Koç, Ç.K., Paar, C. (eds.) CHES 2002. LNCS, vol. 2523, pp. 13–28. Springer, Heidelberg (2003)
6. Bogdanov, A.: Improved side-channel collision attacks on AES. In: Proceedings of 14th International Workshop on Selected Areas in Cryptography, Ottawa, pp. 84–95 (2007)
7. Bogdanov, A.: Multiple-differential side-channel collision attacks on AES. In: Proceedings of 10th Workshop on Cryptographic Hardware and Embedded Systems, Washington, pp. 30–44 (2008)
8. Bogdanov, A., Kizhvatov, I.: Beyond the limits of DPA: combined side-channel collision attacks. IEEE Trans. Comput. **61**(8), 1153–1164 (2012)
9. Ren, Y., Wu, L., Wang, A.: Double sieve collision attack based on bitwise detection. Ksii Trans. Internet Inf. Syst. **9**(1), 296–308 (2015)
10. Moradi, A., Mischke, O., Eisenbarth, T.: Correlation-enhanced power analysis collision attack. In: Proceedings of 12th Workshop on Cryptographic Hardware and Embedded Systems, Santa Barbara, pp. 125–139 (2010)
11. Clavier, C., Feix, B., Gagnerot, G., et al.: Improved collision-correlation power analysis on first order protected AES. In: Proceedings of 13th Workshop on Cryptographic Hardware and Embedded Systems, Nara, pp. 49–62 (2011)
12. Wang, A., Zhang, Y., Tian, W., et al.: Right or wrong collision rate analysis without profiling: Full-automatic collision fault attack. Sci. China (Inf. Sci.) **61**(3), 032101 (2018)
13. Herbst, C., Oswald, E., Mangard, S.: An AES smart card implementation resistant to power analysis attacks. In: Proceedings of 4th International Conference on Applied Cryptography and Network Security, Singapore, pp. 239–252 (2006)

A Novel Image Encryption Scheme Based on Poker Cross-Shuffling and Fractional Order Hyperchaotic System

Zhong Chen[1,2](✉), Huihuang Zhao[1,2], and Junyao Chen[3]

[1] College of Computer Science and Technology, Hengyang Normal University, Hengyang 421002, China
chenzhong@hynu.edu.cn
[2] Hunan Provincial Key Laboratory of Intelligent Information Processing and Application, Hengyang 421002, China
[3] Department of Computer Engineering, Ajou University, Suwon 16499, Korea

Abstract. A novel digital image encryption scheme based on Poker shuffling and fractional order hyperchaotic system is proposed. Poker intercrossing operation has nonlinearity and periodicity, and can form permutation group. Firstly, the plain image is transformed into block image, and each block is confused by Poker shuffling operation with key. Secondly, bit plane of every pixel in each block is confused. Lastly, the scrambled image is encrypted by fractional order hyperchaotic sequence. This encryption scheme provides a secure and efficient key stream, and guarantees a large key space. The experimental results and security analysis imply that the new encryption method has secure encryption effect and property of resisting common attacks.

Keywords: Image encryption algorithm · Poker shuffling operation · Fractional order hyperchaotic system

1 Introduction

1.1 A Subsection Sample

Chaos has been one of the most important research topics in the past decade years [1], and chaos has the following properties: ergodicity and confusion, sensitivity to the initial values, diffusion with a small change, deterministic pseudo-randomness, and so on [2]. Some encryption approaches based on discrete chaotic systems have been proposed during the last decade, including block cipher and stream cipher cryptosystem [3]. Certainly, the more continuous chaotic systems are used to design encryption algorithm [4]. Thus, fractional order chaotic system has the larger key space, including its initial values, parameters and fractional derivative orders. A large amount of literatures have reported many image scrambling algorithms based on space domain. Jiang et al. [5] report encryption method based on the Hilbert curves. Priya et al. [6] use Paillier homomorphic cryptosystem to encrypt the original medical image based on poker shuffling transformation. The paper [7] presents an image scrambling method by using Poker shuffle, which can be controlled dynamically by chaos.

Q. Liu et al. (Eds.): CENet 2020, AISC 1274, pp. 818–825, 2021.
https://doi.org/10.1007/978-981-15-8462-6_94

In this paper, a new hyperchaotic system is presented, and a novel image encryption scheme is proposed based on block Poker shuffling operation and fractional order chaotic system. Firstly, the original image is divided into several blocks, and pixels of each block are scrambled. Secondly, bit planes of each block are confused. Thirdly, the scrambled image performs an XOR operation on fractional order chaotic sequence. Some experiments show that this encryption algorithm is effective, ant this scheme has very large key space.

2 Related Knowledges

2.1 A New Hyperchaotic System and Its Fractional Form

In this paper, a new hyperchaotic system is proposed to generate chaotic sequences as follow:

$$\begin{cases} \dot{x} = ax - yz \\ \dot{y} = xz - by \\ \dot{z} = xy - cz + xw \\ \dot{w} = -xz \end{cases} \tag{1}$$

where x, y, z and w are state variables, and a, b and c are known non-negative constants. In this paper, we mainly generate pseudo-random sequences by using fractional order hyperchaotic system as follows:

$$\begin{cases} D^{q_1}x = ax - yz \\ D^{q_2}y = xz - by \\ D^{q_3}z = xy - cz + xw \\ D^{q_4}w = -xz \end{cases} \tag{2}$$

$q_i, (i = 1, 2, 3, 4)$ is order of the Eq. (2). When order $q_1 = q_2 = q_3 = q_4 = 0.995$, and parameters a = 10, b = 35 and c = 25, the Eq. (2) is hyperchaotic system.

2.2 Poker Shuffling Operation

In general, Poker shuffling is mainly intercrossing operation, and card extracting and card cutting can be considered. In this paper, intercrossing operation is only operation. Without loss of generality, the initial sequence with n numbers is arranged by its nature order, i.e., s = {1, 2, 3, …, n}. When $n = 2k, (k \in N)$, split sequence s into two subsequences $s_1 = \{1, 2, \ldots, k\}$ and $s_2 = \{k+1, k+2, \ldots, 2k\}$, then intercross uniformly s_1 and s_2 with interval 1 to get new sequence $s = \{1, k+1, 2, k+2, 3, \ldots, k, 2k\}$. When $n = 1 + 2k, (k \in N)$, split the odd sequence into two subsequences $s_1 = \{1, 2, \ldots, k+1\}$ and $s_2 = \{k+2, k+3, \ldots, 2k+1\}$, thus the new sequence is $s = \{1, k+2, 2, k+3, 3, \ldots, 2k+1, k+1\}$ can be formed. Keep repeating these steps over and over again, and the new sequence can get messier. Figure 1 displays the cross-shuffling operation of 15 cards. A sequence after intercrossing operation can be considered as a permutation group, and it's period is less than the length of this sequence. Poker cross-shuffling operation can be suitable to encrypt rectangular image, and it's period can be encryption key. Intercrossing operation can rearrange sequence, and has periodicity. Table 1 shows some sequence length and their period. According to Table 1, if a sequence length is 2^n, the period of this sequence is n. To Lena image with size 150 × 150, after n times Poker intercrossing operation, the shuffled image can be shown in Fig. 2.

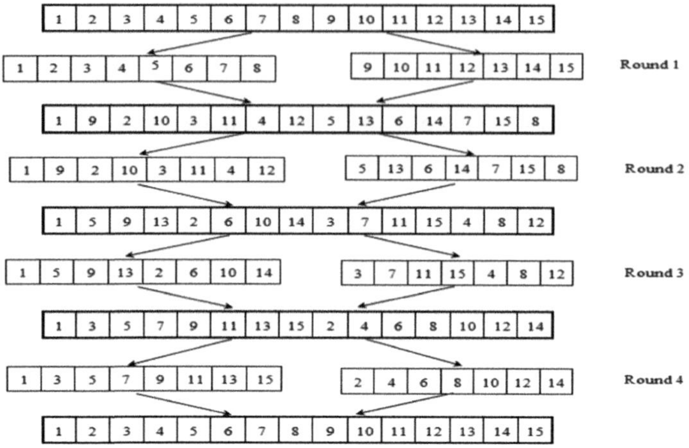

Fig. 1. Flowchart of 4 round of intercrossing operation.

Table 1. Some sequence length and their period.

Number	Period	Number	Period	Number	Period	Number	Period
10	6	80	39	256	8	14400	1320
20	18	90	11	400	18	20000	102
30	28	100	30	512	9	22500	2220
50	21	150	148	2000	7	30000	4940

3 Image Encryption Algorithm

In this paper, the fractional order hyperchaotic system (2) is used to generate pseudo-random sequence. The specific encryption algorithm is as follows:

Step 1: Choose a proper image Lena I and the initial values (x_k, y_k, z_k, w_k) of fractional order hyperchaotic system (2). Step 2: Computing height(h) and width(w) of the image, and do h × w iterations to get chaotic sequences (x_k, y_k, z_k, w_k), $k \in [1, h \times w]$. According to the following relative (3), normalize sequence x_k to [−0.5, 0.5]

$$X_k = 10^{12}x_k - int(10^{12}x_k) - 0.5 \tag{3}$$

Step 3: Divide I into m parts, and get $I_1, I_2, ..., I_m$. Each part $I_i, i \in [1, 2, ..., m]$ is shuffled by using Poker intercrossing operation with keys $(k_i, i = 1, 2, ..., m)$, and obtain shuffling image I'. Step 4: Each bit plane of image I' is confused by using $f(x)$, and one can get an new image I''. Step 5: Perform XOR operation of chaotic sequence (3) from Eq. (2) and image I'', and lastly encrypted image I''' is obtained. Flowchart of the encryption algorithm is displayed in Fig. 3(a).

Conversely, the specific decryption algorithm can also be given as follows:

Step 1: Retain the initial values (x_0, y_0, z_0, w_0) of fractional order Lorenz system (2), and generate chaotic sequences. Step 2: Choose the length of chaotic sequence as $h \times w$, and obtain chaotic sequence after h × w iterations. According to the Eq. (3), one can normalize x_k to $[-0.5, 0.5]$. Step 3: Realize XOR of the encrypted image I''' and chaotic sequence, and one has the image I''. Step 4: Each bit of the image I'' are performed inverse function $f^{-1}(x)$, and we get the image I'. Step 5: Each block is performed by utilizing Poker intercrossing operation with T-keys $(T_i - k_i, i \in [1, 2, ..., m])$ (T_i is period of each block). Last, decrypted image can obtained. Figure 3 (b) shows flow-chart of decryption algorithm.

Fig. 2. Poker shuffling of Lena image (size: 150 × 150) after *n*th inter-crossing operation.

4 Experimental Results and Security Analysis

In this section, we simulate the presented algorithm for standard test Lena image with size 150 × 150. The Lena original image, scrambled pixel of block scrambled image, confused bit image and encrypted image by using above scheme are depicted in Figs. 4 (a)-(d), respectively. To prove the robustness of the presented algorithm, statistical analysis can be performed to demonstrate its superior diffusion and confusion properties. Generally, the standard parameters to gauge an encrypted image are: histogram, entropy, correlation coefficient of adjacent pixels, and so on. The distribution approximating to a uniform distribution, is significantly different from the histogram of the encrypted image in Fig. 5(d). They are much different from the histograms of the Fig. 5(a-b).

4.1 The Correlation Coefficients of Adjacent Pixels

In general, the correlation coefficient is used to evaluate the quality of the proposed encryption algorithm. The correlation coefficient between two adjacent pixels of an encrypted image is calculated by:

$$R_{x,y} = \frac{\sum_{i=1}^{N}[x_i - E(x)][y_i - E(y)]}{N\sqrt{D(x)D(y)}} \tag{4}$$

where $E(x) = \frac{1}{N}\sum_{i=1}^{N} x_i$, $D(x) = \frac{1}{N}\sum_{i=1}^{N}[x_i - E(x)]^2$.

Here x and y are gray values of two adjacent pixels. The correlation distribution of two horizontal adjacent pixels in the original Lena image (150 × 150), block scrambling image, bit plane disturbing image and its encrypted image are shown in Fig. 4.

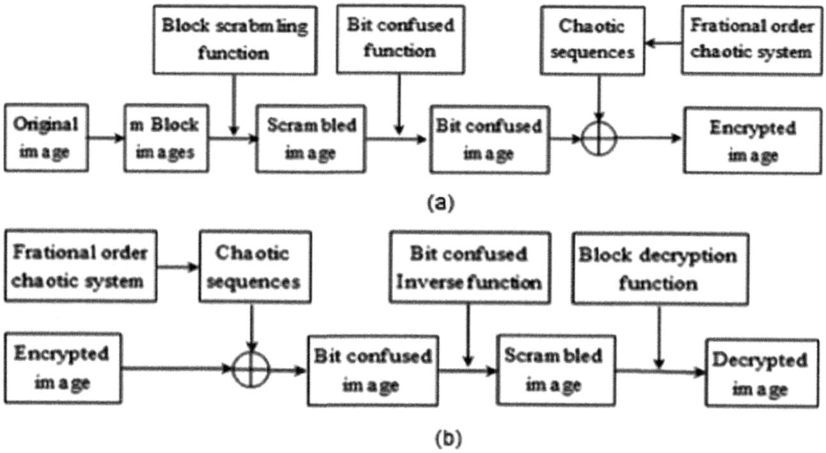

Fig. 3. Encryption and decryption process of image (a) Flowchart of encryption algorithm (b) Flowchart of decryption algorithm.

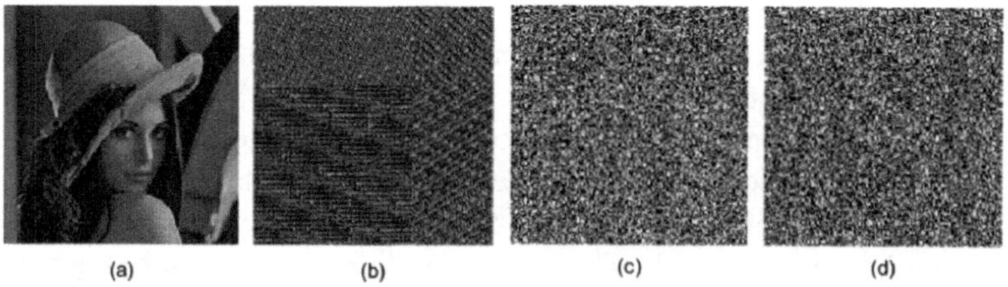

Fig. 4. Encryption process (a) original image, (b) block scrambled image, (c) bit plane scrambled image, (d) encrypted image.

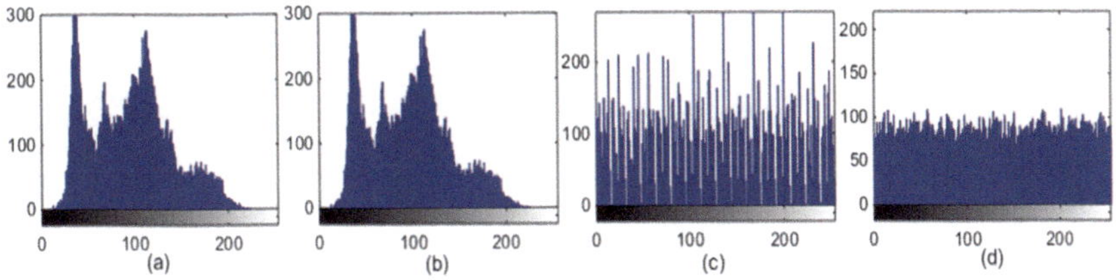

Fig. 5. Histograms (a) original image, (b) block scrambled image, (c) bit scrambled image, (d) encrypted image.

Table 2. Correlation coefficients.

Direction	Horizontal	Vertical	Diagonal
Original image	0.885008	0.861041	0.834218
Block scrambled image	0.006861	0.173399	0.041085
Bit scrambled image	0.003157	0.004107	0.007651
Encrypted image	0.008451	0.001153	0.004421

The similar results for horizontal, vertical and diagonal directions are obtained, displayed in Table 2. The values of correlation coefficients are close to 1 in the original image, whereas the values of encrypted images are close to 0. This implies that the proposed method highly de-correlate the adjacent pixels in the encrypted image.

4.2 Information Entropy

The entropy of an information source can be defined by following equation:

$$H(s) = -\sum_{i=0}^{255} p(s_i)log_2(p(s_i)) \tag{5}$$

where $p(s_i)$ is the probability of symbol s_i in an image. The entropy of an encrypted image can represent the distribution of gray value, and it is significantly less than the ideal value 8. The more uniform the distribution of gray value become, the smaller the entropy is not. The information entropy of the original image is 7.301684, and the encrypted image has 7.991429 entropy value. This means that the information leakage in the proposed encryption scheme is negligible, and the encrypted image is secure against the entropy attack (Fig. 6).

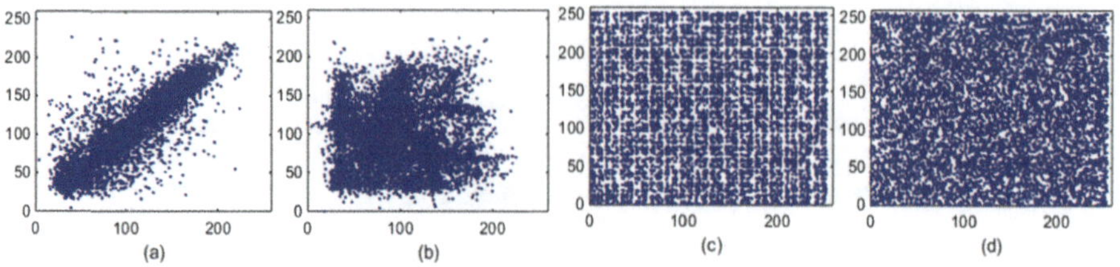

Fig. 6. Adjacent pixels' horizontal correlation (a) original image, (b) block scrambled image, (c) bit scrambled image, (d) encrypted image.

4.3 Key Space

In this algorithm, the initial values (x_0, y_0, z_0, w_0) can be used as secret keys of encryption and decryption in this cryptosystem. Four state variables are defined as double type with 15-digit precision, then the decimal fractions are multiplied by 10^{12}, it is easy to know the key space is $(10^{12})^4 \approx 2^{159}$. Meanwhile, the initial values of parameters a, b, c and orders $q_i(i = 1, 2, 3, 4)$ also can increase the size of key space. Thus the key space of this cryptosystem is enough large to resist the brute-force attack.

4.4 Differential Attacks: NPCR and UACI

In the differential attack, the attackers can only realize a slight change by modifying one pixel of the plain image to detect the changes on the corresponding encrypted image, and try to find a relationship between the original image and the encrypted image. The important arguments of differential attack are the number of pixel change rate (NPCR) and unified average changing intensity (UACI). According to definition PPCR and UACI, the values of NPCR and UACI are 99.595555% and 32.537638%, respectively. In general, the larger NPCR is, the better the encryption; the larger UACI is, the better the encryption is not.

5 Conclusions

This paper presents a novel image encryption algorithm based on Poker shuffling transform and fractional order hyperchaotic system. On the one hand, the original image is scrambled by using Poker inter-crossing operation with period, on the other hand, the bits of each pixel on scrambling image are confused. By the application of fractional order hyperchaotic system, one has the chaotic sequence. Performimg XOR operation of chaotic sequence and scrambled image, one can obtain the encrypted image. The experimental results are simulated, and security analysis shows that the new chaotic encryption algorithm has good encryption effect and resistance to common attacks.

Acknowledgement. This work was supported by the scientific research project of Hengyang Normal University (No. 18D24), Hunan Provincial Natural Science Foundation of China (No. 2020JJ4152), the Science and Technology Plan Project of Hunan Province (No. 2016TP1020), the General Scientific Research Fund of Hunan Provincial Education Department (No. 18A3 33, No. 19A066), the Double First Class University Project of Hunan Province (No. Xiang jiaotong [2018]469), Postgraduate Research and Innovation Projects of Hunan Province (No. Xiangjiaotong [2019]248-998), and Hengyang guided science and technology projects and Application-oriented Special Disciplines (No. Hengkefa [2018]60-31).

References

1. Ma, C., Wang, X.: Hopf bifurcation and topological horseshoe of a novel finance chaotic system. Commun. Nonlinear Sci. Numer. Simul. **17**, 721–730 (2012)
2. Boriga, R., Dăscălescu, A.C., Priescu, I.: A new hyperchaotic map and its application in an image encryption scheme. Signal Process. Image Commun. **29**, 887–901 (2014)
3. Xu, X.: Generalized function projective synchronization of chaotic systems for secure communication. EURASIP J. Adv. Signal Process.14, (2011)
4. Mishra, M., Mankar, V.H.: Chaotic cipher using Arnolds and Duffings map. Adv. Intell. Syst. Comput. **167**, 529–539 (2012)
5. Priya, S., Varatharajan, R., Manogaran, G., Sundarasekar, R., Kumar, P.M.: Paillier homomorphic cryptosystem with poker shuffling transformation based water marking method for the secured transmission of digital medical images. Pers. Ubiquit. Comput. **22**, 1141–1151 (2018). https://doi.org/10.1007/s00779-018-1131-8
6. Pareek, N.K., Patidar, V., Sud, K.K.: Diffusion-Csubstitution based gray image encryption scheme. Digit. Signal Process. **23**, 894–901 (2013)
7. Xu, S., Wang, Y., Wang, J., Min, T.: Cryptanalysis of two chaotic image encryption schemes based on permutation and XOR operations. (2008)

Research on High Speed and Low Power FPGA Implementation of LILLIPUT Cryptographic Algorithm

Juanli Kuang(✉), Rongjie Long, and Lang Li

College of Computer Science and Technology, Hengyang Normal University,
Hengyang 421002, China
kuangjuanli0917@126.com

Abstract. The computing capability of micro encryption devices has been a concerned problem by the people. This paper improves the implementation of LILLIPUT which was proposed in 2016. In this paper, we adopt the method of compromising area and power consumption. And with the utilization of parallel hardware processing and reduced excess hardware, the execution speed of the algorithm is extended with the minimal power consumption. Tested by experiments, the third-order LILLIPUT algorithm of parallel processing is 40% higher than that of the non-parallel processing, with only a small amount of area and power consumption used to significantly improve the computing performance of the micro-equipment.

Keywords: LILLIPUT algorithm · Verilog HDL · Hardware implementation

1 Introduction

LILLIPUT, based on EGFN (extended generalized Feistel structure) [1], is a newly proposed packet cipher algorithm in 2016 with the packet length of 64 and the key length of 80. On the basis of traditional generalized extended Feistel structures Type-1, Type-2 and Type-3 [2–4], through the well-designed EGFN structure, it can not only improve the diffusion performance greatly to offset the slow diffusion of encryption process in Feistel encryption structure and reduce the number of rounds needed for iteration process, but also inherit the basic consistent characteristics of Feistel encryption and decryption structure, and cut down on the area usage of the decryption module.

With the development of Internet of Things, a large number of wireless sensing devices are used extensively. In general, wireless sensors are restricted in power, computing efficiency and communication [5]. How to improve the computing speed of cryptographic algorithms in such devices and reduce the power consumption is a problem worth studying. In the process of the implementation of the cryptographic algorithm, if the algorithm is intended to be executed quickly, it will inevitably increase the power consumption. However, due to the computational requirements of the system, we often cannot reduce the execution speed to lower the power consumption. Therefore, we usually have to balance in the indexes that determine password performance such as through put capacity, dynamics, occupied area, etc. [6]. The idea of space exchange time

Q. Liu et al. (Eds.): CENet 2020, AISC 1274, pp. 826–834, 2021.
https://doi.org/10.1007/978-981-15-8462-6_95

is a choice to achieve the increment of power consumption and the maximal increase in encryption speed, through the minimal area occupation. This paper introduces the research background of LILLIPUT algorithm, and then based on the parallel implementation method [7–9] it proposes an implementation architecture which is verified by Spartan-3 (xc3s700an-5fgg484) devices. Thereafter, the experimental results are analyzed, and in the meantime, a comprehensively overview is summarized. Compared with its basic implementation, the hardware implementation method proposed in this paper balances the area, power consumption and throughput capacity, and obtains the maximal encryption execution speed with a small amount of the area and power consumption.

2 LILLIPUT Algorithm

LILLIPUT encryption structure is shown in Fig. 1. The 64-bit plaintext state can be considered as 16 half bytes, which can be represented as $X_{15}...X_0$. Each round of iterations is called OneRoundEGFN, and the total encryption process is 30 times repeated OneRoundEGFN iterations.

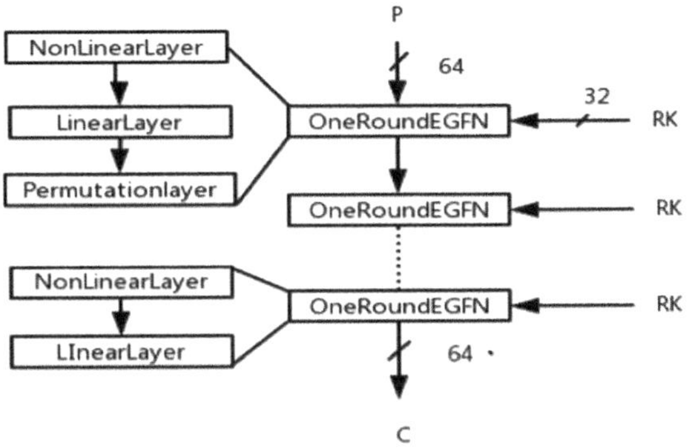

Fig. 1. LILLIPUT encryption process diagram.

The encryption operation in the OneRoundEGFN is shown in Fig. 2, including: nonlinear layer operation (NonLinearLayer), linear layer operation (LinearLayer), and p replacement layer (PermutationLayer). It is worth noting that there is no Permutation operation in the last OneRoundEGFN transformation.

As can be seen from the above Fig. 2, the main operation of the nonlinera layer is F function operation. The F function is defined as F_i, with the value of i from 0 to 7, and then the operation of the F function in the nonlinera layer can be defined as the formula: $F_i = S(X_{7-i} \oplus RK_i)$, within which S represents a half-byte S-box, as shown in the Table 1 below, and RK_i represents a subkey generated from a key extension operation. C language code is implemented as follows:

```
void Function (int state, int key){
state = sbox [state ^ key];}
```

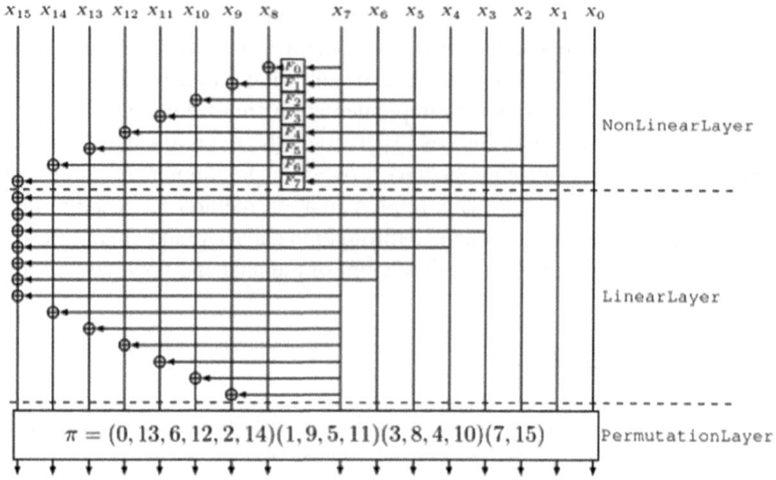

Fig. 2. OneRoundEGFN operation process.

Table 1. S-Box.

$x_{(4)}$	0	1	2	3	4	5	6	7	8	9	A	B	C	D	E	F
$S(x_{(4)})$	4	8	7	1	9	3	2	E	0	B	6	F	A	5	D	C

Operation of P replacement layer: the half byte branch is replaced in accordance with the replacement Table Π, and the specific contents of the replacement table are given in detail in the Table 2. C language implementation is as follows:

```
void PPLayer (int state [16]){
for (int i = 0; i < 16; i ++)
state[i] = p[state[i]]; }
```

Table 2. Π of the sub-group replacement table and its reverse replacement table.

i	0	1	2	3	4	5	6	7	8	9	10	11	12	13	15	16
$\prod(i)$	13	9	14	8	10	11	12	15	4	5	3	1	2	6	0	7
$\prod^{-1}(i)$	14	11	12	10	8	9	13	15	3	1	4	5	6	0	2	7

The process diagram of the key extension algorithm is shown in Fig. 3, which contains two operations: extracting key and updating key status, in which the key transformation operation contains two operations: confusion and diffusion. The 80-bit key is defined as a semi-byte packet $Y_{19}...Y_0$.

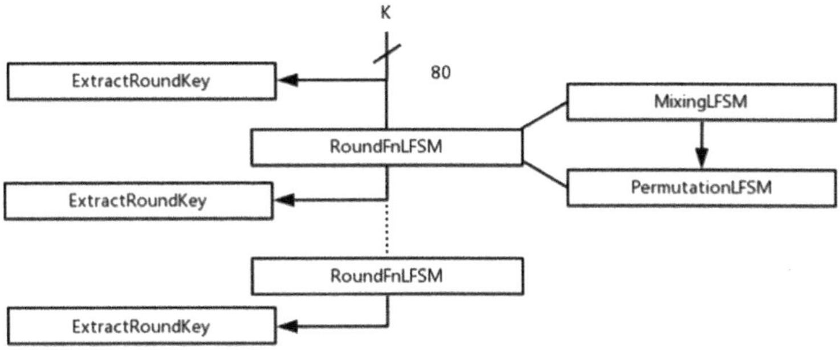

Fig. 3. Key extension process diagram.

Extract the key (ExtractRoundKey) operation extracts the wheel key required for each round of encryption. Firstly it extracts some keys:$Z_{(32)} \leftarrow Y_{18} \| Y_{16} \| Y_{13} \| Y_{10} \| Y_9 \| Y_6 \| Y_3 \| Y_1$ and the state Z is divided into $Z_{31}\ldots\ldots Z_0$ in accordance with bit, then proceeds $RK_j = S(Z_j \| Z_{8+j} \| Z_{16+j} \| Z_{24+j}), j \in [0, 7]$, where S is consistent with the S-box used in the encryption transformation.

The key status update is proceeded by LFSRs, and the implementation process is shown in Fig. 4 below, and four LFSR are respectively named as *L*0 to complete this operation.

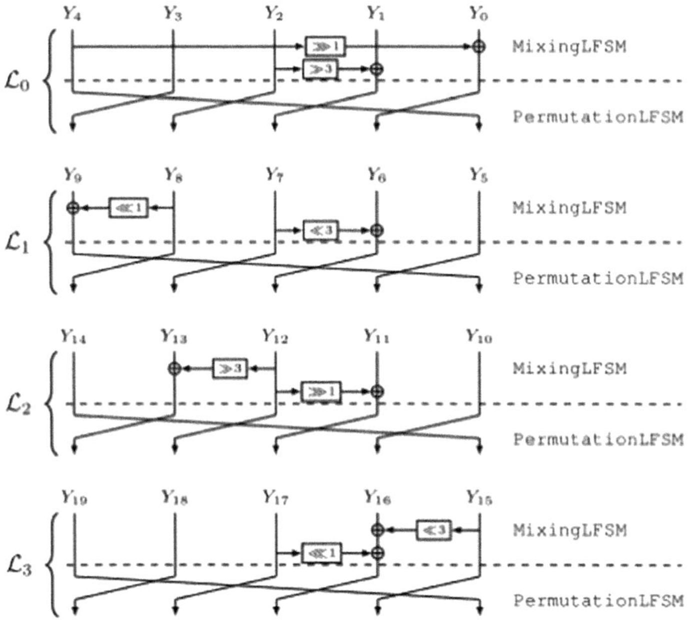

Fig. 4. Key status update process diagram.

Each LFSR operation is described as follows:

$L_0\ :\ Y0 \leftarrow Y0 \oplus (\mathrm{Y4}\ >\ >\ >1)\ Y1 \leftarrow Y1 \oplus (Y2 >\ > 3)$
$L_1\ :\ Y6 \leftarrow Y6 \oplus (Y7 < < 3)\ Y8 \leftarrow Y8 \oplus (Y9 < < < 1)$
$L_2 : Y11 \leftarrow Y11 \oplus (Y12 >\ > 3)\ Y12 \leftarrow Y12 \oplus (Y13 >\ > 3)$
$L_3 : Y16 \leftarrow Y16 \oplus (Y15 < < 3) \oplus (Y17 < < < 1)$

3 High Speed and Low Power Consumption

Parallel implementation is always employed to improve the execution speed of cryptographic algorithms, and such method can greatly reduce the execution time and complexity of the algorithm.

For example, the unparallel implementation of Fig. 5 is a low speed implementation method, and it is also the basic implementation method of block cipher algorithm. If we need to get 10 ciphertexts, by this execution method, nearly 600 execution cycles are needed. But through the parallel implementation mode, such as the implementation mode in Fig. 6, the LILLIPUT algorithm is implemented in parallel order 15, and only 40 cycles are required to obtain the same number of ciphertexts. But the increase in the latter's area occupation is also a fact, with the area surging more than 15 times in parallelization approaches. Although the speed is improved, the area is occupied too much. Such an implementation method obviously cannot meet the practical requirements of micro-embedded devices. So we need to make a trade-off in the area and speed.

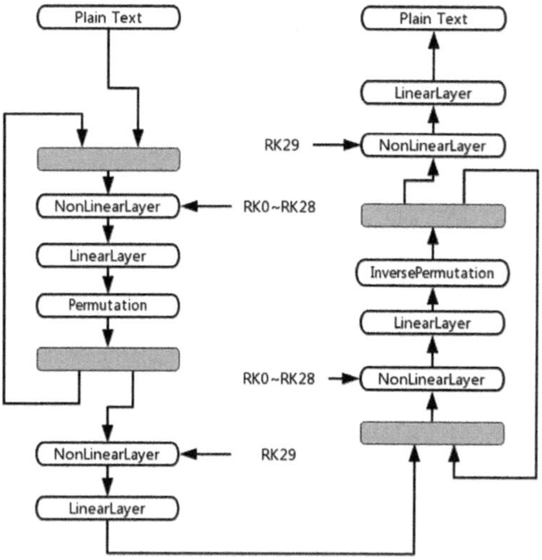

Fig. 5. LILILIPUT implementation process without parallel processing.

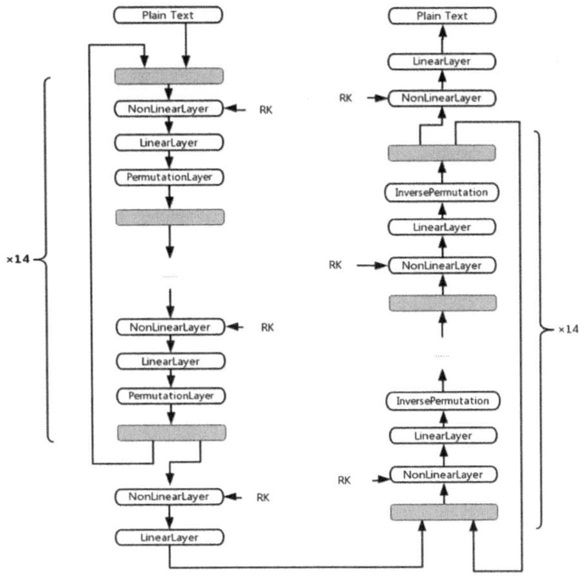

Fig. 6. 15th order parallel processing LILILIPUT implementation process.

To balance the occupancy on the equilibrium area and seek the maximum possible to increase the speed of LILLIPUT algorithm implementation for reducing the power consumption, we propose the parallel implementation method in Fig. 7 below. In this method we adopt the third-order parallel implementation. Compared to the parallel execution mode mentioned in Fig. 6 above, the occupied area will be greatly reduced, while for the common serial implementation, the encryption speed will increase by nearly three times.

Since there is no P replacement layer in the last OneRoundEGFN operation in the encryption process of LILLIPUT algorithm, the basic implementation is implemented separately. While for the implementation method proposed in this paper, we adopt a unified OneRoundEGFN, and then improve the implementation by adding an inverse P replacement layer after the last OnrRoundEGFN operation. Through the improvement, all rounds can be achieved by the unified module. To compare with the basic implementation mode, it slightly improves in the module utilization rate.

In the process of Verilog implementation, we use the main control module to control the increase of wheel counter cnt, key generation, and OneRoundEGFN operation. The code is as follows:

```
always @(posedge clk) begin
always @(posedge clk) begin
cnt <= (cnt^10)? cnt + 1: cnt;
res <= ready ? ((cnt^5' b11111) ? temp: state): res;
k <= ready ? ((cnt^5'b11111) ?out_keys:Key) :k;
ready <= (cnt^10)? 1:0;
end
```

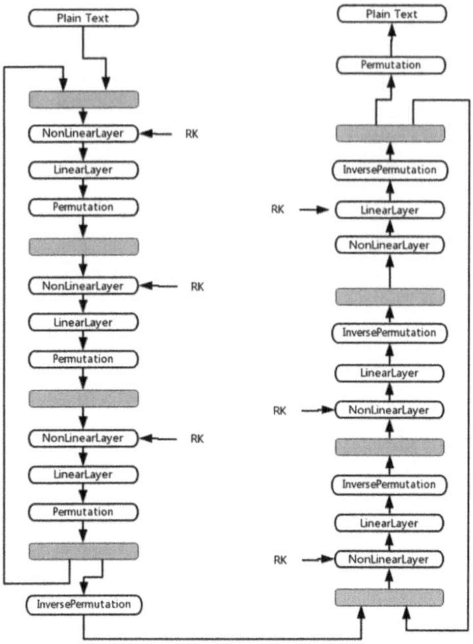

Fig. 7. LILILIPUT implementation process for level 3 parallel processing.

The final encryption result is obtained through an inverse Permutation operation after all the OneRoundEGFN operation is implemented.

```
assign result = {
res[31:28], res[3:0], res [27:24], res[11:8], res[7:4], res[15:12],
res[23:20], res[19:16], res[63:60], res[51:48], res[47:44], res[43:40],
res[35:32], res[59:56], res[39:36], res[55:52]
};
```

4 Implementation and Results Analysis

We respectively implement the Spartan-3(xc3s700an-5fgg484) setup for the three structures referred to above on the Xilinx Design Suite 14.7 application software, and their power consumption, throughput capacity and area are compared. As shown in Fig. 7 below, the algorithm performs a power and throughput capacity line diagram. As we can see, power consumption increases as throughput capacity goes up. The power consumption of the 15th-order parallel execution method is nearly five times higher than that of the non-parallel implementation method. And the power consumption is respectively, 53.44 mW, 74.63 mW, and 223.89 mW. We represent throughput capacity for data sets encrypted every 30 cycles. And the encrypted array groups of unparallel implementation, third-order parallelization implementation, and 15th-order parallelization implementation are respectively 1, 3, 15 groups. We can clearly see that 3 rounds of parallel implementations are relatively low in terms of power consumption growth, while the implementation rate has tripled. There are, of course, many parallel implementation methods, but in terms of the increase in the Fig. 8, third-order is the option of the lowest cost.

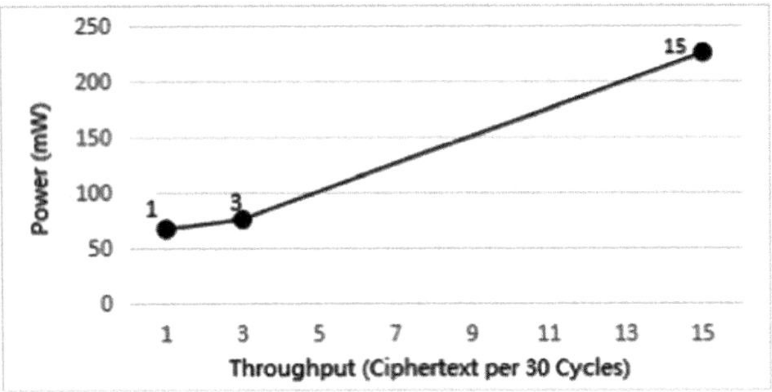

Fig. 8. Power consumption and throughput growth trend.

Figure 9 shows a broken line diagram of the relationship between power consumption and area occupation, and the area increases linearly as the number of wheels in parallel processing increases. As we can see from the graph, the third-order parallel implementation is the most cost-effective to exchange for the unparallel implementation.

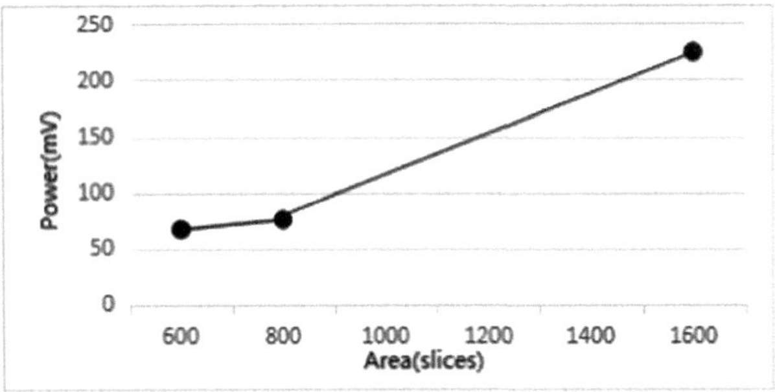

Fig. 9. Power consumption and area growth trends.

5 Conclusion

With the use of unlimited Internet, a large number of embedded devices are deployed in our daily life. And for these micro devices, how to obtain higher computing ability with only limited consumption is a topic worth studying. Hence, this paper proposes a new LILLIPUT cryptographic algorithm for parallel processing of third-order, which balances throughput capacity and area occupation. It can be observed in the results that the proposed third-order parallel processing mode can achieve higher throughput capacity in hardware. Compared with the unparallel mechanism, the latter consumes less power but has a low throughput capacity. The proposed structure consumes 28.39% more power than the unparallel processing, but simultaneously increases throughput capacity by 40%. Although the speed of 15th-order parallel implementation has improved greatly, its power consumption is not appropriate for micro devices.

Acknowledgements. This research is supported by the Hunan Province Special Funds of Central Government for Guiding Local Science and Technology Development No. 2018CT5001, Hunan Provincial Natural Science Foundation of China with Grant No. 2019JJ60004, the Scientific Research Fund of Hunan Provincial Education Department with Grant No. 19A072, Subject group construction project of Hengyang Normal University No. 18XKQ02, Application-oriented Special Disciplines, Double First-Class University Project of Hunan Province (Xiangjiaotong [2018] 469), the Science and Technology Plan Project of Hunan Province No. 2016TP1020.

References

1. Berger, T.P., Francq, J., Minier, M., et al.: Extended generalized Feistel networks using matrix representation to propose a new lightweight block cipher: Lilliput. IEEE Trans. Comput. **65**(7), 2074–2089 (2016)
2. Kumarkushwaha, P., Singh, M., Kumar, P.: A Survey on lightweight block ciphers. Int. J. Comput. Appl. **96**(17), 1–7 (2014)
3. Xue, W., Lai, X.: Impossible differential cryptanalysis of MARS-like structures. Inf. Secur. Iet **9**(4), 219–222 (2015)
4. Berger, T.P., Minier, M., Thomas, G.: Extended generalized Feistel networks using matrix representation. In: International Conference on Selected Areas in Cryptography. vol 8282. Pp. 289–305, Springer, Berlin, Heidelberg (2013)
5. Jing, Q., Vasilakos, A.V., Wan, J., Lu, J., Qiu, D.: Security of the internet of things: perspectives and challenges. Wirel. Netw. **20**(8), 2481–2501 (2014). https://doi.org/10.1007/s11276-014-0761-7
6. Lara-Nino, C.A., Diaz-Perez, A., Morales-Sandoval, M.: Lightweight hardware architectures for the present cipher in FPGA. IEEE Trans. Circuits Syst. I: Regular Papers **64**(9), 2544–2555 (2017)
7. Feizi, S., Nemati, A., Ahmadi, A. et al.: A high-speed FPGA implementation of a bit-slice ultra-lightweight block cipher, RECTANGLE. In: 2015 5th International Conference on Computer and Knowledge Engineering (ICCKE). pp. 206–211, IEEE, Mashhad, Iran (2015)
8. Knežević, M., Nikov, V., Rombouts, P.: Low-Latency Encryption—Is "Lightweight = Light + Wait"?. In: International Conference on Cryptographic Hardware & Embedded Systems. vol. 7428, pp. 426–446, Springer-Verlag (2012)
9. Researcher, I., Suba Sundary, S., Vibhasri, S., Rajalakshmi, D.: FPGA implementation of light weight block cipher Midori. Int. J. Adv. Res. Trends Eng. Technol. **3**(19), 158–164 (2016)

Adversarial Domain Adaptation for Network-Based Visible Light Positioning Algorithm

Luchi Hua[1(✉)], Yuan Zhuang[2], Longning Qi[1], and Jun Yang[1]

[1] Southeast University, Nanjing 210096, China
hualuchi@gmail.com
[2] Wuhan University, Wuhan 430072, China

Abstract. Indoor positioning system (IPS) provides fundamental location-based service in indoor environment where GPS signal cannot reach. Different from WiFi, Bluetooth and other wireless signal based IPSs, visible light positioning (VLP) has been studied in recent years due to its high accuracy and robustness, and free from electromagnetic interference. Therefore, VLP is a promising technique in indoor public places such as hospital, airport, and shopping center. To achieve higher performance, deep learning has been widely investigated for VLP for its high tolerance to noise and studies have shown significant improvement in challenging environment compared to traditional algorithm such as least square. Unfortunately, we found that deep learning, or network-based algorithms are vulnerable to device heterogeneity and setup deviations for different users and environments, thus trained model may have poor performance in test stage. Besides, feature-label pair is often unavailable during testing for further model reinforcement. In this paper, an unsupervised learning method based on adversarial training is proposed for visible light network regressor. In this method, a feature extractor works to align feature distributions from different domains, while a domain classifier differentiates data from training set and test set. Results show that when multi-dimensional attack exists, our adversarial network model can achieve significant improvement than directly applying a trained model to the test set.

Keywords: Indoor positioning · Visible light positioning · Deep learning · Adversarial training

1 Introduction

Indoor positioning system (IPS) provides fundamental location-based service in indoor environment since Global Positioning System (GPS) signals cannot reach or poorly cover indoor space. During recent years, efforts have been made to utilize indoor facilities such as Wireless Fidelity (WiFi) [1], Bluetooth [2], inertial sensors [3], Ultra-Wideband (UWB) [4], etc., with different signal characteristics like Received Signal Strength (RSS), Time-of-Arrival (TOA), Time-Difference-of-Arrival (TDOA), Angle-of-Arrival (AOA), or Channel State Information (CSI) to setup a reliable localization system. IPS applications include providing positioning services for pedestrian in large-scale indoor area such

Q. Liu et al. (Eds.): CENet 2020, AISC 1274, pp. 835–844, 2021.
https://doi.org/10.1007/978-981-15-8462-6_96

as hospital, shopping center, airport and museums, tracking products and navigate robots in factories or warehouses, etc. Many of those applications require sub-meter-level or even centimeter-level accuracy. However, many proposed techniques have difficulties to meet those requirements for suffering from limited signal range, or multipath disturbance. During the last decades, visible light communication (VLC) has attracted much attention in both academic and industry field for numerous advantages such as high output (15 Gbit/s [5]), energy efficiency, and no interference to or from electromagnetic signals [6]. Although VLC is far from being commercialized due to the bottleneck of ununiform networking standards, underdeveloped VLC access point (VAP) collaboration, and limited hardware signal processing speed, studies have successfully utilized visible light signal to setup high-accuracy positioning systems [7]. Another advantage of Visible Light Positioning (VLP) is the full coverage of LED lighting infrastructures indoor, which can be used as the transmitters as well. In recent years, many positioning algorithms such as trilateration, fingerprinting, and fusion methods have been investigated in VLP studies and most methods can achieve accuracy of few centimeters [8–10]. However, many studies have pointed out under physical environment many types of noise and disturbance can hamper the robustness of VLP [11, 12]. Thus, researchers have introduced machine learning methods to VLP system. Especially, efforts have been made to prove deep learning methods can improve VLP performance under challenging conditions. In the study [13], multilayer perceptron (MLP) structure was proposed to map 4 RSSs to the receiver's 2D coordinates. With two hidden layers and a total of 8 hidden units, the system could achieve an average positioning error of 6.39 cm. Another study [14] proposed to use Bayesian regularization to reinforce network performance. Simulation results showed that 3.4 to 4.58 cm 2D error can be achieved $1.8 \times 1.8 \times 2.1$ m^3 area. Although deep neural network methods have shown promising results than traditional algorithms in noisy environment, there still exists some risks for using this technique. One major problem is that many networks-based systems were based on the priority that the training data and test data were from the same distribution. In physical environment, a VLP system would be facing many risks such as user device heterogeneity and the variation of lighting conditions due to long-time working. Moreover, when copying the same layout of LED transmitters in a new area, displacement of deployment (coordinates and orientations of LEDs, parameters of devices) could result in huge difference in the current system performance compared with test performance. Collecting new dataset would be considered with extra cost of labor and time.

Inspired by recent advance in generative adversarial network studies, we propose an adversarial training method for network-based VLP algorithms. The adversarial training network is combined with the MLP regressor for 3D localization using RSSs feature. In the network, a feature extractor works to align feature distributions from different domains, while a domain classifier differentiates data from training set and test set. The proposed adversarial training is unsupervised and it can be utilized in the user end. Compared with directly applying trained network to the test set, this method greatly improves the performance of VLP regressor under the variation of multiple parameters.

2 Generative Adversarial Network

Generative Adversarial Network (GAN) is first proposed by Ian J. Goodfellow in 2014 that enable a net to generate adversarial examples from noise vectors for network reinforcement [15]. Before this some approaches use intractable criteria for deep net model such as using fixed noise distribution to train a generative model [16], or use objective that minimize the difference between two networks' outputs [17]. In GAN mechanism, two networks, namely generator and discriminator, work together yet compete with each other. Given input noise variables $p_z(z)$, generator would work to map this noise to space $G(z;\theta_g)$. Generated examples would be the input along with training examples, given different labels, in the discriminative network. The generator's objective is to minimize the loss $\log(1-D(G(z)))$, while the discriminator tries to maximize the probability of labeling correctly. If examples from training data would be given the label of '1', and '0' for the generated examples, the objective of the whole minimax game is defined as:

$$\min_G \max_D V(D,G) = \mathrm{E}_{\mathrm{X}\sim p\mathrm{data(x)}}[\log D(\mathbf{x})] + \mathrm{E}_{z\sim p_z(z)}[\log(1-D(G(\mathbf{z})))] \quad (1)$$

In each iteration of training GAN, training discriminator and generator would further be separated in two steps. The discriminator is updated by ascending gradient:

$$\nabla_{\theta_d}\frac{1}{m}\sum_{i=1}^{m}\left[\log D\left(\mathbf{x}^{(i)}\right) + \log\left(1-D\left(G\left(\mathbf{z}^{(i)}\right)\right)\right)\right]. \quad (2)$$

Then the generator is updated by descending gradient:

$$\nabla_{\theta_g}\frac{1}{m}\sum_{i=1}^{m}\log\left(1-D\left(G\left(\mathbf{z}^{(i)}\right)\right)\right). \quad (3)$$

To avoid G being updated too often that z would be mapped to the same value of x, GAN usually takes several steps to train D to its optimal then G. In this way, the diversity of training data could be maintained. GAN would finally align $G(z;\theta_g)$ with training data distribution, which is the core idea of our proposed method.

3 Adversarial VLP Network

The proposed adversarial domain adaptation method is combined with VLP regressor to prove that generative adversarial mechanism can be successfully applied to regression problem in VLP applications. Before this, numerous classification applications such as face recognition, handwriting recognition have proved remarkable performance with GAN involved. The proposed network is shown in Fig. 1. Three networks in the form of MLP structures are included. The inputs are the Received Signal Strength of five LEDs, as depicted in Fig. 2. RSSs would first propagate through feature extractor net for high features $F(\cdot;\theta_F)$, then through coordinate regressor to output 3D coordinates $R(\cdot;\theta_R)$. A domain classifier aims to label high features from source domain and target domain with corrected domain label $C(\cdot;\theta_C)$. In this system, data from the source domain would be given label '1' and data from the target domain would be given label '0'. The feature extractor net is similar to the generator in GAN, which aims to minimize the probability that discriminator labels the target data with '0'. In other words, the feature extractor would be trained to fool the domain classifier to misclassify the data from two domains. The domain classifier instead would try to maximize the probability of giving correct labels like the discriminator in GAN.

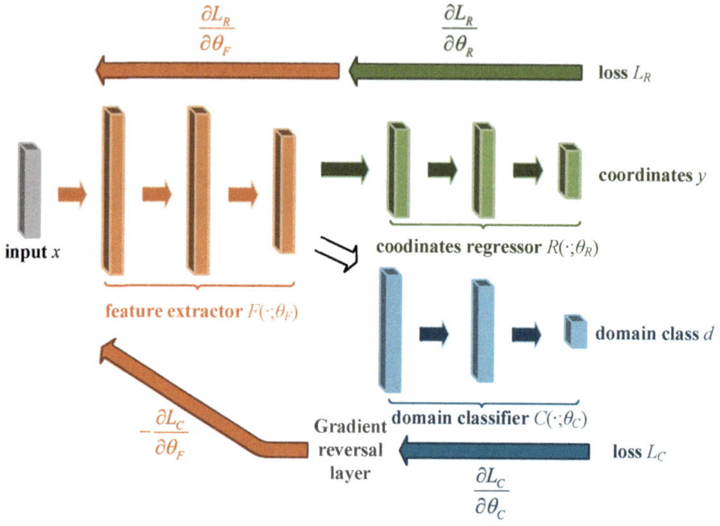

Fig. 1. Architecture of adversarial training network for VLP.

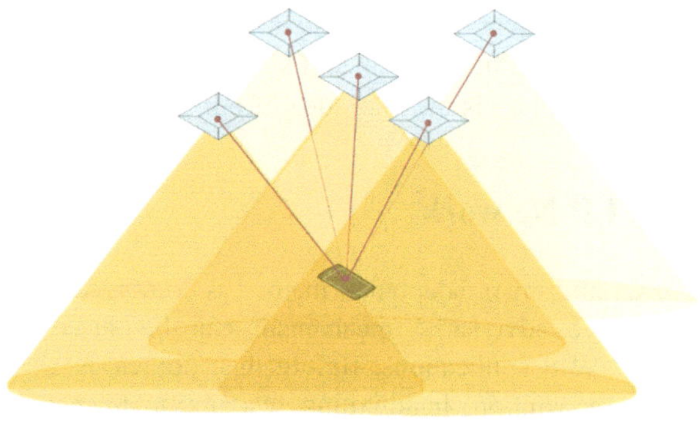

Fig. 2. VLP system layout. The LED light are uniformly distributed. The photodiode is assumed to be attached to a smartphone carrying by the pedestrian.

The objective of updating feature extractor and regressor includes two losses:

$$\left(\hat{\theta}_F, \hat{\theta}_R\right) = \arg\min_{\theta_F,\theta_R} \sum_{i=1..N} L_R(R(F(\mathbf{x}_{si};\theta_F);\theta_R), y_i) - \lambda \sum_{i=1..N} L_C\left(C\left(F(\mathbf{x}_{si};\theta_F);\hat{\theta}_C\right), d_i\right) \quad (4)$$

where θ_F represents the parameters of the feature extractor net F, θ_R refers to the parameters of the regressor net R, $\widehat{\theta}_C$ refers to the fixed parameters of the regressor net C, d_{si} represents domain class labels, x_{si} refers to the ith training data, y_i refers to the coordinates corresponding with the ith training data, λ is the scaler factor that balance the two losses. To achieve the objective, θ_R and θ_F are coorperated to minimize the loss L_R, while θ_F is to maximize the loss L_C. The domain classifier is updated with θ_R and θ_F fixed, therefore the objective would not involve loss L_R:

$$\left(\widehat{\theta}_C\right) = \arg\min_{\theta_C} \lambda \sum_{i=1..N} L_C\left(C\left(F\left(\mathbf{x}_i;\hat{\theta}_F\right);\theta_C\right), d_i\right). \quad (5)$$

where x_i refers to data from both source domain or target domain. The complete training procedures are listed in Table 1.

Table 1. Adversarial domain adaptation algorithm for VLP.

Algorithm Iterative training procedure of our approach

1: training iterations = N
2: **for** t in 1: N **do**:
3: Sample k RSSs-coordinate pairs from source domain S: $\{s_i, y_i\}_{i=1}^{k}$
4: Sample l RSSs from target domain T: $\{t_i\}_{i=1}^{l}$
5: Set F network and R network weights with saved weights
6: Update network using the following objectives:

$$\left(\hat{\theta}_F, \hat{\theta}_R\right) = \arg\min_{\theta_F,\theta_R} \sum_{i=1..N} L_R\left(R\left(F\left(\mathbf{x}_{si};\theta_F\right);\theta_R\right), y_i\right) - \lambda \sum_{i=1..N} L_C\left(C\left(F\left(\mathbf{x}_{si};\theta_F\right);\hat{\theta}_C\right), d_i\right)$$

7: Save F network and R network layer weights
8: Set C network layer weights with saved weights
9: Update network using the following objectives:

$$\left(\hat{\theta}_C\right) = \arg\min_{\theta_C} \lambda \sum_{i=1..N} L_C\left(C\left(F\left(\mathbf{x}_i;\hat{\theta}_F\right);\theta_C\right), d_i\right)$$

10: Save C network layer weights

4 Experiments

The experiment was conducted in the platform of python 3.6 and tensorflow 2.0. We set F net with 3 hidden layers, each with 40 hidden units. The R net consists of 1 hidden layer with 10 units. The C net has 1 hidden layer (32 hidden units) and 1 dropout layer (0.5 dropout rate). The parameter λ is set to 1 in our experiment. Apart from the output layers of R net and C net, we use Exponential Linear Unit (ELU) activation function for each layer. Adam optimizer was adopted for training. The simulation setup includes 5 LED lights as transmitters and 1 photodiode as the receiver, as shown in Fig. 2. We assumed LED lights are modulated by frequency-division multiplexing therefore the power of each signal can be found in the frequency domain. The RSSs were generated in a set of grids based on Lambertian propagation model [6]:

$$P_i = \frac{(m_i + 1)AP_{T_i}}{2\pi} \cdot \frac{\cos^{m_i}(\theta)\cos^M(\psi)T_s(\psi)g(\psi)}{D_i^2}. \tag{6}$$

where P_i is the RSS value, D_i is the distance between the ith LED and the receiver, A is the effective area of the photodetector, P_{Ti} is the transmit power of the LED, θ is the irradiance angle at the source and ψ is the incidence angle at the receiver. $T_s(\psi)$ and $g(\psi)$ represent the gain of the optical filter and the gain of the concentrator, respectively (in our experiment both are set to 1). M represents the Lambertian order of the photodetector and m_i is the Lambertian order of the LED source. More details of simulation setting are listed in Table 2. The parameters are altered to generate target data, which are further seperated by 4:1 into adversarial training data and test data. We considered several combinations of adversarial attacks and compare the performance between non-adversarial regressor training and adversarial training. Details are discussed with the results in the following sections.

Table 2. Parameters setting of VLP simulation

Parameter	Value
Cell dimension (m × m × m)	6 × 5 × 2.8
Power of each LED (W)	10
Position of LEDs (x, y, z) (m)	LED1 (4.5, 0.5, 2.828) LED2 (4.5, 4.5, 2.828) LED3 (2.5, 2.5, 2.828) LED4 (0.5, 4.5, 2.828) LED5 (0.5, 0.5, 2.828)
Effective area of the photodiode (mm^2)	5.2
Lambertian order of each LED m	1
Lambertian order of photodiode	1
The SNR of the system (dB)	30
Receiver's location in source domain (m)	(x: 0 − 5, y: 0 − 6, z: 1 − 1.3), interval: 0.05
Receiver's location in target domain (m)	(x: 0 − 5, y: 0 − 6, z: 1 − 1.3), interval: 0.12

4.1 1-Dimensional Adversarial Attack

The influence of altering only 1 parameter in the channel model was investigate first. We altered Lambertian order of the photodiode M to 1.1 and 1.4 to see the difference. In fact, most VLP works assumed M to be 1, while the parameter M of photodiode may actually be different for different models or within the same model [18]. The simulation results for $M = 1.1$ are shown in Fig. 3. Note that we use positioning error instead of training loss to represent the training curve, due to 'Mean Square Error' (MSE) loss for R net.

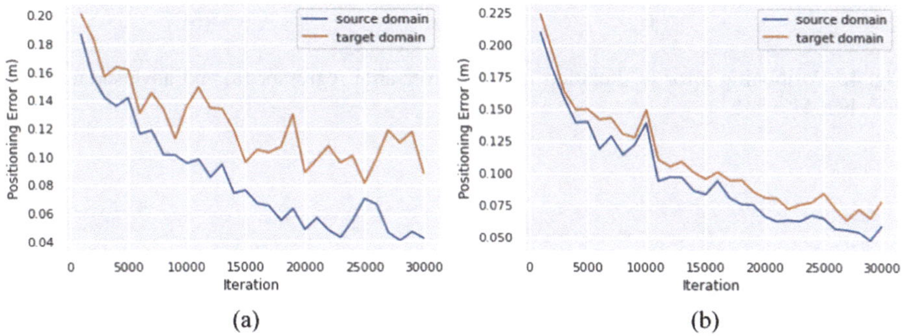

Fig. 3. Positioning error *vs.* Iteration for non-adversarial regressor (a) and adversarial training (b) when $M = 1.1$.

In Fig. 3, two training methods show similar performance while adversarial training achieve slightly better in the accuracy. From the result we can also notice that adversarial training would harm the model performance in source domain, which is comprehensively studied in another literature [19]. The close performance in accuracy mainly results from the weak adversarial attack that the distribution of source data and target data are quite similar. We increased the adversarial power with parameter $M = 1.4$. In Fig. 4, it is clear that the performance of the non-adversarial training greatly degrades while adversarial training still converges. Therefore, it is too risky for trained model being directly applied to different receivers without the prior knowledge of their parameters.

4.2 N-Dimensional Adversarial Attack

We further considered the performance of non-adversarial net when multiple weak adversarial attacks existed. A scenario for this consideration is that when trying to apply the same layout of lighting fixtures to a similar environment. The risk is that the physical locations, orientations for the lights as well as the lighting condition may be slightly different from the testbed. In addition, we considered there existed small difference for the user devices. We tested a group of attacks: 1) $M = 1.4$, $z_{\text{LED2}} = 2.75$; 2) $y_{\text{LED3}} = 1.98$, $z_{\text{LED2}} = 2.78$, $P_{\text{LED1}} = 9.6$, $M = 1.3$.

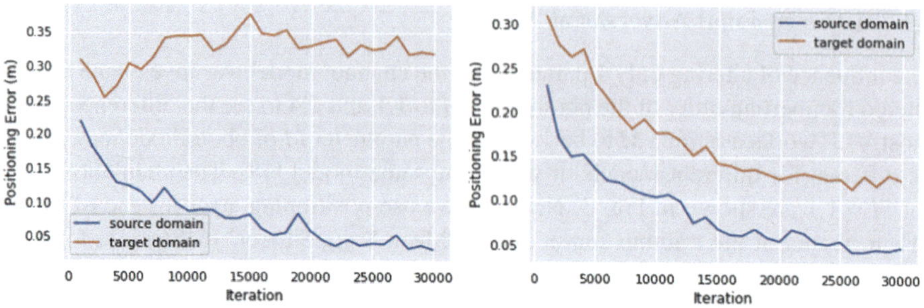

(a) (b)

Fig. 4. Positioning error *vs.* Iteration for non-adversarial regressor (a) and adversarial training (b) when $M = 1.4$.

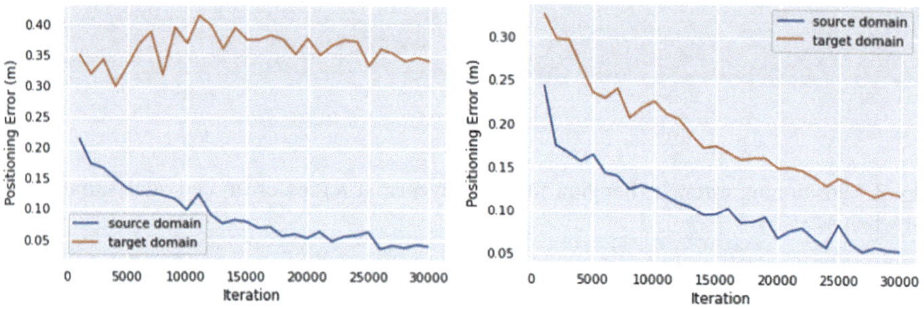

(a) (b)

Fig. 5. Positioning error *vs.* Iteration for non-adversarial regressor (a) and adversarial training (b) when $M = 1.4$, $z_{\mathrm{LED2}} = 2.75$.

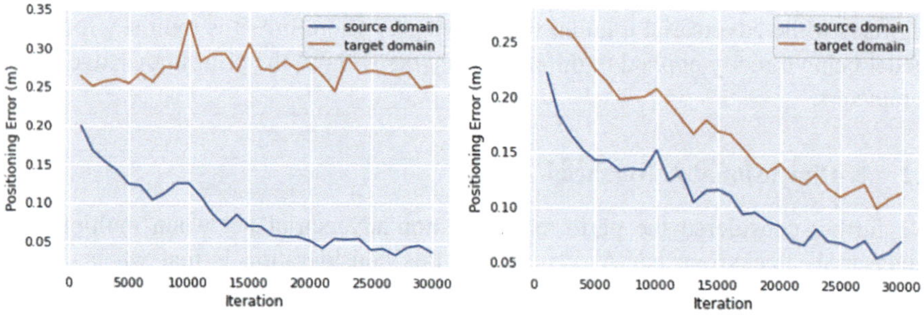

(a) (b)

Fig. 6. Positioning error *vs.* Iteration for non-adversarial regressor (a) and adversarial training (b) when $y_{\mathrm{LED3}} = 1.98$, $z_{\mathrm{LED2}} = 2.78$, $P_{\mathrm{LED1}} = 9.6$, $M = 1.3$.

The performance of model influenced by two-dimensional attack is shown in Fig. 5. With only one extra attack, the original network diverges to higher level of positioning error, and the performance gap between the two training methods become larger. Figure 6 gives the positioning results of 4-dimensional attacks. It can be told that although each adversarial attack is weak, the high complexity of attacks still makes a great impact in the performance of non-adversarial model. The results further prove the significance of adversarial training

5 Conclusions

In this paper, we proposed an unsupervised adversarial training method for VLP regressor network. We have the confidence in this approach to be successfully applied to nonlinear regression problems for localization using other signal features as well as other wireless techniques. The training procedure is simplified for VLP problem, thus for other specific problems one should consider changing hyperparameters and number of steps training classifier according to the data complexity. For VLP network, we investigated different adversarial attacks and their influence on the performance. The performance of original network and adversarial model would both vary with the adversarial attack power and dimension.

References

1. Zhuang, Y., Syed, Z., Li, Y., et al.: Evaluation of two WiFi positioning systems based on autonomous crowdsourcing of handheld devices for indoor navigation. IEEE Trans. Mob. Comput. **15**(8), 1982–1995 (2015)
2. Zhuang, Y., Yang, J., Li, Y., et al.: Smartphone-based indoor localization with bluetooth low energy beacons. Sensors **16**(5), 596 (2016)
3. Kang, W., Han, Y.: SmartPDR: Smartphone-based pedestrian dead reckoning for indoor localization. IEEE Sens. J. **15**(5), 2906–2916 (2014)
4. Müller, P., Wymeersch, H., Piché, R.: UWB positioning with generalized Gaussian mixture filters. IEEE Trans. Mob. Comput. **13**(10), 2406–2414 (2014)
5. Bian, R.T., Iman Haas, H.: 15.73 Gb/s visible light communication with off-the-shelf LEDs. J. Lightwave Technol. **37**(10), 2418–2424 (2019)
6. Zhuang, Y., Hua, L.C., Qi, L.N., Yang, J., et al.: A survey of positioning systems using visible LED lights. IEEE Commun. Surv. Tutorials **20**(3), 1963–1988 (2018)
7. Rehman, S.U., Ullah, S., Chong, P.J.H., et al.: Visible light communication: a system perspective-overview and challenges. Sensors **19**(5), 1153 (2019)
8. Xie, B., Tan, G., Liu, Y.H., et al.: LIPS: a light intensity-based positioning system for indoor environments. ACM Trans. Sens. Netw. **12**(4), 1–27 (2016)
9. Luo, Z.J., Zhang, W.N., Zhou, G.F.: Improved spring model-based collaborative indoor visible light positioning. Opt. Rev. **23**(3), 479–486 (2016)
10. Yang, S.H., Jeong, E.M., Kim, D.R., Kim, H.S.: Indoor three-dimensional location estimation based on LED visible light communication. Electron. Lett. **49**(1), 54–56 (2013)

11. Ding, J.P., Ji, Y.F.: Evolutionary algorithm-based optimisation of the signal-to-noise ratio for indoor visible-light communication utilising white light-emitting diode. IET Optoelectron. **6**(6), 307–317 (2012)
12. Wang, T.Q., Sekercioglu, Y.A., Neild, A.: Position accuracy of time-of-arrival based ranging using visible light with application in indoor localization systems. J. Lightwave Technol. **31** (20), 3302–3308 (2013)
13. Huang, H.Q., Yang, A.Y., Feng, L.H., et al.: Artificial neural-network-based visible light positioning algorithm with a diffuse optical channel. Chin. Opt. Lett. **15**(5), 050601 (2017)
14. Zhang, H.Q., Cui, J.H., Feng, L.H., Yang, A.Y.: High-precision indoor visible light positioning using deep neural network based on the Bayesian regularization with sparse training point. IEEE Photon. J. **11**(3), 1–10 (2019)
15. Goodfellow, I., Pouget-Abadie, J., Mirza, M., Xu, B.: Generative adversarial nets. In: Advances in Neural Information Processing Systems (2014)
16. Gutmann, M., Aapo, H.: Noise-contrastive estimation: a new estimation principle for unnormalized statistical models. In: Proceedings of the Thirteenth International Conference on Artificial Intelligence and Statistics (2010)
17. Schmidhuber, J.: Learning factorial codes by predictability minimization. Neural Comput. **4** (6), 863–879 (1992)
18. Cossu, G., Presi, M., Corsini, R., Choudhury, P.: A visible light localization aided optical wireless system. In: IEEE GLOBECOM Workshops (GC Wkshps) (2011)
19. Javanmard, A., Mahdi, S., Hamed, H.: Precise tradeoffs in adversarial training for linear regression. arXiv preprint arXiv:2002.10477 (2020)

Robustness Detection Method of Chinese Spam Based on the Features of Joint Characters-Words

Xin Tong, Jingya Wang(✉), Kainan Jiao, Runzheng Wang, and Xiaoqin Pan

People's Public Security University of China, Beijing, China
wangjingya@ppsuc.edu.cn

Abstract. Since the current Chinese spam messages often contain artificial adversarial perturbation and can bypass the traditional detection system, this paper proposed an integrated model based on "ALBERT-Capsule Network."On the one hand, the model accepts both raw text and word segmentation text as input data, uses ALBERT and capsule networks to extract character-level features and word-level features respectively, and uses a multi-head output structure to ensure that both the word feature stream and the word feature stream can be meaningfully trained to prevent the neural network from depending on one of its branches. Experiments on the SMS spam dataset and trec06c datasets prove that the model is better than the traditional single-feature models such as Bi-LSTM and TextCNN; On the other hand, the traditional character replacement, pinyin replacement and word splitting and other adversarial methods are used to augment the original training dataset, which further enhances the robustness of the model, so that the model can effectively detect spam messages containing adversarial perturbation.

Keywords: Spam detection · Joint features · Robustness · Adversarial training

1 Introduction

Deep learning theory has greatly promoted the development of natural language processing (NLP). It has achieved amazing results in downstream tasks of text classification, as well as being used in various fields of industries. However, spam messages such as advertisement SMS and phishing emails often contain some adversarial perturbations like artificial sensitive word replacement and keywords splitting, making the distribution of these text data quite different from the normal text. They could by-pass the relevant text classification-based detection system without affecting the reading and understanding of readers. Therefore, network communication interfered, and citizens' property, privacy safety, and public safety were affected.

Based on such background, this paper focuses on how to effectively detect spam and makes the following contributions.

(1) A "ALBERT [1] -capsule network [2] " Chinese spam detection model is proposed, which can simultaneously mine the character-level and word-level features

Q. Liu et al. (Eds.): CENet 2020, AISC 1274, pp. 845–851, 2021.
https://doi.org/10.1007/978-981-15-8462-6_97

of the text, effectively avoiding the problem of information loss caused by the use of monolithic text features. The model also introduces multi-head structures to ensure that both character branch and word branch in this network can be trained efficiently.

(2) Extend the train set by using common adversarial methods to spam messages, such as traditional character transformations, pinyin rewriting, and word splitting, and train the model based on these adversarial methods to improve robustness.
(3) Train and test on publicly available spam datasets and compare them to classical NLP models. The results show that our models has a high performance of respectively 99.45% and 99.55% accuracy in spam SMS detection and spam mail detection, and can effectively defend against adversarial perturbation after an adversarial training.

2 Related Work

Given the dangers of spam messages, many researchers have attempted to explore efficient automatic detection methods, mainly filtering in the early stages by setting up the sensitive word, blocking and allowing a list of the email sender, or other rules. However, there are inadequacies in rules lag and easily bypassing detection. The development of statistic learning provides a solution for intelligent detection. Researchers realized the automatic detection based on the logistic regression [3], K-nearest neighbor's algorithm [4], and Naïve Bayes [5]. Some methods, combined with machine learning and rules, also have a good detection effect [6, 7]. The deep learning theory has further promoted the development of machine learning, and the neural network algorithm has been proved to be significantly better than traditional machine learning methods [8]. The accuracy of the model based on RNN and LSTM on the SpamSMSCollection dataset is as high as 98% [9]. The accuracy can be further improved by using "WordNet," "ConceptNet" and another external knowledge base [10].

The methods mentioned above mainly detect English spam messages, while related detection in the Chinese field is also facing more severe challenges. According to the 2019 Global Spam Report released by Kaspersky, China is currently the world's first source of spam, with the global proportion of spam soaring from 11.69% in 2018 to 21.26% [11]. Simultaneously, the differences between Chinese and English data make the detection methods for English spam not fully applicable to the Chinese field. Based on such predecessors, this paper further attempts to use the hybrid deep learning model with stronger representation ability to recognize Chinese spam messages, taking into account the robustness of the model to prevent artificial perturbations.

3 Model

The structure of the "ALBERT-capsule network" spam message classification model is shown in Fig. 1, which consists of five main components: dual input, character-level features stream, word-level features stream, joint feature, and multi-head output.

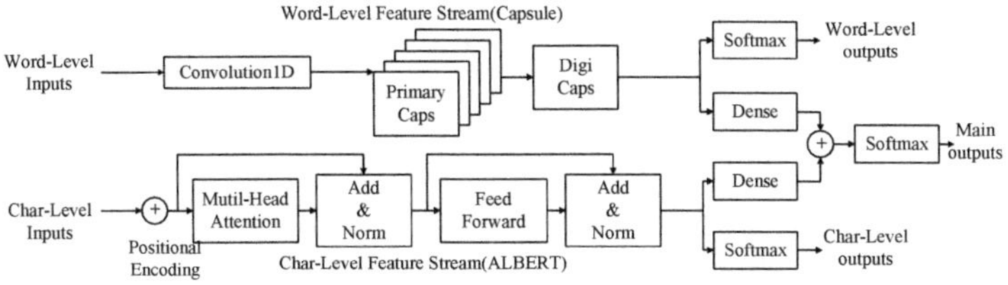

Fig. 1. ALBERT-Capsule network structure.

3.1 Dual Input

Text classification models for Chinese can be broadly divided into two branches: character-feature based models and word-feature based models. The character-level model encodes and processes Chinese characters directly, which will not be limited by the size of the word embedding vocabulary, with the disadvantage of potentially ignoring the grammar and semantics of the text. The word-level model, on the other hand, will first slice the Chinese text in terms of words using the segment technique, and then use the word embedding model to train more meaningful word vectors to reflect higher-level natural language semantic features, but will also face with problems such as out-of-vocabulary and sparse data due to low-frequency words. Also, the effectiveness of word-level models is affected by the quality of the segmentation tools and the existence of clear boundaries for the target words [12]. Therefore, we propose a dual-input mechanism to receive both the raw text and the segmented text, enabling the model to acquire both character-level and word-level features to improve the performance.

3.2 Two-Level Features Stream

Unlike English characters, Chinese characters are inherently capable of expressing specific meanings, and these semantics can be dynamically adjusted to the context, while this multiple meaning problem cannot be overcome by one-hot or word2vec encoding techniques commonly used in English character models. Therefore, we selected the ALBERT model to perform feature extraction of the raw text to obtain richer character-level features. ALBERT is a pre-trained language model based on a self-supervised training approach, similar in structure to the BERT., but has some engineering changes to reduce redundant parameters, making the model less likely to overfit when processing short text data sets such as SMS. For example, by using factorized embedding parameterization, the number of parameters of the embedding layer reduced from $O(\mathbf{V} \times \mathbf{H})$ to $O(\mathbf{V} \times \mathbf{E}+\mathbf{E} \times \mathbf{H})$, where $\mathbf{H} \gg \mathbf{E}$, and $\mathbf{V}$, $\mathbf{H}$ and $\mathbf{E}$ are the size of the one-hot embedding, the token embedding and the new hidden layer, respectively. Besides, setting cross-layer parameter sharing for the attention layer further reduces the number of network parameters, making the model have faster training speed.

However, the Chinese ALBERT model does not perform word segmentation during pre-training. It is a character-level pre-training language model and is not suitable for fine-tuning on the segmented dataset. Therefore, the model chooses to use capsule network to extract word-level features from the segmented text and uses a one-dimensional convolution layer to the output part to process long text more efficiently. The Capsule network was once considered as the replacement of neurons. It used the input-output structure of "Vector in Vector out" to replace the traditional "Vector in Scalar out" form of neurons. Among them, the length of the output vector can represent the existence of the feature, while the value of the vector element is used to represent the sub-feature, so the representational capability is superior to the traditional scalar output.

3.3 Joint Features and Multi-head Output

The model uses a multi-head output structure, which includes one main output and two branch outputs. The main output stitches the data from the word feature stream and the characters feature stream to obtain the joint features of words and characters. To make the model easier to be extended to the multi-classification problem to realize more category recognition of spam messages, the softmax function is used instead of the sigmoid function to predict the message category.

However, if the dense layer of the main output is too wide, the model would become overfitting easily, while on the contrary, the too narrow dense layer will cause loss of information. Therefore, we add separate branch outputs for the ALBERT model and the capsule network to ensure that they can be fully trained and reduce the width of the dense layers, and the loss function is set as the sum of the three losses of output, as shown in Eq. (1), where $\boldsymbol{W}$ is the neural network parameters, $\boldsymbol{\theta}$ used to adjust the importance of the different losses.

$$\text{loss}_{total} = \theta_1 \times \text{loss}(\boldsymbol{W}_{ALBERT}) + \theta_2 \times \text{loss}\left(\boldsymbol{W}_{Capsule}\right) + \theta_3 \times \text{loss}(\boldsymbol{W}_{all}) \tag{1}$$

3.4 Adversarial Training

Adversarial training refers to a training method that makes the neural network adapt to this change by introducing small perturbation in the training dataset, thereby making the model robust to the adversarial samples. It can be implemented by minimizing the loss function shown in Eq. (2). $\boldsymbol{\varepsilon}$ is the perturbation vector, $\boldsymbol{x}$ is a training example.

$$loss = -\log P(y|x + \boldsymbol{\varepsilon}) \tag{2}$$

To make the "ALBERT-Capsule Network" immune to the artificial adversarial perturbation method commonly used in spam messages, the training dataset is randomly extended using traditional Chinese character replacement, pinyin replacement, and word splitting.

4 Experiment Analysis

4.1 Experimental Datasets

To validate the effectiveness of the model proposed in this paper in the task of detecting spam messages, we collected data from public spam SMS dataset and trec06c email dataset for experimentation and used 'jieba' word segmentation tool to preprocess these datasets for training the word-level classifier. The details were shown in Table 1.

Table 1. Datasets details.

Dataset	P&N ratio	Training data	Validation data	Test data
SMS	1:1	18000	2000	2000
Trec06c	1:1	7200	800	2000

4.2 Performance

The detection results of our model and comparison models on the spam SMS and spam email datasets are shown in Table 2. For Bi-GRU, TextCNN, and capsule networks, word segmentation data was used for training. ALBERT is trained by an unsegmented texts, while the ALBERT-capsule network uses both the raw text and the text after the word segmentation for training. Because the benchmark performance of the dataset is high, accuracy, precision and F1 score are almost the same, so the experiment only selects the accuracy as the performance display.

Table 2. Model comparison experiments.

Models	Acc in SMS	Acc in Trec06c
Bi-GRU	98.10%	99.25%
TextCNN	98.00%	98.95%
Capsule	98.30%	99.20%
ALBERT	99.00%	98.75%
ALBERT-Capsule (single output)	98.30%	99.30%
ALBERT-Capsule (ours)	**99.45%**	**99.55%**

From the experimental results, it can be seen that the "ALBERT-Capsule network" proposed in this paper generally exceeds the benchmark method and is superior to other comparative models, and can maintain a detection accuracy of more than 99.4% on both datasets. For the SMS dataset, the short text length after word segmentation causes the word-level model to perform worse than the character-based ALBERT model.

However, the traditional GRU and TextCNN models in the trec06c email dataset with a long text length outperformed ALBERT. The ALBERT-capsule network can comprehensively extract characters-level and word-level features, so it is less affected by the difference between SMS and email dataset and has stable performance.

An in-depth analysis of the impact of different output structures on the performance of ALBERT-capsule networks shows that multi-head output structures can enhance the model effect. The experiment proves that the model tends to enhance the training of simpler capsule networks to improve the joint output if the penalty weights for ALBERT output are small. For example, one of the results showing the accuracy of the model's main output and word-level output in the validation dataset is 98.1% and 98.2%, while the accuracy of the characters-level output is only 50.9%, which is an approximate random guess. It is only when larger penalty weights are selected for the ALBERT output that both branches of the model are more likely to be trained effectively at the same time, and positive impact on the main output, e.g., when a model in the results table achieves 99.45% accuracy on the test dataset, The accuracy of characters-level output and word-level output was 98.45% and 98.25%.

4.3 Robustness Test

The experiment repeatedly extracts 30% of the words from the original SMS training dataset to modify to generate adversarial samples and expands the dataset by 5 times. On this basis, adversarial training is conducted for the previous models, and the experimental results are shown in Table 3.

Table 3. Adversarial Training experiments.

Models	Before adv training		After adv training		
	Adv Acc	Decrease	Org Acc	Adv Acc	Decrease
Bi-GRU	85.65%	12.45%	98.10%	97.00%	1.10%
TextCNN	91.30%	7.25%	98.05%	97.40%	0.65%
Capsule	87.70%	10.60%	97.40%	97.10%	0.30%
ALBERT	**97.15%**	1.82%	98.85%	98.70%	0.15%
ALBERT-Capsule	90.45%	9.00%	**99.25%**	**98.85%**	0.40%

According to the above experiments, SMS dataset with random perturbation can reduce the performance of the conventional models. Simultaneously, compared with the ALBERT model based on characters-level features, the word-level models are more susceptible to misleading. The ALBERT-Capsule Network model suffers from double interference of character-level and word-level. However, the ALBERT-capsule network model using adversarial training can effectively detect spam messages with adversarial perturbations, in the presence of the robustness while still maintaining better detection than other models. It was also found that as the size of the adversarial training dataset increases, the defenses of the model become stronger, but on the normal dataset, the accuracy of the detection system would be negatively affected.

5 Conclusion

The ALBERT-Capsule model based on the features of joint characters-words can effectively detect spam, and further enhance the robustness of the model through the adversarial training technology. In addition, methods such as multi-modal learning can be introduced to enhance the model's effect and detect spam messages that contain multimedia data such as pictures and videos.

Acknowledgement. This study was supported by Research Program of the Ministry of Public Security (2019JZX009, 2018GABJC03), the Scientific research and technological innovation on public security behavior of the People's Public Security University of China (Research on Deep Learning based Computer Vision Model Attack and Defense), the Key scientific research projects of colleges and universities in Henan Province (20B520008) and the research project of Henan police college (HNJY-2019-33, HNJY-2019-39).

References

1. Lan, Z.: Albert: A lite bert for self-supervised learning of language representations. arXiv: 1909.11942 (2019)
2. Sabour, S.: Dynamic routing between capsules. In: Advances in Neural Information Processing Systems, pp. 3856–3866 (2017)
3. Dedeturk, B.K.: Spam filtering using a logistic regression model trained by an artificial bee colony algorithm. Appl. Soft Comput. 106229 (2020)
4. Hassani, Z.A.: Classification method for e-mail spam using a hybrid approach for feature selection optimization. J. Sci. **31**(2), 165–173 (2020)
5. Venkatraman, S.: Spam e-mail classification for the internet of things environment using semantic similarity approach. J. Supercomput. **76**(2), 756–776 (2020)
6. Sedhai, S.: Semi-supervised spam detection in Twitter stream. IEEE Trans. Comput. Soc. Syst. **5**(1), 169–175 (2017)
7. Arif, M.H.: Sentiment analysis and spam detection in short informal text using learning classifier systems. Soft Comput. **22**(21), 7281–7291 (2018)
8. Jain, A.K.: Predicting spam messages using back propagation neural network. Wire. Pers. Communi. **110**(1), 403–422 (2020)
9. Chandra, A.: Spam SMS filtering using recurrent neural network and long short term memory. In: 2019 4th International Conference on Information Systems and Computer Networks (ISCON), pp. 118–122. IEEE (2019)
10. Jain, G.: Spam detection in social media using convolutional and long short term memory neural network. Ann. Math. Artif. Intell. **85**(1), 21–44 (2019)
11. Spam Report 2019. https://securelist.com/spam-report-2019. Accessed 1 Jun 2020
12. Li, X.: Is word segmentation necessary for deep learning of Chinese representations? arXiv: 1905.05526 (2019)

Domain Resolution in LAN by DNS Hijacking

Zongping Yin(✉)

Hefei Technology College, Anhui 230000, China
4751430@qq.com

Abstract. The LAN is often equipped with web servers. At this time, the LAN domain name resolution should be able to access the internal website and be easy to conFig without relying on the LAN export equipment. Based on the DNS protocol, the conventional method is to use dual-domain names, or use a dedicated DNS server, or use bidirectional NAT. These methods are insufficient. Domain name hijacking is a kind of hacker technology, but it can be used to apply domain name resolution in LAN properly. Hijacking can be realized through the DNAT function of net-work devices such as a firewall. The internal website's domain name can be resolved to the IP address of the local area network by the dedicated DNS server, and other domain name resolution requests can be forwarded to the external DNS server. The method has the advantages of convenience, reducing a load of network exit equipment and no increase in cost.

Keywords: LAN · DNS · Domain hijacking · DNAT

1 Traditional Method of Domain Name Resolution in LAN

1.1 Structure of LAN

LAN is the most normal part of the Internet, such as the school network, the enterprise network. The web server can be deployed in LAN. The typical structure of the LAN is shown in Fig. 1. This paper takes this Fig. as the background.

Source Network Address Translation (SNAT) should be conFig. d on the router so that computers in the LAN can access the Internet. The specific method is to change the source address of the IP datagram to the Internet IP address of the router for all traffic entering the router from the LAN IP address [1]. The operation process is shown in Table 1.

In order to access the websites in the LAN from the Internet, Destination Network Address Translation (DNAT) should be conFig. d on the router. Specifically, for the traffic whose destination address is the Internet IP address of the router, change the destination address to the IP address of the web server in the LAN [2]. The operation process is shown in Table 2.

When web servers are deployed in the LAN, domain resolution requirements are: can access the websites in the LAN from the LAN, not to manually conFig. DNS parameters for each client, and reduce dependence on export devices.

Q. Liu et al. (Eds.): CENet 2020, AISC 1274, pp. 852–858, 2021.
https://doi.org/10.1007/978-981-15-8462-6_98

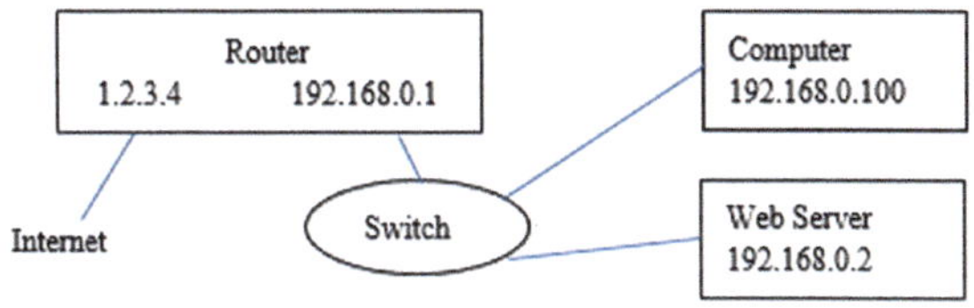

Fig. 1. Typical structure of LAN.

Table 1. Operation process of SNAT.

NO.	Operation process
1	192.168.0.100 sends IP datagram and requests to access server 5.6.7.8 on the Internet
2	The IP datagram arrives at the router and its source address is modified from 192.168.0.100 to 1.2.3.4
3	IP datagram is forwarded on the Internet and finally reaches 5.6.7.8
4	5.6.7.8 return the response IP datagram, target address 1.2.3.4, and forward it to the router
5	Router forwards IP datagram to 192.168.0.100 according to previous records

Table 2. Operation process of DNAT.

NO.	Operation process
1	Internet client requests access www.a.com, the IP to be resolved is 1.2.3.4
2	Internet customers send access requests to 1.2.3.4, and IP datagrams are forwarded to the router
3	On router 1.2.3.4, the destination address of IP datagram is modified to 192.168.0.2
4	IP datagram forwarded to web server 192.168.0.2
5	192.168.0.2 respond to the access request, and the response data is forwarded to the visitor

1.2 Principle of Domain Resolution

The Domain Name System delegates the responsibility of assigning domain names and mapping those names to Internet resources by designating authoritative name servers for each domain. Network administrators may delegate authority over sub-domains of their allocated name space to other name servers [3]. See Fig. 2.

When the client accesses the domain, the domain resolution request is processed locally. If the host file exists, the resolved IP address is returned according to its configuration. If this file does not exist or there is no resolution record, the resolution request is sent to the DNS server specified by this machine for recursive query.

The recursive query [4] process is as follows: the local specified DNS server is usually provided by ISP, or the public DNS server on the Internet, such as Google's public DNS server 8.8.8.8. Generally, there are two DNS servers specified by this machine. If the first one does not respond, the resolution request will be sent to the

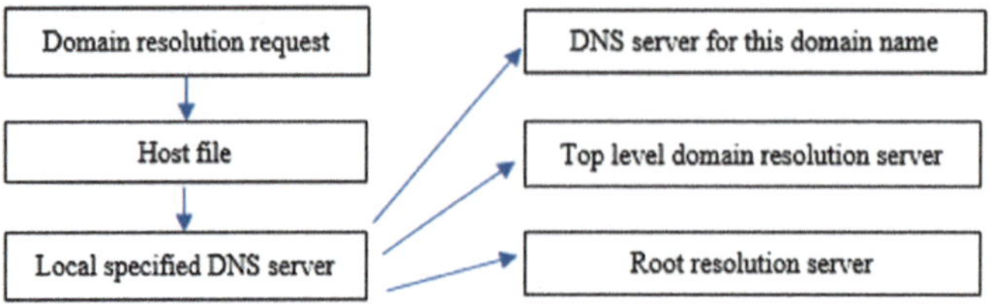

Fig. 2. DNS resolution principle.

second one. After receiving the domain resolution request, the DNS server specified by this computer will process it according to the rules. Usually it will first find in its resolution record, or in its resolution cache. If found, returns the resolved IP address to the client. If not, make an iterative query request to the root resolution server.

The iterative query [5] process is as follows: the root resolution server returns the next level DNS server related to the domain name, such as the resolution server of the top-level domain name such as. Com. Next, the DNS server specified by this computer sends a query request to the resolution server of the top-level domain name, and generally obtains the DNS server of the domain name. The DNS server specified by this computer, then send a query request to the DNS server of this domain name to obtain the resolved IP address. This resolution record will be cached by the DNS server specified by this machine and fed back to the client.

1.3 Methods of Domain Resolution in LAN

When web servers are deployed in the LAN, the domain resolution methods in the LAN are as follows.

Use Two Domains. The website deployed in the LAN is accessed by double domains. One domain points to the LAN IP address of the web server (such as 192.168.0.2), which is used by the intranet. One domain points to the Internet IP address (such as 1.2.3.4) used by destination address translation on the LAN exit router for Internet access.

The disadvantage of this method is that it is inconvenient to remember two domain names.

Use Dedicated DNS Server. The dedicated DNS server [6] is recommended to be conFig. d in the LAN (for example, the IP address is 192.168.0.3), or on the Internet. In the dedicated DNS server, conFig. zone for the domain of the website in the LAN, and conFig. the resolution record in the zone (for example www.a.com), which points to the web server in the LAN (such as 192.168.0.2). Because it also responds to other domain resolution requests, it is allowed to forward the domain resolution requests of other zones to the DNS server provided by ISP or the public DNS server on the Internet, such as Google's public DNS server 8.8.8.8, which will resolve and return the results.

Set the DNS parameters of each computer in the LAN to 192.168.0.3 manually, so that the resolution requests of computers in the LAN will be delivered to the dedicated DNS server, www.a.com will be responded to as 192.168.0.2.

The disadvantage of this method is to manually conFig. the client, which is difficult to operate when there are many computers in the LAN.

Bidirectional NAT in Export Router. Bidirectional NAT refers to the simultaneous configuration of DNAT and SNAT [7], which have been conFig. d in the previous example, the router has only one Internet IP address, which is used for both SNAT sharing Internet surfing, and for DNAT binding internal website.

When the client in the LAN (such as 192.168.0.100) accesses the website in the LAN, and use the same domain (such as www.a.com, resolved to 1.2.3.4), the access request arrives at the exit IP address of the LAN (such as 1.2.3.4), due to the SNAT, the access request is forwarded to the web server in the LAN to realize the website access.

This method is more commonly used. It does not need dual domains or dedicated DNS servers. Table 3 analyzes the traffic passing through the router. All the data of access request and response will pass through the exit router. In some extreme application scenarios, for example, there is a large amount of network teaching video traffic in the campus network, which will increase the load of export equipment, and may form a network performance bottleneck.

Table 3. Router traffic in bidirectional NAT.

NO.	Operation process	Through router
1	Client 192.168.0.100 request resolution www.a.com	Yes
2	Client gets DNS response, IP is 1.2.3.4	Yes
3	Client request access 1.2.3.4	Yes
4	The router forwards the request to the server 192.168.0.2	Yes
5	The server responds to the request, and the target address of the response data is 1.2.3.4	Yes
6	The router forwards the response data to the client 192.168.0.100	Yes

2 Domain Resolution Through Domain Hijacking

This article propose a method of domain name resolution in LAN, which can solve many shortcomings of traditional methods.

2.1 Domain Hijacking

Domain hijacking or domain theft is the act of changing the registration of a domain name without the permission of its original registrant, or by abuse of privileges on domain hosting and registrar software systems [8].

One way to hijack a domain is to modify the DNS server of the domain after the hacker obtains the domain control right, so that the domain related businesses, such as websites and mailboxes, are under the control of the hacker. The second way of domain hijacking is that hackers intercept data packets on domain requests on the network, tamper with them, make them return false resolution IP address, and realize the control of domain related businesses. In this article mode two is studied, which is used for the legitimate purpose of domain resolution in LAN.

2.2 Principle of Domain Resolution by Domain Hijacking

DNS has two types of messages: query and response. Both types have the same format. The query message consists of a header and question records; the response message consists of a header, question records, answer records, authoritative records, and additional records [9].

Both DNS request message and DNS response message do not contain IP address part. DNS message is first transmitted through UDP protocol in the transmission layer and encapsulated into UDP message. And then in the network layer, it is transmitted through IP protocol and encapsulated into IP datagram. In the IP datagram, it has the source IP address part and the destination IP address part.

In domain resolution in LAN, the client's domain resolution request is sent to the local designated DNS server. If the destination address of IP datagram can be modified on the network, the domain resolution request will reach the modified address. If the address is a DNS server, it will respond to the domain resolution request and return the resolution result to the client in the LAN.

2.3 Method of Domain Resolution by Domain Hijacking

Applying the above principles to realize domain resolution in LAN, we need to solve two problems: one is how to modify the destination address of IP datagram on the network, the other is how to respond to all domain resolution requests in DNS server, and the final network structure is shown in Fig. 3.

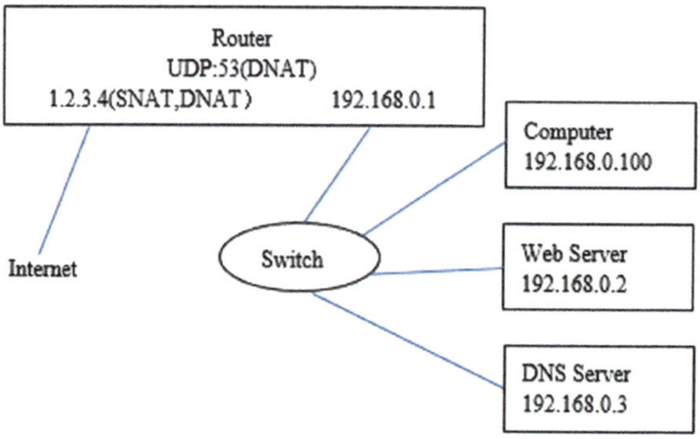

Fig. 3. Structure of domain resolution by domain hijacking.

Modify the Destination Address. By using the function of destination address conversion in network equipment, the destination address can be easily modified, which is possessed by general firewalls and routers. In order to achieve special functions, the content of IP datagram can be modified or its flow direction can be changed through network devices or software methods. There have been relevant research results [10]. ConFig. rules in firewall or router to capture all IP datagrams with DNS request characteristics, and select all UDP traffic with target port 53. The destination address is designated as the DNS server for LAN.

Use a Dedicated DNS Server. The dedicated DNS server shall resolve the domain of the website deployed in the LAN to the IP address of the LAN, and resolve other domains normally. This is the same as the function of the dedicated DNS server mentioned above, which can be conFig. d as described above.

The operation process of domain resolution in LAN by domain hijacking is shown in Table 4.

Table 4. The process of domain resolution in LAN by domain hijacking.

NO.	Operation process	Through router
1	Client 192.168.0.100 requests domain resolution	Yes
2	The resolution request arrives at the router, and the destination address is replaced with 192.168.0.3 and forwarded	Yes
3	The resolution request reaches 192.168.0.3, the response address is 192.168.0.2, and it is forwarded	Yes
4	192.168.0.100 get DNS response, request to visit 192.168.0.2	No
5	Web server 192.168.0.2 responds to the request and delivers it to 192.168.0.100	No

When a user in the LAN visits the website, no matter which DNS server he specifies locally, the domain request will be sent to the dedicated DNS server. It is convenient without modifying the DNS of the client. This dedicated DNS server is only used by this LAN. You still need to conFig. this domain DNS server on the Internet.

If you visit the website in the LAN, the IP address of the LAN of the web server is returned. The subsequent website visit and response will be forwarded through the switch in the LAN. There is no need to go through the exit router, firewall and other devices. So that traffic-jam in these devices can be avoided. When the request and response flows are large, the possibility of network congestion is reduced and the network performance is improved.

This method does not need to modify the network structure or add hardware and software.

This method can also realize the access control function. For example, to block the access to the illegal domain, we only need the domain needs to be established in the dedicated DNS server and resolved to the suggestive web page.

References

1. Source Network Address Translation. https://www.juniper.net/documentation/en_US/contrail19/topics/task/configuration/snat-vnc.html. Accessed 25 July 2018
2. Port forwarding. https://en.wikipedia.org/wiki/Port_forwarding. Accessed 31 May 2020
3. Domain Name System. https://en.wikipedia.org/wiki/Domain_Name_System. Accessed 31 May 2020
4. What Is Recursive DNS. https://www.cloudflare.com/learning/dns/what-is-recursive-dns. Accessed 31 May 2020
5. Difference between iterative and recursive dns query. https://www.slashroot.in/difference-between-iterative-and-recursive-dns-query. Accessed 12 Jan 2013
6. BIND 9 Administrator Reference Manual. https://bind9.readthedocs.io/en/latest/index.html. Accessed 31 May 2020
7. The TCP/IP Guide. http://wwwtcpipguide.com/free/t_IPNATBidirectionalTwoWayInboundOperation.htm. Accessed 20 Sep 2005
8. Domain hijacking. https://en.wikipedia.org/wiki/Domain_hijacking. Accessed 15 Feb 2020
9. Explain DNS message format with neat diagram. https://www.ques10.com/p/10908/explain-dns-message-format-with-neat-diagram-1. Accessed 01 Aug 2016
10. Yin, Z.P.: Method of server using the remote ip address. Comput. Telecommun. **3**, 68–70 (2015)

Study on Distributed Intrusion Detection Systems of Power Information Network

Shu Yu[1(✉)], Taojun[1], and Zhang Lulu[2]

[1] Economic and Trade Department, Anhui Business College, Wuhu 241003, China
603903027@qq.com
[2] School of Computer Science and Engineering, Southeast University, Nanjing 211102, China

Abstract. This paper takes the electric power information network as the model of research, studies the captured data packets and the characteristics of data flow, analyzes the specific characteristics of the audit data from the angle of anomalynetwork traffic, and combines statistical variance method and regression analysis method to put forward an intrusion detection system on bases of distributed Network flow.

Keywords: IDS · Electric power information network · Regression analysis

1 Introduction

This paper gives research and analysis on the common attacks on power information network intrusion, through collected information from the Power Information Network (IDS weekly). It is found that the traditional Power Information Network IDS systems exists problems like long response times, higher false positives. In practical applications, network security system is not responding. In this paper, the network traffic is divided into normal and abnormal traffic flow, analysis characteristics of intrusion in abnormal traffic flow. The traffic processing is based on the current intrusion flow characteristics, filter and test the current accumulation of abnormal traffic flow congestion control features, restore accidentally misread flows. We can construct a flow analysis model and use it to design a suitable Intrusion Detection flow analysis algorithms for Power Information Network. Construct a flow monitoring prototype based on the flow analysis algorithm. According to different states of normal and abnormal flow conditions, combined with the analysis of actual test environment, to establish a real-time internal key server and multiple computer abnormal flow processing network monitoring system.

Q. Liu et al. (Eds.): CENet 2020, AISC 1274, pp. 859–865, 2021.
https://doi.org/10.1007/978-981-15-8462-6_99

2 Modeling of Intrusion Detection Based on the Flow Prediction and the Methods

2.1 Modeling of Network Traffic and Linear Regression Algorithm

Objective things always interrelated and interdependent, the quantity relationship between them is generally divided into two types. One is the deterministic relationship, which is the function relationship; another kind is uncertain relations, called the relation. Because the network flow and the time axis have a certain relationship, we can make flow predictions based on the shift of time axis. Regression analysis prediction method is the analysis of a strain of one or more variables dependencies, and through the already known or settled value to estimate and predict the variable mean stream. Regression analysis prediction method can be divided into linear regression, nonlinear regression analysis.

2.2 Unary Linear Regression Algorithm

Linear regression prediction method means there is a linear relationship between one or more variables and dependent variables (single variable called unary linear regression, two or more variables called multi linear regressions). It is a method of predict the average trends of dependent variables based on the change of variables.

And Unary Linear Regression is a simple and widely used prediction method. so, here I give a brief introduction to the Unary Linear Regression mode: Set X as variables, Y as dependent variables, there is a linear relationship between Y and X: choose prediction model:

$$\{Y = a + bX + \varepsilon \\ \varepsilon \sim N(0, \sigma^2) \tag{1}$$

Here Y is the random error, a, d are the regression coefficients So $Y \sim N(a + bx, \sigma^2)$. We have $E(Y) = a + bx + E(\varepsilon) = a + bx$, μ(X)is the regression function of Y about X, this is a question of Unary Linear Regression.

2.3 Unary Nonlinear Regression Algorithm

In reality we have complex problems, under certain circumstances, we can solve it as a unary linear regression problems though appropriate variable transformations.

For the regression prediction of flow analysis algorithm, the flow is a fluctuation value, its function type is quite similar to the S type curve. So S type curve is used as the predictive model.

Predictive model $y = \frac{1}{a + be^{-x}}$:

$$\begin{cases} \hat{a} = \bar{y} - \bar{x}\hat{b} \\ \hat{b} = \frac{\sum_{i=1}^{n}(x_i - \bar{x})}{\sum_{i=1}^{n}(x_i - \bar{x})^2}(y_i - \bar{y}) \end{cases} \tag{2}$$

Get the value of nonlinear regression mode $y = \frac{1}{a + be^{-x}}$.

3 Test and Analysis of Network Anomaly Detection System

3.1 Linear Regression Algorithm Prediction

We can acquire the network flows by capture the data packets module, and sampling the intervals, by divide the network traffic into segments. And predict flows of other intervals through the network traffic linear regression algorithm, in this way we can get the whole network flow traffic.

Results of the test is in Table 1, the test time is 20 min, with 10 intervals of 2 min a time unit, base on the flow of the first 5 samples we can predict the flows of the next 5 units. So we can get the detecterror rate by comparing with the actual flow samples, as it shows in Fig. 2 and Fig (1).

Table 1. Experiment Result of Linear Regression Algorithm

Time (min)	2	4	6	8	10	12	14	16	18	20
Actual Flow (kb)	517	657	485	412	521	364	431	525	484	455
Predict flow (kb)						449	426	403	380	357
Error rate(%)						19%	1%	30%	27%	27%

From the table we can see that the error rate is pretty big because of the characteristics of linear function, as the time goes, the offset value of prediction is bigger.

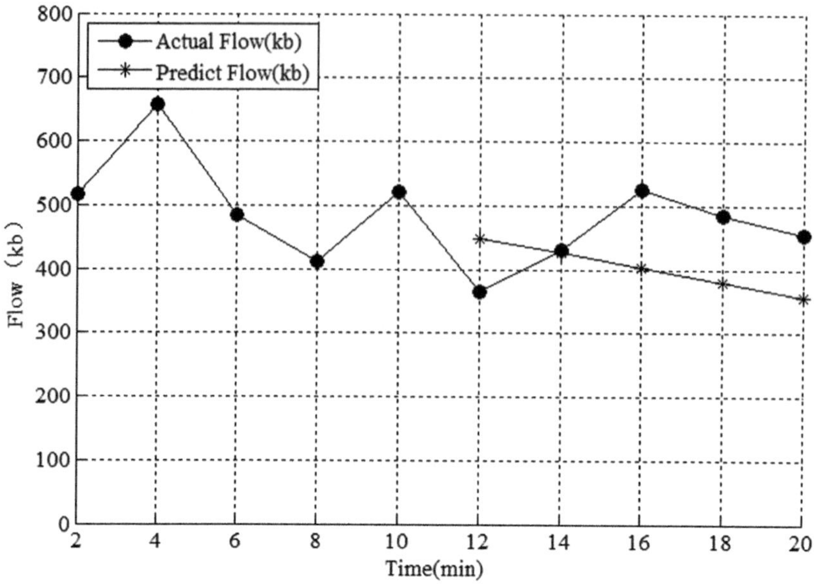

Fig. 1. Experiment result of linear regression algorithm

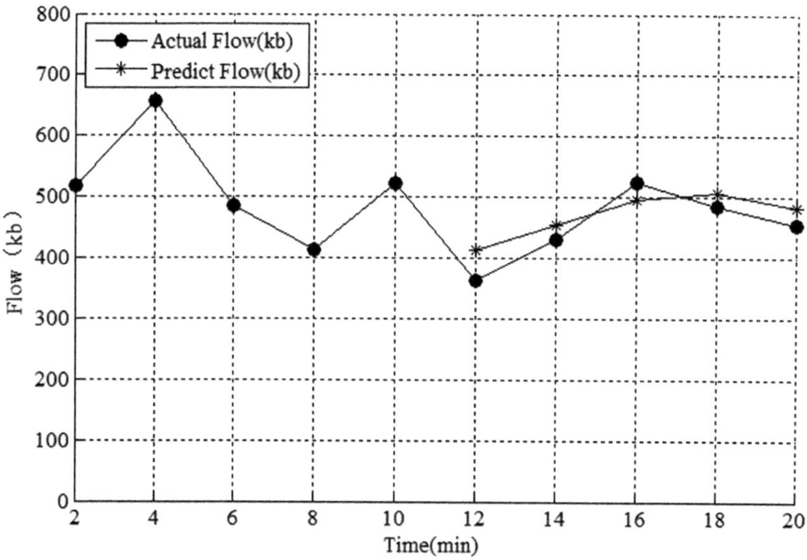

Fig. 2. Experiment result of nonlinear regression algorithm

3.2 Nonlinear Regression Algorithm Prediction

We can acquire the network flows by capture the data packets module, and sampling the intervals, by divide the network traffic into segments. And predict flows of other intervals through the network traffic nonlinear regression algorithm, in this way we can get the whole network flow traffic.

Results of the test is in Table 2, the test time is 20 min, with 10 intervals of 2 min a time unit, base on the flow of the first 5 samples we can predict the flows of the next 5 units. So we can get the detect error rate by comparing with the actual flow samples, as it shows in Fig. 3.

Table 2. Experiment result of nonlinear regression algorithm

Timemin	2	4	6	8	10	12	14	16	18	20
Actual flow (kb)	517	657	485	412	521	364	431	525	484	455
Predict flow (kb)						414	455	497	507	483
Error rate(%)						13%	6%	5%	5%	6%

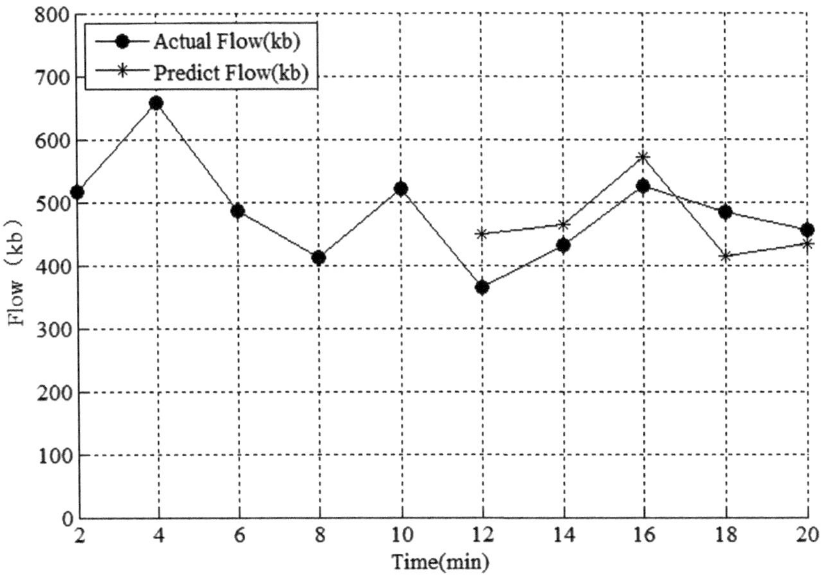

Fig. 3. Experiment result of improved linear regression algorithm

3.3 Result Comparison

Through the comparison of the two methods above, we find in long-term flow prediction the nonlinear regression algorithm is more accurate than linear prediction algorithm, and the linear regression algorithm is more suitable for predicting the flow right after the sampling point.

Thus we conclude a new and improved linear regression network prediction algorithm, Record the actual flow when the interval linear regression algorithm predicts the flow adjacent to the flow of sampling space and record the flow of that time, then it goes on. In this way, although the sampling space is changing, the predicted flow is always followed by the flow of the time when it is sampled. This improved Linear regression Prediction algorithm is more accurate than the pervious linear regression algorithm. As in following Table 3, the data is in Fig. 4.

Table 3. Experiment result of improved linear regression algorithm

Time min	2	4	6	8	10	12	14	16	18	20
Actual flow (kb)	517	657	485	412	521	364	431	525	484	455
Predict flow(kb)						449	464	571	414	433
Error rate(%)						19%	8%	9%	16%	5%

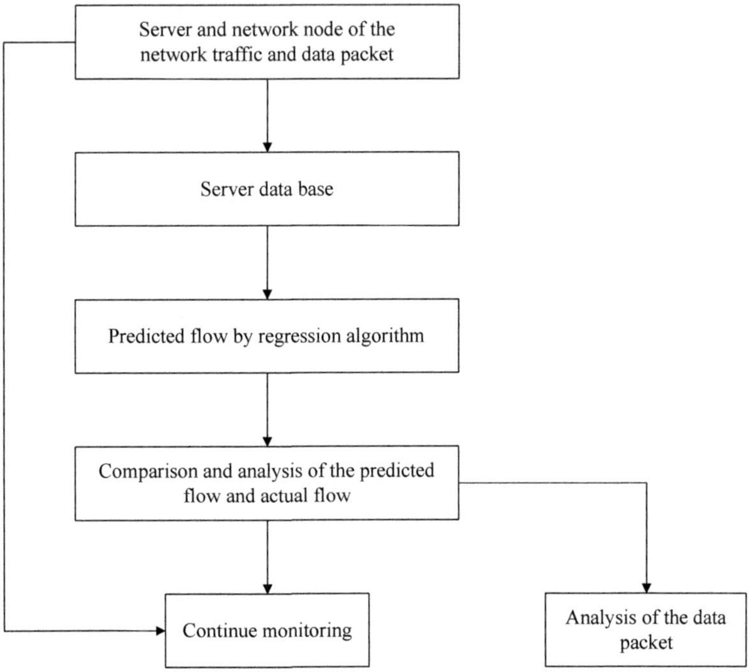

Fig. 4. Network monitoring mode I

The accuracy of improved algorithm is greatly improved, but the predictions are not as accurate as nonlinear regression algorithm.

Because the function of the nonlinear algorithm sampling is S type function, its character is more close to the network flow we predicted, so the value is relatively accurate, but the error will not increase with the extension of time detection point.

3.4 Model Construction

Through the previous analysis of the experimental data, we can construct a network traffic monitoring model suitable for electric power information network.

4 Conclusion

All the current intrusion detection systems of power information network are based on the Anomaly detection or misuse detection technology. But due to insurmountable disadvantage of Anomaly detection and misuse detection technology, these systems not always working well. The model I proposed in this paper have overcame all those shortcomings, it is a good way to solve these problems.

References

1. Bawany, N.Z., Shamsi, J.A.: SEAL: SDN based secure and agile framework for protecting smart city applications from DDoS attacks. J. Netw. Comput. Appl. **145**, 102381 (2019)
2. Conti, M., Lal, C., Mohammadi, R., Rawat, U.: Lightweight solutions to counter DDoS attacks in software defined networking. Wire. Netw. **25**(5), 2751–2768 (2019)
3. Nabil, H.R., Kamoun, S.F.: DDoS flooding attack detection scheme based on F-divergence: Comput. Commun. **35**(11), 1380–1391 (2012)
4. Yu, S., Zhou, W., Doss, R., Jia, W.: Traceback of DDoS attacks using entropy variations. IEEE Trans. Parallel Distrib. Syst. **22**(3), 412–425 (2011)
5. Xiang, Y., Li, K., Zhou, W.: Low-Rate DDoS attacks detection and traceback by using new information metrics. IEEE Trans. Inf. Forensics Secur. **6**(2), 426–437 (2011)
6. Zhang, C.W., Yin, J.P., Cai, Z.P., Chen, W.F.: RRED: Robust RED algorithm to counter low-rate denial-of-service attacks. IEEE Commun. Lett. **14**(5), 489–491 (2010)
7. Ramadhan, G., Kurniawan, Y., Kim, C.S.: Design of TCP SYN Flood DDo S attack detection using artificial immune systems. In: International Conference on Frontiers of Information Technology (2016)
8. Kumar, M., Tripathi, S., Agrawal, N., Singh, S.N.: Growth of premature neonates admitted in a level III neonatal unit. Clin. Epidemiol. Glob. Health **2**(2), 56–60 (2014)
9. Ji, Q.J., Dong, Y.Q.: A load-adaptive active queue management algorithm. J. Softw. **17**(5), 1140–1148 (2006)
10. Buragohain, C., Yoti, M.J., Singh, S., et al.: Anomaly based DDoS attack detection. Int. J. Comput. Applications (2015)
11. Sun, Z.X., Li, Q.D.: Defending DDos attacks based on the source and destination IP address database. J. Softw. **18**(10), 2613–2623 (2007)
12. Minghui, Y., Ruchuan, R.: DDoS detection based on wavelet kernel support vector machine. J. Chin. Univ. Posts Telecommun. **15**(3), 59–94 (2008)
13. LU, N., Wang, Y., Shi, W., Diao, D.: Filtering location optimization for defending against large-scale BDoS attacks. Chin. J. Electron. **26**(2), 435–444 (2017)
14. Li, M. Li M.: An adaptive approach for defending against DDoS attacks (2010)

Credible Identity Authentication Mechanism of Electric Internet of Things Based on Blockchain

Liming Wang[1], Xiuli Huang[2(✉)], Lei Chen[2], Jie Fan[2], and Ming Zhang[1]

[1] State Grid Jiangsu Electric Power Co., Ltd., Nanjing 211106, China
[2] State Grid Key Laboratory of Information & Network Security Global Energy Interconnection Research Institute Co., Ltd., Beijing 102200, China
71737069@qq.com

Abstract. In order to solve the problems of low authentication efficiency and easy tampering of authentication data in the electric IoT system, this paper proposes the electric IoT based on blockchain. First, the authentication model based on blockchain is proposed, and four key issues are analyzed, such as data security storage, user biometrics extraction, power terminal equipment feature extraction, and authentication key generation, then technical solutions are proposed. Secondly, based on the authentication model, the power terminal registration process, user registration process, and authentication process are proposed. The performance analysis of the authentication mechanism is carried out from four aspects: availability, integrity, privacy and efficiency. The analysis results show that the authentication mechanism proposed in this paper has good performance. Finally, the application scenarios and precautions of the authentication mechanism are analyzed.

Keywords: Electric Internet of Things · Authentication mechanism · Blockchain · Biometrics

1 Introduction

With the maturity and development of the Internet of Things technology, the application of the Internet of Things technology in the field of power is gradually increasing. Because power terminals can be controlled and managed remotely, attackers can exploit system vulnerabilities to achieve illegal control of power terminals. In order to ensure the safe access of the power terminal, the trusted service technology has been applied to the authentication work and has achieved good results [1].

In order to improve the security of the authentication technology, the existing research mainly studies from the terminal side security and platform side security. Regarding the research on improving authentication technology on the terminal side, literature [2] proposed a mobile terminal authentication mechanism based on biometrics technology for the problem that identity information is easily tampered with during the authentication process. In reference [3], the NFC (near field communication) technology is integrated with the electronic ID card to solve the problem of easy data

Q. Liu et al. (Eds.): CENet 2020, AISC 1274, pp. 866–875, 2021.
https://doi.org/10.1007/978-981-15-8462-6_100

leakage during the authentication process, and a data security transmission mechanism is proposed. Reference [4] uses SWP encryption technology to improve the security capability of the data processing module of the SIM card, which effectively improves the security of the SIM card. Regarding the issue of identity leakage of mobile terminals, literature [5] and literature [6] respectively proposed key generation algorithm and two-dimensional code encryption technology, which effectively solved the problem of identity leakage of mobile terminals. In terms of research on improving authentication technology on the platform side, literature [7] uses deep learning algorithms to improve the mobile terminal's identity recognition algorithm. Documents [8, 9] realized the safe transmission of mobile terminal data by adopting virtualization technology, and solved the problem of data leakage. Literature [10] proposed a blockchain-based authentication mechanism to solve the problem of centralized failure in cross-domain authentication. Literature [11] and literature [12] respectively adopt quantum key algorithm and generative adversarial network technology to effectively solve the problem that key data is easy to be attacked.

We can see that in the Internet of Things identity authentication, more research results have been achieved. However, the existing identity authentication mechanism generally adopts a trusted center, an Internet of Things terminal, and a user for mutual authentication. In order to solve this problem, this paper proposes a blockchain-based trusted authentication mechanism for electric power Internet of Things, and verifies the performance of this authentication mechanism through performance analysis.

2 Architecture

2.1 Blockchain-Based Authentication Model

With the rapid construction and operation of the electric power internet of things in the electric power field, there are currently many electric power businesses operating on the electric power internet of things. The business on the power Internet of Things can be divided into internal business and external business. Typical internal business includes online national network, power trading, smart supply chain, mobile office, etc. Typical external business includes smart car networking, energy finance, and energy services.

Blockchain technology has the advantages of decentralization and non-tamperable data. Based on this, in order to achieve security authentication between users and power terminals, this paper proposes a user authentication model based on blockchain as shown in Fig. 1, including three levels of user terminals, power terminals, and authentication centers. Among them, the user terminal mainly completes the functions of collecting user information, collecting user instructions, and secure data communication between the user and other devices. The power terminal mainly completes the functions of collecting information in the application scenario, accepting and executing user instructions, and performing secure communication and operation with related equipment. The authentication center mainly realizes the identity authentication of users and power terminals. The authentication server mainly implements the work of receiving, authenticating, and generating communication keys for user and power terminal authentication requests; the blockchain node mainly implements storage of data such as user information, power terminal information, and authentication information of the authentication server.

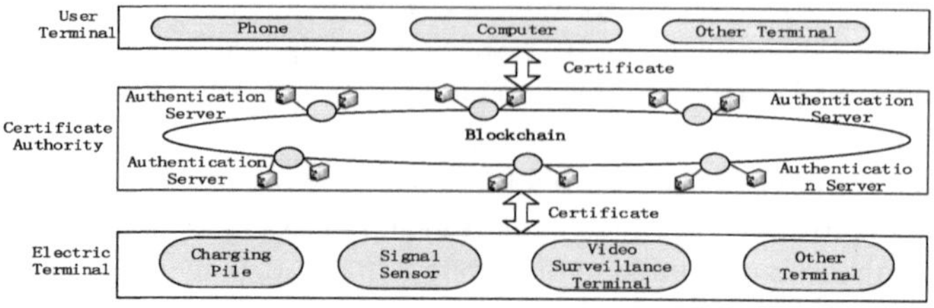

Fig. 1. Blockchain-based authentication model.

Seen from Fig. 1, each blockchain node is connected to two authentication servers and can be deployed to a power company for user and power terminal authentication within the jurisdiction of the power company. Among them, two authentication servers constitute the main and standby architecture, which can effectively solve the problem of single point of failure. In terms of the implementation of blockchain technology, considering that the alliance chain is very suitable for solving the data security problem within an organization, this article chooses the alliance chain technology to build the authentication service function.

2.2 Key Technologies in the Model

In order to achieve security authentication, the following four key issues need to be solved: the data of the blockchain nodes cannot be tampered with and decentralized, user feature information collection and identity information generation, power terminal feature information collection and identity information Generate, generate the authentication key.

Blockchain Technology. In terms of distributed storage, in order to mobilize the enthusiasm of data storage resources, Inter Planetary File System (IPFS) is used. IPFS is the most used storage technology in the current alliance chain. It can achieve rapid storage through an incentive mechanism, and can guarantee the security and privacy of data. In terms of blockchain services, a practical Byzantine fault-tolerant algorithm (PBFT) is used to implement the distributed ledger function. In terms of smart contracts, it adopts two technologies: identity management contract and data management contract. Among them, in terms of identity management contracts, by defining power terminal identity management contracts, user identity management contracts, and authentication center identity management contracts, the creation and management of identity data is realized. In terms of data management contracts, it mainly implements the storage and query of power terminal data, user identity data, and authentication center data.

User Biometric Extraction Technology. In order to ensure that the feature information generated for the user has high security, this article introduces the user's biological characteristics on the basis of the user's name, user password and other basic features. In order to improve the security of the authentication mechanism, the biometrics used in this paper include three elements: fingerprint, iris, and face. It is considered that the collected biometric data is related to the environment where the user is located and the terminal recognition ability. In order to reduce the uncertainty caused by the difference in biometrics to the authentication process, this paper uses the biometrics generation function to standardize the user's biometrics.

Biometric generation functions include feature initialization functions and feature normalization functions. The feature initialization function (FIF) is $FIF(B_i) = (R_i, P_i)$, which means that this function can get two character strings R_i and P_i composed of (0, 1) and by processing a certain biological feature B_i of the user. The feature normalization function (FNF) is $FNF(B_i', P_i) = R_i$, which means that the function can obtain a standardized string R_i by processing a user's certain biological characteristics B_i' and random string P_i. Based on this, each time the user's biometrics are collected, a standardized character string R_i can be generated as standard user biometrics.

The Feature Extraction Technology of Power Terminal. This paper uses a combination of device fingerprint features and device environment features to implement. Device fingerprint characteristics refer to the unique identification of the device during the production process. For example, the network access identifier of the hardware, the MAC address of the network card, etc. Equipment environmental characteristics refer to the characteristics of the equipment during normal operation. For example, the latitude and longitude range where the device is located, relevant device information around the device, etc. Based on this, this paper uses the device's fingerprint D_{fp}^{j}, IP/MAC address D_{im}^{j}, and protocol type D_{p}^{j} to generate characteristic information of the power terminal device.

The Generation Technology of Authentication Key. This paper uses a combination of random numbers and hash functions to implement the technology. First, the certification center generates secret values for each user and power terminal participating in the certification; second, a hash function operation based on the secret values of the certification center and the secret values of the participants can obtain a unique authentication key for each participant. As for the secret value, use a value x_g representing the secret value of authentication center, use a value y_i representing the secret value of user i, and use a value D_j representing the secret value of power terminal equipment j. Based on this, the calculation process of the authentication keys k_{cu} of the certification center and users is $k_{cu} = h(x_g||y_i)$, and the calculation process of the authentication keys k_{cd} of the certification center and the power terminals is $k_{cd} = h(x_g||D_j)$. Among them, $h(\cdot)$ represents the hash function.

3 Authentication Mechanism

The authentication mechanism designed in this paper includes the registration process and the authentication process. The registration process includes two processes: power terminal registration and user registration. The details are described below.

3.1 The Registration Process of Power Terminal

The power terminal registration process includes collecting power terminal feature data, generating a power terminal key, sending registration parameters to a certification center, the certification center generating an authentication key for the power terminal, the certification center calculating the power terminal's authentication characteristics, and converting the power terminal's authentication information Stored to the blockchain node, the blockchain node stores the authentication information of the power terminal in seven steps.

1. Collect power terminal feature data: The collected power terminal feature data includes device fingerprints D^j_{fp}, IP/MAC addresses D^j_{im}, and protocol types D^j_p.
2. Generate the key of the power terminal: the key D^j_B of the power terminal is calculated using the formula $D^j_B = h\left(D^j_{fp}||D^j_{im}||D^j_p\right)$.
3. Send registration parameters to the certification center: The registration parameters that the power terminal needs to send to the certification center include the power terminal characteristic data and the owner's data of the power terminal. The power terminal feature data includes device name D^j_{id}, device password D^j_{pwd}, and device key D^j_B. The data of the owner of the power terminal includes the company to which it belongs and the authority level (query data, modify data, delete data).
4. The authentication center generates an authentication key for the power terminal: the authentication center generates a secret value D_j for the power terminal, and uses the formula $k^j_{cd} = h(x_g||D_j)$ to calculate the authentication key k^j_{cd} for the power terminal.
5. The certification center calculates the authentication characteristics of the power terminal: the certification center uses the formula $A^j_D = h\left(D^j_{id}||D^j_{pwd}||D^j_B||k^j_{cd}\right)$ to calculate the authentication characteristics A^j_D of the power terminal.
6. Store the authentication information of the power terminal to the blockchain node: the authentication center stores the relevant information of the power terminal to the blockchain node, and the stored information includes the characteristic data of the power terminal, authentication characteristics A^j_D, secret value D_j, and owner information.
7. The blockchain node stores the authentication information of the power terminal: To store the authentication information of the power terminal, the identity management contract of the blockchain node is first triggered to create the identity information for the power terminal. Secondly, the data management contract is triggered, and the distributed information of the power terminal is distributed and stored by calling the distributed storage module.

3.2 User Registration Process

The user registration process includes collecting user feature data, generating user keys based on passwords and biometrics, sending user registration parameters to the certification center, the certification center generating authentication keys for users, the certification center calculating the user's authentication characteristics, and converting the user's authentication information Stored to the blockchain node, the blockchain node stores the user's authentication information in seven steps.

1. Collect user characteristic data: The collected user characteristic data includes user name u_{id}^{i}, password u_{pwd}^{i}, and biometrics B_i.
2. Generate the user key based on the password and biometrics: use the u_B^i to represent user key, and use the formula $u_B^i = h(R_i)$ to calculate the user key u_B^i after using the random character string R_i obtained from the biometrics B_i.
3. Send the user's registration parameters to the certification center: The registration parameters that the user needs to send to the certification center include user characteristic data, user's company information and authority information. User characteristic data includes user name u_{id}^{i}, password u_{pwd}^{i}, and key u_B^i. The user's company information and authority information include the company to which the user belongs and the authority level (query data, modify data, delete data).
4. The authentication center generates an authentication key for the user: the authentication center generates a secret value y_i for the user, and uses the formula $k_{cu}^{i} = h\left(x_g || y_i\right)$ to calculate the user's authentication key k_{cu}^{i}.
5. The authentication center calculates the user's authentication characteristics: the authentication center uses the formula $A^i = h\left(u_{id}^{i} || u_{pwd}^{i} || u_B^i || k_{cu}^{i}\right)$ to calculate the user's authentication characteristics A^i.
6. Store the user's authentication information to the blockchain node: The authentication center stores the user's relevant information to the blockchain node, and the stored information includes the user's characteristic data, authentication characteristics A^i, secret value y_i, user's company information and authority information.
7. The blockchain node stores the user's authentication information: To store the user's authentication information, the identity management contract of the blockchain node is first triggered to create identity information for the user. Secondly, the data management contract is triggered, and the user's identity information is distributed and stored by calling the distributed storage module.

3.3 Certification Process

The authentication process mainly includes six steps: user login, user authentication, power terminal login, power terminal authentication, generation of shared key, and communication based on the shared key. The details are described below.

1. User login: The user generates authentication information through the user terminal and applies for login to the authentication center. The user inputs user characteristic information through the user terminal, including the user name $u_{id^*}^{i}$, password $u_{pwd^*}^{i}$, and key $u_{B^*}^{i}$, and sends these characteristic information to the certification center.
2. User authentication: The authentication center authenticates users. After calculating with the formula $A^{i^*} = h\left(u_{id^*}^{i} || u_{pwd^*}^{i} || u_{B^*}^{i} || k_{cu}^{i}\right)$, the certification center takes the A^i

of user u_{id}^{i} from the blockchain and judges whether it is equal to A^{i^*}. If they are equal, the user authentication is successful. Otherwise, user authentication fails.

3. Power terminal login: The authentication center obtains the device name D_{id}^{j} information of the power terminal that the user proposes to access based on the user's request information. The certification center searches for the power terminal based on the device name D_{id}^{j} and sends a broadcast to the power terminal to find the power terminal device with the device name D_{id}^{j}.
4. Power terminal certification: The certification center certifies the authenticity of power terminals. The power terminal collects feature information such as device name $D_{id^*}^{j}$, device password $D_{pwd^*}^{j}$, and device key $D_{B^*}^{j}$. The certification center uses the formula $A_D^{j^*} = h\left(D_{id^*}^{j}||D_{pwd^*}^{j}||D_{B^*}^{j}||k_{cd}^{j}\right)$ to calculate. The certification center takes out A_D^{j} of the power terminal D_{id}^{j} from the blockchain and judges whether it is equal to $A_D^{j^*}$. If they are equal, the power terminal authentication is successful. Otherwise, power terminal authentication fails.
5. Generate shared key: In order to realize the secure communication between the user and the power terminal, the certification center first generates a random number p_{ud}^{ij} and then uses the formula $k_{ud}^{ij} = h\left(p_{ud}^{ij}||x_g\right)$ to generate a shared key k_{ud}^{ij} for the user and the power terminal.
6. Communication based on shared key: The certification center sends the shared key k_{ud}^{ij} to the user and the power terminal, and the user and the power terminal encrypt data communication based on this shared key k_{ud}^{ij}.

4 Performance Analysis

This paper evaluates the authentication mechanism from the aspects of usability, integrity, privacy and authentication efficiency. Among them, the availability of the authentication mechanism is to allow legitimate users to use the corresponding authentication resources. The integrity of the authentication mechanism refers to the fact that the authentication information has not been tampered with. The privacy of the authentication mechanism refers to ensuring the privacy of the authentication data. The efficiency of the mechanism refers to the execution speed and energy consumption scale of the authentication mechanism and the existing authentication mechanism in this paper.

In terms of the availability of the authentication mechanism, from the blockchain-based authentication model architecture, it can be seen that the availability of the authentication mechanism needs to ensure that the user terminal, the certification center, and the power terminal are available to all three parties. On the user terminal side, this paper uses biometrics and password information for authentication, and biometrics extraction uses biometrics generation functions $FIF(B_i) = (R_i, P_i)$ and $FNF(B_i^{'}, P_i) = R_i$ to do the standardization processing, thereby improving the usability of the authentication mechanism. In terms of certification center, the use of distributed storage, blockchain services, smart contracts, authentication server master and other technical means to ensure the availability of certification center. On the power terminal side, device fingerprints and device environment are combined to generate device characteristics, which improves the usability level of the power terminal authentication mechanism.

In terms of the integrity of the authentication mechanism, it mainly analyzes from the single point of failure of the authentication center and data tampering. In terms of the occurrence of single points of failure, the authentication server adopts the main and standby modes, and the data storage adopts blockchain technology, which can effectively solve the problem of single points of failure. In terms of data tampering, the user side adopts a combination of username u_{id}^{i}, password u_{pwd}^{i}, and biometric B_i authentication mode to prevent attackers from forging user identity information. The identity authentication mode combining device fingerprint D_{fp}^{j}, IP/MAC address D_{im}^{j}, and protocol type D_{p}^{j} is used on the power terminal side to prevent the attacker from forging the power terminal information. The use of blockchain technology in the authentication center can effectively prevent attackers from tampering with user data, power terminal data and authentication server data. Therefore, the problem of data tampering is effectively solved.

In terms of the privacy of the authentication mechanism, the privacy of the authentication mechanism is embodied in the four dimensions of power terminal key, user key, authentication key, and shared key. The power terminal key is calculated by $D_{B}^{j} = h\left(D_{fp}^{j}||D_{im}^{j}||D_{p}^{j}\right)$, and the user key is calculated by $u_{B}^{i} = h(R_i)$. The authentication key is divided into the certification key k_{cd} of the certification center and the user, the certification key k_{cd} of the certification center and the power terminal, they are calculated by $k_{cu} = h(x_g||y_i)$ and $k_{cu} = h(x_g||D_j)$, and the shared secret key is calculated by $k_{ud}^{ij} = h\left(p_{ud}^{ij}||x_g\right)$. All of them have better privacy. Therefore, the authentication mechanism in this paper not only has better privacy in the authentication process, but also has better privacy in the communication process after authentication.

In terms of the efficiency of the authentication mechanism, the efficiency of the authentication mechanism is evaluated in terms of execution speed and energy consumption scale. Based on the advantages of decentralization and non-tampering of blockchain technology, this paper effectively solves the problem of mutual distrust among the three parties and improves the efficiency of authentication. In terms of energy consumption scale, each certification process will consume the resources of all parties to the certification. The authentication mechanism in this paper reduces the authentication process, so it reduces the energy consumption of authentication.

5 Application Scenarios

A schematic diagram of a typical authentication mechanism application scenario is shown in Fig. 2, which includes four key authentication elements: user, authentication server, blockchain node, and power terminal. Power users need to go through two key links, registration and authentication, to control remote power terminals through user terminals.

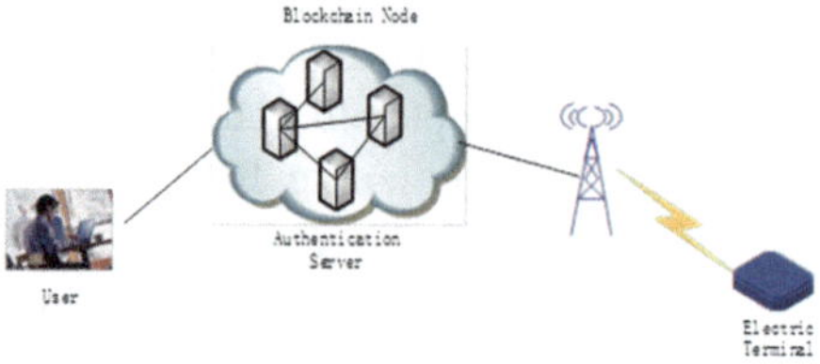

Fig. 2. Schematic diagram of the authentication mechanism to application scenario.

The registration process includes user registration and power terminal registration. The data collected during user registration mainly includes the user name u_{id}^{i}, password u_{pwd}^{i}, and biometrics B_i. The user key u_B^i is obtained using the formula $u_B^i = h(R_i)$ after using the random character string R_i obtained from the biometrics B_i. When registering at the power terminal, the collected data includes the device fingerprint D_{fp}^{j}, IP/MAC address D_{im}^{j}, protocol type D_p^j, and the device key D_B^j is calculated using the formula $D_B^j = h\left(D_{fp}^j || D_{im}^j || D_p^j\right)$. In terms of the relationship between the user and the power terminal, the owner of the power terminal needs to be marked during registration, that is the company to which it belongs, and the authority level (query data, modify data, delete data); also the company to which the user belongs, authority level (query data, Modify data, delete data). In the authentication process, the user first logs in using the user name $u_{id^*}^{i}$, password $u_{pwd^*}^{i}$, and key $u_{B^*}^{i}$, and then the authentication center searches for the power terminal based on the device name D_{id}^{j} and checks the authenticity of the power terminal. Finally, the authentication center uses the formula $k_{ud}^{ij} = h\left(p_{ud}^{ij} || x_g\right)$ to generate the shared key for the user and the power encryption of terminal communication data.

6 Conclusion

In order to solve the problems of low authentication efficiency and low security level in existing authentication systems, this paper proposes a blockchain-based trusted authentication mechanism for electric power Internet of Things. This mechanism not only improves the efficiency of the authentication mechanism, but also implements the secure authentication of trusted identities based on user biometrics and device fingerprint link feature technologies, and enhances the security of the authentication mechanism. Although the authentication mechanism in this paper effectively solves the efficiency and safety problems in the existing authentication system during the registration process and authentication process. However, the authentication mechanism also needs to have more complete functions such as user attribute change, permission change, and related participant information update. The next step is to optimize these functions. On the other hand, the authentication mechanism of this paper does not consider the problem of malicious identity registration of power terminals and users, nor does it deeply study the problem of replay attacks based on time windows. Improve the performance of this authentication mechanism in the actual environment.

Acknowledgements. This work is supported by the science and technology project of State Grid Corporation: Blockchain-Based Research and Typical Scenario Application of Trusted Service Support System for Electric Power Business (5700-201918243A-0-0-00).

References

1. Song, X.R., Zhang, M.: Present situation and trend of the foreign network trusted identity authentication technology and it enlightenment to China. Cyberspace Secur. **9**(2), 6–11 (2018)
2. Ma, X., Zhao, F.G.: Mobile terminal multi source biometric real-time identity authentication system for mobile internet. Video Eng. **4** (2017)
3. Fan, Y., Xv, J., Gao, Y.T.: Research and implementation of e id-based identity authentication system. Netinfo Secur. **3**, 48–53 (2015)
4. Zhang, T., Lin, W.X.: Implementation method of mobile terminal identity authentication and use authorization based on SWP-SIM technology. Technol. Innov. Appl. **9**, 21–22 (2018)
5. Li, Y., Guo, J.W., Du, L.P., et al.: Research on mobile terminal identity authentication scheme based on combined symmetric key algorithm. Netw. Secur. Technol. Appl. **1**, 94–95 (2016)
6. Ma, L.L.: An identity authentication scheme of mobile terminal based on two-dimension code in cloud computing environment. Microelectron. Comput. **33**(1), 140–143 (2016)
7. Sun, Z.W., Zhang, Y.C.: Deep belief network model for mobile terminal identity authentication. Netinfo Secur. **3**, 34–42 (2019)
8. Leng, X.W., Chen, G.P., Jiang, Y., et al.: Data specification and processing in big-data analysis system for monitoring and operation of smart grid. Autom. Electric Power Syst. **42** (19), 169–176 (2018)
9. Xia, Z.Q., Zhao, L., Wang, J., et al.: Research on a privacy protection method for power users based on virtual ring architecture. Netinfo Secur. **18**(2), 48–53 (2018)
10. Dong, G.S., Chen, Y.X., Li, H.W., et al.: Cross-domain authentication credibility based on blockchain in heterogeneous environment. Commun. Technol. **52**(6), 27 (2019)
11. Chen, Z.Y., Gao, D.Q., Wang, D., et al.: Quantum key based optimal data protection model for power business. Autom. Electric Power Syst. **42**(11), 113–121 (2018)
12. Wang, S.X., Chen, H.W., Pan, Z.X., et al.: A reconstruction method for missing data in power system measurement using an improved generative adversarial network. Proc. CSEE **1**, 1–7 (2019)

Trusted Identity Cross-Domain Dynamic Authorization Mechanism Based on Master-Slave Chain

Xiuli Huang[1(✉)], Qian Guo[1], Qigui Yao[1], and Xuesong Huo[2]

[1] State Grid Key Laboratory of Information & Network Security Global Energy Interconnection Research Institute Co., Ltd., Beijing 102200, China
71737069@qq.com
[2] State Grid Jiangsu Electric Power Co., Ltd., Nanjing 211106, China

Abstract. In order to solve the security problems existing when power users access power services in different domains, this paper proposes a dynamic authorization model for power users based on master-slave chain. In this model, the slave Blockchain conducts identity authentication in each autonomous domain, and the main Blockchain undertakes identity authentication among autonomous domains. Moreover, by constructing an attribute access control model, a trusted identity cross-domain authorization mechanism is proposed, and an attribute-based allocation strategy in both single-domain and cross-domain scenarios is designed in detail. Through application analysis and security analysis, it is verified that the trusted identity cross-domain authorization mechanism proposed in this paper can be conveniently applied to existing systems, and has better performance in confidentiality, integrity, and availability.

Keywords: Blockchain · Cross-domain authentication · Authorization mechanism

1 Introduction

With the rapid development of smart grid business, the types and number of power businesses are increasing rapidly. However, in recent years, security threats in the power sector have also been increasing [1]. Under this circumstance, how to build a secure and convenient trusted service architecture has attracted great attention [2]. In order to achieve security management of the power business, the power business is divided into multiple domains for management. Then how to realize the power users' secure access to the power business in different domains has become a key problem. Existing researches on cross-domain authentication are mainly divided into two types: studying new architecture, as well as improving the efficiency of existing authentication mechanism.

In terms of research on new architecture: In [3], the blockchain technology is adopted to improve the credibility evaluation, and the alliance blockchain technology is applied to the risk evaluation mechanism, so as to better improve the cross-domain authentication performance in heterogeneous environment. In view of the problem that

Q. Liu et al. (Eds.): CENet 2020, AISC 1274, pp. 876–882, 2021.
https://doi.org/10.1007/978-981-15-8462-6_101

the existing authentication mechanism relies heavily on centralized authentication centers, Zhou, Z. C. et al. [4] applies the blockchain technology to cross-domain authentication and proposes a new efficient user authentication mechanism. To solve multiple domain authentication between the agent deployment of complex problems, Gao, Y. et al. [5] applies cryptography to each agency's trust evaluation. It effectively reduces communication and computation overhead of cross-domain authentication. In terms of improving the efficiency of existing authentication mechanism: in order to solve the complex interaction between authentication systems, a new information interaction model is proposed in [6], which improved management efficiency in authentication. In view of the authentication process problems existing in the authentication of mobile terminals, Sun, Z. W. et al. [7] adopts deep learning algorithm to improve the authentication model, which improves the efficiency of mobile terminal authentication. In view of the problem of excessive authentication cost in cross-domain authentication, Ma, X. T. et al. [8] improves the existing cross-domain authentication architecture and reduces the cost of storage, communication, computation and other aspects in the process of cross-domain authentication.

Through the analysis of the existing studies, it can be seen that most research achievements have been made on identity authentication in a single domain. However, in the cross-domain authentication, there is a problem of information leakage. To solve this problem, this paper proposes a master-slave chain based cross-domain authorization model for trusted identities of power users. Based on this model, a cross-domain authorization mechanism for trusted identities is proposed. Finally, from the application method and security, the availability and security of the cross-domain authorization mechanism for trusted identities proposed in this paper are analyzed.

2 Trusted Identity Cross-Domain Authorization Model Based on Master-Slave Chain for Power Users

In order to achieve a secure cross-domain authorization mechanism for power users, a master-slave chain-based cross-domain authorization model is proposed in this paper, as shown in Fig. 1. As can be seen from the figure, the model includes power user module and authorization module. Among them, the power user module is composed of wearable devices, mobile phones, custom terminals and audit terminals. The authorization module consists of a master chain and a slave chain. The slave chain refers to the blockchain network composed of a single autonomous domain, to complete the identity authentication of power users in each autonomous domain. It includes the certification cente and business server. The main chain refers to the blockchain network formed by the key nodes of each slave chain, which completes the identity authentication of power users between autonomous domains. As can be seen from Fig. 1, the main chain is composed of certification centers and coordination centers in each autonomous domain.

First, in each slave chain, in order to achieve the authorization management of power users, each power user is assigned a five-tuple of identity, including user classification (U), user role (R), attribute (A), operation (Op), and object (Ob). The relationship of the tuples is: user U has role R, and each role R has certain attribute A,

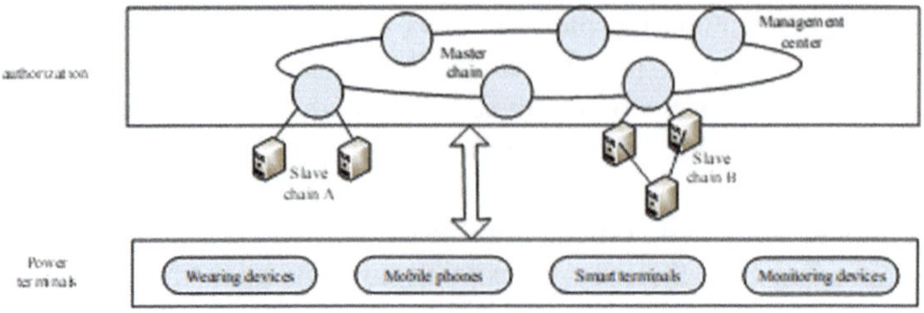

Fig. 1. Trusted identity cross-domain authorization model based on master-slave chain for power users.

in which the operation Op and the object Ob that the current user has are reflected. For example, the role of the user Zhang San is the network maintenance engineer of the power trading system. The attributes of the network maintenance engineer include the authority of remote login, query and modification to the power trading system. Among them, the operation includes remote login, query, modification, etc. The object of operation refers to the maintenance-related functional modules in the power trading system. Secondly, D is used to represent the blockchain node on the main chain. According to the model description, the main chain is composed of the certification center and coordination center of each autonomous domain. Therefore, D0 is used to represent the coordination center, and Di ($0 < I < N + 1$) is used to represent the certification centers for each autonomous domain.

3 Trusted Identity Cross-Domain Authorization Mechanism

The permission of each user is closely related to the user property. If the user's attributes can be managed validly, the user's permissions can be protected validly. Based on this, this paper proposes an attribute-based allocation strategy. The strategy is designed from single domain environment and multi-domain environment.

3.1 Single Domain Attribute-Based Allocation Strategy

In each autonomous domain, there are certification centers and various power business servers. The recognition center mainly completes the management of user identity. As described above, the certification authority is represented by Di. Each power business server mainly completes the service provision, which is denoted by SP in this paper.

For each SP, define a triple including property, operation, object. Among them, the attribute refers to the attribute that the user can have, the operation refers to the operation on the service object, and the object refers to the specific function on which the operation is performed. In order to realize the mapping of user attributes to the specific service object, define $AS\subseteq(A \times S) = (Op \times Ob)$, to represent the specific operation of user attributes A to the service, where A represents the user's property, S represents the current service, and represents the cartesian product of Op and Ob. Op is the operation method that attribute A has on the current service S, and Ob is the specific operation target of the operation Op for service S.

3.2 Cross-Domain Attribute-Based Allocation Strategy

The similarity between attribute-based allocation policies in cross-domain and single-domain is that the user's attributes can determine the user's permissions on a service. Therefore, after the user's attributes are determined, the user's permissions can be determined. In order to determine the user's properties in the cross-domain authentication process, this paper uses the property mapping mechanism to map the user's properties in one domain to the properties in another domain. In the master-slave chain based power user trusted identity cross-domain authorization model, each autonomous domain is independent of each other. In order to implement the property mapping, it is necessary for the coordination center to establish the mapping between the domain properties.

There are two types of direct mapping and indirect mapping when attribute mapping is performed in the coordinator center. The direct mapping means that the properties of two domains have established the mapping method through negotiation. Indirect mapping refers to the fact that no mapping method is established between the properties of two domains. It should be noted that user permissions are determined by attributes. When the indirect mapping does not find an appropriate mapping, the two domains need to negotiate. When an appropriate mapping is found, permission of the two domains should be checked. If at least one of the two domains does not agree to perform the property mapping, the coordination center must stop doing the property mapping, otherwise security issues may arise.

In the direct mapping, the attributes Aa and Ac representing the autonomous domain Di and the autonomous domain Dk are directly mapped. The property Ab and the property Ac representing the autonomous domain Dj and autonomous domain Dk are directly mapped. When autonomous domain Di, autonomous domain Dj, autonomous domain Dk, and all agree to conduct indirect attribute mapping, the property Aa and property Ab of autonomous domain Di and autonomous domain Dj are indirectly mapped using the representation. When the mapping is successful, the property Aa of the autonomous domain Di has the same permissions as the property Ab of the autonomous domain Dj.

3.3 Application Analysis

First, this paper divides authorization information into two types: intra-domain storage and inter-domain storage. The domain stores the information related to the five tuples of the main storage users; Inter-domain stores information about property mappings between major storage domains. Due to the decentralized and tamper-free nature of blockchain technology, this paper applies blockchain to store information within and between domains.

When used specifically, the architecture of the cross-domain authorization mechanism of trusted identity based on attribute mapping is shown in Fig. 2, which includes

Fig. 2. Attribute-base trusted authorization process.

three processes: discovering service, locating local property, and mapping new property.

1. Discovery service: user A needs to use the service provided by A service provider. First, user A looks for the service in the service center. If required service can be found, go to step 2. Otherwise, return with no required service.

2. Location of local properties: in order to access A service, user A first requests to use the service from the local domain. The certification authority in the region sees whether the service is in the region or in another domain. If the service is the service of the local region, the user is directly assigned the corresponding permissions according to the user's attributes.

3. Mapping new attributes: the certification center in this region requests the coordination center to map new attributes. The coordination center looks for the authentication nodes of the two domains and determines the configuration information of the authentication nodes of the two domains. If the authentication authority for both domains allows property mapping, the coordinator center performs the property mapping. Based on the mapping results, user - related properties are returned. The user operates on the services of the target domain based on the properties returned.

4 Security Analysis

Generally speaking, the security level and the ability to resist attacks of an identity authentication system are evaluated in terms of confidentiality, integrity and availability. In order to facilitate security analysis, the prerequisite conditions for security analysis are described as follows: the premise for the attacker to obtain the public-private key is to obtain the user name, private key generation parameters, asymmetric encryption algorithm and other information of the attacked; The attacker may be an external user or an internal user. The confidentiality, integrity and availability of power service security are analysis as follows.

Confidentiality: the slave chain in this article consists of autonomous domains. In the autonomous domain, Kerberos, PKI and other mature authentication mechanisms can be used 4 or authentication. In the process of communication between master and slave chains, technologies such as TLS and IPSec VPN can be adopted to realize the secure encryption of data. In the two interaction processes of slave chain and master slave chain, the secure transmission of user's user name, private key generation parameters, asymmetric encryption algorithm and other information can be ensured.

Therefore, the architecture of this paper can guarantee the confidentiality of user information.

Integrity: from the knowledge of attack and defense, if you want to destroy the integrity of the user's identity information, you can take the way of middleman tampering. To successfully complete a man-in-the-middle attack, an attacker needs to make one of the communicating parties trust the attacker. However, according to the block chain technology, in order to complete the man-in-the-middle attack, the attacker needs to attack at least half of the blockchain nodes, so as to have a chance to obtain the man-in-the-middle identity through the consensus mechanism. This is impossible for an attacker to achieve. Therefore, the user authorization mechanism in this paper has integrity.

Availability: in terms of the availability of the user authentication mechanism, an attacker can compromise the server where the data is stored. However, the blockchain technology is adopted in this paper to prevent the data from being unavailable, which can effectively solve the problem of the destruction of stored data.

5 Application Scenario

The schematic diagram of the power user trusted identity cross-domain authorization application scenario based on master slave chain is shown in Fig. 3. The application scenario includes four parts: user, service center, trusted identity authentication center and data center. First, the user logs in through the terminal device, injects the user's basic information and collects the biometric information. Secondly, wired or wireless network is adopted to request relevant services from the service center. According to the authentication requirements of the data center providing services, the service center requests authentication and authorization from the trusted authentication center. When a user requests an authenticated service for his domain, local authentication is performed on the authentication server. If a user requests an authenticated service for another domain, the trusted identity authority authorizes the user across domains.

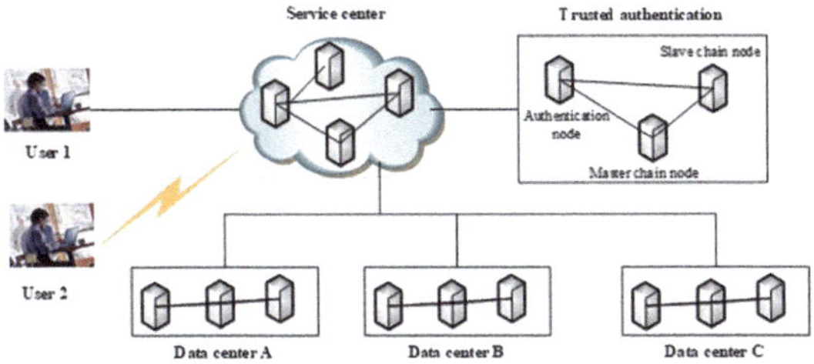

Fig. 3. Scenario.

6 Conclusion and Future

How to realize secure access to power business in different domains has become a key problem that needs to be solved urgently. To solve this problem, this paper proposes a master-slave chain based trusted identity cross-domain authorization model, and proposes a trusted identity cross-domain authorization mechanism by building a property access control model. Through application analysis and security analysis, it is verified that the cross-domain authorization mechanism of trusted identity can be conveniently applied to existing systems and has good confidentiality, integrity and availability. In the future work, by analyzing the current situation of power users' trusted identity cross-domain authorization system, the research results will be optimized, so as to improve the practical value of the research results.

Acknowledgements. This work is supported by the science and technology project of State Grid Corporation: Blockchain-Based Research and Typical Scenario Application of Trusted Service Support System for Electric Power Business (5700-201918243A-0-0-00).

References

1. Wang, Q., Tai, W., Tang, Y., et al.: A review on false data injection attack toward cyber-physical power system. Acta Automatica Sinica **45**(1), 72–83 (2019)
2. Jiang, D.X., Hou, J.N., Han, F., et al.: The research on trusted identity system on the Internet—from identity management to identity service. Inf. Sec. Commun. Priv. **3**, 102–109 (2018)
3. Dong, G.S., Chen, Y.X., Li, H.W., et al.: Cross-domain authentication credibility based on blockchain in heterogeneous environment. Commun. Technol. **6**, 27 (2019)
4. Zhou, Z.C., Li, L.X., Li, Z.H.: Efficient cross-domain authentication scheme based on blockchain technology. J. Comput. Appl. **38**(2), 316–320 (2018)
5. Gao, Y., Ma, W.P., Liu, X.X.: Cross-domain authentication scheme based on trust for service entity. Syst. Eng. Electron. **41**(2), 438–443 (2019)
6. Xie, Y.R., Ma, W.P., Luo, W.: New cross-domain authentication model for information services entity. Comput. Sci. **45**(9), 177–182 (2018)
7. Sun, Z.W., Zhang, Y.C.: Deep belief network model for mobile terminal identity authentication. Netinfo Security **3**, 34–42 (2019)
8. Ma, X.T., Ma, W.P., Liu, X.X.: A cross domain authentication scheme based on blockchain technology. Acta Electronica Sinica **46**(11), 2571–2579 (2018)

Trusted Identity Authentication Mechanism for Power Maintenance Personnel Based on Blockchain

Zhengwen Zhang[1], Sujie Shao[2], Cheng Zhong[1(✉)], Shujuan Sun[1], and Peng Lin[3]

[1] State Grid Xiongan New Area Power Supply Company, Heibei 071000, China
zhongc@126.com

[2] Beijing University of Posts and Telecommunications, Beijing 100876, China

[3] Beijing Vectinfo Technologies Co. Ltd., Beijing 100000, China

Abstract. In order to solve the problem of impersonation of the electronic identity of power maintenance personnel, this paper proposes an trusted identity authentication model based on Blockchain. This model uses plug-in authentication technology to improve the scalability of the identity authentication system. In order to improve the usability of the authentication model, smart devices are named uniformly, and information such as names and corresponding parameters are registered on the distributed data trusted chain through smart contracts. In order to improve the practicability of the model, based on the model architecture, the registration process, login authentication process, and trusted identity card system architecture were designed. In terms of the architecture of the trusted identity system, it includes a Blockchain server, an authentication server, and a mobile terminal. Through the performance analysis, the authentication mechanism proposed in this paper is verified, which effectively solves the problems of poor reliability and centralized storage that easily lead to information leakage in traditional identity authentication.

Keywords: Identity authentication · Blockchain · Reliability

1 Introduction

With the rapid development of smart grid business, new services such as energy internet, charging piles, and electronic malls have developed rapidly. These new businesses have posed greater challenges to the safe and stable operation of the power grid. In order to ensure that power maintenance personnel can participate in maintenance work in different locations, they can log in to the background operation and maintenance system through a secure authentication mechanism, and trusted services have been applied to the identity management of the power grid [1]. In order to prevent the identity of the maintenance personnel from being impersonated, the identity authentication mechanism needs to be further optimized. The existing research on identity authentication can be divided into two aspects: strengthening the standardized management of the identity authentication process and applying new technologies.

Q. Liu et al. (Eds.): CENet 2020, AISC 1274, pp. 883–889, 2021.
https://doi.org/10.1007/978-981-15-8462-6_102

In terms of strengthening the standardized management of the identity authentication process, in order to prevent user identity data from being attacked and improve the security of power data, reference [2] used entropy weight-gray theory to provide risk warning for the security of power grids. In view of the problem of easy leakage of power data assets, reference [3] reconstructed the existing data management system and proposed a power grid data security management system, which effectively improved the power grid data security management capability. To address the problem of data injection attacks, reference [4] proposed a protection strategy for automatic dispatch of power data for each possible attack form. In terms of the application of new technologies, since the power grid security problem is likely to cause power grid damage after destruction, reference [5] used generative confrontation network technology to reconstruct the damaged grid data, reducing the negative impact of missing data on the network. Reference [6] solved the single point of failure in the centralized data protection technology through the improvement of the data management method, and the blockchain technology was applied to the data fusion system to realize the decentralized architecture of data management. In view of the problem of easy leakage of power grid data, reference [7] combined power user data protection with virtualization technology to effectively improve the security of power data.

It can be known from the analysis of existing studies that the identity authentication mechanism in the power system has achieved more research results. However, the scalability of existing research results is not high, and when a new authentication method appears, it cannot be quickly added to the existing authentication system. To solve this problem, this paper proposes a trusted identity authentication model based on blockchain, and designs the registration process, login authentication process, and trusted ID card system architecture. Performance analysis verified the authentication mechanism proposed in this paper could effectively solve the problem of poor scalability in traditional identity authentication.

2 Model Design

2.1 Trusted Identity Authentication Model Based on Blockchain

The trusted identity authentication model based on blockchain proposed in this paper is shown in Fig. 1. This model mainly includes three parts: authentication device, authentication server and blockchain server. The authentication device is generally realized by a mobile terminal, and has plug-in authentication capabilities, and provides registration and authentication services for users. The authentication server is generally implemented by a server with higher security, and provides authentication services for the authentication device. The blockchain server is generally implemented by a distributed blockchain network, and cooperates with the authentication server to complete user identity authentication.

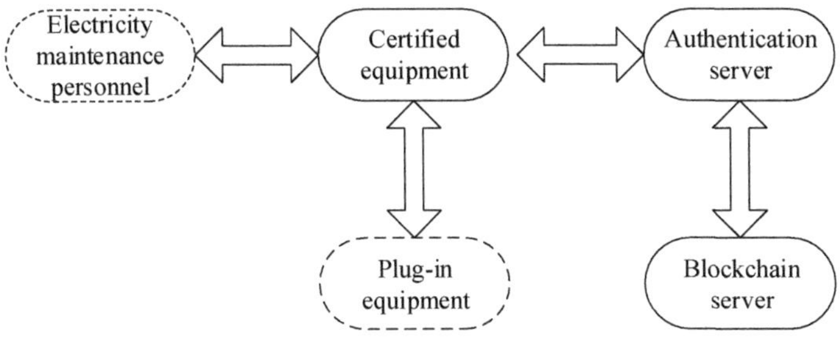

Fig. 1. Trusted identity authentication model based on blockchain.

In order to ensure the scalability of the authentication model, the plug-in design is adopted, so that the authentication device of this model supports various access devices such as computer terminals, mobile terminals, and vehicle-mounted terminals to meet the needs of various application scenarios. Considering the increase in the types and number of access devices, it is easy to make authentication management more complicated. In order to simplify the management of certified devices, this model adopts a unified naming mechanism for devices to achieve unified management of certified devices. The plug-in design and the unified naming management content of the device are described in detail below.

2.2 Plug-in Authentication Capability

Because the blockchain technology has the advantage of strong expansion ability, this paper makes full use of the update mechanism of each blockchain node to realize the plug-in authentication capability of each authentication device. All authentication methods are stored in the smart contract of the blockchain nodes in the form of files. When a new authentication method is needed, it can be easily achieved by adding a smart contract. Authentication methods include fixed authentication and mobile authentication. Among them, fixed authentication mainly includes certificate authentication, fingerprint authentication, iris authentication, mail authentication, short message authentication and other authentication methods. Mobile authentication mainly includes QR code scanning authentication, face recognition authentication, dynamic key authentication, and fingerprint authentication. Therefore, based on the smart contract management mechanism in this paper, the authentication data used in fixed authentication and mobile authentication can be conveniently saved and updated, thereby improving the scalability of the authentication mechanism.

2.3 Unified Naming of Smart Certification Equipment

When each blockchain node receives a new authentication device, it generates a globally unique device name for the current device according to the naming convention. After that, the consensus mechanism and smart contract technology of the blockchain are used to diffuse the name of the current device and device-related information to other nodes of the blockchain, thereby achieving global consistency and global awareness of the device name. Based on this, all nodes of the blockchain can query the relevant information of this device, which is convenient for the management of authentication.

3 Key Process Design

The key process design is carried out based on the user registration process and the login authentication process to describe the use of the identity authentication mechanism in this article.

3.1 Registration Process

The registration process includes two processes: the registration process and the process of generating identity authentication blocks.

The registration process includes four participants: users, authentication equipment, authentication services, and blockchain nodes. The detailed registration process is described below. (1) The user realizes the authenticity verification of the user's identity on the authentication device through the mobile phone dynamic SMS. Among them, the authentication device may be a desktop computer or a mobile terminal. (2) After the user inputs the work unit, authentication authority and other information on the authentication device, the face data is collected through the camera, thereby implementing the user identity authentication mechanism based on face recognition. (3) The authentication device uses data signature technology to send the user's relevant information to the authentication server for registration. (4) After the authentication server receives the user information, it is compared with the user information in the authentication server based on the user's face data. If the user's work unit and authentication authority are the same as those in the authentication server, the user feedback successful registration information.

The stage of generating the identity authentication block includes three participants: user, authentication service, and blockchain node. The detailed process of generating an identity authentication block is described below. (1) The user generates a public and private key pair and sends the public key information to the authentication server. (2) The authentication service sends the user's public key information and identity information to the blockchain node. The blockchain node uses a consensus mechanism to complete the global consensus of user identity information at the blockchain node.

3.2 Login Authentication Process

The user uses the private key to encrypt the user name, password, facial feature value and other information required for login, and sends the encrypted information to the authentication server, and the authentication server performs identity authentication. The authentication server obtains the user's public key and identity information from the blockchain nodes based on the user name information. If the user information in the blockchain is the same as the information sent by the user after the user's public key is used to decrypt the user information, it means that the user's identity is legal and the user is allowed to access resources allowed by their permissions.

Considering that the security level of some power grid systems is relatively high, a time-based authentication mechanism can be adopted to realize the mechanism strategy of one-time login and time-based authentication. It can prevent the occurrence of events due to the loss of the authentication terminal or the use by non-authentication users. First, for the authenticated user, the authentication server generates a timer authentication timer based on the user's identity information. The timing authentication information includes user face recognition information and time interval. Secondly, when the timer is triggered, the authentication terminal automatically collects the user's face information and sends it to the server for comparison. If a recognition error occurs, a prompt box will be returned to prompt the user to use the camera for face recognition. If the recognition is successful, the user continues to use the corresponding power system resources. Otherwise, the authentication service system forces users to log off.

4 Implementation of Trusted ID Card System

By upgrading the existing certification system, this solution can be easily realized. The following describes the implementation scheme of the trusted ID card system.

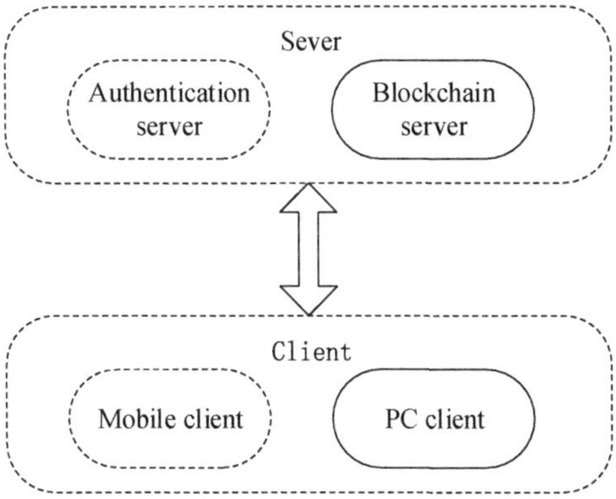

Fig. 2. Implementation framework of trusted identity authentication system.

The implementation framework of the trusted identity authentication system is shown in Fig. 2, which includes two parts, the server and the client. Among them, the server includes authentication server and blockchain server, and the client includes mobile client and computer client. In the implementation process, the mobile client, computer client, and authentication server can make full use of existing resources. The blockchain server is a newly added authentication resource.

The users of this system are mainly power maintenance personnel. According to the position of the user, each user is assigned different permissions, so as to realize the user's permission management. The client side includes fixed devices such as traditional computers and certified firmware, as well as mobile terminals such as mobile phones and customized terminals. Both authentication devices can collect the user's identity information and facial features, thereby providing a multi-dimensional authentication method for identity authentication. Regarding the authentication server, a specific authentication server provides authentication services for power terminals, and cooperates with the blockchain server to implement the security management function of power terminal identity information. When implementing authentication services, a variety of authentication mechanisms are used to achieve the scalability of authentication capabilities. In terms of blockchain servers, it mainly realizes the storage and intelligent update of user identity information, so as to ensure the safe storage and non-tampering of user data.

5 Performance Comparison

In view of the problem of identity information leakage in the traditional identity authentication mechanism, this paper proposes a trusted identity authentication mechanism for power maintenance personnel based on blockchain. In order to verify the advantages of this model, the followings are compared with the traditional identity authentication mechanism from the aspects of identity information security, authentication mechanism scalability, and authentication mechanism performance.

(1) Security of identity information. The traditional authentication method uses a single-point database storage mode. If an attacker successfully attacks a single-point database, all identity information will be obtained, resulting in the leakage of identity information. The model in this paper uses blockchain to store identity information, which belongs to a distributed ledger, and realizes data sharing between alliance nodes through an encryption mechanism. If an attacker attacks one of these nodes, it will not pose a serious threat to the entire identity system.

(2) Scalability of the authentication mechanism. The authentication capability of the traditional authentication mechanism is relatively simple. In order to overcome the problem of poor scalability of the traditional authentication mechanism, this paper uses a plug-in authentication design, and through a unified naming strategy for smart devices, to achieve support for multiple authentication methods. This implementation includes at least the following two advantages. The first is to save investment and achieve rapid access to various new authentication mechanisms. Moreover, if a certain authentication mechanism stores hidden safety hazards, a new and safer identity authentication mechanism can be directly enabled to ensure the safety of the power system.

(3) The performance of the authentication mechanism is more optimized. For the power user, after he is successfully registered at a certain node, his identity information is stored in the blockchain to achieve global sharing. When a power user needs to log in at another node due to work needs, there is no need to repeat registration. In addition, due to the use of multiple authentication mechanisms, it can be ensured that users' authentication results are more secure.

6 Summary

The electric power maintenance work has high requirements for the security authentication of the identity of electric power maintenance personnel. In order to solve the problems that the identity of the maintenance personnel is impersonated and the identity authentication mechanism is not easy to expand, this article uses plug-in authentication and unified naming technology of smart devices, and the name and corresponding parameters are registered on the distributed data trust chain through smart contracts. The identity authentication function based on the blockchain has flexible scalability. Performance analysis verifies that authentication mechanism proposed in this paper effectively solves the single point of failure caused by the traditional identity authentication centralized storage. In the next step, we will strengthen cooperation with the power company and improve the existing power maintenance personnel identity authentication system based on the research results of this paper, so as to realize the social value of the research results of this paper.

Acknowledgments. This work is supported by State Grid Technical Project "Research on key technologies of secure and reliable slice access for Energy Internet services" (5204XQ190001).

References

1. Liao, H.M., Xuan, J.X., Zhen, P., et al.: Overview of ubiquitous power internet of things information security. Electric Power Inf. Commun. Technol. **17**(8), 18–23 (2019)
2. Li, W.J., Li, M., Xing, N.Z., et al.: Risk prediction of power data network based on entropy weight-gray model. J. Beijing Univ. Posts Telecommun. **41**(3), 39–45 (2018)
3. Yang, F., Zhang, Q., Liu, J., et al.: Research on the construction of the state grid sichuan electric power data asset security management system. Electric Power Inf. Commun. Technol. **16**(1), 90–95 (2018)
4. Jiang, R.H.: Protection strategy for data injection attacks in power dispatching automation systems. Autom. Appl. (5), 37 (2018)
5. Wang, S.X., Chen, H.W., Pan, Z.X., et al.: Reconstruction method of missing data in power system measurement using improved generative confrontation network. Proc. CSEE **1**, 1–7 (2019)
6. Yang, T., Zhao, J.J., Zhang, W.X., et al.: Data blockchain generation algorithm for power information physics fusion system. Power Autom. Equipment **38**(10), 74–80 (2018)
7. Xia, Z.Q., Zhao, L., Wang, J., et al.: Research on a power user privacy protection method based on virtual ring architecture. Inf. Network Sec. **18**(2), 48–53 (2018)

Power Data Communication Network Fault Recovery Algorithm Based on Nodes Reliability

Meng Ye[1], Huaxu Zhou[1], Guanjin Huang[1], Yaodong Ju[1], Zhicheng Shao[1], Qing Gong[1], and Meiling Dai[2(✉)]

[1] CSG Power Generation Co. Ltd., Beijing 100070, China
[2] Beijing University of Posts and Telecommunications, Beijing 100876, China
8367688@qq.com

Abstract. In the context of the rapid increase of the number of faults in power data communication networks, in order to ensure the reliability of power data communication services, how to quickly recover the affected power data communication services has become an urgent problem to be solved, when the underlying network fails. In order to solve this problem, this paper first analyzes the two indicators of the affected service from the availability of power service and the return of the power service, and the three indicators of node reliability from the node degree, the node centrality, and the proximity between the nodes. Secondly, the entropy weight method is used to calculate the index weights, and the objective evaluation of each index weight is realized. Finally, the fault recovery algorithm of power data communication network based on node reliability is proposed. In the simulation experiment, compared with the traditional algorithm, the algorithm has achieved good results in terms of fault recovery rate and power data communication network revenue.

Keywords: Power data communication network · Fault recovery · SDN · Reliability · Entropy weight method

1 Introduction

In the context of the rapid construction and application of smart grids, power companies have begun to build a large number of power data communication networks based on software-defined network (SDN) technology, and gradually replace existing traditional power data communication network equipment [1]. The power data communication network based on SDN can divide the traditional power data communication network into the underlying network and the virtual network. Each virtual network provides network services for a power data communication business, realizing flexible scheduling of the power data communication service, service dynamics deployment and fast delivery [2]. However, with the rapid development of power data communication networks, the scale of the network and power data communication services have grown rapidly. Under this background, the number of faults in the power data communication network has also increased rapidly. In order to ensure the reliability of the power data communication service, how to quickly recover the affected

Q. Liu et al. (Eds.): CENet 2020, AISC 1274, pp. 890–898, 2021.
https://doi.org/10.1007/978-981-15-8462-6_103

power data communication service has become an urgent problem to be solved when the underlying network of the power data communication fails.

Generally speaking, there are two types of protection mechanisms and restoration mechanisms for improving the reliability of power services. Among them, the protection mechanism realizes the stable operation of service through resource reservation, but the resource utilization rate is low. For example, literature [3] aims at high service reliability, and allocates an in-use link and a backup link to each service at the same time, thus realizing near real-time switching of power services and satisfying the services with high SLA. The recovery mechanism can realize the rapid recovery of power service on the premise of improving the utilization rate of power data communication network resources. Since the recovery mechanism can effectively use the underlying network resources, it has become the preferred strategy for power companies. Literature [4, 5] studied the service recovery algorithm after the failure of the power data communication network from the perspective of network attack leading to the failure of the power data communication network. This research has achieved good results for the problem of failure recovery caused by the network attack. Literature [6] studied the regional power service recovery method with the goal of minimizing the number of operation switches and maximizing the power service load, and realized the goals of minimizing the loss of the power company and maximizing the recovery speed of the affected power business. Literature [7, 8] aimed at the problem that the smart grid service has higher and higher requirements on the communication network. Taking the power data communication network in the SDN environment as the research object, it proposes a highly real-time power service recovery mechanism to realize the goal of quickly restoring power service and reducing economic losses of power companies in the event of a failure. Reference [9] addresses the issues of service restoration and optimization of power data communication network resource utilization, taking link load and link usage upper limit as important evaluation indicators for service restoration. This dimension achieves the goal of minimizing the recovery time of the affected business.

Through the analysis of the existing research, we can see that for the problem of fault recovery, existing research has made more research results from improving the reliability of the service and the utilization rate of network resources. However, when the reliability of the underlying network allocated to the power service is low, it is easy to cause the power service to be affected by the underlying network failure again. To solve this problem, when reallocating resources for the affected power services, this paper proposes a fault recovery mechanism for allocating highly reliable resources for important services. Through simulation experiments, it is verified that the algorithm of this paper effectively improves the performance of fault recovery rate and power data communication network revenue.

2 Problem Description

The power data communication network includes two parts: the underlying network and the virtual network. The underlying network includes two resources: network nodes and network links. Use $G_D = (N_D, E_D)$ for the underlying network, N_D for the set of network nodes, and E_D for the set of network links. Each network node $n_i^D \in N_D$ includes CPU resources $cpu(n_i^D)$, and each network link $e_j^D \in E_D$ includes bandwidth resources $bw(e_j^D)$. The upper layer virtual network is used to carry a certain kind of power service. If the power service is to operate on the underlying network, the virtual network needs to request resources from the underlying network based on the resource requirements of the power business, mainly including service node resource requests and service link resource requests. Use $G_Q = (N_Q, E_Q)$ for resource allocation request, N_Q for service node resource request set, E_Q for service link resource request set, each service node resource request $n_i^Q \in N_Q$ includes CPU resource request $cpu(n_i^Q)$, and each service link resource request $e_j^Q \in E_Q$ includes bandwidth resource request $bw(e_j^Q)$.

3 Reliability Analysis of Affected Services and Nodes

3.1 Affected Service Analysis

When the underlying network fails, it will affect the power service carried on the failed underlying network resources. In order to reduce the impact of the failure on the power business, the affected power service needs to be restored as soon as possible. Based on the characteristics of the power business, this paper analyzes and restores the affected power service in terms of the remaining time of the service and the economic benefits of the power business. In terms of the remaining time of power services, analyze the remaining time t_i of all services carried on the faulty underlying network. The longer the remaining time of a business, the greater the impact. In terms of the economic benefits of the power business, the benefits brought to the power company by restoring the service carried on the faulty underlying network are analyzed.

In general, the main influencing factors of the benefits brought by restoring the power service include the number of hops n_i^{hop} of the path traveled by the business, the sum of the bandwidth of all links $\sum_{l \in p}^{i} b_i^l$, and the service duration time t_i. Use c_i^r to represent the benefits brought by the power business. The number of hops n_i^{hop} through the path is measured by the number of hops through the link. The more hops the link the service takes, the more resources it uses and the higher the revenue. The sum of the bandwidth of all links $\sum_{l \in p}^{i} b_i^l$ uses the bandwidth to measure. The larger the bandwidth, the more resources are used. The service time t_i is measured by the time required for service. The longer the service time after recovery, the greater the revenue.

3.2 Node Reliability Analysis

In order to allocate high-reliability underlying network resources to important services, the reliability of network nodes is analyzed from three aspects: node degree, node center degree, and proximity between nodes.

In terms of the degree of the node, it is measured by the number of links connected to the current node. Use δ_{ij} to indicate the connection between node n_i^s and node n_j^s.

When $\delta_{ij} = 1$, it indicates that node n_i^s is connected to node n_j^s; when $\delta_{ij} = 0$, it indicates that node n_i^s is not connected to node n_j^s. Based on this, the degree k_i of the network node $n_i^s \in N_S$ is calculated using formula (1). It can be seen from the formula that the larger the value of k_i, the greater the degree of the current node, the more the number of edges with adjacent nodes, and the higher the reliability of the node.

$$k_i = \sum\nolimits_{n_j^s \in N^S} \delta_{n_i^s n_j^s} \tag{1}$$

In terms of node centrality, it is measured by the hop count between a certain node and all other nodes in the underlying network. Use $hops(n_i^s, n_j^s)$ to indicate the number of links from node n_i^s to Node n_j^s. Based on this, the centrality hop_i of the network node $n_i^s \in N_S$ is calculated using formula (2). It can be seen from the formula that the greater the centrality hop_i of the network node $n_i^s \in N_S$ is, the closer the distance of the network node $n_i^s \in N_S$ to all other nodes in the underlying network is, the more likely it becomes the central node of the underlying network.

$$hop_i = \frac{1}{\sum\nolimits_{n_j^s \in N_S} hops(n_i^s, n_j^s)} \tag{2}$$

In terms of the proximity between nodes, the number of links d_{ij} from the shortest path of the current node to other network nodes is used to measure. Use formula (3) to calculate the proximity AP_i of network node $n_i^s \in N_S$, where N represents the number of nodes in the underlying network. It can be seen from the formula that the fewer the number of links d_{ij} from the shortest path of node $n_i^s \in N_S$ to other network nodes, it means that node $n_i^s \in N_S$ is easier to be replaced by other nodes.

$$AP_i = \frac{N-1}{\sum\nolimits_{j=1}^{N} d_{ij}} \tag{3}$$

3.3 Index Weight Solution

The priority of the affected service is analyzed from the three indicators of the number of hops through the path, the sum of the bandwidth of all links, and the service time. The reliability of the node is measured from the three indicators of the degree of the node, the center of the node, and the proximity between the nodes. In order to objectively evaluate the weight of each indicator, the following uses the entropy weight method to calculate the index weight.

The entropy weight method is to measure the weight of each index from the information amount of each index, and calculate the entropy value e_j of the index j through formula (4). In the formula, N represents the number of objects to be evaluated (priority of affected services or node reliability), the decision matrix composed of N indicators to be evaluated is represented by R, and element $r_{ij} \in R$ represents the j-th indicator of object i to be evaluated value.

$$e_j = -\frac{1}{\ln N}\sum\nolimits_{i=1}^{N}\frac{r_{ij}}{\sum\nolimits_{i=1}^{N} r_{ij}}\ln\frac{r_{ij}}{\sum\nolimits_{i=1}^{N} r_{ij}} \tag{4}$$

Based on the entropy value e_j of each indicator, the weight w_j of each indicator is calculated using formula (5), where m represents the number of indicators.

$$w_j = \frac{1 - e_j}{\sum\nolimits_{j=1}^{m}(1 - e_j)} \tag{5}$$

From formula (5), we can see that the constraint conditions satisfied by w_j are $0 \le w_j \le 1$ and $\sum\nolimits_{j=1}^{m} w_j = 1$. The index weight vector composed of m indexes is $W = [w_1, \ldots, w_j, \ldots w_m]$.

3.4 Affected Service and Node Reliability Analysis

Considering the different value ranges of the indicators of the affected services and node reliability, when solving the final values of the objects to be evaluated, the covariance function $u(x) = \frac{x}{\sum x^2}$ is used to balance each evaluation object. Based on the above analysis, use formula (6) to calculate the priority of each service to be restored. Among them, w_n, w_b, and w_t are the weights of the three indicators of the service path hop count, the sum of all link bandwidths, and the service duration determined according to the entropy weight method.

$$c_i = w_n\frac{n_i^{hop}}{\sqrt{\sum\nolimits_{j=1}^{N}(n_i^{hop})^2}} + w_b\frac{\sum\nolimits_{l\in p}^{i} b_i^l}{\sqrt{\sum\nolimits_{j=1}^{N}(\sum\nolimits_{l\in p}^{i} b_i^l)^2}} + w_t\frac{t_i}{\sqrt{\sum\nolimits_{j=1}^{N}(t_i)^2}} \tag{6}$$

Use formula (7) to calculate the priority of each node. Among them, w_k, w_{hop}, and w_{AP} are the weights of the three indicators of node degree, node center degree, and proximity between nodes determined according to the entropy weight method.

$$N_i = w_k\frac{k_i}{\sqrt{\sum\nolimits_{j=1}^{N}(k_i)^2}} + w_{hop}\frac{hop_i}{\sqrt{\sum\nolimits_{j=1}^{N}(hop_i)^2}} + w_{AP}\frac{AP_i}{\sqrt{\sum\nolimits_{j=1}^{N}(AP_i)^2}} \tag{7}$$

4 Algorithm

Based on the analysis results of affected services and node reliability, the fault recovery algorithm of power data communication network based on node reliability (PCNFRA-NR) proposed in this paper includes the following three steps: (1) Based on the analysis results of affected services, sort the affected services; (2) Sort the node reliability based on the results of node reliability; (3) Restore the services one by one to the sorted set of services that need to be restored. Step (3) includes two sub-processes. (A), restore service nodes one by one. Select resources for service nodes that meet the CPU resource requirements to allocate resources; (b), restore service links one by one. Use the shortest path algorithm to restore the service link (Table 1).

Table 1. Algorithm PCNFRA-NR

Power data communication network fault recovery algorithm based on nodes reliability
1. Based on the analysis results of the affected business, sort the affected business;
(1) Use formula (6) to calculate the final influence of each service to be restored;
(2) Based on the influence of each business, sort the service in descending order;
2. Based on the node reliability analysis results, sort the node reliability;
(1) Use formula (7) to calculate the reliability of each node;
(2) Based on the reliability of each node, arrange the nodes in descending order;
3. For the sorted service collections that need to be restored, restore the services one by one;
(1) Restore service nodes one by one;
(2) Restore service links one by one.

5 Performance Analysis

5.1 Simulation Environment

In the experiment, the GT-ITM tool was used to generate the power data communication network topology under the SDN environment [10]. In order to verify the performance of the algorithm under different network scales, the number of network nodes in the experiment was increased from 100 to 700, with an increase of 100 steps. The conditional probability of link failure of the underlying network follows a uniform distribution in the range of (0, 1), and the probability of link prior failures follows a uniform distribution in the range of [0.002, 0.01].

In order to verify the performance of the algorithm in this paper, we compare the algorithm (PCNFRA-NR) with the random recovery algorithm (PCNFRA-R) and the business-level recovery algorithm (PCNFRA-ST) from three dimensions: fault recovery rate, power data communication network revenue, and fault recovery time. Among them, the algorithm PCNFRA-R refers to randomly extracting services from the affected services for recovery; the algorithm PCNFRA-ST refers to selecting the higher-level services from the affected services for priority recovery.

5.2 Performance Analysis

The results of the failure recovery rate, power data communication network revenue, and failure recovery time comparison are shown in Figs. 1 to 3.

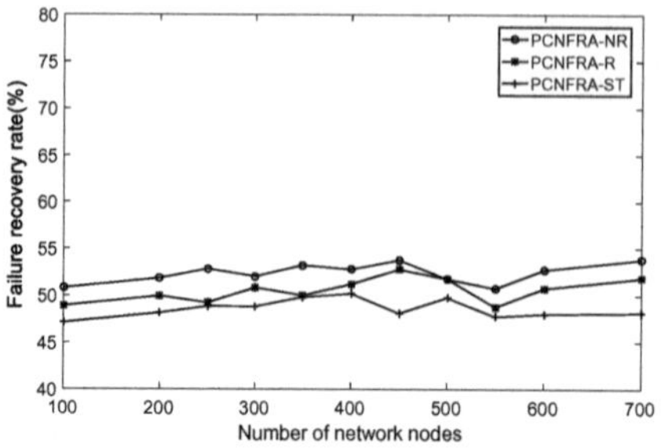

Fig. 1. Comparison of failure recovery rate.

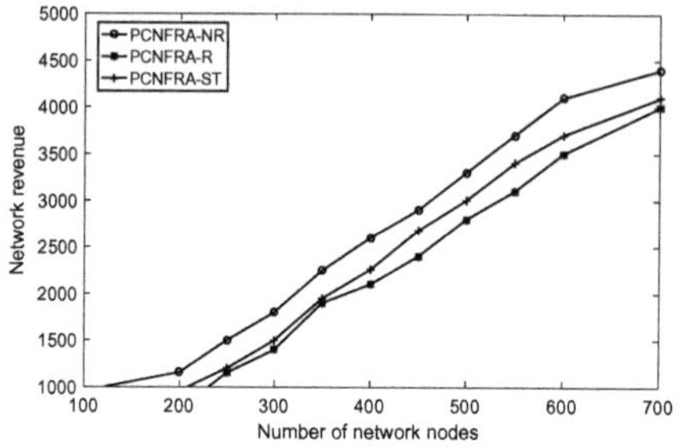

Fig. 2. Comparison of the revenue of the power data communication network.

It can be seen from Fig. 1 that in terms of failure recovery rate comparison, as the network scale increases, the failure recovery rates of the three algorithms are relatively stable, indicating that the three algorithms are more suitable for network environments of different sizes. In terms of specific failure recovery rate comparison, the failure recovery rates of the three algorithms are maintained at around 51%. The algorithm in this paper is slightly higher than the other two algorithms, indicating that the three algorithms have similar failure recovery capabilities.

It can be seen from Fig. 2 that with the increase in the number of network nodes, the revenue of the power data communication network of the three algorithms is increasing, indicating that the larger the network scale, the higher the revenue obtained by restoring services. This conclusion is consistent with the fact that the network scale increases, the number of faults and the number of affected services will increase. In terms of the specific power data communication network revenue comparison of the three algorithms, the results of the algorithm PCNFRA-NR in this paper are significantly higher than the other two algorithms, and the algorithm PCNFRA-R has the smallest power data communication network revenue. This fact shows that in addition to the algorithm PCNFRA-R, the other two algorithms give priority to the recovery of the power service with a higher return rate when recovering the affected business, thereby bringing greater benefits to the power company.

As can be seen from Fig. 3, as the number of network nodes increases, the failure recovery time of the three algorithms is increasing. It shows that when the network scale increases, the number of services that need to be recovered by the three algorithms is increasing, thus increasing the time for fault recovery. Regarding the comparison of the fault recovery time of the three algorithms, the recovery time of the algorithm in this paper is longer than that of the other two algorithms, mainly because the algorithm of this paper needs to analyze the affected services and nodes before recovering the affected services. Therefore, the fault recovery time of this algorithm is longer.

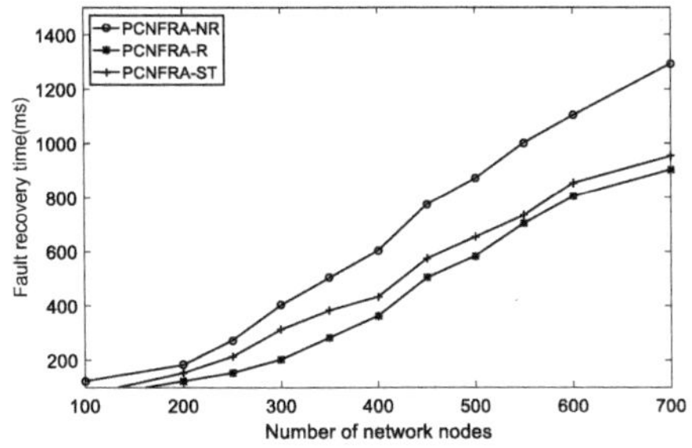

Fig. 3. Comparison of fault recovery time.

Through the analysis of the experimental results of Figs. 1 to 3, we can see that compared with the two existing algorithms, the algorithm in this paper has achieved good results in the two dimensions of fault recovery rate and power data communication network revenue. In addition, because the resources reallocated by the algorithm in this paper to each affected service are highly reliable resources, it also reduces the impact of these services on the underlying network failure again, thereby increasing user satisfaction.

6 Conclusion

In the context of the rapid construction and application of smart grids, power data communication networks based on software-defined network technology have gradually replaced existing traditional power data communication network equipment. The rapid expansion of network scale and power data communication services has led to the number of failures has also increased rapidly. In order to improve the reliability and resource utilization of services, this paper proposes a fault recovery mechanism that allocates high-reliability resources for important services, and allocates high-reliability resources in the underlying network for power services. Through simulation experiments, it is verified that the algorithm of this paper effectively improves the performance of fault recovery rate and power data communication network revenue. Although the algorithm in this paper has achieved good results in terms of failure recovery rate and power data communication network revenue, the failure recovery time of this algorithm is still relatively long. In the next step, on the basis of this research, two measures of the optimization algorithm and operating environment are adopted to further optimize the performance of the power data communication network fault recovery algorithm and provide support for the operation of the power company.

Acknowledgments. This work is supported by China Southern Power Grid Technology Project "Research on application of data network simulation platform" (STKJXM20180052).

References

1. Jain, R., Paul, S.: Network virtualization and software defined networking for cloud computing: a survey. IEEE Commun. Mag. **51**(11), 24–31 (2013)
2. Bari, M.F., Boutaba, R., Esteves, R., et al.: Data center network virtualization: a survey. IEEE Commun. Surveys Tuts. **15**(2), 909–928 (2012)
3. Hong, K., Kyung, Y., Nguyen Tri, M. et al.: A provisioning scheme for guaranteeing Recovery time in WDM mesh networks. In: 7th International conference on Ubiquitous and Future Networks, pp. 585–587. IEEE, Sapporo, Japan (2015)
4. Bie, Z., Lin, Y., Li, G., et al.: Battling the extreme: a study on the power system resilience. Proc. IEEE **105**(7), 1253–1266 (2017)
5. Xu, Y.Q., Li, X.D., Zhang, L.: Distribution network power supply restoration based on multi-agent immune algorithm. J. North China Electric Power Univ. (Natural Science Edition) **37**(2), 15–19 (2010)
6. Ying, H., Yang, J., Liu, F., et al.: An attack-oriented method for restoring distribution area services. J. North China Electric Power Univ. **46**(2), 45–51 (2019)
7. Zhang, J.N., Liu, Z.: Research on key technologies of programmable distribution network communication system based on SDN. Electric Power Inf. Commun. Technol. **13**(5), 51–56 (2015)
8. Hannon, C., Yan, J., Jin, D.: DSSnet: A smart grid modeling platform combining electrical power distribution system simulation and software defined networking emulation. In: Proceedings of the 2016 ACM SIGSIM Conference on Principles of Advanced Discrete Simulation, pp. 131–142. ACM (2016)
9. Liang, J., Shen, J.H.: An optical network shared path protection algorithm based on differentiated service levels. Optical Commun. Res. **39**(2), 19–21 (2013)
10. Medina, A., Matta, I., Byers, J.: On the origin of power laws in Internet topologies. ACM SIGCOMM Comput. Commun. Rev. **30**(2), 18–28 (2000)

Congestion Link Inference Algorithm of Power Data Network Based on Bayes Theory

Meng Ye[1], Huaxu Zhou[1], Yaodong Ju[1], Guanjin Huang[1], Miaogeng Wang[1], Xuhui Zhang[1], and Meiling Dai[2(✉)]

[1] CSG Power Generation Co., Ltd, Beijing 100070, China
[2] Beijing University of Posts and Telecommunications, Beijing 100876, China
8367688@qq.com

Abstract. In order to solve the problems of high false alarm rate and long inference time of the congestion link inference algorithm in power data networks, this paper proposes a congestion link inference algorithm of power data network based on Bayes theory. First, based on the network topology and detection relationship, a detection matrix is constructed, and the detection matrix is simplified using Gaussian Jordan elimination method. In order to reasonably select the detection, the detection information gain is designed to make the judgment. Secondly, in order to accurately infer congested links, a Bayesian model for detecting link associations is constructed based on the detection results, and based on the model, congested link inference is performed using the maximum posterior probability. Compared with the existing algorithms, it is verified that the algorithm in this paper has achieved good results in three aspects: the accuracy of congested link inference, the rate of false positives, and the length of inference.

Keywords: Power data network · Congested link · Bayes · Information gain

1 Introduction

With the gradual expansion of the power data network, more and more important services are running on the power data network, which puts more demands on the reliability of the power data network. In order to improve the reliability of the network, it has become an urgent problem to quickly find the congested links of the network. To solve this problem, active detection technology has been proposed, and has become a key technology to solve this problem [1, 2]. The existing research mainly includes the realization of detection technology and the selection of detection. In terms of the realization of detection technology, literature [3] based on the analysis of the flow table of the detection message to obtain the true path that the detection message passes through, thus providing data support for link detection. Reference [4] adopts the method of storing probe path information in the packet header, which avoids the deployment of relevant flow table rules that record probe information on the switch, and reduces the impact of probe on network performance. In terms of detection selection, literature [5] aims to reduce the complexity of the fault location algorithm, uses path association analysis, and proposes a detection site selection mechanism based

Q. Liu et al. (Eds.): CENet 2020, AISC 1274, pp. 899–907, 2021.
https://doi.org/10.1007/978-981-15-8462-6_104

on path independence, which better reduces the complexity of the fault location algorithm. References [6] and [7] aim to solve the problem of link fault location in optical networks, analyze the link relationship, and formulate a detection and selection strategy for the link fault location process. Reference [8] aims to select the optimal detection site, adopts the source routing method, and increases and optimizes the detection site according to the characteristics of the network topology. Reference [9] aims to solve the problem of resource allocation under network uncertain conditions, and uses Bayesian theory to optimize network detection information, and to determine the detection site during multiple interactions.

It can be known from the analysis of the existing research that the current selection of probes is mainly concerned with the selection of probes, and there is less research on the problem of network congestion caused by the increase in the number of probes. To solve this problem, based on the relationship between network topology and detection, this paper constructs a detection matrix and uses Gaussian Jordan elimination method to simplify the detection matrix. Based on the detection results, the Bayesian model of the detection link is constructed, and the maximum posterior probability is used to infer the congested link. The performance of the algorithm in this paper is verified through experiments.

2 Network Model

2.1 Construct a Routing Matrix for Detecting Paths

The network topology includes network nodes and network links, and $G = (N, E)$ is used to represent the network topology, where $n_i \in N$ represents the network node and $e_j \in E$ represents the network link. Detection refers to the end-to-end path $P_k \in P$ sent from the detection point to the target node. According to the results returned by the detection, the status of the path can be judged. The state of the path can be used to infer the state of the link $e_j \in P_k$ it contains. When probe $P_k \in P$ returns a normal result, it indicates that all links $e_j \in P_k$ included in it are in a normal state. When the result returned by probe $P_k \in P$ is abnormal, it indicates that at least one link among all the links $e_j \in P_k$ contained in it is in an abnormal state. Examples of detection paths are shown in Table 1.

Table 1. Examples of detection paths

Path	Links	State
$P1 : H1 \rightarrow H2$	$e1 \rightarrow e2 \rightarrow e3$	1
$P2 : H1 \rightarrow H3$	$e1 \rightarrow e8 \rightarrow e7 \rightarrow e12$	0
$P3 : H1 \rightarrow H4$	$e1 \rightarrow e11 \rightarrow e10$	1
$P4 : H2 \rightarrow H3$	$e3 \rightarrow e4 \rightarrow e12$	0
$P5 : H2 \rightarrow H4$	$e3 \rightarrow e4 \rightarrow e9 \rightarrow e10$	0
$P6 : H3 \rightarrow H4$	$e12 \rightarrow e9 \rightarrow e10$	0

In order to analyze the relationship between detections, the detection path is expressed as a detection matrix form $M_{kj} \in M$. Each row of the matrix represents a probe, and each column of the matrix represents the links contained in the network. Matrix element $M_{kj} = 1$ indicates that probe k passes link j. Matrix element $M_{kj} = 0$ indicates that probe k does not pass through link j. The routing matrix of the probe path in Table 1 is shown in Fig. 1.

$$
\begin{array}{c}
\begin{array}{cccccccccc} e1 & e2 & e3 & e4 & e7 & e8 & e9 & e10 & e11 & e12 \end{array} \\
M = \left[\begin{array}{cccccccccc}
1 & 1 & 1 & 0 & 0 & 0 & 0 & 0 & 0 & 0 \\
1 & 0 & 0 & 0 & 1 & 1 & 0 & 0 & 0 & 1 \\
1 & 0 & 0 & 0 & 0 & 0 & 0 & 1 & 1 & 0 \\
0 & 0 & 1 & 1 & 0 & 0 & 0 & 0 & 0 & 1 \\
0 & 0 & 1 & 1 & 0 & 0 & 1 & 1 & 0 & 0 \\
0 & 0 & 0 & 0 & 0 & 0 & 1 & 1 & 0 & 1
\end{array} \right] \begin{array}{c} P1 \\ P2 \\ P3 \\ P4 \\ P5 \\ P6 \end{array}
\end{array}
$$

Fig. 1. The routing matrix of the probe path in Table 1

When there are many probes in the network, the routing matrix will be very large. The detection in the network needs to increase the detection traffic in the network, which affects the normal services on the network. In order to reduce the impact of detection on the network, the number of detections needs to be reduced according to the relationship between the detections. When reducing the number of probes, it is necessary to minimize repeated probes to ensure that more link states in the network are detected. Considering that the Gaussian Jordan elimination method can simplify the matrix by row transformation, this paper uses the Gaussian Jordan elimination method to simplify the routing matrix, thereby reducing the number of probes and reducing the negative traffic caused by the probe to the network. Based on the above analysis, the Gaussian Jordan elimination method is used to simplify the detection matrix $M_{kj} \in M$, and the simplified detection matrix is represented as $M_{kj} \in M^{'}$.

2.2 Analysis of Detection Results

Active detection technology uses the method of sending detections one by one to quickly observe the network status. Compared with sending all probes to the network at the same time, the active detection technology can minimize the negative impact on the network. Therefore, this paper uses active detection technology to detect the network status.

In order to reasonably select the detection, the following defines the detection information gain to judge. The calculation method of the information gain $G(P_k)$ of the probe P_k is shown in formula (1). Among them, $H(E|P)$ represents the uncertainty of the network state $E = (S(e_1), S(e_2), \ldots, S(e_n))$ when the detection result is $P = (S(P_1), S(P_2), \ldots, S(P_m))$, which is calculated using formula (2). $S(P_k)$ represents the detection result returned by detection P_k. $S(e_n)$ indicates the state of network link e_n. Therefore, the physical meaning of the information gain $G(P_k)$ of the probe P_k is the reduced value of the uncertainty brought to the network after the probe P_k is sent. Among them, $pr(E,P)$ represents the joint probability distribution of the network state and the detection state. $pr(E,P)$ represents the probability distribution of the network state under the condition that the detection state is known.

$$G(P_k) = H(E|P) - H(E|P \cup \{S(P_k)\}) \tag{1}$$

$$H(E|P) = -\sum_E \sum_P pr(E,P) \log pr(E|P) \tag{2}$$

3 Inferred Model

This paper first builds a Bayesian model of the detection link association based on the detection results. Second, based on the model, the maximum posterior probability is used to infer the congested link. The following describes the method of constructing the link-associated Bayesian model and the algorithm for congested link inference with the maximum posterior probability.

3.1 Constructing Bayesian Model of Association PF Probe Link

After obtaining the detection result and the network link status, based on the relationship between the network detection and the link, a Bayesian model of association of probe link is constructed as shown in Fig. 2.

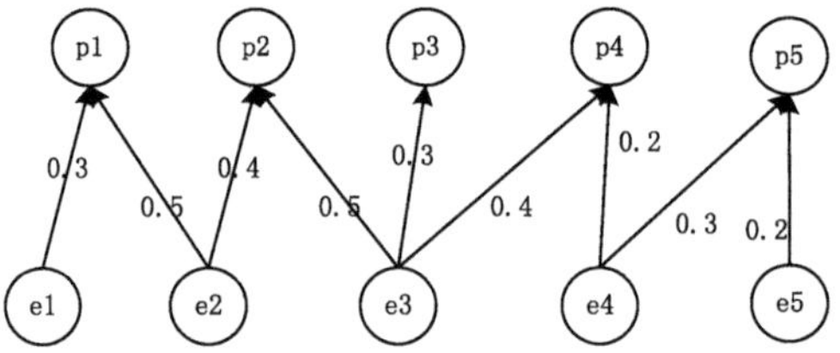

Fig. 2. Bayesian model of association of probe link.

The detection link association Bayesian model includes the parent node e_i, the child node p_j, the connection $pr(e_i|p_j)$ between the parent node and the child node. The parent node $pr(e_i)$ represents priori probability of network link congestion, which can be obtained based on long-term operational data statistics. The child node $pr(p_j)$ indicates the result status of the detection. The connection $pr(e_i|p_j)$ between the parent node and the child node represents the probability that the child node will appear abnormal when the parent node is congested. The goal of this article is to infer the probability of link congestion when the state of the probe result is known. Therefore, using $pr(e_i|p_j)$ to indicate that the detection result is abnormal and the congested link is inferred, which can be calculated using formula (3).

$$pr(e_i|p_j) = \frac{pr(p_j|e_i)pr(e_i)}{\sum_{p_k \in P} pr(p_k|e_i)pr(e_i)} \tag{3}$$

Use formula (3) to calculate the congestion probabilities of all links, and arrange them in descending order to form a set $E_O = \{e_1, e_2, \ldots, e_j\}$ of suspected congested links.

3.2 Maximum Posterior Probability Inference Algorithm

It can be seen from the Bayesian model associated with the probe link that a probe contains multiple links. Therefore, when a certain detection result is abnormal, it may be caused by congestion on one or several links it contains. In order to quickly find the congested link, this paper uses the method of solving the optimal set of suspected congested links. The formula is shown in (4). This formula means that the link set E with the highest congestion probability that can explain the abnormal detection result P is solved. Among them, $1 - \prod_{P_j \in P}(1 - pr(e_i|P_j)pr(P_j))$ means that when link e_i is congested, at least one probe P_j results in abnormal results. $1 - \prod_{e_i \in E}(1 - pr(P_j|e_i))$ indicates that the result of detecting P_j is abnormal because at least one link e_i is congested.

$$\arg\max_{E} C(E,P) = l$$
$$\prod_{e_i \in E}(1 - \prod_{P_j \in P}(1 - pr(e_i|P_j)pr(P_j))) \times \prod_{P_j \in P}(1 - \prod_{e_i \in E}(1 - pr(P_j|e_i))) \tag{4}$$

Take the suspected congested links from E_O in sequence, and calculate the value of formula (4) until the value of formula (4) grows smaller than ε. The set of taken out suspected congested links is the congested link set.

4 Congested Link Inference Algorithm

The congestion link inference algorithm of power data network based on Bayes theory (CLIoB) is presented in Table 2. The algorithm includes four processes: constructing alternative detection sets, actively detecting and obtaining detection results, solving suspected congested link sets, and using the maximum posterior probability to perform congestion link inference.

Table 2. Algorithm CLIoB.

Input: Network topology $G = (N, E)$
Output: Congested link set E_{con}
1. Build an alternative detection set 1) Use Dijkstra shortest path algorithm to construct a detection set P and a detection matrix M; 2) The Gaussian Jordan elimination method is used to simplify the detection matrix M to obtain the simplified detection matrix M'; 2. Actively detect and obtain detection results 1) Use formula (1) to calculate the information gain $G(P_k)$ of each detection in the detection matrix M' and arrange them in descending order; 2) Select the probe with the largest information gain to detect and receive the returned result; 3) Update the information gain of all probes using the probe return results; 4) Select the detection with the maximum information gain to detect until the detection gain reaches the detection threshold; 3. Solve the set of suspected congested links 1) Construct a Bayesian model of the detection link association based on the detection results; 2) Use formula (3) to calculate the congestion probability of all links, and arrange them in descending order to form a set $E_O = \{e_1, e_2, ..., e_j\}$ of suspected congested links; 4. Use maximum posterior probability for congested link inference 1) Take the links from $E_O = \{e_1, e_2, ..., e_j\}$ in turn and use formula (4) to solve the congested link set; 2) When the value growth of formula (4) is less than ε, the set of taken out suspected congested links is the set E_{con} of congested links.

5 Performance Analysis

5.1 Simulation Environment

In order to verify the performance of the algorithm, the BRITE tool [10] was used to generate the network environment in the experiment, and the LLRD 1 model [11] was used to simulate network congestion in the network environment. The priori probability of link congestion is [0.01, 0.003]. To simulate network noise, 0.5% of normal links in the network link are modeled as congested links.

In terms of algorithm evaluation indicators, the algorithm CLIoB in this paper is compared with the traditional algorithm CLIoA (congestion link inference algorithm based on all probe) from the three dimensions of the accuracy rate, false alarm rate, and inference duration. Among them, the algorithm CLIoA refers to sending all probes and inferring congested links based on the probe results.

5.2 Algorithm Comparison

In this paper, the algorithm CLIoB and the algorithm CLIoA are compared in terms of accuracy, false alarm rate, and inferred time. The results are shown in Fig. 3 to 5. Among them, the X axis represents the number of network nodes in the network environment. The performance indicators of the two algorithms are analyzed when the number of network nodes increases from 100 to 500.

In Fig. 3, the Y axis represents the accuracy of congested link inference, which is used to measure the proportion of the number of congested links inferred by the algorithm in the total number of congested links. As can be seen from the Figure, as the network size increases, the performance of both algorithms is relatively stable, and the algorithm in this paper is slightly better than the traditional algorithm.

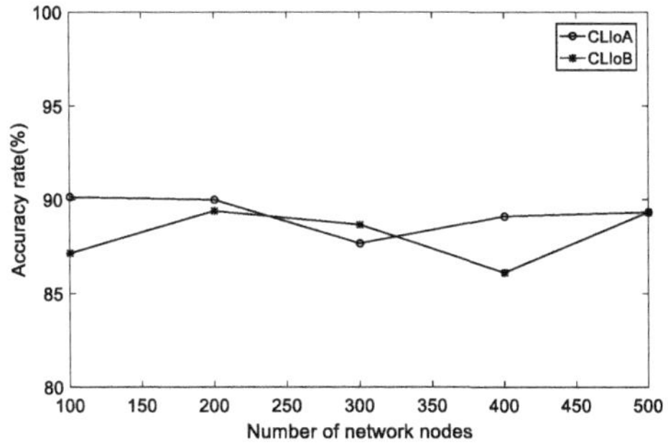

Fig. 3. Comparison of accuracy rate.

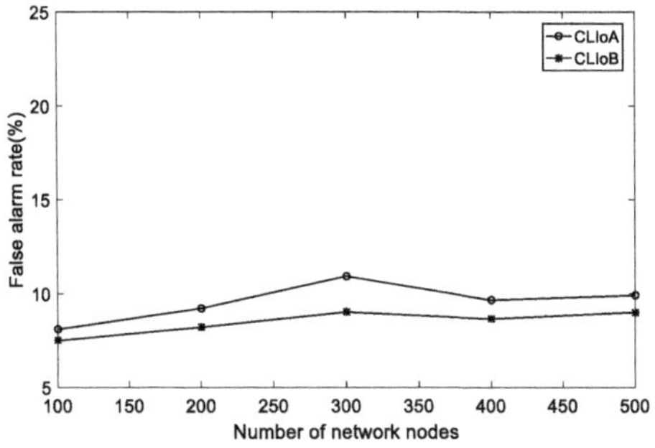

Fig. 4. Comparison of false alarm rate.

In Fig. 4, the Y axis represents the false alarm rate of the inferred algorithm, which refers to the proportion of erroneous congested links inferred by the algorithm in the total congested links. It can be seen from the Figure that as the number of network nodes increases, the false alarm rate of the two algorithms increases slightly. The algorithm in this paper reduces the false alarm rate than the algorithm CLIoA. This is because the algorithm in this paper optimizes the detection. The algorithm CLIoA has a large number of detections, resulting in an increase in the false alarm rate.

In Fig. 5, the inference time of the two algorithms is analyzed, and the Y axis represents the time spent by the inference algorithm. It can be seen from the Figure that as the number of network nodes increases, the inference time of both algorithms grows faster. This is because the size of the network becomes larger and the number of probes in the network increases, resulting in a longer time for algorithm inference. Comparing the two algorithms, the time increase of the algorithm in this paper is relatively gradual, while the inference time of the algorithm CLIoA increases rapidly. This is because the detection scale of the algorithm CLIoA grows rapidly with the growth of the network scale, resulting in a rapid increase in the inference time.

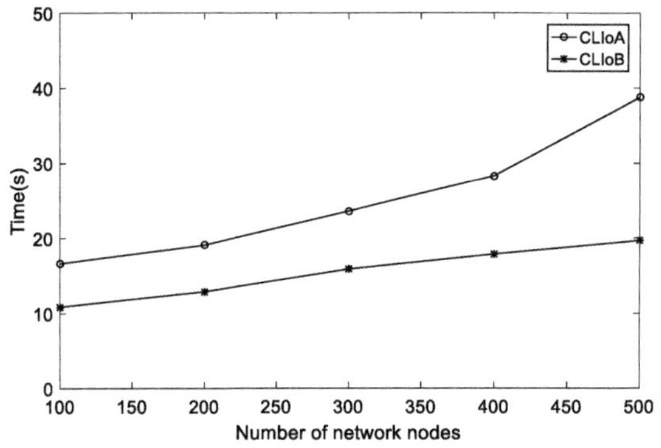

Fig. 5. Comparison of inferred time.

6 Conclusion

In order to ensure the high reliability of power services on the power data network, the accurate positioning of congested links has become an urgent problem to be solved. In order to solve this problem, this paper first constructs a detection matrix based on the relationship between the network topology and detection, and uses the Gaussian Jordan elimination method to simplify the detection matrix. Secondly, based on the detection results, a Bayesian model of the detection link association is constructed, and a Bayesian-based algorithm for inferring congested links in power data networks is proposed. Based on the detection results and Bayesian inference model, the algorithm uses the maximum posterior probability to infer the congested link, and obtains good results in terms of congested link inference performance. In the next step, we will further study the performance monitoring and fault location of the wireless side equipment, so as to achieve end-to-end power business reliability assurance.

Acknowledgments. This work is supported by China Southern Power Grid Technology Project "Research on application of data network simulation platform" (STKJXM20180052).

References

1. Dovrolis, C., Ramanathan, P., Moore, D.: What do packet dispersion techniques measure? In: Proceedings IEEE Infocom 2001. Conference on Computer Communications, vol. 2, pp. 905–914. IEEE (2001)
2. Uludag, S., Lui, K.S., Ren, W., et al.: Secure and scalable data collection with time minimization in the smart grid. IEEE Trans. Smart Grid **7**(1), 43–54 (2016)
3. Handigol, N., Heller, B., Jeyakumar, V., et al.: I know what your packet did last hop: using packet histories to troubleshoot networks. Proc. NSDI **14**, 71–85 (2014)
4. Tammana, P., Agarwal, R., Lee, M.: Cherrypick: tracing packet trajectory in software-defined datacenter networks. In: Proceedings of the 1st ACM Sigcomm Symposium on Software Defined Networking Research, pp. 1–7. ACM (2015)
5. Natu, M., Sethi, A.: Probe station placement for robust monitoring of networks. J. Netw. Syst. Manag. **16**(4), 351–374 (2008)
6. Ali, M.L., Ho, P.H., Tapolcai, J.: SRLG failure localization using nested m-trails and their application to adaptive probing. Networks **66**(4), 347–363 (2015)
7. Xuan, Y., Shen, Y.L., Nguyen, N.P., et al.: Efficient multi-link failure localization schemes in all-optical networks. IEEE Trans. Commun. **61**(3), 1144–1151 (2013)
8. Jeswani, D., Korde, N., Patil, D. et al.: Probe station selection algorithms for fault management in computer networks. In: 2010 Second International Conference on COMmunication Systems and NETworks (Comsnets 2010), pp. 1–9. IEEE (2010)
9. Zheng, A. X., Rish, I., Beygelzimer, A.: Efficient test selection in active diagnosis via entropy approximation. arXiv preprint arXiv: 1207. 1418 (2012)
10. Brite. http://www.cs.bu.edu/brite/
11. Padmanabhan, V.N., Qiu, L., Wang, H, J.: Server-based inference of internet link lossiness. In: IEEE Infocom 2003. Twenty-second Annual Joint Conference of the IEEE Computer and Communications Societies, vol. 1, pp. 145–155. IEEE (2003)

Multimodal Continuous Authentication Based on Match Level Fusion

Wenwei Chen[1], Pengpeng Lv[2], Zhuozhi Yu[1], Qinghai Ou[1], Yukun Zhu[1], Huifeng Yang[2], Lifang Gao[2], Yangyang Lian[2], Qimeng Li[2], Kai Lin[3], and Xin Liu[3(✉)]

[1] State Grid Communication Industry Group, Beijing Zhongdian Feihua Communication Co., Ltd., Beijing 100084, China

[2] State Grid Hebei Electric Power Co., Ltd., Information Communication Branch, Shijiazhuang 051011, China

[3] State Key Laboratory of Networking and Switching Technology, Beijing University of Posts and Telecommunications, Beijing 100876, China

19801116781@163.com

Abstract. With the rapid popularization of mobile devices, power grid system relies on mobile devices more and more. The traditional identity authentication cannot meet the security requirements of power grid system. Although continuous authentication can be achieved through behavioral features, there is still a problem of insufficient authentication accuracy. This paper proposes a method of fusion of gait and touch behavior. The fusion level is match level fusion. Weighted addition fusion is used to fuse gait matching score and touch matching score to improve the accuracy of authentication.

Keywords: Power grid system · Gait · Touch · Matching level fusion

1 Introduction

At present, with the rapid popularization of mobile devices, power grid staff can access the power grid system through mobile devices anytime and anywhere. With the convenience of using mobile devices, the security of mobile devices has also attracted our attention. Only relying on the traditional identity authentication, such as password authentication, cannot guarantee the account security of the staff and the security of the power grid system. Staff information leakage and illegally intruding system has become an important problem to be solved in power grid system. Therefore, how to better ensure the security of power grid system in mobile devices has become a research hotspot.

Identity authentication can be divided into the following three categories: (1) "what does she/he know", for example, authentication by password or pin; (2) "what does she/he have", for example, authentication by token or smart card; (3) "who is she/he" [1], which is authentication by biometrics. Using the first and second types of identity authentication, once the password is leaked or "brutally cracked" [2] and the smart card is lost, the privacy data of the power grid system and the staff will be exposed; the third

Q. Liu et al. (Eds.): CENet 2020, AISC 1274, pp. 908–914, 2021.
https://doi.org/10.1007/978-981-15-8462-6_105

type is more reliable than the first two, the authentication basis is the characteristics of the authenticator and the authentication basis is difficult to copy [3].

With the popularity of mobile devices, at the same time, the built-in sensors in mobile devices are becoming more and more diversified [4]. Common sensors include acceleration sensors, pressure sensors, gyroscope sensors, etc. [5], which makes it possible to collect data of gait feature and touch feature [6]. Patel et al. Proposed a simple authentication method to distinguish legitimate and illegal users by shaking the device, whose input is 2-axis acceleration data [7]. Conti et al. Proposed a biometric authentication method based on the user's movement, which authenticates by the user's movement when receiving or making a call [8]. Yang et al. Improved the authentication accuracy to as low as 2.6% EER by using wearable devices [9]. Touch based authentication provides user authentication in an accessible and convenient way through behavioral data collected by mobile devices [10, 11].

The above results confirm that gait and touch authentication based on mobile devices has certain authentication potential. However, compared with face or fingerprint authentication, the accuracy of gait and touch based authentication is still insufficient. At the same time, although the accuracy of authentication is high enough in some literatures, the experiment is in a strictly controlled environment (the operation of experimental participants is strictly limited), and there is no such controlled operation in a specific power grid business operation scenario. In this study, we propose a behavioral feature fusion scheme based on the grid business scenario. We choose gait and touch behavior fusion, and the fusion level chooses matching level fusion. Through matching level fusion, the accuracy of identity authentication is improved.

2 Behavior Feature Fusion Technology

With the rapid popularization of mobile devices and the rapid development of the Internet, the power grid system increasingly depends on mobile devices, whether based on the traditional way of identity authentication, or using a single behavior feature for identity authentication, cannot meet the security needs of the power grid system. By combining the information fusion technology [12] with the identity authentication technology based on behavior characteristics, not only can the security of user account be detected continuously, but also the accuracy of authentication can be improved.

2.1 Information Fusion Technology

In view of the shortcomings of the method of authentication based on single behavior, this paper proposes a method to integrate multiple behaviors [13], and chooses to integrate gait behavior and touch behavior, and the fusion level is at the matching level. Information fusion is the basis of multimodal biometric technology [14] research. Fusion [15] refers to the synthesis of different forms of information collected through different channels, so as to make the obtained information more comprehensive, so as to achieve more accurate, perfect and effective. Therefore, by fusing different behavior features, the result of fusion can contain useful information of each fusion behavior. In the identification method of biological behavior characteristics, multiple behavior fusion recognition has the following advantages:

(1) The security of multi behavior fusion recognition technology is high, the forgery of multi behavior fusion is more difficult than single behavior, and has higher security than single mode;

(2) The recognition accuracy of multiple behavior fusion recognition technology is higher, and the information of multiple behavior fusion is complementary, which improves the recognition rate and reliability [16].

2.2 Multi-model Feature Fusion Level

According to different fusion levels, multi-modal biological feature fusion can be divided into the following three types: feature-level fusion, matching-level fusion and decision-level fusion. As shown in Fig. 1.

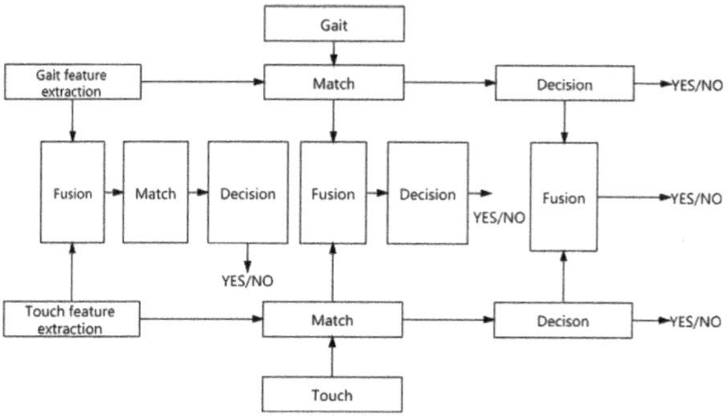

Fig. 1. Fusion framework of all levels.

Level of Fusion Comparison. Different levels of fusion are reflected in different stages of the authentication process. Feature-level fusion fuses unrelated behavioral features. Although the fused features can contain the information amount of each behavioral feature, the feature dimension increases suddenly after fusion, which is easy to bring "dimensional disaster" [17]. Decision-level fusion is only a simple fusion operation of the identification results of various behavioral features, which is often too one-sided and cannot take advantage of two or more behavioral features at the same time. Match-level fusion performs identity recognition by fusing the matching scores of each behavior feature. This method can comprehensively consider the advantages of different behavior features, and the identity authentication results obtained are more relevant and accurate.

3 Gait and Touch Behavior Analysis

3.1 Data Set

We use data sets from other papers [18]. The data set conducted two studies of 30 users to collect five behavioral biological characteristics (gait, touch dynamics, electrocardiogram, eye and mouse movements) in different situations. We chose the gait and touch data sets. Gait data is obtained by letting the user wear 5 different sensors, and instructs the user to walk from the starting point to the end point, then back to the starting point, then jog to the end point, and finally jog back to the starting point. Touch data allows users to play a "discover difference" game on their smartphones.

3.2 Data Preprocessing

Data collection through mobile devices, we can only obtain the original data, the original data contains "noise data" and "bad" data. We use mean filtering to process pressure data, Savitzky–Golay [2] to process acceleration data, and Kalman [2] to process gyroscope data.

3.3 Feature Extraction

Due to the selection of fusion authentication of gait and touch, this paper will extract two types of features, gait and touch.

For touch behavior, we divide features into the following two categories:

(1) Touch click feature: touch click action mainly occurs in the user's click action on mobile devices, combined with specific grid business, mainly occurs in the user's selection of specific functions of grid system;

(2) Sliding features: the subdivision of sliding features can be divided into up and down sliding and left and right sliding. Combined with specific grid business, up and down sliding mainly occurs in browsing information text in grid system, and left and right sliding mainly occurs in browsing pictures.

For gait behavior, we choose the maximum, minimum and average values of velocity, acceleration in three-axis direction (corresponding to x, y, Z axis) as each element of the eigenvector.

3.4 Match Distance

We use the random forest algorithm to obtain the matching distance of the touch operation. Random forest is an integrated method with multiple decision trees. Each tree in the random forest is guided by the data set, and each tree classifies the data and votes in the tree. We use SVM to obtain the matching distance of gait behavior. SVM maps data to a high-dimensional space and attempts to find an optimal hyperplane. In our research, we use RBF kernel SVM, and set the parameters $g = 0.0625$, $nu = 1$.

3.5 Match-Level Fusion

For gait and touch, the matching scores of the gait matching module and the touch matching module are not closely related, so we choose weighted addition fusion. Even if the matching score of one module is wrong, the impact on the overall fusion effect will not be too great. At the same time, due to the different types of gait modules and touch modules, the form of matching scores obtained is also different. Therefore, before fusing values, we only need to normalize the matching score.

Normalization. In this study, the Min-Max method was used to normalize the matching score. Min-Max normalization idea: Convert all the values to be processed into the interval [0, 1], the specific conversion formula is as follows.

Suppose there are n matching scores d_1, d_2, d_3, …, d_n, calculated as follows:

$$d_n = \left(\frac{d_n - d_{min}}{d_{max} - d_{min}}\right), d \in [0, 1] \tag{1}$$

Among them, d_{max} and d_{min} are the maximum and minimum values of matching scores.

Weighted Addition Fusion. Assuming that the number of normalized matching values is M after gait matching and touch matching, the matching values are recorded as: gait matching value: $\{s(O_{gait}|\lambda_i)\}_{i=1,2,\ldots,M}$ and touch matching value: $\{s(O_{touch}|\lambda_i)\}_{i=1,2,\ldots,M}$,

According to the weighted addition fusion algorithm, the fusion formula is as follows:

$$s(O_{gait}O_{touch}|\lambda_i) = \alpha s(O_{touch}|\lambda_i) + \beta s(O_{gait}|\lambda_i), i = 1, 2, \ldots, M \tag{2}$$

Where $\beta = 1\text{-}\alpha$, $s(O_{gait}O_{touch}|\lambda_i)$ is the matching score after fusion. The higher the score, the higher the template matching.

4 Results

In the experiment, we use two indicators to compare the experimental results, false acceptance rate and false rejection rate. The false acceptance rate is the ratio of the number of times the illegal users are identified as legitimate users to the total number of identification times; the false rejection rate is the ratio of the number of times the legitimate users are mistakenly identified as illegal users to the total number of identification times. The experimental results are shown in Fig. 2 or Table 1. Through experimental results, we found that the use of identity authentication combining gait and touch can effectively improve the accuracy of authentication. At the same time, the false acceptance rate is better improved, while the false rejection rate is less improved than the false acceptance rate. We infer that this may be because after the gait and touch are fused and authenticated, the verification conditions are more sufficient and "harsh", resulting in fewer false acceptances and more false rejections.

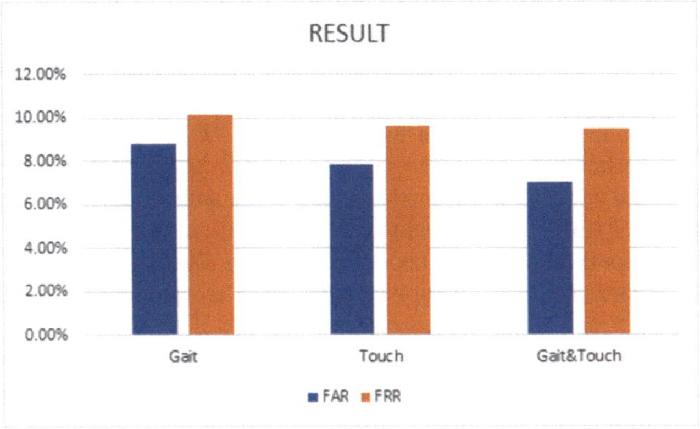

Fig. 2. Experimental results.

Table 1. Experimental results.

Behavior	FAR	FRR
Gait	8.79%	10.17%
Touch	7.85%	9.64%
Gait & Touch	7.04%	9.44%

5 Conclusion

This paper proposes a method of fusing gait behavior and touch behavior, and selects matching-level fusion. Through weighted addition fusion, the fusion matching score is obtained, and a higher authentication accuracy rate is obtained. Compared with continuous authentication using only a single behavioral characteristic, the accuracy of authentication is improved.

Acknowledgment. This paper is supported by the Science and Technology Project of State Grid Corporation of China. Research on Key Technologies of dynamic identity security authentication and risk control in power business.

References

1. Shen, C., Chen, Y.F., Guan, X.H.: Performance evaluation of implicit smartphones authentication via sensor-behavior analysis. Inf. Sci. **430–431**, 538–553 (2018)
2. Derawi, M.O., Nickel, C., Bours, P., Busch, C.: Unobtrusive user-authentication on mobile phones using biometric gait recognition. In: 2010 Sixth International Conference on Intelligent Information Hiding and Multimedia Signal Processing, pp. 306–311 (2010)
3. Nickel, C., Wirtl, T., Busch, C.: Authentication of smartphone users based on the way they walk using k-nn algorithm. In: 2012 Eighth International Conference on Intelligent Information Hiding and Multimedia Signal Processing, pp. 16–20 (2012)

4. Zheng, N., Bai, K., Huang, H., Wang, H.: You are how you touch: user verification on smartphones via tapping behaviors. In: 2014 IEEE 22nd International Conference on Network Protocols, pp. 221–232 (2014). https://doi.org/10.1109/icnp.2014.43
5. Patel, S.N., Pierce, J.S., Abowd, G.D.: A gesture-based authentication scheme for untrusted public terminals. In: Proceedings of the 17th Annual ACM Symposium on User Interface Software and Technology, in: UIST 2004, pp. 157–160. ACM, New York, NY, USA (2004)
6. Yang, J., Li, Y., Xie, M.: Motionauth: motion-based authentication for wrist worn smart devices. In: 2015 IEEE International Conference on Pervasive Computing and Communication Workshops (PerCom Workshops), pp. 550–555 (2015)
7. Muaaz, M., Mayrhofer, R.: Smartphone-based gait recognition: from authentication to imitation. IEEE Trans. Mob. Comput. **16**(11), 3209–3221 (2017)
8. A, AateF., Nappi, M., Ricciardi, S.: I-am: implicitly authenticate me person authentication on mobile devices through ear shape and arm gesture. IEEE Trans. Syst. Man. Cybern. **49** (3), 469–481 (2017)
9. Sandnes, F.E., Zhang, X.: User identification based on touch dynamics. In: Proceedings the 9th International Conference on Ubiquitous Intelligence & Computing and Autonomic & Trusted Computing, pp. 256–263 (2012)
10. Luca, A.D., Hang, A., Brudy, F., Lindner, C., Hussmann, H.: Touch me once and i know it's you!: implicit authentication based on touch screen patterns. In: Proceedings the SIGCHI Conference on Human Factors in Computing Systems, pp. 987–996. Austin, Texas, USA (2012)
11. Sae-Bae, N., Ahmed, K., Isbister, K., Memon, N.: Biometric-rich gestures: a novel approach to authentication on multi-touch devices. In: Proceedings the 2012 ACM Annual Conference on Human Factors in Computing Systems, pp. 977–986 (2012)
12. Li, L., Zhao, X., Xue, G.: Unobservable re-authentication for smartphones. In: Proceedings 20th Annual Network & Distributed System Security Symposium (NDSS), San Diego, USA, (2013)
13. Saravanan, P., Clarke, S., Chau, D., Zha, H.: LatentGesture: active user authentication through background touch analysis. in; Proc. the Second International Symposium of Chinese CHI, Ontario, Canada, pp. 110–113 (2014)
14. Shen, C., Zhang, Y., Guan, X.H., Roy, A.: Maxion, fellow, IEEE performance analysis of touch-interaction behavior for active smartphone authentication. IEEE Trans. Inf. Forensics Secur. **11**(3), 498–513 (2015)
15. Ross, A., Jain, A.: Information fusion in biometrics. Pattern Recogn. Lett. **24**(13), 2115–2125 (2003)
16. Breiman, L.: Random forests. Mach. Learn. **45**(1), 5–32 (2001)
17. Chang, C.C., Lin, C.J.: Libsvm: a library for support vector machines. ACM Trans. Intell. Syst. Technol. **2**(3), 1–27 (2011). https://doi.org/10.1145/1961189.1961199
18. Simon, E., Giulio, L., Andrea, P., Marta, K., Vincent, L., Ivan, M.: When your fitness tracker betrays you: quantifying the predictability of biometric features across contexts. In: 2018 IEEE Symposium on Security and Privacy (2018)

NAS Honeypot Technology Based on Attack Chain

Bing Liu, Hui Shu(✉), and Fei Kang

State Key Laboratory of Mathematical Engineering and Advanced Computing, Zhengzhou 450001, China
shuhui123@126.com

Abstract. With the wide application of network attached storage (NAS), the security problem is becoming more and more serious, together with the increasingly fierce attacks against NAS devices of different manufacturers. In order to better capture various types of attacks against NAS devices and detect security threats in time, this paper proposes a solution named NAS honeypot, mining the potential security threats through modeling and analyzing the NAS threats. Then a NAS honeypot based on device virtualization technology and depth monitoring and attack induction was designed on the basis of the construction NAS attack chain. Experimental results show that the NAS honeypot can effectively capture, record and analyze network attacks against multi-types NAS devices with a strong guiding effect in mastering popular attack method of NAS devices and alleviating the security threats of NAS.

Keywords: NAS · Threat modeling · Attack chain · Honeypot

1 Introduction

In recent years, with the rapid development of network technology and improvement of computer data processing capability, information and data on the network have been increasing explosive, bringing unprecedented demands to the data access, transmission and processing capacity of storage system. NAS has emerged with the increasing demand for storage capacity and reliability. Taking data as the center, it completely separates the storage device from the server and connects a group of computers by standard network topology [1]; clients are allowed to access data directly on the NAS without the server, providing a solution for heterogeneous platforms to use unified storage system [2].

Compared with traditional storage, NAS shows strong advantages, but it also causes new security problems. NAS not only has to assume the threat risk of traditional storage system, but also has the security risks from the network, such as data leakage caused by network transmission. Data resources are the most valuable in the Internet age, but security problems frequently emerge in NAS, a direct storage of data, in recent years. Nearly all manufacturers' devices have not been spared. CVE disclosed more than 300 related public vulnerabilities and almost covered all types. For example, for Gain Privileges: CVE-2018-18471, attackers can execute remote commands as root on the specific NAS of Seagate and NETGEAR [3]; for Bypass Something: CVE-2016-10108,

Q. Liu et al. (Eds.): CENet 2020, AISC 1274, pp. 915–926, 2021.
https://doi.org/10.1007/978-981-15-8462-6_106

Western Digital MyCloud NAS allows an unauthenticated attacker to run remote command as root [4]; for Code Execution: CVE-2020-9054, Multiple ZyXEL NAS devices allow a remote, unauthenticated attacker to execute arbitrary code [5]; for Gain Information: CVE-2018-13291, Synology DiskStation Manager allows remote authenticated users to obtain sensitive information [6]; for Directory Traversal: CVE-2018-13322, Buffalo TS5600D1206 allows attackers to list directory contents [7]; for Overflow: CVE-2018-14749, buffer overflow vulnerabilities in QTS can have unspecified impact on the NAS [8] and so on.

To address the endless security problems of NAS, we carry out research on NAS honeypot technology to capture attacks and enhance the security of NAS. Honeypot technology enables the defense side to clearly understand the security threats they face, and enhances the security protection ability of the actual system by means of technology and management, so it has been extensively and deeply studied in the field of network security. Moore [9] applied the honeypot technology to ransom worm detection, using two services to control the Windows security log, and establishing a graded solution strategy for attack; Saud et al. [10] used NIDS and KFSensor honeypot to detect APT attacks actively, and sent alarm information to the console when the honeypot service is called and run upon request. Anirudh et al. [11] proposed a solution of DoS attack honeypot for IoT devices, using IDS intrusion detection system to process client requests, and comparing information with log library to isolate and guide abnormal requests to honeypot and record abnormal source information. With the continuous development of honeypot technology, it is applied more and more in different scenarios and realized different functions. However, the research on NAS honeypot technology needs to be carried studied further.

In this paper, security threats are comprehensively analyzed and NAS attack chain is constructed through NAS threat modeling. NAS honeypot which is based on virtualization technology, interact with the attacker deeply through the virtual response to the attack, to induce them to launch in-depth attacks, capture and analyze the techniques and avenues of attacks. This paper solves the common NAS honeypot's problems of poor flexibility, poor expansibility and lack of unified deployment and control mechanism. The prototype system is implemented and the effectiveness of the proposed scheme is verified by actual deployment and experiment.

The rest of this paper is as follows: the second part introduces the NAS threat model and attack chain; the third part introduces the design and framework of NAS honeypot; the fourth part describes the experimental test; the fifth part summarizes the whole paper and looks forward to the next step.

2 NAS Threat Model and Attack Chain

The honeypot, which is designed based on the attack chain, can classify captured attacks according to specific attack chains, and intuitively and accurately identify the methods of attack is conducive to targeted security measures. We analyze and identify a variety of threats and separate attack steps of different attacks though threat modeling, and finally form a general attack link that can cover all attacks.

2.1 NAS Threat Model

Threat modeling is simply a structured method for identifying, quantifying and responding to threats. It helps to think about risks through abstract methods after the software design stage and before the software deployment stage. Generally, the process of threat modeling can be divided into 6 steps: identifying assets, creating an architecture overview, decomposing the application, identifying the threats, documenting the threats, and rating the threats [12]. We use the six-stage process to carry out threat modeling analysis for NAS.

First, we identify the assets that need to be protected by decomposing the composition of NAS hardware and software, including NAS devices, firmware, system application, mobile application and data storage, etc. By refining asset identification, we can discover as many types and numbers of threats as possible.

Then, we analyze the architecture, physical deployment, application functions and related technologies of the NAS system, so as to find potential vulnerabilities in the design or implementation of the application. We summarize the system architecture by creating the NAS system architecture diagram, and further identify the trust boundaries of the system, identify the data flow in the system, form the NAS system architecture and data flow diagram (Fig. 1), and then identify the entry points, and focus on the entry points and the data flow across the trust boundary, because the security of transferring data from outside is not trusted. We should focus on the threat analysis of this kind of data. According to Fig. 1, we can confirm entry points that affect system security, including firmware, system application, mobile application, user client, router, cloud and so on.

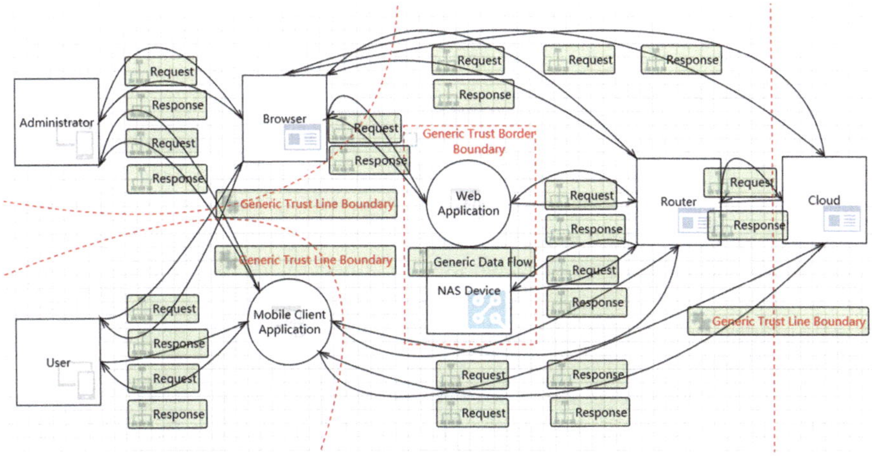

Fig. 1. The architecture and data flow of NAS system.

After that, based on the previous work, the STRIDE model is used to identify threats to NAS devices. STRIDE derived from an acronym for the following six threat categories: Spoofing, Tampering, Repudiation, Information disclosure, Denial of service, and Elevation of privilege. We classify and identify potential security threats according to these six threat categories, such as obtaining the account password of device login, tampering with device configuration, tampering with device log record, obtaining device sensitive information, denial of service attack, remote control of NAS device, etc. Finally, we take a record of threats, take "Remote control of NAS device" as an example (shown in Table 1).

Table 1. The threat record of "Remote control of NAS device".

Threat description	Remote control of NAS device
Threat target	NAS user, NAS communication data, NAS device
Attack techniques	Improper use by NAS users, failure to modify the original password in time, or disclosure of the password in other ways; Attackers can analyze and find the administrator password by sniffing and intercepting the NAS communication data; The control of the device can be obtained by means of brute force, the use of firmware vulnerabilities and bypassing authentication
Countermeasures	Do well in guiding the use of the user's equipment, do well in the password protection of equipment, regularly change the password of equipment; Do well in auditing communication data, do well in communication data encryption; Set up multiple password error locking; Control firmware permissions strictly, and audit the authentication authority

2.2 NAS Attack Chain

NAS attack chain refers to the complete attack path that is composed of a general description of each attack stage of NAS and the attack sequence. Detecting and discovering devices is the first step to achieve NAS attacks. Therefore, the first part of NAS attack chain is NAS Detection. Then, we complete the attack chain through comprehensive analysis of threats identified by threat modeling. The representative high-risk threats are selected and the attack path is symbolized with STRIDE. To facilitate the description, NAS Detection is denoted as DT. Some symbolic descriptions of high-risk threats are shown in Table 2.

Through the symbolic description of threat, we can find that when NAS is discovered, some attackers will choose Spoofing attack (S) or Elevation of Privilege attack (E), and some attackers will choose to bypass the authentication attacks. We define the attack stage of S or E as Penetration Attack, as the second part of the attack chain. If achieving penetration attack or bypassing the authentication successfully, the attacker can carry out further operations. This stage is defined as the third part of the attack chain: Attack and Invade. When an attacker invade the device successfully, he can carry out various malicious attacks on the device. This stage is defined as the fourth part of the attack chain: Malicious Control. NAS attack chain is formed as follows:

Table 2. Symbolic description of threat.

Threat	Description
Tamper with device configuration	DT → T/R DT → S→T/R DT → E→T/R
Denial of service attack	DT → D/R DT → S→D/R DT → E→D/R
Remote control of NAS	DT → S→T/R/I/D/E DT → S→E → T/R/I/D DT → E→S/T/R/I/D

① NAS Detection → ② Penetration Attack → ③ Attack and Invade → ④ Malicious Control, the second stage can be skipped.

3 Design and Framework of NAS Honeypot

Designing honeypot by NAS attack chain can accurately grasp the stage of interaction with attackers, timely respond to attackers' attack actions, improve the probability of attackers attacking the honeypot, and lure attackers to conduct in-depth attacks.

3.1 Design of NAS Honeypot

The first part of the attack chain is NAS Detection. Attackers usually try to find NAS devices through search engines, port scanning, ICMP detection packets and so on, and determine the manufacturers and models of NAS devices according to the fingerprint information of equipment. NAS devices of some manufacturers will leak the firmware versions of devices and other more detailed information. In order to induce attackers to carry out further targeted attacks, the information is chosen back to the attacker in time by virtual response mechanism.

The second part of the attack chain is Penetration Attack. After attacker discovers NAS, it needs a highly simulated NAS honeypot to interact with him. In order to respond to attacks by different attackers against different honeypots, it needs to simulate NAS devices of different manufacturers and models based on virtualization technology. Therefore, it is necessary to establish virtual NAS sets, including NAS honeypots of different manufacturers.

The third part of the attack chain is Attack and Invade. In view of the attack of the attacker at this stage, we mainly consider providing support for subsequent attack classification, and set up specific attack detection units in each NAS honeypot, such as firmware update, configuration tamper, buffer overflow, etc., and the corresponding detection unit can be added according to the characteristics of equipment and detection needs.

The fourth part of the attack chain is Malicious Control. When capturing the attacker's malicious behaviors successfully, how to determine which are known attacks, which are unknown attacks, which are attacked and expanded, and how to record these attacks for better research and analysis. To solve these problems, we need to identify and classify attacks.

3.2 The Framework of NAS Honeypot

According to the design idea in 3.1, we designed and implemented the NAS honeypot. Its architecture is shown in Fig. 2. It including 5 parts: Control Center, Data Center, Virtual Response Center, Virtual NAS Set, and Attack Recognition Center. Data Center mainly collects attack data, classifies and collects the attack data collected by each part, so as to follow-up research and analysis.

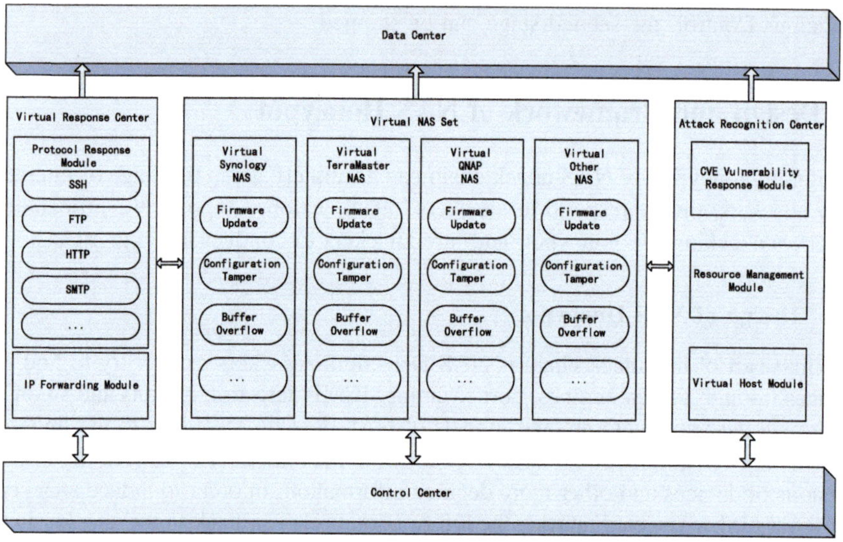

Fig. 2. The framework of NAS honeypot.

Control Center. The control center mainly controls each part of honeypot to respond to the attacker's attack, and connects to an external fingerprint database. In fact, the external fingerprint database is a collection of response packets for each vendor and each protocol. By sending different protocol query packets to different vendor devices, the response packets are extracted, classified and sorted out to form a fingerprint database for response packets for NAS attack detection. When receiving the protocol and port information sent by the Virtual Response Center, Control Center randomly selects the corresponding response packets from the external fingerprint database, returns it to the Virtual Response Center, controls the Virtual NAS Set to activate the corresponding virtual NAS, transmits the network data to the virtual NAS, and control the Virtual Host Module of the Attack Recognition Center connecting to the virtual

NAS. Record the attacker's attack behavior in the virtual NAS, and interact with the Attack Recognition Center to form a complete attack path.

Virtual Response Center. The Virtual Response Center mainly implements the response to the attacker's probe packets, including the Protocol Response Module and the IP Forwarding Module. The Protocol Response Module sets up a variety of protocol response units (such as SSH, FTP, HTTP, SMTP and other protocol types, which can be expanded according to actual needs). When receives the attacker's probe packets, the Virtual Response Center identifies the type of packets by feature extraction. The corresponding protocol response unit transfers the protocol and port information to the Control Center, and sends the response packets generated by Control Center to the attacker. IP forwarding module records the IP information of the attacker, and forwards the IP to the Virtual NAS Set after the virtual NAS is started, so as to deceive the attacker for subsequent attack. The Virtual Response Center sends attacker's detection behavior data to the data center.

Virtual NAS Set. The virtual NAS set is made up of virtual NAS from different manufacturers. When one of the virtual NAS is successfully launched, it becomes a specific NAS honeypot. Based on FIRMADYNE [13], we realize the simulation of every virtual NAS in the Virtual NAS Set.

First of all, we need to obtain firmware for NAS devices. Most NAS vendors constantly upgrade firmware to improve the function and security of the system, and provide firmware downloads on their official website, so we can get the firmware of mainstream NAS vendors through crawler technology.

Secondly, extract file system and kernel from firmware with the extractor that FIRMADYNE developed based on the binwalk API. Due to the difference of firmware structure, the extractor cannot extract all firmware, such as QNAP's firmware, we use the specific firmware extracting program "extract_qnap_fw.sh" [14] invoking PC1 tool to achieve firmware extraction. After all the firmware is extracted, it is normalized and stored in the firmware database.

Finally, we simulate the NAS device based on FIRMADYNE. We define the kernel and libcnvram.so according to the file system extracted from the firmware, and hijack the NVRAM related operations, and then perform the system level simulation based on the QEMU system mode. We analyze the extracted kernel and find that the NAS firmware is mostly ARM and X86 architecture, while FIRMADYNE only supports ARM and MIPS kernel architecture, so we mainly choose the NAS device of ARM architecture for simulation.

Repeat the above steps to simulate the NAS devices of various manufacturers to form the Virtual NAS Set. The Virtual NAS Set has a strong scalability, and can continuously add various virtual NAS devices when the hardware resources are allowed.

Attack Recognition Center. Attack Recognition Center mainly combines attack chain to identify attacker's attack mode and method, including 3 modules: CVE Vulnerability Response Module, Resource Management Module, and Virtual Host Module. CVE Vulnerability Response Module contains a NAS vulnerability database, we get vulnerabilities information from CVE official website, and store vulnerabilities in accordance with the mode of attack chain (① → ②→③ → ④) into the database. When an attacker attacks with a vulnerability in the NAS vulnerability database, the CVE Vulnerability Response Module compares the attack paths. If it is consistent with the way in which the vulnerability is exploited, it is considered to exploit the known vulnerability attack; if the path is inconsistent, it is considered to exploit the known vulnerability to expand the attack. Through resource access control, the Resource Management Module sets files and data accessed by administrators or specific users in the virtual NAS devices. If the controlled resources are accessed, it proves that the attacker has succeeded in invasion. If the attack path is: ① → ②→③, it is regarded as a device attack by password cracking; if the attack path is: ① → ③, an unknown attack is considered to have occurred. The Virtual Host Module mainly sets up the virtual host connected to the virtual NAS device. As a user of the NAS device, it monitors the attacker's penetration attacks through NAS. If the virtual host is attacked, it will be regarded as a successful penetration attack. The attack path is: ① → ② ③ → ④ or ① → ③→④. The Attack Recognition Center sends all the above attack information to the data center.

4 Experiments

4.1 Preparation

In order to detect the capture of attack behavior and the effect of system analysis, the NAS honeypot was built and connected to the Internet to attract attacks against NAS devices. Different attack behaviors are analyzed by capturing the attack behavior of the honeypot and capturing the system traffic. Select a router and connect to a host for traffic monitoring, because of that the NAS devices need to connect to the Internet through routing. And then connect the NAS honeypot to the network.

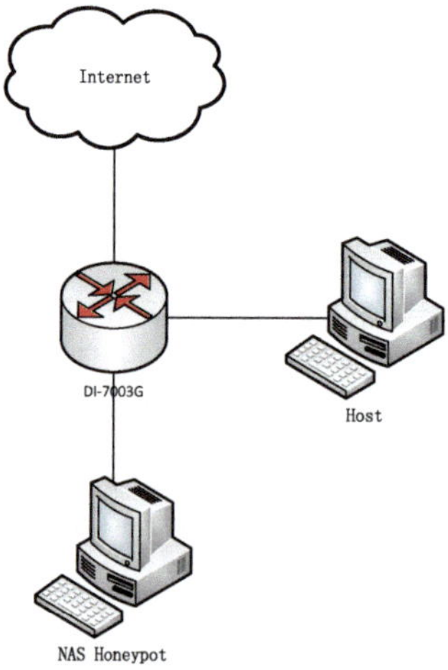

Fig. 3. The deployment of the experimental environment.

The deployment of the experimental environment is shown in Fig. 3. The equipment used is as follows:

(1) Two Windows10 hosts: one for NAS honeypot deployment and one for router traffic monitoring.

(2) One D-Link router, DI-7003G: used to connect the honeypot system to the Internet.

From 8:30 a.m. on March 24, 2020 to 8:30 a.m. on April 24, 2020, the system was deployed for one month. A total of 3053 NAS detection data were collected, among of which 2,247 successfully realized honeypot attacks.

4.2 Results and Analysis

In terms of the source of attack, the 3053 collected data came from 423 IP addresses in 41 countries and regions. Figure 4 shows the countries and regions with the source IP number in the top 10.

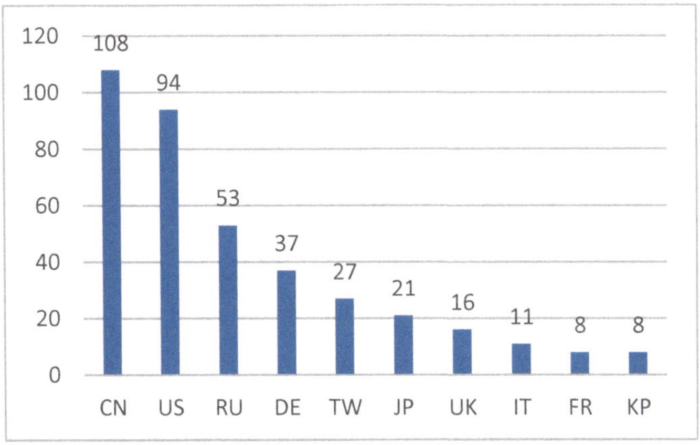

Fig. 4. Countries and regions with the source IP number in the top 10.

In terms of protocol utilization, the detection protocol based on SSH protocol is detected 1291 times, 1434 times based on HTTP protocol, 123 times based on FTP protocol, 16 times based on SMTP protocol, and 189 other probes, and the data statistics are shown in Fig. 5.

In terms of vulnerability utilization, according to the classification of attack types in NAS honeypot data center, 1587 of the 2247 successful attacks have been attacked by known vulnerabilities, 267 of which have expanded attacks, 655 have been attacked by device cracking, and 8 unknown attacks. In 1587 known attacks, gain privileges 183 times, bypass something 169 times, code execution 383 times, gain information 462 times, directory traversal 157 times, overflow 230 times, and data statistics are shown in Fig. 6.

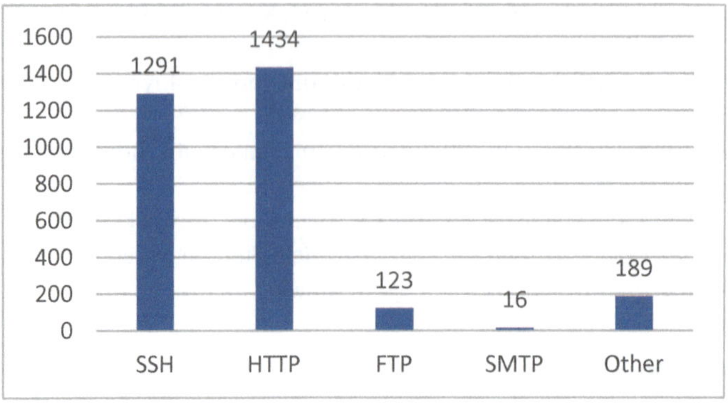

Fig. 5. Statistics of detection data protocol.

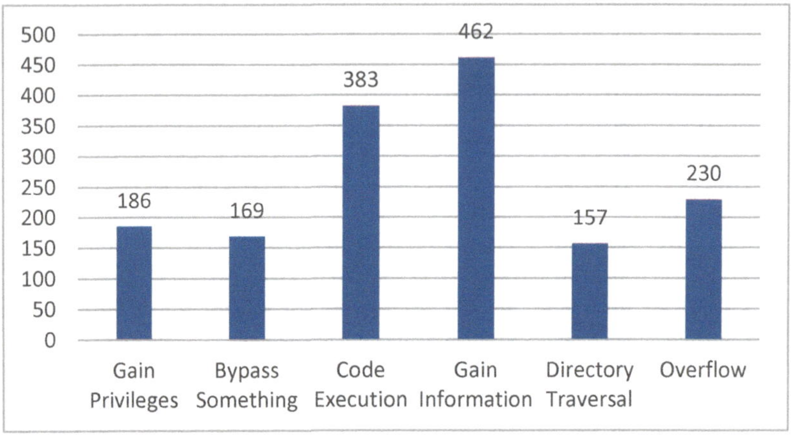

Fig. 6. Statistics of CVE vulnerability exploitation.

Honeypot system currently only provides NAS simulation of 3 vendors, including Synology, TerraMaster, and QNAP. The distribution of 2247 attacks in vendors: Synology 638 times, TerraMaster 529 times, and QNAP 1080 times, as shown in Fig. 7.

Select an example of QNAP attack which is successfully captured to prove the process of NAS honeypot catching, analyzing and recognizing the attack. The attack occurred at 13:00, April 15. The attacker detected the NAS device by sending the ICMP packet, the NAS honeypot system captured the detection data packet, used the QNAP response packet in the external fingerprint database to respond, and launched virtual QNAP NAS. The Virtual Response Layer forwarded attacker IP to the virtual QNAP NAS. The attacker did not engage in a password cracking attack. About 7 min later, the attacker exploited the vulnerability of CVE-2018-14746 to access the folder "import" that we had set up to view only with administrator privileges, and downloaded the 10 documents that were pre-created in the folder, through the way of command injection. Then, the attacker visited other folders without further attack actions, and the

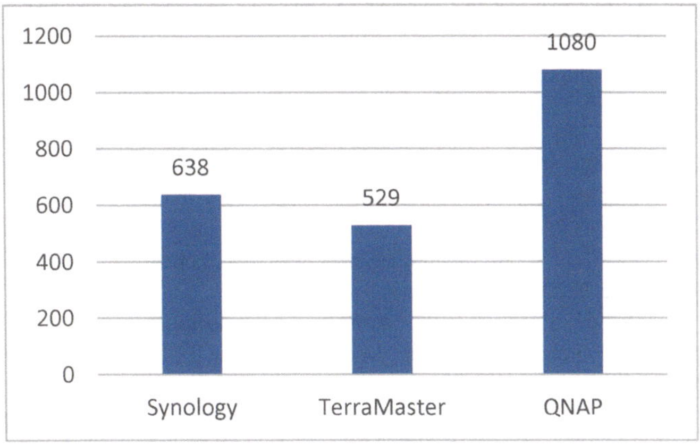

Fig. 7. Statistics of vendor attacks.

virtual NAS captured the above attacks. The NAS honeypot recorded the attack path of the attacker, ① → ③→④, which was consistent with the attack path in the CVE vulnerability database, and we judged that the attack behavior was using the known vulnerabilities to attack. The NAS honeypot maintained the virtual NAS to 18:00. No further attacks had been detected, and the honeypot had been automatically recovered to prepare for the next attack capture.

5 Conclusion

In this paper, NAS honeypot technology is studied based on threat model, and prototype system is implemented. The feasibility and flexibility and expansibility of the prototype system are verified by experiments. It can dynamically adjust the honeypot operation strategy, trap and analyze attacker attacks, and help to fully understand the security threats faced by NAS. Next, we will do further research on the technologies of NAS firmware virtualization for multi architecture, so that the honeypot can simulate more manufacturers and types of NAS devices to capture more unknown attacks.

Acknowledgment. This work was supported by the National Key Research and Development Program of China (2016YFB08011601).

References

1. Jiang, Z.: Network storage server advantage analysis. Modern economic information (03), 214 (2010)
2. Introduction to storage knowledge.https://wenku.baidu.com/view/efabd90e240c844768eaeea9.html. Accessed 20 May 2020
3. CVE-2018-18471. http://cve.mitre.org/cgi-bin/cvename.cgi?name=CVE-2018-18471. Accessed 16 June 2020

4. CVE-2016-10108.http://cve.mitre.org/cgi-bin/cvename.cgi?name=CVE-2016-10108. Accessed 16 June 2020
5. CVE-2020-9054. http://cve.mitre.org/cgi-bin/cvename.cgi?name=CVE-2020-9054. Accessed 16 June 2020
6. CVE-2018-13291. http://cve.mitre.org/cgi-bin/cvename.cgi?name=CVE-2018-13291. Accessed 16 June 2020
7. CVE-2018-13322.http://cve.mitre.org/cgi-bin/cvename.cgi?name=CVE-2018-13322. Accessed 16 June 2020
8. CVE-2018-14749. http://cve.mitre.org/cgi-bin/cvename.cgi?name=CVE-2018-14749. Accessed 16 June 2020
9. Moore, C.: Detecting ransomware with honeypot techniques. In: Cybersecurity and Cyberforensics Conference, Amman, Jordan, pp. 77–81 (2016). https://doi.org/10.1109/ccc.2016.14
10. Saud, Z., Islam, M.H.: Towards proactive detection of advanced persistent threat (APT) attacks using honeypots. In: 8th International Conference on Security of Information and Networks, Sochi, Russia, pp. 154–157 (2015). https://doi.org/10.1145/2799979.2800042
11. Anirudh, M., Thileeban, S. A., Nallathambi, D.J.: Use of honeypots for mitigating DoS attacks targeted on IoT networks. In: International Conference on Computer, Communication and Signal Processing, Chennai, India, pp. 1–4 (2017). https://doi.org/10.1109/icccsp.2017.7944057
12. Threat Modeling. https://docs.microsoft.com/en-us/previous-versions/msp-n-p/ff648644(v=pandp.10). Accessed June 2016
13. Chen, D.D., Egele, M., Woo, M., Brumley, D.: Towards fully automated dynamic analysis for embedded firmware. In: Proceedings of the 23rd Network and Distributed System Security Symposium (NDSS 2016). San Diego, CA: Internet Society, pp. 21–37 (2016)
14. https://github.com/max-boehm/qnap-utils

Key Technology and Application of Customer Data Security Prevention and Control in Public Service Enterprises

Xiuli Huang(✉), Congcong Shi, Qian Guo, Xianzhou Gao, and Pengfei Yu

State Grid Key Laboratory of Information and Network Security, Global Energy Interconnection Research Institute Co. Ltd., Nanjing 210003, China
71737069@qq.com

Abstract. As the value of the big data is revealed, the problem of security prevention and control of customer service data is highlighted. Once the leakage of sensitive customer information will cause serious legal consequences. At present, State Grid Corporation has effectively resisted the risk of data collection, storage, and transmission through internal and external network isolation and data encryption. However, at the data application level, due to frequent data sharing and business collaboration, new risks of leakage continue to emerge in data mining, data interaction between systems, terminal data access, etc. This article first analyzes the risks faced by data security in terminal data access, cross-domain data transfer, and open data sharing. Then, it proposes terminal data access behavior management and control technology, data cross-domain transfer security interaction technology, and data differentiated privacy protection technology in open sharing. Finally, t on the basis of technical research, the core equipment of data protection was developed and tested, and verified the effectiveness of the technology in practice.

Keywords: Data risk · Sensitive data · Privacy protection · Dynamic desensitization

1 Introduction

With the advent of the digital age, the value of data has become prominent, but data is both wealth and responsibility. With the increasing pressure of transformation and upgrading of traditional public service industries in recent years, the realization of data has received extensive attention. Public service enterprises have a large amount of customer service data, and the problem of security prevention and control of customer service data has become prominent.

As the world's largest public service company, State Grid has 460 million customers' electricity usage data. Once leaked, it will cause serious legal consequences. At present, State Grid Corporation has effectively resisted the risk of data collection, storage and transmission through internal and external network isolation and data encryption. However, at the data application level, due to frequent data sharing and business collaboration, new risks of leakage continue to emerge in data mining, data

Q. Liu et al. (Eds.): CENet 2020, AISC 1274, pp. 927–933, 2021.
https://doi.org/10.1007/978-981-15-8462-6_107

interaction between systems, terminal data access, etc. The traditional protection measures have problems such as extensive rigidity and disconnection from application scenarios, which seriously affect the availability of data. How to balance the security and availability of data in data applications has become a major problem affecting the development of the digital economy.

2 Data Security Risk Source Analysis

With data integration and continuous deepening of business collaboration services, data will be openly shared between businesses and across multiple security domains, covering a large number of terminals, master stations and background application services vertically. Sensitive data leakage shows significant multi-source and Complexity, including: risks from heterogeneous access terminals, risks from cross-system data interaction, risks from abnormal access within business systems, and risks of data outgoing privacy leakage for data modeling [1, 2].

Risk of Data Access Leakage on the Terminal Side. The data security risks on the terminal side mainly come from three aspects: 1) the terminal itself has a security hole that the attacker uses to obtain data; 2) Application layer account is stolen and data is stolen illegally; 3) Improperly set user permissions at the application layer, leading to abuse.

Risk of Data Exchange Leakage on the Network Side. In the interaction between users and application data, different users, different business scenarios, and different regions have different data usage requirements and security requirements. If sensitive data cannot be provided and used with the principle of minimizing usage rights, it will lead to the leakage of sensitive data. In the system-level invocation process, traditional static security protection measures are difficult to flexibly adjust to the data sensitivity according to the system level, cannot provide differentiated data security protection methods on demand, and it is easy to leak data in low-level systems.

Data Sharing Risk. When enterprises use big data [3], artificial intelligence and other advanced technologies to mine data value, they need to use all kinds of business data and customer data in batches. If data encryption and desensitization are processed across the board, the availability of data will be affected, and if used directly without processing, it will lead to data leakage.

3 Sensitive Data Protection Technology for Multiple Risk Sources

Aiming at the sources of data security risks, we build customer sensitive data protection systems from the aspects of terminal data access behavior management and control, data cross-domain and cross-system interaction security protection, open shared data differentiation privacy protection [4].

3.1 Security Management of Business Terminal Data Access Behavior

The traditional terminal protection for public user access lacks background knowledge of user data access characteristics, and is limited to vulnerability discovery and protection from the operating system and desktop control layer. There is a lag in discovering the risk of data leakage on the terminal side and weak early warning capabilities. Data access for customers with different roles in the energy service industry such as electric power [5] has typical pattern characteristics, which can be obtained by extracting terminal summary data. Based on the pattern deviation detection model, this paper analyzes the difference between terminal data access characteristics and the platform's global access mode, evaluates terminal data access behavior, and then builds a terminal process access evaluation model for process access risk evaluation to achieve terminal-side warning of high-risk data leakage and blocking.

After collecting and storing the terminal access feature data, the decision model is used for further threat analysis and judgment. The decision model includes the process tree model, the process behavior model, the threat discovery model, the risk discovery model, and the TISE intelligence analysis. The combination and application of different models can support users to operate the security business process.

The process tree is a process relationship representation method, which consists of a parent process, a child process, and a sibling process. Process behavior includes the process of the system from the generation to the end of the life cycle, the system file, registry, network access, memory calls and other operations records. Because there is no session ID between the process and the behavior, the MID, process PID, process MD5, and time are delineated according to the PLC (process life cycle range) as the association between the process and the behavior.

3.2 Business Data Cross-Domain and Cross-System Interaction Security Protection

In the process of complex and dynamic data service interface invocation, in view of the problem of "one-size-fits-all" desensitization flexibility in data service invocation scenarios, this paper proposes a security filtering method based on sensitive data feature mapping and an interactive logic driven data behavior analysis method, and constructed a business scene recognition model based on multi-message analysis, and realized the dynamic differentiated desensitization of sensitive data of data service interface for diverse business scenarios.

Recognition of Sensitive Data Based on Vector Space. A multi-dimensional discriminant feature extraction method based on feature mapping of sensitive data [6] is used to quickly and accurately identify sensitive data from massive exchange data. This method first constructs multi-dimensional feature space vectors of sensitive data, and performs self-learning of feature vector weights and discrimination thresholds. Then extract the discriminating characteristics of sensitive data, pre-analyze the packets called by the data service interface, and extract the information such as user roles and access objects to blur the feature vector dimension of the sensitive data to be discriminated, so that the identification of sensitive data is fast and accurate.

Sensitive data recognition technology for structured data should be able to automatically and accurately identify the name, mobile phone number, ID number, address, bank card, fixed phone, passbook account number, postal code, email address, passport number, business license number, and other field attributes. In addition, for special names, address information, company names and other natural language descriptions of sensitive data, automatic identification of sensitive data is realized through feature extraction and analysis modeling. For the types of sensitive data such as user name, company name and other features that are not clear, or the verification features and regular features are not known in advance, the space vector feature extraction needs to be performed by the corresponding machine learning algorithm.

Differentiation and Desensitization of Sensitive Data for Business Scenarios. Differentiated sensitive information desensitization [7, 8] is a differentiated desensitization based on different business scenarios and different user data permissions, and can be set up in a white list. That is, according to the application scenario accessed by the application system and the user information of the visitor, the preset permission for data use can be accurately identified, and then the desensitization rules under the permission can be matched to achieve differentiated desensitization, as shown in the following Fig. 1. The white list may include user white lists and scene white lists. The user whitelist can specify that whitelist users do not perform desensitization processing when accessing sensitive data, and display real data to them. Scenario whitelist, by desensitizing the data in specific application scenarios by specifying the whitelist.

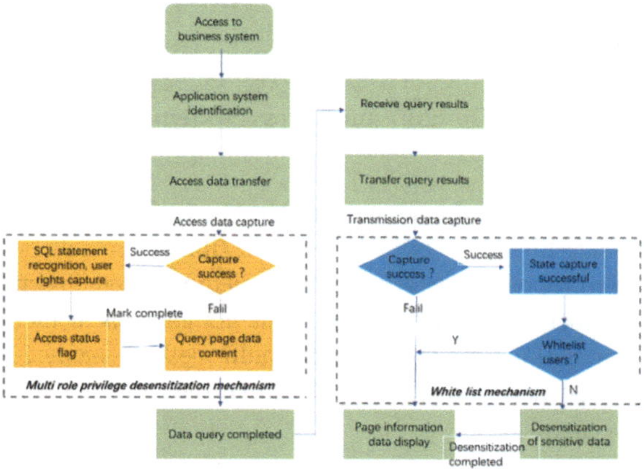

Fig. 1. Differentiated desensitization of sensitive data for business scenarios.

3.3 Differentiated Privacy Protection for Business Data Sharing

The difficulty of customer data privacy protection [9, 10] stems from the diversity of customer data application models, complex data structure features, privacy disclosure scenarios, and protection technology methods. From the analysis of privacy leakage scenarios, there are query privacy, publishing privacy, computing and mining analysis privacy, etc. The effect of privacy protection directly depends on the adaptability of privacy risk scenarios and protection methods. The diversification of the two levels makes the traditional privacy protection methods single, passive and scene-dependent. It is necessary to study the association mechanism between the multiple elements of the two dimensions of risk scenarios and protection technology to realize the fusion and customization of privacy protection methods based on multiple associations.

In the differential association adaptation of privacy protection methods and privacy protection scenarios, combined with customer data privacy grading and privacy leakage risk measurement mechanisms, a bipartite graph matching technology is used to build a multi-association model between privacy scenarios, privacy elements and protection methods. Actively recommend privacy protection technology through random walks in the two-part graph. The constructed triples take the form of <application scenarios, privacy grading, protection strategies and methods> . At the same time, the graph database is used to extract the entity and relationship of the triple data, and the nodes and edges are constructed to generate the customer data privacy protection knowledge graph as shown in Fig. 2.

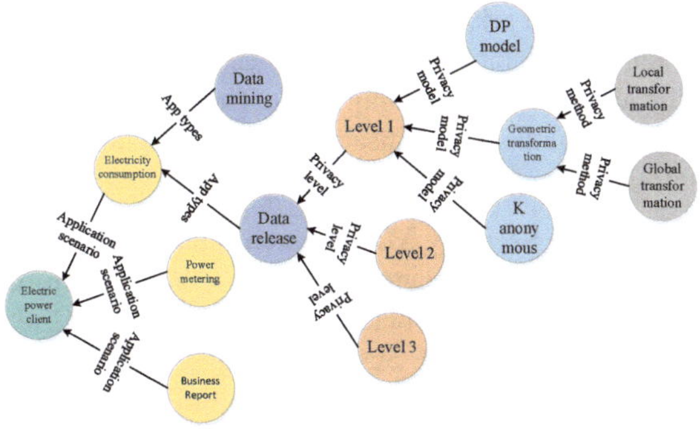

Fig. 2. Diagram of customer data privacy knowledge.

4 Test Verification

Based on customer privacy data sharing security, sensitive data dynamic desensitization, terminal security protection and other technologies, the core equipment of data security protection "network and information security management and control platform" was developed. The platform includes data security protection functions such as terminal security protection management, business sensitive data dynamic desensitization control and privacy data sharing management.

Prototype of Network and Information Security Management and Control Platform. The network and information security management and control platform is a set of ecosystems constructed according to the user's specific business environment, usage habits, and security policy requirements. It integrates data security components such as terminal security protection management, business sensitive data dynamic desensitization management and control, and private data sharing management, as shown in Fig. 3 below.

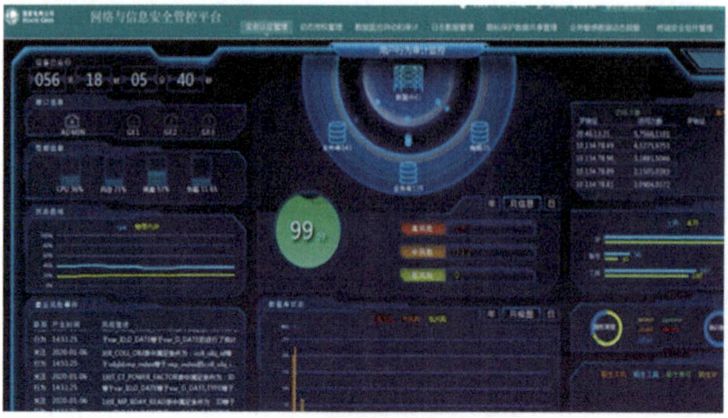

Fig. 3. Network and information security management and control platform.

The sensitive data dynamic desensitization component supports scene-driven differential desensitization, supports lightweight integration and online adjustment of desensitization strategies, and realizes the rapid implementation of desensitization transformation. Differentiated privacy protection component for big data mining, taking into account the high availability of data privacy protection and data clustering, association, classification and other mining tasks. The terminal data security protection component realizes continuous data transaction monitoring and authority management and control capabilities during terminal data access.

Network and Information Security Management and Control Platform Pilot. The network and information security management and control platform has been extensively applied in more than 10 provincial units such as State Grid Big Data Center, State Grid Customer Service Center, State Grid Jiangsu and State Grid Hunan. The application of the platform has significantly improved the security protection capabilities of power customer data in terms of terminal access, cross-domain and cross-system transfer, and big data modeling.

In the pilot application, the network and information security management and control platform has cumulatively protected about 260,000 terminals such as PCs and mobile operation terminals. A total of 8,300 more aggressive accesses on the terminal side and more than 1,200 data transaction operations at the system layer were discovered. It has realized the identification and analysis of more than 1,000 power monitoring instructions, accumulated more than 50 big data analysis tasks, and improved the efficiency of safety protection by about 52%. Through the application of the project results, the overall level of business data security protection has been improved to avoid data security incidents and support customer data privacy security protection.

5 Conclusion

In view of the risk of multi-source data leakage in customer data applications, this article focuses on the practical need to solve the business terminal data security prevention and control, customer data cross-domain and cross-system security interaction, and customer data outsourcing and shared privacy protection issues, and launch theoretical research and technical engineering practice. This paper integrates the application of differential privacy protection, intelligent analysis and other technologies, and proposes solutions. It has good innovation significance and engineering application value, and has verified the effectiveness of the program in practice.

Acknowledgment. This work is supported by the science and technology project of State Grid Corporation of China: Research and application of key technologies of data security sharing and distribution for data center, No. 5700-202090192A-0-0-00.

References

1. Li, Q.M., Zhang, H.: Information security risk assessment technology of cyberspace: a review. Information **15**(11), 677–683 (2012)
2. Li, Q.M., Li, J.: Rough outlier detection based security risk analysis methodology. China Commun. **9**(7), 14–21 (2012)
3. Li, G.J., Cheng, X,Q.: Research status and scientific thinking of big data. Bull. Chinese Acad. Sci. **27**(6), 647–657 (2012)
4. Haque, M.S., Chowdhury, M.U.: A new cyber security framework towards secure data communication for unmanned aerial vehicle (UAV). In: Proceedings 13 Security and Privacy in Communication Networks: Secure Comm 2017 International Workshops, ATCS and SePrIoT, pp. 113–122. Springer International Publishing (2018)
5. Li, J., Zhao, G., Chen, X., et al.: Fine-grained data access control systems with user accountability in cloud computing. In: The 2nd International Conference on Cloud Computing Technology and Science, IEEE Computer Society, pp. 89–96 (2010)
6. Wang, M.F., Liu, P.Y., Zhu, Z.F.: Feature selection method based on TFIDF. Comput. Eng. Des. **28**(23), 5795–5796 (2007)
7. Liu, M.H., Zhang, N., Zhang, Y.Y., et al.: Research on sensitive data protection technology on cloud computing. Telecommun. Sci. **11**, 2–8 (2014)
8. Jiang, R.M.: Data masking system construction plans of telecommunication operator. Inf. Technol. **08**, 132–133 (2014)
9. Liu, J., Xiao, Y., Li, S., et al.: Cyber security and privacy issues in smart grids. IEEE Commun. Surv. Tutorials **14**(4), 981–997 (2012)
10. Cheng, L., Jiang, F.: A-diversity and k-anonymity big data privacy preservation based on micro-aggregation. Netinfo Secur. **3**, 19–22 (2015)

Efficient Fault Diagnosis Algorithm Based on Active Detection Under 5G Network Slice

Zhe Huang[1(✉)], Guoyi Zhang[2], Guoying Liu[1], and Lingfeng Zeng[1]

[1] Shenzhen Power Supply Co. Ltd., Shenzhen 518010, China
13418763020@163.com
[2] Power Control Center of China Southern Power Grid, Guangzhou 510623, China

Abstract. In order to solve the problem of low accuracy of fault diagnosis in network slicing environment, this paper proposes an efficient fault diagnosis algorithm based on active detection. By calculating the business correlation of the underlying network resources, a set of candidate detection nodes is formed. Based on the overlapping characteristics of the detection paths, the set of detection nodes is optimized. Active detection technology is used to actively acquire network characteristics, and a suspected fault set is constructed based on the historical fault probability and detection performance, and the fault is quickly diagnosed through the credibility evaluation of the fault set. By comparing the performance with existing algorithms, it is verified that the algorithm in this paper improves the accuracy of fault diagnosis.

Keywords: 5G network · Network slicing · Fault diagnosis · Active detection

1 Introduction

The era of 4G network has realized the rapid connection between people and smart phones and the rapid development of mobile Internet. However, in the 5G network environment, based on the network characteristics of high-bandwidth and low-latency, the Internet of Everything will be implemented to carry different services such as high-bandwidth, low-latency, and high connection counts [1]. To ensure the reliability and stability of 5G services, when a service fails, fast and accurate fault location technology has become an important research focus.

Fault diagnosis technology can generally be divided into two types: fault diagnosis based on passive monitoring and fault diagnosis based on active detection [2–4]. The former is mainly based on the alarm data and network topology information of the network management system to perform fault inference, which is simple to implement and has little impact on network services [5–7]. At present, studies have focused on passively accepting symptoms and establishing fault diagnosis models for fault diagnosis. There is a problem of low accuracy of fault diagnosis. In order to solve this problem, based on the network characteristics, this paper uses active detection technology to actively obtain network characteristics and quickly diagnose faults, which can effectively solve the problem of low accuracy of fault diagnosis.

Q. Liu et al. (Eds.): CENet 2020, AISC 1274, pp. 934–941, 2021.
https://doi.org/10.1007/978-981-15-8462-6_108

2 Problem Description

According to the different needs of the business for the network, the network can be divided into mobile broadband slices (bearing communication services, Internet services, etc.), massive IoT slices (bearing intelligent agriculture, smart security, etc.), mission-critical IoT slices (bearing smart factories, etc.). This article takes 5G core network as the main research object, and the network topology diagram is shown in Fig. 1. Use $G = (N, E)$ to represent the underlying network topology, where N represents the set of underlying network nodes, including the underlying network node $n_i \in N$. E represents the set of underlying network links, including the underlying link $e_j \in E$. Use $G^V = (N^V, E^V)$ to represent the virtual network topology, where N^V represents the set of virtual network nodes, including virtual network node $n_i^V \in N^V$. E^V represents a set of virtual network links, including virtual link $e_j^V \in E^V$.

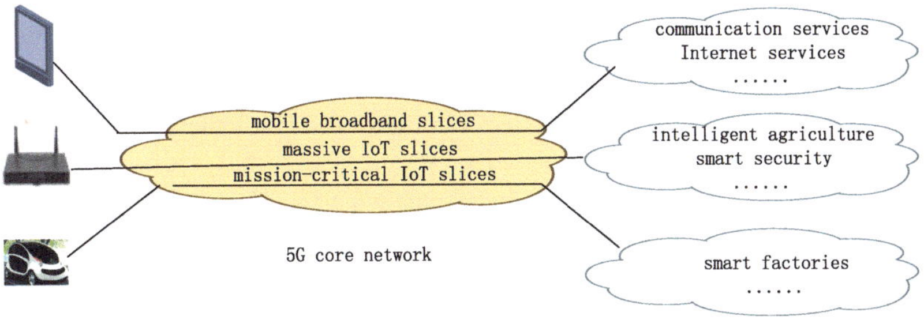

Fig. 1. Schematic diagram of 5G core network.

Active detection technology refers to a technology that judges the network performance of the end-to-end path that the detection passes through by sending end-to-end data packets to certain network nodes on the designated network node and receiving the data packets according to the destination node. Considering that the detection technologies of the wireless network subsystem, the transmission network subsystem and the core network subsystem are similar, this paper takes the detection of the core network subsystem as an example for research. The fault diagnosis technology based on active detection generally includes two processes: detection phase and fault location phase. The two processes are described in detail below.

3 Fault Propagation Model

In the detection phase, it includes two processes: selection of detection station and execution of detection. The detection site is generally a device that deploys specific software and hardware. Therefore, the selection of the detection site is a key part of the detection performance and efficiency. Executing detection generally includes end-to-end detection and process of collecting detection results. The following first describes the selection process of the detection node based on the service correlation of the underlying network resources and the overlapping characteristics of the detection paths, and secondly, the process of performing the detection.

3.1 Business Relevance of Underlying Network Resources

As can be seen from the problem description section, each underlying network resource $G = (N, E)$ will carry multiple virtual networks. Each virtual network $G^V = (N^V, E^V)$ can run multiple 5G services simultaneously. Therefore, if the number of virtual resources on each underlying network resource and the number of services on each virtual resource can be obtained, then the business relevance of the current underlying network resources can be known. The service relevance of the underlying network resources is represented by Ω_{n_i}. The larger the value, the more the number of services on it. z represents the number of virtual networks carried on the underlying network resources. λ_j represents the number of services carried on the j-th virtual network.

$$\Omega_{n_i} = \sum\nolimits_{j=1}^{z} \lambda_j \tag{1}$$

Therefore, by calculating the business correlation Ω_{n_i} of the underlying network resources, important underlying network resources can be selected. Arranged in descending order according to the business relevance Ω_{n_i} of each underlying resource constitutes the initial candidate detection set T_{star}.

3.2 Optimization of Detection Node Set

Because the routing protocol is mainly dynamic routing protocol, there is uncertainty in the detection of passing through the network node n_i, use $p(t_x, n_i)$ to indicate the probability of the network node n_i being detected by the detection t_x, the calculation method is the proportion of the detection that passes the network node n_i in a certain number of detection runs. Use $node(t_x)$ to denote the set of nodes that the detection t_x passes through. If the network node n_j is the node that probes t_x and t_y pass by together, the network node n_j is called the shadow node of probes t_x and t_y. The independent evaluation function $EV(t_x, t_y)$ of probes t_x and t_y is shown in formula (2), where $n_j \in node(t_x) \cap node(t_y)$ represents the intersection of probes t_x and t_y passing through the node.

$$EV(t_x, t_y) = 1 - \cup_{n_j} p(t_x, n_j) \cdot p(t_y, n_j) \tag{2}$$

Put all the underlying network nodes into the set N_S to be detected. First, take the first node n_k^1 in T_{star}, put it into the set of detection nodes T_{end}. Take the nodes from node n_k^1 to all the leaf nodes of the network that have passed the detection from the set to be detected N_S, and mark the detection set of the current node as $T_{n_k^1}$. Secondly, the candidate detection nodes are sequentially taken from the set T_{star} until the set N_S to be detected is empty. The sub-processes of this process are as follows: (1) Use formula (2) to calculate the independence of the detection formed by each node in the set T_{star} and the existing detection set; (2) The network node with the largest sum of the independence of the network nodes as the detection node and which is put into the detection node set T_{end}; (3) All the network nodes through which the detection of the newly put detection node passes are taken out from the detection set N_S until the detection set N_S is empty.

3.3 Perform Probing

The detection matrix is a two-dimensional matrix, and the rows of the matrix are composed of the network nodes passed by the detection and the detection results. The columns of the matrix are composed of network nodes. For example, Table 1 is a schematic diagram of the detection dependence matrix, including 5 detections and 6 network nodes.

Table 1. Schematic diagram of detection dependence matrix.

	N1	N2	N3	N4	N5	N6	Results
t1	0	0	1	0	0	1	0.8
t2	1	0	1	0	0	1	0.7
t3	0	0	0	1	0	0	0.5
t4	0	1	0	0	1	1	0.6
t5	1	0	1	0	0	0	0.9

According to the detection results, a fault diagnosis model can be established based on Bayesian theory. For example, based on the schematic diagram of the detection dependence matrix in Table 1, the schematic diagram of the fault diagnosis model shown in Fig. 2 can be obtained. The upper node in the model represents the network node, the lower node represents the detection node, and the directed line from the upper node to the lower node represents the probability that the detection result of the lower layer is abnormal when the upper node has an abnormality.

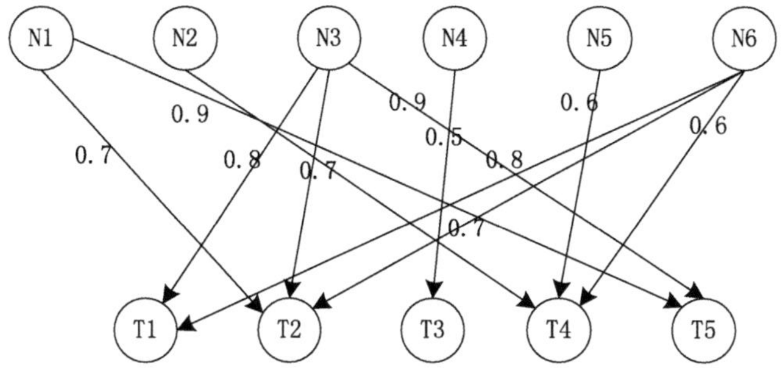

Fig. 2. Schematic diagram of fault diagnosis model.

4 Algorithm

The efficient fault diagnosis algorithm based on active detection under 5G network slice (EFDAoAD) includes the following three key processes: (1) constructing the initial detection set; (2) Optimize the detection set and perform detection; (3) Fault location. In the stage of constructing the initial detection set, based on the mapping relationship, the business correlation of the underlying network resources is calculated to form a set of candidate detection nodes. In optimizing the detection set and performing the detection phase, first, the detection node set is optimized based on the overlapping characteristics of the detection path; second, the detection is sent to obtain a detection result set, and a detection dependency matrix and a fault diagnosis model are constructed. In the fault location stage, the fault set capability is first calculated based on the historical fault probability and detection performance. Second, the credibility evaluation of the fault set is performed to obtain the final fault set. Step (1) and step (2) have been described in detail in the previous section. The step (3) will be described in detail below.

4.1 Build Suspected Fault Sets

When constructing a suspected fault set, a suspected fault set is constructed from the fault node set X based on the historical fault probability and detection capability. Assume that the number of simultaneous failures is k. Construct k suspected faults sets with 1 to k faults. Considering the network dynamics and network noise, this paper analyzes from the two dimensions of historical failure probability and detection performance. Use $f_{n_i}^h$ to represent the historical failure probability of the underlying node n_i, and the value is the number of failures in the time period T. Use DP_{n_i} to indicate the detection performance of the faulty node n_i. The value is the number of detection results corresponding to the faulty node n_i less than 0.5. The larger the value, the greater the number of abnormalities detected by the underlying network nodes in the set. O_{kj} represents the fault set, and $Abl_{O_{kj}}$ represents the explanatory power of the fault set O_{kj}, which is calculated using formula (3). k indicates the number of faulty nodes in the fault set, and j indicates the j-th fault set. α and β represent adjustment factors.

$$Abl_{O_{kj}} = \alpha \sum\nolimits_{n_i \in O_{kj}} f_{n_i}^h + \beta \sum\nolimits_{n_i \in O_{kj}} DP_{n_i} \tag{3}$$

4.2 Reliability Assessment of Fault Sets

In order to improve the performance of fault diagnosis, this paper constructs k types of fault sets, each type of fault set contains j network nodes. In order to select the best fault set from k*j fault node sets, the formula (4) is used to calculate the credibility evaluation of the fault set. Among them, $\prod\nolimits_{t_{n_i} \in T_O} (1 - \prod\nolimits_{f_{n_i} \in O_{kj}} (1 - p(t_{n_i}|f_{n_i})))$ represents the probability that the observed abnormality detection belongs to the fault set O_{kj}, and $\prod\nolimits_{t_{n_i} \in T_{f_{n_i}}}^{f_{n_i} \in O_{kj}} (1 - \prod\nolimits_{f_{n_i} \in O_{kj}} (1 - p(t_{n_i}|f_{n_i})))$ represents the probability of all abnormality detections that the fault set O_{kj} can generate. $p(t_{n_i}|f_{n_i})$ represents the probability that the underlying network node f_{n_i} causes t_{n_i} to be abnormal.

$$CL(O_{kj}) = \frac{\prod_{t_{n_i} \in T_O} (1 - \prod_{f_{n_i} \in O_{kj}} (1 - p(t_{n_i}|f_{n_i})))}{\prod_{t_{n_i} \in T_{f_{n_i}}}^{f_{n_i} \in O_{kj}} (1 - \prod_{f_{n_i} \in O_{kj}} (1 - p(t_{n_i}|f_{n_i})))} \quad (4)$$

5 Performance Analysis

5.1 Simulation Environment

In order to evaluate the performance of the algorithm, this paper uses the Inet topology generator [8] to generate the virtual network and the underlying network topology. The node size of the underlying network has increased from 100 to 500. The number of nodes in the virtual network is uniformly distributed from 5 to 15. The mapping of the virtual network to the underlying network uses a basic resource mapping algorithm. In terms of fault simulation, the network is simulated based on a priori failure probability [0.001, 0.01]. During algorithm analysis, the algorithm EFDAoAD in this paper is compared with the fault diagnosis algorithm based on fault diagnosis model (FDAoFDM). The algorithm FDAoFDM uses the mapping relationship between the underlying network and the virtual network to correlate the service status with the underlying network resources, and build a fault propagation model for fault location.

5.2 Algorithm Comparison

The comparison result of fault diagnosis accuracy is shown in Fig. 3. It can be seen from the figure that the diagnostic accuracy of the two algorithms is less affected by the size of the network. The diagnostic accuracy of the algorithm in this paper is significantly improved compared with traditional algorithms. It shows that the algorithm of this paper has a better optimization effect on the fault diagnosis model.

The comparison result of the fault diagnosis false alarm rate is shown in Fig. 4. It can be seen from the figure that the impact of network size on both algorithms is relatively small, and the false alarm rate of the algorithm in this paper is lower than that of traditional algorithms.

The comparison result of fault diagnosis duration is shown in Fig. 5. It can be seen from the figure that with the increase of the network scale, the diagnosis duration of both algorithms increases rapidly, indicating that the network scale increases, and the amount of data of the fault diagnosis model to be processed increases rapidly, prolonging the time of fault diagnosis.

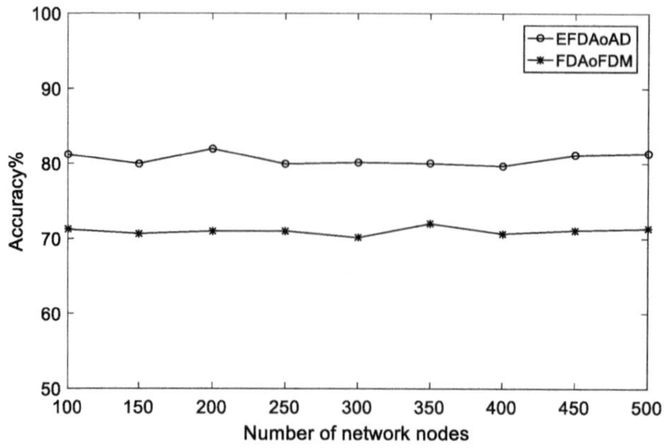

Fig. 3. Comparison of fault diagnosis accuracy.

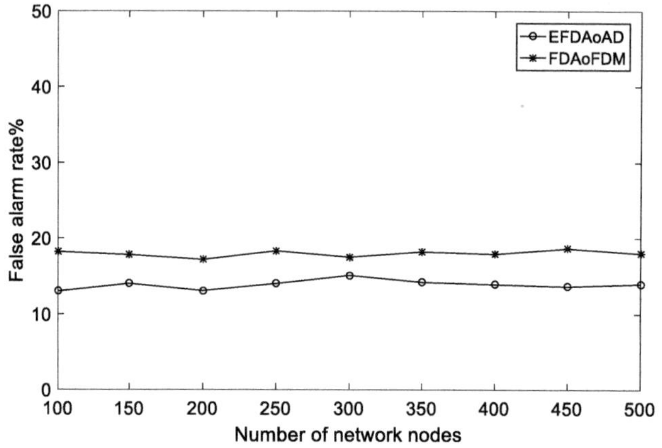

Fig. 4. Comparison of false alarm rate in fault diagnosis.

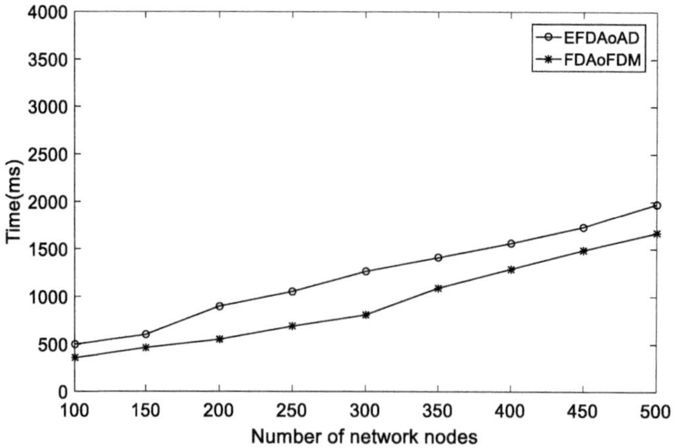

Fig. 5. Comparison of fault diagnosis time.

6 Conclusion

This paper proposes an efficient fault diagnosis algorithm based on active detection. The algorithm constructs a more accurate fault diagnosis model by selecting detection nodes and active detection technology. Through experiments, it is verified that the algorithm of this paper has achieved better performance in fault diagnosis. The algorithm in this paper does not classify the type of service when selecting the detection node. In a real environment, different business types have different priorities and different time limits for troubleshooting. In the next step, the detection sites and fault diagnosis algorithms will be optimized based on business priorities.

References

1. Tang, L., Wei, Y.N., Ma, R.L., et al.: Online learning-based virtual resource allocation for network slicing in virtualized cloud radio access network. J. Electr. Inf. Technol. **41**(7), 1533–1539 (2019)
2. Dusia, A., Sethi, A.S.: Recent advances in fault localization in computer networks. IEEE Commun. Surv. Tutorials **18**(4), 3030–3051 (2016)
3. Wu, B., Ho, P. H., Tapolcai, J., et al.: Optimal allocation of monitoring trails for fast SRLG failure localization in all-optical networks. In: Proceedings of 2010 IEEE Global Telecommunications Conference. Miami, USA, pp. 1–5 (2010)
4. Tang, Y.N., Al-Shaer, E.S., Boutaba, R.: Active integrated fault localization in communication networks. In: Proceedings of the 9th IFIP/IEEE International Symposium on Integrated Network Management. Nice, France, pp. 543–556 (2005)
5. Gontara, S., Boufaied, A., Korbaa, O.A.: Unified approach for selecting probes and probing stations for fault detection and localization in computer networks. In: Proceedings of the 2019 IEEE International Conference on Systems, Man and Cybernetics (SMC), pp. 2071–2076 (2019)
6. Jin, R.F., Wang, B., Wei, W., et al.: Detecting node failures in mobile wireless networks: a probabilistic approach. IEEE Trans. Mob. Comput. **15**(7), 1647–1660 (2016)
7. Yan, C.X., Wang, Y., Qiu, X.S., et al.: Multi-layer fault diagnosis method in the network virtualization environment. In: Proceedings of the 16th Asia-Pacific Network Operations and Management Symposium. Hsinchu, China, pp. 1–6 (2014)
8. Winick, J., Jamin, S.: Inet-3.0: internet topology generator, University of Michigan, Michigan, USA, Technical Report CSE-TR-456-02 (2002)

GPSR Compliance

The European Union's (EU) General Product Safety Regulation (GPSR) is a set of rules that requires consumer products to be safe and our obligations to ensure this.

If you have any concerns about our products, you can contact us on ProductSafety@springernature.com

In case Publisher is established outside the EU, the EU authorized representative is:

Springer Nature Customer Service Center GmbH
Europaplatz 3
69115 Heidelberg, Germany

Batch number: 10755036

Printed by Printforce, the Netherlands

Advances in Intelligent Systems and Computing

Volume 1274

The series "Advances in Intelligent Systems and Computing" contains publications on theory, applications, and design methods of Intelligent Systems and Intelligent Computing. Virtually all disciplines such as engineering, natural sciences, computer and information science, ICT, economics, business, e-commerce, environment, healthcare, life science are covered. The list of topics spans all the areas of modern intelligent systems and computing such as: computational intelligence, soft computing including neural networks, fuzzy systems, evolutionary computing and the fusion of these paradigms, social intelligence, ambient intelligence, computational neuroscience, artificial life, virtual worlds and society, cognitive science and systems, Perception and Vision, DNA and immune based systems, self-organizing and adaptive systems, e-Learning and teaching, human-centered and human-centric computing, recommender systems, intelligent control, robotics and mechatronics including human-machine teaming, knowledge-based paradigms, learning paradigms, machine ethics, intelligent data analysis, knowledge management, intelligent agents, intelligent decision making and support, intelligent network security, trust management, interactive entertainment, Web intelligence and multimedia.

The publications within "Advances in Intelligent Systems and Computing" are primarily proceedings of important conferences, symposia and congresses. They cover significant recent developments in the field, both of a foundational and applicable character. An important characteristic feature of the series is the short publication time and world-wide distribution. This permits a rapid and broad dissemination of research results.

**** Indexing: The books of this series are submitted to ISI Proceedings, EI-Compendex, DBLP, SCOPUS, Google Scholar and Springerlink ****

More information about this series at http://www.springer.com/series/11156

Qi Liu · Xiaodong Liu ·
Tao Shen · Xuesong Qiu
Editors

The 10th International Conference on Computer Engineering and Networks

Set 2

Springer

Editors
Qi Liu
Science and Technology
Nanjing University of Information
Nanjing, Jiangsu, China

Xiaodong Liu
School of Computing
Edinburgh Napier University
Edinburgh, UK

Tao Shen
School of Electrical Engineering
University of Jinan
Jinan, Shandong, China

Xuesong Qiu
Institute of Network Technology
Beijing University of Posts
and Telecommunications
Beijing, Beijing, China

ISSN 2194-5357 ISSN 2194-5365 (electronic)
Advances in Intelligent Systems and Computing
ISBN 978-981-15-8461-9 ISBN 978-981-15-8462-6 (eBook)
https://doi.org/10.1007/978-981-15-8462-6

This Springer imprint is published by the registered company Springer Nature Singapore Pte Ltd.
The registered company address is: 152 Beach Road, #21-01/04 Gateway East, Singapore 189721, Singapore

Preface

This conference proceedings is a collection of the papers accepted by the CENet 2020—the 10th International Conference on Computer Engineering and Networks held on October 16–18 2020, in Xi'an, China.

This proceedings contains the five parts: Part I Artificial Intelligence and Applications (51 papers); Part II Communication System Detection, Analysis and Application (32 papers); Part III Information Security and Cybersecurity (25 papers), Part IV Intelligent System and Engineering (17 papers), Part V Internet of Things and Smart Systems (61 papers) and Part VI Medical Engineering And Information Systems (13 papers).

Each part can be used as an excellent reference by industry practitioners, university faculties, research fellows and undergraduates as well as graduate students who need to build a knowledge base of the most current advances and state of practice in the topics covered by this conference proceedings. This will enable them to produce, maintain and manage systems with high levels of trustworthiness and complexity.

Thanks to the authors for their hard work and dedication as well as the reviewers for ensuring the selection of only the highest-quality papers; their efforts made the proceedings possible.

Contents

Intelligent System and Engineering

Research on Blue Force Simulation System of Naval Surface Ships

Rui Guo(✉) and Nan Wang

Department of Warships Command,
Dalian Naval Academy, Dalian 116018, China
msegr@163.com

Abstract. It is very necessary to carry out simulation research on the naval surface ships of the world's powerful army because of their strong combat effectiveness. At present, it is a scientific and effective method to analyze and study the opponents to build the Blue Force simulation system and carry out the virtual combat between Red Force and Blue Force based on the computer network. On the basis of summarizing the current situation of Blue Force simulation system of naval surface ships, this paper expounds the construction method of Blue Force simulation system from three aspects of system function analysis, structure design and running process. It is considered that: in the aspect of function analysis, it is necessary to face the joint operation simulation, access the real army and real equipment to participate in the military exercise; in the aspect of structure design, it is necessary to highlight the fleet-level computer generated forces system, pay attention to the cooperative operation between ships; in the aspect of running process, it is necessary to use new means such as large data technology to fully analyze the key nodes in the running process. The research conclusion can provide ideas and basis for the construction of Blue Force, and can be extended to other combat simulation system construction.

Keywords: Navy · Surface ships · Blue Force · Simulation system

1 Foreword

The Blue Force, some countries called the "imaginary enemy" force, is the name of the enemy forces in military exercises, used to imitate the opponent's combat thinking, combat principles, equipment performance, and tactical methods, in order to imitate the opponent's mode of fight with "Red Force". It has become the common practice of the world's military powers to hone the troops through the Blue Force [1]. In 1966, Israel formed the world's first regular simulated force of enemy, and its air force greatly improved combat capability by training with it, and the study of the Blue Force has gradually gained attention since then. The U.S. Navy began exploring the Blue Force in 1986, offering war simulation lectures at its Naval War College, and systematically promoting "imaginary enemies" to play related issues. In 1990, the "imaginary enemy" role of the U.S. Navy combat troops also gradually developed from the tactical level to the strategic level, and added a large number of long-range simulated confrontation

Q. Liu et al. (Eds.): CENet 2020, AISC 1274, pp. 945–950, 2021.
https://doi.org/10.1007/978-981-15-8462-6_109

exercises, which had a direction with more simulation, specialization and systematization.

In recent years, the construction of naval surface ship Blue Force has been paid more and more attention, some research institutes have carried out theoretical analysis, simulation research and online virtual confrontation. Some troops have also carried out practical exploration, and achieved many gratifying achievements. However, in general, the construction of naval surface ships Blue Force is lagging behind; there is still a lot of detailed and in-depth work to be done. It is a scientific method to develop simulation system and carry out simulation the confrontation with Red Force based on computer network platform. In this paper, the simulation system of the naval surface ship Blue Force is studied, the simulation method of the system is analyzed from the aspects of function, structure and mode of running.

2 System Function Analyses

The naval surface ship Blue Force simulation system should be able to accomplish the following tasks: First, to provide theoretical basis for fight with rival warships; Second, to provide some virtual battlefield environment and confrontation conditions for joint training and exercises, naval arms training and evaluation; And third, to participate as imaginary enemy in the simulated confrontation exercise solely organized by superior organizations. On the basis of the realization of these functions, the future should also focus on joint combat simulation, and realize the join with actual army as the main goal of construction. Based on this, the basic functions of the simulation system include:

(1) Showing the technical and tactical parameters and operational effectiveness of the Blue Force warships, introducing the typical configuration and formation form in their campaign groups, as well as the generalizing common tactical operational methods;
(2) Generating Blue Force combat scenarios, which include the digitization of operation plan, providing sturdy technology and tools to generate confrontation scenarios, and displaying organization process and war situation;
(3) Multi-scale displaying the simulation process according to the requirements of the mission, such as single weapon handling, single warship operation, and naval fleet combat;
(4) Large data processing, which can collect the key data from programming to results analysis of the whole process, such as real-time position of every force and motion elements, therefore use large data processing methods to analyze;
(5) Realizing the two modes of running, i.e., "automatic simulation" and "man-in-loop simulation", which can not only set specific parameters for the simulation system to operate and obtain important data automatically, but also modify the parameters in real time to control the process;
(6) Replaying and evaluating, which can carry out the simulation of the whole process of confrontation, analyze the key data of the simulation process, assess the gains and losses of Blue Force, and provide data support for the final training summary.

3 System Structure Design

The naval surface ship Blue Force simulation system should be a complete network system based on computer platform, surface ship force and equipment simulation model, sea battlefield database, etc. Using this system, the researchers can simulate the opponent's surface ship operation by means of battle simulation, so as to complete the Red-Blue confrontation based on computer network, and provide theoretical basis for the analysis of military equipment combat effectiveness, as well as capability for commander decision making.

The system can be logically divided into the following sections by task: evaluation and demonstration section, simulation management section, single warship Blue Force generation section, fleet force Blue Force generation section.

(1) Simulation management section. It is mainly used to manage the simulation personnel and combat position settings, develop, modify and adjust into combat and simulation scenarios, produce virtual environment, manage system simulation process, set up "automatic simulation" and "man-in-loop simulation " and other simulation modes, complete the simulation model solution, modify the simulation running parameters, control the simulation resolution parameters, set up the method of ruling the results of the confrontation, achieve docking with other simulation system and so on;
(2) Assessment and presentation section. It is mainly used to record and evaluate all data in the simulation process. The evaluation method can be not only using computer to evaluate automatically, but also organizing experts for collective assessment on the platform. The content shown through two-dimensional and three-dimensional forms can include whole elements and process of the sea battlefield, whole elements and process of single-weapon fight, such as missiles and artillery. This section has the functions of analyzing simulation conclusions and replaying the combat process.
(3) Single ship Blue Force simulation section. It is mainly used to simulate all aspects of warship combat and corresponding weapons, so as to construct a virtual warship Blue Force which can flexibly configure technical and tactical parameters to participate in simulation. There must be more emphasis on coordinated operations between warships, i.e., organic links must be made between single warship and other warships.

- Command & control subsystem, which are used to direct the warship Blue Force decision-making on how to fight at sea;
- Navigation system, which are used to direct warships to navigate and maneuver in the designated sea battlefield;
- Weapons and equipment, which include technical & tactical parameters and usage of weapons and equipment including missiles, artillery, torpedoes, radars, sonar, etc.;
- Carrier-based aircraft subsystem, which are used to command and manage carrier-based fixed-wing aircraft and rotocraft to carry out a variety of operations;

- Air-defense operation subsystem, which are used to fight with air targets by air-defense weapons;
- Anti-ship operation subsystem, which are used to carry out fighting with sea targets with anti-ship missiles, artillery and other weapons;
- Anti-submarine operation subsystem, which are used to carry out fighting with underwater targets by torpedoes, deep bombs and other weapons;
- Electronic warfare subsystem, which are used to carry out electronic warfare with Red Force by a variety of soft and hard weapon;
- Detection and reconnaissance subsystem, which are used to carry out the detection and reconnaissance of enemy by radar, sonar and other weapons.

(4) Fleet Blue Force generation section, which includes the simulation model library, battlefield database, and computer generated force subsystem. The simulation object of the fleet Blue Force should mainly include the U.S. nuclear-powered aircraft carrier group, the amphibious assault ship group, the Japanese light carrier group, the Russian aircraft carrier group, the missile cruiser group and so on, so as to realize the group-level air defense, anti-ship and anti-submarine combat simulation.

- The simulation model library, which provides various supporting models for system simulation and research, including entity model, combat model, evaluation model, visual model, etc.
- The battlefield database, which provides all kinds of data needed for system simulation, including technological & tactical performance data, evaluation data, geographic information, simulation data, combat scenario, sea battlefield environment data, etc.
- The computer generated force subsystem, which can produce the simulation of naval single ship Blue Force, especially the naval fleet force, which mainly includes the intelligent target of submarine, surface ship and carrier-based aircraft.

4 System Running Process

The system running process is closely related to military requirements and experimental methods, and varies with the size of the experiment. The running process of the surface ship Blue Force simulation system is generally as follows:

(1) Military personnel and technical personnel, together with the military needs of the simulation experiment, conduct a collective discussion and analysis of the Blue Force's combat mission and methods. This process is very important, directly affecting the level of subsequent simulation experiments. Military personnel and technical personnel cooperate with each other, the latter provide tools to help the former normalize and format their proposed military needs, while the former instructs the latter to reflect their ideas in as much detail as possible. This process generally needs to be developed many times to achieve satisfactory results.

(2) Military personnel use simulation management subsystem to design combat scenario, determine combat background, the Blue Force weapons, sea battlefield environment, combat process and other content. The combat scenario is the blueprint of the simulation, which reflects the detailed requirements of the military personnel for the simulation experiment. Because of the complexity of the combat process, it is difficult to describe it in a simple form, and the description method used must be flexible to accommodate the diversity of the description of the combat process. However, the battle process between the Red Force and Blue Force must be described in detail in accordance with the fight time, so as to construct the simulation scheme.
(3) Technical personnel enrich and improve all kind of simulation experimental resources in accordance with the combat scenario, form a preliminary combat simulation program. The program mainly includes simulation parameter setting, Red Force configuration, Surface ship Blue Force configuration, operations personnel function, simulation mode, simulation resource list and data processing method.
(4) Technical personnel use the single ship simulation section to establish the simulation model of each warship Blue Force, and load the technical and tactical data. For example, to simulate an "Arleigh Burke" class guided missile destroyer of the U.S. Navy, it should establish simulation models for command & control subsystem, navigation subsystem, weapons and equipment, carrier-based aircraft subsystem, air defense operation subsystem, anti-ship operation subsystem, anti-submarine operation subsystem, electronic warfare subsystem, detection and reconnaissance subsystem. Among them, the warship's Aegis system simulation model is the most critical.
(5) Technical personnel use fleet simulation system, establish fleet-level simulation model, as well as the corresponding sea battlefield environment. For example, to simulate a common U.S. Navy nuclear-powered aircraft carrier group, it is necessary to build simulation models of the various surface ships and carrier-based aircraft, which mainly include Nimitz class nuclear-powered aircraft carriers, Ticonderoga class missile cruisers, Arleigh Burke class missile destroyers, and carrier-based aircrafts on the above-mentioned warships.
(6) Link the director's desk with simulation system, start running and debugging the simulation experiment. When the simulation system is completed, a wide area network for remote off-site simulation can be set up, and other military research institutions can be connected for online exercises. Therefore the system can be connected to the actual military force. In this way, the surface ship Blue Force should be placed in the navy and even the whole army Red-Blue confrontation environment for construction, and it will certainly have obvious results.
(7) The simulation experiment running process and features are displayed by the evaluation demonstration subsystem in two-dimensional and three-dimensional form as needed. The display here is not only to show the real-time situation of the Red Force and Blue Force, but also to show the fighting power index of both sides, in order to analyze their possibility of victory and defeat at any time.

(8) The data of this experiment will be collected and analyzed by the evaluation demonstration subsystem, especially new technologies such as large data technology and artificial intelligence technology should be introduced, and the final result of the evaluation processing results will be formed. The use of artificial intelligence technology can be strengthened, the latest artificial intelligence technology similar to AlphaGo [2] can be introduced to carry out large sample simulation, explore key nodes of the combat process, and then provide scientific solutions for the Red-Blue confrontation of real army.

(9) Military personnel and technical personnel carry out in-depth analysis of the experiment results together. For the purpose of this experiment, they draw technical and tactical conclusions, and feedback to modify and improve the simulation system. In the use of surface ship Blue Force simulation system to carry out combat research, we should adhere to the guiding concept of war design. Only with a scientific design of the war can the Blue Force simulation study be targeted and do more with less.

5 Conclusions

The naval surface ship Blue Force simulation system can complete the Red-Blue confrontation exercise on virtual sea battlefield by simulating the warship force and its combat operations, which can provide a simulated experimental environment for the naval real force exercise, the evaluation of the method of combat and the training of the commander's decision-making ability. At present, the construction of the naval Surface ship Blue Force simulation system has a long way to go, and the gap between the developing requirements of the times is not small, there are still a lot of theoretical and practical works to be done. In the future, the naval surface ship Blue Force simulation system can not only fight virtually with Red Force through the director's department, but also can be connected to the actual military equipment and army for more practical characteristics of the exercise. The functional analysis, structural design and running process proposed in this paper can provide ideas and basis for the construction of the Blue Force simulation system. At the same time, the content of the thesis has certain universality, can be applied to other combat simulation system construction. In future, large data technology, artificial intelligence technology, cloud technology, unmanned system technology and other advanced ideas and technology should also be brought in research, development and construction of the simulation system, which will certainly give the compute-based naval Red-Blue confrontation exercise bring great impetus and promotion.

References

1. Yang, J.K., Zhang, C.Y., Chang, X.F., Zhou, H.: Systematic construction and operation for navy test and training blue force. Mod. Defense Technol. **2**(4), 22–29 (2007)
2. Fu, S.J., Feng, S.P.: The artificial intelligence in AlphaGo. China Air Force **5**(5), 53–56 (2016)

Research on Evaluation of Distributed Enterprise Research and Development (R & D) Design Resources Sharing Based on Improved Simulated Annealing AHP: A Case Study of Server Resources

Yongqing Hu[1], Weixing Su[2(✉)], Yelin Xia[3], Han Lin[2], and Hanning Chen[1,2]

[1] School of Electrical Engineering and Automation, Tiangong University, Tianjin 300000, China
[2] School of Computer Science and Technology, Tiangong University, Tianjin 300000, China
suweixing@tiangong.edu.cn
[3] School of Mechanical Engineering, Tiangong University, Tianjin 300000, China

Abstract. Resource sharing in a distributed environment can improve the utilization of resources. To promote the benign development of resource sharing behavior, a basic evaluation model of resource sharing is proposed, and an evaluation index system of distributed server resource sharing is constructed based on the model in this paper. The judgment matrices are constructed according to the analytic hierarchy process (AHP), and the simulated annealing algorithm (SAA) and the improved simulated annealing algorithm (ISAA) are respectively used to improve the consistency of each judgment matrix. The results show that the average consistency deviation of the judgment matrices optimized by ISAA is reduced 0.0421 and 0.0106 compared to EM and SAA respectively, and the convergence speed is about 77.8% higher than SAA. The standard deviations of the weight differences before and after optimization using ISAA and SAA are 0.0074 and 0.0259, respectively, so the weight fluctuations after ISAA optimization are relatively smaller. The weight distribution corresponding to each index is obtained when the consistencies of the judgment matrices are close to the optimal state, which provides a necessary technical foundation for the evaluation of distributed server resource sharing.

Keywords: Evaluation of distributed R & D design resources · AHP · Improved simulated annealing algorithm

1 Introduction

Enterprise R & D and design resources refer to many hardware resources, software resources, and knowledge resources accumulated by the enterprise over the years. However, in a distributed R & D design environment, resources are geographically

Q. Liu et al. (Eds.): CENet 2020, AISC 1274, pp. 951–959, 2021.
https://doi.org/10.1007/978-981-15-8462-6_110

dispersed, heterogeneous, and diverse, and the efficiency of resource sharing and reuse is low. Unified and useful resource sharing technology is urgently required.

Establishing a comprehensive, multi-level, and fine-grained sharing evaluation system is an essential prerequisite for the healthy development of resource sharing. With the advent of the era of big data, compared with ordinary computers, the server resources have become an essential type of enterprise R & D and design resources because of their higher processing speed and data security.

At present, there are rare reports on the related researches on resource sharing evaluation of distributed servers. AHP [1] is a decision analysis method that layers the complex factors related to decision-making and digitizes each factor's relative importance. It can be used in the complex evaluation and analysis systems, such as natural resources evaluation [2], program decision analysis [3], and risk management evaluation [4]. Due to these advantages of AHP, it can be used as a primary evaluation method for distributed server resource sharing.

However, due to the intense subjectivity of AHP [5], using the improved AHP to determine the relative weight distribution of each index in the evaluation system is currently the focus of research. Zhao et al. [5], Girsang et al. [6], Pereira et al. [7] and Lin et al. [8] proposed using swarm intelligence algorithm, nonlinear model and Bayesian correction method to adjust the judgment matrices. However, the above methods reduce the consistency deviations by modifying the elements in the judgment matrices to obtain each factor's weight, which has a certain "indirectness." When the order of the judgment matrix is n (n > 3), the matrix involves ((“n” ^”2” “-n”))⁄”2” correction elements, which are computationally intensive and require high-performance algorithms.

In response to the above problems, this paper proposes a basic resource sharing evaluation model and establishes a distributed server resource sharing evaluation index system. AHP is used as the primary evaluation method, retains the original judgment matrices, directly adjusts the weight of each index, and reduces optimized variables from ((“n” ^”2” “-n”))⁄”2” to “n”. ISAA is adopted to adjust each index's weight to determine the weight distribution of each index that meets the consistency requirements. The applicability and effectiveness of the evaluation model and improved algorithm are verified through the comparison of experimental results.

2 Evaluation Index System of Distributed Server Resource Sharing Based on AHP

2.1 Evaluation Model of Resource Sharing

In order to promote the virtuous cycle of the resource sharing platform and maximize the use of high-quality resources, it is necessary to carry out distributed resource sharing evaluation based on high user satisfaction. The evaluation index system must be scientific and detailed to ensure the comprehensive and objective evaluation. According to the characteristics of resources and the process of resource sharing behaviors, the overall evaluation of resource sharing can be summarized as: finding resources, obtaining resources, using resources, basic evaluation and post-use evaluation. The above aspects can basically represent the status of each link in the process of resource sharing and realize a comprehensive, multi-dimensional, and fine-grained evaluation of server resources. According to the user experience honeycomb model proposed by Rosenfeld et al. [9], a Desirable-Usable-Valuable-Accessible-Findable (DUVAF) five-dimensional structure model is established for evaluation of distributed resource sharing evaluation in this paper.

Evaluation index systems can be established according to the DUVAF basic model, then the relative weight of each evaluation index can be determined in a certain way, and finally the comprehensive evaluation results of shared resources can be calculated by weighting through expert scores or user surveys.

2.2 Evaluation Index System of Distributed Server Resource Sharing

The steps to determine the relative weight of each evaluation index of distributed server resources based on AHP are as follows:

(1) Construction of hierarchical structure: The target layer A is a comprehensive evaluation of server resources, the criterion layer B is set to a 5-dimensional evaluation criterion according to the DUVAF model, and the index layer C sets corresponding evaluation indexes according to each criterion. The evaluation index system is shown in Table 1:

(2) Construction of judgment matrices and consistency check: The judgment matrix plays a vital role in determining the weight of each factor. Based on subjective experience, the evaluators quantify the relative importance of each factor and expresses it in the form of a matrix. Use the 1-9 scaling method [1] to construct the judgment matrix

$$A = (a_{ij})_{m \times m}, m \text{ is a positive integer} \tag{1}$$

where m is the number of evaluation indicators in the current level and the order of judgment matrix A. a_{ij} represents the relative importance level determined by factor a_i to factor a_j according to the 1-9 scale method. The sorting weight is written as

$$W = (w_1, w_2, w_3 \ldots w_m) \tag{2}$$

where w_m is the normalized eigenvector corresponding to the maximum eigenvalue λ_{max} of the judgment matrix A.

In AHP, the judgment matrices are constructed according to the 1-9 scale method, and their consistency are checked by using the eigenvalue method (EM). If the consistency ratio CR is less than 0.1, the judgment matrix has satisfactory consistency. The weight of each factor and consistency check results of each judgment matrix are shown in Table 2. A is the target judgment matrix; B_1 is the desirable judgment matrix; B_2 is the usable judgment matrix; B_3 is the valuable judgment matrix; B_4 is the accessible judgment matrix; B_5 is the findable judgment matrix. It can be seen that, except for the judgment matrix B_3, the remaining judgment matrices have satisfactory consistency.

Table 1. Server resource sharing evaluation index system.

Target layer	Criterion layer	Index layer	Criterion layer	Index layer
Evaluation of distributed server resource sharing	Desirable B_1	Subjective satisfaction after use C_1	Accessible B_4	Resource acquisition cost C_{14}
		Meet user needs C_2		Ease of access to resources C_{15}
		Economical and practical C_3		Timeliness of resource call C_{16}
		Shared services C_4		
	Usable B_2	run normally C_5		Resource retrieval service navigation C_{17}
		Suitable for current application tasks C_6		
		Downtime during working hours C_7	Findable B_5 (Openness of information)	Instrument informationC_{18}
		Manageability C_8		
	Valuable B_3	Calculating speed C_9		Resource holder information C_{19}
		Reliability C_{10}		
		Interoperability C_{11}		Regulatory documentsC_{20}
		Scalability C_{12}		
		Standby time C_{13}		Other information C_{21}

Table 2. The weight of each factor and *CR* of judgment matrices obtained by EM.

Judgment matrix	w_1	w_2	w_3	w_4	w_5	CR
A	0.4764	0.2778	0.1258	0.0793	0.0407	0.0525
B_1	0.4855	0.2674	0.1555	0.0664		0.0270
B_2	0.5594	0.2688	0.1070	0.0648		0.0347
B_3	0.4668	0.2885	0.1597	0.0537	0.0314	0.1410
B_4	0.6204	0.2012	0.1209	0.0575		0.0428
B_5	0.5041	0.3005	0.1226	0.0727		0.0115

3 AHP Based on Improved Simulated Annealing Algorithm

According to the definition of the judgment matrix, if the judgment matrix is completely consistent, then

$$a_{ij} = \frac{w_i}{w_j} = \frac{1}{a_{ji}}, 1 \le i \le m, 1 \le j \le m \quad (3)$$

After simple deformation,

$$\sum_{i=1}^{m} a_{ij} w_j - m w_i = 0 \quad (4)$$

From the definition of the judgment matrix, the consistency correction problem of the judgment matrix is attributed to the following nonlinear optimization problem according to Eqs. (3) and (4),

$$\begin{cases} \min\left(CIF_{(w_k)}\right) = \frac{\sum_{i=1}^{m} \left| \sum_{k=1}^{m} b_{ij} w_k - m w_i \right|}{n} \\ s.t. w_k > 0, k = 1 \sim m; \sum_{k=1}^{m} w_k = 1 \end{cases} \quad (5)$$

where $CIF_{(w_k)}$ is the consistency index function, w_k is the optimization variable. The smaller the $CIF_{(w_k)}$, the better the consistency of the judgment matrix.

SAA likens the optimization problem to the metal physical annealing process. For the optimization problem, the process of finding the optimal solution corresponds to the physical process of finding the internal energy of the metal object when the temperature drops to the minimum [10]. The algorithm focuses on the global search in the early stage, and focuses on the local search in the later stage. In theory, SAA has a strong local search capability, but the algorithm optimization time is long [11]. GA is a heuristic global optimization search algorithm derived from simulating the natural laws of survival of the fittest in the natural world. It screens offspring according to their fitness to gradually approach the optimal offspring. Its global search ability is strong and the convergence speed is fast, but due to the randomness in the crossover and mutation process, premature convergence is prone to occur and the local search ability is poor [12].

ISAA combines the fast convergence of GA with the precise searchability of SAA. The main steps of the improved algorithm are as follows:

Step 1: Initialize the population, set the population size N, the initial temperature T_0, the end temperature T_{end}, the temperature drop coefficient c, the initial iteration value L, the maximum number of iterations MAXL;

Step 2: If $T_0 > T_{end}$, execute Step 3; otherwise terminate the algorithm and output the optimal solution W_{best};

Step 3: If $L \le$ MAXL, execute Step 4; otherwise, perform the temperature-lowering operation:$T_0 = c*T_0$;

Step 4: Calculate and rank the fitness value f for each individual of the current population, min (f) corresponds to the worst individual W_{worst};

Step 5: Select the parent to cross and mutate to produce the offspring W_{new};

Step 6: Calculate the average fitness value f_{ave} for the parent population, let $\Delta f = f - f_{ave}$, if $\Delta f = f - f_{ave} > 0$, then $W_{worst} = W_{new}$, otherwise, if $\exp(-\Delta f/T_0) \geq$ random (0,1), then $W_{worst} = W_{new}$;

Step 7: If the algorithm termination conditions are met, the algorithm ends, and the optimal solution W_{best} is output; otherwise, the temperature-reducing operation is performed and go to Step 2. The termination condition is set to N consecutive optimal solutions without change.

Applying ISAA to AHP, using Eq. (5) as the fitness function in ISAA, the process of solving its optimal solution corresponds to the process of finding the optimal offspring in ISAA, thus determining the optimal weight of each index in the judgment matrix, to achieve the purpose of correcting the consistency of the judgment matrix.

4 Results Comparison

The consistency deviations of the judgment matrices are corrected by ISAA, and the weights obtained by EM are taken as the basic data, then new weight vectors are generated by randomly generating fluctuations to form new population based on basic data. Random selection is used to select individuals in the population, and the crossover operation uses decimal random crossover. In order to maintain the diversity of the population, the mutation operation needs to be performed with a certain probability during the evolution of the population, and new individuals will be generated after the crossover and mutation. Since the optimization goal is to determine the value of the optimization variable w_k when the minimum value is obtained in Eq. (5), each generation of population regards the individual with the smallest objective function value as the contemporary optimal individual, continuously updates the population. The algorithm will be terminated and the optimal solution can be obtained until the algorithm termination condition is met. In this way, the optimal weight distribution of each evaluation index is obtained, which provides a data basis for solving evaluation problems in the future.

The results of using SAA and ISAA to modify the consistency of judgment matrix A, B_1, B_2, B_3, B_4 and B_5 are shown in Tables 3 and 4.

Judging from the consistency index function of the judgment matrices before and after the improvement, for the six judgment matrices participating in the optimization, the consistency correction results of the judgment matrices using ISAA are significantly better than those of EM and SAA. For the judgment matrix B_3, in terms of operation results, the consistency ratio $CR = 0.1410 > 0.1$ obtained by EM does not have satisfactory consistency; after optimizing by SAA and ISAA, the values of consistency index function are reduced to 0.0497 and 0.0187 respectively, both of which are met the consistency index ($CIF_{(w_k)} < 0.1$). The consistency index obtained by ISAA is lower than that of EM and SAA by 0.1223 and 0.0310 respectively. The remaining five judgment matrices have satisfactory consistency. In order to analyze the effectiveness and sensitivity of the algorithm, the weights of each judgment matrix are used as

Table 3. SAA correction results for the consistency of judgment matrices.

Judgment matrix	w_1	w_2	w_3	w_4	w_5	$CIF_{(w_k)}$	Number of iterations
A	0.4527	0.2859	0.1415	0.0842	0.0357	0.0258	724
B_1	0.4904	0.3032	0.1418	0.0646		0.0098	605
B_2	0.5766	0.2643	0.0989	0.0602		0.0138	682
B_3	0.4050	0.3160	0.1843	0.0507	0.0376	0.0497	740
B_4	0.6236	0.1888	0.1411	0.0466		0.0145	798
B_5	0.5777	0.2447	0.1091	0.0685		0.0073	647

Table 4. ISAA correction results for the consistency of judgment matrices.

Judgment matrix	w_1	w_2	w_3	w_4	w_5	$CIF_{(w_k)}$	Number of iterations
A	0.4889	0.2786	0.1232	0.0773	0.0320	0.0141	155
B_1	0.4901	0.2828	0.1587	0.0684		0.0060	147
B_2	0.5583	0.2726	0.1093	0.0598		0.0079	183
B_3	0.4807	0.2857	0.1447	0.0504	0.0385	0.0187	175
B_4	0.6167	0.1961	0.1182	0.0690		0.0077	107
B_5	0.4978	0.2960	0.1209	0.0853		0.0028	160

optimization variables to participate in the optimization. From the comparison of Tables 2, 3 and 4, it can be seen that the consistency deviations of the judgment matrices obtained by ISAA are lower than that of EM and SAA by an average of about 0.0421 and 0.0106 respectively. ISAA optimization produces the best results.

In terms of convergence speed, the average number of iterations of the ISAA and SAA for the consistency correction of the six judgment matrices are 690 and 155 respectively, and the ISAA convergence speed has increased than SAA by about 77.8%.

In the aspect of weight fluctuations, calculate the differences of the weight of each factor before and after optimization and calculate their standard deviations, so as to quantify the fluctuation of the weight of each factor before and after optimization. The calculation formula is as Eq. (6),

$$\sigma = \frac{\left\{\sum_{i=1}^{m}\left[\left|w_i - w_i^*\right| - \frac{\sum_{i=1}^{m}\left|w_i - w_i^*\right|}{m}\right]\right\}^{0.5}}{m-1} \tag{6}$$

where σ is the standard deviation of the weight difference before and after optimization, w_i is the weight before optimization, w_i^* is the weight after optimization, $||$ means the absolute value. For the above six judgment matrices, the standard deviations of the weight differences between the factors before and after ISAA and SAA optimization are 0.0074 and 0.0259 respectively. It means that the weight changes before and after ISAA optimization are relatively small, which achieves the goal of making the consistency of the judgment matrices close to the optimal under the premise of respecting the subjective will of the evaluators.

5 Conclusion

In this paper, a basic evaluation model of server resources is proposed. The AHP is used to establish a server resource sharing evaluation index system. ISAA is used to optimize the evaluation index weights. The consistency deviations of the judgment matrices are lower than that of EM and SAA by an average of about 0.0421 and 0.0106, and the convergence speed is increased by about 77.8% than that of SAA. In addition, the method in this paper can avoid the possible contradiction between the consistency deviation of the judgment matrix and the subjective will of the evaluators.

Subsequent weighted calculations such as scoring can be used to calculate the overall score of the resources to be evaluated, which provides a certain theoretical and technical basis for evaluation of enterprise shared resources.

Acknowledgement. This paper is supported by the National Key R & D Program of China (No. 2018YFB1701802).

References

1. Saaty, R.W.: The analytic hierarchy process-what it is and how it is used. Math. Model. **9**(3–5), 161–176 (1987)
2. Rabia, S.: Water resource vulnerability assessment in Rawalpindi and Islamabad, Pakistan using analytic hierarchy process (AHP). J. King Saud Univ.-Sci. **28**(4), 293–299 (2016)
3. Fikri, D.: Designing an integrated AHP based decision support system for supplier selection in automotive industry. Expert Syst. Appl. **62**, 273–283 (2016)
4. Natasha, K.: Ranking the indicators of building performance and the users' risk via analytical hierarchy process (AHP): Case of Malaysia. Ecol. Ind. **71**, 567–576 (2016)
5. Zhao, J.: Water resources risk assessment model based on the subjective and objective combination weighting methods. Water Resour. Manage. **30**(9), 3027–3042 (2016)
6. Girsang, A.S., Tsai, C.-W., Yang, C.-S.: Ant algorithm for modifying an inconsistent pairwise weighting matrix in an analytic hierarchy process. Neural Comput. Appl. **26**(2), 313–327 (2014). https://doi.org/10.1007/s00521-014-1630-0
7. Pereira, V., Costa, H.G.: Nonlinear programming applied to the reduction of inconsistency in the AHP method. Ann. Oper. Res. **229**(1), 635–655 (2014). https://doi.org/10.1007/s10479-014-1750-z
8. Lin, C.S.: Bayesian revision of the individual pair-wise comparison matrices under consensus in AHP-GDM. Appl. Soft Comput. **35**, 802–811 (2015)
9. Rosenfeld, L., Morville, P.: Information Architecture for the World Wide Web, 2nd edn. O'Reilly & Associates, USA (2002)
10. Selim, Z.S.: A simulated annealing algorithm for the clustering problem. Pattern Recogn. **24** (10), 1003–1008 (1991)

11. Fanchao, M.: Optimal placement of SaaS components based on hybrid genetic simulated annealing algorithm. J. Softw. **27**(4), 916–932 (2016)
12. Boukhalfa, G.: Genetic algorithm and particle swarm optimization tuned fuzzy PID controller on direct torque control of dual star induction motor. J. Cent. South Univ. **26**(7), 1886–1896 (2019)
13. Zhang, Z.C.: Analysis on decision-making model of plan evaluation based on grey relation projection and combination weight algorithm. J. Syst. Eng. Electr. **29**(4), 789–796 (2018)
14. Meng, C.: Development and application of evaluation index system and model for existing building green-retrofitting. J. Therm. Sci. **28**(6), 1252–1261 (2019)

A Simplified Simulation Method for Measurement of Dielectric Constant of Complex Dielectric with Sandwich Structure and Foam Structure

Yang Zhang(✉), Qinjun Zhao, and Zheng Xu

School of Electrical Engineering, University of Jinan, Jinan 25000, China
cse_liy@ujn.edu.cn

Abstract. Empirical formula and computer simulation are the common two ways to estimate the dielectric constant of dielectric with mixed structure. Empirical formula can be calculated very quickly, but it restricts the shape and volume of the filling material. Simulation is more reasonable theoretically, but it is computationally complex. This paper introduced a simplified simulation method for measurement of dielectric constant of complex dielectric with sandwich structure and foam structure. The experiments showed that its results became near to the results of common simulation and Maxwell-Garnett empirical formula.

Keywords: Dielectric constant · Complex dielectric · Simplified simulation

1 Introduction

Dielectric materials have been widely used in electricity, electronics, and many other fields. In many cases, the dielectric constant of the dielectric is required to be much higher [1–3] or lower [4–6] than any single material. So complex dielectric is introduced. Complex dielectric can be made physically or by chemical reaction. In the physical manufacturing methods, sandwich structure [7–9] and foaming [10–12] are frequently used. These methods mix two or more different materials and the dielectric constant is determined by mixture feature.

In the process of complex dielectric physical generation, it is necessary to predict the dielectric constant of it. Maxwell-Garnett empirical formula is one way, which can be calculated very quickly. However, this empirical formula restricts the shape and volume of the filling material. Computer simulation is another way, which is more reasonable theoretically and computationally complex. This paper attempts to present a simplified computer simulation method. Based on the resonance principle, complex composition will be converted into simple composition with the same volume.

Q. Liu et al. (Eds.): CENet 2020, AISC 1274, pp. 960–967, 2021.
https://doi.org/10.1007/978-981-15-8462-6_111

2 Simplified Simulation Method

The dielectric constant of a composite dielectric can be measure in many ways, such as the transmission/reflection method [13–15], resonant cavity method, and free-space method. In the simulation, these methods and complex dielectrics are virtualized. If a method is chosen, the mixture structure of a complex dielectric and the dielectric constant will correspond one to one. The resonant cavity method is used in this simulation method due to its being realized easily.

Choose a cube resonant cavity for modeling. Because the simulation using the cube resonant cavity not only does not affect the accuracy of the dielectric constant measurement, but also simplifies the derivation process due to the symmetry of the shape. The rectangular resonant cavity can be considered as the case of a short circuit at the end of the rectangular waveguide. At this moment, the electromagnetic wave is all reflected, forming a standing wave in the waveguide. The resonant cavity modes can correspond to the modes in the waveguide, which can be divided into the low or high mode. Figure 1 is the diagram of the first six resonant modes when the cavity is not filled with medium.

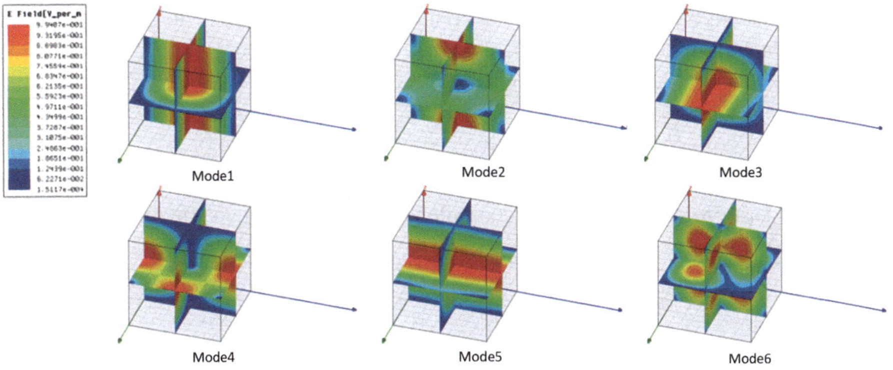

Fig. 1. The first six low resonant modes in a cubic cavity filled with no medium

In Fig. 1, mode 1, 2, 3 are the lowest mode in rectangular waveguide with different orientations, i.e., TE_{10}. Mode 4, 5 are the mode next to the lowest in rectangular waveguide with different orientations.

2.1 Complex Dielectric with Sandwich Structure

Sandwich structure is simple and easy to implement, as shown in Fig. 2.

In many cases there are many layers. This is because the original physical properties of the material are supposed to maintain, and the dielectric constant is supposed to change. In this case, the simulation will consume large computational quantity.

Fig. 2. Complex dielectric with sandwich structure.

In a rectangular cavity, mode TE_{mnl} resonant frequency is calculated below:

$$f_{mn1} = \frac{c}{2\pi\sqrt{\varepsilon\mu}}\sqrt{(\frac{m\pi}{a})^2 + (\frac{n\pi}{b})^2 + (\frac{l\pi}{d})^2} \tag{1}$$

where ε is dielectric constant of the medium, μ is magnetic conductivity of the medium, a is the length of the rectangular cavity, b is the width of the rectangular cavity, d is the height of the rectangular cavity.

Let $a = b = d$ and divide the cavity into i^3 equal parts. Every part can be considered as a little rectangular cavity whose layer number changes to $1/i$ of it before. If $m = n = l = 1$ in little rectangular cavity, then its resonant frequency is

$$f_1 = \frac{c}{2\pi\sqrt{\varepsilon\mu}}\sqrt{(\frac{\pi}{\frac{1}{i}a})^2 + (\frac{\pi}{\frac{1}{i}a})^2 + (\frac{\pi}{\frac{1}{i}a})^2} = \frac{\sqrt{3}ci}{2a\sqrt{\varepsilon\mu}} \tag{2}$$

In this case, $m = n = l =$ i in rectangular cavity and resonant frequency is

$$f_2 = \frac{c}{2\pi\sqrt{\varepsilon\mu}}\sqrt{(\frac{i\pi}{a})^2 + (\frac{i\pi}{a})^2 + (\frac{i\pi}{a})^2} = \frac{\sqrt{3}ci}{2a\sqrt{\varepsilon\mu}} \tag{3}$$

It seems that $f_1 = f_2$. However, f_1 is in mode TE_{111} and f_2 is in mode TE_{iii}. Let the volume of a little rectangular cavity expands to the same size as a rectangular cavity. This Imaginary rectangular cavity has the same number of layers as little rectangular cavity and the same volume size as the rectangular cavity.

In mode TE_{111}, resonant frequency of the rectangular cavity is

$$f_3 = \frac{\sqrt{3}c}{2a\sqrt{\varepsilon\mu}} \tag{4}$$

In mode TE_{111}, resonant frequency of the imaginary rectangular cavity is

$$f_4 = \frac{f_1}{i} = \frac{\sqrt{3}c}{2a\sqrt{\varepsilon\mu}} \tag{5}$$

Then, f3 = f4, which means the resonant frequency of the rectangular cavity is same as the imaginary rectangular cavity. As the imaginary rectangular cavity's medium has a smaller number of layers, its simulation will consume lesser computational quantity.

So when complex dielectric with sandwich structure is simulated to predict the dielectric constant, we can use the imaginary rectangular cavity to reduce the computational quantity. There are two caveats: one is that the shape, particle size, and surface properties of the filling material all affect the properties of the composite dielectric. This means too simple an imaginary rectangular cavity may magnify the effect. Another is that complex dielectric with sandwich structure is not isotropic. There is a complex dielectric which can be isotropic, that is complex dielectric with foam structure.

2.2 Complex Dielectric with Foam Structure

Foam material is a kind of composite dielectric that satisfies isotropy. In order to reduce the dielectric constant, the foaming method is often used to modify materials with smaller dielectric constants to obtain foam materials with smaller dielectric constants. Figure 3 is a dielectric model of foam materials. The simplified simulation model of complex dielectric with foam structure is the same as that of the complex dielectric with a sandwich structure.

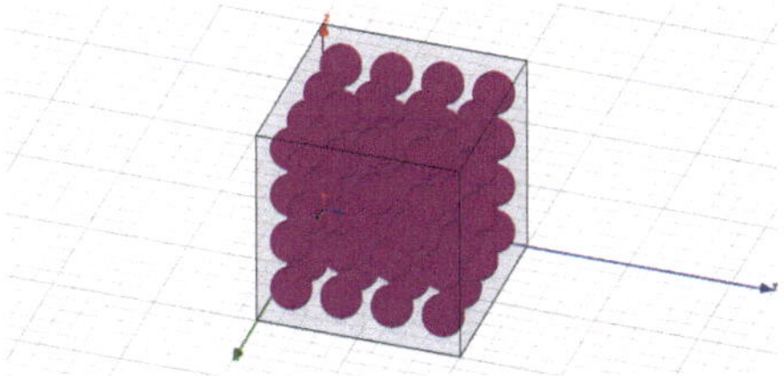

Fig. 3. Complex dielectric with foam structure.

The simplified simulation model of complex dielectric with foam structure is the same as that of the complex dielectric with a sandwich structure. In the case of high-frequency resonance, the rectangular cavity is divided into several little rectangular cavity. One little rectangular cavity is enlarged in imagination, i.e., imaginary rectangular cavity. Then resonant mode changed and the resonant frequency of a rectangular cavity is the same as the imaginary rectangular cavity.

3 Experiments

HFSS is selected as the simulation environment of the model. HFSS is a 3D electromagnetic simulation software based on the electromagnetic field finite element method (FEM) to analyze microwave engineering problems. A cube resonant cavity with a side length of 10 mm is selected in experiments and the complex dielectric is glass with air in it. In experiments resonant frequencies are compared instead of dielectric constants as they correspond one to one.

The experiments include two parts. Firstly, as the first caveat mentioned in Sect. 2.1, the simplest model may lose efficacy. Which is an effective simulation model that is tested in this part? Secondly, the results of Simplified simulation are compared with the Maxwell-Garnett empirical formula.

3.1 Effective Simulation Model

In Simplified simulation of complex dielectric with sandwich structure, the number of layers decreases. However, one-layer simulation is not accurate because of the first caveat. Figure 4 shows how resonant frequencies change with the number of layers in a complex dielectric with a sandwich structure in the lowest 5 models. Note that more layers means layers are thinner, because they are put into a cavity of the same volume.

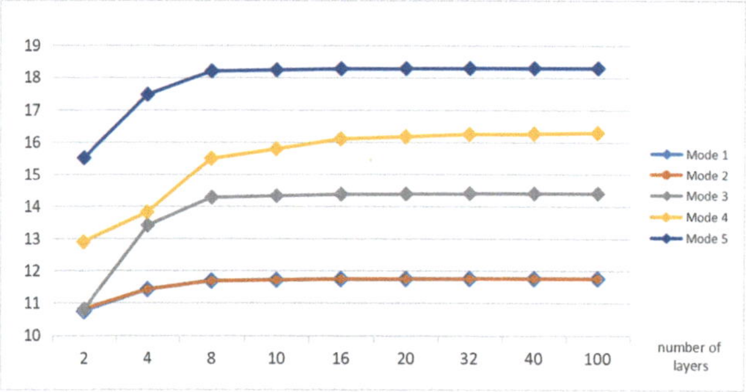

Fig. 4. The resonance frequency varies with the number of layers in a cavity filled with sandwich structure complex dielectric in mode 1 ~ 5.

As can be seen in Fig. 4, resonant frequencies change a lot when layer numbers are small, resonant frequencies level off when layer numbers are large which can be seen as Simplified simulation is little affected by the first caveat. Modes 1, 2, 3 have frequencies with the same value and different directions. Modes 1 and 2 works similarly and Mode 3 works differently. This is because the complex dielectric with the sandwich structure is not isotropous. Finally, it can be seen that the imaginary rectangular cavity with a 16-layer medium is a simple and effective simulation model.

Figure 5 shows how resonant frequencies change with the number of air foams in complex dielectric with foam structure in the lowest 5 models.

As can be seen in Fig. 5, the imaginary rectangular cavity with 8-foam medium is a simple and effective simulation model. Modes 1, 2, 3 work similarly and modes 4, 5 work similarly. This implies that complex dielectric with foam structure is isotropous.

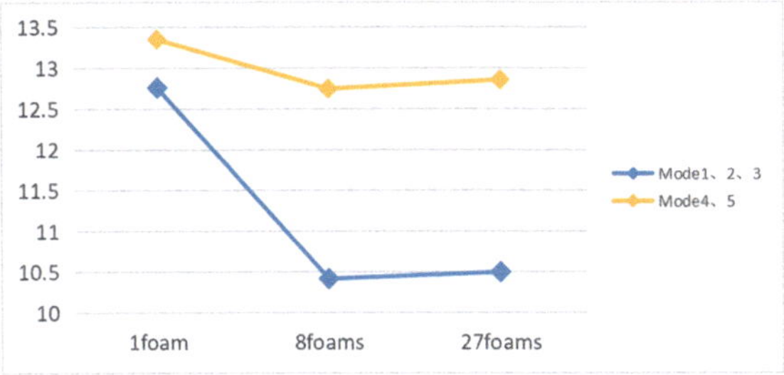

Fig. 5. The resonance frequency varies with the number of foams in a cavity filled with foam structure complex dielectric in mode 1 ~ 5.

3.2 Compared with Maxwell-Garnett Empirical Formula

Maxwell-Garnett empirical formula describes the approximate calculation of the effective dielectric constant of a composite dielectric with a spherical dielectric (ε_1) dispersed in a continuous dielectric (ε_2). So it can be used to calculate the dielectric constant of complex dielectric with a foam structure.

If the volume ratio of the spherical dielectric to the entire dielectric is V, the Maxwell-Garnett empirical formula is expressed as below:

$$\frac{\varepsilon_{eff} - \varepsilon_2}{\varepsilon_{eff} + 2\varepsilon_2} = V \frac{\varepsilon_1 - \varepsilon_2}{\varepsilon_1 + 2\varepsilon_2} \tag{6}$$

After the dielectric constant ε_{eff} is calculated, it can be put into Eq. (1) to calculate empirical formula resonant frequency value f'. The resonance frequencies of 8 foams dielectric were compared with the values calculated according to the empirical formula. Table 1 shows the comparison between the parameters obtained by the 8-foam dielectric in the HFSS environment and those calculated by Maxwell-Garnett empirical formula. R is the radius of foam, f_1 is the simulated resonant frequency in mode 1, f_4 is the simulated resonant frequency in mode 4, f_1' is the resonant frequency calculated by

Table 1. Parameters of 8-foams dielectric obtained by Simplified simulation and empirical formula.

R(mm)	f_1(GHz)	f_4(GHz)	V	ε_{eff}	f_1'(GHz)	f_4'(GHz)
1.8	10003	12241	0.195	4.373	10136	12414
1.85	10093	12352	0.212	483	10241	12543
1.9	10191	12471	0.229	39	10355	12682
1.95	10296	12600	0.248	4.093	10477	12832
2	10409	12738	0.268	3.992	10608	12993

the empirical formula in mode 1, f_4' is the resonant frequency calculated by the empirical formula in mode 4.

The empirical formula has a good approximation when foam volumes are 20% $\sim$ 50% of the dielectric. So results of simplified simulation and empirical formula are compared in this range. Their resonant frequencies in modes 1 and 4 are very close. This proves the validity of simplified simulation to some extent.

4 Conclusion

A simplified and low computational simulation method is proposed to estimate the dielectric constant of dielectric with mixed structure. This method is based on the resonance principle and it has shown the results which became near to the results of the common simulation. Its results are also compared with the Maxwell-Garnett empirical formula. Within the empirical formula's scope of application, their results were similar, too.

Acknowledgement. This work was supported by the Shandong Agricultural machinery equipment R & D innovation plan project under Grant 2018YF011, the High School Science and Technology Project in Shandong Province under Grant J18KA333 and the Key R & D plan of Shandong Province under Grant 2019GGX104026.

References

1. Kim, P., Jones, S.C., Hotchkiss, P.J., et al.: Phosphonic acidmodified barium titanate polymer nanocomposites with high permittivity and dielectric strength. Adv. Mater. **19**(7), 1001–1005 (2007)
2. Kim, P., Doss, N.M., Tillotson, J.P., et al.: High energy density nanocomposites based on surface-modified BaTiO 3 and a ferroelectric polymer. ACS Nano **3**(9), 2581–2592 (2009)
3. Barber, P., Balasubramanian, S., Anguchamy, Y., et al.: Polymer composite and nanocomposite dielectric materials for pulse power energy storage. Materials **2**(4), 1697–1733 (2009)
4. Lee, Y.J., Huang, J.M., Kuo, S.W., et al.: Low-dielectric, nanoporous polyimide films prepared from PEO-POSS nanoparticles. Polymer **46**(23), 10056–10065 (2005)
5. Xi, K., Meng, Z., Heng, L., et al.: Polyimide-polydimethylsiloxane copolymers for low-dielectric-constant and moisture-resistance applications. J. Appl. Polym. Sci. **113**(3), 1633–1641 (2010)
6. Zhang, Y., Yu, L., Su, Q., et al.: Fluorinated polyimide-silica films with low permittivity and low dielectric loss. J. Mater. Sci. **47**(4), 1958–1963 (2012)
7. Seibert, H.F.: Applications for PMI foams in aerospace sandwich structures. Reinf. Plast. **50**(1), 44–48 (2006)
8. Seibert, H.F.: PMI foam cores find further applications. Reinf. Plast. **44**(1), 36–38 (2000)
9. Choi, I., Kim, J.G., Lee, D.G., et al.: Aramid/epoxy composites sandwich structures for low-observable radomes. Compos. Sci. Technol. **71**(14), 1632–1638 (2011)
10. Krause, B., Koops, G., Vegt, N., et al.: Ultralow-k dielectrics made by supercritical foaming of thin polymer films. Adv. Mater. **14**(15), 1041–1046 (2002)

11. Wang, Q., Wang, C., Wang, T.: Controllable low dielectric porous polyimide films templated by silica microspheres: Microstructure, formation mechanism, and properties. J. Colloid Interface Sci. **389**(1), 99–105 (2013)
12. Jiang, L., Liu, J., Wu, D., et al.: A methodology for the preparation of nanoporous polyimide films with low dielectric constants. Thin Solid Films **510**(1–2), 241–246 (2006)
13. Vanzura, E.J., Baker-Jarvis, J.R., Grosvenor, J.H., et al.: Intercomparison of permittivity measurements using the transmission/reflection method in 7-mm coaxial transmission lines. IEEE Trans. Microw. Theor. Tech. **42**(11), 2063–2070 (1994)
14. Ni, E.: An uncertainty analysis for the measurement of intrinsic properties of materials by the combined transmission reflection method. IEEE Trans. Instrum. Meas. **41**(4), 495–499 (1992)
15. Boughriet, A.H., Legrand, C., Chapoton, A.: Noniterative stable transmission/reflection method for low-loss material complex permittivity determination. IEEE Trans. Microw. Theor. Tech. **45**(1), 52–57 (1997)

Research on PSO-MP DC Dual Power Conversion Control Technology

Yulong Huang and Jing Li(✉)

College of Electrical and Information Engineering,
Hunan Institute of Engineering, Xiangtan, China
ljing1999@163.com

Abstract. The power supply continuity is the key factor affecting the normal operation of the load. The complex power quality problems (such as two-frequency ripple superposition, voltage sag and ripple superposition) in the dc power supply system will directly affect the continuous power supply of the load. The internal conversion control technology of the dual power switch can guarantee the continuous power supply of the load. To solve the problem of compound power quality in dc dual power supply system, this paper proposes a dual power conversion control technique based on PSO-MP. In this technique, PSO algorithm is used to perform coarse search and MP algorithm to identify the composite power quality disturbance. The example analysis not only verifies the validity and accuracy of the method proposed in this paper, but also improves the reliability of the double power switch and quickly realizes the conversion between two dc power sources.

Keywords: Supply continuity · DC power supply system · Composite power quality problem · Double power switch · PSO-MP algorithm

1 Introduction

Electrical continuity is the basic function index of the reliability of important load operation. On the one hand, with the continuous improvement of industrial production and department electrification, the load demands more and more on power quality and power supply continuity. On the other hand, with the emergence of new technologies such as power electronics and artificial intelligence, the existing AC power supply system is faced with great challenges in terms of load growth, tight transmission corridor, high line loss and low reliability of power supply in the grid [1]. In recent years, DC power supply system has been widely used in occasions with large proportion of distributed power supply and load due to many advantages such as high transmission efficiency, large transmission capacity, high power quality and strong flexibility of grid connection control [2, 3]. In DC power supply system, if there is a composite power quality problem (such as voltage two-frequency ripple superposition, voltage sag and ripple superposition), it will lead to the normal continuous power supply of load.

In general, the use of dual-power automatic transfer switch has become the best solution to ensure the continuous power supply of the load, and dual-power transfer

Q. Liu et al. (Eds.): CENet 2020, AISC 1274, pp. 968–988, 2021.
https://doi.org/10.1007/978-981-15-8462-6_112

control technology can quickly realize the conversion between the two dc power supply. The core of the dual-power conversion control technology is to identify the power quality disturbance signal of the two-channel DC power supply. If the fault of one circuit is detected, it will be automatically converted to the other circuit according to the conversion program inside the device to ensure the continuous power supply of the load [4].

With the application of DC in various fields, more and more algorithms are used to identify dc energy mass disturbance, including Fourier transform, support vector machine method, k-means clustering algorithm, etc. [5–7]. The Fourier transform can only analyze the frequency spectrum of periodic signals, and it is not suitable for non-stationary signals. The recognition ability of SVM is easily affected by its own parameters. If there are abnormal points in the cluster, k-means clustering algorithm will lead to serious mean deviation, that is, it is sensitive to noise and outlier data.

In recent years, with the intelligent development of dual power supply system, the combined algorithm of matching tracking (MP) algorithm and particle swarm optimization (PSO) algorithm particle swarm optimization matching tracking algorithm (PSO-MP) has been widely used in power quality disturbance signal detection. At present, there have been a lot of researches on the power quality disturbance signal detection technology. In the ac power supply system, [8–10] first MP algorithm is used to change the basic component of an alternating current is extracted, then using the fast Fourier transform (FFT) to find the solution to find the maximum frequency electrical signal and operation, by PSO coarse search and fine search of MP discretization process and matching the best particles complete disturbance signal characteristic parameters extraction. In the DC power supply system, PSO-MP algorithm is also used in literature [11–13] to extract the characteristic parameters of disturbance signal. The algorithm process is similar to the extraction of characteristic parameters of AC disturbance signal, which will not be described here.

Based on the above research results, this paper proposes a dual-power conversion control technology based on particle swarm optimization matching tracking (PSO-MP) algorithm to solve the composite power quality problem in the DC power supply system, and introduces the technology into the dual-power conversion switch. Through the example analysis, the validity and accuracy of the method proposed in this paper are verified, and the reliability of the dual-power transfer switch is improved, and the conversion between the two DC power sources is realized quickly to maximize the continuous power supply of the load.

2 Analysis of Single Power Quality Disturbance Problem

The bus voltage is the only index to show whether the power quality in the system is good or not. In general, several single power quality problems constitute a composite power quality problem. This section analyzes the single power quality problems (such as voltage ripple and voltage sag) existing in the system according to the bus voltage operation status.

2.1 Voltage Ripple

Voltage ripple is an ac component of dc voltage. The voltage ripple mainly comes from the residual AC volume of inverter and filtering of power electronic devices in the DC distribution system. Its frequency is an integer multiple of the switching frequency or input AC voltage frequency and keeps in sync with it. Therefore, the ripple is periodic to a certain extent and is not affected by the fluctuation of load and external nonlinear devices [14, 15]. Dc distortion is shown in Eq. (1):

$$U_{\alpha} = \sqrt{\frac{(U_{\omega 1}^{2} + U_{\omega 2}^{2} + \ldots U_{\omega n}^{2})}{n}} \tag{1}$$

where, DC distortion U is the root mean square value of each AC voltage harmonic component U_{wi} on DC voltage.

The DC distortion coefficient is shown in Eq. (2):

$$THD_{U} = \frac{U_{\alpha}}{U} \times 100\% \tag{2}$$

where, THD_U is the ratio of DC distortion U_α to dc steady-state voltage U.

According to Eqs. (1) and (2), the distortion spectrum quantifies the proportion of DC distortion in amplitude E_m of each frequency component. The amplitude-spectrum formula of the finite discrete amplitude is as follows:

$$E_{m} = E(f_{m}) = E(\omega_{m}) = \sum_{n=0}^{N-1} e[n] \exp\left(-j\frac{2\pi}{N}mn\right) \tag{3}$$

$$e[n] = \frac{1}{N}\sum_{n=0}^{N-1} E_{m} \exp\left(j\frac{2\pi}{N}mn\right) \tag{4}$$

2.2 Voltage Sag

In a DC system, when a large number of distributed power sources are connected, the extreme environment will affect distributed power sources such as wind turbines and photovoltaic cells, resulting in sudden changes in output power and voltage sag [16]. Dynamic voltage sag often occurs in distributed power systems. This paper takes the equivalent circuit of distributed power supply as an example to illustrate the specific process of dynamic voltage sag. Set relevant parameters in the circuit, where: U is dc power supply; C is equivalent capacitance; R is equivalent resistance; L is equivalent inductance (Fig. 1).

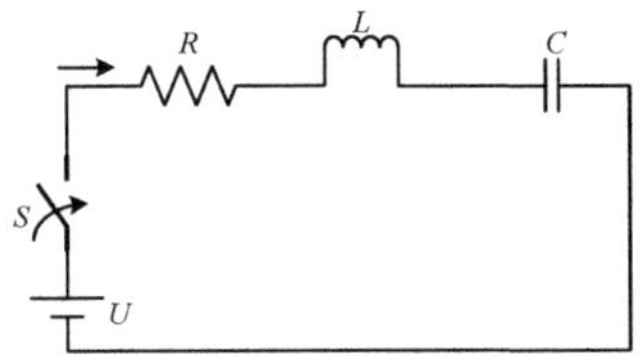

Fig. 1. Distributed power equivalent circuit.

The circuit is expressed as the full response process of the second-order circuit, and the equation with the voltage *Uc* as the output state variable is:

$$LC\frac{d^2u_c}{dt^2} + RC\frac{du_c}{dt} + u_c = U \tag{5}$$

Suppose the voltage solution $U_c = Ae^{pt}$, and substitute into the above equation, the characteristic equation of the response is:

$$LCp^2 + RCp + 1 = U \tag{6}$$

The characteristic root of the general solution of the corresponding homogeneous equation is:

$$p = -\frac{R}{2L} \pm \sqrt{\left[\left(\frac{R}{2L}\right)^2 - \frac{1}{LC}\right]} \tag{7}$$

To accommodate these two values, the voltage *Uc* can be written as:

$$U_C = A_1e^{p_1t} + A_2e^{p_2t} \tag{8}$$

3 Based on PSO-MP DC Dual Power Conversion Control Technology Analysis

Transfer control technology is one of the most important technologies in dual power transfer switch. This section explains the basic principle of dual power transfer switch, the basic idea of MP algorithm, the basic principle of Gabor atomic library, the construction of disturbed atomic library, PSO parameter setting and PSO-MP-based power quality disturbance detection respectively.

3.1 Basic Principle of Dual Power Transfer Switch

Automatic transfer switch (ATSE) is an electronic/electrical switch capable of monitoring, controlling and switching between two or more power sources. Its specific requirements are: can detect when all the dc power supply such as the superposition of two frequency ripple, sag and ripple stacking fault automatically convert to normal standby dc power supply (transformation of DC panel), and one or more of the load circuit from a power conversion to another power, to ensure that the important load continuous electric appliances [17].

ATSE is mainly composed of controller and switch body. The working principle of the controller is to compare the collected electrical signal with the set value inside the microprocessor through the voltage acquisition and conditioning circuit, and judge whether the electrical signal conforms to the normal standard value. The ATSE in this article focuses on the PC level. The system structure of dual power transfer switch is shown in Fig. 2.

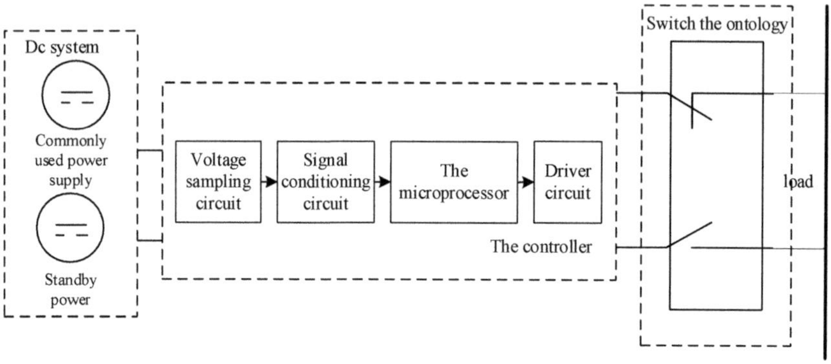

Fig. 2. Double power switch system structure diagram.

3.2 The Basic Idea of MP Algorithm

At the early stage, MP algorithm was mainly used in image and signal processing, and its core was greedy. Dc original signal to X, the dictionary is defined as the signal space is a collection of basic building unit specification, these units mode vector called atom, where $D = \{g_\gamma\}_{\gamma \in \Gamma}$ library for complete atoms. You normalize the atoms, $\|g_\gamma\| = 1$. $g_{\gamma(n)}$ is the dictionary atom. Assuming that the dictionary is complete and redundant, there is no unique way to express the signal from DC as a linear combination of atoms, and the expression is:

$$X = \sum_{n} \alpha_n g_{\gamma(n)} \tag{9}$$

To calculate the coefficient α_n, $g_{\gamma(0)}$ was selected, and the expression was:

$$\alpha_0 = \langle X, g_{\gamma(0)} \rangle = \max_{\gamma \in \tau} \langle X, g_\gamma \rangle \tag{10}$$

Let's divide X into two parts of our definition $R_X^0 = X - \alpha_0 g_{\gamma(0)}$. Through iteration, the residual quantity R^{n+1} of order $n + 1$ can be calculated, and its expression is:

$$R_X^{n+1} = R_X^n - \alpha_{n+1} g_{\gamma(n+1)} \tag{11}$$

Among them, the maximum value can be obtained from γ:

$$\alpha_{n+1} = \left\langle R_X^n, g_{\gamma(n+1)} \right\rangle \tag{12}$$

Let's go from the residual at step n to the $n + 1$. Is orthogonal to $g_{\gamma(\mathrm{n})}$. In other words, the decomposition is carried out in step m, and its expression is:

$$\|X\|^2 = \sum_{n=0}^{m-1} \left\| \left\langle R_X^n, g_{\gamma(n)} \right\rangle \right\|^2 + \left\| R_X^m \right\|^2 \tag{13}$$

Therefore, in the process of calculating the residual energy, each approximate process decreases [18].

3.3 Basic Principles of Gabor Atomic Library

Gabor atomic library is generated by time shift, scaling and modulation of Gaussian window. The real number expression is:

$$g_\gamma(t) = \frac{K_\gamma}{\sqrt{s}} g\left(\frac{t - \tau}{s}\right) \cos(\omega t + \phi) \tag{14}$$

where, $g(t) = 2^{1/4} e^{-\pi t^2}$ is gaussian window function; $\gamma = (s, \tau, \omega, \phi)$ is the parameter set for Gabor. The expression of K_γ is as follows:

$$K_\gamma = 1 \Big/ \sqrt{\left\langle g'_\gamma(t), g'_\gamma(t) \right\rangle} \tag{15}$$

where K is the normalized coefficient of $\|g_\gamma(t)\| = 1$.

The time-frequency information inside the atom is discretized [19], and its discretization expression is:

$$\gamma = (s, \tau, \omega, \phi) = \left(2^j, pa^j \frac{1}{2}, ka^{-j}\omega, i\frac{\pi}{6}\right) \tag{16}$$

where, $0 < j \le \log_2 N$, $0 \le p \le N 2^{j+1}$, $0 < i \le 12$.

3.4 Construction of Power Quality Disturbed Atomic Library and Calculation of Fitness Value

In dc system, how to complete the construction of disturbed atomic library is the most important step to detect dc signal, and also the first step of PSO-MP algorithm. PSO algorithm calculates the fitness value of each atomic library (dc like atomic library, dynamic atomic library and ripple atomic library) to judge the matching degree of particles. This section mainly analyzes the basic idea of PSO algorithm, extraction of DC components, the atomic library of DC-like, dynamic DC and ripple, and parameter setting.

Basic Idea of PSO Algorithm. PSO is an intelligent optimization algorithm, which is inspired by the foraging behavior of birds [19]. The D-dimensional space constitutes a population x of n particles. The ith particle in the d-dimensional solution space can be expressed as: $x_i = (x_{i1}, x_{i2}, \ldots, x_{iD})^T$; $p_i = (p_{i1}, p_{i2}, \ldots, p_{iD})^T$ represents the individual extreme value; $v_i = (v_{i1}, v_{i2}, \ldots, v_{iD})^T$ represents the velocity of the ith particle; $p_g = (p_{g1}, p_{g2}, \ldots, p_{gD})^T$ represents the population extremum of the population.

During each iteration, the particle updates its speed and position through individual extremum and group extremum, i.e.,

$$v_{id}^{k+1} = \omega v_{id}^{k} + c_1 r_1 \left(p_{id}^{k} - x_{id}^{k}\right) + c_2 r_2 \left(p_{gd}^{k} - x_{id}^{k}\right) \tag{17}$$

$$x_{id}^{k+1} = x_{id}^{k} + v_{k+1_{id}} \tag{18}$$

where, C_1 and C_2 are acceleration factors; V_{id} is the particle speed; Omega for inertia weight; Random Numbers r_1 and r_2 are distributed between [0, 1]. In the operation process of the algorithm, the particle speed may be too fast, leading to skipping the optimal solution and limiting its position and speed vid to the interval $[-v_{\max}, v_{\max}]$.

Extraction of DC Components. The phase, frequency and reactive power of the electrical signal are extracted and separated during the disturbance detection of dc energy mass, so as to obtain the residual signal with disturbance component in the original signal.

Suppose the original DC signal to be analyzed is $f(t)$. Since there is no frequency in the DC component signal, So the frequency of the DC component in the Gabor atom is 0, the sampling time is s, the sampling frequency is f_s, and the number of sampling points is N. The DC atomic library is constructed, and atom $g_{1\gamma}(t)$ is defined as

$$g_{1\gamma}(t) = K_{\gamma 1} \cos(\phi_1) \tag{19}$$

where, $K_{\gamma 1}$ is the normalized coefficient of $\|g_{1\gamma}(t)\| = 1$; ϕ_1 is DC component extracted for the matching of atoms. The ϕ_1 discretization, expression is:

$$\phi_1 = \frac{2\pi}{N} \varphi_1,\ \varphi_1 = [0, N-1] \tag{20}$$

The specific diagnostic parameters ϕ_1 of each DC atomic weight are determined according to the different values obtained from Eq. (20). Let,$g'_{1\gamma}(t) = \cos(\phi_1)$ the expression is:

$$K_{\gamma 1} = 1 \Big/ \sqrt{\left\langle g'_{1\gamma}(t), g'_{1\gamma}(t) \right\rangle} \tag{21}$$

The amplitude of the basic DC component can be expressed as:

$$A_1 = \langle f(t), g_{1\gamma}(t) \rangle K_{\gamma 1} \tag{22}$$

The inner product value of DC atoms in the original signal $< f(t),\ g_1(t) >$ are calculated until the DC atoms corresponding to the maximum inner product value are determined as matching DC atoms. When matching DC atoms are determined, the extracted DC component is:

$$f_1(t) = \langle f(t), g_{1\gamma}(t) \rangle g_{1\gamma}(t) \tag{23}$$

When the DC component is separated from the original signal, the residual signal can be obtained, and the expression is as follows:

$$r(t) = f(t) - f_1(t) \tag{24}$$

Construction and Fitness Value Calculation of DC-Like Atomic Library. Voltage sag can be classified as dc-like disturbance in DC distribution system. Atoms in the atomic-like library are defined as:

$$g_{2\gamma}(t) = K_{\gamma 2} \cos(\phi_2) e^{-\rho_2 (t - t_{s2})} [u(t - t_{s2}) - u(t - t_{e2})] \tag{25}$$

where, $K_{\gamma 2}$ is the normalized coefficient of $\|g_{2\gamma}(t)\| = 1$; ρ_2 is attenuation coefficient; ϕ_2 is the initial phase; t_{s2} is the start and end time of disturbance; t_{e2} is the end time of disturbance; $u(t)$ is the step function; $l = [\rho_{2,}\ t_{s2,}\ t_{e2}]$ is a characteristic parameter of a DC-like atomic library. ϕ_2 is discretization can be expressed as:

$$\phi_2 = \frac{2\pi}{N} \varphi_2 \qquad \varphi_2 \in [0, N - 1] \tag{26}$$

Each characteristic parameter is discretized, and the expression is:

$$\rho_2 = f_s \frac{1}{N} r \qquad r \in [0, N - 1] \tag{27}$$

After discretization of t_{s2} and t_{e2}, $t_{s2} = n_{s2}/f_s$ and $t_{e2} = n_{e2}/f_s$ are obtained.

The amplitude of the DC-like atom is expressed as:

$$A_2 = \langle r(t), g_{2\gamma}(t) \rangle K_{\gamma 2} \tag{28}$$

where, $r(t)$ is the disturbance signal in the presence of DC-like atoms.

The fitness value of the DC-like atomic library was calculated, and the acceleration factor was set as $c_1 = c_2 = 2$, and the number of atomic library variables was set as $N = 3$. Take ripple superposition signal (without noise) as an example. Through simulation analysis, the iterative convergence curve of dc like atomic library is shown in Fig. 3.

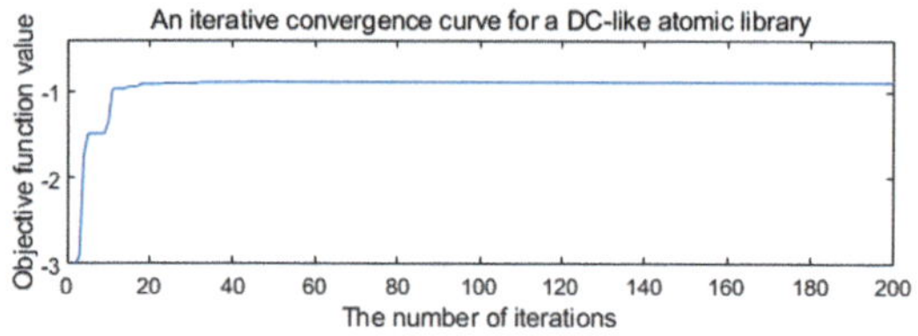

Fig. 3. An iterative convergence curve for a DC-like atomic library.

Construction of Dynamic DC Atomic Library and Calculation of Fitness Value. In a DC distribution system, the atom of dynamic voltage sag is defined as:

$$g_{3\gamma}(t) = K_{\gamma 3}\left(e^{-\rho_3'(t-t_{s3})} - e^{-\rho_3''(t-t_{s3})}\right)[u(t - t_{s3}) - u(t - t_{e3})] \tag{29}$$

where, $K_{\gamma 3}$ is the normalized coefficient of $\|g_{3\gamma}(t)\| = 1$; ρ_3' is the recovery speed of ascent; ρ_3'' is the recovery rate of descent; t_{s3} is the start time of disturbance; t_{e3} is the end time of disturbance; $u(t)$ is the step function; $s = [\rho_3', \rho_3'', t_{s3}, t_{e3}]$ is the characteristic parameter of dynamic DC voltage atomic library.

The amplitude expression of dynamic DC atom is as follows:

$$A_3 = \langle s(t), g_{3\gamma}(t) \rangle K_{\gamma 3} \tag{30}$$

where, $s(t)$ is the residual component of the dynamic DC atom.

The fitness value of dynamic DC atomic library was calculated, and the acceleration factor was set as $c_1 = c_2 = 2$, and the number of atomic library variables was set as $N = 4$. Taking ripple superposition signal (without noise) as an example, through simulation analysis, the iterative convergence curve of dynamic DC atomic library is shown in Fig. 4.

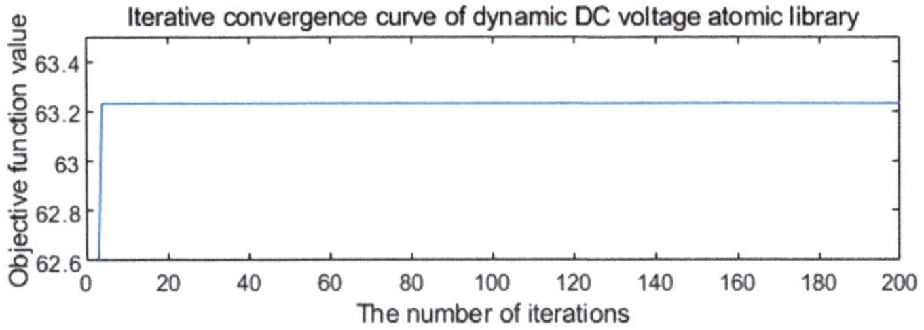

Fig. 4. Dynamic DC atomic library iterative convergence curve.

Construction of Ripple Atomic Library and Calculation of Fitness Value. In a DC distribution system, the voltage ripple is defined as:

$$g_{4\gamma}(t) = K_{\gamma 4}\cos(\omega_4 t + \phi_4)e^{-\rho_4(t-t_{s4})}[u(t - t_{s4}) - u(t - t_{e4})] \tag{31}$$

where, $K_{\gamma 4}$ is the normalized coefficient of $\|g_{4\gamma}(t)\| = 1$; ω_4 is the ripple Angle frequency; ϕ_4 is the initial phase of AC; t_{s4} is the start time of disturbance; t_{e4} is the end time of disturbance; $r = [\omega_{4,}\ \varphi_{4,}\ t_{s4}\ t_{e4}]$ was the characteristic parameter of the ripple atomic library. $u(t)$ is the step function; ρ_4 is attenuation coefficient.

ω_4 is discretized, and the expression is:

$$\omega_4 = \frac{2\pi}{N}\varphi_4 \qquad \varphi_4 \in [1, N] \tag{32}$$

By discretizing the initial phase ϕ_4, the expression is:

$$\phi_4 = \frac{2\pi}{N}\varphi \qquad \varphi \in [0, N-1] \tag{33}$$

where, N is the sampling point of ripple signal.

The amplitude of ripple atoms can be expressed as:

$$A_4 = \langle l(t), g_{4\gamma}(t)\rangle K_{\gamma 4} \tag{34}$$

To calculate the fitness value of ripple atomic library, set the acceleration factor as $c_1 = c_2 = 2$ and the number of atomic library variables $N = 5$. Take ripple superposition signal (without noise) as an example. Through simulation analysis, the iterative convergence curve of ripple atomic library is shown in Fig. 5.

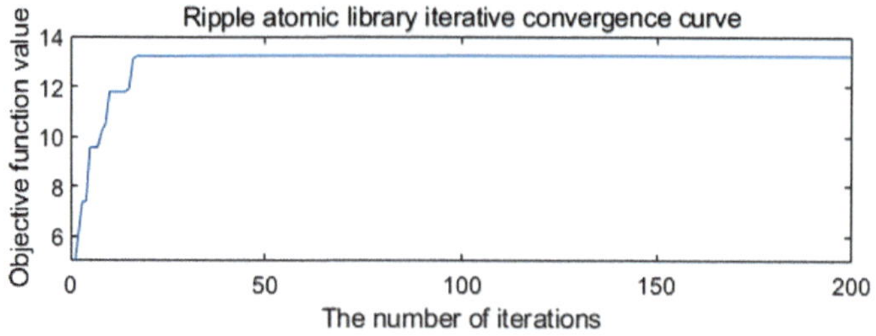

Fig. 5. Ripple atomic library iterative convergence curve.

3.5 Power Quality Disturbance Detection Based on PSO-MP

The specific steps of PSO-MP algorithm include calculation of residual energy, PSO initialization, fitness evaluation, particle update and MP algorithm fine search. The specific steps are as follows:

(1) Calculation of residual energy. MP algorithm was used to match the best atomic library and number the remaining signals. Then the characteristic parameters and residual energy of the initial test were calculated to extract the DC component of the DC signal.
(2) PSO initialization. According to the actual requirements of this paper, parameters were set, the increment constant $c_1 = c_2 = 1$, the maximum inertia factor $\omega_{\max} = 0.9$, the minimum inertia factor $\omega_{\min} = 0.3$, the population size M was 20, the number of iterations was 60, the maximum and minimum values of velocity were 0.8 and -1 respectively, the maximum and minimum values of individual were 6 and 2. At the same time the population is initialized and the initial extremum is searched and iterative optimization is completed.
(3) Fitness evaluation. The fitness values of dc atomic library $\langle r(t), g_{2\gamma}(t) \rangle$, dynamic atomic library $\langle s(t), g_{3\gamma}(t) \rangle$ and ripple atomic library $\langle l(t), g_{4\gamma}(t) \rangle$ in D-dimensional space were calculated through characteristic parameters, parameter setting, population initialization, initial extremum search and iterative optimization. FFT was used to calculate the maximum frequency existing in the DC component residual signal, and the maximum frequency was set to 1.2 times the standard frequency.
(4) Particle update. The best matching particle is obtained through rough search of PSO algorithm, and the particle velocity and position are adjusted by using Eqs. (17) and (18). Characterize the matching feature parameters as ω_{best}, φ_{best}, ρ_{best}, $t_{s\text{best}}$ and $t_{e\text{best}}$. The matching feature parameters were taken into the whole treatment, and $[\omega_{best}]$, $[\varphi_{best}]$, $[\rho_{\text{best}}]$, $[t_{s\text{best}}]$, $[t_{e\text{best}}]$ were obtained. Finally, these feature parameters were discretized to complete the decomposition.

(5) MP algorithm fine search. After the PSO algorithm completes the rough search, MP algorithm completes a set of parameter matching, calculates the residual component and energy of the row according to the relevant information of the characteristic quantity, and then provides other disturbance components through PSO to extract the dc signal disturbance characteristic parameters.
(6) Repeat steps (2)~(5) above until the specified iteration termination condition is met.

4 The Example Analysis

The DC voltage simulation model parameters collected by the controller inside the device are shown in Table 1. To real easy to operate, set the ac grid voltage amplitude value is 253 V, through the voltage sampling circuit and signal conditioning circuit output to 1 V, dc voltage controller internal A/D converter sampling frequency is set to 10 kHz, each cycle collects 200 points, the sampling interval is 100 us, sampling time is 0.4 s, add 40 dB gaussian white noise in dc original disturbance signal.

Table 1. DC voltage simulation model parameters.

Related parameters	The numerical
Dc voltage	1 V
Ac grid voltage	253 V
Support the capacitance	3500 uF
Equivalent inductance	6 nH
Equivalent resistance	0.3 Ω
Filtering inductance	5 mH

4.1 Simulation Analysis of DC Voltage Ripple Superposition

Based on the characteristics of a single voltage ripple signal, a two-frequency ripple superposition signal is constructed in MATLAB simulation software, and the disturbance signal is (Fig 6):

$$f(t) = 1 + A_1 \cos(100\pi t) + A_2 \cos\left(300\pi t + \frac{\pi}{6}\right) + A_3 \cos\left(500\pi t + \frac{\pi}{3}\right) \tag{35}$$

where, $A_1 = \begin{cases} 0.12, & 0.12s \le t \le 0.24s \\ 0, & \text{t for other values} \end{cases}$, $A_2 = \begin{cases} 0.03, & 0.02s \le t \le 0.25s \\ 0, & \text{t for other values} \end{cases}$,

$A_3 = \begin{cases} 0.08 & 0.04s \le t \le 0.16 \\ 0, & \text{t for other values} \end{cases}$.

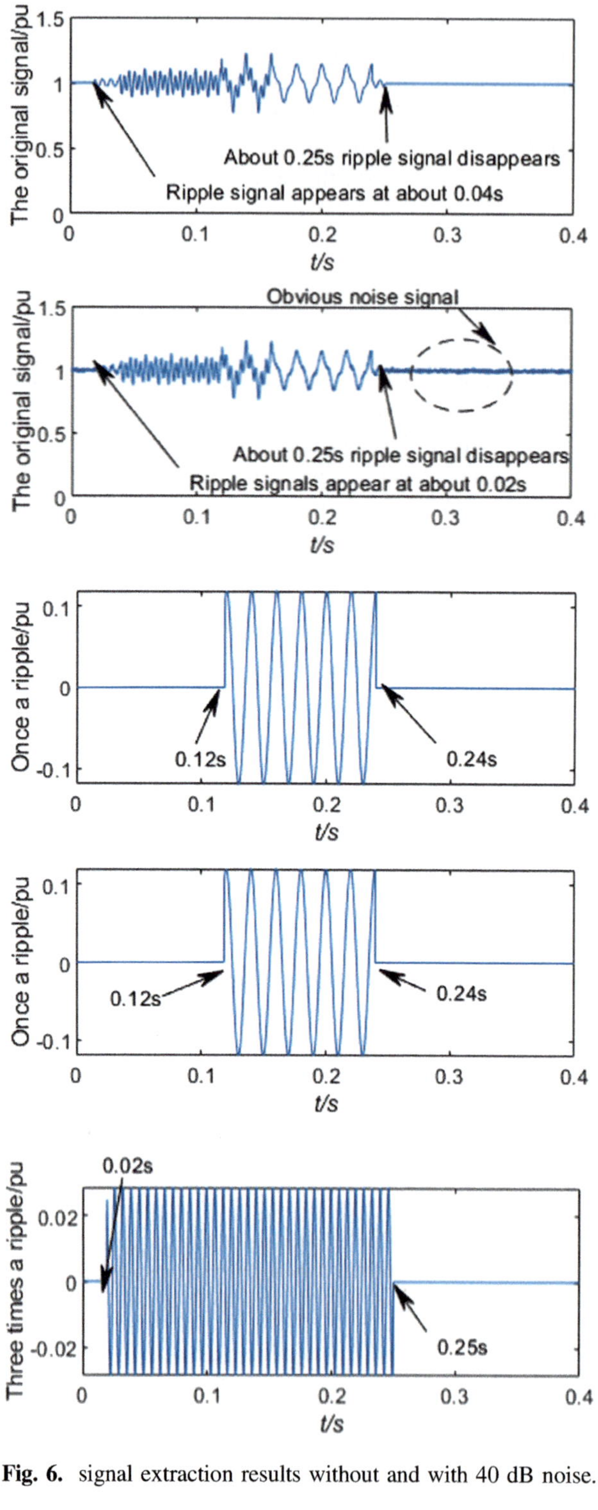

Fig. 6. signal extraction results without and with 40 dB noise.

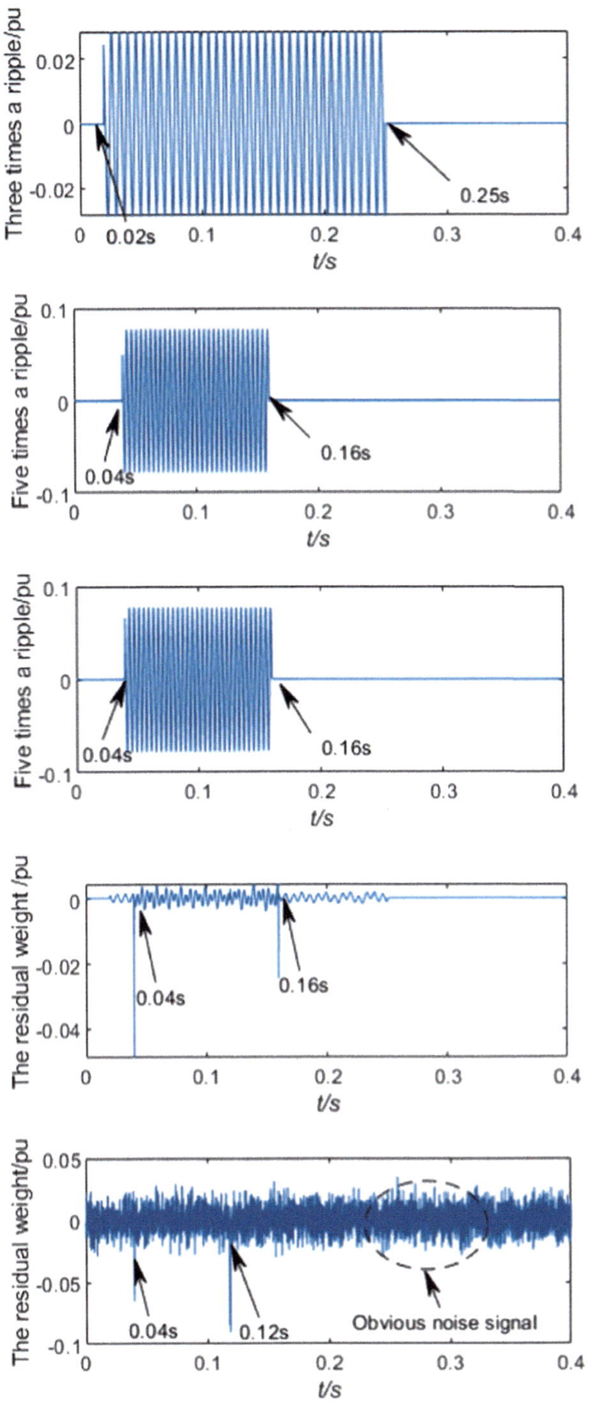

(a) Simulation results without noise. (b) Simulation results with 40 dB noise.

Fig. 6. (*continued*)

Table 2. PSO-MP algorithm's detection results of two-frequency ripple superposition without noise.

Signal types	Basic DC	Once a ripple	First ripple error rate	Three times a ripple	Tertiary ripple error rate	Five times a ripple	Fifth ripple error rate
V	0.9999	0.1196	−0.2929	0.0287	−4.3971	0.078	−2.4771
f//Hz	0	49.9830	−0.0341	150.0013	0.0009	250.0596	0.0238
ϕ/rad	–	0.0185	–	0.5180	−1.06	1.0082	−3.7235
ρ	–	0	–	0	–	0.023	–
load-point	0	1187	–	200	–	398	–
End point	4000	2400	–	2501	–	1604	–
t_s/s	0	0.1187	−1.0895	0.02	−0.0205	0.0398	−0.4411
t_e/s	0.4	0.24	−0.0016	0.2501	0.0343	0.1604	0.2343

Table 3. PSO-MP algorithm detects the superposition of two frequency ripple with 40 dB noise.

Signal types	Basic DC	Once a ripple	First ripple error rate	Three times a ripple	Tertiary ripple error rate	Five times a ripple	Fifth ripple error rate
V	0.9998	0.1198	−0.1959	0.0283	−5.5284	0.0781	−2.3263
f/Hz	0	49.9983	−0.0034	150.0102	0.0068	250.0483	0.0175
ϕ/rad	–	0	–	0.5147	−1.7038	1.011	−3.456
ρ	–	0	–	0	–	0.0793	–
load-point	0	1185	–	200	–	396	–
End point	4000	2399	–	2501	–	1604	–
t_s/s	0	0.1185	−1.2802	0.02	−0.1189	0.0396	−0.9657
t_e/s	0.4	0.2399	−0.0212	0.2501	0.0222	0.1604	0.2288

From the disturbance signal expression, we can see that the amplitude of the original DC signal collected by the controller is 1 V, the amplitude of the first ripple is 0.12 V, the amplitude of the third ripple is 0.03 V, and the amplitude of the fifth ripple is 0.08 V. The detection results of PSO-MP algorithm for two-frequency ripple superposition signals without noise are shown in Table 2, and the detection results of PSO-MP algorithm for two-frequency ripple superposition signals with 40 dB noise are shown in Table 3.

By Table 2 shows that on the basis of not join gaussian white noise, found in the case of atomic library has a good applicability, the error of each characteristic parameters are small, at the same time, the signal component test results of small amplitude can meet the requirements, in each time period can detect the voltage fault, conform to the requirements of the transformation of dc double the power switch.

It can be seen from Table 3 that when 40 dB Gaussian white noise is added to the original signal, the detection error of characteristic parameters directly increases, but the error can still be determined within the allowed range by the data in the table. If the time interval of each component of the signal can be directly superimposed by ripple, meanwhile, with the accumulation of Gaussian white noise, the final residual component will also appear obvious fluctuation. Even in the gaussian white noise environment, the algorithm can meet the requirement of dc dual power switch.

4.2 Simulation Analysis of DC Voltage Sag and Ripple Superposition

Based on the characteristics of a single voltage sag signal, a voltage sag and ripple superposition signal is constructed in MATLAB simulation software, and the disturbance signal is:

$$f(t) = 1 + A_1 + A_2 \cos\left(300\pi t + \frac{\pi}{6}\right) + A_3 \cos(500\pi t) \quad (36)$$

where, $A_1 = \begin{cases} -0.2 & 0.01s \le t \le 0.18s \\ 0, & \text{t for other values} \end{cases}$, $A_2 = \begin{cases} -0.1 & 0.02s \le t \le 0.2s \\ 0, & \text{t for other values} \end{cases}$, $A_3 = \begin{cases} 0.08 & 0.04s \le t \le 0.16s \\ 0, & \text{t for other values} \end{cases}$.

From the disturbance signal expression, we can see that the amplitude of the original DC signal collected by the controller is 1 V, the amplitude of voltage sag is B1 is −0.2 V, the amplitude of the third ripple B3 is −0.1 V, and the amplitude of the fifth ripple is B5 is 0.08 V. Then, the initial residual components of multi-frequency DC ripple signals are extracted according to the DC-like atomic library. The detection results of PSO-MP algorithm for voltage sag and ripple superposition without noise are shown in Table 4, and the detection results of PSO-MP algorithm for voltage sag and ripple superposition signal with 40 dB noise are shown in Table 5.

Table 4. PSO-MP algorithm detects the results of voltage sag and ripple stack without noise.

Signal types	Basic DC	Voltage sag	Transient error rate	Three times a ripple	Tertiary ripple error rate	Five times a ripple	Fifth ripple error rate
V	0.9998	0.1196	−0.2235	0.1012	1.1949	0.08	−0.0356
f//Hz	0	0	–	149.9678	−0.0215	250	0
ϕ/rad	–	0	–	0.5433	4.1464	0	–
ρ	–	0	–	0.1218	–	0	–
load-point	0	99	–	200	–	399	–
End point	4000	1800	–	2000	–	1601	–
t_s/s	0	0.0099	−0.9137	0.02	−0.0146	0.0399	−0.1916
t_e/s	0.4	0.18	−0.0169	0.2	0.0001	0.1601	0.0442

Table 5. PSO-MP algorithm detects dc voltage sag and ripple stack with 40 dB noise.

Signal types	Basic DC	Voltage sag	Transient error rate	Three times a ripple	Tertiary ripple error rate	Five times a ripple	Fifth ripple error rate
V	0.9998	0.1198	−0.0909	0.1034	3.4051	0.0806	0.7224
f//Hz	0	0	−0.0034	149.9734	−0.0178	249.998	−0.0008
ϕ/rad	–	0	–	0.5447	4.0215	0	–
ρ	–	0	–	0.4625	–	0.0701	–
load-point	0	99	–	200	–	400	–
End point	4000	1795	–	2000	–	1601	–
t_s/s	0	0.0099	−0.5739	0.02	-0.0663	0.04	−0.0462
t_e/s	0.4	0.1795	−0.2617	0.2	0.0136	0.1601	0.0517

Table 4 shows that the decomposition results between the two atomic libraries can meet the requirements of detection accuracy on the premise that no Gaussian white noise is added to the original signal. The amplitude of a single voltage sag was calculated by simulation to be 0.995 V, close to 1 V. The error of characteristic parameters was small, and the modified amplitude was 0.9995 × cos(0) = 0.9995 V. The error of characteristic parameters was small, and the atomic library belonging to the detected disturbance component could be correctly identified. Take the common power supply as an example, the instantaneous voltage sag occurs at 0.01 s-0.18 s, the transient amplitude of which is 1.2 V, and the ripple signal occurs between 0.02 s-0.2 s, making the dual-power automatic transfer switch from the common power supply to the standby power supply. Through the analysis of the examples, the voltage faults in the dc dual power system can be detected, and the conversion between the two dc power sources can be better realized Fig 7.

It can be seen from Table 5 that, compared with the detection results of signals without Gaussian white noise, the error percentages of the third and fifth ripple waves are 3.4051% and 0.7224%, respectively, which are both greater than 1.1949% and-0.0356%. When the signal with Gaussian white noise is added, its amplitude is 0.9998 × cos(0) = 0.9998 V, which still meets the accuracy requirements. Even in the gaussian white noise environment, the transient and ripple signals can be detected at 0.01 s-0.18 s and 0.02 s-0.2 s, meeting the conversion requirements of the DC dual-power automatic transfer switch.

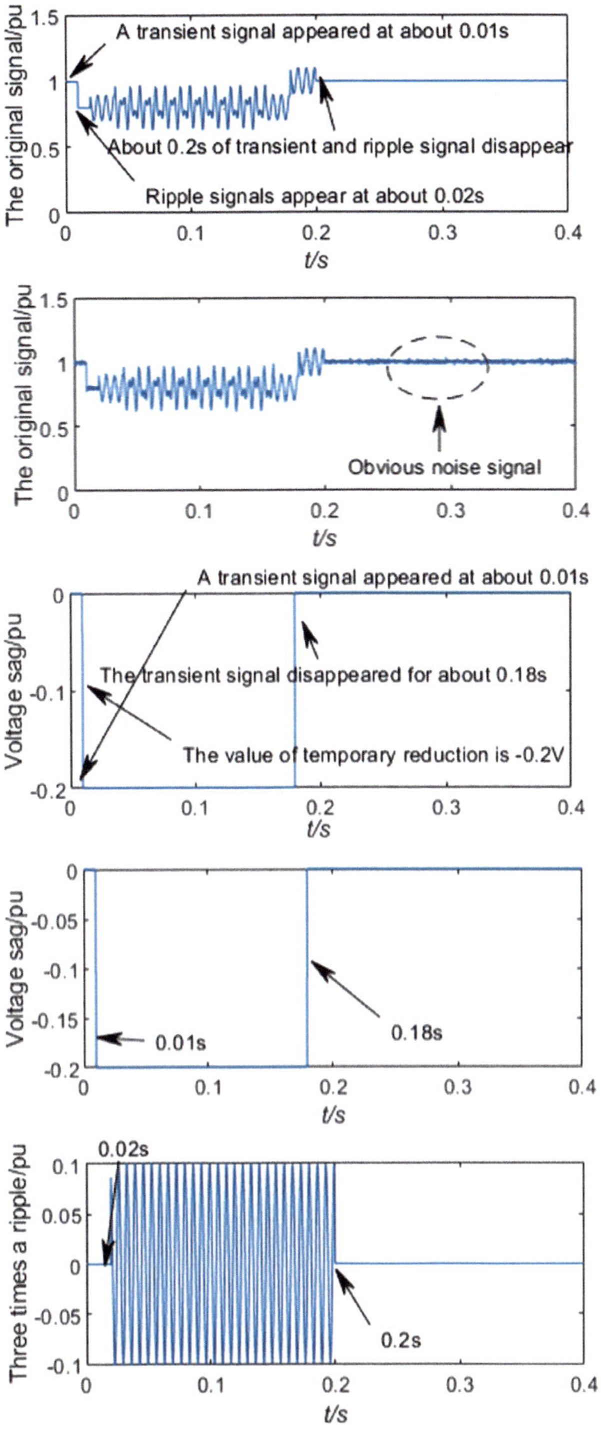

Fig. 7. Signal extraction results without noise and with 40 dB noise.

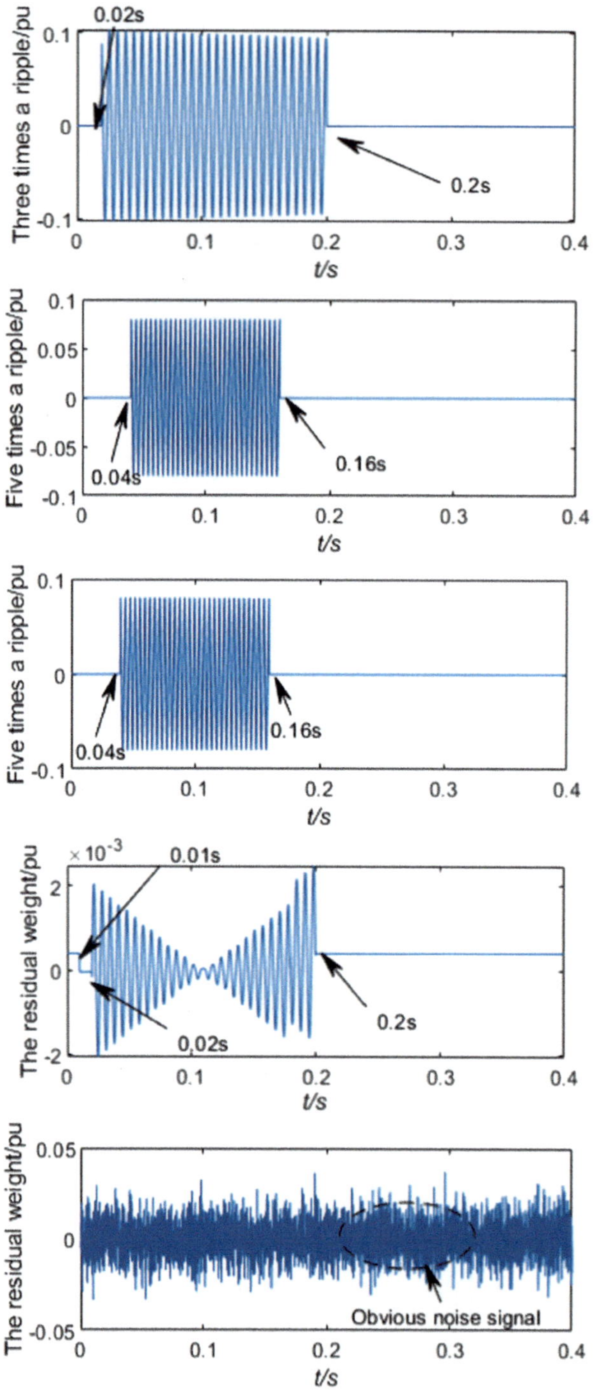

(a) Simulation results without noise (b) Simulation results with 40dB noise

Fig. 7. (*continued*)

5 Conclusion

Dual power transfer control technology is the key of DC dual power transfer switch. In this paper, a dual-power conversion control technology based on pSO-MP is proposed. The main conclusions are as follows:

(1) In the dc dual power transfer switch, when one power supply fails, the PSO-MP algorithm can be used as the starting criterion of switch transfer to quickly realize the automatic transfer between two DC power sources and improve the reliability of the dual power automatic transfer switch.

(2) Most of the DC energy quality disturbance detection algorithms have problems such as low detection accuracy and large calculation amount, while PSO-MP algorithm can better detect the complex power quality (such as two-frequency ripple superposition, transient and ripple superposition) disturbance in the DC power supply system, which not only has fast detection speed but also small calculation amount.

The effectiveness and accuracy of the algorithm are verified by example analysis. At the same time, the algorithm can ensure a certain precision in the case of Gaussian white noise, and has strong anti-interference ability to noise, which has a wide range of engineering application value.

References

1. Xiao, X.N., et al.: Power Quality Analysis And Control. China Electric Power Press, Beijing (2010)
2. Jiang, D.Z., Zheng, H.: Research status and prospect of dc distribution network. Autom. Power Syst. **36**(08), 98–104 (2012)
3. Song, Q., Zhao, B., Liu, W.H., Zeng, R.: Research review of intelligent dc distribution network. Chinese J. Electr. Eng. **33**(25), 9–19+5 (2013)
4. Tian, B., et al.: 400 V/1000 kVA hybrid automatic transfer switch. IEEE Trans. Ind. Electron. **60**(12), 5422–5435 (2013)
5. Li, H.S.: Several fourier algorithms for filtering the influence of attenuation aperiodic components. Modern Electron. Technol. **08**, 87–88 + 100 (2005)
6. Zhao, L.Q., Long, Y.: Power quality composite disturbance classification based on improved SVM. New Technol. Electr. Energy **35**(10), 63–68 (2016)
7. Wang, Z.F., Yang, X., Pan, A.Q., Chen, T.T., Xie, Z.Z.: Voltage deviation prediction based on improved integrated clustering and bp neural network. New Technol. Electr. Energy **37** (05), 73–80 (2016)
8. Qu, Z.W., Hao, W.R., Wang, N.: Application of fast atomic decomposition algorithm in power quality disturbance analysis. Power Autom. Equipment **35**(10), 145–150 (2015)
9. Zhang, Y.J., Gong, Q.W., Li, X., Guan, Q.Y., Jia, J.J., Wang, S.L.: Application of atomic decomposition method based on PSO in inter-harmonic analysis. Power Syst. Protection Control **41**(15), 41–48 (2013)
10. Cui, Z.Q., Wang, N., Jia, Q.Q.: Identification method of power quality composite disturbance parameters based on layered matching tracking algorithm. Power Autom. Equipment **37**(03), 153–159 (2017)
11. Zhu, X.L., Gao, Y.H., Xie, X.Y., Jiao, J.R., Jia, Q.Q.: Power quality disturbance detection method for DC power distribution network based on relevant atomic library. J. Yanshan Univ. **42**(05), 402–408+421 (2016)

12. Jiao, J.R.: Power quality analysis and disturbance detection of dc distribution network. Yanshan University (2017)
13. Li, Y.F., Xie, X.Y., Jiao, J.R., Zhao, T.J., Huang, L.X., Peng, Y.T., Wang X.B.: Detection method and device for DIRECT current energy mass disturbance. Hebei: CN108287279A (2018)
14. Mariscotti, A.: Methods for ripple index evaluation in DC low voltage distribution networks. In: IEEE Instrumentation and Measurement Technology Conference, Warsaw, pp. 1–4 (2007)
15. Liao, J.Q., Zhou, N.C., Wang, Q.G., Li, C.Y., Yang, J.: Definition and correlation analysis of power quality index of DC distribution network. Chinese J. Electr. Eng. **38**(23), 6847–6860 + 7119 (2016)
16. Li, L.L., Yong, J., Liang, S.B., Tian, Q.S., Zeng, L.Q.: Overview of protection of civil low-voltage dc power supply system. J. Electr. Technol. **30**(22), 133–143 (2015)
17. Li, J.: High and Low Voltage Electrical Appliances and Design. Machinery Industry Press, Beijing (2016)
18. Lovisolo, L., da Silva, E.A.B., Rodrigues, M.A.M., Diniz, P.S.R.: Coherent decompositions of power systems signals using damped sinusoids with applications to denoising. In: IEEE International Symposium on Circuits and Systems. Proceedings (Cat. No. 02CH37353), Phoenix-Scottsdale, AZ, USA, V–V (2002)
19. Chen, J.H., Li, X.Y., Deng, D.H., Liao, D.L.: Application review of particle swarm optimization algorithm in power system. Relay **23**, 77–84 (2007)

Design of Lithium Battery Management System for Underwater Robot

Baoping Wang[1], Qin Sun[1], Dong Zhang[2(✉)], and Yuzhen Gong[1]

[1] School of Construction Machinery, Shandong Jiaotong University, Jinan 250357, China
[2] Institute of Automation, Shandong Academy of Sciences, Jinan 250000, China
Jnzd156@163.com

Abstract. At present, the research of the power lithium battery management system which was applied to the underwater robot is still in the initial stage. The lithium battery management has become one of the main contents of the development of the underwater robot technology. When the use time of rechargeable battery is short, various factors will reduce the life of the battery. Due to the problem of materials, the price of the battery is limited. Effective management of lithium battery can improve the use efficiency of the battery. In addition, the safe and effective use of electricity has great significance for the energy of the pool, which can extend the battery life and improve the reliability of the battery. The main functions of the battery management system were introduced in this paper. The hardware management system, temperature sampling, temperature detection circuit and charging module of the lithium battery were mainly designed.

Keywords: Underwater robot · Lithium Battery Management System (BMS) · AUR (Autonomous Underwater Robot)

1 Introduction

For AUR, energy system plays an important role in long-term, deep-sea and complex environment detection. Energy system is an important part of AUR. Power quality directly affects the performance of the whole system and the service life of hardware. As the main form of power driven AUR, battery is an important part of the energy supply of AUR. The selection of battery needs to be considered in many aspects. In addition to energy density and energy/volume ratio, utilization efficiency, ambient temperature, noise, gas leakage during charging and discharging, safety performance of water closed environment, environmental pollution and other factors should also be fully considered. Power management system is an important part of AUR energy system, which is responsible for real-time monitoring and management of AUR energy. In order to construct the AUR precise integrated positioning system in the deep-sea complex environment, multi-sensor device is used [1].

The research of the power lithium battery management system applied to the underwater robot is still in the initial stage, and the lithium battery problem has become

Q. Liu et al. (Eds.): CENet 2020, AISC 1274, pp. 989–995, 2021.
https://doi.org/10.1007/978-981-15-8462-6_113

one of the main reasons limiting the development of the underwater robot technology. When the use time of the rechargeable battery is short, it affects the life of the battery cycle and effectively manages the lithium battery, which is of great significance for improving the use efficiency of the battery. In addition, the safe and effective use of electricity has great significance for the energy of the pool, which can extend the battery life and improve the reliability of the battery. The lithium battery management system of the underwater robot was designed.

2 Main Functions O Power Management System

The structure diagram of power management system is shown in Fig. 1 [2]. It is mainly divided into data collection module, switch control module, power calculation module and emergency processing module. The power management system collects analog input (voltage, current, temperature, etc.) from the monitoring node, calculates the electric quantity, receives the command from the industrial computer via the RS232 interface, preprocesses the data, and controls the switch value (on/off state of the relay, etc.). Then, in case of failure, the emergency processing module is called.

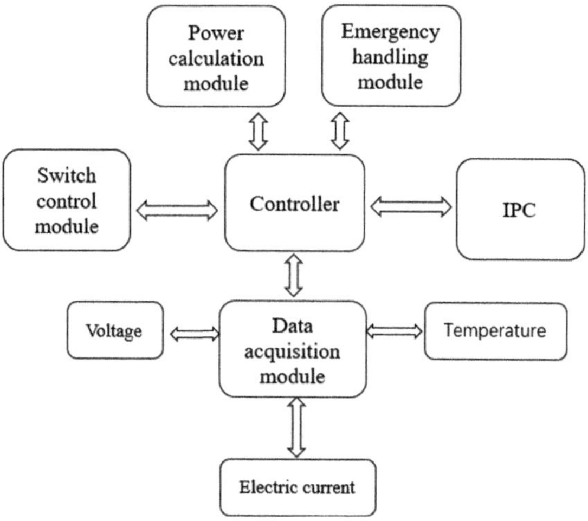

Fig. 1. Schematic diagram of power management system.

Data acquisition module can acquire and preprocess the output voltage, current, temperature around the battery pack and high-power equipment in the AUR studio, and provide the basis for system state analysis [3].

Switch control module can control relay switch simply by driving circuit. In case of short circuit and other faults of one device, switch control module can manage the power supply of each power consumption device without affecting the power supply of other devices.

For power calculation module, the average current per second is the calculation unit of the module. The remaining time of the system can be calculated according to the total battery power, consumed power and current working current, and the operation of the whole system can be referred to.

When the battery power supply is insufficient, the emergency processing module informs the industrial computer to switch over the standby power supply in a short time. The corresponding devices are shut down according to the instructions of the industrial computer to reduce the power consumption. A float is prepared at the same time. When the total current is over-current, the system closes all channels and enters the accident handling mode. In the accident handling mode, the power module combines the current and temperature values of each test point to quickly detect the corresponding fault channel, detect and close the fault channel, and report the status to the industrial computer [4].

3 Design of Lithium Battery Management System

3.1 Hardware Design of Battery Management System

Due to the underwater working environment, the whole system needs to be stored in the waterproof box when designing the battery management system. In addition, the size of the actual underwater robot also needs to be considered. The storage space of ROV will not be large in appearance. The general design scheme needs to meet the reasonable and simple structural requirements, and do not occupy the invalid space. The system has perfect functions and ROV has specific post expansion to ensure safe and long-term operation in water.

The following three basic functions were studied.

(1) Real time collection of battery voltage, charging and discharging current, and working temperature.
(2) Real-time display of battery operation status is provided by user friendly computer interface.
(3) RS232 serial communication function can be connected to PC for hardware debugging of later maintenance.

Figure 2 is the structure diagram of BMS system. It can be seen from the figure that the basic functions of BMS data acquisition, SOC estimation, energy balance, etc. can be realized through the central processing device.

After the system is powered on, the system needs to be initialized first before starting the AUR. At this time, when the devices are closed, the data acquisition module checks the voltage of each node in turn and measures the leakage current. After the exception exceeds the set threshold value, the exception handling subroutine is called to find the fault point, start the buzzer alarm, and wait for the industrial control computer to decide [5].

If the whole power supply system of AUR is normal, instructions can be received from IPC through interruption after startup. Corresponding devices can be turned on according to IPC requirements, and devices which are not used for a long time should

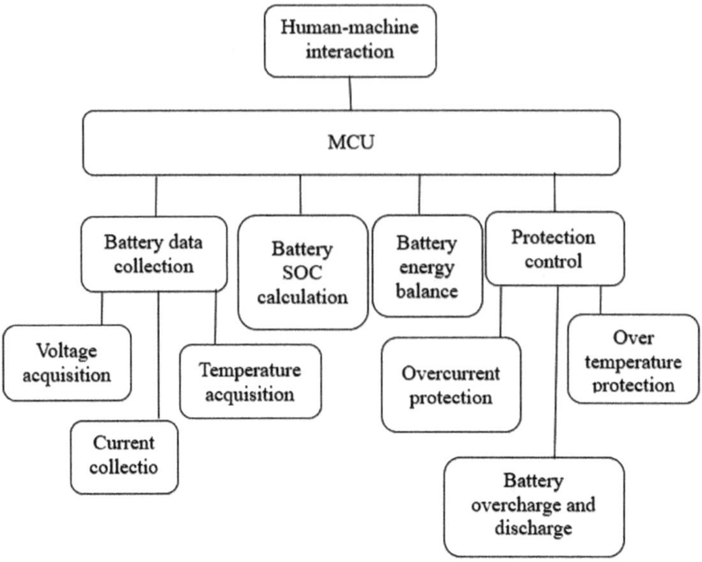

Fig. 2. BMS system structure.

be turned off. The low power consumption requirements of AUR as a whole can be met. The voltage, current and temperature of each test point are monitored in real time, and the subroutine of interrupt processing is also called the exception.

After the system is powered on, the system needs to be initialized first before starting the AUR. At this time, when the devices are closed, the data acquisition module checks the voltage of each node in turn and measures the leakage current. After the exception exceeds the set threshold value, the exception handling subroutine is called to find the fault point, start the buzzer alarm, and wait for the industrial control computer to decide.

If the whole power supply system of AUR is normal, instructions can be received from IPC through interruption after startup. Corresponding devices can be turned on according to IPC requirements, and devices which are not used for a long time should be turned off. As a result, the low power consumption requirements of AUR as a whole can be met. The voltage, current and temperature of each test point are monitored in real time, and the subroutine of interrupt processing is also called the exception.

3.2 Temperature Sampling

Battery temperature is a key parameter to measure whether the battery is in normal use. If the battery temperature exceeds a certain value, the battery will not be recovered. The temperature difference between batteries will lead to imbalance between batteries and shorten battery life. Temperature identification does not need A/D, and can output 9–12-word chip structure. All types of DS18B20 operations can be performed on the data line, and the data line can also be used on the power line. Several DS18B20 have the same functions, sometimes can be connected to a bus. DS18B20 interface design is flexible, including temperature sensor, 64 bit read memory interface, intermediate data high-speed register, TL trigger and redundant cycle test code generator, which is only limited to the low temperature of user group. EPROM is used to store user-defined high temperature alarm TL value and user-defined low temperature alarm TL value. The first internal configuration register can be used to determine the number of digital revolutions of the temperature value. When setting high resolution and long temperature data conversion time, the balance between resolution and conversion time should be considered in practical application. The first and second bytes are temperature information, the third and fourth bytes are TH and TL values, the sixth to eighth bytes are not used, and the ninth byte is CRC code to ensure correct communication. Eds18b20 stores the conversion temperature values of the first and second bytes of the temporary high-speed memory in the 16-bit extended binary addition. Calculate the corresponding temperature: if the sign bit = 0 h, convert the binary number to decimal number directly. If s = 1, first convert the compensation code to the original code [6].

3.3 Design and Operation Principle of Temperature Detection Circuit

The temperature detection system adopts direct power mode. DS18B20 needs to pull up the bus in light memory operation and temperature A/D conversion operation. The maximum pull-up start time is 10 μs. before the installation and operation of the system, the host connects 1 through DS18B20 and reads the serial number. The host sends pulses. Then, when the "O" level is greater than 480 μs, DS18B20 receives the sent response pulse, and the host 18 sends the read 18 command code 33H (the low level is the front side), then sends the pulse (15 μs), and reads the 1 digit of the DS18B20 serial number. Read the 56 bits of the serial number in the same way. The overall flow chart of DS18B20 action of the system is completed in three stages. The system obtains the DS18B20 serial number through repeated operations. For temperature A/D conversion, start all online DS18B20, and read the conversion temperature data of DS18B20 online. The host starts temperature conversion and reads the temperature value. The host writes memory data. In addition, if there are many detection points to measure the temperature, use other ports to expand [7]. The specific circuit diagram is shown in Fig. 3.

3.4 Charging Module Design

A It is necessary to connect a matching constant voltage source and a current limiting power supply bracket when charging a lithium-ion battery equipped with a battery management system. Constant voltage U is equal to U4.2 $\times$ N + loss voltage, where N is the number of cells. The traditional charging current limit of power lithium battery is 0.3C (C is the battery capacity). The initial setting must be carried out before charging. Automatic charging can be carried out in the three stages of pre charging, constant current charging and constant voltage charging.

(1) Initial Settings

The initialization phase is not the beginning of battery charging, but an important step in the whole charging process. In this stage, the intelligent energy management module determines whether it works normally and whether the charging conditions meet the charging requirements.

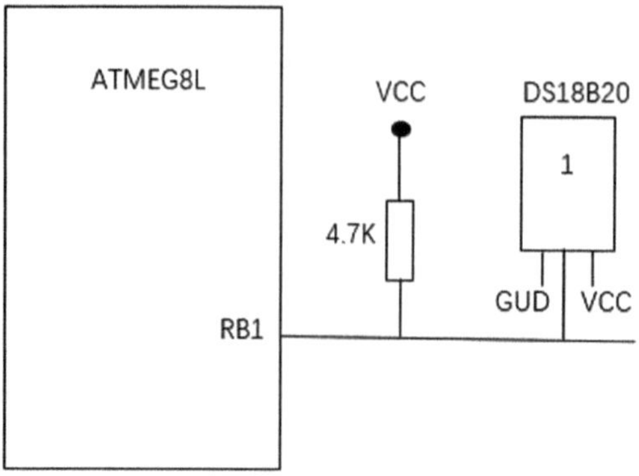

Fig. 3. Temperature detection circuit design.

1) Whether the anode value of external charge source is correct; 2) Whether external charging voltage is installed; 3) The temperature is within the allowable range; 4) Whether the clamping voltage of lithium-ion battery (single battery) may be higher than the minimum charging voltage; 5) Whether the terminal voltage of lithium-ion battery exceeds the battery control.

(2) Precharge

The purpose of precharge is to discharge the battery so as to remember whether the battery is too long or damaged. When the battery terminal voltage is lower than the allowable minimum charging voltage, a small current (usually about 1/10 of the charging current) must be used for precharge. The terminal voltage of lithium-ion battery rises above the minimum allowable charging voltage, which can be converted to the next charging sequence. The principle of precharge is to apply relatively small charging current (about 1/10 of the normal charging current) to the battery through the power adapter controlled by MCU, so that the battery under the minimum charging voltage can reach the minimum allowable charging voltage within a certain period of time. Therefore, a non-rechargeable battery can be used as a battery.

(3) Constant Current Charging

The battery management system requires that the external charging power supply for constant current charging of lithium battery is constant current, and its constant current value is less than the maximum allowable charging current of lithium-ion battery. When the battery power reaches 70%-80% of the total power, the charging time is 2–3 h. When the voltage of the battery unit reaches the set end voltage, the constant current charging ends, and the charging current sharply decreases, and enters the charging maintenance process.

(4) Continue Charging

The battery management system adopts the pulse charging mode in the maintenance charging stage. At this time, the charging circuit is closed by intermittently charging for a certain time t with the same current value as that during a certain charging period. According to the charging current, the battery voltage is higher than the ending voltage. After the charging circuit is cut off, the battery voltage drops again. When the battery voltage returns to the end of the charging voltage, the charging circuit is opened again, the battery is always charged with a certain current value, waiting for the charging circuit to close, and the battery voltage drops. Under the action of pulse charging current, the battery gradually meets the requirements. The voltage at the battery terminals drops slowly. This process continues until the battery voltage recovery time reaches the set value. The batteries are almost full. High power electronic switches are used to protect all types and operating states of the system. Because the use of high power and low on resistance MOSFET includes charge circuit and discharge circuit, the electronic switch device has bidirectional capability, and MOSFET has this capability. In the actual circuit, IFR 4905 is used, the typical on resistance is 20 ms, VX = 55v, IA = 74a.

4 Conclusion

This paper focuses on the functions of data collection module, switch control module, power calculation module and emergency processing module of battery management system. This paper focuses on the design of lithium battery management system, including the hardware management system of lithium battery, temperature application, temperature detection circuit and charging module.

Acknowledgment. The Project is supported by 2019 Shandong Province Key R&D Project (2019JZZY020703), "2019 Youth Innovation and Technology Program" for colleges and universities in Shandong province (2019KJB014), PhD start-up fund of Shandong Jiaotong University (BS201901040) and "Climbing" Research Innovation Team Program of Shandong Jiaotong University (sdjtuc18005).

References

1. Deng, C.: Research and design of lithium battery intelligent management system for underwater robots. Southwest Jiaotong University (2019)
2. Sun, H.L., Zhao, G.D.: SOC estimation of lithium iron phosphate battery based on Elm neural network. Commun. Power Technol. **35**(09), 69–71 (2018)
3. Chang, R.F.: Design and implementation of control system of small underwater robot. Hefei University of Technology (2018)
4. Sun, C.L., Zhao, J.B., Cui, T.Y., et al.: Application status and prospect of AUR power battery technology. Ship Eng. **39**(07), 65–70 (2017)
5. Ren, L.B., Sang, L., Zhao, Q., et al.: Application status and development trend of AUR power battery. Power Technol. **41**(06), 952–955 (2017)
6. Shen, C.: Research on lithium battery management system of underwater robot. Harbin Engineering University (2011)
7. Lei, J.Z.: Research on self-rescue beacon system of autonomous underwater vehicle. Huazhong University of Science and Technology (2008)

Research and Application of Gas Wavelet Packet Transform Algorithm Based on Fast Fourier Infrared Spectrometer

Wanjie Ren[1(✉)], Xia Li[2], Guoxing Hu[1], and Rui Tuo[1]

[1] Shandong Institute of Non-Metallic Materials, Jinan, China
renwanjie66@126.com
[2] Shandong Academy of Pharmaceutical Sciences, Jinan, China

Abstract. In the multi-component analysis of Fourier infrared spectroscopy, the gas mixture of unknown components can be obtained by infrared spectroscopy. This article focuses on the following problems: Under the library spectrum of pure gas, by using the known Pure spectrum, qualitative and quantitative analysis of the mixed spectrum within a certain error range, based on Fourier infrared theory, the corresponding wavelet packet transform algorithm for mixed gas separation is obtained. The experimental results show that gas separation based on Fourier infrared spectrum, the wavelet packet algorithm has the advantages of high accuracy and wide application range. It has a very broad development prospect in the field of environmental gas detection technology.

Keywords: Fast fourier infrared spectrometer · Wavelet packet transform · Computer algorithm

1 Introduction

With the development of economy, gas pollution caused by industry is more and more concerned by the public. Gas pollution not only causes great pressure on environmental protection, but even unconsciously has affected people's lives and caused people's lives and health. Invisible harm.

So far, the application of gas detection systems in China is not extensive, the types are relatively small, and the scope of use is also limited to the detection of gases with known compositions, which can be roughly summarized into two types of technologies: one is based on metal sensors Gas detection technology; another type is based on nonlinear fluorescence spectroscopy and neural network gas detection technology.

Foreign gas detection technology has diversity in detection methods and technologies. Among these detection technologies, spectrum detection technology is a popular detection method. Spectral detection technology can be divided into two kinds of mid-infrared and near-infrared spectrum detection technology. The ion mobility spectrum technology using spectral information is a complementary gas detection technology that combines the two. This technology is mainly used for the detection of poisonous gas and drugs during actual accidents. It is widely used in military warfare and defense fields. In China, sensor-based gas detection technology is rarely used in

Q. Liu et al. (Eds.): CENet 2020, AISC 1274, pp. 996–1001, 2021.
https://doi.org/10.1007/978-981-15-8462-6_114

practice, and reports are rare. Although nitrogen dioxide, ammonia, hydrogen sulfide and other gases can be detected and the detection accuracy is relatively high, an accuracy level of 5 ~ 200 ppb can be achieved Detection, but due to the inherent limitations of this technology's own principles, its scope of application is relatively small, so it has been relatively restricted in use.

Up to now, the detection and analysis of mixed gas based on Fourier infrared is a relatively advanced measurement method at home and abroad. Compared with other gas detection technologies, Fourier infrared detection technology has many advantages. First, the standard detection spectrum of known pure gas In the case of mixed gas, infrared spectrum absorption is performed on the mixed gas, and the coefficients of these pure gas spectra in the mixed gas spectrum are calculated within a certain error range, and then the corresponding content of the gas is obtained. The method has fast response speed and high measurement accuracy. The second infrared, the detection technology has stability. After the algorithm and its machine characteristics are combined, when the fixed gas detection is performed, it will be almost not interfered by other environmental gases. Finally, the monitoring instrument based on Fourier change infrared is small and has a long service life. And it is easy to operate in the process of use, and the later maintenance and repair are relatively simple and convenient. Based on the infrared absorption principle of Fourier infrared spectrometer, a variety of gases such as nitric oxide, nitrogen dioxide, and sulfur oxide are used as the gas to be tested. The experimental results verify the effectiveness and practicability of Fourier infrared spectrometer.

2 Multi-component Analysis of Fourier Infrared Spectroscopy

In multi-component analysis, if a large number of library spectra of pure gases are known, these pure spectra can be used to calculate the coefficients of these pure gases in the mixed gas within a certain error range. The errors in the obtained values are caused by the measurement noise of the spectrum. When performing gas composition analysis, the required characteristic wavelength is selected based on the method of wavelet packet transformation, and the spectral signal is analyzed in the time domain and frequency domain in more detail. The low frequency part has higher frequency resolution and lower. The time resolution has a higher time resolution and a lower frequency resolution in the high-frequency part. Compared with Fourier analysis and simple wavelet analysis, wavelet packet analysis of the signal is more refined, and the high frequency part is further subdivided. And according to the characteristics of the signal to select the frequency band matching the signal spectrum, so the wavelet packet has a higher time-frequency resolution.

Figure 1 is an example of a three-layer wavelet packet decomposition tree. After the signal is decomposed by wavelet packets, the coefficients of each node contain some detailed information of the signal. Therefore, the spectral signal can be analyzed in each frequency band based on the wavelet packet decomposition method. On the one hand, the dispersion of the signal of the sample spectrum in each frequency band is calculated; on the other hand, the wavelength corresponding to the position where the coefficient dispersion of each frequency band is the largest is output.

1) For the collected sample spectrum matrix, average the corresponding spectra of samples of the same substance, and only retain the average spectrum. In the new spectrum matrix obtained, the substance content corresponding to each spectrum is different;
2) Perform wavelet packet decomposition on all the spectra in the new matrix to obtain spectral information in multiple frequency bands;
3) The wavelet coefficients of each spectrum in each frequency band are reconstructed, and all spectral signals in each frequency band are analyzed for dispersion, that is, the standard deviation of the coefficients of the spectrum at each wavelength point in the frequency band is calculated, where the larger the standard deviation indicates the dispersion The greater the degree;
4) Output the wavelength corresponding to the coefficient with the largest dispersion in each frequency band.

$$\sigma = \sqrt{\frac{1}{N}\sum_{i=1}^{N}(x_i - \mu)^2} \tag{1}$$

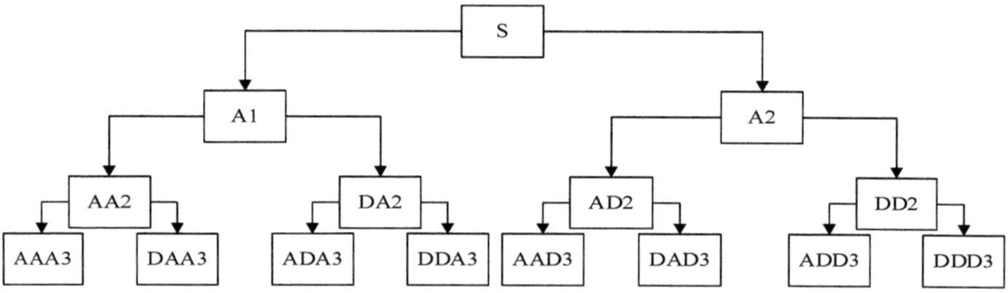

Fig. 1. Three-layer wavelet packet decomposition tree.

3 Experimental Results

This article mainly takes the gas detection experiment as the main content, and actually calculates the measurement data of the mixed gas obtained and the standard experimental data of multiple pure gases to complete the qualitative and quantitative analysis of the gas. The verification and application of the algorithm are mainly divided into for the following 2 Steps:

1) First verify the algorithm with a mixed gas of known coefficients;
2) Perform actual calculations on the measured data of the mixed gas and standard experimental data of various pure gases to complete the qualitative and quantitative analysis of the gas.

3.1 Verification of the Algorithm

Multiply a variety of standard pure gases with the previously set random number, and add a set of random numbers as noise interference on this basis to obtain the data of the mixed gas. After obtaining the spectrum and data of the mixed gas, perform the most Solve and calculate the optimal coefficient vector, calculate the final coefficient matrix, and compare the coefficient obtained by the algorithm with the previously set random number to judge the correctness of the algorithm.

The coefficient vector obtained by the above algorithm is basically consistent with the previously set random number, indicating that the above algorithm is feasible, and the next step will be used for practical applications.

3.2 Practical Application

The qualitative and quantitative analysis of multiple gases in the randomly collected mixed gas is performed by the above algorithm, and the coefficient matrix calculation of the mixed gas is performed using the verified algorithm. The first group of mixed gas spectrum simulation is shown in Fig. 1, pure gas coefficient spectrum. The noise spectra are shown in Figs. 2 and 3, respectively, and the pure gas coefficients of the three groups of mixed gases are shown in Table 1.

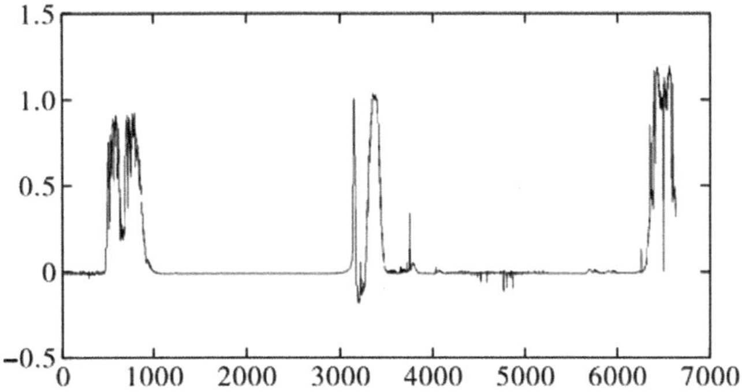

Fig. 2. Mixed gas spectrum.

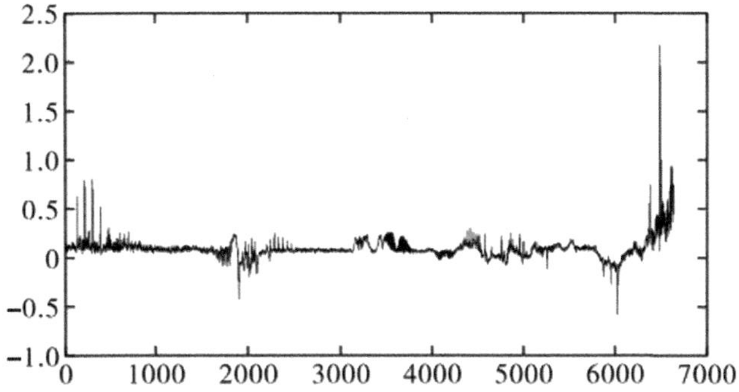

Fig. 3. Pure gas coefficient spectrum.

Table 1. Corresponding coefficients of several pure gases in mixed gas.

Correlation coefficient 1	Correlation coefficient 1	Correlation coefficient 2
0.62	0.63	0.65
0.12	0.13	0.15
5.21	5.23	5.18
6.11	6.10	6.19
9.10	9.08	9.13

4 Conclusion

Suppress and eliminate the noise during the use of the detection system, perform more precise processing on the pure gas spectrum, remove the factors with greater interference during the data acquisition process, the wavelet packet analyzes the signal more finely, and further the high-frequency part Subdivide and select the correlation coefficient that matches the frequency spectrum of the signal according to the characteristics of the signal. Therefore, using this algorithm, you can get more optimized results.

References

1. Rahman, M.A., Khanam, F., Ahmad, M., Uddin, M.S.: Multiclass EEG signal classification utilizing Rényi min-entropy-based feature selection from wavelet packet transformation. Brain Inf. **7**(1), 1–11 (2020). https://doi.org/10.1186/s40708-020-00108-y
2. Networks-Telecommunications.: Findings from Wuhan University provide new insights into telecommunications (Carrier-phase multipath mitigation based on adaptive wavelet packet transform and Tb strategy). Telecommunications Weekly (2020)
3. Chandra, S., Sharma, A., Kumar, S.G.: A comparative analysis of performance of several wavelet based ECG data compression methodologies. IRBM (2020)
4. Othman, A.M., Kotb, H.E., Sabry, Y.M., Khalil, D.: Micro-electro-mechanical system fourier transform infrared spectrometer under modulated-pulsed light source excitation. Appl. Spectro. **74**, 799–807 (2020)
5. Baldazzi, G., Sulas, E., Urru, M., Tumbarello, R., Raffo, L., Pani, D.: Wavelet denoising as a post-processing enhancement method for non-invasive Foetal electrocardiography. Comput. Methods Programs Biomed. **195**, 105558 (2020)
6. Daniel, S., Pär, M., Kim, B., PerErik, L.: Bearing monitoring in the wind turbine drivetrain: a comparative study of the FFT and wavelet transforms. Wind Energy **23**(6), 1381–1393 (2020)
7. Ji, D.H., Zhou, M.Q., Wang, P.C., Yang, Y., Wang, T., Sun, X.Y., Hermans, C., Yao, B., Wang, G.C.: Deriving temporal and vertical distributions of methane in Xianghe Using ground-based Fourier transform infrared and gas-analyzer measurements. Adv. Atmos. Sci. **37**(7), 597–607 (2020)
8. Sensor Research: Report summarizes sensor research study findings from China meteorological administration (Intensity simulation of a Fourier transform infrared spectrometer). Journal of Technology (2020)

9. Ni, Z.Y., Lu, Q.F., Xu, Y.S., Huo, H.Y.: Intensity simulation of a Fourier transform infrared spectrometer. Sensors (Basel, Switzerland) **20**(7), 1833 (2020)
10. Luo, S., Tian, J.H., Liu, Z.M., Lu, Q., Zhong, K., Yang, X.: Rapid determination of styrene-butadiene-styrene (SBS) content in modified asphalt based on Fourier transform infrared (FTIR) spectrometer and linear regression analysis. Measurement **151**, 107204 (2020)

Segment Wear Characteristics of the Frame Saw for Hard Stone in Reciprocating Sawing Mode

Qin Sun[1(✉)], Baoping Wang[1], Zuoli Li[1], and Zhiguang Guan[1,2]

[1] School of Construction Machinery, Shandong Jiaotong University, Jinan 250357, China
206075@sdjtu.edu.cn

[2] Institute of Automation, Shandong Academy of Sciences, Jinan 250000, China

Abstract. The frame saw is not limited by the size of the stone. It can be installed 120 saw blades with a length of 4 m. The thickness of the saw blade is generally less than 3.5 mm, which reduces the generation of waste debris in the stone sawing process. There is no noise and dust pollution in the sawing process, which meets the requirements of high energy, low consumption and environmental protection advocated. However, in the process of sawing hard stone, the flatness of stone surface is poor. The experiment of sawing hard stone with reciprocating frame saw is carried out, and the wear morphology of the saw tooth is analyzed by SEM. The main wear forms of diamond particles are whole crystal, micro-fractured, macro-fractured, flat, and pull out. Among the pulling out diamond grains, the abnormal pulling-out diamond grains are the most, and the depth is more than or close to half of the diameter of the grains. Cracking was found at the interface between diamond grain and matrix. Through the systematic analysis of diamond grain wear, the theoretical basis is provided for the design of new sawing mode.

Keywords: Frame saw · Diamond segment · Wear

1 Introduction

Sawing refers to the process of using different forms of sawing machines to cut the stone materials into rough slabs (semi-finished slabs). Sawing cost accounts for about 50% of the total from block stones to facing slabs [1]. In fact, the material removal relies on of diamond segment grains during the sawing process. Therefore, the grains wear plays an important role in the sawing process. Literature [2] analyzed the difference of sawing movement mode between the circular saw and the frame saw. it is also proposed that the sawing environment of frame saw is worse than that of the circular saw. The reciprocating sawing process of the frame saw is not conducive to heat dissipation and chip removal. In addition, the tailing of the matrix cannot be formed, which reduces the holding force of the matrix on the diamond grain to a certain extent. In order to provide theoretical basis for the design of new sawing modes, the wear characteristics and existing problems of the diamond segment in the process of

Q. Liu et al. (Eds.): CENet 2020, AISC 1274, pp. 1002–1007, 2021.
https://doi.org/10.1007/978-981-15-8462-6_115

sawing hard stone with the frame saw are analyzed by using the observation and research method of the micro morphology of the diamond segment.

2 Main Wear Forms of Diamond Grains

2.1 Main Wear Forms of Diamond Grains

According to literature [3], the wear patterns of diamond grains are divided into six categories, as shown in Fig. 1

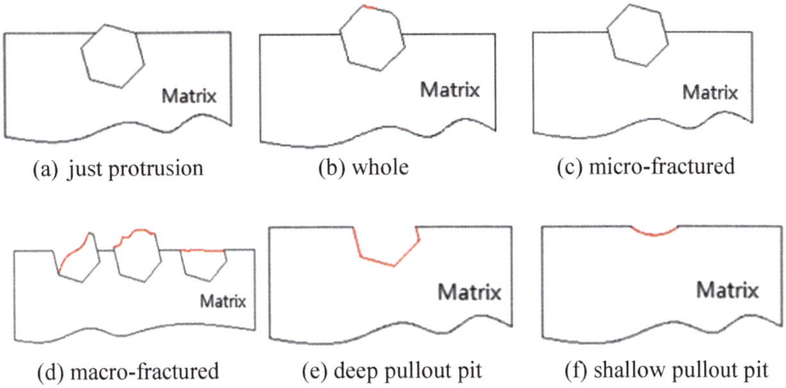

Fig. 1. Wear diagram of 6 typical diamond grains.

According to literature [4], the wear forms of diamond grains are divided into four types, as shown in Fig. 2. (a) shows that the cutting edge of the sharp diamond grain is blunt; (b) shows that the diamond grain is broken, and a new cutting edge is formed in the broken area, and the stone can be cut continuously; (c) shows that the diamond grain is broken seriously, and the cutting height is low; (d) shows that the diamond grain has pulled out.

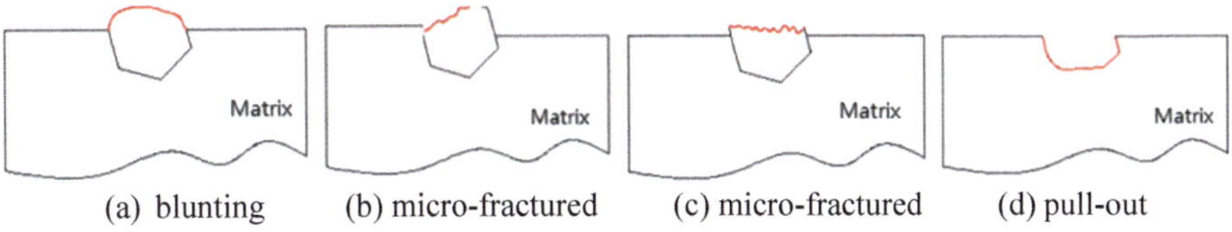

Fig. 2. Wear diagram of four typical diamond grains.

2.2 Wear Process of Diamond Segment

In order to sawing stone continuously, the diamond segment must have self-sharpening ability, that is, the diamond grains from the whole crystal form (not damaged) to the micro-fractured, then to the macro-fractured, finally the diamond grains are pulled out. The new diamond grain grains come out again with the matrix wears. Sawing stone can be realized by the cycle alternative wear form of "relay" type.

3 The Establishment of Experiment Platform for Sawing Hard Stone

The main purpose of the experiment is to study and analyze the problems existing in the process of sawing hard stone with the frame saw machine by observing the micro wear morphology of diamond grains, so as to provide theoretical basis for the design of the new saw machine.

3.1 Saw Machine and Blade

The schematic diagram of reciprocating frame saw is shown in Fig. 3. The saw frame can be installed with a maximum of 120 blades, with a length of 4500 mm. Each saw blade is welded with 26 diamond segments, and the maximum size of stone block is 3200 mm × 2200 mm × 2200 mm.

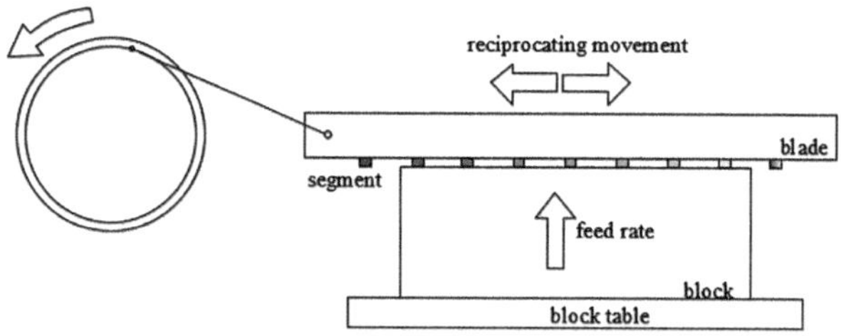

Fig. 3. Sawing diagram of reciprocating frame saw.

Segment are welded on the saw blade in unequal distance, as shown in Fig. 4.

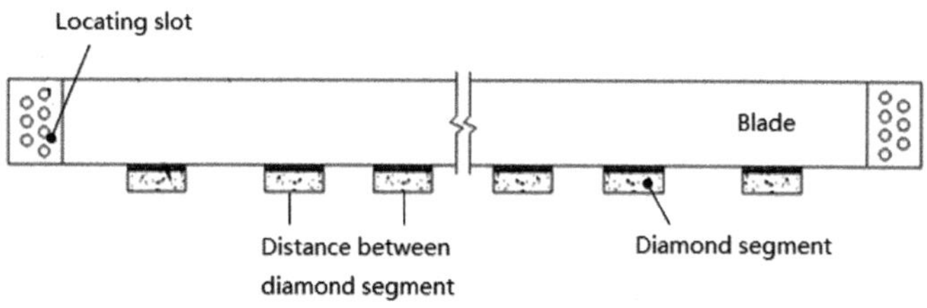

Fig. 4. Sawing diagram of reciprocating frame saw.

3.2 Saw Machine and Blade

The main mineral components, physical and mechanical properties of the workpiece in the test are shown in Table 1.

Table 1. Main characteristics of hard stone.

Mineral components (%)			Physical characteristics				
Quartz	Plagioclase	Orthoclase	Shore hardness (HSD)	Mohs hardness	Compressive strength (MPa)	Density (g/cm^3)	Bending strength (MPa)
29.2	20.35	42.5	85	7.4	92.3	2.56	8.93

3.3 Saw Machine and Blade

As shown in Fig. 5, the micro wear morphology of diamond particles was observed by means of scanning electron microscope (s-2500).

Fig. 5. Testing instrument of diamond segment micromorphology.

4 Analysis of Experimental Results

The main wear forms of diamond grain are whole crystal, micro-fractured, macro-fractured when using reciprocating frame saw for hard stonet, pullout and flat, as shown in Fig. 6.

According to the statistics of the diamond grain exposed on the segment surface, the percentage of whole crystal form and micro-fractured is about 15% and 25%, the percentage of macro-fractured is high, about 28%, the percentage of grinding flat is about 15%, the percentage of pullout is about 10%, and the percentage of grinding blunt is about 7%. Compared with the other five forms of wear, the proportion of grinding blunt is the least. The reason may be that the existence time of blunt wear is short, and the diamond wear grains are finally in the form of pullout or flat. There are

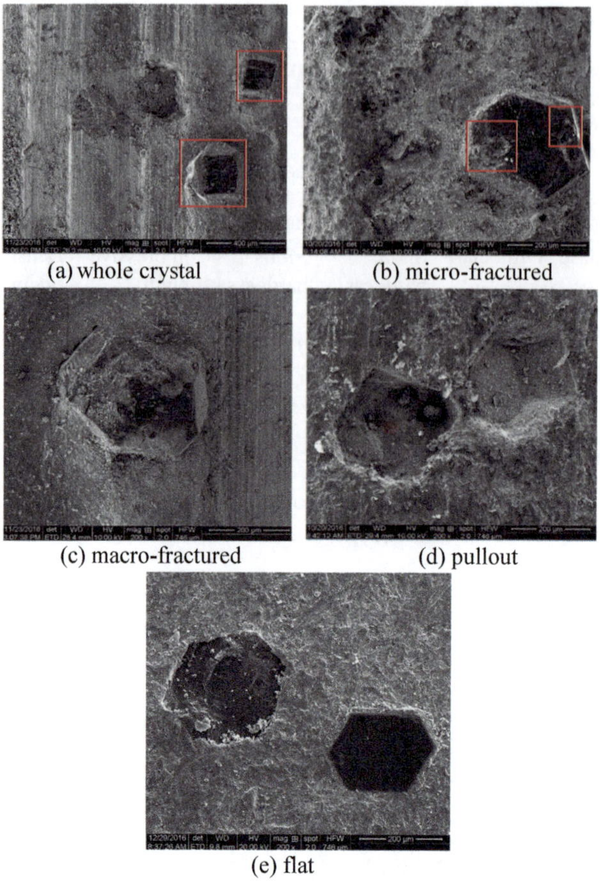

Fig. 6. Typical morphology of diamond grains.

many abnormal desquamated diamond grains among the desquamated diamond grains, and the depth of pullout pit is more than or close to half of the diameter of the grains.

It is found that when the stone is sawed to a certain height, the diamond grain becomes blunt, the height of cutting edge is reduced, the sawing ability is poor, the diamond grain cannot protrude in time, and the self-sharpness is poor.

In addition, cracks were found at the interface between diamond grain and matrix. On the one hand, the reason of cracking may be that the thermal expansion coefficient of diamond and matrix is inconsistent due to the difficulty of heat dissipation of serrations and the temperature rise, and the joint surface will crack; on the other hand, both sides of diamond grains are constantly affected by the sawing force during sawing, and the diamond grains have no protection of matrix tailing, so the holding force of matrix on diamond is low, and the joint surface will crack. The cracking of the interface will accelerate the abnormal pullout of diamond grains before the sawing effect is fully exerted.

The wear of matrix on both sides of diamond is the same in the process of reciprocating sawing stone with frame saw, so tailing cannot be formed. Tailing does not play a decisive role; however, tailing can play an important role in the process of diamond tool sawing stone, which can enhance the holding force of matrix on diamond grains [5, 6].

5 Conclusion

The main conclusions are as follows:

(1) The typical wear morphology of diamond grains is whole crystal (15%), micro-fractured (25%), macro-fractured (28%), pullout (10%) and grinding flat (15%). Blunt diamond grains are less than the above five wear forms (7%). The reason may be that the existence time of blunt wear forms is shorter, and diamond grains finally show the wear forms of pullout and grinding flat. Whole crystal and micro-fractured are the most effective sawing forms of diamond grains.
(2) Experimental results show that the self-sharpening of diamond grains is poor, because the proportion of flat and blunt is more than 20%. The higher the proportion of flat and blunt is, the worse the sawing performance is.
(3) In order to design the new sawing method, it is necessary to solve the problems of heat dissipation and chip removal in the reciprocating sawing method, reduce the proportion of blunt and flat wear forms, and improve the self-sharpening of diamond grains.

Acknowledgments. The Project is supported by "2019 Youth Innovation and Technology Program" for colleges and universities in Shandong province (2019KJB014), PhD start-up fund of Shandong Jiaotong University (BS201901040, BS201901041) and "Climbing" Research Innovation Team Program of Shandong Jiaotong University (SDJTUC18005).

References

1. Zhao, M.: Review of China's stone processing equipment and technology in the past decade. Stone **2**, 25–29 (2016)
2. Konstanty, J.: Theoretical analysis of stone sawing with diamonds. J. Mater. Process. Technol. **123**(1), 146–154 (2002)
3. Wang, C.Y., Fan, J.M., Xu, Z.C.: Study on diamond frame sawing (V)-Characteristics of diamond agglomerate wear. Diam. Grain Tool Eng. **6**, 6–12 (2002)
4. Xi, H.: Mining and processing of facing stone. Beijing: China Construction Industry Press 387, (1986)
5. Tönshoff, H.K., Hillmann-Apmann, H., Asche, J.: Diamond tools in stone and civil engineering industry: cutting principles, wear and applications. Diam. Relat. Mater. **11**(3–6), 736–741 (2002)
6. Özçelik, Y.: The effect of marble textural characteristics on the sawing efficiency of diamond segmented frame saws. Ind. Diam. Rev. 2 (2007)

Dongba Hieroglyphs Visual Parts Extraction Algorithm Based on MSD

Yuting Yang[1] and Houliang Kang[2(✉)]

[1] School of Computer Engineering, Suzhou Vocational University, Suzhou 215000, China
[2] Sports Department, Suzhou Vocational University, Suzhou 15000, China
tudou-yeah@163.com

Abstract. Part-based representations (part-based representations) are widely used in the field of shape matching and classification. We can significantly improve the robustness of the recognition algorithm by using parts, so that the computer can analyze objects from different perspectives, such as global and local. Therefore, we use the theory of part-based representations in the field of shape matching, combined with multi-scale shape decomposition (MSD), to give a multi-scale decomposition based visual part extraction algorithm (MDVPE). The algorithm can accurately separate the visual parts from the feature curves of Dongba hieroglyphs, lay the foundation for designing an efficient Dongba hieroglyphs recognition method, and also provide technical support for researching radicals and other content.

Keywords: Dongba hieroglyphs · Visual parts extraction · Part-based representations · Multi-scale decomposition · Feature curve decomposition

1 Introduction

Dongba character is a very primitive pictograph hieroglyph [1, 2]. As a form of early human pictograph to hieroglyph, it has the characteristics of pictograph with graphic meaning and modern words using simple lines [3, 4]. Extracting feature curves and using local visual parts to represent prominent and visual local features can effectively improve the robustness of shape recognition algorithms. Simultaneously, the segmentation of local features is also crucial for the study of Dongba hieroglyph.

Therefore, we use the theory of part-based representations in the field of shape matching [5, 6], combined with multi-scale shape decomposition (MSD), to give a multi-scale decomposition-based visual part extraction algorithm (MDVPE). This algorithm can accurately separate the visual parts from the feature curves of Dongba hieroglyphs, lay the foundation for designing an efficient Dongba hieroglyph recognition method, and also provide technical support for researching radicals and other content.

Q. Liu et al. (Eds.): CENet 2020, AISC 1274, pp. 1008–1013, 2021.
https://doi.org/10.1007/978-981-15-8462-6_116

2 The Algorithm of MDVPE

Part-based representations (part-based representations) are widely used in the field of shape matching and classification. It refers to the use of intuitionistic local curves containing visual object features to represent local features of shapes [5, 6]. Using parts, we can significantly improve the robustness of the recognition algorithm and help the computer analyze objects from different perspectives, such as global and local.

The concavity and convexity of the curve is the basis of human recognition [7]. To be consistent with the feature extraction habit of human vision, convex arcs or arcs that are nearly convex are generally used to represent the visual parts of objects. Among them, arcs close to convex are composed of convex arcs and arcs of other forms. In order not to be affected by the unevenness of the curve and to accurately extract the visual parts of Dongba characters, we use the MSD algorithm to represent the visual parts by gradually merging the concave arcs at the advanced stage of the hierarchical decomposition. Therefore, multi-scale decomposition-based visual part extraction algorithm of Dongba hieroglyphs can be divided into three steps: First, we use MSD algorithm to decompose the feature curve layer by layer; Second, we get the demarcation points of the visual parts and count the total number of parts in each evolution level; third, we determine the final number of parts based on the structural features of Dongba hieroglyphs.

2.1 The Algorithm of MSD

Multi-scale shape decomposition (MSD) is a shape hierarchy decomposition algorithm based on Discrete Curve Evolution (DCE). The algorithm uses a digital curve composed of line segments to replace the traditional continuous curve, and uses the DCE algorithm to merge line segments with low contributions in the digital curve until the curve converges to a convex polygon. It can be seen that DCE algorithm is the foundation and key content of MSD.

The main idea of Discrete Curve Evolution is: In every evolution step, a pair of consecutive line segments s1, s2 is substituted with a single line segment joining the endpoints of s1 $\cup$ s2. The key property of this evolution is the order of the substitution. The substitution is done according to a relevance measure K [8] given by:

$$K(s_1, s_2) = \frac{\beta(s_1, s_2) l(s_1) l(s_2)}{l(s_1) + l(s_2)}$$

where $\beta(s_1, s_2)$ is the turn angle at the common vertex of segments s1, s2 and l is the length function normalized with respect to the total length of a polygonal curve. It can be proved by the rules on the salience of a limb [9] that the greater the relative arc length and the total turning angle of the curve, the higher its contribution to the shape. In other words, the contribution K is proportional to the values of the relative arc length and the total turning angle. Since the curve obtained at the highest stage of hierarchical decomposition contains the most important parts of the object, we merge the line segments in the curve according to the K value at each stage of evolution until the visual parts gradually appear and the number of parts reaches the specified number. As shown in Fig. 1.

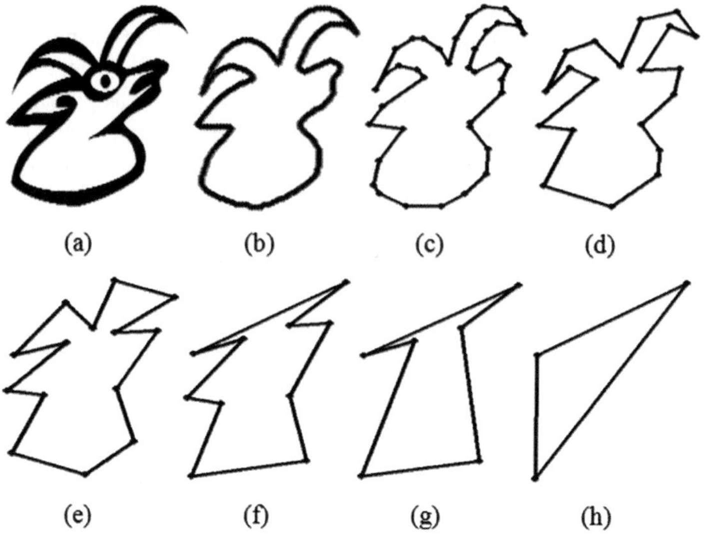

Fig. 1. Merging line segments at each evolution stage.

2.2 Determine the Demarcation Points of Visual Part

In the MSD algorithm, the visual part refers to the maximal convex arc in the feature curve, and the endpoints of maximal convex arc are concave points. Obviously, the higher the evolution stage, the more important concave points in the curve, the more integrated of the visual part can be reflected. Therefore, we start with the convex polygon which is the result of convergence and tracing back all concave points of the maximal convex arc. The way we judge the concave points is as follows and the demarcation points which we extracted at each stage of backtracking evolution are shown in Fig. 2:

For a feature vertex V_i, if its turning angle $\beta_i \in (0, \pi)$, the vertex is convex; if $\beta_i \in (\pi, 2\pi)$, the vertex is concave [10].

Combined with the definition, first we use the cross-multiplication of the adjacent edge vectors to determine the direction of the curve, and then we determine the turning angle based on the angle of the adjacent edge vectors. Finally, we determine the concavity and convexity of the vertex based on turning angle β_i. Therefore, for a given curve, its feature vertices sequence are $P_i(i = 2, 3, \ldots, n-1)$, we select 3 adjacent vertices P_{i-1}, P_i and P_{i+1} where:

$$P_{i-1} = (x_{i-1}, y_{i-1}), P_i = (x_i, y_i), P_{i+1} = \left(x_{i+1}, y_{i+1}\right)$$

Let

$$A = \overrightarrow{P_{i-1}P_i} = (x_i - x_{i-1}, y_i - y_{i-1}), B = \overrightarrow{P_iP_{i+1}} = (x_{i+1} - x_i, y_{i+1} - y_i)$$

And the direction of the curve is:

$$N = A \times B = \overrightarrow{P_{i-1}P_i} \times \overrightarrow{P_iP_{i+1}} = [(x_i - x_{i-1})(y_{i+1} - y_i) - (x_{i+1} - x_i)(y_i - y_{i-1})]k \\ = n_i k$$

Then, the vector angle $\alpha_i(\alpha_i \in [0, \pi])$ which is corresponding to P_i can be calculated by the vertices set of the vector, so that the sorted direction of the curve is clockwise (reverse direction), and the vector angle α_i is:

$$\alpha_i = \begin{cases} \cos^{-1}\left(\frac{A \cdot B}{|A||B|}\right) n_i \geq 0 \\ \pi - \cos^{-1}\left(\frac{A \cdot B}{|A||B|}\right) n_i < 0 \end{cases} \text{in which, } \alpha_i \in [0, \pi]$$

And, in the reverse curve, the turning angle β_i corresponding to the feature vertex $P_i(i = 1, 2, 3, \ldots, n)$ is [10]:

$$\beta_i = \begin{cases} \infty & i = 1 \text{ or } i = n \\ \alpha_i + \pi & n_i > 0 \, and \, i = 2, 3, \ldots, n-1 \\ \alpha_i & n_i < 0 \, and \, i = 2, 3, \ldots, n-1 \end{cases}$$

Therefore, based on the value of turning angle, we can judge the concavity and convexity of the vertex $P_i(i = 1, 2, 3, \ldots, n)$ is:

$$P_i = \begin{cases} \text{convex vertex} & 0 \leq \beta_i < \pi \\ \text{concave vertex} & \pi \leq \beta_i < 2\pi \end{cases}$$

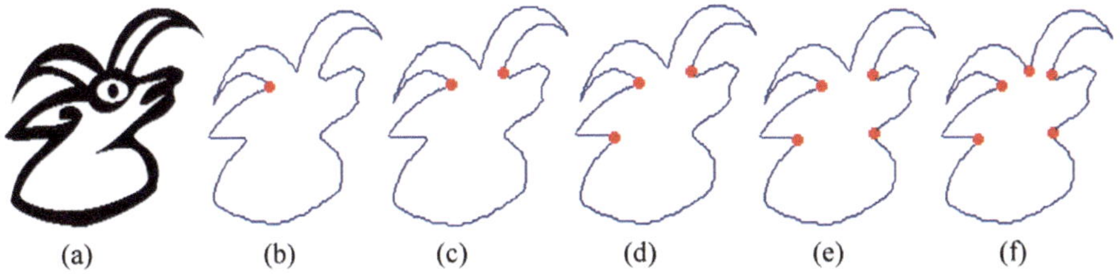

Fig. 2. The demarcation points are shown in the feature curve.

2.3 Extraction of Visual Parts

The demarcation points of maximal convex arc are the feature points on the curve, and are arranged in a clockwise direction. But, the demarcation points we obtained by backtracking are arranged according to the metric value K from large to small, and their order is different from the original order. Therefore, in order to extract visual parts, we should first extract a specified number of demarcation points as required; then, find the original positions of the demarcation points in the curve and arrange them; finally, connect the adjacent demarcation points in pairs to extract the visual parts.

Taking the Dongba character "cow" as an example, first of all, by analyzing the character's structure, we find that the character can contain 5 local features, namely: 2 horns, ears, mouth and upper body, which means that the feature curve should contain 5 demarcation points, As shown in Fig. 2(f); secondly, we find the positions of the 5 demarcation points in the original curve and arrange them; thirdly, we connect the demarcation points in pairs to complete the extraction of the visual parts. The result is shown in Fig. 3(d). Obviously, the number of visual parts is not unique. Based on the character features, we can also extract only 2 or 4 parts. The extraction results are shown in Figs. 3(b) and 3(c). This shows that in practice, we can extract different numbers of visual parts for the same character. The extraction process is independent and does not interfere with each other, and the obtained parts also have good stability.

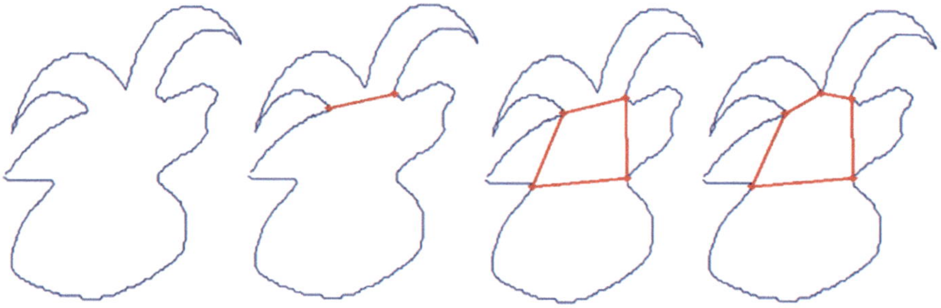

(a) Feature Curve (b) Number=2 (c) Number=4 (d) Number=5

Fig. 3. The segmentation of visual parts.

3 Conclusion

MDVPE is a visual part extraction algorithm for Dongba hieroglyph based on the part representation theory and MSD algorithm. The algorithm continuously merges the line segments with low contributions in the feature curve through the multi-scale decomposition process, so that the maximal convex arc which represents the visual part gradually appears and its boundary is clear. The algorithm can accurately separate the visual parts from the feature curves of Dongba hieroglyphs, lay the foundation for designing an efficient Dongba hieroglyphs recognition method, and also provide technical support for researching radicals and other content.

Acknowledgement. This research was supported by the Talent Introduction Research Fund of Suzhou Vocational University (201905000034) and the Scientific Research Fund of Yunnan Education Department (2019J1152).

References

1. He, L.M.: The research of Dongba culture heritage. Soc. Sci. Yunnan **1**, 83–87 (2004)
2. He, J.G.: The development trend of Dongba culture studies of the Naxi nationality. J. Yunnan Nationalities Univ. **24**(1), 81–84 (2007)
3. Ge, A.G.: Dongba culture review. J. Nationalities Artistic Res. **2**, 71–80 (1999)
4. Fang, G.Y., He, Z.W.: The dictionary of Naxi hieroglyphic. Yunnan People's Publishing House, Kunming (2005)
5. Homan, D.D., Richards, W.A.: Parts of recognition. Cognition **18**, 65–95 (1984)
6. Siddiqi, K., Tresness, K.J., Kimia, B.B.: Parts of visual form: ecological and psychophysical aspects. Perception **25**, 399–424 (1996)
7. Zhou, Y., Liu, J.T., Bai, X.: Research and perspective on shape matching. Acta Autom. Sinica **38**(6), 889–910 (2012)
8. Latecki, L.J., LakaÈmper, R.: Polygon evolution by vertex deletion. In: Processing in International Conference of Scale-Space Theories in Computer Vision, pp. 398–409 (1999)
9. Siddiqi, K., Kimia, B.B.: Parts of visual form: computational aspects. IEEE Trans. Pattern Anal. Mach. Intell. **17**(3), 239–251 (1995)
10. Chen, B.F., Qian, Z.F.: An algorithm for identifying convexity-concavity of a simple polygon. J. Comput. Aided Des. Comput. Graph. **2**(14), 214–217 (2002)

Key Structural Points Extracting Algorithm for Dongba Hieroglyphs

Houliang Kang[1] and Yuting Yang[2](✉)

[1] Sports Department, Suzhou Vocational University, Suzhou 215000, China
[2] School of Computer Engineering, Suzhou Vocational University, Suzhou 215000, China
tudou-yeah@163.com

Abstract. Dongba hieroglyph is a kind of very primitive picture hieroglyphs; it has a characteristic of pictograph to express meaning by using pictures, but also have some features of pictographic, ideographic, self-explanatory and echoism like hieroglyphs. The order of writing makes Dongba hieroglyphs and images essentially different. Extracting the key structural points of Dongba hieroglyphs is very important for studying the writing order and automatically extracting the strokes. Therefore, we combined the structural features of Dongba hieroglyphs and the multiple-scales decomposition theory of shape in computer vision to give the key structure point extraction algorithm. The algorithm uses a discrete curve evolution algorithm to decompose the feature curve by multiple-scales decomposition layer by layer and finally extract the points that can show the key structure of the character. And the experiments show that the algorithm has high accuracy, good stability, and good scaling, rotation and translation invariance.

Keywords: Key structural points · Morphologic change points · DCE · MSD

1 Introduction

Dongba hieroglyph is a kind of very primitive picture hieroglyphs [1, 2]; it has a characteristic of pictograph to express meaning by using pictures, but also have some features of pictographic, ideographic, self-explanatory and echoism like hieroglyphs [3, 4]. In addition, dongba hieroglyphs take a partial or full side view of the object [5, 6], and when we write, we need to follow the writing standards. For example, animal hieroglyphs generally take a side view of the head. When we write, we usually draw the eyes first, then the nose, lips, horns, ears, and then the neck, hands (wings) and toes [3].

The order of writing makes Dongba hieroglyphs and images essentially different. Extracting the key structural points of Dongba hieroglyphs is very important for studying the writing order and extracting the strokes. Therefore, we combined the structural of Dongba hieroglyphs and the MSD theory of shape in computer vision to give the key structure point extraction algorithm. The algorithm uses a discrete curve evolution algorithm to decompose the feature curve by multiple-scales decomposition layer by layer and extracts the points that can show the key structure of hieroglyphs.

Q. Liu et al. (Eds.): CENet 2020, AISC 1274, pp. 1014–1019, 2021.
https://doi.org/10.1007/978-981-15-8462-6_117

2 Key Structural Points Extracting Algorithm for Dongba Hieroglyphs

As a key structural point that can represent the structural characteristics of Dongba pictograph, under ideal conditions, it should contain four properties: (1) Structure points can reflect the key structure of hieroglyphs (for example, morphological changes, joint changes, connections between objects, etc.); (2) The position of the structural point should be accurate; (3) It should be stable under small deformations; (4) It should be invariant under Euclidean transformations, such as rotations and translations;

In order to extract the key structural points of Dongba hieroglyphs which will achieve all the above properties, first we use the chain code based connected domain priority marking algorithm (CDPM) to obtain the feature curve of Dongba hieroglyphics. Then, we combined the discrete curve evolution algorithm (DCE) to remove the redundant points. Finally, by performing multi-scale shape decomposition (MSD) on the feature curve, we extracted the key points which can reflect the structure.

2.1 Extract Feature Curve of Dongba Hieroglyph

CDPM is a feature curve extraction algorithm for Dongba hieroglyphs. It can quickly extract feature curves while maintaining the structural of Dongba hieroglyphs [7]. The basic idea is: first, we determine the extraction direction in counterclockwise, and use the starting point as a reference point. Then, we calculate the weights of the 8 neighboring points adjacent to the reference point according to the direction of the initial reference vector, and select the point with the largest weight as the feature point; Finally, we use the new feature point as a reference point to calculate the new reference vector direction and the weights of the 8 neighboring points adjacent to the reference point. We will repeat the above process until we return to the beginning or complete the traversal of all the terminal points. The feature curves of "Yak" and "Supplication" are shown in Fig. 1. Although the feature curves extracted in Figs. 1(2) and 1(4) only describe the external of the characters, this does not affect the extraction of key structural points.

(1) Hieroglyph (2)Feature Curve (3) Hieroglyph (4) Feature Curve

Fig. 1. Hieroglyph and their feature curve extraction.

2.2 Simplification of Feature Curve

DCE algorithm is a shape hierarchy decomposition algorithm which is suitable for object feature curve simplification. The algorithm uses a polygonal digital curve composed of line segments to replace the traditional continuous curve. Moreover, it has been proved to be a very effective noise point removal algorithm [8]. The main idea of DCE is: In every evolution step, a pair of consecutive line segments s1, s2 is substituted with a single line segment joining the endpoints of s1 ∪ s2. The key property of this evolution is the order of the substitution. The substitution is done according to a relevance measure K [8] given by:

$$K(s_1, s_2) = \frac{\beta(s_1, s_2) l(s_1) l(s_2)}{l(s_1) + l(s_2)}$$

where $\beta(s_1, s_2)$ is the turn angle at the common vertex of segments s1, s2 and l is the length function normalized with respect to the total length of a polygonal curve. It has been proved by the rules on the salience of a limb [9] that the greater the relative arc length and the total turning angle of the curve, the higher its contribution to the shape. That is, the contribution K is proportional to the values of the relative arc length and the total turning angle. As shown in Fig. 2, the feature curve eventually converges into a convex polygon through the evolution of (b) → (c) → (d) → (e) → (f) stage.

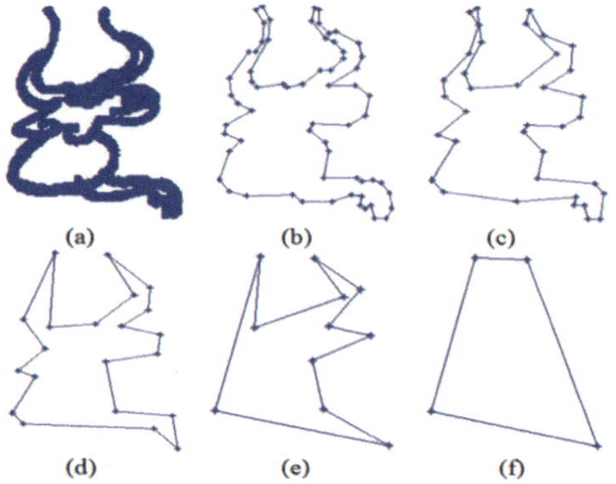

Fig. 2. Different evolution stages of the feature curve.

2.3 Extract the Key Structural Points

The key structural point is the inflection point which is showing the change of the feature curve. According to the MSD theory, by decomposing the feature curve layer by layer, it will eventually converge into a convex polygon. The convex polygon is not only the highest level of shape decomposition, but also contains the most critical convex points in the whole curve. Each concave point traced back from the convex polygon is not only the end point of the maximal convex arc, but also the key structural point of the curve change [8], and this theory is also applicable to the discrete space [9]. Therefore, we transform the problem of extracting key structural points into extracting the endpoints which are the concave points of the maximal convex arc.

Determine the Concave Points in the Feature Curve. The points in the feature curve are divided into two categories: convex points and concave points. The way we judge the concave points in the curve is as follows:

For a feature vertex V_i, if its turning angle $\beta_i \in (0, \pi)$, the vertex is convex; if $\beta_i \in (\pi, 2\pi)$, the vertex is concave [10].

Combined with the definition, first we use the cross-multiplication of the adjacent edge vectors to determine the direction of the curve, and then we determine the turning angle based on the angle of the adjacent edge vectors. Finally, we determine the concavity and convexity of the vertex based on turning angle β_i. Therefore, for a given curve, its feature vertices sequence are $P_i(i = 2, 3, \ldots, n-1)$, we select 3 adjacent vertices P_{i-1}, P_i and P_{i+1} where:

$$P_{i-1} = (x_{i-1}, y_{i-1}), P_i = (x_i, y_i), P_{i+1} = (x_{i+1}, y_{i+1})$$

Let

$$A = \overrightarrow{P_{i-1}P_i} = (x_i - x_{i-1}, y_i - y_{i-1}), B = \overrightarrow{P_iP_{i+1}} = (x_{i+1} - x_i, y_{i+1} - y_i)$$

And the direction of the curve is:

$$\begin{aligned} N = A \times B = \overrightarrow{P_{i-1}P_i} \times \overrightarrow{P_iP_{i+1}} &= [(x_i - x_{i-1})(y_{i+1} - y_i) - (x_{i+1} - x_i)(y_i - y_{i-1})]k \\ &= n_i k \end{aligned}$$

Then, the vector angle $\alpha_i(\alpha_i \in [0, \pi])$ which is corresponding to P_i can be calculated by the vertices set of the vector, so that the sorted direction of the curve is clockwise (reverse direction), and the vector angle α_i is:

$$\alpha_i = \begin{cases} \cos^{-1}\left(\frac{A \cdot B}{|A||B|}\right) & n_i \geq 0 \\ \pi - \cos^{-1}\left(\frac{A \cdot B}{|A||B|}\right) & n_i < 0 \end{cases} \quad \text{and, } \alpha_i \in [0, \pi]$$

And, in the reverse curve, the turning angle β_i corresponding to the feature vertex $P_i(i = 1, 2, 3, \ldots, n)$ is [11]:

$$\beta_i = \begin{cases} \infty & i = 1 \text{ or } i = n \\ \alpha_i + \pi & n_i > 0 \, and \, i = 2, 3, \ldots, n-1 \\ \alpha_i & n_i < 0 \, and \, i = 2, 3, \ldots, n-1 \end{cases}$$

Therefore, based on the value of turning angle, we can judge the concavity and convexity of the vertex $P_i (i = 1, 2, 3, \ldots, n)$ is:

$$P_i = \begin{cases} \text{convex vertex} & 0 \le \beta_i < \pi \\ \text{concave vertex} & \pi \le \beta_i < 2\pi \end{cases}$$

Extract the Key Structural Points. We start from the convex polygon of the convergence result and trace back the maximal convex arc at each stage in the evolution of the feature curve. When we trace back at each stage, if the number of maximal convex arcs in the previous stage and the next stage does not increase, then continue to trace back; If it increases, then we mark the new added endpoint (the key structural point) on the original curve. As shown in Fig. 3, the key structural points of "Yak" and "Supplication" are marked at different stages of evolution.

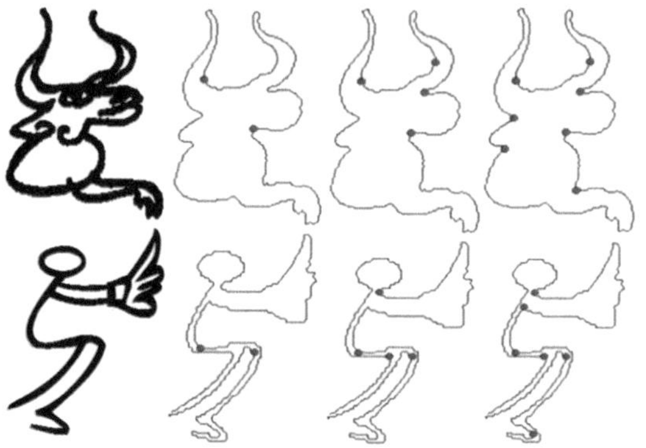

Fig. 3. The extraction process of key structural points in feature curve.

First of all, the maximal convex arc determines the alternation and overlap of convex and concave parts on the object boundary. We use the endpoint of the maximal convex arc as the key point to ensure that it can reflect the key structure of the hieroglyph. Moreover, the higher the curve evolution stage, the more complete the structural features contained in the maximal convex arc, the more accurately the extracted concave points can reflect the key structure.

Secondly, if the endpoint of the maximal convex arc is determined, no matter what stage the discrete curve evolves to, its position will not change. Therefore, it ensures the accuracy of the position of the points.

Thirdly, we use the DCE algorithm to remove the redundant points in the Dongba feature curve. It is a process of constantly refining the structure of Dongba hieroglyph. Because the noise in the feature curve only exists in the early stage of evolution, it will not affect the judgment of the structural point. At the same time, the scaling, rotation, and translation invariance of the DCE algorithm make our extraction algorithm also have the same properties. It shows that our algorithm satisfies the requirement of extracting structural points under ideal conditions.

3 Conclusions

Key structure point extraction algorithm can obtain the points which can describe the structural features of Dongba hieroglyphs. Moreover, it can meet the requirements of ideal structural points. The experiments show that the structure points we extracted are accurate in position and stable in structure, and have the invariance of scaling, rotation and translation. It provides technical support for continuing research on the strokes of Dongba hieroglyphs, as well as assistance for in-depth study of Dongba hieroglyphs.

Acknowledgement. This research was supported by the Talent Introduction Research Fund of Suzhou Vocational University (201905000034) and the Scientific Research Fund of Yunnan Education Department (2019J1152).

References

1. He, L.M.: The research of dongba culture heritage. Soc. Sci. Yunnan **1**, 83–87 (2004)
2. He, J.G.: The development trend of Dongba culture studies of the naxi nationality. J. Yunnan Natl. Univ. **24**(1), 81–84 (2007)
3. Ge, A.: Dongba culture review. J. Natl. Artistic Res. **12**(2), 71–80 (1999)
4. Fang, G.Y., He, Z.W.: The Dictionary of Naxi Hieroglyphic, vol. 12. Yunnan People's Publishing House, Kunming (2005)
5. Hu, R.B., Zhang, X.S., Xu, R.P.: A study on the structure of Naxi Dongba hieroglyphs. J. Zhengzhou Univ. Light Ind. (Soc. Sci.) **14**(2), 94–100 (2013)
6. Wang, L.Y.: Study on Ancient Chinese Characters and Naxi Dongba Hieroglyphs. East China Normal University Press, Shanghai (1998)
7. Yang, Y.T., Kang, H.L.: Research on the extracting algorithm of Dongba hieroglyphic feature curves. J. Graph. **4**, 697–703 (2019)
8. Freeman, H.: Shape description via the use of critical points. Pattern Recogn. **10**, 159–166 (1978)
9. Jan Latecki, L., Lakamper, R.: Convexity rule for shape decomposition based on discrete contour evolution. Comput. Vis. Image Underst. **73**(3), 441–454 (1999)
10. Chen, B.F., Qian, Z.F.: An algorithm for identifying convexity-concavity of a simple polygon. J. Comput. Aided Des. Comput. Graph. **2**(14), 214–217 (2002)
11. Song, X.M., Cheng, C.X.: An analysis and investigation of algorithms for identifying convexity-concavity of a simple polygon. Remote Sens. Land Resour. **9**(13), 25–31 (2011)

Research on a High Step-Up Boost Converter

Jingmei Wu(✉), Ping Ji, Xusheng Hu, and Ling Chen

Wanjiang University of Technology, Ma'anshan 243032, China
61918723@qq.com

Abstract. According to the photovoltaic cell miniature power supply characteristics, this paper proposed a new type of high step-up Boost converter. Compared with the traditional Boost converter, this circuit has three switch tubes, and two switch capacitors are added. A high step-up ratio can be obtained within a range of duty ratios ranging from 0 to 1. Simultaneously, the voltage stress of the power device is reduced, which is beneficial to the selection of switching devices and reduces the loss of the circuit. If the inductor current is continuous, the working principle of the circuit when the duty ratio changes from 0 to 0.5 and 0.5 to 1 is analyzed in detail, and the mathematical model of the converter is established. Finally, the correctness and feasibility of the theory are proved by the Simulink simulation experiment.

Keywords: High step-up ratio · Crisscross control · Boost converter · Voltage stress

1 Introduction

With the increasing improvement of the economy, technology and living standard, energy shortage, and environmental problems are getting worse. Solar energy as new energy has become a focus of attention and development in academia and industry [1–3]. Due to the low output voltage of the photovoltaic cells, the voltage boost of the DC/DC circuit must be passed to satisfy the requirements of the electrical equipment of the rear stage or the grid connection. To improve the system's efficiency, the DC/DC converter must have a small ripple of input current, high gain, and high efficiency [2–10].

Literature [8] proposed the conventional parallel Boost converter has a small voltage gain, and the duty cycle (D) is close to l in high-gain applications, resulting in a broad ripple of input and output current and a reverse severe recovery problem for the diode in the converter. The loss of diode is larger, and EMI interference is caused, which affects the converter's efficiency. Moreover, the high voltage stress of this converter's switching device leads to significant switching off the loss and on loss, further reducing the efficiency of the converter. So the traditional parallel Boost converter is not suitable for photovoltaic power generation with low voltage input and high voltage output. Therefore, in order to make up for the defect of the traditional stagger parallel Boost converter, a novel high step-up ratio Crisscross control Boost converter is proposed in this paper, which can not only achieve a higher voltage gain under the same duty cycle condition but also reduce the voltage stress of the switch tube and work in the range of duty cycle from 0 to 1.

Q. Liu et al. (Eds.): CENet 2020, AISC 1274, pp. 1020–1032, 2021.
https://doi.org/10.1007/978-981-15-8462-6_118

2 The Topology and Operating Principle of Converter

The new high step-up ratioparallel interleaving boost chopper circuit proposed in this paper is shown in Fig. 1, which consists of two inductors, three switches, three diodes and three capacitors. To simplify the theoretical analysis of the circuit and the following assumptions are made. 1st-The switch tube is an ideal element without considering the influence of its parasitic parameters; 2nd-Capacitances C_1, C_2 and C_3 are large enough to ignore the ripple on the capacitances, and the capacitive voltage is considered to remain unchanged in the working process; 3rd-The inductive current of L_1 and L_2 work in CCM; 4th-The switches T_1 and T_2 are controlled by phase shift, and the driving signals of the two are staggered by 180°. The driving signals of T_3 are obtained by the driving signals of T_1andT_2 NOR.

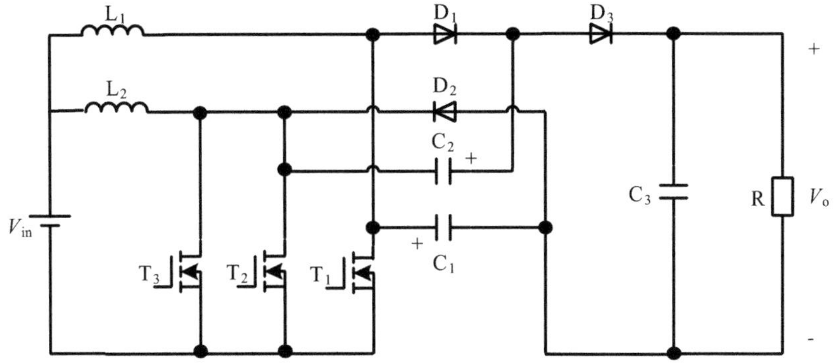

Fig. 1. High Boost ratio stagger Boost circuit.

2.1 The Operating Principle of Converter at 0 < D < 0.5

The switch modal characteristics are shown in Table 1 at $0 < D < 0.5$. In one switching cycle T_s the circuit has 4 switching modes namely mode A, mode B, mode C and mode D. The equivalent circuit is shown in Fig. 2. Figure 3 shows the main working waveform of the circuit at $0 < D < 0.5$.

When $0 < D < 0.5$, T1 conductance time is DT, T2 conductance time is DT, T3 conductance time is (1-2D)T within a Ts period, equivalent to the complementary control, T1 conductance time is DT, T2 conductance time is (1-2D)T.

According to the volt-second balancing principle of inductance and the ampere second balancing principle of capacitance, the relationship between variables in each mode can be expressed as follows:

Table 1. The states of switch at $0 < D < 0.5$.

Modal	T_1	T_2	T3	D_1	D_2	D_3
A	ON	OFF	OFF	OFF	OFF	ON
B	OFF	OFF	ON	ON	ON	OFF
C	OFF	ON	OFF	ON	ON	OFF
D	OFF	OFF	ON	ON	ON	OFF

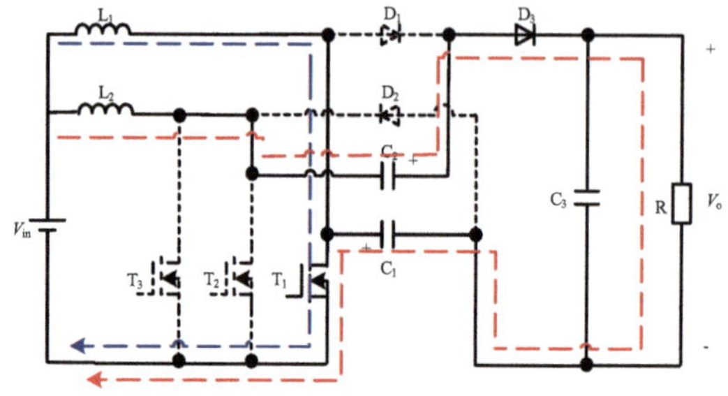

(a) Modal A

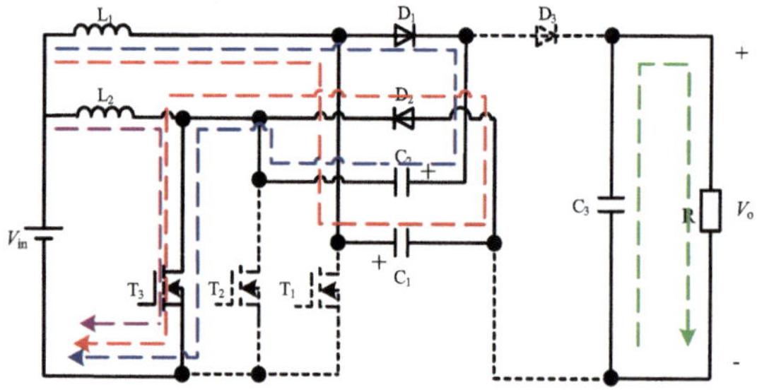

(b) Modal B and Modal D

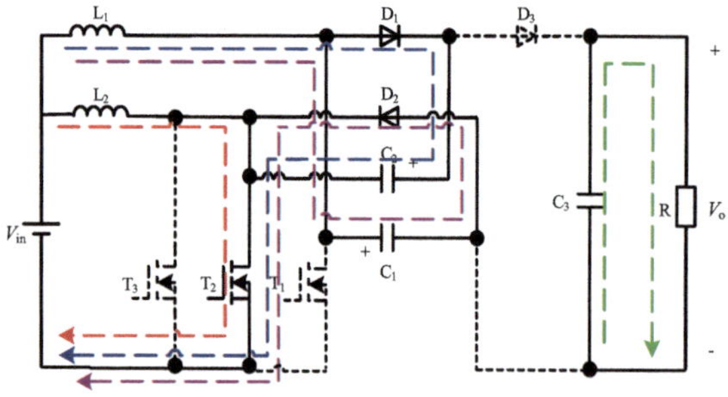

(c) Modal C

Fig. 2. Four states of the circuit at $0 < D < 0.5$.

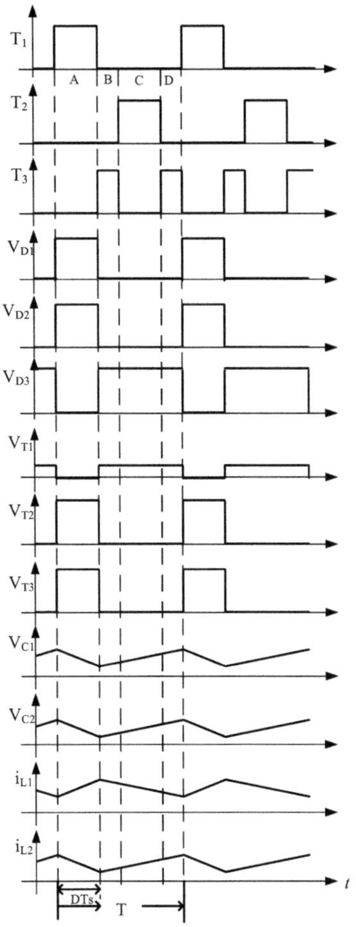

Fig. 3. Key operation waveforms of the circuit at 0 < D < 0.5.

1st-Modal A:

$$\begin{cases} V_{in} = L_1 \frac{di_{L1}}{dt} \\ V_{in} = L_2 \frac{di_{L2}}{dt} - V_{C2} + V_O - V_{C1} \end{cases} \tag{1}$$

2nd-Modal B and Modal D:

$$\begin{cases} V_{in} = L_1 \frac{di_{L1}}{dt} + V_{C2} = L_1 \frac{di_{L1}}{dt} + V_{C1} \\ V_{in} = L_2 \frac{di_{L2}}{dt} \end{cases} \tag{2}$$

3rd- Modal C:

$$\begin{cases} V_{in} = L_1 \frac{di_{L1}}{dt} + V_{C2} = L_1 \frac{di_{L1}}{dt} + V_{C1} \\ V_{in} = L_2 \frac{di_{L2}}{dt} \end{cases} \tag{3}$$

Combining formula (1)–(3) and switching Law get capacitor voltage and ratio of output to input voltage, i.e.,

$$V_{C1} = V_{C2} = \frac{V_{in}}{1 - D} \tag{4}$$

$$\frac{V_O}{V_{in}} = \frac{1 + D}{D(1 - D)} \tag{5}$$

2.2 The Operating Principle of Converter at 0.5 < D < 1

As can be seen from the control mode of the three switch tubes, when 0.5 < D < 1, T_3 is always closed, playing a role of redundancy. So T_3 can be removed. The equivalent circuit is shown in Fig. 4, which is different from 0 < D < 0.5, the control mode after equivalence is still staggered control. In a switching cycle T_s, the circuit has three switching modes and the switching mode characteristics are shown in Table 2. The equivalent circuits of the three switching modes are shown in Fig. 5. Figure 6 shows the main working waveform of the circuit at 0.5 < D < 1.

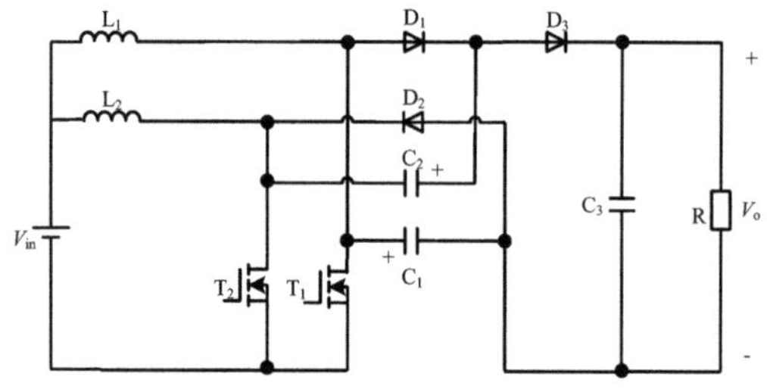

Fig. 4. The equivalent circuit at 0.5 < D < 1.

According to the volt-second balancing principle of inductance and the ampere second balancing principle of capacitance, the relationship between variables in each mode can be expressed as follows:

1st-Modal A:

$$\begin{cases} V_{in} = L_1 \frac{di_{L1}}{dt} \\ V_{in} = L_2 \frac{di_{L2}}{dt} - V_{c2} - V_{c1} + V_o \end{cases} \tag{6}$$

Table 2. The states of switch at 0.5 < D < 1.

Modal	T_1	T_2	T_3	D_1	D_2	D_3
A	ON	OFF	OFF	OFF	OFF	ON
B	OFF	ON	OFF	ON	ON	OFF
C	ON	ON	OFF	OFF	OFF	ON

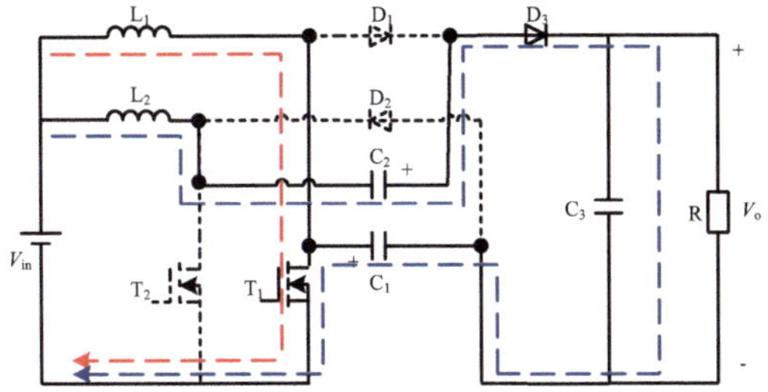

(a) Modal A

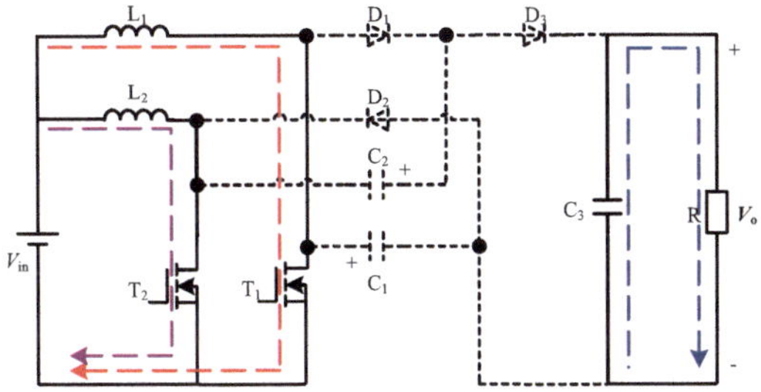

(b) Modal B

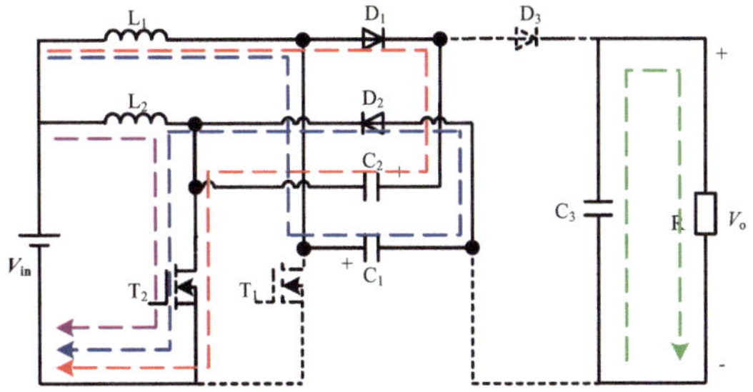

(c) Modal C

Fig. 5. Three states of the circuit at 0.5 < D < 1.

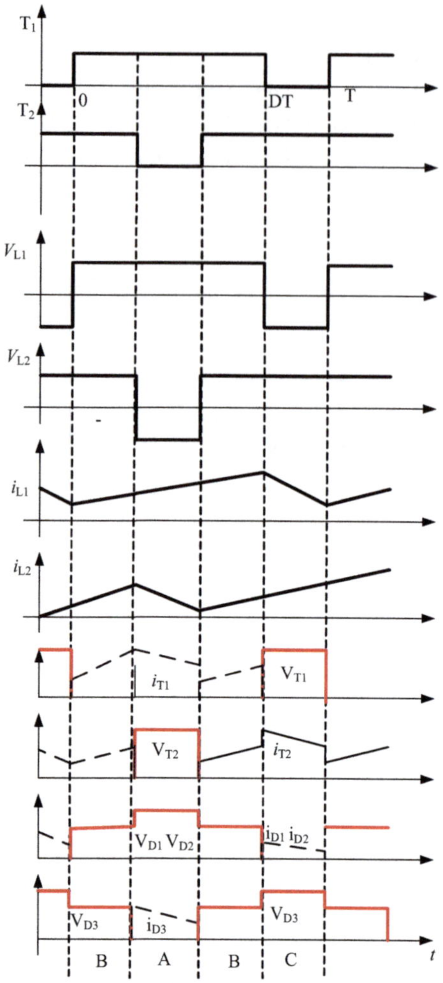

Fig. 6. Key operation waveforms of the circuit at 0.5 < D< 1

2nd-Modal B and Modal D:

$$\begin{cases} V_{in} = L_1 \frac{di_{L1}}{dt} + V_{C2} = L_1 \frac{di_{L1}}{dt} + V_{C1} \\ V_{in} = L_2 \frac{di_{L2}}{dt} \end{cases} \tag{7}$$

3rd- Modal C:

$$\begin{cases} V_{in} = L_1 \frac{di_{L1}}{dt} \\ V_{in} = L_2 \frac{di_{L2}}{dt} \end{cases} \tag{8}$$

Combining formulas (6)–(8) and switching Law get capacitor voltage and ratio of output to input voltage, i.e.

$$V_{C1} = V_{C2} = \frac{V_{in}}{1-D} \tag{9}$$

$$\frac{V_O}{V_{in}} = \frac{3}{(1-D)} \tag{10}$$

Review, due to the complementary control compared with staggered control make the power output of the current fluctuation is bigger, more damage to devices in the circuit, and the staggered parallel topology has inhibition of output current ripple, reduction of output filter and expand the obvious advantages of system power output capacity, so the staggered control is advantageous to the circuit output voltage stable and high quality. But, if there is no T_3, there is a mode of circuit breaking under staggered control in a T_s cycle, which makes the inductance L_1 and L_2 in the circuit not satisfy the ampere-second balance, and the capacitance C_1, C_2 and C_3 don't satisfy the volt-second balance, the circuit doesn't have the effect of boosting the voltage. Therefore, the addition of T_3 can automatically convert staggered control to complementary control. In addition, T_3 doesn't work, but the circuit still works in the staggered control state at $0.5 < D < 1$. According to Eqs. (5) and (10), when the duty cycle is 0.5, the boost ratio of the circuit is the same as 6. Therefore, T_3 enables smooth conversion between complementary control and staggered control and providing a basis for closed-loop control.

In general, the duty cycle ratio of DC-DC converter at steady state operation is greater than 0.5 in PV power generation system, so this paper focuses on the working state of $D > 0.5$.

3 The Mathematical Model and Analysis of Converter Circuit

The mathematical model of the converter is established by using the state space average method and Euler's Formula. The mathematical model is established by the state space average method,which have to satisfy the assumptions, i.e., (1) The frequency of AC small signal and the converter's natural turning frequency should be much less than the switching frequency; (2) The amplitude of the ac component of each variable in the circuit must be much smaller than the corresponding dc component. Because the converter works at high frequency, the circuit can satisfy the above conditions.

Hypothetical state quantity: $x(t) = \begin{bmatrix} i_1(t) & i_2(t) & v_c(t) & v(t) \end{bmatrix}^T i_1(t)$ and $i_2(t)$ is the instantaneous current of inductance L_1 and L_2. $v_c(t)$ is the instantaneous voltage of C_1 and C_2, $C_1 = C_2 = C$. Input quantity: $u(t) = [v_{in}(t)]$. Output quantity: $y(t) = \begin{bmatrix} i_{in}(t) & v_o(t) \end{bmatrix}^T$, $i_{in}(t)$ is the input instantaneous current, $v_o(t)$ is output voltage. D is the on-off duty cycle of the switch.

3.1 The Mathematical Model and Analysis of Converter at 0.5 < D < 1

When 0.5 < D < 1, the equation of state and the output equation of the average variable are listed in sections according to the hypothesis and the above principle, i.e.,

$$\begin{bmatrix} i'_{L1}(t) \\ i'_{L2}(t) \\ v'_c(t) \\ v'_o(t) \end{bmatrix} = \begin{bmatrix} 0 & 0 & \frac{D-1}{L1} & 0 \\ 0 & 0 & \frac{2(1-D)}{L2} & \frac{D-1}{L2} \\ \frac{1-D}{2C} & \frac{D-1}{C} & 0 & 0 \\ 0 & \frac{1-D}{C_3} & 0 & \frac{-1}{RC_3} \end{bmatrix} \begin{bmatrix} i_{L1}(t) \\ i_{L2}(t) \\ v_c(t) \\ v_o(t) \end{bmatrix} + \begin{bmatrix} \frac{1}{L1} \\ \frac{1}{L2} \\ 0 \\ 0 \end{bmatrix} [v_{in}(t)] \tag{11}$$

$$\begin{bmatrix} i_{in}(t) \\ v_o(t) \end{bmatrix} = \begin{bmatrix} 1 & 1 & 0 & 0 \\ 0 & 0 & 0 & 1 \end{bmatrix} \begin{bmatrix} i_{L1}(t) \\ i_{L2}(t) \\ v_c(t) \\ v_o(t) \end{bmatrix} + \begin{bmatrix} 0 \\ 0 \end{bmatrix} [v_{in}(t)] \tag{12}$$

A signal disturbance is introduced into formula (11). The steady-state component and the quadratic component are eliminated. Meanwhile, separate and linearize the perturbation. Then, make the state vector, input vector and output vector correspond to the dc component vector as follows:

$$\begin{aligned} X &= [I_{L1} \quad I_{L2} \quad Vc \quad Vo]^T \\ U &= [V_{in}] \\ Y &= [I_{in} \quad Vo]^T \end{aligned} \tag{13}$$

The static working point of the circuit can be obtained by combining formulas (11)–(13), i.e.,

$$\begin{bmatrix} I_{L1} \\ I_{L2} \\ V_c \\ V_o \\ I_{in} \end{bmatrix} = \begin{bmatrix} \frac{6}{(D-1)^2 R} \\ \frac{3}{(D-1)^2 R} \\ \frac{1}{1-D} \\ \frac{3}{1-D} \\ \frac{9}{(D-1)^2 R} \end{bmatrix} V_{in} \tag{14}$$

When the converter works steady-state, The relationship between voltage and current is shown in formulas (15)–(18).

$$I_{L1} = \frac{6V_{in}}{(D-1)^2 R} = \frac{2V_o}{(1-D)R} \tag{15}$$

$$I_{L2} = \frac{3V_{in}}{(D-1)^2 R} = \frac{V_o}{(1-D)R} \tag{16}$$

$$V_C = V_{C1} = V_{C2} = \frac{V_{in}}{1 - D} \tag{17}$$

$$V_o = \frac{3V_{in}}{1 - D} \tag{18}$$

According to formulas (15)–(18), when $0.5 < D < 1$, the Boost ratio of the converter is 3/(1-D), and the voltage gain is increased by 3 times compared with the traditional Boost circuit. The voltage on capacitance C_1 and C_2 is divided automatically, and the current on inductor L_1 and L_2 is not divided equally, but there is a certain ratio of 2:1.

3.2 The Voltage Stress Analysis of Switch Tubes

According to the analysis of the main working modes of the converter mentioned above, ignore the on-off voltage drop of the diode and the switch tube. The voltage stress of T_1, T_2 and T_3 are $V_{s\text{-}T1}$, $V_{s\text{-}T2}$, $V_{s\text{-}T3}$, i.e.,

$$V_{s-T1} = V_{s-T2} = V_{s-T3} = \frac{V_o}{3} \tag{19}$$

The voltage stress of D_1, D_2 and D_3 are $V_{s\text{-}D1}$, $V_{s\text{-}D2}$, $V_{s\text{-}D3}$, i.e.,

$$V_{s-D1} = V_{s-D2} = V_{s-D3} = \frac{2V_o}{3} \tag{20}$$

The voltage stress of switching devices and diodes in a typical Boost circuit is the output voltage V_o. It can be seen from Eqs. (19) and (20) that the voltage stress of the switching device is 3 times lower than that of a typical Boost converter. The voltage stress of the diode is reduced to 2/3 of that of a typical Boost converter. Therefore, the new converter proposed in this paper can effectively improve the efficiency of the converter.

4 Simulated Analysis

In order to verify the correctness of the theoretical analysis, the above circuit is simulated and verified.The simulation parameters set by the circuit are $V_{in} = 30$ V, $L_1 = L_2 = 0.4$ mH, $C_1 = C_2 = 20$ uF, $C_3 = 200$ uF, f = 20 kHz, R = 20 Ω. Under the mode of $0.5 < D < 1$, D is selected to be 0.7 and the simulation results are shown in Fig. 7.

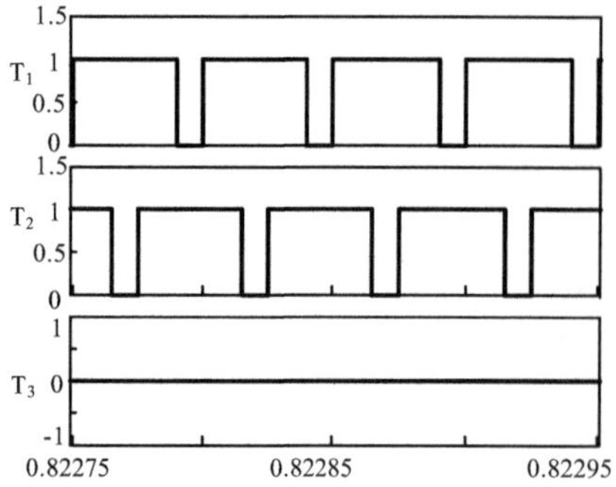

(a) T_1, T_2 and T_3 drive signal waveform

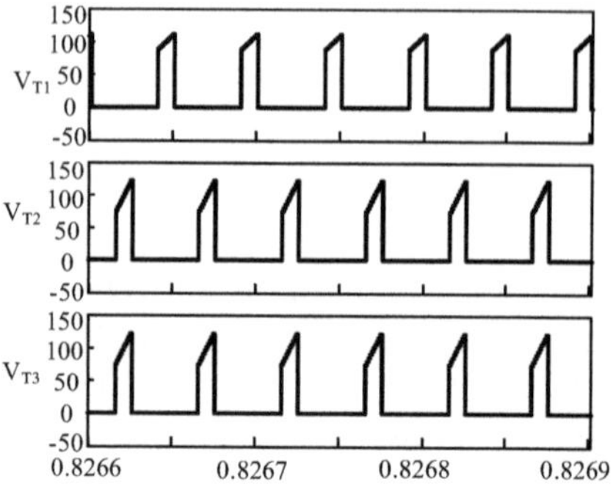

(b) T_1, T_2, T_3 voltage waveform

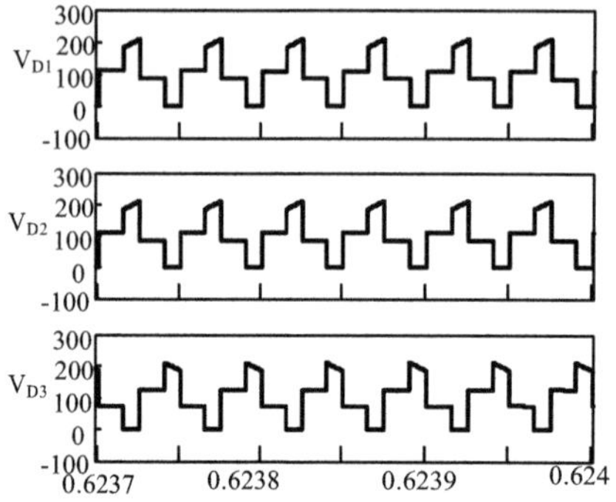

(c) Diode voltage waveform

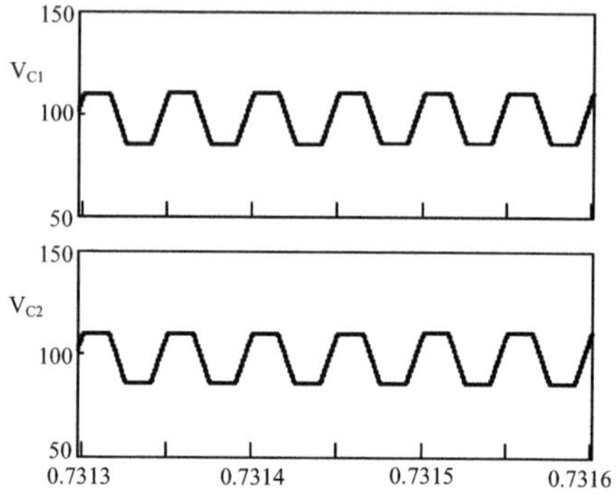

(d) C_1 and C_2 voltage waveform

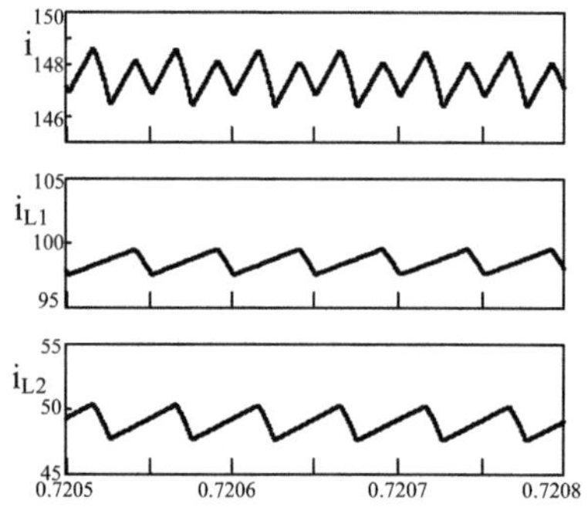

(e) Input current and inductive current waveform s

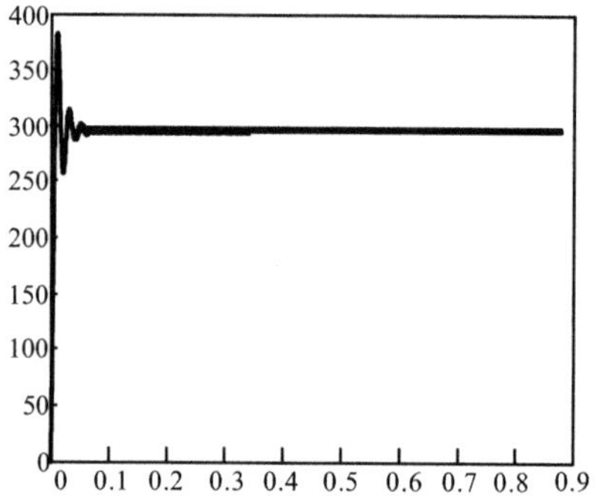

(f) Output voltage waveform

Fig. 7. Circuit simulation result diagram at $0.5 < D < 1$.

Table 3. Comparison of circuit parameters.

Converter parameter	Voltage gain	Switching tube voltage stress	Diode voltage stress
Traditional Boost circuits	1/1-D	V_o	V_o
New type converter	3/1-D	$V_o/3$	$2V_o/3$

From the above simulation diagram and compared to traditional Boost circuits, the voltage fluctuation of the new high boost converter is within 0–100 V. The voltage stress of the switch tube is about one third of the output voltage, and the voltage stress on the diode was also reduced by a third. The performance of the new high Boost converter compared with the steady state performance of the traditional Boost circuit is shown in Table 3. By comparison, the voltage gain of the converter is obviously increased in CCM mode, and the voltage stress of the switching device is effectively reduced, which has a positive effect on the PV cells power generation system.

5 Conclusions

For photovoltaic power generation systems requiring high gain converters and high efficienc, the traditional converter can not meet the requirement. Therefore, a novel high Boost ratio stagger Boost converter is proposed in this paper. It has the following characteristics: (1) High step-up ratio; (2) Input current ripple is small; (3) Simple control method. The circuit topology consists of three switches T_1, T_2 and T_3. The switches T_1 and T_2 are controlled by phase shift, and the driving signals of the two are staggered by 180°. The driving signals of T_3 are obtained by the driving signals of T_1 and T_2 NOR. T_2 and T_3 are in parallel in the circuit, T_3 will interlace control automatically into complementary control at $0 < D < 0.5$, therefore, conversion circuit can work normally in the range of 0 to 1.

Funding Details: Outstanding young talents support program of universities of Anhui Province (Grant no. gxyq2019149), Natural science research project of universities of Anhui Province (Grant no. KJ2019A1275), Natural science research project of universities of Anhui Province (Grant no. KJ2018A0618).

References

1. Wang, L., Hu, X.F.: A high gain Boost converter suitable for fuel cell power generation. J. Power Supply **5**, 6–66 (2014)
2. Zhang, F., Yue, B.B., Zhang, Y.: Maximum power tracking of solar power generation system. Power Source Technol. **39**(11), 2549–2551 (2015)
3. Samuel, V.A., Peter, Z., Regine, M.: Highly efficient single-phase transformer less inverters for grid-connected photovoltaic systems. IEEE Trans. Ind. Electr. **57**(9), 3118–3128 (2010)
4. Wu, Z.D.: Study on new high-step boost converter. Electr. Meas. Techniques **4**(3), 59–61 (2017)

5. Wu, G., Ruan, X.B., Ye, Z.H.: Nonisolated high step-up DC-DC converters adopting switched-capacitor cell. IEEE Trans. Circuits Syst. **62**(1), 383–393 (2015)
6. Brovont, R.: Cuzner DM and CM modeling of non-isolated buck converters for emifilter design. In: Proceedings IEEE Transportation Electrification Conference and Expo, pp. 140–145 (2018)
7. Brovont, D., Pekarek, S.D.: Derivation and application of equivalent circuits to model common-mode current in microgrids. IEEE Emerg. Top. Power Electron. **5**(1), 297–308 (2017)
8. Hu, P., Ji, X.L., Yang, J.: High Step-Up Ratio Interleaved Boost circuit. Telecom. Power Technol. **28**(1), 4–7 (2011)
9. Duan, R.Y., Lee, J.D.: High-efficiency bidirectional DC-DC converter with coupled inductor. IET Power Electron. **5**(1), 115–123 (2012)
10. Miao, S., Wang, F.Q.: A novel buck-boost converter with low stresses on switches and diodes. In: IEEE International Power Electronics and Motion Control Conference Asia, pp. 290–302 (2016)

CAUX-Based Mobile Personas Creation

Mo Li and Zhengjie Liu(✉)

Dalian Maritime University, Dalian 116026, China
644630224@qq.com, liuzhj@dlmu.edu.cn

Abstract. In order to solve some problems occurred when creating mobile personas, especially at data collection and data analysis period, a CAUX (Context-awareness User Experience)-based method is used in this paper. We use CAUX tool to collect objective data such as APP using APP usage information, mobile phone power, Call/SMS information etc., and then doing data visualization and analyze data, display user's mobile phone usage with Gantt chart. Next, summarize behaviors of different users when using mobile APP, and classify according to a certain dimension. Finally, form a set of personas. We create mobile personas with CAUX tools, which compared with traditional method is more concentrate on users'real behaviors. The CAUX-based personas focused on data collection and data analysis, which can effectively solve the problems when creating personas and leads to a better performance when using.

Keywords: Personas · Context-awareness · Data visualization

1 Introduction

With the development of Internet technology and prosperity of mobile devices, User Experience (UX) has been increasingly valued and recognized by the industry community. As a very important research method in the field of user experience, personas has been increasingly used and widely recognized. The personas method was first proposed by Alan Cooper, who defined persona as "a collection of images made up of massive amounts of data from user information."[1] Personas can describe user's needs, characteristics, and interest preferences. Therefore, the personas can assist in the development of products, help developer focus more on user's real needs, and improve user experience of products.

After years of usage, personas method is increasingly considered to be a good way to assist product development [2, 10] but in the research process, it also exposes many problems, especially in the data collection and data processing stages. Following problems are considered as disadvantages when creating personas:

1) Time-consuming and need great effort, traditional personas needs to recruit a number of users, negotiate with users about time and place to do interview. After a long time user interview, researchers can get the user's data, and then do summarize. Finally get the personas. In traditional method, user interviews take a lot of time and researchers' effort.

Q. Liu et al. (Eds.): CENet 2020, AISC 1274, pp. 1033–1039, 2021.
https://doi.org/10.1007/978-981-15-8462-6_119

2) Lack of objectivity. Because of create personas needs data and most of the user's data is obtained through user's feedback. Therefore, traditional methods often cannot obtain user information objectively when understanding user's specific behavior.
3) Authenticity issues, during user interview process, users will adopt a deceptive strategy for certain problems that may cause embarrassment (such as asking students how much time you spent on games every day), thus leads personas lack of authentic.

In response to these problems, many researchers worked hard. The "Data-driven personas" method adopted by Jennifer McGinn and Nalini Kotamraju solves problem of lack of objectivity in the process of establishing personas [3]. Jisu Ann, Soon-gyo Jung, etc. use APG (Automation Personas Generation) to automatically generate personas in social media applications by crawling real behavior data of users on Youtube website [4, 9]. Xiang Zhang and Hans-Frederick Brown etc. solved the problem of authenticity and objectivity by analyzing user's click flow and using machine learning technology to establish personas [5]. Rahimi, Mona etc. tries to create personas by feature gathering [6]. The efforts of the above researchers and other researchers have made contributions to solving problems in establishing personas, but the current methods are mostly for PC field and other fields, less consideration is given in mobile field.

In order to improve the problem of the establishment of mobile personas, the author proposes to use Context Awareness User Experience tool, referred to as CAUX to support creation of mobile personas. This method automatically collects user's APP usage status data and mobile phone status data. Then doing data visualization for semi-automatic analysis, summarizing user behavior, summarizing user characteristics, and finally establishing complete personas based on the user's basic information. The second section of this paper introduces CAUX and research solutions. The third section explores the capabilities that CAUX should have, and proposes visualization design. The fourth section conducts a formal case study, introduces the program, elaborates on the process of creating personas, and presents the final personas result, compare with traditional personas. The fifth section summarizes the full paper, puts forward shortcomings of the research, and looks forward to future improvement direction.

2 Introduction to CAUX and an Overview of the Research Plan

Due to the weakness of traditional data collection tools for collecting user data, such as ignoring context information, our team designed and developed the CAUX tool to trigger data collection actions and collect relevant data by capturing and judging low-level context information. The advantage of CAUX tool in data collection is that it only needs to be negotiated with users when doing recruitment. After that, the data collection work is automatically performed without disturbing the user and realizing effect of remote asynchronous data collection. Meanwhile, the data collected by CAUX is objective behavior of the user, and does not require the user to self-report. Therefore, in solving lack of objectivity and authenticity of the data, CAUX should be a good solution.

In previous research, our group successfully completed case studies such as mobile game behavioral needs analysis and intelligent public transportation software design, but none of them used CAUX to establish personas. Meanwhile, there has been no previous research on how to use CAUX tools to assist in the creation of personas. More experiments are needed to explore the capabilities that CAUX tools should have in personas creation and how to create personas with CAUX.

This article adopts the guiding principle of "learning by doing" [7], and iteratively explores it. Based on the existing CAUX database data, through experiments, try to solve the following problems: What kind of data can support mobile phone characters? Establish, UX researchers should choose which data visualization scheme, how UX researchers should analyze the data, and what dimensions of the final persona should contain. During the experiment, the problem was continuously found and improved, and the method of creating mobile personas supported by CAUX was found, then CAUX was used to conduct case study to obtain personas.

3 Exploration and Promotion of CAUX Capabilities

3.1 Abilities CAUX Should Have

According to Steve Mulder and Ziv Yaar's book [8], personas should contain the following information: goals, behaviors, opinions, demographics. The traditional method obtains the above information through user interviews and user self-reports. The CAUX tool can capture and record the user's behavior more comprehensively and carefully through pre-defined rules. The data that CAUX can collect is shown in Figs. 1, 2, 3 below.

app_id	app_state	app_time	app_running_state	app_name	user_id
1	[默认采集场景（App状态）	2017-03-07 10:57:18	开启服务	Monitor	1001
2	[默认采集场景(App状态)	2017-03-07 10:58:05	退到后台	Monitor	1001

Fig. 1. APP usage information

basic_id	basic_state	time	basic_electric	basic_latitude	basic_lontitude	speed	basic_network	user_id
1	{默认采集场景（基本信息）	2017-06-16 08:45:57	10	5e-324	5e-324	0	"i-LiaoNing"WIFI	1001
2	{条件触发场景（Screen off）	2017-06-16 08:46:30	10	5e-324	5e-324	0	"i-LiaoNing"WIFI	1001

Fig. 2. Phone status information

systm_state_id	system_des	system_time	system_name	system_state	user_id
19612	*基本状态	2017-06-18 13:41:48	电话	您接听了一个来电	1001
19613	*基本状态	2017-06-18 13:41:59	电话	您结束了一个0分10秒钟来电	1001

Fig. 3. Call/SMS information

Through the data collected by CAUX, the scene when the user uses the mobile APP can be restored and then researchers can understand user uses the mobile APP, and what features the user has when using different mobile APPs. Therefore, the author believes that CAUX can better complete the acquisition of "behavior" data in data collection of creating personas.

3.2 Data Visualization

An important part of establishing a persona is data analysis, which classifies users by analyzing their goals, behaviors, opinions, and other information. An important way in data analysis phase is data visualization. In view of CAUX's ability, the data visualization scheme should present the data collected by CAUX: user number, time of data, user's use of APP in chronological order, mobile phone power, which APP the user used, time statistics of the user's use of APP, The user's phone/text message status, the user's demographic information, and the resulting discovery editing area. Through the front-end technology, the data visualization interface is as shown in Fig. 4.

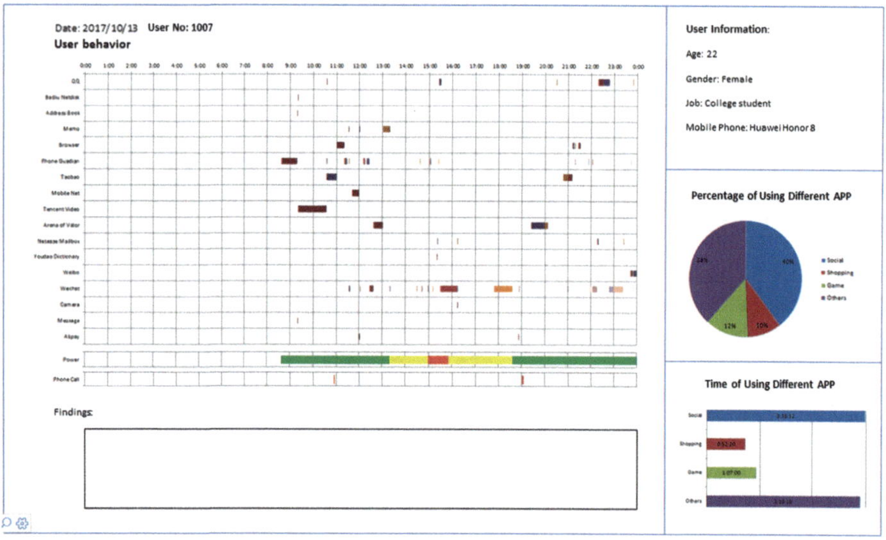

Fig. 4. Data visualization interface

4 Exploring Method of Creating Personas Based on CAUX

This section describes how to get user behavior through data visualization and generalize the behavior, ultimately deducing the personas and the final personas sample.

4.1 Preparation of Experiment

According to the existing data in CAUX database, six different user samples were selected for this experiment. The ratio of male to female was 1:2, and the age range of 21–25 years old, 7 days for each sample, for a total of 13756 data. Generate the final user role by the following dimensions:

1) Type, duration, and time period of the user's daily use of the APP
2) User behavior characteristics, such as whether to use new Apps, user hobbies, unique behaviors, phone power, and the impact of incoming Calls/SMS on users etc.
3) User's type.

4.2 Data Analysis and Methods to Create Personas

The data analysis is based on the visualization interface and relies on the behavioral pattern analysis method. There are two main types of data for analysis: the collected user operation data (APP usage status, APP usage time) and mobile phone status (power, incoming call/sms). Data analysis is performed as follows:

Step 1: Summary of the behavior of a single user for one day. First of all, to determine which types of APP users use, the duration of each type of APP and proportion of time spent using mobile phone every day, to get the basic characteristics of the user, such as social emphasis, weighted games, and emphasis on learning. Secondly, analyze the specific APP behaviors used by users in different fields, including the duration of use, time period, and frequency. Thirdly, analyze impacts of incoming calls/sms and power on user's use of mobile APP.

Step 2: Summary of the behavior of a single user for multiple days. Firstly, summarize the basic features to clarify the specific characteristics of the user. Secondly, construct the behavior characteristics of a single user, analyze the multi-day APP usage behavior of previous step, and generate characteristics of user's daily use of the APP in the form of a timeline. Again, analyze how incoming calls/messages and power affect users. Finally, summarize unique behavior of the user. This gives a user's behavior, after repeat this step several times, the behavior characteristics of six users will be get.

Step 3: Generate personas. First, analyze the basic characteristics of six users and divide them into prototypes of several personas. Next, compare the APP user behavior timeline of 6 users, and summarize different personas use APP based on the usage time period and usage time of APP. Thirdly, analyze the impact of six users on the use of mobile APP in call/sms, and fill in the personas prototype. Finally, review personas, put in photos, demographic information, and give adjectives based on the user's basic characteristics, and get the final personas. A sample is shown in Fig. 5.

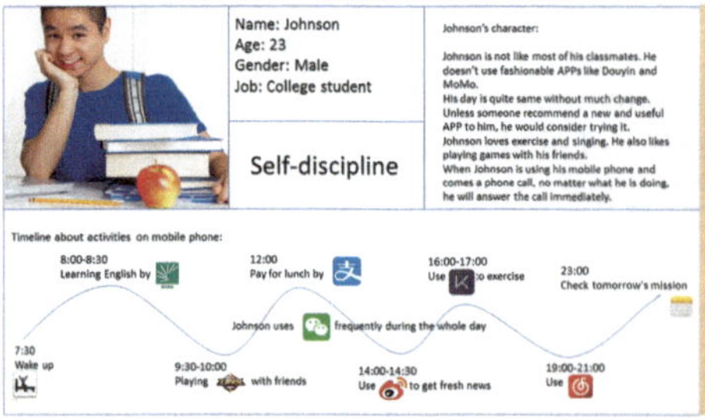

Fig. 5. Personas created by CAUX

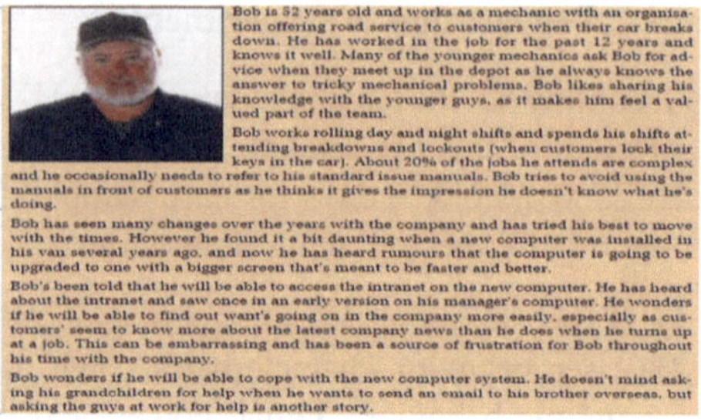

Fig. 6. Traditional personas

4.3 Compare with Traditional Personas

By comparison, it can be found that CAUX-based personas is more concise and direct than traditional personas [8] (showed in Fig. 6) in clearly defining the user's behavior, and can better display user's daily activities. Meanwhile, CAUX-based personas can more accurately summarize user behavior than traditional personas, and can more accurately describe how users use mobile phones for activities in a day. Besides, a comparison with traditional method on UX researchers' feelings and performance when using personas is shown in Table 1.

Table 1. UX researchers' feelings and performance

UX researchers' feelings and performances	CAUX-based	Traditional
Time to understand personas	2 min	5 min
Ease of use	Yes	No
Focused on users' real behavior	Yes	No
Findings when doing product design	12	7
Findings when doing product test	16	11

5 Conclusion

Based on context-aware technology, the author proposes a method of using CAUX tools to assist in creation of personas, that is, using CAUX to automatically collect user data, and then through semi-automatic methods, analyze data based on user behavior, and finally get the personas. Compared with traditional method, CAUX-based personas can save researchers' time and effort cost in data collection phase. Researchers do not need to conduct user interviews, and because automated data collection can lightly disturbing users, the cost of recruiting users can also be reduced. In addition, user data collected by CAUX can reflect user's real behavior and can avoid user fraudulent behavior that may occur in traditional methods. Meanwhile, the objectivity of data can better solve problem that traditional method can only collect user's subjective data.

Personas created with CAUX have certain advantages over traditional methods, but due to limitations in time and cost, there are still some shortcomings. For example, for the user's subjective data collection, such as user's attitude, opinions and other information cannot be well obtained. In the data analysis stage, researchers are required to make a subjective judgment, and it is temporarily impossible to perform automatic and retrograde analysis by means of formulas. In the future, efforts should be made as far as possible in subjective data collection and automated analysis of data to continue to explore the capabilities of CAUX.

References

1. Cooper, A.: About Face 3-The Essentials of Interaction Design. John Wiley & Sons, Inc., New York (2007)
2. Stolterman, E., Chang, Y., Lim, Y.: Personas: From theory to practices. In: Nordic Conference on Human-computer Interaction Building Bridges, pp. 439–442. ACM (2008)
3. Mcginn, J., Kotamraju, N.: Data-driven persona development. In: SIGCHI Conference on Human Factors in Computing Systems, pp. 1521–1524. ACM (2008)
4. An, J., Kwak, H., Jensen, B.J.: Automatic generation of personas using youtube social media data. In: Hawaii International Conference on System Sciences. University of Hawaii. pp. 833–842 (2017)
5. Zhang, X., Brown, H.F., Shankar, A.: Data-driven Personas: Constructing Archetypal Users with Clickstreams and User Telemetry, pp. 5350–5359 (2016)
6. Rahimi, M., Clelandhuang, J.: Personas in the middle: automated support for creating personas as focal points in feature gathering forums. In: ACM/IEEE International Conference on Automated Software Engineering, pp. 479–484. ACM (2014)
7. Hippel, E.V., Tyre, M.J.: How learning by doing is done: problem identification in novel process equipment. Res. Policy **24**(1), 1–12 (2004)
8. Mulder, S., Yaar, Z.: The user is always right: a practical guide to creating and using personas for the web. Inf. Res. Int. Electr. J. **55**(1), 74–76 (2006)
9. Salminen, J., Jung, S.-G., Kwak, H., et al.: Persona perception scale: developing and validating an instrument for human-like representations of Data. In: CHI 2018 Late-Breaking Abstract, pp. 1–6 (2018)
10. Teka, D., Dittrich, Y., Kifle, M.: Adapting lightweight user-centered design with the scrum-based development process. In: SEiA' 2018, pp. 35–42. ACM (2018)

Exploration and Research on CAUX in High-Level Context-Aware

Ke Li[1(✉)], Zhengjie Liu[1], and Vittorio Bucchieri[2]

[1] Dalian Maritime University, Dalian 116026, China
452544364@qq.com, liuzhj@dlmu.edu.cn
[2] Harvard University, Massachusetts, Cambridge, USA
vittorio_bucchieri@harvard.edu

Abstract. With the rapid development and popularity of mobile devices, the aggressive adoption of mobile products poses new challenges to both users and User Experience (UX) researchers: Reducing the disruption to users when collecting subjective data, and recognizing users' subjective data. The goal of this study is to enable the existing context-awareness tools with the ability to collect and recognize high-level context data sought by UX researchers. In this study, the author selects the idle context of users as the research objective, asks users to mark their own data, and summarizes the mapping relationship between the combination of three groups of reliable low-level context information. The Context Awareness User Experience (CAUX) system, which combines context-aware with remote data collection technology, is used for this research. This study shows that when students are both in both a high-level and idle context, if the low-level context combination is used to judge, the accuracy rate is 88%. This method of judging high-level context based on the combination of low-level context information can effectively solve the defects in the perception of high-level contexts by the current remote context-aware tools, thereby solving one of the challenges with remote user experience evaluation.

Keywords: Context-Aware · High-Level context · User experience

1 Introduction

As mobile devices become more popular, their user experience is gaining more attention by researchers. User Experience (UX) refers to a person's emotions and attitudes about using a particular product, system or service. With the development of mobile Internet, the use of communication products has gradually penetrated into all aspects of the social and professional life, and its adoption and engagement have become increasingly diverse and complex.

Traditional research methods based on the field and the laboratory have difficulties to simulate a real-world environment [1], but this intrinsic state has a significant impact on the validity of the collected data [2]. Therefore, the mobile user experience data collection tool is a promising solution for collecting users' data without affecting the realistic context [3].

Q. Liu et al. (Eds.): CENet 2020, AISC 1274, pp. 1040–1047, 2021.
https://doi.org/10.1007/978-981-15-8462-6_120

As research objectives evolve, UX researchers aim to use data collection tools to capture users' subjective feeling. In the traditional method, researchers need to recognize users' context and then collect specific context-data [4]. To meet this goal, researchers have developed a context-aware user experience data collection system (CAUX - Content-Aware User Experience) based on mobile data collection methods. Since this system allows only for preliminary exploration of context-aware data, achieving the collection of high-level, subjective context-aware data is the focus of this paper. The CAUX system includes both a mobile application (the client) and a server. Once this application is installed on a mobile device, the CAUX collects user data according to the user location and interactions with other mobile apps.

The contribution of this paper is to propose a redesign of the processing function of the context-awareness data collection tool by combining the low-level data with the high-level context data labelled by users' tagging.

This study comprises of:

1. An overview of context-awareness tools and research
2. An introduction of high-level context research
3. Recommendations for improving the current context-awareness tool (CAUX)
4. A summary of this research, including the advantages and disadvantages of CAUX, and recommendations for future development.

2 Overview

2.1 Context and Context-Aware Tools

There is currently no consensus on the definition of the term "context" in academia. The concept of context-aware was first proposed by Bill Schilit in 1994. Schilit declared that three important aspects (i.e. where you are, who you are with, and what resources are nearby) of context [5].

As technology evolved, in 2000 Anind Dey gave a more complete definition: Context is any information that can be used to characterize the situation of an entity. An entity is a person, place, or object that is considered relevant to the interaction between a user and an application, including the user and applications themselves [6].

User Experience defines context as comprised by the user, environment, task, society, location, time, device, and infrastructure [7]. The context can be further divided into a Low-Level Context (LCX) and a High-Level Context (HCX). Low-level context can be directly perceived through the collected raw, quantitative data, such as electricity consumption, temperature, timed events, etc. High-level context is more complex and it cannot be directly perceived just by data, as for enjoying a meal, entertaining a friend, reacting to an experience, etc.

User Experience research is increasingly employing context-aware data collection systems. Such systems use sensors to obtain relevant context information from and about the user. Current context-aware systems generally tend to target a specific product or service [8], such as auto vehicle intelligent energy systems or marketing recommendation services. However, these systems are based on low-level context information. Data collection tools for high-level context research are still scarce.

The CAUX tool is a context-aware user experience research tool that combines context-aware with remote data collection technology. The CAUX system includes both a mobile application (the client) and a server. Once this application is installed on a mobile device, the CAUX collects user data according to user location and interactions with other mobile apps. This data is periodically uploaded to the to the server, where UX researchers can manage the collected data. Since the CAUX tool is still in the initial exploration phase, only low-level context can currently be identified. The improvement of its functionality would potentially access and collect also high-level context-aware data. Schematic diagram of UX researchers using the CAUX system for research is shown in Fig. 1.

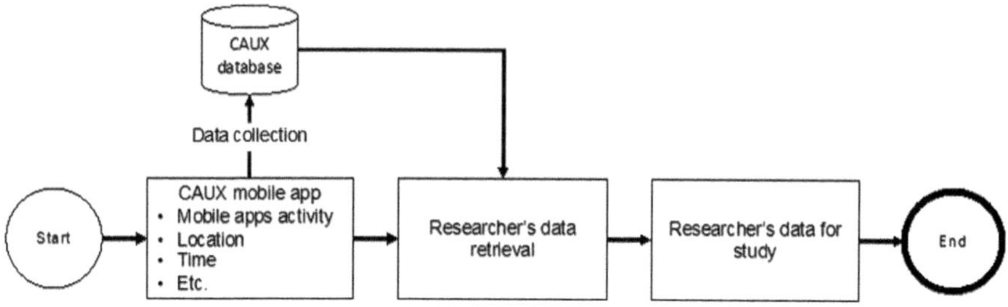

Fig. 1. Schematic diagram of UX researchers using the CAUX system for research

2.2 Research Overview

In the early stages, UX researchers selected college students as target user groups, with their idle context as the high-level context object for research. Once the feasibility of low-level context set inference and the high-level context were verified, the context-aware function of the tool was improved by acquiring a low-level context recognition. The goal to combining the capability enhancement tool for high-level context-aware was achieved.

In order to enhance the CAUX's ability to recognize high-level context, the following data is collected and studied:

- The definition of high-level context defined through user interviews
- The collection of subjective data, such as questionnaires and photography
- The context of users in the idle state (as opposed to changing location)

By studying accurate high-level context and the mapping relationship between low-level and idle context the tool capability can be improved.

3 High-Level Context-Aware Research

During user research, UX researchers often need to interrupt users' activities and ask to provide subjective data through questionnaires or logs, although they would prefer to avoid disturbing the user. For this study, college students were chosen as target user groups, with idle and high-level context as the research objective.

3.1 Data Collection

In this study, 6 college students were recruited according to:

- 1:1 male to female ratio
- Graduate and undergraduate academic level
- Users of Android phones with at least 4 different mobile apps, such as social communication, shopping, music, games, etc.

Users' data marking and collection was carried out for 10 days. A total of 52,859 data points was collected.

3.2 Data Marking and Summary Mapping

After user's data was collected for 7 days, we asked each user to mark the data collected during the idle context. Based on the prior knowledge of the target user group in the idle context, we extract, classify and analyze the common features of the low-level context from these data, analyzed and the mapping of 4 low-level context sets was summarized. As shown in Table 1.

Table 1. Idle state mapping of 4 sets of low-level contexts

LCX combination	App type	Location	Time
1	Music	Outdoor	
2	Music	Indoor	
3	Social communication		Lunch or Dinner
4		Dorm room	Holidays

3.3 Mapping Verification

After collecting data for 10 days, the researcher tagged the data of the last 3 days according to the rules summarized in the above table, while again sending this data to users for context tagging. Table 2 below compares the data marked by the researcher with the data marked by the users.

Table 2. Idle state mapping of accurate 4 sets of low-level contexts.

LCX combination	Number of tags (researcher)	Number of tags (user)	Accuracy
1	72	68	94%
2	1	1	100%
3	74	60	81%
4	30	27	90%

To improve the tool's awareness of the idle context, using the combination of LCX 1, 3, 4 as the law of mapping, the accuracy rate should be about 88%.

4 CAUX High-Level Context-Aware Enhancement

Collection of user data is a critical functionality of the CAUX client. The CAUX mobile client consists of four modules:

1. Data collection
2. Instruction parsing
3. Context detection
4. Communication.

4.1 Functional Structure of the CAUX Client

1. The data collection module collects low-level situation information and corresponding user data according to the events of the mobile applications. Currently, collected user data is divided into two categories: objective and subjective. Subjective data is collected by the user, as shown in Table 3. Objective data is collected by the tool, as shown in Table 4.

Table 3. Subjective data types

Application	Device component or mobile application	User activity
Photo	Camera	Taking a picture
Video	Camera and microphone	Recording video
Audio	Microphone	Recording audio
CAUX questions	Questionnaire on a pop-up window	Typing answers
CAUX log	CAUX mobile client	Typing log information

Table 4. Objective data types

Data type	Details
Basic information of the device	Including model and name, Operating system version, Screen parameters
Battery power	Current battery power
Networking status	Current network status of the phone
Geographic location	Longitude and latitude coordinates and specific address of the location
Time	Current time of the system
Application event	Including the application's opening, closing, switching to the front and back events
Telephone call	Detecting the outgoing event, the incoming call ringing event, the incoming call answering event, the incoming call is not connected to the hang up event, the incoming call is connected to the hang up event, and the call duration can be recorded
SMS sending and receiving	Sending and receiving message events
Music player events	Player playback, pause, stop activity
Screen lock events	The user lights the screen, unlocks the screen, and locks the screen

2. The instruction parsing module parses the instruction file obtained from the communication module by calling the ParseJson utility class to set the trigger condition and to trigger the operation of the next stage.
3. The context detection module detects real-time events triggered by user interactions. Upon each triggered event, the CAUX tool initiates both the subjective data collection module, according to the instruction file, and the objective data collection module.
4. The communication module in the CAUX mobile application periodically communicates to the server the instructions set by the researcher. It also sends the collected user data to the server-side communication module, which stores it in the server-side database.

The workflow of the CAUX client is shown in Fig. 2.

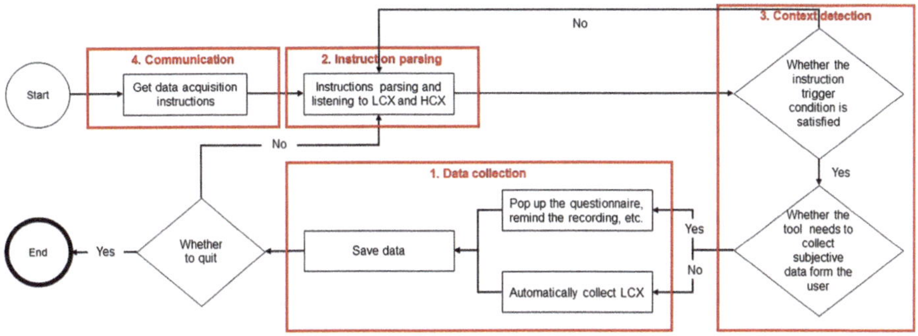

Fig. 2. CAUX client workflow

4.2 Improvement of the Context-Aware of Tools

The existing instruction of the parsing function can only identify a single low-level context, and then make a service call to collect data that is useful to User Experience researchers.

In order to improve the context processing capability of the tool, researchers should improve the instruction for the parsing module by allowing a single low-level context recognition to identify multiple low-level context information.

To use as an identifiable defining condition, the high-level context corresponding to multiple low-level contexts has to be marked by users, initialized and attached to the true value.

This can be done by

1. Adding to the parsing module a plurality of low-level context information
2. Judging whether the high-level context is true
3. Assigning a value to the specified high-level context in the service call
4. Initializing the value of the high-level context

LCXT a, b, c… represents the collected low-level context-data. The value of HCXT a,b,c… indicates the true state of the high-level context.. Here is an example of the instructions for adding high-level context.

if $LCXT_a$ = y and $LCXT_b$ = y and $LCXT_a$ = z/*multiple LCXT are satisfied. */
then $HCXT_A \leftarrow$ true/*Set the high-level context A at this time to true. */
if $HCXT_A$ = true/*Judge whether the high-level context A is true. */
then Data collection/*Collect data set by UX researchers. */

After the improvement, the user data collected is as shown in Fig. 3.

Prior to this improvement, when the software specified by the UX researcher was pushed to the background, the tool recognized the LCX.

Before improvement:
[LCXT(App Event),2018-3-28 6:35:44,back,WeChat,1009]
{LCX(WeChat Back),2018-4-12 6:35:44,50,38.868739,121.528318, 0,MOBILE,325.0,
{Collecting subjective data,Pop up the questionnaire, http：//wjx.top/30829182.aspx}
[LCXT(App Event),2018-3-28 6:37:46,Open,WeChat,1009]
[LCXT(App Event),2018-3-28 6:41:21,back,WeChat,1009]
{LCX(WeChat Back),2018-4-12 6:41:21,50,38.868739,121.528318, 0,MOBILE,325.0,1,
{Collecting subjective data,Pop up the questionnaire, http：//wjx.top/30829182.aspx}

After improvement:
[LCXT(App Event),2018-8-12 09:42:29,back,WeChat,1009]
[LCXT(App Event),2018-8-12 09:42:31,Open,Weibo,1009]
[LCXT(App Event),2018-8-12 09:42:37,back,Weibo,1009]
[LCXT(App Event),2018-8-12 09:42:39,Open,WeChat,1009]
[LCXT(App Event),2018-8-12 09:42:42,back,WeChat,1009]
[LCXT(App Event),2018-8-12 11:22:25,Open,WeChat,1009]
[LCXT(App Event),2018-8-12 11:27:42,back,WeChat,1009]
{HCX(IDLE Context),2018-4-12 11:27:42,50,38.868739,121.528318, 0,MOBILE,325.0,1,
{Collecting subjective data(IDLE Context),Pop up the questionnaire, http：//wjx.top/30829182.aspx}

Fig. 3. Data comparison before and after improvement

Frequently showing the questionnaire to collect subjective data, disturbed the user throughout the day, affecting the user's experience of the mobile device, and especially when the user was in motion. These interruptions led to the collection of poor-quality subjective data, as shown in the red highlight in Fig. 3.

After the tool was improved, it showed the questionnaire only when it detected the user's idle context, which greatly reduced distractions. As shown in the green highlight in Fig. 3, the tool detected the use of communication software at lunch time, determined the idle context, and showed the questionnaire for subjective data collection.

5 Conclusion

This study improved the capability of the CAUX tool to collect and define high-level context-aware data and therefore reduce the users' interruption when collecting users' subjective data. Although the recommended approach cannot yet completely replace human's judgment, it demonstrates improvements that may lead toward that solution, especially with enhanced mobile devices and high-level mapping between low-level and high-level context information.

1. At present, the low-level context types collected by the tools are not comprehensive enough, and are limited by the types of sensors in mobile devices. In future research, more sophisticated sensors could obtain lower-level context information of more dimensions.

2. Presently, the mapping between the high-level and the low-level context of the tool is still in the preliminary exploratory stage. Future efforts will allow to study the mapping between more different high-level and low-level context sets.

We believe that the high-level context-aware ability of the tool will be enhanced so that UX researchers will be able to capture better quality data with minimum disruption to users.

References

1. Christensen, K.P., Mikkelsen, M.R., Nielsen, T.A.S., et al.: Children, mobility, and space: using GPS and mobile phone technologies in ethnographic research. J. Mixed Methods Res. **5**(3), 227–246 (2011)
2. Hassenzahl, M., Tractinsky, N.: User experience-a research agenda. Behav. Inf. Technol. **25**(2), 7 (2006)
3. John, S.I., Rondoni, J., Kukla, C., et al.: A context-aware experience sampling tool. In: CHI' 2003 Extended Abstracts on Human Factors in Computing Systems. ACM (2003)
4. Tu, Z.: Auto-context and its application to high-level vision tasks. IEEE Trans. Pattern Anal. Mach. Intell. **32**(10), 1744–1757 (2010)
5. Schilit, B., Adams, N., Want, R.: Context-aware computing applications. In: Workshop on Mobile Computing Systems & Applications (1994)
6. Dey, A.K.: Understanding and using context. Pers. Ubiquit. Comput. **5**(1), 4–7 (2010)
7. Han, L., Liu, Z.J., Li, H., et al.: A method based on context-awareness for remote user experience data capturing. Chinese J. Comput. **11**, 8 (2015)
8. Nebeling, M., Speicher, M., Norrie, M.: W3Touch: metrics-based web page adaptation for touch. In: ACM SIGCHI Conference on Human Factors in Computing Systems. ACM (2013)

Research on Photovoltaic Subsidy System Based on Alliance Chain

Cheng Zhong[1(✉)], Zhengwen Zhang[1], Peng Lin[2], and Yajie Zhang[1]

[1] State Grid Xiongan New Area Power Supply Company, Heibei 071000, China
zhongc@126.com
[2] Beijing Vectinfo Technologies Co., Ltd., Beijing 100088, China

Abstract. At present, the photovoltaic subsidy system is based on the centralized trusted third-party operation, which has the problems of data tampering, loss and non-traceability. Blockchain technology has the characteristics of decentralization. The application based on blockchain technology does not need the intervention of a trusted third party. Any two nodes can directly trade, making the transaction data not easy to be tampered with, lost, and traceable. In this paper, the blockchain technology is applied in the field of photovoltaic subsidy, and a photovoltaic subsidy system based on the alliance chain is developed, which solves the above shortcomings of the original photovoltaic subsidy system and makes the data in the system real and reliable.

Keywords: Blockchain · Photovoltaic subsidy · Alliance chain

1 Introduction

In 2009, Satoshi Nakamoto proposed the blockchain technology in "Bitcoin: A peer-to-peer electronic cash system". This technology can enable two parties without a foundation of trust to conduct transactions directly without relying on trusted third parties [1]. At present, blockchain technology is highly valued by governments, banks and other institutions. NASDAQ in the United States and 90% of global central banks are actively exploring the application of blockchain. It is speculated that the GDP brought by the blockchain in 2025 will account for about 10% of global GDP [2]. It can be seen that blockchain technology has broad development prospects in the future. Most reliable transactions on the Internet need to rely on trusted third-party agencies, which uses a credit-based model. Credit-based transactions have the following problems:

1) Data is difficult to share. Because the data is only managed by third-party organizations, it is difficult to achieve data sharing.
2) The data is not secure. Data is easy to tamper with: Data information is managed by third-party agencies and may be tampered with by criminals; Data is easy to lose: All data information is stored on third-party organizations. If an unrepairable failure occurs in the third-party organization, all data may be lost; Data is not traceable: The data kept by third-party organizations is not traceable and cannot guarantee the authenticity and reliability of the data.

Q. Liu et al. (Eds.): CENet 2020, AISC 1274, pp. 1048–1055, 2021.
https://doi.org/10.1007/978-981-15-8462-6_121

Blockchain technology emerged to solve these problems [3]. Blockchain technology is based on P2P network operation. Its biggest feature is decentralization, which enables two parties without a foundation of trust to conduct transactions directly. The essence of the blockchain is a globally synchronized distributed ledger [4]. The ledger records all transactions, and each transaction has the characteristics of immutability, loss and traceability, which greatly improves the security of the transaction [5]. Among them, the data cannot be tampered with because the data on the blockchain will be difficult to modify once it is on the chain. If a malicious node wants to tamper with the data on the chain, it needs to gather more than 50% of the computing power of the entire network, which is almost impossible to achieve. The data is not easy to lose because all nodes in the entire network keep a complete transaction record, and a down of a node will not cause data loss [6]. The traceability of data means that the hash of the previous block is stored in the header of each block, and the hash can be used to trace back to the previous block, in this way, you can trace the previous block in turn to find the source of the data [7].

At present, the photovoltaic subsidy system is operated based on a centralized and trusted third party, and there are problems with data security such as easy data tampering, easy loss, and untraceable source. The blockchain technology has the characteristics of de-centralization [8], which can solve the above data security problems. Applications based on blockchain technology can allow any two nodes to conduct transactions directly without the intervention of a third party. This paper designs a photovoltaic subsidy system based on the alliance chain, so that the data generated by photovoltaic subsidies are not easy to tamper with, not easy to lose, and traceable.

2 System Design

2.1 Demand Analysis

In order to achieve the photovoltaic subsidy function between power generation individuals, governments, banks and grid companies, after repeated analysis, the functional modules mainly included in the photovoltaic subsidy system should include four major functional modules: user information management, electricity meter information management, electricity price information management, and photovoltaic subsidy management. The user information management module includes five function points: registration, login, user information query, user information modification, and logout. Considering the security of the system, the registration function can only register users in the individual role of power generation. Users in the government, bank, and grid company can only register through offline application with related staff verifying user identity. The electricity meter information management module includes four functions: querying the meter list, adding new meters, modifying meter information, and deleting meters. This module can be accessed by the users in the individual role of power generation, because only users in this role own the meter. The function of querying the meter list will query the information such as the meter number, the name of the meter, and the amount of power stored in the meter. The electricity price information management module includes two function points for querying electricity prices and modifying electricity prices. Among them, only users in government have the function of modifying electricity prices. The photovoltaic subsidy management module includes two functions as trading electricity and querying transactions, only users with individual role of power generation have the function of trading electricity. Because users of different roles pay different attention to transaction information, the transaction information queried by users of different roles will be different.

From the above, the use case diagram of the photovoltaic subsidy system is shown in Fig. 1. The detailed design of each functional module will be explained in the subsequent chapters.

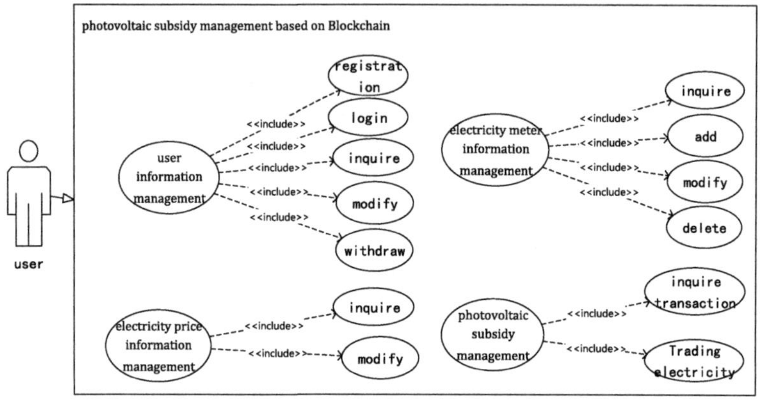

Fig. 1. System use case diagram.

2.2 Analysis and Design of User Information Management Module

The user information management module is used to manage user information. This module includes functions such as registration, login, viewing user information, modifying user information, and logout. The specific process is shown in Fig. 2. All functions of this module interact with the smart contract on the blockchain through web3.js. The write operation will be stored in the blockchain in the form of a transaction, and the read operation will take the corresponding data through the smart contract and return it to the front end.

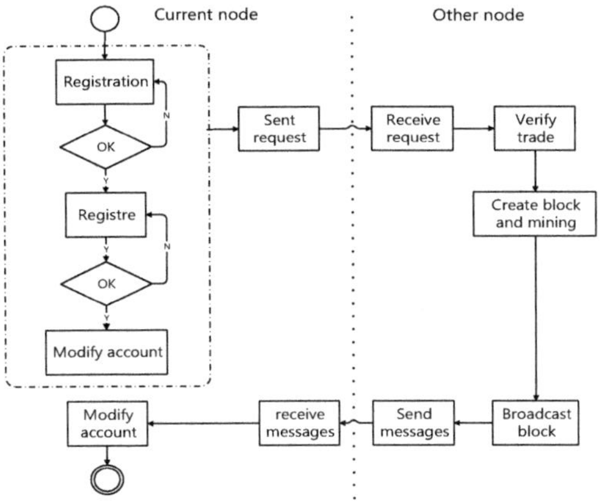

Fig. 2. Flow chart of user information management module.

2.3 Electricity Meter Information Management Module

This module includes functions such as querying the meter, adding a new meter, modifying the meter, and deleting the meter. The specific process is shown in Fig. 3. After the individual power generation user logs in to the system, they will have access to the module, and non-power generation individual users will not have access to the module. All functions of this module interact with the smart contract on the blockchain through web3.js. The write operation will be stored in the blockchain in the form of a transaction, and the read operation will take the corresponding data through the smart contract and return it to the front end.

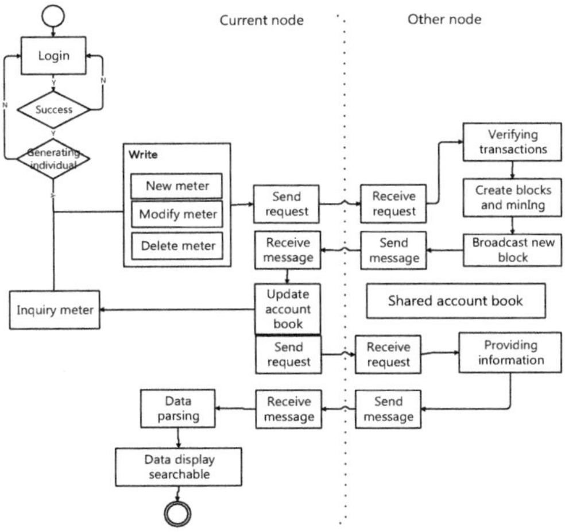

Fig. 3. Flow chart of electricity meter information management module.

2.4 Electricity Price Information Management Module

The electricity price information management module is used for users to manage the electricity price information. This module includes functions such as querying electricity prices and modifying electricity prices. After logging in to the system, users will have access to the module, but non-government users only have the right to view electricity prices. The process for non-government users to query electricity prices is shown in Fig. 4. Government users have the right to modify electricity prices, and government users have the right to modify electricity prices. The process is shown in Fig. 5. All functions of this module interact with the smart contract on the blockchain through web3.js. The write operation will be stored in the blockchain in the form of a transaction, and the read operation will take the corresponding data through the smart contract and return it to the front end.

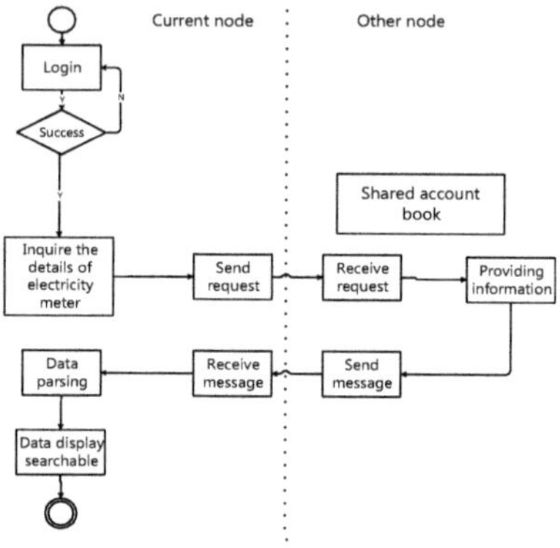

Fig. 4. Flow chart of non-government users querying electricity prices.

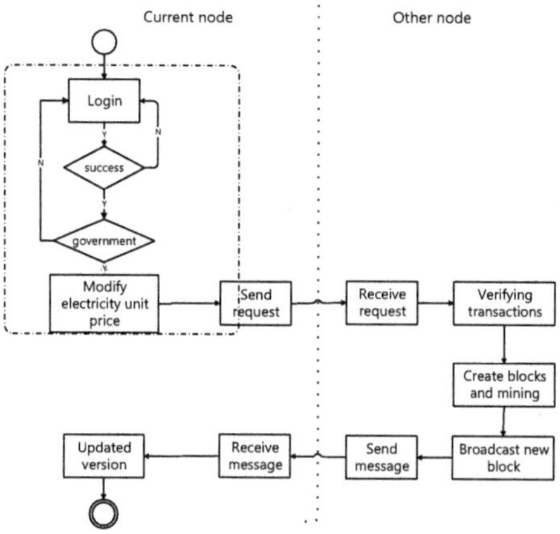

Fig. 5. Flow chart of government users modifying electricity prices.

2.5 PV Subsidy Management Module

The photovoltaic subsidy management module is used to manage the transactions among the four types of users: power generation individuals, grid companies, banks, and governments. This module includes functions such as trading electricity and querying transactions. The specific process is shown in Fig. 6. All functions of this module interact with the smart contract on the blockchain through web3.js. The write operation will be stored in the blockchain in the form of a transaction, and the read operation will take the corresponding data through the smart contract and return it to the front end.

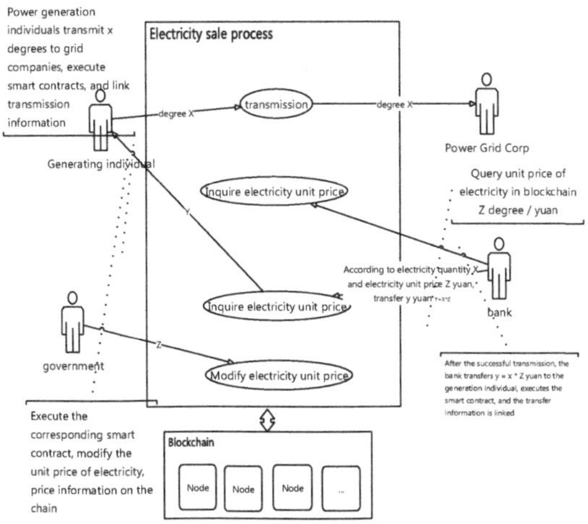

Fig. 6. Flow chart of photovoltaic subsidy management module.

3 System Implementation and Verification

The startup script is executed, the hyperledger blockchain network is started, and the output in the console during the startup process is shown in Fig. 7.

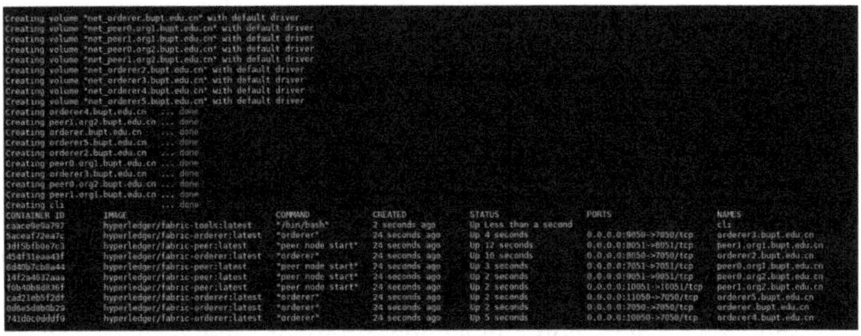

Fig. 7. Super ledger network startup.

Run Ganache to start the Ethereum blockchain network. After the startup is completed, the Ethereum blockchain account tab and log tab are shown in Fig. 8 and Fig. 9.

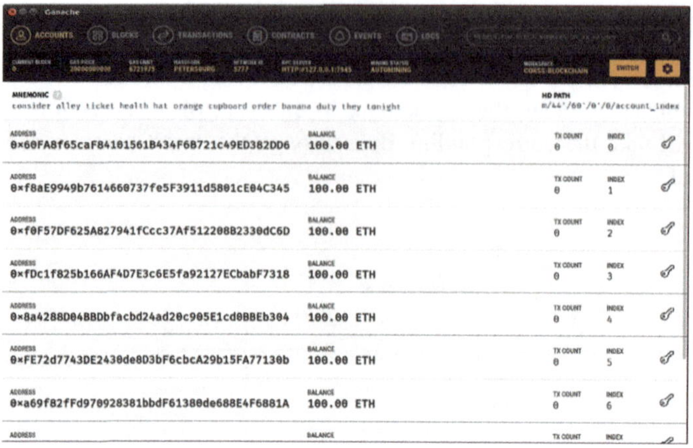

Fig. 8. Ganache account tab.

Fig. 9. Ganache log tab.

Open the overview page of the monitoring management system blockchain, and you can see that the blockchain ID of the super ledger blockchain is BC # 01 and the blockchain ID of the Ethereum blockchain is BC # 02, The joining time is the time when the relay module is started, and the running status of the two blockchain networks is running.

4 Conclusion

In view of the difficulty of data sharing in photovoltaic subsidy system, this paper analyzes the demand of Internet of things data sharing system according to Multi Chain Architecture and cross chain communication protocol, and designs the Internet of things data sharing system, including Internet of things terminal module, blockchain network module and cross chain relay module, which can organize registration, equipment registration and Internet of things on the super ledger blockchain The data is collected and uploaded, cross chain query and cross chain call are made on the Ethereum blockchain client, and cross chain interaction can be viewed on the relationship system of the relay module. According to the system design, each module is coded. It includes the configuration and start of blockchain network, the deployment of smart contracts, the interaction between client programs and smart contracts, and the realization of IOT data sharing system interface and data visualization. The test of each functional module of the Internet of things data sharing system proves the feasibility of cross chain communication of Internet of things data among different blockchains.

Acknowledgments. This work is supported by State Grid Technical Project "Research on key technologies of secure and reliable slice access for Energy Internet services" (5204XQ190001).

References

1. Christidis, K., Devetsikiotis, M.: Blockchains and smart contracts for the Internet of Things. IEEE Access **4**, 2292–2303 (2016)
2. David, Y.: Corporate governance and blockchains. Rev. Financ. **21**(1), 7–31 (2017). Social Science Electronic Publishing
3. Zhang, Y., Wen, J.T.: The IoT electric business model: using blockchain technology for the Internet of Things. Peer Peer Netw. Appl. **10**(4), 983–994 (2017)
4. Kang, J., Yu, R., Huang, X., Maharjan, S.: Enabling localized peer-to-peer electricity trading among plug-in hybrid electric vehicles using consortium blockchains. IEEE Trans. Ind. Inform. **13**(6), 3154–3164 (2017)
5. Knirsch, F., Unterweger, A., Engel, D.: Privacy-preserving blockchain-based electric vehicle charging with dynamic tariff decisions. Comput. Sci. Res. Dev. **33**, 71–79 (2017)
6. Wang, A.P., Fan, J.G., Guo, Y.L.: Application of blockchain in energy interconnection. Electric Power Inf. Commun. Technol. **14**(9), 1–6 (2016)
7. Strugar, D., Hussain, R., Mazzara, M., et al.: On M2M micropayments: a case study of electric autonomous vehicles. Preprint at https://arxiv.org/abs/1804.08964 (2018)
8. Subramanian, H.: Decentralized blockchain-based electronic marketplaces. Commun. ACM **61**(1), 78–84 (2018)

Design of Multi-UAV System Based on ROS

Qingjin Wei[1], Jiansheng Peng[1,2(✉)], Hemin Ye[2], Jian Qin[1], and Qiwen He[1]

[1] School of Physics and Mechanical and Electronic Engineering, Hechi University, Yizhou 546300, China
sheng120410@163.com
[2] School of Electronic Engineering, Guangxi Normal University, Guilin 541004, China

Abstract. Aiming at the problems that the single drone cannot comprehensively check the disaster situation and the single drone suddenly fails and cannot continue the rescue during the execution of the rescue request task. A multi-drone system based on ROS is designed. The hardware part of the system adopts Pixhawk to be responsible for UAV attitude control. Obtain the position of the aircraft on the world coordinate system through GPS. Using the Raspberry Pi 3B as the on-board computer is responsible for dividing each module into nodes. This enables publication and subscription of topics such as sensors, aircraft status and coordinates. This facilitates communication between multiple drones. Eventually, a set of three drones taking off at the same time, a PC capable of changing the target position of multiple drones, and a multi-drones system that can avoid collisions between drones can be realized. Compared with the single drone, the multi-drones system based on ROS can effectively solve the problem that the single drone cannot detect the disaster in all directions.

Keywords: Multi-UAV system · Multi-machine collaboration · Multi-machine communication · Electronic governor

1 Introduction

Unmanned aerial vehicles not only have unparalleled superiority in military strike and defense [1], but also have been widely used in civilian use due to their flexibility and high efficiency [2, 3]. For example, emergency medical treatment, fire disaster relief, geological detection, electric power maintenance and other fields. In 2011, Manathara et al. of the University of Porto in the United States designed a task allocation strategy for a variety of heterogeneous drones, and used heuristic algorithms to solve them, improving resource utilization [4]. In 2016, Kownacki of Bialystok University of Technology in Poland proposed a combination of virtual structure, leaderfollower method, and behavior-based formation method to improve the reliability and throughput of information sharing between aircraft during the formation of UAV formation flight [5]. The European COMETS project has completed the design of heterogeneous multi-UAV, integrated distributed perception and real-time image processing technology, and realized the distributed real-time control system [6]. On

Q. Liu et al. (Eds.): CENet 2020, AISC 1274, pp. 1056–1067, 2021.
https://doi.org/10.1007/978-981-15-8462-6_122

November 14, 2016, Intel Corporation performed a "light show" realized by group flying of 500 drones at the Tmall "Double 11" Global Creative Eco Summit [7].

In response to the problem of multiple UAVs searching for targets in a complex environment, Shaofei Chen of the National University of Defense Technology proposed for the first time an approximate optimal online planning algorithm for multiple UAVs, which makes the computational complexity polynomial with the number of UAVs Relationship rather than the original index relationship [8]. Yu Sinan and others from Beijing University of Aeronautics and Astronautics proposed a method for segmenting an arbitrary convex polygon search area based on the initial position and search area of the UAV, and using the number of turns as the main basis to evaluate the segmentation results [9]. Guangdong Yihang Bailu Company has set a world record of 1,374 successful UAV performances during the rehearsal on April 29 [10].

Because the role of a single drone is limited. For example, to detect a building fire situation, only one location can be detected at the same time, and the situation of disaster detection blind spots occurs, which fails to detect emergencies in time and causes heavy losses. There may also be a single drone in the fire rescue process. The fire was damaged by sudden changes. This led to the inability to continue rescue and seriously affected the rescue efficiency. The multi-unmanned aerial vehicle system based on ROS designed in this paper can comprehensively monitor real-time for this kind of problem and can implement emergency treatment to the emergency happening somewhere. Once the emergency occurs, it can be dealt with. During the rescue process, some drones were damaged, and the remaining drones can also complete the rescue. This solves the problem of unable to continue rescue due to equipment failure.

2 Multi-UAV System Hardware Design

2.1 Multi-UAV System Hardware Configuration

As shown in Fig. 1, the multi-drone system hardware uses Pixhawk as the underlying controller. Pixhawk peripheral access receiver, alarm (buzzer), safety switch, GPS, electronic governor. The receiver receives the control signal of the remote controller. An alarm (buzzer) serves as an information reminder. We use a safety switch to lock the power output of the drone during standby to ensure personnel safety. We carry GPS to obtain the position information of the UAV based on the world coordinate system. We use electronic governors to drive brushless motors. Using the Raspberry Pi 3B as the top controller, it is responsible for monitoring the position information of other drones in the system, obtaining the GPS coordinates of the local machine and publishing them to other drones in the system, and receiving the control data from the PC. After data processing, it is sent to Pixhawk to control the operation of the underlying actuator.

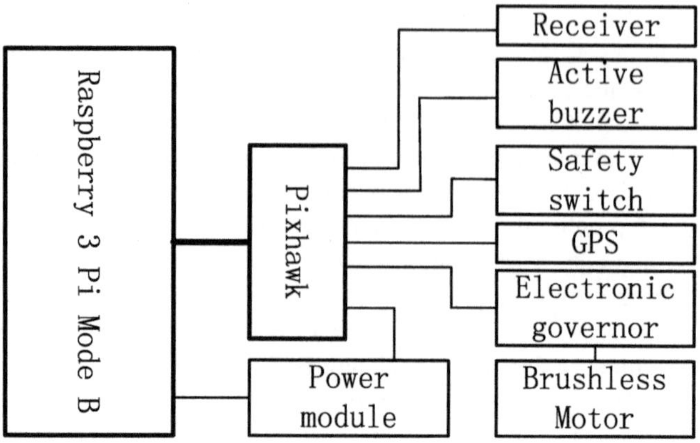

Fig. 1. Multi-UAV system hardware framework.

2.2 Hardware Circuit Design of Electronic Governor

The multi-UAV system only sends control signals to control the power motor, but the power motor needs high voltage and large current. It is necessary to design a motor drive for the power part of the multi-UAV system, that is, the electronic governor. The hardware framework of the electronic governor of the multi-UAV system is shown in Fig. 2, which is mainly divided into four parts: the power supply system, the main control module, the zero-crossing detection module, and the six-arm full-bridge drive circuit module. The function of the zero-crossing detection module is to detect that the rotor of the brushless motor reaches the zero point and trigger the commutation. The main control module is responsible for all working states of the electronic governor. The six-arm full-bridge drive module is responsible for the hardware equipment of motor commutation, and drives the motor to rotate at the same time.

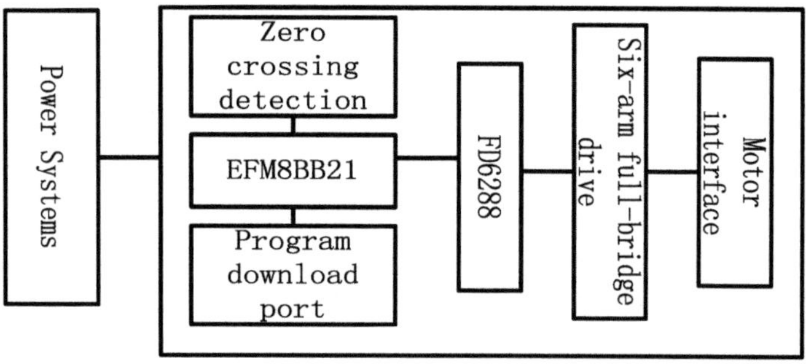

Fig. 2. Hardware frame diagram of electronic governor of multi-drone system.

Commutation Circuit Design. The direction of the magnetic force received by the rotor at different positions is different, so the position of the rotor needs to be determined. By changing the two energized phases, the rotor is continuously rotated. The rotor position is determined by the non-inductive measurement method. The non-inductive measurement can remove the measurement sensor, which greatly reduces the overall component of the system and has a simpler structure. But its shortcomings are also obvious. It has poor controllability when it is started, and needs to be controlled after reaching a certain speed. After the aircraft motor is started, the motor generally does not need to stop rotating, and the poor controllability of the start can be ignored for the UAV system.

The circuit design of the non-inductive measurement mode is shown in Fig. 3. VS1, VS2, and VS3 are respectively connected to the C phase, B phase, and A phase of the motor. CENTER is the voltage at the intermediate junction. A, B, and C are the three-phase voltage obtained by dividing the three phase voltages of the motor through three pairs of voltage dividing circuits formed by R7 and R8, R9 and R10, R11 and R12. The reason for the voltage division is that the main control working voltage is 3.3 V, and the output voltage of the three phases of the motor is the battery voltage (11.1 ~ 16.8 V).

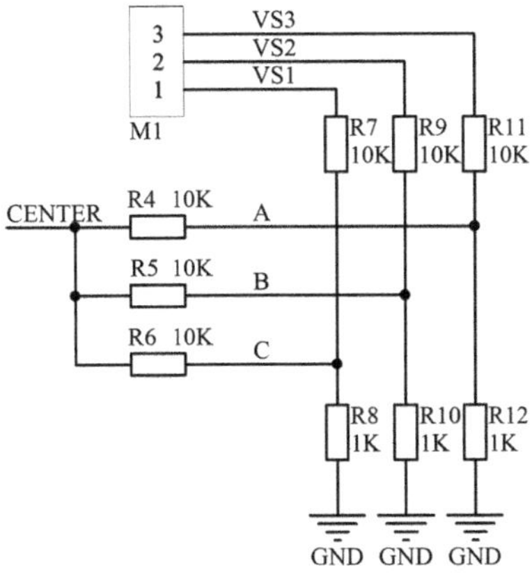

Fig. 3. Zero crossing detection circuit diagram.

Six-Arm Full-Bridge Circuit Design. The six-arm full-bridge drive circuit is shown in Fig. 4. It consists of six N-channel MOS tubes of the model SL150N03Q, which are divided into an upper arm and a lower arm. Its drive control method is to simultaneously open two MOS tubes that are not on the same phase to complete the power-on of the two phases of the power motor.

As can be seen from Table 1, the SL150N03Q gate voltage can be turned on when it is greater than or equal to 2.5 V. The maximum output voltage of the pin of the main control chip EFM8BB21 is 3.3 V, which satisfies the conduction condition of SL150N03Q. When one of the lower arms of the six-arm full bridge is turned on and the corresponding source voltage of the upper arm is the same as the power supply voltage. The gate voltage of the upper arm must be 2.5 V to the power supply voltage to make the upper arm conductive. At this time, the output of the main control chip can no longer meet the requirements. In response to this problem, the motor drive chip

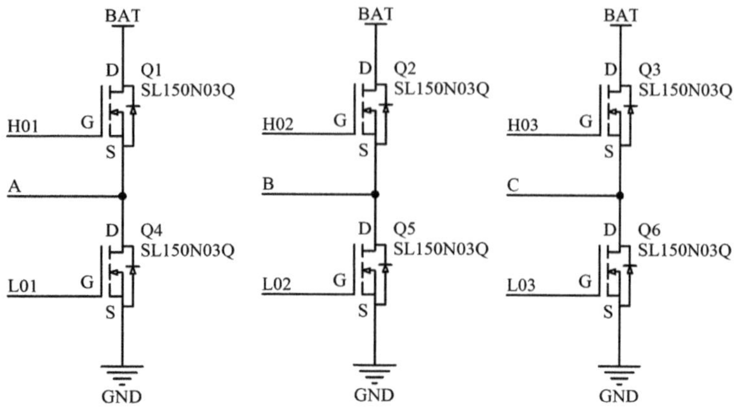

Fig. 4. Six-arm full-bridge drive circuit diagram.

FD6288 is used as an intermediate driver, as shown in Fig. 5. C8 is the power supply filter capacitor. D2, D3 and D4 act as bootstrap diodes. C4, C5, and C9 are used as bootstrap capacitors, and are formed by bootstrap diodes and bootstrap capacitors. The output of the FD6288 is controlled by the main control chip, thereby controlling the on and off of the drive bridge.

PWM (Pulse-Width Modulation) controls the effective on-time of the upper arm, and at the same time controls the conduction of the corresponding lower arm. The energy obtained by the two-phase energization of the motor in the same time is different, so that the speed of the motor can be controlled.

Table 1. SL150N03Q Part of the electrical parameter table.

Parameter	Symbol	Minimum value	Standard value	Maximum	Unit
Drain to source voltage	V_{DSS}	–	–	30	V
Gate to source voltage	V_{GSS}	−20	10	20	V
Drain current (DC)	I_D	–	82	150	A
Maximum threshold voltage	V_{th}	2.5	2.5	–	V
Power consumption	P_D	–	–	166	W
Operating temperature	T_w	−55	–	175	°C

Power Circuit Design. The power supply design of the electronic governor consists of three parts, namely the driving power supply (the voltage depends on the battery voltage, the range is 11.1 ~ 16.8 V), the primary power supply (the voltage is 6 V) and the secondary power supply (the voltage is 3.3 V). The driving power supply is the input voltage, that is, BAT in Fig. 6, which represents the input voltage of the electronic governor, and the main function is to give the driving chip of the driving circuit part. The six-arm full-bridge circuit provides working power and primary power through the 78L06 linear regulator chip. The output voltage of the six-arm full-bridge

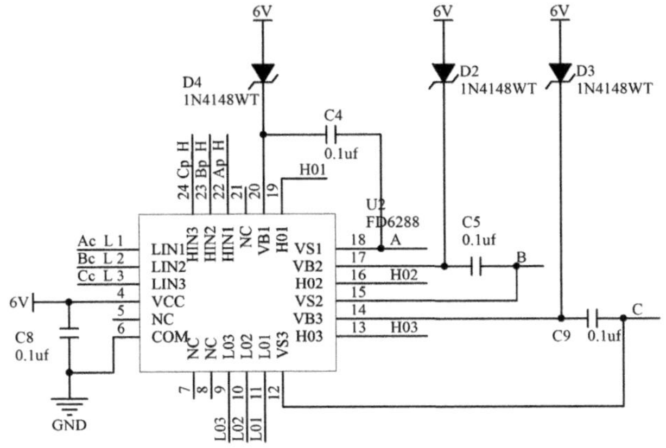

Fig. 5. Intermediate driver circuit diagram.

circuit is 6 V. The circuit design of the primary power supply derived from the drive power supply is shown in Fig. 6. Capacitor E1 plays a role in filtering the power supply, D1 is to prevent reverse connection.

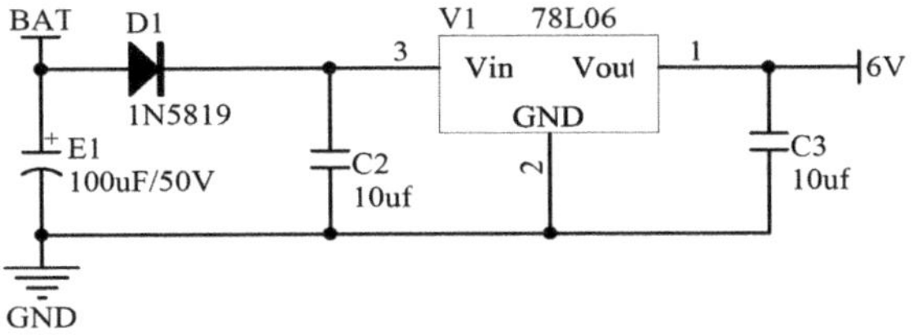

Fig. 6. Primary power supply circuit diagram.

Aiming at the problem of large voltage fluctuation, a two-stage step-down voltage reduction method is adopted to reduce the power supply ripple to provide a good secondary power supply to the main control chip. The role of the primary power supply is to provide an input voltage with less ripple for the secondary power supply, thereby obtaining a stable secondary power supply with extremely low ripple. This effectively avoids the adverse effects on the chip that requires 3.3 V voltage supply due to large driving voltage fluctuations. The circuit is shown in Fig. 7.

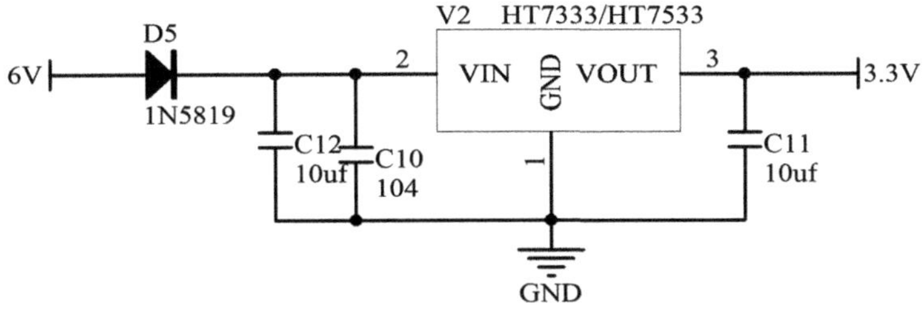

Fig. 7. Secondary power supply circuit diagram.

Main Controller Interface Design. The main control adopts 8-bit processor EFM8BB21, which is an 8051 core and has three-channel PWM output. It can realize PWM speed regulation, a 12-bit precision ADC (Analog-to-Digital Converter) and two comparators can be used for zero-crossing detection. It supports C2 interface simulation and download. C2D and CK (the fifth pin of EFM8BB21) are used as the download program interface, which can effectively prevent the short circuit of the input terminal from burning the main control chip EMF8BB21 through R2 in series with the signal input line. LED1 is a four-pin RGB lamp, and the three control pins are connected to P0.5, P0.6 and P0.7 of EFM8BB21. By identifying LED1 to identify the state of the electronic governor, the main controller interface circuit design is shown in Fig. 8.

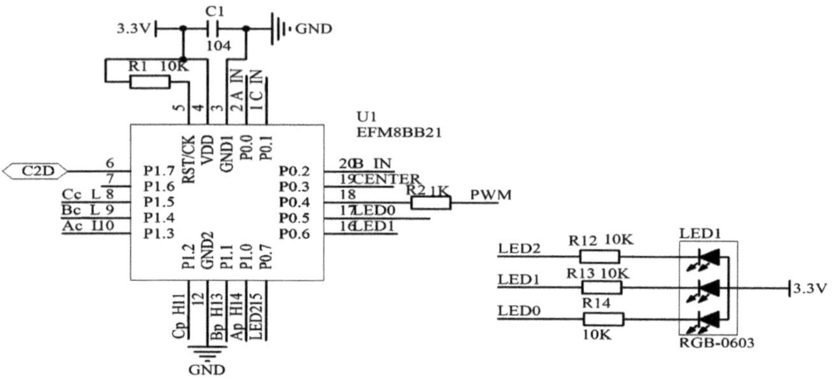

Fig. 8. Master controller interface design.

3 Software Design of Multiple UAV Systems

3.1 Multi-UAV System Software Framework

The software framework builds distributed data processing through a publish-subscribe communication framework provided by ROS. As shown in Fig. 9, the data of all drones in the system interact through the ROS node manager. Each drone is managed by the function package mavros to manage the drone's position control, coordinated control, collision avoidance and independent control. The PC also obtains drone data and sends commands to control multiple drones through the Master.

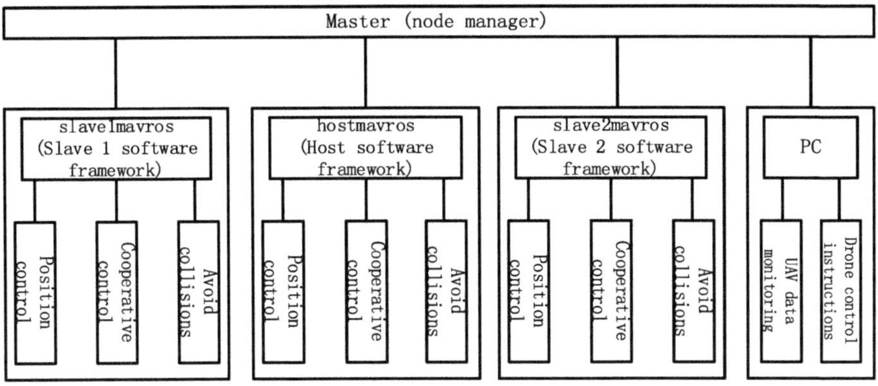

Fig. 9. Multi-UAV system software framework.

3.2 Design of Position Control Software for Multi-UAV System

There are two kinds of position coordinates in this design, which are the world coordinate system and the body coordinate system. The world coordinate system is a three-dimensional coordinate system based on the latitude, longitude and altitude of GPS. The body coordinate system uses the takeoff point of the aircraft as the coordinate origin. The Y axis points directly in front of the drone nose. The X axis points to the positive direction on the right side of the drone. The Z axis is perpendicular to the drone.

Because the difference in latitude and longitude near the equator is 111.3195 km per degree (Earth radius R = 6378.140 km, the perimeter is known from Eq. (1), L = 2π * R = 2 * 3.14159 * 6378.140 = 40075.0355 km, the distance per degree can be expressed as (2) Learned that D = L/360 = 40075.0355/360 = 111.3195 km). Therefore, the world coordinates are suitable for a large range of mobile tasks or go to a clear coordinate point.

$$L = 2\pi * R \tag{1}$$

$$D = L/360 \tag{2}$$

The position control flow is shown in Fig. 10. When the number of satellites received by GPS reaches more than 10 (including 10), the first position control mode switch is performed. If the first switch is unsuccessful, we will continue to switch until we switch to the position control mode before we start to receive position control data from the PC. PC control data is divided into GPS coordinates and body coordinates. According to different control data, the UAV is controlled to move in different ways. In order to prevent the position control error caused by the drone flying, Pixhawk has designed a timeout mode to automatically return to the flight mode before entering the position control mode. The timeout period is set to 500 ms. When the Pixhawk does not receive the position control data again after more than 500 ms, it will determine that the external control is out of control. Therefore, the refresh frequency of the position control data must be set greater than 20 Hz.

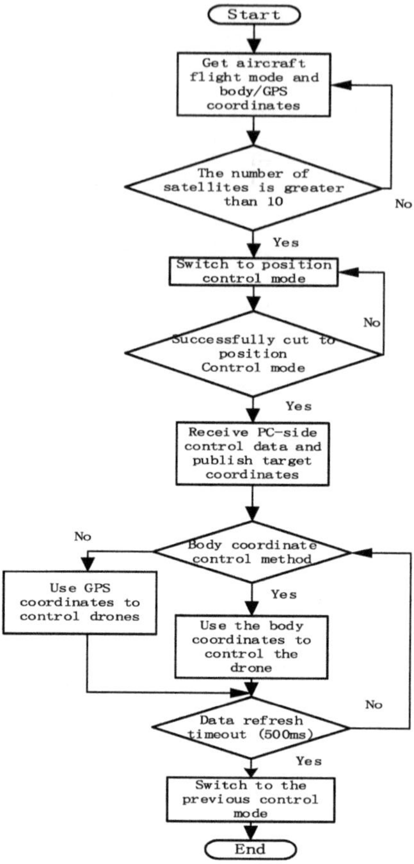

Fig. 10. Flow chart of multi-drone position control.

3.3 Design of Cooperative Control Software for Multiple UAV Systems

Multi-UAV relies on latitude and longitude data to ensure that the GPS stars meet the fixed-point flight requirements when using position control. Therefore, when the aircraft is started, it should be in the fixed-point flight mode. After the GPS searches for enough satellites and the onboard computer (Raspberry Pi) is successfully connected to the flight controller (Pixhawk) through mavros, the position control node is activated to subscribe to the state. The two themes local_position/pose get the status of the aircraft, flight mode, and body coordinates. Then create the flight mode selection service and position control service, and finally set the flight mode to position control mode and poll whether there is new position information according to the set refresh frequency. And through the theme setpoin_position/local to publish the target position to the node manager (responsible for processing the data interaction between nodes), if there is, then refresh the target position. This node is only responsible for receiving location information and publishing target coordinates. The specific execution process is dispatched and controlled by Pixhawk to control the output of the electronic governor to move the aircraft to the target position. When the coordinates of the aircraft body and the coordinates of the target position coincide, it means that the aircraft has reached the target position. The control flow is shown in Fig. 11.

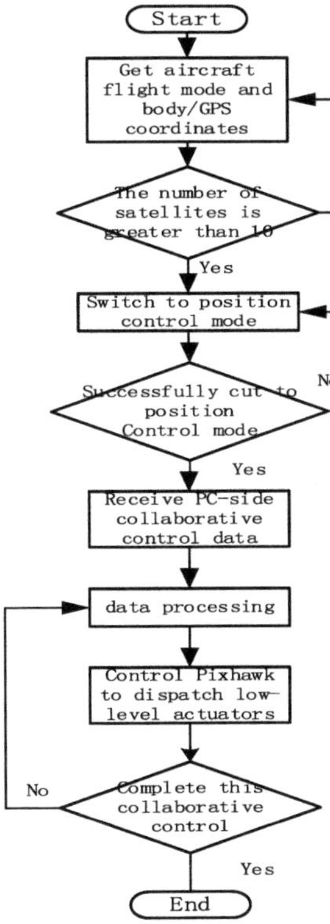

Fig. 11. Collaborative control flow chart.

3.4 Software Design for Multi-UAV System to Avoid Collision

The relative position calculation of multiple drones is based on latitude and longitude and altitude as position coordinates. The horizontal distance and vertical distance between each two UAVs can be calculated through the two-point coordinates. The coordinates of the two points can be expressed as A (ALat, Alon), B (Blat, Blon). Because the difference in latitude differs by one degree. Viewing the earth as a sphere, the horizontal distance between two points can be obtained through its radius combined with latitude and longitude. The two-point distance solution is calculated by substituting the result of Eqs. (3) into (4).

$$C = \sin(ALat) * \sin(BLat) * \cos(ALon - BLon) + \cos(ALat) * \cos(BLat) \tag{3}$$

$$S = R^* \arccos(C) * \pi/180 \tag{4}$$

After the relative position of the unmanned aerial vehicle is obtained, when coordinated to the target point, ensuring that each unmanned aerial vehicle is at least 2 meters apart horizontally and Vertically. The effective model for avoiding collisions is shown in Fig. 12. The central point is the drone. The area within the frame is a collision threat area. The software design for avoiding collisions is shown in Fig. 13. After obtaining the GPS coordinates of the drone in the system, the distance between the drones is calculated according to the GPS coordinates. We judge whether collision warning is reached by distance. If it is reached, the collision avoidance will be started to avoid the drone collision, otherwise the detection of collision avoidance will be ended.

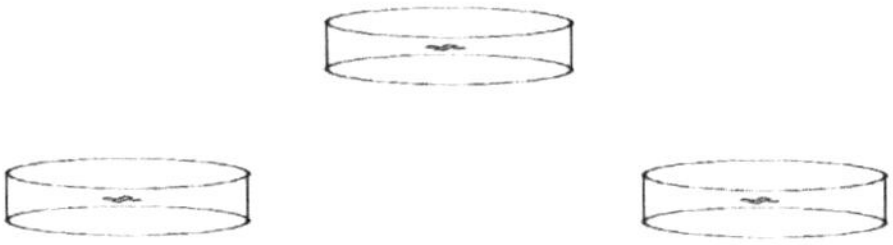

Fig. 12. Model diagram of multi-UAV system to avoid mutual collision.

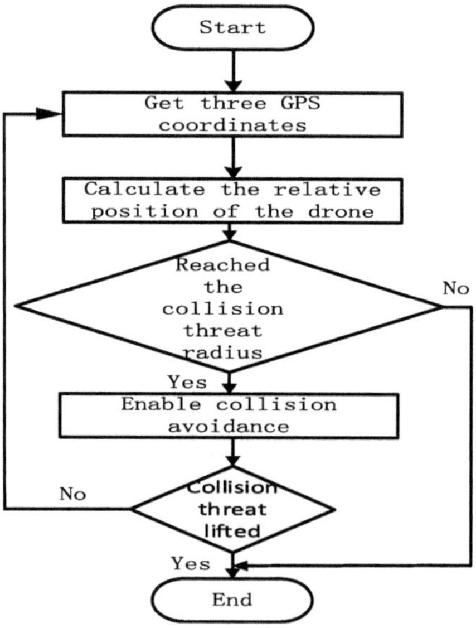

Fig. 13. Multi-UAV system software design flow chart for avoiding collision.

4 Conclusion

This design uses ROS unique distributed communication mechanism combined with Pixhawk, and finally realizes a multi-drones system based on ROS. The system is based on the communication mechanism of the ROS system, which makes each module in the multi-UAV system independent and has certain flexibility. The electronic governor supports multiple signal formats such as PWM, ONEDSHOT125, DSHOT300 and so on, and has high scalability. Pixhawk is used as the flight controller to provide stable and reliable attitude control. The control mode of the system is diverse, supporting two control modes of PC and remote control. You can choose to control one or all drones. It has early warning of collision of drones in the system to ensure safe and reliable cooperation of drones We return flight data through the onboard computer, and can connect to the ground station via UDP network to adjust the parameters of the flight controller, which is highly time-efficient.

Acknowledgement. The authors are highly thankful to the Research Project for Young and Middle-aged Teachers in Guangxi Universities (ID: 2019KY0621), to the Natural Science Foundation of Guangxi Province (No. 2018GXNSFAA281164), This research was financially supported by the project of outstanding thousand young teachers' training in higher education institutions of Guangxi, Guangxi Colleges and Universities Key Laboratory Breeding Base of System Control and Information Processing.

References

1. Li, B., Wang, H.Y., Yang, X.T.: Application and development of UAV system in forest fire prevention. Electron Technol. **44**(05), 15–18 (2015)
2. Chen, M.: Research on the key technology of remote sensing system for light and small UAV. Electron. Technol. Softw. Eng. **10**, 131–132 (2017)
3. Wu, Q., Lei, L.: Technology and development of Multi-UAV measurement and control and information transmission system. Telecommun. Technol. **10**, 107–111 (2008)
4. Manathara, J.G., Sujit, P.B., Beard, R.W.: Multiple UAV coalitions for a search and prosecute mission. J. Intell. Rob. Syst. **62**(1), 125–158 (2011)
5. Zong, Q., Wang, D.D., Shao, S.K., Zhang, B.Y., Han, Y.: Research status and development of multi-UAV cooperative formation flight control research. J. Harbin Inst. Technol. **49**(03), 1–14 (2017)
6. Deng, Q.B.: Research on multi-UAV collaborative mission planning technology. Beijing Institute of Technology (2014)
7. Xu, J.: Group flying characteristics and control analysis of multiple drones. Nanjing University of Science and Technology (2017)
8. Chen, S.F.: Reconnaissance and surveillance mission planning method for UAV cluster system. National University of Defense Technology (2016)
9. Yu, S.N., Zhou, R., Xia, J., Che, J.: Multi-UAV collaborative search area segmentation and coverage. J. Beijing Univ. Aeronaut. Astronaut. **41**(01), 167–173 (2015)
10. Gao, B.: Xi'an UAV performs "garbled code" or signal interference. Science and Technology Daily (2018)
11. Wang, H.: Application research on building IoT environment based on ROS. Hefei University of Technology (2017)

Design of Quadrotor Aircraft System Based on msOS Platform

Yong Qin[1], Jiansheng Peng[1,2(✉)], Hemin Ye[2], Liyou Luo[1], and Qingjin Wei[1]

[1] School of Physics and Mechanical and Electronic Engineering, Hechi University, Yizhou 546300, China
sheng120410@163.com
[2] School of Electronic Engineering, Guangxi Normal University, Guilin 541004, China

Abstract. In order to solve the problems of multi-sensor read data and PID control time interval, data return and flight control mode of quadrotor UAV, a quadrotor UAV control system based on msOS platform was designed. The system uses MPU9250 to obtain UAV attitude data, MS5611 barometer to obtain UAV altitude data, AT7456E as the management chip of the OSD unit, and S-Bus signal as the remote-control signal input. Compared with the traditional four-rotor UAV bare-metal system, the system uses dual-task msOS management sensors to read data and PID control. Finally, an MSOS platform quadrotor aircraft system with OSD data return, attitude, fixed altitude and optical flow hovering flight modes were designed.

Keywords: OSD · Four-Rotor UAV · Barometer altitude setting · Optical flow hover

1 Introduction

With the development of modern technology, more and more modern electronic devices are used in the military and the public. Unmanned aerial vehicle (UAV) is a type of pilotless piloting. Aircraft (UAV) operates through electronic input or through an onboard autonomous flight control management system without the need for flight controller intervention [1]. In the 21st century, unmanned aerial vehicles play a major role in modern military affairs and are widely used in military reconnaissance, surveillance, anti-submarine, and jamming technologies. More and more countries are focusing on the development of unmanned aerial vehicles [2, 3].

Since the early 1980s, drones have become increasingly miniaturized and civilianized [4]. In the 21st century, with the rapid development of microelectronics, some people and teams have developed four-rotor unmanned aerial vehicle controllers (referred to as flight control). Sensors such as GPS and compass may also be installed outside the flight control [5]. Through the combination of these sensors, the attitude and movement of the aircraft can be calculated. In the current research status abroad, people are more concerned about APM flight control, Pixhawk flight control, SPRacingF3 flight control, OpenPilotCC3D flight control, MWC flight control [6].

Q. Liu et al. (Eds.): CENet 2020, AISC 1274, pp. 1068–1079, 2021.
https://doi.org/10.1007/978-981-15-8462-6_123

As early as 2004, the National University of Defense Technology began to design aircraft [7]. In 2006, Shenzhen DJI Innovation Technology Co., Ltd. was founded to inspire the development of civilian drones. DJI's innovative leading technologies and products have been widely used in aerial photography, remote sensing surveying and mapping, forest fire prevention, power line patrol, search and rescue, film and television advertising, etc. Their unmanned aerial vehicles have become the best choice for many aerial photography enthusiasts around the world [8]. With the development and promotion of DJI technology, the number of domestic R&D teams has become more and more. Many companies are engaged in the research and development of UAV flight control, designing more advanced flight control. They also designed peripheral devices such as optical flow modules and vision modules [9].

2 Hardware Design

The hardware circuit determines the implementation of the code. Poor hardware circuit design often affects code debugging. The system design block diagram is shown in Fig. 1. The system block diagram is mainly divided into 3 parts. The first part is the power system. The second part is the sensor circuit. The third part is the function block circuit. There are 3 different voltage outputs in the power supply part. The sensor part has attitude sensor circuit, barometer sensor circuit and optical flow module circuit. The function module circuit is composed of USB interface circuit, S-Bus decoding circuit, STM32 main control circuit and OSD function circuit. This chapter will tell about the hardware circuit design of each function block.

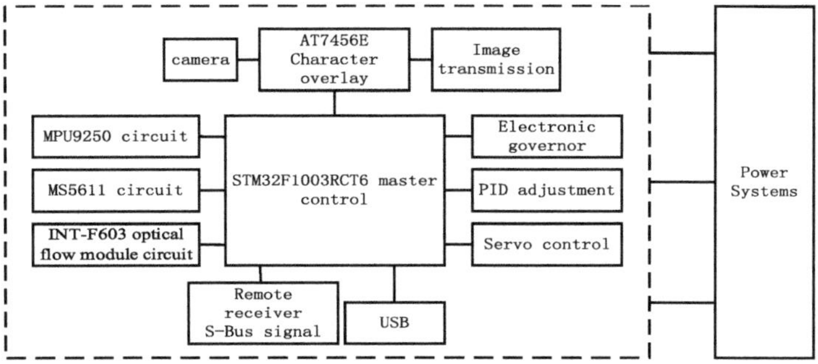

Fig. 1. System hardware framework.

2.1 Power Circuit Design

This design UAV flight control board has three voltage levels, DC3.3 V, DC5.0 V and battery voltage (Bat voltage for short). DC3.3 V is the main control MCU chip. MPU9250 and MS5611 and other chips provide power. DC5.0 V is power supply for AT7456E and remote control receiver. The UAV power supply can choose 3S – 4S power supply, then the Bat voltage is DC12.6 – 16.8 V.

The 3.3 V power circuit diagram using 662 K chip is shown in Fig. 2. Among them, U3 and U4 are 662 K voltage regulator chips. U3 is the power supply for the UAV flight control board. When the external expansion sensor needs 3.3 V power supply, U4 provides external output DC3.3 V. This ensures the feasibility of the development platform to expand the power supply of multiple sensors.

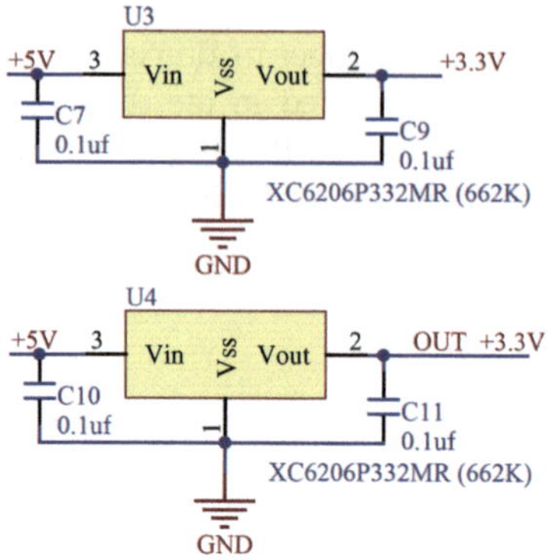

Fig. 2. DC3.3 V voltage regulator circuit diagram.

The ADC detects the battery voltage circuit as shown in Fig. 3. The voltage sampled by the ADC is Vadc = VBat * (R7/(R7 + R6)) = VBat * 10/60 = VBat/6. In this way, it can meet the 3S – 4S power supply.

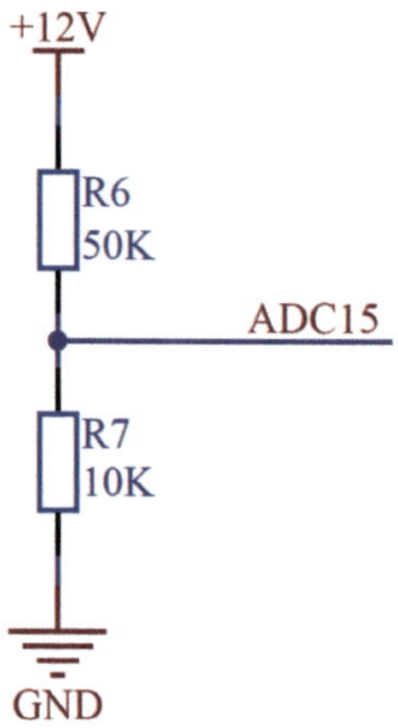

Fig. 3. Circuit diagram of ADC detecting battery voltage.

2.2 Sensor Circuit Design

MPU9250 Attitude Sensor. This design uses the data of DPU channel of MPU9250. The circuit design is shown in Fig. 4. R8 and R9 are pull-up resistors. SDA needs to work in input and output mode in the I^2C bus. The advantage of adding a pull-up resistor on the SDA bus is to set the I/O port corresponding to STM32 to open-drain output mode. The I/O port of the main control chip is readable and writable. This is much more convenient when writing programs.

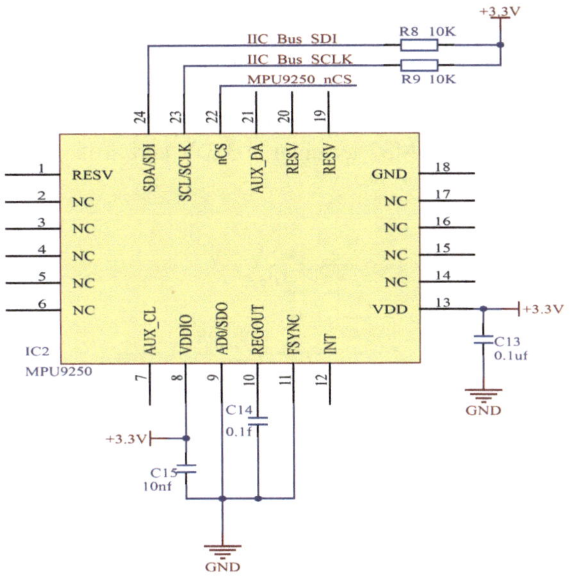

Fig. 4. MPU9250 working circuit diagram.

MS5611 Barometer Sensor. The MS5611 barometer can measure high-precision air pressure and temperature values. The main control chip calculates the sea wave height by reading the air pressure data and temperature data of MS5611, and then compares it with the sea wave height at takeoff to obtain the flight height of the aircraft. The main control chip of this design uses I^2C bus to communicate with MS5611. The design schematic is shown in Fig. 5.

2.3 Function Module Circuit Design

USB Bus Interface Design. O This design uses the USB interface of STM32 to realize multiple functions. The USB interface circuit is shown in Fig. 6. R10 is the pull-up resistor of the USB D+ signal line. This pull-up resistor has two functions. One is the confirmation of the USB working mode: when D+ is connected to a 1.5 K pull-up resistor, the USB works in high-speed mode. Therefore, when the USB is inserted, the

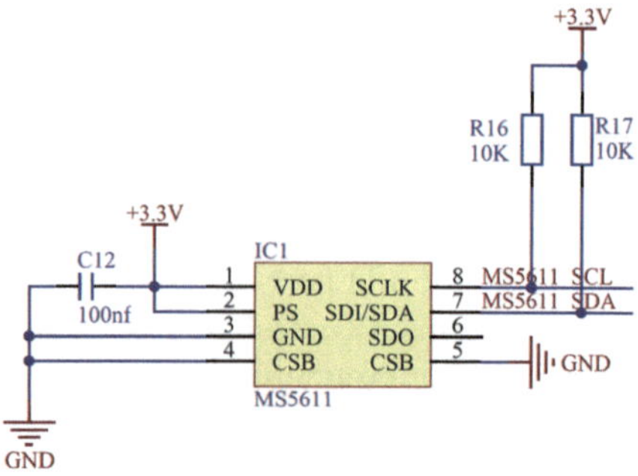

Fig. 5. MS5611 working circuit diagram.

terminal can detect the USB device. The second is to ensure the stability of the data. Because the voltage at the STM32 terminal is 3.3 V here it is pulled up to 3.3 V.

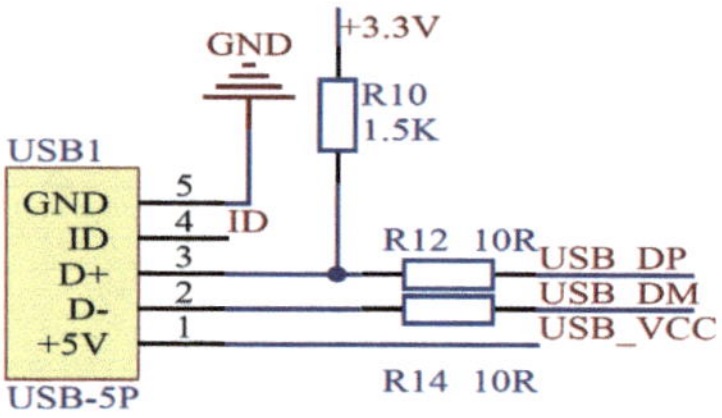

Fig. 6. USB interface circuit diagram.

When doing a USB interface, a circuit for detecting USB access is often needed. Here, partial pressure detection is required. The circuit diagram is shown in Fig. 7.

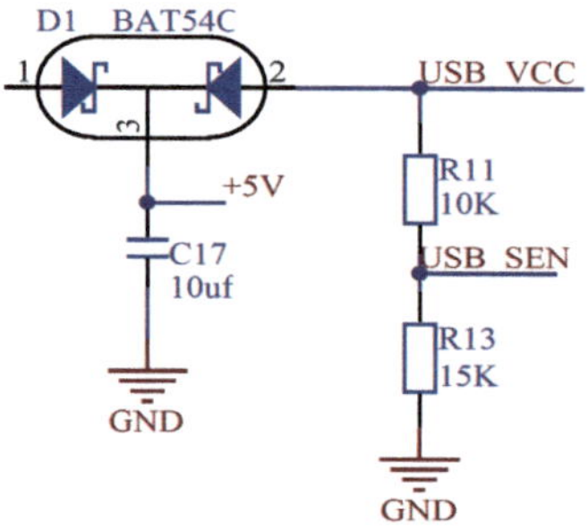

Fig. 7. USB insertion detection circuit diagram.

S-Bus Decoding Design. Because the remote controller receiver uses the hardware circuit to reverse the level of the S-Bus signal output from the code, it cannot restore the true S-Bus signal by software inversion. Therefore, it is necessary to invert the

S-Bus signal level with a hardware circuit at the input terminal of the decoded signal to obtain the true S-Bus signal level. After getting the real S-Bus signal, this real S-Bus signal is input to the RX end of UART1. Then you can use software to decode the data of each channel in the S-Bus signal. S-Bus decoding hardware circuit is shown in Fig. 8. The left side is the plug-in interface, which is used to connect the S-Bus signal output of the receiver. Here you can also power the receiver. When S-BUS_IN is 0, the transistor Q1 is off. UART1_RX is 1. When S-BUS_IN is 1, the transistor Q1 is in a conducting state. UART1_RX is 0. In this way, the signal negation requirement is met.

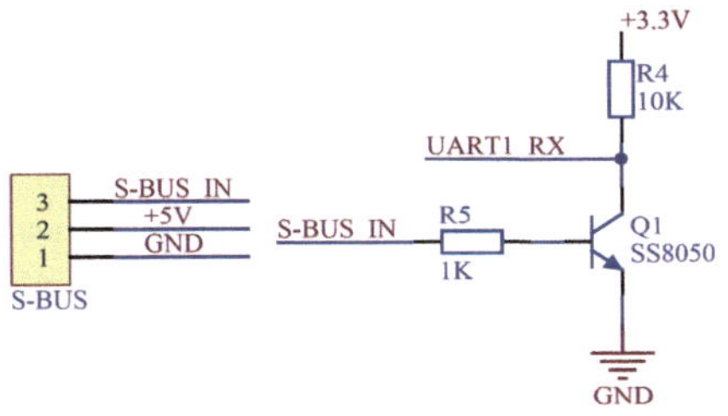

Fig. 8. S-Bus decoding hardware circuit diagram.

Other Extended Function Interface Design. The main control chip of this design OSD module adopts AT7456E. The main control MCU communicates with AT7456E through the SPI bus. At the same time, the hardware SPI1 interface of STM32 is used to communicate with AT7456E. The OSD circuit is shown in Fig. 9.

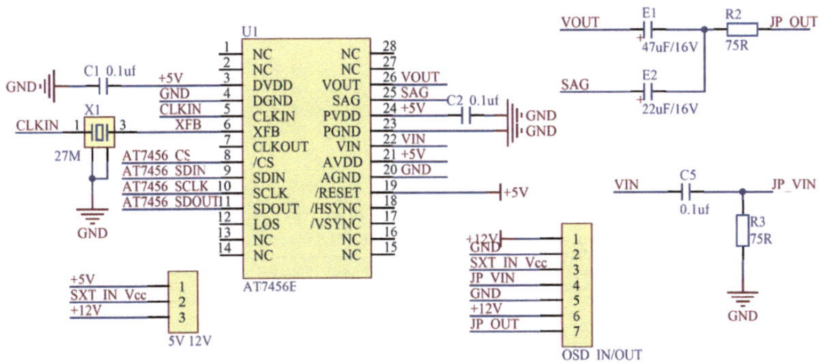

Fig. 9. OSD module circuit diagram.

The circuit diagram is shown in Fig. 10. This design leads to USART2 and USART3 of STM32. The role of USART2 is to debug the program and read the data of the optical flow module. USART3 is used for other sensor expansion interfaces. At the same time, I^2C bus interface circuit is also designed to facilitate the expansion of the sensor.

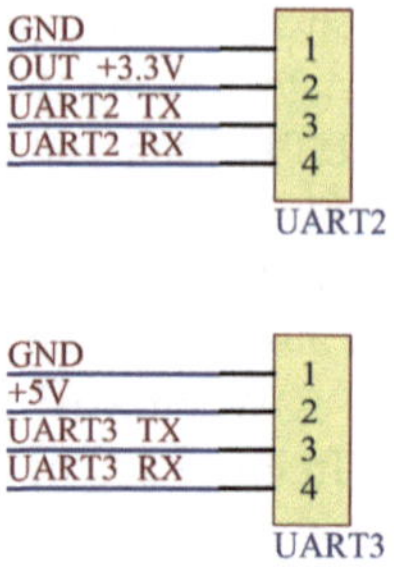

Fig. 10. STM32 serial interface circuit.

The single-chip computer model used in the main control chip of this design is STM32F103RCT6, which has multiple hardware PWMs inside, and its load carrying capacity is also relatively strong. So only need to lead out the PWM output port. STM32 download program can use a variety of methods. The simplest and most practical is SWD mode debug download. This debugging download method requires 4 wires, namely the positive power supply, the negative power supply, SWCLK (clock) and SWDIO (data). The wiring diagram of the MCU chip circuit pin of the four-rotor UAV is shown in Fig. 11.

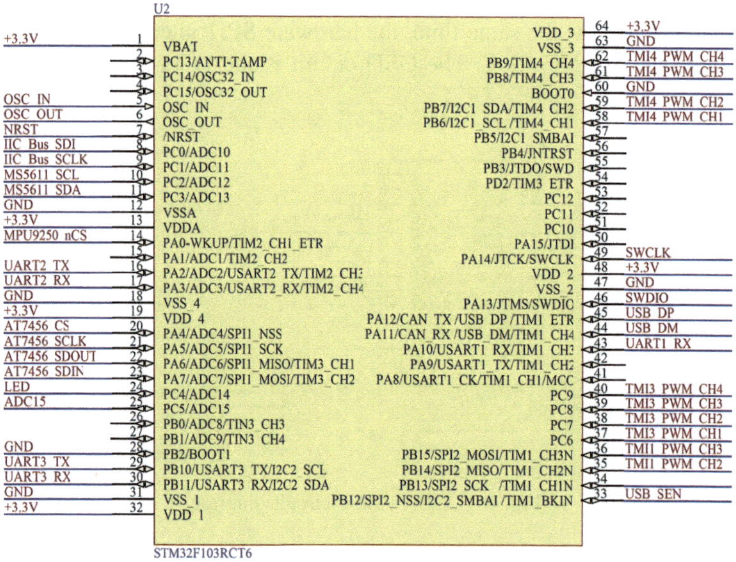

Fig. 11. Four-rotor UAV main control MCU chip circuit diagram.

3 Software Design

3.1 Software System Framework Design

This design uses msOS as the body of the software system. The software design is based on msOS. The software framework of the deeply customized msOS platform quadrotor UAV system is shown in Fig. 12. Originally msOS was an industrial control RTOS written based on STM32R8/RB series single chip microcomputer. Here transplantation and deep customization of msOS-V1.3.3 to STM32F103RC series MCU. In the transplantation and in-depth customization, the function realization circuit of the msOS industrial control part was removed from the hardware circuit. A USB interface circuit and various sensor circuits are added to the hardware circuit. According to the requirements of the UAV control system and the hardware circuit design, the corresponding function realization module program is written on the software system. This constitutes the software system of the quadrotor drone of the msOS platform.

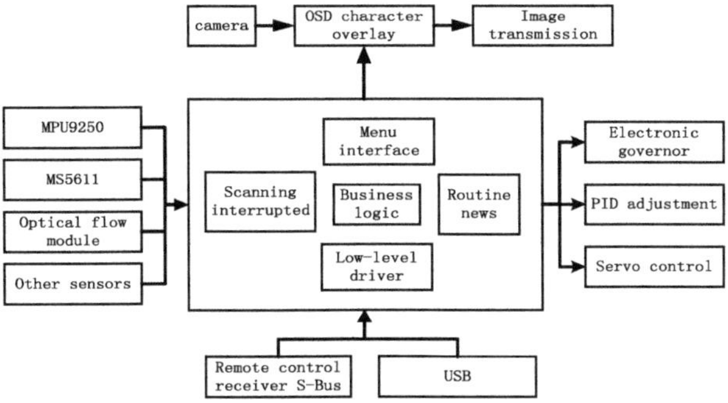

Fig. 12. Software framework of quadrotor UAV system based on msOS platform.

3.2 S-Bus Decoding Software Design

The full name of S-Bus is Serial Bus. It was first introduced by Japanese manufacturer FUTABA. The USART1 configuration of STM32 is shown in Table 1. One frame of S-Bus data has 25 bytes. The starting byte is 0x0F. The 22 bytes in the middle are the data of 16 channels. Each channel is represented by 11-bit binary. The range is 0 to 2047. The next 2 bytes are the flag bit and the end bit [10].

3.3 USB Multifunctional Software Design

This design adds a USB interface, and uses the USB interface to achieve a multi-functional design. The USB here mainly works on the virtual serial port. The first function is to achieve remote control calibration. When in use, power off the drone flight controller completely. Then connect the USB cable to the PC to enter the remote controller calibration mode. Finally, complete the remote controller calibration according to the reminder of the host computer software. The second function is to

Table 1. SART32 USART1 configuration.

Parameter name	Parameter regulations
USART_BaudRate	100000
USART_WordLength	USART_WordLength_8b
USART_Parity	USART_Parity_Even
USART_StopBits	USART_StopBits_2
HardwareFlowControl	USART_HardwareFlowControl_None

update the OSD font. When in use, let the UAV flight controller start with non-USB interface power supply. After the system is initialized. We can connect to the PC with a USB cable to enter the OSD font update mode. Then update the upper computer through the font update. Later, you can also design a USB interface to implement DFU to update the firmware.

3.4 Control Function Software Design

There are 3 flight modes in this UAV development platform. The first is that the mode is an attitude flight mode. The second is MS5611 fixed-height mode. The third is INT-P603 optical flow hovering mode. All three modes use PID algorithm, but the difference is that the sources of error are different. This section will introduce the implementation of three control algorithms.

Attitude Mode Software Design. The attitude mode is named after the PID error comes from the attitude sensor. Before the control is to obtain the attitude and attitude solution. As mentioned in the previous chapter, this design uses the internal DMP channel data of the MPU9250. The data obtained is the four-element data of the 30th power after being solved and filtered. After conversion, the attitude angle is obtained. Among them, Mpu is an IMU structure, which stores all the attitude information of the drone and keeps it up to date.

Fixed-Height Software Design. This system has designed a set of better schemes for acquiring flying height of unmanned aerial vehicles. When the drone is less than 5 m in height, the ultrasonic module is used to obtain height data (the maximum height of the ultrasonic wave can be measured is 6 m). When the high-speed data is greater than 5 m, MS5611 is used to obtain the pressure and temperature data of the current drone environment. It is then converted to altitude and compared with the altitude at takeoff. In this way, the height of the drone can be obtained. The fixed-height mode is realized by using the ascending height data of the drone. The control algorithm is also PID control algorithm. Repeatedly reading the current height of the drone and performing PID control can achieve fixed-height control. The PID control flow chart for fixed height control is shown in Fig. 13.

Software Design of Optical Flow Hovering Mode. Use DMA1 channel 6 of STM32 to receive displacement data sent by optical flow module to USART 2. This reduces the CPU usage and ensures data accuracy. The optical flow control PID control flowchart is

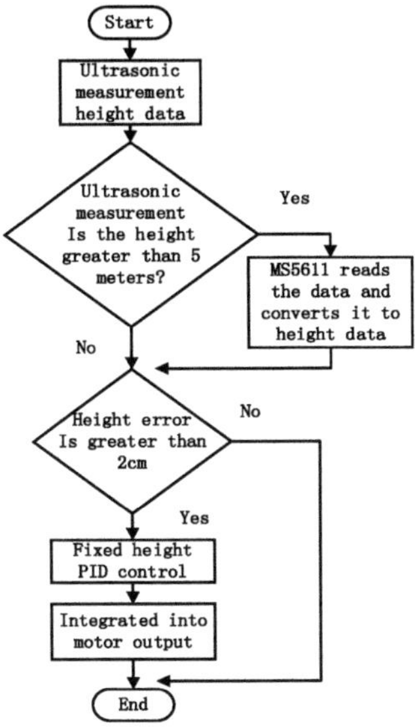

Fig. 13. Flow chart of fixed-height PID control.

shown in Fig. 14. We first determine whether we have received the data from one frame of optical flow module. If it is indeed received a frame of data, the optical flow PID control. Otherwise, the optical flow PID control is ended.

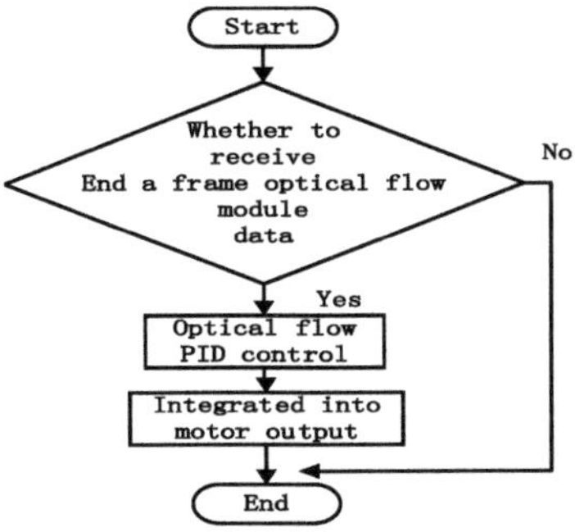

Fig. 14. Optical flow PID control flow chart.

3.5 Design of OSD Data Return Display Software

OSD data return is a relatively good design. The OSD module has been initialized when the system is turned on. After the system is initialized, you can directly call the display function to achieve data return. The main control chip uses the SPI bus to communicate with the OSD module chip AT7456E. It mainly writes the data that needs to be returned to the AT7456E chip, and then the AT7456E chip superimposes the display data on the video screen, and then sends it out through the image transmission. Finally, the camera screen and the returned data can be seen through the receiving device. The OSD module work flow chart is shown in Fig. 15.

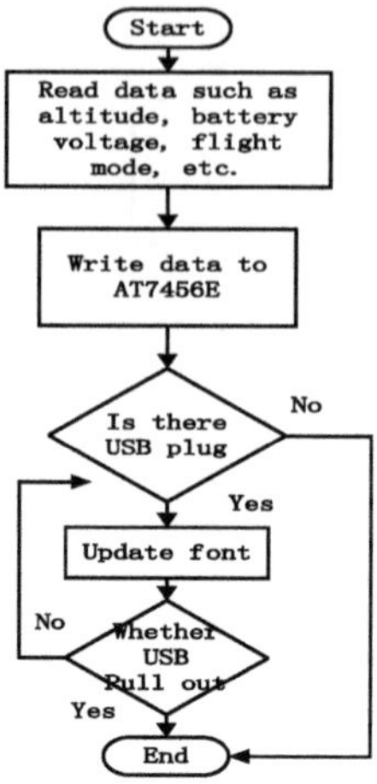

Fig. 15. OSD module data work flow chart.

4 Conclusion

This design msOS four-rotor aircraft system uses the deeply customized msOS as the software system, which has the characteristics of real-time and flexibility. The IMU system uses the internal DMP channel data of the MPU9250 to ensure the accuracy and stability of the data. The UAV flight control board is designed with an OSD circuit to facilitate the real-time data transmission. We use the mature S-Bus protocol as a remote control signal input to support multiple remote control receivers in the market. The system has a fixed-height, optical flow hovering mode, which can achieve outdoor and indoor hovering. The system has a USB function, which can easily update the remote controller calibration data and OSD font. At the same time, it has multiple data interfaces to facilitate secondary development.

Acknowledgement. The authors are highly thankful to the Research Project for Young and Middle-aged Teachers in Guangxi Universities (ID: 2019KY0621), to the Natural Science Foundation of Guangxi Province (No. 2018GXNSFAA281164). This research was financially supported by the project of outstanding thousand young teachers' training in higher education institutions of Guangxi, Guangxi Colleges and Universities Key Laboratory Breeding Base of System Control and Information Processing.

References

1. Zhang, C.X.: Simulation and experimental research on attitude fusion and control algorithm of quadcopter. Taiyuan University of Technology (2017)
2. Chen, H.B., Shi, G.H.: Design of a quadcopter. Lab. Res. Explor. **03**, 41–44 (2013)
3. Xu, Y.Q.: Research on flight control of quadcopter. Xiamen University (2014)
4. Chen, X.Q.: Design and research of four-rotor UAV flight control system. Nanchang Hangkong University (2014)
5. Vishal, G.: Sensor and connection function of UAV. China Integr. Circ. **10**, 46–49 (2017)
6. Li, D.W., Yang, J.: How much do you know about open source flight control. Robot Ind. **03**, 83–93 (2015)
7. Li, X.W.: Flight control and implementation of quadcopter. Southwest University of Science and Technology (2017)
8. Ji, K.: DJI innovation: global UAV leader. Chin. Brand **04**, 72–74 (2015)
9. Deng, Y.M.: Research on multi-rotor UAV optical flow/inertial integrated navigation technology. Nanjing University of Aeronautics and Astronautics (2016)
10. Zhang, Y.C., Sun, C., Lin, K., Qin, S.H.: Design and implementation of S-BUS communication protocol for quadrotor aircraft based on MSP430. Management and Technology of Small and Medium-sized Enterprises (Mid-Term), vol. 04, pp. 187–188 (2018)

Design and Implementation of Traversing Machine System Based on msOS Platform

Qiwen He[1], Jiansheng Peng[1,2(✉)], Hanxiao Zhang[2], and Yicheng Zhan[1]

[1] School of Physics and Mechanical and Electronic Engineering, Hechi University, Yizhou 546300, China
sheng120410@163.com
[2] School of Electrical and Information Engineering, Guangxi University of Science and Technology, Liuzhou 545000, China

Abstract. In order to solve the problem of multi-task simultaneous operation of sensor reading data, remote control data receiving, OSD (On Screen Display) data sending and PID (Proportion Integration Differentiation) attitude control, a traversing machine system based on msOS platform was designed. The system implements a traversing machine system with attitude control, fixed height control, OSD data return, and fast system response speed. First, the structure and flight principle of the traversing aircraft are introduced. Then introduced the hardware design of the system. The hardware system adopts the idea of holistic design. Centralize the interface between the sensor and the function module on the flight control board. Then introduced the software design of the traversing machine system in this paper. Then read various sensor data through msOS. The software executes the cascade PID algorithm to realize the attitude control of the rider. Then introduced the debugging process. Finally, the shortcomings of the traversing machine system based on msOS platform in this paper are summarized and the research prospect is prospected.

Keywords: Traversing machine · msOS · PID algorithm · OSD

1 Introduction

Unmanned aerial vehicle (UAV) aircraft is aerodynamically operated by wireless remote control or the self-control program included in the aircraft itself. Aircraft that do not require a pilot in the UAV and can be recycled and reused [1]. According to the flight mode, the UAV can be divided into a rotorcraft and a fixed wing. The advantage of the gyroplane is that it can rise or fall vertically and can hover in the air. The rotorcraft is easy to control and is suitable for use in complex geographical environments or in small operating spaces. With the development of science and technology, the integration of electronic devices used in the quadcopter is becoming more and more compact. The quadrotor has a wider range of applications and is more and more popular [2].

The traversing machine is a special branch of the first-angle aircraft. It has the characteristics of small size, easy to carry, fast flight speed, flexible operation, low cost

Q. Liu et al. (Eds.): CENet 2020, AISC 1274, pp. 1080–1090, 2021.
https://doi.org/10.1007/978-981-15-8462-6_124

and flexible requirements for the flight site. Most of the traversing machines use a four-rotor structure. Since the beginning of the 21st century, especially in recent years, more and more well-known universities and companies have begun to research and develop the quadrotor flight, so that the quadrotor aircraft will be miniaturized [3]. And the quadcopter is gradually used in various fields. Riders are a special branch of the quadcopter family. Unlike traditional four-rotor aircraft, the traversing machine is a small drone with high speed and short duration. Most players usually buy accessories themselves to assemble them. In terms of flight control, the main difference between the rover and the traditional four-rotor aircraft is reflected in the flight mode and control flexibility. Riders require a variety of flight modes, adapt to various flight environments, and flexible control. Traversing machines can more easily travel through complex environments and terrain. The flight control of the flying machine is divided into open source flight control and commercial flight control.

The full name of Pix is Pixhawk. The main control of Pix is the 32-bit microprocessor of STM32F427. It has a 168 MHz operating frequency and runs fast. It uses the Mission Planner ground station and the software is all open source. Pix has two sets of gyroscope and accelerometer sensors, which are complementary to each other. CC3D flight control is the only heavily used flight control provided by OpenPilot. CC3D is favored by the majority of model friends because of its simple configuration, stable flight control firmware, cheap price, and strong ground station support. This flight control is mainly used in traversing aircraft. Its flight parameters and stability and rich modes are suitable for all types of traversing aircraft.

The Sect. 2 of this article introduces the structure and flying principle of the traversing machine. Section 3 introduces the hardware design of the traversing machine. The design uses STM32F103RX series as the main control. The MPU9250 attitude sensor is responsible for collecting flight attitude data. The MS5611 barometer sensor is responsible for the altitude of the aircraft. S-Bus decoding is responsible for the transmission of remote control data. AT7456E as the OSD processing chip. Section 4 introduces the software design of the traversing machine. The software design uses the msOS system framework to complete more complex tasks. Section 5 introduces the debugging process of the traversing machine. Finally, it further summarizes the shortcomings of the existing methods and prospects.

2 The Structure and Flight Principle of the Ride-Through Aircraft

2.1 Establishment of Body Coordinate System

There are two types of traversing machine structures, one is a cross type perpendicular to each other, and the other is an orthogonal X type. In the choice of models, most of the traversing machines use X-shaped racks. In order to obtain the attitude change of the aircraft, a body coordinate system must first be built around the rider. Take the X-type rack type as an example. The coordinate system uses the cross point of the frame, that is, the center point of the traversing machine, as the origin. Use the angle bisector of the frame to make the X axis and Y axis respectively. The Z axis is perpendicular to the plane of the frame and upward. Establish the body coordinate system in this way, as shown in Fig. 1. The various movement modes and attitude changes of the traversing machine can be decomposed by this body coordinate system.

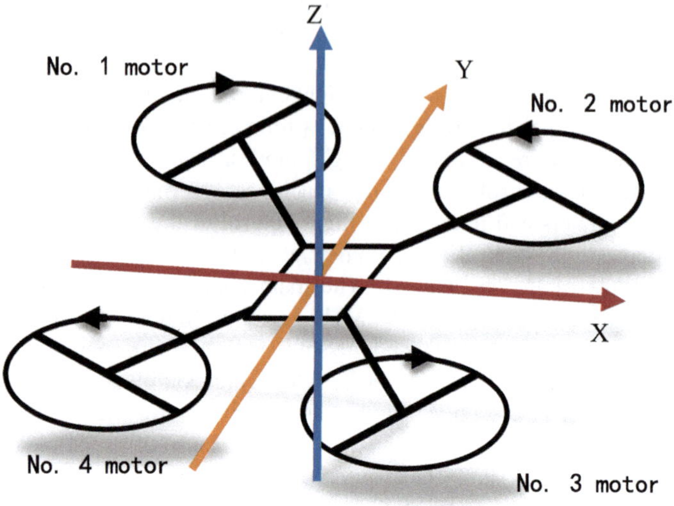

Fig. 1. Four-rotor fuselage coordinate system.

2.2 Analysis of the Motion Model of a Flying Aircraft

Vertical Ascent and Descent. Ideally, when the four propellers produce the same lift and the combined force is greater than the weight of the aircraft itself, the aircraft can be raised vertically, otherwise it can be lowered vertically. Let the lift generated by the propeller be F1, F2, F3, F4. Then their combined force is F. The local gravity acceleration is g, the weight of the traversing machine is m, and the gravity of the aircraft is G. To enable the aircraft to vertically ascend and descend needs to satisfy the conditions of formula (1):

$$\begin{cases} G = mg \\ F = \mathrm{F}_1 + \mathrm{F}_2 + \mathrm{F}_3 + \mathrm{F}_4 \\ \mathrm{F}_1 = \mathrm{F}_2 = \mathrm{F}_3 = \mathrm{F}_4 \end{cases} \tag{1}$$

When the above conditions are met, the output power of four motors is increased at the same time. The increase in the speed of the four rotors increases the total lift and overcomes the weight of the whole machine [4]. That is, when F > G, the quadcopter rises vertically, whereas the total lift F decreases.

Pitch Motion. After the aircraft stabilizes horizontally, it rotates around the X axis. The pitch motion needs to satisfy formula (2).

$$\begin{cases} F_1 = F_2 \\ F_3 = F_4 \end{cases} \tag{2}$$

Because of the change in motor speed, the lift forces F1 and F2 decrease, and F3 and F4 increase. At this time F1 + F2 < F3 + F4. The unbalanced torque caused by the lift force causes the fuselage to rotate clockwise along the X axis, which causes the aircraft to move forward. When F1 + F2 > F3 + F4, the aircraft moves backward.

Roll Motion. The principle of roll motion and pitch motion are the same. It transforms from rotation movement around the X axis to rotation movement around the Y axis. Rolling motion needs to satisfy formula (3):

$$\begin{cases} F1 = F4 \\ F2 = F3 \end{cases} \tag{3}$$

Observe from the front view of the body coordinate system. When F1 + F4 > F2 + F3, the fuselage rotates counterclockwise around the Y axis of the coordinate system, that is, moves to the right. When F1 + F4 < F2 + F3, move to the left.

Yaw Movement. When the rotation speeds of the four motors are the same, the generated reverse torques cancel each other out. At this time, no yaw movement occurs, so that the fuselage reaches a stable state [5]. In order to enable the traverse aircraft to do yaw movement, formula (4) must be satisfied.

$$\begin{cases} \tau_{g1} = \tau_{g3} \\ \tau_{g2} = \tau_{g4} \end{cases} \tag{4}$$

When the clockwise anti-torque received by the aircraft is less than the counter-clockwise anti-torque, the aircraft moves counterclockwise around the Z axis, that is, yaws to the left. The same applies to the right yaw movement.

Horizontal Exercise. The combined force of the lift generated by the four motors is F, and the gravity of the fuselage is G. The force of horizontal movement is generated by the component F5 of the resultant force F in the horizontal direction. The lift is generated by the component F6 of the vertical force of the total force F. The angle of inclination between the fuselage and the horizontal plane is θ [6]. To make the traversing machine move horizontally without generating up and down displacement, the formula (5) must be satisfied.

$$\begin{cases} \begin{cases} F = F_1 + F_2 + F_3 + F_4 \\ F_5 = F \sin\theta \end{cases} \\ \begin{cases} F_6 = F \cos\theta \\ F_6 = G \end{cases} \end{cases} \tag{5}$$

3 Hardware Design

The hardware system block diagram of the ride-through machine is shown in Fig. 2. The hardware system is mainly divided into six parts, namely sensor part, power part, power part, function interface part, main control part and image display part. The power supply part can output voltages of 12 V, 5 V and 3.3 V to supply power and modules of the ride-through system respectively. The sensor part includes MPU9250 attitude sensor and MS5611 barometer sensor. As the data acquisition part of the traversing machine, they are mainly responsible for collecting attitude data and altitude data of the traversing machine. The sensor transmits the collected data to the main control part. The main control part is the core of the entire system. It is responsible for handling the driving of various modules and interfaces. And it receives various information data for processing. Then realize the analysis and calculation of the attitude of the rider. Finally, the calculated control quantity is output in the form of PWM, so as to achieve the purpose of controlling the power of the motor. The functional interface part includes the S-Bus decoding interface and the AT7456 character overlay chip interface. The remote control end transmits the data to the receiver, and the remote control data is obtained through the S-Bus decoding circuit analysis. Then transmit the remote control data to the main control to analyze and execute the remote control commands. The AT7456 character superposition chip performs character processing on the data collected by the main control. Then superimpose the characters on the image data transmitted by the camera. Then transfer the processed image data to the image transmission module. Eventually, the display will show the data status of the aircraft.

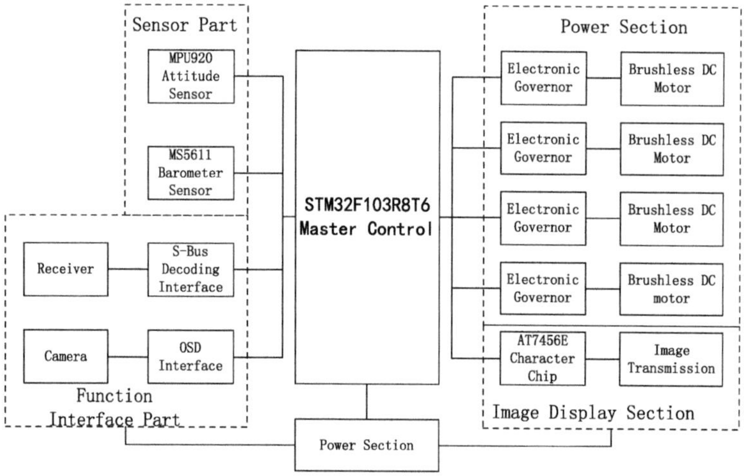

Fig. 2. Block diagram of the traversing machine hardware.

The main control part of the ride-through system is the most important part of the whole system. The main tasks it needs to perform are: acquiring MPU9250 attitude sensor data, acquiring MS5611 barometer sensor data, acquiring receiver control signals, OSD character data processing, aircraft attitude calculation, height setting algorithm, PWM signal output, voltage ADC detection and many more. This ride-through system uses the STM32F103R8T6 chip. The design of the hardware system of the traversing machine system adopts the holistic idea. Integrate the sensors and data processing modules used in the ride-through system into the flight control board. The circuit design of the flight control board including power module, attitude measurement circuit, height measurement circuit and OSD circuit is completed. And according to the actual needs of the traversing machine, a frame conforming to the system of the traversing machine is cut out. Then select the electronic governor and motor that fit the rider. The experimental platform construction of the traversing machine system was finally completed. The 3D rendering of the hardware of the ride-through system is shown in Fig. 3.

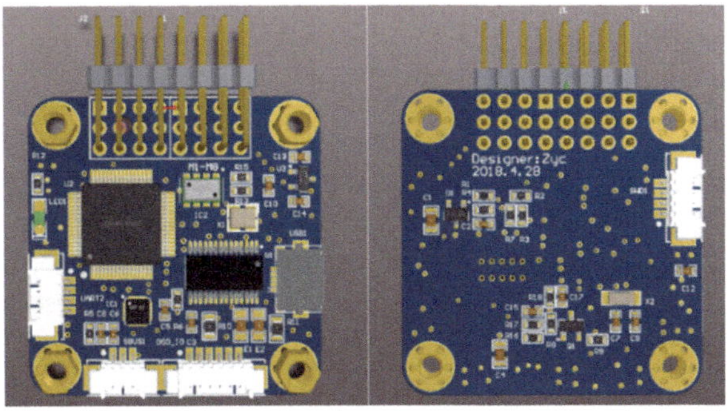

Fig. 3. 3D rendering of the hardware of the ride-through system.

4 Software Design

4.1 MsOS System Framework

All software designs in this article are based on msOS. It can more effectively support the execution of multiple tasks and complete more complex tasks. As an industrial control system, msOS has many industrial control parts that are not needed in the traversing machine system. In order to more effectively use the resources of the MCU, the redundant part of the industrial control drive is removed. The design simplifies and transplants msOS to the traversing machine system. Finally, a software framework based on the msOS platform traversing machine system is formed. The software framework is shown in Fig. 4.

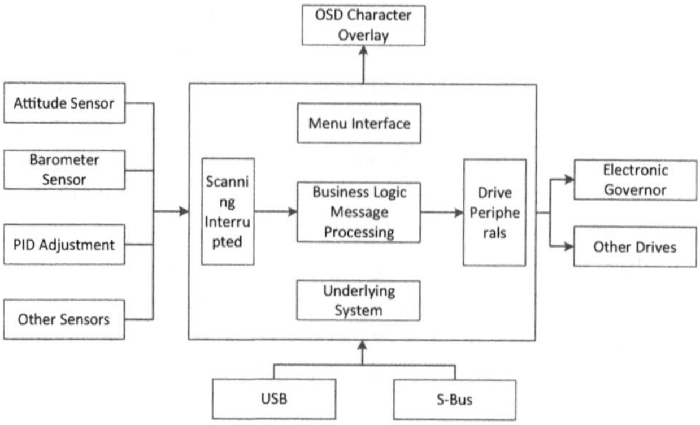

Fig. 4. Software framework of traversing machine system based on msOS platform.

4.2 OSD Display Software Design

In the software design, the processing chip for the OSD function is AT7456E. The main control chip communicates with it through the SPI bus. The ride-through system initializes the OSD module. After the system is initialized, it can directly use the display function to realize the function of data return. The main control chip writes the data needed to be returned to the AT7456E chip and then processes it. Then superimpose the displayed data on the video screen. Finally, the video screen is transmitted on the display screen through the image transmission module.

4.3 S-Bus Decoding Software Design

First, the remote controller sends a remote control signal to the receiver. Secondly, the receiver transmits the received remote control signal to the serial port of the main control chip in the S-Bus protocol. Finally, the remote control data of each channel is obtained through decoding to realize the remote control of the traversing machine. S-Bus is a serial communication protocol used by futaba [7]. Before reading S-Bus data, you need to configure the serial port first. Its configuration parameters are shown in Table 1.

Table 1. S-Bus serial port configuration requirements table.

Baud rate	Data bit	Check digit	Stop bit	Flow control
100 kbps	8 bit	Even parity	2 bit	No flow control

After detecting that the serial port receives a frame of message, it is determined whether the start byte of the received data is 0X0F. When the start byte satisfies the judgment condition, the S-Bus decoding message is thrown into the business logic, and the S-Bus decoding function is executed. When the judgment condition is not satisfied, exit the serial port interrupt judgment.

4.4 Attitude Sensor Software Control Design

Attitude Angle Acquisition. Attitude control is the key to flight control of the traversing system. Only real-time control of the aircraft's attitude angle can ensure the stable flight of the aircraft. In order to ensure the completion of the flight mission, the first thing to do is to collect the attitude angle of the aircraft.

In the case of keeping the reference coordinate system unchanged, the change of the attitude of the aircraft can be regarded as the X, Y, Z axis rotation of the body coordinate system relative to the reference coordinate system to obtain a new coordinate system. The angle of its rotation around the three axes of X, Y and Z is expressed by Euler angle. They are respectively the heading angle ψ, roll angle φ and pitch angle θ. The rotation in three-dimensional space is generally expressed by quaternion. The rotation of the aircraft around the three axes of X, Y and Z is expressed by quaternion. The quaternion of each axis is multiplied by the quaternion multiplication rule to obtain the quaternion of the attitude angle [q0, q1, q2, q4] [8]. Finally, it will be converted into Euler angles. The Euler angle of quaternion is converted as formula (6).

$$\begin{bmatrix} \emptyset \\ \theta \\ \psi \end{bmatrix} = \begin{bmatrix} \arctan \frac{2(q_0 q_1 + q_2 q_3)}{q_0^2 - q_1^2 - q_2^2 + q_3^2} \\ \arcsin[2(q_1 q_2 - q_1 q_3)] \\ \arctan \frac{2(q_0 q_3 + q_1 q_2)}{q_0^2 + q_1^2 - q_2^2 - q_3^2} \end{bmatrix} \tag{6}$$

Rider system uses the internal DMP data of MPU6050. This data can be used to obtain quaternion data of the 30th power after solution and filtering. Putting the quaternion into formula 6 to calculate, we can get the Euler angle of XYZ axis respectively. Angular velocity and acceleration are also obtained through the DMP channel. Directly convert the acquired data units to obtain effective motion posture data. You can get the Euler angle, acceleration and angular velocity data of the XYZ axis of the aircraft and then throw out the PID attitude control message.

Cascade PID Control Algorithm. Compared with single-loop PID, cascade PID connects two PIDs in series. They are the inner ring and the outer ring, as shown in Fig. 5. In the traversing machine system, the angle data is the outer loop input feedback, and the angular velocity data is the inner loop input feedback [9]. Among them, the inner ring control occupies a larger proportion, and the outer ring angle control occupies a smaller proportion. In this way, even if the outer ring data changes significantly, the inner ring can still make the aircraft in a relatively stable state. This is not easy to cause the aircraft to oscillate.

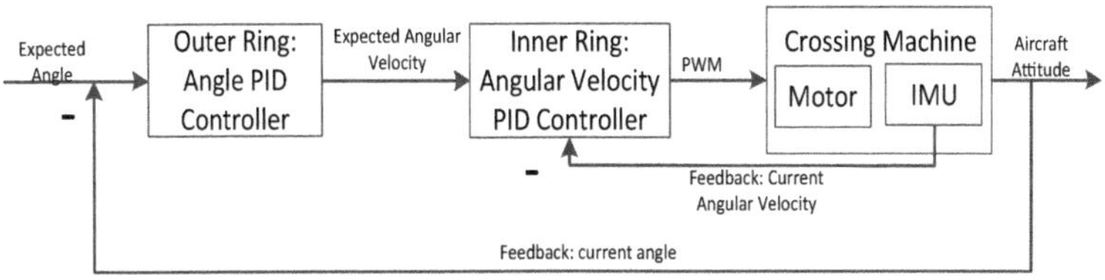

Fig. 5. Block diagram of cascade PID control.

In the inner ring, the main task of P is to correct the traversing machine from the deviation angular velocity to the desired angular velocity. I is to eliminate the angular velocity error left by the P term. D can improve system stability. P in the outer ring can correct angular errors, I eliminates static angle errors, and D improves system stability. After we obtain the output data of the cascade PID in three axes, we integrate them. Then output four combined PWM data to four motors respectively. Then control the speed of the four motors separately. Finally, through the combined changes of the four motor speeds, the attitude control of the aircraft is realized.

Fixed-Height Software Design. It is inconvenient to install an ultrasonic module for height measurement because of the narrowness of the airframe. Therefore, the traversing machine system adopts the method of barometer measurement to obtain altitude data. When the barometer works for a long time, the detected height is small in error deviation. Because some interference will affect the detection of the barometer, the height detected by the barometer will float around the true height in a short time. Therefore, the flight height obtained by acceleration conversion is added. The traversing machine system obtains the acceleration component of the Z axis of the aircraft through attitude calculation. The quadratic integral of this acceleration component can get the flying height of the traversing machine. This data is more accurate in

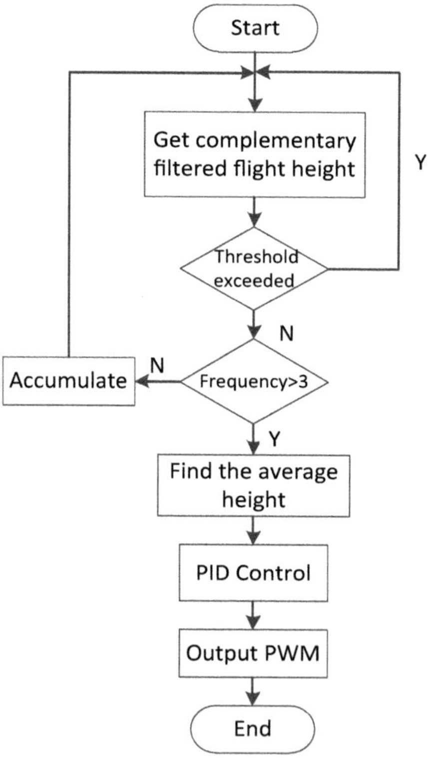

Fig. 6. Flow chart of height setting program.

a short time. But with the extension of the integration time, the data is prone to errors after a long time. Therefore, the above two schemes are complementarily filtered. Refer to the value obtained with the barometer for a long time. Refer to the height of Z axis acceleration conversion in a short time. After complementary filtering, a more accurate height value is obtained. Then control the throttle output value of the four motors through the PID algorithm to achieve the fixed height function. The flow chart of fixed height algorithm is shown in Fig. 6.

5 Commissioning Process

Traversing machine is a special branch of UAV. The main meaning of its existence is to participate in various racing competitions. In the racing game, the traversing machine passes various obstacles and finally reaches the finish line. The winner with the shortest time. In order to check the stability of the traversing machine system based on the msOS platform, arch doors of different sizes were set to test the traversing machine. As shown in Fig. 7.

Fig. 7. Through the machine through the door test chart.

The controller took off 10 m from the obstacle gate. Determine the position of the obstacle door through the screen of the display. Make the traversing machine finally pass through the door to complete a test. The test results are shown in Table 2.

Table 2. Crossing machine barrier door test table.

Barrier door specifications	Successful times	Number of failures	Success rate
1.6 m * 1.6 m	10	0	100%
1.4 m * 1.4 m	10	0	100%
1.2 m * 1.2 m	7	3	70%
1.0 m * 1.0 m	8	2	80%
0.8 m * 0.8 m	7	3	70%
0.6 m * 0.6 m	4	6	40%
0.4 m * 0.4 m	1	9	10%

Table 2 shows the test results of the traversing machine passing through obstacle doors of different specifications. The process of crossing the obstacle gate will not only be affected by the stability of the crossing machine, but also by the operator's technology.

6 Conclusion

The use of the system can well solve the problem of multi-task real-time execution. The system can greatly improve the operation efficiency of the MCU and the stability of the system. In this design, the expected effect is basically achieved. But compared with commodity flight control, the stability of flight is not small. Flight is easily disturbed by external factors. There are still many areas where the system can be optimized. Next, the attitude control algorithm and filtering algorithm of the traversing aircraft can be further optimized to improve the performance of the aircraft.

Acknowledgement. The authors are highly thankful to the Research Project for Young and Middle-aged Teachers in Guangxi Universities (ID: 2019KY0621), to the Natural Science Foundation of Guangxi Province (No. 2018GXNSFAA281164), This research was financially supported by the project of outstanding thousand young teachers' training in higher education institutions of Guangxi, Guangxi Colleges and Universities Key Laboratory Breeding Base of System Control and Information Processing.

References

1. Wang, S.F., Wu, W.: Intelligence research on UAV development. Flying Missile **10**, 32–41 (1998)
2. Yang, J., Zha, X.J., Zhou, H.H.: Research and production of small four-rotor aircraft. Comput. Times **01**, 9–11 (2016)
3. Fang, X., Zhong, B.C.: Research and application of quadrotor aircraft. J. Shanghai Univ. Eng. Sci. **02**, 113–118 (2015)
4. Li, H.B.: Research and design of attitude stability control for quadcopter. Xi'an University of Science and Technology (2017)
5. Wu, Q.B.: Research on reconstruction control method of quadrotor helicopter based on adaptive control. Nanjing University of Aeronautics and Astronautics (2015)
6. Huang, X.: Research on DSP-based cable six-rotor flight control system. Nanchang Hangkong University (2016)
7. Liu, K.: Design and implementation of small quadrotor aircraft. Southwest University of Science and Technology (2017)
8. Xu, Y.C.: Design and simulation of attitude calculation algorithm for quadcopter. Sci. Technol. Horiz. **23**, 17–18 (2016)
9. Xue, L.: Design and implementation of multi-rotor UAV flight control system. Nanjing University of Aeronautics and Astronautics (2016)

Single Image Super-Resolution Based on Sparse Coding and Joint Mapping Learning Framework

Shudong Zhou[1], Li Fang[2], Huifang Shen[2(✉)], Hegao Sun[2], and Yue Liu[1]

[1] Liaoning Technical University, Huludao 125105, China
[2] Quanzhou Institute of Equipment Manufacturing Haixi Institutes, Chinese Academy of Sciences, Quanzhou 362216, China
shenhf@fjirsm.ac.cn

Abstract. Many image super-resolution algorithms can be formulated in a mapping framework based on the natural image prior. Generally, the mapping function with free parameters is learned by minimizing the reconstruction mapping error. In this paper, we obtain a considerable image super-resolution algorithm which gains in accuracy and speed by combining joint mapping learning with fast approximations of sparse coding. A novel "dictionary" training method for single image super-resolution based on feed-forward neural network is proposed. The training algorithm alternates between solving sparse coding problem and learning joint mapping relation problem. The learning process enforces that the sparse coding of a low-resolution image patch can be regarded as the shared latent coordinates for reconstructing its underlying high-resolution image patch with the image high-resolution dictionary. Experiments validate that our learning method shows excellent results both quantitatively and qualitatively.

Keywords: Image super-resolution · Machine learning · Sparse coding · Joint mapping framework · LISTA

1 Introduction

Single image super-resolution (SR) technology has received much attention in image processing and computer vision due to its practical properties, which offers a high-resolution (HR) enlargement of a single low-resolution (LR) image. It plays an important role in the video surveillance, satellite images application and medical images diagnosis. However, image SR technology is an inherently ill-posed problem. It is an effective way to incorporate strong prior information [1].

Recently, there are mainly two families of SR methods based on the prior. The first one formulates the problem within the Bayesian framework. The difference is that which kind of prior is adopted. It can be generally modeled as:

Q. Liu et al. (Eds.): CENet 2020, AISC 1274, pp. 1091–1101, 2021.
https://doi.org/10.1007/978-981-15-8462-6_125

$$X = arg\ \min_X \|Y - DHX\|_2^2 + R(X) \quad (1)$$

where X is the unknown HR image, Y is a degraded measurement image, D represents a down-sampling operator, H is a blurring operator and $R(X)$ is the prior knowledge of the image. Sun *et al.* [2] exploited the gradient profile priors for local image structures. Glasner *et al.* [3] developed the self-similarities of image patches within and across different scales in the same image. Dong *et al.* [4] took advantage of the non-local similarity, sparse representation and autoregressive models in an image. The methods above estimate the final HR image with a maximum a posterior estimation approach.

The other is to learn the relationship between two types of images space, which learns the mapping function from LR to HR image space. It can be modeled as:

$$X = arg\ \min_X \|X - F(Y)\|_2^2 + \lambda R(F) \quad (2)$$

where F represents the mapping function and $R(F)$ is the penalty on the parameters. Freeman *et al.* [5] learned the relationship via a Markov Random Field by belief propagation. Inspired by manifold learning, Chang *et al.* [6] took account of the similarity between the two manifolds in the LR and HR patch spaces. Yang *et al.* [7] extended the method based on sparse representation of HR and LR patch pairs over a dictionary pair. Sparse representation-based image SR methods have been widely used and studied. The main improvements focus on dictionary-learning algorithms, sparse coding methods and the mapping relation between LR and HR image patches. For example, Zhang *et al.* [8] learned compact low- and high-resolution sub-dictionary pairs for multiple feature subspaces. Ahmed *et al.* [9] proposed a coupled dictionary learning with mapping function for clustered training data. Liu *et al.* [10] learned a large dictionary for LR image patches and search for similar patches in it to obtain the sub-dictionary.

The main focus of this paper is on the design of suitable mapping techniques. Yang's method [7] suffers from asymmetry problem between the training step and the test step. Meanwhile, it is too slow for real-time application. This paper will analyze these problems and propose a novel method based on sparse coding and iterative joint mapping learning to alleviate them.

The remainder of this paper is organized as follows. A brief review of sparse coding-based SR is provided in Sect. 2. We present the proposed method in more detail in Sect. 3 and experimental results are shown in Sect. 4. The conclusions and ideas for further work are drawn in Sect. 5.

2 Sparse Representation-Based Super-Resolution

Sparse representation has been widely employed in many areas including machine learning, computer vision, and signal processing, since a signal or an image patch can be well-represented as a sparse linear combination of elements from a learned over-complete dictionary. Coupled dictionaries learning method proposed by Yang et al. [7] is based on this idea. They learn two coupled dictionaries: D_l for LR patches and D_h for

HR ones, which are forced to have the same sparse representation for LR patches as their corresponding HR counterparts.

The resulting optimization problem is:

$$\min_{D_h,D_l,\alpha}\frac{1}{N}\|X_h - D_h\alpha\|_F^2 + \frac{1}{M}\|X_l - D_l\alpha\|_F^2 + \lambda\|\alpha\|_1 \tag{3}$$

where N and M are the dimensionality of HR and LR patches, and α is the sparse coefficient vector. λ is the parameter to balance the sparsity and the fidelity. A large λ corresponds to a belief that α is very sparse. Thus, Eq. (3) can be written as:

$$\min_{D,\alpha}\|X - D\alpha\|_F^2 + \lambda\|\alpha\|_1 \tag{4}$$

where $X = \begin{bmatrix} \frac{1}{\sqrt{N}}X_h \\ \frac{1}{\sqrt{M}}X_l \end{bmatrix}$ and $D = \begin{bmatrix} \frac{1}{\sqrt{N}}D_h \\ \frac{1}{\sqrt{M}}D_l \end{bmatrix}$. The testing phase attempts to recover the HR patches, given that the LR patches, D_h and D_l are known. The HR patches can be easily recovered as $x = D_h\alpha^*$. The sparse codes α^* are obtained as follows:

$$\min_{\alpha^*}\|y - D_l\alpha^*\|_F^2 + \lambda\|\alpha^*\|_1 \tag{5}$$

where y represents the LR patches.

It is worth noting that there is a significant deficiency. The training phase and the testing phase are not symmetrical. At the training phase, the sparse coding is obtained by solving Eq. (4), taking advantage of both LR and HR patches. However, the sparse coding of the testing phase is only inferred using the LR patches by solving Eq. (5). Both constraints on the sparse coding are not equivalent. Thus, the codes obtained at the testing phase are not the desired solution. The experimental result will be affected more or less, which can be observed in Sect. 4. Yang et al. [7] addressed this problem by a coupled dictionary learning formulation via bi-level programming. It can speed up the algorithm, but the calculation of the gradient $\partial L/\partial D_l$ (L represents the reconstruction error) consumes extremely complex computation.

3 Methodology

3.1 Basic Idea

Image SR method based on sparse coding can be described as the problem of regression, where we estimate a mapping $P : Y \to X$ between a LR image $Y \in R^L$ and a HR image $X \in R^H$, but going through a latent sparse coding $\alpha \in R^D$ with $L<H<D$. It is similar to the neural network with a hidden layer. Figure 1 depicts the relationship between SR and neural network. The model is based on two main components:

- The representation process: represent the LR image by sparse coding;
- The mapping process: reconstruct the HR image using the hidden sparse coding as its input.

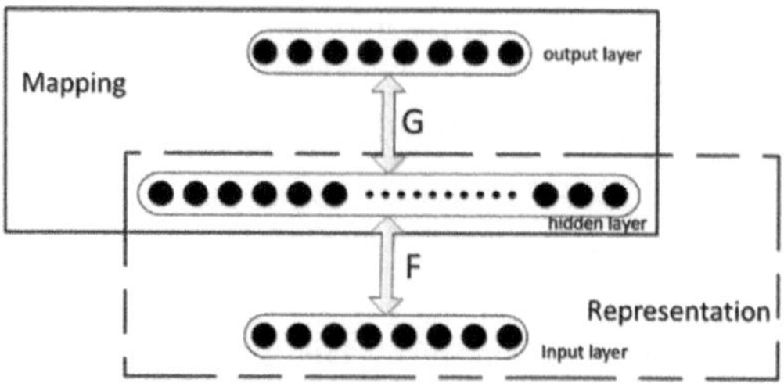

Fig. 1. Depiction of the relationship between image SR and neural network. The input layer is the LR image, the hidden layer is a latent sparse coding of LR image and the output layer represents the HR image.

We attempt to learn the relationship directly with neural network applied to image patches by considering the problem of learning the dictionary pair in a feed-forward neural network using back-propagation method, with respect to the quadratic error between the predicted HR image and the optimal one. That is to say, we regard the dictionary pair as the connection weights in a neural network. The network minimizes the following objective function to estimate the sparse coding F of LR image patches y and the HR image patches x reconstruction operator G:

$$E(F,G) = \sum\nolimits_{n=1}^{N} \|x_n - G(F(y_n))\|_2^2 \tag{6}$$

where $F_{D_l} = arg\ \min_\alpha \sum_{n=1}^{N} \|y_n - D_l\alpha_n\|_2^2 + \lambda\|\alpha_n\|_1$ and $G_{D_h} = D_h^\circ F_{D_l}$, given the training image patch pairs $\{x_n, y_n\}, n = 1, \cdots, N$. We can compute the derivative ∇E_D by back-propagation, where D collectively designates the parameters D_l and D_h. There are several approaches to solve this optimization problem, such as stochastic gradient descent as well as L-BFGS, etc. But they are computationally expensive and the convergence is slow. Instead of computing ∇E_D directly, we find the algorithm based on decoupling parameters from each other can yield more accurate results, which approximate the objective function Eq. (6) by expanding it into two squared terms:

$$\min_{D_h, D_l, \alpha} \rho \sum\nolimits_{n=1}^{N} \|\alpha_n - F_{D_l}(y_n)\|_2^2 + \|x_n - D_h\alpha_n\|_2^2 \quad s.t. \|\alpha\|_1 \le \varepsilon \tag{7}$$

where parameters $\alpha_n, n = 1, \cdots, N$ is the sparse coding, ε reflects the sparsity power, and positive constant ρ is to improve the stability and robustness of the method, which satisfies $\rho^2 = \|D_h\|_F^2$. We iteratively optimize Eq. (7) by alternatingly optimization with respect to α and (D_h, D_l), which is similar to the coordinate descent method.

3.2 Optimization Process

Optimization over the Mapping Relationships. We find that the above two terms in Eq. (7) are irrelevant given that α is a constant. Thus, parameters D_h and D_l can be optimized independently by fixing α. The optimization over D_h is a direct linear regression problem and has a pseudo-inverse solution. The problem becomes:

$$\min_{D_h} \sum_{n=1}^{N} \|x_n - D_h \alpha_n\|_2^2 \tag{8}$$

The solution of this problem is given by the following closed-form expression:

$$D_h = X\alpha^{\dagger} = X\alpha^T \left(\alpha\alpha^T\right)^{-1} \tag{9}$$

where $X = [x_1, \cdots, x_N]$ and $\alpha = [\alpha_1, \cdots, \alpha_N]$. We can also use linear conjugate gradient descent method to minimize the quadratic function taking account of the huge dimension of α and take the dictionary D_h obtained at the previous iteration as initialization in the current iteration. To prevent the columns of D_h to be arbitrarily large, the constraint is introduced as follows:

$$D_h(:,j) = \frac{D_h(:,j)}{max\left\{1, \|D_h(:,j)\|_2^2\right\}} \tag{10}$$

The optimization of D_l can be expressed as follows:

$$\min_{D_l} \sum_{n=1}^{N} \|\alpha_n - F_{D_l}(y_n)\|_2^2 \quad s.t. \; F_{D_l} = \min_{\alpha} \frac{1}{2} \|y_n - D_l\alpha\|_2^2 + \lambda\|\alpha\|_1 \tag{11}$$

However, Eq. (11) may be difficult to solve because it involves the derivative of F_{D_l} over D_l, and the function is non-differentiable at certain points. Bradley *et al.* [11] addressed this problem by introducing a smooth prior (KL-divergence) which can preserve the benefits of sparse prior l_1-norm while allowing efficient computation. Marial *et al.* [12] used heuristics to tackle the problem while the prior is still the l_1-regularization on the premise that it is well initialized. However, it involves too many mathematical derivations which may increase the possibility of manual error.

Indeed, the problem Eq. (11) can be regarded as the sparse coding predictors learning process. Several works have been proposed to learn approximations of sparse coding by training a feed-forward neural network predictor with a fixed depth [13, 14]. That is to say, we can tune the parameter variables in a supervised way by feed-forward neural network. Meanwhile, these methods lead to dramatic reduction of time complexity by reducing the iterations. In this paper, we use the same architecture as [14] allowing for its small approximation error, real-time application and training in easiness.

Here, we briefly discuss learning fast approximation of sparse coding proposed by [14] which is named as Learned iterative shrinkage-thresholding algorithm (LISTA), one of version of fast approximation algorithms, is an improvement on iterative shrinkage algorithm (ISTA). The difference lies in way of obtaining two parameter matrices: ISTA from the basis vectors and LISTA by learning in a supervised way. ISTA iterates the following recursive equation to convergence:

$$Z(k+1) = h_\theta(W_e Y + SZ(k))Z(0) = 0 \tag{12}$$

where Y is an input vector and Z is a sparse coding vector. The other elements in Eq. (12) are defined as follows:

$$W_e = \frac{1}{L} D^T \tag{13}$$

$$S = I - \frac{1}{L} D^T D \tag{14}$$

$$h_\theta(Z) = sign(Z)(|Z| - \theta)_+ \tag{15}$$

where D represents the basis vector, I is an identity matrix, $h_\theta(Z)$ is a component-wise shrinkage function and L is a constant. The ISTA will take hundreds of thousands of iterations to converge to the optimal solution if the dictionary is over-complete. Certainly not in the way we would all like to see. The LISTA encoder can get the same approximation error, taking the same form of Eq. (12), with a fixed number of steps T. It learns the parameters $W = (W_e, S)$ from training set in a supervised way. The architecture of encoders will be represented as $Z = F_W(Y, W)$, where Y are input vectors, W are the parameters to be learned and $F_W(Y, W)$ is defined by repeating Eq. (12) T iterations. Training the encoder to minimize the error with the predicted sparse codes and the optimal ones by feed-forward neural network. The error function is defined as follows:

$$L(W, X) = \|Z^* - F_W(W, X)\|_F^2 \tag{16}$$

Now, let the parameters $W = (W_e, S)$ substitute the parameters D_l and F_{D_l} is replaced by F_W. The parameter Z^* is rewritten as α. We can find that Eqs. (11) and (16) are equal. The optimization over D_l turns to the optimization over W. We compute the derivative $\partial L(W, X)/\partial W$ through the back-propagation method. And then, the parameters $W = (W_e, S)$ are estimated by training on pairs of LR and HR image patches using stochastic gradient descent.

Optimization over Sparse Coding. For fixed W (D_l is substituted by W) and D_h, we have the following minimization over α:

$$\min_\alpha \sum\nolimits_{n=1}^{N} \rho^2 \|\alpha_n - F_W(y_n)\|_2^2 + \|x_n - D_h \alpha_n\|_2^2 \quad s.t. \quad \|\alpha\|_1 \le \varepsilon \tag{17}$$

It can be written as:

$$\min_\alpha \sum\nolimits_{n=1}^{N} \left\| \begin{bmatrix} D_h \\ \rho I \end{bmatrix} \alpha_n - \begin{bmatrix} x_n \\ \rho F_W(y_n) \end{bmatrix} \right\|_2^2 \quad s.t. \quad \|\alpha\|_1 \le \varepsilon \tag{18}$$

where I represents identity matrix. Let $A = \begin{bmatrix} D_h \\ \rho I \end{bmatrix}$ and $b_n = \begin{bmatrix} x_n \\ \rho F_W(y_n) \end{bmatrix}$, Eq. (18) can be written as:

$$\min_{\alpha} \sum_{n=1}^{N} \|A\alpha_n - b_n\|_2^2 + \lambda\|\alpha_n\|_1 \quad (19)$$

Notice that it is a l_1-norm problem, known in statistics as the Lasso [15], which can be solved by fast iterative shrinkage algorithm [16].

3.3 Summary of the Algorithm

The proposed SR method involves two parts: the "dictionaries" learning algorithm (the "dictionaries" refer to the sparse coding parameter W and HR dictionary D_h) and LR image mapping to HR image algorithm. The "dictionaries" learning process is summarized in Algorithm 1. And Algorithm 2 gives the SR reconstruction process.

Algorithm 1. The "dictionaries" learning algorithm

1: **Task:** Learning the parameter W and HR dictionary D_h.
2: **Input:** Training image pairs $\{X_l, X_h\}$. Initial parameter W and HR dictionary D_h.
3: **For** each iteration until convergence:
  1) Fixed α, update W and D_h in Eq. (9) and Eq. (16), respectively.
  2) Fixed W and D_h, update α in Eq. (19).
4: **End**
5: **Output:** W and D_h.

Algorithm 2. The proposed single image SR algorithm

1: **Task:** Estimating HR image of a LR image.
2: **Input:** Parameter W and HR dictionary D_h, and LR test image Y.
3: Obtain the magnified image $\tilde{Y}$ using Bicubic interpolation.
4: **For** each 9×9 patch $\tilde{y}$ of $\tilde{Y}$, in the common sliding-window fashion, with 1 pixel overlap in each direction.
  1) Compute the first- and second-order gradients of the patch y and then perform dimensionality reduction through PCA.
  2) Solve the inference problem with W and $\tilde{y}$ defined as: $\alpha = F_W(\tilde{y})$, where F_W is defined in Eq. (12).
  3) Generate the HR patch $x = D_h\alpha$.
5: **End**
6: Put all the HR patches x into a HR image X.
7: **Output:** SR image X.

4 Experimental Results

In this section, we will discuss some details of the algorithm such as features, some factors which have an influence on the training process, and evaluate the performance by comparing it to other state-of-the-art methods. Since humans are more sensitive to illuminance changes, we adopt the YCbCr color model instead of RGB model, and apply the method to the Y channel only and interpolate the Cb and Cr channels using Bicubic interpolation.

4.1 Experimental Settings and Parameter Analysis

In our experiments, 100,000 h and LR image patch pairs are randomly sampled from training images, which are collected from Yang *et al.* [7]. To effectively deal with the high frequency information, the first- and second-order derivatives of the image patches are used as LR image feature representation. We perform dimensionality reduction to 30 dimensions by PCA, with preserving 99.9% of average energy.

We first train a dictionary D_l using SPArse Modeling Software. The sparsity weight parameter is set as 0.15 and the dictionary size is fixed as 512. Then, FISTA algorithm is used to compute the optimal sparse coding α of the LR image patches. Therefore, we can obtain the initialization of the parameter W and D_h by Eqs. (16) and (8), respectively and update them according to Algorithm 1. In the learning fast approximation of sparse coding process, a crucial issue in the recursive equation Eq. (12) is the choice of the iteration number. We try to set the inference iteration number from 1 to 10, respectively and Fig. 2 give the average values of PSNR and SSIM of 25 test images with different number of iterations. Meanwhile, the time complexity is proportional to the iteration number, which is given by Fig. 3.

It can be seen that the values of PSNR and SSIM increase with the number of iterations, but the improvement is not unlimited, and the running time of the algorithm will increase along with the number of iterations. Therefore, the number of iterations can be selected from the SR effect and the real-time performance of the algorithm according to the actual demand.

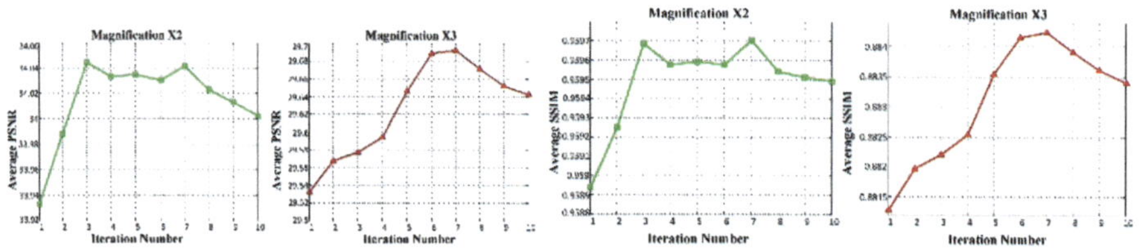

Fig. 2. Average values of PSNR and SSIM of 25 test images with different number of iterations.

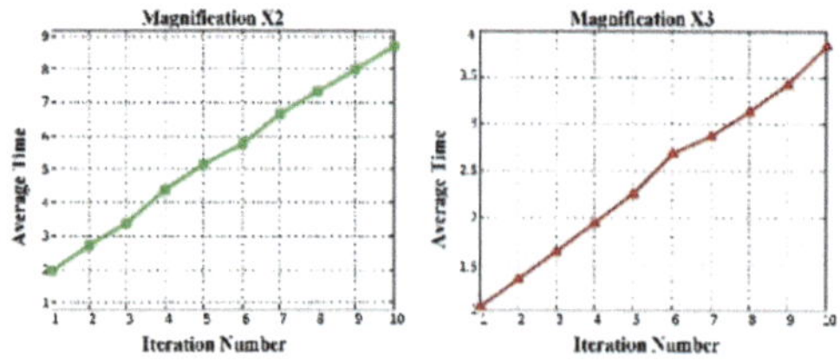

Fig. 3. Average running time of 25 test images with different number of iterations.

4.2 Experimental Results Compared with Other SR Methods

In this section, we will show experimental results both visually and qualitatively and compare them with the other SR algorithms. The methods to be compared share the same trained dictionary of 1024, except for Yang *et al.* [7] with a dictionary of 1022. The size of our HR dictionary D_h is set as 512. Figures 4 and 5 show the visual quality of different methods.

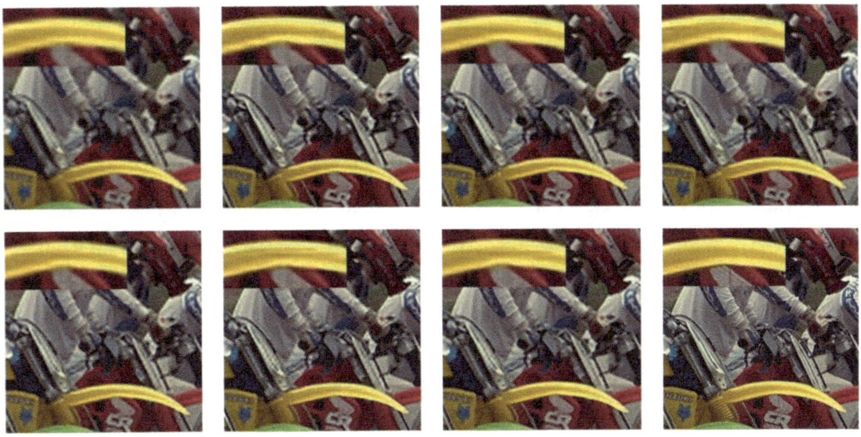

Fig. 4. Results of the bike image by a factor of 3. **Top row:** Bicubic interpolation, Yang *et al.* [7], Zeyde *et al.* [17], Bevilacqua *et al.* [18] **Bottom row:** Timofte *et al.* [19], Peleg *et al.* [20], The proposed method with 7 iteration, and the ground truth HR image.

Fig. 5. Results of the butterfly image by a factor of 2. **Top row:** Bicubic interpolation, Yang *et al.* [7], Zeyde *et al.* [17], Bevilacqua *et al.* [18] **Bottom row:** Timofte *et al.* [19], Peleg *et al.* [20], The proposed method with 7 iteration, and the ground truth HR image.

Table 1. Magnification ×3 performance in terms of PSNR and SSIM per image.

Images	Index	Bicubic	Yang	Zedye	Bevilacqua	Timofte	Peleg	Ours
Bike	PSNR	22.79	23.90	23.87	23.74	23.96	24.20	**24.57**
	SSIM	0.7029	0.7655	0.7643	0.7590	0.7694	0.7884	**0.7939**
Butterfly	PSNR	24.04	25.58	25.94	25.61	25.90	26.75	**27.46**
	SSIM	0.8216	0.8611	0.8770	0.8663	0.8717	0.8999	**0.9082**
Comic	PSNR	23.11	23.90	23.95	23.83	24.04	24.40	**24.45**
	SSIM	0.6988	0.7556	0.7548	0.7484	0.7606	0.7765	**0.7809**
Flowers	PSNR	27.23	28.25	28.43	28.21	28.49	28.82	**29.05**
	SSIM	0.8013	0.8297	0.8375	0.8320	0.8403	0.8474	**0.8506**
Foreman	PSNR	31.18	32.04	33.19	32.87	33.23	33.88	**34.01**
	SSIM	0.9057	0.9132	0.9292	0.9243	0.9300	**0.9374**	0.9372
Leaves	PSNR	23.50	25.03	25.42	25.06	25.36	26.07	**26.61**
	SSIM	0.8045	0.8638	0.8751	0.8627	0.8697	0.8949	**0.9069**
Woman	PSNR	28.56	29.94	30.37	29.89	30.33	30.66	**31.07**
	SSIM	0.8896	0.9037	0.9176	0.9110	0.9169	0.9247	**0.9260**
Parrots	PSNR	27.91	29.06	29.25	29.07	29.35	29.73	**29.77**
	SSIM	0.8786	0.8925	0.9031	0.8997	0.9039	0.9105	**0.9115**
Starfish	PSNR	26.89	27.80	27.94	27.70	28.00	28.64	**28.64**
	SSIM	0.8108	0.8407	0.8460	0.8397	0.8481	0.8632	**0.8651**
Plants	PSNR	31.07	32.28	32.54	32.39	32.61	62.95	**33.28**
	SSIM	0.8674	0.8906	0.9006	0.8979	0.9027	0.9059	**0.9094**

It can be seen from Figs. 4 and 5 that the reconstructed HR images by Bicubic interpolation produce very smooth textures and zigzag edges. Yang's method is very competitive in terms of visual quality compared to Bicubic. However, reconstructed edges by Yang's method have ringing artifacts and "ghost" artifacts for the edges can be found. Our reconstructed edges are much sharper than the other methods which generate the similar result and cannot recover some fine image structure. And Table 1 summarizes the results of 10 testing images and show the PSNR and SSIM values to further verify that the proposed method is superior to the other methods in single image SR reconstruction.

5 Conclusion

In this paper, a novel single image super-resolution based on sparse coding and joint mapping learning framework is proposed, which results in a very efficient optimization. To achieve the real-time image SR, we propose to learn approximations of sparse coding by training a feed-forward neural network predictor with a fixed depth. Experimental results demonstrate the proposed method obtains a trade-off between accuracy and computation time. Since it needs image pretreatment for processing which may lose some necessary information, the future work is to consider how to integrate the preprocessing with the joint mapping learning framework.

References

1. Park, S.C., Min, K.P., Kang, M.: Super-resolution image reconstruction: a technical overview. IEEE Signal Process. Mag. **20**(3), 21–36 (2003)
2. Sun, J., Xu, Z., Shum, H. Y.: Image super-resolution using gradient profile prior. In: Computer Vision and Pattern Recognition, pp. 1–8 (2008)
3. Glasner, D., Bagon, S., Irani, M.: Super-resolution from a single image. In: International Conference on Computer Vision, pp. 349–356 (2009)
4. Dong, W., Zhang, L., Shi, G., Wu, X.: Image deblurring and super-resolution by adaptive sparse domain selection and adaptive regularization. IEEE Trans. Image Process. **20**(7), 1838–1857 (2011)
5. Freeman, W.T., Jones, T.R., Pasztor, E.C.: Example-based super-resolution. IEEE Comput. Graphics Appl. **22**(2), 56–65 (2002)
6. Chang, H., Yeung, D.Y., Xiong, Y: Super-resolution through neighbor embedding. In: Computer Vision and Pattern Recognition, pp. 56–65 (2004)
7. Yang, J., Wright, J., Huang, T.S., Ma, Y.: Image super-resolution via sparse representation. IEEE Trans. Image Process. **19**(11), 2861–2873 (2010)
8. Zhang, K., Tao, D., Gao, X., Li, X., Xiong, Z.: Learning multiple linear mappings for efficient single super-resolution. IEEE Trans. Image Process. **24**(3), 846–861 (2015)
9. Ahmed, J., Shah, M.A.: Single image super-resolution by directionally structured coupled dictionary learning. EURASIP J. Image Video Process. **2016**(1), 1–12 (2016). https://doi.org/10.1186/s13640-016-0141-6
10. Liu, N., Liang, S.: Single image super-resolution using sparse representation on a K-NN dictionary. In: International Conference on Image and Signal Processing, pp. 169–278 (2016)
11. Bradley, D.M., Bagnell, J. A.: Differential sparse coding. In: Proceedings Advances in Neural Information Processing Systems, vol. 21, pp. 113–120 (2008)
12. Mairal, J., Bach, F., Ponce, J.: Task-driven dictionary learning. IEEE Trans. Pattern Anal. Mach. Intell. **34**(4), 791–804 (2012)
13. Kavukcuoglu, K., Ranzato, M., Lecun, Y.: Fast inference in sparse coding algorithms with applications to object recognition. In: Technical Report CBLL-TR-2008-12-01, Computational and Biological Learning Lab, Courant Institute, NYU (2008)
14. Gregor, K., Lecun, Y.: Learning fast approximations of sparse coding. In: International Conference on Machine Learning, pp. 399–406 (2010)
15. Tibshirani, R.: Regression shrinkage and selection via the lasso. J. Roy. Stat. Soc. Ser. B (Methodol.) **58**, 267–288 (1996)
16. Daubechies, I., Defrise, M., Mol, C.E.: A fast iterative shrinkage-thresholding algorithm for linear inverse problems. SIAM J. Imaging Sci. **2**(1), 183–202 (2009)
17. Zeyde, R., Elad, M., Protter, M.: On single image scale-up using sparse-representations. In: International Conference on Curves and Surfaces, pp. 711–730 (2010)
18. Bevilacqua, M., Roumy, A., Guillemot, C., Morel, A.: Low-complexity single-image super-resolution based on nonnegative neighbor embedding. In: Proceedings of the British Machine Vision Conference (2012)
19. Timofte, R., De, V., Gool, L. V.: Anchored neighborhood regression for fast example-based super-resolution. In: IEEE International Conference on Computer Vision, pp. 1920–1927 (2014)
20. Peleg, T., Elad, M.: A statistical prediction model based on sparse representations for single image super-resolution. IEEE Trans. Image Process. **23**(6), 2469–2582 (2014)

Internet of Things and Smart Systems

A Study on the Satisfaction of Consumers Using Ecommerce Tourism Platform

Kun-Shan Zhang[1(✉)], Chiu-Mei Chen[1], and Hsuan Li[2]

[1] University of Zhao Qing, Zhaoqing 526061, China
963677272@qq.com
[2] Peking University, Beijing 100871, China

Abstract. E-commerce tourism is a model derived from the emergence of e-commerce. This study is aimed at the consumers who use e-commerce tourism products. Taking the consumers who purchase tourism products through tourism websites within one year as the survey object, the satisfaction of consumers with e-commerce tourism products transactions is discussed. 227 random samples are selected for the questionnaire survey. Based on the data of satisfaction of technology acceptance dimension, transaction cost dimension and service quality dimension, the research model discusses customer satisfaction with e-commerce platform.

The results show that transaction cost has a significant positive impact on platform service quality perception, and service quality has a significant positive impact on platform use satisfaction. It is suggested that e-commerce tourism platform make use of the advantages of big data to make tourism products and grasp the needs of consumers. In the aspect of setting contract cost, reduce the cost for consumers to unsubscribe, so as to prevent the loss of platform caused by the pressure of unsubscribe cost that consumers fail to purchase products. Increase and simplify complaint channels, automatically identify customer needs, reasonably refund, and reasonably increase the number of customer service platform, so that consumers can solve problems more quickly and efficiently.

Keywords: Travel platform · E-commerce · Consumer satisfaction · Tourism cyber marketing

1 Introduction

Tourism e-commerce is an electronic business model that integrates all aspects of tourism information through the Internet and realizes all links in the whole tourism process. E-commerce tourism marketing is implemented through the Internet, big data and other media means, through the contact and interaction of potential tourism consumers on the Internet, to provide consumers with relevant tourism products.

In recent years, due to the rise of C-trip tourism combined with action payment, the global tourism business model and consumer decision-making process have gradually changed. Hotel and air ticket booking in e-commerce tourism websites seize the opportunity of information asymmetry in the traditional travel agency industry chain in the past. This study focuses on the research and analysis of consumers using traditional

Q. Liu et al. (Eds.): CENet 2020, AISC 1274, pp. 1105–1112, 2021.
https://doi.org/10.1007/978-981-15-8462-6_126

travel agencies and e-commerce tourism products the level of tourism product service involvement will affect the behavior mode of consumers.

Chen Jinping and Liu Zhen (2019) have seen a continuous increase in China's total tourism revenue from 2018 to 2020. It is easy to see the momentum of China's tourism industry and its infinite potential. The promotion of e-commerce technology also drives the traditional tourism industry. By the end of 2018, China's tourism e-commerce websites will soon reach 5000 [1]. Xia Yan (2016) pointed out that the advantages of tourism e-commerce lie in streamlining the payment process of tourism services, realizing product customization and saving the transaction cost of consumers [2].

Due to the rapid development of e-commerce tourism in recent years, low price, fast speed, quality, etc., focusing on one point can become the core competitiveness of the website. From the perspective of the trend, outbound tourism is a tourist. E-commerce tourism websites, such as C-trip, Qunar, Tuniu and Fliggy are the fastest growing markets in recent years. E-commerce tourism uses the Internet as a means to reasonably and efficiently allocate resources through multi-party information interaction, so as to meet the customized needs of tourists as much as possible.

This study is to explore the consumer satisfaction of online tourism products. The scope of the study is to use consumers who use the Internet to order tourism related products within one year as the research object of e-commerce tourism products, so as to know whether the e-commerce tourism platform is in line with consumer satisfaction.

2 Literature Review

2.1 Electronic-Commerce Travel

Tourism e-commerce refers to the use of the most advanced electronic means to operate tourism and its distribution system, which is based on the tourism information database and e-commerce bank. Tourism e-commerce provides an Internet platform for tourism industry. E-commerce activities are essentially economic and trade activities based on the Internet and electronic information technology. The change of transaction platform and transaction form makes the interest relationship between the original unequal operators and consumers more unbalanced and complex [3]. (Younghwa et al. 2003) Consumers are in a weaker position than traditional commerce in e-commerce activities. Characteristics of mobile e-commerce: first, convenience; second, openness; third, the scale of users is huge [4].

2.2 Technology Acceptance

The TAM proposed by Davis (1986) points out that users' willingness and effect to use products are affected by their ideas and attitudes towards technology [5]. The six main aspects of Technology Acceptance architecture are external forces, perceived usefulness, and perceived ease of use, attitudes, intentions and actual use behavior.

The perceived usefulness and perceived ease of use mentioned in tar will affect users' attitudes. If users don't need to spend too much time and energy on the acceptance of new technology, their perception of this technology will usually be more positive (Cheong and amp; Park, 2005) [6]. Davis extended Technology Acceptance according to tar and TPB. Tar and TPB point out that the control and value of attitude and perceived behavior can have a positive impact on the intention of continuous adoption (al Debei 2012) [7]

2.3 Transaction Costs

According to the transaction costs theory proposed by Coase (1937), personal factors will affect his transaction decision-making, that is, the factors of consumers themselves will affect the transaction process. In the context of online shopping, even if the external environment is the same, consumers will make different evaluation decisions because of their own perception and rationality [8].

In the past, although there were literatures that included personal factors in the discussion of repurchase intention (Deng Z, 2010); most of them discussed the influence of personal factors on repurchase intention directly, and seldom discussed the interference effect of personal factors [9].

2.4 Consumer Satisfaction

Customer satisfaction refers to the customer's feelings about the extent to which their reasonable needs or expectations are met. Satisfaction is the feedback of customer satisfaction. Customer satisfaction is not only a psychological state, but also a kind of self-experience. Kolter and Keller (2006) defined satisfaction as the degree of pleasure or disappointment felt by an individual, which originates from the special nature of the product and the expectation of the individual for the product, and is formed by comparing the two [10].

3 Research Design

From December 2019 to January 2020, 230 questionnaires were sent out, 211 questionnaires were recovered, and 201 valid questionnaires were left after eliminating the invalid ones, which will be taken as the survey data.

3.1 Research Framework

Based on the introduction and literature review, this study designed a questionnaire to study the core needs of consumers for e-commerce tourism and the influencing factors of satisfaction through the data collection of the market, so as to carry out empirical analysis. This study uses technology acceptance, transaction cost, service quality and consumer satisfaction to understand the correlation between online tourism websites and consumer variables the architecture is shown in Fig. 1.

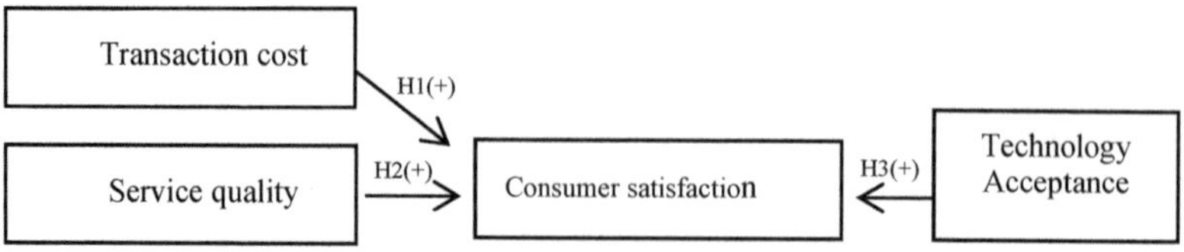

Fig. 1 Research framework.

3.2 Research Hypothesis

H1: Consumers' perception that using e-commerce tourism platform can save transaction costs has a positive impact on the perception of platform service quality.
H2: Consumers' perception of service quality using e-commerce platform has a positive impact on the satisfaction of platform use.
H3: Consumers' acceptance of e-commerce tourism platform technology has a positive impact on the satisfaction of platform use.

4 Data Statistics and Analysis

4.1 Descriptive Statistics

There are three demographic variations in this study, namely gender, age and education level. In the sample data analysis, 26% of them is aged from 26 to 30, 54.9% of them are females, and 61% of them are undergraduates.

Analysis of consumer characteristics, Using e-commerce platform once a year 31.2% and once half a year (28.9% are mostly used, with 2–3 times 36.7%, most of them prefer to buy transportation tickets 32.9%, scenic spot tickets 31.3% and spend between RMB 5,000 and 10,000 is 43.0%, as shown in Table 1.

Table 1. Analysis of consumer characteristics.

variable	grouping	Distribution	(%)	variable	grouping	Distribution	(%)
Times frequency	More than one year	41	19.9	Spend	Under 5,000	38	18.3
	Once a year	64	31.2		5000–10,000	89	43
	Once half a year	60	28.9		10,000–20,000	49	23.9
	Once a month	13	6.5		More than 20000	30	14.8
	2–3 times a month	13	6.5				
	More than three times a month	14	7				
Times of use	1st time	50	24.3	Purchase type	Travel itinerary	24	11.7
	2–3 times	76	36.7		Accommodation reservation	21	10.1
	4–5 times	59	28.8		Transportation ticket	68	32.9
	6–10 times	8	3.9		Scenic spot ticket	64	31.3
	More than 10 times	13	6.3		Restaurant roll	29	14

4.2 Correlation Analysis and Regression Analysis

SPSS 22 was used to analyze Pearson coefficients among variables to observe their correlation. There is a significant positive correlation between the variables of technology acceptance, satisfaction, transaction cost and service quality at the level of 0.01, so they can be used for regression analysis.

From Tables 2 and 3, it can be seen that the model passed the F test (F = 62.197, P = 0.000 < 0.05) when conducting the F test on the model, which means that the model construction is meaningful. Taking transaction cost and technology acceptance as independent variables and service quality as dependent variables for linear regression analysis, the R square value of the model is 0.554, which means that transaction cost and technology acceptance can explain 55.4% of the change of service quality. It is found that the model passes the F-test (F = 62.197, P = 0.000 < 0.05), that is to say, at least one of the transaction costs and technology acceptance will have an impact on service quality.

Table 2. Variance analysis of technology acceptance, transaction cost and service quality.

	Sum of squares	*df*	mean square	*F*	*p*
regression	17.606	2	8.803	62.197	0.000
residual	14.154	100	0.142		
total	31.760	102			

a. dependent variable: Service quality
b. Independent variable: (constant), technology acceptance, transaction cost

Table 3. A collinearity analysis of technology acceptance, transaction cost and service quality.

	Non standardized coefficient		Standardization coefficient	t	p	95%	CI	VIF
	B	Standard error	Beta					
Constant	0.669	0.278	–	2.402	0.018*	0.123	1.214	–
Transaction cost	0.732	0.116	0.689	6.307	0.000**	0.504	0.959	2.680
Technology acceptance	0.063	0.100	0.068	0.625	0.534	−0.134	0.260	2.680

dependent variable: Service quality
* $p < 0.05$ ** $p < 0.01$

The model formula is: service quality = 0.669 + 0.732 * transaction cost + 0.063 * technology acceptance. The final analysis shows that the regression coefficient of transaction cost is 0.732 (t = 6.307, P = 0.000 < 0.01). The value of regression coefficient accepted by science and technology was 0.063 (t = 0.625, P = 0.534 > 0.05).

It can be concluded from the summary analysis that H1 is established, that is, transaction cost will have a significant positive impact on the perception of platform service quality, but technology acceptance will not have an impact on the perception of platform service quality.

4.3 Technology Acceptance, Transaction Cost, Service Quality and Consumer Satisfaction

From Tables 4 and 5, we can see that transaction cost, technology acceptance and service quality are taken as independent variables, while satisfaction is taken as dependent variable for linear regression analysis. The R square value of the model is 0.602, which means that transaction cost, technology acceptance and service quality can account for 60.2% of the change of satisfaction. In the F-test of the model, it was found that the model passed the F-test (F = 49.980, P = 0.000 < 0.05), that is to say, at least one of transaction cost, technology acceptance and service quality would have an impact on satisfaction.

Table 4. Variance analysis of technology acceptance, transaction cost, service quality and consumer satisfaction.

	Sum of squares	*df*	Mean square	*F*	*p*
regression	20.030	3	6.677	49.980	0.000
residual	13.225	99	0.134		
total	33.256	102			

Table 5. A collinearity analysis of technology acceptance, transaction cost, service quality and consumer satisfaction.

	Non standardized coefficient		Standardization coefficient	t	p	95%	CI	VIF
	B	Standard error	Beta					
constant	0.587	0.278	–	2.111	0.037*	0.042	1.132	–
transaction cost	0.133	0.133	0.122	0.995	0.322	−0.129	0.394	3.747
technology acceptance	0.032	0.098	0.034	0.328	0.744	−0.160	0.224	2.691
service quality	0.673	0.097	0.658	6.929	0.000**	0.483	0.864	2.244

Dependent variable: consumer satisfaction

* $p < 0.05$ ** $p < 0.01$

The model formula is: satisfaction = 0.587 + 0.133 * transaction cost + 0.032 * technology acceptance + 0.673 * service quality. The final analysis shows that the regression coefficient of transaction cost is 0.133 ($t = 0.995$, $P = 0.322 > 0.05$). The regression coefficient of science and technology acceptance was 0.032 ($t = 0.328$, $P = 0.744 > 0.05$). The regression coefficient of service quality was 0.673 ($t = 6.929$, $P = 0.000 < 0.01$).

Summary and analysis show that: H3 is established, that is, service quality will have a significant positive impact on the satisfaction of platform use, but transaction cost, and technology acceptance will not have an impact on the satisfaction of platform use.

5 Conclusions and Suggestions

5.1 Conclusions

Due to the popularization of education and the rapid development of science and technology, it is no longer as difficult for consumers to use e-commerce platform for tourism activities. Therefore, the improvement of technology acceptance will not improve consumers' perception of platform service quality. The saving of transaction cost is what every consumer wants to see. Before using the products of e-commerce tourism platform, the biggest difference between online and offline tourism products for consumers is transaction cost. If the transaction cost can be saved, more consumers will choose to purchase products through e-commerce tourism platform. Therefore, e-commerce tourism platform needs more discussions with product developers on the price of products in order to obtain a larger market share.

Service quality is the whole process from purchasing to using products. In this process, consumer satisfaction is gradually formed and improved. Therefore, if the e-commerce tourism platform wants to make consumers get enough satisfaction, it needs to strictly control the service quality of the platform.

5.2 Suggestions

Through the research, it is found that saving transaction cost can help consumers improve their positive perception of the service quality of the platform, while improving the service quality will affect consumers' satisfaction after using the platform.

References

1. Chen, J., Liu, Z.: On the challenges and opportunities of tourism e-commerce-Explore the future development of yuetu tourism platform. Mod. Trade Ind. **40**(25), 24–25 (2019)
2. Cheng, B., Xia, Y.: On the development trend of e-commerce in China, tourism overview (08), 171–172 (2016)
3. Wang, K., Ling, W.: On the development mode of China's mobile e-commerce. J. Mod. Mark. **03**, 178 (2019)

4. Lee, Y., Kozar, K.A., Larsen, K.R.T.: The technology acceptance model: past, present, and future. Commun. Assoc. Inf. Syst. **12**(50), 752–780 (2003)
5. Davis, F.D.: A technology acceptance model for empirically testing new end-user information systems theory and results. Massachusetts Institute of Technology (1986)
6. Cheong, J.H., Park, M.C.: Mobile internet acceptance in Korea. Internet Res. **15**(2), 125–140 (2005)
7. Panagiotopoulos, P., Al-Debei, M.M., Fitzgerald, G., Elliman, T.: A business model perspective for ICTs in public engagement. Gov. Inf. Q. **29**(2), 191–202 (2012)
8. Coase, R.H.: The nature of the firm. Economical **4**(16), 386–405 (1937)
9. Deng, Z., Lu, Y., Wei, K.K., Zhang, J.: Understanding customer satisfaction and loyalty: an empirical study of mobile instant messages in China. Int. J. Inf. Manage. **30**(4), 289–300 (2010)
10. Kolter, P., Keller, K.L.: Marketing Management. Pearson, London (2006)

Research on Construction of Smart Training Room Based on Mobile Cloud Video Surveillance Technologies

Lihua Xiong[1(✉)] and Jingming Xie[2]

[1] Guangdong Polytechnic of Water Resources and Electric Engineering, Guangzhou City 510635, Guangdong Province, China
9720091@qq.com

[2] Guangzhou Panyu Polytechnic, Guangzhou City 511483, Guangdong Province, China

Abstract. Intelligent technologies can effectively improve the management level and safety of training rooms in vocational colleges. This article first analyzes the challenges encountered in the construction of the smart training rooms, then designs a system architecture based on mobile cloud video surveillance technologies for building smart training rooms, analyzes the main software function modules, and discusses video transmission and cameras control. The realization of smart training rooms has made a useful exploration of applying a new generation of information technology to manage training rooms in colleges.

Keywords: Video surveillance · Cloud computing · Raspberry Pi · Smart training room

1 Introduction

A training room is an important place for vocational students to carry out practical activities. A training room may store various types of training equipment, and students from different classes would carry out teaching practice in a same training room, which causes a challenge how to manage a training room and ensure the safety of training equipment and the training room. In May 2019, the Ministry of Education of China issued the "Opinions on Strengthening the Safety of Laboratories in Universities", which proposed the need to complete the laboratory safety responsibility system, improve the laboratory safety management rule and the level of laboratory safety management information, develop the laboratory safety information management system, monitoring and early warning system, and promote the in-depth integration of the information system and security work. Therefore, in addition to strict scientific management ways in a training room, a new generation of information technology is also needed to improve the efficiency and quality of management [1].

In recent years, with the rise of next-generation information technologies such as the Internet of Things, cloud computing, mobile Internet, big data, and artificial intelligence, traditional video surveillance technologies have become networked and intelligent, and are widely applied to build smart campuses and safe communities. Related technologies in the field of video surveillance are booming. Clum explored the

Q. Liu et al. (Eds.): CENet 2020, AISC 1274, pp. 1113–1119, 2021.
https://doi.org/10.1007/978-981-15-8462-6_127

implementation mechanism of task scheduling between mobile devices and cloud computing platforms, realized the saving, migration and recovery of running threads, finally achieved the cloud computing platform's support [2]. Haiwen Wen et al. designed a video surveillance integrated service platform based on SOA architecture and cloud computing technology, using virtualization technology in cloud computing to integrate and manage multiple heterogeneous software and hardware resources in the bottom layer of the platform, and adopting HDFS distributed file system and HBase distributed storage system perform efficient distributed storage management of massive video data, and using the Mapreduce distributed programming framework to achieve distributed parallel processing and resource scheduling of user services [3]. According to the characteristics of the face image in the surveillance video, Wang Hailong et al. first obtained the video face sequence by combining face detection and tracking technology, and then selected all faces based on the partial face image recognition results in the video face sequence, those representative face images in the sequence images are recognized, and finally the face information was comprehensively reflected based on the selected face image recognition results [4].

2 Challenges in the Construction of an Intelligent Training Room

The application of new generation information technology to build an intelligent training room has already been explored. Liu Anmin et al. used technologies of Internet and Internet of Things to establish the intelligent training base of "scientific research teaching and industrial production" based on the combination of localization and cloud in accordance with the sequence of industrial production processes [5]. Huang Lifen et al. established equipment-aware systems in each training room to realize the interconnection of objects and objects, and further connected the equipment-aware network to the campus Internet to realize the interconnection of the Internet of Things, and finally achieved the interconnection of the Internet and the Internet of Things network [6]. However, there are still some challenges in building an intelligent training room in reality as following:

1. Large investment. Building an intelligent training room often requires purchasing a batch of expensive intelligent management equipment and software. An ordinary vocational college has dozens to hundreds of training rooms, so comprehensively carrying out intelligent construction would require a large amount of capital.
2. High scalable requirements. Training rooms in higher vocational colleges are important places to actively carry out vocational technical education practice activities, so they should have high scalability to support vocational education reform.
3. How to relate the construction of intelligent training rooms with the cultivation of information technology talents in higher vocational colleges. Through the way of independent research and development leaded by teachers from information technology major, the research and exploration of the construction of intelligent training rooms can improve practical ability of teachers and students. On the other hand, it is also convenient to maintain the intelligent training rooms and reduce investment costs.

3 System Implementation

3.1 System Architecture Based on Mobile Cloud Video Surveillance Technologies

The system architecture design of intelligent training rooms is mainly divided into the video monitoring suite, the cloud server and the mobile terminal. Among them, the video monitoring suite is deployed to every training room to realize the capture and processing of a training room information. The cloud server is used to improve the storage capacity and processing performance of video transmission. As a central server, it responds to mobile terminal operation requests and interacts with video monitoring suites. Mobile terminals can access the cloud server to obtain training rooms' status information and issue control commands (see Fig. 1).

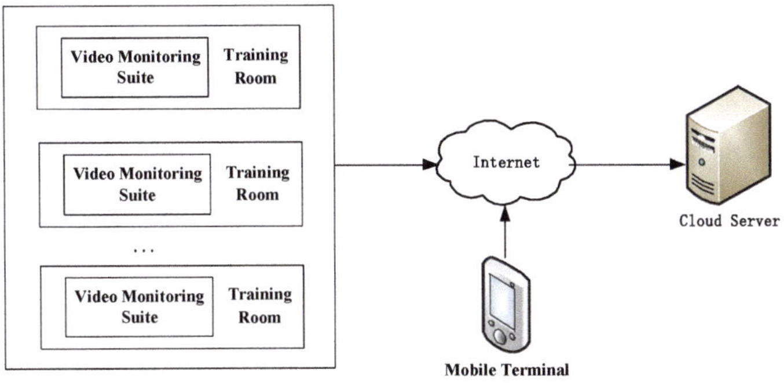

Fig. 1. System architecture based on mobile cloud video surveillance.

Raspberry Pi is a powerful microcomputer that can be used to carry out a variety of Internet of Things application experiments, and has a wide range of applications in the school's IT talent training and scientific experiments. Therefore, in this study, Raspberry Pi was selected as the terminal control center of the video monitoring suite. A Raspberry Pi provides a wealth of GPIO interfaces for connecting various sensors and devices, including smoke sensors, human body infrared sensors, flame sensors, access controllers, servos, USB cameras, etc. (see Fig. 2). In the future, other types of sensors can be expanded according to the needs of the application.

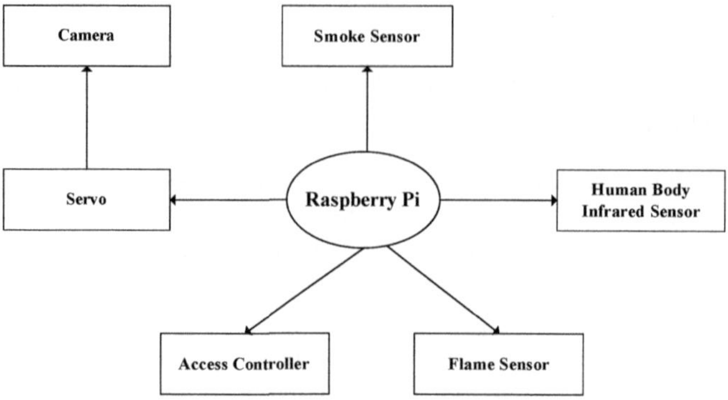

Fig. 2. The video monitoring suite.

3.2 Main Software Function Modules

In the mobile cloud video surveillance system architecture, different software modules need to be separately deployed on the cloud server, the video monitoring suite, and the mobile terminal to achieve functional requirements of a smart training room (see Fig. 3).

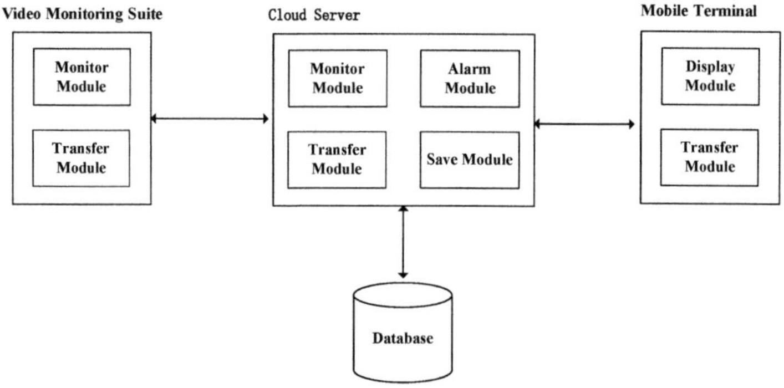

Fig. 3. Main software function modules.

Write a service program in the cloud server to process information in a multi-threaded pool. Its specific functions are as follows:

1. Monitoring Module. It receives information captured by sensors such as flame, smoke, human body infrared, etc., and transmitted by video monitoring suits. It also receives data requests from mobile clients.
2. Transmission Module. It transmits the information captured by sensors such as flame, smoke, human body infrared to a mobile client, and forward the control instructions of the mobile client to a specific video monitoring suite.

3. Save Module. It saves the information captured by the flame, smoke, human body infrared and other sensors transmitted by video monitoring suits into the database, which is convenient for historical data tracking.
4. Alarm Module. When a video monitoring suite captures an alarm event captured by sensors, it immediately sends a short message to an administrator's mobile phone, and sends the instantly captured photo to the administrator's mailbox for storage.

Write a service program in a video monitoring suite to complete the following functions:

1. Monitoring Module. It waits for a control command sent by the cloud server. If the control command is to move the camera, then control the movement of the servo by the order. If the instruction is to open the access control, the electronic access control will be opened after the verification is passed successfully.
2. Transmission Module. It regularly sends sensor status information to the cloud server.

The user interface and control command sending functions are implemented on the mobile client, mainly to showing the video images and sensor data of the training room through the display module. The transmission module is responsible for controlling the movement of a camera and the opening of the electronic access control.

3.3 Video Transmission

There are many solutions to remotely access the images captured by the USB camera on a Raspberry Pi in real time. Writing code from zero base will involve camera data collection, compression, transmission, decoding, etc., which will be a cumbersome process. Another better solution is to install third-party software to achieve video transmission, such as the Motion, Avconv, GStreamer, motionEyeOS, mjpg-streamer or other software. After installing a third-party software, a user can enter the parameter information such as the IP address and port number on the Raspberry Pi in the Chrome browser, and view images of a camera on the Raspberry Pi in real time through the web page, for example, http://raspberryip/parameter.

The majority of USB cameras use UVC drive-free cameras. Generally, there are two transmission formats, YUY2 and MJPG. The former is uncompressed video. No decoding is required, so no decoder is required. The system resources are less occupied, but its disadvantage is that the frame rate is slower. The latter is a JPEG image compression format. The advantage is that the frame rate is high, but its image is not clear enough, and a decoder is required, which will occupy system resources.

The mjpg-streamer is an open source video streaming program developed by Tom Stöveken. It is written for embedded devices with very limited RAM and CPU resources. Its purpose is to enable these devices to achieve fast and high-performance MJPG streaming. It copies a JPEG frame from one or more input plug-ins to multiple output plug-ins through the command line. For example, it gets JPG images from the camera, file system or other input plug-in, and transfers them to Web browser in the form of MJPG through HTTP Device, VLC and other software. You can use SSH reverse tunnel to penetrate the internal network, so that a mobile terminal can access video outside the school. Figure 4 shows the main procedure of playing video collected by a Raspberry Pi uvc camera by using mjpg-streamer technology.

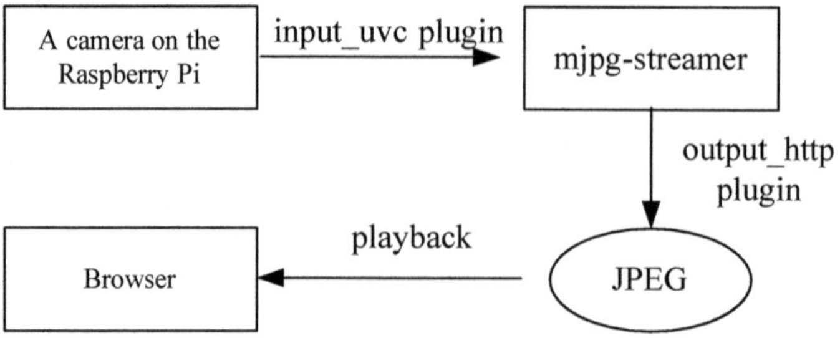

Fig. 4. The main process of browser image playback based on mjpg-streamer technology.

3.4 Control a Camera's Movement

When performing video surveillance, an administrator possibly need to move a camera up, down, left and right, then install a dual g9s servo on the camera, and write the main python code on a Raspberry Pi to control the movement of its servo as follows:

```
//Define the class that controls the servo to move left and right
class xy:
servopin1 = 21//gpio Pin
GPIO. setmode(GPIO.BCM)//Use BCM Mode
GPIO. setup(servopin1, GPIO.OUT, initial = False)//register gpio
P = GPIO. PWM(servopin1, 50)//50 is the electrical frequency Hz
p. start(6)
time. sleep(0.2)
def moveLeft(self, left): Research on Construction of Smart Training Room?
self.p. ChangeDutyCycle(left)
time. sleep(0.2)
def moveRight(self, right):
self. p.ChangeDutyCycle(right)
time. sleep(0.2)
//Define the class that controls the servo to move up and down
class z:
servopin1 = 20
GPIO. setmode(GPIO.BCM)
GPIO. setup(servopin1, GPIO.OUT, initial = False)
P = GPIO.PWM(servopin1, 50)
p. start(2)
time. sleep(0.2)
def moveDown(self, down):
self.p. ChangeDutyCycle(down)
time. sleep(0.2)
def moveUp(self, up):
self.p. ChangeDutyCycle(up)
time. sleep(0.2)
```

4 Conclusion

This paper designs a smart training room system architecture through the application of mobile Internet, cloud computing, Internet of things and other technologies. Experimental deployment has been carried out in the campus training rooms. The experimental results show that the architecture has good ease of use and scalability. The next step for this research will further study how to improve the video image transmission performance in intelligent training rooms and increase the image recognition function of artificial intelligence.

Acknowledgements. This research was supported by the First-class vocational scientific research and social service capacity building project of Guangdong Polytechnic of Water Resources and Electric Engineering, China (Grant No. GX030102y14).

References

1. Ministry of Education of the People's Republic of China. Opinions of the ministry of education on strengthening the safety of university laboratory. http://www.moe.gov.cn/srcsite/A16/s3336/201905/
2. Chun, B.G., Ihm, S., Maniatis, P., Naik, M., Patti, A.: Clonecloud: elastic execution between mobile device and cloud. In: Proceedings of the Sixth Conference on Computer Systems (EuroSys 2011), pp. 301–314. ACM, NY, USA (2011)
3. Han, H.W., Qi, D.Y., Feng, B.: Video surveillance integrated service platform based on cloud computing technologies. Comput. Eng. Des. **34**(5), 1657–1662 (2013)
4. Wang, H.L., Wang, H.B., Wang, R.Y., Wang, H.T., Liu, Q., Zhang, L.Y., Jiang, M.H.: Face recognition method based on video surveillance. Comput. Measure. Control **28**(4), 137–141 (2020)
5. Liu, An, M., Jia, J.M., Tu, Q., Xu, B.: Construction of intelligent training base for higher vocational colleges orienting made in China 2025 plan. Voca. Educ. Newslett. **4**, 11–14 (2019)
6. Huang, L., Liu, D.: Construction of intelligent laboratory management center for higher vocational college based on the internet of things. J. Shunde Poly. **14**(1), 40–44 (2016)

Design of a Real-Time and Reliable Multi-machine System

Jian Zhang[1(✉)], Lang Li[1,2], Qiuping Li[1], Junxia Zhao[1], and Xiaoman Liang[1]

[1] College of Computer Science and Technology, Hengyang Normal University, Hengyang 421002, China
zhangjian2100@126.com

[2] Hunan Provincial Key Laboratory of Intelligent Information Processing and Application, Hengyang 421002, China

Abstract. An IoT (Internet of Things) system is a multi-machine system composed of a master controller and multiple slave controllers. In some automatic control systems such as petroleum and electric power, the main control uses UART (Universal Asynchronous Receiver/Transmitter) to communicate with the slave controller. However, the communication rate of the UART interface is limited and the delay time is long, which affects the system's response to external events. In order to improve the response speed of the system to external events, this paper designs a multi-machine system based on SPI (Serial Peripheral Interface) interface. First, we use the SPI interface of the microcontroller to build a communication network. The circuit structure of the system is simple, and the hardware resource consumption is low. Then, for the problem that the SPI interface cannot detect communication errors and there is no response mechanism, we designed a communication protocol to make up for these defects from the software level. Finally, we use STM32F429IGT6 as the master controller and STM32F103C8T6 as the slave controller to build the experimental platform. The experimental results show that the system is safe and reliable, and can meet the requirements of real-time applications.

Keywords: Real-Time · Multi-Machine system · SPI

1 Introduction

In the past few years, the IoT technology has developed rapidly, and it has been widely used in agricultural production, industrial manufacturing, urban management and personal households, bringing new changes to our life. Smart agriculture uses IoT technology to promote crop growth, which not only reduces production costs, but also improves the quality of agricultural products. Various smart devices at home improve our quality of life. For example, environmental monitoring equipment uses sensors to collect various environmental parameters, provide suggestions for improving our living environment, and bring us a warm and comfortable home. In the future, more IoT devices will be used in our lives to improve our work efficiency and living conditions [1].

Q. Liu et al. (Eds.): CENet 2020, AISC 1274, pp. 1120–1127, 2021.
https://doi.org/10.1007/978-981-15-8462-6_128

The IoT system is usually composed of multiple subsystems, which can be divided into a master controller and slave controllers in terms of function. In addition, in order to provide an interactive interface, some systems are also equipped with a personal computer. The personal computer provides a human-computer interaction interface, which is convenient for manual operation and monitors the running status of the system. The master controller is the control center of the entire system. According to the operation instructions of the personal computer, the master controller schedules and manages the slave controllers to complete the task together. The slave controller receives the control signal of the master controller and completes the corresponding tasks. The UART has a simple structure and is easy to use. The master controller can communicate with the slave controller easily through the UART, which is widely used in industry. In order to solve the problem of high communication cost and complex structure between multiple IoT devices, Ma et al. [2] designed a multi-machine communication protocol based on the UART. In this communication protocol, the slave controller communicates with the master controller in a time-sharing manner to achieve stable and reliable communication between various modules in the system. There are two main types of UART interface: RS232 and RS485. The RS232 interface requires only three signals (RXD, TXD, and GND) to communicate between the two devices. Milic et al. [3] uses the RS232 interface in his temperature measurement and detection system to establish a communication connection for each subsystem. The RS485 interface uses differential signals to transmit data, has strong anti-interference ability, and is suitable for use in a noisy environment [4]. Gu et al. [5] used the RS485 bus to connect multiple smart devices and designed a heat pump hot water monitoring system. The smart electric meter and the smart water meter respectively measure the electric quantity and water quantity of the heat pump, and then send the collected data to the personal computer through the RS485 bus. Hu et al. [6] developed a variable frequency speed control system with multiple asynchronous motors based on the RS485 bus. In this system, the microcontroller STC89C51 controls multiple motors to adjust speed synchronously through the RS485 bus.

However, the communication rate of UART is low, generally no more than 115200 bit/s [7]. In some applications with high real-time requirements, the UART will be difficult to meet the requirements. Compared with the UART interface, the communication rate of the SPI interface is much higher [8]. In order to achieve high-speed data transmission, Wan Wenjie uses the SPI interface to connect ARM processor and FPGA in multi-channel spectrometer [9]. However, there are two defects in the SPI interface. The SPI interface has no data check and response mechanism. In order to solve the problem of high-speed communication between IoT devices, this paper designs a multi-machine system based on the SPI interface. In addition, this paper proposes a communication protocol, which defines the data frame and communication process of the system to ensure the safety and reliability of the communication. We add CRC (Cyclie Redundancy Check) at the end of the data frame to detect whether the data has an error during transmission. In the communication process, the slave controller cannot initiate communication actively, and sends a message to the master controller through the SPI interface. Therefore, in this communication protocol, the master initiates communication to read the slave status. At last, we use STM32 F429IGT6 as the master controller and STM32F103C8T6 as the slave controller to test the system.

2 The SPI Interface

The SPI interface is a universal serial communication interface with low circuit resource consumption and high data transmission rate. The microprocessor can easily communicate with other high-speed devices through the SPI interface. The master controller can communicate with a single slave controller or multiple slave controllers through the SPI interface. The connection of the SPI interface is shown in Fig. 1. The main controller is connected to the slave controller through four signals: clock signal (SCLK), chip select signal (CS), output signal (MOSI) and input signal (MISO).

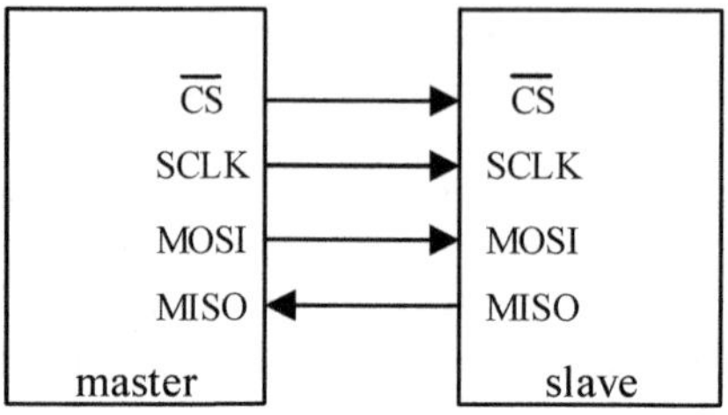

Fig. 1. The connection of the SPI interface.

The CS signal is the enable signal of the slave controller, and the master controller uses the CS signal to select the communication object. When the CS signal is low, the slave controller is enabled; when the CS signal is high, the slave controller is disabled. The SCLK is a clock signal that provides a synchronization signal for data transmission between the master controller and the slave controller. The SPI interface works in master-slave mode, the clock signal is provided by the master controller, and only the master controller can initiate communication. The MOSI signal and MISO signal are data signals. The master controller uses the MOSI signal to send data to the slave controller, and receives the data from the controller through the MISO signal.

3 System Design

3.1 The System Structure

The structure of a real-time and reliable multi-machine system is shown in Fig. 2. The system is mainly composed of a personal computer (PC), a master controller and multiple slave controllers. The main controller has a UART communication port and an SPI communication port, the UART interface is connected to the PC, and the SPI interface is connected to the slave controller. The slave controller has an SPI interface, and the SPI interfaces of all slave controllers are connected in parallel to the master controller. The decoder converts the address signal of the slave controller into a chip select signal, and only the slave controller of the corresponding address will be enabled. According to the slave controller address, the master controller only communicates with one slave controller at a time.

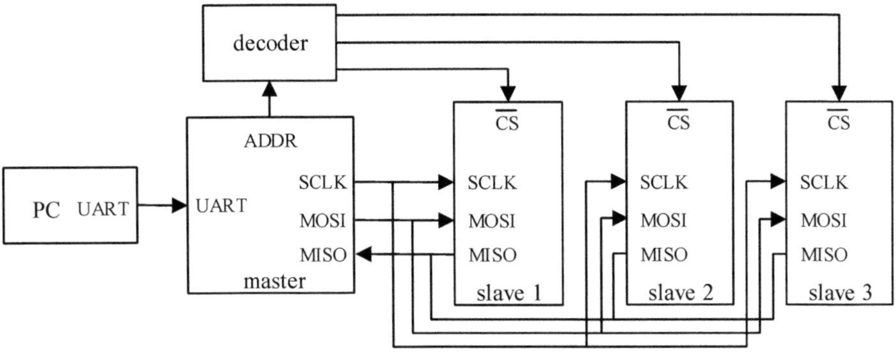

Fig. 2. The system structure.

The PC provides the user with an interactive interface, responsible for downloading various control commands and data, and monitoring the running status of the master controller and slave controllers. The main controller is responsible for the control of the system, and transfers the control signals and data from the PC and itself to the slave controller. The slave controller responds to the control signal of the master controller in real time, and performs secondary operations based on the received data and its own data.

3.2 The Communication Protocol

The master controller will poll all the slave controllers and update the data table, the PC does not need to directly read and write the slave controllers. The PC communicates with the master controller through the UART and queries the data table of the master controller to obtain the status of all slave controllers. At the same time, the master controller receives the operation instructions of the PC and controls the slave controller. The communication between the PC and the main controller is not the main content of this study. We focus on the high-speed communication between the main controller and the slave controller.

The communication rate of the SPI interface can be set to 10 Mbit/s, which improves the speed of the system's response to external events. In order to achieve high-speed and reliable communication between the master controller and the slave controller, we have designed a communication protocol. The data format and data receiving and sending methods are defined in the communication protocol. The format of the data is defined in the communication protocol, and the command and data are encapsulated into a data frame. The main controller and the slave controller communicate in the form of data frames. The SPI interface itself does not have the ability to check data errors. We add a CRC at the end of the data frame to detect whether errors occur during data transmission. The clock signal of the SPI interface is provided by the master controller, and the slave controller cannot initiate communication to send data to the master controller. In order to obtain the status of the slave controller, the master controller needs to send two instructions to the slave controller. The master controller sends the first command to the slave controller, and the slave controller prepares the data according to the command. Then, the master controller sends the second command to the slave controller, and the slave controller sends the data to the master controller at the same time. In addition, the communication protocol stipulates that the slave controller must reply after receiving the command from the master controller. We implement a response mechanism from the software level to ensure the safety and reliability of communications. The communication process is as follows:

(1) The main controller receives control commands from the PC;
(2) The master controller controls the decoder to enable the corresponding slave controller according to the control instructions from the PC;
(3) The main controller encapsulates the control instructions into a data frame with CRC;
(4) The master controller sends data to all slave controllers in broadcast mode, and only enabled slave controllers can receive data;
(5) The slave controller receives data from the master controller in an interrupted manner until the data frame is received;
(6) The slave controller uses CRC to detect whether an error occurs in the data frame during transmission, and then prepares to reply to the master controller according to the detection result;
(7) The main controller sends a command to read the reply from the controller;
(8) If the slave controller replies with an error, the master controller resends the data frame.

The data frame format is shown in Table 1. It is composed of three parts: header, information and the check value. The data frame header is located at the beginning of the data frame, and is a sign that the master controller starts to communicate with the slave controller. The information of the data frame includes the length of the data frame, the command type and the command parameters. The communication protocol supports variable-length data frames. The two bytes of data behind the data frame header are the length information of the data frame, which records the total length of the data frame. We divide the data frame into two types: the command frame and the status frame. The command frame is used by the master controller to set the parameters of the slave controller. The status frame is used by the master controller to read the status of the slave controller. The number of parameters in the data frame can be configured, and the developer can set it according to the specific situation, which improves the flexibility of the system. The last two bytes of the data frame is the CRC value, which is used to detect whether there is an error in the data frame.

Table 1. The SPI communication data frame.

Identifier	Size (Byte)	Starting address
Header	2	0
Length	2	2
Type	2	4
Parameters	n	6
CRC	2	6 + n

4 System Test

The master controller of the system is processor STM32F429IGT6. It has 1 MB Flash and 256 KB SRAM, and its clock frequency is up to 180 MHz. The core circuit board of the master controller is shown as in Fig. 3. The slave controller is the high-performance, low-cost processor STM32F103C8T6. The circuit board is shown in Fig. 4.

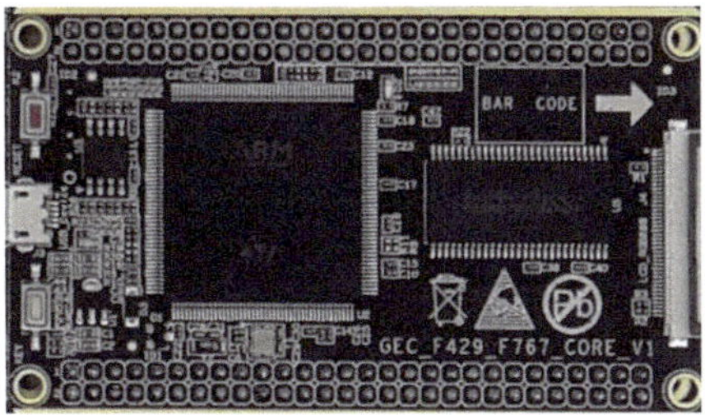

Fig. 3. The core circuit board of the master controller.

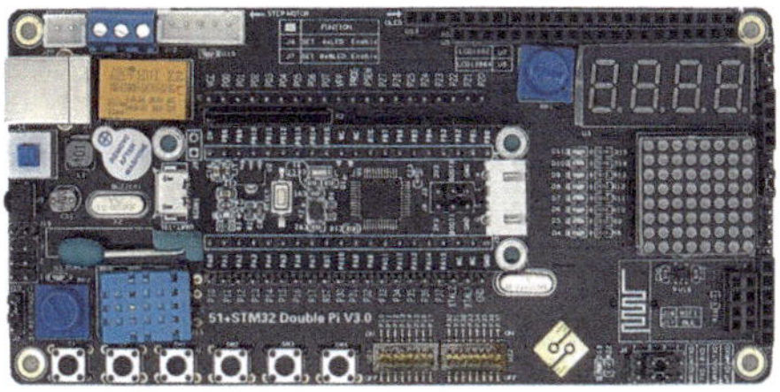

Fig. 4. The circuit board of the slave controller.

In the experiment, the master controller is connected to three slave controllers through the SPI interface, and the three slave controllers collect the temperature, humidity and light in the environment through the sensors. The master controller summarizes the data collected by each slave controller, and sends it to the PC using UART. The test result is shown in Fig. 5. We can see the environmental parameters collected by the three slave controllers on the PC. The first data is temperature, the second data is humidity, and the third data is light.

Fig. 5. The test result.

5 Conclusion

We use the SPI interface to design a real-time and reliable multi-machine system, which provides a solution for some IoT systems that require high response speed. This article introduces the structure and communication protocol of the system. In this system, the SPI ports of the slave controllers are connected in parallel, and the master controller sends messages to all slave controllers in a broadcast manner. I in order to solve the problem of communication conflict, we assign a unique address to each slave controller, and the master controller communicates with the slave controller according to the address signal. The system uses CRC for error checking, and guarantees the safety and reliability of data transmission through the response mechanism and retransmission mechanism. The test platform was built using the PC, the master controller STM3-2F429IGT6 and three slave controllers STM32F103C8T6. The main controller reads the data collected by the three slave controllers and sends them to the PC through the UART interface. The experimental results confirmed that the multi-machine system built with the SPI interface of the microprocessor can run at high speed and reliably.

Acknowledgements. This research is supported by the Science Foundation Project of Hengyang Normal University (18A14), Scientific Research Fund of Hunan Provincial Education Department No. 19C0278, National Natural Science Foundation of China under Grant No. 61572174, Application-oriented Special Disciplines, Double First-Class University Project of Hunan Province (Xiangjiaotong [2018] 469), Hunan Province Special Funds of Central Government for Guiding Local Science and Technology Development No. 2018CT5001, Hunan Provincial Natural Science Foundation of China with Grant No. 2019JJ60004, the Science and Technology Plan Project of Hunan Province No. 2016TP1020, Subject group construction project of Hengyang Normal University No. 18XKQ02, Scientific Research Fund of Hunan Provincial Education Department No. 18C0678.

References

1. Ghanbari, Z., Jafari, N.N., Hosseinzadeh, M., et al.: Resource allocation mechanisms and approaches on the Internet of Things. Cluster Comput. **22**(5), 1253–1282 (2019)
2. Ma, J.P.: A multi-microcontroller communication method based on UART asynchronous serial communication protocol. J. Shangdong Univ. **50**(3), 24–30 (2020)
3. Milic, S.D., Sreckovic, M.Z.: A stationary system of noncontact temperature measurement and hotbox detecting. IEEE Trans. Veh. Technol. **57**(5), 2684–2694 (2008)

4. Tang, X.Q., Li, J.M., She, X.S.: Design and development of RS485 bus interface performance tester. Electr. Measure. Instrum. **56**(07), 142–147 (2019)
5. Gu, H.Q., Yang, Y., Quan, Y., Ma, Y.: Design of bidirectional communication between intelligent instruments and PC computer based on MODBUS portocol. Instrum. Tech. Sensor **12**, 33–35 (2013)
6. Hu, J.G., Luo, Y.W.: Multi-motor communication control by SCM and inverter based on RS485. Mach. Tool Hydraulics **22**, 139–141 (2013)
7. Zhang, C., Tan, Y.S., Yan, S. S.: Method of rate detecting for serial communication signal and information distinguish. Mod. Electron. Tech. (9), 24–25+28 (2002)
8. Kuang, C. Y., Ma, Q., Chen, K. M.: Design and verification of SPI applied to SoC. Mod. Electron. Tech. (24), 149–151+155 (2013)
9. Wang, W.J., Zhou, J.B., Fei, P., Li, Y.P., Qin, H.J., Yu, J.: Implementation of high-speed communication based on SPI bus interface in multichannel energy spectrum analyzer. Nucl. Electron. Detect. Technol. 37(1), 29–32+42 (2017)

A Detection Method of Safety Helmet Wearing Based on Centernet

Bo Wang, Qinjun Zhao, Yong Zhang(✉), and Jin Cheng

School of Electrical Engineering, University of Jinan, Jinan 250022, China
cse_zhangy@ujn.edu.cn

Abstract. In some construction sites, workers often do not wear helmets and cause safety accidents. In order to prevent the occurrence of safety accidents caused by not wearing safety helmets, we propose a detection method of safety helmet wearing based on CenterNet. Input image into fully convolutional network to obtain a heat map, and the peak of the heat map corresponds to the center of the target. The image features on each peak can predict the width and height of the target frame. The network uses dense supervised learning for training. The inference stage is a single forward propagation network without NMS post-processing. Using video capture from the construction site as a part of the data set. Theoretical analysis and experimental results show that when using detection method of safety helmet wearing based on CenterNet, its recognition accuracy and rate meet the requirements of helmet wearing detection.

Keywords: Safety engineering · Helmet identification · Deep learning

1 Introduction

As a personal protective equipment, the helmet is the most common and practical in life, it can be done to effectively prevent and mitigate head injuries for external hazards. When entering or leaving some basic construction sites, security inspectors will check whether the workers are wearing safety helmets one by one. However, this kind of inspection does not guarantee that the works will wear safety helmet all the time during the construction, and the original inspection and supervision method is time-consuming and labor-consuming. In this evolving environment driven by science and technology, in order to ensure that workers can wear safety helmet during the operation, the helmet recognition system for smart factory and smart construction site came into being, which brought safety guarantee for the majority of the workers in the construction sites [1].

Construction industry is a labor-intensive industry with complex working environment and frequent safety accidents [2]. An automatic safety detection method for safety helmet wearing of workers is helpful to improve the safety management level of construction sites. In recent years, the popularity of camera in construction sites and the efficient application of deep learning in speech recognition [3], image recognition and natural language processing provides a new perspective for safety management in construction sites. Instead of traditional manual monitoring, deep learning is used to automatically identify. It is conducive to real-time monitoring in the construction sites,

Q. Liu et al. (Eds.): CENet 2020, AISC 1274, pp. 1128–1137, 2021.
https://doi.org/10.1007/978-981-15-8462-6_129

which not only saves labor costs, but also improves site safety, and lays the foundation for the development of "smart construction site".

The use of image recognition technology for automatic recognition of safety helmets is the latest development direction at present, and domestic and foreign scholars have related research: Bi et al. [4] proposed to divide the image into three types: background, wearing a helmet and not wearing a helmet, and directed input to the convolutional neural network for classification; Fang et al. [5] collected data from workers who did not wear safety helmets correctly at the construction sites by using the method of image recognition based on deep learning, and divided them into 5 classes and 19 subclasses of data sets according to the influencing factors of image recognition effect. The results show that the recognition accuracy of the collected data sets is 90%, it provides a solid foundation for detecting whether workers are wearing helmets correctly. Mneymneh et al. [6] used image information to extract geometric space position information of workers and safety helmet to match, and seted cascader according to feature points to judge whether workers wear safety helmets. However, due to the different working postures and physical conditions of workers, this method has encountered great challenges in practical operation. In 2016, Liu et al. proposed SSD [7] (single shot multibox detector) detection algorithm, which has achieved good results in accuracy and speed. On the basis of YOLO, Redmon also proposed YOLOv2 [8] and YOLOv3 [9] detection algorithms. In the experiment, YOLOv3 detection effect is better. In the coco data set, the mAP reaches 57.9% in 51 ms, and the speed is 3.8 times of RetinaNet. Thus, YOLOv3 can achieve better accuracy and speed. In April 2019, Zhou et al. proposed CenterNet: objects as points [10], and proposed a new anchor free algorithm based on key points. The framework is more convenient, concise and faster.

This paper proposes a safety helmet detection method based on the CenterNet algorithm. Input image into fully convolutional network to obtain a heat map, and the peak of the heat map corresponds to the center of the target. The image features on each peak can predict the width and height of the target frame. The network uses dense supervised learning for training. The inference stage is a single forward propagation network without NMS post-processing.

2 CenterNet Principle

In this paper, we build the model with the target as a point—the center point of the target BBox. Our detector uses key point estimation to find the center point, and return to the other target properties, such as size, 3D position, orientation, and attitude. Our method based on the center point is called: CenterNet. We input the image into a fully convolutional network to obtain a heat map, and the peak of the heat map corresponds to the center of the target. Compared with the BBox-based detector, our model is end-to-end differentiable, simpler, faster and more precise. And Pailla et al. applied CenterNet to aerial image target detection [11].

2.1 Overall Structure

The framework flowchart is shown in Fig. 1.

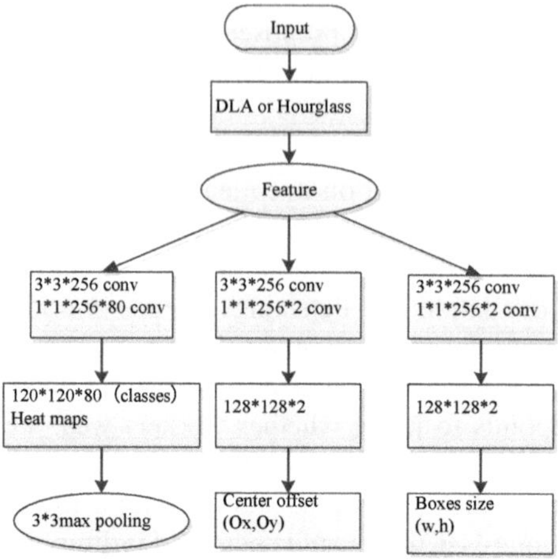

Fig. 1. Framework flow chart.

2.2 Key Point Prediction Network

Set $I \subseteq R^{W \times H \times 3}$ as input image, the goal is to generate a heat map for key point prediction $Y \subseteq [0, 1]^{\frac{W}{R} \times \frac{H}{R} \times C}$, Where R is the output stride (that is the size scaling), and C is the number of output feature maps (the number of key point types, if the recognition category is 80, then $C = 80$). The key point training network refers to the paper of CornerNet and uses Focal Loss normalization. For the key point c of Ground Truth, its coordinate position is $\rho \in R^2$,after the network processing (equivalent to downsampling), the position on the feature map is

$$\tilde{\rho} = \left\lfloor \frac{\rho}{R} \right\rfloor \tag{1}$$

The paper passes the key points of Ground Truth through the Gaussian core

$$Y_{x,y,c} = exp\left(-\frac{\left(x - \tilde{\rho}_x\right)^2 + \left(y - \tilde{\rho}_y\right)^2}{2\sigma_\rho^2} \right) \tag{2}$$

disperse to heat map in $Y \subseteq [0, 1]^{\frac{W}{R} \times \frac{H}{R} \times C}$, The loss function for training is as follows:

$$L_k = -\frac{1}{N} \sum_{xyc} \begin{cases} \left(1 - \hat{Y}_{xyc}\right)^{\alpha} \log\left(\hat{Y}_{xyc}\right) \; if \; Y_{xyc} = 1 \\ \left(1 - Y_{xyc}\right)^{\beta} \left(\hat{Y}_{xyc}\right)^{\alpha} otherwise \\ \log\left(1 - \hat{Y}_{xyc}\right) \end{cases} \tag{3}$$

Among them,α and β is the hyperparameter of focal loss, which is set to 2 and 4 respectively in the experiment; N is the number of key points in an image. Without considering $\left(1 - Y_{xyc}\right)^{\beta}$ on this item first, the formula for converting the above formula is as follows:

$$(1 - P_t)^{\alpha} * \log P_t \tag{4}$$

$$P_t = \begin{cases} \hat{Y}_{xyc}, \; if \; Y_{xyc} = 1 \\ 1 - \hat{Y}_{xyc}, otherwise \end{cases} \tag{5}$$

Among them, $\log P_t$ is the standard cross-entropy loss function; when the value of P_t is large, it indicates that the prediction result has been relatively correct, and $(1 - P_t)^{\alpha}$ is relatively small at this time; and when the value of P_t is small, it indicates that the current sample is hard examples, and the optimizer needs to focus on it, so the corresponding part of $(1 - P_t)^{\alpha}$ is large, which realizes a major function of focal loss. Then let's look at the term of $\left(1 - Y_{xyc}\right)^{\beta}$ again. First, the Y_{xyc} is a value less than 1, and the value of β is 4. Therefore, it does a weighting of less than 1 for all negative samples, and this coefficient can be regarded as 1 for positive samples, so the positive sample The weighting is greater than that of negative samples, which solves the problem of imbalance between positive and negative samples. For the weighted term of negative samples, if only for one negative sample point, more attention will be paid to negative samples away from the center point, and the number of negative samples away from the center point will also be more, so this item will pay more attention to away negative sample of the center point.

Since the image is down-sampled by convolution, the key points of Ground Truth will be biased. In this paper, the prediction of the local offset is added to each key point (the same prediction value is used for all classes) $\hat{O} \subseteq R^{\frac{W}{R} \times \frac{H}{R} \times 2}$. This offset is used L1 loss to train, this operation is only performed at key points, other positions are ignored.

$$L_{off} = \frac{1}{N} \sum\nolimits_{p} \left| \hat{O}_{\tilde{p}} - \left(\frac{p}{R} - \tilde{p} \right) \right| \tag{6}$$

2.3 Target Size Network

Suppose $\left(x_1^k, y_1^k, x_2^k, y_2^k\right)$ is the bounding box k of the target, so its center position is

$$p_k = \left(\frac{x_1^k + x_2^k}{2}, \frac{y_1^k + y_2^k}{2} \right), \tag{7}$$

therefore, it can be estimated that the size of the target

$$S_k = \left(x_2^k - x_1^k, y_2^k - y_1^k\right) \tag{8}$$

increases L1 Loss at the center point:

$$L_{size} = \frac{1}{N}\sum\nolimits_{k=1}^{N} \left|\hat{S}_{p_k} - S_k\right| \tag{9}$$

Instead of normalizing the scale, we use the original pixel coordinates directly. In order to adjust the influence of the loss, it is multiplied by a coefficient, the target loss function of the entire training is:

$$L_{det} = L_k + \lambda_{size} L_{size} + \lambda_{off} L_{off} \tag{10}$$

Among $\lambda_{size} = 1, \lambda_{off} = 1$; the whole network predicts that $C + 4$ values will be output at each location (i.e. keypoint category C, offset x, y, size w, h), and all outputs share a fully convolution backbone.

2.4 Reasoning Process

In the process of reasoning, we extract the peak points of each category on the thermodynamic diagram respectively: compare all the response points on the thermodynamic diagram with the eight adjacent points connected to them, if the response value of this point is greater than or equal to the eight adjacent points, then keep it, and finally we keep all the first 100 peak points that meet the previous requirements.

Let $\hat{P}\sqrt{b^2 - 4ac_c}$ is the set of n center points the detected category c. Each key point of

$$\hat{P} = (\hat{x}_i, \hat{y}_i)_{i=1}^{n} \tag{11}$$

is given in integer coordinates (x_i, y_i). Finally bounding box is generated:

$$\left(\hat{x}_i + \delta\hat{x}_i - \frac{\hat{w}_i}{2}, \hat{y}_i + \delta\hat{y}_i - \frac{\hat{h}_i}{2}, \hat{x}_i + \delta\hat{x}_i + \frac{\hat{w}_i}{2}, \hat{y}_i + \delta\hat{y}_i + \hat{h}_i/2\right) \tag{12}$$

Below are the results predicted offset:

$$(\delta\hat{x}_i, \delta\hat{y}_i) = \hat{O}_{\hat{x}_i, \hat{y}_i} \tag{13}$$

The following results are predicted size:

$$\left(\hat{w}_i, \hat{h}_i\right) = \hat{S}_{\hat{x}_i, \hat{y}_i} \tag{14}$$

Once NMS without subsequent treatment.

3 Experimental Data Set Making

For deep learning detection tasks, the experimental data set is the basic condition. Moreover, there has been no public data set for the detection of helmet wearing, therefore, This article made a set of safety helmet wearing data set, in the course of construction of the data set, includes four main areas: data collection, data preprocessing, data filtering and data marking.

3.1 Data Collection

The data source of the helmet wearing data set mainly comes from two ways: One part is the video surveillance video of the construction site, and the other is to use the crawler code to obtain related pictures from the Internet.

3.2 Data Preprocessing

Because the site video surveillance data is a video format file, Therefore, you need to use the OpenCV development library to convert the avi format video file into a jpg format image frame. The software environment used at this time is OpenCV3.4.4; Visual studio 2017.

3.3 Data Filtering

By framing the video to get a lot of pictures, But a lot of pictures without people and helmets body, all the background image, and this part of the picture is not useful for the research of this article, so delete this part of the picture.

3.4 Data Marking

In the work of establishing a helmet wear detection data set, select labelImg tool to mark the image, the whole process needs to manually mark each picture, and the tagging process is mainly to select the target frame by using software, then mark different targets with different names. The operation interface is shown in Fig. 2.

Fig. 2. Annotation diagram.

4 Experimental Results and Comparative Analysis

The basic process is shown in Fig. 3.

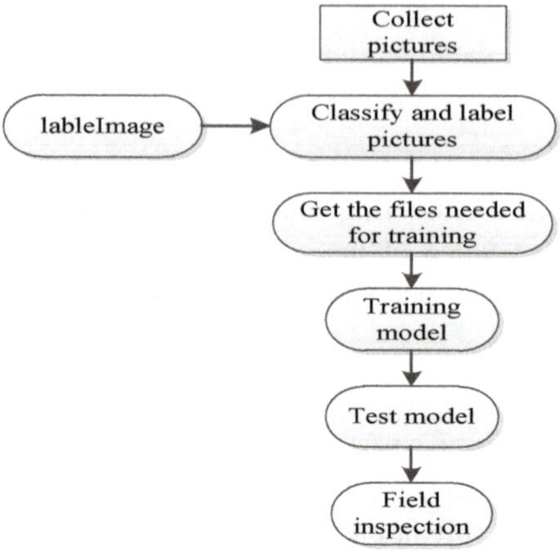

Fig. 3. Experimental flow chart.

4.1 Experimental Scheme

During the experiment, the data set can be divided into training set and test set according to the proportion of 9:1: the first part is the training set, this part of the data set will be used to train the neural network; the second part is the test set, which is the data used to test the performance of the network after the completion of network training, it can test the effect of the trained network, including the main performance indexes such as speed and accuracy. In the training set, positive samples are divided into four classes: A, B, C, and D. Among them, class A is workers wearing yellow helmets; class B is workers wearing white helmets; class C is workers wearing blue

Fig. 4. Sample diagram.

helmets; class D is workers wearing red helmets. However, In this paper, negative samples only set a class E, class E is a worker who does not wear safety helmet as required. That is, the data set is divided into five classes in total, as shown in Fig. 4.

4.2 Experimental Platform and Network Training

Experimental platform. In this paper, the hardware environment is required to be high during the experiment, and GPU is used for calculation. Table 1 shows the hardware configuration of the experiment. At the same time, the experimental environment of the computer is Ubuntu 16.04, CUDA, python, opencv and other common environments, using the DLA framework.

Table 1. Configuration description of experimental hardware environment.

Product name	Model	Number
CPU	Intel (R) Core (TM) i7-8750H CPU @ 2.20 GHz	1
Main board	LNVNB161216	1
RAM	DDR4	1
GPU	Intel (R) UHD Graphics 630 NVIDIA GeForce GTX 1060	2
SSD	500 G	1
Disk	West Blue 1 TB	1

Network Training. Using the weight parameter provided by CenterNet as the initialization parameter of network training. Then, through many parameter tuning tests, we find the parameters suitable for the current network detection. It makes the detection effect of the whole network to reach the best. Some experimental parameters shown in Table 2.

Table 2. Description of network parameters.

Parameter	Parameter value
Learning rate	$2.5 \times 10 - 5$
Epoch	100
Batch size	5
nms	3
lr	$1.25e - 4$

4.3 Experimental Platform and Network Training

The performance of CenterNet for target detection is verified by experiment, it is compared with different algorithms of Faster R-CNN, RetinaNet and YOLO v3. At the same speed, the accuracy of CenterNet is 4 points higher than that of Yolo v3.

Meanwhile, using the Faster R-CNN, RetinaNet and YOLO v3 for comparison, and taking map and the number of recognition frames per second as the evaluation index of detection effect, the results shown in Table 3.

Table 3. Comparison of experimental results.

Algorithm	mAP/%	Frame rate recognition/(f·s − 1)
Faster R-CNN	40.5	5
RetinaNet	37.0	5.4
YOLO v3	31.0	20
CenterNet	39.2	28

4.4 Result Analysis

From the experimental results, Basically, CenterNet is more powerful than SSD, Faster R-CNN and YOLO v3 in performance and speed. In addition, CenterNet requires a lower number of images in the dataset than other algorithms, the frame is smaller, and it is faster.

In this article, I selected 100 images as the data set used for training, and use the monitoring video as the test,the test results are shown in Fig. 5.

Fig. 5. Test result.

By observing Fig. 5, we can find that the accuracy of the test results is very high.

5 Conclusion

This paper proposes a detection method of safety helmet wearing based on the CenterNet algorithm. It can be seen from the results of video detection, while ensuring higher detection accuracy, it can still have a faster detection speed. It can meet the testing requirements of safety helmet wearing.

References

1. Fang, Q., Li, H., Luo, X.C., Ding, L.Y., Luo, H.B., Timothy, M.R., An, W.P.: Detecting non-hardhat-use by a deep learning method from far-field surveillance videos. Autom. Constr **85**, 1–9 (2018)
2. Guo, B.H.W., Goh, Y.M., Wong, K.L.X.: A system dynamics view of a behavior-based safety program in the construction industry. Safe. Sci. **104**, 202–215 (2018)
3. Sun, K., Liu, Z.W., Wu, Y. Q., Guo, D.X.: Software college, Shenyang normal university: a method for deep learning speech recognition based on python. J Shenyang Norm. Univ. (Nat. ence Ed.) (2019)
4. Bi, L., Xie, W., Cui, J.: Identification research on the miner's safety helmet wear based on convolutional neural network. Gold Sci. Technol. **25**(4), 73–80 (2017)
5. Fang, Q., Li, H., Luo, X.C., et al.: Detecting non-hardhat-use by a deep learning method from farfield surveillance videos. Autom. Constr. **85**, 1–9 (2018)
6. Mneymneh, B.E., Abbas, M., Khoury, H.: Automated hardhat detection for construction safety applications. In: Proceedings of the 6th Creative Construction Conference. Primosten, Croatia: Elsevier Ltd., 895 (2017)
7. Wang, Y.Y., Wang, C., Zhang, H.: Combining a single shot multibox detector with transfer learning for ship detection using sentinel-1 SAR images. Remote Sens. Lett. **9**(8), 780–788 (2018)
8. Redmon, J., Farhadi, A.: YOLO9000: better, faster, stronger. In: IEEE Conference on Computer Vision and Pattern Recognition, pp. 6517–6525 (2017)
9. Redmon, J., Farhadi, A.: YOLOv3: an incremental improvement. In: IEEE Conference on Computer Vision and Pattern Recognition, pp. 89–95 (2018)
10. Zhou, X.Y., Wang, D., Philipp, K.: Objects as Points. (2019)
11. Pailla, D.R., Kollerathu, V., Chennamsetty, S. S.: Object detection on aerial imagery using CenterNet (2019)

A Hybrid Sentiment Analysis Method

Hongyu Han, Yongshi Zhang(✉), Jianpei Zhang, Jing Yang, and Yong Wang

Harbin Engineering University, Harbin 150001, China
zhangyongshi@hrbeu.edu.cn

Abstract. Sentiment analysis has attracted a wide range of attentions in the last few years. Supervised-based and lexicon-based methods are two mainly sentiment analysis categories. Supervised-based approaches could get excellent performance with sufficient tagged samples, while the acquisition of sufficient tagged samples is difficult to implement in some cases. Lexicon-based method can be easily applied to variety domains but excellent quality lexicon is needed, otherwise it will get unsatisfactory performance. In this paper, a hybrid supervised review sentiment analysis method which takes advantage of both of the two categories methods is proposed. In training phrase, lexicon-based method is used to learn confidence parameters which used to determine classifier selection from a small-scale labeled dataset. Then training set which is used to train a Naive Bayes sentiment classifier. Finally, a sentiment analysis framework consist of the lexicon-based sentiment polarity classifier and the learned Naive Bayes classifier is constructed. The optimal hybrid classifier is obtained by obtaining the optimal threshold value. Experiments are conducted on four review datasets.

Keywords: Hybrid method · Lexicon-based method · Sentiment analysis

1 Introduction

With the booming development of information technologies, a variety of new and convenient applications appear in people's lives [1]. People can obtain information more conveniently, and unlike the past, users are not simply receiving information, but also can express their feelings more easily and share their experiences [2]. People communicate more and more frequently on social networks. Every day, huge amounts of user generated content (UGC) [3] are created. How to extract valuable information from UGC has attracted a wide range of attentions from researchers, and a lot of sentiment analysis techniques have been proposed. Those presented sentiment analysis methods are mainly divided into two categories: supervised learning and lexicon-based methods.

Supervised learning sentiment analysis methods have been widely studied, machine learning technologies such as Support Vector Machines (SVM), Naive Bayes (NB), K-Nearest Neighbor (KNN), Ensemble Learning, etc. have been utilized with good performances [4]. However, sufficient quantity of labeled samples is indispensable to supervised learning sentiment analysis methods. However, most of the UGC is

Q. Liu et al. (Eds.): CENet 2020, AISC 1274, pp. 1138–1146, 2021.
https://doi.org/10.1007/978-981-15-8462-6_130

unmarked, and the work to obtain enough labeled samples is time consuming and labor intensive [5].

The lexicon-based method which based on sentiment lexicon obtains the final emotional polarity of a comment by accumulating the emotional polarity and strength of each word in the comment. The lexicon-based sentiment analysis method has the characteristics of being widely applied and easy to implement. However, performance of lexicon-based method is heavily dependent on the quality of the sentiment lexicon. The general sentiment lexicon can be used in a variety of fields, but it is clear that its general features limit its performance in specific areas. Domain related lexicons will perform better in its specific field than general sentiment lexicons. Domain-related sentiment lexicons also face the problem of updating emotional polarity strength.

In our proposed method, we propose a hybrid supervised review sentiment analysis method which combines SWN-based method and NB method. Confidence parameters are introduced into the proposed hybrid sentiment analysis method to decide when to use dictionary-based methods and when to use NB methods. Experiments are conduct on datasets which are open access and widely used to test the effectiveness of the proposed method.

2 Literature Review

Machine learning technologies are the most common methods used in supervised learning method, which are successfully used in building sentiment classifiers [6]. A wide range of technologies (SVM, NB, ANN, KNN, etc.) are introduced into sentiment polarity detection. Pang et al. [7] introduced three machine learning techniques (NB, SVM, and maximum entropy classification) into the sentiment analysis problem. Reviews collected from Internet Movie Database (IMDb) are used as the experimental dataset, then experiments were conducted on the dataset to examine the effectiveness of the proposed classifiers. Machine learning methods are found to be superior to human-produced methods. Moraes et al. [8] adopt a standard evaluation context with popular supervised methods for feature selection and weighting in a traditional bag-of-words model, and ANN was used to construct a document-level sentiment classifier. Wang et al. [9] employed ten public sentiment analysis datasets to verify the effectiveness of three ensemble methods (Bagging, Boosting, and Random Subspace) when using five base learners (SVM, NB, KNN, Maximum Entropy, and Decision Tree). Experimental results reveal that the ensemble methods achieve better performances than individual base learners.

Sentiment lexicon is consist of lexical features which convey people's emotions [10, 11]. Each entry in the lexicon can be categorized as positive, negative, or neutral depend on its sentiment orientations and strength [12]. Sentiment lexicon construction has attracted the attentions of some researchers, and they have published several lexicons (SWN [13], LIWC [14], MPQA [15], etc.) which are widely recognized by other scholars. The lexicon-based sentiment analysis method first scores each word in the review, then obtains the review's sentiment score by summing up the scores of all the words in it, and finally categorize reviews as positive, negative, or neutral based on their sentiment scores [16]. A large range of lexicon-based studies have been presented in recent years. Hu et al. [17] assign positive words in the comments to +1, negative words to-1, neutral words to 0, and reverse the sentiment score if negation words appears. Taboada et al. [18] made more fine-grain evaluation for the lexical features, the features are scored in the range of −5 to +5, intensifiers and negators are also considered. The lexicon-based sentiment analysis method is widely used in the absence of sufficient labeled training dataset, and it could get excellent accuracy when the sentiment scores of the reviews are high enough [19].

3 The Proposed Hybrid-Supervised Sentiment Analysis Approach

3.1 Text Pre-processing

Text pre-processing is a necessary process since the original reviews contain lots of useless elements which affect further utilization. We use several processing steps to preprocess the reviews to make them more appropriate to be used in the remaining steps. The text pre-processing steps being used are as follows: Remove URLs, Lemmatization Stop word removal, and Part of speech (POS). More details could be found in the study [19].

3.2 SWN-Based Sentiment Rating Process

We introduce the SWN-based sentiment score process in the following section. First, we make a brief introduction to the SentiWordNet 3.0. Then, SWN-based sentiment score process is introduced in detail. Table 1 shows the composition of the SentiWordNet 3.0.

POS (part-of-speech) and ID uniquely identifies a synset of WordNet [18]. SynsetTerms column reports the synset (terms # sense number). *PosScore* is the positivity strength of the synset, and *NegScore* is the negativity strength of the synset, they are in the range of (0, 1). For example, for the last row in Table 1, "straiten#1 distress#1" is a synset from WordNet, the synset contains two terms, and both of they sense number is 1. Its *PosScore*=0 and *NegScore*=0.625, it recognized as negative. Calculation of the synset's sentiment score is shown below:

$$SynsetScore = PosScore - NegScore \tag{1}$$

If a term appears in n synsets with the same POS, the sentiment strength it can be calculated as follows:

$$TermScore = \left(\sum_{r-1}^{n} SynsetScore(r)/r\right) \Big/ \left(\sum_{r-1}^{n} 1/r\right)) \tag{2}$$

Where r is the term's sense number.

Table 1. Composition of SentiWordNet 3.0 [13].

POS	ID	PosScore	NegScore	SynsetTerms	Gloss
a	1800349	0.625	0.125	pleasant#1	affording pleasure; being in harmony with your taste or likings; "we had a pleasant evening together"; "a pleasant scene"; "pleasant sensations"
a	1586866	0.75	0	pleasant#2	(of persons) having pleasing manners or behavior; "I didn't enjoy it and probably wasn't a pleasant person to be around"
n	530018	0	0.5	modern_dance#1	a style of theatrical dancing that is not as restricted as classical ballet; movements are expressive of feelings
r	236393	0	0.375	dully#2	without luster or shine; "the light shone dully through the haze"; "unpolished buttons glinted dully"
v	2603424	0	0.625	straiten#1 distress#1	bring into difficulties or distress, especially financial hardship

When we calculate the sentiment score of a review, preprocess it at first, then calculate the TermScore of each term #POS in the processed review, and summed them up to get the sentiment score of the review.

3.3 Confidence Parameter

As we found that when SWN-based sentiment scoring process is applied to score the reviews, reviews with high absolute values are more likely being rightly classified than those reviews who have lower absolute values. Two confidence parameters: Positive Confidence Value (PCV) and Negative Confidence Value (NCV) are introduce to the proposed sentiment classifier. For a review, if its sentiment score is greater than PCV, classified it as positive. if its sentiment score is less than NCV, classified it as negative. In the training process, PCV and NCV could be assigned at where the proposed sentiment classifier got the highest accuracy.

3.4 Naive Bayes Sentiment Classifier

Naive Bayes classifier is based on the Bayes' rule:

$$\mathrm{P}(Y|X) = \frac{\mathrm{P}(Y)\mathrm{P}(X|Y)}{\mathrm{P}(X)} \tag{3}$$

where X play no role in selecting Y. In the review sentiment polarity classification problem, the Bayes' rule is as follows:

$$\mathrm{P}(C_k|x_1, x_2, x_3 \ldots, x_n) = \frac{\mathrm{P}(C_k)\mathrm{P}(x_1, x_2, x_3 \ldots, x_n|C_k)}{\mathrm{P}(x_1, x_2, x_3 \ldots, x_n)} \tag{4}$$

where C_k is the sentiment polarity category $\{C_1 = Positive, C_2 = Negative\}$, $x_1, x_2, x_3 \ldots, x_n$ is the feature vector. With the assumption that all features in $x_1, x_2, x_3 \ldots, x_n$ are mutually independent, conditional on the category C_k. Under this assumption, the conditional distribution over the class variable C_k is:

$$\mathrm{P}(C_k|x_1, x_2, x_3 \ldots, x_n) \propto \mathrm{P}(C_k) \prod_{i=1}^{n} \mathrm{P}(x_i|C_k) \tag{5}$$

Where $\propto$ denotes proportionality.

Then, we get the Bayes classifier, is the function that assigns a class label $\hat{y} = C_k$ for some k as follows:

$$\hat{y} = \underset{k \in \{1,2\}}{\arg\max}\, \mathrm{P}(C_k) \prod_{i=1}^{n} \mathrm{P}(x_i|C_k) \tag{6}$$

Terms with specific POS tags appear in both training dataset and the SWN are used as the features.

Algorithm 1: Feature Selection
Input: SWN, Training set T:
Output: Feature set F:
Method:

```
Init F= ∅
For each Synset#POS in SWN:
    For each Term in Synset:
        If Term#POS appear in T:
            Add Term#POS to F
        End If
    End For
End For
Return F
```

3.5 Hybrid Sentiment Analysis Framework

The overview of the hybrid supervised sentiment classification framework is shown in Fig. 1. For a review, text pre-processing is the first process which makes the review get the right format for further use. Then SWN-based sentiment score process is used to assign the review. If the review's sentiment score is greater than NCV and less than PCV, the trained NB sentiment classifier is used to determine which categories the review belongs to. Otherwise, if the review's sentiment score is greater than PCV, classify the review as positive. If the sentiment score is less than NCV, classify the review as negative. In the training process, PCV and NCV could be assigned at where the proposed sentiment classifier got the highest accuracy.

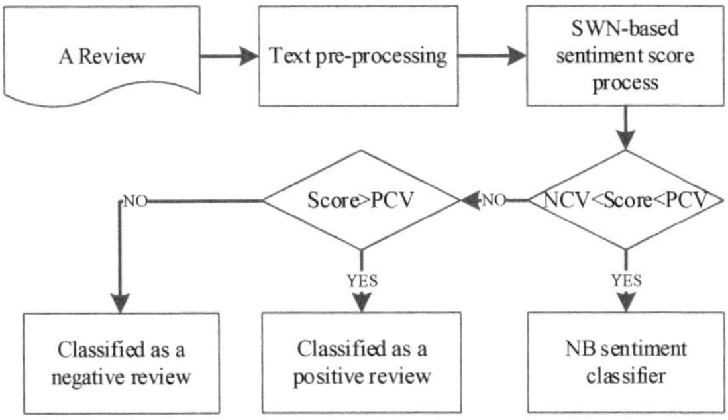

Fig. 1. Hybrid supervised sentiment classification framework.

4 Evaluation, Results, and Discussion

In this section, we first introduce details of the used experimental datasets, then experiments are carried out on these datasets compare with SWN-based, NB, and the proposed method.

Four domain review datasets (Books, DVD, Kitchen, and Electronics) in the Multi-Domain Sentiment Dataset (an open access dataset) which collected from Amazon.com by Blitzer et al. [20] are used to conduct the experiments. One thousand tagged positive and One thousand tagged negative reviews are included in each of the four datasets. Accuracy and F-measure are used to measure the experimental results.

$$Accuracy = \frac{TP + TN}{TP + FP + TN + FN} \tag{7}$$

$$Sensitivity = \frac{TP}{TP + FN} \tag{8}$$

$$Specificity = \frac{TN}{TN + FP} \tag{9}$$

$$F - measure = \frac{2 \times Sensitivity \times Specificity}{Sensitivity + Specificity} \quad (10)$$

where TP/TN is the amount of positive/negative reviews being truly classified, FP/FN is the amount of wrongly classified positive/negative reviews.

During the course of the experiment, first we random selection 500 positive and 500 negative reviews from each of the four datasets as the training set, the other reviews as the corresponding testing set. The training set is used to assign PCV and NCV, and also training a NB-based sentiment classifier, then we can get the sentiment classifier of the proposed method. The performance of SWN-based classifier, NB classifier and the proposed classifier are measured as show in Table 2, the experimental data in the table is the average of five experiments.

Table 2. Polarity classification performance comparison.

	SWN		NB		Hybrid (SWN + NB)	
	F-measure	Accuracy	F-measure	Accuracy	F-measure	Accuracy
Books	48.17%	61.40%	71.61%	72.70%	**74.12%**	**74.32%**
DVD	52.72%	63.60%	77.37%	77.43%	**78.24%**	**78.30%**
Electronics	58.75%	64.80%	76.72%	77.08%	**78.51%**	**78.87%**
Kitchen	59.57%	66.55%	79.03%	79.18%	**80.60%**	**80.66%**

The results indicate that, with the same experimental environment and datasets, compare to SWN-based method the proposed hybrid sentiment method presented overall superior performances on both of the four product review datasets, compare to NB sentiment classifier the proposed hybrid method also could get some improvement. We could see that the proposed hybrid method inherits the advantages of both SWN-based and NB method. The introduced confidence parameters make sure that the proposed hybrid method could get excellent performance. For reviews with sentiment scores greater than PCV or Less than NCV the SWN-based sentiment classifier used to implement the classification process, the phenomenon that review with greater sentiment score usually get a more accurate classification make sure the proposed hybrid method well. For the other reviews whose sentiment scores in range of (NCV, PCV) the NB sentiment classifier used to implement the classification process, this make sure the reviews could be classified properly.

Then we make use Electronics review dataset as the object of study to see what will happened if training dataset size changed. 900 positive and negative reviews are selected as the training set and the rest 100 positive and negative reviews are as the testing set. When the training set size need to change we secondly random select out corresponding number of samples from the above selected training set. Experimental results are measured as Fig. 2. The data used to mark the points in the figure is the average of five experiments.

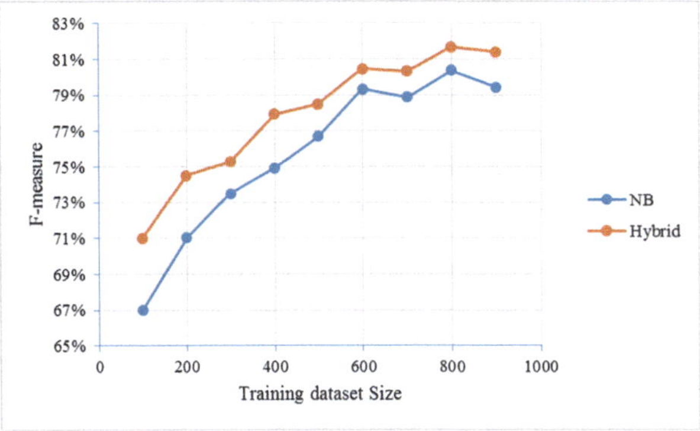

Fig. 2. F-measure of NB classifier and the proposed sentiment classifier following training dataset size changes.

The results indicate that, with the same experimental environment and datasets, as the training set size increases, performances of both NB sentiment classifier and the proposed hybrid sentiment classifier increases correspondingly and the proposed hybrid method perform well than the NB method.

5 Conclusions

In this paper we propose a hybrid supervised review sentiment analysis method which combines SWN-based method and NB method by introducing confidence parameters. Four domain review datasets in the Multi-Domain Sentiment Dataset are used to conduct the experiments. Compare to SWN-based method the proposed hybrid sentiment method presented overall superior performances on both of the four product review datasets, compare to NB sentiment classifier the proposed hybrid method also could get some improvement. Additional we examined the effectiveness of the proposed hybrid method by making use Electronics review dataset as the object of study to see what will happened if training dataset size changed. The proposed method outperforms the NB method along with dataset size changes.

Acknowledgments. This paper is supported by the National Natural Science Foundation of China (61672179, 61370083 and 61402126), the Natural Science Foundation of Heilongjiang Province (F2015030), the Youth Science Foundation of Heilongjiang Province of China (QC2016083) and Heilongjiang Postdoctoral Fund (LBH-Z14071, LBH-Z19015), the China Postdoctoral Science Foundation (2019M651262), and the Youth Fund Project of Humanities and Social Sciences Research of the Ministry of Education of China (20YJCZH172) and the Fundamental Research Funds for the Central Universities (3072020CF0605).

References

1. Yu, N.: Exploring co-training strategies for opinion detection. J. Assoc. Inf. Sci. Technol. **65**(10), 2098–2110 (2014)
2. Zhao, W., Guan, Z., Chen, L., He, X., Cai, D., Wang, B., Wang, Q.: Weakly-supervised deep embedding for product review sentiment analysis. IEEE Trans. Knowl. Data Eng. **30**(1), 185–197 (2018)
3. da Silva, N.F.F., Coletta, L.F.S., Hruschka, E.R., Hruschaka Jr, E.R.: Using unsupervised information to improve semi-supervised tweet sentiment classification. Inf. Sci. **355-356**, 348–365 (2016)
4. Abbasi, A., Chen, H., Thoms, S., Fu, T.: Affect analysis of web forums and blogs using correlation ensembles. IEEE Trans. Knowl. Data Eng. **20**(9), 1168–1180 (2008)
5. Park, S., Lee, W., Moon, I.C.: Efficient extraction of domain specific sentiment lexicon with active learning. Pattern Recogn. Lett. **56**, 38–44 (2015)
6. Onan, A., Korukoğlu, S., Bulut, H.: A multiobjective weighted voting ensemble classifier based on differential evolution algorithm for text sentiment classification. Expert Syst. Appl. **62**, 1–16 (2016)
7. Pang, B., Lee, L., Vaithyanathan, S.: Thumbs up?: sentiment classification using machine learning techniques. In: Proceedings of the 2002 Conference on Empirical Methods in Natural Language Processing (EMNLP 2002), pp. 79–86. ACM, New York City, USA (2002)
8. Moraes, R., Valiati, J.F., Neto, W.P.W.: Document-level sentiment classification: An empirical comparison between SVM and ANN. Expert Syst. Appl. **40**(2), 621–633 (2013)
9. Wang, G., Sun, J., Ma, J., Xu, K., Gu, J.: Sentiment classification: the contribution of ensemble learning. Decis. Support Syst. **57**, 77–93 (2014)
10. Zhang, L., Liu, B.: Sentiment Analysis and Opinion Mining. Springer, Boston, MA, USA (2017)
11. Vashishtha, S., Susan, S.: Fuzzy rule based unsupervised sentiment analysis from social media posts. Expert Syst. Appl. **138**, 112834 (2019)
12. Deng, S., Sinha, A.P., Zhao, H.: Adapting sentiment lexicons to domain-specific social media texts. Decis. Support Syst. **94**, 65–76 (2017)
13. Baccianella, S., Esuli, A., Sebastiani, F.: SentiWordNet 3.0: an enhanced lexical resource for sentiment analysis and opinion mining. In: Proceedings of the Seventh International Conference on Language Resources and Evaluation (LREC 2010). European Language Resources Association (ELRA), Paris, France (2010)
14. Pennebaker, J.W., Boyd, R.L., Jordan, K., Blackburn, K.: The Development and Psychometric Properties of Liwc2015. University of Texas at Austin, TX, USA (2015)
15. Wilson, T., Wiebe, J., Hoffmann, P.: Recognizing contextual polarity: an exploration of features for phrase-level sentiment analysis. Comput. Linguist. **35**(3), 399–433 (2009)
16. Saif, H., He, Y., Fernandez, M., Alani, H.: Contextual semantics for sentiment analysis of Twitter. Inf. Process. Manage. **52**(1), 5–19 (2016)
17. Hu, M., Liu, B.: Mining and summarizing customer reviews. In: Proceedings of the tenth ACM SIGKDD International Conference on Knowledge Discovery and Data Mining, pp. 168–177. ACM, New York City, USA (2004)
18. Taboada, M., Brooke, J., Tofiloski, M., Voll, K., Stede, M.: Lexicon-based methods for sentiment analysis. Computat. Linguistics **37**(2), 267–307 (2011)
19. Han, H., Zhang, J., Yang, J., Shen, Y., Zhang, Y.: Generate domain-specific sentiment lexicon for review sentiment analysis. Multi. Tools Appl. **77**(16), 21265–21280 (2018)
20. Blitzer, J., Dredze, M., Pereira, F.: Biographies, bollywood, boom-boxes and blenders: Domain adaptation for sentiment classification. In: Proceedings of the 45th Annual Meeting of the Association of Computational Linguistics, pp. 440–447. ACL, PA, USA (2007)

Optimizing Layout of Video Surveillance for Substation Monitoring

Yiqi Lu(✉), Yang Zhong, Chuye Hu, and Shaorong Wang

State Key Laboratory of Advanced Electromagnetic Engineering and Technology, Huazhong University of Science and Technology, Wuhan 430074, China
980828237@qq.com

Abstract. An automated video surveillance system plays a vital role in unattended substations. To minimize the camera cost and maximize the monitoring coverage of the substation, efficiently placing the camera is critical. Two optimization models are developed in this paper to satisfy two kinds of requirements of unattended substations monitoring in practice: minimizing cost with full coverage, optimizing coverage with a fixed budget. The second model is mainly proposed for the first time to get the minimum monitoring fitness function as the optimization goal. By introducing piecewise functions in the second model, it meets different monitor-ing requirements in different regions of substations and reduces the over-lapping monitoring area efficiently. The two models are tested using a genetic algorithm to get the optimal layout of the cameras. The validity and practicability of the two models of monitoring in unattended substations are verified by simulation. The results suggest that this model-based optimization approach to cameras layout can be used to improve the efficiency of substation video surveillance systems.

Keywords: Optimization model · Automated video surveillance · Genetic algorithm · Substation monitoring · Set covering problem

1 Introduction

With the continuous development of the smart grid business, substations in various countries are gradually achieving real unattended operation [1]. Automat-ed video surveillance system and the optimized layout of its camera terminals play an essential role in unattended substations and become the physical foundation of artificial intelligence in applying substation safety monitoring. There are many pieces of research on video surveillance and substation safety monitoring. Fan et al. [2] put forward a large network video monitoring platform by integrating various video systems and building the network video monitoring platform to a higher level. Li et al. [3] concentrate on the monitoring system getting video data through the State Grid Unified Video monitoring Platform (SG-UVP) and make a deep research on the video intelligence algorithm to identify and judge critical equipment and place status. Ye et al. [4] propose an air-ground integrated monitoring system based on robot inspection and video surveillance [4]. However, the references mentioned above have not taken the layout of cameras

Q. Liu et al. (Eds.): CENet 2020, AISC 1274, pp. 1147–1157, 2021.
https://doi.org/10.1007/978-981-15-8462-6_131

into account. Several references combine the layout of cameras and the monitoring coverage; for example, BAO uses location models to optimize watchtowers locations for forest fire monitoring [5]. Zhang et al. [6] use an optimization algorithm to establish a network of video surveillance to protect the Golden Camellia. However, these researches have not integrated the camera layout and the actual situation of the substation.

Therefore, this paper proposes a modeling approach to optimize the monitoring coverage of cameras in unattended substations. This approach integrates coverage models and the substation's actual situation into a spatial optimization frame-work and applies the result to unattended substations. The purpose of the approach is to develop a procedure for finding optimal solutions for locating cameras that satisfy the substation's specific requirements and constraints in practice. In the remainder of this paper, Sect. 2 analyzes technical problems associated with unattended substations monitoring; Sect. 3 presents the two proposed optimization models, Sect. 4 uses Matlab to get the results of the two models and Sect. 5 summarizes and concludes this paper.

2 Problem Analysis

There are two key issues involved in the layout of video surveillance system in unattended substations: the actual substation model and the camera view shed.

2.1 Rasterized Model of Substation

When it comes to the actual substation model, it is often constructed through 3D model. However, the 3D modeling process is very complicated. Considering the floor plan of the substation is generally rectangular, and the equipment areas like transformer are also mostly shaped in rectangular. In this paper, we simplify the research of the layout by rasterizing the substation floor plan [7], shown in Fig. 1. And define the jth small raster in the ith row as the substation raster V_{ij}, and define the intersection of each small raster diagonal as the center of the small raster, and record the center of the small area V_{ij} as $O_{V_{ij}}$. The darker areas mean substation electronic equipment like transformer, GIS, etc.

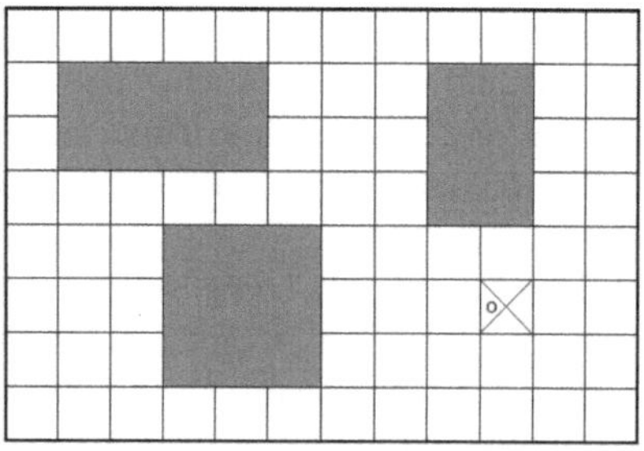

Fig. 1. Rasterized model of Substation

2.2 Camera View Shed Model

When considering the camera view shed, it is necessary to design reasonable camera model in order to describe the camera view shed in unattended video surveillance system. In some former research, 2-D Boolean model with direction optionally is usually used [8], shown in Fig. 2, which can be defined as a four parameter array (C, R, V(t), θ). Where C represents the location of the camera; R represents the sensing radius; V(t) is the sensing direction at time t, and θ is 360° represents the sensing angle of the camera. At time t, the sensing area is a sector, as the sensing direction changes with time, the camera can coverage a circle with radius R. If a target T can be covered by a camera, Eq. (1) should be satisfied,

$$d \leq R \ \beta \leq \frac{\theta}{2} \tag{1}$$

In this paper, we adopt ball cameras with different sensing radius. Then, the sensing angle of the ball camera θ. We develop the model of the ball camera view shed in the raster map—the sharpness of a camera view shed, shown in Fig. 3.

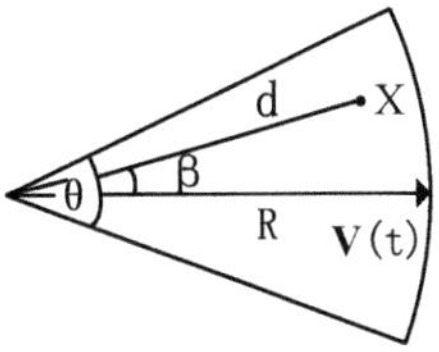

Fig. 2. 2D Boolean model.

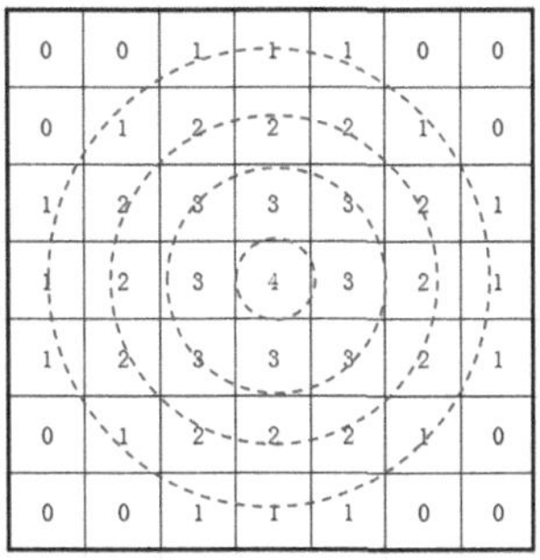

Fig. 3. Sharpness of a ball camera view shed.

We define the sharpness value of the ball camera view shed as s_{ij},in Eq. (2),

$$s_{ij} = \begin{cases} 4 & D \leq k_1 R \\ 3 & k_1 R < D \leq k_2 R \\ 2 & k_2 R < D \leq k_3 R \\ 1 & k_3 R < D \leq R \\ 0 & D > R \end{cases} \tag{2}$$

in which D means the distance between the center of target-located raster V_{ij} to the camera-located raster center; k means the coefficient of the sharpness and $0 < k_1 < k_2 < k_3 < 1$. The sharpness value of the ball camera view shed varies from 0-4. It can be seen that the s_{ij} is relevant to target position. When the target is closer to the camera within the sensing radius, the s_{ij} is higher (not high out of 4); when the target can't be seen from the ball camera (out of the sensing radius), the s_{ij} is 0. Then the sharpness matrix of a ball camera S is shown below:

$$S = \begin{bmatrix} S_{11} & \cdots & S_{1m} & \cdots & S_{1b} \\ \vdots & & \vdots & & \vdots \\ S_{i1} & \cdots & S_{im} & \cdots & S_{ib} \\ \vdots & & \vdots & & \vdots \\ S_{a1} & & S_{am} & & S_{ab} \end{bmatrix}$$

When two or more ball cameras are close to each other, some areas will enter the monitoring range of multiple cameras at the same time. At this time, it is assumed that the sharpness matrix of the ball camera view shed in overlapping areas can be linearly and discretely superimposed [7], shown in Eq. (3).It means there are total Q cameras monitoring the raster V_{ij}.Then the sharpness value of the raster V_{ij} is the sum of each camera's sharpness value contribution.

$$s_{ij} = \sum_{q=1}^{Q} s_{ij}^{q} \tag{3}$$

The two cameras are distributed at the center of two different rasters, and the sharpness matrix of the view shed is the sum of the sharpness matrix of the two cameras, and the method of obtaining the sharpness matrix of multiple cameras is similar.

3 Optimization Camera Location Models

Optimization camera location models must be used to help search for camera locations in order to satisfy a set of requirements regarding monitoring coverage and constraints. There are two typical optimization models will be discussed below.

3.1 Minimum Cost with Full Coverage

When cameras must monitor the entire substation, the location set covering problem or LSCP can be used to minimize the total monitoring equipment cost while guaranteeing that each raster of the substation is covered at least once. In the following equals, there are several indices and constants needed to be explained.

M is the total number of types of ball cameras; N is the total number of candidate locations for building ball cameras. C_m is the cost for building per m-type ball camera. $X_{m,n}$ is a binary decision variable. $S_{m,n}$ is the sharpness matrix of the m-type camera locating at the n-th candidate location.

$$Min\sum_{m=1}^{M}\sum_{n=1}^{N}C_mX_{m,n} \tag{4}$$

subject to

$$a_{ij}\geq 1,\ a_{ij}\in\left\{\sum_{m=1}^{M}\sum_{n=1}^{N}S_{m,n}X_{m,n}\right\} \tag{5}$$

$$X_{m,n}\in\{0,1\},\ m\in\phi_M \qquad n\in\phi_N \tag{6}$$

$$\sum_{m=1}^{M}X_{m,n}=1,\ n\in\phi_N \tag{7}$$

Equation (4) is the object function to find the minimum cost.

Equation (5) ensures that each raster of the substation can be covered in range of cameras at least once.

Equation (5) ensures that each raster of the substation can be covered in range of cameras at least once.

Equation (6) means when the m-type ball camera locating at the nth candidate location, $X_{m,n}=1$; otherwise, $X_{m,n}=0$.

Equation (7) makes sure that each candidate location can only be equipped with one type of ball camera.

3.2 Optimizing Coverage with a Fixed Budget

Usually, what we concerned mostly about substation video surveillance is the equipment failure monitoring like insulation failure [9]. Most failures happen in the electric equipment like transformers, Gis. The location of this focused equipment needed to be payed more attention [10]. Besides, full coverage means more cameras cost that usually can't be achieved. Therefore, this paper proposes an optimizing model in order to make the monitoring sharpness matrix of cameras as consistent as the expected monitoring needs of the substation.

First, we separate the substation into a × b rasters. According to the importance of different areas of the substation, different weight values are given to the rasters which can be defined as substation monitoring importance matrix (simplified as W), shown below. The weight value of each raster (W_{vij}) varies from 0-4. For those most important areas needed to be payed more attention, the weight value of the raster is 4, and so on.

$$W=\begin{bmatrix} W_{V11} & \cdots & W_{V1m} & \cdots & W_{V1b} \\ \vdots & & \vdots & & \vdots \\ W_{Vi1} & \cdots & W_{Vim} & \cdots & W_{Vib} \\ \vdots & & \vdots & & \vdots \\ W_{Va1} & & W_{Vam} & & W_{Vab} \end{bmatrix}$$

In the optimizing coverage model, we introduce monitoring fitness function (simplified as $f_M(d)$) as the optimization object function:

$$Min\ f_M(d) = Min\ \frac{1}{ab}\sum_{i=1}^{a}\sum_{j=1}^{b} d_{ij}^2 \tag{8}$$

in Eq. (8), we introduce piecewise function to define the variable d_{ij}:

$$d_{ij} = \begin{cases} P_1 & \text{when } (s_{ij} - w_{ij} < 0) \\ s_{ij} - w_{ij} s_{ij} - w_{ij} & \text{when } (0 \le s_{ij} - w_{ij} \le 3) \\ P_2 & \text{when } (s_{ij} - w_{ij} > 3) \end{cases} \tag{9}$$

P_1, P_2 are the positive constant. The value of these two constant are usually the larger positive constant.

In Eq. (9) $w_{ij} \in W_m$, W_m is the substation monitoring importance matrix and is an a × b matrix for the a × b rastered substation;

In Eq. (9) $s_{ij} \in S$, S is sum of the sharpness matrix of camera locating at the candidate location and S is an a × b matrix for the a × b rastered substation:

$$S = \sum_{m=1}^{M}\sum_{n=1}^{N} S_{m,n} X_{m,n} \tag{10}$$

In Eq. (10), M is the total number of types of ball cameras; N is the total number of candidate locations for building ball cameras; $S_{m,n}$ is the sharpness matrix of the m-type camera locating at the nth candidate location; $X_{m,n}$ is a binary decision variables.

Equation (9) means: if the sharpness matrix is lower than the substation monitoring importance matrix, the monitoring bestow P_1 value on the difference of the two matrix; if the sharpness matrix is higher over 3 than the substation monitoring importance matrix, it means there are areas over-monitored. Therefore, by the Eq. (9) piecewise function, it can guide the cameras layout as much as possible to satisfy the actual requirements of the substation.

The optimization object function Eq. (8) subject to

$$\sum_{m=1}^{M}\sum_{n=1}^{N} C_m X_{m,n} \le B \tag{11}$$

$$X_{m,n} \in \{0, 1\},\ m \in \phi_M \quad n \in \phi_N \tag{12}$$

$$\sum_{m=1}^{M} X_{m,n} = 1,\ n \in \phi_N \tag{13}$$

Equation (11) means the budget limitation. Equation (12) means when the m-type ball camera locating at the nth candidate location, $X_{m,n} = 1$; otherwise, $X_{m,n} = 0$. Equation (13) makes sure that each candidate location can only be equipped with one type of ball camera.

4 Computing Experiments

The first model (minimum cost with full coverage) is integer programs, the second model (optimizing coverage with a fixed budget) is a typical set coverage problem, and it is also a NP-hard mathematical problem of mixed integer programming, and the larger the area where the substation needs to be equipped with cameras, the larger the solution scale of the problem. Therefore, we use genetic algorithm toolbox in Matlab for solving optimization model in order to get the camera optimal layout.

First, we subdivide the study area into a map of 400 rasters in 20 columns and 20 rows; the solution was 5 m by 5 m. The candidate positions for the cameras are marked in the map, shown in Fig. 4. It is assumed that there 60 candidate camera locations. We arrange these 60 candidate locations by line.

Then, we choose three types of ball cameras as our candidate cameras. The parameter of each ball camera is shown in Table 1.

Table 1. Parameters of different ball cameras.

Type	Sensing radius/m	Cost/yuan per camera
1	15	2500
2	25	3200
3	35	3600

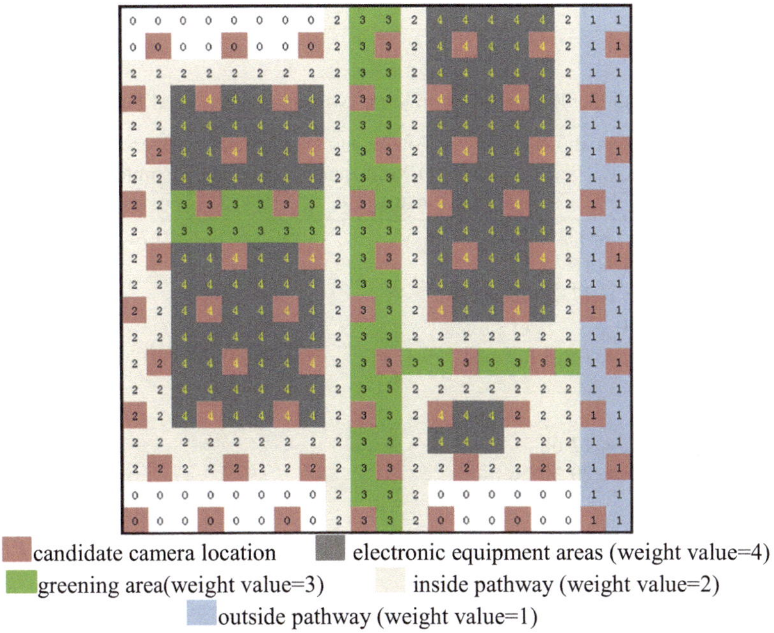

Fig. 4. Rasterized substation with different weight values.

4.1 Results of Minimum Costs with Full Coverage

The optimal location solution to the first model that optimizing coverage with a fixed budget is presented in Table 2. And the total cost to build these cameras is 85600 yuan. There are 43 primary covered rasters (the rasters that are only covered once), 156 rasters that are covered twice, and 201 rasters that are covered thrice. This means that more than 85% of the cells are covered by more than one camera and there is a significant overlap of coverage between the selected cameras.

Table 2. Results of minimum costs with full coverage.

Camera type	Optimal locations (Ci)	Total cost of optimal solution (Yuan)	Coverage (%)	Iterations
1	i = 1 2 3 6 7 8 10 11 12 14 23 28	85600	100	132
2	i = 24 32 35 51 58			
3	i = 18 20 27 30 37 41 46 49 54 60 64			

4.2 Results of Optimizing Coverage with a Fixed Budge

The optimal location solution to the second model that optimizing coverage with a fixed budget is presented in Table 3. In this model, we assume the total budget is 70000 Yuan.

Table 3. Results of optimizing coverage with a fixed budge.

P1	P2	Optimal locations (Type: Ci)	Total cost (10 thousand yuan)	Under-coverage (%)	Over-monitored coverage (%)	Itineration
10	20	Type 1: i = 1 2 7 42 61 70 Type2: i = 16 32 67 Type3: i = 17 28 30 34 38 40 46 60	5.34	23.62	10.74	123
	50	Type1: i = 1 3 5 6 17 34 38 45 48 70 Type2: i = 12 68 Type3: i = 20 30 36 64 60 66	5.3	25.11	7.11	115
	100	Type1: i = 1 2 4 5 6 8 66 70 Type2: i = 16 32 67 Type3: i = 36 43 47 50 53 57	4.92	28.29	1.38	131

(continued)

Table 3. (*continued*)

P1	P2	Optimal locations (Type: Ci)	Total cost (10 thousand yuan)	Under-coverage (%)	Over-monitored coverage (%)	Itineration
100	20	Type1: i = 1 3 7 20 22 24 27 50 52 54 70 Type2: i = 36 38 40 42 Type3: i = 17 30 34 40 46 64 66 68	6.91	6.71	18.62	123
	50	Type1: i = 1 3 6 33 37 42 47 52 56 70 Type2: i = 14 24 28 61 Type3: i = 20 35 40 54 60 66 68	6.3	8.33	18.24	127
	100	Type1: i = 1 2 3 4 5 6 8 41 46 54 67 70 Type2: i = 11 35 63 Type3: i = 15 21 25 38 45 47 50 56 66	6.66	10.66	16.21	131
1000	20	Type1: i = 1 3 5 7 20 22 24 28 34 38 45 48 70 Type2: i = 37 39 45 Type3: i = 15 30 34 38 40 46 60 64 66	5.95	0	23.71	131
	50	Type1: i = 1 3 5 6 12 17 21 27 Type2: i = 12 28 62 Type3: i = 20 30 36 38 40 50 54 56 60	5.6	0	23.54	127
	100	Type1: i = 1 2 3 4 5 6 8 66 70 Type2: i = 16 32 64 Type3: i = 19 22 27 36 43 47 50 53 58	6.45	0	22.89	131

There are two evaluation indexes introduced to evaluate the efficiency of the layout scheme: under-coverage (Uc) and over-monitored coverage (Oc).

$$U_C = \frac{n_c}{a \times b} \times 100\% \tag{14}$$

$$O_C = \frac{n_o}{a \times b} \times 100\% \tag{15}$$

In Eq. (14), n_c is the total number of rasters in which $s_{ij} - w_{ij} < 0$, $w_{ij} \in W_m$, W_m is the substation monitoring importance matrix and is an a × b matrix for the a × b rastered substation;$s_{ij} \in S$, S is sum of the sharpness matrix of camera locating at the candidate location and S is an a × b matrix for the a × b rastered substation. n_o is the total number of rasters in which $s_{ij} - w_{ij} > 3$.

In summary, according to the result, when the parameter (P1, P2) is (10, 100), the over-monitored coverage is the least; when P1 is larger enough, the under-monitored coverage can be reduced to 0. Compared with the first model, this model can adjust the over-monitored coverage by changing the value of P1, P2. The acquisition cost also meets the given constraints, which better reflects the monitoring multiplier function in eliminating the under-monitored areas and as far as possible to achieve the minimum expected monitoring requirements in two aspects, under the condition of limited costs, to achieve the efficient configuration of surveillance cameras and optimize the deployment.

5 Conclusion

This paper develops two optimization models to satisfy two kinds of requirements of unattended substations monitoring in practice: minimizing cost with full coverage, optimizing coverage with a fixed budget. By introducing piecewise functions in the second model, it meets different monitoring requirements in different regions of substations and reduces the overlapping monitoring areas efficiently. The two models are tested using genetic algorithm to get the optimal layout of the cameras. The layout results of actual examples show that the optimization models in this paper can effectively eliminate the over-monitoring areas after introducing the piecewise function as well as getting the largest coverage under different budget limitation, and maximize the realization of the differentiated optimal configuration of limited monitoring resources. It has high practical value in guiding the scientific deployment of smart terminals. At the same time, based on the optimized layout of cameras and the increasingly mature smart monitoring technology, it will further improve the real-time monitoring level of the substation's personal and equipment, thereby greatly reducing personal casualties and the risk of equipment failure. Therefore, the application of intelligent monitoring and intelligent analysis technology in intelligent substations will be the focus of the next stage of research.

References

1. Wang, J.B., Fu, H., Sun, Q., et al.: Research and application of intelligent operation and maintenance control system for substation auxiliary facilities. Bull. Sci. Technol. **36**, 70–73 (2020)
2. Yu, Y.K., Zhang, Z.P., Huang, D. et al.: Application research on management platform of thermal power unit video surveillance system. In: Jiangxi Society for Electrical Engineering Conference, Nanchang (2010)
3. Li, Z.H., Zhang, R.H., Zhang, T.B.: Research on intelligence substation routine inspection system based on video surveillance. Comput. Technol. Dev. **24**, 233–237 (2014)
4. Ye, Q., Cui, D.M., Huang, W., et al.: Research and practice on intelligent inspection technology of space-ground integration in substation. Electric Eng. **15**, 102–103 (2019)

5. Bao, S.T., Xiao, N.C., Lai, Z.H., et al.: Optimizing watchtower locations for forest fire monitoring using location models. Fire Saf. J. **71**, 100–109 (2015)
6. Zhang, K., Huang, Z., Zhang, S.L.: Using an optimization algorithm to establish a network of video surveillance for the protection of Golden Camellia. Ecol. Inf. **42**, 32–37 (2017)
7. Lin, X.B., Jiang, H.X., Hu, J.L., et al.: Layout optimization of video intelligent terminal for substation safety monitoring. Power Syst. Protection Control **47**, 146–151 (2019)
8. Sun, Q., Xu, D., Wu, Z., et al.: Coverage performance analysis of multi-camera networks based on observing reliability model. Optik **125**, 2220–2224 (2014)
9. Zhang, T.H.: Common problems and preventive measures in the operation of 35 kV substation. Modern Chem. Res. **16**, 123–124 (2019)
10. Lu, Y.B., Li, Y.: Application analysis of condition based maintenance of primary electrical equipment. Telecom Power Technol. **37**, 149–150 (2020)

Multi-objective Load Dispatch of Microgrid Based on Electric Vehicle

Zeyu Wang, Zhangyu Lu, Chongzhuo Tan, and Xizheng Zhang(✉)

Hunan Institute of Engineering, Xiangtan 411104, China
z_x_z2000@163.com

Abstract. With the development of electric vehicles (EVs), the random access of large-scale EVs to the grid will become a reality in the future, which will have a significant impact on the security and stability of the power system operation. Therefore, this paper proposes an improved multi-objective particle swarm optimization algorithm to optimize the micro grid load scheduling model based on EV. Microgrid is an autonomous system that can realize self-protection, management and control. It is an effective way to realize active distribution network, which is the transition from traditional power grid to smart grid. At present, micro-grid has become a useful supplement of power grid with its flexible power supply and diversified power generation methods. The model includes operation cost, pollutant treatment cost and load variance. The distributed generation (DG) including diesel engine (DE), micro turbine (MT), photovoltaic (PV) array and wind turbine (WT) is studied in the model. The uncertainty of electric vehicle is modeled by Monte Carlo simulation. The simulation results show the effectiveness of the algorithm.

Keywords: EV · Microgrid · Load dispatch · Monte carlo simulation

1 Introduction

Random access of large-scale electric vehicles to the grid will increase the burden of peak load on power grid, increase the peak valley difference of the load, and affect the normal operation of the grid. But EVs have played a significant role in alleviating environmental pollution and energy shortages.

Scholars use various methods to optimize the load dispatch model of power system. Lu et al. [1] proposed an improved particle swarm optimization algorithm for load dispatch of microgrid model based on EV. Genetic algorithm is used by Zhou et al. to optimize the model in [2]. Xu [3] proposed the binary universal gravitation search algorithm to solve the unit combination optimization of micro power supply in micro grid. In this paper, an improved particle swarm optimization algorithm is used to optimize the model. At the same time, by comparing with other algorithms to highlight the performance of the algorithm, the results show the superiority of the algorithm.

Q. Liu et al. (Eds.): CENet 2020, AISC 1274, pp. 1158–1163, 2021.
https://doi.org/10.1007/978-981-15-8462-6_132

2 Distributed Generation and EV Modeling

2.1 Generation Characteristics of DGs

The Model of DE. The fuel cost [3] function of DE is defined as follows.

$$C_{DE} = a + bP_{DE} + cP_{DE}^2 \tag{1}$$

C_{DE} is the fuel cost of DE (¥), P_{DE} is the output power of DE (kW), a, b, c are coefficients of DE.

The Model of MT. The MT fuel cost [3] function is defined as follows.

$$C_{MT} = \frac{C_{GAS} \times P_{MT}}{LHV \times \eta_{MT}} \tag{2}$$

C_{MT} is the fuel cost of MT(¥), C_{GA} is the price of natural gas (¥(Chinese Yuan)/m^3), P_{MT} is the output power of MT(kW), LHV is the low calorific value of natural gas (kWh/m^3), η_{MT} is the generation efficiency of MT, expression is as follows.

$$\eta_{MT} = x\left(\frac{P_{MT}}{P_R}\right)^3 + y\left(\frac{P_{MT}}{P_R}\right)^2 + z\left(\frac{P_{MT}}{P_R}\right) + d \tag{3}$$

x, y, z, d are MT coefficients, P_R is the rated power of MT (kW).

The Model of PV. The output power model of PV [3] is expressed as.

$$P_{PV} = P_{STC}\frac{G_{ING}}{G_{STC}}[1 + k(T_C - T_{STC})] \tag{4}$$

P_{PV} is the output power of PV(kW), P_{STC} is the rated power of PV cells under STC (kW), G_{ING} is the actual light intensity of PV cells (Lux), G_{STC} is the light intensity of PV cells under STC(Lux), k is the temperature coefficient of PV, T_c is the actual temperature of PV cells(K), T_{STC} is the temperature of PV cells under STC(K).

The Model of WT. The output power model of WT [3] is expressed as.

$$P_{WT} = \begin{cases} 0 & V \le V_{ci} \\ \alpha \times V^3 - P_r & V_{ci} < V < V_r \\ P_r & V_r < V \le V_{co} \\ 0 & V > V_{co} \end{cases} \tag{5}$$

P_{WT} is the output power of WT(kW), V is the actual wind speed (m/s), V_{ci} is the cut-in wind speed (m/s), V_r is the rated wind speed(m/s), V_{co} is the cut-out wind speed (m/s), α, β are MT coefficients.

2.2 EV Charging Model

Starting Charging Time of EVs. The starting charging time model of EV [4] is defined as follows.

$$f_t(x) = \begin{cases} \frac{1}{\sigma_t\sqrt{2\pi}}\exp\left(-\frac{(x+24-\mu_t)^2}{2\sigma_t^2}\right) & 0<x\leq\mu_t-12 \\ \frac{1}{\sigma_t\sqrt{2\pi}}\exp\left(-\frac{(x-\mu_t)^2}{2\sigma_t^2}\right) & \mu_t-12<x\leq 24 \end{cases} \tag{6}$$

Driving Distance of EVs. The probability density function [4] of the driving distance by an electric vehicle is defined as follows.

$$f_s(x) = \frac{1}{\sigma_s x\sqrt{2\pi}}\exp\left(-\frac{(\ln\ x-\mu_s)^2}{2\sigma_s^2}\right) \tag{7}$$

Charging Load of EVs. The starting charging time and charging load of EV are independent of each other, so the charging load can be calculated by random experiment. In this paper, Monte Carlo simulation method is used to calculate the total charging load of electric vehicles.

3 Optimization Model

3.1 Operating Cost of the Microgrid

The operating cost of this system is defined as follows.

$$C_1 = \sum_{i=1}^{N}\sum_{t=1}^{T}\left[F_i(P_{i,t}) + OM_i(P_{i,t})\right] + C_{BAT} + C_{GRD} \tag{8}$$

C_1 is the operation cost of microgrid (¥), N is the total number of DG in microgrid, T is the total amount of periods in scheduling cycle, $F_{i(}P_{i,t)}$ is the fuel cost of the ith DG in period t (¥), $OM_i(P_{i,t})$ is the maintenance cost of the ith DG in period t (¥), C_{BAT} is the depreciation cost of EV batteries (¥), C_{GRD} is the transaction cost between the microgrid and the grid (¥).

3.2 Pollutant Treatment Cost of the Microgrid

Pollutants are generated in the process of DG and the grid. The treatment cost is defined as follows.

$$C_2 = \sum_{t=1}^{T}\sum_{i=1}^{N}\sum_{h=1}^{H}(C_h u_{ih})P_{i,t} + \sum_{t=1}^{T}\sum_{h=1}^{H}\left(C_h u_{grid,h}\right)P_{grid,t} \tag{9}$$

H is the total amount of the pollutant emissions, h is the pollutant emission, C_h is the treatment cost of the hth pollutant (¥/kg), $u_{i,h}$ is the emission coefficient of hth pollutant of ith DG, u_{gridh} is the emission coefficient of hth pollutant of grid, $P_{grid,t}$ is the transmission power between microgrid and the main power grid in period t (kW).

3.3 Load Variance of Grid

Load variance [3] is used to reflect the variance of peak load and valley load in this paper. Its definition is as follows.

$$F = \frac{1}{T}\sum_{t=1}^{T}\left(P_{load,t} + P_{EV,t} - \sum_{i=1}^{N}P_{i,t} - P_{av}\right)^2 \tag{10}$$

F is the load variance of the grid, $P_{load,t}$ is the original load without charging load in period t (kW), $P_{EV,t}$ is the total charging/discharging power of all EVs in period t (kW), P_{av} daily average load (kW).

3.4 Objective Function

In this paper, the operation cost, pollutant treatment cost and load variance need to be considered, in which the operation cost and pollutant treatment cost are related to the cost of microgrid system, so they can be consolidated into one objective. The function of microgrid load optimal dispatch model is defined as follows.

$$\min C = \lambda(C_1 + C_2) + (1-\lambda)F \tag{11}$$

4 Methodology

The PSO algorithm is improved in the selection of inertia weight factor parameters in this paper. The modification strategy [5] is defined as follows.

$$\omega(t) = \begin{cases} 0.9 & t \le \alpha T \\ \dfrac{1}{1+e^{(10t-2T)}} + 0.4 & t > \alpha T \end{cases} \tag{12}$$

t is the current number of iterations in Eq. (12), T is the total number of iterations, α is a predefined constant.

5 Comparative Analysis of the PSO Algorithm

In this section, standard particle swarm optimization (PSO), particle swarm optimization with shrinkage factor (SPSO) [6], chaos Particle Swarm Optimization (CPSO) [7] compared with the improved particle swarm optimization in performance. In the improved PSO algorithm, ω_{max} is 0.9, ω_{min} is 0.6; $c_1 = c_2 = 2$. In standard particle swarm optimization $\omega = 1$, and $c_1 = c_2 = 2$. The parameter of HPSO and CPSO is the recommended value in the corresponding literature. The maximum number of iterations of the four algorithms is 1000. Both particle population size is 600, and the particle dimension is 144. The performance results of these four particle swarm optimization algorithms are compared as shown in Table 1.

Table 1. Comparison results of four different PSO.

	Standard PSO	SPSO	CPSO	Improved PSO
Running times	20	20	20	20
Computation time (s)	42.46	42.38	74.02	44.02
Optimal value	265230.57	240693.85	226956.50	206539.68
Average value	399988.51	323524.86	337786.80	297476.99

The optimal value and average value refer to the minimum value and average value obtained by the four algorithms running independently 20 times in Table 1. The computation time CPSO of four algorithms is larger than that of the other three algorithms. The computation time of standard PSO, HPSO and improved PSO is approximate. Comparing the optimal and average values of the four algorithms. It is found that the values of HPSO, CPSO and improved PSO are smaller than those of standard PSO. It shows that the performance of these three algorithms is better than that of standard PSO.

It is found that the improved PSO algorithm is less than HPSO and CPSO by comparing HPSO, CPSO and improved PSO in the optimal value and average value, which shows that the performance of the improved PSO algorithm is better than the two algorithms.

The optimal value curves of the four algorithms are shown in Fig. 1. As shown in Fig. 1, the fastest convergence is the standard PSO, the optimal values of the other three algorithms are smaller than the standard PSO, and the smallest is the improved PSO. Considering the advantages of the improved PSO, the improved PSO is chosen as the scheduling algorithm.

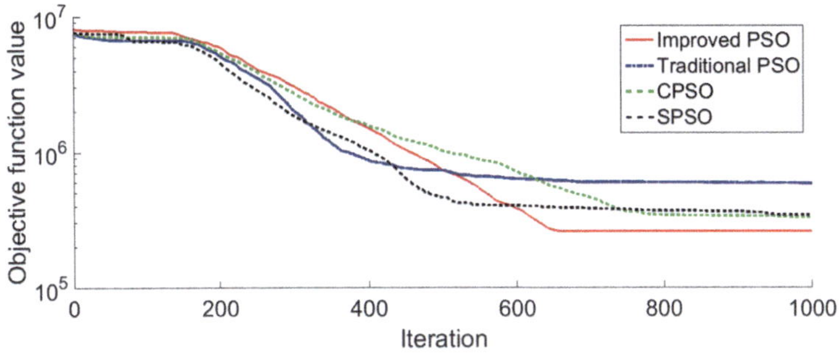

Fig. 1. Convergence curves based on the algorithms.

6 Conclusions

The development of EVs is conducive to reducing the generation of environmental pollutants and alleviating energy shortage. At the same time, large-scale electric vehicles connected to the power grid will lead to the increase of load and affect the stability of the power grid. The economic dispatch of general power grid refers to the lowest operation cost and the least emission. The stability and safety of power grid are also emphasized with the development of society. Therefore, load variance is added to the dispatch model to represent the fluctuation level of power grid, so as to determine the stability of power grid. Simulation results show the effectiveness of the improved PSO.

Acknowledgment. This work was supported by the National Natural Science Foundation of China (61673164), the Natural Science Foundation of Hunan Province (2020JJ6024) and the Scientific Research Fund of Hunan Provincal Education Department (17A048, 19K025).

References

1. Lu, X.H., Zhou, K.L., Yang, S.L., Liu, H.Z.: Multi-objective optimal load dispatch of microgrid with stochastic access of electric vehicles. J. Cleaner Prod. **195**, 187–199 (2018)
2. Zhou, K.L., Shen, C., Ding, S., Yang, S.L.: Optimal load distribution of microgrid based on genetic algorithm. China Manage. Sci. **22**(03), 68–73 (2014)
3. Xu, W.N.: Research on operation optimization method of microgrid. North China Electric Power University (2014)
4. Huang, Y.N., Guo, C.X., Wang, L.C., Baoi, Y.K., Dai, S., Ding, Q.: Group scheduling strategy of electric vehicles considering user satisfaction. Power Syst. Autom. **39**(17), 183–191 (2015)
5. Tian, D.P., Shi, Z.Z., Shi, Z.Z.: MPSO: Modified particle swarm optimization and its applications. Swarm Evol. Comput. **41**, 49–68 (2018)
6. Eberhart, R.C.: Comparing inertia weights and constriction factors in particle swarm optimization. In: Proceedings of the 2000 IEEE Congress on Evolutionary Computation, La Jolla, CA. IEEE, 2000
7. Gao, Y., Xie, S.L.: Chaos particle swarm optimization algorithm. Comput. Sci. **08**, 13–15 (2004)

Local Path Planning Based on an Improved Dynamic Window Approach in ROS

Desheng Feng[1], Lixia Deng[1(✉)], Tao Sun[2], Haiying Liu[1], Hui Zhang[1], and Yang Zhao[1]

[1] School of Electrical Engineering and Automation, Qilu University of Technology (Shandong Academy of Science), Jinan 250353, China
AmandaDeng084@126.com
[2] Academy of Advanced Interdisciplinary Studies, Qilu University of Technology (Shandong Academy of Science), Jinan 250353, China
tutorsun@163.com

Abstract. We consider the problem of robot local path planning using traditional dynamic window approach based ROS. By means of an improved dynamic window approach, we are able to reduce the complexity of the problem and provide a practically efficient procedure for its solution. Improved dynamic window approach based on robot dynamics model is introduced and described in the paper. By abandoning the trajectory that the velocity reduces to zero when it encounters an obstacle to simplify the search speed space. This algorithm improves the real-time performance and reduces the computational complexity of the algorithm.

Keywords: Local path planning · Dynamic window approach · Robot dynamic model · ROS

1 Introduction

The basic path planning problem also known as the find-path problem, can be described as following [1]: suppose that robot is free to move in a three dimensional space amidst a collection of obstacles whose geometry is known to the robot; given an initial placement *S* (start position) and a desired target placement (goal position) of robot, determines an collision-free path of robot from *S* to *G*.

Several methods for solving the basic path planning have developed. As a heuristic methods, A* algorithm uses a best-first search and finds a least-cost path from a given initial node to one goal node (out of one or more possible goals). However, the heuristic function cannot be always right to make search process in an endless loop. Dijkstra algorithm [2] has the defects of high searching time, without consider the target information in global space.

Many intelligent algorithms have been used to avoid obstacles such as fuzzy logic given primarily by Zadeh [3] in 1965, Genetic Algorithm discovered by Bremerman [4] and Neural network used to a wheeled mobile robot navigation in a partially unknown environment [5]. Dynamic window approach proposed by Dieter Fox [6], it differs from previous approaches in that the search for commands controlling the translational

Q. Liu et al. (Eds.): CENet 2020, AISC 1274, pp. 1164–1171, 2021.
https://doi.org/10.1007/978-981-15-8462-6_133

and rotational velocity of the robot is carried out directly in the space of velocities, which greatly improved the real-time performance [6].

In ordinary dynamic window approach, there is a set of speed that can stop before reaching the obstacles in the multiple group constraint speed sampling space. These speeds are not required in situations where safety requirements are not high. But it increased the memory load of navigation system and reduced the real-time performance of navigation. This paper improved dynamic window approach, the calculation of the minimum distance to the obstacle and the breaking distance is abandoned. If there is an obstacle on a trajectory, the track is be abandoned. In extensive experiments, the approach has been found to control our mobile robot model safely based ROS in static and dynamic environment.

2 Motion Equations for a Synchro-Drive Robot

This section describes the fundamental motions equation for synchro-drive mobile robot. The derivation begins with the correct dynamic laws, assuming that the robot's translational and rotational velocity can be controlled independently (with limited torques). Each group of speed ($v_{t,}$ ω_t) in velocity space produces different tracks in a certain period of time and simulates the feasible trajectories in time t. The variations of linear velocity v_t and angular velocity ω_t in the velocity space represents the motion state of the mobile robot. Among all feasible trajectories, the optimal trajectory is obtained by the evaluation function. Therefore, assumes that the mobile robot moves within the interval Δt, the kinematics model of robot:

$$x = x + v_x \Delta t \cos \theta_t - v_y \Delta t \sin \theta_t \tag{1}$$

$$y = y + v_x \Delta t \sin\theta_t - v_y \Delta t \cos\theta \tag{2}$$

$$\theta_t = \theta_t + \omega_t \Delta t \tag{3}$$

3 The Improved Dynamic Window Approach

In the ordinary dynamic window approach the search for commands controlling the robot is carried out directly in the space of velocities. The dynamic of the robot is incorporate into the method by reducing the search space to those velocities which are reachable under the dynamic constraints. In addition to this restriction only velocities are considered which are safe with respect to the obstacles. This pruning of the search space is done in the first step of the algorithm. In the second step the velocity maximizing the objective function is chosen from the remaining velocities [6].

3.1 Search Space

There are infinitely many sets of velocities (vt, ω_t) in the velocity space, and in the actual process, it is necessary to restrict the sampling speed range according to the constraints of the mobile robot and the environment.

Speed constraint of mobile robot:

$$V_{\mathrm{m}} = \{(v, \omega) \, | v \in [v_{min}, v_{max}], \omega \in [\omega_{min}, \omega_{max}]\} \tag{4}$$

Within the time interval of the mobile robot moving in the dynamic window, the speed constraint brought by acceleration and deceleration constraints of the motor is

$$\begin{aligned} V_d = \{(v, \omega) | v \in [v_c - \dot{v}_b \Delta t, v_c + \dot{v}_a \Delta t], \\ \omega \in [\omega_c - \dot{\omega}_b \Delta t, \omega_c + \dot{\omega}_a \Delta t]\} \end{aligned} \tag{5}$$

where v_{c}, ω_{c} represent current speed of mobile robot, $\dot{v}_a, \dot{\omega}_a$ represent the maximum acceleration of the robot, $\dot{v}_b, \dot{\omega}_b$ represents the maximum deceleration of the robot.

Braking distance constraint of mobile robot: when avoiding obstacles in a local environment, the safety of mobile robot should be ensured and under the constraint of the maximum deceleration, the speed can be reduced to 0 m/s before impacts the obstacles, and the braking constraints are shown as following:

$$\begin{aligned} V_d = \Big\{(v, \omega) | v \leq (2d(v, \omega)\dot{v}_b)^{1/2}, \\ \omega \leq (2d(v, \omega)\dot{\omega}_b)^{1/2}\Big\} \end{aligned} \tag{6}$$

where d (v, ω) represents the nearest distance between trajectory (v, ω) and obstacle.

In this paper, the trajectories with obstacle will be abandoned in search space V_{d}. Therefore, there is no need to calculate the minimum distance to the obstacle and the braking distance, so it will improve efficiency of algorithm and reduce computational complexity of system. A large number of simulation experiments show that the improved algorithm is feasible for the environment fully or partially known.

3.2 Objective Function

After having determined the resulting search space V_{d}, a velocity is selected from V_{d}. In order to incorporate the criteria target heading, clearance, and velocity, the maximum of the object function:

$$G(v, \omega) = \sigma(\alpha \cdot angle(v, \omega) + \beta \cdot dist(v, \omega) + \gamma velocity(v, \omega)) \tag{7}$$

Equation (7) is computed over V_{d}. This is done by discretization of the resulting search space, where the target heading *angle* (v, ω) measures the alignment of the robot with target direction, the function *dist* (v, ω) represents the distance to the closest obstacle that intersects with the curvature. If no obstacle is on the curvature this value is set to a large constant, the function *velocity* (v, ω) is used to evaluate the progress of the robot on the corresponding trajectory. It is simply a projection on the translational velocity v.

3.3 Smoothing

All three components of the objective function are normalized to [0, 1].

$$normal\_angle(i) = \frac{angle(i)}{\sum_{i=1}^{n} angle(i)} \tag{8}$$

$$normal\_dist(i) = \frac{dist(i)}{\sum_{i=1}^{n} dist(i)} \tag{9}$$

$$normal\_velocity(i) = \frac{velocity(i)}{\sum_{i=1}^{n} velocity(i)} \tag{10}$$

where n is all the sampling trajectories, i is current trajectory will be evaluated.

4 Implementation and Experimental Results

4.1 Robot System Operation

ROS is an open-source system for robot maintained by worldwide users. It provides rich simulation softwares and a good development environment, supports for multiple programming languages such as C++ and Python [7]. The real physics engine Gazebo can guide into the robot models of user designs and three-dimensional environmental map of physic model, by setting physical parameters to simulate the real world. The publisher of robot (Robotstate Publisher) as a publish-node to release the wheel speed of robot (Odometry), inertial navigation information (IMU), the transfer matrix of each movement joint Tree (Tf Tree) and sweep the point cloud information (Scan) of the tracer. Finally it receives control commands from other nodes to control the movement of robot in the simulation environment.

This experiment also uses ROS as an operating environment for simulation. ROS also provides a graphic parameter display software Rviz. It is the same with the node under the environment of ROS to send and receive messages. Rviz can also show the laser point cloud, map, virtual layer, specified trajectory camera image and so on. And it provides a good interactive structure, which provides various perspectives of observation work. It can also complete the work target that be specified in the interface.

4.2 Simulation Platform

In this paper, the physical model built based the Chinese Academy of Sciences, Institute of Software and it is used as the simulation platform shown in Fig. 1. The A* algorithm is used as the search method of global path, and the Improved Dynamic Window Approach is used as the method of generating a local path. The performance of the algorithm is tested under the following two conditions: 1) The environment maps is completely known. 2) The environment maps partially known (in the process of robot moving, a cube obstacle is artificially added to test the performance of the local path planning algorithm shown in Fig. 2). And completing the following steps:

1) Running the 3D simulation environment in Gazebo.
2) Creating a 3D environment maps of simulation environment in Rviz.
3) Setting goal point of navigation in Rviz.
4) Verifies improved algorithms in two conditions.

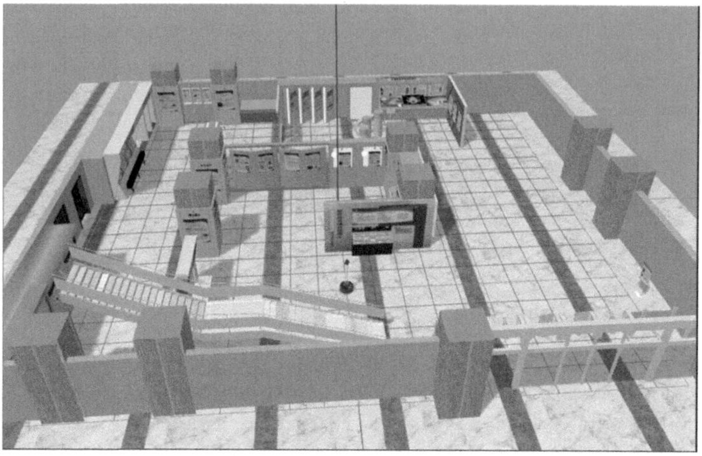

Fig. 1. Physical model of environment.

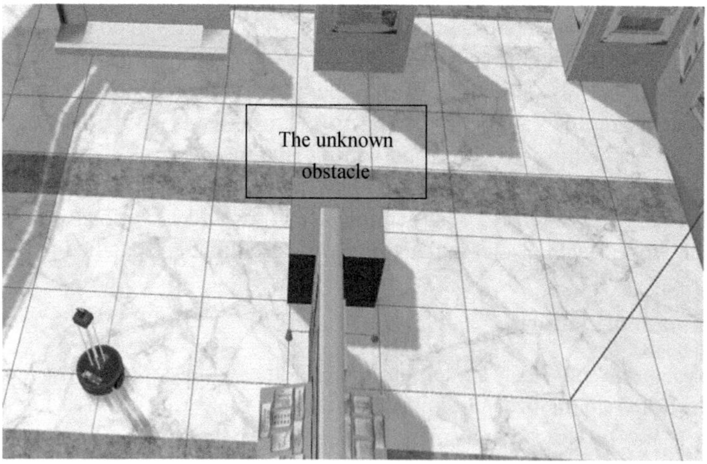

Fig. 2. Simulation environment with a cube obstacle.

4.3 Experimental Results

Based on the Improved Dynamic Window approach to collision avoidance, robot has been operated safely in various environment. It performs well in both fully known and partially known environment maps.

Fully Known Environment. Firstly, creating 3D environmental map and using the navigation system to test the obstacle avoidance performance under static obstacles. After the target pose is specified in the Rviz interface in Fig. 3, the navigation program starts to run and the robot starts to carry out the navigation task, in Fig. 4 can be seen that when there is a obstacle between the target pose and the starting point, the navigation system can successfully calculate the detour point, plan the feasible trajectory and reach the target point stably in Fig. 5.

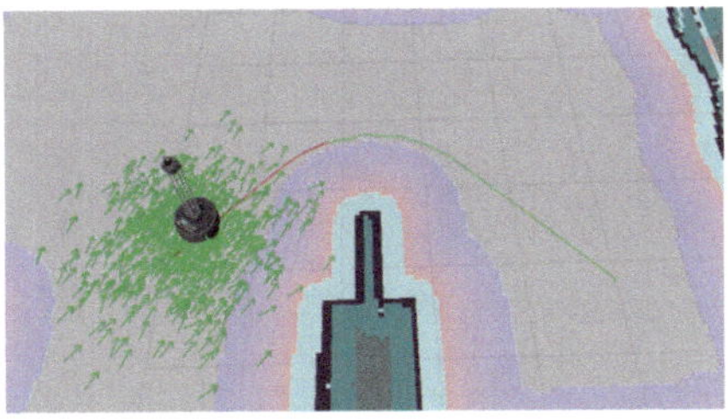

Fig. 3. Specified target pose.

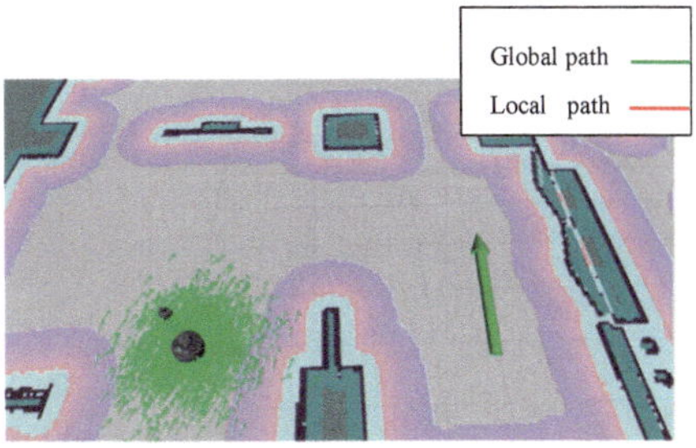

Fig. 4. Planned path of static environment.

Fig. 5. Completed target point.

Partially Known Environment. Adding a cube obstacle to the generated global path to test the dynamic performance of algorithm as the Fig. 2. Running the traditional Dynamic Window Approach and the improved algorithm respectively, and giving values to α, β, γ of 0.2, 0.2 and 2.0. After setting the target point on the right side of the obstacle, it can be seen from the Fig. 6, the velocity space contains trajectories that can reduce the velocity to zero. When robot reached the cube obstacle, the navigation system replanned the paths. And in Fig. 7, the obstacle can be well recognized and the robot planned a feasible trajectory of collision-free. The generated velocity space does not contain trajectories that make the velocity to zero when reaching the dynamic obstacle. In Fig. 8 the robot rounded the obstacle directly to reach the target point.

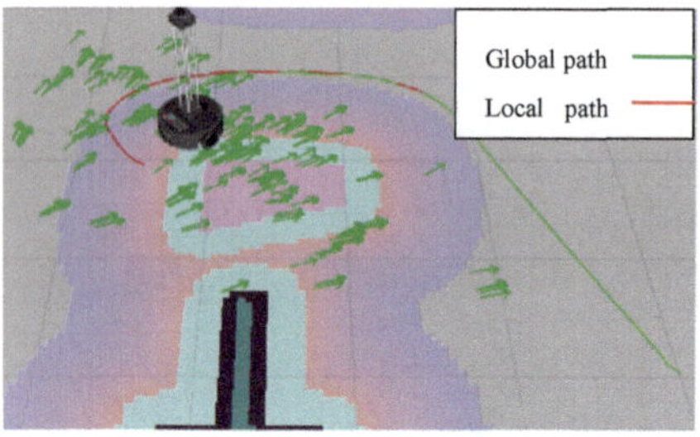

Fig. 6. Paths using traditional algorithm.

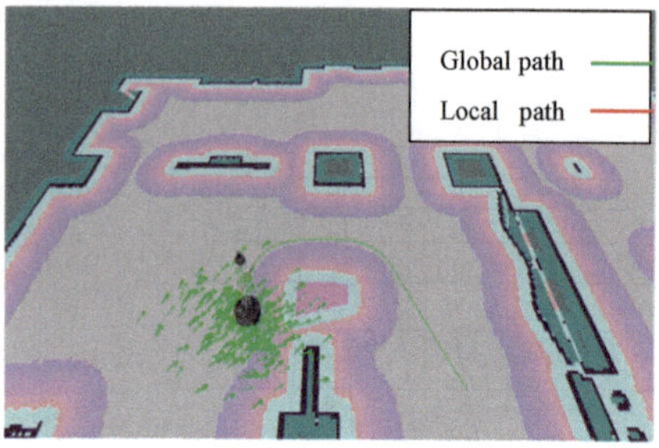

Fig. 7. Paths using improved algorithm.

Fig. 8. Completed goal point with cube obstacle.

The performance of the two algorithms is compared in Table 1. The data in the Table 1 shows that the improved algorithm improves the speed of path searching.

Table 1. Algorithm performance comparing.

Algorithm	Time(s)
Traditional DWA	23
Improved DWA	15

5 Conclusion

In order to improve the real-time performance of the algorithm, an improved algorithm based on dynamic window is proposed in this paper. By abandoning the trajectory that the velocity reduces to zero when it encounters an obstacle to simplify the search speed space. This algorithm improves the real-time performance and reduces the computational complexity of the algorithm. The effectiveness of the improved algorithm is proved by experimental comparison.

Acknowledgement. This work has been supported by the Key Research and Development Program of Shandong Province (2019GGX104079), Natural Science Foundation of Shandong Province (ZR2018QF005), Qilu University of Technology (Shandong Academy of Science) Special Fund Program for International Cooperative Research (QLUTGJHZ2018019), Key Research and Development Program of Shandong Province (2019GGX104091), Natural Science Foundation of Shandong Province (ZR2018LF011).

References

1. Zhu, D., Latombe, J.C.: New heuristic algorithms for efficient hierarchical path planning. IEEE Trans. Robot Autom. **7**(1), 89 (1991)
2. Kadali, R., Huang, B., Rossiter, A.: A data driven subspace approach to predictive controller design. Control Eng. Pract. **11**(3), 261 (2003)
3. Zadeh, L.A.: Fuzzy sets, Fuzzy Logic, & Fuzzy Systems. World Scientific Publishing Co. Inc., Hackensack (1996)
4. Bremermann, H.J.: The evolution of intelligence. The Nervous system as a model of its environment. Technical Report No.1, Contract No. 477(17). Washington, Seattle: Department Mathematics, University (1958)
5. Janglova, D.: Neural networks in mobile robot motion. Int. J. Adv. Rob. Syst. **1**(1), 15–22 (2008)
6. Fox, D., Burgard, W., Thrun, S.: The dynamic window approach to collision avoidance. IEEE Robot. Autom. Mag. **4**(1), 23–33 (2002)
7. Quigley, M., Gerkey, B., Smart, W.D.: Programming robots with ROS. O'Reilly Media, Inc, 1005 Gravenstein Highway North, Sebastopol, CA 95472

An Effective Mobile Charging Approach for Wireless Sensor and Actuator Networks with Mobile Actuators

Xiaoyuan Zhang(✉), Yanglong Guo, Hongrui Yu, and Tao Chen

School of Computer and Information Engineering, Tianjin Normal University, Tianjin, China
zh_xiao_yuan@126.com

Abstract. In wireless sensor and actuator networks (WSANs), actuator nodes require much longer charging time to be fully re-charged compared with sensor nodes. Also, actuators are usually mobile in the network. These features bring new challenges to the charging problem in WSANs. Based on the characteristics of WSANs, a practical mobile charging approach (EMC) for WSANs is proposed. In order to ensure the actuator nodes do not fail and meanwhile minimize the number of failed sensor nodes, the next charging candidate is selected according to both the remaining energy situation of the node and current location. Simulation results show that this proposed approach can guarantee the survival of all actuators. Further, it can effectively reduce the node failure ratio of sensor nodes and achieve the trade-off be-tween the charging delay and the charging cost.

Keywords: WSANs · Mobile actuator · Node failure · Average charging delay

1 Introduction

Wireless Sensor and Actuator Networks (WSANs) is a derivative of Wireless Sensor Networks (WSNs) [1]. Besides sensors, a few actuator nodes with high energy, reliable computing, and communication capabilities join the network. The actuator nodes are organized into an internal network, in which actuators can cooperate and communicate with one another. Moreover, each actuator covers a specific field and is responsible for events from this area in time. Usually, the sensor node transmits the sensing event to the adjacent actuator node. The actuator node processes the event preliminarily and then sends the processed data to the base station (BS) through the internal network composed of the actuator nodes. Relative to WSNs, WSANs owns features as

Actuator nodes can process the sensed events in its coverage area timely quickly;

Sensor nodes are generally immobile, but actuators can move randomly;

The actuator node's battery capacity is much larger than that of the sensor nodes, which implies the actuator nodes need more time duration to become full energy.

Nowadays, WSANs are popular in automated braking, target tracking, facility protection, smart homes, and other fields.

Q. Liu et al. (Eds.): CENet 2020, AISC 1274, pp. 1172–1179, 2021.
https://doi.org/10.1007/978-981-15-8462-6_134

In WSANs, besides sensor nodes, actuator nodes also have energy constraints. Although the energy supplement of WSNs, has been thoroughly studied, the charging methods of WSNs cannot be well applied to WSANs. The main reasons are: firstly, the actuator nodes play an essential role in the whole network. Once the actuator nodes die, the performance of the whole network will be significantly affected. So when designing a charging approach for WSANs, the most crucial issue is to ensure the actuator nodes do not die from energy depletion. Secondly, compared with the sensor, the actuator's time to be fully recharged is much longer. In this situation, we should ensure the actuator nodes do not die and guarantee the number of sensor nodes that die from not being recharged timely be minimized. Thirdly, the actuator nodes are mobile in the network. Hence, the problem that accurately tracks actuators to recharge them effectively should be considered.

2 Network Model

As shown in Fig. 1, the network consists of a BS, a MV, a group of sensor nodes and several actuator nodes. Set $S = \{s_1, s_2, s_3, ..., s_n\}$ represents sensor nodes set, nodes in set $A = \{a_1, a_2, a_3, ..., a_m\}$ are actuator nodes, where m is far less than n. The whole network is divided into several subnets, in which one actuator node is deployed, and each actuator node moves within its own area according to the random waypoint mobility model (RWP) [2]. Sensor nodes are randomly distributed in the whole network and immobile. The BS is located in the network center, which has enough energy and strong communication ability to communicate with MV directly. MV is a strong charging device, which moves in the network at a constant speed v to charge each node in time. Here we assume the energy of MV is enough to charge the nodes in the network.

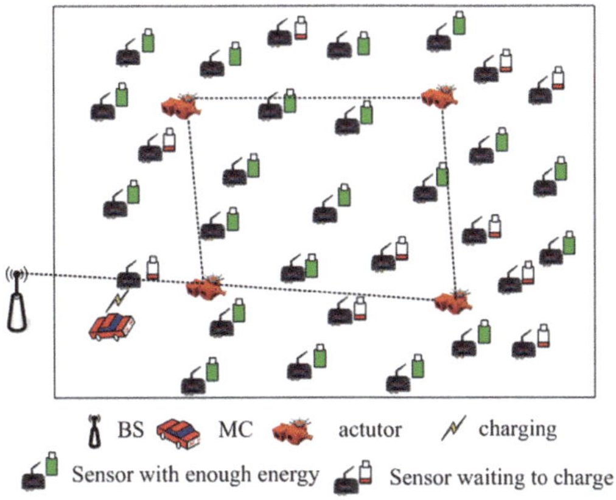

Fig. 1. System architecture.

In this network, each node should continuously monitor its own energy situation. Once it detects that the residual energy is less than the preset critical threshold, it sends a charging message to the BS in time. After receiving the message, the BS forwards the message to MV timely, which is responsible for the selection of the next charging node.

3 Charging Algorithm Designation

3.1 Estimate of Current Node Energy Situation

Affected by the surrounding environment, perception tasks, execution decisions and other factors, the energy consumption rate of each node and the energy consumption of the same node at different time are different. In order to avoid node death happen, it is particularly important to master the energy consumption of nodes in time. To let BS master the current energy situation of each node, each node periodically delivers a message containing its basic information to BS. For example, the message format from the node i can be expressed as <ID_i, RE_{in},t_{in}, *task*>, where ID_i represents the node i, RE_{in} represents its current energy value for the nth record, t_{in} represents the timestamp of this message and task is the task information to be performed.

For node i, let R_i^n represents its current estimation value of energy consumption rate, according to the exponential weighted average method [3], we have:

$$R_i^n = \varphi R_i^{n-1} + (1 - \varphi) r_i^n \tag{1}$$

where φ is the weight coefficient, R_i^{n-1} is node i's energy consumption rate estimation value at the last moment, r_i^n represents true value of energy consumption rate of node i in period $t - 1$ to t, which is calculated as:

$$r_i^n = \frac{RE_{in} - RE_{i(n-1)}}{\Delta t} \tag{2}$$

When $n = 1$, we can set R_i^1 as:

$$R_i^1 = r_i^1 = \frac{RE_{i1} - RE_{i0}}{\Delta t} \tag{3}$$

Take formula (3) into formula (1), and then deduce according to formula (1), we have:

$$R_i^n = \varphi^n \frac{RE_{i1} - RE_{i0}}{\Delta t} + (1 - \varphi)\left(r_i^n + \varphi r_i^{n-1} + \varphi^2 r_i^{n-2} + \varphi^3 r_i^{n-3} + \cdots\cdots + \varphi^{n-1} r_i^1\right) \tag{4}$$

where φ is commonly set as $\varphi \geq 0.9$, so we have $\varphi^{1/1-\varphi} \approx 1/e \approx 0.36$. Thus, for estimating each node's current energy situation, BS only needs to keep the real consumption records of the latest 1/1−φ moments, which can greatly reduce the storage cost of BS, especially for the large-scale network.

3.2 Algorithm Design

***Step* 1**: Calculate the minimum waiting time of each request node:

When one node (e.g. node *j*) is charged as the candidate, MV will calculate the minimum waiting time of other remaining request nodes. If *j* is a sensor node, the minimum waiting time for node *i* named *W*(*j*,*i*) is:

$$W(j,i) = t(MV,j) + t(j,i) + \frac{E - RE_j(t) - R_j \times t(MV,j)}{\mu} \tag{5}$$

When *j* is an actuator node, to reduce the quantity of dead nodes in the process of this actuator's charging, *w*(*j*,*i*) is represented as.

$$w(j,i) = t(MV,j) + t(j,i) + \frac{\alpha E' - RE_j(t) - R_j \times t(MV,j)}{\mu} \tag{6}$$

where *t*(*MV*,*j*) and *t*(*j*,*i*) represent MV's traveling time from MV's current position to node *j* and MV's traveling time from node *j* to node *i* respectively. $\frac{E-RE_j(t)-R_j\times t(MV,j)}{\mu}$ and $\frac{\alpha E'-RE_j(t)-R_j\times t(MV,j)}{\mu}$ denote MV's charging time for the sensor node and the actuator node. *E* and *E'* represents the battery capacity of the sensor node and the actuator node respectively, α is a parameter between 0 and 1, $\alpha E'$ represents the upper charging limit of the actuator. $RE_j(t)$ represents node *i*'s latest remaining energy value, R_j is its energy consumption rate, t_j is the time for node *j* sending the last charging request. μ is the charging rate of MV.

***Step* 2:** Count the number of dead nodes:

When node *j* is charged as the candidate, for the current time *t*, the maximum delay to ensure node *i* do not die is as follows:

$$max_i = \frac{RE_i}{R_i} + t_i - t \tag{7}$$

if $max_i > W(j,i)$, node *i* will not die when node *j* is about to be charged. The total number of dead nodes when *j* is to be recharged is named as $N_{num,i}$, which will be used in formula (10).

***Step* 3:** Calculate the distance between two nodes:

Since the sensor nodes are stationary, the distance between any two sensors can be calculated easily. But for the actuator node, in order to calculate the distance between a sensor and an actuator, the current position coordinates of the actuator must be determined at first. Let (x_g,y_g), (x_p,y_p) represent the coordinates of the starting position G and the next destination position P of the actuator node respectively. C is the current position of the actuator node at the time *t*, we have:

$$x_c = x_g + \frac{V_a(t - t_g)(x_p - x_g)}{\sqrt{(x_p - x_g)^2 + (y_p - y_g)^2}} \tag{8}$$

$$y_c = y_g + \frac{V_a(t - t_g)\left(y_p - y_g\right)}{\sqrt{\left(x_p - x_g\right)^2 + \left(y_p - y_g\right)^2}} \tag{9}$$

where t_g represents the time when the actuator node leaves G. After that, the distance between any two nodes can be calculated easily.

***Step* 4:** Calculate the charging metric:

Now we calculate the charging metric value of each node, which is used for the next charging node selection. For node i, we have:

$$Metric_i = \frac{1}{\beta N_{num,i} + (1 - \beta) distance(MV, i)} \tag{10}$$

where *distance* (*MV,i*) denotes the distance between MV and the node i, β is the weight coefficient. MV selects the next node according to this charging metric value. Usually, the node with the largest metric value is selected as the candidate node. Because the larger the metric value is, the smaller the value of $N_{num,i}$ and *distance* (*MV,i*) are. That's to say, the nodes which cause fewer dead nodes and are closer to the MV are more likely to be charged firstly.

***Step* 5:** Consider the emergency actuators:

In our algorithm, to guarantee emergency actuator nodes to be recharged timely, each actuator node keeps monitoring its own residual energy. Once it detects its residual energy is lower than the critical threshold δ, it sends an emergency message to BS in time. BS transfers the emergency message to MV immediately. Then MV selects this actuator as the next charging node. If there are more than one actuator node in emergency, MV gives priority to the actuator that sends the emergency message first. Maximum charging time of each actuator is adaptively adjusted to ensure that all actuators do not die.

3.3 Actuator Node Tracking Strategy

Since actuator nodes are always in moving status, it is necessary to predict the recharging location of the actuator to recharge it. Suppose actuator i is chosen as the next node for charging, MV will check whether formula (11) hold firstly.

$$\frac{distance(MV, S_i)}{v} < T_{i,s} \tag{11}$$

where $T_{i,S}$ is the retention time of actuator at source point S. If this formula is true, it indicates that when MV reaches the point S, actuator i does not leave here temporarily, so MV chooses to charge the actuator node at point S. If not, MV should move to actuator i's next destination point D directly before the actuator and recharge it when the actuator i arrives there. Once actuator i moves away from the charge range of MV, to continue to recharge it, MV checks whether formula (11) hold.

$$\frac{distance(D_i, S_i)}{v'} + T_{i,D} > \frac{distance(D_i, S_i)}{v} \tag{12}$$

where *v*' is the actuator's moving speed, $T_{i,D}$ is the retention time at point *D*. If formula (12) is true, MV should move to the point *D* to go on recharging actuator *i*. If not, MV has to stay at point *S*, because actuator *i* is not at point *D* when MV arrives there. Only when MV gets the geographical information of the next destination, it will move there to recharge the actuator. When actuator *i* gets away, MV will keep finding its next charging position according to formula (12) to recharge actuator *i* until its remaining energy reaches the upper limit.

4 Performance Evaluation

We use Java to evaluate the performance of EMC proposed in this paper. Compared with First Come First Serve (FCFS) method, NJNP [4] and FEMER, the performance of EMC is evaluated in terms of Sensor Failure Ratio, Average Charging Delay and Charging Cost. Sensor failure ratio is the percentage of dead nodes in total nodes. Average charging latency refers to the average time interval between the time when a node sends a charging request and the time when MV starts to charge for the node. Charging cost is the total travel distance of MV.

In the simulation network, the field is divided into four grids of the same size, with an actuator in each gird. Actuators move randomly according to RWP model in each grid. Table 1 gives the default parameters of the simulation.

Table 1. Default parameters

	Parameters	Value
Sensor node	number	100
	energy	20000 mJ
	Energy consumption in work status	10 mJ/s
	Energy consumption in idle status	1 mJ/s
Actuator node	number	4
	energy	400000 mJ
	Energy consumption in work status	100 mJ/s
	Energy consumption in idle status	10 mJ/s
Other parameters	Charging efficiency μ	200 mJ/s
	MV moving speed v	5 m/s
	Charging thresholdα	0.6
	Emergency threshold of actuator node	0.2
	Time interval between nodes sending Δt	10 s
	Weight coefficient β	0.6

4.1 Varying the Number of Actuators

In simulation, we increase the number of actuator nodes from 1 to 10. As shown in Fig. 2 (a), sensor failure ratios of the four algorithms grow with the growth of the number of actuator nodes. EMC performs better than other three strategies. This is because we set the upper charging threshold for the actuator, which can effectively reduce the failure of sensor nodes caused by the long recharging time of actuators. Secondly, this is because we can predict the current energy situation of the node and choose the next charging node reasonably according to the predicted value.

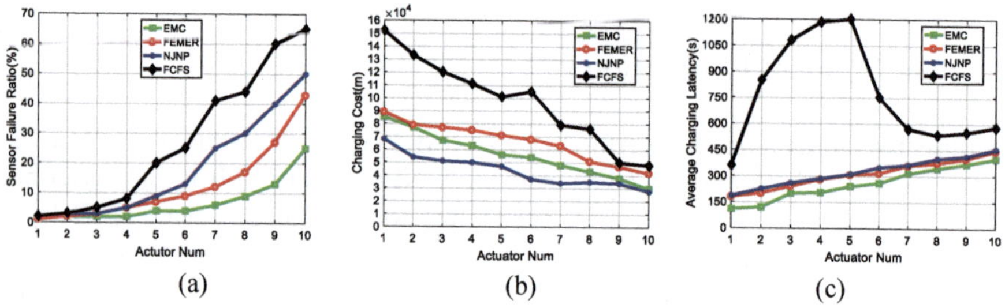

Fig. 2. Different number of actuators.

Figure 2 (b) compares the charging costs of the four schemes. EMC gets low charging cost than FCFS and FEMER. However, the charging cost of our approach is larger than NJNP. The average charging delays of the four schemes are shown in Fig. 2 (c). Except FCFS's charging latency increases first and then decreases as the number of actuator increases, the other three charging strategies all rise up. In EMC, due to the actuator node has a charging upper limit to avoid sensor death and the charging moving distance is considered, so it has obvious advantages in the four algorithms.

4.2 Varying MV's Velocity

Next we vary MV's velocity from 1 to 10 m/s, Fig. 3 (a) shows that the sensor failure ratios of the four charging algorithms all come down, this is because as MV's velocity increasing, its traveling time to the next charging node reduces, which leads to the waiting time reduction of charging nodes. In EMC, because it not only predicts the consumption rate of nodes to ensure the accuracy of the next nodes, but also considers the distance between MV and charging node, the failure ratio in this approach is always lowest.

In Fig. 3 (b), with MV's velocity increases, the charging costs of the four charging algorithms is on the rise, this is because the faster the MV moves, the longer MV has to travel in the same period of time. Since EMC not only set upper charging limit, but also has to track the actuators that move according to RWP model, which result in a relatively large moving distance of the MV, thus EMC have a higher charging cost than FEMER and NJNP.

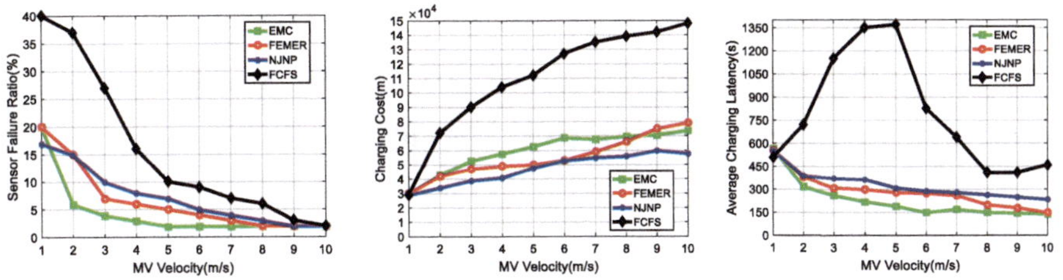

(a) (b) (c)

Fig. 3. Different MV's velocities.

In Fig. 3 (c), FCFS shows a trend of rising first and then declining, and other algorithms decrease with the increase of speed. While EMC takes into account the remaining energy of the next charging node and its distance to MV, so it shows the lowest charging delay among these four approaches. Figures 3 (b) and 3 (c) indicate that our algorithm achieves a good tradeoff between charging cost and charging delay.

5 Conclusion

In this paper, a novel mobile energy supplement approach named EMC is proposed for WSANs. It can select the next charging node according to the predicted current energy condition of each node and its distance to MV. EMC not only effectively solves the high charging cost due to the movement of MV, but ensures all actuators do not fail during charging. Simulation experiment shows that effective of EMC.

References

1. Shen, H., Li, Z.: A Kautz-based wireless sensor and actuator network for real-time, fault-tolerant and energy-efficient transmission. IEEE Trans. Mobile Comput. **15**(1), 1–16 (2016)
2. Hyytia, E., Lassila, P., Virtamo, J.: Spatial node distribution of the random waypoint mobility model with applications. In: To Appear in Transactions on Mobile Computing, pp. 1–15
3. Exponential Weighted Average Method. https://blog.csdn.net/hit1110310422/article/details/81049476, Accessed 15 Jul 2018
4. He, L., Gu, Y., Pan, J., Zhu, T.: On-demand charging in wireless sensor networks: Theories and applications. In: Proceedings IEEE 10th International Conference on Mobile Ad-Hoc and Sensor Systems, Hangzhou, China, pp. 28–36 (2013)

An Overview of Outliers and Detection Methods in General for Time Series from IoT Devices

Bin Sun and Liyao Ma(✉)

University of Jinan, Jinan 250022, China
cse_maly@ujn.edu.cn

Abstract. As internet of things (IoT) devices are booming, a huge amount of data is sleeping without being used. At the same time, reliable and accurate time series analysis plays a key role in modern intelligent systems for achieving efficient management. One reason why the data are not being used is that outliers are preventing many algorithms from working effectively. Manual data cleaning is taking the majority time before one solution could really work on data. Thus, data cleaning, especially fully automated outlier detection is the bottleneck which should be resolved as soon as possible. Previous work has investigated this topic but lacks study on overview from outlier and detection categorization aspects at the same time. This works aims to start covering this topic and to find a direction regarding how to make outlier detection and labelling more automated and general to be suitable for most time series data from IoT devices.

Keywords: Survey · Anomaly novelty detection · Time series · Internet of things

1 Introduction

Time series analysis is widely used in intelligent transport, smart medical assistant, weather forecast, financial systems among other time-dynamic science and engineering topics.

To achieve desired results, a lot of data are needed, which should be clean and in a good quality. However, one big problem is dirty data nowadays. Monitoring and getting data are important, but before any analysis starts, we need clean data. No matter what resources we have, it is nearly always necessary to clean and label outliers in the data. Outliers in raw data are preventing algorithms to achieve their best performance.

In this paper, we try to get an overview of outliers and detection methods to see how to tackle the dirty data.

2 Categorization of Outliers

For data in general, outliers are commonly categorized as three general types: point, contextual and collective [1, 2].

Q. Liu et al. (Eds.): CENet 2020, AISC 1274, pp. 1180–1186, 2021.
https://doi.org/10.1007/978-981-15-8462-6_135

Point outliers are often used while analysing multi-dimensional data which are shown in Fig. 1a [3] which are also known as global outliers due to the fact that the original point-based methods are not considering the local context. Global outliers are usually detected by applying some kind of threshold.

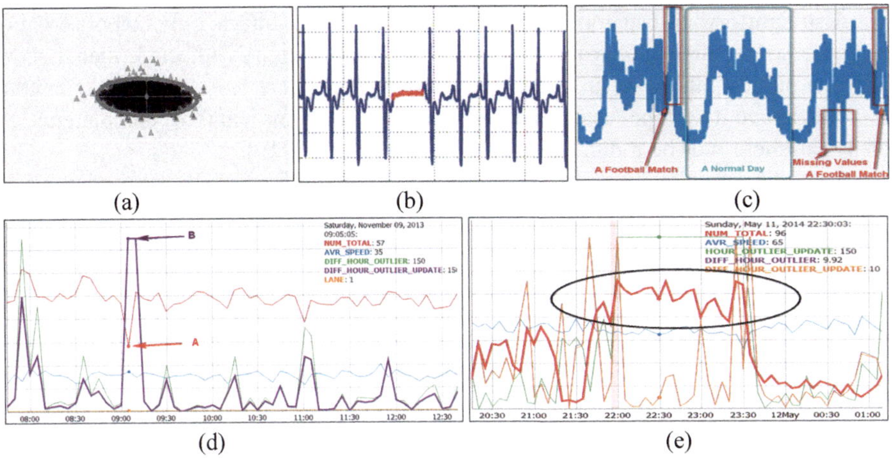

Fig. 1. (a) Point (global) outliers are marked as triangles [3]. (b) A collective outlier in electrocardiographic signal [5]. (c) Two collective outliers due to football matches [6]. (d) An additive outlier is marked as A while a consecutive outlier is marked as B. (e) A long-time range consecutive outlier in time series.

In contract, contextual outliers are useful when observation's context matters. For example, 80 °C is a global outlier and 30 °C is a contextual outlier in Nordic area but normal in India. Contextual outliers are also known as conditional outliers and the definition is a general version of "local outlier" which is defined based on density [4].

A collective outlier contains multiple observations. Any individual observation in this collection is not anomalous either globally or locally. For example, the red-coloured observations in an electrocardiographic signal plot (Fig. 1b [5]). For any red observation, it is similar to its surrounding context and is within the range of the overall signal range. However, when considering the consecutive red observations together, they form a collective outlier. A collective outlier is often called an abnormal subsequence in time series.

Beside silence, another typical collective outlier is irregular peak in time series which is shown in Fig. 1c [6]. It is worth mention that the peak traffic due to football matches is only considered high when compared to the traffic before and after matches but are not too high if compared with all other data as the daily work traffic is also much, so some research may call the peaks or the individual observation points in the peaks (local) outliers [1].

2.1 Categorization of Outliers in Time Series

Considering outliers in time series (OTS) specifically, they are usually categorized from other aspects. Although there is not a fixed/common categorization methodology for OTS, if we summarize current research, some widely accepted categorizations can be described as below.

A straightforward categorization is to see if the outlier affects only one observation or some consecutive observations. If only one observation is affected, then it is an additive outlier, otherwise an innovation outlier [7].

The above two-type categorization contains "patterns" and if the patterns are analysed, there will be a more detailed categorization [8–10].

An additive outlier (Fig. 1d. A) is one outlier that differs much from other observations and affects a single observation but not the others. For example, errors in individual observations may occur when measured data are transmitted to a datacentre in a network.

A level shift outlier is often a result of policy change and will cause all following observations to shift by a constant. This happens very often. For example, a monitoring camera was focusing on one lane of a road but now is catching all traffic in all lanes. Sometimes new monitoring devices for the opposite road direction are also added into the original data stream. When monitoring skin temperature, attribute identifier may remain the same, but the corresponding sensor is moved to another place.

Level shift is one kind of concept drift (or concept change) [11].

A transient outlier (a.k.a. temporary change) impacts its following observations with exponentially decayed impact down to zero.

An innovation outlier will also influence following observations, but the time range and extent might be varying. The influence may be limited within several consecutive observations (in stationary series) or become bigger and bigger (in nonstationary series).

A local trend outlier starts at a particular series point with increasing impact at the beginning which lead to a trend for a while but then dies out. The impact during this time period is irregular. Some research considers local trend outliers together with transient outliers (forming two subcategories) [1, 12].

A seasonal additive outlier regularly impacts its following observations with patterns and occurs seasonally (not only seasons of years but arbitrary periodic components). Each season is affected equally. For example, human body temperature peaks around evening time and road traffic peaks around noon time every day.

A consecutive additive outlier (Fig. 1d. B) or additive outlier patch is similar to the mentioned additive outlier, but the number of observations is different and at least two consecutive observations are deviating and lasting for a limited time period. Those consecutive observations are considered together as one outlier.

One outlier can be one of the outlier types above or combined [7]. In specific domains, there will be some other ways to define and categorize outliers [13].

3 Categorization of Detection Methods

Great work has been done to categorize outlier detection methods. One categorization schema proposed by Pimentel et al. (2014) [14] divides detection methods into five categories: *probability/density-based, distance-based, spectral/refactor-/reconstruction-based, domain-/boundary-based, and information-theory-based methods* and this categorization is widely used [15, 16]. The exact category names differ from research to research.

When a new query instance (a new observation point or a subsequence) occurs, *a probability/density-based method* calculates the probability of the query instance being an outlier according to a (predefined) mathematical distribution model. For a distribution, the calculation result is more or less a (normalized) distance to the centre/centroid. *A distance-based method* calculates some kind of distance from the query instance to its neighbours or other instances. For example, clustering algorithms can be used to calculate its decision metric and compare this value with a threshold (which is often pre-defined) to check if the query instance is beyond the threshold. *A reconstruction-based method* compares the query instance with expected values from normal prediction algorithms [17, 18] and see if it deviates too much. The entire approach includes three steps: decomposition/denoising, reconstruction and residual/error analysis. *A boundary-based method* checks the query instance with a calculated/defined separation to see which part the new instance belongs to. A neat and powerful separation can be calculated by support vector machine (SVM). *An information-theory-based method* calculate contribution of each instance to the overall entropy. The one contributes most is considered as an outlier. Usually, the number of outliers is predefined as k so the top k outliers are identified from k entropy calculation iterations. Some research considers information-theory-based methods are within distance-based method category [19]. Most research has the same or similar categorization schema as the schema described above.

Falcao et al. (2019) [20] divides detection methods into six categories which are similar to Pimentel et al. Some main differences are: (a) density methods are considered as a subcategory within *distance* category; (b) SVM and iForest are categorized as *classification* methods; (c) principal component analysis (PCA) is moved from reconstruction to statistical method category; (d) a new category of *angle-based methods* [21] is used; (e) all the discussed methods are considered as *unsupervised methods*.

If we go further from this point, all methods can fall into three categories: supervised, unsupervised, and semi-supervised methods [22] with a terminology naming style like machine learning. Wang et al. (2019) [23] uses the same three terminologies but all under statistical category. Their overall seven categories include *density-based, statistical, distance-based, clustering-based, graph-based, ensemble-based, and learning-based methods*. Wang et al. discussed both advantages and disadvantages of all categories in detail, especially the ensemble methods often proven to be well-performed [24, 25].

Instead of considering from supervised vs. unsupervised vs. semi-supervised aspect, methods can also be sub-categorized using *parametric vs. non-parametric* aspect [23, 26] or *global vs. local* aspect or *labelling vs. scoring* aspect [1, 26]. A parametric statistical-based method assumes instances follow one of existing statistical distributions or a mixture of them, then estimates distribution's parameters and check if an instance deviates beyond a certain (normalized) distance threshold from the model centre (e.g., 3-sigma away) to decide if it is an outlier. Parametric statistical category also includes *subspace* methods such as spectral and PCA. A typical non-parametric density is kernel density. The aforementioned reconstruction-based methods can also be categorized in this way. A parametric reconstruction-based method example is fbprophet [27] and well-performed non-parametric reconstruction-based methods include autoencoders, generative adversarial network and DWD [28].

More surveys are available for detailed advantages and disadvantages of techniques beyond categorization. For example, Chandola et al. (2009) [11] investigate statistical, classification-based, clustering-based, nearest neighbour-based, information theory-based, spectral techniques etc.

4 Discussion

To the extent of our knowledge and literature, it is hard to find a mapping relationship between outliers and detection methods. Thus, it is needed to conduct a systematic cross-comparison for outliers and detection methods.

In practice, some dataset has univariate time series while others have multivariate data. Though most algorithms can be used for multivariate after some or much modification, the modified algorithm may not work well, for example, vectorized seasonal autoregressive integrated moving average (SARIMA). Besides, some methods cannot handle multi-seasonality or dirty data with missing values [29] or varying intervals among observations. Those dataset attributes should be considered. Also, if the data are not only dirty but also imprecise or uncertain, more precondition should be done [30, 31].

We plant to conduct more systematic study and experimentation on those problems to investigate if there are good automated solutions to solve them.

Acknowledgment. This work is supported by Shandong Provincial Natural Science Foundation No. ZR2018PF009 and Shandong Key Research and Development Program No. 2019JZZY021005.

References

1. Gupta, M., Gao, J., Aggarwal, C., Han, J.: Outlier detection for temporal data. Synth. Lect. Data Min. Knowl. Disc. **5**(1), 1–129 (2014)
2. Zhang, H., Nian, K., Coleman, T.F., Li, Y.: Spectral ranking and unsupervised feature selection for point, collective, and contextual anomaly detection. Int. J. Data Sci. Anal. **9**(1), 57–75 (2018). https://doi.org/10.1007/s41060-018-0161-7
3. Sun, B., Cheng, W., Bai, G., Goswami, P.: Correcting and complementing freeway traffic accident data using mahalanobis distance based outlier detection. Tehnicki Vjesnik-Technical Gazette **24**(5), 1597–1607 (2017)
4. Han, J., Kamber, M.: Data Mining: Concepts and Techniques, 3rd edn. Elsevier, Singapore (2012)

5. Marta, E., Keshav, D., Anant, J.: Anomaly Detection. Learn Machine Learning Algorithms (2020)
6. Sun, B.: Toward Automatic Data-Driven Traffic Time Series Prediction. In: DIVA, Gothenburg, Sweden, vol. 12 (2017)
7. Douglas, M., Cheryl, J., Murat, K.: Introduction to Time Series Analysis and Forecasting, 2nd edn. Wiley-Interscience, Hoboken, New Jersey (2015)
8. Jakaša, T., Andročec, I., Sprčić, P.: Electricity price forecasting-ARIMA model approach. In: 8th International Conference on the European Energy Market. Zagreb, Croatia, pp. 222–225 (2011)
9. Lotto, M., Aguirre, P.E.A., Rios, D., Machado, M.A.A.M., Cruvinel, A.F.P., Cruvinel, T.: Analysis of the interests of Google users on toothache information. PLoS ONE **12**(10), e0186059 (2017)
10. IBM: Outliers-SPSS Modeler 18.1 Document, https://clck.ru/PExAL, Accessed 8 Sep 2017
11. Chandola, V., Banerjee, A., Kumar, V.: Anomaly detection: a survey. ACM Comput. Surv. **41**(3), 15 (2009)
12. Tsay, R.S., Peña, D., Pankratz, A.E.: Outliers in multivariate time series. Biometrika **87**(4), 789–804 (2000)
13. Menezes, R., Oliveira, Á., Portela, S.: Investigating detrended fluctuation analysis with structural breaks. Phys. Stat. Mech. Appl. **518**, 331–342 (2019)
14. Pimentel, M.A.F., Clifton, D.A., Clifton, L., Tarassenko, L.: a review of novelty detection. Sig. Process. **99**, 215–249 (2014)
15. Kanarachos, S., Christopoulos, S.R.G., Chroneos, A., Fitzpatrick, M.E.: Detecting anomalies in time series data via a deep learning algorithm combining wavelets, neural networks and Hilbert transform. Expert Syst. Appl. **85**, 292–304 (2017)
16. Dong, X., Jin, B., Tang, B., Tang, H.: On real-time monitoring on data stream for traffic flow anomalies. In: IEEE International Conference on Parallel & Distributed Processing with Applications, Ubiquitous Computing & Communications, Big Data & Cloud Computing, Social Computing & Networking, Sustainable Computing & Communications, Los Alamitos, pp. 322–329. IEEE Computer Society (2018)
17. Sun, B., Cheng, W., Goswami, P., Bai, G.: Short-term traffic forecasting using self-adjusting k-nearest neighbours. IET Intell. Transp. Syst. **12**(1), 41–48 (2018)
18. Sun, B., Cheng, W., Goswami, P., Bai, G.: Flow-aware WPT k-nearest neighbours regression for short-term traffic prediction. In: 22nd IEEE Symposium on Computers and Communication, Heraklion, Greece, pp. 48–53 (2017)
19. DSMI. Anomaly Detection Toolbox. NTUST (2016)
20. Falcao, F., Zoppi, T., Vieira da Silva, C.B., Santos, A.: Quantitative comparison of unsupervised anomaly detection algorithms for intrusion detection. In: Proceedings of the 34th ACM/SIGAPP Symposium on Applied Computing, Assoc Computing Machinery, New York (2019)
21. Kriegel, H. P., Schubert, M., Zimek, A.: Angle-based Outlier Detection in High-dimensional Data. In: Proceedings of the 14th ACM SIGKDD International Conference on Knowledge Discovery and Data Mining, New York, NY, USA, pp. 444–452 (2008)
22. Hodge, V.J., Austin, J.: A survey of outlier detection methodologies. Artif. Intell. Rev. **22**(2), 85–126 (2004)
23. Wang, H., Bah, M.J., Hammad, M.: Progress in outlier detection techniques: a survey. IEEE Access **7**, 107964–108000 (2019)
24. Ma, L., Sun, B., Li, Z.Y.: Bagging likelihood-based belief decision trees. In: 20th International Conference on Information Fusion, Xi'an, China, pp. 1–6 (2017)

25. Ma, L., Sun, B., Han, C.Y.: Training instance random sampling based evidential classification forest algorithms. In: International Conference on Information Fusion, London, United Kingdom (2018)
26. Zimek, A., Filzmoser, P.: There and back again: Outlier detection between statistical reasoning and data mining algorithms. Data Min. Knowl. Disc. **8**(6), e1280 (2018)
27. Taylor, S., Letham, B.: p*rophet: Automatic Forecasting Procedure.* https://github.com/facebook/prophet, 2020
28. Sun, B., Wei, C., Liyao, M., Prashant, G.: Anomaly-aware traffic prediction based on automated conditional information fusion. In: International Conference on Information Fusion, Cambridge, United Kingdom, pp. 2283–2289 (2018)
29. Sun, B., Ma, L., Cheng, W., Wen, W., Goswami, P.: An improved k-nearest neighbours method for traffic time series imputation. In: Chinese Automation Congress, Jinan, China, (2017)
30. Ma, L., Destercke, S., Wang, Y.: Evidential likelihood flatness as a way to measure data quality: the multinomial case. In: 16th World Congress of the International Fuzzy Systems Association and the 9th Conference of the European Society for Fuzzy Logic and Technology, Gijon, Spain, pp. 313–319 (2015)
31. Ma, L., Sun, B., Han, C.: Learning decision forest from evidential data: the random training set sampling approach. In: 4th International Conference on Systems and Informatics, Hangzhou, China (2017)

A Path Planning Algorithm for Mobile Robot with Restricted Access Area

Youpan Zhang[1], Tao Sun[2], Hui Zhang[1](✉), Haiying Liu[1], Lixia Deng[1], and Yongguo Zhao[3]

[1] School of Electrical Engineering and Automation, Qilu University of Technology (Shandong Academy of Science), Jinan 250353, China
zhanghuifly111@126.com
[2] Academy of Advanced Interdisciplinary Studies, Qilu University of Technology (Shandong Academy of Science), Jinan 250353, China
[3] Institute of Automation, Qilu University of Technology (Shandong Academy of Sciences), Jinan 250014, China

Abstract. In the actual path planning of the mobile robot, the algorithm planning of the robot is almost always to find the shortest path, but if the robot is in a narrow working environment with many corners, it will make the robot too close to the obstacle, and the robot's edge will contact with the obstacle, resulting in the path planning failure. In order to solve the security problem of mobile robot path planning, we put forward a kind of inflation obstacles, refinement algorithm first determine skeleton and area of the mobile robot on the map, and then the refinement algorithm can get skeleton to improve traffic area, connected to the path of starting point and end point, define the passage area. Then by searching the first and second nodes, we can find the optimal path that accords with the safety of the robot. The new path planning algorithm can realize non-collision planning, improve the safety and stability of the robot, and verify its effectiveness and authenticity through simulation results.

Keywords: Mobile robot · Path planning · Qualified pass · Refinement algorithm

1 Introduction

Mobile robot is a comprehensive system integrating environment perception, dynamic decision and planning, behavior control and execution. With the development and progress of science and technology, more and more robots have entered people's life and changed people's lifestyle. As one of the most basic robots, mobile robot has derived a variety of robots with different functions. For example, medical robots used to help doctors make ward rounds, diagnose diseases [1], follow robots to help people shop and take things [2], military robots used for military detection and strike [3], household cleaning robots used for sweeping and washing dishes [4].

The path planning technology of mobile robot is an important research problem of robot control technology. It is an indispensable part of robot navigation and an important sign of robot intelligence. Mobile path planning technology refers to that the

Q. Liu et al. (Eds.): CENet 2020, AISC 1274, pp. 1187–1196, 2021.
https://doi.org/10.1007/978-981-15-8462-6_136

robot independently plans a safe running route according to the sensing information of its own sensors to the environment, and at the same time completes the operation tasks efficiently, which includes the tasks of geographic environment information modeling, path planning, positioning and obstacle avoidance. In the common path planning of mobile robot, it is mainly divided into global path planning method and local path planning method. The global path planning method is to plan an optimal path in a known map by searching the starting position and key position of the robot. For example, raster method [5], topology method [6], Voronoi graph method [7], viewable method [8]. The local path rule refers to that in an unknown map, the robot's position on the unknown map and the surrounding environment can be determined through the information obtained by the robot's sensors, and an optimal path can be planned. Local planning methods include artificial potential field method [9], genetic algorithm [10], ant colony algorithm [11], and fuzzy algorithm.

In the path planning algorithm of mobile robot, most of the algorithms usually focus on how to effectively avoid obstacles to get the optimal path, but rarely involve the safety problem of robot. For example, if the robot is too close to the obstacle, the tail of the robot will touch the obstacle, and the robot will replan the path at a certain position, causing the robot to swing its course. To solve the above problems, this paper proposes a new method of path planning, the algorithm is based on the elaboration and passable area Dijkstra algorithm is put forward, through the expansion of obstacles, refinement algorithm first determine skeleton and area of the mobile robot on the map, and then the refinement algorithm can get traffic area skeleton was improved, the beginning and end of the connected path. Finally, the optimal path for robot safety is found by searching the first and second nodes.

The organization of this article is as follows. The second part introduces the detailed diagram of the passable area of the mobile robot, the third part introduces the extraction of the key nodes of the mobile robot, the fourth part introduces the path selection of the mobile robot, and the fifth part introduces the simulation and verification of the mobile robot.

2 Construction of Mobile Robot Detail Diagram

2.1 Environment Model of Mobile Robot

Firstly, the environment model of mobile robot was built and the robot was placed in a narrow working environment with many obstacles. In this paper, raster method is used to establish a static two-dimensional map of mobile robot, which is used to represent the actual working environment of mobile robot. The path of the mobile robot is divided into passable area and impassable area, and the passable area is divided by the obstacle. We use "1" to represent the passable area, which is the white area in Fig. 1, and "0" to represent the impassable area, which is the black area (obstacle) in Fig. 1. In order to ensure the stability and reliability of the algorithm, a static two-dimensional map with a size of 500 * 500 pixels is constructed, which is more complex than the real environment.

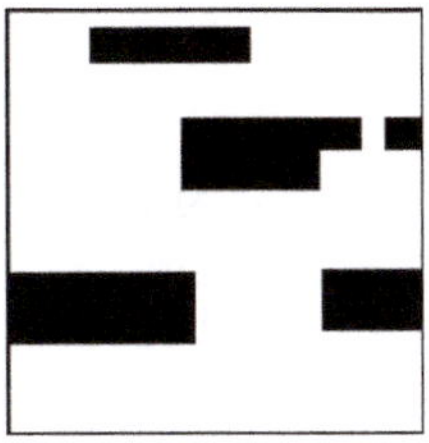

Fig. 1. Environment Youpan Zhang.

2.2 Zhang-Suen Algorithm to Build a Refinement Diagram

We used the zhang-suen algorithm [13] to extract the map of the passable skeleton, and the map is the fixed navigation path of the mobile robot to keep it away from the surrounding obstacles. The extraction principle and method of skeleton in passable area are as follows:

(1) First, we take a passable point P1, denoted as the center grid. Regions with 3 * 3 in the neighborhood around P1 are respectively represented as P2, P3, P4, P5, P6, P7, P8 and P9. Where P2 is directly above P1, the rest of the points are named clockwise around the center grid (Fig. 2).

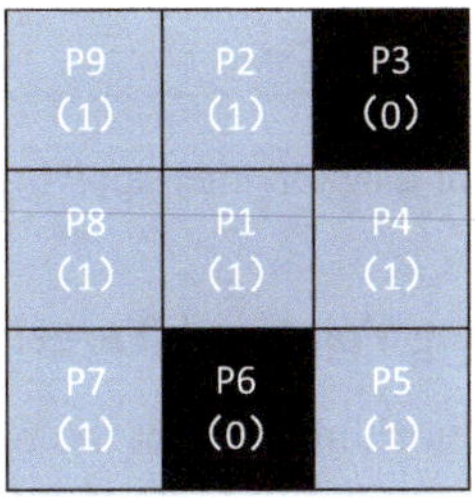

Fig. 2. 3 * 3 grid area.

(2) loop all passable points and delete the points that meet the following conditions:

$$2 < = N(p1) < = 6 \tag{1}$$

$$S(P1) = 1 \tag{2}$$

$$P2 * P4 * P6 = 0 \tag{3}$$

$$P2 * P6 * P8 = 0 \tag{4}$$

where N(p1) refers to the number of passable points in the 8 points adjacent to p1, and S(p1) refers to the number of p2-p9-p2 with paired values of 0 and 1 in sequence

(3) continue to loop all passable points, and delete the points that meet the following conditions

$$2 <= N(P1) <= 6 \tag{5}$$

$$S(P1) = 1 \tag{6}$$

$$P2 * P4 * P8 = 0 \tag{7}$$

$$P2 * P6 * P8 = 0 \tag{8}$$

Loop through the above two steps and iterate until no point in the two steps is marked as deletion, and the output result is the thinning diagram after the extraction of the mobile robot skeleton.

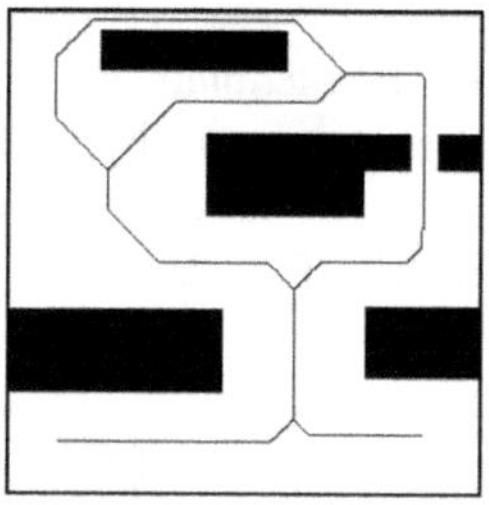

Fig. 3. The refinement diagram after applying Zhang-suen algorithm.

2.3 Detail the Access Starting Point and Target Point

According to the zhang-suen algorithm, we get the refined Fig. 3. Figure 3 is only the basic skeleton of the passable region, but there is no connection between the skeleton and the target point and the starting point. Now we access the starting point and target point of the mobile robot in Fig. 3, as shown in Fig. 4. We improve the zhang-suen algorithm, and obtain the starting point and end point of the skeleton connection based on Fig. 4.

1) If the end point or target point is on the skeleton. If the target point or starting point is on the skeleton, only the skeleton of the passable area needs to be analyzed to find the shortest distance.
2) If the end point or starting point is not on the skeleton. If the target point or starting point is not on the skeleton of the refinement diagram, the algorithm needs to be improved. Firstly, the starting point or target point is regarded as an inner point P1 on the map, and the inner point of the robot should satisfy the following Eq. (9) (Fig. 5):

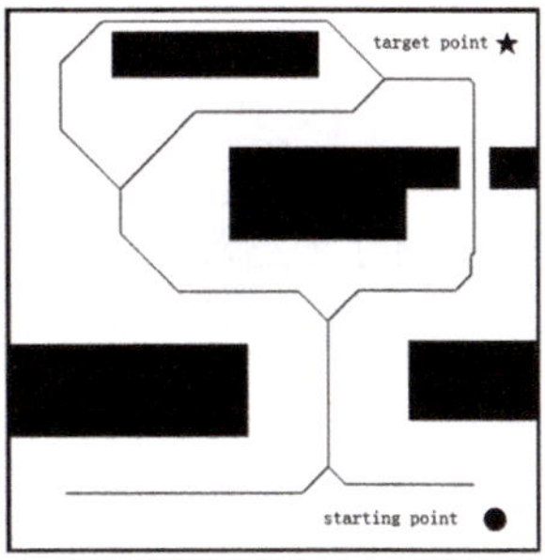

Fig. 4. Adds a refinement diagram of the starting point and the target point.

$$\text{If p1 is an inner point}\begin{cases} P1 = 0 \\ \& \\ P2 + P3 + P4 + P5 + P6 + P7 + P8 + P9 = 8 \end{cases} \tag{9}$$

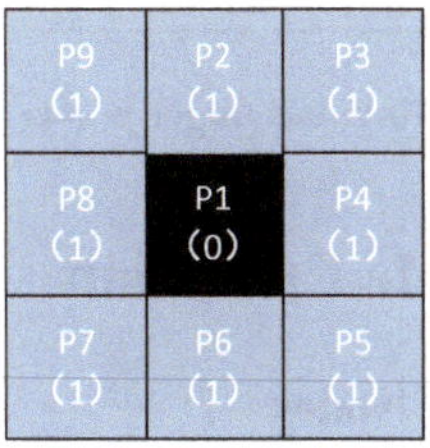

Fig. 5. As the starting point or target point of the inner point.

Then, a circle with the inner point as the center of the circle and a radius of 1 per unit length is established. The radius is expressed by Ri as I = 1, 2, 3, 4……, the points on the circle are represented by Mi. We just look for the point on the circle to the first contact point of the skeleton. When Mi skeleton is connected and Mi + 1 is not 0, that is, Eq. (10),

$$\text{If Mi is the first contact point}\begin{cases} Mi = 0 \\ \& \\ Mi + 1 = 1 \end{cases} \tag{10}$$

The identified point Mi is the shortest contact distance to the skeleton in Fig. 4 (Fig. 6).

Through the above improvement measures, the starting point and the target point of the connection path can be formed. As shown in Fig. 7.

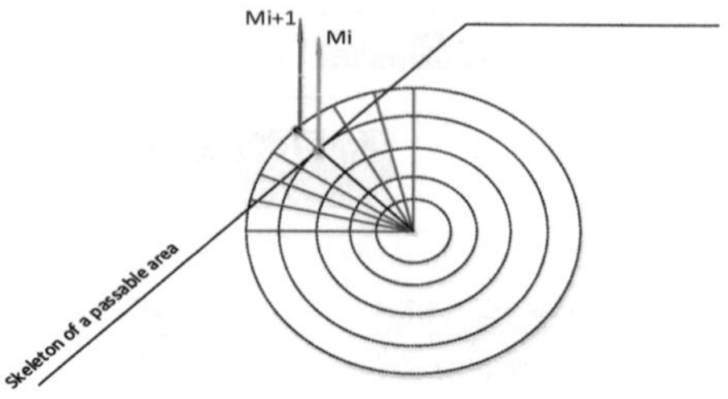

Fig. 6. The solution diagram for the first contact point.

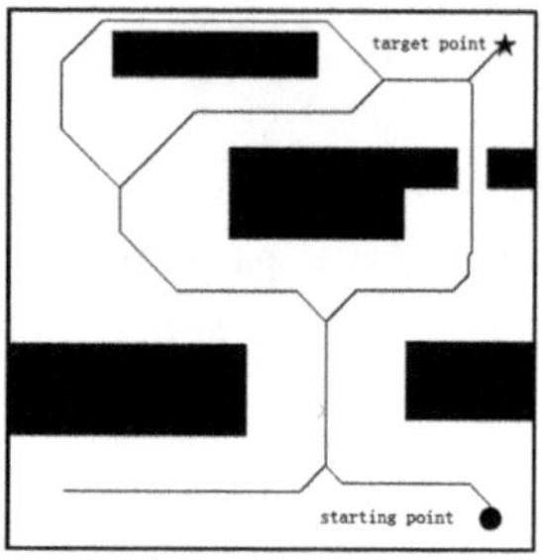

Fig. 7. Join the starting and target points to refine the diagram.

3 Mobile Robot Key Node Extraction

3.1 Extraction of First-Level Nodes

After the starting point and the target point of the mobile robot are connected into the thinning diagram, a thinning grid map Fig. 7 containing the starting point and the target point is obtained. Each line in Fig. 7 is composed of continuous points. Next, we extract the first-level node on the map. The first-level node is defined as the bifurcation point of the improved refinement diagram. The bifurcation point refers to two points in the diagram that can be used by the robot to choose the path.

We analyzed each point on the refined skeleton. If a node P1 is P2 + P3 + P4 + P5 + P6 + P7 + P8 + P9 = 5 in the adjacent 3 * 3 raster map, that is, Eq. (11) is satisfied, then this point P1 is a bipartite point on the refined map. Through the filtering of the algorithm again and again, all the first-level nodes on the refinement graph are selected iteratively, thus obtaining Fig. 8.

$$\text{If p1 is a first level node} \begin{cases} P1 = 0 \\ \& \\ P2 + P3 + P4 + P5 + P6 + P7 + P8 + P9 = 5 \end{cases} \tag{11}$$

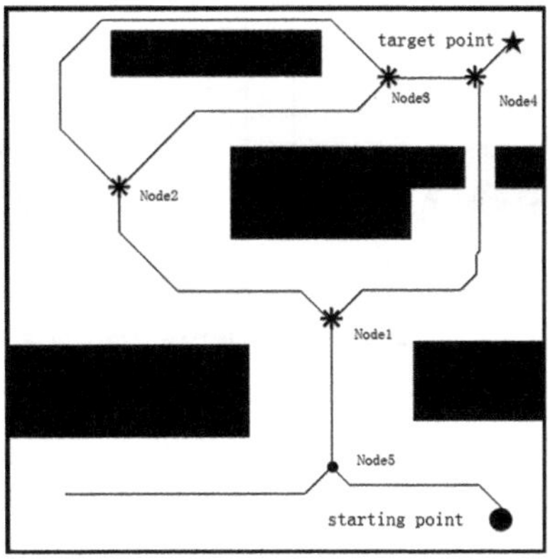

Fig. 8. The first-level nodes.

3.2 Extraction of Second-Level Nodes

When the extraction of the first-level node is completed for the refinement diagram of the mobile robot, the second-level node of the refinement diagram is obtained and then extracted. The point that changes the path direction of the robot.

We again after the detailed analysis of each point on the skeleton, if a node P1 with its adjacent near 3 * 3 points in P2, P4, P6, P8 in only one point for "0", and P2, P3 and P4 + P5, P6 and P7 P8 + + P9 = 6. The first-order nodes on the skeleton are screened again and again by the algorithm. This results in Fig. 9.

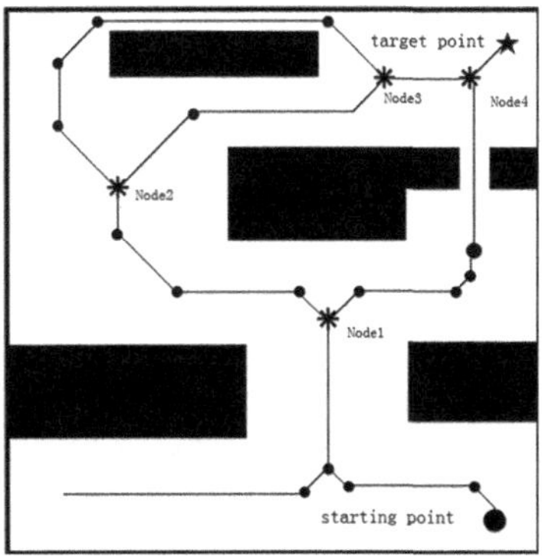

Fig. 9. The second-level nodes.

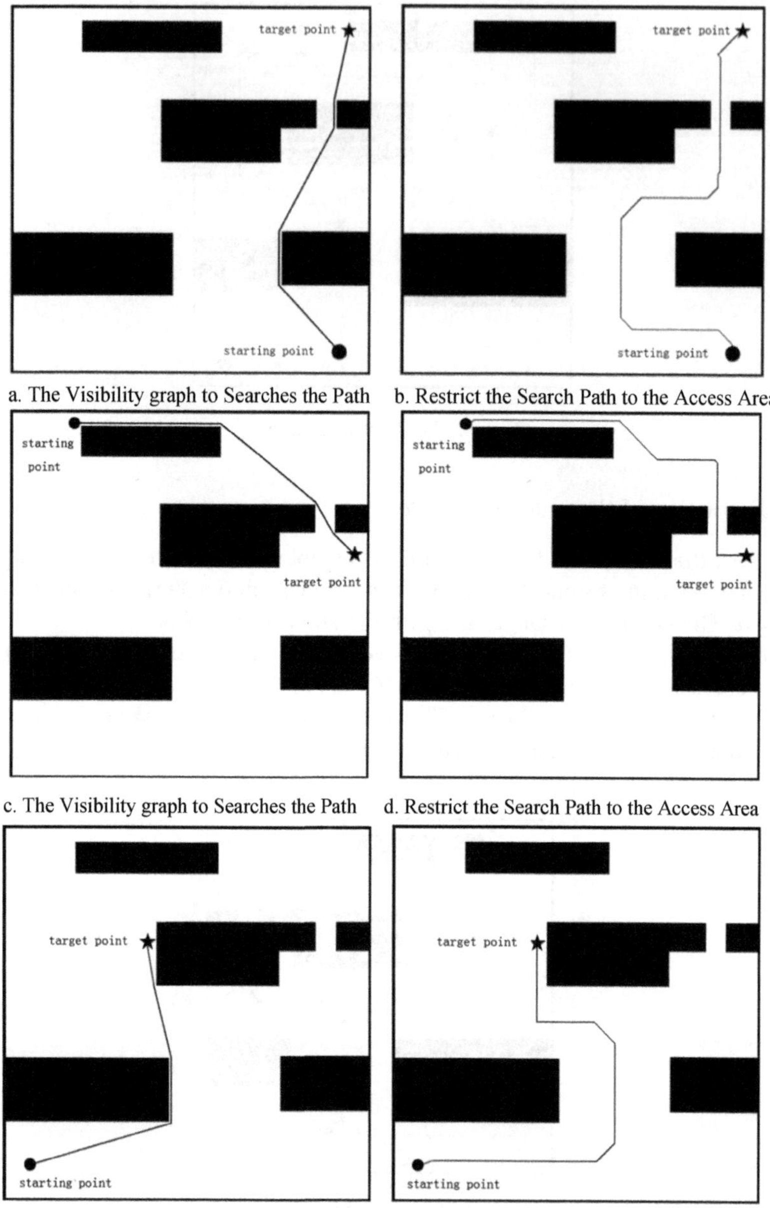

a. The Visibility graph to Searches the Path b. Restrict the Search Path to the Access Area

c. The Visibility graph to Searches the Path d. Restrict the Search Path to the Access Area

e. The Visibility graph to Searches the Path f. Restrict the Search Path to the Access Area

Fig. 10. Comparison of simulation results.

$$\text{If P1 is a second level node}\begin{cases} \text{P1} = 0 \\ \& \\ \text{P2} + \text{P4} + \text{P6} + \text{P8} = 3 \\ \& \\ \text{P2} + \text{P3} + \text{P4} + \text{P5} + \text{P6} + \text{P7} + \text{P8} + \text{P9} = 6 \end{cases} \tag{12}$$

After the second-level nodes of the refined graph are obtained, each first-level node is composed of N second-level nodes, so the weighted value of first-level nodes can be calculated, and the path length D = {d1, d2, d3...Di}. In this paper, the weight of the second-level node is set as the distance between two second-level nodes $|AB| = \sqrt{(x_1 - x_2)^2 + (y_1 - y_2)^2}$, Find the distance between the two nodes.

4 Simulation and Verification

Based on Fig. 8 and Fig. 9, we can find all the connected paths from the starting point to the target point. Through the analysis of First-level and Second-level nodes, we use Dijkstra algorithm to find the combination of the shortest line segments connecting the start and end of the path on the refinement diagram. so as to find the passable path and plan the interpretation. In order to verify the feasibility of the algorithm, we carried out simulation under the experimental platform of MATLAB2017. Three simulation experiments with different starting points and target points were carried out on the traditional Visibility graph, the zhang-suen algorithm in this paper and the algorithm of primary and secondary nodes, respectively. The simulation results are shown in Fig. 10.

5 Conclusion

Traditional path planning algorithms generally focus on how to avoid obstacles or connect to a new target position, but rarely involve the problem that the mobile robot is too close to the obstacle and the robot edge touches the obstacle. Through the analysis of the experimental results, the new algorithm of restricted passage area ensures that the robot avoids touching obstacles to the greatest extent in a narrow passage area with a small part of the robot's path efficiency, so as to achieve the purpose of limiting the robot's movement path. It is especially suitable for mobile robot to work in a narrow passage area with many corners to ensure the safety and stability of the robot to the greatest extent.

Acknowledgement. This work has been supported by the Key Research and Development Program of Shandong Province (Nos. 2019GGX104091, 2019GGX104079), Shandong Provincial Natural Science Foundation (Nos. ZR2018LF011, ZR2018QF005), and Young Doctor Cooperation Foundation of Qilu University of Technology (Shandong Academy of Sciences) (No. 2018BSHZ006), Qilu University of Technology (Shandong Academy of Science) Special Fund Program for International Cooperative Research (QLUTGJHZ2018019).

References

1. Yakimovich, T.; Baerenweiler, R.: System and apparatus for crush prevention for medical robot applications. U.S. Patent Application No 16/486,204 (2020)
2. Xu, J., Yang, X., Chen, Y., et al.: Design of the autonomous following robot system with dual-mode control. Process Automation Instrumentation (2018)
3. Bhat, S., Meenakshi, M.: Military robot path control using RF communication. In: Proceedings of the First International Conference on Intelligent Computing and Communication, pp. 697–704. Springer, Singapore (2017)
4. Jones, J.L., Mack, N.E., Nugent, D.M., et al.: Autonomous floor-cleaning robot. U.S. Patent No 10,420,447 (2019)
5. Nguyen, V.T., Vu, D.T., Park, W.G., et al.: Navier-stokes solver for water entry bodies with moving chimera grid method in 6DOF motions. Comput. Fluids **140**, 19–38 (2016)
6. Kuric, I., Bulej, V., Saga, M., et al.: Development of simulation software for mobile robot path planning within multilayer map system based on metric and topological maps. Int. J. Adv. Robot. Syst. (2017)
7. Wang, J.K., Meng, M.Q.H.: Optimal path planning using generalized Voronoi graph and multiple potential functions. IEEE Trans. Ind. Electron. (2020)
8. Majeed, A., Lee, S.C.: A fast global flight path planning algorithm based on space circumscription and sparse visibility graph for unmanned aerial vehicle. Electronics **7**(12), 375 (2018)
9. Chen, Y.B., Luo, G.C., Mei, Y.S., et al.: UAV path planning using artificial potential field method updated by optimal control theory. Int. J. Syst. Sci. **47**(6), 1407–1420 (2016)
10. Márcio da Silva, A., Jesimar da Silva, A., Toledo, C., et al.: A hybrid multi-population genetic algorithm for UAV path planning. In: Proceedings of the Genetic and Evolutionary Computation Conference, pp. 853–860 (2016)
11. Liu, J., Yang, J., Liu, H., Tian, X., Gao, M.: An improved ant colony algorithm for robot path planning. Soft Comput. **21**(19), 5829–5839 (2016)
12. Wahyunggoro, O., Cahyadi, A.I.: Quadrotor path planning based on modified fuzzy cell decomposition algorithm. Telkomnika **14**(2) (2016)
13. Soni, S., Kaur, S.: To propose an improvement in Zhang-Suen algorithm for image thinning in image processing. Int. J. Sci. Technol. Eng. **3**(I), 7481 (2016)

Weighted Slopeone-IBCF Algorithm Based on User Interest Attenuation and Item Clustering

Peng Shi(✉) and Wenming Yao

The 15th Research Institute of China Electronics Technology Group Corporation, Beijing 100083, China
shipengpku@163.com

Abstract. Aiming at the problem that the traditional collaborative filtering algorithm has low recommendation accuracy and unsatisfactory results when the matrix is sparse, weighted SlopeOne-IBCF algorithm based on user interest attenuation and item clustering is proposed. Firstly, use the weighted Slope One algorithm to predict user ratings and fill in matrix, fuzzy cluster the items in the data set; then combine the interest decay function to modify the target user's historical item score, and determine the category of items that the target user is interested in at the current stage; then integrate the time decay function into the process of calculating the similarity, pay attention to the factors that interest and hobbies change with time; finally, the similarity Top k items that are the highest and are not in the user's historical item set are recommended to the target user. Experiments show that the algorithm in this paper effectively alleviates the problem of poor recommendation and inaccuracy in the case of sparse data, and further improves the accuracy of recommendation.

Keywords: Fuzzy clustering · Weighted-SlopeOne · Collaborative filtering

1 Introduction

Personalized recommendation system [1] is an effective solution to solve the information load and it mainly divided into content-based recommendations [2], association rules-based recommendations [3], and collaborative filtering-based recommendations [4]. Among them, collaborative filtering algorithm is one of the most widely researched and applied and the most successful recommendation algorithm in academia and industry. The collaborative filtering algorithm is a recommendation method based on user behavior, and the user behavior may be browsing, purchasing, and scoring of products in the past. Among them, the logic of the collaborative filtering algorithm is popularly said: "the information that is of interest to people with similar hobbies is also of interest to you" or "the information that is similar to the information you are interested in is also of interest to you", which also is the algorithmic idea of UBCF [5] and IBCF [6] in the collaborative algorithm.

Although the collaborative filtering algorithm has been used in the field of personalized recommendation and it has achieved great success, but with the increase of

Q. Liu et al. (Eds.): CENet 2020, AISC 1274, pp. 1197–1208, 2021.
https://doi.org/10.1007/978-981-15-8462-6_137

the number of users and the rapid growth of the number of projects, the determination of the nearest neighbor requires large-scale calculations and a long time-consuming problem. At the same time, in the huge amount of data, only a few users are actually active in fewer projects, so they are facing the problem of sparse matrix and cold start. Deng [7] proposed a clustering method, combined with a collaborative filtering algorithm, to solve the problem of sparse data rows; Billsus [8] proposed matrix decomposition through the SVD algorithm to decompose the original scoring matrix into a user factor matrix and factor matrix solves the problem of sparse matrix; Pilaszy [9] proposed to use the ALS algorithm to solve the problem of the SVD algorithm that is difficult to deal with missing items. Zhou [10] improved the ALS algorithm and proposed a weighted regular alternating ALS algorithm, which handles user preferences through confidence weights. Wang [11] from the user's perspective, based on the score database, using the K-Means algorithm to cluster users, effectively alleviating the data sparsity problem and improving the recommendation accuracy; Lin [12] proposed singular value decomposition and fuzzy clustering, the CF algorithm improves the recommendation effect by narrowing the search range of the nearest neighbor; Gao [13] combined the Slope One algorithm with the content-based recommendation algorithm to improve the accuracy of the recommendation, but the effect is not ideal and the scalability is poor when the data matrix is sparse; Zhang [14] filled the sparse matrix by using the weighted Slope One algorithm to predict the score in the nearest neighbor set, which relatively eased the matrix The sparse problem improves the recommendation accuracy.

In this paper, based on the weighted SlopeOne collaborative filtering algorithm, the weighted SlopeOne algorithm is used to predict the score, filling the matrix; and then through the fuzzy clustering of items, and introducing a time decay function To quantify the weight of the user's interest, and then to improve the calculation of the item similarity, and finally recommend the Top k item set to the target user.

2 Related Work

2.1 IBCF Algorithm

The basic idea of the IBCF algorithm is to analyze and mine the similarity between items based on the historical evaluation information of all users, and then recommend a set of items with higher similarity to the user based on the user's historical preference information. Suppose that user A likes Item_1 and Item_3, user B likes Item_1, Item_2 and Item_3, and user C likes Item_1. From the historical preferences of these users, it can be considered that item Item_1 is similar to item Item_3. Those who like item Item_1 like item Item_3, based on this Judging that user C may also like Item_3, the recommendation system recommends item Item_3 to user C, as shown in Fig. 1:

The main steps of IBCF algorithm: firstly, format the user evaluation data of the project in the database, and convert it into a user row evaluation matrix $R_{U,I}$ of m rows and n columns, as shown in Table 1, where r_{ij} represents the score of user r_{ij} to I_j; then, according to all users rate the items, calculate the similarity between the items in pairs,

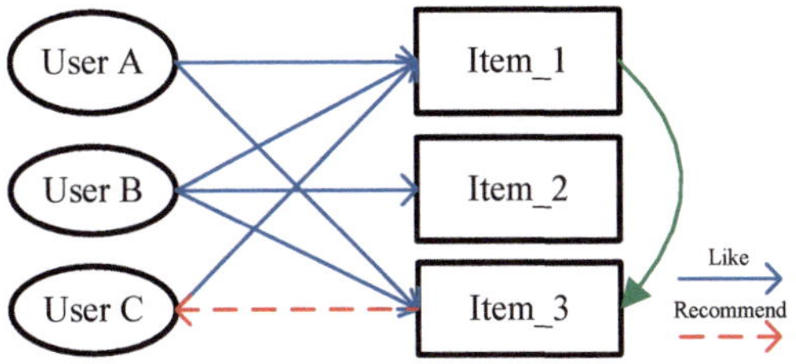

Fig. 1. IBCF algorithm.

Table 1. User item rating matrix $R_{U,I}$.

	I_1	I_2	…	I_n
U_1	r_{11}	r_{12}	…	r_{1n}
U_2	r_{21}	r_{22}	…	r_{2n}
…	…	…	…	….
U_m	r_{m1}	r_{m2}	…	r_{mn}

find the nearest neighbor according to the similarity, predict the preference; finally, generate a recommendation list to the target user.

2.2 Similarity Calculation

The important step of the collaborative filtering algorithm is the calculation of similarity. The current methods of calculating similarity are based on vectors. In fact, it calculates the distance between two vectors. The closer the distance, the greater the similarity.

Pearson Similarity. The Pearson correlation coefficient reflects the degree of linear correlation between two variables, and its value is between [−1, 1]. When the linear relationship between the two variables is enhanced, the correlation coefficient tends to 1 or −1; Due to decentralization and normalization, Pearson similarity can better measure the linear correlation between two variables than other methods; expressed by mathematical formula, Pearson correlation coefficient is equal to the co-variance of two variables divided by two The standard deviation of each variable, the formula is as follows:

$$sim_{a,b}^{Pearson} == \frac{\sum_{i \in U_{a,b}} (r_{ai} - \overline{r_a}) \cdot (r_{bi} - \overline{r_b})}{\sqrt{\sum_{i \in U_{a,b}} (r_{ai} - \overline{r_a})^2} \sqrt{\sum_{i \in U_{a,b}} (r_{bi} - \overline{r_b})^2}} \tag{1}$$

Where a and b are two n-dimensional vectors, $sim_{a,b}^{Pearson}$ is the Pearson similarity of a and b, $U_{a,b}$ is the rating set shared by a and b.Co-variance is a quantity describing the degree of correlation between a and b.

2.3 Weighted Slope One Algorithm

Slope One is a collaborative filtering algorithm based on scoring. It has easy implementation, fast calculation speed, good scalability, and good adaptability to data sparsity; it has other personalities compared with the recommendation algorithm, the Slope One algorithm does not need to calculate the similarity between items, but uses a simple linear regression model $w = f(v) + b$ to predict, where w represents the target user's rating of the item, and f(v) represents the target user Rate items other than this item, b is the average deviation of item i relative to other items (use j for other items), expressed by $Dev_{i,j}$, as shown in Fig. 2, there are two users, user A and user B, of which user A Both Item 1 and Item 2 were evaluated. User B only evaluated Item 1. The above conditions were used to predict User B's rating for Item 2. Using the Slope One algorithm, we know that user B's rating for item 2 is: 2 + (1.5 − 1) = 2.5.

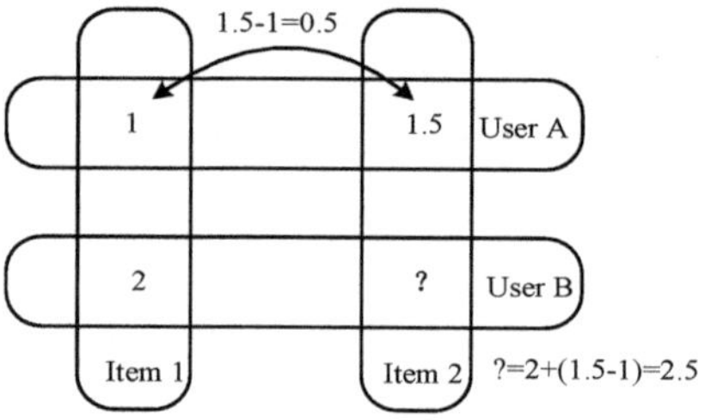

Fig. 2. The idea of slope one algorithm.

Let $Dev_{i,j}$ denote the average rating deviation between the target item and other items. The calculation formula is shown in Eq. 1.

$$Dev_{i,j} = \sum_{u \in S_{i,j}(x)} \frac{r_{u,i} - r_{u,j}}{Num(S_{i,j}(x))} \tag{2}$$

Where x represents the training set, $S_{i,j}(x)$ represents the set of users who have rated item i and other items (use j for other items), and $Num(S_{i,j}(x))$ represents the total number of users in the $S_{i,j}(x)$ set.

Based on the above analysis and on the basis of formula (1) is known, the predicted score of the target user on the target item can be calculated by using $P_{u,i} = Dev_{i,j} + r_{u,j}$,

$$P_{u,i} = \frac{1}{Num(r_i)} \times \sum_{j \in r_i} (Dev_{i,j} + r_{u,j}), r_i = \{j | j \in S(u), i \neq j\} \tag{3}$$

Where r_i represents the target user's set of all rating items, and $Num(r_i)$ represents the number of target user rating items.

The Slope One algorithm does not consider the number of items evaluated by the user but calculates the average when calculating the relative average deviation $Dev_{i,j}$ of the two items, which lacks reasonableness. Therefore, Lemire D proposed the

Weighted Slope One algorithm, which is the weighted Slope One algorithm. The idea of the algorithm is that the greater the number of commodities that are evaluated together with the target commodities, the greater the weight. The algorithm effectively corrects the calculation of the evaluation deviation. The accuracy of the recommendation is further improved. The formula is shown in Eq. (4),

$$P_{u,i} = \frac{\sum\limits_{j \in S(u)} (Dev_{i,j} + r_{u,j})\left|C_{i,j}\right|}{\sum\limits_{j \in S(u)} \left|C_{i,j}\right|} \tag{4}$$

Among it, $P_{u,i}$ represents the rating prediction of user u for item i, and $\left|C_{i,j}\right|$ is the weight of the number of users.

2.4 Fuzzy C-Means Clustering Algorithm

Define any element x in the domain D. If there is a correspondence between $U(x)$ and x, then $U(x)$ is called the fuzzy set on D, and $U(x)$ is called the membership degree of D pair. The membership function is a function that represents the degree to which an object belongs to a set. It is denoted as $U(x)$ and has a value range of [0, 1]. The closer the function value is to 1, the higher the membership degree;

Fuzzy C-means clustering introduces the concept of fuzzy logic. It uses a membership matrix to indicate the degree to which each object belongs to a certain class. The objective function of fuzzy C-means clustering is shown in Eq. (5):

$$J(U,C) = \sum_{s=1}^{N}\sum_{i=1}^{K} (\mathrm{u}_{is})^m \times dist(c_i, x_s)^2 \tag{5}$$

N is the size of the data set, K is the number of cluster centers, and m is the weighted index. That is to say, the objective function $J(U,C)$ of fuzzy clustering is the weighted square sum of each data point to each clustering center. In addition to calculating the clustering center, fuzzy clustering does not directly classify data points into a certain class, but calculates the membership matrix and takes $\{i,u\} = \max_i\{U\}$. The above problem is to find the minimum value of $J(U,C)$ under the condition of membership $\sum\limits_{i=1}^{K} u_{is} = 1$, as shown in Eq. (6).

$$\min\{J(U,C)\} = \min\left\{\sum_{i=1}^{K} (u_{is})^m \times dist(c_i, x_s)^2\right\} \tag{6}$$

We need to transform the above conditional extremum problem into an unconditional extremum problem, introduce n Lagrangian factors, then use the Lagrangian multiplier method to convert the problem into an unconditional extremum problem. Degree matrix U, the process is as follows:

$$F = \sum_{i=1}^{K} (u_{is})^m \times dist_{is}^2 + \lambda(\sum_{i=1}^{K} u_{is} - 1) \quad (7)$$

From formula (7), we can get:

$$\frac{\partial F}{\partial u_{is}} = m \times u_{is}^{m-1} \times dist_{is}^2 - \lambda = 0 \Rightarrow u_{is} = \left(\frac{\lambda}{m}\right)^{\frac{1}{m-1}} \left(\frac{1}{dist_{is}^2}\right)^{\frac{1}{m-1}} \quad (8)$$

$$\frac{\partial F}{\partial \lambda} = (\sum_{i=1}^{K} u_{is} - 1) = 0 \quad (9)$$

$$\sum_{i=1}^{K} (\frac{\lambda}{m})^{\frac{1}{m-1}} (\frac{1}{dist_{is}^2})^{\frac{1}{m-1}} - 1 = 0 \Rightarrow (\frac{\lambda}{m})^{\frac{1}{m-1}} = 1/\sum_{i=1}^{K} (dist_{is}^2)^{\frac{1}{m-1}} \quad (10)$$

Put Eq. (10) into Eq. (8) to get u_{is}, as shown in Eq. (11), thereby obtaining the membership matrix U:

$$u_{is} = 1/\sum_{j=1}^{K} \left(\frac{dist_{is}^2}{dist_{js}^2}\right)^{\frac{2}{1-m}} \quad (11)$$

The solution process of cluster center C is as follows:

$$\frac{\partial J(U,C)}{\partial c_i} = \frac{\sum_{s=1}^{N} \sum_{i=1}^{K} (\mathrm{u}_{is})^m \times \partial dist(c_i, x_s)^2}{\partial c_i} \quad (12)$$

$$dist(c_i, x_s)^2 = \|c_i - x_s\| A = (c_i - x_s)^T A (c_i - x_s)^2 \quad (13)$$

A is the weight, and the formula (12) and the formula (13) are combined to obtain the formula (14):

$$\frac{\partial J(U,C)}{\partial c_i} = \sum_{s=1}^{N} u_{is}^m \frac{\partial (c_i - x_s)^T A (c_i - x_s)}{\partial c_i} = \sum_{s=1}^{N} u_{is}^m (2A(c_i - x_s)) = 0 \quad (14)$$

You can get the clustering center, as shown in Eq. (15):

$$2A\left(\sum_{s=1}^{N} u_{is}^m c_i - \sum_{s=1}^{N} u_{is}^m x_s\right) = 0 \Rightarrow c_i = \frac{\sum_{s=1}^{N} u_{is}^m x_s}{\sum_{s=1}^{N} u_{is}^m} \quad (15)$$

3 Proposed Algorithm Model

The algorithm proposed in this paper is mainly based on the following considerations: the first is to use the weighted Slope One algorithm to process the user scoring matrix; the second is to fuzzy cluster the items to determine the similarity group of the items; the third is calculating the similarity in the cluster set, a time decay function is introduced, which fully takes into account the factors that the user's interest decreases with time. At the same time, calculating the similarity in the cluster set which reduces the calculation scale and calculation time complexity, Fig. 3 shows the algorithm Process.

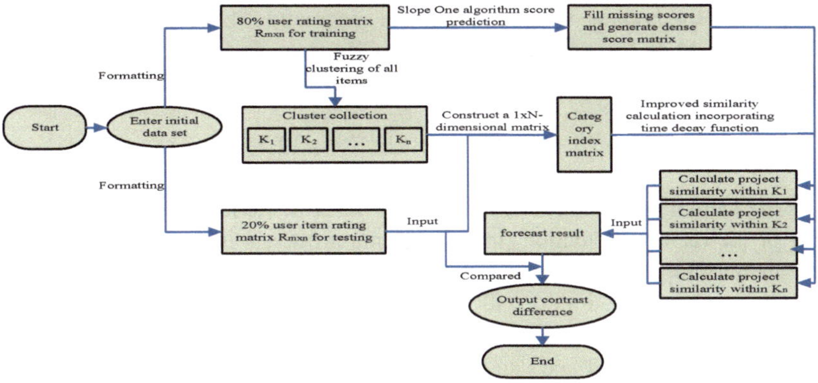

Fig. 3. Algorithm flow chart.

3.1 Fuzzy C-Means Clustering of Itemsets

Clustering items is an effective means to reduce computational complexity and improve system performance. Facing the multi-attribute and multi-features of the project, fuzzy clustering uses mathematical methods to quantitatively determine the intimacy and thinning relationship of the samples. Compared with the hard-clustering algorithm, it can more objectively divide the sample types.

Let N be the size of the item set, K be the number of cluster centers, $U_{N \times K} = \{U_{i,s}\}$ be the membership matrix, $C = \{c_1, c_2, \ldots, c_k\}$ be the list of cluster centers, and Eq. (16) be the objective function for fuzzy clustering of items,

$$J(U, C) = \sum_{s=1}^{N} \sum_{i=1}^{K} (U_{is})^m \times \left(sim_{i_s, c_i}^{Pearson} \right)^2 \tag{16}$$

Where $sim_{i_s, c_i}^{Pearson}$ is the similarity between the item s and the i-th cluster center using the Pearson similarity method, the algorithm flow is as follows:

Input: number of clusters K and item vector table;

Output: the membership matrices and cluster centers of samples for various classes;

Step 1: Initialize each cluster center;

Step 2: Repeat the follows, until the objective function converges.
Step 3: Use the current clustering center to calculate the membership function according to the following formula and update the membership matrix U;
Step 4: Use the current membership function to update and calculate cluster center C;

3.2 Calculation of User Interest Degree and Item Similarity Incorporating Time Forgetting Weight

As a biological individual, people's interest in things tends to shift with time. Therefore, the traditional collaborative filtering algorithm does not consider the factor of loss of interest over time when calculating the similarity, so it is difficult for the algorithm to break through the bottleneck and improve the accuracy of the recommendation results.

Using the Matlab curve fitting toolbox CFtool to fit the Ebbinghaus forgetting curve, we can get the forgetting fitting function (17):

$$\mathrm{f}(x) = 34.92 \times x^{-02028} + 12.71 \tag{17}$$

$f(x)$ is the memory retention rate, the range is 0.0–1.0, the larger of the $f(x)$ value, the higher of the memory retention (interest retention rate); x is the time (days) after the initial memory input. It can be seen from the characteristics of the power function that the memory retention rate $f(x)$ will gradually decrease with the increase of time.

Using the above forgetting fitting function can track the user's interest changes, which can be constructed and can further improve the prediction accuracy;

$$T_{dev} = t_{un} - t_{ui} \tag{18}$$

In Eq. (18), T_{dev} is the difference between the number of days of the latest evaluation and the initial evaluation of the user, t_{un} is the time of the latest evaluation of user u, and t_{ui} is the time of the first evaluation of the project by user u;

$$\mathrm{f}(T_{dev}) = 34.92 \times (T_{dev} + 1)^{-02028} + 12.71 \tag{19}$$

Where $\mathrm{f}(T_{dev})$ is the retention rate of user u's interest in the project.

By calculating the user's $\mathrm{f}(T_{dev})$ for all historical items, and then calculating and accumulating the retained interest values of the items contained in each class, find the cluster set Q in which the user is interested at this stage; finally, the evaluation values of all items in the interest cluster are updated, Eq. (20) is the improved method of similarity.

$$sim_{I_i,I_j}^{Pearson\_time} = \frac{\sum_{u=1}^{n} \left(r_{ui} \times \mathrm{f}(T_{dev}) - \overline{r_{I_i}}\right) \cdot \left(r_{uj} \times \mathrm{f}(T_{dev}) - \overline{r_{I_j}}\right)}{\sqrt{\sum_{u=1}^{n} \left(r_{ui} \times \mathrm{f}(T_{dev}) - \overline{r_{I_i}}\right)^2} \sqrt{\sum_{u=1}^{n} \left(r_{uj} \times \mathrm{f}(T_{dev}) - \overline{r_{I_j}}\right)^2}} \tag{20}$$

3.3 Generate a Recommendation Set

Since fuzzy clustering was performed on the item set in Sect. 3.1, a cluster set was generated, and the similarity between all recommended items and the historical items of the target user was calculated by using the improved Pearson similarity calculation method in Sect. 3.2 within the cluster set. The target user's rating of the recommended items can be obtained using Eq. (21), find the items with the highest similarity to the items, and select the Top N items as the recommendation set.

$$P_{u,l} = \overline{r_l} + \frac{\sum_{i=1}^{N} (r_{u,l_i} - \overline{r_{l_i}}) \times sim(l, l_i)}{\sum_{i=1}^{N} |sim(l, l_i)|} \tag{21}$$

In Eq. (21), $sim(l, l_i)$ represents the similarity between items l and l_i, N represents the number of nearest neighbors with a high similarity to item i, $\overline{r_l}$ represents the average rating of the user on item i, and $\overline{r_{l_i}}$ represents the average rating of the user on item l_i.

4 Experiment and Result Analysis

4.1 Experimental Environment and Data Set Selection

Experimental environment: macOS system, hardware configuration parameters: CPU is 2.3 GHz quad-core Intel Core i5, memory is 8 GB 2133 MHz, experimental language environment is Python3.7; ml-20 m was collected by the GroupLens research team at the University of Minnesota, and describes the 5-star rating and free text labeling activities of the movie recommendation service. It contains 20000263 ratings and 465564 tags from 27278 movies. Movies.csv contains movie ID, movie name, movie category/genre; Rating.csv contains user ID, movie ID, rating, and timestamp.

4.2 Evaluation Standard

In the standard evaluation of the algorithm model in this paper, Mean Absolute Error (MAE) is used to measure the effectiveness of recommendation results. The mean absolute error (MAE) measures the accuracy of the algorithm by calculating the average deviation between the predicted value of the user score and the actual value of the user score. For example, if the predicted score vector of n items is $\{P_1, P_2, \ldots, P_n\}$, and the actual score vector is $\{T_1, T_2, \ldots, T_n\}$, then the algorithm is expressed as:

$$MAE = \frac{\sum_{i=1}^{n} |P_i - T_i|}{n} \tag{22}$$

4.3 Comparison and Analysis of Experimental Results

The first experiment is set up to verify the accuracy of the model under different cluster numbers, and observe the influence of the number of clusters on the experiment at intervals of 4 in the interval of cluster numbers of clusters [0, 30]. The results are shown in Fig. 4 As shown, when the number of clusters is less than 16, the MAE value is in a decreasing state, and the best is achieved when the number of clusters is 10, and when it is greater than 10, the MAE value shows a slowly rising trend, indicating that the number of clusters When it is 16, it is the best state of the algorithm model in this paper.

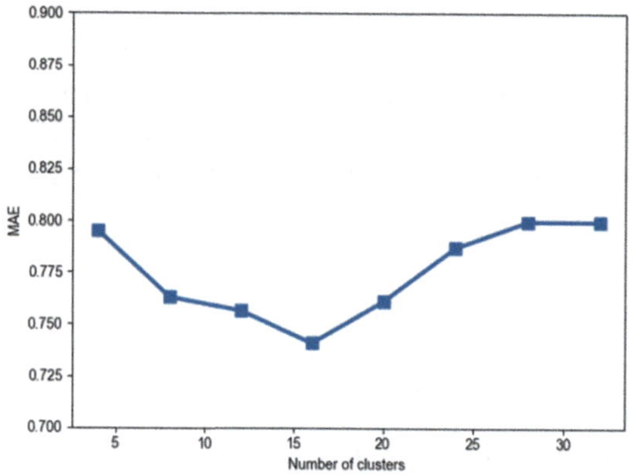

Fig. 4. MAE under different clustering numbers.

The second experiment is conducted comparative experiments on different numbers of nearest neighbors to verify the effectiveness and stability of the algorithm model in this paper. Compare the algorithm of this paper with the traditional item-based CF, Slope One algorithm (SO), cluster-based Weighted-Slope One algorithm (IC-WSO) and the weighted Slope One algorithm with user similarity (US-WSO) under the condition of 16 clusters. As can be seen from Fig. 5, as the number of nearest neighbors gradually increases, the average absolute error of the Slope One algorithm is not limited by the number of nearest neighbors, so the average absolute error of the text algorithm and the other three algorithms decreases. The trend then gradually approached a flat state. It means that the nearest neighbors are not as good as possible, but the weight of influence on MAE becomes smaller after reaching the optimal number.

After experiments and comparison with other algorithms, the effectiveness and stability of the text algorithm are verified. Finally, through Fig. 6 we can see that the algorithm in this paper does improve the recommendation effect, and to some extent, it has effectively improved the recommendation accuracy compared to the previous algorithm.

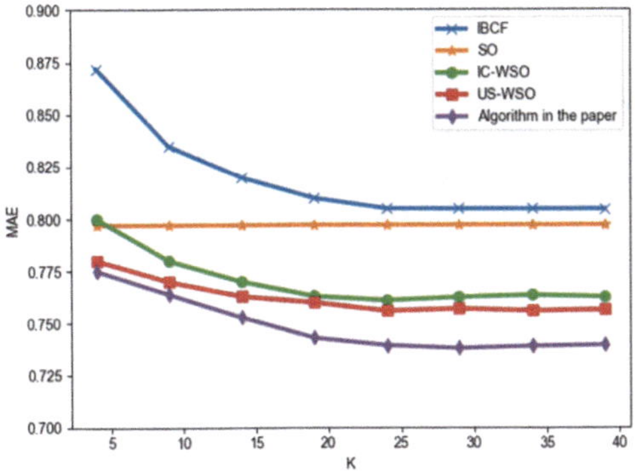

Fig. 5. MAE changes with K (nearest neighbor).

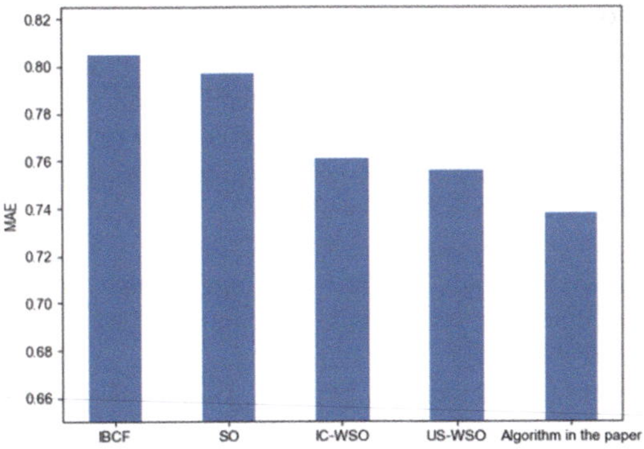

Fig. 6. Comparison between those algorithms.

5 Conclusion

This article analyzes the ideas and defects of the traditional Slope One algorithm, and describes and references some breakthroughs and achievements made by scholars on this research. It aims at the accuracy and stability of the traditional Slope One algorithm in prediction and recommendation. For further optimization, first use the weighted SlopeOne algorithm to score prediction and fill the scoring matrix, then fuzzy cluster the items, and then use the target user's historical items combined with the interest forgetting function to determine the current user's preference category, in the cluster The IBCF algorithm is used to find the optimal set of recommended items. Finally, the algorithm model of this paper is compared and analyzed with other classic algorithm models through experiments to verify the effectiveness of the algorithm model of this paper, and to a certain extent, improve the efficiency and accuracy of recommendation rate.

References

1. Amatriain, X., Basilico, J.: Past, present, and future of recommender systems: an industry perspective. In: RecSys 2016: Proceedings of the 10th ACM Conference on Recommender Systems, pp. 211–214. ACM, NY, USA (2016)
2. Wang, L., Meng, X., Zhang, Y.: Context-aware recommender systems. J. Softw. **23**(1), 1–20 (2012)
3. Chen, P., Chen, C., Hong, Y.: A collaborative filtering recommendation algorithm combining association rules. J. Chin. Comput. Syst. **37**(02), 287–292 (2016)
4. Breese, J.S., Hecherman, D., Kadie, C.: Empirical analysis of predictive algorithms for collaborative filtering. In: Proceedings of the 14th Conference on Uncertainty in Artificial Intelligence (UAI 1998), pp. 43–52. Morgan Kaufmann Publishers Inc., CA, USA (1998)
5. Yin, F., Wang, Z., Tan, W., Xiao, W.: Sparsity-tolerated algorithm with missing value recovering in user-based collaborative filtering recommendation. J. Inf. Comput. Sci. **10**(15), 4939–4948 (2013)
6. Sarwar, B., Karypis, G., Konstan, J., Riedl, J.: Item-based collaborative filtering recommendation algorithms. In: WWW 2001: Proceedings of the 10th International Conference on World Wide Web, pp. 285–295. ACM, NY, USA (2001)
7. Deng, A., Zuo, Z., Zhou, Y.: Collaborative filtering recommendation algorithm based on item clustering. J. Chin. Comput. Syst. **25**(9), 1665–1670 (2004)
8. Billsus, D., Pazzani, M. J.: Learning collaborative information filters. In: ICML 1998: Proceedings of the Fifteenth International Conference on Machine Learning, pp. 46–54. Morgan Kaufmann Publishers Inc., CA, USA (1998)
9. Pilaszy, I., Zibriczky, D., Tikk, D.: Fast ALS-based matrix factorization for explicit and implicit feedback datasets. In: Proceedings of the fourth ACM Conference on Recommender Systems, pp. 71–78. ACM, NY, USA (2010)
10. Zhou, Y., Wilkinson, D., Schreiber, R., Pan, R.: Large-scale parallel collaborative filtering for the netflix prize. In: Proceedings of the 4th international conference on Algorithmic Aspects in Information and Management, pp. 337–348. Springer, Heidelberg (2008)
11. Wang, Z., Yu, X., Feng, N.: An improved collaborative movie recommendation system using computational intelligence. J. Vis. Lang. Comput. **25**(6), 667–675 (2014)
12. Lin, J., Yan, X., Huang, B.: Collaborative filtering recommendation algorithm based on SVD and fuzzy clustering. Appl. Comput. Syst. **25**(11), 156–163 (2016)
13. Gao, M., Wu, Z.: Incorporating personalized contextual information in item-based collaborative filtering recommendation. J. Softw. **5**(7), 729–736 (2010)
14. Zhang, Z., Tang, X., Chen, D.: Applying user- favorite- item- based similarity into Slope One scheme for collaborative filtering. In: 2014 World Congress on Computing and Communication Technologies, pp. 203–205. IEEE, NJ (2014)

Simulation Analysis of Output Characteristics of Power Electronic Transformers

Zhiwei Xu, Zhimeng Xu(✉), and Weicai Xie

Hunan Provincial Key Laboratory of Wind Generator and Its Control, Hunan Institute of Engineering, Xiangtan 411101, China
1010884740@qq.com

Abstract. Output current is an important parameter of the power electronic transformer. In order to study the control strategy of the output current of the power electronic transformer, this paper, aiming at ac-dc-ac type power electronic transformer, carries out the corresponding control strategy simulation of the power electronic transformer in MATLAB/SIMULINK, including: PWM control, parallel operation, SVPWM control and current tracking hysteresis comparison mode. The simulation results show that the output current of the power electronic transformer can present different output characteristics by using different control strategies, and the output characteristics can be optimized to some extent.

Keywords: Power electronic transformer · Output current · PWM · SVPWM · Parallel operation

1 Preface

Power Electronic Transformer (PET) is also known as electronic power transformer [1], and is also known as solid state transformer (Solid State Transformer, SST) and flexible transformer (Flexible Transformer, FT). It is a new type of transformer that introduces power electronics technology and high-frequency transformer to achieve energy conversion energy transfer [2]. This new transformer not only has the functions of a traditional power transformer, but also has the advantages of small size, light weight, safety and environmental protection. In addition, it also has special functions such as improving power quality and reactive power compensation [3, 4].

With the rapid development of high-power devices and power electronic technology, power electronic transformers have received extensive attention [5]. Literature [6] proposed a power electronic transformer topology suitable for high-power power systems. The control strategy is analyzed to provide a good reference; literature [7] proposes a control strategy for a permanent magnet wind power generation system based on matrix power electronic transformers, which achieves a good energy conversion of the wind power generation system; literature [8] design A new grid-connected device structure for power electronic transformers is proposed, and the corresponding grid-connected device control strategy is proposed to achieve fast and accurate voltage conversion.

Q. Liu et al. (Eds.): CENet 2020, AISC 1274, pp. 1209–1218, 2021.
https://doi.org/10.1007/978-981-15-8462-6_138

This article first introduces the topology of PET, selects AC-DC-AC PET with DC link as the research object, establishes its typical general mathematical model, and then performs control simulation analysis on it, and finally verifies PET through simulation modeling The feasibility of output current control strategy.

2 Topological Structure and Working Principle of Power Electronic Transformer

The structure of power electronic transformer includes two basic elements: power electronic converter and high frequency transformer. According to the presence or absence of DC link, power electronic transformers can be divided into two categories: AC-AC type and AC-DC-AC type [9, 10]. Figure 1 is a circuit diagram of an AC-DC-AC type power electronic transformer. This article is based on the AC-DC-AC type power electronic transformer.

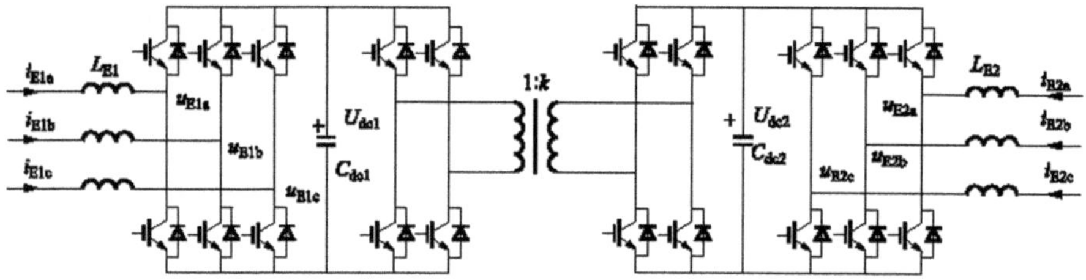

Fig. 1. Topology of ac-dc-ac power electronic transformer.

The working principle of the power electronic transformer is shown in Fig. 2. The primary side is connected to the power supply. The original power electronic converter rectifies the input current of the power supply into DC, and then converts it into high frequency AC, which is coupled to the secondary through the primary winding of the high frequency transformer. at this time, the secondary winding of the high-frequency transformer is induced by the law of electromagnetic induction, and this induced electromotive force is applied to the secondary power electronic converter connected to the secondary side of the high-frequency transformer. Inverse transformation of the power frequency to obtain the required power frequency AC power output [11, 12].

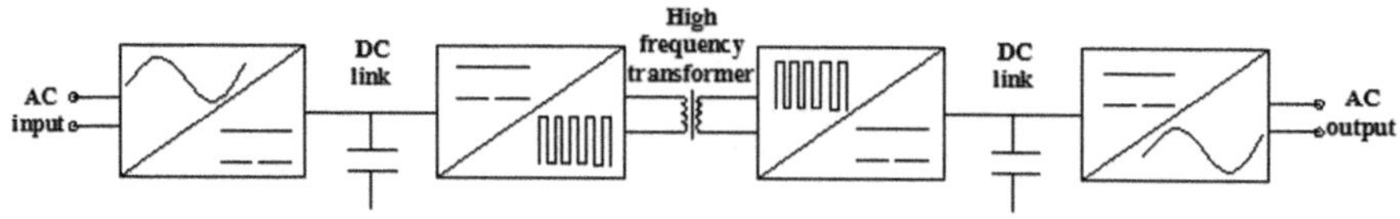

Fig. 2. Principle of power electronic transformer working.

3 Mathematical Model of Power Electronic Transformer

The PET studied is a typical three-phase PET implementation.

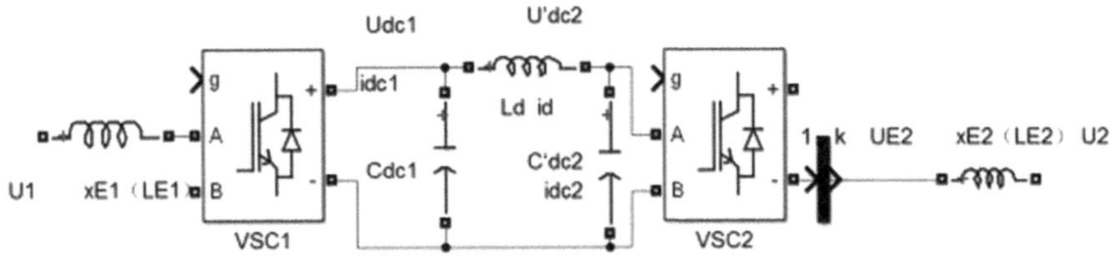

Fig. 3. Equivalent physical model.

For the AC-DC-AC type power electronic transformer, it can be equivalent to the form shown in Fig. 3, According to the basic theory of the circuit and the simplified physical model, the PET mathematical model can be obtained as shown in (1)–(8). i_{d1} and i_{d2} are the current flowing through the main and auxiliary power electronic converters, U_{dc1} and U'_{dc2} are the voltage across the main and auxiliary power electronic converters.

$$\begin{cases} L_{E1}\dfrac{di_{E1a}}{dt} = u_{1a} - u_{E1a} \\ L_{E1}\dfrac{di_{E1b}}{dt} = u_{1b} - u_{E1b} \\ L_{E1}\dfrac{di_{E1c}}{dt} = u_{1c} - u_{E1c} \end{cases} \tag{1}$$

$$\begin{cases} L_{E2}\dfrac{di_{E2a}}{dt} = u_{2a} - u_{E2a} \\ L_{E2}\dfrac{di_{E2b}}{dt} = u_{2b} - u_{E2b} \\ L_{E2}\dfrac{di_{E2c}}{dt} = u_{2c} - u_{E2c} \end{cases} \tag{2}$$

Considering the transient process of the high-frequency transformer, L_d is the equivalent inductance, U_{dc2}, C_{dc2} is the conversion amount on the primary side U'_{dc2}, C'_{dc2}.

$$\begin{cases} C_{dc1}\dfrac{dU_{dc1}}{dt} = i_{dc1} \\ C'_{dc2}\dfrac{dU'_{dc2}}{dt} = i_{dc2} \\ L_d\dfrac{di_d}{dt} = U_{dc1} - U'_{dc2} \end{cases} \tag{3}$$

$$\begin{cases} i_{dc1} = i_{d1} - i_d = \sum\limits_{j=a,b,c} i_{E1j} d_{E1j} - i_d \\ i_{dc2} = i_{d2} + i_d = \sum\limits_{j=a,b,c} k i_{E2j} d_{E2j} + i_d \end{cases} \quad (4)$$

$$\begin{cases} d_{E1a} = (m_{E1}/\mathbf{2}) \cos(\omega \mathbf{t} - \theta_{E1}) + \mathbf{1/2} \\ d_{E1b} = (m_{E1}/\mathbf{2}) \cos[\omega t - (\mathbf{2\pi/3}) - \theta_{E1}] + \mathbf{1/2} \\ d_{E1c} = (m_{E1}/\mathbf{2}) \cos[\omega t + (\mathbf{2\pi/3}) - \theta_{E1}] + \mathbf{1/2} \end{cases} \quad (5)$$

$$\begin{cases} d_{E2a} = (m_{E2}/\mathbf{2}) \cos(\omega \mathbf{t} - \theta_{E2}) + \mathbf{1/2} \\ d_{E2b} = (m_{E2}/\mathbf{2}) \cos[\omega t - (\mathbf{2\pi/3}) - \theta_{E2}] + \mathbf{1/2} \\ d_{E2c} = (m_{E2}/\mathbf{2}) \cos[\omega t + (\mathbf{2\pi/3}) - \theta_{E2}] + \mathbf{1/2} \end{cases} \quad (6)$$

$$\begin{cases} u_{E1a} = (m_{E1} U_{dc1}/\mathbf{2}) \cos(\omega \mathbf{t} + \theta_{E1}) \\ u_{E1b} = (m_{E1} U_{dc1}/\mathbf{2}) \cos[\omega t - (\mathbf{2\pi/3}) + \theta_{E1}] \\ u_{E1c} = (m_{E1} U_{dc1}/\mathbf{2}) \cos[\omega t + (\mathbf{2\pi/3}) + \theta_{E1}] \end{cases} \quad (7)$$

$$\begin{cases} u_{E2a} = (k m_{E2} U_{dc2}/\mathbf{2}) \cos(\omega \mathbf{t} + \theta_{E2}) \\ u_{E2b} = (k m_{E2} U_{dc2}/\mathbf{2}) \cos[\omega t - (\mathbf{2\pi/3}) + \theta_{E2}] \\ u_{E2c} = (k m_{E2} U_{dc2}/\mathbf{2}) \cos[\omega t + (\mathbf{2\pi/3}) + \theta_{E2}] \end{cases} \quad (8)$$

k is the high-frequency transformer transformation ratio, and θ is the phase angle of the control signal.

4 Simulation Analysis

A simulation model of AC-DC-AC was built in SIMULINK, and the research was conducted by using different control methods. The simulation parameters are set as follows: three-phase power supply phase-to-phase voltage $V = \mathbf{380\ v}$, phase A phase angle $\alpha = \mathbf{30}^\circ$, frequency $\omega = \mathbf{50\ Hz}$; double pulse, pulse width $t = \mathbf{25}^\circ$, high frequency transformer frequency $f = \mathbf{500\ Hz}$.

4.1 PWM Control

Figure 4 shows the PWM control model of the power electronic transformer. the modulation signal is the amplitude $A = \mathbf{1}$. Frequency $f = \mathbf{100\ \pi}$ and phase are sequentially sinusoidal signals with 120° difference; using two-level PWM converter, three-phase six-pulse bridge, carrier frequency $f_Z = \mathbf{90} * \mathbf{50\ Hz}$; filter inductance and capacitance $L = \mathbf{1} * \mathbf{10^{-3}H}$, $C = \mathbf{10.13214} * \mathbf{10^{-3}F}$, simulation duration $T_1 = \mathbf{0.2\ s}$.

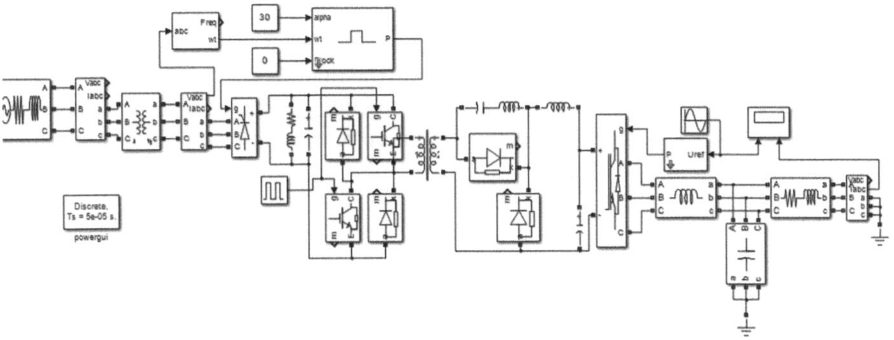

Fig. 4. PWM control model for power electronic transformer.

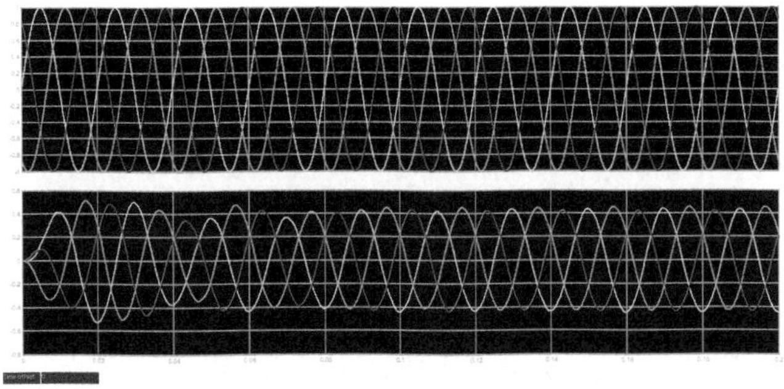

Fig. 5. Modulation signal and output current waveform.

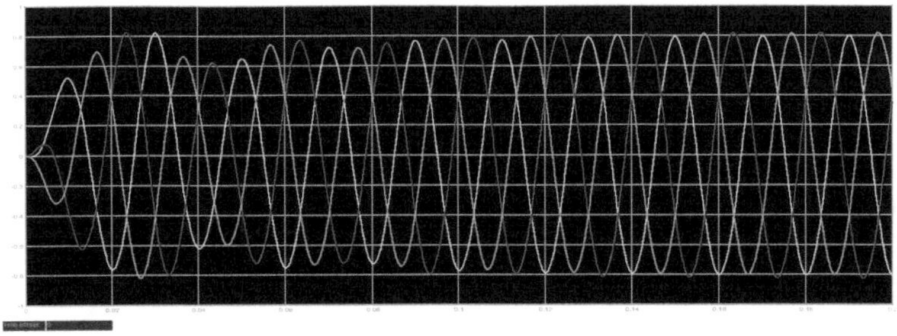

Fig. 6. Increase the output current waveform of the sampling point.

The modulation signal and output current waveform are shown in Fig. 5. The upper waveform is the modulation signal, and the lower waveform is the PET output current waveform. At the left and right of **0.1 s**, the output current waveform is stable. Increasing the carrier frequency and sampling point, as shown in Fig. 6, obviously increases the carrier frequency, and the output current waveform becomes smoother.

4.2 PET Running in Parallel

The key to running PET in parallel is to achieve a reasonable distribution of load between PET and to achieve current sharing. On the basis of the above PET, a PET model with the same parameter settings is incorporated. The two PETs share a three-phase power supply. The control method is still the above-mentioned PWM control method. The other parameter settings are unchanged, and the simulation results are as follows. As shown in Fig. 7.

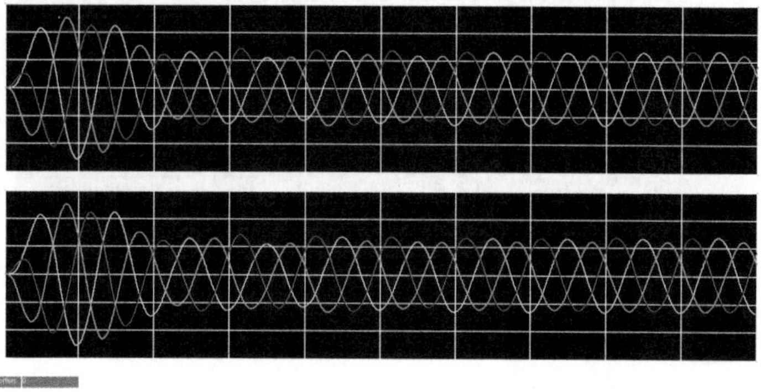

Fig. 7. Pet parallel running output current waveform.

The two upper and lower waveforms in Fig. 7 are the output current waveforms of two PETs respectively. The output current waveforms of the two PETs are the same. There is no circulating current between them, and the current sharing control is well implemented, which greatly improves the reliability and flexibility of power supply.

4.3 SVPWM Control

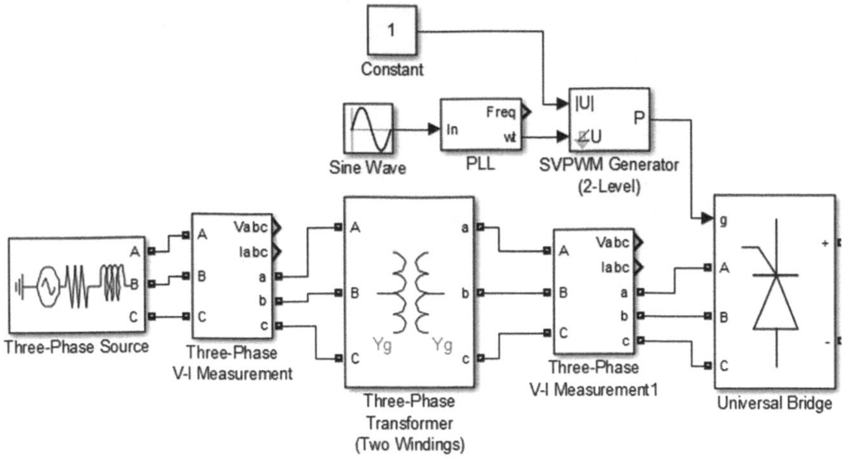

Fig. 8. SVPWM control module.

The SVPWM control module is shown in Fig. 8. SVPWM amplitude phase control and switching mode 1. Initial phase $A_1 = \mathbf{0}^{\circ}$, initial frequency $f_1 = \mathbf{50\ Hz}$, PWM Frequency $f_{\mathbf{pwm}} = \mathbf{33 * 50\ Hz}$, duration of simulation $T_2 = \mathbf{0.05\ s}$. As shown in Fig. 9.

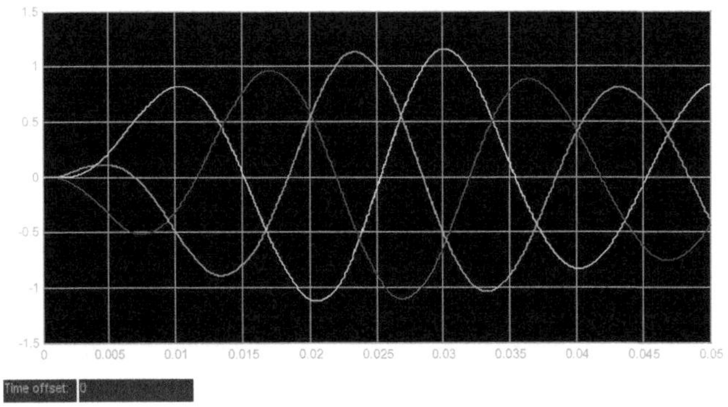

Fig. 9. SVPWM controls the output current waveform.

The PLL will output by default without being controlled by the waveform, and then observe the phase waveform, as shown in Fig. 10. The phase waveform is a sawtooth wave with a minimum value of 0, a maximum value of 2 pi, and a frequency of 50 Hz. From the simulation results, the output current waveform is stable, well controlled, and the current distortion rate is small.

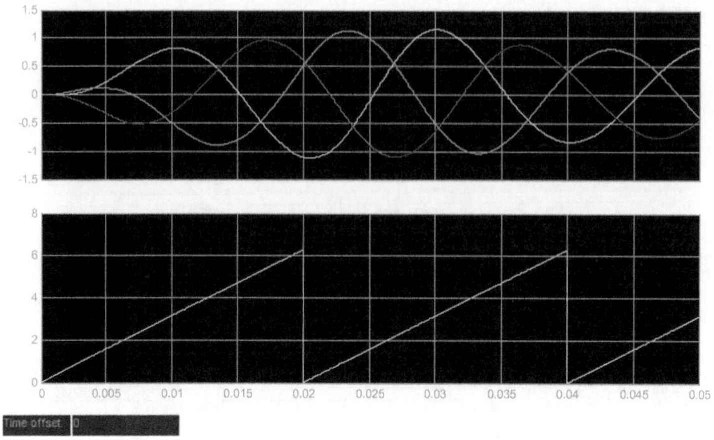

Fig. 10. Output current and phase waveform.

4.4 Closed-Loop Current Tracking Hysteresis Comparison

Power electronic transformers based on the current tracking hysteresis comparison method directly use the given three-phase current command signal to make a difference with the output current signal on the inverter side, The value is passed to the hysteresis controller, the PWM signal is obtained, and the inverter is controlled by the PWM module [13–15]. Figure 11 shows the simulation model of PWM current tracking control using hysteresis comparison. The three-phase current command signal differs by 120° in sequence, and the filter inductance $L_1 = \mathbf{5.5\ mH}, L_2 = \mathbf{0.1\ mH}$, filter capacitor $C = \mathbf{20}\ \mu\mathbf{F}$ and series resistance $R = \mathbf{3\ \Omega}$ are selected. The switching state of the hysteresis controller is 0.01, −0.01, and the simulation duration $T_3 = \mathbf{0.2\ s}$, the output current waveform shown in Fig. 12. From the simulation results, the output current is stable, the three-phase symmetry.

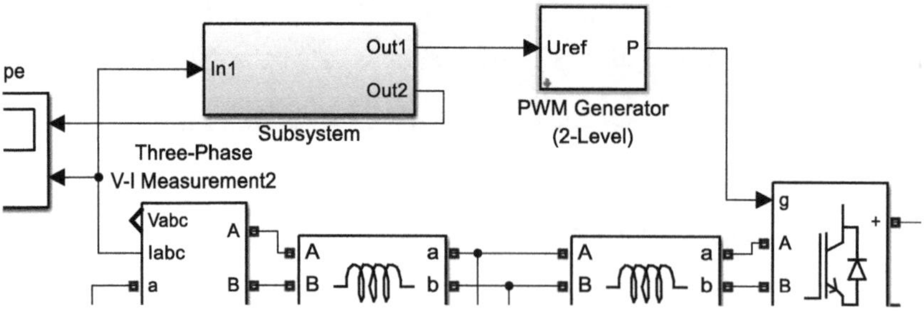

Fig. 11. Current tracking hysteresis module.

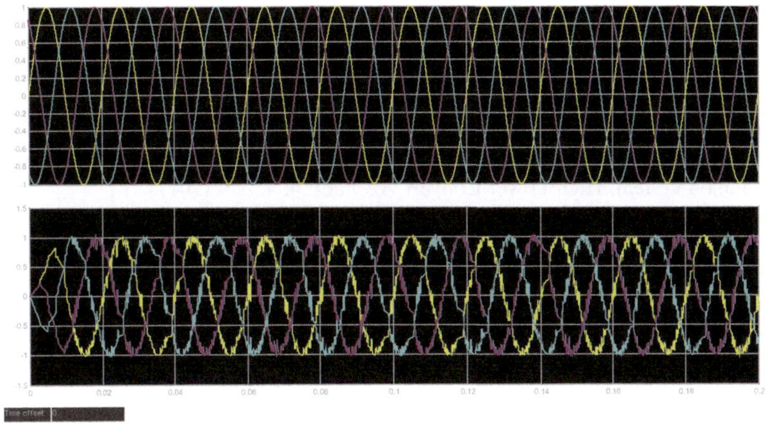

Fig. 12. Output current waveform under modulation signal and current tracking control.

5 Conclusion

In this paper, under the MATLAB/SIMULINK environment, the AC-DC-AC type power electronic transformer output current control strategy is studied. Among them, PWM control can make the output current waveform smooth by increasing the carrier frequency; double PET Parallel operation can effectively achieve current sharing, reduce the load of each PET, and avoid fault strikes; SVPWM control can stabilize the current output, low distortion rate, and good control effect; the current tracking hysteresis comparison method belongs to closed-loop control and can also be well controlled The output current is stable, the three phases are symmetrical and there is little fluctuation. The above four control strategies have optimized the output current waveform to a certain extent, thus verifying the feasibility of the control strategy.

Acknowledgment. The work in this paper was supported by the Hunan Youth Education Department Outstanding Youth Project (18B381); the wind power generator and the Open Research Fund of the Key Laboratory of Hunan Province (FLFD1902); Key projects in Hunan Province Department of Education (18A348).

References

1. Wang, D.: Electronic power transformer of distribution system. Huazhong University of Science and Technology (Bosch Paper) (2006)
2. Zhao, Z.M., Feng, G.H., Yuan, L.Q., et al.: Development and key technologies of electric energy routers. Chin. J. Electr. Eng. **37**(13), 3823–3834 (2017)
3. Li, Z.X., Gao, F.Q., Zhao, C., et al.: Review of power electronic transformer technology research. Chin. J. Electr. Eng. **38**(5), 1274–1289 (2018)
4. Pan, S.F., Zhao, J.F.: A summary of power electronic transformers and their development. Jiang Su Electr. Eng. **22**(6), 52–54 (2003)

5. Guan, J.P., Xu, Y.H.: A review of the research on the application of power electronic transformers in wind power generation systems. New Technologies of Electrical Energy (2018)
6. Zhou, Y.C., Yu, F.: Overview of high-power power electronic transformers. Marine Electr. Technol. **38**(4), 1–6 (2018)
7. Wang, H., Su, M.: Matrix power electronic transformer for permanent magnet wind power generation system. High Power Convert. Technol. **4**, 45–49 (2017)
8. Dai, X.: Research on microgrid control based on power electronic transformer grid-connected device. Hangzhou Dianzi University (Master thesis) (2014)
9. Huang, Y.Y.: Research on electronic power transformer. Wuhan: Huazhong University of Science and Technology (Master Thesis) (2004)
10. Deng, W.H., Zhang, B., Hu, Z.B.: Research on power electronic transformer circuit topology and control strategy. Power Syst. Autom. (20), 40–44+48 (2003)
11. Zhang, X.D.: Power electronic transformer and its application in power system. Shandong University (Master thesis) (2012)
12. Li, Z.: Application research of power electronic transformer. Coal Mine Electromech. **40**(3), 56–59 (2019)
13. Dong, M., Luo, A.: Design and control method of inverter in photovoltaic grid-connected power generation system. Autom. Power Syst. **20**, 97–102 (2006)
14. Wang, W., Luo, A., Xu, X.Y., Fang, H.H., Shuai, Z.K., Li, Z., Meng, J.L.: Active hysteretic filter double hysteresis space vector discrete control method. Chin. J. Electr. Eng. **33**(12), 10–17+180 (2013)
15. Zhao, Y.: Research on double closed-loop control of PWM inverter based on current hysteresis tracking. Electron. Technol. **29**(12), 51–54 (2016)

Virtual Synchronous Control Strategy for Frequency Control of DFIG Under Power-Limited Operation

Yunkun Mao(✉), Guorong Liu, Lei Ma, and Shengxiang Tang

Hunan Institute of Engineering, Xiangtan 411104, China
493564970@qq.com

Abstract. Large scale application of wind power in power system will reduce the system equivalent inertia and primary frequency control ability. This paper proposes a frequency coordination control scheme for DFIG under power-limited operation to increase the amount of system inertia via virtual synchronization control, so DFIG can provide a transient frequency support for system. The released reserve power caused by decrease of pitch angle can compensate the output power sag, and reduce steady state frequency deviation. The characteristics of virtual synchronization control and the pitch angle control of DFIG are combined to effectively suppress the system frequency fluctuation caused by load change. Simulation results show that the proposed coordinated control strategy can make full use of the wind turbine's own load reduction operation and virtual synchronous generator technology to effectively improve the frequency stability and inertia support effect of the grid-connected system, which is conducive to the safe and stable operation of the power system.

Keywords: Wind generation · Virtual synchronization control · Virtual inertia · Pitch angle · Primary frequency regulation

1 Introduction

In order to meet the sustainable development needs of resources and the environment, China has vigorously developed new energy power generation in recent years, and the installed capacity of wind power continues to grow rapidly. The proportion of new energy in the power grid in some regions has exceeded 50%. While the proportion of new energy permeability increases continuously, the inertia response and primary frequency regulation ability of the power grid are also declining, which brings risks to the frequency stability and recovery ability of the power grid under high-power shortage shocks. Therefore, increasing the inertia of the system, enhancing the frequency regulation ability of the wind turbine and improving the stability of the system have become the research hotspots, and are receiving more and more attention worldwide.

Many theoretical studies focus on inertial control and primary frequency regulation control of wind turbines. Some scholars have proposed the virtual inertia method [1–3], that is, when the frequency changes, let the wind turbine quickly release the kinetic

Q. Liu et al. (Eds.): CENet 2020, AISC 1274, pp. 1219–1228, 2021.
https://doi.org/10.1007/978-981-15-8462-6_139

energy stored in the wind turbine to the power grid, or absorb energy from the power grid to increase the kinetic energy, so as to simulate the inertial constant and transient frequency response characteristics similar to synchronous machines. Some scholars have proposed to use the power backup method, which can allow wind turbines to better participate in primary frequency regulation. For example, the control method of overspeed deceleration operation of the wind farm by modifying the power curve is described, and the frequency regulation is completed by using active/frequency droop control [4–6]. A load-reduction control scheme combining overspeed and pitch angle control is proposed, so that the wind turbine can use pitch angle/frequency sag control for primary frequency regulation [7–9].

In order to make the wind turbine simulate the frequency regulation characteristics of synchronous generator, virtual synchronous generator technology has been widely concerned. Taking the traditional mathematical model of synchronous generator as a reference and simulating its rotational inertia and damping characteristics, the concept of virtual synchronous machine was first proposed [10]. It is proposed to build a traditional synchronous generator model in a three-phase inverter to achieve electromagnetic characteristics, rotational inertia, frequency regulation and voltage regulation characteristics similar to traditional synchronous machines [11, 12]. The virtual synchronous machine control strategy has been initially applied.

By analyzing the research of the above scholars, this paper proposes a virtual synchronization control strategy for DFIG under de-loading operation mode. First, the operating characteristics of the DFIG are analyzed. Secondly, the wind turbine is put into the load limiting operation mode through pitch control. On this basis, the virtual synchronous control strategy is introduced into the wind turbine for control, and the active output is added through the inertia response link to achieve the purpose of improving the system inertial support and primary frequency regulation. Finally, a system simulation model is built on the Matlab/Simulink platform to verify the correctness and effectiveness of the proposed control strategy, thus providing theoretical and experimental basis for large-scale wind power consumption.

2 Analysis of the Operating Characteristics of DFIG

Figure 1 is the structure diagram of DFIG. The system consists of doubly-fed induction motor, gearbox, wind turbine, power converter, grid-connected transformer, etc. DFIG is a wound-rotor asynchronous generator whose rotor circuit can be controlled by an external power device to achieve variable speed operation. The stators of doubly-fed induction motor are directly connected to the power grid, and the rotors are also connected to the power grid through AC converters. Both the stators and the rotors are involved in the feeding [13]. Double-fed wind turbines achieve maximum wind energy capture through motor rotor side converters to improve energy efficiency.

Figure 2 shows the power characteristic curve of the DFIG. Each curve corresponds to a different wind speed. When the rotation speed is constant, different wind speeds cause the wind turbine to output different power. When the wind speed v changes within a certain range, adjust wind turbine speed ω in time, so that wind turbine blade tip speed ratio λ is the optimal blade tip speed ratio λ to track the Pmppt

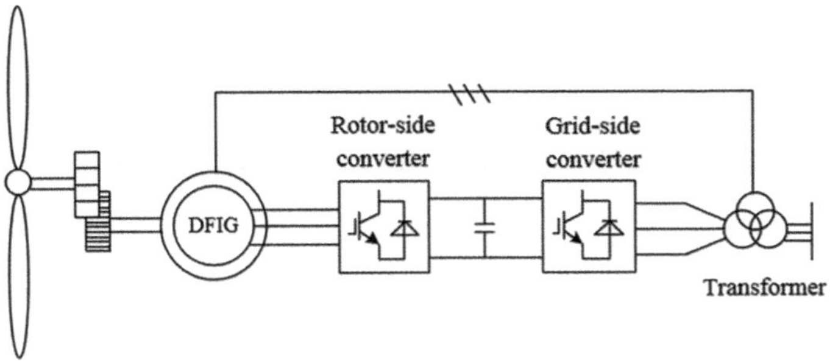

Fig. 1. Structure diagram of DFIG.

curve, and at the same time the generator shaft obtains the maximum input mechanical power [14], so that wind turbine can maintain the best efficiency within a certain range.

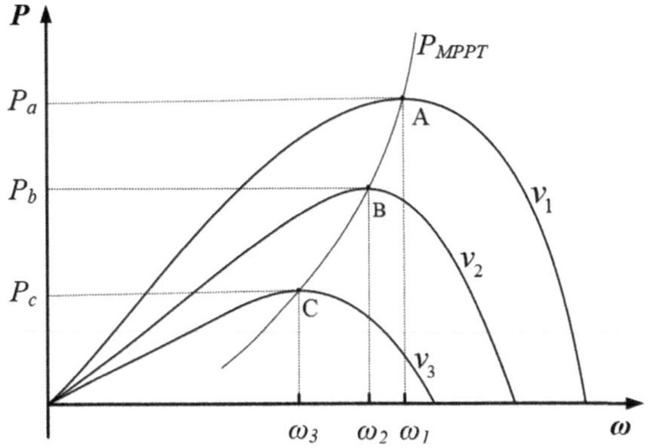

Fig. 2. Power characteristic curve of the DFIG.

However, the weakness of this strategy is that it does not provide active support for the grid to participate in frequency regulation and that the turbine speed ω is decoupled from the grid frequency. Loss of inertia response capability. Especially when the frequency dip voltage flicker and other faults occur in the power system, the performance of DFIG unit itself to withstand voltage and frequency fluctuations Poor, easy to large-scale off-grid when the system is disturbed, triggering a chain reaction and expanding the scope of the accident.

Inertial response of traditional generator set is inherent and uncontrollable, and its primary frequency regulation is accomplished by adjusting the mechanical power of the prime mover through the governor according to the static frequency characteristics of the system. DFIG needs to realize both inertia and primary frequency regulation to provide frequency support for the system, so that it has similar frequency response characteristics to traditional synchronous generator sets, in order to effectively solve the problem of inertia loss and frequency instability caused by large-scale wind power grid-connected.

3 DFIG Integrated Frequency Regulation Strategy

3.1 Pitch Angle Control

Without considering the configuration of energy storage, in order to realize the coordinated control of inertia response and primary frequency regulation of virtual synchronous control, the wind turbine needs to perform load reduction operation for primary frequency regulation, and the operating characteristics of wind turbine after load reduction also need to be applicable to virtual synchronous generator technology.

There are two schemes for load shedding operation of wind turbine, one is rotor speed control, the other is pitch control. Load shedding operation is to give up the maximum capture of wind energy, so that the wind turbine can maintain a certain level of reserve capacity, so as to participate in the primary frequency regulation of the system. In this paper, the pitch control scheme is used to reduce the load, so that the wind turbine can also provide power reserve at high wind speed.

Figure 3 shows the relationship between power and speed at different pitch angles. At a certain rotor speed, the greater the pitch angle, the lower the power of the wind turbine. The principle of pitch control is to increase the pitch angle in advance to make the wind turbine run at the sub optimal power point, so as to reserve the active power for the system frequency adjustment. When the system frequency is reduced, the pitch angle is reduced, the output of the wind turbine is increased, and the frequency response of the system is supported.

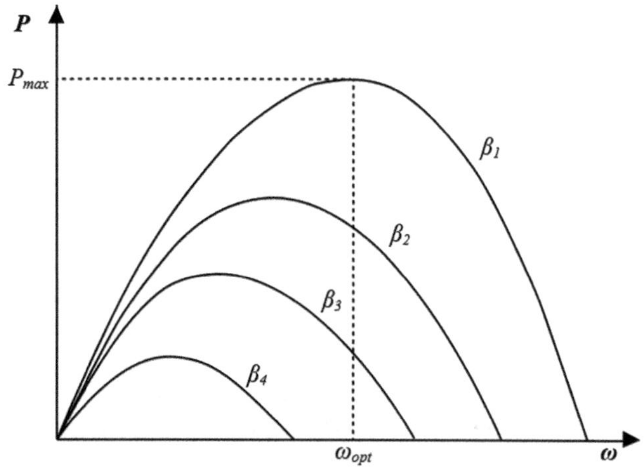

Fig. 3. Power-rotor speed diagram.

Figure 4 shows de-loading operation by pitch angle control schematic diagram The initial value of the pitch angle is calculated by the power (1 − d%) PMPPT (where d% is the load reduction level) after the load is reduced, and the pitch angle is reserved. The relationship between frequency deviation and pitch angle is introduced to enable wind turbine pitch angle to respond to the change of system frequency, thereby adjusting wind turbine output to support system frequency. The pitch angle limiting module [15] is added to prevent the pitch angle from exceeding the allowable range of wind turbine.

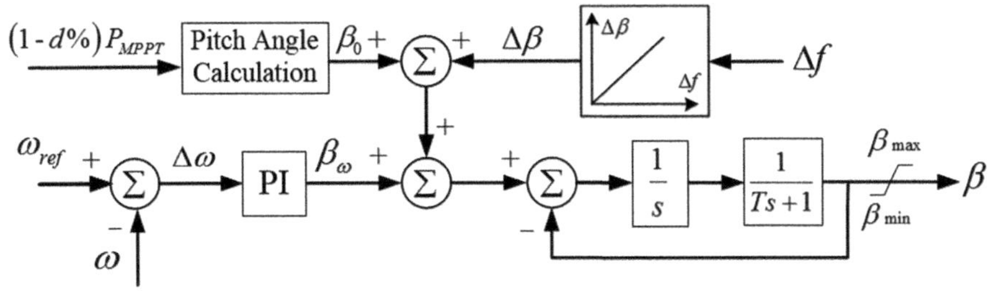

Fig. 4. De-loading operation by pitch angle control block diagram.

3.2 Virtual Synchronization Control

The system inertia has a certain buffering effect on the grid frequency disturbance, and the inertia response capability of the synchronous generator is simulated by virtual synchronous generator (VSG) control. The classic second-order model of the synchronous generator is used as a reference to establish the VSG mathematical model, which is composed of mechanical and electromagnetic characteristics. The mechanical characteristics are expressed by the rotor motion Eq. (1)

$$\begin{cases} 2H\frac{d\Delta\omega}{dt} = P_m - P_e - D(\omega - \omega_N) \\ \frac{d\delta}{dt} = \omega - \omega_N \end{cases} \tag{1}$$

In the formula, H is the inertial time constant; D is the damping coefficient; P_m and P_e are the mechanical power and electromagnetic power of the grid-side converter; ω is the rotor angular velocity; ω_N is the system rated angular velocity; δ is the rotor power angle.

The electromagnetic part is modeled with the stator electrical equation as the reference object, as shown in Eq. (2).

$$\dot{E}_{0\mathrm{i}} = E_0 \sin\theta = \dot{U}_\mathrm{i} + \dot{I}_\mathrm{i}(r_\mathrm{s} + jx_\mathrm{s}) \tag{2}$$

In the formula, $\dot{E}_{0\mathrm{i}}$, $\dot{U}_\mathrm{i}$, $\dot{I}_\mathrm{i}$ are the three-phase induced electromotive force, stator terminal voltage and stator current; θ is the phase angle; r_s is the stator armature resistance; x_s is the line equivalent impedance.

In the control algorithm, power control is the core link of VSG, and the instantaneous electromagnetic power output by the change of rotor kinetic energy can be expressed as [16].

$$P_e(t) = -\frac{P_N H}{f_N^2} f(t) \frac{df(t)}{dt} \tag{3}$$

In the formula, $P_e(t)$ is the instantaneous electromagnetic power; P_N is the rated power of the synchronous machine; f_N is the rated frequency of the system; $f(t)$ is the instantaneous frequency of the system.

Because the absolute value of the system frequency change exceeds a certain fixed value, low-frequency load shedding action can be caused, and its relative value is very small, so $f(t) \approx f_N$, then Eq. (3) can be simplified to

$$P_e(t) \approx -\frac{H}{f_N} \frac{df(t)}{dt} P_N \tag{4}$$

In order to prevent the momentary inertia support power output from causing torque imbalance and causing severe torque shock to DFIG, in the VSG inertia response design, an additional first-order inertia module with adjustable time constant K was introduced [17], thereby slowing the torque mutation, as shown in Fig. 5.

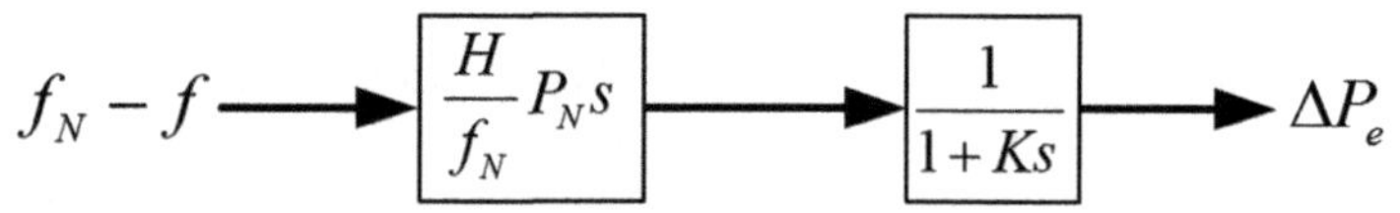

Fig. 5. Inertia support control of VSG schematic diagram.

When VSG control is introduced in DFIG grid-connected control, the wind turbines exhibit inertial response characteristics to the grid by establishing a link between the power variation and the internal potential. When disturbed, the frequency change of power grid is suppressed and the system stability is improved.

3.3 The Integrated Control Strategy for Primary Frequency Control

The virtual synchronous control strategy of wind turbine based on load shedding mentioned in this paper is mainly designed for primary frequency regulation and inertia response of power system. The control method is divided into three parts, as shown in Fig. 6.

(1) The load shedding level of the wind turbine is set according to the operating status of wind turbine.
(2) The pitch angle of wind turbine is adjusted according to the load reduction level and the rotational speed of the wind turbine, and the electromagnetic power is adjusted by setting P_{ref} by PI controller and VSG inertia response module.

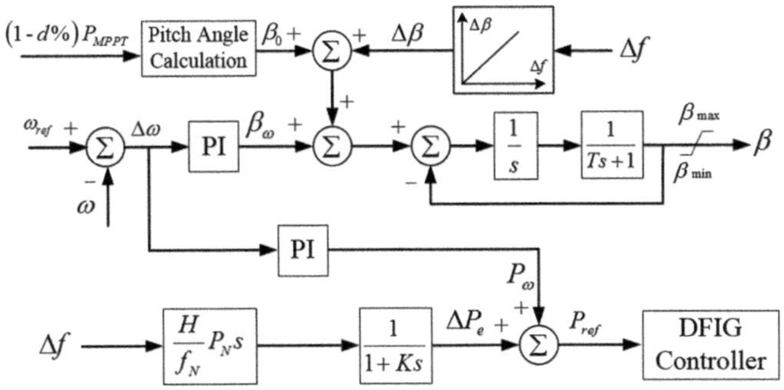

Fig. 6. VSG inertial support and primary frequency control structure diagram.

(3) When the system frequency changes, the VSG inertia response control module can increase the active power, and wind turbine adjusts the pitch angle, so that the whole control structure simulates the frequency response characteristics similar to the traditional synchronizer. Under the joint action of the two parts, the system inertia response and primary frequency regulation ability are improved.

4 Simulation Analysis

This article establishes the simulation system shown in Fig. 7 in MATLAB/Simulink. The simulation system includes 4 synchronous generators with a capacity of 50 MW and 20 DFIG with a capacity of 2 MW. L_1 and L_2 are system loads with capacities of 100 MW and 80 MW respectively. C_1 and C_2 are reactive power compensation devices.

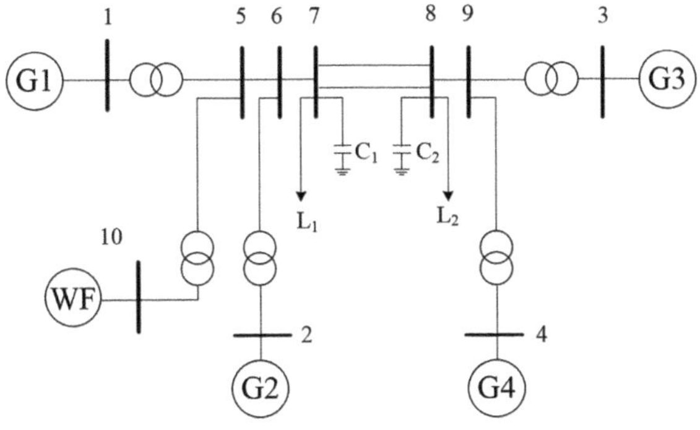

Fig. 7. Structure of the simulation system

Assuming that the wind speed is set at a fixed value of 10 m/s, the wind turbine initially maintains the operating state of 10% load reduction, and the load L_1 suddenly

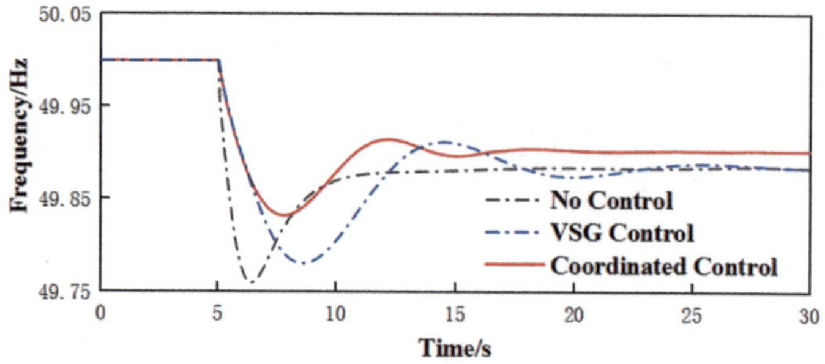

(a) Grid frequency

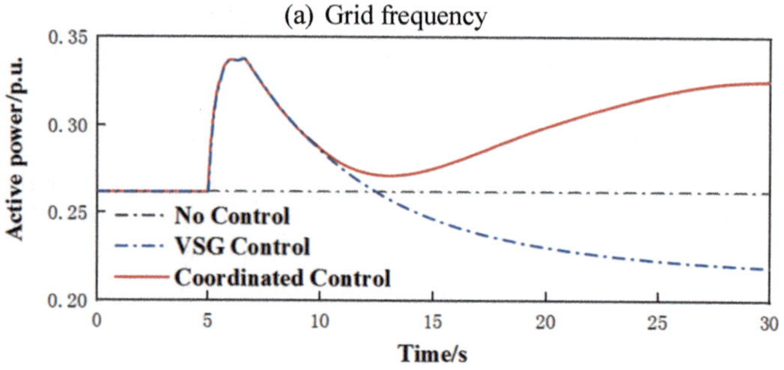

(b) Active power of DFIG

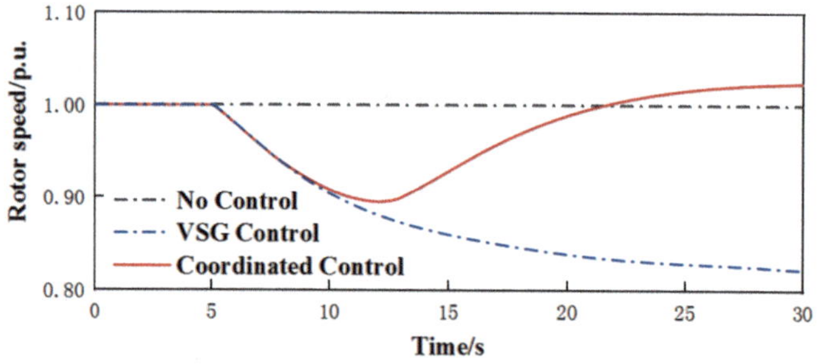

(c) Rotor speed of DFIG

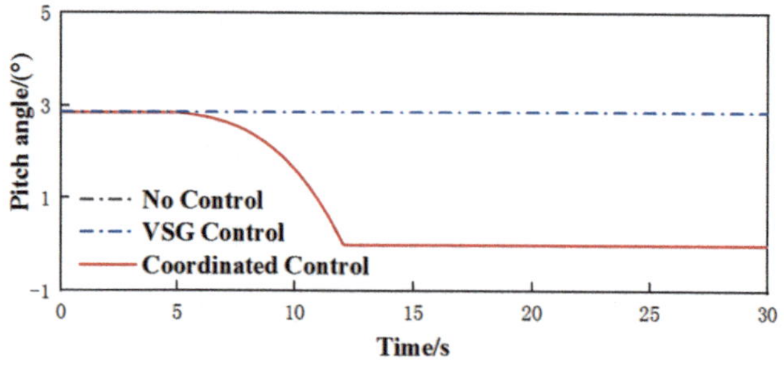

(d) Pitch angle of DFIG

Fig. 8. Comparison of dynamic response of system after sudden increase of load L_1.

increases from 100 MW to 110 MW at 5 s, resulting in a decrease in the system frequency. The simulation results are shown in Fig. 8.

Figure 8 compares the dynamic response of the system frequency, active output of the wind turbine, rotor speed, and pitch angle in DFIG without additional control, VSG control, and frequency coordination control. It can be seen from Fig. 8 that when there is no additional control of the wind turbine, the rate and amplitude of the frequency decrease is the largest, and there is almost no response to the system frequency change. Using VSG control can make the wind turbine to suppress the maximum frequency deviation to a certain extent, and make the lowest frequency 49.76 Hz to 49.78 Hz. During the frequency regulation process, wind turbine can play a certain role in supporting the system by reducing the speed to release kinetic energy. However, due to the limited kinetic energy provided by wind turbines, DFIG cannot continuously share the unbalanced power of the system. However, under the coordinated frequency control of DFIG load reduction operation and VSG technology, DFIG can not only maintain the inertial support of the system, but also participate in the system frequency adjustment. Compared with DFIG without control, the maximum frequency offset is The reduction of 0.24 Hz to 0.17 Hz, a reduction of 29.2%, can be seen that the frequency characteristics of the system have been significantly improved. Among them, the active spare capacity reserved for DFIG load reduction operation can provide continuous active power support for the system, and reduce the power change rate at the initial stage of the system frequency drop, which is conducive to the recovery of the system frequency.

5 Conclusion

Based on the proposed load reduction control scheme for DFIG, this paper further proposes a virtual synchronous control method for wind turbines under limited power operation. This method uses the pitch method to reduce the load of the wind turbine, thereby generating a certain amount of reserve capacity for the system's primary frequency regulation, and combined with the virtual synchronous generator technology to support the system frequency. Simulation results show that the proposed control strategy can suppress the system frequency fluctuations caused by wind speed changes and grid load mutations. Compared with traditional virtual inertial control methods, it can effectively improve the system's inertia response and primary frequency regulation capability.

References

1. Kayikci, M., Milanovic, J.V.: Dynamic contribution of DFIG-based wind plants to system frequency disturbances. IEEE Trans. Power Syst. **24**(2), 859–867 (2009)
2. Morren, J.: Wind turbines emulating inertia and supporting primary frequency control. IEEE Trans. Power Syst. **21**(1), 433–434 (2006)
3. Keung, P.: Kinetic energy of wind-turbine generators for system frequency support. IEEE Trans. Power Syst. **24**(1), 279–287 (2008)

4. Ramtharan, G., Ekanayake, J.B., Jenkins, N.: Frequency support from doubly fed induction generator wind turbines. IET Renew. Power Gener. **1**(1), 3–9 (2007)
5. Vidyanandan, K.V.: Primary frequency regulation by deloaded wind turbines using variable droop. IEEE Trans. Power Syst. **28**(2), 837–846 (2012)
6. Erlich, I., Wilch, M.: Primary frequency control by wind turbines. In: Power & Energy Society General Meeting, pp. 1–8. IEEE (2010)
7. Dreidy, M.: Inertia response and frequency control techniques for renewable energy sources: a review. Renew. Sustain. Energy Rev. **69**, 144–155 (2017)
8. Teninge, A., Jecu, C., Roye, D., Bacha, S., Duval, J., Belhomme, R.: Contribution to frequency control through wind turbine inertial energy storage. IET Renew. Power Gener. **3** (3), 358–370 (2010)
9. Ghosh, S.: Doubly fed induction generator (DFIG)-based wind farm control framework for primary frequency and inertial response application. IEEE Trans. Power Syst. **31**(3), 1861–1871 (2015)
10. Beck, H.: Virtual synchronous machine. In: 9th International Conference on Electrical Power Quality and Utilisation, pp. 1–6. IEEE (2007)
11. Alipoor, J., Miura, Y., Ise, T.: Power system stabilization using virtual synchronous generator with alternating moment of inertia. IEEE J. Emerg. Sel. Top. Power Electron. **3**(2), 451–458 (2014)
12. Wang, S., Hu, J., Yuan, X., Sun, L.: On inertial dynamics of virtual-synchronous-controlled DFIG-based wind turbines. IEEE Trans. Energy Convers. **30**(4), 1691–1702 (2015)
13. Liu, Y., Jiang, L., Wu, Q.H., Zhou, X.: Frequency control of DFIG-based wind power penetrated power systems using switching angle controller and AGC. IEEE Trans. Power Syst. **32**(2), 1553–1567 (2016)
14. El Itani, S.: Short-term frequency support utilizing inertial response of DFIG wind turbines. In: 2011 IEEE Power and Energy Society General Meeting, pp. 1–8. IEEE (2011)
15. Moutis, P.: Primary load-frequency control from pitch-controlled wind turbines. In: 2009 IEEE Bucharest PowerTech, pp. 1–7. IEEE (2009)
16. Qin, X.: Functional orientation discrimination of inertia support and primary frequency regulation of virtual synchronous generator in large power grid. Autom. Electr. Power Syst. **42**(9), 36–43 (2018)
17. Kheshti, M., Ding, L., Nayeripour, M., Wang, X., Terzija, V.: Active power support of wind turbines for grid frequency events using a reliable power reference scheme. Renew. Energy **139**, 1241–1254 (2019)

Design on Underwater Fishing Robot in Shallow Water

Zhiguang Guan[1,2](✉), Dong Zhang[2], and Mingxing Lin[3]

[1] School of Construction Machinery, Shandong Jiaotong University, Jinan 250357, China
guanzhiguang@sdjtu.edu.cn
[2] Institute of Automation, Shandong Academy of Sciences, Jinan 250000, China
[3] School of Mechanical Engineering, Shandong University, Jinan 250061, China

Abstract. Underwater fishing robots have been more and more widely used in ocean breeding nowadays. They have become the most effective and potential tools in underwater fishing. A small-scale underwater fishing robot has been developed, which has the functions of sensing and detecting. This robot can measure the conductivity, temperature, depth (CTD) and robot's attitude in real time, and at the same time the controller transmits them to the mother ship through CAN bus. In this paper, firstly, the whole structure of the self-designed robot is introduced, then the control system is designed, finally, a physical experiment is carried out. The experiment shows that the robot can meet the design requirement.

Keywords: Underwater fishing robot · Manipulator · Thruster · CAN bus

1 Introduction

The ocean occupies 71% of the earth's surface area, which stores rich resources such as biological and mineral resources. In 2018, the State Council approved the construction of a comprehensive experimental zone for Replacing Old Growth Drivers with New Ones in Shandong Province. So modern marine industry is focused cultivation and development as one of the "top ten" industries. Shandong Province is adjacent to the Yellow Sea and the Bohai Sea, a series of marine industries especially the sea cucumber breeding zones have been formed in the coastal areas in recent years. The total area of aquaculture in China in 2019 was 6954.3 thousand hectares, with a total output of 64.5 million tons. However, at present, aquaculture fishing still faces with many serious problems. For example, sea cucumber, abalone need divers to carry oxygen masks to submerge into the sea floor to fish them, which will endanger the diver's life because of the strong sea pressure and long-term underwater work. Using underwater robot to fish sea cucumber and abalone automatically can reduce work intensity and improve work efficiency. It is a future trend to use underwater robot to replace artificial sea cucumber fishing. The current domestic aquaculture mainly concentrated in the coastline within 10 nautical miles, water depth of about 30 m of the sea, farther from the coastline is not yet developed, so the market of underwater fishing robots has great development potential. The control system of the robot and the

Q. Liu et al. (Eds.): CENet 2020, AISC 1274, pp. 1229–1235, 2021.
https://doi.org/10.1007/978-981-15-8462-6_140

real-time identification and location of sea cucumber are the keys to its success. The research on the acquisition of objects by underwater robots abroad mainly focuses on the United States and Canada. Underwater fishing robot CHINOOK was designed by SEAMOR Company, whose maximum depth can reach 300 m underwater. The robot can use different module to complete a certain work. In China PI Zhifeng designed a two-stage structure sea cucumber fishing robot. The sea cucumber fishing device and sea cucumber collecting device were designed and realized separately, and connected by a straw and an umbilical cord cable.

2 Design of Underwater Fishing Robot Structure

The prototype designed in this topic mainly consists of manipulator, buoyancy tank, frame, thrusters and control tank, which is shown in Fig. 1.

Fig. 1. Prototype.

2.1 Manipulator

Manipulator is used to fish sea cucumber and abalone, whose efficiency is closely related to underwater work. In the underwater robot, an open chain structure is adopted, which can effectively reduce the weight and volume of the manipulator. The driven motor is placed on the frame, so it can reduce the load of the large arm and improve the fishing weight. Manipulator structure is shown in Fig. 2.

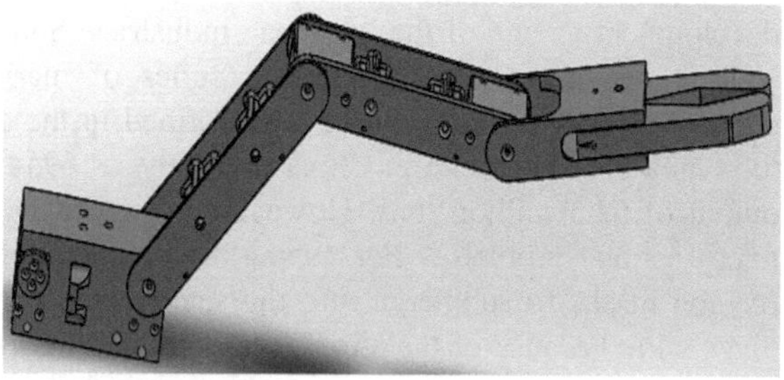

Fig. 2. Manipulator structure.

2.2 Buoyancy Tank

Buoyancy tank can increase buoyancy by increasing the displacement of robot, which is an important source of buoyancy for underwater robot. The main function of the underwater robot is fishing, so it needs to carry a relatively stable buoyancy device. The buoyancy device adopts a cylindrical metal aluminum alloy. Aluminum alloy can reduce the weight of robot, and has good corrosion resistance. The two buoyancy tanks are placed on the upper part symmetrically.

2.3 Frame

The frame carries the weight and working load of the underwater robot. The strength and quality of the frame affects various structural parameters of the robot. In this paper, the robot structure based on nylon material is established by considering the characteristics of various structural materials. The positioning assembly of the prototype is connected by punching holes and tapping threads on the nylon frame, and the weight reduction holes are punched in a large area to reduce the overall quality of the UNDERWATER robot and improve its dynamic performance. The frame structure of the prototype is shown in Fig. 3.

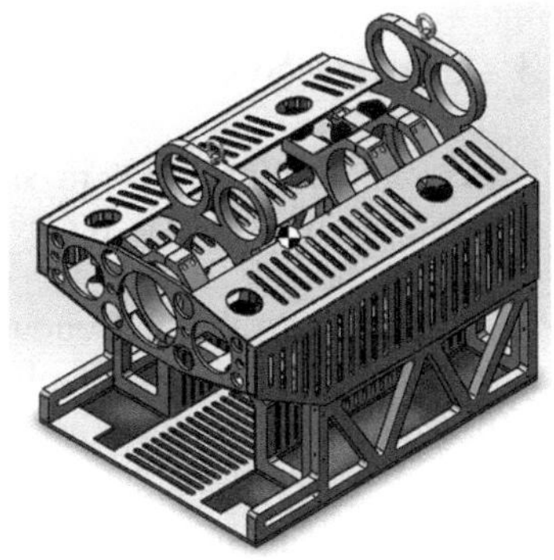

Fig. 3. Frame.

2.4 Thruster

Thruster supplies navigation power of the underwater robot. All the functions of the underwater robot, such as diving, floating, forward, backward and turning are controlled by the speed of thruster. The number and arrangement of thrusters have many types. This prototype adopts a four-axis vertical thruster and two-axis horizontal thruster. The four thrusters are arranged in a vector form in the same plane, which enhances the mobility of the robot.

The force and torque of the four thrusters can be shown in (1).

$$\begin{bmatrix} T \\ M \end{bmatrix} = \begin{bmatrix} \sum_{i=1}^{2} T_i \\ \sum_{i=1}^{2} M_i \end{bmatrix} = \begin{bmatrix} (T_1 + T_2 + T_3 + T_4)\cos\alpha \\ (T_1 - T_2 - T_3 + T_4)\sin\alpha \\ 0 \\ 0 \\ 0 \\ c(T_1 - T_2 + T_3 - T_4)\cos\alpha \end{bmatrix} \tag{1}$$

where, T_i is the force of ith thruster; M_i is the Torque of ith thruster.

3 Control System Design of Underwater Robot

Underwater fishing robot in shallow water is an intelligent experimental platform, whose control system is relatively complex. In order to make the underwater robot stable and reliable sailing, the control system needs to collect some sensors' signal, such as attitude, CTD (Conductivity, Temperature, Depth) and so on. the control system of the underwater robot is composed by two parts. Both the underwater information and command are transmitted each other through CAN bus.

3.1 Measurement System

Detection is an important information acquisition system for underwater fishing robot. The attitude sensor mainly helps the robot to carry out its own stability control and make the robot work stably and normally. The robot can monitor underwater information through CTD sensor and transmit to the control center, so that the operator can understand the underwater environment and deal with the situation accordingly.

Attitude sensor is used to detect the robot's own posture. In order to make the robot work stably, the robot's attitude should be adjusted in real-time. MPU6050 is adopted to collect the attitude information of the underwater fishing robot. The sensor chip integrates three-axis Angle and three-axis acceleration sensor. It supports various communication protocols, such as IIC and SPI. Conductivity, temperature and depth information can be detected using integral CTD sensor. The temperature range of seawater is −5-40 °C, which belongs to the normal temperature measurement range. When CTD measures the temperature, the output resistance changes with the temperature. According to output resistance, the seawater temperature can be calculated. The conductivity is obtained by the resistance between two electrodes in seawater. AC source is added to the electrodes as the excitation power. The measured conductance electrode can be equivalent to a network circuit combining capacitance and resistance, which is shown in the Fig. 4.

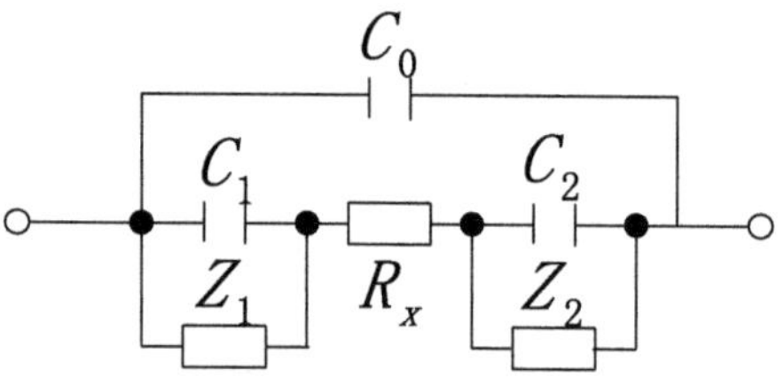

Fig. 4. Equivalent network.

where, Z_1 and Z_2 are the polarization impedances of the two electrodes; C_1 and C_2 represent the double-layer capacitance; R_x is the resistance of seawater; C_0 is the capacitance of the seawater.

The depth of the robot can be measured through the pressure of the seawater. The relationship between seawater pressure and the depth is (2):

$$h = \frac{P}{\rho g} \tag{2}$$

where: P is seawater pressure; g is the gravitational acceleration at the measurement point; ρ is density of seawater.

3.2 Control System

The underwater fishing robot is composed of the monitoring and operating system in the mother ship and the robot itself, which is shown in Fig. 5. The two parts are connected by a special zero-buoyancy cable. This system adopts two ARM micro-controller to realize real-time communication between the mother ship and the robot. The display screen of the monitoring device displays the image information collected by the underwater robot as well as the parameters such as depth, salinity and temperature. At the same time, the rocker control method is more convenient, reliable and stable.

The underwater control includes a controller, binocular vision system, temperature sensor, salinity sensor and depth sensor. The controller is integrated with a gyroscope. The binocular vision system includes two sets of cameras. The binocular vision system is connected with the underwater controller through USB, the CTD sensor is connected with the analog input port of the controller, and the thrusters are connected with the PWM pins of the controller. When the robot is working, the rocker controls the vertical propeller mounted on the robot to make the robot dive to the set depth. During the diving process, the controller collects the depth and gyroscope signals in real time to control the operation stability of the robot. As the robot reaches to setting depth, the binocular vision system begins to work and locates the target. When positioning the target, the rocker is out of control, the underwater robot will find the target automatically, and drive the manipulator to catch the target. After completing all the corresponding actions, the underwater controller will notify the mother ship controller to make the rocker work again. When the robot fails, the control system can detect the fault location according to the sensor signal and send the fault code to the mother ship.

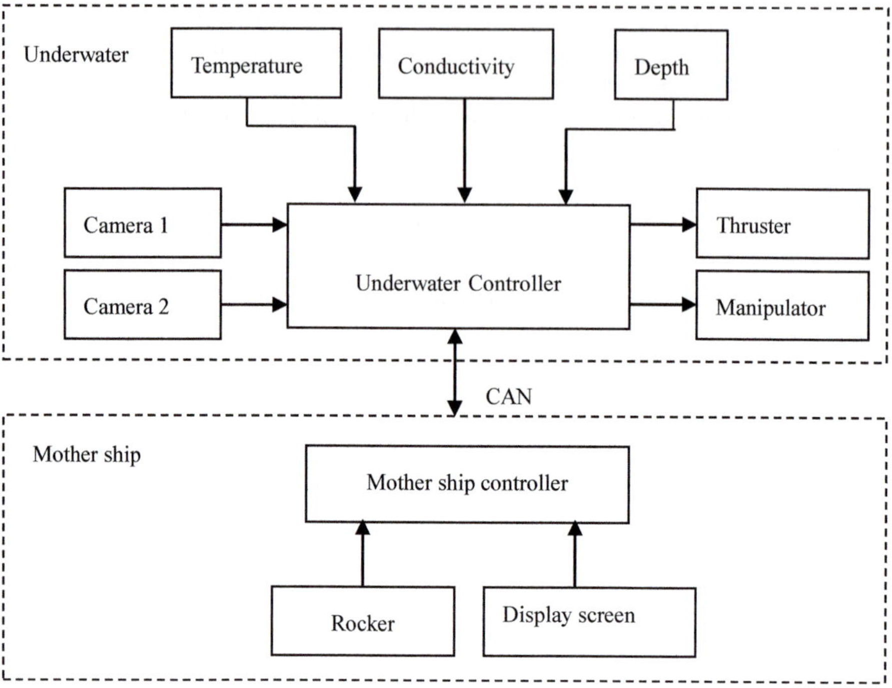

Fig. 5. Structure of control system.

4 Experiment

After the prototype is assembled, it is tested in a lake. Firstly, the buoyancy balance and the gyroscope data are calibrated. Then, rolling, turning, diving and floating functions are tested respectively, which is shown in Fig. 6. The experiment shows that the prototype can satisfy with the design requirement.

a) Rolling b) Turning c) Diving

Fig. 6. Experiment.

5 Conclusion

The development of the underwater robot fishing can improve the production efficiency and reduce the labor cost. The user uses the terminal recker to control underwater robot and monitior seawater envirement. The experiment shows that the prototype can satisfy with the design requirement.

Acknowledgment. This work is supported by Major Science and Technology Innovation Project of Shandong Province (2019JZZY020703, 2019JZZY020712), Science and Technology Support Plan for Youth Innovation in Universities of Shandong Province colleges, and universities (2019KJB014), Shandong Jiaotong University "Climbing" Research innovation Team Program (SDJTUC1805).

References

1. Wen, X.P.: Research on system characteristics analysis and control methods of underwater vehicle. Harbin Engineering University (2012)
2. Guan, Z.G., Zhang, D., Lin, M., et al.: Mechanical analysis of remotely operated vehicle. In: 15th International Conference on Ubiquitous Robots, pp. 857–862 (2018)
3. Guan, Z.G., Zhang, D., Ma, R.L. et al.: Control system design of remotely operated vehicle. In: 4th International Conference on Control, Automation and Robotics, pp. 446–450 (2018)
4. Wu, J.M., Chen, D.J.: Trajectory following of a tethered underwater robot with multiple control techniques. J. Offshore Mech. Arctic Eng. Trans. ASME **141**, 0511041–0511049 (2019)
5. Wang, Z., Lin, M.X., Ban, C.Q.: Research on hydrodynamics analysis and double loop integral sliding mode control of 4-joint underwater manipulator. In: 14th International Conference on Ubiquitous Robots and Ambient Intelligence, pp. 728–733 (2017)
6. Liu, S.J., Liu, R., Zheng, H., et al.: The system design of an autonomous underwater vehicle. Mach. Des. Manuf. **05**, 233–236 (2020)
7. Saruchi, S.A., Zamzuri, H., Zulkarnain, N., Wahid, N., Ariff, M.H.M.: Composite nonlinear feedback with disturbance observer for active front steering. Indonesian J. Electr. Eng. Comput. Sci. **7**(2), 434–441 (2017)
8. Lu, T., Lan, W.Y.: Composite nonlinear feedback control for strict-feedback nonlinear systems with input saturation. Int. J. Control **92**, 1–8 (2018)

Structure Design of a Cleaning Robot for Underwater Hull Surface

Qin Sun[1(✉)], Zhiguang Guan[1,2], and Dong Zhang[2]

[1] School of Construction Machinery, Shandong Jiaotong University, Jinan 250357, China
206075@sdjtu.edu.cn

[2] Institute of Automation, Shandong Academy of Sciences, Jinan 250000, China

Abstract. A cleaning robot for underwater hull surface is designed, which mainly includes adsorption mode, track driving mode, driving mechanism and cleaning mechanism. The double track moving mode has the advantages of large contact area with the hull surface, good adaptability to the wall surface, large magnetic adsorption capacity, strong bearing capacity and stable movement. The form of rear drive was adopted. When the track passes through the obstacle, the magnetic track that the robot fits on the hull surface is lengthened, so that the track can better adapt to the fluctuation of the hull surface. Washing tools are divided into soft material cleaning tools and hard material cleaning tools. Only the soft material cutter is close to the hull surface, which can avoid the damage of the hull surface. Electromagnet adsorption mode was adopted, which is easy to realize rapid movement. In addition, the working range can be increased with the rotary mechanism, and the working efficiency can be greatly improved with the rotary cleaning head.

Keywords: Hull surface · Cleaning robot · Track

1 Introduction

There are many organisms in the marine environment, and the part below the ship waterline becomes the main attachment of marine organisms. Fuel consumption increases by 10% and ship speed decreases by 0.33% for every 10 μm increase in hull surface roughness [1]. For civil ships, it may only slow down the speed and increase the operating cost. For warships, it will affect the speed and maneuverability of warships, and even change the outcome of the war. Therefore, regular cleaning is required. Scraping the bottom is very hard because the bottom of the ship is generally large and the biological attachment is very strong, especially for underwater ship bottom cleaning. For "frogman", the workload is unimaginable. Instead of frogman diving, the cleaning climbing robot automatically cleans the hull of the port, wharf and anchorage, which can greatly improve the cleaning efficiency of the ship and reduce the labor cost.

The first wall climbing robot designed by A. Nishi of Japan in 1966. In addition, UA600, a typical ship derusting climbing robot developed by Japan's Pushang Institute of technology, mainly used vacuum adsorption and wheel mechanism for moving, which had limited obstacle surmounting ability and was not suitable for uneven hull

Q. Liu et al. (Eds.): CENet 2020, AISC 1274, pp. 1236–1242, 2021.
https://doi.org/10.1007/978-981-15-8462-6_141

surface. NASA and the Robotics Institute of Carnegie Mellon University have developed the wall climbing robot M3500, which is based on the M2000 prototype and optimized part of the structure. The traveling mechanism adopts articulated connection, which greatly improves the adaptability of the wall. However, robots are still heavy and need further lightweight design; at the same time, robot manufacturing and maintenance costs are too high for most ship enterprises to bear. Flow company has also developed a ship rust removal robot, hydrocat, which can remove rust on the flat wall with a large arc and has certain wall adaptability. However, due to its operation on the uneven wall, it may be overturned due to vacuum failure. The wall climbing robot developed by kamat company of Germany adopts the pneumatic motor driving mode with large driving torque, which effectively solves the driving difficulty of tracked robot. The effective cooperation of driving and walking mechanism makes the robot move flexibly and rust removal efficiency high [2–4].

In China, Xi'an Tianhe Haifang Intelligent Technology Co., Ltd., Harbin Engineering University, Dalian Maritime University, Shenzhen Institute of advanced technology, Chinese Academy of Sciences, Zhejiang University, Tsinghua University, etc. have also studied the ship embroidery robot.

Based on the review of the research status of wall climbing robot at home and abroad, the current wall climbing robot is mainly used for ship derusting. However, the underwater bottom cleaning and ship derusting conditions are different. At present, most ship derusting robots are driven by negative pressure adsorption, propulsion in water or wheel motor, which has poor adsorption stability in water and low cleaning efficiency.

2 Overall Structure Design

The main task of the cleaning robot is to remove the marine dirt on the surface of the underwater hull. As a transport mechanism, the robot should first have good adsorption and movement ability, be able to reliably adsorb on the hull surface, and be able to move on the hull surface by itself. At the same time, considering the particularity of the working environment and nature of the operation, the robot should have the ability of surmounting obstacles, loading and keeping the right walking posture.

The overall structure of the cleaning robot is shown in Fig. 1. It mainly includes cleaning mechanism, walking mechanism, adsorption mechanism, etc. Electromagnet adsorption mode makes the robot easy to move quickly. The rotary mechanism 5 can increase the working range, and the rotary washing cutter head 1 can improve the working efficiency. The function of the vacuum recovery device is to recover the garbage effectively and avoid the secondary pollution of the ocean.

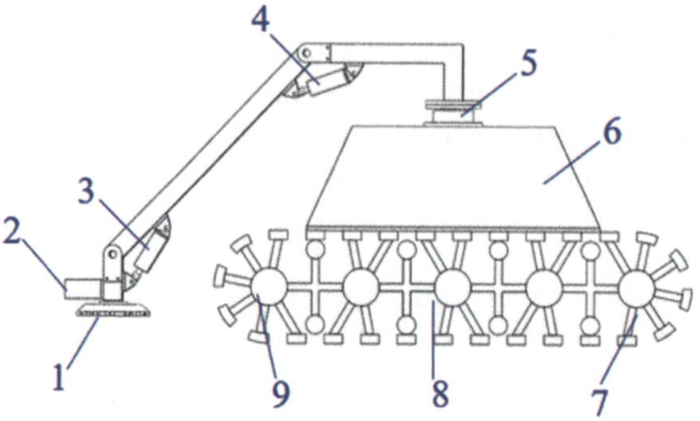

1–Cleaning tool head; 2–Tool head motor; 3, 4–Oil cylinder; 5–Turning mechanism; 6–Recovery box; 7–Rear wheel; 8–Track; 9–Front wheel

Fig. 1. General schematic diagram of robot.

3 Main Structure Design

3.1 Mobile Scheme

The most commonly moving modes of the cleaning robot for underwater hull surface are wheel type and track type. The wheel type has the advantages of fast-moving speed, flexible control, and easy steering, but the contact area between the wheel and the wall is small, so it is difficult to maintain a certain adsorption force. The track type has strong adaptability to the hull surface, large contact area, strong adsorption force and fast-moving speed, but it is not easy to turn and transfer to the hull surface. Considering the characteristics of all aspects comprehensively, track cleaning robot can not only produce large adsorption force, but also have fast movement speed, strong obstacle surmounting ability and load capacity, and relatively simple control, so it has outstanding advantages and is suitable for the movement of the ship surface.

There are many types of tracks, as shown in Fig. 2. Double track mode is employed in this paper. Compared with other track structures, the structure is simple and flexible, which can meet the requirements of hull wall. The moving function of the robot is completed by the track. Because the track is hinged, it has certain flexibility, can adapt to the curvature change of the hull surface, and can cross the obstacles such as the weld.

According to the working environment and requirements of the cleaning robot, in order to ensure the reliability and safety of the cleaning robot, the double track moving mode was designed, as shown in Fig. 3. Compared with the wheel type, the double track moving mode is more complex in structure, but its contact surface is large, and its center of gravity is low. The double track moving mode is stable, which can increase the load and is easy to carry the work tools.

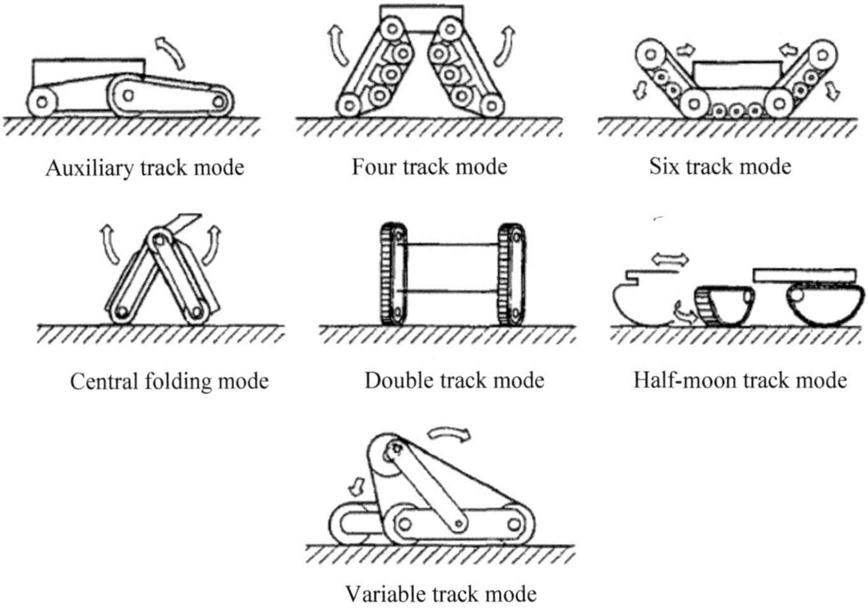

Fig. 2. Track type [5].

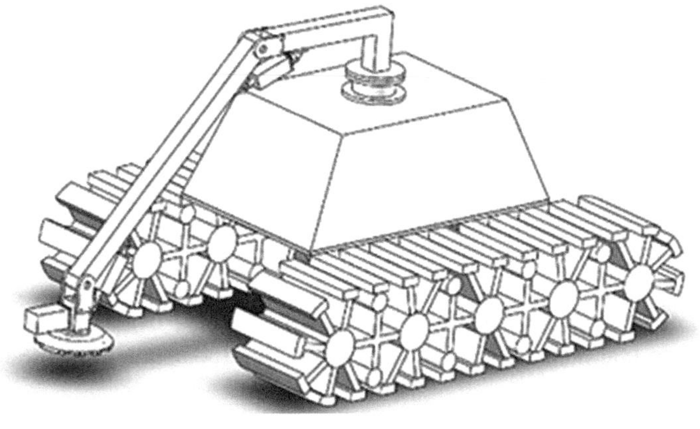

Fig. 3. Double track mode.

3.2 Adsorption Mode

According to the actual situation of its working wall and object, it is required that the cleaning robot can be reliably adsorbed on the hull surface and move flexibly, so the cleaning robot must have good adsorption ability. The adsorption methods of wall mobile robot include vacuum adsorption, magnetic adsorption, thrust adsorption and bionic adsorption. The electromagnet adsorption mode is selected, which is easy to realize the separation and closing between the magnet and the wall, and easy to realize the rapid movement.

3.3 A Subsection Sample

Because of the different driving modes of the track moving mechanism, the tension distribution in the track is also different. The tension distribution in the track of front drive and rear drive is shown in Fig. 4. T_0 is the initial tension of the track and F_{ZT} is the traction on the driving wheel.

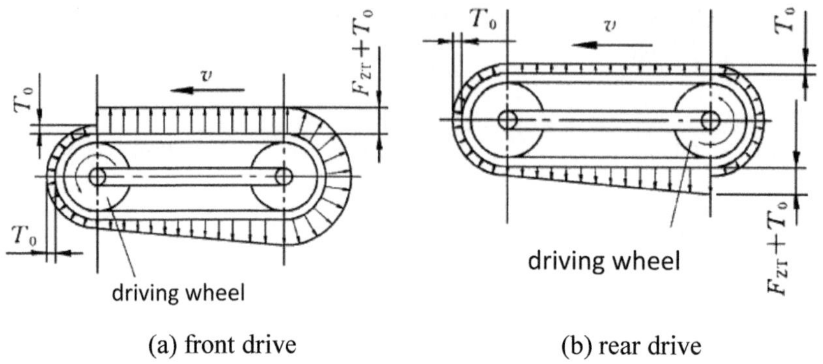

Fig. 4. Different tension distributions in tracks due to the robotic different driving modes.

It can be seen from Fig. 4a that when the driving device is in front, most of the tracks bear large traction when driving, which affects the service life of the tracks. It also causes the tracks to extend violently, resulting in the formation of the so-called "track abdomen" at the front and lower tracks during driving, so that the tracks may fall off when the robot turns. However, in the form of rear drive (see Fig. 4b), the above problems will not occur due to the short section of high tensile stress in the track [6].

The method of rear drive is employed. The more important advantage is that it improves the adaptability of the wall climbing robot. When the track passes through the obstacle, the front driven wheel will turn more Δh angle, so that the extended part of the track moves to the contact section with the wall surface through the driven wheel, as shown in Fig. 5. In this way, the magnetic track of the robot and the wall surface is lengthened, so that the track can better adapt to the fluctuation of the wall surface.

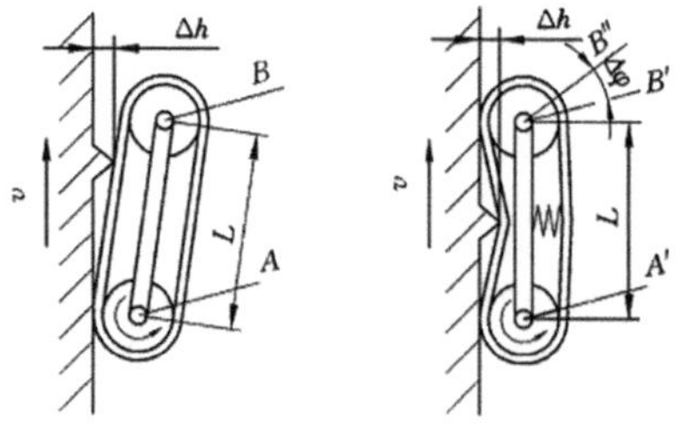

Fig. 5. The kinematic characteristics of the front driven chain wheel of the robot.

3.4 Driving Mechanism

The commonly driving modes of cleaning robot are hydraulic driving, pneumatic driving, electric driving and mechanical driving. Electric drive is employed in this paper.

3.5 Cleaning Mechanism

The cleaning mechanism of the robot consists of a driving motor, a reducer and a disc brush. The motor drives the disc brush to rotate, the cutter follows the disc to do circular movement, and the cutter angle is automatically adjustable, as shown in Fig. 6.

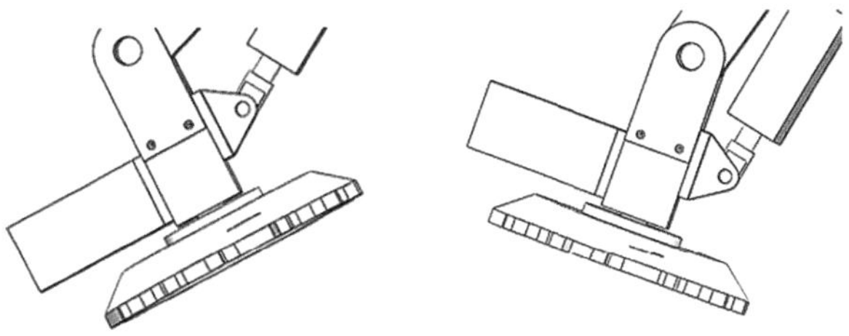

Fig. 6. Cleaning mechanism.

The cleaning tools are fixed by the way of slot positioning. The tool has two different materials, soft material cleaning tool and hard material cleaning tool. It can be seen from the Fig. 7. that the height of the two kinds of cutters is different. The soft material cutters are close to the hull surface, which can better protect the hull from damage and complete the removal of algae, while the hard material cleaning cutters mainly complete the cleaning of hard materials such as shells.

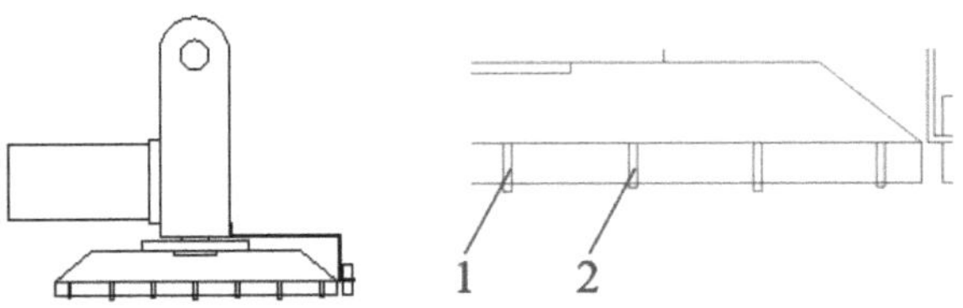

1–soft material tool; 2–hard material tool

Fig. 7. General schematic diagram of robot.

3.6 Other Mechanism

A vacuum recovery device was designed to recover the cleared garbage timely and effectively, and also avoid the secondary pollution of the ocean.

4 Conclusion

The structure of the cleaning robot is designed, including the overall scheme, the moving scheme, the adsorption mode, the track driving mode, the driving mechanism, the cleaning mechanism, etc.

The robot has the following advantages:

(1) The double magnetic track has a large contact area with the hull surface. The double magnetic track drive can produce a large magnetic adsorption force. This driving mode has strong bearing capacity and stable movement.
(2) The hull surface adaptability of the cleaning robot is improved by adopting the rear driving mode. When the track passes through the obstacle, the magnetic track of the robot and the wall is lengthened, so that the track can better adapt to the fluctuation of the wall.
(3) The cleaning tools are soft material cleaning tools and hard material cleaning tools. The soft material cutter clings to the hull surface to complete the removal of algae and other soft materials, which can also avoid damage to the hull surface, while the hard material cutter mainly completes the cleaning of hard materials such as shells.

Acknowledgment. The Project is supported by 2019 Shandong Province Key R&D Project (2019JZZY020703), “2019 Youth Innovation and Technology Program” for colleges and universities in Shandong province (2019KJB014), PhD start-up fund of Shandong Jiaotong University (BS201901040, BS201901041) and “Climbing” Research Innovation Team Program of Shandong Jiaotong University (sdjtuc18005).

References

1. Wu, J.G., Liu, D., Wang, X.M., et al.: Hydrodynamic analysis and experimental research of underwater machine for ship wall cleaning. Ship Eng. **40**(03), 91–97 (2018)
2. Wang, X.C.: Research on the technology of ship derusting climbing robot. South China University of Technology (2016)
3. Ross, B., Bares, J., Chris, F.A.: Semi-autonomous robot for striping paint from large vessels. Int. J. Robot. Res. **22**(7–8), 617–626 (2003)
4. Coste-Maniere, E., Simmons, R.: Architecture, the backbone of robotic systems. In: Proceedings 2000 ICRA. IEEE International Conference on Robotics and Automation. Symposia Proceedings, San Francisco, USA, vol. 1, pp. 67–72 (2000)
5. Ren, M.H.: Research on composite adsorption method of underwater hull surface cleaning robot. Harbin Engineering University (2009)
6. Tong, J.G., Ma, P.S., Chen, J.M.: Study on the adaptability of tracked wall climbing robot. J. Shanghai Jiaotong Univ. (07), 89–92 (1999)

Design of Control System for Tubeless Wheel Automatic Transportation Line

Qiuhua Miao[1(✉)], Zhiguang Guan[1,2], Dong Zhang[2], and Tongjun Yang[3]

[1] School of Construction Machinery, Shandong Jiaotong University, Jinan 250357, China
miaoqiuhua@126.com
[2] Institute of Automation, Shandong Academy of Sciences, Jinan 250000, China
[3] Jinan Branch of China Unicom Software Research Institute, Jinan 250000, China

Abstract. At present, the heavy truck tubeless workshop adopts manual or low automatic way to transport the wheels, it is necessary to design a set of control system of wheel automatic transportation line which can run safely and stably in workshop environment, instead of the existing low automation transportation mode. The paper introduces an automatic transportation line using PLC and WinCC technology, the main content includes the introduction of the overall structure of the automatic transportation line, and the composition and function of automatic production line are described in detail, the selection of PLC, step motor to drive, etc. Use PLC technology to design control programs, apply configuration software technology to design the monitoring system, and connect PLC and configuration software to form a control system through Ethernet communication. Finally, the paper describes the experimental results. The system has the characteristics of high automation, data visualization, simple operation and stable operation. It is a safe, stable, efficient and low labor cost tubeless wheel transportation mode.

Keywords: Automated transportation line · Configuration software · PLC · Tube-less wheels

1 Introduction

The tubeless wheels used in heavy trucks have the characteristics of large diameter and heavy weight. At present, in the transportation link of the wheel production line, the wheel handling adopts manual handling or partial automation. The transportation efficiency is low, the hidden danger is big, and the workers have high work intensity and need to use a large number of handling tools, so the transportation cost is high. Enterprises need to improve production efficiency and reduce production and operation costs in order to adapt to the development and reform of industrial modernization. This research can help enterprises to eliminate the traditional manual handling mode and realize the automation transformation in the process of wheel transportation.

Q. Liu et al. (Eds.): CENet 2020, AISC 1274, pp. 1243–1249, 2021.
https://doi.org/10.1007/978-981-15-8462-6_142

2 Overall Structure and Workflow

The wheel automatic transportation line is composed of inspection workshop, manual maintenance line, sorting line, return line, main line, turnover line and overturn fixture. The total length of transportation line is 100 m and the working beat of the transportation line is about 19.7 s, and the wheels can automatically turn 180° at the end of the transportation line. The overall structure diagram is shown in Fig. 1.

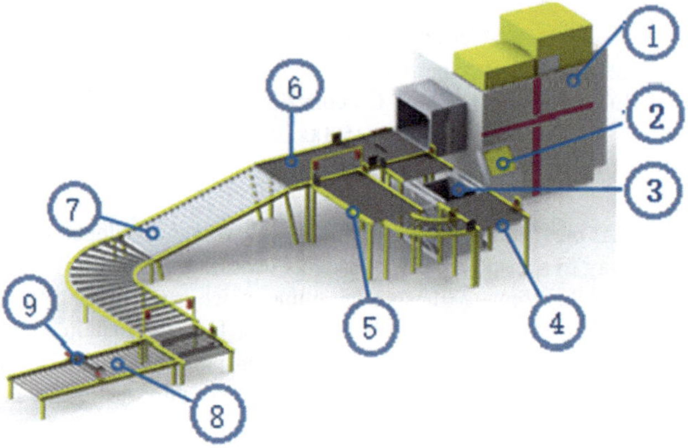

1. Inspection workshop; 2. Host computer; 3. Maintenance operation station; 4. Manual maintenance line; 5. Return line; 6. Sorting line; 7. Main line; 8. Overturn line; 9. Overturn fixture

Fig. 1. Overall composition diagram of the transportation line.

As shown in the figure above, the wheels enter the inspection workshop first, the inspection workshop inspects the wheels to determine whether they are qualified, and then the sorting line is responsible for sending the unqualified products to the manual maintenance operation station, the unqualified wheels are sent to return the line after passing the manual maintenance, the qualified wheels are sent to the overturn line through the main line, and then they are turned over by the overturn fixture before entering the wheels storage area.

The overturn fixture is an important tool on the automatic transportation line. Before the wheel is put into storage area, it is necessary to turn over the wheel. The overturn fixture has a more complex mechanical structure and its 3D model is shown in Fig. 2. The overturn fixture is at the end of the transportation line and which is composed of two parts, one part is the rotating shaft, which is installed on the base and driven by the stepping motor and is responsible for 180° rotation; the other part is the clamping arm, which is installed on the rotating shaft and driven by the cylinder and is responsible for clamping the wheel. The overturn fixture is made of 45 steel. On the premise of ensuring safety, the mechanical mechanism adopts hollow structure in order to reduce its own weight. The contact part of the clamping arm and the wheel is circular arc type, the diameter of the circular arc is the same as the diameter of the wheel. The rotating shaft is driven by the stepping motor, which can precisely control the rotation angle of the shaft when used in combination with the reducer. It is possible to exceed the rotation angle during the operation of the rotating shaft, which may cause dangerous situations. In order to ensure safety, hardware limit switches are installed on both sides of the rotating shaft. If the limit is exceeded, the operation of the rotating shaft will be stopped. In the overall work flow, when the wheel reaches the position of turning fixture, the clamping arm stretches out first to clamp the wheel. After the action is completed, the motor drives the shaft to make the arm slowly and stably complete the rotation work when the wheel is clamped. After turning the arm in place, release and reset, then reset the rotating shaft, and wait for the next work.

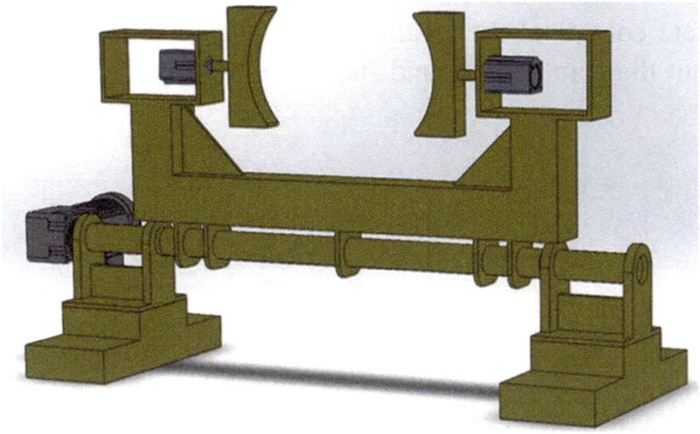

Fig. 2. 3D model of overturn fixture.

3 Design of Control System

PLC is a microprocessor-based control device, which uses programmable memory to control the production process of various types of output equipment through digital or analog input and output. PLC is the core equipment of the whole control system, which is responsible for input and output, logic calculation and communication. S7-200 smart PLC and host computer WinCC configuration software are selected as the control system of the tubeless wheel automatic transportation line. The configuration software has the functions of real-time displaying the number of wheel transmission, alarming, controlling the transmission line, monitoring the wheel transmission process and so on. The control system ensures that there will be no collision in the process of wheel transmission.

3.1 Hardware Design of Control System

In this control system, the stepper motor needs to be controlled by step driver, so the control system needs PLC which can output high speed pulse. The output PLC of relay has strong load capacity, but the output contact life of relay is limited, so it cannot be used for high-frequency action load. The output mode of transistor is suitable for high frequency output and sensitive response. The output contact has no limit of service life.

To sum up, transistor output PLC is selected, and CPU model is ST60, which can meet the design requirements without adding expansion module.

Step motor needs to be used with step driver, which can convert electrical pulse into mechanical angular displacement. The stepping motor operates at a fixed step angle, which is the fixed angle for the motor output shaft to rotate after the controller sends a pulse signal. By controlling the number of input pulses and the frequency of signal, the speed, acceleration and rotation angle of the motor output shaft can be controlled, and the precise motion control can be realized. The system uses 130 three phase stepping motor, so the driver is selected with 220 V input and 380 V output. To sum up, the stepping driver of model ZD-3HE22150 is selected, the input signal of the driver is connected with PLC, and PLC sends high-speed pulse signal to control the driver, so as to realize accurate control of the rotation speed and rotation angle of the shaft. The connection circuit diagram of PLC and step driver is shown in Fig. 3.

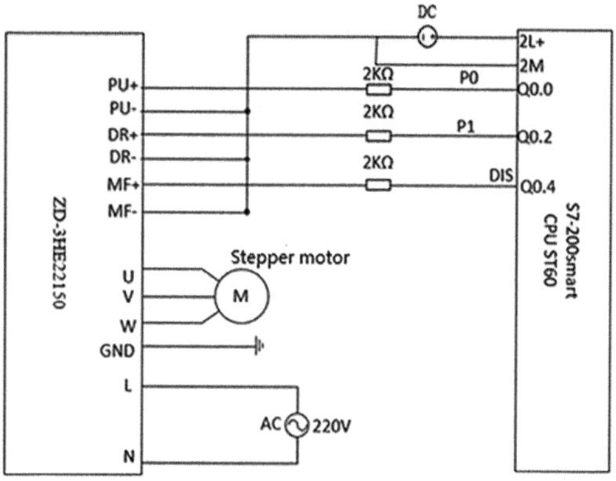

Fig. 3. Connection circuit diagram of PLC and step driver.

In this system, the clamping arms are driven by the cylinder. In this section, the cylinder is selected by calculation, and the corresponding solenoid valve control cylinder is selected. The clamping arm part clamps the wheel by extending the left and right pneumatic arms at the same time and outputting certain pressure. The inner side of the arm is arc-shaped, and the diameter of the arc is 508 mm. The cylinder needs to be controlled by solenoid valve. When the solenoid coil is powered on, it will generate electromagnetic force to push the valve core. In this design, the solenoid valve is used for simple control of the cylinder. In the system, two position three-way solenoid valve LDDE2308 is selected.

In this project, proximity switch is mainly used for position detection and counting, and photoelectric switch is selected after integrating various factors. There are many types of photoelectric switches, such as diffuse reflection type, opposite reflection type and so on. The tested object is a metal wheel with smooth surface and high reflective rate. Considering economic factors and installation convenience, diffuse reflective photoelectric switch is selected. The specific model is E3F-DS50B1, and the detection distance is 50 cm adjustable.

Ethernet communication is selected as the communication mode, which can directly connect the host computer and the lower computer through the Ethernet interface with the network cable. Finally CSM 1277 Ethernet switch is selected to connect PLC and WinCC, which can be used to connect multiple devices.

3.2 Software Design of Control System

The software of the system includes WinCC configuration software and PLC control program. The smart version of STEP-7 software is used to design PLC the control program. The program is mainly divided into the main line part and the turning section

Table 1. I/O address assignment.

Signal name	I/O
Start button	I0.0
Stop button	I0.1
Sorter return to position photoelectric switch	I0.2
Sorter limiter up to position photoelectric switch	I0.3
Sorter limit in place photoelectric switch	I0.4
Sorter push in position photoelectric switch	I0.5
Manual maintenance completion signal	I0.6
Sorter limiter down to position photoelectric switch	I0.7
Sorting line in place signal	I1.1
Main line distribution in place signal	I1.2
Turn over line entrance in place signal	I1.3
Reverser back to position photoelectric switch	I1.4
Tipper detection in place signal	I1.5
Flipper in place photoelectric switch	I1.6
Pulse	Q0.0
Direction	Q0.1
Enable	Q0.2
Sorting line motor	Q0.3
Sort push motor	Q0.4
Sort return motor	Q0.5
Sort stop down motor	Q0.6
Main line distribution motor	Q0.7
Cylinder exhaust	Q1.1
Reverse line motor	Q1.2
Cylinder charging	Q1.3
Main line motor	Q1.4
Workshop test start	Q1.5
Sorter limiter is the up motor	Q1.6

part in function. The two parts of the program play their respective roles and influence each other, and form the control program of the automatic line together to realize the control function. I/O address assignment is shown in Table 1.

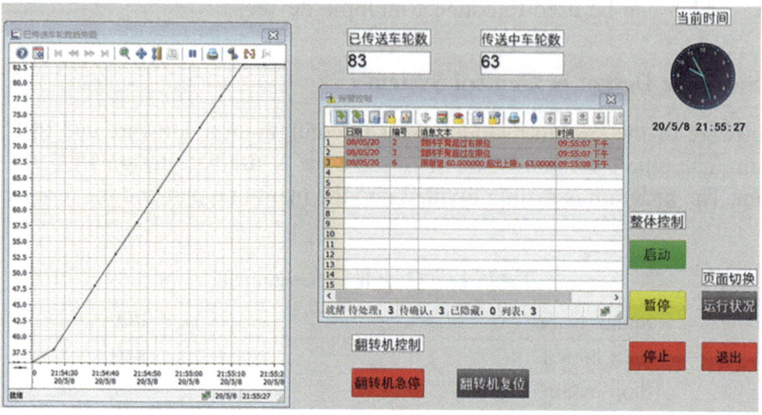

Fig. 4. Control interface.

Configuration software can communicate with PLC and other equipment for process control, data acquisition and other work. It is used to design industrial automation monitoring system software, widely used in various industries related fields. Siemens WinCC is selected for the design of configuration software. The software version is 7.4sp1, which can be run in 64 bit environment.

The user interface consists of three process screens, which are login interface, control interface and monitoring interface. The login interface is the starting page. When using, the user first enters the login interface. Click the "login" button in the login interface to type in the user name and password. After the user has typed correctly, the login succeeds. Then click the "enter the system" button to enter the control interface. Click the "logout" button to exit the current user. You can switch the user operation or exit the login to prevent irrelevant personnel from operating the software. Click the "exit system" button to exit the configuration software. The control interface is shown in Fig. 4. There are three buttons "start", "pause" and "stop" in the control interface to control the transportation line. In addition to the control, this interface also has the functions of real-time display of the number of transmitted wheels, display of the change trend of the number of transmitted wheels and alarm.

4 Field Experiment

The control system of wheel automatic transportation line has been used for one year, the operation of the whole system is stable. The system has the characteristics of convenient control management and high efficiency of wheel transportation and the system realizes the unmanned transportation process, reduces labor cost and improves working conditions of workers. Picture of field experiment is shown in Fig. 5.

Fig. 5. Field experiment.

References

1. Han, G.R., Chen, X.B.: The design of full-automatic handing system of automobile hub based on PLC technology, robot and visual system. Mod. Manuf. Eng. **9**, 52–58 (2017)
2. Chen, R.H., Shi, Z.M.: Design of workpiece transport system based on Q series PLC and CC-Link. Manuf. Autom. (7) (2019)
3. Lin, Z.X., Huang, R.Q., Zhang, Z.S.: Design of electro-pneumatic control system for pad printing press automatic transfer device. J. Anqing Norm. Univ. (Nat. Sci. Edn.) **25**(2), 85–90 (2019)
4. Shen, X.L.: Design of conveyor control system based on PLC. Technol. Innov. Appl. **30**, 86–88 (2018)

Research on Positioning Algorithm of Indoor Mobile Robot Based on Vision/INS

Tongqian Liu, Yong Zhang(✉), Yuan Xu, Wanfeng Ma, and Jidong Feng

School of Electrical Engineering, University of Jinan, Jinan 250022, China
cse_zhangy@ujn.edu.cn

Abstract. In order to meet the requirements of positioning accuracy of indoor mobile robot navigation system, this paper merges vision and inertial navigation system (INS) to improve robot positioning accuracy. Aiming at the low frequency of the visual navigation system (VNS) and the high frequency of the INS, a multi-frequency Kalman filter algorithm is proposed, and two different measurement equations are designed. The measurement Eq. (1) updates the position information of the inertial navigation after sampling in the INS, and the measurement Eq. (2) updates the position deviation of the mobile robot after sampling in the VNS. Finally, the accurate position information of the mobile robot is estimated by the inertial navigation position updated by the measurement equation one minus the optimal error updated by the measurement equation two. Realize the effective fusion of the location information of visual matching estimation and INS location information. The experimental results show that the multi-frequency Kalman filter algorithm is improved on the basis of the traditional filtering method, and the integrated navigation method proposed in this paper further improves the positioning accuracy of the INS.

Keywords: INS · Vision · Multi-frequency Kalman filter

1 Introduction

In recent years, with the continuous progress of automation integration technology, communication and electronic technology, and high-precision sensor technology, mobile robot navigation technology has developed rapidly [1]. The requirements for the positioning accuracy of mobile robots are also getting higher and higher [2]. In order to meet the positioning accuracy, different sensors are gradually used in mobile robot positioning systems. Inertial navigation system is a completely autonomous navigation system, which has the advantages of not relying on any external information, good concealment, working around the clock [3]. However, since inertial navigation is a kind of inferential navigation, its errors will accumulate with time, and its navigation errors will diverge seriously after long-term work of the mobile robot. Therefore, inertial navigation systems need other navigation systems to assist in use [4]. The advantages of using vision for mobile robot navigation are mainly reflected in the low cost, small size, strong electromagnetic interference resistance and good concealment of vision sensors. Inertial navigation and visual navigation have good

Q. Liu et al. (Eds.): CENet 2020, AISC 1274, pp. 1250–1255, 2021.
https://doi.org/10.1007/978-981-15-8462-6_143

complementarity and autonomy [5]. Navigation technology combining the two is one of the promising directions in indoor navigation. In order to achieve high-precision navigation of mobile robots, this paper proposes a technology that combines visual navigation information and inertial navigation information [6].

2 Vision/INS Localization Strategy

In this chapter, we will study the Vision/INS localization strategy. The overall localization strategy is shown in Fig. 1. Set the sampling frequency f^I of the IMU to frequency one and the sampling frequency f^V (among them, $f^I = N * f^V, N = 3, 4, 5$) of the vision system to frequency two. First, the IMU sends the measured raw data to the computer for calculation, and obtains the position P_o^I and speed V_o^I of the robot. The camera obtains the original image and obtains the position P_o^V and speed V_o^V of the mobile robot after processing by the computer. Then, input the positions P_o^I and P_o^V into the filter for filtering to obtain the filtered δP_O^I and $\hat{P}_o^I$.

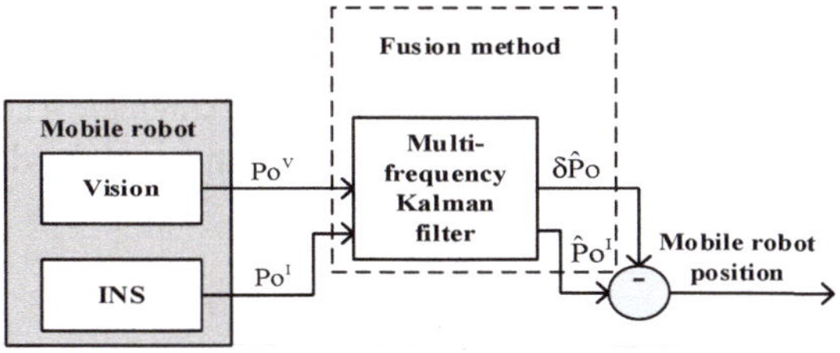

Fig. 1. Vision/INS fusion positioning strategy.

In the data fusion positioning stage, at frequency one, the system first filters the INS position P_o^I and velocity V_o^I navigation information; at frequency two, the difference δP between the filtered position $\hat{P}_o^I$ and the VNS position P_o^V is used as the input of the filter 2 to achieve Update to achieve an optimal estimate of the position error δP. Finally, subtract the optimal estimate δP of the position error from $\hat{P}_o^I$ to get the optimal estimate of the position. The data fusion algorithm is shown in Fig. 2.

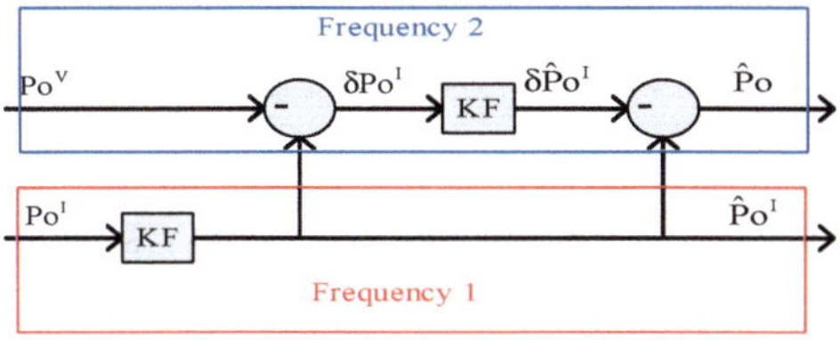

Fig. 2. The data fusion algorithm.

Assuming that the acceleration is constant during the filtering period, the equations of the velocity and displacement difference as the system input are discretized, and the state equations of the Kalman filter 1 and 2 are obtained as follows:

$$\begin{bmatrix} P_{E,k} \\ V_{E,k} \\ P_{N,k} \\ V_{N,k} \end{bmatrix} = \begin{bmatrix} 1 & T^I & 0 & 0 \\ 0 & 1 & 0 & 0 \\ 0 & 0 & 1 & T^I \\ 0 & 0 & 0 & 1 \end{bmatrix} \begin{bmatrix} P_{E,k-1} \\ V_{E,k-1} \\ P_{N,k-1} \\ V_{N,k-1} \end{bmatrix} + \omega_{k-1} \tag{1}$$

$$\begin{bmatrix} \delta P_{E,k} \\ \delta V_{E,k} \\ \delta P_{N,k} \\ \delta V_{N,k} \end{bmatrix} = \begin{bmatrix} 1 & T^V & 0 & 0 \\ 0 & 1 & 0 & 0 \\ 0 & 0 & 1 & T^V \\ 0 & 0 & 0 & 1 \end{bmatrix} \begin{bmatrix} \delta P_{E,k-1} \\ \delta V_{E,k-1} \\ \delta P_{N,k-1} \\ \delta V_{N,k-1} \end{bmatrix} + \omega_{k-1}^V \tag{2}$$

where $\delta P_{E,k}$ is the eastward position error of mobile robot and $\delta P_{N,k}$ is the northward position error of mobile robot at time K. $\delta V_{E,k}$ is the eastward velocity error of mobile robot and $\delta V_{N,k}$ is the northward velocity error of mobile robot, and white Gaussian noise $\omega_k \sim N(0, Q_k)$, $\omega_k^V \sim N(0, Q_k^V)$.

In filter 1, the inertial navigation system measurements a and b are used as the input values; in filter two, in order to obtain better filtering effects, the inertial navigation system measurement and visual navigation system measurement The difference values q and f are used as input values. The observation equations of the discrete form of the filter can be obtained as:

$$Y_k = \begin{bmatrix} \tilde{P}_{E,k}^I \\ \tilde{P}_{N,k}^I \end{bmatrix} = \begin{bmatrix} 1 & 0 & 0 & 0 \\ 0 & 0 & 1 & 0 \end{bmatrix} \begin{bmatrix} P_{E,k}^I \\ V_{E,k}^I \\ P_{N,k}^I \\ V_{N,k}^I \end{bmatrix} + v_k \tag{3}$$

$$Y_k^V = \begin{bmatrix} \delta\tilde{P}_{E,k}^I \\ \delta\tilde{P}_{N,k}^I \end{bmatrix} = \begin{bmatrix} 1 & 0 & 0 & 0 \\ 0 & 0 & 1 & 0 \end{bmatrix} \begin{bmatrix} \delta P_{E,k}^I \\ \delta V_{E,k}^I \\ \delta P_{N,k}^I \\ \delta V_{N,k}^I \end{bmatrix} + v_k^V \tag{4}$$

where $(P_{E,k}^I, P_{N,k}^I)$ is the position information provided by INS, and white Gaussian noise $v_k \sim N(0, R_k)$, $v_k \sim N(0, R_k^V)$.

3 Test

In the test phase, this paper validates the proposed algorithm through actual test experiments. The experiment used a robot, an INS system, and a binocular camera. The equipment used in the actual test is shown in Fig. 3.

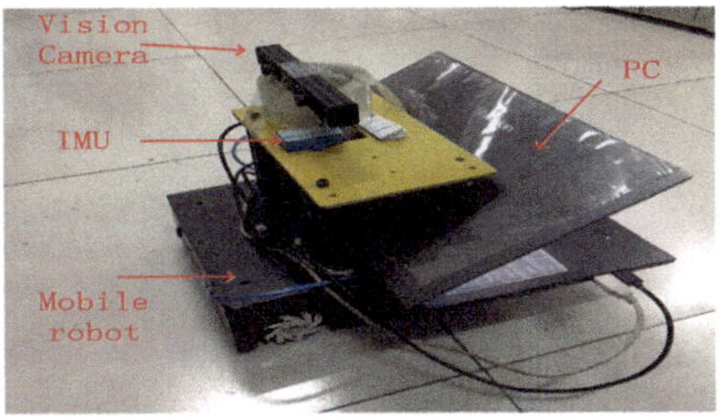

Fig. 3. The mobile robot and devices.

Figure 4 shows the reference path and estimated position trajectory. The inertial navigation system has accumulated errors, so the continuous working stability is poor, and the vision system is more stable. Integrated navigation system has higher accuracy than single sensor. Table 1 lists the position error comparisons. Figure 5 shows a comparison of the North position error, and Fig. 6 shows a comparison of the East position error. As can be seen from the graphs and table, the error of the proposed method is smaller.

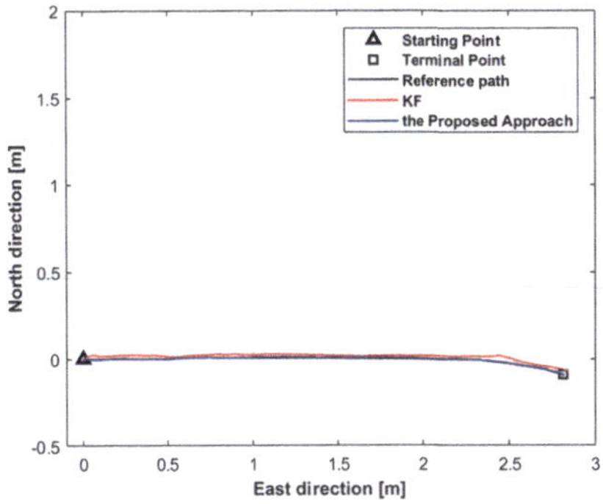

Fig. 4. The track of INS and vision.

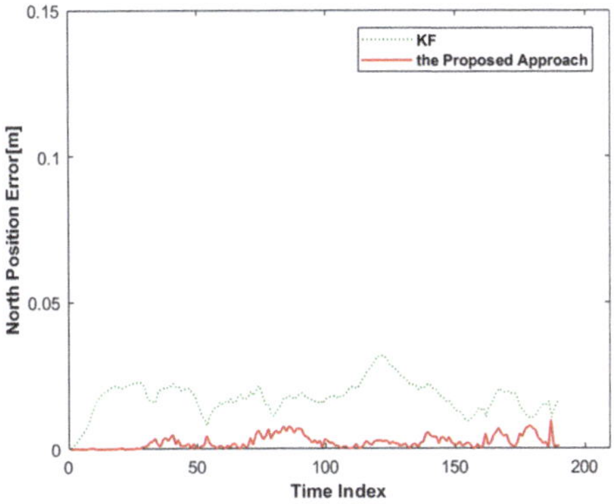

Fig. 5. North position error.

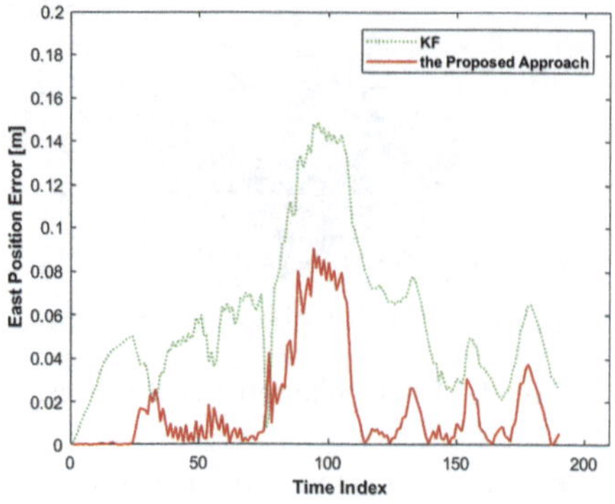

Fig. 6. East position error.

Table 1. The comparison of the localization.

Model	East direction mean [m]	North direction mean [m]
KF	0.0596	0.0177
The proposed approach	0.0174	0.0023

4 Conclusion

On the basis of Kalman filtering, this paper proposes a multi-frequency Kalman filtering algorithm for the problem of inconsistent sampling frequencies of inertial navigation and visual navigation to reduce the position error of mobile robots. For two different frequencies, two filters are used and data fusion and position estimation are realized. Finally, the optimal position estimate of the mobile robot is calculated by subtracting the position error of the optimal estimate from the corrected position of the INS system.

Acknowledgement. This work was supported in part by the National Natural Science Foundation of China under Grants 61803175.

References

1. Hu, S.: Analysis on the development direction and application of computer information technology. China Comput. Commun. (2017)
2. Shi, F., Liu, W., Wang, X.: Design of indoor laser radar navigation system. Infrar. Laser Eng. **44**, 3570–3575 (2015)
3. Baranski, P., Strumillo, P.: Enhancing positioning accuracy in urban terrain by fusing data from a GPS receiver, inertial sensors, stereo-camera and digital maps for pedestrian navigation. Sensors **12**, 6764–6801 (2012)

4. Jian, T., Chen, Y., Niu, X., et al.: LiDAR scan matching aided inertial navigation system in GNSS-denied environments. Sensors **15**(7), 16710–16728 (2015)
5. Kim, J.M., Leeghim, H.: INS/Multi-vision integrated navigation system based on landmark. Korean Soc. Aeronaut. Space Sci. **45**(8), 671–677 (2017)
6. Kim, Y., Hwang, D.H.: Vision/INS integrated navigation system for poor vision navigation environments. Sensors **16**, 1672 (2016)

Design of Smart Electricity Meter with Load Identification Function

Jia Qiao, Yong Zhang(✉), and Lei Wu

School of Electrical Engineering, University of Jinan, Jinan 250022, China
cse_qiaoj@ujn.edu.cn

Abstract. With the development of the electronic industry, smart meters are more and more required by people. Compared with traditional mechanical meters, smart electricity meters have more functions, such as multi-rate metering, intelligent interaction, data transmission, safe electricity, etc. NILM algorithm, also known as Non-intrusive load monitoring algorithm, can analyze the power consumption data through PC to show more detailed power consumption to users, which can greatly improve the power consumption experience of users. Based on AT89C51 microcontroller, ADE7755 energy measurement chip, LCD1602 display module, etc., this paper designed the overall hardware circuit of smart meter; In the software part, it not only realized basic ability of smart meters, but also realized the load recognition function by using the NILM algorithm.

Keywords: Smart electricity meter · AT89C51 microcontroller · NILM algorithm

1 Introduction

Since the initiative to build a global energy internet, the functional requirements of smart grid have been further improved [1]. Smart meters play an important role in the interactive service system between users and the power grid, which has a great impact on the intelligent construction of the entire power grid. Therefore, it is extremely urgent to research smart meters with more advanced functions.

Non-intrusive load monitoring technology was first proposed by Hart [1], which can realize load identification function. Load identification is the first step of interactive services, just calculate the total electricity user data sent to the user. However, if the non-intrusive load monitoring (NILM) algorithm can be used for load detection, the calculation and detection of the power consumption and power consumption behavior of each electrical appliance of the user can be realized. It will have a great impact on the electricity consumption of the entire society, and realize more convenient and intelligent interactive services, which will greatly help the construction of the global smart grid. For users, if they can know the use status and power consumption status of each type of electrical appliances, they will be able to better allocate the use time of each type of electrical appliances, which will help to save electricity and ease the user's voltage. Alleviate the use of electrical power in the entire society and achieve green development of energy conservation and emission reduction.

Q. Liu et al. (Eds.): CENet 2020, AISC 1274, pp. 1256–1262, 2021.
https://doi.org/10.1007/978-981-15-8462-6_144

In this paper, the hardware circuit, software programming and load recognition algorithm of upper computer are carried out.

2 Hardware Circuit

2.1 Overall Hardware Circuit

The main components of the designed hardware circuit are: master control module, power metering module, display module, storage module, key module, clock module and communication module, as shown in Fig. 1. The main control module of intelligent electricity meter selected AT89C51 MCU to control the whole system. The power metering chip ADE7755 was used to measure and analyze the load power consumption; LCD1602 was used for digital display; The storage module selected SD card to store data; DS1302 was used to segment the time. The communication module selected MAX485 to realize communication with PC terminal.

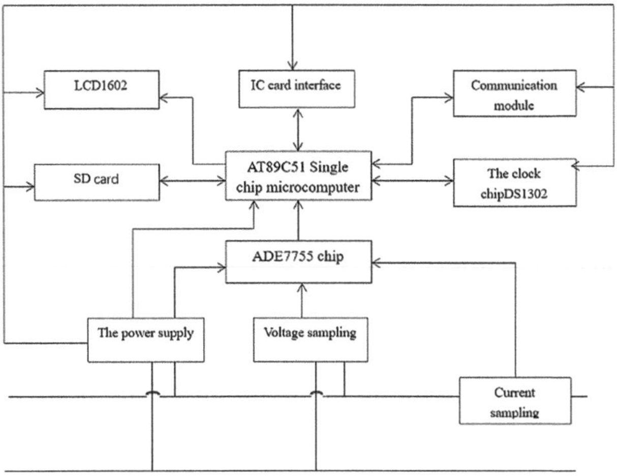

Fig. 1. Overall hardware circuit.

2.2 Power Metering Module

In the process of measuring the user's electricity using the ADE7755 chip, current transformers and voltage transformers can convert voltage and current into analog signals, and then transfer the analog signals into the energy measurement chip into digital signals, and then perform digital integration operations, output pulse signal whose frequency is proportional to the energy data. The signals are then passed microcontroller for processing, followed by operations such as time judgment, data transmission, and liquid crystal display. The hardware circuit of the power metering module is shown in Fig. 2.

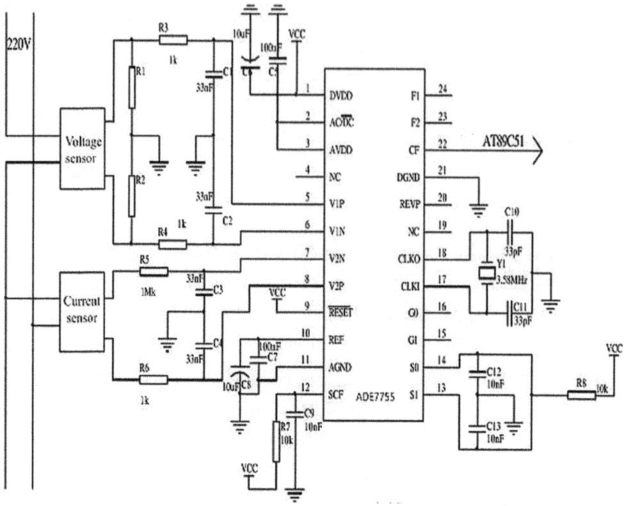

Fig. 2. Hardware circuit design of power metering module.

2.3 LCD Display Module

LCD1602, a character-based liquid crystal display module, is selected in this design, which is not only easy to operate but also highly cost-effective [2]. The microcontroller converts the pulse signal into digital signal and inputs it into the LCD1602 display chip, which is displayed on the LCD through the pre-programmed C language control program. Its circuit is shown in Fig. 3.

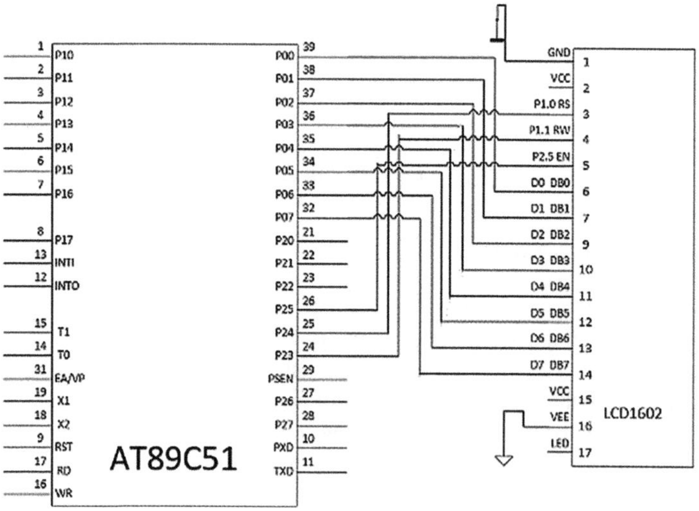

Fig. 3. Hardware circuit design of LCD display module.

3 Software Programming

3.1 Overall Software Programming

The data source of the helmet wearing data set mainly comes from two ways: One part is the video surveillance video of the construction site, and the other is to use the crawler code to obtain related pictures from the Internet.

The smart meter software is mainly based on AT89C51 s microcontroller to achieve data acquisition, peripherals, chip control, data processing and other functions. Specifically, it can be divided into ADE7755 driver module, LCD 1602 chip driver, display module, RS485 data transmission, SPI communication and other module program design. The ADE7755 module program uses timer interruption to read the power data calculated by DSP in ADE7755, and sends the data results to the MCU through SPI. The MCU reads the data and sends the data package to the upper computer via RS485 bus. Another function undertaken by MCU is to support LCD display and menu switching functions while reading and sending data [3]. The overall software programming flowchart is shown in the Fig. 4.

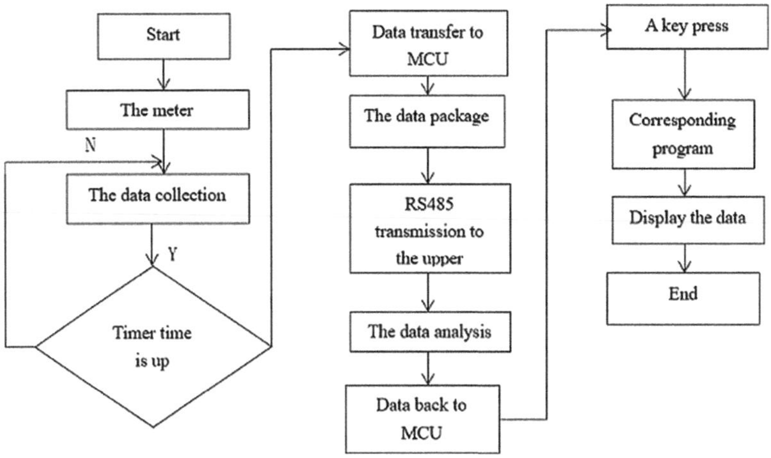

Fig. 4. Flowchart of overall software programming.

3.2 Data Acquisition Subroutine

After the electricity meter is initialized, the data acquisition subroutine flow chart is shown in Fig. 5.

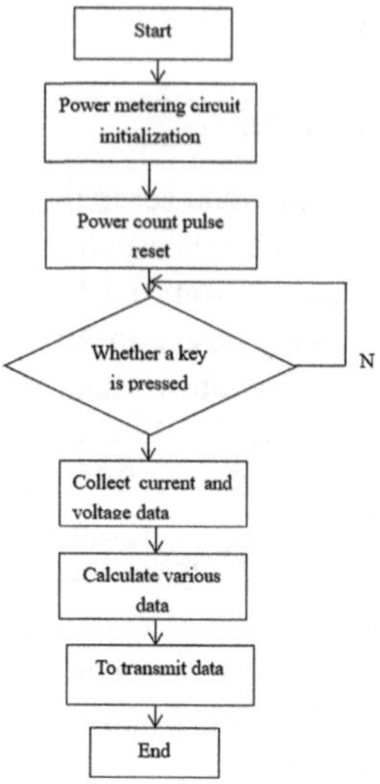

Fig. 5. Flowchart of data acquisition subroutine.

3.3 Key Subroutine

The key of intelligent electricity meter is realized by programming scanning. When the key is pressed, the main program will start. If the program finally determines that a key has been pressed, the key value will be scanned, and one key corresponds to one function operation [4]. For operations that need to be combined, the key values of all pressed keys are added up and then entered into a register to complete the corresponding function. Key subroutine flowchart is shown in Fig. 6.

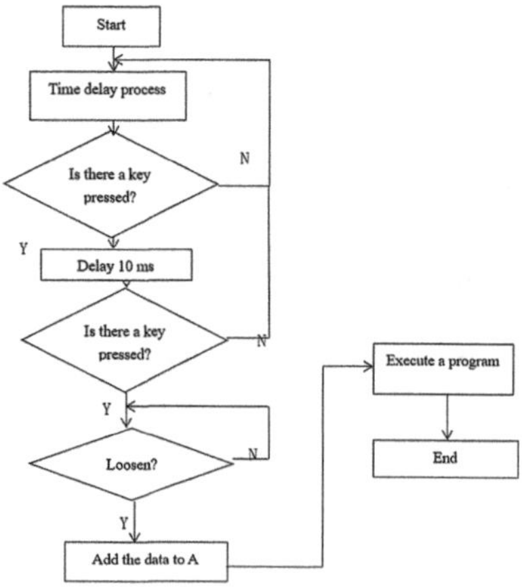

Fig. 6. Key subroutine flow chart.

4 Realization of Load Identification Function

4.1 Selection of Load Characteristics

The ADE7755 power metering chip used in this design can accurately measure the active power, reactive power and power factor of the user's electrical appliances. Five characteristics, namely current, voltage, active power, reactive power and power factor, are taken as load identification features [5]. Power factor is defined as the ratio of active power and apparent power, by the phase difference between voltage and current φ decision.

The formula for calculating active power is:

$$P = U \cdot I \cdot \cos \varphi \tag{1}$$

The formula for calculating reactive power is:

$$\mathrm{Q} = U \cdot I \cdot \sin \varphi \tag{2}$$

4.2 SVM for Load Identification

Support Vector Machines (SVM) is a dichotomy model, which means to find a support vector in the training set and obtain the optimal partition hyperplane. The principle is to maximize the sample interval. Given the training sample set, which can be based on the training sample set, it is concluded that a classification hyperplane in sample space, the sample was divided into different classes, but is not only a partitioning hyperplane training samples can be classified correctly. So, how should make the division of the optimal hyperplane is an important part of the support vector machine [6].

In this paper, 106 electrical appliances are selected, each of which has 5 characteristics (current, voltage, active power, reactive power, and power factor). There are 6 types of electrical appliances (fan, kettle, incandescent lamp, microwave oven, laptop and energy-saving lamp). 70% of them are randomly selected as training sets and the rest 30% as test sets. Using the LibSVM toolbox as the experimental platform, the recognition accuracy reaches 80.7%, and the recognition results are shown in Fig. 7.

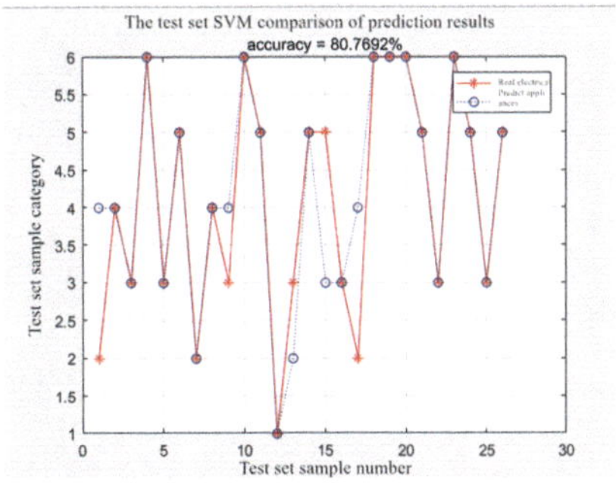

Fig. 7. The SVM for load identification results.

Figure 7 is the comparison diagram of the prediction of the final results, in which the red line represents the real electrical appliance category, the blue line represents the electrical appliance category predicted by vector machine, the X-axis is the sample number of the test set, and the Y-axis is the electrical appliance category. It can be seen that the red line basically coincides with the blue line, and only a small number of samples, such as the 13th sample, have errors in prediction, with the total accuracy reaching more than 80%. This indicates that the load recognition function of smart electricity meters can be basically realized by SVM vector machine.

5 Conclusion

In this paper, the hardware circuit, software programming and load identification example verification experiments are carried out for the smart meter with load identification function. The experimental results show that the load identification accuracy of this design reaches more than 80%, and the load identification function of intelligent electricity meter is basically realized. Since many electrical features are selected, the accuracy of SVM algorithm is not very high. Choosing more appropriate parameters as electrical identification features and improving SVM algorithm is the direction to further improve the identification accuracy in the future.

References

1. Rong, C.K.: Design of remote electricity meter based on MSP430 single chip microcomputer. Electron. World (08), 28–29 (2013)
2. Cheng, Y.: Research on online parameter identification method of household appliance load. China Excell. Master Dissert. Full-Text Database (02), 57-–57 (2013)
3. Cai, G.C., Cheng, L.J.: SVM based non-invasive load identification. J. Lingnan Norm. Univ. (06), 46—51 (2018)
4. Hart, G.W.: Nonintrusive appliance load monitoring. Proc. IEEE **80**(12), 1870–1891 (1992)
5. Chang, H.H., Lin, C.L., Lee, J.K.: Load identification in nonintrusive load monitoring using steady-state and turn-on transient energy algorithms. In: The 14th International Conference on Computer Supported Cooperative Work in Design, pp. 27–32 (2010)
6. Sun, D.S.: Classification and regression methods of support vector machines. Central South University, Changsha (2004)

Entropy Enhanced AHP Algorithm for Heterogeneous Communication Access Decision in Power IoT Networks

Yao Wang[1(✉)], Yun Liang[1,2], Hui Huang[1,2], and Chunlong Li[1,2]

[1] Global Energy Interconnection Research Institute Co., Ltd., Nanjing 210000, China
wangyao@geiri.sgcc.com.cn

[2] Electric Power Intelligent Sensing Technology and Application State Grid Corporation Joint Laboratory, Beijing 102209, China

Abstract. In this paper, the importance of the coverage, security, transmission delay, service rate and cost in each heterogeneous wireless network to the power business is comprehensively analyzed from the perspective of business preference of power distribution communication and the objective conditions of 5G heterogeneous network. The weight distribution of each network attribute for different business is more accurately obtained by using the entropy enhanced AHP and method. The performance ranking of each network provides a scientific, objective, accurate and optimized selection scheme for the power distribution communication service. Compared with other algorithms, the proposed algorithm shows better performance in term of blocking rate.

Keywords: Power IoT networks · 5G · Heterogeneous model · Access mode

1 Introduction

With the development of science and technology, power IoT networks have become the inevitable development direction of traditional power grid. Compared with the traditional power grid, power IoT networks will achieve the overall, accurate and real-time information acquisition of the power grid, improve the physical performance of the power grid and establish a perfect information interaction platform. The use of information data can provide auxiliary decision support and control management scheme for power grid practitioners [1]. Intelligent distribution grid is a heterogeneous network coverage, multi-mode service terminal environment, multiple services and multiple communication options, which has the problem of how to choose. For a long time, network selection mechanism is the most basic and critical problem in the heterogeneous integrated public telecommunication network [2]. In the context of power IoT networks distribution network communication, the design of network selection mechanism becomes more and more difficult due to the more diverse and complex business requirements and network environment [3]. Nowadays, the communication methods used in the intelligent power distribution include: optical fiber communication, power line carrier (PLC), broadband wireless access technology (5G, 4G), etc.

Q. Liu et al. (Eds.): CENet 2020, AISC 1274, pp. 1263–1270, 2021.
https://doi.org/10.1007/978-981-15-8462-6_145

In 5G heterogeneous fusion network scenario, network access decision-making algorithm is also an important aspect of network resource management. At present, there are many researches on network decision-making mechanism based on multi-attribute.

In document [4], grey relational analysis and TOPSIS are combined to select multi-attribute network. In document [5], simple weighting method saw based on SINR and AHP is used to select network. Analytic hierarchy process belongs to the subjective weighting method, and the results are in line with the generally accepted scheme, but it is easy to achieve the original intention of using mathematical methods to obtain accurate results because of its arbitrary subjectivity. Generally, the entropy method is used to revise again. The entropy method mainly calculates the weight of the evaluation index to be considered according to the relevant information provided by each evaluation index itself and a certain mathematical model, which has absolute objectivity, but often results in unsatisfactory results [6]. This paper proposes an analytic hierarchy process and entry evaluation method based on business preference.

The rest of this paper is organized as follows: In Sect. 2, we introduce the main analysis process of AHP-E algorithm in the scenario of power 5G heterogeneous communication. Section 3 gives the simulation results and analysis results, and Sect. 4 gives some conclusions.

2 Analysis of AHP-E Algorithm

In this section, we build hierarchical relationships between business, performance requirements, and network technology. The target layer is a certain power business, such as meter reading, fault warning, etc. The criterion layer is service QoS requirements, such as coverage, security, transmission delay, service rate, and use cost. The scheme layer is the communication mode in the heterogeneous model of power IoT networks.

2.1 Single Level Sorting and Consistency Test

The decision matrix from the scheme level to the criterion level and from the criterion level to the target level is a single level ranking matrix. The decision matrix between network attributes reflects the demand preference of communication services. The industry experts give the quantitative scale to the comparison results of the two, and then arrange a priority through mathematical operation, that is to find out the characteristic quantity of the decision matrix.

$$\mathrm{A} = (a_{ij})_{m\times n} = \begin{bmatrix} a_{11} & a_{12} & a_{13} & \cdots & a_{1n} \\ a_{21} & a_{22} & a_{23} & \cdots & a_{2n} \\ a_{31} & a_{32} & a_{33} & \cdots & a_{3n} \\ \vdots & \vdots & \vdots & \vdots & \vdots \\ a_{m1} & a_{m2} & a_{m3} & \cdots & a_{mn} \end{bmatrix} \tag{1}$$

where a_{ij} denotes the value of the jth parameter of QoS of every network.

1) We normalize the column vectors of the multi-objective parameter matrix A to get the matrix X. x_{ij} is the element in matrix X:

$$x_{ij} = \frac{a_{ij}}{\sum_{i=1}^{n} a_{ij}}, j = 1, 2, \ldots, n \tag{2}$$

2) By summing the rows of matrix X, the column vector $\bar{\mathrm{Y}}$ is obtained. $\bar{y}_i$ is the element in the column vector $\bar{\mathrm{Y}}$:

$$\bar{y}_i = \sum_{j=1}^{n} x_{ij}, i = 1, 2, \ldots, n \tag{3}$$

3) By normalizing column vector $\bar{Y}$, we get the eigenvector Y. y is the element in Y which is the weight of the QoS parameter:

$$y_i = \frac{\bar{y}_i}{\sum_{i=1}^{n} \bar{y}_i}, i = 1, 2, \ldots, n \tag{4}$$

Then

$$\mathrm{Y} = [y_1, y_2, \ldots, y_n]^T \tag{5}$$

4) We calculate the largest eigenvalue λ_{max} of the matrix A. The calculation formula is as follow:

$$\lambda_{max} = \frac{1}{n} \sum_{i=1}^{n} (Ay)_i / y_i \tag{6}$$

The necessary step after each level of ranking is to test the consistency of the decision matrix. The consistency here refers to the logical consistency of the judgment thinking. The specific steps are as follows:

1) The first step is to calculate the consistency index:

$$\mathrm{CI} = (\lambda_{max} - \mathrm{n})/(\mathrm{n} - 1) \tag{7}$$

2) The second step is to calculate the consistency ratio:

$$\mathrm{CR} = \mathrm{CI}/\mathrm{RI} \tag{8}$$

When CR is less than 0.1, it is considered that the matrix A satisfies the consistency requirement. Otherwise, the matrix must be adjusted and consistency checked again until the matrix meets the requirements. The RI values are shown in Table 1, where n is the order of matrix.

Table 1. The value of random consistency index RI.

n	1	2	3	4	5	6	7	8	9	10
RI	0.00	0.00	0.52	0.89	1.12	1.26	1.36	1.41	1.46	1.49

2.2 Hierarchy Total Sort

The single order of target layer in known criterion layer C is $Y^{(1)} = \left[y_1^{(1)}, y_2^{(1)}, \ldots, y_m^{(1)}\right]^T$ The single order of n schemes in scheme layer p to C_i criterion is $Y^{(1)} = \left[y_{1i}^{(1)}, y_{2i}^{(1)}, \ldots, y_{mi}^{(1)}\right]^T (i = 1, 2, \ldots m)$. The total ranking is the ranking of the lowest layer relative to the highest layer, that is, the ranking of the scheme layer relative to the target layer. The weight of single layer elements is synthesized from the top to the bottom, and the calculation is carried out according to the following formula. Finally, the total ranking of the p layer is $Y^{(0)} = \left[Y_1^{(0)}, Y_2^{(0)}, \ldots, Y_m^{(0)}\right]^T$.

$$Y_n^{(0)} = \sum\nolimits_{j=1}^{M} Y_j^{(1)}, Y_{nj}^{(2)} \tag{9}$$

It is also necessary to evaluate the consistency of the total ranking matrix. The consistency index of the p layer elements for the C_j single ranking is CI_j, and the corresponding average random consistency index is RI_j, so the random consistency ratio of the total ranking is

$$CR = \frac{\sum_{j=1}^{M} Y_j^{(1)} * CI_j}{\sum_{j=1}^{M} Y_j^{(1)} * RI_j} \tag{10}$$

When $CR > 0.1$, it indicates that the matrix needs to be modified if it does not meet the requirements; when $CR < 0.1$, the results meet the requirements of consistency, which can be used as the weight vector.

2.3 Analysis of Entropy Algorithm

Entropy Algorithm is a simple and intuitive objective weighting method. Entropy is a physical quantity to measure the amount of information in information theory. The larger the amount of information, the smaller the uncertainty of the event, the smaller the entropy, and vice versa. The certainty of the event can also be determined from the perspective of the probability of the event occurrence, so the first step of the entropy method is to calculate the probability of the event occurrence. Generally, the attributes of each network are listed in (3.10) matrix form:

$$B = \left(b_{ij}\right)_{m\times n} = \begin{bmatrix} b_{11} & b_{12} & b_{13} & \cdots & b_{1n} \\ b_{21} & b_{22} & b_{23} & \cdots & b_{2n} \\ b_{31} & b_{32} & b_{33} & \cdots & b_{3n} \\ \vdots & \vdots & \vdots & \vdots & \vdots \\ b_{m1} & b_{m2} & b_{m3} & \cdots & b_{mn} \end{bmatrix} \tag{11}$$

The element B_{ij} of matrix B represents the i-th attribute value of the j-th network. The attribute values have different units and measurement methods. If you want to put them together for processing, you need to standardize them first. Set $b_{i\_max} = \max\left(b_{i1}, \ldots, b_{ij}\right)$, $b_{i\_min} = \min\left(b_{i1}, \ldots, b_{ij}\right)$ for each attribute value. For the completion quality of communication services and user requirements, the faster the service rate is, the better the security is. Therefore, the standardized processing of this type of attribute scores the sum of the maximum and the minimum. The formula is $B_{ij} = \frac{b_{ij}}{b_{i\_max} + b_{i\_min}}$.

For both transmission delay and use cost, it is expected to be as small as possible, and its standardized processing method is $B_{ij} = \frac{b_{i\_min} - b_{ij}}{b_{i\_max} + b_{i\_min}}$. therefore, the standardized attribute matrix can be obtained as follows:

$$R = \left(r_{ij}\right)_{m\times n} = \begin{bmatrix} r_{11} & r_{12} & r_{13} & \cdots & r_{1n} \\ r_{21} & r_{22} & r_{23} & \cdots & r_{2n} \\ r_{31} & r_{32} & r_{33} & \cdots & r_{3n} \\ \vdots & \vdots & \vdots & \vdots & \vdots \\ r_{m1} & r_{m2} & r_{m3} & \cdots & r_{mn} \end{bmatrix} \tag{12}$$

According to the definition and principle of entropy, when there are N possible mutually independent states in the scheme of service selection network, and the probability of each state is $p_j (j = 1, 2, \ldots, n)$, the information entropy selected for this service is

$$E(i) = -K \sum\nolimits_{j=1}^{n} r_{ij} ln r_{ij} \tag{13}$$

Where, i is the network attribute, j is the candidate network, r_{ij} is the standardized value of the network attribute, K is a constant, simplified calculation, k = 1. According to the definition of entropy, the probability of occurrence of the i-th scheme under the j-th attribute is:

$$p_j = r_{ij} / \sum\nolimits_{i=1}^{m} r_{ij}, i = 1, 2, \ldots m; j = 1, 2, \ldots n. \tag{14}$$

According to the above formula, the entropy value e_j of the j-th attribute is calculated. For the j-th attribute, the greater the influence of the change degree of its value on the merits and demerits of the scheme, the smaller the calculated entropy value is, and vice versa. Then, the weight of the j-th attribute to the influence degree of the overall attribute on the scheme is calculated as follows:

$$w_j = (1 - e_j)/(n - \sum_{j=1}^{n} e_j) \tag{15}$$

2.4 Network Sorting

Through AHP and entropy method, the weights of subjective and objective are $W^{(0)}$ and w_j.

$$w = \alpha * W^{(0)} + (1 - \alpha) * w_j \tag{16}$$

Where α is a real number greater than 0 and less than 1.

After calculating the weight of network attribute from the previous formula, the total value of network performance can be calculated in the way of product sum:

$$S(j) = \sum_i w_i r_{ij} \tag{17}$$

Where w_i is the weight of network attribute, and r_{ij} is the standardized value of attribute i in the jth network. The total performance value of the selected network can be obtained in turn, and the optimized network sequencing can be obtained, which provides the applicable basis for the selection of communication modes of different services.

3 Simulation Results and Analysis

At present, the choice of communication mode in each power system is empirical and fixed. According to the literature review, the information collection class basically uses power line carrier and wireless public network, while the management and detection class mostly uses wireless private network support. Such rough allocation is difficult to ensure the high quality of service completion, and it is easy to cause load imbalance.

When the traffic is saturated, there is also a multi access method based on network load balancing, and the representative one is the maximum load balance, MLB) algorithm: scan to the optional access system, calculate the resource requirements estimation and current network load of the access systems, select the network with the lowest ratio of consumed resources to available resources of the network after the users access the system, and achieve the purpose of load balancing between different networks.

The process of simulation steps simulates the simplified process of generating communication service in power system:

1) According to the requirements of communication services and corresponding bandwidth delay, 20 initial services are randomly generated to select network access according to three algorithms;
2) In a fixed period of time, 40 new arrivals are randomly generated: the high priority services adopt queue jumping mode, and the services within the same priority

follow FIFO first come first serve mode, and the network is selected to join the service queue according to three algorithms respectively;

3) Get the network selection result from step 2, get the index to measure the algorithm performance through the method described below, and draw the curve for analysis.

In this paper, quality of service (QoS) and service blocking rate are selected as the performance indicators of the algorithm. For data type services, the blocking rate is calculated by (1- outflow flow/inflow flow); for control type services, the blocking rate of access delay is calculated by the ratio of the number of stopped services at a certain time to the total number of services. Then calculate the average value, and then calculate the relationship between the traffic blocking rate and the number of services, as shown in Fig. 1 simulation results.

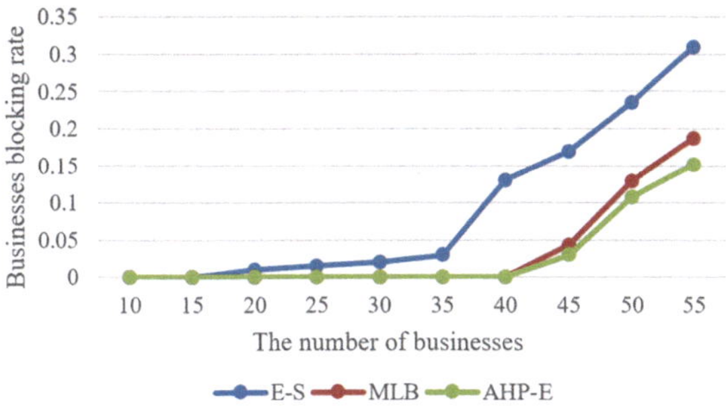

Fig. 1. The relationship between businesses blocking rate and the number of business.

Figure 1 shows that power grid users (services) access the communication network through different heterogeneous network selection algorithms, and the E-S performance is the worst, because the decision of network access is only based on experience, not combined with the actual situation of the network, so the service blocking rate is high. MLB algorithm ensures the load balance of the network and makes the operation of the power grid more stable. AHP-E really takes into account the characteristics and requirements of the business and guarantees the service quality of the business. Therefore, the performance of AHP-E is optimal.

4 Conclusion

This paper aims to optimize the decision-making of service access technology of 5G heterogeneous network integration of power IoT networks and improve the service quality of power business. AHP-E network selection algorithm is put forward, and it is proved that the improved network selection algorithm can play an accurate guiding role in all aspects of intelligent distribution grid business. In order to further verify the performance of the algorithm, the service blocking rate and the service QoS are taken as the performance indicators to evaluate the service satisfaction, and the load balancing algorithm and the empirical selection scheme are compared and analyzed. It is verified that the AHP-E network selection scheme proposed in this paper has better application effect in the distribution grid communication system.

Acknowledgement. This work is funded by industrial Internet innovation and development project in 2019 (Industrial Internet Testbed based on Ubiquitous Electric Internet of Things", No. SGGR000DCJS 1901068); State Grid Corporation of China (Research and Application of Wireless Sensor Self Powered Technology Based on Micro Energy Collection, No. SGXJXT 00JFJS2000053).

References

1. Matinkhah, S.M., Shafik, W.: Smart grid empowered by 5G technology. In: 2019 Smart Grid Conference (SGC), pp. 1–6. IEEE, USA (2019)
2. Saxena, N., Roy, A., Kim, H.S.: Efficient 5G small cell planning with eMBMS for optimal demand response in smart grids. IEEE Trans. Ind. Inform. **13**(3), 1471–1481 (2017)
3. Wang, Q., Zhao, F., Chen, T.: A base station DTX scheme for OFDMA cellular networks powered by the smart grid. IEEE Access **6**, 63442–63451 (2018)
4. Ansari, H.T., PremKumar, S., Saminadan, V.: Heterogeneous network modeling for smart grid technology. In: 2016 International Conference on Communication and Signal Processing (ICCSP), pp. 2336–2339. IEEE, New Jersey (2016)
5. Knirsch, F., Engel, D., Frincu, M., Prasanna, V.: Model-based assessment for balancing privacy requirements and operational capabilities in the smart grid. In: 2015 IEEE Power & Energy Society Innovative Smart Grid Technologies Conference (ISGT), pp. 1–5. IEEE, New Jersey (2015)
6. Chou, H.: A heterogeneous wireless network selection algorithm for smart distribution grid based on chi-square distance. In: 2018 10th International Conference on Communications, Circuits and Systems (ICCCAS), pp. 325–330. IEEE, New Jersey (2018)

An Off-line Handwritten Numeral Recognition Method

Yuye Zhang[1(✉)], Xingxiang Guo[1], Yuxin Li[2], and Shujuan Wang[1]

[1] Qingdao Branch, Navy Aeronautical and Astronautical University, Qingdao 266041, China
amigo_yezi@sohu.com

[2] Department of Foreign Languages, Shandong Normal University, Jinan 250000, China

Abstract. Off-line handwritten numeral recognition is a pattern recognition problem of the images of ten numbers. To improve the recognition efficiency, the number's image's character dimension should be decreased. In order to improve the recognition veracity, the character mode instability resulting from different writing styles and habits should be considered. The article proposed a numbers recognition method which combined with the statistical characteristics and structural features of numbers. Firstly, the principal component analysis (PCA) method was adopted to extract the numeral image's statistical characteristics. The numeral recognition will be realized through analysis of the reconstruction error of the model, which reconstructed by the principal components. To further determine the type of numeral, the structural features of width and height rate should be added. Finally, through experiments on the numeral image identification, the reliability and accuracy of this method of digital recognition were verified. The deficiency of this method in real-time recognition was analyzed.

Keywords: Offline handwritten numeral recognition · Principal component analysis · Structural features · Statistical characteristics

1 Introduction

Off-line digital recognition cannot make use of dynamic information such as time and stroke order obtained by online recognition, so the system implementation is difficult [1]. The most critical part of handwritten numeral recognition is a feature extraction of numeral characters. At present, the characteristics of handwritten numbers can be divided into statistical and structural features. The statistical feature is to determine the statistical rule of the spatial distribution of each class of characters from 0 to 9 by using the character sample library. The structural feature includes the construction of Numbers such as endpoints, intersections, contours, etc. The two types of characteristics have their advantages. The statistical characteristics can describe the essential characteristics of Numbers, which is suitable for a given training set with little difference. The structural features can accurately describe the numbers' detailed features

Q. Liu et al. (Eds.): CENet 2020, AISC 1274, pp. 1271–1278, 2021.
https://doi.org/10.1007/978-981-15-8462-6_146

and have a high recognition rate for the Numbers with more standard writing. Two kinds of features can be combined to make better digital recognition.

This paper proposes a recognition method that can combine the statistical characteristics and structural characteristics of characters. Firstly, the statistical characteristics of digital character samples are extracted by the principal component analysis method, and the character recognition is carried out by error analysis of the principal component reconstruction model. To further improve the accuracy of character recognition, the feature of the width-to-height ratio structure is added for character matching recognition.

2 The Statistical Characteristics of Numbers Are Extracted by Principal Component Analysis

The character recognition process based on statistical feature extraction mainly has two stages: the training stage and the recognition stage. In the training stage, the acquired sample information is given to the classifier for training after preprocessing and feature extraction, so that the classifier can acquire the recognition ability on the basis of recognition. In the recognition stage, the information to be recognized will be classified and recognized by the classifier after the same preprocessing and feature extraction. This method uses PCA (Principal Component Analysis) [2–4] to complete the above two stages [5–7].

2.1 Training Stage

PCA is a multivariate statistical analysis method which selects a few important variables through linear transformation. The process of character feature extraction using PCA method is as follows: preprocessed digital character sample image pixels are arranged into N-dimensional column vector x_i, then K samples constitute $N \times K$ matrix X, and its covariance matrix is

$$\mathrm{C_X} = \frac{1}{\mathrm{K}}\sum_{\mathrm{i=1}}^{\mathrm{K}} (\mathrm{x_i} - \overline{\mathrm{x}})(\mathrm{x_i} - \overline{\mathrm{x}})^{\mathrm{T}}.$$

In this formula

$$\overline{x} = \frac{1}{K}\sum_{i=1}^{K} x_i$$

The N eigenvalues of the covariance matrix C_x are arranged in descending order $\lambda_1 \geq \lambda_2 \geq \cdots \geq \lambda_N$, and their corresponding eigenvectors $\mu_1, \mu_2, \cdots \mu_N$ constitute a basis of the eigenspace. If the first d(<<N) basis vectors can describe the main information of x_i, then the subspace spanned by the d basis vectors forms the characteristic space of the description of x_i.

For n-dimensional pattern classes, the d(<<N) dimensional eigenspaces are obtained by solving the eigenvectors of the covariance matrix of the training sample set. Since the description ability of each component corresponding to the basis vector to the character pattern features decreases in turn, the first d basis vectors of each handwritten character can be used to estimate the reconstructed model. The following equation is used for reconstruction:

$$\tilde{x} = \sum_{i=1}^{d} \mu_i^T (x - \bar{x}) \mu_i + \bar{x} \tag{1}$$

For training samples of 0 to 9 Numbers, a reconstruction model of 10 types of patterns is obtained.

2.2 The Recognition Stage

After a test sample is input, the corresponding reconstruction model is estimated respectively in the feature space of each model, and then the reconstruction model of the test sample is subtracted from the reconstruction model of each sample model, and the reconstruction error is obtained as follows:

$$E(x) = \|x - \tilde{x}\|^2 = \left\| \sum_{i=1}^{d} \mu_i^T (x - \bar{x}) \mu_i - (x - \bar{x}) \right\|^2 \tag{2}$$

There, $\tilde{x}$ is the estimation of the reconstruction of the original input data x, and $E(x)$ is the error between the reconstruction model and the actual original input. If the reconstruction error relative to a certain model is minimal, it indicates that the sample most conforms to the description of the principal component of the corresponding pattern, and the corresponding model is the recognition result.

For handwritten number recognition, the sample x to be identified is given if r satisfies

$$E^{(r)}(x) = \min_{l=0,1,\cdots,9} \{E^{(l)}(x)\} \tag{3}$$

Then the sample to be identified belongs to the type pattern.

3 Extract the Aspect Ratio Feature of Handwritten Numerals

Considering that some people's writing habit is to write flat characters, so that except for “1”, the remaining 9 digits can be written with the width of the word greater than the height, so we cannot directly take the aspect ratio of handwritten digits as the structural features of the number sample, but take 5 positions from the top to the bottom of the number, and extract the character aspect ratio features respectively.

In the process of preprocessing, there is a step to normalize the image size of all digital characters. This step is very important, it is related to the accuracy of the digital structure features.

Steps:

(1) For the image after a given tilt correction, firstly search the height h and width w of the calculated number, find the center position of the handwritten number $\left(\frac{w}{2}, \frac{h}{2}\right)$, set a certain size, divide the number horizontally and vertically 5 equal parts (ensure to cover the number), and form a 5×5 uniform area.
(2) Check the presence of black pixels in 5 grid areas in each row from top to bottom. If the horizontal width of black pixels exceeds half of the width of the grid, add up a width, with a maximum width of 5 and a minimum of 1. Since the height of each row is consistent, five aspect ratio features of the digital sample are obtained in this way. The feature template is represented by a vector l with length of 5.

4 Offline Handwritten Numeric Character Feature Matching Recognition

After the Principal Component Analysis (PCA) number identification step, m (experiments found $m \leq 3$) categories $w_1, w_2, \cdots w_m$ that are closest to the number to be identified can be screened out:

Then, according to the method in Sect. 3, the aspect ratio feature is extracted to obtain the 5-dimensional vector: $X_i = [x_{i1}, x_{i2}, \ldots, x_{i5}]^T$. For the identified number X, its feature template is: $X = [x_1, x_2, \ldots, x_5]^T$. Computer the distance of $d(X_i, X)$, if some i existed to make $d(X, X_i) < d(X, X_j)$, $j = 1, 2, \cdots, m$, $i \neq j$, then $X \in \omega_i$.

5 Experimental Design, Results and Analysis

In order to verify the feasibility of the method in this paper, it is necessary to conduct experiments with the help of handwritten number sample library and analyze the experimental results. In general, the number of sample base is 5–10 times of the number of features, while the structural feature template in this method only has 5 values. When extracting statistical features, samples with different written features are also needed, and at least 20 samples are given for each number. Ten people were asked to write 0–9 twice, so as to get 200 training samples with significant differences in the shape, thickness and gray level of character strokes. Five other people were asked to write 0–9 twice, and 100 test samples were obtained. Each sample is preprocessed and normalized to a 64×64 pixel gray image.

In the experiment, the character image preprocessing method in Sect. 1 is firstly adopted to improve the consistency of character morphology by refining and expanding characters. Then, the method of principal component analysis was used to extract statistical features from samples of 10 Numbers, and the first d principal components were selected in turn according to the size of the contribution rate. Finally, using the experimental data obtained by the principal component analysis to identify the Numbers, the Numbers that are easy to misjudge are identified by adding the structural features of the aspect ratio to analyze the recognition rate results.

The recognition rate is defined as:

$$= (\text{total number of correctly identified characters/total number of test samples}) \times 100\%.$$

5.1 Digital Experiment of Principal Component Analysis

Step 1: the training phase

The principal component analysis method was used to extract statistical features from training samples of 0 to 9 Numbers, and the numerical model reconstructed from the first d principal components of 0 to 9 pattern classes was obtained. Taking the number "6" as an example, the reconstruction model diagram of the first d principal components is given. The number of principal components d is 3, 5, 7 and 10 respectively. It can be seen from the result that the larger d is, the more information the model image contains.

Step 2: identification phase

The test samples are estimated respectively in the feature space of each mode corresponding to the reconstruction model of each mode, and then the reconstruction error is obtained by subtracting from the reconstruction model of each mode. The smaller the error is, the more similar it is to the corresponding digital pattern class.

Arbitrary given 0–9 digital test samples of the image, for preprocessing. In the feature space of number 6, take the number of principal components equal to 10 as an example. The reconstruction model images constructed respectively are shown in Fig. 1.

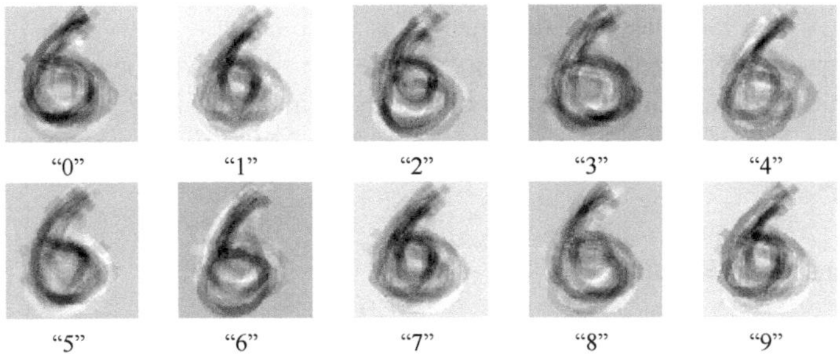

Fig. 1. The reconstruction model images constructed by 10 principal components

The gray scale of the number's background image changes significantly. This is actually because the reconstructed model image calculated by the algorithm program is not suitable for direct display (the image obtained is the one with all black eyes). Therefore, the image obtained here is the gray scale stretching operation of the reconstructed model by using mat2gray command in MATLAB, and a series of images listed in the above table are obtained.

According to formula (2), the model is calculated by subtracting the reconstruction model of the 10 randomly selected Numbers (test samples) from the reconstruction model of the training samples, and identify the obtained model error value.

Here, we only analyze the sample set of "6". When d = 10, the test samples of any ten Numbers were subtracted from the reconstruction model of the "6" sample set, and the subtraction results of the test samples of ten Numbers and the reconstruction model of the "6" sample set were obtained as shown in Fig. 2.

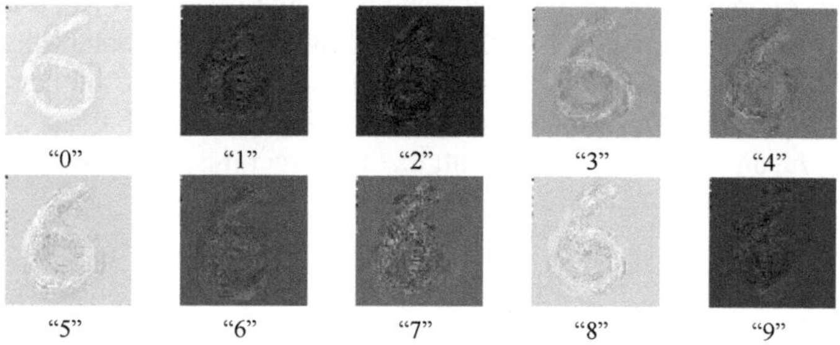

Fig. 2. The subtraction results of the test samples of ten Numbers and the reconstruction model of the "6" sample set

The images are also drawn in grayscale. The actual results are more uniform, and it is difficult to distinguish the difference between them.

The error value is obtained by modulating the image in Fig. 2. The smaller the error value is, the more similar the shape of the test sample number is to the number "6".

It is found that the reconstruction model error value of test sample number "6" is the lowest, which indicates that this method is reasonable.

The next smallest error is "4", then "5", the rest of the Numbers have a relatively large error. After analyzing the image of the test sample in Fig. 2, it is not difficult to find that the Numbers 4 and 5 written have the most similar stroke radians to the Numbers 6.

5.2 Analyze the Results

It can be seen from the experimental results that when only three principal components are taken to reconstruct the data, a high recognition rate $\eta = 84\%$ is obtained, indicating that the three-dimensional feature space spaned by the first three principal components has described most features of characters. When 5 principal components are selected, the reconstruction model is more accurate and the recognition rate increases to $\eta = 89\%$.The number of principal components was further increased to 7, and the recognition rate increased slightly to $\eta = 91\%$. When the principal component number increases to 10, the recognition rate does not change. This shows that the main information describing character features is concentrated in the first five principal components of the character, then increasing the number of principal components has little effect on improving the character recognition rate, but will greatly increase the computational cost of the recognition process.

Taking the experimental results obtained when taking 7 principal components as an example, it is found that the recognition rate of the number "1" is the lowest ($\eta = 70\%$), followed by the number "7" and "9" ($\eta = 80\%$).The main reason is that the stroke of the character "1" is too simple, with only one vertical stroke, so the features of the stroke structure are not obvious, and the structural features can be extracted by PCA is limited. While "7" are similar to "1" and "9" these two digital structure characteristics, also leads to its relatively low recognition rate. For this reason, the character's aspect ratio structure feature is added to identify.

Aiming at the problems of principal component analysis (PCA) in identifying Numbers, the method in Sect. 4 is adopted to identify Numbers in combination with the features of width-to-height ratio. It was found that the number recognition rate increased from 91% to 96%, in which the number "1" was recognized by all Numbers, because its aspect ratio feature was the most significant. The recognition rate of "7" and "9" is also greatly improved, because the structural features are different from "1", and the structural features make the difference between "9" and "6" more significant.

6 The Methods Need to Be Improved in This Paper

The method adopted in this paper is mainly aimed at improving the problem of simple statistical feature recognition, the experimental analysis steps are programmed in Matlab software, only to verify the recognition effect, no requirements for real-time algorithm. And a good algorithm must have excellent recognition efficiency, so we need to use professional signal processing software to improve real-time reprogramming.

References

1. Hanmandlu, M., Ramana Murthy, O.V.: Fuzzy model based recognition of handwritten numerals. Pattern Recogn. **40**, 1840–1854 (2007)
2. Wang, Y.W., Liu, J.: A new method for slant correction in recognition of handwritten numerals. Comput. Eng. **11**(30), 128–130 (2004)
3. Jiang, W., Lu, Z.Y., Li, J.: Skew correction of handwritten Uyghur character based on Hough transform. Microcomput. Appl. **32**(8), 29–31 (2013)
4. Zhang, G.H., Wan, J.L.: Offline handwritten numeral recognition based on principal component analysis. Comput. Eng. **9**(18), 219–221 (2007)
5. Chen, X., Zhu, R., Wang, Z.Y.: Handwritten digits recognition based on fused convolutional neural network model. Comput. Eng. **43**(11), 187–192 (2017)

6. Ye, X.B., Qin, H.F., Lü, Y.L.: Handwritten digits recognition based on multivariate normal distribution. Comput. Era **10**, 11–15 (2018)
7. Wang, Z.Q., Liu, P.F., Hao, F.: Application of handwritten digit recognition technology based on deep learning. Digit. Technol. Appl. **36**(11), 78–79 (2018)

3D Point Cloud Multi-target Detection Method Based on PointNet++

Jianheng Li[1], Bin Pan[1(✉)], Evgeny Cherkashin[2], Linke Liu[1], Zhenyu Sun[1], Manlin Zhang[1], and Qinqin Li[1]

[1] Liaoning Shihua University, Fushun 113001, China
panbin@lnpu.edu.cn
[2] ISDCT SB RAS, Ac. Lavrentieva ave. 17, Novosibirsk 630090, Russia

Abstract. 3D object detection is an important research direction in the fields of computer vision and pattern recognition in recent years. This technology can provide important technical support for unmanned driving and intelligent robots. Aiming at the challenges of object detection caused by the sparseness of 3D point clouds in outdoor scenes, this paper designs a 3D point cloud multi-target detection method based on pointnet++. The method first preprocesses the collected original point cloud; after obtaining the point cloud of the region of interest, the point cloud is clustered, and then the 3D target detection is performed by pointnet++ to obtain the object category. Finally, get the size and orientation of the target object through the 3D boundingbox. In order to verify the effectiveness of the method in this paper, the point cloud data of real outdoor scenes were collected using lidar, and a sample set was produced for network training. The final results verify that the method can achieve higher detection accuracy and meet the requirements of real-time performance.

Keywords: 3D object detection · PointNet++ · Point clound · Deep learning

1 Introduction

Object detection is one of the traditional tasks in the field of computer vision. Unlike image recognition, object detection not only recognizes objects on the image, but also gives the corresponding category of the object. It also needs to give the object through the minimum bounding box Location information [1]. According to the different output results required for target detection, the method of using RGB images for target detection to output the type of the object and the smallest bounding box on the image is generally referred to as 2D target detection. The detection using RGB images, RGB-D depth images, and laser point clouds to output object types, and the object's length, width, height, and rotation angle in three-dimensional space is called 3D target detection. With the development of deep learning in recent years, the two-dimensional detection technology has become very mature, such as FasterRCNN [2, 3] and MaskRCNN [4] of the RPN series, and YOLOv1-YOLOv3 [5] of the OneShot series.

In recent years, 3D object detection for outdoor scenes has been widely used, such as driverless and home intelligent robots. However, ordinary 2D detection in application scenarios such as unmanned driving, intelligent robots, and augmented reality

Q. Liu et al. (Eds.): CENet 2020, AISC 1274, pp. 1279–1290, 2021.
https://doi.org/10.1007/978-981-15-8462-6_147

cannot provide all the information needed to perceive the environment. It can only provide the position information of the target in the two-dimensional picture. Currently, the mainstream sensors used in outdoor scenes are camera and lidar. The more commonly used 3D target detection methods are classified according to different data types, including monocular images, multi-view images (stereo images) and point clouds (RGB-D images) [1, 6]. Compared with the two-dimensional image information obtained by the camera, the three-dimensional point cloud obtained by the lidar can provide more reliable depth and contour information of the three-dimensional object. In recent years, it has been gradually applied to target detection. However, the laser radar unevenly samples the three-dimensional space, resulting in the farther and thinner the obtained point cloud, which poses a greater challenge to the accuracy of target detection.

Due to the irregularity and high sparseness of the point cloud, it is difficult for CNN to process it. Common 3D detection methods based on point clouds either use a mature 2D detection framework to project the point cloud to a bird's eye view for processing or discretize the sparse point cloud is converted into a common 3D voxel format to process the point cloud data [5, 7, 8]. However, none of these methods can achieve good results, which can easily lead to loss of information or extremely high computational costs. However, the above methods will lose point cloud information during processing, and there are bottlenecks, especially the detection of some small objects. In response to this problem, [9] first divided the three-dimensional space by voxel grids, then extracted the features of each voxel grid point by a learning method, and finally performed detection. However, this process is time-consuming and difficult to meet the needs of real-time detection of outdoor scenes. Reference [10] proposed PointNet, which can directly learn the features of the original point cloud, and has excellent performance on datasets such as ModelNet. However, PointNet only considers the global features of the point cloud and ignores the local features. To this end, PointNet++ is further proposed to integrate local and global point cloud features to improve its performance on the data set [11]. This article applies PointNet++ to multi-target detection of 3D point clouds in outdoor scenes and obtains the bounding box of 3D objects.

2 Related Work

2.1 Object Detection on 2D Images

Object detection on 2D images has been developed for a long time [12]. The detection ideas are generally divided into suspicious area block selection, feature extraction and model training. The core part is how to determine the suspicious area block. Specifically, first determine the image area that may contain the target. That is to find all objects of interest in the image, including two sub-tasks of object localization and object classification, and determine the category and position of the object at the same time. Before 2012, target detection mainly used manual features such as Haar, HOG (gradient histogram), LBP (local binary mode), and machine learning methods such as AdaBVM (Support Vector Machine), DPM (Deformable Part Model) (called the Feature-B method) [13, 14]. In 2012, Alex and others improved the CNN (Convolutional Neural Networks) network and trained a large neural network DCNN (Deep Convolutional Neural Networks) on the basis of CNN. DCNN won the ILSVRC-2012 competition with a huge advantage of 10% lower error rate than the second place, and the CNN network was again valued. In the same year, J.R.R. Uijlings et al. [15] used selective search to select useful areas from the entire image and extract features for detection, laying the foundation for the subsequent CNN-derived TWO-STATE method. Since then, the speed and accuracy of two-dimensional target detection have improved significantly, and CNN-based detection methods have become main stream [16].

2.2 3D Object Detection from 2D Images

The appearance of a 3D object can be reflected by the brightness, posture, and shape in the 2D image. At present, there are many algorithms for detecting 3D object bounding boxes based on 2D pictures. [8, 17] use geometric constraints between 3D and 2D bounding boxes to recover 3D object poses. [12, 18, 19] exploited the similarity between 3D objects and CAD models. Chen et al. [20, 21] Formulate the 3D geometric information of the object as an energy function to score a predefined 3D box. Due to the lack of depth information, these works can only generate rough 3D inspection results, and may be substantially affected by changes in appearance.

2.3 3D Object Detection Based on Point Cloud

Existing state-of-the-art point cloud-based 3D object detection methods have proposed various methods for learning recognition features from sparse point clouds and generating 3D bounding boxes. [1, 14, 22–24] Project the point cloud to a bird's eye view, and use 2DCNN to learn the point cloud function generated by the 3D box. Song et al. [13] and Zhou et al. [9] grouped these points into voxels and used 3DCNN to learn the features of the voxels to generate 3D boxes. However, due to the data quantization, the bird's-eye view projection and voxelization will lose information, and 3DCNN is both too high in memory and low in computational efficiency. [6, 25] Generate 2D suggestions from images using mature 2D detectors and reduce the size of 3D points in each cropped image area. Then, use PointNet [10, 11] to learn the point cloud features of the 3D box estimation. However, proposal generation based on 2D images may fail in some challenging situations, which can only be observed well from 3D space. Such failures cannot be recovered through the 3D box estimation step. In contrast, we proposed the application of PointNet++ to multi-target detection of 3D point clouds in outdoor scenes, which is both efficient and accurate.

3 Our Method

Outdoor scenes are often complicated. The point cloud data obtained from the lidar contains various other objects besides the target of interest [26]. First, the point cloud needs to be pre-processed to obtain the point cloud of the region of interest; then, the point cloud processed by PointNet++ for classification detection is a single object point cloud, and the point cloud within the region of interest needs to be clustered to distinguish different objects; Then, the distinguished objects are input to the PointNet++ network to obtain the detection category result; finally, the boundingbox generation algorithm is used to obtain the size and orientation of the three-dimensional object [7, 11]. The overall framework is shown in Fig. 1.

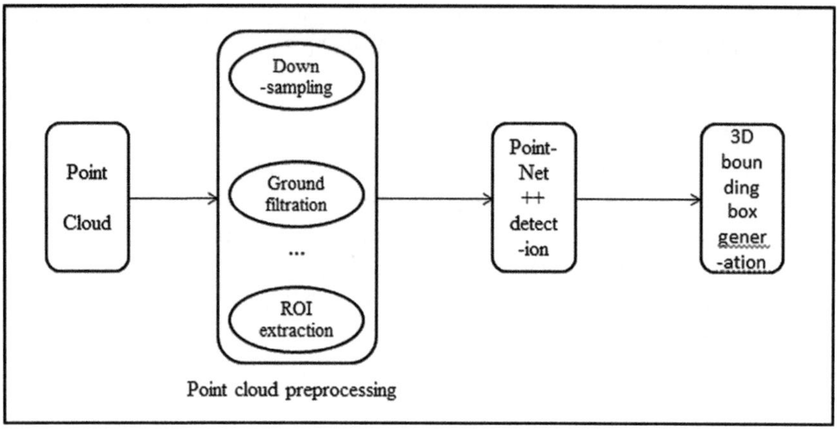

Fig. 1. The overall framework for 3D point cloud multi-target detection method based on PointNet++.

3.1 Point Cloud Preprocessing

The original point cloud data obtained by using lidar in outdoor scenes is a set of points with irregular structure and sparse distribution. It contains point clouds inside the road of interest and non-interest point clouds outside the road. The original point cloud obtained by the outdoor scene lidar is shown in Fig. 2. Training and inferring point cloud data with powerful CNNs is not easy. In order to obtain the point cloud in the region of interest and reduce the calculation of subsequent steps, the point cloud needs to be pre-processed [21, 27]. The point cloud preprocessing process is shown in Fig. 3.

Downsampling. Lidar's original point cloud has a large number of points. Even the 16-line lidar has more than 20,000 original point clouds. It is extremely time-consuming to fully process these points in subsequent steps. Therefore, the original point cloud needs to be down-sampled. At the same time, in order to maintain the original contour information of the down-sampled point cloud objects, we use voxel lattice filtering for downsampling. The voxel lattice size is 0.1 m.

Ground Filtering. In order to obtain the point cloud of each object through subsequent clustering, the point clouds belonging to an object should be close to each other and have a clear distance boundary with other objects [20]. As shown in Fig. 2, the point cloud on the ground connects the point cloud of each object on the ground, which is not conducive to subsequent clustering and segmentation. In addition, the ground can

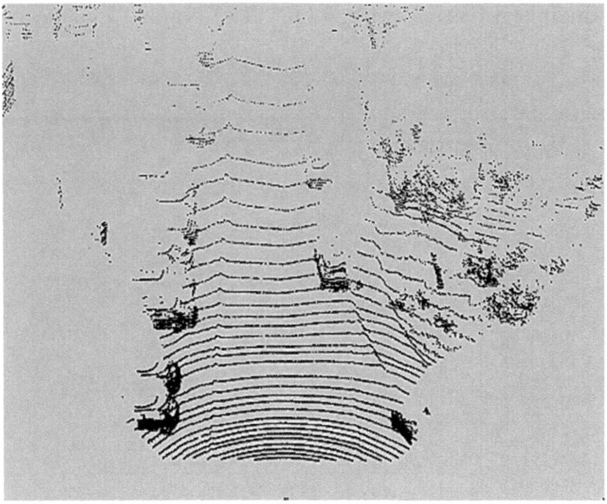

Fig. 2. Outdoor scene lidar acquisition of the original point cloud.

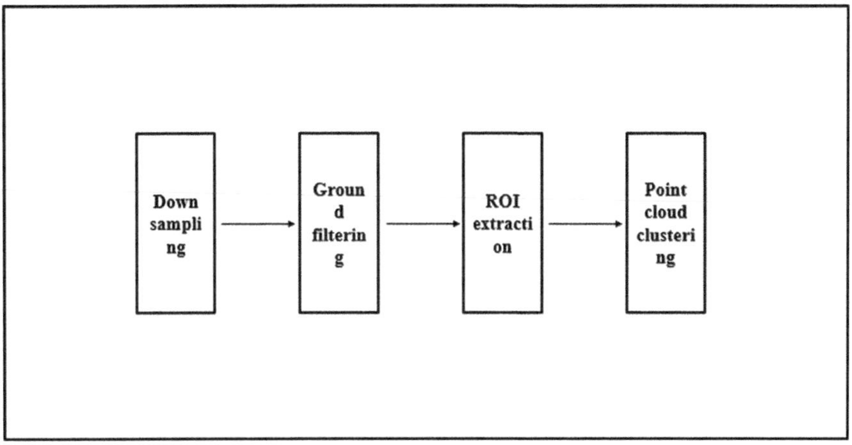

Fig. 3. Point cloud preprocessing process.

generally be approximated as a plane. Therefore, this paper uses the sampling consistency method to filter the ground point cloud by using the plane model.

ROI Extraction. The point cloud obtained through the above processing still includes point clouds other than roads. This article uses LiDAR to build a map to obtain roadside point clouds to filter out point clouds outside the roadside, and finally to obtain the point clouds of the objects to be detected on the road.

Point Cloud Clustering. After pre-processing of the point cloud, according to the characteristics of the lidar point cloud, the point cloud in the road is segmented by improving the Euclidean Cluster method [10]. The original Euclidean Cluster method is shown in Fig. 4.

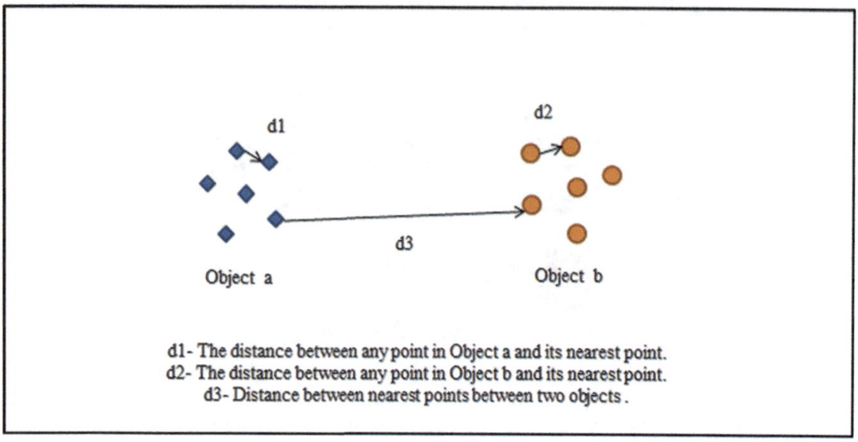

Fig. 4. Euclidean Cluster method.

Set a distance threshold that separates two categories is named d_t. Suppose the point cloud of an object *a* is composed of dots in the figure, the distance between any point in the object and its nearest point is showed as $d_1 < d_t$. The point cloud of another object *b* consists of triangular points, the distance between any point in this object and its nearest point is expressed as $d_2 < d_t$. And the distance between the closest points between these two objects is written as $d_3 > d_t$, then divide all the point clouds into 2 categories according to circles and triangles. The point cloud data obtained by the lidar, from near to far, changes from dense to sparse. The point clouds of three objects at different distances are shown in Fig. 5.

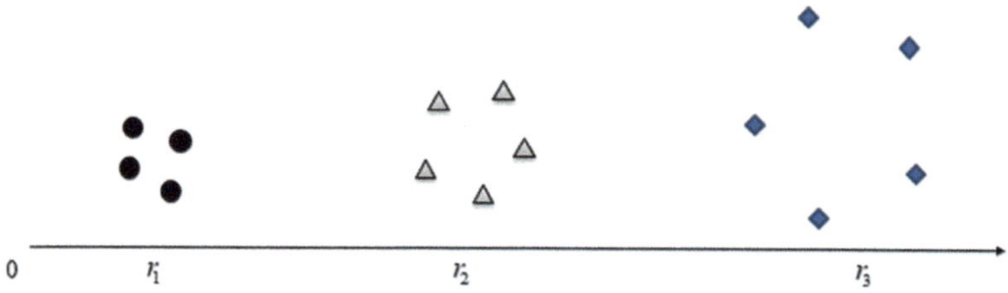

Fig. 5. Lidar gets three object point clouds at different distances.

Suppose that the circular point cloud, triangular point cloud, and rhombus point cloud are three objects with a distance of r_1, r_2, and r_3 from the origin, and $r_1 < r_2 < r_3$. The longer the distance, thc grcater the distance between points within the same object. In this way, it is difficult to obtain a good segmentation effect according to the

Euclidean Cluster method. To this end, the distance threshold d of the two categories is adjusted to a variable threshold according to the distance r

$$d = \begin{cases} 0.6, r \leq 3 \\ 0.2r, r > 3 \end{cases} \tag{1}$$

where r is the distance from a point to the origin. When $r \leq 3$ m, set a fixed threshold of 0.6 m, if $r > 3$, the threshold increases with r.

3.2 PointNet++

PointNet++ Classification Network. PointNet is the first deep learning framework to directly process unstructured point cloud data. PointNet's network structure can fit arbitrary continuous set functions and can be used as a general continuous set function estimator, as shown in the following formula [11].

$$f(x_1, x_{2,}, \cdots, x_n) = \gamma(\max_{i=1,\cdots,n} \{h(x_i)\}) \tag{2}$$

where $\{x_1, x_{2,}, \cdots, x_n\}$ is an unordered point set, $x_i \in R^d$. Set function $f : x \rightarrow R$ represents the mapping of a point set to a vector, which has invariance to the arrangement of input points, and can approximate any continuous set function; γ and h are continuous functions, here stands for a multilayer perceptron network.

PointNet++ is an improvement on the PointNet, adding hierarchical structure on the basis of PointNet and use PointNet to extract different scale features at each layer. Pointnet++ network hierarchy consists of many set abstraction layers. The set abstraction layer input is $N * (d + C)$, where N is the number of input points, d is the coordinate dimension, and C is the feature dimension. The output is $N' * (d + C')$, where N' is the number of output points, d is the same coordinate dimension, and C' is the new feature dimension. Each group of set abstraction layers of the network mainly includes three parts: Sampling layer, Grouping layer and PointNet layer. The sampling layer uses FPS (farthest point sampling) algorithm to sample a limited number of points from the initial point cloud for subsequent processing; the grouping layer uses each sampling point as the center of the sphere, and finds a special number of K points that fall within the range of the sphere with a specific radius to form a group; the PointNet layer treats each group obtained by the Grouping layer as a "local point cloud", and uses the PointNet network to calculate the feature of this local point cloud. After extracting features through the set abstraction layer, the point cloud-to-point cloud classification results are obtained through fully connected network integration [10, 11, 25].

The difficulty of laser point cloud detection is mainly sparse. This requires comprehensive consideration of the global and local characteristics of each point cloud object to obtain comprehensive object contour information [4, 28]. For outdoor scenarios, the PointNet++ network structure used in this article is shown in Fig. 6. Due to the sparseness of the point cloud, only a set abstraction layer is used here to extract the local features of the input 3D point cloud and set a larger radius parameter (due to the sparse point cloud, too small radius parameters cannot obtain better local features); Then use the PointNet layer to get global features; and finally output the classification results of point cloud objects through a fully connected network.

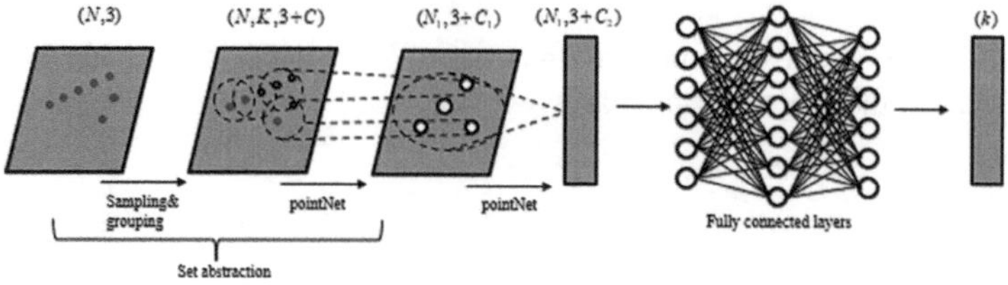

Fig. 6. PointNet++ structure.

3.3 Generation of 3D Bounding Box

Different from two-dimensional detection, in addition to category detection, three-dimensional objects need to know the length, width, height, and orientation information of these objects. Based on the characteristics of Lidar point clouds, this paper designs a method for generating a three-dimensional boundingbox of an object [12, 27]. The process is shown in Fig. 7.

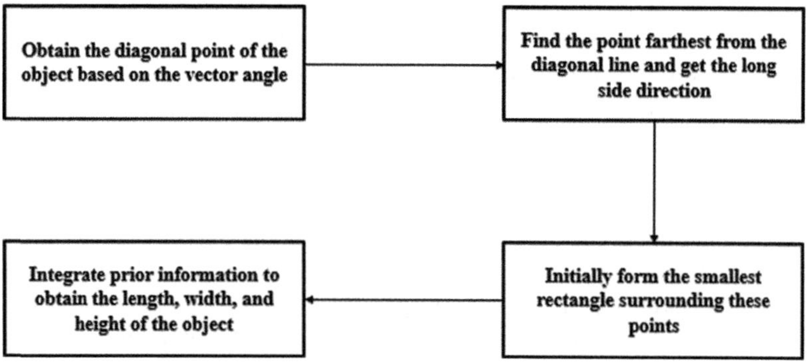

Fig. 7. 3D object bounding box generation process.

Obtain the Diagonal Point of the Object Based on the Vector Angle. The projection of a car's point cloud in a top view is shown in Fig. 8(a). The coordinate origin of the point cloud is the origin of the lidar. The coordinates of each point are averaged to obtain the estimated center point of the object. The vector corresponding to the center point is v_c. Find the two points in the point cloud in the clockwise and counter-clockwise directions of v_c and the two angles with the largest angle between v_c, to get the points (represented by triangle points) for the vectors v_a and v_b.

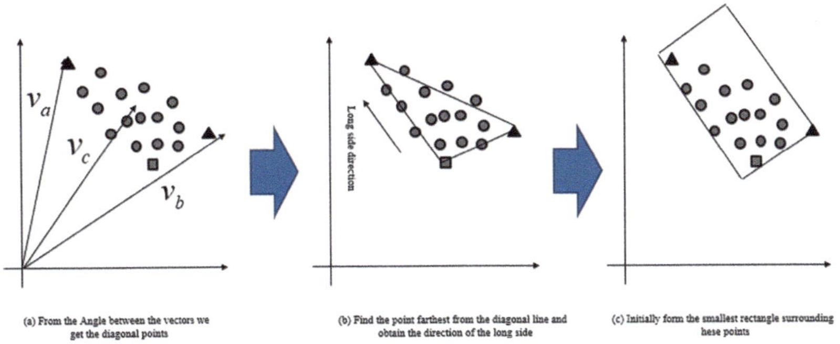

Fig. 8. 3D object bounding box generation

Find the Point Farthest from the Diagonal Line and Get the Long Side Direction. Find the point farthest from the diagonal line, as shown by the diamond point in Fig. 8(b). Connect the point furthest from the diagonal line with the two diagonal points, and take the long side direction as the object's orientation. Gets the edge points of the long side direction and the vertical long side direction.

Initially Form the Smallest Rectangle Surrounding These Points. In the long side and vertical long side directions and the opposite directions of these two directions, get the point with the most edge. A rectangle that minimally surrounds these points is initially formed, as shown in Fig. 8(c).

Integrate Prior Information to Obtain the Length, Width, and Height of the Object. Because the point cloud of the same object at different positions in space is fixed, the length and width of each calculation are inconsistent. However, the size of outdoor objects of interest can generally be obtained a priori. For example, the length and width of a car on an outdoor road is generally about 4 m × 2 m. Therefore, the stable object length and width can be obtained by integrating the prior information and the detected rectangular information.

4 Experimental Results and Analysis

In order to verify the effectiveness of the proposed method, we evaluated the performance of the algorithm. First, we use 16-line velodyne lidar to collect data on outdoor roads, and intercept point cloud data within 20 m from the lidar. Samples were made based on cars, trucks, pedestrians, cyclists, and 5 other categories, as shown in Table 1. Except for cars, trucks, pedestrians, and cyclists, the types of other objects on outdoor roads are different, so other categories need to make more samples to train the network to ensure detection robustness.

The number of points of objects in the outdoor scene collected by Lidar varies, and the number of points of each sample input by the PointNet++ network is fixed. Therefore, the point cloud of each object needs to be normalized, and the number of points of each object is unified to 1024. In addition, the number of sampling center

Table 1. Data sets of ours.

Categories	Car	Truck	Pedestrians	Cyclist	Others
Number	7534	2438	3921	2137	16027

points in the setabstraction layer is set to 128, and each center point gathers 32 points selected within 1.5 m around it, and then performs subsequent local feature extraction. Randomly scramble the data in Table 1, select 4/5 of each category as the training set, and the remaining 1/5 as the test set. The training results are shown in Table 2.

Table 2. Results of ours method.

Categories	Car	Truck	Pedestrians	Cyclist	Others
AC/%	94.3	88.2	91.4	89.1	97.5
Recall/%	96.8	82.7	95.1	79.2	97.3

It can be seen from Table 2 that the classification results obtained by training with the PointNet++ network structure have higher accuracy. Among them, the two categories of trucks and cyclists appear less in the actual measurement scenes. Compared with other categories, the samples obtained are relatively small, so the accuracy and recall rate are slightly affected. The data of these two categories will continue to be collected in the future. Improve accuracy and recall.

5 Conclusion

The 3D target detection algorithm based on point cloud data in outdoor scenes has been a research hotspot in driverless perception technology. With the development and maturity of deep neural network technology, the method of 3D target detection using neural networks has begun to show great promise [9, 25]. Advantage. Aiming at the sparseness and real-time problems of lidar point cloud object detection in outdoor scenes, this paper uses PointNet++ network and designs corresponding methods. This article uses the point cloud data collected by the vehicle's 16-line lidar as the research basis to study how to quickly and accurately detect the position, size and direction of obstacles in the environment based on the point cloud data, providing a reliable source for vehicle tracking and path planning. information. The results show that this method can effectively recognize three-dimensional objects of interest within a range of 20 m, and has a high detection accuracy rate. At the same time, it can meet the real-time requirements.

There are still unresolved problems in this paper. The small number of samples in the data set limits the improvement of the algorithm performance, and only uses the 3D coordinate information of the point cloud, but does not utilize other information, and will be further studied in subsequent work.

Acknowledgment. This work was supported by The National Natural Science Foundation of China under Grant Nos. 61602228, 61572290; and the Revitalization Talents Program of LiaoningProvince under Grant No. XLYC1807266.

References

1. Ku, J., Mozifian, M., Lee, J., Harakeh, A., Waslander, S.L.: Joint 3D proposal generation and object detection from view aggregation. In: 2018 IEEE/RSJ International Conference on Intelligent Robots and Systems (IROS), pp. 1–8. IEEE (2018)
2. Girshick, R.: Fast R-CNN. In: Proceedings of the IEEE International Conference on Computer Vision, pp. 1440–1448 (2015)
3. Ren, S., He, K., Girshick, R., Sun, J.: Faster R-CNN: towards real-time object detection with region proposal networks. In: Advances in Neural Information Processing Systems, pp. 91–99 (2015)
4. He, K., Gkioxari, G., Dollár, P., Girshick, R.: Mask R-CNN. In: Proceedings of the IEEE International Conference on Computer Vision, pp. 2961–2969 (2017)
5. Ali, W., Abdelkarim, S., Zidan, M., Zahran, M., El Sallab, A.: Yolo3D: end-to-end real-time 3D oriented object bounding box detection from lidar point cloud. arXiv preprint, arXiv:1808.02350 (2018)
6. Xu, D., Anguelov, D., Jain, A.: Pointfusion: deep sensor fusion for 3D bounding box estimation. In: Proceedings of the IEEE Conference on Computer Vision and Pattern Recognition, pp. 244–253 (2018)
7. Beltran, J., Guindel, C., Moreno, F.M., Cruzado, D., Garcia, F., De La Escalera, A.: Birdnet: a 3D object detection framework from lidar information. In: 2018 21st International Conference on Intelligent Transportation Systems (ITSC), pp. 3517–3523. IEEE (2018)
8. Li, B., Ouyang, W., Sheng, L., Zeng, X., Wang, X.: Gs3D: an efficient 3D object detection framework for autonomous driving. In: Proceedings of the IEEE Conference on Computer Vision and Pattern Recognition, pp. 1019–1028 (2019)
9. Zhou, Y., Tuzel, O.: Voxelnet: end-to-end learning for point cloud-based 3D object detection. In: Proceedings of the IEEE Conference on Computer Vision and Pattern Recognition, pp. 4490–4499 (2018)
10. Qi, C.R., Su, H., Mo, K., Guibas, L.J.: Pointnet: deep learning on point sets for 3D classification and segmentation. In: Proceedings of the IEEE Conference on Computer Vision and Pattern Recognition, pp. 652–660 (2017)
11. Qi, C.R., Yi, L., Su, H., Guibas, L.J.: Pointnet++: deep hierarchical feature learning on point sets in a metric space. In: Advances in Neural Information Processing Systems, pp. 5099–5108 (2017)
12. Zhu, M., Derpanis, K.G., Yang, Y., Brahmbhatt, S., Zhang, M., Phillips, C., Lecce, M., Daniilidis, K.: Single image 3D object detection and pose estimation for grasping. In: 2014 IEEE International Conference on Robotics and Automation (ICRA), pp. 3936–3943. IEEE (2014)
13. Song, S., Xiao, J.: Deep sliding shapes for a modal 3D object detection in RGB-D images. In: Proceedings of the IEEE Conference on Computer Vision and Pattern Recognition, pp. 808–816 (2016)
14. Yang, B., Liang, M., Urtasun, R.: HDNET: Exploiting HD maps for 3D object detection. In: Conference on Robot Learning, pp. 146–155 (2018)
15. Uijlings, J.R., Van De Sande, K.E., Gevers, T., Smeulders, A.W.: Selective search for object recognition. Int. J. Comput. Vis. **104**, 154–171 (2013)

16. Krizhevsky, A., Sutskever, I., Hinton, G.E.: ImageNet classification with deep convolutional neural networks. In: Advances in Neural Information Processing Systems, pp. 1097–1105 (2012)
17. Mousavian, A., Anguelov, D., Flynn, J., Kosecka, J.: 3D bounding box estimation using deep learning and geometry. In: Proceedings of the IEEE Conference on Computer Vision and Pattern Recognition, pp. 7074–7082 (2017)
18. Chabot, F., Chaouch, M., Rabarisoa, J., Teuliere, C., Chateau, T.: Deep manta: a coarse-to-fine many-task network for joint 2D and 3D vehicle analysis from monocular image. In: Proceedings of the IEEE Conference on Computer Vision and Pattern Recognition, pp. 2040–2049 (2017)
19. Mottaghi, R., Xiang, Y., Savarese, S.: A coarse-to-fine model for 3D pose estimation and sub-category recognition. In: Proceedings of the IEEE Conference on Computer Vision and Pattern Recognition, pp. 418–426 (2015)
20. Chen, X., Kundu, K., Zhang, Z., Ma, H., Fidler, S., Urtasun, R.: Monocular 3D object detection for autonomous driving. In: Proceedings of the IEEE Conference on Computer Vision and Pattern Recognition, pp. 2147–2156 (2016)
21. Chen, X., Kundu, K., Zhu, Y., Berneshawi, A.G., Ma, H., Fidler, S., Urtasun, R.: 3D object proposals for accurate object class detection. In: Advances in Neural Information Processing Systems, pp. 424–432 (2015)
22. Chen, X., Ma, H., Wan, J., Li, B., Xia, T.: Multi-view 3D object detection network for autonomous driving. In: Proceedings of the IEEE Conference on Computer Vision and Pattern Recognition, pp. 1907–1915 (2017)
23. Yang, B., Luo, W., Urtasun, R.: Pixor: real-time 3D object detection from point clouds. In: Proceedings of the IEEE Conference on Computer Vision and Pattern Recognition, pp. 7652–7660 (2018)
24. Liang, M., Yang, B., Wang, S., Urtasun, R.: Deep continuous fusion for multi-sensor 3D object detection. In: Proceedings of the European Conference on Computer Vision (ECCV), pp. 641–656 (2018)
25. Qi, C.R., Liu, W., Wu, C., Su, H., Guibas, L.J.: Frustum pointnets for 3D object detection from RGB-D data. In: Proceedings of the IEEE Conference on Computer Vision and Pattern Recognition, pp. 918–927 (2018)
26. Jiang, M., Wu, Y., Zhao, T., Zhao, Z., Lu, C.: Pointsift: A sift-like network module for 3D point cloud semantic segmentation. arXiv preprint arXiv:1807.00652 (2018)
27. Tombari, F., Di Stefano, L.: Hough voting for 3D object recognition under occlusion and clutter. IPSJ Trans. Comput. Vis. Appl. **4**, 20–29 (2012)
28. Guo, Y., Bennamoun, M., Sohel, F., Lu, M., Wan, J.: 3D object recognition in cluttered scenes with local surface features: a survey. IEEE Trans. Pattern Anal. Mach. Intell. **36**, 2270–2287 (2014)

Distance Learning System Design in Edge Network

Wei Pei(✉), Junqiang Li, Bingning Li, and Rong Zhao

XSCC, Xi'an 710000, China
pw2580@163.com

Abstract. With the development of technology, the bandwidth of public networks continues to increase, and the means of information exchange is constantly enriched. Especially the impact of COVID-19, distance education has gradually become popular, and many problems have also been exposed. The fact that the current distance education users are widely dispersed, the number of users is increasing, the network development is unbalanced, and the access methods are diverse, this paper designs a distance education system suitable for edge network environments. The system uses a B/S architecture, uses cloud computing technology to build a data center, and uses edge computing to build edge servers, enabling educational resources to move forward to users, reducing network latency, shortening response time, and improving user experience. This article outlines the system's architecture, the composition of the functional modules of each part, the working method, the core technology adopted, and the expected effect.

Keywords: Edge network · Edge computing · Data center · Distance learning system

1 Introduction

The sudden outbreak of COVID-19 promoted distance education's rapid development, and it also brought a sharp increase in the number of users. The apparent increase in the amount of data generated by terminal devices, coupled with the uneven development of regional networks, the network environment that users are in Different network speeds, receiving methods, and network stability have brought different levels of challenges to existing distance education systems based on core networks or cloud data centers (after this collectively referred to as data centers) [1].

Users are located on the edge of the public telecommunications network, often called the edge network. It has a wide geographical distribution, a complex network environment, and a variety of terminals. The limited bandwidth data center reception mode is challenging to meet the explosive data transmission between the user terminal and the data center demand, and lead to increased network load, which results in high network latency and severely affects user experience [2]. This paper used edge computing technology to construct a distance education system suitable for this condition.

Q. Liu et al. (Eds.): CENet 2020, AISC 1274, pp. 1291–1297, 2021.
https://doi.org/10.1007/978-981-15-8462-6_148

2 System Structure

This system relies on cloud computing technology to build a data center in the core network for processing and responding to user needs [3], storing educational resources, and providing educational services. Establishing edge servers on the user side provides limited user authentication [4], caching educational resources, reducing network latency, and Shorten response time to improve user experience. Based on factors such as network conditions, user request frequency, user preferences, hot resource levels, and other factors, the data center and the edge servers formulate communication and update the cache strategy to improve the cache hit rate and reduce the load of a large amount of data demand on limited bandwidth. At the same time, the edge servers, from time to time, send the user group characteristics data, network status, access terminal statistics and other data served by it to the data center for statistical analysis to optimize the system operation mechanism. The technology stack generally adopts the B/S architecture, uses HTML5, VueJS framework, etc. to realize the rapid development of the front-end manual interactive interface and the asynchronous call of data, and builds a service program based on NodeJS and Python, which makes the front-end and back-end development languages consistent and reduces the difficulty of system development [5]. The system architecture is shown in Fig. 1.

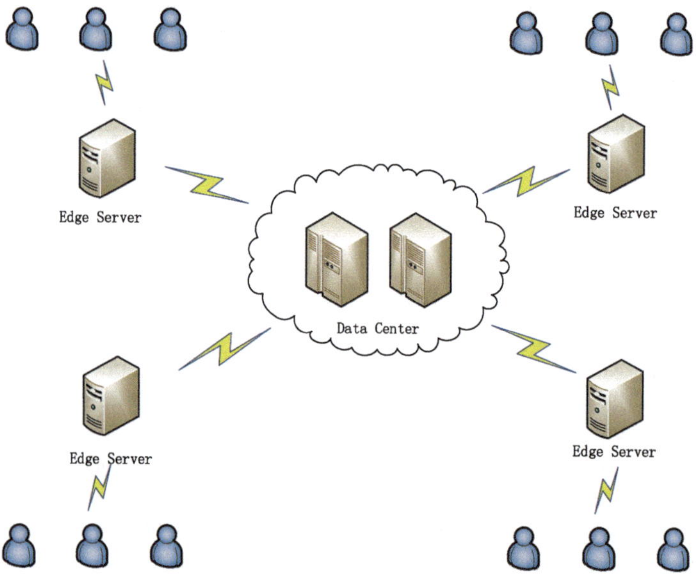

Fig. 1. System architecture diagram.

3 Data Center

Cloud computing uses a large number of computers to form a shared resource pool [6], uses virtualization technology to provide scalable storage, distributed computing, and other resources, and provides services to users through the network. At present, cloud computing has gradually matured, with on-demand services, low cost, and high scalability, and has a reliable service provider [7]. This system builds a data center on the core network and relies on cloud computing technology to achieve high elasticity of core storage and computing capabilities to cope with the explosive service demand growth while reducing system maintenance and operating costs. The data center includes 7 modules: user management, resource management, label maintenance, demand response, hotspot assessment, edge server management, and security management (Fig. 2).

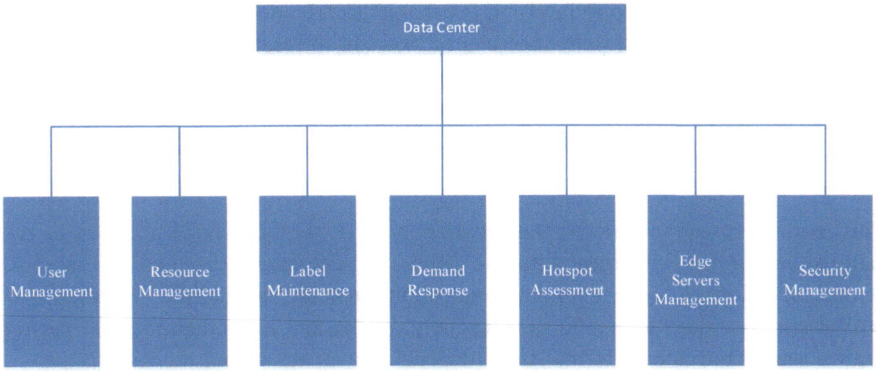

Fig. 2. Data center modules.

3.1 User Management

This module is responsible for managing and maintaining the basic information (ID, name, age, mobile phone number, etc.) and characteristic information (resident, hobbies, knowledge level, hierarchical stage, behavior data, etc.) of registered users. And it is also responsible for the edge server interacts, obtains the user information collected by the edge server, and performs correlation analysis and mining on it, tags the user, obtains the user's sensitive content prediction, and provides guidance for the cache push of the edge servers.

3.2 Resource Management

This module is responsible for the management and maintenance of all resources of the data center, including the addition, deletion, modification and inspection of resources and the maintenance of basic information (ID, name, size, storage location, etc.). It uses multi-dimensional labels to describe resource coverage, interest preferences, etc., to provide data support for association with user labels.

3.3 Label Maintenance

This module is responsible for managing and maintaining the preset label model of the system, including user labels and resource labels, and the correlation coefficient between different labels. At the same time, according to changes in data resources, social hot spots, etc., it constantly adjusts the number of labels and their relationship, and evaluates the fine-grained labels based on the cache hit rate to support the system to analyze the rationality of the label hierarchy model.

3.4 Demand Response

This module is responsible for processing user requests submitted by the edge servers, mainly using Web Service to respond, text-based small amounts of data are interacted in JSON data format, and large amounts of data are interacted by streaming. And the user's request characteristics and browsing features are recorded for optimizing network system and cache resource configuration.

3.5 Hot Spot Assessment

This module is responsible for statistical analysis of the number of accesses to educational resources, combined with the labels of accessing users and the labels of resources, to form a hot evaluation of educational resources, determine whether it will push cache to a specific area or push cache globally, and its storage location perform evaluations to provide performance bottleneck alerts, such as hot blocks [8, 9].

3.6 Edge Servers Management

This module is responsible for managing all edge servers, collecting the edge server status, cache hit rate, load, transmission delay, etc., and modeling and analyzing these data, and using Negative Selection Algorithm (NSA) to provide optimization services deployment strategy.

3.7 Security Management

This module is responsible for data center security protection, including data encryption processing, secure transmission protocols, firewall configuration strategies, network access rights management, intrusion detection, etc., to prevent leakage of user information and tagged portrait descriptions in the data center, causing adverse effects to improve information security in a cloud environment.

4 Edge Servers

An edge network is a non-technically described network, which refers to the edge of the network that is physically close to the source of the data, with specific multi-user access, unstable network connections, and complex user terminals. As the cost of caching equipment decreases, infrastructure providers provide virtualized infrastructure on the edge network to reduce user request response delays and improve various service qualities. This system builds edge servers on the edge network and relies on edge computing technology to implement limited data processing, data caching, and information mining functions to enhance the user experience of the entire system. The edge server includes five modules: user management, demand response, feature analysis, cache management, and security management (Fig. 3).

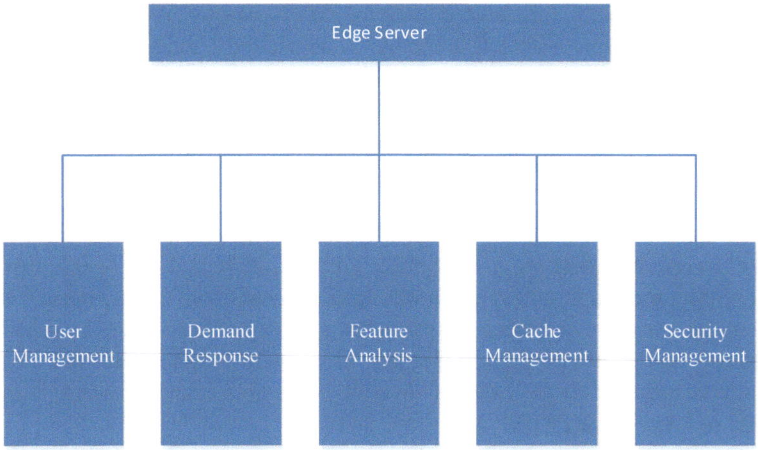

Fig. 3. Edge server modules.

4.1 User Management

This module is responsible for managing and maintaining the basic information (ID, name, age, mobile phone number, etc.) and characteristic information (resident, hobbies, knowledge level, hierarchical stage, behavior data, etc.) of the users registered on the server. And it is responsible for interacting with the data center and uploading the collected user information.

4.2 Demand Response

This module is responsible for processing the submitted user requests, and mainly uses Web Service to respond. Text-based small amounts of data are interacted in JSON data format. Large amounts of data are interacted using stream push. Priority is given to using local cache resources. At the same time, it records the user's request characteristics and browsing characteristics, used to optimize the network system and cache resource configuration.

4.3 Feature Analysis

This module is responsible for using data crawler technology to collect user behavior habits, attention preferences, etc. with the permission of the user. It digs out the user's deep features, establishes the user's label portrait, and associates with the resource label. At the same time, it establishes an evolution model of content relevance, predicts notes on highly sensitive resources, and guides data caching and information push.

4.4 Cache Management

This module is responsible for managing the cache content based on the user's characteristic analysis results, resource popularity, label association, etc., using integrated cache and peak-shaving cache technology to adjust the local cache data in time to achieve the optimal configuration of the cache content of the edge server and calculate the cache hit rate. The indicator data is pushed to the data center.

4.5 Security Management

This module is responsible for the security protection of user sensitive information in the edge server, including data encryption processing, secure transmission protocol, firewall configuration strategy, network access permission management, intrusion detection, etc. At the same time, based on the security defense model of the average field game, comprehensive consideration of network equipment and computing resource consumption, it provides an overall protection strategy against the limited resources of edge servers [10].

5 Conclusion

The system combines the characteristics of edge computing and cloud computing to achieve intelligent caching of educational resources based on user behavior characteristics, reducing the impact of network delay on user experience, has strong scalability and stability, and can respond well to distance education's explosive growth of demand. And at the same time, considering the limited resources of the edge network, it provides a targeted overall protection strategy to ensure the safety of user data.

References

1. Ma, J., Liu, C., Lin, S.J.: Research on mobile edge cache technology supporting mobile big data. Commun. Technol. **52**(06), 1379–1384 (2019)
2. Zeydan, E., Bastug, E., Bennis, M., et al.: Big data caching for networking: moving from cloud to edge. IEEE Commun. Mag. **54**(9), 36–42 (2016)

3. Gong, C., Liu, J., Zhang, Q., et al.: The characteristics of cloud computing. In: Parallel Processing Workshops (ICPPW) 2010 39th International Conference on IEEE, pp. 275–279 (2010)
4. Fu, Y.Q., Li, D.S.: Application-driven network delay measurement and optimization technology in edge computing environment. J. Comput. Res. Dev. **55**(3), 512–523 (2018)
5. Piao, L.: Easy to Understand NodeJS, pp. 22–55. Posts and Telecommunications Press, Beijing (2013)
6. Buyya, R., Broberg, J., Goscinski, A.M., (eds.): Cloud Computing: Principles and Prardigms. Wiley, Hoboken (2010)
7. Durao, F., Carvalho, J.F.S., Fonseka, A., et al.: A systematic review on cloud computing. arXiv preprint arXiv:1609.01967 (2016)
8. Zhang, H.Y.: Research on intelligent awareness in information and communication engineering. BUPT School of Information and Communication Engineering, Beijing (2018)
9. Wang, L.J., Shen, J.: Data-intensive service provision based on partical swarm optimization. Int. J. ComPut. Intell. Syst. **2**(1), 330–339 (2018)
10. Zhang, J., Chen, B., Zhao, Y., et al.: Data security and privacy-preserving in edge computing paradigm: survey and open issues. IEEE Access **6**, 18209–18237 (2018)

Marine Intelligent Distributed Temperature and Humidity Collection System Based on Narrow-Band IoT Architecture

Congliang Hu(✉), Huaqing Wan, and Wei Ding

719 Research Institute, Wuhan, China
hucongliang0015@163.com

Abstract. In recent years, the rapid development of the shipping industry at sea has become an essential part of China's economic development. Large ships carry more items, and changes in temperature and humidity in the closed cabin are especially crucial for preserving items. To grasp the change of temperature and humidity in the cabin in time, an intelligent distributed temperature and humidity collection system based on narrowband Internet of things is designed. The system combines microprocessor module (STM32), sensor module, and narrowband Internet of things to realize a real-time collection of temperature and humidity in the cabin and alarm function when the change exceeds the threshold. The mobile APP is used to realize remote monitoring and data collection monitoring. narrowband internet of things; temperature and humidity collection; STM32; remote control.

Keywords: Narrow-band Internet of Things · Temperature and humidity collection · STM32 · Remote control

1 Introduction

Marine transport is becoming more and more critical with the development of economic globalization. Comparing with other kinds of transport, it is safer at a lower cost. Statistics show that 80% of bulk cargos in the world are transported by ship at present [1]. Because of the longtime marine transport and many cargos in transport, it is especially vital to well store cargos and knows the change of temperature and humidity in a closed cabin. Temperature and humidity sensors shall be installed in the place of cargo storage for real-time monitoring. Data shall be transmitted to users' mobile phones by the communication device.

Traditional data transmission on most ships is wired. As the ships are in complex steel structures, it is of considerable significance to develop the traditional internal communication architecture and apply narrow-band IoT technology to the communication system on ships [2]. The specific benefits include [3] low cost: comparing with wired communication, there is no need to excavate cable trench and place cable for wireless communication, which can save human resources and material resources in a large quantity; short construction period: it is challenging to place wires for a wired communication network in complex application environment on ships whereas wireless

Q. Liu et al. (Eds.): CENet 2020, AISC 1274, pp. 1298–1307, 2021.
https://doi.org/10.1007/978-981-15-8462-6_149

network requires an IoT architecture, which shortens the construction period, saves both time and labor in later maintenance; great flexibility: wired network is limited by cable layout and reserved ports while wireless network is not influenced by space, user increase, and port, which makes the status of equipment on ship monitored in any place at any time; high reliability: several switching nodes are needed to realize wired data transmission while reliability can be strengthened by employing the direct point-to-point data transmission between nodes and data gateway.

In modern technology, temperature and humidity sensor is fixed in the place tested. Bluetooth or WIFI connects mobile terminal, and manpower is not needed for checking test data, which significantly saves manpower resources and cost. When using Bluetooth, a short distance between the user and the place tested is required, limiting the application of the technology. In navigation, WIFI might not be accessible, or the coverage rate of mobile data is low. That approach is also infeasible because of the high cost. In this design, NB-IoT is used to monitor the temperature and humidity in the air. Distance is not a limiting factor, and the cost is low. Combining microprocessor STM32 and sensor module, the cabin's temperature and humidity can be remotely monitored and saved in real-time. In this way, the user can know the change of data.

2 Overview Scheme of System Design

The distributed temperature and humidity collection system based on NB-IoT architecture is comprised of the following modules: temperature and humidity measurement module, microprocessor STM32F103ZET6, power circuit, A/D conversion circuit, alarm circuit, data storage module, LED screen, NB-IoT module, mobile phone APP and system architecture, as shown in Fig. 1.

This system collects data by using module of temperature and humidity sensor. It connects the ADC pin of microprocessor STM328S103F2 by A/D conversion circuit to collect data. A/D conversion circuit transforms the analog signal from sensor to be the equivalent digital signal that can be recognized by computer. Then the digital signal is sent to the central microprocessor STM32F103ZET6; the central microprocessor is responsible for controlling the overall system, storing and transmitting data collected. It is the control center of the whole system. When temperature or humidity exceeds the threshold, the alarm system will send alarm signal to remind user to take measures. LED screen is the display of the whole system and displays data of temperature and humidity. Data storage module uses MicroSD card to store the data collected by sensor module in real time. NB-IoT is the communication module that sends data of temperature and humidity to mobile phone APP for convenience of user. Power circuit supplies power to central processor, sensor module and NB-IoT module.

Figure 1 shows the data collection structure of a node. Large ships are often filled with cargos in the cabins. Each cabin needs a sensor to collect data and sends them to user. Figure 2 shows multi-node transmission network topology. A BC95 module is placed in each cabin to monitor temperature and humidity. The data monitored by the

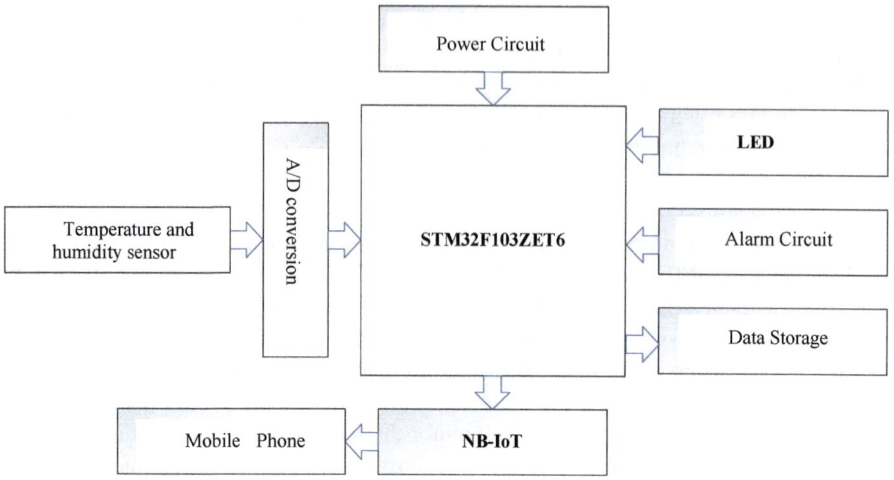

Fig. 1. Block diagram of system structure.

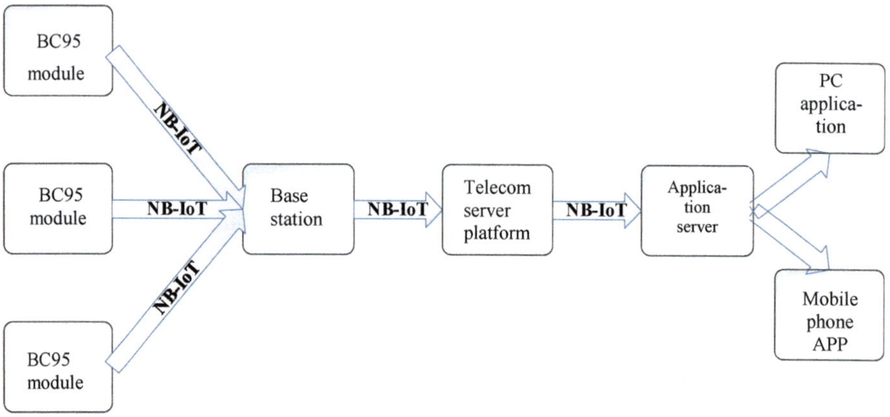

Fig. 2. Multi-node transmission network topology.

module can be transmitted by NB-IoT of telecom company to the service platform of telecom remotely. The external development kit supplied by NB-IoT network platform receives the temperature and humidity monitored by the module on application server platform. Then data can be displayed on PC application program and mobile APP. User can monitor the data collected in real time.

3 NB-IoT

The IoT communication used in traditional monitoring system can be divided into short-distance communication and long-distance communication [4]. Common traditional wireless communication technologies include ZigBee, WIFI, Bluetooth and 2G. With development of the society, the two ways of communication can no longer meet the demand of people. Traditional communication technology often fails to address both power consumption and distance. The NB-IoT developed in the latest years mainly targets the equipment with low power consumption, low complexity, low cost and high transmission delay tolerance. It has become the most active wireless communication system [5]. As a low-power Wide Area Network (LPWAN), NB-IoT is developed to connect equipment and service by cellular communication frequency band. The frequency band can be shared by LTE system to solve the contradiction between distance and power consumption. Favored by network operators, it has become the backbone force of wireless communication technology.

NB-IoT is a wireless access technology based on cellular network and wireless communication technology based on operator. The bandwidth of radio frequency is 180 kHz. Low-power WAN connection is supplied in licensed frequency spectrum. It is different from the short-distance technology in non-licensed frequency spectrum. 3GPP design goal in the 13th version is the typical goal of MTC: long battery life of equipment; low complexity of equipment to guarantee low cost and support equipment in large quantity; in addition, coverage range is wide to reach the equipment in basement and other challenging places [8]. Therefore, NB-IoT technology enjoys the following superiorities: (1) Low power consumption: NB-IoT can keep equipment online all the time; lower power consumption by decreasing unnecessary commands in PSM state and the target is 5Wh battery life 10 years in 164dbMCL. (2) Low complexity: The target is to offer equipment of low complexity and support batch application. At the same time, it contributes to lowering cost. (3) Wide coverage: The target is to expand 20 dB comparing with traditional GPRS equipment, which is to reach the targeted maximum coupling loss (MCL) 164 dB. Based on the MCL, upward and downward application layer shall support data rate 160 dps at least. (4) Wide access: The target is to support 52547 devices in a sector area of community. This target is set for 40 warehouses. Family density is based on London Hypothesis (1517 families and community distance 1732 m in each square kilometer) [5]. Transmission speed of NB-IoT is low; the upward and downward peak speed does not exceed 250kbit/s [7]. Therefore, NB-IoT is suitable for small flow and can be directly deployed in cellular or LTE (long-term evolution) network to expand deployment and lower cost. It can be applied widely, including smart home, intelligent transportation, smart power grid and wireless sensor network [6].

4 Hardware Design of the System

4.1 Introduction of Microprocessor

In our system, the model of central microprocessor is STM32F103ZET6. STM32F series is mid-low end 32-bit ARM microprocessor. The series of chips is produced by STMicroelectronics (ST) and its core is Cortex-M3. STM32F103ZET6 is a microprocessor of high performance and low cost that uses RISC (Reduced Instruction Set Computer) command, containing 8 timers, 2 DMA controllers, 3 AD conversion modules, 1 DA conversion module and 1 high-speed memory (flash memory of 512 kHz and SPAM of 64 K bytes); at the same time, 2 interfaces and SPI memory card interface are conFig. d for connection.

4.2 NB-IoT Data Transmission Module

The NB-IoT data transmission module used in NB-IoT hygro-thermograph system for ship is an engineering module BC95 developed by Quectel Wireless Solutions Co., Ltd. As the first home-made engineering module in standard NB-IoT R13, it was put into market at the end of 2016. BC95 is a NB-IoT wireless communication module of high performance and low power in dimensions 19.9 mm × 23.6 mm × 2.2 mm, which maximally meets the demand of terminal device for modules of small size. BC95 module is compatible with M95 module in the series of mobile communication module GSM/GPRS, which is convenient for users to design and upgrade products rapidly and flexibly. Besides, LCC packaging is used for easy soldering. Standard SMT equipment can help realize fast production of modules. Supplying reliable connection to clients, it is suitable for modern production of automation, large scale and low cost. Depending on its small size, ultra-low power consumption and ultra-wide range of working temperature, BC95 has become an ideal choice in NB-IoT application field [9].

When central processor MCU communicates with BC95 module by UART interface, serial port level transformation chip RS232 is added between MCU and module in

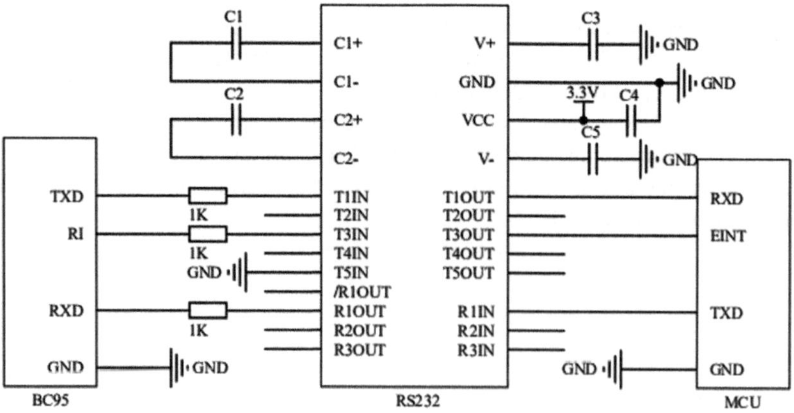

Fig. 3. Circuit of serial port communication.

design because the voltage domain of signal is different between MCU and module. As shown in Fig. 2, a resistor in value larger than 1 KΩ is added between BC95 module and RS232 to decrease current of serial port (Fig. 3).

4.3 Data Collection Module

As large ships carry a large quantity of cargos, users need to be timely informed of the temperature and humidity in the cabins, so it is necessary to install digital temperature and humidity sensor in every cabin. To complete data collection, the sensors work with other modules, such as microprocessor.

The model of temperature and humidity sensor selected is DHT11, which can measure temperature 0 °C–50 °C and humidity 20–90%. It is a compound temperature and humidity sensor that outputs calibrated digital signal. Ultra-high reliability and outstanding long-term stability of products can be guaranteed by using special digital module technology and temperature & humidity sensing technology. Containing a resistor-type humidity part and a NTC temperature measurement part, sensor connects a high-performance 8-bit single chip microcomputer. Therefore, the product has many superiorities, including excellent quality, immediate response, strong capacity of resisting disturbance and high performance-cost ratio. Every DHT11 sensor is calibrated in ultra-precise humidity calibration laboratory. Calibration coefficients are stored in OTP memory in program form. The calibration coefficients are used for signal detection in sensor. Serial interface of single wire system makes system integration convenient and rapid. The ultra-small size and ultra-low power consumption make it the optimal choice for various applications and even the most rigorous application scenario.

4.4 Power Module

The system needs to monitor temperature and humidity in cabin in long term, so reliable power supply is needed. Power circuit can guarantee the normal work of the whole system by supplying effective current, voltage and power to different modules. Therefore, the design of power module is also important. Corresponding to different components selected in power circuit of the system, there are 4 levels of voltage supply: 12 V DC voltage, 5 V DC voltage, 3.6 V DC voltage and 3.3 V DC voltage. The whole circuit panel needs power supply and we use 12 V DC voltage as power supply; 5 V DC voltage is for sensor module; 3.6 V DC voltage is for communication module; 3.3 V DC voltage is for central processor STM32F103ZET6 and other auxiliary circuits of control system. 12 V lithium battery supplies 12 V voltage to the whole circuit panel directly without processing. Other three levels of voltage are supplied from 12 V lithium battery after voltage reduction.

4.5 Data Storage Module

The data of temperature and humidity measured are in large quantity. Users can observe change of temperature and humidity in cabin depending on the data, so the data shall be stored by a storage device with large capacity. MicroSD card is used as storage device in the system. It not only has large capacity (above 32 GB) but also supports SPI interface for convenient movement, which meets different demands. MicroSD card system can store the data collected in real time. When communication fails, the data that cannot be transmitted will be stored.

5 Software Design of the System

Software design of NB-IoT intelligent temperature & humidity hygro-thermograph system is comprised of two parts: data monitoring design and microcontroller software design.

5.1 Data Monitoring Design

Java language is used for this design. As an object-oriented program design language for designing cross-platform application software, it combines Java program design language and Java platform issued in May 1995. With excellent universality, high efficiency, platform portability and safety, Java has been widely used in personal PC, data center, game console, scientific super computer, mobile phone and Internet. At the same time, there is a global largest Java developer community. In the industrial environment of global cloud computing and mobile internet, Java presents its outstanding superiority and promising future. Eclipse software development platform is used. It is a software development platform with open source code. Only containing a group of frames and a group of service, the platform constructs development environment by using plug-in parts.

5.2 Software Design of Microprocessor

Modularized programming idea is used for microprocessor software design. Generally, it contains main program, data collection program, data storage program and data transmission program. The software design is completed in KeiluVision5.0 based on C language. KeiluVision5.0 is a software development tool issued by ARM company in October 2013. It targets various embedded processors, containing compiler, assembler, project manager, debugger and real-time operation system. KeiluVision5.0 can match ARM parts perfectly.

The work flow is shown in Fig. 4. The first step is system initialization, including basic setting (such as clock and serial port) initialization and NB-IoT module initialization. As for NB-IoT module initialization, there are three steps: The first step is to judge whether the communication between MCU and NB-IoT module is normal and conFig. parameters; the second step is to judge whether BC95 module is in working mode, whether SIM information reading is normal, whether there is signal, whether it has been registered to network and whether it is in link status; the third step is to judge whether socket-udp transmission channel is created. The system initialization guarantees normal work status of all devices.

Server checks time and then power will be supplied. Sensors begin collecting signal. A/D conversion circuit converts signal to be digital signal. After data are collected, they will be stored and transmitted to server. If the transmission fails, the system will record the data and wait for successful communication. We need to know temperature and humidity in cabin to prevent the data exceeding the threshold we set beforehand. Threshold setting is determined by the cargo transported in ship. When the data exceed or are lower than the thresholds set beforehand, the data will be recorded and alarm will be given. If all data are normal, the system will continue working normally.

Besides, users can know the change of data on APP. Thanks to multi-node transmission, temperature and humidity in specified cabin can be known. In case of abnormal data, APP will send alarm for users to know and take action.

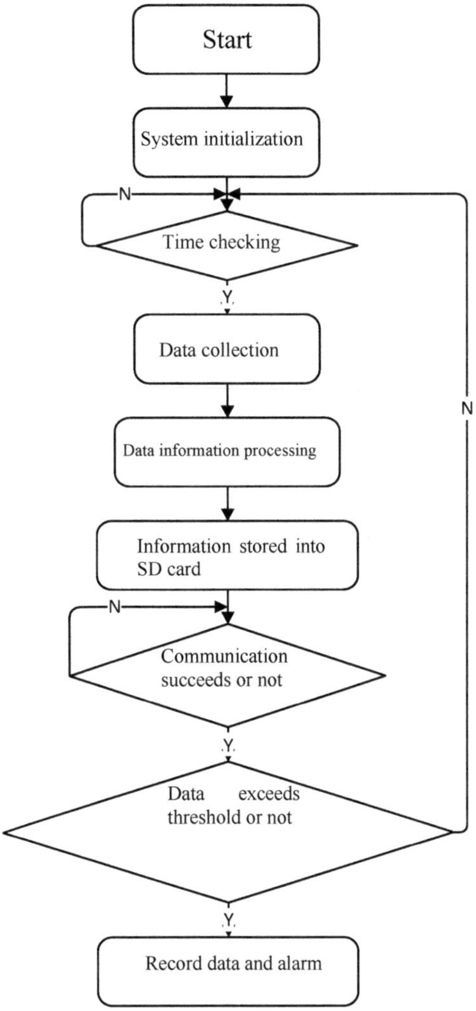

Fig. 4. Work process of the system.

6 System Test Analysis

NB-IoT intelligent distributed temperature & humidity collection system receives the data collected by sensors. For system test, user sets thresholds beforehand; then data will be collected and compared with the thresholds. Firstly start system monitoring interface and wait for the initialization of the system and external interfaces. After power is supplied to monitoring nodes, the receiving end can be accessed. Data are collected per 5 min. The test lasts for about 1.5 h. Taking the data of temperature and humidity for example, the temperature graph and humidity graph are shown in Fig. 5 and Fig. 6, respectively. In the Fig. s, temperature and humidity fluctuate slightly during the test. The temperature is kept at about 25 °C and humidity about 46%, which sufficiently reflects the stability and reliability of data collection. In this case, data are in normal range and the system can work normally.

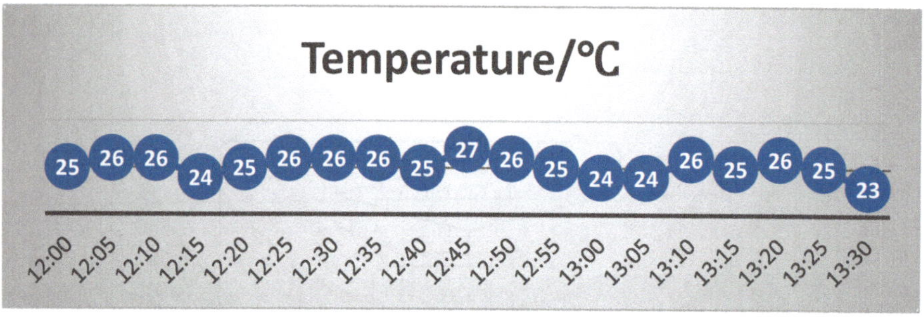

Fig. 5. Temperature data.

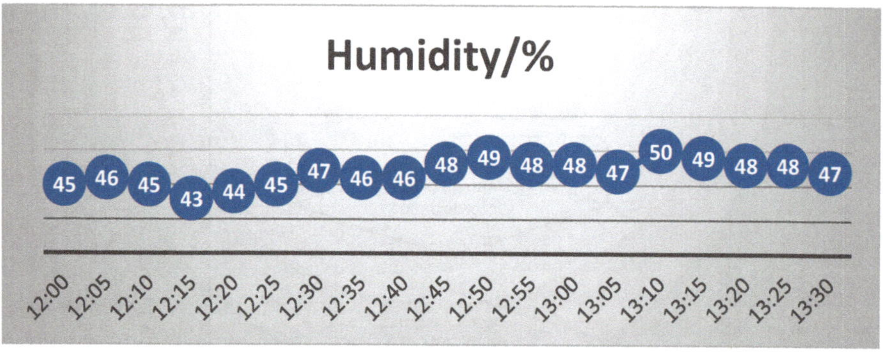

Fig. 6. Humidity data.

7 Conclusion

This paper introduces a data monitoring system with communication means NB-IoT to monitor the temperature and humidity in cabin effectively. Hardware design and software design are mainly expounded. The hardware design is based on central processor STM32F103ZET6 and NB-IoT module BC95. Temperature and humidity sensor module is used to monitor the temperature and humidity in cabin. The intelligent distributed temperature and humidity collection system based on narrow-band IoT architecture has a few superiorities, including low power consumption, fast networking and high stability. It can effectively monitor temperature and humidity in cabin in real time, solving the problem of small monitoring range and high communication cost.

Acknowledgment. This work is supported in part by national major research and development project, marine nuclear platform technology, equipment development and applying demonstrating project (No. 2017YFC0307800).

References

1. Yang, J.W., Wang, H.Z.: Analysis of the actual role of Internet of Things in ship safety monitoring system. Ship Sci. Technol. **41**(16), 202–204 (2019)
2. Chen, W.: Research on the application of data communication technology in mobile communication system of ship. Comput. Commun. **03**, 11–14 (1999)
3. Meng, L.C.: Research on the application of NB-IoT in ship communications system. JiMei University (2019)
4. Li, C.Y.: New technology of narrow-band Internet of Things (NB-IOT) and application. Telecom World **23**, 61–62 (2017)
5. Ratasuk, R., Mangalvedhe, N., Zhang, Y., et al.: Overview of narrow-band IoT in LTE Rel-13. In: 2016 IEEE conference on standards for communications and networking (CSCN), pp. 1–7(2016)
6. Wang, M.: Design and realization of monitoring system based on NB-IoT. University of Chinese Academy of Sciences (Shenyang Institute of Computing Technology, Chinese Academy of Sciences) (2019)
7. Oh, S.M., Shin, J.S.: An efficient small data transmission scheme in the 3GPP NB-IoT system. IEEE Commun. Lett. **21**(3), 660–663 (2016)
8. Zhang, H., Li, J., Wen, B., et al.: Connecting intelligent things in smart hospitals using NB-IoT. IEEE Internet Things J. **5**(3), 1550–1560 (2018)
9. Gozalvez, J.: New 3GPP standard for IoT [mobile radio]. IEEE Veh. Technol. Mag. **11**(1), 14–20 (2016)
10. Guo, X.L., Wu, R.J., Zhang, T.X.: Design and realization of temperature monitoring system based on NB-IoT. J. Hebei North Univ. (Nat. Sci. Ed.) **34**(05), 29–32 (2018)

Design of Intelligent Clothes Hanger System Based on Rainfall Data Analysis

Binbin Tao[1,2](✉), Jing Zhang[2], Xusheng Hu[2], and Jingmei Wu[2]

[1] Maanshan Engineering Technology Research Center for Wireless Sensor Network and IntelliSense, Maanshan 243031, China
282753853@qq.com

[2] Wanjiang Institute of Technology, Maanshan 243031, China

Abstract. Rainfall plays a vital role in guiding clothes drying. This paper proposes an intelligent clothes hanger system based on the analysis of rainfall data. The system consists of an electric mechanism, weather monitoring module, wireless communication module, central control module, network server, and user APP. The weather data monitored by the local weather monitoring module is uploaded to the network server in real-time, users can monitor and control the clothes hanger through the APP. Besides, the web server constructs a two-dimensional rainfall map based on the weather data provided by the clothes hangers in different geographic locations. Using the weight matrix to perform convolution operation on the rainfall map, the weighted average of the rainfall conditions at any point and its surrounding points is realized. The result is used as the rainfall risk prediction of the area corresponding to this point, used to guide drying. Finally, the feasibility of the algorithm is verified by Labview simulation.

Keywords: Clothes drying · Rainfall data analysis · Weighted average · Rainfall prediction · Labview simulation

1 Overview

With the development of Internet of Things technology, smart home products have brought many conveniences to people's lives. The weather forecast only predicts the probability of rainfall in districts and counties. However, their area is large, and rainfall has significant regional characteristics [1]. According to the weather forecast to guide clothes' drying, they cannot meet people's needs, so smart clothes hangers were invented. The widely studied smart clothes hangers mainly use light monitoring, raindrop detection, humidity monitoring, and electric retractable devices. According to the real-time weather conditions, it controls whether to withdraw the clothes. The rain cover can be opened when it rains to prevent the clothes from getting wet. This type of intelligent clothes hanger only completes shrinking the clothes hanger, covering the rain cloth, reminding the user and other functions after detecting rain. To a certain extent, it can reduce the negative impact of adverse weather. However, because the treatment process is not predictable, it is impossible to promptly remind the user to take back the clothes before the rain.

Q. Liu et al. (Eds.): CENet 2020, AISC 1274, pp. 1308–1318, 2021.
https://doi.org/10.1007/978-981-15-8462-6_150

In order to solve the above problems, this paper has designed an intelligent clothes hanger. The system includes clothes hanger, network server, and user APP (see Fig. 1). The clothes hanger comprises the STM32F103 microcontroller, environment sensor module, wireless network module, electric mechanism, motor and motor drive module, and so on. According to the changes in the environment in real-time, the clothes hanger can automatically complete unwinding actions, gathering clothes and avoiding the rain. The local environmental data is uploaded to the network server through the wireless communication module. The server divides the data information upload-ed by all the clothes hanger according to their location. This can make a detailed regional weather map. By analyzing the weather map, it can predict the short-term trend of weather changes. This is very beneficial for guiding users to deal with sudden weather changes.

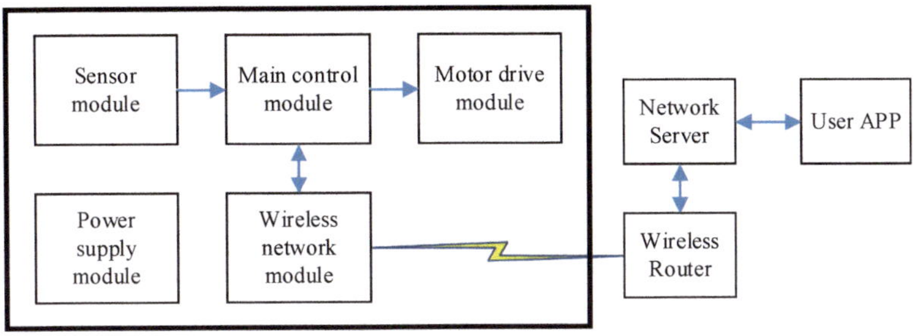

Fig. 1. System hardware block diagram.

2 Mechanical Structure Design

Conventional clothes hangers are often inconvenient to use, because of their large span. The clothes hanger designed in this paper can be stretched in both directions. Therefore, it only needs a small operating space to hang and retract the clothes. At the same time, a rain shield which can be moved left and right is added near the window sill, which can also be used as a drying platform in normal times. The mechanical mechanism is divided into three parts, including: support rails, cross rails and rain shield. The support rails are two hard rails on the left and right sides. One end of the support rails fixed on the wall to support the clothes hanger. The cross rails are placed horizontally between the support rails. They are equipped with sliders at both ends. A certain length of wire rope is connected between the cross rails and fixed to the wall. The slider of the far-end cross rail is pulled by the belt, and its movement can drive the other cross rails to move forward and backward on the support rails. There are some hook sliders on the cross rail. The hook slider is used to hang clothes. They can be dragged left and right on the cross rail. The traction method is the same as the cross rail. This structure enables the cross bar and the hook slider to move by control, it can realize the unfolding and gathering of the clothes. The rain shield slide rails are installed above the support rails, and the rain shield can be moved left and right by the belt.

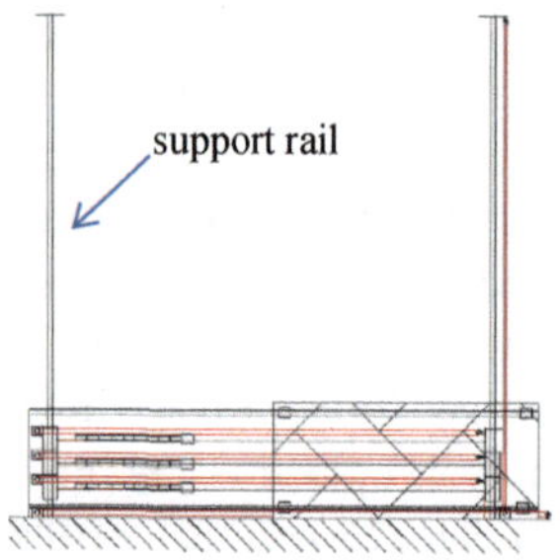

Fig. 2. Collapsed state.

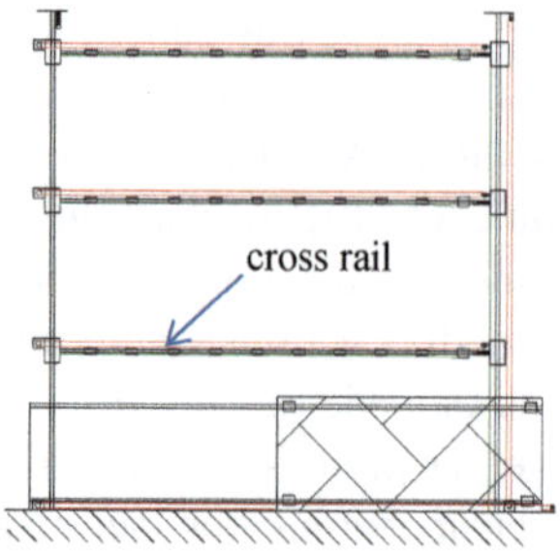

Fig. 3. Expanded state.

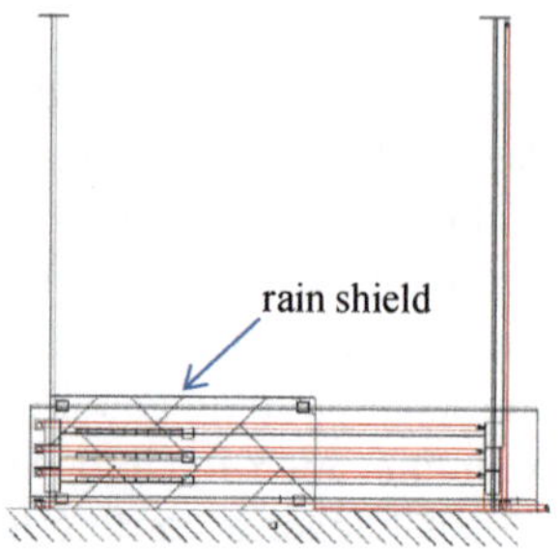

Fig. 4. Rain cover state.

There are three states of the clothes hanger:

State 1: When the user prepares to hang or retract the laundry, the clothes hanger is in the state of Fig. 2. In this state, the clothes are gathered on the left side and close to the user, and the rainproof plate is on the right side. It is convenient for users to hang or retract clothes from the left half of the clothes hanger.

State 2: After the user hangs the clothes, the cross rail moves forward, the clothes hook slider on the cross rail is pulled apart, and the clothes hanger is in the state of Fig. 3.

State 3: If the clothes are not suitable for drying, the clothes are automatically gathered, and the rain shield is moved above the clothes to cover them. The clothes hanger is in the state of Fig. 4.

3 Hardware Design

The hardware circuit part of the system consists of power system, environmental monitoring sensor group, wireless network module, stepper motor driver, limit switch and STM32 microcontroller. The modules are widely used in the design, which greatly reduce the design difficulty and ensure the system performance.

3.1 Power System

The system power supply scheme is shown in Fig. 5.

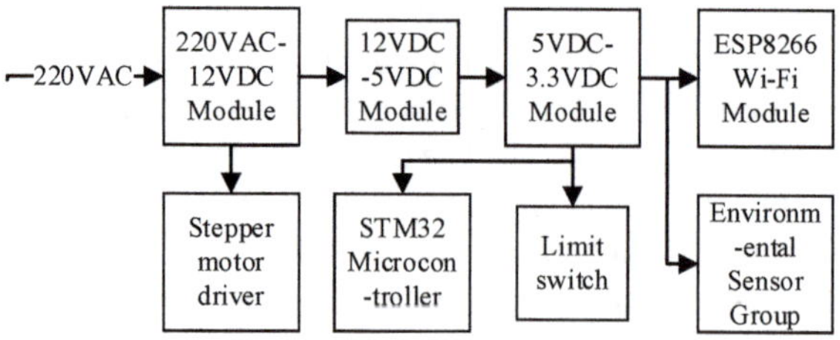

Fig. 5. System power supply scheme.

The main power is supplied by AC220V, and the AC power is converted into 12 V DC power through the AC-DC module. DC12 V can be used for the 12 V stepper motor driver. Then DC12 V is converted into DC3.3 V through the DC-DC module, which is supplied to environmental monitoring sensor group, wireless network module, limit switch and STM32 microcontroller.

3.2 Environmental Monitoring Sensor Group

The system needs to monitor the main environmental factors to guide drying. These environmental factors mainly include: temperature, humidity, light intensity and rainfall. The environmental monitoring sensor group includes: GY-39 module and raindrop sensor. GY-39 is a low-cost meteorological station module that can measure light intensity, temperature, humidity, air pressure, and altitude, and supports serial transmission. However, there is no necessary connection between humidity and rainfall [5]. So it is necessary to cooperate with the raindrop sensor to detect rain. It monitors rain by monitoring the conductivity of the sensor's surface. When rain drops on, the conductivity and the output voltage will increase. When the rainwater evaporates, the conductivity recovers and the output voltage decreases. This paper uses the voltage comparator LM393 to compare the output voltage of the raindrop sensor with the threshold value and convert the analog voltage value into a digital output.

The Schematic diagram of the environmental monitoring sensor group is shown in Fig. 6. The GY-39 module is directly connected to the serial port of the microprocessor, and the LM393 is connected to the microprocessor I/O port.

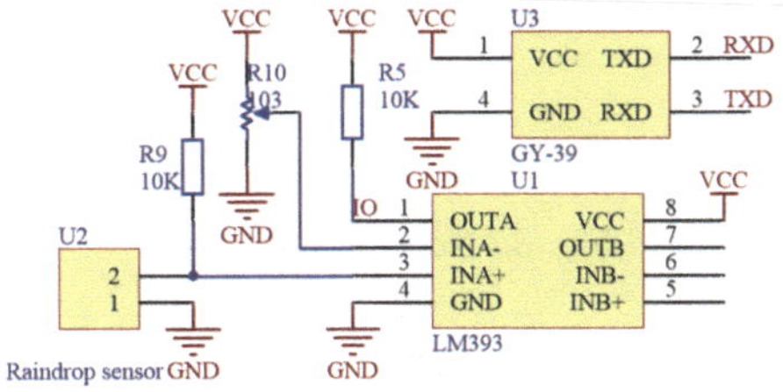

Fig. 6. Schematic diagram of the environmental monitoring sensor group.

3.3 Wireless Network Module

The main function of the wireless network module is to connect a wireless router, and establish a data link with a network server to realize the functions of data transmission and command reception. The wireless communication module uses ESP8266, which is a wireless Wi-Fi module that supports Station mode, AP mode and Station + AP mode. It supports transparent transmission. Wi-Fi data transmission can be achieved by sending AT commands through the serial port. The designer does not need to touch the TCP/IP protocol, which greatly simplifies the design difficulty of wireless networks [6]. As shown in Fig. 7, the ESP8266 module uses a serial port to connect to the serial port of the microcontroller.

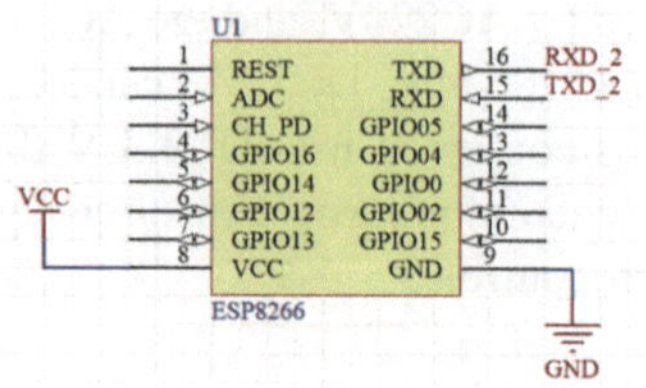

Fig. 7. Schematic diagram of wireless communication module.

The microcontroller of the clothes hanger is connected to the network server through the router by ESP8266. After matching database, user's APP and the clothes hanger can realize data interaction, then the user can check the weather status and control the clothes hanger by the APP (Fig. 8).

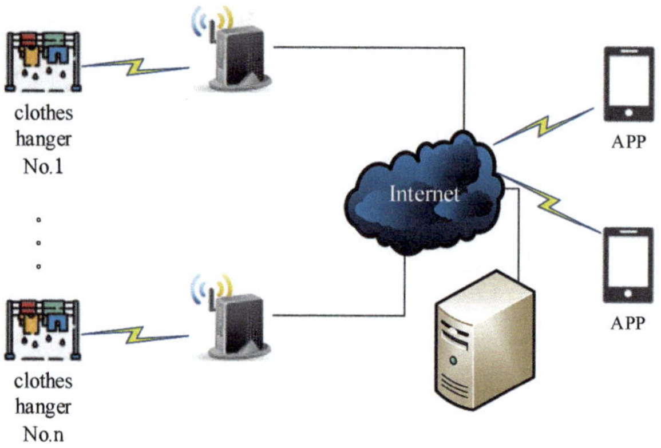

Fig. 8. Network topology.

3.4 Stepping Motor Drive Circuit

The unfolding and gathering of hangers and the movement of the rain shield are powered by 42 stepper motors. This paper uses a two-phase hybrid stepper motor driver chip THB6032MQ to drive the stepper motor. The chip uses a full-bridge MOSFET, the maximum withstand voltage is 50 VDC, the peak current is 4 A, and the maximum support is 32 subdivisions.

As shown in Fig. 9, when the working mode of the chip is configured, only two signals of CLK and CW are needed to complete the control of the stepper motor. And the rotation speed and rotation angle of the stepper motor are controlled by the frequency and the number of CLK. The rotation direction of the stepper motor is controlled by the CW signal. This control method is simple and effective.

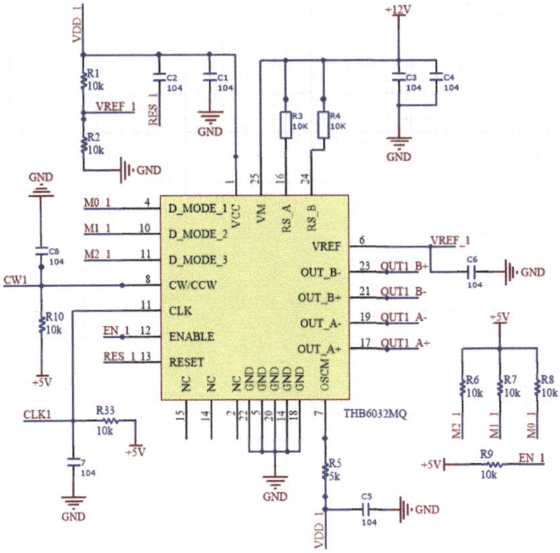

Fig. 9. Schematic diagram of stepper motor drive circuit.

3.5 Limit Switch

The stepping motor drives each part of the clothes hanger through the belt. During the movement, it is necessary to detect whether each part reaches the limit position. The motor rotation must be stopped immediately after reaching the limit position. Otherwise, the system will be damaged because the motor is blocked. The limit detection of the motor is realized by the limit switch (Fig. 10). Each motor requires two limit switches for positive and negative limit. The limit switch is in the off state when the moving mechanism does not reach the limit position. Due to the action of the pull-up resistor, the terminal M1_K1 is in a high state. When the limit switch is touched, the switch is closed. The terminal M1_K1 is short-circuited to ground. Therefore, it can be effectively judged whether the limit position is reached according to the state of the terminal level.

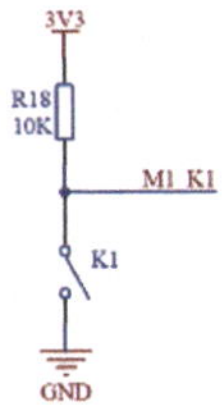

Fig. 10. Wiring diagram of limit switch.

3.6 STM32 Microcontroller

STM32F103c8t6 microprocessor is a high-performance, low-cost, low-power embedded application design ARM core controller, which has up to 8 timers, 3 USART serial ports and 37 independent I/O ports [6]. Its internal resources can meet system requirements.

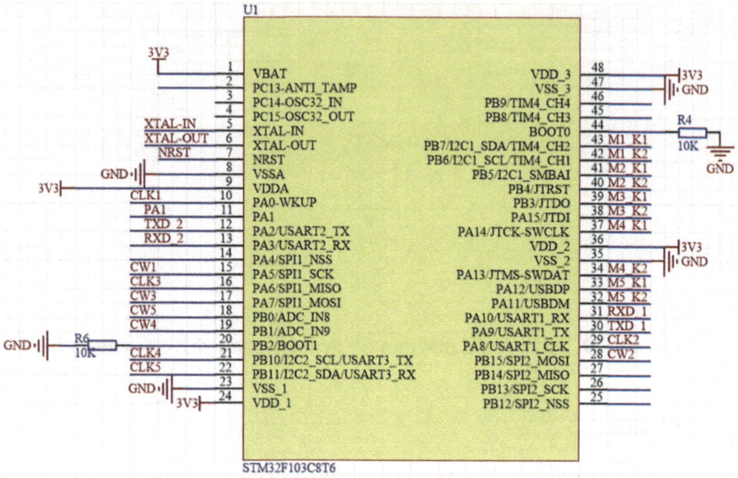

Fig. 11. STM32 microcontroller wiring diagram.

The pin assignments are shown in Fig. 11: Two USART serial ports communicate serially with the environmental monitoring module GY-39 and the wireless communication module ESP8266; the PA1 port is connected to a raindrop sensor; multi-channel clock signal CLKx and direction signal CWx are respectively connected to the stepper motor driver to control the stepper motors; multiple Mx_K1 and Mx_K2 connection limit switches are used to detect the positive and negative limit of the motor.

4 Software Design

The software design includes three parts: the control program of the main control module, the network server design and the user APP design.

4.1 A Subsection Sample

The main control module of the clothes hanger is mainly used for motor control, environmental monitoring and network communication. The control program logic is shown in Fig. 12.

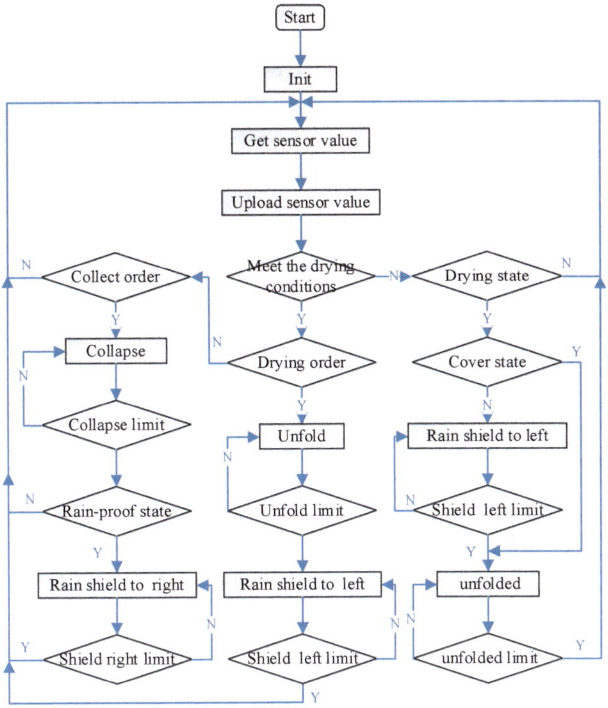

Fig. 12. Main control module program logic diagram.

The control system starts to obtain environmental information and upload the data to the network server in real time after power-on initialization. There are three actions according to the current environmental status and control instructions:

1. When the drying conditions are satisfied and the drying order is received, the clothes hanger is unfolded and the rain shield is moved to the right side;
2. When the drying conditions are met and the withdraw order is received, the clothes hanger gathers and the rain shield moves to the right side;
3. When it is not suit for drying (rainy, night, etc.), but it is in the drying state. The clothes hanger gathers and the rain shield moves to the left side to be in the rain-proof state.

4.2 Rainfall Warning Algorithm

The rainfall situation is an important guiding role for the drying of the clothes. It is not enough to completely solve the problem by the raindrop sensor. Rainfall is regional, and the scope and movement of rain clouds determine the scope and trend of rainfall. The further away from the area where is raining, the less risk of rain. Therefore, this paper proposes an analysis method for realizing rainfall warning based on the rainfall situation uploaded by many clothes hangers in different geographical locations. The specific steps are as follows:

Step 1: Divide the map into grids of 1 square kilometer to form a two-dimensional map array.

Step 2: The server collects the rainfall of the clothes hangers at different positions and stores them in the map array to form the rainfall-map array N. The value of each bit of the array indicates whether the corresponding area is raining.

Step 3: Averaging multiple data that needs to be stored in the same array position, and then judging whether there is rain by the threshold value.

Step 4: Make the weight matrix according to the principle that the longer the distance, the smaller the influencing factors.

Step 5: Use the weight matrix to convolve the rainfall-map array to realize the weighted average of the rainfall conditions of all points to form a rainfall warning map.

Step 6: According to the rain warning map, the user and the clothes hanger are warned.

4.3 User APP

The role of the user APP is to receive data and remote control. When the clothes hanger is used for the first time, the user needs to use the APP to scan the serial number on the clothes hanger product to bind the user to the clothes hanger. The user APP needs to select the address and direction of the clothes hunger during the registration process. The location information of the clothes hangers is useful for later data analysis.

5 Labview Simulation

This paper uses LabVIEW to simulate the rainfall warning algorithm. First create a 30 * 30 two-dimensional array of rainfall conditions N. Each number in array N corresponds to the rainfall situation within one square kilometer of the array position, where “1” indicates that it is raining and “0” indicates that there is no rainfall. Then fill the data with the two-dimensional rainfall array N to get the rainfall distribution. As shown in Fig. 13-1, some points in the array are in the rain state (red square), and no rain is detected in the remaining positions (green square). The values in the weight array A (Eq. 1) are set by the distance from the center. Use the weight array A to perform a convolution operation on the rainfall array N to obtain the rainfall risk array P. Then obtain the calculation results by surface fitting, as shown in Fig. 13-2. The area nearby the rain area is regarded as a high-risk area (red area). This method realizes the rainfall warning with the accuracy of the area range represented by the data in the array. The user can be reminded to take back the clothes before the rain occurs.

$$A = \begin{bmatrix} 0.0625 & 0.125 & 0.25 & 0.125 & 0.0625 \\ 0.125 & 0.25 & 0.5 & 0.25 & 0.125 \\ 0.25 & 0.5 & 1 & 0.5 & 0.25 \\ 0.125 & 0.25 & 0.5 & 0.25 & 0.125 \\ 0.0625 & 0.125 & 0.25 & 0.125 & 0.0625 \end{bmatrix} \quad (1)$$

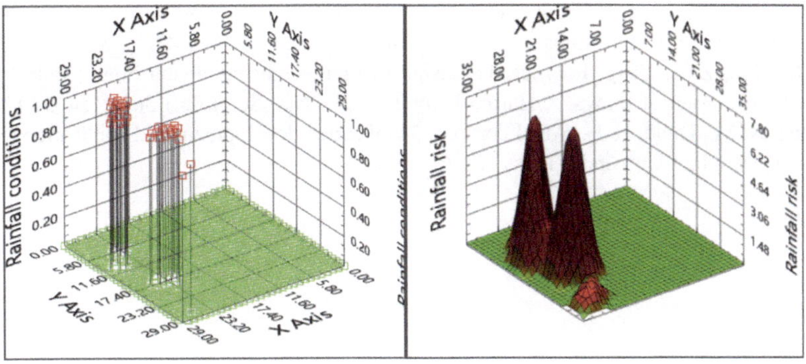

(1) (2)

Fig. 13. Main control module program logic diagram.

6 Conclusion

This paper designs a freely retractable clothes hanger in both vertical and horizontal directions. Only a small operation space is needed to complete the hanging and collection of clothes. And it proposes a method of using the weather data by clothes hangers to establish a rain array and calculate the rain risk. The simulation proves the effectiveness of the algorithm, and achieves the purpose of early warning before rainfall occurs. It makes up for the lack of weather forecasts and ordinary smart clothes hangers.

Due to the many heights and directions of users installing clothes hangers and the complicated weather data, this paper only considers the data provided by raindrop sensor for rainfall warning. In order to more effectively predict the rainfall area, in the future, other meteorological factors and clothes hangers installation factors can be considered. It will provide individualized drying solutions for each user.

Funding Details. School-level scientific research projects of Wanjiang Institute of Technology (Grant no. WG18032ZD), Outstanding young talents support program of universities of Anhui Province (Grant no. gxyq2019149), Natural science research project of universities of Anhui Province (Grant no. KJ2019A1275).

References

1. Tao, X.C., Zhang, X., Wang, X.Q.: Analysis on characteristics of rainfall distribution in shanghai. Water Purificat. Technol. **38**(10), 25–28 (2019)
2. Wang, F.J.: Design and production of intelligent clotheshorese based on STC89C52RC MCU. Microcomput. Appl. **35**(10), 157–160 (2019)
3. Zhang, S.W.: Design of intelligent laundry rack control system based on single chip microcomputer. Value Eng. **38**(3), 125–127 (2019)
4. Shi, J.Z., Liang, J.: Design of intelligent clothes hanger system based on 51 single chip microcomputer. Electr. Test **11**, 37–38 (2019)

5. Liu, X., Ma, Y., Ling, Y., et al.: Analysis of two cyclone heavy rain processes with different low-level humidity conditions in Qingdao. Meteorol. Sci. Technol **47**(5), 818–829 (2019)
6. Li, M., Ma, Z.C., Yang, P.Y., et al.: Design of lightweight intelligent home control system based on Internet of things. J. Harbin Univ. Comm. (Nat. Sci. Edn.) **36**(1), 54–57 (2020)
7. Zhang, X. M., Duan, H. L.: ARM cortex-m3 embedded development and application (STM32 series). Tsinghua University Press (2016)

Smart Shoes for Obstacle Detection

Wenzhu Wu, Ning Lei, and Junquan Tang(✉)

Chongqing Medical and Pharmaceutical College, Chongqing 401331, China
tangjunquan84@qq.com

Abstract. Objective To develop Smart Shoes with obstacle detection system for visually impaired people. **Method** The obstacle detection system adopts a triple-axis accelerometer to collect feet's acceleration data, and uses an ultrasonic sensor to detect obstacles. The system is controlled by STM32L432KC (a microcontroller from STMicroelectronics), and powered by a Lithium-ion battery that can be recharged either by a charger or by walking. A gait events recognition algorithm is proposed to detect the motion state of feet. Obstacles are detected only when users are walking with foot in stance (ST) phase. Moreover if a fall is detected, the Smart Shoes will connect to the cellphone and call the emergency contacts. **Results** The overall recognition ratio of the gait events was 90.9%, the ratio of walking, jiggling and fall (simulated) were 91%, 88.5% and 100% respectively. The detection resolution of Smart Shoes depends on the ultrasonic sensor. User's average obstacles detection distance are all above 50 mm, and for each user the detection distance is proportional to the obstacles' dimension. **Conclusion** Experimental results indicate that the Smart Shoes performs stably in real-time, and has high detection accuracy with low false-alarm rate.

Keywords: Smart shoes · Obstacle detection · Visually impaired people · Gait events recognition

1 Introduction

Globally, at least 2.2 billion people have a vision impairment or blindness [1], and the number keeps on increasing. China accounts for at least 20% of world's blind population. Vision loss will limit this group of people's mobility severely. On the other hand, falls are the number one cause of injury and injury-related deaths in people over the age of 65. In China, falls caused injury makes up to 34.84% injury-related deaths in elderly people over 65 [2]. In a study of influencing factors of falls in Chinese elderly, vision loss and improper light condition counts for 2.52% and 2.38%; while uneven ground and obstacles counts for 2.45%, 2.28% in the factors of falls respectively [3].

Fall prevention is essential. To overcome the limitation caused by visual impairment, the elderly and visual impaired people turn towards assistive devices to get more independent mobility. Currently, the assistive mobility aid devices can be divided into four categories [4–9]: (1) guide crutches, (2) Electronic Travel Aids (ETA), (3) movable guiding robots, (4) wearable aid devices; every device has a certain value of popularization and application. However, most obstacle avoidance and guide devices cannot automatically switch the obstacle detection state according to the user's gait.

Q. Liu et al. (Eds.): CENet 2020, AISC 1274, pp. 1319–1326, 2021.
https://doi.org/10.1007/978-981-15-8462-6_151

When a user is talking with others face to face, or is stationary in a closed space (such as in a room or in a vehicle), the obstacle detection system need to be turned off, otherwise there will be falsely alarms.

In this paper, authors designed a Smart Shoe which can recognize gait events and only detect obstacles when users are in stance (ST) phase, besides if users fall down, the Smart Shoe can connect to their cellphone and call the emergency contacts. The Smart Shoes proposed in this paper can be used as an auxiliary tool for elderly and visual impaired people to walk more independently.

2 Design

2.1 System Hardware Design

In order to detect low obstacles, an ultrasonic sensor embeds in the tip of shoe, which can continuously transmit and receive sound waves to detect the obstacles in front of the user. The other modules are placed under the insole. Under the heel a Piezoelectric transducer consist of Polyvinylidene Fluoride (PVDF) is employed. A PVDF film can transform foot's pressure into electricity which can be collected to recharge the battery. A vibration motor is placed under the forefoot, the control board and the rechargeable battery is under the arch of the foot.

The structure of the smart shoes is shown in Fig. 1.

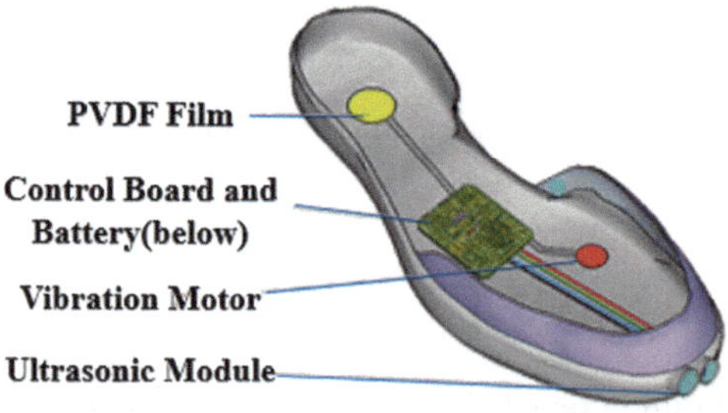

Fig. 1. Structure of smart shoes for obstacle detection.

The system of the smart shoes employs STM32L432KC, an ultra-low-power and high performance microcontroller, as CPU, which is suitable for battery powered system. The shoes adopts the ultrasonic ranging module HC-SR04 to detect obstacles, it provides up to 400 mm non-contact measurement with the ranging accuracy of 3 mm. The ultrasonic module has certain dirt resistance and temperature compensation, also it is not affected by light and smoke, which means it can be used in various environment [10]. An ADXL335 triple-axis accelerometer is used to collect the motion data of foot. The system is powered by a rechargeable Lithium-ion battery that can be recharged by a charger or by walking. Walking energy harvesting module uses the MAX17710 to collect electricity power from a PVDF film (Fig. 2).

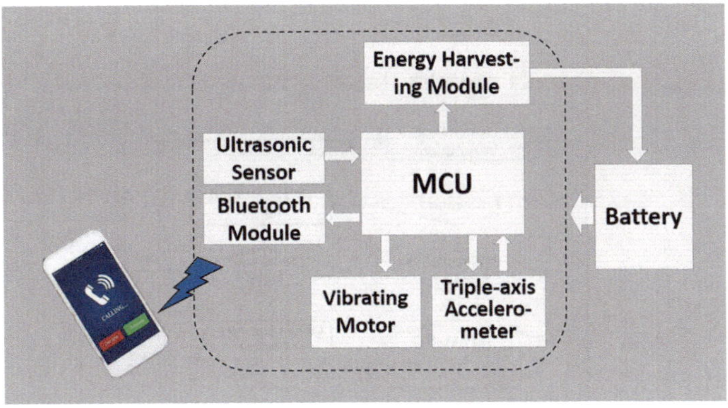

Fig. 2. System diagram of the smart shoes.

An LED is employed to indicate the system status: the LED lamp keeps on during system initialization, the LED lamp flashes slowly when collecting motion data and detecting obstacles. If the battery power is less than 5%, the LED flashes quickly and prompts charging. Figure 3 depicts the pictures of the real smart shoes.

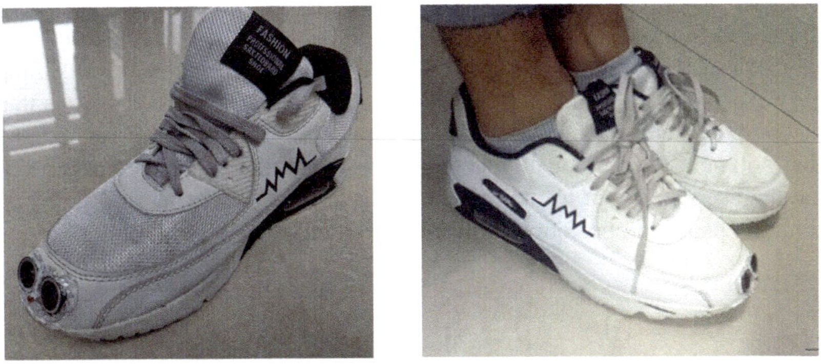

a. Appearance of the smart shoes b. Smart shoes on feet

Fig. 3. Pictures of the smart shoes.

2.2 System Software Design

The frequency components of almost all acceleration signals generated by human activities are below 20 Hz [11]. In order to ensure the real-time performance of the device and reduce the amount of data, the system adopts a 50 Hz sampling frequency to collect motion acceleration data. Fall signals are the slant angles calculated from acceleration data. Gait events signals are the root mean square of the three axis acceleration data. The flow chart of the system software is shown in Fig. 4.

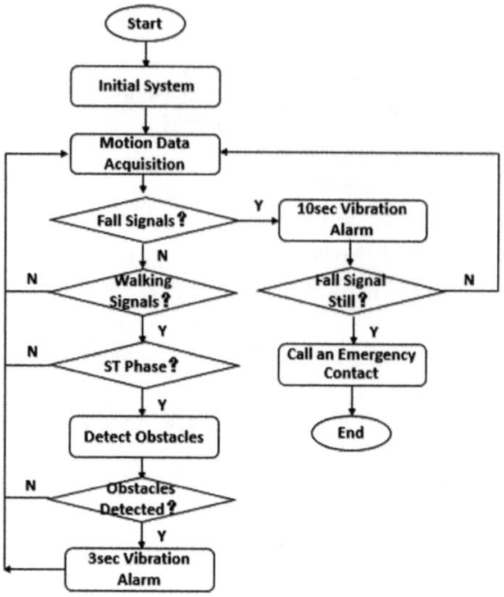

Fig. 4. Flowchart of the system software.

2.3 Gait Event Algorithm

As showed in Fig. 5, one gait cycle can be divided into two phases, stance (ST) and swing (SW) [12]. For the Smart Shoes are designed to help the visually impaired people walk more safely, the shoes are intended to detect obstacles only when the users are walking. There are two conditions that the system should not alarm: i) the users' feet stay still or have some small jitter (e.g. the user are standing, sitting or in a vehicle), so that the users need not to turn off the obstacle detection system. ii) in SW phase, since in this phase the tip foot will face the ground for a short time.

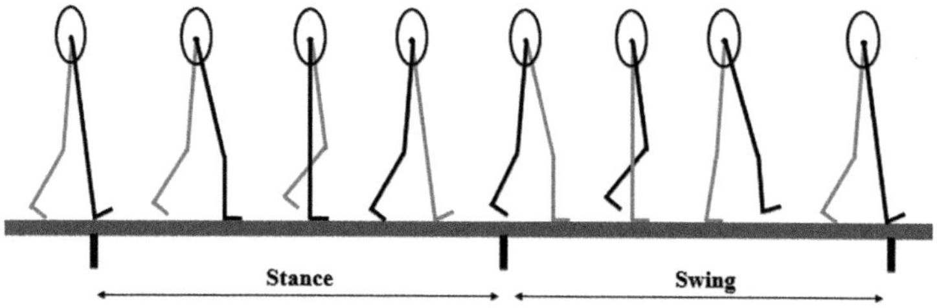

Fig. 5. A gait circle can be divided into two phases: stance (ST) and swing (SW).

In order to avoid the false detection in above situation, we proposed a gait events detection algorithm. This algorithm detect gait events by analyzing the foot's acceleration features. There are 4 different gait events can be identified: SW, ST, fall, stay still or with some small jitter, the MCU will enable different function accordingly. The correspondence between gait events and system function is listed in Table 1.

Table 1. Correspondence between gait events and system function.

Gait events	System function	
	Obstacle detection	Alarm
Stay still or jitter	Off	–
SW phase	Off	–
ST phase	On	Vibrate
Fall	Off	Vibrate, Call the Emergency

As indicated in Table 1, the system detects obstacles only when the user's movement is featured as walking and the foot is in ST phase. If the user is walking but the foot is in SW phase, the ultrasonic sensor does not detect obstacles. When an obstacle is successfully detected, the MCU sends a signal to enable the vibration to alert the user to be aware of the obstacle. When a fall is detected, the Bluetooth module will send a signal to the APP in cellphone, and the APP will call the emergency contact automatically.

3 Methods

A total of 6 users with low vision whose feet-function normally, 3 males and 3 females, aging from 50 to 71 were recruited in the study. And all the users volunteered for the study.

To assess the effectiveness of Smart Shoes in gait events recognition and obstacles detection, two studies were designed. The studies took place in the campus of CQMPC, Chongqing, China.

The first study tested the accuracy of gait events recognition. The testing required the users to accomplish three sets of actions: 1) walk on flat ground for 50 m, upgrade for 15 m, down grade for 15 m; 2) shake their feet for 10 s without walking; 3) lay down on a cushion in three positions (supine, lateral, prone) to stimulate falls. Feet's motion acceleration data was collected by the smart shoes and sent to a computer. A total of 845 gait events signals were collected, there are 723 walking signals (457 signals from walking on flat ground, 142 signals from walking upgrade, 124 signals from walking down grade), 104 shaking feet signals, 18 falls simulating signals.

The second study tested the sensitivity of obstacles detection algorithm, subjects in different sizes and shapes were placed on the ground. All the obstacles are frequently encountered in daily life. The users wore the Smart Shoes walk towards the subjects. The distance between the user and the obstacle was measured.

4 Results and Discussion

The overall recognition ratio of the gait events was 90.9%, the ratio of walking, jiggling and fall were 91%, 88.5% and 100% respectively. The results of gait event recognition are showed in Table 2.

Table 2. Results of gait events recognition.

Gait events	Gait signals	Missed	False	Accuracy
Walking	723	65	0	91%
Jiggling	104	0	12	88.5%
Fall (simulated)	18	0	0	100%
Total	845	65	12	90.9%

As illustrated in Table 2, 65 walking signals were missed to be identified, most of the signals are the first several steps after starting to walk and the last several steps before stopping. 12 jiggling signals were falsely identified as walking, these falsely identified signals are all from one user who shook his feet sharply. Further research about this need to be discussed.

User's obstacles detection distance varies in some ranges. The average detection distances are above 50 mm. The results of the test are showing in Table 3.

Table 3. Distance results of different obstacles.

Obstacles	Width (diameter)/cm	Height/cm	Alarm Distance/cm	Alarm Distance/cm (Average)
Wall	150	270	160–221	186
Chair	330	50	158–207	184
Person (Front Body)	15(leg) ~ 42 (shoulder)	160	120–158	144
Pole (lengthways)	6	60	80–119	105
Pole (sideways)	60	6	60–97	88
Stair	150	4	56–99	85
Stone	8	6.5	53–94	81
Thread (horizontal braced)	0.8	7	42–76	62
Thread ((horizontal braced)	0.4	7	33–62	51

As illustrated in Table 3, for each user the detection distance is proportional to the obstacles' dimension. If the dimension of the obstacle is reaching the limit resolution of the ultrasonic module, the detection distance is significantly reduced. Detection distance among users various, probably because of individual difference in response time.

5 Conclusion

This paper proposes Smart Shoes for obstacle detection with gait recognition algorithm. The Smart Shoes can recognize different gait events and only detect obstacles when users are walking with foot in ST phase, and if the users fall down, the Smart Shoe will connect to the cellphone and call the emergency contacts. Experimental results indicate that the Smart Shoes performs stably in real-time, and has high detection accuracy with low false-alarm rate.

In order to improve the accuracy and enhance the performance of the Smart Shoes, further research and improvement can be made in the following aspects:

(1) For the ultrasonic sensor can measure the distance, a voice broadcasting function can be added on the APP. Users can know the distances of obstacles through audio broadcast equipment.
(2) High-power and high-sensitivity ultrasonic sensors can be used to increase the detection range and improve the resolution.
(3) A GPS module should be added in the smart shoes. The users can share their location information with families. It is especially useful in badly injured cases caused by fallings that users are incapable of answering phone calls.

References

1. World Health Organization. Visual Impairment and Blindness statistics: Fact sheet (282) (2019)
2. Chinese Center for Disease Control and Prevention Center for Prevention and Control of Chronic Noncommunicable Diseases, National Health and Family Planning Commission Statistical Information Center. Death Cause Surveillance Data Set in China (2014). Popular Science Press, Beijing (2015)
3. Yang, Y.l., Zhang, H.Y., Liao, T., Ma, Y.X., Han, L.: Visualization analysis on influencing factors of falls in Chinese elderly. Shanghai Nurs. **20**(04), 19–24 (2020)
4. Fang, R.J., Zhu, W.B.: Design of guiding blind cane based on GPS positioning and ultrasonic detection. Comput. Measur. Control **19**(05), 1154–1157 (2011)
5. Rodríguez, A., Yebes, J.J., Alcantarilla, P.F., Bergasa, L.M., Almazán, J., Cela, A.: Assisting the visually impaired: obstacle detection and warning system by acoustic feedback. Sensors **12**(12), 17476–17496 (2012)
6. Zhu, J.H., Hu, J., Zhang, M., Chen, Y.N., Bi, S.: A fog computing model for implementing motion guide to visually impaired. Simul. Model. Pract. Theor. **101** (2020)
7. Dakopoulos, D., Bourbakis, N.G.: Wearable obstacle avoidance electronic travel aids for blind: a survey. IEEE Trans. Syst. Man Cybern. **40**(1), 25–35 (2010)
8. Yu, J.H., Chung, H.I., Hahn, H.S.: Walking assistance system for sight impaired people based on a multimodal transformation technique. In: Proceedings of ICROS-SICE International Joint Conference, pp. 1639–1643 (2009)
9. Bertozzi, M., Broggi, M.: GOLD: a parallel real-time stereo vision system for generic obstacle and lane detection. IEEE Trans. Image Proc. **7**(1), 62–81 (1998)
10. ElecFreaks. HC-SR04 datasheet. www.Elecfreaks.com

11. Antonsson, E.K., Mann, R.W.: The frequency content of gait. J. Biomech. **18**(1), 39–47 (1985)
12. Caldas, R., Mundt, M., et al.: A systematic review of gait analysis methods based on inertial sensors and adaptive algorithms. Gait Postur. **57**, 204–210 (2017)

Research on Differential Power Analysis of Lightweight Block Cipher LED

Yi Zou[1,2(✉)], Lang Li[1,2], Hui-huang Zhao[1,2], and Ge Jiao[1]

[1] College of Computer Science and Technology, Hengyang Normal University, Hengyang 421002, China
51674074@qq.com

[2] Human Provincial Key Laboratory of Intelligent Information Processing and Application, Hengyang 421002, China

Abstract. LED algorithm is a new lightweight encryption algorithm proposed in CHES 2011, which is used for IOT to protect the communication security of RFID tags and smart cards. It has been found that it is possible to retrieve secret key of algorithm by appropriate analysis of the devices's power consumption, and the differential power analysis is the most powerful attack. This paper proposed a method of differential power analysis on LED algorithm. Combined with hardware circuit, it designed and implemented differential power analysis of the LED encryption system. The experimental results show that method which is proposed in this paper was cracked the 64 bits key of LED. Thus, without protection, the cipher LED would be difficult to resist differential power analysis. and according this, We provide a general analysis method of differential power analysis on other lightweight cryptographic algorithms.

Keywords: Lightweight block cipher · Differential power analysis · LED

1 Introduction

The Internet of things are more and more deeply into our life in recent years, and Internet security has aroused the attention of all cryptography researchers. The Ultra-lightweight algorithm-LED was proposed in 2011 [1], The cipher algorithm is mainly for encryption research and development of IOT smart card which is resource-constrained, that is a kind of ultra-lightweight symmetric encryption algorithm.

Although the time of Lightweight block cipher proposed is shorter, but with the development of the Internet of things, its application is more and more. therefore, the research of its security is indispensable. Along with the continuous improvement of cryptographic algorithm, the traditional method of attack by mathematical means are more and more difficulty. Under this background, the bypass attack research gradually rise, the common methods of bypass attack is included: Power Analysis Attack [1–3], Timing Attack [4], the Electromagnetic Analysis [5], Cache Attack [6], Fault Attack [7], and the most threatening in the bypass Attack should be belong to Power Attack. Kocher and Jaffe proposed the DPA technology in 1998, Kocher said around one thousand power curve, they can be decoded on most of the smart card at the present market. The methods of Power attack were Simple Power Analysis (SPA), Differential Power Analysis (DPA) and High Order Power consumption Analysis (HO - PA).

Q. Liu et al. (Eds.): CENet 2020, AISC 1274, pp. 1327–1334, 2021.
https://doi.org/10.1007/978-981-15-8462-6_152

In this paper, we proposed a new method of differential power analysis on LED-64 algorithm. the other chapters are arranged as follow: the Second part was introduced the principle of LED cipher; the Third part was simply described the consumption model of differential power analysis; the Fourth part, the process was described in detail, which is about the new method of differential power analysis on LED-64 algorithm; the Fifth part is the experiment and analysis result of the new differentical power analysis; Finally, the work of this paper was summarized.

2 LED Algorithm

LED is a lightweight block cipher algorithm with a substitution permutation network (SPN) structure, and uses 64-bit input plaintext, 64-bit or 128-bit key to generate 64-bit cipher text in 32 rounds. In this paper, the target is the 64-bit key, expressed as LED-64, the key space of the encryption system for 264. Each round operation includes four basic steps: AddConstant, SubCell, ShiftRow, MixColumn. Computing with intermediate result of each round and round keys to xor, called AddRoundKey, the result is as the next input.

2.1 Definition

The symbol and its meaning of instructions used in LED - 64 algorithm are as follows:

- P: 64-bit plain text
- C: 64-bit cipher text
- Ki: 64-bit key
- State: 64-bit state
- ||: Concatenation Operator
- $\oplus$: XOR Operator

2.2 The Encryption Algorithm

LED-64 encryption algorithm consists of 32 rounds of SPN iterative structure, encryption process is shown in Fig. 1. Now supposed $P = (X_0, X_1, X_2,\ldots, X_{15})$, is 64 bits plaintext, its data processing is as follows:

- AddConstant AddConstant transformed the intermediate state and round constant, which make exclusive xor operation.

  $State_i{:}\{0,1\}^{64} \longrightarrow Y{:}\{0,1\}^{64}$
  $Y = State_i \oplus RC_{i+1}$ $(i = 0,1,\ldots,31)$ (while $i = 0$, $State_i = P$)

- SubCell Calculated by sbox, the intermediate state was transformed by nonlinear transform, which is transformed from a nibble to the other nibble, and stored in e_res; the e_res value will be assigned to the result.

  $S{:}\ \{0,1\}^{64} \longrightarrow \{0,1\}^{64}$
  $Y = Y_{15}||Y_{14}||\ldots||Y_1||Y_0 \longrightarrow Z - Z_{15}||Z_{14}||\ldots||Z_1||Z_0$
  $Z_i = S(Y_i)$ $(i = 0,1,\ldots,15)$

- ShiftRow Row i of the state is rotated i distance to the left, for i = 0,1,2,3. The relationship between the number of rows and the number of bits in the state matrix is shown in Table 1.

Table 1. The relationship between the number of rows and bits

ROW	0	1	2	3
Shift bits	0	4	8	12

$Z = Z_{15}||Z_{14}||\ldots||Z_1||Z_0 \longrightarrow U = U_{15}||U_{14}||\ldots||U_1||U_0$
$U_0\ ||U_1\ ||U_2\ ||U_3 = Z_0||Z_1||Z_2||Z_3; U_4\ ||U_5\ ||U_6\ ||U_7 = Z_5||Z_6||Z_7||Z_4;$
$U_8\ ||U_9\ ||U_{10}||U_{11} = Z_{10}||Z_{11}||Z_8||Z_9; U_{12}||U_{13}||U_{14}||U_{15} = Z_{15}||Z_{12}||Z_{13}||Z_{14}$

- MixColumn The output is the multiplication results of Intermediate state and the matrix.

$U:\{0,1\}^{64} \longrightarrow State':\{0,1\}^{64}$
$U = U_{15}||U_{14}||\ldots||U_1||U_0$

$$State' = U \times \begin{pmatrix} 4 & 2 & 1 & 1 \\ 8 & 6 & 5 & 6 \\ B & E & A & F \\ 2 & 2 & F & B \end{pmatrix}$$

- AddRoundKey The output of AddRoundKey is xor with the intermediate state and roundkey.

$K:\{0,1\}^{64}$
$State':\{0,1\}^{64} \longrightarrow Statei:\{0,1\}^{64}$
$State_i = State' \oplus K_i (i = 0,1,\ldots,31)$ (while i = 0, Statei = P)

3 Differential Power Attack Model

According to the principle of CMOS logic gates inverter, while the embedded encryption system is running, the data in the system is computing, and the capacity is charging and discharging, then the power is generating power. At this time, in circuit level, the load capacity is charging or discharging; In register level, the internal state of register is tripping; in instruction level, the data is changing when the instructions are carried. The power consumption of encryption system is decided to the effect of the processing data.

For hardware implementation of encryption systems, usually using hamming distance to calculate the power consumption, because the hamming distance model not only pay attention to be processed the data itself, also considering the changes of data is processed, it is a kind of more in line with the power consumption of CMOS circuit power consumption situation simulation model. Hamming distance refers to the number of corresponding position of the different characters of two strings, or a string into the next state the number of characters in the string needed. Model is based on a hypothesis: internal state bits from 0 -> 1 and 1 -> 0 flip when the leakage power consumption is the same, while ignoring the logic element static power. The power consumption model expressed by formula 1.

$$P_j = aH(D \oplus R) \oplus L \oplus \omega \tag{1}$$

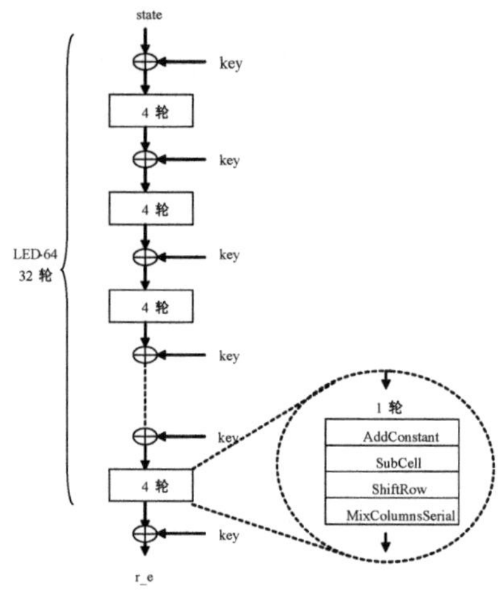

Fig. 1. Structure of LED-64

In this paper, by using the hamming distance model, it analyzes the power consumption of the encryption system. During the processing of the data In LED, the power point is setting at the operation point between the data is changing and unchanging. In LED algorithm, after eight rounds of AddRoundKey, AddConstant, SubCell, ShiftRow, MixColumn, the state is xor with the key at last, then the output is the cipher text, and there are 41 power points in all.

4 The Process of Differential Power Attack

4.1 The Principle

The principle of differential power attack is as follow:

First, Data Collecting. The target of differential power attack is the encryption key. To the same key, the attacker needs to input enough encryption plaintext Pi (0 < i<= N), using the simulation platform, it acquired and saved power consumption Si[j] in the process of encryption system (i is the sequence number of pain text, j is the j-th sampling points, if the number of sampling points is M, then 0 < j<= M; it would be obtained a two-dimensional array of Si[j], it's the j-th sampling points power, while plaintext is Pi.

Second, Data Analyzing. Construct a Distinguishing function that is the D function, D is a bit with the value of 0 or 1. D is related to the cipher text Ci and the related key bits (4 bits of the sub key), or to the plaintext Pi and the related key bits. For example, the D can be estimated as STATE1-1, while STATE1-1 is the first bit of output after 1st round S box, at this time, the D value is a power point time, known as follow:

$$STATE_i = SHIFT(S(STATE_0 \oplus K) \oplus RC_0)\begin{pmatrix} 4 & 2 & 1 & 1 \\ 8 & 6 & 5 & 6 \\ B & E & A & F \\ 2 & 2 & F & B \end{pmatrix}$$

STATE0, RCi are known, K (64 bits) is unknown, with the analysis of the LED algorithm, it shows that the STATE0 is actually the 16th bit after S box selection, corresponding to the 4th S box, is related with the input bits of S box, the value of STATE0 depends on STATE0 which is the known value, RCi and the unknown K (0-3). the value space of K (0-3) is 24 = 16, exhaustive it, and we can get 16 groups of the values for D.

4.2 The Steps of Power Attack on LED

The steps of LED-64 are as follows:

1) Given N plaintext Pi ($1 \leq i \leq N$), and obtained the power curve of the LED encryption process, Got the power matrix which is recorded the actual power dissipation, the size is N × 41.
2) To each round of S box, the input of the 1st S box is State′ ⊕ Ki, 1-4, State′ is the intermediate state, in the first round is Pi, Ki, 1-4 is 1 to 4 bit of the key; and the others, there are 15 Sbox, the input of K is 4 bits of the remaining 60 bits key; the output of S-box is $S(P_i \oplus K_{i,1-4})$; while the value of variable D is $P_i \oplus K_{i,1-4} \oplus S(P_i \oplus K_{i,1-4})$. $K_{i,1\text{-}4}$ may be 24 values, corresponding to 16 D values, therefore, given an known plaintext, it could be obtained N * 16 values the D matrix.
3) To the value of D, the $STATE_{1\text{-}1}$ is a power point, while the value of $STATE_{1\text{-}1}$ is 0 or 1, the power dissipation is different, therefore, let the D value to estimate the actual value of $STATE_{1\text{-}1}$. The power samples corresponding to different plaintext inputs are divided into two sets by D value:
$S_0 == \{S_i[j] | D = 0\}; S_1 = \{S_i[j] | D = 1\}$
4) Computing the difference with the arithmetic mean values of two groups of $S_i[j]$:

$$\Delta[j] = \frac{1}{|S_1|}\sum_{S_i[j]\in s_1} S_i[j] - \frac{1}{|S_0|}\sum_{S_i[j]\in s_0} S_i[j]$$

$|S_1|$, $|S_0|$ represents the number of elements in the set S_0 and S_1, evermore, Have to meet the equation: $|S_0| + |S_1| = N$.

While D is the time for attack, and N is large enough, basing on correlation of mathematical statistical, then the results of selecting guess key are three cases as following:

1. It is equal to classify the collection with a random function while at non D value time, N is infinity, $\Delta[j] = 0$, then the guess is the correct key.
2. It is equal to classify the collection with a random function while at D value time, N is infinity, $\Delta[j] = 0$, then the guess is the error key.
3. the power difference will be shown while it deals with different data at D times, N is infinity, $\Delta[j] \neq 0$, the guess is the correct key.

And so on, for other bits of K, to construct D function 15 times, according with the above method, it would crack the 64 bits key, the value space is 24×16, and then, it can get the key quickly by using exhaustion.

5 Experiment and Analysis

The experiment platform includes SAKURA-G FPGA, oscilloscope and PC. The bitstreams of LED encryption algorithm, which was implemented by verilog HDL, was downloaded on FPGA; the oscilloscope is used with PicoScope oscilloscope and its supporting software; and the data was processed by MATLAB, to analysis the attack waveform.

The LED encryption algorithm was described by hardware description language (such as Verilog HDL, VHDL and other languages), and was taken the logic simulation by Modelsim (seen as Fig. 2). After successful verificating, we downloaded the bit-streams of LED encryption algorithm, which is described in this paper, on FPGA board;

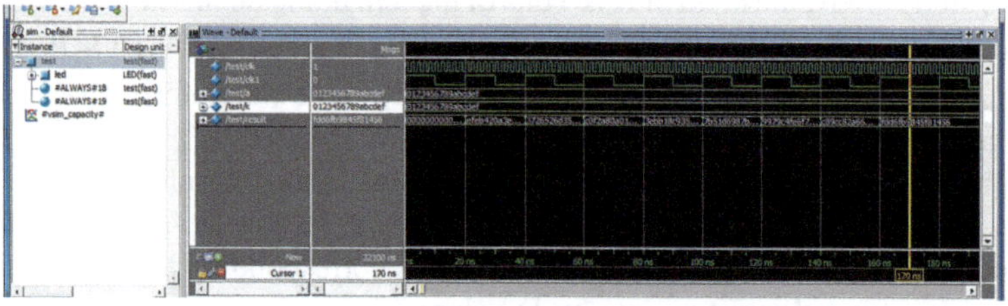

Fig. 2. Simulation results of LED by Modelsim

PicoScope oscilloscope and Picoscope 6.0 software platform is used to get the power consumption files and the intermediate variable values; and then, the waveform about the attack was analysised by using MATLAB 7.0 with the related datas. Figure 3 is the results of the attack by collecting 1000 plaintext, the original sub key is: sub-key = 'h0123456789abcdef, through the power model proposed in this paper, repeated 16 times to crack the 64 bits key.

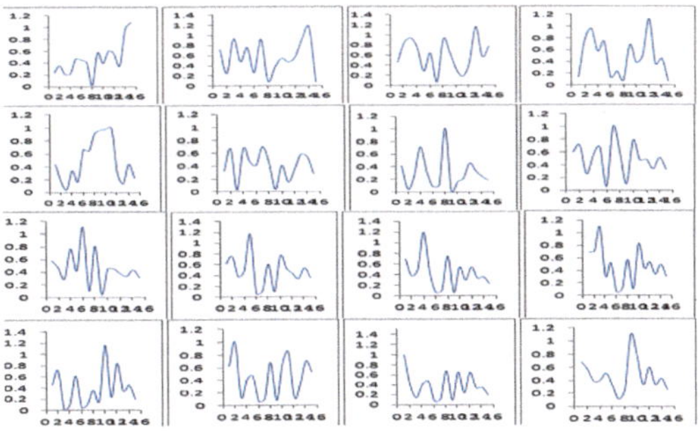

Fig. 3. The results of the attack

The D value is the estimated value of STATE1-1, according to the above study, the corresponding sub key for the K (1-4), the value space is 16, by X-axis. To crack the 64 bits subkey, repeat 16 times, the corresponding values are accumulated in the value space; the Y-axis is represented by the deviation of D between two groups of power curve. The corresponding peak value of the waveform is the correct value of the sub key, which is 00H,01H,02H,03H,04H,05H,06H,07H,08H,09H,0AH,0BH,0CH,0DH, 0EH,0FH; Compared with the original value of sub key, it is correct. The results show that the proposed model is effective. And it was shown that there would be security risks without any defense on LED encryption system.

6 Conclusion

This paper analyzed the principle of power consumption on CMOS chip, used power model of the Hamming distance, and proposed a new method of differential power analysis for lightweight cipher LED, and verified the correctness and effectiveness of the proposed scheme. We cracked the sub key of LED encryption algorithm. The result is shown that there would be security risks without any defense on LED encryption system. The next step, we will focus on that how to achieve and enhance the security of lightweight encryption algorithm, such as LED.

Acknowlegements. This work was supported by the National Natural Science Foundation of China under Grant (No. 61572174 and No. 61503128), Open fund project of the Science and Technology Plan Project of Hunan Province (No. 2017IIPAYB01), the Science and Technology Plan Project of Hunan Province (No. 2016TP1020), the Hunan Province Special Funds of Central Government for Guiding Local Science and Technology Development No. 2018CT5001, Hunan Provincial Natural Science Foundation of China (Grant No. 2017JJ2010), the Scientific Research Fund of Hunan Provincial Education Department (16C0226) and Young Backbone Teacher of Heng Yang Normal University.

References

1. Kocher, P., Jaffe, J., Jun, B.: Differential power analysis. In: Wiener, M. (ed.) CRYPTO 1999. LNCS, vol. 1666, pp. 388–397. Springer, Heidelberg (1999). https://doi.org/10.1007/3-540-48405-1_25
2. Akkar, M.-L., Bevan, R., Dischamp, P., Moyart, D.: Power analysis, what is now possible…. In: Okamoto, T. (ed.) ASIACRYPT 2000. LNCS, vol. 1976, pp. 489–502. Springer, Heidelberg (2000). https://doi.org/10.1007/3-540-44448-3_38
3. Messerges, T.S., Dabbish, E.A., Sloan, R.H.: Power analysis attacks of modular exponentiation in smartcards. In: Koç, Çetin K., Paar, C. (eds.) CHES 1999. LNCS, vol. 1717, pp. 144–157. Springer, Heidelberg (1999). https://doi.org/10.1007/3-540-48059-5_14
4. Kocher, P.C.: Timing Attacks on Implementations of Diffie-Hellman, RSA, DSS, and Other Systems. In: Koblitz, N. (ed.) CRYPTO 1996. LNCS, vol. 1109, pp. 104–113. Springer, Heidelberg (1996). https://doi.org/10.1007/3-540-68697-5_9
5. Gandolfi, K., Mourtel, C., Olivier, F.: Electromagnetic Analysis: Concrete Results. In: Koç, Ç.K., Naccache, D., Paar, C. (eds.) CHES 2001. LNCS, vol. 2162, pp. 251–261. Springer, Heidelberg (2001). https://doi.org/10.1007/3-540-44709-1_21
6. Osvik, D.A., Shamir, A., Tromer, E.: Cache attacks and countermeasures: the case of AES. In: Pointcheval, D. (ed.) CT-RSA 2006. LNCS, vol. 3860, pp. 1–20. Springer, Heidelberg (2006). https://doi.org/10.1007/11605805_1
7. Boneh, D., DeMillo, R.A., Lipton, R.J.: On the importance of checking cryptographic protocols for faults. In: Fumy, W. (ed.) EUROCRYPT 1997. LNCS, vol. 1233, pp. 37–51. Springer, Heidelberg (1997). https://doi.org/10.1007/3-540-69053-0_4
8. Li, W., Gu, D., Zhao, C.: Security analysis of the LED lightweight cipher in internet of things. Chin. J. Comput. **35**(3), 434–445 (2012)
9. Jiang, H., Mao, Z.: Advanced DES algorithm against differential power analysis and its hardware implementation. Chin. J. Comput. **27**(3) (2004)
10. Liu, H., Wang, T., Zhao, X.: Research on correlation power analysis attack against PRESENT. Comput. Sci. **38**(11), 40–42 (2011)
11. Li, L., Du, G., Zeng, T., et al.: Research on the PRINCE algebraic attack. Math. Practice Theory **45**(5), 153–159 (2015)
12. Zou, Y., Li, L., Jiao, G.: Differential fault analysis of PRINCE lightweight cryptographic algorithm. Comput. Sci. **44**(6A), 377–379 (2017)
13. Zou, Y., Li, L., Jiao G.: FPGA optimal implementation of PRINCE against power analysis attacks. In: ISCC 2017, December 2017

Research on Network Optimization and Network Security in Power Wireless Private Network

Yu Chen(✉), Kun Liu, and Ziqian Zhang

NARI Group Corporation/State Grid Electric Power Research Institute, Nanjing, China
chenyul@sgepri.sgcc.com.cn

Abstract. In view of the problems of electric power wireless communication and to serve the power consumer better, on the basis of the basic principles of TD-LTE's network structure, traffic model of power service is analyzed in detail, and in the actual case, through the data processing method and comparison analysis method, the technical parameters, theoretical analysis and field test results of 1.8 G and 230 M LTE systems are compared, and the ability of power wireless private network to support power service is verified. The analysis and testing provide a technology support for electric power communication access network construction and lays the foundations for popularization and application of LTE power wireless private network.

Keywords: TD-LTE · Wireless private network · Hybrid networking · Traffic model · Bandwidth analysis

1 Introduction

With the development of the construction of smart grid, the types of power communication services are becoming more and more abundant, network coverage and construction scale are expanding year by year [1–3]. In order to support the core tasks and objectives of smart grid construction, it is necessary to construct a terminal access network that has a wide access and flexibility [4, 5].

Fiber access network has many advantages, such as high bandwidth, anti-interference, anti-damage, security, reliability and so on. It is the best access method of user experience [6]. However, there are many problems in fiber-optic access network, such as the difficulty of laying, the long construction period, the long time of fault recovery, the high cost of investment, and so on, which can not meet the needs of mass access or mobile access [7]. Wireless technology system becomes a necessary supplement to optical fiber access.

At present, we mainly use the way of public network leasing to realize [8, 9], this approach has the following problems, one is that many intranet service applications can not be extended to the wireless terminal of the outer network, the second is that applications with high bandwidth requirements, such as pictures and videos, will generate expensive traffic costs, the third is restricted by the national network information

Q. Liu et al. (Eds.): CENet 2020, AISC 1274, pp. 1335–1344, 2021.
https://doi.org/10.1007/978-981-15-8462-6_153

security factors, video surveillance and other services can not access the intranet, there is a lot of inconvenience in the application of the business. In addition, with the development of clean energy substitution and electric energy substitution, distributed photovoltaic, electric vehicle charging piles and other business will be explosive growth [10, 11]. Power wireless private network is one of the most important technologies to realize access and acquisition of power network terminal services, therefore, more and more attention has been paid to the establishment of power wireless communication broadband private network based on special authorized frequency band [12, 13]. In order to ensure that the construction of power wireless private network can meet the current and even future needs of power business applications, in this paper, the traffic model of electric power business is obtained at the present stage and on this basis, the power service support capability of power wireless special network is analyzed and verified.

2 Architecture of Power Wireless Private Network

2.1 Base Station

The base station is generally located in the power supply building [14], including network management platform, monitoring center, data center, etc. The network management platform is mainly responsible for network state monitoring, fault diagnosis and alarm. At the same time, it can integrate the existing electric power information management, and form a unified dispatching command system on the basis of various multimedia means and GIS (Geographic Information System) technology, mainly including dispatching command center, field emergency command and dispatch system, videophone scheduling system and monitoring system.

2.2 Core Network

The core network is directly connected to the base station [15]. It mainly provides the connection for the user, manages the user and carries the service, including responsible for terminal authentication, terminal IP address management, mobility management, etc. The core network of the power wireless private network system can provide basic services including distribution automation, load management, power information collection, emergency repair and maintenance, dispatching command and visual management of mobile assets, etc.

2.3 Terminal

Terminal equipment module is the general name of remote terminal modules such as user data acquisition, power monitoring and dispatching, power video transmission, etc. It is the executive unit of information collection and monitoring and dispatching in Internet of things terminal, such as collector, concentrator, control switch and so on. Thc wireless terminal (also referred to as the UE) and the base station are air interfaces that provide communication between the terminal device and the base station.

3 Bandwidth Analysis

We use Okrmuram-Hata model to calculate the coverage ability of power wireless private network system. The model is established according to the measured data, which provides complete data and is widely used in 150–1920 MHz frequency band. The basic transmission loss formulas of Okumuram-Hata model in different propagation environments are as follows

$$\begin{aligned}\Delta = 69.55 + 26.16 \times \log_{10}(f_c) - 13.82 \times \log_{10}(h_t) \\ + (44.9 - 6.55 \times \log_{10}(h_t)) \times \log_{10}(d) \\ -((1.1 \times \log_{10}(f_c) - 0.7) \times h_r - 1.56 \times \log_{10}(f_c) + 0.8) + C\end{aligned} \quad (1)$$

where f_c is the working frequency, h_t is the effective antenna height transmitted by the base station, h_r is the effective antenna height received by the terminal, and d is the overlay distance, and C is the environmental compensation factor, the value is as follows

$$\begin{cases} city: \; C = 0 \\ suburbs: \; C = -2 \times \left(\log_{10}(f_c/28)\right)^2 - 5.4 \end{cases} \quad (2)$$

When the path loss is achieved, the link budget is then analyzed. Assume that the transmission power of the signal is p (dBm), Then the sensitivity of the receiver is $p - \Delta$ (dBm). If the bandwidth of the signal is π (dB−Hz), the noise coefficient of the receiver is ζ (dB), and the corresponding noise power of the receiver is $\pi + N_0 + \zeta$ (dBm), where N_0 (dBm/Hz) denotes the noise power spectral density (−174 dBm/Hz). It is known that the minimum receiving SNR of the receiver is $\Gamma = (p - \Delta) - (\pi + N_0 + \zeta)$, which is also the minimum SNR that requires the receiving device to be able to work.

Since power infrastructure mainly depends on the uplink data transmission, the capacity of uplink is analyzed here. In the uplink, the uplink signal transmitted by the user device can take up all of its transmit power, $p = P_{UL}$. The corresponding receiving SNR on the receiving end is

$$\Gamma = (P_{UL} - \Delta) - (\pi + N_0 + \zeta) \quad (3)$$

And then use the modified Shannon formula

$$r = \pi \cdot \log_2(1 + \beta \cdot \Gamma) \text{ (bit/s)} \quad (4)$$

The calculation of the channel capacity is carried out, where β is the compensation parameter used to compensate for the theoretical formula and the actual situation.

To simplify the analysis, a cell is divided into three regions according to SNR, as shown in Fig. 1. The three regions are 20 dB regions with SNR above 20 dB, 10 dB region with SNR between 10 dB and 20 dB, and 0 dB region with SNR between 0 dB and 10 dB. In urban environment, the corresponding radius of the three regions are 0.75 km, 1.45 km and 2.8 km, respectively. In the suburban environment, the radius of the three regions are 1.65 km, 3.2 km and 6.1 km respectively. The area percentage and spectral efficiency of each region are given by simulation, as shown in Table 1.

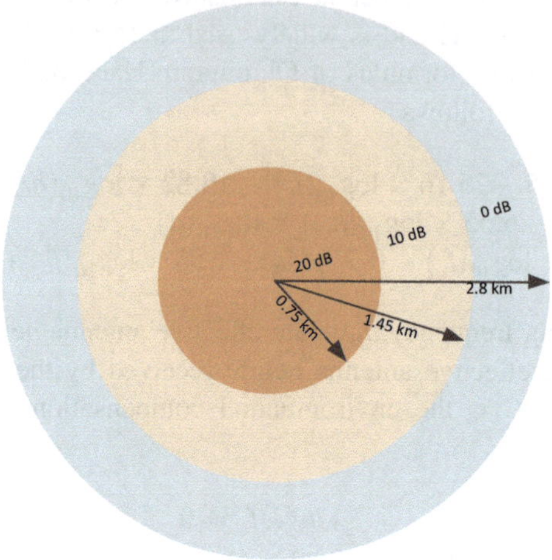

Fig. 1. Plot diagram of district division

Table 1. Area percentage and spectral efficiency of each region within the cell

	Area percentage %	Spectral efficiency bits/s/Hz
20 dB	$\alpha_1 = \begin{cases} 7.2, & urban \\ 7.3, & \text{suburban} \end{cases}$	$r_1 = 4.9$
10 dB	$\alpha_2 = \begin{cases} 19.6, & urban \\ 20, & suburban \end{cases}$	$r_2 = 2.5$
0 dB	$\alpha_3 = \begin{cases} 73.2, & city \\ 72.7, & suburban \end{cases}$	$r_3 = 0.82$

The average transmission rate per RB can be expressed as

$$\bar{r} = \sum_{n=1}^{3} r_n b\alpha_n = r_1 b\alpha_1 + r_2 b\alpha_2 + r_3 b\alpha_3 \text{ (bits/RB)} \tag{5}$$

$b = 144$ is the number of available resources per RB, r_i is the spectral efficiency of the region in Table 1, and α_i is the area percentage of the region in Table 1. Bring the data from Table 1 to get the average number of bits transmitted by a single RB $\bar{r} \approx 200$

bits/RB. It is worth noting that the average rate is basically the same because the percentage of area between the city and the suburb is similar. The transmission rates that each RB can support can be represented as

$$\bar{r} = 200 \times 0.6/0.001 = 120 \text{ kbps} \tag{6}$$

Table 2 shows the number of power terminals to be accommodated in each cell by 2020 and the requirements of different power infrastructure services for wireless communication transmission rates. For wireless meter reading, there are currently two solutions, one is that 1 poll per day, 300 KB per poll, including positive and negative active power, voltage and current power, daily freezing data, etc., and another is that 1 h 1 poll, 28 kB each time, additional 1 day 1 poll, 8 KB each time. For cities, an average of 92 m are hung by a single concentrator, and 53 m by a single concentrator in suburban areas.

Table 2. The demand of wireless communication rate for power basic service

Type of traffic	Single terminal transmission rate	Number of terminals in cell	Total rate demand (kbps)
92 m hanging under concentrator	300 KB/day	485	13.5
92 m hanging under concentrator	28 KB/hour + 8 KB/day	485	30.5
53 m hanging under concentrator	182 KB/day	763	12.85
53 m hanging under concentrator	20 KB/hour + 6 KB/day	763	34.33
Load control	6 MB/month	urban: 365 suburban: 277	urban: 6.8 suburban: 5.1
Distribution automation	800 MB/month	urban: 83 suburban: 84	urban: 205 suburban: 207
Distributed photovoltaic	100 MB/month	urban: 10 suburban: 250	urban: 3.1 suburban: 77.1
Charging pile	150 MB/month	urban: 150 suburban: 235	urban: 69.5 suburban: 108.8

According to Table 2, these four types of basic power services require approximately 315 kbps (cities) and 433 kbps (suburbs) for transmission rates, as a result, at least 3 RBs and 4 RBs need to be reserved for urban and suburban areas, respectively. Considering the overhead of uplink control channel, at least 1.4 MHz bandwidth is required for TD-LTE system, which is exactly the minimum bandwidth of existing LTE system.

It can be seen from Table 2 that power distribution automation requires the highest transmission rate in four types of services. But with the progress of science and technology, the collection of electricity information will become particularly important.

In general, the minimum bandwidth requirement for a power wireless private network is 1.4 MHz if only basic power services are considered. When the user information acquisition interval is reduced to 5 min, the minimum bandwidth is 3 MHz. When it is necessary to transmit the data of distribution automation with high precision, the minimum bandwidth is 5 MHz. The minimum bandwidth required to support other extended services, such as video conferencing, emergency repair, etc., is 10 MHz.

4 Road Survey

4.1 Test Location Selection

The total area of the test area is 927.68 km^2. At present, the area has been fully covered by 230 M wireless private network, and 14 base stations have been set up. Through careful planning, and considering the system capacity, coverage and other problems, the average radius of the base station is set to 4.8 km.

The contrast test of LTE1.8 G and 230 M is within the 5 km coverage radius of a base station. A representative wireless communication scene is selected for field measurement, and the throughput, delay and diffraction capability are compared and tested. In the north, a variety of sheltered environments, such as visual field transmission, buildings, bridges and trees, were selected for community coverage testing.

4.2 Relationship Between Distance and Throughput

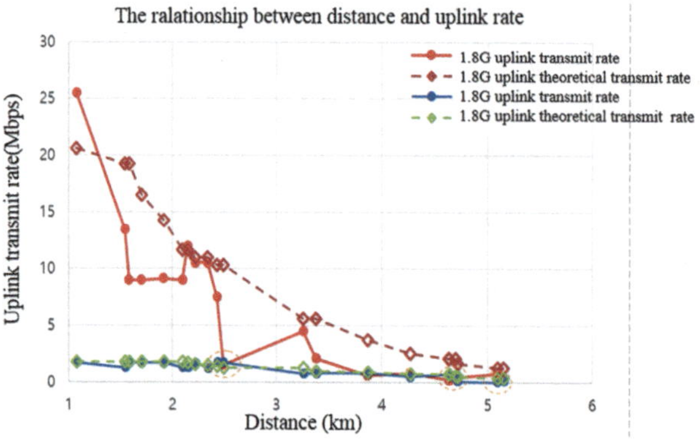

Fig. 2. Comparative diagram of the relationship between test distance and uplink rate

The uplink transmit rate can reach 10 Mbps in the 1.8 G wireless private network when the distance between the base station and test point is about 2 km, and the downlink transmit rate can reach 12.5 Mbps. It can support the application of video conference. The uplink transmit rate can reach 2 Mbps when the distance between the base station and test point is about 3.5 km, and the downlink transmit rate can reach 5 Mbps. It can support the application of power data monitoring. The uplink transmit rate can reach

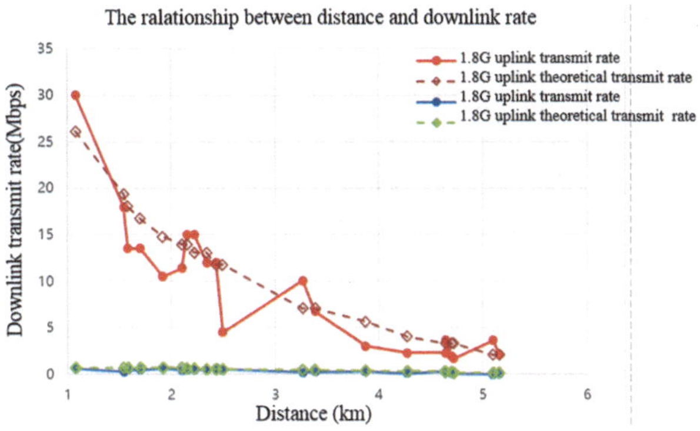

Fig. 3. Comparative diagram of the relationship between test distance and downlink rate

0.5 Mbps when the distance between the base station and test point is about 5 km, and the downlink transmit rate can reach 0.8 Mbps. It can support the application of power data monitoring (Figs. 2 and 3).

The uplink transmit rate can reach 1.5 Mbps in the 230 M wireless private network when the distance between the base station and test point is about 2 km, and the downlink transmit rate can reach 0.5 Mbps. The uplink transmit rate can reach 800 kbps when the distance between the base station and test point is about 3.5 km, and the downlink transmit rate can reach 300 kbps. It can support the application of power data monitoring.

4.3 Relationship Between Distance and Delay

The bidirectional average delay of 1.8 G is within 50 ms, and the packet loss rate is "0". One-way transmission delay of 230 M is about 140 ms (Fig. 4).

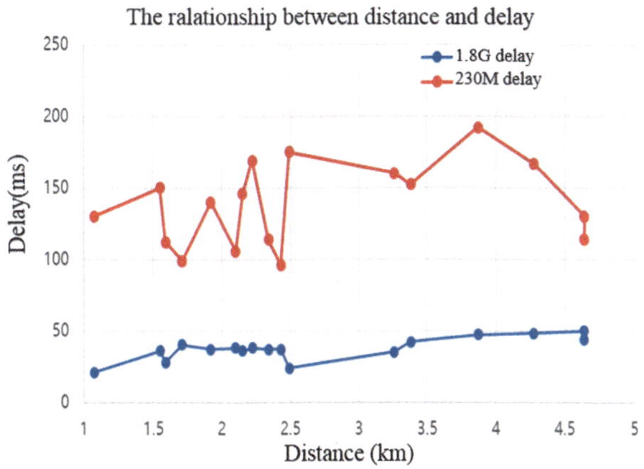

Fig. 4. Comparative diagram of the relationship between test distance and delay

4.4 Relationship Between Distance and Traffic Support Capability

By comparing the technical parameters of 1.8 G and 230 M, the theoretical analysis and field test results show that, in addition to being slightly superior to the former in terms of cell coverage radius, the 230 M system is much weaker than the 1.8 G system in terms of throughput, delay, service support capability, and the smoothness of subsequent technology evolution, and can only support static state. Small amount of basic power business. Therefore, 1.8 G is more suitable for urban area, and 230 M has a certain coverage advantage in suburbs and rural areas, which is about 1.6 times of 1.8 GHz, suitable for the construction of special network of small granular area with fewer service terminals (Figs. 5 and 6).

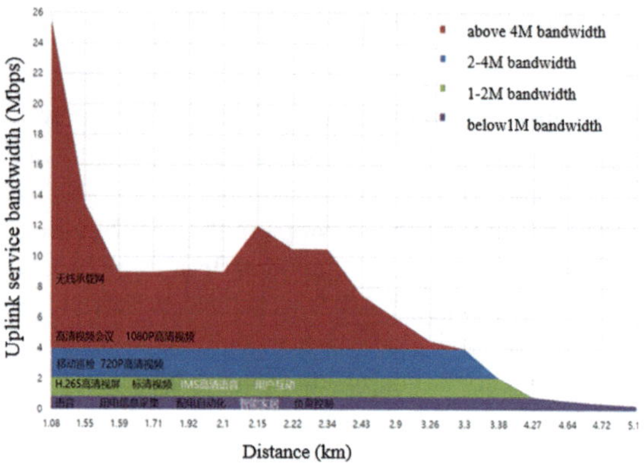

Fig. 5. Distance and traffic support capability diagram of 1.8 G

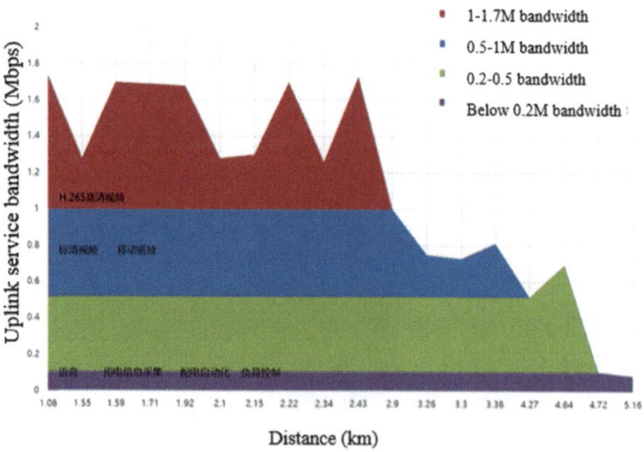

Fig. 6. Distance and business support capability diagram of 230 M

In addition, through theoretical simulation analysis, we can see that the minimum bandwidth of power wireless private network is 1.4 MHz under the condition of considering only the basic power services. When the information collection interval is reduced to 5 min, the minimum bandwidth is 3 MHz. The minimum bandwidth required is 5 MHz when high precision data transmission is required for distribution automation, and 10 MHz is the minimum bandwidth required to support other extended services, such as videoconferencing, emergency repair, etc.

5 Conclusion

On the basis of the basic principle of TD-LTE network structure, this paper analyzes the traffic model, and takes the actual case as the background, the technical parameters of 1.8 G and 230 M system are compared by data processing method and comparative analysis method. The theoretical analysis and test results verify the ability of power wireless private network to support power service. By comparing the technical parameters of 1.8 G and 230 M, the theoretical analysis and test results show that, in addition to being slightly superior to the former in terms of cell coverage radius, the 230 M system is much weaker than the 1.8 G system in terms of throughput, delay, service support capability and so on. The analysis and test provide technical support for the unified construction of power terminal communication access network and lay a solid foundation for the popularization and application of LTE power wireless private network.

References

1. Yao, J.M.: Random access technology of electric dedicated LTE network based on power priority. Autom. Electr. Power Syst. **40**(10), 127–131 (2016)
2. Cao, J.P., Liu, J.M., Li, X.Z.: A power wireless broadband technology scheme for smart power distribution and utilization network. Autom. Electr. Power Syst. **37**(11), 76–80 (2013)
3. Yu, J., Liu, J.S., Cai, S.L.: Performance simulation on TD-LTE electric power wireless private network. Guangdong Electr. Power **30**(1), 39–45 (2017)
4. Sun, S.W., Chen, Y.: Research on LTE power wireless private network for service coverage. Electr. Power Inf. Commun. Technol. **13**(4), 6–10 (2015)
5. Yu, J., Liu, J.S., Cai, S.L.: Research on LTE wireless network planning in electric power system. Electr. Power Inf. Commun. Technol. **10**(2), 7–11 (2016)
6. Gao, X., Zhu, J., Chen, Y.: Research on multi-services bearing solution of LTE power wireless private network. Electr. Power Inf. Commun. Technol. **12**(12), 26–29 (2014)
7. Liu, J.C., Quan, H.D.: RF interference cancellation based on multi-tap delay and orthogonal combination in multipath channel. J. Electron. Inf. Technol. **39**(3), 654–661 (2017)
8. Zhang, L.J., Liu, Z.H., Zhang, H.B.: Wireless secure enhancement with cooperative jamming. J. Commun. **38**(2), 183–195 (2017)

9. Wang, J., Zhao, H.Z., Tang, Y.X.: Quick adaptive self-interference cancellation at RF domain in full duplex systems. J. Univ. Electron. Sci. Technol. China **46**(4), 505–512 (2017)
10. Cao, J.P., Liu, J.M.: A two-stage double-threshold local spectrum sensing algorithm research for the power private communication network. Proc. CSEE **35**(10), 2471–2479 (2015)
11. Li, B.L., Zhou, J., Ren, X.H.: Research on key issues of TD-LTE network optimization. Telecom Eng. Tech. Stand. **1**, 57–61 (2015)
12. Gesbert, D., Shafi, M., Smith, P., Shiu, D., Naguib, A.: From theory to practice: an overview of MIMO space-time coded wireless systems. IEEE J. Sel. Areas Commun. (2004). Special Issue on MIMO systems, guest edited by the authors
13. He, S., Huang, Y., Yang, L., et al.: Coordinated multiuser procoding for maximizing weighted sum energy efficiency. IEEE Trans. Sig. Process. **62**(3), 741–751 (2014)
14. Wu, J., Zhang, Y., Zukerman, M., et al.: Energy-efficient base-stations sleep-mode techniques in green cellular networks: a survey. IEEE Commun. Surv. Tutor. **17**(2), 803–826 (2015)
15. Garcia, V., Zhou, Y., Shi, J.: Coordinated multipoint transmission in dense cellular networks with user-centric adaptive clustering. IEEE Trans. Wirel. Commun. **13**(8), 4297–4308 (2014)

Review and Enlightenment of the Development of British Modern Flood Risk Management System

Zhi Cong Ye[1], Yi Chen[1(✉)], Jun Zhou Ma[2], and Hui Liu[1]

[1] School of Architecture and Urban Planning, Nanjing Tech University, Nanjing 211800, China
njut_chenyi@126.com
[2] State Grid Nanjing Power Supply Company, Nanjing 210024, China

Abstract. Flood disaster is one of the most serious natural disasters since the 20th century. Flood risk management has become an important means to deal with flood problems. The UK's flood management system has undergone nearly a century of development, and its evolutionary experience is worth thinking about. This paper focuses on the development of the British flood management system based on the relationship between central and local authorities and between water and planning agencies. The development process of British flood management system is summarized, from "flood disaster prevention system" to "flood disaster management system", and finally to "flood risk management system". By comparing the development history of the British flood management system with the current situation of flood management in China, this paper suggests that the concept should be the same, and the organization should be tailored to local conditions, and measures should be broadened.

Keywords: Flood risk management · UK · Modern · Non-structural measures

1 Introduction

Since the 20th century, flood disasters have been one of the most serious natural disasters, causing huge economic losses to all countries in the world. All regions are facing severe flood threat [1]. For the past century, the concept of flood risk management has gradually gained popularity. It refers to a series of management measures and emergency strategies formulated and implemented in a planned and organized way by the central and local governments and relevant units to mitigate the threats and damages brought by sudden flood events [2]. Risk management measures are divided into two broad categories: structural measures and non-structural measures. The former refers to the development of strict building technical specifications, the improvement of construction methods and the construction of public disaster reduction engineering structures as a preventive means to reduce risks, the latter refers to the use of social structural methods to enhance disaster reduction capacity, through non-engineering and non-technical planning, education Wait for the disaster reduction [3].

Q. Liu et al. (Eds.): CENet 2020, AISC 1274, pp. 1345–1355, 2021.
https://doi.org/10.1007/978-981-15-8462-6_154

The United Kingdom has superior geographical conditions and few types of natural disasters. Floods are the main types of disasters [2], which are mostly manifested by coastal flood disasters, river flooding and urban infighting [4]. In general, the formation of the UK's flood risk management system can be understood as a process of ongoing conflict between water and planning, between local and central authorities, and a coordinated response to conflict. The development track of a century, in which the evolution of thought and value deserves our serious consideration and excavation.

Domestic research on flood risk is fruitful, but there are fewer related studies involving a larger-scale political perspective [5], and there is also a lack of vertical collation in a single country [6]. The British political system is close to China and belongs to the type of central power concentration. However, it advocates local autonomy, especially the "localism" thought that it has risen in recent years, which is worthy of our critical reference. In summary, this article will trace and analyze the development of the UK flood risk management system by collating the flood disaster related policies and literature since the mid-to-late 19th century. This will not only have a better understanding of the current institutional mechanisms and measures of disaster management in the UK, but also to some extent to make up for the gaps in the vertical research in relevant fields, and to provide some help for better reference to foreign experience.

2 Mid-Nineteenth Century to the 1960s: Structural Measures Dominated by Water Sector

The process of the initial formation of a flood disaster prevention system in the UK is strongly influenced by two factors: the need for agricultural development and the tradition of local autonomy. Since 1930, England and Wales have established regional catchment departments, and after 1945 the United Kingdom began to accelerate agricultural intensification, further creating the need for effective drainage and protection of high-yield flooded agricultural areas in the southeast [7]. The tradition of British local autonomy was challenged by the accumulation of "negative externalities" since the Industrial Revolution, which led to the promulgation of the Public Health Act in 1848, which initially involved many issues related to urban drainage [8]. After nearly a century, the Land Drainage Act, promulgated in 1930, officially made up for the vacancy in the UK's Independent Drainage Act. Although the local behavior was restricted by law, the 1930 Act also clearly stated that the problem of land drainage is mainly regarded as a local problem, and the management power of the land drainage department is not mandatory [4].

The Land Drainage Act of 1930 constructed a stable administrative organization structure [4] (Fig. 1). At the local level, the main water sector includes the Local Land Drainage Committees (LLDCs) and the Internal Drainage Committees (IDBs), whose parent organization is the Ministry of Agriculture, Fisheries and Food (MAFF), and the main planning department, the District Council (DCs), is responsible for Planning work in the floodplain. In addition, the water department also manages the land drainage method, designing and supervising the construction of flood control projects for major rivers. In the field of water affairs, the superiors have greater power to bind the lower levels, and the superiors enjoy a great voice in both legislation and engineering construction. In the planning area, there is a lack of supervision in the water sector, and communication between the two departments is not smooth. For example, the water department may have to try to balance the hydrological function and development pressure of the river, while the entire floodplain is transformed and built by the local council, and flood control is not necessarily the main target.

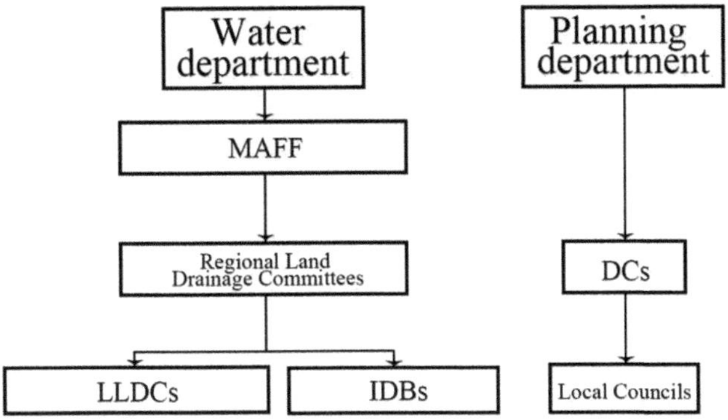

Fig. 1. Organizational structure of 1930. Source: Authors sorted by reference [4]

Structural measures are more popular in the UK than non-structural measures. This is mainly because the water department has partial construction power in the floodplain, but it does not have regulatory power. Therefore, structural measures will be taken as its main solution, and non-structural incentives such as the central government's Grant-aid System are also Structural engineering fits well. Non-structural measures in the UK include legislation and Grant-aid System, as well as land use planning. Due to the high dispersion of drainage responsibilities and the economic depression caused by the two world wars, the local government did not have sufficient resources and capacity to solve the drainage problems that need attention at the time. Therefore, the central government provided resources in the form of subsidies to become local flood control and drainage. The main driving force of construction [4]. Because the central intervention is in an auxiliary position and does not form a unified guiding principle, it can only be maintained on the surface, and it is difficult to deeply influence local management. This makes the local water department and the planning department lack effective coordination from superiors, causing a series of problems.

3 Late 1960s to 1994: First Attempt to Coordinate Water and Planning Departments

In the 1960s, North American flood control theory had an important impact on the UK. In 1942, American geographer Gilbert White first proposed the need to focus on non-structural programs and coordinate public behavior [9]. The government then gradually established a flood management system [3] combining structural and non-structural measures. The past studies of flood disasters in the UK were mainly influenced by natural geographers, and the success of the United States in flood theory in the 1950s and 1960s made this research model led by economic and social geographers highly respected by the United Kingdom. The Chicago model became the main reference example for flood research in the UK [9].

Britain has not been under the "shadow of North America". Scholars and officials soon discovered that there are differences between the UK and the US in terms of political system and social tradition, so some experiences are not fully instructive. They concluded that the UK is not facing a severe flood threat like the United States [4]. Therefore, the British government has obvious choices when absorbing the experience of non-structural measures in North America: the reform measures related to national policies have been met with a cold shoulder, and the government has turned to improve its original non-structural measures, namely, the planning department and the water department. Coordination to better play the role of land development control.

After critically drawing on the American experience, the UK quickly put the work of departmental coordination on the agenda. The government notices issued in 1961, 1969 and 1981 emphasized that the planning department and the water department must be in close contact to ensure development. The decision was taken into account when flooding occurred [4]. In 1970, the government established the Department of the Environment (DoE) to deliver housing, planning, and environmental protection to the department. Later, in 1976, the new Land Drainage Law was promulgated. Based on the 1930 Act, it made more detailed regulations on land development control in the floodplain. The improved organizational structure is shown in Fig. 2. MAFF conducts overall supervision of land drainage related matters and performs functions through its regional land drainage committee, while local governments exercise planning authority under the supervision of DoE. In addition to structures or adjacent structures in rivers or rivers, the water department has no direct control over the development of major rivers, floodplains in major rivers or flood-hazard areas. The Water Department has two main functions in the process of coordination between the two parties: providing advice to the planning department on flood risk and flood impacts in the development area, and ensuring that development does not affect the effective management and operation of the drainage system under its jurisdiction. The bill avoids the practice of passive protection. If there is a risk of flooding, the government does not directly "prohibit development", but delays planning decisions, tries to reduce or eliminate this risk, and puts the interests of the whole people at the center of development, not because of protection. Abandoned development.

While the United Kingdom improved its organizational structure, some of the non-structural measures showed a disconnect from the organizational structure. In the aspect of flood warning system, there is a lack of self-help early warning plan for "affected potential communities". It is difficult to achieve protection for the whole of the UK by relying solely on the top-down warning system. At the same time, the statutory body is missing, and the responsibility for issuing flood warnings involves water, Local departments and police (the actual person in charge on many occasions), the mutual urging work between public agencies greatly reduces the speed and accuracy of early warning propagation. In terms of flood insurance, compared to the United

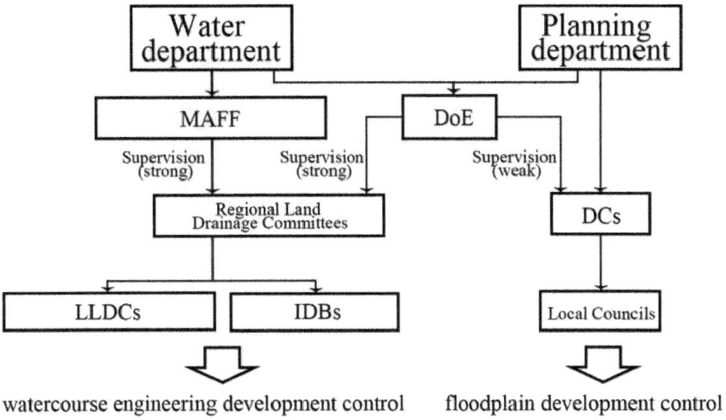

Fig. 2. Organizational structure of 1976. Source: Authors sorted by reference [4]

States's "government-led, private commercial insurance companies to cooperate, and individuals in the community involved in the buying [10]" legal system, The UK is based on a signing with the insurance industry in the 1960s. This "Gentleman Agreement" stipulates that flood insurance provided by private insurance companies will not be directly interfered by the government, and the government will make flood risk insurable by building flood defense facilities [2]. The British central government maintains the stability of insurance ecology indirectly, and the community residents are also in a passive position in the insurance system. The failure to incorporate the insurance system into the system has led to many problems: on the one hand, there is no relevant institution to manage the flood insurance industry, on the other hand, the lack of guidance from national policies makes it lose contact with other flood area management methods to a large extent and unable to play a better effect.

On the positive side, in constant practice and exploration, the UK is beginning to realize how balancing the relationship between protection and development is the key to disaster management. In the planning field, the government tries to ensure the long-term development of the region to the greatest extent, and is not superstitious about pure ecological protection. In the technical field, the assessment method of flood risk has evolved from the past cost-benefit analysis to the service level analysis, marking the focus on social development. The economic perspective turns to a humanistic perspective that guarantees citizen security [4].

4 1995–2010: The Second Institutional Coordination under the Sustainable Development Theory

In the 1990s, countries generally accepted the concept of sustainable development. In 1995, the United Kingdom enacted the Environmental Act. At the request of the Act, a new government non-sectoral public agency: Environment Agency (EA) was established to promote the goal of achieving sustainable development. The rapid spread of the concept of sustainable development in the UK stems from the constant changes in the environment. Three severe floods occurred in 1998, 2000 and 2002, causing huge economic losses to the UK. These floods have a striking feature: they do not occur in major floodplains, but are scattered throughout the country [11]. Such flood problems cannot be dealt with solely by the functional scope of the water department, and large-scale spatial planning for floods needs to be effectively implemented.

In 1996, the British government formally established a non-departmental agency, EA, to integrate drainage related matters. In 2004, it immediately carried out departmental-level reforms. In the past, the functions of MAFF and DoE and other related departments were integrated to form an environment, food and Ministry of Rural Affairs (Defra). Defra develops a national policy, and the EA, the Internal Drainage Committee, and the local sector are responsible for implementation. EA uses its regulatory responsibilities and designs defense programs for "main rivers" through its regional agencies, while IDB and local departments design plans for "ordinary waterways". In addition, EA is the designated agency to provide and disseminate flood warnings, and also serves as an advisory position when local authorities review new developments in floodplains (Fig. 3).

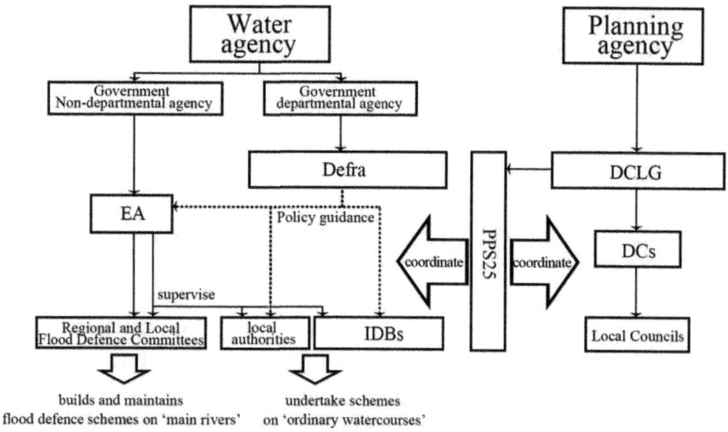

Fig. 3. Organizational structure in 2006. Source: Authors sorted by reference [7, 12]

To strengthen the relationship between planning and flood management, in 2001 the UK issued Planning Policy Guide 25: Development and Flood Risk (PPG25) as the first planning guide for floods in England [11]. In 2004, the United Kingdom made major changes to the planning regulations, and the planning area also made new arrangements for coordination with the water sector. In 2006, the "Planning Policy Statement 25: Development and Flood Risk" (PPS25) was released, replacing the previous PPG25. PPS25 describes its efforts in absorbing water theory and experience in three areas: integrating flood risk into planning zoning, coordinating risk assessment with planning levels, and ensuring that flood risk management is reflected in the full phase of planning [12]. The success of the policy in the planning area has made up for the lack of a policy framework. At the same time, Defra, as a government water agency, has played a role in coordinating the role of water non-sectoral institutions (such as EA) and space planning agencies. However, this kind of negotiation has a limited nature and cannot be related to grassroots society [13]. Flood risk management has been transformed into a movement that requires the concerted efforts of all sectors of society. The government needs to establish a broader consultative organization to build a framework of action for all interest groups.

5 2011-Present: Integration and Decentralization of Powers and Responsibilities Under the Trend of Localism

After 2010, the UK began to implement the concept of "localism". The concept recognizes the role of the central government in ensuring fairness, and it also recognizes the primacy of local governments in providing public services on behalf of the central government, that is, the central government decentralizes the localities, and the local authorities simultaneously decentralize the communities and individuals [14]. The performance of "localism" in the field of flood risk management is that all circles recognize that the government should play an important role in providing and guiding venture capital, but the community should also obtain more licenses and participate in the risks they face, and take the initiative. Choose to be more flexible in reflecting the local situation.

A severe summer flood occurred in 2007, costing more than £3.2 billion to the UK. Sir Pitt published a report on the flood in 2008. In this report, he suggested that beneficiaries should be better integrated with payers to create a more efficient and responsive organization. The government attaches great importance to Pitt's report. In 2010, the Flood and Water Management Act was introduced. The bill required Defra to issue a national-level policy document in the field of flood risk management as the core of the implementation of the bill. So in 2011, the UK National Flood and Coastal Erosion Risk Management Strategy (FCERM) was prepared by EA based on Defra's advice, which provides an overall picture of all future risk management efforts to address flood and coastal erosion in England. The framework marks an important milestone in the national strategy. The management organization mode of FCERM is shown in Fig. 4 [15]. Developed by Defra to develop FCREM policies, EA has a strategic and supervisory function, and it also works with the Bureau of Meteorology to provide flood forecasts and warnings through the National Flood Forecasting Center, the functions of local leading flood authorities (LLFAs) include the development of local flood risk management Strategy, as well as managing flood risks in surface water, groundwater and ordinary waterways, in addition to DCs, IDBs can also manage flood risk in ordinary waterways, EA has also established Regional Flood and Coastal Commissions (RFCCs) to play a key role in coordinating FCERM.

On this basis, the UK has established a mature system of measures for "risk management system + flood rescue operations". FCERM, issued in 2011, has built a circular flood risk management model: understanding risk-management possibilities–helping people manage their risks–avoiding inappropriate development–improving flood forecasting, warning and post-disaster recovery–further understanding risk [16].

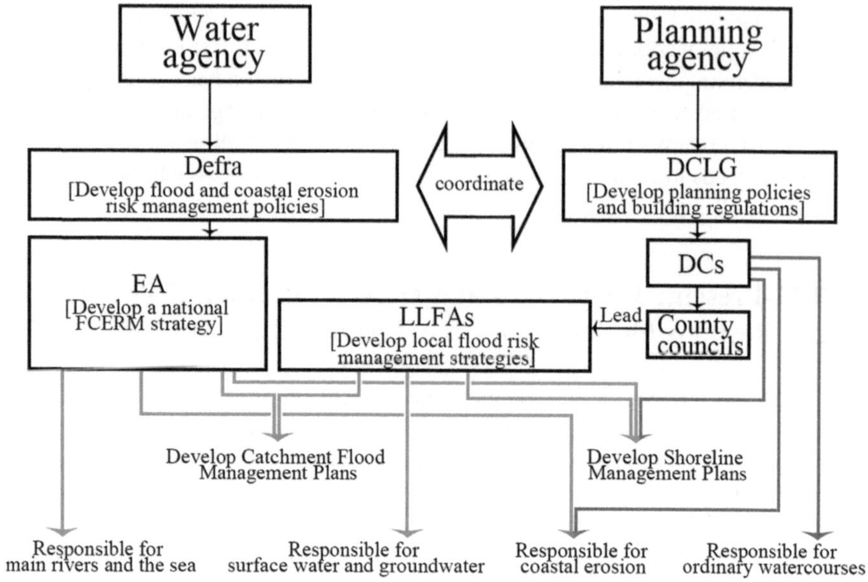

Fig. 4. Organizational structure of 2011. Source: Authors sorted by reference [18]

FCERM pointed out that local protection and investment plans cannot be subject to national budgets, reforming the way national funds are allocated, and allowing other sources of funding to function to balance the protection of relationships between inland and coastal areas, cities and rural areas. It is worth noting that FCERM accepts the concept of "resilience" and regards "risk" as the logical focus of management in management methods. It no longer treats floods with "disasters". FCERM has almost no use of "disaster". This term accepts the natural attributes of the flood, that is, its positive and negative benefits [15]. In 2013, Defra established a system called the "Flood Rescue National Augmentation Project" to form the concept of "National Flood Relief Action". The main goal of the Flood Relief National Augmentation Project is to make more effective use of existing flood rescue capabilities through a comprehensive multi-agency flood rescue operation (FRCO). The system includes the Civil Emergency Law, the Flood and Water Management Act, the National Flood Emergency Framework, the Competency Plan, and the Pitt Review [17, 18]. After decades of development, the UK has established a relatively complete flood risk management system for the central, local and community, as well as individuals.

6 Revelation and Experience

The development of the flood management system is a process of continuous feedback. Guided by ideas and principles, the government has established an organizational structure and corresponding measures system. The organizational structure guides the system to develop toward a certain tendency, and the continuous maturity of the system of measures will also aggravate the solidification of the organizational structure. When the region is exposed to flood disasters, it is exposed to weaknesses in the system. These experiences continue to converge to promote a new consensus in society, and thus evolve. After nearly a century of development, the UK's flood control system has evolved from a "flood disaster prevention system" to a "flood disaster management system" and has finally been refined into today's "flood risk management system" (Table 1).

Table 1. Development history of the British flood management system

Stage of development	Idea/principle	Organizational structure	Measure system
1930	Economic development orientation	Local autonomy and central soft intervention	Structural measures dominate the system of measures
1976	Coordinating development and protection	Local departments coordinate organizational relations	Top-down non-structural measures
2006	Sustainable development orientation	National policy coordination and organizational relations	Non-structural measures coordinated by departments
2011	Resilience development orientation	National policies guide community participation	All sectors of society together constitute the system of measures

China has entered the stage of developing and improving the "flood risk management system". In terms of the principle, China has changed from "control flood" to "management flood" to ensure the sustainable development of social economy. In terms of organizational structure, the current system mainly includes two parts: the flood control project management system and the flood control emergency management system. In addition to traditional engineering flood control, the system also includes preventive and early warning mechanisms, emergency response mechanisms and loss compensation mechanisms [1]. Comparing China's current system with the United Kingdom, we can see that although there are differences in social traditions and development stages, there are still many experiences worth learning from:

(1) Concept: The Same Way

Both the UK and China have accepted the goal of sustainable development. In order to achieve this goal, the United Kingdom chose the reform direction of "localism", while China developed the "social social governance system with Chinese characteristics", both of which are adapted to the path of national tradition. The types of flood problems in the UK are relatively simple, and the capital market is relatively mature. The potential for local communities to be excavated is large. China has a vast territory, and the development gap between the regions is large. There are many types of floods, so they are facing flood disasters. There must be strong central support, but no one-size-fits-all policy.

(2) Organization: Adapting to Local Conditions

However, the history of flood management in the UK shows that future flood problems will tend to be "variable, complex and difficult to control". China's flood disaster management has a high center of gravity. Various powers and resources are mainly concentrated in higher-level governments such as the central government. The enthusiasm and actual ability of local government emergency management are relatively insufficient. The communication of information is very poor in both vertical and horizontal directions [1]. This has led to China's "flood risk management system" as a "flood disaster relief system" in many cases, and has not really "reduced the possibility of disaster losses". For some developed regions or cities, it is recommended to carry out reforms and adjustments to the corresponding organizational structure to enhance the corresponding decision-making power of local governments. The central government provides an efficient resource platform and regulatory framework for disaster prevention and relief.

(3) Measures: broaden channels

Many of China's current non-structural measures are still in the early stages, and there are many aspects that can be improved. Drawing on the experience of the United Kingdom, we should increase the protection and scope of the government, and at the same time fully guide the local potential to join the flood control. The current "risk management + rescue operation" system in the UK is compared with the current system of "flood prevention project management + flood control emergency management" in China. The former fully absorbs non-structural measures, while the latter still lacks, such as lack of flood insurance. There are faults between the system, planning and actual management and rescue. The current departmental reform is the first step in resolving the inadequacy of various measures, and the follow-up work should be carried out in an orderly manner.

References

1. Hong, W.: Research on flood disaster risk management system. Wuhan University (2012)
2. Li, S., Zhai, G., Wu, Y.: Basic experience of urban flood risk management in the UK. Int. Urban Plann. **26**(04), 32–36 (2011)
3. Zhou, L.: From structural disaster reduction to unstructured disaster reduction: a new trend of international disaster reduction policy. China Adm. **12**, 94–100 (2013)
4. Handmer, J.W. (ed.): Flood Hazard Management: British and International Perspectives. Geo Books, Norwich (1987)
5. Shi, F.: Flood risk management in the UK. Water Conservancy Hydropower Express **24**, 1–4 (2006)
6. Che, W., Yan, P., Zhao, Y., Frank, T.: Development and analysis of international modern rainwater management system. China Water Wastewater **30**(18), 45–51 (2014)
7. Werritty, A.: Sustainable flood management: oxymoron or new paradigm? Area **38**(1), 16–23 (2006)

8. Wang, Y., Tan, Z.: Review and enlightenment of the development of modern and contemporary planning system in the UK——based on the perspective of land development rights. Int. Urban Plann. **34**(02), 94–100 + 135 (2019)
9. Parker, D.J., Penning-Rowsell, E.C.: Flood hazard research in Britain. Prog. Geogr. **7**(2), 182–202 (1983)
10. Zili, R.: Research on the legal system of american flood insurance——concurrently on the enlightenment of its reform to China. Tsinghua Law **6**(01), 122–135 (2012)
11. Howe, J., White, I.: Like a fish out of water: The relationship between planning and flood risk management in the UK. Plann. Pract. Res. **19**(4), 415–425 (2004)
12. Department for communities and local government. planning policy statement 25: Development and Flood Risk (2006)
13. Green, C.: Competent authorities for the flood risk management plan–reflections on flood and spatial planning in England. J. Flood Risk Manage. **10**(2), 195–204 (2017)
14. Sun, H., Tan, R.: On the characteristics and path choice of British "New Localism". J. Natl Sch. Adm. **06**, 115–119 + 128 (2016)
15. Environment agency, Department for environment, food and rural affairs. The national flood and coastal erosion risk management strategy for England (2011)
16. Environment Agency. Thames Catchment Flood Management Plan (2009)
17. Department for environment, food and rural affairs. Flood rescue national enhancement project: flood rescue concept of operations (2010)
18. Department for Environment, Food and Rural Affairs. The national flood emergency framework for England (2014)

Research on 5G in Electric Power System

Long Liu(✉), Wei-wei Kong, Shan-yu Bi, and Guang-yu Hu

NARI Group Corporation/State Grid Electric Power Research Institute,
Nanjing, China
401945335@qq.com

Abstract. The collection, bearing, analysis and application of the basic electric power data from the ubiquitous Internet of things have put forward new and higher requirements for the data collection, bearing, analysis and application, and higher requirements for communication technologies and methods is raised. In view of this situation, the integration and development means of 5G technology and power system business are put forward, and the application of typical business scenarios in various power systems under 5G technology is analyzed. A typical accurate load control service is selected for testing, and the test results show that the delay is improved. The application analysis of 5G in electric power system supports the Internet of things application, provides the technical support, and lays a solid foundation for the popularization and application of 5G technology.

Keywords: VoLTE · Smart grid · Road test · Voice quality · Wireless private network

1 Introduction

In recent years, the National Network Company has actively promoted the construction of smart grid [1–3]. With the rapid development of distributed energy access, electric vehicle service, power information collection, distribution automation, two-way interaction between users and so on, explosive growth in communication demand for all types of power grid equipment, power terminals and electricity customers, and there is an urgent need for real-time, stable, reliable and efficient new communication technologies [4–6].

The user experience rate of 5G is 10 times that of 4G, the peak rate is 20Gbps, and the empty port delay is 1 ms [7]. The application scenarios of 5G include mass couplet, low delay and high reliability. China is the leader of 5G industry, and it promote 5G from the mobile Internet of things to the all things are connected to the Internet of things [8]. The development requirements of State Grid Company are highly in line with the 5G application scenario, and it has inherent resource advantages in base station site, communication network, business application and so on [9]. Therefore, how to deeply and systematically explore the integration and development means of 5G technology and power system in this environment, and promote the sustainable and stable development of power system in our country, is one of the key points of work.

Q. Liu et al. (Eds.): CENet 2020, AISC 1274, pp. 1356–1363, 2021.
https://doi.org/10.1007/978-981-15-8462-6_155

2 Overview of 5G

5G refers to the fifth generation mobile communication technology [10]. The main core technology is to improve the performance and meet the requirements of diversification by perfecting the technology. 5G has broad application scene and development prospect in electric power, and the main characteristics of 5G are low cost, reliable, low delay and architecture of open technical. The characteristics of 5G are corresponding to the basic requirements of the power system, as shown in the following table [11] (Table 1).

Table 1. Characteristics of 5G communication and the demands of the power systems

The characteristics of 5G	Demand for electricity
High speed	Massive data transmission
High power capacity	Interconnection of all things
High reliability	Power system reliability
Low delay	Flexible response and Cooperative Control
Low energy consumption	Battery life guarantee

There is complementarity between the characteristics of 5G communication and the characteristics and requirements of power system. For power system, 5G has a wide range of applications in mass measurement, precision control, broadband communication and so on [12], as shown in the figure (Fig. 1).

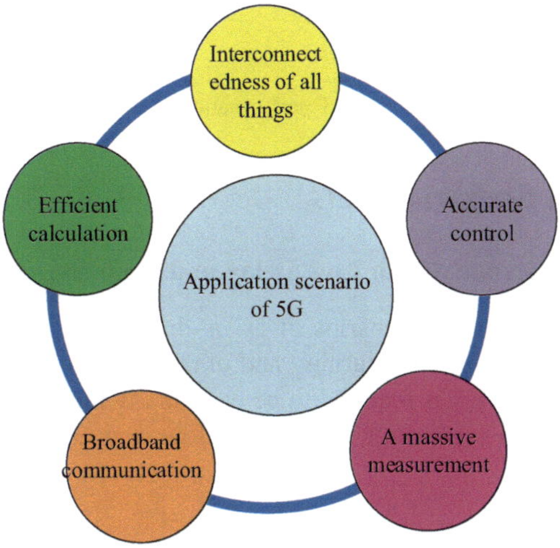

Fig. 1. Application scenarios of 5G communication in the power systems

The communication system of 5G is mainly composed of macro base station, micro base station and core network [13]. The core network is mainly responsible for the control system and the transmission of data information, macro base station is mainly responsible for connecting to the core network through optical fiber or microwave, the information of base stations and users in different areas is mainly transmitted through wireless communication.

The performance of 5G network is better than that of 4G [14]. The following figure shows the comparison of 5G and 4G (Fig. 2).

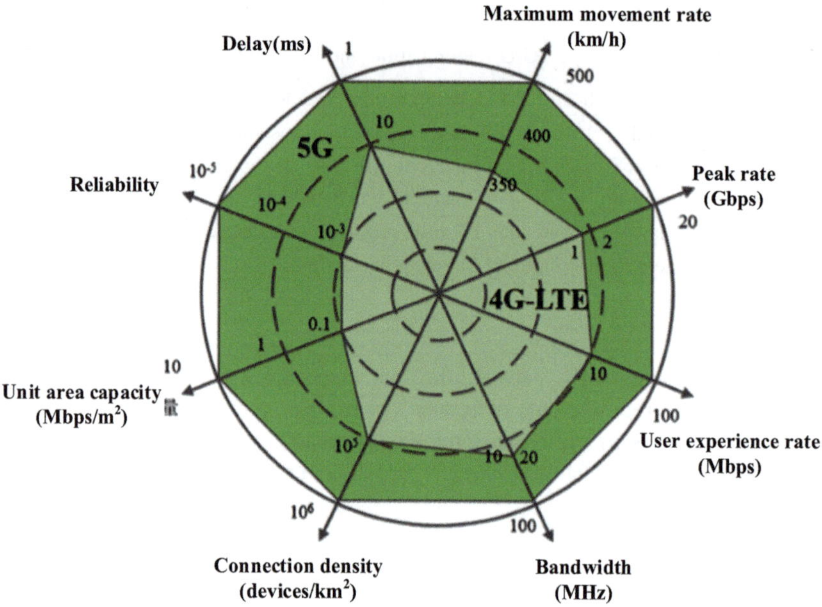

Fig. 2. Comparison of performance between 5G and 4G

3 Application Scenario of 5G

3.1 Application Scenario with Low Delay and High Reliability

One of the main application scenarios of 5G is driverless cars, because in driverless cars, cars need to make braking, turning and other judgment behavior in a very short period of time, so the delay is required to be particularly high. The precise load control in a typical application scenario is described below (Fig. 3).

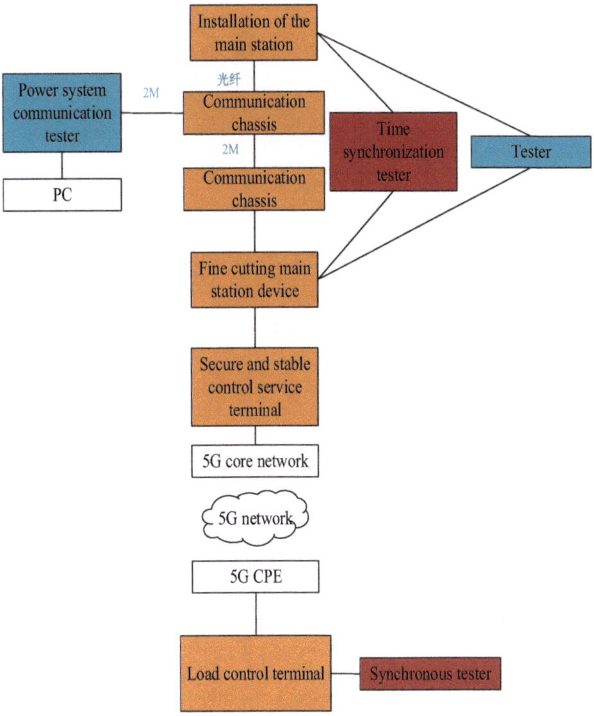

Fig. 3. The structure diagram of precise load control

3.2 Application Scenario with Wide Coverage and Large Connections

The following is an introduction to the collection of electricity information. The power information acquisition system is composed of a main station, a remote and local channel, a concentrator and a collector.

In the electricity collection, although application of the electricity collection in our country are widely used, many of the electricity meters only upload the daily data instead of the data for half an hour, so it is inconvenient for the data analysis. Because of the high communication rate of 5 G, it brings convenience to the analysis of a large number of data collected by electricity (Fig. 4).

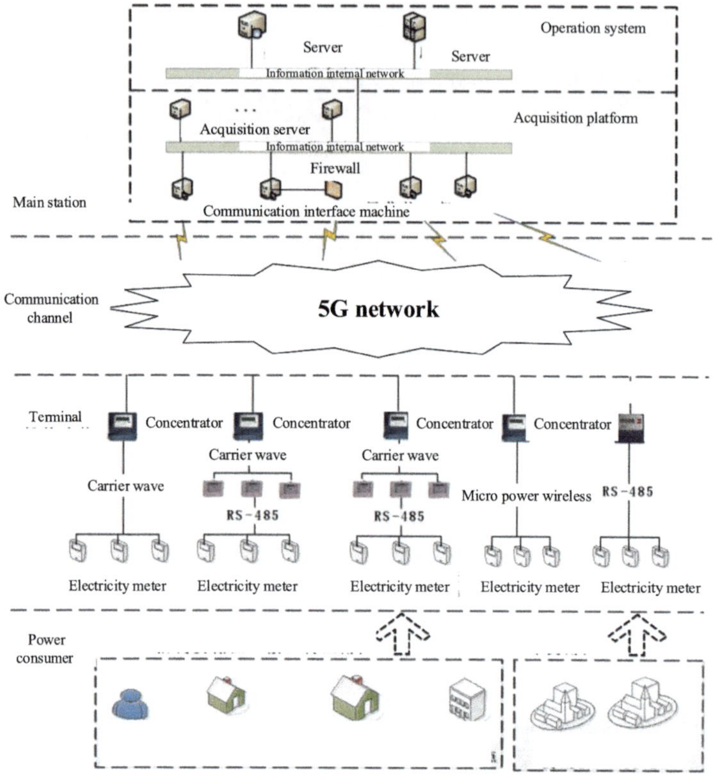

Fig. 4. The structure diagram of electricity information collection

4 Verification Test of 5G

4.1 Background of Test

In this paper, the precision control service is selected as an example, adopt the method of combining the actual environment with the test terminal, a 5G verification test environment is built, and the time delay of the precision control under the 5G network is detected, and the basis for subsequent fine control service access is provided.

4.2 Scheme of Test

Using 5G network, the master station and sub-station are built in the computer room, and it uses CPE to access the load control terminal, then the delay test of precision load control is carried out. The network topology diagram is shown in the following figure (Fig. 5).

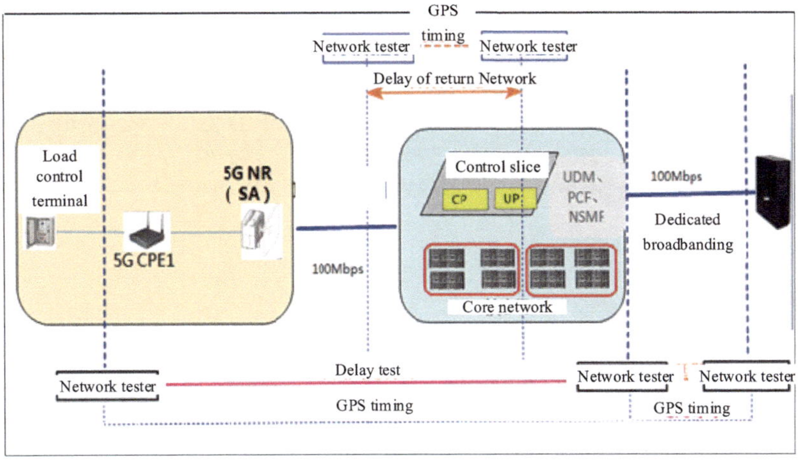

Fig. 5. Test topology diagram

1) CPE connects distribution terminals via Ethernet ports, the CPE is inserted into the USIM card and turned on into the idle state;
2) Correctly configure the parameters of CPE, and complete the activation process of the core network;

The on-site test environment is shown in the following figure (Fig. 6).

Fig. 6. Field test diagram

4.3 Test Results

Four test points are selected for this test. The relevant test data are as follows (Table 2).

The average time delay of this verification in the 5G wireless environment is 35 ms, in 4G environment, the average delay is 40 ms. The farther away the distance, the worse the signal and the longer the delay. The outdoor delay is shorter than indoor delay. The delay in 5G environment is 5 ms less than that in 4G environment. The low-delay characteristic of 5G is verified by this experiment.

Table 2. 5G Test results

Test position	Scene	Distance/meter	RSRP/dBm	SINR	5G Delay/ms	4G Delay/ms
Testpoint1	Outdoor	150	−75	36.8	37.5	44.30
Testpoint2	Indoor	300	−73	29	38.35	45.40
Testpoint3	Outdoor	350	−72	35	32.79	34.20
Testpoint4	Outdoor	450	−83	34	34.11	40

5 Conclusion

This paper puts forward a higher requirement for the load, collection, application and analysis of power data in the power network. The service requires high communication technology, but 5G can be realized, so it is necessary to verify the application ability of 5G in power system. In this paper, the integration and development of the 5G technology in the power system are analyzed systematically, and the application of the typical scenarios in power systems under the 5G technology is analyzed. Precise load control is selected for testing, and the test results show that the delay is reduced. The application analysis of 5G in power system supports the application of power Internet of things in the future. This research provides technical support and lays a solid foundation for the popularization and application of 5G technology.

References

1. Xia, X., Zhu, X., Mei, C., Li, W., Fang, H.: Research and practice on 5G slicing in power Internet of Things. Mob. Commun. **43**(1), 63–69 (2019)
2. You, X.H., Pan, Z.W., Gao, X.Q., et al.: The 5G mobile communication: the development trends and its emerging key techniques. Sci. Sinica **44**(5), 551 (2014)
3. Liu, G., Jiang, D.: 5G: vision and requirements for mobile communication system towards year 2020. Chin. J. Eng 1–8 (2016)
4. Ma, Z., Zhang, Z.Q., Ding, Z.G., et al.: Key techniques for 5G wireless communications: network architecture, physical layer, and MAC layer perspectives. Sci. China **58**(4), 41301–41321 (2015)
5. Qiu, J., Ding, G., Wu, Q., et al.: Hierarchical resource allocation framework for hyper-dense small cell networks. IEEE Access **4**(99), 8657–8669 (2017)
6. Yao, J.: Random access technology of electric dedicated LTE network based on power priority. Autom. Electr. Power Syst. **40**(10), 127–131 (2016)
7. Cao, J., Liu, J., Li, X.: A power wireless broadband technology scheme for smart power distribution and utilization network. Autom. Electr. Power Syst. **37**(11), 76–80 (2013)
8. Yu, J., Liu, J., Cai, S.: Performance simulation on TD-LTE electric power wireless private network. Guangdong Electr. Power **30**(1), 39–45 (2017)
9. Yu, J., Liu, J., Cai, S.: Research on LTE wireless network planning in electric power system. Electr. Power Inf. Commun. Technol. **10**(2), 7–11 (2016)
10. Wu, J., Zhang, Y., Zukerman, M., et al.: Energy-efficient base-stations sleep-mode techniques in green cellular networks: a survey. IEEE Commun. Surve. Tutor. **17**(2), 803–826 (2015)

11. Cao, Y., Wu, L., Li, Z., et al.: Application project and security research in wireless power emergency communication network. Telecommun. Netw. Technol. **7**(7), 86–90 (2012)
12. Carvalho, G.H.S., Woungang, I., Anpalagan, A., et al.: Qo S-aware energy-efficient joint radio resource management in multi-RAT heterogeneous networks. IEEE Trans. Veh. Technol. **65**(8), 6343–6365 (2016)
13. Li, W., Chen, B., Wu, Q., et al.: Applied research of TD-LTE power wireless broadband private network. Telecommun. Electr. Power Syst. **33**(241), 82–87 (2012)
14. Hamidi, M.M., Edmonson, W.W., Afghah, F.: A non-cooperative game theoretic approach for power allocation in intersatellite communication. In: 2017 IEEE International Conference on Wireless for Space and Extreme Environments (WISEE), pp. 13–18 (2017)

Research on Interference of LTE Wireless Network in Electric Power System

Shanyu Bi(✉), Junyao Zhang, Weiwei Kong, Long Liu, and Pengpeng Lv

NARI Group Corporation/State Grid Electric Power Research Institute, Nanjing, China
bishanyu@sgepri.sgcc.com.cn

Abstract. In view of the problems of electric power wireless communication and to serve the power consumer better, On the basis of the basic principles of TD-LTE's network structure, interference in the wireless network optimization of power system are analyzed in detail, and put forward a method of interference remediation for power system. Based on the actual cases, the feasibility of interference remediation is verified by means of data processing and comparative analysis. The interference analysis provides a technology support for electric power communication access network construction and late operation and lays the foundations for popularization and application of LTE power wireless private network, and the security of the power network is guaranteed.

Keywords: TD-LTE · Barrage jamming · Intermodulation interference · Network security · Spurious emission · Network optimization

1 Introduction

As one of the important technologies of power communication network, wireless communication has been widely used in national network marketing, transportation inspection and other systems [1, 2]. At present, many provincial and municipal districts have carried out pilot applications of power private networks based on different wireless communication technologies, including WiMAX, McWILL, 230 MHz LTE, and data transmission stations [3]. Business applications include distribution automation, power information collection, monitoring of transmission and transformation status and so on [4]. At the same time, some problems in power wireless network are gradually exposed. The interference problem will affect the performance of the network, therefore, it is urgent to study the interference analysis of wireless network optimization technology in power system. In this paper, the interference in the power system is introduced in detail, and different kinds of interference is analyzed in detail.

Q. Liu et al. (Eds.): CENet 2020, AISC 1274, pp. 1364–1373, 2021.
https://doi.org/10.1007/978-981-15-8462-6_156

2 Process for Interference Analysis

For wireless communication networks, the premise of ensuring service quality is to use a clean spectrum, this band is not used or interfered by other systems [5]. Otherwise, the performance of the interference system and the terminal will be affected negatively [6].

The screening of interference cells is to screen out the high interference cells that need to be dealt with according to certain conditions [7]. Interference analysis is the third step in the overall process of interference trouble shooting, and is a very important step. Interference analysis determines the next working direction of site interference detection [8]. Correct interference analysis is helpful to improve the overall efficiency of interference checking. The flow chart of interference analysis is shown in Fig. 1.

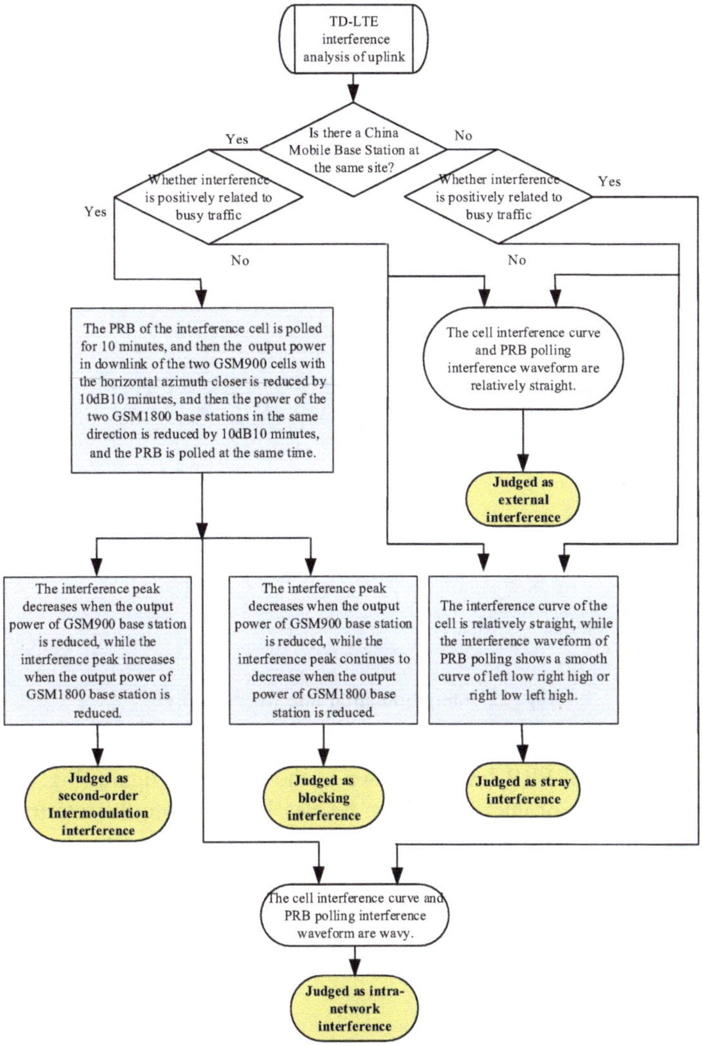

Fig. 1. Interference process of uplink in TD-LTE.

3 Analysis and Treatment of Intermodulation Interference

3.1 Analysis of Intermodulation Interference

Intermodulation interference is generally caused by the fact that the Intermodulation signal transmitted by nearby radio equipment falls into the frequency band received by the TD-LTE base station [9]. At present, the Intermodulation interference is mainly the second-order Intermodulation interference produced by China Mobile GSM900 system, which interferes with the F-band of TD-LTE. In addition, in other places, because the frequency band used in GSM1800 system reaches 1870 MHz, the third or fifth order Intermodulation interference will also fall in the F band of TD-LTE. The characteristics of interference are as follows.

The average interference level of the cell is closely related to the traffic of 2G. The busier the traffic of 2G is, the greater the interference of TD-LTE is. The smaller the isolation between the antenna of 2G cell and the antenna of TD-LTE cell, the more serious the interference is.

The interference curve of a cell is shown in the following figure (Fig. 2).

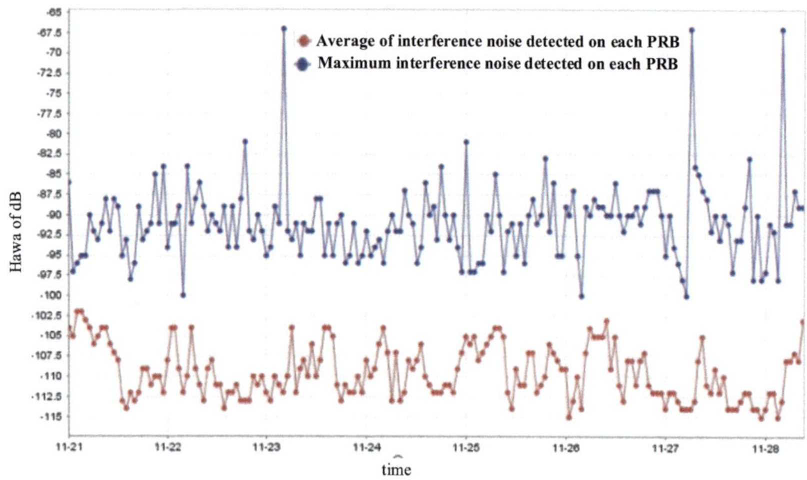

Fig. 2. Intermodulation interference curve.

The interference of PRB level is shown in the following figure (Fig. 3).

The characteristics of interference can be clearly seen from the interference of the cell. The interference in the early hours of the morning is the least, because traffic in the early hours of the morning is lower. According to the parameters of information, there are GSM900 base stations in the same sector of the site, and there are many frequency points. The frequency point of BCCH is 69, and the frequency point of TCH is 13/10. The calculated second order intermodulation and second harmonics are shown in the following Table 1.

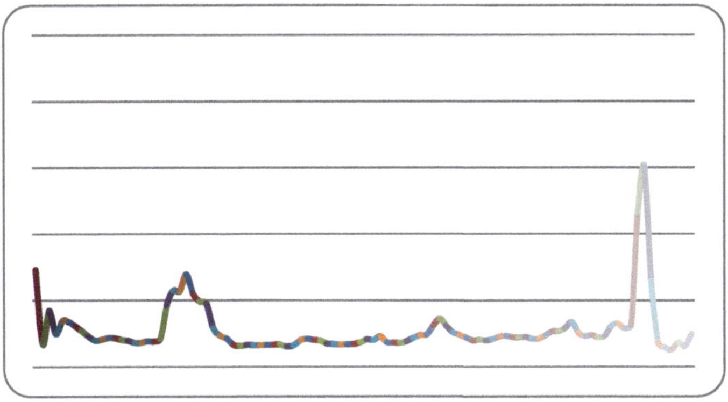

Fig. 3. Intermodulation interference in Class PRB.

Table 1. Two order intermodulation and two harmonic

Second order intermodulation & frequency point of second harmonic	Interference with the number of PRB in F band
Second-order Intermodulation generated by Frequency points 69 and 10	PRB25 ~ 27
Second-order Intermodulation generated by Frequency points 69 and 13	PRB28 ~ 31
Second Harmonic produced by Frequency Point 69	PRB91 ~ 93

Therefore, it can be judged from the above table that the sector of TD-LTE is interfered by the second order Intermodulation of GSM900 cell.

Because the antenna is about the same height, the site can be initially determined to be subjected to blocking interference from the GSM900/1800 base station. And because of the frequent occurrence of BCCH frequency points, the second harmonic interference is the most obvious, which is of course related to the low traffic volume in the cell.

3.2 Treatment of Intermodulation Interference

There are two ways to regulate the interference of Intermodulation. First, the antenna of the interference source base station and the antenna of the interfered base station are transformed from horizontal isolation to vertical isolation, the isolation degree can generally be increased by more than 10 dB. Second, the horizontal distance between the antenna of interference source base station and the antenna of the interfered base station is more than 2 m, or in the case of vertical isolation, the antenna of the interference source base station can be replaced with the antenna with higher second-order intermodulation suppression system. At present, the antenna with intermodulation index of 100 dBm and 43 dBm can be replaced.

4 Analysis and Treatment of Blocking Interference

4.1 Analysis of Blocking Interference

Blocking interference is generally caused by a strong signal transmitted by a nearby radio device received by the device of the TD-LTE. The blocking interference found at this stage is mainly caused by China Mobile GSM900 base station and closer base station. The characteristics of interference are as follows.

The average interference level of the cell is related to the traffic of the interference source, and the busier the traffic of the interference source is, the greater the interference is. The smaller the antenna isolation between the interference base station and the TD-LTE cell, the more serious the interference. Of course, it is impossible to know the isolation degree of the antenna between the systems only through the information of the working parameters, but the isolation degree of the antenna can be roughly understood from the antenna height and the horizontal azimuth angle of the antenna. The characteristic of PRB interference is that there is an obvious protruding before PRB10, and there is no obvious interference waveform behind the raised PRB (Fig. 4).

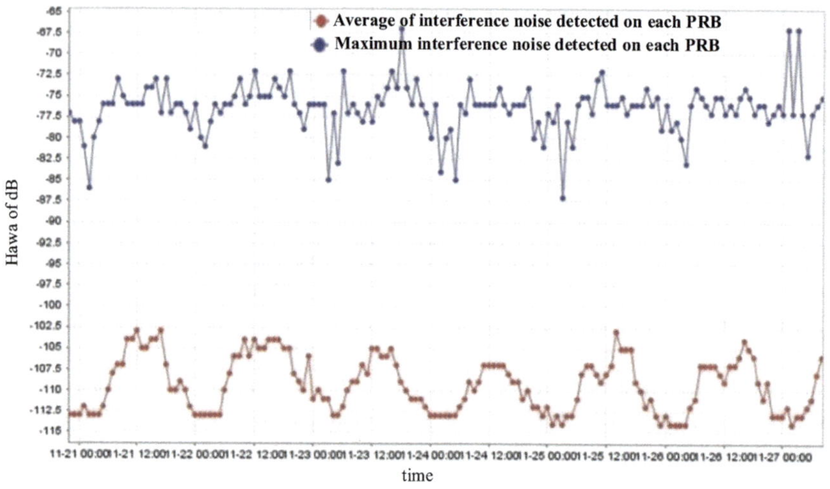

Fig. 4. Block interference curve.

The interference of PRB level is shown in the following figure (Fig. 5).

The interference characteristics of the cell can be clearly seen from the interference of the cell. The interference in the early hours of the morning is the least, because traffic in the early hours of the morning is lower. From the interference of PRB, we can see that there is a large uplink interference in the cell around PRB1. According to the information of working parameters, there is a GSM900/1800 base station in the same sector, and the antenna of the base station is the same as that of the LTE base station. Therefore, the site can be initially determined to be subjected to blocking interference from the GSM900 base station.

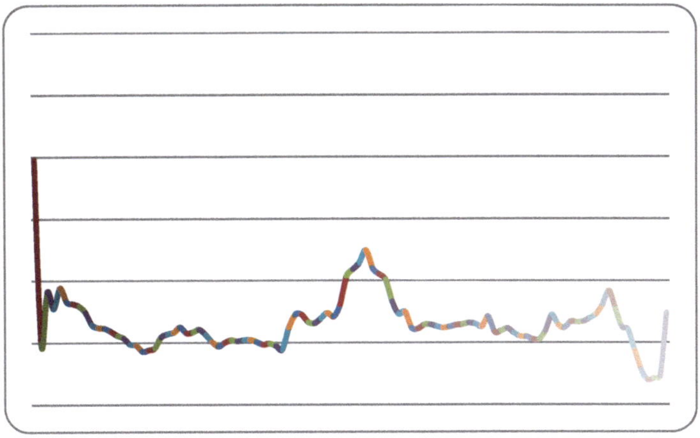

Fig. 5. Block interference in class PRB

4.2 Treatment of Blocking Interference

There are three ways to deal with blocking interference.

First, filters in the corresponding frequency band can be installed on interfered TD-LTE base stations. It should be noted that the filter installed in the RRU, common mode with the TD-SCDMA in the A band must be 2010–2025 MHz compatible.

Second, increase the isolation between the two systems, for example, raise the antenna height of the interference source base station or the interfered base station, change it from horizontal isolation to vertical isolation.

Third, replace the interfered RRU with the RRU of more anti-blocking. For example, replace it with RRU produced after 2012, and its anti-blocking ability is developed and produced according to the latest 3GPP specification, compared with the previous RRU, the resistance to stopper is obviously enhanced, so the number of blocking interference sites is small.

5 Analysis and Treatment of Stray Interference

5.1 Analysis of Stray Interference

Stray interference is the interference caused by stray transmission outside the transmission band of one system falling into the receiving band of another system. Stray interference directly affects the receiving sensitivity of the system. If the stray falls into the receiving frequency band of the system, the receiver system of the interfered system can not filter out the stray signal, therefore, the filter must be added to the input port of the transmitter to control the stray interference.

In the process of actual investigation, it is found that the stray interference mainly comes from three aspects. One is the stray interference from China Mobile GSM1800MHz base station, second, the 1G FDD-LTE base station of China Telecom at present, third, the TD-LTE base station in E band is vulnerable to the stray interference of WLAN AP.

The diagram of interference curve is as follows (Fig. 6).

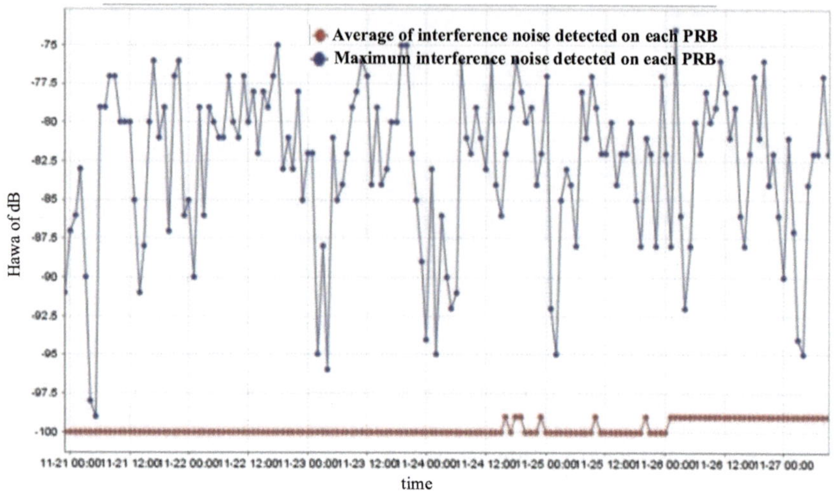

Fig. 6. Stray interference curve.

From the diagram, the interference curve of the cell affected by stray interference is relatively straight, and the fluctuation is generally about 1 dB.

The interference of PRB level is shown in the following figure.

From the interference curve, we can obviously see that the interference of the cell has obvious busy characteristics, and there are many interference peaks in the interference polling waveform of PRB.

5.2 Treatment of Stray Interference

There are two kinds of treatment methods for stray interference.

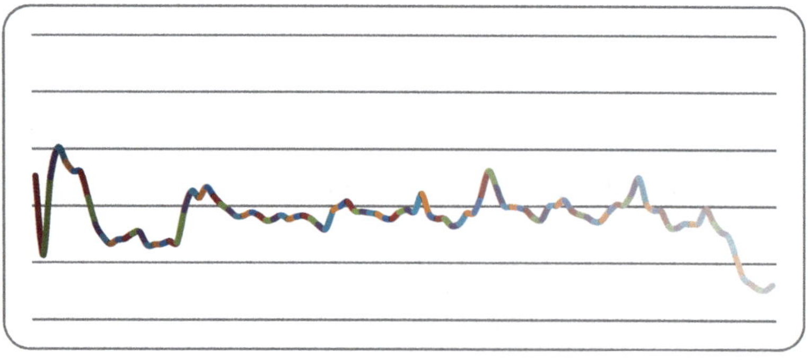

Fig. 7. Interference curve in class PRB

First, by increasing the isolation between the antenna of the base station and the base station of the interference source, the purpose of reducing the interference can be achieved, horizontal isolation can generally be changed to vertical isolation. The second is to reduce stray interference by adding bandpass filter to the base station of interference source (Fig. 7).

6 Interference Analysis of Power Wireless Private Network

There are two main ways to find interference. One is to statistics the uplink interference through the base station side, the other is to collect the downlink interference through the terminal side. Taking Y city as an example, the interference index in uplink of the last week is extracted, and the granularity is the one day. The uplink interference can only be extracted on the base station side, and the downlink interference can only be collected on the terminal side. Uplink interference is ideal when the low noise is below-120 dbm, and can be determined as interference when it is higher than-120 dbm. When the uplink interference is higher than-110dbm, it is considered that the interference begins to deteriorate, and when it is higher than-105 dbm, it is considered that the interference is more serious (Fig. 8 and Table 2).

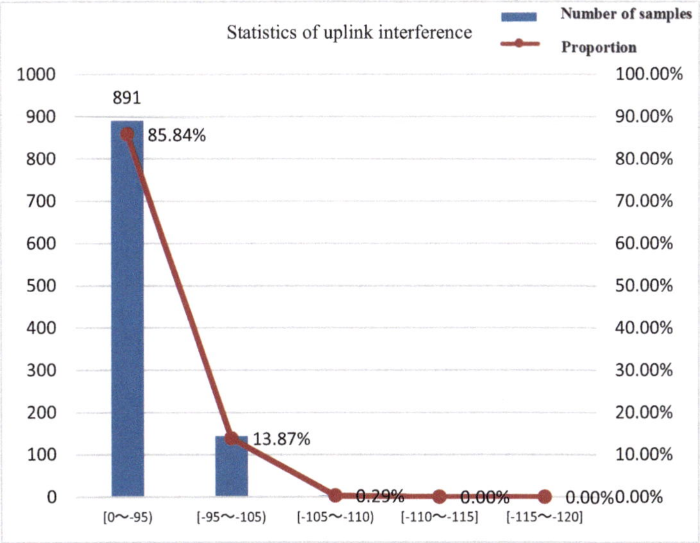

Fig. 8 Statistics of uplink interference

Table 2. Uplink interference statistics

RSSI	Number of samples	Proportion
[0 ~ −95)	891	85.84%
[−95 ~ −105)	144	13.87%
[−105 ~ −110)	3	0.29%
[−110 ~ −115]	0	0.00%
[−115 ~ −120]	0	0.00%

From the above statistics, we can see that, the proportion of interference above −95 dbm is very high, reaching 85.84%, it shows that the interference in the frequency band is very serious and needs to be checked.

7 Conclusion

This paper is based on the basic principle of TD-LTE network structure, the frequency sweep test, interference analysis, alarm processing and coverage optimization in wireless network optimization of power system are analyzed in detail. In this paper, an optimization method in TD-LTE wireless private network is proposed. The network optimization scheme provides technical support for the construction of power network, and lays a solid foundation for the popularization and application of power wireless private network.

References

1. Qiu, J., Ding, G., Wu, Q., et al.: Hierarchical resource allocation framework for hyper-dense small cell networks. IEEE Access **4**(99), 8657–8669 (2017)
2. Chen, X., Jiang, Y.Z., Wang, X., et al.: Research on the development of energy internet from the perspective of internet. Autom. Electr. Power Syst. **41**(9), 2–11 (2017)
3. Gao, F., Zeng, R., Qu, L., et al.: Research on identification of concept and characteristics of energy internet. Electr. Power 1–6 (2018)
4. Xu, X.L.: Time-sensitive network technology and its application in industrial network. Telecommun. Netw. Technol. **5**, 1–5 (2018)
5. Craciunas, S.S., Oliver, R.S.: Combined task and network-level scheduling for distributed time-triggered systems. Real-Time Syst. **52**(2), 161–200 (2016)

6. Sun, G.D.: Research and Simulation of Clock Synchronization and Scheduling Algorithm in Time-Sensitive Networks. Beijing University of Posts and Telecommunications (2018)
7. Bahnasse, A., Louhab, F.E., Oulahyane, H.A., et al.: Novel SDN architecture for smart MPLS traffic engineering-diff serv aware management. Fut. Gener. Comput. Syst. **11**(2), 212–219 (2018)
8. Li, B.L., Zhou, J., Ren, X.H.: Research on key issues of TD-LTE network optimization. Telecom Eng. Tech. Stand. **1**, 57–61 (2015)
9. Zheng, L., Tse, D.N.C.: Diversity and multiplexing: a fundamental tradeoff in multiple antenna channels. IEEE Trans. Inf. Theor. **49**, 1073–1096 (2003)

Research on Time-Sensitive Technology in Electric Power Communication Network

Yu-qing Yang(✉)

NARI Group Corporation/State Grid Electric Power Research Institute, Nanjing, China
yujia@sgepri.sgcc.com.cn

Abstract. There are many problems in distributed energy access, such as many access points, different access distances and miscellaneous access protocols, which results in low reliability of synchronous operation of the whole system and poor reliability of clock, which further affects the efficiency and quality of centralized regulation and control. Therefore, the technical scheme based on time sensitive network is put forward, and the Ethernet equipment based on time sensitive network technology is developed, all the sampling data in the distributed energy control task are under the same time section, and the time sequence of the control task can be measured and predicted. It provides important technical support for the establishment of efficient and low-cost source network load storage and control center. The results can be popularized and applied to the national source network storage and control center, and has a broad application prospect.

Keywords: Time-sensitive · Smart grid · Distributed energy · Switch · Protocol stack

1 Introduction

The continuous growth of world energy consumption and the increasing environmental pollution have promoted the wide application of distributed power supply [1–3]. The virtual power plant, the energy storage cloud and the micro-network provide a practical solution for distributed power operation management through advanced communication technology and control strategy, it can give full play to the benefits of distributed power supply, improve the reliability of power supply of users, and solve the problem of control and scheduling of distributed power supply [4, 5].

However, a large number of uncontrollable distributed power sources, such as wind power, photovoltaic and so on, are greatly affected by nature, and there is great randomness and volatility in the actual output, if not guided and controlled in time, the reliability of power supply will be reduced [6]. Secondly, due to the fact that the electric energy can not be stored in large quantities, the high permeability distributed power supply needs to be sent and used immediately after it is connected to the power grid, if the power load of the corresponding capacity is not enabled in time, the wind and light can only be abandoned, resulting in the waste of a large number of renewable energy. Finally, there are many problems in distributed energy access, such as many

Q. Liu et al. (Eds.): CENet 2020, AISC 1274, pp. 1374–1381, 2021.
https://doi.org/10.1007/978-981-15-8462-6_157

access points, different access distances, miscellaneous access protocols and so on, resulting in low reliability of synchronous operation of the whole system [7, 8].

Therefore, this paper puts forward a technical scheme based on time-sensitive network (TSN), and develops an Ethernet device based on TSN technology. All the sampled data in the distributed energy control task are under the same time section. Measurement and predictability of the time at which the control task is realized. The result can be applied to the national regulation and control center, and has a wide application prospect.

2 Time-Predictability Model

Through system analysis and functional requirement analysis of new distributed energy, energy storage, controllable load and other application scenarios in smart grid [9], it is clear how power information and communication technology can better support the regulation and effective utilization of new distributed energy.

2.1 Differentiated Requirements of Distributed Energy Service

1) Analysis of Technical Scene

Four kinds of distributed energy control technologies such as micro-network, active power distribution network, virtual power plant and energy storage cloud are used as research objects [10], and various technical scenes are analyzed, and the functional system of various technical scenes is described.

2) Analysis of Regulation and Control Process

Around the main goal of distributed energy operation and management, the regulation and control process of each function is established, the relationship between functional modules is clarified, and the degree of interdependence and influence between functions is described.

3) Analysis of interaction process of information system

According to the framework of functional system and control process, the application system is designed, and it provides conditions and reference for analysis of communication requirement.

4) Demand Analysis of Communication Service

On the basis of establishing the interaction process of information system, the categories of communication service between each system are further clarified, and the QoS template of communication service requirements is given in the form of format.

5) Time Constraint

Based on the QoS parameters of communication service, the parameters related to transmission time are abstracted, and the time constraint function is constructed. In the course of the follow-up study, the time constraint function is not only used as the time quality description method of multi-agent bidirectional real-time communication service in distributed energy control scenario, but also as the objective function of communication network planning, design and implementation.

2.2 Temporal Dynamic Characteristics of Distributed Energy

1) Selection of Scenes

There are many distributed energy scenarios, the first task is to select the most appropriate and representative scenarios as the research object [11]. The selection principles of scenarios include importance, criticality, universality, comprehensiveness and realizability.

2) Characteristics of Regulation and Control Process

The control process is a process in which multiple information entities interact, and the more appropriate feature description method is the time series model. The basic attributes of distributed energy management control can be effectively analyzed by using time series to describe the characteristics of the control process.

3) Network Event Mapping

At the physical level, time series describes the regulation and control process of distributed energy management control [12], however, at the information level, time series describes the transmission events of all kinds of messages in the communication network and the characteristics of their transmission events. Therefore, the time series describing the characteristics of the regulation process is mapped to the communication network, and the time series describing the characteristics of the network event is formed. In the process of mapping, the related theories of information system are the basis of research.

4) Correlation and Concurrency Analysis of Time Series

Through the feature mapping of the principle of information system, the characteristics of the control process are represented as the time series of network events. According to the distributed energy regulation process, this project studies the relationship between all kinds of time series. By analyzing the time-scale correlation of the time series, the event correlation of the control process can be obtained.

5) Analysis of the Influence of Task Execution Process

Based on the relevance and concurrency of events in the regulatory process, the influence of task execution process in the process of regulation and control is comprehensively studied.

3 Scheduling Control Strategy of Source Net Load

The core of the coordinated control of the source network load is that the resources on the three sides of the power grid, power supply and load can be effectively and fully utilized based on time prediction. On the net side, the control system can be managed in time. On the source side, distributed power supply can participate in time. On the load side, the timely response of flexible load is realized. The application of source network load control system is based on time sensitive technology to realize timely and accurate perception of docking element information. By collecting the real-time operation data of each load point in the whole network, the switching state information, the network topology information, the operating condition of the distributed power supply and the charge state information of the energy storage unit, etc., and according to the constraint conditions, the global optimal control strategy of active power and the global optimal control strategy of reactive power are obtained. The control implementation of the source network load is shown in the Fig. 1.

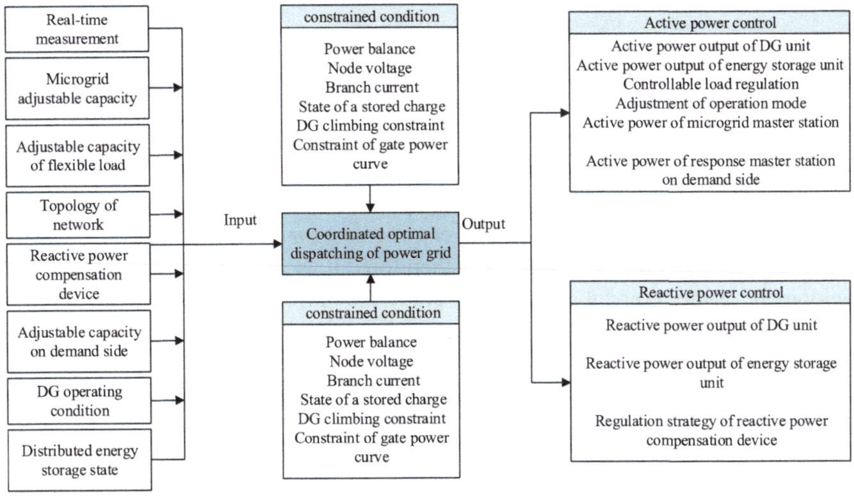

Fig. 1. General idea of coordinated optimal control of storage load in source network

4 Transformation Model of Unified Protocol for Mass Data

On the basis of intelligent metering technology such as remote reading of measured value, visual market price signal, monitoring load curve and so on. The source network load control center realizes intelligent energy management and intelligent distribution automation by aggregating demand-side resources and distributed energy. The source network load control center realizes intelligent energy management and intelligent distribution automation by aggregating demand-side resources and distributed energy. However, the implementation of these technologies requires massive data exchange, and the typical communication system structure is shown in the following Fig. 2.

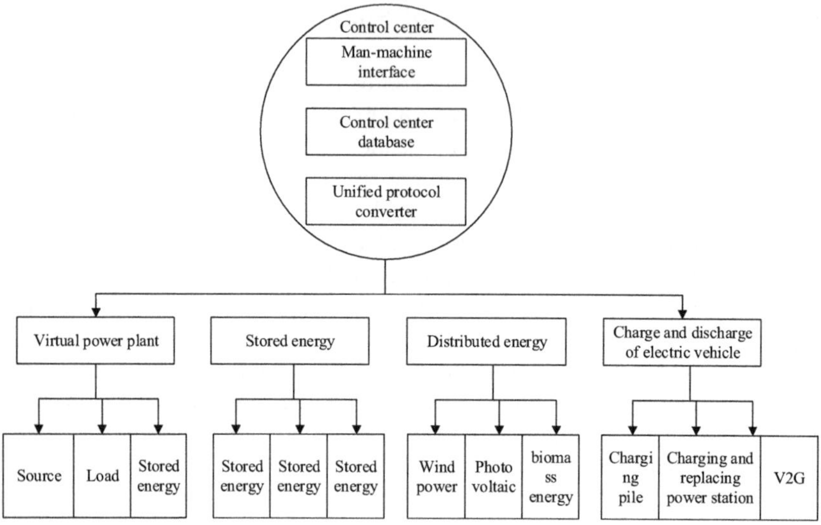

Fig. 2. Communication system structure

The core part of the source network load control center is the public information model database based on IEC61970. The database contains all the information needed for the system to run. The communication between the control center and the control center is to face a large number of different protocols such as IEC61850, the massive information conversion between these communication protocols and IEC61970 is realized by unified protocol converter, therefore, it is the key of the research and development of the protocol converter to study the unified protocol conversion model of mass data based on the network management technology.

The protocol stack of time-sensitive network is analyzed, and the key mechanism of TSN is studied, including priority time-aware scheduling technology and synchronous transmission cycle and time synchronization technology, to ensure clock synchronization of each device node connected to the network and to achieve a microsecond precision error. According to the protection bandwidth and frame preemption technology and traffic shaping technology, assign a specific time slot to time-sensitive critical data with higher priority through a time-aware scheduler, and it is required that all nodes in the network must give priority to ensuring the passage of important data frames within the specified time nodes. Through network management and configuration, the dynamic configuration of network parameters is realized. In order to meet the changes of equipment nodes and data requirements, a time deterministic industrial Ethernet data transmission technology system based on TSN is proposed, and a standard data communication platform for distributed energy load is established.

5 The Development of Data Communication Equipment for Time-Sensitive Network

5.1 Research on Key Technology of Prototype and Realization of Prototype

In addition to the hardware application of chips that support TSN functionality, TSN switches rely more on protocol support. Based on the current mature switch technology platform, the TSN function protocol is implemented through C language and runs in the form of independent process. Based on the analysis of existing platform architecture and TSN protocol function, the functional interface of TSN protocol is abstracted, and the adaptation processing is carried out in combination with the existing software architecture, so as to realize product development quickly.

Based on the function of switch hardware chip, through independent monitoring and management of the process, dynamically identify and extract logical link feature information according to configuration, analyze and process the link information, and record the abnormal situation and report the alarm.

5.2 Development of TSN Switch

On the basis of comprehensive analysis of technical maturity and economic benefit, the function and performance of the device are analyzed, and the software and hardware implementation technology of the key technology is studied, and carry on schematic diagram, PCB design, software code writing and debugging, complete the development of the device prototype, and carry on the complete test (Figs. 3 and 4).

1) Design of Hardware

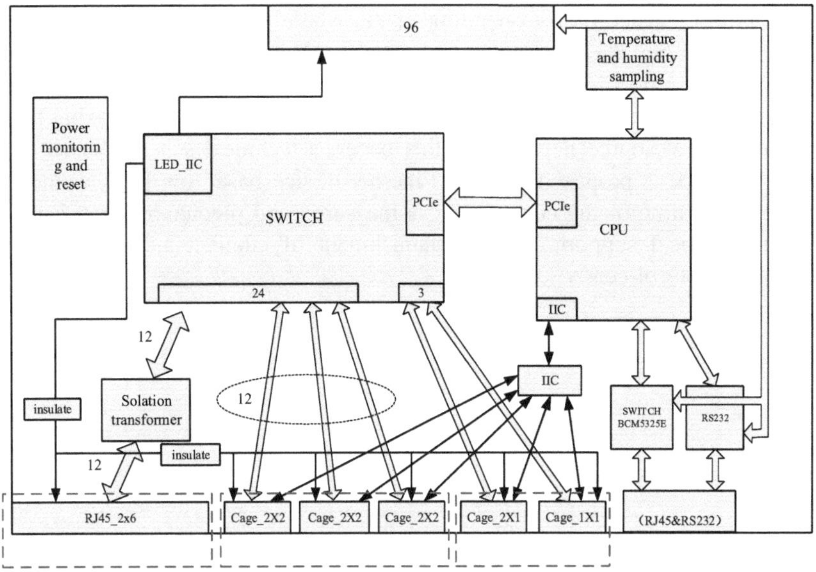

Fig. 3. Design scheme of hardware

2) Design of Software

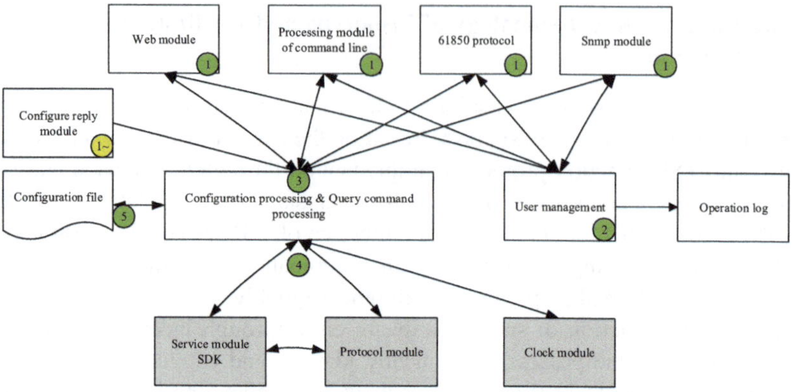

Fig. 4. Software design scheme

The system is based on Linux operating system, switch application function is implemented in a multi-process manner, the application function and the protocol module carry out data interaction with the driving interface of the chip through the adaptation management module.

6 Conclusion

There are many problems in distributed energy access, such as many access points, different access distances, miscellaneous access protocols and so on, which results in low reliability of synchronous operation of the whole system and poor reliability of clock, which further affects the efficiency and quality of centralized regulation and control. The operation level of time system in distributed energy and the synchronization of data acquisition have become the bottleneck of further improving the level of load control in source network. In this paper, a technical scheme based on time sensitive network is proposed, and an Ethernet device based on TSN technology is developed. The time of the control task is measured and predictable, and it provides important technical support for the establishment of efficient and low-cost source network load control center.

References

1. Xia, X., Zhu, X., Mei, C., Li, W., Fang, H.: Research and practice on 5G slicing in power Internet of Things. Mobile Commun. **43**(1), 63–69 (2019)
2. You, X.H., Pan, Z.W., Gao, X.Q., et al.: The 5G mobile communication: the development trends and its emerging key techniques. Sci. Sinica **44**(5), 551 (2014)
3. Liu, G., Jiang, D.: 5G: vision and requirements for mobile communication system towards year 2020. Chin. J. Eng. **2016**, 1–8 (2016)
4. Ma, Z., Zhang, Z.Q., Ding, Z.G., et al.: Key techniques for 5G wireless communications: network architecture, physical layer, and MAC layer perspectives. Sci. China **58**(4), 41301–41321 (2015)
5. Qiu, J., Ding, G., Wu, Q., et al.: Hierarchical resource allocation framework for hyper-dense small cell networks. IEEE Access **4**(99), 8657–8669 (2017)
6. Chen, X., Jiang, Y.Z., Wang, X., et al.: Research on the development of energy internet from the perspective of Internet. Autom. Electr. Power Syst. **41**(9), 2–11 (2017)
7. Gao, F., Zeng, R., Qu, L., et al.: Research on identification of concept and characteristics of energy Internet. Electr. Power, 1–6 (2018)
8. Xu, X.: Time-sensitive network technology and its application in industrial network. Telecommun. Netw. Technol. **5**, 1–5 (2018)
9. Craciunas, S.S., Oliver, R.S.: Combined task and network-level scheduling for distributed time-triggered systems. Real Time Syst. **52**(2), 161–200 (2016)
10. Sun, G.D.: Research and simulation of clock synchronization and scheduling algorithm in time-sensitive networks. Beijing University of Posts and Telecommunications (2018)
11. Nayak, N.G., Dürr, F., Rothermel, K.: Time-sensitive software-defined network (TSSDN) for real-time applications. In: The International Conference, Brest, France, pp. 193–202 (2016)
12. Bahnasse, A., Louhab, F.E., Oulahyane, H.A., et al.: Novel SDN architecture for smart MPLS traffic engineering-diff serv aware management. Future Gener. Comput. Syst. **11**(2), 212–219 (2018)

Research on High Reliability Planning Method in Electric Wireless Network

Zewei Tang[1(✉)] and Chengzhi Jiang[2]

[1] School of Management and Engineering, Nanjing University, Nanjing 210093, China
tzw@nju.edu.cn
[2] Nanjing Institute of Technology, Nanjing 211167, China

Abstract. Aimed at the problems existing in the planning of power wireless communication, in order to better serve smart grid and carry out the application of time-sharing long-term evolution technology in power communication, a highly reliable planning method for wireless private network is proposed. It improves the survival of the user's service by using overlapping coverage of base station signals, and the overlapping cells are merged to avoid the same frequency interference. By optimizing the deployment of the organization of the base station, in the case of the same number of base stations, the average inter station distance of overlapping coverage is increased, and the network reliability is improved on the premise of ensuring network performance. The planning scheme solves the problem of mutual interference caused by the same frequency overlapping coverage between base stations.

Keywords: TD-LTE · Network planning · High reliability · Interference · Electric wireless network

1 Introduction

To support the core tasks and objectives of the smart grid, the power system needs to build a terminal access network with wide access and flexibility [1]. Optical fiber access is the best access method for users, but there are some problems, such as difficult laying, long construction cycle, long fault recovery time, high investment cost and so on [2, 3]. The reliability and security of communication network are very important in power system. Power monitoring system should be isolated from data network and external information network of electric enterprises. Therefore, the wireless private network is the necessary path for the construction of the power system terminal access network [4, 5].

Wireless network planning is the foundation of wireless construction [6, 7]. The coverage of the power wireless private network shall meet the needs of automatic power distribution, power utilization information acquisition, power transmission and transformation status monitoring [8, 9], etc. In power application, if the base station fails due to sudden power outage, equipment failure and cable damage, shutdown maintenance must be arranged. And under normal circumstances, the traffic accessed by the base station will not be maintained, which will have a serious impact on the

Q. Liu et al. (Eds.): CENet 2020, AISC 1274, pp. 1382–1392, 2021.
https://doi.org/10.1007/978-981-15-8462-6_158

power system [10–13]. Therefore, the reliability of the service should be strengthened, and it must ensure that the key service is covered by two or more base stations [14]. One service is covered by multiple base stations, which leads to self-interference [15]. In the public network, interference avoidance is realized in the self-interference region through hetero-frequency network. However, due to the shortage of frequency resources in power system, hetero-frequency network may not be implemented, therefore, it is necessary to study the high reliability planning method under the same frequency.

2 Novel Networking Structure

Each base station in the power wireless private network includes a remote radio frequency unit (RRU), the baseband unit (BBU), connected to the optical fiber between the remote radio frequency unit and the baseband unit [16]. A baseband unit can usually support multiple remote radio frequency units. In this section, it first describes the structure of the traditional cellular network, and then introduces two improved structures with low self-interference and high reliability.

2.1 Traditional Cellular Networking

According to the general description of physical layer, the cell of the base station is recorded as $C_{N^{ID}}$, where N^{ID} is a physical cell identification, $N^{ID} = 3N^{ID(1)} + N^{ID(2)}$, Where $N^{ID(2)}$ is the primary synchronous sequence identifies (PSSI). ID represents the identification of the cell of a base station, and the regular value is $\{0,1,2\}$. The direction of the three cells of the base station refers to the direction angle of the three cells whose PSSI is 0, 1 and 2, the conventional cellular communication system is deployed as shown in Fig. 1.

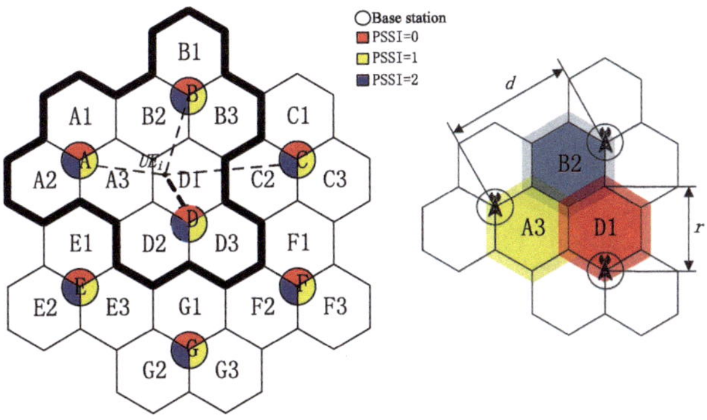

Fig. 1. Traditional cellular network structure.

The coverage radius of the base station is recorded as r, r is also the diameter of the cell. The distance between base stations is recorded as d, the area covered is recorded as S, the number of base stations in the area is recorded as N, Cell A1 represents the first cell of base station A.

The user i in the base station D of cellular networking is recorded as UE_i. The received signal-to-noise ratio is

$$SINR_i = \frac{\rho(D, i)}{\sum_{\substack{k \in \Psi \\ k \neq D}} \rho(k, i) + P_N} \tag{1}$$

2.2 Network Structure 1

In LTE network, by increasing the density of base station, the same terminal can receive more than one signal from the base station to avoid service failure caused by the failure of base station. In a single coverage network, the overlap of coverage can be increased by adding the same configuration of the base station. The existing base station is called the backbone base station. The new base station is called disaster recovery base station, represented by set Φ, the signal-to-noise ratio (SNR) at UE_i after adding the station is as follows.

$$SINR_i = \frac{\rho(D, i)}{\sum_{\substack{k \in \Psi \\ k \neq D}} \rho(k, i) + \sum_{j \in \Phi} \rho(j, i) + P_N} \tag{2}$$

In that case of not changing the original site, the site selection method for doubling redundant coverage is shown in Fig. 2.

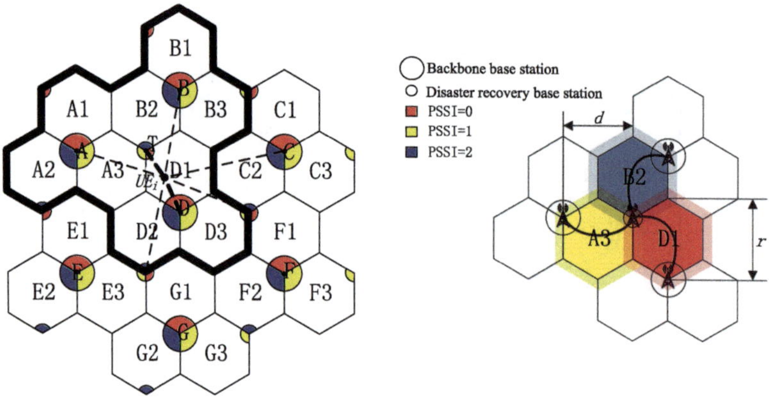

Fig. 2. Network structure 1.

Because the backbone base station D is merged with the adjacent disaster recovery base station T cell, the signal-to-noise ratio of UE_i is

$$SINR_i = \frac{\rho(D,i) + \rho(T,i)}{\sum_{\substack{k \in \Psi \\ k \neq D}} \rho(k,i) + \sum_{\substack{j \in \Phi \\ j \neq T}} \rho(j,i) + P_N} \tag{3}$$

Formula (3) is obviously larger than (2), so SINR is improved and the self-interference is reduced.

The calculation method of one station spacing in networking structure is shown as follows. The relationship of the distance between disaster recovery base station and backbone base station d_D and cell radius r is as follows.

$$d_D = \sqrt{3}r/2 \tag{4}$$

The number of network base stations is $N_D = 2N$, and the relationship between the coverage area S and the number of base stations N_D is.

$$S = 9\sqrt{3}N_D r^2/16 = 3\sqrt{3}N_D d_D^2/4 \tag{5}$$

The networking structure is obtained by adding N disaster recovery base stations in a cellular network composed of N backbone base stations,

The distance between stations in uniform distribution is shown in the following formula.

$$\bar{d} = 3r/2\sqrt{2} = \sqrt{3/2}d_D \tag{6}$$

Because the failure of base station is due to the power outage, the dug out optical cable is dug up and so on, increasing the distance between stations is beneficial to increase the independence between disaster tolerant base stations and backbone base stations.

2.3 Network Structure 2

The second method of networking structure is shown in Fig. 2.

For the user UE_i of the base station D in Fig. 3, the signal-to-noise ratio of UE_i is

$$SINR_i = \frac{\rho(D,i) + \rho(T,i)}{\sum_{\substack{k \in \Psi \\ k \neq D}} \rho(k,i) + \sum_{\substack{j \in \Phi \\ j \neq T}} \rho(j,i) + P_N} \tag{7}$$

Formula (7) it is obviously larger than (2), so the SINR is improved and the self-interference is reduced.

There are two station spacing d_1 and d_2 in the repeat area. Due to symmetry, the distance between the first backbone base station and the adjacent second disaster recovery base station, as well as the distance between the second backbone base station

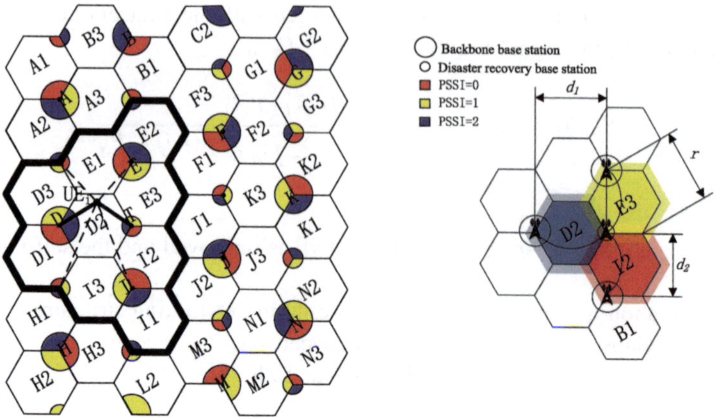

Fig. 3. Network structure 2.

and the adjacent first disaster recovery base station, are recorded as d_1, $d_1 = r$. The distance between the first backbone base station and the first disaster recovery base station, and the distance between the second backbone base station and the second disaster recovery base station are recorded as d_2, $d_2 = \sqrt{3}r/2$.

Since the number ratio of d_1 and d_2 is 1:2, the average station spacing is

$$\begin{aligned} d_{D'} &= \left(1+\sqrt{3}\right)r/3 \\ &= \left(1+\sqrt{3}\right)2d_D/3\sqrt{3} \\ &= \left(1+\sqrt{3}\right)2\sqrt{2}\bar{d}/9 \end{aligned} \tag{8}$$

Under the condition that the coverage overlap and anti-interference ability are the same as the structure-1, the average distance between the base stations is increased by 5.2%, and the distance between the stations is increased by 85.9%, which increases the reliability of disaster recovery.

3 The Process of Network Planning

In the above two networking structures, the former's networking scheme is to add disaster recovery base station to the single coverage cellular network, which is suitable for the reinforcement of the built network. The latter is a kind of networking structure with larger station spacing, which provides better disaster recovery effect in different places and needs to change the traditional cellular network structure, so it can be used in the new network. Two kinds of networking structures and the particularity of power service are integrated into the process of wireless network planning, as shown in Fig. 4.

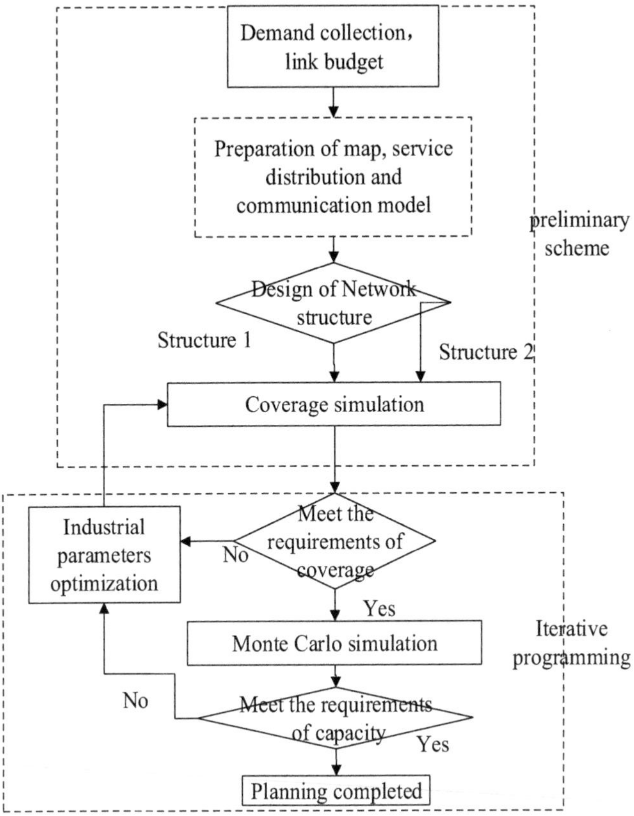

Fig. 4. Flow chart of network planning.

4 Simulation Analysis

The simulation adopts atoll, a professional planning simulation software for wireless communication. City Y is a small city, with an area of 132 km^2, the overall elevation is flat, the terrain of the city fluctuates within 3 m, and 2 base stations have been built. There is a need for full coverage of 58 km^2 of urban areas and to ensure the reliability of service in central and industrial areas. The installed terminal is a fixed CPE terminal with an edge rate of 256 kbps, the link budget results are as follows (Table 1):

Based on the results of the link budget, the cell radius of the service at the rate of 256 kbps is about 1.36 km, 5 new backbone base stations are built in the urban area, and 4 disaster recovery stations are deployed in the central urban and industrial areas. A total of 11 base stations were built. The base station distribution and coverage simulation during co-channel networking are shown in the following figure. The results of the simulation are shown in Table 2 (Fig. 5 and Table 3).

Table 1. Link budget.

CPE-256kbps link budget table	
Antenna gain of base station (dBi)	17.0 dBi
Processing gain of base station (dB)	0.0 dB
Loss of feeders and joints (dB)	0.5 dB
Number of RB in uplink	12
Thermal noise power spectral density (dBm/Hz)	−173.98
Demodulation bandwidth thermal noise power(dBm)	−110.6 dBm
Noise coefficient of receiver (dB)	3.0 dB
Demodulation bandwidth receiver noise power (dB)	−107.6 dB
Demodulation threshold (dB)	−5.8 dB
Sensitivity of receiver (dBm)	−113.4 dBm
Interference margin (dB)	3.0 dB
Body loss (dB)	0.0 dB
The minimum received level of the demodulated bandwidth (dBm)	−110.4 dBm
Transmission power of terminal (dBm)	23.0 dBm
Antenna gain of terminal (dBi)	0.0 dBi
Feeder and joint loss of terminal (dB)	0.0 dB
Requirements for coverage probability	95.0%
Fading variance of shadow	10.0 dB
The margin of shadow fading (dB)	11.7 dB
Penetration loss (dB)	0.0 dB
Antenna height of terminal (m)	1.5 m
Antenna height of base station (m)	45.0 m
Center frequency (MHz)	1795 MHz
Maximum allowable path loss (dB)	138.2 dB
The radius of coverage in uplink	1.36 km

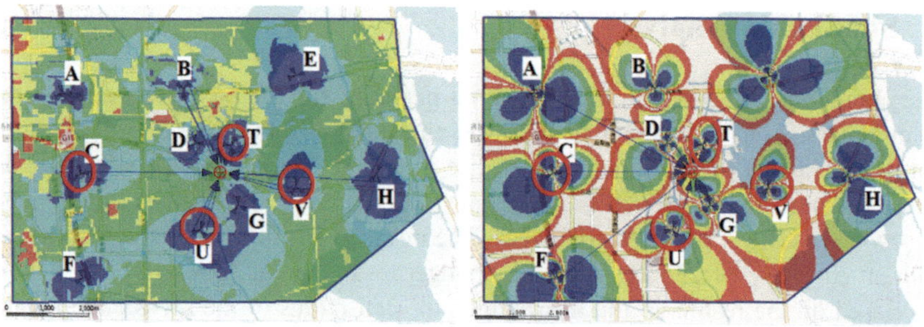

Fig. 5. Simulation (RSRP & SINR) of overlap coverage with co-channel interference.

Table 2. Results of the coverage simulation.

RSRP (dBm)	Area of coverage (km^2)	Fraction of coverage (%)	SINR (dB)	Area of coverage (km^2)	Fraction of coverage (%)
≥−75	10.3755	17.4	≥9	9.0549	15.2
≥−95	50.9343	85.6	≥3	24.5682	41.3
≥−115	59.0764	99.3	≥-3	47.1537	79.2

Table 3. Results of the UE_i single point simulation.

The strongest cell	The distance of propagation (m)	Path loss (dB)	SINR (dB)	RSRP (dBm)
D	961	139.1	−5.47	−102.1
T	670	139.6	–	−102.6
G	847	151.2	–	−116.6
U	1459	157.6	–	−123.0

The proportion of RSRP >−115 dBm is 99.3% and that of SINR >−3 dB is 79.2%. From the simulation results, it can be seen that when the disaster recovery base station and the backbone base station cover a terminal together, it will cause serious co-frequency self-interference.

The table shows the static simulation results of user UE_i, in which RSRP = −102.1, SINR = −5.47, indicates that the user in this location can not be connected to the network due to self-interference.

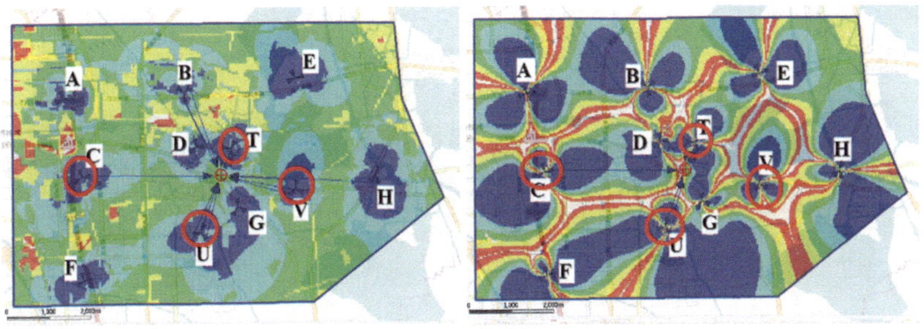

Fig. 6. Simulation (RSRP & SINR) of Overlap Coverage with SFN Cell merging.

In Fig. 6, the coverage simulation after the scheme proposed in this paper is applied. The simulation results are shown in Table 4, in which the coverage area of RSRP >−115 dBm is 99.3%, and that of SINR >−3 dB is 93.1%.

Table 4. Results of the coverage simulation

RSRP (dBm)	Area of coverage (km^2)	Fraction of coverage (%)	SINR (dB)	Area of coverage (km^2)	Fraction of coverage (%)
≥ -75	10.3635	17.4	≥ 9	26.7093	44.9
≥ -95	50.9598	85.6	≥ 3	40.4271	67.9
≥ -115	59.0841	99.3	≥ -3	55.4436	93.1

Table 5. Results of the UE_i single point simulation

The strongest cell	The distance of propagation (m)	Path loss (dB)	SINR (dB)	RSRP (dBm)
D/T merged cell	961 (D) 6709 (T)	139.1	1.51	−101.4
U/G merged cell	847 (G) 1459 (U)	151.2	–	−114.4
V	1718	168.1	–	−130.6

From the simulation results, it can be seen that due to the merger of interference cells, the problem of self-interference is significantly reduced, and the coverage of SINR is increased by 13.9%, which meets the coverage requirements of power wireless special network.

Table 5 shows the static simulation of user UE_i in the same location, where the RSRP = −101.4, SINR = 1.51, signal-to-noise ratio (SNR) is improved by about 7 dB.

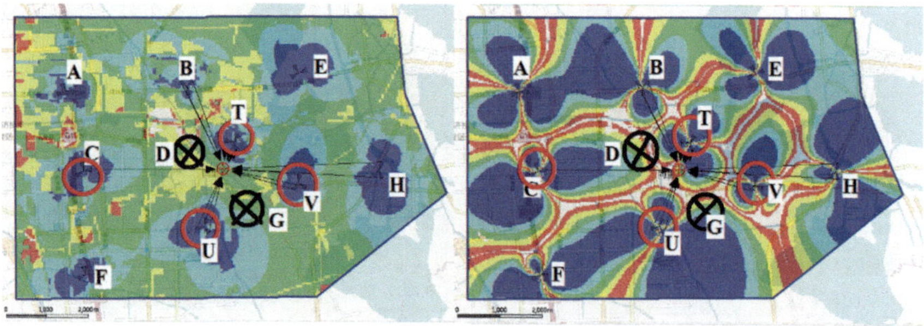

Fig. 7. Simulation (RSRP & SINR) of coverage with eNodeB failure.

Table 6. Results of the coverage simulation.

RSRP (dBm)	Area of coverage (km^2)	Fraction of coverage (%)	SINR (dB)	Area of coverage (km^2)	Fraction of coverage (%)
≥−75	8.0613	13.5	≥9	23.3397	39.2
≥−95	49.6422	83.4	≥3	42.2874	71
≥−115	58.8942	98.9	≥-3	56.4336	94.8

Table 7. Results of the UEi single point simulation.

The strongest cell	The distance of propagation (m)	Path loss (dB)	SINR (dB)	RSRP (dBm)
D/T merged cell	670 (T)	139.5	3.95	−101.99
U/G merged cell	1459 (U)	158	–	−120.49
V	1718	169.8	–	−133.37

Figure 7 shows the coverage simulation results of the proposed scheme in the event of base station failure in the key area. Suppose station D and G fail, and they are marked with ⊗ in the figure. Table 6 shows the result of simulation, in which the coverage area of RSRP > −115dbm is 98.9%, and that of SINR >−3 dB is 94.8%, the results are similar to those in Table 4 without failure. The results show that due to the existence of disaster tolerant base station, the field strength and signal quality in the coverage area do not change significantly, which meets the coverage requirements of the power network and achieves the purpose of disaster recovery.

Table 7 shows the static simulation results of user UE_i in the same location, where RSRP = −101.99, SINR = 3.95, it shows that users can connect to the network.

5 Conclusion

In this paper, a high reliability planning method for power wireless private networks with low self-interference is proposed based on SFN technology. It improves the survival of the user's service by using overlapping coverage of base station signals, and the overlapping cells are merged to avoid the same frequency interference. By optimizing the deployment of the organization of the base station, in the case of the same number of base stations, the average inter station distance of overlapping coverage is increased, and the network reliability is improved on the premise of ensuring network performance. The scheme solves the mutual interference of the same frequency between the base stations of the existing technology, which will lead to mutual interference, and it also solves problems of the frequency resource constraints which may be unable to implement.

References

1. Sun, S.W., Chen, Y.: Research on business-oriented covering LTE power wireless private network. Electr. Power Inf. Commun. Technol. **13**(4), 6–10 (2015)
2. Yao, J.M.: Random access technology of electric dedicated LTE network based on power priority. Autom. Electr. Power Syst. **40**(10), 127–131 (2016)
3. Cao, J.P., Liu, J.M., Li, X.Z.: A power wireless broadband technology scheme for smart power distribution and utilization network. Autom. Electr. Power Syst. **37**(11), 76–80 (2013)
4. Yu, J., Liu, J.S., Cai, S.L.: Performance simulation on TD-LTE electric power wireless private network. Guangdong Electr. Power **30**(1), 39–45 (2017)
5. Cao, J.P., Liu, J.M.: A two-stage double-threshold local spectrum sensing algorithm research for the power private communication network. Proc. CSEE **35**(10), 2471–2479 (2015)
6. Yan, J.M., Xu, J.B., Ni, M., et al.: Impact of communication system interruption on power system wide area protection and control system. Autom. Electr. Power Syst. **40**(5), 17–24 (2016)
7. Sun, W., Lu, W., Li, Q.Y., et al.: Reliability confidence interval prediction for communication link of wireless sensor network in smart grid. Autom. Electr. Power Syst. **41**(4), 29–34 (2017)
8. Han, J.H., Duan, Z.L.: Optimal base station selection based on dynamic programming for reprogramming in mine wireless network. J. Commun. **38**(3), 7–15 (2017)
9. Hamza, A.S., Khalifa, S.S., Hamza, H.S., et al.: A survey on inter-cell interference coordination techniques in OFDMA based cellular networks. IEEE Commun. Surv. Tutor. **15**(4), 1642–1670 (2013)
10. Bjomson, E., Zakhour, R., et al.: Cooperative multicell precoding: Rare region characterization and distributed strategies with instantaneous and statistical CSI. IEEE Trans. Signal Process. **58**(8), 4298–4310 (2010)
11. Garcia, V., Zhou, Y., Shi, J.: Coordinated multipoint transmission in dense cellular networks with user-centric adaptive clustering. IEEE Trans. Wirel. Commun. **13**(8), 4297–4308 (2014)
12. Shi, Y., Zhang, J., Letaief, K.B., et al.: Large-scale convex optimization for ultra-dense Cloud-RAN. IEEE Wirel. Commun. **22**(3), 84–91 (2015)
13. George, K.: Green network planning of single frequency networks. IEEE Trans. Broadcast. **56**(4), 541–550 (2010)
14. Chen, H., Zhong, X., Wang, J.: Optimal deployment of assisting cells for multimedia push in single frequency network. J. Commun. Netw. **19**(2), 114–123 (2017)
15. GPP. TS 36.201 LTE physical layer: General description. 3GPP Technical Specification (2017)
16. Li, C., Wen, C., Bin, W., Xin, Z.: System-level simulation methodology and platform for mobile cellular systems. IEEE Commun. Mag. **49**(7), 148–155 (2011)

Task Allocation Method for Power Internet of Things Based on Two-Point Cooperation

Jun Zhou[1], Yajun Shi[1], Qianjun Wang[2](✉), Zitong Ma[2], and Can Zhang[3]

[1] State Grid Jiangsu Electric Power Co., Ltd., Nanjing Gaochun District Power Supply Branch, Nanjing 210000, China
[2] Beijing University of Posts and Telecommunications, Beijing 100876, China
wqianjuna@sina.com
[3] State Grid Jiangsu Power Co., Ltd., Nanjing Power Supply Branch, Nanjing 210000, China

Abstract. Edge computing calculates computing tasks on computing resources close to the data source, which can effectively reduce the latency of the computing system, reduce data transmission bandwidth, and ease the pressure on the cloud computing center. However, with the explosive growth of business terminals, the capacity of a single edge node is limited, and it is difficult to meet all business requirements at the same time. Therefore, a task allocation method for power Internet of Things based on two-point cooperation is proposed. First, a task allocation model based on two-point cooperation was established to minimize the average task completion delay while meeting business resource requirements. Then the ECTA-MPSO (Edge Collaborative Task Allocation based on Modified Particle Swarm Optimization) algorithm is proposed, which solves the problem that the task allocation scheme easily falls into a local optimum. Simulation results show that the average delay decreases by 32.8% and 12% respectively compared with benchmark and GA algorithm.

Keywords: Task allocation · Power Internet of Things · Edge cooperation

1 Introduction

Electricity Internet of Things effectively integrates communication infrastructure resources and power system infrastructure resources, realizes the interconnection of everything in the power system, comprehensive state awareness and efficient information processing. The Power Internet of Things provides various services, such as video surveillance, induction detection, intelligent operation and maintenance, and equipment inspection. With the construction of the electric power Internet of Things and the continuous development of business, business terminals are showing explosive growth, and the amount of data processing is also increasing. As a result, under the cloud computing model, the pressure on the Internet of Things transmission increases, and the processing load on the cloud center increases.

Edge computing has been applied in the powerful IoT as an extended solution for cloud computing to solve the proplem. The powerful IoT architecture based on edge

Q. Liu et al. (Eds.): CENet 2020, AISC 1274, pp. 1393–1402, 2021.
https://doi.org/10.1007/978-981-15-8462-6_159

computing is shown in Deploy edge nodes with computing and storage functions at the edge of the network, such as wireless access points, routers, SDN switches and edge servers. Power business terminals are connected to edge nodes via wired, Wi-Fi, micro-power wireless, 4G/5G and low-power WAN. By placing computing tasks on edge nodes, they can reduce network transmission and cloud loads, reduce business processing and data transmission delays, and meet increasingly stringent business delay requirements.

However, compared to the cloud center, the resources of edge nodes (such as computing and storage) are still very limited. With the increase in the number of business applications that a single edge node needs to access and process, it is difficult for edge nodes with limited resources to simultaneously meet the different needs of multiple and distinct power IoT services. In order to solve the above problems, the task computing method of the edge node cooperation can be adopted. The work in [1, 2] shows that edge node cooperative computing is better than offloading tasks to a single edge node, while being able to balance the load between edge servers. Cao, X. et al. and Yang, C. S. et al. [3, 4] proposed cooperative computing of edge nodes, where the main concern is how to perform optimal task allocation and resource allocation. Reduce task completion delay or terminal energy consumption, and improve user perception performance. Sahni, Y. et al. [5] focuses on network traffic scheduling while considering task scheduling issues, and models joint issues to minimize overall completion latency.

Edge cooperation mainly studies how to allocate and complete task requests initiated by business terminals in the edge network to provide satisfactory services to business terminals. To address the problems of insufficient network edge resources, there are currently two types of edge collaboration research: 1) One is edge collaboration, with latency as the optimization goal. For service scenarios requiring high latency. 2) Edge collaboration technology with energy consumption as the optimization goal. It mainly considers that some edge devices have limited battery capacity due to their portability.

Due to its importance in IoT applications, latency has often become a key starting point for research on edge collaboration technologies. Kao, Y. H. et al. [6] studies the task assignment problem with task dependencies, minimizes service delay under resource cost constraints, and finally provides an approximate polynomial algorithm Hermes to solve this problem. problem. Aiming at the computing load problem in fog computing networks, Xiao, Y. et al. [7] proposed a fog node collaboration strategy for fog load problems, which translates the load problem into a workload Allocation issues, minimizing service delays. Given the power efficiency, a parallel optimization framework based on Alternating Direction Multiplier (ADMM) can be used to solve this problem and improve the network performance of fog computing.

We proposes a task allocation mechanism based on the cooperation of edge computing to minimize the average task completion delay under the constraints of business resource requirements. More specifically, the main contributions of this article are as follows:

1) In order to make better use of edge node resources and reduce business completion delay, a task allocation model based on two-point cooperation is proposed. In the model, business request and resource models and time delay models were established respectively, and then collaborative computing models were built, aiming to shorten the average task completion delay to the greatest extent under the condition of meeting business resource requirements.
2) The above model is transformed into an integer non-linear programming problem, and a ECTA-MPSO algorithm is proposed for solving.

2 Task Allocation Model

2.1 Business Request and Resource

Assume that the number of user terminals UE and edge nodes EN in the network are N, M. EN is an edge device with computing and storage capabilities. $\mathcal{N} = \{1, 2, \ldots, N\}$ and $\mathcal{M} = \{1, 2, \ldots, M\}$ are used to denote UE set and EN set, respectively.

The task request of the UE is cooperatively completed on the edge node. Considering that multiple EN cooperation will bring certain communication resource overhead, we adopts two-point cooperation method. The decision of the cooperative node is represented by R, where the task of UE i is completed by EN set $R_i = \{r_{i1}, r_{i2} \in \mathcal{M}\}$. Default r_{i1} is the access point of UE i.

The task set requested by UE i is $\mathcal{T}_i$, where the subtask j is represented by $w_{ij} = \left(c_{ij}, e_{ij}, d_{ij}, t_{ij}, \lambda_{ij}\right)$. c_{ij} represents the computing resource requirement, e_{ij} represents the storage resource requirement, d_{ij} represents the amount of input data, t_{ij} represents the calculation delay if the resource meets the situation, and represents the ratio of the amount of data in the calculation result to the amount of input data. There is no timing dependency between subtasks, and they can be completed independently. Because edge nodes have heterogeneous resources, we uses containers and virtualization technologies to support the implementation of EN resource allocation. The amount of resources required for subtasks is represented by the number of virtual resource units.

Assume that all UE requests are sent simultaneously at a certain time. Let the remaining resources of EN k be (C_k, E_k), where C_k and E_k respectively indicate the number of virtual computing units and the number of virtual storage units. The value of task allocation decision $X = \left\{x_{ijk}\right\}$ is specified as follow,

$$x_{ijk} = \left\{ \begin{matrix} 1 & \textit{If task j of UE i is allocated to EN k} \\ 0 & \textit{else} \end{matrix} \right\} \tag{1}$$

A subtask can only be executed by one EN in R_i, so it has the following constraints,

$$\sum\nolimits_{k \in \mathcal{M}} x_{ijk} = 1 \tag{2}$$

$$k \in R_i, \forall x_{ijk} = 1 \tag{3}$$

EN needs to meet the computing and storage resource requirements of the sub-tasks allocated to this node. $Sum_{R,X,k}^{comp}$ and $Sum_{R,X,k}^{sto}$ represent the amount of EN's computing and storage resources that should be satisfied under cooperative node decision X and task allocation decision R. Therefore, the constraints is as follow,

$$Sum_{R,X,k}^{comp} = \sum\nolimits_{i \in N} \sum\nolimits_{j \in T_i} c_{ij} x_{ijk} \le C_k, k \in \mathcal{M} \tag{4}$$

$$Sum_{R,X,k}^{sto} = \sum\nolimits_{i \in \mathcal{N}} \sum\nolimits_{j \in T_i} e_{ij} x_{ijk} \le E_k, k \in \mathcal{M} \tag{5}$$

2.2 Delay

Each UE will access the nearest edge node. Let access point of UE i be EN $u_i \in \mathcal{M}$, so the UE set associated with EN k can be represented as $\mathcal{N}_k = \{i : i \in \mathcal{N}, u_i \in \mathcal{M}\}$.

The bandwidth resource of EN k is B_k Hz. We defaults that the UEs associated with EN allocate the EN bandwidth resources evenly. The signal-to-noise ratio of UE i and EN u_i is $y_i = \frac{p_i h_{i,u_i}}{\sigma^2}$, where p_i represents UE i transmission power, h_{i,u_i} represents UE i and EN u_i channel gain, and σ^2 represents additive Gaussian white noise power. Therefore, when the UE i accesses the EN u_i, the uplink data transmission rate of the UE can be obtained from the Shannon theorem.

$$v_i = \frac{B_{u_i}}{|\mathcal{N} u_i|} log_2(1 + y_i) \tag{6}$$

Similar to X, because the downlink bandwidth of the UE i is much higher than the uplink bandwidth, and the amount of data of the calculation result is small, the downlink transmission delay of the calculation result from EN u_i, to the UE i is ignored.

Let the data transmission rate from EN k to EN k' be $v_{k,k'}$.

2.3 Collaborative Computing

The tasks of UE i are computed cooperatively by $R_i = \{r_{i1}, r_{i2} \in \mathcal{M}\}$. r_{i1} and r_{i2} each calculate a part of them $r_{i1} = u_i, r_{i1} \ne u_i$. The calculation results are aggregated at one of the nodes and finally returned to the access point EN k of UE i. The cooperation methods are shown in Fig. 1, in which the access point completes the sub-task T1 and T2, and another neighbor EN completes the sub-task T3, and the calculated results finally converge to the access point.

$T_{i,r_{i1},r_{i2}}^{finish}$ represents the cooperative task completion delay of UE i, which include the communication delay of task input data sent from EN to EN, the computing delay on EN r_{i1} and r_{i2}, and the delay of results merge and returned to EN u_i.

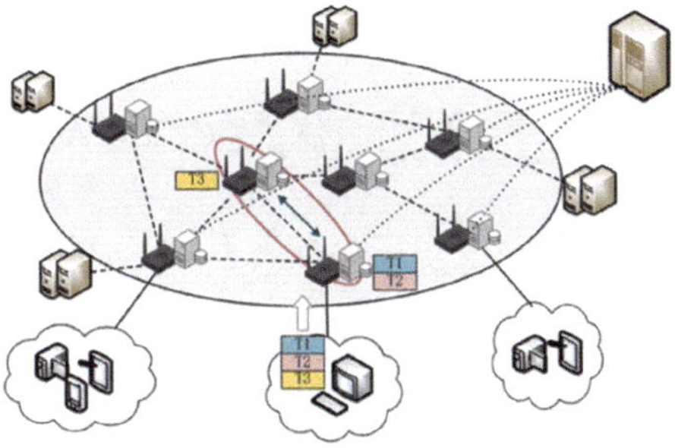

Fig. 1. Cooperation method.

First, the EN u_i completes some sub-tasks of the UE, and another part sends them to the neighbor EN r_{i2} for computing. After the sub-task computing on the EN r_{i2} is completed, the results are returned to the EN u_i for merging.

The computing delays of the sub-tasks on EN A and EN B are $T_{i,u_i}^{comp} = \sum_{j \in T_i, x_{ijk}=1} t_{ij}$, $T_{i,r_{i2}}^{comp} = \sum_{j \in T_i, x_{ijk}=1} t_{ij}$.

The data transmission delay from EN u_i to EN r_{i2} is $T_{u_i,r_{i2}}^{comm} = \sum_{j \in T_i, x_{ijr_{i2}}=1} \frac{d_{ij}}{V_{u_i,r_{i2}}}$, and the transmission delay of the computing result back to u_i is $T_{r_{i2},u_i}^{comm} = \sum_{j \in T_i, x_{ijr_{i2}}=1} \frac{\lambda_{ij} d_{ij}}{V_{r_{i2},u_i}}$, so the UE task cooperation completion delay is

$$T_{i,r_{i1},r_{i2}}^{finish} = \max\left\{T_{i,u_i}^{comp}, T_{u_i,r_{i2}}^{comm} + T_{i,r_{i2}}^{comm} + T_{r_{i2},u_i}^{comm}\right\}) \tag{7}$$

2.4 Problem

The total delay from the time the UE i sends the task request to the time it receives the calculation result is

$$T_i = T_{i,u_i}^{up} + T_{i,r_{i1},r_{i2}}^{finish} + T_{i,u_i}^{down} = T_{i,u_i}^{up} + T_{i,r_{i1},r_{i2}}^{finish} \tag{8}$$

T_{i,u_i}^{down} represents the delay of returning the calculation result from EN to the UE. As described above, the delay is small and ignored, and the computing delay of execution decision and the transmission delay of returning decision data are ignored. Therefore, the average completion delay of tasks for all UEs is

$$\overline{T} = \frac{1}{N} \sum_{i \in \mathcal{N}} T_i \tag{9}$$

This paper describes the task allocation problem as minimizing the average completion delay of tasks. The task assignment decision of the UE is represented by $W = (R, X)$, $R = \{R_i, i \in \mathcal{N}\}$. Therefore, the task assignment problem in this article is described as follows,

$$P1: Minimize\,\overline{T} \tag{10}$$

$$s.t \sum\nolimits_{i \in N} \sum\nolimits_{j \in \mathcal{T}_i} c_{ij} x_{ijk} \leq C_k, k \in \mathcal{M} \tag{C1}$$

$$\sum\nolimits_{i \in \mathcal{N}} \sum\nolimits_{j \in \mathcal{T}_i} e_{ij} x_{ijk} \leq E_k, k \in \mathcal{M} \tag{C2}$$

$$R_i = \{r_{i1}, r_{i2} \in \mathcal{M}\}, i \in \mathcal{N} \tag{C3}$$

$$k \in R_i, \forall x_{ijk} = 1, i \in \mathcal{N}, j \in \mathcal{T}_i, \tag{C4}$$

$$\sum\nolimits_{k \in \mathcal{M}} x_{ijk} = 1, i \in \mathcal{N}, j \in \mathcal{T}_i \tag{C5}$$

$$x_{ijk} \in \{0, 1\}, i \in \mathcal{N}, j \in \mathcal{T}_i, k \in \mathcal{M} \tag{C6}$$

3 ECTA-MPSO Algorithm

3.1 Problem Encoding

Assume the particle population size is A, the lth particle is represented as D dimensional position vector, denoted as $S^l = (s_1^l, s_2^l, \ldots, s_D^l)$, the optimal position of the lth particle so far is the individual extreme value, denoted as $PS_{best}^l = (p_1^l, p_1^l, \ldots, p_D^l)$, and the optimal position of the whole particle swarm so far is the global optimal value $GS_{best} = (g_1, g_1, \ldots, g_D)$. The velocity of the particle is expressed as $V^l = (v_1^l, v_2^l, \ldots, v_D^l)$.

We adopted the discrete coding strategy to generate candidate particles, and each particle represented the cooperation scheme and task assignment scheme of the edge nodes. The particle l after the t iteration was expressed as

$$S^l(t) = (R^l(t), Z^l(t)) = \left(r_1^l(t), r_2^l(t), \ldots, r_N^l(t), z_{11}^l(t), z_{12}^l(t), \ldots, z_{N|\mathcal{T}_N|}^l(t)\right) \tag{11}$$

Where $r_i^l(t) \in \mathcal{M}$ represents the edge nodes selected by the ith terminal node to computing with u_i, and $z_{ij}^l(\text{t}) \in \{1, 2\}$ represents the subtask j of the UE i executed on $r_i^l(\text{t})$ or u_i. For the sake of convenience, let's unify the symbols

$$S^l(t) = (s_1^l(t), s_2^l(t), \ldots, s_D^l(t)) \tag{12}$$

where $D = N + \sum_{i=1}^{N} |\mathcal{T}_i|$.

3.2 Fitness Function

The fitness function is used to evaluate particle quality. There are constraints in the problem description, but evolutionary algorithm is an unconstrained search technology, and certain constraint processing techniques must be combined in the process of solving the constraint optimization problem. In the problem coding, the constraints C3–C6 have been satisfied. For inequality constraints C1 and C2, this paper adopts constraint violation degree. The degree of constraint violation of a particle is defined as

$$G(S) = \sum\nolimits_{k \in \mathcal{M}} \max\left\{Sum_{S,k}^{comp} - C_k, 0\right\} + \max\left\{Sum_{S,k}^{sto} - C_k, 0\right\} \quad (13)$$

$G(S)$ represents the sum of constraint violations of all EN's computing and storage resources. When the particle is within the feasible region, $G(S) = 0$, that is, all the satisfied particles constitute the feasible region of the search space. When x is not in the feasible region, $G(S) > 0$.

Define the particle fitness function as

$$Fitness(S) = \frac{1}{\overline{T}(S) + G(S)} \quad (14)$$

3.3 Update Strategy

$$v_i^l(t+1) = v_i^l(t) + c_1 r_1 \left(p_i^l(t) - s_i^l(t)\right) + c_2 r_2 \left(G_i^l(t) - s_i^l(t)\right) \quad (15)$$

$$S^l(t+1) = S^l(t) \oplus v_i^l(t+1) \quad (16)$$

Because task assignment is a discrete problem, after updating particles according to the above formula, the position values of the examples will become unnatural values. Therefore, these examples need to be discretized and out of bounds. This article uses the absolute value, rounding, and remainder method.

$$R^l(t) = \{ \begin{matrix} [|R^l(t)|] & 0 \le |R^l(t)| \le N \\ [|R^l(t)|]\%N & else \end{matrix} \quad (17)$$

$$Z^l(t) = \{ \begin{matrix} [|Z^l(t)|] & 0 \le |Z^l(t)| \le 1 \\ [|Z^l(t)|]\%2 & else \end{matrix} \quad (18)$$

4 Experimental Results and Analysis

In this section, the proposed algorithm is simulated and its performance is verified. At the same time, we compare the two allocation methods GA and benchmark with the proposed method. GA uses genetic algorithm to solve the above P1 to get the task allocation scheme. In the Benchmark method the tasks sent by the UE are only completed by the access point independently. And there is no cooperation between the edge nodes.

4.1 Simulation Settings

It is assumed that the simulation environment is an area with a radius of 1 km and 1 km. The area contains 10 EN, 50 UEs, and the positions of EN and UE are randomly generated in the area. The CPU frequency (GHz) and storage space size (GB) of each edge node obey normal distribution. Set a virtual computing resource unit to be 0.1 GHz and a virtual storage unit to be 0.5 GB. The calculation of sub-tasks, the number of storage virtual resource unit requirements, and the calculation delay follow a Poisson distribution. The mean values of Poisson distribution are $\lambda_1 = 8$, $\lambda_2 = 10$, $\lambda_3 = 40$ (Table 1).

Table 1. Parameter list.

Parameter	Value
Band with B_k of EN k	20 MHz
transmission power p_i of UE i	[20,30] dBm
Gaussian white noise power σ^2	2×10^{-13} W
Subtask input data size d_{ij}	[0.05, 1] MB
The data size ratio of the output and input λ_{ij}	[0.01,0.1]
Population size Y	30
Constraint violation factor γ	0.95

4.2 Simulation Results

In order to verify the superiority of cooperative computing in ECTA-MPSO algorithm, the benchmark algorithm and GA algorithm are selected for comparison, In benchmark method the tasks sent by the UE are only completed by the access point independently. And there is no cooperation between the edge nodes. GA is a method to search for the optimal solution by simulating the natural evolution process, and iteratively update genes through a cross mutation strategy.

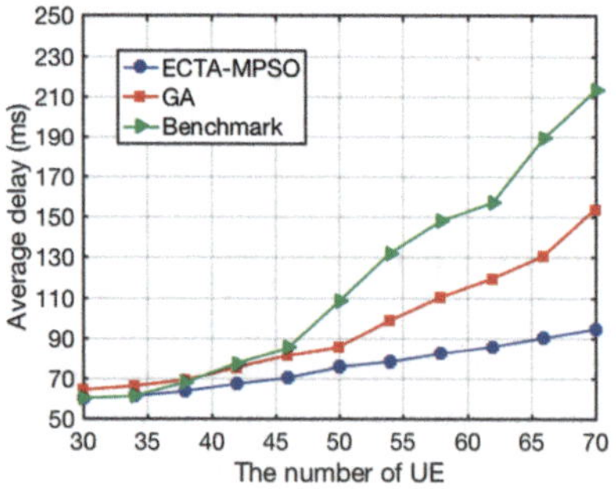

Fig. 2. Average delay of task allocation method with different number of UE.

Figure 2 shows the average delay comparison of three task allocation mechanisms, ECTA-MPSO, Benchmark and GA under different UE's numbers. With the increase of the number of UE, the average delay increases gradually, and ECTA-MPSO has the best effect. In benchmark, the increase of UE makes it difficult for some EN to meet the business demand at the same times, so tasks begin to queue for execution, which has a great impact on the completion delay. GA has no memory, and the effect is worse than ECTA-MPSO.

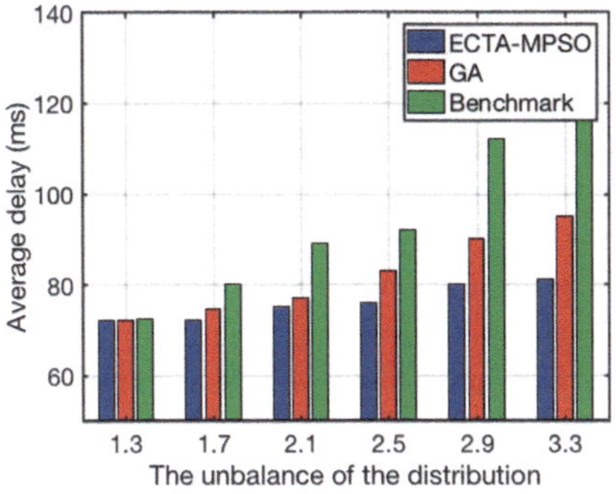

Fig. 3. Average delay of task allocation whit different UE distribution unbalance degree.

Figure 3 shows the average delay of the three task allocation algorithm under different UE distribution unbalance degree. The distribution unbalance degree is represented, $\sigma = \sqrt{\frac{1}{N}\sum_{k=1}^{N}\left(|\mathcal{N}_k| - \mu\right)^2}$, $\mu = \frac{\sum_{k=1}^{N}|\mathcal{N}_k|}{N}$. It can be seen that the average delay of the three task allocation algorithm increase with the σ. The average delay of ECTA- MPSO is lower than that of GA and benchmark.

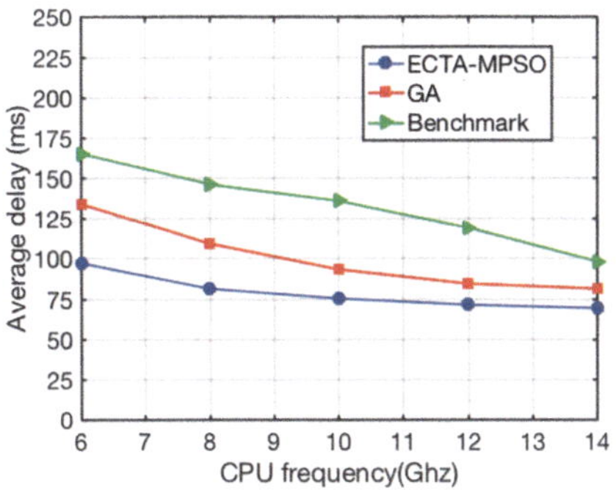

Fig. 4. Average delay of task allocation method with different CPU frequencies.

Figure 4 respectively show the average delay of the three task allocation mechanisms under different CPU frequency. It can be seen that the average delay of the three algorithm gradually decreases with the increase of resources. The average delay of ECTA-MPSO has been kept to the minimum, because when EN load is too high and the remaining resources are unable to meet the business demand, some subtasks are diverted to other idle EN to meet the business resource demand through cooperation. In Benchmark, tasks need to be queued due to limited EN resources, adding an additional queuing delay. In GA, task allocation scheme is not as good as ECTA-MPSO.

5 Conclusion

The resources of edge nodes in the power Internet of Things scene are limited, so it is necessary to meet all the increasing business processing demands in a timely through the cooperation of edge nodes. For the issue we propose a task allocation method for power Internet of Things based on two-point cooperation. First, a task allocation model is established to minimize the average task completion delay while meeting business resource requirements. Then the ECTA-MPSO algorithm is proposed, which solves the problem that the task allocation scheme easily falls into a local optimum.

Acknowledgment. This work is supported by State Grid Jiangsu Electric Power Co., Ltd. Science and Technology Project "Research and Application of Online Intelligent Monitoring and Security Protection Technology for Distribution Equipment IoT Terminals Based on Edge Computing" (J2019065).

References

1. Tran, T.X., Hajisami, A., Pandey, P., et al.: Collaborative mobile edge computing in 5G networks: new paradigms, scenarios, and challenges. IEEE Commun. Mag. **55**(4), 54–61 (2017)
2. Jia, M., Cao, J., Liang, W.: Optimal cloudlet placement and user to cloudlet allocation in wireless metropolitan area networks. IEEE Trans. Cloud Comput. **5**(4), 725–737 (2015)
3. Cao, X., Wang, F., Xu, J., et al.: Joint computation and communication cooperation for energy-efficient mobile edge computing. IEEE Internet Things J. **6**(3), 4188–4200 (2019)
4. Yang, C.S., Pedarsani, R., Avestimehr, A.S.: Communication-Aware Scheduling of Serial Tasks for Dispersed Computing (2018)
5. Sahni, Y., Cao, J., Yang, L.: Data-aware task allocation for achieving low latency in collaborative edge computing. IEEE Internet Things J. **6**, 3512–3524 (2018)
6. Kao, Y.H., Krishnamachari, B., Ra, M.R., et al.: Hermes: latency optimal task assignment for resource-constrained mobile computing. In: IEEE Conference on Computer Communications. IEEE (2015)
7. Xiao, Y., Krunz, M.: QoE and power efficiency tradeoff for fog computing networks with fog node cooperation. In: IEEE INFOCOM 2017-IEEE Conference on Computer Communications, pp. 1–9. IEEE (2017)

Heterogeneous Sharing Resource Allocation Algorithm in Power Grid Internet of Things Based on Cloud-Edge Collaboration

Bingbing Chen[1], Guanru Wu[2], Qinghang Zhang[3](✉), and Xin Tao[2]

[1] State Grid Jiangsu Power Co., Ltd., Nanjing Power Supply Branch, Nanjing 210000, China

[2] State Grid Jiangsu Electric Power Co., Ltd., Nanjing Gaochun District Power Supply Branch, Nanjing 210000, China

[3] Beijing University of Posts and Telecommunications, Beijing 100876, China
zhangqh@bupt.edu.cn

Abstract. The power Internet of things (IOT), which integrates edge computing, has become a research hotspot because of its characteristics of edge intelligence, wide interconnection and real-time decision-making. However, with the construction and business application development of the power Internet of things, problems such as insufficient processing capacity of edge devices and insufficient utilization of distributed edge resources are increasingly prominent, affecting the high-quality service provision of the power Internet of things. To this end, this paper uses cloud-side collaboration and service provision ideas to propose a cloud-based collaboration-based power IoT heterogeneous shared resource allocation method, through resource optimization to improve the quality of power IoT service provision. From the perspective of resource service matching and integration, the algorithm distinguishes service types, and allocates different cloud-based collaborative resources according to different service provision types to meet the corresponding real-time processing needs of services and ensure optimal allocation of edge resources. Simulation experiments show that the algorithm can give priority to ensuring the efficient use of edge resources on the basis of ensuring that the resource requirements of the business request are fully met, and achieve the purpose of improving the quality of power Internet of Things service provision.

Keywords: Power Internet of Things · Cloud-edge collaboration · Resource allocation · Edge computing · Service provision

1 Introduction

With its ability to connect a large number of intelligent devices across a wide geographical area, the IOT has become a key infrastructure of energy Internet carrying multiple advanced applications [1, 2]. With the gradual development of power IoT business applications, the cloud computing model is facing problems that it is difficult to effectively support some requirements new power Internet of Things applications with low latency and on-site control. Edge computing, with its characteristics of data

Q. Liu et al. (Eds.): CENet 2020, AISC 1274, pp. 1403–1413, 2021.
https://doi.org/10.1007/978-981-15-8462-6_160

processing and analysis and calculation close to the terminal [3, 4], shortens the response time of business applications, reduces network load, and ensures the personalized terminal security, effectively solves the above-mentioned problems in cloud computing, and becomes a key technology for the development of the Internet of Things.

However, with the comprehensive construction of power IoT integrated with edge computing and the deepening development and update of business applications, problems such as insufficient processing capacity of edge devices and insufficient utilization of distributed edge resources under multi-service access have become increasingly prominent, affecting the high power IoT Quality service provided. To this end, it is necessary to further optimize the resource allocation of the electric power Internet of Things, consider the spatial and temporal distribution of business and the characteristics of business access, accurately match business needs and resource allocation, optimize the use of network resources and the carrying capacity of services, improve resource utilization and reduce margin Insufficient side resources, unbalanced resource distribution, and insufficient utilization impact the business, thereby improving the overall performance of the business.

At present, resource allocation based on edge computing is becoming a research hotspot in academia. Reference [5] presents a strategy for the allocation of computational resources using deep reinforcement learning in mobile edge computing networks. Reference [6] presents a multi-user Wi-Fi-based MEC architecture. Reference [7] proposes a novel decentralized resource management technique and accompanying technical framework for the deployment of latency-sensitive IoT applications on edge devices. Reference [8] presents the offloading system model and presents an innovative architecture, called “MVR”.

To a certain extent, the above research has solved the current problem of insufficient computing power and other problems faced by the Internet of Things from the perspective of resource allocation, but the impact of the differentiation of service types in resource allocation is not considered, and the collaborative use of edge resources is not enough. Therefore, based on the idea of cloud-edge collaboration and service provision, this paper proposes a heterogeneous sharing resource allocation algorithm in power grid Internet of things based on cloud-edge collaboration, the algorithm distinguishes service types, allocates different cloud-based collaborative resources according to different service provision types, meets the corresponding real-time processing needs of services, and ensures optimal allocation of edge resources. The characteristics of this algorithm to differentiate service types make it more flexible and extensible in the process of resource allocation decision, and provide an effective solution for dynamic resource allocation.

2 System Framework

The framework of heterogeneous shared resources allocation based on cloud edge collaboration proposed in this paper is shown in Fig. 1, which mainly includes three layers: cloud computing layer, edge layer and terminal layer.

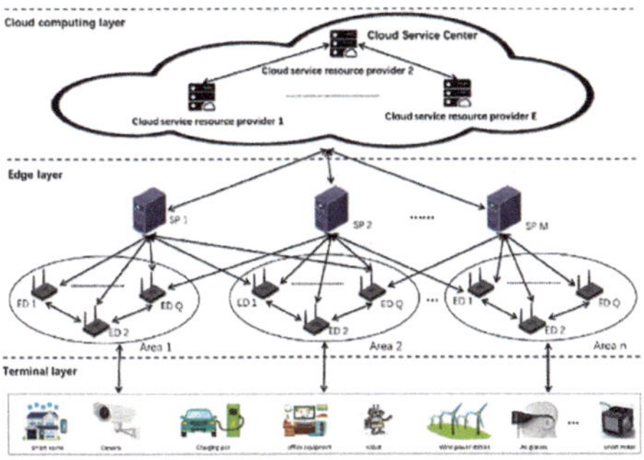

Fig. 1. Allocated framework of heterogeneous shared resources of power Internet of things based on cloud edge collaboration.

There are multiple power IoT business terminal devices at the terminal layer, which can generate different service type data, such as video monitoring, status monitoring, charging piles and other business data. There are n edge regions in the edge layer, there are Q edge devices in an edge area, represented by the set $\text{Area}_i = \{ED_1, ED_2, \ldots, ED_Q\}$, $1 \leq i \leq n$, represents the j-th edge device in the i-th edge area. One edge device accesses multiple service terminal devices and processes the service request data obtained from the service terminal devices. The cloud computing layer uses the cloud service center to provide processing data resources and business capabilities. Considering the limited capacity of the edge devices in the edge layer, and making the edge devices process the business with high real-time requirements as much as possible, therefore, if the resource of the edge layer is insufficient or the real-time requirement of the business is not high, the corresponding business can be sent to the cloud service center for processing to optimize the use of the resource of the edge layer.

The resource allocation system includes n service providers, and each service provider supports a class of power IoT services, which can be represented by SP_k, $k = 1, 2, 3, \ldots, n$. In order to improve the efficiency of resource allocation and ensure that the resource requirements of the business request are fully satisfied, the service is allocated to the edge device supporting this type of service, and the edge area where the service terminal is located has priority. Considering the differentiated resource requirements of computing data intensive service, storage type service and delay sensitive service, the tasks of class n service terminals are classified. In this paper, the tasks of received request resources are divided into three levels: real-time task, quasi real-time task and non-real-time task. Real-time tasks and quasi-real-time tasks are preferentially allocated to the edge area where the business terminal is located, otherwise they are allocated to adjacent edge areas. If neither can meet the resource and business requirements, they are sent to the cloud service center, where the cloud service center allocates resources and processes them. For non-real time tasks, if the area where the business terminal is located cannot meet the resource and business requirements, the task resource request is directly sent to the cloud service center, and the cloud service center allocates resources and processes them.

3 Cloud Edge Shared Resource Allocation Algorithm Based on Service Partition

3.1 Resource Allocation Model

Based on the results of system framework analysis, the cloud edge shared resource allocation proposed in this paper is divided into two parts: edge device shared resource allocation and cloud service center shared resource allocation. For business requests, resources should be allocated in the edge devices first. If they cannot be met, resources should be allocated in the cloud service center, and the opportunity to enter the resource allocation stage of the cloud service center should be differentiated according to the service level.

In the shared resource allocation of edge devices, in addition to considering the service level, it is necessary to further consider the support of edge devices for different types of services.

Let $SP_{i,j}^{k}$ indicate whether the edge device $ED_{i,j}$ supports the service SP_k, $SP_{i,j}^{k} = 1$ means support, $SP_{i,j}^{k} = 0$ means not support. Then the set of edge devices in the i-th region that can perform class k services can be expressed as:

$$\mathrm{EDSP_i^k} = \left\{\mathrm{ED_{i,j}} | \mathrm{SP_{i,j}^k} = 1, 1 \leq \mathrm{j} \leq \mathrm{Q}\right\} \tag{1}$$

Let Q^k represent the size of set $EDSP_i^k$, T_{ED}^k is $\mathrm{Q}^k \times \mathrm{Q}^k$ matrix, $T_{ED}^k[\mathrm{a}][\mathrm{b}]$ represents the transmission delay from the a-th edge device to the b-th edge device in $EDSP_i^k$, $1 \leq a, b \leq \mathrm{Q}^k$.

The set of edge regions that can perform class k services is expressed as:

$$ASP^k = \left\{Area_i | SP_{i,j}^k = 1, 1 \leq i \leq N, 1 \leq \mathrm{j} \leq \mathrm{Q}\right\} \tag{2}$$

Let N^k represent the size of set ASP^k, T_A^k is $\mathrm{N}^k \times \mathrm{N}^k$ matrix, $T_A^k[\mathrm{c}][\mathrm{d}]$ represents the transmission delay from the c-th edge device to the d-th edge device in ASP^k, $1 \leq c, d \leq \mathrm{N}^k$.

The service request VB property of the service terminal is defined as a binary set $\{\mathrm{R}, \beta\}$, Where R represents the number of resources required by the business request and β represents the delay threshold of the business request. $T\_CVB, ED_{i,j}$ represents the completion time of business request VB on edge device $ED_{i,j}$. The calculation formula can be described as follows:

$$\mathrm{T\_C}\left(\mathrm{VB}, \mathrm{ED_{i,j}}\right) = \mathrm{T\_W}\left(\mathrm{VB}, \mathrm{ED_{i,j}}\right) + \mathrm{T\_EX}\left(\mathrm{VB}, \mathrm{ED_{i,j}}\right) \tag{3}$$

$T\_W\left(VB, ED_{i,j}\right)$ represents the waiting time of resource request VB on edge device $ED_{i,j}$. $T\_EX\left(VB, ED_{i,j}\right)$ represents the execution time of resource request VB on edge device $ED_{i,j}$.

$$T\_EX(VB, ED_{i,j}) = \frac{CR(VB)}{abt(ED_{i,j})} \tag{4}$$

Among them, $abt(ED_{i,j})$ represents the processing capacity of edge device $ED_{i,j}$, and $CR(VB)$ represents the computing resource demand of business request VB.

The cost of business request VB executing on edge device $ED_{i,j}$ can be expressed as follows:

$$cost_{VB,i,j} = T\_EX(VB, ED_{i,j}) * P_{i,j} \tag{5}$$

Where $P_{i,j}$ is the unit time price of edge device $ED_{i,j}$ resources.

For real-time or quasi-real-time services, it is necessary to ensure the minimum resource cost as much as possible on the basis of meeting time constraints. For non-real-time services, only the minimum resource cost needs to be considered. Therefore, the objective function of shared resource allocation of edge devices can be expressed as follows:

$$\begin{aligned} & RAgoal(VB, ED_{i,j}) = mincost_{VB,i,j} \\ & s.t. \qquad T\_C(VB, ED_{i,j}) \leq \beta \end{aligned} \tag{6}$$

In the allocation of shared resources of cloud service center, considering the powerful fusion computing ability of cloud computing, only considering the service level, all cloud service resource providers support all types of services by default. There are multiple cloud service resource providers in the cloud service center. Each cloud service resource provider manages the resource status of its cloud service physical data center through a unified resource control unit. Since each cloud service resource provider may operate multiple geographically dispersed cloud service physical data centers, in the process of cloud service shared resource allocation, when allocating resources to real-time or quasi-real-time services, in addition to meeting resource requirements, it is necessary to consider the delay factor caused by the transmission distance between the control unit and the physical data center, which is not necessary for non-real-time services.

3.2 Resource Allocation Algorithm

When the edge device $ED_{i,j}$ receives the resource service request VB sent by the service terminal and performs resource allocation, First, the edge device judges the service type of VB request, assuming that the resource request belongs to SP_k service type, and then processes it according to the level of SP_k business request.

According to the system framework for real-time, quasi-real-time and non-real-time classification of business requests, let EG_k be the level of SP_k business request, $EG_k \in \{1, 2, 3\}$, $EG_k = 1$, $EG_k = 2$ and $G_k = 3$ represents real-time, quasi-real-time, and non-real-time business requests, respectively. For these three types of service request levels, combined with the resource allocation and service bearing principles corresponding to the business levels in the system framework, the cloud-side shared resource allocation algorithm is designed into four situations, as shown in Algorithm 1.

The first case: $ED_{i,j}$ confirms whether the resource request required by VB can be satisfied. The judgment conditions are as follows:

$$\left(CED^{u}_{i,j} + R \leq CED_{i,j}\right) \cap \left(T\_EX\left(VB, ED_{i,j}s\right) \leq \beta\right) \tag{7}$$

Among them, $CED^{u}_{i,j}$ represents the number of resources that $ED_{i,j}$ has been occupied currently, $CED_{i,j}$ represents the total resources of $ED_{i,j}$. If the conditions are met, $ED_{i,j}$ allocates the required resources to VB and VB is processed in $ED_{i,j}$. Otherwise, enter the second or fourth situation. Real-time and quasi-real-time services enter the second situation, and non-real-time services directly enter the fourth situation.

The second case: $ED_{i,j}$ cannot satisfy the resource request required by VB, however, the edge device set in the edge area $Area_i$ where it can perform k-type services can meet the resource requirements of VB, that is, one or more edge devices in the $EDSP^k_i$ can satisfy the resource request required by the VB. Considering that there are multiple edge devices to meet the needs of VB, the judgment conditions are as follows:

$$\left(\mu \leq \sum_{l=1}^{Q^k} CED_{i,l}\right) \cap \left(EG_k = 1 \cup EG_k = 2\right) \tag{8}$$

Among them, $= \sum_{l=1}^{Q^k} CED^{u}_{i,l} + R$, $ED_{i,l} \in EDSP^k_i$. If the condition is satisfied, it indicates that the resource request required by VB can be satisfied in the edge area $Area_i$. Otherwise, enter the third situation. When the condition shown in (8) is met, it is necessary to find the edge device that specifically confirms to undertake the service request in the edge device set $EDSP^k_i$. Algorithm 2 determines the best set of edge devices to carry the service request through a dynamic search method, and completes the corresponding resource allocation. At this time, if the number of selected edge devices is 1, the device cannot be $ED_{i,j}$. If multiple edge devices are selected, they can include $ED_{i,j}$.

The third case: $ED_{i,j}$ where the edge area $Area_i$ cannot satisfy the resource request required by VB, but other edge devices in the adjacent edge area that can perform the service of type k can meet the resource request, that is, one or more edge devices in other edge areas other than $Area_i$ in ASP^k can satisfy the resource request required by VB. The judgment conditions are as follows:

$$\left(\mu > \sum_{l=1}^{Q^k} CED_{i,l}\right) \cap \left(EG_k = 1 \cup EG_k = 2\right) \tag{9}$$

If the condition shown in (9) is met, the condition (8) is used to judge each edge area other than $Area_i$ in ASP^k, and using Algorithm 2 to find the best edge device set in the corresponding edge region to meet the resource requirements of VB. If all areas in ASP^k cannot carry VB, then enter the fourth case.

The fourth case: The cloud service center provides the resources required for the business request and processes the business request. At this time, there are two types of service levels, one is the real-time or quasi-real-time service entered in the third case, and the other is the non-real-time service directly entered in the first case.

In order to optimize resource allocation as much as possible, for real-time or quasi real-time services, service request processing delay needs to be improved. Considering the distance difference and resource occupancy status difference between each cloud service resource provider control unit and the specific cloud service physical data center it manages, the service allocation is optimized.

Assume that there are E cloud service resource providers in the cloud service center, let CC_e represent the total resource of the e-th cloud service resource provider, $1 \leq e \leq E$. CC_e^u represents the number of resources currently the e-th cloud service resource provider is occupied. Assuming that each cloud service resource provider has n_e cloud service physical data centers, the distance from the data center to the cloud service resource provider control unit can be expressed by D_cud.

$$D\_cud = \begin{bmatrix} d_{1,1} & \cdots & d_{1,\varphi} \\ \vdots & \ddots & \vdots \\ d_{E,1} & \cdots & d_{E,\varphi} \end{bmatrix} \tag{10}$$

$\varphi = \max\{n_e | e = 1, 2, 3 \ldots E\}$ and $d_{e,g} \in D\_cud$,

$$d_{e,g} = \begin{cases} Distance\ from\ the\ e-th\ control\ unit\ to\ the\ g-th\ data\ center & g \leq n_e \\ -1 & other \end{cases} \tag{11}$$

The time delay constraint is transformed into the distance threshold between the acceptable control unit and the data center, which is represented by ε_{VB}. In algorithm 1, the optimal data center satisfying the distance threshold constraint is found to allocate the resources needed by VB, and VB is confirmed to be processed. Let $\arg min_h \left[\delta_{1,h} - \varepsilon_{VB}\right]$ represents the minimum difference between the distance $d_{e,g}$ of the h control units to the data center and the distance threshold ε_{VB}.

The flow chart of cloud edge shared resource allocation algorithm based on service partition is shown in Fig. 2.

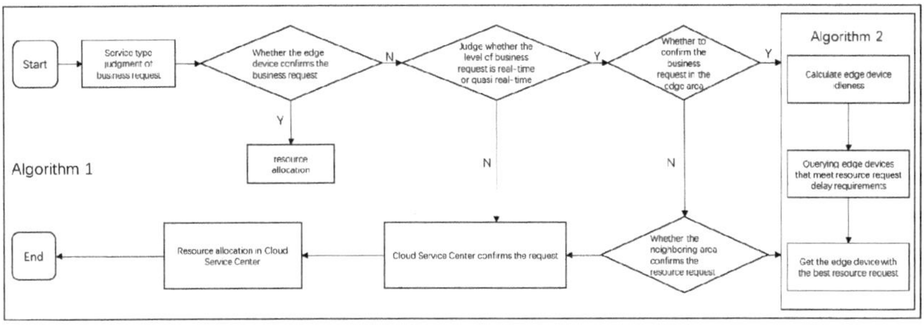

Fig. 2. The flow chart of cloud edge shared resource allocation algorithm based on service partition.

The algorithm steps are as follows:

Algorithm 1: cloud edge shared resource allocation algorithm based on service partition

Initialization: resource requirement R of business request VB, delay threshold β, acceptable distance threshold ε_{VB}, and cloud service resource search function $CS(R)$ that meets resource requirements

1) if$(CED_{i,j}^{u} + R \le CED_{i,j}) \cap (T_EX(VB, ED_{i,j}) \le \beta)$ then
2) $CED_{i,j}^{u} \leftarrow CED_{i,j}^{u} + R$
3) else if $(\mu \le \sum_{l=1}^{Q^k} CED_{i,l}) \cap (EG_k = 1 \cup EG_k = 2)$ then
4) p ← 0
5) go to Algorithm 2
6) Obtain ED ID ξ and allocate resource
7) else if $(\mu > \sum_{l=1}^{Q^k} CED_{i,l}) \cap (EG_k = 1 \cup EG_k = 2)$ then
8) for $s \leftarrow 1$ to N do
9) if $s \ne i \cap (Area_s \in ASP^k)$ then
10) p ← 1
11) end if
12) if $p = 1 \cap (\sum_{l=1}^{Q^k} CED_{s,l}^{u} + R \le \sum_{l=1}^{Q^k} CED_{s,l})$ then
13) Go to Algorithm 2
14) x=1
15) end if
16) end for
17) Use formula (6) to get the appropriate device ξ to confirm the resource demand of business request.
18) Obtain ED ID ξ and allocate resource
19) if $x \ne 1 \cap (\sum_{e=1}^{E} CC_e^{u} + R \le \sum_{e=1}^{E} CC_e)$ then
20) $\hat{e} \leftarrow CS(R)$
21) $\delta \leftarrow d_{e,1:n_{\hat{e}}}$
22) h ← sizeof(δ)
23) if $\arg min_h [\delta_{1,h} - \varepsilon_{VB}] \le 0$ then
24) $CC_{\hat{e}}^{u} \leftarrow CC_{\hat{e}}^{u} + R$
25) end if
26) end if
27) else if $(CED_{i,j}^{u} + R \le CED_{i,j}) \cap EG_k = 3$ then

Algorithm 2: A real-time or quasi-real-time resource allocation algorithm

Initialization: the number of edge devices Q^k in the i-th area that can allocate resources for the service type SP_k, Resource requirement R of business request VB, delay threshold β, The usable capacity $CED_{i,j}^{a}$ of the *l-th* edge device

1) for $l \leftarrow 1$ to Q^k do
2) $CED_{i,l}^{a} \leftarrow CED_{i,l} - CED_{i,l}^{u}$
3) if $CED_{i,l}^{a} \ge R$ then
4) $y_l \leftarrow 1$
5) end if
6) end for
7) for $l \leftarrow 1$ to Q^k do
8) if $y_l = 1 \cap p = 0$ then
9) if $T_{ED}^{k}[j][l] + T\_C(VB, ED_{i,l}) > \beta$ then
10) $y_l \leftarrow 0$
11) end if
12) else if $y_l = 1 \cap p = 1$ then
13) if $T_A^{k}[i][s] + T\_C(VB, ED_{s,l}) > \beta$ then
14) $y_l \leftarrow 0$
15) end if
16) end if
17) end for
18) for all $y_l = 1$, Use formula (6) to get the appropriate deviceξto confirm the resource demand of business request.
19) Return ξ

```
    if $\sum_{e=1}^{E} CC_e^u + R \leq \sum_{e=1}^{E} CC_e$ then
      $\hat{e} \leftarrow CS(R)$
      $CC_{\hat{e}}^u \leftarrow CC_{\hat{e}}^u + R$
    end if
return
```

4 Simulation Experiment

In order to verify the feasibility and efficiency of the proposed scheme, the resource allocation method proposed in this paper is evaluated. The simulation results show that compared with Full-local and DTMO, the resource allocation efficiency of this method in edge computing network is greatly improved.

Assuming that the number of edge devices is 100, the service type is 10, and there are 10 edge devices in each edge area. The number of real-time and quasi real-time business requests accounts for 60% of the total number of business requests in each time slot. The number of non-real time business requests accounts for 40% of the total number of business requests in each time slot. With the increase in the number of business requests, the comparison between full local, DTMO and the resource allocation method based on service partition proposed in this paper is shown in Fig. 3 and Fig. 4. It can be seen from Fig. 3 that executing all business requests at the edge layer does not guarantee that the resource requirements of all requests are confirmed, Full-local business request resource allocation satisfaction rate is lower than DTMO and resource allocation method based on service division. The resource allocation method based on service partition has the highest business request resource allocation satisfaction rate.

Figure 5 shows the comparison results of Full-local, DTMO and Resource allocation methods based on service division on the satisfaction rate of resource allocation of business requests when the number of business requests is fixed and the number of edge devices increases from 50 to 500. The results show that with the growth of edge devices, the efficiency of Full-local is greatly affected. In contrast, due to the service classification of business requests and distributed heterogeneous resource allocation according to service types, the resource allocation method based on service division has better performance than Full-local and DTMO, and effectively diverts business requests, which has certain decision flexibility and effectively allocates resources. As can be seen from the results, as the number of edge devices increases, the degree of resource allocation for business requests is higher. Because the more edge devices, the more edge resources can be provided to complete the request, which significantly improves the efficiency of resource allocation and utilization of edge resources. Therefore, this method provides an effective technical solution for the heterogeneous resource sharing of the power Internet of things.

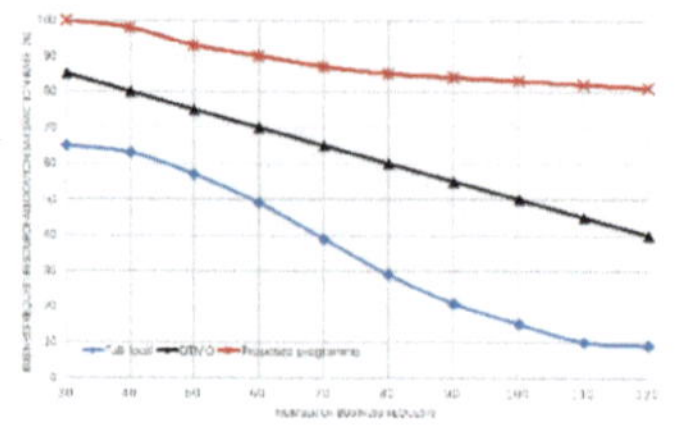

Fig. 3. Satisfaction rate of business request resource allocation under different business request numbers.

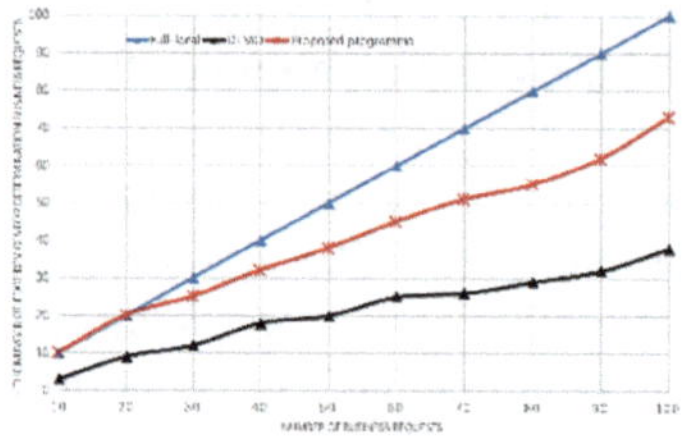

Fig. 4. The number of edge devices for determining business requests under different business requests.

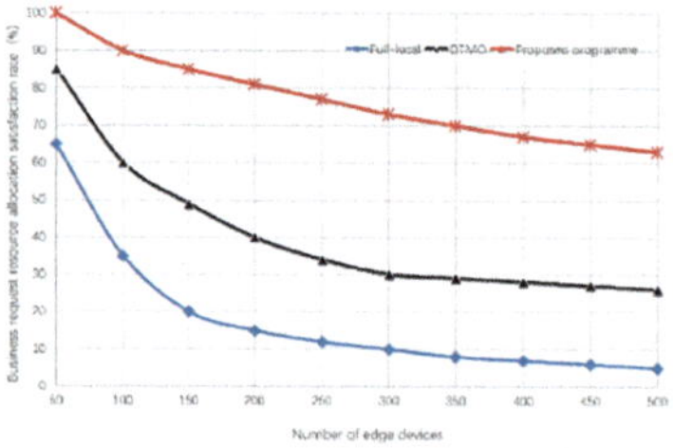

Fig. 5. Business request resource allocation satisfaction rate under different number of edge devices.

5 Conclusion

This paper studies the problem of resource allocation of cloud-edge collaboration in the power IoT. Using cloud-edge collaboration and service provision ideas, this paper proposes a method of heterogeneous resource allocation based on cloud edge collaboration in the power IoT, which improves the quality of service provision of power IoT by optimizing resource allocation. From the perspective of resource service matching integration, this method distinguishes service categories, allocates different cloud edge collaboration resources according to different service providing types, meets the corresponding real-time processing needs of services, and ensures the optimal allocation of edge resources. The experimental results show that the solution proposed in this paper has obvious effect in realizing resource allocation of business request. With the increase of the number of business requests, the satisfaction rate of business request resource allocation is maintained at a high level, and with the increase of edge devices, it effectively guarantees utilization of edge resources.

Acknowledgment. This work is supported by State Grid Jiangsu Electric Power Co., Ltd. Science and Technology Project "Research and Application of Online Intelligent Monitoring and Security Protection Technology for Distribution Equipment IoT Terminals Based on Edge Computing" (J2019065).

References

1. Alrowaily, M., Lu, Z.: Secure edge computing in IoT systems: review and case studies. In: 2018 IEEE/ACM Symposium on Edge Computing (SEC), Seattle, WA, pp. 440–444 (2018)
2. Song, Y., Yau, S.S., Yu, R., Zhang, X., Xue, G.: An aproach to QoS-based task distribution in edge computing networks for IoT applications. In: 2017 IEEE International Conference on Edge Computing (EDGE), Honolulu, HI, pp. 32–39 (2017)
3. Huang, Y., Zhang, J., Duan, J., Xiao, B., Ye F., Yang, Y.: Resource allocation and consensus on edge blockchain in pervasive edge computing environments. In: 2019 IEEE 39th International Conference on Distributed Computing Systems (ICDCS), Dallas, TX, USA, pp. 1476–1486 (2019)
4. Chen, S., et al.: Internet of things based smart grids supported by intelligent edge computing. IEEE Access **7**, 74089–74102 (2019)
5. Yang, T., Hu, Y., Gursoy, M. C., Schmeink, A., Mathar, R.: Deep reinforcement learning based resource allocation in low latency edge computing networks. In: 2018 15th International Symposium on Wireless Communication Systems (ISWCS), Lisbon, pp. 1–5 (2018)
6. Dab, B., Aitsaadi, N., Langar, R.: A novel joint offloading and resource allocation scheme for mobile edge computing. In: 2019 16th IEEE Annual Consumer Communications & Networking Conference (CCNC), Las Vegas, NV, USA, pp. 1–2 (2019)
7. Avasalcai, C., Tsigkanos, C., Dustdar, S.: Decentralized resource auctioning for latency-sensitive edge computing. In: 2019 IEEE International Conference on Edge Computing (EDGE), Milan, Italy, pp. 72–76 (2019)
8. Wei, X., et al.: MVR: an architecture for computation offloading in mobile edge computing. In: 2017 IEEE International Conference on Edge Computing (EDGE), Honolulu, HI, pp. 232–235 (2017)

A Load Balancing Container Migration Mechanism in the Edge Network

Chen Xin[1], Dongyu Yang[1], Zitong Ma[2], Qianjun Wang[2](✉), and Yang Wang[3]

[1] State Grid Jiangsu Electric Power Co., Ltd., Nanjing Gaochun District Power Supply Branch, Nanjing 210000, China
[2] Beijing University of Posts and Telecommunications, Beijing 100876, China
wqianjuna@sina.com
[3] State Grid Jiangsu Power Co., Ltd., Nanjing Power Supply Branch, Nanjing 210000, China

Abstract. Aiming at the problem of difference in business busyness between edge nodes caused by obvious uneven distribution of service requests in the edge network, this paper proposes a container migration mechanism for load balancing. First of all, the timing of container migration is determined based on the classical static resource utilization threshold model. Then, a container migration model of load balancing combined migration cost is established to minimize the impact of container migration while balancing the load of edge network. Finally, the priority of container migration is calculated from the perspective of resource correlation, and a migration algorithm based on the improved ant colony algorithm is designed for container migration in the context of the power Internet of things. The simulation results show that the container migration mechanism proposed in this paper can not only improve the load balancing degree of edge network but also reduce the cost of container migration.

Keywords: Edge computing · Container migration · Load balancing · ACO · Power Internet of Things

1 Introduction

Over the past decade, cloud computing, which enables users to access services universally on demand, has become a popular paradigm [1]. As a highly aggregated service computing, cloud computing centralizes the computing, storage, and network management functions of data centers, freeing users from many detailed specifications. However, with the development of interconnected devices and sensors, massive data transmission and processing leads to a significant increase in cloud center computing and storage load along with network transmission pressure, making it hard to meet the processing delay of businesses [2].

By deploying the intelligent equipment to provide services on the edge side closer to the user terminal, edge computing can effectively alleviate cloud center computing pressure while ensuring the delay of power Internet of Things business [3]. As can be

Q. Liu et al. (Eds.): CENet 2020, AISC 1274, pp. 1414–1422, 2021.
https://doi.org/10.1007/978-981-15-8462-6_161

seen from Fig. 1, the architecture of edge computing under the scenario of the Internet of Things mainly concludes three parts, namely the end layer, the edge layer and the cloud layer, where end users communicate with edge nodes via WIFI or 4G/5G, and send task requests to edge nodes instead of the cloud platform.

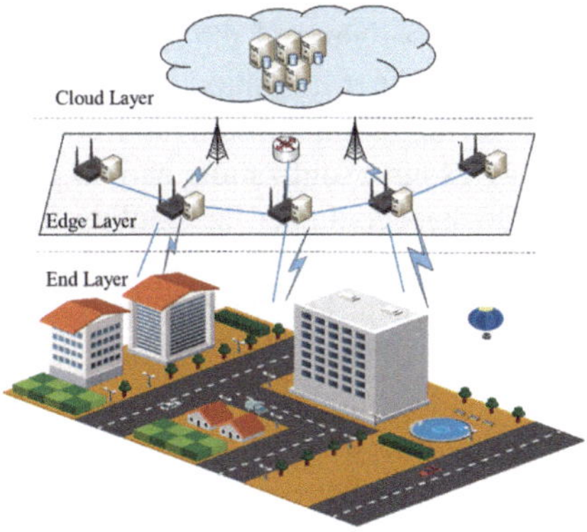

Fig. 1. The architecture of edge computing under the scenario of the Internet of Things.

However, the power Internet of Things based on edge computing still faces lots of problems. Due to the finiteness of the resource of edge nodes and the uneven distribution of business requests in the edge network, there are obvious differences in the degree of business between edge nodes. At the same time, various types of business in power Internet of Things and different business resource needs make the utilization rate of resources of edge nodes unbalanced. Therefore, it is of great significance to optimize the resource utilization of edge nodes and guarantee the overall performance of the network in the case of unbalanced load in the edge network.

With the development of virtualization technology, container migration has become a key technology to balance the load between edge nodes and improve the utilization of edge node resources. In contrast to virtual machines, the container is lightweight and supports migration [4, 5]. However, container migration inevitably has a migration cost [6–8]. As a result, it is necessary to make an effective migration mechanism to optimize the resource utilization of the edge network on the premise of meeting the multi-type business needs of the power Internet of things [9, 10].

In summary, in view of the problem that due to the uneven distribution of business requests in time and space under the scenario of power Internet of Things, some edge nodes are overloaded thus resulting in a sharp increase in computing delay and a decline in business service quality while the others cannot be fully utilized, this paper established a joint load balancing and migration cost container migration model, and proposes a load balancing container migration mechanism in the edge network.

The structure of the paper is as follows. Section 2 introduces the research status at home and abroad. Section 3 proposes a container based migration decision mechanism. Section 4 describes the algorithm to solve the container migration problem in detail. Section 5 uses MATLAB tools to compare the performance of the container migration decision mechanism and verify its effectiveness. Section 6 summarizes the full text and conducts future technology research prospects.

2 The Related Work

Containers are considered a lightweight and efficient virtualization technology [11]. Concerning to the problem of container migration, researchers at present mainly focus on two directions, namely the migration strategy and concrete migration implementation. Aiming at balancing resource utilization and improving application performance, [12] proposed a Docker container scheduling based on the ant colony optimization algorithm. Zhao, et al. [13] proposed a load balancing model in cloud environment, and designed a heuristic load balancing scheduling method for virtual machines through virtual machine layout and dynamic virtual machine scheduling. Based on time delay and power consumption, Tang, et al. [4] modeled the container migration strategy in the environment of fog computing as a multidimensional Markov decision process space, and improved the deep reinforcement learning algorithm to realize fast decision-making. Based on the characteristics of hierarchical storage in Docker container, a framework was proposed in [14] to support efficient and real-time migration of services between edge nodes, which greatly reduced the total migration time and user-perceived service interruption.

3 Container Migration Mechanism

3.1 Container Migration Trigger

In the edge network, $\mathcal{N}=\{n_1,n_2,\cdots,n_J\}$ represents the set of edge nodes and $\mathcal{C}=\{c_1,c_2,\cdots,c_I\}$ represents the set of containers, where J and I are the number of edge nodes and containers respectively. The set of types of resources is $\mathcal{R}=\{r_1,r_2,\cdots,r_K\}$, where K is the number of resources. For each edge node $n_j \in \mathcal{N}$, let $\overrightarrow{W_j} = \left(W_j^1, W_j^2, \cdots, W_j^K\right)$ denote its resource capacity of all types and $\overrightarrow{u_j} = \left(u_j^1, u_j^2, \cdots, u_j^K\right)$ denote its resource utilization. For each container $c_i \in \mathcal{C}$, let $\overrightarrow{d_i} = \left(d_i^1, d_i^2, \cdots, d_i^K\right)$ represents the requirement of container c_i for all types of resources.

Define the two-dimensional container deployment decision matrix $\mathrm{X}=\left[x_{i,j}\right]_{I\times J}$ for container-to-edge node mapping, where the decision variable $x_{i,j}$ represents whether container c_i has been deployed on edge node n_j. Let $x_{i,j} = 1$ if c_i is placed on n_j, and $x_{i,j} = 0$ else.

The utilization rate of resource r_k on edge node n_j can be calculated as follows.

$$u_j^k = \frac{\sum_{c_i \in \mathcal{C}} x_{i,j} d_i^k}{W_j^k} \tag{1}$$

Given the upper threshold T_k of resource utilization, the timing of containers migration is:

$$\exists u_j^k \geq T_k \quad \forall n_j \in \mathcal{N}, r_k \in \mathcal{R} \tag{2}$$

3.2 Container Migration Model

The size of the container's CPU and memory is a direct factor in migration downtime and resource loss. The smaller the CPU and memory utilization, the less downtime the container migrates, and therefore the less impact it has on performance and business QOS. Therefore, based on migration time, the migration probability $pro_{i,j}$ of container c_i on edge node n_j can be expressed as:

$$pro_{i,j} = \omega_{cpu} \frac{d_i^{k_{cpu}}}{\sum\limits_{c_i \in C} x_{i,j} d_i^{k_{cpu}}} + \omega_{mem} \frac{d_i^{k_{mem}}}{\sum\limits_{c_i \in C} x_{i,j} d_i^{k_{mem}}} \tag{3}$$

Define the two-dimensional container deployment decision matrix $\widehat{X} = \left[\widehat{x_{i,j}}\right]_{I \times J}$ for container-to-edge node mapping after migration and the migration decision variables $y_{j,j'}^i$. If $x_{i,j} = 1$ and $\widehat{x_{i,j}} = 1$, then there is $y_{j,j'}^i = 1$, indicating that c_i has migrated from n_j to $n_{j'}$, $y_{j,j'}^i = 0$ otherwise.

The migration cost of c_i mainly refers to the network delay caused by the transmission of data between two edge nodes.

$$Cost_{mig}^i = \frac{\Lambda_i}{B_{j,j'}} \tag{4}$$

Therefore, we can get the total migration cost as follows:

$$Cost_{mig}^{total} = \sum_{c_i \in C} y_{j,j'}^i Cost_{mig}^i \tag{5}$$

For each type of resources $r_k \in \mathcal{R}$, the resource utilization equilibrium after migration is defined as the variance of resource utilization on all edge nodes.

$$U_{balance}^k = \sum_{n_j \in \mathcal{J}} \frac{\left(\widehat{u_j^k} - \overline{\widehat{u_j^k}}\right)^2}{J} \tag{6}$$

where $\widehat{u_j^k}$ denotes the utilization of resource r_k on n_j, $\overline{\widehat{u_j^k}}$ is the mean utilization of resource r_k of all edge nodes and the total resource utilization equilibrium of edge network is the sum of the resource utilization equilibrium of all types of resources.

$$U_{balance}^{total} = \sum_{r_k \in \mathcal{R}} U_{balance}^{k} \tag{7}$$

To sum up, the migration model under the scenario of power Internet of Things aiming for load balancing can be described as follows.

$$\begin{aligned} &\min\left\{\theta U_{balance}^{total} + \gamma Cost_{mig}^{total}\right\} \\ s.t.\ \ & x_{i,j}, \widehat{x_{i,j}}, y_{j,j'}^{i} \in \{0,1\}, \forall c_i \in \mathcal{C}, n_j \in \mathcal{N} \end{aligned} \tag{8}$$

4 Container Migration Algorithm

4.1 Heuristic Information

The heuristic factor can be expressed as $\eta_{i,j}$, which represents the desirability of migrating the container c_i to n_j. In the ACO algorithm, they are combined with pheromones to construct a migration decision scheme.

Based on the CMM scheme proposed in this paper, heuristic information is mainly calculated according to the migration cost of moving the container c_i to n_j.

For the migration costs incurred by migrating containers c_i to n_j, the heuristic information $\eta_{i,j}$ can be expressed as:

$$\eta_{i,j} = B_{j',j} \tag{9}$$

4.2 Pheromone Update

The pheromone can be represented as $\tau_{i,j}$, representing the pheromone value that migrates container c_i to n_j. And we define the initial pheromone value to be 1.

After selecting a new mapping tuple, the ant updates the pheromone level of the traversal mapping based on the following pheromone update rule:

$$\tau_{i,j} = (1-\rho)\tau_{i,j} + \Delta\tau_{i,j} \tag{10}$$

where $\rho \in [0,1]$ is the local pheromone evaporating parameter and $\Delta\tau_{i,j}$ is the increment of additional pheromones, and it can be calculated as follows.

$$\Delta\tau_{i,j} = \begin{cases} \dfrac{1}{\theta U_{balance}^{total} + \gamma Cost_{mig}^{total}} & if\ \widehat{x_{i,j}} = 1 \\ 0 & if\ \widehat{x_{i,j}} = 0 \end{cases} \tag{11}$$

4.3 Pseudo-Random-Proportion Rule

Ants tend to choose tuples with the largest pheromone and heuristic information. Therefore, ants will select mapping tuples to traverse according to the following pseudo-random proportion rules:

$$p_{i,j} = \begin{cases} \frac{[\tau_{i,j}]^{\alpha}\cdot[\eta_{i,j}]^{\beta}}{\sum_{n_u \in \Theta_l(i)} [\tau_{i,u}]^{\alpha}\cdot[\eta_{i,u}]^{\beta}}, & \textit{if } n_j \in \Theta_l(i) \\ 0, & \textit{otherwise} \end{cases} \tag{12}$$

where $p_{i,j}$ represents the probability that c_i selects n_j as its destination edge node. α and β are the weights of pheromone update and heuristic information respectively. $\Xi_l(i)$ is a set of effective edge nodes of Ant_l that satisfy the constraints, and is defined as follows:

$$\Xi_l(i) = \left\{ n_j \middle| \sum_{c_s \in C} x_{s,j} d_s^k + d_i^k \leq T \times W_j^k, \forall r_k \in \mathcal{R} \right\} \tag{13}$$

5 Simulation and Analysis

5.1 Simulation Parameters

In the simulation, it is assumed that the heterogeneous edge network environment is a rectangular region of 10 km × 5 km, the number of edge devices is 10 and the number of containers varies from 30 to 210. Consider the resource types of edge nodes: compute, memory, and storage resources. Among them, the CPU core number of edge devices is 32, the value range of memory and storage resource capacity is [16, 32], [100, 300] GB respectively, and the value range of transmission rate between edge devices is [100, 300] Mb/s. The resource requirements of the container are randomly generated, and the values of the CPU resource requirements, memory resource requirements and storage resource requirements of the container range from [1, 3] core, [1, 4], [5, 15] GB, respectively. The service delay constraints of the business carried by the container are randomly set from 50 ms to 500 ms.

5.2 Simulation Results

Figure 2 compares the load changes of each edge node before and after container migration. It can be seen that, after the implementation of the CMM mechanism, the load of the edge nodes in the edge network is relatively balanced, and the resource utilization of the idle edge nodes is optimized while the execution pressure of individual busy edge nodes and the possibility of unsatisfied service QOS demand is reduced.

Figure 3 shows the load balancing degree of edge network changes before and after container migration. The load balancing degree includes the utilization balancing degree of different edge nodes to any resource. The smaller the value of load balancing

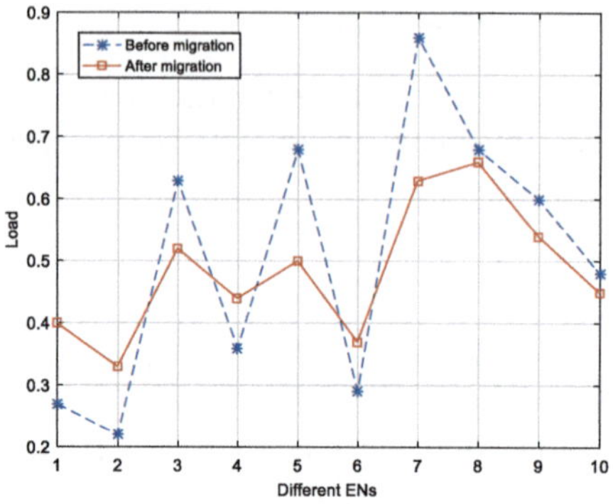

Fig. 2. Load comparison diagram of each edge node before and after container migration.

degree is, the more balanced the load of each edge node in the edge network is. As can be seen from the figure, that with the increase of the number of containers, the CMM mechanism can ensure the load balance of the edge network while improving the system migration cost.

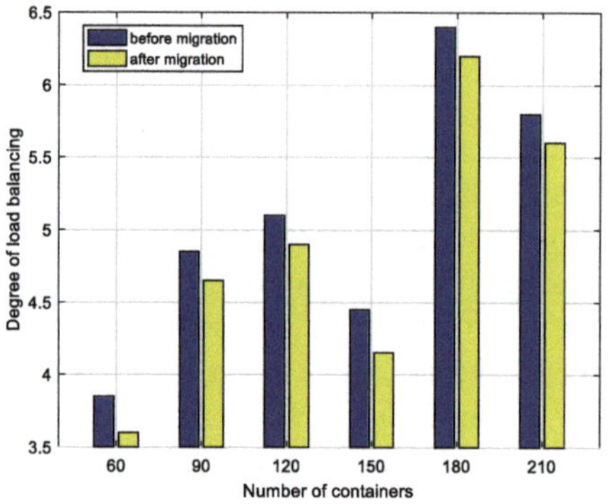

Fig. 3. Comparison of load balancing degree before and after container migration.

Figure 4 compares the value of container migration costs under the three algorithms. In general, with the rising number of container, the edge nodes resource usage rate goes up and the edge network load difference significantly, which leading to more containers need to be migrated to the relatively free edge node so as to ensure the time delay of container business constraints. Based on the global optimal algorithm, the load balancing container migration mechanism in the edge network proposed in this paper can minimize the impact of container migration while balancing the load of the edge network.

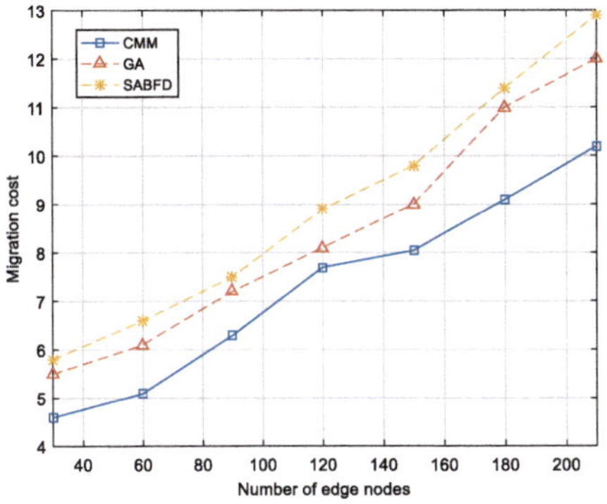

Fig. 4. Comparison of the container migration cost.

6 Conclusion

In the scenario of power Internet of Things, some edge nodes are overloaded due to the uneven distribution of business requests in time and space, resulting in a sharp increase in computing delay and a decline in business service quality while the remaining edge nodes have idle processing power and resources cannot be fully utilized. In view of this, a container migration model of combined load balancing and migration cost is established, and a container migration mechanism CMM for edge network load balancing is proposed in this paper. Among them, CMM decomposes the problem of container migration into three aspects: determining the timing of migration triggering, establishing a container migration model and designing a migration algorithm based on the ACO algorithm. Simulation results show that the container migration mechanism proposed in this paper improves the edge network load while reducing the influence of container migration on users.

Acknowledgment. This work is supported by State Grid Jiangsu Electric Power Co., Ltd. Science and Technology Project "Research and Application of Online Intelligent Monitoring and Security Protection Technology for Distribution Equipment IoT Terminals Based on Edge Computing" (J2019065).

References

1. Botta, A., et al.: Integration of cloud computing and internet of things: a survey. Future Generation Computer Systems **56**, 684–700 (2016)
2. Schulz, P., Matthe, M., Klessig, H., et al.: Latency critical IoT applications in 5G: perspective on the design of radio interface and network architecture. IEEE Commun. Mag. **55**(2), 70–78 (2017)
3. Pan, J., Mcelhannon, J.: Future edge cloud and edge computing for internet of things applications. IEEE Internet Things J. **PP**(99), 1 (2017)
4. Tang, Z., Zhou, X., Zhang, F., et al.: Migration modeling and learning algorithms for containers in fog computing. IEEE Trans. Serv. Comput. **12**(5), 712–725 (2018)

5. Ma, L., Yi, S., Carter, N., et al.: Efficient live migration of edge services leveraging container layered storage. IEEE Trans. Mob. Comput. **18**(9), 2020–2033 (2018)
6. Puthal, D., Obaidat, M.S., Nanda, P., et al.: Secure and sustainable load balancing of edge data centers in fog computing. IEEE Commun. Mag. **56**(5), 60–65 (2018)
7. Kurve, A., Griffin, C., Miller, D.J., et al.: Game theoretic iterative partitioning for dynamic load balancing in distributed network simulation (2012)
8. Jia, M., Liang, W., Xu, Z., et al.: Cloudlet load balancing in wireless metropolitan area networks. In: IEEE Infocom the IEEE International Conference on Computer Communications. IEEE (2016)
9. Ma, X., Zhang, S., Li, W., et al.: Cost-efficient workload scheduling in cloud assisted mobile edge computing. In: 2017 IEEE/ACM 25th International Symposium on Quality of Service (IWQoS). ACM (2017)
10. Tewksbury, S.K., Hornak, L.A.: Communication network issues and high-density interconnects in large-scale distributed computing systems. IEEE J. Sel. Areas Commun. **6**(3), 587–609 (1988)
11. Rodrigues, T.G., Suto, K., Nishiyama, H., et al.: Hybrid method for minimizing service delay in edge cloud computing through VM migration and transmission power control. IEEE Trans. Comput. **66**(5), 810–819 (2017)
12. Kaewkasi, C., Chuenmuneewong, K.: Improvement of container scheduling for docker using ant colony optimization. In: 2017 9th International Conference on Knowledge and Smart Technology (KST). IEEE (2017)
13. Zhao, J., Hu, L., Ding, Y., et al.: A heuristic placement selection of live virtual machine migration for energy-saving in cloud computing environment. PLoS ONE **9**(9), e108275 (2014)
14. Willis, D., Dasgupta, A., Banerjee, S.: Paradrop: a multi-tenant platform to dynamically install third party services on wireless gateways. In: Proceedings ACM International Workshop Mobility Evolving Internet Architecture, pp. 43–48. ACM (2014)

Research on Key Agreement Security Technology Based on Power Grid Internet of Things

Weidong Xia[1], Ping He[2], Yi Li[3], Qinghang Zhang[3(✉)], and Han Xu[2]

[1] State Grid Jiangsu Electric Power Co., Ltd., Nanjing Gaochun District Power Supply Branch, Nanjing 210000, China
[2] State Grid Jiangsu Power Co., Ltd., Nanjing Power Supply Branch, Nanjing 210000, China
[3] Beijing University of Posts and Telecommunications, Beijing 100876, China
zhangqh@bupt.edu.cn

Abstract. With the rapid development of information technology, the Power grid Internet of Things technology has been integrated into all aspects of life. The intelligent network formed by the Power grid Internet of Things that interconnects people, things and things has greatly promoted the intelligent development in various fields. However, due to the large number of IoT devices and the limited resources of many devices, it has brought huge challenges to the protection of IoT information data and device management. This paper proposes a lightweight and secure IoT authentication key agreement protocol for the problems in the IoT environment. The protocol adds identity authentication based on the ECDH algorithm to prevent man-in-the-middle attacks. The identity authentication part is designed based on the dynamic password technology with hash as the main operation, and realizes the mutual authentication of the dynamic password based on the identification. The entire protocol has good security and can effectively resist man-in-the-middle, impersonation, eavesdropping, and replay attacks, ensuring the security of IoT device communications. At the same time, the protocol resource overhead is small, and it can adapt to resource-limited IoT devices. In addition, the authentication mechanism combined with cloud-edge-end collaboration can efficiently manage many IoT devices.

Keywords: Authentication key agreement · Dynamic password · Edge computing · Lightweight · Power grid Internet of Things

1 Introduction

As the third revolution of the information technology industry [1], the Power grid Internet of Things technology has greatly promoted the intelligent development of all walks of life. At present, regardless of whether it is artificial intelligence, intelligent transportation, Internet of vehicles, smart home and other fields, we can see the important role played by the Power grid Internet of Things technology.

Q. Liu et al. (Eds.): CENet 2020, AISC 1274, pp. 1423–1430, 2021.
https://doi.org/10.1007/978-981-15-8462-6_162

However, the rapid development of Power grid Internet of Things technology also brings many security issues. A large amount of data is generated due to the access of a large number of devices, and there is no lack of a large amount of private data. Faced with security issues such as data eavesdropping, data tampering, data forgery, man-in-the-middle attacks, and so on. In addition, due to limited resources and restrictions on related device management, many IoT sensor devices have open communication channels. This allows attackers to easily tamper, forge, or replay transmitted data, resulting in very serious security problems and huge losses. Therefore, according to the characteristics of the Power grid Internet of Things environment, how to protect the data transmitted by the device is still a security problem to be solved urgently.

As the first key agreement protocol, the Diffie-Hellman protocol enables the communication parties to negotiate a shared temporary session key for building a secure channel on an open channel. Unfortunately, the Diffie-Hellman protocol cannot resist man-in-the-middle attacks due to lack of authentication. In the follow-up development, the researchers added the identity authentication function based on the DH protocol. In view of the limited resources of a large number of IoT devices, based on the research of related authentication key agreement technology, a lightweight IoT security authentication key agreement protocol is proposed to adapt to the IoT environment.

In addition, in the face of massive IoT devices, the traditional centralized management mechanism centered on cloud servers will bring a huge burden to the server. Mass device authentication and key agreement will cause delays and network congestion, and cannot achieve efficient management of mass devices. Based on the edge computing technology, a cloud-edge-end collaborative authentication mechanism is built. The authentication and key agreement tasks are delegated to the edge gateway-level device, and the edge gateway device is responsible for managing the IoT devices in the area. So as to solve the problems caused by the large number of devices in the Power grid Internet of Things environment.

2 Protocol

2.1 Identity-Based Dynamic Password Authentication Protocol

Registration Process. The entire registration process is very simple. Users only need to register new terminal devices through the registration platform in a secure environment. During the device registration process, the user needs to enter the relevant device information (including device area number, device type number, device number, etc.) and preset password *pw*, and then the system will generate a unique device identification *ID* from the device information, and then generate The random number N_i is calculated as $C_i = H(ID, pw \oplus N_i)$. The device side will save the *ID*, *pw* and C_i, and transmit the device C_i and *ID* to the server of the registration platform through a secure channel for inspection and storage. The registration server inquires through the device *ID*, and if it is already registered, returns relevant information; if the device *ID* is not registered, the server saves C_i and *ID* and returns the registration success information. The schematic diagram of the device registration process is shown in Fig. 1.

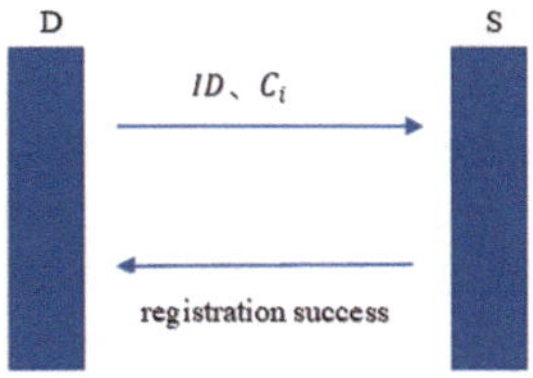

Fig. 1. Device registration flow chart.

Authentication Process. Based on identity identification and implemented with improved one-time password authentication technology as the core. The entire authentication process of this protocol is based on hash functions to calculate and provide two-way identity authentication, which occupies less equipment resources and communication and is more secure and efficient. The flow of the entire authentication process is shown in Fig. 2.

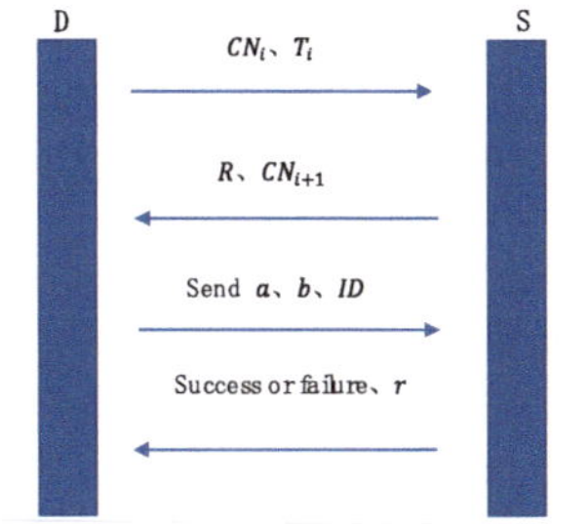

Fig. 2. Two-way authentication process.

- First, the device initiates an identity authentication request to the server and generates a random challenge number CN_i and a time stamp T_i, and sends the ID and CN_i to the server together.
- After receiving the information, the authentication server (edge gateway) first judges whether the time stamp T_i is valid. If it is valid, it looks up the information according to the ID and verifies whether the user exists. If it does not exist, it sends a failure message; otherwise, it finds the corresponding C_i according to the ID., And calculate $R = H(ID, C_i \oplus CN_i)$ as a response, and generate a random number CN_{i+1}, and send it to the device together.
- The user calculates $R' = H(ID, C_i \oplus CN_i)$ based on the information saved by himself, and compares the data. If the data comparison is the same, the authentication to the server is successful, and if it is different, the authentication request is resent.
- The user generates a random number N_{i+1}, and calculates and generates a new password $C_{i+1} = H(ID, PW \oplus N_{i+1})$, and calculates $C_{i+1} = H(ID, PW \oplus N_{i+1})$, calculate $D_i = H(ID, C_i \oplus CN_{i+1})$, calculate $a = C_{i+1} \oplus (D_{i+1} + D_i)$, calculate $b = C_i \oplus D_{i+1}$, and send (a, b, ID) to the server.

- When the edge gateway, that is, the authentication server, receives the authentication data transmitted by the device and performs related operations, calculate $D_{i+1} = b \oplus C_i$, and calculate $D_i = H(ID, C_i \oplus CN_{i+1})$, calculate $C_{i+1} = \mathrm{a} \oplus (D_{i+1} + D_i)$, calculate $D'_{i+1} = H(ID, \mathrm{C}_{\mathrm{i}+1})$, and put D'_{i+1} is compared with D_{i+1}. If the comparison is different, the authentication fails, otherwise the authentication succeeds. The server will update the database information and replace the original C_i with C_{i+1} to complete the password update.
- The authentication server calculates $r = H(ID, D_{i+1})$, and sends Success and r to the user to provide secondary authentication.
- After receiving r, the device terminal calculates $r' = H(ID, D_{i+1})$, and compares it with the received r. If the comparison is consistent, the authentication is passed and key negotiation begins. (And use $ak = D_{i+1}$ as the shared authentication material in the subsequent key negotiation stage).

2.2 Lightweight Key Agreement Protocol Based on Identity Authentication

The entire key agreement protocol is based on the authentication protocol designed above to verify the identity of the two parties, and then use the ECDH protocol to exchange keys to generate a temporary session key. For the entire key agreement protocol, the core is mainly composed of two stages: negotiation exchange and negotiation verification. The overall negotiation process of the lightweight key agreement protocol is shown in Fig. 3.

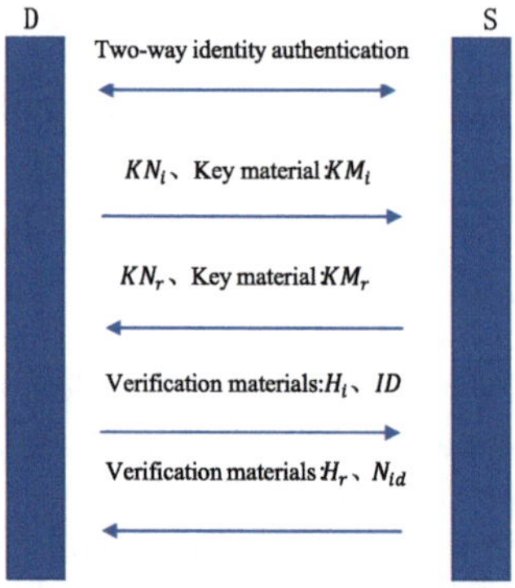

Fig. 3. Key agreement protocol process.

- First, the two parties negotiate for two-way identity authentication (that is, the authentication protocol described above). If the authentication is passed, shared authentication material $ak = D_{i+1}$ will be generated, which will be used in the subsequent key agreement phase to negotiate the two parties Perform authentication to protect against man-in-the-middle attacks.

- After the two-way identity authentication is passed, it enters the key agreement exchange stage. The device generates a random number KN_i, generates a random integer a, and calculates $KM_i = a \cdot G$, (G is the base point of the elliptic curve). Send KN_i, KM_i to the server.
- After the two-way identity authentication is passed, the server also enters the key negotiation and exchange phase, generates a random number KN_r, generates a random integer b, calculates $KM_r = b \cdot G$, receives the key material and other data from the device, and sends it to the device KM_r, KN_r.
- After receiving the key material, the device and the server calculate the shared key $K = k_i = a \cdot KM_r = k_r = b \cdot KM_i$. Calculate the temporary session key $Key = H(ak, K \oplus KN_i \oplus KN_r)$.
- The device side also calculates the verification material $H_i = H(ID, ak \oplus Key)$, and sends it to the server side for verification together with the ID.
- The server receives the verification materials sent from the device and calculates $H_i' = H(ID, ak \oplus Key)$ to compare with H_i. If they are different, the verification fails, and the negotiation is re-negotiated. If the two are the same, the verification is passed, and the Generate a random number N_{id}, and calculate $H_r = H(N_{id}, ak \oplus Key)$, then send N_{id}, H_r to the client together, and the server enters the secure communication stage.
- The client calculates $H_r' = H(N_{id}, ak \oplus Key)$ and compares it with H_r. If the two are consistent, the verification is passed. After the verification is passed, the secure communication stage is entered.

3 Protocol Analysis

3.1 Security Analysis

Anti-eavesdropping Attack. First, the result of hash operation based on PW C_i is a very important material, and subsequent materials such as D_i are related to it, which is related to the security of the entire authentication protocol. The protection of C_i is based on the secure channel of the registration platform, which is protected by encryption, and the attacker simply cannot get any important information from it. In the subsequent authentication process, only the Hash value and the random number are transmitted, and there is no algebraic association before and after the data sequence, and the identity verification is related to C_i and C_{i+1}. So no matter how the attacker eavesdrops because it does not know C_i, then it cannot pass the verification, nor can it infer any data and private information from the exchanged data. Therefore, this authentication scheme can prevent eavesdropping.

Anti-replay Attack. In the authentication scheme designed in this paper, the timestamp and random number are included in the sent message data, the challenge and response data are random, and the password will be dynamically updated after each identity authentication is passed, so the last time of data is already invalid. Therefore, in the case of time stamp verification, data randomness, and dynamic update of passwords, it can well resist replay attacks.

Impersonation Attack. Due to the plain text transmission of relevant data (such as *ID*) during the authentication process, the attacker deceives the edge gateway by pretending to be a legitimate user to impersonate the identity of the legitimate user and forging related hash data. However, because the private information is not known, the corresponding challenge information cannot be accurately answered, and the identity verification of the edge gateway cannot be passed at all. At the same time, the impersonation server is also unable to pass the authentication of the client, so the design scheme of this article can resist impersonation attacks.

Anti-man-in-the-middle Attack. OTP for traditional one-way authentication is vulnerable to man-in-the-middle attacks. In this solution, both parties are authenticated using two-way challenge/response authentication, and C_i participates in the generation of related data. Therefore, the attacker can neither pass authentication nor construct legal data, so it can effectively resist man-in-the-middle attacks.

Key Security. Only relevant hash data is transmitted during the entire authentication process. Password and other data have never appeared in the message, so there is no threat of key eavesdropping. Has high key security.

3.2 Performance Evaluation

The authentication key agreement protocol proposed in this paper, on the basis of ensuring protocol security, providing two-way identity authentication and key agreement functions, has less overhead for computing and communication resources and obvious advantages. T_{PM}: calculate the time of an elliptic curve scalar multiplication operation, T_{PA}: calculate the time of an elliptic curve scalar addition operation, T_{HP}: calculate the time of a hash function mapped to a point, T_H: calculate the time of a one-way hash function, T_I: Calculate the time of a modular inverse operation; T_{SE}: Calculate the time of a symmetric encryption and decryption (AES). The following Table 1 shows the comparisons of our proposed scheme and the previous schemes on performance evaluation.

Table 1. Performance evaluation of schemes.

Schemes	Data exchange	User	Server
Islam et al.	816 *bits*	$3T_{PM}+2T_{PA}+3T_H$	$4T_{PM}+2T_{PA}+T_{HP}$
Yang et al.	752 *bits*	$4T_{PM}+2T_{PA}+T_{HP}$	$4T_{PM}+2T_{PA}+T_{HP}$
Chou et al.	912 *bits*	$3T_{PM}+3T_H$	$3T_{PM}+5T_H$
Ours	1072 *bits*	$2T_{PM}+5T_H$	$2T_{PM}+4T_H$

It can be clearly seen from the above table that the overall performance of the authentication key agreement protocol designed in this paper is significantly better than other schemes, especially the computing resource occupies little, and the communication resource occupies less, so the entire authentication key agreement protocol can meet A large number of devices with limited resources in the IoT environment. At the same time, the entire protocol provides two-way identity authentication between the device and the server. It has higher security than most protocols. The entire protocol is safe, efficient and lightweight.

4 Conclusion

The lightweight security authentication key agreement protocol proposed in this paper can improve the efficiency of authentication and key agreement, and reduce the calculation and storage overhead while ensuring the safety of the device. For the current privacy data of a large number of Power grid Internet of Things devices cannot be guaranteed, there are problems of clear text transmission, low shared key security, and limited device resources. This protocol can solve these problems well. And in comparison with other solutions, this protocol has better advantages in terms of security and performance.

Acknowledgment. This work is supported by State Grid Jiangsu Electric Power Co., Ltd. Science and Technology Project "Research and Application of Online Intelligent Monitoring and Security Protection Technology for Distribution Equipment IoT Terminals Based on Edge Computing" (J2019065).

References

1. Wang, X.H.: Internet of things information security solution. Sci. Technol. Inf. **12**(16), 12 (2014)
2. Lu, Y.: Research and Implementation of Trusted Authentication and Automatic Access Technology for Internet of Things Terminals. Nanjing University of Posts and Telecommunications (2019)
3. Yang, J.H., Chang, C.C.: An ID-based remote mutual authentication with key agreement scheme for mobile devices on elliptic curve cryptosystem. Comput. Secur. **28**(3–4), 138–143 (2009)
4. Xu, Q.: Research on Unified Authentication Platform for Dynamic Passwords Based on BS Architecture. Zhejiang University of Technology (2012)
5. Xiao, Y.F.: Study on Diffie-Hellman protocol key interaction system. Comput. Knowl. Technol. **14**(03), 34–36 (2018)
6. Xu, D.W., Chen, W.: Digital signature and encryption algorithm based on elliptic curve. Comput. Eng. **37**(04), 168–169 + 189 (2011)
7. Shi, J.L.: Improvement and Application of Dynamic Password S/Key Authentication Protocol. Hefei University of Technology (2010)
8. Cui, W., Zhang, G.S., Li, H.: Research on a lightweight dynamic key agreement for internet of things identity authentication protocol. Comput. Appl. Res. **33**(2), 531–533 (2016)
9. Zhou, Y., Yu, X.S.: Exploration of issues related to Internet of Things security technology. China Manage. Inf. **20**(04), 151–152 (2017)
10. Wu, C.K., Zhang, L., Li, J.L.: IoT device trust system architecture and lightweight identity authentication scheme design. Inf. Netw. Secur. **09**, 16–20 (2017)
11. Diffie, W., Hellman, M.: New directions in cryptography. IEEE Trans. Inf. Theor. **22**(6), 644–654 (1976)

12. Igor, E.S.: Security of polynomial transformations of the Diffie-Hellman key. Finite Fields Appl. **10**, 123–131 (2003)
13. Debiao, H., Jianhua, C., Jin, H.: An ID-based client authentication with key agreement protocol for mobile client–server environment on ECC with provable security. Inf. Fusion **13** (3), 223–230 (2012)
14. Liao, Y.P., Wang, S.S.: A secure dynamic id based remote user authentication scheme for multi-server environment. Comput. Standar. Interf. **31**(1), 24–29 (2009)
15. Yoon, E.J., Yoo, K.Y.: Robust ID-based remote mutual authentication with key agreement scheme for mobile devices on ECC. In: IEEE International Conference on IEEE Cse. IEEE (2009)
16. Chou, C.H., Tsai, K.Y., Lu, C.F.: Two ID-based authenticated schemes with key agreement for mobile environments. J. Supercomput. **66**(2), 973–988 (2013)
17. Wu, W.: Research on Internet of Things Identity Authentication and Privacy Protection Technology in Edge Computing Environment. Xidian University (2019)
18. Huang, R.J., Wu, X.P., Li, H.C.: Identity authentication scheme based on identity identification encryption. J. Netw. Inf. Secur. **2**(06), 32–37 (2016)

Design and Implementation of a Relay Transceiver with Deep Coverage in Power Wireless Network

Jinshuai Wang(✉), Xunwei Zhao, LiYu Xiang, Lingzhi Zhang, and Gaoquan Ding

State Grid Information and Communication Industry Group Co. Ltd., Beijing 102211, China
1053349427@qq.com

Abstract. With the promotion of the power Internet of things strategy of State Grid Corporation of China, using wireless communication technology to realize the interconnection of everything in multi business scenarios has become a research focus. However, there are still blind coverage areas such as basement, dense urban area and building shadow area in the existing power wireless private network, which can not achieve full communication coverage. For this reason, this paper designs a relay transceiver for deep coverage of power wireless network, which adds relay nodes in the weak coverage area of the network, and evaluates its coverage ability by defining its importance, which effectively improves the coverage of power wireless communication system. The experimental results show that the relay transceiver designed in this paper can realize the deep coverage of the power wireless network and improve the reliability of the power wireless communication network.

Keywords: Power wireless network · Depth coverage · Relay transceiver · Network connectivity

1 Introduction

In 2020, State Grid Corporation of China will start a new journey to build an international leading energy Internet enterprise with Chinese characteristics, and the construction of power Internet of things is the core task to achieve this strategic goal. Advanced grid communication technology can ensure the smooth realization of smart grid construction goals. In order to promote the construction of diversified smart grid, it is difficult to rely on optical fiber access network. Only by using terminal communication can the construction of smart grid covering a wide range of areas be realized [1]. The application of power wireless private network technology in the field of power communication is conducive to the stable access of terminal communication to the network of power supply enterprises, so as to promote the long-term development of power grid enterprises [2].

However, with the application environment of power wireless communication network becoming more and more diversified and complex, part of the scenes (such as

Q. Liu et al. (Eds.): CENet 2020, AISC 1274, pp. 1431–1438, 2021.
https://doi.org/10.1007/978-981-15-8462-6_163

basement, dense buildings, etc.) in practical application are often interfered by signal, which directly leads to the decline of communication quality, the deterioration of received signal-to-noise ratio, and even the paralysis of communication system [3]. Therefore, wireless network deep coverage is very important in network optimization.

In order to solve this problem, based on the analysis of the existing power business demand for wireless communication, this paper designs a deep coverage scheme of power wireless signal based on relay transceiver, which can improve the network connectivity of the coverage area of power wireless network, reduce the communication network instability caused by weak coverage, cell handover failure and other problems.

2 The Demand of Power Business for Wireless Communication

With the development of Internet of things, access terminals need to meet the business requirements of ubiquitous access, real-time acquisition, comprehensive perception and intelligent computing. At present, power wireless network is widely used in multi service scenarios of power grid, mainly involving the following three scenarios [4].

(1) Emergency communication. When the power system is damaged by natural disasters, its stability will inevitably be affected, but wireless communication technology has a very strong anti disaster ability, good stability, and can be deployed quickly, so wireless communication can be used as an emergency communication technology [5]. When there are some faults in the optical cable that can not be dealt with quickly, wireless communication can be used in emergency to ensure the orderly operation of the whole system.
(2) Remote access and extension. In some areas of the existing power system, the distance between the power supply station and the main power substation is relatively long. If the technology of laying optical cable is adopted, not only the coverage is relatively small, but also the cost of laying the line is large, so a large amount of manpower, material resources and capital needs must be invested. If we use the wireless network technology to cover the local area, we can avoid these problems and reduce the unnecessary financial loss [6]. At the same time, we can ensure that the main substation and the regional power supply station support communicate more stably and reliably.
(3) User's meter reading. Wireless network technology can be used to establish a complete radiation communication system, so that the system can be used to effectively monitor the user's power consumption, ensure the data security, and facilitate the accurate control of power consumption.

3 Principle and Architecture Design of Coverage Enhancement for Power Wireless Relay Transceiver

3.1 Depth Coverage Enhancement Principle

Different from the deep coverage problem of wireless public network, there are few mobile terminals due to the fixed location of power system terminals. Therefore, the deployment of wireless relay transceivers in the coverage blind area of electric power wireless network can effectively improve the connectivity of the network, thus improving the service carrying capacity of the network.

Definition: network connectivity of n nodes $\partial[G_n]$:

$$\partial[G_n]=\frac{1}{n\cdot p_{SNR}} \tag{1}$$

Where P_{SNR} is the average signal-to-noise ratio of the network, defined as follows:

$$p_{SNR}=\frac{\sum\limits_{i\neq j\in N} p_{ij}}{n-1} \tag{2}$$

Where P_{ij} is the signal-to-noise ratio between any two nodes in the network, substituting (2) into (1) can obtain:

$$\partial[G_n]=\frac{n-1}{n\sum\limits_{i\neq j\in N} p_{ij}} \tag{3}$$

In order to improve the connectivity of the network, a relay node $(n + 1)$ is added in the weak coverage area of the network. The definition of importance $I(n + 1)$ is as follows:

$$I(n+1)=\frac{\partial[G_{n+1}]}{\partial[G_n]}-1 \tag{4}$$

The higher the importance I is, the stronger the enhanced coverage ability of relay node $(n + 1)$ is.

3.2 Overall Architecture Design

The relay transceiver designed in this paper is mainly composed of analog RF, digital front-end and baseband digital processing part. And Fig. 1 shows its architecture.

Among them, the RF part is the analog RF part, which is composed of antenna, coupler, band-pass/low-pass filter, power amplifier and other modules. This part has the functions of digital to analog/analog-to-digital conversion, up and down conversion of signal and interference filtering. The architecture design is shown in Fig. 2.

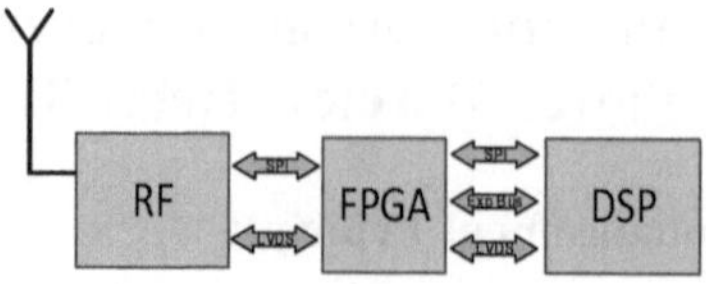

Fig. 1. Schematic diagram of relay transceiver architecture.

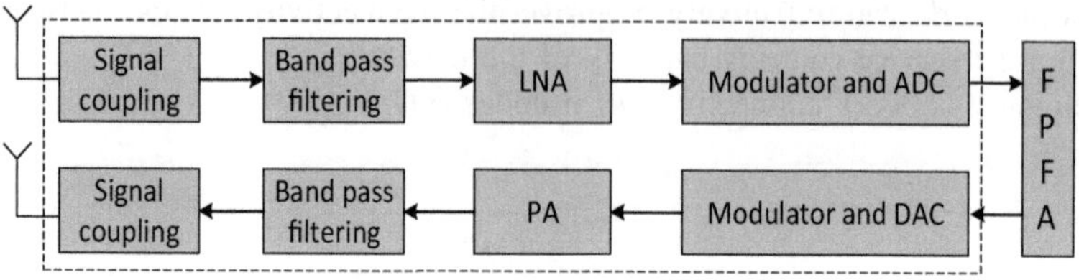

Fig. 2. Architecture design of analog RF circuit.

FPGA part mainly completes the function of digital front-end, which is composed of receiving channel, transmitting channel, clock management, interference elimination and other modules. It mainly completes the interference suppression, real-time acquisition of sending and receiving signals and other functions. The architecture design is shown in Fig. 3.

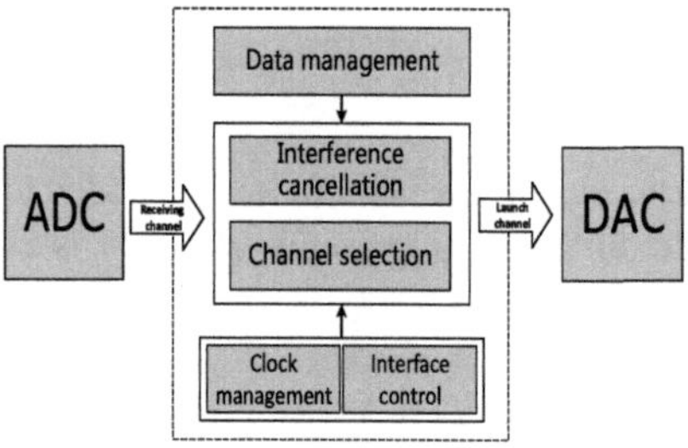

Fig. 3. FPGA architecture design.

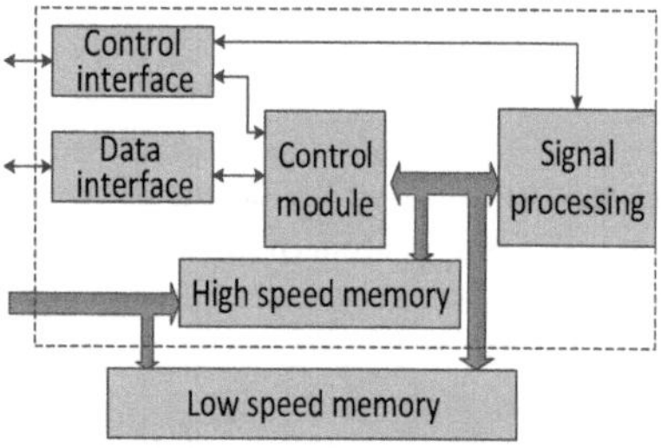

Fig. 4. Architecture of baseband processing part.

The DSP part mainly completes the baseband processing of the signal, including the control and algorithm processing, and can realize the power control of the receiver and the transmitter as well as the synchronization of the receiving and transmitting signals of the system.

3.3 Baseband Processing Design

In the design scheme of the transceiver, the baseband processing part chooses DSP as the wireless baseband processor because it undertakes the processing of the received data. As shown in Fig. 4, the architecture of baseband processing part is shown.

In the baseband processing module, from the perspective of functional modules, it is mainly divided into three parts: processing unit, storage unit and interface unit. The storage module is divided into high-speed storage and low-speed storage. The processing unit is composed of control module and signal processing module

(1) Control module: the control module mainly completes the main protocol processing and control functions, and can access almost all modules, interfaces and addresses in the chip. It can complete the processing function of the basic protocol stack and the control and management function.
(2) Signal processing module: it mainly completes various functions of physical layer protocol stack and signal processing related operation. From the architecture diagram, it can be seen that the external data is sent to the signal processing module and stored in the storage unit through the interface module. The signal processing module will access the storage unit and process the corresponding data.

4 Testing and Verification

4.1 Test Plan Structure

In order to test and verify the transceiver performance of the wireless repeater designed in this paper, and avoid uncertainties as much as possible, and still be able to complete the test normally and smoothly, the test scheme shown in Fig. 5 is adopted in this paper.

The test signal mainly includes monophonic signal and 4 m wide-band signal. The purpose is to verify the performance of the system, such as frequency point configuration, signal power, interference suppression and isolation ability, by observing the position, energy and interference spurious of the corresponding signal on the coupling network side.

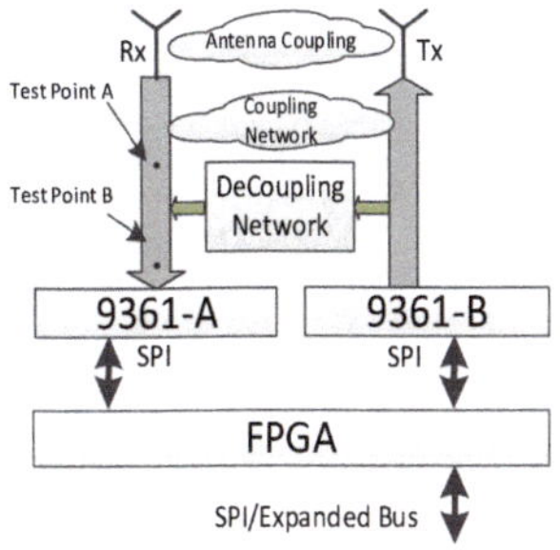

Fig. 5. Test scheme structure.

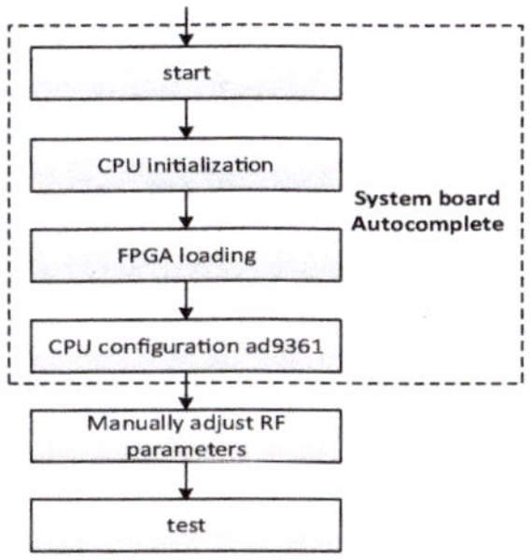

Fig. 6. System test flow chart.

The signal is injected from the receiving end of the uplink channel, processed by FPGA, D/A conversion, power amplifier etc., sent from the RF port, and then processed by the power amplifier, A/D conversion and FPGA of the downlink channel to obtain the repeater signal. Finally, the data of the receiver is sampled from 1024 sampling points. And the waveform or spectrum chart is drawn by MATLAB. So as to verify whether the repeater can realize the reproduction and transmission of the received signal.

4.2 Test Process

According to the test plan designed in the previous section, the test flow adopted in this paper is shown in Fig. 6.

The system testing process is divided into three parts:

(1) System startup and automatic configuration: this part mainly includes board power on, DSP startup, FPGA logic loading and system configuration process.
(2) Adjust RF configuration parameters: this phase is mainly to adjust the corresponding individual registers after the basic configuration of the previous phase, such as frequency point, sampling rate, test mode, test single tone frequency configuration, etc.
(3) Test: conduct performance test according to the test plan.

4.3 Verification Results

In this section, the monophonic signal and broadband signal are respectively poured into the system from the receiving end through the signal generator. The signal spectrum of the two measured signals at the transmitting end is mapped with MATLAB. Figure 7 and Fig. 8 are the setting interface of the signal generator and the spectrum diagram of the bandwidth signal displayed on the spectrometer.

Fig. 7. Signal generator setting interface.

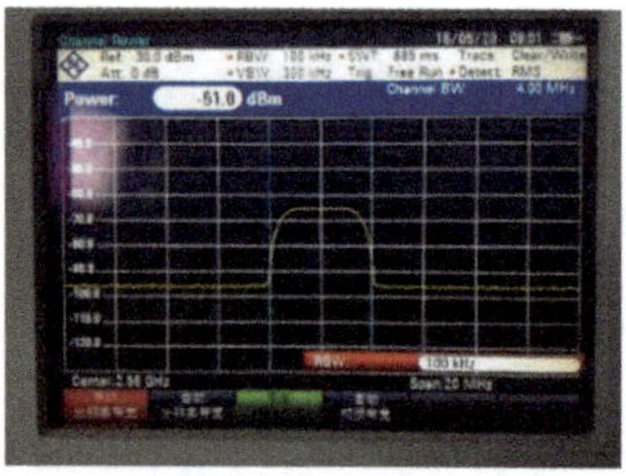

Fig. 8. 4 M broadband signal spectrum.

At present, the working frequency point of the relay transceiver is 2555 M. In order to observe the system's ability to suppress the image signal, this test uses the offset signal with the frequency point at 2560 M. Figure 9 and Fig. 10 are the frequency spectrums of single tone and broadband signals collected by the transmitter respectively.

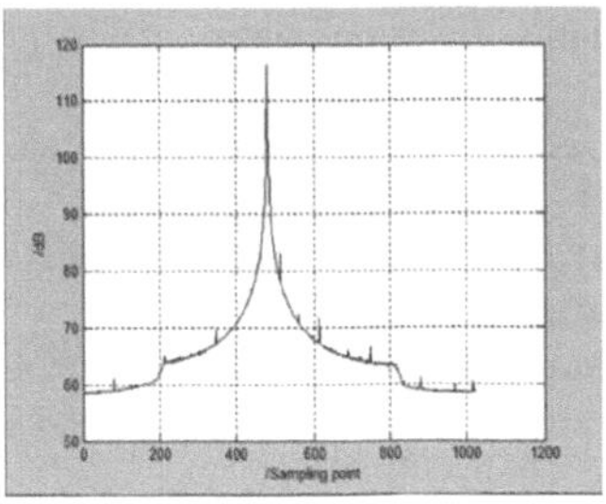

Fig. 9. Spectrum of single tone signal at transmitter.

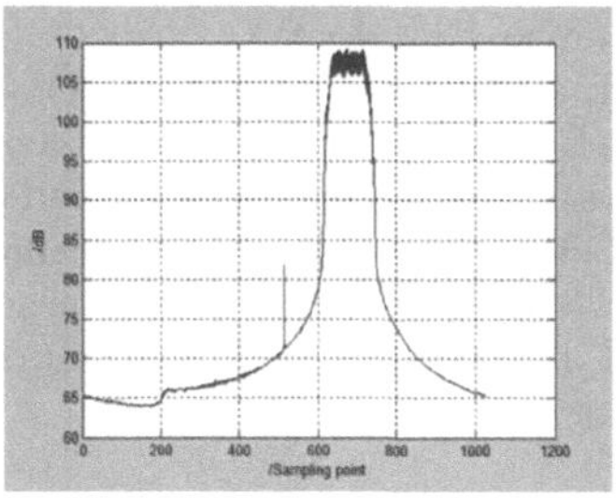

Fig. 10. Broadband signal spectrum of transmitter.

It can be seen from the Fig. that the transmitter can reproduce the signal of the receiver very well. From the spectrum diagram of the single tone signal, it can be seen that the image suppression of the system can reach 44 dB. And the image of the broadband signal is basically submerged in the bottom noise. However, due to the existence of local oscillator, mixing and noise, the signal at the transmitter also has some interference.

The test scheme of this paper verifies that the relay transceiver can realize the signal reproduction and forwarding well, and has better performance in interference suppression.

5 Conclusion

Based on the demand of electric power business for electric power wireless network, this paper designs a solution based on wireless relay transceiver, aiming at the blind coverage area of electric power wireless network in basement, dense urban area, building shadow area and so on. Firstly, this paper analyzes the demand of electric power business for electric power wireless network. Aiming at the blind coverage area of electric power wireless network in basement, dense urban area, building shadow area, etc., which can not achieve full coverage, a solution based on wireless relay transceiver is designed. From the verification results, it can be seen that the wireless relay transceiver designed in this paper can support the deep coverage of power wireless signal, meet the real-time and high reliable communication requirements of power consumption information collection and intelligent power consumption services, which is of great significance for the development of smart grid.

References

1. Zhu, J.J.: Research on Resource Allocation Scheme in Wireless Heterogeneous Network Environment. North China Electric Power University (Beijing) (2019)
2. Li, Z., Chen, K.Y.: Coverage analysis of electric power communication wireless private network. Inf. Commun. **09**, 168–170 (2019)
3. Lu, M., Mao, P., Zhao, P.Y.: Study on coverage planning of electric power wireless private network. China New Commun. **21**(11), 48–50 (2019)
4. Zhang, C.: Application of wireless communication technology in power communication private network. Digital Commun. World **06**, 167 (2019)
5. Li, T.R.: Research and Analysis of Intelligent Electricity Meter Network Based on Internet of Things. Nanjing University of Posts and telecommunications (2017)
6. Gao, L.: Research on LTE power wireless private network based on service coverage. China New Commun. **20**(16), 218 (2018)

A Multi-domain Virtual Network Embedding Approach

Yonghua Huo[1], Chunxiao Song[1], Yi Cao[2], Juntao Zheng[3], and Jie Min[3(✉)]

[1] The 54th Research Institute of CETC, Shijiazhuang 050000, China
[2] Military Representative Office of General Military Equipment Development Department, Shijiazhuang 050000, China
[3] State Key Laboratory of Networking and Switching Technology, Beijing University of Posts and Telecommunications, Beijing 100876, China
minj6@bupt.edu.cn

Abstract. As a key technology to promote the development of network applications, network virtualization decouples traditional network service providers into service providers and infrastructure providers, and separates the physical network into multiple virtual end-to-end networks to meet the different needs of the upper tenants. In application scenarios represented by edge computing, virtual networks usually need to be carried by multiple different infrastructure networks in different geographic locations. In order to ensure the security of the internal network topology, each infrastructure provider has incomplete information sharing between them, and the embedding is also restricted by region and resource type. Therefore, this paper considers the impact of the substrate network topology on virtual network embedding from multi-angle, and designs a multi-domain virtual network embedding method based on particle swarm optimization. In this paper, a calculation index based on embedding cost and link interference is designed, and a particle generation mechanism based on node resource richness in particle swarm optimization is proposed to reduce the computing time of the algorithm. Simulation results prove the superiority of the embedding algorithm proposed in this paper in terms of embedding cost and computing time.

Keywords: Multi-domain virtual network embedding · PSO · Network topology · Virtual network request division

1 Introduction

Over the past decade, the Internet has grown exponentially, and it has faced severe technical challenges in terms of security, mobility, and QoS. In recent years, network virtualization has aroused widespread concern in academia and industry, and is regarded as one of the most promising technologies for the Internet in the future. Network virtualization technology divides traditional network into infrastructure providers (InP) and service providers (SP), allowing multiple heterogeneous virtual networks to coexist on shared infrastructure [1]. It guaranteed better flexibility and manageability.

Q. Liu et al. (Eds.): CENet 2020, AISC 1274, pp. 1439–1446, 2021.
https://doi.org/10.1007/978-981-15-8462-6_164

The most critical technology in network virtualization is the virtual network embedding method [2] (Virtual Network Embedding, VNE). Virtual network embedding is to embed the nodes and links of each virtual network into the corresponding substrate nodes and links. In application scenarios represented by edge computing, virtual networks often need to be carried by multiple infrastructure networks in different geographic locations. A variety of single-domain virtual network embedding methods previously proposed are not applicable to multi-domain virtual network embedding. We need to design efficient multi-domain virtual network embedding methods to reduce embedding cost and obtain the best embedding scheme.

This paper designs a multi-domain virtual network embedding algorithm based on Particle Swarm Optimization (PSO) algorithm [3]. The main contributions of this paper are as follows: (1) The algorithm refers to the continuous PSO algorithm and adopts the discrete PSO algorithm, optimized according to the characteristics of the virtual network embedding. (2) This paper designs a new calculation index, adds consideration to link interference, uses constant value to adjust the impact of embedding cost and link interference on this index, which considers complex network from multi-angle. (3) a particle generation mechanism based on node resource richness in particle swarm optimization is proposed to helps particles iterate to the best position quickly, reduce the computing time of the algorithm.

The remainder of this paper is organized as follows. In Sect. 2, we discuss some related works. In Sect. 3, we introduces the process of the multi-domain virtual network embedding algorithm based on PSO. In Sect. 4, we compare our algorithm with other schemes numerically. We conclude the paper in Sect. 5.

2 Related Work

In the previous research, the multi-domain virtual network embedding method has achieved some results. In reality we consider different schemes to solve the embedding problem in different situations.

One solution is not to use VNP for coordination and communication, but to perform distributed mapping between InP based on a common protocol. We classify it as a distributed algorithm. Chowdhury et al. introduced the concept of PolyViNE in [4]. PolyViNE is a policy-based embedding architecture, which completes the embedding by forwarding network requests to the remaining InPs after completing the mapping within the InP domain through a global distributed algorithm. The distributed consistent bidding mechanism proposed by Esposito et al. [5] maximizes the efficiency of the substrate network.

The other is to add a virtual network provider (Virtual Network Provider, VNP), that is a global controller, between SP and InP, responsible for acting as an intermediary, statistics and analysis of the information of the substrate network. Such algorithms are classified as centralized algorithms. In literature [6], Vaishnavi et al. transformed the goal of the algorithm into embed to the entire substrate network which knows network resources, computing power, etc., treating each InP as a pseudo node. In literature [7], a multi-domain virtual network embedding method with limited information disclosure was proposed, which use the flow matrix to seek better scheme in a wider range. Literature [8] applies the ant colony algorithm. The ant colony secretes and updates the pheromone based on the cost of link embedding, and learn each other according to the feedback information.

3 Algorithm Design

3.1 Problem Model

In actual application scenarios, considering the amount of resources and types of resources required by a virtual network request, an infrastructure provider may not be able to fulfill the network requirements, and often requires cooperation between multiple infrastructure providers to complete the embedding. The most important thing is that, for the security of the internal network topology of each domain only exposes limited information for virtual network embedding.

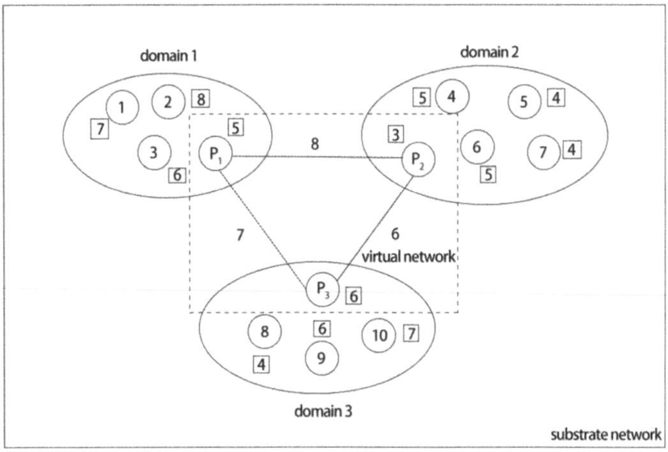

Fig. 1. Multi-domain virtual network embedding.

As shown in Fig. 1, we embed virtual nodes and virtual links into the substrate network composed of multiple domains to form a virtual network. The solid line rectangle represents the substrate network environment with three domains, and the dotted rectangle represents the virtual network requests for three virtual nodes and three virtual links. The number in the box represents the amount of resources owned or required by the node. The number next to the link represents the amount of bandwidth required. We need to meet the service requirements and complete multi-domain virtual network embedding efficiently while considering the global embedding cost and computing time in the limited substrate network information.

This article introduces a multi-domain virtual network embedding algorithm based on the network topology, considering the network topology and the characteristics of the multi-domain network, as well as the reference indicators for the multi-domain virtual network embedding problem. The main notations are provided in Table 1.

Table 1. Notations.

n^s	Substrate nodes
n^v	Virtual nodes
l^s	Substrate links
l^v	Virtual links
$C()$	Cost of the virtual node or virtual link
$P()$	Cost of the substrate node or substrate link
SC	Total cost of the embedding plan
D^v	The set of domains which virtual nodes can be embedded
D^s	The domain which the substrate node belongs
D_S^V	The domain to which the substrate node embedded by a virtual node belongs
t	computing time of the algorithm

Next, the constraints of the problem and the objective function are given.

First, the amount of CPU resources required by the virtual node is less than the amount of CPU resources of the substrate node embedded:

$$C(n^v) \leq C(n^s) \tag{1}$$

Similarly, the required bandwidth of the virtual link is less than the bandwidth of the substrate link embedded:

$$C(l^v) \leq C(l^s) \tag{2}$$

In addition, the substrate node to which the virtual node is embedded belongs to the specified domain:

$$D_S^V \in D^V \tag{3}$$

Our objective function is to minimize the embedding cost of the virtual network:

$$\min SC = \sum C(n^v) \times P(n^s) + \sum C(l^v) \times P(l^s) \tag{4}$$

At the same time, the computing time of the algorithm should be reduced as much as possible:

$$\min t \tag{5}$$

3.2 Flow Control Algorithm Based on PSO

We designed a calculation index based on the embedding cost and link interference in the algorithm, referring to the literature [9], considering complex networks, as a new embedding cost instead of a measure which is simply resource occupancy multiply price. The formula as follows:

$$LI = \sum_{n^v} \frac{\sum Far}{Degree} \tag{6}$$

$$R = a \times LI + b \times SC \tag{7}$$

LI is expressed as link interference and is defined as the average length of the link connected by the node. *SC* is formula (4). By controlling the two constant values *a* and *b* to control the influence ratio. A new calculation index R is denoted.

In addition, the algorithm in this paper refers to the idea of discrete PSO algorithm for multi-domain virtual network embedding [10]. At the same time, reference [11] designed a particle generation mechanism based on node resource to reduce algorithm computing time. The calculation formula of node resource richness is as follows:

$$NR(n_i^s) = C(n_i^s) \times \sum_{l \in L(n_i^s)} C(l) \tag{8}$$

NR is expressed as the resource richness of a node, which is defined as the amount of node resources multiplied by the total bandwidth of the connected link. In many embedding methods, the optimal scheme often adopts a node with high node resource richness, so the higher the richness, the higher the probability that the node is selected. In particle swarm optimization, each particle has two parameters. They are location and speed:

$$X_i = (x_i^1, x_i^2, x_i^3 \cdots x_i^D) \tag{9}$$

$$V_i = (v_i^1, v_i^2, v_i^3 \cdots v_i^D) \tag{10}$$

Location vectors are no longer expressed as location in D-dimensional space. Each location represents D virtual nodes embedding scheme. Each velocity vector represents a modification direction of the embedding scheme.

The algorithm records its own optimal location X_{pb} and global optimal location X_{gb} in iterations, and the global optimal location output after multiple iterations is regarded as the best embedding scheme. The formulas as follows:

$$V_{i+1} = \alpha V_i \oplus \beta(X_{pb} \ominus X_i) \oplus \gamma(X_{gb} \ominus X_i) \tag{11}$$

$$X_{i+1} = X_i \otimes V_{i+1} \tag{12}$$

$$其中，\alpha + \beta + \gamma = 1 \tag{13}$$

The three constants values α, β, and γ are all from 0 to 1, which are used to control the influence of the two optimal solutions and previous speed on the new speed. ⊕ is to select one of the three components according to the probability. The Θ operator used to compare the differences between two embedding results. The ⊗ symbol is refer to the velocity vector, and adjust the regeneration of the scheme.

After design of the algorithm, we define the PSO-based multi-domain virtual network embedding algorithm process in Fig. 2:

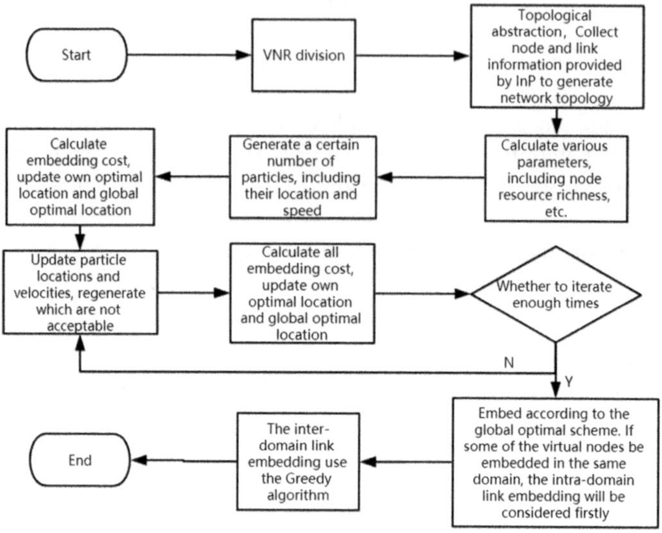

Fig. 2. Multi-domain virtual network embedding algorithm.

4 Simulation

4.1 System Model

We use JAVA language (IDE: Eclipse) to implement it. The virtual network request, the substrate network parameters and the algorithm parameters shown as follows (Table 2):

Table 2. Parameters

Parameters	Value
Number of substrate domains	5
Number of substrate nodes per domain	12
Initial resources of each substrate node	10000–20000
Price of each substrate node	3–6
Initial bandwidth resources of each substrate link	1000–2000
Price of each intra-domain substrate link	1
Length of each substrate link	500–1000
Number of virtual nodes	14
Node resource requirement of each virtual node	2000–4000
Candidate domains of each virtual node	2
The bandwidth requirement of each virtual link	200–400
Number of particles	20
Iteration time of PSO	50
α, β, γ	0.2, 0.4, 0.4
a, b	1, 0.2

4.2 Simulation Results

In order to reflect its superiority in embedding cost and computing time, one of the comparison algorithms we choose is a random multi-domain virtual network embedding algorithm VNE-Random. Another comparison algorithm is the algorithm proposed in [12], which adopted PSO. We call it VNE-PSO. The differences with the algorithm in this paper are the generation mechanism, the selection of embedding nodes and the calculation of embedding costs. In addition, the algorithm in this paper is represented by VNE-PSO-I.

The experiment shown in Fig. 3 uses the number of iterations as the controllable variable, and evaluates the algorithm by calculating the average embedding cost. VNE-PSO-R is an algorithm that does not use the node resource richness generation mechanism in this paper. It is obvious that the algorithm in this paper can be used in fewer number of iterations to achieve an embedding scheme that is close to the least cost, which reduce the computing time.

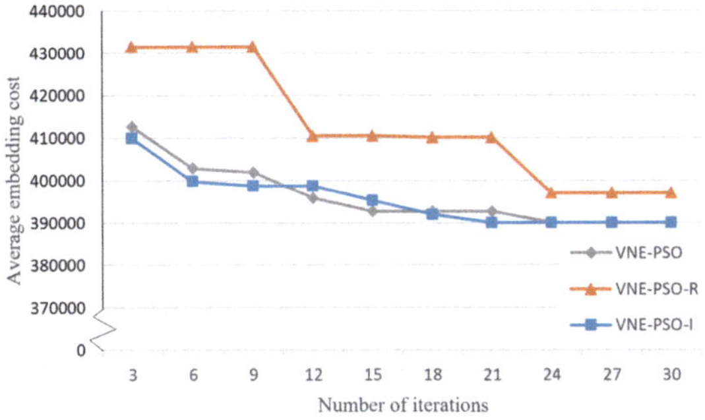

Fig. 3. Average embedding cost influenced by number of iterations.

The experiments shown in Fig. 4 and Fig. 5 evaluate the algorithm performance by separately controlling the number of substrate domains and virtual nodes. As the number of domains and the number of virtual nodes increase, the average embedding cost is increasing. In this paper, the average embedding cost is always better than the random algorithm. Therefore, it can be concluded that it can quickly obtain the optimal or sub-optimal embedding scheme in complex situation.

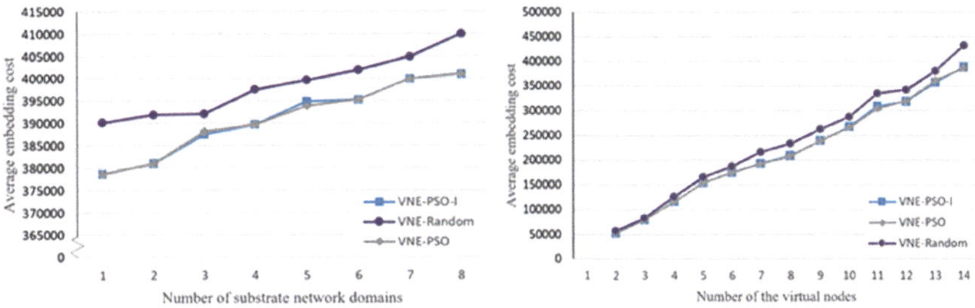

Fig. 4. Average embedding cost influenced by number of substrate domains and virtual nodes.

5 Conclusion

This paper studies the research background and research status of multi-domain virtual network embedding, designs a PSO-based multi-domain virtual network embedding algorithm and simulate it. The results show that the algorithm can perform multi-domain virtual network embedding well, and has superiority on embedding cost and computing time.

References

1. Feamster, N., Gao, L., Rexford, J.: How to lease the Internet in your spare time. ACM SIGCOMM Comput. Commun. Rev. **37**(1), 64 (2007)
2. Andreas, F., et al.: Virtual network embedding: a survey. IEEE Commun. Surv. Tutor. **15**, 1888–1906 (2013)
3. Kennedy, J., Eberhart, R., et al.: Particle swarm optimization. In: Proceedings of IEEE International Conference on Neural Networks, Perth, Australia, vol. 4, pp. 1942–1948 (1995)
4. Chowdhury, M., Samuel, F., Boutabar, R.: PolyViNE: policy-based virtual network embedding across multiple domains. In: Proceedings of the second ACM SIGCOMM Workshop on Virtualized Infrastructure Systems and Architectures, New Delhi, pp. 49–56. ACM, India (2013)
5. Esposito, F., Paolad, D., Matta, I.: A general distributed approach to slice embedding with guarantees. In: IFIP Networking Conference [s.l.]: [s.n.], pp. 1–9 (2013)
6. Vaishnavi, I., Guerzoni, R., Trivisonno, R.: Recursive, hierarchical embedding of virtual infrastructure in multi-domain substrates. In: IEEE Conference on Network Softwarization, London, pp. 1–9. IEEE, UK (2015)
7. David, D., Amr, R., Panagiotis, P.: Multi-provider virtual network embedding with limited information disclosure. IEEE Trans. Netw. Serv. Manag. **12**(2), 188–201 (2015)
8. Cao, W., Wang, H., Liu, L.: A modified ant colony optimization algorithm for virtual network embedding. In: Algorithms and Architectures for Parallel Processing, pp. 68–78 (2014)
9. Cao, H.T., et al.: Novel node-ranking approach and multiple topology attributes-based embedding algorithm for single-domain virtual network embedding. IEEE Internet Things J. **5**, 108–120 (2018)
10. Kennedy, J., Eberhart, R.: A discrete binary version of the particle swarm algorithm. IEEE Int. Conf. Syst. Man Cybernet. **5**, 4104–4108 (1997)
11. Zhang, Z.B., Cheng, X., Su, S., Wang, Y.W., Shuang, K., Luo, Y.: A unified enhanced particle swarm optimization-based virtual network embedding algorithm. Int. J. Commun. Syst. **26**(8), 1054–1073 (2013)
12. Ni, Y.J., Huang, G.Y., Wu, S., Li, C.X., Zhang, P.Y., Yao, H.P.: A PSO based multi-domain virtual network embedding approach. China Commun. **16**(04), 105–119 (2019)

An Intent-Based Network Slice Orchestration Method

Jie Min(✉), Ying Wang, and Peng Yu

State Key Laboratory of Networking and Switching Technology,
Beijing University of Posts and Telecommunications, Beijing 100876, China
{minj6,wangy,yupeng}@bupt.edu.cn

Abstract. Considering a large number of end-to-end services, in order to orchestrate and manage the deployment requests of edge computing servers to provide better intelligent services, in this paper, we adopt network slice. In the regard of network slice, the most important challenge is how to deploy the node and link resource requirements in the request to the underlying network with limited resources. Current researches ignores the needs of multi-domain services. In addition, the greedy method of orchestration is likely to cause node resource constraints, resulting in problems such as low acceptance rate and low revenue. Therefore, in this paper, we propose an intent-based multi-domain anti-fragmentation request orchestration method, including expression method of service requirements, scheduling of services based on priorities, selection of candidate nodes, and multi-domain orchestration algorithm. The experimental results show that our approach has a good performance in improving the acceptance rate and overall revenue.

Keywords: Network slice · Multi-domain · Anti-fragmentation · Intent

1 Introduction

As a technology promotes the development of future network, edge computing has received widespread attention. Edge computing [1] provides high-quality services to devices at a close distance on the edge of network to meet various application requirements. To implement edge computing, the deployment of edge devices are required. Therefore, we take network slice (NS) to orchestrate requests. NS decouples traditional network service providers into service providers (SP) and infrastructure providers (InP), making InP focus on underlying facility management, while SP leases resources from InP to satisfy tenants.

The most critical thing of NS is orchestration. Existing researches have the following challenges: (1) In traditional network, administrators with professional knowledge realize requirements raised by tenants to manage the network. However, tenant without such skills is prone to the problem of demand representation errors. (2) Most approaches only consider single domain orchestration, ignoring the need for multi-domain service in real scenarios. Some methods take service division to handle multi-domain requests, rather than from a global perspective. (3) Pursuit of maximizing physical resources in node ranking may lead to excessive consumption of node's resources, reducing the acceptance rate.

Q. Liu et al. (Eds.): CENet 2020, AISC 1274, pp. 1447–1455, 2021.
https://doi.org/10.1007/978-981-15-8462-6_165

In this paper, we propose an intent-based multi-domain anti-fragmentation orchestration approach (IMAF). The main contributions can be summarized as follows: (1) We define a description method based on intent for NS requirements description, so that users can describe their needs without grasping specific network configuration details. (2) We propose and implement a request scheduling method that calculates the priority for each network slice request (NSR) and a anti-fragmentation selection method for nodes in NSR. (3) We design and implement a network slice service orchestration algorithm that considers the positional relationship between nodes and each domain in a request based on a topology abstraction method.

2 Related Work

2.1 Intent-Based Service Request Representation

ONF proposed intent [2] to convert user's demand to underlying instructions. Existing intent description methods can be divided into graphical, self-defined and natural. Graphical PGA [3] intuitively expresses service function strategies and detects potential conflicts. However, the focus of [3] is not on how tenants express NSRs. Natural [4] converts key-value pairs extracted from needs into network configuration commands, but relies on natural language processing technology. Self-defined NEMO [5] can flexibly define requests in SDN scenario. However, when using NEMO, it is complicated for tenants to describe each element in detail. Compared with above researches, our approach is self-defined to express the service by abstracting the network elements involved in multi-domain NS scenario.

2.2 Network Slice Orchestration Methods

According to the application range of NSRs, orchestration algorithms can be divided into: single-domain-oriented approach and multi-domain-oriented approach.

Li et al. [6] is an exact algorithm that classifies requests and specifies different linear programming optimization goals for them: load balancing, minimizing delay, minimizing resource consumption. Since NS orchestration problem proved to be NP-hard, exact algorithm may not be a good way to solve problem in polynomial time. Based on greedy algorithm, Zhang et al. [7] proposed two node ranking indicators considering resource and consumption. However, it only considers resource cost of nodes. Moreover, greedy method is easy to cause node resource fragmentation.

Based on heuristic algorithm, Ni et al. [8] separates requests by nodes, selects candidate nodes according to resource requirements and alternative domains, then uses an improved PSO algorithm for final solutions. However, dividing NSR and mapping sub-requests one by one is not considered from a global perspective. Houidi, et al. [9] realizes the multi-domain NSR orchestration with three steps: resource matching, splitting and embedding. In splitting step, the authors adopt exact algorithm to get results, which may result in the same problem as [6].

Hua et al. [10] 's topology abstraction method only exposes edge switches and provides full connectivity among them. In this way, we can hide the internal details of each domain to view end-to-end routing issue from a global perspective.

Different from traditional greedy method used in node ranking, we take revenue, complexity, location and other resource requirements into account to improve performance. In this paper, we first define a simple and intuitive description method for NSR. Then we design an anti-fragmentation node selection method and propose a node ranking heuristic algorithm to get better solutions.

3 Network Slice Orchestration Problem Description

3.1 Substrate Network and NSR Model

Substrate Network (SN) can be modeled by weighed undirected graph $G_s = (N_s, L_s)$, where N_s represents set of substrate nodes, L_s represents set of substrate links. Each $n_s \in N_s$ has available CPU capacity $CPU(n_s)$ and domain attribute $Domain(n_s)$. Each $l_s \in L_s$ owns bandwidth capacity $BW(l_s)$ and delay $Delay(l_s)$. P_s composed of $P_s(i,j)$ is a set of all loop-free paths in the underlying network, where $i \in n_s$, $j \in n_s$.

NSR is modeled by $G_r = \left(N_r, L_r, T_a, T_d, T_{life}\right)$, where N_r is set of NSR nodes and L_r is set of NSR links, T_a represents the arrive time of NSR, T_d represents its departure time, T_{life} is its lifetime. Each $n_r \in N_r$ has CPU requirement $CPU(n_r)$, and each $l_r \in L_r$ has bandwidth requirement $BW(l_r)$ and delay constraint $Delay(l_r)$.

3.2 Problem Formulation

In the following section, we present the mathematical programming model of NS orchestration problem. Our target is to increase acceptance rate and total revenue.

Objectives:

$$\text{Maxmize} \quad Revenue(NSR_i) = \sum\nolimits_{n_r \in N_r} CPU(n_r) + \sum\nolimits_{l_r \in L_r} BW(l_r) \tag{1}$$

$$\text{Maxmize} \quad Acceptance\,Rate = \lim\nolimits_{T \to \infty} \frac{\sum_{t=0}^{T} NSR_{acc}}{\sum_{t=0}^{T} NSR_{tot}} \tag{2}$$

where NSR_{acc} declares the number of accepted NSR, NSR_{tot} indicates the number of total NSR.

Constraints:

$$CPU(n_s) \geq CPU(n_r), \forall n_s \in N_s, \forall n_r \in N_r \tag{3}$$

$$BW(l_s) \geq BW(l_v), \forall l_s \in P_s, \forall l_r \in L_r, P_s = Map(l_r) \tag{4}$$

$$Delay(l_s) \leq Delay(l_v), \forall l_s \in P_s, \forall l_r \in L_r, P_s = Map(l_r) \tag{5}$$

Equation (3) ensures that available CPU capacity of node n_s can meet the CPU requirement of node n_r. Equations (4)–(5) guarantee that available bandwidth resource of l_s can satisfy the bandwidth requirement of l_r, and the delay of l_s is within the delay constraint of l_r. Moreover, relation between n_s and n_r is one-one mapping.

4 The Proposed Approach

In this section, we detail our IMAF. First, we present our self-defined intent description method for tenants. Second, we propose NSR scheduling approach based on calculated priority before orchestration. Third, we detail the candidate nodes selection algorithm. Last, based on above results and inter-domain requirements of each node, we propose the multi-domain NSR orchestration algorithm.

4.1 Intent Specification of NSR

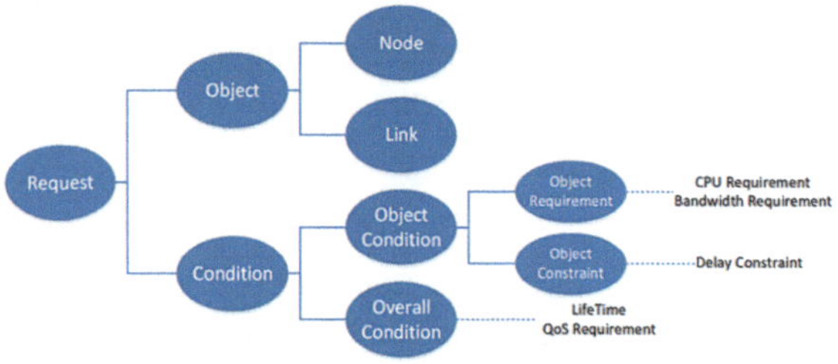

Fig. 1. Network elements abstraction.

```
<NSR> ::= "create NSR " NSR_id ":" (<Objects>|<Conditions>)
<Objects> ::= <Object>{"\n"<Object>}
<Object> ::= "add "(<Node> | <Link>){("," | ",\n")(<Node> | <Link>)}
<Node> ::=   "node id: "node_id
<Link> ::= "link id: "link_id | "src_node: "src_id | "dest_node: " dest_id
<Conditions> ::= <Condition>{("\n")<Condition>}
<Condition> ::= "add "(<Object Condition> | <Overall Condition>) {("," | ",\n")(<Object
Condition> | <Overall Condition>)}
<Object Condition> ::= <Node Condition> | <Link Condition>
<Node Condition> ::= ("node: "node_id) + ("cpu:"cpu_demand)
<Link Condition> ::= ("link: "link_id) + ("bw:"bw_demand) | ("latency: "latency)
<Overall Condition> ::= "lifetime: " lifetime | ("QoS: " <QoS>{","<QoS>})
<QoS> ::= "l-latency" | "h-cpu"
```

Fig. 2. Definition of NSR intent.

In order to simplify the complexity of tenant describing service requirements, reduce the difficulty of network management, we propose an intent-based expression method of multi-domain NS requirements. We first abstract and classify the elements involved in this problem. In this paper, the elements in NSRs are grouped into two categories: object and condition, as shown in Fig. 1.

Object is element to be orchestrated in NSR, which consists of Node and Link (logical connectivity relationships between nodes). Condition is divided into object condition and overall condition. Object condition include object requirement (e.g. requirement of node CPU/link bandwidth) and object constraint (e.g. constraint of link delay). Overall condition is the restriction on the entire NSR (e.g. lifetime or QoS).

As shown in Fig. 2, we describe NSR in BNF from the tenant's perspective:

4.2 Scheduling Method of NSR Based on Calculated Priority

When pre-processing NSR, we design corresponding priority for each NSR, so that the NSR with highest priority P can be orchestrated preferentially.

$$P(NSR_i) = \frac{QoS(NSR_i) * avg\_Revenue(NSR_i)}{Complexity(NSR_i)} \tag{6}$$

Where $avg\_Revenue(NSR_i)$ is the revenue generated by NSR_i per timeunit in its life time. $QoS(NSR_i)$ is QoS value for current NSR.$Complexity(NSR_i)$ measures the orchestration cost of NSR_i by considering elements to be managed in NSR_i.

$$QoS(NSR_i) = \alpha * CPU(N_r) + \beta * BW(L_r) + \gamma * Delay(L_r) \tag{7}$$

$$Complexity(NSR_i) = (|N_r| + |L_r|) * domains(NSR_i) \tag{8}$$

$$avg\_Revenue(NSR_i) = \frac{Revenue(NSR_i)}{T_{life}(NSR_i)} \tag{9}$$

Where α, β, γ range from 0 to 1, and $\alpha + \beta + \gamma = 1$. $CPU(N_r)$ is total CPU demands of all nodes in N_r. $Delay(L_r)$ and $BW(L_r)$ are respectively delay constraints and bandwidth demands of all links in L_r, and $domains(NSR_i)$ is the number of domains involved in NSR_i.

4.3 Selection of Candidate Substrate Nodes

After receiving a NSR, in order to reduce the time cost of orchestration process, we screens the set of candidate substrate nodes in each domain for the next operation.

Greedy may lead to node resource fragmentation. For example, an NSR contains three nodes a, b, c, which requires resources of 8, 3 and 7 respectively. And a SN with two nodes A, B, whose capacity are 10 and 8. In traditional greedy algorithm, we can get: $Map(a) \to A, Map(c) \to B$, but A and B don't have enough resource to deploy b.

In Algorithm 1, we propose the following two schemes: (1) Select substrate node that reserves the expected value of CPU requirement predicted based on past CPU requirement condition. (2) Select the substrate node that has the closest CPU capacity.

Algorithm 1 Anti-Fragmentation Candidate Nodes Selection Algorithm

Input: NSR, SN

Output: The set of candidate substrate nodes for each node in NSR Ψ

add $CPU(N_r)$ to historical data set Ω;
calculate next possible CPU demand avg_cpu by quadratic exponential smoothing, standard deviation $deviation$ based on Ω, and the fluctuation range of CPU demand $\Phi = [avg\_cpu - deviation, avg\_cpu + deviation]$;
for $n_r \in N_r$ **do**
  for unmarked $n_s \in N_s$ and $domain(n_s) = domain(n_r)$ and $CPU(n_s) \in \Phi$ **do**
    if $\sum BW(l_r) = 0$ where l_r is adjacent to n_r **then**
      if $\Psi(n_r).\text{size} < \text{Defined\_Size}$ **then** $\Psi(n_r).add(n_s)$;
      else Replace node with more adjacent bandwidth in $\Psi(n_r)$ with n_s;
    else if adjacent bandwidth resources of n_s are better than those of n_r;
      if $\Psi(n_r).\text{size} < \text{Defined\_Size}$ **then** $\Psi(n_r).add(n_s)$;
      else choose better node corresponding $adjR(n_s) = a * |adj\_link(n_s)| + (1 - a) * adj\_bw(n_s)$;
  if $\Psi(n_r).\text{size} < \text{Defined\_Size}$ **then**
    for unmarked $n_s \in N_s$ and $domain(n_s) = domain(n_r)$ **do**
      if $\sum BW(l_r) = 0$ where l_r is adjacent to n_r **then**
        Select node whose $CPU(n_s)$ is closer to $CPU(n_r)$ with fewer adjacent bandwidth resources;
      else Select node whose $CPU(n_s)$ is closer to $CPU(n_r)$ with sufficient adjacent bandwidth resources;

4.4 Multi-domain Orchestration Algorithm Description

After getting candidate sets, we perform pruning for SN, then classify nodes into three types according to their surrounding links and take different strategies:

- Isolated nodes: nodes without surrounding links. To reserve sufficient resources for links around other types of nodes, we select the node with fewer available adjacent bandwidth resources among substrate node candidate set.
- Inter nodes: nodes with more surrounding inter-domain links than or equal to intra-domain links. Since resource of edge node is easy to become a bottleneck in such scenario, and the delay of inter-domain link is much larger than that of intra-domain links, we think more about inter-domain links around them.
- Intra nodes: nodes with fewer inter-domain links than intra-domain links. In contrast to inter nodes, we focus more on the intra-domain links' demands.

Algorithm 2 Multi-domain Orchestration Algorithm

Input: NSR, SN, Candidate Set of Nodes
Output: Orchestration Result of NSR

```
for isolated nodes do
  Select candidate node with less available surrounding bandwidth resources;
for inter nodes do
  Get shortest candidate-edge paths that meet requirements in cache;
Compare these paths proportionally based on connection to adjacent domains;
for intra nodes do
  if exists neighbor-domain nodes and inter nodes in adjacent nodes then
    Get shortest intra-edge paths to get node closer to accepted adjacent nodes;
  else if only intra nodes in adjacent nodes
    Get a better shortest intra-neighbor path in cache or through Dijkstra;
for each link do
  if no shortest path in cache meets the requirement then
    if inter link then
      Adopt inter-domain path composed of shortest paths calculated above;
    else Dijkstra Algorithm;
    Update Cache;
```

5 Evaluation

5.1 Related Settings

In this paper, we compared 4 algorithms: our IMAF, Greedy, NRM-VNE [7] and RCR-VNE. Relevant settings of parameters are as Table 1. Moreover, $\alpha = 0.3$, $\beta = 0.3$, $\gamma = 0.4$ when calculating NSR's priority. The arrival of NSRs follows Poisson distribution, with an average rate of 4 NSRs per timeunit (i.e. 100 ms).

Table 1. Network topology parameters.

Parameters	SN	NSR
Number of domains	4	[2, 4]
Number of edge nodes per domain	4	/
Number of nodes	30 * 4	[4, 12]
CPU capacity/requirement of node	[100,300]	[1, 10]
Bandwidth capacity/requirement of intra-domain link	[100, 200]	[1, 10]
Bandwidth capacity/requirement of inter-domain link	[100, 200]	[15,20]
Delay of intra-domain link	[1, 10]	[10, 25]
Delay of inter-domain link	[10, 30]	[30, 50]
Connect ratio between nodes	50%	50%

5.2 Performance Evaluation

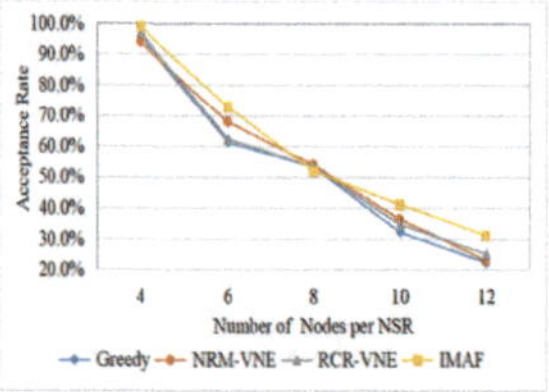

Fig. 3. Acceptance rate.

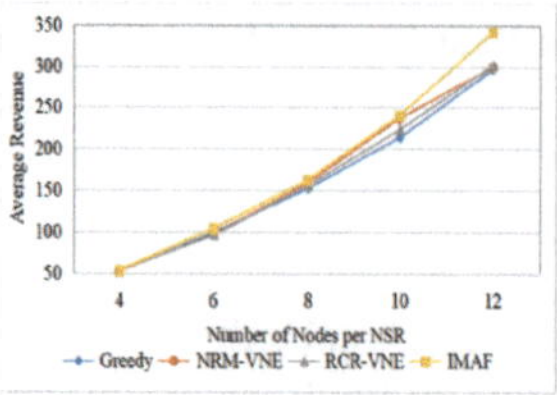

Fig. 4. Average revenue.

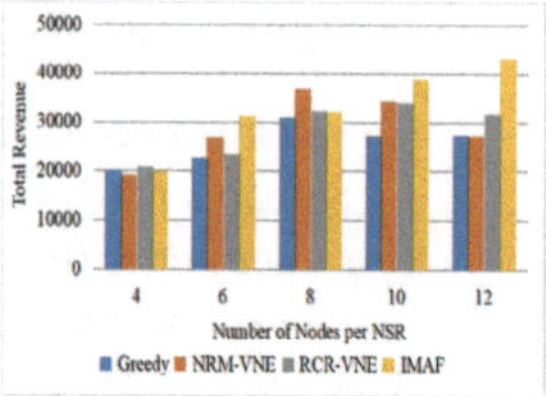

Fig. 5. Total revenue.

Figure 3 demonstrates how the overall acceptance rate changes with the scale of NSR. The experiment is set to run 100 timeunits. Overall, acceptance rate is negatively correlated with network scale. Compared with other algorithms, IMAF has an average increase in acceptance rate of 5%, because node selection method we have adopted improves the availability of node's CPU resources, thereby increasing the success rate of node deployment. Meanwhile, our multi-domain orchestration strategy makes the processing of requests easier and faster, resulting in the increase in acceptance rate.

Figure 4 shows the average revenue of NSR under different strategies. As the request scale is larger, the average NSR revenue is higher, and IMAF has a higher average revenue than other algorithms. This means that the NSR scheduling strategy is effective. Figure 5 is the situation of total revenue generated by adopting four strategies, when there are 4 nodes in each NSR, according to the acceptance rate of each approach, the difference in total revenue is not obvious. As average revenue increases and acceptance rate changes, the total revenue generated by IMAF increases linearly.

6 Conclusion

In this paper, we propose a multi-domain NSR orchestration algorithm IMAF with anti-fragmentation node ranking method and intent-based requirement description method. Moreover, we verified its feasibility and compare with three baseline algorithms to prove that our algorithm increase the acceptance rate and revenue.

References

1. Zhao, Z., et al.: Deploying edge computing servers for large-scale IoT: a diversity aware approach. IEEE Internet Things J. **PP**(99), 1–8 (2018)
2. Intent NBI-definition and principles, The Open Networking Foundation (ONF) Tech. Rec. TR-523, October 2016 (2016)
3. Prakash, C., et al.: PGA: Using graphs to express automatically reconcile network policies. In: ACM on Special Interest Group on Data Communication, pp. 29–42 (2015)

4. Jacobs, A., et al.: Refining network intents for self-driving networks. In: Proceedings of ACM SIGCOMM 2018 Workshop on Self-Driving Networks, pp. 15–21 (2018)
5. NEMO homepage. http://www.hickoryhill-consulting.com/nemo/
6. Li, Z., et al.: A self-adaptive virtual network embedding algorithm based on software-defined networks. IEEE Trans. Netw. Serv. Manage. **16**(1), 362–373 (2019)
7. Zhang, P., et al.: Virtual network embedding based on computing, network, and storage resource constraints. IEEE Internet Things J. **5**(5), 3298–3304 (2018)
8. Ni, Y., et al.: A PSO based multi-domain virtual network embedding approach. China Commun. **16**(4), 105–119 (2019)
9. Houidi, I., et al.: Virtual network provisioning across multiple substrate networks. Comput. Netw. **55**(4), 1011–1023 (2011)
10. Hua, J., et al.: Topology-preserving traffic engineering for hierarchical multi-domain SDN. Comput. Netw. **140**, 62–77 (2018)

Service Offloading Algorithm Based on Depth Deterministic Policy Gradient in Fog Computing Environment

Biao Zou(✉), Jian Shen, Zhenkun Huang, Sijia Zheng, Jialin Zhang, and Wei Li

State Grid General Aviation Company Limited, Beijing 102200, China
274232887@qq.com

Abstract. In the fog computing environment, in order to solve the problem of large service offload delay in large-scale network, this paper first designs a layered service offload management architecture of the Internet of things in large-scale fog computing environment. Based on the management architecture, four service unloading modes are proposed according to the task characteristics. Secondly, based on the theory of deep deterministic policy gradient (DDPG), a decision model of service unloading based on DDPG is constructed. Finally, aiming at minimizing the service offload delay, a service offload algorithm based on DDPG is proposed. In the experiment, the algorithm is compared with the traditional algorithm, which proves that the algorithm has great advantages in the delay and success rate of service unloading.

Keywords: Fog computing · Service offload · Storage · Depth deterministic policy gradient

1 Introduction

Under the background of the rapid development of 5G network, the application scope of Internet of things is more and more wide, providing more new services for people's work and life. However, some of the Internet of things services are sensitive to computing delay, which puts forward higher requirements for the performance of the Internet of things. Due to the advantages of low delay, fog computing is gradually used in the Internet of things as a supplement to cloud computing technology. In this context, how to make full use of fog computing resources to provide low latency response services for Internet of things services has become an important research content. In reference [1, 2], aiming at solving the energy consumption of nodes in fog computing, game theory and intelligent optimization are adopted. The distributed computing service offloading algorithm is put forward in the theory, which can reduce the energy consumption of fog computing nodes. In reference [3], aiming at solving the delay of multiple computing tasks, a service offload algorithm is proposed to reduce the delay of task execution and reduce the energy consumption of edge nodes. In reference [4], aiming at reducing the energy consumption, task execution delay and task payment cost of fog computing nodes, a deep reinforcement learning algorithm is used to

Q. Liu et al. (Eds.): CENet 2020, AISC 1274, pp. 1456–1465, 2021.
https://doi.org/10.1007/978-981-15-8462-6_166

propose an optimized service unloading algorithm, which has achieved good results in experiments. In reference [5], the intelligent optimization algorithm is used to model and solve the problem with the goal of solving the delay of HD video signal processing and the goal of minimizing energy consumption and low delay. In reference [6], based on the research background of dense and complex network environment, a joint allocation algorithm of service unloading and resource scheduling in network intensive environment is proposed, which solves the resource allocation problem in dense network environment and reduces the task calculation delay. In reference [7], a service scheduling algorithm is proposed to reduce the competition time of user resources, optimize the allocation mechanism of user resources, and reduce the calculation delay of user tasks.

From the existing research, we can see that the existing research has achieved good results in the aspects of service offload delay, energy consumption and so on. However, in the fog computing environment, the service type, the type of edge nodes and other factors are seldom considered, and the problem of service offloading in the overall network environment is not considered, especially the factors such as large scope of network environment, dynamic changes and so on. Therefore, a new algorithm of service offload is urgently needed, which can minimize the delay of service offload in the scenarios of network dynamic change, network scale, service type and so on. In order to solve this problem, this paper first designs a large-scale hierarchical Internet of things architecture in the fog computing environment, then according to the characteristics of the task, proposes four service unloading modes, and finally, with the goal of minimizing the service unloading delay, designs a service unloading algorithm based on the depth deterministic policy gradient. Through experiments, the advantages of this algorithm are verified.

2 Network Architecture and Service Unloading Mode

2.1 Network Architecture

In order to solve the problem that large-scale network leads to complex offloading of fog computing services, this paper first divides large-scale network into multiple control areas, so as to reduce the complexity of management. Based on this, the proposed layered service offload management architecture of the Internet of things is shown in Fig. 1. In the Figure, the layered IOT service unloading management architecture from top to bottom includes cloud service master layer, regional SDN control layer, edge node layer, data layer and device layer.

2.2 Computing Service Uninstall Mode

In order to effectively reduce the delay of service offloading and study the dynamics of network topology as a variable, this paper divides service offloading into four ways: idle device offloading, edge node offloading, cloud offloading and local offloading. Idle device offloading refers to offloading services to the computing nodes in the device layer that meet the task requirements. Edge node offloading refers to offloading services

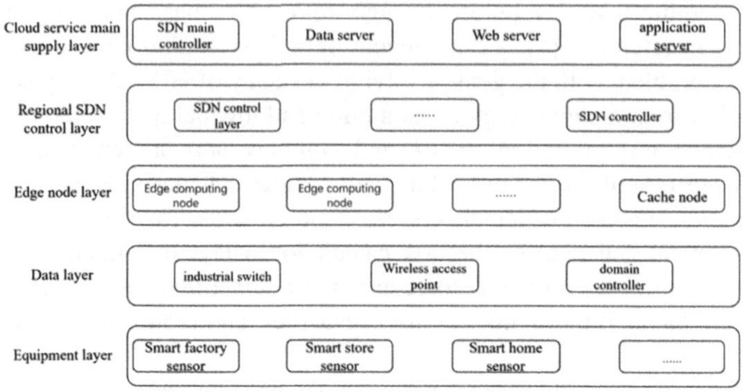

Fig. 1. Management structure of service unloading of layered Internet of Things.

to the edge nodes in the edge node layer. Cloud offloading refers to offloading services to the corresponding servers in the cloud service master layer. Local offloading refers to offloading services to the computing nodes in the device layer that meet the task requirements Count the nodes. The situation of idle device unloading and local unloading is more suitable for the environment with small computing task demand. The edge node offloading comparison is suitable for the environment with large computing task demand and time-delay sensitivity. The cloud offload comparison is suitable for the environment with large computing task demand and insensitive time delay.

3 Problem Description

3.1 Internet of Things System

In terms of network environment, the network environment of the Internet of Things is divided into N cell areas, which are represented by $\{Cell_1, Cell_2, \ldots Cell_N\}$. Cloud computing services are represented by R. Within each cell area $Cell_i$, there are m sensor terminals (denoted by $\mathcal{D} = \{D_0, D_1, \ldots, D_m\}$), n fog computing nodes (using $\mathcal{M} = \{M_1, M_2, \ldots, M_n\}$).

In computing tasks, use K_p represents a service request, and each task request includes p subtask K_p. Use $K_p = (Q_p, S_p, T_m)$. Where, Q_p represents the computing resources required by the task, S_p represents the transmission resources required by the task, T_m is the constraint of task to delay. Task delay consists of computation delay and transmission delay. For the four service unloading modes, only the local service unloading mode does not need transmission delay, and the other three service unloading modes need to consider transmission delay.

$$R_{D2X} = B \log_2\left(1 + \frac{P_{D_i} d_{D,X}{}^{-\delta} h_{D,X}^2}{N_0}\right), X \in \mathcal{M} \cup R \cup \mathcal{D}, X \neq D_i \tag{1}$$

3.2 Service Uninstall Mode

Representation of Four Services Uninstallation. For calculation subtask K_p. The four service uninstallation modes are as follows: (1) the idle device uninstallation mode uses $a_{D_i,D_j,K_p} = 1, j \neq i$ means. (2) local uninstall mode using $a_{D_i,D_i,K_p} = 1$; (3) the unloading mode of the fog calculation node uses $a_{D_i,M_i,K_p} = 1$. (4) Cloud offload mode using $a_{D_i,M_i,K_p} = 1$.

Task Delay of Four Service Offloads. For calculation subtask K_p. The task delays of the four services are described respectively below. The task delay of idle device unloading mode is expressed by formula (2), where T_p^D is the execution time of the task on the idle device, f_{D_j} represents the computing power of idle equipment. The local unloading mode is represented by formula (3), where T_p^1 indicates the execution time of the task on the local device, f_{D_i} represents the computing power of the local device. The unloading mode of the fog calculation node is represented by formula (4), where T_p^M is the execution time of the task on the edge device, f_{M_i} represents the computing power of edge devices. The cloud offload mode is represented by formula (5), where T_p^R represents the execution time of the task on the cloud device, f_R represents the computing power of cloud devices.

$$T_p^D = \frac{Q_p}{f_{D_j}} + \frac{S_p}{R_{D2D}}, D_j \in \mathcal{D}, j \neq i \tag{2}$$

$$T_p^1 = \frac{Q_p}{f_{D_i}} \tag{3}$$

$$T_p^M = \frac{Q_p}{f_{M_i}} + \frac{S_p}{R_{D2M}}, M_i \in \mathcal{M} \tag{4}$$

$$T_p^R = \frac{Q_p}{f_R} + \frac{S_p}{R_{D2R}} \tag{5}$$

To sum up, task K_p the total calculation delay can be calculated by formula (6).

$$T_{K_p} = a_{D_i,D_i,K_p} T_p^1 + a_{D_i,M_i K_p} T_p^M + a_{D_i,R,K_p} T_p^R + a_{D_i,D_j,K_p} T_p^D \tag{6}$$

4 Service Offload Algorithm

4.1 Decision Model of Service Unloading Based on DDPG

In this paper, when DDPG algorithm is used to solve the optimal service unloading strategy, Actor-Critic architecture is used. Actor refers to the policy network, including

Actor_M network and Actor_T network, among them, the main function of the Actor_T network is to generate training data sets, and the main function of the Actor_M network is to train the parameters of the optimized network. Critic refers to the evaluation network, including Critic_M network and Critic_T network, where the main function of Critic_T network is to generate training data set, and the main function of Critic_M network is to train and optimize network parameters.

Deep reinforcement learning refers to the use of deep neural networks to learn action strategies based on traditional reinforcement learning. In general, the reinforcement learning model contains four main elements: environment, states, strategy π, and action a. The agent in the reinforcement learning model obtains the reward r as the evaluation standard, and optimizes the above four main elements. For convenience of description, at time t, s_t is used to represent the state at time t, and s_{t+1} is used to represent the state at t + 1. Among them, s_{t+1} is the next state that the Agent chooses the action a_t according to the policy π and executes in the environment. In order to evaluate the effectiveness of the strategy, the reward of the agent's action using the strategy in the environment is defined as the return r_t. The key elements are described in detail below.

4.2 Elements Modeling of Deep Reinforcement Learning

In terms of state space, the state of the fog computing nodes and idle device terminals around the sensor device is combined into the state space. For fog computing nodes, use $F_R^M = \{F_R^{M_1}, F_R^{M_2}, \ldots, F_R^{M_n}\}$ to represent the remaining resource set of each edge node in the area, where $F_R^{M_i} = F_m^{M_i} - \sum_{p=1}^{n} a_{D_i,M_iK_p} Q_p$. For idle device terminals, use $F_R^D = \{F_R^{D_1}, F_R^{D_2}, \ldots, F_R^{D_n}\}$ to represent the remaining resource set of each idle device in the area, where $F_R^{D_j} = F_m^{D_j} - \sum_{p=1}^{n} a_{D_i,D_j,K_p} Q_p$. Based on this, the formula (7) is used to represent the state space, where $T_{all}^{\mathcal{K}_p}$ represents the total delay of the service, and the calculation method is $T_{all}^{\mathcal{K}_p} = \sum_{p=1}^{n} T_{K_p}$.

$$St = \left(F_R^M, F_R^D, T_{all}^{\mathcal{K}_p}\right) \tag{7}$$

In terms of action space, the service offload decision is defined as formula (8), where $F_A^M = \{F_A^{M_1}, F_A^{M_2}, \ldots, F_A^{M_n}\}$, $F_A^{M_i} = a_{D_i,M_iK_p} Q_P$; $F_A^D = \{F_A^{D_1}, F_A^{D_2}, \ldots, F_A^{D_n}\}$, $F_A^{D_i} = a_{D_i,D_j,K_p} Q_P$.

$$\mathfrak{D}_{K_p} = \{\mathfrak{D}_1, \mathfrak{D}_2, \ldots, \mathfrak{D}_n; F_A^M, F_A^D\} \tag{8}$$

Regarding the reward function, the reward function is defined as formula (9), where $T_1^{\mathcal{K}_p}$ represents the time required when the service $\mathcal{K}_p$ is all executed locally, and $T_{all}^{\mathcal{K}_p}$

represents that the service task takes the action a_t in the state s_t After the execution time.

$$r(s_t, a_t) = T_1^{\mathcal{K}_p} - T_{all}^{\mathcal{K}_p} \tag{9}$$

In terms of the value function, formula (10) is used to calculate the value $V_\pi(s)$ function, which represents the expectation of the reward r of the strategy π in the initial states. Among them, E(•) represents the expected operation.

$$V_\pi(s) = \mathbb{E}[r|s_0 = s, \pi] \tag{10}$$

Based on this, the action value function Q_π (s, a) can be calculated using formula (11), γ represents the attenuation factor, and the value range is 0 to 1, and s' represents the next state after s is taken as a. In order to simplify the calculation of the action value function, it is known from the Bellman equation that formula (12) can be obtained according to formula (11). Among them, μ represents the strategy generated by the Actor network.

$$Q_\pi(s, a) = \mathbb{E}[r(s, a) + \gamma\mathbb{E}[V_\pi(s')]] \tag{11}$$

$$Q^\mu(s_t, a_t) = E[r(s_t, a_t) + \gamma Q^\mu(s_{t+1}, \mu(s_{t+1}))] \tag{12}$$

4.3 Deep Deterministic Strategy Gradient

Strategy gradient is a gradient-based optimization search algorithm. In this paper, the gradient update parameters of Critic_M network and Actor_M network are θ^{Q_i} and θ^{μ_i}. The gradient update parameters of Critic_T network and Actor_T network are $\theta^{Q_i'}$ and $\theta^{\mu_i'}$. The calculation process of the loss function Loss function is defined as formula (13). Among them, y_i^m is calculated using formula (14).

$$Loss = \frac{1}{m}\sum\nolimits_i \left(y_i^m - Q_i\left(s_i^m, a_i^m|\theta^{Q_i}\right)\right)^2 \tag{13}$$

$$y_i^m = r_i^m + \gamma Q_i'\left(s_i^{m+1}, \mu_i'\left(s_i^{m+1}|\theta^{\mu_i'}\right)|\theta^{Q_i'}\right) \tag{14}$$

Based on this, the gradient calculation method is shown in formula (15), where function J is calculated using formula (16) and is used to measure the performance of strategy μ. ρ^β represents the distribution function of state s. Therefore, the meaning of formula (16) is that when the state space is s, the expected value of $Q^\mu(s, \mu(s))$ is calculated according to the distribution of ρ^β.

$$\nabla_{\theta^{\mu_i}} J_i \approx \frac{1}{m}\sum_i \left(\nabla_{a_i^m} Q_i\left(s_i^m, a_i^m|\theta^{Q_i}\right)|_{a_i^m=\mu_i(s_i^m)} \cdot \nabla_{\theta^{\mu_i}} \mu\left(s_i^m|\theta^{\mu_i}\right)\right) \tag{15}$$

$$J(\mu) = \mathbb{E}_{s \sim \rho^{\beta}}[Q^{\mu}(s, \mu(s))] \quad (16)$$

When the Actor network interacts with the environment, in order to prevent the correlation of the conversion data sequence in time, resulting in the problem that the neural network cannot converge, this article stores the conversion data sequence in the empirical pool, based on mini-batch random sampling, so as to realize the data Time is not relevant.

4.4 Service Offload Algorithm Based on DDPG

Based on the above analysis, the service offloading algorithm based on deep deterministic policy gradient in Fog computing environment (SOoDDPG) proposed in this paper is shown in Table 1. It includes two processes: model initialization and solving the optimal service offload strategy for each area of the network environment. The model initialization steps include steps 1 to 7 in Table 1, mainly to initialize the four network models and key parameters in the Actor-Critic architecture. The steps to solve the optimal service offload strategy in each area of the network environment include steps 8 to 25 in Table 1, mainly to solve the optimal service offload strategy within the number of iterations.

5 Performance

5.1 Environment Settings

In order to verify the performance of the algorithm in this paper, the network environment of fog computing was developed using Python language. Based on the idea of dividing regions in this paper, the network environment is divided into 25 units. The minimum distance between each unit is set to 10 m. In each unit, set 2 idle devices and 2 fog computing nodes. Among them, the calculation frequency of idle devices is 200 and 150. The computing frequency of the fog computing node is 650 and 600. The transmission power of the device is subject to an even distribution between (5, 38). In terms of services that need to be uninstalled, 100 services that need to be uninstalled are randomly generated, and the even distribution of computing resource services (100, 800) required by each service.

In terms of parameter settings, the minimum batch in the experience pool is set to 500, the Actor_M network learning rate is set to 0.0001, the Critical_M network learning rate is set to 0.001, and the Gaussian noise power during transmission delay calculation is set to −114 dbm, memory. The pool capacity is set to 50000, the maximum number of iterations is set to 5000, and the number of explorations is set to 10. Considering that the problem studied in [3] is similar to this paper, and the proposed cooperative algorithm (DCTO, distributed cooperative task offloading) based on multiple edge nodes achieves better offloading of computing services of intelligent terminals. In the experimental part, the algorithm SOoDDPG is compared with the algorithm DCTO.

Table 1. Service unloading algorithm based on DDPG.

Algorithm: Service offload algorithm based on DDPG

For every area of the network environment i, $i \in N$
    Initialize critic_M network $Q_i(s_i, a_i; \theta^{Q_i})$ with random parameterθ^{Q_i}
    Initialize actor_M network $\mu_i(s_i; \theta^{\mu_i})$ with random parameterθ^{μ_i}
    Initialize critic_T network $\mu_i(s_i; \theta^{\mu_i})$ with random parameterθ^{μ_i}
    Initialize actor_T network $\mu_i'(s_i; \theta^{\mu_i'})$ with random parameter$\theta^{\mu_i'}$
    Initialize replay buffer D_i
End of initialization
    For each region i of the network environment, $i \in N$
    Solve the optimization strategy N_e in the iteration number threshold N_e
        Action search in Gaussian space $\mathcal{N}_i$
        Get initial state s_i^1
        Search within t
            Use formula (8) to calculate the states s_i^t
            Use formula (10) to calculate incomer$_i^t$ and update status s_i^{t+1}
            Set $(s_i^t, a_i^t, r_i^t, s_i^{t+1})$ stored in replay buffer
            From $D_i$randomly selected Mini_ Batch $(s_i^t, a_i^t, r_i^t, s_i^{t+1})$
            Update the critic_M network using formulas (14) and (15)
            Update the critic_M network using formulas (16) and (17)
            Update critic_T network and actor_T network:
            $\theta^{Q_i'} \leftarrow \tau\theta^{Q_i} + (1-\tau)\theta^{Q_i'}$ (18)
            $\theta^{\mu_i'} \leftarrow \tau\theta^{\mu_i} + (1-\tau)\theta^{\mu_i'}$ (19)
            Update state space $s_i^t = s_i^{t+1}$
        End of search
    End of solution optimization strategy
Optimal strategies are solved in all regions

5.2 Algorithm Comparison

The two algorithms are compared in terms of average data transfer rate, service uninstallation success rate, and task execution time. In terms of the average data transmission rate, the two dimensions are compared from the maximum distance between adjacent units and the number of devices connected to each computing node.

The effect of the maximum distance between adjacent units on the average data transmission rate is shown in Fig. 2. Among them, the X axis indicates that the maximum distance between adjacent units varies from 100 m to 1000 m. The Y axis represents the average data transfer rate (bps). It can be seen from the Fig. that as the maximum distance between adjacent units increases, the average data transmission rate under both algorithms increases. This is because the network scale increases, the mutual interference in the network environment decreases, and the data transmission rate increases. In various environments, the average data transmission rate of the algorithm SOoDDPG in this paper is larger than the average data transmission rate of the algorithm DCTO. This shows that the algorithm in this paper can optimize the service offload, reduce the interference between network devices, and thus improve the data transmission rate.

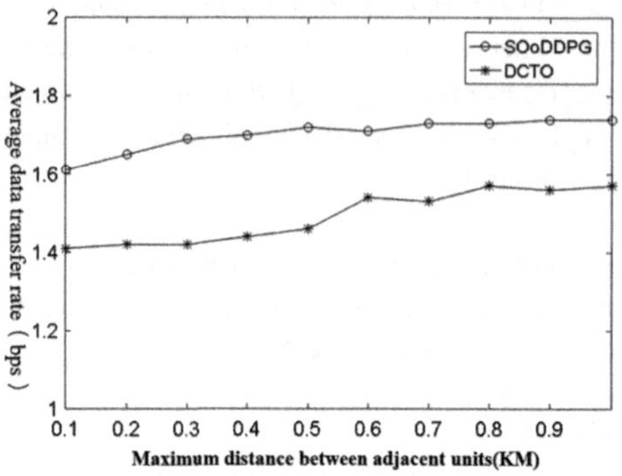

Fig. 2. Influence of maximum distance between adjacent cells on average data transmission rate.

The effect of the number of devices connected to each computing node on the average data transmission rate is shown in Fig. 3. Among them, the X axis indicates that the number of devices connected to each computing node varies from 1 to 8. The Y axis represents the average data transfer rate (bps). As can be seen from the Figure, as the number of devices connected to each node increases, the average data transmission rate under both algorithms is rapidly decreasing, indicating that the increase in the number of devices can easily increase the load on the computing node, thereby reducing the average data transmission rate. However, the algorithm SOoDDPG has achieved better results than the algorithm DCTO in the environment of different numbers of devices.

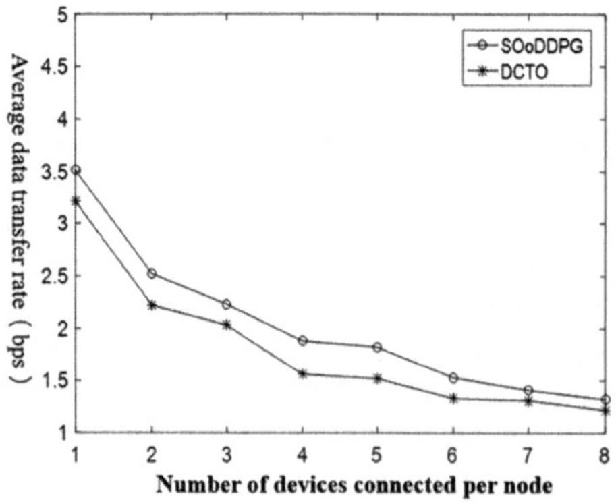

Fig. 3. The impact of the number of devices connected to each node on the average data transmission rate.

6 Conclusion

Under the background of rapid application of Internet of things technology, in order to make full use of fog computing resources and provide low latency response service for Internet of things business, service offload has become an important research content. In this context, in order to solve the problem of large service offloading delay in large-scale network, this paper designs a large-scale layered Internet of things service offloading management architecture in the fog computing environment. A service offloading algorithm based on depth deterministic policy gradient is proposed. Through experiments, the advantages of this algorithm in service offloading are verified. From the research and analysis of this paper, we can see that when the mobile speed of the Internet of things terminal equipment increases, the SDN control of each region needs to cooperate with each other to ensure the stability and reliability of the service unloading of the terminal equipment. In the next step, based on the research results of this paper, we will further study the cooperation mechanism between multiple units, so as to solve the problem of service unloading in high mobile environment.

Acknowledgement. This work was supported by the State Grid General Aviation Co., Ltd. technology Program 2400/2019-44006B.

References

1. Chen, X., Jiao, L., Li, W.Z., et al.: Efficient multi-user computation offloading for mobile-edge cloud computing. ACM Trans. Netw. **24**(5), 2795–2808 (2016)
2. Tao, X.Y., Ota, K., Dong, M.X., et al.: Performance guaranteed computation offloading for mobile-edge cloud computing. IEEE Wirel. Commun. Lett. **6**(6), 774–777 (2017)
3. Chen, M., Hao, Y.X.: Task offloading for mobile edge computing in software defined ultra-dense network. IEEE J. Sel. Areas Commun. **36**(3), 587–597 (2018)
4. Xue, N., Huo, J., Zeng, S.Q., Wang, S., Huang, T., et al.: Algorithm of task unloading and resource scheduling for MEC based on DRL. J. Beijing Univ. Posts Telecommun. 1–6. https://doi.org/10.13190/j.jbupt.2019-155
5. Trinh, H., Calyam, P., Chemodanov, D., et al.: Energyaware mobile Fog computing and routing for low-latency visual data processing. IEEE Trans. Multimedia **20**(10), 2562–2577 (2018)
6. Zhang, H.B., Li, H., Chen, S.X., et al.: Service offload and resource optimization based on mobile fog computing in super dense networks. J. Electron. Inf. **41**(5), 1194–1201 (2019)
7. Alnoman, A., Anpalagan, A.: A dynamic priority service provision scheme for delay-sensitive applications in fog computing. In: The 29th Biennial Symposium on Communications, Toronto, Canada, pp. 1–5 (2018)

Server Deployment Algorithm for Maximizing Utilization of Network Resources Under Fog Computing

Wei Du, Hongbo Sun, Heping Wang, Xiaobing Guo, and Biao Zou(✉)

State Grid General Aviation Company Limited, Beijing 102200, China
274232887@qq.com

Abstract. In an edge computing network environment, in order to improve the utilization of network resources, this paper proposes a server deployment algorithm for maximizing network resource utilization under edge computing. First, the problem is modeled from four aspects: network topology, user service requests, storage and computing resource allocation, and service node traffic usage matrix. Secondly, an objective function for minimizing network traffic is constructed, and a server deployment algorithm for maximizing network resource utilization under edge computing is proposed to solve. The algorithm uses the idea of slack constraints to solve the optimal server deployment strategy. Use tabu genetic search algorithm to find the optimal storage node and compute node deployment location. Use Lagrange operator to solve the best set of service nodes. In the experimental part, it is verified that the proposed algorithm achieves better results in reducing network traffic.

Keywords: Edge computing · Resource allocation · Resource utilization · Tabu genetic algorithm

1 Introduction

With the rapid application of 5G networks in production and life, the response speed of various services to computing resources and storage resources has put forward higher and higher requirements [1]. Under this background, fog computing technology has been proposed and has been rapidly developed [2]. How to deploy and allocate fog computing resources has become an important research problem. Existing studies mainly use intelligent algorithms, automatic perception theory, optimization and other methods to solve the problems of low resource utilization and long execution time of user tasks in fog computing. Literature [3] proposed a resource demand prediction algorithm based on genetic algorithm based on the goal of resource demand in a dynamic environment, which solved the problem of low server resource utilization [4]. In order to maximize the utilization of computing resources, based on the application-aware theory, the server is deployed to the edge nodes with greater resource requirements, thereby improving the server utilization. Reference [5] aimed at minimizing user energy consumption, and designed a task offloading mechanism based on data partitioning, which better reduced

Q. Liu et al. (Eds.): CENet 2020, AISC 1274, pp. 1466–1474, 2021.
https://doi.org/10.1007/978-981-15-8462-6_167

the transmission power of tasks in the network. Reference [6] aimed at minimizing resource energy consumption and user task execution time, designed a joint optimization objective function of resource allocation and task offloading, and obtained the optimal solution through an approximate algorithm. Reference [7] aims to increase the amount of tasks in the fog computing server, and uses server utilization as a constraint, periodically adding tasks to the server, thereby increasing the number of tasks in the fog computing network. Reference [8] takes the task execution time limit as a constraint, and based on the energy sensing theory, proposes a scheduling algorithm that maximizes the utilization rate of virtual machine resources, which effectively reduces the energy consumption of virtual machines.

From the analysis of existing studies, we can see that in terms of resource allocation of fog computing, existing studies have achieved good results in terms of task execution time, number of tasks, and server resource utilization. However, existing studies have not conducted joint research on computing resources, storage resources, and network resources, and have neglected the relationship between various constraints, resulting in uneven use of resources in fog computing. To solve this problem, this paper models the problem from four aspects: network topology, user service request, storage and computing resource allocation, and service node's traffic usage matrix. An objective function for minimizing network traffic is constructed, and a server deployment algorithm for maximizing network resource utilization under fog computing is proposed for solution. In the experimental part, it is verified that the algorithm in this paper has achieved good results in solving the optimal server deployment scheme and comparing the network traffic under different service request numbers.

2 Problem Description

Currently, the fog computing network architecture based on SDN technology is the mainstream architecture of fog computing. The network devices in this architecture include three devices: controller, repeater, and remote server. The controller is responsible for the configuration and management of the repeater. Transponders are divided into network nodes, computing nodes, and storage nodes, and have three functions: transmission, calculation, and storage. The process of users using services mainly includes the process of using storage services and the process of using computing services. When a user uses a storage service, the storage node closest to the user checks whether there are user resources; if not, requests resources from a remote server; and returns the resources to the user after obtaining the resources. When a user uses a computing service, the computing node closest to the user provides the computing function for the user. The following models the problem from four aspects: network topology, user service request, storage and computing resource allocation, and service node traffic usage matrix.

In terms of network topology, $G = \{V, E\}$ represents network topology, V represents node set, $S_A \subset V$ represents node with storage capacity, and c_s^a represents storage capacity. $S_B \subset V$ represents the node with computing power, and the computing capacity of $S_B \subset V$ represents the c_s^b. $T \subset V$ represents the user node, $o \subset V$ represents the source service node that meets the user requirements. E is the link set.

In terms of user service requests, $k^a \in K_A$ is used to represent content service requests, and its capacity is o_{k^a}. Use $k^b \in K_B$ to express the calculation service request, and its calculation amount using o_{k^b}. The demand matrix is represented by $M : \{m_t^k\}_{(k,t)\in K\times T}$, where m_t^k represents the rate requirement of user $t \in T$ for service k.

In terms of storage and computing resource allocation, $H : \{h_s^k\}$ is used to indicate storage and computing resource allocation identification, where h_s^k is a binary variable. When $h_s^k = 1$, the identification service node s can satisfy the service k of user t. When $h_s^k = 0$, it indicates that the service node s cannot satisfy the service k of user t. Therefore, the capacity limit of the storage node is expressed as formula (1), and the capacity limit of the computing node is expressed as formula (2).

$$\sum\nolimits_k h_s^k o_{k^a} \le c_s^a, \forall s \in S_A \tag{1}$$

$$\sum\nolimits_k h_s^k o_{k^b} \le c_s^b, \forall s \in S_B \tag{2}$$

In terms of the traffic usage matrix of the service node, when multiple service nodes cooperate to satisfy a user request, $\rho_{s,t}^k$ is used to represent the proportion of service k of the user t carried by the service node s. Therefore, the service node selection matrix is expressed as $P : \{\rho_{s,t}^k\}$. At this time, the traffic $x_{s,t}^k$ of the service k of the user t carried by the service node s is expressed as formula (3). This formula indicates that the flow rate is equal to the requested rate * the proportion of services provided * the storage/calculation strategy, and $\sum_s \rho_{s,t}^k h_s^k = 1$.

$$x_{s,t}^k = m_t^k \rho_{s,t}^k h_s^k \tag{3}$$

The calculation method of the load x_s of the service node s is shown in formula (4). Therefore, use $X : \{x_{s,t}^k\}, \forall s \in S$ to indicate the flow usage matrix.

$$x_s = \sum\nolimits_{k,t} x_{s,t}^k \tag{4}$$

When traffic passes through multiple links, the traffic proportion of the link between the service node s and the user t is expressed by $r_l^{s,t} \in [0, 1]$. When the capacity of the link l is c_l, $x_l = \sum_{k,s,t} x_{s,t}^k r_l^{s,t} \le c_l$. This formula means: The flow $x_{s,t}^k$ of the service flow k from the user t to the service node s * the proportion of the flow through the link l $r_l^{s,t}$. Based on the above analysis, this paper defines the function of the total network traffic as formula (5). Among them, $D_{s,t}$ means that the average hop count of service node s and user node t is $D_{s,t}$, so the objective function of minimizing network traffic is formula (6). Constraints a and b indicate that the storage capacity and computing capacity of the allocated service cannot exceed the capabilities of the service node; constraint c indicates that user t's demand for service request k can be served by multiple service nodes at the same time; constraint d indicates that the network link. The traffic carried cannot exceed the upper limit of the capacity of the network link.

$$F = \sum\nolimits_{s \in S} \sum\nolimits_{t \in T} D_{s,t} (\sum\nolimits_{k^a \in K_A} x_{s,t}^{k^a} + \sum\nolimits_{k^b \in K_B} x_{s,t}^{k^b}) \tag{5}$$

$$G = \min \sum\nolimits_{s \in S} \sum\nolimits_{t \in T} D_{s,t} (\sum\nolimits_{k^a \in K_A} x_{s,t}^{k^a} + \sum\nolimits_{k^b \in K_B} x_{s,t}^{k^b}) \tag{6}$$

s.t.

$$\sum\nolimits_k h_s^k o_{k^a} \le c_s^a, \forall s \in S_A \tag{6-a}$$

$$\sum\nolimits_k h_s^k o_{k^b} \le c_s^b, \forall s \in S_B \tag{6-b}$$

$$\sum\nolimits_s \rho_{s,t}^k h_s^k = 1 \tag{6-c}$$

$$x_l \le c_l \tag{6-d}$$

3 Server Deployment Algorithm

It can be seen from the constraints that the constraints (a) and (b) solve the problem of storage node and computing node deployment location, and the constraints (c) and (d) solve the problem of optimal service node set. If they are solved at the same time, it is more difficult.

The server deployment algorithm for maximizing network resource utilization under edge computing (SDAoMNRU) proposed in this paper is shown in Table 1. To simplify the solution of the problem, this paper divides the problem into three steps. (1), Step 1 first relax the service allocation and flow capacity constraints, only use the constraints (a) and (b) to solve the optimal storage node and computing node deployment location. This problem belongs to the optimal strategy search problem. The tabu-genetic search algorithm is used to achieve the global optimal search. The tabu-genetic search algorithm will be described in the next Sect. (2) In step 2, based on the optimal storage node and computing node deployment location, relax the storage capacity and computing capacity restrictions, and use constraints (c) and (d) to solve the optimal set of service nodes. In this case, the function is convex and can be calculated using the Lagrange operator. (3) Use the total network flow formula (5) to calculate the network flow.

Table 1. Server deployment algorithm for maximizing network resource utilization under fog computing.

Input: Network topology, $G = \{V, E\}$ Request for service K_A, K_B
Output: Network traffic
Step 1: Solve the best storage node and computing node deployment plan
1) Model the objective function formula (6) for minimizing network traffic as an optimization problem containing only constraints (a) and (b);
2) Use the tabu genetic search algorithm to search for the best storage node and computing node deployment scheme;
3) According to the best storage node and computing node deployment plan, deploy storage nodes and computing nodes in the network.
Step 2: Solve the best set of service nodes
1) Model the objective function formula (6) for minimizing network traffic as an optimization problem containing only constraints (a) and (b);
2) Use Lagrange operator for calculation to solve the best set of service nodes.
Step 3: Solve network traffic
1) Update the objective function based on the storage node, computing node, and optimal service node set deployed in the network;
2) Use the total network flow formula (5) to calculate and output the network flow.

4 Storage Node and Computing Node Deployment Solutions

In the server deployment algorithm proposed by this paper to maximize network resource utilization under fog computing, the optimal storage node and computing node deployment scheme is solved based on the tabu genetic search algorithm. The tabu genetic algorithm [9] combines the tabu search algorithm with the genetic algorithm, which can significantly improve the quality of the algorithm solution. It has been successfully applied to the study of optimization problems and has achieved good results.

First, according to the chromosome coding strategy, N chromosomes are randomly generated. Using the objective function F_i as the fitness function, the chromosomes are judged and optimized through selection, crossover, and mutation operators until the number of iterations specified by the algorithm is reached. Finally, the optimal result is taken as the best storage node and computing node deployment plan. Among them, the selection operator uses the market selection method for selection. The crossover operator uses a two-point crossover method for crossover operations. The mutation operator uses a random mutation method to perform mutation operations. During the mutation operation, if the newly generated allocation scheme does not belong to the allowed network node, it needs to be regenerated until the network topology requirements are met.

In the initial population, select the L chromosomes with the highest fitness value, construct a tabu table of length L, and set the value to the fitness value of the chromosome. The fitness value with the largest fitness value in the chromosome is assigned to the craving level of the tabu list and replaces the average fitness value of the parental chromosome population. In terms of chromosome renewal, when the fitness value of the new chromosome is greater than the craving level or the new chromosome is not on the taboo list, these chromosomes are accepted as the next generation chromosome. After that, select the chromosome update contraindication table with the best fitness value from the updated chromosome set.

In terms of chromosome coding strategies, the goal of this section is to solve the optimal storage node and computing node deployment plan, so this article uses the deployment plan to construct chromosomes. Suppose that n user requests are represented as K_A, K_B, and $\sum_{a \in R} |N^a|$ is used to represent the storage node and computing node deployment schemes that provide services for the request. The $\sum_{a \in [1,j-1]} |N^a| + i$ element represents the service node where the jth task of the ith user is located (Table 2).

Table 2. Taboo genetic search for the optimal storage node and computing node deployment scheme.

	Input: Network topology $G = \{V, E\}$,Request for service K_A, K_B
	Output: The best storage node and computing node deployment plan
1.	Parameter initialization: randomly generate N initial populations, genetic algebra T, crossover probability P_c , mutation probability P_m , $count = 0$;
2.	Use the fitness function formula (5) to calculate the fitness value of the chromosome;
3.	**If** ($count > T$) go to step 7;
4.	Build a taboo list and calculate the level of desire;
5.	Perform selection, crossover, and mutation operations: use a random selection mechanism to perform selection operations; use crossover probability pc to perform crossover operations; use mutation probability pm to perform mutation operations;
6.	Use the fitness function formula (5) to calculate the fitness value of contemporary chromosomes;
7.	$count++$ go to step 3
8.	Use the chromosome strategy with the maximum fitness value F_i as the optimal storage node and computing node deployment scheme.

5 Performance

5.1 Environment

The experimental environment includes two aspects: the generation of network topology environment and the setting of key technical parameters. In order to generate a network topology environment that conforms to the business environment, this paper uses S64 [10], a typical representative of the business network topology, to simulate the network topology environment. In terms of setting key technical parameters, the routing protocol in network traffic uses OSPF protocol, and the scale of storage request and calculation request is set to obey the uniform distribution of [1 MB, 5 MB].

In order to verify the performance of the algorithm, this paper compares the algorithm SDAoMNRU with the algorithm SDAoMF (server deployment algorithm based on minimum flow). Among them, algorithm SDAoMNRU uses the minimum traffic mechanism to provide network resources for services. The performance analysis includes two processes: solving the optimal server deployment scheme and comparing the network traffic under different service requests. Among them, the solution to the optimal server deployment scheme is to plan the number of storage nodes and computing nodes based on different service request arrival rates. The comparison of network traffic under different service requests is to verify the advantages and disadvantages of algorithm SDAoMNRU and algorithm SDAoMF in network traffic consumption.

5.2 Algorithm Comparison

When solving the optimal server deployment scheme, the maximum number of storage nodes and the maximum number of computing nodes are 70 and 20 respectively. The experimental results are shown in Fig. 1. The x-axis represents the arrival rate of service requests, and the y-axis represents the optimal number of service deployments. It can be seen from the figure that the number of storage nodes and computing nodes increases with the increase of service request arrival rate. When the service request arrival rate reaches 20 per second, the number of storage nodes and computing nodes tend to be stable.

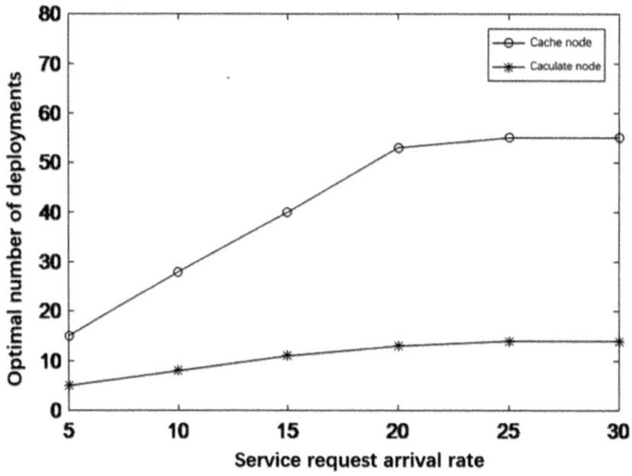

Fig. 1. Solving the optimal server deployment scheme.

In order to compare the network traffic of algorithm SDAoMNRU and algorithm SDAoMF. When running the algorithm, the two algorithms use the same number of storage nodes and computing nodes for analysis. The experiment compares the network traffic that the two algorithms spend to satisfy the service under different service request arrival rates.

The experimental results are shown in Fig. 2. The X axis represents the number of service requests and the Y axis represents the network traffic. It can be seen from the figure that with the increase of the number of service requests, the network traffic overhead of both algorithms is increasing. Compared with the algorithm SDAoMF, the growth of network traffic overhead in this algorithm SDAoMNRU is relatively flat. It shows that the algorithm in this paper selects better computing and storage nodes when providing services for service requests, so as to reduce the overhead of network traffic.

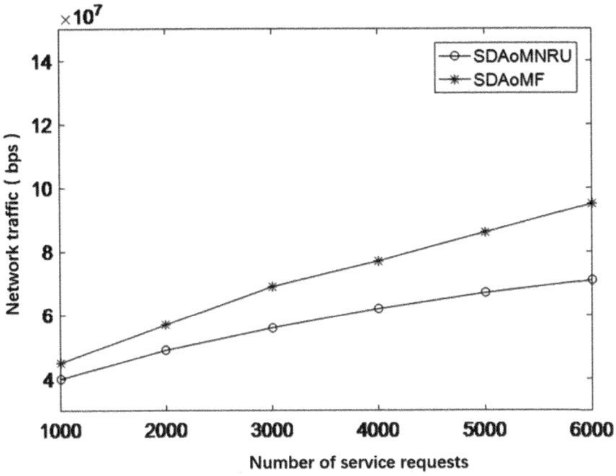

Fig. 2. Comparison of network traffic under different service requests.

6 Conclusion

Fog computing network is a key technology that 5 g network can be applied rapidly in production and life. In order to maximize the utilization of network resources in fog computing network, this paper proposes a server deployment algorithm to maximize the utilization of network resources in fog computing. The algorithm uses the idea of relaxing constraints to solve the optimal server deployment strategy. In the experimental part, the algorithm has achieved good results in two aspects: optimal server deployment and network traffic minimization. Although this paper improves the utilization of network resources, the fast response of fog computing server is a key technical index for 5 g services with low delay and high efficiency. In order to solve this problem, the next work will further optimize the processing and response speed of fog computing server for time delay sensitive services based on the research results of this paper.

Acknowledgement. This work was supported by the State Grid General Aviation Co., Ltd. technology Program 2400/2019-44006B.

References

1. Lee, H.S., Lee, J.W.: Task offloading in heterogeneous mobile cloud computing: modeling, analysis, and cloudlet deployment. IEEE Access **6**, 14908–14925 (2018)
2. Shi, W.S., Cao, J., Zhang, Q., et al.: Edge computing: vision and challenges. IEEE Internet Things J. **3**(5), 637–646 (2016)
3. Tseng, F.H., Wang, X.F., Chou, L.D., et al.: Dynamic resource prediction and allocation for cloud data center using the multi-objective genetic algorithm. IEEE Syst. J. **12**(2), 1688–1699 (2018)
4. Fan, Q., Ansari, N.: Application aware workload allocation for edge computing-based IoT. IEEE Internet Things J. **5**(3), 2146–2153 (2018)
5. Wang, Y., Sheng, M., Wang, X., et al.: Mobile-edge computing: partial computation offloading using dynamic voltage scaling. IEEE Trans. Commun. **64**(10), 4268–4282 (2016)
6. Chen, M.H., Dong, M., Liang, B.: Joint offloading decision and resource allocation for mobile cloud with computing access point. In: 2016 IEEE International Conference on Acoustics, Speech and Signal Processing (ICASSP), pp. 3516–3520. IEEE (2016)
7. Song, Y., Yau, S.S., Yu, R., et al.: An Approach to QoS-based task distribution in edge computing networks for IoT applications. In: 2017 IEEE International Conference on Edge Computing (EDGE). IEEE Computer Society (2017)
8. Shrawankar, U., Hatwar, P.: Approach towards v m management for green computing. In: IEEE International Conference on Computational Intelligence & Computing Research. IEEE (2015)
9. Glover, F., Kelly, J.P., Laguna, M.: Genetic algorithms and tabu search: hybrids for optimization. Comput. Oper. Res. **22**(1), 111–134 (1995)
10. Choi, N., Guan, K., Kilper, D.C., et al.: In-network caching effect on optimal energy consumption in content-centric networking. In: 2012 IEEE International Conference on Communications (ICC), pp. 2889–2894. IEEE (2012)

Power Data Network Resource Allocation Algorithm Based on TOPSIS Algorithm

Huaxu Zhou[1], Meng Ye[1], Guanjin Huang[1], Yaodong Ju[1], Zhicheng Shao[1], Qing Gong[1], and Linna Ruan[2(✉)]

[1] CSG Power Generation Co., Ltd., Beijing 100070, China
[2] Beijing University of Posts and Telecommunications, Beijing 100876, China
29200011@qq.com

Abstract. In order to identify the key nodes in the power data service and assign high reliability network resources, this paper proposes a power data network resource allocation algorithm based on TOPSIS algorithm, which consists of five parts: calculating the importance index of communication network resources, calculating the importance index of the service node, calculating the importance evaluation of the service node based on the TOPSIS algorithm, allocating resources to the service node in turn, and allocating resources to the service link. In the simulation experiment, the success rate of resource allocation, resource utilization, and resource allocation performance of key nodes are compared. It is verified that the algorithm cannot decrease success rate and resource utilization, and key service nodes are allocated better resources and improve the performance of resources.

Keywords: Power data network · Resource allocation · Topsis algorithm · Resource utilization

1 Introduction

With the rapid development of smart grids in research and application, the types of smart grid services are becoming more and more abundant, meeting the increasingly diverse needs of the people. Under this background, the development of smart grid business has put forward more requirements for the power data network. How to allocate key power data network resources to important smart grid business has become an important research content. In previous studies, from the perspective of network virtualization technology, literature [1] proposed a resource allocation mechanism that satisfies the QoS of power services to maximize the utility of the power data network. Reference [2] analyzes the business needs of the power data network and models it, and proposes a topology planning optimization mechanism for the power data network, which effectively solves the optimization problems of communication network construction and planning. Reference [3] analyzed the potential problems of power data network from the perspective of power grid business security defense, and put forward key recommendations for the construction of power data network in terms of safe and stable operation of power business. Reference [4] analyzes the characteristics of power business flow, analyzes the optimization algorithm of multi-service routing based on

Q. Liu et al. (Eds.): CENet 2020, AISC 1274, pp. 1475–1484, 2021.
https://doi.org/10.1007/978-981-15-8462-6_168

information entropy theory, and proposes a quantum genetic algorithm to solve the multi-objective maximum constraint problem, and realizes the load balancing and stable operation of power business. In order to ensure the service quality of power business, literature [5] and literature [6] based on the optimization theory, realized the optimal scheduling of power data network, thus ensuring the stable operation of network business.

From the analysis of existing studies, we can see that more research results have been achieved in the related research of power data network business resource allocation and service quality assurance. However, from the existing literature [7, 8], we can see that the power data business contains some key business nodes. If these business nodes fail, it will have a very significant impact on the power data business. Therefore, identifying kcy nodes in the power data business and allocating highly reliable network resources to it is a key issue that needs to be resolved urgently.

To solve this problem, this paper first identifies important nodes in the power business, evaluates the importance of communication network resources. Secondly, proposes a power data network resource allocation algorithm based on the TOPSIS algorithm, which includes: allocate resources to important business nodes, and use the shortest path algorithm to allocate link resources. Through simulation experiments, it is verified that the algorithm of this paper allocates better resources to important business nodes without reducing the success rate and resource utilization indicators, and improves the resource performance of important business nodes.

2 Problem Description

The resource allocation of power data network includes two processes: resource allocation of network nodes and resource allocation of network links. This article uses $G = (N, E)$ to denote the network resources of the power data network. Among them, N represents a network node set composed of network nodes $n_i \in N$, and E represents a network link set composed of network links $e_j \in E$. At the time of resource allocation, the network node $n_i \in N$ contains the CPU attribute $cpu(n_i)$. The network link $e_j \in E$ contains the bandwidth attribute $bw(e_j)$.

For the formal description of smart grid business, use $G_R = (N_R, E_R)$ to denote smart grid business. Among them, N_R represents the business node set composed of smart grid business nodes $n_i^R \in N_R$, and E_R represents the business set of smart grid business links composed of smart grid business links $e_j^R \in E_R$. When a resource is requested, the smart grid service node $n_i^R \in N_R$ contains the CPU attribute $cpu(n_i^R)$. Smart grid service link $e_j^R \in E$ contains bandwidth attribute $bw(e_j^R)$.

3 Business Node Importance Evaluation Index

The importance of power data network business nodes is not only related to their position in the network topology, but also closely related to the business carried by the business nodes. Based on this, this paper builds the importance of the evaluation system of business nodes from three aspects: node type, node load, and network topology characteristics.

In terms of node categories, they generally include different categories such as power plants, power supply companies, and dispatch centers. Nodes of the same category also include different scales. For example, some dispatching stations are 220 kV dispatching stations, and some dispatching stations are 500 kV dispatching stations. Based on this, in terms of node categories, this paper will classify the two indicators of node level and node size and divide them into five levels. In terms of node load, it is mainly related to the user type of the load. For example, the level of user power service in military and political aspects is higher than that of general industrial and economic power service. And the importance of general industrial and economic electricity business is higher than that of ordinary residents. Based on this, in terms of node load, this paper will classify the two indicators from load level and load size, and divide it into five levels. In terms of network topology characteristics, it mainly includes two aspects: the degree of connection between nodes and other nodes, and the degree of proximity of nodes to other nodes. Among them, the degree of connection between a node and other nodes is calculated using formula (1). The formula indicates that the more edges k_i of a node, the higher the degree of connection with other nodes. N represents the total number of business nodes.

$$D_{1i} = \frac{k_i}{N - 1} \tag{1}$$

In terms of the proximity of a node to other nodes, it is calculated using formula (2), which indicates that the fewer the number of links d_{ij} on the shortest path traversed by a node to other nodes, the higher its proximity to other nodes.

$$D_{2i} = \frac{N - 1}{\sum_{j=1}^{N} d_{ij}} \tag{2}$$

Based on business node importance evaluation index, this paper constructs the decision matrix $X_a = (x_{ij})_{N\times m}$ of the a-th index as follows. Among them, N represents the total number of nodes, m represents the number of indicators of evaluation dimension a. Because the dimensions and data types of each indicator are not uniform, it is not convenient for further analysis and decision-making. This paper uses formula (3) to normalize each index to obtain the decision matrix $R_a = (r_{ij})_{N\times m}$.

$$r_{ij} = \frac{x_{ij} - \min\{x_{ij}|1 \le i \le N\}}{\max\{x_{ij}|1 \le i \le N\} - \min\{x_{ij}|1 \le i \le N\}} \tag{3}$$

4 The Importance of Business Nodes Based on Topsis Algorithm

In order to evaluate the three-dimensional indicators of node type, node load, and network topology characteristics, this paper uses TOPSIS algorithm. Among them, the TOPSIS algorithm ranks the indicators to be evaluated based on idealized goals [9]. The idealized goals constructed in this paper include positive ideal goal solutions and negative ideal goal solutions as shown in formula (4) and formula (5).

$$R_{aj}^{+} = \{r_1^{\max}, \ldots, r_m^{\max}\} \tag{4}$$

$$R_{aj}^{-} = \{r_1^{\min}, \ldots, r_m^{\min}\} \tag{5}$$

Based on this, the distance from each node to the positive ideal target solution and the negative ideal target solution can be obtained by formulas (6) and (7).

$$L_{aj}^{+} = \sqrt{\sum\nolimits_{j=1}^{m} (r_{ij} - r_j^{\max})^2}, i = 1, 2, \ldots .N \tag{6}$$

$$L_{aj}^{-} = \sqrt{\sum\nolimits_{j=1}^{m} (r_{ij} - r_j^{\min})^2}, i = 1, 2, \ldots .N \tag{7}$$

Based on the distance from each node to the positive ideal target solution and the negative ideal target solution, use formula (8) to calculate the closeness of the a-th index of node i.

$$T_{ai} = \frac{L_{ai}^{-}}{L_{ai}^{-} + L_{ai}^{+}}, i = 1, 2, \ldots, N \tag{8}$$

After obtaining the decision matrix consisting of three categories: node category, node load, and network topology, how to summarize the attributes of the three aspects requires first solving the vector weight W_j of the three indicators. From the literature [10], we can use formula (9) to solve the vector weight W_j of the three indicators based on the elements in the decision matrix. E_j uses formula (10) for calculation, which is based on information entropy to calculate the weight values of different indicators.

$$w_j = \frac{1 - E_j}{3 - \sum_{j=1}^{n} E_j} \tag{9}$$

$$E_j = \frac{1}{\ln m} \sum\nolimits_{i=1}^{m} \frac{r_{ij}}{\sum_{i=1}^{m} r_{ij}} \ln \frac{r_{ij}}{\sum_{i=1}^{m} r_{ij}} \tag{10}$$

Based on the index weights, use formula (11) to solve the final closeness value of each node. The greater the closeness, the more important of the node in the network.

$$T_{x_i} = \sum_{j=1}^{N} w_j \times T_{ij} \tag{11}$$

5 Evaluation of the Importance of Communication Network Resources

In the long-term operation of the power data network, power companies have accumulated a large amount of resource allocation data. By analyzing these data, important node resources in the power data network can be obtained. In order to make full use of these data, the statistics of power data network resource allocation are used in this paper to identify important resources in the power data network. Based on this, the diagonal element value $a_{ii} \in M_k^N$ of the $n \times n$ matrix M_k^N represents the CPU resource demand allocated to the power service node.

$$M_k^N = \begin{bmatrix} a_{11} & 0 & \ldots & 0 \\ 0 & a_{22} & \ldots & 0 \\ \ldots & \ldots & \ldots & \ldots \\ 0 & 0 & \ldots & a_{nn} \end{bmatrix} \tag{12}$$

$$a_{ii} = \begin{cases} cpu(n_j^R) & n_i = M_N(n_j^R), n_j^R \in G_R \\ 0 & other \end{cases}$$

Among them, $n_i = M_N(n_j^R)$ indicates that n_i allocates CPU resources to the power business n_j^R. For the link $e_j^R \in E_R$ of the power service n_j^R, if resources are allocated by the path $p^s(i,j)$ of the power data network, the element b_{ij} of the matrix M_k^L can be used to represent the association relationship between the network node $n_i \in N$ and the network node $n_j \in N$ in the power data network, using the formula (13) to calculation, where $hops(p^s(i,j))$ represents the number of links included in the path $p^s(i,j)$ of the power data network.

$$b_{ij} = \begin{cases} \frac{degree(n_i^v)}{hops(p^s(i,j))} & p^s(i,j) = M_L(l^v), l^v \in VN_k \\ 0 & other \end{cases} \tag{13}$$

After the power data network allocates resources to k power services, the resource usage of matrix M_k^N and matrix M_k^L can be calculated by matrix summation formula (14) and formula (15).

$$S^N = \sum_{i=1}^{k} M_k^N \tag{14}$$

$$S^L = \sum_{i=1}^{k} M_k^L \tag{15}$$

Considering that the elements of the matrix respectively represent the CPU resources of the nodes and the degree of association between the nodes, they do not belong to the same dimension. Equation (16) is used to normalize and sum matrix elements In matrix S^{NL}, diagonal element $M_{ii} \in S^{NL}$ represents the importance of network node $n_i \in N$ in the power data network, and non-diagonal element $M_{ij} \in S^{NL}$ represents the degree of association between network node $n_i \in N$ and network node $n_j \in N$ in the power data network.

$$S^{NL} = \frac{S_{ij}^{N}}{\sum_{i=1}^{n}\sum_{j=1}^{n} S_{ij}^{N}} + \frac{S_{ij}^{L}}{\sum_{i=1}^{n}\sum_{j=1}^{n} S_{ij}^{L}} \tag{16}$$

6 Algorithm

The power data network resource allocation algorithm based on the TOPSIS algorithm RAAoTA proposed in this paper is composed of five steps, which consist of the calculation of the communication network resource importance index, the calculation of the business node importance index, the calculation of the TOPSIS algorithm-based business node importance evaluation, the allocation of resources for the business node in turn, and the business link allocation resources. Details as follows.

1) Calculate the importance of communication network resources, including the following two steps: (a), after the power data network allocates resources to k power services, the resource usage of matrix M_k^N and matrix M_k^L can be calculated by formula (14) and formula (15). (b), use formula (16) to normalize the matrix elements and obtain the final resource importance analysis matrix S^{NL}.

2) Calculate the evaluation index of business node importance, including the following two steps: (a), construct a business node importance evaluation system from three aspects: node type, node load, and network topology characteristics. (b), use formula (3) to normalize each index to obtain the normalized decision matrix $R_a = (r_{ij})_{N \times m}$.

3) Calculate the importance index of business nodes based on TOPSIS algorithm, including the following six steps: (a), use formula (4) and formula (5) to construct positive ideal target solution and negative ideal target solution. (b), use formula (6) and formula (7) to obtain the distance from each node to the positive ideal target solution and the negative ideal target solution. (c), based on the distance of each node to the positive ideal target solution and the negative ideal target solution, use formula (8) to calculate the closeness of the a-th index of node i. (d), use formula (9) to solve the vector weight W_j of the three indicators based on the elements in the decision matrix. (e), based on the index weights, use formula (11) to solve the final closeness value of each node. (f), sort nodes in descending order based on the closeness value.

4) Assign resources to the service nodes in sequence, including the following two steps: (a), find the network node that meets the $cpu(n_i^R)$ requirements of service node in matrix S^{NL}, and allocate the network node resources with the largest sum of $m_{ii} \in S^{NL}$

and $m_{ij} \in S^{NL}$ to the service node. (b), if there is no network node that meets the $cpu(n_i^R)$ demand of service node in matrix S^{NL}, resource allocation fails.

5) Allocate resources for business links, including the following two steps: (a), among the network nodes that have been allocated business node resources, find the power data network link that meets the demand $e_j^R \in E_R$ of the power grid business link $e_j^R \in E_R$ based on the shortest path algorithm. (b) If the power data network link resource that satisfies the service link $bw(e_j^R)$ requirement is not found in the matrix S^{NL}, the resource allocation fails.

7 Performance Analysis

In the simulation experiment, the GT-ITM tool is used to generate the power data network topology and smart grid business topology [11]. The generated power data network topology has 100 communication network node resources, and the number of network node and network link resources is evenly distributed among [45, 95]. The number of business nodes of the generated smart grid business topology is evenly distributed among [5, 12], and the number of resources of business nodes and business links is evenly distributed among [2, 12].

In order to verify the performance of the algorithm in this paper, the algorithm RAAoTA in this paper is compared with the algorithm basic resource allocation algorithm (BRAA) that does not consider the importance of business. The comparative dimensions include resource allocation success rate, resource utilization rate, and key business node allocation resource performance. Among them, resource allocation success rate refers to the success rate of power data network allocating resources for smart grid business within a period of time. Resource utilization refers to the ratio of used power data network resources to total resources after the power data network allocates resources for smart grid business within a period of time. When comparing the performance of key business nodes in allocating resources, select the 5% of the nodes with the largest closeness among the power business nodes as important nodes. In terms of the performance of key node allocation resources, the sum of $m_{ii} \in M$ and $m_{ij} \in M$ of the allocated power data network nodes is used. The experimental results are shown in Figs. 1 to 3.

Figure 1 shows the comparison results of the resource allocation success rate of the two algorithms. It can be seen from the figure that the success rate of resource allocation of the two algorithms is maintained at about 77%. The resource allocation success rate of the algorithm BRAA, which does not consider business importance, is slightly higher than that of the algorithm RAAoTA in this paper, but the overall level of the resource allocation success rate of the algorithm in this paper is slightly different from that of the algorithm BRAA. Figure 2 shows the results of resource utilization comparison between the two algorithms. It can be seen from the figure that the resource utilization rate of the two algorithms is maintained at about 56%. The resource allocation algorithm BRAA, which does not consider the importance of business, has a slightly higher resource utilization rate than the algorithm RAAoTA in this paper, but the overall resource utilization level of the algorithm in this paper differs slightly from

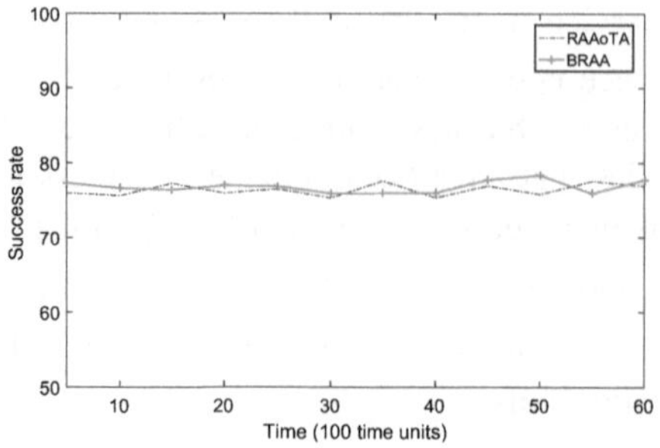

Fig. 1. Comparison of success rate of resource allocation.

that of the algorithm BRAA. Figure 3 shows the comparison results of the resource allocation performance of the key business nodes of the two algorithms. It can be seen from the figure that the algorithm BRAA, which does not consider the importance of business, maintains the performance of resource allocation of key business nodes at about 280, and the performance of the key business node of the algorithm RAAoTA of this paper maintains at about 310, indicating that the algorithm in this paper allocates optimized power data network resources to important nodes of the power business.

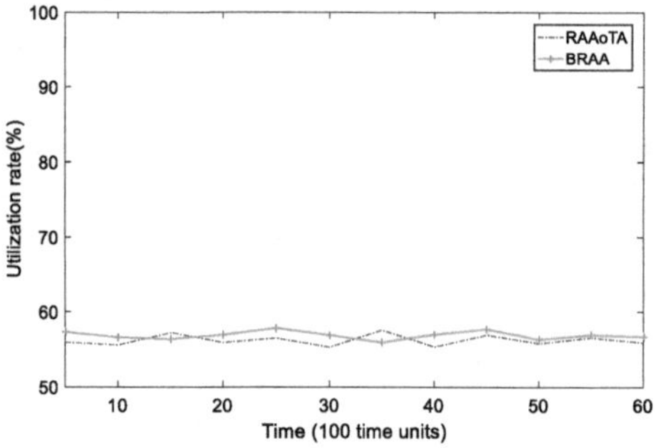

Fig. 2. Comparison of resource utilization rate.

It can be seen from the analysis of Figs. 1 to 3 that the performance of resource allocation success rate and resource utilization of the algorithm RAAoTA in this paper is lower than that of the algorithm BRAA, but it still belongs to the acceptance range, and has less impact on the resource allocation of the power business. However, the algorithm of this paper has achieved good results in the performance of key business nodes in allocating resources, which is helpful to improve the performance of important nodes in the power business.

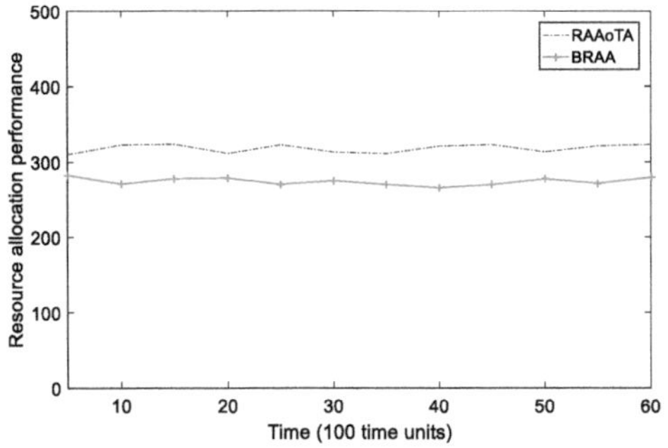

Fig. 3. Comparison of resource allocation performance of important business nodes.

8 Conclusion

The types of smart grid business are becoming more and more abundant, meeting the increasingly diverse needs of the people. Under this background, allocating key power data network resources to important smart grid business nodes has become an important research content. To solve this problem, this paper firstly identifies the important nodes in the power business, and secondly evaluates the importance of communication network resources. Based on this, a power data network resource allocation algorithm based on TOPSIS algorithm is proposed. Through simulation experiments, it is verified that the algorithm of this paper allocates better performance resources to the key nodes of the power business without reducing the success rate and resource utilization indicators, and improves the performance of the key business nodes. In the next step, based on historical operating data and equipment redundancy information, the reliability factors of communication network resources are studied to ensure that important nodes can obtain highly reliable resources.

Acknowledgments. This work is supported by China Southern Power Grid Technology Project "Research on application of data network simulation platform" (STKJXM20180052).

References

1. Li, M., Xu, Z.F., Xu, C.Z., Nian, A.J.: QoS-driven resource allocation mechanism for power utility network to maximize utility. Comput. Syst. Appl. **27**(7), 265–271 (2018)
2. Zhou, J., Hu, Z.W., Liu, G.J., et al.: Business demand analysis and topology planning of power backbone communication network. Opt. Commun. Res. **02**, 15–18 (2016)
3. Zhu, S.S., Huang, Y.B., Zhu, Y.F., et al.: Research on key technologies for information security protection of industrial control systems. Electric Power Inf. Commun. Technol. **11** (11), 106–109 (2013)
4. Cui, L.M., Sun, J.Y., Li, S.J., et al.: An entropy-based uniform distribution algorithm for power data network business resources. Power Grid Technol. **41**(9), 3066–3073 (2017)
5. Zeng, Y., Li, W.J., Chen, Y.Y., et al.: Congestion avoidance algorithm of power dispatching data network based on business priority. Power Syst. Protect. Control **42**(2), 49–55 (2014)

6. Cui, Z.H.: Research on Power Data Network Optimization Strategy to Achieve Load Balancing. Tianjin University, Tianjin (2014)
7. Liu, D.C., Ji, X.P., Wang, B., et al.: Topological vulnerability analysis and countermeasures of power data network based on complex network theory. Power Grid Technol. **39**(12), 3615–3621 (2015)
8. Jiang, Z.P., Zhang, D.L., Wang, L., et al.: Evaluation method for the importance of command network nodes under multidimensional constraints. J. PLA Univ. Sci. Technol. (Natural Science Edition) **16**(3), 294–298 (2015)
9. Patil, S.K., Kant, R.: A fuzzy AHP-TOPSIS framework for ranking the solutions of knowledge management adoption in supply chain to overcome its barriers. Expert Syst. Appl. **41**(2), 679–693 (2014)
10. Liu, J., Liu, S.F., Zhou, X.Z., et al.: Research on multi-attribute decision making based on similarity relationship. Syst. Eng. Electronic Technol. **33**(5), 1069–1072 (2011)
11. Zegura, E.W., Calvert, K.L., Bhattacharjee, S.: How to model an Internet work. In: IEEE Infocom, pp. 594–602 (1996)

Application of Dynamic Management of 5G Network Slice Resource Based on Reinforcement Learning in Smart Grid

Guanghuai Zhao(✉), Mingshi Wen, Jiakai Hao, and Tianxiang Hai

State Grid Beijing Electric Power Company, Beijing 100031, China
sg_zhaoguanghuai@163.com

Abstract. With the rapid development of power grid, the types of power services are becoming more and more diversified, resulting in different service demands. As one of the important technologies of 5G, network slice is used to accommodate different services on the same physical network. After a brief analysis of smart grid background, this paper makes a deep research on the network slice. Each service in smart grid has its own requirements of bandwidth, reliability and delay tolerance. In order to ensure the QoS of Smart Grid, a dynamic optimization scheme of network slicing resources based on reinforcement learning is proposed. This algorithm adjusts the network slice resource dynamically, we can predict the traffic by considering the change of traffic in future network slice, and then deduce the partition of network resource in the future. The reinforcement learning algorithm is then used to make the state of network resource partitioning at future moments influence the current partitioning policy to get the best current policy. Based on this algorithm, the fast response to the change of network demand can be guaranteed in the process of resource allocation, and it is verified by simulation.

Keywords: Smart grid · 5G · Network slice · Resource allocation

1 Introduction

With the development of 5G network, it can not only bring better bandwidth experience, but also shoulder another important mission: enabling vertical industry [1]. The ultra-high bandwidth, ultra-low delay and ultra-large scale connection will change the operation mode and operation mode of the core business of the vertical industry, overall promote of traditional vertical industry operating efficiency and intelligent level of decision-making. As one of the key technologies of 5G, network slicing has been paid more and more attention. Based on intel virtualization technology, a 5G physical network is logically cut into virtual end to end networks. The network slices are isolated from each other, and the congestion, overload and configuration adjustment of any network slice will not affect other network slices. 5G network slicing enables the operators to build agile and flexible networks to cater to multiple use-cases in different industry verticals [2].

The explosive growth of communication demand from all kinds of grid equipment, power terminals and power users has forced the world power grid to transform from the

Q. Liu et al. (Eds.): CENet 2020, AISC 1274, pp. 1485–1492, 2021.
https://doi.org/10.1007/978-981-15-8462-6_169

traditional grid to the smart grid. Due to the different needs of grid use cases, an ultra-reliable and low End-to-End (E2E) delay, flexible and low-cost network is needed [3]. And the 5G network slice just has corresponding ability to match.

According to the classification of 5G application scenarios, power services can be divided into three categories: mobile application services, control services and information collection services [4]. Each kind of power service has different quality of service requirements, which leads the division of network slices to change. In the slice-based network architecture, the quality of slicing directly affects the network performance, so it is very important to optimize the slicing resources dynamically.

Aiming at the dynamic optimization of network slicing resources, some scholars have proposed a semi-static resource allocation scheme based on proportional fairness algorithm. This scheme enables a more equitable allocation of resources among network slices [5]. Some scholars believe that through the statistical analysis of the traffic, the traffic distribution characteristics of the whole network can be obtained, and then the basic slices can be constructed in advance according to the traffic distribution. After that, the load and demand of the real-time traffic are analyzed, and the construction results are sent to the switching node through the OpenFlow protocol [6]. In paper [7], a proportional fair resource allocation scheme based on quality of service is proposed, which weighs the fairness between slices and the needs of users. It achieves higher user spectrum efficiency while ensuring fairness. However, the above slicing algorithms are optimized according to the current traffic, without considering the impact of future network traffic changes. In fact, the future network traffic needs to be taken into account when dynamically optimizing network slicing resources, because if the future network traffic changes are taken into account in the decision-making, it is equivalent to introducing the prediction function into the slicing strategy. So that the partition results can respond more quickly to the changes in the needs of the future network. In order to solve this problem, we propose a dynamic optimization scheme of network slice resources based on reinforcement learning (DOSoNS).

2 Analysis of DOSoNS Algorithm

2.1 Reinforcement Learning

Reinforcement learning (RL) is one of the main methods in the field of machine learning and intelligent control in recent years. RL determines a set of behaviors that agents should take in the environment by maximizing cumulative utility. The method of calculating the cumulative utility of RL does not depend on the past behavior, but on the future state, that is to say, the future state will affect the current state choice. Through reinforcement learning, an agent can know what action should be taken in a particular state.

The idea of RL is much like Markov processes, as shown in Fig. 1. It defines four tuples {S, A, Psa, R}. Among them, S is the current state of intelligent agent, A is the action that the agent takes. PSA is the probability distribution of an agent doing action a in state s and then moving to another state. R is the utility function for each state. Furthermore, RL defines mappings between states to action, π: S $\rightarrow$ A, is known as strategies.

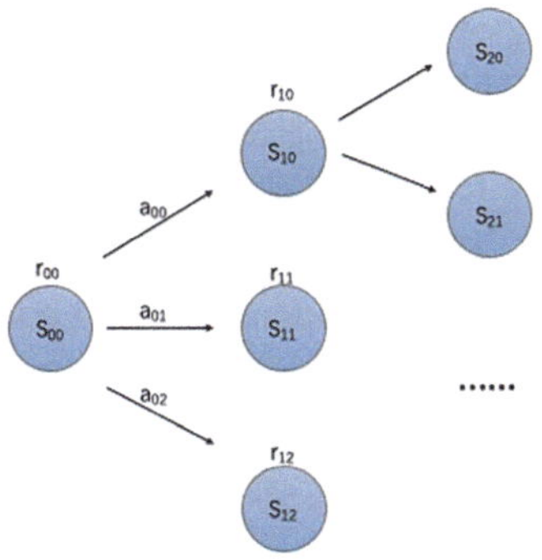

Fig. 1. Markov process.

Reinforcement learning obtains the optimal strategy by defining and optimizing the value function, the most common form of value function such as formula (1).

$$U^{\pi}(s) = E_{\pi}\left(\sum\nolimits_{j=0}^{\infty} \gamma^{j} r_{j} | s_{0} = s\right) \tag{1}$$

It can be seen that this is the weighted sum expectation of a set of utility functions, in which γ is called the conversion factor, which describes the importance of the future utility to the current utility.

With the definition of the value function, the optimal strategy becomes the maximum value function, that is:

$$\pi^{*} = \arg\max_{\pi} U^{\pi}(s), \forall s \in S \tag{2}$$

We can improve the existing strategy according to the following principle: if the other behaviors of the strategy π remain unchanged, only the action a in the state s is changed to a', and a new strategy π' is obtained. If the value function $U^{'} > U$, then the policy π is better than the strategy π'. We can get the optimal policy π^* through the dynamic programming algorithm.

2.2 Some Definition in Algorithm

(1) Link state and node state. First of all, it is necessary to discretize the CN resources, that is, to divide the core network links and core network node resources into resource slices. In dynamic optimization, a resource slice is taken as the smallest unit of change.

(2) Predict the state of future links and nodes. If the dynamic optimization algorithm has a certain prediction function, then the partition result of CN can respond to the change of network demand more quickly, so it is necessary to consider the possible partition state of the core network in the future. Therefore, we also need to carry out traffic analysis to predict the state of CN at various times in the future.

(3) Link utility function, node utility function and total utility function. The utility function describes several indicators. Due to the fact that links and nodes are often concerned about different indicators, link utility functions and node utility functions need to be defined respectively.

Taking the link utility function as an example, assuming that there are n types of traffic in the network, the link utility function in a certain state can be defined as: $U_m = \sum_{j=0}^{n} U_{mj}$. Here U_{mj} is the utility function of the j-th network slice in this link. The sub-utility function of the j-th network slice can be defined as follows: first, the link utilization should be an appropriate value, because too high link utilization will lead to congestion and packet loss and low link utilization will lead to a waste of resources. Therefore, when defining the utility function, a reference link utilization β can be given. The smaller the actual link utilization deviates from β, the higher the sub-utility function of the network slice. In addition, for a business, the importance of different slices can be different, so the more important slices, the higher the utility function.

Similarly, the utility function U_n of a node can be defined. Therefore, under a certain time t, the total utility function can be defined as $U_t = U_m + U_n$, and the maximization objective function at time t_0 is:

$$U_{total} = u_{t_0} + \gamma^1 \cdot U_{t_1} + \gamma^2 \cdot U_{t_2} + \cdots + \gamma^T \cdot U_{t_T} = \sum_{k=0}^{T} \gamma^k \cdot U_{t_k} \tag{3}$$

Here, γ is converted factor which describes the importance of future state to the current strategy.

2.3 Algorithm Steps

The algorithm includes input and output. Input refers to the historical data packets of each link and node within a period of time. The output refers to how the network links and network nodes should be divided at this time. The description of the algorithm is as follows:

a. Content prediction. Based on the data requests in the core network in the past period of time, the traffic of all kinds of service packets in each link in the future discrete time is predicted.
b. Discretization of resources. The resources of each link and network node are divided into resource slices, and one resource slice is used as the smallest partition unit in the resource allocation in the future.
c. Maintain two state matrices to describe the link state and node state of the core network respectively. It is assumed that there are k slices in the network, and the initial state of the link S_a is given. At this time, the resource division of the link is $a_i = (b_1, b_2, \ldots, b_n)$, where b_k is the number of resource slices allocated by the k-th network slice, and $\sum_{i=0}^{n} b_i$ is the total number of resource slices of the link. The link resource partition matrix $B_{m \times n}$ of the whole network can be obtained, in which each row vector describes the resource partition of each link. Define $S_a = B_{m \times n}$ (m links). Given the initial state of the node S_n, the resource division of the j th node at this time is $n_j = (c_1, c_2, \ldots, c_n)$, where c_k is the number of resource slices allocated by

the k-th network slice, and $\sum_{i=0}^{n} c_i$ is the total number of resource slices of the network nodes, the node resource partition matrix $C_{k \times n}$ of the whole network can be obtained, in which each row vector describes the resource partition of each node. Define $S_n = C_{k \times n}$ (k nodes).

d. Define the utility function. We have got the formula as (3).

e. Assuming that the time t_0 needs to be optimized, if the link state is S_a and the node state is S_n, the corresponding link utility function and node utility function can be obtained. Under a certain action, the link state and node state will be transferred to their adjacent states S_a' and S_n' at t1 time. Both S_a' and S_n' are collections in which the elements are adjacent to the current state.
The link resources of the core network are often more abundant than the node resources, and different services have different requirements for network nodes. So after transferring to a new set of states Sa' and Sn', you need to replace the Sa' into the Sn' for verification. Check whether the new link resource partition meets the node partition requirements at this time. If so, the utility functions of Sa' and Sn' in the new state can be obtained, and then proceed to the next state transition. If it is not satisfied, the state pair is marked as invalid, that is, the transition node is deleted from the state transition diagram. The link state Sa' falls back to the state Sa, and retransfers to other states in the Sa' set, and then continues to verify, so that we can get the network state of T moments in the future, then calculate their respective utility functions according to the predicted packet traffic, influence the current decision with a certain discount rate, and get the corresponding effect value of this group of policies.

f. Maximize the objective function U_{total}. The dynamic programming algorithm is used to improve the strategy, so as to converge to the optimal strategy, and the resource division of links and nodes between the current time and the future moment can be obtained.

3 Simulation Results and Analysis

We compare the performance of the three algorithms: algorithm 1 is a proportion-based dynamic optimization scheme for network slices, which divides the slicing resources proportionally according to the current traffic; algorithm 2 is a fair static network slicing allocation scheme, which allocates network resources fairly to each slice and does not adjust according to the changes of network traffic. Algorithm 3 is a dynamic optimization scheme of 5G network slice resources based on reinforcement learning. The following experiment simulates the 5G core network, in which the core network nodes are assumed to be distributed collaboration. First of all, ensure that the network resources of the core network remain unchanged, by increasing the amount of requested data in the network, compare the resource utilization of the three algorithms. Then ensure the core network resources and the amount of requested data remain unchanged, by changing the value of the conversion factor γ, investigate the utilization of network resources. Suppose there are four core gateway nodes, the number of resources provided by each node is generated according to the uniform distribution with 40 as the center, the resources of the CN node include computing resources, storage resources, etc. The links between the two CN nodes are generated with a certain probability P, and the number of resources per link is generated according to the uniform distribution with 55 as the center. Suppose the operator constructs two CN slices at the same time, and the number of request packets for each slice is randomly generated within a certain range. Its value is the sum of the amount of data of each link connected to the node, and a total of T groups of data can be generated to simulate the demand of each slice in the future T time.

Figure 2 and Fig. 3 describe the average node resource utilization and link average resource utilization of each algorithm in the 5G core network. We can get when the maximum number of content requests of each CN slice increases, that is, the total amount of data in the core network increases, the resource utilization of the three algorithms increases, and the resource utilization of the two dynamic adjustment algorithms is higher than that of the static partition algorithm. This is mainly because the dynamic resource allocation results better match the network requirements, so that the core network resources are better utilized.

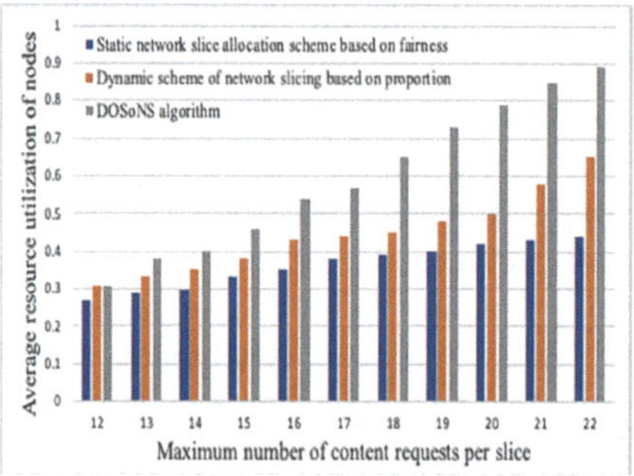

Fig. 2. Node resource utilization of each algorithm at present.

Figure 4 and Fig. 5 describe the average node resource utilization and link average resource utilization of each algorithm in the 5G core network in the future. The average resource utilization of the dynamic optimization scheme of 5G network slice resources based on reinforcement learning is the highest, the static partition algorithm based on fairness is the second, and the proportion-based partition algorithm is the last, among which the dynamic optimization scheme of network slices based on proportion is the most unstable. Intuitively, the increase in the maximum number of content requests of each CN slice will mainly bring about changes in two aspects: first, it increases the total amount of data in the core network, so that resources are more fully utilized; second, the change range of traffic demand of each CN slice is increased. This means that when the traffic in the network changes, the proportion-based partition algorithm needs to adjust the network partition results repeatedly, while the continuous adjustment of core network resources will consume a lot of time and resources. therefore, the resource partition result is slow to respond to the future network demand changes and is not stable. Therefore, when the result of network partition does not match the traffic in the future, it will lead to low resource utilization.

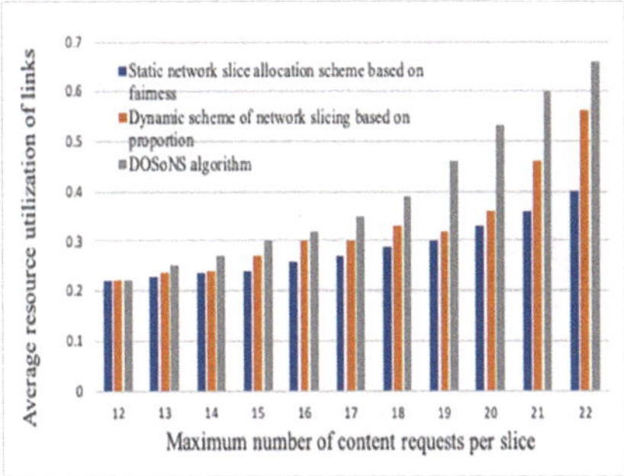

Fig. 3. Link resource utilization of each algorithm at present.

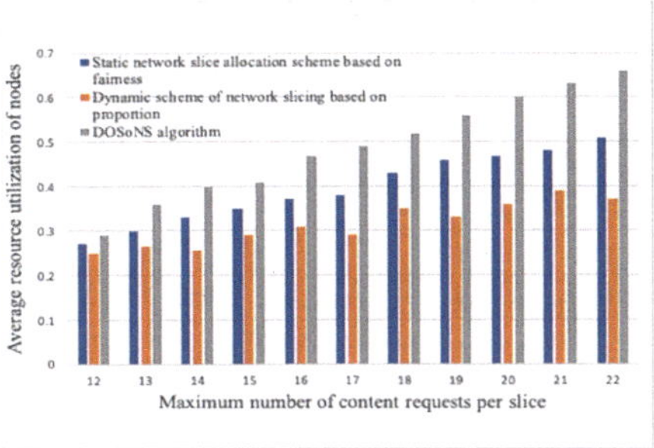

Fig. 4. Node resource utilization of each algorithm for future.

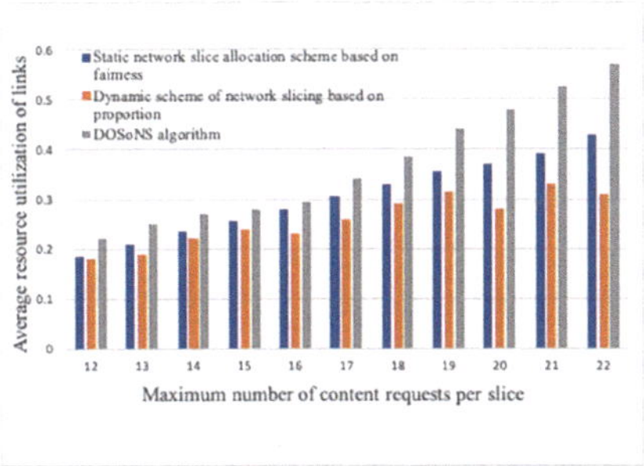

Fig. 5. Link resource utilization of each algorithm for future.

4 Conclusion

Aiming at how to dynamically adjust 5G core network slicing resources to better meet the needs of smart grid, we propose a dynamic optimization scheme of 5G network slicing resources based on reinforcement learning. The algorithm first predicts the state of the core network in the future, and then affects the current decision according to the state of the core network in the future, so that the slicing results can respond to the changes in the needs of the network more quickly and improve the overall performance of the network.

Acknowledgement. This work is supported by 2020 State Grid Beijing Municipality Science and Technology project "Research and application of ubiquitous business agile access technology in 5G integrated smart grid".

References

1. Xia, X., Zhu, X., Mei, C., Li, W., Fang, H.: Research and practice on 5G slicing in power internet of things. Mobile Commun. **43**(1), 63–69 (2019)
2. Smart Energy International, China to invest $77.6 bn in smart grid infrastructure, https://www.smart-energy.com/regional-news/north-america/smart-grid-china-northeast-group/, 04 Aug 2016
3. Hassebo, A., Mohamed, A.A., Dorsinville, R., Ali, M.A.: 5G-based converged electric power grid and ICT infrastructure. In: 2018 IEEE 5G World Forum (5GWF), pp. 33–37. IEEE, NJ, USA (2018)
4. Wang, Z., Meng, S., Sun, L., Ding, H., Wu, S., Yang, D., Li, Y., Wang, X., Xi, L.: Slice management mechanism based on dynamic weights for service guarantees in smart grid. In: 2019 9th International Conference on Information Science and Technology (ICIST), pp. 391–396. IEEE, NJ, USA (2019)
5. Su, X., Gong, J.J., Zeng, J.: Radio resource allocation for 5G network slicing. Electronic Eng. Product World **24**(4), 30–32 (2017)
6. Zhou, H., Chang, Z., Yang, W., Guo, J.: An algorithm for arranging 5G network slices. Telecommun. Sci. **33**(8), 130–137 (2017)
7. Wang, W., Xu, Z., Tian, Z.: QoS-based 5G slice resource allocation. Res. Opt. Commun. **207** (03), 59–63 (2018)

Improved Genetic Algorithm for Computation Offloading in Cloud-Edge-Terminal Collaboration Networks

Dequan Wang[1(✉)], Ao Xiong[1], Boxian Liao[1], Chao Yang[2], Li Shang[3], Lei Jin[2], and Xiaolei Tian[4]

[1] State Key Laboratory of Networking and Switching Technology, Beijing University of Posts and Telecommunications, Beijing 100876, China
dequanwang@bupt.edu.cn
[2] Information and Communication Branch, State Grid Liaoning Electric Power Co., Ltd, Shenyang 110000, China
[3] State Grid Hebei Information and Telecommunication Branch, Shijiazhuang 050000, China
[4] State Grid Liaoning Electric Power Supply Co. Ltd., Shenyang 110000, China

Abstract. The massive use of Internet of Things (IoT) mobile devices (MDs) and the increasing demand for their computing have created huge challenges for the current development of the IoT. Mobile edge computing (MEC) and cloud computing provide a scheme for these problems. In the process of offloading, MDs and servers are facing difficulties such as high consumption and high latency. So it is necessary to reasonably offload computing tasks to MDs, edge servers, or cloud servers. In view of this situation, the research direction of this article is how to reduce the power consumption of the device and the server while ensuring that the delay requirements of different tasks are met. First we formulate the proposed problem as a nonlinear combinatorial optimization problem, then propose the cloud-edge-terminal collaboration offloading algorithm based on improved genetic algorithm (IGA). Finally, the characteristics of the algorithm are studied by simulation and compared with other algorithms to verify the performance of the algorithm.

Keywords: Mobile edge computing · Internet of Things · Genetic algorithm

1 Introduction

In recent years, many new types of mobile devices (MDs) have appeared and are widely used, and the number and complexity of computing tasks generated on MDs have increased dramatically [1]. So computing tasks processed on MDs can no longer meet user needs. Through mobile edge computing (MEC) or cloud computing, using computation offloading to offload computing tasks to an edge server or cloud server can effectively solve this problem [2]. However, a key issue in computation offloading is how to offload it optimally [3]. We should try our best to reduce the calculation time consumption, transmission time consumption and energy consumption of all tasks.

Q. Liu et al. (Eds.): CENet 2020, AISC 1274, pp. 1493–1499, 2021.
https://doi.org/10.1007/978-981-15-8462-6_170

The increasing number of computing tasks, MDs and edge servers make computation offloading face unprecedented challenges [4].

Recently, some researchers have started to pay attention to the computation offloading in different scenarios. Shi, W. et al. proposed a simple cloud offloading decision strategy to minimize energy consumption based on computing communication ratio and network environment [5]. Sardellitti, S. et al. studied the MECO problem in a MIMO multi-base station system connected to a public cloud server, and proposed an energy-efficient iterative algorithm based on continuous convex approximation as a scheme [6]. Chen, X. et al. studied the problem of computer offloading in a multi-channel wireless interference environment and used a game theory approach [7].

Note that although there have been many works on computation offloading, most of these works are based on simple computing task scenarios, such as the scenario of a single mobile device and non-concurrent tasks. Considering the limitations of the existing work, this paper studies the computation offloading problem with multiple tasks, multiple MDs, and multiple base stations, which takes into account all the above situations. Besides, we take the cloud serve into consideration.

The contributions of this paper are listed as follows:

- We consider the computation offloading in the MEC scenario of multiple computing tasks, multiple MDs, multiple small cell base stations, multiple micro base stations and a cloud serve. Then we present a system model that contains the elements considered above.
- In order to solve the problem of constrained optimization, we propose the cloud-edge-terminal collaboration offloading algorithm based on improved genetic algorithm (IGA).
- We theoretically analyze and verify the proposed algorithm, and carry out simulation experiments. We have simulated different numbers and capability of MDs and edge servers. From the large amount of data results, our proposed algorithm has superior performance.

The organizational structure of the rest of the article is as follows: we will introduce the system model in the second section, analyze the problem and propose the algorithm in the third section, and finally in the fourth and fifth sections introduce simulation experiments and conclusions.

2 System Model

Our system model is shown in Fig. 1. It contains a cloud server, several macro base stations and micro base stations with edge servers, and a large number of MDs connected by wireless.

We denote the MDs set as $U = \{1, 2, \ldots, j, \ldots, U\}$, U_j as the j-th MD, edge servers set as $B = \{1, 2, \ldots, i, \ldots, B\}$, B_i as the i-th edge service, and cloud server as P. The total number of tasks for all MDs is $X = \{1, 2, \ldots, x, \ldots, X\}$, and the x-th task is $D_x = \{d_x, c_x, t_x, t_x^{(\max)}\}$, where d_x represents the data size of the task, c_x represents the calculation cycle required by the task, t_x represents the time when the task is generated,

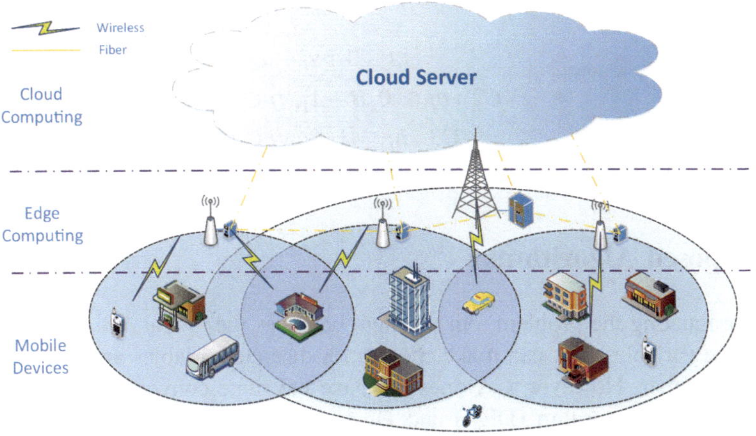

Fig. 1. System model.

and $t_x^{(\max)}$ represents the maximum delay that D_x can tolerate. The connection between U_j and the base station is denoted as h_j, and $h_j \in [1,\ B]$ indicates that U_j connected to B_{h_j}. The offloading position of D_x is expressed as a_x, $a_x \in [0, B+1]$. When $a_x = 0$, D_x is offloaded to U_j. When $1 \le a_x \le B$, D_x is offloaded to B_{a_x}. Otherwise, D_x is offloaded to P.

Time consumption is expressed as T, which is composed of transmission time consumption and calculation time consumption. The transmission time consumption can be calculated by Shannon's formula, and the calculation time consumption can be calculated according to c_x and CPU frequency. The energy consumption is expressed as E, which can be calculated based on the calculation time consumption and CPU power. T and E are weighted to obtain the total consumption by the following formula.

$$\phi = \alpha_t \zeta\left(T, t_x^{(\max)}\right) T + \alpha_e E \tag{1}$$

where α_t, α_e are the weighting coefficient of time consumption and energy consumption respectively, and $\alpha_t + \alpha_e = 1$. Through the expression (1), the total consumption calculated by D_x on U_j, B_i and P is , $\phi_i^{(b)}$ and $\phi^{(p)}$. The total consumption of all MDs, base stations and cloud servers is

$$\Psi = \sum_{j \in U} \sum_{i \in B} \left(I_{\{a_j=0\}} \phi_j^{(u)} + I_{\{0<a_j<=B\}} \phi_i^{(b)} + I_{\{a_j=B+1\}} \phi^{(p)}\right) \tag{2}$$

where $I_{\{\eta\}}$ is an indicator function. When η is true, $I_{\{\eta\}} = 1$; otherwise, $I_{\{\eta\}} = 0$. According to the system model, the problem can be formulated as

$$\begin{aligned} & \min \Psi \\ & s.t.\ C1 : \alpha_t + \alpha_e = 1 \\ C2 : & a_j \in [0, B+1], \forall j \in U \\ C3 : & h_j \in [1, B], \forall j \in U \end{aligned} \tag{3}$$

3 Proposed Algorithms

After formulating the problem, our problem becomes the global optimization problem of discrete variablcs. GA can make choices for discrete variables and get approximate optimal schemes. However, GA does not have the same convergence effect as differential evolution algorithm (DEA) and DEA can only deal with the optimization of continuous variables. Therefore, we improved GA based on the idea of DEA.

3.1 Initialization Improvements

From a probabilistic perspective, the consumption incurred by offloading a computing task generated by an MD to a base station connected to it is less than the consumption incurred by offloading to a base station not connected to it. Therefore, when initializing the task to which base station to offload, we initialize most tasks to offload to the base station to which it is connected. The number of tasks in the current cycle is X, the connection decision of all MDs is $\hat{H} = \{h_j | j \in U\}$, the offload decision is $\hat{A} = \{a_x | x \in X\}$, U_j connection decision is h_j, and the offloading decision of the x-th arriving task in the current cycle is a_x. In order to facilitate writing and expression, $\{\hat{A}, \hat{H}\}$ is represented by $\hat{I}$. The first initialization method is as follows. Initialization method of $\hat{H}$ is $h_j = rand([1, B])$. 30% and 70% initialization method of $\hat{A}$ is $a_x = rand(\{0, rand([1, B]), B+1\})$ and $a_x = rand(\{0, h_j, B+1\})$ separately. Where $rand([1, B])$ indicates that a number is randomly selected from $1, 2, \ldots, B$. $rand(\{0, rand([1, B]), B+1\})$ indicates that one is randomly selected from 0, $rand([1, B])$ or $B+1$. For convenience of expression, the initialization method is collectively referred to as IM.

3.2 Improvement of Algorithm Steps

The mutation operation is basically the same as the simple genetic algorithm (SGA), except that the mutation of SGA takes random values, and we improve it to use IM assignment. *MP* and *CP* are mutation operators and crossover operators, respectively. The value range is $(0, 1)$. At the k iteration, $\hat{P}^{(k)}$ is generated from $\hat{I}^{(k)}$ through the mutation operation.

$$h_j^{(p)} = \begin{cases} result\ of\ MI, & if\ rand(0,1) \le MP \\ h_j \quad , & otherwise \end{cases} \tag{4}$$

$$a_x^{(p)} = \begin{cases} \textit{result of MI}, & \textit{if } rand(0,1) \le MP \\ a_x \quad , & \textit{otherwise} \end{cases} \tag{5}$$

$rand(0,1)$ means that a decimal is randomly taken between 0 and 1. The crossover operation of IGA is to replace some genes in the parent with the genes of the mutated offspring. Through the cross operation, $\hat{Q}^{(k)}$ is generated from $\hat{I}^{(k)}$ and $\hat{P}^{(k)}$.

$$h_j^{(q)} = \begin{cases} h_j^{(p)}, & \textit{if } rand(0,1) \le CP \textit{ or } j = j_{rand} \\ h_j \ , & \textit{otherwise} \end{cases} \tag{6}$$

$$a_x^{(q)} = \begin{cases} a_x^{(p)}, & \textit{if } rand(0,1) \le CP \textit{ or } j = j_{rand} \\ a_x \ , & \textit{otherwise} \end{cases} \tag{7}$$

We randomly picked a fixed $j_{rand} = rand([1, U])$ in advance to make at least one-dimensional data different from the parent. Through selection operations, $\hat{I}^{(k+1)}$ is generated from $\hat{I}^{(k)}$ and $\hat{Q}^{(k)}$.

$$\hat{I}^{(k+1)} = \begin{cases} \hat{Q}^{(k)}, & \textit{if } \Psi(\hat{Q}^{(k)}) < \Psi(\hat{I}^{(k)}) \\ \hat{I}^{(k)}, & \textit{otherwise} \end{cases} \tag{8}$$

The process of the cloud-edge-terminal collaboration offloading algorithm based on IGA is shown in Algorithm 1, and its tasks are all the tasks accumulated in a cycle. The population size is Z, an individual is $\hat{I}$, the sum of all $\hat{I}$ is I and the optimal individual is I'.

4 Simulation Results

In order to verify the validity of experiments, we conducted experiments on the convergence of the algorithm. Fig. 2 shows the convergence speed of IGA. As can be seen from the figure, when the number of MDs is 30, 60 and 90, IGA converges after approximately 700, 1300 and 1900 iterations, respectively. The overall global consumption after convergence is 67.5, 146.5, and 201.6. The increase in the number of MDs has led to an increase in computing tasks. In the case of multiple experiments, the global minimum consumption of 60 and 90 for MDs is approximately 2 and 3 times that of 30.

Figure 3 shows the consumption comparison between IGA and SGA. In order to clarify that the difference in cost is caused by the main body of the algorithm, the initialization process of IGA and SGA are both used in IM. In order to make the contrast difference more obvious, we increase the processing time, so IGA cycle time is set to 2 s. For the same tasks, IGA has lower consumption than SGA, and the difference becomes more and more obvious as the number of MDs increases.

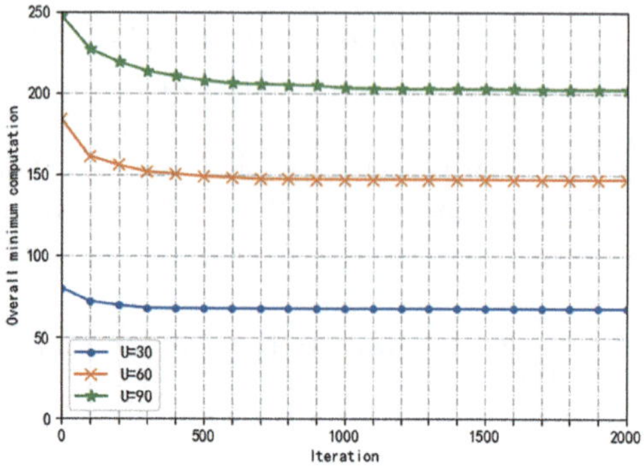

Fig. 2. Comparison of IGA's convergence speed.

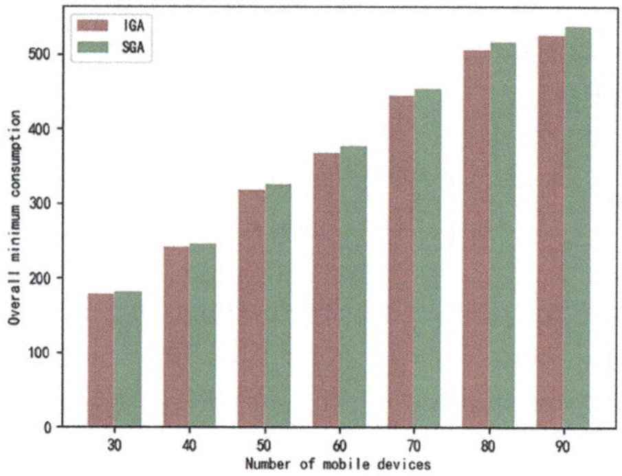

Fig. 3. Comparison between IGA and SGA.

5 Conclusion

In this paper, we study the computation offloading in the MEC scenario of multiple computing tasks, multiple MDs, multiple small cell base stations and multiple micro base stations. The computational goal of this problem is to meet the constraints of computational delay while minimizing the overall consumption of all tasks. First we formulate the proposed problem as a nonlinear combinatorial optimization problem. After that, we proposed the cloud-edge-terminal collaborative offloading algorithm based on IGA as our scheme. Finally, numerical results prove that our proposed algorithm performs better than SGA.

Acknowledgment. This work is supported by the Science and Technology Project of State Grid Corporation of China: Research and Application of Key Technologies in Virtual Operation of Information and Communication Resources.

References

1. Sun, W., Liu, J., Zhang, H.: When smart wearables meet intelligent vehicles: challenges and future directions. IEEE Wirel. Commun. **24**(3), 58–65 (2017)
2. Yu, Y.: Mobile edge computing towards 5G: vision, recent progress, and open challenges. China Commun. **13**(2), 89–99 (2016)
3. Yang, L., Cao, J., Cheng, H., et al.: Multi-user computation partitioning for latency sensitive mobile cloud applications. IEEE Trans. Comput. **64**(8), 2253–2266 (2015)
4. Mao, Y., You, C., Zhang, J., et al.: A survey on mobile edge computing: the communication perspective. IEEE Commun. Surveys Tuts. **19**(4), 2322–2358 (2017)
5. Shi, W., Zhang, J., Zhang, R.: Share-based edge computing paradigm with mobile-to-wired offloading computing. IEEE Commun. Lett. **23**(11), 1953–1957 (2019)
6. Sardellitti, S., Scutari, G., Barbarossa, S.: Joint optimization of radio and computational resources for multicell mobile-edge computing. IEEE Trans. Signal Inf. Process. Netw. **1**(2), 89–103 (2014)
7. Chen, X., Jiao, L., Li, W., et al.: Efficient multi-user computation offloading for mobile-edge cloud computing. IEEE/ACM Trans. Netw. **24**(5), 2795–2808 (2016)

A Computation Offloading Scheme Based on FFA and GA for Time and Energy Consumption

Jia Chen[1], Qiang Gao[1], Qian Wu[1], Zhiwei Huang[1], Long Wang[1], Dequan Wang[2(✉)], and Yifei Xing[2]

[1] Shenzhen Power Supply Co, Ltd., Shenzhen 518000, China
[2] State Key Laboratory of Networking and Switching Technology, Beijing University of Posts and Telecommunications, Beijing 100876, China
dequanwang@bupt.edu.cn

Abstract. With the development of Internet of Things (IoT) technology, the types and the volume of business have been increasing rapidly. The existing centralized cloud processing model is hard to meet the requirements of delay-sensitive and compute-intensive services. So, mobile edge computing and cloud computing are introduced to realize a cloud-edge-terminal collaboration network architecture. However, there still exist problems such as high energy consumption and long delay among devices and servers. To overcome these challenges, a cloud-edge-terminal collaboration offloading scheme based on first fit algorithm (FFA) and genetic algorithm (GA) is proposed, which combines two allocation modes. On one hand, for delay-sensitive tasks, FFA is designed to quickly offload tasks. On the other hand, for dense tasks, GA is designed to accurately offload tasks. To make the best of the advantages and avoid the disadvantages of FFA and GA, we adopt the method of using two algorithms alternately, and restrict the rules of the alternations. At the end, the characteristics of the algorithm are studied by simulation and compared with other algorithms to verify the performance of the algorithm.

Keywords: Mobile edge computing · Internet of Things · First fit algorithm · Genetic algorithm

1 Introduction

Due to the rapid development of Internet of Things (IoT) and 5G, smart cities are increasingly valued in improving the quality of urban life [1]. And with the continuous advancement of computer science, the calculation amount of emerging algorithms is increasing [2]. However, most mobile devices (MDs) have low computing power and small battery capacity, which makes MDs unable to complete delay-sensitive tasks. Through mobile edge computing (MEC) or cloud computing, computation offloading can effectively solve this problem [3]. Therefore, MEC and cloud computing have attracted more and more attention, and their research is also increasing [4].

Recently, some researchers have started to pay attention to the computation offloading in different scenarios. Yang, L. et al. designed a framework that supports

Q. Liu et al. (Eds.): CENet 2020, AISC 1274, pp. 1500–1506, 2021.
https://doi.org/10.1007/978-981-15-8462-6_171

single-user adaptive partitioning, and the framework decided whether to offload [5]. Zheng, J. et al. studied the multi-user MECO problem in a dynamic environment, in which the activity of mobile users and the gain of wireless channels were changed over time. Then an effective multi-agent random learning algorithm was proposed as a scheme to reduce the system-wide computational consumption [6]. Zhang, J. et al. studied the allocation of trade-offs between energy consumption and transmission delay of MDs in single and multi-base station networks, and proposed an energy-aware computation offloading scheme by jointly optimizing communication and computing resource allocation [7].

However, in most current researches, MDs and edge servers only handle one task in a single thread, ignoring that MDs and edge servers are mostly multitasking operating systems and dynamically allocate resources. Considering the limitations of the existing work, this paper studies the computation offloading problem with multiple tasks, multiple MDs, and multiple base stations, which takes into account all the above situations. Besides, we take the cloud serve and dynamic allocation factors into consideration.

The contributions of this paper are listed as follows:

- In order to reduce task computing time and energy consumption, we use the cloud-edge-terminal collaborative offloading system model computing the total consumption.
- To offload the task quickly, first fit algorithm (FFA) is designed to execute delay-sensitive tasks. For tasks that are computationally intensive, we use genetic algorithm (GA) for offloading to approach the optimal offloading decision. To make use of the advantages of FFA and GA and complement each other, we designed a cloud-edge-terminal collaboration offloading scheme based on FFA and GA (COS-FG).
- In the simulation, we consider a practical scenario that hundreds of MDs and tens of base stations are deployed. From the large amount of data results, our proposed algorithm has superior performance.

The organizational structure of the rest of the article is as follows: we will introduce the system model in the second section, analyze the problem and propose the algorithm in the third section, and finally in the fourth and fifth sections introduce simulation experiments and conclusions.

2 System Model

Figure 1 shows our system model that contains a cloud server, several macro base stations and micro base stations with edge servers, and a large number of MDs connected by wireless. The right side of the figure is the task resource allocation diagram for MDs and edge servers. The queuing system of MDs and edge servers is M/M/C queues. For task resource allocation, MDs and edge servers allocate fewer computing resources for later tasks, and use the maximum parallel computing resource cap (MPCRC) and parallel task cap (PTC) as the criteria. In the task queue of MDs and edge servers, tasks that are less than or equal to MPCRC are allocated to the fixed and largest computing resources, and tasks that are greater than MPCRC and less than or equal to the PTC are allocated to fewer computing resources. Tasks that are greater than the PTC are suspended, then wait for the previous computation tasks to end and resources to be released, and finally computing resources will be allocated. Allocating more computing resources to the first-come task can effectively increase the utilization of system resources and speed up the processing speed of the task. The cloud server allocates equal amounts of computing resources to all computation tasks.

Let $U = \{1, 2, \ldots, j, \ldots, U\}$, $B = \{1, 2, \ldots, i, \ldots, B\}$ and P denote the sets of MDs, edge servers and cloud server. We denote U_j as the j-th MD and B_i as the i-th edge service. The total number of tasks for all MDs is $X = \{1, 2, \ldots, x, \ldots, X\}$, and the x-th task is $D_x = \{d_x, c_x, t_x, t_x^{(\max)}\}$, where d_x represents the data size of the task, c_x represents the calculation cycle required by the task, t_x represents the time when the task is generated, and $t_x^{(\max)}$ represents the maximum delay that D_x can tolerate.

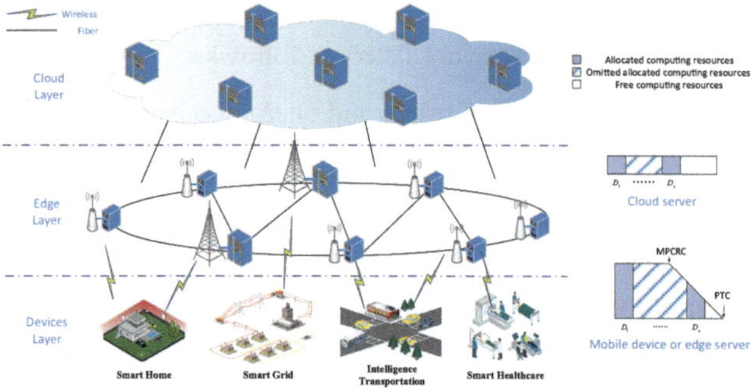

Fig. 1. Cloud-edge-terminal collaborative offloading system model.

Time consumption is expressed as T, which is composed of transmission time consumption and calculation time consumption. The transmission time consumption can be calculated by Shannon's formula, and the calculation time consumption can be calculated according to c_x and CPU frequency. The energy consumption is expressed as E, which can be calculated based on the calculation time consumption and CPU power. T and E are weighted to obtain the total consumption by the following formula.

$$\phi = \alpha_t T + \alpha_e E \tag{1}$$

where α_t, α_e are the weighting coefficient of time consumption and energy consumption respectively, and $\alpha_t + \alpha_e = 1$. Through the expression (1), the total consumption calculated by D_x on U_j, B_i and P is $\phi_j^{(u)}$, $\phi_i^{(b)}$ and $\phi^{(p)}$. The total consumption of all MDs, base stations and cloud servers is

$$\Psi = \sum_{j \in U} \sum_{i \in B} \left(I_{\{a_j = 0\}} \phi_j^{(u)} + I_{\{0 < a_j <= B\}} \phi_i^{(b)} + I_{\{a_j = B+1\}} \phi^{(p)} \right) \tag{2}$$

where $I_{\{\eta\}}$ is an indicator function. When η is true, $I_{\{\eta\}} = 1$; otherwise, $I_{\{\eta\}} = 0$. According to the system model, the problem can be formulated as

$$\begin{aligned} & \min \Psi \\ s.t. C1 &: \alpha_t + \alpha_e = 1 \\ C2 &: a_j \in [0, B+1], \forall j \in U \\ C3 &: h_j \in [1, B], \forall j \in U \end{aligned} \tag{3}$$

3 Proposed Algorithms

For delay-sensitive tasks, we use FFA as our fast algorithm, which can quickly make offload decisions for tasks. For tasks that are computationally intensive, we use GA as our accurate algorithm, achieving the minimum consumption of all tasks that arrive within a period of time.

3.1 FFA

FFA arranges the free area of the storage space in the increasing order of its starting address. We use the FFA to first select the offloading location of the computation task in the order of U, B, P, *Queue*. Here, B indicates the base station to which U is connected. *Queue* represents the priority queue, which stores several base stations. When the calculation time of the task at the current position is less than the maximum delay that the task can tolerate, the task is offloaded to the current position; otherwise, it moves to the next position and repeats the previous judgment. When no location is satisfied, the location with the lowest consumption is selected for offloading.

3.2 GA

GA is a computational model that simulates the natural evolution and biological mechanism of Darwin's biological evolution theory. It is a method to search for the optimal scheme by simulating natural evolution. We use GA as our accurate algorithm, which can achieve the minimum consumption of all tasks. Since we are using a general-purpose GA, the related processes used are not repeated here.

3.3 COS-FG

To make the best of the advantages and avoid the disadvantages of each algorithm, we adopt the method of using two algorithms alternately, and restrict the rules of the alternations. We set GA to run once to continue Q s when using GA to offload. GA will offload all tasks within this period at the end of the duration. When using FFA to offload, FFA offload it immediately after the task is generated. The maximum processing time of FFA is *TMAX*. We use the *penalty*(c) function to calculate the penalty value of the tasks. Then add up the penalty values of the tasks processed by FFA to get the cumulative penalty value. When the cumulative penalty value of a pending computation task exceeds *PMAX*, the computation task is no longer processed FFA.

$$penalty(c) = 1.060083^{\frac{c}{c(mean)}} - 1.010738 \tag{4}$$

When the $\frac{c}{c(mean)}$ value is 0.2, 1 and 1.8, the $penalty(c)$ value is about 0.001, 0.05 and 0.1, respectively. When *PMAX* is set to 1 and all tasks are of average size, the execution times ratio of GA to FFA is 1:20.

How to set *Q*, *TMAX* and *PMAX* to minimize the consumption will be verified through experiments.

Based on the above discussion, we propose a cloud-edge-terminal collaboration offloading scheme based on FFA and GA (COS-FG). The process of COS-FG is Algorithm 2. If the system starts to run, it will first implement GA, and then update *Queue* and *TLIMIT*. As the computation task comes, count the penalty value of each task. Unless the cumulative penalty value or the number of loops reaches the upper limit, Algorithm 2 is always executed. After the cumulative penalty value or the number of loops reaches the upper limit, if the system continues to execute, the process of Algorithm 1 and Algorithm 2 will be repeated.

Algorithm 2 COS-FG

Input: The number of tasks D_x X, the upper limit of the task tolerance $PMAX$, the time limit of OOAFFA $TMAX$, the execution time for AOAGA T, the current time t, and the current cumulative penalty value y.

Output: Computation task offloading decisions and MD connection decisions

while [System continues to run] **do**
  Wait for one cycle to execute GA
  Update *Queue* with the result of FFA
  $TLIMIT = t + TMAX$
  $y = 0$
  while $t < TLIMIT$ **and** $y < PMAX$ **do**
    Execute FFA
    $y = y + penalty(c_x)$
  end while
end while

4 Simulation Results

Figure 2 shows the trade-offs between FFA and GA from a calculation time perspective. We will consider the limitation of the calculation time individually, so *PMAX* is set to infinity. In order to verify the relationship between *Q* and *TMAX*, we set the sum of *Q* and *TMAX* to 5 s. In the figure, as the *Q* continues to increase, the general trend of the overall consumption is continuously increasing, but it is slightly lower when the $Q = 0.4$. This shows that this time is when the sum of the advantages of FFA and GA is maximized. GA allocates computation tasks closer to the optimal scheme,

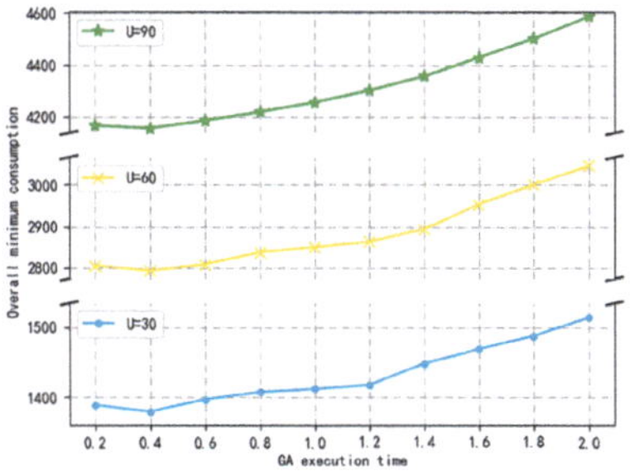

Fig. 2. Overall minimum consumption of GA execution time.

so its overall consumption will be lower. And FFA need not wait for allocation of computation tasks, so their respective advantages will be maximized during $Q = 0.4$. To prove that COS-FG's offloading decision is better, we compare it with FFA only and GA only. Figure 3 is comparison diagrams of COS-FG, FFA and GA consumption under the conditions of $\alpha_t = 0.5$. $\alpha_t = 0.5$ is a compromise between time and energy consumption. It can be seen from the figure that for any number of MDs, the time consumption and energy consumption of COS-FG are lower than FFA and GA, and it becomes more and more obvious as the number of MDs increases.

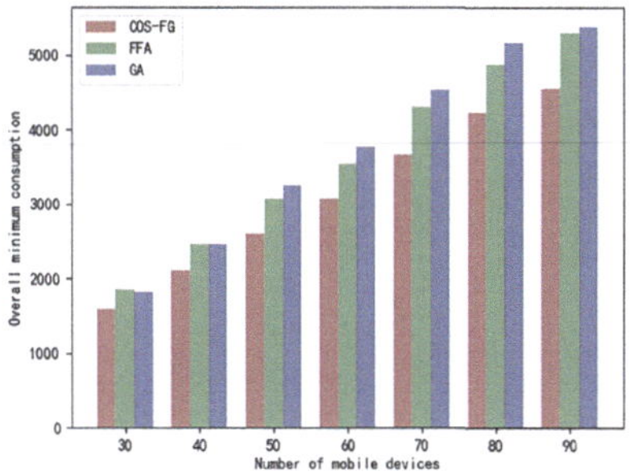

Fig. 3. Comparison of total consumption.

5 Conclusion

To better utilize the communication and computing resource in cloud-edge-terminal collaboration network with multiple IoT tasks, MDs, small cell base stations and micro base stations, a cloud-edge-terminal collaboration offloading scheme based on FFA and GA (COS-FG) is proposed. It consists of two algorithms, FFA and GA, which can offload task quickly and accurately respectively. FFA is designed to quickly offload tasks and GA is designed to accurately offload tasks. We combine these two algorithms and set optimal cycle mode according to the requirement of tasks. Simulation results clearly show the superiority of the proposed COS-FG in saving energy consumption and reducing processing time.

Acknowledgment. This work is supported by the project of Research and Application of Key Technologies of Trusted Cloud-Network Collaborative Resource Chain Management (09000-0KK52190088).

References

1. Ke, Z., Supeng, L., Yejun, H., et al.: Cooperative content caching in 5G networks with mobile edge computing. IEEE Wirel. Commun. **25**(3), 80–87 (2018)
2. Patel, M., Naughton, B., Chan, C., et al.: Mobile-edge Computing Introductory Technical White Paper. White Paper, Mobile-edge Computing (MEC) industry initiative (2014)
3. Li, H., Shou, G., Hu, Y., et al.: Mobile edge computing: progress and challenges. In: 2016 4th IEEE International Conference on Mobile Cloud Computing, Services, and Engineering (MobileCloud). IEEE (2016)
4. Mao, Y., You, C., Zhang, J., et al.: A survey on mobile edge computing: the communication perspective. IEEE Commun. Surveys Tuts. **19**(4), 2322–2358 (2017)
5. Yang, L., Cao, J., Yuan, Y., et al.: A framework for partitioning and execution of data stream applications in mobile cloud computing. ACM SIGMETRICS Perform. Eval. Rev. **40**(4), 23–32 (2013)
6. Zheng, J., Cai, Y., Wu, Y., et al.: Stochastic computation offloading game for mobile cloud computing. In: 2016 IEEE/CIC International Conference on Communications in China (ICCC), pp. 1–6. IEEE (2016)
7. Zhang, J., Hu, X., Ning, Z., et al.: Energy-latency tradeoff for energy-aware offloading in MEC networks. IEEE IoT J. **5**(4), 2633–2645 (2017)

IoT Terminal Security Monitoring and Assessment Model Relying on Grey Relational Cluster Analysis

Jiaxuan Fei(✉), Xiangqun Wang, Xiaojian Zhang, Qigui Yao, and Jie Fan

Global Energy Interconnection Research Institute Co. Ltd., State Grid Information and Network Security Key Laboratory, Nanjing 210003, China
444965979@qq.com

Abstract. IoT (Internet of Things) terminal security monitoring and evaluation encompasses technique and administration. There are several uncertainties in the monitoring and assessment process of various types of IoT terminals, which cannot be fully quantified. Therefore, it is difficult to achieve completely objective safety risk evaluation. For this to happen, the study proposes an IoT terminal security monitoring and assessment model based on grey relational cluster analysis. First, by integrating experts' knowledge, the conditional probability matrix (CPM) of cluster analysis are explained, which lays the foundation for establishing the security monitoring and assessment model. Then, through the grey relational cluster algorithm, the subjective judgment information of the experts on the threat degree of the target information system is synthesized as prior information. At the same time, through the observation node of objective evaluation information, the safety threat levels are synergized to realize the continuity and accumulation of safety evaluation. Ultimately, simulation examples verify the rationality and effectiveness of the pattern.

Keywords: IoT security · Security monitoring · Grey relational cluster analysis · Network operators · Quantitative evaluation

1 Introduction

As computer technology and the Internet develop [1], illegal intruders and hackers have widely used new attacks with system security flaws [2, 3]. Moreover, the security risks and threats faced by information system security are gradually eliminated. People have been concerned about the security of information systems. Network information security risk evaluation is an effective approach to solve information system security problems, covering FTA, AHP, and FCE. This approach was used by safety assessors [4, 5]. Nevertheless, so far, the effect of human considerations and managerial measures on information systems has not been fully considered [6, 7]. At the same time, network information security evaluation covers technology and administration. There are many uncertain considerations in the evaluation, which can't be fully quantified [8, 9]. Therefore, it's difficult to achieve a completely objective network information security risk evaluation [10].

Q. Liu et al. (Eds.): CENet 2020, AISC 1274, pp. 1507–1513, 2021.
https://doi.org/10.1007/978-981-15-8462-6_172

In the study, the subjective and objective security evaluation information is synthesized, and a model of IoT terminal security monitoring and assessment on basis of grey relational clustering analysis is developed. For starters, the group assessment approach on basis of grey relational cluster analysis makes full use of each assessor's experience and knowledge to assess the target information system. It compensates for the singleness of the assessor's personal judgment to a large extent. Second, similar to the neural networks, cluster analysis can completely exhibit the human reasoning processes. The security assessment based on circular arrays can explain the security evaluation course quantitatively and exhibit the continuity and accumulation of the safety evaluation.

2 Grey Relational Cluster Analysis Algorithm

Definition 1. Preset $F : R_n \to R$, and there exists an n-dimensional weight vector related to F, $w_i \in [0,1], 1 \leq i \leq n$ and $\sum_{i=1}^{n} w_i = 1$ for making

$$F(a_1, a_2, \cdots, a_n) = \sum_{i=1}^{n} w_i b_i \tag{1}$$

Among them, b represents the ith largest factor of the array $(a_1, a_2, \cdots, a_n)$. F is dubbed n-dimensional grey relational cluster analysis. Grey relational cluster analysis is an analysis between maximal analysis and minimal analysis.
If $w = (1, 0, 0, \cdots, 0)$

$$F(a_1, a_2, \cdots, a_n) = \max(a_1, a_2, \cdots, a_n) = b_1 \tag{2}$$

In fuzzy calculations, grey relational cluster analysis is equivalent to "or" analysis:
If $w = (0, 0, 0, \cdots, 1)$

$$F(a_1, a_2, \cdots, a_n) = \max(a_1, a_2, \cdots, a_n) = b_n \tag{3}$$

Grey relational cluster analysis is equivalent to "sum" analysis in fuzzy analysis: If $w = (1/n, 1/n, 1/n, \cdots, 1/n)$

$$F(a_1, a_2, \cdots, a_n) = \frac{1}{n} \sum_{i=1}^{n} a_i \tag{4}$$

Grey relational cluster analysis is equivalent to the arithmetic mean analysis. In grey relational cluster analysis, the determination of the weight vector is directly associated with the dataset size. For ensuring the fairness and rationality of the assessment outcomes, the Gaussian distribution was discretized for defining the weight vector of positions. In this approach, the degree of freedom is placed in the position where the weighted value is relatively small, thereby effectively eliminating the negative influence of the emotional factor on the assessment course.

Set μ represents the expectation given to $(1, 2, \cdots, n)$ in the $w = (1/n, 1/n, \cdots, 1/n)$; σ represents the SD of μ and $(1, 2, \ldots, \mathrm{n})$ in w. Therefore, there are

$$\mu_n = \frac{1}{n}\frac{n(n+1)}{2} = \frac{n+1}{2}\ ;\ \sigma_n = \sqrt{\frac{1}{n}\sum_{i=1}^{n}(i-\mu_n)^2}$$
$$\omega' = \frac{1}{\sqrt{2\pi}\sigma_n}e^{-\frac{(i-\mu_n)^2}{2\sigma_n^2}}\ ;\ \omega = \frac{\omega'}{\sum_{i=1}^{n}\omega'} \tag{5}$$

In order to assess n information systems, the assessment group set is $D = (d_1, d_2, \cdots, d_n)$, in which $d_k (k = 1, 2, \cdots, m)$ indicates the kth assessor. The subjective judgment (SJ) in the form of assessment: the utility value is $u^{(k)} = \left(u_1^k, u_2^k, \cdots, u_n^k\right)^T$, the fuzzy language assessment value is $S = (s_0, s_1, \cdots, s_T)$, and the fuzzy complementary judgment matrix is $p_{(k)} = \left(p_{ij}^{(k)}\right)_{n\times n}$. Thus, $u^{(k)}$ comes from the information corresponding to these judgments.

1) Convert $P_{(k)}$ into $u^{(k)}$.

$$u_i^{(k)} = \left(\sum\nolimits_{j=1}^{n} p_{ij}^{(k)} + \frac{n}{2} - 1\right)\Big/ n(n-1), i = 1, 2, \cdots .n \tag{6}$$

2) Convert the S into the $u^{(k)}$. The fuzzy language assessment value c can be made to correspond to the utility value and be described in natural language. For instance,

$$\begin{aligned} S = \{s_0 = level_1 = 0, s_1 = level_2 = 0.1, s_2 = level_3 = 0.3, s_3 = level_4 = 0.5, \\ s_4 = level_5 = 0.7, s_5 = level_6 = 0.9, s_6 = level_7 = 1\} \end{aligned} \tag{7}$$

Therefore, $u^{(k)}$ has a formula below:

$$u_i^{(k)} = S_i^{(k)} \Big/ \sum\nolimits_{i=1}^{n} S_i^{(k)}, i = 1, 2, \cdots .n \tag{8}$$

The target population level assembled by grey relational cluster analysis and threatening the SJ of the i-th piece of information is:

$$u_i = OWA_w\left(u_i(1), u_i(2), \cdots, u_i^{(n)}\right), i = 1, 2, \cdots, n \tag{9}$$

The assessment group $u = (u_1, u_2, \cdots, u_n)^T$ as the information SJ is:

$$u_i = u_i \Big/ \sum\nolimits_{i=1}^{n}, i = 1, 2, \cdots, n \tag{10}$$

Cluster analysis is a reasoning algorithm suitable for simple connection spatial cluster analysis. In the multi-tree communication algorithm, it is assumed that at node X, there exist m child nodes $(Y_1, Y_2, \cdots, Y_n)$ and n parent nodes $(Z_1, Z_2, \cdots, X_n)$. Assuming that Bel is a posterior probability distribution, the evidence information obtained from the Y and Z. $M_{X|Z} = P(X = x|Z = z)$ represents the probability of occurrence of event x in parent node Z in child node X in case z. Since X is discrete, $\lambda(x)$ and $\pi(x)$ are actual vectors. Their elements are associated with each discrete value:

$$\begin{aligned} \lambda(x) &= [\lambda(X = x), \lambda(X = x_2), \cdots, \lambda(X = x_l)] \\ \pi(x) &= [\pi(X = x), \pi(X = x_2), \cdots, x(X = x_l)] \end{aligned} \quad (11)$$

The inference of cluster analysis is centered on a single node. λ can be obtained from the Y and π can be obtained from the Z. Bel, λ and π on the node are computed to trigger the update of the neighboring node. The update process is below.

Step 1. Update $Bel(x) = \alpha\lambda(x)\pi(x)$, in which α is the normalization factor, so there are

$$\sum Bel(x) = 1, \lambda(x) = \prod \lambda_{y_j}(x), \pi(x) = \prod \pi_{z_i} M_{X|Z} \quad (12)$$

Step 2. Renew the contract from bottom to top:

$$\lambda_x(z) = \lambda(x) M_{X|Z} \quad (13)$$

Step 3. Renew the contract from top to bottom:

$$\pi_y(x) = \alpha\pi(x) \prod_{k \neq j} \lambda_{y_j}(x) \quad (14)$$

3 IoT Terminal Security Monitoring and Assessment Model

3.1 Analysis of Assessment Factors Affecting the Threat Level of Security

Network information safety risk can be regarded as an impact on capital. For simplifying the pattern, only considerations below are considered: the effect on capital, the frequency of capital threats, and the vulnerability of capital f. The threat level is TL. Thus, a network information safety evaluation pattern based on circular arrays is established.

The state of the variables in the model is collected below.

$$TL = \{high, medium, low\} C = \{big, middle, small\} T = \{high, medium, low\} \quad (15)$$

The CPM exhibits the expert view of the relationship between related nodes, becoming expert knowledge. For instance, (◇ represents the probability of small, medium, and large capital losses) if TL is high, ◇will be 10%, 30%, and 60%; if TL is medium, ◇ will be 40%, 40%, and 20%; if TL is low, ◇ will be 60%, 30%, and 10%. t and f can be interpreted similarly to the above description, as exhibited in Table 1.

Table 1. Conditional probability matrix of inference rules.

TL	$P(C\|TL)$	$P(t\|TL)$	$P(f\|TL)$
High	Small, medium and large	Low, medium and high	Not serious, ordinary and serious
Medium Low	$\begin{pmatrix} 0.1 & 0.3 & 0.6 \\ 0.4 & 0.4 & 0.2 \\ 0.6 & 0.3 & 0.1 \end{pmatrix}$	$\begin{pmatrix} 0.1 & 0.5 & 0.4 \\ 0.4 & 0.3 & 0.3 \\ 0.6 & 0.2 & 0.2 \end{pmatrix}$	$\begin{pmatrix} 0.8 & 0.1 & 0.1 \\ 0.6 & 0.3 & 0.1 \\ 0.1 & 0.45 & 0.45 \end{pmatrix}$

3.2 Analysis of Abnormal Network Access Behavior of IoT Terminal

This article takes all the traffic generated by the terminal equipment as the object of investigation, and comprehensively considers the physical layer of the terminal equipment, network traffic and protocol behavior characteristics to establish network information security. The specific implementation route is shown in Fig. 1.

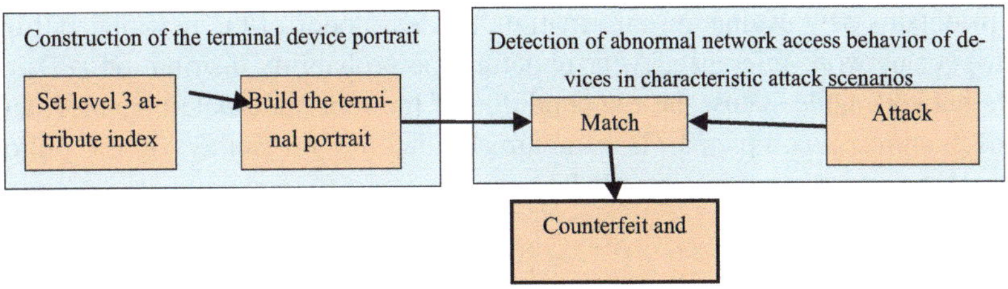

Fig. 1. Implementation roadmap of abnormal network access behavior detection.

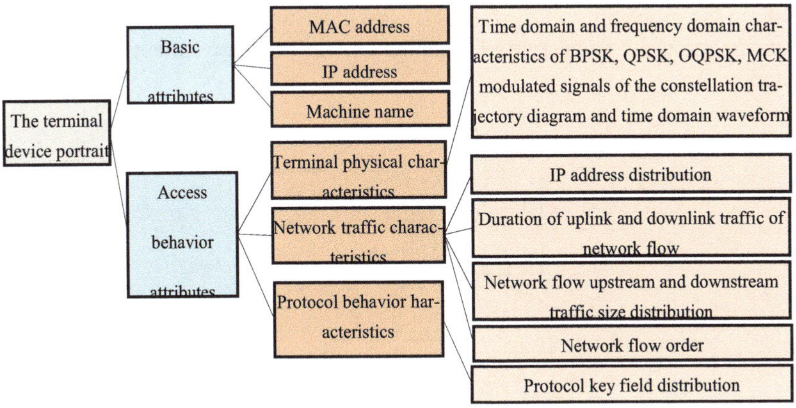

Fig. 2. Network information security content.

The core content of network information security includes basic attributes and access behavior attributes, as shown in Fig. 2.

4 Analysis of Examples and Results

Suppose the target TL is high, medium, and low respectively. The threat assessment information provided by the four assessors is

$$\begin{array}{c} U_1 = (0.2, 0.6, 0.2)\ ;\ U_2 = (0.3, 0.3, 0.4) \\ U_3 = (0.7, 0.33, 0.6)\ ;\ U_4 = (0.37, 0.3, 0.33) \end{array} \tag{16}$$

The weight vector of the grey relational cluster algorithm can be obtained:

$$w = (0.155, 0.345, 0.155, 0.345) \tag{17}$$

And the TL evaluation of the assessment team is

$$U{=}(0.247, 0.367, 0.387) \tag{18}$$

After initializing cluster analysis, the evaluation system is waiting. When the system obtains new evaluation information, the leaf nodes of the network will update and trigger network inference. After updating the probability distribution (PD) of the whole network node status, the PD condition of the root node state is gained, and the TL evaluation is completed. It is assumed that the possibility of the following influencing factors has been recorded:

$$\lambda_c = [0 \quad 0 \quad 1]\,\lambda_t = [0 \quad 1 \quad 0]\lambda_f = [0 \quad 1 \quad 0] \tag{19}$$

Instance: Assuming that TL is generated according to the group assessment approach on basis of grey relational cluster analysis, the prior information of cluster analysis is $\Pi(TL)$. The evaluation outcomes are exhibited in Fig. 3. That indicates an increase in the probability of medium TL. The other two levels are reduced likely. The assessment group's subjective TL judgment information affects the evaluation outcomes.

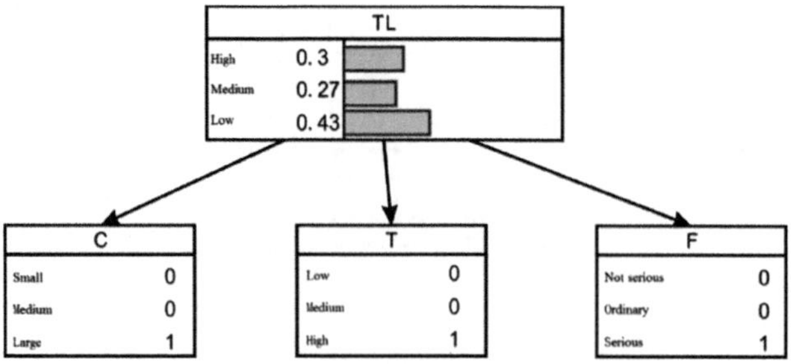

Fig. 3. Assessment result of example.

The common prior information encompasses the prior information needing to be set when the algorithm is started, and the result of the previous cycle in the calculation cycle of the algorithm.

5 Conclusion

In this study, based on the systematic analysis of IoT terminal security threat considerations, according to subjective *TL* judgment information and objective situation information, an IoT terminal security monitoring and assessment model based on cluster analysis and grey relational cluster algorithm is established. It is verified that the model more aligns with the actual course of network information safety evaluation and can more accurately exhibit the true *TL*. The algorithm instance proves the effectiveness of the approach and provides a novel idea for network information security monitoring and assessment.

Acknowledgment. The work is supported by the science and technology project of State Grid Corporation of China: "Research on SG-eIoT Basic Protection Architecture and Terminal Layer Security Monitoring and Protection Technology" (SGGR0000XTJS1900274).

References

1. Pan, L., Li, T.Y.: Dynamic information security evaluation model in mobile Ad Hoc network. J. Comput. Appl. **35**(12), 3419–3423 (2015)
2. Chen, Y.Q., Wu, X.P., Fu, Y., et al.: Network security evaluation based on stochastic game and network entropy. J. Beijing Univ. Posts Telecommun. **37**(s1), 92–96 (2014)
3. Dong, J.Q.: The power struggles of cyber space and information security management in South Korea and its inspirations to China's information governance. Inf. Sci. **4**, 153–157 (2016)
4. Navare, J., Gemikonakli, O.: Governance and risk management of network and information security: the role of public private partnerships in managing the existing and emerging risks. Commun. Comput. Inf. Sci. **92**, 170–177 (2010)
5. Cheng, Y.X., Jiang, W., Xue, Z. et al.: Multi-objective network security evaluation based on attack graph model. Journal of Computer Research and Development (s2), 23–31 (2012)
6. Xu, C., Liu, X., Wu, J., et al.: Software behavior evaluation system based on BP neural network. Comput. Eng. **9**, 149–154 (2014)
7. Samet, H., Rashidi, M.: Early warning system based on power system steady state and dynamic security assessment. J. Electrical Syst. **11**(3), 249–257 (2015)
8. Hong, Y., Li, P.: Information security defense mechanism based on wireless sensor network correlation. Journal of Computer Applications 33(2), 423–425,467 (2013)
9. Zhao, L., Xue, Z.: Synthetic security assessment for incomplete interval-valued information system. High Technol. Lett. **18**(2), 160–166 (2012)
10. Zou, K., Xiang, S., Zhangzhong, Q.Y., et al.: Model construction and empirical study on smart city information security risk assessment. Library Inf. Serv. **7**, 19–24 (2016)

Research on Congestion Control Over Wireless Network with Delay Jitter and High Ber

Zulong Liu[1], Yang Yang[1(✉)], Weicheng Zhao[2], Meng Zhang[3], and Ying Wang[1]

[1] State Key Laboratory of Networking and Switching Technology, Beijing University of Posts and Telecommunications, Beijing 100876, China
yyang@bupt.edu.cn

[2] China Academic of Electronics and Information Tech, Beijing 100000, China

[3] Xinhua Power Investment Co., Ltd., Beijing 100000, China

Abstract. Most current researches use packet loss as a prerequisite for congestion to design congestion control algorithms. Wi-Fi, 4G, satellite networks and other wireless networks have been widely used. Given that it is easily influenced by the natural environment, the transmission of wireless network features in large delay jitter and high random bit error rate (BER). In this network environment, the performance of traditional congestion control algorithm is poor. This paper proposes an improved congestion control algorithm NBBR. The algorithm is based on a feedback idea to adjust the sending rate of the sending end. Link capacity is used to adjust the transmission rate and the size of the congestion window to shield the transmission rate reduction caused by high bit error rate in wireless communication environment. Different delay measurement strategies are determined by the degree of delay jitter, which can reflect the network condition more accurately and improve the utilization rate of the network. Finally, by comparing the simulation experiment results, it is proved that the proposed algorithm can maintain higher throughput in the wireless network with higher delay jitter and higher BER.

Keywords: Congestion control · Wireless network · Jitter · Throughput

1 Introduction

With the application of wireless network more and more widely, congestion control has become the focus of research. Congestion occurs when the user's demand for network resources exceeds the capacity provided by the network itself. Without proper congestion control algorithms to control the amount of data sent, the network may crash directly.

The congestion control algorithm adjusts the amount of data sent to the network mainly through the congestion window. Here we adjust the transmission rate by determining the link capacity and the data transferred in the link. This method can solve the problem that traditional congestion control algorithm based on packet loss is sensitive to packet loss rate.

Q. Liu et al. (Eds.): CENet 2020, AISC 1274, pp. 1514–1521, 2021.
https://doi.org/10.1007/978-981-15-8462-6_173

Firstly, we calculate the propagation time in the link by calculating the jitter of the round trip time. The bottleneck bandwidth is obtained by continuously probing the maximum throughput during transmission. Then the link capacity is determined by the propagation time and bottleneck bandwidth. Finally, the transmission rate is determined by the relationship between link capacity and transmission data.

2 Related Work

In the development of modern networks, as the network situation becomes more and more diverse, wireless networks are also more and more widely used. Congestion control algorithms have also become the focus of continuous research. The earliest congestion control protocol is TCP Tahoe [2]. This method reduces the window rate by a larger margin and lower bandwidth utilization. TCP Reno algorithm improves TCP Tahoe and proposes a fast recovery mechanism [3]. Scholz, D., Jaeger B. and others proposed an improved TCP Veno [5] algorithm, which could increase throughput up to 80% in a typical wireless access network with 1% random packet loss rate.

With the continuous improvement of network transmission capacity, Sangtae Ha, Injong Rhee et al. proposed CUBIC algorithm [6]. The linear window growth function is modified into a cubic function to improve the scalability of TCP over fast and long distance networks. In 2016, Google proposed BBR [1] algorithm based on bottleneck bandwidth idea. BBR has a smaller delay than CUBIC [4, 7] with similar throughput. However, BBR has poor performance in wireless links with serious delay jitter [10]. Brakmo, L. S., Peterson, L. L. et al. proposed TCP Vegas [8] algorithm, which judged the congestion state of the network and controlled the sending rate of the sender according to the change of RTT. Wei, D. X., Jin, C. et al. proposed FASTTCP [9] that combines multiplication growth and linear growth to improve network performance. The delay based congestion control protocol is less fair and less applied in practice.

3 Congestion Control Model Based on Link Capacity

The N-BBR algorithm uses the bottleneck bandwidth and the minimum round-trip delay to adjust the sending rate. Update the value of BltBw (bottleneck bandwidth) and RT_{prop} (Round-trip propagation time) through continuous feedback

The calculation formulas of PacingRate and Cwnd in N-BBR are respectively:

$$\text{PacingRate} = \text{PacingGain} * \text{BltBw} * RT_{prop} \tag{1}$$

$$\text{cwnd} = \text{cwnd}_{\text{gain}} * \text{BltBw} * RT_{prop} \tag{2}$$

Among them, the Pacing said send rate gain factor, used for sending rate adjustment algorithm under different running condition; CwndGain represents the congestion window gain factor, which is used to adjust the sending window under the running state of the algorithm.

3.1 Bottleneck Bandwidth

Set the time of sending packet X to be t1 and the time of receiving confirmation to be t2. And in this period of time to confirm the number of packages received for n_{ack}. Then the real-time rate collected when packet X is answered is:

$$\text{BtlBw} = \frac{n_{ack}}{t2 - t1} \quad (3)$$

3.2 Network Delay Jitter

When the detected RTT is not considered queuing, the delay is satisfied as follows:

$$\text{RTT} = RT_{prop} + \mu \quad (4)$$

RT_{prop} is the inherent delay propagation time. is the communication link jitter, affected by the network environment in the process of data transmission.

Calculating RTT is the standard deviation of $\sqrt{RTT_{dev}}$, use it to judge the time delay. The following formula is the weighted average delay:

$$RTT_{smo} = \alpha * RTT_{smo} + (1 - \alpha) * \text{RTT} \quad (5)$$

The following is the calculation method of delay variance:

$$RTT_{dev} = 1 - \beta * RTT_{dev} + \beta * abs(RTT_{smo} - RTT) \quad (6)$$

- RT_{prop}: When $RTT_{dev} > \gamma * RT_{prop}$ think jitter is larger, make $RT_{prop} \approx RT_{prop}$. When $RTT_{dev} < \gamma * RT_{prop}$, make $RT_{prop} \approx T_{min}$, ($\gamma$ = 0.3).
- Solution for slow convergence after the wireless link changes: Since the distribution of round-trip time can be adjusted to approximate normal distribution [11, 12].

$$\text{P(RTT)} = \frac{1}{\sqrt{2\pi RTT_{dev}}} e^{-\frac{(RTT - RTT_{smo})^2}{2RTT_{dev}}} \quad (7)$$

When the round-trip time $|\text{RTT} - RTT_{smo}| > 3\sqrt{T_{dev}}$, it immediately enters the stage of detecting RT_{prop} and BtlBw. This method can also effectively solve the problem of slow convergence of BBR when the link changes.

3.3 Congestion Control Strategy

The N-BBR congestion control strategy designed in this paper is divided into four states: startup state, Drain state, ProbeBW state ProbeRTT state.

Startup State. BltBW, RTT_{min}, RT_{prop}, RTT_{smo}, RTT_{dev} are global variables and the life cycle of a transmission connection is the same. These variables are updated every time a sample measurement is made. The update methods in Fig. 1, 2 and 3 are the same. Formulas (3), (5) and (6) propose calculation methods for BltBW, RTT_{smo}, and RTT_{dev}, respectively. RTT_{min} simply makes a comparison and then updates, which is relatively simple. RT_{prop} determines the measurement method based on the degree of delay jitter.

The details of the operation at this stage are shown in Fig. 1. This stage is similar to slow start, but not as blind as slow start. If BltBW increases by less than 25% over 3 rounds, it enters the Drain state, otherwise it continues to the Start state. The purpose of this state is to quickly increase the transmission rate to an appropriate level. Although it is not as aggressive as a slow start during detection, it also outputs too much data in the link in order to measure the maximum bottleneck bandwidth.

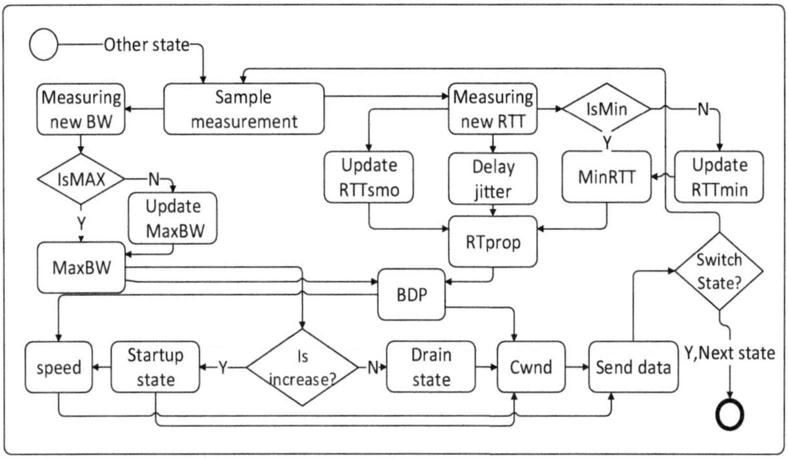

Fig. 1. Startup state running details.

Drain State. In the start state, we mentioned that extra data will be generated in the link, which will cause queuing. The main purpose of this stage is to reduce the transmission rate and consume excess data in the link. The gain coefficient at this stage is $ln2/2$, and the transmission rate and the size of the congestion window are calculated according to formulas (1) and (2).

The operation details are shown in Fig. 2. First, the global variables are updated. The key to Drain state is to judge the relationship between inflight and BDP. Inflight is a data packet that has been sent but has not yet been received. If inflight > BDP indicates that there is excess queuing data in the link, continue to implement the strategy at this stage. Otherwise, it means that there is no queuing situation in the link. The task at this stage has been completed and the ProbeBW state is entered.

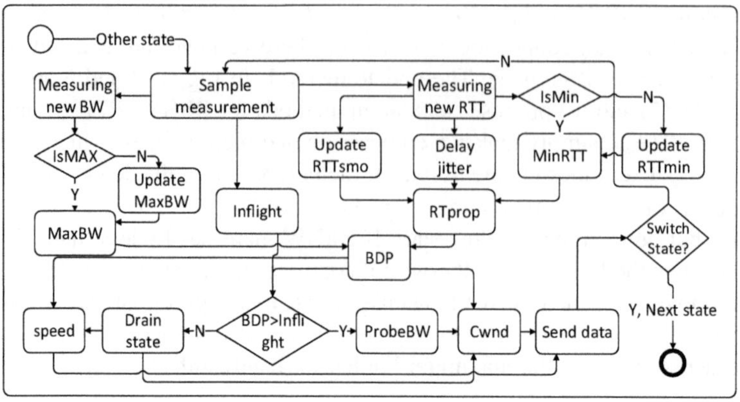

Fig. 2. Drain state running details.

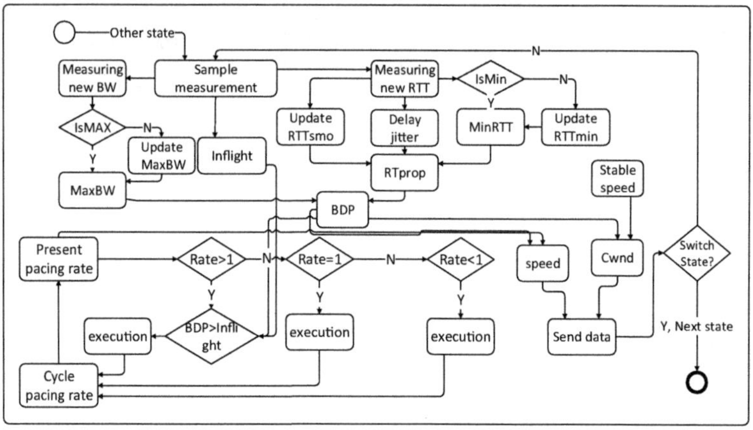

Fig. 3. ProbeBW state running details.

ProbeBW State. When entering this stage, a maximum bottleneck bandwidth has been measured, and there is no queuing phenomenon in the Drain state link. In the entire data transmission process, most of the time should be in ProbeBW state. As shown in Fig. 3, in a ProbeBW state loop, poll the following gain coefficients, [1.25, 0.75, 1, 1, 1, 1, 1, 1]. During the first RTT time, the gain coefficient gain is increased to 1.25 to detect whether there is excess bandwidth that can be used. In the second RTT, reducing the gain to 0.75 consumes the extra data packets sent in the previous stage, so that there is no queued data in the link. In the next 6 RTT time, set gain to 1 and continue to send data at this stage. In this way, every 8 RTTs are used as a cycle to detect the bandwidth. This stage maintains a relatively uniform transmission rate and can be fine-tuned by polling the gain factor.

ProbeRTT State. This state will be entered only when the delay jitter is satisfied: $RTT_{dev} < \gamma * RT_{prop}$. No smaller RTT value is estimated for more than 10 s, then enter the RTT detection state. Reduce the amount of packets sent, and measure a more accurate RTT value, at least 200 ms or a packet back and forth to exit this state. Different states are entered depending on whether the bandwidth is full. If it is not full, enter the Startup state, otherwise enter the ProbeBW state.

4 Simulation Analysis

The proposed congestion control algorithm is designed and implemented on NS3 network simulation platform. In order to verify the performance of the algorithm proposed in the article, it is compared with the two widely used algorithms of CUBICB6 and BBRB1. The specific simulation process is:

(1) The transmission performance is compared under different delay jitter;

(2) Comparison of transmission performance under different packet loss rates;

(3) Comparison of convergence between BBR and NBBR;

(4) When both delay jitter and packet loss rate are high, the transmission performance is compared;

The bandwidth is set to 10Mbps and the bit error rate is set to 0.001%. In addition, the round trip time of links is normally distributed with a mean value of 100 ms and a standard deviation of 40 ms. The performance of the congestion control algorithm was tested when the delay jitter was serious, and the experimental results were shown in Fig. 4.

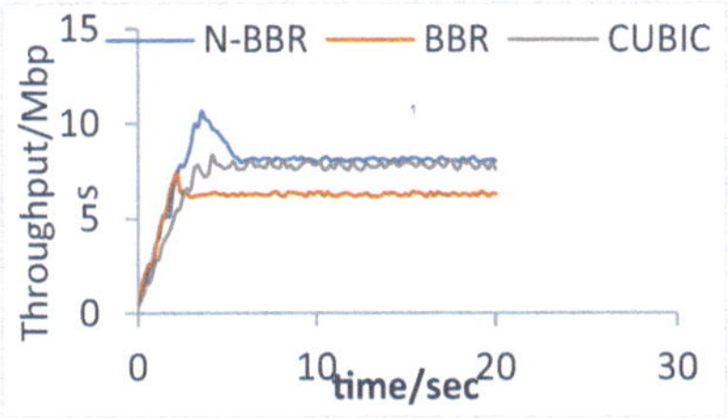

Fig. 4. Throughput of different congestion control algorithms when round-trip delay jitter is serious.

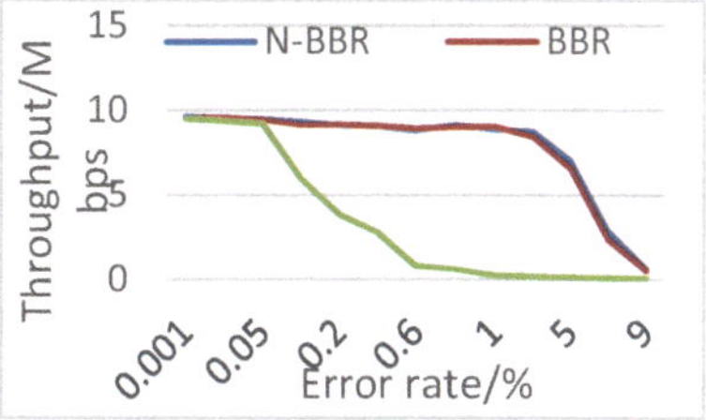

Fig. 5. Experimental data results of different congestion control algorithms transmitted under different bit error rates.

The communication bandwidth in the topological link is set to 10Mbps, round-trip delay is set to 100 ms, and standard deviation is set to 2 ms. Ber value respectively 0.001%, 0.01%, 0.05%, 0.1%, 0.2%, 0.4%, 0.6%, 0.8%, 1%, 3%, 5%, 7%, 9%. Throughput of different congestion control algorithms was measured under different bit error rates, and the experimental results were shown in Fig. 5.

The bandwidth in the topology link is set to 10Mbps, the bit error rate is set to 0.001%, and the round-trip delay value is 100 ms. At 10 s, the round trip time of the communication link was changed to 150 ms to observe the convergence speed of N-BBR and BBR. The experimental results are shown in Fig. 6.

The bandwidth in the topological link was set to 10Mbps, the bit error rate was set to 0.1%, the round-trip delay normal distribution was changed, the mean value was 100 ms, and the standard deviation was 40 ms. When bit error rate and delay jitter are both high, the performance of the algorithm is tested. The experimental results are shown in Fig. 7.

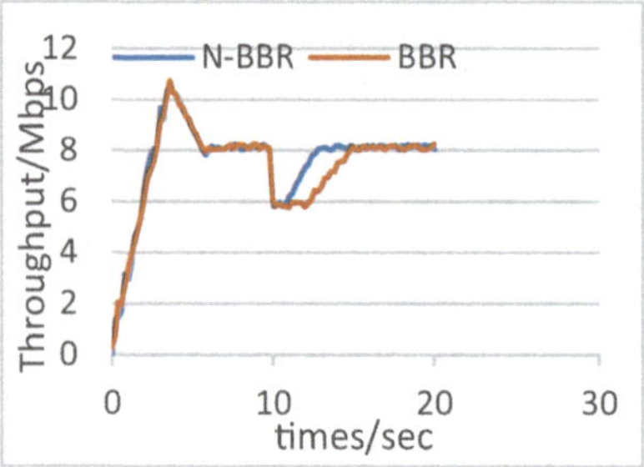

Fig. 6. Comparison of convergence between BBR and N-BBR.

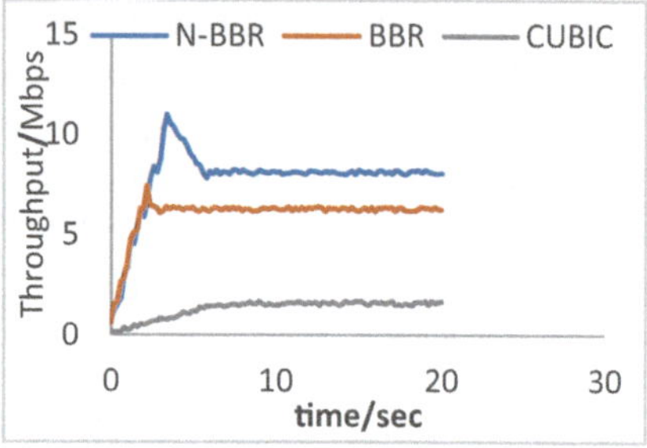

Fig. 7. Comparison of transmission efficiency of different congestion control algorithms with high bit error rate and severe delay jitter.

According to the experimental data, when the delay jitter reaches 40%, the throughput of N-BBR improves by nearly 15.5% compared with BBR. Because the delay has little impact on CUBIC, the throughput is basically the same as N-BBR efficiency. But with a packet loss rate of 1%, CUBIC's transmission performance dropped by more than 80%, and the higher the packet loss rate, the worse it performed. However, N-BBR began to decrease significantly when the packet loss rate was 5%. When the propagation delay changes, the convergence speed time of N-BBR is 32.1% higher than that of BBR.

5 Conclusion

The new congestion control algorithm NBBR proposed in this paper uses bottleneck bandwidth and round-trip delay to adjust the data transmission rate and congestion window size. Different measurement methods are used to estimate the propagation delay of link by evaluating the degree of delay jitter. In the network environment with high bit error rate and serious delay jitter, compared with the traditional algorithm, the new congestion control algorithm proposed in this paper has a greater improvement in transmission performance.

Acknowledgment. This work is supported by National Key Research and Development Program of China (2019YFB2103200), the Fundamental Research Funds for the Central Universities (500419319 2019PTB-019), Open Subject Funds of Science and Technology on Information Transmission and Dissemination in Communication Networks Laboratory (SKX182010049), the Industrial Internet Innovation and Development Project 2018& 2019 of China.

References

1. Cardwell, N., et al.: BBR: congestion-based congestion control. ACM Queue **60**(2), 58–66 (2016)
2. Jacobson, V.: Congestion avoidance and control. ACM Special Interest Group on Data Commun. **18**(4), 314–329 (1988)
3. Padhye, J., et al.: Modeling TCP Reno performance: a simple model and its empirical validation. IEEE ACM Trans. Netw. **8**(2), 133–145 (2000)
4. Scholz, D., et al.: Towards a deeper understanding of TCP BBR congestion control. In: 2018 IFIP Networking Conference (IFIP Networking) and Workshops, pp. 1–9 (2018)
5. Fu, C.P., Soung, C.L.: TCP veno: TCP enhancement for transmission over wireless access networks. IEEE J. Sel. Areas Commun. **21**(2), 216–228 (2003)
6. Sangtae, H., Rhee, I., Xu, L.S.: CUBIC: a new TCP-friendly high-speed TCP variant. Oper. Syst. Rev. **42**(5), 64–74 (2008)
7. Li, F., et al.: TCP CUBIC versus BBR on the highway. In: Passive and Active Network Measurement, pp. 269–280 (2018)
8. Brakmo, L.S., Larry, L.P.: TCP Vegas: end to end congestion avoidance on a global Internet. IEEE J. Sel. Areas Commun. **13**(8), 1465–1480 (1995)
9. Wei, D.X., et al.: FAST TCP: motivation, architecture, algorithms, performance. IEEE ACM Trans. Netw. **14**(6), 1246–1259 (2006)
10. Atxutegi, E., et al.: On the use of TCP BBR in cellular networks. IEEE Commun. Mag. **56** (3), 172–179 (2018)
11. Zhao, W.F., et al.: Transmission control protocol based on statistic process control. Int. J. Adv. Comput. Technol. **5**(5), 1206–1214 (2013)
12. Fontugne, R., Johan, M., Kensuke, F.: An empirical mixture model for large-scale RTT measurements. In: International Conference on Computer Communications, pp. 2470–2478 (2015)

Reinforcement Learning Based QoS Guarantee Traffic Scheduling Algorithm for Wireless Networks

Qingchuan Liu[1(✉)], Ao Xiong[1], Yimin Li[1], Siya Xu[1], Zhiyuan An[2], Xinjian Shu[2], Yan Liu[2], and Wencui Li[2]

[1] Key Laboratory of Networking and Switching Technology, Beijing University of Posts and Telecommunications, Beijing 100876, China
lqc_201996@bupt.edu.cn

[2] Communication & Dispatching Center State Grid Henan Information & Telecommunication Company, Zhengzhou 450000, China

Abstract. With the rapid development of wireless network, the amount of heterogeneous services increases, which have significant differences in QoS requirements. However, the traditional service management methods realized by artificial distinction are difficult to satisfy the QoS requirements of services with high bandwidth and burstiness. Therefore, it has great significance to better use the limited network resources and design suitable QoS guarantee mechanisms for wireless network. In this paper, a reinforcement learning based QoS guarantee traffic scheduling algorithm RQTS is proposed to optimize the utilization rate of wireless network. First, the traffic scheduling optimization model is constructed based on Lyapunov theory, which sets the optimal system utilization as the objective and all the restrictions are transformed into queue stability problems. Then, the QoS routing method RQTS is adopted to solve the optimization problem through traffic access control and transmission path control, the effects of different transmission channels, transmission rates and transmission path delays of the system are examined. Simulation results show that, the proposed mechanism can satisfy the QoS requirements and optimize the weighted system utilization.

Keywords: Traffic scheduling · Resource allocation · Reinforcement learning · Lyapunov

1 First Introduction

In the harsh smart grid environment, electromagnetic interference, equipment noise and multi-path transmission make QoS guarantee a great challenge in the wireless sensor network of intelligent power communication network [1]. There are many achievements in the research of smart grid QoS routing protocol, but it is only applicable to specific applications, such as electricity price and network operation quality monitoring [2–4]. The distributed optimization algorithm proposed in [5] increases the rate of each flow iteratively until all flows converge to the optimal rate to maximize the throughput. In addition, in reference [6], an opportunity scheduling algorithm with packet delay

Q. Liu et al. (Eds.): CENet 2020, AISC 1274, pp. 1522–1532, 2021.
https://doi.org/10.1007/978-981-15-8462-6_174

guarantee is proposed, which can effectively reduce the end-to-end delay. But these algorithms considered little about reliability and bandwidth [7, 8]. Most of the existing network traffic scheduling algorithms only consider the absolute priority of the primary user (PU), not the relative priority of the secondary user (SU), which cannot provide differentiated QoS services for heterogeneous requirements [9]. To solve the above problems, a reinforcement learning based QoS guarantee traffic scheduling algorithm for wireless network (RQTS) is proposed. First of all, aiming at the optimization of system utilization with the QoS requirements of priority services, this paper makes the operation optimization of wireless network as a Lyapunov drift optimization problem. Then, through traffic access control and transmission path control to interpret the influence of different transmission channels, transmission rate and transmission path on dependability, data rate and system delay. RQTS achieves the premise of satisfying the requirements of QoS and comprehensive optimization of network operation.

2 Problem Formulation

The wireless network system model involved in this paper is shown in Fig. 1. We use $\mathbf{N}$ to represent the nodes set of wireless network, then the number of nodes in the network is $|\mathbf{N}| = N$; $\mathbf{F}$ is the set of all traffic flows in the network, then the number of traffic flows is $|\mathbf{F}|$. $\mathbf{J}$ represents a channel set. The time-varying capacity of these channels is $\mathbf{C}_n(t) = [C_n^1(t), C_n^2(t), \cdots C_n^j(t)]$. $\mathbf{C} = \{a, b, c, \cdots k\}$ represents all service classes in intelligent power communication network. In wireless networks, the goal is to optimize the weighted system utilization of each traffic flow between nodes under limit of the data rate, delay and reliability requirements of the traffic. Then the optimization objective function can be abstracted as:

$$\begin{aligned} \max U(\mathbf{F}_n) = \sum_{c\in\mathbf{C}} \sum_{i\in\mathbf{F}_c} & \omega_c * U_n^c(\beta_i, \tau_i, \alpha_i) \\ s.t.\ \mathrm{C1:}\ & \beta_i(t) \le C_n^j(t)^{Res} \\ \mathrm{C2:}\ & \sum_{\overrightarrow{mn}\in\mathbf{R_i}} \overline{\tau}_{\overrightarrow{mn}} \le \tau_i \\ \mathrm{C3:}\ & \alpha_i \ge \alpha_c^{\min}, c \in \mathbf{C} \end{aligned} \tag{1}$$

Where, $\mathbf{F}_n$ represents the collection of all service flows in all service classes on the node. C_1 is the transmission rate limit. Where $C_n^j(t)^{Res}$ is the residual capacity of the selected channel j for the node n. C_2 is the transmission delay limit. And $\mathbf{R}_i$ is the entire path of the traffic from the source node to the destination node, and $\overline{\tau}_{\overrightarrow{mn}}$ is the average single-hop delay from the node n to the node n. τ_i indicates end-to-end latency limit for service flows. C_3 is the reliability requirement, where α_i is the reliability of the service flow i in the service class c and $\alpha_c^{\min}$ is the minimum reliability requirement for the priority class c.

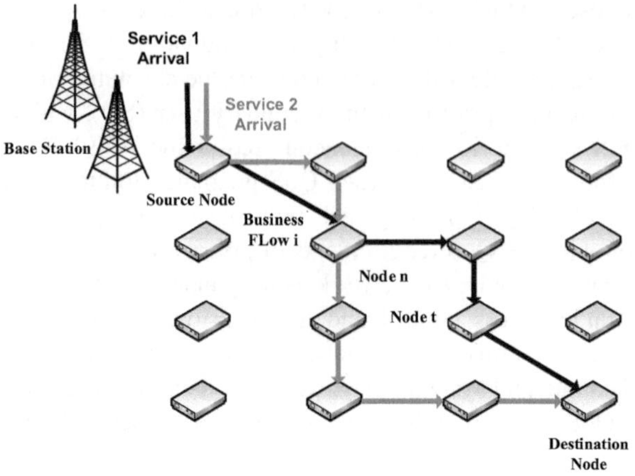

Fig. 1. The wireless network system model.

3 RQTS Algorithm

3.1 Lyapunov Drift Penalty Reduction Function

In this paper, the Lyapunov drift optimization algorithm is used to optimize the utilization rate of the system. First, all the restrictions are transformed into queue stability problems, and then a distributed control algorithm is designed to make the average rate of all physical and virtual queues stable and maximize system utilization.

First, define $Q_n^c(t)$ as the service flow queue backlog with priority c in the node n at the time slot t. Next, to satisfy the constraint C_1, a virtual queue $P_n^c(t)$ is defined to observe the output residual capacity of node n connected channel j. Finally, a virtual queue $Z(t)$ is defined to control the total inflow and to prevent causing excessive traffic queue delay, that is, to meet the delay limit C_2 C_3. In order to control the transmission route, a routing variable $p_{n,s}^j(t)$ is set to indicate whether the path chooses a channel $n \rightarrow s$ for transmission. $\mu_{n,s}^c(t)$ indicates the $n \rightarrow s$ outflow data of a service flow with priority class c in a time slot t; $\mu_{t,n}^{c*}(t)$ represents the actual $t \rightarrow n$ inflow data of a service flow with priority class c in a time slot t; and $\beta_i(t)$ determines the flow control decision for a service i in a time slot t, that is, the new traffic that actually reaches the network. The data arrival rate for the node n is $\lambda_n^c(t)$, that is $\lambda_n^c(t) = \sum_{t=1}^{N} \mu_{t,n}^{c*}(t) + \sum_{i \in \mathbf{F}_c} \beta_i(t)$. The rate at which traffic actually enters the network is $\beta_i^*(t)$ which satisfies $\beta_i^*(t) = \beta_i(t) - d_n^c(t) \leq \beta_i(t)$. ψ_m and ψ_n represents all accessible channel collections for nodes n and nodes m. $\mathbf{\Pi}_n$ is the collection of all neighbor nodes of a node n.

Defined $\boldsymbol{\Theta}(t) = (\mathbf{Q}(t), \mathbf{P}(t), \mathbf{Z}(t))$ as a queue backlog vector, and the Lyapunov drift is $\Delta(\boldsymbol{\Theta}_n(t))$. According to *Lyapunov optimization theory*, in order to minimize system utilization, the drift-plus-penalty functions should be considered, which are:

$$\begin{aligned} &\Delta(\boldsymbol{\Theta}_n(t)) - VE\{U_n^c(\mathbf{F}_n^c)|\boldsymbol{\Theta}_n(t)\} \\ &\leq B - VE\{U_n^c(\mathbf{F}_n^c)\} + \frac{1}{2}E\left\{ \begin{array}{l} Q_n^c(t) * \left(\lambda_n^c(t) - \sum\limits_{s \in \boldsymbol{\Pi}_u} p_{n,s}^j(t) * \mu_{n,s}^c(t)\right) \\ + P_n^c(t) * \left(C_n^j(t)^{Res} - \beta_i^*(t)\right) \\ + Z_n^c(t) * \left(\lambda_n^c(t) - \sum\limits_{s \in \boldsymbol{\Pi}_n} \mu_{n,s}^{c*}(t)\right) \end{array} \right\} \end{aligned} \tag{2}$$

3.2 RQTS Algorithmic Analysis

This paper designs a QoS guaranteed traffic scheduling algorithm RQTS to minimize the drift penalty reduction function. This optimization problem is decomposed into several sub-problems to solve, which include traffic admission control and transmission path selection. By decomposing the constraints, RQTS separately optimizes the rightmost items of the formula (2), and a joint optimal solution is obtained.

Priority-Based Traffic Access Control. The second sub-problem further determines how much traffic is allowed to access the channel, that is, traffic access control. Network controllers need to make traffic access control decisions $\boldsymbol{\Omega}(t)$ by confirming resource availability and channel service metrics to ensure the quality of service of current streams. In decision making, a node chooses the access amount of each stream $\beta_i^*(t)$ to minimize the remaining capacity of the link. This subproblem can be expressed as:

$$\begin{aligned} \min_{\beta_i^*(t)} : &\sum_{c \in \mathbf{C}} \sum_{i \in \mathbf{F}_n^c} P_n^c(t)\left[C_n^j(t)^{Res} - \beta_i^*(t)\right] - V \sum_{c \in \mathbf{C}} U_n^c(\mathbf{F}_n^c) \\ &s.t.\, 0 \leq \beta_i^*(t) \leq \beta_c^{\max} \end{aligned} \tag{3}$$

To improve system utilization, reserve the highest priority of SU for users with higher latency and lower bandwidth requirements. In traffic access control, the remaining bandwidth $C_n^j(t)^{Res}$ provided by the channel j is examined in priority order to see if it meets the requirements of the traffic flow i. If $\beta_i \leq C_n^j(t)^{Res}$, access the traffic flow i; if $\beta_i > C_n^j(t)^{Res}$, consider discarding a portion of the data $\beta_i * (\alpha_i - \alpha_c^{\min})$ based on minimum reliability requirements $\alpha_c^{\min}$. Then compare the remaining capacity $C_n^j(t)^{Res}$ of the channel j to see if it can meet the new tolerable rate $\beta_i^{\lim}$. If it is satisfied, traffic flow i that discard data are allowed to access the channel $\beta_i^{\lim}$.

Transmission Path Control. The purpose of transmission path control is to control end-to-end transmission delay. On the one hand, the transmission path of each hop needs to be selected, on the other hand, the total number of hops from source node to destination node needs to be controlled to meet the traffic delay limit of C_3. Then the optimal objective function is:

$$\min : \sum_{c \in \mathbf{C}} \sum_{i \in \mathbf{F}_c} \left\{ \begin{array}{c} Q_n^c(t) * \left(\lambda_n^c(t) - \sum_{s \in \mathbf{\Pi}_u} p_{n,s}^j(t) * \mu_{n,s}^c(t) \right) \\ + Z_n^c(t) * \left(\lambda_n^c(t) - \sum_{s \in \mathbf{\Pi}_u} p_{n,s}^j(t) * \mu_{n,s}^{c*}(t) \right) \\ - VU_n^c(\beta_i, \tau_i, \alpha_i) \end{array} \right\} \tag{4}$$

An algorithm based on reinforcement learning is used to solve the path control problem. The problem is represented as a Markov decision process (MDP), which is characterized by four tuples (S, A, P, R), and the Q-Learning algorithm in reinforcement learning is used to solve path control problem. S is a finite set of states. In this problem, the state is expressed as a node in a wireless network, and the number of states in the network is $|\mathbf{N}| = N$; A is a finite set of actions. Actions $a \in A$ is changes of routing variables $p_{n,s}^j(t)$. $\chi - greedy$ is a greedy rule. Specifically, the action of current is determined by randomly selecting the action with a certain probability χ, or by selecting the action who has the largest Q-value with the probability $1 - \chi$. R is the reward. This algorithm aims to improve the utilization rate of the weighted system, so the instantaneous reward is defined as the following equation:

$$R(s, a) = \begin{cases} -U_{be} + U_{af} & |\mathbf{R}_i| \leq |\mathbf{R}_i^L| \\ r_c & \text{else} \end{cases} \tag{5}$$

Where r_c is the reward when the selected transmission path is longer than the longest path that the traffic can tolerate, so the value is a negative number with a larger absolute value to avoid exceeding the delay limit and select valid actions. The reward $-U_{be} + U_{af}$ for effective action is to make the selected path approach the optimal goal, where U_{be} is the cost value before the action is implemented and U_{af} is the cost after the action. In the process of optimizing the objective function (4), the transmission delay limit C_3: $\tau(\mathbf{R}_i) \leq \tau_i$, that is, transmission path in RQTS cannot be longer than the longest path $\mathbf{R}_i^L$ that the traffic i can tolerate, that is, the longest path $|\mathbf{R}_i| \leq |\mathbf{R}_i^L|$.

3.3 RQTS Algorithmic Design

In summary, the RQTS proposed in this paper decomposes the optimization problem with multiple constraints into three subproblems. First, control traffic access to determine the flow rate of the access channel; and then, determine the transmission path of the traffic from the source node to the destination node. The flow of RQTS is shown in Table 1.

Table 1. RQTS Algorithm.

Input: Arrived service flow set $\mathbf{F}_n^c$ at node *n*

Output: $\mathbf{R}_i^*$, $\beta_i^*(t)$

For each episode:
  Classify data traffic set $\mathbf{F}$ into $|\mathbf{C}|$ classes;
  Assign ω_c to S_c;
  For $c = 1 : |\mathbf{C}|$
    Select traffic flow $i \in |\mathbf{F}_n^c|$, $c \in \mathbf{C}$
    Calculate $\beta_i^*(t)$
    Calculate the max length $|\mathbf{R}_i^L|$ under hip constraint
    Introduce $p_{n,s}^j(t)$ and use j^*, $\beta_i^*(t)$ to find all feasible path set $\mathbf{R}_i$
    While $steps < \mathrm{T}_p$ do
      Find state and Construct action set $A_c = \{a, a \in A, Q(s,a) > 0\}$
      If $A_c = \varnothing$ then
        Randomly determine action
      Else
        Determine action $a \in A_c$ according to $\chi - greedy$
      End if
      Calculate reward and $R(s,a)$ and update Q(s,a)
    End while
  End for
End
Return $\mathbf{R}_i^*$, $\beta_i^*(t)$, j^*

The algorithm minimizes the objective function (1) by scheduling as much data as possible to reduce the cached data in the queue. At the same time, priority-based scheduling mechanisms are used to maximize latency utility by giving higher priority to service flows with strict latency requirements. At the beginning of the MDP, the state of the transmission path is selected randomly. That is, the initial position is random and therefore not optimized. MDP then enters a different state. Each state transition generates positive or negative returns. Positive/negative rewards result in an increase/decrease in the overall objective function. Finally, we have reached some states of MDP where the immediate reward function R often fails to generate any positive rewards. Finally, the optimal transmission path is reached, because any action to be adopted no longer results in a decrease in the target function. Therefore, the goal of MDP is to find a series of state-action pairs (paths) that ultimately reach the optimal transmission path efficiently. Moreover, for each priority class c, when the cumulative transmission delay of the traffic i exceeds the end-to-end maximum delay limit τ_i of the traffic $i(\tau(\mathbf{R}_i) > \tau_i)$, the node n discards the data of the traffic i to avoid delays in the transmission of other packets.

4 Simulation Result

4.1 Parameter Setting

In this section, four priority classes of SU are selected for simulation, and reliability, delay and data rate are taken as performance parameters. Adjust the number of flows in each priority class and set the number of channels to at most 10. The number of sensor nodes in the network is 64, which is deployed according to the grid. The specific parameters are shown in Table 2:

Table 2. Simulation parameters of wireless network.

User type	Arrival rate	β min	τ max	α min	Priority	Weight ω	Service type
Service 1	20	18	0.75	99	1	0.4	Protection
Service 2	40	36	2	–	2	0.3	Video
Service 3	30	27	1	90	3	0.2	Information monitoring
Service 4	10	9	–	90	4	0.1	Marketing

4.2 Analysis of Simulation Results

In order to verify the effectiveness of RQTS proposed in this paper, several other advanced scheduling algorithms need to be selected as the comparison scheme. In this paper, we choose the random access strategy (RSS) and a delay based adaptive dynamic programming (DADP) strategy for performance comparison. RSS is not priority sensitive, and DADP is similar to RQTS, which adjusts the scheduling strategy by learning the behavior of PU and SU, while RQTS proposed in this paper makes the real-time scheduling strategy by making Lyapunov drift optimization theory.

In the first scenario, assume that there are four flows in each priority class, and the data arrival rate of each SU flow is shown in Table 1. In this scenario, the performance parameters to be considered include the average delay and throughput of each service class. The simulation results are shown in Fig. 2. As can be seen from Fig. 2, the average delay of services with higher priority level is smaller. The reason is that services with higher priority will be given priority. With the increase of the number of channels, the average delay of all services decreases, but the average transmission delay of services with higher real-time requirements decreases more obviously. However, when the number of channels increases to 7, the delay of each priority class tends to be stable. This is because for the data arrival intensity set in scenario 1, 7 channels can basically meet the transmission requirements of all priority classes. As can be seen from Fig. 3, under the stable working state of the system, RQTS proposed

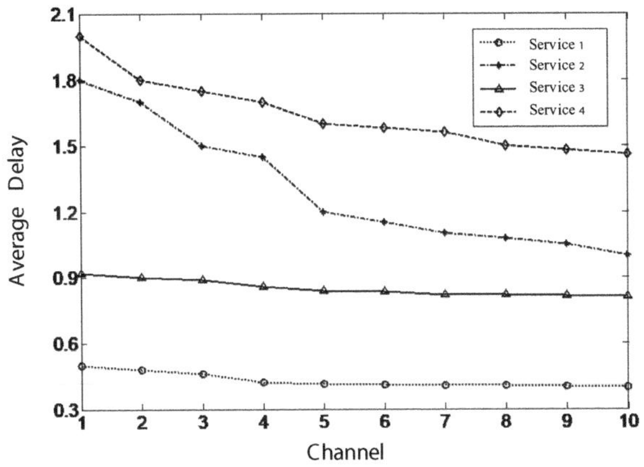

Fig. 2. Average delay of each service in the change of channel number.

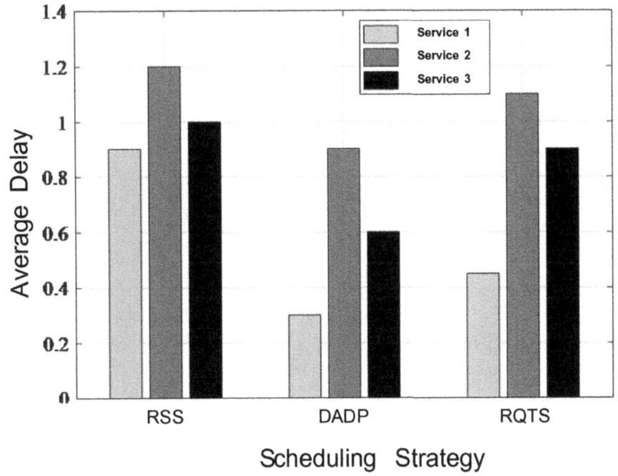

Fig. 3. Average delay of each service under different scheduling strategies.

in this paper improves the system delay of each priority class compared with RSS. This is because in the priority system, the services with strict delay requirements are set with higher priority, which can give priority to the transmission of data. Compared with the RSS with non-priority, its average transmission delay is lower. However, due to the adaptive scheduling oriented to delay optimization, the optimization target of DADP is the average delay of higher priority services, so compared with DADP, the delay performance of RQTS is slightly worse.

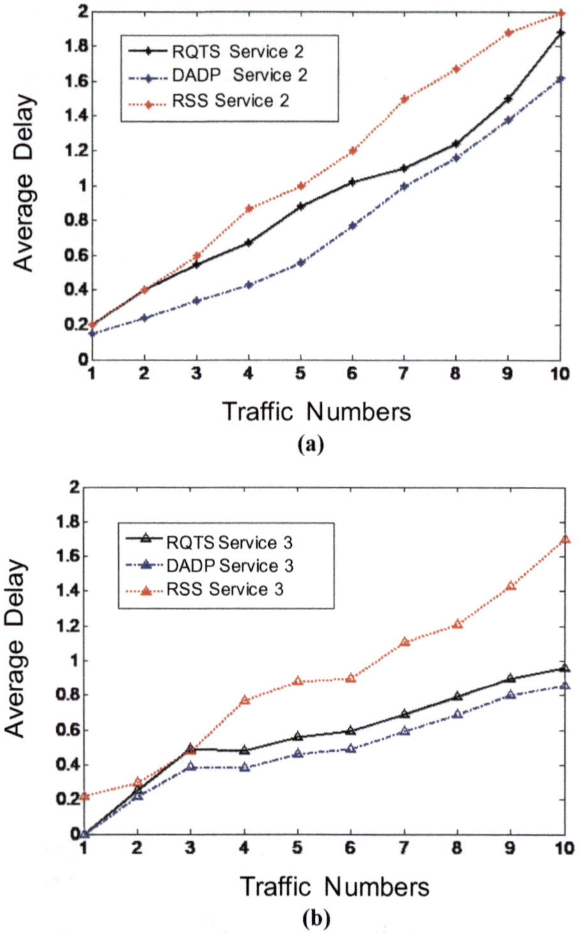

Fig. 4. (a) Average delay of service 2 under the change of traffic numbers. (b) Average delay of service 3 under change of traffic numbers.

In the second scenario, increase the number of service 3 flows to more clearly observe the impact of QoS oriented scheduling policies on the system. In this scenario, it is assumed that the number of service flows of service 2 is still 4, but the number of service flows of service 3 is variable. As can be seen from Fig. 4 (a-b), when the number of service 3 flows reaches 9, the delay of traffic class using RQTS tends to the delay boundary value. When using RSS, the delay of each service increases rapidly with the increase of traffic flow. This is because in a non-prioritized system, all services get the same service opportunity. In the priority system, the delay of each priority class using DADP and RQTS is relatively stable. Until the number of traffic flows of service 3 increases to 7, the delay of service 2 begins to increase exponentially, which is due to the high bandwidth requirements.

5 Conclusion

In this paper, according to the differentiated QoS requirements of service in wireless sensor network of intelligent power communication network, the optimal traffic scheduling model is established to maximize the utilization of weighted system. Based on the theory of Lyapunov optimization, a wireless network QoS guarantee traffic scheduling algorithm RQTS is designed to solve the optimization problem. The algorithm adopts the idea of distributed routing, and realizes the control of reliability, data rate and delay through flow access control and transmission path control. The simulation results show that the proposed mechanism can improve the QoS of the network, and can still provide high-performance communication services for high priority services when the traffic of low priority is large.

Acknowledgment. This work was supported by the Science and Technology Project of State Grid Henan Electric Power Co., Ltd. (5217Q018003C) and the Fundamental Research Funds for the Central Universities (2019RC07).

References

1. Felemban, E., Lee, C.G., Ekici, E., Boder, R., Vural, S.: MMSPEED: multipath multi speed protocol for QoS guarantee of reliability and timeliness in wireless sensor networks. IEEE Trans. Mob. Comput. **5**(6), 738–754 (2006)
2. Al-Anbagi, I., Erol-Kantarci, M., Mouftah, H.T.: An adaptive QoS scheme for WSN-based smart grid monitoring. In: 2013 IEEE International Conference on Communications Workshops (ICC), pp. 1046–1051 (2013)
3. Li, H., Zhang, W.: QoS routing in smart grid. In: IEEE Globecom, pp. 1–6 (2010)
4. Sun, W., Yuan, X., Wang, J., Han, D., Zhang, C.: Quality of service networking for smart grid distribution monitoring. In: IEEE Smart Grid Communication, pp. 373–378 (2010)
5. Shi, Y., et al.: A distributed optimization algorithm for multi-hop cognitive radio networks. In: IEEE Infocom, pp. 1292–1300 (2008)
6. Neely, M.J.: Opportunistic scheduling with worst case delay guarantees in single and multi-hop networks. In: IEEE Infocom, pp. 1728–1736 (2011)

7. Shah, G.A., Gungor, V.C., Akan, O.B.: A cross-layer design for QoS support in cognitive radio sensor networks for smart grid applications. In: IEEE ICC, pp. 1378–1382 (2012)
8. Ghalib, A.S., Vehbi, C.G., Ozgur, B.A.: A cross-layer QoS-aware communication framework in cognitive radio sensor networks for smart grid applications. IEEE Trans. Industr. Inf. **9**(3), 1477–1485 (2013)
9. Deshpande, J.G., Kim, E., Thottan, M.: Differentiated services QoS in smart grid communication networks. Bell Labs Tech. J. **16**(3), 61–81 (2011)

Network Traffic Prediction Method Based on Time Series Characteristics

Yonghua Huo[1], Chunxiao Song[1], Sheng Gao[1], Haodong Yang[2], Yu Yan[2], and Yang Yang[2(✉)]

[1] The 54th Research Institute of CETC, Shijiazhuang 050000, China
[2] State Key Laboratory of Networking and Switching Technology, Beijing University of Posts and Telecommunications, Beijing 100876, China
yyang@bupt.edu.cn

Abstract. With the continuous development of computer networks in recent years, the scale and types of services carried by the network are increasing. Accurate traffic prediction results provide the main support and reference basis for network operation and maintenance functions such as network attack detection. Since network traffic has certain dynamics, continuity, and long correlation and self-similar characteristics, the artificial intelligence method is generally used for network traffic prediction. Among them, the recurrent neural network has a short-term memory performance and has a good prediction effect for time series data such as network traffic. However, when the time series span is relatively long, the problem of gradient disappearance or gradient explosion may occur, so further optimization is required. In this paper, we propose a network traffic prediction method based on parameter pre-training of clockwork neural network. This method is first based on CW-RNN, and then introduces the differential evolution algorithm to pre-train the clock parameters. At the same time, the differential evolution algorithm is further improved by changing its crossover factor and mutation factor to improve the accuracy of its convergence. Through computer simulation, the flow prediction method proposed in this paper can obtain accurate prediction results.

Keywords: Internet traffic prediction · Differential evolution algorithm · Recurrent neural network

1 Introduction

As the scale of computer networks expands, there are more and more types of services carried by the Internet, and the nature of network traffic becomes more complicated. Since the Internet is a complex nonlinear system, in order to ensure that users can achieve stable and reliable data transmission and reasonable network resource allocation, it is necessary to deeply understand the new characteristics of network traffic and achieve accurate predictions. Network traffic prediction is of great significance for improving network service quality (QoS).

In recent years, researchers have discovered that due to the characteristics of short-term memory, the recurrent neural network (RNN) has a good predictive effect on time

Q. Liu et al. (Eds.): CENet 2020, AISC 1274, pp. 1533–1541, 2021.
https://doi.org/10.1007/978-981-15-8462-6_175

series data such as network traffic. However, RNN has great restrictions on the memory of historical data. For example, when the span of the time series is relatively long and the number of model layers is relatively large, the problem of gradient disappearance or gradient explosion may occur, so further optimization is required.

In view of the above problems, this paper proposes a traffic prediction method based on clock cycle recurrent neural network (Clockwork RNN, CW-RNN) and improved differential evolution algorithm. First, the basic model of CW-RNN is used, and then the improved differential evolution algorithm is introduced to improve the clock cycle parameters, so that the selection of the clock cycle is more intelligent. Experiments verify that this method can provide better data fitting and more accurate results for network traffic prediction.

The rest of this paper is organized as follows. Section 2 sorts out some related works, Sect. 3 introduces a traffic prediction method based on CW-RNN parameter pre-training. Simulation results are discussed in Sect. 4. Finally, Sect. 5 concludes the paper.

2 Related Work

At present, there are many methods for flow prediction, which can be roughly divided into two categories: traditional flow prediction method such as FARIMA [1], artificial intelligence method mainly including neural network [2, 3].

Katris, C. et al. [4] compared some forecasting model building procedures like ARIMA, FARIMA and Artificial Neural Networks (ANN) models with Internet traffic data. The experiment shows that, predictions may further be improved if we recognize the non-linear structure, existing sometimes in the time series, with the help of ANN.

Feng, J. et al. [5] proposed Deep Traffic Predictor (DeepTP) to forecast traffic demands from spatial-dependent and long-period cellular traffic, which can be divided into two components: a general feature extractor, and a sequential module.

Xiang, L. et al. [6] proposed a new hybrid network traffic prediction method based on the combination of the covariation orthogonal prediction and the artificial neural network prediction to capture the burstiness in the network traffic and the accuracy of the new prediction method has been effectively improved.

Wei, Y. et al. [7] introduced a prediction algorithm based on traffic decomposition. The complex correlation structure of the network history traffic is decomposed according to different protocols and then predicted with wavelet method separately. Simulation results show that the combination method can achieve higher prediction accuracy rather than that without traffic decomposition.

However, these methods often have difficulty adapting to the flow characteristics when dealing with continuous-span flows and the prediction results are inaccurate. Based on this, this paper proposes a network traffic prediction model, which is based on the CW-RNN model and improved differential evolution algorithm to pre-train the clock parameters.

3 Traffic Prediction Model Based on CW-RNN Parameter Pre-training

3.1 Overall Process

This paper refers to the overall framework of CW-RNN and proposes a traffic prediction method based on CW-RNN parameter pre-training. The improved differential evolution algorithm is used to pre-train the clock parameters of the CW-RNN algorithm, and the number of clock cycles obtained from the training is taken as the clock cycle of the CW-RNN. In addition, we also improve the hidden module activation rules of CW-RNN for traffic prediction. The specific structure of CW-RNN for internet traffic prediction is shown in the Fig. 1.

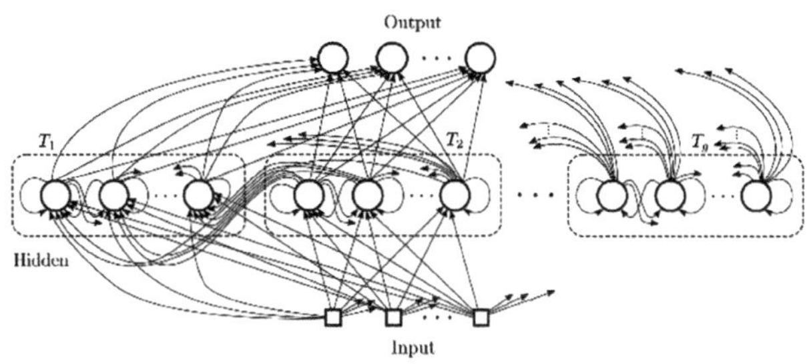

Fig. 1. The structure of clockwork RNN.

3.2 Improved Differential Evolution Algorithm

In this paper, on the basis of CW-RNN combined with improved differential evolution algorithm to train clock cycle T_n, so that more modules can be trained in the hidden layer. To a certain extent, the performance of the differential evolution algorithm has a great relationship with the setting of control parameters. In the differential evolution algorithm, it mainly includes three control parameters: population size NP, scaling factor F and cross coefficient CR.

In the process of the standard differential evolution algorithm, the value of the scaling factor F is generally a fixed real number, and the value interval is generally [0, 2]. This fixed value method will cause F to not adapt well to the execution process of the model. If the value of F is too large, the efficiency of the model in seeking the optimal solution will become lower. On the other hand, if the value of F is too small, the diversity of the population will decrease, which will cause the model to fall into a local optimal solution during the solution process. Experimental research found that the scaling factor F should be larger at the beginning to ensure the diversity of the population, and should be smaller in the later stage to ensure its convergence. Similarly, the crossover factor CR should be smaller at the beginning to ensure the stability of the algorithm, and it should be larger in the later stage to ensure the diversity of the crossover.

The overall process of the improved differential evolution algorithm is as follows:

Population Initialization. Initially, an initial population with a population size of NP is initially generated in the search space $[X_{\min}, X_{\max}]$ of the solution of the clock cycle. The vector dimension of each individual in the population is D, and the jth dimension of each individual $\vec{X}_i$ in the initial population can be expressed as:

$$\mathrm{X}_{i,j} = X_{minj} + rand(0,1) * \left(X_{maxj} - X_{minj}\right) \tag{1}$$

Among them, $X_{\min j}$ is the lower bound of the value of $X_{i,j}$, and $X_{\max j}$ is the upper bound of the value of $X_{i,j}$. The fitness function of the differential evolution algorithm designed for this flow prediction experiment is: the maximum value of $f = \sum y(i)$, where

$$y(i) = \begin{cases} 1 & if(zi/xj = 0) \\ 0 & else \end{cases} \tag{2}$$

Mutation Operation. During the evolution process, the improved differential evolution algorithm will generate the corresponding mutation vector in each generation to perform mutation operations on the previous generation. The principle is to add a disturbance vector to the base vector. The mutation strategy of the algorithm in the mutation operation is as follows:

$$\vec{V}_i = \vec{X}_{r1} + F*(\vec{X}_{r2} - \vec{X}_{r3}) \tag{3}$$

Among them, $\vec{V}_i$ represents the mutation vector of the current individual, and $\vec{X}_{r1}$, $\vec{X}_{r2}$, and $\vec{X}_{r3}$ are randomly selected vectors in the sample (Among them, $\vec{X}_{r1}$ is the basis vector, $\vec{X}_{r2}$ and $\vec{X}_{r3}$ are different vectors in the population, and the difference is the disturbance vector). The control parameter F is a scaling factor. Formula (4) is used to improve the differential evolution algorithm, we can obtain

$$F = F_{\min} + \frac{2(F_{\max} - F_{\min})}{t} \tag{4}$$

Where t represents the current number of iterations.

Crossover Operation. During the evolution process, the improved differential evolution algorithm will perform crossover operation in each generation, and combine the current individual vector $\vec{X}_i$ and mutation vector $\vec{V}_i$ into the test vector $\vec{U}_i$ by performing the binomial crossover method.

$$U_{i,j} = \begin{cases} V_{i,j} & if\ rand(0,1) \le CR, or j = I \\ X_{i,j} & else \end{cases} \tag{5}$$

Where I represents an integer randomly generated from [1, D], and its meaning is to ensure that at least one dimension of the test vector $\vec{U}_i$ comes from the mutation vector

$\vec{V}_i$, and the control parameter CR is the crossover coefficient. Formula (6) is used to improve the differential evolution algorithm, we can obtain

$$CR = CR_{\max} - \frac{2(CR_{\max} - CR_{\min})}{t} \tag{6}$$

Where t represents the current number of iterations.

Selection operation. The overall idea of the differential evolution algorithm is to adopt a greedy selection strategy with local optimization as the global optimization. Among the test vector $\vec{U}_i$ and the current individual vector $\vec{X}_i$, the one with better fitness value will be selected into the next generation. When the training objective function is a maximum problem, the individual vector entering the next generation is

$$\vec{X}_i = \begin{cases} \vec{U}_i & if\, f(\vec{U}_i) > f(\vec{X}_i) \\ \vec{X}_i & else \end{cases} \tag{7}$$

Where $f(\vec{X})$ is the objective equation for maximization.

3.3 Improved CW-RNN

This paper uses CW-RNN as the basis and improves the hidden module activation rules for traffic prediction. With the structural design of module-division, each module of the CW-RNN can process data at different time cycles to enhance the memory effect. Each module has a different update rate according to different clock cycles. Modules with short clock cycle time will be updated quickly at a high frequency, which are mainly responsible for short-term memory; and modules with long clock cycle time will be updated slowly, which are responsible for the long-term memory.

However, CW-RNN has some disadvantages in traffic forecasting. The CW-RNN increase the training speed by only selecting the modules whose period can be divided by current epoch to activate, and other modules keep the previous hidden layer output value unchanged. In this way, not every output can receive all corresponding long-term and short-term hidden layer output because each time only partial modules of neurons are activated. In fact, the long-term memory maintained by a slow module is often kept at a certain value. The output of the hidden layer may retain the same value for a long period, especially for slower modules. Though the long-term memory can be learned, the information it learned is only from a certain fixed moment before, and the information of its surrounding time is not transmitted to the current moment because the slow module's value remains unchanged for a long time.

Since the traffic is data with obvious cyclic trend, the output corresponding to each moment is affected by the traffic data of a certain period before, rather than the traffic data in a fuzzy period. Therefore, we improved the hidden module activation rules that will update the hidden layer output of each module according to the state of the hidden layer before the corresponding time period. At each moment, all modules are updated to determine the output to improve the accuracy of the traffic forecast. Through this improved activation rule, it can be guaranteed that every time the information is transmitted in the forward direction, all modules in the hidden layer will be activated, and the information will be transmitted to the output result. In this way, it can be guaranteed that the output result of each step is a synthesis of the cycle memory information stored in each different cycle reserve pool. This can greatly improve the accuracy of the final prediction results for data with a more obvious periodic trend, such as network traffic.

4 Experiment

In this experiment, we use the traffic data set downloaded from the MAWI website. This data set is the network traffic of a backbone network from Japan to the United States, and the time interval is every 5 min. The flow data set used in this experiment is shown in Fig. 2, and its time span is from February 7, 2013 to May 4, 2015. The total experimental data is 12,800.

In order to measure whether the prediction effect of the model is accurate, in addition to observing the intuitive prediction effect graph, we also use two commonly used quantitative indicators for evaluation: root mean square error (RMSE) and mean absolute error (MAE) to compare the results.

Figure 2 and 3 respectively show the prediction results. Figure 2 is the overall prediction using the traffic prediction method proposed in this article, and Fig. 3 is the overall prediction using the original CW-RNN traffic prediction method.

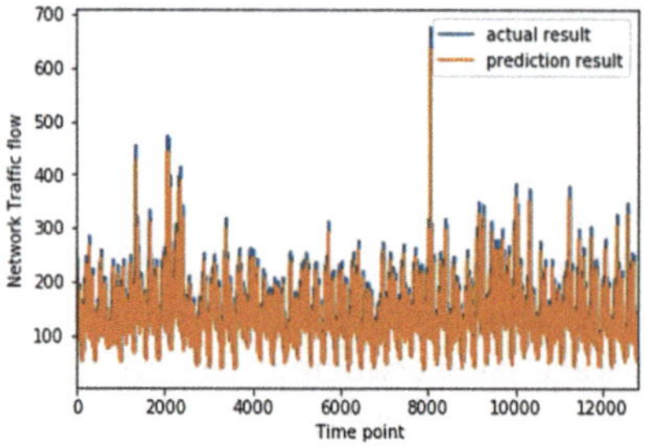

Fig. 2. The overall prediction results using our method.

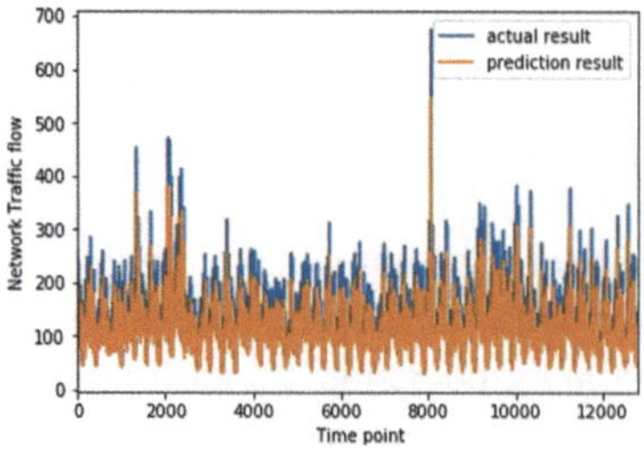

Fig. 3. The overall prediction results using original CW-RNN.

As can be seen, the overall prediction effect of both algorithms is good. However, at some details and the position of the peak, the prediction effect of the traffic prediction method proposed in this paper is better than that of the original CW-RNN traffic prediction method. In addition, it can also be seen on the graph that at 2200 min, 8100 min and other individual points are prone to large errors, the other overall prediction effects are good. In order to facilitate careful observation, we intercept the prediction results of the two algorithms in about 4200 min and zoom in to observe the prediction. Figure 4 is the detailed prediction graph of the traffic prediction method proposed in this article, and Fig. 5 is of the original CW-RNN prediction method.

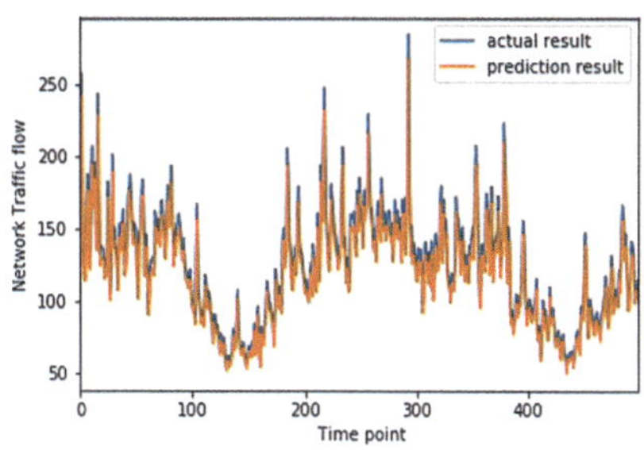

Fig. 4. The detailed prediction results using our method.

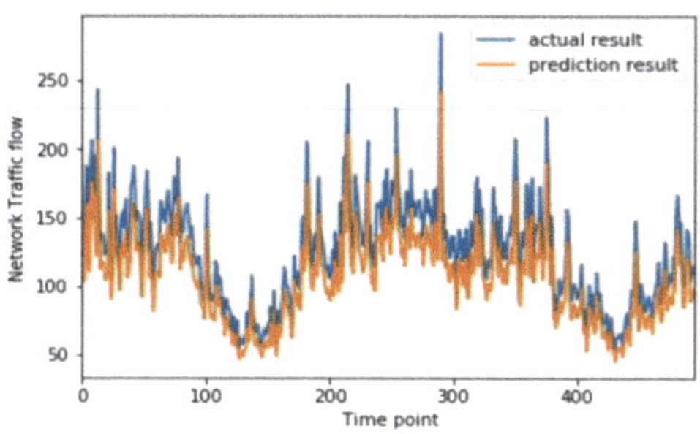

Fig. 5. The detailed prediction results using original CW-RNN.

From the above figure, we can see that from the details, the traffic prediction method proposed in this paper has a better prediction effect. The original CW-RNN traffic prediction method has no big error in the overall prediction effect, but it does not achieve a good fitting effect at many time points such as the peak.

In order to make the obtained data more credible, we conducted three experiments and averaged the quantitative index results obtained three times. Table 1 shows the average RMSE and average MAE results. For the results of RMSE, the prediction effect of the model proposed in this paper is improved by 10.57% compared with the comparison algorithm; for the results of MAE, the prediction effect of the model proposed in this paper is improved by 10.35% compared to the comparison algorithm.

Table 1. The RMSE and MAE of each algorithm.

	RMSE	MAE
Improved CW-RNN	0.03731	0.01869
CW-RNN	0.04172	0.02085

It can be seen that the prediction method proposed in this paper has a better prediction effect. The main reason is that with the structure of the traffic prediction method proposed in this paper, there are more modules participating in the training, which can store more historical information, better capture the timing relationship on the data, and achieve a better prediction effect.

5 Conclusion

In order to improve the accuracy of traffic prediction, this paper proposes an improved differential evolution network traffic prediction model based on CW-RNN. The model first uses the basic model of CW-RNN, and then introduces an improved differential evolution algorithm to improve the clock cycle parameters, making the selection of clock cycles more flexible and accurate. Through experimental simulation, it is proved that the traffic prediction method based on CW-RNN parameter pre-training proposed in this paper can get better prediction effect in dealing with network traffic prediction.

Acknowledgment. This work is supported by National Key Research and Development Program of China (2019YFB2103200), the Fundamental Research Funds for the Central Universities (500419319 2019PTB-019), Open Subject Funds of Science and Technology on Information Transmission and Dissemination in Communication Networks Laboratory (SKX182010049), the Industrial Internet Innovation and Development Project 2018 & 2019 of China.

References

1. Wang, P., Zhang, S., Chen, X.: SFARIMA: a new network traffic prediction algorithm. In: 2009 First International Conference on Information Science and Engineering, pp. 1859–1863. Nanjing (2009)
2. Wang, X., Zhang, C., Zhang, S.: Modified elman neural network and its application to network traffic prediction. In: 2012 IEEE 2nd International Conference on Cloud Computing and Intelligence Systems, pp. 629–633. Hangzhou (2012)
3. Koutník, J., et al.: A clockwork RNN. Comput. Sci. **18**(2), 1863–1871 (2014)
4. Katris, C., Daskalaki, S.: Prediction of internet traffic using time series and neural networks. In: International work-conference on Time Series (2014)
5. Feng, J., Chen, X., Gao, R., Zeng, M., Li, Y.: DeepTP: an end-to-end neural network for mobile cellular traffic prediction. IEEE Network **32**(6), 108–115 (2018)

6. Xiang, L., Ge, X., Liu, C., Shu, L., Wang, C.: A new hybrid network traffic prediction method, 2010 IEEE Global Telecommunications Conference Globecom 2010, pp. 1–5. Miami, FL (2010)
7. Wei, Y., Wang, J., Wang, C. et al.: Network traffic prediction by traffic decomposition. 2012 Fifth International Conference on Intelligent Networks and Intelligent Systems, pp. 158–161. IEEE (2012)

E-Chain: Blockchain-Based Energy Market for Smart Cities

Siwei Miao[1], Xiao Zhang[2], Kwame Omono Asamoah[3], Jianbin Gao[2,4], and Xia Qi[3](✉)

[1] China Electric Power Research Institute, Beijing 100000, China
[2] National Dispatching Center, State Grid Corporation of China, Beijing 100000, China
[3] School of Computer Science and Engineering, University of Electronic Science and Technology of China, Chengdu 610000, China
qixia@uestc.edu.cn
[4] School of Resources and Environment, University of Electronic Science and Technology of China, Chengdu 610000, China

Abstract. Electricity is the most critical input product for most businesses in the world nowadays. The utilization of electricity has assisted in the discovery of breakthrough disclosures and advances and has in time, become the most crucial constraint in the development of economies. With the introduction of distributed energy resources in microgrids, energy users are gradually turning into prosumers i.e., users who simultaneously produce and consume energy. This has brought about peer-to-peer (P2P) energy markets where prosumers can trade their surplus energy. However, there are privacy and security issues associated with such markets. Prosumers find it insecure to trade their power in an untrusted and nontransparent environment. In a P2P energy market, an intermediary oversees the transaction process between parties on the market and this poses a threat to the privacy of the prosumers. Consequently, a unified and robust energy trading market is required for prosumers to trade their energy. To address these security and privacy challenges, we leverage the properties of the blockchain innovation to propose a secure energy trading market where prosumers can trade their energy without compromising their privacy.

Keywords: Microgrid · Blockchain · Prosumers · Smart city · Energy market

1 Introduction

The notion of smart cities has gained enormous attention of researchers in recent times [1], which is a critical component of the long run change of human settlements. In smart cities, there is an interconnection of electrical systems through sensing gadgets and actuators with extensive networking and computing capabilities. The traditional power grid has been the main source of electricity in smart cities and industries over the past years [2]. The grid comprises of various components like synchronous machines, power transformers, transmission lines, transmission substations, conveyance lines, dissemination substations and proficient sorts of loads.

Q. Liu et al. (Eds.): CENet 2020, AISC 1274, pp. 1542–1549, 2021.
https://doi.org/10.1007/978-981-15-8462-6_176

A collaborative sharing of energy supply will help curb this problem by improving the balance of energy supply among prosumers in microgrids [3]. However, energy buying and selling is mainly done on large-scale basis and it is normally between utility companies and consumers. There is therefore, the need for new mechanisms for trading energy on a small-scale basis [4]. Peer-to-peer (P2P) energy trading is anticipated to be one of the foremost vital components of next-generation power frameworks.

However, there are common privacy and security issues associated with trading energy in a P2P manner. First, it is insecure for prosumers to trade power in an untrusted and nontransparent energy market. Secondly, prosumers with surplus energy are sometimes reluctant to participate as energy suppliers in energy market because of their concerns about their privacy. Furthermore, in a P2P energy market, there is an intermediary to verify and approve transactions between prosumers. The intermediary can cause problems such leakage of private information about prosumers and delay in transactions when there is a point of failure with the system of the intermediary [5]. This work presents a blockchain based remedy coupled with smart contracts for designing an immutable energy market for trading of electrical power among prosumers.

2 Preliminaries

This part explains the tools that we employed in our Blockchain Energy Market for Smart Cities. In this section, we describe the blockchain as a service and we further give a brief explanation of the microgrid technology.

2.1 Blockchain

A blockchain is a disseminated database which consists of a continuous growing list of records linked together using cryptography [5]. By design, the blockchain architecture is impervious to data modification. The list of records are connected together through chains on blocks [6].

A basic comparison for understanding blockchain innovation is a Google Doc. When a file is created and shared with a group of people, the file is disseminated rather than replicated or transferred [5]. This presents a decentralized chain which allows everybody access to the file at the same time. No one is bolted out anticipating changes from another party, whereas all adjustments to the file are being listed in real-time, making alterations totally transparent [8].

In a blockchain environment, there is no need for a centralized control or actor for data maintenance [7]. Whereas conventional monetary standards are distributed by central banks, bitcoin has no central power. Instead, the bitcoin blockchain is sustained by a network of individuals referred to as miners [8]. These "miners," often referred to as "nodes" on the network, are people running purpose-built computers that are actually involved in solving complex mathematical problems in order to make a transaction successful [12].

2.2 Microgrid

A microgrid consists of a number of interconnected loads and dispersed energy assets inside a clearly characterized electrical confines that acts as a single controllable substation with regard to the grid [9]. Generally speaking, hydropower generating stations and grid supply the homes and businesses in a neighborhood. A microgrid can be set up by including innovations such as solar panels, large-capacity batteries and electric vehicles. In the event that the microgrid can be integrated with the area's primary power network, it is normally referred to as a hybrid microgrid.

A micorgrid permits communities to be more energy independent and in a few cases, more environmentally friendly. Microgrid technologies can supply homes and businesses in parallel with the hydro grid. If the main grid goes out, the microgrid can continue to supply power to homes and businesses for a while, keeping customers comfortable. When the sun is shining and less power is being used, the microgrid can supply the whole neighborhood or part of it and if the microgrid generates more electricity than the neighborhood needs, the surplus can be fed back into the hydro grid.

3 Related Works

Wang, H. G. et al. [11] presented ID-based proxy re-signcryption scheme in. Their scheme allows a semi-trusted entity called the "proxy" to convert the signcryption addressed to a "delegator" to those that can be de-signcrypted to a "delegatee". This process was achieved by using some special information given by the "delegator" called "rekey". The proxy is not allowed to know the secret keys of the delegator or the delegatee or the plaintext during the conversation.

Li, Z. T. et al. [13] proposed a consortium blockchain for secure energy trading in industrial internet of things. Blockchain technology was utilized to secure an energy trading mechanism, which can be adapted in general, scenarios of p2p energy trading by getting rid of a trusted intermediary.

Biswas, S. et al. [10] presented a scalable blockchain framework for secure transactions in IoT. This work dealt with the issue of scalability of the ledger on a blockchain network and the rate of transaction when IoT is integrated with blockchain technology. The network architecture proposed in this work was divided into two, a local peer network and a global blockchain network. Transactions between devices on the same local peer network are processed locally whilst transactions between devices on different local peer networks are processed on the global network.

4 System Architecture

In this part, we show a graphical representation of our proposed system and give a brief description of the various entities on the system and the functions of the said entities. Figure 1 shows the architecture of our proposed system.

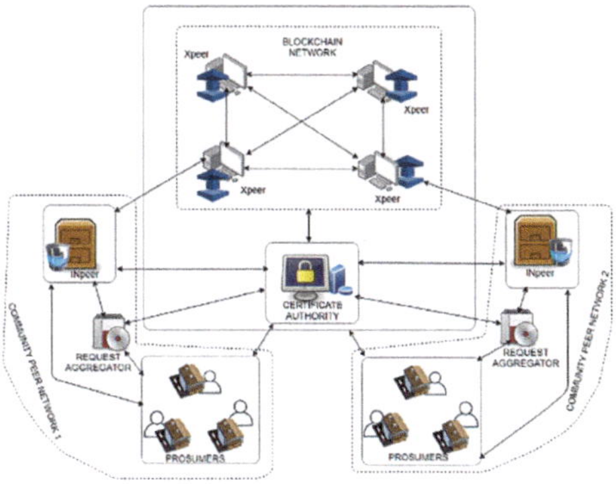

Fig. 1. System architecture with various entities.

4.1 Community Network

This network directly interfaces with the main blockchain network. A community implements a community peer network which consists of all prosumers and consumers within a specific location, request aggregator and an *INpeer*. The importance of this network is to separate transactions between parties within the same community from transactions between parties from different communities. Transactions within a community are processed by the *INpeer* and the transactions between parties in different communities are processed by the *Xpeers* on the main blockchain network. We implement this principle by separating our system into two parts as shown in Fig. 1.

Prosumers. Prosumers produce energy with their energy harvesting devices and sell it to customers on the platform and also use part of the energy which they produce. A prosumer or a consumer cannot join the system unless it has an association with a community.

INpeers. An *INpeer* serves as a local peer for the community and interacts with an *Xpeer* in the main blockchain network on behalf of the community which it represents. Each *INpeer* maintains a ledger where internal transactions are processed and stored. An *INpeer* also maintains a list of keys and other vital credentials of all the prosusmers connected to it. All external transactions are forwarded by the *INpeer* to its corresponding *Xpeer* on the blockchain network where they are processed and added to the main blockchain upon consensus.

The network was structured in this way in order to enhance the scalability of the ledger of the *Xpeers* and also to increase the rate of transaction of the peers on the system. The *INpeers* have been divided into *INpeer_*1, *INpeer_*2 to *INpeer_*N. *INpeer_1* represents the main instance of the network with other *INpeers* serving as a secondary instance distributed geographically to cater for the problem of single of point of failure. It is important to note that there can be two or more *INpeers* on a community network but we represent them with a single *INpeer* on our architecture for clarity purposes.

Request Aggregator. The main function of the request aggregator is to fetch transactions from the energy request pool and convert into a block based on the aggregating algorithm's batch instructions. Copies of the certificates generated by the CA are stored by the request aggregator and these are used in validating the requests it receives from prosumers. It must be noted that, the request aggregator on our system is not a miner. The function of the request aggregator is restricted to fetching transactions and generating blocks from them.

Energy Request Pool. This is a pool of unprocessed energy requests from prosumers with energy deficit and energy surplus. These requests contain the credentials of prosumers as well as the amount of power they want to purchase.

Blockchain Network. The blockchain network is of a network of all the *Xpeers* on the system and the certificate authority.

Certificate Authority. The CA is a fully trusted server on the blockchain network that provides keys, signatures and certificates to all the entities on the system. It is a key entity on the system because there is no other entity on the system which provides such functionalities. All the entities on the system request for their encryption keys and signatures from the CA and in addition that, the CA provides credential validation.

Xpeers. On the blockchain network, all the peers on the network are connected together and each peer keeps its own ledger with related smart contracts. An *INpeer* interacts with a corresponding peer (*Xpeer*) on the blockchain network on behalf of the entities (authenticated by the CA). It must be noted that the working principles of the core blockchain network remains the same. The *Xpeers* communicates with other peers on the blockchain network which may be working as *Xpeers* for other communities. As demonstrated in Fig. 1, *Xpeer* _1 acts as an external peer on the blockchain network and communicates with *INpeer*_1 in community 1.

Smart Contract. Smart contract can be defined as a computerized contract which characterizes the terms and conditions of a transaction between two entities. It is executed within the frame of chaincode based on business model and resource definitions. The code allows the mutual settlement of contractual terms, and ensures that there is no alteration in every recording in the network. hence, trust is assured and high-priced duplication is eliminated. We implement smart contract in our work as they are defined for global blockchain networks.

5 Experiment

We carried out experiments and some valuable parameters were measured. We tested our proposed system on a private ethereum blockchain network. Ethereum is a programmable blockchain platform, which makes use of the robust nature of Solidity. We made use of two machines having these specifications for emulating the network topology. The first machine had a 2.2 GHz intel core i7 processor and an 8 GB RAM. The second machine had a 2.4 GHz intel core i5 processor and a 16 GB RAM.

We represented the community network with the first machine and the blockchain network with the second machine. A virtual Fedora container-based technique was used in running the network peers. An application was designed in Python 3.7.2 using the Charm-Crypto 0.50 framework, which connects the peers and continuously generates transaction messages to *INpeer* through smart contract. We allocated one aggregator and one *INpeer* to a community network and 4 *Xpeers* with 2 controlling communities for the blockchain network. The values were adjusted according to experimental demands.

5.1 Experimental Results

In evaluating the performance of our system, we performed two experiments. The first experiment was focused on the time it takes to append a signature on a transaction by an *INpeer* or *Xpeer* and the time it takes to verify a signature on a transaction received by an *INpeer* or *Xpeers* from the certificate authority. The number of transactions were set within the intervals of 10, 20, 30, …, 100 and the delay it takes for a successful signing and verification to be accomplished was measured. We simulated the variations obtained and the corresponding results are shown in Fig. 2. Our experiment revealed that appending a signature is more computationally intensive than verifying a signature on a transaction.

The process involved in locating prosumers in the same community is faster than that of prosumers in different communities. This means that all things being equal, transactions between prosumers in the same community will be faster than that of transactions between prosumers in different communities. Therefore, the effect on weight of block against increasing number of transactions from prosumers was analyzed and the results are shown in Fig. 3.

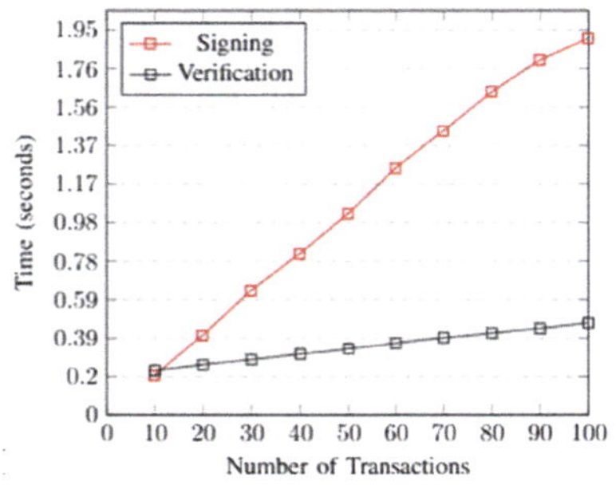

Fig. 2. Verification and signing of blocks.

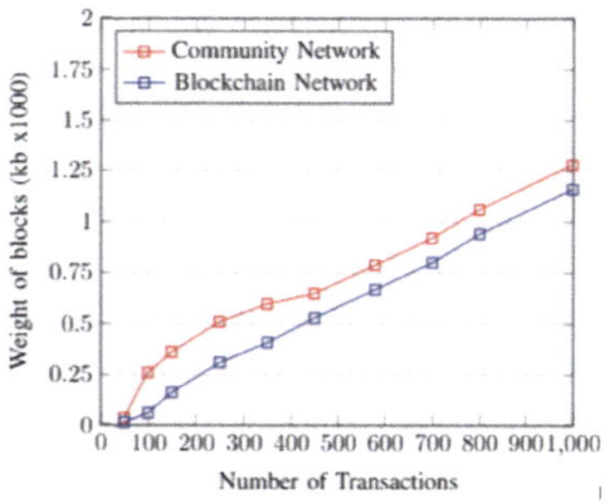

Fig. 3. Weight of blocks vs number of transactions.

The red curve depicts the weight of blocks created by the *INpeer* on a community network whilst the blue curve represents the weight of blocks created by *Xpeer* on the blockchain network. We should note that the probability of a transaction taking place between prosumers in the same community is 0.7. The weight of blocks for the *Xpeer* increases but not as the rate of the *INpeer*. This is because the number of transactions on the community network is relatively higher than number of transactions of the blockchain network.

5.2 Discussions

In this section, we discuss the security and privacy of our proposed system.

Privacy. The privacy of our proposed system is derived from the blockchain were each prosumer uses a unique public key in communicating with a peer on the network. A peer on the network communicates with another peer using a unique a public key. This prevents an attacker from tracking a prosumer or a node on the network.

Security. The security provided by our system is generally attributed to the use of blockchain. Each transaction, which takes place on the network, contains the hash of the data and this ensures integrity. All transactions on the system are encrypted by the use of asymmetric encryption methods which guarantees confidentiality. Recall that, an *INpeer* on a community network keeps up a key list of all the prosumers connected to it. This gives access control for the prosumers in such a way that, only the transactions for which the embedded keys match the key list of an *INpeer* can be processed.

Distributed Denial of Service DDoS Attack. In order for an attacker to launch a DDoS attack, it is necessary to compromise majority of the prosumers on the network. The compromised prosumers issue a larger number of transactions to a targeted *INpeer* on the network. Remember that, an *INpeer* either processes a transaction (i.e., if it finds a buyer or seller on the community network) or forwards it to the blockchain network. An *INpeer* looks for a buyer or seller only if the keys in the transaction match a key in its key list. The keys in the transactions that are part of the DDoS in our system would not match a key in the key list of an *INpeer*. Hence such transactions are dropped and will not have an impact on the targeted *INpeer*.

Data Unforgeability. The decentralized nature of our blockchain energy market coupled with digitally-signed transactions ensures that a malicious user cannot pose as a legitimate prosumer or corrupt the network, as that would mean that, the malicious user forged a digital signature or gained control over majority of the system resources. This is because only transactions with verifiable digital signatures are processed by the *INpeers* on the system.

6 Conclusion

In this work, we have presented a unified P2P energy market for secure energy trading between prosumers in smart cities. We employed microgrids where each microgrid forms a local energy market and these community energy markets are connected together to allow prosumers in different communities to trade energy with each other. We utilized the blockchain innovation in designing our system to guarantee the privacy of prosumers and the security of transactions. This platform can guarantee security of transactions even in large scale while protecting the privacy of prosumers in the P2P environment. We next intend to extend our idea by linking different cities and countries on a single blockchain energy market for prosumers to trade energy.

Acknowledgement. This work was supported by the science and technology project of State Grid Corporation of China, "Research on the key technologies of Energy Internet Mobile and Interconnection Security".

References

1. Albino, V., et al.: Smart cities: definitions, dimensions, performance, and initiatives. J. Urban Technol. **22**(1), 3–21 (2015)
2. Gao, J.B., et al.: Gridmonitoring: secured sovereign blockchain based monitoring on smart grid. IEEE Access **6**, 9917–9925 (2018)
3. Gómez-Expósito, A., et al.: Electric energy systems: analysis and operation. CRC Press, United States (2017)
4. Liu, N., et al.: Energy-sharing model with price-based demand response for microgrids of peer-to-peer prosumers. IEEE Trans. Power Syst. **32**(5), 3569–3583 (2017)
5. Gao, J.B., et al.: A blockchain-SDN-enabled Internet of vehicles environment for fog computing and 5G networks. IEEE Internet Things J. **7**(5), 4278–4291 (2019)
6. Sifah, E.B., et al.: Blockchain based monitoring on trustless supply chain processes. In: Proceedings of IEEE International Conference Internet Things (iThings) IEEE Green Computer Commununication (GreenCom) IEEE Cyber Physical Social Computer (CPSCom) IEEE Smart Data (SmartData), pp. 1189–1195 (2018)
7. Xia, Q., et al.: BBDS: blockchain-based data sharing for electronic medical records in cloud environments. Information **8**(2), 44 (2017)
8. Obour, A., et al.: A secured proxy-based data sharing module in IoT environments using blockchain. Sensors **19**(5), 1235 (2019)
9. Bakken, D., et al.: Smart grids: clouds, communications, open source, and automation. CRC Press, United States (2014)
10. Biswas, S., et al.: A scalable blockchain framework for secure transactions in IoT. IEEE Internet Things J. **6**(3), 4650–4659 (2018)
11. Hugie, W., et al.: ID-based proxy re-signcryption scheme. In: 2011 IEEE International Conference on Computer Science and Automation Engineering, vol. 2, pp. 317–321 (2011)
12. Sabounchi, M., Wei, J.: Towards resilient networked microgrids: blockchain-enabled peer-to-peer electricity trading mechanis. In: 2017 IEEE Conference on Energy Internet and Energy System Integration, EI2 2017-Proceedings (2017)
13. Li, Z.T., et al.: Consortium blockchain for secure energy trading in industrial internet of things. IEEE Trans. Industr. Inf. **14**(8), 3690–3700 (2017)

Deep Reinforcement Learning Cloud-Edge-Terminal Computation Resource Allocation Mechanism for IoT

Xinjian Shu[1], Lijie Wu[1], Xiaoyang Qin[1], Runhua Yang[1], Yangyang Wu[1], Dequan Wang[2(✉)], and Boxian Liao[2]

[1] Communication & Dispatching Center, State Grid Henan Information & Telecommunication Company, Zhengzhou 450000, China
[2] State Key Laboratory of Networking and Switching Technology, Beijing University of Posts and Telecommunications, Beijing 100876, China
dequanwang@bupt.edu.cn

Abstract. With the development of Internet of Things (IoT), the types and the volume of IoT services have been increasing rapidly. Mobile edge computing (MEC) and cloud computing has recently emerged as a promising paradigm for meeting the increasing computational demands of IoT. More and more computation offloading algorithms of MEC and cloud computing have appeared. However, existing computation offloading algorithms cannot have particularly good performance in various scenarios. In this regard, we proposed a cloud-edge-terminal collaborative computation offloading algorithm based on Asynchronous Advantage Actor-Critic. It uses Asynchronous Advantage Actor-Critic to make the task choose one of the two algorithms that has better performance in their respective scenarios, and achieves the complementarity of the advantages and disadvantages of the two algorithms. Finally, the characteristics of the algorithm are investigated by simulation and compared with other algorithms to verify the algorithm's performance.

Keywords: Mobile edge computing · Cloud computing · Internet of Things · Asynchronous Advantage Actor-Critic

1 Introduction

Nowadays, Internet of Things (IoT) is playing an important role in improving the quality of urban life, involving manufacturing, transportation, construction, mining and energy. In IoT scenarios, a large amount of data is generated in equipment monitoring and maintenance, which needs to be stored, processed and analyzed [1]. The traditional processing method for data is to transfer the data to the centralized cloud [2]. However, lots of IoT services need real-time computing, such as safety assistance driving and autonomous driving. Therefore, cloud-edge-terminal collaborative computation offloading schemes have attracted more and more attention and have already been effectively applied in various IoT scenes with the help of other techniques like deep learning and reinforcement learning [3].

Q. Liu et al. (Eds.): CENet 2020, AISC 1274, pp. 1550–1556, 2021.
https://doi.org/10.1007/978-981-15-8462-6_177

A new cloud-edge-terminal offloading framework based on sequential neural learning network was constructed to automatically discover the common patterns of various applications in different scenarios, so as to infer the optimal offloading strategy of the machine learning application [4]. Methods based on reinforcement learning were applied to minimize the energy consumption by changing states, accumulating experience and changing actions without obtaining global information [5]. Some researches focus on a multi-user scenario in cloud-edge-terminal architecture. An adaptive sequential offloading scheme for multi-cell MEC was proposed in [6], in which the number of offloaded users was adjusted to avoid unexpected queueing delays. Authors in [7] used genetic algorithm and adjusted the multi-site offloading problem. Literature [8] studied the mobile edge computation offloading problem in a multiple input multiple output multi-base station system connected to a public cloud server. Thus, most of the current computing modes of IoT devices and edge servers are multithreaded and dynamically allocating resources [9].

Note that although there have been many works on computation offloading, most of these works have advantages for a single scene or are only balanced for most scenes. In this regard, we propose a cloud-edge-terminal collaborative computation offloading algorithm based on Asynchronous Advantage Actor-Critic (CETCO-A3C), which uses two algorithms with their own advantages. The main contributions of this paper are as follows:

1. In order to take full use of cloud, edge nodes and intelligent terminals, as well as reducing task computing time and energy consumption, greedy algorithm and particle swarm optimization are used for computation offloading
2. According to actions and state transition of offloading decision, we give the markov decision process. Then a cloud-edge-terminal collaborative offloading scheme based on Asynchronous Advantage Actor-Critic is designed.
3. We simulated different numbers and capabilities of IoT devices. Due to the large amount of data results, the proposed algorithm performs well.

The organizational structure of the rest of the article is as follows: we will analyze the problem and propose the algorithm in the second section, introduce simulation experiments in the third section and give the conclusion in the fourth section.

2 Proposed Algorithms

We selected two commonly used computation offloading algorithms and used a deep reinforcement learning (DRL) algorithm called A3C to make the task choose an algorithm to offload to ensure that the task's energy consumption was reduced while meeting the task's needs.

2.1 Computation Offloading Algorithm

At present, there are many algorithms in the field of computation offloading. We choose greedy algorithm and PSO. The core idea of the greedy algorithm is to greedyly choose the current optimal offloading position. The advantage is that it can get better results quickly. The disadvantage is that it does not consider other tasks, so the result is not the global optimal. PSO is a heuristic algorithm. The advantage is that it can obtain the offloading result that is approximately optimal. The disadvantage is that the calculation complexity is high so the result cannot be obtained quickly. The advantages and disadvantages of the greedy algorithm and the PSO complement each other, so that it can better meet the task needs.

2.2 Markov Decision Process

The optimization problem of computation offloading can be modeled as a Markov decision process, which includes state, action, reward, etc.

State. The result of offloading each task is whether greedy algorithm is used or PSO. Let the state be $S = \{s_1, s_2, \ldots, s_X\}$, where $s_i = \{0, 1, 2\}$. When the task is not offloaded, the state is $s_i = 0$, greedy algorithm offloading state is $s_i = 1$, and PSO offloading state is $s_i = 2$. The initial state is that all task states are 0. Because tasks are offloaded in order, when $s_i = 0$, $\{s_{i+1}, \ldots, s_X\}$ is 0.

Action. Let action be $A = \{a_1, a_2, \ldots, a_X\}$, of which $a_i = \{1, 2\}$. At $a_i = 1$, tasks perform actions that use greedy algorithm, and at $a_i = 2$, tasks perform actions that use PSO.

Reward. For each step s in the process of finding the best state, the agent performs a possible action in the state s and will get a reward r. When the task moves from state a to state s through action s', the reward of DRL is

$$R(s, s', a) = \Psi(s) - \Psi(s') \tag{1}$$

2.3 Cloud-Edge-Terminal Collaborative Offloading Scheme Based on A3C

The single thread of the A3C algorithm is shown in Algorithm 3. We define the stochastic policy in the state s as $\pi(s)$, which determines the agent's action. $\pi(s, a)$ represents the probability of choosing a under s, so its formula is

$$\pi(s, a) = P(a|s) \tag{2}$$

π is the strategy we want to learn. It is a function of s that returns the probability of all actions. In DRL, policy s has value function $V(s)$ and action value function $Q(s, a)$. The formula for $V(s)$ is

$$V(s) = E_{\pi(s)}(r + \gamma V(s')) \tag{3}$$

Where γ represents the reward attenuation factor. where $\gamma \in [0, 1]$ is a discount rate.$\gamma = 1$ is allowed only in episodic task. The formula represents the return that can be obtained by the current state s, which is the sum of the next state s' can obtain return and the reward r obtained during the state transition. The formula for $Q(s, a)$ is

$$Q(s, a) = r + \gamma V(s') \tag{4}$$

At this time, we can define a new function $AD(s, a)$, called advantage function, whose formula is

$$AD(s, a) = Q(s, a) - V(s) \tag{5}$$

It expresses how good the action s is in the state a. If the action a is better than average, then the advantage function is positive, otherwise it is negative.

Under the policy based reinforcement learning method, we approximate $\pi(s)$. At this time, $\pi(s)$ can be expressed as a function containing the parameter θ, namely $\pi_\theta(s)$. In order to get a better $\pi_\theta(s)$, we need to update it. Before optimizing $\pi_\theta(s)$, some metrics are needed to measure it. We define a function $J(\pi)$, which represents the average return of a strategy from the initial state s_0. The formula is

$$J(\pi) = E_{\rho^{s_0}}[V(s_0)] \tag{6}$$

The gradient derived from θ can be expressed as

$$\nabla_\theta J(\pi) = E_{s \sim \rho^{s_0}, a \sim \pi(s)}[AD(s, a) \cdot \nabla_\theta \log \pi(a|s)] \tag{7}$$

The proof process is in the appendix of [10]. Accumulate actor's local gradient update is

$$\begin{aligned} d\theta \leftarrow & d\theta + \nabla_{\theta'} log\pi(a_i|s_i; \theta')(R - V(s_t, \theta'_v)) \\ & + \delta \nabla_{\theta'} H(\pi(s_i; \theta')) \end{aligned} \tag{8}$$

Cumulative critic's local gradient update is

$$d\theta_v \leftarrow d\theta_v + \partial(R - V(s_t, \theta'_v))^2 / \partial\theta'_v \tag{9}$$

Algorithm 3 Asynchronous Advantage Actor-Critic

Input: Global shared parameter vectors θ and θ_v, global shared counter $T = 0$, thread-specific parameter vectors θ' and θ'_v.

Output: The policy π and the value V.

Initialize thread step counter $t \leftarrow 1$
repeat
  Reset gradients: $d\theta \leftarrow 0$ and $d\theta_v \leftarrow 0$
  Synchronize thread-specific parameters $\theta' = \theta$ and $\theta'_v = \theta_v$
  $t_{start} = t$
  Get state s_t
  repeat
    Perform a_t according to policy $\pi(a_t \mid s_t; \theta')$
    Receive reward r_t and new state s_{t+1}
    $t \leftarrow t + 1$
    $T \leftarrow T + 1$
  until terminal s_t **or** $t - t_{start} == t_{\max}$
  if s_t is a terminal state **then**
    $R = 0$
  else
    $R = V(s_t, \theta'_v)$
  end if
  for $i \in \{t_1, \ldots, t_{start}\}$ **do**
    $R \leftarrow r_i + \gamma R$
    Calculate the accumulate gradient $d\theta$ for actor network by (8)
    Calculate the accumulate gradient $d\theta_v$ for actor network by (9)
  end for
  Asynchronous updating of θ using $d\theta$ and of θ_v using $d\theta_v$
until $T \geq T_{\max}$

3 Simulation Results

In this section, we evaluate the performance of cloud-edge-terminal collaborative offloading scheme based on A3C and give some comparison results between our proposed scheme and other available methods.

Figure 1 depicts the training results of long term reward. It can be seen from the figure that with the advancement of reinforcement learning process, each round of implementation results are closer to the ideal reward value. When the number of training steps reaches about 600, the total reward starts to converge to about −17.62. In order to express more data, the reward is the value of all agents.

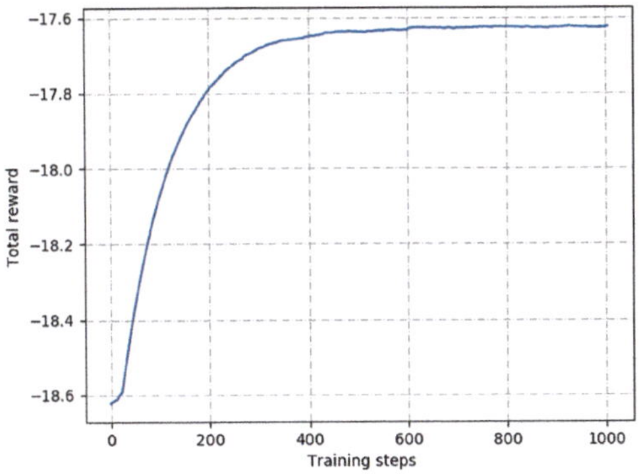

Fig. 1. Long-term reward.

To prove that CETCO-A3C's offloading decision is better, we compare it with greedy algorithm and PSO. Figure 2 is comparison of consumption of CETCO-A3C, greedy algorithm and PSO under the conditions of $\gamma = 0.9$. It can be seen from the figure that for any number of IoT devices, the overall minimum consumption of CETCO-A3C are lower than greedy algorithm and PSO, and the overall minimum consumption of PSO are lower than greedy algorithm.

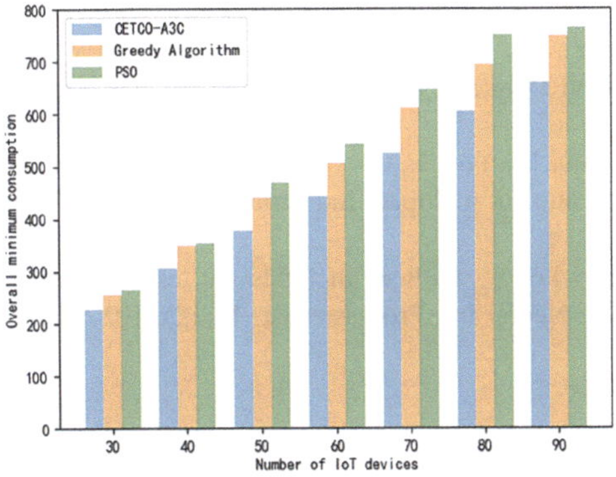

Fig. 2. Comparison of total consumption.

4 Conclusion

To better utilize the computing resource in cloud-edge-terminal collaboration network with multiple IoT tasks, IoT devices, edge nodes and cloud, CETCO-A3C is proposed. It contains three algorithms: greed algorithm, PSO and A3C. Among them, greed algorithm and PSO make choices about the offloading position of the task, and A3C chooses which offloading algorithm to use for the task. Simulation results clearly show the superiority of the proposed CETCO-A3C in saving energy consumption and reducing processing time.

Acknowledgment. This work was supported by the Science and Technology Project of State Grid Henan Electric Power Co., Ltd. (5217Q020002R).

References

1. Yang, C.W., Huang, Q.Y., Li, Z.L., et al.: Big data and cloud computing: innovation opportunities and challenges. Int. J. Digit. Earth **10**(1), 13–53 (2017)
2. Xu, J., Chen, L., Ren, S.: Online learning for offloading and autoscaling in energy harvesting mobile edge computing. IEEE Trans. Cogn. Commun. Netw. **3**(3), 361–373 (2017)
3. Du, J.B., Zhao, L.Q., Feng, J., et al.: Computation offloading and resource allocation in mixed fog/cloud computing systems with Min-Max fairness guarantee. IEEE Trans. Commun. **66**(4), 1594–1608 (2018)
4. Huang, K.S.Y., Zhang, J., Hu, J., Chen, X., Li, J.: Cost-driven offloading for DNN-based applications over cloud, edge and end devices. IEEE Trans. Ind. Inf. (2019)
5. Liu, Y., Yu, H., Xie, S., Zhang, Y.: Deep reinforcement learning for offloading and resource allocation in vehicle edge computing and networks. IEEE Trans. Veh. Technol. **68**(11), 11158–11168 (2019)
6. Deng, M., Tian, H., Lyu, X.: Adaptive sequential offloading game for multi-cell MEC. In: 2016 23rd International Conference on Telecommunications (ICT), pp. 1–5 (2016)
7. Goudarzi, M., Zamani, M., Toroghi Haghighat, A.: A genetic-based decision algorithm for multisite computation offloading in mobile cloud computing. Int. J. Commun. Syst. **30**(10), e3241 (2017)
8. Sardellitti, S., Scutari, G., Barbarossa, S.: Joint optimization of radio and computational resources for multicell mobile-edge computing. IEEE Trans. Signal Inf. Process. Netw. **1**(2), 89–103 (2015)
9. Song, F., Ai, Z., Zhou, Y., You, I., Choo, K.R., Zhang, H.: Smart collaborative automation for receive buffer control in multipath industrial networks. IEEE Trans. Industr. Inf. **16**(2), 1385–1394 (2020)
10. Sutton, R.S., et al.: Policy gradient methods for reinforcement learning with function approximation. Adv. Neural Inf. Process. Syst. **12**(12), 1057–1063 (1999)

Low-Energy Edge Computing Resource Deployment Algorithm Based on Particle Swarme

Jianliang Zhang, Wanli Ma, Yang Li(✉), Honglin Xue, Min Zhao, Chao Han, and Sheng Bi

State Grid Shanxi Information and Telecommunication Company, Taiyuan 030000, China
sxzyfx@163.com

Abstract. In the edge computing environment, in order to reduce the energy consumption of the entire network on the premise of meeting user needs, this paper proposes a low-energy edge computing resource deployment algorithm based on Particle Swarm. First, an edge computing service model based on SDN is designed, which includes three types of devices and three types of management processes. Secondly, the two requirements of user content service request and calculation service request are analyzed, and three energy consumption models of the entire network are constructed: storage energy consumption, calculation energy consumption, and transmission energy consumption. Finally, based on the main idea of the particle swarm optimization algorithm, the solution method is modeled, and a low-energy edge computing resource deployment algorithm based on particle swarm optimization is proposed. In the experimental part, the algorithm in this paper is compared with the traditional algorithm in different service request arrival rate environments. It is verified that the algorithm in this paper reduces the number of storage nodes and computing nodes and saves energy consumption on the entire network.

Keywords: Edge computing · Resource allocation · Particle swarm optimization · Energy consumption

1 Introduction

In the 5G network environment, in order to reduce the delay of user services, edge computing technology has been successfully applied to solve this problem. With the growth of user services, the demand for edge computing resources is increasing. How to reduce the energy consumption of the entire network as much as possible under the premise of meeting user needs has become an urgent problem to be solved [1, 2]. Existing researches mainly use methods such as virtual machine migration, game theory and optimization theory to solve the problems of low business execution efficiency and energy consumption reduction. For example, literature [3] propose a sustainable service plan for fog server nodes to improve the efficiency of user task execution, which reduced the energy consumption of fog nodes and the execution time of user tasks. Literature [4] In order to solve the server performance in edge computing

Q. Liu et al. (Eds.): CENet 2020, AISC 1274, pp. 1557–1564, 2021.
https://doi.org/10.1007/978-981-15-8462-6_178

cannot meet the needs of high response speed and low latency services in the manufacturing industry, based on the rapid migration of virtual machines and network traffic control technology, a business response model with high speed transmission and fast processing is presented, which better solves the problem of low performance of edge computing servers. Literature [5] takes the user task delay requirement as a constraint, combines the task calculation and communication modeling, and proposes a scheduling mechanism that minimizes the user task delay, which reduces the user task calculation and communication delay. In [6], the user task execution time limit is used as a constraint, and an optimization algorithm for unloading user task calculation requirements is proposed based on game theory, which better solves the problem of low user task execution efficiency. In [7], with the goal of reducing resource consumption, a resource allocation mechanism for cloud service providers and edge service providers is proposed. Under the condition of meeting user task execution delay constraints, the total resource overhead is reduced and verified the scalability and reliability of the distribution mechanism. Literature [8] designed the information transmission bandwidth resource allocation mechanism in the mobile edge computing network with the goal of maximizing the task transmission rate, realized the bandwidth resource allocation scheme that provides corresponding services according to the specific needs of users, and met the different needs of different users.

Through the analysis of the existing research, we can see that the current research mainly solves different problems such as delay, rate, and resource amount, and ignores the constraint relationship between various resources. Therefore, this paper designs the resource allocation scheme under the constraints of energy consumption and bandwidth with the joint goal of minimizing the energy consumption of the entire network and the occupation of network bandwidth. The contributions of this paper include: designing an SDN-based edge computing service model, designing an objective function that minimizes the energy consumption of the entire network, proposing a particle swarm-based low-energy edge computing resource deployment algorithm, and verifying that the algorithm reduces the number of storage nodes and computing nodes, and saves the energy consumption of the entire network.

2 Edge Computing Service Model Based on SDN

At present, edge computing technology based on SDN has become the main technology and development trend. In the edge computing architecture of SDN, it includes three devices: controller, repeater, and remote server. The controller adopts the primary and secondary redundancy mode to manage the entire network resources. Under the management of the controller, the repeater realizes the calculation, storage, and transmission of data across the entire network. Repeaters include three types: network nodes, storage nodes and computing nodes. Among them, the storage node provides network transmission function and data storage function, and the computing node provides network transmission function and business calculation function.

The service use process under the SDN-based edge computing architecture mainly includes three processes: control process, user use of storage resource process, and user use of computing resource process. The following are introduced separately.

(1) Control process: The controller interacts with the repeater and the remote server to register and configure the resource status; (2) User uses storage resource process: the user initiates a content access request to the closest storage node; the storage node checks to see if there is a content resource required by the user; if not, request the content resource from the remote server and store the returned content resource; the storage node returns the requested resource to the user; (3) The process of using computing resources by the user: the user makes a computing request to the closest computing node, which calculates the computing task of the user and returns the result to the user.

3 Problem Description

Through the analysis of the service usage process under the SDN-based edge computing architecture, it can be seen that in order to meet user needs, it is necessary to increase the number of storage nodes and computing nodes deployed in the network. This paper mainly studies how to minimize the energy consumption of the whole network on the premise of meeting user needs.

Under the SDN-based edge computing architecture, service types include content service requests and computing service requests. Among them, the content service request is represented by $k^a \in K_A$, and the computing service request is represented by $k^b \in K_B$. In terms of content service requests, λ_{k^a} is used to represent the number of service requests that arrive within time t, and the size of the stored content of each content service request is represented by o_{k^a}. In terms of computing service requests, λ_{k^b} is used to denote the number of service requests arriving in time t, the task execution time of each service request is expressed in o_{k^b}, and the amount of communication required during the service is expressed in c_{k^b}.

In the edge computing network environment, energy consumption includes storage energy consumption, computing energy consumption, and transmission energy consumption. The storage energy consumption is mainly consumed by storage nodes, which is denoted by $E_{k^a}^{ca}$, and the calculation method is as formula (1). Among them, p_{ca} represents the average storage energy consumption, unit $J/(bit \cdot s)$

$$E_{k^a}^{ca} = \lambda_{k^a} o_{k_a} p_{ca} \tag{1}$$

The computing energy consumption is mainly consumed by computing nodes, including two types of dynamic computing energy consumption $E_{k^b}^{active}$ and static computing energy consumption $E_{k^b}^{static}$. The calculation method for dynamic calculation of energy consumption is shown in formula (2). p_{active} represents the average power consumption of the virtual machine copy performing the computing task under the calculation dynamics, in $J/(bit \cdot s)$. λ_{k^b} represents the number of service requests arriving in time t. o_{k^b} indicates the duration of task execution for service requests. The calculation method for static calculation of energy consumption is shown in formula (3). m_{k^b} represents the number of VMs in a static environment. Among them, p_{static} represents the average power consumption of the virtual machine copy that performs the computing task at static, the unit is J/s.

$$E_{k^b}^{active} = \lambda_{k^b} o_{k^b} p_{active} \tag{2}$$

$$E_{k^b}^{static} = m_{k^b} p_{static} \tag{3}$$

Transmission energy consumption is mainly consumed by network nodes, including content transmission energy consumption and calculation transmission energy consumption. The content transmission energy consumption $E_{k^a}^{tr}$ is calculated using formula (4). o_{k^a} represents the size of the content requested by the storage service, and p_{link} represents the energy consumption parameter of the link, in J/bit. p_{node} represents the energy consumption parameter of the network node, unit J/bit. d_{k^a} represents the average number of hops from the content service request to the service node. The transmission energy consumption $E_{k^b}^{tr}$ is calculated using formula (5). c_{k^b} represents the amount of communication required during the calculation service. d_{k^b} represents the average number of hops from the calculation service request to the service node.

$$E_{k^a}^{tr} = \lambda_{k^a} o_{k^a} \left[p_{link} d_{k^a} + p_{node}(d_{k^a} + 1)\right] \tag{4}$$

$$E_{k^b}^{tr} = \lambda_{k^b} c_{k^b} \left[p_{link} d_{k^b} + p_{node}(d_{k^b} + 1)\right] \tag{5}$$

Based on the above analysis, the total energy consumption is defined as formula (6).

$$F = \left(\sum\nolimits_{k^a \in K_A} E_{k^a}^{ca} + \sum\nolimits_{k^a \in K_A} E_{k^a}^{tr}\right) + \left(\sum\nolimits_{k^b \in K_B} E_{k^b}^{active} + \sum\nolimits_{k^b \in K_B} E_{k^b}^{static} + \sum\nolimits_{k^b \in K_B} E_{k^b}^{tr}\right) \tag{6}$$

Therefore, in this paper, the objective function to minimize energy consumption is defined as formula (7). Among them, n_{k^a} represents the optimal number of storage nodes, n_{k^b} represents the optimal number of computing nodes, N represents the maximum number of storage nodes, and M represents the maximum number of computing nodes.

$$\min F(n_{k^a}, n_{k^b}) \tag{7}$$

$$\text{s.t. } 1 \le n_{k^a} \le N, 1 \le m_{k^b} \le M$$

4 Resource Deployment Algorithms

4.1 Particle Swarm Optimization Algorithm

Through the analysis of the intelligent search algorithm, we can see that the particle swarm optimization algorithm is an efficient global random search algorithm [9], which has been successfully applied to solve optimization problems [10]. Therefore, this paper uses particle swarm optimization to derive the optimal number of storage nodes n_{k^a} and the optimal number of computing nodes n_{k^b}.

The main idea of the particle swarm optimization algorithm is: describe the solution of the network resource that needs to be deployed as the position of a particle; after that, the particle moves in these solution spaces in a direction determined by the best historical position of the particle X_{pb}, and the best historical position of the neighborhood X_{gb}. Therefore, the key content includes position, speed, position update, and speed update. The details are described below.

1) Position: Use the network resource deployment plan as the particle position. Assuming that the deployment scheme of the i-th network resource includes n tasks that need to be completed, the particle position of the solution is expressed as $X_i = [x_i^1, x_i^2, \ldots, x_i^n]$. Among them, the element x_i^j represents the number of the network resource that the j-th task acquires the resource.
2) Speed: The optimization method of the network resource deployment scheme is taken as the moving speed of the particles. Assuming that the deployment scheme of the i-th network resource is $X_i = [x_i^1, x_i^2, \ldots, x_i^n]$, the particle moving speed of the solution is expressed as $V_i = [v_i^1, v_i^2, \ldots, v_i^n]$. Among them, the element v_i^j takes values of 1 and 0. When $v_i^j = 0$ indicates that the current task needs to update the acquired network resource number. When $v_i^j = 1$ means that the current task does not need to update the acquired network resource number.
3) Position update: update the position based on the position at the previous time and the speed at the current time, the calculation method is as formula (8). Formula (8) indicates that the position X_{i+1} is the product operation $\otimes$ of the position X_i at the previous time and the speed V_{i+1} at the current time. Among them, the product operation $\otimes$ refers to the resource used by the task $v_i^j = 0$ in the speed V_{i+1}, which needs to be adjusted in the position X_i.

$$X_{i+1} = X_i \otimes V_{i+1} \tag{8}$$

4) Speed update: update the speed based on the position and speed at the previous time, the calculation method is as formula (9). Formula (9) indicates that the speed V_{i+1} is the sum operation + of the three values of $P_1 V$, $P_2(X_{pb} \Theta X_i)$ and $P_3(X_{gb} \Theta X_i)$. Among them, P1, P2, and P3 are three constants, indicating the probability of whether the three values are updated, and $P_1 + P_2 + P_3 = 1$. The sum operation + refers to the addition operation of the corresponding element under the specified probability. The operator between $X_{pb} \Theta X_i$ and $X_{gb} \Theta X_i$ refers to the subtraction operation, used to compare the similarities and differences of the

positions of two particles. When the elements at the same position of the two particles are the same, the result of the subtraction operation is 1; when the elements at the same position of the two particles are different, the result of the subtraction operation is 0.

$$V_{i+1} = P_1 V_i \oplus P_2(X_{pb} \Theta X_i) \oplus P_3(X_{gb} \Theta X_i) \quad (9)$$

4.2 Resource Deployment Algorithms

Based on the above description, the low-energy edge computing resource deployment algorithm based on Particle Swarm (RDAoPS) proposed in this paper. The algorithm includes five processes: parameter initialization, calculating initial position, particle velocity and particle position update, updating global optimal initial position and individual optimal initial position, and judging whether the end condition is reached.

5 Performance

To verify the performance of the algorithm in this paper, US64 [11] is used to simulate the network topology environment. Among them, US64 is a typical topology of a commercial network, which can effectively describe the characteristics of network nodes in the commercial network. In terms of performance parameters of network nodes, set the power consumption of static computing resources to 50 W, the energy consumption of storage nodes to J/bit, and the energy consumption of link resources to J/bit. The energy consumption of the network node is set to J/bit. In terms of the number of service requests, the number of service requests arriving per second is set to 1 to 25. The maximum number of storage nodes is set to 80, and the maximum number of computing nodes is set to 20.

In order to verify the advantages and disadvantages of this algorithm in solving the deployment number of storage nodes and computing nodes, the RDAoPS algorithm in this paper is compared with the traditional RDAoRN algorithm (Resource Deployment Algorithm based on Request Number). Among them, the RDAoRN algorithm is to deploy storage nodes and computing nodes according to the characteristics of the network topology and the number of service requests. This paper verifies the number of storage nodes and computing nodes of the two algorithms under different service request arrival rate environments. The comparison results see Fig. 1 and Fig. 2.

In Fig. 1, the X axis represents the increase in service request arrival rate from 5 to 30 per second, and the Y axis represents the number of storage nodes. As can be seen from the figure, as the service request arrival rate increases, the number of storage nodes increases under both algorithms. This is because service demand increases, which need more storage nodes to be deployed to meet the service demand. From the comparison of the results of the two algorithms, it can be seen that when the service request arrival rate of this algorithm reaches 20 per second, the number of deployed storage nodes tends to be stable. This shows that the storage node location deployed in this article is more reasonable.

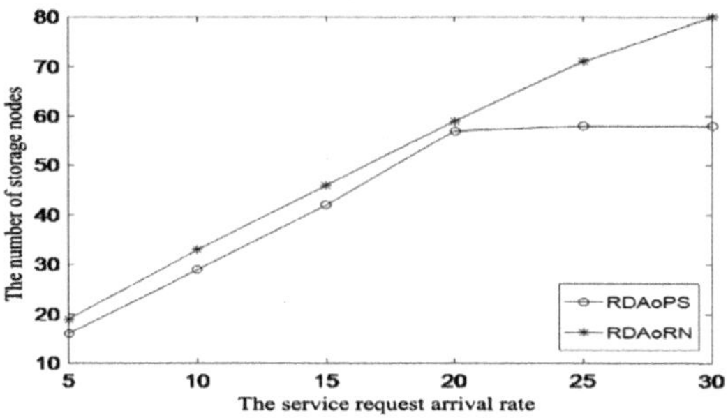

Fig. 1. Comparison of the number of storage nodes

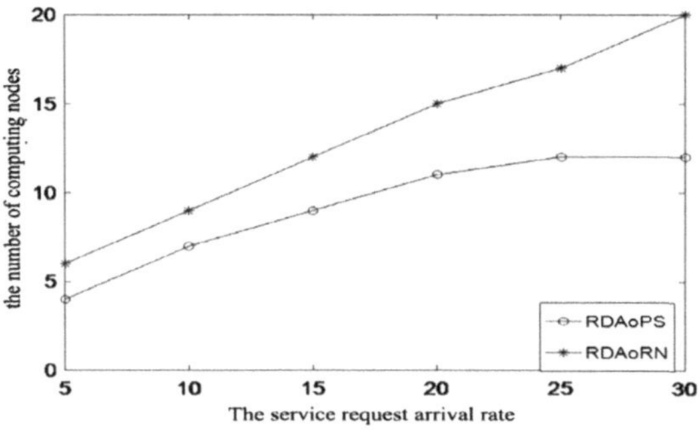

Fig. 2. Comparison of number of computing nodes.

In Fig. 2, the X axis also represents the service request arrival rate, and the Y axis represents the number of computing nodes. As can be seen from the graph, as the service request arrival rate increases, the number of computing nodes increases under both algorithms. This is because service demand increases, which need more computing nodes to be deployed to meet the service demand. From the comparison of the results of the two algorithms, it can be seen that when the service request arrival rate of this algorithm reaches 25 per second, the number of deployed computing nodes tends to be stable. This shows that the location of the computing node deployed in this article is more reasonable.

6 Conclusion

The edge computing technology has been successfully applied to the 5G network environment, which improves the execution efficiency of user services. Because edge computing resources are affected by factors such as mobility, power and environment, how to reduce the energy consumption of edge computing devices has become an urgent problem to be solved. To solve this problem, this paper designs an edge computing service model based on SDN, proposes a low-energy edge computing resource deployment algorithm based on particle swarm, and verifies through experiments that this algorithm reduces the number of storage nodes and computing nodes, and saves the entire network energy consumption. In the next step, based on the research in this paper, we will classify the edge network resources and user service requests to meet the needs of business diversity in the 5G network environment, and further improve the practicality and usability of the algorithm in this paper.

Acknowledgments. This work is supported by the Technical Project of State Grid Shanxi Power Grid (52051C19006R).

References

1. Tong, L., Li, Y., Gao, W.: A hierarchical edge cloud architecture for mobile computing. In: IEEE INFOCOM 2016-The 35th Annual IEEE International Conference on Computer Communications, pp. 1–9. IEEE (2016)
2. Ahmad, A., Paul, A., Khan, M., et al.: Energy efficient hierarchical resource management for mobile cloud computing. IEEE Trans. Sustain. Comput. **2**(2), 100–112 (2017)
3. Mishra, S.K., Puthal, D., Rodrigues, J.J.P.C., et al.: Sustainable service alposition using a metaheuristic technique in a fog server for industrial applications. IEEE Trans. Industr. Inf. **14**(10), 4497–4506 (2018)
4. Rodrigues, T.G., Suto, K., Nishiyama, H., et al.: Hybrid method for minimizing service delay in edge cloud computing through VM migration and transmission power control. IEEE Trans. Comput. **66**(5), 810–819 (2017)
5. Meng, X., Wang, W., Zhang, Z.: Delay-constrained hybrid computation offloading with cloud and fog computing. IEEE Access **5**, 21355–21367 (2017)
6. Chen, X.: Decentralized computation offloading game for mobile cloud computing. IEEE Trans. Parallel Distrib. Syst. **26**(4), 974–983 (2015)
7. Xu, J., Palanisamy, B., Ludwig, H., et al.: Zenith: utility-aware resource alposition for edge computing. In: IEEE International Conference on Edge Computing. IEEE Computer Society (2017)
8. Ito, Y., Koga, H., Iida, K.: A bandwidth alposition scheme to meet flow requirements in mobile edge computing. In: IEEE International Conference on Cloud Networking. IEEE (2017)
9. Kennedy, J., Eberhart, R.: Particle swarm optimization. In: Proceedings of ICNN'95-International Conference on Neural Networks, vol. 4, pp. 1942–1948. IEEE (1995)
10. Holland, J.: Adaptation in Natural and Artificial Systems. MIT Press, Cambridge (1992)
11. Choi, N., Guan, K., Kilper, D.C., et al.: In-network caching effect on optimal energy consumption in content-centric networking. In: 2012 IEEE International Conference on Communications (ICC), pp. 2889–2894. IEEE (2012)

Efficient Fog Node Resource Allocation Algorithm Based on Taboo Genetic Algorithm

Yang Li(✉), Wanli Ma, Jianliang Zhang, Jian Wu, Junwei Ma, and Xiaoyan Dang

State Grid Shanxi Information and Telecommunication Company, Taiyuan 030000, China
sxzyfx@163.com

Abstract. In the fog computing network, in order to improve user task execution efficiency under resource constraints, this paper proposes an efficient fog node resource allocation algorithm based on Taboo genetic algorithm. First, the resource constraint problem is modeled as a communication resource constraint and a computing resource constraint problem, and the performance constraint problem is modeled as a task execution delay constraint problem. Secondly, the advantages of taboo genetic algorithm are analyzed, and key processes such as chromosome coding, fitness function, selection process, taboo crossing process, and taboo mutation process are designed. Finally, an efficient fog node resource allocation algorithm based on Taboo genetic algorithm is proposed. In the experimental part, from the two aspects of the number of different tasks and the number of different fog nodes, it is verified that the algorithm in this paper significantly reduces the execution time of user tasks.

Keywords: Fog computing · Fog nodes · Resource allocation · Taboo genetic algorithm · Chromosome

1 Introduction

The fog wireless access network architecture is a key technology for 5G networks to achieve large-scale applications, effectively overcoming the problem of long computing delays in cloud computing technology [1]. In the fog wireless access network architecture, fog nodes have computing and communication resources and can provide users with fast computing and communication services. However, in the wireless environment, the resource capacity of the fog node is limited, and the stability of the fog node itself is also affected by factors such as the wireless environment and power supply. Therefore, how to allocate the resources of fog nodes has become a research focus. At present, there are studies that mainly use intelligent algorithms and multi-objective optimization methods to solve problems such as resource utilization and task execution efficiency. For example, literature [2] aims at minimizing the cost of public transportation networks, and proposes a resource balancing algorithm between micro cloud servers and fog node servers based on genetic algorithms to achieve an increase in the utilization rate of public transportation network resources. Literature [3] In order to improve the execution speed of tasks in the medical big data network, a distributed

Q. Liu et al. (Eds.): CENet 2020, AISC 1274, pp. 1565–1573, 2021.
https://doi.org/10.1007/978-981-15-8462-6_179

fog node coordinated resource allocation algorithm based on particle swarm algorithm is proposed, which realizes the distributed processing of user tasks and significantly improves the execution efficiency of the algorithm. In literature [4], takes user task execution efficiency and fog resource cost as joint optimization goals, and proposes an efficient task scheduling algorithm based on the heuristic algorithm, which better solves the problems of slow task execution and high fog node resource consumption. Literature [5] proposed an approximate optimization algorithm with green energy consumption as the goal, which achieved a balanced optimization of resource allocation and carbon emissions, and reduced the negative impact of server operation on the environment. In literature [6], with the goal of minimizing the cost of fog nodes, an optimization model for the deployment location and number of fog nodes is proposed, which effectively reduces the resource overhead of fog nodes. Literature [7] aims at minimizing the completion time of user tasks, modeling and solving the problem of fog node resource allocation as a mixed integer nonlinear programming problem, which effectively improves the execution speed of user tasks.

Through the analysis of the existing research, it can be seen that the fog node resource allocation has achieved more research results in terms of time constraints, energy consumption constraints and other aspects. However, the communication resources of fog nodes also play a key role in the efficiency of user task execution. Therefore, when allocating fog node resources, it is necessary to consider how to improve the efficiency of communication resources. How to efficiently allocate the resources of fog nodes to users under the constraints of communication resources still needs further research. In order to solve this problem, this paper models the problem of fog node resource allocation as the problem of minimizing the service delay of users designs the key processes such as chromosome coding, fitness function, selection process, taboo crossing process, and taboo mutation process, and proposes an efficient fog node resource allocation algorithm based on taboo genetic algorithm. Experiments verify that the algorithm in this paper significantly reduces the execution time of user tasks.

2 Problem Description

Fog wireless access network architecture, including m fog nodes $Fog = \{Fog_1, Fog_2, \ldots, Fog_i, \ldots, Fog_m\}$ and n users $T = \{T_1, T_2, \ldots, T_j, \ldots, T_n\}$. Each fog node Fog_i has a communication resource G_i and a computing resource H_i. Each user T_j has multiple tasks, which need to apply for the communication resources D_j j and computing resources O_j of the fog node. Each fog node can execute tasks of multiple users at the same time. However, considering the independence of task execution, each task can only be executed on one fog node. Use l_{ij} to represent the relationship between the fog node and the user task. When $l_{ij} = 1$, it means that part of the resources of the fog node R_i are allocated to the task T_j. When $l_{ij} = 0$, the fog node Fog_i has not allocated any resources to the task T_j. In resource allocation, resource constraints and performance constraints need to be satisfied. It will be described separately below.

In terms of resource constraints, including formula (1), formula (2) two constraints Formula (1) indicates that the sum of the communication resources D_j allocated to all tasks T_j by each fog node Fog_i cannot be greater than the amount of communication resources G_i it has; formula (2) indicates that the sum of calculation resources O_j allocated to all tasks T_j by each fog node Fog_i cannot be greater than the amount of calculation resources H_i it has.

$$\sum\nolimits_{j=1}^{n} l_{ij} D_j \leq G_i \tag{1}$$

$$\sum\nolimits_{j=1}^{n} l_{ij} O_j \leq H_i \tag{2}$$

In terms of performance constraints, including a constraint of formula (3). Formula (3) indicates that the execution delay of each task is less than or equal to its delay threshold T_{T_j}, that is, each task can be completed within the required time. Among them, $t_{T_j} = \max_{i=1}^{m} t_i$ means that the execution time of all tasks for each user is determined by the task with the longest execution time among all fog nodes. Therefore, the delay of each fog node Fog_i executing its task is calculated by formula (4).

$$t_{T_j} \leq T_{T_j} \tag{3}$$

$$t_i = \sum\nolimits_{j=1}^{n} l_{ij} \cdot t_{ij} = \sum\nolimits_{j=1}^{n} l_{ij} \cdot \left(\frac{D_j}{G_i} + \frac{O_j}{H_i}\right), i \in \{1, 2, \ldots, m\}, j \in \{1, 2, \ldots, n\} \tag{4}$$

Based on the above analysis, the objective function of the fog node's resource allocation problem is to minimize the service delay of n users, as shown in formula (5), and includes two delays, resource constraints and performance constraints.

$$F = \min \sum\nolimits_{j=1}^{n} t_{T_j} \tag{5}$$

$$\text{s.t. } \sum\nolimits_{j=1}^{n} l_{ij} D_j \leq G_i, \sum\nolimits_{j=1}^{n} l_{ij} O_j \leq H_i, t_{T_j} \leq T_{T_j}$$

3 Taboo Genetic Algorithm

Taboo genetic algorithm is a multi-objective optimization request algorithm combining genetic algorithm and taboo search algorithm [8]. Compared with genetic algorithm, taboo genetic algorithm improves the crossover process and mutation process to make it have a memory function, which can effectively prevent the premature phenomenon of traditional genetic algorithm and obtain a better optimal solution.

The taboo genetic algorithm includes six processes: chromosome coding, initial population generation, fitness function construction, selection process, taboo crossover process and taboo mutation process. Among them, the process of initial population generation uses a random generation method to generate N individuals, and the fitness function is calculated by the objective function in formula (5). The four processes of chromosome coding, selection process, taboo crossing process, and taboo mutation process are described in detail below.

3.1 Chromosome Coding

Assuming that n users contain p tasks in total, which are represented by the set $G\{G^j|j \in R, R = [1,p\}$. A chromosome is used to represent the allocation plan of the p tasks, and the length of the chromosome is $\sum_{a\in R}|N^a|$. The value of the $\sum_{a\in[1,j-1]}|N^a| + i$ element of the chromosome represents the number $R_j^i \in \Omega(T_j^i)$ of the fog node that allocates resources to the task T_j^i. Among them, the task T_j^i represents the i-th task of the j-th user. $\Omega(T_j^i)$ represents the set of fog nodes that meet the requirements of task T_j^i. Based on this, $[\sum_{c\in[1,a-1]}|N^c| + 1, \sum_{b\in[1,a]}|N^b|]$ represents the fog node set that allocates resources for all tasks of the a-th user.

3.2 Selection Process

In order to ensure that individuals with better fitness values can be selected, so as to ensure the optimization of each solution, when calculating of selected individuals, this paper adopts the strategy of combining no-remaining method and roulette selection method. First, when calculating the survival quantity of each individual, the calculation method of the survival quantity N_i of the $i(1 \leq i \leq N)$-th individual is shown in formula (6). Second, the quantity of $N - \sum_{j=1}^{N}[N_j]$ individuals generated is calculated by the roulette selection method.

$$[N_i] = \frac{N \cdot F_i}{\sum_{j=1}^{N} F_j} \tag{6}$$

3.3 Taboo Crossover Process

In order to avoid the loop problem of local solutions in the search for the optimal solution, a taboo table is constructed, which contains fitness values of L chromosomes. The average fitness of the parents of the population can be calculated using the desirability level of the taboo table. Among them, the desire level value is the target value of the best solution obtained in the last iteration calculation. The taboo crossover process in this paper includes three processes: cross generating new chromosomes, calculating the fitness value of new chromosomes, and updating the taboo table.

In terms of cross-generating new chromosomes, two-point cross strategy is used to mutate the two chromosomes. That is, two intersections are randomly generated, and the sequence values in the intersections of the two parents are interchanged, thereby generating two new chromosomes. When a new chromosome is generated, it is necessary to determine whether there is the same fog node number in the same user task within each chromosome. If there is, the duplicate fog node number needs to be updated. This is because multiple tasks of the same user cannot be allocated resources by the same fog node. If duplication occurs, this article uses the random replacement method to select a non-duplicate fog node from the fog node set for replacement.

In calculating the fitness value of the new chromosome, the fitness function is used to calculate the fitness value of the new chromosome. Compare the fitness value of the new chromosome with the taboo table and its desirability level. When the fitness value of the new chromosome is greater than or equal to the desirability level of the taboo table, accept the new chromosome. Otherwise, check whether the new chromosome belongs to the taboo table set. If not, accept the new chromosome. If yes, discard the new chromosome and select the chromosome with the highest unused fitness value from the parent chromosome to replace the current chromosome.

In the term of updating the taboo table, we use the new chromosome to update the chromosome in the taboo table, form a new chromosome with length of L, and calculate the desirability level of the new taboo table.

3.4 Taboo Mutation Process

The taboo mutation process includes three processes of generating new chromosomes by mutation, calculating the fitness value of new chromosomes, and updating the taboo table. Among them, the process of calculating the fitness value of the new chromosome and updating the taboo table is the same as the taboo crossover process, and will not be described again. The process of generating new chromosomes by mutation adopts the method of randomly selecting a single task node in the chromosome for mutation. When mutating a single task node, you need to ensure that the mutated value is still in the $\Omega(n_j^i)$ set. If two tasks that do not belong to the $\Omega(n_j^i)$ set or the same user are allocated resources by the same fog node, the mutated value needs to be updated until the two tasks that belong to the M set or the same user are not allocated resources by the same fog node.

4 Resource Allocation Algorithm

The Efficient Fog Node Resource Allocation Algorithm Based on Taboo Genetic Algorithm (RAo TG) proposed in this paper is shown in Table 1. The algorithm includes ten processes: parameter initialization, calculating fitness values, judging whether the number of iterations meets the threshold, constructing a taboo table and *desirability* level, performing a selection process, performing a taboo crossover process, performing a taboo mutation process, calculating the fitness value of a new chromosome, updating parameters and resource allocation. Among them, the fitness value F_i of the chromosome is calculated by formula (5).

Table 1. The efficient fog node resource allocation algorithm based on taboo genetic algorithm.

Input: Fog node set $Fog = \{Fog_1, Fog_2, ..., Fog_i, ..., Fog_m\}$, task set $G\{G^j \mid j \in R, R = [1, p\}$,
Output: Optimal fog node resource allocation subset R'
1. parameter initialization: Initialized parameters include crossover probability P_c, mutation probability P_m, genetic algebra *T*, population size N, $count = 0$.
2. calculating fitness values: Use formula (5) to calculate the fitness value of the chromosome F_i.
3. judging whether the number of iterations meets the threshold: if ($count > T$) {Go to step 10.}
4. constructing a taboo table and *desirability* level: Construct a taboo table of length L and use the optimal fitness value of all chromosomes as the level of *desirability*.
5. performing a selection process: Chromosomes are selected using the strategy of no-remaining method and roulette selection method.
6. performing a taboo crossover process: Use the crossover probability P_c to perform the taboo crossover process.
7. performing a taboo mutation process: Use mutation probability P_m to perform the taboo mutation process.
8. calculating the fitness value of a new chromosome: Use formula (5) to calculate the fitness value F_i of the new chromosome.
9. updating parameters: $count = count + 1$, go to step 3.
10. resource allocation: Based on the chromosomal strategy with the largest fitness value F, fog nodes are used to allocate resources to tasks, and a resource allocation set R′ is generated.

5 Performance Analysis

5.1 Environment

In order to verify the performance of the algorithm in this paper, iFogSim [9] tool was used in the experiment to simulate the fog wireless access network environment. In terms of simulation algorithm execution efficiency, the instruction number index is used for simulation. The completion time of user tasks is subject to the uniform distribution of [1000 ms, 1500 ms]. The number of tasks of the user is subject to a uniform distribution of [10, 400], and the number of instructions in the task is subject to a uniform distribution $[2 * 10^9, 10 * 10^9]$. The number of fog nodes is subject to a uniform distribution of [50, 300], and the speed of each fog node's execution of task instructions per second is subject to a uniform distribution of $[1 * 10^9, 1.5 * 10^9]$.

In terms of taboo genetic algorithm parameter setting, cross probability $P_c = 0.7$, mutation probability $P_m = 0.1$, genetic algebra $T = 30$, population size $N = 30$, taboo table length L. Considering that the algorithm RAoTI (Resource Allocation algorithm based on Task scheduling and Image placement) [7] solves the problem similar to the algorithm in this paper, and has achieved good results in the time performance of task allocation, so in the experiment, the algorithm of this paper RAoTG compared with the algorithm RAoTI.

5.2 Algorithm Comparison

The experiment includes two aspects: the comparison of execution time under different numbers of tasks and the comparison of execution time under different numbers of fog nodes.

First of all, in terms of the execution time comparison under different task numbers, when the number of fog nodes is 100, the execution time comparison under different task numbers see Fig. 1.

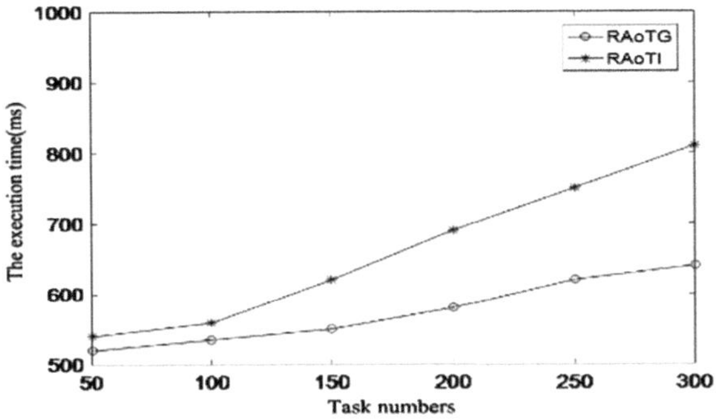

Fig. 1. The execution time comparison under different task numbers.

The X axis represents the number of tasks, increasing from 50 to 300, with a growth step of 50. The Y axis represents the task execution time. It can be seen from the figure that as the number of tasks increases, the execution time of the two algorithms increases. This is because as the number of tasks increases, more and more tasks cannot obtain resources, resulting in more tasks waiting for resources. From the performance analysis of the two algorithms, it can be seen that when the algorithm of this paper performs resource allocation, the execution time of the task increases relatively slowly, indicating that the algorithm of this paper allocates more optimized resources for user tasks.

Secondly, in terms of the comparison of the execution time under different numbers of fog nodes, when the number of tasks is 200, the comparison of execution time under different numbers of fog nodes see Fig. 2.

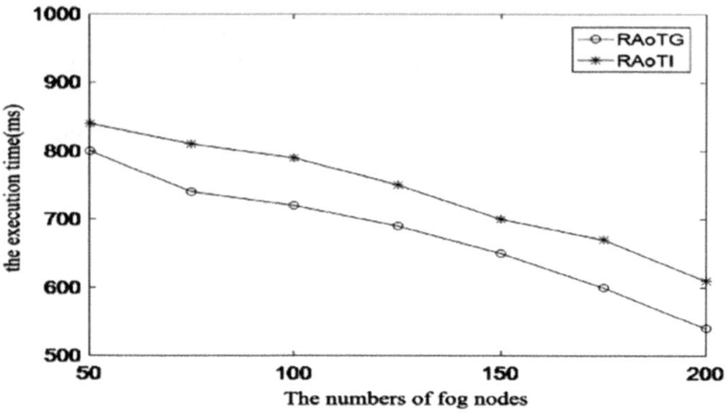

Fig. 2. The comparison of the execution time under different numbers of fog nodes.

The Y axis represents the task execution time. It can be seen from the graph, as the number of fog nodes increases, the execution time of tasks under both algorithms is declining. It shows that as the number of fog nodes increases, the number of tasks executed on each fog node decreases and the task execution time decreases. From the performance analysis of the two algorithms, it can be seen that under the algorithm of this paper, the task execution time decreases faster, which shows that the algorithm of this paper allocates more optimized resources for user tasks than the algorithm RAoTI, thus significantly reducing the execution time of user tasks.

6 Conclusion

The foggy wireless access network architecture overcomes the problem of large computing delays in cloud computing technology, and is a key technology for 5G networks to achieve large-scale applications. In order to improve the execution efficiency of user tasks in fog wireless access network environment under resource constraints, this paper models the problem of fog node resource allocation as minimizing the service delay of n users. The advantages of taboo genetic algorithm are analyzed, and the problem in this paper is modeled as the key process of taboo genetic algorithm.

An efficient fog node resource allocation algorithm based on taboo genetic algorithm is proposed. Experimental results show that the algorithm in this paper significantly reduces the execution time of user tasks. In the next step, on the basis of the research in this paper, we will increase the reliability constraints of resources to ensure the stability of user task execution.

Acknowledgments. This work is supported by the Technical Project of State Grid Shanxi Power Grid (52051C19006R).

References

1. Du, J., Zhao, L., Feng, J., et al.: Computation offloading and resource allocation in mixed fog/cloud computing systems with min-max fairness guarantee. IEEE Trans. Commun. **66**(4), 1594–1608 (2018)
2. Ye, D., Wu, M., Tang, S., et al.: Scalable fog computing with service offloading in bus networks. In: 2016 IEEE 3rd International Conference on Cyber Security and Cloud Computing (CSCloud), pp. 247–251. IEEE (2016)
3. He, X.L., Ren, Z.Y., Shi, C.H., et al.: A cloud and fog network architecture for medical big data and its distributed computing scheme. J. Xi'an Jiaotong Univ. **50**(10), 71–77 (2016)
4. Pham, X.Q., Huh, E.N.: Towards task scheduling in a cloud-fog computing system. In: 2016 18th Asia-Pacific Network Operations and Management Symposium (APNOMS), pp. 1–4. IEEE (2016)
5. Do, C.T., Tran, N.H., Pham, C., et al.: A proximal algorithm for joint resource allocation and minimizing carbon footprint in geo-distributed fog computing. In: 2015 International Conference on Information Networking (ICOIN), pp. 324–329. IEEE (2015)
6. Zhang, W., Lin, B., Yin, Q., et al.: Infrastructure deployment and optimization of fog network based on microdc and lrpon integration. Peer-to-Peer Netw. Appl. **10**(3), 579–591 (2017)
7. Zeng, D., Gu, L., Guo, S., et al.: Joint optimization of task scheduling and image placement in fog computing supported software-defined embedded system. IEEE Trans. Comput. **65**(12), 3702–3712 (2016)
8. Glover, F., Kelly, J.P., Laguna, M.: Genetic algorithms and tabu search: hybrids for optimization. Comput. Oper. Res. **22**(1), 111–134 (1995)
9. Gupta, H., Vahid Dastjerdi, A., Ghosh, S.K., et al.: iFogSim: a toolkit for modeling and simulation of resource management techniques in the internet of things, edge and fog computing environments. Softw.: Pract. Exp. **47**(9), 1275–1296 (2017)

Network Reliability Optimization Algorithm Based on Service Priority and Load Balancing in Wireless Sensor Network

Pengcheng Ni[1(✉)], Zhihao Li[1], Yunzhi Yang[1], Jiaxuan Fei[2], Can Cao[1], and Zhiyuan Ye[1]

[1] Anhui Jiyuan Software Co. Ltd., SGITG, Hefei 230088, China
18326134067@163.com

[2] Global Energy Interconnection Research Institute Co. Ltd., Nanjing 210000, China

Abstract. The rapid growth of the wireless sensor network scale and the number of services presents new challenges to the reliability of the network. To solve this problem, this paper proposes a network reliability optimization algorithm based on service priority and load balancing. The algorithm includes three steps: selecting the virtual network to be migrated, migrating the virtual network, and evaluating whether the life cycle of all physical nodes meets the threshold. The strategy of judging whether the life cycle of all physical nodes meets the threshold is adopted to effectively prevent the problem that some physical node resources are overused due to migration. In the simulation experiment, the algorithm in this paper is compared with the traditional algorithm, and it is verified that the algorithm in this paper has achieved good results in terms of effective node indicators.

Keywords: Wireless sensor network · Sensor · Network reliability · Quantum communication

1 Introduction

Wireless sensor networks have been applied to various fields such as industry, medical treatment, transportation, and environment [1]. Because virtualization technology can effectively improve the utilization rate of network resources and the reliability of network services, wireless virtualization sensor network technology based on virtualization has been proposed and has been applied in a large range [2]. Regarding the problem of sensor network resource allocation and network reliability optimization, existing research can be divided into two types: traditional network environment and network virtualization environment. In the traditional network environment, the two strategies of routing protocol optimization and rapid fault recovery are mainly adopted [3–5]. The network virtualization environment includes two types of resource allocation policy: optimal resource allocation to improve utilization rate and resource allocation considering survivability [6–8].

Q. Liu et al. (Eds.): CENet 2020, AISC 1274, pp. 1574–1580, 2021.
https://doi.org/10.1007/978-981-15-8462-6_180

When allocating resources, it is easy to waste resources by adopting reserved resources. In addition, due to the imbalance of resource allocation, it is easy to cause some basic nodes to exhaust their energy due to the long-term bearing of the virtual network. Therefore, the virtual resources carried on the basic nodes need to be balanced to ensure the stable operation of the basic nodes. To solve this problem, based on the analysis of business priority and load balancing theory, this paper optimizes the network reliability, thereby improving the quality of service.

2 Problem Description

The physical network topology is represented by an undirected graph $G(N, E)$. Among them, N represents the set of physical network nodes, and E represents the set of physical network links. For network node $n_i \in N$, it includes three attributes: CPU computing power $C(n_i)$, node position $P(x_i, y_i)$, and remaining energy E_i^{left}. The attribute of physical link $l_{ij} \in L$ is bandwidth $B(l_{ij})$. The virtual network topology is represented by an undirected graph $G^v(N^v, E^v)$. Among them, N^v represents the set of virtual network nodes, and E^v represents the set of virtual network links. The attribute included in the virtual network node $n_i^v \in N^v$ is CPU computing power $C_{re}(n_i^v)$, and the attribute included in the virtual network link $l_{ij}^v \in L^v$ is bandwidth $B_{re}(l_{ij}^v)$. Using $\{G_1^v, G_2^v, \ldots, G_M^v\}$ to represent M virtual network requests, the delay limits are $\{Del_1^{re}, Del_2^{re}, \ldots, Del_K^{re}\}$ and $Del_1^{re} < Del_2^{re} < \ldots < Del_k^{re}$. Use $C^{init}(n_i)$ for initial CPU computing power, $C^{used}(n_i)$ for used CPU computing power, $B^{init}(l_{ij})$ for initial bandwidth, and $B^{left}(l_{ij})$ for remaining bandwidth.

3 Modeling

3.1 Life Cycle of Physical Network Nodes

This paper uses formula (1) to calculate the remaining life cycle T_{life}^i of the physical network node $n_i \in N$. Among them, E_{cost}^i represents the total energy consumption of network nodes, and E_{left}^i represents the remaining energy of network nodes.

$$T_{life}^i = E_{left}^i / E_{cost}^i \tag{1}$$

The energy consumption E_{send} for sending k-bit data is calculated using formula (2), and the energy consumption $E_{receive}$ for receiving k-bit data is calculated using formula (3). *pre* represents the indicator vector. When $d < d_0$, $pre = 2$. When $d \geq d_0$, $pre = 4$, $d_0 = \sqrt{\varepsilon_{fs}/\varepsilon}$. ε_{fs} represents the energy attenuation coefficient, and ε represents the multipath attenuation coefficient. d_{ij}^{pre} represents the Euclidean distance between physical node $n_i \in N$ and its next hop $n_i \in N$, which is calculated using formula (4). E_0 represents the RF energy consumption coefficient.

$$E_{send} = k(E_0 + \varepsilon d_{ij}^{pre}) \tag{2}$$

$$E_{receive} = kE_0 \tag{3}$$

$$d_{ij} = P_i - P_j = \sqrt{(x_i - x_j)^2 + (y_i - y_j)^2} \tag{4}$$

Based on the above analysis, the calculation process of E_{cost}^{i} can be obtained as shown in formula (5).

$$E_{cost}^{i} = E_{send} + E_{receive} = k(E_0 + \varepsilon d_{ij}^{indic}) + kE_0 = k * (2E_0 + \varepsilon d_{ij}^{indic}) \tag{5}$$

3.2 Physical Network Load Balancing

Use η_{reload} to indicate the load balancing rate of the physical network, and calculate it using formula (6). Among them, $E_{left}^{\max}$ and $E_{left}^{\min}$ represent the maximum and minimum values of the remaining energy in all physical nodes, respectively. It can be seen from formula (6) that the load balancing rate η_{reload} of the physical network is a number greater than 1, and the closer the value is to 1, the more balanced the remaining energy of all physical network nodes. Use η_{reload}^{last} to indicate the load balancing rate of the physical network in the last resource allocation. When $\eta_{reload} < \eta_{reload}^{last}$, it means that the current resource allocation strategy can better achieve load balancing of network resources.

$$\eta_{reload} = E_{left}^{\max} / E_{left}^{\min} \tag{6}$$

3.3 Virtual Network Delay Limit

In order to associate the delay limit of the virtual network with the attributes of the physical network, this paper converts the delay limit T_x into the hop number H_x of the physical network, and calculates it using formula (7). Among them, T_x^{ave} represents the average value of the data processing duration and transmission duration of each link.

$$H_x \leq \frac{Del_k^{re}}{T_x^{ave}} \tag{7}$$

4 Algorithm

Network reliability optimization algorithm based on service priority and load balancing in wireless sensor network (NROA-SPoLB) is shown in Table 1. The algorithm includes three steps: selecting the virtual network to be migrated, migrating the virtual network, and evaluating whether the life cycle of all physical nodes meets the

Table 1. Algorithm NROA-SPoLB.

Input: physical network topology $G(N,E)$, virtual network $\{G_1^v, G_2^v, ..., G_M^v\}$

Output: optimized physical network topology $G^{'}(N^{'}, E^{'})$

1. Select the virtual network to be migrated

(1) For each physical network node $n_i \in N$, use formula (1) to calculate its life cycle T_{life}^i.

(2) Put physical nodes with T_{life}^i less than the life cycle threshold φ_{life}^i into set Θ.

(3) Put the services on each physical node in set Θ into the set Ω of the virtual network to be migrated.

(4) Use formula (6) to calculate the load balancing coefficient of the physical network and assign it to η_{reload}^{last}.

2. Migrate the virtual network

(1) The virtual network G_x^v in the set Ω is used to calculate the delay requirement H_x of the virtual network using formula (7).

(2) Sort in ascending order according to H_x to get set Ω_{ord}.

(3) For each virtual network in set Ω_{ord}, use the following process to migrate.

(a) For the current virtual network G_x^v, the Dijkstra algorithm is used to solve the shortest path, and the path hop number H_x^D is calculated.

(b) Determine whether H_x^D is less than H_x. If it is not satisfied, the migration fails and the migration is not carried out, go to step (a).

(c) Calculate the load balancing coefficient η_{reload} and determine whether $\eta_{reload} < \eta_{reload}^{last}$ is satisfied. If satisfied, jump to step (e).

(d) Select the node with the next shortest path for mapping, and return to step (b).

(e) Map the delay-sensitive virtual network to the current path, and return to step (a).

3. Evaluate whether the life cycle of all physical nodes meets the threshold

(1) Determine whether the optimization threshold T is exceeded, if it has been exceeded, end.

(2) For each physical node $n_i \in N$, evaluate whether the life cycle meets the threshold φ_{life}^i. If they meet, algorithm ends. If not satisfied, put node into the set Ω_r to be optimized.

(3) If set Ω_r is not empty, return to step 2.

threshold. In step 1, by selecting the physical node to be optimized, all virtual services of the physical node that do not meet the life cycle are migrated. In step 2, for a virtual service that needs to be migrated, the shorter its delay, the more it needs to be migrated first, thereby obtaining a shorter link and meeting the delay requirement. In step 3, threshold T of optimization times is set to limit the number of times the algorithm is executed to prevent over-calculation.

5 Performance Analysis

5.1 Simulation Environment

In order to verify the performance of the algorithm, the simulation language is written in java language in the experiment. The CPU resources of each physical network node are subject to an even distribution of (40, 50). The bandwidth resources of each physical network link are subject to an even distribution of (20, 30). For each virtual network, the number of virtual network nodes follows the uniform distribution of (2, 4), and the CPU resources of each virtual network node follow the uniform distribution of (1, 5). The bandwidth resources of each virtual network link follow the uniform distribution of (1, 3). Considering that the maximizing recovery of available nodes algorithm (MRANA) is a typical network reliability optimization algorithm, this paper compares the algorithm NROA-SPoLB with the algorithm MRANA. Among them, the MRANA algorithm selects all unavailable nodes and uses the shortest path method to remap the virtual network on it.

5.2 Algorithm Comparison

When the number of physical network nodes is 200, the comparison result of the number of virtual networks on the effective nodes is shown in Fig. 1.

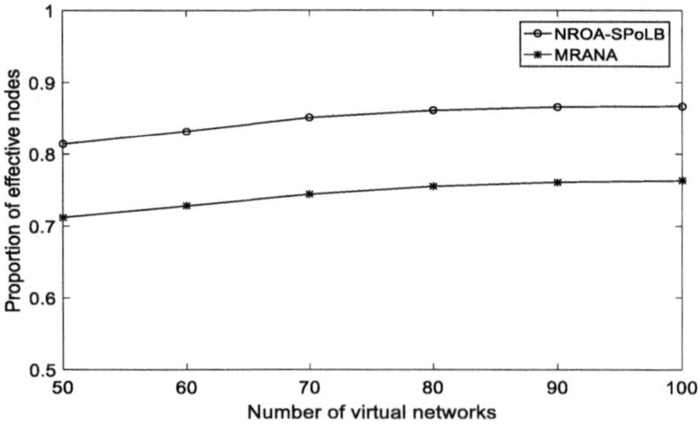

Fig. 1. The effect of the number of virtual networks on the proportion of effective nodes.

As can be seen from the figure, as the number of virtual networks increases, the proportion of effective nodes under both algorithms increases, indicating that the number of virtual networks increases, and both algorithms can optimize more physical network nodes. Comparing the two algorithms, the network optimization ability of the algorithm in this paper is stronger, which shows that the algorithm of this paper can better optimize the network resources.

When the number of virtual networks is 70, the comparison result of the effect of physical network size on effective nodes is shown in Fig. 2.

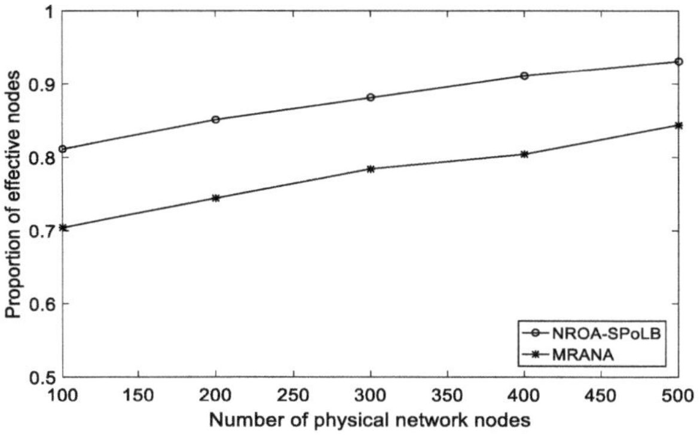

Fig. 2. The effect of physical network size on the proportion of effective nodes.

As can be seen from the figure, as the number of physical network nodes increases, the proportion of effective nodes under both algorithms increases, indicating that the number of physical network nodes increases, and the physical network resources that can be selected by the virtual network increase rapidly, thereby better implementing the network resource optimization. From the comparison of the two algorithms, the network optimization ability of the algorithm in this paper is relatively strong.

6 Conclusion

How to improve the reliability of wireless sensor networks has become an urgent problem to be solved. To solve this problem, this paper proposes a network reliability optimization algorithm based on business priority and load balancing. In the simulation experiment part, it is verified that the algorithm of this paper has achieved good results in terms of effective node index. Considering that some services require high latency, an algorithm to improve the efficiency of service execution is urgently needed. In the next step, based on the research results of this paper, we will further study the network reliability optimization technology for delay-sensitive services.

Acknowledgments. This work is supported by science and technology project of State Grid Corporation headquarters (End-to-End Security threat analysis and accurate protection technology of Ubiquitous power Internet of things: 5700-201958466A-0-0-00).

References

1. Ezdiani, S., Acharyya, I.S., Sivakumar, S., et al.: Wireless sensor network softwarization: towards WSN adaptive QoS. IEEE Internet Things J. **4**(5), 1517–1527 (2017)
2. Khan, I., Belqasmi, F., Glitho, R., et al.: Wireless sensor network virtualization: a survey. IEEE Commun. Surv. Tutor. **18**(1), 553–576 (2015)
3. Ren, X.L., Chen, Y.: A routing protocol optimized for data transmission delay in wireless sensor networks. Comput. Appl. **40**(1), 196–201 (2019)
4. Wang, J.H., Chen, Y.D., Chen, M.Y.: Research on real-time and reliability optimization of WSNs in intelligent distribution network based on fuzzy cognitive graph. J. Sens. Technol. **29** (2), 213–219 (2016)
5. Wang, G., Wang, R.Y.: Self-powered wireless sensor network routing algorithm based on cluster head optimization. Comput. Appl. **6**(1721), 1725–1736 (2018)
6. Delgado, C., Canales, M., Ortín, J., et al.: Joint application admission control and network slicing in virtual sensor networks. IEEE Internet Things J. **5**(1), 28–43 (2017)
7. Soualah, O., Aitsaadi, N., Fajjari, I.: A novel reactive survivable virtual network embedding scheme based on game theory. IEEE Trans. Netw. Serv. Manag. **14**(3), 569–585 (2017)
8. Shahriar, N., Chowdhury, S.R., Ahmed, R., et al.: Virtual network survivability through joint spare capacity allocation and embedding. IEEE J. Sel. Areas Commun. **36**(3), 502–518 (2018)

5G Network Resource Migration Algorithm Based on Resource Reservation

Guoliang Qiu[1], Guoyi Zhang[2(✉)], Yinian Gao[1], and Yujing Wen[1]

[1] Shenzhen Power Supply Co. Ltd., Shenzhen 518010, China
[2] Power Control Center of China Southern Power Grid, Guangzhou 510623, China
13418763020@163.com

Abstract. In order to solve the problem of low success rate of virtual network resource allocation, this paper proposes a virtual network resource migration algorithm based on resource reservation under 5G network slicing. In order to improve the utilization of network resources through virtual network migration, this paper proposes a prediction method of the underlying network link resource demand to measure the resource demand for a certain link in the future. The urgency calculation method of the underlying link is designed, and a greedy migration algorithm is adopted to realize the resource migration on the underlying link. This paper proposes a virtual network resource migration algorithm based on resource reservation, and compares it with traditional algorithms through simulation experiments. It is verified that the proposed algorithm achieves good results in terms of virtual network mapping success rate and underlying network resource utilization.

Keywords: Network slicing · Resource allocation · Resource migration · Resource reservation · Resource utilization

1 Introduction

With the rapid development and application of 5G technology, the demand for network resources in all walks of life is increasing [1]. In order to meet the demands of network resources by increasing the utilization rate of network resources, network slicing technology has gradually become a key solution. In the network slicing environment, the traditional basic network is divided into the underlying network and the virtual network. The underlying network provides network resources for the virtual network. By applying resources from the underlying network, the virtual network constructs the virtual network and carries various services to provide services for users. In order to efficiently use the underlying network resources, virtual network resource mapping has become a key research content [2]. Existing researches mainly use methods such as migration technology, intelligent algorithms, and optimization theory to solve the problems of low resource utilization and low mapping success rate in resource allocation [3–6].

However, in resource migration, existing studies have mainly considered load balancing, and have not considered the reservation of important resources, which has

Q. Liu et al. (Eds.): CENet 2020, AISC 1274, pp. 1581–1589, 2021.
https://doi.org/10.1007/978-981-15-8462-6_181

led to the high utilization rate of some important underlying network resources, affecting the success rate of virtual network mapping. In order to solve this problem, this paper proposes a prediction method for the resource requirements of the underlying network link, designs an urgency calculation method for the migration of the underlying link, and proposes a virtual network resource migration algorithm based on resource reservation under network slicing.

2 Problem Description

2.1 Network Description

In the network slicing environment, the traditional network is divided into the underlying network $G_S = (N_S, E_S)$ and the virtual network $G_V = (N_V, E_V)$. The underlying network $G_S = (N_S, E_S)$ includes an underlying node $n_i^s \in N_S$ and an underlying link $e_j^s \in E_S$, and provides CPU resources $cpu(n_i^s)$ and bandwidth resources $bw(e_j^s)$ for the virtual network, respectively, for quickly constructing the virtual network. The virtual network $G_V = (N_V, E_V)$ includes a virtual node $n_i^v \in N_V$ and a virtual link $bw(e_j^v)$, and applies for the CPU resource $cpu(n_i^v)$ of the virtual node and the bandwidth resource $bw(e_j^v)$ of the virtual link by sending a virtual network request to the underlying network. In order to allocate resources to the virtual network, the underlying network uses a virtual network mapping algorithm. The virtual network mapping algorithm generally includes two processes: node mapping and link mapping. Node mapping is represented by $n_i^v \downarrow n_i^s$, which means that the underlying node $n_i^s \in N_S$ allocates the virtual node $n_i^v \in N_V$ with the CPU resource $cpu(n_i^s)$ that satisfies its CPU constraint $cpu(n_i^v)$. Link mapping is represented by $e_j^v \downarrow p_j^s$, which means that the underlying link $p_j^s \in E_S$ allocates the virtual link $e_j^v \in E_V$ with a bandwidth resource $bw(e_j^s)$ that satisfies its bandwidth constraint $bw(e_j^v)$. Among them, p_j^s is formed by the bottom link, which represents the set of bottom links through which the end-to-end connection between the bottom nodes mapped by the two endpoints of the virtual link $e_j^v \in E_V$ passes.

2.2 Migration Description

Regarding how to migrate to achieve the global optimal resource allocation strategy, the next part will be studied. This section mainly describes the concept of resource migration in virtual networks.

The virtual node migration is represented by $M_N : N^S \rightarrow N^S$, which refers to the migration of virtual node $n_i^V \in N^V$ from the underlying node n_i^s to the underlying node n_j^S, which is specifically defined as formula (1). The constraint condition indicates that the CPU resources of the two bottom nodes before and after the migration must meet the CPU resource requirements of the virtual network node.

$$M_N((n_i^S) \to (n_j^S)), F_N(n_i^V) = n_i^S \in N^S, F_N(n_i^V) = n_j^S \in N^S \tag{1}$$

$$\text{S.t. } cpu(n_i^V) \le cpu(n_i^S), cpu(n_i^V) \le cpu(n_j^S)$$

The virtual link migration is indicated by $M_E : P^S \to P^S$, which means that the virtual link $e^V = (n_i^V, n_j^V) \in E^V$ is migrated from the underlying path to the new underlying path. The definition of link migration is formula (2). The constraint condition indicates that the bandwidth resources of the two underlying paths before and after migration must meet the bandwidth resource requirements of the virtual network link.

$$M_E(P_{old}^S \to P_{new}^S) \tag{2}$$

$$\text{S.t. } bw(P_{old}^S) \ge bw(e^V), bw(P_{new}^S) \ge bw(e^V)$$

3 Modeling

3.1 Find Important Link Resources

In order to improve the utilization of network resources through virtual network migration, the idea adopted in this paper is to reduce the number of virtual networks mapping failures, thereby improving the utilization rate of the underlying network resources.

Generally speaking, the reasons for the virtual network mapping failure include the underlying nodes that do not meet the CPU resource requirements of the virtual nodes and the underlying links that do not meet the virtual link bandwidth resource requirements. Therefore, the main reason for the virtual network mapping failure is the shortage of CPU resources of the underlying nodes and the bandwidth resources of the underlying links. If you can find frequently used bottom node resources and bottom link resources, migrate the allocated resources on them or expand the resources of them, it is very helpful to improve resource utilization.

Through research and analysis of virtual network resource allocation, it can be seen that the frequently used bottom nodes and bottom links are related to their location in the network. This article is based on link prediction technology [7, 8] to measure the demand for a certain link in the future. The link bandwidth resource requirement between node n_i^s and node n_j^s is denoted by $LN_{n_i^s n_j^s}$ and calculated using formula (3). This formula fully considers the centrality of nodes and the characteristics of common neighbors between nodes to measure the link demand between node n_i^s and node n_j^s. The larger the value, the greater the demand for the link. Among them, the result of $\frac{k_{n_i^s}}{N_s - 1}$ represents the centrality of the current node in the network. $k_{n_i^s}$ represents the degree of the node, and N_s represents the number of nodes in the underlying network. This formula means that the centrality of a node is measured by the degree of the node.

$S_{n_i^s n_j^s} = |\Gamma(n_i^s) \cap \Gamma(n_j^s)|$ represents the common neighbor of node n_i^s and node n_j^s. $\Gamma(n_i^s)$ represents the set of nodes connected to node n_i^s.

$$LN_{n_i^s n_j^s} = S_{n_i^s n_j^s} + \left(\frac{\sum_{x \in S_{n_i^s n_j^s}} \frac{k_x}{N_s - 1}}{k_x} + \frac{\sum_{x \in S_{n_i^s n_j^s}} \frac{k_x}{N_s - 1}}{k_y}\right) \quad (3)$$

3.2 Migration Urgency Evaluation

For ease of description, the important underlying link is represented by $e_j^s \in E_S$. The two nodes of the link are represented by $n_{e_{j-l}}^s$ and $n_{e_{j-r}}^s$, which respectively represent the left node $n_{e_{j-l}}^s$ and the right node $n_{e_{j-r}}^s$ of the link $e_j^s \in E_S$. Through the analysis of the working process of the virtual network mapping, it can be seen that the virtual resources carried on the important underlying links include node type, link type, and node link type. Among them, the node type refers to that the end node of the bottom link carries virtual node resources. The link type refers to carrying virtual link resources on the underlying link. Node link type refers to resources that carry virtual nodes and virtual links on the end nodes and links of the bottom link, respectively.

In the node-type mapping, there are two cases: (1), the virtual node is carried on the left (or right) node of the link; (2), the virtual node is carried on both the left and right nodes of the link. In both cases, link resources are not used, so migration is not considered. In link-type mapping, there are the following two cases: (1), the link belongs to the left (or right) end link of the underlying path of the virtual link mapping; (2), the link belongs to the intermediate link of the underlying path of the virtual link mapping. In both cases, migration is required. That is, the current underlying link is set to be unavailable, and the shortest path is again found for the virtual link carried on it. On the node link type mapping, there are the following two situations: (1), the left (or right) node of the link carries the virtual node, and the current link carries the virtual link; (2), the link both the left and right nodes carry virtual nodes at the same time, and the current link carries virtual links. In case (1), the virtual node to be carried is first migrated out, and then the current underlying link is set to be unavailable, and the shortest path is again searched for the virtual link which is carried on the current underlying link. In case (2), the current resource allocation belongs to the optimal resource, and the migration cost is relatively large. In this case, migration is not performed.

Through the analysis of migration ideas, we can see that resource migration is an NP problem. In order to solve this problem, the greedy migration algorithm proposed in this paper is as follows: calculate the migration urgency of the underlying link that needs to migrate resources, and move one by one according to the urgency from low to high. Migrating from low to high one by one is to ensure that the most urgent bottom link is migrated last, which can effectively reduce resources being occupied repeatedly. The migration urgency of the underlying link is denoted by $Urg_{e_j^s}$ and calculated using formula (4). Among them, $\sum_{n_{e_{j-l}}^s} cpu(n_k^v) + \sum_{n_{e_{j-r}}^s} cpu(n_k^v)$ represents the amount of node resources carried; $\sum_{n_{e_j^s}^s} bw(e_p^v)$ represents the amount of link resources carried.

From formula (4), it can be seen that the greater the amount of resources carried, the greater the cost of migration, and the lower the migration urgency.

$$Urg_{e_j^s} = \alpha \frac{1}{\sum_{n^s_{e_{j-l}}} cpu(n_k^v) + \sum_{n^s_{e_{j-r}}} cpu(n_k^v)} + \beta \frac{1}{\sum_{n^s_{e_j^s}} bw(e_p^v)} \tag{4}$$

4 Algorithm

The virtual network resource migration algorithm based on resource reservation (VNRMorr) under the network slice proposed in this paper is shown in Table 1. The algorithm includes two processes: search for important link resources and migration of important link resources.

Table 1. Virtual network resource migration algorithm based on resource reservation

Input: the underlying network $G_S = (N_S, E_S)$, the virtual network $G_V = (N_V, E_V)$ Output: the underlying network after migration $G_S^{'} = (N_S, E_S)$
Step 1: Find important link resources 1) Use formula (3) to calculate the bandwidth resource requirement $LN_{n_i^s n_j^s}$ of each link and arrange them in descending order; 2) Find the shortest path of the link node pair of the previous M, and label the shortest path; 3) Reserve resources for the N links with the largest number of shortest paths through; Step 2: Migration of important link resources 1) Use formula (4) to calculate the urgency of link resource migration and arrange them in descending order; 2) Migrating link resources one by one: (A) Determine whether node resource migration is required, and if necessary, find the bottom node that meets the CPU resource requirements for the current node in the bottom network again; (B) Set the current underlying link to be migrated to be unavailable, and use the shortest path algorithm to allocate the underlying link resources for the virtual link of the virtual node after migration and the virtual node connected before migration.

5 Performance Analysis

5.1 Simulation Environment

This article uses the GT-ITM [9] tool to generate the network environment. The network environment includes the underlying network and the virtual network. In terms of the topology of the underlying network, the number of underlying nodes of the underlying network is 100, and the underlying links are formed by interconnecting the underlying nodes with a probability of 0.5. The CPU resources of the bottom node and the bandwidth resources of the bottom link follow the uniform distribution of [20, 50]. In terms of the topology of the virtual network, the number of virtual nodes follows the uniform distribution of [2, 8], and virtual links are formed by connecting virtual nodes to each other with a probability of 0.5. The CPU resources and bandwidth resources requested by the virtual network follow the uniform distribution of [1, 8]. In terms of simulating virtual network requests, the arrival of each virtual network request follows the Poisson distribution at intervals of 1.5 time units. The life cycle of each virtual network request is 20 time units. A total of 6000 time units were run in the experiment.

In terms of algorithm performance comparison, the algorithm VNRMoRR and the algorithm VNRMoRU (virtual network resource migration algorithm for reducing utilization) are compared in three dimensions from the virtual network mapping success rate, the average utilization rate of the underlying network links, and the average utilization rate of the underlying network nodes. The algorithm VNRMoRU judges the urgency of migration by the level of resource utilization. In terms of evaluation indicators, the virtual network mapping success rate refers to the ratio of the number of successfully mapped virtual network requests to the total number of virtual network requests. The average utilization rate of the underlying network link refers to the proportion of the amount of the underlying link bandwidth allocated to the virtual network in the total amount of the underlying link bandwidth. The average utilization of the underlying network nodes refers to the proportion of the number of CPUs of the underlying nodes allocated to the virtual network in the total number of CPUs of the underlying nodes.

5.2 Algorithm Comparison

In order to evaluate the impact of virtual network resource migration on the performance of the mapping algorithm, the virtual network mapping algorithm [10] is used for mapping. The algorithm first generates 100 virtual network requests and allocates resources, then executes the migration algorithm, and finally generates 100 virtual network requests and allocates resources, and calculates three indicators of the virtual network mapping success rate, the average utilization rate of the underlying network links, and the average utilization rate of the underlying network nodes. The algorithm is executed 5 times in total, and the experimental results are shown in Figs. 1, 2 and 3.

As can be seen from Fig. 1, in five experiments, the success rate of the virtual network mapping of the two algorithms is relatively stable. It shows that both algorithms have achieved an increase in the success rate of virtual network mapping. From the performance analysis of the two algorithms, we can see that the virtual network

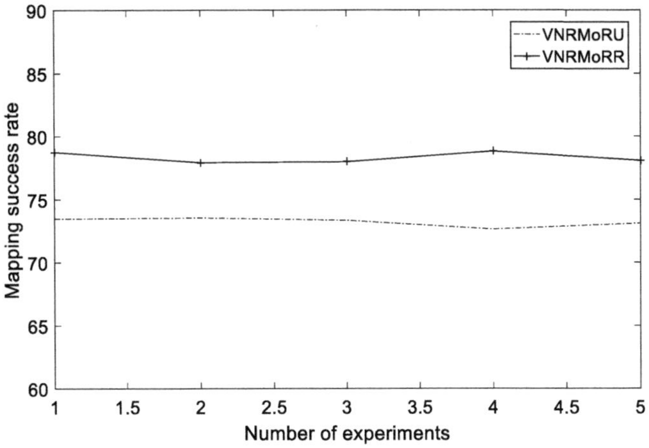

Fig. 1. Analysis of the success rate of virtual network mapping

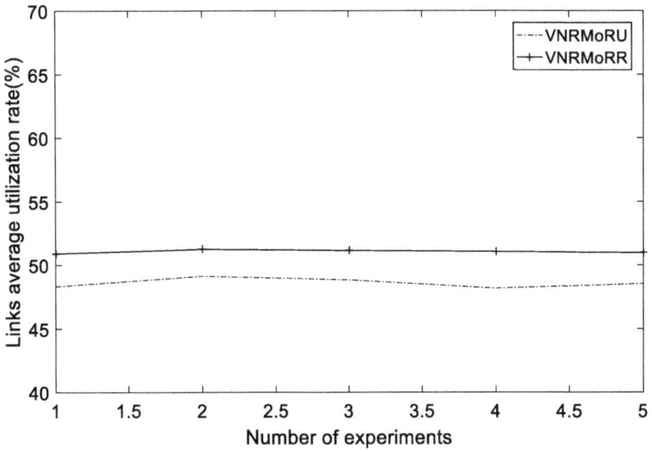

Fig. 2. Analysis of the average utilization rate of the underlying network links

mapping success rate of the algorithm VNRMoRR is about 6.5% higher than that of the algorithm VNRMoRU.

It can be seen from Fig. 2 that in the five experiments, the average utilization of the underlying network links of the two algorithms is relatively stable. From the performance analysis of the two algorithms, we can see that the average utilization rate of the underlying network link of the algorithm VNRMoRR in this paper is about 8.2% higher than that of the algorithm VNRMoRU.

As can be seen from Fig. 3, in the five experiments, the average utilization values of the underlying network nodes of the two algorithms are relatively stable. From the performance analysis of the two algorithms, we can see that the average utilization rate

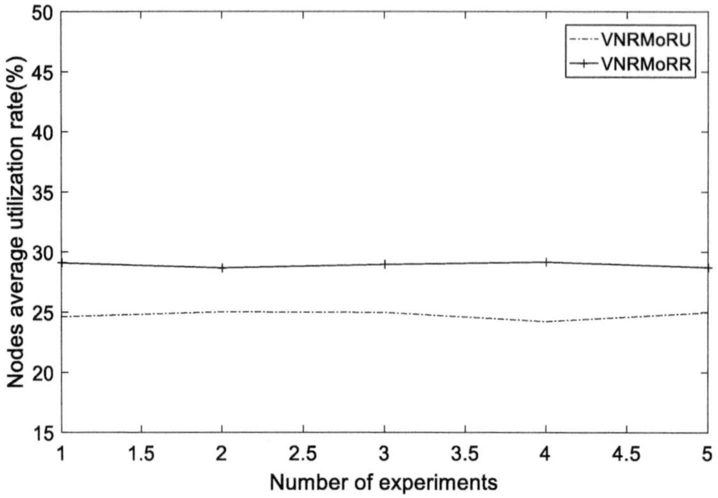

Fig. 3. Analysis of the average utilization rate of the underlying network nodes

of the underlying network nodes of the algorithm VNRMoRR in this paper is about 11.5% higher than that of the algorithm VNRMoRU.

Through analysis of the three indicators of the virtual network mapping success rate, the average utilization rate of the underlying network links, and the average utilization rate of the underlying network nodes in Figs. 1, 2 and 3, it can be seen that the migration algorithm proposed in this paper improves the virtual network mapping success rate and resource utilization. It shows that the migration algorithm in this paper can migrate the virtual network on the critical urgent resource when migrating resources, thus providing the required resources for the new virtual network request.

6 Conclusion

In order to improve the utilization of the underlying network resources and the success rate of virtual network mapping in a network slicing environment, the idea adopted in this paper is to reduce the number of virtual networks mapping failures, thereby improving the utilization rate of the underlying network resources. Based on this, this paper proposes a link demand prediction calculation method to measure the future demand for a link. In order to further enhance the application value of the algorithm in this paper, in the next step, on the basis of this study, we will increase the energy consumption limitation of the underlying network resources, so as to reduce the energy consumption of the underlying network resources on the premise of satisfying the virtual network resource allocation.

References

1. Aijaz, A.: Hap-SliceR: a radio resource slicing framework for 5G networks with haptic communications. IEEE Syst. J. **12**(3), 2285–2296 (2018)
2. Fischer, A., Botero, J.F., Beck, M.T., et al.: Virtual network embedding: a survey. IEEE Commun. Surv. Tutor. **15**(4), 1888–1906 (2013)
3. Soto, P., Botero, J.F.: Greedy randomized path-ranking virtual optical network embedding onto EON-based substrate networks. In: 2017 IEEE Colombian Conference on Communications and Computing (COLCOM), Colombia, pp. 1–6. IEEE (2017)
4. Chowdhury, S.R., Ahmed, R., Shahriar, N., et al.: ReViNE: reallocation of virtual network embedding to eliminate substrate bottlenecks. In: Integrated Network and Service Management, Portugal. IEEE (2017)
5. Zhang, Y., Zhu, Y., Yan, F., et al.: Energy-efficient radio resource allocation in software-defined wireless sensor networks. IET Commun. **12**(3), 349–358 (2017)
6. Parwez, M.S., Rawat, D.B.: Resource allocation in adaptive virtualized wireless networks with mobile edge computing. In: 2018 IEEE International Conference on Communications (ICC), Kansas City, pp. 1–7 (2018)
7. Lv, L.: Link prediction of complex networks. J. Univ. Electron. Sci. Technol. China **39**(5), 651–661 (2010)
8. Sarukkai, R.R.: Link prediction and path analysis using markov chains. Comput. Netw. **33** (1–6), 377–386 (2000)
9. Zegura, E.W., Calvert, K.L., Bhattacharjee, S.: How to model an internet work. In: IEEE Infocom, pp. 594–602 (1996)
10. Yu, M., Yi, Y., Rexford, J., Chiang, M.: Rethinking virtual network embedding: Substrate support for path splitting and migration. ACM SIGCOMM Comput. Commun. Rev. **38**(2), 17–29 (2008)

Virtual Network Resource Allocation Algorithm Based on Reliability in Large-Scale 5G Network Slicing Environment

Xiaoqi Huang[1], Guoyi Zhang[2(✉)], Ruya Huang[1], and Wanshu Huang[1]

[1] Shenzhen Power Supply Co. Ltd., Shenzhen 518010, China
[2] Power Control Center of China Southern Power Grid, Guangzhou 510623, China
13418763020@163.com

Abstract. In the 5G network slicing environment, in order to solve the problem of low utilization rate when a large-scale underlying network allocates resources to virtual networks, this paper proposes a virtual network resource allocation algorithm based on reliability in large-scale network environment. First, the reliability of the virtual network is modeled from the importance of the virtual network nodes. Second, the reliability of the underlying network is analyzed from the perspective of the community introverted character and the community relationship value, and the underlying network community division algorithm is proposed. Community reliability and node reliability are analyzed in terms of community and node reliability. Finally, a virtual network resource allocation algorithm based on reliability in large-scale network environment is proposed. In the experimental part, from the two aspects of the underlying network revenue and virtual network mapping success rate, it is verified that the proposed algorithm achieves good results in large-scale virtual network resource allocation.

Keywords: Network slicing · Network resource allocation · Large-scale · Reliability

1 Introduction

In the 5G network environment, the wireless network speed reaches more than 1 Gbps, which poses a greater challenge to the core network. Network slicing technology uses network virtualization technology to divide the traditional network into the underlying network and the virtual network, which significantly improves the networking flexibility of the network, facilitates the rapid deployment of 5G services, and also significantly improves the utilization of network resources [1]. After network slicing, how to allocate resources has become a research hotspot. Because resource allocation is affected by the constraints of the underlying network resources, the virtual network requirements on resource capacity and response speed, etc., the virtual network resource allocation problem has been proved to be an NP problem [2–7].

Q. Liu et al. (Eds.): CENet 2020, AISC 1274, pp. 1590–1598, 2021.
https://doi.org/10.1007/978-981-15-8462-6_182

However, with the rapid development of 5G networks, the scale of the core network is also increasing. How to allocate virtual network resources in a large-scale environment has become an urgent problem to be solved. To solve this problem, this paper models the reliability of the virtual network and the reliability of the underlying network, and proposes an underlying network community division algorithm and a large-scale virtual network resource allocation algorithm based on reliability.

2 Network Modeling

The underlying network G_S is composed of a underlying node N_S and a underlying link E_S, denoted by $G_S = (N_S, E_S)$. Underlying node N_S has CPU resource $cpu(n_i^s)$, and underlying link E_S has bandwidth resource $bw(e_j^s)$. The virtual network G_V is composed of a virtual node N_V and a virtual link E_V, denoted by $G_V = (N_V, E_V)$. Virtual node N_V has CPU resource $cpu(n_i^v)$, and virtual link E_V has bandwidth resource $bw(e_j^v)$.

This paper defines the process of the underlying network allocating resources to the virtual network as the virtual network mapping process. Among them, the process that the underlying node allocates CPU resources to the virtual node is defined as $n_i^v \downarrow n_i^s$, which means that the virtual node $n_i^v \in N_V$ is mapped onto the underlying node $n_i^s \in N_S$. The process that the underlying link allocates bandwidth resources to the virtual link is defined as $e_j^v \downarrow p_j^s$, which means that the virtual link $e_j^v \in E_V$ is mapped to the underlying path $p_j^s \in E_S$. The underlying path p_j^s is formed by connecting one or more underlying links $e_j^s \in E_S$. When the underlying node allocates resources to the virtual node, the amount of CPU resources allocated to the virtual node must be greater than or equal to the amount of CPU resources requested by the virtual node. When the underlying link allocates resources to the virtual link, the bandwidth resources allocated to the virtual link must be greater than or equal to the number of bandwidth resources requested by the virtual link.

In order to ensure the reliability of the virtual network, this paper analyzes the reliability from three aspects: virtual network reliability analysis, underlying network reliability analysis, and virtual network resource allocation process, and performs resource allocation based on the characteristics of network reliability.

3 Reliability Analysis

3.1 Virtual Network

Importance Analysis of Virtual Nodes. The importance of a node with a larger CPU resource amount is high. The CPU resource amount refers to the CPU resource amount $cpu(n_i^v)$ that the current virtual node has. The central value of the virtual node is a measure of the current virtual node's central position in the virtual network. The more the virtual node is in the center of the virtual network, indicating that the current virtual node has a greater impact on the reliability of the virtual network. The calculation method of the central value $NC(n_i^v)$ of the virtual node is shown in formula (1), where

$hops(n_i, n_j)$ represents the number of end-to-end links between n_i^v and n_j^v. Therefore, the central value $NC(n_i^v)$ of the virtual node is the derivative of the sum of the hops from the current node to all nodes in the virtual network.

$$NC(n_i^v) = \frac{1}{\sum_{n_j^v \in N_V} hops(n_i^v, n_j^v)} \tag{1}$$

The adjacent link bandwidth resource is a measure of the bandwidth resource of the link connected to the current virtual node. The calculation method is formula (2), where $E(n_i^v)$ represents the link set of the virtual link connected to virtual node n_i^v. This formula indicates that the more link resources connected to a virtual node, the more important the current virtual node.

$$AL(n_i^v) = \sum_{e_j^v \in E(n_i^v)} bw(e_j^v) \tag{2}$$

Through the above analysis, the importance of virtual nodes can be calculated using formula (3).

$$IMPORT(n_i^v) = (cpu(n_i^v) + AL(n_i^v)) \times NC(n_i^v) \tag{3}$$

Virtual Network Importance Ranking. When the virtual network request $G = \{G_1^V, G_2^V, \ldots, G_M^V\}$ has M virtual networks, formula (3) is used to calculate the importance of virtual nodes in the M virtual network requests, and each virtual network request is summed according to the node importance. After that, the M virtual network requests are sorted in descending order according to the sum of the importance of the nodes to obtain $G^{des} = \{G_1^V, G_2^V, \ldots, G_M^V\}$.

3.2 Underlying Network

Underlying Network Analysis. According to the complex network theory [8], the network has the characteristics of a community. RS of the community restraint is used to judge the division of the community, and it is calculated using formula (4). $\delta(n_i^s, n_j^s)$ indicates whether nodes n_i^s and n_j^s belong to the same community. When $\delta(n_i^s, n_j^s) = 1$, it means that nodes n_i^s and n_j^s belong to the same community. When $\delta(n_i^s, n_j^s) = 0$, it means that nodes n_i^s and n_j^s do not belong to the same community. $E(n_i^s)$ represents the set of links formed by the underlying links connected to the underlying node n_i^s. $E2E(n_i^s)$ represents the set of links included in the path between node i and other nodes. $\sum_{e_j^s \in E2E(n_{ij}^s)} bw(e_j^s)$ represents the sum of the bandwidth resources of the link with node i and node j. $\sum_{n_i^s \in N_k^s} \sum_{e_j^s \in E2E(n_i^s)} bw(e_j^s)$ represents the sum of link bandwidth resources from all nodes to other nodes. $\sum_{e_j^s \in E(n_i^s)} bw(e_j^s)$ and $\sum_{e_i^s \in E(n_j^s)} bw(e_i^s)$ represent the sum of the bandwidth resources of all sides connected by node i and node j, respectively.

$$RS_k = \frac{\sum_{ij}\left(\sum_{e_j^s \in E2E(n_{ij}^s)} bw(e_j^s) - \frac{\sum_{e_j^s \in E(n_i^s)} bw(e_j^s) \times \sum_{e_i^s \in E(n_j^s)} bw(e_i^s)}{\sum_{n_i^s \in N_k^s} \sum_{e_j^s \in E2E(n_{ij}^s)} bw(e_j^s)}\right)\delta(n_i^s, n_j^s)}{\sum_{n_i^s \in N_k^s} \sum_{e_j^s \in E2E(n_{ij}^s)} bw(e_j^s)} \tag{4}$$

The community relationship value OR of node i is calculated using formula (5), where $E(n_i^s)$ represents the link set formed by the underlying links connected to the underlying node n_i^s. $k_{n_i^s}$ represents the degree of node n_i^s.

$$OR_{n_i^s} = \frac{E(n_i^s)}{k_{n_i^s}(k_{n_i^s} - 1)} \tag{5}$$

Algorithm for Dividing the Underlying Network Community. The underlying network community division algorithm proposed in this paper is shown in Table 1. The algorithm includes two steps: initial division of communities based on the value of community relations, and optimization of the results of community division based on the restraint of the communities. When the community is initially divided based on the community relationship value, the community relationship value is used to determine whether the node and neighboring nodes belong to the same community.

Reliability Analysis of Communities and Nodes

Community reliability CR is calculated using formula (6). Among them, O_i represents the ith community in the community collection. The association relationship between O_i and O_j is $RC_{ij} = \{(u, v) \in E, u \in O_i, v \in O_j\}$, which means that communities O_i and O_j are connected through edges u and v. Then the reliability of O_i and O_j is $v(O_i, O_i) = |RC_{ij}|$, where $|RC_{ij}|$ represents the number of edges connected by community O_i and O_j. Therefore, $RC_{O_i} = \cup_{O_j \in O_i^j} v(O_i, O_i)$ represents the number of edges of all communities O_i^j connected to the community O_i. $\sum_{O_j \in O} |RC_{O_j}|$ represents the number of connected edges between all communities.

$$CR_{O_i} = \frac{|RC_{O_i}|}{\sum_{O_j \in O} |RC_{O_j}|} \tag{6}$$

The node reliability $CNR(n_i^s)$ is calculated using formula (7). Among them, the first half $\sum_{k=1}^{|O|} \delta(n_i^s, k) CR(O_k)$ represents the reliability of the community to which the node belongs, and the second half $BR(n_i^s)$ represents the reliability of the node itself. In the first half, $|O|$ indicates the number of clubs. $\delta(n_i^s, k)$ indicates whether node n_i^s belongs to community k, if yes, $\delta(n_i^s, k) = 1$, otherwise, $\delta(n_i^s, k) = 0$. CR_{O_k} indicates the reliability of community O_k. In the second half, $BR(n_i^s) = \frac{\prod_{k=1}^{|O|} CR_{O_k}}{\sum_j^{j \in \lambda_i^{inside}} e_{ij}}$. Among them,

Table 1. Underlying network community division algorithm.

Input: $G_S = (N_S, E_S)$
Output: Community O
Step 1: Initially divide the community based on the community relationship value OR
1) For each underlying node, use formula (5) to calculate the value of its association relationship;
2) According to the value of the community relationship, the underlying nodes are arranged in descending order to form a node set;
3) Take out the nodes in the node set and all neighbor nodes in sequence;
4) Determine whether the neighbor node of the current node has been put into other communities, if not, put it into the community where the current node is.
Step 2: Optimize the results of community division based on community restraint RS
1) For each community, take out the node n_i^s one by one;
2) Take out all the neighbor nodes of this node, and add the nodes that are not in the community of n_i^s to the set ON;
3) Put the nodes in the set ON into the community of n_i^s one by one, and use the formula (4) to calculate the community restraint RS;
4) When the value ΔRS of change after node $n_j^s \in ON$ is put into community of n_i^s is greater than the specified threshold, put it into community of n_i^s.

$\prod_{k=1}^{|O|} CR_{O_k}$ represents the reliability product of all communities directly connected by node n_i^s. $\sum_j^{j \in \lambda_i^{inside}} e_{ij}$ represents the sum of the edge bandwidth values directly connected to the node. The larger the value, the higher the utilization rate of the node and the less reliable it is.

$$CNR(n_i^s) = \sum_{k=1}^{|O|} \delta(n_i^s, k) CR_{O_k} \times BR(n_i^s) \tag{7}$$

4 Virtual Network Resource Allocation Algorithm

The virtual network resource allocation algorithm based on reliability in large-scale network environment (VNRAoR) proposed in this paper is shown in Table 2. The algorithm includes three steps: dividing communities and reliability analysis, reliability

Table 2. Algorithm VNRAoR.

Input: $G_S = (N_S, E_S), G = \{G_1^V, G_2^V, ..., G_M^V\}$
Output: Resource allocation plan
Step 1: Divide communities and reliability analysis
1) For the underlying network $G_S = (N_S, E_S)$, use the community division algorithm to divide the community;
2) Use formula (6) to calculate the reliability of the community and arrange them in descending order;
3) Use formula (7) to calculate the nodes in the community and arrange them in descending order;
Step 2: Reliability ranking of virtual network requests
1) For M virtual network requests $G = \{G_1^V, G_2^V, ..., G_M^V\}$, use formula (3) to calculate the importance of the virtual nodes;
2) For each virtual network request, sum the importance of the nodes it contains, and arrange them in descending order to get set $G^{des} = \{G_1^V, G_2^V, ..., G_M^V\}$;
Step 3: Assign resources to virtual network requests one by one
1) Take the first virtual network request from the virtual network request set $G^{des} = \{G_1^V, G_2^V, ..., G_M^V\}$;
2) Virtual node resource allocation: find the high-reliability underlying nodes that meet the requirements of virtual nodes in the high-reliability underlying network community. When the nodes cannot meet the requirements, search from the second high-reliability underlying network community until all allocated;
3) Virtual link resource allocation: use the shortest path algorithm to allocate virtual link resources.

ranking of virtual network requests, and allocating resources one by one for virtual network requests. When dividing communities and reliability analysis, the underlying network is divided into communities, and the reliability of the community and its internal nodes are judged. When sorting the reliability of the virtual network request, first calculate the importance of the virtual node, and secondly sort the virtual nodes based on the importance of all virtual nodes in the virtual network request.

5 Performance Analysis

5.1 Simulation Environment

In order to verify the performance of the algorithm in this paper, the GT-ITM tool [9] is used to generate the underlying network and virtual network. The CPU resources and bandwidth resources of the underlying network follow the uniform distribution of [25, 50]. The CPU resources and bandwidth resources of the virtual network are subject to the uniform distribution of [1, 5] and [1, 10], respectively. The underlying network includes 100 underlying network nodes. The nodes of the virtual network obey the uniform distribution of [3, 10]. The two nodes of the underlying network and the virtual network are connected with a probability of 0.5.

In order to analyze the performance of the algorithm in this paper, the algorithm VNRAoR and the algorithm VNRAoO (Virtual network resource allocation algorithm based on order) were compared. Algorithm VNRAoO allocates resources one by one according to the arrival order of virtual network requests. In terms of comparative indicators, it analyzes from two aspects: underlying network revenue and virtual network mapping success rate.

5.2 Performance Analysis

In terms of underlying network revenue, the experimental results are shown in Fig. 1. It can be seen from the figure that as the algorithm runs, the underlying network revenue of both algorithms tends to be stable. The algorithm VNRAoR in this paper increases the underlying network revenue by 8.5%. This shows that the algorithm in this paper allocates the underlying network resources for more virtual network requests.

It can be seen from the Fig. 2 that the success rate of the virtual network mapping of the two algorithms tends to be stable as the algorithm runs. In this paper, the virtual network mapping success rate of algorithm VNRAoR is increased by 9.8% compared with algorithm VNRAoO, and more underlying network resources are allocated to the virtual network.

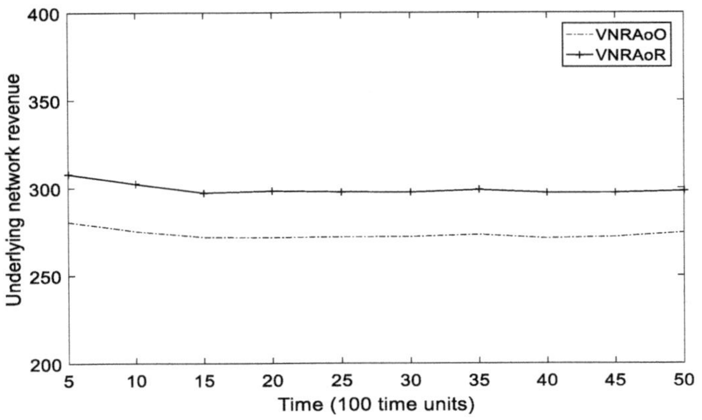

Fig. 1. Underlying network revenue.

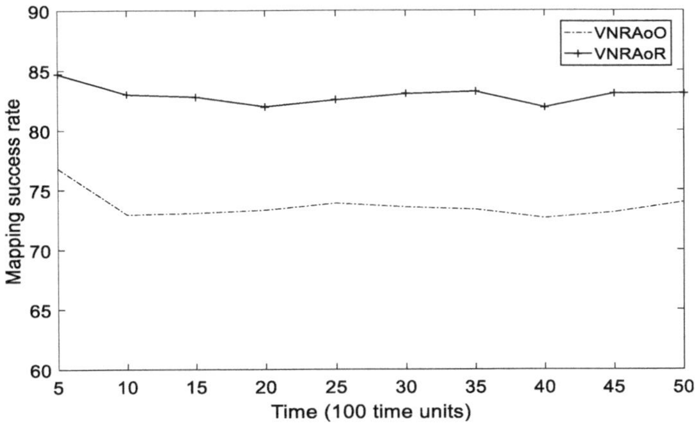

Fig. 2. Virtual network mapping success rate.

6 Conclusion

In order to solve the problem of low utilization rate when the large-scale underlying network allocates resources to the virtual network, this paper proposes a large-scale virtual network resource allocation algorithm based on reliability under network slicing. This paper models the importance of virtual network nodes, the community restraint of the underlying network and the relationship value of the community, and proposes an algorithm for dividing the community of the underlying network and a large-scale virtual network resource allocation algorithm based on reliability. In the next step, on the basis of the research in this paper, the time efficiency of resource allocation will be studied as a new constraint to meet the needs of time-sensitive virtual network services.

References

1. Hap-SliceR, A.A.: A radio resource slicing framework for 5G networks with haptic communications. IEEE Syst. J. **12**(3), 2285–2296 (2018)
2. Fischer, A., Botero, J.F., Beck, M.T., et al.: Virtual network embedding: a survey. IEEE Commun. Surv. Tutor. **15**(4), 1888–1906 (2013)
3. Soto, P., Botero, J.F.: Greedy randomized path-ranking virtual optical network embedding onto EON-based substrate networks. In: 2017 IEEE Colombian Conference on Communications and Computing (COLCOM), Colombia, pp. 1–6. IEEE (2017)
4. Chowdhury, S.R., Ahmed, R., Shahriar, N., et al.: ReViNE: reallocation of virtual network embedding to eliminate substrate bottlenecks. In: Integrated Network and Service Management, Portugal. IEEE (2017)
5. Zhang, Y., Zhu, Y., Yan, F., et al.: Energy-efficient radio resource allocation in software-defined wireless sensor networks. IET Commun. **12**(3), 349–358 (2017)
6. Guan, W., Wen, X., Wang, L., et al.: A service-oriented deployment policy of end-to-end network slicing based on complex network theory. IEEE Access **6**, 19691–19701 (2018)
7. Guo, L., Ning, Z., Song, Q., et al.: A QoS-oriented high-efficiency resource allocation scheme in wireless multimedia sensor networks. IEEE Sens. J. **17**(5), 1538–1548 (2016)
8. Newman, M.E.J., Girvan, M.: Finding and evaluating community structure in networks. Phys. Rev. E **69**(2), 026113 (2004)
9. Zegura, E.W., Calvert, K.L., Bhattacharjee, S.: How to model an Internet work. In: IEEE Infocom, pp. 594–602 (1996)

Resource Allocation Algorithm of Power Communication Network Based on Reliability and Historical Data Under 5G Network Slicing

Yang Yang[1], Guoyi Zhang[2](✉), Junhong Weng[1], and Xi Wang[1]

[1] Shenzhen Power Supply Co. Ltd., Shenzhen 518010, China
[2] Power Control Center of China Southern Power Grid, Guangzhou 510623, China
13418763020@163.com

Abstract. In the 5G network slicing environment, in order to solve the problem of low success rate of virtual network mapping in the existing research, this paper proposes a resource allocation algorithm of power communication network based on reliability and historical data under 5G network slicing. First, the reliability of the virtual network node is analyzed from three aspects: the CPU resources of the virtual node, the connected link resources, and the centrality of the node. The reliability of the underlying network node is analyzed from three aspects: the reliability matrix of the underlying node, the CPU allocation history matrix of the underlying node, and the underlying link allocation history matrix. Secondly, based on the virtual network model and the underlying network model, a resource allocation algorithm of power communication network based on reliability and historical data is proposed. In simulation experiments, it is verified that the algorithm in this paper effectively improves the revenue of the underlying network and the mapping success rate of the virtual network.

Keywords: Network slicing · Resource allocation · Reliability · Resource utilization

1 Introduction

With the gradual commercialization of 5G technology in the power communication network, the power business has put forward more requirements on the power communication network [1]. Network slicing technology uses virtualization, SDN and other technologies to divide the traditional basic network into the underlying network and virtual network. The underlying network provides computing resources and bandwidth resources for the virtual network. Virtual networks carry different power services. After network slicing, the network resource utilization rate of the power communication network has been improved, and the development and operation of new power services have become more flexible [2]. There have been studies to solve the problems of low resource utilization and low network reliability through the use of intelligent algorithms and optimization algorithms [3–8].

However, there have been studies that have not fully evaluated the importance of the underlying network when allocating resources, resulting in an increased failure rate

Q. Liu et al. (Eds.): CENet 2020, AISC 1274, pp. 1599–1607, 2021.
https://doi.org/10.1007/978-981-15-8462-6_183

due to lack of resources when requesting virtual network mappings. In order to solve this problem, this paper analyzes the reliability of virtual network nodes from three aspects: CPU resources of virtual nodes, connected link resources, and the centrality of nodes. A resource allocation algorithm for power communication network based on reliability and historical data is proposed.

2 Problem Description

In the network slicing environment, the power communication network is divided into the underlying network $G_S = (N_S, E_S)$ and the virtual network $G_V = (N_V, E_V)$. The underlying node $n_i^s \in N_S$ uses its own CPU resource $cpu(n_i^s)$ to provide the virtual node $n_i^v \in N_V$ with the CPU resource $cpu(n_i^v)$ that satisfies its request, denoted by $n_i^v \downarrow n_i^s$. The underlying link $e_j^s \in E_S$ uses its own bandwidth resource $bw(e_j^s)$ to provide the virtual node $n_i^v \in N_V$ with the bandwidth resource $bw(e_j^v)$ that satisfies its request, denoted by $e_j^v \downarrow p_j^s$. Among them, the underlying path $p_j^s \in E_S$ indicates that a single virtual link may require one or more underlying links to provide bandwidth resources.

The underlying network overhead refers to the sum of the underlying network resources used to allocate resources to the virtual network within the time period t, and is calculated using formula (1), where $hop(e_j^v)$ refers to the underlying link to which the virtual link e_j^v is mapped set. The underlying network revenue refers to the sum of CPU resources and bandwidth resources allocated by the underlying network to the virtual network within the time period t, and is calculated using formula (2).

$$C_t^S = \sum\nolimits_{n_i^v \in N_V} cpu(n_i^v) + \sum\nolimits_{e_i^v \in E_V} hop(e_j^v) \times bw(e_j^v) \tag{1}$$

$$R_t^S = \sum\nolimits_{n_i^v \in N_V} cpu(n_i^v) + \sum\nolimits_{e_j^v \in E_V} bw(e_j^v) \tag{2}$$

3 Reliability Analysis of Virtual Network Nodes

The CPU resource of the virtual node is determined by the business, and is denoted by $cpu(n_i^v)$. The more CPU resources required by the power business, the more important the current virtual node. Connected link resources refer to the sum of the bandwidth resources of all links connected to the current virtual node (the link set is denoted by $E(n_i)$), which is calculated using formula (3).

$$AL(n_i) = \sum\nolimits_{e_j \in E(n_i)} bw(e_j) \tag{3}$$

The centrality of the node is calculated using formula (4). Among them, $hops(n_i, n_j)$ represents the number of links included in the end-to-end path between n_i^v and n_j^v.

$$NC(n_i^v) = \frac{1}{\sum_{n_j^v \in N_V} hops(n_i^v, n_j^v)} \tag{4}$$

Based on the above analysis, the reliability of the virtual network node is calculated using formula (5).

$$IMPORT(n_i^v) = (cpu(n_i^v) + AL(n_i^v)) \times NC(n_i^v) \tag{5}$$

4 Reliability Analysis of the Underlying Network Nodes

4.1 Reliability Matrix of Underlying Nodes

The reliability of the underlying network nodes is mainly related to three factors: resource utilization rate $RU(n_i^s)$, failure rate $FN(n_i^s)$, and distance $UH(n_i^s)$ to the neighboring virtual node corresponding to the underlying node. Among them, the higher the resource utilization rate $RU(n_i^s)$ of the underlying node, the greater the probability of failure of the underlying node. The failure rate $FN(n_i^s)$ of the underlying node is an important factor of the reliability of the underlying node. If the underlying node has a higher frequency of failure before, the probability of failure in the future operation is also higher. The distance $UH(n_i^s)$ to the neighboring virtual node corresponding to the underlying node is used to measure the distance between the current underlying node and the neighbor virtual node corresponding to the underlying node. If the distance is closer, it indicates that this underlying node helps to save network link resources. Use formula (6) to calculate $UH(n_i^s)$. Among them, $\mu(n_i^s)$ represents the set of underlying nodes that the virtual network has mapped before selecting the current node n_i^s as the available underlying node.

$$UH(n_i^s) = \sum_{n_j^s \in \mu(n_i^s)} hop(n_i^s, n_j^s) \tag{6}$$

Through the above analysis, the reliability of the underlying node can be calculated using formula (7). Among them, the parameters κ and λ are used to adjust the weight between the first part of the resource amount and the second half of the resource performance.

$$RELIAB(n_i^s) = \kappa \frac{cpu(n_i^s) + AL(n_i^s)}{UH(n_i^s)} \times \lambda \frac{1}{FN(n_i^s) \times RU(n_i^s)} \tag{7}$$

Based on the calculation method of the reliability of the underlying nodes, the reliability matrix M_{node} of the underlying nodes can be constructed, and the matrix element $r_{ii} \in M_{node}$ is the reliability of each underlying node calculated using formula (7).

4.2 The CPU Allocation History Matrix of the Underlying Node

The CPU allocation history matrix M_{CPU} of the underlying node is shown in formula (8). The value of element $a_{ii} \in M_{CPU}$ represents the sum of the CPU resources allocated by node $n_i^s \in N_S$ to the virtual node during time period t.

$$M_{CPU} = \begin{bmatrix} a_{11} & 0 & 0 & 0 & 0 & 0 \\ 0 & a_{22} & 0 & 0 & 0 & 0 \\ \cdots & \cdots & \cdots & \cdots & \cdots & \cdots \\ 0 & 0 & 0 & a_{ii} & 0 & 0 \\ \cdots & \cdots & \cdots & \cdots & \cdots & \cdots \\ 0 & 0 & 0 & 0 & 0 & a_{nn} \end{bmatrix} \quad (8)$$

4.3 The Underlying Link Allocation History Matrix

The link allocation history matrix M_{LINK} of the underlying node is shown in formula (9). The value of element $b_{ij} \in M_{LINK}$ represents the sum of the bandwidth allocated by the underlying path $P(n_i^s, n_j^s)$ for the virtual link in the time period t divided by the number of the underlying links contained in the underlying path $P(n_i^s, n_j^s)$.

$$M_{LINK} = \begin{bmatrix} 0 & b_{12} & \cdots & b_{1j} & \cdots & b_{1n} \\ b_{21} & 0 & \cdots & b_{2j} & \cdots & b_{2n} \\ \cdots & \cdots & \cdots & \cdots & \cdots & \cdots \\ b_{i1} & b_{i2} & \cdots & 0 & \cdots & b_{in} \\ \cdots & \cdots & \cdots & \cdots & \cdots & \cdots \\ b_{n1} & b_{n2} & \cdots & b_{nj} & \cdots & 0 \end{bmatrix} \quad (9)$$

4.4 Reliability History Matrix of the Underlying Nodes

In order to use the reliability matrix M_{node} of the underlying node, the CPU allocation history matrix M_{CPU} of the underlying node, and the underlying link allocation history matrix M_{LINK} to construct the reliability history matrix M_{RELIAB} of the underlying node, first use the min-max normalization method scales the value of each element in the matrix to the range of [0, 1], thereby obtaining three new matrices M'_{CPU}, M'_{LINK} and M'_{node}. Use formula (10) to calculate the reliability history matrix M_{RELIAB} of the underlying nodes.

$$M_{RELIAB} = M'_{CPU} + M'_{LINK} + M'_{node} \quad (10)$$

5 Algorithm

The resource allocation algorithm of power communication network based on reliability and historical data (RAAoRH) proposed in this paper is shown in Table 1. The algorithm includes four steps: calculating the reliability of virtual nodes, ranking virtual nodes, allocating resources for virtual nodes, and allocating resources for virtual links.

Table 1. Algorithm RAAoRH.

Input: $G_S = (N_S, E_S)$, $G_V = (N_V, E_V)$, Reliability history matrix M_{RELIAB} of the underlying node

Output: list G_V of mappings

1. Calculate the reliability of the virtual node: For each virtual network $G_V = (N_V, E_V)$ in the virtual network request, use formula (5) to calculate the reliability $IMPORT(n_i^v)$ of the virtual network node $n_i^v \in N_V$.

2. Virtual node sorting: Sort the virtual nodes in N_V in descending order based on $IMPORT(n_i^v)$ to get a new set $N_V^{'}$;

3. Allocate resources for virtual nodes: allocate resources to n_i^v in $N_V^{'}$ in order;

1) From the underlying network, select the underlying node with the largest $m_{ij} \in M_{RELIAB}$ ($i = j$) and the CPU resource meets the demand of $cpu(n_i^v)$, and allocate resources for the first n_i^v. If there is no underlying node that meets this condition, resource allocation fails and ends.

2) For the other n_i^v in set $N_V^{'}$, select the underlying node with the largest $m_{ij} \in M_{RELIAB}$ ($i \neq j$) and the CPU resource meets the requirements of $cpu(n_i^v)$, and allocate resources for n_i^v. If there is no underlying node that meets this condition, resource allocation fails and ends.

4. Allocate resources for virtual links: The k-shortest path algorithm is used to allocate the underlying link resources that satisfy the $bw(e_j^v)$ constraint to the virtual link $e_j^v \in E_V$ in E_V. If there is no underlying node that meets this condition, resource allocation fails and ends.

6 Performance Analysis

6.1 Simulation Environment

To analyze the performance of the algorithm in this paper, the GT-ITM tool [9] is used to generate the network topology environment. The network topology consists of the underlying network and the virtual network. The underlying network contains 100 underlying nodes, and the underlying nodes are connected to each other with a probability of 0.5. The resources of the underlying node and the underlying link follow the uniform distribution of [30, 60]. The virtual nodes of the virtual network obey the uniform distribution of [2, 8], and the virtual links are connected to each other with a probability of 0.5. In resource allocation, the virtual network request includes 2000 virtual networks, and the arrival time interval of each virtual network request is 2 time units, and the life cycle of the virtual network is 10 time units.

In the algorithm comparison, the two indicators of underlying network revenue and mapping success rate are used. Among them, the underlying network revenue refers to the sum of resources allocated by the underlying network to the virtual node resources and link resources of the virtual network. The mapping success rate refers to the ratio of the number of virtual network requests that successfully obtain the underlying network resources divided by the total number of virtual network requests.

6.2 Reliability Matrix Analysis

In order to choose the appropriate reliability matrix construction method, the algorithm performance of this paper is better. When constructing the reliability matrix, the number of virtual network mappings (N-VNM) that have been mapped is used. Next, the reliability matrix with N-VNM values of 500, 700, 900, 1100, 1300, and 1500 is applied to resource allocation, and the algorithm is compared from the two indicators of underlying network revenue and mapping success rate.

The experimental results of underlying layer network revenue and mapping success rate (Fig. 1 and Fig. 2) show that as the algorithm runs, under different N-VNM values, the underlying network revenue and mapping success rate both tend to stabilize. When

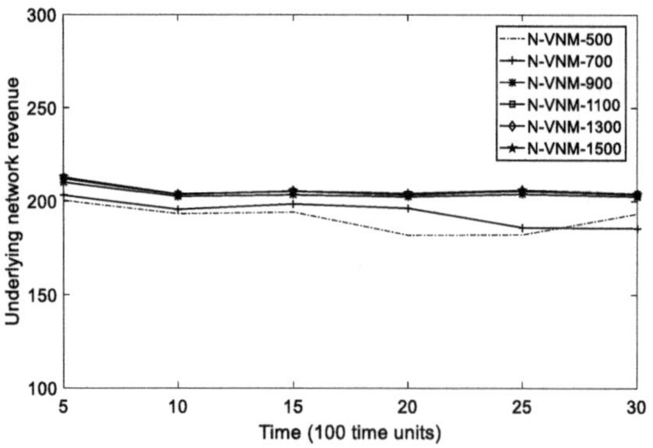

Fig. 1. Underlying network revenue.

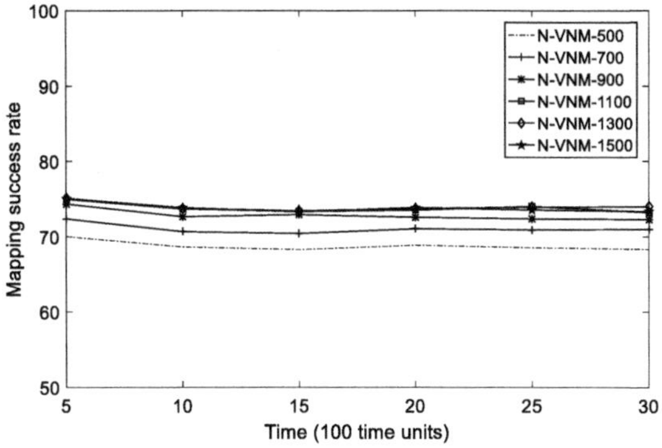

Fig. 2. Mapping success rate.

the N-VNM value is 1100, the underlying network revenue and mapping success rate converge to a better effect. When comparing the following algorithms, the reliability matrix constructed when the N-VNM value is 1100 is used for analysis.

6.3 Algorithm Comparison

In order to verify the performance of the algorithm RAAoRH in this paper, the algorithm of this paper is compared with the algorithm RAAoRO (resource allocation algorithm based on order). Based on the order of virtual network requests, algorithm RAAoRO allocates resources to virtual network requests according to the first-come-first-served strategy.

In terms of underlying network revenue, the underlying network revenue of the algorithm RAAoRH in this paper is 11.4% higher than that of the algorithm RAAoRO. In terms of mapping success rate, the mapping success rate of the algorithm RAAoRH in this paper is 7.4% higher than that of the algorithm RAAoRO. From the analysis of the experimental results, it can be seen that the algorithm in this paper allocates more reasonable resources to the virtual network request, thereby improving the underlying network revenue and the success rate of mapping (Figs. 3 and 4).

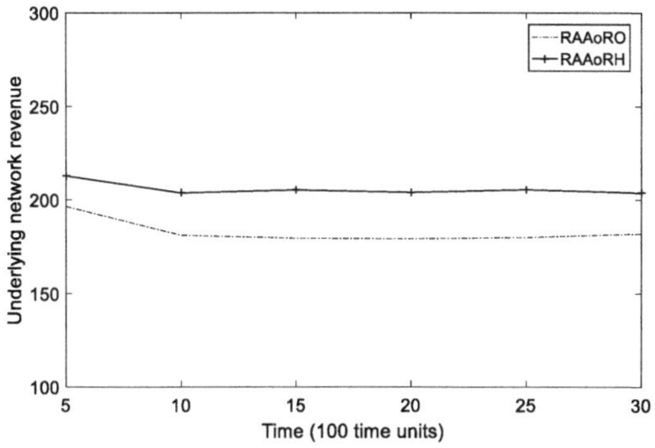

Fig. 3. Comparison of underlying network revenue.

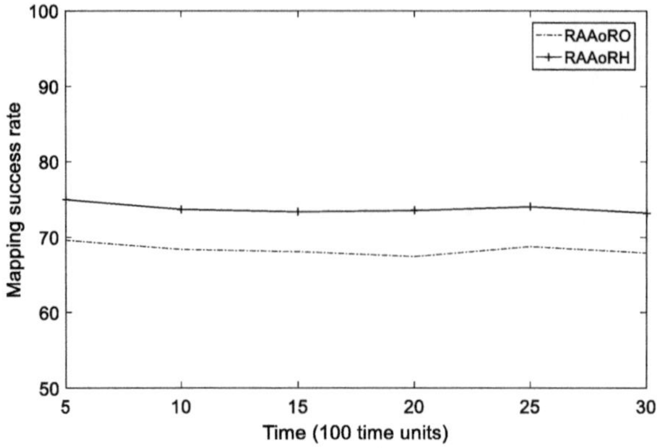

Fig. 4. Comparison of mapping success rate.

7 Conclusion

Network slicing technology can fully improve the utilization rate of existing network resources. How to allocate existing network resources to virtual networks has become a research focus. In order to solve the problem of low success rate of virtual network mapping in the existing research, this paper proposes a resource allocation algorithm of power communication network based on reliability and historical data under network slicing. First, the reliability of the virtual network nodes and the reliability of the underlying network nodes are modeled. Secondly, based on the virtual network model and the underlying network model, the power communication network resource allocation algorithm based on reliability and historical data is proposed. Finally, the performance of the algorithm in this paper is verified in terms of the underlying network revenue and the success rate of mapping. In the next step, based on the research results of this paper, we will further optimize the correlation between node resource allocation and link resource allocation, thereby optimizing the effect of the resource allocation algorithm.

References

1. Hap-SliceR, A.A.: A radio resource slicing framework for 5G networks with haptic communications. IEEE Syst. J. **12**(3), 2285–2296 (2018)
2. Fischer, A., Botero, J.F., Beck, M.T., et al.: Virtual network embedding: a survey. IEEE Commun. Surv. Tutor. **15**(4), 1888–1906 (2013)
3. Lira, V., Tavares, E., Oliveira, M., et al.: Virtual network mapping considering energy consumption and availability. Computing **101**, 1–31 (2018)
4. Zheng, X., Tian, J., Xiao, X., et al.: A heuristic survivable virtual network mapping algorithm. Soft. Comput. **23**(5), 1453–1463 (2018)
5. Chowdhury, S.R., Ahmed, R., Khan, M.M.A., et al.: Dedicated protection for survivable virtual network embedding. IEEE Trans. Netw. Serv. Manag. **13**(4), 913–926 (2016)

6. Md, M., Nashid, S., Reaz, A., et al.: Multi-path link embedding for survivability in virtual networks. IEEE Trans. Netw. Serv. Manag. **13**(2), 253–266 (2016)
7. Chowdhury, S.R., Ahmed, R., Shahriar, N., et al.: Revine: reallocation of virtual network embedding to eliminate substrate bottlenecks. In: 2017 IFIP/IEEE Symposium on Integrated Network and Service Management (IM), pp. 116–124. IEEE (2017)
8. Raza, M.R., Fiorani, M., Rostami, A., et al.: Dynamic slicing approach for multi-tenant 5G transport networks. IEEE/OSA J. Opt. Commun. Netw. **10**(1), 77–90 (2018)
9. Zegura, E.W., Calvert, K.L., Bhattacharjee, S.: How to model an Internet work. In: IEEE Infocom, pp. 594–602 (1996)

5G Slice Allocation Algorithm Based on Mapping Relation

Qiwen Zheng[1], Guoyi Zhang[2(✉)], Minghui Ou[1], and Jian Bao[1]

[1] Shenzhen Power Supply Co. Ltd., Shenzhen 518010, China
[2] Power Control Center of China Southern Power Grid, Guangzhou 510623, China
13418763020@163.com

Abstract. In the 5G network slicing environment, in order to improve the utilization of underlying network resources, this paper proposes a resource allocation algorithm of power communication network based on mapping relation under 5G network slicing. The successful virtual network resource allocation case is modeled as a knowledge underlying to provide data support for the new virtual network resource allocation problem. According to the topology characteristics of the virtual network, the importance of the virtual node is analyzed to determine the priority of the virtual node when the virtual network resource is allocated. In the experimental part, it is verified that the algorithm in this paper has achieved good results in terms of virtual network mapping success rate and underlying network resource utilization.

Keywords: Network slicing · Power communication network · Resource allocation · Virtual network mapping · Resource utilization

1 Introduction

The transmission rate of 5G networks reaches more than 1 Gbps per second, which poses challenges to the construction and operation of power communication networks. In order to meet the communication needs of 5G networks, network slicing technology has become a key technology to improve the utilization rate of power communication network resources [1]. After using network slicing technology, the traditional power communication network is divided into underlying network and virtual network. Therefore, how the underlying network allocates resources to the virtual network has become a key research content [2]. Existing studies have solved the problems of high energy consumption, low mapping success rate, and low resource utilization by using neural networks, genetic algorithms, and deep learning methods [3–9].

Through the analysis of the existing research, it can be seen that the existing research on virtual network resource allocation mainly focuses on improving the utilization rate of underlying network resources, and has achieved good results. However, there is a lack of analysis of the mapping relationship of virtual nodes in resource allocation, which leads to the need to further improve the resource utilization of the underlying network. This paper models the successful virtual network resource allocation case as a knowledge base, analyzes the importance of virtual nodes according to

Q. Liu et al. (Eds.): CENet 2020, AISC 1274, pp. 1608–1616, 2021.
https://doi.org/10.1007/978-981-15-8462-6_184

the topological characteristics of the virtual network, and finally builds a node mapping model based on Bayesian network, and a resource allocation algorithm of power communication network based on mapping relationship is presented.

2 Problem Description

The underlying network is represented by $G_S = (N_S, E_S)$. Among them, N_S represents the underlying node, provides CPU resources $cpu(n_i^s)$ for the virtual nodes of the virtual network. E_S represents the underlying link and provides bandwidth resources $bw(e_j^s)$ for the virtual links of the virtual network. The virtual network is represented by $G_V = (N_V, E_V)$, where N_V represents a virtual node, which is used to carry the computing function of the power business by applying for CPU resource $cpu(n_i^v)$ from the underlying network. E_V represents a virtual link. By applying bandwidth resource $bw(e_j^v)$ from the underlying network, it is used to carry the communication function of the power service.

Virtual network mapping includes two processes: node mapping and link mapping. The node map is represented by $n_i^v \downarrow n_i^s$, which represents the underlying node $n_i^s \in N_S$ allocating CPU resources to the virtual node $n_i^v \in N_V$. When performing node mapping, the amount of resources allocated by the underlying node to the virtual node needs to meet the CPU resource requirement $cpu(n_i^v)$ proposed by the virtual network request. Link mapping is represented by $e_j^v \downarrow p_j^s$, which represents the underlying path $p_j^s \in E_S$ to allocate link resources to the virtual link $e_j^v \in E_V$. During link mapping, the link resources allocated by the underlying link to the virtual link need to meet the bandwidth resource requirement $bw(e_j^v)$ proposed by the virtual network request. The success rate of the virtual network mapping is calculated using formula (1). Q_{win}^V represents the number of virtual network requests that successfully obtained resources at time t, and $Q^V(t)$ represents the number of virtual network requests that arrived at time t.

$$Q_{win}^V = \lim_{T \to \infty} \frac{\sum_{t=0}^{T} Q_{win}^V(t)}{\sum_{t=0}^{T} Q^V(t)} \tag{1}$$

3 Network Modeling

3.1 Underlying Node Mapping Association Matrix

When constructing a virtual network resource allocation case, the matrix M of $n \times n$ is used to represent the association relationship between the allocated virtual networks. The matrix element $m_{ij} \in M$ represents the relationship between the underlying node $n_j^s \in N_S$ and the underlying node $n_i^s \in N_S$ after the virtual node $n_i^v \in N_V$ of the virtual network is successfully mapped to the underlying node $n_i^s \in N_S$, in which virtual node $n_j^v \in N_V$ adjacent to the virtual node $n_i^v \in N_V$ is successfully mapped to the underlying node $n_j^s \in N_S$. $m_{ij} \in M$ is calculated by dividing the sum of the bandwidth values (assigned to the virtual link by the end-to-end path $P(n_i^s, n_j^s)$ between the underlying node $n_i^s \in N_S$ and the underlying node $n_j^s \in N_S$ within the specified time period T) by the number of links of the end-to-end path $P(n_i^s, n_j^s)$.

3.2 Virtual Network Feature Analysis

The amount of CPU resources of a virtual node refers to the remaining amount of CPU resources of the current virtual node, which is indicated by $cpu(n_i^v)$. The amount of link resources of a virtual node refers to the sum of the amount of bandwidth resources of the virtual links connected to the current virtual node, which is expressed by $AL(n_i^v)$, and the calculation method is as formula (2). Among them, $E(n_i)$ represents the set of virtual links directly connected to the virtual node n_i^v.

$$AL(n_i^v) = \sum\nolimits_{e_j \in E(n_i^v)} bw(e_j) \tag{2}$$

The position of the virtual node in the virtual network is measured by the distance from the current virtual node to all other virtual nodes of the virtual network, which is denoted by $NC(n_i^v)$ and calculated using formula (3). Where, $Dis(n_i^v, n_j^v)$ represents the number of links included in the end-to-end path from the virtual node n_i^v to the virtual node n_j^v.

$$NC(n_i^v) = \frac{1}{\sum\nolimits_{n_j^v \in N_V} Dis(n_i^v, n_j^v)} \tag{3}$$

Based on the above analysis, the importance of virtual node $n_i^v \in N_V$ is defined as $IMPORT(n_i^v)$, and it is calculated using formula (4).

$$IMPORT(n_i^v) = (cpu(n_i^v) + AL(n_i^v)) \times NC(n_i^v) \tag{4}$$

3.3 Node Mapping Model Based on Bayesian Network

The key issue of virtual network resource allocation is to select appropriate underlying nodes for virtual nodes. Therefore, the association between virtual nodes and physical nodes is the key to solving the problem. In order to associate virtual nodes with underlying nodes, this paper proposes a node mapping model based on Bayesian networks, as shown in Fig. 1.

Suppose that virtual network request $N_V = \{n_i^v, n_2^v, \ldots, n_i^v, \ldots, n_m^v\}$ includes m virtual nodes. The underlying network $N_S = \{n_i^s, n_2^s, \ldots, n_i^s, \ldots, n_m^s\}$ includes m underlying nodes. Then, the Bayesian map calculation of virtual node n_m^v is shown in formula (5). Among them, $P(n_m^s | n_1^s, \ldots, n_{m-1}^s)$ represents the conditional probability

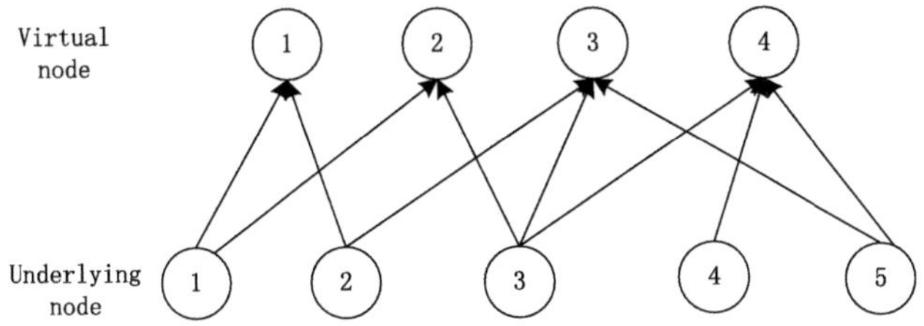

Fig. 1. Node mapping model based on Bayesian network

that virtual node n^v_m is mapped to the underlying node n^s_m under the condition $n^s_1, \ldots, n^s_{m-1}$. Among them, $n^s_1, \ldots, n^s_{m-1}$ represents the underlying node set mapped by the virtual node $n^v_1, \ldots, n^v_{m-1}$.

$$P(n^v_m) = P(n^s_m | n^s_1, \ldots, n^s_{m-1}) \ldots P(n^s_2 | n^s_1) P(n^s_1) \tag{5}$$

If the mapping of virtual node n^v_m is only related to its directly connected virtual nodes, then formula (5) can be simplified to formula (6). Among them, $pa(n^v_{m-pa})$ represents the parent node in the Bayesian network of the virtual node n^v_{m-pa} that has been successfully allocated resources and is connected to the virtual node n^v_m.

$$P(n^v_m) = P(n^s_m | pa(n^v_{m-pa})) \tag{6}$$

When calculating the probability of formula (6), if the number of the $pa(n^v_{m-pa})$ is less than or equal to 2, the underlying node mapping association relationship matrix can be used to quickly solve. Otherwise, it is $pa(n^v_m) = \{n_1^{pa,s}, n_2^{pa,s} \ldots, n_j^{pa,s}\}$. Formula (6) becomes Formula (7), and its computational complexity will increase rapidly and become very complicated.

$$P(n^v_m) = P(n^s_m | n_1^{pa,s}, n_2^{pa,s} \ldots, n_j^{pa,s}) = \frac{P(n^s_m, n_1^{pa,s}, n_2^{pa,s} \ldots, n_j^{pa,s})}{P(n_1^{pa,s}, n_2^{pa,s} \ldots, n_j^{pa,s})} \tag{7}$$

The derivation process of formula (7) is as follows.

$$
\begin{aligned}
P(n_m^v) &= P(n_m^s | n_1^{pa,s}, n_2^{pa,s} \ldots, n_j^{pa,s}) \\
&= \frac{P(n_m^s, n_1^{pa,s}, n_2^{pa,s} \ldots, n_j^{pa,s})}{P(n_1^{pa,s}, n_2^{pa,s} \ldots, n_j^{pa,s})} \\
&= \frac{P(n_m^s, n_j^{pa,s}) P(n_1^{pa,s}, n_2^{pa,s} \ldots, n_{j-1}^{pa,s} | n_m^s, n_j^{pa,s})}{P(n_j^{pa,s}) P(n_1^{pa,s}, n_2^{pa,s} \ldots, n_{j-1}^{pa,s})} \\
&= P(n_m^s | n_j^{pa,s}) \frac{P(n_1^{pa,s}, n_2^{pa,s} \ldots, n_{j-1}^{pa,s} | n_m^s, n_j^{pa,s})}{P(n_1^{pa,s}, n_2^{pa,s} \ldots, n_{j-1}^{pa,s})} \\
&= P(n_m^s | n_j^{pa,s}) \frac{P(n_1^{pa,s}, n_2^{pa,s} \ldots, n_{j-1}^{pa,s} | n_m^s)}{P(n_1^{pa,s}, n_2^{pa,s} \ldots, n_{j-1}^{pa,s})}
\end{aligned} \tag{8}
$$

Due to $P(n_1^{pa,s}, n_2^{pa,s} \ldots, n_{j-1}^{pa,s} | n_m^s) = \frac{P(n_1^{pa,s}, n_2^{pa,s} \ldots, n_{j-1}^{pa,s}, n_m^s)}{P(n_m^s)}$, so,

$$
\begin{aligned}
P(n_m^v) &= P(n_m^s | n_j^{pa,s}) \times \frac{1}{P(n_m^s)} \times \frac{P(n_1^{pa,s}, n_2^{pa,s} \ldots, n_{j-1}^{pa,s}, n_m^s)}{P(n_1^{pa,s}, n_2^{pa,s} \ldots, n_{j-1}^{pa,s})} \\
&= \frac{1}{P(n_m^s)} \times (P(n_m^s | n_j^{pa,s}) \times \frac{P(n_1^{pa,s}, n_2^{pa,s} \ldots, n_{j-1}^{pa,s}, n_m^s)}{P(n_1^{pa,s}, n_2^{pa,s} \ldots, n_{j-1}^{pa,s})})
\end{aligned} \tag{9}
$$

Due to $P(n_m^v) = \frac{1}{P(n_m^s)^{m-1}} \prod_{j=1}^m P(n_m^s | n_j^{pa,s})$, so, the formula (7) is reduced to formula (10).

$$
P(n_m^v) = \frac{1}{P(n_m^s)^{m-1}} \prod_{j=1}^m P(n_m^s | n_j^{pa,s}) \tag{10}
$$

4 Resource Allocation Algorithm

The resource allocation algorithm of power communication network based on mapping relation (RAAoMR) proposed in this paper is shown in Table 1. The algorithm includes four processes: calculating the importance and ranking of virtual nodes, generating a breadth-first search tree based on virtual node relationships, allocating resources for virtual nodes, and allocating resources for virtual links.

Table 1. Resource allocation algorithm

Input: $G_S = (N_S, E_S)$, $G_V = (N_V, E_V)$, Underlying node mapping association matrix M

Output: The result G_V of resource allocation

1. Calculate the importance and sort of virtual nodes: Use formula (4) to calculate the importance $IMPORT(n_i^v)$ of $n_i^v \in N_V$, and arrange the virtual nodes in descending order according to the importance, to obtain the sorted virtual node set $N_V^{'}$;

2. Generate a breadth-first search tree based on the virtual node relationship: take the first virtual node n_i^v of the virtual node set $N_V^{'}$ as the root node, and generate a breadth-first search tree $Tree(n_i^v)$;

3. Allocate resources for virtual nodes: allocate resources layer by layer for the virtual nodes of breadth-first search tree $Tree(n_i^v)$:

a) Sort the virtual nodes in descending order based on $IMPORT(n_i^v)$;

b) For each virtual node, calculate $P(n_m^v)$ based on formula (10), and select the underlying node that has the largest $P(n_m^v)$ and meets the CPU requirements of n_m^v to allocate resources for it. If there is no underlying node that meets the requirements, resource allocation fails. The algorithm ends.

4. Allocate resources for virtual links: Use the k-shortest path algorithm to allocate underlying link resources to meet the needs of A for each virtual link.

5 Performance Analysis

5.1 Simulation Environment

To verify the performance of the algorithm in this paper, the experiment used GT-ITM tool [10] to generate the network environment. The network environment includes underlying network and virtual network. The underlying network contains 200 underlying nodes, and the nodes are connected to each other with a probability of 0.5. The algorithm RAAoMR and the algorithm RAAoRO (resource allocation algorithm of power communication network based on order) are compared from the three aspects of the virtual network mapping success rate, the average utilization rate of the underlying network link, and the average utilization rate of the underlying network nodes.

5.2 Algorithm Comparison

It can be seen from the figure that with the running of the algorithm, the success rate of virtual network mapping under both algorithms tends to be stable. The mapping success rate of the algorithm RAAoMR in this paper is about 9.5% higher than that of the algorithm RAAoRO, indicating that the algorithm in this paper allocates more reasonable resources for virtual network requests (Fig. 2).

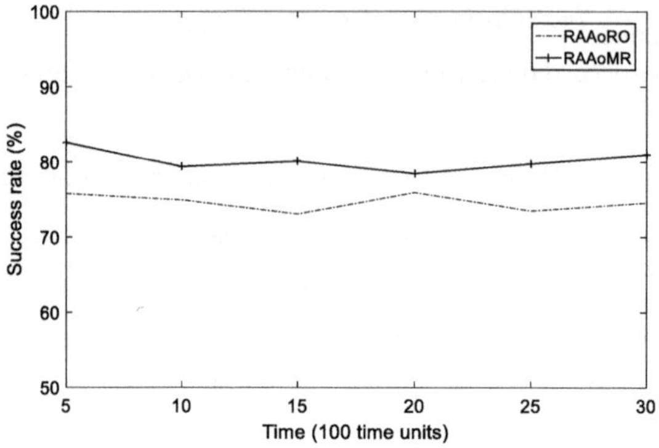

Fig. 2. Analysis of the success rate of virtual network mapping

It can be seen from the figure that with the operation of the algorithm, the average utilization rate of the underlying network link and the average utilization rate of the underlying network nodes under the two algorithms are relatively stable. In this paper, the average utilization rate of the underlying network link of the algorithm RAAoMR is higher than the algorithm RAAoRO by about 6.2%, and the average utilization rate of the underlying network node is higher than the algorithm RAAoRO by about 9.2%, indicating that the algorithm in this paper allocates more node resources and link resources to virtual network requests (Figs. 3 and 4).

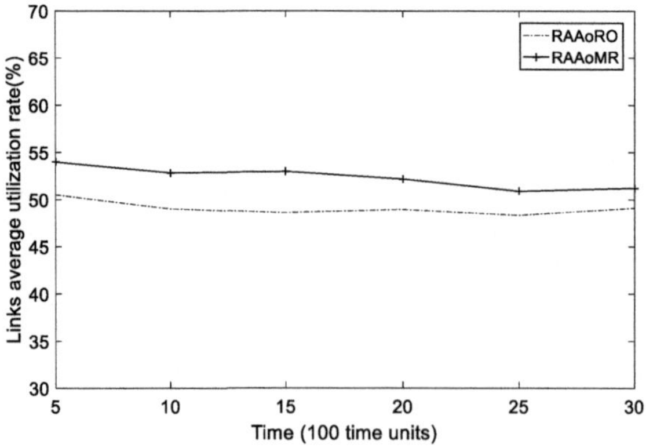

Fig. 3. Analysis of average utilization rate of underlying network links

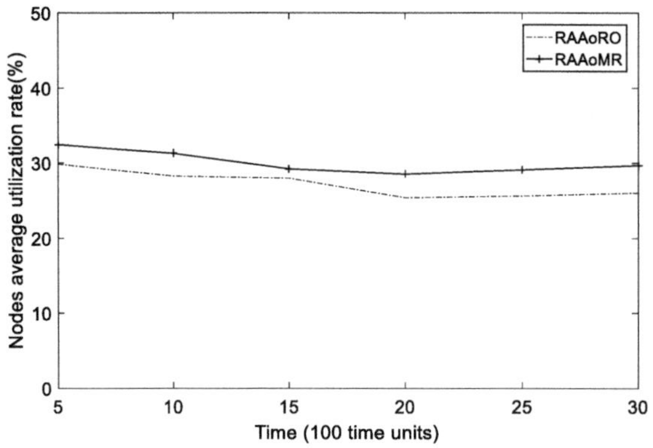

Fig. 4. Analysis of average utilization rate of underlying network nodes

6 Conclusion

This paper models the successful case of virtual network resource allocation as a knowledge underlying and provides data support for the new virtual network resource allocation problem. Experiments verify that the algorithm in this paper has achieved good results in terms of virtual network mapping success rate and underlying network resource utilization. In the next step, based on the research results of this paper, the energy consumption limit of the underlying network will be added to the constraints of resource allocation, so as to further improve the scope of application of the research results of this paper.

References

1. Aijaz, A.: Hap-SliceR: a radio resource slicing framework for 5G networks with haptic communications. IEEE Syst. J. **12**(3), 2285–2296 (2018)
2. Fischer, A., Botero, J.F., Beck, M.T., et al.: Virtual network embedding: a survey. IEEE Commun. Surv. Tutor. **15**(4), 1888–1906 (2013)
3. Jahani, A., Khanli, L.M., Hagh, M.T., et al.: Green virtual network embedding with supervised self-organizing map. Neurocomputing **351**, 60–76 (2019)
4. Jahani, A., Khanli, L.M., Hagh, M.T., et al.: EE-CTA: energy efficient, concurrent and topology-aware virtual network embedding as a multi-objective optimization problem. Comput. Stand. Interfaces **66**, 1–17 (2019)
5. Zhang, P., Yao, H., Li, M., et al.: Virtual network embedding based on modified genetic algorithm. Peer-to-Peer Netw. Appl. **12**(2), 481–492 (2019)
6. Dehury, C.K., Sahoo, P.K.: DYVINE: fitness-based dynamic virtual network embedding in cloud computing. IEEE J. Sel. Areas Commun. **37**(5), 1029–1045 (2019)
7. Dolati, M., Hassanpour, S.B., Ghaderi, M., et al.: Deep ViNE: virtual network embedding with deep reinforcement learning. In: IEEE INFOCOM 2019-IEEE Conference on Computer Communications Workshops (INFOCOM WKSHPS), pp. 879–885. IEEE (2019)
8. Mijumbi, R., Serrat, J., Gorricho, J.L., et al.: A path generation approach to embedding of virtual networks. IEEE Trans. Netw. Serv. Manag. **12**(3), 334–348 (2015)

9. Melo, M., Sargento, S., Killat, U., et al.: Optimal virtual network embedding: node-link formulation. IEEE Trans. Netw. Serv. Manag. **10**(4), 1–13 (2013)
10. Zegura, E.W., Calvert, K.L., Bhattacharjee, S.: How to model an internet work. In: IEEE Infocom, pp. 594–602 (1996)

High-Reliability Virtual Network Resource Allocation Algorithm Based on Service Priority in 5G Network Slicing

Huicong Fan(✉), Jianhua Zhao, Hua Shao, Shijia Zhu, and Wenxiao Li

State Grid Hebei Economic Research Institute, Shijiazhuang 050000, China
jyy_fanhc@163.com

Abstract. In the context of 5G network slicing, in order to solve the problem of low service reliability, a high-reliability virtual network resource allocation algorithm based on service priority is proposed in this paper. The algorithm includes four steps: priority ranking of virtual network request and virtual nodes, reliability ranking of underlying nodes, resource allocation for virtual nodes, and resource allocation for virtual links. The priority ranking of virtual network request and virtual nodes mainly include two processes of ordering the virtual network based on service priority and ordering the virtual node based on centrality. In the step of allocating resources to virtual nodes, the main work is to allocate resources to virtual nodes based on the resources and reliability of the underlying nodes. In the simulation experiment, the algorithm in this paper is compared with the traditional algorithm, which verifies that the algorithm in this paper has achieved good results in terms of key business reliability indicators.

Keywords: 5G network · Network slicing · Resource allocation · Quality of experience (QoE)

1 Introduction

With the increasing construction and application of 5G networks, virtualized evolved packet core (vEPC) technology has become a key technology of 5G core networks [1]. In the vEPC technology environment, the network slicing architecture based on virtualization has become the main architecture of the 5G core network. Under the network slicing architecture, the original basic core network is divided into the underlying network and the virtual network. Among them, the underlying network is responsible for the construction of the basic network, and the network virtualization technology is used to divide the basic network resources into multiple independent virtual resources. The virtual network is responsible for renting network resources from the underlying network to carry 5G services and provide services to end users. Under this background, how to allocate the underlying network to the virtual network has become a critical problem that needs to be solved urgently [2].

Q. Liu et al. (Eds.): CENet 2020, AISC 1274, pp. 1617–1625, 2021.
https://doi.org/10.1007/978-981-15-8462-6_185

From the main problems solved by the existing research, the existing research can be divided into two aspects of improving resource utilization and improving network reliability. In terms of improving resource utilization, there have been many studies, mainly to optimize resource allocation algorithms to improve resource utilization. For example, document [3] improves the utilization of link resources by proposing an optimized link resource allocation algorithm. Literature [4] uses a greedy algorithm to solve the problem of low utilization of the underlying network resources. Literature [5] introduced the popular deep learning algorithm into the field of resource allocation, utilized the self-learning ability of the deep learning algorithm, and improved the resource utilization rate by improving the adaptive ability of the algorithm. In terms of improving network reliability, it mainly solves the problem that virtual network services are affected after the underlying network fails. For example, literature [1] uses network coding theory to propose a network link protection and recovery mechanism, which effectively improves the reliability of network links. Literature [6] proposed a single-region network failure recovery mechanism, which effectively solved the problem of service recovery under single-region network failure. Reference [7] marks some of the underlying network resources as backup resources, and uses the heuristic algorithm to solve the problem of low reliability of the underlying network resources.

Existing research has achieved more results. However, the existing research mainly solves the problem of the reliability of the underlying network from the perspective of resource backup, while ignoring the basic characteristics of different services that require different reliability of the network. This paper allocates resources based on the service priority of the virtual network, so as to allocate high-reliability resources to the high-priority virtual network, thereby improving the reliability of the virtual network service.

2 Problem Description

2.1 Network Architecture

The network architecture under the 5G network slice is shown in Fig. 1. As can be seen from the figure, the 5G core network contains multiple vEPCs. These vEPCs provide various business application services for 5G wireless networks. vEPCs are connected to each other through a network link. The network functions included in each vEPC include a Home Subscriber Server (HSS), Mobility Management Entity (MME), Packet Data Network Gateway (PGW), and Serving Gateway (SGW). Considering that the internal resources of each vEPC can be quickly migrated through virtualization technology, so if a single vEPC internal resource can meet the fault recovery, there is no need to consider using the resources of other networks.

This article mainly studies the problem that failure recovery cannot be achieved through vEPC's own backup resources when a single or multiple vEPC failure occur. Therefore, this article mainly solves the problem of how to improve the reliability of the network in the event of a failure of the vEPC and the network link.

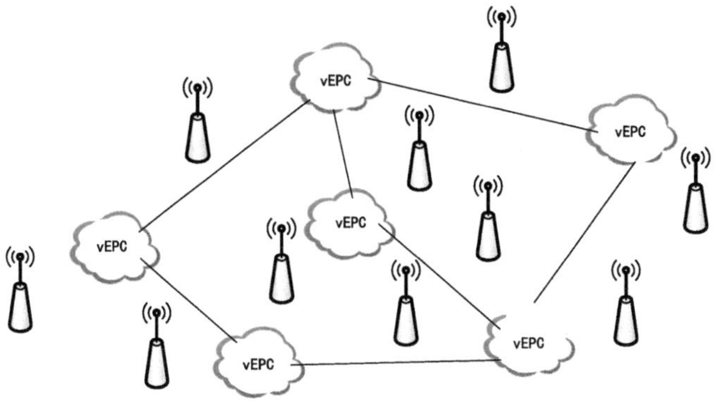

Fig. 1. Network architecture under 5G network slice.

2.2 Virtual Network Mapping Problem Description

The following describes the problem of virtual network mapping from three aspects: underlying network, virtual network, and virtual network mapping. In terms of the underlying network, the underlying network is modeled as an undirected weighted graph $G^S = (N^S, E^S)$, where N^S and E^S represent the underlying node set and the underlying link set, respectively. The attribute set possessed by the underlying node $n_i^s \in N^s$ is denoted by $\eta_{n_i}^S = \{\sigma_{n_i}^S, \tau_{n_i}^S, Loc_{n_i}^S\}$, and includes computing resource attribute $\sigma_{n_i}^S$, storage resource attribute $\tau_{n_i}^S$, and location attribute $Loc_{n_i}^S$. The attribute set possessed by the underlying link $l_{ij}^s \in L^s$ is denoted by $\eta_{e_j}^S$ and contains the link bandwidth attribute $B_i(l_{ij}^s)$.

In terms of virtual networks, the virtual network is modeled as an undirected weighted graph $G^V = (N^V, E^V)$, where N^V and E^V represent the set of virtual nodes and the set of virtual links, respectively. The attribute set possessed by the virtual node $n_i^v \in N^v$ is denoted by $\eta_{n_i}^V = \{\sigma_{n_i}^V, \tau_{n_i}^V, Loc_{n_i}^V\}$, and includes a computing resource attribute $\sigma_{n_i}^V$, a storage resource attribute $\tau_{n_i}^V$, and a location attribute $Loc_{n_i}^V$. The attribute set possessed by the virtual link $l_{ij}^v \in L^v$ is denoted by $\eta_{e_j}^V$, and includes the link bandwidth attribute $B(l_{ij}^v)$.

In terms of virtual network mapping, when the virtual network requests the underlying network resources, the underlying network allocates resources to the virtual network. Each virtual network has a period from the application of resources to the release of resources. This document is called the life cycle of the virtual network, and is denoted by Del_{re}^v. The process of allocating resources from the underlying network to the virtual network is called virtual network mapping. Use $M_N : (N^V \to N^S, E^V \to P^S)$ to represent, where P^S represents the underlying path of the virtual link mapping, including multiple connected underlying links.

3 Network Reliability Model

3.1 Reliability Model of Virtual Network

The reliability model of the virtual network is modeled from two aspects: the service priority of the virtual network and the priority of the virtual node. The service priority of the virtual network is used to ensure that critical services can be allocated with high-reliability resources. The priority of the virtual node is used to ensure the reliability of key virtual nodes, thereby improving the reliability of the virtual network.

In terms of the service priority of the virtual network, P_v^i is used to indicate the service priority of the virtual network. This article mainly analyzes from three aspects: business type, quality of experience (QoE), and the amount of resources requested. The type of business is represented by P_i, which is divided into key business and general business. When $P_i = 1$, the current business is a key business. When $P_i = 0$, the current business is a general business. Service uses QoE_i to indicate that the value of QoE_i is related to the specific network characteristics. This article sets its value to a real number greater than zero. The requested resource amount is represented by Re_i, and the value of Re_i is determined by the node resources and link resources requested by the virtual network. This paper sets the sum of node resources and link resources requested by the virtual network as the amount of resources requested by the virtual network. Based on the above analysis, the service priority P_v^i of the virtual network is calculated using formula (1). Among them, α, β and γ represent the weighting factors of service type, service QoE, and the amount of requested resources, respectively.

$$P_v^i = \alpha P_i + \beta QoE_i * \gamma \mathrm{Re}_i \tag{1}$$

In terms of the priority of virtual nodes, the evaluation is mainly based on the centrality of virtual nodes. Because the more central virtual node, the more routing data through it. In this paper, $CC(n_i^v)$ is used to indicate the centrality of virtual node n_i^v, and formula (2) is used for calculation. Among them, $n_j^v \in \varphi(n_i^v)$ represents other virtual nodes except node n_i^v in the node set of the virtual network. d_{ij} represents the number of links in the shortest path between virtual node n_i^v and virtual node n_j^v.

$$CC(n_i^v) = \frac{1}{\sum_{n_j^v \in \varphi(n_i^v)} d_{ij}} \tag{2}$$

3.2 Reliability Model of the Underlying Network

It can be seen from the network architecture model under the 5G network slice that the underlying network nodes contain many types of resources, and the reliability of the underlying network nodes is critical to the reliability of the virtual network. The higher the reliability of the underlying network nodes, the higher the reliability of the underlying network.

In order to evaluate the reliability of the underlying network nodes, this paper evaluates the historical failure rate of the underlying network nodes and the replacement rate of the underlying network nodes. Among them, the historical failure rate of the underlying network node is evaluated using the number of failures in the time period T. The greater the number of failures, the less reliable the current underlying network nodes. The replacement rate of the underlying network node is measured by the hop count index of the rerouting strategy, and is calculated using formula (3). Among them, $J(n_i^s, n_j^s)$ represents the mutual replacement rate of the underlying network node n_i^s and node n_j^s, and $N(n_i^s)$ represents the set of underlying network nodes directly connected by the underlying network node n_i^s within one hop. $|*|$ represents the number of nodes contained in the set. Therefore, the value of $J(n_i^s, n_j^s)$ represents the probability that node n_i^s and node n_j^s can replace each other. When $J(n_i^s, n_j^s) = 1$, it means that node n_i^s and node n_j^s are directly connected.

$$J(n_i^s, n_j^s) = \frac{|N(n_i^s) \cap N(n_j^s)|}{|N(n_i^s) \cup N(n_j^s)|} \tag{3}$$

Based on formula (3), the alternative probability of the bottom node n_i^s is defined as $Rp_{n_i^s}$, and the formula (4) is used for calculation. It can be seen from the formula that the larger the value of $Rp_{n_i^s}$, the greater the probability that the current node n_i^s is replaced by other nodes. At this time, when the node n_i^s fails, it can be quickly replaced by the adjacent node, so as to quickly recover from the failure and ensure the quality of the virtual network carried on it.

$$Rp_{n_i^s} = \sum\nolimits_{n_{ia}^s, n_{ib}^s \in N(n_i^s)} J(n_{ia}^s, n_{ib}^s) \tag{4}$$

4 Algorithm

The high-reliability virtual network resource allocation algorithm based on service priority (HVNRAoSP) proposed in this paper is shown in Table 1. The algorithm includes four steps: priority ranking of virtual network request and virtual nodes, reliability ranking of underlying nodes, resource allocation for virtual nodes, and resource allocation for virtual links. (1) In the priority ranking of virtual network request and virtual nodes, it mainly includes two processes of sorting the virtual network based on service priority and sorting the virtual nodes based on centrality. (2) In the reliability sorting step of the bottom nodes, the main work is to sort the bottom nodes based on the reliability of the bottom nodes. (3) In the step of allocating resources for virtual nodes, the main work is to allocate resources for virtual nodes based on the resources and reliability of the underlying nodes. (4) In the step of allocating resources for the virtual link, the main work is to allocate the link resources of the shortest path to the virtual link based on the resource constraints of the virtual link.

Table 1. Algorithm HVNRAoSP.

Input: underlying network $G^S = (N^S, E^S)$, virtual network $G^V = (N^V, E^V)$

Output: resource allocation strategy $M_N : (N^V \rightarrow N^S, E^V \rightarrow P^S)$

1. Virtual network request and priority ranking of virtual nodes

(1) For each virtual network $G_i^V \in G^V$, use formula (1) to calculate the priority P_v^i and arrange it in descending order to obtain a new virtual network set G_{ord}^V;

(2) For each virtual node $n_i^v \in N_i^v$ of the virtual network G_i^V, use formula (2) to calculate its centrality $CC(n_i^v)$, and arrange it in descending order to obtain a new virtual node set N_i^{v-ord};

2. Reliability ranking of the underlying nodes

(1) For each underlying node $n_i^s \in N^s$ of the underlying network, use formula (4) to calculate its reliability $Rp_{n_i^s}$;
(2) Based on the reliability $Rp_{n_i^s}$ of the bottom node n_i^s, arrange the bottom node in descending order to obtain a new set N_{ord}^s;

3. Allocate resources for virtual nodes

(1) For each virtual network G_i^V in the virtual network set G_{ord}^V, allocate resources in sequence;

(2) For each virtual node n_i^v in the node set N_i^{v-ord} of virtual network G_i^V, select the bottom node that meets the resource constraints of n_i^v and the highest reliability $Rp_{n_i^s}$ from the set N_{ord}^s, and allocate resources for the virtual node n_i^v;

4. Allocate resources for virtual links

(1) Use Dijkstra's algorithm to find the shortest path for each virtual node n_i^v;

(2) Determine whether the link resource on the current path meets the resource constraints of the virtual link. If it meets, allocate the link resource for the next virtual node n_j^v; if it does not, use the Dijkstra algorithm to look for the sub-optimal shortest path for node n_i^v.

5 Performance Analysis

In the performance analysis part, similar to the existing research, the GT-ITM tool [8] is used to generate the underlying network and virtual network topology. The number of underlying network nodes increased from 100 to 500, and the virtual network nodes obey the uniform distribution of [3, 10]. For the underlying network and the virtual network, any two nodes are connected with a probability of 0.3. The underlying network nodes and link resources follow the uniform distribution of [25, 50], and the virtual network nodes and link resources follow the uniform distribution of [1, 5]. In terms of business types of the virtual network, 20% of the services in the virtual network are set as key services, and the rest of the services are non-critical services. When simulating the failure of the underlying network node, the 20% of node with the highest failure rate of the underlying nodes is set as the failed node.

This paper compares the algorithm HVNRAoSP with the virtual network resource allocation algorithm under random resource backup (VNRAoRRB). The algorithm VNRAoRRB uses the same amount of underlying network resources as the algorithm in this paper. To verify the performance indicators of the two algorithms, the critical business reliability is used for evaluation. Critical business reliability is measured by the proportion of the number of critical business operations that are normally running in the network node failure environment to the total number of critical business operations.

The experimental results of the impact of the scale of the underlying network on the performance of the algorithm are shown in Fig. 2. The X axis represents five network environments that simulate the increase of the number of nodes in the underlying network from 100 to 500, and is used to analyze the impact of the size of the underlying network on the performance of the algorithm. It can be seen from the figure that with the increase of the scale of the underlying network, the reliability indicators of key services under the two algorithms are relatively stable, indicating that the two algorithms have achieved good performance results under different scales of the underlying network. From the comparison of the two algorithms, the reliability of the key business

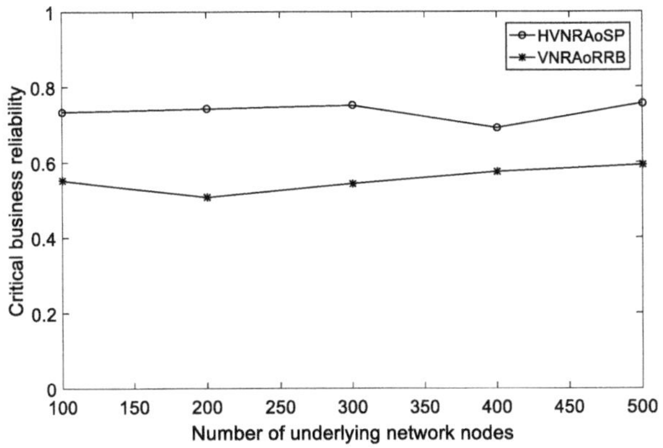

Fig. 2. The impact of the size of the underlying network on algorithm performance.

under this algorithm is higher, which shows that the algorithm of this paper allocates more optimized bottom network resources to the key business, thereby improving the reliability of the key business.

The experimental results of the impact of the underlying node failure rate on algorithm performance are shown in Fig. 3. The X axis indicates that the failure rate obeys (0.01%, 0.05%), (0.05%, 0.1%), (0.1%, 0.5%), (0.5%, 0.1%) (1%, 1.5%) (sequentially numbered 1 to 5) are used to analyze the impact of the underlying network node failure probability on algorithm performance.

As can be seen from the figure, as the failure rate of the underlying network nodes increases, the reliability of key services under both algorithms is decreasing. When the failure rate of the underlying network node increases to 0.1% or more, the impact of the failure of the underlying network node on the reliability of critical services increases rapidly, indicating that the increased incidence of the underlying network failure affects more critical services. From the comparison of the two algorithms, we can see that the algorithm in this paper improves the reliability of critical services better than the algorithm VNRAoRRB.

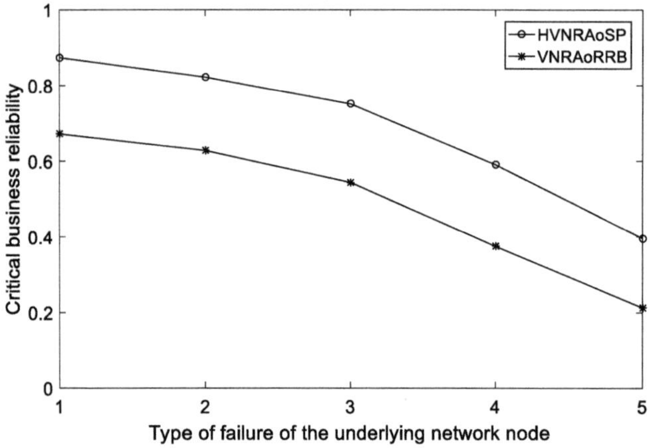

Fig. 3. The impact of bottom node failure rate on algorithm performance.

6 Conclusion

With the increase in the number and types of 5G network services, some services have put higher and higher requirements on the reliability of the network. In the existing resource allocation research, it mainly solves the problem of resource utilization improvement, and ignores the reliability requirements of the service on the network. In order to solve the problem of low service reliability, this paper proposes a high reliability virtual network resource allocation algorithm based on service priority. In the simulation experiment part, the algorithm of this paper is compared with the traditional algorithm in terms of key business reliability indicators, which verifies that the algorithm of this paper has achieved good experimental results. As the scale of 5G networks increases, the probability of network failures is increasing. In order to ensure the reliability of high-priority services, rapid service recovery in the event of a network failure has been called an urgent problem. In the next step, based on the research results of this paper, we will further study the rapid failure recovery mechanism of high-priority services.

References

1. Peng, M., Li, Y., Jiang, J., et al.: Heterogeneous cloud radio access networks: a new perspective for enhancing spectral and energy efficiencies. IEEE Wirel. Commun. **21**(6), 126–135 (2014)
2. Tang, J., Tay, W.P., Quek, T.Q.S.: Cross-layer resource allocation in cloud radio access network. In: 2014 IEEE Global Conference on Signal and Information Processing (GlobalSIP), pp. 158–162. IEEE (2014)
3. Mijumbi, R., Serrat, J., Gorricho, J.L., et al.: A path generation approach to embedding of virtual networks. IEEE Trans. Netw. Serv. Manag. **12**(3), 334–348 (2015)
4. Lira, V., Tavares, E., Oliveira, M., et al.: Virtual network mapping considering energy consumption and availability. Computing **101**(8), 937–967 (2018)
5. Dolati, M., Hassanpour, S.B., Ghaderi, M., et al.: DeepViNE: virtual network embedding with deep reinforcement learning. In: IEEE INFOCOM 2019-IEEE Conference on Computer Communications Workshops (INFOCOM WKSHPS), pp. 879–885. IEEE (2019)
6. Yousaf, F.Z., Loureiro, P., Zdarsky, F., et al.: Cost analysis of initial deployment strategies for virtualized mobile core network functions. IEEE Commun. Mag. **53**(12), 60–66 (2015)
7. Zheng, X., Tian, J., Xiao, X., et al.: A heuristic survivable virtual network mapping algorithm. Soft. Comput. **23**(5), 1453–1463 (2018). https://doi.org/10.1007/s00500-018-3152-7
8. Zegura, E.W., Calvert, K.L., Bhattacharjee, S.: How to model an internetwork. In: Proceedings of IEEE INFOCOM 1996. Conference on Computer Communications, vol. 2, pp. 594–602. IEEE (1996)

Design of Real-Time Vehicle Tracking System for Drones Based on ROS

Yong Xu[1], Jiansheng Peng[1,2(✉)], Hemin Ye[2], Wenjian Zhong[1], and Qingjin Wei[1]

[1] School of Physics and Mechanical and Electronic Engineering, Hechi University, Yizhou 546300, China
sheng120410@163.com
[2] School of Electronic Engineering, Guangxi Normal University, Guilin 541004, China

Abstract. When using vehicles to hunt down criminal vehicles, the difficulty of chasing criminals is increased due to factors such as vision, surrounding environment and road conditions. In order to solve this problem, we have designed a real-time vehicle tracking system based on ROS. The system uses Pixhawk as the flight control platform in the hardware part. Its internal integrated attitude control module and altitude control module are mainly responsible for controlling the attitude of the aircraft to achieve the effect of stable flight of the aircraft. The GPS module is responsible for acquiring coordinate position data, so that the aircraft can achieve the fixed point effect. This system solves the problem that a processor has insufficient resources when processing relatively large data. Compared with using vehicles to chase criminal vehicles, this system can solve the problem of increasing the difficulty of chasing criminals due to factors such as visual field, surrounding environment and road conditions. This system greatly improves the pursuit efficiency.

Keywords: ROS · Target detection · Color segmentation · Tracking

1 Introduction

At present, the process of chasing criminal vehicles is still using vehicles. Due to the influence of the visual field, surrounding environment, road conditions and other conditions during the pursuit, the difficulty of pursuing the criminal is increased. As a brand-new high-tech product, the multi-rotor UAV has the characteristics of small size, good maneuverability, high flexibility, good concealment and no space constraints.

In 2005, the University of British Columbia, Canada developed the Hockey Tracking System project. Based on the analysis of the video sequence of the ice hockey game, this project finds the respective positions of the athletes and draws a chart of their corresponding positions on the court. The chart can search the position information of any athlete at any time. In addition, some related characteristic points can also be generated by the team on the ice of the stadium. Maintaining the changing homography between each frame can help the coach discover the key information in the game and make an effective adjustment strategy [1, 2]. In 2004, Carnegie Mellon

Q. Liu et al. (Eds.): CENet 2020, AISC 1274, pp. 1626–1637, 2021.
https://doi.org/10.1007/978-981-15-8462-6_186

University used binocular vision, module matching, feature tracking, color segmentation, artificial icon recognition, current relative displacement SFM real estimation, terrain reconstruction, target recognition and tracking on the unmanned helicopter YAMAHA R50 platform [3]. In 2005, MIT used module matching on the unmanned helicopter X-CELL60 platform to achieve target detection [4]. In 2004, Chiba University used binocular vision, SFM, SLAM and optical flow methods to achieve autonomous landing and relative pose estimation on the quadrotor platform [5].

In 2003, Beijing University of Aeronautics and Astronautics used artificial icon recognition and binocular vision to achieve autonomous landing and target tracking path planning on an unmanned helicopter platform [6]. In 2005, Tsinghua University used artificial icon recognition, SFM to achieve autonomous landing and target tracking on the unmanned helicopter Raptor-30 platform [7]. In 2008, Zhejiang University used artificial icon recognition, binocular vision, homography matrix and mean drift to realize relative pose estimation, autonomous landing and target tracking in the unmanned helicopter HIROBO Freya90 [8]. Northwestern Polytechnical University uses digital map scene matching, manual icon detection and runway recognition to achieve autonomous landing and target tracking on unmanned helicopters and fixed-wing UAV platforms [9]. In 2009, Shanghai University used artificial icon detection and binocular vision to achieve autonomous landing, target tracking and obstacle avoidance on an unmanned helicopter platform [10].

In this design, the OpenCV library is used to process the image data in real time on the platform of the multi-rotor UAV to achieve real-time tracking of the vehicle. The design uses the ROS system, which has distributed communication and a point-to-point network. The system can divide the huge data processing on different computers or processors, and then combine the data processed by different computers through the publication, subscription and service of topics to realize the communication and data interaction between multiple computers or processors. It can well solve the performance problems of a single processor. Therefore, a real-time vehicle tracking system based on ROS is designed.

2 Overall Design of Real-Time UAV Tracking Vehicle System

2.1 UAV Real-Time Tracking Vehicle System Design

The overall system framework of system design is shown in Fig. 1. Raspberry Pi as the core processor. Pixhwak as a flight controller. The receiver receives the data transmitted by the remote control. The electronic governor controls the motor rotation and speed regulation. GPS obtains position coordinate information. The USB camera collects image data.

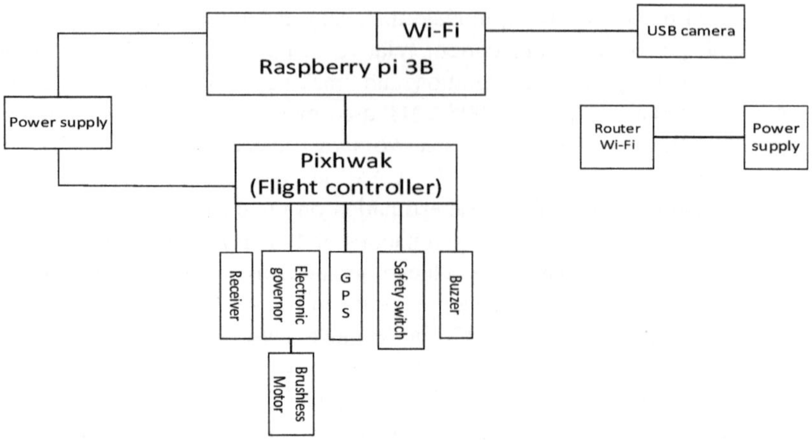

Fig. 1. UAV real-time tracking vehicle system framework.

2.2 Realization Method of UAV Real-Time Tracking Vehicle System

The flight control platform part is composed of the aircraft power part, the power supply part, and the flight controller part. We installed a USB camera and a raspberry pi 3B core processor on the successfully tested flight control platform. The raspberry pi 3B processes the USB camera data, and the execution program realizes the purpose of keeping track of the target color vehicle. Based on the OpenCV vision library, the functions of opening the USB camera, reading the camera image, setting the target color threshold, searching for the target color area, searching for the maximum contour in the target color area, and searching for the maximum contour centroid coordinates are realized. Then, through the error between the coordinates of the center of mass of the target vehicle and the center coordinates of the image, the movement of the aircraft is controlled to eliminate the errors. By keeping the center of mass of the target at the center of the image, the tracking effect is achieved.

3 UAV Real-Time Tracking Vehicle System Hardware Design

3.1 UAV Real-Time Tracking Vehicle System Hardware Composition

The hardware structure of UAV real-time tracking vehicle system is shown in Fig. 2. The hardware consists of raspberry pi 3B processor, Pixhawk (flight controller), USB camera, receiver, GPS module, electronic governor module, brushless motor, safety switch module and buzzer module.

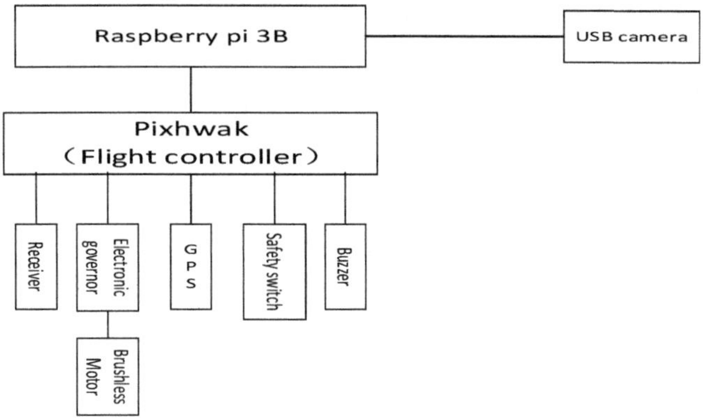

Fig. 2. Hardware block diagram of UAV real-time tracking vehicle system.

3.2 UAV Tracks the Motor Part of the Vehicle System in Real Time

As shown in Fig. 3, the principle of rotation of the brushed motor is to energize the carbon brush, and the current flows from the carbon brush to the commutator. This causes the coil to generate a magnetic field that interacts with the magnetic field of the stator. Like repels, opposite attracts. The commutator changes the direction of the current to produce a changing magnetic field that causes the rotor to rotate. As shown in Fig. 4, the brushless motor changes the conduction sequence of the three wires to realize a magnetic field that changes the coil, thereby causing the motor to rotate. Turn on sequentially from AB to CB to change the direction of the magnetic field to achieve the rotation of the motor.

Fig. 3. Hardware block diagram of UAV real-time tracking vehicle system.

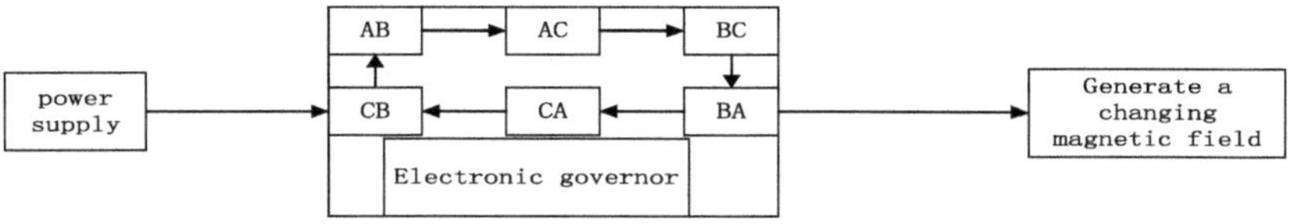

Fig. 4. Block diagram of variable magnetic field generated by brushless motor.

3.3 UAV Real-Time Tracking Vehicle System Electronic Governor Part

The rotation of the motor needs to continuously change the conduction sequence of the three phases. Due to the use of DC power, there is no converter inside the motor that can sequentially change the three-phase conduction sequence. Therefore, an external converter is needed to change the three-phase conduction sequence in order to achieve the effect of driving the motor. This converter is called an electronic governor. It can not only change the three-phase conduction sequence, but also can adjust the speed of the motor through the PWM sent by the flight controller to achieve the effect of rotating the motor and adjusting the speed. As shown in Fig. 5.

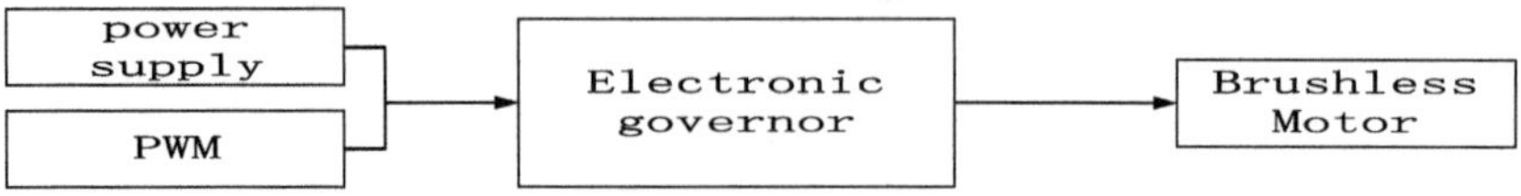

Fig. 5. Block diagram of ESC drive motor.

3.4 UAV Real-Time Tracking Vehicle System Power Supply Part

The most basic part of the entire flight control platform is a stable input power supply. The stable input power plays a vital role in the normal operation of the entire control system. The input power requirements for the flight controller and raspberry pi 3B processor are both 5V3A. The power supply of the electronic governor is a battery, and the electronic governor supplies power to the motor. Receiver power and GPS power are directly connected from the Pixhawk aircraft. The power supply block diagram is shown in Fig. 6.

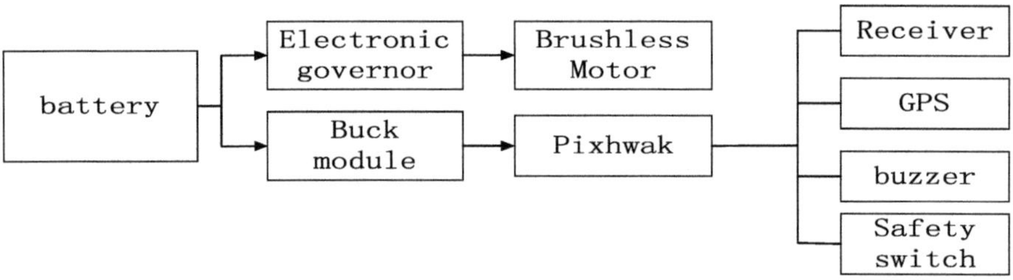

Fig. 6. Power supply block diagram.

The power circuit is shown in Fig. 7. The working principle of the entire buck: First, we set the output voltage by adjusting the resistance of the output terminal and the feedback terminal. When the resistance is set and the output voltage is set accordingly. The feedback voltage is obtained after the output terminal passes through the voltage dividing resistor. The feedback voltage is compared with the reference voltage (0.8 V). When the feedback voltage being smaller than the reference voltage, we use the comparator to keep the MOS transistor on and increase the duty cycle. When the feedback voltage exceeding the set value, we turn off the MOS tube and reduce the duty cycle. Until the output voltage is equal to the set voltage.

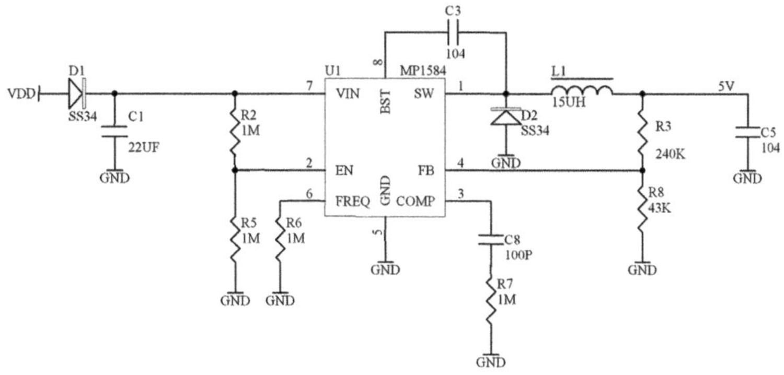

Fig. 7. MP1584 power circuit.

4 UAV Real-Time Tracking Vehicle System Software Design

The overall block diagram of the software design is shown in Fig. 8. The ROS system serves as the underlying processing system of the software part. The USB camera is responsible for acquiring image data. The OpenCV library provides a machine vision library. GPS obtains position coordinate information. The target detection processing part acquires target contour information and centroid. The target tracking processing part acquires the deviation of the centroid of the target from the center of the image and acquires position control information. The mode control processing section performs flight mode control. After processing the location data tracked by the target, the ROS system publishes location topics to the aircraft to subscribe and execute the corresponding location information.

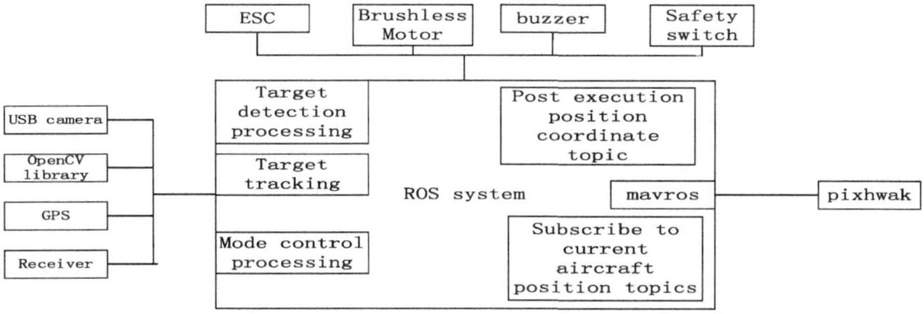

Fig. 8. Overall framework of software.

4.1 Design of UAV Real-Time Tracking Vehicle System Target Detection Software

Target detection is the processing of image data. Filter the target vehicle from the image. The UAV real-time tracking vehicle system uses a color image segmentation method. The target vehicle area is selected by the difference between the color of the target vehicle and the background color. First, grayscale the image. Secondly, the target area is filtered through the threshold of the target vehicle color to obtain the target binary image, and then the contour of the target color area is extracted, and then the target centroid is extracted after the target contour is obtained, and the target detection is finally completed. The system block diagram of the entire target detection part is shown in Fig. 9.

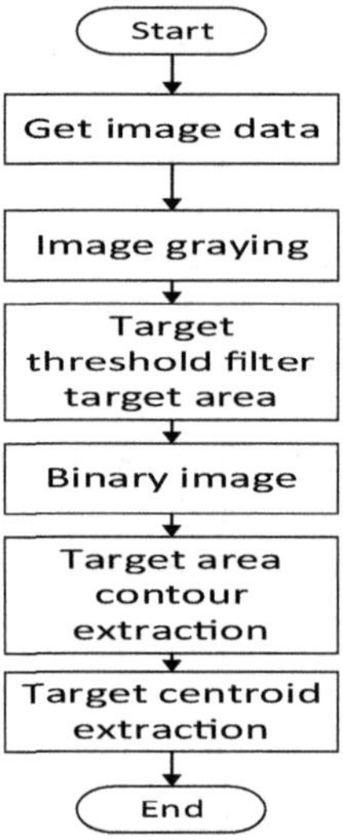

Fig. 9. Block diagram of target detection process.

Image Color Segmentation of Target Detection. This system uses the color difference between the target vehicle color and the background to obtain the target color vehicle through image color segmentation. Image segmentation generally has RGB, HSV and other methods. The system uses the HSV method for image segmentation. The setting value of this method is relatively continuous, and the effect is better than RGB, and it is easier to detect the target color. The flowchart of color segmentation is shown in Fig. 10.

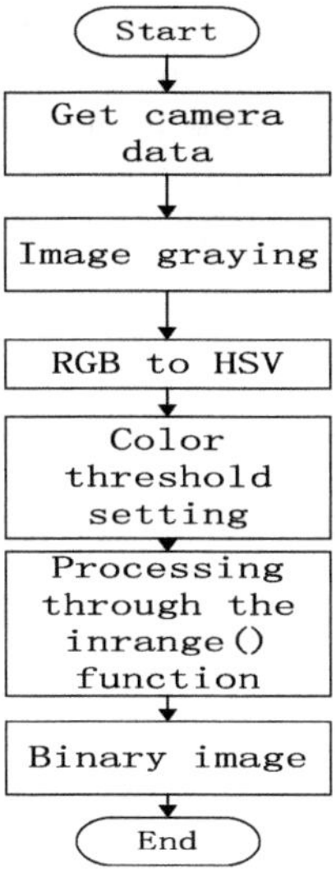

Fig. 10. Flow chart of color segmentation.

Target Contour Extraction. The system uses the findContours() function of the OpenCV library to extract the target contour. After getting the outline, we draw the outline. Drawing the outline to reflect whether the obtained outline is correct. This part draws the contour through the drawContours() function, and finally outputs the contour drawing. Figure 11 is a flowchart of the contour extraction part.

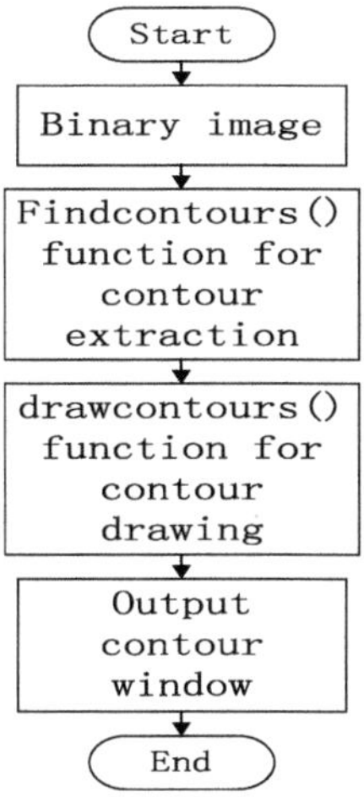

Fig. 11. Contour extraction flowchart.

Target Centroid Extraction for Target Detection. After obtaining the outline of the target area, the target tracking of the system uses the deviation of the center of the target vehicle area and the center of the image to obtain the error, thereby eliminating the error to achieve the tracking effect. Therefore, it is critical to determine the center of mass of the target area to track the target.

The centroid formula is shown in the following formula (1):

$$Xc = \frac{My}{m} \quad Yc = \frac{Mx}{m} \quad m = \sum\nolimits_{i=1}^{n} mi \quad My = \sum\nolimits_{i=1}^{n} miXi \quad Mx = \sum\nolimits_{i=1}^{n} miYi \tag{1}$$

where Xc is the abscissa of the centroid coordinate. Yc is the ordinate of the centroid coordinate. My and Mx are the static moments of X axis and Y axis respectively. m is the total mass of particles. In image processing, the first-order matrix is related to shape and size. So by transforming into a matrix form, the first moment of the matrix is used to find the centroid.

The first-order matrix formula is as follows Eqs. (2), (3), and (4):

$$\mathrm{m00} = \sum\nolimits_{x=y=-r}^{r} x^0 y^0 I(x,y) = \sum\nolimits_{x=y=-r}^{r} I(x,y) \tag{2}$$

$$\mathrm{m01} = \sum\nolimits_{x=y=-r}^{r} x^0 y^1 I(x,y) = \sum\nolimits_{x=y=-r}^{r} x * I(x,y) \tag{3}$$

$$\mathrm{m10} = \sum\nolimits_{x=y=-r}^{r} x^1 y^0 I(x,y) = \sum\nolimits_{x=y=-r}^{r} y * I(x,y) \tag{4}$$

The centroid formula (5) obtained by the first-order matrix:

$$C = \left(\frac{\mathrm{m10}}{\mathrm{m00}}, \frac{\mathrm{m01}}{\mathrm{m00}}\right) \tag{5}$$

This formula corresponds to the original formula of the centroid. m01 corresponds to Mx which is the static moment of X axis. m10 corresponds to the static moment of the Y axis of My. m00 corresponds to m which is the sum of masses of particles.

4.2 Design of Target Tracking Software for UAV Real-Time Tracking Vehicle System

This system uses PID control algorithm and aircraft position control to eliminate this deviation, keeping the target in the center of the image, and achieving the effect of the aircraft tracking the target vehicle. Part of the target tracking process is shown in Fig. 12.

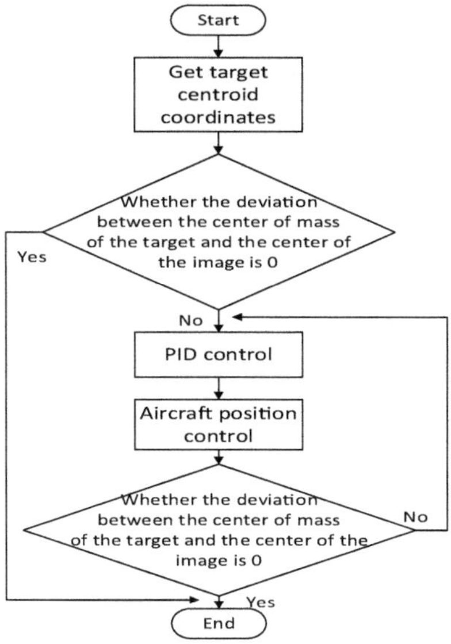

Fig. 12. Block diagram of target tracking system.

PID Control Algorithm for Target Tracking. PID control algorithm is a more commonly used feedback control algorithm, which can quickly eliminate the errors and interferences. The PID control algorithm formula is shown in the following formula (6):

$$\mathrm{U}(t) = Kp(err(t) + \frac{1}{\mathrm{Ti}}\int err(t)dt + \frac{\mathrm{Td*derr(t)}}{\mathrm{dt}}) \tag{6}$$

P is a proportional term and is the core of the entire PID control algorithm. Kp is the scale factor. When an error occurring, the scale item will respond immediately. The larger the scale factor, the faster the response. However, if the scale factor is too large, it will cause high frequency oscillation and system instability. I is the integral term. Ti is the integration time constant. The role of I is to eliminate the steady-state error produced by P. D is the differential term. Td is the differential time constant. The role of D is to advance the regulation, increasing the degree of damping of the system, suppressing the oscillation generated by P, increasing the stability of the system and reduce the regulation time of the system. So I and D play the role of assisting P. The purpose is to eliminate the influence of P.

As shown in Fig. 13. The error between the coordinates of the image center and the coordinates of the center of mass of the target input into the PID controller for error processing. The result of the processing sent to the aircraft position control. In this way, the aircraft can be controlled to execute the flight, reducing the error, and then judging whether the error is eliminated by feedback comparison. Looping control until the error is eliminated.

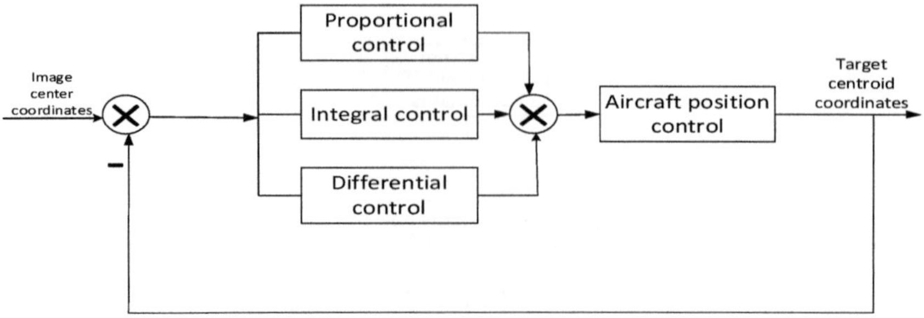

Fig. 13. Control system block diagram.

Position Control of Target Tracking. This system controls the position of the aircraft to perform target tracking. The information output by GPS is relative to the geodetic coordinate system, taking the earth's center of mass as the origin and the local ENU coordinate system. The coordinate system is based on the takeoff point of the aircraft as the origin. The Y axis points to the North Pole. The X axis points east. The Z axis is vertically upward. The location control block diagram of target tracking is shown in Fig. 14.

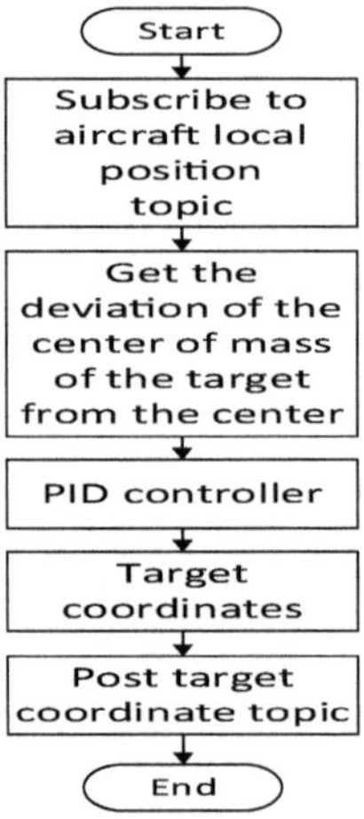

Fig. 14. Block diagram of target tracking position control.

5 Conclusion

Based on the design of ROS-based UAV real-time tracking vehicle system, in the hardware circuit design, the power supply part is the primary link for the normal operation of the entire system. Therefore, the system uses a switching power supply for power supply, which improves the stability of the power supply and enables the circuit to operate stably. We use OpenCV library functions for image data processing and use the image data processing data to complete the corresponding projects. When tracking the target, it is very important to deal with the coordinate position of the tracking target. This system uses PID to deal with errors, greatly improving the accuracy of target tracking. In this system design, the tracking effect of slower vehicles can be achieved. However, the target detection part is not stable enough, and is greatly affected by light. The tuning of PID control parameters has not achieved a good result.

Acknowledgements. The authors are highly thankful to the Research Project for Young and Middle-aged Teachers in Guangxi Universities (ID: 2019KY0621), to the Natural Science Foundation of Guangxi Province (No. 2018GXNSFAA281164). This research was financially supported by the project of outstanding thousand young teachers' training in higher education institutions of Guangxi, Guangxi Colleges and Universities Key Laboratory Breeding Base of System Control and Information Processing.

References

1. Meingast, M., Geyer, C., Sastry, S.: Vision based terrain recovery for landing unmanned aerial vehicles. In: 43rd IEEE Conference on Decision and Control (CDC) (2004)
2. Heble, H., Cameron, C.: OTAS: Oxford aerial tracking system. Robot. Auton. Syst. **55**(9), 661–666 (2007)
3. Ng, A.Y., Kim, H.J., Jordan, M.I., et al.: Inverted autonomous helicopter flight via reinforcement learning (2003)
4. Lookingbill, A., Lieb, D., Stavens, D., et al.: Learning activity-based ground models from a moving helicopter platform (2005)
5. Ollero, A., Maza, I.: Multiple heterogeneous unmanned aerial vehicles (2007)
6. Qiu, L.W., Song, Z.S., Shen, W.Q.: Computer vision algorithm for autonomous landing of unmanned helicopters. J. Beijing Univ. Aeronaut. Astronaut. **29**(2), 99–102 (2003)
7. Liu, S.Q.: Research and implementation of vision-based autonomous landing method for unmanned helicopter. Institute of Software, Chinese Academy of Sciences, Beijing (2005)
8. Pan, S.L., Wang, X.J., Shen, W.Q., et al.: Visually guided simulation of autonomous landing system for unmanned helicopter. Aerosp. Control **26**(2), 63–67 (2008)
9. Qiu, L.W., Song, Z.S., Shen, W.Q.: Research on computer vision technology for landing control of unmanned helicopters. J. Aeronaut. **24**(4), 351–354 (2003)
10. Shan, H.Y.: Research on flight control technology of quadrotor unmanned helicopter. Nanjing University of Aeronautics and Astronautics, Jiangsu (2008)

Medical Engineering and Information Systems

Study of Cold-Resistant Anomalous Viruses Based on Dispersion Analysis

Hongwei Shi[1,3], Jun Huang[2,3(✉)], Ming Sun[5], Yuxing Li[1], Wei Zhang[4], Rongrong Zhang[1], Lishen Wang[1], Tong Xu[3], and Xiumei Xue[4]

[1] Suqian College, Suqian, Jiangsu, China
[2] Jiangsu University, Zhenjiang, Jiangsu, China
steedhuang@ujs.edu.cn
[3] Suqian Institute of Industrial Technology, Suqian, Jiangsu, China
[4] The Affiliated Suqian Hospital of Xuzhou Medical University, Suqian, Jiangsu, China
[5] The Affiliated Drum Tower Hospital of Nanjing University, Nanjing, Jiangsu, China

Abstract. This paper deals with the recognition of a relationship between the cold-resistant viruses by fluorescent staining, especially the recognition of Hurst characteristic of cold-resistant viruses based on the time series of differential normalized fluorescence indices, and derived consensus gene position maps. The complex covariance is calculated to find the kin relationship between the different coronavirus clade. Differential Normalized Fluorescence Indices (DNFI) is one of the most commonly used indexes to extract virus information by fluorescent staining medical images, widely used in virus classification and growth evaluation. In this paper, a novel method based on Hurst of time series of differential normalized fluorescence indices derived gene sequence is proposed, which considers the whole time series of differential normalized fluorescence indices generated various gene maps and is simple and practical.

Keywords: Cold-resistant · Anomalous viruses · DNFI · SARS-COV-2

1 Introduction

Low-temperature abnormal viruses are among the essential health-threatening viruses in the world today, the virus can hide in ice for many years [1], which makes it prone to worldwide spread. In the winter season, the cold-resistant abnormal virus is the main virus. The timely and accurate extraction of the characteristics of the cold-resistant abnormal virus is the basis of prevention [2] and is an essential factor related to national border security and social economic stability.

The traditional regional examination method based on section gene observation cannot meet the need to acquire the characteristics of low temperature-resistant abnormal viruses with lengthy gene compositions [3]. For example, some French experts and Indian experts made a mistake mismatching the COVID-19 gene with the HIV gene section. As such [4], we need to look at the entire gene sequence as a whole.

Q. Liu et al. (Eds.): CENet 2020, AISC 1274, pp. 1641–1648, 2021.
https://doi.org/10.1007/978-981-15-8462-6_187

With the rapid development of fluorescent dyeing technology [5], fluorescent dyeing image has been widely used in low temperature-resistant abnormal virus monitoring. In the early stage, identifying low-temperature-resistant abnormal viruses was carried out mainly by using single-phase fluorescence image data. Due to the complexity and diversity of virus types [6], there was an obvious spectrum overlap between different viruses. It is difficult to achieve an ideal classification accuracy because of the "wrong classification" and "missing classification" when viruses are classified by using single-time fluorescent image data [7]. To study the relationship of different virus traits, we need to look into the gene sequence that is mainly derived from DNFI analysis of medical image from Reverse transcription-polymerase Chain Reaction (RT-PCR) test machine [8]. An intron is a non-coding segment of DNA in a gene that separates its adjacent exons [9]. Introns are sequences that block t linear expression of genes. Introns may contain "old codes" parts of genes that have been disabled by evolution. Because introns have no meaning for the translated product's structure, they accumulate more mutations than exons. Introns play an essential role in Alternative splicing [10], and a gene can therefore produce many different proteins. In the final analysis, it is the same piece of DNA spliced together, sometimes as an Exon, sometimes as an intron [11].

A particular type of Intron called a self-splicing intron (ribozyme) can be cut away by its actions to leave the mRNA. The ratio of introns to exons varies from species to species. Viruses have fewer introns; humans have much more [12].

However, unlike "junk DNA", which is the sequence outside the genes, introns have not been found to have any functional DNA, but may be involved in gene regulation and alternative splicing regulation. Nevertheless, extensive mutations can occur if the mRNA [13] fragments corresponding to the introns are not removed.

Splicing is not unique, so exons can only be seen in adult mRNA. Even with the bioinformatics method, it is difficult to predict exons' exact location because each gene formation process is highly related to each other; the later depends of the former, forming the long-range dependence, which leads to the Fractal phenomenon [14].

DNFI is one of the most commonly used indexes to extract virus information by fluorescent staining medical image, which is widely used in virus classification and growth evaluation. DNFI time series data can accurately reflect the natural information of fluorescent dyes (exons, introns, spacers, and proteins), effectively weaken the phenomenon of "isomer" and play an important role in the study of virus classification. It can be applied to the identification of cold-resistant abnormal viruses. In the domain of virus recognition with DNFI time series, the most mature method is decision tree classification [15], which only uses a few characteristic bands in the time series, but does not consider the whole sequence.

2 Detail Method

The aim of this paper is to propose a novel method to identify low temperature resistant abnormal viruses based on Hurst's values of DNA sequence, which takes full advantage of the natural characteristics of the gene sequences of low temperature resistant abnormal viruses that distinguish them from other viruses, through the Hurst dispersion of DNFI time series, the difference between the cold-resistant abnormal virus and other virus strains was enhanced, and the high-precision identification of relationship of cold-resistant abnormal virus was realized.

In order to achieve the above-mentioned goal, this article goes through the following technical plan:

Hurst's cold-resistant anomaly virus relation identification method based on the time series of DNFI derived gene sequence includes the following steps:

Step 1: Acquire the fluorescent staining image sequence in the growth cycle of the cold-resistant abnormal virus, and construct the DNFI time sequence to form the DNFI time sequence curve with time as the horizontal coordinate and DNFI as the vertical coordinate;
Step 2: Obtaining cold-resistant abnormal virus gene sample data through field investigation or historical gene data library;
Step 3: Based on the cold-resistant abnormal virus samples, the corresponding Pixel DNFI time series curves are obtained, and the quasi-symmetric Hanning windowed moving average of all sample Pixel DNFI time series is performed, the reference curve of cold-resistant Abnormal Virus DNFI time series was formed;
Step 4: Based on the reference curve of DNFI time series of cold-resistant abnormal virus, Hurst value, that is, subtract the reference curve from the reference curve of DNFI time series of all pixels in the test area, the DNFI time series conversion curve of each gene is calculated according to the reference motif;
Step 5: Based on the conversion curve of gene sequence, the complex covariance real and imaginary of the conversion curve corresponding to the sample pixel are calculated respectively using the data of the cold-resistant abnormal virus samples, then, the maximum and minimum of both are obtained by statistical analysis as the threshold of kin recognition, and the automatic determination of threshold is realized;
Step 6: Using the conversion curve obtained in Step 5, calculate the Hurst value, and judge by the threshold value obtained in Step 5, when the differences are in the threshold range, the gene section is determined to be a cold-resistant abnormal virus, and then the distribution map of cold-resistant abnormal virus is formed;

Further, in the first step, the growth cycle of the cold-resistant abnormal virus is 30 min, ensuring as far as possible at least 30 s of a piece of fluorescent dyeing data, and the data need to be calibrated by radiation and corrected by the control group before the construction of the DNFI time series, then the DNFI is calculated by the fluorescent band and the white light band, and then the DNFI time series is formed.

Further, in Step 4, the Pixel DNFI time series is represented as

$$\mathrm{Vi} = (\mathrm{v1, i,\ v2, i, \ldots v8,\ i}) \tag{1}$$

$$Vi \in VI(V1, I, V2, I, \ldots V8, I) \tag{2}$$

The reference curve of DNFI time series of cold-resistant viruses is

$$\text{Vref} = (\text{v1}, \text{ref}, \ \text{v2}, \text{ref}, \ldots \text{v8}, \text{ref}) \tag{3}$$

$$\text{Vref} \in \text{VREF (V1, Ref, V2, REF,} \ldots \text{V8, Ref)} \tag{4}$$

According to the formula,

$$\text{Ti} = \text{Vi} - \text{Vref} = (\text{v1}, \text{i} - \text{v1, ref, v2}, \text{i} - \text{v2, ref,} \ldots \text{v8, i} - \text{v8, ref}) \tag{5}$$

Further, in Step 5, the cold-resistant abnormal virus sample data is representative, that is, the dynamic range of Hurst's sectional dispersion value obtained from the sample gene can represent the dynamic range of the transition region of anomalous virus series in the whole test regions.

3 Detail Calculations

Obtain the fluorescent staining data of the cold-resistant abnormal virus during its growth cycle, and construct the DNFI time series. The DNFI is obtained by using the Fluorescent Band (FB) and the White Band (WB) of the fluorescent dyeing data, and is calculated by formula (1).

$$DNFI = \frac{FB - WB}{FB + WB} \tag{6}$$

In Formula: FB is the fluorescence band reflection value, WB is the white light band reflection value.

Hurst value is calculated by formula below:

$$CH = \frac{1 + \log \frac{IDC(m)}{IDC(1)} \times \log \frac{Level(1)}{Level(m)}}{2}, \ IDC = \frac{E(\xi - E\xi)^k}{\mu} \tag{7}$$

Where k = 2, M is the last point on the x-axis of the graph, E is the mean.

Figure 1 A flowchart of anomaly virus identification method based on a Hurst sequence of confirmed gene sequence from DNFI time series, as we can see the right side is fetching operation from the database, the left side is the calculation procedure.

Figure 2 The Hurst map of the SARS-COV-2 strain [16], it shows that the point before 2.36, which is spike protein is extra higher than normal, this is where human ACE2 is bounded, it is positive charged, and the copper or charged other material may be used to inhibit it.

Figure 3 The Hurst map of the Bat strain [17], it shows that the point way before 2.36, which is non- structure protein is extra higher than normal, they can be targets of pharmaceutical agents like protease inhibitors etc., it helps us looking for more Chinese or Western medicines [18].

Figure 4 The Hurst map of the Pangolin strain, it shows that the point way after 2.36, which is ORF10 that does not have any similar proteins in a huge repository in

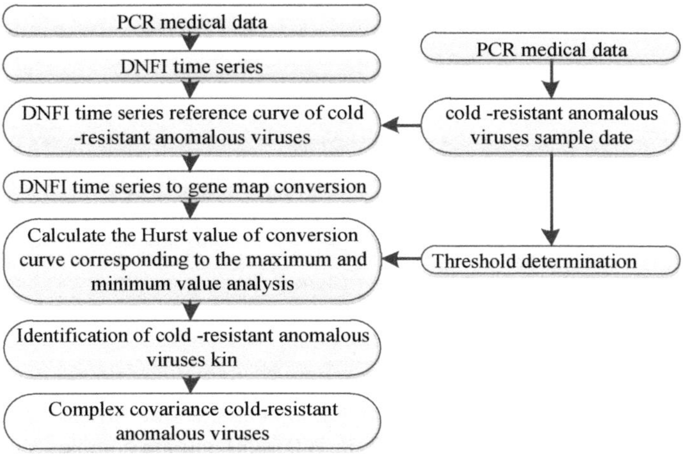

Fig. 1. A flowchart of anomaly virus identification method.

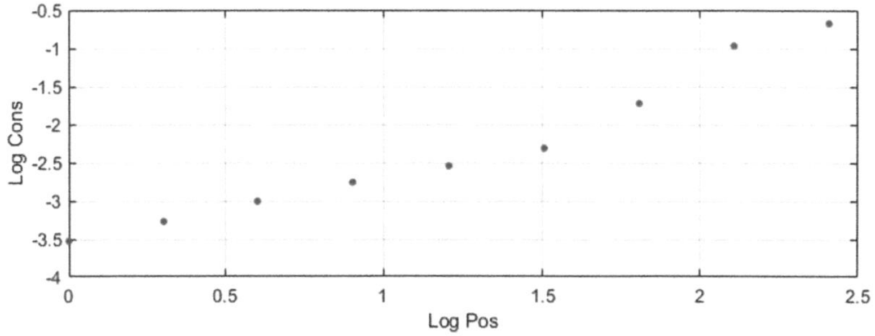

Fig. 2. Covid 19 genomic trace hurst = 0.93.

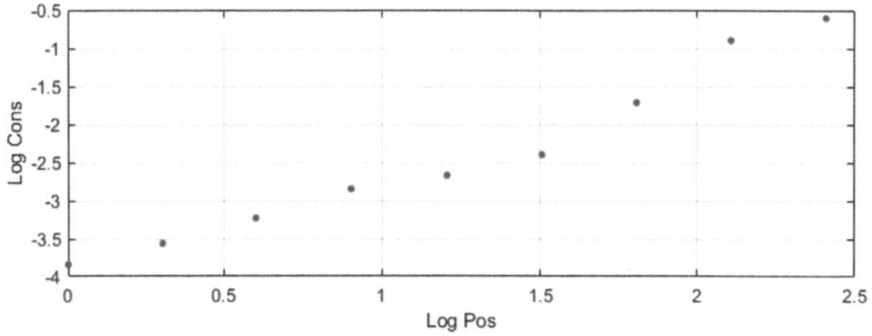

Fig. 3. Bat21 genomic trace hurst = 0.97.

NCBI. ORF10 is a short protein or peptide of length 38 residues. This unique protein can be utilized to detect the virus more quickly than PCR based methods.

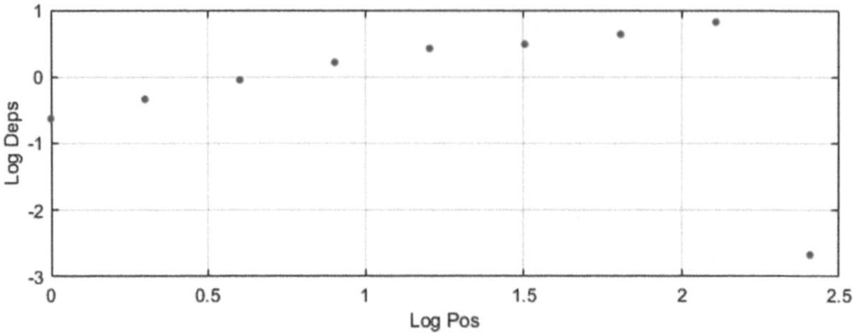

Fig. 4. Pangolin genomic trace hurst = 0.96.

The Table 1 shows the Pearson correlation coefficient of the local Hurst value of the genome of human strain and bat strain is 0.6859 and the complex sub correlation coefficient is 0.7411. No matter which measurement method is used, it shows that there is still a distance between bat and human, and the intermediate host has not been found.

Table 1. Fluorescence staining images.

	Span	Human	Bat	Hypo cov		Hypo var human		Hypo var bat	
		Slope	Slope	Real	Imag	Real	Imag	Real	Imag
1	2	0.8412	0.9093	0.147	0.000	0.000	0.341	0.000	0.432
2	4	0.8721	1.1028	0.074	0.000	0.000	0.310	0.000	0.238
3	8	0.8308	1.2638	0.027	0.000	0.000	0.351	0.000	0.077
4	16	0.7034	0.5884	0.360	0.000	0.000	0.479	0.000	0.753
5	32	0.7747	0.9124	0.175	0.000	0.000	0.407	0.000	0.429
6	64	1.9614	2.2761	0.729	0.000	0.779	0.000	0.935	0.000
7	128	2.4977	2.7024	1.791	0.000	1.316	0.000	1.361	0.000
8	256	0.9762	0.9706	0.076	0.000	0.000	0.206	0.000	0.370
AVG	64	1.182	1.341			a	b	c	d
Cor	0.975			0.238	0.000	0.000	0.199	0.000	0.284
Std		0.817	0.862	0.142	0.000	0.000	0.172	0.000	0.116
Cov	0.6859			0.067	0.000	0.000	0.208	0.000	0.021
				0.465	0.000	0.000	0.331	0.000	0.653
				0.270	0.000	0.000	0.260	0.000	0.281
				0.789	0.000	0.688	0.000	0.904	0.000
				1.548	0.000	1.509	0.000	1.589	0.000
				0.145	0.000	0.000	0.093	0.000	0.225
			Hypo	2.376	0.000	1.690	0.292	1.839	0.339
Cov	0.7411		Var			1.7152		1.8695	
Hypo Covariance between Dispersion Slope of Genomic Concensus Position of Human and Bat virus									

4 Conclusion

We have calculated the complex covariance between virus between Chinese patient, American bat and Malasia pangolins, showing the relationship among these cold-resistant viruses by fluorescent staining genesequence, especially the recognition of Hurst characteristic of derived consensus gene position maps, based on the concept of time series of differential normalized fluorescence indices, and the complex covariance is carefully calculated to find the kin relationship between the different coronavirus traits, and suggestions for finding better drug, detection and sanitization methods are given as well.

References

1. Van, D.N., Bushmaker, T., Morris, D., Holbrook, M.G., Gamble, A., Williamson, B.N., Tamin, A., Harcourt, J.L., Thornburg, N.J., Gerber, S.I., et al.: Aerosol and surface stability of SARS-CoV-2 as compared with SARS-CoV-1. N. Engl. J. Med. **382**, 1564–1567 (2020)
2. Aemro, K., Andrés, F., Marie, P., Joel, R.C., Amos, B.S., Joseph, S.: Identifification of a human immunodefificiency virus type 1 envelope glycoprotein variant resistant to cold inactivation. J. Virol. **83**(9), 4476–4488 (2009)
3. Samuel, D.C., Inder, S., Carlos, A.A., Amy, L.D., Patrick, B.P., Benjamin, D.D.: Real-time detection of COVID-19 epicenters within the United States using a network of smart thermometers. medrxiv.org (2020)
4. Luc, M., Jin, Y., Rong C.: New property of Chen Rong's DNA: occurrence of electromagnetic Signal (in Chinese) (2010)
5. Takara Bio companies: real-time qPCR for COVID-19 research (2020)
6. Justin, D.B.: Group size and nest spacing attect buggy greek virus infection in nesting house sparrow. PLoS One **6**(9), e25521–e25528 (2011)
7. Alba, G., John, S., Yun, Z., Richard, H.S., Bjoern, P., Alessandro, S.: A sequence homology and bioinformatic approach can predict candidate targets for immune responses to SARS-CoV-2. Cell Host Microbe **27**(4), 671–680.e2 (2019)
8. Guo, J.J., Sun, X.L., Yan, Q.L.: Establishment of TaqMan real-time PT-PCR for detection of flavivirus. J. Virol. Methods **177**(1), 75–79 (2011)
9. Pavlos, G.P., Karakatsanis, L.P., Iliopoulos, A.C., Pavlos, E.G., Xenakis, M.N., Peter, C., Jamie, D., Monos, D.S.: Measuring complexity, nonextensivity and chaos in the DNA sequence of the major histocompatibility complex. Phys. A: Stat. Mech. Appl. **438**, 188–209 (2015)
10. Dimitri, S.M.: Deconvoluting the most clinically relevant region of the human genome. ARUP LABORATORIES, Pathology Grand Rounds, 20 September 2018
11. Zhang, T., Wu, Q.F., Zhang, Z.G.: Probable pangolin origin of SARS-CoV- 2 associated with the COVID-19 outbreak. Curr. Biol. **30**(8), 1578 (2020). 20
12. Pross, A.: Toward a general theory of evolution: extending darwinian theory to inanimate matter. J. Syst. Chem. **2**, 1 (2011)
13. Wang, J.W., Yamoto, T., Zhao, T.C.: The epidemic situation and coping strategy of Novel coronavirus pneumonia at the initial stage of transmission in Japan (in Chinese) (2020)
14. Liu, Q., Liu, Z., Dalakas, E.: Prevalence of porcine endogenous retrovirus in Chinese pig breeds and in patients treated with a porcine liver cell-based bioreactor. World J. Gastroenterol. **11**(30), 4727–4730 (2005)

15. Wong, C.W.: Tracking the evolution of the SARS coronavirus using high-throughput, high-density resequencing arrays. Genome Res. **14**(3), 398–405 (2004)
16. Deslandes, A., Berti, V., Tandjaoui-Lambotte, Y., Chakib, A.M.C., Zahar, J.R., Brichler, S., Cohen, Y.: SARS-COV-2 was already spreading in France in late December. Int. J. Antimicrobial Agents **55**(6), 106006 (2019)
17. Jun, S.H., Wang, L.S.: Hurst dispersion graph for genomic characterization of 2019 novel corona virus how far from bat to human (2019). https://doi.org/10.13140/rg.2.2.21575.24488
18. Anthony, S., Johnson, C., Greig, D., Kramer, S.: Global patterns in coronavirus diversity. Virus Evol. **3**(1), vex012 (2017)

A Pilot Study on the Music Regulation System of Autistic Children Based on EEG

Xiujin Zhu[1], Sixin Luo[2], Xianping Niu[3], Tao Shen[1], Xiangchao Meng[4], Mingxu Sun[1](✉), and Xuqun Pei[5]

[1] University of Jinan, Jinan 250022, China
[2] Dezhou Special Education School, Dezhou 253000, China
[3] Special Education Center of Zhoucun District, Zibo 255300, China
[4] Jinan Children's Hospital, Jinan 250022, China
cse_sunmx@ujn.edu.cn
[5] Jinan Central Hospital, Jinan 250013, China

Abstract. At present, the traditional music therapy for autistic children requires professional music therapists to carry out auxiliary therapy on the side of the patients, judge the emotional state of the children, and manually play the corresponding music for treatment according to the current emotional state. Traditional music therapy not only requires professional therapist to keep an eye on it and rely on professional experience to observe and judge the effect of music therapy on autistic children, but also makes mistakes in judgment because autistic children do not show their inner emotional activities. The system uses the EEG acquisition equipment of EMOTIVEPOC+14 channel to collect the brain waves of autistic children. The median negative emotions of normal people correspond to the impulsive outward emotions of autistic children, calm emotions, and restrained autistic emotions to conduct preliminary experiments. After the self-EEG signal is collected, the detail component threshold is denoised by wavelet decomposition, the SVM algorithm is used to classify the positive, neutral and negative emotions. According to the principle of playing homogeneous music, autistic children finally have a calm state, so as to achieve the purpose of music intervention therapy. The system visually displays the identified EEG signals to the interface, which can feedback the emotional state in real time, so that the effect of music therapy for autistic children can be systematically evaluated. In this paper, the EEG data of normal people are used to verify the feasibility of the system, and the classification accuracy is 88.8%.

Keywords: Autistic children · Brain-computer interface · Emotion recognition · Music regulation

1 Introduction

Autism is a serious neuro developmental disorder with social communication disorder, speech development disorder and language communication ability deficiency as typical symptoms, accompanied by stereotyped behavior. More and more scholars have paid attention to the intervention and treatment of autistic children. As the etiology and pathogenesis of autism have not been fully elucidated, and there is no special and

Q. Liu et al. (Eds.): CENet 2020, AISC 1274, pp. 1649–1657, 2021.
https://doi.org/10.1007/978-981-15-8462-6_188

effective treatment [1]. At present, comprehensive intervention methods such as special education and psychological and behavioral therapy are often used in clinical treatment. Among them, music intervention therapy is more and more widely used. The basic principle of music therapy is to use the law of people's response to music to regulate people's physiology, psychology and behavior, to change disordered physiological reactions and maladjusted behaviors, and establish new and appropriate responses [2].

The application of EEG technology in the auxiliary treatment of autistic children plays an auxiliary role in judging the emotional state of autistic children and greatly reduces the workload of therapist. And the real-time music therapy system can timely feedback the emotional state of autistic children to the system to judge and observe the emotional changes of autistic children, visualize the emotional state, and systematically evaluate the treatment effect. For music therapy of autistic children, music is divided into positive music, negative music and neutral music, and the emotional state of autistic children is divided into introverted and autistic type, explicit impulsive type and calm type by Jin et al. [3]. Before music therapy, therapists play homogenous music with the same emotional type according to the emotional state of autistic children, and gradually transition to heterogeneous music contrary to their emotional type, with the ultimate goal of stabilizing their emotional state and finally playing neutral music. The system requires autistic children to wear EEG caps to identify the emotional state of autistic children in real time. According to the method of Jin et al. [3], homogenous music is played corresponding to autistic emotional state, and then it gradually transitions to heterogeneous music. When it is finally detected that their emotional state tends to calm down, neutral music is played. In the whole process, the system will detect and judge the emotional state of autistic children every 10 s, and if there is a change, it will transition to change the type of music, and the therapist only needs to supervise it.

Emotion recognition based on EEG is a common and effective recognition method. The main steps of emotion recognition using EEG include: emotional induction, EEG signal acquisition, EEG signal preprocessing, feature extraction, emotion pattern learning and classification, etc. [4]. In this system, the emotion are divided into three types: positive, neutral and negative, corresponding to the impulsive explicit emotion, calm emotion and introverted autistic emotion of autistic children respectively. The nature is marked mainly by the therapist to judge the emotional state of autistic children. Because the EEG signal is weak, there will be a lot of interference noise, which requires us to remove interference artifacts and preprocess the EEG signal. Luo et al. [5], through wavelet decomposition and reconstruction, eliminated Gaussian noise to complete preprocessing. The separated signals extracted by noise reduction source separation show a weak correlation, while the target separation signal has a strong correlation with the source signal. The results show that the ophthalmic artifact can be eliminated effectively. In the aspect of feature extraction and classification, the features are mainly divided into time domain features, frequency domain features, time-frequency features and other features [6]. Pham et al. [7] used EmotivEPOC headphones to collect EEG signals, used 2–30 Hz bandpass filter to divide EEG data into four frequency bands, used fast Fourier transform (FFT) with 5 s non-overlapping window to calculate the power of each frequency band of each channel and its average power as a feature, and finally used Adboost classifier to classify positive and negative emotions with an accuracy of 92.8%. In this paper, we extract the frequency band energy features and use SVM classifier to classify, and the accuracy is more than 85%.

Because the experimental agreement with autistic children has not been signed at present, we first use EmotivEPOC+ equipment to collect data from normal people, and preprocess the data for feature extraction and emotion classification (see Fig. 1).

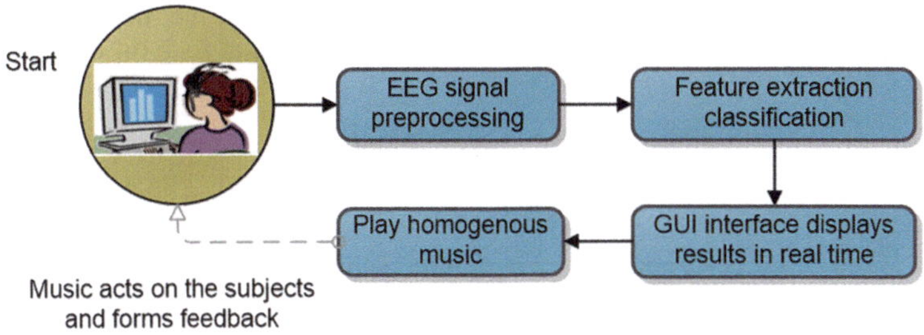

Fig. 1. Block diagram of music regulation system for autistic children based on EEG signals.

2 Experiment

2.1 Subjects and Emotional Induction

Because the EEG data of autistic children are not yet available in the experimental stage, the data of five adult men were collected to correspond to the impulsive explicit emotion, calm emotion and introverted autistic mood of autistic children according to the positive, neutral and negative emotion of normal people. The five subjects were all graduate students with an average age of 24.6 years. Each of the subjects had no mental illness, had a good mental state before collecting emotional data, and had normal vision and hearing. To ensure the external environment will not cause physical discomfort, the laboratory temperature should be appropriate, and the light should be soft. The subjects normally sat next to the computer and kept the position around the 70 cm with the video playback screen.

In the emotion-induced collection, 15 positive, moderate and negative emotion-induced videos were selected, each type had 5 videos, these video clips were selected by themselves on the network, and ten people watched the video and got the potency by scoring. The video that induces positive emotion consists of video clips that make people feel up, the background music is cheerful and radical (potency: 8.41 ± 1.72). The video that induces neutral emotion is composed of landscape video. Some relaxing sounds such as rain and sea tides are used as background music (potency: 5.21 ± 1.24). The video that induces negative emotion consists of disaster movie clips and breakup movie clips, and the background music is sad (potency: 2.42 ± 1.32). The video clips are all 1 min, and the format resolution is the same.

2.2 Experiment Process

For each subject, after the start of recording, there is 3 s of preparation time, then relax and begin to record 15 s of eye-opening baseline data, after recording, there will still be 3 s of preparation time, then relax and begin to record 15 s of eye-closing data, a total of 36 s. Then the video is played randomly, and the video is often 60 s. After that, the emotional type is marked according to the video type. During this period, relax for 2 min, so repeatedly, and finally collect 20 groups of the positive, neutral and negative emotions data. The EEG signal recorded in each experiment was 96 s (see Fig. 2).

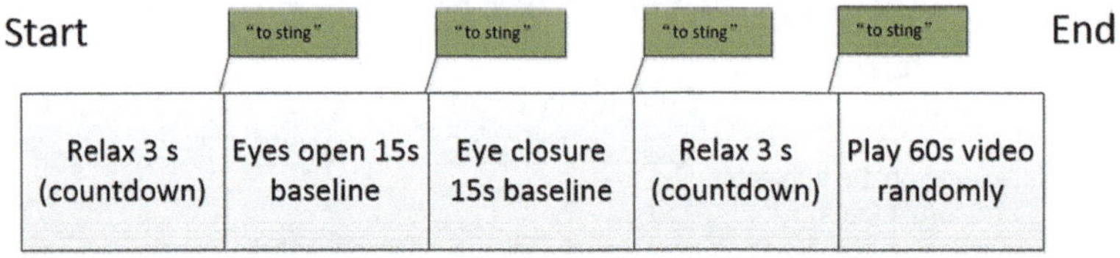

Fig. 2. Flow chart of mood induction.

2.3 Data Acquisition and Preprocessing

The music regulation system based on EEG signals of autistic children uses EMOTIVEPOC+14 channel mobile Brainwear. The EEG device has 14 electrodes located in AF3, F7, F3, FC5, T7, P7, O1, O2, P8, T8, FC6, F4, F8 and AF4 respectively (see Fig. 3). The device does not need to wet the hat to transmit EEG signals. The sampling rate is 128 Hz and the bandwidth is 0.2–45 Hz.

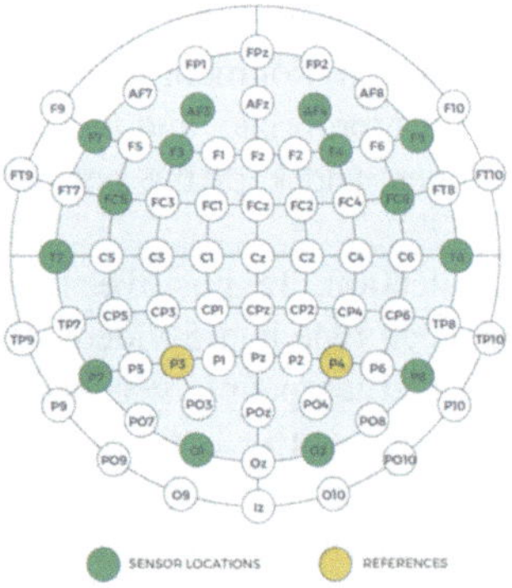

Fig. 3. Device channel diagram.

When collecting and recording data, we only verify the feasibility of the algorithm, so we used the software EmotivPRO provided by EPOC+ to record (if you need real-time identification, you need to call EPOC's SDK development package). Through this software, we recorded the data and induced video tags according to the emotions played.

According to the device's official website, headphones have a large number of signal processing and filtering functions, which can remove power noise and harmonic frequencies. So when processing the signal, we can aim at other interference signals.

In the original EEG, there are mainly breathing, skin electrical, ECG and other low-frequency noise and high-frequency noise generated by EMG, the most important is the eye electrical artifacts. For the processing of artifacts, wavelet packet decomposition reconstructed signal, the signal for further analysis and extraction. We use db4 basis function of the signal multi-scale decomposition, wavelet decomposition later, the need for the details of the component part of the wavelet coefficients do threshold processing, complete wavelet threshold denoising. Wavelet soft threshold denoising algorithm is used in this paper. After filtering and wavelet decomposition and reconstruction, it can be found that the sharp noise signal and ophthalmogram artifacts were basically eliminated (see Fig. 4).

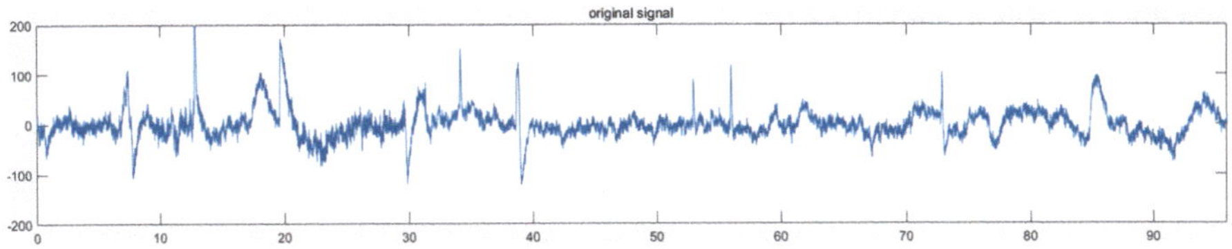

(a) original signal before processing

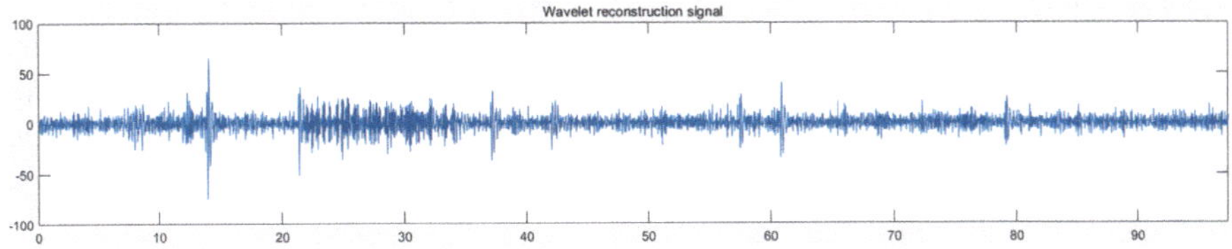

(b) The processed wavelet reconstructed signal

Fig. 4. (a) Original signal before processing (b) The processed wavelet reconstructed signal.

After the data preprocessing was completed, the 36 s baseline data was deleted, and we segmented the remaining 60 s EEG data of each subject in each experiment. Candra et al. [8] found that when the sample segmentation length is 3–12 s, the classification and recognition accuracy is the highest when using the pre-processed 32-channel EEG signals in DEAP database to study two kinds of emotions in valence and arousal dimensions. In this paper, we use a window with a length of 6 s and an overlap rate of 0 for sample segmentation. After segmentation, for each experiment, we get 10 samples, the data length of each sample is 6 s, that is, 768 (6 s $\times$ 128 Hz) data points, and there is no overlap between adjacent samples. In this way, we got 600 samples for each subject's 60 experiments. Feature extraction and classification.

3 Feature Extraction and Classification

3.1 Feature Extraction of Frequency Band Energy

According to the research of brain neuroscience and psychology, the four rhythms of Delta (1 Hz–3 Hz), Theta (4 Hz–7 Hz), Alpha (8 Hz–12 Hz) and Beta (13 Hz–30 Hz) are closely related to human physiological activities [9]. In the study of emotional EEG, frequency band energy is generally used as a classification feature, and the value of energy under stress is generally greater than that under calm emotion. From a physiological point of view [10], when people are in a state of stress, the brain is in a tense state of high load, which increases the interaction between brain neurons, so the energy of EEG signals is higher in the state of excitement or stress.

For the reconstructed signal, we use a series of Butterworth bandpass filter to decompose it into theta, alpha, beta and gamma four bands, and then use the Fast Fourier transform to calculate the energy value of each data point and add it, the calculated energy and as a feature. Each experiment corresponds to a 56-dimensional (14 channels × 4 powers) feature sample. For the $X_i(n)$ of the $\boldsymbol{i}$ band, the corresponding energy formula is:

$$\mathrm{B}_{\mathrm{band}} = \sum_{k=1}^{N} |X_i(k)|^2 \tag{1}$$

where $X_i(k)$ is the result of the fast Fourier transform corresponding to the signal $X_i(n)$, N is the length of the fast Fourier transform, and here the length of the signal $X_i(n)$ is equal to 768.

For each sample, we calculated the band energy sum of each frequency band of the sample as a feature, and finally get a 56-dimensional (14 channels × 4 powers) eigenvector. The following picture shows the energy characteristics of the AF3 channel of one of the experimenters (see Fig. 5).

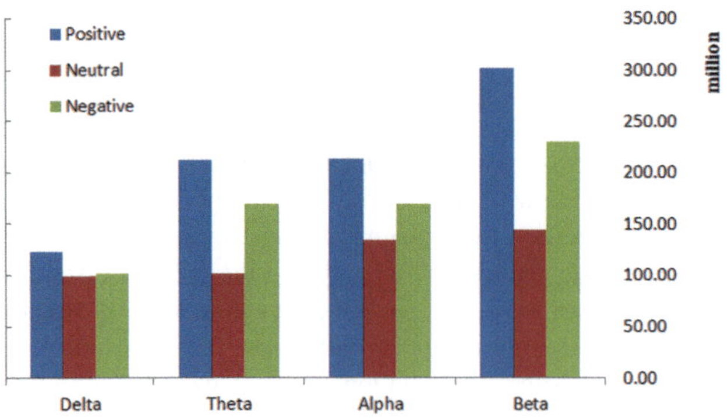

Fig. 5. One experimenter showed the energy characteristics of each AF3 channel of a sample.

Through the characteristic chart, it can be found that the energy sum of positive emotion is the largest, while that of neutral emotion is the lowest, and that of negative emotion is moderate, and the distinction between Beta band is the most obvious.

3.2 Emotion Classification

SVM (Support Vector Machine) is a machine learning algorithm based on statistical learning developed in the mid-1990s. In this algorithm, we marked all the data with points in the N (N is the total number of features) dimension space, and looked for a hyperplane that can divide the training sample points. In the case of linear separability, there are one or more hyperplanes in which the training samples are completely separated. The goal of SVM is to find an optimal hyperplane with the maximum distance from the data points to ensure the highest classification accuracy [10]. In this paper, the SVM classifier is used to classify the positive and negative emotions of the feature samples.

According to the feature samples, theta, alpha, beta, gamma and all feature samples were classified by SVM classifier, and they were divided into 30% test samples and 70% training samples (Table 1).

Table 1. Classification accuracy of four frequency bands.

Experimenter	Delta (%)	Theta (%)	Alpha (%)	Beta (%)	ALL (%)
S01	70.12	73.27	80.28	83.25	88.55
S02	71.36	75.23	81.25	82.16	89.42
S03	70.22	73.20	81.37	86.25	88.62
S04	71.75	75.48	80.34	83.25	87.50
S05	72.24	72.16	80.75	82.26	89.88
Average	71.14	73.87	80.80	83.43	88.80

The results show that when using the energy of four bands for feature classification, the highest recognition rate of Beta band can reach 83.43%. When using the energy of four bands and 56-dimensional feature vectors, the classification accuracy is the highest, up to 88.80%.

4 Software Design

After verifying the feasibility of the music regulation system algorithm of autistic children based on EEG signals, the software needs to be designed. The software has not been tested by patients, and data collection is not completed by the software. EPOC + headphones need to apply for the SDK development kit from the official if they need to transmit the EEG signals in real time. Use Python to run. Through the SDK development package, the control data is transmitted in real time, and the data is continuously stored in the .csv file. Matlab reads this .csv file every 10 s, and the result is in real time to identify emotions. The software design flow chart is shown below (see Fig. 6).

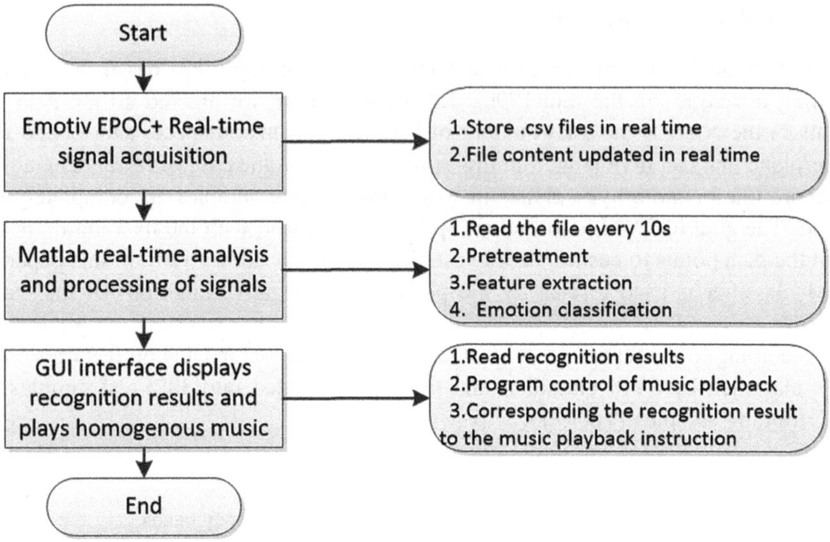

Fig. 6. Software flow chart.

5 Conclusion

In this paper, the positive and negative emotions of 5 male correspond to the impulsive explicit emotions, calm emotions, and restrained autistic emotions of autistic children. Using the EEG data of 5 male to verify the feasibility of the music therapy system for autistic children based on EEG signals. Using wavelet threshold method to remove electrooculogram artifacts, extract the energy features of the 4 frequency bands and finally classify the positive and negative emotions with the SVM classifier. Combine traditional music therapy methods of autistic children with emotional recognition of EEG signals, play homogenous music with the same emotional type according to the determined emotional state, and gradually transition to heterogeneous music with the opposite emotional type, and finally play Neutral music stabilizes people's emotional state, and achieves the purpose of real-time recognition through software design. After obtaining the EEG data of the autistic children in the later period, the software is tested and applied to the rehabilitation regulation of autistic children. Applying the EEG signal emotion recognition technology to the rehabilitation treatment of autistic children can record the effect of music treatment of autistic children through a visual interface system, which is of great significance for the rehabilitation treatment of autistic children.

Acknowledgement. This work is supported by Key R & D projects of Shandong Province under grant 2019GNC106093, Shandong Agricultural machinery equipment R & D innovation plan project under grant 2018YF011, and Key R & D projects of Shandong Province under grant 2019JZZY021005.

References

1. Liu, Z.H., Zhang, L.H., Yin, X., Li, Z.L., Feng, S.H.: A clinical study of music therapy in the treatment of children's autism. Pediatr. TCM **7**(06), 17–19 (2011)
2. Qiu, T.L.: Music therapy intervention principles and methods of autism children. J. Yichun Coll. **34**(01), 143–146 (2012)
3. Jin, Y., Wang, J.R., Li, L.Q., Huang, Z.M., Xiao, Y.T., Lu, H.Y., Zhou, H.S.: Construction and application of visual music therapy system for special children. Chin. J. Spec. Educ. (05), 7–12 (2008)
4. Nie, D., Wang, X.W., Duan, R.N., Lv, B.L.: A survey on EEG based emotion recognition. Chin. J. Biomed. Eng. **31**(04), 595–606 (2012)
5. Luo, Z.Z., Jin, S., Li, Y.D.: Denoising method of EEG signal based on tangent function of denoising source separation. J. Huazhong Univ. Sci. Technol. (Nat. Sci. Edn.) **46**(12), 60–64 (2018)
6. Jenke, R., Peer, A., Buss, M.: Feature extraction and selection for emotion recognition from EEG. IEEE Trans. Affect. Comput. **5**(3), 327–339 (2014)
7. Pham, T.D., Tran, D.: Emotion recognition using the emotiv EPOC device. In: International Conference on Neural Information Processing, pp. 394–399. Springer, Heidelberg (2012)
8. Candra, H., Yuwono, M., Chai, R.: Investigation of window size in classification of EEG-emotion signal with wavelet entropy and support vector machine **2015**(5), 7250–7253 (2015)
9. Li, X., Cai, E.J., Tian, Y.X., Sun, X.Q., Fan, M.D.: An improved electroencephalogram feature extraction algorithm and its application in emotion recognition. J. Biomed. Eng. **34** (04), 510–517+528 (2017)
10. Liu, C.Y., Li, W.Q., Bi, X.J.: Emotional feature extraction and classification based on EEG signals. Chin. J. Sens. Actuat. **32**(01), 82–88 (2019)

EEG Characteristics Extraction and Classification Based on R-CSP and PSO-SVM

Xue Li, Yuliang Ma(✉), Qizhong Zhang, and Yunyuan Gao

Institute of Intelligent Control and Robotics, Hangzhou Dianzi University, Hangzhou 310018, China
mayuliang@hdu.edu.cn

Abstract. In order to improve the EEG recognition accuracy and real-time performance, a classification and recognition method for optimizing the penalty factor C and kernel parameter g of Support Vector Machine (SVM) based on Particle Swarm Optimization (PSO) algorithm is proposed in this paper. Firstly, the Regularization Common Spatial Pattern (R-CSP) was used for EEG feature extraction. Secondly, the penalty factor and the kernel function were optimized by the proposed PSO algorithm. Finally, the constructed SVM classifiers were trained and tested by the two class EEG data of right foot and right hand movements. The experimental results show that the recognition rate for EEG classification of the PSO-SVM is average 2.2% higher than the non-parameter-optimized SVM classifier, and it is significantly higher than the traditional LDA classifier, which proves the feasibility and higher accuracy of the algorithm.

Keywords: EEG signal · Regularization Common Spatial Pattern (R-CSP) · SVM · PSO

1 Introduction

Brain computer interface (BCI) is a communication and control interface between the human brain and a computer or other device based on EEG [1]. BCI technology refers to a new type of human-computer interaction method. It enables people to output control signals through computers and other electronic devices without the participation of peripheral nervous system and muscle tissue, thus further communicating with the external environment [2]. EEG-based brain computer interface technology, in recent years, has become a hot research topic in this field and gradually developed into a rising multi-disciplinary cross technology.

The key technology of BCI is how to quickly and effectively extract EEG characteristics and improve the accuracy of recognition. Various methods such as linear discriminant analysis (LDA), K-nearest neighbor analysis, artificial neural network, etc., has been proposed. Support vector machine (SVM) is a kind of technique used in pattern classification and nonlinear regression, which was first proposed in 1995. The idea is to transform the vector map into a higher dimensional space, and get an optimal classification surface through calculation, so that the samples are linearly separated. kor

Q. Liu et al. (Eds.): CENet 2020, AISC 1274, pp. 1658–1667, 2021.
https://doi.org/10.1007/978-981-15-8462-6_189

Kor Shoker of Cardiff University extracted feature parameters of the event-related synchronization (ERS) and event-related desynchronization (ERD) [3] when they studied the mu rhythm signal of the hands' movements. Inspired by the fact that SVM is more adept at solving practical problems such as small samples, nonlinearity and high dimensionality, they used SVM to classify the introduced parameters, and the accuracy reached 83.5%. However, it led to the uncertainty of the selection in the optimal kernel function of SVM, due to the randomness and non-stationarity of EEG signals during the use of SVM, and the lack of a priori knowledge about the distribution of signals [4].

The traditional EEG feature classification algorithms of SVM usually adopt the empirical data when selecting the penalty parameters and the kernel parameters, while ignoring the importance of parameter optimization which enhances the classification effect of SVM. In recent years, particle swarm optimization (PSO) algorithm has been combined with SVM to solve the selection of SVM kernel function and improve the classification accuracy. Maali et al. used particle swarm optimized SVM to recognize sleep dysfunction such as apnea [5]. Subasi et al. adopted PSO to optimize SVM for the classification of EMG signals, which obtained good results in the diagnosis and recognition of neuromuscular diseases [6].

The effectiveness of SVM classification depends mainly on the choice of kernel function and parameters. According to the characteristics of EEG signals, PSO optimized SVM is used to classify the EEG signals of motor imagery in this paper, the kernel function and penalty parameters of SVM are optimized by particle swarm, which improves the classification accuracy of EEG signals.

2 Data Description and R-CSP Feature Extraction

2.1 Experimental Data

The experimental data are based on the public data of the third Brain computer interface Contest in 2005. In this paper, the EEG data of motor imagery (Data set IVa) provided by neurology department of Berlin research center in Germany is used to verify the algorithm.

The experimental data were obtained from 5 healthy subjects (aa, al, av, aw, ay). The subjects were aged between 24–35 years old, during the experiment, the subjects were asked to sit in a chair and relax. According to the target of the display, the subjects made corresponding actions, including the left hand movement imagination, the right hand movement imagination and the right foot movement imagination. The target cue duration was 3.5 s, the sampling frequency was 100 Hz and the interval was 1.75–2.25 s, which was used for the rest of the subjects. The experimental data were collected from 118 consecutive EEG signals recorded 280 times. Table 1 shows the number of training trials per subject and the number of tests.

Table 1. Sheet of experimental data.

Number	#tr	#te
aa	168	112
al	224	56
av	84	196
aw	56	224
ay	28	252

2.2 Feature Extraction of EEG Signals Based on R-CSP

As an effective feature extraction algorithm, Common Spatial Pattern (CSP) algorithm is widely used in the feature extraction of motor imagery EEG. CSP uses the principle of simultaneous diagonalization of matrices to find a set of spatial filters, making the variance of one class of signals reached the maximum, and another class reached minimum to achieve the purpose of classification [7]. However, CSP shows an obvious deficiency, it is easily influenced by various factors such as the emotional and physical condition of the subjects. In this paper, the regularized CSP algorithm is used to extract the features of EEG signals and the regularization parameters β and γ are introduced. The experimental data of the main and auxiliary subjects are utilized reasonably [8]. The rational use of the experimental data of the auxiliary subjects can not only reduce the individual differences in the classification of EEG signals, but also reduce the classification errors caused by the small number of main subject samples. Since the regularization algorithm is suitable for the case of fewer samples, $\beta = \gamma = 0$ is taken when it contains large samples, which is equivalent to the traditional CSP.

Suppose the EEG signal is E_c^i, $c \in \{1, 2\}$ represents the EEG signal of the i-th sample in category C, $N \times T$ dimensions, where N is the number of channels, and T is the number of sampling points for each channel in each experiment. In this paper, R-CSP is used to obtain the spatial filter W with $N \times Q$ dimensions, where $Q = 2\alpha$. The feature vector is obtained by $X = W^T E$, and X is transformed to obtain the final Q-dimensional feature vector Y, where the q-th feature vector is expressed as follows:

$$y_q = log\left(\frac{var(x_q)}{\sum_{q=1}^{Q} var(x_q)}\right) \quad (1)$$

where x_q is the q-th row of X, and $var(x_q)$ is the variance of x_q.

3 Classification and Recognition Based on PSO-SVM

3.1 Support Vector Machine

Support vector machine (SVM) is a kind of classification technology proposed by Vapnic et al. at the end of the 20th century, which solves the problem of pattern recognition such as small samples, nonlinear and high dimension [10]. SVM usually uses the kernel function $K(x, y)$ instead of the inner product operation in the optimal classification [11]. It transforms the nonlinear problem into a linear classification problem by lifting the dimension method when solving the nonlinear classification problem. This article mainly aims at the classification of the two kinds of motor imagery of right hand and right foot. The set of samples is set to be (x_i, y_i), $i = 1, 2, \cdots, l, x \in R^N$ and $y_i \in \{+2, +1\}$ is the vector of the X category identification meet as follows:

$$y_i[(w \cdot x_i) + b] - 1 \geq 0, i = 1, 2, \cdots, l. \tag{2}$$

It is transformed into a dual problem represented by training data through Lagrange multiplier method, then the optimization objective function is transformed as follows:

$$minQ(a) = \frac{1}{2}\sum\nolimits_{i,j=1}^{l} a_i a_j y_i y_j \cdot K\left(x_i, x_j\right) - \sum\nolimits_{i=1}^{l} a_i \tag{3}$$

which meets the constraints: $\sum\nolimits_{i=1}^{l} a_i y_i = 0, 0 \leq a_i \leq c$, where a_i is the Lagrange multiplier for each constraint corresponding to M, and c is the penalty parameter of the sample classification.

In order to solve the above problems, it is necessary to select the appropriate kernel function, in this paper, the kernel function of RBF is selected as follows:

$$K(x, x_i) = exp(-|x - x_i|^2)/\sigma^2 \tag{4}$$

And then the optimal classification function is obtained as follows:

$$f(x) = sgn(\sum\nolimits_{i=1}^{l} a_i^* y_i K(x, x_i) + b^*) \tag{5}$$

where a^*, b^* are the determined optimal classification surface parameters which can be obtained by any support vector.

3.2 Particle Swarm Optimization

Particle Swarm Optimization (PSO) is an evolutionary optimization algorithm proposed by Kennedy and Eberhart in 1995 [12]. It uses an iterative method to find the optimal solution, more precisely, PSO determines the position of the individual particle at the next moment according to the individual's current optimal position and the population's optimal position.

Suppose that N particles are included in the D-dimensional space, the position and velocity of the i-th particle at t moment are as follows: $x_i^j = \left(x_{i1}^t, x_{i2}^t, \ldots\ldots, x_{id}^t\right), v_i^t = \left(v_{i1}^t, v_{i2}^t, \ldots, v_{id}^t\right)$, where $i = 1, 2, \cdots, N$. Each particle i adjusts its speed and position continuously compared to the global extremes p_{gd}^t and individual extreme value p_{id}^t. The Optimal parameters of SVM are obtained by iterative optimization. The formula for iterative update of particle velocity and position is as follows:

$$v_{id}^{t+1} = v_{id}^{t} + c_1 r_1 \left(p_{id}^{t} - x_{id}^{t}\right) + c_2 r_2 \left(p_{gd}^{t} - x_{gd}^{t}\right) \tag{6}$$

$$x_{id}^{t+1} = x_{id}^{t} + v_{id}^{t+1} \tag{7}$$

where the position $x_{id}^{t} \in [L_d, U_d]$, L_d, U_d are the lower and upper limit of D dimensional space. The speed is $v_{id}^{t} \in \left[v_{min,d}, v_{max,d}\right]$, $v_{min,d}, v_{max,d}$ are the lower and upper limit of particle velocity. r_1, r_2 are random numbers evenly distributed at (0,1). c_1, c_2 are constants, called learning factors. Shi and Eberhart add the inertia weight parameter w to the original PSO, controlling the search scope and reducing the importance of the upper speed limit. The formula (6) is converted as follows:

$$v_{id}^{t+1} = wv_{id}^{t} + c_1 r_1 \left(p_{id}^{t} - x_{id}^{t}\right) + c_2 r_2 \left(p_{gd}^{t} - x_{id}^{t}\right) \tag{8}$$

3.3 PSO Optimization SVM Algorithm

In practical applications, in order to obtain a higher classification accuracy of SVM classifier, it is necessary to optimize the parameters of SVM. Particle swarm optimization algorithm is very suitable to solve the nonlinear problem in high dimensional space because of its nature of parallelism and the independence of the objective optimization function. In this paper, the global search ability of PSO is used to optimize the penalty parameter C and the kernel function radius g in the process of SVM modeling to obtain a more accurate SVM classifier with better classification effect.

In the *D*-dimensional space, *M* particles constantly update their own position and speed according to the formulas (7) and (8). The optimal parameters C and g of SVM are obtained by iterative optimization. The implementation steps of support vector machine based on particle swarm optimization algorithm are as follows:

1) Initialization: In the *D*-dimensional parameter space, the *M* particles are initialized, including the initial parameters c_1, c_2 of the particle swarm, the initial velocity of each particle and the optimal position of each particle and the global optimal position.
2) Calculate the fitness: The fitness function of the particle swarm is selected to calculate the particle fitness function.
3) Adjustment: According to the fitness value of the particle to adjust the individual optimal position and the global optimal position of the particle.
4) Update: Update the particle state according to formulas (7) and (8) and obtain the new parameters of SVM.
5) Judgment: When the particle fitness function value is met or the maximum number of iterations is reached, the iteration is terminated and then output, otherwise return to step 3) to continue the calculation.
6) Classification: If the terminate condition is satisfied, the optimal parameters are obtained. Re-training SVM as the final classifier for training and classification prediction.

After the above six steps, PSO is used to optimize the SVM penalty parameter C and the kernel function number radius g, getting the smallest error of C and g, which are used for the classification of EEG signals of motor imagery.

4 Experiment and Result Analysis

In this paper, we used the EEG data of motor imagery Data set Iva and adopted the regularized CSP to extract the features of the right hand and right foot, and then the PSO-SVM was used to classify the extracted feature vector. The initial parameters of the PSO algorithm are set to: $w = 0.8$, $c_1 = 1.5$, $c_2 = 1.7$, the number of particles is set to 50 and the maximum number of iterations is set to 100. The iterative process is shown in Fig. 1 with al as the subject. The iterative curve of feature classification is obtained when the number of experimental samples is 100. With the increase of the number of iterations, the accuracy rates tend to be the best. When the terminate condition is satisfied, the accuracy is about 90%, and the optimal parameters are c = 4.5294, g = 0.01. Support vector machine classifier can quickly find the best penalty parameter C and kernel parameter g. PSO is a parallel global search based on population with fewer parameters to adjust but faster convergence speed, which reflects the obvious advantage of PSO-SVM in this paper.

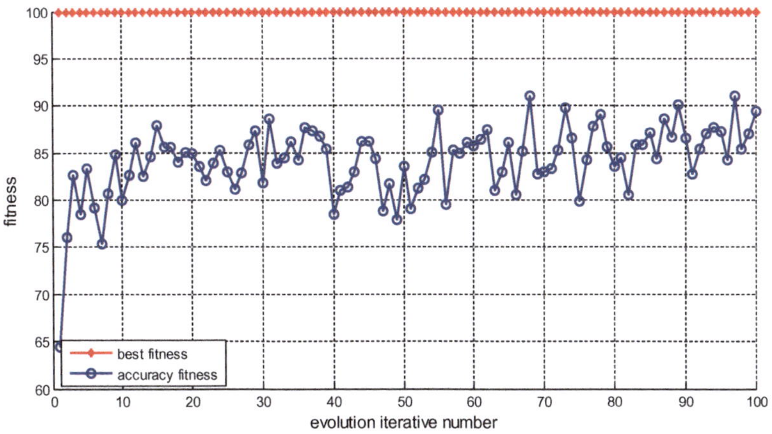

Fig. 1. Curve of PSO searching for the optimal parameters.

In this experiment, aa, al, av, aw, ay were used as the main subjects, and the other four were auxiliary subjects. The number of samples is chosen from large to small. It is equivalent to the traditional CSP algorithm when the regularization parameters are $\beta = 0, \lambda = 0$. Reasonable selection of two regular parameters to achieve the best feature classification effect, when the number of aa, al samples are 200 and 100 respectively, the regularized parameters are both set to 0, when the number of av and aw samples are 80 and 56 respectively, the optimal regularization parameters $\beta = 0, \gamma = 0.01$ are selected for feature extraction. First, the R-CSP algorithm is used for feature extraction, and then PSO-SVM and SVM are used for classification and recognition respectively. In Figs. 2 and 3, the number of samples is 100 and aa is the subject. They show the correctness of the test set in SVM and PSO-SVM respectively. As can be seen from the graph, the classification effect of PSO optimization is better than that before optimization.

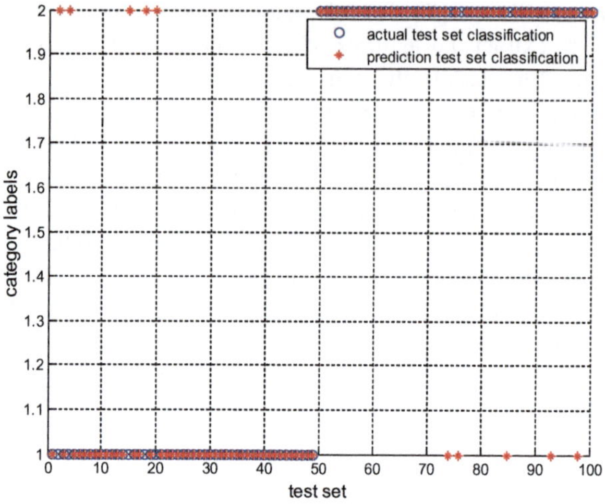

Fig. 2. Classification accuracy schematic of SVM before optimization.

Tables 2 and 3 are classification of the experimental data using SVM and PSO-SVM respectively when aa, al, av, aw as the main subjects, the others as the auxiliary subjects. The data analysis shows that the SVM classification of PSO optimized chosen the different number of samples is significantly improved compared with the traditional SVM, and the recognition rate of PSO-SVM is about 2.2% higher than that of traditional SVM classification, which shows that PSO-SVM obtains the optimal parameters and improves the performance of SVM. Comparing the results of PSO-SVM with traditional SVM and LDA, taking one as the main subject, and the others as the auxiliary subjects. The data from the first 4 experiments are adopted in each experiment and the regularization parameters are set to $\beta = 0, \gamma = 0.01$. As shown in Table 4, PSO-SVM classification method obtains the highest classification and recognition rate compared with the traditional LDA and SVM. All of the results show the obvious advantages of PSO-SVM in the classification of EEG signals, which effectively improves the accuracy of EEG classification.

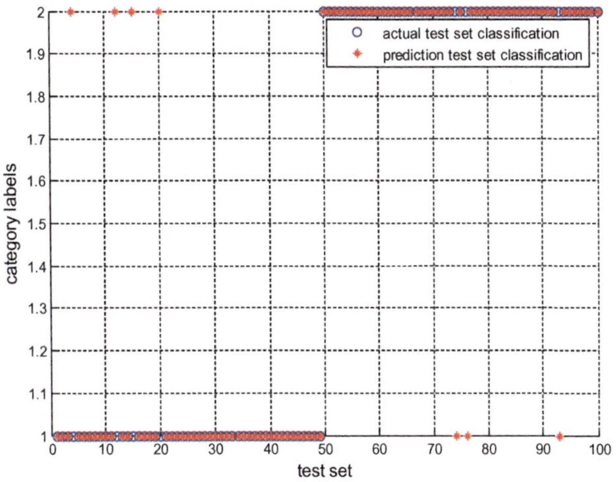

Fig. 3. Classification accuracy schematic of PSO-SVM after optimization.

Table 2. SVM classification data before optimization.

Subject	aa (100)			al (200)			av (80)			aw (56)		
Training sample set	52	54	56	101	108	110	42	44	46	30	32	34
Test sample set	48	46	44	99	92	90	38	36	34	26	24	22
Accurate identification	95.83	95.65	97.72	92.55	88.04	87.78	81.58	77.8	73.53	88.46	87.51	90.91
Average recognition	96.4			89.46			77.63			88.96		

Table 3. PSO-SVM classification data after optimization.

Subject	aa (100)			al (200)			av (80)			aw (56)		
Training sample set	52	54	56	101	108	110	42	44	46	30	32	34
Test sample set	48	46	44	99	92	90	38	36	34	26	24	22
Accurate identification	97.92	97.83	95.45	93.94	90.22	91.11	81.58	80.5	79.42	92.31	91.67	90.91
Average recognition	97.4			91.79			80.5			91.63		

Table 4. Classification accuracy comparison data of PSO-SVM, SVM and LDA.

Subject	aa	al	av	aw	ay
PSO-SVM accuracy (%)	100	98.33	94.26	97.61	100
SVM accuracy (%)	98.48	87.32	86.33	89.78	94.81
LDA accuracy (%)	94.64	96.43	87.86	95.36	98.45

5 Conclusion

Aiming at the classification and recognition of EEG signals of motor imagery, this paper proposes to use PSO to optimize the penalty factor C and the kernel function radius g of SVM to improve the classification accuracy. The PSO-SVM classification method overcomes the data limitation of the traditional SVM parameter selection method, which not only reduces the time consumption, but also finds the optimal parameters quickly. In the process of PSO optimization, the adaptive particles are more in line with the actual situation. The optimal parameters of the support vector machine can be found quickly and accurately by dynamically balancing the global search and local search abilities, which improves the classification efficiency and accuracy. The PSO-SVM classification method has higher classification accuracy than the traditional classification methods.

Acknowledgement. This research was funded by the Natural Science Foundation of Zhejiang Province grant number LY18F030009 and the National Natural Science Foundation of China grant number 61971168, 61372023.

References

1. Hassan, M., Wendling, F.: Electroencephalography source connectivity: aiming for high resolution of brain networks in time and space. IEEE Signal Process. Mag. **35**(3), 81–96 (2018)
2. Friman, O., Volosyak, I., Graser, A.: Multiple channel detection of steady-state visual evoked potentials for brain-computer interfaces. IEEE Trans. Rehabil. Eng. **54**(4), 742–750 (2007)
3. Ming, D., Wang, K., He, F., et al.: Study on physiological information detection and application evoked by motor imagery: review and prospect. Instrumentation **35**(9), 1921–1931 (2014)
4. Liu, B., Wei, M.R., Luo, C.: Research progress on BCI based on EEG. Comput. Knowl. Technol. **07**(10), 1493–1495 (2014)
5. Maali, Y., Al-Jumaily, A.: A novel partially connected cooperative parallel PSO-SVM algorithm: study based on sleep apnea detection. In: Proceedings of the 2012 IEEE Congress on Evolutionary Computation (CEC), Brisbane, Australia, pp. 1–8 (2012)
6. Subasi, A.: Classification of EMG signals using PSO optimized SVM for diagnosis of neuromuscular disorders. Comput. Biol. Med. **43**(5), 576–586 (2013)
7. Yang, Y.H., Lu, W.Y., He, M.Y., et al.: Feature extraction based on WPD and CSP in brain computer interface. Instrumentation **33**(11), 2560–2565 (2012)
8. Lu, H., Eng, H.L., Guan, C., et al.: Regularized common spatial pattern with aggregation for EEG classification in small-sample setting. Biomed. Eng. IEEE Trans. Rehabil. Eng. **57**(12), 2936–2946 (2010)
9. Mishuhina, V., Jiang, X.: Feature weighting and regularization of common spatial patterns in EEG-based motor imagery BCI. IEEE Signal Process. Lett. **25**(6), 783–787 (2018)
10. Vapnik, V.: The nature of Statistical Learning Theory, pp. 123–179. Springer, New York (1995)
11. Gu, W.C., Chai, B.R., Teng, Y.P.: Research on support vector machine based on particle swarm optiminzation. Trans. Beijing Inst. Technol. **24**(7), 705–709 (2014)

12. Yang, P., Tsai, J., Chou, J.: Best selection for the parameters of fractional-order particle swarm optimizer. In: IEEE International Conference on Systems, Man, and Cybernetics (SMC), Banff, AB, 3302–3304 (2017)
13. Shen, M., Zhang, Q., Lin, L., Sun, L.: a novel approach for analyzing EEG signal based on SVM. In: 2016 International Conference on Information System and Artificial Intelligence (ISAI), Hong Kong, pp. 310–313 (2016)
14. Nivedha, R., Brinda, M., Vasanth, D., Anvitha, M., Suma, K. V.: EEG based emotion recognition using SVM and PSO. In: 2017 International Conference on Intelligent Computing, Instrumentation and Control Technologies (ICICICT), Kannur, pp. 1597–1600 (2017)
15. Park, Y., Chung, W.: Optimal channel selection using covariance matrix and cross-combining region in EEG-based BCI. In: 7th International Winter Conference on Brain-Computer Interface (BCI), Gangwon, Korea (South), pp. 1–4 (2019)
16. He, C., Li, J. L., Xing, J. C.: A new model for structural damage assessment using adaptive mutation particle swarm optimization and support vector machine. In: 2018 Chinese Control and Decision Conference, Shenyang, pp. 6711–6714 (2018)
17. Gao, F.R., Wang, J.J., Xi, X.G., She, Q.S., et al.: Gait recognition for lower extremity electromyographic signals based on PSO-SVM method. J. Electron. Inf. Technol. **37**(5), 1154–1159 (2015)
18. Azimirad, V., Hajibabzadeh, M., Shahabi, P.: A new brain-robot interface system based on SVM-PSO classifier. In: 2017 Artificial Intelligence and Signal Processing Conference (AISP), Shiraz, pp. 124–128 (2017)
19. Guan, C., Robinson, N., Handiru, V.S., Prasad, V. A.: Detecting and tracking multiple directional movements in EEG based BCI. In: 2017 5th International Winter Conference on Brain-Computer Interface (BCI), Sabuk, pp. 44–45 (2017)

Motor Imagery EEG Feature Extraction Based on Fuzzy Entropy with Wavelet Transform

Tao Yang, Yuliang Ma(✉), Ming Meng, and Qingshan She

Institute of Intelligent Control and Robotics,
Hangzhou Dianzi University, Hangzhou 310018, China
mayuliang@hdu.edu.cn

Abstract. Due to the nonlinear characteristics of EEG signals and the rhythm characteristics of motor imagery, the low recognition rate of using single feature extraction algorithm, a feature extraction method based on wavelet transform and fuzzy entropy is presented in this paper. The EEG signals are decomposed to three levels by the wavelet transform, according to the ERS/ERD phenomena during motor imagery, the alpha rhythm and beta rhythm signal can be extracted by the algorithm of fuzzy entropy. Finally, the motor imagery EEG signals are classified by a support vector machine classifier. BCI Competition IV Datasets1 has been used to conduct the experiment, the experimental results show that the feature extraction method combining wavelet transform and fuzzy entropy is much better than the ways of using single fuzzy entropy, sample entropy, or others, and its highest recognition rate is 90.25%.

Keywords: Electroencephalograph signal · Fuzzy entropy · Wavelet transform · Feature extraction · Support vector machine

1 Introduction

Brain-computer Interface (BCI) is a new technology that integrates multiple disciplines such as artificial intelligence, computer science and information, biomedical engineering and neuroscience [1]. This technology has broad applications prospects in the fields of medical treatment, rehabilitation of the disabled and cognitive science of the brain.

BCI is a non-muscle communication system, and it allows the intention of brain connect with environment directly [2]. Feature extraction is a key problem in BCI system, and the result is directly related to the design of classifier and recognition rate. Feature extraction is done for making pattern recognition easier by extracting the most representative features from the EEG signals. The EEG signals are non-linear, so there are many nonlinear analysis methods applied in analyzing the EEG signals. With the development of the relevant theory, Approximate Entropy Characteristic (ApEN) [3], Sample Entropy (SampEn) [4, 5] and etc. have appeared one after another, and they are used as the measure of complexity. Acharya et al. [6] used ApEN, SampEn to extract the features of epilepsy electro-encephalogram and the normal EEG signals, characterizing the features of epileptic seizure in EEG signals efficiently. Chen et al. [7] improved SampEn and defined fuzzy entropy, and used fuzzy entropy to the feature

Q. Liu et al. (Eds.): CENet 2020, AISC 1274, pp. 1668–1678, 2021.
https://doi.org/10.1007/978-981-15-8462-6_190

extraction of the EMG signals with success. Meanwhile, according to the phenomenon that the event related desynchronization and the event related synchronization (ERD/ERS) [8] occur with the motor imaginary of unilateral limb, the feature of alpha rhythm (8–15 Hz) and beta rhythm (15–30 Hz) is most obvious. Thus, it is necessary to resolve the EEG signals into different frequency bands.

Wavelet transform [9] is a common time-frequency analysis method. This method has multiresolution characteristics and favorable resolution in time and in frequency domain. We can get the information of EEG signals in different frequency bands by wavelet decomposition. But wavelet transform fails to reflect the nonlinear characteristics of EEG signals. Fuzzy Entropy [10, 11] is a nonlinear analysis method which inherits the advantages of ApEN and SampEn and has the similar physically meaningful. This method is sensitive to the change of signal complexity. It is more suitable for observations on signals that reflecting the nonlinear characteristics of EEG signals by measuring complexity of EEG signals.

In view of the nonlinear characteristics of EEG signals and the rhythms of motor imaginary, we introduce a feature extraction method combining wavelet transform with fuzzy entropy. This method use the time-frequency characteristics extracted by wavelet decomposition and the nonlinear characteristics extracted by analysis of fuzzy entropy, and we can get the feature vectors that reflect the time-frequency characteristics and the nonlinear characteristics of EEG signals. Using Support Vector Machine to classify the EEG signals of motor imaginary, the highest recognition rate is 90.25%. The experimental results show that this feature extraction method is much better than the ways of using single fuzzy entropy, sample entropy, or others.

2 Wavelet Transform

The EEG signals are time variant non-stationary signals, and the traditional method cannot distinguish the frequency component within a certain time range and transient features in detail. Wavelet transform is a leap of Fourier Transform, and Wavelet Decomposition Coefficients can reflect energy distribution of signals in time and in frequency domain. So the features of EEG signals in time and in frequency domain can be reflected by energy of wavelet decomposition coefficients [12]. Let $f(t)$ denote the discrete EEG signals acquired, so discrete wavelet transform of $f(t)$ is defined as

$$C_{j,k}(f,\psi_{j,k}) = \langle f, \psi_{j,k}(t)\rangle = 2^{-\frac{j}{2}}\sum\nolimits_{-\infty}^{+\infty} f(t)\psi^*(2^{-j}t-k)dt \quad j,k \in Z \tag{1}$$

Wherein: $\psi_{j,k}(t)$ is a wavelet series; j is the decomposition scale of discrete wavelet; k is the displacement parameter, which is the component of signals at a certain scale 2^j. To improve the effectiveness of wavelets analysis, we use orthogonal wavelet transform to realize the wavelet transform in EEG signals.

Mallat algorithm is used to realize EEG signal decomposition and reconstruction, and Fig. 1 shows decomposition and reconstruction on discrete wavelet.

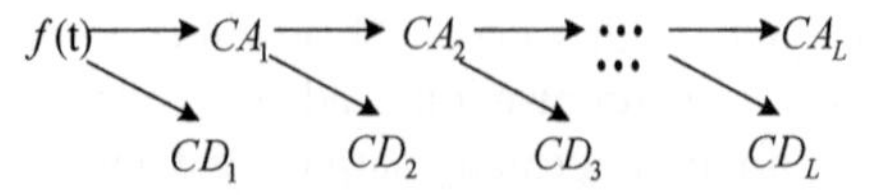

(a) Discrete wavelet decomposition.

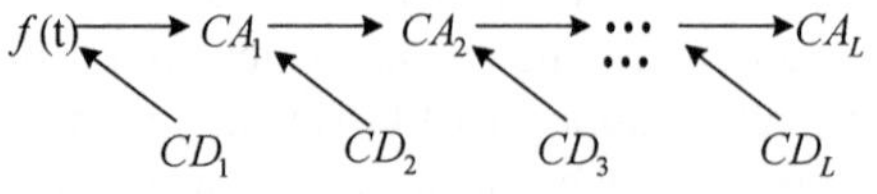

(b) Discrete wavelet reconstruction.

Fig. 1. Decomposition and reconstruction on discrete wavelet.

As shown in Fig. 1(a), the signals are decomposed at larger scale first, and then decomposing the low-frequency part of signals. This was repeated until acquiring the frequency band information at different levels. Its corresponding formulas and coefficients of the levels are:

$$f(t) = f_L^A(n) + \sum_{j=1}^{L} f_j^D(n) = A_1 + D_1 = A_2 + D_2 + D_1 = \ldots \tag{2}$$

$$\begin{cases} c_k^{j-1} = \langle f, \phi_{j-1,k} \rangle = \sum_{n \in Z} \langle f, \phi_{j,k} \rangle = \sum_{n \in Z} \overline{h}_{n-2k} c_n^j \\ d_k^{j-1} = \langle f, \psi_{j-1,k} \rangle = \sum_{n \in Z} \langle f, \psi_{j,k} \rangle = \sum_{n \in Z} \overline{g}_{n-2k} c_n^j \end{cases} \tag{3}$$

Wherein: L is decomposition level; c_k^{j-1} is the coefficient of approximation component; d_k^{j-1} is the coefficient of detail component; h represents the high pass filter; g represents low pass filter; $\phi_{j,k}$ is scaling function; $\psi_{j,k}$ is wavelet function.

If f_x is the sampling frequency of signals, the frequency ranges of sub-bands that correspond to each component are $[0, f_x/2^{L+1}]$, $[f_x/2^{L+1}, f_x/2^L]$, $[f_x/2^L, f_x/2^{L-1}]$, …, $[f_x/2^2, f_x/2]$. The corresponding approximation coefficients and the detailed coefficients [13] are cA_L, cD_L, cD_{L-1}, …, cD_1. During the wavelet transform, the EEG signals are transformed into different frequency bands after passing the high pass filter and the low pass filter. The low-frequency part of output is approximation information, and the high-frequency part of output is detail information.

Wavelet reconstruction is the inverse process of wavelet decomposition. The process of wavelet reconstruction is shown as 1(b), and reconstruction formulas are:

$$\begin{aligned} c_k^j &= \langle f, \phi_{j,k} \rangle = \sum_{n \in Z} h_{n-2k} \langle f, \phi_{j-1,k} \rangle + \sum_{n \in Z} g_{n-2k} \langle f, \psi_{j-1,k} \rangle \\ &= \sum_{k \in Z} h_{n-2k} c_n^{j-1} + \sum_{k \in Z} g_{n-2k} d_n^{j-1} \; j, k \in Z \end{aligned} \tag{4}$$

3 Theory of Fuzzy Entropy

The vector similarity of the approximate entropy and sample entropy is measured by the Heaviside function which is a binary classifier, and the formula of function is defined as:

$$\theta(k) = \begin{cases} 1 & z \geq 0 \\ 0 & z < 0 \end{cases} \tag{5}$$

Heaviside function is a binary function, and it makes a final judgment through the threshold. If certain conditions are satisfied, the result is a certain class. Otherwise, the result is another class. But there is not a clear boundary of classes in actual application, so it is not able to judge the classes of inputs correctly. In order to solve the problem of the approximate entropy and sample entropy, fuzzy set is introduced into the definition of fuzzy entropy. Then it is able to characterize input/output relations by fuzzy sets theory.

The core of fuzzy entropy is using the means algorithm and membership function so that the fuzzification of the vector similarity can be realized. Membership function in fuzzy theory are used to replace the criteria of hard thresholds. For example, we can know whether an element belongs to a set and the extent that the element belongs to the set. Fuzzy set theory uses a value in the close interval [0, 1], and the larger value means higher membership. Otherwise, the smaller value means lower membership. Exponential function is used as the fuzzy function in definition. The value of fuzzy entropy is smooth and continuous because of the continuity of exponential function. And the shape of fuzzy function which is determined by exponential function decides the vector similarity so that the fuzzification of similarity is ensured.

The definition algorithm of fuzzy entropy is:

(1) Let the sampling sequence with N sampling points of signals be $\{u(i)\text{: } 1 \leq i \leq N\}$, and this sequence forms a set of m-dimensional vectors in its order as followed:

$$x_i^m = \{u(i), u(i+1), \ldots, u(i+m-1)\} - u_0(i), i = 1, 2, \ldots, L-m \tag{6}$$

Wherein: $u(i)\,, u(i+1)\,, \ldots, u(i+m-1)$ are m continuous node that are point i to point $i+m-1$ in u, and $u_0(i)$ is the mean value of them:

$$u_0(i) = \frac{1}{m}\sum_{j=0}^{m-1} u(i+j) \tag{7}$$

(2) d_{ij}^m is the distance of vector x_i^m and x_i^m, and its value means the maximum difference between the corresponding element of two vectors. The definition of d_{ij}^m is:

$$d_{ij}^m = d[x_i^m, x_j^m] = \max_{k \in (0, m-1)} \{|u(i+k) - u_0(i) - (u(j+k) - u_0(j))|\} \tag{8}$$

Wherein: the value-taking range of i, j meet the formula: $i, j = 1 \sim N - m$ and $j \neq i$.

(3) Defining the similarity D_{ij}^{m} of vector x_i^m and x_j^m by fuzzy entropy function $u(d_{ij}^{m}, n, r)$:

$$D_{ij}^{m} = \mu(d_{ij}^{m}, n, r) = exp(-(d_{ij}^{m})/r) \tag{9}$$

Wherein: fuzzy entropy function $\mu(d_{ij}^{m}, n, r)$ is a exponential function. In this function, n and r represent the gradient and width of fuzzy function boundaries.

(4) Function ϕ^n is defined as:

$$\phi^n(n, r) = \frac{1}{n-m}\sum_{i=1}^{N-m}\left(\frac{1}{N-m-1}\sum_{j=1, j\neq i}^{N-m} D_{ij}^{m}\right) \tag{10}$$

(5) Repeating steps (1) to (4), new set of $m + 1$ dimensional vectors is formed in sequence's order, and fuzzy similarity of this vector are calculated. The functional definition is:

$$\phi^{n+1}(n, r) = \frac{1}{n-m}\sum_{i=1}^{N-m}\left(\frac{1}{N-m-1}\sum_{j=1, j\neq i}^{N-m} D_{ij}^{m+1}\right) \tag{11}$$

(6) The fuzzy entropy is defined as:

$$FuzzyEn(m, n, r) = \lim_{N\to\infty}\left[ln\phi^n(n, r) - ln\phi^{n+1}(n, r)\right] \tag{12}$$

According to the steps above, when N which is the length of sequence is limited, the value of fuzzy entropy is defined as:

$$FuzzyEn(m, n, r, N) = ln\phi^n(n, r) - ln\phi^{n+1}(n, r) \tag{13}$$

In the above formulas, m is embedding dimensionality. r is the similarity tolerance and N is the length of sequence. Their values affect the value of fuzzy entropy, and it is a very important step that choosing m and r. If r is too large, there will be too much losses of detail information. And if r is too small, the influence of noise will increase. Thus, let $m = n = 2$ and let $r = 0.2SD$ in which SD is the standard deviation of time sequence for enough detail information and better conditional probability estimation.

4 Experimental Data Acquisition

BCI Competition IV Datasets1 [14] which is provided by Berlin research center has been used to conduct the experiment as the experimental data. This data set recorded EEG signals of seven healthy subjects(a-g) when they had motor imaginary. There were 14 sets of data including 7 sets of calibration data and 7 sets of test data. There are 4 sets of calibration data from subject a, b, f and g in this article. There are three types of motor imaginary including left hand, right hand and single foot or double feet. Each subject performed two types of these tasks according to the marked words in screen. Single experiment continued 8 s: there were 2 s for subjects to prepare quietly when a cross symbol was displayed in the screen, 4 s for task execution and 2 s for an end task with a blackout display of screen, as shown in Fig. 2. Specifically, subject a and f conducted the left hand and foot motor imaginary, and subject b and g conducted the left hand and right hand motor imaginary. The electrode configuration for EEG signals acquisition used international standard lead 10/20 system. Data acquisition is from lead of C3 and C4, and there were 100 Hz sampling frequency.

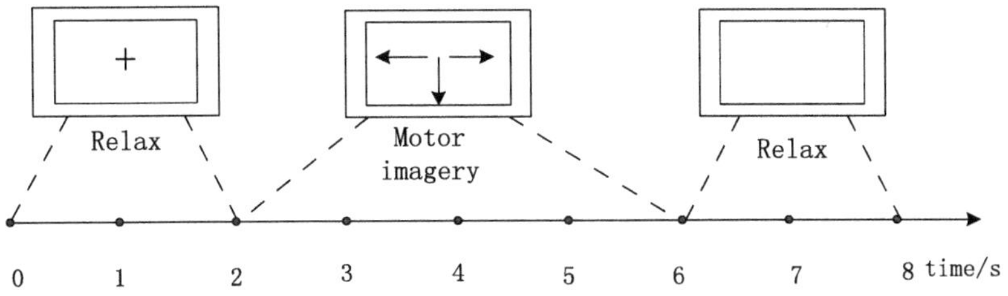

Fig. 2. Motor imagery sequence diagram of 2008 BCI competition IVData Set1.

5 Motor Imaginary EEG Feature Extraction Based on Wavelet Fuzzy Entropy

5.1 Feature Extraction Based on Wavelet Transform and Fuzzy Entropy

Wavelet transform can describe the time-frequency characteristics of EEG signals, but it cannot reflect the nonlinear characteristics of EEG signals. Fuzzy entropy inherits the advantages of ApEN and SampEn, and it reflect the nonlinear characteristics of EEG signals by measuring complexity of signals. Thus, wavelet transform and fuzzy entropy were used for feature extraction to get higher recognition rate of EEG signals, and the feature extraction procedure is shown as Fig. 3. First, the preprocessing using adaptive threshold denoising [15] were carried out for the data from channel C3 and C4. Then EEG signals were decomposed into different frequency bands by wavelet transform ERD/ERS occur with the motor imaginary. When the left-hand motor imaginary was executed, motor cortex in the contralateral functional area of brain was activated so that metabolism and blood flow increased in this area. Thereupon, synchronous neuronal activity was broken. Meanwhile, the amplitudes of α and β band fell sharply, and the fuzzy entropy of EEG signals was increased. Conversely, the fuzzy entropy of EEG signals that acquired from corresponding electrodes in the same side of brain. So fuzzy entropy feature extraction is only to signals of α and β band that was extracted by wavelet transform, and the number of wavelet transform levels is decided by useful EEG signal contribution and sampling frequency.

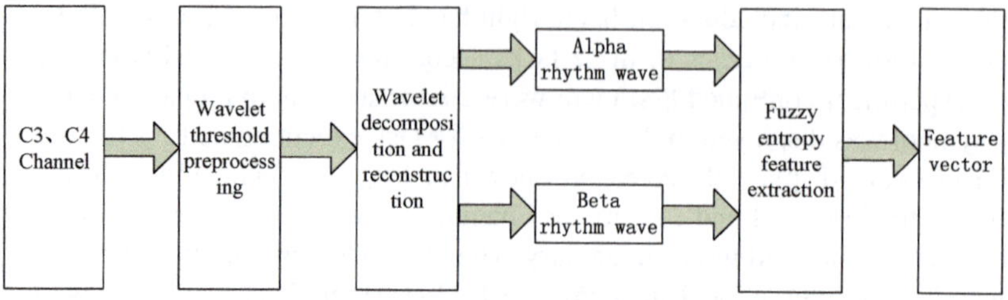

Fig. 3. Flow chart of EEG signal feature based on wavelet fuzzy entropy.

According to ERD/ERS, alpha rhythm(8-15 Hz) and beta rhythm(15-30 Hz) are useful EEG signal contribution relating to motion. The data from channel C3 and C4 was decomposed into different frequency bands as shown in Table 1 by wavelet transform for three layers. D2 and D3 were in the corresponding range of frequency band β and α. Frequency band D2 and D3 were reconstructed so that we could get EEG signals in frequency bands of alpha and beta rhythm. Alpha and beta rhythm of C3 and C4 channels with the left hand and foot motor imagery of subject f were shown as Fig. 4.

Table 1. Different frequency bands with wavelet decomposition.

Signal	A3	D3	D2	D1
$f(Hz)$	0–6.25	6.25–12.5	12.5–25	25–50

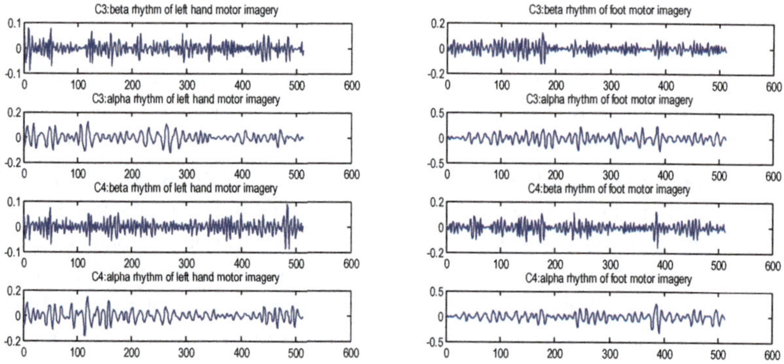

(a) Left hand motor imagery. (b) Foot motor imagery.

Fig. 4. Alpha and beta rhythm of C3 and C4 channels with the left hand and foot motor imagery of subject.

We used the fuzzy entropy method for feature extraction from the reconstructed signals so that each subject corresponded to a four-dimensional vector. The feature extraction results of subject f and g are shown in the lower views as the comparison.

The left hand and feet motor imagery was performed by subject f, and the left- and right-hand motor imagery was performed by subject g. Figure 5 is showing the results of fuzzy entropy feature extraction from subject f and g. And two classes of signal features which extracted by single fuzzy entropy had a greater aliasing. It is shown as Fig. 6 and Fig. 7 that the result of features which extracted by wavelet fuzzy entropy, and Fig. 6(a) and Fig. 6(b) were the results of fuzzy entropy feature extraction of beta and alpha rhythm waves respectively that generated when the left hand and foot motor imagery was performed by subject f. Figure 7(a) and Fig. 7(b) were the results of fuzzy entropy feature extraction of beta and alpha rhythm waves respectively that generated when the left and right hand motor imagery was performed by subject g. The results of feature extraction by wavelet fuzzy entropy and single fuzzy entropy were obtained to conduct comparative analysis, and the features extracted by wavelet transform for selecting frequency band and fuzzy entropy for feature extraction are more separable for better pattern recognition rather than that extracted by single fuzzy entropy method.

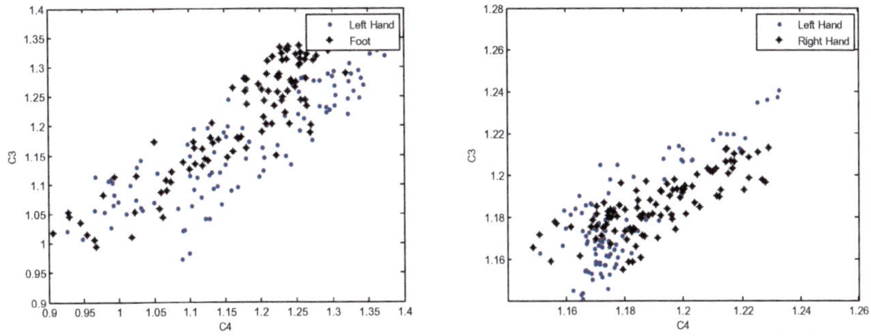

(a) Fuzzy entropy feature of subject f (b) Fuzzy entropy feature of subject g

Fig. 5. Fuzzy entropy chart of subjects f and g.

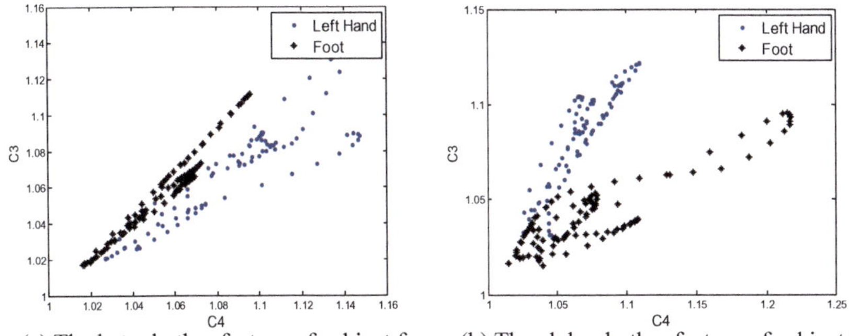

(a) The beta rhythm feature of subject f. (b) The alpha rhythm feature of subject f.

Fig. 6. Alpha and beta rhythm waves of subject f.

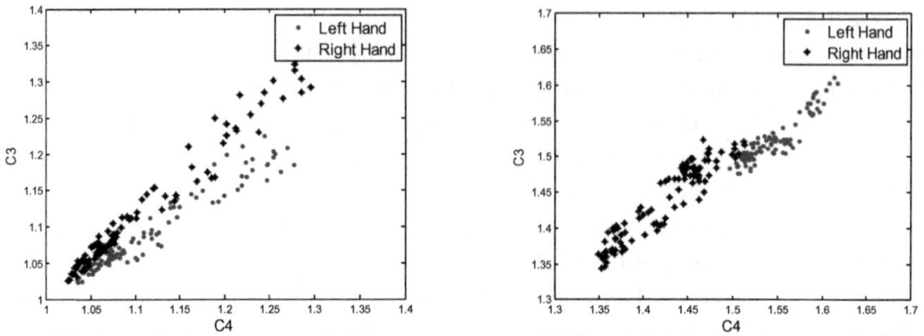

(a) The beta rhythm feature of subject g. (b) The alpha rhythm feature of subject g.

Fig. 7. Alpha and beta rhythm waves of subject.

5.2 Classification by SVM

Support Vector Machine (SVM) [16] is a machine learning method based on statistical learning theory. Its basic idea is that transforming a nonlinear mapping to a higher dimensional vector space according to the kernel function defined and gaining an optimal classification surface by computing. In consequence, the samples could be linearly separated. SVM has the sound theoretical foundation so that it is a good method for solving the problems such as nonlinear problem, small sample size and the local extremum problem. SVM has good study capability generalization ability. This paper used SVM to pattern recognition for features of EEG signals, and radial basis function was selected as the kernel function of SVM.

5.3 Analysis of Experiment Result

512 sample points including 400 points in motor imaginary stage and 56 points respectively before and after it were selected for feature extraction. According to the ERS/ERD phenomena during motor imagery, data in frequency band α and β of subject a, b, f and g performed feature extraction by wavelet fuzzy entropy method, and each subject corresponded to 200 four-dimensional feature vectors. In order to compare the results of fuzzy entropy method and wavelet fuzzy entropy method produced in this paper, feature extraction was performed by two methods respectively and classification was performed by SVM, and the results are shown in the Table 2. There are classification rates of subject a, b, g and g in the two pictures. According to the 3/4, 2/4 and 1/4 quantities in the box plots, the over half classification correctness rates based on wavelet fuzzy entropy were greater than 84%. The average rate of recognition of subject f reached 90.25% which was the highest among them. The experimental results show that the feature extraction method combining wavelet transform and fuzzy entropy is much better than the ways of using single fuzzy entropy. Therefore, the wavelet fuzzy entropy is more suitable for feature analysis of two-classes motor imagery EEG signals.

Table 2. Comparison of fuzzy entropy and wavelet fuzzy entropy classification accuracy.

Average accuracy/subject	a	b	f	g
Fuzzy entropy	78.64	77.56	79.68	78.82
Wavelet fuzzy entropy	86.47	83.78	90.12	87.92

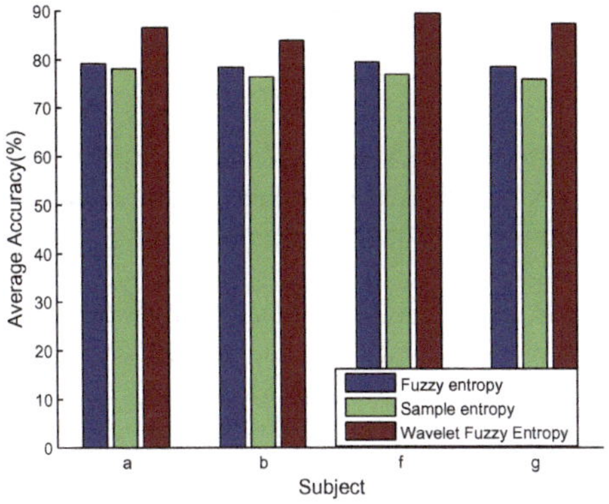

Fig. 8. Comparison of wavelet fuzzy entropy, fuzzy entropy and sample entropy classification accuracy.

To further verify feature extraction effect of wavelet fuzzy entropy, comparing it with traditional sample entropy method. As shown in the Fig. 8, feature extraction was performed by wavelet fuzzy entropy, fuzzy entropy and sample entropy respectively and classification was performed by SVM. As can be seen from a comparison of these results, classification correctness rates of wavelet fuzzy entropy is obviously higher than classification rates of traditional sample entropy method.

In the experiments, wavelet fuzzy entropy method and single fuzzy entropy method were compared as the methods for feature extraction. As can be seen from the analysis of the classification rates of four subjects, the feature extraction effect of wavelet fuzzy entropy method is better under the same experimental conditions. Moreover, it can be seen that fuzzy entropy is obviously better than traditional sample entropy for feature extraction by comparing fuzzy entropy with sample entropy method. This comparison better illustrates that feature analysis method based on wavelet fuzzy entropy is better for characterizing the features of two-classes motor imagery EEG signals.

6 Conclusion

A feature extraction method based on wavelet transform and fuzzy entropy is presented in this paper, and the motor imagery EEG signals after feature extraction are classified by a support vector machine classifier. The experimental results show that the feature extraction method combining wavelet transform and fuzzy entropy is much better than the ways of using single fuzzy entropy, sample entropy, or others. And it had a higher recognition rate. This method is limited to the two-class case, and only channel C3 and C4 are selected. The next research will focus on multi-channel and multi-class classification to develop multi-command BCI system.

Acknowledgement. This research was funded by the National Natural Science Foundation of China (Nos. 61871427 and 61372023).

References

1. Lotte, F., Roy, R. N.: Chapter7-brain–computer interface contributions to neuroergonomics. In: Neuroergonomics, pp. 43–48. Academic Press, Cambridge (2019)
2. Hughes, M.A.: Engineering brain-computer interfaces: past, present and future. J. Neurosurg. Sci. **58**(2), 117–123 (2014)
3. Namazi, H., Aghasian, E., Ala, T.S.: Fractal-based classification of electroencephalography (EEG) signals in healthy adolescents and adolescents with symptoms of schizophrenia. Technol. Health Care: Off. J. Eur. Soc. Eng. Med. **27**(3), 233–241 (2019)
4. Ma, M., Guo, L., Su, K., Liang, D.: Classification of motor imagery EEG signals based on wavelet transform and sample entropy. In: IEEE 2nd Advanced Information Technology, Electronic and Automation Control Conference (IAEAC), pp. 905–910. IEEE, NJ (2017)
5. Hasnaoui, L.H., Djebbari, A.: Discrete wavelet transform and sample entropy-based EEG dimensionality reduction for electroencephalogram classification. In: 2019 International Conference on Advanced Electrical Engineering (ICAEE), pp. 1–6. IEEE, NJ (2019)
6. Acharya, U.R., Molinari, F., Sree, S.V., Chattopadhyay, S., Ng, K.H., Suri, J.S.: Automated diagnosis of epileptic EEG using entropies. Biomed. Signal Process. Control **7**(4), 401–408 (2012)
7. Chen, W., Wang, Z., Xie, H., Yu, W.: Characterization of surface EMG signal based on fuzzy entropy. IEEE Trans. Neural Syst. Rehabil. Eng. **15**(2), 266–272 (2007)
8. Zhou, Y., Wang, T., Feng, H., Jiang, Z.: ERD/ERS analysis for motor imaginary EEG. Beijing Biomed. Eng. **23**(4), 263–268 (2004)
9. Faust, O., Acharya, U.R., Adeli, H., Adeli, A.: Wavelet-based EEG processing for computer-aided seizure detection and epilepsy diagnosis. Seizure **26**, 56–64 (2015)
10. Li, M., Wang, R., Yang, J., Duan, L.: An improved refined composite multivariate multiscale fuzzy entropy method for MI-EEG feature extraction. Comput. Intell. Neurosci. **2019**(2), 1–12 (2019)
11. Tian, J., Luo, Z.: Motor imagery EEG feature extraction based on fuzzy entropy. J. Huazhong Univ. Sci. Technol. (Nat. Sci. Edn.) **S1**, 92–94+98 (2013)
12. Sadiq, M.T., Yu, X., Yuan, Z., Zeming, F., Rehman, A.U., Ullah, I., Li, G., Xiao, G.: Motor imagery EEG signals decoding by multivariate empirical wavelet transform based framework for robust brain-computer interfaces. IEEE Access **7**, 1 (2019)
13. Shi, J.: Research on signal processing of motor imagery EEG data and P300 stimulus presentation paradigm. Zhejiang University (2012)
14. Blankertz, B., Dornhege, G., Krauledat, M., Müller, K.R., Curio, G.: The non-invasive Berlin brain-computer interface: fast acquisition of effective performance in untrained subjects. NeuroImage **37**(2), 539–550 (2007)
15. Ma, Y., Xu, M., She, Q., Gao, Y., Sun, Y., Yang, J.: De-noising method of the EEG based on adaptive threshold. Chin. J. Sens. Actuat. **10**, 1368–1372 (2014)
16. Adankon, M. M., Cheriet, M.: Support vector machine. In: Encyclopedia of Biometrics, pp. 1303–1308. Springer, USA (2009)

An Automatic White Balance Algorithm via White Eyes

Yuanyong Feng[1], Weihao Lu[1], Jinrong Zhang[2], and Fufang Li[1(✉)]

[1] School of Computer Science and Cyber Engineering, Guangzhou University, Guangzhou 510006, China
liff@gzhu.edu.cn

[2] School of Mechanical and Electrical Engineering, Guangzhou University, Guangzhou 510006, China

Abstract. Aiming to tackle the shortage of traditional white balance algorithms on face image, an eye white based algorithm is proposed under the assumption of stability of eye white in YCrCb color space. It first compares the eye white parts of input image with standard ones under normal white light source, to produce a gain coefficient, then corrects the color cast to the image. The experiments show that our method overcomes traditional counterparts both in subjective vision viewpoint and objective evaluation while retains good applicability and simple implementation.

Keywords: White balance · Color correction · Face recognition · White eyes

1 Introduction

As smart phone becomes more and more widespread, it is more and more common to take selfies to record people's daily life [1]. Such photos are often affected by non-white light sources. As we know, the color of an object is determined by the light source and the object itself. When the spectral composition of the light source varies, the object color in image varies, especially for white objects. Therefore, in order to allow machines collect images accurately under various light sources, a white balance adjustment is required.

There are many general white balance algorithms, including gray world algorithm [2–4], perfect reflection algorithm [5], histogram stretching algorithm [6, 7], etc. The gray world algorithm assumes that the average value of the three RGB components will tend to be equal when shooting any pair of images. The disadvantage is that when a large amount of monochrome appears in the image, the white balance effect will be greatly reduced, and the image color may even be distorted [8]. The perfect reflection algorithm assumes that the brightest point in the image is the white point, and calculates the image color correction coefficient from the white point. This algorithm has high accuracy, but when the image does not have an obvious white point, the method deteriorates. The effect of the histogram stretching algorithm to face is also not effective [9].

While general white balance algorithms do not work on face images, many researchers found that, human eye has color constancy under varying light sources [10, 11]. This reminds us color adjustment according to eye white should work, as it

Q. Liu et al. (Eds.): CENet 2020, AISC 1274, pp. 1679–1686, 2021.
https://doi.org/10.1007/978-981-15-8462-6_191

exists in most of face images while they are least affected by environmental factors. According to this feature, combined with eye extraction, an algorithm is proposed to restore face image to the white light environment by comparing the values of the white color components in different light environments. The experimental results show that the algorithm has a good white balance effect on the face image with color deviation.

2 Face Recognition

With the development of face recognition technology for many years, the face recognition algorithm has made remarkable progress. After detecting the face, it can accurately extract facial features. In the past few decades, most scholars have proposed many face recognition algorithms, and Turk and Pentland have proposed eigenface method to realize face recognition [12], Penev [13] et al. Proposed local feature analysis method, which can be divided into two categories: statistical method and geometric method [14]. In this paper, combined with the face key point detection function in the third-party library Dlib, the face is recognized and 68 key points are obtained from the face, among which the eye area is between the key point 36 and the key point 47, as shown in Fig. 1 after the capture, It is helpful to cut off the part of eyes accurately to reduce the error of skin to extract the RGB mean of eye white.

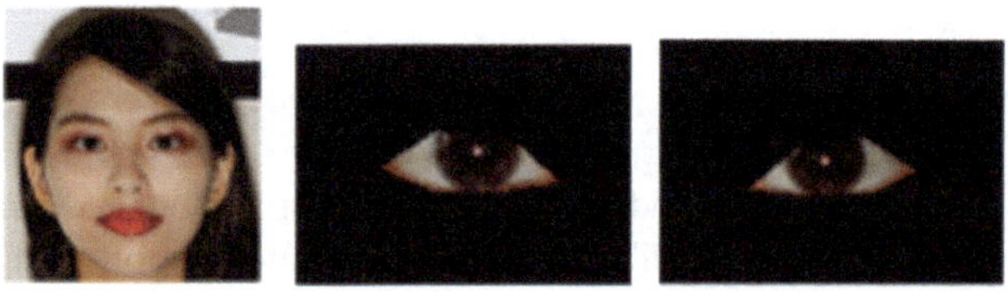

Fig. 1. Eye amputation. left: face; center: left eye; right: right eye.

3 Automatic Face White Balance Algorithm Based on White Eyes

3.1 Capture the White Area of the Eyes

The effect of the correction image of the white balance algorithm depends on the RGB mean of the eye white. To achieve this goal, we need to distinguish the eye area and the eye white area (as is shown in Fig. 1).

In paper [15], an algorithm based on CbCr Gauss distribution is proposed to segment whites from the whole picture. The algorithm is complex. This paper designs a simple method to separate whites and irises.

Color the pixel points white (R, G, B are all 255), and the segmentation diagram is as shown in Fig. 2 and the result sample is as shown in Fig. 3.

The above method uses a fixed value of 180 to distinguish the whites and the eyes. The calculation is simple, but it can only be used in some cases, and the accuracy

1. Read in the color image and cut off the eye part by face recognition, and get the R, G, B component values of each pixel.
2. Traversing RGB color space, if the RGB component sum of pixel points is greater than 180, it is regarded as eye white.

Fig. 2. Fixed area eye white truncation algorithm.

Fig. 3. White eye segmentation.

cannot be guaranteed. It is helpful to solve the problems of small application range and low accuracy (or instability) by using dynamic distinguishing value. In this paper, according to the characteristics of large differences in RGB components and values of eye whites and eyeballs, the eye pixels are traversed and sorted according to RGB components and values. The pixels with the largest sum value and the pixels with the smallest sum value are calculated as RGB components and mean values of eye whites and eyeballs respectively. Then, the mean value s of these two values is taken as the distinguishing value of eye whites and eyeballs. The calculation formula of S is as follows,

$$\mathrm{S} = \frac{5}{n}\left(\sum\nolimits_{0}^{a\%*n} m_i + \sum\nolimits_{(1-a\%)*n}^{n} m_i\right) \tag{1}$$

Here, i's is the eye pixel point, m_i is the sum of RGB components and n is the total number of all pixels in the eye. If the coefficient a is set too low, it is easy to appear that the extracted part cannot represent the mean value of eye white and eyeball, resulting in too many or too few interceptions; if the coefficient a is set too high, when the difference between eye white and eyeball pixels is large, a% of the pixels will contain both eye white and eyeball, resulting in inaccurate interceptions. When a value is 10, the white area is stable.

The improved procedure is as shown in Fig. 4,

Color the pixel points white (R, G, B are all 255), and the segmentation diagram is as shown in Fig. 5.

From the test results, we can see that the improved algorithm is more detailed and accurate than the original algorithm.

1. Read in the color image and cut off the eye part by face recognition, and get the R, G, B component values of each pixel.
2. Find S according to formula (1).
3. Traversing RGB color space, if the sum of RGB components of pixel points is greater than S, it is recognized as the segmented eye white part.

Fig. 4. Dynamic discriminative eye white truncation algorithm.

Fig. 5. White eye segmentation via dynamic Alg.

3.2 Eye White Balance Algorithm

According to the results of eye white truncation, this paper proposes a white balance algorithm for human eye white. The algorithm detail is as shown in Fig. 6.

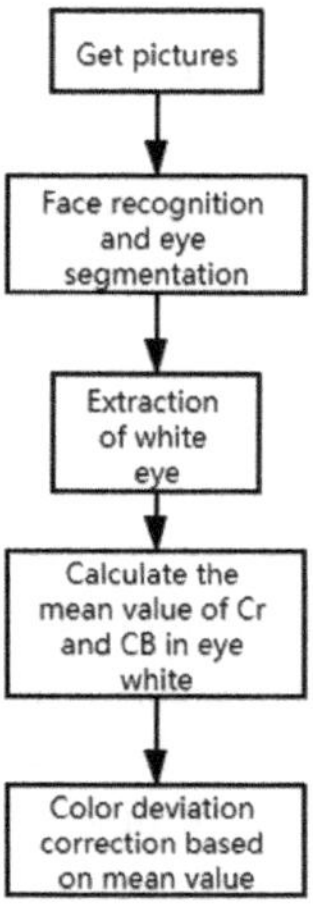

Fig. 6. Flow chart of automatic white balance based on eye white.

After obtaining the pixel points of eye white, the RGB components of all eye white pixels are statistically averaged according to the following formula ($\bar{R}$, $\bar{G}$, $\bar{B}$),

$$\bar{R} = \sum\nolimits_{0}^{n} R_i/n, \quad \bar{G} = \sum\nolimits_{0}^{n} G_i/n, \quad \bar{B} = \sum\nolimits_{0}^{n} B_i/n \tag{2}$$

Among them, $\bar{R}$, $\bar{G}$, $\bar{B}$, respectively, are the mean values of the red, green and blue components, R_i, G_i, B_i are the pixel values of the ith pixel in the eye white area, and n is the total number of pixel points in the eye white area. Transform the data of eye white image from RGB color space to YCrCb color space [16],

$$\begin{aligned}\bar{Y} &= 0.2990 * \bar{R} + 0.5870 * \bar{G} + 0.1140 * \bar{B} \\ \overline{Cr} &= 0.5000 * \bar{R} - 0.4187 * \bar{G} - 0.0813 * \bar{B} + 128 \\ \overline{Cb} &= -0.1687 * \bar{R} - 0.3313 * \bar{G}0.5000 * \bar{B} + 128\end{aligned} \tag{3}$$

where $\bar{Y}$ is the average value of the brightness of the white eye, $\overline{Cr}, \overline{Cb}$ is the average value of the color component.

The white balance of the photos can be achieved by the contrast value of the eye white, but the quality depends on whether the Cr and Cb values of the eye white are within a certain range. In order to verify this feature, 1277 pictures in vggface2 picture data [17] set are verified.

Step 1: face recognition of these images.
Step 2: calculate the mean value of Cr and Cb components of each picture.
Step 3: statistical analysis shows that 95% confidence interval of Cr value is [120, 166], 95% confidence interval of Cb value is [103, 147], and the statistical results are as shown in Fig. 7.

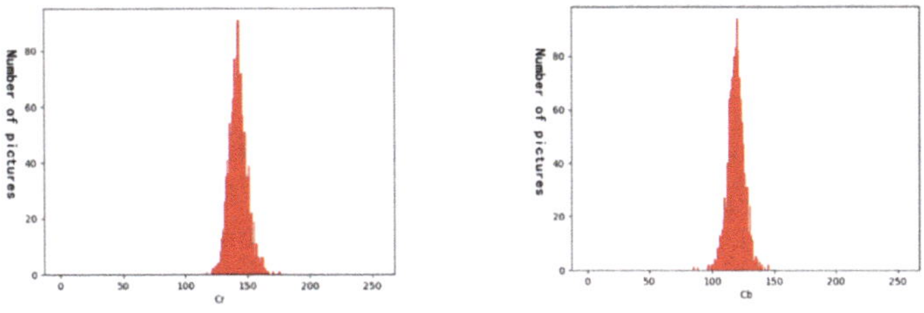

Fig. 7. Cr and Cb value in white eyes.

It can be seen from the statistical chart that most values of Cr and Cb are in a certain range. The algorithm proposed in this paper is to correct the white balance of the image according to the above characteristics. Among them, the average value of Cr and Cb is 142 and 125. When 137 and 128 are taken as the standard values of Cr and Cb, the white balance effect is the most stable. Compare the average value of the picture with the standard value to calculate the gain coefficients K1 and K2 of Cr and Cb color components.

$$K_1 = \overline{Cr}/137, \quad K_2 = \overline{Cb}/128 \tag{4}$$

If K_1 is greater than 1, it means the original image is red; if K_2 is greater than 1, it means the original image is blue.

By using the correction coefficient obtained in formula (4) to correct all pixels in the original image, the color deviation correction of the face image can be completed.

$$Cr' = Cr/K_1, \quad Cb' = Cb/K_2 \tag{5}$$

4 Experimental Results and Analysis

4.1 Effect Evaluation Based on Image Color Similarity

In order to verify the effectiveness of the algorithm, the partial color image is restored to the color under the white light source, and 100 face images under more than 100 environments are found from Baidu pictures and daily selfie photographs. The effects of our eye white balance and other methods are compared, and the results are as shown in Fig. 8.

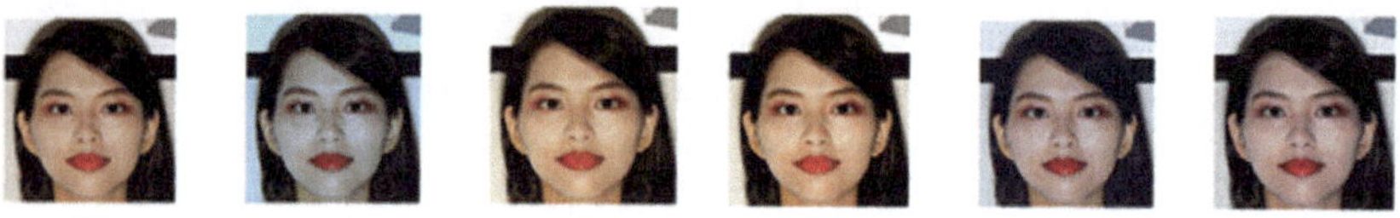

(a) (b) (c) (d) (e) (f)

Fig. 8. White balance rendering: (a) original picture in partial color light, (b) Gray world, (c) Perfect reflection, (d) Histogram Stretch, (e) Algorithm in this paper, (f) Original picture in white light.

Subjectively, compared with other white balance algorithms, the image generated by the white balance algorithm in this paper is closest to the original image in white light. The image color contrast formula is introduced to compare the image similarity. The formula is as follows,

$$\mathrm{X} = \sqrt{(R1 - R2)^2 + (G1 - G2)^2 + (B1 - B2)^2} \tag{6}$$

Among them, R1, G1, B1 and R2, G2, B2 are the RGB mean of the two images respectively. This algorithm is aimed at the white balance of the face, so only the face is compared when comparing the effect.

4.2 Visual Effect Comparison Based on Human Eye Observation

Cross Contrast. In addition, in order to compare the effect of the algorithm more intuitively, this paper uses the algorithm and histogram white balance algorithm to process 50 pictures containing human faces, and uses the intuitive feeling of human beings to compare the two groups of pictures for 2500 times. Among them, 1283 pictures obtained by this algorithm are considered to be closer to the real effect. Finally, it is concluded that the accuracy of this algorithm is 2.6% higher than that of histogram white balance algorithm.

Comparison of the Same Original. In addition, the algorithm and histogram white balance algorithm are used to carry out white balance processing on 1028 face pictures. The pictures are taken from vggface2 data set, and some pictures are shown in Fig. 9.

Fig. 9. Dataset picture.

Two groups of pictures are obtained by comparing two kinds of white balance algorithms with people's intuitive feelings. Among them, 580 pictures are considered to have better white balance effect. Compared with histogram algorithm, the correction accuracy of this algorithm is improved by 12.8%.

5 Conclusion

In this paper, we propose an automatic white balance algorithm based on eye white, which is proved to be more advantageous in face white balance by experiments. The disadvantage of this algorithm is that it can only adjust the white balance of the face image, which is less applicable than other algorithms. The next work is to reduce the influence of different people's eye whites on the white balance effect, so as to have a better white balance effect.

Acknowledgement. This work is supported in part by the National Natural Science Foundation of China (No. 61472092) and Guangdong College Student Innovation Project (No. S201911078074).

References

1. Zhou, Q.: Research on the application of smart phones in mobile learning. Softw. Guide (Educ. Technol.) **7**, 89–90 (2011)
2. Lam, E. Y.: Combining gray world and retinex theory for automatic white balance in digital photography. In: Proceedings of the Ninth International Symposium on Consumer Electronics, pp. 134–139. IEEE, NJ (2005)

3. Shi, R.: Research and implementation of automatic white balance algorithm. Inf. Technol. **36**(03), 85–88+93 (2012)
4. Lukac, R.: New framework for automatic white balancing of digital camera images. Signal Process. **88**(3), 582–593 (2008)
5. Fierro, M., Ha, H., Ha, Y.: An automatic color correction method inspired by the Retinex and opponent colors theories. In: International Symposium on Optomechatronic Technologies, pp. 316–321. IEEE, NJ (2009)
6. Jin, W., He, G., He, W., Mao, Z.: A 12-bit 4928 × 3264 pixel CMOS image signal processor for digital still cameras. Integration **59**, 206–217 (2017)
7. Wei, C., He, G.: Research on white balance algorithm based on histogram. Microelectron. Comput. **35**(06), 75–78 (2018)
8. Shen, L., Zhuo, L.: Wavelet Coding and Network Video Transmission. Science Press, Beijing, (2005)
9. Li, B.: Research on color constancy calculation. Beijing Jiaotong University (2009)
10. Zhao, P., Wang, W., Chen, W.: Research on layered color correction algorithm. Comput. Eng. Appl. **51**(06), 158–162 (2015)
11. Yang, H.: An automatic white balance method based on basic color system. J. Zhejiang Univ. Sci. Technol. **27**(01), 42–47 (2015)
12. Turk, M., Penland, A.: Eigenfaces for recognition. J. Cogn. Neurosci. **3**(1), 71–86 (1991)
13. Zhao, W.: Robust image based 3d face recognition. University of Maryland (1999)
14. Tang, D.: Research on image feature extraction and matching technology in face recognition. Dalian Maritime University (2013)
15. Kuang, W., Mao, K., Huang, J., Li, H.: Fatigue driving detection based on gaussian eye white model. Chin. J. Image Graph. **21**(11), 1515–1522 (2018)
16. ISO/IEC 10918-5: Information Technology - Digital Compression and coding of continuous-tone still images: JPEG File Interchange Format (JFIF). ITU-T (2013)
17. Cao, Q., Shen, L., Xie, W., Parkhi, O.M., Zisserman, A.: VGGFace2: a dataset for recognizing faces across pose and age. In: 13th IEEE International Conference on Automatic Face & Gesture Recognition, pp. 67–74. IEEE, NJ (2018)

Experimental Study on Mechanical Characteristics of Lower Limb Joints During Human Running

Lingyan Zhao[1], Shi Zhang[2], Lingtao Yu[2], Kai Zhong[2], and Zhiguang Guan[1(✉)]

[1] School of Construction Machinery, Shandong Jiaotong University, Jinan 250357, China
{206012, guanzhiguang}@sdjtu.edu.cn
[2] School of Mechanical and Electric Engineering, Harbin Engineering University, Harbin 150001, China

Abstract. With the vigorous development of marathon, running has been favored by people. While running brings people health, joint injury is also a major hidden danger in this sport. In order to study the characteristics of the joints of the lower limbs of the human body including the joint angle and joint torque during running, especially the effect of running speed on the above mechanical characteristics, this paper recruited 20 testers to complete periodic running at a running speed of 6 km/s–14 km/s. The Functional Assessment of Biomechanics System (FAB system) was used to collect kinematics and dynamic parameter data. On basis of the seven-body rigid body model, the Lagrangian method was used to establish the dynamic model of the human lower limbs to obtain the joint moments of the various links of the lower limbs. Through the above model and combined with experimental research, this paper concludes: running speed has different effects on hip, knee, and ankle flexion angles and extension angles. From the perspective of joint torque, the hip and ankle joints of the human lower limbs are significantly affected by the changes in running speed during running and are prone to injury. For runners, running performance can be improved by increasing swing leg torque and reducing swing leg inertia. This research has guiding significance for improving athletes' level, designing and improving sports equipment and developing exoskeleton robots.

Keywords: Joint injury · Lower limb · Gait analysis · Seven rigid body model · Mechanical properties

1 Introduction

Gait is a characteristic of walking and running, and gait control is a complex process that requires the nervous system to cooperate with the muscles, bones and other systems [1]. In the absence of pathological interference, gait is characterized by coordination, but under the interference of disease, the coordination and periodicity of gait will be affected accordingly. With the improvement of medical standards, gait

Q. Liu et al. (Eds.): CENet 2020, AISC 1274, pp. 1687–1693, 2021.
https://doi.org/10.1007/978-981-15-8462-6_192

measurement has attracted much attention, and various gait measurement systems have emerged accordingly. The gait measurement system combining the optical motion capture system with the force plate was once considered as a standard measurement system by some researchers [2]. The system uses an optical motion capture system to perform image enhancement, spatial filtering, and other image processing techniques on the collected image information to calculate eigenvalues, which, combined with the plantar pressure data obtained by the force plate, help obtain gait-related data [3–5]. Although the system can measure human joint motion information with high accuracy, the expensive equipment cost has discouraged many scholars [6]. To this end, Kanika et al. [7] designed a passive marker optical gait analysis system, which measures gait parameters including joint angle and walking speed by measuring 5 reflective markers. The advantage of this method is that the system is simple and it doesn't need too much time to set the flag. But it can obtain the information of only 5 joint angles. So it's impossible to perform systematic analysis of gait by using this system. With the development of micro-electromechanical technology, various inertial motion capture sensors are used in gait measurement systems to form wearable gait measurement systems. This type of system obtains the information such as the rotation angle and the acceleration of each limb joint through sensors. After the corresponding coordinate conversion, the speed, displacement, joint angle and other information of the movement can be calculated [8]. Ferrari et al. compare the gait information obtained based on the optical capture system with the gait information obtained based on the inertial sensor system. The results showed that two systems had an error of less than 5% in step size [9]. Analyzing the above gait measurement system, this paper takes the speed as the starting point and uses the inertial sensor-based FAB system to measure the joint angle of the human lower extremity joint, and then analyzes the mechanical properties of the human lower extremity joint.

2 Biomechanical Model of Human Lower Limb Gait Movement

Running is a rather frequent behavior in the daily life of the human body. It is a periodic movement completed by the central nervous system acting on each muscle group, being coordinated with each joint at the same time. The characteristics of each cycle are described by gait [10]. The support phase and the swing phase of the running process form a complete gait cycle, and the specific stages are divided as shown in Fig. 1. Studies have shown [11] that during normal walking and running, sagittal motion (e.g., flexion and extension) is the main form of the joints motion of the lower extremities. The movements in the coronal plane (e.g., internal rotation, external rotation) and horizontal plane (e.g., adduction, abduction) are not obvious, so this paper only focuses on the movement model of the lower limb joints and plantar area in the sagittal plane. As shown in Fig. 2, the lower extremity system model is simplified into a seven-rigid body model including the pelvis, hip joint, thigh fibula, knee joint, calf fibula, ankle joint, and metatarsal bone.

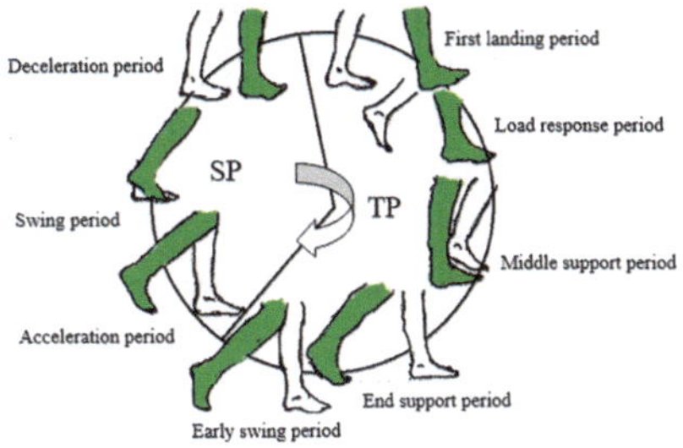

Fig. 1. Gait cycle division.

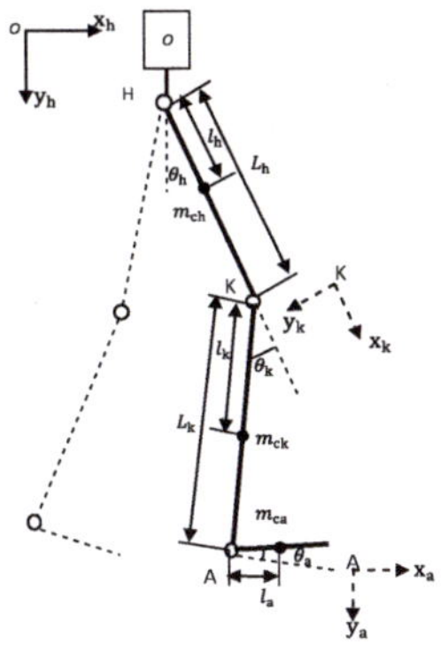

Fig. 2. Human lower extremity system model.

By establishing mathematical models, classical mechanics theory is used to explain human motion characteristics. This paper focuses on analyzing the dynamic equations during the support period. According to the characteristics of the model in this paper and the feasibility of the solution, the Lagrange equation is selected to establish the biomechanical model of human motion. The Lagrange equation has the form:

$$\frac{d}{dt}(\frac{\partial L}{\partial q_j}) - \frac{\partial L}{\partial q_j} = Q_j(j = 1, 2, \ldots, n) \tag{1}$$

3 Experiments and Analysis of Human Lower Limb Joint Stress Characteristics

In this paper, the FAB system developed by NEODYN of Canada is used to collect and measure the kinematics information. The FAB system integrates a system for measuring static human posture and dynamic human motion capture, and it is based on sensors such as accelerometers and gyroscopes. Subsequently, the collected multi-cycle data is successively subjected to cubic spline interpolation [12] and filtering [13] to obtain smoother data of the same dimension for subsequent analysis.

3.1 Research on Characteristics of Joint Rotation Angle of Human Lower Limbs with Different Running Speeds

Figure 3 shows the average curve of hip, knee, and ankle rotation angles of 20 testers at four speeds of 6 km/h, 8 km/h, 10 km/h, and 12 km/h, as the pictures show:

(1) In the support phase, the hip joint is in a flexion state during the gait cycle of 0% to 40%, and it is in the extension state during the gait cycle of 40% to 60%. At the beginning of the gait-the moment when the supporting leg and the heel started to touch the ground-the hip flexion angle of the tester was 16°. At 40% of the gait cycle, the hip joint angle is 0° and the swing leg reaches the maximum swing

angle. At this time, the supporting leg and the trunk are on the same vertical line. After that, the center of gravity of the person moves forward, the swinging legs fall back, and the extension angle of the hip joint gradually increases. At the end of the single support period-at 50% of the gait cycle- the hip extension angle reaches the maximum of 6.1°. Within the swing phase, the support legs began to gradually leave the ground, at this time mainly manifested as flexion. The flexion angle gradually increases within 60%–80% of the gait cycle, and reaches the maximum flexion value of 24.8° at 80% of the gait cycle. Later, during the gait cycle, the hip flexion angle decreased slowly and remained near 20°. As the running speed increases, the flexion angle of the hip joint does not change much when the heel of the supporting leg touches the ground, but the maximum flexion angle and the maximum extension angle increase in size and phase advances. The increase in flexion and extension angles is used to provide more power and speed up to the running, while the flexion and extension phase advance are the reason for the increase in cadence.

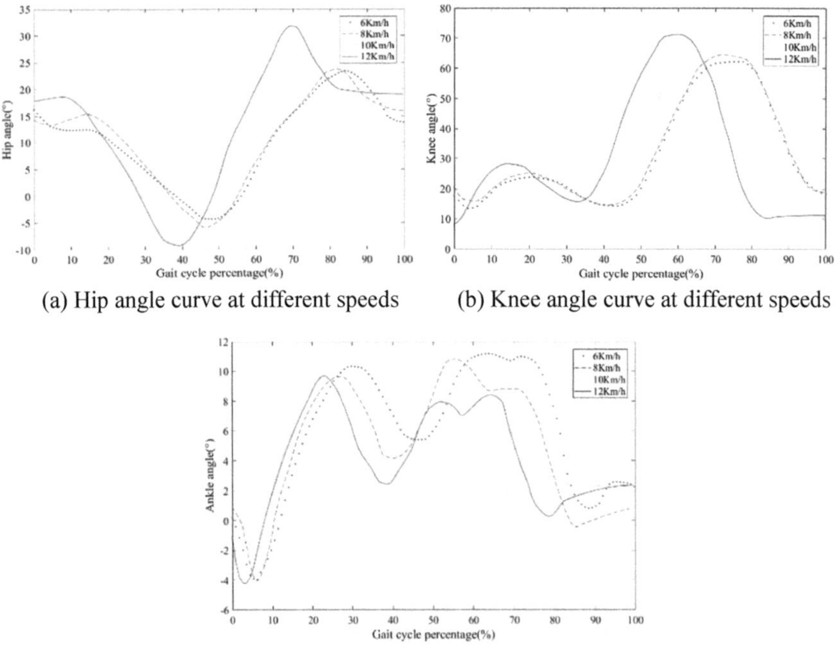

(a) Hip angle curve at different speeds (b) Knee angle curve at different speeds

(c) Ankle angle curve at different speeds

Fig. 3. Changes in hip, knee and ankle joint angles at different running speeds.

(2) The knee joint showed flexion throughout the gait cycle, and there were two maximum flexion angles, which appeared at the end of the first double support period and the swing period, respectively. At 20% of the gait cycle, the knee reaches the first maximum flexion angle of 23.5°. At 40% of the gait cycle, the

knee flexion angle decreased to a minimum of 13.8°. After that, the knee joint entered the second double support period, and the flexion angle gradually increased until it entered the swing period. Within the swing phase, the knee angle gradually increases, and the maximum value of the second flexion angle is 60° at 70% of the gait cycle. In the second half of the swing phase, the knee joint angle gradually decreases until the end of the swing phase, and the knee joint angle reaches a minimum value of about 10°. Similar to the hip joint, as the running speed increases, the knee joint angle does not change much when the supporting leg touches the ground. The maximum values of the two flexion angles used for pedaling and reducing the moment of inertia of the swinging legs change significantly, respectively, to 31.6° and 70°. The maximum knee extension angle remains unchanged at 10°.

(3) During the entire gait cycle, the ankle joint mainly exhibits dorsiflexion. In the support phase, the ankle joint has a fast plantar flexion when the heel of the supporting leg touches the ground, and the maximum plantar flexion angle is 4° at this time, which is used to cushion the impact of the ground. At 30% of the gait cycle, that is, when the swinging amplitude of the swinging leg is at its maximum, the ankle joint reaches the maximum dorsiflexion angle of 10°. Within the swing phase, the dorsiflexion angle gradually increases from the beginning of the swing to 80% of the gait cycle, the maximum value is about 10°, and it is accompanied by fluctuations. This is because the ankle joint of the swinging leg controls muscle contraction through the hook on the toe to reduce the rotational inertia of the swinging leg, and the fluctuation is reflected in the unstable phenomenon during the swinging process. In the second half of the swing phase, the flexion angle of the ankle joint gradually decreased, and eventually remained stable in preparation for touchdown. As the running speed increases, the ankle joint angle does not change much.

3.2 Research on Joint Torque of Human Lower Limb with Different Running Speed

Based on the kinetic equation established based on the Lagrange equation, combined with the kinematics parameters, ground reaction forces, and body inertia parameters of the lower limbs of the human body, the moments of the lower limbs of the human body can be calculated. Figure 4 shows the average curve of the joint torque of the tester at different running speeds.

It can be seen from Fig. 4:

(1) In a complete exercise cycle, the hip joint mainly exhibits flexion torque and there will be two large torque peaks, one in the support period and one in the swing period. The peak torque during the support period is greater than the peak during the swing period. At the end of the first double support period, the hip joint torque reaches a maximum value of 140 Nm, which is the process of supporting the legs and is used to provide power for the lower limbs. Subsequently, the hip flexion moment gradually decreases until it reaches a minimum value of 40 Nm. Then it began to show a stretched state, and the torque gradually increased to reach the

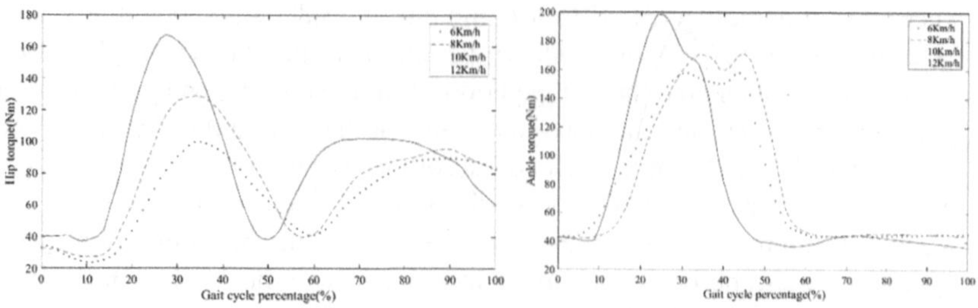

Fig. 4. Hip and ankle torque curves at different speeds.

second torque peak value of 80 Nm. This torque is used to control the backward swing of the leg and maintain body balance. As the running speed increases, the two torque peaks increase to 170 Nm and 100 Nm, respectively, and there is a phase advance situation, which leads to increasing the running speed while also bringing the risk of increased hip injury.

(2) Compared with the hip torque, the ankle torque is too large, which causes the ankle to be the most vulnerable link in running. The ankle joint shows dorsiflexion torque within the support phase, and the flexion torque increases slowly from the moment when the supporting leg heel touches the ground. The maximum dorsiflexion moment in the single support phase reached 170 Nm, after which the support foot began to gradually leave the ground. At this time, the ankle joint is converted from dorsiflexion to plantar flexion, and the angle of dorsiflexion gradually decreases until the end of the support phase. In the subsequent swing phase, the flexion torque of the ankle joint remains stable at around 40 Nm. As the running speed increases, the ankle torque increases during the support period, and the maximum flexion torque increases to 200 Nm. Since there is no angular change in the ankle joint during the swing period, the ankle joint torque does not change much.

4 Conclusion

In this paper, the FAB system is used to study the mechanical properties of the lower extremity joints, including joint rotation angle and joint torque, in a complete gait cycle. At the same time, we focus on the impact of running speed on the above mechanical characteristics, and draw the following conclusions:

(1) The hip joint transitions from the flexed state to the extended state in the support phase, and the flexed state in the swing phase. As the running speed increases, the maximum flexion angle and extension angle of the hip joint increase. The knee joint showed flexion throughout the gait cycle. As the running speed increases, the maximum value of the flexion angles for pedaling and reducing the moment of inertia of the swinging legs increases significantly, and the maximum knee extension remains unchanged. The ankle joint is dorsiflexed throughout the gait cycle, and is less affected by the running speed.

(2) The hip joint mainly exhibits flexion torque and two large torque peaks during the complete gait cycle. As the running speed increases, the peaks of the two waves increase, and the phases of the peaks move forward. Compared with the hip joint torque, the ankle joint torque is larger, and the running speed has a greater influence on the flexion torque during the support period.

The curves of joint angle and joint moment obtained by experiments in this paper can be applied to the design of exoskeleton joints. However, this article only considers the sagittal plane motion of the human lower limbs, and does not consider the coronal and horizontal plane motions. The above factors will be taken into account in future research.

Acknowledgment. This work is supported by Major Science and Technology Innovation project of Shandong Province (2019JZZY020703), Science and Technology Support Plan for Youth Innovation in Universities of Shandong Province (2019KJB014), PhD Startup Fund of Shandong Jiaotong University (BS2019020) and Shandong Jiaotong University "Climbing" Research innovation Team Program (SDJTUC1805).

References

1. Ma, X.H.: Body eleven rigid body biomechanical model and biomechanical study of running exercise. Harbin Engineering University (2018)
2. Passmore, E., Sangeux, M.: Improving repeatability of setting volume origin and coordinate system for 3D gait analysis. J. Gait Posture **39**(2), 831–833 (2014)
3. Xue, Y. X.: Lower limb rehabilitation training evaluation method based on gait analysis. Guangdong University of Technology (2019)
4. Armand, S., Sangeux, M., Baker, R.: Optimal markers' placement on the thorax for clinical gait analysis. Gait Posture **39**(1), 147–153 (2014)
5. Zhou, H., Hu, H.: Upper limb motion estimation from inertial measurements. Int. J. Inf. Technol. **13**(1), 1–14 (2007)
6. Maenaka, K.: MEMS inertial sensors and their applications. In: Source: Proceedings of INSS 2008-5th International Conference on Networked Sensing Systems, pp. 71–73 (2008)
7. Chandra, P., Kanika, G., Anshul, M., Rajesh, K., Vijay, L.: Passive marker based optical system for gait kinematics for lower extremity. Proc. Comput. Sci. **45** (2015)
8. Kim, J.N., Ryu, M.H., Yang, Y.S., et al.: Detection of gait event and supporting leg during overground walking with mediolateral swing angle. Appl. Mech. Mater. **145**, 567–573 (2011)
9. Ferrari, A., Rocchi, L., Noort, J.V.D., et al.: Toward the use of wearable inertial sensors to train gait in subjects with movement disorders. In: Converging Clinical and Engineering Research on Neurorehabilitation, pp. 937–940. Springer, Heidelberg (2013)
10. Perry, J., Burnfield, J.M., Cabico, L.M.: Gait analysis: normal and pathological function. Slack (2010)
11. Song, L.M.: Study on biological factors of gait stability. Tianjin University of Science and Technology (2017)
12. Wu, J.M., Liu, Y.Y., Zhu, C.G., Zhang, X.L.: Multilevel integro cubic spline quasi-interpolation. J. Image Graph. **22**(03), 327–333 (2017)
13. Zhu, Z.H., Sun, J.H., Wang, W.S., Li, N., He, H.: Study of wavelet threshold de-noising algorithm based on Grubbs rule. Northwest Hydropower (s1), 45–48 (2011)

Study on the Gait Motion Model of Human Lower Limb Joint

Lingyan Zhao[1(✉)], Shi Zhang[2], Lingtao Yu[2], Kai Zhong[2], and Guan Zhiguang[1]

[1] School of Construction Machinery, Shandong Jiaotong University, Jinan 250357, China
{206012, guanzhiguang}@sdjtu.edu.cn
[2] School of Mechanical and Electric Engineering, Harbin Engineering University, Harbin 150001, China

Abstract. Accurately simulating the biomechanical parameters of human lower extremity movement is very important to analyze the biomechanical characteristics of human lower limbs. In this paper, the joints of the lower limbs of the human body are simplified to a seven-rigid body model in the sagittal plane, and the kinematics of the lower limbs of the human body and the dynamic model of the supporting period are established by the D-H method and the Lagrange method. In order to verify the above model, this paper builds a simulation model of human rigid body based on SimMechanics. Finally, comparing the experimental data of FAB measurement with the simulation data, it is found that the kinematics and dynamics models used in this paper are accurate enough to complete the measurement of the mechanical properties of the human lower limbs.

Keywords: Biomechanical characteristics · Human lower limbs · Mechanical properties

1 Introduction

In recent years, sports biomechanics has gradually attracted people's attention. In [1], Jacquelin Perry detailed the research methods of pathological gait and normal gait. Now researchers believe that the gait measurement system that combines the optical motion capture system and the force plate is the standard measurement system [2]. In order to accurately measure human biomechanical information, various gait measurement systems came into being. With the development of micro-electromechanical technology, various inertial motion capture sensors are used in gait measurement systems to form wearable gait measurement systems. Anna et al. [3] used inertial sensors to quantitatively assess gait information, and conducted clinical evaluations of patients with ankle replacements. Gao et al. [4] designed a gait measurement system based on multi-MEMS inertial sensors, which can collect data such as lower limb swing amplitude and foot acceleration. The system is inexpensive and easy to wear, but it can only establish a gait database for specific pathologies. In addition, a gait measurement system based on plantar area pressure has also been developed. It collects the

Q. Liu et al. (Eds.): CENet 2020, AISC 1274, pp. 1694–1701, 2021.
https://doi.org/10.1007/978-981-15-8462-6_193

pressure data of various parts of the plantar area during walking or running, and makes pressure distribution maps, and then measures the biomechanical characteristics of the human foot [5]. The plantar area pressure measurement system can be used clinically for disease identification [6] and correction [7]. Combined with experimental data, a model for analyzing human biomechanics needs to be established. In order to solve the above problems, the kinematics of the lower limbs of the human body and the dynamic model of the supporting period are established in this paper, and the simulation of the lower limbs of the human body is realized by SimMechanics. This data is compared with the experimental data collected by the Functional Assessment of Biomechanics System (a system developed by NEODYN of Canada is used to collect and measure the kinematics information) to prove the feasibility and accuracy of the biomechanical model proposed in this paper.

2 Human Lower Limb Biomechanical Model

2.1 Human Lower Limb Gait Model

Studies have shown [8] that during normal walking and running, sagittal motion (e.g., flexion and extension) is the main form of the joints motion of the lower extremities. The movements in the coronal plane (e.g., internal rotation, external rotation) and horizontal plane (e.g., adduction, abduction) are not obvious, so this paper only focuses on the movement model of the lower limb joints and plantar area in the sagittal plane. In this paper, the angle change of each joint in the sagittal plane is defined as: the angle of the hip joint is the angle between the extension line of the thigh fibula and the vertical ground reference line. The positive is the flexion angle, and the negative is the extension angle. The knee joint accompanies the hip joint movement refering to the angle between the thigh fibula and the calf fibula. The situation closer to the thigh reflects the flexion angle, otherwise it reflects the extension angle. The ankle joint's angle between the plantar area and a reference line parallel to the horizontal plane, the back extension angle is positive, and the plantar area flexion angle is negative. As shown in Fig. 1, the lower extremity system model is simplified into a seven-rigid body model including the pelvis, hip joint, thigh fibula, knee joint, calf fibula, ankle joint, and metatarsal bone.

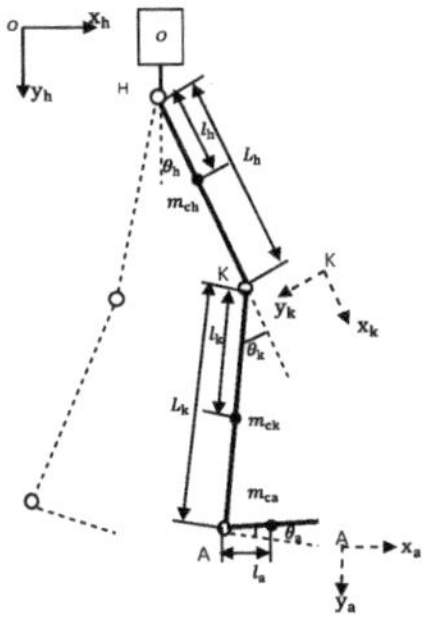

Fig. 1. Human lower extremity system model.

2.2 Lower Limb Kinematics Model Based on D-H Method

For the seven-rigid-body model shown in Fig. 1, the right lower limb is taken as the research object. H, K, and A represent the hip, knee, and ankle joints respectively. m_{ch}, m_{ck}, and m_{ca} are the center of mass of the thigh fibula, calf fibula, and metatarsal. L_h, L_k, and L_a are the length of the thigh fibula, calf fibula, and metatarsal, respectively. l_h, l_k, and l_a are the lengths from the center of mass of the thigh fibula, calf fibula, and metatarsal to the proximal end respectively, and θ_h, θ_k, and θ_a are the rotation angles of the hip, knee, and ankle joints, respectively.

Select the hip joint H as the origin. For the convenience of derivation, this paper assumes:

$$\begin{cases} q_1 = \theta_h \\ q_2 = \theta_k \\ q_3 = 90^\circ + \theta_a \end{cases} \tag{1}$$

The transformation matrix of the two adjacent coordinates of the hip, knee and ankle joints is:

$$T_H^0 = R_{z,q_1} T_{x,l_1} = \begin{bmatrix} cq_1 & -sq_1 & 0 \\ sq_1 & cq_1 & 0 \\ 0 & 0 & 1 \end{bmatrix} \begin{bmatrix} 1 & 0 & L_h \\ 0 & 1 & 0 \\ 0 & 0 & 1 \end{bmatrix} = \begin{bmatrix} cq_1 & -sq_1 & L_h cq_1 \\ sq_1 & cq_1 & L_h sq_1 \\ 0 & 0 & 1 \end{bmatrix} \tag{2}$$

$$T_K^1 = R_{z,q_2} T_{x,l_2} = \begin{bmatrix} cq_2 & -sq_2 & 0 \\ sq_2 & cq_2 & 0 \\ 0 & 0 & 1 \end{bmatrix} \begin{bmatrix} 1 & 0 & L_k \\ 0 & 1 & 0 \\ 0 & 0 & 1 \end{bmatrix} = \begin{bmatrix} cq_2 & -sq_2 & L_k cq_2 \\ sq_2 & cq_2 & L_k sq_2 \\ 0 & 0 & 1 \end{bmatrix} \tag{3}$$

$$T_A^2 = R_{z,q_3} T_{x,l_3} = \begin{bmatrix} cq_3 & -sq_3 & 0 \\ sq_3 & cq_3 & 0 \\ 0 & 0 & 1 \end{bmatrix} \begin{bmatrix} 1 & 0 & L_a \\ 0 & 1 & 0 \\ 0 & 0 & 1 \end{bmatrix} = \begin{bmatrix} cq_3 & -sq_3 & L_a cq_3 \\ sq_3 & cq_3 & L_a sq_3 \\ 0 & 0 & 1 \end{bmatrix} \tag{4}$$

$$\begin{aligned} T_1 = T_H^0 T_K^1 &= \begin{bmatrix} cq_1 & -sq_1 & L_h cq_1 \\ sq_1 & cq_1 & L_h sq_1 \\ 0 & 0 & 1 \end{bmatrix} \begin{bmatrix} cq_2 & -sq_2 & L_k cq_2 \\ sq_2 & cq_2 & L_k sq_2 \\ 0 & 0 & 1 \end{bmatrix} \\ &= \begin{bmatrix} c(q_1+q_2) & -s(q_1+q_2) & L_k c(q_1+q_2) + L_h cq_1 \\ s(q_1+q_2) & c(q_1+q_2) & L_k s(q_1+q_2) + L_h sq_1 \\ 0 & 0 & 1 \end{bmatrix} \end{aligned} \tag{5}$$

$$T_2 = T_1 T_A^2 = \begin{bmatrix} c(q_1+q_2) & -s(q_1+q_2) & L_k c(q_1+q_2)+L_h cq_1 \\ s(q_1+q_2) & c(q_1+q_2) & L_k s(q_1+q_2)+L_h sq_1 \\ 0 & 0 & 1 \end{bmatrix} \begin{bmatrix} cq_3 & -sq_3 & L_a cq_3 \\ sq_3 & cq_3 & L_a sq_3 \\ 0 & 0 & 1 \end{bmatrix}$$
$$\begin{bmatrix} c(q_1+q_2+q_3) & -s(q_1+q_2+q_3) & L_a c(q_1+q_2+q_3)+L_k c(q_1+q_2)+L_h cq_1 \\ s(q_1+q_2+q_3) & c(q_1+q_2+q_3) & L_a s(q_1+q_2+q_3)+L_k s(q_1+q_2)+L_h sq_1 \\ 0 & 0 & 1 \end{bmatrix} \quad (6)$$

In the formula: $c(q_1) = \cos(q_1), s(q_1) = \sin(q_1)$.

From the above formula, the positions of the joint points H, K, A and the center of mass of the lower limbs of the human body can be obtained, and the speed, acceleration and other indicators of each position can be obtained by deducing each position.

2.3 Lower Limb Dynamics Model Based on Lagrangian Method

The establishment of dynamic equations of human motion is a key step in the use of biomechanical theoretical analysis methods. By establishing mathematical models, classical mechanics theory is used to explain human motion characteristics. During the running process of the human body, the activities of the lower limb legs during the supporting period are crucial. The four processes of supporting the leg's landing, cushioning, kicking, and leaving the ground are the core issues of human running. Therefore, this paper focuses on analyzing the dynamic equations during the support period. According to the characteristics of the model in this paper and the feasibility of the solution, the Lagrange equation is selected to establish the biomechanical model of human motion. The Lagrange equation has the form:

$$\frac{d}{dt}\left(\frac{\partial L}{\partial q_j}\right) - \frac{\partial L}{\partial q_j} = Q_j (j = 1, 2, \ldots, n) \quad (7)$$

In the formula, q_j represents the generalized coordinate of the system; n is the degree of freedom of the system; L is the difference between the kinetic energy and potential energy of the system, and the generalized force corresponding to the generalized coordinate q_j is Q_j.

According to Fig. 2, a simplified model of the three-body dynamics of the lower limbs of the human body is established. $m_h g$, $m_k g$, and $m_a g$ represent the equivalent gravity at the center of mass of the thigh, calf, and plantar area, and N and f represent the ground support reaction force and friction force received by the center of mass of the plantar area. The process of establishing a three-degree-of-freedom model of human lower limbs using Lagrange's equation is as follows:

(1) System kinetic energy:

$$\begin{aligned} T =& \frac{1}{2} m_h (l_h \dot{\theta}_1)^2 + \frac{1}{2} m_k \left[(L_h \dot{\theta}_1)^2 + l_k^2 (\dot{\theta}_1 + \dot{\theta}_2)^2 + 2 L_h l_k (\dot{\theta}_1^2 + \dot{\theta}_1 \dot{\theta}_2) \cos\theta_2 \right] + \frac{1}{2} m_a \left[L_k^2 (\dot{\theta}_1 + \dot{\theta}_2)^2 \right] \\ &+ \frac{1}{2} m_a \left[+ 2 L_h l_a (\dot{\theta}_1^2 + \dot{\theta}_1 \dot{\theta}_2 + \dot{\theta}_1 \dot{\theta}_3) \cos(\theta_2 + \theta_3) + + 2 L_k l_a (\dot{\theta}_1^2 + \dot{\theta}_2^2 + 2 \dot{\theta}_1 \dot{\theta}_2 + \dot{\theta}_1 \dot{\theta}_3 + \dot{\theta}_2 \dot{\theta}_3) \cos\theta_3 \right] \\ &+ \frac{1}{2} m_a \left[(L_h \dot{\theta}_1)^2 + 2 L_h l_k (\dot{\theta}_1^2 + \dot{\theta}_1 \dot{\theta}_2) \cos\theta_2 + l_a^2 (\dot{\theta}_1 + \dot{\theta}_2 + + \dot{\theta}_3)^2 \right] \end{aligned} \quad (8)$$

(2) System potential:

$$\begin{aligned} P =& m_h g l_h (1 - \cos\theta_1) + m_k g \{ L_h (1 - \cos\theta_1) + l_k [1 - \cos(\theta_1 + \theta_2)] \} + m_a g \{ L_h (1 - \cos\theta_1) \\ &+ L_k [1 - \cos(\theta_1 + \theta_2)] + l_a [1 - \cos(\theta_1 + \theta_2 + \theta_3)] \} \end{aligned} \quad (9)$$

(3) Lagrange function: $L = T - P$
(4) System dynamics equation: The Lagrange function is used to derivate the generalized coordinates θ_1, $\dot{\theta}_1$, θ_2, $\dot{\theta}_2$, θ_3, $\dot{\theta}_3$ and other variables in order to obtain the corresponding generalized force.

Add the external forces N and f to the Lagrange equation above to obtain the joint torque model as follows:

$$M_i = Q_i + f L_i \cos\theta_i + N L_i \sin\theta_i (i = 1, 2, 3) \quad (10)$$

3 Experiments and Results

Combined with the above model, this paper uses SimMechanics to complete the modeling of the human body's multi-rigid body model, and compares it with the data collected by the FAB system. The result is shown in the Figs. 2 and 3:

From Fig. 2: the data collected through experiments, and then the calculated joint torque is similar to the simulated torque, and the shape of the torque curve is consistent.

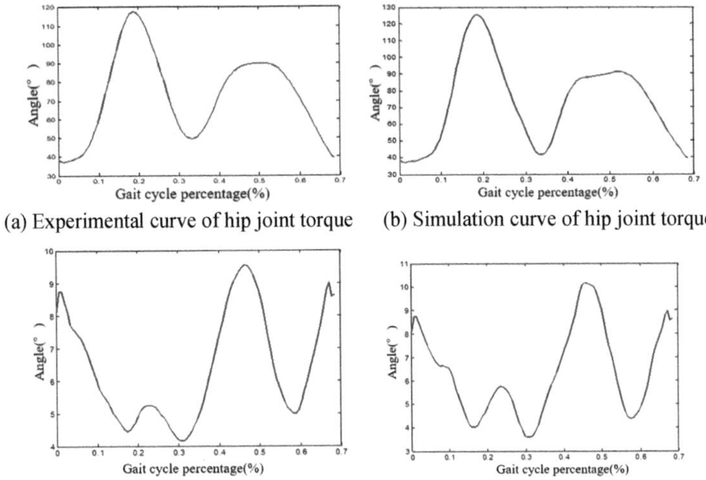

(a) Experimental curve of hip joint torque (b) Simulation curve of hip joint torque

(c) Experimental curve of ankle joint torque (d) Simulation curve of ankle joint torque

Fig. 2. Comparison of experimental and simulation curves of joint torque

However, there are some differences in that the simulation curve has less fluctuation, and the value of the simulation curve will be smaller than the actual torque value. Considering the complexity of the human body, this difference is within the allowable range.

From Fig. 3, the experimental curve is similar to the simulation curve and there are differences: The curve of the actual measured hip joint angle data is not very smooth, and there will be a small range of fluctuations. The value of the simulation curve will be slightly larger than the experimental curve value, and the trajectory of the simulation curve will be smoother. The simulation curve of the knee joint angle is larger than the actual value, and the whole is in accordance with the law. There is a slight deviation between the ankle rotation curve and the actual curve. This is due to the flexibility of the foot, and there will be a certain error. This can prove the correctness of the mathematical model of the human joint rotation and can be used to simulate human motion.

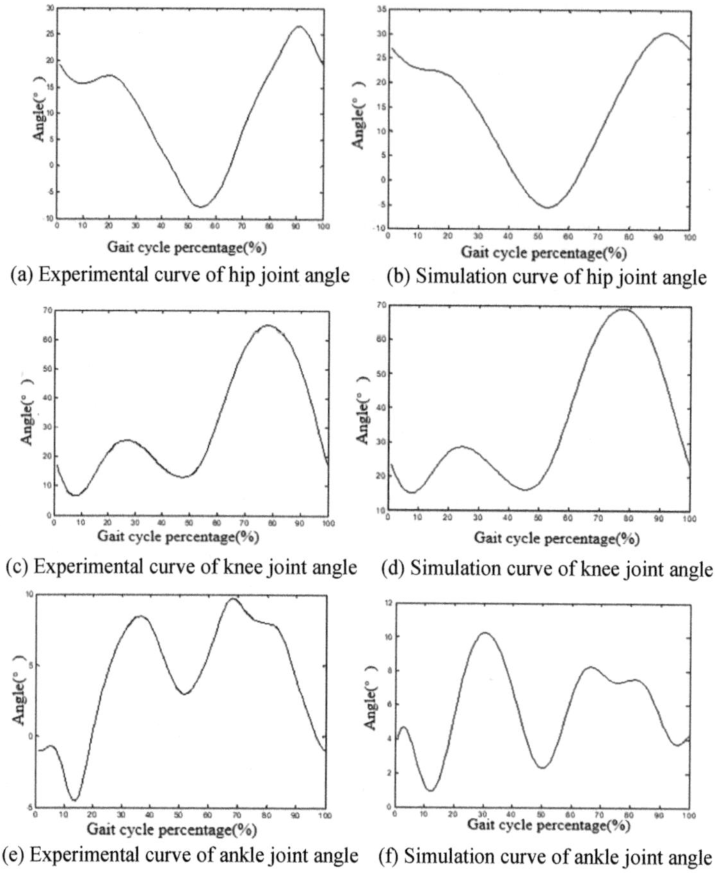

(a) Experimental curve of hip joint angle (b) Simulation curve of hip joint angle

(c) Experimental curve of knee joint angle (d) Simulation curve of knee joint angle

(e) Experimental curve of ankle joint angle (f) Simulation curve of ankle joint angle

Fig. 3. Comparison of experimental and simulation curves of joint rotation.

4 Conclusion

In order to further explore the biomechanics of human lower extremity movement, on the basis of simplifying the human lower extremity into seven rigid bodies, the D-H method and Lagrangian method were used to establish the human lower extremity kinematics and dynamic model of the supporting period. Simultaneously, Sim was used to simulate the above model, and the obtained simulation data was compared with the experimental data collected by the FAB system. The results show that the above model can accurately simulate the biomechanics of human lower limb movement. The results show that the above model can accurately simulate the biomechanics of human lower limb movement.

Acknowledgment. This work is supported by Major Science and Technology Innovation project of Shandong Province (2019JZZY020703), Science and Technology Support Plan for Youth Innovation in Universities of Shandong Province (2019KJB014), PhD Startup Fund of Shandong Jiaotong University (BS2019020) and Shandong Jiaotong University "Climbing" Research innovation Team Program (SDJTUC1805).

References

1. van der Linden, M.: Gait Analysis, Normal and Pathological Function, 2 edn. J. Perry, J.M. Burnfield, Slack Inc., Physiotherapy, vol. 97, no. 2, ISBN 978-1-55642r-r766-4 (2010)
2. Elyse, P., Morgan, S.: Improving repeatability of setting volume origin and coordinate system for 3D gait analysis. J. Gait Posture **39**(2), 831–833 (2014)
3. Sant'Anna, A., Wickström, N., Eklund, H., et al.: Assessment of gait symmetry and gait normality using inertial sensors: in-lab and in-situ evaluation. In: International Joint Conference on Biomedical Engineering Systems and Technologies, pp. 239–254. Springer, Heidelberg (2012)
4. Gao, C. L.: Gait analysis and research based on multi-MENS inertial sensors. Zhejiang University of Technology (2018)
5. Abdul Razak, A.H., Zayegh, A., Begg, R.K., Wahab, Y.: Foot plantar pressure measurement system: a review. Sensors **12**(7), 9884–9912 (2012)
6. Ma, Y.P.: Early warning system design of foot pressure for detection of human disease. Electron. Test **Z1**, 12–15 (2018)
7. Ellen, L., Rodrigo, V., Fabio, A.B., Diego, O., Lucas, S., Lilian, T., Bucken, G.: Continuous use of textured insole improve plantar sensation and stride length of people with parkinson's disease: a pilot study. Gait Posture **58**, 495–497 (2017)
8. Song, L.M.: Study on biological factors of gait stability. Tianjin University of Science and Technology (2017)

CWAM: Analysis and Research on Operation State of Medical Gas System Based on Convolution

Lida Liu, Song Liu, and Yanfeng Xu(✉)

Shandong Raone Intelligent Technology Co., Ltd., Jinan 250000, China
rosun@raonetech.com, xuyanfeng@sdrxhb.cn

Abstract. The medical gas system is a system engineering directly related to the life and property safety of patients and doctors. In order to ensure the safety of hospital patients' gas use and the regular operation of related medical equipment, this paper introduces the analysis model (Convolution Weight Analysis Model, CWAM for short) of medical oxygen operation state in the medical gas system. It obtains the operation state regulation of the tank through CWAM model. Through the analysis of the operation status of the medical oxygen tank, we can adequately grasp the abnormal operation status, timely prevent, and ensure the regular operation of the hospital oxygen supply system.

Keywords: Medical gas · Convolution Weight · Normalization

1 Background and Current Situation

Medical gases are gases used for medical treatment, including oxygen, carbon dioxide, nitrogen, laughing gas, argon, positive air, negative pressure air, etcetera [1]. It is a single or mixed component gas centrally supplied by medical piping systems for patient treatment, diagnosis, prevention, or drive surgical tools, for the treatment of critically ill patients in hospitals and for a variety of medical devices, whose safety is directly related to the safety of patients and physicians [2].

Due to the expansion of the use and construction scale of various advanced medical devices, modern hospitals need to use various medical gasses and adopt a centralized supply mode. The medical gas system refers to a complete system including gas source subsystem, monitoring and alarm subsystem, and equipped with terminal facilities such as valves and terminal components for the supply of medical gas. Therefore, monitoring and managing the medical gas system is a significant systematic project [3]. Due to the close relationship between medical gas and patient's life, the medical gas system of centralized supply and management is called life support system in the international scope [4], which is also the only special mechanical and electrical system named from the height of life support in the hospital mechanical and electrical supply system, and its importance is self-evident [5].

With the progress of medical technology and the construction of the modern hospital, the medical gas system has already developed into a systematic project integrating machinery, electronics, pipeline transportation, centralized control, construction equipment construction, and medical equipment application [6]. The medical

Q. Liu et al. (Eds.): CENet 2020, AISC 1274, pp. 1702–1709, 2021.
https://doi.org/10.1007/978-981-15-8462-6_194

gas system research is scattered, and it is not able to compare every link of medical gas system construction, and then discuss its safety comprehensively and systematically. Therefore, ensuring the safety of the medical gas system is a problem worthy of deep thought, which is also related to the medical safety and quality of the hospital.

2 Problem Definition and Analysis

2.1 Research Contents

The medical gas system is the life support system of the hospital, and the safe operation of the system is an essential guarantee for life support, in which the safe supply of medical oxygen gas is also particularly important. At present, there are many oxygen storage tanks in most hospitals. If there is leakage or explosion in the oxygen storage tanks, it will cause huge safety problems of life and property. To monitor the operation status of the oxygen storage tank, under the normal data, observe the pressure of the oxygen storage tank, the liquid level pressure of the tank and the effective range of the normal operation of the pressure after the tank gasification. During the process of oxygen filling and oxygen releasing, the pressure of the tank changes violently, the liquid level pressure of the tank changes slowly and the pressure changes most violently after the gasification. When abnormal operation occurs, the pressure data of the oxygen storage tank will change abnormally. This paper attempts to determine the regularity of the tank's pressure, the liquid level pressure of the tank, and the internal operation of the pressure after gasification, whether there are periodicity and volatility.

2.2 Data Sampling

The operation state of the medical oxygen storage system is described by several signals which can be continuously measured in real time. Define that the multidimensional system U contains N data signals, expressed as

$$f = \{f_1(t), \ldots, f_j(t), \ldots, f_N(t)\} \tag{1}$$

where $f_n(t), n \in \{1, \ldots, n, \ldots, N\}$ is the signal value (reading) of n-th signal at time t.
Definition

$$U^m = \{u_1^m(t), \ldots u_n^m(t), \ldots u_N^m(t)\} \tag{2}$$

Is the sampling data array, in which the parameter m is the sequence number of the sampling window and the parameter n is the sequence number of the sampling signal. Set the data obtained from sampling is discrete, so the data after sampling processing is

$$u_n^m(t) = \{u_n^m(t_0), u_n^m(t_1), u_n^m(t_2), \ldots, u_n^m(t_k), u_n^m(t_K)\} \tag{3}$$

$$t_1 = t_0 + \Delta t; t_2 = t_0 + 2\Delta t; \ldots, t_k = t_0 + k\Delta t; \ldots, t_K = t_0 + K\Delta t \tag{4}$$

Among them, t_0 is the initial time of the data sampling window, t_K is the end time of the data sampling window, Δt is the data sampling interval, and K is the width of the data sampling window.

Finally, the sampling data of signal n in window m can be expressed as

$$\begin{aligned} u_n^m(t) &= u_n^m(t_0) \\ &= f_n^m(t_0), u_n^m(t_1) = f_n^m(t_1), u_n^m(t_2) \\ &= f_n^m(t_2), \ldots, u_n^m(t_k) = f_n^m(t_k), \ldots, u_n^m(t_K) \\ &= f_n^m(t_K) = f_n^m(t_0), f_n^m(t_0 + \Delta t), f_n^m(t_0 + 2\Delta t), \ldots, \\ & f_n^m(t_0 + k\Delta t), \ldots, f_n^m(t_0 + K\Delta t) \end{aligned} \tag{5}$$

Definition and Description of Formula Symbols

Definition: the semaphore is N, and the signal representative parameter is j.

Definition: parameter n is the serial number of the sampling signal, representing parameter i.

Definition: X* is the standardized symbol.

Definition:

$$y_{j(j-1)}(n) = \sum_{i=-\infty}^{\infty} x(i)k(n-i) = x(n) * k(n) \tag{6}$$

Represents convolution of two semaphores, wherein

$$y(j) = x(i) \tag{7}$$

$$y(j-1) = k(i) \tag{8}$$

Definition: $W_{j(j-1)}$ represents the convolution weight of the semaphore.

3 Algorithm Model

CWAM model consists of standardization algorithm, convolution algorithm, weight analysis algorithm and data fitting algorithm.

3.1 Standardized Algorithm

Because different types of indicators often have different dimensions and dimensional units, such a situation will affect the results of data analysis. In order to eliminate the dimensional impact between indicators [7], data standardization is needed to solve the comparability between data indicators.

The model uses min-max standardization algorithm, also known as deviation standardization, which is a linear transformation of the original data, so that the result value is mapped between [0–1]. The conversion function is as follows:

$$X^{*} = \frac{x - min}{max - min} \tag{9}$$

where max is the maximum value of sample data and min is the minimum value of sample data.

3.2 Convolution Algorithm

Convolution is the sum of two variables multiplied in a certain range [8]. If the convolution variables are sequences x(n) and k(n), the convolution results:

$$y(n) = \sum\nolimits_{i=-\infty}^{\infty} x(i)k(n-i) = x(n) * k(n) \tag{10}$$

where asterisk * represents convolution. When the sequence n = 0, the sequence h(−i) is the result of the reversal of the sequence i of h(i); the reversal of the sequence causes h(i) to flip 180 degrees with the vertical axis as the center, so the calculation method of the sum after multiplication is called convolution sum, which is called convolution for short.

3.3 Weight Analysis Algorithm

The weight analysis is based on the data convolution to get the convolution sum between the two indexes and find out the proportion relationship between the two indexes.

$$W_{j(j-1)}(i) = \frac{y_{j(j-1)}(n)_{max}}{\sum\nolimits_{i=1}^{i} y_{j(j-1)}(n)_{max}} \tag{11}$$

where j is the index signal and $y_{j(j-1)}(n)_{max}$ is the maximum value in the convolution sequence of two signals.

3.4 Data Fitting Algorithm

Data fitting refers to the convolution and weight function relationship between fitting indexes on the basis of weight analysis.

$$F(i) = \sum\nolimits_{i=1}^{n} W_i * y_j(i)\ (i \in 1\ldots n;\ j \in 1\ldots N) \tag{12}$$

where W_j is the weight data between two signals, $y_j(i)$ is the convolution sequence between two signals.

4 Experimental Verification

4.1 Data Sources

The experimental verification data are taken from three monitoring indicators of a hospital's medical oxygen tank, including the pressure of the tank, the liquid level pressure of the tank and the pressure after gasification. The specific index data collection location is shown in Fig. 1.

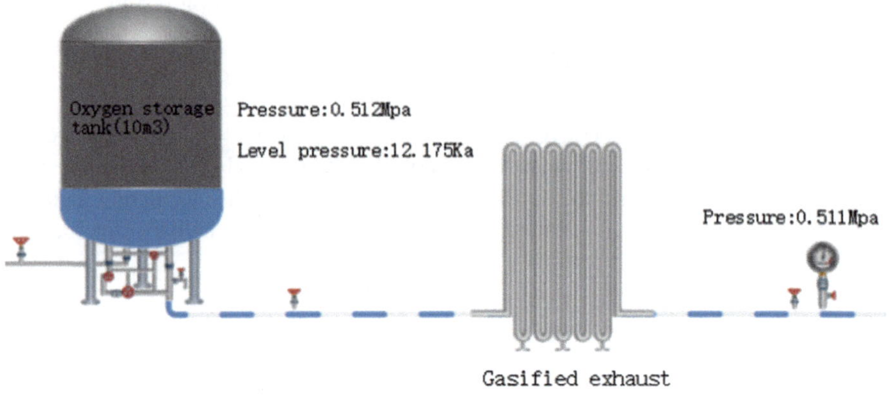

Fig. 1. Index diagram of medical oxygen tank.

4.2 Experimental Process

Standardization. The data is linearly transformed by min-max standardization algorithm to map the result value to [0–1]. The algorithm can map the original data to the same dimension, effectively improving the convergence speed and accuracy of the model [9].

Figure 2 is the data after linear standardization, including tank pressure, tank liquid level pressure and tank gasification pressure. The red dot line is tank pressure, the green * is tank liquid level pressure, and the blue *o* is tank gasification pressure.

Convolution. The data is convoluted by convolution algorithm. Firstly, the tank pressure and tank liquid level pressure are convoluted to get the red dot line relationship sequence in the figure; secondly, the tank pressure and the pressure after tank gasification are convoluted to get the relationship sequence of green * in the figure below; finally, the tank liquid level pressure and the pressure after tank gasification are convoluted to get the relationship sequence of blue *o* in the figure below, such as Fig. 3.

Weight Analysis. Observe data standardization convolution sequence data, use formula 1 to get the weight relationship between convolution sequences.

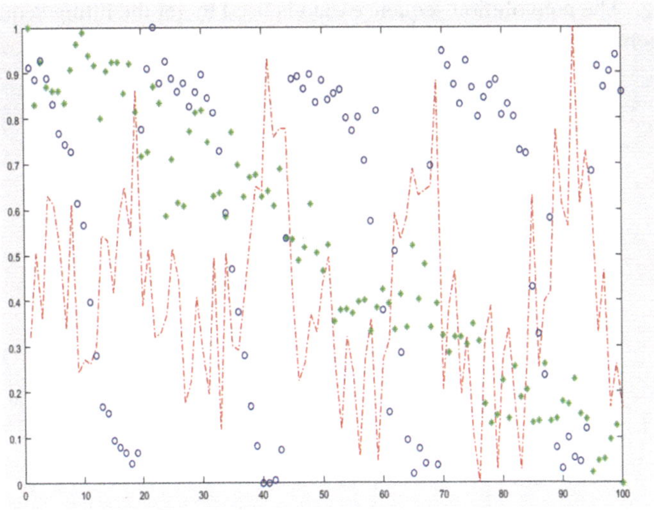

Fig. 2. Standardized data of medical oxygen index.

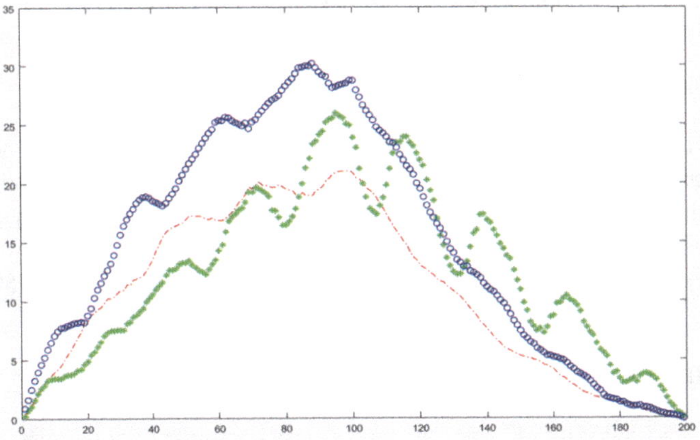

Fig. 3. Convolution sequence data of medical oxygen index.

$$W_i = \frac{y_i(n)_{max}}{\sum_{i=1}^{i}\sum y_i(n)_{max}} \tag{13}$$

According to formula 1, the weight of convolution action relationship can be calculated as follows: the convolution action weight of tank pressure and tank level pressure is 0.27; the convolution action weight of tank pressure and tank pressure after gasification is 0.338; the convolution action weight of tank pressure and tank level pressure after gasification is 0.392.

Data Fitting. The convolution sequence data is used to get the fitting sequence data of Fig. 4 through the fitting algorithm.

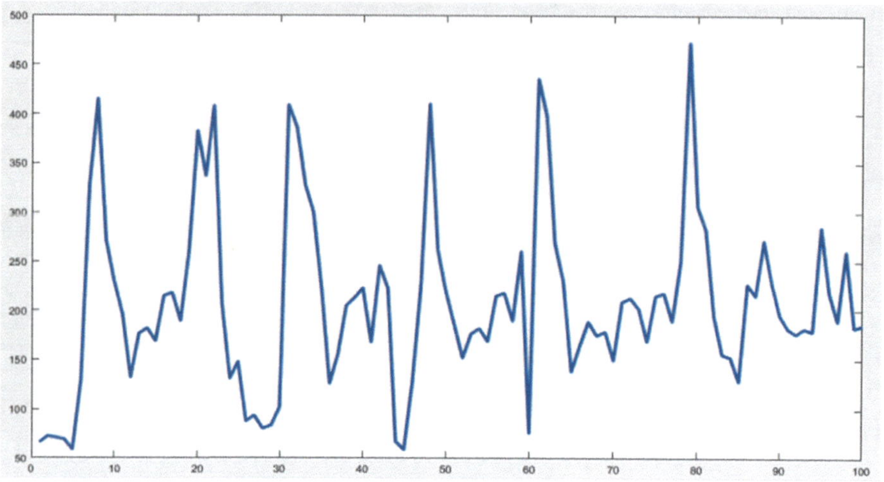

Fig. 4. Fit data of medical oxygen index.

4.3 Data Analysis and Summary

Through cwam analysis of the tank pressure, tank liquid level pressure and pressure after tank gasification, Fig. 4 data sequence is obtained. According to the analysis in Fig. 4, it can be verified that the internal operation state of the medical oxygen tank has certain periodicity, first slowly rising, and then slowly falling. It has certain volatility in the periodic operation law, and the effective operation range interval is [50, 500].

5 Conclusion

Medical gas system is the life support system of the hospital, and the safe operation of the system is an important guarantee of life support. In this paper, CWAM convolution weight analysis model is composed of standardization algorithm, convolution algorithm, weight analysis algorithm, data fitting algorithm, etcetera, through the calculation of this model, the operation state law of medical oxygen system can be observed intuitively. The internal operation state of medical oxygen tank has certain periodicity, and has certain volatility in the periodic operation law. Through the analysis of the operation status of the medical oxygen tank, we can effectively grasp the abnormal operation status, timely prevent and ensure the normal operation of the hospital oxygen supply system.

Acknowledgement. This project was selected as the first batch of technological innovation projects in Shandong Province in 2020 and received financial support.

References

1. Kunyu, X.M., Fan, C., Zheng, P., Wu, S.K.: Practice and exploration of hospital gas monitoring and alarm system. Mod. Hospit. **20**(01), 79–82+86 (2020)
2. Shi, X.: Design example of medical gas centralized supply system in modern hospital. Chin. Hospit. Archit. Equip. **20**(10), 41–44 (2019)
3. Wei, X.Y., Huang, B.Q.: Brief talk on the design of medical gas system in new hospital. Chin. Hospit. Archit. Equip. **21**(02), 49–51 (2020)
4. Pan, S.W., Jin, Y.S.: Discussion on safety analysis and optimization measures of medical gas system in a hospital. Chin. Med. Equip. J. **40**(06), 67–69+95 (2019)
5. Zhang, F.J., Guo, L.J., Zhang, R., Zhou, F.P.: Design and implementation of medical gas real-time monitoring platform. Comput. Appl. Softw. **36**(04), 91–98+185 (2019)
6. Yuan, M., Li, M., Shi, C., Huong, J., Sun, L.: Design of real-time monitoring and early warning system for medical oxygen. J. Nantong Univ. (Nat. Sci. Edn.) **18**(01), 16–20 (2019)
7. Zhang, W.K., Yu, K., Wu, X.F.: Entity recommendation based on normalized similarity measure of metagraph. J. Shandong Univ. (Eng. Sci.) **50**(02), 66–75 (2020)
8. Yu, B., Yuan, B.B., Zhuang, K.X., Qian, J., Ni, X.Y., Zhang, M.R.: Learning recommendation system based on convolutional neural network. J. Fujian Comput. **36**(04), 55–57 (2020)
9. Liu, J.Y., Zhang, K., Wang, G.H.: Comparative study on data standardization methods in comprehensive evaluation. Digit. Technol. Appl. **36**(06), 84–85 (2018)

Yi Zhuotong Intelligent Security Management Platform for Hospital Logistics

Lida Liu, Song Liu, and Fengjuan Li(✉)

Shandong Raone Intelligent Technology Co., Ltd., Jinan 250000, China
rosun@raonetech.com, lifengjuan@sdrxhb.cn

Abstract. The hospital logistics management is an integral part of hospital management, and safety management is the most important part of hospital logistics management. Intelligent and efficient standardized management mode is an essential guarantee for the safety production of hospital logistics. Therefore, intelligent security management of hospital logistics has gradually become the core of hospital management. The realization of this platform, on the one hand, strengthens the hospital to the logistics jurisdiction power equipment security monitoring management, enhances the Early Warning and prevention ability, improves the level of safety management in the hospital.

Keywords: Hospital logistics · Security · Intelligence · Management

1 Background to the Question

With the rapid development of the economy and the continuous improvement of modern medical technology, people's demand for medical treatment is increasing, the number of hospitals is increasing, and hospitals' scale is expanding [1]. In the medical profession's rapid development, hospitals have combined their characteristics to complete the hospital information system, the electronic medical record system, the operation anesthesia monitoring system, the imaging system, the examination system, and the office automation system. The quality of medical work and the level of hospital management have been improved. However, as an important supporting link of the hospital, the logistic service is always on the edge of intelligent construction, restricting the development and management level of the Hospital Logistic Service [2].

The safe operation of the hospital logistics business provides high-quality medical services to patients, and it is also the foundation of the smooth operation of the hospital. At present, the hospital's logistics management is generally difficult to ensure security, heavy responsibility, many types of equipment, wide distribution, aging staff structure and Low education level, the workload is increasing day by day, and the clinical service requirements are high [3].

Q. Liu et al. (Eds.): CENet 2020, AISC 1274, pp. 1710–1717, 2021.
https://doi.org/10.1007/978-981-15-8462-6_195

2 Current Situation of Hospital Logistics

2.1 Status

In 2015, an elevator failure occurred in a hospital in Yongchuan, Chongqing. The reason for the elevator failure was that the instantaneous voltage fluctuation caused the elevator balance sensor to be abnormal, which caused the elevator to stop running after the top flushing. People in the elevator are trapped for 30 min.

In 2006, the Department of Radiology of Chaohu First People's Hospital caused a fire due to electric sparks, and more than 20 houses and related equipment were turned to ashes. With the burning of medical equipment, the damage is estimated at over 20 million yuan.

From 2011 to 2016, there were 524,000 electric fires in China, causing 3,261 deaths and more than 2,000 injuries, with direct economic losses of 9.2 billion yuan.

Based on the above cases, the following problems exist in hospital logistics management:

Equipment Maintenance Is Not in Place. Hospital logistics involves the management and maintenance of power supply and distribution, water supply and drainage, central air conditioning, medical gas, boilers, elevator equipment, etc. The maintenance work is generally undertaken by the electrician class and maintenance class of the General Affairs Department. However, the employees generally lack of theoretical knowledge and practical experience. When the equipment has a serious failure, it is helpless, which seriously affects the normal and orderly operation of the hospital, and even leads to more serious safety accidents [4].

Aging of Logistics Personnel. Hospital logistics generally have the status of aging, technical faults, lack of successors, etc., relying on empirical management methods, resulting in incomplete data, weak management of infrastructure and equipment; maintenance without planning; lack of statistical analysis of information data [5]; The professional skills of the personnel in the logistics department are not high and lack professionalism. Secondly, there is a blind spot in the manual inspection equipment, and it is impossible to find the operation failure of the equipment in time.

Outsourcing Unit Lacks Supervision. Logistics services in many hospitals are now undergoing social reforms and implementing logistics service outsourcing. On the one hand, it reduces operating costs, improves the efficiency of hospital logistics [6], and promotes the development of hospital logistics reform; at the same time, due to the large number of service outsourcing, frequent staff turnover, and difficulty in communication and supervision, most of the personnel in outsourcing units have not been systematically trained. The outsourcing unit acts as a third party, and the hospital management system does not constrain the service personnel. After the problem occurs, communication is not smooth, and there may be safety accidents.

Energy Use Is Unscientific and Wasteful. The hospital's energy consumption index is relatively high, 1.6 to 2 times that of general public buildings. In the process of energy use, there is a lack of awareness of energy conservation, energy-saving management measures are generally lacking and missing, public resources are overused and consumed, and energy waste is serious. On the other hand, the hospital has a large building area, scattered energy consuming equipment, and many end energy consumption nodes. There are often situations where the local energy consumption is unknown, and there are loopholes in energy management.

2.2 National Policies

In response to the above problem, the state has issued a series of policy documents:

(1) "Notice of the State Council's Work Safety Commission on the Comprehensive Management of Electrical Fires" encourages social units to apply electrical fire monitoring technology to improve the monitoring, early warning, and disposal capabilities of electrical products and their circuit operation status. Encourage the public to report potential electrical safety hazards, and form a strong atmosphere in which all people pay attention to participate in the prevention and control of electrical fires.

(2) The "Thirteenth Five-Year Plan for the Development of Special Equipment Safety and Energy Conservation" strengthens the building of scientific and technological support capabilities. Carry out research on key technologies for the safety, energy conservation and environmental protection of special equipment based on big data. Using advanced technologies such as big data, Internet of Things, cloud computing, mobile Internet, etc., with the intelligent collection, transmission and storage, analysis, processing, and mining applications of data and information as the main lines, we will cultivate new formats for intelligent testing of special equipment.

(3) "Opinions of the General Office of the State Council on Strengthening the Work of Elevator Quality and Safety" uses big data, Internet of Things and other information technologies to build an elevator safety public information service platform, and establish an elevator quality and safety evaluation system based on failure rate and service life, Gradually establish a quality and safety traceability system for the entire life cycle of elevators to achieve problem investigation, accountability, and social supervision.

3 Platform Design and Implementation

In response to the above problems combined with national policy documents, this paper designs and implements a smart security management platform for hospital logistics.

3.1 Platform Architecture Design

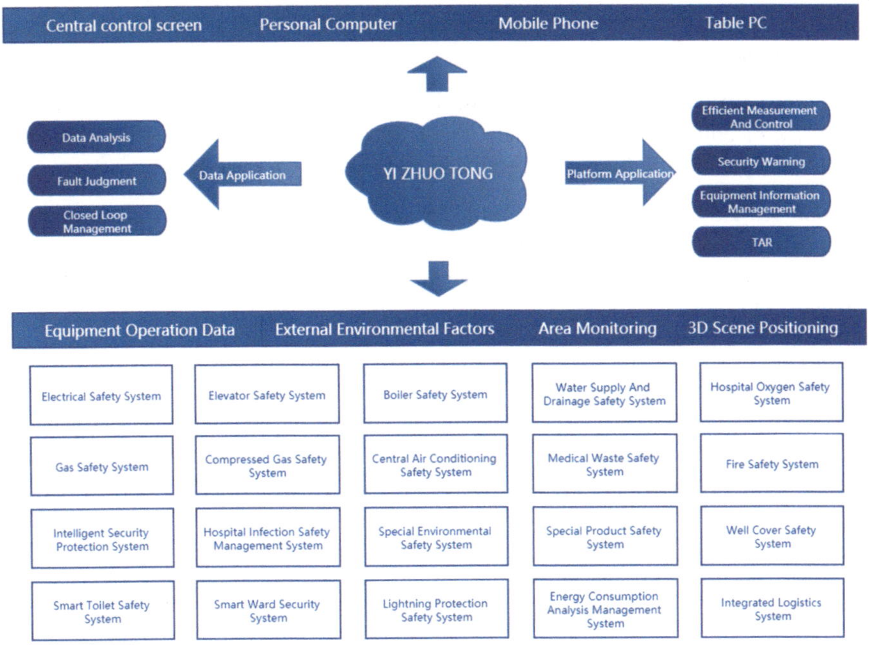

Fig. 1. Platform architecture design.

YI Zhuotong safety intelligent management platform includes electrical safety intelligent management system, elevator safety intelligent management system, boiler safety intelligent management system, water supply and drainage safety intelligent management system, medical oxygen safety intelligent management system, gas safety intelligent management system, compressed gas gas safety Intelligent management system, central air conditioning safety management system, medical waste safety management system, fire safety intelligent management system, hospital security intelligent management system, hospital sense safety management system, special environment safety management system, special product safety management system, manhole cover safety intelligent management System, smart toilet safety management system, smart ward safety management system, lightning protection safety intelligent management system, energy consumption analysis management system, logistics integrated service management system (Fig. 1).

Platform functions mainly include efficient measurement and control, safety warning, equipment information management, and TAR closed-loop management.

Efficient Measurement and Control. Build a real-time intelligent perception platform for the Internet of Things. The data can be uploaded once every 1–3 s at the fastest, realizing all-round real-time acquisition of equipment operation faults or abnormal information, and quickly grasping equipment fault parts and causes.

Safety Warning. Use data visualization to perform three-dimensional measurement and control of equipment operating data and environmental factors. Alarm permissions and methods can be set according to the fault level and staff level requirements, effectively reducing the intensity of equipment patrols and security risks for management personnel.

Equipment Information Management. Establish an equipment account database to keep track of equipment maintenance, repair records and laws in real time to predict equipment operation failures and problems and extend equipment life.

TAR Closed-Loop Management. Build a systematic, three-dimensional, and dynamic mathematical model, use edge algorithms and data analysis to dynamically sense the operating safety of the device, and intervene in advance to achieve device health status management before the device fails.

3.2 Platform Technology Application

Naked Eye VR Technology. Yi Zhuotong platform takes a panoramic picture of the hospital scene. Through the dynamic perception of the equipment operation data, the three-dimensional display of the operation status of the on-site equipment [7], the abnormal operation of the equipment platform can accurately locate the fault location, real-time view of the cause of the equipment failure, greatly improved Staff productivity.

Full V Hawkeye Technology. Important equipment and hidden places inside the hospital, especially power supply, water supply, gas supply and other dangerous places. Smart cameras can monitor the scene of the equipment in real time. When the equipment fault alarm information is pushed to the platform, the camera is actively triggered to take pictures of the fault site After uploading the background, carry out photo information detection, analysis, processing, and comparison to confirm the location of the fault and the actual cause.

Physical Water Quality Sensing Technology. The self-developed physical water quality sensing technology can achieve rapid monitoring of water quality indicators. Cod/ammonia nitrogen and other indicators are quickly collected and uploaded to the Yi Zhuotong platform. Compared with the original water quality online monitoring system, the number of equipment used is small, the data monitoring time is fast, and the cost low investment, mobile intelligent management of sewage water quality through mobile phone app.

Data Security Encryption Technology. The world's leading original algorithm is used to encrypt the data [8]. Under non-homomorphic conditions in cloud computing, homomorphic functions are realized, and the performance is expected to be 100 times faster than existing technologies. Double encrypted data backup on the client and server to ensure the security of data stored in Yi Zhuotong's cloud.

At the same time, build a Yi Zhuotong IoT card monitoring platform to monitor the flow trends, signal strength and operation status of customers' IoT cards to ensure the stability and security of information transmission.

3.3 Implementation Effect

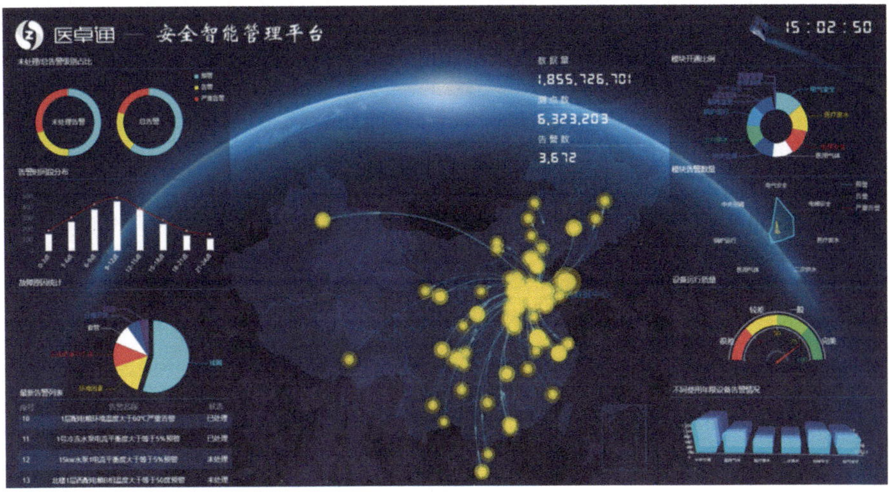

Fig. 2. System screenshot.

Improve Equipment Inspection Efficiency. The traditional patrol inspection method requires personnel to work on shift inspections. The main work performed by the staff is equipment patrol inspection. The work tasks are relatively simple and mechanical, and there are loopholes in the supervision of patrols [9] (Fig. 2).

The Yi Zhuotong platform terminal focuses on the parts that were difficult to inspect in the past. The inspection plan has been changed from twice daily to weekly inspection, which reduces operating costs and strengthens the inspection security through the use of the platform. The Yi Zhuotong platform can determine the fault location and cause more quickly and accurately.

Improve Personnel Management Efficiency. The hospital general management staff is relatively small compared with medical staff, especially with the social outsourcing of the department's logistics business, the existing management staff has been unable to meet the regular safety inspections of all the equipment in the hospital area, mainly through regular visits to various departments to understand the property work Check the situation [10].

Through the Yi Zhuotong platform, you can know the cause and location of the failure, and the platform's warranty dispatch management function can intelligently dispatch the ticket, and the ticket management follows the progress of the property staff's equipment maintenance (no need to repeat the previous phone confirmation). The assessment of the work of the property has significantly improved the work efficiency of the previous staff.

Reduce Equipment Maintenance Costs. Hospital equipment maintenance is mainly to manage and maintain the equipment regularly every year. During the equipment maintenance process, the equipment parts submitted by the staff are replaced. The hospital cannot arrange personnel to go to the site to check [11], and the purchase price of equipment parts is difficult to control.

The Yi Zhuotong platform combines the expert library, knowledge base, sample library and runtime library, and uses data modeling and deep learning algorithms to achieve real-time monitoring of equipment operation health status, online automatic diagnosis of equipment failures, and equipment life analysis Forecasting, prompting staff to perform "demand" maintenance on specific equipment. Through the intelligent comprehensive comparison and analysis of the operating data of different brands, different models and different years of equipment, we found differences in the stability, reliability, quality and performance of the equipment, and then selected the future equipment purchase Provide high-quality decision support.

4 Conclusion

Based on the research and application of Internet of things, big data, artificial intelligence and other modern information technologies, the intelligent security management platform of Yi Zhuotong carries out three-dimensional measurement and control on the operation state and environmental factors of hospital equipment and facilities, when abnormal operation of equipment and facilities occurs, the platform realizes fault location in visual scene through real-time data collection, dynamic analysis and complex calculation, and according to the fault level and the level of personnel respectively provide SMS, telephone and other multi-way classification early warning. Based on the original mathematical model and big data analysis to establish a multidimensional database, and the potential failure of equipment and facilities to predict, early intervention for predictive maintenance, to achieve the closed-loop intelligent management of hospital safety in a more accurate, visual and intelligent way.

Acknowledgement. This article selected 2020 Shandong Province first batch technology innovation project and obtains the fund support.

References

1. Qiu, Q.Z.: Analysis on the refined management of hospital logistics cost. China Townsh. Enterp. Acc. (04), 144–145 (2020)
2. Qian, Z.H.: Discussion on the practice of hospital logistics management innovation. Chin. Hospit. Archit. Equip. **21**(03), 78–79 (2020)

3. Li, T., Yang, H.L., Chen, Z.Y., Li, M.: Research on the current situation and application trend of hospital logistics information management. Technol. Innov. Appl. (05), 188–189 (2020)
4. Wu, S.H.: Study on standardization of logistics equipment management in modern hospital. Chin. Hospi. Archit. Equip. **21**(01), 76–78 (2020)
5. Zhu, Y.C., Qiu, P., Ruan, P.Q.: Thinking of constructing modern hospital logistics management system. J. Tradit. Chin. Med. Manage. **27**(24), 184–185 (2019)
6. Zhao, G.B.: Construction of hospital network security information system based on artificial intelligence. Inf. Technol. Inf. (02), 205–207 (2020)
7. Zhan, G.L.: Application of AI in hospital medical equipment. Wirel. Internet Technol. **17** (02), 157–158 (2020)
8. Wu, N., Wei, W.: Hospital information management system based on big data analysis. Mod. Electron. Tech. **41**(21), 33–36 (2018)
9. Xu, N.: Research on the security of Network database under the background of big data—a case study of hospital. Netw. Secur. Technol. Appl. (05), 77–78 (2020)
10. Cui, J.H.: Application of Internet of things technology in fire safety work. Mod. Electron. Tech. (05), 133–134 (2020)
11. Tang, X.B., Wang, Y.N.: Big data: the technical underpinning of public health system reform. Beijing Daily (2020)

Research on Tibetan Medicine Entity Recognition and Knowledge Graph Construction

Luosanggadeng, Nima Zhaxi(✉), Renzeng Duojie, and Suonan Jiancuo

Tibet University, Lhasa, China
niqiongda@163.com

Abstract. Tibetan medicine entity recognition is a primary task of medical entity recognition. Building a Tibetan medicine knowledge graph is the preliminary work of medical big data research. In this paper, Bi-directional Long Short-Term Memory (BiLSTM-CRF) is used for automatic recognition of Tibetan medicine entities. Then we construct a Tibetan medicine knowledge graph based on Nodejs. Entity recognition in medical documents and electronic medical lay a solid foundation for the research of medical big data.

Keywords: Tibetan medicine · Entity recognition · Knowledge graph

1 Introduction

Tibetan medicine, known as "Sowariba", is concentrated in the Yalong River Valley and the Dzongkha mountains of the Tibetan plateau of the Tibetan farming and pastoral areas. Tibetan medicine is widely spread in such provinces as Tibet, Qinghai, Sichuan, Gansu, and Yunnan. It plays a vital role in protecting the lives and health of Tibetan people as well as preventing and curing diseases. Tibetan medicine and pharmacy are a treasure of Chinese civilization, embodying the great wisdom of the Chinese people and nation.

With the continuous development and accumulation of medical information, medical literature, electronic medical records, health records, and other data, mining the potential information is helpful for doctors in decision-making, disease analysis etc. Medical Entity Recognition (MER) is an important task for information extraction and artificial medical intelligence. Currently, many Chinese and British people are focused on medical entity recognition research. In the early stage, rule-based methods were used. Hettne et al. [1] employed a dictionary-based method to extract drug names. Hu et al. [2] constructed several dictionaries for the entity according to the training set. Long et al. [3] constructed a medical dictionary with semantic information by using network resources and then recognized the disease named entity by combining the dictionary-based method with the conditional random field. Later, medical entity recognition based on the machine learning method appeared. Wei et al. [4] inputted the named entities recognized by each model into Support Vector Machine (SVM) to combine the results when establishing the disease named entity recognition and

Q. Liu et al. (Eds.): CENet 2020, AISC 1274, pp. 1718–1724, 2021.
https://doi.org/10.1007/978-981-15-8462-6_196

standardization system. Jiang et al. [5] developed a new hybrid clinical entity extraction system in the i2b2/VA competition 2010. They integrated the heuristic rules module with named entity recognition module. Liu et al. [6] added character features, part of speech features, dictionary features, and word clustering features to the CRF algorithm and then designed different feature templates for experiments, and applied part of speech features, dictionary features, and word clustering features. Liang et al. [7] proposed a new Chinese medicine entity recognition method cascade-type, which aimed at combining the sentence classifier based on SVM with Chinese medicine entity recognition based on CRF. Then, a deep neural network was used in t medical entity recognition. Almgren et al. [8] proposed a character-based deep two-way recurrent neural network for medical named entity recognition, which trained in an end-to-end way, and performed boundary detection and classification simultaneously. Xu et al. [9] proposed a medical named entity recognition model based on bi-directional long memory and conditional random field (Bi-LSTM-CRF). Experimental results showed that this method was better than the traditional method.

While there is no Tibetan medical entity recognition research now. In this paper, we research the Tibetan medicine entity recognition and Tibetan medicine functions and classifications relation extraction to construct the Tibetan medicine knowledge graph.

2 Tibetan Medicine Entity Recognition Based on BiLSTM-CRF

Named entity recognition (NER) is a basic task in natural language processing, which is to identify named references from text and lay the foundation for relationship extraction. Compared with traditional entity recognition, medical entity recognition (MER) mainly focuses on such entities as disease, symptom, examination, and treatment. Tibetan medical entities are complex in word formation, various in writing forms, different in character length, difficult to distinguish between borders, and often have a variety of references. This paper mainly focuses on the automatic recognition of Tibetan medicine names in Tibetan medical entities.

Recurrent Neural Network (RNN) is a feedforward neural network with sequence modeling. In natural language processing, word sequences are usually modeled. In this case, each input is specific and the basic RNN model is difficult to learn long dependency, which will suffer from the problem of gradient disappearance. So, it is challenging to train the RNN model, which leads to the development of long-term, short-term memory (LSTM) to solve these shortcomings to some extent.

This paper proposes a deep bidirectional LSTM model-based character for medical named entity recognition. The main reference is literature [10, 11], which have a CRF output layer.

The first layer of the model is the look-up layer. Each word xi in the sentence is mapped from a one-hot vector to a low-dimensional dense word vector (character embedding) $x_i \in R^d$ by using the embedding matrix of pre-training or random initialization. d is the dimension of embedding. Before entering to next layer, set dropout to ease overfitting.

The second layer of the model is the bidirectional LSTM layer, which can automatically extract sentence features. The char_embedding sequence ($X_1, X_2, \ldots, X_N$) of each word of a sentence is taken as the input of each time step of bidirectional LSTM. Then, the hidden state sequence ($H_{1f}, H_{2f}, \ldots, H_{Nf}$) outputted by the forward LSTM and the hidden state sequence ($H_{1r}, H_{2r}, \ldots, H_{Nr}$) outputted by the reverse LSTM are spliced according to the positions $HT = [H_{TF}; H_{TR}] \in R_M$ to obtain the complete hidden state sequence $(H_1, H_2, \ldots, H_N) \in R^{N \times M}$. After set dropout, a linear layer is connected, and the hidden state vector is mapped from m dimension to K dimension. K is the number of labels in the label set, so that the sentence feature extracted automatically is recorded as matrix $P = (P_1, P_2, \ldots, P_N) \in R^{N \times K}$.

The third layer of the model is the CRF layer, which is used for sequence annotation at the sentence level. The CRF layer parameter is a $(K + 2) \times (K + 2)$ matrix A. A_{ij} represents the transfer score from the i label to j label and when labeling a position, the previously labeled label can be used. The reason for adding 2 is to add a starting state for the first part of the sentence and an ending state for the end of the sentence. If we record a tag sequence $y = (Y_1, Y_2, \ldots, Y_N)$, is equal to the sentence length, then the model will give a score of Y for the tag of sentence X.

$$score(x, y) = \sum_{i=1}^{n} P_{i,y_i} + \sum_{i=1}^{n+1} A_{y_{i-1},y_i} \tag{1}$$

It can be seen that the score of the whole sequence is equal to the sum of the scores of each position. The score of each position is obtained by two parts, one is determined by the output Pi of LSTM, and the other is determined by the transfer matrix A of CRF. Then we can use softmax to get the normalized probability:

$$P(y|x) = \frac{\exp(score(x, y))}{\sum_{y'} \exp(score(x, y'))} \tag{2}$$

The structure of the model is shown in Fig. 1:

The experimental results are as follows (Table 1):

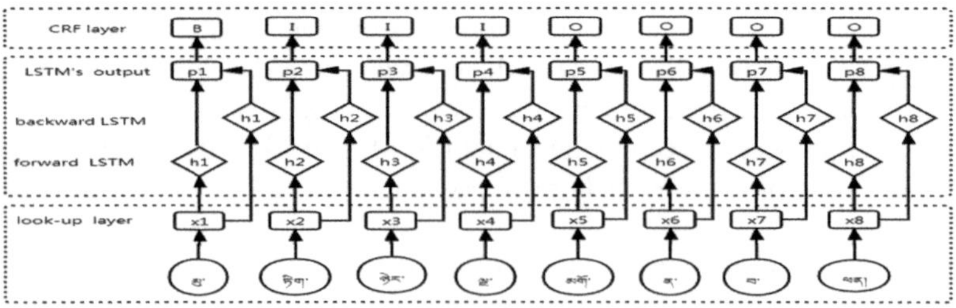

Fig. 1. Character-based BiLSTM-CRF model framework.

Table 1. Comparison of CRF and BiLSTM-CRF model results.

Model	F1
CRF	0.83
BiLSTM-CRF	0.89

3 Tibetan Medicine Knowledge Graph Construction

Please In the era of big data, knowledge graph and data visualization can present data in a structured and visual way and establish a knowledge system based on keyword centered, to show the relationship between data. The medical knowledge graph is the cornerstone of intelligent medical application, which can provide a knowledge basis for machine reading to understand medical texts, intelligent consultation, and intelligent diagnosis [12].

At present, the Tibetan medicine knowledge graph is in the initial stage in terms of scale, standardization, systematization, and formalization. The accurate description of the complex Tibetan medical knowledge is a major challenge to construct the medical knowledge graph. As before-mentioned, this paper uses natural language processing and text mining technology to build a Tibetan medicine knowledge graph in the way of man-machine combination, hoping to provide some reference for medical knowledge graph construction and application.

Based on Tibetan medicine entity recognition, this paper uses the name, efficacy, composition of Tibetan medicine as the relationship to sort out. As shown in Table 2:

Table 2. Tibetan medicine entity name effect, component correspondence

སྨན་མིང་། medicine	སྨན་སྦྱོར། Ingredients	ཕན་ནུས། efficacy	བསྟེན་ཚུལ་དང་ཐུན་ཚད། dosage	ཉར་ཚགས་བྱེད་སྟངས། storage
གོ་ཡོ་ཉེར་ལྔ།	གོ་ཡོ་དཀར་པོ། སེ་འབྲུ། ཤིང་ཚ། སུག་སྨེལ། ཡུང་བ། བྲེ་ག།	འབས་བུ་སྐྲངས་སམ་དཀར་དམར་གྱུང་བ་འཛོག། མཁལ་མའི་ཟྟོད་སོར་དབྱི་མི་ག་གཟེར་ལ་བསྟུགས།	ཉིན་རེར་ཐེངས་༢ ༣དང་། ཐེངས་རེར་རིལ་བུ༦ ༥ཚུ་མཁོལ་ཟྟོད་འཇམ་གྱིས་གཏོང་དགོས།	ཁ་བཙད་དེ་བསིལ་གྲིབ་ལྡན་པའི་སྣམ་སར་འཛོག་དགོས།
ཨ་རུ་བཅུ་པ།	ཨ་རུ་ར། མཁལ་ཞོ་དཀར་པོ། སུག་སྨེལ། ཤུག་ཚེར་ཁཎྜ་བཅས་ལས་སྦྱོན་ཚུལ་བཞིན་བྱས་ཏེ་བཟོས་པ་ཡིན།	མཁལ་མ་བསྐྱེད་འགྲམས་རྒྱུ་སྒྲི་ཏྟེད་པ་འཁོར། བཤལ་འབྱེས་མཁལ་ཚད་མ་ཡུས་སེལ་བ་བྱེད།	ཉིན་རེ་ཐེངས་༢དང་། ཐེངས་རེར་རིལ་བུ་༤ ༡༢ཞིབ་མོར་བས་ནས་རྒྱུ་འཁྱག་ཟྟོད་འཇམ་གྱིས་གཏོང་དགོས།	ཁ་བཙད་དེ་བརྟན་ལ་གཞག་དགོས
སྤོས་དཀར་བཅུ་པ།	སྤོས་དཀར། ཨ་རུ། སྣེ་ཏྲིས། པ་ཤ་ཀ། ཟྣག་ཤུན་ཁཎྜ་བཅས་ལས་གྲུབ།	དྲེག་གསུམ་རྒྱ་སེར་ཀོ་ལེར་ཚ་འབྱུག་འཛོམས།	ཉིན་རེར་ཐེངས་༢ ཐེངས་རེར་ལ་རིལ་བུ་༦ ༥པར་ཞིབ་མོར་བསྐུར་ནས་བཞེས་རྒྱ།	བསིལ་གྲིབ་ལྡན་པའི་སྣམ་སར་ཁ་དམ་པོ་བརྒྱབ་ནས་འཇོག་དགོས།

Finally, each entity is structured as the following structured format, and then used as the data source for the next step of building the knowledge graph (Fig. 2).

```
{"Medication":
    {   "Id": 1900,
        "Name": "གཙོ་བོ་ཉེར་ལྔ།",
        "category":"སྙོ་བའི་སྨན་སྡེ།",
        "Efficacy": "ཚད་པ་ཆེ་ཆུང་སྙོ་དང་རུས་ལ་ཞེན། སྙོ་ཆེ་འགོག་དཀའ་དབུགས་ཐུང་གྱིན་མི་འཛེར།"
    }
}
```

Fig. 2. Structured data format of Tibetan medicine.

Finally, we divide the complex data into drug name data, classification data, and efficacy data, and constructs the knowledge graph as shown in Figs. 3 and 4, based on Nodejs platform, using the B/S architecture. The function of searching by drug name is realized. The extended interface of other medical entity relationship modeling is reserved in the system. The system visualizes the name, category, and efficacy information of Tibetan medicine in a rich and intuitive form, which is helpful and of reference value to relevant researchers.

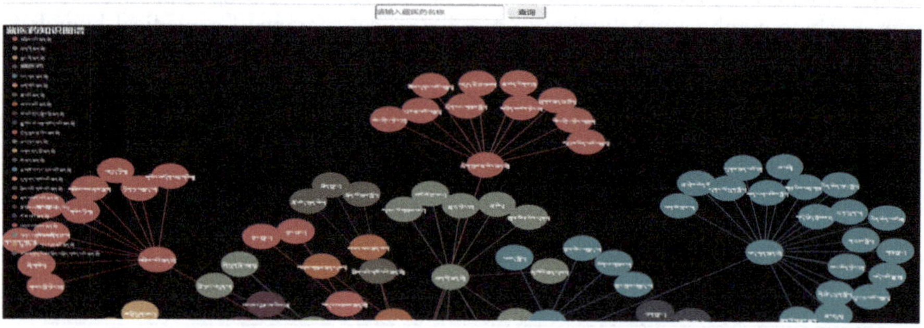

Fig. 3. Classification map of Tibetan medicine.

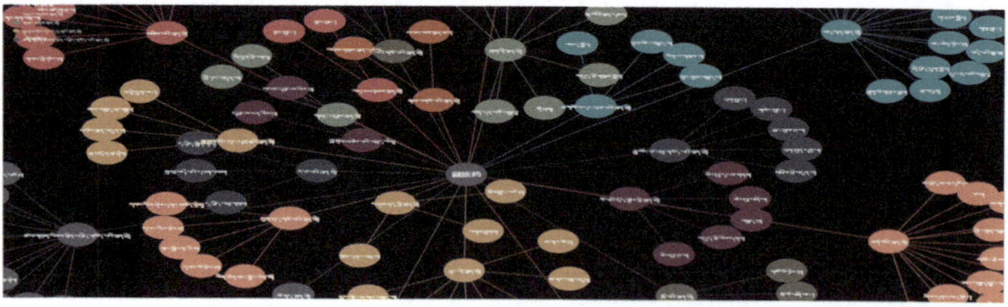

Fig. 4. Tibetan medicine efficacy relationship map.

4 Conclusions

In this paper, Tibetan medicine entity recognition based on BiLSTM-CRF is constructed by using Tibetan medicine medical documents and electronic medical records as data sources. Besides, Tibetan medicine knowledge map is established. Compared with the CRF model BiLSTM-CRF, the F1 value is significantly higher. However, this paper only recognizes the Tibetan medicine entity. We will further study the recognition of medical entity, including the recognition of the disease, symptom, treatment, and other entities. Besides, the knowledge graph constructed in this paper only contains the entity relationship related to Tibetan medicine. Next, we will study the construction of the knowledge map of the whole Tibetan medical system.

Acknowledgments. This work has been done within the project "Construction of Computer and Tibetan Information Technology National Team and Key Laboratory" of the Education Department of Tibet Autonomous Region (zjcz [2018] No. 81); research on Tibetan automatic segmentation and part of Speech Tagging Based on the integration of statistics and rules of the Natural Science Fund of Tibet Autonomous Region (Project No. xz2017zrg-08); foundation project of the University of Tibet's in school Cultivation Fund Research on Tibetan Intelligent Question Answering System Based on the neural network (Project No. zdczjh19-20); Study on the construction of knowledge graph of Tibetan Medicine (Project No. lzj2020004).

References

1. Hettne, K.M., Stierum, R.H., Schuemie, M.J., et al.: A dictionary to identify small molecules and drugs in free text. Bioinformatics **25**(22), 2983–2991 (2009)
2. Hu, J., Shi, X., Liu, Z., et al.: HITSZ CNER: a hybrid system for entity recognition from Chinese clinical text. In: CEUR Workshop Proceedings. Chengdu, China: the Technical Committee on Language and Knowledge Computing of the Chinese Information Processing Society of China, pp. 25–30 (2017)
3. Pang, G.Y., Xu, Y., et al.: Disease named entity recognition based on CRF and dictionary. Microcomput. Appl. **21**, 51–53 (2017)
4. Wei, Q., Chen, T., Xu, R., et al.: Disease named entity recognition by combining conditional random fields and bidirectional recurrent neural networks. Database **baw140**, 1–8 (2016)
5. Jiang, M., Chen, Y., Liu, M., et al.: A study of machine-learning-based approaches to extract clinical entities and their assertions from discharge summaries. J. Am. Med. Inf. Assoc. **18** (5), 601–606 (2011)
6. Liu, K., Hu, Q., Liu, J., et al.: Named entity recognition in Chinese electronic medical records based on CRF. In: 2017 14th Web Information Systems and Applications Conference (WISA), pp. 105–110. IEEE Press, Piscataway (2017)
7. Liang, J., Xian, X., He, X., et al.: A novel approach towards medical entity recognition in Chinese clinical text. J. Healthc. Eng. **4898963**, 1–16 (2017)
8. Almgren, S., Pavlov, S., Mogren, O.: Named entity recognition in Swedish health records with character-based deep bidirectional LSTMs. In: Proceedings of the Fifth Workshop on Building and Evaluating Resources for Biomedical Text Mining (BioTxtM2016), pp. 30–39. University of Manchester UK, Osaka (2016)

9. Xu, K., Zhou, Z., Hao, T., et al.: A bidirectional lstm and conditional random fields approach to medical named entity recognition. In: International Conference on Advanced Intelligent Systems and Informatics, pp. 355–365. Springer, Berlin (2017)
10. Huang, Z., Xu, W., Yu, K.: Bidirectional LSTM-CRF models for sequence tagging. arXiv preprint arXiv:1508.01991 (2015)
11. Lample, G., Ballesteros, M., Subramanian, S., et al.: Neural architectures for named entity recognition. In: Proceedings of NAACL-HLT, pp. 260–270 (2016)
12. Dema, A., Yang, Y.F., et al.: On the construction of Chinese medical knowledge graph CMeKG. Chin. Inf. J. (2019)

Spatial Distribution Characteristics and Optimization Strategies of Medical Facilities in Kunming Based on POI Data

Xin Shan[1], Jian Xu[1(✉)], Yunfei Du[2], Ruli Wang[1], and Haoyang Deng[1]

[1] School of Architecture and Planning, Yunnan University, Kunming 650504, China
xujian@ynu.edu.cn
[2] Beijing Rail Transit Branch of Grandland Group, Beijing 100871, China

Abstract. Based on the POI data of medical facilities and residential quarters in Kunming in 2018, using GIS spatial analysis methods such as Kernel Density Estimation, Average Nearest Distance, and Direction Distribution were used to study the spatial distribution characteristics of Kunming medical facilities and propose optimization strategies. The results show that: (1) From the perspective of spatial structure, the spatial distribution of medical facilities in Kunming is relatively uneven, with the main urban in the city as a "single-center" cluster, there are four characteristics have emerged: 1) Urban crowding 2) The suburbs are empty 3) Old city crowded 4) New city zone is empty. (2) From the degree of regional agglomeration, the concentration of medical facilities in Kunming is high. The concentration of Special Hospitals are higher than that of Grade A Class Three Hospitals, General Hospitals and Community Hospitals. (3) From the perspective of spatial development direction, medical facilities develop along the "northwest southeast" direction, which is consistent with the direction of spatial expansion of residential quarters and the direction of urban expansion.

Keywords: POI · Medical facilities · Spatial distribution · GIS · Kunming

1 Introduction

China's Urbanization Rate (the proportion of urban population to total population) has exceeded 60 percent in 2020, it means the tendency and process of the mass population gathering towards cities. In 2020, a public health event has aroused international attention that the NCP (Novel Coronavirus Pneumonia) broke up in Wuhan. Because of the limit of the population flow and the transportation of relief supplies, urban medical facilities still suffer from severe shortage of medical resources, uneven allocations and ill-suited combinations of sanitation facilities, and other problems. In daily urban environment, we should use urban medical facilities spatial distribution characteristics wisely and form the grading system of urban medical services network that ensure the system has the flexible with reasonable function reorganization to improve the quality and level of urban public health services. We can use this method to face outbreaks of public health events of international concern such as SARS (severe acute respiratory

Q. Liu et al. (Eds.): CENet 2020, AISC 1274, pp. 1725–1733, 2021.
https://doi.org/10.1007/978-981-15-8462-6_197

syndrome), NCP (Novel Coronavirus Pneumonia) or the other malignant infectious disease.

The study of urban medical facilities in foreign countries began at the end of the 19th century [1], and under the influence of humanism and post-modernism theory, the location selection and layout mode, the fairness and accessibility of medical facilities became a focus of issue [2–4]. Since the 20th century, urban planning has integrated various disciplines, and the perspective of medical facilities has begun to turn to the formation mechanism of spatial difference [5, 6]. China's research on urban medical facilities focuses on spatial location selection, system construction and formation mechanism, externality research and so on [7–9]. With the development and application of big data, scholars combined with visual analysis software such as GIS to explore the fairness and accessibility of medical facilities [10–13]. Yu Canning analyzed the road status of Kunming by using spatial syntax analysis and make suggestions for improving the layout of medical facilities [14]. Liu Jing used GIS and Voronoi Polygonto, geographic accessibility and other research methods to analyze the balance of medical facilities layout in Beijing's secondary medical system [15]. The current research pays more attention to the use of big data in the spatial layout of urban medical facilities in order to provide a scientific basis for the rational layout of medical facilities.

Urban planning originated from human response to health demands, the relationship between disease and city has been the focus of modern urban planning. Especially at the current time of the outbreak of the NCP (Novel Coronavirus Pneumonia). Kunming is one of the core cities in western China, Kunming is also a provincial capital which has a high primacy ratio [16]. Kunming's medical service facilities not only serve the city, but also radiation throughout Yunnan Province and even South Asia and Southeast Asia. The research on medical service facilities in Kunming can clarify the spatial distribution characteristics of various hospitals and used to propose urban medical facilities space optimization plan, to provide a basis for the improvement and optimization of Kunming's infrastructure and urban construction. It can also Meet Kunming's medical resource management and meet the health needs of all residents in the region, then set up a goal to make Kunming a healthy city for Southeast Asia.

2 Spatial Distribution Characteristics of Medical Facilities in Kunming

2.1 POI Basic Data

In order to quantify the data of medical facilities in Kunming, this research uses the online map open interface in 2018 to obtain and screen out 1466 POI data of medical facilities in Kunming, including 37 Grade A Class Three Hospitals, 317 General Hospitals, 711 Special Hospitals, and 401 Community Hospitals. As shown in Fig. 1.

Because the distribution of urban residents is crucial to the study of the layout of medical facilities, and the population of housing center is an important indicator for measuring various needs, this paper takes the community as the smallest settlement unit, and obtains a total of 3279 POI data from residential areas. As shown in Fig. 2.

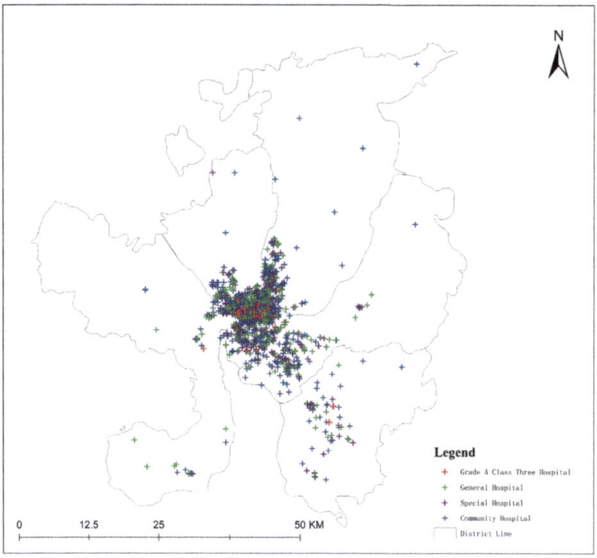

Fig. 1. Spatial distribution of hospitals in Kunming.

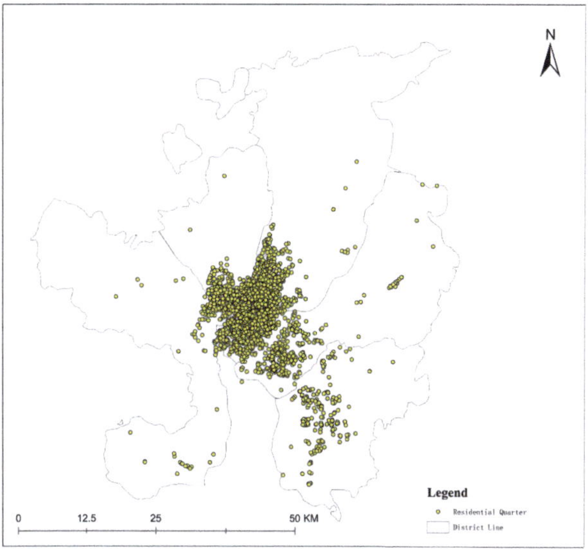

Fig. 2. Spatial distribution of residential quarter in Kunming.

2.2 Structural Characteristics of Spatial Distribution of Medical Facilities Based on Kernel Density Estimation

This paper uses GIS spatial analysis method: Kernel Density Estimation, statistics hospital point elements, used to analyze the spatial distribution of medical facilities in Kunming, and summarize its structural characteristics.

The research based on the Kernel Density Estimation map of the spatial distribution of medical facilities in Kunming (Fig. 3) shows that the spatial distribution of medical facilities in Kunming is uneven, there are four characteristics have emerged: 1) Urban crowding 2) The suburbs are empty 3) Old city crowded 4) New city zone is empty. Medical facilities which has obvious scale advantages are centered around the Dongfeng Square in Kunming's main urban area, mostly in the old urban area within the Third Ring Road. The distribution density is much higher than in new urban areas and suburbs. It indicates that the medical facilities in the center of the city are relatively adequate, but in other areas of the city (like Chenggong New District, urban suburbs), medical facilities are relatively sparsely distributed.

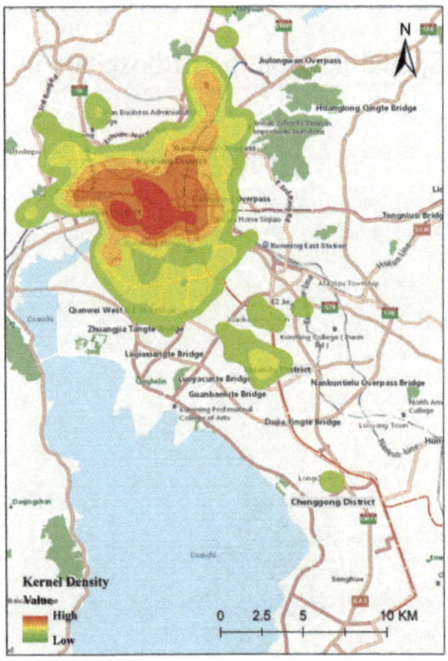

Fig. 3. Kernel density of medical service facilities in Kunming.

From the perspective of different types of medical facilities (Fig. 4), the Grade Three A Hospitals form a gathering center on Renmin West Road in the Second Ring Road, and the Grade Three A Hospitals form subcenters in Dongfeng Square in Wuhua District and Caiyun South Road in Chenggong District; General Hospitals form three gathering centers in the Second Ring Road (near Dongfeng Square, Yunnan University and Daguan Road), one sub-center (near Jinxing Overpass); the Special Hospitals form a gathering center in the second ring (near Dongfeng Square); the Community Hospitals form a multi-center gathering in the central city area. There is no gathering center in Chenggong New District.

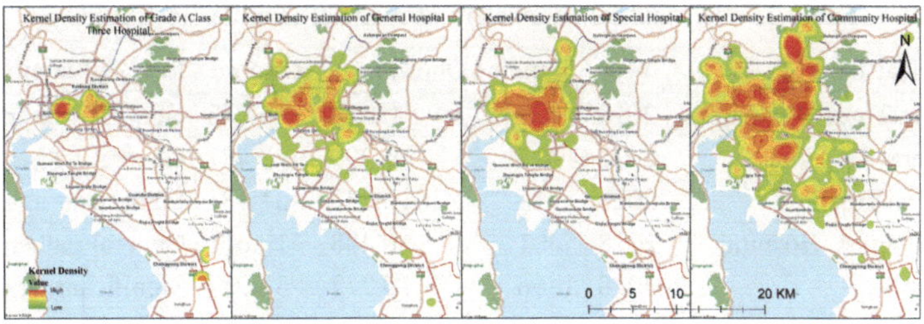

Fig. 4. Kernel density of different types of medical service facilities in Kunming.

To view the situation as a whole, Kunming City, which faces Kunming City, Yunnan Province, and even South Asia and Southeast Asia, has large-scale medical facilities that are mostly serviced in the downtown area. The high concentration of medical resources and patients not only affects the fairness of the use of medical resources and the efficiency of medical services, but also is not conducive to the dispersal of the population and industries in the main urban area, and it has also increased the pressure on traffic, environment and energy in the main urban area.

2.3 Clustering Characteristics of Medical Facilities Based on Average Nearest Distance

In this paper, the Average Nearest Distance tool in GIS software is used to compare the shortest distance between points in a medical facility in space and the distance between the nearest points in a theoretical model to summarize the degree of spatial distribution. The study summarizes the spatial performance of path dependence and agglomeration effects of medical implementation layout (Table 1). This is to provide cities with efficient, convenient and timely medical conditions, and to provide conditions for the exchange of information, technology and complementary of medical resources. At the same time, it can also provide a basis for the independence and closedness of health units when special public health emergencies occur (Table 2).

Table 1. POI in the study area.

	Medical Facilities				Residential Quarter
	Grade A class Three Hospital	General Hospital	Special Hospital	Community Hospital	
POI/One	37	317	711	401	3279
Percentage/%	2.5	21.6	48.5	27.4	100

According to the Average Nearest Distance Index of medical facilities in Kunming, the NNI of different types of medical facilities in Kunming are all less than 1, indicating that all types of hospitals always present a spatial state of cluster distribution, especially

Table 2. NNI and its parameters of medical facilities in Kunming.

Medical facilities	Year	Samples/One	NNI	Z-Score	Level of significance
Grade Three A Hospital	2018	37	0.428	−6.648	General gathering
General Hospital	2018	317	0.399	−20.481	General gathering
Special Hospital	2018	711	0.330	−34.181	Significant aggregation
Community Hospital	2018	401	0.488	−19.611	General gathering
Overall	2018	1466	0.277	−52.957	Significant aggregation

in the main urban area. Among them, the medical facilities of the specialist hospitals have shown significant aggregation. The Average Nearest Distance Index of Special Hospitals are always smaller than those of Grade Three A Hospital, General Hospitals, and Community Hospitals. This shows that the concentration of the spatial layout of Special Hospitals is stronger than that of Grade Three A Hospital, General Hospitals and Community Hospitals. General Hospitals and Community Hospitals present general aggregation. Studies have shown that the concentration phenomenon of medical facilities in Kunming is obvious, and the level of division of labor among medical institutions is unclear, the overloaded operation of General Hospitals and underutilization of service resources in a large number of primary level of health institutions.

2.4 Features of Medical Facilities Development Direction Based on Direction Distribution

Use GIS software to perform Direction Distribution analysis for the Top Three Hospitals, General Hospitals, Special Hospitals, Community Hospitals, and Residential Quarter of Kunming. It is used to compare the central, discrete and directional trends of the spatial distribution of medical facilities and residential communities, and to analyze the characteristics of the spatial development direction of medical facilities in Kunming (Fig. 5)

From the perspective of the expansion of the Direction Distribution: the main axes of the expansion of the Grade A Three Hospitals, General Hospitals, and Special Hospitals are the same, it's always along the "northwest-southeast" direction of the city, basically consistent with the expansion of the Residential Quarters, also consistent with the expansion of the city. The expansion angles of the Grade A Three Hospitals, General Hospitals, and Special Hospitals are 141.6°, 141.6°, and 156.1°, compared with the Residential Quarters expansion direction, it is more southwest. The direction of the Community Hospital's main axis of expansion is along the city's "South-North" direction, and There is a certain deviation from the direction of the main axis of the residential quarters expansion.

Kunming medical facilities and Residential Quarters are consistent in the direction of spatial expansion. The reason is that the main urban area of Kunming has expanded, and the system of medical treatment, residence and education in the northern urban area of Kunming is relatively perfect matching function. The new campus and university

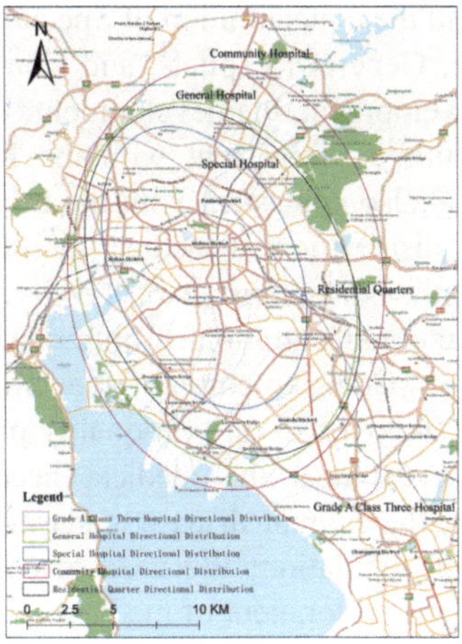

Fig. 5. Direction distribution of medical facilities and residential quarters in Kunming.

moved into Chenggong New District in the southeast direction. Under the high concentration and demand of residential quarters and research land, the development of medical facilities, especially general hospitals and Top Three Hospitals, has been promoted.

3 Discussion

POI, as geospatial big data, has the characteristics of timely updating and easy access, which can make up for the lack of investigation and research on the analysis of traditional medical facilities. Through the study of POI in medical facilities can accurately reflect the problems of medical institutions in urban layout, and this has a positive effect on urban planning. But in the actual urban medical facility space layout, medical institutions are highly dependent on the convenience of urban traffic. The influence of density of road network on hospital layout is not fully considered in this study. In the future, it is necessary to further consider the research of density of road network on the spatial distribution pattern of medical facilities.

4 Conclusions

This study is based on POI data, taking Kunming as an example, comprehensively using multiple GIS spatial analysis methods, and discussing the spatial distribution characteristics of medical facilities with Grade A Three Hospitals, General Hospitals, Special Hospitals, and Community Hospitals as the research objects. The main conclusions are as follows: (1) From the spatial structure, the spatial distribution of medical facilities in Kunming is uneven, there are four characteristics have emerged:1) Urban crowding 2) The suburbs are empty 3) Old city crowded 4) New city zone is empty; (2) From the perspective of spatial concentration, the concentration of medical facilities in Kunming is significant, and the concentration of Special Hospitals is higher than that of Grade A Three Hospitals, General Hospitals, and Community Hospitals. (3) From the perspective of spatial development, the development of medical facilities along the "northwest-southeast" direction is consistent with the Residential Quarters' spatial expansion direction, which is also consistent with the urban expansion direction.

According to the spatial distribution of medical facilities in Kunming, this can be optimized and adjusted with the characteristics of the city. (1) Adjust and optimize the layout structure. Aiming at the concentration of high-quality medical resources in the main urban area, adjust the spatial structure of the Grade A Three Hospitals and General Hospitals to reduce the overlap of high-quality medical resources; Adjust the layout of Special Hospitals to meet the medical needs of different populations according to the changes in the human disease spectrum, population structure, and medical demand of the local population; We need to combine the characteristics of urban structure to promote the development of medical facilities from a "single center" to a "multi-center" space structure; We need to speed up the improvement of medical facilities in the new area and reduce the pressure on the main urban area. (2) Activate stock resource and balance resources. Divert the advantage resources, and at the same time, improve the utilization efficiency of medical facilities to activate the stock resource. High quality medical facilities should focus on improving the overall use efficiency and technical services, and guide the transfer and development of medical resources within the Third Ring Road to the outside of the Third Ring Road, as well as to the new area and suburb areas; at the same time, we need to support the development of qualified Specialized Hospitals into large-scale and high-level medical institutions, and guide social capital to develop high-end medical services; We also need to strengthen the role of community hospitals in screening and shunting patients, and relieve the pressure of hospital treatment. (3) According to the characteristics of the city, a scientific and flexible network system should be established. We should further analyze the characteristics of urban space and population, refine the level, scale, radiation range and characteristics of medical facilities, and build a network system of urban medical facilities. It is used to provide flexible and scientific contingency measures and mechanisms for emergencies.

Funding. This research was funded by National Natural Science Foundation of China, grant numbers 51878591.

References

1. Cheng, S.Q., Qi, X.H., Jin, X.X., Li, D.M., Lin, H.: Progress in domestic and foreign study on spatial layout of public service facilities. Trop. Geogr. **36**, 122–131 (2016)
2. Cheol-joo, C.: An equity-efficiency trade-off model for the optimum location of medical care facilities. Soc.-Econ. Plann. Sci. **32**, 99–112 (1998)
3. Love, D., Lindquist, D.: The geographical accessibility of hospitals to the aged: a geographic information systems analysis within Illinois. Health Serv. Res. **29**, 629–651 (1995)
4. Kontodimopoulos, N., Nanos, P., Niakas, D.: Balancing efficiency of health services and equity of access in remote areas in Greece. Health Policy **76**, 49–57 (2007)

5. Jones, A.P., Bentham, G.: Emergency medical service accessibility and outcome from road traffic accidents. Public Health **109**, 169–177 (1995)
6. Alexis, J.C., Chris, B., Robert, R.: A spatial analysis of variations in health access: linking geography, socio-economic status and access perceptions. Int. J. Health Geogr. **10**, 2187–2198 (2011)
7. Lu, X.L., Hou, Y.X., Lin, W., Shen, Q.: Functional optimization of emergency medical service centre of small towne based on facility location theory: a case study of Tengzhou city in Shandong province. Econ. Geogr. **31**, 1119–1123 (2011)
8. Cao, Y., Zhen, F.: Medical facilities service evaluation and planning response, Nanjing. Planners **34**, 93–100 (2018)
9. Peng, B.F., Shi, Y.S., Shan, Y., Chen, D.L.: The spatial impacts of class 3A comprehensive hospitals on peripheral residential property prices in Shanghai. Sci. Geogr. Sinica **35**, 860–866 (2015)
10. Li, Z., Gao, Y.Y., Cui, W.Y., Deng, K.: Study on the accessibility of urban medical facilities layout and their elderly-oriented development—take Hefei city as an example. Sci. Geogr. Sinica **35**, 860–866 (2015)
11. Xie, X.H., Wang, R.Z., Wen, D.H., Zhang, Z.Y.: Evaluating the medical facilities layout based on gis: an application of Xiang'an district. J. Geo-Inf. Sci. **17**, 317–328 (2015)
12. Li, H., Pan, X. F., Yang, L., Wang, L., Wan, B.: The determination of the service radius of healthcare facility based on taxi GPS trajectory. Bull.Surv. Mapp. 78–83 (2018)
13. Hou, S.Y., Jiang, H.T.: An analysis on accessibility of hospitals in Changchun based on urban public transportation. Geogr. Res. **33**, 915–925 (2014)
14. Yu, C.N., Li, Z.Y., Wang, X.Y., Long, Y., Tian, J.H.: Research on medical facilities fairness of spatial layout in Kunming city. J. Hum. Settle. West China **34**, 76–81 (2019)
15. Liu, J., Zhu, Q.: Research of equalizing layout of public service facilities: take health facilities of central six districts of beijing for example. Urban Dev. Stud. **23**, 6–11 (2016)
16. Shan, X., Xu, J., Liu, Y. X., Sui, X., He, X.: A research on spatial distribution pattern of urban catering industry in Kunming based on POI data. J. Kunming Univ. Sci. Technol. (Nat. Sci.) 115–120 (2019)

Study on the Abnormal Expression MicroRNA Network of Pancreatic Cancer

Bo Zhang(✉), Lina Pan, and HuiPing Shi

School of Computer Engineering,
Jinling Institute of Technology, Nanjing, China
Zhangb@jit.edu.cn

Abstract. With the continuous progress of biotechnology and gene technology, scientists have found a large number of microRNA-related data through experiments. Among them, analyzing the characteristics, structure and function of microRNA by various methods is one of the most important applications in bioinformatics. To survey the mechanisms involving gene alteration and miRNAs in pancreatic cancer, we focused on transcriptional factors as a point of penetration to build an abnormal expression regulatory network. In this network, we found that some pathways with differentially expressed elements (genetic and miRNA) showed some self-adaption relations, such as SMAD4. A total of 32 genes and 88 microRNAs with 199 directed edges were identified. The evidence have shown that the expression of microRNAs varies with the severity of pancreatic cancer deterioration.

Keywords: Pancreatic cancer · Network · Transcription factor · MicroRNA

1 Introduction

Pancreatic cancer (PC) is a highly lethal disease with few effective treatment options. Patients with pancreatic cancer have been identified as having a chance to develop a vertical relationship, usually an genetic pattern [1]. Emerging evidence suggests that the ability of tumors to grow and reproduce depends on a small number of cells in tumors, called cancer stem cells. leukemia suppressor (LIF), a key factor mediating signal transduction between pancreatic cancer cells and stellate cells, has systematically validated its feasibility as a therapeutic target and biomarker for pancreatic cancer [2]. The expression profile of microRNAs has entered cancer diagnosis as a biomarker for diagnosis and prognosis [3]. Evaluate the tumorigenesis, progression and response to treatment of cancer patients, as well as the genetic mechanism of genetic cancer. The effects of regulatory mechanisms on diseases play an important role in the application of microRNAs in cancer diagnosis, treatment and genetic control [4].

After the establishment of abnormal expression network in pancreatic cancer, it was found that genes and miRNA interact to form a regulatory relationship. The genes and miRNA in the pathway are expressed abnormally in pancreatic cancer, such as 80–90% of SMAD4 gene abnormalities in all pancreatic cancer [5]. SMAD4 gene inactivation and shorter correlated with overall survival (risk ratio, 1.92; 95% confidence interval, 1.20–3.05; P = 0.006). Pancreatic cancer is driven by the accumulation of mutations

Q. Liu et al. (Eds.): CENet 2020, AISC 1274, pp. 1734–1740, 2021.
https://doi.org/10.1007/978-981-15-8462-6_198

and genetic changes, the activation of oncogenes and the inactivation of tumor suppressor genes in the course of pathogenesis. However, the relationship of these genes was regulated in pancreatic cancer remains unclear [6].

Transcription factor (TF) and microRNA (microRNA) have been found to play key roles in the development of cancer. However, the role of TF and microRNAs in the relationship of pancreatic cancer remains unclear [7]. TFs are special proteins that are expressed positively and negatively during gene transcription [8]. Bioinformatics is used to explain the relationship between miRNA, target genes and miRNA transcription factors, and to distinguish related diseases. The research focus of bioinformatics has shifted to how to interpret these data and discover potential knowledge from them. It is difficult to study the relationship between miRNA and host genes [9].

MiRNA, TFS, and their downstream target genes form a regulatory network. In this network, there is a positive correlation between miRNA and transcription factors in biological process, and it is inclined that one miRNA can be regulated by multiple transcription factors at the same time [10, 11]. These correlations indicate that these target genes play an important role in the regulation of miRNA, such as the correlation between "stress" and "metabolism" [12]. Finally, the relationship between miRNA transcription factors and target genes, transcription factors regulate target genes through miRNA. I can also find that the feedforward loop composed of transcription factors and miRNAs, miRNAs and target genes plays an important role in gene regulation [13].

In the present study, we investigated the Abnormal expression network of miRNAs, targets of miRNAs, TFs, host gene of miRNAs in pancreatic cancer. According to the abnormal network data of cancer, I can find that miRNA is not isolated, it can only regulate the development of disease with gene regulation. MiRNA, target genes, transcription factors are all closed-loop regulation, through the study of these closed-loop, I can find the occurrence, development, transformation of cancer Transfer mechanism. In comparison, I found that although they shared genes and miRNAs, they had different regulatory pathways. The network is so complex that we don't clearly identify pathways associated with pancreatic cancer. We compare the similarities and differences of these pathways in the network.

2 Materials and Methods

The microRNAs pathway in this data method is not a simple gene pathway, involving a single type of regulator, but ignoring the transcription factors and microRNAs involved in transcriptional and post-transcriptional regulation. It is a method that can reflect all interactions between microRNAs and genes. The data in this paper can be used to study the complex regulators of cells System. Based on these regulatory relationships, we establish abnormal expression networks. At the same time, the relationship between microRNAs and target genes, microRNAs and host genes, microRNAs and transcription factors was used to organize the abnormal expression data.

In this paper, text mining data algorithm and manual collection are combined to collect and collate the experimental verification of the relationship between transcription factors regulating microRNAs, microRNAs regulating target genes, and microRNAs and their host genes from authoritative resources. We collect data flows such as Fig. 1 (Structural Diagram of Data Collection System).

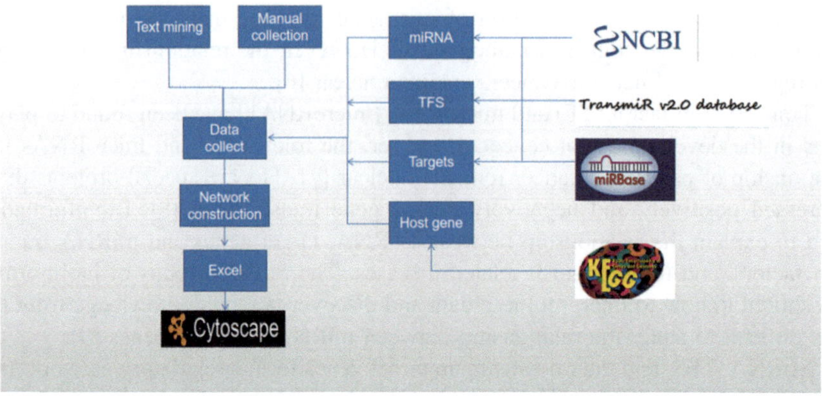

Fig. 1. Structural diagram of data collection system.

I use the proposed text mining algorithm to find the regulatory relationship between miRNA and gene in PubMed database. The miRNA gene regulation is direct, we call it "miRNA target gene" relationship. The other cases of miRNA gene regulation correspond to the rule is indirect, or it is not clear from the text whether it is directly involved in the regulation, so we establish the regulatory network. KEGG pathway database mainly through gene abnormal expression pathway, the data is visual data, miRNA does not directly affect, but through gene regulation. The role of miRNA can be divided into two layers. The first layer is to inhibit target genes, and the second layer is to regulate mRNA to regulate gene expression. So far, there are more than 20 miRNA and target gene databases. The principle is the same, based on the binding mechanism of miRNA and target gene. Transcription factors and miRNA play an important role in transcriptional regulation and post transcriptional regulation of genes. They have become an important research field in bioinformatics. Previous prediction results have reached academic requirements, so bioinformatics research using experimental transcription factors to control miRNA is still the mainstream.

The method of text mining and manual mining to collect abnormal expression miRNA, abnormal expression gene, target gene, transcription factor and host gene. Three kinds of regulation were constructed: TF regulates miRNA; miRNA regulates its target genes; host gene contains miRNA. Based on the regulatory relationship between gene and miRNA, the relationship between abnormal expression gene and miRNA was extracted and stored in the set. Establish the data collection and expression diagram as shown in Fig. 1, obtain the relevant data from various databases, and process the data to obtain the abnormal expression network diagram. In this data method, the miRNA pathway is not a simple gene pathway. It involves a single type of regulator. It ignores the transcription and post transcriptional regulation of transcription factors and miRNA. It is a method that can reflect all the interactions between miRNA and genes. We can use this data to study the complex regulatory mechanisms of cells (Tables 1 and 2).

Table 1. Abnormal genes in pancreatic cancer

Gene	Gene ID	Gene	Gene ID	Gene	Gene ID
PRPF31	26121	DNM3	26052	NR6A1	2649
PANK1	53354	C9orf3	84909	ZRANB2	9406
HOXB3	3213	MCM7	4176	TLN2	83660
EGFL7	51162	SKA2	348235	NDUFAF3	DALRD3
RTL1	388015	SKA2	348235	MIR196A1 (196a-1)	406972
ARPP21	10777	NFYC (30c-1)	4802	HOXB7 (196a-1)	3217
PDE2A	5138	C6orf155 (30c-1)	79940	HOXC5 (196a-2)	3222
COPZ1	22818	C9orf5	23731	DNM2	1785
MIR155HG	114614	MIR375	494324	DNM3	26052
DLEU2	10301	ATAD2 (548d-1)	29028	TRPM3	80036
IFT80	57560	PITPNC1	26207	MIR17HG	407975
DLEU2	10301	MIR17HG	407975	TMEM49	81671
PRPF31	26121	ABLIM2	84448	IGF-2	3481

Table 2. Abnormal microRNAs in pancreatic cancer

miRNA	miRNA	miRNA	miRNA
let-7d	miR-132	miR-181d	miR-212
let-7f-1	miR-139	miR-186	miR-213
miR-100	miR-142-5p	miR-190	miR-214
miR-100-1	miR-143	miR-191	miR-220
miR-100-2	miR-146a	miR-192	miR-221
miR-103	miR-148a	miR-196a	miR-222
miR-103-2	miR-148b	miR-199a-1	miR-223
miR-106a	miR-150	miR-199a-2	miR-23a
miR-107	miR-155	miR-200a	miR-23b
miR-10a	miR-15a	miR-200b	miR-24-1
miR-10b	miR-15b	miR-200c	miR-24-2
miR-125a	miR-16-1	miR-203	miR-25
miR-125b	miR-17-5p	miR-204	miR-27a
miR-125b-1	miR-181a	miR-205	miR-29b-2
miR-126	miR-181b-1	miR-20a	miR-301
miR-127	miR-181b-2	miR-21	miR-301a
miR-128b	miR-181c	miR-210	miR-30c
miR-32	miR-34c	miR-421	miR-520 h
miR-345	miR-372	miR-424	miR-548d
miR-34a	miR-375	miR-429	miR-92-1
miR-34b	miR-376a	miR-483-3p	miR-92-2
miR-95	miR-96	miR-99a	

3 Results and Discussion

To study the abnormal expression data of pancreatic cancer, establish abnormal network through data standardization, and determine the regulatory relationship (transcription, host, targeting) between the two according to the directional edges of the cancer data [16]. In addition to these determinant relationships, there may be nodes, because the research is not in-depth or data mining is not enough to cause node isolation. Except that a small number of host genes are not abnormally expressed in pancreatic cancer, the genes and microRNAs expressed in other nodes are abnormal [17].

We found that there are 32 genes and 88 microRNAs in the abnormal network of pancreatic cancer, of which 199 are related to their regulation, which can be called Pathway. TP53 gene regulates 12 kinds of microRNAs (hsa-mir-107, hsa-mir-125b, hsa-mir-143, hsa-mir-155, hsa-mir-192, hsa-mir-200a, hsa-mir-200b, hsa-mir-200c, hsa-mir-29b-2), covering, hsa-mir-34b, hsa-mir-34c), the target genes of these microRNAs (TP53, KRAS, APC, CDKN2A, BB2, BB2, etc.).

The core transcription networks are TP53, Smad4, CDKN2A, target microRNAs, microRNAs and their host genes.

After the establishment of abnormal expression network in pancreatic cancer, it was found that genes and microRNAs interact to form regulatory relationships such as KRAS microRNA-155 APC, SMAD4 microRNA-17 TGFBR2, TP53 microRNA-25 MCM7. Genes and microRNAs in the pathway are abnormally expressed in pancreatic cancer [18]. For example, KRAS gene abnormalities occur in 80%–90% (Smit VT et al.) of all pancreatic cancer patients. These genes are a sign of cancer derivation. SMAD4 gene inactivation was significantly associated with shorter overall survival (risk ratio, 1.92; 95% confidence interval, 1.20–3.05; $P = 0.006$). The median survival of patients with Smad4 gene inactivation was 11.5 months, compared with 14.2 months without Smad4 gene inactivation [17]. CDKN2A, also known as multiple tumor suppressor genes, can form enzymatic complexes with CDK4. If the gene is deleted or mutated, cells will gain infinite proliferation function leading to canceration [18].

3.1 Comparison and Analysis of Differentially Expressed miRNAs

Through the establishment of abnormal expression network in pancreatic cancer, we also found loop regulation network and co-regulation network in abnormal expression pathway, such as Fig. 2. By visualizing the data in the database, we found some useful regulation such as microRNA-155 KRAS microRNA-142; TFS regulation and expression of multiple microRNAs, such as KRAS microRNA-155, APC microRNA-155, and SMAD4 microRNA-155.

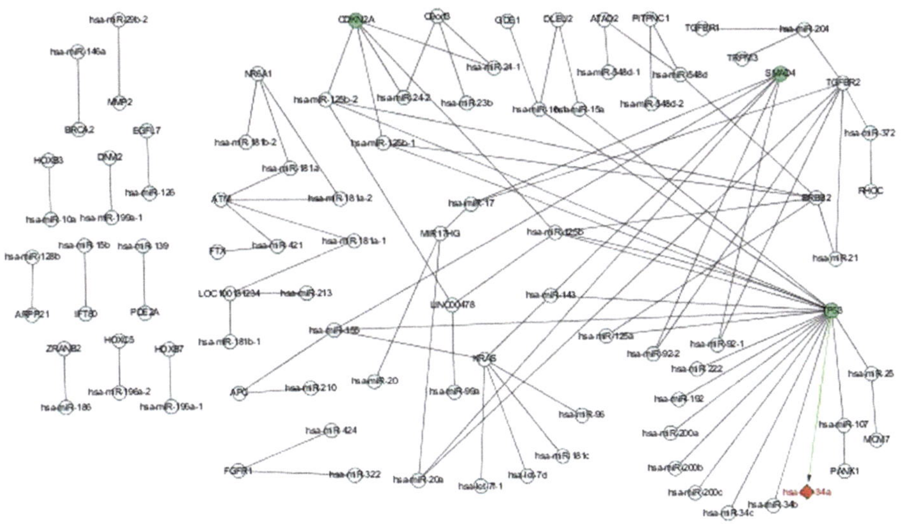

Fig. 2. Abnormal expression network of pancreatic cancer.

3.2 Analysis of Host Genes and miRNAs in Pancreatic Cancer

We found that microRNAs can map outward to regulate genes, such as microRNA-17 SMAD4 and microRNA-17 TGFBR2; a single microNA can be mapped by multiple transcription factors to form a regulatory relationship. For example, SMAD4 microRNA-125, TGFBR2 microRNA-125, TP53 microRNA-155; a target gene is regulated by multiple microRNAs, such as microRNA-222 TP53, microRNA-192 TP53, microRNA-34B TP53, microRNA-25 TP53 (Table 3).

Table 3. Circular regulatory pathway

miRNA	Gene	miRNA	Gene
miR-125b	TP53	miR-17	SMAD4
miR-17	SMAD4	miR-17	TGFBR2
MiR-20a	SMAD4	Let-7	KRAS
miR-20a	TGFBR2	miR-155	KRAS
miR-20a	MIR17HG	miR-155	TP53

References

1. Yip, D., Karapetis, C., Strickland, A.: Chemotherapy and radiotherapy for inoperable advanced pancreatic cancer. Cochrane Syst. Rev. (2009)
2. Heidt, D.G., et al.: Identification of pancreatic cancer stem cells. **130**(2), 194–195 (2006)
3. Wang, X., et al.: The regulatory roles of miRNA and methylation on oncogene and tumor suppressor gene expression in pancreatic cancer cells. Biochem. Biophys. Res. Commun. **425**(1), 51–57 (2012)
4. Nguyen, H., Lai, T., Kim, D.R.: miRCancerdb: a database for correlation analysis between microRNA and gene expression in cancer (2018)
5. Mangal, M., Sagar, P., Singh, H., Raghava, G.P., Agarwal, S.M.: NPACT: naturally occurring plant-based anti-cancer compound-activity-target database. Nucleic Acids Res. (2013)
6. Metzler-Zebeli, B.U., et al.: Interactions between metabolically active bacteria and host gene expression at the cecal mucosa in pigs of diverging feed efficiency. J. Animal Sci. **96**(6), 2249–2264 (2018)
7. Rodriguez, A., Griffiths-Jones, S., Ashurst, J.L., et al.: Identification of mammalian microRNA host genes and transcription units. Genome Res. **14**, 1902–1910 (2004)
8. Cao, G., Huang, B., Liu, Z., et al.: Intronic miR-301 feedback regulates its host gene, ska2, in A549 cells by targeting MEOX2 to affect ERK/CREB pathways. Biochem. Biophys. Res. Commun. **396**, 978–982 (2010)
9. Wang, J., Lu, M., Qiu, C., et al.: TransmiR: a transcription factor-microRNA regulation database. Nucleic Acids Res. **38**, D119–D122 (2009)
10. Kozomara, A., Griffiths-Jones, S.: miRBase integrating microRNA annotation and deep-sequencing data. Nucleic Acids Res. **39**, D152–D157 (2011)
11. Kanehisa, M., Goto, S.: KEGG: Kyoto encyclopedia of genes and genomes. Nucleic Acids Res. **28**, 27–30 (2000)
12. Mocellin, S., Shrager, J., Scolyer, R., et al.: Targeted Therapy Database (TTD): a model to match patient's molecular profile with current knowledge on cancer biology. PLoS One **5**, e11965 (2010)
13. Chekmenev, D.S., Haid, C., Kel, A.E.: P-Match: transcription factor binding site search by combining patterns and weight matrices. Nucleic Acids Res. **33**, W432–W437 (2005)
14. Fujita, P.A., Rhead, B., Zweig, A.S., et al.: The UCSC genome browser database: update 2011. Nucleic Acids Res. **39**, D876–D882 (2011)
15. Jiang, Q., Wang, Y., Hao, Y., et al.: miR2Disease: a manually curated database for microRNA deregulation in human disease. Nucleic Acids Res. **37**, D98–D104 (2009)
16. Smit, V.T., Boot, A.J., Smits, A.M., Fleuren, G.J., Cornelisse, C.J., et al.: KRAS codon 12 mutations occur
17. Blackford, A., Serrano, O.K., Wolfgang, C.L., Parmigiani, G., Jones, S.: SMAD4 gene mutations are associated with poor prognosis in pancreatic cancer. Clin. Cancer Res. Off. J. **15**(14), 4674 (2009)
18. Bartsch, D.K., Sinafrey, M., Lang, S., Wild, A., Gerdes, B.: CDKN2A germline mutations in familial pancreatic cancer. Ann. Surg. **236**(6), 730 (2002)

Modeling RNA Secondary Structures Based on Stochastic Tree Adjoining Grammars

Sixin Tang(✉), Huihuang Zhao, and Jie Jiang

College of Computer Science and Technology, Hengyang Normal University, Hengyang 421002, China
tangsix@163.com

Abstract. This paper presents a new method for modeling RNA secondary structures based on stochastic tree adjoining grammars. In order to get better predict results, we shows using Stochastic Tree Adjoining Grammars to inference RNA secondary structures. firstly, we expound the tree adjoining grammars and stochasitc model and the operations of TAG tree. Secondly, discusses the key problems for RNA secondary structure prediction using tree adjoining grammars modeling, finally we design a tree adjoining grammars model to predict RNA structure, The experiment choose 8 species sequence from EMBL database, and experimental verification of the validity of the model, the experimental results shows that the tree adjoining grammars in the prediction of RNA sequence structure has long-range correlation, can improve the prediction accuracy.

Keywords: Tree Adjoining Grammars · RNA · Structure prediction

1 Introduction

1.1 A Subsection Sample

RNA secondary structure prediction is a hot topic in computational molecular biology. As the spatial structure prediction algorithm relies on the precise model of molecular folding, which does not exist yet, the spatial structure prediction at this stage is still very immature in technology. But the spatial structure of largely depends on the secondary structure, secondary structure prediction algorithm after 20 years of research, has made great progress', and has been confirmed by practice in many ways, therefore, research of secondary structure, to explore the spatial structure, function and promote protein structure prediction is of great significance.

RNA secondary structure prediction is usually carried out according to the following principles: minimum free energy; the maximum number of base pairs; unity of structure and function [1]. Under the guidance of these principles, due to the complexity of secondary structures, people try all possible means to consider the problem from different angles, hoping to find a better method. Therefore, there are a variety of methods, and the general trend is to learn from each other, make use of all kinds of useful information as much as possible, and make the calculated results closer to the real situation under the control of the computational complexity. Among the many

Q. Liu et al. (Eds.): CENet 2020, AISC 1274, pp. 1741–1749, 2021.
https://doi.org/10.1007/978-981-15-8462-6_199

kinds of prediction algorithms, two kinds of algorithms are generally accepted. One is based on the comparative sequence analysis [2]. For the predicted target sequence, the structure of the homologous sequence is modified according to the difference between the sequences when the exact secondary structure of the homologous sequence is known. Because the structure of homologous sequence is conservative, this method has high accuracy and low time complexity. However, this kind of method is limited by the comparative structure and has great limitations when used.

Another kind of method is the method of ab initio prediction [3], which is typically based on the algorithm of minimum free energy. As early as 1981, Zuker et al. [4] proposed a secondary structure prediction algorithm with the lowest molecular free energy after folding, and the time complexity of this algorithm was $O(n^3)$. This algorithm has a great limitation in the accuracy of prediction. The reason is that there is a kind of structural domain called “false junction” widely in the secondary structure, and many important functional sites are often located in the false junction region. The prediction algorithm of et al. can only deal with the traditional secondary structure, but cannot deal with the false junction structure. In other words, the prediction results of Zuker et al.’s [5]. algorithm do not contain false knots. If the real structure contains false knots, the prediction results cannot be correct. The formal language method was initially used to predict the secondary structure of RNA. The context-free grammar was suitable for modeling the long-term correlation, but this method could not model the pseudonode. Various other grammars have been proposed for this purpose [6].

2 Tree Adjoining Grammars

2.1 The Model of Tree Adjoining Grammars

Context-free grammar has long been used to predict the secondary structure of RNA because it can model sequences of long range correlation [7]. However, it has the disadvantage of high computational complexity and inability to model the crossover structure. Tree adjoining grammar is more expressive than context-free grammar [8]. The context-free grammar is linear, and the connection relation between characters is from left to right dimension (called chain), which can be considered as the depth of 1 tree adjoining grammar. If the one-dimensional connection relation is extended to multiple dimensions, the chain can be extended to become a tree, and the grammar with such tree connection relation between characters is called tree adjoining grammar. Therefore, lexical tree adjoining grammar is better than context-free grammar in terms of the ability to express various linguistic phenomena in natural language. In recent years, tree adjoining grammar has been widely used, such as: machine translation, automatic summarization, fingerprint image recognition, etc. [9].

The TAG grammar can be defined as a 5-tuple: G = (VN, VT, S, I, A). Where: VN is a non-terminal; VT is the terminal. The specific characters in a language; S is the starting symbol ($S \in VN$); I is the initial tree; A is the additional tree. The set of I union A is the basic tree in TAG. In LTAG, all the basic trees are lexical trees. Figure 1 shows the composition of the initial tree and the additional tree.

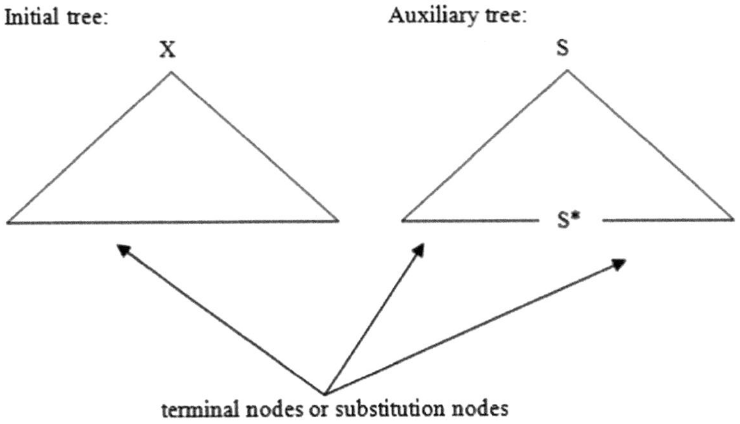

Fig. 1. The composition of the initial tree and the additional tree.

Initial tree (I tree) is short for "I tree", which means a simple structure without recursive structure, such as stem ring structure in RNA secondary structure, unpaired single chain, etc. The initial tree definition is as follows:

(1) All non-leaf nodes are non-terminal characters;
(2) All leaf nodes are terminators or non-terminators used for replacement operations.

The node where the non-terminal of the replacement operation is marked "↓" is called the replacement node.

Auxiliary trees are called "A trees" for short. It is used to represent a recursive structure that is an addition to a basic structure that usually modifies, such as a modified site base. The form of the attached tree is defined as:

(1) All non-leaf nodes are non-terminal;
(2) Except for the foot node, all leaves are terminal or non-terminal used for replacement operation;
(3) The symbol of the foot node of the tree is the same as that of the root node.

2.2 Operation of TAG

TAG models sequences with long range correlation and cross-pairing through substitution and adjunction, which makes it very suitable for analyzing RNA sequences. A(Adjoin) in TAG means "adjoining". Roughly, it is A combination of lexicalized trees. When some specific lexical trees of the sequence to be analyzed are selected, TAG uses a new algorithm nameed Earley algorithm [10] to find all the spanning trees of the sentence. This is the process of performing a series of adjacency operations on the selected tree. These operations can be divided into two types according to the adjacent objects and positions: replace and add.

In Fig. 2, T is a derived tree which generates the RNA sequences "gucacg" and represents the secondary structure {(1, 5), (2, 4), (3, 6)}. Note that the structure

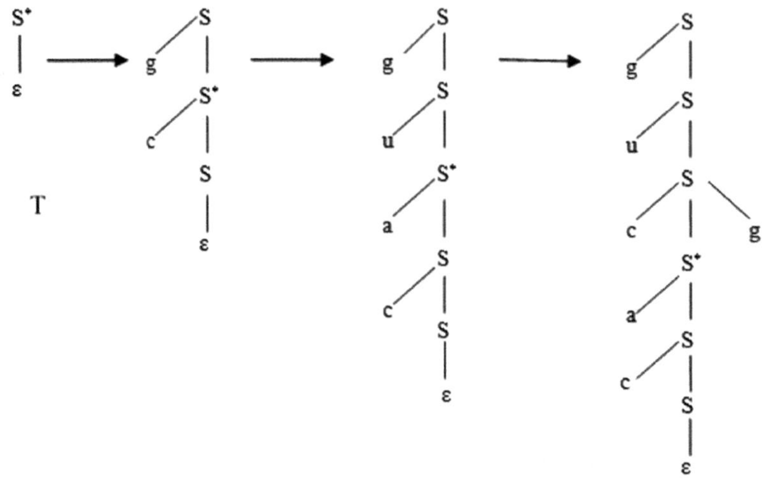

Fig. 2. A derivation of a RNA secondary structure in Example 1.

represented by T has crossing dependenciey which cannot be modeled by any context-free grammar.

TAG is a tree generation system rather than a string generation system, but the tree generated by TAG can be used to analyze and interpret the string language of the target language. In a TAG, the tree in the target language is generated through the derivation of the tree. The "tree" in the target language and the generation process of these "trees" are described below. If A tree is formed from any two trees in the set of I union A, it is called A derived tree. The process of deriving a derivation tree is called the "derivation process." in this process, two operations are used: one is to insert and the other is to substitute.

(1) Substitution

The operation of merging the root node of one i-tree with the replacement node of another tree to produce a new tree is called substitution. The condition for the substitution is that the root of one I tree is the same nonterminal as the replacement of another tree (either an I tree or an A tree). The replacement node is a special leaf node, and the replacement node is a non-terminal. Because leaf nodes are specific words in the analysis sentence, and specific words are terminators in the grammar, they can be non-terminators only when leaf nodes are replacement nodes. This allows it to have a terminal (that is, a concrete word) as a child by substitution.

(2) Adjunction

The operation of inserting an A tree into an I tree at A non-endpoint is called adding. The addition operation can be performed if the root and foot of A tree are the same as A node in I. Since the root and foot nodes of tree A are the same, other structures of tree I will not be affected after adding operations.

3 STAG Models RNA Structures

3.1 Modeling RNA Structure Prediction by STAG

Stochastic Tree Adjoining Grammars (STAG) is a probabilistic model that adds a probability to each production of TAG [11]. Just like the three problems mentioned in the hidden markov model [12], STAG also inevitably needs to answer how to solve these three questions, which are respectively: 1) How to input an observation sequence under a known model, and how to calculate its probability, namely, how to calculate P (O| λ), where the O represents the observable sequence, which is actually the observable discrete sequence of table states for a discrete sequence, that is, how to evaluate the score of the existing structure, which is equivalent to the criterion for judging whether it is a reasonable result; The second problem is to calculate which state sequence has the highest probability (that is, the hidden sequence with the greatest possibility) under the existing model λ and observation sequence Q. The third problem involves the training of the model. How to use the information in the training set to solve a reasonable result for the whole model.

For the first problem, the Stochastic Tree adjoining grammar model takes the X sequence set of table state and the hidden state Y data set given by the condition, calculates the conditional probability value of a certain Y to X through the following formula (1), which is equivalent to solving the final scoring standard.

$$P_{\lambda}(Y|X) = \frac{e^{\lambda \times F(y,x)}}{Z(x)} \tag{1}$$

where, λ is equivalent to the parameters already set for the model; Z(x) is a normalization factor, which represents all the cases where Y sequence set appears, so the probability of all Y sets is a necessary event, that is, it represents that if x is an observable sequence set, then one of the sets of sequence set Y must appear. F(y, x) is a characteristic function, often take a Boolean value.

For the second problem, we use the maximum likelihood method to deal with it. By taking the derivative of formula (3–1), the problem of finding the most likely marker sequence for a given observation sequence can be simplified to the following formula (2).

$$Y = \arg\max P_{\lambda}(y|x) = \arg\max \lambda \times F(y,x) \tag{2}$$

For a given input observation sequence x, find a marker sequence y that maximizes $P_{\lambda}(y|x)$. This generally use Viterbi algorithm for computing. We use maximum likelihood method to train a STAG model. Therefore, for a training set TM = $\{(x_k; y_k)\}$, M represents the existence of M sample sets in the training set, and the value range of k is 1 < k < M, so we can get formula (3).

$$L_{\lambda} = \sum_{K} \log P_{\lambda}(y_k|x_k) = \sum_{K} [\lambda \times F(y_k, x_k) - \log Z_{\lambda}(x_k)] \tag{3}$$

Using the gradient descent method, let the gradient of L_λ is zero, we can get:

$$\Delta L_\lambda = \sum_K [F(y_k, x_k) - E_{p\lambda(Y|x)} F(Y, x_k)] \tag{4}$$

The third problem is parameter training, that is, how to train the parameters under the existing model. The general STAGs approach is to construct forward and backward method similar to the hidden markov models [13]. The formula (4) is the result of taking partial derivatives with respect to each one, so that the optimal value of each parameter can be fixed, so that these parameters can be used to deal with those new sequences.

3.2 Steps of STAG to Predict RNA Structure

Modeling of RNA secondary structure by STAGs can be summarized into three Steps.

Step 1: convert RNA sequences and structures into mathematical symbols. For an RNA sequence, find y_i, the matching tag corresponding to each base x_i on the sequence, and the problem of RNA structure prediction is transformed into a classification problem in machine learning. First, the paired label y_i is simplified to 0–1–2 digital character set, which 0 means the base is not paired, 1 means the stem structure in the non-pseudoknot formed by pairing the base with other bases, and 2 means the stem structure in the pseudoknot formed by pairing the base with other bases.

Step 2: select the core child node (hc) of this node, and judge the relationship between this node and its siblings (predicate relationship, modification relationship, connection relationship).

(1) model input, including: (1) RNA sequence, it is a collection of base characters. Namely Σ = {A, C, G, U} generate a stochaistic sequence, (2) the matching probability of RNA bases, because the RNA base pairing has regularity, such as A-U and G-C matching probability is very big, but A-G, G-U matching probability is small, so can be set up according to the experience: P (A, U) = 0.9, P (A, G) = 0.1, etc.
(2) model output: a series of information labels of base pairing types: $y_1, y_2, \ldots, y_n$. And base pair label y_{ij}. The value of the former is any value in the set {0, 1, 2}. The latter refers to the pairing of any two bases in RNA. y_{ij} means the pairing of base I and base j, and 0 means no pairing. All y_{ij} sets are stored in a two-dimensional triangular matrix.

Step 3: in the derived tree, find A main path P from R to leaf node A. Create a central tree Ts and copy the nodes in P to Ts (R3 is the root of Ts and A3 is the leaf of Ts). Then copy the son of each non-connected node (copy only the node, not the subtree root of the node) to the corresponding node in Ts. In the Ts tree, all connection nodes are merged, then in the Ts tree, A is the specific lexical node, and all other leaf nodes are replacement nodes.

Figure 3 shows the flow chart of Modeling RNA Structure by STAG.

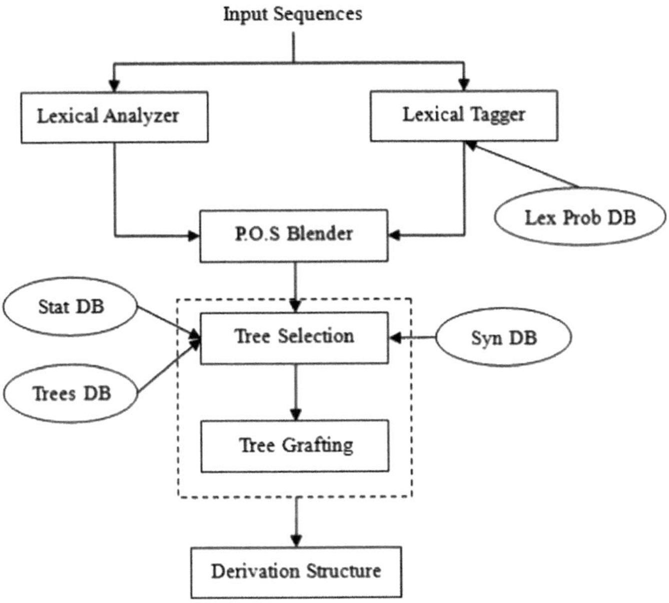

Fig. 3. The flow chart of Modeling RNA Structure by TAG.

3.3 Experimental and Prediction Results

We chose in EMBL database [14] includes virus comes, the archaea and the eubacteria, cyanellae, cytoplasm, and a variety of tRNA mitochondria sequence, a total of 675. On sample selection and use in tRNA we have constructed a call MT10CY10 training sample set, the sample set was drawn at random in the cytoplasm and mitochondria class 15 each for a total of 30 series sample set as our teacher. In the process of sampling samples in the experiment, we adopted the random number generation method of Matlab to conduct sampling, and the test sample set was processed by other tRNA data mentioned above (excluding the data of the test sample set).

Sn (Sensitivity) and Specification are adopted as the evaluation index of prediction accuracy in this study. The evaluation method is to use the TAG prediction model to conduct comparison experiments on Mfold and Pfold. Among them, Pfold is a prediction algorithm of RNA secondary structure based on SCFG model, and Mfold is a prediction algorithm of RNA secondary structure based on the minimum free energy in thermodynamics. The experimental results are shown in Table 1.

Figure 4 shows the comparison of experimental results in Table 1. It can be seen from the experimental results that the prediction accuracy of various length sequences, especially those with false junction structure, is improved to some extent by using the convolutional neural network prediction model with pooling layer. It indicates that convolutional neural network has a good effect on extracting long-range correlation of RNA sequences.

Table 1. Experimental results of STAG, Mfold and Pfold model comparison.

Dataset	Mfold		Pfold		STAG	
	Sens.	Spec.	Sens.	Spec.	Sens.	Spec.
ARCHAE	63.70	58.45	66.75	62.34	69.12	68.45
CY	59.54	55.65	61.21	58.66	62.36	61.56
CYANELCHLORO	68.45	61.22	70.23	64.53	79.20	75.19
EUBACT	62.14	58.76	63.45	59.76	73.14	62.34
MT	60.41	60.20	64.15	62.35	69.98	62.90
PARTIII	65.74	67.34	66.89	69.43	67.09	63.87
VIRUS	62.75	65.46	63.45	65.43	76.75	74.15
Total	63.12	59.56	65.97	64.78	69.80	65.68

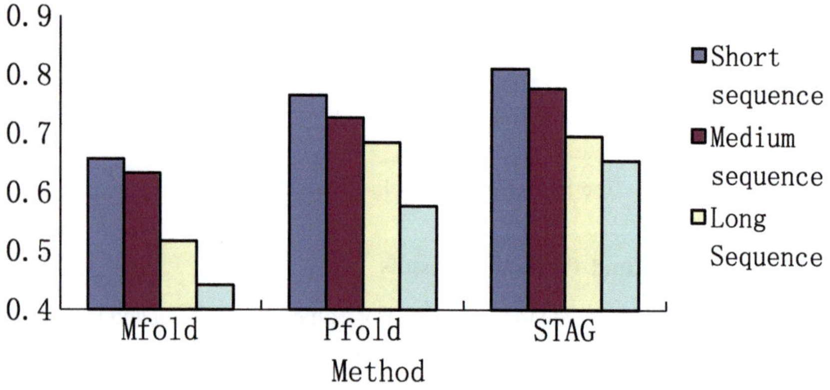

Fig. 4. Comparison of experimental results.

4 Conclusion

In this paper, we present a method to predict the secondary structure of RNA using Stochastic Tree Adjoining Grammars model. The theoretical basis is that TAG can well deal with the long-distance dependence of bases in biological sequences, so it is suitable for structural prediction of RNA sequences with long range correlation, especially when RNA sequences contain false junctions, and the accuracy of prediction is directly affected by the treatment of long range correlation. Through experimental comparison, this method uses Stochastic Tree Adjoining Grammars model to effectively improve the prediction accuracy of RNA structure prediction model, indicating that this method has good application value.

Acknowledgement. This work was supported by the Application-oriented Special Disciplines, Double First-Class University Project of Hunan Province (Xiangjiaotong [2018] 469), Hunan Provincial Natural Science Foundation of China (2020JJ4152), and Hengyang technology innovation guidance projects (Hengkefa [2020]-5), the Science and Technology Plan Project of Hunan Province (No. 2016TP1020), Postgraduate Research and Innovation Projects of Hunan Province (No. Xiangjiaotong [2019]248-998). Hengyang guided science and technology projects and Application-oriented Special Disciplines (No. Hengkefa [2018]60-31) and Postgraduate Scientific Research Innovation Project of Hunan Province (No. CX20190998).

References

1. Joshi, A.K., Levy, L., Takahashi, M.: Tree adjunct grammar. J. Comput. Syst. Sci. **10**(1), 136–163 (1975)
2. Shieber, S.M., Schabes, Y.: Generation and synchronous tree adjoining grammars. Comput. Intell. **7**(4), 220–228 (2007)
3. Hutchison, L.A.D.: Pika parsing: parsing in reverse solves the left recursion and error recovery problems. arXiv preprint arXiv:2005.06444 (2020)
4. Widder, S.J.: Syntactically-constrained paraphrase generation with tree-adjoining grammar (2020)
5. Nawrocki, E.P.: Structural RNA homology search and alignment using covariance models. Washington University School of Medicine (2009)
6. Come, E., Oukhellou, L., Denoeux, T.: Learning from partially supervised data using mixture models and belief functions. Pattern Recogn. **42**(3), 334348 (2009)
7. Tanzera, A., Hofackerab, I.L., Lorenz, R.: RNA modifications in structure prediction-status quo and future challenges. Methods **39**(10), 23–38 (2018)
8. Jenkins, A.M., Waterhouse, R.M., Muskavitch, M.A.T.: Long non-coding RNA discovery across the genus anopheles reveals conserved secondary structures within and beyond the Gambiae complex. BMC Genom. **16**(1), 337–350 (2015)
9. Quadrini, M., Merelli, E., Tesei, L.: Alignment tree for RNA pseudoknots. BMC Bioinf. **11**(4), 337–343 (2018)
10. Tang, S., Zhou, Y., Zou, S.: The RNA secondary structure prediction based on the lexicalized stochastic grammar model. Comput. Eng. Sci. **3**(31), 128–131 (2009)
11. Griffiths-Jones, S., Bateman, A., Marshall, M., Khanna, A., Eddy, S.R.: Rfam: an RNA family database. Nucleic Acids Res. **31**(1), 429–441 (2003)
12. Quadrini, M., Tesei, L., Merelli, E.: An algebraic language for RNA pseudoknots comparison. BMC Bioinf. **20**(4), 161 (2019)
13. Bellaousov, S., Mathews, D.H.: ProbKnot: fast prediction of RNA secondary structure including pseudoknots. RNA **16**(10), 1870–1880 (2010)
14. Kato, Y., Seki, H., Kasami, T.: On the generative power of grammars for RNA secondary structure. IEICE Trans. Inf. Syst. **88**(1), 53–64 (2005)

Author Index

Q. Liu et al. (Eds.): CENet 2020, AISC 1274, pp. 1751–1757, 2021.
https://doi.org/10.1007/978-981-15-8462-6